The Prokaryotes

Third Edition

KU-262-171

The **Prokaryotes**

A Handbook on the Biology of Bacteria

Third Edition

Volume 3: Archaea. Bacteria: Firmicutes, Actinomycetes

MARTIN DWORKIN (Editor-in-Chief), STANLEY FALKOW, EUGENE ROSENBERG,
KARL-HEINZ SCHLEIFER, ERKO STACKEBRANDT (Editors)

 Springer

Editor-in-Chief
Professor Dr. Martin Dworkin
Department of Microbiology
University of Minnesota
Box 196
University of Minnesota
Minneapolis, MN 55455-0312
USA

Editors
Professor Dr. Stanley Falkow
Department of Microbiology
and Immunology
Stanford University Medical School
299 Campus Drive, Fairchild D039
Stanford, CA 94305-5124
USA

Professor Dr. Eugene Rosenberg
Department of Molecular Microbiology
and Biotechnology
Tel Aviv University
Ramat-Aviv 69978
Israel

Professor Dr. Karl-Heinz Schleifer
Department of Microbiology
Technical University Munich
80290 Munich
Germany

Professor Dr. Erko Stackebrandt
DSMZ- German Collection of Microorganisms
and Cell Cultures GmbH
Mascheroder Weg 1b
38124 Braunschweig
Germany

URLs in *The Prokaryotes*: Uncommon Web sites have been listed in the text. However, the following Web sites have been referred to numerous times and have been suppressed for aesthetic purposes: www.bergeys.org; www.tigr.org; dx.doi.org; www.fp.mcs.anl.gov; www.ncbi.nlm.nih.gov; www.genome.ad.jp; www.cme.msu.edu; umbbd.ahc.umn.edu; www.dmsz.de; and www.arb-home.de. The entirety of all these Web links have been maintained in the electronic version.

Library of Congress Control Number: 91017256

Volume 3
ISBN-10: 0-387-25493-5
ISBN-13: 978-0387-25493-7
e-ISBN: 0-387-30743-5
Print + e-ISBN: 0-387-33489-0
DOI: 10.1007/0-387-30743-5

Volumes 1–7 (Set)
ISBN-10: 0-387-25499-4
ISBN-13: 978-0387-25499-9
e-ISBN: 0-387-30740-0
Print + e-ISBN: 0-387-33488-2

Printed on acid-free paper.

© 2006 Springer Science+Business Media, LLC
All rights reserved. This work may not be translated or copied in whole or in part without the written permission of the publisher (Springer Science+Business Media, LLC, 233 Spring Street, New York, NY 10013, USA), except for brief excerpts in connection with reviews or scholarly analysis. Use in connection with any form of information storage and retrieval, electronic adaptation, computer software, or by similar or dissimilar methodology now known or hereafter developed is forbidden.
The use in this publication of trade names, trademarks, service marks, and similar terms, even if they are not identified as such, is not to be taken as an expression of opinion as to whether or not they are subject to proprietary rights.

Printed in Singapore.   (BS/KYO)

9 8 7 6 5 4 3 2 1

springer.com

Preface

Each of the first two editions of *The Prokaryotes* took a bold step. The first edition, published in 1981, set out to be an encyclopedic, synoptic account of the world of the prokaryotes—a collection of monographic descriptions of the genera of bacteria. The Archaea had not yet been formalized as a group. For the second edition in 1992, the editors made the decision to organize the chapters on the basis of the molecular phylogeny championed by Carl Woese, which increasingly provided a rational, evolutionary basis for the taxonomy of the prokaryotes. In addition, the archaea had by then been recognized as a phylogenetically separate and distinguishable group of the prokaryotes. The two volumes of the first edition had by then expanded to four. The third edition was arguably the boldest step of all. We decided that the material would only be presented electronically. The advantages were obvious and persuasive. There would be essentially unlimited space. There would be no restrictions on the use of color illustrations. Film and animated descriptions could be made available. The text would be hyperlinked to external sources. Publication of chapters would be seriati—the edition would no longer have to delay publication until the last tardy author had submitted his or her chapter. Updates and modifications could be made continuously. And, most attractively, a library could place its subscribed copy on its server and make it available easily and cheaply to all in its community. One hundred and seventy chapters have thus far been presented in 16 releases over a six-year period. The virtues and advantages of the online edition have been borne out. But we failed to predict the affection that many have for holding a bound, print version of a book in their hands. Thus, this print version of the third edition shall accompany the online version.

We are now four years into the 21st century. Indulge us then while we comment on the challenges, problems and opportunities for microbiology that confront us.

Moselio Schaechter has referred to the present era of microbiology as its third golden age—the era of "integrative microbiology." Essentially all microbiologists now speak a common language. So that the boundaries that previously separated subdisciplines from each other have faded: physiology has become indistinguishable from pathogenesis; ecologists and molecular geneticists speak to each other; biochemistry is spoken by all; and—mirabile dictu!—molecular biologists are collaborating with taxonomists.

But before these molecular dissections of complex processes can be effective there must be a clear view of the organism being studied. And it is our goal that these chapters in *The Prokaryotes* provide that opportunity.

There is also yet a larger issue. Microbiology is now confronted with the need to understand increasingly complex processes. And the modus operandi that has served us so successfully for 150 years—that of the pure culture studied under standard laboratory conditions—is inadequate. We are now challenged to solve problems of multimembered populations interacting with each other and with their environment under constantly variable conditions. Carl Woese has pointed out a useful and important distinction between empirical, methodological reductionism and fundamentalist reductionism. The former has served us well; the latter stands in the way of our further understanding of complex, interacting systems. But no matter what kind of synoptic systems analysis emerges as our way of understanding host–parasite relations, ecology, or multicellular behavior, the understanding of the organism as such is sine qua non. And in that context, we are pleased to present to you the third edition of *The Prokaryotes*.

Martin Dworkin
Editor-in-Chief

Foreword

The purpose of this brief foreword is unchanged from the first edition; it is simply to make you, the reader, hungry for the scientific feast that follows. These four volumes on the prokaryotes offer an expanded scientific menu that displays the biochemical depth and remarkable physiological and morphological diversity of prokaryote life. The size of the volumes might initially discourage the unprepared mind from being attracted to the study of prokaryote life, for this landmark assemblage thoroughly documents the wealth of present knowledge. But in confronting the reader with the state of the art, the Handbook also defines where more work needs to be done on well-studied bacteria as well as on unusual or poorly studied organisms.

This edition of *The Prokaryotes* recognizes the almost unbelievable impact that the work of Carl Woese has had in defining a phylogenetic basis for the microbial world. The concept that the ribosome is a highly conserved structure in all cells and that its nucleic acid components may serve as a convenient reference point for relating all living things is now generally accepted. At last, the phylogeny of prokaryotes has a scientific basis, and this is the first serious attempt to present a comprehensive treatise on prokaryotes along recently defined phylogenetic lines. Although evidence is incomplete for many microbial groups, these volumes make a statement that clearly illuminates the path to follow.

There are basically two ways of doing research with microbes. A classical approach is first to define the phenomenon to be studied and then to select the organism accordingly. Another way is to choose a specific organism and go where it leads. The pursuit of an unusual microbe brings out the latent hunter in all of us. The intellectual challenges of the chase frequently test our ingenuity to the limit. Sometimes the quarry repeatedly escapes, but the final capture is indeed a wonderful experience. For many of us, these simple rewards are sufficiently gratifying so that we have chosen to spend our scientific lives studying these unusual creatures. In these endeavors many of the strategies and tools as well as much of the philosophy may be traced to the Delft School, passed on to us by our teachers, Martinus Beijerinck, A. J. Kluyver, and C. B. van Niel, and in turn passed on by us to our students.

In this school, the principles of the selective, enrichment culture technique have been developed and diversified; they have been a major force in designing and applying new principles for the capture and isolation of microbes from nature. For me, the "organism approach" has provided rewarding adventures. The organism continually challenges and literally drags the investigator into new areas where unfamiliar tools may be needed. I believe that organism-oriented research is an important alternative to problem-oriented research, for new concepts of the future very likely lie in a study of the breadth of microbial life. The physiology, biochemistry, and ecology of the microbe remain the most powerful attractions. Studies based on classical methods as well as modern genetic techniques will result in new insights and concepts.

To some readers, this edition of the *The Prokaryotes* may indicate that the field is now mature, that from here on it is a matter of filling in details. I suspect that this is not the case. Perhaps we have assumed prematurely that we fully understand microbial life. Van Niel pointed out to his students that—after a lifetime of study—it was a very humbling experience to view in the microscope a sample of microbes from nature and recognize only a few. Recent evidence suggests that microbes have been evolving for nearly 4 billion years. Most certainly those microbes now domesticated and kept in captivity in culture collections represent only a minor portion of the species that have evolved in this time span. Sometimes we must remind ourselves that evolution is actively taking place at the present moment. That the eukaryote cell evolved as a chimera of certain prokaryote parts is a generally accepted concept today. Higher as well as lower eukaryotes evolved in contact with prokaryotes, and evidence surrounds us of the complex interactions between eukaryotes and prokaryotes as well as among prokaryotes. We have so far only scratched the surface of these biochemical interrelationships. Perhaps the legume nodule is a pertinent example of nature caught in the act of evolving the "nitrosome," a unique nitrogen-fixing organelle. Study of prokaryotes is proceeding at such a fast pace that major advances are occurring yearly. The increase of this edition to four volumes documents the exciting pace of discoveries.

To prepare a treatise such as *The Prokaryotes* requires dedicated editors and authors; the task has been enormous. I predict that the scientific community of microbiologists will again show its appreciation through use of these volumes—such that the pages will become "dog-eared" and worn as students seek basic information for the

hunt. These volumes belong in the laboratory, not in the library. I believe that a most effective way to introduce students to microbiology is for them to isolate microbes from nature, i.e., from their habitats in soil, water, clinical specimens, or plants. *The Prokaryotes* enormously simplifies this process and should encourage the construction of courses that contain a wide spectrum of diverse topics. For the student as well as the advanced investigator these volumes should generate excitement.

Happy hunting!

Ralph S. Wolfe
Department of Microbiology
University of Illinois at Urbana-Champaign

Contents

Preface   **v**

Foreword by Ralph S. Wolfe   **vii**

Contributors   **xxix**

Volume 1

1.   *Essays in Prokaryotic Biology*

1.1   How We Do, Don't and Should Look at Bacteria and Bacteriology   3
CARL R. WOESE

1.2   Databases   24
WOLFGANG LUDWIG, KARL-HEINZ SCHLEIFER and ERKO STACKEBRANDT

1.3   Defining Taxonomic Ranks   29
ERKO STACKEBRANDT

1.4   Prokaryote Characterization and Identification   58
HANS G. TRÜPER and KARL-HEINZ SCHLEIFER

1.5   Principles of Enrichment, Isolation, Cultivation, and Preservation of Prokaryotes   80
JÖRG OVERMANN

1.6   Prokaryotes and Their Habitats   137
HANS G. SCHLEGEL and HOLGER W. JANNASCH

1.7   Morphological and Physiological Diversity   185
STEPHEN H. ZINDER and MARTIN DWORKIN

1.8   Cell-Cell Interactions   221
DALE KAISER

1.9   Prokaryotic Genomics   246
B. W. WREN

1.10   Genomics and Metabolism in *Escherichia coli*   261
MARGRETHE HAUGGE SERRES and MONICA RILEY

1.11   Origin of Life: RNA World versus Autocatalytic Anabolism   275
GÜNTER WÄCHTERSHÄUSER

1.12   Biotechnology and Applied Microbiology   284
EUGENE ROSENBERG

1.13   The Structure and Function of Microbial Communities   299
DAVID A. STAHL, MEREDITH HULLAR and SEANA DAVIDSON

2.   Symbiotic Associations

2.1   Cyanobacterial-Plant Symbioses   331
DAVID G. ADAMS, BIRGITTA BERGMAN, S. A. NIERZWICKI-BAUER,
A. N. RAI and ARTHUR SCHÜßLER

2.2   Symbiotic Associations Between Ciliates and Prokaryotes   364
HANS-DIETER GÖRTZ

2.3   Bacteriocyte-Associated Endosymbionts of Insects   403
PAUL BAUMANN, NANCY A. MORAN and LINDA BAUMANN

2.4   Symbiotic Associations Between Termites and Prokaryotes   439
ANDREAS BRUNE

2.5   Marine Chemosynthetic Symbioses   475
COLLEEN M. CAVANAUGH, ZOE P. MCKINESS, IRENE L.G. NEWTON and
FRANK J. STEWART

3.   Biotechnology and Applied Microbiology

3.1   Organic Acid and Solvent Production   511
PALMER ROGERS, JIANN-SHIN CHEN and MARY JO ZIDWICK

3.2   Amino Acid Production   756
HIDEHIKO KUMAGAI

3.3   Microbial Exopolysaccharides   766
TIMOTHY HARRAH, BRUCE PANILAITIS and DAVID KAPLAN

3.4   Bacterial Enzymes   777
WIM J. QUAX

3.5   Bacteria in Food and Beverage Production   797
MICHAEL P. DOYLE and JIANGHONG MENG

3.6   Bacterial Pharmaceutical Products   812
ARNOLD L. DEMAIN and GIANCARLO LANCINI

3.7   Biosurfactants   834
EUGENE ROSENBERG

3.8   Bioremediation   850
RONALD L. CRAWFORD

3.9   Biodeterioration   864
JI-DONG GU and RALPH MITCHELL

3.10   Microbial Biofilms   904
DIRK DE BEER and PAUL STOODLEY

Index   939

Volume 2

1.   Ecophysiological and Biochemical Aspects

1.1   Planktonic Versus Sessile Life of Prokaryotes   3
KEVIN C. MARSHALL

1.2   Bacterial Adhesion   16
ITZHAK OFEK, NATHAN SHARON and SOMAN N. ABRAHAM

1.3   The Phototrophic Way of Life   32
JÖRG OVERMANN and FERRAN GARCIA-PICHEL

1.4   The Anaerobic Way of Life   86
RUTH A. SCHMITZ, ROLF DANIEL, UWE DEPPENMEIER and
GERHARD GOTTSCHALK

1.5   Bacterial Behavior   102
JUDITH ARMITAGE

1.6   Prokaryotic Life Cycles   140
MARTIN DWORKIN

1.7   Life at High Temperatures   167
RAINER JAENICKE and REINHARD STERNER

1.8   Life at Low Temperatures   210
SIEGFRIED SCHERER and KLAUS NEUHAUS

1.9   Life at High Salt Concentrations   263
AHARON OREN

1.10   Alkaliphilic Prokaryotes   283
TERRY ANN KRULWICH

1.11   Syntrophism among Prokaryotes   309
BERNHARD SCHINK and ALFONS J.M. STAMS

1.12   Quorum Sensing   336
BONNIE L. BASSLER and MELISSA B. MILLER

1.13   Acetogenic Prokaryotes   354
HAROLD L. DRAKE, KIRSTEN KÜSEL and CAROLA MATTHIES

1.14   Virulence Strategies of Plant Pathogenic Bacteria   421
BARBARA N. KUNKEL and ZHONGYING CHEN

1.15   The Chemolithotrophic Prokaryotes   441
DONOVAN P. KELLY and ANNE P. WOOD

1.16   Oxidation of Inorganic Nitrogen Compounds as an Energy Source   457
EBERHARD BOCK and MICHAEL WAGNER

1.17   The H_2-Metabolizing Prokaryotes   496
EDWARD SCHWARTZ and BÄRBEL FRIEDRICH

1.18   Hydrocarbon-Oxidizing Bacteria   564
EUGENE ROSENBERG

1.19   Cellulose-Decomposing Bacteria and Their Enzyme Systems   578
EDWARD A. BAYER, YUVAL SHOHAM and RAPHAEL LAMED

1.20   Aerobic Methylotrophic Prokaryotes   618
MARY E. LIDSTROM

1.21   Dissimilatory Fe(III)- and Mn(IV)-Reducing Prokaryotes   635
DEREK LOVLEY

1.22   Dissimilatory Sulfate- and Sulfur-Reducing Prokaryotes   659
RALF RABUS, THEO A. HANSEN and FRIEDRICH WIDDEL

1.23   The Denitrifying Prokaryotes   769
JAMES P. SHAPLEIGH

1.24   Dinitrogen-Fixing Prokaryotes   793
ESPERANZA MARTINEZ-ROMERO

1.25   Root and Stem Nodule Bacteria of Legumes   818
MICHAEL J. SADOWSKY and P. H. GRAHAM

1.26   Magnetotactic Bacteria   842
STEFAN SPRING and DENNIS A. BAZYLINSKI

1.27   Luminous Bacteria   863
PAUL V. DUNLAP and KUMIKO KITA-TSUKAMOTO

1.28   Bacterial Toxins   893
VEGA MASIGNANI, MARIAGRAZIA PIZZA and RINO RAPPUOLI

1.29   The Metabolic Pathways of Biodegradation   956
LAWRENCE P. WACKETT

1.30   Haloalkaliphilic Sulfur-Oxidizing Bacteria   969
DIMITRY YU. SOROKIN, HORIA BANCIU, LESLEY A. ROBERTSON and
J. GIJS KUENEN

1.31   The Colorless Sulfur Bacteria   985
LESLEY A. ROBERTSON and J. GIJS KUENEN

1.32   Bacterial Stress Response   1012
ELIORA Z. RON

1.33   Anaerobic Biodegradation of Hydrocarbons Including Methane   1028
FRIEDRICH WIDDEL, ANTJE BOETIUS and RALF RABUS

1.34   Physiology and Biochemistry of the Methane-Producing Archaea   1050
REINER HEDDERICH and WILLIAM B. WHITMAN

Index   1081

Volume 3

A: Archaea

1.   The Archaea: A Personal Overview of the Formative Years   3
RALPH S. WOLFE

2.   Thermoproteales   10
HARALD HUBER, ROBERT HUBER and KARL O. STETTER

3.   Sulfolobales   23
HARALD HUBER and DAVID PRANGISHVILI

4.   Desulfurococcales   52
HARALD HUBER and KARL O. STETTER

5.   The Order Thermococcales   69
COSTANZO BERTOLDO and GARABED ANTRANIKIAN

6.   The Genus *Archaeoglobus*   82
PATRICIA HARTZELL and DAVID W. REED

7.   Thermoplasmatales   101
HARALD HUBER and KARL O. STETTER

8.   The Order Halobacteriales   113
AHARON OREN

9.   The Methanogenic Bacteria   165
WILLIAM B. WHITMAN, TIMOTHY L. BOWEN and DAVID R. BOONE

10.   The Order Methanomicrobiales   208
JEAN-LOUIS GARCIA, BERNARD OLLIVIER and WILLIAM B. WHITMAN

11.   The Order Methanobacteriales   231
ADAM S. BONIN and DAVID R. BOONE

12.   The Order Methanosarcinales   244
MELISSA M. KENDALL and DAVID R. BOONE

13. Methanococcales   257
WILLIAM B. WHITMAN and CHRISTIAN JEANTHON

14. Nanoarchaeota   274
HARALD HUBER, MICHAEL J. HOHN, REINHARD RACHEL and
KARL O. STETTER

15. Phylogenetic and Ecological Perspectives on Uncultured
Crenarchaeota and Korarchaeota   281
SCOTT C. DAWSON, EDWARD F. DELONG and NORMAN R. PACE

B: *Bacteria*

1. *Firmicutes (Gram-Positive Bacteria)*

1.1. *Firmicutes with High GC Content of DNA*

1.1.1 Introduction to the Taxonomy of Actinobacteria   297
ERKO STACKEBRANDT and PETER SCHUMANN

1.1.2 The Family Bifidobacteriaceae   322
BRUNO BIAVATI and PAOLA MATTARELLI

1.1.3 The Family Propionibacteriaceae: The Genera *Friedmanniella*,
Luteococcus, *Microlunatus*, *Micropruina*, *Propioniferax*,
Propionimicrobium and *Tessarococcus*   383
ERKO STACKEBRANDT and KLAUS P. SCHAAL

1.1.4 Family Propionibacteriaceae: The Genus *Propionibacterium* 400
ERKO STACKEBRANDT, CECIL S. CUMMINS and JOHN L. JOHNSON

1.1.5 The Family Succinivibrionaceae   419
ERKO STACKEBRANDT and ROBERT B. HESPELL

1.1.6 The Family Actinomycetaceae: The Genera *Actinomyces*, *Actinobaculum*,
Arcanobacterium, *Varibaculum* and *Mobiluncus*   430
KLAUS P. SCHAAL, ATTEYET F. YASSIN and ERKO STACKEBRANDT

1.1.7 The Family Streptomycetaceae, Part I: Taxonomy   538
PETER KÄMPFER

1.1.8 The Family Streptomycetaceae, Part II: Molecular Biology 605
HILDGUND SCHREMPF

1.1.9 The Genus *Actinoplanes* and Related Genera   623
GERNOT VOBIS

1.1.10 The Family Actinosynnemataceae   654
DAVID P. LABEDA

1.1.11 The Families Frankiaceae, Geodermatophilaceae, Acidothermaceae
and Sporichthyaceae   669
PHILIPPE NORMAND

1.1.12   The Family Thermomonosporaceae: *Actinocorallia, Actinomadura, Spirillospora* and *Thermomonospora*
REINER MICHAEL KROPPENSTEDT and MICHAEL GOODFELLOW   682

1.1.13   The Family Streptosporangiaceae
MICHAEL GOODFELLOW and ERIKA TERESA QUINTANA   725

1.1.14   The Family Nocardiopsaceae
REINER MICHAEL KROPPENSTEDT and LYUDMILA I. EVTUSHENKO   754

1.1.15   *Corynebacterium*—Nonmedical
WOLFGANG LIEBL   796

1.1.16   The Genus *Corynebacterium*—Medical
ALEXANDER VON GRAEVENITZ and KATHRYN BERNARD   819

1.1.17   The Families Dietziaceae, Gordoniaceae, Nocardiaceae and Tsukamurellaceae
MICHAEL GOODFELLOW and LUIS ANGEL MALDONADO   843

1.1.18   The Genus *Mycobacterium*—Nonmedical
SYBE HARTMANS, JAN A.M. DE BONT and ERKO STACKEBRANDT   889

1.1.19   The Genus *Mycobacterium*—Medical
BEATRICE SAVIOLA and WILLIAM BISHAI   919

1.1.20   *Mycobacterium leprae*
THOMAS M. SHINNICK   934

1.1.21   The Genus *Arthrobacter*
DOROTHY JONES and Ronald M. KEDDIE   945

1.1.22   The Genus *Micrococcus*
MILOSLAV KOCUR, WESLEY E. KLOOS and KARL-HEINZ SCHLEIFER   961

1.1.23   *Renibacterium*
HANS-JÜRGEN BUSSE   972

1.1.24   The Genus *Stomatococcus*: *Rothia mucilaginosa*, basonym *Stomatococcus mucilaginosus*
ERKO STACKEBRANDT   975

1.1.25   The Family Cellulomonadaceae
ERKO STACKEBRANDT, PETER SCHUMANN and HELMUT PRAUSER   983

1.1.26   The Family Dermatophilaceae
ERKO STACKEBRANDT   1002

1.1.27   The Genus *Brevibacterium*
MATTHEW D. COLLINS   1013

1.1.28   The Family Microbacteriaceae
LYUDMILA I. EVTUSHENKO and MARIKO TAKEUCHI   1020

xvi   Contents

1.1.29   The Genus *Nocardioides*   **1099**
             JUNG-HOON YOON and YONG-HA PARK

Index   **1115**

Volume 4

1.   *Firmicutes (Gram-Positive Bacteria)*

1.2   *Firmicutes with Low GC Content of DNA*

1.2.1   The Genera *Staphylococcus* and *Macrococcus*   5
            FRIEDRICH GÖTZ, TAMMY BANNERMAN and KARL-HEINZ SCHLEIFER

1.2.2   The Genus *Streptococcus*—Oral   76
            JEREMY M. HARDIE and ROBERT A. WHILEY

1.2.3   Medically Important Beta-Hemolytic Streptococci   108
            P. PATRICK CLEARY and QI CHENG

1.2.4   *Streptococcus pneumoniae*   149
            ELAINE TUOMANEN

1.2.5   The Genus Enterococcus: Taxonomy   163
            LUC DEVRIESE, MARGO BAELE and PATRICK BUTAYE

1.2.6   Enterococcus   175
            DONALD J. LEBLANC

1.2.7   The Genus *Lactococcus*   205
            MICHAEL TEUBER and ARNOLD GEIS

1.2.8   The Genera *Pediococcus* and *Tetragenococcus*   229
            WILHELM H. HOLZAPFEL, CHARLES M. A. P. FRANZ, WOLFGANG LUDWIG,
            WERNER BACK and LEON M. T. DICKS

1.2.9   Genera *Leuconostoc*, *Oenococcus* and *Weissella*   267
            JOHANNA BJÖRKROTH and WILHELM H. HOLZAPFEL

1.2.10 The Genera *Lactobacillus* and *Carnobacterium*   320
            WALTER P. HAMMES and CHRISTIAN HERTEL

1.2.11 *Listeria monocytogenes* and the Genus *Listeria*   404
            NADIA KHELEF, MARC LECUIT, CARMEN BUCHRIESER, DIDIER CABANES,
            OLIVIER DUSSURGET and PASCALE COSSART

1.2.12 The Genus *Brochothrix*   473
            ERKO STACKEBRANDT and DOROTHY JONES

1.2.13 The Genus *Erysipelothrix*   488
            ERKO STACKEBRANDT, ANNETTE C. REBOLI and W. EDMUND FARRAR

1.2.14 The Genus *Gemella*   507
MATTHEW D. COLLINS

1.2.15 The Genus *Kurthia*   515
ERKO STACKEBRANDT, RONALD M. KEDDIE and DOROTHY JONES

1.2.16 The Genus *Bacillus*—Nonmedical   526
RALPH A. SLEPECKY and H. ERNEST HEMPHILL

1.2.17 The Genus *Bacillus*—Insect Pathogens   559
DONALD P. STAHLY, ROBERT E. ANDREWS and ALLAN A. YOUSTEN

1.2.18 The Genus *Bacillus*—Medical   605
W. EDMUND FARRAR and ANNETTE C. REBOLI

1.2.19 Genera Related to the Genus *Bacillus*—*Sporolactobacillus,*
Sporosarcina, Planococcus, Filibacter and *Caryophanon*   627
DIETER CLAUS, DAGMAR FRITZE and MILOSLAV KOCUR

1.2.20 An Introduction to the Family Clostridiaceae   650
JÜRGEN WIEGEL, RALPH TANNER and FRED A. RAINEY

1.2.21 Neurotoxigenic Clostridia   674
CESARE MONTECUCCO, ORNELLA ROSSETTO and MICHEL R. POPOFF

1.2.22 The Enterotoxic Clostridia   693
BRUCE A. MCCLANE, FRANCISCO A. UZAI,
MARIANO E. FERNANDEZ MIYAKAWA, DAVID LYERLY and
TRACY WILKINS

1.2.23 *Clostridium perfringens* and Histotoxic Disease   748
JULIAN I. ROOD

1.2.24 The Genera *Desulfitobacterium* and *Desulfosporosinus*: Taxonomy   766
STEFAN SPRING and FRANK ROSENZWEIG

1.2.25 The Genus *Desulfotomaculum*   782
FRIEDRICH WIDDEL

1.2.26 The Anaerobic Gram-Positive Cocci   790
TAKAYUKI EZAKI, NA (MICHAEL) LI and YOSHIAKI KAWAMURA

1.2.27 The Order Haloanaerobiales   804
AHARON OREN

1.2.28 The Genus *Eubacterium* and Related Genera   818
WILLIAM G. WADE

1.2.29 The Genus *Mycoplasma* and Related Genera (Class Mollicutes)   831
SHMUEL RAZIN

1.2.30 The Phytopathogenic Spiroplasmas   897
JACQUELINE FLETCHER, ULRICH MELCHER and ASTRI WAYADANDE

1.3   *Firmicutes with Atypical Cell Walls*

1.3.1   The Family Heliobacteriaceae   943
MICHAEL T. MADIGAN

1.3.2   *Pectinatus*, *Megasphaera* and *Zymophilus*   957
AULI HAIKARA and ILKKA HELANDER

1.3.3   The Genus *Selenomonas*   974
ROBERT B. HESPELL, BRUCE J. PASTER and FLOYD E. DEWHIRST

1.3.4   The Genus *Sporomusa*   983
JOHN A. BREZNAK

1.3.5   The Family Lachnospiraceae, Including the Genera *Butyrivibrio*,
Lachnospira and *Roseburia*   994
MICHAEL COTTA and ROBERT FORSTER

1.3.6   The Genus *Veillonella*   1014
PAUL KOLENBRANDER

1.3.7   Syntrophomonadaceae   1033
MARTIN SOBIERJ and DAVID R. BOONE

2.   *Cyanobacteria*

2.1   The Cyanobacteria—Isolation, Purification and Identification   1045
JOHN B. WATERBURY

2.2   The Cyanobacteria—Ecology, Physiology and Molecular Genetics   1066
YEHUDA COHEN and MICHAEL GUREVITZ

2.3   The Genus *Prochlorococcus*   1091
ANTON F. POST

Index   1103

Volume 5

3.   *Proteobacteria*

Introduction to the Proteobacteria   3
KAREL KERSTERS, PAUL DE VOS, MONIQUE GILLIS, JEAN SWINGS,
PETER VAN DAMME and ERKO STACKEBRANDT

3.1.   *Alpha Subclass*

3.1.1   The Phototrophic Alphaproteobacteria   41
JOHANNES F. IMHOFF

3.1.2   The Genera *Prosthecomicrobium* and *Ancalomicrobium*   65
GARY E. OERTLI, CHERYL JENKINS, NAOMI WARD, FREDERICK A. RAINEY,
ERKO STACKEBRANDT and JAMES T. STALEY

3.1.3   Dimorphic Prosthecate Bacteria: The Genera *Caulobacter*,
Asticcacaulis, *Hyphomicrobium*, *Pedomicrobium*, *Hyphomonas*
and *Thiodendron*   72
JEANNE S. POINDEXTER

3.1.4   The Genus *Agrobacterium*   91
ANN G. MATTHYSSE

3.1.5   The Genus *Azospirillum*   115
ANTON HARTMANN and JOSE IVO BALDANI

3.1.6   The Genus *Herbaspirillum*   141
MICHAEL SCHMID, JOSE IVO BALDANI and ANTON HARTMANN

3.1.7   The Genus *Beijerinckia*   151
JAN HENDRICK BECKING

3.1.8   The Family Acetobacteraceae: The Genera *Acetobacter*, *Acidomonas*,
Asaia, *Gluconacetobacter*, *Gluconobacter*, and *Kozakia*   163
KAREL KERSTERS, PUSPITA LISDIYANTI, KAZUO KOMAGATA and
JEAN SWINGS

3.1.9   The Genus *Zymomonas*   201
HERMANN SAHM, STEPHANIE BRINGER-MEYER and GEORG A. SPRENGER

3.1.10   The Manganese-Oxidizing Bacteria   222
KENNETH H. NEALSON

3.1.11   The Genus *Paracoccus*   232
DONOVAN P. KELLY, FREDERICK A. RAINEY and ANN P. WOOD

3.1.12   The Genus *Phenylobacterium*   250
JÜRGEN EBERSPÄCHER and FRANZ LINGENS

3.1.13   *Methylobacterium*   257
PETER N. GREEN

3.1.14   The Methanotrophs—The Families Methylococcaceae and
Methylocystaceae   266
JOHN P. BOWMAN

3.1.15   The Genus *Xanthobacter*   290
JÜRGEN WIEGEL

3.1.16   The Genus *Brucella*   315
EDGARDO MORENO and IGNACIO MORIYÓN

3.1.17   Introduction to the Rickettsiales and Other Intracellular Prokaryotes   457
DAVID N. FREDRICKS

3.1.18   The Genus *Bartonella*   467
MICHAEL F. MINNICK and BURT E. ANDERSON

3.1.19   The Order Rickettsiales   493
XUE-JIE YU and DAVID H. WALKER

3.1.20   The Genus *Coxiella*   529
ROBERT A. HEINZEN and JAMES E. SAMUEL

3.1.21   The Genus *Wolbachia*   547
MARKUS RIEGLER and SCOTT L. O'NEILL

3.1.22   Aerobic Phototrophic Proteobacteria   562
VLADIMIR V. YURKOV

3.1.23   The Genus *Seliberia*   585
JEAN M. SCHMIDT and JAMES R. SWAFFORD

3.2.   Beta Subclass

3.2.1   The Phototrophic Betaproteobacteria   593
JOHANNES F. IMHOFF

3.2.2   The *Neisseria*   602
DANIEL C. STEIN

3.2.3   The Genus *Bordetella*   648
ALISON WEISS

3.2.4   *Achromobacter, Alcaligenes* and Related Genera   675
HANS-JÜRGEN BUSSE and ANDREAS STOLZ

3.2.5   The Genus *Spirillum*   701
NOEL R. KRIEG

3.2.6   The Genus *Aquaspirillum*   710
BRUNO POT, MONIQUE GILLIS and JOZEF DE LEY

3.2.7   *Comamonas*   723
ANNE WILLEMS and PAUL DE VOS

3.2.8   The Genera *Chromobacterium* and *Janthinobacterium*   737
MONIQUE GILLIS and JOZEF DE LEY

3.2.9   The Genera *Phyllobacterium* and *Ochrobactrum*   747
JEAN SWINGS, BART LAMBERT, KAREL KERSTERS and BARRY HOLMES

3.2.10   The Genus *Derxia*   751
JAN HENDRICK BECKING

3.2.11   The Genera *Leptothrix* and *Sphaerotilus*   758
STEFAN SPRING

3.2.12   The Lithoautotrophic Ammonia-Oxidizing Bacteria   778
HANS-PETER KOOPS, ULRIKE PURKHOLD, ANDREAS POMMERENING-RÖSER,
GABRIELE TIMMERMANN and MICHAEL WAGNER

3.2.13   The Genus *Thiobacillus*   812
LESLEY A. ROBERTSON and J. GIJS KUENEN

3.2.14   The Genera *Simonsiella* and *Alysiella*   828
BRIAN P. HEDLUND and DAISY A. KUHN

3.2.15   *Eikenella corrodens* and Closely Related Bacteria   840
EDWARD J. BOTTONE and PAUL A. GRANATO

3.2.16   The Genus *Burkholderia*   848
DONALD E. WOODS and PAMELA A. SOKOL

3.2.17   The Nitrite-Oxidizing Bacteria   861
AHARON ABELIOVICH

3.2.18   The Genera *Azoarcus*, *Azovibrio*, *Azospira* and *Azonexus*   873
BARBARA REINHOLD-HUREK and THOMAS HUREK

Index   893

Volume 6

3.   *Proteobacteria*

3.3.   *Gamma Subclass*

3.3.1   New Members of the Family Enterobacteriaceae   5
J. MICHAEL JANDA

3.3.2   Phylogenetic Relationships of Bacteria with Special Reference to
Endosymbionts and Enteric Species   41
M. PILAR FRANCINO, SCOTT R. SANTOS and HOWARD OCHMAN

3.3.3   The Genus *Escherichia*   60
RODNEY A. WELCH

3.3.4   The Genus *Edwardsiella*   72
SHARON L. ABBOTT and J. MICHAEL JANDA

3.3.5   The Genus *Citrobacter*   90
DIANA BORENSHTEIN and DAVID B. SCHAUER

3.3.6   The Genus *Shigella*   99
YVES GERMANI and PHILIPPE J. SANSONETTI

3.3.7   The Genus *Salmonella*   123
CRAIG D. ELLERMEIER and JAMES M. SLAUCH

3.3.8   The Genus *Klebsiella*   159
SYLVAIN BRISSE, FRANCINE GRIMONT and PATRICK A. D. GRIMONT

3.3.9   The Genus *Enterobacter*   197
FRANCINE GRIMONT and PATRICK A. D. GRIMONT

3.3.10 The Genus *Hafnia*   215
MEGAN E. MCBEE and DAVID B. SCHAUER

3.3.11 The Genus *Serratia*   219
FRANCINE GRIMONT and PATRICK A. D. GRIMONT

3.3.12 The Genera *Proteus*, *Providencia*, and *Morganella*   245
JIM MANOS and ROBERT BELAS

3.3.13 *Y. enterocolitica* and *Y. pseudotuberculosis*   270
ELISABETH CARNIEL, INGO AUTENRIETH, GUY CORNELIS,
HIROSHI FUKUSHIMA, FRANÇOISE GUINET, RALPH ISBERG,
JEANNETTE PHAM, MICHAEL PRENTICE, MICHEL SIMONET,
MIKAEL SKURNIK and GEORGES WAUTERS

3.3.14 *Yersinia pestis* and Bubonic Plague   399
ROBERT BRUBAKER

3.3.15 *Erwinia* and Related Genera   443
CLARENCE I. KADO

3.3.16 The Genera *Photorhabdus* and *Xenorhabdus*   451
NOEL BOEMARE and RAYMOND AKHURST

3.3.17 The Family Vibrionaceae   495
J. J. FARMER, III

3.3.18 The Genera *Vibrio* and *Photobacterium*   508
J. J. FARMER, III and F. W. HICKMAN-BRENNER

3.3.20 The Genera *Aeromonas* and *Plesiomonas*   564
J. J. FARMER, III, M. J. ARDUINO and F. W. HICKMAN-BRENNER

3.3.21 The Genus *Alteromonas* and Related Proteobacteria   597
VALERY V. MIKHAILOV, LYUDMILA A. ROMANENKO and
ELENA P. IVANOVA

3.3.22 Nonmedical: *Pseudomonas*   646
EDWARD R. B. MOORE, BRIAN J. TINDALL, VITOR A. P. MARTINS DOS SANTOS,
DIETMAR H. PIEPER, JUAN-LUIS RAMOS and NORBERTO J. PALLERONI

3.3.23 *Pseudomonas aeruginosa*   704
TIMOTHY L. YAHR and MATTHEW R. PARSEK

3.3.24 Phytopathogenic Pseudomonads and Related
Plant-Associated Pseudomonads   714
MILTON N. SCHROTH, DONALD C. HILDEBRAND and
NICKOLAS PANOPOULOS

3.3.25 *Xylophilus*   741
ANNE WILLEMS and MONIQUE GILLIS

3.3.26   The Genus *Acinetobacter*   746
KEVIN TOWNER

3.3.27   The Family Azobacteriaceae   759
JAN HENDRICK BECKING

3.3.28   The Genera *Beggiatoa* and *Thioploca*   784
ANDREAS TESKE and DOUGLAS C. NELSON

3.3.29   The Family Halomonadaceae   811
DAVID R. ARAHAL and ANTONIO VENTOSA

3.3.30   The Genus *Deleya*   836
KAREL KERSTERS

3.3.31   The Genus *Frateuria*   844
JEAN SWINGS

3.3.32   The Chromatiaceae   846
JOHANNES F. IMHOFF

3.3.33   The Family Ectothiorhodospiraceae   874
JOHANNES F. IMHOFF

3.3.34   *Oceanospirillum* and Related Genera   887
JOSÉ M. GONZÁLEZ and WILLIAM B. WHITMAN

3.3.35   *Serpens flexibilis*: An Unusually Flexible Bacterium   916
ROBERT B. HESPELL

3.3.36   The Genus *Psychrobacter*   920
JOHN P. BOWMAN

3.3.37   The Genus *Leucothrix*   931
THOMAS D. BROCK

3.3.38   The Genus *Lysobacter*   939
HANS REICHENBACH

3.3.39   The Genus *Moraxella*   958
JOHN P. HAYS

3.3.40   *Legionella* Species and Legionnaire's Disease   988
PAUL H. EDELSTEIN and NICHOLAS P. CIANCIOTTO

3.3.41   The Genus *Haemophilus*   1034
DORAN L. FINK and JOSEPH W. ST. GEME, III

3.3.42   The Genus *Pasteurella*   1062
HENRIK CHRISTENSEN and MAGNE BISGAARD

3.3.43   The Genus *Cardiobacterium*   1091
SYDNEY M. HARVEY and JAMES R. GREENWOOD

3.3.44   The Genus *Actinobacillus*   1094
JANET I. MACINNES and EDWARD T. LALLY

3.3.45   The Genus *Francisella*   1119
FRANCIS NANO and KAREN ELKINS

3.3.46   Ecophysiology of the Genus *Shewanella*   1133
KENNETH H. NEALSON and JAMES SCOTT

3.3.47   The Genus *Nevskia*   1152
HERIBERT CYPIONKA, HANS-DIETRICH BABENZIEN,
FRANK OLIVER GLÖCKNER and RUDOLF AMANN

3.3.48   The Genus *Thiomargarita*   1156
HEIDE N. SCHULZ

Index   1164

Volume 7

3.   *Proteobacteria*

3.4   *Delta Subclass*

3.4.1   The Genus *Pelobacter*   5
BERNHARD SCHINK

3.4.2   The Genus *Bdellovibrio*   12
EDOUARD JURKEVITCH

3.4.3   The Myxobacteria   31
LAWRENCE J. SHIMKETS, MARTIN DWORKIN and HANS REICHENBACH

3.5.   *Epsilon Subclass*

3.5.1   The Genus *Campylobacter*   119
TRUDY M. WASSENAAR and DIANE G. NEWELL

3.5.2   The Genus *Helicobacter*   139
JAY V. SOLNICK, JANI L. O'ROURKE, PETER VAN DAMME and ADRIAN LEE

3.5.3   The Genus *Wolinella*   178
JÖRG SIMON, ROLAND GROSS, OLIVER KLIMMEK and ACHIM KRÖGER

4.   *Spirochaetes*

4.1   Free-Living Saccharolytic Spirochetes: The Genus *Spirochaeta*   195
SUSAN LESCHINE, BRUCE J. PASTER and ERCOLE CANALE-PAROLA

4.2   The Genus *Treponema*   211
STEVEN J. NORRIS, BRUCE J. PASTER, ANNETTE MOTER and
ULF B. GÖBEL

4.3   The Genus *Borrelia*   235
MELISSA J. CAIMANO

4.4   The Genus *Leptospira*   294
BEN ADLER and SOLLY FAINE

4.5   Termite Gut Spirochetes   318
JOHN A. BREZNAK and JARED R. LEADBETTER

4.6   The Genus *Brachyspira*   330
THADDEUS B. STANTON

5.   *Chlorobiaceae*

5.1   The Family Chlorobiaceae   359
JÖRG OVERMANN

6.   ***Bacteroides and Cytophaga Group***

6.1   The Medically Important *Bacteroides* spp. in Health and Disease   381
C. JEFFREY SMITH, EDSON R. ROCHA and BRUCE J. PASTER

6.2   The Genus *Porphyromonas*   428
FRANK C. GIBSON and CAROLINE ATTARDO GENCO

6.3   An Introduction to the Family Flavobacteriaceae   455
JEAN-FRANÇOIS BERNARDET and YASUYOSHI NAKAGAWA

6.4   The Genus *Flavobacterium*   481
JEAN-FRANÇOIS BERNARDET and JOHN P. BOWMAN

6.5   The Genera *Bergeyella* and *Weeksella*   532
CELIA J. HUGO, BRITA BRUUN and PIET J. JOOSTE

6.6   The Genera *Flavobacterium*, *Sphingobacterium* and *Weeksella*   539
BARRY HOLMES

6.7   The Order Cytophagales   549
HANS REICHENBACH

6.8   The Genus *Saprospira*   591
HANS REICHENBACH

6.9   The Genus *Haliscomenobacter*   602
EPPE GERKE MULDER and MARIA H. DEINEMA

6.10   *Sphingomonas* and Related Genera   605
DAVID L. BALKWILL, J. K. FREDRICKSON and M. F. ROMINE

6.11   The Genera *Empedobacter* and Myroides   630
CELIA J. HUGO, BRITA BRUUN and PIET J. JOOSTE

6.12   The Genera *Chryseobacterium* and *Elizabethkingia*   638
JEAN-FRANÇOIS BERNARDET, CELIA J. HUGO and BRITA BRUUN

6.13   The Marine Clade of the Family Flavobacteriaceae: The Genera
Aequorivita, Arenibacter, Cellulophaga, Croceibacter, Formosa,
Gelidibacter, Gillisia, Maribacter, Mesonia, Muricauda, Polaribacter,
Psychroflexus, Psychroserpens, Robiginitalea, Salegentibacter,
Tenacibaculum, Ulvibacter, Vitellibacter and Zobellia   677
JOHN P. BOWMAN

6.14   Capnophilic Bird Pathogens in the Family Flavobacteriaceae:
Riemerella, Ornithobacterium and Coenonia   695
PETER VAN DAMME, H. M. HAFEZ and K. H. HINZ

6.15   The Genus *Capnocytophaga*   709
JARED R. LEADBETTER

6.16   The Genera *Rhodothermus*, *Thermonema*, *Hymenobacter*
and *Salinibacter*   712
AHARON OREN

7.   *Chlamydia*

7.1   The Genus *Chlamydia*—Medical   741
MURAT V. KALAYOGLU and GERALD I. BYRNE

8.   *Planctomyces and Related Bacteria*

8.1   The Order Planctomycetales, Including the Genera *Planctomyces*,
Pirellula, *Gemmata and Isosphaera* and the Candidatus Genera
Brocadia, *Kuenenia* and *Scalindua*   757
NAOMI WARD, JAMES T. STALEY, JOHN A. FUERST, STEPHEN GIOVANNONI,
HEINZ SCHLESNER and ERKO STACKEBRANDT

9.   *Thermus*

9.1   The Genus *Thermus* and Relatives   797
MILTON S. DA COSTA, FREDERICK A. RAINEY and M. FERNANDA NOBRE

10.   *Chloroflexaceae and Related Bacteria*

10.1   The Family Chloroflexaceae   815
SATOSHI HANADA and BEVERLY K. PIERSON

10.2   The Genus *Thermoleophilum*   843
JEROME J. PERRY

10.3   The Genus *Thermomicrobium*   849
JEROME J. PERRY

10.4   The Genus *Herpetosiphon*   854
          NATUSCHKA LEE and HANS REICHENBACH

11.   Verrucomicrobium

11.1   The Phylum Verrucomicrobia: A Phylogenetically Heterogeneous
          Bacterial Group   881
          HEINZ SCHLESNER, CHERYL JENKINS and JAMES T. STALEY

12.   Thermotogales

12.1   Thermotogales   899
          ROBERT HUBER and MICHAEL HANNIG

13.   Aquificales

13.1   Aquificales   925
          ROBERT HUBER and WOLFGANG EDER

14.   Phylogenetically Unaffiliated Bacteria

14.1   Morphologically Conspicuous Sulfur-Oxidizing Eubacteria   941
          JAN W. M. LA RIVIÈRE and KARIN SCHMIDT

14.2   The Genus *Propionigenium*   955
          BERNHARD SCHINK

14.3   The Genus *Zoogloea*   960
          PATRICK R. DUGAN, DAPHNE L. STONER and HARVEY M. PICKRUM

14.4   Large Symbiotic Spirochetes: *Clevelandina, Cristispira, Diplocalyx,*
          Hollandina and *Pillotina*   971
          LYNN MARGULIS and GREGORY HINKLE

14.5   *Streptobacillus moniliformis*   983
          JAMES R. GREENWOOD and SYDNEY M. HARVEY

14.6   The Genus *Toxothrix*   986
          PETER HIRSCH

14.7   The Genus *Gallionella*   990
          HANS H. HANERT

14.8   The Genera *Caulococcus* and *Kusnezovia*   996
          JEAN M. SCHMIDT and GEORGI A. ZAVARZIN

14.9   The Genus *Brachyarcus*   998
          PETER HIRSCH

14.10   The Genus *Pelosigma*   1001
PETER HIRSCH

14.11   The Genus *Siderocapsa* (and Other Iron- and Maganese-Oxidizing
Eubacteria)   1005
HANS H. HANERT

14.12   The Genus *Fusobacterium*   1016
TOR HOFSTAD

14.13   Prokaryotic Symbionts of Amoebae and Flagellates   1028
KWANG W. JEON

Index   1039

Contributors

Sharon L. Abbott
Microbial Diseases Laboratory
Berkeley, CA 94704
USA

Aharon Abeliovich
Department of Biotechnology Engineering
Institute for Applied Biological Research
Environmental Biotechnology Institute
Ben Gurion University
84105 Beer-Sheva
Israel

Soman N. Abraham
Director of Graduate Studies in Pathology
Departments of Pathology, Molecular Genetics
   and Microbiology, and Immunology
Duke University Medical Center
Durham, NC 27710
USA

David G. Adams
School of Biochemistry and Microbiology
University of Leeds
Leeds LS2 9JT
UK

Ben Adler
Monash University
Faculty of Medicine, Nursing and Health
   Sciences
Department of Microbiology
Clayton Campus
Victoria, 3800
Australia

Raymond Akhurst
CSIRO Entomology
Black Mountain
ACT 2601 Canberra
Australia

Rudolf Amann
Max Planck Institute for Marine Microbiology
D-28359 Bremen
Germany

Burt E. Anderson
Department of Medical Microbiology and
   Immunology
College of Medicine
University of South Florida
Tampa, FL 33612
USA

Robert E. Andrews
Department of Microbiology
University of Iowa
Iowa City, IA 52242
USA

Garabed Antranikian
Technical University Hamburg-Harburg
Institute of Technical Microbiology
D-21073 Hamburg
Germany

David R. Arahal
Colección Española de Cultivos Tipo (CECT)
Universidad de Valencia
Edificio de Investigación
46100 Burjassot (Valencia)
Spain

M. J. Arduino
Center for Infectious Diseases
Centers for Disease Control
Atlanta, GA 30333
USA

Judith Armitage
Department of Biochemistry
Microbiology Unit
University of Oxford
OX1 3QU Oxford
UK

Ingo Autenrieth
Institut für Medizinische Mikrobiologie
Universitatsklinikum Tuebingen
D-72076 Tuebingen
Germany

Hans-Dietrich Babenzien
Leibniz-Institut für Gewässerökologie und
  Binnenfischereiim Forschungsverbund
  Berlin
12587 Berlin
Germany

Werner Back
Lehrstuhl für Technologie der Brauerei I
Technische Universität München
D-85354 Freising-Weihenstephan
Germany

Margo Baele
Department of Pathology
Bacteriology and Poultry Diseases
Faculty of Veterinary Medicine
Ghent University
B-9820 Merelbeke
Belgium

Jose Ivo Baldani
EMBRAPA-Agrobiology
Centro Nacional de Pesquisa de Agrobiologia
Seropedica, 23851-970
CP 74505 Rio de Janeiro
Brazil

David L. Balkwill
Department of Biomedical Sciences
College of Medicine
Florida State University
Tallahassee, FL 32306-4300
USA

Horia Banciu
Department of Biotechnology
Delft University of Technology
2628 BC Delft

Tammy Bannerman
School of Allied Medical Professions
Division of Medical Technology
The Ohio State University
Columbus, OH 43210
USA

Bonnie L. Bassler
Department of Molecular Biology
Princeton University
Princeton, NJ 08544-1014
USA

Linda Baumann
School of Nursing
Clinical Science Center
University of Wisconsin
Madison, WI 53792-2455
USA

Paul Baumann
Department of Microbiology
University of California, Davis
Davis, CA 95616-5224
USA

Edward A. Bayer
Department of Biological Chemistry
Weizmann Institute of Science
Rehovot 76100
Israel

Dennis A. Bazylinski
Department of Microbiology, Immunology and
  Preventive Medicine
Iowa State University
Ames, IA 55001
USA

Jan Hendrick Becking
Stichting ITAL
Research Institute of the Ministry of
  Agriculture and Fisheries
6700 AA Wageningen
The Netherlands

Robert Belas
The University of Maryland Biotechnology
  Institute
Center of Marine Biotechnology
Baltimore, MD 21202
USA

Birgitta Bergman
Department of Botany
Stockholm University
SE-106 91 Stockholm
Sweden

Kathryn Bernard
Special Bacteriology Section
National Microbiology Laboratory
Health Canada
Winnipeg R3E 3R2
Canada

Jean-François Bernardet
Unité de Virologie et Immunologie
  Moléculaires
Institut National de la Recherche
  Agronomique (INRA)
Domaine de Vilvert
78352 Jouy-en-Josas cedex
France

Costanzo Bertoldo
Technical University Hamburg-Harburg
Institute of Technical Microbiology
D-21073 Hamburg
Germany

Bruno Biavati
Istituto di Microbiologia Agraria
40126 Bologna
Italy

Magne Bisgaard
Department of Veterinary Microbiology
Royal Veterinary and Agricultural University
1870 Frederiksberg C
Denmark

William Bishai
Departments of Molecular Microbiology and
    Immunology, International Health, and
    Medicine
Center for Tuberculosis Research
Johns Hopkins School of Hygiene and Public
    Health
Baltimore, MD 21205-2105
USA

Johanna Björkroth
Department of Food and Environmental
    Hygiene
Faculty of Veterinary Medicine
University of Helsinki
FIN-00014 Helsinki
Finland

Eberhard Bock
Institute of General Botany
Department of Microbiology
University of Hamburg
D-22609 Hamburg
Germany

Noel Boemare
Ecologie Microbienne des Insectes et
Interactions Hôte-Pathogène
UMR EMIP INRA-UMII
IFR56 Biologie cellulaire et Porcessus
    infectieux
Université Montpellier II
34095 Montpellier
France

Antje Boetius
Max-Planck-Institut für Marine Mikrobiologie
D-28359 Bremen
Germany

Adam S. Bonin
Portland State University
Portland OR 97207
USA

David R. Boone
Department of Biology
Environmental Science and Engineering
Oregon Graduate Institute of Science and
    Technology
Portland State University
Portland, OR 97207-0751
USA

Diana Borenshtein
Massachusetts Institute of Technology
Cambridge, MA 02139-4307
USA

Edward J. Bottone
Division of Infectious Diseases
The Mount Sinai Hospital
One Gustave L. Levy Place
New York, NY 10029
USA

Timothy L. Bowen
Department of Microbiology
University of Georgia
Athens, GA 30602
USA

John P. Bowman
Australian Food Safety Centre for Excellence
School of Agricultural Science
Hobart, Tasmania, 7001
Australia

John A. Breznak
Department of Microbiology and Molecular
    Genetics
Michigan State University
East Lansing, MI 48824-1101
USA

Stephanie Bringer-Meyer
Institut Biotechnologie
Forschungszentrum Jülich
D-52425 Jülich
Germany

Sylvain Brisse
Unité Biodiversité des Bactéries Pathogènes
    Emergentes
U 389 INSERM
Institut Pasteur
75724 Paris
France

Thomas D. Brock
Department of Bacteriology
University of Wisconsin-Madison
Madison, WI 53706
USA

Robert Brubaker
Department of Microbiology
Michigan State University
East Lansing, MI 48824
USA

Andreas Brune
Max Planck Institute for Terrestrial
   Microbiology
Marburg
Germany

Brita Bruun
Department of Clinical Microbiology
Hillerød Hospital
DK 3400 Hillerød
Denmark

Carmen Buchrieser
Laboratoire de Génomique des
   Microorganismes Pathogènes
Institut Pasteur
75724 Paris
France

Hans-Jürgen Busse
Institut für Bakteriology, Mykologie, und
   Hygiene
Veterinärmedizinische Universität Wien
A-1210 Vienna
Austria

Patrick Butaye
CODA-CERVA-VAR
1180 Brussels
Belgium

Gerald I. Byrne
Department of Medical Microbiology and
   Immunology
University of Wisconsin—Madison
Madison, WI 53706
USA

Didier Cabanes
Department of Immunology and Biology of
   Infection
Molecular Microbiology Group
Institute for Molecular and Cellular Biology
4150-180 Porto
Portugal

Melissa Caimano
Center for Microbial Pathogenesis
and
Department of Pathology
and
Department of Genetics and Development
University of Connecticut Health Center
Farmington, CT 06030-3205
USA

Ercole Canale-Parola
Department of Microbiology
University of Massachusetts
Amherst, MA 01003
USA

Elisabeth Carniel
Laboratoire des *Yersinia*
Institut Pasteur
75724 Paris
France

Colleen M. Cavanaugh
Bio Labs
Harvard University
Cambridge, MA 02138
USA

Jiann-Shin Chen
Department of Biochemistry
Virginia Polytechnic Institute and
   State University—Virginia Tech
Blacksburg, VA 24061-0308
USA

Zhongying Chen
Department of Biology
University of North Carolina
Chapel Hill, NC 27514
USA

Qi Cheng
University of Western Sydney
Penrith South
NSW 1797
Australia

Henrik Christensen
Department of Veterinary Microbiology
Royal Veterinary and Agricultural University
Denmark

Nicholas P. Cianciotto
Department of Microbiology and Immunology
Northwestern University School of Medicine
Chicago, IL
USA

Dieter Claus
Deutsche Sammlung von Mikroorganismen
D-3300 Braunschweig-Stockheim
Germany

P. Patrick Cleary
Department of Microbiology
University of Minnesota Medical School
Minneapolis, MN 55455
USA

Yehuda Cohen
Department of Molecular and Microbial
  Ecology
Institute of Life Science
Hebrew University of Jerusalem
91904 Jerusalem
Israel

Matthew D. Collins
Institute of Food Research
Reading Lab, Early Gate
UK

Guy Cornelis
Microbial Pathogenesis Unit
Université Catholique de Louvain and
Christian de Duve Institute of Cellular
  Pathology
B1200 Brussels
Belgium

Pascale Cossart
Unité des Interactions Bactéries-Cellules
INSERM U604
Institut Pasteur
75724 Paris
France

Michael Cotta
USDA-ARS North Regional Research
  Center
Peoria, IL 61604-3902
USA

Ronald L. Crawford
Food Research Center
University of Idaho
Moscow, ID 83844-1052
USA

Cecil S. Cummins
Department of Anaerobic Microbiology
Virginia Polytechnic Institute and State
  University
Blacksburg, VA 24061
USA

Heribert Cypionka
Institut für Chemie und Biologie des Meeres
Fakultät 5, Mathematik und
  Naturwissenschaften
Universität Oldenburg
D-26111 Oldenburg
Germany

Milton S. da Costa
M. Fernanda Nobre
Centro de Neurociências e Biologia Celular
Departamento de Zoologia
Universidade de Coimbra
3004-517 Coimbra
Portugal

Rolf Daniel
Department of General Microbiology
Institute of Microbiology and Genetics
37077 Göttingen
Germany

Seana Davidson
University of Washington
Civil and Environmental Engineering
Seattle, WA 98195-2700
USA

Scott C. Dawson
Department of Molecular and Cellular
  Biology
University of California-Berkeley
Berkeley, CA 94720
USA

Dirk de Beer
Max-Planck-Institute for Marine Microbiology
D-28359 Bremen
Germany

Jan A.M. de Bont
Department of Food Science
Agricultural University
6700 EV Wageningen
The Netherlands

Maria H. Deinema
Laboratory of Microbiology
Agricultural University
6703 CT Wageningen
The Netherlands

Jozef de Ley
Laboratorium voor Microbiologie en
  Microbiële Genetica
Rijksuniversiteit Ghent
B-9000 Ghent
Belgium

Edward F. DeLong
Science Chair
Monterey Bay Aquarium Research Institute
Moss Landing, CA 95039
USA

Arnold L. Demain
Department of Biology
Massachusetts Institute of Technology
Cambridge, MA 02139
USA

Uwe Deppenmeier
Department of Biological Sciences
University of Wisconsin
Milwaukee, WI 53202
USA

Paul de Vos
Department of Biochemistry, Physiology and
  Microbiology
Universiteit Gent
B-9000 Gent
Belgium

Luc Devriese
Faculty of Veterinary Medicine
B982 Merelbeke
Belgium

Floyd E. Dewhirst
Forsyth Dental Center
140 Fenway
Boston, MA 02115
USA

Leon M. T. Dicks
Department of Microbiology
University of Stellenbosch
ZA-7600 Stellenbosch
South Africa

Michael P. Doyle
College of Agricultural and Environmental
  Sciences
Center for Food Safety and Quality
  Enhancement
University of Georgia
Griffin, GA 30223-1797
USA

Harold L. Drake
Department of Ecological Microbiology
BITOEK, University of Bayreuth
D-95440 Bayreuth
Germany

Patrick R. Dugan
Idaho National Engineering Laboratory
EG & G Idaho
Idaho Falls, ID 83415
USA

Paul V. Dunlap
Department of Molecular
Cellular and Developmental Biology
University of Michigan
Ann Arbor, MI 48109-1048
USA

Olivier Dussurget
Unité des Interactions Bactéries-Cellules
INSERM U604
Institut Pasteur
75724 Paris
France

Martin Dworkin
University of Minnesota Medical School
Department of Microbiology
Minneapolis, MN 55455
USA

Jürgen Eberspächer
Institut fur Mikrobiologie
Universitat Hohenheim
D-7000 Stuttgart 70
Germany

Paul H. Edelstein
Department of Pathology and Laboratory
  Medicine
University of Pennsylvania Medical
  Center
Philadelphia, PA 19104-4283
USA

Wolfgang Eder
Lehrstuhl für Mikrobiologie
Universität Regensburg
93053 Regensburg
Germany

Karen Elkins
CBER/FDA
Rockville, MD 20852
USA

Craig D. Ellermeier
Department of Microbiology
University of Illinois
Urbana, IL 61801
and
Department of Molecular and Cellular
  Biology
Harvard University
Cambridge, MA 02138
USA

Lyudmila I. Evtushenko
All-Russian Collection of Microorganisms
Institute of Biochemistry and Physiology of the
  Russian, Academy of Sciences
Puschino
Moscow Region, 142290
Russia

Takayuki Ezaki
Bacterial Department
Gifu University Medical School
40 Tsukasa
Machi Gifu City
Japan

Solly Faine
Monash University
Faculty of Medicine, Nursing and Health
  Sciences
Department of Microbiology
Clayton Campus
Victoria, 3800
Australia

J. J. Farmer, III
Center for Infectious Diseases
Centers for Disease Control
Atlanta, GA 30333
USA

W. Edmund Farrar
Department of Medicine
Medical University of South Carolina
Charleston, SC 29425
USA

Mariano E. Fernandez Miyakawa
California Animal Health and Food Safety
  Laboratory
University of California, Davis
San Bernardino, CA 92408
USA

Doran L. Fink
Edward Mallinckrodt Department of Pediatrics
  and Department of Molecular Microbiology
Washington University School of Medicine
St. Louis, Missouri 63110
USA

Jacqueline Fletcher
Department of Entomology and Plant
  Pathology
Oklahoma State University
Stillwater, OK
USA

Robert Forster
Bio-Products and Bio-Processes Program
Agriculture and Agri-Food Canada
Lethbridge Research Centre
Lethbridge T1J 4B1
Canada

M. Pilar Francino
Evolutionary Genomics Department
DOE Joint Genome Institute
Walnut Creek, CA 94598
USA

Charles M. A. P. Franz
Institute of Hygiene and Toxicology
BFEL
D-76131 Karlsruhe
Germany

David N. Fredricks
VA Palo Alto Healthcare System
Palo Alto, CA 94304
USA

J. K. Fredrickson
Pacific Northwest National Laboratory
Richland, Washington 99352
USA

Bärbel Friedrich
Institut für Biologie/Mikrobiologie
Homboldt-Universität zu Berlin
Chaussesstr. 117
D-10115 Berlin
Germany

Dagmar Fritze
Deutsche Sammlung von Mikroorganismen
D-3300 Braunschweig-Stockheim
Germany

John A. Fuerst
Department of Microbiology and
  Parasitology
University of Queensland
Brisbane
Queensland 4072
Australia

Hiroshi Fukushima
Public Health Institute of Shimane
  Prefecture
582-1 Nishihamasada, Matsue
Shimane 690-0122
Japan

Jean-Louis Garcia
Laboratoire ORSTOM de Microbiologie des
    Anaérobies
Université de Provence
CESB-ESIL
13288 Marseille
France

Ferran Garcia-Pichel
Associate Professor
Arizona State University
Tempe, AZ 85281
USA

Arnold Geis
Institut für Mikrobiologie
Bundesanstalt für Milchforschung
D-24121 Kiel
Germany

Caroline Attardo Genco
Department of Medicine
Section of Infectious Diseases
    and Department of Microbiology
Boston University School of Medicine
Boston, MA 02118
USA

Yves Germani
Institut Pasteur
Unité Pathogénie Microbienne Moléculaire
and
Réseau International des Instituts Pasteur
Paris 15
France

Frank C. Gibson
Department of Medicine
Section of Infectious Diseases
and '
Department of Microbiology
Boston University School of Medicine
Boston, MA 02118
USA

Monique Gillis
Laboratorium voor Mikrobiologie
Universiteit Gent
B-9000 Gent
Belgium

Stephen Giovannoni
Department of Microbiology
Oregon State University
Corvallis, OR 97331
USA

Frank Oliver Glöckner
Max-Planck-Institut für Marine Mikrobiologie
D-28359 Bremen
Germany

Ulf B. Göbel
Institut für Mikrobiologie und Hygiene
Universitaetsklinikum Chariteacute
Humboldt-Universitaet zu Berlin
D-10117 Berlin
Germany

José M. González
Department de Microbiologia y Biologia
    Celular
Facultad de Farmacia
Universidad de La Laguna
38071 La Laguna, Tenerife
SPAIN

Michael Goodfellow
School of Biology
Universtiy of Newcastle
Newcastle upon Tyre NE1 7RU
UK

Friedrich Götz
Facultät für Biologie
Institut für Microbielle Genetik
Universität Tübingen
D-72076 Tübingen
Germany

Hans-Dieter Görtz
Department of Zoology
Biologisches Institut
Universität Stuttgart
D-70569 Stuttgart
Germany

Gerhard Gottschalk
Institut für Mikrobiologie und Genetik
Georg-August-Universität Göttingen
D-37077 Göttingen
Germany

P. H. Graham
Department of Soil, Water, and Climate
St. Paul, MN 55108
USA

Paul A. Granato
Department of Microbiology and Immunology
State University of New York Upstate Medical
    University
Syracuse, NY 13210
USA

Peter N. Green
NCIMB Ltd
AB24 3RY Aberdeen
UK

James R. Greenwood
Bio-Diagnostics Laboratories
Torrance, CA 90503
USA

Francine Grimont
Unite 199 INSERM
Institut Pasteur
75724 Paris
France

Patrick A. D. Grimont
Institut Pasteur
75724 Paris
France

Roland Gross
Institut für Mikrobiologie
Johann Wolfgang Goethe-Universität
Frankfurt am Main
Germany

Ji-Dong Gu
Laboratory of Environmental Toxicology
Department of Ecology & Biodiversity
and
The Swire Institute of Marine Science
University of Hong Kong
Hong Kong SAR
P.R. China
and
Environmental and Molecular Microbiology
South China Sea Institute of Oceanography
Chinese Academy of Sciences
Guangzhou 510301
P.R. China

Françoise Guinet
Laboratoire des *Yersinia*
Institut Pasteur
75724 Paris
France

Michael Gurevitz
Department of Botany
Life Sciences Institute
Tel Aviv University
Ramat Aviv 69978
Israel

H. M. Hafez
Institute of Poultry Diseases
Free University Berlin
Berlin
German

Auli Haikara
VTT Biotechnology
Tietotie 2, Espoo
Finland

Walter P. Hammes
Institute of Food Technology
Universität Hohenheim
D-70599 Stuttgart
Germany

Satoshi Hanada
Research Institute of Biological Resources
National Institute of Advanced Industrial
    Science and Technology (AIST)
Tsukuba 305-8566
Japan

Hans H. Hanert
Institut für Mikrobiologie
Technische Univeristät Braunschweig
D-3300 Braunschweig
Germany

Michael Hannig
Lehrstuhl für Mikrobiologie
Universität Regensburg
D-93053 Regensburg
Germany

Theo A. Hansen
Microbial Physiology (MICFYS)
Groningen University
Rijksuniversiteit Groningen
NL-9700 AB Groningen
The Netherlands

Jeremy M. Hardie
Department of Oral Microbiology
School of Medicine & Dentistry
London E1 2AD
UK

Timothy Harrah
Bioengineering Center
Tufts University
Medford, MA 02155
USA

Anton Hartmann
GSF-National Research Center for
    Environment and Health
Institute of Soil Ecology
Rhizosphere Biology Division
D-85764 Neuherberg/Muenchen
Germany

Sybe Hartmans
Department of Food Science
Agricultural University Wageningen
6700 EV Wageningen
The Netherlands

Patricia Hartzell
Department of Microbiology, Molecular
  Biology, and Biochemistry
University of Idaho
Moscow, ID 83844-3052
USA

Sydney M. Harvey
Nichols Institute Reference Laboratories
32961 Calle Perfecto
San Juan Capistrano, CA 92675
USA

John P. Hays
Department of Medical Microbiology and
  Infectious Diseases
Erasmus MC
3015 GD Rotterdam
The Netherlands

Reiner Hedderich
Max Planck Institute für Terrestriche
  Mikrobiologie
D-35043 Marburg
Germany

Brian P. Hedlund
Department of Biological Sciences
University of Nevada, Las Vegas
Las Vegas, NV 89154-4004
USA

Robert A. Heinzen
Department of Molecular Biology
University of Wyoming
Laramie, WY 82071-3944
USA

Ilkka Helander
VTT Biotechnology
Tietotie 2, Espoo
Finland

H. Ernest Hemphill
Department of Biology
Syracuse University
Syracuse, NY 13244
USA

Christian Hertel
Institute of Food Technology
Universität Hohenheim
D-70599 Stuttgart
Germany

Robert B. Hespell
Northern Regional Research Center, ARS
US Department of Agriculture
Peoria, IL 61604
USA

F. W. Hickman-Brenner
Center for Infectious Diseases
Centers for Disease Control
Atlanta, GA 30333
USA

Donald C. Hildebrand
Department of Plant Pathology
University of California-Berkeley
Berkeley, CA 94720
USA

Gregory Hinkle
Department of Botany
University of Massachusetts
Amherst, MA 01003
USA

K. H. Hinz
Clinic for Poultry
School of Veterinary Medicine
D-30559 Hannover
Germany

Peter Hirsch
Institut für Allgemeine Mikrobiologie
Universität Kiel
D-2300 Kiel
Germany

Tor Hofstad
Department of Microbiology and
  Immunology
University of Bergen
N-5021 Bergen
Norway

Michael J. Hohn
Lehrstuhl für Mikrobiologie
Universität Regensburg
D-93053 Regensburg
Germany

Barry Holmes
Central Public Health Laboratory
National Collection of Type Cultures
London NW9 5HT
UK

Wilhelm H. Holzapfel
Federal Research Centre of Nutrition
Institute of Hygiene and Toxicology
D-76131 Karlsruhe
Germany

Harald Huber
Lehrstuhl für Mikrobiologie
Universität Regensburg
D-93053 Regensburg
Germany

Robert Huber
Lehrstuhl für Mikrobiologie
Universität Regensburg
D-93053 Regensburg
Germany

Celia J. Hugo
Department of Microbial, Biochemical and
   Food Biotechnology
University of the Free State
Bloemfontein
South Africa

Meredith Hullar
University of Washington
Seattle, WA
USA

Thomas Hurek
Laboratory of General Microbiology
University Bremen
28334 Bremen
Germany

Johannes F. Imhoff
Marine Mikrobiologie
Institut für Meereskunde an der Universität
   Kiel
D-24105 Kiel
Germany

Ralph Isberg
Department of Molecular Biology and
   Microbiology
Tufts University School of Medicine
Boston, MA 02111
USA

Elena P. Ivanova
Senior Researcher in Biology
Laboratory of Microbiology
Pacific Institute of Bioorganic Chemistry of the
   Far-Eastern Branch of the Russian Academy
   of Sciences
690022 Vladivostok
Russia

Rainer Jaenicke
6885824 Schwalbach a. Ts.
Germany
and
Institut für Biophysik und Physikalische
   Biochemie
Universität Regensburg
Regensburg
Germany
and
School of Crystallography
Birbeck College
University of London
London, UK

J. Michael Janda
Microbial Diseases Laboratory
Division of Communicable Disease Control
California Department of Health Services
Berkeley, CA 94704-1011
USA

Holger W. Jannasch
Woods Hole Oceanographic Institution
Woods Hole, MA 02543
USA

Christian Jeanthon
UMR CNRS 6539–LEMAR
Institut Universitaire Europeen de la Mer
Technopole Brest Iroise
29280 Plouzane
France

Cheryl Jenkins
Department of Microbiology
University of Washington
Seattle, WA 98195
USA

John L. Johnson
Department of Anaerobic Microbiology
Virginia Polytechnic Institute and State
   University
Blacksburg, VA 24061
USA

Dorothy Jones
Department of Microbiology
University of Leicester, School of Medicine
Lancaster LE1 9HN
UK

Piet J. Jooste
Department of Biotechnology and Food
    Technology
Tshwane University of Technology
Pretoria 0001
South Africa

Edouard Jurkevitch
Department of Plant Pathology and
    Microbiology
Faculty of Agriculture
Food & Environmental Quality Services
The Hebrew University
76100 Rehovot
Israel

Clarence I. Kado
Department of Plant Pathology
University of California, Davis
Davis, CA 95616-5224
USA

Dale Kaiser
Department of Biochemistry
Stanford University School of Medicine
Stanford, CA 94305-5329
USA

Murat V. Kalayoglu
Department of Medical Microbiology and
    Immunology
University of Wisconsin—Madison
Madison, WI 53706
USA

Peter Kämpfer
Institut für Angewandte Mikrobiologie
Justus Liebig-Universität
D-35392 Gießen
Germany

David Kaplan
Department of Chemcial and Biological
    Engineering
Tufts University
Medford, MA 02115
USA

Yoshiaki Kawamura
Department of Microbiology
Regeneration and Advanced Medical
    Science
Gifu University Graduate School of
    Medicine
Gifu 501-1194
Japan

Ronald M. Keddie
Craigdhu
Fortrose
Ross-shire IV 10 8SS
UK

Donovan P. Kelly
University of Warwick
Department of Biological Sciences
CV4 7AL Coventry
UK

Melissa M. Kendall
Department of Biology
Portland State University
Portland, OR 97207-0751
USA

Karel Kersters
Laboratorium voor Mikrobiologie
Department of Biochemistry
Physiology and Microbiology
Universiteit Gent
B-9000 Gent
Belgium

Nadia Khelef
Unité des Interactions Bactéries-Cellules
INSERM U604
Institut Pasteur
75724 Paris
France

Kumiko Kita-Tsukamoto
Ocean Research Institute
University of Tokyo
Tokyo 164
Japan

Oliver Klimmek
Johann Wolfgang Goethe-Universität
    Frankfurt
Institut für Mikrobiologie
D-60439 Frankfurt
Germany

Wesley E. Kloos
Department of Genetics
North Carolina State University
Raleigh, NC 27695-7614
USA

Miloslav Kocur
Czechoslovak Collection of Microorganisms
J.E. Purkyně University
662 43 Brno
Czechoslovakia

Paul Kolenbrander
National Institute of Dental Research
National Institute of Health
Bethesda, MD 20892-4350
USA

Kazuo Komagata
Laboratory of General and Applied
  Microbiology
Department of Applied Biology and
  Chemistry
Faculty of Applied Bioscience
Tokyo University of Agriculture
Tokyo, Japan

Hans-Peter Koops
Institut für Allgemeine Botanik
Abteilung Mikrobiologie
Universität Hamburg
D-22069 Hamburg
Germany

Noel R. Krieg
Department of Biology
Virginia Polytechnic Institute
Blacksburg, VA 24061-0406
USA

Achim Kröger
Institut für Mikrobiologie
Biozentrum Niederursel
D-60439 Frankfurt/Main
Germany

Reiner Michael Kroppenstedt
Deutsche Sammlung von Mikroorganismen
  und Zellkulturen
D-3300 Braunschweig
Germany

Terry Ann Krulwich
Department of Biochemistry
Mount Sinai School of Medicine
New York, NY 10029
USA

J. Gijs Kuenen
Department of Biotechnology
Delft University of Technology
2628BC Delft
The Netherlands

Daisy A. Kuhn
Department of Biology
California State University
Northridge, CA 91330
USA

Hidehiko Kumagai
Division of Applied Sciences
Graduate School of Agriculture
Kyoto University
Kitashirakawa
606 8502 Kyoto
Japan

Barbara N. Kunkel
Department of Biology
Washington University
St. Louis, MO 63130
USA

Kirsten Küsel
Department of Ecological Microbiology
BITOEK, University of Bayreuth
D-95440 Bayreuth
Germany

David P. Labeda
Microbial Genomics and Bioprocessing
  Research Unit
National Center for Agricultural Utilization
  Research
Agricultural Research Service
U.S. Department of Agriculture
Peoria, IL 61604
USA

Edward T. Lally
Leon Levy Research Center for Oral Biology
University of Pennsylvania
Philadelphia, Pennsylvania, 19104-6002
USA

Bart Lambert
Plant Genetic Systems N.V.
J. Plateaustraat 22
B-9000 Ghent
Belgium

Raphael Lamed
Department of Molecular Microbiology and
  Biotechnology
George S. Wise Faculty of Life Sciences
Tel Aviv University
Ramat Aviv 69978
Israel

Giancarlo Lancini
Consultant, Vicuron Pharmaceutical
21040 Gerenzano (Varese)
Italy

Jan W. M. la Rivière
Institut für Mikrobiologie
Universität Göttingen
D-3400 Göttingen
Germany

Jared R. Leadbetter
Environmental Science and Engineering
California Institute of Technology
Pasadena, CA 91125-7800
USA

Donald J. LeBlanc
ID Genomics
Pharmacia Corporation
Kalamazoo, MI 49001
USA

Marc Lecuit
Unité des Interactions Bactéries-Cellules
INSERM U604
Institut Pasteur
75724 Paris
France

Adrian Lee
School of Microbiology & Immunology
University of New South Wales
Sydney, New South Wales
2052 Australia

Natuschka Lee
Lehrstuhl für Mikrobiologie
Technische Universität München
D-85350 Freising
Germany

Susan Leschine
Department of Microbiology
University of Massachusetts
Amherst, MA 01003-5720
USA

Na (Michael) Li
Division of Biostatistics
School of Public Health
University of Minnesota
Minneapolis, MN 55455
USA

Mary E. Lidstrom
Department of Chemical Engineering
University of Washington
Seattle, WA 98195
USA

Wolfgang Liebl
Institut für Mikrobiologie und Genetik
Georg-August-Universität
D-37077 Göttingen
Germany

Franz Lingens
Institut fur Mikrobiologie
Universitat Hohenheim
D-7000 Stuttgart 70
Germany

Puspita Lisdiyanti
Laboratory of General and Applied
  Microbiology
Department of Applied Biology and
  Chemistry
Faculty of Applied Bioscience
Tokyo University of Agriculture
Tokyo, Japan

Derek Lovley
Department of Microbiology
University of Massachusetts
Amherst, MA 01003
USA

Wolfgang Ludwig
Lehrstuhl für Mikrobiologie
Technische Universität München
D-85350 Freising
Germany

David Lyerly
TechLab, Inc.
Corporate Research Center
Blacksburg VA 24060-6364
USA

Janet I. Macinnes
University of Guelph
Guelph N1G 2W1
Canada

Michael T. Madigan
Department of Microbiology
Mailcode 6508
Southern Illinois University
Carbondale, IL 62901-4399
USA

Luis Angel Maldonado
School of Biology
Universidad Nacional Autonoma de Mexico
  (UNAM)
Instituto de Ciencias del Mar y Limnologia
Ciudad Universitaria CP
04510 Mexico DF
Mexico

Jim Manos
The University of Maryland Biotechnology
  Institute
Center of Marine Biotechnology
Baltimore, MD 21202

Lynn Margulis
Department of Botany
University of Massachusetts
Amherst, MA 01003
USA

Kevin C. Marshall
School of Microbiology
University of New South Wales
Kensington
New South Wales 2033
Australia

Esperanza Martinez-Romero
Centro de Investigacion sobre Fijacion de
  Nitrogeno
Cuernavaca Mor
Mexico

Vitor A. P. Martins dos Santos
Gesellschaft für Biotechnologische Forschung
Division of Microbiology
Braunschweig D-38124
Germany

Vega Masignani
IRIS, Chiron SpA
53100 Siena
Italy

Paola Mattarelli
Istituto di Microbiologia Agraria
40126 Bologna
Italy

Carola Matthies
Department of Ecological Microbiology
BITOEK, University of Bayreuth
D-95440 Bayreuth
Germany

Ann G. Matthysse
Department of Biology
University of North Carolina
Chapel Hill, NC 27599
USA

Megan E. McBee
Biological Engineering Division
Massachusetts Institute of Technology
Cambridge, MA
USA

Bruce A. McClane
Department of Molecular Genetics and
  Biochemistry
University of Pittsburgh School of Medicine
Pittsburgh, PA 15261
USA

Zoe P. McKiness
Department of Organic and Evolutionary
  Biology
Harvard University
Cambridge, MA 02138
USA

Ulrich Melcher
Department of Biochemistry and Molecular
  Biology
Oklahoma State University
Stillwater, OK
USA

Jianghong Meng
Nutrition and Food Science
University of Maryland
College Park, MD 20742-7521
USA

Valery V. Mikhailov
Pacific Institute of Bioorganic Chemistry
Far-Eastern Branch of the Russian Academy of
  Sciences
690022 Vladivostok
Russia

Melissa B. Miller, Ph.D.
Department of Pathology and Laboratory
  Medicine
University of North Carolina at Chapel Hill
Chapel Hill, NC 27599
USA

Michael F. Minnick
Division of Biological Sciences
University of Montana
Missoula, MT 59812-4824
USA

Ralph Mitchell
Laboratory of Microbial Ecology
Division of Engineering and Applied
  Sciences
Harvard University
Cambridge, MA 02138
USA

Cesare Montecucco
Professor of General Pathology
Venetian Institute for Molecular Medicine
35129 Padova
Italy

Edward R. B. Moore
The Macaulay Institute
Environmental Sciences Group
Aberdeen AB158QH
UK
and
Culture Collection University of Göteborg
  (CCUG)
Department of Clinical Bacteriology
University of Göteborg
Göteborg SE-416 43
Sweden

Nancy A. Moran
University of Arizona
Department of Ecology and Evolutionary
  Biology
Tucson, AZ 85721
USA

Edgardo Moreno
Tropical Disease Research Program
  (PIET)
Veterinary School, Universidad Nacional
Costa Rica

Ignacio Moriyón
Department of Microbiology
University of Navarra
32080 Pamplona
Spain

Annette Moter
Institut für Mikrobiologie und Hygiene
Universitaetsklinikum Chariteacute
Humboldt-Universität zu Berlin
D-10117 Berlin
Germany

Eppe Gerke Mulder
Laboratory of Microbiology
Agricultural University
6703 CT Wageningen
The Netherlands

Yasuyoshi Nakagawa
Biological Resource Center (NBRC)
Department of Biotechnology
National Institute of Technology and
  Evaluation
Chiba 292-0818
Japan

Francis Nano
Department of Biochemistry & Microbiology
University of Victoria
Victoria V8W 3PG
Canada

Kenneth H. Nealson
Department of Earth Sciences
University of Southern California
Los Angeles, CA 90033
USA

Douglas C. Nelson
Department of Microbiology
University of California, Davis
Davis, CA 95616
USA

Klaus Neuhaus
Department of Pediatrics, Infection, Immunity,
  and Infectious Diseases Unit
Washington University School of Medicine
St. Louis, MO 63110
USA

Diane G. Newell
Veterinary Laboratory Agency (Weybridge)
Addlestone
New Haw
Surrey KT1 53NB
UK

Irene L. G. Newton
Department of Organismic and Evolutionary
  Biology
Harvard University
Cambridge, MA 02138
USA

S.A. Nierzwicki-Bauer
Department of Biology
Rensselaer Polytechnic Institute
Troy, NY
USA

M. Fernanda Nobre
Departamento de Zoologia
Universidade de Coimbra
3004-517 Coimbra
Portugal

Philippe Normand
Laboratoire d'Ecologie Microbienne
UMR CNRS 5557
Université Claude-Bernard Lyon 1
69622 Villeurbanne
France

Steven J. Norris
Department of Pathology and Laboratory
  Medicine and Microbiology and Molecular
  Genetics
University of Texas Medical Scvhool at
  Houston
Houston, TX 77225
USA

Howard Ochman
Department of Biochemistry and Molecular
  Biophysics
University of Arizona
Tucson, AZ 85721
USA

Gary E. Oertli
Molecular and Cellular Biology
Unviersity of Washington
Seattle, WA 98195-7275
USA

Itzhak Ofek
Department of Human Microbiology
Tel Aviv University
69978 Ramat Aviv
Israel

Bernard Ollivier
Laboratoire ORSTOM de Microbiologie des
    Anaérobies
Université de Provence
CESB-ESIL
13288 Marseille
France

Scott L. O'Neill
Department of Epidemiology and Public
    Health
Yale University School of Medicine
New Haven, CT 06520-8034
USA

Aharon Oren
Division of Microbial and Molecular
    Ecology
The Institute of Life Sciences
and
Moshe Shilo Minerva Center for Marine
    Biogeochemistry
The Hebrew University of Jerusalem
91904 Jerusalem
Israel

Jani L. O'Rourke
School of Microbiology and Immunology
University of New South Wales
Sydney, NSW 2052
Australia

Jörg Overmann
Bereich Mikrobiologie
Department Biologie I
Ludwig-Maximilians-Universität München
D-80638 München
Germany

Norman R. Pace
Department of Molecular, Cellular and
    Developmental Biology
Unversity of Colorado
Boulder, CO 80309-0347
USA

Norberto J. Palleroni
Rutgers University
Department of Biochemistry and
    Microbiology
New Brunswick 08901-8525
New Jersey
USA

Bruce Panilaitis
Department of Chemcial and Biomedical
    Engineering
Tufts University
Medford, MA 02155
USA

Nickolas Panopoulos
Department of Plant Pathology
University of California-Berkeley
Berkeley, CA 94720
USA

Yong-Ha Park
Korean Collection for Type Cultures
Korea Research Institute of Bioscience &
    Biotechnology
Taejon 305-600
Korea

Matthew R. Parsek
University of Iowa
Iowa City, IA 52242
USA

Bruce J. Paster
Department of Molecular Genetics
The Forsyth Institute
Boston, MA 02115
USA

Jerome J. Perry
3125 Eton Road
Raleigh, NC 27608-1113
USA

Jeannette Pham
The CDS Users Group
Department of Microbiology
South Eastern Area Laboratory Services
The Prince of Wales Hospital Campus
Randwick NSW 2031
Australia

Harvey M. Pickrum
Proctor and Gamble Company
Miami Valley Laboratories
Cincinnatti, OH 45239
USA

Dietmar H. Pieper
Gesellschaft für Biotechnologische Forschung
Division of Microbiology
Braunschweig D-38124
Germany

Beverly K. Pierson
Biology Department
University of Puget Sound
Tacoma, WA 98416
USA

Mariagrazia Pizza
IRIS, Chiron SpA
53100 Siena
Italy

Jeanne S. Poindexter
Department of Biological Sciences
Barnard College/Columbia University
New York, NY 10027-6598
USA

Andreas Pommerening-Röser
Institut für Allgemeine Botanik
Abteilung Mikrobiologie
Universität Hamburg
D-22069 Hamburg
Germany

Michel R. Popoff
Unité des Toxines Microbiennes
Institut Pasteur
75724 Paris
France

Anton F. Post
Department of Plant and Environmental
   Sciences
Life Sciences Institute
Hebrew University
Givat Ram
91906 Jerusalem
Israel

Bruno Pot
Laboratorium voor Microbiologie en
   Microbiële Genetica
Rijksuniversiteit Ghent
B-9000 Ghent
Belgium

David Prangishvili
Department of Mikrobiology
Universitity of Regensburg
D-93053 Regensburg
Germany

Helmut Prauser
DSMZ-German Collection of
   Microorganisms and Cell Cultures GmbH
D-38124 Braunschweig
Germany

Michael Prentice
Bart's and the London School of Medicine and
   Dentistry
Department of Medical Microbiology
St. Bartholomew's Hospital
London EC1A 7BE
UK

Ulrike Purkhold
Lehrstuhl für Mikrobiologie
Technische Universität München
D-80290 Munich
Germany

Wim J. Quax
Department of Pharmaceutical Biology
University of Groningen
Groningen 9713AV
The Netherlands

Erika Teresa Quintana
School of Biology
Universtiy of Newcastle
Newcastle upon Tyne NE1 7RU
UK

Ralf Rabus
Max-Planck-Institut für Marine Mikrobiologie
D-28359 Bremen
Germany

Reinhard Rachel
Lehrstuhl für Mikrobiologie
Universität Regensburg
D-93053 Regensburg
Germany

A. N. Rai
Biochemistry Department
North-Eastern Hill University
Shillong 793022
India

Frederick A. Rainey
Department of Biological Sciences
Louisiana State University
Baton Rouge, LA 70803
USA

Juan-Luis Ramos
Estación Experimental del Zaidin
Department of Biochemistry and Molecular
   and Cell Biology of Plants
Granada E-18008
Spain

Rino Rappuoli
IRIS Chiron Biocine Immunobiologie
Research Institute Siena
53100 Siena
Italy

Shmuel Razin
Department of Membrane and Ultrastructure
   Research
The Hebrew University-Hadassah Medical
   School
Jerusalem 91120

Annette C. Reboli
Department of Medicine
Hahneman University Hospital
Philadelphia, PA 19102
USA

David W. Reed
Biotechnology Department
Idaho National Engineering and
    Environmental Laboratory (INEEL)
Idaho Falls, ID 83415-2203
USA

Hans Reichenbach
GBF
D-3300 Braunschweig
Germany

Barbara Reinhold-Hurek
Laboratory of General Microbiology
Universität Bremen
Laboratorium für Allgemeine Mikrobiologie
D-28334 Bremen
Germany

Markus Riegler
Integrative Biology School
University of Queensland
Australia

Monica Riley
Marine Biological Lab
Woods Hole, MA 02543
USA

Lesley A. Robertson
Department of Biotechnology
Delft University of Technology
2628 BC Delft
The Netherlands

Edson R. Rocha
Department of Microbiology and Immunology
East Carolina University
Greenville, NC 27858-4354
USA

Palmer Rogers
Department of Microbiology
University of Minnesota Medical School
Minneapolis, MN 55455
USA

Lyudmila A. Romanenko
Senior Researcher in Biology
Laboratory of Microbiology
Pacific Institute of Bioorganic Chemistry of the
    Far-Eastern Branch of the Russian Academy
    of Sciences
Vladivostoku, 159
Russia

M. F. Romine
Pacific Northwest National Laboratory
Richland, WA 99352
USA

Eliora Z. Ron
Department of Molecular Microbiology and
    Biotechnology
The George S. Wise Faculty of Life Sciences
Tel Aviv University
Ramat Aviv
69978 Tel Aviv
Israel

Julian I. Rood
Australian Bacterial Pathogenesis Program
Department of Microbiology
Monash University
Victoria 3800
Australia

Eugene Rosenberg
Department of Molecular Microbiology &
    Biotechnology
Tel Aviv University
Ramat Aviv
69978 Tel Aviv
Israel

Frank Rosenzweig
Division of Biological Sciences
University of Montana
Missoula, MT 59812-4824
USA

Ornella Rossetto
Centro CNR Biomembrane and Dipartimento
    di Scienze Biomediche
35100 Padova
Italy

Michael J. Sadowsky
Department of Soil, Water, and Climate
University of Minnesota
Minneapolis, MN 55455
USA

Hermann Sahm
Institut Biotechnologie
Forschungszentrum Jülich
D-52425 Jülich
Germany

Joseph W. St. Gemer, III
Department of Molecular Microbiology
Washington University School of Medicine
St. Louis, MO 63110
USA

James E. Samuel
Department of Medical Microbiology and
  Immunology
College of Medicine
Texas A&M University System Health Science
  Center
College Station, TX, 77843-1114
USA

Philippe J. Sansonetti
Unité de Pathogénie
Microbienne Moléculaire
Institut Pasteur
75724 Paris
France

Scott R. Santos
Department of Biochemistry & Molecular
  Biophysics
University of Arizona
Tucson, AZ 85721
USA

Beatrice Saviola
Departments of Molecular Microbiology and
  Immunology
Johns Hopkins School of Hygiene and Public
  Health
Baltimore, MD 21205-2105
USA

Klaus P. Schaal
Institut für Medizinische
Mikrobiologie und Immunologie
Universität Bonn
D-53105 Bonn
Germany

David B. Schauer
Biological Engineering Division and Division
  of Comparative Medicine
Massachusetts Institute of Technology
Cambridge, MA 02139
USA

Siegfried Scherer
Department für Biowißenschaftliche
  Grundlagen
Wißenschaftszentrum Weihenstephan
Technische Universität München
D-85354 Freising, Germany

Bernhard Schink
Fakultät für Biologie der Universität Konstanz
D-78434 Konstanz
Germany

Hans G. Schlegel
Institut für Mikrobiologie der Gessellschaft
  für Strahlen- und Umweltforschung mbH
Göttingen
Germany

Karl-Heinz Schleifer
Lehrstruhl für Mikrobiologie
Technische Universität München
D-85354 Freising
Germany

Heinz Schlesner
Institut für Allgemeine Mikrobiologie
Christian Albrechts Universität
D-24118 Kiel
Germany

Michael Schmid
GSF-Forschungszentrum für Umwelt und
  Gesundheit GmbH
Institut für Bodenökologie
D-85764 Neuherberg
Germany

Jean M. Schmidt
Department of Botany and Microbiology
Arizona State University
Tempe, AZ 85287
USA

Karin Schmidt
Institut für Mikrobiologie
Georg-August-Universität
D-3400 Göttingen
Germany

Ruth A. Schmitz
University of Göttingen
D-3400 Göttingen
Germany

Hildgund Schrempf
FB Biologie/Chemie
Universität Osnabrück
49069 Osnabrück
Germany

Milton N. Schroth
Department of Plant Pathology
University of California-Berkeley
Berkeley, CA 94720
USA

Heide N. Schulz
Institute for Microbiology
University of Hannover
D-30167 Hannover
Germany

Peter Schumann
DSMZ-German Collection of Microorganisms
    and Cell Cultures GmbH
D-38124 Braunschweig
Germany

Arthur Schüßler
Institut Botany
64287 Darmstadt
Germany

Edward Schwartz
Institut für Biologie/Mikrobiologie
Homboldt-Universität zu Berlin
D-10115 Berlin
Germany

James Scott
Geophysical Laboratory
Carnegie Institution of Washington
Washington, DC 20015
USA

Margrethe Haugge Serres
Marine Biological Lab
Woods Hole, MA 02543
USA

James P. Shapleigh
Department of Microbiology
Cornell University
Wing Hall
Ithaca, NY 14853-8101
USA

Nathan Sharon
The Weizmann Institute of Science
Department of Biological Chemistry
IL-76100 Rehovoth
Israel

Lawrence J. Shimkets
Department of Microbiology
The University of Georgia
Athens, GA 30602-2605
USA

Thomas M. Shinnick
Center for Infectious Diseases
Centers for Disease Control
Atlanta, GA 30333
USA

Yuval Shoham
Department of Food Engineering and
    Biotechnology
Technion—Israel Institute of Technology
Haifa 32000
Israel

Jörg Simon
Johann Wolfgang Goethe-Universität Frankfurt
Campus Riedberg
Institute of Molecular Biosciences
Molecular Microbiology and Bioenergetics
D-60439 Frankfurt
Germany

Michel Simonet
Départment de Pathogenèse des Maladies
    Infectieuses et Parasitaires
Institut de Biologie de Lille
59021 Lille
France

Mikael Skurnik
Department of Medical Biochemistry
University of Turku
20520 Turku
Finland

James M. Slauch
Department of Microbiology
College of Medicine
University of Illinois
and
Chemical and Life Sciences Laboratory
Urbana, IL 61801
USA

Ralph A. Slepecky
Department of Biology
Syracuse University
Syracuse, NY 13244
USA

C. Jeffrey Smith
Department of Microbiology and
    Immunology
East Carolina University
Greenville, NC 27858-4354
USA

Martin Sobierj
Department of Biology
Environmental Science and Engineering
Oregon Graduate Institute of Science and
    Technology
Portland State University
Portland, OR 97291-1000
USA

Pamela A. Sokol
Department of Microbiology and Infectious
    Diseases
University of Calgary Health Science Center
Calgary T2N 4N1
Canada

Jay V. Solnick
Department of Interanl Medicine (Infectious
    Diseases) and Medical
Microbiology and Immunology
University of California, Davis
School of Medicine
Davis, CA 95616
USA

Dimitry Yu. Sorokin
Department of Biotechnology
Delft University of Technology
2628 BC Delft
The Netherlands
and
S.N. Winogradsky Institute of Microbiology
117811 Moscow
Russia

Georg A. Sprenger
Institut Biotechnologie
Forschungszentrum Jülich
D-52425 Jülich
Germany

Stefan Spring
Deutsche Sammlung von Mikroorganismen und
    Zellkulturen
D-38124 Braunschweig
Germany

Erko Stackebrandt
Deutsche Sammlung von Mikroorganismen und
    Zellkulturen
D-38124 Braunschweig
Germany

David A. Stahl
University of Washington
Seattle, WA
USA

Donald P. Stahly
Department of Microbiology
University of Iowa
Iowa City, IA 52242
USA

James T. Staley
Department of Microbiology
University of Washington
Seattle, WA 98105
USA

Alfons J.M. Stams
Laboratorium voor Microbiologie
Wageningen University
NL-6703 CT Wageningen
The Netherlands

Thaddeus B. Stanton
PHFSED Research Unit
National Animal Disease Center
USDA-ARS
Ames, IA 50010
USA

Daniel C. Stein
Department of Cell Biology and Molecular
    Genetics
University of Maryland
College Park, MD 20742
USA

Reinhard Sterner
Universitaet Regensburg
Institut fuer Biophysik und Physikalische
    Biochemie
D-93053 Regensburg
Germany

Karl O. Stetter
Lehrstuhl für Mikrobiologie
Universität Regensburg
D-93053 Regensburg
Germany

Frank J. Stewart
Department of Organic and Evolutionary
    Biology
Harvard University
Cambridge, MA 02138
USA

Andreas Stolz
Institut für Mikrobiologie
Universität Stuttgart
70569 Stuttgart
Germany

Daphne L. Stoner
Idaho National Engineering Laboratory
EG & G Idaho
Idaho Falls, ID 83415
USA

Paul Stoodley
Center for Biofilm Engineering
Montana State University
Bozeman, MT 59717-3980
USA

James R. Swafford
Department of Botany and Microbiology
Arizona State University
Tempe, AZ 85287
USA

Jean Swings
Laboratorium voor Microbiologie
Department of Biochemistry
Physiology and Microbiology
BCCM/LMG Bacteria Collection
Universiteit Gent
Gent
Belgium

Mariko Takeuchi
Institute for Fermentation
Osaka 532-8686
Japan

Ralph Tanner
University of Oklahoma
Norman, OK, 73019-0390
USA

Andreas Teske
Department of Marine Sciences
University of North Carolina at Chapel Hill
Chapel Hill, NC 27599
USA

Michael Teuber
ETH-Zentrum
Lab Food Microbiology
CH-8092 Zürich
Switzerland

Gabriele Timmermann
Institut für Allgemeine Botanik
Abteilung Mikrobiologie
Universität Hamburg
D-22069 Hamburg
Germany

Brian J. Tindall
Deutsche Sammlung von Mikroorganismen und
   Zellkulturen
Braunschweig D-38124
Germany

Kevin Towner
Consultant Clinical Scientist
Public Health Laboratory
University Hospital
Nottingham NG7 2UH
UK

Hans G. Trüper
Institut für Mikrobiologie und Biotechnologie
D-53115 Bonn
Germany

Elaine Tuomanen
Department of Infectious Diseases
St. Jude Children's Research Hospital
Memphis, TN 38105-2394
USA

Francisco A. Uzal
California Animal Health and Food Safety
   Laboratory
University of California, Davis
San Bernardino, CA 92408
USA

Peter Van damme
Laboraroorium voor Microbiologie
Faculteit Wetenschappen
Universiteit Gent
B-9000 Gent
Belgium

Antonio Ventosa
Department of Microbiology and
   Parasitology
Faculty of Pharmacy
University of Sevilla
41012 Sevilla
Spain

Gernot Vobis
Centro Regional Universitario Bariloche
Universidad Nacional de Comahue
Barioloche 8400, Rio Negro
Argentina

Alexander von Graevenitz
Department of Medical Microbiology
University of Zürich
GH-8028 Zürich
Switzerland

Günther Wächtershäuser
80331 Munich
Germany

Lawrence P. Wackett
Department of Biochemistry, Molecular
   Biology
and
Biophysics and Biological Process Technology
   Institute
University of Minnesota
St. Paul, MN, 55108-1030
USA

William G. Wade
Department of Microbiology
Guy's Campus
London, SE1 9RT
UK

Michael Wagner
Lehrstuhl für Mikrobielle Ökologie
Institut für Ökologie und Naturschutz
Universität Wien
A-1090 Vienna
Austria

David H. Walker
Department of Pathology
University of Texas Medical Branch
Galveston, TX 77555-0609
USA

Naomi Ward
The Institute for Genomic Research
Rockville, MD 20850
USA

Trudy M. Wassenaar
Molecular Microbiology and Genomics
  Consultants
55576 Zotzenheim
Germany

John B. Waterbury
Woods Hole Oceanographic Institution
Woods Hole, MA 02543
USA

Georges Wauters
Université Catholique de Louvain
Faculté de Médecine
Unité de Microbiologie
B-1200 Bruxelles
Belgium

Astri Wayadande
Department of Entomology and Plant
  Pathology
Oklahoma State University
Stillwater, OK
USA

Alison Weiss
Molecular Genetics, Biology and Microbiology
University of Cincinnati
Cincinnati, OH 45267
USA

Rodney A. Welch
Medical Microbiology and Immunology
University of Wisconsin
Madison, WI 53706-1532
USA

William B. Whitman
Department of Microbiology
University of Georgia
Athens, GA 30605-2605
USA

Friedrich Widdel
Max-Planck-Institut für Marine Mikrobiologie
D-28359 Bremen
Germany

Jürgen Wiegel
University of Georgia
Department of Microbiology
Athens, GA 30602
USA

Robert A. Whiley
Queen Mary, University of London
London E1 4NS
UK

Tracy Whilkins
TechLab, Inc.
Corporate Research Center
Blacksburg VA 24060-6364
USA

Anne Willems
Laboratorium voor Mikrobiologie
Universiteit Gent
B-9000 Gent
Belgium

Carl R. Woese
Department of Microbiology
University of Illinois
Urbana, IL 61801
USA

Ralph S. Wolfe
Department of Microbiology
University of Illinois
Urbana, IL 61801

Ann P. Wood
Division of Life Sciences
King's College London
London WC2R 2LS
UK

Donald E. Woods
Department of Microbiology and Infectious
  Diseases
University of Calgary Health Science Center
Calgary T2N 4N1
Canada

B. W. Wren
Department of Infectious and Tropical
  Diseases
London School of Hygiene and Tropical
  Medicine
London WC1E 7HT
UK

Timothy L. Yahr
University of Iowa
Iowa City, IA 52242
USA

Atteyet F. Yassin
Institut für Medizinische
Mikrobiologie und Immunologie
Universität Bonn
D-53105 Bonn
Germany

Jung-Hoon Yoon
Korean Collection for Type Cultures
Korea Research Institute of Bioscience and
    Biotechnology
Yuson, Taejon 305-600
Korea

Allan A. Yousten
Biology Department
Virginia Polytechnic Institute and State
    University
Blacksburg, VA 24061
USA

Xue-Jie Yu
University of Texas Medical Branch
Galveston, TX
USA

Vladimir V. Yurkov
Department of Microbiology
University of Manitoba
Winnipeg R3T 2N2
Canada

Georgi A. Zavarzin
Institute of Microbiology
Academy of Sciences of the USSR
117312 Moscow
Russia

Mary Jo Zidwick
Cargill Biotechnology Development Center
Freshwater Building
Minneapolis, MN 55440
USA

Stephen H. Zinder
Department of Microbiology
Cornell University
272 Wing Hall
Ithaca, NY 14853
USA

Archaea

Prokaryotes (2006) 3:3–9
DOI: 10.1007/0-387-30743-5_1

CHAPTER 1

The Archaea: A Personal Overview of the Formative Years

RALPH S. WOLFE

"—a new scientific truth does not triumph by convincing its opponents and making them see the light, but rather because its opponents eventually die, and a new generation grows up that is familiar with it." Max Planck

The year 2002 marks the 25th anniversary of the discovery of the Archaea. Data to support this discovery were presented in the *Proceedings of the National Academy of Science* of October 1977 (Fox et al., 1977). However, most people learned about methanogens, "a third form of life," from the front pages of their newspapers on Thursday, November 3, 1977, the day the *PNAS* issue was available. The National Science Foundation (NSF) and the National Aeronautics and Space Administration (NASA) supported the research of Carl Woese and were pleased to sponsor and receive publicity for the press release he presented. When discussing the importance of the discovery with reporters prior to the press release, Woese had difficulty communicating in scientific terms that they could understand, until he hit upon the phrase, "a third form of life." They could relate to this phrase, and it became prominent in their articles. Methanogens, though prokaryotes, were only distantly related to other prokaryotes.

The immediate response of the scientific community to the press release was negative with disbelief and much hostility, especially among microbiologists. Scientists were suspicious of scientific publication in newspapers, and only a very few were familiar with the use of 16S rRNA oligonucleotides to define relationships among organisms. Among the phone calls that I received the morning of November 3, the one by S. E. Luria was the most civil and free of four-letter words. Luria was a Professor of Microbiology, when I joined the Department at Illinois in 1953 and had later moved to MIT. Luria: "Ralph, you must dissociate yourself from this nonsense, or you're going to ruin your career!" "But, Lu, the data are solid and support the conclusions: they are in the current issue of *PNAS*." Luria: "Oh yes, my issue just arrived." "If you would like to discuss the paper after you have had a chance to look at it, give me a ring." He did not call again.

I wanted to crawl under something and hide. Fortunately I was able to escape the hostility and left graduate students to cope, because my wife and I were leaving for Philadelphia to help celebrate her father's 90th birthday. We collected a few newspapers in airports. Carl was on the front page of the *New York Times* with his Adidas-clad feet prominently displayed on his desk. Much later he would say, "You know, Adidas never offered me a contract." In Philadelphia, I explained in dismay to my father-in-law what my colleague had done, and his response was: "You know, in my long life I have observed something, if you don't overstate your case, no one will listen." I felt better. However, in hindsight, the press release polarized the scientific community, and the majority refused to read the literature, delaying acceptance of the Archaebacteria for perhaps a decade. The discovery of the Archaea resulted from the intersection of two independent lines of research; the culmination of the informational-macromolecule line was the research of Carl Woese, whereas the biochemistry of methanogenesis was pursued by my research group.

Informational Macromolecules

This line began with the publications of Sanger and coworkers (Sanger and Tuppy, 1951; Sanger and Thompson, 1953), who showed each amino acid in the two chains of insulin occupies a precise position in the protein molecule. This discovery had enormous implications for genetics and for the emerging area of molecular biology. By the 1960s, insulin molecules from various animal species had been sequenced, and it was apparent that insulin from different species possessed variations in the sequence of certain amino acids. Zukerkandl and Pauling (1965) pointed out that these differences could be used to determine the relationship among the molecules, and hence, the organisms from which they were obtained; macromolecules which showed only minor sequence changes were closely related, whereas those with larger differences were more distantly related. The next macromolecules to be sequenced were other small proteins, for example, cytochromes, but the limitations of protein molecules to study relatedness among organisms soon became apparent.

In the early 1960s, Carl Woese's study of the ribosome convinced him that this structure was highly conserved. He reasoned that because the ribosome is of ancient origin, is universally distributed, and is functionally equivalent in all living cells, it would be the ideal structure to use for study of evolution. In addition, the ribosome has only one function, translation of the genetic code into the amino acid sequence of a protein, and may be somewhat "insulated" from the variables of phenotypic encounters. Then, Sanger et al. (1965) published a method for a two-dimensional (2-D) fractionation of radioactive nucleotides, which could be used to sequence RNA. Woese's seminal insight was to recognize that ribosomal RNA was the ideal molecule to follow evolution to very ancient events. He modified the Sanger RNA sequencing technique and explored the use of ribosomal 5S, 16S, and 23S rRNA. He chose 16S rRNA as a "statistical ensemble" of 1540 monomers that would allow investigations of organisms to reach the very root of the tree of life. Carl Woese was alone in developing the use of ^{32}P-labelled oligonucleotides from T_1 endonuclease digests of 16S rRNA to generate a similarity coefficient that could be used to determine the relatedness of organisms (Sogin et al., 1972; Pechman and Woese, 1972).

I was fascinated by this new approach, for as a graduate student I had been prejudiced against "bug sorting" and its tenuous results by my professor, and I resolved never to get involved in taxonomy with its constant reshuffling and renaming of species. But here was something I could believe in, for it had a sense of permanence. I was especially impressed by results of Woese's initial study of the genus, *Bacillus* (Woese et al., 1976). I became a believer. Today with greatly improved technology, data can be generated in one day that took Woese months to do in 1969. By 1976, Woese had obtained the 16S rRNA similarity coefficients of 60 microbes, representing a wide diversity.

Biochemistry of Methanogenesis

This line had an early beginning with the experiments of Alessandro Volta in 1776 which showed that the gas produced by decaying vegetable residues in sediments was combustible; its identity as methane came a century later, and definitive experiments with methanogenic bacteria came with the isolation of *Methanobacillus omelianskii* (Barker, 1936; Barker, 1940). For

over 20 years after 1936, Barker's laboratory was the leader in studying methanogens, and many observations led him to conclude that these organisms, though morphologically diverse, had a common physiology. In 1961, I was attracted to and began to investigate the unexplored biochemistry of methanogenesis. An enzymically active cell extract had not been prepared. Because of the formidable challenges in cultivating these organisms, knowledge of their biochemistry lagged decades behind other organisms.

Barker's culture of *M. omelianskii* was mass cultured, and the first cell-free formation of methane was obtained (Wolin et al., 1963). In collaboration with Bryant and Wolin, Barker's culture of *M. omelianskii* was found to be a symbiotic association of two organisms, i.e., interspecies hydrogen transfer had been discovered (Bryant et al., 1967). The methanogenic organism from the mixed culture oxidized hydrogen and reduced carbon dioxide to methane. My laboratory developed a technology for mass culture of the organism on hydrogen and carbon dioxide (Bryant et al., 1968), which was scaled up to the 200-liter level. With kilogram quantities of cells available, the first unique coenzyme of methanogenesis was discovered and named "coenzyme M" (McBride and Wolfe, 1971). Its structure was determined, and its methylated active form was synthesized (Taylor and Wolfe, 1974a). The unusual coenzyme was found to be a "vitamin" required by a methanogen from the cow's rumen, *Methanobacterium ruminantium* (Taylor et al., 1974b). To assay this vitamin, this organism was grown via a new procedure, the Balch modification of the Hungate technique, involving use of a pressurized atmosphere of hydrogen and carbon dioxide (Balch and Wolfe, 1976). After exhaustive analyses, CoM was found nowhere else in nature. A second unique coenzyme (Eirich et al., 1978), the blue-green fluorescent deazaflavin, F_{420}, was found to be the coenzyme for formic dehydrogenase (Tzeng et al., 1975). By 1976, evidence for other unique coenzymes was in hand in my laboratory.

Discovery of the Archaea

The simple method of growing methanogenic cells in an anaerobic pressurized atmosphere of hydrogen and carbon dioxide was pivotal to the discovery of the Archaea, for hydrogen and carbon dioxide could be replenished with safe containment of high levels of ^{32}P and without exposure of cells to oxygen or contamination. After analysis of the 2-D chromatographic pattern of the T_1 endonuclease digestion fragments from the 16S RNA of the first methanogen, I asked about the results. Woese was puzzled, for the pattern was unlike any he had seen; he could only conclude that: "Somehow we must have isolated the wrong RNA." So the experiment was repeated with special care, and this time, his response was: "Wolfe, these methanogens are not bacteria." "Of course they are Carl; they look like bacteria." "They are not related to any bacteria I've seen." Because Woese had spent nearly 10 years alone developing the method and analyzing the 2-D chromatographic patterns of T_1 endonuclease digestion patterns of ^{32}P-labelled 16S rRNA from 60 different bacteria, he was easily able to discern that methanogens were different! But what should the group be called?

In November 1977, Woese and Fox proposed that ribosomal RNA sequence characterization could be used to define three "aboriginal lines of descent" (Woese and Fox, 1977). One line, the typical bacteria, was designated "eubacteria." "Archaebacteria" was proposed as a name for the methanogen line, and the term "urkaryotes" was proposed for the cytoplasmic component of eukaryotic cells. So the methanogens were Archaebacteria. The name Archaebacteria was suggested by David Nanney. In a way, the term was unfortunate because the case was being made that the methanogens were "old" (i.e., represented a very ancient divergence in evolution), so that they were no more related to bacteria than to eukaryotes. On the other hand, they were bacteria. (Over a decade later, Woese, Kandler, and Wheelis would propose the name "Archaea" for the Archaebacteria [Woese et al., 1990]. By that time, much evidence had accumulated showing the Archaea clearly belonged on the eukaryotic line of descent, and it would be less confusing if the word "bacteria" in Archaebacteria was deleted.)

All of the available methanogens in pure culture were obtained and subjected to 16S rRNA analysis. The results were startling (Fox et al., 1977); none of the species was related to typical bacteria, but they fell into natural groups. Additionally, the unusual biochemistry of methanogens now assumed importance. Methanogens are the only organisms that produce methane as their metabolic product, and they do so by a biochemistry that employs unique enzymes and coenzymes. So, the initial concept that methanogens were different rested on 16S rRNA analysis and was supported by unique biochemistry. While these studies were underway, it occurred to me that, if methanogens were so different in these aspects, they should show differences in other areas. So, I wrote to Otto Kandler in November 1976 explaining that Carl Woese had determined methanogens to be only distantly related to bacteria and asked him whether he would be interested in determining the structure

of the cell walls of methanogens, which we could send to him. He was enthusiastic about the project for his laboratory had been studying the cell walls of diverse bacteria including *Halococcus morrhuae* (Schleifer and Kandler, 1972; Steber and Schleifer, 1975) and *Methanosarcina* (Kandler and Hippe, 1977) which were shown to lack peptidoglycan. This letter would have an effect far beyond its simple question, for Kandler found a gold mine and became convinced that the Archaebacterial concept represented a major turning point for prokaryotic phylogeny. In 1977, he made a trip to Urbana, Illinois, to obtain the most direct information. He became convinced that German microbiologists should begin serious work on the Archaebacteria. His laboratory soon showed that species of *Methanobacterium* lacked typical peptidoglycan (murein) cell walls, known to be a characteristic of all bacteria. Other species of methanogens had no peptidoglycan derivatives at all (Kandler and König, 1978), and when the project was complete, Kandler's laboratory had shown that there were as many variations in the cell walls of methanogens as in all bacteria combined! This was a breakthrough—a third pillar to support the concept that the methanogens were different.

The year 1978 was eventful in fine tuning the concept that methanogenic bacteria (10 species) were so distantly related to typical bacteria that they should be considered as a distinct phylogenetic group, the Archaebacteria. Tornabene et al. (1978) investigated the lipids of *Methanobacterium thermoautotrophicum* and found that phytanyl-glycerol ethers and squalenes were major lipid components. A fourth pillar to support the Archaebacterial concept was in place. This unusual lipid composition was similar to that of *Halobacterium cutirubrum* reported by Kates et al. (1965) and Tornabene et al. (1969). However, the discovery of these unusual lipids in *Halobacterium* had not created wide interest at the time of publication, for their possession was largely perceived to be a consequence of adaptation to growth in saturated salt environments. Similarly, organisms previously isolated from thermoacidic aquatic environments, *Sulfolobus* and *Thermoplasma* summarized by Brock (1978), had been shown to contain only negligible amounts of ester-linked lipids (Langworthy et al., 1972; DeRosa et al., 1975; DeRosa et al., 1976). Cell walls of *Sulfolobus* had been shown by Brock et al. (1972) to be deficient in peptidoglycan, and of course, species of mycoplasma were known to have only a protein covering. Again it was presumed that perhaps the lipids and lack of peptidoglycan were related (somehow) to growth in hot acidic environments. Then Woese, Magrum, and Fox brought it all together in an article entitled "Archaebacteria" (Woese et al., 1978), and expanded the concept to include the methanogens, extreme halophiles, and thermoacidophiles, the common factors being: 1) the possession of characteristic ribosomal RNAs and tRNAs; 2) the absence of murein cell walls; and 3) the presence of ether-linked lipids in phytanyl chains.

In 1978, Zillig and coworkers began to report results of experiments on the remarkably similar component patterns of the DNA-dependent RNA polymerases of Archaebacteria and eukaryotes (Zillig et al., 1978; Zillig et al., 1979; Stetter et al., 1980; Sturm et al., 1980). In addition, the Archaebacterial polymerases showed characteristic eukaryotic resistance against the inhibitors rifampicin, streptolydigin, and α-amanitin. Because the DNA-dependent RNA polymerases of eubacteria have the standard four-component composition β'βασ, it was assumed that all prokaryotes would be similar. However, the Archaebacterial RNA polymerases contained 7 to 12 different components, showing a pattern similar to *Saccharomyces cerevisae*. The results summarized by Zillig et al. (1985) were so astounding that many viewed them as conclusive evidence, alone, that Archaebacteria indeed represented a third domain of life. Another pillar to support the Archaebacteria was in place.

In 1979, an interesting observation was made in the laboratory of Friedrick Klink by M. Kessel. He found that in vitro protein synthesis by halobacterial preparations was inhibited by diphtheria toxin (Kessel and Klink, 1980). These studies were extended to 18 Archaebacteria (Kessel and Klink, 1982). The state of the field was summarized by Klink (1985). The Archaebacteria have an elongation apparatus in the ribosome that is distinct from those of typical bacteria and eukaryotes. All Archaebacteria contain a structural domain in elongation factor-2 (EF-2) that renders it a substrate for diphtheria toxin. Archaebacterial EF-2 is highly specific for Archaebacterial ribosomes and does not work with ribosomes from bacteria or eukaryotes. These experiments attracted wide interest because they exploded the perceived dogma that all prokaryotes are insensitive to the diphtheria toxin; so another pillar to support the Archaebacterial concept was in place.

However, the concept of the Archaebacteria, outside of a very few believers, was not embraced by the microbiological community. So in 1978, I suggested to William Balch, a graduate student in my laboratory, that we gather together in a review article all of the evidence in support of methanogens as Archaebacteria. He did a magnificent job dedicating his considerable talents to the task and working closely with Woese. With the dedicated help of Marvin Bryant, a new

classification of methanogens based on 16S rRNA was included in the article. The review displayed eloquently the power of 16S rRNA to show the relatedness of organisms, i.e., a poorly understood group of methanogens as diverse as all bacteria combined was finally organized into a phylogeny that made sense. Marv Bryant, the world authority on methanogens, pronounced the results "beautiful, beautiful." The four types of cell walls, the C_{20} diphytanyl glycerol diethers, the C_{40} dibiphytanyl diglycerol tetraethers, and $C_{30}H_{30}$ squalene of the neutral lipid fraction were Archaebacterial. The structure and function of coenzyme M and coenzyme F_{420} had been determined (Taylor and Wolfe, 1974a; Taylor et al., 1974b; Eirich, L. D., et al., 1978; Tzeng et al., 1975) and were clearly unique to the biochemistry of methanogenesis. Although initially the uniqueness of these coenzymes supported the Archaebacterial concept, it soon appeared that they were not found at the time in known species of extreme halophiles or thermoacidophiles and so were a general characteristic of methanogens only. In addition to ribosomal RNA, transfer RNAs of methanogens were found to be unique; they do not contain the universal common arm sequence. The review, "Methanogens: Reevaluation of a unique biological group," appeared in the June 1979 issue of *Microbiological Reviews* (Balch et al., 1979) and was received with respect by the microbiological community. The work attracted the interest of a number of new investigators to the Archaebacteria.

Because Rolf Thauer had close contact with the German Forschungsgemeinschaft, Kandler asked him in March 1978 to think about a proposal for funds to initiate a Schwerpunkt-program for research on methanogens and other Archaebacteria. Thauer agreed to head the Schwerpunkt, but to avoid controversy, they decided not to use the term "Archaebacteria" in their grant application, since this term was not well accepted at the time. The DFG-Schwerpunkt-program, "Methanogene Bakterien," was funded in 1979. Despite the title, research on other Archaebacteria was encouraged. The initial investigators included J. Andreesen (Göttingen), A. Bacher (Garching), G. Diekert (Marburg/Stuttgart), G. Fuchs (Marburg/Ulm), G. Gottschalk (Göttingen), O. Kandler (München), A. Klein (Marburg), H. König (Regensberg/Ulm), P. Scherer (Jülich/Weihenstephan), P. Schönheit (Marburg/Berlin), R. Thauer (Marburg), J. Winter (Regensberg). Kandler was the instigating force in the initiation of this effort, which stands in stark contrast to the rejection of or ambivalence toward the Archaebacterial concept by microbiologists in other countries, who were quite content with the entrenched taxonomy of the time.

The Schwerpunkt paved the way for the first international workshop on Archaebacteria at München-Martinsried, June 27–July 1, 1981. W. Zillig, O. Kandler, and K. Stetter were involved in setting up the workshop, which was supported by the Stiftung Volkswagenwerk and the Max-Planck-Gesellschaft. The workshop was devoted to all areas of Archaebacterial research, with emphasis on molecular and biochemical research as well as geochemical, paleontological, and taxonomic areas. Over 50 workers attended, and it was the first time that many workers in the field had a chance to meet personally. Participants were enthusiastic about the success of the workshop. It was clear that the Schwerpunkt was not only focusing the efforts of German microbiologists on Archaebacteria but was having an international impact as well. It was also clear that German microbiologists had assumed a leadership role in the study of the Archaebacteria.

The proceedings of the 1st International Workshop were published in May 1982 (Kandler, 1982). The 17 articles covering cell walls, the ribosome, translation apparatus, transcription, lipids, membranes, coenzymes, CO_2 fixation, anaerobic reduction of molybdenum, phylogeny, new species, and 5S ribosomal RNA well documented the beautiful flowering of the Archaebacterial concept that had occurred in just five years! To celebrate the breakthrough of the Archaebacterial concept after the workshop, Kandler and his wife Trudy took Woese and Wolfe for a hike in the Bavarian Alps to the top of "Hohe Hiss" 1850 m, 80 km south of Munich. Woese and especially Wolfe were not in top physical shape, but with some huffing and puffing, they reached the top via a well-graded path. The photograph (Fig. 1) was taken by Trudy Kandler at the summit. After we returned to the car, the Kandlers showed us some beautiful Bavarian churches, and then, because it was a hot afternoon, we stopped at an open-air beer garden for

Fig. 1. (left to right) Carl Woese, Ralph Wolfe, and Otto Kandler celebrate the success of the 1st international workshop on Archaebacteria in the Bavarian Alps, July 1, 1981.

a glass of beer under a large linden tree. The beer was delivered to us in large graceful glasses, and after "Prost" (to your health in English), we each had a sip of the cold beer. Otto Kandler seemed to have consumed about a third of the beer in his glass, when he pronounced, "Ah, the first sip is the best!" I have always remembered the volume of a German "sip" on a hot afternoon.

In 1985, owing to the fast-growing field of Archaebacterial research, international symposia were again organized in Germany. The second International Workshop on Biology and Biochemistry of Archaebacteria was organized by O. Kandler and K. O. Stetter and sponsored by the Deutsche Forschungsgessellschaft. A workshop on Molecular Genetics of Archaebacteria was organized by W. Zillig and A. Böck and sponsored by the European Molecular Biology Organization (EMBO) and the Max-Planck-Gesellschaft (München). These workshops were held back-to-back June 23–July 1, 1985, in München. Proceedings of the workshops were published in a volume entitled *Archaebacteria '85* (Kandler and Zillig, 1986). The book contains 65 reprints of papers and 38 extended poster abstracts and thoroughly documents the astounding progress made in all aspects of the Archaebacteria in four years (1981–1985). Again, German microbiologists took a leadership role and were ahead of the international community of microbiologists in fully accepting evidence for the three domains of life as indisputable; in contrast, misunderstandings, doubts, and hesitation prevented many microbiologists in other countries from embracing the Archaebacterial concept. It would be perhaps another 10 years before the concept was fully accepted.

It should be reemphasized that the road was difficult and experimentally demanding, especially in the early years, when Carl Woese's laboratory alone generated the primary data. To obtain the radioactive nucleotides for 2-D fractionation, the microbial cells had to be grown in media from which phosphate had been removed and to which high levels of ^{32}P-phosphate were added. The trick was to get cells to take up enough radioactive phosphate before they were killed by radiation. The procedure worked well for organisms with a short generation time but was less suitable for slow-growing microbes. The results of the T_1 endonuclease digestion and subsequent fractionation procedures to identify the sequence of each oligonucleotide, i.e., 5-mers to 18-mers, provided data for only about 25% of the 16S rRNA. Yet the whole system for obtaining the similarity coefficient (S_{AB}) for each organism to construct dendrograms of relationship among organisms was developed by Woese within these limitations. The S_{AB} was the method of choice until cloning and sequencing of the 16S rRNA genes became available. The first to appear was the 16S rRNA gene from *Escherichia coli* (Brosius et al., 1978), and the first archaeal 16S rRNA sequence appeared in 1983 (Gupta et al., 1983). There was a long lag before the complete 16S rRNA sequences became useful, because for comparative studies, a significant number of sequences needed to be determined. Thus, the method of choice now is to clone rDNA from an organism and obtain the complete sequence of the gene encoding 16S rRNA. The increasing number of complete genome sequences of organisms has made analysis relatively simple compared to the initial procedures used by Woese to define the tree of life.

The second edition of *The Prokaryotes* (four volumes published in 1992) recognized the almost unbelievable impact that the work of Carl Woese had in defining the Archaea as well as a 16S rRNA phylogenetic basis for the microbial world. In the 1990s, microbiology textbooks in the United States began to present phylogeny based on 16S rRNA. In 1997, twenty years after the discovery, the textbook *Brock, Biology of Microorganisms* (Madigan et al., 1977) completely endorsed the Archaea and the universal phylogenetic tree based on 16S rRNA. However, the vast inertia invested in the morphological approach to taxonomy and phylogeny, since the time of Linnaeus and later Darwin by the biological community, makes changing the course of such a huge ship very difficult. This is illustrated dramatically in textbooks of general biology from the 1990s that include perhaps a paragraph or so on the Archaebacteria with very little or nothing on the use of monomer sequences of rRNA to reveal a molecular tree of life. This approach trickles down from college textbooks to high school textbooks, perpetuating the teaching of biology based on morphology. However, in the United States, the 6th edition of the general text *Biology* by Raven and Johnson (2002) refers to taxonomy and phylogeny based on morphology as the "traditional approach" and then proceeds with a well-presented three-page treatment of the domains of life: Eubacteria, Archaea, and Eukarya. So, all is not lost, just delayed. Remember the quote from Max Planck!

Acknowledgments. I thank Otto Kandler and Gary Olsen for kindly providing factual information.

Literature Cited

Balch, W. E., and R. S. Wolfe. 1976. New approach to the cultivation of methanogenic bacteria: 2-mercaptoethanesulfonic acid (HS-CoM)-dependent growth of Methanobacterium ruminantium in a pressurized atmosphere. Appl. Environ. Microbiol. 32:781–791.

Balch, W. E., G. E. Fox, L. J. Magrum, C. R. Woese, and R. S. Wolfe. 1979. Methanogens: Reevaluation of a unique biological group. Microbiol. Rev. 43:260–296.

Barker, H. A. 1936. Studies upon the methane-producing bacteria. Arch. Mikrobiol. 7:420–438.

Barker, H. A. 1940. Studies upon the methane fermentation. IV: The isolation and culture of Methanobacillus omelianskii. Ant. v. Leeuwenhoek 6:201–220.

Brock, T. D., K. M. Brock, T. R. Belly, and R. L. Weiss. 1972. Sulfolobus: A new genus of sulfur-oxidizing bacteria living at low pH and high temperature. Arch. Mikrobiol. 84:54–68.

Brock, T. D. 1978. Thermophilic Organisms and Life at High Temperatures. Springer-Verlag. Berlin, Germany.

Brosius, J., M. L. Palmer, P. J. Kennedy, and H. F. Noller. 1978. Complete nucleotide sequence of a 16S ribosomal RNA gene from Escherichia coli. Proc. Natl. Acad. Sci. USA 75:4801–4805.

Bryant, M. P., E. A. Wolin, M. J. Wolin, and R. S. Wolfe. 1967. Methanobacillus omelianskii, a symbiotic association of two species of bacteria. Arch. Mikrobiol. 59:20–31.

Bryant, M. P., B. C. McBride, and R. S. Wolfe. 1968. Hydrogen-oxidizing methane bacteria. I. Cultivation and methanogenesis. J. Bacteriol. 95:1118–1123.

DeRosa, M., A. Gambacorta, and J. D. Bu'Lock. 1975. Extremely thermophilic acidophilic bacteria convergent with Sulfolobus acidocaldarius. J. Gen. Microbiol. 86:156–164.

DeRosa, M., A. Gambacorta, and J. D. Bu'Lock. 1976. The Caldariella group of extreme thermoacidophilic bacteria: Direct comparison of lipids in Sulfolobus, Thermoplasma, and the MT strains. Phytochemistry 15:143–145.

Eirich, L. D., G. D. Vogels, and R. S. Wolfe. 1978. Proposed structure for coenzyme F420 from Methanobacterium. Biochemistry 17:4583–4593.

Fox, G. E., L. J. Magrum, W. E. Balch, R. S. Wolfe, and C. R. Woese. 1977. Classification of methanogenic bacteria by 16S ribosomal RNA characterization. Proc. Natl. Acad. Sci. USA 74:4537–4541.

Gupta, R., J. M. Lanter, and C. R. Woese. 1983. Sequence of the 16S ribosomal RNA from Halobacterium volcanii, an archaebacterium. Science 221:656–659.

Kandler, O., and H. Hippe. 1977. Lack of peptidoglycan in the cell walls of Methanosarcina barkeri. Arch. Microbiol. 113:57–60.

Kandler, O., and H. König. 1978. Chemical composition of the peptidoglycan-free cell walls of methanogenic bacteria. Arch. Microbiol. 118:141-152.

Kandler, O. (Ed.) 1982. Proceedings of the first international workshop on Archaebacteria. Zbl. Bakt. Hyg., 1. Abt. Orig. C3(2):171–345.

Kandler, O., and W. Zillig (Eds.) 1986. Archaebacteria '85. Gustav Fischer Verlag. Stuttgart, Germany.

Kates, M., L. S. Yengoyan, and P. S. Sastry. 1965. A diether analog of phosphatidyl glycerophosphate in Halobacterium cutirubrum. Biochim. Biophys. Acta 98:252–268.

Kessel, M., and F. Klink. 1980. Archaebacterial elongation factor is ADP-ribosylated by diphtheria toxin. Nature (London) 287:250–251.

Kessel, M., and F. Klink. 1982. Identification and comparison of eighteen Archaebacteria by means of the diphtheria toxin reaction. Zentralbl. Bakt. Mikrobiol. Hyg. Abt. I, Orig. C3:140–148.

Klink, F. 1985. Elongation factors. *In:* C. R. Woese and R. S. Wolfe (Eds.)The Bacteria. Academic Press. New York, NY. VIII:379–410.

Langworthy, T. A., P. F. Smith, and W. R. Mayberry. 1972. Lipids of Thermoplasma acidophilum. J. Bacteriol. 112:1193–1200.

Madigan, M. T., J. M. Martinko, and J. Parker. 1977. Brock, Biology of Microorganisms. Prentice-Hall. Upper Saddle River, NJ.

McBride, B. C., and R. S. Wolfe. 1971. A new coenzyme of methyl transfer, coenzyme M. Biochemistry 10:2317–2324.

Pechman, K. J., and C. R. Woese. 1972. Characterization of the primary structural homology between the 16S ribosomal RNAs of Escherichia coli and Bacillus megaterium by oligomer cataloging. J. Molec. Evol. 1:230–240.

Raven, P. H., and G. B. Johnson. 2002. Biology. McGraw-Hill. New York, NY.

Sanger, F., and H. Tuppy. 1951. The amino-acid sequence in the phenylalanyl chain of insulin. Biochem. J. 49:481–490.

Sanger, F., and E. O. P. Thompson. 1953. The amino-acid sequence in the glycyl chain of insulin. Biochem. J. 53:353–374.

Sanger, F., G. G. Brownlee, and B. G. Barrell. 1965. A two-dimensional fractionation procedure for radioactive nucleotides. J. Molec. Biol. 13:373–398.

Schleifer, K. H., and O. Kandler. 1972. Peptidoglycan types of bacterial cell walls and their taxonomic implications. Bacteriol. Rev. 36:407–477.

Sogin, S. J., M. L. Sogin, and C. R. Woese. 1972. Phylogenetic measurement in prokaryotes by primary structural characterization. J. Molec. Evol. 1:173–184.

Steber, J., and K. H. Schleifer. 1975. Halococcus morrhuae: A sulfated heteropolysaccharide as the structural component of the bacterial cell wall. Arch. Microbiol. 105:173–177.

Stetter, K. O., J. Winter, and R. Hartlieb. 1980. DNA dependent RNA polymerase of the archaebacterium Methanobacterium thermoautotrophicum. Zbl. Bakt. Hyg., I Abt. Orig. C1:201–218.

Sturm, S., V. Schönefeld, W. Zillig, D. Janekovic, and K. O. Stetter. 1980. Structure and function of the DNA dependent RNA polymerase of the archaebacterium Thermoplasma acidophilum. Zbl. Bakt. Hyg., I Abt. Orig. C1:12–25.

Taylor, C. D., and R. S. Wolfe. 1974a. Structure and methylation of coenzyme M. J. Biol. Chem. 249:4879–4885.

Taylor, C. D., B. C. McBride, R. S. Wolfe, and M. P. Bryant. 1974b. Coenzyme M, essential for growth of a rumen strain of Methanobacterium ruminantium. J. Bacteriol. 120:974–975.

Tornabene, T. G., M. Kates, E. Gelpi, and J. Oró. 1969. Occurrence of squalene, di and tetrahydrosqualenes and vitamin MK8 in an extremely halophilic bacterium, Halobacterium cutirubrum. J. Lipid Res. 10:294–303.

Tornabene, T. G., R. S. Wolfe, W. E. Balch, G. Holzer, G. E. Fox, and J. Oró. 1978. Phytanyl-glycerol ethers and squalene in the archaebacterium Methanobacterium thermoautotrophicum. J. Molec. Evol. 11:259–266.

Tzeng, S. F., R. S. Wolfe, and M. P. Bryant. 1975. Factor-420-dependent pyridine nucleotide-linked hydrogenase system of Methanobacterium ruminantium. J. Bacteriol. 121:184–191.

Woese, C. R., M. Sogin, D. Stahl, B. J. Lewis, and L. Bonen. 1976. A comparison of the 16S ribosomal RNAs from mesophilic and thermophilic Bacilli: Some modifications in the Sanger method for RNA sequencing. J. Molec. Evol. 7:197–213.

Woese, C. R., and G. E. Fox. 1977. Phylogenetic structure of the prokaryotic domains: The primary kingdoms. Proc. Natl. Acad. Sci. USA 74:5088-5090.

Woese, C. R., L. J. Magrum, and G. E. Fox. 1978. Archaebacteria. J. Molec. Evol. 11:245–252.

Woese, C. R., O. Kandler, and M. L. Wheelis. 1990. Towards a natural system of organisms. Proposal for the domains Archaea, Bacteria, and Eucarya. Proc. Natl. Acad. Sci. USA 87:4576–4579.

Wolin, E. A., M. J. Wolin, and R. S. Wolfe. 1963. Formation of methane by bacterial extracts. J. Biol. Chem. 238:2882–2886.

Zillig, W., K. O. Stetter, and M. Tobien. 1978. DNA dependent RNA polymerase from Halobacterium halobium. Eur. J. Biochem. 91:193–199.

Zillig, W., K. O. Stetter, and D. Janekovic. 1979. DNA dependent RNA polymerase from the archaebacterium Sulfolobus acidocaldarius. Eur. J. Biochem. 96:597–604.

Zillig, W., K. D. Stetter, R. Schnabel, and M. Thomm. 1985. DNA-dependent RNA polymerases of the Archaebacteria. In: C. R. Woese and R. S. Wolfe (Eds.) The Bacteria. Academic Press. New York, NY. VIII:499–524.

Zukerkandl, E., and L. Pauling. 1965. Molecules as documents of evolutionary history. J. Theor. Biol. 8:357–366.

Prokaryotes (2006) 3:10–22
DOI: 10.1007/0-387-30743-5_2

CHAPTER 2

Thermoproteales

HARALD HUBER, ROBERT HUBER AND KARL O. STETTER

Introduction

The order Thermoproteales (Zillig et al., 1981) is one of the three orders of the archaeal phylum Crenarchaeota (Woese et al., 1990). This branch is further represented by the orders Sulfolobales (Stetter, 1989) and Desulfurococcales (Huber and Stetter, 2001b). Members of the Thermoproteales are rod-shaped extreme thermophiles or hyperthermophiles, which grow either as anaerobes or facultative anaerobes. Under autotrophic conditions, they gain energy by oxidation of hydrogen, using sulfur, thiosulfate, sulfite, oxygen, selenate, and arsenate as electron acceptors. Alternatively, they grow by several types of respiration, using sulfur, oxygen, nitrate, nitrite, arsenate, ferric iron, selenate, selenite, L-cystine, and oxidized glutathione as electron acceptors, or by fermentation of organic substrates. Following the classification listed in the new edition of *Bergey's Manual of Systematic Bacteriology* (Huber and Stetter, 2001a), the order Thermoproteales comprises two validly published families, the Thermoproteaceae (Zillig and Stetter, 1982a) and the Thermofilaceae (Burggraf et al., 1997). The members of these two families can be quite easily distinguished by their morphology: The Thermoproteaceae are rods of at least 0.4 µm in diameter, and the Thermofilaceae are very thin rods of only 0.15–0.35 µm.

The term "sulfur-dependent" archaebacteria (now Archaea) was used formerly to designate the branch which is now the phylum Crenarchaeota. However, several members of the Thermoproteales (and the Desulfurococcales) can replace sulfur by various inorganic or organic compounds for energy formation and are unable to metabolize sulfur or are even inhibited in the presence of sulfur. Therefore, the designation "sulfur-dependent" Archaea ("Archaebacteria") should no longer be used.

In the former editions of *Bergey's Manual of Systematic Bacteriology* (Zillig, 1989) and *The Prokaryotes* (Huber and Stetter, 1992), the Thermoproteales represented, together with the order Sulfolobales, one of the two main lineages of the phylum Crenarchaeota. The order had been comprised of two families (the Thermoproteaceae [Zillig et al., 1981] and the Desulfurococcaceae [Zillig et al., 1982b]), and the genera *Pyrodictium* (Stetter, 1982; Stetter et al., 1983), *Thermodiscus* (Stetter and Zillig, 1985; Stetter, 2001), and *Staphylothermus* (Fiala et al., 1986) had been tentatively placed within the Thermoproteales on the basis of their physiological properties (Huber and Stetter, 1992). Meanwhile, mainly based on 16S rRNA sequence comparisons, these genera are placed either within the Pyrodictiaceae (*Pyrodictium*) or Desulfurococcaceae (*Thermodiscus* and *Staphylothermus*) within the novel order Desulfurococcales (Huber and Stetter, 2001b).

Phylogeny

Based on 16S rDNA sequence data, the order Thermoproteales is a group of the crenarchaeotal branch of the Archaea (Woese et al., 1990). It forms a separate cluster and is distinct from the already mentioned Sulfolobales and Desulfurococcales (Fig. 1). The representatives of the Thermoproteales exhibit phylogenetic similarities between 0.87 and 0.99 to members within the order and 0.77–0.89 (distance matrix values without correction) to all other species of the Crenarchaeota. These values indicate that the order is clearly defined by 16S rDNA gene sequence comparisons and also the two described families are obvious from these results (Fig. 1). Problems arise with members of the genera *Thermoproteus* and *Pyrobaculum*, since they do not form separate clusters in the phylogenetic trees. However, since the two genera are well defined by physiological properties, the present taxonomy is suitable for reidentification of the different species.

Taxonomy

The order Thermoproteales (including the family Thermoproteaceae) was first defined by Zillig and Stetter in 1982 (effective publication: Zillig

Fig. 1. Phylogenetic tree based on 16S rRNA sequences. The tree was calculated using the maximum parsimony program, which is included in the ARB package (Ludwig and Strunk, 2001). Scale bar: 10 estimated exchanges within 100 nucleotides. Red lines = Thermoproteales; blue lines = Desulfurococcales; yellow lines = Sulfolobales; and green lines = Euryarchaeota.

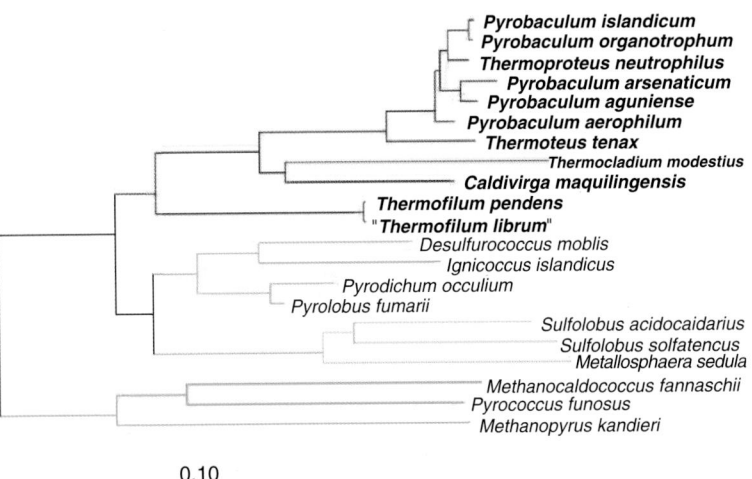

0.10

et al., 1981). The description of both taxa was emended by Burggraf et al. (1997). Today, the Thermoproteales are comprised of two families: the Thermoproteaceae containing four validly published genera and the Thermofilaceae containing a single genus (Huber and Stetter, 2001a).

The genus *Thermoproteus* (Zillig et al., 1981), the type genus of the family Thermoproteaceae and the order Thermoproteales, harbors two validly described species: *T. tenax* (Zillig and Stetter, 1982a) and *T. neutrophilus* (Zillig, 1989; Table 1). The type strain for *T. tenax* is strain Kra1^T (ATCC 35583^T [American Type Culture Collection, Rockville, MD, United States] and DSM 2078^T [Deutsche Sammlung von Mikroorganismen und Zellkulturen GmbH, Braunschweig, Germany]) and for *T. neutrophilus* strain V24Sta^T (DSM 2338^T and JCM 9278^T [Japan Collection of Microorganisms, Wako, Japan]). Furthermore, *T. uzoniensis* was published by Bonch-Osmolovskaya et al. (1990) and validated recently (deposited strain: Z-605^T, DSM 5263^T; Bonch-Osmolovskaya et al., 2001). The three species are characterized by optimal growth temperatures between 85 and 90°C, an anaerobic mode of life, and the capability of sulfur respiration with complex organic substrates. They can be distinguished by their pH optima and the possibility of being a host for four different rod-shaped viruses.

In contrast to *Thermoproteus*, members of the genus *Pyrobaculum* (Huber et al., 1987; Huber and Stetter, 2001c) are characterized by temperature optima between 90 and 100°C and anaerobic or facultatively anaerobic growth. They can respire sulfur, thiosulfate, oxygen, nitrate, nitrite, ferric iron, selenate, selenite, arsenate, L-cystine or oxidized glutathione. At present, five species are described (Table 1): *P. islandicum* (type strain GEO3^T, DSM 4184^T; Huber et al., 1987),

P. organotrophum (type strain H10^T, DSM 4185^T; Huber et al., 1987), *P. aerophilum* (type strain IM2^T, DSM 7523^T; Völkl et al., 1993), *P. arsenaticum* (type strain PZ6*^{T}, DSM 13514^T and ATCC 700994^T; Huber et al., 2000b), and *P. oguniense* (type strain TE7^T; DSM 13380^T and JCM 10595^T; Sako et al., 2001; Table 1). *Pyrobaculum islandicum*, *P. organotrophum*, and *P. arsenaticum* are obligate anaerobes, while the two other species are facultative anaerobes, which can in addition use oxygen as an electron acceptor. *Pyrobaculum arsenaticum* is characterized by the ability to grow on arsenate, which serves as electron acceptor forming arsenite as metabolic end product. Furthermore, its genomic DNA has a significantly higher G+C content than that of the other members of the genus (58.3 mol%). *Pyrobaculum oguniense* exhibits the lowest temperature optimum of all *Pyrobaculum* species (between 90 and 94°C) and is the only species that does not grow at 100°C. In DNA-DNA hybridization experiments, *P. islandicum* and *P. organotrophum* show a similarity of 65%, indicating that they are closely related to each other, although they can be clearly distinguished by physiological properties, such as metabolic pathways. No significant DNA similarity has been obtained between the other *Pyrobaculum* species including members of the genus *Thermoproteus* (Sako et al., 2001). However, 16S rDNA-based phylogenetic studies revealed that *Thermoproteus neutrophilus* clusters with the *Pyrobaculum* species (Fig. 1). The phylogenetic similarities of the 16S rDNA sequences of the organisms within the cluster are quite high (at least 98.5%).

The genera *Caldivirga* (Itoh et al., 1999) and *Thermocladium* (Itoh et al., 1998b) are both represented by a single species. *Caldivirga maquilingensis* (type strain IC-167^T; JCM 10307^T and MCC-UPLB 1200^T [Microbial Culture Collection of the Museum of Natural History,

Table 1. Morphological and physiological characteristics of the validly described species of the *Thermoproteales*.

	Temp (°C) [optimum]	pH range [pH optimum]	NaCl-range (%)	G+C-content (mol%)	Cell size/motility	Aerobic/anaerobic	Metabolism	Electron acceptors	Electron donors	References
Thermoproteus tenax DSM 2078[T]	70–98 [88]	2.5–6.0 [5.5]	n.d.	55.5	0.4 × 1–80 μm	Anaerobic	Facultative lithoautotrophic	S^0, malate	H_2 and single and complex organics	Zillig et al., 1981
Thermoproteus neutrophilus DSM 2338[T]	~97 [85]	5.0–7.5 [6.5]	n.d.	56.2	0.4 × 1–40 μm	Anaerobic	Facultative lithoautotrophic	S^0	H_2 and organic components	Zillig, 1989
Thermoproteus uzoniensis DSM 5263[T]	74–102 [90]	4.6–6.8 [5.6]	n.d.	56.5	0.3–0.4 × 1–20 μm nonmotile	Anaerobic	Obligate heterotrophic	S^0	Peptides	Bonch-Osmolovskaya et al., 1990
Pyrobaculum islandicum DSM 4184[T]	74–102 [100]	5.0–7.0 [6.0]	0–0.8	46	0.5 × 1–8 μm flagellated (bipolar)	Anaerobic	Facultative lithoautotrophic	S^0, $S_2O_3^{2-}$, SO_3^{2-}, oxidized glutathione, L-cystine, and Fe^{3+}	H_2 and complex organic compounds	Huber et al., 1987
Pyrobaculum organotrophum DSM 4185[T]	78–102 [100]	5.0–7.0 [6.0]	0–0.8	47	0.5 × 1–8 μm flagellated (peritrichous)	Anaerobic	Obligate heterotrophic	S^0, L-cystine, and oxidized glutathione	Complex organic compounds	Huber et al., 1987
Pyrobaculum aerophilum DSM 7523[T]	75–104 [100]	5.8–9.0 [7.0]	0–3.6 (1.5)	52	0.6 × 3–20 μm flagellated (monopolar)	Aerobic/anaerobic	Facultative lithoautotrophic	Nitrate, nitrite, Fe^{3+}, O_2, arsenate, selenate, and selenite	H_2, $S_2O_3^{2-}$, and complex organic compounds	Völkl et al., 1993
Pyrobaculum arsenaticum DSM 13514[T]	68–100	n.d.	0–3.0	58.3	0.7 × 3–7 μm flagellated	Anaerobic	Facultative lithoautotrophic	S^0, $S_2O_3^{2-}$, arsenate, and selenate	H_2 and complex organic compounds	Huber et al., 2000b
Pyrobaculum oguniense DSM 13380[T]	70–97 [90–94]	5.4–7.4 [6.3–7.0]	0–1.5 (0)	48	0.6–1.0 × 2–10 μm flagellated (monopolar)	Aerobic/anaerobic	Obligate heterotrophic	O_2, S^0, $S_2O_3^{2-}$, L-cystine, and oxidized glutathione	Complex organic compounds	Sako et al., 2001
Thermocladium modestius JCM 10088[T]	45–82 [75]	2.6–5.9 [4.2]	0–1	52	0.4–0.5 × 5–20 μm nonmotile	Anaerobic (microaero-tolerant)	Obligate heterotrophic	S^0, $S_2O_3^{2-}$, SO_4^{2-} and L-cystine	Complex organic components	Itoh et al., 1998a
Caldivirga maquilingensis JCM 10307[T]	60–92 [85]	2.3–6.4 [3.7–4.2]	0–0.75	43	0.4–0.7 × 3–50 μm nonmotile	Anaerobic (microaero-tolerant)	Obligate heterotrophic	S^0, $S_2O_3^{2-}$, SO_4^{2-}	Complex organic components	Itoh et al., 1999
Thermofilum pendens DSM 2475[T]	70–95 [88]	4.0–6.5 [5.5]	0.1–2	57.4	0.17–0.35 × 5–100 μm nonmotile	Anaerobic	Obligate heterotrophic	S^0	Peptides	Zillig et al., 1983

Abbreviations: DSM, Deutsche Sammlung von Mikroorganismen; [T], type strain; n.d., not determined; and JCM, Japan Collection of Microorganisms.

University of the Philippines, Los Banos, the Philippines]) grows heterotrophically under anaerobic or microaerobic conditions. Optimal growth is obtained at weakly acidic pH (3.7–4.2) at 85°C. By 16S rDNA sequence comparisons, the closest relative of *Caldivirga maquilingensis* is *Thermocladium modestius*, although it clearly represents a different genus. *Thermocladium modestius* (type strain IC-125^T; JCM 10088^T) grows optimally at 75°C, under anaerobic or microaerobic conditions. It is an obligate heterotroph using sulfur, thiosulfate or L-cystine as electron acceptor. Furthermore, the G+C content of the genomic DNA is 52 mol% and therefore significantly different from *Caldivirga maquilingensis* (43 mol%).

The Thermofilaceae represent the second family within the Thermoproteales, which consists so far of only one genus, *Thermofilum* (Zillig et al., 1983). The characteristic feature of the *Thermofilum* strains is rod-shaped cells with diameter of only 0.15–0.35 μm and length of 1–>100 μm. The organisms are obligate anaerobes which grow by sulfur respiration with complex organic substrates. They are hyperthermophilic with temperature optima between 85 and 90°C. So far, only one species is validly published: *Thermofilum pendens* (type strain Hrk5^T; DSM 2475^T and ATCC 35544^T). Furthermore, "*Thermofilum librum*" was mentioned in several publications, but is so far not validly published and is not listed in the last edition of *Bergey's Manual of Systematic Bacteriology* (Zillig and Reysenbach, 2001).

Habitat

Members of the Thermoproteales have been found exclusively in biotopes of volcanic activity with temperatures up to 100°C and even higher. Representatives of the genera *Thermoproteus*, *Pyrobaculum* and *Thermofilum* have been isolated from springs, water holes, mud holes, and soils of continental solfataric fields with low salinity and acidic to neutral pH values (1.7–7) and found in many places worldwide, including Pisciarelli Solfatara (Italy), Ribeira Quente (the Azores), the Krafla area (Iceland), Yellowstone National Park (United States), Uzon Caldera (Kamchatka), Indonesia, and New Zealand. *Caldivirga* has been obtained from an acidic hot spring in the Philippines and *Thermocladium* from several solfataric areas in Hokkaido, Akita, and Fukushima, Japan. *Pyrobaculum islandicum* was isolated from an outflow of superheated water of an overpressure valve at the Krafla geothermal power plant (Iceland). As an exception, *Pyrobaculum aerophilum* is the only species obtained from a marine hydrothermal system, located at the Maronti Beach, Ischia Island, Italy.

Isolation

Enrichment

Thermoproteus species can be enriched in anaerobic low-salt medium (Allen, 1959), supplemented with elemental sulfur, with a gas phase of H_2/CO_2 (80:20, v/v) at incubation temperatures around 85°C. Alternatively, peptides, amino acids, glycogen, starch, or acetate can serve as carbon sources during sulfur respiration. Members of the genus *Pyrobaculum* can be selectively enriched using incubation temperatures of around 100°C. With the exception of *P. aerophilum* (which needs a half strength seawater medium for enrichment and cultivation), all *Pyrobaculum* species grow in low salt media (pH ~6), which are supplemented with complex organic substrates, like peptides or yeast extract. *Pyrobaculum arsenaticum* can be selectively enriched using arsenate as sole electron acceptor, while *P. islandicum* can be grown in a medium without organic components. For *P. aerophilum* and *P. oguniense*, enrichment under microaerobic conditions is successful. A highly selective enrichment method for strains of the extremely thin *Thermofilum* cells involves passage through a 0.2 μm sterile filter and incubation of the filtrate at 85°C with yeast extract or cell extract of *Thermoproteus tenax*, and elemental sulfur, with a gas phase of H_2/CO_2 (80:20, v/v).

Isolation Procedures

Anaerobic or microaerobic culture media are inoculated with 0.5–1 ml of environmental samples. The culture attempts are examined for growth by phase microscopy over a period of up to four weeks. The isolates are obtained from the enrichment cultures by serial dilutions (three times) or by plating. The plates can be solidified by 20% starch or by 0.6–0.8% Gelrite, and colonies are visible within 1–2 weeks.

Selected single cell isolation using the "optical tweezers trap":

For a safe, efficient and fast isolation of organisms from mixed cultures, including the Thermoproteales, a novel, plating-independent isolation procedure was developed (Huber et al., 1995). The method is based on separation of a single cell from enrichment cultures by the use of a laser microscope and subsequent growth of this cell. The "optical tweezers trap" (also called the "laser trap") consists of a computer-controlled inverted microscope. This microscope is equipped with a continuously operating neodymium-doped yttrium aluminum garnet laser (Nd:YAG laser) with an emission wavelength of 1064 nm, which is near infrared, and a maximum output power of 2.5 Watts. By the use of a high-numerical-aperture oil immersion objective (magnification 100X), the laser can be focused to a spot size of less than 1 μm in diameter. The strong intensity of the laser light makes optical trapping and manipulation in three dimensions of single μm-size cells possible (Ashkin and Dziedzic, 1987a; Ashkin et al., 1987b). The micromanipulation of the cells is performed by keeping the laser beam at a fixed position and moving the motor-driven mechanical stage.

The cell separation unit consists of a rectangular micro-slide as an observation and separation chamber (inside dimension: 0.1×1 mm^2; length: 10 cm), which is connected by a tube to the needle of a 1-ml syringe (Huber et al., 1995; Huber, 1999; Fig. 2a). A cutting line separates the micro-slide into two compartments (Fig. 2a). After sterilization of the cell separation unit, about 90% of the micro-slide is filled with sterile medium. The mixed culture is then filled into the remaining volume of the micro-slide. Under 1000-fold magnification, a single cell is selected and is optically trapped in the laser beam by activation of the laser (Fig. 2a). Within 3–10 min, this cell can be separated at least 6 cm from the mixed culture into the sterile compartment by moving the microscopic stage (Huber, 1999; Fig. 2b). The micro-slide is gently broken at the cutting line, and the single cell is flushed into sterile medium ("selected cell cultivation technique"). Depending on the strain, 20–100% of the separated cells from logarithmically growing enrichment cultures generate pure cultures within only 1–5 days. In addition, the combination of the "selected cell cultivation technique" and the use of the membrane potential-sensitive probe 1,3-dibutylbarbituric acid trimethine oxonol (DiBAC$_4$(3)) can enhance the isolation of novel organisms either from enrichment cultures or directly from natural samples (Beck and Huber, 1997).

Identification

Members of the Thermoproteales are rod-shaped cells about 0.15–0.7 μm in diameter and 1–~100 μm in length (Table 1). They do not divide by septa formation, but snapping division has been observed (Horn et al., 1999). Spores are not formed. The cell envelope S-layer is composed of protein or glycoprotein subunits in hexagonal dense packing; no muramic acid is present. The DNA-dependent RNA polymerase is resistant to rifampicin and streptolydigin. Lipids contain glycerol ethers of polyisoprenoid C$_{40}$ alcohols and a lesser amount of C$_{20}$ alcohols (De Rosa et al., 1986; Langworthy and Pond, 1986). The G+C contents of genomic DNA range from 43 to 58 mol% (Table 1). The Thermoproteales are anaerobic to facultatively anaerobic organisms with optimal growth temperatures between 75 and 100°C and a pH range from 2 to 8 with an optimum around 5 to 6.5.

Cells of *Thermoproteus* are straight, nonmotile rods, with pili attached laterally and/or terminally. Branching forms are occasionally found. When cells are observed at room temperature under aerobic conditions, spherical bodies ("golf clubs") can be observed, mostly at the terminal end of the rods (Zillig et al., 1981; Fig. 3a, b). In

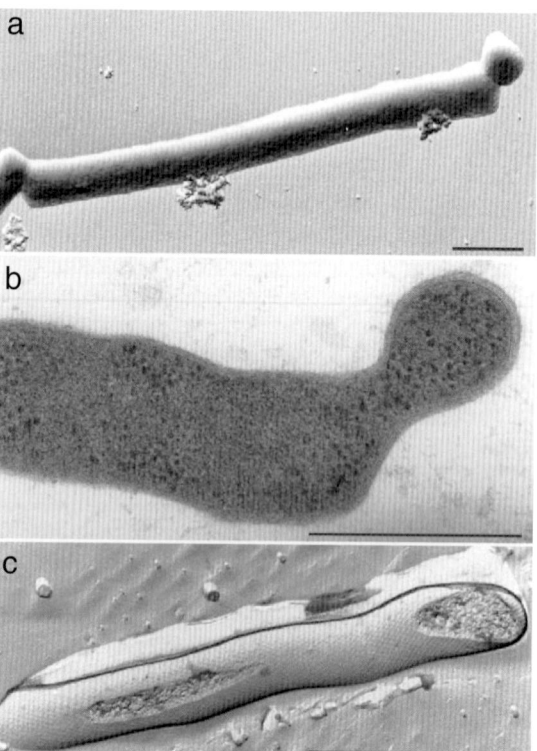

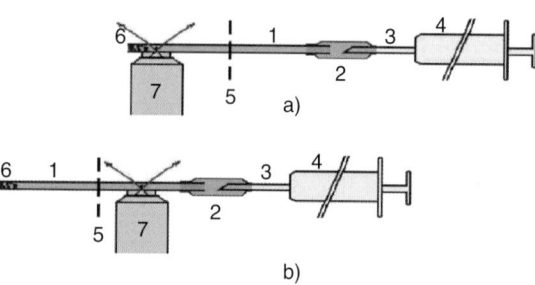

Fig. 2. A schematic drawing showing the isolation of a single cell from an enrichment culture by use of a laser microscope. a) A single cell is optically trapped within the focus of the laser beam. b) The single cell is separated from the mixed culture into the sterile compartment. 1: Microslide; 2: tube; 3: needle; 4: syringe; 5: cutting line; 6: mixed culture; and 7: objective.

Fig. 3. Electron micrographs of cells of *Thermoproteus tenax*: a) Pt-shadowed; b) thin section; and c) freeze-etched. Bar: 1 μm for a) and c) and 0.5 μm for b).

contrast, they cannot be detected under anaerobic conditions at growth temperature, using a phase contrast microscope with a heatable stage (Horn et al., 1999). The cell envelopes of *Thermoproteus tenax* and *Thermoproteus neutrophilus* exhibit proteins arranged in p6-symmetry and a lattice constant of approx. 30 nm (Messner et al., 1986). Analysis of the lattice orientation, together with the number and distribution of lattice faults on intact cells, provides a strong indication that the S-layers of both organisms have a shape determining function (Messner et al., 1986). The layer of *Thermoproteus tenax* can be seen as a helical structure consisting of a right-handed, two-stranded helix, with the individual chains running parallel (Wildhaber and Baumeister, 1987; Fig. 3c). For *T. tenax*, four different temperate viruses of the *Lipothrixviridae* and *Bacilloviridae* have been identified (Zillig et al., 1996). While the *Lipothrixviridae* are flexible filaments, the *Bacilloviridae* are stiff rods. New types of menaquinones have been described within the genus *Thermoproteus* (Thurl et al., 1985). The DNA-dependent RNA polymerase shows the BAC-type and is resistant to rifampicin and streptolydigin. Furthermore, two D-glycerate-3-phosphate dehydrogenases have been characterized (Hensel et al., 1987).

Cells of *Pyrobaculum* are cylinder-shaped, rigid rods with nearly rectangular ends (Fig. 4). About 1–10% of the rods form terminal spherical bodies at the end of the exponential growth phase, similar to the "golf clubs" of *Thermoproteus* (Zillig et al., 1981). In the stationary phase, the terminal spheres of *Pyrobaculum aerophilum* tend to enlarge, and several rods convert completely into spheres during further incubation. Cultures of *P. aerophilum* contain about 50% coccoid cells at high nitrate concentrations or pH values above 8.0. *Pyrobaculum* species are flagellated and show motility, which can be strongly enhanced by heating a microscopic slide to about 90°C. The cell envelope of *Pyrobaculum organotrophum* is composed of two distinct hexagonally arrayed crystalline protein layers. Fibrils appear to be sandwiched between the two layers. The outer layer has a p6 symmetry and a

lattice spacing of 20.6 nm (Phipps et al., 1991). In contrast, the cell wall of *P. aerophilum* and *Pyrobaculum islandicum* consists of one S-layer with a p6 symmetry and is connected via a spacer to the cytoplasmic membrane (Phipps et al., 1990; Völkl et al., 1993). An additional surface coat of fibrillar material covers the S-layer (Phipps et al., 1990; Rieger et al., 1997). A novel type of cell-to-cell connection in the form of thin tubes approximately 15–20 nm in diameter was identified in *P. aerophilum* (Rieger et al., 1997). The complex lipids of the *Pyrobaculum* species consist mainly of phosphoglycolipids of types I and II, while aminophospholipids are absent (Trincone et al., 1992; Völkl et al., 1993). The core lipids are mainly composed of acyclic and cyclic glycerol diphytanyl glycerol tetraethers with one to four pentacyclic rings. In addition, traces of glycerol diphythanyl glycerol diethers are present (Trincone et al., 1992; Völkl et al., 1993). Fully saturated menaquinones have been identified in *P. islandicum* and *P. organotrophum* (Tindall, 1989; Tindall et al., 1991). In *P. islandicum*, a highly thermostable glutamate dehydrogenase was identified, which requires NAD^+ as a coenzyme for L-glutamate deamination (Kujo and Ohshima, 1998). Furthermore, a dissimilatory siroheme-sulfite-reductase-type protein was detected, which has an $\alpha_2\beta_2$ structure and contains high-spin siroheme, non-heme iron and acid-labile sulfide (Molitor et al., 1998). In addition, *P. islandicum* contains two new tRNA nucleosides, 3-hydroxy-*N*-[[(9-β-D-ribofuranosyl-9*H*-purine-6-yl)amino]carbonyl]norvaline and 3-hydroxy-*N*-[[(9-β-D-ribofuranosyl-9*H*-2-methyl-thiopurine-6-yl)amino]carbonyl]norvaline (Reddy et al., 1992). In *P. aerophilum*, a subtilisin-type protease ("aerolysin") with activity between 80–130°C was identified, and cytochrome oxidases with novel prenylated hemes as cofactors were present (Lübben and Morand, 1994; Völkl et al., 1994). *Pyrobaculum islandicum* and *P. organotrophum* are resistant to streptomycin, phosphomycin, chloramphenicol, penicillin G and vancomycin (100 µg/ml). In the presence of rifampicin (100 µg/ml), a temporary growth inhibition occurs.

Cells of *Thermocladium* (Itoh et al., 1998b) are straight or slightly curved rods (Table 1), which are nonmotile and multiply by branching and budding. In addition, bent cells, branched cells forming T-shapes or Y-shapes, and extensively branched cells occur. At least five tetraether core lipids containing cyclopentane rings in the isoprenyl chains have been identified. The organism is insensitive to chloramphenicol, kanamycin and oleandomycin, but sensitive to rifampicin.

Cells of *Caldivirga* (Itoh et al., 1999) are straight or slightly curved rods, occasionally bent

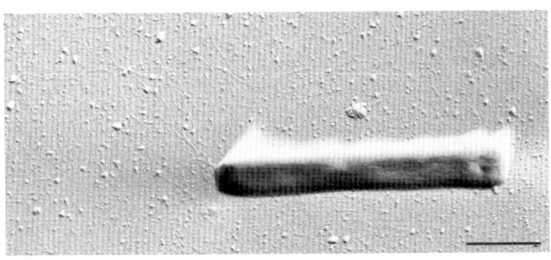

Fig. 4. Electron micrograph of a Pt-shadowed *Pyrobaculum aerophilum* cell with monopolar flagellation. Bar: 1 µm.

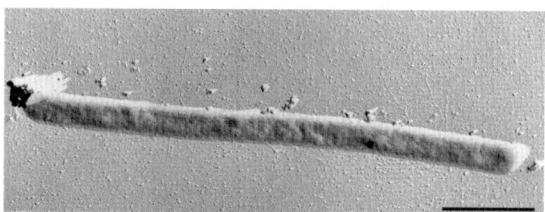

Fig. 5. Electron micrograph of a single Pt-shadowed cell of *Thermofilum pendens*. Bar: 1 μm.

or branched. They are highly variable in length (usually 3–20 μm), but occasionally reach up to 50 μm (Table 1). Furthermore, globular bodies are observed, located at the cell poles, sometimes also laterally. Pili are attached to the cells, and no motility is observed. *Caldivirga* possesses at least five cyclized glycerol-bisphythanyl-glycerol tetraethers. The organism is insensitive to chloramphenicol, kanamycin, streptomycin and oleandomycin, but sensitive to rifampicin, erythromycin and novobiocin.

Cells of *Thermofilum* (Zillig et al., 1983) are extremely thin rods (Fig. 5) and can be easily overlooked in the microscope. The length varies between 1 and 100 μm, but the diameter of the cells is only 0.17–0.35 μm. Cells are only rarely branched or with sharp bends. Tetraether lipids in combination with very low amounts of 2,3-di-*O*-phythanyl-*sn*-glycerol are typical for this genus.

Cultivation

Media for Cultivation

Basal Culture Medium for *Thermoproteus, Thermofilum, Pyrobaculum islandicum, Pyrobaculum organotrophum* and *Pyrobaculum oguniense*
(Modified Medium of Allen, 1959)

$(NH_4)_2SO_4$	1.30 g
KH_2PO_4	0.28 g
$MgSO_4 \cdot 7H_2O$	0.25 g
$CaCl_2 \cdot 2H_2O$	0.07 g
$FeCl_3 \cdot 6H_2O$	0.02 g
$MnCl_2 \cdot 4H_2O$ (10 mg/ml)	180 μl
$Na_2B_4O_7 \cdot 10H_2O$ (10 mg/ml)	450 μl
$ZnSO_4 \cdot 7H_2O$ (10 mg/ml)	22 μl
$CuCl_2 \cdot 2H_2O$ (10 mg/ml)	5 μl
$Na_2MoO_4 \cdot 2H_2O$ (10 mg/ml)	3 μl
$VoSO_4 \cdot 5H_2O$ (10 mg/ml)	3 μl
$CoSO_4 \cdot 7H_2O$ (10 mg/ml)	1 μl
Yeast extract (Bacto Difco)	0.20 g
Resazurin	1.0 mg
Elemental sulfur	2.0 g

Dissolve components in and dilute up to 1 liter with distilled water.

For *P. islandicum* and *P. organotrophum*, add 0.05% peptone (Bacto Difco), adjust pH to 6.0, and grow in gas phase of N_2 (300 kPa). Both

species can also be cultivated in 1/20 MG-CB-medium (see below).

For *P. oguniense*, replace sulfur by 0.1% $Na_2S_2O_3 \cdot 5H_2O$, adjust pH to 6.5–7.5, and grow aerobically in gas phase of air or anaerobically in N_2 (200 kPa).

For *Thermofilum*, add 20 ml of 10% extract (see below) of *Thermoproteus tenax* or *Methanothermus fervidus* per liter of medium, adjust pH to 6.5, and grow in gas phase of N_2/CO_2 (300 kPa).

For *Thermoproteus tenax*, adjust pH to 5.5, and grow in gas phase of N_2/CO_2 (300 kPa).

For *T. neutrophilus*, adjust pH to 6.5, and grow in a gas phase of H_2/CO_2 (300 kPa).

Culture Medium for *Thermoproteus uzoniensis*
(Bonch-Osmolovskaya et al., 1990)

NH_4Cl	0.33 g
KCl	0.33 g
$MgCl_2 \cdot 2H_2O$	0.33 g
$CaCl_2 \cdot 2H_2O$	0.33 g
KH_2PO_4	0.33 g
Peptone	2.5 g
Yeast extract	0.1 g
Elemental sulfur	10.0 g
Trace element solution (see below)	1.0 ml
Resazurin	1.5 mg

Dissolve components in and dilute up to 1 liter with distilled water. Adjust pH to 6.5 and grow in gas phase of N_2/CO_2 (300 kPa).

Trace Element Solution
(Pfennig and Lippert, 1966)

Nitrilotriacetic acid	500 mg
$FeSO_4 \cdot 7H_2O$	200 mg
H_3BO_3	30 mg
$CoCl_2 \cdot 6H_2O$	20 mg
$ZnSO_4 \cdot 7H_2O$	10 mg
$MnCl_2 \cdot 4H_2O$	3 mg
$Na_2MoO_4 \cdot 2H_2O$	3 mg
$NiCl_2 \cdot 6H_2O$	2 mg
$CuCl_2 \cdot 2H_2O$	1 mg

Dissolve components in and dilute up to 1 liter with distilled water. Adjust pH to 6.5 with KOH.

Culture Medium for *Pyrobaculum aerophilum*
(BS medium; Völkl et al., 1993)

NaCl	5.89 g
$MgCl_2 \cdot 6H_2O$	2.82 g
$NaHCO_3$	2.20 g
Na_2SO_4	0.41 g
$CaCl_2 \cdot 2H_2O$	0.39 g
NH_4Cl	0.25 g
KCl	0.15 g
$(NH_4)_2Fe(SO_4)_2 \cdot 6H_2O$ (0.2%)	1.0 ml
KJ solution (0.05%)	0.1 ml
BS stock solution	10.0 ml
Trace elements I	2.5 ml
Trace minerals II	1.0 ml
Trace vitamin solution	1.0 ml
Resazurin (0.1%)	1.0 ml
KNO_3	1.0 g

Dissolve components in and dilute up to 1 liter with distilled water. Adjust pH to 7.0, and grow in gas phase of H_2/CO_2 (300 kPa).

BS Stock Solution

KH_2PO_4	7.0 g
Na_2CO_3	1.2 g
NaBr	1.3 g
H_3BO_3	0.38 g
$SrCl_2 \cdot 6H_2O$	0.19 g

Dissolve components in and dilute up to 1 liter with distilled water.

Trace Elements I

KBr	4.0 g
$SrCl_2 \cdot 6H_2O$	2.85 g
H_3BO_3	1.1 g
$Na_2HPO_4 \cdot 2H_2O$	0.5 g
$Na_2SiO_3 \cdot 5H_2O$	0.2 g
NaF	0.12 g

Dissolve components in and dilute up to 1 liter with distilled water. Adjust pH to 7.0.

Trace Minerals II

Nitrilotriacetic acid	15.0 g
$MgSO_4 \cdot 7H_2O$	30.0 g
$MnSO_4 \cdot H_2O$	5.0 g
NaCl	10.0 g
$Ni(NH_4)_2(SO_4)_2 \cdot 6H_2O$	2.8 g
$CoSO_4 \cdot 7H_2O$	1.8 g
$ZnSO_4$	1.8 g
$FeSO_4 \cdot 7H_2O$	1.0 g
$CaCl_2 \cdot 2H_2O$	1.0 g
$KAl(SO_4)_2 \cdot 12H_2O$	0.18 g
$CuSO_4 \cdot 5H_2O$	0.1 g
H_3BO_3	0.1 g
$Na_2MoO_4 \cdot 2H_2O$	0.1 g
$Na_2WO_4 \cdot 2H_2O$	0.1 g
Na_2SeO_4	0.1 g

Dissolve components in and dilute up to 1 liter with distilled water. Adjust pH to 6.5 with KOH.

Trace Vitamin Solution

(Balch et al., 1979)

Vitamin B_6 (pyridoxine) hydrochloride	10 mg
Vitamin B_1 (thiamine) hydrochloride	5 mg
Vitamin B_2 (riboflavin)	5 mg
Nicotinic acid (niacin)	5 mg
Pantothenic acid (DL Ca-pantothenate)	5 mg
PABA (p-aminobenzoic acid)	5 mg
Lipoic (thioctic) acid	5 mg
Biotin	2 mg
Folic acid	2 mg
B12 (crystalline)	0.1 mg

Dissolve components in and dilute up to 1 liter with distilled water.

Culture Medium for *Pyrobaculum arsenaticum*, *Pyrobaculum islandicum* and *Pyrobaculum organotrophum* (1/20 MG-CB-medium)

$MgCl_2 \cdot 6H_2O$	1215 mg
NaCl	900 mg
$NaHCO_3$	500 mg
$CaCO_3$	50 mg

KCl	17 mg
NH_4Cl	12 mg
$CaCl_2 \cdot 2H_2O$	7 mg
$K_2HPO_4 \cdot 3H_2O$	7 mg
$(NH_4)_2Fe(SO_4)_2 \cdot 6H_2O$	0.1 mg
Trace vitamin solution (see above)	10 ml
Trace mineral solution (see below)	10 ml
Resazurin	5 mg

Dissolve components in and dilute up to 1 liter with distilled water. Adjust pH to 6.5–7 with HCl, and grow in gas phase of H_2/CO_2 (300 kPa).

Trace Mineral Solution

(a modification of that described by Balch et al., 1979)

$MgCl_2 \cdot 6H_2O$	2.48 g
$MnCl_2 \cdot 4H_2O$	0.53 g
NaCl	1.0 g
$FeCl_2 \cdot 4H_2O$	0.72 g
$CoCl_2 \cdot 6H_2O$	0.18 g
$CaCl_2 \cdot 2H_2O$	0.1 g
$ZnCl_2$	0.09 g
$CuCl_2 \cdot 2H_2O$	6.8 mg
KCl	3.15 mg
$AlCl_3$	5.6 mg
H_3BO_3	0.01 g
$Na_2MoO_4 \cdot 2H_2O$	0.01 g
$Na_2WO_4 \cdot 2H_2O$	0.01 g
Na_2SeO_4	0.01 g
$NiCl_2 \cdot 6H_2O$	0.14 g

Dissolve components in and dilute up to 1 liter with distilled water. Adjust to pH 1.5 with HCl.

Culture Medium for *Thermocladium* and *Caldivirga*

(TCD medium; Itoh et al., 1998b)

$(NH_4)_2SO_4$	1.30 g
KH_2PO_4	0.28 g
$MgSO_4 \cdot 7H_2O$	0.25 g
$CaCl_2 \cdot 2H_2O$	0.07 g
Combined trace element solution	1.0 ml
Yeast extract	0.5 g
Resazurin	1.0 mg
Sulfur powder	10.0 g

Dissolve components in and dilute up to 1 liter with distilled water. Adjust pH to 4-4.5 and grow in gas phase of H_2/CO_2 or N_2 (100 kPa).

Combined Trace Element Solution

$FeCl_3 \cdot 6H_2O$	20.00 g
$Na_2B_4O_7 \cdot 10H_2O$	4.50 g
$MnCl_2 \cdot 4H_2O$	1.80 g
$ZnSO_4 \cdot 7H_2O$	0.22 g
$CuCl_2 \cdot 2H_2O$	0.05 g
$VoSO_4 \cdot 5H_2O$	0.04 g
$Na_2MoO_4 \cdot 2H_2O$	0.03 g
$CoSO_4 \cdot 7H_2O$	0.01 g

Dissolve components in and dilute up to 1 liter with distilled water. Adjust pH to 1.5.

For growth of *Thermocladium modestius*, add 1 ml of archaeal cell-extract solution from *Sulfolobus acidocaldarius*, *Methanosarcina barkeri*, *Halobacterium salinarum* or *Thermoplasma acidophilum*.

For growth of *Caldivirga maquilingensis*, add 10 ml of trace vitamin solution (see above).

Preparation of Cell Extracts

Ten g of cells (wet weight) are suspended in 40 ml of distilled water and disrupted by a French press. The volume of the homogenate is adjusted to 100 ml, the homogenate is centrifuged, and the supernatant is passed through a sterile filter (0.2-μm pore width) into a sterile serum bottle. After gassing the filtered supernatant with nitrogen, oxygen is reduced by the addition of 1 ml of Na_2S solution (2.5%). The extracts (designated "10% extracts") are stored at 4°C.

Cultivation in Liquid Media

All anaerobic members of the Thermoproteales are cultivated using the anaerobic technique of Balch and Wolfe (1976). For small volumes, the organisms are grown in stoppered and pressurized 28-ml serum tubes (boric silicate glass; Schott, Mainz, Germany) or in 120-ml serum bottles (Bormioli, Italy) made of "type III" glass. One liter of the medium is flushed with nitrogen for 20 min in a stoppered glass bottle. Then, 0.5 g of $Na_2S \cdot 9H_2O$, dissolved in 1 ml of medium, are added with a syringe and the medium is adjusted to the desired pH with H_2SO_4 or HCl. The medium is dispensed in an anaerobic chamber. The culture vessels are stoppered, and the gas phase is exchanged three times. Finally, the desired gas phase is pressurized.

Cultivation on Plates

Microcolonies (diameter about 0.5 mm) of *Thermoproteus* and *Thermofilum* are obtained on 5% polyacrylamide gels (Zillig et al., 1981) at 85°C in an atmosphere of 100 kPa N_2 plus 20 kPa H_2S and 20 kPa CO_2 in an anaerobic jar. *Pyrobaculum islandicum* can be plated under anaerobic conditions, using a stainless-steel anaerobic jar (Balch et al., 1979) and plates of medium solidified by 20% starch and containing thiosulfate instead of sulfur. After anaerobic incubation for about seven days at 85°C, tiny, greenish-black colonies will be formed. *Pyrobaculum aerophilum* can be cultivated on plates of medium solidified by 0.6% Gelrite, and these plates are incubated in a stainless-steel anaerobic jar at 85°C. (Note: If the medium is of low ionic strength, 0.4% $MgSO_4$ should be added for a better solidification of the Gelrite). Round grayish-yellow colonies will be visible after about two weeks. High plating efficiencies approaching 100% can be obtained on Gelrite plates at 92°C under aerobic conditions and an incubation time of only four days (Völkl et al., 1993). For *Thermocladium modestius* and *Caldivirga maquilingensis*, no plating procedures have been reported.

Preservation

For short-time storage, most Thermoproteales isolates can be stored at 4°C for several months. In contrast, *P. aerophilum* should be transferred once a week, when stored at 4°C. Long-term storage is best achieved in culture media containing 5% dimethyl sulfoxide and in the gas phase over liquid nitrogen (which is around –140°C). No loss of cell viability of *Pyrobaculum-* and *Thermoproteus* relatives was observed after storage of about ten years.

Physiology

Within the Thermoproteales, only extreme thermophiles or hyperthermophiles are known with optimal growth temperatures between 75 and 100°C. With the exception of *Thermocladium modestius*, no Thermoproteales organisms have been observed to grow below 60°C (Table 1). Almost all Thermoproteales grow best at slightly acidic pH values (pH 4–6; Table 1). Since the natural habitats of all Thermoproteales except *Pyrobaculum aerophilum* are terrestrial solfataric fields, optimal growth is obtained in media of low ionic strength (Table 1). All members of the Thermoproteales are able to grow under strictly anaerobic culture conditions. In addition, *Pyrobaculum aerophilum*, *Thermocladium modestius* and *Caldivirga maquilingensis* propagate in the presence of low oxygen concentrations (e.g., *Pyrobaculum aerophilum* up to 6% oxygen), while *Pyrobaculum oguniense* can grow under aerobic conditions.

Facultative and obligate heterotrophs are known within the Thermoproteales (Table 1). Under autotrophic conditions, molecular hydrogen is used as the electron donor, and sulfur is reduced to hydrogen sulfide (H_2S). Beside sulfur, *Pyrobaculum arsenaticum* can use arsenate and thiosulfate as alternative electron acceptors. *Pyrobaculum aerophilum* grows autotrophically with hydrogen and thiosulfate, arsenate, nitrate, oxygen or selenate (Völkl et al., 1993; Huber et al., 2000a).

Heterotrophic growth occurs by sulfur respiration using various organic substrates (e.g., glucose, starch, glycogen, fumarate, peptides, and several amino acids [genus *Thermoproteus*]) and gelatin and complex organic substrates (e.g., beef extract, peptone, tryptone, and yeast extract [genera *Thermofilum*, *Caldivirga* and *Thermocladium*]). Besides sulfur, thiosulfate or sulfate (*Caldivirga*) or L-cystine (*Thermocladium*) can serve as an alternative electron acceptor. *Pyrobaculum oguniense* and *Pyrobaculum organotrophum* (both obligate heterotrophs) and

Pyrobaculum islandicum grow anaerobically on peptone, yeast extract, meat extract and cell homogenates of bacteria (e.g., *Thermotoga maritima*) or archaea (e.g., *Staphylothermus marinus*) in the presence of sulfur, L(–)-cystine or oxidized glutathione or on yeast extract combined with hydrogen and ferric iron (Kashefi and Lovley, 2000). DL-Lanthionine, tetrathionate, thiosulfate, sulfite, dimethyl sulfone, fumarate, or sulfate is not used as electron acceptor. No growth occurs on glycogen, starch, maltose, galactose, glucose, casamino acids, formamide, methanol, ethanol, formate, malate, propionate, L(+)lactate or acetate. *Pyrobaculum aerophilum* grows microaerophilically in the presence of complex organic substrates (and propionate and acetate) by nitrate reduction (producing nitrite and traces of nitric oxide [NO]) and by nitrite reduction (forming N_2 and traces of N_2O and NO as final products). Nitrous oxide (N_2O) is not used as an alternative electron acceptor. Growth of *Pyrobaculum aerophilum* and *Pyrobaculum oguniense* was inhibited by elemental sulfur. *Pyrobaculum arsenaticum* grows anaerobically by respiration of arsenate, selenate, thiosulfate or elemental sulfur.

In *Pyrobaculum islandicum*, *Pyrobaculum organotrophum* and *Thermoproteus tenax* the complete oxidation of organic compounds to CO_2 proceeds via the citric acid cycle (Selig and Schönheit, 1994). Furthermore, in *Thermoproteus tenax*, two glyceraldehyde phosphate (GAP)-dehydrogenases have been found, one specific for NAD^+, the second for $NADP^+$. In *Thermoproteus neutrophilus*, a reductive citric acid cycle for carbon assimilation was found (Schäfer et al., 1986). Glycogen was present in *Thermoproteus* (König et al., 1982), while trehalose was absent (Nicolaus et al., 1988).

Thermofilum pendens has unusual growth factors. For growth, the cell debris or a polar lipid fraction of *Thermoproteus tenax* must be added to the medium (Zillig et al., 1983). These fractions can be substituted by a similar fraction obtained from methanogens (Stetter, 1986) but not by a fraction from *Thermoplasma* (Zillig et al., 1983) or by combined fractions from different bacteria (Stetter, 1986).

Genetics

A 713-bp intron, which cyclizes upon excision, was identified in the 16S rRNA gene of *Pyrobaculum aerophilum*. The intron contains an open reading frame whose protein translation shows no statistically significant homology with any known protein sequence (Burggraf et al., 1993). Two introns occur within the single 23S rRNA-

encoding gene of *Pyrobaculum organotrophum*. The RNA products also circulate after excision from the 23S rRNA and are stable in the cell (Dalgaard and Garrett, 1992). Furthermore, an intron was detected in the 16S rRNA genes of *Thermoproteus neutrophilus* and *Thermoproteus* spp. (Itoh et al., 1998a), while two introns were located in this gene in *Caldivirga maquilingensis* (Itoh et al., 1999). However, the taxonomic significance of these introns is still unclear.

The sequencing and annotation of the complete 2.2 Mb genome of *Pyrobaculum aerophilum* by three groups under the leadership of the California Institute of Technology were published very recently (Fitz-Gibbon et al., 2002). This sequence has been deposited in the GenBank database (accession number).

Ecology

Representatives of the Thermoproteales thrive in neutral to slightly acidic biotopes, covering temperatures from 45 up to 104°C. With the exception of *Pyrobaculum aerophilum*, they are typical inhabitants of high-temperature continental solfataric areas. They have been isolated from such biotopes nearly world-wide. They often share these habitats with other thermophilic and hyperthermophilic organisms such as members of the orders Sulfolobales, Desulfurococcales, Thermotogales and Aquificales. Such hot solfataric soils often consist of two dominating different zones. The upper oxidized zone is often rich in ferric iron and therefore colored orange. In a depth between a few centimeters and about 50 cm, depending on the volcanic activity and emission of reducing volcanic gases (e.g., H_2S and H_2) below these oxidized layers, the anaerobic zone (often black-colored due to the presence of ferrous sulfide) begins. The chemical composition of the solfataric fields is highly variable and depends on the site. Often iron, followed in quantity by arsenic and antimony, is found in these habitats (Huber et al., 2000a).

In their ecosystem, the Thermoproteales are important organisms within complex food webs. They can function as primary producers and/or as consumers of organic material. As primary producers, they use oxygen, elemental sulfur, sulfate, thiosulfate, sulfite, and nitrate as electron acceptors for growth and molecular hydrogen as the electron donor in these energy-yielding reactions. Recently, it was shown that heavy metal compounds are also suitable electron acceptors for several members of *Pyrobaculum* that can reduce ferric iron to form magnetite (Stetter, 1999). By coexistence with anaerobic iron oxidiz-

Fig. 6. Organotrophic growth of *Pyrobaculum arsenaticum* at 95°C in the presence of 10 mM arsenate and 0.1% thiosulfate (gas phase 300 kPa H_2/CO_2 = 80:20, v/v). a) Yellow-orange precipitate is realgar (As_2S_2) and b) uninoculated control.

ers such as *Ferroglobus placidus*, a hot iron cycle can be postulated, which may have existed already within the ecosystems of early life forms on earth (Stetter, 2000). Also, autotrophic arsenate and/or selenate reduction with hydrogen as the electron donor occurs under anaerobic conditions by *Pyrobaculum arsenaticum* and *Pyrobaculum aerophilum* (Huber et al., 2000b).

The obligate heterotrophs, growing by different types of respiration or fermentation, are consumers of the organic matter. For anaerobic respiration, sulfur, sulfurous compounds and nitrate are used as electron acceptors. *Pyrobaculum arsenaticum* grows heterotrophically in the presence of thiosulfate or arsenate. Interestingly, when both acceptors are supplied simultaneously, a precipitate (identified as realgar [As_2S_2]) is formed during growth and is visible as yellow-orange flocs in the culture medium (Fig. 6). Arsenate-reducing hyperthermophiles occur in high densities in these natural environments. Pisciarelli Solfatara sediment has been found to contain at least 10^7 viable cells of arsenate reducers per gram (Huber et al., 2000a). Considering these results, hyperthermophilic arsenate reducers may be important members in their high-temperature environment and may be a critical part of a possible combined arsenic and sulfur cycle (Huber et al., 2000b).

Biotechnology

A continuously increasing number of heat stable enzymes from hyperthermophiles is being checked for industrial and research applications. This includes extracellular enzymes (e.g., amylases, lipases, proteases, or xylanases) and intracellular enzymes (e.g., dehydrogenases, oxidoreductases, and DNA polymerases). The enzymes are usually not extracted from their "parental" hyperthermophiles, but are produced from genetically engineered mesophilic production strains. The total genome sequence of *Pyrobaculum aerophilum* is now available, offering the opportunity to use its genes as tools for different biotechnological applications (Fitz-Gibbon et al., 2002).

Acknowledgments. The authors thank R. Rachel for electron microscopy and Brian Hedlund for critically reading the manuscript.

Literature Cited

Allen, M. B. 1959. Studies with Cyanidium caldarium, an anomalously pigmented chlorophyte. Arch. Mikrobiol. 32:270–277.

Ashkin, A., and J. M. Dziedzic. 1987a. Optical trapping and manipulation of viruses and bacteria. Science 235:1517–1520.

Ashkin, A., J. M. Dziedzic, and T. Yamane. 1987b. Optical trapping and manipulation of single cells using infrared laser beams. Nature 330:769–771.

Balch, W. E., and R. S. Wolfe. 1976. New approach to the cultivation of methanogenic bacteria: 2-mercaptoethanesulfonic acid (HS-CoM)-dependent growth of Methanobacterium ruminantium in a pressurized atmosphere. Appl. Environ. Microbiol. 32:781–791.

Balch, W. E., G. E. Fox, L. J. Magrum, C. R. Woese, and R. S. Wolfe. 1979. Methanogens: Reevaluation of a unique biological group. Microbiol. Rev. 43:250–296.

Beck, P., and R. Huber. 1997. Detection of cell viability in cultures of hyperthermophiles. FEMS Microbiol. Lett. 147:11–14.

Bonch-Osmolovskaya, E. A., M. L. Miroshnichenko, N. A. Kostrikina, N. A. Chernych, and G. A. Zavarzin. 1990. Thermoproteus uzoniensis sp. nov., a new extremely thermophilic archaebacterium from Kamchatka continental hot springs. Arch. Microbiol. 154:556–559.

Bonch-Osmolovskaya, E. A., M. L. Miroshnichenko, N. A. Kostrikina, N. A. Chernych, and G. A. Zavarzin. 2001. Validation of publication of new names and new combinations previously effectively published outside the IJSEM. List No. 82. Int. J. Syst. Evol. Microbiol. 51:1619–1620.

Burggraf, S., N. Larsen, C. R. Woese, and K. O. Stetter. 1993. An intron within the 16S ribosomal RNA gene of the archaeon Pyrobaculum aerophilum. Proc. Natl. Acad. Sci. USA 90:2547–2550.

Burggraf, S., H. Huber, and K. O. Stetter. 1997. Reclassification of the crenarchaeal orders and families in accordance with 16S rRNA sequence data. Int. J. Syst. Bacteriol. 47:657–660.

Dalgaard, J. Z., and R. A. Garrett. 1992. Protein-coding introns from the 23S rRNA-encoding gene form stable circles in the hyperthermophilic archaeon Pyrobaculum organotrophum. Gene 121:103–110.

De Rosa, M., A. Gambacorta, and A. Gliozzi. 1986. Structure, biosynthesis, and physiochemical properties of archaebacterial lipids. Microbiol. Rev. 50:70–80.

Fiala, G., K. O. Stetter, H. W. Jannasch, T. A. Langworthy, and J. Madon. 1986. Staphylothermus marinus sp. nov. represents a novel genus of extremely thermophilic submarine heterotrophic archaebacteria growing up to 98°C. Syst. Appl. Microbiol. 8:106–113.

Fitz-Gibbon, S., H. Ladner, U. J. Kim, K. O. Stetter, M. I. Simon, and J. H. Miller. 2002. Genome sequence of the hyperthermophilic crenarchaeon Pyrobaculum aerophilum. Proc. Natl. Acad. Sci, USA 99:984–989.

Hensel, R., S. Laumann, J. Lang, H. Heumann, and F. Lottspeich. 1987. Characterization of two D-glyceraldehyde-3-phosphate dehydrogenases from the extremely thermophilic archaebacterium Thermoproteus tenax. Eur. J. Biochem. 170:325–333.

Horn, C., B. Paulmann, G. Kerlen, N. Junker, and H. Huber. 1999. In vivo observation of cell division of anaerobic hyperthermophiles by using a high-intensity dark-field microscope. J. Bacteriol. 181:5114–5118.

Huber, H., and K. O. Stetter. 2001a. Order I: Thermoproteales. In: G. Garrity (Ed.) Bergey's Manual of Systematic Bacteriology, 2nd ed. Springer-Verlag. New York, NY. 1:170.

Huber, H., and K. O. Stetter. 2001b. Order II: Desulfurococcales. In: G. Garrity (Ed.) Bergey's Manual of Systematic Bacteriology, 2nd ed. Springer-Verlag. New York, NY. 1:179–180.

Huber, H., and K. O. Stetter. 2001c. Genus III: Pyrobaculum. In: G. Garrity (Ed.) Bergey's Manual of Systematic Bacteriology, 2nd ed. Springer-Verlag. New York, NY. 1:174–177.

Huber, R., J. K. Kristjansson, and K. O. Stetter. 1987. Pyrobaculum gen. nov., a new genus of neutrophilic, rod-shaped archaebacteria from continental solfataras growing optimally at 100°C. Arch. Microbiol. 149:95–101.

Huber, R., and K. O. Stetter. 1992. The order Thermoproteales. In: A. Balows, H. G. Trüper, M. Dworkin, W. Harder, and K.-H. Schleifer (Eds.) The Prokaryotes, 2nd ed. Springer-Verlag. New York, NY. 677–683.

Huber, R., S. Burggraf, T. Mayer, S. M. Barns, P. Rossnagel, and K. O. Stetter. 1995. Isolation of a hyperthermophilic archaeum predicted by in situ RNA analysis. Nature 367:57–58.

Huber, R. 1999. Die Laserpinzette als Basis für Einzelzellkultivierungen. Biospektrum 5:289–291.

Huber, R., H. Huber, and K. O. Stetter. 2000a. Towards the ecology of hyperthermophiles: biotopes, new isolation strategies and novel metabolic properties. FEMS Microbiol. Rev. 24:615–623.

Huber, R., M. Sacher, A. Vollmann, H. Huber, and D. Rose. 2000b. Respiration of arsenate and selenate by hyperthermophilic archaea. Syst. Appl. Microbiol. 23:305–314.

Itoh, T., K. Suzuki, and T. Nakase. 1998a. Occurrence of introns in the 16S rRNA genes of members of the Thermoproteus. Arch. Microbiol. 170:155–161.

Itoh, T., K. Suzuki, and T. Nakase. 1998b. Thermocladium modestius gen. nov., sp. nov., a new genus of rod-shaped, extremely thermophilic crenarchaeote. Int. J. Syst. Bacteriol. 48:879–887.

Itoh, T., K. Suzuki, P. C. Sanchez, and T. Nakase. 1999. Caldivirga maquilingensis gen. nov., sp. nov., a new genus of rod-shaped crenarchaeote isolated from a hot spring in the Philippines. Int. J. Syst. Bacteriol. 49:1157–1163.

Kashefi, K., and D. R. Lovley. 2000. Reduction of Fe(III), Mn(IV), and toxic metals at 100°C by Pyrobaculum islandicum. Appl. Env. Microbiol. 66:1050–1056.

König, H., R. Skorko, W. Zillig, and W. D. Reiter. 1982. Glycogen in the thermoacidophilic archaebacteria of the genera Sulfolobus, Thermoproteus, Desulfurococcus and Thermococcus. Arch. Microbiol. 132:297–303.

Kujo, C., and Ohshima, T. 1998. Enzymological characteristics of the hyperthermostable NAD-dependent glutamate dehydrogenase from the archaeon Pyrobaculum islandicum and effects of denaturants and organic solvents. Appl. Environ. Microbiol. 64:2152–2157.

Langworthy, T. A., and J. L. Pond. 1986. Membranes and lipids of thermophiles. In: T. Brock (Ed.) Thermophiles: General, Molecular, and Applied Microbiology. John Wiley. New York, NY. 107–135.

Lübben, M., and K. Morand. 1994. Novel prenylated hemes as cofactors of cytochrome oxidases. J. Biol. Chem. 269:21473–21479.

Ludwig, W., and O. Strunk. 2001. ARB: A software environment for sequence data.

Messner, P., D. Pum, M. Sára, K. O. Stetter, and U. Sleytr. 1986. Ultrastructure of the cell envelope of the archaebacteria Thermoproteus tenax and Thermoproteus neutrophilus. J. Bacteriol. 166:1046–1054.

Molitor, M., C. Dahl, I. Molitor, U. Schäfer, N. Speich, R. Huber, R. Deutzmann, and H. G. Trüper. 1998. A dissimilatory sirohaem-sulfite-reductase-type protein from the hyperthermophilic archaeon Pyrobaculum islandicum. Microbiology 144:529–541.

Nicolaus, B., A. Gambacorta, A. L. Basso, R. Riccio, M. De Rosa, and W. D. Grant. 1988. Trehalose in archaebacteria. Syst. Appl. Microbiol. 10:215–217.

Pfennig, N., and K. D. Lippert. 1966. Über das Vitamin B_{12}-Bedürfnis phototropher Schwefelbakterien. Arch. Mikrobiol. 55:245–256.

Phipps, B. M., H. Engelhardt, R. Huber, and W. Baumeister. 1990. Three-dimensional structure of the crystalline protein envelope layer of the hyperthermophilic Pyrobaculum islandicum. J. Struct. Biol. 103:152–163.

Phipps, B. M., R. Huber, and W. Baumeister. 1991. The cell envelope of the hyperthermophilic archaebacterium Pyrobaculum organotrophum consists of two regularly arrayed protein layers: Three-dimensional structure of the outer layer. Molec. Microbiol. 5:253–265.

Reddy, D. M., P. F. Crain, C. G. Edmonds, R. Gupta, T. Hashizume, K. O. Stetter, F. Widdel, and J. A. McCloskey. 1992. Structure determination of two new amino acid-containing derivatives of adenosine from tRNA of thermophilic bacteria and archaea. Nucl. Acids Res. 20:5607–5615.

Rieger, G., K. Müller, R. Hermann, K. O. Stetter, and R. Rachel. 1997. Cultivation of hyperthermophilic archaea in capillary tubes resulting in improved preservation of fine structures. Arch. Microbiol. 186:373–379.

Sako, Y., T. Nunoura, and A. Uchida. 2001. Pyrobaculum oguniense sp. nov., a novel facultatively aerobic and hyperthermophilic archaeon growing at up to 97 °C. Int. J. Syst. Bacteriol. 51:303–309.

Schäfer, S., C. Barkowski, and G. Fuchs. 1986. Carbon assimilation by the autotrophic thermophilic archaebacterium Thermoproteus neutrophilus. Arch. Microbiol. 146:301–308.

Selig, M., and P. Schönheit. 1994. Oxidation of organic compounds to CO_2 with sulfur or thiosulfate as electron acceptor in the anaerobic hyperthermophilic archaea

Thermoproteus tenax and Pyrobaculum islandicum proceeds via the citric acid cycle. Arch. Microbiol. 162:286–294.

Stetter, K. O. 1982. Ultrathin mycelia-forming organisms from submarine volcanic areas having an optimum growth temperature of 105°C. Nature 300:258–259.

Stetter, K. O., H. König, and E. Stackebrandt. 1983. Pyrodictium gen. nov., a new genus of submarine disc-shaped sulphur reducing archaebacteria growing optimally at 105°C. Syst. Appl. Microbiol. 4:535–551.

Stetter, K. O., and W. Zillig. 1985. Thermoplasma and the thermophilic sulfur-dependent archaebacteria. *In:* C. Woese and R. S. Wolfe (Eds.) The Bacteria. Academic Press. New York, NY. 8:100–201.

Stetter, K. O. 1986. Diversity of extremely thermophilic archaebacteria. *In:* T. D. Brock (Ed.) Thermophiles: General, Molecular, and Applied Microbiology. John Wiley. New York, NY. 40–74.

Stetter, K. O. 1989. Order III: Sulfolobales. *In:* J. T. Staley, M. P. Bryant, N. Pfennig, and J. G. Holt (Eds.) Bergey's Manual of Systematic Bacteriology. William and Wilkins. Baltimore, MD. 2250.

Stetter, K. O. 1999. Extremophiles and their adaptation to hot environments. FEBS Lett. 452:22–55.

Stetter, K. O. 2000. Sulfur-containing high-temperature biotopes and their microorganisms.

Stetter, K. O. 2001. Genus VII: Thermodiscus. *In:* G. Garrity (Ed.) Bergey's Manual of Systematic Bacteriology, 2nd ed. Springer-Verlag. New York, NY. 1:189–190.

Thurl, S., J. Butrow, and W. Schäfer. 1985. New types of menaquinones from the thermophilic archaebacterium Thermoproteus tenax. Biol. Chem. Hoppe-Seyler 366:1079–1083.

Tindall, B. J. 1989. Fully saturated menaquinones in the archaebacterium Pyrobaculum islandicum. FEMS Microbiol. Lett. 60:251–254.

Tindall, B. J., V. Wray, R. Huber, and M. D. Collins. 1991. A novel, fully saturated cyclic menaquinone in the archaebacterium Pyrobaculum organotrophum. Syst. Appl. Microbiol. 14:218–221.

Trincone, A., B. Nicolaus, G. Palmieri, M. De Rosa, R. Huber, G. Huber, K. O. Stetter, and A. Gambacorta. 1992. Distribution of complex and core lipids within new hyperthermophilic members of the archaea domain. Syst. Appl. Microbiol. 15:11–17.

Völkl, P., R. Huber, E. Drobner, R. Rachel, S. Burggraf, A. Trincone, and K. O. Stetter. 1993. Pyrobaculum aerophilum sp. nov., a novel nitrate-reducing hyperther-

mophilic archaeum. Appl. Env. Microbiol. 59:2918–2926.

Völkl, P., P. Markiewicz, K. O. Stetter, and J. H. Miller. 1994. The sequence of a subtilisin-type protease (aerolysin) from the hyperthermophilic archaeum Pyrobaculum aerophilum reveals sites important to thermostability. Protein Sci. 3:1329–1340.

Wildhaber, I., and W. Baumeister. 1987. The cell envelope of Thermoproteus tenax: Three-dimensional structure of the surface layer and its role in shape maintenance. EMBO J. 6:1475–1480.

Woese, C. R., O. Kandler, and M. L. Wheelis. 1990. Towards a natural system of organisms: Proposal for the domains Archaea, Bacteria and Eukarya. Proc. Natl. Acad. Sci. USA 87:4576–4579.

Zillig, W., K. O. Stetter, W. Schäfer, D. Janekovic, S. Wunderl, I. Holz, and P. Palm. 1981. Thermoproteales: A novel type of extremely thermoacidophilic anaerobic archaebacteria isolated from Icelandic solfataras. Zbl. Bact. Hyg., I. Abt. Orig. C. 2:205–227.

Zillig, W., and K. O. Stetter. 1982a. Validation of the publication of new names and new combinations previously effectively published outside the IJSB. List No. 8. Int. J. Syst. Bacteriol. 32:266–268.

Zillig, W., K. O. Stetter, D. Prangishvili, W. Schäfer, S. Wunderl, D. Janekovic, I. Holz, and P. Palm. 1982b. Desulfurococcaceae, the second family of the extremely thermophilic, anaerobic, sulfur-respiring Thermoproteales. Zbl. Bact. Hyg., I. Abt. Orig. C 3:304–317.

Zillig, W., A. Gierl, G. Schreiber, W. Wunderl, D. Janekovic, K. O. Stetter, and H.-P. Klenk. 1983. The archaebacterium Thermofilum pendens represents a novel genus of the thermophilic, anaerobic sulfur respiring Thermoproteales. Syst. Appl. Microbiol. 4:79–87.

Zillig, W. 1989. Order II: Thermoproteales. *In:* J. T. Staley, M. P. Bryant, N. Pfennig, and J. G. Holt (Eds.) Bergey's Manual of Systematic Bacteriology. William and Wilkins. Baltimore, MD. 2240–2244.

Zillig, W., D. Prangishvili, C. Schleper, M. Elferink, I. Holz, S. Albers, D. Janekovic, and D. Götz. 1996. Viruses, plasmids and other genetic elements of thermophilic and hyperthermophilic Archaea. FEMS Microbiol. Rev. 18:225–236.

Zillig, W., and A.-L. Reysenbach. 2001. Genus I: Thermofilum. *In:* G. Garrity (Ed.) Bergey's Manual of Systematic Bacteriology, 2nd ed. Springer-Verlag. New York, NY. 1:178–179.

Prokaryotes (2006) 3:23–51
DOI: 10.1007/0-387-30743-5_3

CHAPTER 3

Sulfolobales

HARALD HUBER AND DAVID PRANGISHVILI

Introduction

Within the crenarchaeotal branch of the Archaea (Woese and Fox, 1977; Woese et al., 1990), three orders have been described so far: the Desulfurococcales (Huber and Stetter, 2001b), the Thermoproteales (Zillig et al., 1981), and the Sulfolobales (Stetter, 1989). Members of the Sulfolobales are well-defined and distinguished from the other orders by morphological, physiological and molecular characters. Cells are regular to irregular cocci, which occur usually singly or in pairs and exhibit cell diameters from about 1.0 up to 5 μm. All members of the order are extreme thermophiles to hyperthermophiles with optimal growth temperatures between 65 and 90°C. An important common property is their pH optimum of around pH 2. They grow aerobically, facultatively anaerobically, or anaerobically. Under autotrophic conditions they gain energy by oxidation of elemental sulfur, thiosulfate, sulfidic ores, or molecular hydrogen. Carbon dioxide (CO_2) is used as a carbon source. Alternatively, organotrophic growth occurs by aerobic respiration or anaerobic sulfur respiration or by fermentation of organic substrates. Following the classification listed in the new edition of *Bergey's Manual of Systematic Bacteriology* (Huber and Stetter, 2001c), the order Sulfolobales comprises one family, the Sulfolobaceae (Stetter, 1989), with six genera and around 20 species. Since the last edition of *The Prokaryotes* (Segerer and Stetter, 1992), several representatives have been renamed and one genus no longer exists ("*Desulfurolobus*"; Zillig et al., 1986). In addition, nucleic acid hybridization studies and 16S rRNA sequencing suggest that most Sulfolobales are genetically rather unrelated, even the phenotypically quite similar species included in the genus *Sulfolobus* (Fuchs et al., 1996; Trevisanato et al., 1996). The members of the genera can be distinguished by their genomic DNA G+C content, the ability of growing aerobically and/or anaerobically, and the use of different electron acceptors. However, probably as a consequence of their common, extreme biotope, the members of the Sulfolobales exhibit high physiological and morphological similarities. Therefore only a few properties can be used to classify and describe the different genera and species. Some problems have occurred since several type strains of *Sulfolobus* species deposited in culture collections (e.g., *S. acidocaldarius* and *S. solfataricus*) proved to be unable to oxidize elemental sulfur, although this capability was used to define *Sulfolobus* spp. in their original descriptions. The reason for this is unknown, but meanwhile closely related strains that exhibit this physiological property have been isolated. Compared to other archaeal groups, members of the Sulfolobales are quite easy to handle and to cultivate, and so a huge number of molecular investigations has been carried out and even genetic systems have been established. Furthermore, complete genome sequences are available for *S. solfataricus* and *S. tokodaii* (the sequence data for *S. acidocaldarius* and several others are not available at the moment).

Phylogeny

On the basis of 16S rRNA sequence comparisons, the order Sulfolobales represents one lineage of the crenarchaeotal branch of the Archaea (Woese et al., 1990; Fuchs et al., 1996; Fig. 1). It forms a separate cluster and is distinct from the two other described orders of the archaeal phylum Crenarchaeota: the Desulfurococcales and the Thermoproteales. The representatives of the Sulfolobales exhibit phylogenetic similarities of 0.85–1.00 to members within the order and 0.88–0.77 to all other species of the Crenarchaeota (Fuchs et al., 1996; H. Huber, unpublished observation). However, the phylogenetic analysis reveals several differences in the present taxonomy of the Sulfolobales. This is especially true for the genus *Sulfolobus*. Four groups can be identified within this genus on the basis of 16S rDNA sequences (Fig. 1). They exhibit phylogenetic distances of 7–15% from each other (H. Huber, unpublished data), indicating that they are at least different genera. The first group is composed of *S. acidocaldarius* and

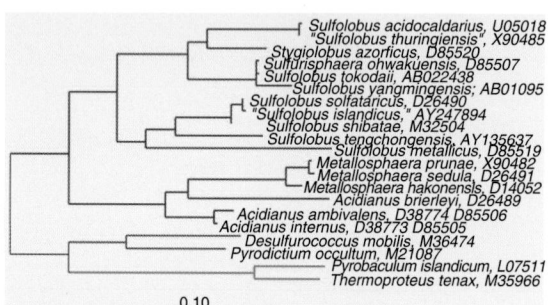

Fig. 1. Phylogenetic tree based on 16S rRNA sequences. The tree was calculated using the maximum likelihood program, which is included in the ARB package (Ludwig and Strunk, 2001). Scale bar: 10 estimated exchanges within 100 nucleotides. Red lines = Sulfolobales; blue lines = Desulfurococcales; and green lines = Thermoproteales.

the undescribed species "*S. thuringiensis*." The second group encompasses *S. tokodaii*, *S. yangmingensis* and *Sulfurisphaera ohwakuensis*. These three organisms represent the same genus, *Sulfurisphaera*, and as a consequence, the two "*Sulfolobus*" species should be reclassified as *Sulfurisphaera* species. The third group includes *S. solfataricus*, *S. shibatae*, the recently published *S. tengchongensis*, and the undescribed "*S. islandicus*," while the fourth phylogenetic group is represented by *S. metallicus*. However, since there are few significant physiological characteristics for the description, the definition of new genera is based nearly exclusively on sequence data. Nevertheless, it can be expected that further reclassification of members of the genus *Sulfolobus* or other members of the Sulfolobales will occur.

Phylogenetic investigations based on analyses of 23S rRNA sequences were carried out by Trevisanato et al. (1996). The resulting phylogeny was consistent with the 16S rRNA tree, although not all members of the Sulfolobales have been included in this study.

Taxonomy

The first member of the order was isolated by Brierley in Yellowstone National Park, Wyoming, United States, and preliminarily described in his thesis (Brierley, 1966). Full description was published in 1973 (Brierley and Brierley, 1973a), one year after the genus *Sulfolobus* was established (Brock et al., 1972).

Another organism, tentatively named "*Caldariella acidophila*," was isolated from the Solfatara crater, Naples, Italy (DeRosa et al., 1974). Although phenotypically convergent, it was considered to be phylogenetically distinct from *Sulfolobus* (DeRosa et al., 1975). Subsequent

studies revealed that the DNA-dependent RNA polymerases of *Sulfolobus acidocaldarius*, Brierley's organism, and strains of "*Caldariella*" were very similar. Consequently, both Brierley's organism and "*Caldariella*" were described as members of the genus *Sulfolobus* and named "*Sulfolobus brierleyi*" and "*S. solfataricus*," respectively (Zillig et al., 1980).

In 1984, facultatively anaerobic strains of *Sulfolobus*-related organisms capable of both sulfur-oxidation and sulfur-reduction were isolated (Segerer et al., 1985) and described as *Acidianus infernus* (Segerer et al., 1986). With regard to the organism's capacity for facultative reduction/oxidation of sulfur, the genus name *Acidianus* refers to Ianus, an ancient Roman god with two faces looking in opposite directions. Closely related isolates with similar features were independently obtained by Zillig and coworkers (Zillig et al., 1985). Strain Lei10, tentatively named "*Sulfolobus ambivalens*," was later described as *Desulfurolobus ambivalens* (Zillig et al., 1986) and in 1996 reclassified as *Acidianus ambivalens* (Fuchs et al., 1996). In addition, *Sulfolobus brierleyi* was recognized to be another facultative sulfur-reducer with the same DNA G+C content of about 31 mol%, which is typical for *Acidianus*. This finding led to an emended description of Brierley's organism as a member of the genus *Acidianus*, *Acidianus brierleyi* comb. nov. (Segerer et al., 1986).

Then in 1989, the order Sulfolobales and the family Sulfolobaceae were defined by Stetter in the third edition of *Bergey's Manual of Systematic Bacteriology* (Stetter, 1989). At present, the Sulfolobales are comprised of one family and six validly described genera (Huber and Stetter, 2001c). In addition to the 16 validated species, three ("*S. islandicus*," "*S. thuringiensis*"and "*S. tengchongensis*") are tentatively described (Xiang et al., 2003). Furthermore, a huge number of mostly unclassified species has been used in different investigations.

Several members of the Sulfolobales have been reclassified during the last years and even today, the classification is sometimes not in line with the phylogenetic data obtained, e.g., by 16S rRNA based sequence analysis. In addition, problems arise with strains which are unavailable or not deposited in public culture collections. Notably, for the description of newly described species (or genera) to be valid, the type strains have to be deposited in two independent official culture collections, located in two different countries. Therefore, in this article only validly described species will be discussed in detail.

In many biochemical and molecular studies of Sulfolobales, isolates assigned to one of the validly described species with little verification were used. However, at least in some cases, these isolates differ significantly from the type strains, rendering their taxonomic status questionable (see "Biochemistry").

The following genera and species have been assigned to the order Sulfolobales:

a) Genus *Sulfolobus* (type genus of the family Sulfolobaceae and the order Sulfolobales) includes *S. acidocaldarius* (type species), *S. acidocaldarius* type strain 98-3^T (American Type Culture Collection [ATCC 33909^T], Deutsche Sammlung von Mikroorganismen und Zellkulturen [DSM 639^T]; Brock et al., 1972); *S. metallicus* (type strain Kra23^T, DSM 5389^T; Huber and Stetter, 1991a); *S. shibatae* (type strain B12^T, DSM 6482^T; Grogan et al., 1990); *S. solfataricus* (type strain ATCC 35091^T, DSM 1616^T; Brock et al., 1972; Zillig et al., 1980); *S. tokodaii* (validated, although only deposited in one public culture collection; type strain 7^T, Japan Collection of Microorganisms [JCM 10545^T]; Suzuki et al., 2002); and *S. yangmingensis* (validated, although not deposited in a public culture collection; type strain YM1^T; Jan et al., 1999). b) Genus *Acidianus* includes *A. infernus* (type species), *A. infernus* type strain So-4a^T (DSM 3191^T; Segerer et al., 1986); *A. ambivalens* type strain Lei1^T (DSM 3772^T; Zillig et al., 1986; Fuchs et al., 1996); and *A. brierleyi* type strain DSM 1651^T (Brierley and Brierley, 1973a; Zillig et al., 1980; Segerer et al., 1986). c) Genus *Metallosphaera* includes *M. sedula* (type species), *M. sedula* type strain TH2^T (DSM 5348^T; Huber et al., 1989); *M. hakonensis* type strain HO1-1^T (ATCC 51241^T, DSM 7519^T, IAM 14250^T, JCM 8857^T; Takayanagi et al., 1996; Kurosawa et al., 2003); and *M. prunae* type strain Ron12/II^T (DSM 10039^T; Fuchs et al., 1995). d) Genus *Stygiolobus* includes *St. azoricus* (type species), *St. azoricus* type strain FC6^T (DSM 6296^T; Segerer et al., 1991). e) Genus *Sulfurisphaera* includes *S. ohwakuensis* (type species), *S. ohwakuensis* type strain TA-1^T (Institute of Fermentation, Osaka, Japan [IFO 15161^T]; Kurosawa et al., 1998). f) Genus *Sulfurococcus* includes *S. mirabilis* (type species), *S. mirabilis* type strain INMI AT-59^T (Golovacheva et al., 1987a; Golovacheva et al., 1987b); and *S. yellowstonensis* type strain Str6kar^T (Karavaiko et al., 1994). These last two strains have no deposition numbers and both are preserved at the Institute of Microbiology, Russian Academy of Sciences, Moscow, Russia!

The main characteristics for distinguishing these genera are summarized in Table 1. Members of the genus *Sulfolobus* are obligate aerobes, which can oxidize a variety of sulfur containing compounds. Most strains are facultative heterotrophs. *Acidianus* strains exhibit a facultative anaerobic mode of life. Under aerobic conditions they oxidize sulfur compounds, H$_2$, and ferrous iron, while they grow anaerobically by S°-H$_2$ autotrophy, producing H$_2$S as the metabolic end product. The G+C content of the genomic DNA is around 31%. *Metallosphaera* cells gain energy by oxidation of sulfur compounds, molecular hydrogen and especially sulfidic ores. All strains are able to grow on complex organic substrates. The DNA G+C content is around 45 mol% (the G+C content of *M. hakonensis* was corrected to 46.2 mol%). The major characteristic of *Stygiolobus azoricus* is its obligate chemolithoautotrophic growth under anaerobic conditions by reduction of elemental sulfur with H$_2$ as electron donor (S°-H$_2$ autotrophy). Hydrogen sulfide (H$_2$S) is produced as sole metabolic end product. The DNA G+C content of *Stygiolobus azoricus* is 38 mol%. *Sulfurisphaera ohwakuensis* is a facultative anaerobic organotroph, which grows anaerobically on elemental sulfur and on complex substrates. However, aerobic growth on sulfur compounds is not described. The G+C content of the genomic DNA is 33 mol%. *Sulfurococcus* species are characterized by an obligate aerobic growth, which occurs on sulfur compounds and sugars or complex organic compounds. The G+C content of their genomic DNAs is around 45 mol%.

Habitats

Members of the Sulfolobales have been isolated almost exclusively from continental solfataric fields, where they thrive within hot, highly acidic water- and mud-holes. In addition, representatives have been obtained from acidic hot soils (Fliermans and Brock, 1972) and smoldering slag heaps (Fuchs et al., 1995).

Two different niches, defined by redox state and pH, are present within solfataric fields. The first is an upper, oxygen-containing zone of rusty color (indicating the presence of ferric iron), which is highly acidic (pH ≤ 2) owing to biotic and abiotic oxidation processes (Brock, 1978b; Stetter et al., 1986; Huber et al., 2000). In this zone, an iridescent, oily glimmering layer containing high concentrations (up to 10^8 cells/ml) of irregular, lobed cells typical for Sulfolobales is commonly seen floating on grayish, boiling mud. The second is a deeper zone of around pH 4, which is anaerobic because of the presence of reducing volcanic exhalations (e.g., hydrogen sulfide) and blackish-gray because of the formation of heavy metal sulfides (Stetter et al., 1986; Huber et al., 2000). Whereas growth of the Thermoproteales is restricted to the reduced depth of solfataras, Sulfolobales species exhibit a broader metabolic diversity (see below) and therefore inhabit both zones.

Table 1. Differentiation of the Sulfolobales species.

Species	Temp (°C) (opt. °C)	pH-range (opt.)	Relation to oxygen	Metabolism	Substrates	G+C content (mol %)	References
Sulfolobus acidocaldarius	55–85 (70–75)	1–6 (2–3)	Aerobic	Mixotrophic or heterotrophic	Complex organic compounds; sugars (e.g., glucose, ribose, sucrose, xylose); and oxidation of molecular hydrogen	37	Brock et al., 1972
Sulfolobus metallicus	50–75 (65)	1–4.5 (2–3)	Aerobic	Obligate lithoautotrophic	Oxidation of sulfidic ores (pyrite, sphalerite, and chalcopyrite); and oxidation of elemental sulfur	38	Huber and Stetter, 1991
Sulfolobus shibatae	n.d.–86 (81)	n.d. (3)	Aerobic	Mixotrophic or heterotrophic	Complex organic compounds; sugars (e.g., glucose, ribose, sucrose, xylose, maltose, and starch); oxidation of molecular hydrogen and elemental sulfur	35	Grogan et al., 1990
Sulfolobus solfataricus	50–87 (85)	2–5.5 (3–4.5)	Aerobic	Mixotrophic or heterotrophic	Complex organic compounds; sugars (e.g., glucose, ribose, sucrose, and xylose); and oxidation of molecular hydrogen	35	Zillig et al., 1980
Sulfolobus tokadaii	70–85 (80)	2–5 (2.5–3)	Aerobic	Chemolithotrophic to chemoheterotrophic	Complex organic compounds; amino acids; and oxidation of elemental sulfur	33	Suzuki et al., 2002
Sulfolobus yangmingensis	65–95 (80)	2–6 (4)	Aerobic	Mixotrophic or heterotrophic	Complex organic compounds; sugars (e.g., glucose, ribose, sucrose, xylose, and maltose); and amino acids (all except cysteine); and oxidation of elemental sulfur	42	Jan et al., 1998
Acidianus infernus	65–95 (85–90)	1–5.5 (2)	Facultative anaerobic	Obligate lithoautotrophic	Oxidation of sulfidic ores (pyrite, sphalerite, and chalcopyrite); oxidation of molecular hydrogen and elemental sulfur; and S^0-H_2 autotrophy	31	Segerer et al., 1986
Acidianus ambivalens	n.d.–87 (80)	1–3.5 (2.5)	Facultative anaerobic	Obligate lithoautotrophic	Oxidation of elemental sulfur; and S^0-H_2 autotrophy	33	Zillig et al., 1986 Fuchs et al., 1996
Acidianus brierleyi	45–75 (70)	1–6 (1.5–2)	Facultative anaerobic	Facultative chemolithoautotrophic	Oxidation of sulfidic ores (pyrite, sphalerite, chalcopyrite); oxidation of molecular hydrogen and elemental sulfur, complex; organic compounds; and S^0-H_2 autotrophy	31	Zillig et al., 1980 Segerer et al., 1986

Organism	Temperature range (optimum)	pH range (optimum)	Oxygen relationship	Metabolism	Substrates	G+C (%)	References
Metallosphaera sedula	50–80 (75)	1.0–4.5 (2–3)	Aerobic	Facultative chemolithoautotrophic	Oxidation of sulfidic ores (pyrite, sphalerite, and chalcopyrite); oxidation of molecular hydrogen and elemental sulfur; complex; and organic compounds	45	Huber G. et al., 1989
Metallosphaera hakonensis	50–80 (70)	1–4 (3)	Aerobic	Facultative chemolithoautotrophic	Oxidation of metal sulfides (FeS); oxidation of elemental sulfur; tetrathionate and H_2S; complex; and organic compounds	46	Takayanagi et al., 1996; Kurosawa et al., 2003
Metallosphaera prunae	55–80 (75)	1–4.5 (2–3)	Aerobic	Facultative chemolithoautotrophic	Oxidation of sulfidic ores (pyrite, sphalerite, and chalcopyrite); oxidation of molecular hydrogen and elemental sulfur; complex; and organic compounds	46	Fuchs et al., 1995
Stygiolobus azoricus	57–89 (80)	1–5.5 (2.5–3)	Anaerobic	Obligate chemolithoautotrophic	S^0-H_2 autotrophy	38	Segerer et al., 1991
Sulfurisphaera ohwakuensis	63–92 (84)	1–5 (2)	Facultative anaerobic	Mixotrophic or heterotrophic	Oxidation of complex; organic compounds; reduction of elemental sulfur with H_2 and complex organic compounds	33	Kurosawa et al., 1998
Sulfurococcus mirabilis	50–86 (70–75)	1–5.8 (2–2.6)	Aerobic	Facultative chemolithoautotrophic	Oxidation of sulfidic ores (pyrite, sphalerite, and chalcopyrite); oxidation of elemental sulfur; complex; organic compounds; and sugars and amino acids	~44	Golovachewa et al., 1987a, b; Karavaiko et al., 1995
Sulfurococcus yellowstonensis	40–80 (60)	1–5.5 (2–2.6)	Aerobic	Facultative chemolithoautotrophic	Oxidation of sulfidic ores (pyrite, sphalerite, and chalcopyrite); oxidation of elemental sulfur and ferrous iron; complex; organic compounds; and sugars (e.g., glucose, fructose, sucrose, and rhamnose)	45	Karavaiko et al., 1995

Abbreviation: n.d., not determined.

There is also evidence that members of the genus *Acidianus* grow not only in continental solfatara fields but in other habitats as well (e.g., two strains of *A. infernus* have been isolated from geothermally heated submarine hydrothermal systems on the beach at Vulcano, Italy; Segerer et al., 1986).

Sulfolobus species are widespread and can be isolated from solfatara fields in a great number of geographically different locations, including the United States (Brock et al., 1972; Brock, 1978a), China (Xiang et al., 2003), Iceland (Brock, 1978a; Huber and Stetter, 1991a), Italy (DeRosa et al., 1974; Brock, 1978a; Zillig et al., 1980), Japan (Furuya et al., 1977; Yeats et al., 1982; Jan et al., 1999; Suzuki et al., 2002), New Zealand (Bohlool, 1975), and the West Indies (Brock, 1978a). In addition, strains were obtained from smoldering slag heaps at Ronneburg, Germany (Fuchs et al., 1995) and Kvarntorp, Sweden (H. Huber and O. Holst, unpublished observation). A recent study investigated the distribution of different strains of one species, "*S. islandicus*" (Whitaker et al., 2003). Although exhibiting identical 16S rRNA sequences, a small but significant level of genetic differentiation among the populations from Yellowstone National Park, Lassen Volcanic Park (Iceland), Uzon Caldera, and Mutnovsky Volcano was detectable.

Members of the genus *Acidianus* were isolated from solfataric springs in Naples, Italy (Segerer et al., 1986), the Yellowstone National Park, Wyoming, United States (Brierley and Brierley, 1973a; Segerer et al., 1986), Iceland (Segerer et al., 1986; Zillig et al., 1986), the Azores (Segerer et al., 1986), Java, Indonesia (Huber et al., 1991b), and the Southern Kurils, Russia (A. Segerer, unpublished observation).

Metallosphaera species have been found within Neapolitan solfataras (Huber et al., 1989), smoldering slag heaps at Ronneburg, Germany (Fuchs et al., 1995), and hot springs in Hakone, Japan (Takayanagi et al., 1996).

The only member of the genus *Stygiolobus* known so far, *St. azoricus*, was exclusively isolated from samples taken near Furnas and Ribeira Quente, Azores (Segerer et al., 1985; Segerer et al., 1991). It is still unclear whether this reflects geographic isolation.

Sulfurisphaera ohwakuensis has been obtained from acidic hot springs in Ohwaku Valley, Hakone, Japan (Kurosawa et al., 1998), while several *Sulfurococcus* species were isolated from a crater of the Uzon volcano in Kamchatka, Russia, and from a thermal spring in Yellowstone National Park, United States (Golovacheva et al., 1987a; Karavaiko et al., 1994).

Isolation

Nearly all members of the Sulfolobales can be cultivated within various low ionic strength media (see the section Medium for Cultivation in this Chapter) in the presence of elemental sulfur and at a pH around 3 (most strains will even grow at pH 1.5). The incubation temperature depends on the species but is usually above 75°C. Depending on the organism, aerobic or anaerobic conditions have to be applied (see media, section Medium for Cultivation). Several species are also able to grow on (complex) organic substrates.

Different representatives of the Sulfolobales quite often exhibit similar physiological features. Therefore, a selective enrichment of members of one genus or even one species is very hard to achieve.

Selective Enrichment of *Sulfolobus* and *Sulfurococcus* Species

Sulfolobus species can be enriched at 85°C in aerobic mineral medium supplemented with 0.1% (w/v) of yeast extract (exception *S. metallicus*). The high incubation temperature prevents growth of *Metallosphaera* strains and *Acidianus brierleyi*. *Acidianus infernus*, *A. ambivalens* and *Sulfurisphaera ohwakuensis* will not grow in this medium owing to the absence of sulfur or sulfide. *Stygiolobus azoricus* cannot grow under aerobic conditions. However, *Sulfurococcus* species will also be obtained under these enrichment conditions.

Selective Enrichment of Members of the Genus *Metallosphaera*

Metallosphaera sedula strains TH2 and TH4 have been selectively enriched with 0.33% (w/v) of pyrite, chalcopyrite, or ore mixture "G1N" as sole substrate at 65°C and pH 2 (Huber et al., 1989). The same conditions are suitable for *M. prunae* and *M. hakonensis*. A third *M. sedula* strain, SP3a, was enriched in the absence of ore on elemental sulfur in combination with yeast extract. The latter conditions are, however, not selective for *Metallosphaera* species.

Selective Enrichment of *Acidianus*, *Sulfurisphaera* and *Stygiolobus* Species

Highly anaerobic growth conditions (Balch et al., 1979) in medium supplemented with 0.02% (w/v) yeast extract in combination with about 0.5% (w/v) sulfur and an atmosphere consisting of H_2 and CO_2 (e.g., 80:20 [v/v], 300 kPa) allow the growth of facultatively or strictly anaerobic members of the Sulfolobales, to the exclusion of *Sulfolobus, Metallosphaera* and *Sulfurococcus*.

Acidianus brierleyi is selectively enriched at 65°C, a temperature allowing only slow growth of the other species, with subsequent transfer into aerobic medium containing 0.1% (w/v) yeast extract, but no sulfur. *Acidianus infernus* and *A. ambivalens* are selectively enriched at 85°C, with subsequent transfer into aerobic culture medium containing 0.01% (w/v) yeast extract and 0.5% (w/v) sulfur. *Sulfurisphaera* grows under aerobic conditions heterotrophically.

Stygiolobus azoricus has been obtained from three anaerobically grown enrichment cultures (incubated at 85°C) by thrice-repeated end-point dilution. As a characteristic, no growth is obtained after transfer to aerobic culture media.

Identification

Sulfolobales cells are coccoid, about 0.8–2 µm (rarely up to 5 µm) in diameter. Characteristically, they are highly irregular in shape and often strongly lobed or edged (Fig. 2A–C). The organisms stain Gram negative. All members of the Sulfolobales are extreme acidophiles, not growing above pH 5.5–6. They exhibit temperature optima between 60 and 90°C (Table 1).

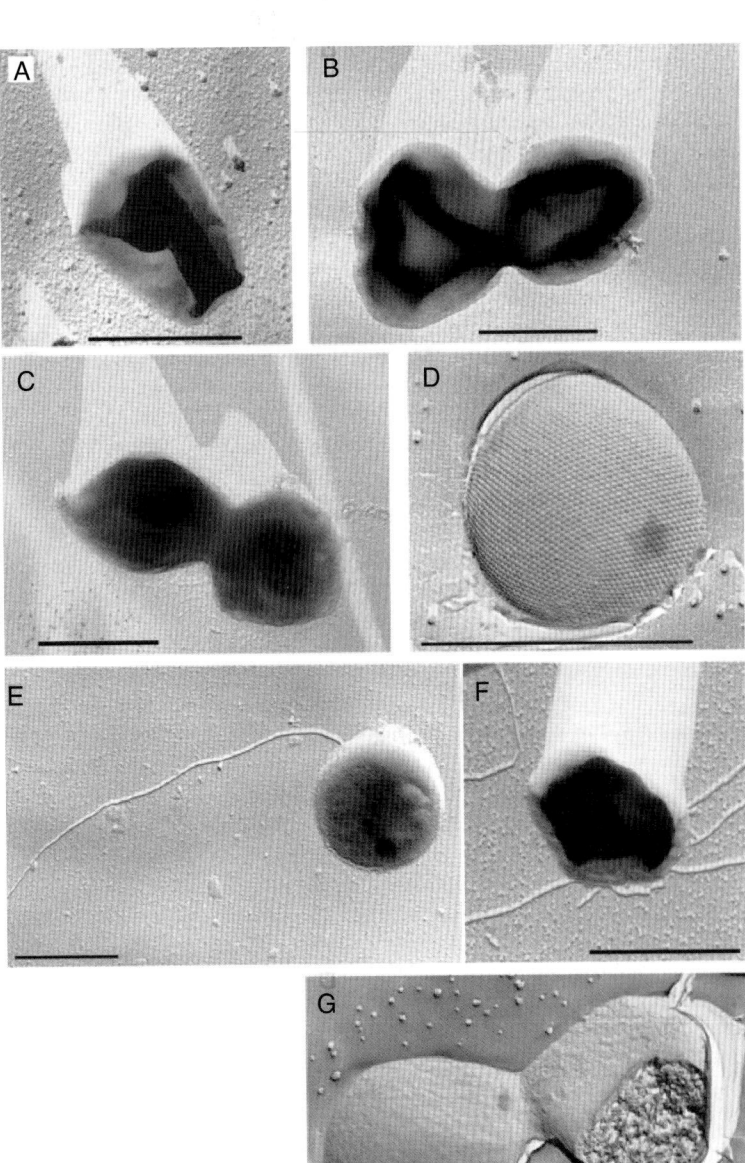

Fig. 2. Electron micrograph of platinum-shadowed cells of A) *Sulfolobus acidocaldarius*; B) *Sulfolobus metallicus*; C) *Acidianus infernus*; bars for (A–C) are each 1 µm; D) electron micrograph of a freeze-etched cell of *Metallosphaera sedula*; bar is 1 µm; E) *Metallosphaera sedula* cell with flagellum; platinum-shadowed; bar is 1 µm; F) *Metallosphaera prunae* cells with flagella; platinum-shadowed; bar is 1 µm; and G) electron micrograph of freeze-etched cells of *Acidianus infernus*; bar is 1 µm.

With the exception of *Stygiolobus azoricus*, they are capable of aerobic growth on elemental sulfur, sulfur compounds, or organic material. In contrast, the extremely thermophilic to hyperthermophilic members of the order Thermoproteales are only moderately acidophilic, almost neutrophilic, and they exhibit exclusive rod-shaped morphologies (Huber and Stetter, 2001a; Huber et al., 2002b). Representatives of the Desulfurococcales are also almost neutrophiles and most of them grow under anaerobic conditions. They exhibit cell shapes from discs to irregular cocci (Huber and Stetter, 2001b; Huber and Stetter, 2002a). The extremely acidophilic members of the order Thermoplasmatales, which have been shown to occur also in solfataric springs (Segerer et al., 1988), can also be easily distinguished from the representatives of the Sulfolobales: They either lack a cell wall or possess an S-layer with p4-symmetry (genus *Picrophilus*), are strictly heterotrophic, and are incapable of sulfur oxidation (Huber and Stetter, 2001d).

Within the Sulfolobales, three distinct physiological groups are evident, allowing a rough but fast classification of an unknown strain (Table 1). There are 1) strictly aerobic sulfur-oxidizers, represented by members of the genera *Sulfolobus*, *Metallosphaera* and *Sulfurococcus*; 2) facultatively aerobic sulfur-oxidizers/reducers, represented by *Acidianus* and *Sulfurisphaera* species; and 3) the strictly anaerobic sulfur-reducing *Stygiolobus azoricus*. Further distinguishing features of the genera and species are listed in Table 1.

Sulfolobus cells are coccoid and differ from members of *Metallosphaera* by the highly irregular, lobed shape and the lower G+C content of their genomic DNA. The validly described species of the genus are phenotypically very similar to each other. They are distinguished by their different optimal and maximal growth temperatures, pH optimum, metabolism (Table 1), the DNA-dependent RNA polymerases, and mainly the 16S rRNA sequences (Zillig et al., 1979; Zillig et al., 1980; Fuchs et al., 1996; Huber and Stetter, 2001a). The cell wall is an S-layer (Weiss, 1974) located about 20 nm from the cytoplasmic membrane. It is attached to the membrane via as yet uncharacterized spacer proteins. The S-layer glycoprotein subunits are arranged on a two-dimensional crystalline lattice with p3 symmetry. The distance between the subunit complexes is around 20–21 nm. The three-dimensional structure of the S-layers from *S. acidocaldarius*, *S. solfataricus* and *S. shibatae* has been investigated in detail by electron crystallography (Taylor et al., 1982; Deatherage et al., 1983; Prüschenk and Baumeister, 1987; Lembcke et al., 1991; Lembcke et al., 1993). Their fine structures are remark-

ably similar to each other and to the S-layer of *A. brierleyi* (Baumeister et al., 1991). They differ from the S-layers of genera of other crenarchaeal families and orders; therefore, at least in this phylum, "S-layer structures are of (limited) usefulness as taxonomic features" (Baumeister and Lembcke, 1992). Pilus- and pseudopodium-like appendages were found on cells attached to solid surfaces, e.g., sulfur crystals (Weiss, 1973; McClure and Wyckoff, 1982). They were also present in cells grown in liquid culture containing yeast extract (DeRosa et al., 1974). Flagella and flagella-like appendages have been described (Weiss, 1973; Typke et al., 1988; Grogan, 1989) and have also been observed in related species and genera (Fuchs et al., 1996; R. Rachel et al., unpublished observation; Fig. 2E, F). The cell membrane consists of isopranyl ether lipids (Langworthy, 1985; Langworthy and Pond, 1986; DeRosa and Gambacorta, 1988), some of which appear to be characteristic of Sulfolobales. These include: 1) a diglycerol tetraether containing a C_{40} biphytanyl chain and two C_{20} phytanyl chains (DeRosa et al., 1983); and 2) a glycerol-alkylnonitol tetraether, in which a branched C_9 nonitol trivially named "calditol" substitutes for a glycerol (DeRosa et al., 1980a). The membrane lipids are mainly composed of tetraethers containing up to four pentacyclic rings. The quantitative distribution of di- and tetraethers varies considerably with different species and growth parameters, e.g., temperature (DeRosa et al., 1980b), redox state (Langworthy and Pond, 1986), and nutrition (Langworthy, 1977). Further lipid compounds typical for Sulfolobales are unique sulfur-containing quinones. A benzo-[β]-thiophen-4, 7-quinone named "caldariellaquinone" (DeRosa et al., 1977) and other unusual quinones (sulfolobus-quinone and tricyclic quinone) have been detected in some species.

Within the group of strict aerobes, members of the genus *Metallosphaera* are readily identified by their slightly lobed, coccoid cell shape and the G+C content of their genomic DNA (around 46 mol%; Table 1), which is the highest among the Sulfolobales. Growth occurs on sulfidic ores, on metal sulfides (characteristic for *M. hakonensis*: growth on FeS), and by oxidation of molecular hydrogen (not determined for *M. hakonensis*). No significant DNA-DNA homology was detected between *Metallosphaera* species and any other member of the Sulfolobales (Huber et al., 1989). The cell wall consists of an S-layer of glycoprotein subunits with p3 symmetry and a center-to-center distance of 21 nm (Fuchs et al., 1995; Fig. 2D). Flagella have been found, exhibiting a diameter of about 13 nm (Fig. 2E, F). Genome sizes of around 1.89 Mb have been determined for *Met-*

allosphaera sedula (Baumann et al., 1998), while the genomes of members of the genus *Sulfolobus* are significantly larger (Baumann et al., 1998; Kawarabayasi et al., 2001; She et al., 2001b). *Metallosphaera sedula* and *M. prunae* can be distinguished by the possession of a fibrillar surface coat and a significantly higher O_2 concentration requirement for growth on molecular hydrogen for *M. prunae* (Fuchs et al., 1995). No data for these characteristics are available for *M. hakonensis*.

Both *Sulfurococcus* species are reported to show a characteristic cell division by binary fission and budding; whether this feature differs from the mode of cell division in other members of the Sulfolobales is still an open question. All other characteristics are very similar to the *Metallosphaera* species, like morphology, metabolism, and G+C content of genomic DNA. Since no 16S rRNA sequences are available, the phylogenetic position of the *Sulfurococcus* species within the Sulfolobales is uncertain (the strains are not available!). Both organisms can be distinguished by their temperature optima and the ability to oxidize ferrous iron (Table 1).

In addition to their unique mode of energy conservation by either reduction or oxidation of elemental sulfur, a low DNA G+C content of ~31 mol% is characteristic for members of the genus *Acidianus*. Within this group, identification of *A. brierleyi* can easily be done by its growth temperature, which is the lowest known for any member of the Sulfolobales (Table 1). The strictly lithotrophic species *A. infernus* and *A. ambivalens* have, on the other hand, quite similar morphology, physiology, and genomic DNA G+C content (Segerer et al., 1986; Zillig et al., 1986). They are best distinguished by their maximal growth temperature (Table 1). *Acidianus brierleyi* does not share significant DNA homology with *A. infernus* and *A. ambivalens*. No DNA homology was found with any other member of the Sulfolobales (Segerer et al., 1986; Segerer et al., 1991). The S-layer also exhibits p3 symmetry (Fig. 2G).

On the basis of 16S rRNA sequence comparisons, *Sulfurisphaera ohwakuensis* is phylogenetically very closely related to the *Sulfolobus* species *S. tokodaii* and *S. yangmingensis*. However, like the *Acidianus* species, it is a facultative anaerobic organism, although in contrast to them, it does not grow aerobically on sulfur compounds.

Stygiolobus azoricus is well characterized by its strictly anaerobic mode of life in combination with an obligate chemolithotrophic metabolism ($S°$-H_2 autotrophy). It shares no significant DNA homology with other members of the Sulfolobales (Segerer et al., 1991).

Cultivation

Medium for Cultivation of All Sulfolobales Members Basal Mineral Medium (Allen, 1959; modified by Brock et al., 1972)

$(NH_4)2SO_4$	1.30 g
KH_2PO_4	0.28 g
$MgSO_4 \cdot 7H_2O$	0.25 g
$CaCl_2 \cdot 2H_2O$	70 mg
$FeCl_3 \cdot 6H_2O$	20 mg
$MnCl_2 \cdot 4H_2O$	1.8 mg
$Na_2B_4O7 \cdot 10H_2O$	4.5 mg
$ZnSO_4 \cdot 7H_2O$	0.22 mg
$CuCl_2 \cdot 2H_2O$	0.05 mg
$Na_2M_0O_4 \cdot 2H_2O$	0.03 mg
$VOSO_4 \cdot 2H_2O$	0.03 mg
$CoSO_4$	0.01 mg

Dissolve ingredients and dilute up to 1 liter with distilled water. Adjust pH to 2.0–3.0.

Although for a few members of the Sulfolobales (e.g., *Sulfurococcus*) slight modifications of this medium were applied for cultivation, all representatives can be grown in this medium.

Cultivation in Liquid Media

For heterotrophic growth, media are supplemented with organic substrates, for example, yeast extract (1 g/liter) or glucose (2 g/liter). For lithotrophic growth, media are supplemented with elemental sulfur (5–10 g/liter). Additional supplementation with low amounts (0.05–0.2 g/liter) of yeast extract or meat extract will provide significantly (about 10 times) better growth of most strains than is obtained by strictly autotrophic cultivation. Metal-mobilizing strains may be grown lithoautotrophically on a mixture of sulfidic ores (e.g., pyrite [Grube Bayerland, Germany], sphalerite [Grube Lüderich, Germany], chalcopyrite [Norway], and pitch blend [Grube Höhenstein, Germany]; Huber et al., 1986; Huber et al., 1989). A detailed chemical analysis of such an ore mixture is described in Huber et al. (1989).

The basal salt medium is sterilized by autoclaving. Organic components, elemental sulfur, or sulfidic ores are added afterwards from sterile stocks. Sterile sulfur is obtained by Tyndall's fractional sterilization procedure; sulfidic ores can be sterilized by dry incubation at 140°C for 15 h.

Aerobic cultivation is routinely carried out in long (e.g., 270 × 18 mm), aluminum-capped test tubes containing 10 or 15 ml of medium. Although the cultures have to be shaken, the comparatively small surface of medium provides microaerobic rather than fully aerobic growth conditions, the former being preferred by most strains. Alternatively, aerobic cultivation can occur in 100-ml Erlenmeyer flasks with shaking. Each flask contains 30 ml of medium and is equipped with an air cooler (e.g., a 1-ml pipette tightly fitted through a rubber stopper). This procedure is especially recommended for lithoautotrophic strains on ores, but is also suitable for growing other Sulfolobales species.

Anaerobic media are prepared by using the technique of Balch et al. (1979). The media are reduced by the addition of Na_2S (0.75 g/liter). Cultivation is performed in rubber-stoppered 100-ml serum bottles containing 20 ml of medium plus sulfur and an atmosphere composed of H_2 and CO_2 (e.g., 80:20, 300 kPa). The bottles are sterilized by fractional sterilization and, if desired, supplemented with sterile organic substrate after the final cooling.

Members of the Sulfolobales may be fermented in conventional fermenters. However, special attention has to be paid to the very high corrosivity of both the medium and the metabolic products (H_2SO_4 and H_2S).

Cultivation on Plates

Sulfolobales cells may be plated onto medium solidified by Gellan gum K9A40 (Gelrite) (Zillig et al., 1986; Lindström and Sehlin, 1989), starch (R. Skorko, personal communication), or polysilicate (Stetter and Zillig, 1985). The use of agar is not recommended (Brock et al., 1972).

An overlay technique is used to plate facultatively and obligately heterotrophic strains and lithotrophic strains capable of efficiently growing by oxidation of soluble sulfur compounds (e.g., tetrathionate). A plating procedure based on Gelrite (Kelco Inc., San Diego, CA, United States) yielding plating efficiencies close to one is outlined by Lindström and Sehlin (1989). Gelrite suspension (1.6 g per liter in distilled water) is autoclaved (20 min, 121°C) and subsequently kept at ≥70°C after autoclaving. Prewarmed double-strength mineral medium containing 20 mM tetrathionate, 2 mM NaCl, 0.2% yeast extract or Casamino Acids™, and 16 mM $MgCl_2$ is added to the Gelrite solution, mixed, poured into Petri dishes, and allowed to solidify. The soft gel contains half the concentration of Gelrite and is prepared similarly: double-strength medium containing the desired titer of bacteria is added to the liquid gel (0.8 g of Gelrite per liter), mixed, and poured onto the supporting gel. Plates can be incubated at 65°C in a sealed plastic bag or if available in stainless steel cylinders up to temperatures of 90°C to minimize evaporation.

Plating of obligately sulfur-dependent strains is still troublesome. Also in this case, a top layer technique rather than mechanical plating is advised. The following procedure was successfully used to obtain single colonies of *Acidianus infernus*.

Boil medium without sulfur and $(NH_4)_2SO_4$, containing 12% (w/v) of starch (Stärke Gel, Serva) and pour after dissolution of the polymer into Petri dishes. Allow these supporting gels to solidify and dry overnight at 4°C. To prepare one top layer, dissolve 20 mg of sulfur in a minimum amount of $(NH_4)_2S$ solution (20%; Merck, Darmstadt) to give a polysulfide solution, and dissolve 0.5 g of starch by boiling in 5 ml of medium, pH 5.5. After cooling the starch solution to about 45°C, add the highly alkaline polysulfide solution and a sufficient amount of 9N H_2SO_4 to give a final pH of 3. After combining the acidified starch medium and polysulfide, colloidal sulfur forms immediately. Vigorously mix this soft gel at once with the bacteria, pour onto the supporting gel, and allow to solidify at 4°C. Incubate plates microaerobically at 75°C in a pressure cylinder (Balch et al., 1979) or a desiccator containing a humid atmosphere of about 60% air and 40% CO_2 (v/v). Small, often slimy, ochreous-brown colonies will develop after 7 or more days of incubation.

Another plating procedure involving Gellan gum K9A40 (Scott) with colloidal sulfur derived from thiosulfate described by Zillig et al. (1986) was successfully used to plate *Acidianus ambivalens* under anaerobic conditions in an atmosphere consisting of H_2 and CO_2.

Preservation

Members of the Sulfolobales maintain an intracellular pH of about 5.5 during growth (Lübben and Schäfer, 1989). To avoid damage to cells during storage caused by penetrating extracellular acid, addition of sterile $CaCO_3$ to raise the pH of all fully grown cultures to about 5.5 is recommended. Short-term storage is possible at 4°C. Although anaerobically grown strains are still sufficiently viable after one year of storage, maintenance requires transfer every 2–3 months. Anaerobically grown cells, e.g., *Acidianus infernus* cells, are significantly more resistant to storage at 4°C than are aerobically grown ones. Therefore, facultatively aerobic strains should be stored anaerobically.

Since oxygen is toxic to resting *Sulfolobus* cells (Grogan, 1989), storing cells under nitrogen significantly increases viability. Thus, storage under nitrogen at pH ~5.5 appears to be the most successful short-term storage procedure for aerobic members of the Sulfolobales.

Although not described for all strains, long-term storage is best achieved by freezing to around −140°C (in the gas phase over liquid nitrogen) cells suspended in culture media containing 5% dimethyl sulfoxide. No loss of cell viability is usually observed after storage of several years (H. Huber and D. Prangshvili, personal observation).

Physiology

As indicated above, all Sulfolobales species are extremely acidophilic thermophiles. With the exception of *Stygiolobus azoricus*, *Sulfurisphaera ohwakuensis* and a few *Sulfolobus* strains, all species can oxidize elemental sulfur to sulfuric acid according to the equation:

$$S_8 + 12O_2 + 8H_2O \rightarrow 8H_2SO_4 \qquad (1)$$

Modifications of this mode of energy conservation have been described. Oxygen can be replaced as an electron acceptor by MoO_4^{2-} (Brierley and Brierley, 1982) and Fe^{3+} (Brock and Gustafson, 1976a). Sulfur can be replaced as an energy source by S^{2-} (Brock, 1978a), $S_4O_6^{2-}$ (Wood et al., 1987), as well as by Fe^{2+} (Brierley and Brierley, 1973a; Brock et al., 1976b). Owing to their ability to oxidize sulfide, some strains are able to mobilize metals from sulfidic ores by formation of soluble sulfates (Brierley, 1978; Huber et al., 1986; Huber et al., 1989; Huber and Stetter, 1991a; Karavaiko et al., 1994; Fuchs et al., 1995; see also the section Applications). In addition, growth by oxidation of molecular hydrogen ("Knallgas reaction") is possible for many Sulfolobales (Huber et al., 1992). However, significant differences in the optimal oxygen concentrations for this metabolism were observed in the investigated strains.

Several species and strains are able to grow on CO_2 as sole carbon source (Shivvers and Brock, 1973; Wood et al., 1987; Huber et al., 1989; Huber and Stetter, 1991a; Segerer et al., 1991). Assimilation of carbon dioxide was first studied with *A. brierleyi*. In the beginning it was thought that the fixation occurs via a reductive carboxylic acid pathway (Kandler and Stetter, 1981). However, by measuring enzymatic activities, key enzymes of the 3-hydroxypropionate cycle were detected in autotrophically grown cells, which led to the conclusion that a modified 3-hydroxypropionate pathway operates in *A. brierleyi* (Ishii et al., 1997). In detailed studies, the enzymes of this pathway were identified also in other autotrophically grown cells of Sulfolobales species (Menendez et al., 1999; Hügler et al., 2003a) and finally the key enzyme for carboxylation (acetyl-CoA/propionyl-CoA carboxylase) was identified as the main CO_2 fixation enzyme for the autotrophic members of the Sulfolobales (Hügler et al., 2003b).

Several representatives of the Sulfolobales (e.g., *A. brierleyi* and some *Sulfolobus* or *Metallosphaera* species) are facultative heterotrophs, capable of growing on various complex organic substrates like yeast extract, meat extract, bacterial extracts, peptone, tryptone, or Casamino Acids™. *Sulfolobus* species also grow well on various mono-, oligo-, and even polymeric sugars and amino acids (Brock et al., 1972; DeRosa et al., 1975; Brock, 1978a; Grogan, 1989). Many strains of Sulfolobales may grow mixotrophically (Wood et al., 1987). Interestingly, the type strains of *S. acidocaldarius* DSM 639 and *S. solfataricus* DSM 1616 have become incapable of oxidizing sulfur, as evidenced by observations in independent laboratories (Marsh et al., 1983; Huber et al., 1989). They apparently have mutated into obligately heterotrophic strains.

Aerobic growth of strictly or facultatively aerobic members of the Sulfolobales is often enhanced at low oxygen tensions and with an atmosphere enriched with 5% (v/v) CO_2 (Brock, 1978a; Stetter and Zillig, 1985; H. Huber, personal observations), indicating that the organisms are microaerophiles. This finding is consistent with their sensitivity to H_2O_2 and the aerophobic response of *Sulfolobus* species (Grogan, 1989).

Representatives of the genera *Acidianus* and *Sulfurisphaera* can both oxidize and reduce molecular sulfur (Segerer et al., 1985; Zillig et al., 1985; Kurosawa et al., 1998). Anaerobic growth occurs via $S°$-H_2 autotrophy (Fischer et al., 1983) and large amounts of H_2S are formed (equation 2):

$$8H_2 + S_8 \rightarrow 8H_2S \qquad (2)$$

This mode of energy conservation operates also in the strictly anaerobic *Stygiolobus azoricus*.

Biochemistry and Molecular Biology

A huge number of biochemical and molecular biology studies has been performed with various strains of Sulfolobales. However, these investigations have not been extended to all members.

RESPIRATORY CHAIN Within the Sulfolobales, fairly complete respiratory chains and individual respiratory catalysts have been characterized mainly from *S. acidocaldarius*, *S. metallicus*, *S. tokodaii* and *A. ambivalens* by identifying their components using absorption spectroscopy of intact cells, membranes, or detergent-solubilized complexes (e.g., Anemüller et al., 1985; Lübben et al., 1986; Wakagi and Oshima, 1987; Iwasaki et al., 1995a; Iwasaki et al., 1995b; Gomes et al., 1998a; Bandeiras et al., 2003; see also the review by Schäfer et al., 1999). The so far simplest aerobic respiratory chain was detected in *A. ambivalens*: it is composed of a type-II reduced nicotinamide adenine dinucleotide (NADH) dehydrogenase (Gomes et al., 2001), a succinate:quinone oxidoreductase (Gomes et al., 1999; Lemos et al., 2001), caldariella quinone, and a single terminal quinol oxidase of the aa_3 type (Giuffre et al., 1997; Purschke et al., 1997).

In contrast, *S. acidocaldarius* has a much more intricate respiratory chain: it is a split electron transport chain. Two different terminal oxidase complexes exist, the so-called "SoxABCD complex" (Lübben et al., 1992; Castresana et al., 1995) and the "SoxM supercomplex" (Lübben et al., 1994; Castresana et al., 1995; see review by Schäfer et al., 1999). In *S. metallicus*, the major NADH dehydrogenase appears to be the type-II enzyme with unique features: the flavin group is covalently bound and contains a flavin mononucleotide instead of a flavin dinucleotide (Bandeiras et al., 2003). Furthermore, the spectroscopic analysis of the respiratory chain revealed the presence of a novel type of iron sulfur cluster and indicated the presence of several respiratory enzymes (Gomes et al., 1998b). So far, the presence of the respiratory NADH dehydrogenase, the essential component of bacterial respiratory chains (Complex I), has not been detected in Sulfolobales (Schäfer et al., 1999; She et al., 2001b).

Sulfolobales species contain unique types of quinones that are part of the respiratory chain of Sulfolobales. Beside caldariella quinone (CQ; DeRosa et al., 1977; Collins and Langworthy, 1983), benzo-[β]-thiophen-4,7-quinone, called "sulfolobus quinone" (SQ), was found in considerable amounts in *A. infernus* and *A. ambivalens*, but only in traces in *S. acidocaldarius* (Thurl et al., 1986). Whereas aerobically grown cells of *A. ambivalens* contain only about one-third of the SQ amount found in anaerobically grown cells, *A. infernus* contains no SQ when grown aerobically and no CQ when grown anaerobically. As in eukaryotes, the aromatic portion of CQ is derived from tyrosine. *Sulfolobus* strain MT 4 contains a tricyclic benzo[1,2-β; 4,5-β']dithiophen-4,8-quinone (Lanzotti et al., 1986), which is absent in *Metallosphaera sedula* (Huber et al., 1989). In *S. solfataricus*, the ratio between *Sulfolobus* quinone and caldariella quinone was reported to vary considerably with growth temperature (Nicolaus et al., 1992).

Activity of a coenzyme A-acylating 2-oxoacid:ferredoxin oxidoreductase was detected in cell-free extracts of *S. acidocaldarius* (Kerscher et al., 1982). A ferredoxin probably containing two $[4Fe-4S]^{2+(2+,1+)}$ clusters per molecule, which had an apparent molecular mass of 12.7 kDa, was purified from this organism (Kerscher et al., 1982), and its amino acid sequence has been determined (Minami et al., 1985). In addition, [3Fe-4S] clusters in the form of di-cluster seven-iron ferredoxins were found within the Sulfolobales. Their ferredoxins are small, monomeric proteins with β-sheet structure that contain one $[3Fe-4S]^{+/0}$ center and one $[4Fe-4S]^{2+/+}$ center (Teixeira et al., 1995; Gomes et al., 1998a). Parameters that affect the linear cluster

formation and disassembly in *S. acidocaldarius* have been investigated (Jones et al., 2002).

Further elements of the respiratory chain of *S. acidocaldarius* consist of different *a*- and *b*- type cytochromes (Anemüller et al., 1985; Anemüller and Schäfer, 1989). In contrast, it is still an open question whether enzymes or analogs to the cytochrome bc_1 complex (or the closely related b_6f complex) can be found in the membranes of Sulfolobales. At present there is only indirect evidence for the existence of archaeal homologues to these complexes. Genes encoding Rieske proteins and *b*-type cytochromes have been detected in the genomes of different *Sulfolobus* species (e.g., Kawarabayasi et al., 2001; She et al., 2001b). A novel type of iron cluster, which may act as a functional of Rieske iron-sulfur proteins, was found in *S. metallicus* (Gomes et al., 1998b).

Membrane-bound ATPases have been isolated from *S. acidocaldarius* DSM 639 (Lübben and Schäfer, 1987a; Lübben et al., 1987b; Schäfer and Meyering-Vos, 1992a; Schäfer and Meyering-Vos, 1992b), *S. solfataricus* (Hochstein and Stan-Lotter, 1992), *S. tokodaii* and *A. ambivalens* (Hinrichs et al., 1999). The A_1 ATPase of *S. acidocaldarius* released from the plasma membrane has an apparent molecular mass of 380 kDa and consists of four subunits of 65, 51, 20 and 12 kDa (Lübben et al., 1987b). The membrane-bound ATPase from *S. tokodaii* has a different apparent molecular mass of 360 kDa and consists of subunits of 69, 54 and 28 kDa, respectively (Wakagi and Oshima, 1985; Konishi et al., 1987).

SULFUR METABOLISM The sulfur oxidation pathway has been studied with *A. ambivalens* as the model organism (recently reviewed in Kletzin et al., 2004). A soluble sulfur oxygenase reductase (SOR) catalyzes the initial step of S° oxidation by performing a sulfur disproportionation reaction with sulfite, thiosulfate and hydrogen sulfide as the products (Kletzin, 1989). The colorless SOR, both native and recombinant, showed the presence of a low-potential mononuclear iron site (E_m = –270 mV). Electron microscopy showed that the enzyme forms a hollow, (bacterio-) ferritin-like sphere of 15.5 nm in diameter which is most probably composed of 24 identical subunits made of dimer building blocks (Urich et al., 2004).

The product thiosulfate is oxidized to tetrathionate by a novel type of quinone-containing, thiosulfate:quinone oxidoreductase purified from the membrane fraction (Müller et al., 2004). The 102-kDa glycoprotein was composed of two subunits with apparent molecular masses of 28 and 16 kDa, which were previously thought to be part of the terminal oxidase (DoxD and DoxA; Purschke et al., 1997). The enzyme has an optimum activity at 85°C and pH 5 (Müller et al., 2004). Thiosulfate-dependent oxygen consumption inhibitable by addition of cyanide indicates that this reaction is one of the entry points of electrons into the electron transport chain. The fate of the tetrathionate is not clear at present.

The other SOR product (sulfite) is oxidized by two different pathways (Zimmermann et al., 1999). A membrane-bound sulfite:acceptor oxidoreductase activity was found. It is also oxidized by soluble enzymes involving an adenylylsulfate adenylyltransferase and an adenosine phosphosulfate reductase leading to energy conservation by substrate-level phosphorylation. The enzymes have not yet been purified in both cases.

Sulfur reduction was studied by Laska and Kletzin (2000) and Laska et al. (2003) in *A. ambivalens*. The purification and biochemical and molecular characterization of the membrane-bound hydrogenase and sulfur reductase from cells grown under anaerobic conditions were reported.

Sugar Degradation Glucose metabolism in *Sulfolobus solfataricus* strain MT 4 and *S. acidocaldarius* was investigated and shown to occur exclusively via a modified Entner-Doudoroff (ED) pathway not involving phosphorylated intermediates and producing pyruvate (DeRosa et al., 1984; Bartels 1989; Selig et al., 1997). This pyruvate is finally almost completely converted to CO_2; no three-carbon fermentation products and no significant amounts of acetate are formed. The CO_2 is generated via acetyl-CoA (by pyruvate-ferredoxin oxidoreductase) and the tricarboxylic acid (TCA) cycle. The net ATP yield of this nonphosphorylated ED pathway is zero, but ATP can be generated from reduced pyridine nucleotides oxidized by molecular oxygen in the respiratory chain (Schönheit and Schäfer, 1995). However, on the basis of analysis of the *S. solfataricus* genome, it was proposed that instead of reduced nicotinamide adenine dinucleotide phosphate (NAD[P]H), ferredoxin may act as the main physiological electron carrier (She et al., 2001b). Several metabolic enzymes involved in catabolism, intermediary metabolism, and anabolism have been investigated, partially or totally purified, and characterized. These include β-galactosidase (Buonocore et al., 1980), pyridine-dependent glucose dehydrogenase (Giardina et al., 1986), NAD^+-dependent alcohol dehydrogenase (Rella et al., 1987), malic enzyme (Bartolucci et al., 1987; Guagliardi et al., 1988), aspartate aminotransferase (Marino et al., 1988), and propylamine transferase (Cacciapuoti et al., 1986) from *S. solfataricus* strain MT 4, citrate synthase from *S. acidocaldarius* and *S. solfataricus* (Großebüter and Görisch, 1985; Löhlein-Werhan et al., 1988), and isocitrate dehydrogenase (Danson and Wood, 1984) and malate dehydrogenase (Görisch et al., 1985; Hartl et al., 1987) from *S. acidocaldarius*.

Sugar Storage and Accumulation As occurs in other Archaea (e.g., *Pyrobaculum aerophilum* or *Thermoplasma acidophilum*), the cells of the Sulfolobales accumulate trehalose (Nicolaus et al., 1988; Martins et al., 1997). The function of trehalose as a possible osmolyte within the Sulfolobales (which are organisms that grow in media with very low salt concentrations) is highly unlikely (Martins et al., 1997) and remains unknown.

In *Sulfolobus shibatae*, glycogen of an average chain length of only seven glucose molecules was isolated and in complexes with four proteins, one of which was identified as a glucosyl-transferase using both ADP-glucose and UDP-glucose as substrates (König et al., 1982).

Solute Transport Proteins A whole range of sugar ATP-binding cassette (ABC) transporters has been found in *S. solfataricus* (Albers et al., 1999; Elferink et al., 2001). These transporters fall into two groups: 1) the glucose, arabinose, trehalose-systems, which show similarity to the sugar ABC transporters of bacteria, and 2) the cellobiose and maltose and maltotriose transporters, which exhibit the highest similarity to bacterial di- and oligopeptide transporters.

Uridine-5′-Monophosphate Biosynthesis Screening of different nucleic acid precursors for their ability to incorporate into nucleic acids of heterotrophically growing *S. acidocaldarius* indicated predominance of the de novo over the "salvage" pathway of pyrimidine nucleotide biosynthesis (Muskhelishvili et al., 1990). The result was confirmed by genetic studies (Grogan and Gunsalus, 1993).

Sulfolobicins Several strains of "*S. islandicus*" produce proteinaceous toxins with molecular mass of 20 kDa, termed "sulfolobicins," which kill cells of other strains of the same species as well as strains of *S. solfataricus* and *S. shibatae*, but not *S. acidocaldarius* (Prangishvili et al., 2000). They were associated with the producer cells as well as with cell-derived S-layer-coated spherical membrane vesicles 90–180 nm in diameter and not released from cells in soluble form (Prangishvili et al., 2000).

Chromatin and Chromosome Structure In different strains of *Sulfolobus* a reverse gyrase was found, an enzyme that introduces positive superturns into DNA at the expense of ATP and is specific for hyperthermophilic archaea (reviewed by Forterre et al. [1996] and Forterre [2000]). *Sulfolobus* strains also contain a type II topoisomerase which relaxes both positive and negative superturns and exhibits a strong decatenase activity (reviewed by Forterre et al., 1996). It is a heterotetramer composed of two subunits, one of which is homologous to the factor Spo11 required for initiation of meiotic recombination in the Eukarya (Bergerat et al., 1997).

Among several DNA binding proteins found in Sulfolobales, the most abundant are proteins of two families. One of them is the Sso7d family (recently renamed the "Sul7d family" by White and Bell, 2002) of small architectural proteins, specifically found only in *Sulfolobus* strains. Proteins were reported to stabilize the double helix from denaturation (Baumann et al., 1994), to promote the annealing of complementary strands (Guagliardi et al., 1997), and to show ATPase-dependent rescue of aggregate proteins (Guagliardi et al., 2000). Binding of these proteins causes unwinding of the helix and negative supercoiling (Agback et al., 1998; Lopez-Garcia et al., 1998; Mai et al., 1998; Robinson et al., 1998) as well as DNA bending and compaction (Napoli et al., 2002). The role of these proteins in vivo is not understood. It has been suggested that they are involved in homeostatic control of DNA superhelicity in Sulfolobales (Lopez-Garcia et al., 1998).

Members of another family of abundant DNA-binding proteins of *Sulfolobus* are known which induce negative supercoiling. They are named "Ssh10," "Sso10b," and "Sac10b," depending on the species of origin (Lurz et al., 1986; Forterre et al., 1999; Xue et al., 2000). This protein, found also in many other Archaea and some Eukarya, has been recently named "Alba" (for "acetylation lowers binding affinity") because it is found acetylated in vivo and is deacetylated by the *Sulfolobus* homolog of the eukaryal histone deacetylase Sir2 (Bell et al., 2002). Deacetylation of Alba reduces its affinity for DNA and its ability to repress transcription and suggests similarities in modulation of chromatin activity by acetylation in Archaea and Eukarya. A model for the Alba-DNA interaction consistent with the available structural data has been suggested (Wardleworth et al., 2002).

DNA TRANSCRIPTION, REPLICATION, REPAIR AND RECOMBINATION Studies on transcription in *Sulfolobus* species started with a purification of an RNA polymerase (Zillig et al., 1979). Complexity of the enzyme, consisting of at least 10 subunits, was much higher than of bacterial RNA polymerases and resembled that of RNA polymerase II of Eukarya. Subsequent bioinformatic and biochemical studies on *Sulfolobus* transcription machinery have confirmed that it resembles the core components of the eukaryal RNA polymerase II apparatus and showed that it uses also bacterial-like gene-specific transcription regulators (reviewed by Bell and Jackson, 2001). Briefly, RNA polymerase together with a TATA-binding protein (TBP) and transcription factor B (TFB), homologues of the eukaryal transcription factors TBP and TFIIB, respectively, is sufficient to direct initiation of transcription on promoters that contain a TATA box and transcription factor B recognition element (BRE), reminiscent of eukaryal RNA polymerase II promoters.

As mentioned above (see the section Chromatin and Chromosome Structure in this Chapter), in clear parallel to eukaryotes, *Sulfolobus* species possess the capacity to modulate transcription through covalent modification of a key chromatin protein. Neither of the two chromatin proteins, Sul7d (Napoli et al., 2002) or Alba (Wardleworth et al., 2002), has in its native state an effect on a reconstituted in vitro transcription. However, acetylation of the latter protein significantly lowers its affinity for DNA and results in repression of transcription (Bell et al., 2002).

The existence of a global gene regulatory system in *S. solfataricus* involving activation is suggested by the coordinated transcription of three physically unlinked glycosyl hydrolase genes in response to change in carbon source (Haseltine et al., 1999). However, elements of this system remain to be elucidated.

The presence of multiple forms of DNA polymerases in *Sulfolobus* cells was first demonstrated biochemically (Prangishvili, 1986) and later confirmed by results of genome sequencing. Three enzymes, PolB1, PolB2 and PolB3, belong to family B of DNA polymerases (She et al., 2001b). The replicative polymerase B1 has been most intensively studied (Klimczak et al., 1985; Rossi et al., 1986; Bukhrashvili et al., 1989; Chinchaladze et al., 1989; Elie et al., 1989; reviewed by Böhlke et al., 2002). *Sulfolobus* possesses also a DNA polymerase belonging to the Y family, which apparently is specialized in the replication of aberrant DNA templates and is able to bypass certain DNA lesions in an error-free manner (Kulaeva et al., 1996; Boudsocq et al., 2001; Gruz et al., 2001; Kokoshka et al., 2002; Potapova et al., 2002). The structural basis of lesion bypass has been investigated by crystallographic approaches utilizing PolY from different strains of *S. solfataricus* (Ling et al., 2001; Silvian et al., 2001; Zhou et al., 2001).

The activity of PolB1 is stimulated and its processivity is enhanced by the sliding clamp, proliferating cell nuclear antigen (PCNA), a heterotrimer of three distinct subunits, PCN1, 2 and 3 (Dionne et al., 2003). The finding is in marked contrast to a report by De Felice et al. (1999) that indicated that high concentrations of individual PCNA1 and PCNA3 could stimulate the polymerase. The heterotrimeric PCNA besides DNA polymerase stimulates the activities of DNA ligase I and the flap endonuclease. A model for clamp loading of PCNA has been suggested by unique subunit-specific interactions between components of the clamp loader, the replication factor C (Dionne et al., 2003).

Studies on the single-stranded DNA-binding protein (SSB) of *S. solfataricus* have suggested that it represents a simplest form of SSB protein studied to date and demonstrated striking similarity of its structure to that of human replication protein A (Wadsworth and White, 2001; Kerr et al., 2003).

Repair of double-stranded breaks in DNA in *Sulfolobus* appears similar to its eukaryal equivalent. In the pathway of homologous recombination, the recombinase that causes stand exchange, RadA, of *S. solfataricus* is significantly more similar in sequence to eukaryal Rad51 than to the eubacterial RecA protein (Sandler et al., 1996) and has the same DNA binding and pairing preferences as its homologues (Seitz and Kowalczykowski, 2000). The processing of DNA ends at the initiation step of homologous recombination in *S. acidocaldarius* involves four proteins Rad50, Mre11, NurA and HerA, the genes of which constitute an operon; the latter two proteins are prototypes of new families of nucleases and helicases, correspondingly (Elie et al., 1997; Constantinesco et al., 2002; Constantinesco et al., 2004). Two distinct Holliday junction resolvases, Hjc and Hje, with differing substrate specificities have been described for *S. solfataricus* (Kvaratskhelia et al., 2000). Hjc has no recognizable branch migration motor for the Holliday junction. Its activity in vitro is stimulated by the chromatin protein Sul7d (Kvaratskhelia et al., 2002). The three-dimensional structure of Hjc revealed extensive structural homology with a superfamily of type II restriction enzymes (Bond et al., 2001).

In nucleotide excision repair in *S. solfataricus*, a structure-specific endonuclease is involved which is homologous to eukaryal nuclease XPF and has the substrate specificity remarkably similar to the human XPF but displays an unexpected requirement for a functional interaction with the sliding clamp PCNA (Roberts et al., 2003; see above).

TRANSLATION For translation initiation in *S. solfataricus*, apparently different mechanisms are used. This was indicated by the finding of a Shine-Dalgarno sequence upstream of the genes inside operons but not generally for the first gene in an operon or isolated genes (Tolstrup et al.,

2000). This finding was later confirmed biochemically (Benelli et al., 2003). Shine-Dalgarno-dependent initiation, resembling the pathway prevalent in extant bacteria, apparently operates on distal cistrons of polycistronic mRNAs, whereas "leaderless" initiation, reminiscent of the eukaryotic pathway, operates on monocistronic mRNAs and on opening cistrons of polycistronic mRNA (Benelli et al., 2003).

In *Sulfolobus* species, small RNAs were found and functionally characterized. Almost all of these are related to the plethora of small RNAs that form the core of the complex ribonucleoprotein machinery responsible for eukaryotic ribosome biosynthesis and mediate diverse, complex and essential processes such as intron excision, RNA modification and editing, protein targeting, etc. (reviewed by Tang et al. [2002] and Omer et al. [2003]).

Translation inhibition by many specific antibiotics in cell free extracts of several Sulfolobales species was investigated by Sanz et al. (1994). The analysis showed that the ribosomes of Sulfolobales were the most refractory to inhibitors of protein synthesis known so far.

CELL CYCLE REGULATION As revealed by epifluorescence microscopy and flow cytometry, the chromosomal DNA in *S. solfataricus* is differently organized in exponentially growing and stationary cells, and it is conceivable that DNA-binding proteins of different classes are involved in this reorganization (reviewed by Bernander, 2000). Within the *Sulfolobus* cell cycle there was found a G2 period, similar to that in the eukaryotic cell cycle, during which cells increase in size (until they reach the desired mass or size for segregating the genomes) and then divide (Hjort and Bernander, 2001). Synchronous cell cycle progression in *S. acidocaldarius* could be generated by daunomycin treatment (Hjort and Bernander, 2001) or by dilution of stationary phase culture into fresh medium (Hjort and Bernander, 1999). Abrupt temperature shifts were shown to induce cell cycle arrest and chromosomal DNA degradation (Hjort and Bernander, 1999).

HEAT SHOCK Little is known on the general heat-shock response in Sulfolobales. It was shown that in "*S. islandicus*," the control of DNA topology during stress relies primarily on the physical effect of temperature on topoisomerase activities and on the geometry of DNA itself (Lopez-Garcia and Forterre, 1999).

In *S. shibatae*, acquired thermotolerance at lethal temperatures ($\geq 90°C$) correlates with the increased synthesis of two heat-shock proteins known as "TF55 α and β" (Trent et al., 1990; Trent et al., 1994; Kagawa et al., 1995). These proteins are isolated from cells as subunits of

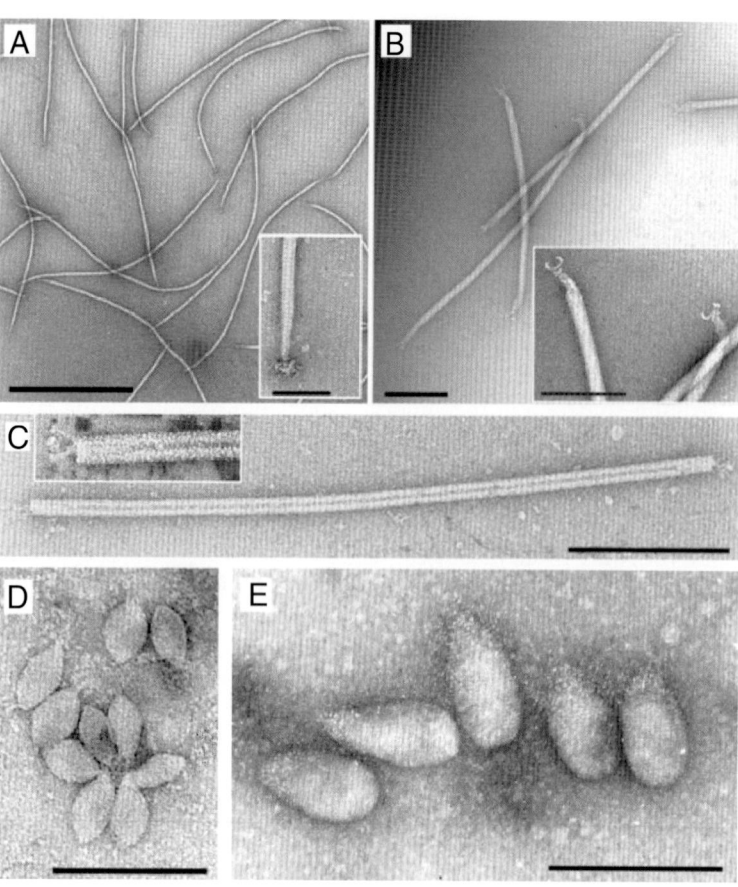

Fig. 3. Electron micrographs of representatives of different families of viruses of Sulfolobales, negatively stained with uranyl acetate: A) Lipothrixvirus SIFV of *Sulfolobus* (reproduced from Arnold et al. [2000b], with permission). B) Lipothrixvirus AFV1 of *Acidianus* (reproduced from Bettstetter et al. [2003], with permission). C) Rudivirus SIRV1 of Sulfolobus. D) Fusellovirus SSV2 of Sulfolobus. E) "Guttavirus" of *Sulfolobus* (courtesy of Wolfram Zillig). Bars are 200 nm; for insets in A), B), and C), bars are 100 nm.

double-ring complexes called "rosettasomes" that share structural and functional features with chaperonins (Trent et al., 1991). They constitute up to 4% of total cellular protein and in vitro can generate a filamentous structure (Trent et al., 1997). Under all conditions, rosettasomes were shown to be membrane-associated, and their function as a membrane skeleton that interacts with lipids and/or membrane-associated proteins to impact the permeability and stability of the cytoplasmic membrane has been suggested (Trent et al., 2003).

Extrachromosomal Genetic Elements

Extrachromosomal genetic elements described for members of the order Sulfolobales include about a dozen of viruses and about two dozen of plasmids, conjugative and cryptic.

Viruses

The unusual features of viruses of Sulfolobales warranted the establishment of novel virus fam-

ilies: spindle-shaped enveloped Fuselloviridae, enveloped lipid-containing filamentous *Lipothrixviridae*, and unenveloped, stiff-rod-shaped Rudiviridae (Prangishvili, 1999b; reviewed by Prangishvili et al., 2001). The fourth family, the *Guttaviridae*, has been suggested but not yet approved for the bearded-droplet-shaped virus (Arnold et al., 2000a). The structure of virus particles from each of the families is shown in Fig. 3A–E and virus properties are summarized in Table 2. In addition to these cultivated viruses, a plethora of virus-like particles was found to be produced by different strains of Sulfolobales (reviewed in Prangishvili, 2003).

Viruses of *Sulfolobus* are exceptional also in characteristics of virus-host interactions. In contrast to nearly all double-stranded (ds)DNA prokaryotic viruses which eventually kill the host cell during release of progeny virions, *Sulfolobus* viruses establish a productive infection without killing or lysing the cell, a so-called "carrier state." Only in one case of the *Sulfolobus* virus SSV1 has there been observed induction of virus production in the lysogenic strain but without subsequent cell lysis. In all cases, the infected cells continue to replicate themselves as well as the virus, although cellular growth is slowed

Sulfolobales 39

Table 2. Viruses of Sulfolobales.

Families	Species	Host	ds DNA (kb)	Peculiarities of genome	References
Fuselloviridae	SSV1	*S. shibatae, S. solfataricus,* "*S. islandicus*"	Circular (15.5)	Positively supercoiled	Schleper et al., 1992[a] Palm et al., 1991[b]
	SSV2	*S. solfataricus,* "*S. islandicus*"	Circular (14.8)	N.F.	Stedman et al., 2003[b]
Lipothrixviridae	SIFV	"*S. islandicus*"	Linear (42.8)	N.F.	Arnold et al., 2000a[a,b]
	AFV1	*Acidianus* spec.	Linear (20.8)	Ends resemble telomeres	Bettstetter et al., 2003[a,b]
Rudiviridae	SIRV1	"*S. islandicus*"	Linear (32.3)	Covalently closed ends, 2 kb-ITRs	Prangishvili et al., 1999[a] Peng et al., 2001[b]
	SIRV2	"*S. islandicus*"	Linear (35.5)	Covalently closed ends, 1.6 kb-ITRs	Prangishvili et al., 1999[a] Peng et al., 2001[b]
"*Guttaviridae*"	SNDV	*Sulfolobus* spec.	Circular (20)	Extensively methylated	Arnold et al., 2000b[a]

Abbreviations: ITRs, inverted terminal repeats; and N.F., none found.
[a]Reference to virus description.
[b]Reference to nucleotide sequence.

down, suggesting inhibition of some specific cell functions.

Double-stranded DNA genomes of several viruses have been sequenced. These include circular genomes of the fuselloviruses SSV1 and SSV2 (Palm et al., 1991; Stedman et al., 2003), linear genomes of lipothrixviruses SIFV (Arnold et al., 2000b) and AFV1 (Bettstetter et al., 2003) and linear genomes of rudiviruses SIRV1 and SIRV2 (Peng et al., 2001). About 90% of all open reading frames (ORFs) on sequenced genomes show no significant matches to sequences in public databases, suggesting that viruses of Sulfolobales have exceptional biochemical solutions for biological functions. Genomes share some homologous genes, and they yield good sequence matches with genes present in the host chromosomes. They include genes encoding glycosyl transferases, CopG family proteins, integrases, dUTPases and Holliday junction resolvases as well as proteins of unknown function (reviewed in Prangishvili and Garrett, 2004); some of these enzymes have been biochemically characterized (Muskhelishvili et al., 1993; Prangishvili et al., 1998; Birkenbihl et al., 2001). None of the viruses share homologous ORFs with viruses of other prokaryotes.

Specific features of genome organization, as well as strategies for DNA replication, suggest phylogenetic relationships between the rudiviruses of *Sulfolobus* and the large eukaryal DNA viruses: poxviruses, the African swine fever virus, and *Chlorella* viruses (Blum et al., 2001; Peng et al., 2001). Sequence patterns at the ends of the linear genome of the lipothrixvirus AFV1 are reminiscent of the telomeric ends of linear eukaryal chromosomes and suggest that a primitive telomeric mechanism operates in this virus (Bettstetter et al., 2003).

Plasmids

Twelve cryptic circular plasmids (5.3–7.6 kb in size) have been isolated from different strains of *Sulfolobus* and *A. ambivalens* (Zillig et al., 1998). Among them five plasmids have been sequenced: pRN1 (Keeling et al., 1996), pRN2 (Keeling et al., 1998), pSSVx (Arnold et al., 1999), and pHEN7 (Peng et al., 2000) from "*S. islandicus*" strains and pDL10 (Kletzin et al., 1999) from *A. ambivalens*. Sequence comparison revealed that all five plasmids belong to the same family (Peng et al., 2000) and do not resemble known bacterial plasmids. The plasmids share three conserved ORFs arranged in identical order and orientation on all plasmids. The corresponding proteins of the plasmid pRN1 have been heterologously expressed and bioche-

mically characterized. Two of them (6.5 and 9.5 kDa) are sequence-specific DNA-binding proteins, and the smaller is believed to participate in regulating the copy number (Lipps et al., 2001a; Lipps et al., 2001b). The third protein is a unique multifunctional enzyme with primase, DNA polymerase, and DNA-dependent ATPase activity (Lipps et al., 2003). The N-terminal half of the protein harbors a novel domain which is able to polymerize DNA without the need of a separate priming enzyme. There is no sequence similarity to other known DNA polymerases and primases, suggesting that this protein constitutes a new DNA polymerase family.

Several families of conjugative plasmids hosted by a *Sulfolobus* sp. from Japan and different strains of "*S. islandicus*" have been described (Schleper et al., 1995; Prangishvili et al., 1999a). Mechanism of conjugation has not been elucidated. Plasmids contain only two ORFs with some sequence similarity to genes essential for conjugation in the Bacteria, *traG* and *trbE* (She et al., 1998; Stedman et al., 2000).

Genetics

Genomic Sequences

Nucleotide sequences of the 3 Mb genome of *S. solfataricus* strain P2 (She et al., 2001b; the sequence and annotation are also available on the Homed at LBMGE website (http://www-archbac.u-psud.fr, www.ncbi.nlm.nih.gov/entrez/query.fcgi?db=nucleotide=search=AE006641) and of the 2.7 Mb genome of *S. tokodaii* (Kawarabayasi et al., 2001) the sequence and annotation are also available on the National Insititute of Technology and Evaluation website (http://www.bio.nite.go.jp) have been determined. In progress is sequencing of the genomes of *S. acidocaldarius*, "*S. islandicus*" strain HVE10/4 (R. A. Garrett, personal communication), *A. ambivalens* (A. Kletzin, personal communication) and *A. brierleyi* (Q. She, personal communication).

Genome Stability and Mutants

Two main types of mobile elements identified on the *S. solfataricus* genome are the autonomous insertion sequence (IS) elements (Brügger et al., 2002) and the nonautonomous miniature inverted repeat elements (Redder et al., 2001). They are considered to be mobilized via transposases encoded by IS elements, constitute about 10% of the genome, and are spread out along it, however, mainly clustered in two areas separated by putative replication origin and termination regions. Such an abundance of mobile elements gives the *S. solfataricus* genome a high level of plasticity. Transposon mutagenesis apparently is a dominant mechanism of mutation in this organism. Transposition of several distinct families of IS elements into the β-galactosidase gene and at the chromosomal *pyrEF* locus was shown to produce spontaneous lacS⁻ and uracyl auxotrophic mutants, correspondingly, with high frequencies (Schleper et al., 1994; Martusewitsch et al., 2000).

There is strong evidence that DNA can enter and leave chromosomes of at least some Sulfolobales, facilitated by an integrase (reviewed by She et al., 2002). This was first demonstrated for the virus SSV1 which encodes an integrase facilitating insertion into the tRNA^{Arg} gene in the chromosome of the host, *S. shibatae* (Muskhelishvili et al., 1993). During insertion, the integrase gene is partitioned into two fragments which flank the linearized virus DNA in the chromosome. Both gene fragments carry perfect direct repeats, and excision involves recombination at these direct repeats. Similarly partitioned integrase genes occur in chromosomes of other members of the Sulfolobales. For example, in *S. solfataricus*, one borders an inserted plasmid (Peng et al., 2000) and another a large (67 kb) integron-like segment (She et al., 2001a).

High genome plasticity is not typical for all Sulfolobales. *Sulfolobus acidocaldarius* has few active IS elements (Grogan et al., 2001; R. A. Garrett, personal communication). According to mutational analysis of the *pyrE* and *pyrF* genes, *S. acidocaldarius* has one of the lowest genomic error rates yet measured (Grogan et al., 2001). Moreover, it was shown that within the low mutant frequency, deletions make up an unusually low proportion of spontaneous mutations (Grogan and Hansen, 2003b).

Several *lacS⁻* and uracyl auxotrophic mutants of *S. acidocaldarius* and *S. solfataricus*, as well as thermosensitive mutants of *S. acidocaldarius*, have been isolated and characterized (Grogan, 1991; Kondo et al., 1991; Grogan and Gunsalus, 1993; Schleper et al., 1994; Bernander et al., 1998; Martusewitsch et al., 2000; Bartolucci et al., 2003). As a result of disruption of the gene encoding a secreted α-amylase of *S. solfataricus*, a mutant strain was produced unable to grow on starch, glycogen or pullulan as sole carbon and energy source (Worthington et al., 2003).

ing conjugative plasmids and virus DNAs are used for construction of vectors. A derivative of the conjugative plasmid pNOB8 containing a *lacS* gene as a selectable marker has been used for transformation of *lacS⁻* mutant of *S. solfataricus* and complementation of *lacS* activity (Elferink et al., 1996). DNA of the virus SSV1 provides a basis for ongoing vector construction. A shuttle vector has been constructed by introducing an *E. coli* plasmid into the nonessential region of the viral genome (Stedman et al., 1999). Upon transfection, this vector spreads efficiently throughout the culture. Development of other recombinant SSV1-based shuttle vectors has resulted in complementation of uracil auxotrophic and β-galactosidase mutants of *Sulfolobus* (Jonuscheit et al., 2003).

A stable *E. coli/S. solfataricus* shuttle vector was constructed also by introduction of the replication sequence of the virus SSV1 into an *E. coli* plasmid (Cannio et al., 1998). Using this vector as a cloning vehicle, functional complementation of *lacS⁻* mutants defective in β-glycosidase activity was demonstrated (Bartolucci et al., 2003). The same vector was used for the gene transfer and expression in *S. solfataricus* of two genes of thermophilic bacteria (Contursi et al., 2003). The selectable system used in these experiments has been developed by adaptation of the hygromycin B phosphotransferase from *E. coli* to high temperatures, allowing resistance to the antibiotic hygromycin and selection in *Sulfolobus* (Cannio et al., 2001).

A plasmid of *Pyrococcus abyssi*, pGT5, has also been used as a basis for transformation protocols for *Sulfolobus* (Aravalli and Garrett, 1997). A vector containing its replication origin and an *adh* gene from *S. solfataricus* could propagate in both *Pyrococcus furiosus* and *S. solfataricus*. The system uses as selection markers toxic alcohols such as butanol that are detoxified by the plasmid-encoded *Sulfolobus adh* gene and the selection provided is weak. The potential of the mobile intron from *Desulfurococcus mobilis*, inserted into an *E. coli* plasmid, to generate a new type of vector was demonstrated in *S. acidocaldarius* (Aagaard et al., 1996). A gene replacement method was used to disrupt an α-amylase coding sequence by insertion of a modified allele of the *S. solfataricus lacS* gene and to produce mutant recombinant strain (Worthington et al., 2003).

Tools for Genetic Manipulation

Transformation protocols reported for *Sulfolobus* are not yet fully suitable and require further improvement. In some cases, efficiently spread-

Restriction-Modification System

Restriction-modification system, designated "*Sua*I," with a recognition sequence GGCC, has been biochemically characterized for *S. acidocal-*

darius (Prangishvili et al., 1985). It uses N^4 methylation of the inner C to protect the recognition sequence from restriction (Grogan, 2003a).

Ecology

Representatives of the Sulfolobales thrive in acidic biotopes with original temperatures from at least 50°C to about 100°C. They are typical inhabitants of high-temperature continental solfataric areas and have been isolated from such biotopes nearly worldwide. They share these habitats with other thermophilic and hyperthermophilic organisms, including members of the orders Desulfurococcales, Thermoproteales, Thermoplasmatales, and Aquificales. They can be enriched from the aerobic and anaerobic zones of solfataric soils, which can be easily identified owing to the different colors caused by oxidized or reduced iron. Depending on the volcanic activity and exhalation of reducing volcanic gases (e.g., H_2S and H_2), the anaerobic zone starts at a depth of a few centimeters up to about 50 cm below the oxidized layers (Huber et al., 2000). In marine hydrothermal systems, members of the Sulfolobales seem to be extremely rare.

In their ecosystems, members of the Sulfolobales can function as primary producers and/or as consumers of organic material. For the chemolithoautotrophic primary producers, oxygen and elemental sulfur (for anaerobes) are suitable electron acceptors for growth. Different sulfur compounds or molecular hydrogen serve as the electron donors. The end product of their metabolism is H_2SO_4, H_2O or H_2S; in addition, the organic cell material is formed. The heterotrophs consume organic matter by growth via respiration or fermentation with single organic compounds (e.g., sugars and proteins) or complex organic substrates (e.g., yeast extract, peptone, and meat extract). Therefore, the Sulfolobales are important for the carbon cycle in acidic high temperature habitats.

Biotechnology

By means of sulfide oxidation, several isolates of Sulfolobales are able to extract metal ions from sulfidic ores ("bioleaching"; Brierley and Murr, 1973b; Brierley, 1978; Marsh et al., 1983; Huber et al., 1986; Huber et al., 1989; Huber and Stetter, 1991a; Fuchs et al., 1995). Whereas some strains only mobilize metal ions weakly, others, e.g., *Metallosphaera* species and *Sulfolobus metallicus*, are very efficient ore leachers (Huber et al., 1986; Huber et al., 1989; Huber and Stetter, 1991a; Fuchs et al., 1995). Thus, they may be suitable for reactor or in situ leaching of geothermally heated ore deposits.

Immobilized cells of *Sulfolobus* strain MT 4 were used for studying production of glucose and 2-keto-3-deoxygluconate at self-sterilizing reactor temperatures (Drioli et al., 1982; Drioli et al., 1986; Nicolaus et al., 1986; Catapano et al., 1988).

Sulfolobales strains have also been investigated for use in the removal of inorganic and organic sulfur from coal and oil (Kargi and Robinson, 1985; Kargi, 1987; Chen and Skidmore, 1988).

Acknowledgments. The authors thank G. Gmeinwieser for technical support and R. Rachel for electron microscopy.

Literature Cited

Aagaard, C., I. Leviev, R. N. Aravalli, P. Forterre, D. Prieur, and R. A. Garrett. 1996. General vectors for archaeal hyperthermophiles: strategies based on a mobile intron and a plasmid. FEMS Microbiol. Rev. 18:93–104.

Agback, P., H. Baumann, S. Knapp, R. Ladenstein, and T. Hard. 1998. Architecture of non-specific protein-DNA interactions in the Sso7-DNA complex. Nature Struct. Biol. 5:579–584.

Albers, S.-V., M. Elferink, R. L. Charlebois, C. Sensen, A. J. Driessen, and W. Konings. 1999. Glucose transport in the extremely thermoacidophilic Sulfolobus solfataricus involves a high-affinity membrane-integrated binding protein. J. Bacteriol. 181:4285–4291.

Allen, M. B. 1959. Studies with Cyanidium caldarium, an anomalously pigmented chlorophyte. Arch. Mikrobiol. 32:270–277.

Anemüller, S., M. Lübben, and G. Schäfer. 1985. The respiratory system of Sulfolobus acidocaldarius, a thermoacidophilic archaebacterium. FEBS Lett. 193:83–87.

Anemüller, S., and G. Schäfer. 1989. Cytochrome aa3 from the thermoacidophilic archaebacterium Sulfolobus acidocaldarius. FEBS Lett. 244:451–455.

Aravalli, R. N., and R. A. Garrett. 1997. Shuttle vectors for hyperthermophilic archaea. Extremophiles 1:183–191.

Arnold, H. P., Q. She, H. Phan, K. Stedman, D. Prangishvili, I. Holz, J. K. Kristjansson, R. A. Garrett, and W. Zillig. 1999. The genetic element pSSVx of the extremely thermophilic crenarchaeon Sulfolobus is a hybrid between a plasmid and a virus. Molec. Microbiol. 34:217–226.

Arnold, H. P., U. Ziese, and W. Zillig. 2000a. SNDV, a novel virus of the extremely thermophilic and acidophilic archaeon Sulfolobus. Virology 272:409–416.

Arnold, H. P., W. Zillig, U. Ziese, I. Holz, M. Crosby, T. Utterback, J. F. Weidmann, J. Kristjanson, H.-P. Klenk, K. E. Nelson, and C. Fraser. 2000b. A novel lipothrixvirus, SIFV, of the extremely thermophilic crenarchaeon Sulfolobus. Virology 267:252–266.

Balch, W. E., G. E. Fox, L. J. Magrum, C. R. Woese, and R. S. Wolfe. 1979. Methanogens: reevaluation of a unique biological group. Microbiol. Rev. 43:260–296.

Bandeiras, T. M., C. A. Salgueiro, H. Huber, C. M. Gomes, and M. Teixeira. 2003. The respiratory chain of the thermophilic archaeon Sulfolobus metallicus: studies on the type-II NADH dehydrogenase. Biochim. Biophys. Acta 1557:13–19.

Bartels, M. 1989. Glucoseabbau über einen modifizierten Entner-Doudoroff Weg be idem thermoacidophilen Archaebacterium Sulfolobus acidocaldarius [PhD thesis]. Medizinische Fakultät, Universität Lübeck. Lübeck, Germany.

Bartolucci, S., R. Rella, A. Guagliardi, C. A. Raia, A. Gambacorta, M. DeRosa, and M. Rossi. 1987. Malic enzyme from the archaebacterium Sulfolobus solfataricus. Purification, structure, and kinetic properties. J. Biol. Chem. 262:7725–7731.

Bartolucci, S., M. Rossi, and R. Cannio. 2003. Characterization and functional complementation of a nonlethal deletion in the chromosome of a beta-glycosidase mutant of Sulfolobus solfataricus. J. Bacteriol. 185:3948–3957.

Baumann, H., S. Knapp, T. Lundback, R. Ladenstein, and T. Hart. 1994. Solution structure and DNA-binding properties of a small thermostable protein from the archaeon Sulfolobus solfataricus. Nature Struct. Biol. 1:808–819.

Baumann, C., M. Judex, H. Huber, and R. Wirth. 1998. Estimation of genome sizes of hyperthermophiles. Extremophiles 2:101–108.

Baumeister, W., S. Volker, and U. Santarius. 1991. The three-dimensional structure of the surface protein of Acidianus brierleyi determined by electron crystallography. Syst. Appl. Microbiol. 14:103–110.

Baumeister, W., and G. Lembcke. 1992. Structural features of archaebacterial cell envelopes. J. Bioenerg. Biomembr. 24:567–575.

Bell, S. D., and S. P. Jackson. 2001. Mechanism and regulation of transcription in archaea. Curr. Opin. Microbiol. 4:208–213.

Bell, S. D., C. H. Botting, B. N. Wardleworth, S. P. Jackson, and M. F. White. 2002. The interaction of Alba, a conserved archaeal chromatin protein, with Sir2 and its regulation by acetylation. Science 296:148–151.

Benelli, D., E. Maone, and P. Londei. 2003. Two different mechanisms for ribosome/mRNA interation in archaeal translation initiation. Molec. Microbiol. 50:635–643.

Bergerat, A., B. De Massy, D. Gadelle, P. C. Varoutas, A. Nicolas, and P. Forterre. 1997. An atypical topoisomerase II from Archaea with implications for meiotic recombination. Nature 386:414–417.

Bernander, R., A. Poplawski, and D. W. Grogan. 1998. Altered patterns of cellular growth, morphology, replication and division in conditional-lethal mutants of the thermophilic archaeaon Sulfolobus acidocaldarius. Microbiology 146:749–757.

Bernander, R. 2000. Chromosome replication, nucleoid segregation and cell division in Archaea. Trends Microbiol. 8:278–283.

Bettstetter, M., X. Peng, R. A. Garrett, and D. Prangishvili. 2003. AFV1, a novel virus infecting hyperthermophilic archaea of the genus Acidianus. Virology 315:68–79.

Birkenbihl, R. P., K. Neef, D. Prangishvili, and B. Kemper. 2001. Holliday junction resolving enzymes of archaeal viruses SIRV1 and SIRV2. J. Molec. Biol. 309:1067–1076.

Blum, H., W. Zillig, S. Mallock, H. Domdey, and D. Prangishvili. 2001. The genome of the archaeal virus SIRV1 has features in common with genomes of eukaryal viruses. Virology 281:6–9.

Böhlke, K., F. M. Pisani, M. Rossi, and G. Antranikian. 2002. Archaeal DNA replication. Extremophiles 6:1–14.

Bohlool, B. B. 1975. Occurrence of Sulfolobus acidocaldarius, an extremely thermophilic acidophilic bacterium, in New Zealand hot springs. Isolation and immunofluorescence characterization. Arch. Microbiol. 106:171–174.

Bond, C. S., M. Kvaratskhelia, D. Richard, M. F. White, and W. N. Hunter. 2001. Structure of Hjc, a holliday junction resolvase, from Sulfolobus solfataricus. Proc. Natl. Acad. Sci. USA 98:5509–5514.

Boudsocq, F., S. Iwai, F. Haoka, and R. Woodgate. 2001. Sulfolobus solfataricus P2 DNA polymerase IV (Dpo4): an archaeal Din-B-like DNA polymerase with lesion-bypass properties akin to eukaryotic polη. Nucleic Acids Res. 29:4607–4616.

Brierley, J. A. 1966. Contribution of Chemolithoautotrophic Bacteria to the Acid Thermal Waters of the Geysir Springs Group in Yellowstone National Park [PhD thesis]. Montana State University. Bozeman, MT. 58–60.

Brierley, C. L., and J. A. Brierley. 1973a. A chemoautotrophic and thermophilic microorganism isolated from an acid hot spring. Can. J. Microbiol. 19:183–188.

Brierley, C. L., and L. E. Murr. 1973b. Leaching: use of a thermophilic and chemoautotrophic microbe. Science 179:488–490.

Brierley, C. L. 1978. Bacterial leaching. CRC Crit. Rev. Microbiol. 6:207–262.

Brierley, C. L., and J. A. Brierley. 1982. Anaerobic reduction of molybdenum by Sulfolobus species. Zbl. Bakteriol. Parasitenkd. Infektionskr. Hyg., Abt. 1 Orig. C3:289–294.

Brock, T. D., K. M. Brock, R. T. Belly, and R. L. Weiss. 1972. Sulfolobus: a new genus of sulfur-oxidizing bacteria living at low pH and high temperature. Arch. Mikrobiol. 84:54–68.

Brock, T. D., and J. Gustafson. 1976a. Ferric iron reduction by sulfur- and iron-oxidizing bacteria. Appl. Environ. Microbiol. 32:567–571.

Brock, T. D., S. Cook, S. Peterson, and J. L. Mosser. 1976b. Biochemistry and bacteriology of ferrous iron oxidation in geothermal habitats. Geochim. Cosmochim. Acta 40:493–500.

Brock, T. D. 1978a. Thermophilic Microorganisms and Life at High Temperatures. Springer-Verlag. New York, NY. 117–179.

Brock, T. D. 1978b. Thermophilic Microorganisms and Life at High Temperatures. Springer-Verlag. New York, NY. 386–418.

Brügger, K., P. Redder, Q. She, F. Confalonieri, Y. Zivanovic, and R. A. Garrett. 2002. Mobile elements in archaeal genomes. FEMS Microbiol. Lett. 206:131–141.

Bukhrashvili, I., D. Chinchaladze, A. Levina, G. Nevinsky, O. Lavrik, and D. Prangishvili. 1989. Comparison of initiating abilities of primers of different lengths and composition in polymerization reactions catalyzed by DNA polymerases from extremely thermophilic archaebacteria. Biochim. Biophys. Acta 1008:102–107.

Buonocore, V., O. Sgambati, M. DeRosa, E. Esposito, and A. Gambacorta. 1980. A constitutive β-galactosidase from the extreme thermoacidophilic archaebacterium Caldariella acidophila: properties in the free state and in immobilized cells. J. Appl. Biochem. 2:390–397.

Cacciapuoti, G., M. Porcelli, M. Carteni-Farina, A. Gambacorta, and V. Zappia. 1986. Purification and characterization of propylamine transferase from Sulfolobus solfataricus, an extreme thermophilic archaebacterium. Eur. J. Biochem. 161:263–271.

Cannio, R., P. Contursi, and S. Bartolucci. 1998. An autonomously transforming vector for Sulfolobus solfataricus. J. Bacteriol. 180:3237–3240.

Cannio, R., P. Contursi, W. Rossi, and S. Bartolucci. 2001. Thermoadaptation of a mesophilic hygromycin B phosphotransferase by directed evolution in hyperthermophilic archaea: selection of a stable genetic marker for DNA transfer into Sulfolobus solfataricus. Extremophiles 3:153–159.

Castresana, J., M. Lübben, and M. Saraste. 1995. New archaebacteria genes coding for redox proteins: implications for the evolution of aerobic metabolism. J. Molec. Biol. 250:202–210.

Catapano, G., G. Iorio, E. Drioli, and M. Filosa. 1988. Experimental analysis of a cross-flow membrane bioreactor with entrapped whole cells: influence of transmembrane pressure and substrate feed concentration on reactor performance. J. Membr. Sci. 35:325–338.

Chen, C.-Y., and D. R. Skidmore. 1988. Attachment of Sulfolobus acidocaldarius cells in coal particles. Biotechnol. Prog. 4:25–30.

Chinchaladze, D., D. Prangishvili, A. Scamrov, R. Beabealashvili, N. Dyatkina, and A. Krayevsky. 1989. Nucleoside 5′-triphosphates modified at sugar residues as substrates for DNA polymerase from the thermoacidophilic archaebacterium Sulfolobus acodocaldarius. Biochim. Biophys. Acta 1008:113–115.

Collins, M. D., and T. A. Langworthy. 1983. Respiratory quinone composition of some acidophilic bacteria. Syst. Appl. Microbiol. 4:295–304.

Constantinesco, F., P. Forterre, and C. Elie. 2002. NurA, a novel 5′-3′ nuclease gene linked to rad50 and mre11 homologs of thermophilic archaea. EMBO Rep. 3:40–52.

Constantinesco, F., P. Forterre, E. V. Koonin, L. Aravind, and C. Elie. 2004. A bipolar archaeal helicase, HerA, and its potential functional association with Rad50, Mre11 and Nur11 proteins. Nucleic Acids Res. 32:1439–1447.

Contursi, P., R. Cannio, S. Prato, G. Fiorentino, M. Rossi, and S. Bartolucci. 2003. Development of a genetic system for hyperthermophilic Archaea: Expression of a moderate thermophilic bacterial alcohol dehydrogenase gene in Sulfolobus solfataricus. FEMS Microbiol. Lett. 218:115–120.

Danson, M. J., and P. A. Wood. 1984. Isocitrate dehydrogenase of the thermoacidophilic archaebacterium Sulfolobus acidocaldarius. FEBS Lett. 172:289–293.

Deatherage, J. F., K. A. Taylor, and L. A. Amos. 1983. Three-dimensional arrangement of the cell-wall protein of Sulfolobus acidocaldarius. J. Molec. Biol. 167:823–852.

De Felice, M., C. W. Sensen, R. L. Charlebois, M. Rossi, and F. M. Pisani. 1999. Two DNA polymerase sliding clamps from the thermophilic archaeon Sulfolobus solfataricus. J. Molec. Biol. 291:47–57.

DeRosa, M., A. Gambacorta, G. Millonig, and J. D. Bu'Lock. 1974. Convergent characters of extremely thermophilic acidophilic bacteria. Experientia 30:866–868.

DeRosa, M., A. Gambacorta, and J. D. Bu'Lock. 1975. Extremely thermophilic acidophilic bacteria convergent with Sulfolobus acidocaldarius. J. Gen. Microbiol. 86:156–164.

DeRosa, M., S. DeRosa, A. Gambacorta, L. Minale, R. H. Thomson, and R. D. Worthington. 1977. Caldariellaquinone, a unique benzo-b-thiophen-4,7-quinone from Caldariella acidophila, an extremely thermophilic and acidophilic bacterium. J. Chem. Soc. Perkin Trans. 1:653–657.

DeRosa, M., S. DeRosa, A. Gambacorta, and J. D. Bu'Lock. 1980a. Structure of calditol, a new branched-chain nonitol, and the derived tetraether lipids in thermoacidophilic archaebacteria of the Caldariella group. Phytochemistry 19:249–254.

DeRosa, M., E. Esposito, A. Gambacorta, B. Nicolaus, and J. D. Bu'Lock. 1980b. Effects of temperature on ether lipid composition of Caldariella acidophila. Phytochemistry 19:827–831.

DeRosa, M., A. Gambacorta, B. Nicolaus, B. Chappe, and P. Albrecht. 1983. Isoprenoid ethers: backbone of complex lipids of the archaebacterium Sulfolobus solfataricus. Biochim. Biophys. Acta 753:249–256.

DeRosa, M., A. Gambacorta, B. Nicolaus, P. Giardina, E. Poerio, and V. Buonocore. 1984. Glucose metabolism in the extreme thermoacidophilic archaebacterium Sulfolobus solfataricus. Biochem. J. 224:407–414.

DeRosa, M., and A. Gambacorta. 1988. The lipids of archaebacteria. Prog. Lipid Res. 27:153–157.

Dionne, I., R. K. Nookala, S. P. Jackson, A. J. Doherty, and S. D. Bell. 2003. A heterotrimeric PCNA in the hyperthermophilic archaeon Sulfolobus solfataricus. Molec. Cell 11:275–282.

Drioli, E., G. Iorio, M. DeRosa, A. Gambacorta, and B. Nicolaus. 1982. High-temperature immobilized-cell ultrafiltration reactors. J. Membr. Sci. 11:365–370.

Drioli, E., G. Iorio, G. Catapano, M. DeRosa, and A. Gambacorta. 1986. Capillary membrane reactors: performances and applications. J. Membr. Sci. 27:253–261.

Elferink, M., C. Schleper, and W. Zillig. 1996. Transformation of the extremely thermophilic archaeon Sulfolobus solfataricus via a self-spreading vector. FEMS Microbiol. Lett. 137:31–35.

Elferink, M., S.-V. Albers, W. N. Konings, and A. J. Driessen. 2001. Sugar transport in Sulfolobus solfataricus is mediated by two families of binding protein-dependent ABC transporters. Molec. Microbiol. 39:1494–1503.

Elie, C., A. M. DeRecondo, and P. Forterre. 1989. Thermostable DNA polymerase from the archaebacterium Sulfolobus acidocaldarius: purification, characteization and immunological properties. Eur. J. Biochem. 178:619–626.

Elie, C., M.-F. Baucher, C. Fondrat, and P. Forterre. 1997. A protein related to eucaryal and bacterial DNA-motor proteins in the hyperthermophilic archaeon Sulfolobus acidocaldarius. J. Molec. Evol. 45:107–114.

Fischer, F., W. Zillig, K. O. Stetter, and G. Schreiber. 1983. Chemolithoautotrophic metabolism of anaerobic extremely thermophilic archaebacteria. Nature (Lond.) 301:511–513.

Fliermans, C. B., and T. D. Brock. 1972. Ecology of sulfur-oxidizing bacteria in hot acid soils. J. Bacteriol. 111:343–350.

Forterre, P., A. Bergerat, and P. Garcia-Lopez. 1996. The unique DNA topology and DNA topoisomerases of hyperthermophilic archaea. FEMS Microbiol. Lett. 18:237–248.

Forterre, P., F. Gonfalonieri, and S. Knapp. 1999. Identification of the gene encoding archaeal-specific DNA-binding proteins of the Sac10b family. Molec. Microbiol. 32:669–670.

Forterre, P. 2000. A hot story from comparative genomics: reverse gyrase is the only hyperthermophile-specific protein. Trends Genet. 18:236–238.

Fuchs, T., H. Huber, K. Teiner, S. Burggraf, and K. O. Stetter. 1995. Metallosphaera prunae, sp. nov., a novel metal-mobilizing, thermoacidophilic Archaeum, isolated from a uranium mine in Germany. Syst. Appl. Microbiol. 18:560–566.

Fuchs, T., H. Huber, S. Burggraf, and K. O. Stetter. 1996. 16S rDNA-based phylogeny of the archaeal order Sulfolobales and reclassification of Desulfurococcus ambivalens as Acidianus ambivalens comb. nov. Syst. Appl. Microbiol. 19:56–60.

Furuya, T., T. Nagumo, T. Itoh, and H. Kaneko. 1977. A thermophilic acidophilic bacterium from hot springs. Agric. Biol. Chem. 41:607–612.

Giardina, P., M.-G. DeBiasi, M. DeRosa, A. Gambacorta, and V. Buonocore. 1986. Glucose dehydrogenase from the thermoacidophilic archaebacterium Sulfolobus solfataricus. Biochem. J. 239:517–522.

Giuffre, A., C. M. Gomes, G. Antonini, E. D'Itri, M. Teixeira, and M. Brunori. 1997. Functional properties of the quinol oxidase from Acidianus ambivalens and the possible role of its electron donor: studies in the membrane integrated and purified enzyme. Eur. J. Biochem. 250:383–388.

Golovacheva, R. S., K. M. Valieho-Roman, and A. V. Troitskii. 1987a. Sulfurococcus mirabilis gen. nov., sp. nov., a new thermophilic archaebacterium with the ability to oxidize sulfur. Mikrobiologiya 56:100–107.

Golovacheva, R. S., I. G. Zhukova, T. P. Nikultseva, and D. N. Ostrovinskii. 1987b. Some properties of Sulfurococcus mirabilis, a new thermoacidophilic archaebacterium. Mikrobiologiya 56:281–287.

Gomes, C. M., A. Faria, J. C. Carita, J. Mendes, M. Refalla, P. Chicau, H. Huber, K. O. Stetter, and M. Teixeira. 1998a. Di-cluster, seven-iron ferredoxins from hyperthermophilic Sulfolobales. J. Biochem. Inorg. Chem. 3:499–507.

Gomes, C. M., H. Huber, K. O. Stetter, and M. Teixeira. 1998b. Evidence for a novel type of iron cluster in the respiratory chain of the archaeon Sulfolobus metallicus. FEBS Lett. 432:99–102.

Gomes, C. M., R. S. Lemos, M. Teixeira, A. Kletzin, H. Huber, K. O. Stetter, G. Schäfer, and S. Anemüller. 1999. The unusual iron sulfur composition of the Acidianus ambivalens succinate dehydrogenase complex. Biophys. Biochim. Acta 1411:134–141.

Gomes, C. M., T. M. Bandeiras, and M. Teixeira. 2001. A new type-II NADH dehydrogenases from the archaeon Acidianus ambivalens: characterization and in vitro reconstitution of the respiratory chain. J. Bioenerg. Biomembr. 33:1–8.

Görisch, H., T. Hartl, W. Groebüter, and J. J. Stezowski. 1985. Archaebacterial malate dehydrogenases: The enzymes from the thermoacidophilic organisms Sulfolobus acidocaldarius and Thermoplasma acidophilum show A-side stereospecifity for NAD(+). Biochem. J. 226:885–888.

Grogan, D. W. 1989. Phenotypic characterization of the archaebacterial genus Sulfolobus: Comparison of five wild-type strains. J. Bacteriol. 171:6710–6719.

Grogan, D., P. Palm, and W. Zillig. 1990. Isolate B12, which harbours a virus-like element, represents a new species of the archaebacterial genus Sulfolobus, Sulfolobus shibatae, sp. nov. Arch. Microbiol. 154:594–599.

Grogan, D. 1991. Selectable mutant phenotypes of the extremely thermophilic archaeabacterium Sulfolobus acidocaldarius. J. Bacteriol. 173:7725–7727.

Grogan, D., and R. P. Gunsalus. 1993. Sulfolobus acidocaldarius synthesizes UMP via a standard de novo pathway: results of biochemical-genetic study. J. Bacteriol. 175:1500–1507.

Grogan, D., G. T. Carver, and J. W. Drake. 2001. Genetic fidelity under harsh conditions: analysis of spontaneous mutation in the thermoacidophilic archaeon Sulfolobus acidocaldarius. Proc. Natl. Acad. Sci. USA 98:7928–7933.

Grogan, D. 2003a. Cytosine methylation by the SuaI restriction-modification system: implications for genetic fidelity in a hyperthermophilic archaeon. J. Bacteriol. 185:4657–4661.

Grogan, D., and J. E. Hansen. 2003b. Molecular characteristics of spontaneous deletions in the hyperthermophilic archaeaon Sulfolobus acidocaldarius. J. Bacteriol. 185:1266–1272.

Großebüter, W., and H. Görisch. 1985. Partial purification and properties of citrate synthases from the thermoacidophilic archaebacteria Thermoplasma acidophilum and Sulfolobus acidocaldarius. Syst. Appl. Microbiol. 6:119–124.

Gruz, P., F. Pisani, M. Shimizu, M. Yamada, I. Hayashi, K. Morikawa, and T. Nohmi. 2001. Synthetic activity of Sso DNA polymerase Y1, an archaeal DinB-like DNA polymerase, is stimulated by processivity factors proliferating cell nuclear antigen and replication factor C. J. Biol. Chem. 276:47394–47401.

Guagliardi, A., M. Moracci, G. Manco, M. Rossi, and S. Bartolucci. 1988. Oxalacetate decarboxylase and pyruvate carboxylase activities, and effect of sulfhydryl reagents in malic enzyme from Sulfolobus solfataricus. Biochim. Biophys. Acta 957:301–311.

Guagliardi, A., A. Napoli, M. Rossi, and M. Ciaramella. 1997. Annealing of complementary DNA strands above the melting point of the duplex promoted by an archaeal protein. J. Molec. Biol. 267:841–848.

Guagliardi, A., L. Cerchia, M. Moracci, and M. Rossi. 2000. The chromosomal protein Sso7d of the crenarchaeon Sulfolobus solfataricus rescues aggregated proteins in an ATP hydrolysis-depenedent manner. J. Biol. Chem. 275:31813–31818.

Hartl, T., W. Großebüter, H. Görisch, and J. J. Stezowski. 1987. Crystalline NAD/NADP-dependent malate dehydrogenase; the enzyme from the thermoacidophilic archaebacterium Sulfolobus acidocaldarius. Biol. Chem. Hoppe-Seyler 368:259–267.

Haseltine, C., R. Montalvo-Rodriguez, E. Bini, A. Carl, and P. Blum. 1999. Coordinate transcriptional control in the hyperthermophilic archaeaon Sulfolobus solfataricus. J. Bacteriol. 181:3920–3927.

Hinrichs, M., G. Schäfer, and S. Anemüller. 1999. Functional characterization of an extremely thermophilic ATPase in membranes of the chrenarchaeon Acidianus ambivalens. Biol. Chem. 380:1063–1069.

Hjort, K., and R. Bernander. 1999. Changes in cell size and DNA content in Sulfolobus cultures during dilution and temperature shift experiments. J. Bacteriol. 181:5669–5675.

Hjort, K., and R. Bernander. 2001. Cell cycle regulation in the hyperthermophilic crenarchaeon Sulfolobus acidocaldarius. Molec. Microbiol. 40:225–234.

Hochstein, L. I., and H. Stan-Lotter. 1992. Purification and properties of an ATPase from Sulfolobus solfataricus. Arch. Biochem. Biophys. 295:153–160.

Huber, G., H. Huber, and K. O. Stetter. 1986. Isolation and characterization of new metal-mobilizing bacteria. Biotech. Bioengin. Symp. 16:239–251.

Huber, G., C. Spinnler, A. Gambacorta, and K. O. Stetter. 1989. Metallosphaera sedula gen. and sp. nov. represents a new genus of aerobic, metal-mobilizing, thermoacidophilic archaebacteria. Syst. Appl. Microbiol. 12:38–47.

Huber, G., and K. O. Stetter. 1991a. Sulfolobus metallicus sp. nov., a novel strictly chemolithotrophic thermophilic archaeal species of metal-mobilizers. Syst. Appl. Microbiol. 14:372–378.

Huber, G., R. Huber, B. Jones, G. Lauerer, A. Neuner, A. Segerer, K. O. Stetter, and E. T. Degens. 1991b. Hyperthermophilic archaea- and eubacteria occurring within Indonesia hydrothermal areas. Syst. Appl. Microbiol. 14:397–404.

Huber, G., E. Drobner, H. Huber, and K. O. Stetter. 1992. Growth by aerobic oxidation of molecular hydrogen in Archaea—a metabolic property so far unknown for this domain. Syst. Appl. Microbiol. 15:502–504.

Huber, R., H. Huber, and Stetter, K. O. 2000. Towards the ecology of hyperthermophiles: Biotopes, new isolation strategies and novel metabolic properties. FEMS Microbiol. Rev. 24:615–623.

Huber, H., and K. O. Stetter. 2001a. Order I: Thermoproteales. In: G. Garrity (Ed.) Bergey's Manual of Systematic Bacteriology, 2nd ed. Springer-Verlag. New York, NY. 1:170.

Huber, H., and K. O. Stetter. 2001b. Order II: Desulfurococcales. In: G. Garrity (Ed.) Bergey's Manual of Systematic Bacteriology, 2nd ed. Springer-Verlag. New York, NY. 1:179–180.

Huber, H., and K. O. Stetter. 2001c. Order III: Sulfolobales. In: G. Garrity (Ed.) Bergey's Manual of Systematic Bacteriology, 2nd ed. Springer-Verlag. New York, NY. 1:198.

Hügler, M., H. Huber, K. O. Stetter, and G. Fuchs. 2003a. Autotrophic CO2 fixation pathways in archaea (Crenarchaeota). Arch. Microbiol. 179:160–173.

Hügler, M., R. S. Krieger, M. Jahn, and G. Fuchs. 2003b. Characterization of acetyl-CoA/propionyl.CoA carboxylase in Metallosphaera sedula. Carboxylating enzyme in the 3-hydroxypropionate cycle for autotrophic carbon fixation. Eur. J. Biochem. 270:736–744.

Ishii, M., T. Miyake, T. Satoh, H. Sugiyama, Y. Oshima, T. Kodama, and Y. Igarashi. 1997. Autotrophic carbon dioxide fixation in Acidianus brierleyi. Arch. Microbiol. 166:368–371.

Iwasaki, T., K. Matsuura, and T. Oshima. 1995a. Resolution of the aerobic respiratory system of the thermoacidophilic archaeon, Sulfolobus sp. Strain 7. I: The archaeal therminal oxidase supercomplex is a functional fusion of respiratory complexes III and IV with no c-type cytochromes. J. Biol. Chem. 270:30881–30892.

Iwasaki, T., T. Wakagi, and T. Oshima. 1995b. Resolution of the aerobic respiratory system of the thermoacidophilic archaeon, Sulfolobus sp. Strain 7. III: The archaeal novel respiratory complex II (succinate:caldariellaquinone oxidoreductase complex) inherently lacks heme group. J. Biol. Chem. 270:30902–30908.

Jan, R.-J., J. Wu, S.-M. Chaw, C.-W. Tsai, and S.-D. Tsen. 1999. A novel species of thermoacidophilic archaeon, Sulfolobus yangmingensis sp. nov. Int. J. Syst. Bacteriol. 49:1809–1816.

Jones, K., C. M. Gomes, H. Huber, M. Teixeira, P. Wittung-Stashede. 2002. Formation of a linear [3Fe-4S] cluster in a seven-iron ferredoxin triggered by polypeptide unfolding. J. Biol. Inorg. Chem. 7:357–362.

Jonuscheit, M., E. Martusewitsch, K. M. Stedman, and C. Schleper. 2003. A reporter gene system for the hyperthermophilic archaeon Sulfolobus solfataricus based on a selectable and integrative shuttle vector. Molec. Microbiol. 48:1241–1252.

Kagawa, H. K., J. Osipiuk, N. Maltsev, R. Overbeek, E. Quaite-Randall, A. Joachimiak, and J. Trent. 1995. The 60 kDa heat shock proteins in the hyperthermophilic archaeon Sulfolobus shibatae. J. Molec. Biol. 253:712–725.

Kandler, O., and K. O. Stetter. 1981. Evidence for autotrophic CO2 assimilation in Sulfolobus brierleyi via a reductive carboxylic acid pathway. Zbl. Bakteriol. Parasitenkd. Infektionskr. Hyg., Abt. 1 Orig. C2:111–121.

Karavaiko, G. I., O. V. Golyshina, A. V. Troitskii, K. M. Valieho-Roman, R. S. Golovacheva, and T. A. Pivovarova. 1994. Sulfurococcus yellowstonii sp. nov., a new species of iron- and sulphur-oxidizing thermoacidophilic archaebacteria. Microbiology 63:379–387.

Kargi, F., and J. M. Robinson. 1985. Biological removal of pyritic sulfur from coal by the thermophilic organism Sulfolobus acidocaldarius. Biotechnol. Bioengin. 27:41–49.

Kargi, F. 1987. Biological oxidation of thianthrene, thioxanthene and dibenzothiophene by the thermophilic organism Sulfolobus acidocaldarius. Biotechnol. Lett. 9:478–482.

Kawarabayasi, Y., Y. Hino, H. Horikawa, K. Jin-no, M. Takahashi, M. Sekine, S. Baba, A. Ankai, H. Kosugi, A. Hosoyama, S. Fukui, Y. Nagai, K. Nishijima, R. Otsuka, H. Nakazawa, M. Takamiya, Y. Kato, T. Yoshizawa, T. Tanaka, Y. Kudoh, J. Yamazaki, N. Kushida, A. Oguchi, K. Aoki, S. Masuda, M. Yanagii, M. Nishimura, A. Yamagishi, T. Oshima, and H. Kikuchi. 2001. Complete genome sequence of an aerobic thermoacidophilic crenarchaeon, Sulfolobus tokodaii strain 7. DNA Res. 8:123–140.

Keeling, P. J., H.-P. Klenk, R. K. Singh, O. Feeley, C. Schleper, and W. Zillig. 1996. Complete nucleotide sequence of the Sulfolobus islandicus multicopy plasmid pRN1. Plasmid 35:141–144.

Keeling, P. J., H.-P. Klenk, R. K. Singh, M. E. Schenk, C. W. Sensen, W. Zillig, and W. F. Doolittle. 1998. Sulfolobus islandicus plasmid pRN1 and pRN2 share distant but common evolutionary distance. Extremophiles 2:391–393.

Kerr, I., D., R. I. M. Wadsworth, L. Cubeddu, W. Blankenfeldt, J. H. Naismith, and M. F. White. 2003. Insights into ssDNA resognition by the OB fold from a structural and thermodynamic study of Sulfolbus SSB protein. EMBO J. 22:2561–2570.

Kerscher, L., S. Nowitzki, and D. Oesterhelt. 1982. Thermoacidophilic archaebacteria contain bacterial-type ferredoxins acting as electron acceptors of 2-oxoacid:ferredoxin oxidoreductases. Eur. J. Biochem. 128:223–230.

Kletzin, A. 1989. Coupled enzymatic production of sulfite, thiosulfate, and hydrogen sulfide from sulfur: Purification and properties of a sulfur oxigenase reductase from the facultatively anaerobic archaebacterium Desulfurolobus ambivalens. J. Bacteriol. 171:1638–1643.

Kletzin, A., A. Lieke, T. Ulrich, R. L. Cherlebois, and C. W. Sensen. 1999. Molecular analysis of pDL10 from Acidianus ambivalens reveals a family of related plasmids from extremely thermophilic and acidophilic archaea. Genetics 152:1307–1314.

Kletzin, A., T. Urich, F. Müller, T. M. Bandeiras, and C. M. Gomes. 2004. Dissimilatory oxidation and reduction of elemental sulfur in thermophilic Archaea. J. Bioenerg. Biomembr. 36:77–91.

Klimczak, L. J., F. Grummt, and K. J. Burger. 1985. Purification and characterization of DNA polymerase from the archaebacterium Sulfolobus acidocaldarius. Nucleic Acids Res. 13:5269–5282.

Kokoshka, R. J., K. Bebenek, F. Boudsocq, R. Woodgate, and T. A. Kunkel. 2002. Low fidelity DNA synthesis by a y family DNA polymerase. J. Biol. Chem. 277:19633–19638.

Kondo, S., A. Yamagishi, and T. Oshima. 1991. Positive selection for uracyl auxotrophs of the sulfur-dependent thermophilic archaeabacterium Sulfolobus acidocaldarius by use of 5-fluoroorotic acid. J. Bacteriol. 173:7698–7700.

König, H., R. Skorko, W. Zillig, and W.-D. Reiter. 1982. Glycogen in thermoacidophilic archaebacteria of the genera Sulfolobus, Thermoproteus, Desulfurococcus, and Thermococcus. Arch. Microbiol. 132:297–303.

Konishi, J., T. Wakagi, T. Oshima, and M. Yoshida. 1987. Purification and properties of the ATPase solubilized from membranes of a thermoacidophilic archaebacterium, Sulfolobus acidocaldarius. J. Biochem. 102:1379–1387.

Kulaeva, O. I., E. V. Koonin, J. P. McDonald, S. K. Randall, N. Rabinovich, J. F. Connaughton, A. S. Levine, and R. Woodgate. 1996. Identification of a DinB/UmuC homolog in the archaeon Sulfolobus solfataricus. Mut. Res. 357:245–253.

Kurosawa, N., Y. H. Itoh, T. Iwai, A, Sugai, I. Uda, N. Kimura, T. Horiuchi, and T. Itoh. 1998. Sulfurisphaera ohwakuensis gen. nov., sp. nov., a novel extremely thermophilic acidophile of the order Sulfolobales. Int. J. Syst. Bacteriol. 48:451–456.

Kurosawa, N., Y. H. Itoh, and T. Itoh. 2003. Reclassification of Sulfolobus hakonensis Takayanagi et al. 1996 as Metallosphaera hakonensis comb. nov. based on phylogenetic evidence and DNA G+C content. Int. J. Syst. Evol. Microbiol. 53:1607–1608.

Kvaratskhelia, M., and M. F. White. 2000. Two holliday junction resolving enzymes in Sulfolobus solfataricus. J. Molec. Biol. 297:923–932.

Kvaratskhelia, M., B. N. Wardleworth, C. S. Bond, J. M. Fogg, D. M. J. Lilley, and M. F. White. 2002. Holliday junction resolution is modulated by archaeal chromatin components in vitro. J. Biol. Chem. 277:2992–2996.

Langworthy, T. A. 1977. Comparative lipid composition of heterotrophically and autotrophically grown Sulfolobus acidocaldarius. J. Bacteriol. 130:1326–1330.

Langworthy, T. A. 1985. Lipids of archaebacteria. In: C. R. Woese and R. S. Wolfe (Eds.) The Bacteria, Volume 8: Archaebacteria. Academic Press. Orlando, FL. 459–497.

Langworthy, T. A., and J. L. Pond. 1986. Archaebacterial ether lipids and chemotaxonomy. Syst. Appl. Microbiol. 7:253–257.

Lanzotti, V., A. Trincone, A. Gambacorta, M. DeRosa, and E. Breitmaier. 1986. ^{2}H and ^{13}C NMR assignment of benzothiophenquinones from the sulfur-oxidizing archaebacterium Sulfolobus solfataricus. Eur. J. Biochem. 160:37–40.

Laska, S., and A. Kletzin. 2000. Improved purification of the membrane-bound hydrogenase and sulfur-reductase complex from thermophilic Archaea using -aminocaroic acid-containing chromatography buffers. J. Chromatogr. B 737:151–160.

Laska, S., F. Lottspeich, and A. Kletzin. 2003. Membrane-bound hydrogenase and sulfur-reductase of the hyperthermophilic and acidophilic archaeon Acidianus ambivalens. Microbiology 149:2357–2371.

Lembcke, G., R. Dürr, R. Hegerl, and W. Baumeister. 1991. Image analysis and processing of an imperfect two-dimensional cystal: the surface layer of the archaebacterium Sulfolobus acidocaldariusi re-investigated. J. Microscopy 161:263–278.

Lembcke, G., W. Baumeister, E. Beckmann, and F. Zemlin. 1993. Cyro-electron microscopy of the surface protein of Sulfolobus shibatae. Ultramicroscopy 49:397–406.

Lemos, R. S., C. M. Gomes, and M. Teixeira. 2001. Acidianus ambivalens Complex II typifies a novel family of succinate dehydrogenases. Biochem. Biophys. Res. Commun. 281:141–150.

Lindström, E. B., and H. M. Sehlin. 1989. High efficiency plating of the thermophilic sulfur-dependent archaebacterium Sulfolobus acidocaldarius. Appl. Environ. Microbiol. 55:3020–3021.

Ling, H., F. Bousdoucq, R. Woodgate, and W. Young. 2001. Crystal structure of a Y-family DNA polymerase in action: a mechanism for error-prone and lesion-bypass replication. Cell 107:91–102.

Lipps, G., P. Ibanez, T. Stroessenreuther, K. Hekimian, and G. Kraus. 2001a. The protein ORF80 from the acidophilic and thermophilic archaeon Sulfolobus islandicus binds highly specifically to double-stranded DNA and represents a novel type of basic leucine zipper protein. Nucleic Acids Res. 29:4973–4982.

Lipps, G., M. Stegert, and G. Krauss. 2001b. Thermostable and site-specific DNA binding of the gene produc of ORF56 from the Sulfolobus islandicus plasmid pRN1, a putative archaeal plasmid copy control protein. Nucleic Acids Res. 29:904–913.

Lipps, G., S. Röther, C. Hart, and G. Krauss. 2003. A novel type of replicative enzyme harbouring ATPase, promase and DNA polymerase activities. EMBO J. 22:2516–2525.

Löhlein-Werhan, G., P. Goepfert, and H. Eggerer. 1988. Purification and properties of an archaebacterial enzyme: citrate synthase from Sulfolobus solfataricus. Biol. Chem. Hoppe-Seyler 369:109–113.

Lopez-Garcia, P., S. Knapp, R. Ladenstein, and P. Forterre. 1998. In vitro DNA binding of the archaeal protein Sso7d induces negative supercoiling at temperatures typical for thermophilic growth. Nucleic Acids Res. 26:2322–2328.

Lopez-Garcia, P., and P. Forterre. 1999. Control of DNA topology during thermal stress in hyperthermophilic archaea: DNA topoisomerase levels, activities and induced thermotolerance during heat and cold shock. Molec. Microbiol. 33:766–777.

Lübben, M., S. Anemüller, and G. Schäfer. 1986. Investigations of the bioenergetic system of Sulfolobus acidocaldarius DSM 639. Syst. Appl. Microbiol. 7:425–426.

Lübben, M., and G. Schäfer. 1987a. A plasma-membrane associated ATPase from the thermoacidophilic archaebacterium Sulfolobus acidocaldarius. Eur. J. Biochem. 164:533–540.

Lübben, M., H. Lünsdorf, and G. Schäfer. 1987b. The plasma membrane ATPase of the thermoacidophilic archaebacterium Sulfolobus acidocaldarius. Purification and immunological relationships to F[1]-ATPases. Eur. J. Biochem. 167:211–219.

Lübben, M., and G. Schäfer. 1989. Chemiosmotic energy conversion of the archaebacterial thermoacidophile Sulfolobus acidocaldarius: oxidative phosphorylation and the presence of an F_0-related N, N′-dicyclohexylcarbodiimide-binding proteolipid. J. Bacteriol. 171:6106–6116.

Lübben, M., B. Kolmerer, and M. Saraste. 1992. An archaebacterial terminal oxidase combines core structures of two mitochondrial respiratory complexes. EMBO J. 11:805–812.

Lübben, M., S. Arnaud, J. Castresana, A. Warne, S. P. J. Albracht, and M. Saraste. 1994. A second terminal oxidase in Sulfolobus acidocaldarius. Eur. J. Biochem. 224:151–159.

Ludwig, W., and O. Strunk. 2001. ARB: A software environment for sequence data. (http://www.arb-home.de/arb/ documentation.html{www.arb-home.de}).

Lurz, R., M. Grote, J. Dijk, R. Reinhardt, and B. Dobrinski. 1986. Electron microscopic study of DNA complexes with proteins from the archaebacterium Sulfolobus acidocaldarius. EMBO J. 5:3715–3721.

Mai, V. Q., X. Chen, R. Hong, and L. Huang. 1998. Small abundant DNA binding proteins from the thermoacidophilic archaeon Sulfolobus shibatae constrane negative DNA supercoils. J. Bacteriol. 180:2560–2563.

Marino, G., G. Nitti, M. I. Arnone, G. Sannia, A. Gambacorta, and M. DeRosa. 1988. Purification and characterization of aspartate aminotransferase from the thermoacidophilic archaebacterium Sulfolobus solfataricus. J. Biol. Chem. 263:12305–12309.

Marsh, R. M., P. R. Norris, and N. W. LeRoux. 1983. Growth and mineral oxidation studies with Sulfolobus. In: G. Rossi and A. E. Torma (Eds.) Recent Progress in Biohydrometallurgy. Associazione Mineraria Sarda. Iglesias, Italy. 71–81.

Martins, L. O., R. Huber, H. Huber, K. O. Stetter, M. S. DaCosta, and H. Santos. 1997. Organic solutes in hyperthermophilic Archaea. Appl. Environ. Microbiol. 63:896–902.

Martusewitsch, E., C. W. Sensen, and C. Schleper. 2000. High spontaneous mutation rate in the hyperthermophilic archaeon Sulfolobus solfataricus is mediated by transposable elements. J Bacteriol. 182:2574–2581.

McClure, M. L., and R. W. G. Wyckoff. 1982. Ultrastructural characteristics of Sulfolobus acidocaldarius. J. Gen. Microbiol. 128:433–437.

Menendez, C., Z. Bauer, H. Huber, N. Gad'on, K. O. Stetter, and G. Fuchs. 1999. Presence of acetyl coenzyme A (CoA) carboxylase and propionyl-CoA carboxylase in autotrophic Crenarchaeota and indication for operation of a 3-hydroxypropionate cycle in autotrophic carbon fixation. J. Bacteriol. 181:1088–1098.

Minami, Y., S. Wakabayashi, K. Wada, H. Matsubara, L. Kerscher, and D. Oesterhelt. 1985. Amino acid sequence of a ferredoxin from the thermoacidophilic archaebacterium, Sulfolobus acidocaldarius. Presence of an N(6)-monomethyllysine and phyletic consideration of archaebacteria. J. Biochem. 97:745–753.

Müller, F., T. Bandeiras, T. Urich, M. Teixeira, C. M. Gomes, and A. Kletzin. 2004. Coupling of the pathway of sulphur oxidation to dioxygen reduction: characterization of a novel membrane-bound thiosulphate:quinone oxidoreductase. Nucleic Acids Res. 53:1147–1160.

Muskhelishvili, G., M. Karseladze, and D. Prangishvili. 1990. Incorporation of exogenous precursors into nucleic acids of the extremely thermophilic acidophilic archaebacterium Sulfolobus acidocaldarius. Biochemistry (USSR) 55:517–524.

Muskhelishvili, G., P. Palm, and W. Zillig. 1993. SSV1-encoded site-specific recombination system in Sulfolobus shibatae. Molec. Gen. Genet. 237:334–342.

Napoli, A., Y. Zivanovic, C. Bocs, C. Buhler, M. Rossi, P. Forterre, and M. Ciaramella. 2002. DNA bending, compaction and negative supercoiling by the architectural protein Sso7d of Sulfolobus solfataricus. Nucleic Acids Res. 30:2656–2662.

Nicolaus, B., A. DeSimone, L. Del Piano, P. Giardina, and L. Lama. 1986. Production of 2-keto-3-deoxygluconate by immobilized cells of Sulfolobus solfataricus. Biotechnol. Lett. 8:497–500.

Nicolaus, B., A. Gambacorta, A. L. Basso, R. Riccio, M. DeRosa, and W. D. Grant. 1988. Trehalose in archaebacteria. Syst. Appl. Microbiol. 10:215–217.

Nicolaus, B., A. Trincone, L. Lama, G. Palmieri, and A. Gambacorta. 1992. Quinone compositionj in Sulfolobus acidocaldarius grown in different conditions. Syst. Appl. Microbiol. 15:18–20.

Omer, A., S. Ziesche, W. A. Decatur, M. J. Fournier, and P. P. Dennis. 2003. RNA-modifying machines in archaea. Molec. Microbiol. 48:617–629.

Palm, P., C. Schleper, B. Grampp, S. Yeats, P. McWilliam, W.-D. Reiter, and W. Zillig. 1991. Complete nucleotide sequence of the virus SSV1 of the archaebacterium Sulfolobus shibatae. Virology 185:2242–250.

Peng, X., I. Holz, W. Zillig, R. A. Garrett, and Q. She. 2000. Evolution of the family of pRN plasmids and their integrase-mediated insertion into the chromosome of the crenarchaeon Sulfolobus solfataricus. J. Molec. Biol. 303:449–454.

Peng, X., H. Blum, Q. She, S. Mallok, K. Brügger, R. A. Garrett, W. Zillig, and D. Prangishvili. 2001. Sequences and replication of genomes of the archaeal rudiviruses SIRV1 and SIRV2: Relationships to the archaeal lipothrixvirus SIFV and some eukaryal viruses. Virology 291:226–234.

Potapova, O., N. D. F. Grindley, and C. M. Joyce. 2002. The mutation specificity of the Dbh lesion bypass polymerase and its implications. J. Biol. Chem. 277:28157–28166.

Prangishvili, D., R. P. Vashakidze, M. G. Chelidze, and I. Y. Gabriadze. 1985. A restriction endonucelase SuaI from the thermoacidophilic archaebacterium Sulfolobus acidocaldarius. FEBS Lett. 192:57–60.

Prangishvili, D. 1986. DNA-dependent DNA polymerases from the thermoacidophilic archaebacterium Sulfolobus acidocaldarius. Molec. Biol. (USSR) 20:477–488.

Prangishvili, D., H.-P. Klenk, G. Jakobs, A. Schmiechen, C. Hanselman, I. Holz, and W. Zillig. 1998. Biochemical and physiological characterization of the dUTPase from the archaeal virus SIRV. J. Biol. Chem. 273:6024–6029.

Prangishvili, D., S.-V. Albers, I. Holz, H. P. Arnold, K. Stedman, T. Klein, H. Singh, J. Hiort, A. Schweier, J. K. Kristjansson, and W. Zillig. 1999a. Conjugation in Archaea: Frequent occurrence of conjugative plasmids in Sulfolobus. Plasmid 40:190–202.

Prangishvili, D., H. P. Arnold, U. Ziese, D. Goetz, I. Holz, and W. Zillig. 1999b. A novel virus family, the Rudiviridae: structure, virus-host interactions and genome variability of Sulfolobus viruses SIRV1 and SIRV2. Genetics 153:1387–1396.

Prangishvili, D., I. Holz, E. Stieger, S. Nickell, J. Kristjansson, and W. Zillig. 2000. Sulfolobicins, specific proteinaceous toxins produced by strains of the extremely thermophilic archaea of the genus Sulfolobus. J. Bacteriol. 182:2985–2988.

Prangishvili, D., K. M. Stedman, and W. Zillig. 2001. Viruses of the extremely thermophilic archaeon Sulfolobus. Trends Microbiol. 9:39–42.

Prangishvili, D. 2003. Evolutionary insights from studies on viruses from hot habitats. Res. Microbiol. 154:289–294.

Prangishvili, D., and R. A. Garrett. 2004. Exceptionally diverse morphotypes and genomes of crenarchaeal hyperthermophilic viruses. Biochem. Soc. Trans. 32:204–208.

Prüschenk, R., and W. Baumeister. 1987. Three-dimensional structure of the surface protein of Sulfolobus solfataricus. Eur. J. Cell Biol. 45:185–191.

Purschke, W., C. L. Schmidt, A. Petersen, S. Anemüller, and G. Schäfer. 1997. The terminal quinol oxidase of the hyperthermophilic archaeon Desulfurolobus ambivalens exhibits unusual subunit structure and gene organization. J. Bacteriol. 179:1344–1353.

Redder, P., Q. She, and R. A. Garrett. 2001. Non-autonomous mobile elements in the crenarchaeon Solfolobus solfataricus. J. Molec. Biol. 306:1–6.

Rella, R., C. A. Raia, M. Pensa, F. M. Pisani, A. Gambacorta, M. DeRosa, and M. Rossi. 1987. A novel archaebacterial NAD+-dependent alcohol dehydrogenase: Purification and properties. Eur. J. Biochem. 167:475–479.

Roberts, J. A., S. D. Bell, and M. F. White. 2003. An archaeal XPF repair endonuclease dependent on a heterotrimeric PCNA. Molec. Microbiol. 48:361–371.

Robinson, H., Y. G. Gao, B. S. McCrary, S. P. Edmondson, J. W. Shriver, and A. H. J. Wang. 1998. The hyperthermophilic chromosomal protein Sac7d sharply kinks DNA. Nature 392:202–205.

Rossi, M., R. Rella, M. Pensa, S. Bartolucci, M. DeRosa, A. Gambacorta, C. A. Raia, and N. Dell'Aversano Orabona. 1986. Structure and properties of a thermophilic and thermostable DNA polymerase isolated from Sulfolobus solfataricus. Syst. Appl. Microbiol. 7:337–341.

Sandler, S. J., L. H. Satin, H. S. Samra, A. J. Clark. 1996. RecA-like genes from three archaean species with putative protein products similar to Rad51 and Dmc1 proteins of the yeast Saccharomyces cerevisiae. Nucleic Acids Res. 24:2125–2132.

Sanz, J. L., G. Huber, H. Huber, and R. Amils. 1994. Using protein synthesis inhibitors to establish the phylogenetic relationships of the Sulfolobales order. J. Molec. Evol. 39:528–532.

Schäfer, G., and M. Meyering-Vos. 1992a. F-type or V-type? The chimeric nature of archaebacterial ATP synthase. Biochim. Biophys. Acta 1101:232–235.

Schäfer, G., and M. Meyering-Vos. 1992b. The plasma membrane ATPase of archaebacteria: A chimeric energy converter. Ann. NY Acad. Sci. 671:293–309.

Schäfer, G., M. Engelhard, V. Müller, V. 1999. Bioenergetics of the Archaea. Microbiol. Molec. Biol. Rev. 63:570–620.

Schleper, C., K. Kubo, and W. Zillig. 1992. The particle SSV1 from the extremely thermophilic archaeon Sulfolobus is

a virus: Demonstration of infectivity and of transfection with viral DNA. Proc. Natl. Acad. Sci. USA 89:7645–7649.

Schleper, C., R. Röder, T. Singer, and W. Zillig. 1994. An insertion element of the extremely thermophilic archaeaon Sulfolobus solfataricus transposes into the endogenous β-galactosidase gene. Molec. Gen. Genet. 243:91–96.

Schleper, C., I. Holz, D. Janekovic, J. Murphy, and W. Zillig. 1995. A multicopy plasmid of the extremely thermophilic archaeaon Sulfolobus effects its transfer to recipients by mating. J. Bacteriol. 177:4417–4426.

Schönheit, P., and T. Schäfer. 1995. Metabolism of hyperthermophiles. World J. Microbiol. Biotechnol. 11:26–57.

Segerer, A., K. O. Stetter, and F. Klink. 1985. Two contrary modes of chemolithotrophy in the same archaebacterium. Nature (Lond.) 313:787–789.

Segerer, A., A. Neuner, J. K. Kristjansson, and K. O. Stetter. 1986. Acidianus infernus gen. nov., sp. nov., and Acidianus brierleyi comb. nov.: facultatively aerobic, extremely acidophilic thermophilic sulfur-metabolizing archaebacteria. Int. J. Syst. Bacteriol. 36:559–564.

Segerer, A., T. A. Langworthy, and K. O. Stetter. 1988. Thermoplasma acidophilum and Thermoplasma volcanium sp. nov. from solfatara fields. Syst. Appl. Microbiol. 10:161–171.

Segerer, A., A. Trincone, M. Gahrtz, and K. O. Stetter. 1991. Stygiolobus azoricus gen. and sp. nov. represents a novel genus of anaerobic, extremely thermoacidophilic archaea of the order Sulfolobales. Int. J. Syst. Bacteriol. 41:495–501.

Segerer, A. H., and K. O. Stetter. 1992. The order Sulfolobales. In: A. Balows, H. G. Trüper, M. Dworkin, W. Harder, and K.-H. Schleifer (Eds.) The Prokaryotes, 2nd ed. Springer-Verlag. New York, NY. 684–701.

Seitz, E. M., and S. C. Kowalczykowski. 2000. The DNA binding and pairing preferences of the archaeal RadA protein demonstrate a universal characteristic of DNA strand exchange proteins. Molec. Microbiol. 37:555–560.

Selig, M., K. B. Xavier, H. Santos, and P. Schönheit. 1997. Comperative analysis of Embden-Meyerhof and Entner-Doudoroff glycolytic pathways in hyperthermophilic archaea and the bacterium Thermotoga. Arch. Microbiol. 167:217–232.

She, Q., H. Phan, R. A. Garrett, S.-V. Albers, K. M. Stedman, and W. Zillig. 1998. Genetic profile of pNOB8 from Sulfolobus: The first conjugative plasmid from an archaeon. Extremophiles 2:417–425.

She, Q., X. Peng, W. Zillig, and R. A. Garrett. 2001a. Gene capture events in archaeal chromosomes. Nature 409:478.

She, Q., R. K. Singh, F. Confalonieri, Y. Zivanovic, G. Allard, M. J. Awayez, C. C.-Y. Chan-Weiher, I. G. Clausen, B. A. Curtis, A. De Moors, G. Erauso, C. Fletcher, P. M. K. Gordon, I. Heikamp-de Jong, A. C. Jeffries, C. J. Kozera, N. Medina, X. Peng, H. P. Thi-Ngoc, P. Redder, M. E. Schenk, C. Theriault, N. Tolstrup, R. Charlebois, W. F. Doolittle, M. Duguet, T. Gaasterland, R. A. Garrett, M. A. Ragan, C. W. Sensen, and J. van der Oost. 2001b. The complete genome of the crenarchaeon Sulfolobus solfataricus P2. Proc. Natl. Acad. Sci. USA 98:7835–7840.

She, Q., K. Brügger, and L. Chen. 2002. Archaeal integrative genetic elements and their impact on genome evolution. Res. Microbiol. 153:325–332.

Shivvers, D. W., and T. D. Brock. 1973. Oxidation of elemental sulfur by Sulfolobus acidocaldarius. J. Bacteriol. 114:706–710.

Silvian, L. F., E. A. Toth, P. Pham, M. F. Goodman, and T. Ellenberger. 2001. Crystal structure of a DinB family DNA polymerase from Sulfolobus solfataricus. Nature Struct. Biol. 8:984–989.

Stedman, K. M., C. Schleper, E. Rumpf, and W. Zillig. 1999. Genetic requirements for the function of the archaeal virus SSV1 in Sulfolobus solfataricus: construction and testing of virual shuttle vectors. Genetics 152:1397–1405.

Stedman, K. M., Q. She, H. Phan, I. Holz, H. Singh, D. Prangishvili, R. A. Garrett, and W. Zillig. 2000. The pING family of conjugative plasmids from the extremely thermophilic archaeon Sulfolobus islandicus demonstrates modes of genomic variation and conjugation in crenarchaeota. J. Bacteriol. 182:7014–7020.

Stedman, K. M., Q. She, H. Phan, H. P. Arnold, I. Holz, R. A. Garrett, and W. Zillig. 2003. Biological and genetic relationships between fuselloviruses infecting the extremely thermophilic archaeon Sulfolobus: SSV1 and SSV2. Res. Microbiol. 154:295–302.

Stetter, K. O., and W. Zillig. 1985. Thermoplasma and the thermophilic sulfur-dependent archaebacteria. *In:* C. R. Woese and R. S. Wolfe (Eds.) The Bacteria, Volume 8: Archaebacteria. Academic Press. Orlando, FL. 85–170.

Stetter, K. O., A. Segerer, W. Zillig, G. Huber, G. Fiala, R. Huber, and H. König. 1986. Extremely thermophilic sulfur-metabolizing archaebacteria. Syst. Appl. Microbiol. 7:393–397.

Stetter, K. O. 1989. Order III: Sulfolobales ord. nov. *In:* J. T. Staley, M. P. Bryant, N. Pfennig, and J. G. Holt (Eds.) Bergey's Manual of Systematic Bacteriology. Williams and Wilkins. Baltimore, MD. 3:2250.

Suzuki, T., T. Iwasaki, T. Uzawa, K. Hara, N. Nemoto, T. Ueki, A. Yamagishi, and T. Oshima. 2002. Sulfolobus tokodaii sp. nov. (f. Sulfolobus sp. strain 7), a new member of the genus Sulfolobus isolated from Beppu Hot Springs, Japan. Extremophiles 6:39–44.

Takayanagi, S., H. Kawasaki, K. Sugimori, T. Yamada, A. Sugai, T. Ito, K. Yamasato, and M. Shioda. 1996. Sulfolobus hakonensis sp. nov., a novel species of acidothermophilic archaeon. Int. J. Syst. Bacteroil. 46:377–382.

Tang, T. H., T. S. Rozdenstvensky, B. Clouet d'Orval, M.-L. Bortolin, H. Huber, B. Charpentier, C. Branlant, J.-P. Bachellerie, J. Brosius, and A. Hüttenhofer. 2002. RNomics in Archaea reveals a further link between splicing of archaeal introns and RNA processing. Nucleic Acids Res. 30:921–930.

Taylor, K. A., J. F. Deatherage, and L. A. Amos. 1982. Structure of the S-layer of Sulfolobus acidocaldarius. Nature (Lond.) 299:840–842.

Teixeira, M., R. Batista, A. P. Campos, C. Gomes, J. Mendes, I. Pacheco, S. Anemüller, and W. R. Hagen. 1995. A seven-iron ferredoxin from the thermoacidophilic archaeon Desulfurococcus ambivalens. Eur. J. Biochem. 227:322–327.

Thurl, S., W. Witke, I. Buhrow, and W. Schäfer. 1986. Quinones from archaebacteria. II. Different types of quinones from sulphur-dependent archaebacteria. Biol. Chem. Hoppe-Seyler 367, pp:191–197.

Tolstrup, N., C. W. Sensen, R. A. Garrett, and I. G. Clausen. 2000. Two different and highly organized mechanisms of translation initiation in the archaeon Sulfolobus solfataricus. Extremophiles 4:175–179.

Trent, J. D., J. Osipiuk, and T. Pinkau. 1990. Acquired thermotolerance and heat shock in the extremely thermophilic archaebacterium Sulfolobus sp. B12. J. Bacteriol. 172:1478–1484.

Trent, J. D., E. Nimmesgern, J. S. Wall, F.-U. Hartl, and A. L. Horwich. 1991. A moleculare chaperone from a thermophilic archaebacterium is related to the eukaryotic protein t-complex polypeptide-1. Nature 354:490–493.

Trent, J. D., M. Gabrielsen, B. Jensen, J. Neuhard, and J. Olsen. 1994. Acquired thermotolerance and heat-shock proteins in thermophiles from the three domains. J. Bacteriol. 176:6148–6152.

Trent, J. D., H. K. Kagawa, T. Yaoi, E. Olle, and N. J. Zaluzec. 1997. Chaperonine filaments: the archaeal cytoskeleton? Proc. Natl. Acad. Sci. USA 94:5383–5388.

Trent, J. D., H. K. Kagawa, C. D. Paavola, R. A. McMillan, J. Howard, L. Janke, C. Lavin, T. Embaye, and C. E. Henze. 2003. Intracellular localization of a group II chaperonin indicates a membrane-related function. Proc. Natl. Acad. Sci. USA 100:15589–15594.

Trevisanato, S. I., N. Larsen, A. H. Segerer, K. O. Stetter, and R. A. Garrett. 1996. Phylogenetic analysis of the archaeal order of Sulfolobales based on sequences of 23S rRNA genes and 16S/23S rDNA spacers. Syst. Appl. Microbiol. 19:61–65.

Typke, D., M. Nitsch, A. Möhrle, R. Hegerl, M. Alam, D. Grogan, and J. Trent. 1988. Image analysis and processing of an imperfect two-dimensional cystal: the surface layer of the archaebacterium Sulfolobus acidocaldariusi re-investigated. Inst. Phys. Conf. Ser. 93(3):379–380.

Urich, T., T. Bandeiras, S. S. Leal, R. Rachel, T. Albrecht, P. Zimmermann, C. Scholz, M. Teixeira, C. M. Gomes, and A. Kletzin. 2004. The sulphur oxygenase reductase from Acidianus ambivalens is a multimeric protein containing a low-potential mononuclear non-haem iron centre. Biochem. J. 381:137–146.

Wadsworth, R. I. M., and M. F. White. 2001. Identification and properties of the crenarchaeal single-stranded DNA binding protein from Sulfolobus solfataricus. Nucleic Acids Res. 29:914–920.

Wakagi, T., and T. Oshima. 1985. Membrane-bound ATPase of a thermoacidophilic archaebacterium, Sulfolobus acidocaldarius. Biochim. Biophys. Acta 817:33–41.

Wakagi, T., and T. Oshima. 1987. Energy metabolism of a thermoacidophilic archaebacterium, Sulfolobus acidocaldarius. Orig. Life 17:391–399.

Wardleworth, B. N., R. J. Russel, S. D. Bell, G. L. Taylor, and M. F. White. 2002. Structure of Alba: Aarchaeal chromatin protein modulated by acetylation. EMBO J. 17:4654–4662.

Weiss, R. L. 1973. Attachment of bacteria to sulphur in extreme environments. J. Gen. Microbiol. 77:501–507.

Weiss, R. L. 1974. Subunit cell wall of Sulfolobus acidocaldarius. J. Bacteriol. 118:275–284.

Whitaker, R. J., D. W. Grogan, and J. W. Taylor. 2003. Geographic barriers isolate endemic populations of hyperthermophilic Archaea. Science 301:976–978.

White, M. F., and S. D. Bell. 2002. Holding together: chromatin in the Archaea. Trends Genet. 18:621–626.

Woese, C. R., and G. E. Fox. 1977. Phylogenetic structure of the prokaryotic domain: the primary kingdoms. Proc. Natl. Acad. Sci. USA 74:5088–5099.

Woese, C. R., O. Kandler, and M. L. Wheelis. 1990. Towards a natural system of organisms: proposal for the domains

Archaea, Bacteria and Eucarya. Proc. Natl. Acad. Sci. USA 87:4576–4579.

Wood, A. P., D. P. Kelly, and P. R. Norris. 1987. Autotrophic growth of four Sulfolobus strains on tetrathionate and the effect of organic nutrients. Arch. Microbiol. 146:382–389.

Worthington, P., V. Hoang, F. Perez-Pomares, and P. Blum. 2003. Targeted disruption of the alpha-amylase gene in the hyperthermophilic archaeaon Sulfolobus solfataricus. J. Bacteriol. 185:482–488.

Xiang, X., X. Dong, and L. Huang. 2003. Sulfolobus tengchongensis sp. nov., a novel thermoacidophilic archaeon isolated from a hot spring in Tengchong, China. Extremophiles 7:493–498.

Xue, H., R. Guo, Y. Wen, D. Liu, and L. Huang. 2000. An abundant DNA binding protein from the hyperthermophilic archaeaon Sulfolobus shibatae affects DNA supercoiling in a temperature-dependent fashion. J. Bacteriol. 182:3929–3933.

Yeats, S., P. McWilliam, and W. Zillig. 1982. A plasmid in the archaebacterium Sulfolobus acidocaldarius. EMBO J. 1:1035–1038.

Zhou, B. L., J. D. Pata, and T. A. Steitz. 2001. Crystal structure of a DinB family error-prone DNA polymerase catalytic fragment reveals a classic polymerase catalytic domain. Molec. Cell 8:427–437.

Zillig, W., K. O. Stetter, and D. Janekovic. 1979. DNA-dependent RNA polymerase from the archaebacterium Sulfolobus acidocaldarius. Eur. J. Biochem. 96:597–604.

Zillig, W., K. O. Stetter, W. Schulz, H. Priess, and I. Scholz. 1980. The Sulfolobus-"Caldariella" group: taxonomy on the basis of the structure of DNA-dependent RNA polymerases. Arch. Microbiol. 125:259–260.

Zillig, W., K. O. Stetter, W. Schäfer, D. Janekovic, S. Wunderl, I. Holz, and P. Palm. 1981. Thermoproteales: a novel type of extremely thermoacidophilic anaerobic archaebacteria isolated from Icelandic solfataras. Zbl. Bakteriol. Parasitenkd. Infektionskr. Hyg., Abt. 1 Orig. C2:200–227.

Zillig, W., S. Yeats, I. Holz, A. Böck, F. Gropp, M. Rettenberger, and S. Lutz. 1985. Plasmid-related anaerobic autotrophy of the novel archaebacterium Sulfolobus ambivalens. Nature (Lond.) 313:789–791.

Zillig, W., S. Yeats, I. Holz, A. Böck, M. Rettenberger, F. Gropp, and G. Simon. 1986. Desulfurolobus ambivalens, gen. nov., sp. nov., an autotrophic archaebacterium facultatively oxidizing or reducing sulfur. Syst. Appl. Microbiol. 8:197–203.

Zillig, W., H. P. Arnold, I. Holz, D. Prangishvili, A. Schweier, K. M. Stedman, Q. She, H. Phan, R. A. Garrett, and J. K. Kristjansson. 1998. Genetic elements in the extremely thermophilic archaeon Sulfolobus. Extremophiles 2:131–140.

Zimmermann, P., S. Laska, and A. Kletzin. 1999. Two modes of sulfite oxidation in the extremely thermophilic and acidophilic archaeon Acidianus ambivalens. Arch. Microbiol. 172:76–82.

Prokaryotes (2006) 3:52–68
DOI: 10.1007/0-387-30743-5_4

CHAPTER 4

Desulfurococcales

HARALD HUBER AND KARL O. STETTER

Introduction

The order Desulfurococcales (Huber and Stetter, 2001b) is one of three orders of the archaeal phylum Crenarchaeota (Woese et al., 1990), which also includes Thermoproteales (Huber and Stetter, 2001a) and Sulfolobales (Stetter, 1989). Cells of members of the Desulfurococcales are regular to irregular cocci, discs or dishes, which occur singly, in pairs, short chains, or grapelike aggregates. Diameters of coccoid cells vary from about 0.5 up to 15 μm, while disc-shaped cells frequently exhibit ultra-flat areas only about 0.1 to 0.2 μm thick. All members of the order Desulfurococcales are hyperthermophiles with optimal growth temperatures between 85 and 106°C. They grow anaerobically, facultatively anaerobically, or aerobically. Under autotrophic conditions, they gain energy by oxidation of hydrogen using elemental sulfur, thiosulfate, nitrate, or nitrite as electron acceptor, and they use CO_2 as a carbon source. Alternatively, organotrophic growth occurs by aerobic respiration or anaerobic sulfur respiration or by fermentation of organic substrates. Following the classification listed in the new edition of *Bergey's Manual of Systematic Bacteriology* (Huber and Stetter, 2001b), the order Desulfurococcales comprises two validly published families, the Desulfurococcaceae (Zillig et al., 1982; Burggraf et al., 1997) and the Pyrodictiaceae (Burggraf et al., 1997). The members of these two families can be physiologically distinguished by their optimal and maximal growth temperatures: the Desulfurococcaceae harbor organisms which grow optimally between 85 and 95°C and exhibit temperature maxima up to 102°C. In contrast, the Pyrodictiaceae are characterized by optimal growth temperatures between 95 and 106°C and temperature maxima between 108 and 113°C. Furthermore, several members of this family form a network of hollow tubules (cannulae).

The term "sulfur-dependent" archaebacteria (now Archaea) was formerly used to designate the branch which is now the phylum Crenarchaeota. However, several members of the Desulfurococcales are obligate organotrophs and are unable to use elemental sulfur or sulfur compounds for energy formation; some strains and species are even inhibited by the presence of elemental sulfur. Therefore, the designation "sulfur-dependent" Archaea ("Archaebacteria") should not be used anymore.

In the former edition of *The Prokaryotes* (Huber and Stetter, 1992), only the family Desulfurococcaceae (Zillig et al., 1982) with the genus *Desulfurococcus* and the genera *Pyrodictium* (Stetter, 1982; Stetter et al., 1983), *Thermodiscus* (Stetter and Zillig, 1985; Stetter, 2001), and *Staphylothermus* (Fiala et al., 1986) had been mentioned and placed within the order Thermoproteales because of their physiological properties (Huber and Stetter, 1992). Meanwhile, mainly on the basis of 16S rRNA sequence comparisons, these genera have been placed either within the Pyrodictiaceae (*Pyrodictium*), Desulfurococcaceae (*Thermodiscus* and *Staphylothermus*) or within the novel order Desulfurococcales (Huber and Stetter, 2001b). Furthermore, several new genera have been described.

Phylogeny

On the basis of 16S rDNA sequence comparisons, the order Desulfurococcales represents one lineage of the crenarchaeotal branch of the Archaea (Woese et al., 1990; Fig. 1). It forms a separate cluster and is distinct from the two other orders described so far, the Sulfolobales and the Thermoproteales. The representatives of the Desulfurococcales exhibit phylogenetic similarities of 0.89–1.00 to members within the order and 0.89–0.79 to all other species of the Crenarchaeota. These values indicate that the order is clearly defined by 16S rRNA gene sequence comparisons. The two described families, the Desulfurococcaceae and the Pyrodictiaceae, are also obvious from these results (Fig. 1). The phylogenetic position of *Acidilobus aceticus* is not absolutely clear; however, in most calculations it branches within the family of the

Fig. 1. Phylogenetic tree based on
16S rRNA sequences. The tree was
calculated using the maximum
likelihood program, which is
included in the (ARB package;
Ludwig and Strunk, 2001). Scale
bar: 10 estimated exchanges
within 100 nucleotides. Red lines
= Desulfurococcales; blue lines
= Thermoproteales; yellow lines =
Sulfolobales; and green lines =
Euryarchaeota.

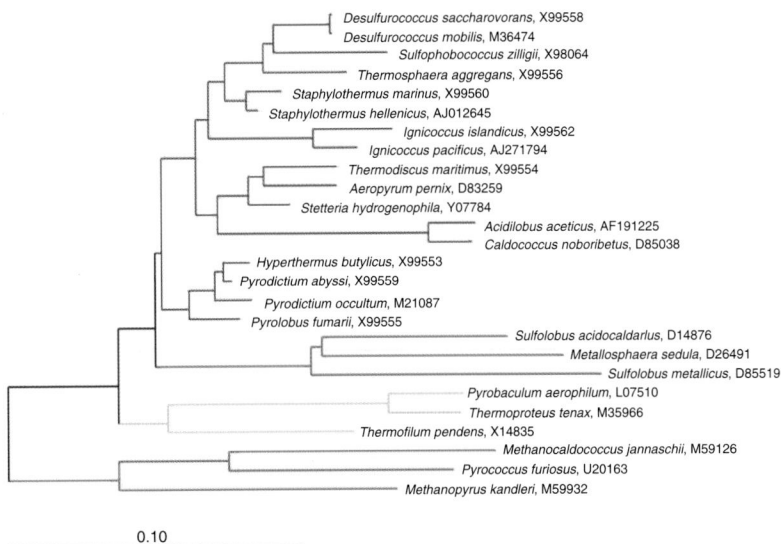

Desulfurococcaceae (Fig. 1). All members of the
order, particularly the Pyrodictiaceae, are characterized by very short lineages in the corresponding phylogenetic trees (Fig. 1). Since the
organisms form a cluster around the root of the
16S rRNA-based phylogenetic tree, hyperthermophiles appear to be primitive (Stetter, 1995).

Phylogenetic investigations based on analyses
of partial 23S rRNA sequences were carried out
by Kjems et al. (1992). The resulting phylogeny,
although less comprehensive, was generally consistent with the 16S rRNA tree.

Taxonomy

The order Desulfurococcales was defined
recently by Huber and Stetter (2001b) in the last
edition of *Bergeys Manual of Systematic Bacteriology*. However, this name had already been proposed in the last edition of *The Prokaryotes*
(Huber and Stetter, 1992). The description of the
family Desulfurococcaceae (Zillig et al., 1982)
was emended by Burggraf et al. (1997). In the
same paper, the family Pyrodictiaceae was validly described, although the name "Igneococ-
cales" was proposed in this publication to
designate the order. At present, the Desulfurococcales are comprised of two families: the Desulfurococcaceae harbor nine validly published
genera, while the Pyrodictiaceae are represented
by three genera (Prokofeva et al., 2000; Huber
and Stetter, 2001b). Furthermore, the 16S rRNA
gene sequence of "*Caldococcus noboribetus*" has
been published, although a description of this
organism has not (Aoshima et al., 1996a).

The genus *Desulfurococcus* (Zillig et al.,
1982), the type genus of the family Desulfurococ-

caceae and the order Desulfurococcales, harbors
three validly described species: *D. mucosus* (Zillig et al., 1982), *D. mobilis* (Zillig et al., 1982) and
D. amylolyticus (Bonch-Osmolovskaya et al.,
1988; validated: 2001: ISJEM 51, 1619–1620),
and the so far not validly described "*D. saccharovorans*" (Stetter, 1986; Table 1). The type
strain for *D. mucosus* is strain 07^T (American
Type Culture Collection [ATCC 35584^T], Deutsche Sammlung von Mikroorganismen und
Zellkulturen [DSM 2162^T] and Japan Collection
of Microorganisms [JCM 9187^T]), the type strain
for *D. mobilis* is strain Hvv3^T (ATCC 35582^T,
DSM 2161^T and JCM 9186^T), and for *D. amylolyticus* is strain Z-533^T (DSM 3822^T and JCM
9188^T). The species are characterized by optimal
growth temperatures between 85 and 90°C, an
anaerobic mode of life, and the capability of sulfur respiration with complex organic substrates
or fermentation of peptides. They can be distinguished by their substrates used and their
motility.

Within the family of the Desulfurococcaceae,
seven other genera have been described:

1) *Aeropyrum* (Sako et al., 1996; type species:
A. pernix; type strain K1^T, JCM 9820^T), 2) *Ignicoccus* (Huber et al., 2000a; type species: *I. islandicus*; type strain Kol8^T, DSM 13165^T; further
species: *I. pacificus*; type strain LPC33^T, DSM
13166^T), 3) *Staphylothermus* (Fiala et al., 1986;
type species: *S. marinus*; type strain F1^T, DSM
3639^T and ATCC 49053^T; further species: *S. hellenicus*; type strain P8^T, DSM 12710^T and JCM
10830^T; Arab et al., 2000), 4) *Stetteria* (Jochimsen
et al., 1997; type species: *S. hydrogenophila*; type
strain 4ABC^T, DSM 11227^T), 5) *Sulfophobococcus* (Hensel et al., 1997; type species: *S. zilligii*;
type strain K1^T, DSM 11193^T and JCM 10309^T),

Table 1. Characteristics of the Desulfurococcales species.

Species	Cell size / Cell surface / Flagellation	Temp (°C) (opt. °C)	pH range (pH opt.)	NaCl conc. (opt. %)	Relation to oxygen	Biotopes	Metabolism	Energy yielding reactions	G+C content (mol %)	References
Desulfurococcus mucosus	Cocci, Ø 0.3–2.0 µm, sometimes up to 10 µm; Mucoid layer of neutral sugars; Not flagellated	76–93 (85)	4.5–7.0 (6.0)	n.d. (around 0)	Anaerobic	Continental	Mixotrophic or heterotrophic	Sulfur respiration or fermentation using complex organic components	51	Zillig et al., 1982
Desulfurococcus mobilis	Cocci, Ø 0.5–1.0 µm; S-layer, p4 symmetry; Polytrichous flagellated	78–87 (8.5)	n.d. (6.0)	n.d. (around 0)	Anaerobic	Continental	Mixotrophic or heterotrophic	Sulfur respiration or fermentation using complex organic components	51	Zillig et al., 1982
Desulfurococcus amylolyticus	Cocci, Ø 0.7–1.5 µm; S-layer, symmetry?; Not flagellated	68–97 (90–92)	5.7–7.5 (6.4)	n.d. (around 0)	Anaerobic	Continental	Heterotrophic S^0 stimulates growth	Fermentation using peptides, amino acids, starch, or glycogen	42	Bonch-Osmolovskaya et al., 1988
Aeropyrum pernix	Cocci, Ø 0.8–1.0 µm; S-layer, p4 symmetry; Motile without flagella (?)	70–100 (90–95)	5.0–9.0 (7.0)	1.8–7.0 (3.5)	Aerobic	Marine	Heterotrophic	Aerobic respiration with O_2 of complex organic compounds	67 (56)	Sako et al., 1996; Kawarabayashi et al., 1999
Ignicoccus islandicus	Cocci, Ø 1.2–3 µm; Outer sheath; Polytrichous flagellated	70–98 (90)	3.8–6.5 (5.8)	0.3–5.5 (2.0)	Anaerobic	Marine	Obligate lithoautotrophic	S^0-H_2 autotrophy	41	Huber et al., 2000a
Ignicoccus pacificus	Cocci, Ø 1.0–2.0 µm; Outer sheath; Polytrichous flagellated	75–98 (90)	4.5–7.0 (6.0)	1.0–5.0 (2.0)	Anaerobic	Marine	Obligate lithoautotrophic	S^0-H_2 autotrophy	45	Huber et al., 2000a
Staphylothermus marinus	Cocci, Ø 0.5–1.5 µm, up to 100 cells in aggregates; S-layer, p4 symmetry; Not flagellated	65–98 (92)	4.5–8.5 (6.5)	1.0–3.5 (1.5)	Anaerobic	Marine	Heterotrophic S^0-required	Fermentation of complex organic compounds	35	Fiala et al., 1986
Staphylothermus hellenicus	Cocci, Ø 0.8–1.3 µm, in large aggregates; n.d.; Single flagellum	70–90 (85)	4.5–7.0 (6.0)	2.0–8.0 (4.0)	Anaerobic	Marine	Heterotrophic S^0-required	Fermentation of complex organic compounds	38	Arab et al., 2000

Species	Morphology	Temperature (°C) range (opt.)	pH range (opt.)	NaCl (%) range (opt.)	O₂	Habitat	Nutrition	Energy metabolism	G+C (mol%)	Reference
Stetteria hydrogenophila	Cocci, Ø 0.5–1.5 µm; n.d.; Single flagellum	68–102 (95)	4.5–7.0 (6.0)	0.5–0.6 (2.0–4.0)	Anaerobic	Marine	Mixotrophic, H₂ and S⁰ (S₂O₃²⁻) required	S⁰, S₂O₃²⁻ respiration with H₂ and complex organic compounds	65	Jochimsen et al., 1997
Sulfophobococcus zilligii	Cocci, Ø 3–5 µm; n.d.; Tuft of filaments	70–95 (85)	6.5–8.5 (7.5)	0–0.2	Anaerobic	Continental	Obligate heterotrophic	Growth by fermentation, only on yeast extract	55	Hensel et al., 1997
Thermodiscus maritimus	Dish to disc-shaped cocci; Ø 0.3–3 µm; n.d.; Not flagellated	75–98 (90)	5.0–7.0 (5.5)	1.0–4.0 (2.0)	Anaerobic	Marine	Obligate heterotrophic	Sulfur respiration and fermentation of complex organic compounds	49	Stetter, 2001
Thermosphaera aggregans	Cocci, Ø 0.2–0.8 µm in aggregates; Amorphous layer; Up to eight flagella	65–90 (85)	5.0–7.0 (6.5)	0–0.7 (0.4)	Anaerobic	Continental	Obligate heterotrophic	Fermentation of complex organic compounds	46	Huber et al. 1998
Pyrodictium occultum	Discs and dishes, Ø 0.3–2.5 µm × 0.2 µm; S-layer, p6 symmetry; Formation of a network	85–100 (105)	4.5–7.2 (5.5)	0.2–12 (1.5)	Anaerobic	Marine	Chemolithoautotrophic to Mixotrophic	S⁰, S₂O₃²⁻–H₂ autotrophy	62	Stetter, 1982; Stetter et al., 1983
Pyrodictium abyssi	Discs, Ø 1–2 µm; S-layer, p6 symmetry; Formation of a network	80–110 (97)	4.7–7.1 (5.5)	0.7–4.2 (2)	Anaerobic	Marine	Obligate heterotrophic	Fermentation of complex organic compounds	60	Pley et al., 1991
Pyrodictium brockii	Discs and dishes, Ø 0.3–2.5 µm × 0.2 µm; S-layer, p6 symmetry; Formation of a network	n.d. (105)	4.5–7.2 (5.5)	0.2–12 (1.5)	Anaerobic	Marine	Chemolithoautotrophic to Mixotrophic	S⁰, S₂O₃²⁻–H₂ autotrophy or growth on complex organic compounds	62	Stetter et al., 1983
Hyperthermus butylicus	Cocci, Ø 1.5 µm; S-layer, p6 symmetry; Not flagellated	72–108 (95–106)	n.d. (7.0)	n.d. (1.7)	Anaerobic	Marine	Obligate heterotrophic	Fermentation of peptides	56	Zillig et al., 1990
Pyrolobus fumarii	Cocci, Ø 0.7–2.5 µm; S-layer, p4 symmetry; Not flagellated	90–113 (106)	4.0–6.5 (5.5)	1.0–4.0 (1.7)	Anaerobic/microaerobic	Marine	Obligate chemolithoautotrophic	Nitrate, S₂O₃²⁻ or O₂ reduction with H₂	53	Blöchl et al., 1997
Acidilobus aceticus	Cocci, Ø 1–2 µm; S-layer with osmophilic layer–; Not flagellated	60–92 (85)	2.0–6.0 (3.8)	n.d. (0)	Anaerobic	Continental	Obligate heterotrophic	Fermentation of complex organic compounds	54	Prokofeva et al., 2000

Symbols and abbreviations: –; Ø = cell diameter; and n.d., not determined.

6) *Thermodiscus* (Stetter, 2001; type species: *T. maritimus*; type strain S2[T], DSM 15173[T] and JCM 11597[T]), and 7) *Thermosphaera* (Huber et al., 1998; type species: *T. aggregans*; type strain M11TL[T], DSM 11486[T]).

In addition to *I. islandicus* and *I. pacificus*, a third strain of *Ignicoccus* was isolated recently from the Kolbeinsey Ridge, north of Iceland (strain KIN4/I). This strain is the host of *Nanoarchaeum equitans*, the first representative of a new phylum of Archaea (Huber et al., 2002).

Furthermore, the genus *Acidilobus* (Prokofeva et al., 2000; type species: *A. aceticus*; type strain 1904[T], DSM 11585[T]) can be attached to this family, although on the basis of 16S rDNA sequence similarities, it is not closely related to the other genera. The closest relative to *Acidilobus* is "*Caldococcus noboribetus*" (Aoshima et al., 1996a); however, this organism was never validly described or deposited in culture collections.

The main characteristics for distinguishing these genera are summarized in Table 1. *Aeropyrum*, the only obligate aerobe within the family, gains energy by respiration of organic substrates. *Ignicoccus* is an obligate chemolithoautotroph (S°-H$_2$ autotrophy) and produces exclusively H$_2$S as the metabolic end product. *Staphylothermus* cells grow in aggregates and gain energy by fermentation of complex organic substrates in the presence of elemental sulfur. They exhibit a low G + C content of their genomic DNA (35–38 mol%). The major characteristic of the genus *Stetteria* is its growth by sulfur respiration with complex organic substrates and its strict requirement of molecular hydrogen. *Sulfophobococcus*, as implied by the name, is inhibited by elemental sulfur. *Thermodiscus* is characterized by disk-shaped cells, which carry out sulfur respiration with complex organic substrates. *Thermosphaera* forms short chains or aggregates and is inhibited not only by elemental sulfur but also by molecular hydrogen. *Acidilobus* exhibits a pH optimum at 3.8 and is therefore, along with "*Caldococcus*," the only acidophile within the group. Furthermore, all genera are clearly defined by the results of 16S rDNA sequence comparisons implemented in the corresponding phylogenetic trees (e.g., Fig. 1).

The Pyrodictiaceae represent the second family within the order Desulfurococcales. It consists of three genera, *Pyrodictium* (Stetter et al., 1983), *Hyperthermus* (Zillig et al., 1991), and *Pyrolobus* (Blöchl et al., 1997). The genus *Pyrodictium*, the type genus of the family, harbors three species: *P. occultum* (the type species for the genus; Stetter et al., 1983), *P. abyssi* (Pley et al., 1991), and *P. brockii* (Stetter et al., 1983). The type strain of *P. occultum* is strain PL-19[T] (DSM 2709[T]), of *P. abyssi*, strain AV2[T] (DSM 6158[T]), and of *P. brockii*, strain S1[T] (DSM 2708[T]). The characteristic feature of the *Pyrodictium* species is the formation of a network of hollow cannulae, in which the coccoid or disk-shaped cells are embedded. The organisms are obligate anaerobes with temperature optima between 97 and 105°C. The species can be distinguished by their growth substrates and cell envelope composition.

Hyperthermus butylicus (Zillig et al., 1990; type strain DSM 5456[T]), the only representative of the genus, is an obligate anaerobe that grows by fermentation of proteolysis products, forming organic acids, butanol, and CO$_2$.

Pyrolobus fumarii (Blöchl et al., 1997; type strain 1A[T], DSM 11204[T]) is a facultatively anaerobic, obligately chemolithoautotrophic, irregular coccus which exhibits the highest temperature maximum of all life forms on earth, 113°C.

Habitats

Members of the Desulfurococcales have been found worldwide in biotopes of volcanic activity on land, in shallow marine areas, and in the deep sea. Representatives of the genera *Desulfurococcus*, *Sulfophobococcus*, *Thermosphaera* and *Acidilobus* have been isolated from hot springs, water and mud holes, and soils of continental solfataric fields with low salinity and acidic to slightly alkaline pH values, such as occur in the Askja-, Hveravellir-, and Hveragerdi areas in Iceland, the Uzon Caldera, Valley of Geysers, and Mendeleev Volcano (on Kamchatka and Kunashir Island, Russia), and the Yellowstone National Park (United States). They are adapted to these low ionic strength biotopes as indicated by their NaCl optima (Table 1).

Strains of the genera *Aeropyrum*, *Ignicoccus*, *Staphylothermus*, *Stetteria*, *Thermodiscus*, *Pyrodictium* and *Hyperthermus* have been obtained from shallow marine high temperature sediments, springs, and venting waters (original temperatures up to 112°C) at Kodakara-Jima Island (Japan), Kolbeinsey Ridge (north of Iceland, depth 105 m), Vulcano Island (Italy), Paleohori Bay (Milos, Greece), and Sao Miguel Island (Azores). *Ignicoccus*, *Staphylothermus*, *Pyrodictium* and *Pyrolobus* have been isolated from deep-sea hydrothermal systems and black smoker fragments at the East Pacific Rise (9–11° N and 21° N, depth 2500 m), the Guaymas Basin hot vent area (27° N, depth 2000 m), and the Mid-Atlantic Ridge (TAG site [geographical landmark]: 26° N, depth 3650 m). *Pyrodictium* has been also found in the active zone of an erupting submarine volcano (Macdonald Seamount, Indonesia; Huber et al., 1990).

Isolation

Enrichment

All representatives of the Desulfurococcales are hyperthermophiles; therefore, enrichment at temperatures of at least 85°C is highly selective. Representatives of the genera *Desulfurococcus*, *Sulfophobococcus*, *Thermosphaera* and *Acidilobus*, which thrive in continental solfataric areas, can be enriched in low-salt media (see Media for Cultivation) prepared by the use of the anaerobic techniques of Balch and Wolfe (1976). The media are supplemented with elemental sulfur and/or various kinds of complex or defined organic substrates with a gas phase of N_2, H_2/CO_2, and N_2/CO_2 (80:20, v/v, 200–300 kPa) at incubation temperatures around 85°C. Depending on the species, peptides, amino acids, glycogen, starch, glucose, or acetate can serve as carbon sources during sulfur respiration or fermentation. *Acidilobus* can be selectively enriched in media with pH 3.

Members of the genera *Aeropyrum*, *Ignicoccus*, *Staphylothermus*, *Stetteria*, *Thermodiscus*, *Pyrodictium*, *Hyperthermus* and *Pyrolobus* grow in artificial seawater based media (see Media for Cultivation). They are adapted to the ionic strength of their natural habitats, which is expressed by optimal growth at NaCl concentrations between 2 and 3%. For *Aeropyrum* enrichment at 95°C in an aerobic medium in the presence of molecular oxygen is successful. *Pyrodictium*, *Hyperthermus* and *Pyrolobus* can be selectively obtained by incubation at temperatures of around 105°C. A highly selective enrichment method for *Pyrolobus* is the use of nitrate as the sole electron acceptor in addition. *Ignicoccus* can be obtained in a medium without organic components.

Isolation Procedures

Anaerobic, microaerobic, or aerobic culture media are inoculated with 0.5–1 ml (or grams) of the environmental samples. The culture attempts should be checked by phase-contrast microscopy over a period of up to several weeks for growth. The isolates are obtained from the enrichment cultures by serial dilution (three times) or by plating. The plates can be solidified by 0.6–0.8% Gelrite and colonies are visible after 4–14 days. However, successful plating is not reported for every member of the Desulfurococcales. A recently developed isolation technique that is safe, fast and suitable for organisms unable to grow on solidified media is the use of the "optical tweezers" trap (also see Thermoproteales in this Volume; Huber et al., 1995).

Identification

Members of the Desulfurococcales are regular to irregular coccoid to disc- or dish-shaped cells with diameters from 0.2 to 15 μm (Table 1). They do not divide by septa formation, and spores are not formed. All members stain Gram negative. In most species the cell envelope is composed of an S-layer, i.e., glycoprotein subunits regularly arranged on a two-dimensional lattice, with four- or sixfold symmetry. Alternatively, an outer membrane or an amorphous layer can occur; muramic acid is always absent. The DNA-dependent RNA polymerase is of the "BAC" type and the transcription is resistant to rifampicin and streptolydigin. The major core lipids are glycerol dialkylglycerol tetraethers and 2,3-di-*O*-phytanyl-*sn*-glycerol. The G+C-contents of genomic DNA range from 35 to 67 mol% (Table 1). The Desulfurococcales are anaerobic, facultatively anaerobic or aerobic organisms with optimal growth temperatures between 85 and 106°C and therefore harbor the organisms with the highest temperature maxima known so far (108–113°C; see also Physiology). With the exception of *Acidilobus* and "*Caldococcus*," they grow in a pH range of 4.5–9 with an optimum around 5–7.0.

Cells of *Desulfurococcus* are regular spheres usually between 0.5 and 1 μm in diameter (Fig. 2a), although cells of up to 10 μm may occur. *Desulfurococcus mucosus* is surrounded by a mucoid layer, which covers the envelope and consists of neutral sugars and a small fraction of amino sugars (Stetter and Zillig, 1985). In contrast, *D. mobilis* is devoid of this mucous polymer. Instead, as indicated by the name, it possesses a bundle of up to seven flagella (diameter 12.5 nm) attached at one pole of the cell (Fig. 2b). *Desulfurococcus* species possess an S-layer with fourfold symmetry and a center-to-center distance of 18 nm (investigated for *D. mobilis*: Wildhaber et al., 1987). It is built up of protein complexes in the shape of a cross. The S-layer encloses a periplasmic space about 30 nm wide. An intron was found in the 23S rRNA sequence of *D. mobilis* (Larsen et al., 1986). The G+C content of the genomic DNA is 51 mol% for *D. mobilis* and *D. mucosus*, while it is only 42 for *D. amylolyticus* (Table 1).

Aeropyrum pernix cells are irregular cocci with diameters of 0.8 to 1 μm, sometimes exhibiting sharp edges. In contrast to their original description, cells are highly motile and possess flagella (Fig. 3a). In the genome, two flagellin genes were detectable (Thomas et al., 2001). *Aeropyrum pernix* has pili-like appendages. The cell envelope consists of an S-layer with fourfold symmetry and a center-to-center distance of

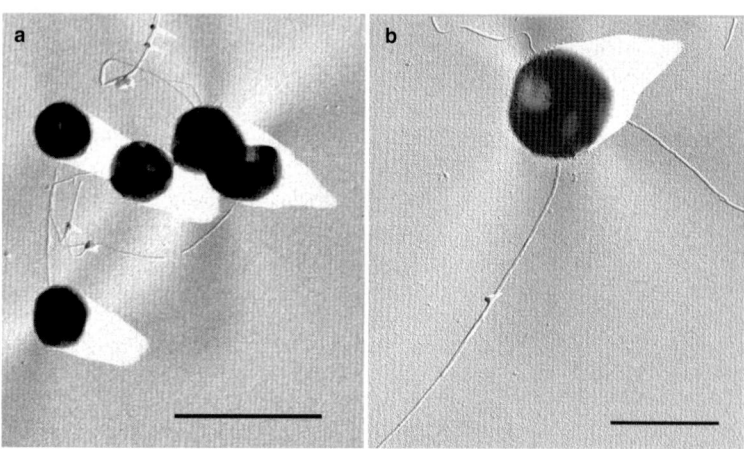

Fig. 2. Electron micrograph of a) cells of *Desulfurococcus mobilis*; platinum-shadowed (bar = 2 μm); and b) a *Desulfurococcus mobilis* cell with flagella; platinum-shadowed (bar = 1 μm).

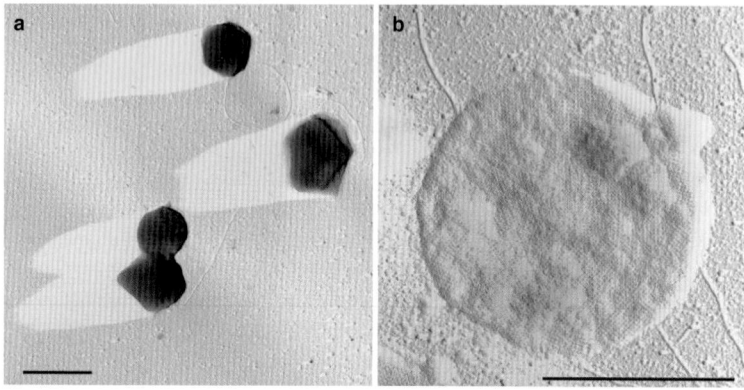

Fig. 3. Electron micrograph of a) *Aeropyrum* cells; platinum-shadowed; and b) an *Aeropyrum* cell envelope, showing the S-layer lattice and flagella; platinum-shadowed. Bars are each 1 μm.

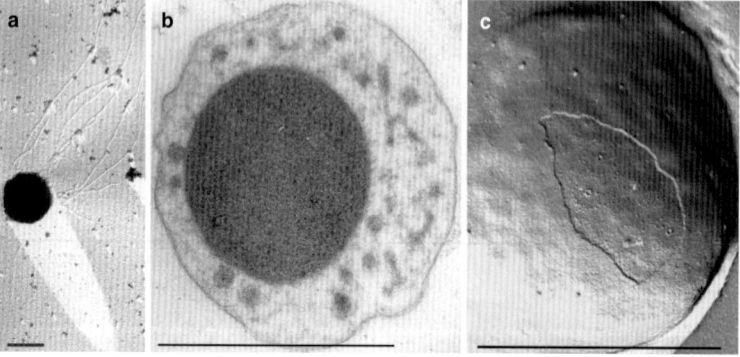

Fig. 4. Electron micrographs of cells of *Ignicoccus islandicus*: a) platinum-shadowed; b) thin sectioned; and c) freeze-etched. Bars are each 1 μm.

18 nm (R. Rachel, unpublished observation; Fig. 3b). The original G+C content of the genomic DNA was determined to be 67 mol%; however, sequencing of the whole genome revealed only 56 mol% (Kawarabayashi et al., 1999). The core lipids of *A. pernix* consist of di-esterterpanyl (C25, C25) glycerol ether lipids. Five components of polar lipids were detected; phosphoglycolipid (archaetidyl [glucosyl] inositol, containing inositol and glucose) and phospholipid (archaetidyl inositol) are predominant (Morii et al., 1999). *Aeropyrum pernix* possesses two family B DNA polymerases (Cann et al.,

1999), and the 16S–23S rRNA operon contains three introns (Nomura et al., 1998).

All *Ignicoccus* cells are slightly irregular cocci with diameters of 1–3 μm. They are monopolar polytrichously flagellated (Fig. 4a). They exhibit a unique cell architecture within the Archaea: a cytoplasmic membrane; a periplasmic space with a variable width of 20 up to 400 nm containing membrane-bound vesicles; and an outer membrane, approximately 10 nm wide (Fig. 4b). The outer membrane contains three different types of particles: numerous tightly, irregularly packed single particles about 8 nm in size; pores with

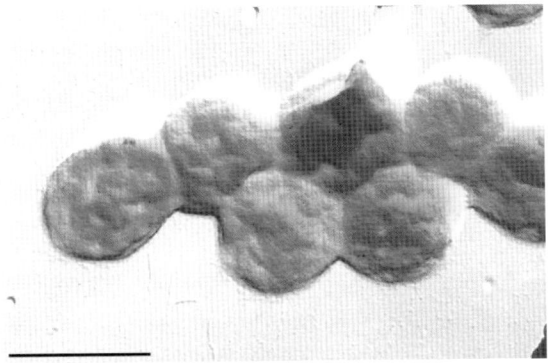

Fig. 5. Electron micrograph of *Staphylothermus marinus* cells; platinum-shadowed. Bar = 1 μm.

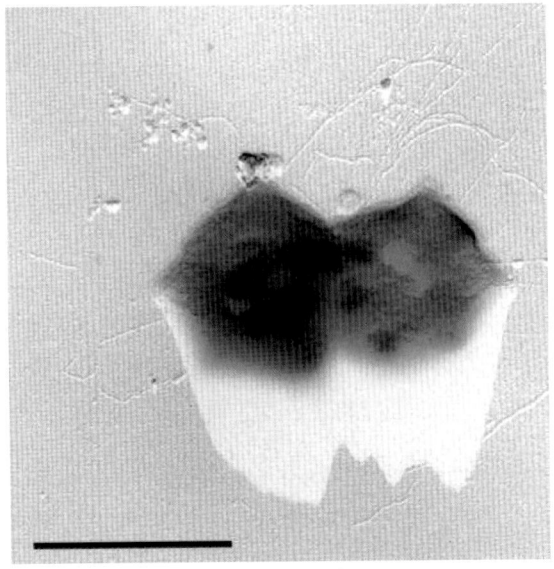

Fig. 6. Electron micrograph of a *Thermodiscus maritimus* cell in division with numerous appendages; platinum-shadowed. Bar = 1 μm.

diameters of 24 nm surrounded by tiny particles arranged on a ring with a diameter of 130 nm; and clusters of up to 8 particles, each 12 nm in size (Fig. 4c). Freeze-etched cells exhibit a smooth surface, without a regular pattern, indicating the absence of an S-layer (Fig. 4c); frequently, fracture planes through the outer membrane are found (Rachel et al., 2002). The G+C content, of the two described species are 41 and 45 mol%, respectively (Table 1). The core lipids of *I. islandicus* are acyclic 2,3-di-*O*-phytanyl-*sn*-glycerol and glyceroldialkyl glycerol tetraether in a relative ratio of 1 : 1 (Huber et al., 2000a).

Cells of both *Staphylothermus* species are regular to slightly irregular cocci, about 0.8–1.3 μm in diameter. They often form aggregates (Fig. 5), sometimes containing up to 100 cells. In the presence of high concentrations of yeast extract (0.2%) in the medium, *S. marinus* forms giant cells of up to 15 μm in diameter. They contain at least one dark granulum (about 1 μm in diameter), which most likely consists of glycogen. The cell envelope of *Staphylothermus* consists of an almost 70 nm wide periplasm and an S-layer of a poorly ordered meshwork of branched, filiform morphological subunits, named "tetrabrachion." Tetrabrachion is a glycoprotein tetramer that forms a parallel, four-stranded α-helical rod, separated at one end into four strands (Peters et al., 1995). An extremely thermostable protease is tightly bound in stoichiometric amounts to this structure (resistant to heat inactivation up to 125°C in the stalk-bound form; Mayr et al., 1996). Both *Staphylothermus* species (*S. marinus* and *S. hellenicus*) exhibit a low G+C content of the genomic DNA of 35 and 38 mol%, respectively.

Stetteria cells are irregular to disc-shaped cocci with a diameter of 0.5–1.5 μm. The cells possess a single flagellum and form cytoplasmic protrusions up to 2 μm in length, especially during the stationary phase. The cell envelope consists of an S-layer, probably with sixfold symmetry (Jochimsen et al., 1997). *Stetteria* exhibits a G+C content of the genomic DNA of 65 mol%, which is the highest of all members of the Desulfurococcales (Table 1).

Cells of *Sulfophobococcus* are regular to slightly irregular cocci, 2–5 μm in diameter. They occur singly or in aggregates and possess a tuft of filaments, whose function remains to be clarified (Hensel et al., 1997). The membrane contains dibiphytanylglycerol tetraethers without pentacyclic rings. As polar head groups, *myo*-inositol phosphate and mono- and diglycosyl units are present. No respiratory lipoquinones were detectable. The G+C content of the genomic DNA is 55 mol% (Table 1).

As indicated by the name, *Thermodiscus* cells are irregular dish to disc-shaped cocci, varying in diameter from about 0.3 to 3 μm (Fig. 6). The discs are sometimes less than 0.2 μm thick (Stetter, 1986). Flagella are absent, but pili-like structures, about 10 nm thick and sometimes connecting two cells, are present. The cell envelope is composed of protein subunits, most likely an S-layer, with unknown symmetry (R. Rachel, unpublished observation). The G+C content of the DNA is 49 mol%.

Thermosphaera forms regular cocci of 0.2–0.8 μm in diameter, which occur singly, in pairs, in short chains, and in grapelike aggregates (Fig. 7a). The aggregates exhibit a weak bluish-green fluorescence at 436 nm. The aggregates are very stable and cannot be disintegrated by enzymatic treatment (e.g., by cellulase, proteinase K or

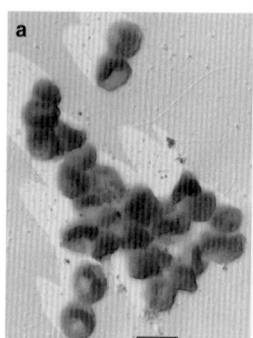
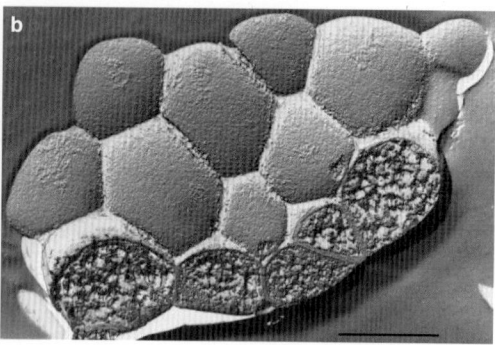

Fig. 7. Electron micrograph of an aggregate of a) *Thermosphaera aggregans* cells; platinum-shadowed; and b) freeze-etched *Thermosphaera aggregans* cells. Bars are each 1 μm.

trypsin), sonification, or mechanical stirring. The cells possess up to eight flagella. The cell envelope contains a cytoplasmic membrane covered by an amorphous layer of unknown composition about 10–15 nm thick; there is no evidence for a regularly arrayed surface layer protein (Huber et al., 1998; Fig. 7b). Acyclic and cyclic glycerol diphytanyl tetraethers with one to four pentacyclic rings were identified as core lipids. In addition, traces of glycerol diphytanyl diethers are present (Huber et al., 1998). The G+C content of the genomic DNA is 46 mol% (Table 1).

As a characteristic, *Pyrodictium* grows as a three-dimensional network of cells and extracellular hollow tubules (the cannulae; Fig. 8a and b); these cannulae interconnect the cells (Stetter, 1982; König et al., 1988; Rieger et al., 1995). In liquid cultures, the network forms flakes of up to 10 mm in diameter, visible by the naked eye, or, in the case of *P. abyssi*, tiny white balls, about 1 mm in size. The cells are dish- to disc-shaped with diameters between 0.3 and 2.5 μm, frequently exhibiting ultra-flat areas, which are only 0.1–0.2 μm thick (Fig. 8c). The cannulae are about 25 nm in diameter and are composed of glycoprotein subunits in a helical array (Rieger et al., 1995); they often aggregate into large bundles. The genes of three proteins forming the cannulae (*canA*, *canB* and *canC*) have been sequenced (Mai, 1998a). The cannulae are up to 150 μm long and elongate at 1.0–1.5 μm/min from the proximal end (Horn et al., 1999). As can be visualized under anaerobic conditions at 95°C, using a heatable phase contrast microscope, *Pyrodictium* cells divide by binary fission (Horn et al., 1999). The cell envelope consists of a cytoplasmic membrane, a 35-nm wide periplasmic space (Rachel, 1999), and a "zigzag"-shaped surface-layer protein. The subunits of the surface layer are arranged on a p6 lattice with a center-to-center distance of 21 nm (Dürr et al., 1991; Fig. 8d). Precipitates of zinc and sulfur (most likely zinc sulfide) occur on the cell surface and on the cannulae (Stetter et al., 1983). In addition to the cannulae, flagella with a diameter of about 10 nm are present (Rieger et al., 1995). A heat-inducible molecular chaperone, called the "thermosome," was found in *Pyrodictium* (Phipps et al., 1993); this chaperone makes up about 80% of all soluble proteins at 108°C (Phipps et al., 1991). In *P. brockii*, sulfur respiration is proposed to be accomplished by an electron-transport chain composed of a membrane-bound hydrogenase of the nickel-iron-sulfur type, a quinone, a cytochrome *c*, and an unknown sulfur-reducing entity (Pihl and Maier, 1991; Pihl et al., 1992). A protein complex with a molecular mass of 520 kDa was purified from *P. abyssi*; the complex catalyzed the reduction of elemental sulfur to H_2S by molecular hydrogen (Dirmeier et al., 1998). *Pyrodictium abyssi* cells also possess an ATPase complex; the head piece (AF_1) consists of five polypeptides and the AF_0 component of a proteolipid (Dirmeier et al., 2000). The complex is able to restore the formation of ATP in membrane vesicles in vitro. The sulfur-reducing complex and the ATPase are both membrane-bound and are extremely heat stable, exhibiting a temperature optimum of 100°C. *Pyrodictium abyssi* possesses a protein-serine/threonine phosphatase, which is active between 40 and 110°C and has a temperature optimum at 90°C (Mai et al., 1998b). The genomic DNAs of all *Pyrodictium* species exhibit high G+C contents (between 60 and 62 mol%; Table 1).

Cells of *Hyperthermus* are irregular cocci of about 1.5 μm in diameter. They often possess pili-like projections on the cell surface; no network is formed. The S-layer exhibits a hexagonal symmetry with a center-to-center distance of the complexes of 25 nm (Baumeister et al., 1990). The RNA polymerase contains an unsplit B component. The membrane lipids are mainly *bis*-isopranyl tetraether lipids containing two or more cyclopentane rings. The G+C content of the genomic DNA is 56 mol%.

Pyrolobus cells are regular to irregular lobed cocci, about 0.7–2.5 μm in diameter (Fig. 9a). Like in *Hyperthermus*, no network is formed. In contrast to the other members of the Pyrodicti-

Fig. 8. Electron micrographs of cells of *Pyrodictium abyssi*. a) Stained with uranyl acetate; bar = 2 µm; b) visualized by scanning electron microscopy; bar = 2 µm; c) thin section after freeze substitution; bar = 0.2 µm; and d) freeze-etched; bar = 1 µm.

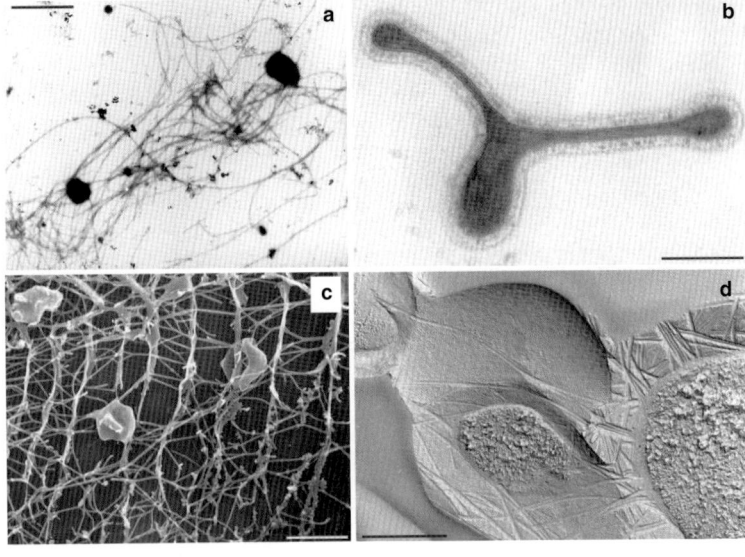

Fig. 9. Electron micrographs of cells of *Pyrolobus fumarii*: a) thin section; and b) freeze-etched. Bars are 1 µm.

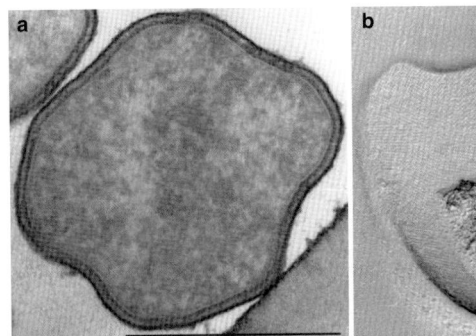
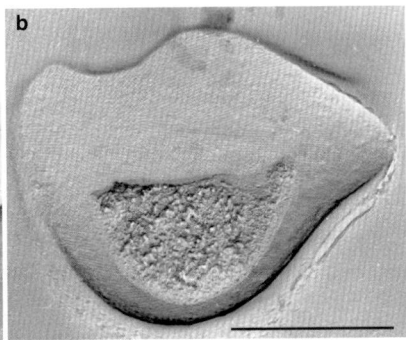

aceae, the cells are covered by an S-layer with p4 (!) symmetry and a center-to-center distance of 19 nm (Blöchl et al., 1997; Fig. 9b). The protein complexes exhibit a central depression, most likely a pore. The periplasmic space is about 45 nm wide. The major core lipids are glycerol dialkylglycerol tetraether and traces of 2,3-di-*O*-phytanyl-*sn*-glycerol. Crude extracts of *Pyrolobus* cells show a strong crossreaction with antibodies prepared against the thermosome of *Pyrodictium occultum*. Furthermore, a membrane-associated hydrogenase with an optimum reaction temperature of 119°C was found in cells grown on molecular hydrogen and nitrate. The G+C content of the genomic DNA is 53 mol%.

Acidilobus cells are regular to irregular cocci, about 1–2 µm in diameter, which are not flagellated. The cell envelope consists of an S-layer, which is covered by an osmophilic layer (Prokofeva et al., 2000). The G+C content of the genomic DNA is 54 mol%. "*Caldococcus*" cells are 1–3 µm in diameter. Besides the 16S rRNA gene, an isocitrate dehydrogenase, which showed high similarities to bacterial isocitrate dehydro-

genases, was sequenced and expressed (Aoshima et al., 1996b).

Cultivation

Media for Cultivation

Medium for *Desulfurococcus amylolyticus* (Bonch-Osmolovskaya et al., 1988)

NaCl	0.50 g
MgCl$_2$ · 6H$_2$O	0.70 g
KH$_2$PO$_4$	0.33 g
NH$_4$Cl	0.33 g
CaCl$_2$ · 2H$_2$O	0.44 g
KCl	0.33 g
NaHCO$_3$	0.80 g
Elemental sulfur	10.0 g
Trace elements (SL 10)	1.0 ml
Trace vitamin solution (see below)	10.0 ml
Yeast extract	0.2 g
Casein hydrolysate	5.0 g
Resazurin	1 mg
Na$_2$S · 9H$_2$O	0.5 g
Distilled water ad	1 liter

Adjust pH to 6.2–6.8 and incubate cultures in a gas phase of N$_2$/CO$_2$ (80:20) under pressure (200 kPa).

Trace Element Solution (SL 10)

HCl (25%; 7.7M)	10 ml
$FeCl_2 \cdot 4H_2O$	1.5 g
$ZnCl_2$	70 mg
$MnCl_2 \cdot 4H_2O$	100 mg
H_3BO_3	6 mg
$CoCl_2 \cdot 6H_2O$	190 mg
$CuCl_2 \cdot 2H_2O$	2 mg
$NiCl_2 \cdot 6H_2O$	24 mg
$Na_2MoO_4 \cdot 2H_2O$	36 mg
Distilled water ad	1 liter

Trace Vitamin Solution (Balch et al., 1979)

B_6 (pyridoxine) hydrochloride	10 mg
B_1 (thiamine) hydrochloride	5 mg
B_2 (riboflavin)	5 mg
Nicotinic (niacin) acid	5 mg
Pantothenic acid (DL Ca-pantothenate)	5 mg
PABA (p-aminobenzoic acid)	5 mg
Lipoic acid	5 mg
Biotin	2 mg
Folic acid	2 mg
B_{12} (crystalline)	0.1 mg
Distilled water ad	1 liter

Standard Medium for *Desulfurococcus mucosus*, *D. mobilis* and *D. saccharovorans* (Brock, 1972; Zillig et al., 1982)

$(NH_4)_2SO_4$	1.30 g
KH_2PO_4	0.28 g
$MgSO_4 \cdot 7H_2O$	0.25 g
$CaCl_2 \cdot 2H_2O$	70 mg
$FeCl_3 \cdot 6H_2O$	20 mg
$MnCl_2 \cdot 4H_2O$	18 mg
$Na_2B_4O_7 \cdot 10H_2O$	4.5 mg
$ZnSO_4 \cdot 7H_2O$	0.22 mg
$CuCl_2 \cdot 2H_2O$	0.05 mg
$Na_2MoO_4 \cdot 2H_2O$	0.03 mg
$VOSO_4 \cdot 2H_2O$	0.03 mg
$CoSO_4$	0.01 mg
Yeast extract (Bacto Difco)	2.0 g
Resazurin	1.0 mg
Elemental sulfur	5.0 g
$Na_2S \cdot 9H_2O$	0.5 g
Distilled water ad	1 liter

Adjust pH to 5.5 and incubate cultures in a gas phase of N_2 under pressure (200 kPa).

For culture of *Desulfurococcus saccharovorans*, add per liter 5 g of glucose and reduce the concentration of yeast extract to 0.2 g. Adjust pH to 7.0 and incubate cultures in a gas phase of H_2/CO_2 (80:20) under pressure (200 kPa).

Culture Medium for *Acidilobus aceticus* (Basal Medium)

KCl	330 mg
$(NH_4)Cl$	330 mg
KH_2PO_4	330 mg
$MgCl_2 \cdot 6H_2O$	330 mg
$CaCl_2 \cdot 2H_2O$	330 mg
Na_2S	500 mg
Starch	5.0 g
Yeast extract (Difco)	0.1 g
Resazurin	1 mg
Trace-elements SL-10	1 ml

Trace vitamin solution (Balch et al., 1979)	1 ml
$Na_2S \cdot 9H_2O$	700 mg
Distilled water ad	1 liter

Adjust pH to 3.0 with 5M H_2SO_4 and incubate cultures in a gas phase of CO_2.

Culture Medium for *Aeropyrum pernix*

Bacto yeast extract (Difco)	1.0 g
Trypticase peptone (BBL)	1.0 g
$Na_2S_2O_3 \cdot 5H_2O$	1.0 g
Synthetic seawater ad	1 liter

Adjust pH to 7.0 and incubate cultures in air. The synthetic seawater known as "JS" consists of Jamarine S synthetic sea salts (Jamarine Laboratory, Osaka, Japan).

Culture Medium for *Pyrodictium*, *Pyrolobus fumarii*, *Hyperthermus butylicus*, *Ignicoccus*, *Staphylothermus marinus* and *Thermodiscus maritimus* (1/2 SME-medium)

NaCl	13.85 g
$MgSO_4 \cdot 7H_2O$	3.50 g
$MgCl_2 \cdot 6H_2O$	2.75 g
KH_2PO_4	0.50 g
NH_4Cl	0.50 g
$CaCl_2 \cdot 2H_2O$	0.38 g
KCl	0.33 g
NaBr	50 mg
H_3BO_3	15 mg
$SrCl_2$	7.5 mg
KI	0.05 mg
Trace mineral solution (see below)	10.0 ml
Resazurin	1 mg
$Na_2S \cdot 9H_2O$	0.5 g
Distilled water ad	1 liter

This medium is based on the originally described "SME" medium (Stetter et al., 1983), modified by Pley et al. (1991), and described as "1/2 SME-medium" by Blöchl et al. (1997).

For culture of *Pyrodictium occultum*, add per liter 30 g of elemental sulfur, adjust pH to 5.5, and incubate in a gas phase of H_2/CO_2 (80:20) under pressure (300 kPa).

For culture of *Pyrodictium brockii*, add per liter 30 g of elemental sulfur and 2 g of yeast extract (Bacto Difco), adjust pH to 5.5, and incubate in a gas phase of H_2/CO_2 (80:20) under pressure (300 kPa).

For culture of *Pyrodictium abyssi*, add per liter 1 g of yeast extract (Bacto Difco) and 1 g of $Na_2S_2O_3$, adjust pH to 5.5, and incubate in a gas phase of H_2/CO_2 (80:20) under pressure (300 kPa).

For culture of *Pyrolobus fumarii*, replace NH_4Cl with 1 g/liter of $NaNO_3$, adjust pH to 5.5, and incubate in a gas phase of H_2/CO_2 (80:20) under pressure (300 kPa).

For culture of *Hyperthermus butylicus*, add per liter 2.5 mg of KI, 6 g of tryptone (Bacto Difco), and 10 g of elemental sulfur, adjust pH to 7.0, and incubate in a gas phase of H_2/CO_2 (80:20) under pressure (200 kPa).

For culture of *Ignicoccus islandicus* and *I. pacificus*, omit trace element solution and add per liter 10 g of elemental sulfur, adjust pH to 5.5, and incubate in a gas phase of H_2/CO_2 (80:20) under pressure (300 kPa).

For culture of *Staphylothermus marinus*, add per liter 1 g of yeast extract (Bacto Difco), 5 g of peptone (Bacto Difco), and 30 g of elemental sulfur, adjust pH to 6.5, and incubate in a gas phase of N_2 under pressure (300 kPa).

For culture of *Staphylothermus hellenicus* and *Stetteria hydrogenophila*, add per liter 1 g of yeast extract (Bacto Difco), 20 g of elemental sulfur, and 20 g of NaCl, adjust pH to 5.5, and incubate in agas phase of H_2/CO_2 (80:20) under pressure (300 kPa).

For culture of *Thermodiscus maritimus*, add per liter 0.2 g of yeast extract (Bacto Difco) and 30 g elemental sulfur, adjust pH to 5.5, and incubate in a gas phase of H_2/CO_2 (80:20) under pressure (300 kPa).

Trace Mineral Solution (modified from Balch et al., 1979)

$MgSO_4 \cdot 7H_2O$	3.0 g
NaCl	1.0 g
$MnSO_4 \cdot H_2O$	0.5 g
$Ni(NH_4)(SO_4)_2 \cdot 6H_2O$	280 mg
$ZnSO_4 \cdot 7H_2O$	180 mg
$CoCl_2 \cdot 6H_2O$	180 mg
$FeSO_4 \cdot 7H_2O$	100 mg
$CaCl_2 \cdot 2H_2O$	100 mg
$KAl(SO_4)_2 \cdot 12H_2O$	18 mg
$CuSO_4 \cdot 5H_2O$	10 mg
H_3BO_3	10 mg
$Na_2MoO_4 \cdot 2H_2O$	10 mg
$Na_2WO_4 \cdot 2H_2O$	10 mg
Na_2SeO_4	10 mg
Distilled water ad	1 liter

The solution is adjusted with H_2SO_4 to pH 1.0.

Culture Medium for *Sulfophobococcus zilligii*

EDTA	25 mg
$CaCl_2 \cdot 2H_2O$	66 mg
$MgCl_2 \cdot 2H_2O$	31 mg
KCl	31 mg
$FeSO_4 \cdot 7H_2O$	51.5 mg
$ZnCl_2$	2.1 mg
$MnSO_4$	2.3 mg
$Na_4B_4O_7 \cdot 10H_2O$	1.8 mg
Glycine	1.5 g
Na_2CO_3	230 mg
Yeast extract	1.0 g
Resazurin	1 mg
Titanium citrate	1–2 mM
Distilled water ad	1 liter

Adjust pH to 7.5 and incubate cultures in a gas phase of N_2/CO_2 (80:20) under pressure (200 kPa).

Culture Medium for *Thermosphaera aggregans* (Huber et al., 1998)

$MgCl_2 \cdot 6H_2O$	2.2 g
$NaHCO_3$	1.0 g
NaCl	0.9 g
KCl	17 mg
$(NH_4)Cl$	12.5 mg
$CaCl_2 \cdot 2H_2O$	7.0 mg
$K_2HPO_4 \cdot 3H_2O$	7.0 mg
$Fe(III)Cl_3$	0.05 mg
Trace vitamin solution (see above; Balch et al., 1979)	10 ml
Bacto yeast extract (Difco)	1.0 g
Bacto peptone (Difco)	1.0 g
Resazurin	0.02 mg
$Na_2S \cdot 9H_2O$	0.5 g
Distilled water ad	1 liter

Adjust pH to 6.5 and incubate cultures in a gas phase of N_2/CO_2 (80:20) under pressure (300 kPa).

Culture Medium for "*Caldococcus noboribetus*" (Aoshima et al., 1996a)

NaCl	0.1 g
$CaCl_2 \cdot 2H_2O$	0.1 g
$FeSO_4 \cdot 2H_2O$	0.05 g
Bacto yeast extract (Difco)	0.1 g
Casamino acids	0.1 g
Bacto peptone (Difco)	0.8 g
Trace vitamin solution (see above; Balch et al., 1979)	10 ml
Trace mineral solution (see below; Aoshima et al., 1996a)	10 ml
Resazurin	1 mg
Elemental sulfur	2.0 g
$Na_2S \cdot 9H_2O$	0.5 g
Distilled water ad	1 liter

Adjust pH to 2.7 and incubate cultures in a gas phase of H_2/CO_2 (80:20) under pressure (200 kPa).

Trace Mineral Solution (Aoshima et al., 1996a)

$MnSO_4 \cdot 2H_2O$	0.5 g
$ZnSO_4 \cdot 7H_2O$	180 mg
$CoSO_4$	180 mg
$CuSO_4 \cdot 5H_2O$	10 mg
H_3BO_3	10 mg
$Na_2MoO_4 \cdot 2H_2O$	10 mg
Distilled water ad	1 liter

Cultivation in Liquid Media

Aerobic cultivation (for *Aeropyrum pernix*) is carried out in screw-capped tubes (e.g., 180 × 18 mm) containing 10 ml of medium using normal air as gas phase or overpressure (200 kPa). Microaerobic culture medium (for *Pyrolobus fumarii*) does not contain resazurin and sodium sulfide. After solubilization of the salts, the medium is gassed for 45 min with N_2 (1 liter/ min), heated to 65°C, evacuated three times, and finally pressurized with N_2 (200 kPa). All anaerobic members of the Desulfurococcales are cultivated using the anaerobic technique of Balch and Wolfe (1976). For small volumes, the organisms are grown in stoppered and pressurized 28-ml serum tubes (borosilicate glass; Schott, Mainz, Germany) or in 120-ml serum bottles (Bormioli, Italy) made of "type III"-glass. One liter of the corresponding medium is flushed with nitrogen or N_2/CO_2 for 20 min in a stoppered glass bottle. Then dissolved oxygen is removed by the addition of 0.5 g of $Na_2S \cdot 9H_2O$, and the pH is adjusted with H_2SO_4 or HCl. The medium is dispensed in an anaerobic chamber and finally the desired gas phase is pressurized by gas exchange at a gas station.

Cultivation on Plates

Growth on solidified media is described for several members of the Desulfurococcales. Colonies of *Pyrolobus fumarii* and *Pyrodictium abyssi* are obtained on anaerobic media solidified by 0.8% Gelrite (Kelco, San Diego, CA, United States) in the presence of nitrate or thiosulfate as electron acceptor, respectively. The plates are incubated using a stainless steel anaerobic jar (Balch et al., 1979) with an atmosphere of H_2/CO_2 (200 kPa) plus H_2S (5 kPa). After about one week, white colonies (about 1 mm in diameter) will be formed (Pley et al., 1991; Blöchl et al., 1997). For the cultivation of *Hyperthermus*, 0.8% Gelrite plates without electron acceptors are soaked for two min with 5 ml of a saturated solution of $S°$ in 1 M $(NH_4)_2S$ and then rinsed with water. Colloidal sulfur precipitates in the surface by acidification with 5–10 ml of 1 M H_2SO_4 for 2 min. After washing with water, the plates are equilibrated overnight with culture medium. The dried plates can be inoculated and incubated as described above. Amber colonies surrounded by clear halos appear in the sulfur layer after 2–3 days (Zillig et al., 1990). *Staphylothermus hellenicus* can be plated on its normal medium, solidified with 1% Phytagel (Sigma) and 2–3% sodium alginate (Roth) according to Kovacs and Rakhely (1996). For growth of *Acidilobus aceticus*, the culture medium (supplemented with 0.05% $MgSO_4$) is solidified by 0.8% Gelrite. Smooth white colonies, about 1 mm in diameter, appear after one week of incubation (Prokofeva et al., 2000).

Preservation

Most Desulfurococcales strains can be stored under anaerobic conditions at 4°C or at room temperatures for several months (or even up to two years). Although not described for all strains, long-term storage is best achieved in culture media containing 5% dimethyl sulfoxide and frozen in the gas phase over liquid nitrogen (around –140°C). No loss of cell viability is usually observed after storage of several years.

Physiology

Within the Desulfurococcales, only hyperthermophilic organisms are known, exhibiting optimal growth temperatures between 85 and 106°C. No representative can grow below 65°C (Table 1). Therefore, the Desulfurococcales is the order that harbors the most thermophilic microorganisms known so far. *Pyrolobus fumarii* exhibits the highest temperature maximum of all life forms on earth, 113°C. It is unable to propagate at a temperature of 90°C or below. Cultures of *Pyrodictium* and *Pyrolobus* species have been shown to survive even autoclaving at 121°C for one hour (Blöchl et al., 1997; Stetter, 1999; Stet-

ter, 2002). In agreement with this, members of the Desulfurococcales have been exclusively enriched and isolated from terrestrial or marine high temperature biotopes. The pH optima of all organisms are neutral to slightly acidic (Table 1). With the exception of *Aeropyrum pernix*, all members of the Desulfurococcales are able to grow under strictly anaerobic culture conditions. In addition, *Pyrolobus fumarii* can alternatively propagate in the presence of low oxygen concentrations.

The metabolic properties of the Desulfurococcales range from strict heterotrophs to obligate chemolithoautotrophs (Table 1). Under autotrophic conditions, molecular hydrogen is used as the electron donor and usually sulfur (or $S_2O_3^{2-}$) is reduced to H_2S ($S°$-H_2 autotrophy, carried out by *Pyrodictium* and *Ignicoccus*). *Ignicoccus islandicus* produces up to 25 μmol of H_2S per ml of culture medium at the end of the log phase and *Ignicoccus pacificus* is able to convert 10 g of elemental sulfur (dissolved in 1 liter of culture medium) quantitatively to H_2S within two days, although cell propagation stops after about 24 hours (H. Huber, unpublished observation). Besides thiosulfate, *Pyrolobus fumarii* can use nitrate or oxygen as electron acceptor, forming either ammonium or water as the metabolic end product (Blöchl et al., 1997). Also, CO_2 serves as sole carbon source for all autotrophs. *Pyrodictium* contains ribulose 1,5-bisphosphate carboxylase activity, indicating the presence of the Calvin cycle, while no significant activities for the key enzymes of the known three CO_2-fixation pathways have been detected in *Ignicoccus* (M. Hügler et al., 2003).

Two different forms of respiration are found within the order Desulfurococcales. Sulfur respiration by using various organic substrates occurs within the genera *Desulfurococcus*, *Stetteria* and *Thermodiscus*. Usually complex organic compounds such as beef extract, peptone, yeast extract, casamino acids, carbohydrates or peptides serve as substrates for growth. Additionally, H_2S and CO_2 are formed as end products; in the case of *Stetteria*, acetate and ethanol were also detected. Besides elemental sulfur, this organism can use thiosulfate as an alternative electron acceptor (Jochimsen et al., 1997).

Aerobic respiration of various complex proteinaceous compounds (e.g., yeast extract, peptone and tryptone) is carried out by *Aeropyrum pernix*. Surprisingly, its growth is highly stimulated in the presence of thiosulfate, and no H_2S could be detected, indicating that the organism has a thiosulfate-oxidizing system for energy generation (Sako et al., 1996).

Desulfurococcus, *Staphylothermus*, *Thermosphaera*, *Hyperthermus* and *Sulfophobococcus* grow, depending on the strain, by fermentation of a broad variety of complex organic compounds like yeast extract, peptides, amino acids, starch, glycogen or pectin. Possible fermentation products include H_2, CO_2, acetate, isovalerate, *n*-butanol, propionic acid and phenylacetic acid (Zillig et al., 1982; Zillig et al., 1990; Fiala et al., 1986; Huber et al., 1998). Although the name of the family and order implies the opposite, growth of *Thermosphaera aggregans* and *Sulfophobococcus zilligii* is completely inhibited in the presence of elemental sulfur (Hensel et al., 1997; Huber et al., 1998).

The enzymes involved in acetate formation were studied in *Desulfurococcus amylolyticus* and *Hyperthermus butylicus* (Schönheit and Schäfer, 1995). They use, as do all other Archaea, one enzyme: acetyl-CoA synthetase (ADP-forming). In *Desulfurococcus* and *Staphylothermus*, extracellular proteases have been isolated and characterized (Leuschner and Antranikian, 1995).

Genetics

The sequence of the whole genome of *Aeropyrum pernix* was published a few years ago (Kawarabayashi et al., 1999). The genome is 1,670 kb and in total nearly 2,700 open reading frames (ORFs). *Aeropyrum pernix* was the first Crenarchaeote and first aerobic member of the Archaea for which the complete genome sequence has been determined. It is still the only member of the Desulfurococcales whose complete genome is sequenced. In the first publication, more than 50% of the ORFs showed no significant similarity to sequences in databases. One single 16S–23S rRNA operon was identified, while two 5S rRNA genes and 47 tRNA genes were found. Furthermore, fewer "eukaryotic genes" were found than in the sequenced Euryarchaeotes (Faguy and Doolittle, 1999). Sequence comparisons among the assigned ORFs suggested that a considerable number of ORFs were generated by sequence duplication. The sequence data are available on the Aeropyrum pernix K1 data base (http://www.mild.nite.go.jp).

Ecology

Representatives of the Desulfurococcales thrive in acidic to slightly alkaline biotopes with original temperatures from at least 60°C to more than 100°C. They are typical inhabitants of high-temperature continental solfataric areas and of submarine hydrothermal systems and have been isolated from such biotopes nearly worldwide. They often share these habitats with other thermophilic and hyperthermophilic organisms, including members of the orders Sulfolobales, Thermoproteales, Thermotogales, and Aquificales. They can be enriched from the anaerobic zone of solfataric soils, which are usually black-colored due to the presence of ferrous sulfide. Depending on the volcanic activity and exhalation of reducing volcanic gases (e.g., H_2S and H_2), this zone starts at a depth of a few centimetres up to about 50 cm below the oxidized layers (Huber et al., 2000b). In marine hydrothermal systems, members of the Desulfurococcales have been found in all depths, from shallow to deep-sea biotopes. Usually they thrive in heated sediments or directly in the chimney walls of "black smokers."

In their ecosystems, the Desulfurococcales can function as primary producers and/or as consumers of organic material. For primary producers (e.g., *Ignicoccus*, *Pyrodictium* and *Pyrolobus*) elemental sulfur, thiosulfate, nitrate, oxygen and sulfite are suitable electron acceptors for growth, and molecular hydrogen serves as the electron donor. The end product of their metabolism is H_2S, H_2O or ammonium; in addition the organic cell material is formed. The obligate heterotrophs consume organic matter by growth via different types of respiration or fermentation. For anaerobic respiration, sulfur and sulfurous compounds are used as electron acceptors, and H_2S is formed as one metabolic end product besides different organic molecules. Fermentation occurs with several compounds (e.g., proteins), and a great variety of organic acids and alcohols are formed by the different strains, making these Desulfurococcales important members of the carbon cycle in high temperature habitats.

Very recently, a member of the Desulfurococcales (*Ignicoccus* strain KIN4/I) was identified as a symbiotic partner of a representative of a novel phylum of Archaea, "*Nanoarchaeum equitans*" (Huber et al., 2002).

Although several 16S rDNA sequences related to the known Crenarchaeota were obtained from low temperature biotopes, e.g., in freshwater and marine sediments, terrestrial soils, anaerobic digesters, deep subsurface paleosols (for an overview, see Phylogenetic and Ecological Perspectives on Uncultured Crenarchaeota and Korarchaeota in this Volume), none of these organisms had been cultivated so far and nothing is known about their morphology or physiology. In addition, all these environmental sequences are very distantly related to members of the Desulfurococcales.

Biotechnology

In general, heat stable enzymes from hyperthermophiles are of increasing interest for industrial applications and research. Besides polymer-degrading extracellular enzymes (e.g., amylases, lipases and proteases), metabolic enzymes may serve as potent biocatalysts in technological processes.

Acknowledgments. The authors thank G. Gmeinwieser for technical support, R. Rachel for electron microscopy, and Brian Hedlund for critically reading the manuscript.

Literature Cited

Aoshima, M., Y. Nishibe, M. Hasegawa, A. Yamagishi, and T. Oshima. 1996a. Cloning and sequencing of a gene encoding 16S ribosomal RNA from a novel hyperthermophilic archaebacterium NC12. Gene 180:183–187.

Aoshima, M., A. Yamagishi, and T. Oshima. 1996b. Eubacteria-type isocitrate dehydrogenase from an Archaeon: Cloning, sequencing, and expression of a gene encoding isocitrate dehydrogenase from a hyperthermophilic archaebacterium, Caldococcus noboribetus. Arch. Biochem. Biophys. 336:77–85.

Arab, H., H. Völker, and M. Thomm. 2000. Thermococcus aegaeicus sp. nov. and Staphylothermus hellenicus sp. nov., two novel hyperthermophilic archaea isolated from geothermally heated vents off Palaeochori Bay, Milos, Greece. Int. J. Syst. Evol. Microbiol. 50:2101–2108.

Balch, W. E., and R. S. Wolfe. 1976. New approach to the cultivation of methanogenic bacteria: 2- mercaptoethanesulfonic acid (HS-CoM)-dependent growth of Methanobacterium ruminantium in a pressurized atmosphere. Appl. Environ. Microbiol. 32:781–791.

Balch, W. E., G. E. Fox, L. J. Magrum, C. R. Woese, and R. S. Wolfe. 1979. Methanogens: Reevaluation of a unique biological group. Microbiol. Rev. 43:250–296.

Baumeister, W., U. Santarius, S. Volker, R. Dürr, G. Lembcke, and H. Engelhardt. 1990. The surface protein of Hyperthermus butylicus: Three-dimensional structure and comparison with other archaebacterial surface proteins. Syst. Appl. Microbiol. 13:105–111.

Blöchl E., R. Rachel, S. Burggraf, D. Hafenbradl, H. W. Jannasch, and K. O. Stetter. 1997. Pyrolobus fumarii, gen. and sp. nov., represents a novel group of Archaea, extending the upper temperature limit for life to 113°C. Extremophiles 1:14–21.

Bonch-Osmolovskaya, E. A., A. I. Slesarev, M. L. Miroshnichenko, T. P. Svetlichnaya, and V. A. Alekseev. 1988. Characteristics of Desulfurococcus amylolyticus n. spec. a new extremely thermophilic archaebacterium isolated from thermal springs of Kamchatka and Kunashir Island. Microbiologiya 57:94–101.

Brock, T. D., K. M. Brock, R. T. Belly, and R. L. Weiss. 1972. Sulfolobus: A new genus of sulfur-oxidizing bacteria living at low pH and high temperature. Arch. Microbiol. 84:56–68.

Burggraf, S., H. Huber, and K. O. Stetter. 1997. Reclassification of the crenarchaeal orders and families in accordance with 16S rRNA sequence data. Int. J. Syst. Bact. 47:657–660.

Cann, I. K. O., S. Ishino, N. Nomura, Y. Sako, and Y. Ishino. 1999. Two family B DNA polymerases from Aeropyrum

pernix, an aerobic hyperthermophilic crenarchaeote. J. Bacteriol. 181:5984–5992.

Dirmeier, R., M. Keller, G. Frey, H. Huber, and K. O. Stetter. 1998. Purification and properties of an extremely thermostable membrane-bound sulfur-reducing complex from the hyperthermophilic Pyrodictium abyssi. Eur. J. Biochem. 252:486–491.

Dirmeier, R., G. Hauska, and K. O. Stetter. 2000. ATP synthesis at 100°C by an ATPase purified from the hyperthermophilic archaeon Pyrodictium abyssi. FEBS Lett. 467:101–104.

Dürr, R., R. Hegerl, S. Volker, U. Santarius, and W. Baumeister. 1991. Three-dimensional reconstruction of the surface protein of Pyrodictium brockii: Comparing two image processing strategies. J. Struct. Biol. 106:181–190.

Faguy, D. M., and F. W. Doolittle. 1999. Genomics: Lessons from the Aeropyrum pernix genome. Curr. Biol. 9:883–886.

Fiala, G., K. O. Stetter, H. W. Jannasch, T. A. Langworthy, and J. Madon. 1986. Staphylothermus marinus sp. nov. represents a novel genus of extremely thermophilic submarine heterotrophic archaebacteria growing up to 98°C. Syst. Appl. Microbiol. 8:106–113.

Hensel, R., K. Matussek, K. Michalke, L. Tacke, B. J. Tindall, M. Kohlhoff, B. Siebers, and J. Dielenschneider. 1997. Sulfophobococcus zilligii gen. nov., spec. nov. a novel hyperthermophilic archaeum isolated from hot alkaline springs of Iceland. Syst. Appl. Microbiol. 20:102–110.

Horn, C., B. Paulmann, G. Kerlen, N. Junker, and H. Huber. 1999. In vivo observation of cell division of anaerobic hyperthermophiles by using a high-intensity dark-field microscope. J. Bacteriol. 181:5114–5118.

Huber, R., P. Stoffers, J. L. Cheminee, H. H. Richnow, and K. O. Stetter. 1990. Hyperthermophilic archaebacteria within the crater and open-sea plume of erupting Macdonald Seamount. Nature 345:179–182.

Huber, R., and K. O. Stetter. 1992. The order Thermoproteales. In: A. Balows, H. G. Trüper, M. Dworkin, W. Harder, and K.-H. Schleifer (Eds.) The Prokaryotes, 2nd ed. Springer-Verlag. New York, NY. 677–683.

Huber, R., S. Burggraf, T. Mayer, S. M. Barns, P. Rossnagel, and K. O. Stetter. 1995. Isolation of a hyperthermophilic archaeum predicted by in situ RNA analysis. Nature 367:57–58.

Huber, R., D. Dyba, H. Huber, S. Burggraf, and R. Rachel. 1998. Sulfur-inhibited Thermosphaera aggregans sp. nov., a new genus of hyperthermophilic archaea isolated after its prediction from environmentally derived 16S rRNA sequences. Int. J. Syst. Bact. 48:31–38.

Huber, H., S. Burggraf, T. Mayer, I. Wyschkony, R. Rachel, and K. O. Stetter. 2000a. Ignicoccus gen. nov., a novel genus of hyperthermophilic, chemolithoautotrophic Archaea, represented by two new species, Ignicoccus islandicus sp. nov. and Ignicoccus pacificus. sp. nov. Int. J. Syst. Evol. Microbiol. 50:2093–2100.

Huber, R., H. Huber, and K. O. Stetter. 2000b. Towards the ecology of hyperthermophiles: biotopes, new isolation strategies and novel metabolic properties. FEMS Microbiol. Rev. 24:615–623.

Huber, H., and K. O. Stetter. 2001a. Order I: Thermoproteales. In: G. Garrity (Ed.) Bergey's Manual of Systematic Bacteriology, 2nd ed. Springer-Verlag. New York, NY. 1:170.

Huber, H., and K. O. Stetter. 2001b. Order II: Desulfurococcales. In: G. Garrity (Ed.) Bergey's Manual of Systematic Bacteriology, 2nd ed. Springer-Verlag. New York, NY. 1:179–180.

Huber, H., M. J. Hohn, R. Rachel, T. Fuchs, V. C. Wimmer, and K. O. Stetter. 2002. A new phylum of Archaea represented by a nanosized hyperthermophilic symbiont. Nature 417:63–67.

Hügler, M., H. Huber, K. O. Stetter, and G. Fuchs. 2003. Autotrophic CO$_2$ Fixation Pathways in Archaea (Crenarchaeota). Arch Microbiol 179(3):160–173.

Jochimsen, B., S. Peinemann-Simon, H. Völker, D. Stüben, R. Botz, P. Stoffers, P. R. Dando, and M. Thomm. 1997. Stetteria hydrogenophila, gen. nov. and sp. nov., a novel mixotrophic sulfur-dependent crenarchaeote isolated from Milos, Greece. Extremophiles 1:67–73.

Kawarabayashi, Y., Y. Hino, H. Horikawa, S. Yamazaki, Y. Haikawa, K. Jin-no, M. Takahashi, M. Sekine, S. Baba, A. Ankai, H. Kosugi, A. Hosoyama, S. Fukui, Y. Nagai, K. Nishijima, H. Nakazawa, M. Takamiya, S. Masuda, T. Funahashi, T. Tanaka, Y. Kudoh, J. Yamazaki, N. Kushida, A. Oguchi, K. Aoki, K. Kubota, Y. Nakamura, N. Nomura, Y. Sako, and H. Kikuchi. 1999. Complete genome sequence of an aerobic hyperthermophilic crenarchaeon, Aeropyrum pernix K1. DNA Res. 6:83–101 and 145–152.

Kjems, J., N. Larsen, J. Z. Dalgaard, R. A. Garrett, and K. O. Stetter. 1992. Phylogenetic relationships amongst the hyperthermophilic archaea determined from partial 23S rRNA gene sequences. Syst. Appl. Microbiol. 15:203–208.

König, H., P. Messner, and K. O. Stetter. 1988. The fine structure of the fibers of Pyrodictium occultum. FEMS Microbiol. Lett. 49:207–212.

Kovacs, K. L., and G. Rakhely. 1996. Plating hyperthermophilic Archaea on solid surface. Analyt. Biochem. 243:181–183.

Larsen, N., H. Leffers, J. Kjems, and R. A. Garrett. 1986. Evolutionary divergence between the ribosomal RNA operons of Halococcus morrhuae and Desulfurococcus mobilis. Syst. Appl. Microbiol. 7:49–57.

Leuschner, C., and G. Antranikian. 1995. Heat-stable enzymes from extremely thermophilic and hyperthermophilic microorganisms. World J. Microbiol. Biotechnol. 11:95–114.

Ludwig, W., and O. Strunk. 2001. ARB: A software environment for sequence data.

Mai, B. 1998a. In vitro Untersuchungen zum extrazellulären Netzwerk von Pyrodictium abyssi TAG11 (PhD thesis). University of Regensburg. Regensburg, Germany.

Mai, B., G. Frey, R. V. Swanson, E. J. Mathur, and K. O. Stetter. 1998b. Molecular cloning and functional expression of a protein-serine/threonine phosphatase from the hyperthermophilic archaeon Pyrodictium abyssi TAG11. J. Bacteriol. 180:4030–4035.

Mayr, J., A. Lupas, J. Kellermann, C. Eckerskorn, W. Baumeister, and J. Peters. 1996. A hyperthermostable protease of the subtilisin family bound to the surface layer of the archaeon Staphylothermus marinus. Curr. Biol. 6:739–749.

Morii, H., H. Yagi, H. Akatsu, N. Nomura, Y. Sako, and Y. Koga. 1999. A novel phosphoglycolipid archaetidyl (glucosyl) inositol with two sesterterpanyl chains from the aerobic hyperthermophilic archaeon Aeropyrum pernix K1. Biochim. Biophys. Acta 1436:426–436.

Nomura, N., Y. Sako, and A. Uchida. 1998. Molecular characterization and postselecting fate of three introns

within the single rRNA operon of the hyperthermophilic archaeon Aeropyrum pernix K1. J. Bacteriol. 180:3635–3643.

Peters, J., M. Nitsch, B. Kühlmorgen, R. Golbik, A. Lupas, J. Kellermann, H. Engelhardt, J.-P. Pfander, S. Müller, K. Goldie, A. Enge, K. O. Stetter, and W. Baumeister. 1995. Tetrabrachion: A filamentous archaebacterial surface protein assembly of unusual structure and extreme stability. J. Molec. Biol. 245:385–401.

Phipps, B. M., A. Hoffmann, K. O. Stetter, and W. Baumeister. 1991. A novel ATPase complex selectively accumulated upon heat shock is a major cellular component of thermophilic archaebacteria. EMBO J. 10:1711–1722.

Phipps, B. M., D. Typke, R. Hegerl, S. Volker, A. Hoffmann, K. O. Stetter, and W. Baumeister. 1993. Structure of a molecular chaperone from a thermophilic archaebacterium. Nature 361:475–477.

Pihl, T. D., and R. J. Maier. 1991. Purification and characterization of the hydrogen uptake hydrogenase from the hyperthermophilic archaebacterium Pyrodictium brockii. J. Bacteriol. 173:1839–1844.

Pihl, T. D., L. K. Black, B. A. Schulman, and R. J. Maier. 1992. Hydrogen-oxidizing electron transport components in the hyperthermophilic archaebacterium Pyrodictium brockii. J. Bacteriol. 174:137–143.

Pley, U., J. Schipka, A. Gambacorta, H. W. Jannasch, H. Fricke, R. Rachel, and K. O. Stetter. 1991. Pyrodictium abyssi sp. nov. represents a novel heterotrophic marine archaeal hyperthermophile growing at 110°C. Syst. Appl. Microbiol. 14:245–253.

Prokofeva, M. I., M. L. Miroshnichenko, N. A. Kostrikina, N. A. Chernyh, B. B. Kuznetsov, T. P. Tourova, and E. A. Bonch-Osmolovskaya. 2000. Acidilobus aceticus gen. nov., sp. nov., a novel anaerobic thermoacidophilic archaeon from continental hot vents in Kamchatka. Int. J. Syst. Bact. 50:2001–2008.

Rachel, R. 1999. Fine structure of hyperthermophilic prokaryotes. In: J. Seckbach (Ed.) Enigmatic Microorganisms and Life in Extreme Environments. Kluwer. Dordrecht, The Netherlands. 277–289.

Rachel, R., I. Wyschkony, S. Riehl, and H. Huber. 2002. The ultrastructure of Ignicoccus: Evidence for a novel outer membrane and for intracellular vesicle budding in an archaeon. Archaea 1:9–18.

Rieger, G., R. Rachel, R. Hermann, and K. O. Stetter. 1995. Ultrastructure of the hyperthermophilic archaeon Pyrodictium abyssi. J. Struct. Biol. 115:78–87.

Sako, Y., N. Nomura, A. Uchida, Y. Ishida, H. Morii, Y. Koga, T. Hoaki, and T. Maruyama. 1996. Aeropyrum pernix gen. nov., sp. nov., a novel aerobic hyperthermophilic archaeon growing at temperatures up to 100°C. Int. J. Syst. Bact. 46:1070–1077.

Schönheit, P., and T. Schäfer. 1995. Metabolism of hyperthermophiles. World J. Microbiol. Biotechnol. 11:26–57.

Stetter, K. O. 1982. Ultrathin mycelia-forming organisms from submarine volcanic areas having an optimum growth temperature of 105°C. Nature 300:258–260.

Stetter, K. O., H. König, and E. Stackebrandt. 1983. Pyrodictium gen. nov., a new genus of submarine disc-shaped sulphur reducing archaebacteria growing optimally at 105°C. Syst. Appl. Microbiol. 4:535–551.

Stetter, K. O., and W. Zillig. 1985. Thermoplasma and the thermophilic sulfur-dependent archaebacteria. In: C. Woese, and R. S. Wolfe (Eds.) The Bacteria. Academic Press. New York, NY. 8:100–201.

Stetter, K. O. 1986. Diversity of extremely thermophilic archaebacteria. *In:* T. D. Brock (Ed.) Thermophiles: General, Molecular, and Applied Microbiology. Wiley and Sons. New York, NY. 39–74.

Stetter, K. O. 1989. Order III: Sulfolobales. *In:* J. T. Staley, M. P. Bryant, N. Pfennig, and J. G. Holt (Eds.) Bergey's Manual of Systematic Bacteriology. William and Wilkins. Baltimore, MD. 2250.

Stetter, K. O. 1995. Microbial life in hyperthermal environments. ASM News 61:285–290.

Stetter, K. O. 1999. Hyperthermophiles: Isolation, classification, and properties. *In:* K. Horikoshi and W. D. Fasman (Eds.) Extremophiles: Microbial Life in Extreme Environments. Wiley. New York, NY. 1–24.

Stetter, K. O. 2001. Genus VII: Thermodiscus. *In:* G. Garrity (Ed.) Bergey's Manual of Systematic Bacteriology, 2nd ed. Springer-Verlag. New York, NY. 1:189–190.

Stetter, K. O. 2002. Mikroorganismen an extremen Standorten. *In:* Rundgespräche der Kommision für Ökologie, Bd. 23: Bedeutung der Mikroorganismen für die Umwelt. Pfeil. Munich, Germany. 123–136.

Thomas, N. A., S. L. Bardy, and K. F. Jarrell. 2001. The archaeal flagellum: A different kind of prokaryotic motility structure. FEMS Microbiol. Rev. 25:147–174.

Wildhaber, I., U. Santarius, and W. Baumeister. 1987. Three-dimensional structure of the surface protein of Desulfurococcus mobilis. J. Bacteriol. 169:5563–5568.

Woese, C. R., O. Kandler, and M. L. Wheelis. 1990. Towards a natural system of organisms: Proposal for the domains Archaea, Bacteria and Eukarya. Proc. Natl. Acad. Sci. USA 87:4576–4579.

Zillig, W., K. O. Stetter, D. Prangishvili, W. Schäfer, S. Wunderl, D. Janekovic, I. Holz, and P. Palm. 1982. Desulfurococcaceae, the second family of the extremely thermophilic, anaerobic, sulfur-respiring Thermoproteales. Zbl. Bakt. Hyg., I. Abt. Orig. C 3:304–317.

Zillig, W., I. Holz, D. Janekovic, H.-P. Klenk, E. Imsel, J. Trent, S. Wunderl, V. H. Forjaz, R. Coutinho, and T. Ferreira. 1990. Hyperthermus butylicus, a hyperthermophilic sulfur-reducing archaebacterium that ferments peptides. J. Bacteriol. 172:3959–3965.

Zillig, W., I. Holz, and S. Wunderl. 1991. Hyperthermus butylicus gen. nov., sp. nov., a hyperthermophilic, anaerobic, peptide-fermenting, facultatively H₂S-generating archaebacterium. Int. J. Syst. Bact. 41:169–170.

Prokaryotes (2006) 3:69–81
DOI: 10.1007/0-387-30743-5_5

CHAPTER 5

The Order Thermococcales

COSTANZO BERTOLDO AND GARABED ANTRANIKIAN

Among the hyperthermophilic archaea, representatives of order Thermococcales form the most numerous group to date. Members of this group are the most frequently isolated hyperthermophiles. They are heterotrophic and as such regarded as the major constituents of organic matter within marine hot water ecosystems (Canganella et al., 1997). They belong to the branch of Euryarchaeota that contains the methanogens, the genus Thermoplasma, and the extremely halophilic archaea. The Thermococcales order is actually represented by three genera: Pyrococcus (Fiala and Stetter, 1986), Thermococcus (Achenbach-Richter et al., 1988) and the newly described Paleococcus (Takai et al., 2000). Phylogenetic analysis based on 16 rDNA sequences indicates that the Paleococcus strains are members of an ancient lineage of Thermococcales that diverged prior to the formation of the genera Pyrococcus and Thermococcus (Takai et al., 2000). These three genera include at present 38 species: 2 belonging to the genus Paleococcus, 6 belonging to the genus Pyrococcus, and 30 to the genus Thermococcus. The optimal growth temperature is 95–100°C for members of the genus Pyrococcus and 80–90°C for those of the genus Thermococcus. Pyrococcus strains have been isolated only from marine hydrothermal vents, whereas species belonging to the genus Thermococcus have been isolated also from terrestrial fresh water (Ronimus et al., 1997), marine solfataric ecosystems, deep-sea hydrothermal vents (Stetter, 1996) and offshore oil wells (Takahata et al., 2001). Representatives of the order Thermococcales have coccoid cells with or without flagella; they are obligate anaerobic organotrophic thermophiles with a fermentative metabolism using peptides, polysaccharides, or other sugars as carbon sources. Elemental sulfur is either stimulatory or necessary for the growth of these microorganisms. Molecular hydrogen that is produced during fermentation reduces elemental sulfur to H_2S (Schönheit and Schäfer, 1995). Most of Thermococcales are neutrophiles growing optimally at pH 6.0–7.0; only the two species Thermococcus alcaliphilus (Keller et al., 1995) and T. acidoam-inovorans (Dirmeier et al., 1998) are able to grow optimally at pH 9.0.

Ecology

The Thermococcales are generally found in natural biotopes that are typical for thermophilic microorganisms. They were originally discovered in terrestrial and submarine hot vents and they were then found also in deep subsurface environments. For example, Thermococcus celer (Zillig et al., 1983), T. litoralis (Britton et al., 1995) and Pyrococcus sp. were discovered in an offshore oil production platform in the North Sea (Stetter, 1996), and T. litoralis (Neuner et al., 1990) was isolated from a continental oil well (Paris Basin, France). This fact probably indicates the indigenous origin of hyperthermophilic archaea in the deep subsurface biosphere. Another strain, T. sibiricus, was isolated from a high temperature oil reservoir in Western Siberia, a location far remote from both the ocean and volcanic areas. The sites where these microorganisms are found can appear to be unusual, but these microorganisms might have been deposited with the original sediment and survived over geologic time by metabolizing buried organic matter (Miroshnichenko et al., 1998; Miroshnichenko et al., 2001).

On the other hand, members of the genus Pyrococcus seem to be isolated from only marine environments and belong to a particular ecological niche (Morikawa et al., 1994). As a result of the hydrostatic pressure at deep-sea vents, geothermally heated seawater remains liquid at temperatures up to 400°C (Stetter, 1996; Duffaud et al., 1998). When the hot fluid that is enriched in polymetal sulfides and gasses is mixed with cold (2°C) seawater, minerals precipitate and form so-called "chimneys," or "smokers" (Fiala and Stetter, 1986). The temperature of the hot fluid that is emitted from the chimney can even reach 300°C (Holden and Baross 1993; Kwak et al., 1995; Gonzalez et al., 1998; Cambon-Bonavita et al., 2003). Members of the genus Pyrococcus are common inhabitants of this ecologic environment together with other hyper-thermophiles belonging to the genera of *Pyrodictium*, *Pyrobaculum*, *Pyrolobus* and *Methanopyrus*. The two strains *Pyrococcus abyssi* and *Paleococcus horikoshii* were isolated from deep-sea vents in the North Fiji Basins, South Pacific Ocean (Erauso et al., 1993), and Okinawa Trough, Japan (Gonzalez et al., 1998), respectively. *Pyrococcus furiosus* and *Pyrococcus woesei* have been discovered along the marine solfataras of the island Vulcano (Italy). The novel barophilic archaeon belonging to the genus *Palaeoococcus* was collected from a deep-sea hydrothermal vent chimney at the Myojin Knoll in the Ogasawara-Bonin-Arc, Japan (Takai et al., 2000).

Isolation

For the isolation of hyperthermophilic archaea, complex anaerobic media are usually used which are prepared according to the Hungate technique (Blamey et al., 1999). After taking samples with special syringes, they should be inoculated into serum vials (25–100 ml) or Hungate tubes (10 ml) containing media under an anaerobic atmosphere of CO_2 or N_2. The 1000 ml of medium should also contain 0.2–1% of elemental sulfur and 10–100 μl of resazurin (1 mg/ml) as a redox indicator. Before inoculation, the medium should be reduced with sodium sulfide (Na_2S; 0.05%). The sample may be transported at ambient temperature and kept at 4°C. Most Thermococcales may survive for a long time at cold or ambient temperature. Using special equipment, the samples taken from the deep sea can be transferred to the laboratory under pressure, and enrichment cultures can be prepared under high pressure and temperature. Samples from the deep sea can be collected by employing a manned submersible. In many cases and to increase the cell concentration, marine samples have to be concentrated many fold using a sterilized cross-flow device equipped with a 50-kDa cut-off membrane. For the growth and isolation of marine Thermococcales, the medium MB (Bacto Marine Broth, Difco) can be used and contains, in general, the following components per liter: bactopeptone, 5 g; bactoyeast extract, 1 g; Fe(III) citrate, 0.1 g; NaCl, 19.45 g; $MgCl_2$, 5.9 g; $NaSO_4$, 3.24 g; $CaCl_2$, 1.8 g; KCl, 0.55 g; Na_2CO_3, 0.16 g; KBr, 0.08 g; $SrCl_2$, 34 mg; H_3BO_3, 22 mg; Na-silicate, 4 mg; NaF, 2.4 mg; $(NH_4)NO_3$, 1.6 mg; and Na_2HPO_4, 8 mg; the final pH is adjusted to 7.6–7.8. If the complete medium from Difco is used, 37.4 g must be added to 1 liter of water. It is important to filter the medium using normal filter paper to prevent the possible precipitation of iron. The medium that is normally

prepared for the cultivation of *Pyrococcus* species is as follows: KH_2PO_4, 0.5 g; $NiCl_2 \cdot 6H_2O$, 2 mg; trace element solution (Balch et al., 1979), 10 ml; sulfur, 30 g; yeast extract, 1 g; peptone, 5 g; and resazurin, 1 mg. The pH has to be adjusted to 6.4–6.5. After boiling the above-mentioned media under N_2 atmosphere, they should be cooled on ice and transferred under N_2 atmosphere to 10-ml Hungate tubes and sterilized by autoclaving. Before use, the medium must be reduced with a sterile neutral solution of $Na_2S \cdot 9H_2O$ (0.5 g/liter).

To obtain single colonies on plates, the same media can be used including Gelrite (0.8%), and the plates should be stored in anaerobic jars under a N_2 atmosphere at the desirable temperature.

Cultivation

Thermococcales are receiving increasing interest from academia and industry because they provide a unique source of stable biocatalysts and other products such as archaeal lipids and compatible solutes. However, until recently only low cell concentrations (10^7–10^8 cells/ml) could be obtained, making application studies very difficult. The main reason for this has to be ascribed to difficulties related to the production and purification of large quantities of biocatalysts and cell components. Special equipment is also needed to cultivate some strains under high pressure and temperature. Innovative bioreactor design to improve biomass yield is required. Because the accumulation of toxic compounds is thought to be responsible for low biomass yields, dialysis fermentation of *Paleococcus woesei* (Blamey et al., 1999) and *P. furiosus* has been performed for effective removal of low-molecular-mass components from the fermentation broth. Unlike many other heterotrophic hyperthermophiles, significant growth of *P. furiosus* is not dependent on the presence of elemental sulfur (S^0). When dialysis membrane reactors were applied, a dramatic increase in cell yields was achieved (Krahe et al., 1996). The cultivation of the hyperthermophilic archaeon *Pyrococcus furiosus* (growth at 90°C) resulted in cell yields of 2.6 g $\cdot$ liter^{-1}. For *P. furiosus* the optimum stirrer speed was 1,800 rpm, and neither hydrogen nor the metabolic products were found responsible for the comparatively low cell yield. The fermentation processes can be scaled-up from 3 liters to over 30 liters (up to 300 liters). The pilot plant scale offers the possibility of transferring the fermentation performance to a larger industrial scale. In recent experiments it was shown that even the results of the 1-liter dialysis reactor can be reproduced in 30-liter reactors using

Fig. 1. Transmission electron micrographs of a) *Thermococcus aggregans* (from Canganella et al., 1998), b) *Thermococcus siculi* (from Grote et al., 1999), negatively stained with 0.3% phosphotungstic acid.

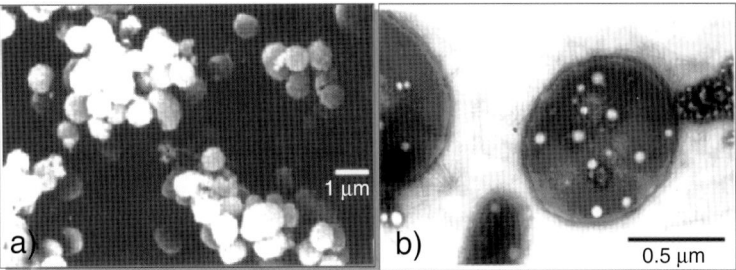

external dialysis modules. On the other hand, a method was described for growing *P. furiosus* in 600-liter fermentors (Verhagen et al., 2001), which resulted in the production of 500 g of cells (wet weight).

As already mentioned, Thermococcales, isolated from deep sea vents, are able to grow not only at high temperature but also at hydrostatic pressure around 20–30 MPa. Owing to practical difficulties, the growth and the metabolism of barophilic Thermococcales have been poorly investigated. It has been demonstrated that the upper growth temperature is extended at least 3°C when cells of *P. abyssi* are cultivated at 20 MPa. Similar higher thermotolerance and upward shift in the optimal temperature were observed also for *P. endeavori* at 22 MPa (Erauso et al., 1993). More detailed studies with *Thermococcus peptonophilus* (Canganella et al., 1997) and *T. barophilus* (Marteinsson et al., 1999) have also demonstrated that both species are barophilic. In fact, in both cases the growth rate at the optimal growth temperature was higher under in situ hydrostatic pressure than under lower pressure.

Identification and Morphology

In general, in phase contrast microscopy, all known species of Thermococcales appear as spherical cells, mostly as diploid forms constricted to various degrees owing to the duration of the division process throughout the whole generation time. When the cells of Thermococcales are compared to those of Sulfolobales and certain Thermoproteales, the Thermococcales cells are more round, slightly irregular, and their size varies from 0.5 to 2.5 μm. In the final stage of division of Thermoccocales cells, the two daughter cells are connected by a thin string of cytoplasm enclosed by a membrane and S-layer. In general, cells show a Gram-negative reaction. Unlike all other strains, *T. aggregans* cells form chains, and cell aggregates are particularly characteristic of this strain. However, the formation of aggregates is not a general characteristic but occurs only after

cultivation of the strain on yeast-tryptone extract (Canganella et al., 1998; Fig. 1a).

Electron microscopy of Thermococcales cells reveals the presence of monopolar polytrichous flagella in most of the species. For example, *P. furiosus* (Fiala and Stetter, 1986), *P. woesei* (Zillig et al., 1987), *P. abyssi* (Erauso et al., 1993), *P. horikoshii* (Gonzalez et al., 1998), *T. acidoaminovorans* (Dirmeier et al., 1998), *T. peptonophilus* (Gonzalez et al., 1995) and *T. chitinophagus* (Huber et al., 1996) are all characterized by the presence of a tuft of polar flagella. Nevertheless, this is not a rule. In other strains, the flagella have a fimbriate arrangement as observed for *T. siculi* (Grote et al., 1999; Fig. 1b). In the case of *T. alcaliphilus* (Keller et al., 1995), only a single flagellum is present and in the case of *T. sibiricus*, the flagella are completely absent (Miroshnichenko et al., 2001).

The cytoplasmic membrane (5–10 nm) of *T. chitonophagus* is covered by a bilayered cell envelope with an inner periplasmic space (15–20 nm) and an external, densely stained layer (5 nm), probably corresponding to a surface-layer protein (Huber et al., 1996). The described organization (structure) of the cytoplasmic membrane is more or less similar among the other members of Thermococcales. Thermococcales have the typical archaeal cytoplasmic membrane lipids (ether lipids), and some members also have simple diether lipids (mainly made up of one or two phospholipids) and trace amounts of tetraethers. The presence of two rare acyclic and cyclic glycerol diphytanyl tetraethers has been reported in *T. chitinophagus* (Huber et al., 1996). However, more detailed studies have been carried out on the lipid structure of *T. hydrothermalis* isolated from a deep-sea hydrothermal vent. On the basis of acid methanolysis and spectroscopic studies, the polar lipids (amounting to 4.5% [w/w] of the dry cells) included diphytanyl glycerol diethers and dibiphytanyldiglycerol tetraethers in a 45:55 ratio. No cyclopentane ring was present in the tetraethers. From the neutral lipids (0.4% [w/w] of the dry cells), four di- and tri-unsaturated acyclic tetraterpenoid hydrocarbons and low amounts of di- and tetraethers (occurring in free form) were identified. All are structurally related to lycopane. The presence of these hydrocarbons provides some evidence that lycopane, widely distributed in oceans, could be derived, at least partially, from the hydrocarbons synthesized by some hyperthermophilic archaea. Analysis of the uninoculated culture medium indicates that fatty acid derivatives and some steroid and triterpenoid compounds identified in the lipidic extract of the archaea probably originate from the culture medium (Lattuati et al., 1998).

Physiology and Metabolism

Most of Thermococcales species are obligate anaerobic organothrophic thermophiles that prefer to utilize polymeric substrates like proteins and carbohydrates (preferentially oligo- and polysaccharides) as carbon and energy sources. Elemental sulfur is required in some cases for the growth and is used as an electron acceptor to remove reducing equivalents that are produced during fermentation. However, these physiological characteristics are not the rule for all members of the order Thermococcales and some differences can be observed in the three genera *Thermococcus*, *Pyrococcus* and *Paleococcus* (Selig et al., 1997; Table 1).

Species of the genus *Pyrococcus* are heterotrophic and sulfur-reducing microorganisms. Growth is observed when complex organic substrates such as yeast extract, peptone, tryptone, meat extract, and peptides are used (Fiala and Stetter, 1986). *Pyrococcus furiosus* and *P. woesei* can also grow on various carbohydrates such as starch (Biller et al., 2002), glycogen, pullulan (Blumentals et al., 1990; Costantino et al., 1990), cellobiose and pyruvate (Kengen et al., 1993). *Pyrococcus glycovorans* grows on proteinaceous substrate and different carbohydrates and, in addition, it is able to use glucose as carbon source, a feature that appears to be unique in hyperthermophiles (Barbier et al., 1999). *Pyrococcus abyssi* and *P. horikoshii* are unable to grow on carbohydrates (Erauso et al., 1993; Gonzalez et al., 1998). Unlike the results reported in the literature, in many cases some members of Thermococcales can even grow on modified media in the absence of sulfur. Members belonging to the genus *Thermococcus* appear to grow mainly on media containing complex proteinaceous substrates such as yeast extract or tryptone as the sole carbon and energy source. Some species like *Thermococcus peptonophilus*, *T. alcaliphilus* and *T. zilligii* are unable to grow on amino acid mixtures (Gonzalez et al., 1995; Keller et al., 1995). There are few reports on the successful cultivation of hyperthermophiles on defined minimal media. A minimal defined medium has been reported for the cultivation of *Thermococcus acidoaminovorans*, which uses defined amino acids as the sole energy source (Dirmeier et al., 1998). *Thermococcus aggregans* and *T. aegaeicus* strains instead are able to use carbohydrates as substrates. *Thermococcus aggregans* is able to grow on starch and maltose and *T. aegaeicus* also utilizes starch but under a N_2/CO_2 atmosphere (Canganella, et al., 1998). Significant growth on maltose and slow growth on cellobiose were observed for *T. hydrothermalis* (Godfroy et al., 1997; Gruyer et al., 2002) and *T. fumicolans* (Godfroy et al., 1996). Interestingly, *T. chitinophagus* represents the only species able to grow on chitin as a carbon source (Huber et al., 1996). With the exception of *T. stetteri* (Miroshnichenko et al., 1989), *T. profundus* (Kobayashi et al., 1994; Kwak et al., 1995) and *T. waiotapuoensis* (Gonzalez et al., 1999), *Thermococcus* strains are stimulated by addition of sulfur, but sulfur is not absolutely required (Arab et al., 2000). The recently identified archaeon *Palaeococcus* sp. possesses most of the morphological and physiological properties typical of Thermococcales. It, however, displays an absolute requirement for either elemental sulfur or ferrous iron (Fe^{2+}). The requirement for iron represents an ancient characteristic of early microbial metabolism, in light of geochemical data suggesting the properties of habitats occupied by microorganisms belonging to the order Thermococcales (Takai et al., 2000).

The generation time of the members of Thermococcales is the shortest among Archaea. The time range is between 25 min for *T. peptonophilus* and 70 min for *T. stetteri*. The doubling time of the strains belonging to the members of the genus *Pyrococcus* is around 30–35 min (Table 1).

Sugar and Peptide Degradation Pathways

As already described, many Thermococcales show heterotrophic growth on a variety of carbohydrates. This suggests that oligosaccharides with varying degrees of polymerization are transported into the cell and are subsequently hydrolyzed to glucose. Various studies have focused both on the transport of the saccharides into the cell and on the pathways that are used to degrade the glucose (Schönheit and Schäfer, 1995). For *Thermococcus litoralis*, a transport system for both maltose and trehalose has been described that probably represents an ATP-binding cassette (ABC) transporter (Horlacher et al., 1998; Xavier et al., 1999). The trehalose-maltose binding protein, TMBP, and the ATPase subunit, MalK, have been functionally expressed in *Escherichia coli* (Greller et al., 1999; Greller et al., 2001; Diederichs et al., 2000). These binding

Table 1. Morphological and physiological characteristics of strains belonging to the order Thermococcales.

Species	G+C content (mol%)	Growth temperature (°C)		pH		NaCl concentration (%)		Carbon sources[a]	Growth on amino acids	Sulfur effects	References
		range	opt.	range	opt.	range	opt.				
Palaeococcus ferrophilus DMJ 10246(T)	53.5	60–88	83	4.0–8.0	6.0	n.d.	n.d.	Complex substrates	n.d.	E	Takai et al., 2000
Pyrococcus furiosus DSM 3638	38	70–103	100	5–9	7	0.5–5	2	Complex substrates, maltose, starch, pyruvate, and casamino acids	Yes	R	Fiala and Stetter, 1986
Pyrococcus abyssi CNCMI-1302	45	67–100	96	4–8.5	6.8	0.7–5	3	Complex substrates, maltose, starch, pyruvate, and casamino acids	Yes	E	Erauso et al., 1993
Pyrococcus woesei DSM 3773	37.5	70–105	100–103	n.d.	n.d.	n.d.	3	Yeast extract, tryptone, glycogen, and gellan	Yes	E	Zillig et al., 1987
Pyrococcus kodakaraeinsis DSM 3773	38	65–100	95	5–9	7	1–5	3	Complex substrates, and peptides	Yes	E	Morikawa et al., 1994
Pyrococcus endeawori (ES4)	55	80–110	98	4–8	7.0	n.d.	2.8	Casamino acids	Yes	E	Holden and Baross, 1993
Pyrococcus horikoshii JCM 9974	44	80–102	98	5–8	7.0	1–5	2.4	Complex substrates	Yes	E	Gonzalez et al., 1998
Pyrococcus glycovorans (A1585T)	47	75–104	95	2.5–9.5	7.5	2–6	3	Complex substrates, and glucose	Yes	E	Barbier et al., 1999
Thermococcus celer DSMZ 2476	57	Up to 93	88	n.d.	5.8	n.d	4	Peptides stimulated by sucrose	Yes	E	Zillig et al., 1983
Thermococcus litoralis DSMZ 5474	38	65–95	88	6.2–8.5	7.2	1.8–6.5	2.5	Peptides, and pyruvate	No	E	Neuner et al., 1990; Britton et al., 1995
Thermococcus stetteri DSMZ 5262	50	60–85	75	5.7–7.2	6.5	1–4	2.5	Starch, pectin, and peptides	No	R	Miroshnichenko et al., 1989
Thermococcus profundus DT5432 52.2	52.2	50–90	80	4.5–8.5	7.5	1–6	2	Pyruvate, starch, and maltose	n.d.	R	Kobayashi et al., 1994
Thermococcus peptonophilus JCM 9653	52	60–100	85	4–8	6	1–5	3	Peptides	n.d.	E	Gonzalez et al., 1995
Thermococcus aggregans DSMZ 10597	42	60–94	88	4.6–7.9	7	n.d.	2	Complex substrates, dextrose, and maltose	n.d.	E	Canganella et al., 1998
Thermococcus pacificus DSMZ 10394	53.3	70–95	80/88	6–8	6.5	1–5	2–3.5	Complex substrates	n.d.	E	Miroshnichenko et al., 1998

Table 1. *Continued*

Species	G+C content (mol%)	Growth temperature (°C)		pH		NaCl concentration (%)		Carbon sources	Growth on amino acids	Sulfur effects	References
		range	opt.	range	opt.	range	opt.				
Thermococcus guaymasiensis DSMZ 11113	46	56–90	88	5.6–8.5	7.2	1–5	2–3.5	Casein, dextrose, and maltose	n.d.	E	Canganell et al., 1998
Thermococcus gorgonarius DSMZ 10395	50.6	68–95	80/88	3.4–9	7	2–10	3	Complex substrate	n.d.	E	Miroshnichenko et al., 1998
Thermococcus hydrothermalis CNCM I1319	58	53–100	85	2–10	7	Sea salt	4	Casein, peptides, and maltose	Yes	E	Godfroy et al., 1997
Thermococcus zilligii DSMZ 2770	46.2	n.d.	75	n.d.	7.4	n.d.	25	Casein	n.d.	E	Ronimus et al., 1997
Thermococcus acidaminovorans DSMZ 11906	49	56–93	85	5–9.5	9	1–6	2–3	Peptides	Yes	E	Dirmeier et al., 1998
Thermococcus aegeicus DSMZ 12767	45.5	50–90	88	4–9	6	0.5–6.5	2.7	Starch	No	E	Arab et al., 2000
Thermococcus barophilus CNCMI 1946	37	75–95	85	4.5–9.5	7	1–4	3	Yeast extract, and peptone	No	E	Marteinsson et al., 1999
Thermococcus barossii DSMZ 9535	60	60–92	82.5	3–9	6.5/7.5	1–4	2	Tryptone, yeast extract, and malto-oligosaccharides			Duffaud et al., 1998
Thermococcus chitinophagus DSM 10152	46.5	60–93	85	3.5–9	6.7	0.8–8	2	Complex substrates, and chitin	No	E	Huber et al., 1996
Thermococcus siculi DSM 12349	55.8	50–93	85	5.0–9.0	7	0.1–0.4	0.2	Peptides	Yes	E	Grote et al., 1999
Thermococcus alcaliphilus DSM 10322	42.4	54–91	85	6.5–10.5	9.0	0.1–0.6	0.2–0.3	Peptides	Yes	E	Keller et al., 1995
Thermococcus waiotapuensis DSM 12768	50.4	60–90	85	5–8	7	Up to 13.9	5.4	Complex substrates, starch, maltose, and pyruvate	Yes	R	Gonzalez et al., 1999
Thermococcus fumiculans CIP 104690	54–55	73–103	90	4.5–9.5	8.5	0.6–4	1.3–2.6	Complex substrates, and pyruvate	Yes	E	Godfroy et al., 1996
Thermococcus sibiricus DSMZ 12597	38.4	40–88	78	5.8–9	7.5	0.5–7	1.8–2	Peptides	n.d.	E	Miroshnichenko et al., 2001
Thermococcus atlanticus CIP-107420T	49.8	70–95	85	4–9	7	1.5–4.6	30	Peptides	No	E	Cambon-Bonavita et al., 2003

Abbreviations: opt., optimal; n.d., no data available; E, elemental sulfur stimulates growth; and R, elemental sulfur is required for growth.
aComplex substrates: yeast extract, tryptone, peptone.

protein-dependent transport systems exhibit an unusually high affinity for the sugar, with a K_m in the submicromolar range. While glycolysis in thermophilic bacteria proceeds in a conventional way, glucose catabolism by Thermococcales (as well as in other hyperthermophilic archaea) differs from the canonical pathways, involves novel enzymes, and shows unique control. In an effort to understand the metabolism of cellobiose in *P. furiosus*, a binding protein-dependent ABC transport system for oligosaccharides was discovered. The 70-kDa protein is responsible for the uptake of cellobiose and most other α-glucosides (Koning et al., 2001).

Two major pathways are known to be involved in the degradation of glucose in prokaryotes: the Embden-Meyerhof (EM) and the Entner-Doudoroff (ED) pathways. In general, they differ in the key enzyme acting on glucose or glucose-6-phosphate and in the subsequent aldolytic cleavage of the intermediates fructose-1,6 biphosphate (EM) and 2-keto-3-deoxy-6-phosphogluconate (ED). So far, sugar degradation has been analyzed in representative species of the genera *Thermococcus* and *Pyrococcus*. The analysis included 1) determination of ^{13}C-labeling patterns by ^{1}H- and ^{13}C-NMR spectroscopy of fermentation products derived from pyruvate after fermentation of specifically ^{13}C-labeled glucose by cell suspensions, 2) identification of intermediates of sugar degradation after conversion of ^{14}C-labeled glucose by cell extracts, and 3) measurements of enzyme activities in cell extracts (Schönheit and Schäfer, 1995; De Vos et al., 1998). It has been established that in the three Thermococcales, *P. furiosus*, *T. celer* and *T. litoralis*, glycolysis appears to occur via a modified EM pathway. This pathway is unusual because the hexose kinase and phosphofructokinase steps are dependent on ADP rather than ATP, and a novel tungsten-containing enzyme termed "glyceraldehyde-3-phosphate:ferredoxin oxidoreductase" (GAPOR) replaces the expected glyceraldehyde-3-phosphate dehydrogenase (GAPDH) and phosphoglycerate kinase. In contrast, other thermophilic Archaea (like *Sulfolobus solfataricus* and *Thermoplasma acidophilum*) degrade glucose via the ED pathway, and hyperthermophilic bacteria belonging to the order Thermotogales degrade glucose via conventional forms of the EM and ED pathways (Schönheit and Schäfer, 1995; De Vos et al., 1998). A final step in sugar fermentation is the conversion of acetyl-CoA into acetate that produces ATP. Also in this crucial step, the Thermococcales are unique because they have a single enzyme, an ADP-dependent acetyl-CoA synthase, while in bacteria this reaction is catalyzed by two different enzymes, phosphate acetyltransferase and acetate kinase (Fig. 2). In addition, *P.*

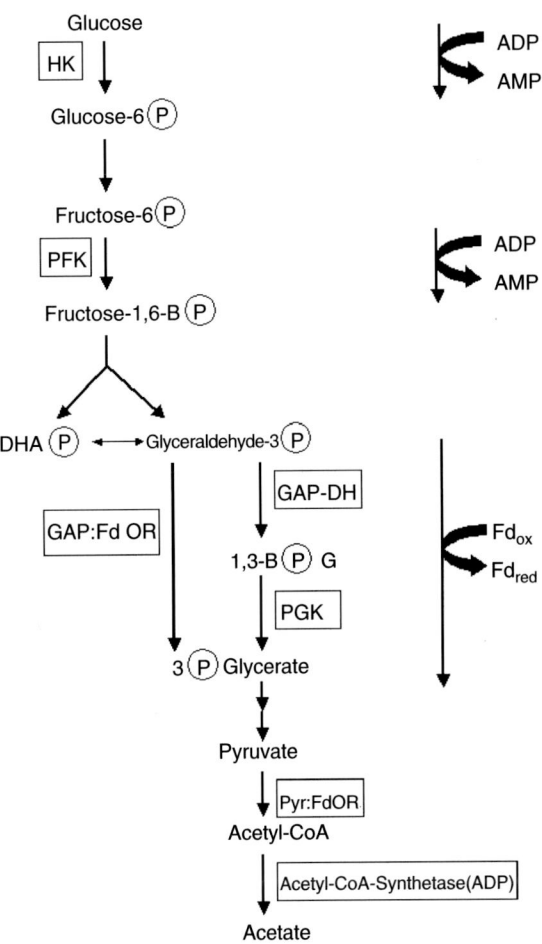

Fig. 2. Embden-Meyerhof-type glycolytic pathway in the genera *Pyrococcus* and *Thermococcus*. The phosphoryl-donor specificities (ADP-AMP) of hexokinase (HK) and 6-phosphofructokinase (PFK) and the enzyme proposed for glyceraldehyde-3-phosphate oxidation are indicated (GAP:FdOR, glyceraldehyde-3-phosphate:ferredoxin oxidoreductase; GAP-DH, glyceraldehyde-3-phosphate dehydrogenase; PGK, phosphoglycerate kinase; DHA-P, dihydroxyacetonephosphate; 1,3-BPG, 1,3-biphosphoglycerate; Fd*ox*, oxidized ferredoxin; Fd*red*, reduced ferredoxin; and Pyr:FdOR, pyruvate:ferredoxin oxidoreductase).

furiosus has the capacity to convert pyruvate into alanine, which acts as an alternative electron sink. This reaction involves the combined activity of both alanine aminotransferase and glutamate dehydrogenase (Kengen et al., 1994).

The operation of a new glycolytic pathway was demonstrated in nongrowing cells of *Thermococcus zilligii* by isotopic enrichment analysis of the end products derived from fermentation of ^{13}C-labeled glucose. The new pathway involves the formation of formate, derived from C-1 in glucose, via cleavage of a six-carbon carboxylic acid. The operation of a novel glycolytic strategy in *T. zilligii* with two branches diverging at the level of glucose-6-phosphate was demonstrated (Ronimus et al., 2001). Glucose is phosphorylated by an ADP-dependent hexokinase to glucose-6-phosphate, which is subsequently degraded by two glycolytic branches: an EM-type glycolytic pathway and a new route where formate is produced by a reaction involving cleavage of the C-1 carboxylic group of a six-carbon compound to yield formate and a pentose phosphate. By analogy with the pyruvate-formate-lyase reaction, it was suggested that the six-carbon compound is a β-ketoacid, such as 2-keto-3-deoxy-6-phosphogluconate, derived from 6-phosphogluconate. The contribution of the novel glycolytic branch was twice as high as that of the EM-type pathway when cells were grown on tryptone, and the inverse relationship was found for cells grown in the presence of glucose. This is the first report of a glycolytic pathway involving the formation of formate from C-1 in glucose. It is noteworthy that the most atypical member of the Thermococcales, *T. zilligii*, possesses also this unusual glycolytic feature (Ronimus et al., 1999; Ronimus et al., 2001).

The pathways of peptide metabolism have been well studied in *P. furiosus*. Amino acid catabolism in *P. furiosus* is thought to involve four distinct 2-keto acid oxidoreductases that convert transaminated amino acids into their corresponding coenzyme A (CoA) derivatives (Blamey and Adams, 1993; Mai and Adams, 1994; Mai and Adams, 1996; Heider et al., 1996). These CoA derivatives, together with acetyl-CoA produced from glycolysis via pyruvate, are then transformed to their corresponding organic acids by two acetyl-CoA synthetases unique to archaea, with concomitant substrate-level phosphorylation to form ATP. Alternatively, it has been postulated that depending on the redox balance of the cell, 2-keto acids are decarboxylated to aldehydes and then oxidized to form carboxylic acids by a second tungsten-containing enzyme, aldehyde:ferredoxin oxidoreductase (AOR; Mai and Adams, 1996). A third enzyme of this type, termed "formaldehyde:ferredoxin oxidoreductase" (FOR), is thought to be involved in the catabolism of basic amino acids (Roy et al., 1999; Adams et al., 2001).

Molecular Biology

The DNA G+C content of members of Thermococcales varies from 37.5 mol% for *Pyrococcus woesei* to 60 mol% for *Thermococcus barossii*. The abyssal strains *P. abyssi*, *P. horiskoshii* and *P. glycovorans* are characterized by a higher G+C (44–47 mol%) content than that of the coastal strains *P. furiosus* and *P. woesei* (38 mol%). A G+C content higher than 40 mol% is a typical feature of most *Thermococcus* species, but also in this genus, a distinction can be made. According to Godfroy et al. (1997), the strains of the genus *Thermococcus* can be divided on the basis of their G+C contents into the following two groups: 1) a group of strains with high G+C content (50–58 mol%), including the strains from shallow marine environment, *T. celer* and *T. stetteri*, and three deep-sea species *T. profundus*, *T. peptinophilus* and *T. funiculans*; and 2) a group of strains with low G+C content (38–47 mol%), including deep-sea strains of *T. chitinophagus*, *T. alcaliphilus* and *T. barophilus*, an organism from a shallow marine environment, *T. litoralis*, and the microorganism from a terrestrial high temperature oil reservoir, *Thermococcus sibiricus* (Table 1).

Phylogenetic studies based on 16S rDNA analysis reveal that *Pyrococcus* and *Palaeococcus* strains are clustered separately from *Thermococcus* species. However, the topology of the dendogram based on neighbor-joining algorithms clusters the Thermococcus species in a different number of branches, thus indicating that at the phylogenetic level, diversity in the same genus within high temperature environments is remarkable (Fig. 3).

The recent advances in genome projects have provided a considerable amount of data, which enables genomes of distant organisms to be compared in a comprehensive and integrative way. Comparisons of closely related species constitute a complementary approach crucial to the understanding of genome evolution. At the genomic level, these comparisons provide a unique opportunity to understand the mechanisms that determine chromosomal organization and evolution. At the proteomic level, this powerful strategy can be used to assess the genuine extent of gene losses and gains that lead to the observed divergence of coding capacity. Comparison of the genomes of Thermococcales has been already performed on three closely related species: *Pyrococcus abyssi* (Cohen et al., 2003; Genoscope website http:\\www.genoscope.cns.fr), *Pyrococcus horikoshii* (Kawarabayasi et al., 1998; Kawarabayasi et al., 2001; The National Institute of Technology and Evaluation(NITE) website http:\\www.bio.nite.go.jp) and *Pyrococcus furiosus* (Maeder et al., 1999; Lecompte et al., 2001; Environmental Genome Project sponsored by the National Institute of Environmental Health Sciences http:\\www.genome.utah.edu; ORNL website http:\\www.ornl.gov). Several genome features of *P. abyssi*, *P. horikoshii* and *P. furiosus* affirm the close relationship among the three species, including similar G+C content and RNA elements, like rRNA and tRNA. At the genomic level, the comparison reveals that a differential conservation among four regions of the *Pyrococcus* chromosomes correlates with the location of genetic elements mediating DNA reorganization. At the proteomic level, the closer proximity of

Fig. 3. Phylogenetic tree of the order Thermococcales as derived from neighbor-joining analysis of 16S rRNA. The tree was constructed by maximum-likelihood analysis using the program CLUSTAL W (Higgins et al., 1996). The 16S rRNA gene sequences were all obtained from GenBank. The accession number of the sequences is indicated within the brackets. Bar indicates one substitution per 100 nucleotides.

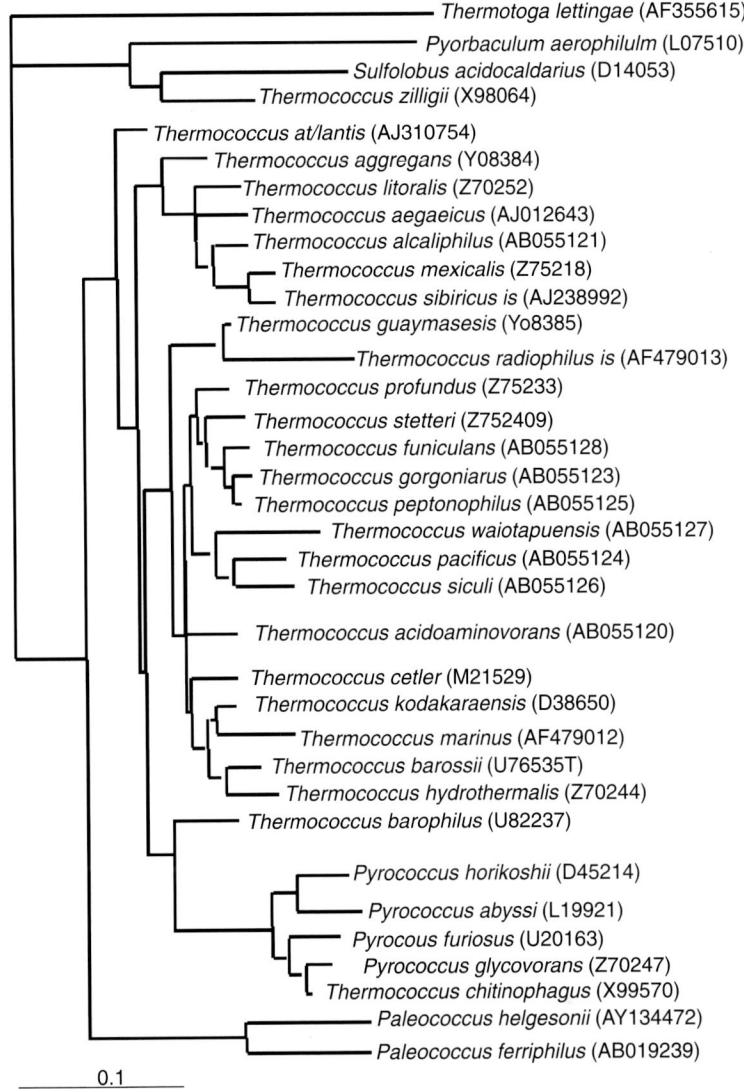

P. abyssi and *P. horikoshii* is affirmed by their average amino acid identity (77%) and their chromosomal organization. Nevertheless, the evolutionary distance between *P. abyssi* and *P. horikoshii* is not negligible relative to *P. furiosus* because the average amino acid identities are also high between *P. furiosus* and the two other species (72% with *P. abyssi* and 73% with *P. horikoshii*). The comparison of the three *Pyrococcus* species sheds light on specific selection pressure acting both on their coding capacities and on their evolutionary rates. The two independent methods, the "reciprocal best hits" approach and a new distance-ratio analysis, allow detection of the false orthology relationships within the *Pyrococcus* lineage. Such analyses reveal a high amount of differential gains and losses of genes since the three closely related species diverged. The resulting polymorphism is probably linked to an adaptation of these free-living organisms to differential environmental constraints.

Enzymology

Because members of Thermococcales grow at high temperature, their enzymes are highly thermoactive and thermostable. A large number of enzymes show no significant loss of activity after several hours at 100°C and are even active at temperatures that exceed the optimal growth temperature of the organism from which they were isolated (Bertoldo and Antranikian, 2002). These properties make hyperthermophiles very attractive for new biotechnological applications. To date, a large number of extracellular and intracellular enzymes have been characterized. They include the extracellular enzymes like amylases, pullulanases, α-glucosidases and proteases but also intracellular enzymes like dehydrogenases, oxidoreductases and DNA polymerases (Table 2). Interestingly, the ability of hyperthermophiles to produce cellulases, xylanases and pectinases seems to be very limited,

JOHN MOORES UNIVERSITY
AVRIL ROBARTS LRC
TITHEBARN STREET
LIVERPOOL L2 2ER

Table 2. Properties of enzymes from Thermococcales with potential biotechnological implications.

Enzymes	Organism[a]	Enzyme properties			Remarks
		Optimal temperature	Optimal pH	Mw (kDa)	
α-Amylase	Pyrococcus furiosus (100)	100	6.5–7.5	129	Purified/cloned/intracellular
		100	7.0	68	Purified/cloned/extracellular
	Pyrococcus kodakaraensis	90	6.5	49.5	Purified/cloned/extracellular
	Pyrococcus woesi (100)	100	5.5	68	Purified/extracellular
	Thermococcus celer (85)	90	5.5	–	Crude extract
	Thermococcus profundus DT5432 (80)	80	5.5	42	Purified/cloned/"Amy S"
	Thermococcus profundus (80)	80	4.0–5.0	42	Purified/"Amy L"
	Thermococcus aggregans (85)	95	6.5	–	Cloned
Pullulanase type II	Pyrococcus woesei (100)	100	6.0	90	Purified/cloned/cell associated
	Thermococcus celer (85)	90	5.5	–	Crude extract
	Thermococcus litoralis (90)	98	5.5	119	Purified/extracell./glycoprotein
	Thermococcus hydrothermalis (80)	95	5.5	128	Purified/extracell./glycoprotein
Pullulan-hydrolase type III	Thermococcus aggregans (85)	100	6.5	83	Purified/cloned
CGTase	Thermococcus sp. (75)	100	2.0	83	Purified
α-Glucosidase	Thermococcus strain AN1 (80)	130	–	63	Purified/extracell./glycoprotein
	Thermococcus hydrothermalis (80)	–	–	–	Cloned
	Pyrococcus furiosus (100)	–	5.0–6.0	125	Pur./extracellular
	Pyrococcus woesei (100)	100	5.0–5.5	90	Pur./cloned/intracellular
Chitinase	Thermococcus chitinophagus (80)	70	7.0	70	Purified
	Pyrococcus kodakaraensis (95)	85	5.0	135	Purified/cloned
	Thermococcus chitinophagus (80)	70	7.0	70	Purified
Exoglucanase	Pyrococcus furiosus (100)	100	6.0	35.9	Cloned
β-Glycosidase	Pyrococcus furiosus (100)	102–105	–	230/58	Purified/cloned
Serine protease	Pyrococcus furiosus (100)	85	6.3	124/19	Protease I/purified
		–	–	105/80	Pyrolysin/pur./cloned
	Thermococcus aggregans (75)	90	7.0	–	Crude extract
	Thermococcus celer (85)	95	7.5	–	Crude extract
	Thermococcus litoralis (90)	95	9.5	–	Crude extract
	Thermococcus stetteri (75)	85	8.5	68	Pur./cloned
Thiol protease	Pyrococcus kodakaraensis (95)	110	7	44	Purified
DNA ligase (Pfu)	Pyrococcus furiosus (100)	90	6.0	35.9	Cloned
DNA polymerase	Thermococcus litoralis (90)	70–80	8.8	98.9	Cloned/purified
DNA polymerase	Thermococcus strain 9°N-7 (90)	70–80	–	89.6	Cloned/purified
DNA polymerase	Thermococcus aggregans (90)	80	7.5	90	Cloned/purified
DNA polymerase	Pyrococcus furiosus (100)	72–78	8.0–9.0	90.1	Cloned/Purified
DNA polymerase	Pyrococcus kodakaraensis (95)	75	7.5	–	Purified

Symbol: –, not determined.
[a]Values in parentheses give the optimal growth temperature for each organism in °C.

and only in few cases the formation of cellobio-hydrolase has been reported. Among the intracellular enzymes, the DNA polymerase and DNA ligase from *Pyrococcus* sp. (Southworth et al., 1996) and the DNA polymerase from *Thermococcus* strains especially have attracted further interest because of their potential in commercial applications to PCR. For reviews on the enzymology of hyperthermophilic archaea and their potential applications refer to the following articles (Sunna et al., 1997; Niehaus et al., 1999; Lévêque et al., 2000; Bertoldo and Antranikian, 2002; Bohlke et al., 2002).

Literature Cited

Achenbach-Richter, L., R. Gupta, W. Zillig, and C. R. Woese. 1988. Rooting the archaebacterial tree: The pivotal role of Thermococcus celer in archaebacterial evolution. Syst. Appl. Microbiol. 10:231–240.

Adams, M. W., J. F. Holden, A. L. Menon, G. J. Schut, A. M. Grunden, C. Hou, A. M. Hutchins, F. E. Jenney Jr., C. Kim, K. Ma, G. Pan, R. Roy, R. Sapra, S. V. Story, and M. F. Verhagen. 2001. Key role for sulfur in peptide metabolism and in regulation of three hydrogenases in the hyperthermophilic archaeon Pyrococcus furiosus. J. Bacteriol. 183(2):716–724.

Arab, H., H. Volker, and M. Thomm. 2000. Thermococcus aegaeicus sp. nov. and Staphylothermus hellenicus sp. nov., two novel hyperthermophilic archaea isolated from geothermally heated vents off Palaeochori Bay, Milos, Greece. Int. J. Syst. Evol. Microbiol. 50 Pt 6:2101–2108.

Balch, W. E., G. E. Fox, L. J. Magrum, C. R. Woese, and R. S. Wolfe. 1979. Methanogens: Reevaluation of a unique biological group. Microbiol. Rev. 43(2):260–296.

Barbier, G., A. Godfroy, J. R. Meunier, J. Querellou, M. A. Cambon, F. Lesongeur, P. A. Grimont, and G. Raguenes. 1999. Pyrococcus glycovorans sp. nov., a hyperthermophilic archaeon isolated from the East Pacific Rise. Int. J. Syst. Bacteriol. 49 Pt 4:1829–1837.

Bertoldo, C., and G. Antranikian. 2002. Starch-hydrolyzing enzymes from thermophilic archaea and bacteria. Curr. Opin. Chem Biol. 2:151–160.

Biller, K. F., I. Kato, and H. Markl. 2002. Effect of glucose, maltose, soluble starch, and CO₂ on the growth of the hyperthermophilic archaeon Pyrococcus furiosus. Extremophiles 6(2):161–166.

Blamey, J. M., and M. W. Adams. 1993. Purification and characterization of pyruvate ferredoxin oxidoreductase from the hyperthermophilic archaeon Pyrococcus furiosus. Biochim. Biophys. Acta 1161(1):19–27.

Blamey, J., M. Chiong, C. Lopez, and E. Smith. 1999. Optimization of the growth conditions of the extremely thermophilic microorganisms Thermococcus celer and Pyrococcus woesei. J. Microbiol. Meth. 38(1–2):169–175.

Blumentals, I. I., S. H. Brown, R. N. Schicho, A. K. Skaja, H. R. Costantino, and R. M. Kelly. 1990. The hyperthermophilic archaebacterium, Pyrococcus furiosus. Development of culturing protocols, perspectives on scaleup, and potential applications. Ann. NY Acad. Sci. 589:301–314.

Bohlke, K., F. M. Pisani, M. Rossi, and G. Antranikian. 2002. Archaeal DNA replication: Spotlight on a rapidly moving field. Extremophiles 6(1):1–14.

Britton, K. L., P. J. Baker, K. M. Borges, P. C. Engel, A. Pasquo, D. W. Rice, F. T. Robb, R. Scandurra, T. J. Stillman, and K. S. Yip. 1995. Insights into thermal stability from a comparison of the glutamate dehydrogenases from Pyrococcus furiosus and Thermococcus litoralis. Eur. J. Biochem. 229(3):688–695.

Cambon-Bonavita, M. A., F. Lesongeur, P. Pignet, N. Wery, C. Lambert, A. Godfroy, J. Querellou, and G. Barbier. 2003. Extremophiles, thermophily section, species description Thermococcus atlanticus sp. nov., a hyperthermophilic Archaeon isolated from a deep-sea hydrothermal vent in the Mid-Atlantic Ridge. Extremophiles 7(2):101–109.

Canganella, F., J. M. Gonzalez, M. Yanagibayashi, C. Kato, and K. Horikoshi. 1997a. Pressure and temperature effects on growth and viability of the hyperthermophilic archaeon Thermococcus peptonophilus. Arch. Microbiol. 168(1):1–7.

Canganella, F., W. J. Jones, A. Gambacorta, and G. Antranikian. 1997b. Biochemical and phylogenetic characterization of two novel deep-sea Thermococcus isolates with potentially biotechnological applications. Arch. Microbiol. 167(4):233–238.

Canganella, F., W. J. Jones, A. Gambacorta, and G. Antranikian. 1998. Thermococcus guaymasensis sp. nov. and Thermococcus aggregans sp. nov., two novel thermophilic archaea isolated from the Guaymas Basin hydrothermal vent site. Int. J. Syst. Bacteriol. 48 Pt 4:1181–1185.

Cohen, G. N., V. Barbe, D. Flament, M. Galperin, R. Heilig, O. Lecompte, O. Poch, D. Prieur, J. Querellou, R. Ripp, J. C. Thierry, J. Van der Oost, J. Weissenbach, Y. Zivanovic, and P. Forterre. 2003. An integrated analysis of the genome of the hyperthermophilic archaeon Pyrococcus abyssi. Molec. Microbiol. 47(6):1495–1512.

Costantino, H. R., S. H. Brown, and R. M. Kelly. 1990. Purification and characterization of an alpha-glucosidase from a hyperthermophilic archaebacterium, Pyrococcus furiosus, exhibiting a temperature optimum of 105 to 115 degrees C. J. Bacteriol. 172(7):3654–3660.

De Vos, W. M., S. W. Kengen, W. G. Voorhorst, and J. Van der Oost. 1998. Sugar utilization and its control in hyperthermophiles. Extremophiles 2:201–205.

Diederichs, K., J. Diez, G. Greller, C. Muller, J. Breed, C. Schnell, C. Vonrhein, W. Boos, and W. Welte. 2000. Crystal structure of MalK, the ATPase subunit of the trehalose/maltose ABC transporter of the archaeon Thermococcus litoralis. EMBO J. 19(22):5951–5961.

Dirmeier, R., M. Keller, D. Hafenbradl, F. J. Braun, R. Rachel, S. Burggraf, and K. O. Stetter. 1998. Thermococcus acidaminovorans sp. nov., a new hyperthermophilic alkalophilic archaeon growing on amino acids. Extremophiles 2(2):109–114.

Duffaud, G. D., O. B. d'Hennezel, A. S. Peek, A. L. Reysenbach, and R. M. Kelly. 1998. Isolation and characterization of Thermococcus barossii, sp. nov., a hyperthermophilic archaeon isolated from a hydrothermal vent flange formation. Syst. Appl. Microbiol. 21(1):40–49.

Erauso, G., Reysenbach, A. L., Godfroy, A., Meunier, J. R., Crump, B., Partensky, F., Baross, J. A., Marteinsson, V., Barbier, G., Pace, N. R., and Prieur, D. 1993. Pyrococcus abyssi sp. nov., a new hyperthermophilic archaeon isolated from a deep-sea hydrothermalvent. Arch. Microbiol. 160:338–349.

Fiala, G., and K. O. Stetter. 1986. Pyrococcus furiosus sp. nov. represents a novel genus of marine heterotrophic archaebacteria growing optimally at 100°C. Arch. Microbiol. 145:56–61.

Godfroy, A., J. R. Meunier, J. Guezennec, F. Lesongeur, G. Raguénès, A. Rimbault, and G. Barbier. 1996. Thermococcus fumicolans sp. nov., a new hyperthermophilic archaeon isolated from a deep-sea hydrothermal vent in the north Fiji Basin. Int. J. Syst. Evol. Microbiol. 46(4):1113–1119.

Godfroy, A., F. Lesongeur, G. Raguenes, J. Querellou, E. Antoine, J. R. Meunier, J. Guezennec, and G. Barbier. 1997. Thermococcus hydrothermalis sp. nov., a new hyperthermophilic archaeon isolated from a deep-sea hydrothermal vent. Int. J. Syst. Bacteriol. 47:622–626.

Gonzalez, J. M., C. Kato, and K. Horikoshi. 1995. Thermococcus peptonophilus sp. nov., a fast-growing, extremely thermophilic archaebacterium isolated from deep-sea hydrothermal vents. Arch. Microbiol. 164(3):159–164.

Gonzalez, J. M., Y. Masuchi, F. T. Robb, J. W. Ammerman, D. L. Maeder, M. Yanagibayashi, J. Tamaoka, and C. Kato. 1998. Pyrococcus horikoshii sp. nov., a hyperthermophilic archaeon isolated from a hydrothermal vent at the Okinawa Trough. Extremophiles 2(2):123–130.

Gonzalez, J. M., D. Sheckells, M. Viebahn, D. Krupatkina, K. M. Borges, and F. T. Robb. 1999. Thermococcus waiotapuensis sp. nov., an extremely thermophilic archaeon isolated from a freshwater hot spring. Arch. Microbiol. 172(2):95–101.

Greller, G., R. Horlacher, J. DiRuggiero, and W. Boos. 1999. Molecular and biochemical analysis of MalK, the ATP-hydrolyzing subunit of the trehalose/maltose transport system of the hyperthermophilic archaeon Thermococcus litoralis. J. Biol. Chem. 274(29):20259–20264.

Greller, G., R. Riek, and W. Boos. 2001. Purification and characterization of the heterologously expressed trehalose/maltose ABC transporter complex of the hyperthermophilic archaeon Thermococcus litoralis. Eur. J. Biochem. 268(14):4011–4018.

Grote, R., L. Li, J. Tamaoka, C. Kato, K. Horikoshi, and G. Antranikian. 1999. Thermococcus siculi sp. nov., a novel hyperthermophilic archaeon isolated from a deep-sea hydrothermal vent at the Mid-Okinawa Trough. Extremophiles 3(1):55–62.

Gruyer, S., E. Legin, C. Bliard, S. Ball, and F. Duchiron. 2002. The endopolysaccharide metabolism of the hyperthermophilic archeon Thermococcus hydrothermalis: Polymer structure and biosynthesis. Curr. Microbiol. 44(3):206–211.

Heider, J., X. Mai, and M. W. Adams. 1996. Characterization of 2-ketoisovalerate ferredoxin oxidoreductase, a new and reversible coenzyme A-dependent enzyme involved in peptide fermentation by hyperthermophilic archaea. J. Bacteriol. 178(3):780–787.

Higgins, D. G., J. D. Thompson, and T. J. Gibson. 1996. Using CLUSTAL for multiple sequence alignments. Meth. Enzymol. 266:383–402.

Holden, J. F., and J. A. Baross. 1993. Enhanced thermotolerance and temperature-induced changes in protein composition in the hyperthermophilic archaeon ES4. J. Bacteriol. 175(10):2839–2843.

Horlacher, R., K. B. Xavier, H. Santos, J. DiRuggiero, M. Kossmann, and W. Boos. 1998. Archaeal binding protein-dependent ABC transporter: Molecular and biochemical analysis of the trehalose/maltose transport system of the hyperthermophilic archaeon Thermococcus litoralis. J. Bacteriol. 180(3):680–689.

Huber, R., J. Stöhr, S. Hohenhaus, R. Rachel, S. Burggraf, H. W. Jannasch, and K. O. Stetter. 1996. Thermococcus chitonophagus sp. nov., a novel, chitin-degrading, hyperthermophilic archaeum from a deep-sea hydrothermal vent environment. Arch. Microbiol. 64:255–264.

Kawarabayasi, Y., M. Sawada, H. Horikawa, Y. Haikawa, Y. Hino, S. Yamamoto, M. Sekine, S. Baba, H. Kosugi, A. Hosoyama, Y. Nagai, M. Sakai, K. Ogura, R. Otsuka, H. Nakazawa, M. Takamiya, Y. Ohfuku, T. Funahashi, T. Tanaka, Y. Kudoh, J. Yamazaki, N. Kushida, A. Oguchi, K. Aoki, and H. Kikuchi. 1998. Complete sequence and gene organization of the genome of a hyperthermophilic archaebacterium, Pyrococcus horikoshii OT3. DNA Res. 5(2):55–76.

Kawarabayasi, Y. 2001. Genome of Pyrococcus horikoshii OT3. Meth. Enzymol. 330:124–134.

Keller, M., F. J. Braun, R. Dirmeier, D. Hafenbradl, S. Burggraf, R. Rachel, and K. O. Stetter. 1995. Thermococcus alcaliphilus sp. nov., a new hyperthermophilic archaeum growing on polysulfide at alkaline pH. Arch. Microbiol. 164(6):390–395.

Kengen, S. W., E. J. Luesink, A. J. Stams, and A. J. Zehnder. 1993. Purification and characterization of an extremely thermostable beta-glucosidase from the hyperthermophilic archaeon Pyrococcus furiosus. Eur. J. Biochem. 213(1):305–312.

Kengen, S. W., F. A. de Bok, N. D. van Loo, C. Dijkema, A. J. Stams, and W. M. de Vos. 1994. Evidence for the operation of a novel Embden-Meyerhof pathway that involves ADP-dependent kinases during sugar fermentation by Pyrococcus furiosus. J. Biol. Chem. 269(26):17537–17541.

Kobayashi, T., Y. S. Kwak, T. Akiba, T. Kudo, and K. Horikoshi. 1994. Thermococcccus profundus sp. nov., a new hyperthermophilic archaeon isolated from a deep-sea hydrothermal vent. Syst. Appl. Microbiol. 17:232–236.

Koning, S. M., M. G. Elferink, W. N. Konings, and A. J. Driessen. 2001. Cellobiose uptake in the hyperthermophilic archaeon Pyrococcus furiosus is mediated by an inducible, high-affinity ABC transporter. J. Bacteriol. 183(17):4979–4984.

Krahe, M., G. Antranikian, and H. Märkl. 1996. Fermentation of extremophilic microorganisms. FEMS Microbiol. Rev. 18:271–285.

Kwak, Y. S., T. Kobayashi, T. Akiba, K. Horikoshi, and Y. B. Kim. 1995. A hyperthermophilic sulfur-reducing archaebacterium, Thermococcus sp. DT1331, isolated from a deep-sea hydrothermal vent. Biosci. Biotechnol. Biochem. 59(9):1666–1669.

Lattuati, A., J. Guezennec, P. Metzger, and C. Largeau. 1998. Lipids of Thermococcus hydrothermalis, an archaea isolated from a deep-sea hydrothermal vent. Lipids 33(3):319–326.

Lecompte, O., R. Ripp, V. Puzos-Barbe, S. Duprat, R. Heilig, J. Dietrich, J. C. Thierry, and O. Poch. 2001. Genome evolution at the genus level: Comparison of three complete genomes of hyperthermophilic archaea. Genome Res. 11(6):981–993.

Lévêque, E., S. Janeek, and H. B. Belarbi. 2000. Thermophilic archaeal amylolytic enzymes. Enz. Microb. Technol. 1:3–14.

Maeder, D. L., R. B. Weiss, D. M. Dunn, J. L. Cherry, J. M. Gonzalez, J. DiRuggiero, and F. T. Robb. 1999. Divergence of the hyperthermophilic archaea Pyrococcus furiosus and P. horikoshii inferred from complete genomic sequences. Genetics 152(4):1299–1305.

Mai, X., and M. W. Adams. 1994. Indolepyruvate ferredoxin oxidoreductase from the hyperthermophilic archaeon Pyrococcus furiosus. A new enzyme involved in peptide fermentation. J. Biol. Chem. 269(24):16726–16732.

Mai, X., and M. W. Adams. 1996. Purification and characterization of two reversible and ADP-dependent acetyl coenzyme A synthetases from the hyperthermophilic archaeon Pyrococcus furiosus. J. Bacteriol. 178(20):5897–5903.

Marteinsson, V. T., J. L. Birrien, A. L. Reysenbach, M. Vernet, D. Marie, A. Gambacorta, P. Messner, U. B. Sleytr, and D. Prieur. 1999. Thermococcus barophilus sp. nov., a new barophilic and hyperthermophilic archaeon isolated under high hydrostatic pressure from a deep-sea hydrothermal vent. Int. J. Syst. Bacteriol. 49 Pt 2:351–359.

Miroshnichenko, M. L., E. A. Bonch-Osmolovskaya, A. Neuner, N. A. Kostrikina, N. A. Chernych, and V. A. Alekseev. 1989. Thermococcus stetteri sp. nov., a new extremely thermophilic marine sulfur-metabolizing archaebacterium. Syst. Appl. Microbiol. 12:257–262.

Miroshnichenko, M. L., G. M. Gongadze, F. A. Rainey, A. S. Kostyukova, A. M. Lysenko, N. A. Chernyh, and E. A. Bonch-Osmolovskaya. 1998. Thermococcus gorgonarius sp. nov. and Thermococcus pacificus sp. nov.: Heterotrophic extremely thermophilic archaea from New Zealand submarine hot vents. Int. J. Syst. Bacteriol. 48 Pt 1:23–29.

Miroshnichenko, M. L., H. Hippe, E. Stackebrandt, N. A. Kostrikina, N. A. Chernyh, C. Jeanthon, T. N. Nazina, S. S. Belyaev, and E. A. Bonch-Osmolovskaya. 2001. Isolation and characterization of Thermococcus sibiricus sp. nov. from a Western Siberia high-temperature oil reservoir. Extremophiles 5(2):85–91.

Morikawa, M., Y. Izawa, N. Rashid, T. Hoaki, and T. Imanaka. 1994. Purification and characterization of a thermostable thiol protease from a newly isolated hyperthermophilic Pyrococcus sp. Appl. Environ. Microbiol. 60(12):4559–4566.

Neuner, A., H. W. Jannasch, S. Belkinn, and K. O. Stetter. 1990. Thermococcus litoralis sp. nov.: S new species of extremely thermophilic marine archaebacterium. Arch. Microbiol. 153:205–207.

Niehaus, F., C. Bertoldo, M. Kahler, and G. Antranikian. 1999. Extremophiles as a source of novel enzymes for industrial application. Appl. Microbiol. Biotechnol. 51(6):711–729.

Ronimus, R. S., A. Reysenbach, D. R. Musgrave, and H. W. Morgan. 1997. The phylogenetic position of the Thermococcus isolate AN1 based on 16S rRNA gene sequence analysis: A proposal that AN1 represents a new species, Thermococcus zilligii sp. nov. Arch. Microbiol. 168(3):245–258.

Ronimus, R. S., J. Koning, and H. W. Morgan. 1999. Purification and characterization of an ADP-dependent phosphofructokinase from Thermococcus zilligii. Extremophiles 3(2):121–129.

Ronimus, R. S., E. de Heus, and H. W. Morgan. 2001. Sequencing, expression, characterisation and phylogeny of the ADP-dependent phosphofructokinase from the hyperthermophilic, euryarchaeal Thermococcus zilligii. Biochim. Biophys. Acta 1517(3):384–391.

Roy, R., S. Mukund, G. J. Schut, D. M. Dunn, R. Weiss, and M. W. Adams. 1999. Purification and molecular characterization of the tungsten-containing formaldehyde ferredoxin oxidoreductase from the hyperthermophilic archaeon Pyrococcus furiosus: The third of a putative five-member tungstoenzyme family. J. Bacteriol. 181(4):1171–1180.

Schönheit, P., and T. Schäfer. 1995. Metabolism of hyperthermophiles. World J. Microbiol. Biotechnol. 11:26–57.

Selig, M., K. B. Xavier, H. Santos, and P. Schonheit. 1997. Comparative analysis of Embden-Meyerhof and Entner-Doudoroff glycolytic pathways in hyperthermophilic archaea and the bacterium Thermotoga. Arch. Microbiol. 167(4):217–232.

Southworth, M. W., H. Kong, R. B. Kucera, J. Ware, H. W. Jannasch, and P. F. B. 1996. Cloning of thermostable DNA polymerases from hyperthermophilic marine Archaea with emphasis on Thermococcus sp. 9°N-7 and mutations affecting 3'-5' exonuclease activity. PNAS 93:5281–5285.

Stetter, K. O. 1996. Hyperthermophiles in the history of life. CIBA Found. Symp. 202:1–10; discussion 11–8.

Sunna, A., M. Moracci, M. Rossi, and G. Antranikian. 1997. Glycosyl hydrolases from hyperthermophiles. Extremophiles 1(1):2–13.

Takahata, Y., T. Hoaki, and T. Maruyama. 2001. Starvation survivability of Thermococcus strains isolated from Japanese oil reservoirs. Arch. Microbiol. 176(4):264–270.

Takai, K., A. Sugai, T. Itoh, and K. Horikoshi. 2000. Palaeococcus ferrophilus gen. nov., sp. nov., a barophilic, hyperthermophilic archaeon from a deep-sea hydrothermal vent chimney. Int. J. Syst. Evol. Microbiol. 50 Pt 2:489–500.

Verhagen, M. F., A. L. Menon, G. J. Schut, and M. W. Adams. 2001. Pyrococcus furiosus: Large-scale cultivation and enzyme purification. Meth. Enzymol. 330:25–30.

Xavier, K. B., R. Peist, M. Kossmann, W. Boos, and H. Santos. 1999. Maltose metabolism in the hyperthermophilic archaeon Thermococcus litoralis: Purification and characterization of key enzymes. J. Bacteriol. 181(11):3358–3367.

Zillig, W., I. Holtz, D. Janecovic, W. Schäfer, and W. D. Reiter. 1983. The archaebacterium Thermococcus celer represents a novel genus within the thermophilic branch of the archaebacteria. Syst. Appl. Microbiol. 4:88–94.

Zillig, W., I. Holz, H. P. Klenk, J. Trent, S. Wunderl, D. Janekovic, E. Imsel, and B. Haas. 1987. Pyrococcus woesei, sp. nov., an ultra-thermophilic marine Archaebacterium, representing a novel order, Thermococcales. Syst. Appl. Microbiol. 9:62–70.

Prokaryotes (2006) 3:82–100
DOI: 10.1007/0-387-30743-5_6

CHAPTER 6

The Genus *Archaeoglobus*

PATRICIA HARTZELL AND DAVID W. REED

Archeal Sulfate Reducers

The domain Archaea contains two genera, thermophilic *Archaeoglobus* and thermoacidophilic *Caldivirga*, which obtain energy by reducing oxidized sulfur compounds to H_2S under anaerobic conditions. Both *Archaeoglobus fulgidus* and *Caldivirga maquilingensis* thrive at 85°C, the optimal temperature for growth (Itoh et al., 1999).

Phylogeny

The genus *Archaeoglobus* is classified in the Archaeoglobaceae family, Archaeoglobales order, Archaeoglobi class, Euryarchaeota phylum, of the domain Archaea (Huber and Stetter, 2001). *Archaeoglobus* is a chemolithoautotrophic or chemoorganotrophic microorganism using sulfate, sulfite or thiosulfate as electron acceptor, with the formation of hydrogen sulfide as the end product.

Habitat

Dissimilatory sulfate-reducers inhabit aquatic and terrestrial sediments and play an essential role in the biogeochemical sulfur cycle. Although oxygen is the most abundant terrestrial element on earth and more electronegative than sulfur, sulfate (an oxidized form of sulfur) is a major anion in anoxic environments in which sulfur predominates (such as seawater) and serves as the electron acceptor for many organisms including *Archaeoglobus*. Sulfate can be reduced by sulfate-reducers to hydrogen sulfide (H_2S), which can be assimilated by other organisms as a substrate for growth.

The *Archaeoglobus* genus contains, to date, four cultured species, *A. fulgidus*, *A. profundus*, *A. veneficus* and *A. lithotrophicus*. The type strain *A. fulgidus* VC-16, the first *Archaeoglobus* species identified, is a chemolithoautotroph that grows preferentially as a chemoorganotroph. It was isolated initially from near anoxic shallow and abyssal submarine hydrothermal vents off

the coast of Italy (Stetter et al., 1987) and later from hot oil field waters in the North Sea (Beeder et al., 1994). In addition to *Archaeoglobus fulgidus* strain VC-16, strains 7324 and Z have been partially characterized.

The chemolithoheterotroph *A. profundus* obligately requires H_2 as an electron donor (Burggraf et al., 1990). *Archaeoglobus profundus* lacks the carbon monoxide dehydrogenase necessary for CO_2 fixation into acetyl-CoA and hence is unable to grow autotrophically (Vorholt et al., 1995). Like *A. fulgidus*, *A. profundus* also was isolated from a marine hydrothermal system. *Archaeoglobus fulgidus* strain Z was isolated from Vulcano (Zellner et al., 1989) and strain 7324 from hot oil field waters off the coast of Norway (Beeder et al., 1994). *Archaeoglobus fulgidus* and related species, such as *A. veneficus*, have been identified in mid-Atlantic ridge hydrothermal vents (Reysenbach et al., 2000) and in association with methanogens in the petroleum hydrocarbon-rich Guaymas Basin off the coast of Mexico (Teske et al., 2002). *Archaeoglobus veneficus*, a black smoker isolate, does not use sulfate as an electron acceptor but can convert sulfite into sulfide (Huber et al., 1997). It is the fastest growing species of the genus *Archaeoglobus* with a doubling time of one hour. *Archaeoglobus lithotrophicus* was isolated from fluids in a North Sea oil field and is an obligate chemolithoautotroph with a doubling time of about two hours (Stetter et al., 1993; Vorholt et al., 1995). *Archaeoglobus lithotrophicus* grows on $H_2 : CO_2$ (80 : 20 ratio) and sulfate, but has lower levels of F_{420} (a coenzyme and a 7,8-didemethyl-8-hydroxy-5-deazariboflavin derivative) and lower activities of F_{420}-dependent dehydrogenases than strain VC-16 has. *Archaeoglobus fulgidus* has also been identified in terrestrial environments including hot springs in Yellowstone National Park (Barns et al., 1994) and terrestrial oil wells (P. Hartzell, unpublished results) where sulfate is an important component. Because sulfate reducers have a high affinity for hydrogen gas, the capacity for sulfate reduction likely allows *A. fulgidus* to compete favorably against hydrogenotrophic methanogens in

Table 1. Characteristics of species from the genus *Archaeoglobus*.

Physiology	*A. fulgidus*	*A. profundus*	*A. veneficus*	*A. lithotrophicus*
Size (μm); shape	0.4–1.2; regular to irregular coccoid	1.3; irregular coccoid	0.5–1.2; irregular coccoid (pairs)	0.4–1.2; regular to irregular coccoid
Flagella	Monopolar polytrichous	Not observed	Polar	Peritrichous
Autofluorescence (coenzyme F_{420})	+++	+++	+++	+
Growth range (°C)	60–95	65–90	65–85	55–87
Electron donor	H_2, lactate, pyruvate, formate	H_2	H_2, pyruvate, acetate, formate	H_2
Electron acceptor	Sulfate, sulfite, thiosulfate	Sulfate, sulfite, thiosulfate	Sulfite, thiosulfate	Sulfate, sulfite, thiosulfate
Carbon source	CO_2, lactate, pyruvate, formate	Lactate, pyruvate, acetate	CO_2, pyruvate, acetate, formate	CO_2
NaCl range (%)	0.5–4.5	0.9–3.6	0.5–4	0.6–4.8
pH range	5.5–8.0	4.5–7.5	6.5–8.0	6.0–8.5
Mol G + C (%)	46 (T_m) 48.5 (sequence)	41 (T_m)	45 (T_m)	40 (T_m)
16S rRNA gene accession number	X05567	AF322392	AF418181	AJ299218

sulfate-rich environments. Although *A. fulgidus* is the focus of this chapter, a summary of the characterization of the other species is included in Table 1.

Archaeoglobus fulgidus

Archaeoglobus fulgidus strain VC-16 is the most extensively characterized member of the genus *Archaeoglobus*. The genome for this strain (DSM4304) has been sequenced at The Institute for Genome Research, *Archaeoglobus fulgidus* DSM4304 Genome page). Initially *A. fulgidus* was shown to be in the euryarchaeotal branch between the Methanococcales and Thermococcales on the basis of 16S rRNA sequencing (Achenbach-Richter et al., 1987). However, reevaluation of this phylogenetic position showed that *Archaeoglobus* branches between the Methanobacteriales and Methanomicrobiales—extreme halophiles (Woese et al., 1991). Our analysis positions the *Archaeoglobus* genus branching most closely with the *Methanococcales* (e.g., *Methanocaldococcus*), between the *Thermococcales*/*Methanopyrales* and *Methanobacteriales* (e.g., *Methanothermus*), Euryarchaeota, as shown in Fig. 1.

The 16S rRNA link between *Archaeoglobus* and methanogens is supported by biochemical data. *Archaeoglobus fulgidus* fluoresces at 420 nm because, like the methanogens, it contains the deazaflavin cofactor F_{420} (Gorris et al., 1991). *Archaeoglobus fulgidus* expresses activity for each of the enzymes and cofactors associated with methanogenesis, but it lacks the components such as 2-mercaptoethane sulfonic acid coenzyme M and the nickel tetrapyrrole cofactor F_{430} that are required for the terminal step in

methane (CH_4) production (Möller-Zinkhan et al., 1989). Genes encoding the methyl-CoM reductase are absent. In *Archaeoglobus*, the methanogenic cofactors are used in the reverse direction during the oxidation of lactate.

Identification

The name *Archaeoglobus* means "ancient spheres." As shown in Figs. 2 and 3, *A. fulgidus* cells are 0.4–1.2 μm wide, with a triangular, dimpled-shaped morphology (Stetter, 1988). *Archaeoglobus fulgidus* cells are polytrichously flagellated at one end (Zellner et al., 1989). The three dimensional structure of *A. fulgidus* is dome-shaped, similar to the structure of *Halobacterium volcanii* and *Sulfolobus solfataricus*. Like other archaea, *A. fulgidus* lacks a peptidoglycan layer; however, the envelope consists of a single layer (S-layer) of hexagonally arranged monomeric peptide units (Kessel et al., 1990). Some of the units are symmetric while others deviate considerably from symmetry. Unlike for *H. volcanii* and *S. solfataricus*, there is no evidence of protein-protein interconnectivity in the hexameric complexes. The S-layer is in close contact with the membrane, providing a narrow periplasmic space and some structural support. Indeed partial removal of the membrane-supporting base weakens the surface layer, making it susceptible to shearing forces.

Archaeoglobus fulgidus membrane has phytanyl ether-linked lipids which are arranged as a bilayer rather than a more stable monolayer membrane found in many hyperthermophiles (Kessel et al., 1990). This is surprising considering that *A. fulgidus* cells grow optimally at 83°C, above the classification for hyperthermophiles (>80°C). *Archaeoglobus fulgidus* produces a

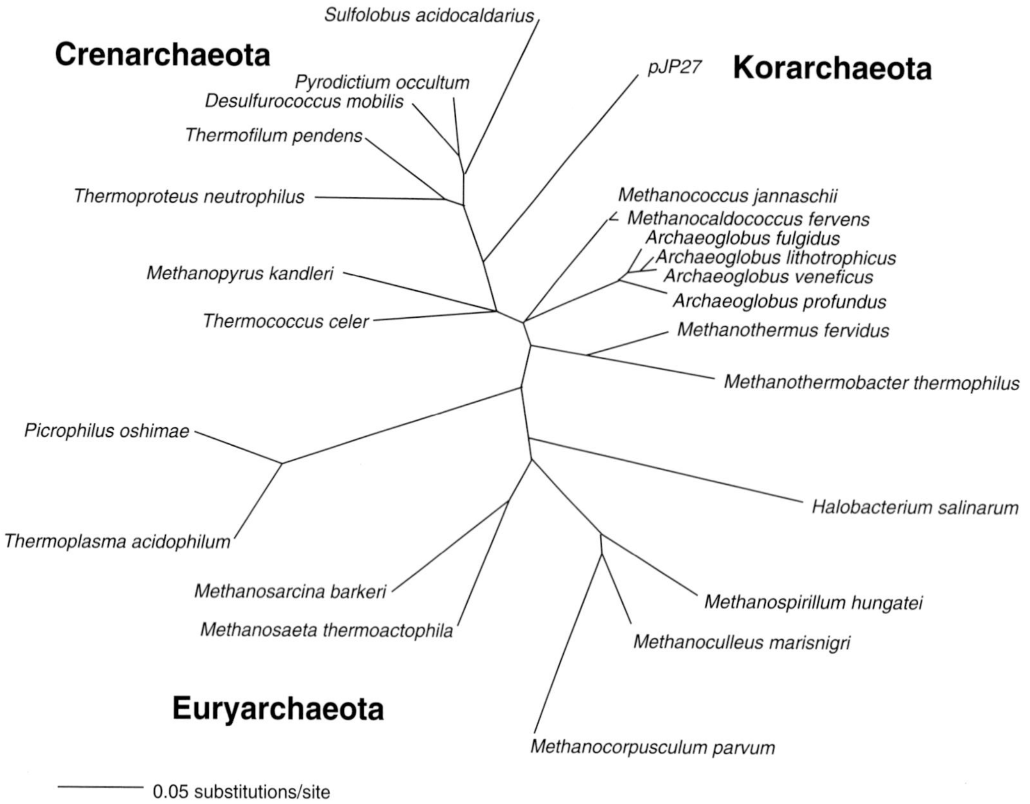

Crenarchaeota

Sulfolobus acidocaldarius

Pyrodictium occultum
Desulfurococcus mobilis
Thermofilum pendens

Thermoproteus neutrophilus

Methanopyrus kandleri

Thermococcus celer

Picrophilus oshimae

Thermoplasma acidophilum

Methanosarcina barkeri

Methanosaeta thermoactophila

Euryarchaeota

pJP27 **Korarchaeota**

Methanococcus jannaschii
Methanocaldococcus fervens
Archaeoglobus fulgidus
Archaeoglobus lithotrophicus
Archaeoglobus veneficus
Archaeoglobus profundus
Methanothermus fervidus

Methanothermobacter thermophilus

Halobacterium salinarum

Methanospirillum hungatei

Methanoculleus marisnigri

Methanocorpusculum parvum

—— 0.05 substitutions/site

Fig. 1. *Archaeoglobus fulgidus* groups near a subset of methanogens in the Euryarchaeota. Phylogenetic relationships of archaeal 16S rRNA sequences as determined by Jukes-Cantor distance analysis. The scale bar represents the average number of nucleotide substitutions per site. Alignments were performed with known taxonomic sequences obtained from the Ribosomal Database Project II (Maidak et al., 2000) and GenBank (Benson et al., 2000; Wheeler et al., 2000) databases using the BioEdit Sequence Alignment Editor freeware (version 5.0.9; Department of Microbiology, North Carolina State University, Raleigh, NC). Sequences were manually corrected using the MacClade software package (version 3.0; Sinauer Associates, Inc., Sunderland, MA) to ensure only homologous nucleotides were compared between sequences. The edited alignments were evaluated with maximum parsimony, maximum likelihood and distance methods using the PAUP package (version 4.0b10; Sinauer Associates, Inc., Sunderland, MA). Phylogenetic inference and evolutionary distance calculations were generated using the Jukes-Cantor distance model (gamma parameter = 2.0). Nucleotide sequence accession numbers are provided: *Archaeoglobus fulgidus*, *Archaeoglobus lithotrophicus*, *Archaeoglobus profundus*, *Archaeoglobus veneficus*, *Desulfurococcus mobilis*, *Halobacterium salinarum*, *Methanocaldococcus jannaschii*, *Methanocorpusculum parvum*, *Methanoculleus marisnigri*, *Methanopyrus kandleri*, *Methanosaeta thermoacetophila*, *Methanosarcina barkeri*, *Methanospirillum hungatei*, *Methanothermobacter thermophilus*, *Methanothermus fervidus*, pJP27, *Picrophilus oshimae*, *Pyrodictium occultum*, *Sulfolobus acidocaldarius*, *Thermococcus celer*, *Thermofilum pendens*, *Thermoplasma acidophilum*, and *Thermoproteus neutrophilus*.

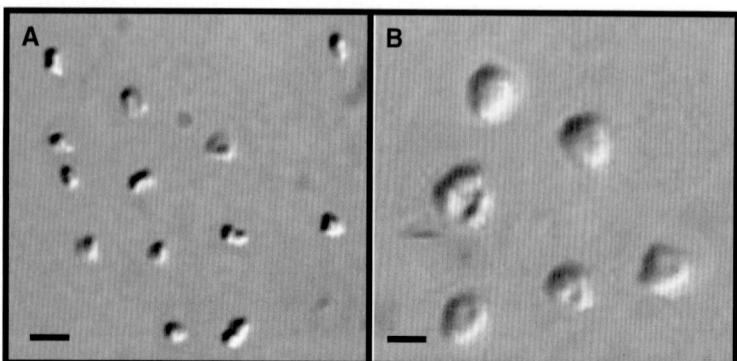

Fig. 2. *Archaeoglobus fulgidus* cells. Cells of *A. fulgidus* are typically irregular in shape, having a triangular or dimpled appearance. A: Bar = 2 μm. B: Bar = 0.5 μm. Photos were taken with a Nikon FXA equipped with diffraction interference contrast.

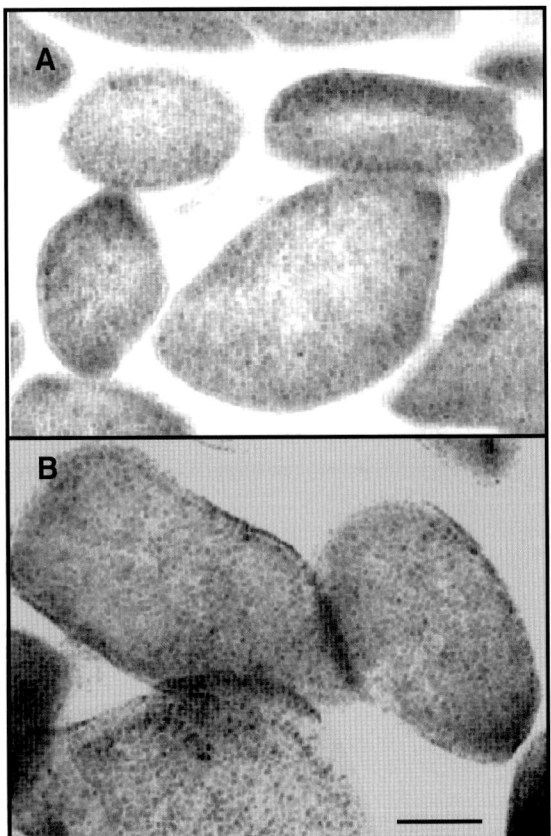

Fig. 3. Thin sections of *A. fulgidus* cells. Cells were embedded in LR White and examined using a JEOL transmission electron microscope at an acceleration magnification of 80 kV (Pagala et al., 2002). Bar = 0.5 μm.

3-hydroxy-3-methylglutaryl CoA (HMG-CoA) reductase (AF1736), an important intermediate for synthesis of cholesterol and lipids, suggesting that these membrane components are produced via mevalonate (Kim et al., 2000).

Isolation

Archaeoglobus fulgidus can be selectively enriched and maintained under anaerobic conditions in sulfate-thiosulfate-lactate (STL, described below) medium at 83°C (Stetter et al., 1987; Hartzell et al., 1999). Lactate can be used as the sole electron and carbon source for *A. fulgidus*. *Archaeoglobus profundus* can be enriched with medium containing acetate and $H_2 : CO_2$ (80 : 20) (Burggraf et al., 1990). Solutions are made anoxic by boiling media and sparging with oxygen-free N_2 gas prior to adding to serum vials. Vials are stoppered, sealed with aluminum crimps and autoclaved. The medium is reduced with 0.5 mM sodium sulfide (Na_2S)

and 1 mM sodium thiosulfate ($Na_2S_2O_3$) prior to inoculation.

Pure isolates can be obtained by serially diluting cells into solid medium under anaerobic conditions. A 0.5-ml aliquot from each dilution is mixed with 5 ml of sulfate-thiosulfate-lactate (STL) medium in melted Gelrite (2%), added to anaerobic tubes, stoppered and spun over an ice bath using a tube roller which produces a thin layer of agar in the tube (Rodabough et al., 1995). Tubes are incubated upright at 83°C, where colonies form in the roll tube agar and are easily seen after about 7–10 days because cells produce H_2S, which forms a black FeS precipitate in the colonies (Fig. 4, Panel B).

For large-scale growth, 40-liter glass carboys can be used (Hartzell et al., 1999; Fig. 4, Panel A). A glass (or enamel-protected) fermentor is used instead of stainless steel because the production of H_2S during growth would cause pitting (from surface oxidation) of metal fermentors. Tris-maleate (20 mM at pH 6.1) can replace 1,4-piperazine bis-(2-ethanesulfonic acid) PIPES in large-scale growth. Higher cell densities are obtained if temperature is maintained at 74–80°C, rather than 83°C, and pH is maintained near 7.2–7.8 (Reed and Hartzell, 1999). The lag period for growth in a carboy is typically 15–20 hours. Typically a 3.6-hour doubling time results in a final cell density of about 5–6×10^8 cells/ml (0.8 g/liter wet weight).

Cultivation

The *A. fulgidus* sulfate-thiosulfate-lactate (STL) base growth medium contains (per liter) 100 ml of 10X salt solution, 20 ml of 1 M PIPES (piperazine-*N*,*N*′-bis[2-ethanesulfonic acid]) buffer at pH 7.0, 1 ml of 1000X trace element stock solution, 1 ml of 0.2% $(NH_4)_2Fe(SO_4) \cdot 2H_2O$, 0.5 g of yeast extract, and 7 ml of 30% sodium lactate (Balch et al., 1979). The pH of the base medium is adjusted to 6.7 with 1 M NaOH or KOH.

The 10X salt solution (1 liter) is prepared using the following three solutions, dissolved separately prior to combining. Solution I contains 74 g of $MgSO_4$, 3.4 g of KCl, 27.5 g of $MgCl_2 \cdot 6H_2O$, 2.5 g of NH_4Cl, 178 g of NaCl and 5 ml of 0.2% resazurin (the redox indicator) dissolved in 800 ml of nanopure (np) H_2O. Solution II contains 1.4 g of $CaCl_2 \cdot 2H_2O$ dissolved in 100 ml of npH_2O. Solution III contains 1.37 g of K_2HPO_4 dissolved in 100 ml of npH_2O.

The 1000X trace element stock (per liter) contains 29 g of $Na_2EDTA \cdot 2H_2O$ (sodium ethylenediamine tetraacetate) dissolved in npH_2O and adjusted to pH 8 with KOH. The following

Fig. 4. Growth of *Archaeoglobus* in 40-liter carboys and serum vials. Panel A: A 40-liter glass carboy heated to 80°C with a heating belt can be used to grow *A. fulgidus*. Sterile, anoxic gas is delivered to the carboy via neoprene tubing and waste gas is removed with constant flushing. Panel B: A colony of *A. fulgidus* in Gelrite agar in a stoppered tube appears black due to precipitation of FeS. Panel C: Serum bottles hold 50–100 ml of medium. Unreduced medium is pink (right bottle) due to oxidation of resazurin, which is colorless when reduced (middle bottle). A growing culture of *A. fulgidus* has a greenish appearance. Cells are transferred to fresh medium using a sterile, N_2-flushed syringe.

are added to the solution: 5 g of $MnSO_4 \cdot 6H_2O$, 1.8 g of $CoCl_2 \cdot 6H_2O$, 1 g of $ZnSO_4 \cdot 7H_2O$, 0.11 g of $NiSO_4 \cdot 6H_2O$, 0.1 g of $CuSO_4 \cdot 5H_2O$, 0.1 g of H_3BO_3, 0.1 g of $KAl(SO_4)_2 \cdot 12H_2O$, 0.1 g of $Na_2MoO_4 \cdot 2H_2O$, 0.1 g of $Na_2WO_4 \cdot 2H_2O$, 0.05 g of Na_2SeO_3, and 0.01 g of V_2O_5.

For a defined medium, 10 ml of Wolfe's vitamin mixture can be added in place of yeast extract (Balch et al., 1979). Wolfe's vitamin mixture contains (per liter) 10 mg of pyridoxine, 5 mg each of thiamine HCl, riboflavin, nicotinic acid, DL-calcium pantothenate, *p*-aminobenzoic acid, and lipoic acid, 2 mg each of biotin and folic acid, and 0.1 mg of cyanocobalamin.

Preservation

Fresh cultures of *A. fulgidus* can be prepared from frozen cell pellets (−80°C), liquid stocks, or a colony from a roll tube. For routine manipulations, *A. fulgidus* can be grown in 50 ml (in a 100-ml serum vial) of STL medium for 1–2 days, then stored at room temperature or 4°C. Liquid stocks should be transferred to fresh medium every 2–3 weeks although growth can be obtained from cells stored at 4°C after 2–3 years. Typically, a 5% (vol) inoculum of cells is transferred to fresh medium using a N_2-flushed sterile syringe and anaerobic techniques.

Genetics

The 2.178 Mbp single circular chromosome of *Archaeoglobus fulgidus* VC-16 has been sequenced and annotated (Klenk et al., 1997; The Institute for Genome Research, *Archaeoglobus fulgidis* DSM4303 Genome page). It is estimated that there are 2436 open reading frames (ORFs) that cover ≈92% of the genome with about 1.1 ORF per kb. Analysis of the genome reveals extensive gene duplication, which likely contributes to the *A. fulgidus* genome being larger than the *Methanocaldococcus jannaschii* genome. The duplicated genes are not identical and likely have slightly different functions. Hence, gene duplication may have provided an evolutionary advantage as a mechanism for increasing physiological diversity. Of the 2436 ORFs, 651 are predicted to encode conserved proteins of unknown function, 428 of which are also found in the genome of *M. jannaschii*. Another 639 have no match in the databases.

The average gene is 822 bp, making the average protein about 30,000 daltons. Predicted polypeptides of *A. fulgidus* range from M_r 1939 to 266,571. The average gene size is similar to that of *M. jannaschii*, but smaller than the average eubacterial gene. While overall the genome has a DNA base composition of 48.5 mol% G+C, some lateral transfer of DNA may have occurred because three regions have low G+C content (<39 mol%) and two have high G+C content (>53 mol%). Low G+C regions contain genes that encode enzymes involved in lipopolysaccharide synthesis, while the high G+C regions contain genes for large ribosomal proteins, heme biosynthesis, and transporters. There is evidence of lateral transfer of DNA from *A. fulgidus* to eukaryotic cells. The malic enzyme of *Entamoeba histolytica*, which decarboxylates malate to pyruvate, appears to have originated with *A. fulgidus* (Field et al., 2000).

While a typical *A. fulgidus* gene initiates with ATG, 22% begin with GTG and 2% begin with TTG. No inteins have been found in *A. fulgidus* although, like in other Archaea, five of the 46 tRNAs carry 15–62-bp introns. The introns are thought to be removed by the intron excision enzyme EndA. Many small nonmessenger (SN) RNAs have been characterized from *A. fulgidus* (Tang et al., 2002). Some of these SN RNAs originated from rRNA splicing, but many were complementary to or overlapping with an ORF. While the development of genetic tools in *A. fulgidus* is in its infancy, one plasmid, pGS5, has been isolated from *A. profundus* strain AV18. The plasmid pGS5 is 2.8 kb and negatively supercoiled (Lopez-Garcia et al., 2000). No phage has been found for members of the genus *Archaeoglobus*.

Archaeoglobus fulgidus contains 40 ATP-binding subunit members of the superfamily of ATP-binding cassette (ABC) transporters for arsenite, Fe^{+2}, Fe^{+3}, proline, ribose, lactate, sulfate, hexuronate, and Cu^{+2}, to name a few. As might be expected for a marine organism, *A.*

fulgidus produces a number of proteins that regulate ionic homeostasis. These include transporters for Na^+, Fe^{+2}, Fe^{+3}, NO_3^-, and SO_4^{-2}, as well as transporters to eliminate toxic compounds such as copper, arsenite and cyanate.

DNA replication in *A. fulgidus* involves DNA polymerases that are related to the Δ polymerase of eukaryotes. Unlike other Archaea, *A. fulgidus* also encodes a homolog of the ε subunit of *E. coli* PolIII, which is involved in proofreading. Several DNA repair enzymes also exist, including a eukaryotic-like Rad25, exodeoxynuclease III, reverse gyrase, and topoisomerase VI. *Archaeoglobus fulgidus* contains a novel type of uracil DNA glycosylase, which is involved in repair of uracil that arises in DNA after hydrolytic deamination of cytosine (Sandigursky and Franklin, 2000; Chung et al., 2001). The product of *nfi* is a deoxyinosine-specific endonuclease which can repair DNA lesions that occur after spontaneous deamination of deoxyadenosine (Liu et al., 2000). Although the *nfi* gene encodes a homolog of endonuclease V from *E. coli*, it is a unique protein in that it recognizes only deoxyinosine, in contrast with the *E. coli* enzyme.

A gene encoding a homolog of the bacterial RecA and eukaryotic Rad51 has been identified from the *A. fulgidus* genome. The *A. fulgidus* *radA* gene was expressed in *E. coli*, and the biochemical properties of purified RadA were compared with RadA proteins from other Archaea. Like the *Desulfurococcus amylolyticus* RadA protein, the *A. fulgidus* RadA is stable at high temperatures. RadA catalyzes DNA-stimulated ATP hydrolysis, D-loop formation and DNA strand exchange at 60–70°C (McIlwraith et al., 2001). The 37-kDa RadA protein does not bind double-stranded (ds)DNA, but forms filaments on single-stranded (ss)DNA. In solution, RadA forms ring-like structures.

Genes encoding homologs of bacterial-like transcription activators, such as Lrp, MerR, TetR, and PhoU, and sensor-kinase response regulators (two-component family) are abundant in *Archaeoglobus* (Klenk et al., 1997). In contrast, homologs of transcription factors (such as σ^{54}, MetJ, NusB, and Rho terminator) have not been found (Kyrpides and Ouzounis, 1999). As is the case for other Archaea, *A. fulgidus* has genes encoding eukaryotic-like initiation factors TFIIB, TFIID, and TFIIEα, small subunits of RNA polymerase (RPB5, RPB6, RPB8, RPB10, and RPB12), the large universal RNA polymerase subunits, and archeal histones. *Archaeoglobus fulgidus* differs from methanogens whose genomes have been sequenced in that it encodes a homolog of the eukaryotic TATA-binding protein (TBP)-interacting protein 49 and Silent Information Regulator (Sir2, the nicotine adenine dinucleotide [NAD]-dependent protein deacetylase that is a eukaryotic transcriptional silencer). In *A. fulgidus*, the Sir2 protein may play a role in cell-cycle progression and genome integrity. Eukaryotic-like transcription factor domains, such as the MADS box or homeoboxes, have not been found.

Ecology and Environmental Stress

Extremophiles have characteristics that allow them to thrive under environmental conditions that are lethal to most known cultured organisms. In particular, hyperthermophiles have developed cellular structures to withstand adverse conditions of heat. Enzymes can be stabilized at high temperature by adopting particular protein conformation, increased substrate turnover, molecular chaperones and/or stabilizing molecules.

A comparison of the stability of the enzymes from *A. fulgidus* suggests that different methods are employed by the cell to stabilize different enzymes. An enzyme such as the D-lactate dehydrogenase seems to have stabilization imprinted in the primary amino acid structure (e.g., recombinant protein half-life of 105 minutes at 83°C was similar to that of the native protein) whereas the native reduced nicotine adenine dinucleotide (NADH) oxidase is much more thermally stable than the recombinant form, suggesting stabilizing components may be important (Reed and Hartzell, 1999; Reed et al., 2001). The L-malate dehydrogenase carries a deletion of a specific loop at the dimer-dimer contact region, which may ensure that the dimeric form is maintained under high temperature (Madern et al., 2001), again suggesting that primary sequence dictates structural stability. The catalase-peroxidase uses turnover for stability. The catalase has a very high turnover number (the highest reported) even though stability is rather low in the presence of peroxide (Kengen et al., 2001).

Substrates, chaperones and solutes also play a critical role in stabilizing structures at high temperatures. Although *A. fulgidus* does not contain a *dnaK* (HSP70) homolog, it does have two genes that encode α and β subunits of the TCP-1 chaperone family (Klenk et al., 1997). Solutes also play a role in thermal stabilization of cellular components. For example the N^5, N^{10} methenyl tetrahydromethanopterin (H₄MPT) cyclohydrolase from *A. fulgidus* requires relatively high concentrations of K_2HPO_4 (1 M) for optimal thermostability at 90°C (Klein et al., 1993). The N^5, N^{10} methylene H₄MPT reductase is rapidly inactivated when incubated at 80°C; however, activity is stabilized in the presence of either substrate (F_{420} [0.2 mM] or methylene-H₄MPT [0.2 mM]) or molecules such as albumin (1 mg/

ml) or KCl (0.5 M; Schmitz et al., 1991). Compatible organic solutes such as trehalose or α-glutamate, which are important for mesophile stability, accumulate in thermophilic or hyperthermophilic archaea. In addition to these, the heat stable solute diglycerol phosphate (1,1′-diglyceryl phosphate) appears to be unique to *A. fulgidus* (Lamosa et al., 2000). Diglycerol phosphate production by *A. fulgidus* increases in concentration as the salinity and temperature of the growth medium increase (Martins et al., 1997; Lamosa et al., 2000). Diglycerol phosphate acts by protecting proteins against denaturation as opposed to refolding denatured proteins. Diglycerol phosphate remains stable at 95°C for at least 3 hours and can stabilize enzymes from mesophilic organisms at thermophilic temperatures (50–60°C). The *A. fulgidus* inositol-1-phosphate synthase (AF1794) is a class II aldolase that catalyzes the irreversible conversion of D-glucose 6-phosphate to L-*myo*-inositol 1-phosphate (Chen et al., 2000). This inositol (a ring-containing compound) is a common hyperthermophilic archaeon osmolyte and may also be an important response solute to stress associated with supraoptimal temperature or high salinity.

Although *A. fulgidus* is a strict anaerobe, it has the ability to survive exposure to oxygen (Beeder et al., 1994; LaPaglia and Hartzell, 1997). While some *A. fulgidus* enzymes, such as the D-lactate dehydrogenase, are not adversely affected by oxygen (Reed and Hartzell, 1999), others, such as formyl methanofuran dehydrogenase, are rapidly inactivated in the presence of oxygen (Schmitz et al., 1991). One mechanism of protection and recovery from oxygen exposure may be provided by the enzymatic reduction of oxygen in the cellular environment. *Archaeoglobus fulgidus* has several NADH oxidases (Nox) that are expressed constitutively and are located in the periplasmic space (between the membrane and the S-layer), which may establish a first line of defense against O_2 (Pagala et al., 2002). One NADH oxidase, the 47-kDa NoxA2 (AF0395), carries out bivalent oxygen reduction with the evolution of H_2O_2 (Reed et al., 2001) and H_2O. NoxA2 is stable in oxygen and at high temperatures (83°C) for up to 35 hours (Reed et al., 2001). Taken together, these findings hint that NoxA2 may be involved in detoxification of oxygen under conditions of oxidative stress.

Archaeoglobus fulgidus also encodes proteins that are homologous to *Desulfovibrio* proteins that have been shown to reduce O_2 to H_2O (Abreu et al., 2000b). These include a flavoprotein rubredoxin and rubredoxin/oxygen oxidoreductase. The *A. fulgidus* genome does not contain a gene predicted to encode a canonical superoxide dismutase (SOD) but instead appears to have an independent mechanism to protect itself from toxic oxygen products such as superoxide ($O_2 \cdot^-$) and H_2O_2. A small (14.5 kDa), soluble, blue-colored, iron-containing protein named "neelaredoxin" (Nlr), which was previously identified in *Desulfovibrio gigas*, has been identified in *A. fulgidus* and is able to reduce and dismutate $O_2 \cdot^-$ efficiently (Abreu et al., 2000a; Abreu et al., 2001; Abreu et al., 2002). The products of the reaction of $O_2 \cdot^-$ and neelaredoxin are H_2O_2 or H_2O and O_2. The reductase activity may allow the cell to eliminate $O_2 \cdot^-$ by a reduced nicotinamide adenine dinucleotide phosphate (NAD[P]H)-dependent pathway, whereas the dismutation activity allows the cell to detoxify $O_2 \cdot^-$ independently of the cell redox status. The electron donor for neelaredoxin-mediated reduction of $O_2 \cdot^-$ to H_2O_2 involves an NADH oxidase present in *A. fulgidus* extracts (Abreu et al., 2000a).

The cytotoxic H_2O_2 generated by NoxA2 and Nlr-dependent reactions can be catabolized by the catalase-peroxidase (*perA* gene-AF2233; Kengen et al., 2001). The catalase activity strongly exceeds the peroxidase activity, and combined NoxA2 and catalase activity can decrease the oxygen content to half (O_2 + $2NADH \leftrightarrow H_2O_2 + 2NAD^+$; $H_2O_2 \leftrightarrow H_2O + \frac{1}{2}O_2$, respectively). The catalase is active only under normal physiological conditions (pH 5.5–7 and 40–80°C), whereas the peroxidase ($AH_2 + H_2O_2 \leftrightarrow A + 2H_2O$) functions optimally during stages that might be considered environmental stress (70–90°C and pH 4–5), when a decrease in oxygen by the catalase only (incomplete oxygen removal) might be inadequate for cell survival. Although the recombinant peroxidase enzyme exhibits high thermostability, rapid inactivation occurs in the presence of H_2O_2 unless a proton donor is present.

Many microorganisms, including the archaea, form biofilms as a common stress response mechanism. *Archaeoglobus fulgidus* forms a morphologically variable biofilm containing protein, polysaccharide, and metals in response to environmental stresses induced by extremes of pH and temperature, nutrient changes, and exposure to metals, antibiotics, xenobiotics, or oxygen (LaPaglia and Hartzell, 1997). The cells within the biofilm have increased tolerance to an otherwise toxic environmental condition, suggesting that cells may produce biofilms to concentrate cells in a protective barrier and, in some cases, to store nutrients.

Metabolism

Archaeoglobus fulgidus grows well using D-lactate, L-lactate or pyruvate, but not acetate, as a source of carbon and energy (Fig. 5). *Archaeoglobus fulgidus* can also use oxaloacetate, formate, or CO_2/H_2 as electron and carbon donor (Stetter et al., 1987). *Archaeoglobus fulgidus* strain VC-16 is reported to grow on substrates including glucose, formamide, casamino acids, peptone, casein, meat extract, yeast extract, cell homogenates, methanol, ethanol, formate, starch and gelatin; however, growth rates are minimal compared to those on the previously mentioned substrates (Stetter et al., 1987; Stetter, 1988). Additional reports show that glucose is not utilized by *A. fulgidus* VC-16 and that neither an uptake-transporter nor a catabolic pathway is identified at the genetic level for glucose (Klenk et al., 1997), which is not surprising considering most sulfate-reducing bacteria typically are unable to utilize glucose. Although growth on acetate has not been demonstrated, the genes for acetate utilization, such as acetyl CoA synthetase, are present in *A. fulgidus* VC-16. Moreover, genes encoding oxidoreductase enzymes that could generate energy from nitrate, nitrite and polysulfide also are present. The presence of multiple enzymes for β-oxidation and ferredoxin-dependent oxidoreductases in the genome suggests that under appropriate conditions, *A. fulgidus* uses hydrocarbons and organic acids as a source of carbon. An acyl-CoA ligase related to FadD may play a role in the import of long-chain fatty acids across the membrane (Klenk et al., 1997).

Archaeoglobus fulgidus is able to synthesize the amino acids, vitamins and cofactors, purines and pyrimidines that it needs to sustain growth, but it lacks biochemical pathways that are common to many bacteria. For example, most if not all of the genes for the pentose-phosphate, Entner-Doudoroff, glycolysis and gluconeogenesis pathways are absent (Klenk et al., 1997). However, there is evidence that *A. fulgidus* uses atypical enzymes to carry out certain reactions within these pathways. For example, *A. fulgidus* does not have a fructose 1,6-bisphosphatase (FBPase), yet it produces an inositol monophosphatase (IMPase) that has FBPase activity (Rashid et al., 2002). Interestingly, *A. fulgidus* does not have typical bacterial polysaccharide biosynthetic machinery, but it does encode glycosyl transferases, suggesting the ability to make polysaccharides. The products of the ≈78 amino acid biosynthetic genes appear to form pathways that are similar to those found in *Bacillus subtilis*. Surprisingly, the genes for synthesis of biotin and pyridoxine are missing from the genome even though both compounds can be detected in cell extracts (Noll and Barber, 1988). Hence, these components may also be produced by an alternative pathway.

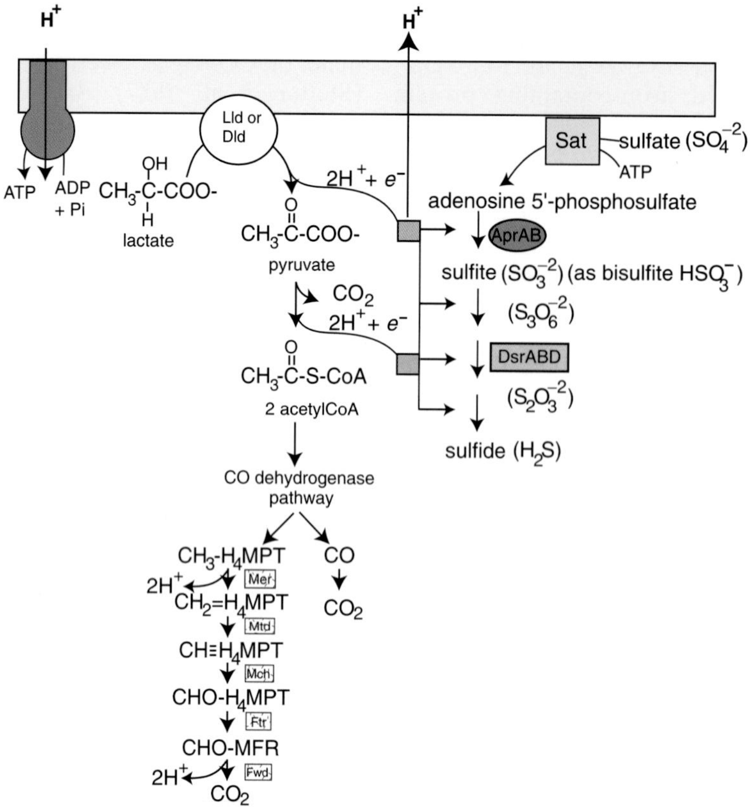

Fig. 5. Energy production using lactate as a source of carbon and energy. Lactate is converted to pyruvate by D- or L-lactate dehydrogenase. Electrons from catabolism of lactate are used to reduce sulfate to sulfide, and release of protons generates a proton motive force (PMF) that is used to make ATP via ATPase. Sat = sulfate adenylyltransferase (ATP sulfurylase; AF1667; 53 kDa); AprAB = adenylylsulfate reductase subunits A (AF1670; 73 kDa) and B (AF1669; 17 kDa); and DsrABD = sulfite reductase (α = AF0423, 47.5 kDa, β = AF0424, 41.8 kDa, and γ = AF0425, 8.8 kDa). Mer = N^5,N^{10}-methylene tetrahydromethanopterin reductase (*mer-1* = AF1066, *mer-2* = AF1196); Mtd = N^5,N^{10}-methylene tetrahydromethanopterin dehydrogenase (AF0714); Mch = N^5,N^{10}-methenyltetrahydromethanopterin cyclohydrolase; Ftr = formylmethanofuran H4MPT formyltransferase (*ftr-1* = AF2073 and *ftr-2* = AF2207); FwdA-G = formylmethanofuran dehydrogenase (A= AF1930, B-1 = AF1650, B-2= AF1929, C = AF1931, D-1 = AF1651, D-2 = AF1928, E= AF0177, F= AF1644, and G= AF1649). *Archaeoglobus fulgidus* also encodes Mtr (N^5-methyltetrahydromethanopterin:coenzyme M methyltransferase [AF0009; not shown]).

Archaeoglobus fulgidus lacks a complete citric acid cycle. Key metabolic intermediates are produced from a partial cycle to synthesize molecules such as amino acids and to generate reducing power (reduced nicotinamide adenine dinucleotide phosphate [NADPH]). For example, enzymes required for reductive formation of succinate from oxaloacetate and oxidative formation of 2-oxoglutarate (α-ketoglutarate) from oxaloacetate and acetyl-CoA are present (Steen et al., 2001). *Archaeoglobus fulgidus* has an NADP$^+$-dependent isocitrate dehydrogenase (AF0647) that catalyzes the formation of 2-oxoglutarate, CO$_2$ and NAD$^+$ from D-isocitrate, although the enzyme also has specificity for NAD$^+$, resulting in low 2-oxoglutarate oxidation (Steen et al., 1997).

The L-malate dehydrogenase (AF0855) from *A. fulgidus* catalyzes the reversible reduction of oxaloacetate to L-malate with NADH (Langelandsvik et al., 1997). The rate of reduction of oxaloacetate with NADH is tenfold greater than malate oxidation with NAD$^+$, suggesting a preferred biosynthetic role. Interestingly, the enzyme shows higher amino acid sequence homology to lactate dehydrogenases than to other malate dehydrogenases, although L-malate and L-lactate dehydrogenases belong to the same family of NAD-dependent enzymes (Madern et al., 2001).

Archaeoglobus fulgidus uses inorganic molecules or amino acids as a source of nitrogen. No genes for N$_2$ fixation have been found. NADP$^+$-specific glutamate dehydrogenase (EC 1.4.1.4)

Enzyme	Subunit structure (kDa)	pH optimum	Temp. optimum (°C)	Prosthetic groups	Substrate	K_m	Specific activity
$F_{420}H_2$ quinone oxidoreductase	7 peptides (56, 45, 41, 39, 37, 33, and 32)	8.0	65	FAD, non-heme iron, acid-labile sulfur, F_{420}	2,3-dimethyl-1,4-naphthoquinone / F_{420}	190 µM / 50 µM	500 U/mg
N^5, N^{10} methylene H_4MPT reductase	200 kDa homomultimer (35)	7.1	65	ND	CH_2H_4MPT / $F_{420}H_2$	16 µM / 4 µM	450 U/mg
methylene H_4MPT dehydrogenase	140 kDa homomultimer (32)	5.5	NA	ND	N^5, N^{10} methylene H_4MPT / F_{420}	17 µM / 13 µM	5000 U/mg
N^5, N^{10} methenyl H_4MPT cyclohydrolase	Homodimer (39)	8.5	85	ND	$CH{\equiv}H_4MPT^+$	220 µM	11300 U/mg
formyl methanofuran H_4MPT formyltransferase	125 kDa homomultimer (30)	5.0	70	ND	formyl methanofuran / H_4MPT	32 µM / 17 µM	14.4 U/mg (total cell extract)
formyl methanofuran dehydrogenase	Probably 6 peptides	NA	NA	ND	NA	10 µM	19 U/mg
ATP sulfurylase	Homotrimer (55)	8.0	90	ND	APS / PPI	0.17 mM / 0.13 mM	480 U/mg
APS reductase	160 kDa $\alpha_2\beta$ trimer	8.0	85	FAD, nonheme iron, labile sulfide	AMP	1 mM	2.39 U/mg
sulfite reductase	218 kDa $\alpha_2\beta_2$ tetramer	NA	85	siroheme, non-heme Fe, acid labile sulphide	NA	NA	NA
ferric reductase	Homodimer (18)	7.0	88	FMN	NADH / NADPH	61 µM / 80 µM	4935 U/mg / 3505 U/mg

NA, not available; ND, not detected.

[a]Mostly strain VC16 (Kunow et al., 1993).

[b]Links catabolism to anabolism (Kunow et al., 1993).

[c]Carbon monoxide (Möller-Zinkhan et al., 1989).

CO$_2$ and 1.5 moles of H$_2$S, and lactate-grown cells contain high amounts of F$_{420}$ (0.6 nmoles/mg protein), indicating utilization of the reverse-methanogenesis related enzymes.

A subset of Embden-Meyerhof pathway enzymes has been detected in extracts of starch-grown cells (Labes and Schönheit, 2001). Present were phosphoglucose isomerase, which converts α-D-glucose 6-phosphate to β-D-fructose 6-phosphate, and fructose-1, 6-diphosphate aldolase, which produces D-glyceraldehyde 3-phosphate from β-D-fructose 1,6-diphosphate. Conventional reactions are carried out by the triose-phosphate-isomerase (dihydroxyacetone-phosphate conversion to D-glyceraldehyde 3-phosphate), phosphoglycerate mutase (3-phospho-D-glycerate conversion to 2-phospho-D-glycerate), enolase (2-phospho-D-glycerate to phosphoenol-pyruvate) and pyruvate kinase (phosphoenol-pyruvate to pyruvate). Although glyceraldehyde-3-phosphate dehydrogenase activity was present, phosphoglycerate kinase activity was not detected. Hence, D-glyceraldehyde 3-phosphate-dependent oxidation does not proceed by 1,3-bisphosphate D-glycerate. Glyceraldehyde-3-phosphate: ferredoxin oxidoreductase (GAP:FdOR) is believed to be the enzyme responsible for D-glyceraldehyde 3-phosphate oxidation to 3-phospho-D-glycerate. The GAP:FdOR enzyme was not present in lactate-grown cells but was present in starch-grown cells.

ATP does not serve as a phosphoryl donor for the ADP-dependent hexokinase (ADP-HK) and for the 6-phosphofructokinase (ADP-PFK). The K$_m$ for ADP and glucose with ADP-HK is 0.2 and 2.6 mM, respectively, and the K$_m$ for ADP and fructose 6-phosphate with ADP-PFK is 1 and 3.3 mM, respectively. ATPase activity is high in starch-grown cells to supply the required ADP. Overall specific activities of ADP-HK, ADP-PFK, GAP:FdOR, and PK are significantly higher in starch-grown cells than in lactate-grown cells, indicating induction of these enzymes during starch catabolism.

The pyruvate:ferredoxin oxidoreductase converts pyruvate to acetyl-CoA. The conventional enzymes involved in conversion of acetyl-CoA to acetate (phosphate acetyltransferase and acetate kinase) have not been detected by enzymatic assay (Möller-Zinkhan and Thauer, 1990), nor are there genes encoding homologs of these enzymes in the genome (Klenk et al., 1997). *Archaeoglobus fulgidus* does however make an ADP-forming acetyl coenzyme A synthetase (ACD), a novel enzyme found in Archaea and eukaryotic protists that is involved in acetate formation and energy production. ACD converts acetyl-CoA to acetate in a single step (acetyl-CoA + ADP + Pi acetate + ATP + CoA; Musfeldt and Schönheit, 2002). ACD also can utilize acyl-CoA esters to produce ATP and the corresponding acid. Two ACD ORFs (AF1211 and AF1938) of *A. fulgidus* each generate homodimers (ACD I and ACD II, respectively) of about 140 kDa composed of two identical 70-kDa subunits. ACD I has high substrate specificity toward acetate (k$_{cat}$/K$_m$ 0.4 μM^{-1}s^{-1}) and acetyl-CoA (k$_{cat}$/K$_m$ 9.2 μM^{-1}s^{-1}), whereas ACD II has higher specificity for phenylacetate (k$_{cat}$/K$_m$ 0.0012 μM^{-1}s^{-1}) and phenylacetyl-CoA (k$_{cat}$/K$_m$ 0.14 μM^{-1}s^{-1}) but not acetyl-CoA (k$_{cat}$/K$_m$ <0.000046 μM^{-1}s^{-1}). Physiological roles for these enzymes in *A. fulgidus* have been proposed on the basis of their similarity with the *Pyrococcus furiosus* enzymes. ACD I is predicted to play a role in utilization of acetyl-CoA derived from carbohydrates whereas ACD II is involved in the degradation of aryl-CoA esters derived from aromatic 2-keto acids. Both enzymes are thought to participate in degradation of branched chain acyl-CoAs derived from branched chain 2-keto acids (Musfeldt and Schönheit, 2002).

Acetate is an end product of starch metabolism. Hence, although the acetyl-CoA/CO-dehydrogenase (which converts acetyl-CoA to CO$_2$ via CO) is present, it must be inactive. The F$_{420}$-dependent enzymes are linked to the acetyl-CoA/CO-dehydrogenase pathway; their apparent absence in cells grown on starch suggests that they limit the rate of acetyl-CoA/CO-dehydrogenase activity. The production of F$_{420}$ appears to be substrate regulated. This is substantiated by the limited production of F$_{420}$ and F$_{420}$-dependent dehydrogenases for *A. fulgidus* VC-16 grown on CO$_2$ and for the chemolithoautotroph *A. lithotrophicus* (Vorholt et al., 1995).

Archaeoglobus fulgidus cannot degrade cellulose (glucose β [1–4] linkage; Bayer et al., 1999) because it lacks a known glycosyl hydrolase gene. However, it has two ORFs that contain cohesin domains and one with a dockerin domain, which are characteristic of proteins in cellulosome complexes used by cellulolytic microorganisms to hydrolyze plant cell wall polysaccharides. It is unclear what function(s) the putative cohesin and dockerin-containing proteins play in *A. fulgidus*.

CARBON FIXATION *Archaeoglobus fulgidus* is able to oxidize lactate to three carbon dioxide molecules via pyruvate and acetyl-CoA using the reverse methanogen C$_1$ and carbon monoxide dehydrogenase pathways (Möller-Zinkhan and Thauer, 1990). As mentioned previously, in *A. fulgidus* the enzymes and cofactors associated with the reduction of CO$_2$ to methane by methanogens are used to oxidize the methyl group from acetyl-CoA to CO_2 (Möller-Zinkhan et al., 1989). In addition, *A. fulgidus* has an acetyl-CoA decarbonylase/synthase (ACDS) multienzyme complex that catalyzes the reversible cleavage and synthesis of acetyl-CoA from CO_2 (Dai et al., 1998). The ACDS complex is the key enzyme component of the acetyl-CoA catabolism pathway and is structurally and functionally similar to the ACDS complex found in methanogens. The C_1-carrier cofactor in these reactions is the tetrahydromethanopterin (H_4MPT, a folate analog) as in methanogens (Gorris et al., 1991); the sulfate-reducing bacteria use tetrahydrofolate as a C_1 carrier. The H_4MPT is typically associated with hydrogenotrophs whereas acetotrophs typically contain sarcinapterin (Gorris et al., 1991). The ACDS complex in *A. fulgidus* is also able to utilize tetrahydrosarcinapterin (H_4SPt) and *Methanosarcina* ferredoxin in catalysis of acetyl-CoA synthesis (Dai et al., 1998).

During lactate catabolism, acetyl-CoA is cleaved into two C_1-units, one of which is oxidized to CO_2 via CO with the ACDS complex (acetyl-CoA + H_4MPT + 2Fd_{ox} $\leftrightarrow$ CH_3-H_4MPT + CO_2 + 2Fd_{red}). This reaction is reversible during autotrophic growth and catalyzes the overall synthesis of acetyl-CoA according to the following reaction: CO_2 + 2 Fd_{red}(Fe^{2+}) + 2H^+ + CH_3H_4SPt + CoA $\leftrightarrow$ acetyl-CoA + H_4SPt + 2Fd_{ox}(Fe^{3+}) + H_2O (Fd = ferredoxin; CH_3-H_4SPt and H_4SPt denote N^5-methyl-tetrahydrosarcinapterin and tetrahydrosarcinapterin, respectively). The ACDS enzyme comprises a significant level (5%) of the total soluble cell protein, suggesting that it may be required to ensure a flux of electrons for sulfate reduction. The corrinoid, 5' methyl-benzimidazolyl cobamide, found in sulfate-reducers but not methanogens, represents a significant component in *A. fulgidus* due to its involvement in this reaction (Dai et al., 1998).

The other C_1-unit of acetyl-CoA is transferred to H_4MPT and electrons are transferred to the 5-deazaflavin coenzyme F_{420} by the N^5, N^{10}-methylene H_4MPT reductase (CH_3-H_4MPT + F_{420} CH_2 = H_4MPT + $F_{420}H_2$; Schmitz et al, 1991; Kunow et al., 1993). The reduced 5-deazaflavin coenzyme ($F_{420}H_2$) is probably oxidized by the membrane-bound $F_{420}H_2$ quinone oxidoreductase, with electrons transferred to menaquinone (Kunow et al., 1994). $F_{420}H_2$ quinone oxidoreductase is encoded by the 11-gene *fqo* gene cluster (*fqo* J, K, M, L, N, A, B, C, D, H, I, F) and is involved in energy conservation (Bruggemann et al., 2000). The amino acid sequences of all other genes (except *fqo*F) show homology to NADH-quinone oxidoreductases from prokaryotes and eukaryotes, suggesting that the F_{420} H_2-dependent and the NADH-dependent enzymes are

functional equivalents. The 39-kDa protein product of the *fqo*F gene is responsible for oxidation of F_{420}. The C_1-unit is oxidized to CO_2 sequentially by the methylene H_4MPT dehydrogenase, transferring electrons to coenzyme F_{420} 5,10-methylene-H_4MPT (CH_2 = H_4MPT) + F_{420} + H^+ $\leftrightarrow$ 5, 10-methenyl-H_4MPT (CH $\equiv$ H_4MPT$^+$) + $F_{420}H_2$; Kunow et al., 1993; Schwörer et al., 1993), N^5, N^{10}-methenyl-H_4MPT cyclohydrolase (5, 10-methenyl) (CH $\equiv$ H_4MPT$^+$ + H_2O $\leftrightarrow$ 5-formyl-H_4MPT (CHO-H_4MPT) + H^+; Klein et al., 1993), formyl methanofuran-H_4MPT formytransferase (5-formyl-H_4MPT (CHO-H_4MPT) + methanofuran (MFR) $\leftrightarrow$ formyl-MFR CHO-MFR + H_4MPT; Schwörer et al., 1993), and formyl methanofuran dehydrogenase (formyl-MRF (CHO-MFR + X + H_2O $\leftrightarrow$ CO_2 + MFR + XH_2; electron acceptor has not been identified; Schmitz et al., 1991).

Most anaerobes in the Archaea capture (fix) CO_2 via the acetyl CoA or reverse citric acid cycle pathways. Hence, finding two genes that are predicted to encode a ribulose 1,5-bisphosphate carboxylase/oxygenase (rubisco)-like protein in the *A. fulgidus* genome was surprising. *Methanococcus jannaschii* also is predicted to encode a rubisco protein, which shares 41 and 45% identity, respectively, with rubisco form 1 and 2 from *A. fulgidus*. Although the Archaea rubisco proteins share only 33% identity with the form II subunit from *Rhodospirillum rubrum* and 41% identify with the form 1 subunit from *Synechococcus*, most of the critical active site residues are present in the Archaeal proteins. The *M. jannaschii* rubisco, expressed in *E. coli*, formed a homodimer that could convert ribulose bisphosphate and CO_2 to 3-phosphoglyceric acid at a reasonable, albeit lower, rate compared with its bacterial and eukaryotic counterparts. Oxygen inhibited the reaction reversibly, although it appears to bind tightly to the enzyme. The physiological role of rubisco in either *A. fulgidus* or *M. jannaschii* is unknown, but hints to its function have come from studies with the *Chlorobium tepidum* rubisco. Like the archaeal proteins, the *C. tepidum* rubisco lacks some of the typical active site residues that are critical for function. Moreover, there is evidence that the *C. tepidum* rubisco plays a role in oxidative stress and/or sulfur metabolism (Hanson and Tabita, 2001). The similarity between the *Chlorobium* and *Archaeoglobus* rubisco subunits hints that they may have similar functions.

Sulfate Reduction   Dissimilatory sulfate-reducing prokaryotes use sulfate, thiosulfate or sulfite as the terminal electron acceptor during anaerobic respiration. Elemental sulfur ($S°$) is inhibitory to growth of *A. fulgidus* when other electron acceptors are present. *Archaeoglobus fulgidus* contains the complete conserved dissimilatory sulfate reduction pathway as characterized in sulfate-reducing bacteria (SRB). Phylogenic analysis of the ATP sulfurylases, adenosine 5'-phosphosulfate (APS) reductases, and sulfite reductases suggests that the proteins became specialized for assimilatory and dissimilatory purposes prior to divergence of Archaea and Bacteria (Dahl and Trüper, 2001).

Sulfate reduction in SRB occurs in the cytoplasm and the enzymes that carry out these reactions associate with unknown membrane-associated electron carriers. During dissimilation, sulfate must first be activated by ATP to accept electrons, because reduction of sulfate to sulfite alone is not energetically favorable. Sulfate is activated to adenylsulfate by ATP sulfurylase, the product of the *sat* gene. Sulfate adenylate transferase (Sat) is the key enzyme in this activation for both dissimilatory and assimilatory sulfate reduction. The Sat protein is similar to homo-oligomeric adenosine triphosphate (ATP) sulfurylases from sulfur-oxidizing bacteria and sulfate assimilating organisms (Sperling et al., 1998). When expressed and purified from *E. coli*, the ATP sulfurylase is a homodimer. The ATP sulfurylase (sulfate adenylyltransferase) catalyzes the reaction of sulfate at the expense of ATP to generate APS. Inorganic pyrophosphate (PP_i) is released during this reaction. APS is reduced to AMP and sulfite (SO_3^{-2}) by the *aprAB* gene products, which encode adenylylsulfate reductase. In the final six-electron transfer, sulfite is reduced to sulfide by the products of the *dsrABD* genes encoding sulfite reductase. Typically, sulfide is released as volatile hydrogen sulfide (H_2S) gas. Although *A. fulgidus* contains *cysC*, which encodes adenylylsulfate 3-phosphotransferase (APS kinase), this enzyme is not involved in dissimilatory sulfate reduction (Sperling et al., 1998).

The ATP sulfurylase (MgATP-sulfate adenylyltransferase, EC 2.7.7.4) forms a homotrimer in *A. fulgidus*, although the recombinant form is a homodimer (Dahl et al., 1990; Sperling et al., 2001). The reaction is thermodynamically unfavorable and there is evidence that activation of ATP sulfurylase by a potent inorganic pyrophosphatase (EC 3.6.1.1) helps to pull the reaction in the forward direction (Dahl et al., 1990). The recombinant form requires 0.5 M NaCl for thermal stability and has an optimal stability at >100°C, whereas the native enzyme is most stable at 90°C. The recombinant protein is tenfold less active than the native enzyme. These differences are thought to be due to incorrect folding or incorrect posttranslational modification (Sperling et al., 2001).

The native electron donor to the APS reductase (adenylylsulfate reductase, EC1.8.99.2) is unknown, but may be a reduced menaquinone (Tindall et al., 1989). The APS reductase structural genes (*aprA* and *aprB*) form an operon with the ATP sulfurylase and an unknown 13.6-kDa cytoplasmic protein (Speich et al., 1994; Sperling et al., 2001). The APS reductase is thermostable and comprises about 1.5% of the total soluble cellular protein in *A. fulgidus* (Speich and Trüper, 1988; Dahl and Trüper, 2001). The enzyme has 1 mole of FAD bound with 8 moles of nonheme iron and 6 moles of labile sulfide per enzyme. Two distinct iron-sulfur centers exist, one of which reacts with the substrates AMP and sulfite (Lampreia et al., 1991). The two centers, which transfer two electrons from the surface of the protein to FAD, differ significantly in reduction potential (–60 and –500 mV; Fritz et al., 2002b). The reductase has been crystallized and its structures in the two-electron reduced state and with sulfite bound to FAD at 1.6-Å and 2.5-Å resolution, respectively, were determined (Fritz et al., 2002b). Sulfite binds to FAD to form a covalent FAD-N5 sulfite adduct (Fritz et al., 2002a). The architecture of the active site suggests that catalysis involves a nucleophilic attack of the N5 atom of FAD on the sulfur of APS. The structure of the APS reductase is very similar to that of fumarate reductase and it has been suggested that they arose from a common ancestor (Fritz et al., 2002a).

The final step of sulfate respiration, the sulfite reductase (EC 1.8.99.1) reaction, is the central energy-conserving step. The native sulfite reductase represents 0.5% of the total soluble cellular protein (Dahl et al., 1993). The green-colored enzyme contains siroheme and non-heme iron atoms and acid labile sulfide that make six [4Fe-4S] clusters, two of which bind siroheme and four of which bind ferredoxin amino acid motifs. The β subunit lacks a single cysteine residue in one of the two cluster motifs, suggesting that only the α subunit binds the siroheme-[Fe_4S_4] complex, and chemical analyses show the presence of only two sirohemes per enzyme molecule. The subunit genes are transcribed in tandem and the peptides have 25.6% amino acid sequence identity, indicating that they may have arisen by a duplication event.

ALTERNATIVE ACCEPTORS AND METAL REDUCTION *Archaeoglobus fulgidus* can use electron acceptors besides sulfate, sulfite or thiosulfate for respiration. For example it can use the humic analog, anthraquinone-2,6-disulfonate (AQDS), as an electron acceptor (Lovely et al., 2000). Although other organisms use hydrogen as the electron donor in this reaction, in *A. fulgidus* only lactate serves as the electron donor in

AQDS reduction. The humics contain extracellular quinone moieties that can serve as electron acceptors that may in turn shuttle electrons to Fe(III) for reduction.

Iron oxidation has been well studied in organisms such as *Thiobacillus* and *Sulfolobus*, which can oxidize ferrous (Fe^{+2}) to ferric (Fe^{+3}) iron for energy, albeit very little energy is gained because Fe^{3+}/Fe^{2+} is +0.77V at pH 2 and $1/2O_2/H_2O$ is +0.82V. While dissimilatory iron (Fe^{3+}) reduction is poorly understood, it is becoming more apparent that a diversity of microorganisms can reduce (and respire) iron. Hence iron (Fe^{3+}) reduction has been hypothesized to be a universal event (Vargas et al., 1998). Although organisms such as *Pyrobaculum islandicum* and *Thermotoga maritima* can grow using H_2 as electron donor and Fe(III) as electron acceptor, these conditions have not been found to support the growth of *A. fulgidus* (Vargas et al., 1998). Neither can *A. fulgidus* reduce Au(III) under conditions that favor formation of colloidal gold by *P. islandicum* (Kashefi et al., 2001).

Archaeoglobus fulgidus contains highly thermostable Fe^{+3}-EDTA reductase activity in its soluble protein fraction (Vadas et al., 1999). The ferric reductase is closely related to the NAD(P)H:FMN oxidoreductases that are associated with monooxygenases involved in oxidative degradation of aromatic or hydrocarbon compounds (Vargas et al., 1998). Dissimilatory ferric reductases catalyze the reduction of Fe^{+3} in energy metabolism, where iron is taken up as complexed Fe^{+3} and reduced to Fe^{+2} using NADPH as the electron donor during assimilation of iron into the cell. The *A. fulgidus* ferric reductase is functionally most similar to the enzymes used in assimilation; however, the high specific activity and high abundance in the cell (0.75% soluble protein) suggest that the enzyme may be involved in dissimilatory iron reduction. It is speculated that a hydrogenase on the outside of the cell membrane would generate a proton gradient and hence a PMF to generate ATP. Electrons would transfer to an NADH dehydrogenase inside the cell to generate NADH. The ferric reductase would then regenerate NAD and serve as the terminal electron acceptor in a process not unlike sulfate reduction.

The *A. fulgidus* ferric reductase (*fer* gene-AF0830) appears to be a novel enzyme because it shares amino acid sequence similarity with a family of NADPH:FMN oxidoreductases but not with ferric reductases. Ferric reductase uses both NADH and NADPH as electron donors to reduce Fe^{+3}-EDTA and other complexes. Uncomplexed Fe^{+3} is not utilized as an acceptor (Vadas et al., 1999). The ferric reductase reduces FMN and FAD, but not riboflavin, with NADPH, which classifies the enzyme as a NADPH flavin oxidoreductase. The crystal structure of recombinant *A. fulgidus* ferric reductase shows the enzyme is homologous to the FMN-binding protein from *Desulfovibrio vulgaris*, with a fold type similar to the flavin-binding domain of the ferredoxin reductase superfamily (Chiu et al., 2001). The *A. fulgidus* ferric reductase is further distinguished from the ferredoxin reductase superfamily by the absence of a Rossmann fold domain that is used to bind the NADPH. Instead, ferric reductase uses its single domain to provide both the flavin and the NADPH binding sites. Potential binding sites for ferric iron complexes are identified near the cofactor binding sites. This enzyme lacks an aromatic residue stacked against the isoalloxazine ring, which typically protects the flavin from O_2 oxidation, suggesting anaerobic development.

Summary

The biochemistry, physiology, and genome sequence of *A. fulgidus* have contributed much to our understanding of the relationship of Archaea with Bacteria and Eukarya. Proteins from *A. fulgidus*, such as Sir2 (Min et al., 2001), adenylylsulfate reductase (Roth et al., 2000), MinD (Cordell and Lowe, 2001), Sm1 and Sm2 (small nuclear ribonucleoprotein particles; Toro et al., 2002), and splicing endosome (Li and Abelson, 2000), whose homologs have been difficult to crystallize from other organisms, have yielded tractable structural data. Structural information from novel enzymes, such as ferric reductase and the IMPase/FBPase, has provided insight into the possible origins and evolution of particular active site conformations.

Literature Cited

Aalen, N., I. H. Steen, N. K. Birkeland, and T. Lien. 1997. Purification and properties of an extremely thermostable NADP(+)-specific glutamate dehydrogenase from Archaeoglobus fulgidus. Arch. Microbiol. 168:536–539.

Abreu, I. A., L. M. Saraiva, J. Carita, H. Huber, K. O. Stetter, D. Cabelli, and M. Teixeira. 2000a. Oxygen detoxification in the strict anaerobic archaeon Archaeoglobus fulgidus: Superoxide scavenging by neelaredoxin. Molec. Microbiol. 38:322–334.

Abreu, I. A., L. M. Saraiva, J. Carita, H. Huber, K. O. Stetter, D. Cabelli, and M. Teixeira. 2000b. Oxygen detoxification in the strict anaerobic archaeon Archaeoglobus fulgidus: Superoxide scavenging by neelaredoxin. Molec. Microbiol. 38:322–334.

Abreu, I. A., L. M. Saraiva, C. M. Soares, M. Teixeira, and D. E. Cabelli. 2001. The mechanism of superoxide scavenging by Archaeoglobus fulgidus neelaredoxin. J. Biol. Chem. 276:38995–39001.

Abreu, I. A., A. V. Xavier, J. LeGall, D. E. Cabelli, and M. Teixeira. 2002. Superoxide scavenging by neelaredoxin: Dismutation and reduction activities in anaerobes. J. Biol. Inorg. Chem. 7:668–674.

Achenbach-Richter, L., K. O. Stetter, and C. R. Woese. 1987. A possible biochemical missing link among archaebacteria. Nature 327:348–349.

Balch, W. E., G. E. Fox, L. J. Magrum, C. R. Woese, and R. S. Wolfe. 1979. Methanogens: Reevaluation of a unique biological group. Microbiol. Rev. 43:260–296.

Barns, S. M., R. E. Fundyga, M. W. Jeffries, and N. R. Pace. 1994. Remarkable archaeal diversity detected in a Yellowstone National Park hot spring environment. Proc. Natl. Acad. Sci. USA 91:1609–1613.

Bayer, E. A., P. M. Coutinho, and B. Henrissat. 1999. Cellulosome-like sequences in Archaeoglobus fulgidu: An enigmatic vestige of cohesin and dockerin domains. FEBS Lett. 463:277–280.

Beeder, J., R. K. Nilsen, J. T. Rosnes, T. Torsvik, and T. Lien. 1994. Archaeoglobus fulgidus isolated from hot North Sea oil field waters. Appl. Environ. Microbiol. 60:1227–1231.

Benson, D. A., I. Karsch-Mizrachi, D. J. Lipman, J. Ostell, B. A. Rapp, and D. L. Wheeler. 2000. GenBank. Nucleic Acids Res. 28:15–18.

Bruggemann, H., F. Falinski, and U. Deppenmeier. 2000. Structure of the $F_{420}H_2$:quinone oxidoreductase of Archaeoglobus fulgidus identification and overproduction of the $F_{420}H_2$-oxidizing subunit. Eur. J. Biochem. 267:5810–5814.

Burggraf, S., H. J. Jannasch, B. Nicolaus, and K. O. Stetter. 1990. Archaeoglobus profundus sp. nov., represents a new species within the sulfate-reducing archaebacteria. Syst. Appl. Microbiol. 13:24–28.

Chen, L., C. Zhou, H. Yang, and M. F. Roberts. 2000. Inositol-1-phosphate synthase from Archaeoglobus fulgidus is a class II aldolase. Biochemistry 39:12415–12423.

Chiu, H. J., E. Johnson, I. Schroder, and D. C. Rees. 2001. Crystal structures of a novel ferric reductase from the hyperthermophilic archaeon Archaeoglobus fulgidus and its complex with NADP(+). Structure 9:311–319.

Chung, J. H., M. J. Suh, Y. I. Park, J. A. Tainer, and Y. S. Han. 2001. Repair activities of 8-oxoguanine DNA glycosylase from Archaeoglobus fulgidus, a hyperthermophilic archaeon. Mutat. Res. 486:99–111.

Cordell, S. C., and J. Lowe. 2001. Crystal structure of the bacterial cell division regulator MinD. FEBS Lett. 492:160–165.

Dahl, C., H. Koch, O. Keuken, and H. G. Trüper. 1990. Purification and characterization of ATP sulfurylase from the extremely thermophilic archaebacterial sulfphate-reducer, Archaeoglobus fulgidus. FEMS Microbio. Lett. 67:27–32.

Dahl, C., N. M. Kredich, R. Deutzmann, and H. G. Trüper. 1993. Dissimilatory sulphite reductase from Archaeoglobus fulgidus: Physico-chemical properties of the enzyme and cloning, sequencing and analysis of the reductase genes. J. Gen. Microbiol. 139:1817–1828.

Dahl, C., and H. G. Trüper. 2001. Sulfite reductase and APS reductase from Archaeoglobus fulgidus. Meth. Enzymol. 331:427–441.

Dai, Y. R., D. W. Reed, J. H. Millstein, P. L. Hartzell, D. A. Grahame, and E. DeMoll. 1998. Acetyl-CoA decarbonylase/synthase complex from Archaeoglobus fulgidus. Arch. Microbiol. 169:525–529.

Field, J., B. Rosenthal, and J. Samuelson. 2000. Early lateral transfer of genes encoding malic enzyme, acetyl-CoA synthetase and alcohol dehydrogenases from anaerobic prokaryotes to Entamoeba histolytica. Molec. Microbiol. 38:446–455.

Fritz, G., T. Buchert, and P. M. Kroneck. 2002a. The function of the [4Fe-4S] clusters and FAD in bacterial and archaeal adenylylsulfate reductases. J. Biol. Chem. 277:26066–26073.

Fritz, G., A. Roth, A. Schiffer, T. Buchert, G. Bourenkov, H. D. Bartunik, H. Huber, K. O. Stetter, P. M. Kroneck, and U. Ermler. 2002b. Structure of adenylylsulfate reductase from the hyperthermophilic Archaeoglobus fulgidus at 1.6-A resolution. Proc. Natl. Acad. Sci. USA 99:1836–1841.

Gorris, L. G. M., A. Voet, and C. Vanderdrift. 1991. Structural characteristics of methanogenic cofactors in the non-methanogenic archaebacterium Archaeoglobus fulgidus. Biofactors 3:29–35.

Hanson, T. E., and F. R. Tabita. 2001. A ribulose-1,5-bisphosphate carboxylase/oxygenase (RubisCO)-like protein from Chlorobium tepidum that is involved with sulfur metabolism and the response to oxidative stress. Proc. Natl. Acad. Sci. USA 98:4397–4402.

Hartzell, P. L., J. Millstein, and C. L. LaPaglia. 1999. Biofilm formation in a hyperthermophilic archaeon. Meth. Enzymol. 310:335–349.

Huber, H., H. Jannasch, R. Rachel, T. Fuchs, and K. O. Stetter. 1997. Archaeoglobus veneficus sp nov, a novel facultative chemolithoautotrophic hyperthermophilic sulfite reducer, isolated from abyssal black smokers. Syst. Appl. Microbiol. 20:374–380.

Huber, H., and K. O. Stetter. 2001. Archaeoglobus. In: D. R. Boone, R. W. Castenholz, and G. M. Garrity (Eds.) Bergey's Manual of Systematic Bacteriology. Springer-Verlag. New York, NY. 1:349–352.

Itoh, T., K. Suzuki, P. C. Sanchez, and T. Nakase. 1999. Caldivirga maquilingensis gen. nov., sp. nov., a new genus of rod-shaped crenarchaeote isolated from a hot spring in the Philippines. Int. J. Syst. Bacteriol. 49:1157–1163.

Kashefi, K., J. M. Tor, K. P. Nevin, Kand D. R. Lovley. 2001. Reductive precipitation of gold by dissimilatory Fe(III)-reducing bacteria and archaea. Appl. Environ. Microbiol. 67:3275–3279.

Kengen, S. W. M., F. J. Bikker, W. R. Hagen, W. M. de Vos, and J. van der Oost. 2001. Characterization of a catalase-peroxidase from the hyperthermophilic archaeon Archaeoglobus fulgidus. Extremophiles 5:323–332.

Kessel, M., S. Volker, U. Santarius, R. Huber, and W. Baumeister. 1990. Three-dimensional reconstruction of the surface protein of the extremely thermophilic archaebacterium Archaeoglobus fulgidus. Syst. Appl. Microbiol. 13:207–213.

Kim, D. Y., C. V. Stauffacher, and V. W. Rodwell. 2000. Dual coenzyme specificity of Archaeoglobus fulgidus HMG-CoA reductase. Protein Sci. 9:1226–1234.

Klein, A. R., J. Breitung, D. Linder, K. O. Stetter, and R. K. Thauer. 1993. N^5,N^{10}-Methenyltetrahydromethanopterin cyclohydrolase from the extrememly thermophilic sulfate reducing Archaeoglobus fulgidus: Comparison of its properties with those of the cyclohydrolase from the extremely thermophilic Methanopyrus kandleri. Arch. Microbiol. 159:213–219.

Klenk, H. P., R. A. Clayton, J. F. Tomb, O. White, K. E. Nelson, K. A. Ketchum, R. J. Dodson, M. Gwinn, E. K.

Hickey, J. D. Peterson, D. L. Richardson, A. R. Kerlavage, D. E. Graham, N. C. Kyrpides, R. D. Fleischmann, J. Quackenbush, N. H. Lee, G. G. Sutton, S. Gill, E. F. Kirkness, B. A. Dougherty, K. McKenney, M. D. Adams, B. Loftus, S. Peterson, C. I. Reich, L. K. NcNeil, J. H. Badger, A. Glodek, L. Zhou, R. Overbeek, J. D. Gocayne, J. F. Weidman, L. McDonald, T. Utterback, M. D. Cotton, T. Spriggs, P. Artiach, B. P. Kaine, S. M. Sykes, P. W. Sadow, K. P. D'Andrea, C. Bowman, C. Fujii, S. A. Garland, T. M. Mason, G. J. Olsen, C. M. Fraser, H. O. Smith, C. R. Woese, and J. C. Venter. 1997. The complete genome sequence of the hyperthermophilic, sulphate-reducing archaeon Archaeoglobus fulgidus. Nature 390:364–370.

Kunow, J., B. Schwörer, K. O. Stetter, and R. K. Thauer. 1993. A F$_{420}$-dependent NADP reductase in the extremely thermophilic sulfate-reducing Archaeoglobus fulgidus. Arch. Microbiol. 160:199–205.

Kunow, J., D. Linder, K. O. Stetter, and R. K. Thauer. 1994. F$_{420}$H$_2$: Quinone oxidoreductase from Archaeoglobus fulgidus: Characterization of a membrane-bound multisubunit complex containing FAD and iron-sulfur clusters. Eur. J. Biochem. 223:503–511.

Kunow, J., D. Linder, and R. K. Thauer. 1995. Pyruvate:ferredoxin oxidoreductase from the sulfate-reducing Archaeoglobus fulgidus: Molecular composition, catalytic properties, and sequence alignments. Arch. Microbiol. 163:21–28.

Kyrpides, N. C., and C. A. Ouzounis. 1999. Transcription in archaea. Proc. Natl. Acad. Sci. USA 96:8545–8550.

Labes, A., and P. Schönheit. 2001. Sugar utilization in the hyperthermophilic, sulfate-reducing archaeon Archaeoglobus fulgidus strain 7324: Starch degradation to acetate and CO$_2$ via a modified Embden-Meyerhof pathway and acetyl-CoA synthetase (ADP-forming). Arch. Microbiol. 176:329–338.

Lamosa, P., A. Burke, R. Peist, R. Huber, M. Y. Liu, G. Silva, C. Rodrigues-Pousada, J. LeGall, C. Maycock, and H. Santos. 2000. Thermostabilization of proteins by diglycerol phosphate, a new compatible solute from the hyperthermophile Archaeoglobus fulgidus. Appl. Environ. Microbiol. 66:1974–1979.

Lampreia, J., G. Fauque, N. Speich, C. Dahl, I. Moura, H. G. Trüper, and J. J. G. Moura. 1991. Spectroscopic Studies on APS Reductase isolated from the hyperthermophilic sulfate-reducing archaebacterium Archaeglobus fulgidus. Biochem. Biophys. Res. Com. 181:342–347.

Langelandsvik, A. S., I. H. Steen, N. K. Birkeland, and T. Lien. 1997. Properties and primary structure of a thermostable L-malate dehydrogenase from Archaeoglobus fulgidus. Arch. Microbiol. 168:59–67.

LaPaglia, C., and P. L. Hartzell. 1997. Stress-induced Production of Biofilm in the Hyperthermophile Archaeoglobus fulgidus. Appl. Environ. Microbiol. 63:3158–3163.

Li, H., and J. Abelson. 2000. Crystal structure of a dimeric archaeal splicing endonuclease. J. Molec. Biol. 302:639–648.

Liu, J., B. He, H. Qing, and Y. W. Kow. 2000. A deoxyinosine specific endonuclease from hyperthermophile, Archaeoglobus fulgidus: A homolog of Escherichia coli endonuclease V. Mutat. Res. 461:169–177.

Lopez-Garcia, P., P. Forterre, J. van der Oost, and G. Erauso. 2000. Plasmid pGS5 from the hyperthermophilic archaeon Archaeoglobus profundus is negatively supercoiled. J. Bacteriol. 182:4998–5000.

Lovely, D. R., K. Kashefi, M. Vargas, J. M. Tor, and E. L. Blunt-Harris. 2000. Reduction of humic substances and Fe(III) by hyperthermophilic microorganisms. Chem. Geol. 169:289–298.

Madern, D., C. Ebel, H. A. Dale, T. Lien, I. H. Steen, N. K. Birkeland, and G. Zaccai. 2001. Differences in the oligomeric states of the LDH-like L-MalDH from the hyperthermophilic archaea Methanococcus jannaschii and Archaeoglobus fulgidus. Biochemistry 40:10310–10316.

Maidak, B. L., J. R. Cole, T. G. Lilburn, C. T. Parker, P. R. Saxman Jr., J. M. Stredwick, G. M. Garrity, B. Li, G. J. Olsen, S. Pramanik, T. M. Schmidt, and J. M. Tiedje. 2000. The RDP (Ribosomal Database Project) continues. Nucleic Acids Res. 28:173–174.

Mander, G. J., E. C. Duin, D. Linder, K. O. Stetter, and R. Hedderich. 2002. Purification and characterization of a membrane-bound enzyme complex from the sulfate-reducing archaeon Archaeoglobus fulgidus related to heterodisulfide reductase from methanogenic archaea. Eur. J. Biochem. 269:1895–1904.

Martins, L., R. Huber, H. Huber, K. Stetter, M. Da Costa, and H. Santos. 1997. Organic solutes in hyperthermophilic archaea. Appl. Environ. Microbiol. 63:896–902.

McIlwraith, M. J., D. R. Hall, A. Z. Stasiak, A. Stasiak, D. B. Wigley, and S. C. West. 2001. RadA protein from Archaeoglobus fulgidus forms rings, nucleoprotein filaments and catalyses homologous recombination. Nucleic Acids Res. 29:4509–4517.

Min, J., J. Landry, R. Sternglanz, and R. M. Xu. 2001. Crystal structure of a SIR2 homolog-NAD complex. Cell 105:269–79.

Möller-Zinkhan, D., G. Borner, and R. K. Thauer. 1989. Function of methanofuran, tetrahydromethanopterin, and coenzyme F$_{420}$ in Archaeoglobus fulgidus. Arch. Microbiol. 152:362–368.

Möller-Zinkhan, D., and R. K. Thauer. 1990. Anaerobic lactate oxidation to 3 CO$_2$ by Archaeoglobus fulgidus via the carbon-monoxide dehydrogenase pathway: Demonstration of the acetyl-CoA carbon-carbon cleavage reaction in cell extracts. Arch. Microbiol. 153:215–218.

Musfeldt, M., and P. Schönheit. 2002. Novel type of ADP-forming acetyl coenzyme-A synthetase in hyperthermophilic archaea: Heterologous expression and characterization of isoenzymes from the sulfate reducer Archaeoglobus fulgidus and the methanogen Methanococcus jannaschii. J. Bacteriol. 184:636–44.

Noll, K. M., and T. S. Barber. 1988. Vitamin contents of archaebacteria. J. Bacteriol. 170:4315–4321.

Pagala, V. R., J. Park, D. W. Reed, and P. L. Hartzell. 2002. Cellular localization of D-lactate dehydrogenase and NADH oxidase from Archaeoglobus fulgidus. Archaea 1:678–689.

Rashid, N., H. Imanaka, T. Kanai, T. Fukui, H. Atomi, and T. Imanaka. 2002. A novel candidate for the true fructose 1,6-bisphosphatase in archaea. J. Biol. Chem. 277:30649–30655.

Reed, D. W., and P. L. Hartzell. 1999. The Archaeoglobus fulgidus D-lactate dehydrogenase is a Zn^{2+} flavoprotein. J. Bacteriol. 181:7580–7587.

Reed, D. W., J. Millstein, and P. L. Hartzell. 2001. H$_2$O$_2$-forming NADH oxidase with diaphorase (cytochrome) activity from Archaeoglobus fulgidus. J. Bacteriol. 183:7007–7016.

Reysenbach, A. L., K. Longnecker, and J. Kirshtein. 2000. Novel bacterial and archaeal lineages from an in situ growth chamber deployed at a mid-Atlantic Ridge hydrothermal vent. Appl. Environ. Microbiol. 66:3798–806.

Rodabough, A., M. S. Foster, and E. C. Neiderhoffer. 1995. Plating techniques for extremely thermophilic methanogens. *In:* Sowers, K. R., Schreier (Eds.) Archaea: A Laboratory Manual—Methanogens. Cold Spring Harbor Press. Plainview, NY.

Roth, A., G. Fritz, T. Buchert, H. Huber, K. O. Stetter, U. Ermler, and P. M. H. Kroneck. 2000. Crystallization and preliminary X-ray analysis of adenylylsulfate reductase from Archaeoglobus fulgidus. Acta Crystallographica Sect. D, Biol. Cryst. 56:1673–1675.

Sandigursky, M., and W. A. Franklin. 2000. Uracil-DNA glycosylase in the extreme thermophile Archaeoglobus fulgidus. J. Biol. Chem. 275:19146–19149.

Schmitz, R. A., D. Linder, K. O. Stetter, and R. K. Thauer. 1991. N^5,N^{10}-methylenetetrahydromethanopterin reductase (coenzyme F_{420}-dependent) and formylmethanofuran dehydrogenase from the hyperthermophile Archaeoglobus fulgidus. Arch. Microbiol. 156:427–434.

Schwörer, B., J. Breitung, A. R. Klein, K. O. Stetter, and R. K. Thauer. 1993. Formylmethanofuran: Tetrahydromethanopterin formyltransferase and N^5,N^{10}-methylenetetrahydromethanopterin dehydrogenase from the sulfate-reducing Archaeoglobus fulgidus: Similarities with the enzymes from methanogenic Archaea. Arch. Microbiol. 159:225–232.

Speich, N., and H. H. Trüper. 1988. Adenylylsulphate reductase in a dissimilatory sulphate-reducing archaebacterium. J. Gen. Microbiol. 134:1419–1425.

Speich, N., C. Dahl, P. Heisig, A. Klein, F. Lottspeich, K. O. Stetter, and H. G. Trüper. 1994. Adenylylsulphate reductase from the sulphate-reducing archaeon Archaeoglobus fulgidus: Cloning and characterization of the genes and comparison of the enzyme with other iron-sulphur flavoproteins. Microbiology 140:1273–1284.

Sperling, D., U. Kappler, A. Wynen, C. Dahl, and H. G. Trüper. 1998. Dissimilatory ATP sulfurylase from the hyperthermophilic sulfate reducer Archaeoglobus fulgidus belongs to the group of homo-oligomeric ATP sulfurylases. FEMS Microbiol. Lett. 162:257–264.

Sperling, D., U. Kappler, H. G. Trüper, and C. Dahl. 2001. Dissimilatory ATP sulfurylase from Archaeoglobus fulgidus. Meth. Enzymol. 331:419–427.

Steen, I. H., T. Lien, and N. K. Birkeland. 1997. Biochemical and phylogenetic characterization of isocitrate dehydrogenase from a hyperthermophilic archaeon, Archaeoglobus fulgidus. Arch. Microbiol. 168:412–420.

Steen, I. H., H. Hvoslef, T. Lien, and N. K. Birkeland. 2001. Isocitrate dehydrogenase, malate dehydrogenase, and glutamate dehydrogenase from Archaeoglobus fulgidus. Meth. Enzymol. 331:13–26.

Stetter, K. O., G. Lauerer, M. Thomm, and A. Neuner. 1987. Isolation of extremely thermophilic sulfate reducers: Evidence for a novel branch of archaebacteria. Science 236:822–824.

Stetter, K. O. 1988. Archaeoglobus fulgidus gen. nov., sp. nov.: A new taxon of extremely thermophilic archaebacteria. Syst. Appl. Microbiol. 10:172–173.

Stetter, K. O., H. Huber, E. Blochl, M. Kurr, R. D. Eden, M. Fielder, H. Cash, and L. Vance. 1993. Hyperthermophilic archaea are thriving in deep North Sea and Alaskan oil-reservoirs. Nature 365:743–745.

Tang, T. H., J. P. Bachellerie, T. Rozhdestvensky, M. L. Bortolin, H. Huber, M. Drungowski, T. Elge, J. Brosius, and A. Huttenhofer. 2002. Identification of 86 candidates for small non-messenger RNAs from the archaeon Archaeoglobus fulgidus. Proc. Natl. Acad. Sci. USA 99:7536–7541.

Teske, A., K. U. Hinrichs, V. Edgcomb, A. de Vera Gomez, D. Kysela, S. P. Sylva, M. L. Sogin, and H. W. Jannasch. 2002. Microbial diversity of hydrothermal sediments in the Guaymas Basin: Evidence for anaerobic methanotrophic communities. Appl. Environ. Microbiol. 68:1994–2007.

Tindall, B. J., K. O. Stetter, and M. D. Collins. 1989. A novel, fully saturated menaquinone from the thermophilic sulphate-reducing Archaebacterium archaeoglobus fulgidus. J. Gen. Microbiol. 135:693–696.

Toro, I., J. Basquin, H. Teo-Dreher, and D. Suck. 2002. Archaeal Sm proteins form heptameric and hexameric complexes: Crystal structures of the Sm1 and Sm2 proteins from the hyperthermophile Archaeoglobus fulgidus. J. Molec. Biol. 320:129–142.

Vadas, A., H. G. Monbouquette, E. Johnson, and I. Schroder. 1999. Identification and characterization of a novel ferric reductase from the hyperthermophilic Archaeon Archaeoglobus fulgidus. J. Biol. Chem. 274:36715–36721.

Vargas, M., K. Kashefi, E. L. Blunt-Harris, and D. R. Lovley. 1998. Microbiological evidence for Fe(III) reduction on early Earth. Nature 395:65–67.

Vorholt, J., J. Kunow, K. O. Stetter, and R. K. Thauer. 1995. Enzymes and coenzymes of the carbon monoxide dehydrogenase pathway for autotrophic CO_2 fixation in Archaeoglobus lithotrophicus and the lack of carbon monoxide dehydrogenase in the heterotrophic A. profundus. Arch. Microbiol. 163:112–118.

Wheeler, D. L., C. Chappey, A. E. Lash, D. D. Leipe, T. L. Madden, G. D. Schuler, T. A. Tatusova, and B. A. Rapp. 2000. Database resources of the National Center for Biotechnology Information. Nucleic Acids Res. 28:10–14.

Woese, C. R., L. Achenbach, P. Rouviere, and L. Mandelco. 1991. Archaeal phylogeny: Reexamination of the phylogenic position of Archaeoglobus fulgidus in light of certain composition-induced artifacts. Syst. Appl. Microbiol. 14:364–371.

Zellner, G., E. Stackebrandt, H. Kneifel, P. Messner, U. B. Sleytr, E. C. Demacario, H. P. Zabel, K. O. Stetter, and J. Winter. 1989. Isolation and characterization of a thermophilic, sulfate reducing archaebacterium, Archaeoglobusfulgidus strain Z. Syst. Appl. Microbiol. 11: 151–160.

Prokaryotes (2006) 3:101–112
DOI: 10.1007/0-387-30743-5_7

CHAPTER 7

Thermoplasmatales

HARALD HUBER AND KARL O. STETTER

Introduction

The order Thermoplasmatales (Reysenbach, 2001) is represented by facultatively anaerobic, thermoacidophilic, autotrophic or heterotrophic organisms that are unique among the Archaea both by their morphology and by their phylogenetic position. So far, the order harbors three families, each represented by one genus: the Thermoplasmaceae (genus *Thermoplasma*; Darland et al., 1970), the Picrophilaceae (genus *Picrophilus*; Schleper et al., 1995), and the recently described Ferroplasmaceae (genus *Ferroplasma*; Golyshina et al., 2000), formerly named *Ferromonas metallovorans*.

Thermoplasma spp. are devoid of a cell wall or envelope. For that reason, the genus *Thermoplasma*, which was for a long time (until 1995) the only member of the group, was first considered to be associated with the (bacterial) mycoplasmas (Darland et al., 1970; Masover and Hayflick, 1981). However, results of 16S rRNA sequence analyses revealed that *Thermoplasma* was a member of the archaeal domain (Woese and Fox, 1977; Woese et al., 1980; Woese et al., 1990). Although the sequences were quite unique, they clustered within the kingdom Euryarchaeota (Woese et al., 1990) and most calculations placed them between the Methanobacteriales and the Archaeoglobales. Furthermore, the affiliation to the Archaea was clear from a number of biochemical and molecular features (Stetter and Zillig, 1985; Langworthy and Smith, 1989). However, several characteristics were more crenarchaeotal than euryarchaeotal, like the physiology (Darland et al., 1970; Belly et al., 1973; Stetter and Zillig, 1985; Segerer et al., 1988) and the composition of the DNA-dependent RNA polymerase (Sturm et al., 1980; Zillig et al., 1982). In contrast, the requirement for polypeptide synthesis and the degree of stability of the ribosomal subunit association (Londei et al., 1986) supported the relationship to the Euryarchaeota, and the structure of the RNA polymerase of *Thermoplasma* was taken to be possibly of no phylogenetic significance (Yang et al., 1985). The lack of a cell wall as well as some other properties resembling those of mycoplasmas, e.g., the shape of the colonies, was therefore due to convergent evolution of these entirely unrelated groups of organisms. At present, there is no doubt that *Thermoplasma*, together with *Picrophilus* and *Ferroplasma*, represents a separate order within the domain Archaea.

Phylogeny

Based on 16S rRNA sequence data, the order Thermoplasmatales is a member of the euryarchaeotal branch of the Archaea (Woese et al., 1990). It forms an isolated cluster which branches in most calculation programs between the Methanobacteriales and the Methanomicrobiales/Halophiles (Fig. 1). However, the 16S rRNAs of all members show an unusual nucleotide sequence (for *Thermoplasma*, see Woese et al., 1980) and a high number of base exchanges in comparison to all other Archaea known so far. As a consequence the Thermoproteales exhibit low phylogenetic similarities (between 0.6 and 0.73 to all other Euryarchaeota and 0.58 to 0.67 to the Crenarchaeota). These values indicate that at the moment, no closer relatives of the members of the Thermoplasmatales are described. Within the order, the representatives of the different genera exhibit phylogenetic similarities between 0.86 and 0.89 (with the exception of the two *Thermoplasma* species, which show an identity of 98.6%).

Taxonomy

The order Thermoplasmatales (including the family Thermoplasmaceae) was first defined in the latest edition of *Bergey's Systematic Bacteriology* (Reysenbach, 2001). It is comprised of three different families, each represented by one single genus. So far, the genus *Thermoplasma* (Darland et al., 1970), the type genus for the family Thermoplasmaceae and the order Thermoplasmatales, harbors two described species: *T. acidophilum* and *T. volcanium* (Segerer et al., 1988). The type strain for *T. acidophilum* is strain 122-1B2^T (ATCC 25905^T and DSM 1728^T;

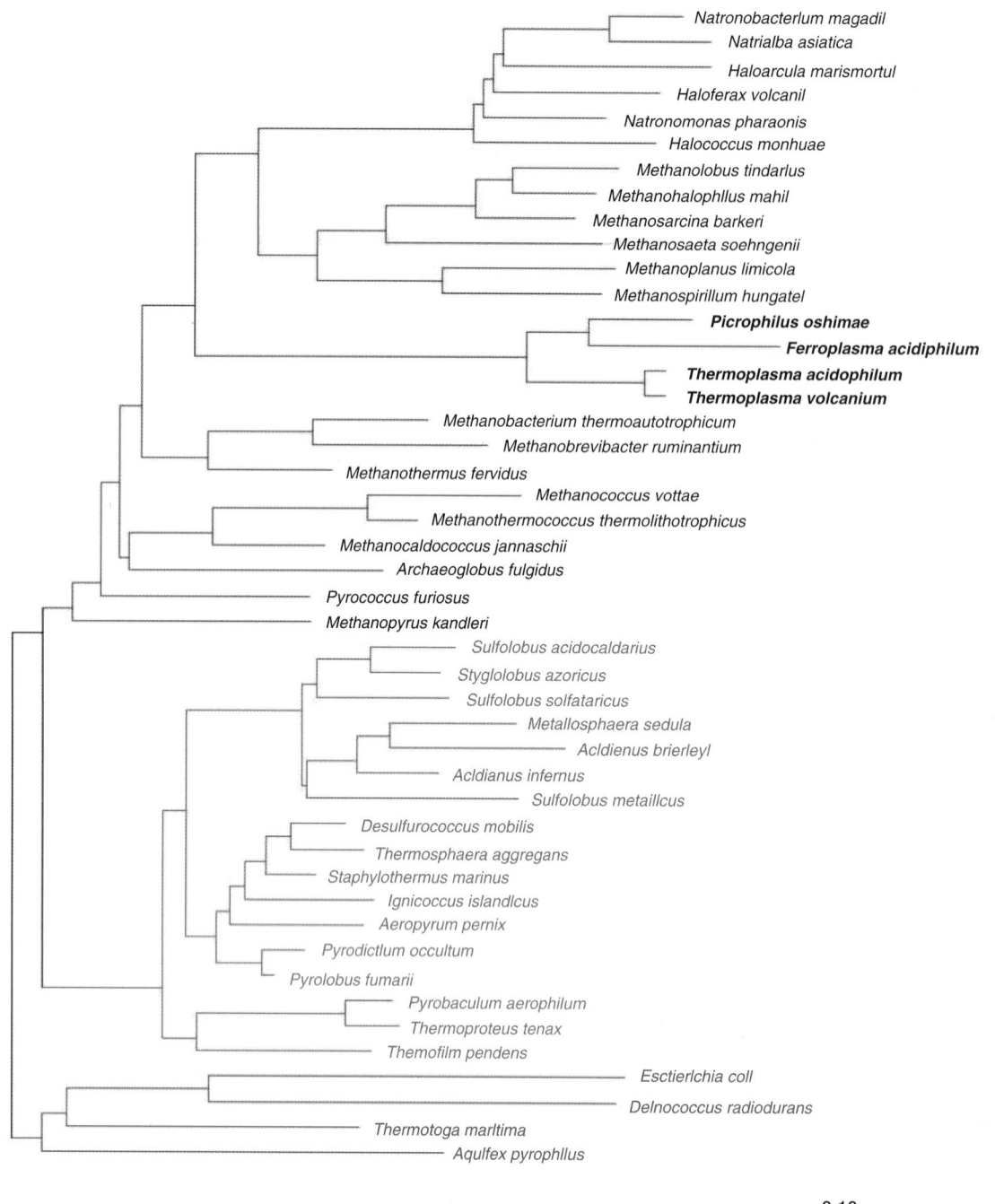

Fig. 1. Phylogenetic tree based on 16S rRNA sequences. The tree was calculated using the neighbor-joining program with Jukes-Cantor correction, which is included in the ARB package (Technische Universität, München; Ludwig and Strunk, 1997). Scale bar: 10 estimated exchanges within 100 nucleotides. Red lines = Euryarchaeota; Blue lines = Crenarchaeota; and Green lines = Bacteria.

American Type Culture Collection [ATCC], Rockville, MD, USA; Deutsche Sammlung von Mikroorganismen und Zellkulturen GmbH [DSMZ], Braunschweig, Germany) and for *T. volcanium* strain GSS1*ᵀ (ATCC 51530ᵀ and DSM 4299ᵀ). The latter presents the DNA homology group 1 of *T. volcanium*. Additionally, two further DNA homology groups within *T. volcanium* have been established: strains KD3 (representing DNA homology group 2) and KO2 (representing DNA homology group 3), which are available from the DSMZ (DSM 4300 and 4301, respectively). They do not hybridize significantly with organisms from the other groups

(Segerer et al., 1988), indicating that the organisms are genomically unrelated (Schleifer and Stackebrandt, 1983). However, owing to the lack of distinctive phenotypic features, a description of separate species was not carried out (Segerer et al., 1988). The representatives of the three homology groups are thus still treated as the single taxon, *T. volcanium*. The two *Thermoplasma* species can be distinguished by the G+C content of their genomic DNA and by DNA-DNA hybridization (Table 1). *Thermoplasma acidophilum* exhibits a G+C content of 46 mol% (Christiansen et al., 1975; Searcy and Doyle, 1975b), whereas the G+C content of *T. volcanium* is ~38 mol%. There is no significant DNA homology between the two species.

The genus *Picrophilus* represents the second family of the Thermoplasmatales, the Picrophilaceae (Schleper et al., 1996). Two *Picrophilus* species are described: *P. oshimae* (type strain: KAW 2/2[T], DSM 9789[T]) and *P. torridus* (type strain KAW 2/3[T], DSM9790[T]). However, since in the original papers (Schleper et al., 1995; Schleper et al., 1996) only poor information is given for the second species (*P. torridus*), their differentiation is not very clear. It is stated that *P. torridus* grows "significantly faster" than *P. oshimae*, contains no plasmids (which is however also true for some *P. oshimae* strains) and has a different DNA restriction pattern, although it resembles that of *P. oshimae*. Furthermore the 16S rRNA shows 3% difference within the first 250 positions. However, the sequence of *P. torridus* is still not available in the databases.

The third family, the Ferroplasmaceae, harbors one genus, *Ferroplasma*, with two species, *F. acidiphilum* type strain Y[T] (DSM 12658[T];

Golyshina et al., 2000) and "*F. acidarmanus*" (Edwards et al., 2000). However, the latter is so far not validly described. The two species can be distinguished by the ability of "*F. acidarmanus*" to grow heterotrophically on yeast extract, whereas *F. acidiphilum* is unable to use yeast extract as sole energy source. Furthermore, the pH optima and range differ significantly (1.2 for "*F. acidarmanus*," range 0–2.5, and 1.7 for *F. acidiphilum*, range 1.3–2.2; Table 1). In 16S rRNA sequence analysis, no differences between both species were obtained.

Habitat

The first representative of the genus *Thermoplasma*, *T. acidophilum*, was isolated aerobically from a coal refuse pile in Indiana, in the United States (Darland et al., 1970). Further isolates were also obtained from self-heated smoldering coal refuse piles in the United States and from water samples at these locations (Belly et al., 1973; Brock, 1978). Although smouldering coal piles were obviously colonized within a short time after ignition (Belly et al., 1973; Brock, 1978) because of their anthropogenic origin, it appeared unlikely that they represent the primary habitat of the organism. A first natural habitat of *Thermoplasma* was reported in 1982, the occurrence of *Thermoplasma* in a Japanese hot spring (Ohba and Oshima, 1982). However, no further details had been published. A broad screening for *Thermoplasma* in numerous solfataric fields in Italy (Naples; Figs. 2–4; and Vulcano Island; Figs. 5 and 6), the United States (Yellowstone National Park, Wyoming), Iceland

Table 1. Differentiation of the species within the order Thermoplasmatales.

Properties	Thermoplasma acidophilum	Thermoplasma volcanium	Picrophilus oshimae	Picrophilus torridus	Ferroplasma acidiphilum	"Ferroplasma acidarmanus"
Morphology, size of cells	Pleomorphic, 0.2–5 μm	Pleomorphic, 0.2–5 μm	Irregular cocci, 1–1.5 μm	Irregular cocci, 1–1.5 μm	Pleomorphic, 1–3 μm long, 0.3–1 μm wide	Pleomorphic
Flagella	+	+	+	+	−	n.d.
Autotrophy	−	−	−	−	+ (needs vitamin solution)	+ and heterotrophic
Relation to oxygen	Facultatively aerobic	Facultatively aerobic	Aerobic	Aerobic	Aerobic	Aerobic
Optimal growth temperature (°C)	59	60	60	60	35	~37
Temperature range (°C)	45–63	33–67	47–65	47–65	15–45	n.d.
Optimal pH	1–2	2	0.7	0.7	1.7	1.2
pH range	0.5–4	1–4	0–3.5	0–3.5	1.3–2.2	0–2.5
G+C content of genomic DNA (mol%)	46	38–40	36	n.d.	36.5	n.d.

n.d. = no data available.

Fig. 2. "Solfatara" crater near Naples, Italy.

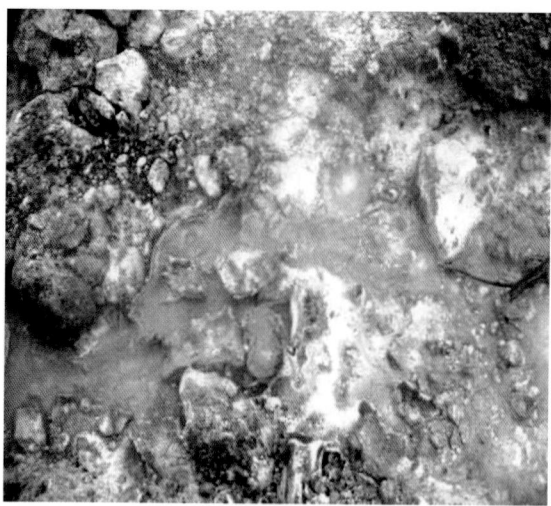

Fig. 4. Sampling site at "Pisciarelli Solfatara" with *Cyanidium caldarum*: pH 2.0; temperature 55°C.

Fig. 3. The small, highly active solfataric area at "Pisciarelli Solfatara," Naples, Italy.

Fig. 5. The island Vulcano, Italy, with the active solfataric area (in the center) and the Vulcanello (in the background).

(Krisuvik), the Azores (Furnas), and Indonesia (Java) demonstrated that aerobic and anaerobic zones of continental volcanic areas are the natural habitat of these organisms (Segerer et al., 1988). In these studies, an additional isolate was obtained from a warm, acidic, tropical swamp in Java (Segerer et al., 1988). Although there was evidence that *Thermoplasma* can also thrive within marine hydrothermal systems, like the submarine solfatara field close to the beach on Vulcano Island, Italy (Segerer et al., 1988), no

isolates from deep-sea "black smoker" vents have been obtained so far (A. Segerer, unpublished observation).

Both *Picrophilus* strains were isolated from geothermal solfataric soils and springs in Hokkaido, northern Japan (Schleper et al., 1996). Therefore, they share in general the same biotopes as *Thermoplasma*, although their pH minima and optima are significantly lower. Since both *Picrophilus* species are inhibited by 0.2 M NaCl, they may be restricted to terrestrial geothermal environments.

Strains of *Ferroplasma* were isolated from a bioreactor, which was operated with gold-containing arsenopyrite/pyrite ore concentrate from Kazakhstan (Golyshina et al., 2000) and from a pyrite ore-body at Iron Mountain, California, United States (Edwards et al., 2000). In addition, 16S rRNA genes were detected in an acidic geothermal pool on the Caribbean island

Fig. 6. Sampling at the solfataric fields at Vulcano, Italy.

of Montserrat (Burton and Norris, 2000). Since both isolated organisms grow at temperatures around 37°C and have been isolated from highly distant locations, they may occur in many sulfidic ore-containing mines and heaps on earth.

Isolation

Enrichment

Several procedures to obtain *Thermoplasma* strains have been described. They can be enriched by aerobic incubation of samples in Darland's culture medium (see recipe below). Although low pH and high temperature are highly selective, overgrowth of rod-shaped, sporeforming, bacterial contaminants has been observed in many cases (Belly et al., 1973; Segerer et al., 1988), and good results were only obtained by adding appropriate antibiotics (e.g., vancomycin) to the cultures. Alternatively, the medium must be adjusted to pH 1 to prevent bacterial growth (Belly et al., 1973).

Alternatively, water samples were passed through a 0.45 μm filter followed by a passage through a 0.22 μm filter. Subsequent incubation of the latter in culture medium yielded pure cultures of *Thermoplasma* (Belly et al., 1973).

A highly selective and convenient procedure for the enrichment of *Thermoplasma* is incubation of samples in anaerobic media (Balch et al., 1979) in the presence of elemental sulfur (Segerer et al., 1988). Darland's medium, pH 2, was supplemented with 0.4% (w/v) of sulfur and incubated with an aliquot of the sample in rubber-stoppered serum bottles containing either a N_2 or N_2/CO_2 (80:20, v/v) atmosphere. The gas phase should be devoid of H_2 to prevent growth of facultatively or strictly anaerobic representatives of the order Sulfolobales, e.g., *Acidianus* or *Stygiolobus* strains. In enrichment cultures obtained by using this procedure, no contaminants were detected by light microscopy or plating (Searcy and Doyle, 1975a).

For the enrichment of *Picrophilus*, an aerobic growth medium described by Smith et al. (1975) was used (see below). Yeast extract (0.1%, w/v) and glucose (1%, w/v) served as carbon sources. Incubation was carried out at pH 1 at 60°C.

Strains of *Ferroplasma* were enriched in a modified 9K-medium (see below; Silverman and Lundgren, 1959; Golyshina et al., 2000) at pH 1.6–1.9 and at incubation temperatures between 28 and 30°C. Alternatively, enrichment in a medium according to Edwards et al. (1998) with a pH of 1 and at 37°C was described, using pyrite as energy source (Edwards et al., 2000). Since yeast extract is essential for the growth of both strains, this (0.02%, w/v) must be added to the culture media.

Isolation Procedures

For the isolation of *Thermoplasma*, Darland's culture medium (see below) is inoculated with samples (5% inoculum) and incubated at 59°C. *Thermoplasma* growth occurs within 2 days to 3 weeks. Isolation is achieved by subsequent plating, which is conveniently performed under aerobic conditions. Therefore, it is recommended that enriched cultures obtained by using the anaerobic enrichment procedure (see above) should be transferred once into aerobic liquid medium before plating. Alternatively, pure cultures can be obtained by thrice-repeated serial dilutions. However, the fastest and most secure isolation procedure is the separation by the use of optical tweezers (Huber et al., 1995).

Picrophilus strains were isolated by plating on 12.5% starch plates (pH 1) containing yeast extract and glucose. Colonies were obtained after 6 days at an incubation temperature of 60°C. Both strains of *Picrophilus* were isolated by serial dilutions. So far, nothing is published about growth on solidified media.

Ferroplasma has been isolated from a bioreactor and from an enrichment culture out of a pyrite ore-body by the use of serial dilutions. Growth on solidified media is so far not documented.

Identification

All representatives of the order are extremely acidophilic, growing optimally at pH < 2. The lack of a cell wall is characteristic for the members of the genera *Thermoplasma* and *Ferroplasma*, while *Picrophilus* possesses an outer S-layer. *Thermoplasma* and *Picrophilus* are further characterized by their thermophily, exhibiting temperature optima around 60°C. All organisms stain Gram negative and lack resting stages.

The diameter of *Thermoplasma* and *Ferroplasma* cells, ranging from 0.2 to roughly 5 μm (Table 1), is highly variable. The same is true for the cell shape: filamentous, disc, and club (Fig. 7), and coccoid forms occur concomitantly in the same culture, making it difficult to ensure purity of the strain by microscopic investigations. Filamentous cells are especially abundant in the early exponential growth phase, while coccoid forms predominate in the stationary phase. Buds, about 0.3 μm in width, are often associated with mother cells. In *Ferroplasma*, the buds appear tubular or vesicular in shape, tending to form septation annuli. The tubular extrusions range from 85 to 142 nm in diameter and up to 1 μm in length (Golyshina et al., 2000). *Thermoplasma* and *Ferroplasma* cells are readily discernible from all other thermoacidophiles by the unique feature of lacking a true cell wall (Fig. 8). They

are surrounded by a single triple-layer membrane about 4–10 nm thick. *Thermoplasma* shows a pale yellowish-green fluorescence under ultraviolet (UV) radiation in the fluorescence microscope (Segerer et al., 1988). Despite the lack of a cell wall, the cells are flagellated and motile (Black et al., 1979; Segerer et al., 1988). Usually, monopolar, monotrichous flagellation is found, but sometimes multiflagellated cells also were observed (Segerer et al., 1988). On solid media, colonies of *Thermoplasma* are usually small (about 0.5 mm in diameter) and are either colorless or brownish. As in the *Mycoplasmas*, fully grown colonies resemble a fried egg in shape (Darland et al., 1970; Belly et al., 1973).

Cells of *Picrophilus* are irregular cocci, 1 to 1.5 μm in diameter (Table 1). In exponentially growing cultures, many cells are present in incompletely divided division forms of two or three individuals. Zones of low electron density were found after thin sectioning, resembling vacuoles, which were however not separated from the cytoplasm by a membrane. A 40-nm thick S-layer, which is situated on top of the cytoplasmic membrane, distinguishes *Picrophilus* from the other members of the Thermoplasmatales. The S-layer has a tetragonal symmetry (center-to-center distance about 20 nm) and consists of an outer dense and inner, almost empty stratum consisting of widely spaced pillars that anchor the surface layer in the membrane. No flagella or pili have been observed (Schleper et al., 1995).

All members of the Thermoplasmatales are resistant to cell wall inhibitors (like ampicillin or vancomycin) and to streptomycin. While *Picrophilus* is sensitive to chloramphenicol and

Fig. 7. Electron micrograph of cells of *Thermoplasma* acidophilum DSM 1728^T, Pt-shadowed. Bar 1 μm.

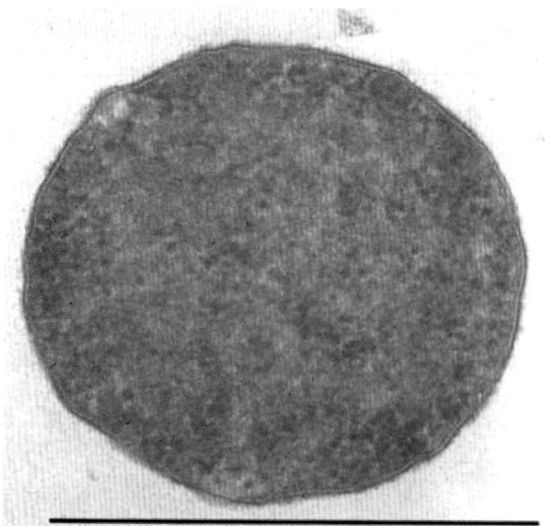

Fig. 8. Electron micrograph of a thin section of *Thermoplasma* acidophilum. Bar 1 μm. (The photograph was kindly provided by H. Engelhardt.)

novobiocin, the other organisms are resistant to these antibiotics.

Many molecular investigations were carried out with *Thermoplasma*, especially with *T. acidophilum*. It turned out that it contains a number of unique and characteristic chemical compounds. An unusual mannose-rich, 152-kDa glycoprotein is located in the cell membrane and is thought to form a hydroskeleton via a network of sugar residues surrounding the cell, thus contributing to the remarkable rigidity of the membrane (Yang and Haug, 1979). The membrane lipids contain predominantly acyclic, mono- and bi- pentacyclic C_{40} biphytanyl diglycerol tetraethers along with small amounts of C_{20} phytanyl glycerol diethers (Langworthy, 1977; Langworthy, 1985; Langworthy et al., 1982; Langworthy and Pond, 1986). The structure of a polar tetraether lipid was fully established and shown to be a lipoglycan (MW 5,300), consisting of 24 mannosyl residues and one glucosyl residue bound to a diglyceryltetraether (Mayberry-Carson et al., 1974; Smith, 1980). Several novel neutral glycolipids, consisting of caldarchaeol (dibiphytanyl-diglycerol tetraethers) and monosaccharide residues (gulose and glucose) on one side or both sides of the core lipids, were described recently for *T. acidophilum* (Uda et al., 1999). *Picrophilus* has similar lipids, with the exception that a β-glycosyl residue is present in the major lipid component. In contrast, *Ferroplasma* contains no tetraethers. The main phospholipid is archaetidyl glycerol. Furthermore, archaetidic acid and dimers of both ether lipids were found in small amounts (Schleper et al., 1995).

The DNA-dependent RNA polymerase of *Thermoplasma* consists of seven subunits and is resistant to rifampicin, streptolydigin and α-amanitin (Sturm et al., 1980). In the corresponding enzyme of *Picrophilus*, several components were absent (e.g., A″ and E) and the enzyme does not crossreact in immunodiffusion assays with antibodies against the *Thermoplasma* enzyme (Schleper et al., 1995). The respiratory chain of *Thermoplasma* contains at least one *b*-type cytochrome (Belly et al., 1973; Holländer, 1978; Searcy and Whatley, 1982) and several quinones (Holländer et al., 1977; Collins and Langworthy, 1983). Among the quinones, a characteristic compound called "thermoplasmaquinone" was found, which was not detected in other Archaea. In addition, menaquinone and methionaquinone are present (Shimada et al., 2001). Coenzyme F_{420} is present in about 1% of the amount typical for methanogens (Lin and White, 1986). The DNA is associated with a small basic histone called "Hta" that exhibits partial amino acid sequence homology to the HU-1 protein of *Escherichia coli* and to the calf thymus histones H3 and H2A

(Searcy, 1975a; Searcy and DeLange, 1980; DeLange et al., 1981. *Thermoplasma acidophilum* possesses a protein that resembles the human ubiquitin (Wolf et al., 1993). Furthermore, it contains proteasomes, which function as threonine proteases in ATP-dependent proteolysis (Seemüller et al., 1995). Structural investigations on the thermosome, which represents the archaeal chaperonin, were carried out by electron cryo-microscopy and by high resolution analysis (2.6 Å) of protein crystals (Nitsch et al., 1997; Ditzel et al., 1998). Based on these results, the molecule is composed of two stacked eight-membered rings of alternating α- and β-subunits. Studies on the mechanism of its ATPase activity revealed that the thermosome had unique allosteric properties (Gutsche et al., 2000).

For *Picrophilus*, bioenergetic studies with liposomes indicate an intrinsic instability of the cytoplasma membrane at higher pH values, resulting in a loss of viability and cell integrity above pH 4 (van de Vossenberg et al., 1998).

Cultivation

Medium for *Thermaplasma* Species

Darland's medium is suitable for the growth of all strains of *Thermoplasma*.

Darland's Medium (Darland et al., 1970)

KH_2PO_4	3.00 g
$MgSO_4 \cdot 7H_2O$	1.02 g
$CaCl_2 \cdot 2H_2O$	0.25 g
$(NH_4)_2SO_4$	0.20 g
Yeast extract	1.00 g
Glucose $\cdot$ H_2O	10.0 g

The mineral base of the medium is dissolved in one liter of double-distilled water and the pH is adjusted to around 2 with 10% (v/v) H_2SO_4.

For aerobic cultivation, 30-ml aliquots are distributed into 100-ml Erlenmeyer flasks equipped with an air cooler (e.g., a 1-ml pipette tightly fitted through a rubber stopper) and autoclaved. Yeast extract, glucose, and (if necessary) meat extract are added separately from sterile stock solutions. Reduction of the glucose content to 0.5% (w/v) does not significantly affect growth and cell yield. For reasons of convenience, we prepare a stock mixture of 30% (w/v) glucose and 6% (w/v) yeast extract (plus, if desired, 3% [w/v] meat extract), which is sterilized by passage through a 0.2-μm filter membrane. From this stock, 0.5-ml are added to each 30 ml of mineral base medium.

For anaerobic cultivation of *Thermoplasma* strains, the medium must be supplemented with about 0.4% (w/v) elemental sulfur. It is distributed as 15-ml aliquots into 100-ml serum bottles. The atmosphere (~150 kPa) may consist of N_2, N_2/CO_2 (80:20 v/v), or H_2/CO_2 (80:20 v/v). Serum bottles containing elemental sulfur must be sterilized by Tyndall's fractional sterilization procedure.

Aerobic and anaerobic media for the strains of *T. volcanium* group 3 (see "Identification") should be supplemented with 0.025–0.05% (w/v) meat extract in addition to yeast extract and glucose.

Medium for *Picrophilus* Species

This medium was originally described by Smith et al. (1975) according to E. A. Freundt for the cultivation of *Thermoplasma acidophilum* (Schleper et al., 1995). It is in principle similar to Darland's medium, with the exception of the absence of KH_2PO_4.

$(NH_4)_2SO_4$	0.20 g
$MgSO_4 \cdot 7H_2O$	0.50 g
$CaCl_2 \cdot 2H_2O$	0.25 g
Yeast extract	1.00 g

The pH is adjusted to around 2 with 10 N H_2SO_4. After autoclaving, glucose is added to a final concentration of 1%.

MEDIUM FOR *FERROPLASMA* SPECIES A modified 9K medium (Silverman and Lundgren, 1959) is used for the cultivation of *Ferroplasma* (Golyshina et al., 2000).

$MgSO_4 \cdot 7H_2O$	0.40 g
$(NH_4)_2SO_4$	0.20 g
KCl	0.10 g
K_2HPO_4	0.10 g
$FeSO_4 \cdot 7H_2O$	25.0 g
Yeast extract	0.20 g
Trace elements	1.0 ml

The pH is adjusted to 1.7 with 10% (v/v) H_2SO_4. Trace elements are described by Segerer and Stetter (1992b). For the growth of both *Ferroplasma* strains, ferrous sulfide can be replaced by pyrite.

Cultivation on Plates

For growth on solidified media, best results are obtained for members of the Thermoplasmatales using 10–12% starch (e.g., Stärke Gel, Serva). Only poor results were obtained by plating *Thermoplasma* spp. onto agar or Gelrite plates (Searcy and Doyle, 1975a). It is recommended to adjust the pH to about 3 just before pouring the plates. High and reproducible yields (plating efficiency ~90 to 100%) were obtained by incubating the plates in a pressure cylinder (Balch et al., 1979) microaerobically, in a humid atmosphere consisting of roughly 60% air and 40% CO_2 (v/v). Small, colorless to brownish colonies showing the "fried egg" appearance typical of *Thermoplasma* emerged within 7 or more days.

Cultivation Conditions

Strains of *Thermoplasma* (*Picrophilus* and *Ferroplasma*) can be cultivated in a wide variety of glass bottles and fermentors. Special attention has to be paid to the construction of fermentors because of the high corrosivity of the medium due to the low pH, elevated temperature, and (in case of anaerobic cultivation) high amounts of H_2S produced during growth. Therefore, only fermentors containing high-quality steel are recommended.

All *Thermoplasma* strains investigated thus far are best grown at 57–59°C and at around pH 2, whereas for *Picrophilus*, the optimal pH is 0.7. Optimal growth for *Ferroplasma* strains occurs between pH 1.2 ("*F. acidarmanus*") and 1.7 (*F. acidiphilum*), with temperature optima of 40 and 35°C, respectively.

Preservation

Cultures of *Thermoplasma* are significantly more resistant to storage procedures when grown anaerobically. The reason for this phenomenon, which is also observed with *Acidianus* spp. (Segerer et al., 1986a), is unknown. Adjustment to pH ~5 with sterile $CaCO_3$ prior to storage has a further positive influence on long-term viability, presumably because *Thermoplasma* actively maintains an internal pH of about 5 (Searcy, 1976). Possible damage to cells during storage caused by the penetration of acids is thus avoided. Therefore, for storage, the use of anaerobically grown cultures at pH ~5 and 4°C is recommended. Although viability of such cultures is sufficient even after 12 months of storage, maintenance transfer every 2–3 months is advised. Best results for long-term storage are achieved in culture media containing 5% DMSO (dimethyl sulfoxide) and storage in the gas phase over liquid nitrogen (around −140°C; H. Huber and K. O. Stetter, unpublished observation).

Viable stock cultures for *Picrophilus* can be obtained by suspending cells in basal salt medium (pH 4.5) containing 20% glycerol. These suspensions can be kept at −70°C. Neutralization of the culture medium for short-term storage is not recommended because *Picrophilus* lyses above pH 5 (Schleper et al., 1995).

Nothing is stated in the original papers on preservation results for *Ferroplasma* (Golyshina et al., 2000; Edwards et al., 2000). However, it should be possible to store these strains in fresh medium supplemented with 5% DMSO over liquid nitrogen.

Physiology

The members of the genus *Thermoplasma* are obligately heterotrophic, facultatively anaerobic thermoacidophiles that require the presence of yeast extract or similar extracts for growth. At growth-limiting concentrations of yeast extract, the carbohydrates sucrose, glucose, mannose, galactose, and fructose were found to stimulate growth significantly (Belly et al., 1973; Brock, 1978). However, no growth occurs on sugars alone or on peptone, tryptone, casamino acids, various amino acids, and alcohols. As shown by Smith et al. (1975), the growth factor(s) present in yeast extract is most likely a basic oligopeptide consisting of 8–10 amino acids. Although *Thermoplasma* has an absolute requirement for yeast extract (Langworthy and Smith, 1989), this organism also grows in the presence of meat extract or bacterial extracts (Segerer et al., 1988), suggesting that the same or similar growth factors are present in those extracts. Therefore, the nutrition of *Thermoplasma* in its natural habitat is most likely based on the products of decomposing cells of organisms sharing the biotope, e.g., *Acidianus brierleyi* (Brierley and Brierley, 1973; Segerer et al., 1986a), *Bacillus acidocaldarius* (Darland and Brock, 1971), *Cyanidium caldarium* (Geitler and Ruttner, 1936), or *Dactylaria gallopava* (Tansey and Brock, 1973). Additionally, *Thermoplasma* was also found to grow poorly on an extract of coal refuse material (Bohlool and Brock, 1974). However, it is not clear from this study whether the coal itself provides all necessary nutrients or not.

Glucose degradation was investigated in T. acidophilum. Although contradictory results were obtained, degradation appears to take place via a modified Entner-Doudoroff pathway involving nonphosphorylated intermediates, rather than via the pentose phosphate pathway or glycolysis (Searcy and Whatley, 1984; Budgen and Danson, 1986). By using D-[U-^{14}C]-glucose as tracer, CO_2 and acetic acid were detected as metabolic products, in addition to the respiration of cell extracts by key intermediates of the citric acid cycle and the presence of the enzymes malate dehydrogenase and citrate synthase. This suggests the operation of a Krebs cycle (Searcy and Whatley, 1984; Grossebüter and Görisch, 1985; Grossebüter et al., 1986).

Thermoplasma spp. grow as facultative anaerobes on molecular sulfur by sulfur respiration, forming large amounts of H_2S (Segerer et al., 1986b; Segerer et al., 1988). This feature was in the beginning unknown, and the organism was considered to be a strict aerobe (Darland et al., 1970; Langworthy and Smith, 1989). Low cell densities are obtained when the cells are grown anaerobically without sulfur, indicating the presence of further unknown electron acceptor(s). Sulfur is not a prerequisite for aerobic cultivation. When *Thermoplasma* was grown aerobically in the presence of sulfur, no formation of H_2SO_4 was found (Brock, 1978; Segerer et al., 1988), a feature typical for members of the genus *Acidianus* (Segerer et al., 1986a). In addition, no growth occurs on ferrous iron.

The growth temperatures range from about 45–67°C for T. acidophilum and from about 33–67°C for T. volcanium. The optimum growth temperature is around 59°C. Both species grow within a pH range of 0.5–4, with an optimum around pH 2 (Table 1), but growth is very slow at both extremes.

Cells lyse at neutral pH, indicating that *Thermoplasma* has an absolute requirement for protons. They cannot be replaced by other monovalent or divalent ions. Nevertheless, the internal pH is near neutrality (Hsung and Haug, 1975; Searcy, 1976). The cells do not lyse in distilled water or during heating up to 100°C. However, cells are rapidly disintegrated in the presence of sodium dodecyl sulfate (SDS).

The representatives of *Picrophilus* are obligately aerobic heterotrophs. They cannot grow by fermentation or by chemolithotrophic pathways, like sulfur respiration. Yeast extract (0.1–0.5%) serves as an energy source, yielding cell densities up to 5×10^8 cells/ml. A slight stimulation is achieved by addition of 1% glucose, sucrose or lactose. No growth occurs on these sugars alone, on starch, or on casamino acids. Growth is inhibited by the addition of relatively small amounts of NaCl (0.2 M). *Picrophilus* strains are thermophilic, exhibiting temperature optima around 60°C. No growth occurs at 40°C and below and at 67°C or above. The optimal pH is 0.7, and no growth is obtained at pH 3.5 or above, indicating that the organisms are "hyperacidophilic." Even at pH 0, cell division occurs (Schleper et al., 1995).

In contrast to the other genera, *Ferroplasma* harbors also chemolithoautotrophic organisms. *Ferroplasma acidophilum* is able to use CO_2 as a carbon source and ferrous iron or pyrite as an energy source. Ferric ion (Fe^{3+}) is the end product of the oxidation. In addition, Mn^{2+} can be oxidized. No growth occurs on other sulfidic ores or reduced sulfur compounds like elemental sulfur, thiosulfate or tetrathionate. No growth could be observed on organic substrates, although the addition of yeast extract is essential for cell propagation (Golyshina et al., 2000). It was detected that yeast extract can be replaced by a vitamin solution. "*Ferroplasma acidarmanus*" is able to grow in addition heterotrophically on yeast extract as sole energy source. It grows between pH 0 and 2.5, with an optimum at 1.2 (*F. acidiphilum* pH 1.7–2.2, optimum 1.7; Edwards et al., 2000). Both organisms are mesophiles with a temperature optimum around 37°C (Table 1).

Genetics

The whole genomes of *Thermoplasma acidophilum* and *Thermoplasma volcanium* were sequenced recently (Ruepp et al., 2000; Kawashima et al., 2000). The genomes have sizes of only 1,565 kb and 1,585 kb, respectively, being sone of the smallest among free-living organisms. In *T. acidophilum*, 1,509 ORFs were identified, and about 16% have so far no database match. Each of the three ribosomal RNA genes is present in one copy, but the three genes are dispersed in the genome. Analyses of the data revealed that *T. acidophilum* is a typical member of the Euryarchaeota, although the highest number of ORFs (17%) was most similar to proteins of *Sulfolobus solfataricus*. The complete sequence also showed that *Thermoplasma acidophilum* is a typical Archaeon and therefore most likely not a direct ancestor of the eukaryotic cytoplasm (Ruepp et al., 2000). In general, it turned out that two classes of genes can be distinguished: the "housekeeping" genes reflect generally the phylogenetic origin, while the "lifestyle" genes (mostly genes related to metabolism) are influenced by the specific environment (Ruepp et al., 2000). In *T. volcanium*, 1,524 genes were identified. The main goal of this genome sequencing was the investigation of a correlation between higher growth temperature and genomic organization (Kawashima et al., 2000).

Ecology

Representatives of the Thermoplasmatales thrive in highly acidic biotopes, covering temperatures from 20 up to 60°C. While *Thermoplasma* and *Picrophilus* are typical inhabitants of heated solfataric areas, recent results document that *Ferroplasma* seems to be widely distributed in pyrite-dominated ore-containing habitats at temperatures around 37°C (Bond et al., 2000; Vásquez et al., 1999). So far, these biotopes were thought to be dominated by bacteria, like *Thiobacillus ferrooxidans* (now *Acidithiobacillus ferrooxidans*; Kelly and Wood, 2000), *Leptospirillum ferrooxidans*, or *Acidiphilum*. However, especially for lower pH regions in mines, ore bodies, or drainage waters, *Ferroplasma* seems to be more important than the other organisms.

Literature Cited

Balch, W. E., G. E. Fox, L. J. Magrum, C. R. Woese, and R. S. Wolfe. 1979. Methanogens: Reevaluation of a unique biological group. Microbiol. Rev. 43:260–296.

Belly, R. T., B. B. Bohlool, and T. D. Brock. 1973. The genus Thermoplasma. Ann. NY Acad. Sci. 225:94–107.

Black, F. T., E. A. Freundt, O. Vinther, and C. Christiansen. 1979. Flagellation and swimming motility of Thermoplasma acidophilum. J. Bacteriol. 137:456–460.

Bohlool, B. B., and T. D. Brock. 1974. Immunofluorescence approach to the study af the ecology of Thermoplasma acidophilum in coal refuse material. Appl. Microbiol. 28:11–16.

Bond, P. L., G. K. Druschel, and J. F. Banfield. 2000. Comparison of acid mine drainage microbial communities in physically and geochemically distinct ecosystems. Appl. Microbiol. 66:4962–4971.

Brierley, C. L., and J. A. Brierley. 1973. A chemolithoautotrophic and thermophilic microorganism isolated from an acidic hot spring. Can. J. Microbiol. 19:183–188.

Brock, T. D. 1978. Thermophilic Microorganisms and Life at High Temperatures. Springer-Verlag. New York.

Budgen, N., and M. J. Danson. 1986. Metabolism of glucose via a modified Entner-Doudoroff pathway in the thermoacidophilic archaebacterium Thermoplasma acidophilum. FEBS Lett. 196:207–210.

Burton, N. B., and P. R. Norris. 2000. Microbiology of acidic, geothermal springs of Montserrat: Environmental rDNA analysis. Extremophiles 4:315–320.

Christiansen, C., E. A. Freundt, and F. T. Black. 1975. Genome size and deoxyribonucleic acid base composition of Thermoplasma acidophilum. Int. J. Syst. Bacteriol. 25:99–101.

Collins, M. D., and T. A. Langworthy. 1983. Respiratory quinone composition of some acidophilic bacteria. Syst. Appl. Microbiol. 4:295–304.

Darland, G., T. D. Brock, W. Samsonoff, and S. F. Conti. 1970. A thermophilic acidophilic Mycoplasma isolated from a coal refuse pile. Science 170:1416–1418.

Darland, G., and T. D. Brock. 1971. Bacillus acidocaldarius sp. nov., an acidophilic, thermophilic sporeforming bacterium. J. Gen. Microbiol. 67:9–15.

DeLange, R. J., L. C. Williams, and D. G. Searcy. 1981. A histone-like protein (HTa) from Thermoplasma acidophilum. II: Complete amino acid sequence. J. Biol. Chem. 256:905–911.

Ditzel, L., J. Löwe, D. Stock, K. O. Stetter, H. Huber, R. Huber, and S. Steinbacher. 1998. Crystal structure of the thermosome, the archael chaperonin and homolog of CCT. Cell 93:125–138.

Edwards, K. J., M. O. Schrenk, R. Hamers, and J. F. Banfield. 1998. Microbial oxidation of pyrite: Experiments using microorganisms from an extreme acidic enviroment. Am. Mineral. 83:1444–1453.

Edwards, K. J., P. L. Bond, T. M. Gihring, and J. F. Banfield. 2000. An archaeal iron-oxidizing extreme acidophile important in acid mine drainage. Science 287:1796–1799.

Geitler, L., and F. Ruttner. 1936. Die Cyanophyceen der Deutschen Limnologischen Sunda-Expedition. Arch. Hydrobiol. Suppl. XIV:308–481.

Golyshina, O. V., T. A. Pivovarova, G. I. Karavaiko, T. F. Kondrat'eva, E. R. B. Moore, W.-R. Abraham, H. Lünsdorf, K. N. Timmis, M. M. Yakimov, and P. N Golyshin. 2000. Ferroplasma acidiphilum gen. nov., sp. nov., an acidophilic, autotrophic, ferrous-iron-oxidizing, cell-wall-lacking, mesophilic member of the Ferroplasmaceae fam. nov., comprising a distinct lineage of the Archaea. Int. J. System. Evol. Microbiol. 50:997–1006.

Grossebüter, W., and H. Görisch. 1985. Partial purification and properties of citrate synthases from the thermoacidophilic archaebacteria Thermoplasma acidophilum and Sulfolobus acidocaldarius. Syst. Appl. Microbiol. 6:119–124.

Grossebüter, W., T. Hartl, H. Görisch, and J. J. Stezowski. 1986. Purification and properties of malate dehydrogenase from the thermoacidophilic archaebacterium Thermoplasma acidophilum. Biol. Chem. Hoppe-Seyler 367:457–463.

Gutsche, I., O. Mihalache, and W. Baumeister. 2000. ATPase cycle of an archaeal chaperonin. J. Molec. Biol. 300:187–196.

Holländer, R., G. Wolf, and W. Mannheim. 1977. Lipoquinones of some bacteria and mycoplasmas, with consideration an their functional significance. Ant. v. Leeuwenhoek 43:177–185.

Holländer, R. 1978. The cytochromes of Thermoplasma acidophilum. J. Gen. Microbiol. 108:165–168.

Hsung, J. C., and A. Haug. 1975. Intracellular pH of Thermoplasma acidophila. Biochim. Biophys. Acta 389:477–482.

Huber, R., S. Burggraf, T. Mayer, S. M. Barns, P. Rossnagel, and K. O. Stetter. 1995. Isolation of a hyperthermophilic archaeum predicted by in situ RNA analysis. Nature 367:57–58.

Kawashima, T., N. Amano, H. Koike, S. Makino, S. Higuchi, Y. Kawashima-Ohya, K. Watanabe, M. Yamazaki, K. Kanehori, T. Kawamoto, T. Nunoshiba, Y. Yamamoto, H. Aramaki, K. Makino, and M. Suzuki. 2000. Archael adaptation to higher temperatures revealed by genomic sequence of Thermoplasma volcanium. Proc. Natl. Acad. Sci. USA 97:14257–14262.

Kelly, D. P., and A. P. Wood. 2000. Reclassification of some species of Thiobacillus to the newly designated genera Acidithiobacillus gen. nov., Halobacillus gen. nov. and Thermithiobacillus gen. nov. Int. J. Syst. Evol. Microbiol. 50:511–516.

Langworthy, T. A. 1977. Long-chain diglycerol tetraethers from Thermoplasma acidophilum. Biochim. Biophys. Acta 487:37–50.

Langworthy, T. A., T. G. Tornabene, and G. Holzer. 1982. Lipids of archaebacteria. Zentralbl. Bakteriol. Parasitenkd. Infektionskr. Hyg. Abt. 1, Orig. Reihe C3:228–244.

Langworthy, T. A. 1985. Lipids of archaebacteria. In: C. R. Woese and R. S. Wolfe (Eds.) The Bacteria. Academic Press. Orlando, FL. 8:459–497.

Langworthy, T. A., and J. L. Pond. 1986. Archaebacterial ether lipids and chemotaxanomy. Syst. Appl. Microbiol. 7:253–275.

Langworthy, T. A., and P. F. Smith. 1989. Group IV: Cell wall-less archaeobacteria. In: J. T. Staley, M. P. Bryant, N. Pfennig, and J. G. Holt (Eds.) Bergey's Manual of Systematic Bacteriology. Williams and Wilkins. Baltimore, MD. 3:2233–2236.

Lin, X.-L., and R. H. White. 1986. Occurrence of coenzyme F_{420} and its γ-monoglutamyl derivative in nonmethanogenic archaebacteria. J. Bacteriol. 168:444–448.

Londei, P., S. Altamura, P. Cammarano, and L. Petrucci. 1986. Differential features of ribosomes and of poly(U)-programmed cell-free systems derived from sulphur-dependent archaebacterial species. Eur. J. Biochem. 157:455–462.

Ludwig, W, and O. Strunk. 1997. ARB: A software environment for sequence data.

Masover, G., and L. Hayflick. 1981. The genera Mycoplasma, Ureaplasma, and Acholeplasma, and associated organisms (Thermoplasmas and Anaeroplasmas). In: M. P. Starr, H. Stolp, H. G. Trüper, A. Balows, and H. G. Schlegel (Eds.). Springer-Verlag. Berlin, 2:2247–2270.

Mayberry-Carson, K. J., T. A. Langworthy, W. R. Mayberry, and P. F. Smith. 1974. A new class of lipopolysaccharide from Thermoplasma acidophilum. Biochim. Biophys. Acta 360:217–229.

Nitsch, M., M. Klumpp, A. Lupas, and W. Baumeister. 1997. The thermosome: Alternating alpha and beta-subunits within the chaperonin of the archaeon Thermoplasma acidophilum. J. Molec. Biol. 267:142–149.

Ohba, M., and T. Oshima. 1982. Some biochemical properties of the DNA synthesizing machinery of acidothermophilic archaebacteria isolated from Japanese hot springs. In: O. Kandler (Ed.) Archaebacteria. G. Fischer Verlag. Stuttgart, Germany. 353.

Reysenbach, A.-L. 2001. Order "Thermoplasmatales" ord. nov. In: G. Garrity (Ed.) Bergey's Manual of Systematic Bacteriology, 2nd ed. Springer-Verlag. New York, NY. 1:35.

Ruepp, A., W. Graml, M.-L. Santos- Martinez, K. K. Koretke, C. Volker, H. W. Mewes, D. Frishman, S. Stocker, A. N. Lupas, and W. Baumeister. 2000. The genome sequence of the thermoacidophilic scavenger Thermoplasma acidophilum. Nature 407:508–513.

Schleifer, K.-H., and E. Stackebrandt. 1983. Molecular systematics of prokaryotes. Ann. Rev. Microbiol. 37:143–187.

Schleper, C., G. Puehler, I. Holz, A. Gambacorta, D. Janekovic, U. Santarius, H.-P. Klenk, and W. Zillig. 1995. Picrophilus gen. nov., fam. nov.: A novel aerobic, heterotrophic, thermoacidophilic genus and family comprising archaea capable of growth around pH 0. Int. J. Syst. Bacteriol. 177:7050–7059.

Schleper, C., G. Pühler, H.-P. Klenk, and W. Zillig. 1996. Picrophilus oshimae and Picrophilus torridus fam. nov., gen. nov., sp. nov., two species of hyperacidophilic, thermophilic, heterotrophic, aerobic archaea. Int. J. Syst. Bacteriol. 46:814–816.

Searcy, D. G. 1975a. Histon.e-like protein in the prokaryote Thermoplasma acidophilum. Biochim. Biophys. Acta 395:535–547.

Searcy, D. G., and E. K. Doyle. 1975b. Characterization of Thermoplasma acidophilum deoxyribonucleic acid. Int. J. Syst. Bacteriol. 25:286–289.

Searcy, D. G. 1976. Thermoplasma acidophilum: Intracellular pH and potassium concentration. Biochim. Biophys. Acta 451:278–286.

Searcy, D. G., and R. J. DeLange. 1980. Thermoplasma acidophilum histone like protein: Partial amino acid sequence suggestive of homology to eukaryotic histones. Biochim. Biophys. Acta 609:197–200.

Searcy, D. G., and F. R. Whatley. 1982. Thermoplasma acidophilum cell membrane: Cytochrome b and sulfate-stimulated ATPase. Zentralbl. Bakteriol. Parasitenkd. Infektionskr. Hyg. Abt. 1, Orig. Reihe C3: 245–247.

Searcy, D. G., and F. R. Whatley. 1984. Thermoplasma acidophilum: Glucose degradative pathways and respiratory activities. Syst. Appl. Microbiol. 5:30–40.

Seemüller, E., A. Lupas, D. Stock, J. Löwe, R. Huber, and W. Baumeister. 1995. Proteasome from Thermoplasma acidophilum: A threonine protease. Science 268:579–582.

Segerer, A., A. Neuner, J. K. Kristjansson, and K. O. Stetter. 1986a. Acidianus infernus gen. nov., sp. nov., and Acidianus brierleyi comb. nov.: Facultatively aerobic, extremely acidophilic, thermophilic sulfur-metabolizing archaebacteria. Int. J. Syst. Bacteriol. 36:559–564.

Segerer, A., K. O. Stetter, and F. Klink. 1986b. Novel facultatively aerobic sulfur-dependent archaebacteria. In: O. Kandler and W. Zillig (Eds.) Archaebacteria. G. Fischer Verlag. Stuttgart, Germany. 430.

Segerer, A., T. A. Langworthy, and K. O. Stetter. 1988. Thermoplasma acidophilum and Thermoplasma volcanium sp. nov. from solfatara fields. Syst. Appl. Microbiol. 10:161–171.

Segerer, A. H., and K. O. Stetter. 1992a. The genus Thermoplasma. *In:* A. Balows, H. G. Trüper, M. Dvorkin, W. Harder, and K.-H. Schleifer (Eds.) *The Prokaryotes,* 2nd ed. Springer-Verlag. New York, NY. 712–718.

Segerer, A. H., and K. O. Stetter. 1992b. The order Sulfolobales. *In:* A. Balows, H. G. Trüper, M. Dvorkin, W. Harder, and K.-H. Schleifer (Eds.) *The Prokaryotes,* 2nd ed. Springer-Verlag. New York, NY. 684–701.

Shimada, H., Y. Shida, N. Nemoto, T. Oshima, and A. Yamagishi. 2001. Quinone profiles of Thermoplasma acidophilum HO-62. J. Bacteriol. 183:1462–1465.

Silverman, M. P., and D. G. Lundgren. 1959. Studies on the chemoautotrophic iron bacterium Ferrobacillus ferrooxidans. 1: An improved medium and harvesting procedure for securing high cell yields. J. Bacteriol. 77:642–647.

Smith, P. F., T. A. Langworthy, and M. R. Smith. 1975. Polypeptide nature of growth requirement in yeast extract for Thermoplasma acidophilum. J. Bacteriol. 124:884–892.

Smith, P. F. 1980. Sequence and glycosidic bond arrangement of sugars in lipopolysaccharide from Thermoplasma acidophilum. Biochim. Biophys. Acta 619:367–373.

Stetter, K. O., and W. Zillig. 1985. Thermoplasma and the sulfur-dependent archaebacteria. *In:* C. R. Woese and R. S. Wolfe (Eds.) The Bacteria. Academic Press. Orlando, FL. 8:85–170.

Sturm, S., V. Schönefeld, W. Zillig, D. Janekovic, and K. O. Stetter. 1980. Structure and function of the DNA-dependent RNA polymerase of the archaebacterium Thermoplasma acidophilum. Zentralbl. Bakteriol. Parasitenkd. Infektionskr. Hyg. Abt. 1, Orig. Reihe C1:12–25.

Tansey, M. R., and T. D. Brock. 1973. Dactylaria gallopava, a cause of avian encephalitis, in hot spring effluents, thermal soils and self-heated coal waste piles. Nature 242:202–203.

Uda, I., A. Sugai, K. Kon, S. Ando, Y. H. Itoh, and T. Itoh. 1999. Isolation and characterization of novel neutral glycolipids from Thermoplasma acidophilum. Biochim. Biophys. Acta 1439:363–370.

van de Vossenberg, J. L. C. M., A. J. M. Driessen, W. Zillig, and W. N. Konings. 1998. Bioenergetics and cytoplasmic membrane stability of the extremely acidophilic, thermophilic archaeon Picrophilus oshimae. Extremophiles 2:67–74.

Vásquez, M., E. R. B. Moore, R. T. Espejo. 1999. Detection by polymerase chain reaction-amplification and sequencing of an archaeon in a commercial-scale copper bioleaching plant. FEMS Microbiol. Lett. 173:183–187.

Woese, C. R., and G. E. Fox. 1977. Phylogenetic structure of the prokaryotic domain: The primary kingdoms. Proc. Natl. Acad. Sci. USA 74:5088–5090.

Woese, C. R., J. Maniloff, and L. B. Zablen. 1980. Phylogenetic analysis of the mycoplasmas. Proc. Natl. Acad. Sci. USA 77:494–498.

Woese, C. R., O. Kandler, and M. L. Wheelis. 1990. Towards a natural system of organisms: Proposal for the domains Archaea, Bacteria and Eucarya. Proc. Natl. Acad. Sci. USA 87:4576–4579.

Wolf, S., F. Lottspeich, and W. Baumeister. 1993. Ubiquitin found in the archaebacterium Thermoplasma acidophilum. FEBS Lett. 326:42–44.

Yang, L. L., and A. Haug. 1979. Purification and partial characterization of a prokaryote glycoprotein from the plasma membrane of Thermoplasma acidophilum. Biochim. Biophys. Acta 556:265–277.

Yang, D., B. P. Kaine, and C. R. Woese. 1985. The phylogeny of archaebacteria. Syst. Appl. Microbiol. 6:251–256.

Zillig, W., R. Schnabel, J. Tu, and K. O. Stetter. 1982. The phylogeny of archaebacteria, including novel anaerobic thermoacidophiles, in the light of RNA polymerase structure. Naturwissenschaften 69:197–204.

Prokaryotes (2006) 3:113–164
DOI: 10.1007/0-387-30743-5_8

CHAPTER 8

The Order Halobacteriales

AHARON OREN

Introduction

Halophilic Archaea of the order *Halobacteriales* are found in hypersaline environments in which salt concentrations exceed 150–200 g/liter. They inhabit salt lakes such as the Great Salt Lake, Utah, the Dead Sea, and other hypersaline water bodies such as the crystallizer ponds of solar salterns in which sea water is evaporated for the production of salt. Additional habitats in which these halophilic Archaea often develop include salted fish and hides preserved by treatment with salt. They also occur in certain fermented food products in which molar concentrations of NaCl are added as part of the manufacturing process, such as Thai fish sauce.

Microbial deterioration of salted hides and salted fish by members of the *Halobacteriales* is clearly visible by the pink-red color of the developing colonies (Clayton and Gibbs, 1927; Harrison and Kennedy, 1922; Lochhead, 1934; Shewan, 1971; Vreeland et al., 1998a). Many of the early studies on the red halophilic Archaea were initiated in an attempt to understand the cause of the damage to fish and hides treated with salt for their preservation. The account by Klebahn, 1919 on the bacteria causing the red discoloration of salted cod probably presents the first accurate description of halobacteria, at the time designated as "*Bacillus halobius* ruber." That study was followed by the isolation of "*Pseudomonas salinaria*" from the red discoloration of cured codfish (Harrison and Kennedy, 1922). This culture is now lost, but the organism is probably very similar to later isolates known as *Halobacterium salinarum*. Another classic early study on the red halophilic bacteria that grow on salted fish is that of Petter, 1931. The phenomenon of bacterial degradation of salted hides was documented by Lochhead, 1934. During the 1930s also the first studies appeared of the microbiology of solar salterns and salt lakes, describing the importance of red halophilic bacteria in these ecosystems (Baas-Becking, 1931; Hof, 1935). Many of the early studies on the *Halobacteriales* were summarized in Larsen's classic essay on "the halobacteria's confusion to

biology" (Larsen, 1973). Tindall, 1992 presented an in-depth discussion on the properties, nomenclature, and taxonomic affiliations of some of the early isolates.

The mode of adaptation of the *Halobacteriales* to life at salt concentrations at or near NaCl saturation has been the subject of in-depth studies. In contrast to most other halophilic or halotolerant microorganisms that keep intracellular ionic concentrations low, cells of the halophilic Archaea of the order *Halobacteriales* contain molar concentrations of ions, especially K^+ and Cl^-, within the cells (Christian and Waltho, 1962; Matheson et al., 1976; Oren, 1999a; Pérez-Fillol and Rodriguez-Valera, 1986). The maintenance of high intracellular salt concentrations requires unique adaptations of the enzymatic machinery to be able to function in the presence of high salt, adaptations that make the cells of most species strictly dependent on the continuous presence of high salt concentrations for the maintenance of structural integrity and viability (Dennis and Shimmin, 1997; Ebel et al., 1999; Eisenberg, 1995; Eisenberg and Wachtel, 1987; Eisenberg et al., 1992; Lanyi, 1974).

Much of the interest of the scientific community in the genus *Halobacterium* and other representatives of the *Halobacteriales* was triggered by the discovery of the retinal pigments, bacteriorhodopsin and halorhodopsin. The structure of bacteriorhodopsin and its function as an outward proton pump were disclosed in the early 1970s (Danon and Stoeckenius, 1974; Oesterhelt and Stoeckenius, 1971). The function of halorhodopsin as an inward chloride pump was recognized in 1982 (Schobert and Lanyi, 1982). Both proteins enable the direct use of light energy by the cells for energy transduction. Additional retinal proteins serve as light sensors to direct the phototactic machinery of cells of motile species to an optimal light environment (Bogomolni and Spudich, 1982; Spudich and Bogomolni, 1988; Spudich et al., 1995). The retinal pigments rapidly became popular models for in-depth biochemical and biophysical studies (Lanyi and Váró, 1995; Oesterhelt, 1995). In addition, the unique properties of these pigments also present promising features for future biotechnological applications (Birge, 1995; Chen and Birge, 1993; Hong, 1986; Oesterhelt et al., 1991).

Different aspects of the biology of the *Halobacteriales* have been reviewed in the past. In addition to review papers cited in the sections relating to specific subjects, much useful information can be found in reviews by Bayley and Morton, 1978; Grant and Ross, 1986; Kushner, 1985; Larsen, 1962; Tindall and Trüper, 1986; Trüper and Galinski, 1986; the reviews on physiology and metabolism by Dundas, 1977 and by Hochstein, 1988 and on the genetics of the group by Pfeifer, 1988; the ecological survey by Oren, 1994a; the taxonomic treatment in *Bergey's Manual* (Grant and Larsen, 1989); and the treatises on the group in the earlier editions of *The Prokaryotes* (Larsen, 1981; Tindall, 1992).

Phylogeny

The question "Are extreme halophiles actually 'bacteria'?" (Magrum et al., 1978) had to be answered negatively when it was recognized in the late 1970s that *Halobacterium salinarum* and related halophiles belong to the newly recognized domain Archaea. This finding again increased the interest in the *Halobacteriales*, as of all archaeal groups, they are the most easily handled in the laboratory (Shand and Perez, 1999a).

The *Halobacteriales* (Grant and Larsen, 1989) form a branch within the Euryarchaeota, branching off close to the *Methanomicrobiales/Methanosarcinales*. At the time of writing (June 1999) a single family was recognized within the order *Halobacteriales*: the *Halobacteriaceae*. Presently this family is divided into 14 genera with 35 validly described species (Table 1).

Table 1. The genera and species within the order Halobacteriales.

Family	Genus	Species
Halobacteriaceae		
	Halobacterium[T]	*Halobacterium salinarum*[T]
	Halobaculum	*Halobaculum gomorrense*[T]
	Halorubrum	*Halorubrum saccharovorum*[T]
		Halorubrum sodomense
		Halorubrum lacusprofundi
		Halorubrum coriense
		Halorubrum distributum
		Halorubrum vacuolatum
		Halorubrum trapanicum
	Haloarcula	*Haloarcula vallismortis*[T]
		Haloarcula marismortui
		Haloarcula hispanica
		Haloarcula japonica
		Haloarcula argentinensis
		Haloarcula mukohataei
		Haloarcula quadrata
	Natronomonas	*Natronomonas pharaonis*[T]
	Halococcus	*Halococcus morrhuae*[T]
		Halococcus saccharolyticus
		Halococcus salifodinae
	Natrialba	*Natrialba asiatica*[T]
		Natrialba magadii
	Natronobacterium	*Natronobacterium gregoryi*[T]
	Halogeometricum	*Halogeometricum borinquense*[T]
	Natronococcus	*Natronococcus occultus*[T]
		Natronococcus amylolyticus
	Haloferax	*Haloferax volcanii*[T]
		Haloferax gibbonsii
		Haloferax denitrificans
		Haloferax mediterranei
	Natrinema	*Natrinema pellirubrum*[T]
		Natrinema pallidum
	Haloterrigena	*Haloterrigena turkmenica*[T]
	Natronorubrum	*Natronorubrum bangense*[T]
		Natronorubrum tibetense

[T] = type genus of the family or type species of the genus.

Fig. 1. Phylogenetic tree of the *Halobacteriaceae*, prepared by T.J. McGenity (Department of Biological Sciences, University of Essex, U.K.). Distances were calculated by the method of Jukes and Cantor, 1969, and the tree was produced by the method of Fitch and Margoliash, 1967. Bootstrap percentages above 75% are shown at the nodes.

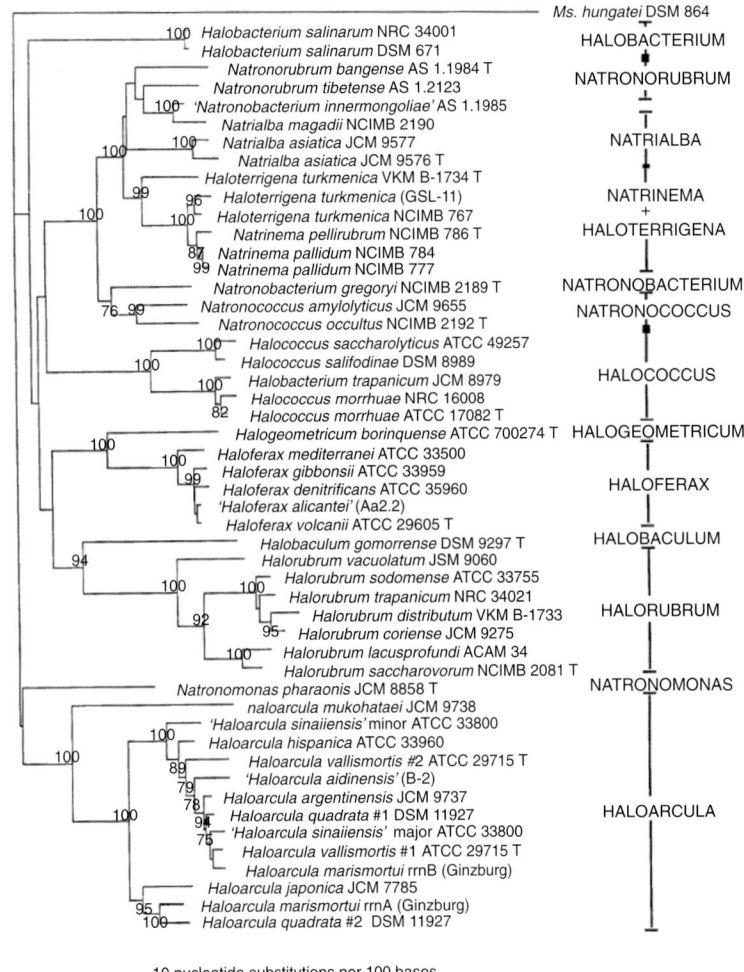

10 nucleotide substitutions per 100 bases

Figure 1 presents a phylogenetic tree of the family, based on 16S rDNA sequence comparisons.

Taxonomy

The order *Halobacteriales* contains a variety of morphological types, from rods and cocci to flat extremely pleomorphic types (Mullakhanbhai and Larsen, 1975), perfectly square flat cells (Kessel and Cohen, 1982; Oren, 1999b; Oren et al., 1996c; Parkes and Walsby, 1981; Romanenko, 1981; Stoeckenius, 1981; Walsby, 1980), and triangular and trapezoid cells such as displayed by *Haloarcula japonica* (Hamamoto et al., 1988; Horikoshi et al., 1993; Takashina et al., 1990). The micrographs shown in Figs. 2–4 present a selection of the morphological types encountered within the group.

Classification and identification of the genera and species belonging to the order *Halobacteriales*, family *Halobacteriaceae* are currently based on a polyphasic approach (Oren et al., 1997a), involving properties such as cell morphology, growth characteristics, chemotaxonomical traits (notably the presence or absence of specific polar lipids), and nucleic acid sequence data. The properties of the validly described genera and species are summarized in Tables 2–6. More detailed information can be found in the individual species descriptions (see the literature citations at the end of each table). Useful taxonomic information can be found online, as well (Euzéby, 1999).

In recent years the comparison of 16S rDNA sequences has led to a new classification, including the creation of a number of new genera, the splitting of existing genera, and in a number of cases also the unification into one genus of two species classified earlier in separate genera (see e.g., Kamekura, 1998; Kamekura, 1999; Kamekura and Seno, 1993). Comparison of 16S rRNA nucleotide sequences led to the insight that the alkaliphiles do not form a defined phylogenetic group, but many alkaliphilic species are interspersed between their neutrophilic relatives

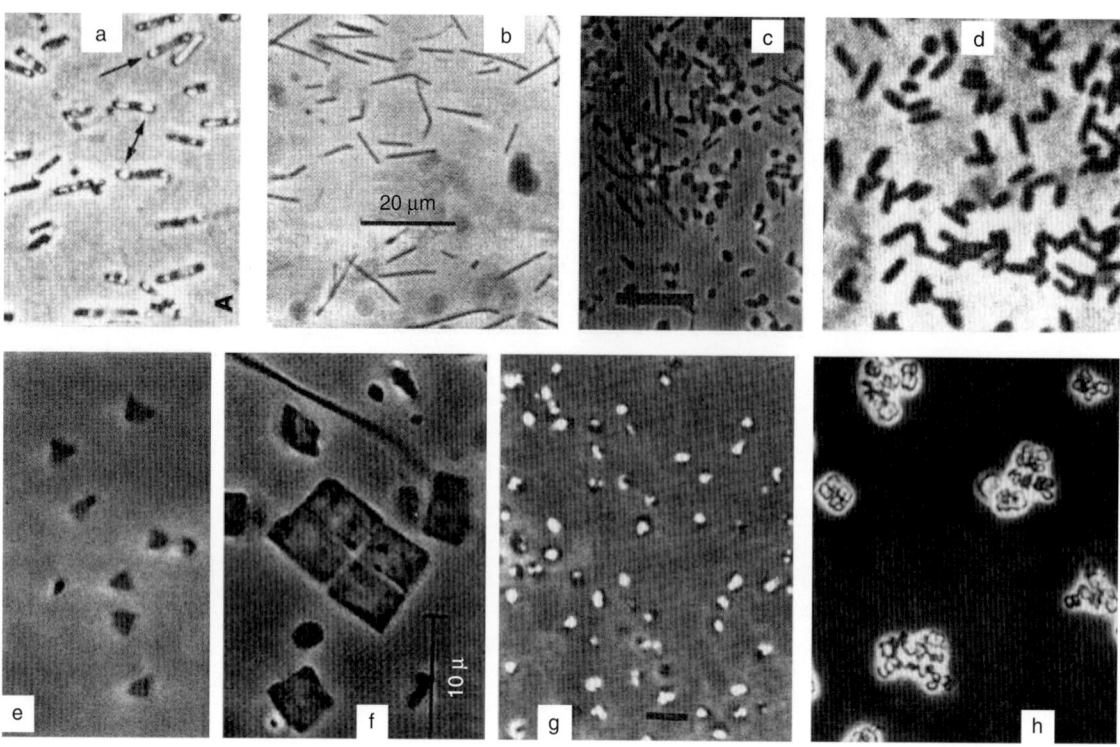

Fig. 2. Some representatives of the *Halobacteriales*: *Halobacterium salinarum* (a), *Natronobacterium gregoryi* (b), *Halorubrum lacusprofundi* (c), *Natronomonas pharaonis* (d), *Haloarcula japonica* (e), Walsby's square bacterium (f), *Halogeometricum borinquense* (g), and *Natronococcus occultus* (h). Phase contrast. Gas vesicles are visible in (a) and (g). Reproduced from Horne and Pfeifer, 1989, Tindall et al., 1984, Franzmann et al., 1988, Soliman and Trüper, 1982, Horikoshi et al., 1993, Stoeckenius, 1981, Mwatha and Grant, 1993, and Tindall et al., 1984, respectively, with permission from the publishers.

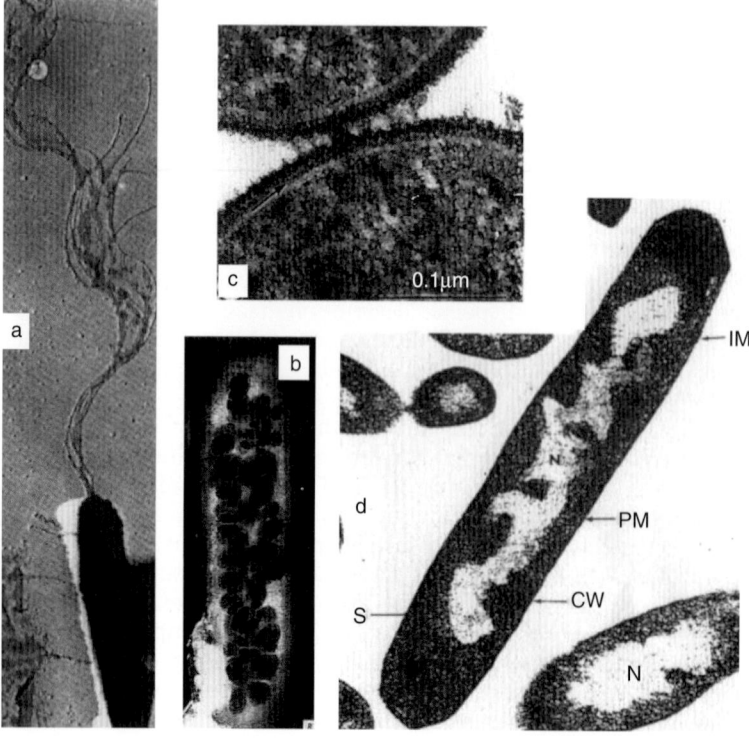

Fig. 3. Electron micrographs of *Halobacterium salinarum*: negatively stained preparations, showing flagella (a) and gas vesicles (b), and thin sections (c, d), showing details of the cell envelope structure and the arrangement of the nuclear material. Reproduced from Alam and Oesterhelt, 1984a, Houwink, 1956, Usukura et al., 1980, and Cho et al., 1967, respectively, with permission from the publishers.

Fig. 4. Electron micrographs of different representatives of the *Halobacteriales*: thin sections of *Haloferax volcanii* (a) and *Haloferax mediterranei* (containing large amounts of poly-β-hydroxyalkanoate) (b), transmission (c) and scanning electron micrographs (d) of *Halococcus salifodinae*, and negatively stained cells of Walsby's square bacterium (e). Reproduced from Mullakhanbhai and Larsen, 1975; Rodriguez-Valera and Lillo, 1992b; Denner et al., 1994; and Parkes and Walsby, 1981, respectively, with permission from the publishers.

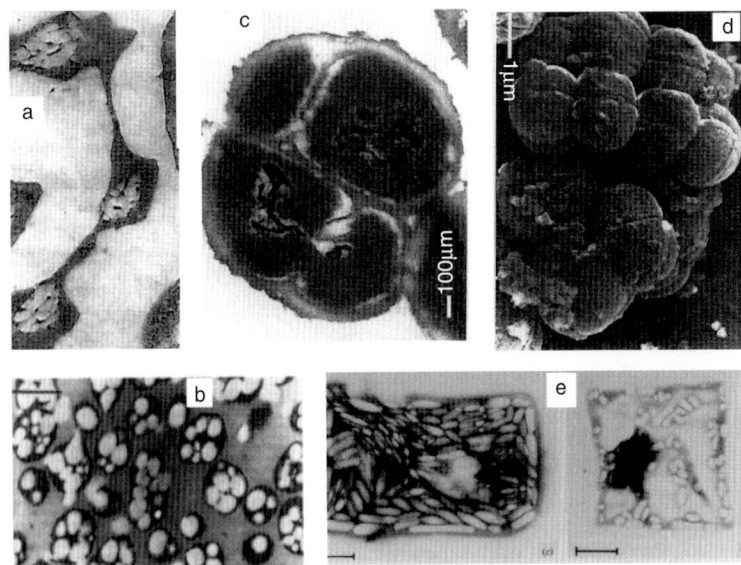

in the phylogenetic tree. The present-day emphasis on 16S-rRNA sequence data for the determination of the phylogenetic position of the different members of the halophilic Archaea has reduced the relative importance of the polar lipid composition, as certain diagnostic glycolipids may be found in more than one genus, and members of certain genera may contain different glycolipids (for an in-depth discussion see Kamekura, 1999). Figure 1 presents a phylogenetic tree of the *Halobacteriaceae*, based on 16S rDNA sequence comparisons. Also the 23S rRNA genes have been used in some taxonomic studies, and the analysis of the available data suggested that the halobacterial genes diverged over a relatively short time (Lodwick et al., 1994).

Lost and contaminated strains, as well as improper strain designations, have caused much confusion in the past. Since the in-depth discussion and critical assessment of strain histories and identities by Tindall, 1992, many of the old controversies have been resolved. However, considerable confusion still exists on the true nature of certain strains (see e.g., the notes to Table 2). Thus, different strains designated *Halorubrum saccharovorum* possess different glycolipids (Tindall, 1990b). The importance of the use of type strains in taxonomic and other (e.g. biochemical and genetic) studies should be stressed here once more. In case of doubt, molecular biological techniques now enable easy tests for the authenticity of the strains used. Random amplified polymorphic DNA analysis (RAPD), using oligonucleotide probes of arbitrary sequence, was used to compare closely related strains (Martinez-Murcia and Rodriguez-Valera, 1994) and was found to be a powerful tool to sort out

problems of strain identity (Martinez-Murcia et al., 1995a). Another useful way to compare strains and resolve questions about their authenticity is to determine protein profiles with SDS polyacrylamide electrophoresis gels (Hesselberg and Vreeland, 1995).

A number of species *incertae sedis* exist that are awaiting to be described either as new species or as representatives of recognized taxa: "*Haloarcula aidinensis*" (Peijin and Yi, 1996; Peijin et al., 1995; Yi et al., 1995), "*Haloarcula californiae*" (Javor et al., 1982), "*Natronobacterium innermongoliae*" (Tian et al., 1997), and "*Haloferax alicantei*" (Holmes et al., 1997). There are also yet uncultured types, known either by their distinct morphology or from 16S rDNA sequences amplified from brine samples. Thus, the gas vesicle containing square nonmotile Archaea first described by Walsby, 1980 and since observed and studied in many hypersaline environments (Guixa-Boixareu et al., 1996; Kessel and Cohen, 1982; Oren, 1999b; Oren et al., 1996c; Parkes and Walsby, 1981; Romanenko, 1981; Stoeckenius, 1981; Torrella, 1986) still defy the microbiologists' attempts to culture them. They probably contain S-DGD-1 as their major glycolipid (Oren et al., 1996c) and are therefore probably unrelated to the motile square isolate 801030/1 obtained from the same brine pool in the Sinai peninsula (Egypt) in which the gas vesicles containing square bacteria were first observed (Alam et al., 1984). This strain was recently described as a new species of the genus *Haloarcula*, *Haloarcula quadrata* (Oren et al., 1999). Random amplification of 16S rDNA sequences from saltern crystallizer ponds in Spain and Israel consistently yielded one phylotype appearing at the highest frequency. This

Table 2. Properties of the species of the genera *Halobacterium*, *Halobaculum*, and *Halorubrum*.

Genus	Halobacterium	Halobaculum	Halorubrum						
Species Basonym	salinarum[T] Halobacterium salinarium, Pseudomonas salinaria[a]	gomorrense[T] Halobaculum gomorrense	saccharovorum[T] Halobacterium saccharovorum	sodomense Halobacterium sodomense	lacusprofundi Halobacterium lacusprofundi	coriense Halorubrobacterium coriense	distributum Halobacterium distributum	vacuolatum Natronobacterium vacuolatum (vacuolata)	trapanicum Halobacterium trapanicum
Morphology	Rod	Rod	Rod	Rod	Rod	Pleomorphic rod	Pleomorphic rod	Rod	Rod
Cell size	$5-10 \times 0.5-1\mu m$	$0.5-1 \times 5-10\mu m$	$0.5 \times 2.5-5\mu m$	$0.5 \times 2.5-5\mu m$	up to $12\mu m$	$0.5-1 \times 0.5-5\mu m$	$0.8-1 \times 2.7-7\mu m$	$0.5-0.7 \times 1.5-3\mu m$	$0.7-1 \times 1.5-3\mu m$
Motility	+	±	+	+	d	+	+	+	-
Gas vesicles	-	-	-	-	-	-	-	+	-
NaCl optimum (M)	4-5	1.5-2.5	3.5-4.5	1.7-2.5	2.5-3.5	2.2-2.7	NR	3.5	NR
NaCl range (M)	3.0-5.2	1.0-2.5	1.5-5.2	0.5-4.3	1.5-5.2	2.0→3.2	NR	2.6-5.1	NR
Mg optimum (M)	>0.005	0.6-1.0	>0.005	0.6-1.2	0.02-0.6	>0.005	NR	<0.001	NR
pH optimum	Neutral	6-7	Neutral	Neutral	Neutral	Neutral	9.5	Neutral	Neutral
Temp. optimum (°C)	35-50	40	50	40	31-37	NR	37-45	35-40	NR
Nitrite from nitrate	-	+	+[b]	±	+ or ±	NR	+	+	+
Gas from nitrate	-	-	-	-	-	NR	-	-	-
Anaerobic growth on nitrate	-	-	-	-	NR	NR	NR	NR	NR
Anaerobic growth on L-arginine	+	+	+	+	+	+	+	+	+
Acids from carbohydrates	-	-	NR	-	+	+	-	NR	NR
Growth on single carbon sources[c]	-	-	-	-	+	-	-	NR	NR
Indole from tryptophan	-	+[d]	-	+	-	NR	-	-	-
Starch hydrolysis	-	-	NR	NR	-	NR	NR	-	NR
Tween 80 hydrolysis	+	-	-	NR	-	NR	+	-	-
Gelatin hydrolysis	-	-	-	-	-	NR	-	NR	NR
Gasein hydrolysis	+	-	-	NR	-	NR	NR	NR	NR
Pigmentation	Red/purple	Red	Red	Red/purple	Red	Red	Red	Red	Red
Major glycolipids	S-TGD, S-TeGD	S-DGD-1	S-DGD-3(?) or S-DGD-1(?)[e]	S-DGD-3	S-DGD-3	S-DGD-3	S-DGD-3	—	S-DGD-5
Presence of PGS	+	-	+	+	+	+	+	-	+

	1	2	3	4	5	6	7	8	9
G+C of DNA	66–70.9 (major) 57–60 (minor)	70	69.1–72.1 (major) 54.8–56.5 (minor)	68	65.3–65.8 (major) 54.6–65.3 (minor)	NR	63.6–70.8	62.7	64.3
16S rRNA sequence	M38380, M11583, D14127	L37444	U17364, X82167	D13379, X82169	U17365, X82170	L00922	D63572	D87972	X82168
Type strain	ATCC 33171 DSM 3754 JCM 8978 NCIMB 764 NRC 34002 VKM B-1769	DSM 9297	ATCC 29252 CCM 2887 DSM 1137 NCIMB 2081 VKM B-1747	ATCC 33755 DSM 3755 NCIMB 2197 VKM B-1771	ACAM 34 DSM 5036 NCIMB 12997 VKM B-1753	ACM 391 JCM 9275	NCIMB 13203 VKM B-1733[f]	NCIMB 13189, JCM 9060	NCIMB 13488[g]
Source of isolation	Salted cow hide	Dead Sea, Israel	Saltern, California	Dead Sea, Israel	Deep Lake, Antarctica	Saltern, Australia	Saline soil, USSR	Lake Magadi, Kenya	Solar salt, Italy
Reference	1	2	3	4	5	6	7	8	9

Data were derived from the original species descriptions and from Grant and Larsen (1989) and Tindall (1992). Data on anaerobic growth on arginine were derived in part from Oren (1994b) and Oren and Litchfield (1999).

+: all strains tested positive; –: all strains tested negative; ±: weak reaction; d: some strains positive; NR: not reported.

[T] type species of the genus.

[a] The species *Halobacterium salinarum* includes strains formerly named *Halobacterium halobium* and *Halobacterium cutirubrum* (Ventosa and Oren, 1996). For additional properties of the group see Colwell etal. (1979).

[b] Franzmann etal. (1988) reported a positive reaction.

[c] Positive growth on single carbon sources may in certain cases require low concentrations of additional growth factors (to be satisfied e.g., by addition of 0.005% yeast extract).

[d] Franzmann etal. (1988) reported a negative reaction.

[e] Controversial data have been published on the structure of the sulfated diglycosyl diether lipids present in different species of *Halorubrum* (Kamekura, 1998; Lanzotti etal. 1998).

[f] Zvyagintseva etal. (1996) proposed strain VKM B-1739 as the new type strain of *Halorubrum distributum* to replace VKM B-1733. However, no compelling reasons for the change were brought forward, and therefore strain VKM B-1733 should remain the type strain of the species (Oren etal., 1997b).

[g] Considerable confusion has existed in the past on strain identities and the assignment of a type strain. For details see Tindall (1992) and Grant etal. (1998a).

References:
1—Harrison and Kennedy, 1922; Elazari-Volcani, 1957.
2—Oren etal., 1995.
3—Tomlinson and Hochstein, 1976; McGenity and Grant, 1995.
4—Oren, 1983a; McGenity and Grant, 1995.
5—Franzmann etal., 1988; McGenity and Grant, 1995.
6—Nuttall and Dyall-Smith, 1995; Kamekura and Dyall-Smith, 1995; Oren and Ventosa, 1996.
7—Zvyagintseva and Tarasov, 1987; Kamekura and Dyall-Smith, 1995; Kostrikina etal., 1990; Oren and Ventosa, 1996; Oren etal., 1997b.
8—Mwatha and Grant, 1993; Kamekura etal., 1997.

Table 3. Properties of the species of the genera *Haloarcula* and *Natronomonas*.

Genus	*Haloarcula*							*Natronomonas*
Species	*vallismortis*[T]	*marismortui*	*hispanica*	*japonica*	*argentinensis*	*mukohataei*[a]	*quadrata*	*pharaonis*[T]
Basonym	*Halobacterium vallismortis*	*Halobacterium marismortui*						*Natronobacterium pharaonis*
Morphology	Pleomorphic rod	Flat pleomorphic	Pleomorphic rod	Flat pleomorphic	Flat pleomorphic	Rod	Flat pleomorphic to square	Rod
Cell size	0.6–1 × 3–5 μm	1–2 × 2–3 μm	0.3 × 0.5–1 μm	0.2–0.5 × 2–5 μm	0.3 × 1 μm	0.5 × 2 μm	2–3 μm	0.8 × 2–3 μm
Motility	+	– or ±	+	+	+	+	+	+
Gas vesicles			–	–	–	–	–	+
NaCl optimum (M)	3.5–4.3	3.4–3.9	2.6	3.4	2.5–3.0	3.0–3.5	3.4–4.3	3.5
NaCl range (M)	NR	1.7–5.1	2.0–5.2	2.5–5.1	2.0–4.5	2.5–4.5	2.7–4.3	2.0–5.2
Mg optimum (M)	NR	>0.05–0.1	>0.005	>0.04	0.1	0.003–0.3	>0.05–0.1	<0.01
pH optimum	7.4–7.5	Neutral	Neutral	7.0–7.5	Neutral	Neutral	6.5–7.0	9.5–10.0
Temp. optimum (°C)	40	40–50	35–40	42	40	40	53	45
Nitrite from nitrate	+	+	+	+	NR	NR	+	±[b]
Gas from nitrate	+	+	+	+	NR	NR	+	–
Anaerobic growth on nitrate	+	+	NR	NR	NR	NR	+	–
Anaerobic growth on L-arginine	–	–	NR	NR	NR	NR	–	NR
Acids from carbohydrates	+	+	+	+	+	+	+	–
Growth on single carbon sources[c]	+	+	+	+	NR	NR	+	–
Indole from tryptophan	+	–	d	+	NR	NR	–	NR
Starch hydrolysis	+	+[d]	+	–	NR	NR	+	–
Tween 80 hydrolysis	–	NR	+	NR	NR	NR	–	–
Gelatin hydrolysis	–	–	+	–	NR	NR	–	+
Gasein hydrolysis		NR	d		NR	NR	–	–
Pigmentation	Red	Red	Red	Red	Red	Red	Red	Red
Major glycolipids	TGD-2	TGD-2	TGD-2	TGD-2	TGD-2	S-DGD-1(?), DGD-1(?)	TGD-2	
Presence of PGS	+	+	+	+	+	+	+	–
G+C of DNA	64.7	62.7 (major) 54.7 (minor)	62.7	63.3	62	65	60.1	61.2 (major) 51.9 (minor)

	1	2	3	4	5	6	7	8
16S rRNA sequence[e]	U17593 (A), D50581 (B)	X61688, X61689	U68541	D28872	D50849	D50850	AB010964, AB010965	D87971
Type strain	ATCC 29715 CCM 3404 DSM 3756 NCIMB 2082 VKM B-1791	ATCC 43049	ATCC33960 DSM 4426 NCIMB 2187 VKM B-1755	JCM 7785 NCIMB 13157	JCM 9737	JCM 9738	DSM 11927	ATCC 35678 CCM 3872 DSM 2160 JCM 8858 NCIMB 2260 VKM B-1749
Source of isolation	Salt pools, Death Valley, California	Dead Sea, Israel	Saltern, Spain	Saltern, Japan	Salt flats, Argentina	Salt flats, Argentina	Sabkha, Sinai, Egypt	Wadi Natrun, Egypt
Reference	1	2	3	4	5	6	7	8

Data were derived from the original species descriptions and from Grant and Larsen (1989) and Tindall (1992). Data on anaerobic growth on arginine were derived in part from Oren (1994b) and Oren and Litchfield (1999).

+: all strains tested positive; −: all strains tested negative; ±: weak reaction; d: some strains positive; NR: not reported.

[T]type species of the genus.

[a]In view of the different glycolipid content, lack of the characteristic *Haloarcula* signature bases in the 16S rDNA sequence, and other properties in which *Haloarcula mukohataei* differs from the other representatives of the genus *Haloarcula*, reclassification in another, newly to be created, genus may be desirable (Kamekura 1999).

[b]Conflicting reports: Soliman and Trüper (1982) reported a negative reaction, while Tindall etal. (1984) stated a positive reaction.

[c]Positive growth on single carbon sources may in certain cases require low concentrations of additional growth factors (to be satisfied e.g., by addition of 0.005% yeast extract).

[d]Conflicting reports exist.

[e]*Haloarcula* species typically have more than one heterologous 16S rRNA: two were identified in *Haloarcula marismortui* (Mylvaganam and Dennis, 1992), *Haloarcula argentinensis* possesses at least four genes, and *Haloarcula mukohataei* contains at least three different genes (Thara etal. 1997).

References:

1—Gonzalez etal. 1978; Torreblanca etal. 1986.
2—Elazari-Volcani, 1957; Ginzburg etal. 1970; Oren etal. 1988; Oren etal. 1990.
3—Juez etal. 1986.
4—Takashina etal. 1990.
5—Ihara etal. 1997.
6—Ihara etal. 1997.
7—Oren etal. 1999.
8—Soliman and Trüper, 1982; Tindall etal. 1984; Kamekura etal. 1997.

Species incertae sedis classified within the genus *Haloarcula* are:

—*Haloarcula sinaiiensis*, isolated from a brine pool in the Sinai peninsula, Egypt (Javor etal., 1982). The strain has been deposited as ATCC 33800 and NCIMB 2268, and its 16S rRNA gene sequences are D14129 (major gene) and D14130 (minor gene).
—*Haloarcula californiae*, isolated from Guerrero Negro, Baja California, Mexico (Javor etal., 1982). The strain has been deposited as ATCC 33799 and NCIMB 2267.
—*Haloarcula aidinensis*, isolated from a salt lake in Xinyang, China (Peijin and Yi, 1996; Peijin etal. 1994; Yi etal., 1995), deposited as AS 1.2042, with 16S rDNA sequence AB000563.

Table 4. Properties of the species of the genera *Halococcus*, *Natrialba*, *Natronobacterium*, and *Halogeometricum*.

Genus	Halococcus			Natrialba	Natronobacterium		Halogeometricum
Species / Basonym	morrhuae[T] / Sarcina morrhuae	saccharolyticus	salifodinae	asiatica[T]	magadii / Natronobacterium magadii	gregoryi[T]	borinquense[T]
Morphology	Coccus	Coccus	Coccus	Rod	Rod	Rod	Pleomorphic flat
Cell size	0.8–1.5μm	0.8–1.5μm	0.8–1.2μm	0.5 × 1–5μm	0.7–0.9 × 2–4μm	0.5–0.07 × 10–15μm	1–2 × 1–3μm
Motility	–	–	–	+	+	–	+
Gas vesicles	–		–	–	+	–	+
NaCl optimum (M)	3.5–4.5	4.3	3.4–4.3	3.5–4.0	3.5	3.0	3.4–4.3
NaCl range (M)	2.5–5.2	2.5–5.2	2.6–5.2	2.0–5.2	2.0–5.2	2.0–5.2	1.4–5.2
Mg optimum (M)	NR	>0.2	NR	>0.005	<0.01	<0.01	0.04–0.08
pH optimum	Neutral	6–8	6.8–9.5	6.6–7.8	9.5	9.5	7
Temp. optimum (°C)	30–45	37–40	40	35–40	37–40	37	40
Nitrite from nitrate	+	+	+	+	+	–	+
Gas from nitrate	–	d	–	–	–	–	+
Anaerobic growth on nitrate	–	–	–	–	–	–	+
Anaerobic growth on L-arginine	NR	NR	NR	–	NR	NR	–
Acids from carbohydrates	–	+	+	d	–	NR	+
Growth on single carbon sources[a]	–	+	NR	NR	–	–	+
Indole from tryptophan	+	+	NR	+	NR	NR	+
Starch hydrolysis	d	–	NR	d	–	–	+
Tween 80 hydrolysis	d	–	NR	NR	NR	NR	–
Gelatin hydrolysis	d	d	+	NR	+[b]	+[b]	NR
Gasein hydrolysis	–	–	NR	+	NR	NR	+
Pigmentation	Red	Red	Red	White	Red	Red	Red

	1	2	3	4	5	6	7
Major glycolipids	S-DGD-1	S-DGD-1	S-DGD-1	–	–	Unknown	–
Presence of PGS	–	–	–	NR	–	–	–
G+C of DNA	61–66	59.5	62	60.3–63.1	63.0 (major), 49.5 (minor)	65.0	59
16S rRNA sequence	X00662, D11106	AB004876	AD004877	D14123	X72495, D14124	D87970	AF002984
Type strain	ATCC 17082 CCM 537 DSM 1307 NCIMB 787 VKM B-1772	ATCC 49257 CCM 4147 DSM 5350 NCIMB 12873	ATCC 51437, DSM 8989	JCM 9576	ATCC 43099 CCM 3739 DSM 3739 JCM 8861 NCIMB 2190 VKM B-1751	ATCC 43098 CCM 3738 DSM 3393 JCM 8860 NCIMB 2189 VKM B-1750	ATCC 700274
Source of isolation	Dead Sea, Israel	Saltern, Spain	Salt mine, Austria	Beach sand, Japan	Lake Magadi, Kenya	Lake Magadi, Kenya	Saltern, Puerto Rico
Reference	1	2	3	4	5	6	7

Data were derived from the original species descriptions and from Grant and Larsen (1989) and Tindall (1992). Data on anaerobic growth on arginine were derived in part from Oren (1994b) and Oren and Litchfield (1999).

+: all strains tested positive; –: all strains tested negative; ±: weak reaction; d: some strains positive; NR: not reported.

[T]type species of the genus.

[a]Positive growth on single carbon sources may in certain cases require low concentrations of additional growth factors (to be satisfied e.g., by addition of 0.005% yeast extract).

[b]Mwatha and Grant (1993) reported a negative reaction.

References:

1—Kocur and Hodgkiss, 1973; Montero et al., 1988.
2—Montero et al., 1989.
3—Denner et al., 1994.
4—Kamekura and Dyall-Smith, 1995.
5—Tindall et al., 1984; Kamekura et al., 1997.
6—Tindall et al., 1984.
7—Montalvo-Rodríguez et al., 1998.

A species incertae sedis has been classified within the genus Natronobacterium: "Natronobacterium innermongoliae," isolated from a soda lake in Mongolia (Tian et al., 1997). The strain has been deposited as AS 1.1985, and its 16S rDNA gene sequence is AF009601.

Table 5. Properties of the species of the genera *Natronococcus* and *Haloferax*.

Genus	Natronococcus		Haloferax			
Species	occultus[T]	amylolyticus	volcanii[T]	gibbonsii	denitrificans	mediterranei
Basonym			Halobacterium volcanii		Halobacterium denitrificans	Halobacterium mediterranei
Morphology	Coccus	Coccus	Pleomorphic flat	Pleomorphic rod	Pleomorphic	Pleomorphic
Cell size	1–2 μm	1–2 μm	1–3 × 2–3 μm	0.4 × 0.5–2.5 μm	0.8–1.5 × 2–3 μm	0.5–1 × 2–3 μm
Motility	–	–	–	–	–	±
Gas vesicles	–	–	–	–	–	+
NaCl optimum (M)	3.4–3.8	2.5–3.4	1.7–2.5	2.5–4.3	2–3	2.9
NaCl range (M)	1.4–5.2	1.4–5.2	1.0–4.5	1.5–5.2	1.5–4.5	NR
Mg optimum (M)	<0.01	<0.01	0.2	>0.02–0.04	NR	0.02–0.04
pH optimum	9.5	9.0	Neutral	6.5–7.0	6.7	6.5
Temp. optimum (°C)	35–40	40–45	40	35–40	50	40[a]
Nitrite from nitrate	+	+	+	d	+	+
Gas from nitrate	–	NR	–	NR	+	+
Anaerobic growth on nitrate	–	NR	–[b]	NR	+	+
Anaerobic growth on L-arginine	NR	NR	–	–	–	–
Acids from carbohydrates	NR	NR	+	+	+	+
Growth on single carbon sources[c]	–	NR	+	+	+	+
Indole from tryptophan	NR	NR	+	+	–	+
Starch hydrolysis	–	+	–	–	–	+
Tween 80 hydrolysis	NR	NR	–	+	–	+
Gelatin hydrolysis	+	–	–	±	+	+
Gasein hydrolysis	NR	NR	–	±	NR	NR
Pigmentation	Red	Red	Red	Red	Red	Weakly pink

	1	2	3	4	5	6
Major glycolipids	—	—	S-DGD-1	S-DGD-1	S-DGD-1	S-DGD-1
Presence of PGS	—	—	—	—	—	—
G+C of DNA	64.0 (major), 55.7 (minor)	63.5	63.4 (major), 55.3 (minor)	61.8	64.2	59.1–62.2
16S rRNA sequence	Z28378	D43628	K00421	D13378	D14128	D11107
Type strain	ATCC 43101 CCM 3871 DSM 3396 JCM 8859 NCIMB 2192 VKM B-1752	JCM 9655	ATCC 29605 CCM 2852 DSM 3757 JCM 8879 NCIMB 2012 VKM B-1768	ATCC 33959 DSM 4427 NCIMB 2188 VKM B-1756	ATCC 35960 DSM 4425 NCIMB 13115 VKM B-1754	ATCC 33500 CCM 3361 DSM 1411 NCIMB 2177 VKM B-1748
Source of isolation	Lake Magadi, Kenya	Lake Magadi, Kenya	Dead Sea, Israel	Saltern, Spain	Saltern, California	Saltern, Spain
Reference	1	2	3	4	5	6

Data were derived from the original species descriptions and from Grant and Larsen (1989) and Tindall (1992). Data on anaerobic growth on arginine were derived in part from Oren (1994b) and Oren and Litchfield (1999).

+: all strains tested positive; −: all strains tested negative; ±: weak reaction; d: some strains positive; NR: not reported.

[T] type species of the genus.

[a] Shand and Perez (1999) reported an optimum temperature of the type strain of 55°C in complex medium and 53°C in defined medium.

[b] Franzmann etal. (1988) reported a positive reaction.

[c] Positive growth on single carbon sources may in certain cases require low concentrations of additional growth factors (to be satisfied e.g., by addition of 0.005% yeast extract).

References:

1—Tindall etal., 1984; McGenity and Grant, 1993.
2—Kobayashi etal., 1992; Kanai etal., 1995.
3—Mullakhanbhai and Larsen, 1975; Torreblanca etal., 1986.
4—Juez etal., 1986.
5—Tomlinson etal., 1986; Tindall etal., 1989.
6—Rodriguez-Valera etal., 1983a; Torreblanca etal., 1986.

A species incertae sedis has been classified within the genus Haloferax: "Haloferax alicantei," isolated from a saltern in Spain (Holmes etal., 1997); its 16S rDNA sequence has been deposited as M33803, M33804 and M33805.

Table 6. Properties of the species of the genera *Natrinema*, *Haloterrigena*, and *Natronorubrum*.

Genus	*Natrinema*		*Haloterrigena*	*Natronorubrum*	
Species	*pellirubrum*[T]	*pallidum*	*turkmenica*[T]	*Bangense*[T]	*tibetense*
Morphology	Rod or pleomorphic	Rod	Coccus	Pleomorphic	Pleomorphic
Cell size	0.6–1.0 × 1–4µm	0.7–1.0 × 1.5–6µm	1.5–2µm	NR	NR
Motility	–	–	–	–	–
Gas vesicles			–	NR	NR
NaCl optimum (M)	3.4–4.3	3.4–4.3	NR	3.8	3.4
Mg optimum (M)	NR	NR	NR	NR	NR
pH optimum	7.2–7.8	7.2–7.6	Neutral	9.5	9.0
Temp. optimum (°C)	NR	37–40	45	45	45
Nitrite from nitrate	+	+	+	–	–
Gas from nitrate	–	–	–	–	–
Anaerobic growth on nitrate	NR	NR	NR	NR	NR
Anaerobic growth on L-arginine	NR	NR	+	NR	NR
Acids from carbohydrates	–	–	NR	NR	NR
Growth on single carbon sources	NR	NR	–	+	+
Indole from tryptophan	–	–	–	+	+
Starch hydrolysis	–	–	NR	–	–
Tween 80 hydrolysis	NR	NR	–	–	–
Gelatin hydrolysis	+	+	NR	–	+
Casein hydrolysis	NR	NR	NR	–	–
Pigmentation	Light red—orange	Pale orange	Red	Red	Red
Major glycolipids	unidentified	unidentified	S_2-DGD-1	–	–
Presence of PGS	–	+	+	–	–
G+C of DNA	69.9 (major), 60.0 (minor)	NR	59.2–60.2	59.9	60.1
16S rRNA sequence	AJ002947	AJ002949	AB004878	Y14028	AB005656
Type strain	NCIMB 786	NCIMB 777	ATCC 51198 NCIMB 13204 VKM B-1734	AS 1.1984	AS 1.2123
Source of isolation	Salted hide	Salted cod	Saline soil, USSR	Alkaline salt lake, Tibet	Alkaline salt lake, Tibet
Reference	1	1	2	3	3

Data were derived from the original species descriptions.
+: all strains tested positive; –: all strains tested negative; ±: weak reaction; NR: not reported.
[T]type species of the genus.

References:
1—McGenity et al., 1998.
2—Zvyagintseva and Tarasov, 1987; Ventosa et al., 1999.
3—Xu et al., 1999.

phylotype has a low similarity to the 16S rDNA sequences of any of the genera defined thus far and is only remotely related to the genus *Haloferax*, its closest relative (Benlloch et al., 1995; Benlloch et al., 1996; Rodríguez-Valera et al., 1999). The organism harboring this 16S rDNA gene still awaits isolation.

Sensitivity to antibiotics and other antibacterial substances

Sensitivity to antibiotics and other antibacterial substances is often used in taxonomic studies in which strains are characterized or compared. Being Archaea, members of the *Halobacteriales* typically are resistant to such Bacteria-specific antibiotics as penicillin, ampicillin, cycloserine, kanamycin, neomycin, polymyxin, and streptomycin (Bonelo et al., 1984a; Hilpert et al., 1981; Pecher and Böck, 1981). Regarding their sensitivity to chloramphenicol, conflicting information exists: although most reports state resistance to chloramphenicol, Pecher and Böck, 1981 found inhibition already at 31 µg/ml, and they also reported some sensitivity to tetracycline and aureomycin (both with a minimal inhibitory concentration of 62 µg/ml). *Halobacterium salinarum* was reported to be inhibited by chloramphenicol and erythromycin at concentrations > 100 µg/ml and to be susceptible to tetracycline at > 8 µg/ml (Chow and Mark, 1980). Sensitivity to high concentrations of chloramphenicol (minimal inhibitory concentrations up to 750–1500 µl/ml in most species tested and 141 µg/ml in *Haloferax mediterranei*) was also reported by Bonelo et al., 1984a. Some species may be inhibited by high concentrations (>150 µg/ml) of erythromycin (Bonelo et al., 1984a; Pecher and Böck, 1981).

Most halophilic Archaea are sensitive to novobiocin and bacitracin. Novobiocin is a DNA gyrase inhibitor (Holmes and Dyall Smith, 1991; Sioud et al., 1988) and acts on the same target in the Archaea as in sensitive Bacteria. *Halococcus saccharolyticus* presents a notable exception, being resistant to novobiocin inhibition (Montero et al., 1989), but *Halococcus morrhuae* is sensitive (Hunter and Millar, 1980). Bacitracin inhibits incorporation of the high-molecular-weight saccharide into the cell wall glycoprotein of non-coccoid halophilic Archaea (Wieland et al., 1980; Wieland et al., 1982) and may also inhibit lipid biosynthesis in these organisms (Basinger and Oliver, 1979).

Members of the *Halobacteriales* were found to be sensitive to a number of additional antibiotics and other antibacterial compounds:

• Anisomycin, a protein synthesis inhibitor of eukaryal ribosomes, also inhibits protein synthesis of all members of the *Halobacteriales* tested,

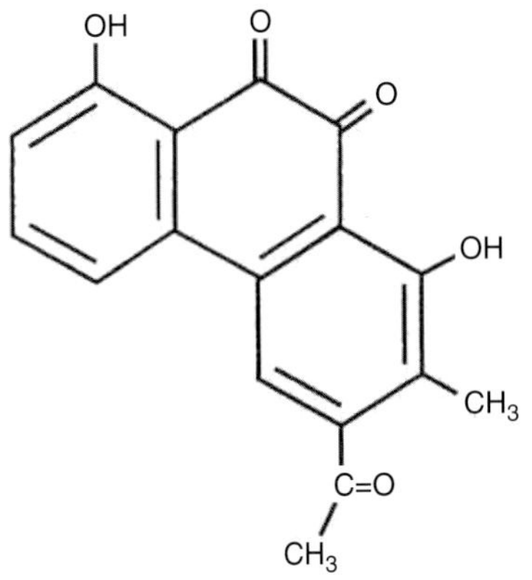

Fig. 5. Structure of the antibiotic haloquinone.

both *in vivo* and *in vitro* (Pecher and Böck, 1981).

• Cerulenin, a potent inhibitor of the synthesis of the straight-chain fatty acids by the fatty acid synthetase complex, inhibits *Halobacterium salinarum* (and possibly other members of the order as well) (Dees and Oliver, 1977). This finding suggests that the production of straight-chain fatty acids is essential even in an organism possessing archaeal type lipids. It is now known that fatty acids are used to acylate certain membrane proteins (Pugh and Kates, 1994).

• *Halobacterium salinarum* was found sensitive to haloquinone, an antibiotic substance produced by *Streptomyces venezuelae* subsp. *xanthophaeus* (Fig. 5), affecting DNA synthesis (Ewersmeyer-Wenk et al., 1981).

• Rifampicin was reported to be inhibitory to many members of the *Halobacteriaceae* (Bonelo et al., 1984a; Hilpert et al., 1981; Pecher and Böck, 1981). The target of rifampicin action in the halophilic Archaea is probably not the DNA-dependent RNA polymerase, but the cell membrane upon which it exerts a detergent effect, causing cell lysis (Pfeifer, 1988).

• The DNA polymerase inhibitor aphidicolin prevents cell division and often causes the formation of elongated cells (Forterre et al., 1984; Schinzel and Burger, 1984).

• Certain antitumor drugs that act on topoisomerase II (adriamycin, daunorubicin, etoposide) and drugs that act on actomyosin (cytochalasin B and D) or on tubulin (vincristine, podophyllotoxin, nocodazole) inhibit *Halobacterium salinarum* (Sioud et al., 1987).

• Inhibition of halophilic Archaea by coumarin and quinolone antibacterial compounds presented evidence for the presence of DNA gyrase-like enzymes (Sioud et al., 1988). The quinolone compound ciprofloxacin inhibited most *Haloferax* and *Haloarcula* species at concentrations of 25–60 µg/ml. *Halobacterium* cells proved less sensitive. At sublethal concentrations swelling and elongation of the cells were observed. Sensitivity to ciprofloxacin and other quinolone derivatives (norfloxacin, perfloxacin) was decreased at increased magnesium concentrations (Oren, 1996a; Sioud et al., 1988). *Haloferax volcanii* became more resistant when the magnesium concentration in the growth medium was increased. Alkaliphiles such as *Natronomonas pharaonis* are very sensitive (Oren, 1996a).

At least certain species are sensitive to gardimycin (inhibiting cell wall biosynthesis), to virginiamycin (acting on the protein synthesis machinery), to monensin and lasalocid (which target the cell membranes) (Hilpert et al., 1981), and to amphotericin B, vibriostat O/129, sulfafurazole, and josamycin (Tindall, 1992). The usefulness of these antibiotics and antibacterials for differentiation purposes in taxonomic studies has not yet been systematically evaluated.

Bile acids at low concentrations cause lysis of the cell envelopes of non-coccid halophilic Archaea and are potent growth inhibitors (Dussault 1956a; Dussault 1956b; Kamekura and Seno, 1991; Kamekura et al., 1988).

Surface layers

The non-coccoid representatives of the *Halobacteriales* possess an S-layer cell wall, whose main constituent is a high molecular weight glycoprotein. This glycoprotein cell wall is responsible for maintaining the native cell shape. Electron micrographs show a hexagonal pattern of protein subunits (Houwink, 1956), and high-resolution electron microscopy enabled a detailed three-dimensional reconstruction of the wall structure of *Haloferax volcanii* (Kessel et al., 1988a; Kessel et al., 1988b) and *Haloarcula japonica* (Nishiyama et al., 1992). In *Halobacterium salinarum* the high molecular weight glycoprotein makes up 40–50% of the wall protein, the remainder consisting of 15–20 proteins of smaller size (Mescher and Strominger, 1976a; Mescher and Strominger, 1976b; Mescher et al., 1974).

The primary structure of the structural cell wall glycoprotein of *Halobacterium salinarum* has been elucidated. Gene cloning and sequencing yielded detailed information on the structure of the S-layer and its biosynthesis (Lechner and Sumper, 1987; Lechner and Wieland, 1989; Lechner et al., 1985; Mescher, 1981; Paul and Wieland, 1987; Paul et al., 1986; Sumper, 1987; Wieland, 1988; Wieland et al., 1980; Wieland et al., 1982). The S-layer glycoprotein (molecular mass about 120 kDa) consists of a 87 kDa core protein rich in acidic amino acids, containing acidic and neutral saccharide chains attached in a manner resembling animal proteoglycans (Lechner and Sumper, 1987; Lechner and Wieland, 1989; Kandler and König, 1993; Sumper, 1987).

- A single large (around 10 kDa) acidic glycosaminoglycan composed of 10–15 repeats of a branched sulfated ($2\ SO_4^{2-}$) pentasaccharide, containing galacturonic acid, is bound to the sub NH_2-terminal asparagine via a special direct asparaginyl GalNAc N-linkage (Paul and Wieland, 1987; Paul et al., 1986). This saccharide complex is considered as the main shape-maintaining component of the S-layer. When its synthesis is inhibited by bacitracin, the rod-shaped cells are converted to spheres (Lechner and Wieland, 1989; Mescher and Strominger, 1976a; Mescher and Strominger, 1976b).

- Low molecular weight tetrasaccharide units (about 10 copies per molecule), composed of 2–3 sulfated hexuronic acid residues (GlcUA, IdUA) attached to a glucose residue, are bound by a direct N-linkage to asparagine (Lechner and Wieland, 1989).

- Collagen-like O-linked glucosyl-(α-1→2)-galactoside disaccharide units (about 12–15 per molecule) are bound via threonines clustered on the protein core domain 755–774 on top of the cell membrane, close to the postulated transmembranal COO^- terminal domain (Paul and Wieland, 1987; Wieland, 1988).

Figure 6 presents a model of the *Halobacterium salinarum* cell wall glycoprotein.

The primary structure of the protein backbone and the mode of glycosylation may vary among the species. The glycosylation pattern of *Haloferax volcanii* involves both N- and O-glycosidic bonds [glucosyl-(1→2)-galactose disaccharides O-linked to threonine residues], in a pattern that differs from that of *Halobacterium salinarum* (Sumper et al., 1990). The S-layer protein of *Haloferax volcanii* has seven N-glycosylation sites, versus 12 in *Halobacterium salinarum* (including the sub NH_2-terminal, negatively charged repeating unit saccharide which is absent in *Haloferax volcanii*). The N-linked oligosaccharides of *Haloferax volcanii* are (β-1→4)-linked repeating glucose residues attached via asparaginyl-glucose linkages (Mengele and Sumper, 1992). It has been suggested that the difference in charge density between the extremely halophilic *Halobacterium salinarum* and *Haloferax volcanii*, a species designated as a moderate halophile with high magnesium tolerance (Mullakhanbhai and Larsen, 1975), may be related to their difference in salt requirement (Mengele and Sumper, 1992). *Halorubrum distributum* was reported to produce thick-walled cyst-like structures, probably representing resting stages (Kostrikina et al., 1991). No information is available as yet on the cell wall structure of these cysts as compared to that of cells in growing cultures.

Fig. 6. Model of the cell wall glycoprotein of *Halobacterium salinarum*,
based on data presented by Lechner
and Wieland, 1989. GalNAc = N-
acetylgalactosamine; GalUA = galacturonic acid; Glc = glucose; GlcNAc =
N-acetylglucosamine; GlcUA = glucu-
ronic acid; IdUA = iduronic acid.

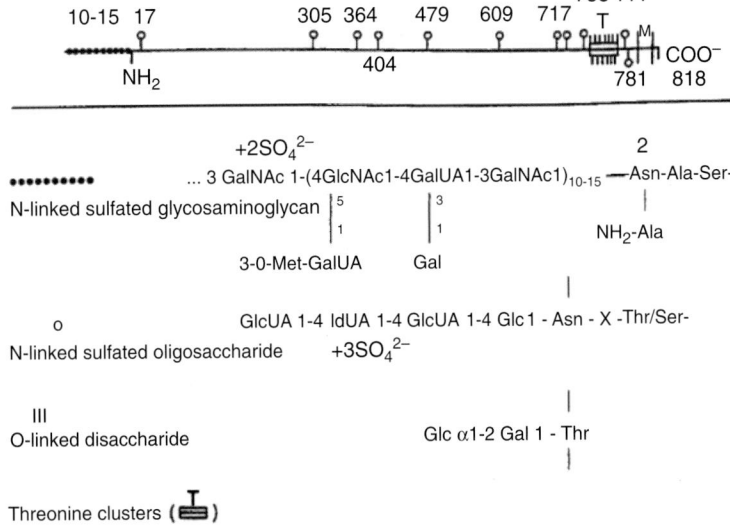

... 3 GalNAc 1-(4GlcNAc1-4GalUA1-3GalNAc1)$_{10-15}$ —Asn-Ala-Ser-

N-linked sulfated glycosaminoglycan

3-0-Met-GalUA   Gal   NH$_2$-Ala

o
N-linked sulfated oligosaccharide   +3SO$_4^{2-}$

GlcUA 1-4 IdUA 1-4 GlcUA 1-4 Glc 1 - Asn - X -Thr/Ser-

III
O-linked disaccharide   Glc α1-2 Gal 1 - Thr

Threonine clusters

The glycoprotein cell wall requires high NaCl concentrations for stability. Similar to most other proteins of halophilic Archaea, the wall protein denaturates when suspended in distilled water. As a result, cells lyse in the absence of salt due to the denaturation and dissolution of the cell wall (Kushner, 1964; Mohr and Larsen, 1963). In addition to high NaCl concentrations, relatively high concentrations of magnesium or other divalent cations are required to maintain the structural stability of the glycoprotein cell wall. When suspended in media lacking magnesium, even in the presence of saturated NaCl concentrations, cells of many species become rounded with the formation of spheroplasts. This phenomenon was documented in studies of *Haloarcula marismortui*, *Haloferax volcanii* (Cohen et al., 1983), and *Haloarcula japonica* (Horikoshi et al., 1993; Nakamura et al., 1992). Concentrations as high as 20–50 mM magnesium may be required in these species to protect the intact morphology of the cell wall. Some magnesium may even be required for the structural stability and morphology of alkaliphilic rods. A not yet formally described rod-shaped species, isolated from a soda lake in India, becomes rounded when suspended at magnesium concentrations below 2 mM (Upasani and Desai, 1990). Spheroplast formation can often be induced by EDTA treatment, and this property has proved of much use in genetic transformation protocols (Dyall-Smith, 1999).

Halococcus species possess a thick sulfated heteropolysaccharide cell wall that does not require high salt concentrations to maintain its rigidity (Steensland and Larsen, 1971). The structure of the polysaccharide wall of *Halococcus morrhuae* has been elucidated in part (Fig. 7). Sulfate groups are linked to hydroxyl groups in

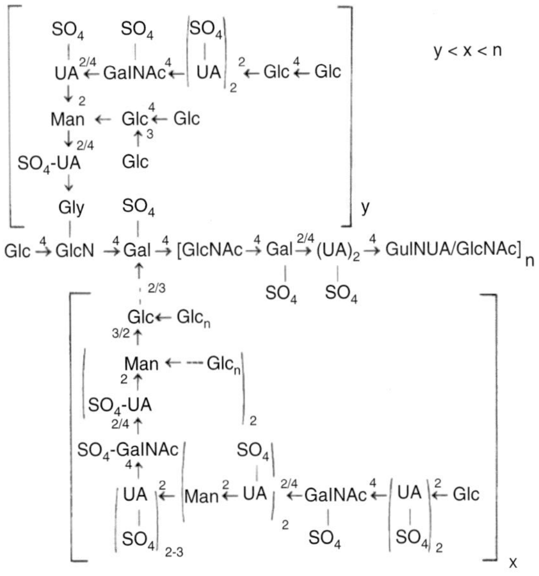

Fig. 7. The tentative structure of the cell wall polymer of *Halococcus morrhuae* (after Schleifer et al., 1982). Gal = galactose; GalNAc = N-acetylgalactosamine; Glc = glucose; GlcNAc = N-acetylglucosamine; Gly = glycine; GulNUA = N-acetylgulosaminuronic acid; Man = mannose; UA = uronic acid.

positions 2 or 3 or both of uronic acids, galactose and galactosamine residues. Glucose, galactose, galacturonic acid and all amino sugars are 1→4 glycosidically linked to the cell wall polymer. Part of the glucose, galactose, and to a lesser extent mannose residues possess more than two glycosidic linkages and represent possible branching points. Glycine residues may play a role in connecting glycan strands through pep-

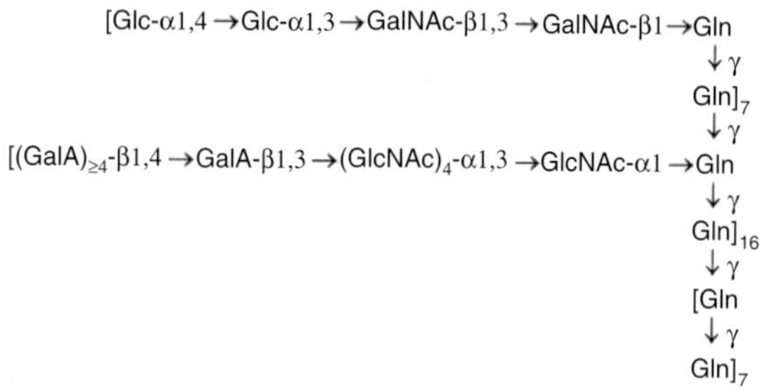

[Glc-α1,4 →Glc-α1,3 →GalNAc-β1,3 →GalNAc-β1 →Gln
                                                    ↓γ
                                                  Gln]$_7$
                                                    ↓γ
[(GalA)$_{\geq4}$-β1,4 →GalA-β1,3 →(GlcNAc)$_4$-α1,3 →GlcNAc-α1 →Gln
                                                    ↓γ
                                                  Gln]$_{16}$
                                                    ↓γ
                                                   [Gln
                                                    ↓γ
                                                  Gln]$_7$

Fig. 8. The tentative structure of the cell wall polymer of *Natronococcus occultus* (after Niemetz et al., 1997). GalA = galactonic acid; GalNAc = N-acetylgalactosamine; Glc = glucose; GlcNAc = N-acetylglucosamine; Gln = glutamine.

tide linkages between the amino groups of glucosamine and the carboxyl group of an uronic acid or gulosaminuronic acid (Schleifer et al., 1982; Steber and Schleifer, 1975; Steber and Schleifer, 1979).

The coccoid *Natronococcus occultus* also has a thick cell wall that retains its shape in the absence of salt. Its structure differs greatly from that of the cell wall polymer of *Halococcus*, and it consists of repeating units of a poly(L-glutamine) glycoconjugate (Niemetz et al., 1997). Figure 8 presents a model of this structure.

Several species of halophilic Archaea, notably of the genus *Haloferax*, excrete massive amounts of exopolysaccharides (Paramonov et al., 1998; Parolis et al., 1996; Severina et al., 1989; Severina et al., 1990). The structure of the exopolysaccharide of *Haloferax mediterranei* was recently identified as consisting of repeating units of:

-4)-β-D-glc*p*NAcA-(1→6)-α-D-Man*p*-(1→4)

-β-D-Glc*p*NAcA-3-O-SO$_3^-$) · (1→

(Parolis et al., 1996). The polysaccharide excreted by *Haloferax gibbonsii* has a different structure:

-4)-β-D-Man*p*-(1→4)-β-D-Man*p*-(1→4)

-α-D-Gal*p*-(1→3)-β-D-Gal*p*(1→

```
       3                    2
       ↑                    ↑
       1                    1
α-D-Gal p-(1→2)        α-D-Glc p

-α-L-Rha p
```

(Paramonov et al., 1998).

In accordance with the variations in the surface layer structure, the halophilic Archaea display a considerable antigenic diversity with limited cross-reaction between the species. Antigenic fingerprinting may thus be useful not only for rapid identification but also for studying the species composition of natural communities

(Conway de Macario et al., 1986). Classic immunological techniques may be hampered by the fact that antibodies may not or only poorly react in the presence of salt concentrations approaching saturation. However, antibodies against halobacterial flagella raised in chicken were shown to interact with the antigen also at 4 M NaCl (Alam and Oesterhelt, 1984a). Also typing the surface layers with different lectins binding to specific sugar residues has been employed in the characterization of halophilic Archaea (Gilboa-Garber et al., 1998).

Polar lipids

Being Archaea, the members of the *Halobacteriales* possess lipids based on branched 20-carbon (phytanyl) and sometimes also 25-carbon (sesterterpanyl) chains bound to glycerol by ether bonds. A great variety of neutral and polar lipids, the last including phospholipids, sulfolipids, and glycolipids, can be encountered in the different representatives of the group. The types of polar lipids present have been used as an important characteristic in the taxonomic classification of strains. Until recently the presence of certain glycolipids was used as a diagnostic trait characteristic for many of the genera recognized within the *Halobacteriaceae* (Torreblanca et al., 1986). In addition, the polar lipid composition of biomass collected from such environments as the Dead Sea or saltern crystallizer ponds has been used as a chemotaxonomic tool to obtain information on the type(s) of halophilic Archaea present in natural communities (Oren, 1994c; Oren and Gurevich, 1993; Oren et al., 1996c). During recent years, however, the increased emphasis on the use of 16S rDNA sequence comparisons in the classification of halophilic Archaea has led to the creation of a number of new genera, and many species had to be reclassified. In the present classification (see also Tables 2–6) members of one genus may contain greatly different glycolipids, and the presence of a certain glycolipid may now be shared by representatives of more than one genus. Details on the lipid composition of halophilic Archaea, the biosynthetic pathways involved in the synthesis of the lipids, and the analytical methods used in their characterization may be found in several reviews (Kamekura and Kates, 1988; Kamekura, 1993; Kamekura, 1998; Kates, 1993; Kates, 1996; Kates and Kushwaha, 1995; Kates and Moldoveanu, 1991). One- or two-dimensional thin-layer chromatographic procedures are useful for the rapid characterization of polar lipid patterns in halophilic Archaea (Norton et al., 1993; Tindall, 1990a; Tindall, 1990b; Torreblanca et al., 1986). Fast-atom-bombardment mass spectrometry has also been suggested as a rapid means of

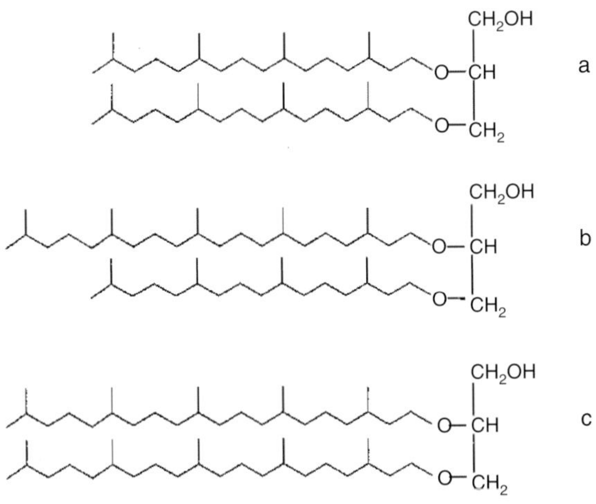

Fig. 9. Core lipids of halophilic Archaea of the family *Halobacteriaceae*: 2,3-di-*O*-phytanyl-*sn*-glycerol (C_{20},C_{20}) (a), 2-*O*-sesterterpanyl-3-*O*-phytanyl-*sn*-glycerol (C_{25},C_{20}) (b), and 2,3-di-*O*-sesterterpanyl-*sn*-glycerol (C_{25},C_{25}) (c).

identifying polar lipids of the *Halobacteriaceae* (Fredrickson et al., 1989a; Klöppel and Fredrickson, 1991). For a detailed structure elucidation, techniques such as mass spectrometry and nuclear magnetic resonance spectroscopy are among the methods used.

The diether core lipid that forms the basis for most of the polar lipid structures present in the halophilic Archaea of the family *Halobacteriaceae* is 2,3-di-*O*-phytanyl-*sn*-glycerol, further abbreviated as C_{20},C_{20} (Fig 9a). Certain species, however, contain in addition the asymmetric 2-*O*-sesterterpanyl-3-*O*-phytanyl-*sn*-glycerol (C_{25},C_{20}) in different amounts (Fig. 9b). The C_{25},C_{20} core lipid has been detected in many of the alkaliphilic types and in the neutrophilic *Natrialba asiatica* (Kamekura and Dyall-Smith, 1995; Matsubara et al., 1994), the genera *Natrinema* (McGenity et al., 1998) and *Halococcus*, and in "*Halobacterium halobium* IAM13167," which may be a *Halobacterium salinarum* strain (Morita et al., 1998). Sometimes 2,3-di-*O*-sesterterpanyl-*sn*-glycerol (C_{25},C_{25}) (Fig. 9c) is encountered as a minor component as well (De Rosa et al., 1982; De Rosa et al., 1983). Thin-layer chromatographic procedures have been developed to separate the C_{20},C_{20} and C_{25},C_{20} lipid species (Ross et al., 1981). Different haloalkaliphilic Archaea contain C_{20},C_{20} and C_{25},C_{20} in different proportions, and strains can be found in which as much as 89% or as little as 0.1% of the lipids are based on the C_{25},C_{20} core lipid structure (Tindall, 1985). The ratio between C_{20},C_{20} and C_{25},C_{20} lipids may depend also on growth conditions, as shown in the alkaliphiles *Natronobacterium gregoryi* and *Natrialba magadii*, in which increasing medium salinity leads to

an increased proportion of C_{25},C_{20} (Morth and Tindall, 1985a; Morth and Tindall, 1985b; Tindall et al., 1991). In most cases the hydrophobic chains are fully saturated, but the occurrence of unsaturated phytanyl ("phytenyl") side chains has been documented in *Halorubrum lacusprofundi* from Deep Lake, Antarctica, a strain able to grow at temperatures down to 4 °C (Franzmann et al., 1988; Fredrickson et al., 1989b). Introduction of double bonds in the carbon chains is probably important in the regulation of membrane fluidity in this cold-adapted species. In *Halobacterium salinarum*, double bonds appear transiently during the biosynthesis of the phytanyl chains (Moldoveanu and Kates, 1988; Fredrickson et al., 1989b).

Polar lipid structures are all derivatives of the C_{20},C_{20} or C_{25},C_{20} core lipids. All halophilic Archaea known thus far contain the diether derivatives of phosphatidyl glycerol (PG) and the methyl ester of phosphatidyl glycerophosphate (Me-PGP) (Fig. 10a,b). The presence of the methyl ester group in the Me-PGP structure was recognized only relatively recently (Kates, 1996; Kates et al., 1993), though it was already suggested from fast bombardment mass spectrometry data published in 1989 (Fredrickson et al., 1989b). The ratio between the amount of PG and Me-PGP present may depend on growth conditions. In *Natronococcus occultus* grown at increased salt concentrations the relative amount of PG(C_{20},C_{20}) decreases with a concomitant increase in Me-PGP(C_{25},C_{20}), while PG(C_{25},C_{20}) and Me-PGP(C_{20},C_{20}) remain unchanged (Tindall et al., 1991). PG and Me-PGP are the only phospholipids identified in most neutrophilic representatives of the *Halobacteriales*. Other unidentified phospholipids have been detected in the genus *Natrinema* (McGenity et al., 1998). In *Natronococcus occultus* a phospholipid with a cyclic phosphate group has been identified: 2,3-di-*O*-phytanyl-*sn*-glycero-1-phosphoryl-3'-*sn* glycerol-1,2-cyclic phosphate (Lanzotti et al., 1989) (Fig. 10c). Additional, yet unidentified phospholipids are present in some haloalkaliphiles (De Rosa et al., 1988; Morth and Tindall, 1985a; Tindall, 1985).

Phosphatidylglycerosulfate (PGS, Fig. 10d) is present in many neutrophilic species (Hancock and Kates, 1973). Its absence in certain genera (*Haloferax*, *Natrialba*, *Halobaculum*, *Halococcus*, *Halogeometricum*) is a useful diagnostic feature in the classification of the *Halobacteriaceae*, and its presence or absence may vary even of different species within a single genus, as the case of *Natrinema* shows (Table 6). The alkaliphilic members characterized thus far all lack PGS.

A variety of glycolipids has been identified in different members of the *Halobacteriales*. Di-,

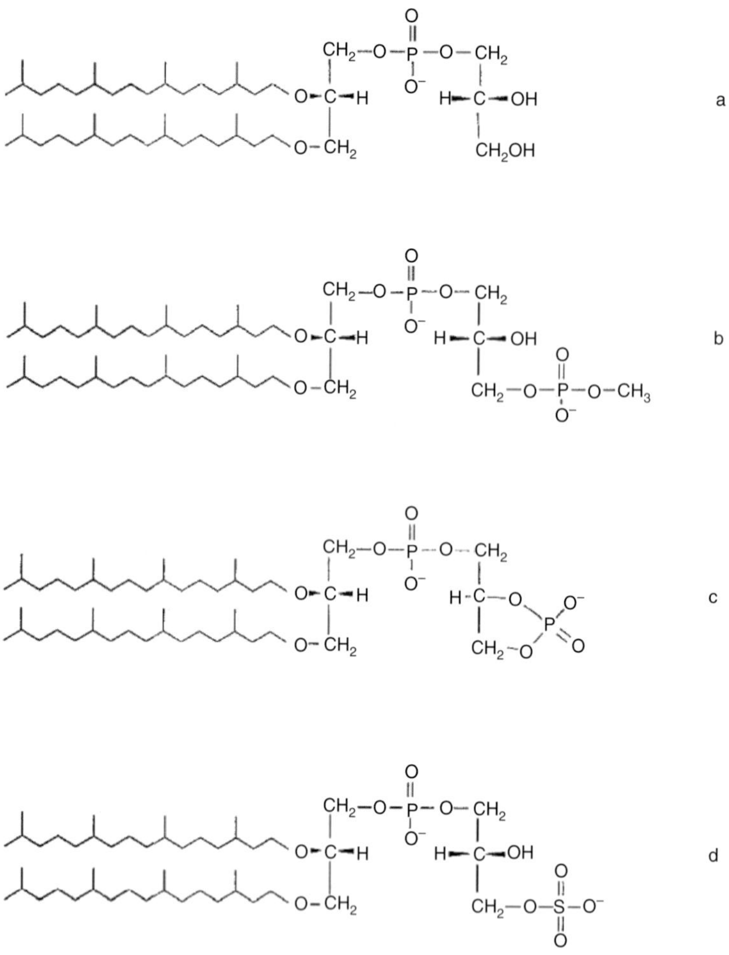

Fig. 10. Structure of the isoprenoyl diether derivatives of phosphatidyl glycerol (PG) (a), the methyl ester of phosphatidyl glycerophosphate (Me-PGP) (b), the cyclic PGP of *Natronococcus occultus* (c) and phosphatidyl glycerosulfate (PGS) (d).

tri-, and tetraglycosyl diether lipids occur, some of them carrying sulfate groups on the sugar moieties. Most haloalkaliphilic members lack substantial amounts of glycolipids, but there are exceptions: the alkaliphilic strain SSL1, isolated from a soda lake in India, possesses DGD-4 (Upasani et al., 1994). Figure 11 presents some of the glycolipids that have been identified in different representatives of the group (unless indicated otherwise, all sugar residues are in the pyranose conformation):

• DGD-1, 1-*O*-[β-D-mannose-(1′→2′)-α-D-glucose]-2,3-di-*O*-phytanyl-*sn*-glycerol (Fig. 11a), is found in minor amounts in *Haloferax* species.

• DGD-2, a minor diglyceride lipid of unknown structure, containing mannose and glucose, is found as a minor component in *Haloarcula* species.

• DGD-4, documented from the haloalkaliphilic strain SSL1, isolated from an Indian soda lake (Upasani et al., 1994). Two possible structures have been proposed for this glycolipid: 1-*O*-[β-1′→6′-glucose-(1′→1′)-α-D-glucose]-2,3-di-*O*-phytanyl-*sn*-glycerol and 1-*O*-[α-1′→6′-glucose-(1′→1′)-β-D-glucose]-2,3-di-*O*-phytanyl-*sn*-glycerol (Fig. 11b,c). It remains to be ascertained which of these is the correct structure.

• S-DGD-1, 1-*O*-[α-D-mannose-(6′-SO₃H)-(1′→2′)-α-D-glucose]-2,3-di-*O*-phytanyl-*sn*-glycerol (Fig. 11d), being the major glycolipid in the genus *Haloferax* (Kushwaha et al., 1982a; Kushwaha et al., 1982b). It is found also as sole or minor glycolipid in the genera *Halobaculum* (Oren et al., 1996c) and *Halococcus* and was reported also from *Halorubrum saccharovorum* (Lanzotti et al., 1988). S-DGD-1 was found also as the major or sole glycolipid in the archaeal biomass of the Dead Sea (Oren and Gurevich, 1993) and saltern crystallizer ponds in Eilat, Israel (Oren, 1994c). Lipid analysis of field samples also suggested that S-DGD-1 may be the main glycolipid of the gas vesicles containing square Archaea (Oren et al., 1996c).

• S-DGD-3, 1-*O*-[α-D-mannose-(2′-SO₃H)-α-D-(1′→4′)-glucose]-2,3-di-*O*-phytanyl-*sn*-glycerol (Fig. 11e), is the glycolipid of several *Halorubrum* species, including *Halorubrum sodomense* (Trincone et al., 1990) and *Halorubrum lacusprofundi* (Tindall, 1990a).

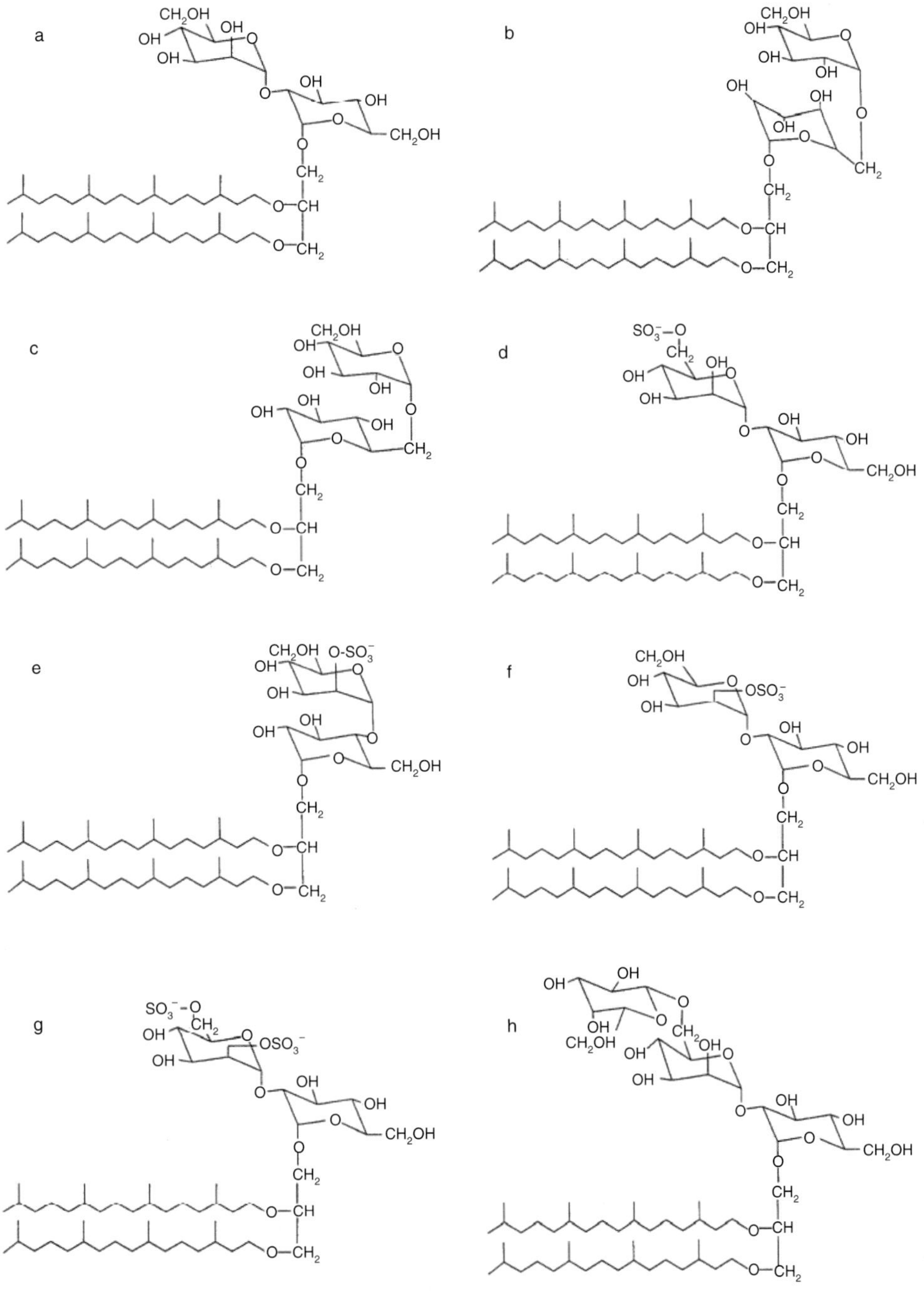

Fig. 11. Structure of some glycolipids found in members of the *Halobacteriales*: DGD-1 (a), proposed structures for DGD-4 (b,c), S-DGD-1 (d), S-DGD-3 (e), S-DGD-5 (f), S₂-DGD-1 (g), TGD-1 (h), TGD-2 (i), S-TGD-1 (j), TeGD (k), and S-TeGD (l).

• S-DGD-5 ([1-*O*-mannose-(2'-SO₃H)-α-D-(1'→2')-glucose]-2,3-di-*O*-phytanyl-*sn*-glycerol) (Fig. 11f) is found in *Halorubrum trapanicum* (Trincone et al., 1993).

• S₂-DGD-1, a bis-sulfated glycolipid, 1-*O*-[α-D-mannose-(2',6'-SO₃H)-α-D-(1'→2')-glucose]-2,3-di-*O*-phytanyl- or phytanyl sesterterpenyl-*sn*-glycerol (Fig. 11g), is found in *Natrialba asiatica* (Kates, 1996; Matsubara et al., 1994; Onishi et al., 1985).

• TGD-1, 1-*O*-[β-D-galactose-(1'→6')-α-D-mannose-(1'→2')-α-D-glucose]-2,3-di-*O*-phytanyl-*sn*-glycerol (Fig. 11h), is a minor glycolipid of *Halobacterium salinarum* (Smallbone and Kates, 1981).

• TGD-2, 1-*O*-[β-D-glucose-(1'→6')-α-D-mannose-(1'→2')-α-D-glucose]-2,3-di-*O*-phytanyl-*sn*-glycerol (Fig. 11i), is the sole or major glycolipid of most *Haloarcula* species (Evans et al., 1980).

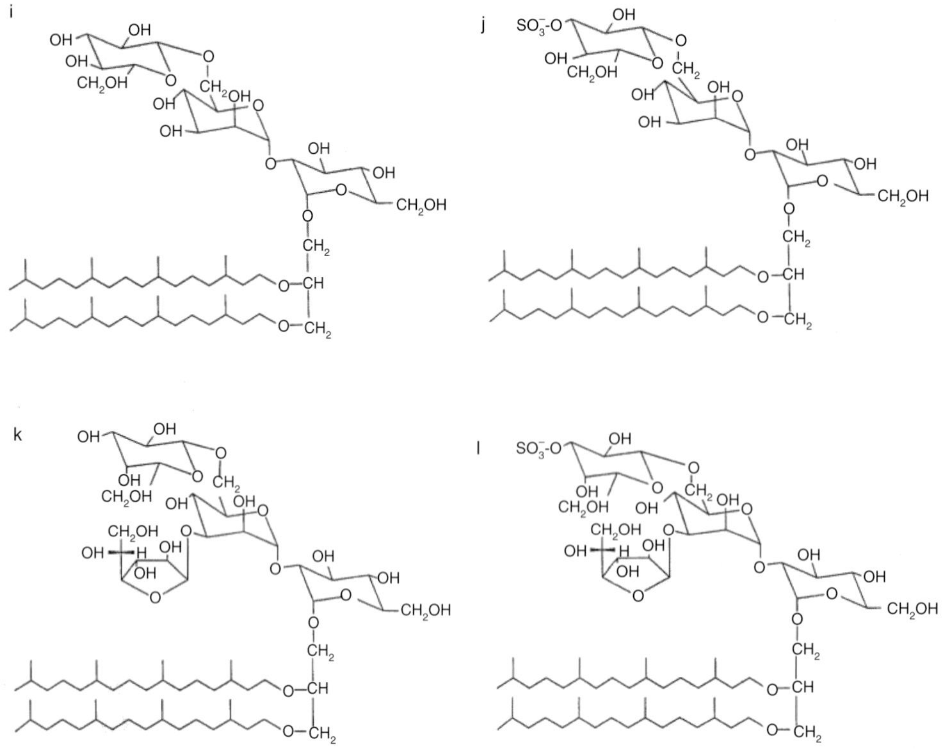

Fig. 11. *Continued*

- S-TGD-1,   1-O-[β-D-galactose-(3'-SO₃H)- (1'→6')-α-D-mannose-(1'→2')-α-D-glucose]-2,3-di-O-phytanyl-*sn*-glycerol (Fig. 11j), is found in the genus *Halobacterium* (Kates, 1978; Kates and Deroo, 1973).
- TeGD,   1-O-[β-D-galactose-(1'→6')-α-D-mannose-(3'←1')-α-D-galactofuranose)-(1'→2')-α-D-glucose]-2,3-di-O-phytanyl-*sn*-glycerol (Fig. 11k), is a minor glycolipid of *Halobacterium salinarum* (Smallbone and Kates, 1981).
- S-TeGD,   1-O-[β-D-galactose-(3'-SO₃H)- (1'→6')-α-D-mannose-(3'←1')-α-D-galactofuranose)-(1'→2')-α-D-glucose]-2,3-di-O-phytanyl-*sn*-glycerol (Fig. 11l), is found in *Halobacterium* (Smallbone and Kates, 1981).

While considerable knowledge has accumulated on the nature of the glycolipids present in different members of the *Halobacteriales*, some controversies remain. Confusion still exists on the correct structures of the glycolipids that occur in different members of the genus *Halorubrum*. Kamekura, 1998 stated that S-DGD-3 is the main glycolipid in the genus *Halorubrum*. However, Lanzotti et al., 1988 reported that the *Halorubrum saccharovorum* glycolipid is identical to S-DGD-1 from *Haloferax*. To add to the confusion, different isolates designated as *Halorubrum saccharovorum* were found to contain different glycolipids (Tindall, 1990b).

Additional glycolipids occur within the halophilic Archaea. An as yet unidentified halophile was isolated from a saltern in Guerrero Negro, Baja California, Mexico. It is able to grown anaerobically in the presence of nitrate and possesses two novel unidentified phosphoglycolipids in addition to PG, Me-PGP, a diglycosyl diether and a sulfated diglycosyl diether (Tindall et al., 1987).

Glycolipids have become valuable taxonomic markers in the classification of halophilic Archaea, as the type(s) of glycolipids present are constant for any species. A single exception has been reported thus far: a *Halococcus* strain (incorrectly designated *Sarcina marina*) was found to contain an unidentified triglycosyl diglyceride when grown at 34°C, but a diglycosyl diglyceride became the major component when the cells were grown at 20°C (Hunter et al., 1981). The ratios in which the different polar lipids occur in the membrane may vary according to the growth conditions: when grown at very high salt concentrations *Haloferax mediterranei* showed an increase in S-DGD-1 with a decrease in PG (Kamekura and Kates, 1988; Kushwaha et al., 1982b). *Natronococcus occultus* showed an increase in Me-PGP(C₂₅,C₂₀) and a decrease in PG(C₂₀,C₂₀) with increasing medium salinity. In addition, an increase was found in the amount of glycine betaine associated with the cell mem-

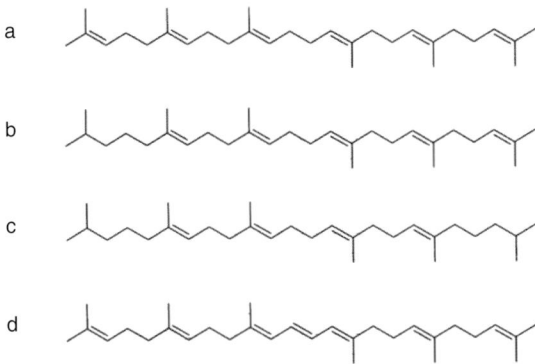

Fig. 12. Some neutral lipids found in the *Halobacteriales*: squalene (a), dihydrosqualene (b), tetrahydrosqualene (c), and dehydrosqualene (d).

brane. The content of Me-PGP(C_{20},C_{20}) and PG(C_{25},C_{20}) was independent of the salt concentration at which the cells were grown (Nicolaus et al., 1989).

Neutral lipids

Neutral lipids may represent about 10% of the total lipid content of halophilic Archaea of the order *Halobacteriales* (Kamekura and Kates, 1988). The following types have been reported:
- C_{20} isoprenoid lipids: geranylgeraniol.
- Neutral phytanyl ethers of glycerol: DL-*O*-phytanyl-*sn*-glycerol and 2,3-di-*O*-phytanyl-*sn*-glycerol (Kushwaha and Kates, 1978).
- C_{30}-isoprenoid compounds: squalene, dihydrosqualene, tetrahydrosqualene, and dehydrosqualene (Kushwaha et al., 1972; Kushwaha et al., 1974) (Fig. 12). Squalenes may make up about 36% of the total neutral lipids of *Halobacterium salinarum* (Kushwaha et al., 1972; Kushwaha et al., 1974). Occurrence of squalene derivatives has been reported also in *Haloferax volcanii* (Mullakhanbhai and Francis, 1972).
- Indole has been indicated as a neutral lipid in *Halobacterium salinarum* and *Halococcus* sp. (Kushwaha et al., 1977).
- Carotenoids and quinones also make up part of the neutral lipids, and these are discussed separately.

Information on the biosynthetic pathways of different lipids mentioned above has been presented by Kamekura and Kates, 1988 and Kushwaha and Kates, 1979.

In addition to the isoprenoid hydrophobic chains characteristic of the archaeal domain, straight-chain fatty acids can be found in members of the *Halobacteriales* in limited amounts. The presence of a fatty acid synthetase has been demonstrated in *Halobacterium salinarum*. Activities measured were low, and the enzyme was inhibited by salt at the concentrations reported to occur intracellularly (Pugh et al., 1971). The straight-chain fatty acids produced are not incorporated into the membrane lipids but serve to acylate membrane proteins to make them more hydrophobic (Pugh and Kates, 1994).

Carotenoid pigments

Most representatives of the *Halobacteriales* are brightly red-orange, colored by a high content of carotenoid pigments in their cell membrane. *Natrialba asiatica*, a species lacking substantial amounts of carotenoids, is a rare exception to the rule (Kamekura and Dyall-Smith, 1995). However, white mutants of species such as *Halobacterium salinarum* can easily be isolated. In the dark, such mutants grow as well as the red wild type, but when incubated at high light intensities approaching those of full sunlight, the white mutants are outcompeted by the pigmented parent strain, demonstrating the role of the carotenoid pigments in protecting the cells against light damage (Dundas and Larsen, 1962). Carotenoid pigments in *Halobacterium salinarum* were also claimed to protect against UV radiation and aid in photoreactivation (Wu et al., 1983), but the data presented are not fully convincing, as little difference in survival can be seen in the red wild-type strain and the colorless mutant.

Most carotenoids of the halophilic Archaea are C_{50} straight-chain derivatives of α-bacterioruberin. The most abundant carotenoid species encountered are α-bacterioruberin, monoanhydrobacterioruberin, and bisanhydrobacterioruberin (Kelly et al., 1970; Kushwaha et al., 1974; Kushwaha et al., 1975; Fig. 13a,b,c). Several other derivatives have been found in minor amounts, such as the dodecaene C_{50} carotenoids 3′,4′-dihydromonoanhydrobacterioruberin (Fig. 13d), haloxanthin (a 3′,4′-dihydromonoanhydrobacterioruberin derivative with a novel peroxide end group) (Fig. 13e), and 3′,4′-epoxymonoanhydrobacterioruberin of *Haloferax volcanii* (Rønnekleiv et al., 1995) (Fig. 13f).

C_{40} carotenoids are generally found in small amounts, lycopene and β-carotene being the most abundant (Kushwaha et al., 1982a; Tindall, 1992). Among the other types identified are lycopersene, *cis*- and *trans*-phytoene, *cis*- and *trans*-phytofluene, neo-α-carotene, and neo-β-carotene. The low concentrations of these compounds suggest that they may serve as precursors for the synthesis of retinal, lycopene, and the C_{50} red carotenoid pigments of the bacterioruberin group. A recent report on the occurrence of major amounts of the C_{40} carotenoids 3-hydroxyechinenone and *trans*-astaxanthin in

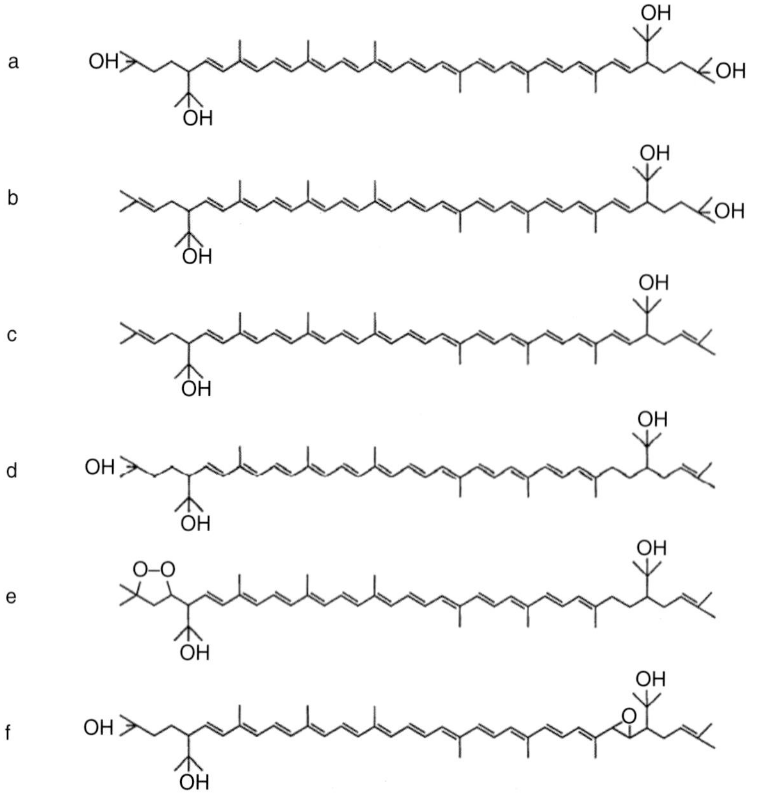

Fig. 13. Structure of carotenoid pigments found in members of the *Halobacteriales*: α-bacterioruberin (a), monoanhydrobacterioruberin (b), bis-anhydrobacterioruberin (c), 3′,4′-dihydromonoanhydrobacterioruberin (d), haloxanthin (e), and 3′,4′-epoxymonoanhydrobacterioruberin (f).

Halobacterium salinarum, Haloarcula hispanica, and *Haloferax mediterranei* (Calo et al., 1995) remains to be confirmed. It was stated that in *Halobacterium salinarum* 24% by weight of pigment is 3-hydroxyechinenone and 11% is *trans*-astaxanthin. However, identification was based on HPLC analysis only.

The pigment content of the cells may depend on their nutritional status (Gochnauer et al., 1972; Kushwaha and Kates, 1979), and also the salinity may play a role: certain *Haloferax* species are pigmented when grown at low salinity (e.g., 15%), while at higher salt concentrations (e.g., 25%) they may be almost colorless (Kushwaha et al., 1982a; Rodriguez-Valera et al., 1980a). *Haloferax mediterranei*, a species that at high salinity is only weakly pigmented, produced massive amounts of bacterioruberin pigments when incubated at 5% salt, a concentration too low to support growth (D'Souza et al., 1997).

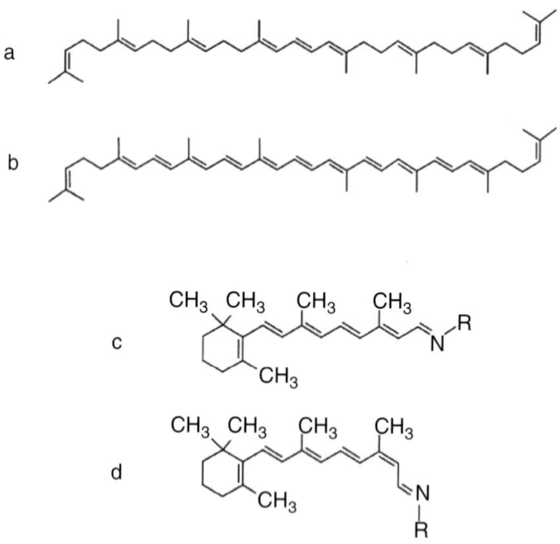

Respiratory quinones

The major respiratory quinones in the *Halobacteriales* are two menaquinones with 8 isoprenoid units: MK-8 (Fig. 15a) and MK-8(H₂) (Fig. 15b) (Collins et al., 1981; Tindall and Collins, 1986a). Quinones are present in large amounts and may amount to about 9% of the total neutral lipid content of the cells (Kamekura and Kates,

Fig. 14. C_{40} carotenoids phytoene (a) and lycopene (b), and the all-*trans* form (c) and the 13-*cis* form (d) of retinal, found in bacteriorhodopsin, halorhodopsin, and the sensory rhodopsins.

1988a). A close examination of the quinones of *Halococcus morrhuae* showed that the saturation in MK-8(H₂) involves the terminal double bond [MK-8(VIII-H₂)] (Tindall and Collins, 1986). The relative amounts of MK-8 and MK-8(VIII-H₂) may depend on the growth conditions (Tindall,

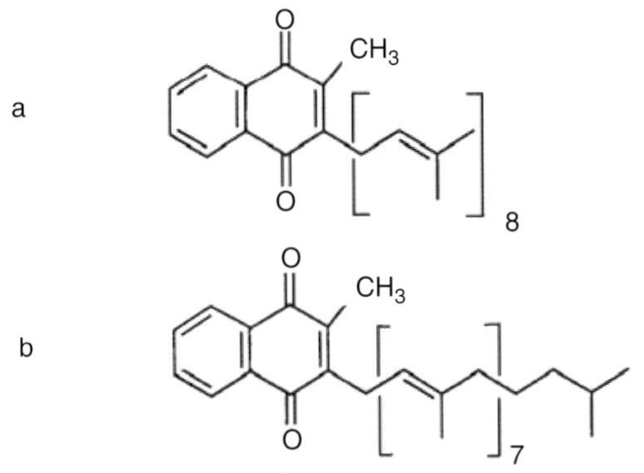

Fig. 15. Structure of respiratory quinones found in members of the *Halobacteriales*: 2-methyl-3-octaprenyl-1, 4-naphtoquinone (menaquinone-8, MK-8) (a); 2-methyl-3-VIII-dihydro-octaprenyl-1,4-naphthoquinone (dihydromenaquinone 8, MK-8[VIII-H_2]) (b).

1992; Tindall et al., 1991). Novel types of methylated menaquinones were detected in *Natronobacterium gregoryi*, which contains a range of octaprenyl menaquinones, monomethylated, and dimethylated menaquinones (Collins and Tindall, 1987).

RETINAL PIGMENTS Four retinal-containing proteins have been identified in representatives of the *Halobacteriales*: bacteriorhodopsin—an outward light-driven proton pump, halorhodopsin—an inward light-driven chloride pump, and two sensory rhodopsins, involved in light-sensing for phototaxis. The presence of the unique retinal proteins has greatly stimulated the interest of the scientific community in the halophilic Archaea, and during the almost three decades that have passed since the structure and function of bacteriorhodopsin were discovered, more publications have been written on the retinal pigments than on all other aspects of the biology of the *Halobacteriales* combined. For more in-depth information on the biology, biochemistry, and biophysics of the retinal proteins specialized reviews should be consulted (e.g., Bickel-Sandkötter et al., 1996; Lanyi, 1993a; Lanyi and Váró, 1995; Oesterhelt, 1995; Skulachev, 1993; Stoeckenius and Bogomolni, 1982; and Wagner and Linhardt, 1988).

The purple pigment bacteriorhodopsin was first identified in *Halobacterium salinarum* (Oesterhelt and Stoeckenius, 1971). This 27-kDa protein, carrying retinal as a prosthetic group (Fig. 14c–d), is located in the cell membrane in certain species, localized in patches called purple membrane. The protein serves as an outward proton pump. Upon excitation by light of the proper wavelength (absorption maximum 570 nm), protons are extruded from the cyto-

plasm to the outside of the cell (for reviews see Lanyi, 1993a; Lanyi, 1993b; Lanyi, 1997). The pathway of the proton transfer through the molecule has now been elucidated at high resolution (Luecke et al., 1998). The proton gradient thus formed is used to drive energy-requiring processes in the cell, including the phosphorylation of ADP to ATP (Danon and Stoeckenius, 1974). Under the proper conditions, *Halobacterium salinarum* produces the pigment in large quantities, imparting a purple color to the cultures. Formation of bacteriorhodopsin does not seem to be widespread among the Halobacteriales. Another species able to synthesize large amounts of bacteriorhodopsin, probably also located in its purple membrane, is *Halorubrum sodomense* (Oren, 1983a). The high content of bacteriorhodopsin detected in the bacterial community in the Dead Sea in 1980–82 (Oren, 1983b; Oren and Shilo, 1981) may be attributed to that species. Smaller amounts of retinal protein proton pumps, sometimes designated archaerhodopsins and cruxrhodopsins, were found in yet unidentified isolates from Australia (Mukohata et al., 1988; Mukohata et al., 1991) and different *Haloarcula* species (*Haloarcula vallismortis*, *Haloarcula argentinensis*) (Kitajima et al., 1996; Mukohata, 1994; Sugiyama et al., 1994; Tateno et al., 1994). In addition, retinal pigments were found in unidentified strains obtained from the Gavish Sabkha (Sinai, Egypt) (Stoeckenius et al., 1985). The gene encoding for bacteriorhodopsin was found even in an isolate that never expresses it (Kamekura et al., 1998). Studies of the bacteriorhodopsin gene cluster of *Halobacterium salinarum* have shown that expression of the protein is induced by low oxygen tension and by light (Betlach et al., 1986; Shand and Betlach, 1991). Presence of bacteriorhodopsin, which enables cells to use light energy directly, is of obvious ecological advantage. It was shown that light can relieve energy starvation at low nutrient concentrations in cells containing the pigment (Brock and Petersen, 1976), and in chemostat cultures, light enhanced growth rate and growth yields (Rodriguez-Valera et al., 1983b; Rogers and Morris, 1978). *Halobacterium salinarum* cells containing bacteriorhodopsin also are able to grow anaerobically in the light.

The second retinal protein, also first discovered in *Halobacterium salinarum*, is halorhodopsin. The structure of this protein is quite similar to that of bacteriorhodopsin, but it acts as a chloride pump: excitation by light (absorption maximum 580 nm) causes the inward transport of chloride ions from the medium (Schobert and Lanyi, 1982). Inward chloride transport is important in maintaining the proper ionic balance and is essential for cell growth. Other properties of the pigment can be found in a number of reviews

(Lanyi, 1986; Lanyi et al., 1990; Oesterhelt, 1995). No systematic survey has been performed as yet of the presence of halorhodopsin among the different genera and species within the *Halobacteriales*, but it is probably more widely distributed than bacteriorhodopsin. Halorhodopsin was in several haloalkaliphilic Archaea (Bivin and Stoeckenius, 1986), and the halorhodopsin of *Natronomonas pharaonis* has been studied in detail (Lanyi et al., 1990).

Halobacterium salinarum contains two sensory rhodopsins involved in light sensing for phototaxis. Sensory rhodopsin I is a green light receptor (light to which the cells are attracted), and sensory rhodopsin II, also termed phoborhodopsin, is a blue light receptor (light that acts as a repellent). In-depth characterizations of the sensory rhodopsin II of *Halobacterium salinarum* (Seidel et al., 1995; Zhang et al., 1996) and of *Natronomonas pharaonis* (Hirayama et al., 1992; Tomioka and Sasabe, 1995) have been published. The subject of sensory rhodopsins in halophilic Archaea has been reviewed by Bogomolni and Spudich, 1982 and by Bickel-Sandkötter et al., 1996.

PTERINS, SULFHYDRYL COMPOUNDS AND POLYAMINES Two unusual pterins have been reported to occur in *Haloarcula marismortui*: sulfohalopterin-2 (Fig. 16) and phosphohalopterin-1 (Lin and White, 1987; Lin and White, 1988). No information is available as yet on their function in the cell.

Glutathione was found to be absent in cells of halophilic Archaea, tested species including *Haloarcula marismortui*, *Haloferax volcanii*, *Halorubrum saccharovorum*, and "*Haloarcula californiae*." γ-Glutamylcysteine is present instead (Newton and Javor, 1985).

Polyamines, which are useful chemotaxonomic markers in many groups of prokaryotes, are of little or no taxonomic value in the halophilic Archaea. Concentrations of polyamines encountered are always very low. Agmatine, putrescine, spermidine, and spermine have been detected in different representatives (Carteni-Farina et al., 1985; Hamana et al., 1985; Hamana et al., 1995; Kamekura et al., 1986; Kneifel et al., 1986). The polyamines may be derived to a large extent from the growth medium, and the cellular polyamine composition is influenced by the type of medium employed.

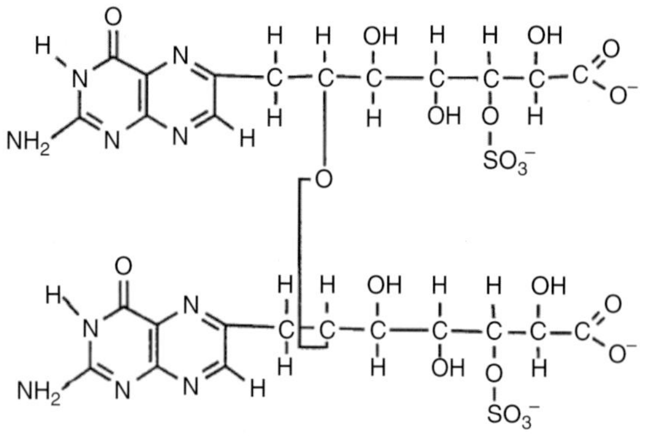

FLAGELLA Many members of the *Halobacteriales* are motile. They were shown to possess right-handed helical flagella (Alam and Oesterhelt, 1984; Alam et al., 1984; Houwink, 1956; Fig. 3a). Clockwise rotation results in forward movement of the cells, counterclockwise rotation in backward movement (Alam and Oesterhelt, 1984; Marwan et al., 1987; Marwan et al., 1991). The flagellin of which the flagella are built is a sulfated glycoprotein (Wieland et al., 1985).

GAS VESICLES Gas vesicles are found in a number of species: *Halobacterium salinarum* (Fig. 2a, Fig. 3b), *Haloferax mediterranei*, *Natrialba vacuolata*, *Halogeometricum borinquense* (Fig. 2g), and the flat square, not yet cultivated cells discovered by Walsby, 1980 (Fig. 2f, 4f). Expression of the potential of gas vesicle formation may be salt dependent: *Haloferax mediterranei* produces gas vesicles only at salt concentrations exceeding 17% (Englert et al., 1990; Röder and Pfeifer, 1996). The property is also easily lost by mutation. The structural and genetic aspects of the gas vesicles and the mechanisms involved in the regulation of their formation have been described in depth (DasSarma and Arora, 1997; Offner et al., 1998; Pfeifer and Englert, 1992; Pfeifer et al., 1997). The presence of gas vesicles and the ability to float to the water surface are advantageous for an aerobic bacterium living in salt saturated brines in which the solubility of oxygen is low. The advantage of possessing gas vesicles became clear during competition experiments between a *Halobacterium salinarum* mutant with a low amount of deficient gas vesicles and the wild type. In shaken culture both strains grew equally well. However, in static cultures cells of the wild type floated and dominated, probably because they competed successfully for oxygen, which was in short supply. In shallow static cultures, the mutant won the competition, an effect that was explained by its smaller protein burden, i.e., the wild type wasted much energy to produce gas vesicles, unnecessary under the conditions employed (Beard et al., 1997). The idea that gas vesicles also may be effective as a shield against harmful UV radiation was not confirmed in laboratory experiments (Simon, 1980).

Fig. 16. Structure of sulfohalopterin-2, an unusual pterin compound found in *Haloarcula marismortui*.

7S RNA A stable 7S RNA has been detected in several members of the *Halobacteriales*. In *Halobacterium salinarum* this molecule is 304 nucleotides long; in other species it may be somewhat larger (Moritz and Goebel, 1985; Moritz et al., 1985). The molecule does not make part of the ribosome, and its function in the cell is still unknown.

Habitat

Halophilic Archaea of the order *Halobacteriales* can be found in a wide variety of hypersaline ecosystems. Aquatic environments include natural salt lakes, such as the Great Salt Lake, Utah, and the Dead Sea, and saltern crystallizer ponds (Borowitzka, 1981; Javor, 1983; Javor, 1989; Oren, 1983c; Oren et al., 1988; Oren 1994a; Post, 1977; Post, 1981). The major factors determining their distribution in nature are the total salt concentration present, the ionic composition of the salts, and the availability of nutrients. Most species will not grow at total salt concentrations below 2.5–3 M; when suspended in solutions containing less than 1–2 M salt, cells will be irreversibly damaged and many species will lyse. Therefore, the presence of a sufficiently high salt concentration and absence of major salinity fluctuations are prerequisites for the development of halophilic Archaea of the order *Halobacteriales*.

In addition to the concentration of Na$^+$ (required in molar concentrations by all known representatives of the *Halobacteriales*), the concentration of divalent cations, especially magnesium and calcium, in the environment is of considerable ecological importance. Besides thalassohaline environments that originated by evaporation of sea water and thus reflect its ionic composition dominated by Na$^+$ and Cl$^-$, there are also athalassohaline water bodies with greatly different ionic compositions. Thus, the Dead Sea is dominated by divalent cations (presently around 1.9 M Mg^{2+} and 0.4 M Ca^{2+}, in addition to about 1.7 M Na$^+$ and 0.14 M K$^+$).

Red halophilic Archaea are also abundant in hypersaline soda lakes such as Lake Magadi (Kenya) (Tindall et al., 1980; Tindall et al., 1984), the Wadi Natrun lakes (Egypt) (Soliman and Trüper, 1982), and soda lakes in China (Wang and Tang, 1989) and India (Upasani and Desai, 1990). Alkaliphilic representatives of the *Halobacteriales* may impart a red color to such lakes. These environments are characterized by a salinity at or close to halite saturation and by high concentrations of carbonates, resulting in pH values around 10–11, often leading to the precipitation of trona (NaHCO$_3$·Na$_2$CO$_3$·2H$_2$O) (Grant and Tindall, 1986).

Acidophilic types of halophilic Archaea have not yet been reported. The Dead Sea with a pH of around 6.0 is probably the most acidic environment in which mass development of halophilic Archaea has been observed (Oren, 1983c; Oren et al., 1988; Oren and Gurevich, 1993). A pH of 6 approximately coincides with the lower boundary of the range of pH values that support growth of members of the *Halobacteriales*. As more acidic hypersaline environments seem to be rare or altogether nonexistent, the halophilic Archaea appear to be well adapted to the whole pH range occurring in hypersaline brines in nature.

Most halophilic Archaea have rather high temperature optima, in the range between 35 and 50°C and sometimes even higher (Shand and Perez, 1999). The ability to grow at high temperatures may be an adaptation to the often relatively high temperatures of salt lakes in tropical areas where the high rates of evaporation easily lead to the formation of hypersaline environments. However, cold-adapted strains also occur. Deep Lake, Antarctica is a hypersaline lake in which the water temperature varies according to the season between below 0 and +11.5°C. *Halorubrum lacusprofundi*, a halophilic Archaeon isolated from this lake, grows optimally at 31–37°C, but slow growth does occur down to temperatures as low as 4°C (Franzmann et al., 1988).

Halophilic Archaea also have been isolated from saline soils. Though soil environments are generally more prone to salinity fluctuations following rainfall events than hypersaline brines in salt lakes are, members of the *Halobacteriales* were found in saline soils of arid areas such as at the Mediterranean coast of Spain (Quesada et al., 1982) and in soils of plains and high mountains of the former USSR (Kulichevskaya et al., 1992; Zvyagintseva and Tarasov, 1988). These soils yielded cultures of species such as *Halorubrum distributum*, *Haloferax mediterranei*, *Haloferax volcanii*, and *Haloterrigena turkmenica*.

Members of the *Halobacteriaceae* also develop on salted food, hides, and other products to which high salt concentrations are added for preservation (Browne, 1922; Clayton and Gibbs, 1927; Harrison and Kennedy, 1922; Lochhead, 1934; Rodriguez-Valera, 1992; Shewan, 1971). Salted fish may develop a red patina, called "the pink," caused by development of *Halobacterium* and *Halococcus* colonies. The economic damage caused by the halophilic Archaea has triggered much of the early research on these organisms. The problem may be caused to a large extent by the use of crude solar salt. During crystallization of halite from NaCl-saturated brines, cells are often trapped within liquid inclusions inside the growing crystals, and these may remain viable for months and even years (Dussault, 1958; Norton and Grant, 1988). Different methods have been suggested to kill halophilic Archaea in contaminated solar salt, including the use of dilute acid (Kushner et al., 1965), peracetic acid vapor or ethylene oxide (Tasch and Todd, 1973; Tasch and Todd, 1974), and bile acids (Rodriguez-Valera, 1992).

To what extent the longevity of cells within salt crystals extends to hundreds of millions of years is still a point of controversy. Viable halobacteria have been recovered from salt collected from British salt mines from the Triassic (195–225 million years B.P.) and Permian (225–270 million years B.P.) periods (Norton et al., 1993). Vreeland and coworkers (Vreeland and Rosenzweig, 1999; Vreeland et al., 1998b) recently recovered a variety of halophilic Archaea from a salt mine of the Salado formation (Permian) near Carlsbad, New Mexico, including some novel types. Some of these strains may be able to degrade cellulose, a property not earlier documented within the *Halobacteriales*. However, it is difficult to assess with certainty whether these bacteria were indeed trapped within the crystals at the time of the formation of the salt deposits or whether they are of more recent origin and may have entered the salt during disturbances of the salt layer, either natural or caused by human activity. Another case in which red halophilic Archaea have been recovered from ancient geological strata is the isolation of *Halorubrum distributum* from an Upper-Devonian oilfield (Zvyagintseva et al., 1998). It is not clear whether the cells found within the salt crystals and in the deep oil deposits remained viable in a state of "suspended animation" or whether they were slowly metabolizing and possibly even growing *in situ*. Critical discussions of the intriguing findings of viable halophilic Archaea within ancient salt deposits were given by Vreeland and Powers, 1999 and by Grant et al., 1998b.

Isolation

A variety of media has been recommended for the growth of different members of the *Halobacteriales*. Table 7 gives a representative selection. Additional information on cultivation can be found in the original species descriptions (see the references given in Tables 2–6) and in reviews (DasSarma et al., 1995b; Eimhjellen, 1965; Gibbons, 1969; Larsen, 1981; Rodriguez-Valera, 1995; Tindall, 1992; Weber et al., 1982). Also the homepage of the Deutsche Sammlung von Mikroorganismen und Zellkulturen mbH (http://www.dmsz.de) gives access to a wealth of information. Another useful website providing descriptions of growth media and laboratory procedures for use with the halophilic Archaea was prepared by Dyall-Smith, 1999.

Growth media often resemble the ionic composition of the environment from which the organism was isolated originally. Media used differ greatly in total salt concentration, ionic composition (e.g., high magnesium concentrations of up to 0.8 M for species isolated from the Dead Sea), and pH (9.5 and higher for the alkaliphilic species, using media very low in divalent cation concentrations). Members of the *Halobacteriales* are grown generally in complex media containing high concentrations of yeast extract, casamino acids, and similar sources of nutrients (see also Table 7). Typical doubling times in such media are from 3–4 h for *Haloferax volcanii* to 8–12 h for *Halobacterium salinarum*. The use of media with high concentrations of peptides and amino acids reflects environments such as salted fish and hides from which many isolates were obtained. It may be argued that the earlier assumption that the members of the group have a very limited metabolic potential was to a large extent because of the restricted range of compounds tested as potential substrates.

Several brands of peptone, notably Bacto-peptone (Difco), are unsuitable for the cultivation of members of the *Halobacteriales* as they cause lysis of the cells. The toxic factor present in Bacto-peptone was identified as bile acids (Kamekura et al., 1988), known since 1956 to cause lysis (when present at very low concentrations) of halophilic Archaea (Dussault, 1956a; Dussault, 1956b). Addition of starch to the medium may bind and neutralize toxic components such as bile acids (Oren, 1990a). Sugars may stimulate growth of many species. When adding sugars, proper buffering may be required to avoid acidification of the medium to values inhibitory for growth. Though light can be used as an energy source in species containing bacteriorhodopsin, no absolute requirement for light has been demonstrated for any strain, and all known members of the *Halobacteriales* grow well in the dark.

Recovery of colonies of halophilic Archaea from samples collected from nature may be enhanced by the addition of natural brine from the sampling site and a whole cell extract of *Halobacterium salinarum* as a source of stimulatory growth factors (Wais, 1988). For the selective isolation of Archaea from natural sources, inclusion of antibiotics such as penicillin or ampicillin has been recommended.

Few procedures have been described for the selective isolation of specific genera and species belonging to the *Halobacteriales*. *Halobacterium* can be selectively enriched under anaerobic conditions in medium containing L-arginine (Oren and Litchfield, 1999). *Halococcus* species may be

Table 7. A selection of media used for the cultivation of neutrophilic and alkaliphilic members of the order *Halobacteriales*.

Ingredient (g/liter)[a]	Medium no.						
	1	2	3	4	5	6	7
NaCl	250	200	175	125	125	250	200
$MgSO_4 \cdot 7H_2O$	20	20	—	—	160	2.5	—
$MgCl_2 \cdot 6H_2O$	—	—	20	50	—	2	1
KCl	2	2	—	5	—	—	—
K_2SO_4	—	5	5	—	—	—	—
$CaCl_2 \cdot 2H_2O$	0.1	0.1	0.1	0.13	0.1	3	—
Na_3Citrate $\cdot 3H_2O$	3	3	—	—	—	3	1
Na glutamate	—	1	—	—	—	2.5	5
Na_2CO_3[b]	—	—	—	—	—	5	5
Yeast extract	10	5	5	5	1	—	5
Casein hydrolysate	7.5	5	—	5	1	15	—
Tryptone	—	—	—	—	2	—	—
Starch	—	—	—	—	—	—	—
$FeCl_3 \cdot 4H_2O$ (mg/liter)	36	36	—	—	—	—	36
$MnCl_2 \cdot 4H_2O$ (mg/liter)	0.36	0.36	—	—	—	—	0.36
Final pH (to be adjusted with NaOH or HCl)	7.2–7.4	7.0–7.4	6.8	6.8	6.8	9.0–9.5	9.5–10.0
Reference	Payne etal., 1960	Tindall and Collins, 1986	Based on Rodriguez-Valera etal., 1983a	Mullakhanbhai and Larsen, 1975	Oren etal., 1995	Soliman and Trüper, 1982	Tindall, 1992

a—For solid media, 20g agar/liter should be added.
b—Sterilized separately.

Media no. 1 and 2 are suitable for the growth of *Halobacterium* and other neutrophilic extremely halophilic representatives of the group. Medium no. 3 may be used for species that require less salt, such as many *Halorubrum* species. Types that require high concentrations of magnesium, such as Dead Sea isolates, can be grown in medium no. 4 (*Haloferax volcanii*) or 5 (*Halorubrum sodomense, Halobaculum gomorrense*). Media no. 6 and 7 were designed for the growth of alkaliphilic species. Additional recipes for media, including defined media, were given by Dyall-Smith (1999), Eimhjellen (1965), Gibbons (1969), Tindall (1992), and Larsen (1981).

selectively isolated by suspension of the sample in medium with a salt concentration sufficiently low to kill all other known neutrophilic halophilic Archaea, followed by cultivation in a suitable high-salinity medium. Viable *Halococcus* cells could be recovered even from sea water: 2–35 *Halococcus* colonies were obtained from 5-liter portions of Mediterranean sea water sampled 5 km off the coast of Spain (Rodriguez-Valera et al., 1979).

Chemically defined media have been designed for a number of species. Members of the genera *Haloferax* and *Haloarcula* grow on inorganic media amended with a suitable single carbon and energy source (Javor, 1984; Rodriguez-Valera et al., 1980a). In certain cases small amounts of vitamins or other growth factors may still be needed. A synthetic medium supporting good growth of *Haloferax volcanii* contains glycerol, succinate, NH₄Cl, thiamine and biotin. Also urea, histidine and glutamate can serve as single nitrogen sources in this species (Kauri et al., 1990). Defined media for more fastidious types such as different *Halobacterium salinarum* strains may contain 10–15 amino acids, nucleotides (adenylic acid, uridylic acid, and/or cytidylic acid) and other compounds (Dundas et al., 1963; Grey and Fitt, 1976; Onishi et al., 1965; Weber et al., 1982).

For solid media, higher than usual agar concentrations should be used, as the high salt concentration of the medium interferes with the solidification of the agar. A concentration of 20 g agar/liter generally gives satisfactory results. For the preparation of agar media for the haloalkaliphiles the agar should be sterilized separately from the sodium carbonate and the other alkaline components of the media. When streaking halophilic Archaea on agar plates, account should be taken that some species may be mechanically fragile, and cells are easily damaged during handling. Drying out of agar plates with the formation of salt crystals on the surface of the agar may present a serious problem in view of the often long incubation times required (up to several weeks) for colonies to appear. Incubation and storage of Petri dishes in plastic bags are then recommended.

Identification

Isolates may be identified according to the properties listed in Tables 2–6. Properties important in the identification of members of the *Halobacteriales* are based on the published list of minimal standards for the description of new taxa within the order (Oren et al., 1997a). The list of properties to be included in such descriptions includes colonial size and shape, cell morphology, motility, pigmentation, Gram stain reaction, salt concentration required to prevent cell lysis, optimum NaCl and MgCl₂ concentrations for growth, the range of salt concentrations enabling growth, temperature and pH ranges for growth, ability of anaerobic growth in the presence of nitrate, reduction of nitrate to nitrite, formation of gas from nitrate, anaerobic growth in the presence of L-arginine, production of acids from a range of carbohydrates, ability to grow on a range of single carbon sources, catalase and oxidase activities, formation of indole, starch hydrolysis, gelatin hydrolysis, casein hydrolysis, Tween 80 hydrolysis, sensitivity to different antimicrobial compounds, characterization of polar lipids, including the types of glycolipids present and the presence or absence of PGS, G+C content of DNA, and 16S rRNA nucleotide sequence information. For the description of new species within existing genera, also DNA-DNA hybridization experiments with related species are to be performed. Further traits recommended to be included in descriptions of new taxa are electron microscopy, anaerobic growth in the presence of dimethylsulfoxide or trimethylamine N-oxide, phosphatase activity, urease activity, β-galactosidase activity, lysine decarboxylase activity, ornithine decarboxylase activity, presence of glycoprotein in the cell envelope, presence of poly-β-hydroxyalkanoate, presence of plasmids, and the electrophoretic pattern of cellular proteins. Procedures and approaches used for the performance of the different tests are described in the minimal standards document (Oren et al., 1997a). The Gram stain is best carried out after fixation of the cells with acetic acid (Dussault, 1955). However, the value of the Gram stain in differentiation of members of the order and family is limited. The catalase and oxidase reactions of all known species are positive.

DNA-DNA hybridization tests are still considered essential to determine whether a new isolate assigned to an existing genus represents a new species or may belong to one of the recognized species within that genus (DNA-DNA hybridization values higher that 70%) (Oren et al., 1997a). DNA-16S rRNA hybridization has been used also in taxonomic studies dealing with the group (Ross and Grant, 1985).

Preservation

Cultures may be kept on agar slopes at 4°C, to be subcultured every 3–6 months. However, because of genetic instability of certain strains (Pfeifer et al., 1981), routine subculturing is not recommended, and long-term storage by freezing or drying is preferred.

The vast majority of described species may be dried under vacuum. A concentrated cell suspension (20 μl) is dropped onto a predried, sterile skimmed milk plug, which rapidly absorbs excess moisture. The tube is then placed under vacuum. The combined effect of high salinity and binding of water to the skimmed milk plug appears to have a protective function. This method is used by the Deutsche Sammlung von Mikroorganismen und Zellkulturen mbH (DSMZ) for routine preservation of members of the *Halobacteriales* (Tindall, 1991; Tindall, 1992). Dried preparations are best stored at 4–8°C. Sakane et al., 1992 recommended liquid drying (L-drying, vacuum drying from the liquid state without freezing). The procedure was successfully used for the preservation of *Halobacterium salinarum*, *Haloferax volcanii*, *Haloarcula vallismortis*, *Halococcus morrhuae*, *Halorubrum saccharovorum*, and *Natronomonas pharaonis*. In this procedure, cells are suspended in 0.1 M K-phosphate buffer pH 7, containing 10% Na-glutamate, 1.5% adonitol, 2% sorbitol, 15% NaCl, and 0.05% Na-thioglycolate. Portions of 0.1 ml suspension in sterile ampoules provided with cottonwool plugs are dried in vacuo, and then the ampoules are sealed. A detailed protocol was presented by Dyall-Smith, 1999.

Storage in a frozen state presents another option for the preservation of cultures of halophilic Archaea. A glass capillary method is used by the DSMZ for freezing cells in liquid nitrogen. As described by Tindall, 1992, cell pellets of late exponential growth phase cells are collected by low-speed centrifugation and suspended in fresh growth medium containing 5% (w/v) dimethylsulfoxide. Sterile glass capillaries are filled with the suspension and stored in liquid nitrogen. Cells also may be stored at –60 to –80°C in 10% or 20% glycerol (w/v) (Hochstein, 1988; Jones et al., 1984). When resuspended in growth medium for revival of the cultures the medium should be properly buffered for those strains that produce acids from glycerol. Dyall-Smith, 1999 has provided detailed protocols.

Physiology

The *Halobacteriales* are aerobic heterotrophic prokaryotes. Aerobic degradation of carbon sources is based on the tricarboxylic acid cycle, if necessary in combination with the glyoxylate cycle, and respiratory electron transport involving a chain of cytochromes. All the enzymes of the tricarboxylic acid cycle have been demonstrated in *Halobacterium salinarum* (Aitken and Brown, 1969), and NMR spectroscopy following ^{13}C-labeling showed very rapid cycling of carbon through the pathway (Ghosh and Sonawat,

1998). Enzymes of the glyoxylate cycle are present as well in *Halobacterium salinarum* (Aitken and Brown, 1969), and a functional glyoxylate cycle has been demonstrated in *Haloferax volcanii* (Serrano et al., 1998) and in *Natronococcus occultus* (Kevbrina and Plakunov, 1992).

A variety of respiratory cytochromes and other electron transport proteins involved in respiration has been characterized in different members of the *Halobacteriales*. In *Halobacterium salinarum*, b-type cytochromes are present in the largest quantities, and four different b-type cytochromes have been identified (Hallberg-Gradin and Colmsjö, 1989). Low levels of a c-type cytochrome and cytochrome o and a_1 are present as well (Cheah, 1970; Hallberg and Baltcheffsky, 1979; Hallberg and Baltscheffsky, 1981).

An interesting blue copper-containing redox protein has been described as a peripheral membrane protein in *Natronomonas pharaonis* and was termed halocyanin (Scharf and Engelhard, 1993). The respiratory electron transport chain of *Natronomonas* thus consists of a cytoplasmic cytochrome c (–142 mV), a membrane-bound cytochrome bc (redox potentials of the heme B and heme C groups –117 and –44 mV, respectively), the copper heme B protein halocyanin (+128 mV), and the membrane-bound cytochrome ba_3 with a heme A_S group (+358 mV) (Scharf et al., 1997).

Though *Halobacterium salinarum*, the type species of the type genus of the *Halobacteriaceae*, is unable to grow on sugars, different carbohydrates can be used by a variety of other species. Carbohydrate utilization was first demonstrated in *Halorubrum saccharovorum* (Tomlinson and Hochstein, 1972a; Tomlinson and Hochstein, 1976). The glycolyic Embden-Meyerhof pathway is not functional in the *Halobacteriaceae* as hexokinase and phosphofructokinase activities are low and strongly salt-inhibited (Rawal et al., 1988a). Instead, breakdown of glucose by *Halorubrum saccharovorum* and probably other members of the group follows a modified Entner-Doudoroff pathway in which the phosphorylation step is postponed. Glucose is oxidized by gluconate to 2-keto-3-deoxygluconate and then phosphorylated to 2-keto-3-deoxy-6-phosphogluconate, which is split into pyruvate and glyceraldehyde-3-phosphate (Tomlinson et al., 1974).

Use of carbohydrates is often associated with the production of acids, as oxidation of such substrates is incomplete (Hochstein, 1978; Hochstein, 1988, see also Tables 2–6). When grown on glucose, *Halorubrum saccharovorum* was shown to excrete acetate and pyruvate (Tomlinson and Hochstein, 1972b; Tomlinson et al., 1978). On other sugars, aldonic acids may be formed: galactonic acid is produced by *Halorubrum saccharo-vorum* in the presence of galactose (Hochstein et al., 1976), lactobionic acid is made from lactose (Tomlinson et al., 1978), and also arabinose, ribose, and xylose are oxidized to the corresponding aldonic acids (Hochstein, 1978; Hochstein, 1988). Acetate, pyruvate, and D-lactate were identified in cultures of a variety of *Haloferax* and *Haloarcula* species grown in the presence of glycerol (Oren and Gurevich, 1994).

Although the Embden-Meyerhof pathway is generally not involved in the degradation of sugars in the halophilic Archaea, many of the enzymes of the pathway are present and function in gluconeogenesis (Sonawat et al., 1990). The fructose-1,6-bisphosphate aldolase involved in the pathway has been characterized (Krishnan and Altekar, 1991). When grown on fructose, *Haloarcula vallismortis* may use the pathway also for dissimilatory purposes, degrading fructose via fructose-1-phosphate, which is then further phosphorylated to fructose-1,6-bisphosphate by an ATP-dependent fructose 1-phosphotransferase (Altekar and Rangaswamy, 1990).

In addition to the enzymes of the modified Entner-Doudoroff pathway and gluconeogenesis, the methylglyoxal bypass (dihydroxyacetone phosphate → methylglyoxal → D-lactate) may be operative in the halophilic Archaea. Methylglyoxal synthase, the first enzyme of the pathway, was found in all species tested except *Halobacterium salinarum*, whereas glyoxalase I was found in all species tested (Oren and Gurevich, 1995).

Glycerol, produced as an osmotic solute by the halophilic green alga *Dunaliella*, may be one of the main substrates supporting growth of halophilic Archaea in salt lakes (Borowitzka, 1981; Javor, 1984; Oren, 1993a; Oren, 1995a; Oren and Shilo, 1985). Production of D-lactate, acetate and pyruvate from glycerol also could be demonstrated in natural communities of halophilic Archaea from the Dead Sea and from saltern crystallization ponds amended with micromolar concentrations of glycerol (Oren and Gurevich, 1994). The acetate was utilized very poorly by the community, and acetate turnover-times in hypersaline environments were shown to be very long—in the order of weeks to months (Oren, 1995b). In a hypersaline cyanobacterial mat, red halophilic Archaea were found (including *Haloarcula japonica* and *Halorubrum distributum*), and it was suggested that these organisms primarily used tricarboxylic-acid-cycle intermediates excreted by *Microcoleus chthonoplastes*, the main component of the cyanobacterial mat (Zvyagintseva et al., 1995a).

In addition to simple substrates such as amino acids, sugars, and organic acids, certain poly-

meric substances can be degraded by halophilic Archaea. Many species of *Halobacteriales* produce exoenzymes such as proteases, lipases, DNAses and amylases (see also Tables 2–6).

A number of unusual substrates such as aliphatic and aromatic hydrocarbons is used at least by certain representatives of the group. Red halophilic Archaea were isolated from a salt marsh in the south of France from an enrichment culture with the C_{20} hydrocarbon eicosane (Bertrand et al., 1990). One of the isolates degraded a wide variety of compounds, including even- and odd-carbon-number saturated hydrocarbons (tetradecane, hexadecane, eicosane, heneicosane), saturated isoprenoid alkanes (pristane), aromatic hydrocarbons (acenaphthene, phenanthrene, anthracene, 9-methylanthracene), and also long-chain fatty acids such as palmitic acid. No taxonomic description of the isolate has been published as yet. Halophilic Archaea have been found in association with oilfield brines (Zvyagintseva et al., 1995b). Also, hydrocarbon-degrading halophilic Archaea (strains taxonomically close to *Halobacterium salinarum*, *Haloferax volcanii* and *Halorubrum distributum*) were isolated from hypersaline brines from an oil deposit in Tatarstan (Kulichevskaya et al., 1991). A *Haloferax* isolate obtained from soil contaminated with oil brine was shown to grow on different aromatic compounds, including benzoate, cinnamate, and 3-phenylpropionate, as sole carbon and energy sources (Emerson et al., 1994). Degradation of 3-phenylpropionate involves a gentisate 1,2-dioxygenase (Fu and Oriel, 1998; Fu and Oriel, 1999).

In view of the fact that the *Halobacteriales* are all strictly heterotrophic bacteria, the finding of ribulose-1,5-bisphosphate carboxylase (RuBisCo) activity in certain *Haloferax* and *Haloarcula* species came as a surprise (Altekar and Rajagopalan, 1990; Rawal et al., 1988b). The RuBisCo of *Haloferax mediterranei* was characterized in further detail. The enzyme is a hexadecamer of 500 kDa apparent molecular mass, consisting of 54 and 14 kDa subunits (Rajagopalan and Altekar, 1994). It cross-reacts with anti-spinach RuBisCo antibodies (Altekar and Rajagopalan, 1990). In contrast to most other proteins of halophilic Archaea the *Haloferax* RuBisCo does not possess an excess of acidic amino acid residues (Rajagopalan and Altekar, 1994). Additional enzymes of the Calvin cycle, including phosphoribulokinase, have been identified in RuBisCo-containing halophilic Archaea (Rawal et al., 1988a), but autotrophic growth was never demonstrated. A correlation has been suggested between the presence of RuBisCo activity and the ability to accumulate poly-β-hydroxyalkanoate as storage polymer, shown in *Haloferax mediterranei* (Lillo and

Rodriguez-Valera, 1990; Rodriguez-Valera and Lillo, 1992), *Haloferax volcanii*, and *Haloarcula marismortui* (Kirk and Ginzburg, 1972). However, the nature of the connection between the two phenomena remains unclear. No information is available as yet on the contribution of CO_2 fixed via the RuBisCo to the total carbon assimilated during growth.

Additional modes of CO_2 fixation, not involving RuBisCo, have been indicated in different members of the halophilic Archaea. *Halobacterium salinarum* cells containing bacteriorhodopsin showed a light-induced stimulation of CO_2 uptake (Danon and Caplan, 1977). The process was suggested to involve a reductive carboxylation of propionate (Danon and Caplan, 1977), biosynthesis of glycine (Javor, 1988), or biosynthesis of pyruvate (Rajagopalan and Altekar, 1991). Light-stimulated CO_2 incorporation by members of the *Halobacteriales* was observed also in a bacteriorhodopsin-containing community of halophilic Archaea (probably involving *Halorubrum sodomense*) in the Dead Sea (Oren, 1983b).

Anaerobic metabolism

Due to its low solubility in salt-saturated brines, oxygen may easily become a limiting factor for development of halophilic Archaea. A few representatives of the group may overcome oxygen limitation by producing gas vesicles, enabling them to float toward the air-water interface. In addition, many halophilic Archaea are able to grow anaerobically. Modes of anaerobic growth documented within the *Halobacteriales* include the use of alternative electron acceptors such as nitrate, dimethylsulfoxide, trimethylamine N-oxide or fumarate, fermentation of arginine, and light-driven anaerobic growth in cells containing bacteriorhodopsin.

The ability to reduce nitrate is widespread among the members of the *Halobacteriales* (see also Tables 2–6). A few species (e.g., *Haloferax denitrificans*, several members of the genus *Haloarcula*, *Halogeometricum borinquense*) are able to grow anaerobically using nitrate as electron acceptor. Nitrate is then reduced to gaseous products, generally N_2, but also N_2O formation has been observed in several species (Hochstein and Tomlinson, 1985; Mancinelli and Hochstein, 1986; Tomlinson et al., 1986). The dissimilatory nitrate reductases of *Haloferax denitrificans* and *Haloferax mediterranei* have been characterized (Hochstein, 1991). The ecological relevance of anaerobic growth on nitrate has never been ascertained. In view of the low concentrations of nitrate generally encountered in hypersaline brines and the apparent lack of regeneration of nitrate by nitrification at high salt concentra-

tions, the process can be expected to occur only to a limited extent in nature (Oren, 1994a).

Other alternative electron acceptors for respiration in many species are dimethylsulfoxide (DMSO), trimethylamine N-oxide (TMAO), and fumarate. Reduction of DMSO and TMAO was found to be coupled with growth in several species. *Halobacterium salinarum*, *Haloferax mediterranei*, *Haloarcula marismortui*, and *Haloarcula vallismortis* grew anaerobically in the presence of DMSO or TMAO, but *Haloferax volcanii* grew anaerobically in the presence of DMSO but not TMAO (Oren and Trüper, 1990). The ecological importance of DMSO reduction is not clear. However, TMAO may be available as an electron acceptor in salted fish, being present in often high concentrations within the fish tissues as an osmotic solute. Fumarate-driven anaerobic growth was reported in certain *Halobacterium salinarum* strains, in *Haloferax denitrificans*, and in *Haloferax volcanii* (Oren, 1991). Likewise, elemental sulfur and thiosulfate have been suggested to serve as potential electron acceptors (Tindall and Trüper, 1986), but little information is available on the nature of the process.

Fermentation of L-arginine to citrulline can drive anaerobic growth in *Halobacterium salinarum* (Hartmann et al., 1980; Oesterhelt, 1982). An in-depth genetic analysis of the genes involved in the arginine fermentation pathway has been performed (Bickel-Sandkötter et al., 1996; Ruepp and Soppa, 1996). Anaerobic growth on arginine does not seem to be widespread among the Archaea; a wide variety of neutrophilic strains tested belonging to genera other than *Halobacterium* gave negative results (Oren, 1994b; Oren and Litchfield, 1999). As a result, a specific enrichment procedure for members of the genus *Halobacterium* was developed, based on their ability to grow anaerobically in the presence of L-arginine (Oren and Litchfield, 1999).

Light can be used as an energy source to drive anaerobic growth in *Halobacterium salinarum*, provided the cells contain the light-driven proton pump, bacteriorhodopsin (Hartmann et al., 1980; Oesterhelt, 1982; Oesterhelt and Krippahl, 1983). As the biosynthesis of bacteriorhodopsin from β-carotene is oxygen-dependent, either trace concentrations of oxygen should be present or retinal or retinaloxime should be supplied to the medium to enable sustained light-driven anaerobic growth (Oesterhelt and Krippahl, 1983).

OSMOTIC ADAPTATION To cope with the high osmotic pressure of their hypersaline medium, the cytoplasm of the halophilic Archaea of the order *Halobacteriales* contains molar concentrations of ions, especially K^+ and Cl^-, with often considerable concentrations of Na^+ being reported as well. This strategy of "salt-in" to balance "salt-out," though it adapts the cellular components to function in the presence of high ionic concentrations, is found in only a small group of obligatory anaerobic halophilic Bacteria (order Haloanaerobiales, families Haloanaerobiaceae and Halobacteroidaceae). All other halophilic microorganisms known, prokaryotic as well as eukaryotic, use organic solutes to balance their cytoplasm osmotically with the salt concentrations found in their environment. The mechanisms enabling the establishment of the huge potassium concentration gradient over the cytoplasmic membrane and also the often large sodium gradient present, as well as the special adaptations of the intracellular enzymatic machinery to the presence of high ionic concentrations, have been described in Life at High Salt Concentrations in Volume 2.

It was recently shown that some haloalkaliphilic Archaea, in addition to possessing high intracellular salt concentrations, may contain an organic osmotic solute as well: 2-sulfotrehalose. When grown in defined media lacking yeast extract, this novel compound was found inside *Natronococcus occultus* at concentrations of up to 0.9 M. However, in rich media containing yeast extract the use of inorganic ions is preferred over synthesis of 2-sulfotrehalose (Desmarais et al., 1997). The role of 2-sulfotrehalose accumulation by haloalkaliphilic Archaea is still far from clear.

Genetics

During the past 15 years considerable progress has been made in the study of the genetics of the *Halobacteriales*. Mutant strains have been characterized, and different types of gene transfer systems have been developed, based on cell fusion, transformation, or transfection. Reviews on the genetics of the group were presented by Doolittle et al., 1992, Pfeifer, 1988, and Schalkwyk, 1993. A wealth of technical information, including detailed protocols for genetic experiments with the halophilic Archaea, is available online (Dyall-Smith, 1999).

Chromosomes and plasmids

It was reported as early as 1963 that, during buoyant density centrifugation in CsCl gradients, the DNA of many species of the *Halobacteriales* is resolved into a major ("FI DNA") band and a minor ("FII DNA" or "satellite DNA") band of lower G+C content (Joshi et al., 1963). The FII fraction accounts for 11–36% of the total DNA

(Joshi et al., 1963; Moore and McCarthy, 1969). Alternative methods for the determination of the G+C composition of the DNA, such as T_m and HPLC techniques, do not provide information on whether two types of DNA are present.

The more A+T-rich FII DNA consists of a heterogeneous collection of covalently closed circular DNA (cccDNA) fractions, but there also may be "islands" of FII DNA within the chromosome (Pfeifer, 1988). The FII islands on the chromosome are often the sites of insertion sequences, which are present in high numbers in *Halobacterium salinarum*, and are responsible for the high frequency of spontaneous mutations found in this species. The genome of *Halobacterium salinarum* ("*halobium*") NRC-1 contains a variety of transposable elements that give rise to frequent DNA rearrangements (DasSarma, 1993; Ng et al., 1998).

In *Halobacterium salinarum* ("*halobium*") NRC817, cccDNA is a heterogeneous collection of molecules ranging in size from 60 to 180 kb. The main species was termed pHH1 (150 kb) and is present in 6–8 copies per genome (Pfeifer, 1988). Large plasmids ("megaplasmids") are found in many representatives of the *Halobacteriales* (Gutiérrez et al., 1986; Ng et al., 1998). Many strains have between 2 and 6 large plasmids with sizes between 150 and 450 kb (Gutiérrez et al., 1986). A large dynamic replicon of 191 kb from *Halobacterium salinarum* ("*halobium*") NRC-1 that can be designated either as a megaplasmid or as a minichromosome was recently characterized (Ng et al., 1998). In addition to major cccDNAs, minor cccDNAs of variable sizes and quantities are present in most strains (Pfeifer, 1988). Detailed information on the structure of a number of plasmids from halophilic Archaea was supplied by DasSarma, 1995.

The function of the plasmids in the halophilic Archaea is largely unknown. However, in certain cases the property of gas vesicle production may be plasmid-linked (Simon, 1978; Weidinger et al., 1979).

Organization of the genome

The isolation of mutants, both spontaneous and induced, has opened the way toward the elucidation of the genomic organization in a number of species of the *Halobacteriales*.

In *Halobacterium salinarum*, phenotypic markers such as changes in pigmentation and colony appearance have proved useful in genetic analyses. Properties such as the red pigmentation caused by presence of bacterioruberins, presence of purple membrane with bacteriorhodopsin, and content of gas vesicles were shown to mutate spontaneously with a high frequency (10^{-2} for gas vesicle formation, 10^{-4} for the production of bacterioruberin and purple membrane) (Pfeifer, 1988).

The genome of *Halobacterium salinarum* contains many repeated sequences, which are highly mobile and are arranged in both clustered and dispersed fashion within the chromosome (Sapienza and Doolittle, 1982). These sequences represent insertion sequences, which by spontaneous insertions, deletions and rearrangements in the genome are responsible for the high mutation frequencies (Charlebois and DasSarma, 1995; Pfeifer, 1986). The genomic instability of *Halobacterium salinarum* seems to be unique and is not a common feature of all halophilic Archaea (Pfeifer, 1988).

Mutagenesis can be induced chemically also, and procedures such as mutagenesis by ethyl methanesulfonate or N-methyl-N′-nitro-N-nitrosoguanidine (Bonelo et al., 1984b; Charlebois et al., 1989) were proven effective for the isolation of auxotrophic mutants.

Analysis of large numbers of mutants has shown that many metabolic pathways are organized in operons. Thus, all of the 29 tryptophan auxotrophic alleles characterized in *Haloferax volcanii* map are in one of two positions in the genome. On the other hand, some other biosynthetic pathways controlled by operons in *Escherichia coli* are probably not organized in operons in *Haloferax volcanii*: 23 histidine auxotrophic mutants mapped in six unlinked positions (Doolittle et al., 1992).

Introns have been found in several genes. Thus, the tryptophan (trp) tRNA gene of *Haloferax volcanii* has an intervening sequence of 105 bp. Another unusually structured gene is the 5S rRNA of *Halococcus morrhuae*, which is abnormally long (231 bp, containing a 108 bp insertion in the putative helix 5) (Luehrsen et al., 1981).

No complete genomic sequences have been published for any of the members of the *Halobacteriaceae* at the time of writing (June 1999), but considerable information on the genome structure is already available for several representatives of the group (Charlebois, 1999). Physical maps of the *Haloferax volcanii* and *Halobacterium salinarum* chromosomes have been constructed (Charlebois, 1995a; Charlebois, 1995b; Charlebois et al., 1989; Charlebois et al., 1991; Doolittle et al., 1992). In the case of *Halobacterium salinarum* genetic studies are complicated because of the instability of the genome, which undergoes spontaneous rearrangements at a high frequency (Pfeifer, 1988; Pfeifer et al., 1981). However, other species such as *Haloferax volcanii* and *Haloferax mediterranei* possess stable genomes, more amenable to analysis (López-García et al., 1995). For *Haloferax volcanii*, a large number of ethyl methanesulfonate-induced auxotrophic mutants has been located (Cohen

et al., 1992). The total size of the genome is 3.7 million bp, as appears from the physical map: 2,920 kbp on the chromosome, and plasmids have 690, 442, 86 and 6.4 kbp (Charlebois et al., 1991; Doolittle et al., 1992).

Genetic manipulation of halophilic Archaea

A number of techniques for the genetic manipulation of members of the *Halobacteriales* are now available. These include gene transfer by mating (Mevarech and Werczberger, 1985; Ortenberg et al., 1999; Rosenshine et al., 1989) and transformation with native plasmids, with chromosomal DNA, with shuttle vectors grown in *Haloferax volcanii* or in *Escherichia coli*, with cosmid DNA from *Escherichia coli*, with restriction fragments of cosmid DNA, and with double- and single-stranded M13 sequencing templates (Doolittle et al., 1992). An overview of the techniques, including detailed laboratory protocols, was given by Dyall-Smith, 1999.

Studies with auxotrophic mutants of *Haloferax volcanii* showed that cells can exchange genetic information by mating, following formation of cytoplasmic bridges. As a result, mixing of different auxotrophs led to the formation of prototrophic recombinants (Mevarech and Werczberger, 1985; Ortenberg et al., 1999; Rosenshine et al., 1989). Direct contact of the cells is necessary, to be achieved, e.g., by filtration on nitrocellulose filters or within cell pellets during centrifugation. Electron micrographs show that *Haloferax volcanii* cells grown in colonies are often connected by cytoplasmic bridges (Mullakhanbhai and Larsen, 1975; see also Fig. 4a). This phenomenon may be related to the cell fusion during mating. Both plasmid and chromosomal DNA may cross in both directions. Plasmids and chromosomal markers are transferred in tight linkage and at similar frequencies (Ortenberg et al., 1999; Tchelet and Mevarech, 1994). The process is not inhibited by DNAse (Rosenshine and Mevarech, 1991). In addition, unidirectional interspecies gene transfer between *Haloferax volcanii* and *Haloferax mediterranei* has been demonstrated, involving fusion of cells and transfer of plasmids. Thus, *Haloferax mediterranei* cells were obtained that carry the *Haloferax volcanii* hydroxymethylglutaryl-CoA reductase gene that bestows resistance to mevinolin (Tchelet and Mevarech, 1994).

Transformation was first shown in *Haloferax volcanii*, using plasmid pHV2 (Charlebois et al., 1987). Transformation can be achieved by addition of DNA and low molecular weight polyethylene glycol to spheroplasts formed upon removal of magnesium and other divalent cations from the medium by EDTA. The spheroplasts are subsequently regenerated on agar plates (Dyall-Smith, 1999). Such transformation has been achieved thus far in *Haloferax volcanii*, *Halobacterium salinarum*, *Haloarcula hispanica*, and *Haloarcula vallismortis* (Cline et al., 1989a; Cline et al., 1989b; Cline et al., 1995; Doolittle et al., 1992).

Different plasmid shuttle vectors carrying one or more selectable markers have been developed for genetic studies of halophilic Archaea, such as gene disruption experiments and construction of mutant strains (Cline and Doolittle 1992; Das-Sarma, 1995a; Doolittle et al., 1992; Dyall-Smith and Doolittle, 1994; Holmes and Dyall-Smith, 1990; Holmes et al., 1991; Holmes et al., 1994; Lam and Doolittle, 1989). Transposon mutagenesis has been achieved in *Haloarcula hispanica*, using a transposon based on the ISH_28-insertion sequence from *Halobacterium salinarum* and the mevinolin-resistance marker of *Haloferax volcanii* (Dyall-Smith and Doolittle, 1994). Transfection was shown as early as 1987, when it was demonstrated that *Halobacterium salinarum* spheroplasts can be transfected with DNA from its bacteriophage ΦH (Cline and Doolittle, 1987).

The recent cloning and characterization of the β-galactosidase gene *bgaH* of "*Haloferax alicantei*" (Holmes and Dyall-Smith, 1999; Holmes et al., 1997) promises its future use as a reporter gene in gene expression studies of halophilic Archaea, similar to the use of the *lacZ* gene in *Escherichia coli*. The gene can be transferred to *Haloferax volcanii* (β-galactosidase-negative) using shuttle plasmid pMDS20 (Holmes et al., 1994).

Ecology

Halophilic Archaea of the order *Halobacteriales* can be found in a wide variety of hypersaline ecosystems, and they often occur there in very high densities. In salt lakes such as the Great Salt Lake, the Dead Sea, and saltern crystallizer ponds, numbers of 10^7 to 10^8 cells per ml and higher are not unusual (Borowitzka, 1981; Javor, 1983; Javor, 1989; Oren, 1983c; Oren, 1990b; Oren, 1994a; Post, 1977; Post, 1981). Owing to the bright red, orange, or purple coloration of most representatives of the group and to the often extremely high community densities (among other factors the result of lack of predators able to thrive at these extremely high salt concentrations), often the presence of the dense communities of halophilic Archaea in hypersaline environments can be observed with the unaided eye. Lakes with salt concentrations approaching saturation may show red hues, such as documented for the north arm of the Great Salt Lake (Post, 1977; Post, 1981), the Dead Sea (Arahal et al., 1996; Kaplan and Friedmann, 1970; Oren, 1983c; Oren et al., 1988; Oren, 1993a) and hypersaline alkaline lakes such as Lake Magadi, Kenya (Grant and Tindall, 1986). Red colored brines are also typically present during the final stages of the evaporation of sea water in solar saltern ponds (Borowitzka, 1981; Javor, 1989; Jones et al., 1981; Oren, 1994a). Though the presence of the β-carotene-rich green halophilic alga *Dunaliella salina* also may contribute to the red coloration of the brines, the red halophilic Archaea generally are responsible for most of the color of the saltern brines (Oren and Dubinsky, 1994; Oren et al., 1992).

Considerable differences exist between the different genera and species within the *Halobacteriaceae* with respect to salt requirement and tolerance (see also Tables 2–6). In contrast to types such as *Halobacterium salinarum*, which will not grow below 3 M NaCl, many representatives, especially those belonging to the genus *Haloferax*, thrive at much lower salt concentrations. Thus, *Haloferax volcanii*, an organism isolated from the Dead Sea, was described as a moderate rather than an extreme halophile, growing optimally at about 2.5 M NaCl (Mullakhanbhai and Larsen, 1975). *Haloferax mediterranei* can grow down to 1.7 M NaCl and survive below 0.85 M (D'Souza et al., 1997). Accordingly, in a study of the microbial communities in saltern ponds of different salinities, *Haloferax* species were recovered with the highest frequency from ponds of intermediate salinities (Rodriguez-Valera et al., 1985).

The concentration of divalent cations, especially magnesium and calcium, in the environment is of considerable ecological importance. Thus, the Dead Sea is dominated by divalent cations (presently around 1.9 M Mg^{2+} and 0.4 M Ca^{2+}, in addition to about 1.7 M Na^+ and 0.14 M K^+). Halophilic Archaea isolated from the Dead Sea are characterized by a relatively low requirement for Na^+ and an extraordinarily high tolerance toward and often also a requirement for high Mg^{2+} concentrations, such as was shown for *Haloferax volcanii* (Mullakhanbhai and Larsen, 1975), *Halorubrum sodomense* (Oren, 1983a), and *Halobaculum gomorrense* (Oren et al., 1995). Strains tolerant to high magnesium concentrations are not limited in their distribution to the Dead Sea, but can be isolated from thalassohaline environments such as saltern crystallizer ponds (Javor, 1983; Javor, 1984). In this respect it should be noted that as a result of the precipitation of halite in these ponds highly increased magnesium concentrations may be encountered. Most halophilic Archaea that live in thalassohaline environments of neutral pH also require relatively high concentrations of

Mg^{2+} as compared to most other prokaryotes, and many species need concentrations of at least 50–100 mM Mg^{2+} for optimal growth. Suboptimal magnesium concentrations often cause loss of the native cell shape and formation of coccoid cells which may continue to grow at suboptimal rates (Brown and Gibbons, 1995) or lose their viability altogether (Cohen et al., 1983). The specific requirements for and tolerances toward total ionic strength and divalent cations in different members of the *Halobacteriales* have been compared by Edgerton and Brimblecombe, 1981, defining the growth boundaries of a thalassohaline strain such as *Halobacterium salinarum* with those of the Dead Sea isolate *Haloferax volcanii*. The approach used can be adapted easily to the characterization of the ecological niches of other members of the group.

A number of alkaliphilic halophilic archaeal species have been isolated from hypersaline soda lakes: *Halorubrum vacuolatum* (Mwatha and Grant, 1993), *Natronomonas pharaonis* (Soliman and Trüper, 1982), *Natrialba magadii*, *Natronobacterium gregoryi*, *Natronococcus occultus* (Tindall et al., 1984), and *Natronococcus amylolyticus* (Kanai et al., 1995). These species grow optimally at pH values between 9 and 10 and are unable to grow at neutral pH. In accordance with the low solubility of the divalent cations Mg^{2+} and Ca^{2+} at high pH, the alkaliphilic halophiles do not require high magnesium concentrations to grow. Optimal growth is achieved at Mg^{2+} concentrations as low as 1 mM, and magnesium concentrations above 10 mM might be inhibitory (Tindall et al., 1980).

At the high salt concentrations required for the growth of the *Halobacteriales* predators are generally absent. Protozoa rarely occur above 20–25% salt, and they were never shown to contribute much to the reduction of the community sizes of halophilic Archaea in nature. However, lysis by bacteriophages may be an important factor in the regulation of the community sizes of halophilic Archaea in their natural environments. Bacteriophages have been shown to attack *Halobacterium salinarum* (Daniels and Wais, 1990; Torsvik and Dundas, 1974; Wais et al., 1975) and other species (Bath and Dyall-Smith, 1998; Nuttall and Dyall-Smith, 1993b). Generally these phages contain DNA only, but phage ΦCh1 of the alkaliphilic *Natrialba magadii* contains both DNA and RNA (Witte et al., 1997). Similarly to their hosts, the bacteriophages of the halophilic Archaea require high NaCl concentrations to maintain structural integrity. Both lysogenic and virulent phages occur (Pfeifer, 1988). The possible importance of bacteriophages in controlling the community sizes of the *Halobacteriaceae* in hypersaline water bodies appears from the high number of virus-like particles—one to two

orders higher than the number of bacteria present—that can be observed by electron microscopic examination of brines from saltern crystallizer ponds (Guixa-Boixareu et al., 1996) and the Dead Sea (Oren et al., 1997c). A study of bacteriophages in a brine pool by Wais and Daniels, 1985 showed that lysis of halophilic Archaea by bacteriophages may be triggered by a decrease in water salinity as a result of dilution with rain water. Halophilic phages have been implicated also as the possible cause of sudden drops in the community size of halophilic Archaea in the Dead Sea (Oren et al., 1997c).

Different species of the *Halobacteriaceae* may compete for nutrients and other resources by excreting halophilic bacteriocins, termed "halocins," protein antibiotics that inhibit the growth of other related species (Meseguer and Rodriguez-Valera, 1986; Rodriguez-Valera et al., 1982). Halocin production appears to be widespread among representatives of the group (Meseguer et al., 1986; Torreblanca et al., 1994). Different modes of action have been suggested for different halocins, and the information presently available has recently been summarized by Shand et al., 1999. Most known halocins act on the cell membrane of the sensitive strains (Meseguer et al., 1991). Halocin H4, excreted by *Haloferax mediterranei*, causes membrane permeability changes and ionic imbalance in *Halobacterium salinarum* (Meseguer and Rodriguez-Valera, 1986). Although the ability to excrete halocins may be expected to have considerable ecological advantage, no data are available as yet that prove that halocins are excreted by natural communities of halophilic Archaea in concentrations sufficient to inhibit the development of competitor strains, thus substantiating their ecological role.

The niche of the halophilic Archaea of the order *Halobacteriales* in nature is quite distinct from that of the halophilic or halotolerant aerobic Bacteria. A number of known Bacteria (e.g. *Halomonas elongata* and other *Halomonas* species) are able to grow at salt concentrations as high as those required by the Archaea. However, in environments with salt concentrations above 20–25% halophilic Bacteria never seem to occur in high numbers, and they are outcompeted by the Archaea. Competition experiments in the laboratory showed that in addition to salinity, temperature is a key factor determining the outcome of the competition, higher temperatures favoring the development of Archaea (Rodriguez-Valera et al., 1980b). The dominance of the halophilic Archaea in saltern ponds of salinities exceeding 250 g/liter was demonstrated by measuring heterotrophic activities in the presence of specific inhibitors such as low concentrations of bile salts, which cause lysis of the halophilic Archaea, or specific antibiotics, which inhibit either one of the groups. Using this technique little overlap was found between the salinity range in which halophilic Bacteria were responsible for the heterotrophic activity (up to about 200–250 g/liter salt) and the higher salinity range in which the activity could be attributed to the *Halobacteriales* (Oren, 1990b; Oren, 1990c). Finally, molecular characterization of the microbial communities in saltern ponds of different salt concentrations in Spain, involving a study of rDNA restriction-fragment length polymorphism of PCR-amplified 16S rDNA, showed a dominance of halophilic Archaea in the crystallizer ponds. However, already at 133 g/liter salt, archaeal 16S rDNA sequences were different from those found between 218 and 308 g/liter salinity (Martinez-Murcia et al., 1995b), suggesting that a yet-to-be-characterized specialized group of halophilic Archaea may be present at the lower salinity range.

Applications

The halophilic Archaea of the order *Halobacteriales* have a number of useful applications, and potential new applications in biotechnological processes are being investigated (Galinski and Tindall, 1992; Rodriguez-Valera, 1992; Ventosa and Nieto, 1995).

The positive effect of the presence of dense communities of red halophilic Archaea in saltern crystallizer ponds has been recognized for a long time. By trapping solar radiation, they raise the temperature of the brine and the rate of evaporation, thereby increasing salt production (Davis, 1974; Javor, 1989). To improve salt production in salterns that do not develop a sufficiently dense archaeal community, fertilization with nutrients such as proteose-peptone and tryptone has been suggested to increase the red color of the ponds (Jones et al., 1981).

Production of fermented fish sauce in the Far East (e.g., nam pla in Thailand) involves participation of halophilic Archaea. The product is traditionally made by adding two parts of fish and one part marine salts. The mixture is covered with concentrated brine (4.4–5.1 M NaCl) and left to ferment for about a year. Red halophilic Archaea (identified as *Halobacterium* and *Halococcus*) reach their maximum density in the liquor after 3 weeks and persist throughout the fermentation period. The halobacterial proteases probably take part in the fermentation process (Thongthai and Siriwongpairat, 1990; Thongthai and Suntinalert, 1991; Thongthai et al., 1992).

Different species of halophilic Archaea synthesize potentially useful products such as bacteriorhodopsin, exopolysaccharides, poly-β-hydroxyalkanoate, and others. Halophilic Archaea have distinct advantages in biotechnological processes as cultivation is relatively easy, risk of contamination is minimal, and culture-size can be upscaled to the use of large fermentors (e.g., 70-liter cultures were grown by Kushner, 1966). However, because of the low solubility of gases in concentrated brines the supply of sufficient amounts of oxygen may cause some problems (Shand and Perez, 1999). In addition, the aggressive nature of the salts should be taken into account when planning the construction of large fermenters with metal parts exposed to the medium.

Bacteriorhodopsin, the light-driven proton pump of *Halobacterium salinarum*, has considerable biotechnological potential (Oesterhelt et al., 1991). It may be used as a biological material for information processing. Other potential uses suggested include conversion of sunlight to electricity, ATP generation, desalination of sea water, use in chemo- and biosensors, and ultrafast light detection. The photochromic effect, shifts between the purple ground state of the molecule (the "B state") and the yellow "M state," may be used for information storage, including holographic storage (König, 1988), and may lead to the development of powerful computer memories and processors (Birge, 1995). Bacteriorhodopsin offers many advantages since it is a very stable molecule, functioning well between 0–45°C and pH 1–11 (Chen and Birge, 1993), is easy to immobilize on solid substrates, and produces very reproducible photoelectric signals (König, 1988). Holographic bacteriorhodopsin films are suitable for the construction of computer memories enabling parallel processing, and the developing technology might lead to a new generation of computers (Hong, 1986). To turn bacteriorhodopsin into a light sensor, it is spread in a thin film sandwiched between an electrode and an electrically conductive gel. Changes in the shape of the molecule create a displacement of charge, generating an electrical signal.

The extracellular polysaccharide produced by *Haloferax* may have biotechnological potential (Antón et al., 1988; Rodriguez-Valera et al., 1991). The bacterium produces up to 3–8 g/liter of an acidic heteropolysaccharide with a high apparent viscosity at relatively low concentrations and resistance to extremes of salt concentrations, temperature, and pH. The structure of this polymer was recently elucidated (Parolis et al., 1996). *Haloferax volcanii* and *Haloferax gibbonsii* also produce exopolysaccharides (Paramonov et al., 1998; Severina et al., 1990). Such polymers may be used to modify rheological properties of aqueous systems, for viscosity stabilization as thickening agents, gelling agents and emulsifiers, and may find applications in microbially enhanced oil recovery (Ventosa and Nieto, 1995). Here a salt-resistant surfactant is advantageous as high salinity brines are encountered often associated with oil deposits. Whole cell preparations may be used, as the lipids liberated upon lysis of halophilic Archaea may act as surfactants to improve the oil-carrying properties (Post and Al-Harjan, 1988).

Haloferax mediterranei cells may contain considerable amounts of poly-β-hydroxyalkanoate (PHA) (Fernandez-Castillo et al., 1986; Rodriguez-Valera et al., 1991) (Fig. 3b). PHA is used for the production of biodegradable plastics. Though halophilic Archaea are not yet being used commercially for PHA production, they have certain obvious advantages over an organism such as *Ralstonia eutropha*, which is already being exploited for the purpose. *Haloferax mediterranei* can be grown on a cheap substrate such as starch. Moreover, downstream processing and purification of the product should be relatively simple, as the cells are easily lysed in water (Ventosa and Nieto, 1995). Also the high genomic stability of the organism and the reduced danger of contamination are clear assets. PHA production is maximal when grown on sugars (glucose or starch) and in the presence of low phosphate concentrations (Lillo and Rodriguez-Valera, 1990; Rodriguez-Valera and Lillo, 1992). PHA production has been demonstrated in certain other halophilic Archaea as well, such as *Haloferax volcanii* and *Haloarcula marismortui* (Kirk and Ginzburg, 1972; Rajagopalan and Altekar, 1991).

A number of additional potential uses of halophilic Archaea have been suggested:

• The glycerol diether lipids may find application as a food additive, to serve as a non-caloric fat substitute (Collins, 1977; Post and Collins, 1982).

• Exoenzymes such as amylases, amyloglucosidases, proteases, and lipases that function at high salinity may be useful in biotechnological processes requiring degradation of macromolecules in the presence of high salt concentrations (Chaga et al., 1993; Ventosa and Nieto, 1995).

• Hydrocarbon degrading halophilic Archaea (Bertrand et al., 1990; Emerson et al., 1994; Kulichevskaya et al., 1991) may prove useful in the bioremediation of oil spills.

• An 84-kDa protein from *Halobacterium salinarum* has been used as an antigen to detect antibodies against the human c-*myc* oncogene product in the sera of cancer patients, and therefore the protein may be useful for the detection of certain forms of cancer (Ben-Mahrez et al., 1988; Ben-Mahrez et al., 1991).

• Site-specific endonucleases of halophilic Archaea (Schinzel and Burger, 1986) may find uses in molecular biological research.

Literature Cited

Aitken, D. M., and A. D. Brown. 1969. Citrate and glyoxylate cycles in the halophil, Halobacterium salinarum. Biochim. Biophys. Acta 177:351–354.

Alam, M., and D. Oesterhelt. 1984. Morphology, function and isolation of halobacterial flagella. J. Mol. Biol. 176:459–475.

Alam, M., M. Claviez, D. Oesterhelt, and M. Kessel. 1984. Flagella and motility behaviour of square bacteria. EMBO J. 3:2899–2903.

Altekar, W., and R. Rajagopalan. 1990. Ribulose bisphosphate carboxylase activity in halophilic Archaebacteria. Arch. Microbiol. 153:169–174.

Altekar, W., and V. Rangaswamy. 1990. Induction of a modified EMP pathway for fructose breakdown in a halophilic archaebacterium. FEMS Microbiol. Lett. 69:139–144.

Antón, J., I. Meseguer, and F. Rodriguez-Valera. 1988. Production of an extracellular polysaccharide by Haloferax mediterranei. Appl. Env. Microbiol. 54:2381–2386.

Arahal, D. E., F. E. Dewhirst, B. J. Paster, B. E. Volcani, and A. Ventosa. 1996. Phylogenetic analyses of some extremely halophilic Archaea isolated from Dead Sea water, determined on the basis of their 16S rRNA sequences. Appl. Environ. Microbiol. 62:3779–3786.

Baas-Becking, L. G. M. 1931. Historical notes on salt and salt manufacture. Sci. Monthly 32:434–446.

Basinger, G. W., and J. D. Oliver. 1979. Inhibition of Halobacterium cutirubrum lipid biosynthesis by bacitracin. J. Gen. Microbiol. 111:423–427.

Bath, C., and M. L. Dyall-Smith. 1998. His1, an archaeal virus of the Fuselloviridae family that infects Haloarcula hispanica. J. Virol. 72:9392–9395.

Bayley, S. T., and R. A. Morton. 1978. Recent developments in the molecular biology of extremely halophilic bacteria. Crit. Rev. Microbiol. 6:151–205.

Beard, S. J., P. K. Hayes, and A. E. Walsby. 1997. Growth competition between Halobacterium salinarum strain PHH1 and mutants affected in gas vesicle synthesis. Microbiology UK 143:467–473.

Benlloch, S., A. J. Martínez-Murcia, and F. Rodríguez-Valera. 1995. Sequencing of bacterial and archaeal 16S rRNA genes directly amplified from a hypersaline environment. Syst. Appl. Microbiol. 18:574–581.

Benlloch, S., S. G. Acinas, A. J. Martínez-Murcia, and F. Rodríguez-Valera. 1996. Description of prokaryotic biodiversity along the salinity gradient of a multipond solar saltern by direct PCR amplification of 16S rDNA. Hydrobiologia 329:19–31.

Ben-Mahrez, K., D. Thierry, I. Sorokine, A. Danna-Muller, and M. Kohiyama. 1988. Detection of circulating antibodies against c-myc protein in cancer patient sera. Br. J. Cancer 57:529–534.

Ben-Mahrez, K., I. Sorokine, D. Thierry, T. Kawasumi, S. Ishii, R. Salmon, and M. Kohiyama. 1991. An archaebacterial antigen used to study immunological human response to c-myc oncogen product. In: F. Rodriguez-Valera (Ed.) General and applied aspects of halophilic microorganisms. Plenum Press. New York, NY. 367–372.

Bertrand, J. C., M. Almallah, M. Aquaviva, and G. Mille. 1990. Biodegradation of hydrocarbons by an extremely halophilic archaebacterium. Lett. Appl. Microbiol. 11:260–263.

Betlach, M. C., D. Leong, and H. W. Boyer. 1986. Bacterio-opsin gene expression in Halobacterium halobium. Syst. Appl. Microbiol. 7:83–89.

Bickel-Sandkötter, S., W. Gärtner, and M. Dane. 1996. Conversion of energy in halobacteria: ATP synthesis and phototaxis. Arch. Microbiol. 166:1–11.

Birge, R. R. 1995. Protein-based computers. Sci. Am. March:66–71.

Bogomolni, R. A., and J. L. Spudich. 1982. Identification of a third rhodopsin-like pigment in phototactic Halobacterium halobium. Proc. Natl. Acad. Sci. USA 79:6250–6254.

Bonelo, G., A. Ventosa, M. Megias, and F. Ruiz-Berraquero. 1984a. The sensitivity of halobacteria to antibiotics. FEMS Microbiol. Lett. 21:341–345.

Bonelo, G., M. Megias, A. Ventosa, J. J. Nieto, and F. Ruiz-Berraquero. 1984b. Lethality and mutagenicity in Halobacterium mediterranei caused by N-methyl-N'-nitro-N-nitrosoguanidine. Curr. Microbiol. 11:165–170.

Borowitzka, L. J. 1981. The microflora: Adaptations to life in extremely saline lakes. Hydrobiologia 81:33–46.

Bivin, D., and W. Stoeckenius. 1986. Photoactive retinal pigments in haloalkaliphilic archaebacteria. J. Gen. Microbiol. 132:2167–2177.

Brock, T. D., and S. Petersen. 1976. Some effects of light on the viability of rhodopsin-containing halobacteria. Arch. Microbiol. 109:199–200.

Brown, H. J., and N. E. Gibbons. 1995. The effect of magnesium, potassium, and iron on the growth and morphology of red halophilic bacteria. Can. J. Microbiol. 1:486–494.

Browne, W. W. 1922. Halophilic bacteria. Proc. Exper. Biol. Med. 19:321–322.

Calo, P., T. de Miguel, C. Sieiro, J. B. Velazquez, and T. G. Villa. 1995. Ketocarotenoids in halobacteria: 3-hydroxy-echinenone and trans-astaxanthin. J. Appl. Bacteriol. 79:282–285.

Carteni-Farina, M., M. Porcelli, G. Cacciapouti, M. De Rosa, A. Gambacorta, W. D. Grant, and H. N. M. Ross. 1985. Polyamines in halophilic archaebacteria. FEMS Microbiol. Lett. 28:323–327.

Chaga, G., J. Porath, and T. Illíni. 1993. Isolation and purification of amyloglucosidase from Halobacterium sodomense. Biomed. Chromatogr. 7:256–261.

Charlebois, R. L., W. L. Lam, S. W. Cline, and W. F. Doolittle. 1987. Characterization of pHV2 from Halobacterium volcanii and its use in demonstrating transformation of an archaebacterium. Proc. Natl. Acad. Sci. USA 84:8530–8534.

Charlebois, R. L., J. D. Hofman, L. C. Schalkwyk, W. L. Lam, and W. F. Doolittle. 1989. Genome mapping in halobacteria. Can. J. Microbiol. 35:21–29.

Charlebois, R. L., L. C. Schalkwyk, J. D. Hofman, and W. F. Doolittle. 1991. Detailed physical map and set of overlapping clones covering the genome of the archaebacterium Haloferax volcanii DS2. J. Mol. Biol. 222:509–524.

Charlebois, R. L. 1995a. Appendix 3. Physical and genetic map of the genome of Halobacterium volcanii DS2. In: S. DasSarma and E. M. Fleischmann (Eds.) Archaea: A laboratory manual. Halophiles. Cold Spring Harbor Laboratory Press. Cold Spring Harbor, NY. 231–235.

Charlebois, R. L. 1995b. Appendix 4: Physical and genetic map of the genome of Halobacterium sp. GRB. In: S. DasSarma, and E. M. Fleischmann (Eds.) Archaea: A laboratory manual. Halophiles. Cold Spring Harbor Laboratory Press. Cold Spring Harbor, NY. 237–239.

Charlebois, R. L., and S. DasSarma. 1995. Appendix 7: Insertion elements of halophiles. In: S. DasSarma, and E. M. Fleischmann (Eds.) Archaea: A laboratory manual. Halophiles. Cold Spring Harbor Laboratory Press. Cold Spring Harbor, NY. 253–255.

Charlebois, R. L. 1999. Evolutionary origins of the haloarchaeal genome. In: A. Oren (Ed.) Microbiology and biogeochemistry of hypersaline environments. CRC Press. Boca Raton, FL. 309–317.

Cheah, K. S. 1970. The membrane-bound ascorbate oxidase system of Halobacterium halobium. Biochim. Biophys. Acta 205:148–160.

Chen, Z., and R. R. Birge. 1993. Protein-based artificial retinas. Trends Biotechnol. 11:292–300.

Cho, K. Y., C. H. Doy, and E. H. Mercer. 1967. Ultrastructure of the obligate halophilic bacterium Halobacterium halobium. J. Bacteriol. 94:196–201.

Chow, K.-C., and K.-K. Mark. 1980. Antibiotic susceptibility of Halobacterium cutirubrum. Microbios Lett. 15:117–122.

Christian, J. H. B., and J. A. Waltho. 1962. Solute concentrations within cells of halophilic and non-halophilic bacteria. Biochim. Biophys. Acta 65:506–508.

Clayton, W., and W. E. Gibbs. 1927. Examination for halophilic microorganisms. Analyst 52:395–397.

Cline, S. W., and W. F. Doolittle. 1987. Efficient transfection of the archaebacterium Halobacterium halobium. J. Bacteriol. 169:1341–1344.

Cline, S. W., W. L. Lam, R. L. Charlebois, L. C. Schalkwyk, and W. F. Doolittle. 1989a. Transformation methods for halophilic archaebacteria. Can. J. Microbiol. 35:148–152.

Cline, S. W., L. C. Schalkwyk, and W. F. Doolittle. 1989b. Transformation of the archaebacterium Halobacterium volcanii with genomic DNA. J. Bacteriol. 171:4987–4991.

Cline, S. W., and W. F. Doolittle. 1992. Transformation of members of the genus Haloarcula with shuttle vectors based on Halobacterium halobium and Haloferax volcanii plasmid replicons. J. Bacteriol. 174:1076–1080.

Cline, S. W., F. Pfeifer, and W. F. Doolittle. 1995. Transformation of halophilic Archaea. In: S. DasSarma and E. M. Fleischmann (Eds.) Archaea: A laboratory manual. Halophiles. Cold Spring Harbor Laboratory Press. Cold Spring Harbor, NY. 197–204.

Cohen, S., A. Oren, and M. Shilo. 1983. The divalent cation requirement of Dead Sea halobacteria. Arch. Microbiol. 136:184–190.

Cohen, A., W. L. Lam, R. L. Charlebois, W. F. Doolitte, and L. C. Schalkwyk. 1992. Localizing genes on the map of the genome of Haloferax volcanii, one of the archaea. Proc. Natl. Acad. Sci. USA 89:1602–1606.

Collins, N. F. 1977. A preliminary investigation of the lipid of halophilic bacteria as a food additive. M. Sc. Thesis Utah State University. Logan, UT.

Collins, M. D., H. N. M. Ross, B. J. Tindall, and W. D. Grant. 1981. Distribution of isoprenoid quinones in halophilic bacteria. J. Appl. Bacteriol. 50:559–565.

Collins, M. D., and B. J. Tindall. 1987. Occurrence of menaquinones and some novel methylated menaquinones in the alkaliphilic, extremely halophilic archaebacterium Natronobacterium gregoryi. FEMS Microbiol. Lett. 43:307–312.

Colwell, R. R., C. D. Litchfield, R. H. Vreeland, L. A. Kiefer, and N. E. Gibbons. 1979. Taxonomic study of red halophilic bacteria. Int. J. Syst. Bacteriol. 29:379–399.

Conway de Macario, E., H. König, and A. J. L. Macario. 1986. Immunologic distinctiveness of archaebacteria that grow in high salt. J. Bacteriol. 168:425–427.

Daniels, L. L., and A. C. Wais. 1990. Ecophysiology of bacteriophage S5100 infecting Halobacterium cutirubrum. Appl. Environ. Microbiol. 56:3605–3608.

Danon, A., and W. Stoeckenius. 1974. Photophosphorylation in Halobacterium halobium. Proc. Natl. Acad. Sci. USA 71:1234–1238.

Danon, A., and S. R. Caplan. 1977. CO_2 fixation by Halobacterium halobium. FEBS Lett. 74:255–258.

DasSarma, S. 1993. Identification and analysis of the gas vesicle cluster on an unstable plasmid of Halobacterium halobium. Experientia 49:482–486.

DasSarma, S. 1995. Natural plasmids and plasmid vectors of halophiles. In:S. DasSarma and E. M. Fleischmann (Eds.) Archaea: A laboratory manual. Halophiles. Cold Spring Harbor Laboratory Press. Cold Spring Harbor, NY. 241–250.

DasSarma, S., E. M. Fleischmann, and F. Rodriguez-Valera. 1995. Appendix 2: Media for halophiles. In: S. DasSarma, and E. M. Fleischmann (Eds.) Archaea: A laboratory manual. Halophiles. Cold Spring Harbor Laboratory Press. Cold Spring Harbor, NY. 225–230.

DasSarma, S., and P. Arora. 1997. Genetic analysis of the gas vesicle gene cluster in haloarchaea. FEMS Microbiol. Lett. 153:1–10.

Davis, J. S. 1974. Importance of microorganisms in solar salt production. In: A. L. Coogan (Ed.) Proceedings of the 4th symposium on salt. Northern Ohio Geological Society. Cleveland, OH. 1:369–372.

Dees, C., and J. D. Oliver. 1977. Growth inhibition of Halobacterium cutirubrum by cerulenin, a potent inhibitor of fatty acid synthesis. Biochem. Biophys. Res. Commun. 78:36–44.

Denner, E. B. M., T. J. McGenity, H. -J. Busse, W. D. Grant, G. Wanner, and H. Stan-Lotter. 1994. Halococcus salifodinae sp. nov., an archaeal isolate from an Austrian salt mine. Int. J. Syst. Bacteriol. 44:774–780.

Dennis, P. P., and L. C. Shimmin. 1997. Evolutionary divergence and salinity-mediated selection in halophilic archaea. Microbiol. Mol. Biol. Rev. 61:90–104.

De Rosa, M., A. Gambacorta, B. Nicolaus, H. N. M. Ross, W. D. Grant, and J. D. Bu'lock. 1982. An asymmetric archaebacterial diether lipid from alkaliphilic halophiles. J. Gen. Microbiol. 128:344–348.

De Rosa, M., A. Gambacorta, B. Nicolaus, N. M. Ross, and W. D. Grant. 1983. A $C_{25:25}$ diether core lipid from archaebacterial haloalkaliphiles. J. Gen. Microbiol. 129:2333–2337.

De Rosa, M., A. Gambacorta, W. D. Grant, V. Lanzotti, and B. Nicolaus. 1988. Polar lipids and glycine betaine from haloalkaliphilic archaeobacteria. J. Gen. Microbiol. 134:205–211.

Desmarais, D., P. E. Jablonski, N. S. Fedarko, and M. F. Roberts. 1997. 2-Sulfotrehalose, a novel osmolyte in haloalkaliphilic archaea. J. Bacteriol. 179:3146–3153.

Doolittle, W. F., W. L. Lam, L. C. Schalkwyk, R. L. Charlebois, S. W. Cline, and A. Cohen. 1992. Progress in developing the genetics of the halobacteria. In: M. J. Danson, D. W. Hough, and G. G. Lunt (Eds.) Archaebacteria: biochemistry and biotechnology: Biochemical Society Symposium no. 58. Biochemical Society. High Holburn, London . 73–78.

D'Souza, S. E., W. Altekar, and S. F. D'Souza. 1997. Adaptive response of Haloferax mediterranei to low concentrations of NaCl (less than 20%) in the growth medium. Arch. Microbiol. 168:68–71.

Dundas, I. D., and H. Larsen. 1962. The physiological role of the carotenoid pigments of Halobacterium salinarium. Arch. Mikrobiol. 44:233–239.

Dundas, I. D., V. R. Srinivasan, and H. O. Halvorson. 1963. A chemically defined medium for Halobacterium salinarium strain 1. Can. J. Microbiol. 9:619–624.

Dundas, I. E. D. 1977. Physiology of Halobacteriaceae. Adv. Microb. Physiol. 15:85–120.

Dussault, H. P. 1955. An improved technique for staining red halophilic bacteria. J. Bacteriol. 70:484–485.

Dussault, H. P. 1956a. Study of red halophilic bacteria in solar salt and salted fish: I. Effect of Bacto-oxgall. J. Fish. Res. Bd. Canada 13:183–194.

Dussault, H. P. 1956b. Study of red halophilic bacteria in solar salt and salted fish: II. Bacto-oxgall as a selective agent for differentiation. J. Fish. Res. Bd. Canada 13:195–199.

Dussault, H. P. 1958. The fate of red halophilic bacteria in solar salt during storage. In: B. P. Eddy (Ed.) The microbiology of fish and meat curing brines: Proceedings of the 2nd international symposium on food microbiology. Her Majesty's Stationery Office. London, 13–19.

Dyall-Smith, M. L., and W. F. Doolittle. 1994. Construction of composite transposons for halophilic archaebacteria (Archaea). Can. J. Microbiol. 40:922–929.

Dyall-Smith, M. L. 1999. The Halohandbook: Protocols for halobacterial genetics. Version 2.9.

Ebel, C., P. Faou, B. Franzetti, B. Kernel, D. Madern, M. Pascu, C. Pfister, S. Richard, and G. Zaccai. 1999. Molecular interactions in extreme halophiles—the solvation–stabilization hypothesis for halophilic proteins. In: A. Oren (Ed.) Microbiology and biogeochemistry of hypersaline environments. CRC Press. Boca Raton, FL. 227–237.

Edgerton, M. E., and P. Brimblecombe. 1981. Thermodynamics of halobacterial environments. Can. J. Microbiol. 27:899–909.

Eimhjellen, K. 1965. Isolation of extremely halophilic bacteria. In: H. G. Schlegel (Ed.) Anreicherungskultur und Mutantenauslese: Supplementsheft 1. Zentralbl. Bakteriol. Parasitenkd. Infektionskr. Hyg. I Abt. Fischer Verlag. Stuttgart, 126–137.

Eisenberg, H., and E. J. Wachtel. 1987. Structural studies of halophilic proteins, ribosomes, and organelles of bacteria adapted to extreme salt concentrations. Ann. Rev. Biophys. Biophys. Chem. 16:69–92.

Eisenberg, H., M. Mevarech, and G. Zaccai. 1992. Biochemical, structural, and molecular genetic aspects of halophilism. Adv. Prot. Chem. 43:1–62.

Eisenberg, H. 1995. Life in unusual environments: progress in understanding the structure and function of enzymes from extreme halophilic bacteria. Arch. Biochem. Biophys. 318:1–5.

Elazari-Volcani, B. 1957. Genus XII. Halobacterium. In: R. S. Breed, E. G. D. Murray, and N. R. Smith (Eds.) Bergey's manual of determinative bacteriology, 7th ed. Williams & Wilkins. Baltimore, MD. 207–212.

Emerson, D., S. Chauhan, P. Oriel, and J. A. Breznak. 1994. Haloferax sp. D1227, a halophilic Archaeon capable of growth on aromatic compounds. Arch. Microbiol. 161:445–452.

Englert, C., M. Horne, and F. Pfeifer. 1990. Expression of the major gas vesicle protein gene in the halophilic archaebacterium Haloferax mediterranei is modulated by salt. Mol. Gen. Genet. 222:225–232.

Evans, R. W., S. C. Kushwaha, and M. Kates. 1980. The lipids of Halobacterium marismortui, an extremely halophilic bacterium in the Dead Sea. Biochim. Biophys. Acta 619:533–544.

Ewersmeyer-Wenk, B., H. Zähner, B. Krone, and A. Zeeck. 1981. Metabolic products of microorganisms: 207. Haloquinone, a new antibiotic active against halobacteria. I. Isolation, characterization and biological properties. J. Antibiot. 34:1531–1537.

Fernández-Castillo, R. F., F. Rodriguez-Valera, J. Gonzalez-Ramos, and F. Ruiz-Berraquero. 1986. Accumulation of poly (β-hydroxybutyrate) by halobacteria. Appl. Env. Microbiol. 51:214–216.

Fitch, W. M., and E. Margoliash. 1967. Construction of phylogenetic trees: a method based on mutation distances as estimated from cytochrome c sequences is of general applicability. Science 155:279–284.

Forterre, P., C. Elie, and M. Kohiyama. 1984. Aphidicolin inhibits growth and DNA synthesis in halophilic archaebacteria. J. Bacteriol. 159:800–802.

Franzmann, P. D., E. Stackebrandt, K. Sanderson, J. K. Volkman, D. E. Cameron, P. L. Stevenson, T. A. McMeekin, and H. R. Burton 1988. Halobacterium lacusprofundi sp. nov., a halophilic bacterium isolated from Deep Lake, Antarctica. Syst. Appl. Microbiol. 11:20–27.

Fredrickson, H. L., J. W. de Leeuw, A. C. Tas, J. van der Greef, G. F. LaVos, and J. J. Boon. 1989a. Fast atom bombardment (tandem) mass spectrometric analysis of intact polar ether lipids extractable from the extremely halophilic archaebacterium Halobacterium cutirubrum. Biomed. Environ. Mass Spectrom. 18:96–105.

Fredrickson, H. L., W. I. C. Rijpstra, A. C. Tas, J. van der Greef, G. F. LaVos, and J. W. de Leeuw. 1989b. Chemical characterization of benthic microbial assemblages. In: Y. Cohen, and E. Rosenberg (Eds.) Microbial mats: Physiological ecology of benthic microbial communities. American Society for Microbiology. Washington DC, 455–468.

Fu, W., and P. Oriel. 1998. Gentisate 1,2-dioxygenase from Haloferax sp. D1227. Extremophiles 3:45–53.

Fu, W., and P. Oriel. 1999. Degradation of 3-phenylpropionic acid by Haloferax sp. D1227. Extremophiles 2:439–446.

Galinski, E. A., and B. J. Tindall. 1992. Biotechnological prospects for halophiles and halotolerant microorganisms. In: R. A. Herbert, and R. J. Sharp (Eds.) Molecular biology and biotechnology of extremophiles. Chapman and Hall. New York, NY. 76–114.

Ghosh, M., and H. M. Sonawat. 1998. Kreb's cycle in Halobacterium salinarum investigated by ^{13}C nuclear magnetic resonance spectroscopy. Extremophiles 2:427–433.

Gibbons, N. E. 1969. Isolation, growth and requirements of halophilic bacteria. Methods Microbiol. 3B:169–184.

Gilboa-Garber, N., H. Mymon, and A. Oren. 1998. Typing of halophilic Archaea and characterization of their cell surface carbohydrate by use of lectins. FEMS Microbiol. Lett. 163:91–97.

Ginzburg, M., L. Sachs, and B. Z. Ginzburg. 1970. Ion metabolism in a Halobacterium: Part I. Influence of age of culture on the intracellular concentrations. J. Gen. Physiol. 55:187–207.

Gochnauer, M. B., S. C. Kushwaha, M. Kates, and D. J. Kushner. 1972. Nutritional control of pigment and isoprenoid compound formation in extremely halophilic bacteria. Arch. Mikrobiol. 84:339–349.

Gonzalez, C., C. Gutierrez, and C. Ramirez. 1978. Halobacterium vallismortis sp. nov., an amylolytic and carbohy-

drate metabolizing extremely halophilic bacterium. Can. J. Microbiol. 24:710–715.

Grant, W. D., and H. Larsen. 1989. Extremely halophilic archaeobacteria. Order Halobacteriales ord. nov. In: J. T. Staley, M. P. Bryant, N. Pfennig, and J. G. Holt (Eds.) Bergey's manual of systematic bacteriology. 3:Williams & Wilkins. Baltimore, MD. 2216–2233.

Grant, W. D., and H. N. M. Ross. 1986. The ecology and taxonomy of halobacteria. FEMS Microbiol. Rev. 39:9–15.

Grant, W. D., and B. J. Tindall. 1986. The alkaline saline environment. In: R. A. Herbert, and G. A. Codd (Eds.) Microbes in extreme environments. Academic Press. London, 25–54.

Grant, W. D., A. Oren, and A. Ventosa. 1998a. Proposal of strain NCIMB 13488 as neotype of Halorubrum trapanicum: request for an opinion. Int. J. Syst. Bacteriol. 48:1077–1078.

Grant, W. D., R. T. Gemmell, and T. J. McGenity. 1998b. Halobacteria: The evidence for longevity. Extremophiles 2:279–287.

Grey, V. L., and P. S. Fitt. 1976. An improved synthetic growth medium for Halobacterium cutirubrum. Can. J. Microbiol. 22:440–442.

Guixa-Boixareu, N., J. I. Caldéron-Paz, M. Heldal, G. Bratbak, and C. Pedrós-Alío. 1996. Viral lysis and bacterivory as prokaryotic loss factors along a salinity gradient. Aquat. Microb. Ecol. 11:215–227.

Gutiérrez, M. C., M. T. García, A. Ventosa, J. J. Nieto, and F. Ruiz-Berraquero. 1986. Occurrence of megaplasmids in halobacteria. J. Appl. Bacteriol. 61:67–71.

Hallberg, C., and H. Baltscheffsky. 1979. Partial purification of membrane-bound b-type cytochrome from Halobacterium halobium. Acta Chem. Scand. B 33:600–601.

Hallberg, C., and H. Baltscheffsky. 1981. Solubilization and separation of two b-type cytochromes from a carotenoid mutant of Halobacterium halobium. FEBS Lett. 125:201–204.

Hallberg-Gradin, C., and A. Colmsjö. 1989. Four different b-type cytochromes in the halophilic archaebacterium, Halobacterium halobium. Arch. Biochem. Biophys. 272:130–136.

Hamamoto, T., T. Takashina, W. D. Grant, and K. Horikoshi. 1988. Asymmetric cell division of a triangular halophilic archaebacterium. FEMS Microbiol. Lett. 56: 221–224.

Hamana, K., M. Kamekura, H. Onishi, T. Akazawa, and S. Matsuzaki. 1985. Polyamines in photosynthetic eubacteria and extreme-halophilic archaebacteria. J. Biochem. 97:1653–1658.

Hamana, K., H. Hamana, and T. Itoh. 1995. Ubiquitous occurrence of agmatine as the major polyamine within extremely halophilic archaebacteria. J. Gen. Appl. Microbiol. 41:153–158.

Hancock, A. J., and M. Kates. 1973. Structure determination of the phosphatidylglycerosulfate (diether analog) from Halobacterium cutirubrum. J. Lipid Res. 14:422–429.

Harrison, F. C., and M. E. Kennedy. 1922. The red discoloration of cured codfish. Trans. Roy. Soc. Canad. Sct. III 16:101–152.

Hartmann, R., H.-D. Sickinger, and D. Oesterhelt. 1980. Anaerobic growth of halobacteria. Proc. Natl. Acad. Sci. USA 77:3821–3825.

Hesselberg, M., and R. H. Vreeland. 1995. Utilization of protein profiles for the characterization of halophilic bacteria. Curr. Microbiol. 31:158–162.

Hilpert, R., J. Winter, W. Hammes, and O. Kandler. 1981. The sensitivity of archaebacteria to antibiotics. Zbl. Bakt. Hyg. I. Abt. Orig. C. 2:11–20.

Hirayama, J., Y. Imamoto, Y. Shichida, N. Kamo, H. Tomioka, and T. Yoshizawa. 1992. Photocycle of phoborhodopsin from haloalkaliphilic bacterium (Natronobacterium pharaonis) studied by low-temperature spectrophotometry. Biochemistry 31:2093–2098.

Hochstein, L. I., B. P. Dalton, and G. Pollock. 1976. The metabolism of carbohydrates by extremely halophilic bacteria: identification of galactonic acid as a product of galactose metabolism. Can. J. Microbiol. 22:1191–1196.

Hochstein, L. I. 1978. Carbohydrate metabolism in the extremely halophilic bacteria: The role of glucose in the regulation of citrate synthase activity. In: S. R. Caplan, and M. Ginzburg (Eds.) Energetics and structure of halophilic microorganisms. Elsevier. Amsterdam, 397–412.

Hochstein, L. I., and G. A. Tomlinson. 1985. Denitrification by extremely halophilic bacteria. FEMS Microbiol. Lett. 27:329–331.

Hochstein, L. I. 1988. The physiology and metabolism of the extremely halophilic bacteria. In: F. Rodriguez–Valera (Ed.) Halophilic bacteria. CRC Press. Boca Raton, FL. II:67–83.

Hochstein, L. I. 1991. Nitrate reduction in the extremely halophilic bacteria. In: F. Rodriguez–Valera (Ed.) General and applied aspects of halophilic microorganisms. Plenum Press. New York, NY. 129–137.

Hof, T. 1935. Investigations concerning bacterial life in strong brines. Rec. Trav. Bot. Néerl. 32:92–171.

Holmes, M. L., and M. L. Dyall-Smith. 1990. A plasmid vector with a selectable marker for halophilic archaebacteria. J. Bacteriol. 172:756–761.

Holmes, M. L., and M. L. Dyall-Smith. 1991. Mutations in DNA gyrase results in novobiocin resistance in halophilic archaebacteria. J. Bacteriol. 173:642–648.

Holmes, M. L., S. D. Nuttall, and M. L. Dyall-Smith. 1991. Construction and use of halobacterial shuttle vectors and further studies on Haloferax DNA gyrase. J. Bacteriol. 173:3807–3813.

Holmes, M., F. Pfeifer, and M. Dyall-Smith. 1994. Improved shuttle vectors for Haloferax volcanii including a dual-resistance plasmid. Gene 146:117–121.

Holmes, M. L., R. K. Scopes, R. L. Moritz, R. J. Simpson, C. Englert, F. Pfeifer, and M. L. Dyall-Smith. 1997. Purification and analysis of an extremely halophilic β-galactosidase from Haloferax alicantei. Biochim. Biophys. Acta 1337:276–286.

Holmes, M. L., and M. L. Dyall-Smith. 1999. Cloning, sequence and heterologous expression of bgaH, a beta-galactosidase gene of "Haloferax alicantei." In: A. Oren (Ed.) Microbiology and biogeochemistry of halophilic microorganisms. CRC Press. Boca Raton, FL. 265–271.

Hong, F. T. 1986. The bacteriorhodopsin model membrane system as a prototype molecular computing element. Biosystems 19:223–236.

Horikoshi, K., R. Aono, and S. Nakamura. 1993. The triangular halophilic archaebacterium Haloarcula japonica strain TR-1. Experientia 49:497–502.

Horne, M., and F. Pfeifer. 1989. Expression of two gas vacuole protein genes in Halobacterium halobium and other related species. Mol. Gen. Genet. 218:437–444.

Houwink, A. L. 1956. Flagella, gas vacuoles, and cell-wall structure in Halobacterium halobium: An electron microscope study. J. Gen. Microbiol. 15:146–150.

Hunter, M. I. S., and S. J. W. Millar. 1980. Effect of wall antibiotics on the growth of the extremely halophilic coccus, Sarcina marina NCMB 778. J. Gen. Microbiol. 120:255–258.

Hunter, M. I. S., T. L. Olawoye, and D. A. Saynor. 1981. The effect of temperature on the growth and lipid composition of the extremely halophilic coccus, Sarcina marina. Ant. v. Leeuwenhoek 47:25–40.

Ihara, K., S. Watanabe, and T. Tamura. 1997. Haloarcula argentinensis sp. nov. and Haloarcula mukohataei sp. nov., two new extremely halophilic archaea collected in Argentina. Int. J. Syst. Bacteriol. 47:73–77.

Javor, B., C. Requadt, and W. Stoeckenius. 1982. Box-shaped halophilic bacteria. J. Bacteriol. 151:1532–1542.

Javor, B. J. 1983. Planktonic standing crop and nutrients in a saltern ecosystem. Limnol. Oceanogr. 28:153–159.

Javor, B. 1984. Growth potential of halophilic bacteria isolated from solar salt environments: carbon sources and salt requirements. Appl. Env. Microbiol. 48:353–360.

Javor, B. 1988. CO_2 fixation in halobacteria. Arch. Microbiol. 149:433–440.

Javor, B. 1989. Hypersaline environments, microbiology and biogeochemistry. Springer-Verlag. Berlin, 362–369.

Jones, A. G., C. M. Ewing, and M. V. Melvin. 1981. Biotechnology of solar saltfields. Hydrobiologia 82:391–406.

Jones, D., P. A. Pell, and P. H. A. Sneath. 1984. Maintenance of bacteria on glass beads at −60°C to −70°C. In: B. E. Kirsop, and J. J. S. Snell (Eds.) Maintenance of microorganisms: A manual of laboratory methods. Academic Press. London, 35–40.

Joshi, J. G., W. R. Guild, and P. Handler. 1963. The presence of two species of DNA in some halobacteria. J. Mol. Biol. 6:34–38.

Juez, G., F., Rodriguez-Valera, A. Ventosa, and D. J. Kushner. 1986. Haloarcula hispanica spec. nov. and Haloferax gibbonsii spec. nov., two new species of extremely halophilic archaebacteria. Syst. Appl. Microbiol. 8:75–79.

Jukes, T. H., and C. R. Cantor. 1969. Evolution of protein molecules. In: H. N. Munro (Ed.) Mammalian protein metabolism. III:Academic Press. New York, NY. 21–132.

Kamekura, M., S. Bardocz, P. Anderson, R. Wallace, and D. J. Kushner. 1986. Polyamines in moderately and extremely halophilic bacteria. Biochim. Biophys. Acta 880:204–208.

Kamekura, M., and M. Kates. 1988. Lipids of halophilic archaebacteria. In: F. Rodriguez–Valera (Ed.) Halophilic bacteria. Vol. II. CRC Press. Boca Raton, FL. 25–54.

Kamekura, M., D. Oesterhelt, R. Wallace, P. Anderson, and D. J. Kushner. 1988. Lysis of halobacteria in bacto-peptone by bile acids. Appl. Env. Microbiol. 54:990–995.

Kamekura, M., and Y. Seno. 1991. Lysis of halobacteria with bile acids and proteolytic enzymes of halophilic archaeobacteria. In: F. Rodriguez–Valera (Ed.) General and applied aspects of halophilic microorganisms. Plenum Press. New York, NY. 359–365.

Kamekura, M. 1993. Lipids of extreme halophiles. In: R. H. Vreeland, and L. I. Hochstein (Eds.) The biology of halophilic bacteria. CRC Press. Boca Raton, FL. 135–161.

Kamekura, M., and Y. Seno. 1993. Partial sequence of the gene for a serine protease from a halophilic archaeum Haloferax mediterranei R4, and nucleotide sequences of 16S rRNA encoding genes from several halophilic archaea. Experientia 49:503–513.

Kamekura, M., and M. L. Dyall-Smith. 1995. Taxonomy of the family Halobacteriaceae and the description of two new genera Halorubrobacterium and Natrialba. J. Gen. Appl. Microbiol. 41:333–350.

Kamekura, M., M. L. Dyall-Smith, V. Upasani, A. Ventosa, and M. Kates. 1997. Diversity of alkaliphilic halobacteria: proposals for transfer of Natronobacterium vacuolatum, Natronobacterium magadii, and Natronobacterium pharaonis to Halorubrum, Natrialba, and Natronomonas gen. nov., respectively, as Halorubrum vacuolatum comb. nov., Natrialba magadii comb. nov., and Natronomonas pharaonis comb. nov., respectively. Int. J. Syst. Bacteriol. 47:853–857.

Kamekura, M. 1998. Diversity of extremely halophilic bacteria. Extremophiles 2:289–295.

Kamekura, M., Y. Seno, and H. Tomioka. 1998. Detection and expression of a gene encoding a new bacteriorhodopsin from an extreme halophile strain HT (JCM 9743) which does not possess bacteriorhodopsin activity. Extremophiles 2:33–39.

Kamekura, M. 1999. Diversity of members of the family Halobacteriaceae. In: A. Oren (Ed.) Microbiology and biogeochemistry of hypersaline environments. CRC Press. Boca Raton, FL. 13–25.

Kanai, H., T. Kobayashi, R. Aono, and T. Kudo. 1995. Natronococcus amylolyticus sp. nov., a haloalkaliphilic archaeon. Int. J. Syst. Bacteriol. 45:762–766.

Kandler, O., and K. König. 1993. Cell envelopes of archaea: structure and chemistry. In: M. Kates, D. J. Kushner, and A. T. Matheson (Eds.) The biochemistry of Archaea. Elsevier. Amsterdam, 223–259.

Kaplan, I. R., and A. Friedmann. 1970. Biological productivity in the Dead Sea: Part 1. Microorganisms in the water column. Israel J. Chem. 8:513–528.

Kates, M., and P. W. Deroo. 1973. Structure determination of the glycolipid sulphate from the extreme halophile Halobacterium cutirubrum. J. Lipid Res. 14:438–445.

Kates, M. 1978. The phytanyl ether-linked polar lipids and isoprenoid neutral lipids of extremely halophilic bacteria. Prog. Chem. Fats Lipids 15:301–342.

Kates, M., and N. Moldoveanu. 1991. Polar lipid structure, composition and biosynthesis in extremely halophilic bacteria. In: F. Rodriguez-Valera (Ed.) General and applied aspects of halophilic microorganisms. Plenum Press. New York, NY. 191–198.

Kates, M. 1993. Membrane lipids of extreme halophiles: biosynthesis, function and evolutionary significance. Experientia 49:1027–1036.

Kates, M., N. Moldoveanu, and L. C. Stewart. 1993. On the revised structure of the major phospholipid of Halobacterium salinarium. Biochim. Biophys. Acta 1169:46–53.

Kates, M., and S. C. Kushwaha. 1995. Isoprenoids and polar lipids of extreme halophiles. In: S. DasSarma, and E. M. Fleischmann (Eds.) Archaea: A laboratory manual. Halophiles. Cold Spring Harbor Laboratory Press. Cold Spring Harbor, NY. 35–54.

Kates, M. 1996. Structural analysis of phospholipids and glycolipids in extremely halophilic archaebacteria. J. Microbiol. Meth. 25:113–128.

Kauri, T., R. Wallace, and D. J. Kushner. 1990. Nutrition of the halophilic archaebacterium, Haloferax volcanii. Syst. Appl. Microbiol. 13:14–18.

Kelly, M., S. Norgård, and S. Liaaen-Jensen. 1970. Bacterial carotenoids: Part XXXI. C_{50} carotenoids 5. Carotenoids of Halobacterium salinarium, especially bacterioruberin. Acta Chem. Scand. 24:2169–2182.

Kessel, M., and Y. Cohen. 1982. Ultrastructure of square bacteria from a brine pool in southern Sinai. J. Bacteriol. 150:851–860.

Kessel, M., E. L. Buhle, Jr., S. Cohen, and U. Aebi. 1988a. The cell wall structure of a magnesium-dependent halobacterium, Halobacterium volcanii CD-2, from the Dead Sea. J. Ultrastruct. Mol. Struct. Res. 100:94–106.

Kessel, M., I. Wildhaber, S. Cohen, and W. Baumeister. 1988b. Three-dimensional structure of the regular surface glycoprotein layer of Halobacterium volcanii from the Dead Sea. EMBO J. 7:1549–1554.

Kevbrina, M. V., and V. K. Plakunov. 1992. Acetate metabolism in Natronococcus occultus. Microbiology 61:534–538.

Kirk, R. G., and M. Ginzburg. 1972. Ultrastructure of two species of Halobacterium. J. Ultrastr. Res. 41:80–94.

Kitajima, T., J. Hirayama, K. Ihara, Y. Sugiyama, N. Kamo, and Y. Mukohata. 1996. Novel bacterial rhodopsins from Haloarcula vallismortis. Biochem. Biophys. Res. Commun. 220:341–345.

Klebahn, H. 1919. Die Schädlinge des Klippfisches. Mitt. Inst. Allg. Botanik Hamburg 4:11–69.

Klöppel, K.-D., and H. L. Fredrickson. 1991. Fast atom bombardment mass spectrometry as a rapid means of screening mixtures of ether-linked polar lipids from extremely halophilic archaebacteria for the presence of novel chemical structures. J. Chromatogr. 562:369–376.

Kneifel, H., K. O. Stetter, J. R. Andreesen, J. Weigel, H. König, and S. M. Schoberth. 1986. Distribution of polyamines in representative species of archaebacteria. Syst. Appl. Microbiol. 7:241–245.

Kobayashi, T., M. Kanai, T. Hayashi, R. Akiba, R. Akaboshi, and K. Horikoshi. 1992. Haloalkaliphilic maltotriose-forming α-amylase from the archaebacterium Natronococcus strain Ah-3b. J. Bacteriol. 174:3439–3444.

Kocur, M., and W. Hogkiss. 1973. Taxonomic status of the genus Halococcus Schoop. Int. J. Syst. Bacteriol. 23:151–156.

König, H. 1988. Archaeobacteria. In: H. J. Rehm, and G. Reed (Eds.) Biotechnology. Verlag Chemie. Weinheim, 6B:699–728.

Kostrikina, N. A., I. S. Zvyagintseva, and V. I. Duda. 1991. Cytological pecularities of some extremely halophilic soil archaeobacteria. Arch. Microbiol. 156:344–349.

Krishnan, G., and W. Altekar. 1991. An unusual class I (Schiff base) fructose-1,6-bisphosphate aldolase from the halophilic archaebacterium Haloarcula vallismortis. Eur. J. Biochem. 195:343–350.

Kulichevskaya, I. S., E. I. Milekhina, I. A. Borezinkov, I. S. Zvyagintseva, and S. S. Belyaev. 1991. Oxidation of petroleum hydrocarbons by extremely halophilic archaebacteria. Microbiology 60:596–601.

Kulichevskaya, I. S., I. S. Zvyagintseva, A. L. Tarasov, and V. K. Plakunov. 1992. The extremely halophilic archaeobacteria from some hypersaline ecotops. Microbiology 46:51–56.

Kushner, D. J. 1964. Lysis and dissolution of cells and envelopes of an extremely halophilic bacterium. J. Bacteriol. 87:1147–1156.

Kushner, D. J., G. Mason, and N. E. Gibbons. 1965. Simple method for killing halophilic bacteria in contaminated solar salt. Appl. Microbiol. 13:288–288.

Kushner, D. J. 1966. Mass-culture of red halophilic bacteria. Biotechnol. Bioeng. 8:237–245.

Kushner, D. J. 1985. The Halobacteriaceae. *In:* C. R. Woese, and R. S. Wolfe (Eds.) The bacteria-a treatise on structure and function. Archaebacteria Academic Press. Orlando, FL. VIII:171–214.

Kushwaha, S. C., E. L. Pugh, J. K. G. Kramer, and M. Kates. 1972. Isolation and identification of dehydrosqualene and C_{40} carotenoid pigments in Halobacterium cutirubrum. Biochim. Biophys. Acta 260:492–506.

Kushwaha, S. C., M. B. Gochnauer, D. J. Kushner, and M. Kates. 1974. Pigments and isoprenoid compounds in extremely and moderately halophilic bacteria. Can. J. Microbiol. 20:241–245.

Kushwaha, S. C., J. K. G. Kramer, and M. Kates. 1975. Isolation and characterization of C_{50} carotenoid pigments and other polar isoprenoids from Halobacterium cutirubrum. Biochim. Biophys. Acta 398:303–313.

Kushwaha, S. C., M. Kates, and J. K. G. Kramer. 1977. Occurrence of indole in cells of extremely halophilic bacteria. Can. J. Microbiol. 23:826–828.

Kushwaha, S. C., and M. Kates. 1978. 2,3-Di-O-phytanyl-sn-glycerol and prenols from extremely halophilic bacteria. Phytochemistry 17:2029–2030.

Kushwaha, S. C., and M. Kates. 1979. Effect of glycerol on carotenogenesis in the extreme halophile, Halobacterium cutirubrum. Can. J. Microbiol. 25:1288–1291.

Kushwaha, S. C., G. Juez-Pérez, F. Rodriguez-Valera, M. Kates, and D. J. Kushner. 1982a. Survey of lipids of a new group of extremely halophilic bacteria from salt ponds in Spain. Can. J. Microbiol 28:1365–1372.

Kushwaha, S. C., M. Kates, G. Juez, F. Rodriguez-Valera, and D. J. Kushner. 1982b. Polar lipids of an extremely halophilic bacterial strain (R-4) isolated from salt ponds in Spain. Biochim. Biophys. Acta 711:19–25.

Lam, W. L., and W. F. Doolittle. 1989. Shuttle vector for the archaebacterium Halobacterium volcanii. Proc. Natl. Acad. Sci. USA 86:5478–5482.

Lanyi, J. K. 1974. Salt-dependent properties of proteins from extremely halophilic bacteria. Bacteriol. Rev. 38:272–290.

Lanyi, J. K. 1986. Halorhodopsin: a light-driven chloride ion pump. Ann. Rev. Biophys. Biophys. Chem. 15:11–28.

Lanyi, J. K. 1990. Halorhodopsin, a light-driven electrogenic chloride-transport system. Physiol. Rev. 70:319–330.

Lanyi, J. K., A. Duschl, G. W. Hatfield, K. May, and D. Oesterhelt. 1990. The primary structure of a halorhodopsin from Natronobacterium pharaonis. J. Biol. Chem. 265:1253–1260.

Lanyi, J. K. 1993a. Ion transport rhodopsins (bacteriorhodopsin and halorhodopsin): structure and function. *In:* M. Kates, D. J. Kushner, and A. T. Matheson (Eds.) The biochemistry of Archaea (Archaebacteria). Elsevier. Amsterdam, 189–207.

Lanyi, J. K. 1993b. Proton translocation mechanism and energetics in the light-driven pump bacteriorhodopsin. Biochim. Biophys. Acta 1183:241–261.

Lanyi, J. K., and G. Váró. 1995. The photocycles of bacteriorhodopsin. Israel J. Chem. 35:365–385.

Lanyi, J. K. 1997. Mechanism of ion transport across membranes. Bacteriorhodopsin as a prototype for proton pumps. J. Biol. Chem. 272:31209–31212.

Lanzotti, V., B. Nicolaus, A. Trincone, and W. D. Grant. 1988. The glycolipid of Halobacterium saccharovorum. FEMS Microbiol. Lett. 55:223–228.

Lanzotti, V., B. Nicolaus, A. Trincone, M. De Rosa, W. D. Grant, and A. Gambacorta 1989. A complex lipid with

a cyclic phosphate from the archaebacterium Natronococcus occultus. Biochim. Biophys. Acta 1001:31–34.

Larsen, H. 1962. Halophilism. *In:* I. C. Gunsalus, and R. Y. Stanier (Eds.) The bacteria. A treatise on structure and function. IV:Academic Press. New York, NY. 297–342.

Larsen, H. 1973. The halobacteria's confusion to biology. Ant. v. Leeuwenhoek 39:383–396.

Larsen, H. 1981. The family Halobacteriaceae. *In:* M. P. Starr, H. Stolp, H. G. Trüper, A. Balows, and H. G. Schlegel (Eds.) *The Prokaryotes:* A handbook on habitats, isolation, and identification of bacteria. I:Springer-Verlag. Berlin, 985–994.

Lechner, J., F. Wieland, and M. Sumper. 1985. Biosynthesis of sulfated saccharides N-glycosidically linked to the protein via glucose. J. Biol. Chem. 260:860–866.

Lechner, J., and M. Sumper. 1987. The primary structure of a procaryotic glycoprotein: Cloning and sequencing of the cell surface glycoprotein gene of halobacteria. J. Biol. Chem. 262:9724–9729.

Lechner, J., and F. Wieland. 1989. Structure and biosynthesis of prokaryotic glycoproteins. Ann. Rev. Biochem. 58:173–194.

Lillo, J. G., and F. Rodriguez-Valera. 1990. Effects of culture conditions on poly-β-hydroxybutyric acid production by Haloferax mediterranei. Appl. Environ. Microbiol. 56:2517–2521.

Lin, X., and R. H. White. 1987. Structure of sulfohalopterin-2 from Halobacterium marismortui. Biochemistry 26:6211–6217.

Lin, X., and R. H. White. 1988. Distribution of charged pterins in nonmethanogenic archaebacteria. Arch. Microbiol. 150:541–546.

Lochhead, A. G. 1934. Bacteriological studies on the red discoloration of salted hides. Canad. J. Res. 10:275–286.

Lodwick, D., T. J. McGenity, and W. D. Grant. 1994. The phylogenetic position of the haloalkaliphilic archaeon Natronobacterium magadii, determined from its 23S ribosomal RNA sequence. Syst. Appl. Microbiol. 17:402–404.

López-García, P., A. St. Jean, R. Amils, and R. L. Charlebois. 1995. Genomic stability in the archaeae Haloferax volcanii and Haloferax mediterranei. J. Bacteriol. 177:1405–1408.

Luecke, H., H.-T. Richter, and J. K. Lanyi. 1998. Proton transfer pathways in bacteriorhodopsin at 2.3 Angstrom resolution. Science 280:1934–1937.

Luehrsen, K. R., D. E. Nicholson, D. C. Eubanks, and G. E. Fox. 1981. An archaebacterial 5S rRNA contains a long insertion sequence. Nature 293:755–756.

Magrum, L. J., K. R. Luehrsen, and C. R. Woese. 1978. Are extreme halophiles actually "bacteria"? J. Mol. Evolut. 11:1–8.

Mancinelli, R., and L. I. Hochstein. 1986. The occurrence of denitrification in extremely halophilic bacteria. FEMS Microbiol. Lett. 35:55–58.

Martinez-Murcia, A. J., and F. Rodriguez-Valera. 1994. Random amplified polymorphic DNA of a group of halophilic archaeal isolates. Syst. Appl. Microbiol. 17:395–401.

Martínez-Murcia, A. J., I. F. Boán, and F. Rodríguez-Valera. 1995a. Evaluation of the authenticity of haloarchaeal strains by random-amplified polymorphic DNA. Lett. Appl. Microbiol. 21:106–108.

Martínez-Murcia, A. J., S. C. Acinas, and F. Rodríguez-Valera. 1995b. Evaluation of prokaryotic diversity by restrictase digestion of 16S rDNA directly amplified from hypersaline environments. FEMS Microbiol. Ecol. 17:247–256.

Marwan, W., M. Alam, and D. Oesterhelt. 1987. Die Geißelbewegung halophiler Bakterien. Naturwissenschaften 74:585–590.

Marwan, W., M. Alam, and D. Oesterhelt. 1991. Rotation and switching of the flagellar motor assembly in Halobacterium halobium. J. Bacteriol. 173:1971–1977.

Matheson, A. T., G. D. Sprott, I. J. McDonald, and H. Tessier. 1976. Some properties of an unidentified halophile: growth characteristics, internal salt concentration, and morphology. Can. J. Microbiol. 22:780–786.

Matsubara, T., N. Iida-Tanaka, M. Kamekura, N. Moldoveanu, I. Ishizuka, H. Onishi, A. Hayashi, and M. Kates. 1994. Polar lipids of a non-alkaliphilic extremely halophilic archaebacterium strain 172: a novel bissulfated glycolipid. Biochim. Biophys. Acta 1214:97–108.

McGenity, T. J., and W. D. Grant. 1993. The haloalkaliphilic archaeon (archaebacterium) Natronococcus occultus represents a distant lineage within the Halobacteriales, most closely related to the other haloalkaliphilic lineage (Natronobacterium). Syst. Appl. Microbiol. 16:239–243.

McGenity, T. J., and W. D. Grant. 1995. Transfer of Halobacterium saccharovorum, Halobacterium sodomense, Halobacterium trapanicum NRC 34021 and Halobacterium lacusprofundi to the genus Halorubrum gen. nov. as Halorubrum saccharovorum comb. nov., Halorubrum sodomense comb. nov., Halorubrum trapanicum comb. nov., and Halorubrum lacusprofundi comb. nov. Syst. Appl. Microbiol. 18:237–243.

McGenity, T. J., R. T. Gemmell, and W. D. Grant. 1998. Proposal of a new halobacterial genus Natrinema gen. nov., with two species Natrinema pellirubrum nom. nov. and Natrinema pallidum nom. nov. Int. J. Syst. Bacteriol. 48:1187–1196.

Mengele, R., and M. Sumper. 1992. Drastic differences in glycosylation of related S-layer glycoproteins from moderate and extreme halophiles. J. Biol. Chem. 267:8182–8185.

Mescher, M. F., J. L. Strominger, and S. W. Watson. 1974. Protein and carbohydrate composition of the cell envelope of Halobacterium salinarium. J. Bacteriol. 120:945–954.

Mescher, M. F., and J. L. Strominger. 1976a. Purification and characterization of a prokaryotic glycoprotein from the cell envelope of Halobacterium salinarium. J. Biol. Chem. 251:2005–2014.

Mescher, M. F., and J. L. Strominger. 1976b. Structural (shape-maintaining) role of the cell surface glycoprotein of Halobacterium salinarium. Proc. Natl. Acad. Sci. USA 73:2687–2691.

Mescher, M. 1981. Glycoproteins as cell-surface structural components. Trends Biochem. Sci. 6:97–99.

Meseguer, I., and F. Rodriguez-Valera. 1986. Effect of halocin H4 on cells of Halobacterium halobium. J. Gen. Microbiol. 132:3061–3068.

Meseguer, I., F. Rodriguez-Valera, and A. Ventosa. 1986. Antagonistic interactions among halobacteria due to halocin production. FEMS Microbiol. Lett. 36:177–182.

Meseguer, I., M. Torreblanca, and F. Rodriguez-Valera. 1991. Mode of action of halocins H4 and H6: are they effective against the adaptation to high salt environments? In: F. Rodriguez-Valera (Ed.) General and applied aspects of halophilic microorganisms. Plenum Press. New York, NY. 157–164.

Mevarech, M., and R. Werczberger. 1985. Genetic transfer in Halobacterium volcanii. J. Bacteriol. 162:461–462.

Mohr, V., and H. Larsen. 1963. On the structural transformations and lysis of Halobacterium salinarium in hypotonic and isotonic solutions. J. Gen. Microbiol. 31:267–280.

Moldoveanu, N., and M. Kates. 1988. Biosynthetic studies of the polar lipids of Halobacterium cutirurbum. Biochim. Biophys. Acta 960:161–182.

Moldoveanu, M., M. Kates, C. G. Montero, and A. Ventosa. 1990. Polar lipids of non-alkaliphilic Halococci. Biochim. Biophys. Acta 1046:127–135.

Montalvo-Rodríguez, R., R. H. Vreeland, A. Oren, M. Kessel, C. Betancourt, and J. López-Garriga. 1998. Halogeometricum borinquense gen. nov., sp. nov., a novel halophilic Archaeon from Puerto Rico. Int. J. Syst. Bacteriol. 48:1305–1312.

Montero, C. G., A. Ventosa, F. Rodriguez-Valera, and F. Ruiz-Berraquero. 1988. Taxonomic study of non-alkaliphilic halococci. J. Gen. Microbiol. 134:725–732.

Montero, C. G., A. Ventosa, F. Rodriguez-Valera, M. Kates, N. Moldoveanu, and F. Ruiz-Berraquero. 1989. Halococcus saccharolyticus sp. nov., a new species of extremely halophilic non-alkaliphilic cocci. Syst. Appl. Microbiol. 12:167–171.

Moore, R. L., and B. J. McCarthy. 1969. Characterization of the deoxyribonucleic acid of various strains of halophilic bacteria. J. Bacteriol. 99:248–254.

Morita, M., N. Yamaguchi, T. Eguchi, and K. Kakinuma. 1998. Structural diversity of the membrane core lipids of extreme halophiles. Biosci. Biotechnol. Biochem. 62:596–598.

Moritz, A., and W. Goebel. 1985. Characterization of the 7S RNA and its gene from halobacteria. Nucl. Acid Res. 13:6969–6979.

Moritz, A., B. Lankat-Buttgereit, H. J. Gross, and W. Goebel. 1985. Common structural features of the genes for two stable RNAs from Halobacterium halobium. Nucl. Acid Res. 13:31–43.

Morth, S., and B. J. Tindall. 1985a. Variation of polar lipid composition within haloalkaliphilic archaebacteria. Syst. Appl. Microbiol. 6:247–250.

Morth, S., and B. J. Tindall. 1985b. Evidence that changes in the growth conditions affect the relative distribution of diether lipids in haloalkaliphilic archaebacteria. FEMS Microbiol. Lett. 29:285–288.

Mukohata, Y., Y. Sugiyama, K. Ihara, and M. Yoshida. 1988. An Australian halobacterium contains a novel proton pump retinal protein: Archaerhodopsin. Biochem. Biophys. Res. Commun. 151:1339–1345.

Mukohata, Y., K. Ihara, K. Uegaki, Y. Miyashita, and Y. Sugiyama. 1991. Australian halobacteria and their retinal-protein ion pumps. Photochem. Photobiol. 54:1039–1045.

Mukohata, Y. 1994. Comparative studies on ion pumps of the bacterial rhodopsin family. Biophys. Chem. 50:191–201.

Mullakhanbhai, M. F., and G. W. Francis. 1972. Bacterial lipids: I. Lipid composition of a moderately halophilic bacterium. Acta Chem. Scand. 26:1399–1410.

Mullakhanbhai, M. F., and H. Larsen. 1975. Halobacterium volcanii sp. nov., a Dead Sea halobacterium with a moderate salt requirement. Arch. Microbiol. 104:207–214.

Mwatha, W. E., and W. D. Grant. 1993. Natronobacterium vacuolata, a haloalkaliphilic archaeon isolated from Lake Magadi, Kenya. Int. J. Syst. Bacteriol. 43:401–404.

Mylvaganam, S., and P. P. Dennis. 1992. Sequence heterogeneity between the two genes encoding 16S rRNA from the halophilic archaebacterium Haloarcula marismortui. Genetics 130:399–410.

Nakamura, S., R. Aono, S. Mizutani, T. Takashina, W. D. Grant, and K. Horikoshi. 1992. The cell surface glycoprotein of Haloarcula japonica TR-1. Biosci. Biotechnol. Biochem. 56:996–998.

Newton, G. L., and B. Javor. 1985. γ-glutamylcysteine and thiosulfate are the major low-molecular-weight thiols in halobacteria. J. Bacteriol. 161:438–441.

Ng, W. V., S. A. Ciufo, T. M. Smith, R. E. Bumgarner, D. Baskin, J. Faust, B. Hall, C. Loretz, J. Seto, J. Slagel, L. Hood, and S. DasSarma. 1998. Snapshot of a large dynamic replicon in a halophilic archaeon: Megaplasmid or minichromosome? Genome Res. 8:1131–1141.

Nicolaus, B., V. Lanzotti, A. Tricone, M. De Rosa, W. D. Grant, and A. Gambacorta. 1989. Glycine betaine and polar lipid composition in halophilic archaebacteria in response to growth in different salt concentrations. FEMS Microbiol. Lett. 59:157–160.

Niemetz, R., U. Kärcher, O. Kandler, B. J. Tindall, and H. König. 1997. The cell wall polymer of the extremely halphilic archaeon Natronococcus occultus. Eur. J. Biochem. 249:905–911.

Nishiyama, Y., T. Takashina, W. D. Grant, and K. Horikoshi. 1992. Ultrastructure of the cell wall of the triangular halophilic archaebacterium Haloarcula japonica strain TR-1. FEMS Microbiol. Lett. 99:43–48.

Norton, C. F., and W. D. Grant. 1988. Survival of halobacteria within fluid inclusions in salt crystals. J. Gen. Microbiol. 134:1365–1373.

Norton, C. F., T. J. McGenity, and W. D. Grant. 1993. Archaeal halophiles (halobacteria) from two British salt mines. J. Gen. Microbiol. 139:1077–1081.

Nuttall, S. D., and M. L. Dyall-Smith. 1993a. Ch2, a novel archaeon from an Australian solar saltern. Int. J. Syst. Bacteriol. 43:729–734.

Nuttall, S. D., and M. L. Dyall-Smith. 1993b. HF1 and HF2: novel bacteriophages of halophilic archaea. Virology 197:678–684.

Oesterhelt, D., and W. Stoeckenius. 1971. Rhodopsin-like protein from the purple membrane of Halobacterium halobium. Nature 233:149–152.

Oesterhelt, D. 1982. Anaerobic growth of halobacteria. Meth. Enzymol. 88:417–420.

Oesterhelt, D., and G. Krippahl. 1983. Phototrophic growth of halobacteria and its use for isolation of photosynthetically-deficient mutants. Ann. Microbiol. (Inst. Pasteur) 134B:137–150.

Oesterhelt, D., C. Bräuchle, and A. Hampp. 1991. Bacteriorhodopsin: A biological material for information processing. Quart. Rev. Biophys. 24:425–478.

Oesterhelt, D. 1995. Structure and function of halorhodopsin. Israel J. Chem. 35:475–494.

Offner, S., U. Ziese, G. Wanner, D. Typke, and F. Pfeifer. 1998. Structural characteristics of halobacterial gas vesicles. Microbiology UK 144:1331–1342.

Onishi, H., M. E. McCance, and N. E. Gibbons. 1965. A synthetic medium for extremely halophilic bacteria. Can. J. Microbiol. 11:365–373.

Onishi, H., T. Kobayashi, S. Iwao, and M. Kamekura. 1985. Archaebacterial diether lipids in a non-alkalophilic, non-pigmented extremely halophilic bacterium. Agric. Biol. Chem. 49:3053–3055.

Oren, A., and M. Shilo. 1981. Bacteriorhodopsin in a bloom of halobacteria in the Dead Sea. Arch. Microbiol. 130:185–187.

Oren, A. 1983a. Halobacterium sodomense sp. nov., a Dead Sea halobacterium with an extremely high magnesium requirement. Int. J. Syst. Bacteriol. 33:381–386.

Oren, A. 1983b. Bacteriorhodopsin mediated CO_2 photoassimilation in the Dead Sea. Limnol. Oceanogr. 28:33–41.

Oren, A. 1983c. Population dynamics of halobacteria in the Dead Sea water column. Limnol. Oceanogr. 28:1094–1103.

Oren, A., and M. Shilo. 1985. Factors determining the development of algal and bacterial blooms in the Dead Sea: a study of simulation experiments in outdoor ponds. FEMS Microbiol. Ecol. 31:229–237.

Oren, A., P. P. Lau, and G. E. Fox. 1988. The taxonomic status of "Halobacterium marismortui" from the Dead Sea: a comparison with Halobacterium vallismortis. Syst. Appl. Microbiol. 10:251–258.

Oren, A. 1990a. Starch counteracts the inhibitory action of Bacto-peptone and bile salts in media for the growth of halobacteria. Can. J. Microbiol. 36:299–301.

Oren, A. 1990b. The use of protein synthesis inhibitors in the estimation of the contribution of halophilic archaebacteria to bacterial activity in hypersaline environments. FEMS Microbiol. Ecol. 73:187–192.

Oren, A., and H. G. Trüper. 1990. Anaerobic growth of halophilic archaeobacteria by reduction of dimethylsulfoxide and trimethylamine N-oxide. FEMS Microbiol. Lett. 70:33–36.

Oren, A., M., Ginzburg, B. Z. Ginzburg, L. I. Hochstein, and B. E. Volcani. 1990. Haloarcula marismortui (Volcani) sp. nov., nom., rev., an extremely halophilic bacterium from the Dead Sea. Int. J. Syst. Bacteriol. 40:209–210.

Oren, A. 1991. Anaerobic growth of halophilic archaeobacteria by reduction of fumarate. J. Gen. Microbiol. 137:1387–1390.

Oren, A., N. Stambler, and Z. Dubinsky. 1992. On the red coloration of saltern crystallizer ponds. Int. J. Salt Lake Res. 1:77–89.

Oren, A. 1993a. Ecology of extremely halophilic microorganisms. In: R. H. Vreeland. and L. I. Hochstein (Eds.) The biology of halophilic bacteria. CRC Press. Boca Raton, FL. 25–53.

Oren, A., and P. Gurevich. 1993. Characterization of the dominant halophilic archaea in a bacterial bloom in the Dead Sea. FEMS Microbiol. Ecol. 12:249–256.

Oren, A. 1994a. The ecology of the extremely halophilic archaea. FEMS Microbiol. Rev. 13:415–440.

Oren, A. 1994b. Enzyme diversity in halophilic archaea. Microbiología SEM 10:217–228.

Oren, A. 1994c. Characterization of the halophilic archaeal community in saltern crystallizer ponds by means of polar lipid analysis. Int. J. Salt Lake Res. 3:15–29.

Oren, A., and Z. Dubinsky. 1994. On the red coloration of saltern crystallizer ponds: II. Additional evidence for the contribution of halobacterial pigments. Int. J. Salt Lake Res. 3:9–13.

Oren, A., and P. Gurevich. 1994. Production of D-lactate, acetate, and pyruvate from glycerol in communities of halophilic archaea in the Dead Sea and in saltern crystallizer ponds. FEMS Microbiol. Ecol. 14:147–156.

Oren, A. 1995a. The role of glycerol in the nutrition of halophilic archaeal communities: A study of respiratory electron transport. FEMS Microbiol. Ecol. 16:281–290.

Oren, A. 1995b. Uptake and turnover of acetate in hypersaline environments. FEMS Microbiol. Ecol. 18:75–84.

Oren, A., and P. Gurevich. 1995. Occurrence of the methylglyoxal bypass in halophilic Archaea. FEMS Microbiol. Lett. 125:83–88.

Oren, A., P. Gurevich, R. T. Gemmell, and A. Teske. 1995. Halobaculum gomorrense gen. nov., sp. nov., a novel extremely halophilic archaeon from the Dead Sea. Int. J. Syst. Bacteriol. 45:747–754.

Oren, A. 1996a. Sensitivity of selected members of the Halobacteriaceae to quinolone antimicrobial compounds. Arch. Microbiol. 165:354–358.

Oren, A., and A. Ventosa. 1996b. A proposal for the transfer of Halorubrobacterium distributum and Halorubrobacterium coriense to the genus Halorubrum as Halorubrum distributum comb. nov. and Halorubrum coriense comb. nov., respectively. Int. J. Syst. Bacteriol. 46:1180–1180.

Oren, A., S. Duker, and S. Ritter. 1996. The polar lipid composition of Walsby's square bacterium. FEMS Microbiol. Lett. 138:135–140.

Oren, A., A. Ventosa, and W. D. Grant. 1997a. Proposed minimal standards for description of new taxa in the order Halobacteriales. Int. J. Syst. Bacteriol. 47:233–238.

Oren, A., M. Kamekura, and A. Ventosa. 1997b. Confirmation of strain VKM B-1733 as the type strain of Halorubrum distributum. Int. J. Syst. Bacteriol. 47:231–232.

Oren, A., G. Bratbak, and M. Heldal. 1997c. Occurrence of virus-like particles in the Dead Sea. Extremophiles 1:143–149.

Oren, A. 1999a. Prokaryotic life at high salt concentrations. In: M. Dworkin, S. Falkow, E. Rosenberg, K.-H. Schleifer, and E. Stackebrandt (Eds.) The Prokaryotes. Springer-Verlag. New York, NY.

Oren, A. 1999b. The enigma of square and triangular halophilic archaea. In: J. Seckbach (Ed.) Enigmatic microorganisms and life in extreme environmental habitats. Kluwer Academic Publishers. Dordrecht, 337–355.

Oren, A., and C. D. Litchfield. 1999. A procedure for the enrichment and isolation of Halobacterium. FEMS Microbiol. Lett. 173:353–358.

Oren, A., A. Ventosa, M. C. Gutiérrez, and M. Kamekura. 1999. Haloarcula quadrata sp. nov., a square, motile Haloarcula species from a brine pool in Sinai (Egypt). Int. J. Syst. Bacteriol. 49:1149–1155.

Oren, A. 2000. Estimation of the contribution of halobacteria to the bacterial biomass and activity in solar salterns by the use of bile salts. FEMS Microbiol. Ecol. 73:41–48.

Ortenberg, R., R. Tchelet, and M. Mevarech. 1999. A model for the genetic exchange system of the extremely halophilic archaeon Haloferax volcanii. In: A. Oren (Ed.) Microbiology and biogeochemistry of hypersaline environments. CRC Press. Boca Raton, FL. 331–338.

Paramonov, N. A., L. A. S. Parolis, H. Parolis, I. F. Boán, J. Antón, and F. Rodríguez-Valera. 1998. The structure of the exocellular polysaccharide produced by the archaeon Haloferax gibbonsii (ATCC 33959). Carbohydr. Res. 309:89–94.

Parkes, K., and A. E. Walsby. 1981. Ultrastructure of a gas-vacuolate square bacterium. J. Gen. Microbiol 126:503–506.

Parolis, H., L. A. S. Parolis, I. F. Boán, F. Rodríguez-Valera, G. Widmalm, M. C. Manca, P.-E. Jansson, and I. W. Sutherland. 1996. The structure of the exopolysaccharide produced by the halophilic archaeon Haloferax

mediterranei strain R4 (ATCC 33500). Carbohydr. Res. 295:147–156.

Paul, G., F. Lottspeich, and F. Wieland. 1986. Asparaginyl-N-acetylgalactosamine. Linkage unit of halobacterial glycosaminoglycan. J. Biol. Chem. 261:1020–1024.

Paul, G., and F. Wieland. 1987. Sequence of the halobacterial glycosaminoglycan. J. Biol. Chem. 262:9587–9593.

Payne, J. I., S. N. Sehgal, and N. E. Gibbons. 1960. Immersion refractometry of some halophilic bacteria. Can. J. Microbiol. 6:9–15.

Pecher, T., and A. Böck. 1981. In vivo susceptibility of halophilic and methanogenic organisms to protein synthesis inhibitors. FEMS Microbiol. Lett. 10:295–297.

Peijin, Z., X. Yi, X. Changsong, M. Yunqing, and L. Hongdi. 1994. New species of Haloarcula. Acta Microbiol. Sin. (in Chinese) 34:89–95.

Peijin, Z., and X. Yi. 1996. Phylogenetic position of a halophilic archaeon Haloarcula aidinensis B-2, determined from its 16S rRNA sequence. Japan Collection of Microorganisms.

Pérez-Fillol, M., and F. Rodriguez-Valera. 1986. Potassium ion accumulation in cells of different halobacteria. Microbiología SEM 2:73–80.

Petter, H. F. M. 1931. On bacteria of salted fish. Proc. Konink. Akad. Wet. Amsterdam. Ser. B 34:1417–1423.

Pfeifer, F., G. Weidinger, and W. Goebel. 1981. Genetic variability in Halobacterium halobium. J. Bacteriol. 145:375–381.

Pfeifer, F. 1986. Insertion elements and genome organization of Halobacterium halobium. Syst. Appl. Microbiol. 7:36–40.

Pfeifer, F. 1988. Genetics of halobacteria. In: F. Rodriguez-Valera (Ed.) Halophilic bacteria. CRC Press. Boca Raton, FL. II:105–133.

Pfeifer, F., and C. Englert. 1992. Function and biosynthesis of gas vesicles in halophilic Archaea. J. Bioenerg. Biomembr. 24:577–585.

Pfeifer, F., K. Krüger, R. Röder, A. Mayr, S. Ziesche, and S. Offner. 1997. Gas vesicle formation in halophilic archaea. Arch. Microbiol. 167:257–268.

Post, F. J. 1977. The microbial ecology of the Great Salt Lake. Microbial. Ecol. 3:143–165.

Post, F. J. 1981. Microbiology of the Great Salt Lake north arm. Hydrobiologia 81:59–69.

Post, F. J., and N. F. Collins. 1982. A preliminary investigation of the membrane lipid of Halobacterium halobium as a food additive. J. Food Biochem. 6:25–38.

Post, F. J., and F. A. Al-Harjan. 1988. Surface activity of halobacteria and potential use in microbially enhanced oil recovery. Syst. Appl. Microbiol. 11:97–101.

Pugh, E. L., M. K. Wassef, and M. Kates. 1971. Inhibition of fatty acid synthetase in Halobacterium cutirubrum and Escherichia coli by high salt concentrations. Can. J. Biochem. 49:953–958.

Pugh, E. L., and M. Kates. 1994. Acylation of proteins of the archaebacteria Halobacterium cutirubrum and Methanobacterium thermoautotrophicum. Biochim. Biophys. Acta 1196:38–44.

Quesada, E., A. Ventosa, F. Rodriguez-Valera, and A. Ramos-Cormenzana. 1982. Types and properties of some bacteria isolated from hypersaline soils. J. Appl. Bacteriol. 53:155–161.

Rajagopalan, R., and W. Altekar. 1991. Products of non-reductive CO_2 assimilation in the halophilic archaebacterium Haloferax mediterranei. Indian J. Biochem. Biophys. 28:65–67.

Rajagopalan, R., and W. Altekar. 1994. Characterisation and purification of ribulose-bisphosphate carboxylase from heterotrophically grown halophilic archaebacterium, Haloferax mediterranei. Eur. J. Biochem. 221:863–869.

Rawal, N., S. M. Kelkar, and W. Altekar. 1988a. Alternative routes of carbohydrate metabolism in halophilic archaebacteria. Indian J. Biochem. Biophys. 25:674–686.

Rawal, N., S. M. Kelkar, and W. Altekar. 1988b. Ribulose 1,5-bisphosphate dependent CO_2 fixation in the halophilic archaebacterium, Halobacterium mediterranei. Biochem. Biophys. Res. Commun. 156:451–456.

Röder, R., and F. Pfeifer. 1996. Influence of salt on the transcription of the gas-vesicle gene of Haloferax mediterranei and identification of the endogenous transcriptional activator. Microbiology UK 142:1715–1723.

Rodriguez-Valera, F., F. Ruiz-Berraquero, and A. Ramos-Cormenzana. 1979. Isolation of extreme halophiles from seawater. Appl. Env. Microbiol. 38:164–165.

Rodriguez-Valera, F., F. Ruiz-Berraquero, and A. Ramos-Cormenzana. 1980a. Isolation of extremely halophilic bacteria able to grow in defined organic media with single carbon sources. J. Gen. Microbiol. 119:535–538.

Rodriguez-Valera, F., F. Ruiz-Berraquero, and A. Ramos-Cormenzana. 1980b. Behaviour of mixed populations of halophilic bacteria in continuous cultures. Can. J. Microbiol. 26:1259–1263.

Rodriguez-Valera, F., G. Juez, and D. J. Kushner. 1982. Halocins: Salt dependent bacteriocins produced by extremely halophilic rods. Can. J. Microbiol. 28:151–154.

Rodriguez-Valera, F., G. Juez, and D. J. Kushner. 1983a. Halobacterium mediterranei spec. nov., a new carbohydrate-utilising extreme halophile. Syst. Appl. Microbiol. 4:369–381.

Rodriguez-Valera, F., J. J. Nieto, and F. Ruiz-Berraquero. 1983b. Light as an energy source in continuous cultures of bacteriorhodopsin-containing halobacteria. Appl. Env. Microbiol. 45:868–871.

Rodriguez-Valera, F., A. Ventosa, G. Juez, and J. F. Imhoff. 1985. Variation of environmental features and microbial populations with salt concentrations in a multi-pond saltern. Microb. Ecol. 11:107–115.

Rodriguez-Valera, F., J. A. G. Lillo, J. Antón, and I. Meseguer. 1991. Biopolymer production by Haloferax mediterranei. In: F. Rodriguez-Valera (Ed.) General and applied aspects of halophilic microorganisms. Plenum Press. New York, NY. 373–380.

Rodriguez-Valera, F. 1992. Biotechnological potential of halobacteria. In: M. J. Danson, D. W. Hough, and G. G. Lunt (Eds.) Archaebacteria: biochemistry and biotechnology: Biochemical Society Symposium no. 58. Biochemical Society. High Holburn, London. 135–147.

Rodriguez-Valera, F., and J. A. G. Lillo. 1992. Halobacteria as producers of polyhydroxyalkanoates. FEMS Microbiol. Rev. 103:181–186.

Rodriguez-Valera, F. 1995. Cultivation of halophilic Archaea. In: S. DasSarma, and E. M. Fleischmann (Eds.) Archaea: A laboratory manual. Halophiles. Cold Spring Harbor Laboratory Press. Cold Spring Harbor, NY. 13–16.

Rodríguez-Valera, F., S. Acinas, and J. Antón. 1999. Contribution of molecular techniques to the study of microbial diversity in hypersaline environments. In: A. Oren (Ed.) Microbiology and biogeochemistry of hypersaline environments. CRC Press. Boca Raton, FL. 27–38.

Rogers, P. J., and C. A. Morris. 1978. Regulation of bacteriorhodopsin synthesis by growth rate in continuous cul-tures of Halobacterium halobium. Arch. Microbiol. 119:323–325.

Romanenko, V. I. 1981. Square microcolonies in the surface water film of the Saxkoye lake. Mikrobiologiya (in Russian) 50:571–574.

Rønnekleiv, M., M. Lenes, S. Norgård, and S. Liaaen-Jensen. 1995. Three dodecaene C_{50}-carotenoids from halophilic bacteria. Phytochemistry 39:631–634.

Rosenshine, I., R. Tchelet, and M. Mevarech. 1989. The mechanism of DNA transfer in the mating system of an archaebacterium. Science 245:1387–1389.

Rosenshine, I., and M. Mevarech. 1991. The kinetic of the genetic exchange process in Halobacterium volcanii mating. In: F. Rodriguez-Valera (Ed.) General and applied aspects of halophilic microorganisms. Plenum Press. New York, NY. 265–270.

Ross, H. N. M., M. D. Collins, B. J. Tindall, and W. D. Grant. 1981. A rapid procedure for the detection of archaebacterial lipids in halophilic bacteria. J. Gen. Microbiol. 123:75–80.

Ross, H. N. M., and W. D. Grant. 1985. Nucleic acid studies on halophilic archaebacteria. J. Gen. Microbiol. 131:165–173.

Ruepp, A., and J. Soppa. 1996. Fermentative arginine degradation in Halobacterium salinarium (formerly Halobacterium halobium): Genes, gene products, and transcripts of the arcRACB gene cluster. J. Bacteriol. 178:4942–4947.

Sakane, T., I. Fukuda, T. Itoh, and A. Yokota. 1992. Long-term preservation of halophilic archaebacteria and thermoacidophilic archaebacteria by liquid drying. J. Microbiol. 16:281–287.

Sapienza, C., and W. F. Doolittle. 1982. Unusual physical organization of the Halobacterium genome. Nature 295:384–389.

Schalkwyk, L. C. 1993. Halobacterial genes and genomes. In: M. Kates, D. J. Kushner, and A. T. Matheson (Eds.) The biochemistry of Archaea (Archaebacteria). Elsevier. Amsterdam, 467–496.

Scharf, B., and M. Engelhard. 1993. Halocyanin, an archaebacterial blue copper protein (type I) from Natronobacterium pharaonis. Biochemistry 32:12894–12900.

Scharf, B., R. Wittenberg, and M. Engelhard. 1997. Electron transfer proteins from the haloalkaliphilic archaeon Natronobacterium pharaonis: Possible components of the respiratory chain include cytochrome bc and a terminal oxidase cytochrome ba_3. Biochemistry 36:4471–4479.

Schinzel, R., and K. J. Burger. 1984. Sensitivity of halobacteria to aphidicolin, an inhibitor of eukaryotic α-type DNA polymerases. FEMS Microbiol. Lett. 25:187–190.

Schinzel, R., and K. J. Burger. 1986. A site-specific endonuclease activity in Halobacterium halobium. FEMS Microbiol. Lett. 37:325–329.

Schleifer, K. H., J. Steber, and H. Mayer. 1982. Chemical composition and structure of the cell wall of Halococcus morrhuae. Zbl. Bakt. Hyg. I. Abt. Orig. C. 3:171–178.

Schobert, B., and J. K. Lanyi. 1982. Halorhodopsin is a light-driven chloride pump. J. Biol. Chem. 257:10306–10313.

Seidel, R., B. Scharf, M. Gautel, K. Kleine, D. Oesterhelt, and M. Engelhard. 1995. The primary structure of sensory rhodopsin II: A member of an additional retinal protein subgroup is coexpressed with its transducer, the halobacterial transducer of rhodopsin II. Proc. Natl. Acad. Sci. USA 92:3036–3040.

Serrano, J. A., M. Camacho, and M. J. Bonete. 1998. Operation of glyoxylate cycle in halophilic archaea: Presence of malate synthase and isocitrate lyase in Haloferax volcanii. FEBS Lett. 434:13–16.

Severina, L. O., I. A. Usenko, and V. K. Plakunov. 1989. Biosynthesis of an exopolysaccharide by the extreme halophilic archaebacterium, Halobacterium mediterranei. Microbiology 58:441–445.

Severina, L. O., I. A. Usenko, and V. K. Plakunov. 1990. Biosynthesis of an exopolysaccharide by the extreme halophilic archaebacterium, Halobacterium volcanii. Microbiology 59:292–295.

Shand, R. F., and M. C. Betlach. 1991. Expression of the bop gene cluster of Halobacterium halobium is induced by low oxygen tension and by light. J. Bacteriol. 173:4692–4699.

Shand, R. F., and A. M. Perez. 1999. Haloarchaeal growth physiology. In: J. Seckbach (Ed.) Enigmatic microorganisms and life in extreme environments. Kluwer Academic Publishers. Dordrecht, 414–424.

Shand, R. F., L. B. Price, and E. M. O'Connor. 1999. Halocins: Protein antibiotics from hypersaline environments. In: A. Oren (Ed.) Microbiology and biogeochemistry of hypersaline environments. CRC Press. Boca Raton, FL. 295–306.

Shewan, J. M. 1971. The microbiology of fish and fishery products—a progress report. J. Appl. Bacteriol. 34:299–315.

Simon, R. D. 1978. Halobacterium strain 5 contains a plasmid which is correlated with the presence of gas vacuoles. Nature 273:314–317.

Simon, R. D. 1980. Interactions between light and gas vacuoles in Halobacterium salinarium strain 5: Effect of ultraviolet light. Appl. Environ. Microbiol. 40:984–987.

Sioud, M., G. Baldacci, P. Forterre, and A.-M. de Recondo. 1987. Antitumor drugs inhibit the growth of halophilic archaebacteria. Eur. J. Biochem. 169:231–236.

Sioud, M., O. Possot, C. Elie, L. Sibold, and P. Forterre. 1988. Coumarin and quinolone action in archaebacteria: Evidence for the presence of a DNA gyrase-like enzyme. J. Bacteriol. 170:946–953.

Skulachev, V. P. 1993. Bioenergetics of extreme halophiles. In: M. Kates, D. J. Kushner, and A. T. Matheson (Eds.) The biochemistry of Archaea (Archaebacteria). Elsevier. Amsterdam, 25–40.

Smallbone, B. W., and M. Kates. 1981. Structural identification of minor glycolipids in Halobacterium cutirubrum. Biochim. Biophys. Acta. 665:551–558.

Soliman, G. S. H., and H. G. Trüper. 1982. Halobacterium pharaonis sp. nov., a new extremely halophilic bacterium with a low magnesium requirement. Zbl. Bakt. Hyg., I. Abt. Orig. C. 3:318–329.

Sonawat, H. M., R. Srivasta, S. Swaminathan, and G. Govit. 1990. Glycolytic and Entner-Doudoroff pathways in Halobacterium halobium: Some new observations based on ^{13}C NMR spectroscopy. Biochem. Biophys. Res. Commun. 173:358–362.

Spudich, J. L., and R. A. Bogomolni. 1988. Sensory rhodopsins of halobacteria. Ann. Rev. Biophys. Biophys. Chem. 17:193–215.

Spudich, J. L., D. N. Zacks, and R. A. Bogomolni. 1995. Microbial sensory rhodopsins: photochemistry and function. Israel J. Chem. 35:495–513.

Steber, J., and K. H. Schleifer. 1975. Halococcus morrhuae: A sulfated heteropolysaccharide as the structural component of the bacterial cell wall. Arch. Microbiol. 105:173–177.

Steber, J., and K. H. Schleifer. 1979. N-glycyl-glucosamine, a novel constituent in the cell wall of Halococcus morrhuae. Arch. Microbiol. 123:209–212.

Steensland, H., and H. Larsen. 1971. The fine structure of the extremely halophilic cocci. Kong. Norske Vidensk. Selsk. Skr. 8:1–5.

Stoeckenius, W. 1981. Walsby's square bacterium: Fine structure of an orthogonal procaryote. J. Bacteriol. 148:352–360.

Stoeckenius, W., and R. A. Bogomolni. 1982. Bacteriorhodopsin and related pigments of halobacteria. Ann. Rev. Biochem. 51:587–616.

Stoeckenius, W., D. Bivin, and K. McGinnis. 1985. Photoactive pigments in halobacteria from the Gavish Sabkha. In: G. M. Friedman and W. E. Krumbein (Eds.) Hypersaline ecosystems: The Gavish Sabkha. Springer-Verlag. Berlin, 288–295.

Sugiyama, Y., N. Yamada, and Y. Mukohata. 1994. The light-driven pump, cruxrhodopsin-2 in Haloarcula sp. arg-2 (bR^+, hR^-), and its coupled ATP formation. Biochim. Biophys. Acta 1188:287–292.

Sumper, M. 1987. Halobacterial glycoprotein biosynthesis. Biochim. Biophys. Acta 906:69–79.

Sumper, M., E. Berg, R. Mengele, and I. Strobel. 1990. Primary structure and glycosylation of the S-layer protein of Haloferax volcanii. J. Bacteriol. 172:7111–7118.

Takashina, T., T. Hamamoto, K. Otozai, W. D. Grant, and K. Horikoshi. 1990. Haloarcula japonica sp. nov., a new triangular halophilic archaebacterium. Syst. Appl. Microbiol. 13:177–181.

Tasch, P., and B. Todd. 1973. Halophilic bacteria susceptibility to peracetic acid vapor and ethylene oxide. Appl. Microbiol. 25:205–207.

Tasch, P., and B. Todd. 1974. Halophile bacteria: Experimental control and its ecological significance. In: A. L. Coogan (Ed.) 4th Symposium on salt. 1:Northern Ohio Geological Society. Cleveland, OH. 373–376.

Tateno, M., K. Ihara, and Y. Mukohata. 1994. The novel ion pump rhodopsins from Haloarcula form a family independent from both the bacteriorhodopsin and archaerhodopsin families/tribes. Arch. Biochem. Biophys. 315:127–132.

Tchelet, R., and M. Mevarech. 1994. Interspecies genetic transfer in halophilic archaebacteria. Syst. Appl. Microbiol. 16:578–581.

Thongthai, C., and M. Siriwongpairat. 1990. The sequential quantitation of microorganisms in traditionally fermented fish sauce (nam pla). In: P. J. A. Reilly, R. W. H. Parry, and L. E. Barile (Eds.) Post-harvest technology, preservation and quality of fish in southeast Asia. International Foundation for Science. Stockholm, 51–59.

Thongthai, C., and P. Suntinanalert. 1991. Halophiles in Thai fish sauce (nam pla). In: F. Rodriguez-Valera (Ed.) General and applied aspects of halophilic microorganisms. Plenum Press. New York, NY. 381–388.

Thongthai, C., T. J. McGenity, P. Suntinanalert, and W. D. Grant. 1992. Isolation and characterization of an extremely halophilic archaeobacterium from traditionally fermented Thai fish sauce (nam pla). Lett. Appl. Microbiol. 14:111–114.

Tian, X., X. Yu, H. Liu, and P. Zhou. 1997. New species of Natronobacterium. Acta Microbiol. Sin. (in Chinese) 37:1–6.

Tindall, B. J., A. A. Mills, and W. D. Grant. 1980. An alkalophilic red halophilic bacterium with a low magnesium requirement. J. Gen. Microbiol. 116:257–260.

Tindall, B. J., H. N. M. Ross, and W. D. Grant. 1984. Natronobacterium gen. nov. and Natronococcus gen. nov. two new genera of haloalkaliphilic archaebacteria. Syst. Appl. Microbiol. 5:41–57.

Tindall, B. J. 1985. Qualitative and quantitative distribution of diether lipids in haloalkaliphilic archaebacteria. Syst. Appl. Microbiol. 6:243–246.

Tindall, B. J., and M. D. Collins. 1986. Structure of 2-methyl-3-VIII-dihydrooctaprenyl-1,4-naphthoquinone from Halococcus morrhuae. FEMS Microbiol. Lett. 37:117–119.

Tindall, B. J., and H. G. Trüper. 1986. Ecophysiology of the aerobic halophilic archaebacteria. Syst. Appl. Microbiol 7:202–212.

Tindall, B. J., G. A. Tomlinson, and L. I. Hochstein. 1987. Polar lipid composition of a new halobacterium. Syst. Appl. Microbiol. 9:6–8.

Tindall, B. J., G. A. Tomlinson, and L. I. Hochstein. 1989. Transfer of Halobacterium denitrificans (Tomlinson, Jahnke, and Hochstein) to the genus Haloferax as Haloferax denitrificans comb. nov. Int. J. Syst. Bacteriol. 39:359–360.

Tindall, B. J. 1990a. Lipid composition of Halobacterium lacusprofundi. FEMS Microbiol. Lett. 66:199–202.

Tindall, B. J. 1990b. A comparative study of the lipid composition of Halobacterium saccharovorum from various sources. Syst. Appl. Microbiol. 13:128–130.

Tindall, B. J. 1991. Cultivation and preservation of members of the family Halobacteriaceae. World J. Microbiol. Biotechnol. 7:95–98.

Tindall, B. J., B. Amendt, and C. Dahl. 1991. Variations in the lipid composition of aerobic, halophilic archaeobacteria. In: F. Rodriguez-Valera (Ed.) General and applied aspects of halophilic microorganisms. Plenum Press. New York, NY. 199–205.

Tindall, B. J. 1992. The family Halobacteriaceae. In: A. Balows, H. G. Trüper, M. Dworkin, W. Harder, and K.-H. Schleifer (Eds.) The prokaryotes: A handbook on the biology of bacteria. Ecophysiology, isolation, identification, applications. I:Springer-Verlag. New York, NY. 768–808.

Tomioka, H., and H. Sasabe. 1995. Isolation of photochemically active archaebacteria photoreceptors, pharaonis phoborhodopsin from Natronobacterium pharaonis. Biochim. Biophys. Acta 1234:261–267.

Tomlinson, G. A., and L. I. Hochstein. 1972a. Isolation of carbohydrate metabolizing, extremely halophilic bacteria. Can. J. Microbiol. 18:698–701.

Tomlinson, G. A., and L. I. Hochstein. 1972b. Studies on acid production during carbohydrate metabolism by extremely halophilic bacteria. Can. J. Microbiol. 18:1973–1976.

Tomlinson, G. A., T. K. Koch, and L. I. Hochstein. 1974. The metabolism of carbohydrates by extremely halophilic bacteria: Glucose metabolism via a modified Entner-Doudoroff pathway. Can. J. Microbiol. 20:1085–1091.

Tomlinson, G. A., and L. I. Hochstein. 1976. Halobacterium saccharovorum sp. nov., a carbohydrate-metabolizing, extremely halophilic bacterium. Can. J. Microbiol. 22:587–591.

Tomlinson, G. A., M. P. Strohm, and L. I. Hochstein. 1978. The metabolism of carbohydrates by extremely halophilic bacteria: The identification of lactobionic acid as a product of lactose metabolism by Halobacterium saccharovorum. Can. J. Microbiol. 24:898–903.

Tomlinson, G. A., L. L. Jahnke, and L. I. Hochstein. 1986. Halobacterium denitrificans sp. nov., an extremely halophilic denitrifying bacterium. Int. J. Syst. Bacteriol. 36:66–70.

Torreblanca, M., F. Rodriguez-Valera, G. Juez, A. Ventosa, M. Kamekura, and M. Kates. 1986. Classification of non-alkaliphilic halobacteria based on numerical taxonomy and polar lipid composition, and description of Haloarcula gen. nov., and Haloferax gen. nov. Syst. Appl. Microbiol. 8:89–99.

Torreblanca, M., I. Meseguer, and A. Ventosa. 1994. Production of halocin is a practically universal feature of archaeal halophilic rods. Lett. Appl. Microbiol. 19:201–205.

Torrella, F. 1986. Isolation and adaptive strategies of haloarculae to extreme hypersaline habitats. Abstracts of the fourth international symposium on microbial ecology, Ljubljana. 59.

Torsvik, T., and I. D. Dundas. 1974. Bacteriophage of Halobacterium salinarium. Nature 248:680–681.

Trincone, A., B. Nicolaus, L. Lama, M. De Rosa, A. Gambacorta, and W. D. Grant. 1990. The glycolipid of Halobacterium sodomense. J. Gen. Microbiol. 136:2327–2331.

Trincone, A., E. Trivellone, B. Nicolaus, L. Lama, E. Pagnotta, W. D. Grant, and A. Gambacorta. 1993. The glycolipid of Halobacterium trapanicum. Biochim. Biophys. Acta 1210:35–40.

Trüper, H. G., and E. A. Galinski. 1986. Concentrated brines as habitats for microorganisms. Experientia 42:1182–1187.

Upasani, V., and S. Desai. 1990. Sambar Salt Lake: Chemical composition of the brines and studies on haloalkaliphilic archaebacteria. Arch. Microbiol. 154:589–593.

Upasani, V. N., S. G. Desai, N. Moldoveanu, and M. Kates. 1994. Lipids of extremely halophilic archaeobacteria from saline environments in India: a novel glycolipid in Natronobacterium strains. Microbiology UK 140:1959–1966.

Usukura, J., E. Yamada, F. Tokunaga, and T. Yoshizawa. 1980. Ultrastructure of purple membrane and cell wall of Halobacterium halobium. J. Ultrastr. Res. 70:204–219.

Ventosa, A., and J. J. Nieto. 1995. Biotechnological applications and potentialities of halophilic microorganisms. World J. Microbiol. Biotechnol. 11:85–94.

Ventosa, A., and A. Oren. 1996. Halobacterium salinarum nom. corrig., a name to replace Halobacterium salinarium (Elazari-Volcani) and to include Halobacterium halobium and Halobacterium cutirubrum. Int. J. Syst. Bacteriol. 46:347–347.

Ventosa, A., M. C. Gutiérrez, M. Kamekura, and M. L. Dyall-Smith. 1999. Proposal to transfer Halococcus turkmenicus, Halobacterium trapanicum JCM 9743 and strain GSL-11 to Haloterrigena turkmenica gen. nov., comb. nov. Int. J. Syst. Bacteriol. 49:131–136.

Vreeland, R. H., S. Angelini, and D. G. Bailey. 1998a. Anatomy of halophile induced damage to brine cured cattle hides. J. Am. Leather Chem. Assoc. 93:121–131.

Vreeland, R. H., A. F. Piselli, Jr., S. McDonnough, and S. S. Meyers. 1998b. Distribution and diversity of halophilic bacteria in a subsurface salt formation. Extremophiles 2:321–331.

Vreeland, R. H., and D. Powers. 1999. Considerations for microbial sampling of crystals from ancient salt formations. *In:* A. Oren (Ed.) Microbiology and biogeochemistry of hypersaline environments. CRC Press. Boca Raton, FL. 53–73.

Vreeland, R. H., and W. D. Rosenzweig. 1999. Survival of halophilic bacteria in ancient salts: Possibilities and potentials. *In:* J. Seckbach (Ed.) Enigmatic microorganisms and life in extreme environments. Kluwer Academic Publishers. Dordrecht, 389–398.

Wagner, G., and R. Linhardt. 1988. The retinal proteins of halobacteria. *In:* F. Rodriguez-Valera (Ed.) Halophilic bacteria. II:CRC Press. Boca Raton, FL. 85–104.

Wais, A. C., M. Kon, R. E. MacDonald, and B. D. Stollar. 1975. Salt-dependent bacteriophage infecting Halobacterium cutirubrum and Halobacterium halobium. Nature 256:314–315.

Wais, A. C., and L. L. Daniels. 1985. Populations of bacteriophage infecting Halobacterium in a transient brine pool. FEMS Microbiol. Ecol. 31:323–326.

Wais, A. C. 1988. Recovery of halophilic archaebacteria from natural environments. FEMS Microbiol. Ecol. 53:211–216.

Walsby, A. E. 1980. A square bacterium. Nature 283:69–71.

Wang, D., and Tang, Q. 1989. Natronobacterium from soda lakes of China. *In:* T. Hattori, Y. Ishida, Y. Maruyama, R. Y. Morita, and A. Uchida (Eds.) Recent advances in microbial ecology. Japan Scientific Societies Press. Tokyo, 68–72.

Weber, H. J., S. Sarma, and T. Leighton. 1982. The Halobacterium group: microbiological methods. Meth. Enzymol. 88:369–373.

Weidinger, G., G. Klotz, and W. Goebel. 1979. A large plasmid from Halobacterium halobium carrying information for gas vacuole formation. Plasmid 2:377–386.

Wieland, F., W. Dompert, G. Bernhardt, and M. Sumper. 1980. Halobacterial glycoprotein saccharides contain covalently linked sulphate. FEBS Lett. 120:110–114.

Wieland, F., J. Lechner, and M. Sumper. 1982. The cell wall glycoprotein of Halobacterium: structural, functional and biosynthetic aspects. Zbl. Bakt. Hyg. I Abt. Orig. C 3:161–170.

Wieland, F., G. Paul, and M. Sumper. 1985. Halobacterial flagellins are sulfated glycoproteins. J. Biol. Chem. 260:15180–15185.

Wieland, F. 1988. The cell surfaces of halobacteria. *In:* F. Rodriguez-Valera (Ed.) Halophilic bacteria. CRC Press. Boca Raton, FL. II:55–65.

Witte, A., U. Baranyi, R. Klein, M. Sulzner, C. Luo, G. Wanner, D. H. Krüger, and W. Lubitz. 1997. Characterization of Natronobacterium magadii phage ΦCh1, a unique archaeal phage containing DNA and RNA. Mol. Microbiol. 23:603–616.

Wu, L., K. Chow, and K. Mark. 1983. The role of pigments in Halobacterium cutirubrum against UV irradiation. Microbios Lett. 24:85–90.

Xu, Y., P. Zhou, and X. Tian. 1999. Characterization of two novel haloalkaliphilic archaea Natronorubrum bangense gen. nov., sp. nov. and Natronorubrum tibetense gen. nov., sp. nov. Int. J. Syst. Bacteriol. 49:261–266.

Yi, X., L. Hongdi, and Z. Peijin. 1995. Nucleotide sequence of the 16S rRNA from an Archaea, Haloarcula aidinensis strain B2. Acta Microbiol. Sin. (in Chinese) 35:77–85.

Zhang, W., A. Brooun, M. M. Mueller, and M. Alam. 1996. The primary structures of the Archaeon Halobacterium salinarium blue light receptor sensory rhodopsin II and its transducer, a methyl-accepting protein. Proc. Natl. Acad. Sci. USA 93:8230–8235.

Zvyagintseva, I. S., and A. L. Tarasov. 1988. Extreme halophilic bacteria from saline soils. Microbiology 56:664–668.

Zvyagintseva, I. S., L. M. Gerasimenko, N. A. Kostrikina, E. S. Bulygina, and G. A. Zavarzin. 1995a. Interaction of halobacteria and cyanobacteria in a halophilic cyanobacterial community. Microbiology 64:209–214.

Zvyagintseva, I. S., S. S. Belyaev, I. A. Borzenkov, N. A. Kostrikina, E. I. Milechina, and M. V. Ivanov. 1995b. Halophilic archaebacteria from the Kalamkass oilfield. Microbiology 64:83–87.

Zvyagintseva, I. S., E. B. Kudryashova, and E. S. Bulygina. 1996. Proposal of a new type strain of Halobacterium distributum. Microbiology 65:352–354.

Zvyagintseva, I. S., N. A. Kostrikina, and S. S. Belyaev. 1998. Detection of halophilic archaea in an Upper Devonian oil field in Tatarstan. Microbiology 67:688–691.

Prokaryotes (2006) 3:165–207
DOI: 10.1007/0-387-30743-5_9

CHAPTER 9

The Methanogenic Bacteria

WILLIAM B. WHITMAN, TIMOTHY L. BOWEN AND DAVID R. BOONE

The methanogenic bacteria are a large and diverse group that is united by three features: 1) They form large quantities of methane as the major product of their energy metabolism. 2) They are strict anaerobes. 3) They are members of the domain Archaea, or archaebacteria, (see Chapter 1) and only distantly related to the more familiar classical bacteria or eubacteria. Like the photosynthetic eubacteria, the methanogenic bacteria are related to each other primarily by their mode of energy metabolism but are very diverse with respect to their other properties.

Methanogenic bacteria obtain their energy for growth from the conversion of a limited number of substrates to methane gas. The major substrates are $H_2 + CO_2$, formate, and acetate. In addition, some other C-1 compounds such as methanol, trimethylamine, and dimethylsulfide and some alcohols such as isopropanol, isobutanol, cyclopentanol and ethanol are substrates for some methanogens. All of these substrates are converted stoichiometrically to methane. In this regard, the metabolism of the methanogens is strikingly different from that of the so-called "minimethane" producers, which are other anaerobic microorganisms that produce very small amounts of methane as a consequence of side reactions of their normal metabolism (for an overview of the minimethane producers, see Rimbault et al., 1988). Another feature that distinguishes the methanogens from the minimethane producers is that the methanogens are obligate methane-producers, that is, they only grow under conditions where methane is formed.

The list of substrates for growth of methanogens may be divided into three groups (Table 1). In the first group, the energy substrate (electron donor) is H_2, formate, or certain alcohols and the electron acceptor is CO_2, which is reduced to methane. The ability to utilize H_2 as an electron donor for CO_2 reduction is almost universal among methanogens. Likewise, many methanogens also utilize formate, but the ability to utilize alcohols is less common (Bleicher et al., 1989; Zellner and Winter, 1987a). Some methanogens also utilize carbon monoxide as an electron donor, but growth is very slow (Daniels et al.,

1977). CO_2 reduction is the major source of methane in certain habitats such as the rumen. In other environments, such as the sediments of freshwater lakes and certain bioreactors, only about one-third of the methane is formed from CO_2 reduction. However, this reaction is still very important for maintaining the very low concentrations of H_2 and formate typical of these anaerobic habitats and facilitating the process of interspecies electron transfer.

In the second group, the energy substrate is one of a variety of methyl-containing C-1 compounds, which can serve as substrates for a few taxa of methanogens. Usually these compounds are disproportionated. Some molecules of the substrate are oxidized to CO_2. The electron acceptors are the remaining methyl groups, which are reduced directly to methane (Table 1). Although dimethylselenide and methane thiol also serve as substrates for methanogenesis, these substrates do not support growth (Kiene et al., 1986). Methanogenesis from C-1 compounds is common where methyl-containing C-1 compounds are abundant. In marine sediments, trimethylamine may be formed from choline, glycine betaine, or trimethylamine oxide. In the large intestine of mammals, methanol may be formed from the anaerobic transformation of the methoxy groups of pectin. Dimethylsulfide is also common in anaerobic environments where it is formed from both methionine and the osmoregulant dimethylsulfoniopropionate.

In the third group, acetate is the major source of methane, but the ability to catabolize this substrate is limited to species of *Methanosarcina* and *Methanosaeta* ("*Methanothrix*"). Acetate is present in many environments, and methane synthesis proceeds by an aceticlastic reaction, in which the methyl carbon of acetate is reduced to methane and the carboxyl carbon is oxidized to CO_2. Methanogenesis from acetate is common in anoxic freshwater sediments where the catabolism of acetate by other anaerobes is limited by the availability of alternate electron acceptors such as sulfate or nitrate.

Methane synthesis is the major source of energy for growth of methanogens. Thus, meth-

Table 1. Reactions and standard changes in free energies for methanogenesis.[a]

Reaction	$\Delta G^{\circ\prime}$ (kJ/mol of methane)
$4\ H_2 + CO_2 \rightarrow CH_4 + 2\ H_2O$	−135.6
$4\ \text{Formate} \rightarrow CH_4 + 3\ CO_2 + 2\ H_2O$	−130.1
$4\ \text{2-Propanol} + CO_2 \rightarrow CH_4 + 4\ \text{Acetone} + 2\ H_2O$[b]	−36.5
$2\ \text{Ethanol} + CO_2 \rightarrow CH_4 + 2\ \text{Acetate}$[c]	−116.3
$\text{Methanol} + H_2 \rightarrow CH_4 + H_2O$	−112.5
$4\ \text{Methanol} \rightarrow 3\ CH_4 + CO_2 + 2\ H_2O$	−104.9
$4\ \text{Methylamine} + 2\ H_2O \rightarrow 3\ CH_4 + CO_2 + 4\ NH_4^+$	−75.0
$2\ \text{Dimethylamine} + 2\ H_2O \rightarrow 3\ CH_4 + CO_2 + 2\ NH_4^+$	−73.2
$4\ \text{Trimethylamine} + 6\ H_2O \rightarrow 9\ CH_4 + 3\ CO_2 + 4\ NH_4^+$	−74.3
$2\ \text{Dimethylsulfide} + 2\ H_2O \rightarrow 3\ CH_4 + CO_2 + H_2S$	−73.8
$\text{Acetate} \rightarrow CH_4 + CO_2$	−31.0

[a]The standard changes in free energies were calculated from the free energy of formation of the most abundant ionic species at neutral pH. Thus, "CO_2" is $HCO_3^- + H^+$ and formate is $HCOO^- + H^+$.
[b]Other secondary alcohols utilized include 2-butanol, 1,3-butanediol, and cyclopentanol.
[c]Other primary alcohols utilized include 1-propanol and 1-butanol. From Kiene etal. (1986), Nagle and Wolfe (1985), and Widdel (1986).

Table 2. Distinctive features found in various members of the domain Archaea (archaebacteria).

Capability of extreme thermophily in some groups
Lipids composed of glycerol ethers of isoprenoids and tertraethers are common
Stereochemistry of lipids is 2,3-*sn* glycerol
Cell walls composed of protein, glycoprotein, or pseudomurein; murein is absent
Antibiotic sensitivity differs from that of eubacteria
Unique modes of energy metabolism in some groups; i.e., bacteriorhodopsin-driven photosynthesis, methanogenesis

anogenesis may be viewed as a form of anaerobic respiration where CO_2, the methyl groups of C-1 compounds, or the methyl carbon of acetate is the electron acceptor. However, the standard change in free energy ($\Delta G^{\circ\prime}$) during most methanogenic reactions is very small, and the amount of ATP produced per mole of methane is probably close to or less than one (Table 1). In most natural habitats where the concentrations of H_2 and formate are very low, the change in free energy may even be much smaller.

The second distinctive feature of methanogens is their extreme sensitivity to oxygen. Thus, the methanogens are very strict anaerobes, and they are generally present in nature only in anoxic environments. For instance, the half-time for survival of one species of *Methanosarcina* is reported to be only 4 minutes in air-equilibrated medium (Zhilina, 1972). In part, some of the sensitivity of methanogens is probably due to the oxygen lability of many of the enzymes involved in methanogenesis. Other mechanisms are probably important as well. For instance, the methanococcal enzyme acetohydroxy acid synthase—an enzyme in the pathway of branched-chain amino acid biosynthesis—is unusually sensitive to oxygen (Xing and Whitman, 1987). Likewise, many methanogens are unable to utilize oxidized sulfur compounds. Because the most widely used sulfur source, sulfide, reacts chemically with oxygen, anaerobiosis is also required to protect the sulfur source in the medium. Paradoxically, when cultures of some species are washed and resuspended in unreduced medium in the absence of substrates, their viability in unaffected by atmospheric levels of oxygen for at least 30 hours (Kiener and Leisinger, 1983).

The third distinctive feature of the methanogens is that they are archaebacteria (Jones et al., 1987; Woese, 1987). Other archaebacteria include the extreme halophiles and the sulfur-dependent extreme thermophiles. Some properties the methanogens share with other archaebacteria are listed in Table 2 and are described in more detail in subsequent sections. However, the methanogens are different from other archaebacteria because they are abundant in environments of moderate temperature, pH, and salinity. Although some extremely halophilic or extremely thermophilic methanogens have been described, most species are found in moderate environments where they are often closely associated with anaerobic eubacteria and eukaryotes. Likewise, the methanogens are unlike other archaebacteria in that they contain large amounts of coenzymes essential for methane synthesis (see below). Some of these, such as coenzyme F_{420} and coenzyme M, may serve as distinctive biomarkers.

Habitats of Methanogenic Bacteria

Methanogenic bacteria are abundant in habitats where electron acceptors such as O_2, NO_3^-, Fe^{3+} and SO_4^{2-} are limiting. Common habitats for methanogens are anaerobic digestors, anoxic sediments, flooded soils, and gastrointestinal tracts. Methanogens are generally absent from the water column of unstratified lakes and rivers because convection currents rapidly aerate the deep waters. However, the diffusion of O_2 between the layers of stratified lakes is often too slow to maintain oxic conditions in the lower layers. Similarly, the physical structure of sediments limits dispersive mechanisms, so deeper sediments are often anoxic and harbor methanogens. Soil environments may be divided into the vadose zone, which is not saturated with water, and the zone below the water table, which is saturated. In the vadose zone, O_2 diffusion is rapid and anoxic vadose zones are extremely rare, although anoxic microenvironments may occur. Thus, the activity of methanogens in the vadose zone is limited. O_2 diffusion through water is slower than through soil, so the water-saturated zone is often anoxic. Because air-saturated water at 20°C contains less than 0.3 mM O_2, even small amounts of organic pollution cause groundwater to become anoxic and suitable for development of methanogenic populations.

In axenic culture, methanogenic bacteria are extremely sensitive to small amounts of O_2. However, in natural habitats, the activities of other organisms protect methanogens in apparently oxic environments. For instance, methanogens have been isolated from large dental caries and subgingival plaque in the human mouth (Belay et al., 1988; Brusa et al., 1987). Methanogens may survive in such environments because O_2-uptake by aerobic and euryoxic bacteria creates anoxic microenvironments.

In anoxic environments, the presence of NO_3^-, Fe^{3+}, and SO_4^{2-} inhibits methanogenesis by allowing other organisms to outcompete methanogens for reduced substrates. For instance, in the presence of sulfate, sulfate-reducing bacteria utilize H_2 at concentrations lower than the minimum concentration which can be utilized by methanogens (Kristjansson et al., 1982; Lovley, 1985). Presumably, the ability of the sulfate-reducing bacteria to outcompete the methanogens is a direct consequence of the more-positive reduction potential of SO_4^{2-} compared to that of CO_2.

In environments with sufficient quantities of sulfate, hydrogen sulfide is the predominant reduced product, and the major fate of biodegradable organic carbon is oxidation to CO_2. If sulfate becomes limiting, methane replaces hydrogen sulfide as the reduced product, and the organic carbon is disproportionated to CO_2 and methane. The fate of the methyl group of acetate is an indicator of whether methanogenesis is a dominant catabolic pathway in an ecosystem. If acetate is catabolized by the sulfate-reducing bacteria, CO_2 is formed from the methyl group of acetate. In a methanogenic system, acetate is catabolized by the aceticlastic reaction, which forms methane from the methyl group. Thus, methanogenic degradation of $[2-^{14}C]$-acetate leads to $^{14}CH_4$, whereas sulfidogenic degradation leads to $^{14}CO_2$, and the ratio of these labeled gaseous products indicates whether methanogenesis predominates (Winfrey and Zeikus, 1979). However, this test fails in some thermophilic ecosystems where acetate is oxidized to CO_2 prior to formation of methane (see below).

Sulfidogenesis normally dominates in estuarine, marine, and hypersaline sediments, where sulfate diffuses from overlying water. Beneath areas of high productivity, such as kelp forests, sulfate may be limiting in the deep organic sediments. Under these conditions, aceticlastic and H_2- or formate-using methanogens develop. Even in surface layers where sulfate is in excess, some methanogenesis may occur from methylated compounds such as trimethylamine and dimethylsulfide. Trimethylamine is often found in marine sediments, where it is formed from betaine glycine or other related osmoprotectants which are produced by marine plants and bacteria to balance the osmolarity of their cytoplasm with that of the seawater. Trimethylamine is not easily utilized by sulfidogenic bacteria, but it is rapidly fermented by methanogens to methane, CO_2, and ammonia. Thus, trimethylamine has been termed a "noncompetitive" substrate for methanogens. Trimethylamine-degrading methanogens from marine environments are all in the family Methanosarcinaceae, and all methanogens that have been isolated to date from hypersaline environments use trimethylamine as catabolic substrate.

Interspecies Electron Transfer and Obligate Syntrophy

Because of their limited substrate range, methanogens depend on fermentative bacteria to convert a wide range of organic compounds into methanogenic substrates. In environments where organic matter is completely degraded to methane and CO_2, the methanogenic precursors are predominantly acetate, formate, and H_2 + CO_2. The organic matter is initially fermented mainly to volatile organic acids, H_2, and CO_2. Methanogens can directly catabolize H_2 + CO_2, formate, and acetate, but longer-chain volatile organic acids (with three or more carbon atoms) such as propionate and butyrate must be metabolized to one or more of these methanogenic precursors by a specialized group of microbes called syntrophs. These syntrophs form methanogenic substrates only in the presence of vanishingly low concentrations of H_2 or formate (for reviews see Boone and Mah, 1988; Wolin and Miller, 1987).

During the catabolism of sugars and amino acids, acetate production is more favorable for fermentative bacteria because an extra ATP is produced from acetyl CoA, which is derived from pyruvate oxidation. However, if acetate is formed, the cells require an alternative mechanism to reoxidize the NADH generated during the fermentation. Many fermentative bacteria have NADH-linked hydrogenases or formate dehydrogenases, but the thermodynamics for NADH oxidation are unfavorable except at very low concentrations of H_2 or formate. For instance, the H_2 concentration must be less than about 1 μM and the formate concentration less than 100 μM for these substances to be produced from NADH. Thus, in pure cultures, H_2 or formate production from NADH is inhibited by end-product accumulation. However, when these bacteria are grown in coculture with H_2 and formate-utilizing methanogens, the concentrations of H_2 and formate remain low, and they become important products for fermentative bacteria. In such methanogenic cocultures, the fermentative bacteria produce more acetate and less reduced products such as propionate, butyrate, lactate, and ethanol. Although the fermentative bacteria can generally grow without methanogens, additional energy is obtained from phosphoroclastic acetate production when methanogens are present. Because the activities of the methanogens are not required by the fermentative bacteria, this type of interspecies electron transfer is called nonobligate interspecies electron transfer.

A second type of interspecies electron transfer cannot proceed without the activities of H_2 or formate-utilizing species. A specialized group of bacteria called obligate syntrophs oxidize compounds such as propionate, longer-chain volatile organic acids, and aromatic compounds. The obligate syntrophs must dispose of the electrons by the reduction of protons to H_2 or of CO_2 to formate. They lack alternative fermentative reactions and cannot produce other reduced organic compounds. When the concentrations of H_2 and formate are high, end-product inhibition prevents the oxidation of the syntrophic substrates. Thus, propionate and butyrate oxidation are accomplished by obligate syntrophy of fatty acid-oxidizing bacteria and methanogens. The activities of CO_2-reducing methanogens keep the concentrations of H_2 and formate low, allowing the exergonic oxidation of fatty acids by the syntrophs.

Until recently it was assumed that H_2 is the major precursor for CO_2-reduction to methane, although many of the CO_2-reducing methanogens can use H_2 or formate equally well. The butyrate-oxidizing bacterium *Syntrophomonas wolfei* can produce either H_2 or formate, so the relative importance of these substances as interspecies electron carriers may depend on the enzyme kinetics of their production and degradation or the rate of diffusion between cells.

Although the diffusion coefficient of formate is only one-fifth as large as that of H_2, the formate concentration may be one hundred times greater than that of H_2. Under this condition, formate may be responsible for the bulk of interspecies electron transfer in some environments (Boone et al., 1989; Thiele and Zeikus, 1988).

HABITATS OF SPECIAL INTEREST. When organic matter is completely catabolized to methane and CO_2, the major substrates of methanogens are usually acetate, formate, and $H_2 + CO_2$. However, in some environments the growth of aceticlastic methanogens and obligate syntrophs is too slow to maintain a large population in the system. For instance, in the rumen and colon, acetate accumulates to concentrations of 50 to 100 mM. Although this is well above the concentration required for aceticlastic growth of methanogens such as *Methanosarcina*, these organisms do not catabolize significant quantities of acetate because their growth rate on this substrate is too slow to maintain the population in a rapid-turnover ecosystem. However, when methylamine or methanol is present, the cell numbers of *Methanosarcina* in the rumen may reach 10^5 to 10^6 per milliliter because these substrates support a faster growth rate. Propionate and butyrate are also present in the rumen at significant concentrations, but the slowly growing propionate- and butyrate-degrading organisms are not found in abundance. In the rumen, a wide range of CO_2-reducing methanogens may be found, including *Methanobrevibacter ruminantium*, *Methanobacterium formicicum*, and *Methanomicrobium mobile*. *Methanobrevibacter* species are the most commonly found CO_2-reducing methanogen in nonruminant intestinal tracts. *Methanosphaera* species have also been isolated from colonic environments; they only grow by using H_2 to reduce methanol to methane.

Methanogenic bacteria form mutualistic associations with rumen, freshwater and marine ciliates and amoebae, growing as ecto- or endosymbionts. The physical associations of these microbes with protozoa may improve the efficiency of interspecies H_2- or formate-transfer. The methanogenic partners in the endosymbioses have been classified as *Methanoplanus endosymbiosus*, *Methanobrevibacter* species, and *Methanobacterium formicicum*. It is quite possible that in some habitats such as sapropel (aquatic sludge rich in organic matter), the majority of the methanogens are involved in these symbioses (Lee et al., 1987; van Bruggen et al., 1985).

In some environments, such as hot springs and solfataras (volcanic fissures that emit steam and other gases), the predominant substrate is geothermal H_2 rather than decaying organic matter. In such environments, methanogens do not rely on the activities of other organisms for provision of their substrates. The waters near these sources are often thermal as well, and thermophilic methanogens have been isolated from hot springs in Yellowstone National Park (Zeikus et al., 1980) and in Iceland (Huber et al., 1982; Lauerer et al., 1986), and from submarine thermal vents (Jones et al., 1983a).

Atmospheric Methane

At its current atmospheric concentration of 1.7 ppm, methane is the second most abundant carbon-containing gas, and its atmospheric concentration is increasing at a rate of about 1% per year (for a review, see Cicerone and Oremland, 1988). Because methane is a major greenhouse gas, its sources and atmospheric chemistry are of considerable interest. Each year, about 400–640 $\times 10^{12}$ g of methane are released into the atmosphere. Estimates based upon the isotopic composition of the atmospheric methane suggest that about 74% of this methane is derived from recent microbiological activity. The total amount of methane that is produced may be far greater, and in some important habitats, microbial methane oxidation is known to be very significant. Thus, the methane-producing bacteria are an important component of the earth's carbon cycle.

The major sources of atmospheric methane include: enteric fermentations in animals; wetlands such as bogs, swamps, and rice paddies; landfills; and termites. Little methane is obtained from marine systems where competition with the sulfate-reducing bacteria and anaerobic methane-oxidizing organisms limits its production and release, respectively.

Isolation

Methanogenic bacteria are extremely sensitive to oxygen, and strict anaerobiosis and prereduced media are required for their isolation. Two general strategies are employed: 1) Enrichment techniques are generally very successful because few other microorganisms are capable of utilizing the major methanogenic substrates under anaerobic conditions. In addition, the sulfate- and sulfur-reducing bacteria, the denitrifying bacteria, and the photosynthetic bacteria may be further excluded by omitting sulfate, sulfur, and nitrate from the enrichment medium and performing the incubations in the dark. While the chemolithotrophic acetogenic bacteria may also be enriched under these conditions, they may be eliminated in subsequent steps. 2) Antibiotics can be employed as selective inhibi-

tors of the eubacteria. Since the methanogens are archaebacteria, many of their enzyme systems are unaffected by a wide variety of common antibiotics.

Enrichments are performed in media that simulate the source of the sample and mimic the environmental pH, salt concentration, and temperature. The presence of methanogens in the enrichment is determined by gas chromatographic analysis of the headspace for methane. Consumption of $H_2 + CO_2$ or gas production from acetate or methylamines is a less reliable indicator of methanogenic activity. For example, acetogenic bacteria will also consume $H_2 + CO_2$ gas under the same conditions as the methanogens (see Chapter 21). Likewise, in enrichments containing sediments or large amounts of other organic material, gas may be produced by fermentative bacteria. If the samples are believed to contain high numbers of methanogens, they are first serially diluted in medium, and after incubation, the highest dilution with methanogenic activity is processed further. Mineral medium is used to obtain autotrophic methanogens. Although most of the heterotrophic contaminants will be eliminated upon successive transfers, low levels of some contaminants may persist due to growth on exudates of the methanogen or on lysed cells. Because many methanogens have nutritional requirements for low levels of acetate, amino acids, volatile fatty acids, or vitamins, a complex enrichment medium is often more suitable. For very fastidious methanogens, the enrichment medium may also be amended with 30% rumen fluid or sludge extract (Mah and Smith, 1981). In this case, the heterotrophic contaminants must be eliminated by colony isolation on solid medium. Antibiotics may also reduce the numbers of contaminants. However, because of the rapid selection of antibiotic resistant contaminants, they are seldom sufficient alone for the isolation of pure cultures.

Antibiotics are especially useful when used in conjunction with enrichment techniques. Because some antibiotics are toxic to certain methanogens (Böck and Kandler, 1985), it is also useful to test their effects on the enrichment culture. Success has been obtained with a variety of antibiotics, used either singly or in combination (Table 3). However, to avoid the selection of resistant contaminants, the antibiotics are only utilized for one or two transfers. In addition, the antibiotics must be omitted from the medium when determining the purity of the isolated methanogen.

When a successful enrichment has been obtained, the methanogen may be observed by fluorescence microscopy (Doddema and Vogels, 1978). For optimal fluorescence, the excitation

Table 3. The use of antibiotics for isolation of methanogens.

Antibiotic (µg/ml)	Source of sample	Kind of methanogen isolated
Vancomycin (100)	Marine sediment	*Methanosarcina acetivorans*[a]
Marine sediment		*Methanoccoccoides methylutens*[b]
Sewage sludge		*Methanosaeta soehngenii*[c]
Sewage sludge		"*Methanothrix*" sp.[d]
Rabbit feces		"*Methanosphaera cunicuh*"[e]
Vancomycin (200)	Sewage sludge	*Methanosarcina* sp.[f]
Vancomycin (500)	Sewage sludge	*M. mazei*[g]
Cephalothin (1.7) and clindamycin (6.7)	Human feces	*Methanobrevibacter smithii*[h]
Human feces		*Methanosphaera stadtmaniae*[i]
Rabbit feces		"*M. cunicuh*"[e]
Groundwater		*Methanobacterium bryantii*[j]
Cycloserine (100) and penicillin G (2,000)	Sewage sludge	*Methanosarcina thermophila*[k]
Lake sediment		*Methanocorpusculum labreanum*[l]
Penicillin G (2,000)	Sewage sludge	*Methanobacterium* sp.[m]
Cyanobacterial mat		*Methanococcus halophilus*[n]
Penicillin G (1,000 units/ml)	Protozoan coculture	*Methanoplanus endosymbiosus*[o]
Protozoan coculture		*Methanobacterium formicicum*[p]
Penicillin G (50)	Pasture mud	*M. thermoaggregans*[q]
Penicillin G (30–100) and vancomycin (50)	Marine sediment	*Methanococcus thermolithotrophicus*[r]
Marine sediment		*Methanosarcina frisia*[s]
Cattle manure		*Methanobacterium thermoalcaliphilum*[t]
Kanamycin (100)	Sewage sludge	*Methanosaeta concilii*[u]
Vancomycin (150), penicillin G (150), kanamycin (150), and tetracycline (100)	Drilling-waste swamp	*Methanoplanus limicola*[v]
Penicillin G (200), erythromycin (200), and streptomycin sulfate (200)	Salt-marsh sediment	*Methanococcus* spp.[w]

References are: [a]Sowers et al., 1984a; [b]Sowers and Feny, 1983; [c]Huser et al., 1982; [d]Ahring and Westermann, 1984; [e]Biavati et al., 1988; [f]Touzel et al., 1985; [g]Touzel and Albagnac, 1983; [h]Miller and Wolin, 1982; [i]Miller and Wolin, 1985; [j]Godsy, 1980; [k]Zinder and Mah, 1979, [l]Zhao et al., 1989; [m]Zhao et al., 1986; [n]Zhilina, 1983; [o]van Bruggen et al., 1986; [p]van Bruggen et al., 1984; [q]Blotevogel and Fischer, 1985; [r]Huber et al., 1982, [s]Blotevogel et al., 1986; [t]Blotevogel et al., 1985; [u]Patel, 1984; [v]Wildgruber et al., 1982; [w]Whitman et al., 1986.

wavelength should be between 350 and 420 nm. Under these conditions, most species autofluoresce a blue-green color due to an abundance of coenzyme F_{420}. However, the intensity of the fluorescence varies greatly, and it may not be observed following growth of *Methanosaeta* on acetate (Zehnder et al., 1980). The fluorescence may also rapidly fade within a few seconds or be absent in older and inactive cells. In spite of these limitations, fluorescence microscopy is useful for determining if more than one type of methanogen is present in an enrichment or for checking the relative purity of the culture.

The isolation of single colonies on solid medium is usually necessary for obtaining axenic cultures of methanogens. Solid medium is prepared by the roll-tube technique or in petri plates in an anaerobic glove box. One advantage of roll tubes is that colonies of methanogens may be frequently visualized through the wall of the tube by their fluorescence. In any case, the anaerobic techniques must be stringent enough to ensure that the plating efficiency is high, since, otherwise, most of the colonies may contain contaminants that are less oxygen-sensitive than the methanogen. In a minimal medium, it may not be possible to obtain isolated colonies because

the methanogen may require a nutrient produced by a contaminating organism. For instance, in media lacking acetate, this nutrient may be produced by the fermentation of amino acids by a contaminating bacterium. Under such conditions, if acetate is required for growth of the methanogen, colonies will contain a mixture of both organisms.

In addition to the normal microbiological tests for axenic cultures, certain growth tests are useful to establish culture purity. Growth should not be observed in complex medium prepared without the substrates for methanogenesis. The medium is prepared with the same salt and trace nutrient composition as the enrichment medium, but 0.2–0.4% trypticase peptones or casamino acids and 0.2% glucose are added. Yeast extract, a source of B-vitamins, frequently contains small amounts of formate and acetate and should be avoided. Growth in a complex medium without a methanogenic substrate indicates the presence of a heterotrophic contaminant. The plating efficiency of the methanogenic culture in the enrichment medium should also be greater than 50%. If it is much lower, it is possible that growth is dependent upon crossfeeding between the methanogen and another organism.

Techniques for Culturing Methanogenic Bacteria

The methodology for culturing methanogenic bacteria is based upon the pioneering work of Hungate on anaerobic rumen bacteria (1969). Anaerobiosis is achieved by the replacement of air with oxygen-free gases and the addition of a reducing agent. The culture vessels are made of glass and sealed with butyl rubber stoppers. Oxygen-permeable rubber and plastic are avoided whenever possible. Media are dispensed in an anaerobic glove box, under a stream of oxygen-free gas, or with a glass syringe. Commercial gases are scrubbed to remove traces of contaminating oxygen. Details of the procedures are described in the primary literature (Balch and Wolfe, 1976; Macy et al., 1972) and in several recent reviews (Ljungdahl and Wiegel, 1986; Mah and Smith, 1981).

Enrichment Media for Growth of Methanogens

A large number of different types of media have been formulated for the growth of methanogens. Many are described in the original species descriptions cited elsewhere in this chapter. Therefore, only a few general media will be described. Although most methanogens will grow in one of these media, optimal growth may require modification of the concentrations of some of the components or special additions.

Medium 1 for Freshwater, Sewage, and Intestinal Species

Its composition is similar to media described earlier (Balch et al., 1979; Mah and Smith, 1981). Its composition is (per liter of medium):

Yeast extract	2.0 g
Trypticase peptones	2.0 g
Salt solution A (see below)	10 ml
Phosphate solution (200 g/l of $K_2HPO_4 \cdot 3H_2O$)	2 ml
Resazurin solution (0.5 g/l in water)	2 ml
Sodium acetate solution (136 g/l of Na acetate $\cdot$ 3H_2O)	10 ml
Trace element solution (see below)	10 ml
Vitamin solution (see below)	10 ml
$NaHCO_3$	5.0 g
Cysteine/hydrochloride	0.5 g
Sulfide solution (see below)	20 ml

Salt solution A is composed of (per liter):

NH_4Cl	100 g
$MgCl_2 \cdot 6H_2O$	100 g
$CaCl_2 \cdot 2H_2O$	40 g

After dissolving the salts, the pH is adjusted to 4 with HCl.

The trace element solution is modified from Wolin et al. 1963 and is composed of (per liter):

Nitrilotriacetic acid	1.5 g
$Fe(NH_4)_2(SO_4)_2 \cdot 6H_2O$	0.2 g
Na_2SeO_3	0.2 g
$CoCl_2 \cdot 6H_2O$	0.1 g
$MnSO_4 \cdot 2H_2O$	0.1 g
$Na_2MoO_4 \cdot 2H_2O$	0.1 g
$Na_2WO_4 \cdot 2H_2O$	0.1 g
$ZnSO_4 \cdot 7H_2O$	0.1 g
$AlCl_3 \cdot 6H_2O$	0.04 g
$NiCl_2 \cdot 6H_2O$	0.025 g
H_3BO_3	0.01 g
$CuSO_4 \cdot 5H_2O$	0.01 g

To prepare the trace element solution, dissolve the nitrilotriacetic acid in 800 ml of water and adjust the pH to 6.5 with KOH. Then dissolve the minerals in order, adjust the pH to 7.0, and bring the volume to 1 liter.

The vitamin solution is modified from Bryant et al. 1971 and is composed of (per liter):

p-Aminobenzoic acid	10 mg
Nicotinic acid	10 mg
Calcium pantothenate	10 mg
Pyridoxine hydrochloride	10 mg
Riboflavin	10 mg
Thiamine hydrochloride	10 mg
Biotin	5 mg
Folic acid	5 mg
α-Lipoic acid	5 mg
Vitamin B_{12} (stored in dark at 5°C)	5 mg

The sodium sulfide solution is prepared in anoxic water under N_2 gas. A 10 mM solution of NaOH, 110 ml, is boiled under a stream of N_2 gas until the volume is reduced to 100 ml. It is then allowed to cool under a stream of N_2 gas in a fume hood. Because sodium sulfide is very toxic, precautions must be taken to avoid contact with the solid and the solution, as well as inhalation of the vapor. To remove sulfur oxides from the surface, a large crystal of $Na_2S \cdot 9H_2O$ (about 3 g) is washed in 50 ml of water for a few seconds. The crystal is blotted dry on a paper towel and weighed. About 2.5 g should remain and is immediately added to the anaerobic solution of NaOH. The sulfide solution is then dispensed anaerobically into tubes or bottles, pressurized with 100 kPa of N_2, and autoclaved. If an anaerobic glove box is available, the tubes or bottles and their stoppers should be placed in the glove box 24 hours before use to remove adsorbed O_2. The stoppers must fit tightly and should not be greased. They may be lubricated with anaerobic water to facilitate insertion and twisting into place. After autoclaving, the sulfide solution may be stored for up to two weeks or until the solution becomes cloudy. For organisms like Methanococcus, best results are obtained if the sulfide solution is stored in the anaerobic chamber immediately after autoclaving.

Medium 2 for Methanogens Isolated from Marine Environments

It is modified from Romesser et al. 1979, and it is composed of (per liter):

Salt solution B (see below)	500 ml
Phosphate solution (14 g/l of $K_2HPO_4 \cdot 3H_2O$)	10 ml
Trace element solution	10 ml

Vitamin solution	10 ml
Iron stock solution (see below)	5 ml
Sodium acetate solution (136 g/l of sodium acetate · 3H$_2$O)	10 ml
Resazurin solution (0.5 g/l in water)	2 ml
NaCl solution (293 g/l)	75 ml
Yeast extract	2 g
NaHCO$_3$	5 g
Cysteine hydrochloride	0.5 g
Sulfide solution	20 ml

Salt solution B is composed of (per liter):

CaCl$_2$ · 2H$_2$O	0.28 g
KCl	0.67 g
NH$_4$Cl	1.00 g
MgCl$_2$ · 6H$_2$O	5.50 g
MgSO$_4$ · 7H$_2$O	6.90 g

To prepare the iron stock solution, add 0.2 g of Fe(NH$_4$)$_2$(SO$_4$)$_2$ · 6H$_2$O to a screw-top bottle. Then add 0.1 ml of concentrated HCl followed by 100 ml of H$_2$O. The solution should be replaced every four weeks.

Medium 3 for Moderately Halophilic Methanogens

It contains the same components as Medium 1 except that 88 g/l of NaCl is added.

As described above, the media do not contain a substrate for methanogenesis. For growth on H$_2$ + CO$_2$ (80:20 vol/vol), the gas mixture is added after dispensing and sterilizing the medium, and the culture tubes are pressurized to 100 kPa above atmospheric pressure. After inoculation, the culture tubes are pressurized to 240 kPa. For growth on other substrates, the gas atmosphere is 100 kPa of N$_2$ + CO$_2$ (80:20 vol/vol), and the NaHCO$_3$ concentration is reduced to 2 g/l. For growth on acetate, 50 ml of the sodium acetate solution (1 M) is added. For growth on trimethylamine/hydrochloride, 12 ml of a 1 M solution (95.5 g/l) is added. For growth on methanol, 4 ml of methanol is added per liter. For growth on 2-propanol, 1-propanol, ethanol, 2-butanol, and 1-butanol, the alcohols are added at a final concentration of 20–30 mM (Widdel, 1986). For growth on formate, 4 g/l of sodium formate is added. To control the pH in batch cultures, a formic-acid reservoir may also be added to the culture tube (Schauer and Whitman, 1989). Regardless of the substrate, care must be taken to avoid exceeding the pressure limit of the culture vessels. For instance, 10 ml of medium containing 50 mM acetate will generate nearly 15 ml of gas if the acetate is completely converted to CO$_2$ and CH$_4$. Therefore, the headspace of the culture vessel must be large enough to prevent an excessive increase in pressure, which could lead to an explosion. To minimize this danger, the culture vessels should be handled in wire baskets or metal cans. During growth on H$_2$ + CO$_2$, the opposite problem occurs. Because the methanogens consume 5 moles of gas for every

mole of CH$_4$ produced, the vessel will quickly develop a negative pressure. Also, as the partial pressure of CO$_2$ decreases, the medium will become strongly alkaline, which may inhibit growth and cause cell lysis. To minimize these problems, the volume of the headspace should be at least five times the volume of the medium, and the culture vessel should be pressurized periodically throughout the growth period. For better control of the pH, the headspace may also be repressurized with H$_2$ + CO$_2$ (75:25 vol/vol) to fully replenish the CO$_2$ consumed. For the thermophilic species, attention must also be given to the expansion of gas at higher temperatures.

All three media may be prepared by the same method. All the components are combined except the gaseous substrates and the reducing agents, cysteine hydrochloride and the sulfide solution. The medium is then brought to a boil under a stream of N$_2$ + CO$_2$ gas (80:20 vol/vol) and immediately removed from the heat. Solid cysteine hydrochloride is added, and the medium is allowed to cool under a stream of N$_2$ + CO$_2$. During this time, the oxygen indicator resazurin changes from blue to pink to colorless. The change from blue to pink occurs upon the pH-dependent formation of resorufin. The change from pink to clear occurs upon the reduction of resorufin to dihydroresorufin. When the medium has cooled to about 50°C, it can be dispensed anaerobically to the culture tubes or bottles under a stream of O$_2$-free N$_2$ + CO$_2$ gas or in an anaerobic glove box (Ljungdahl and Wiegel, 1986; Mah and Smith, 1981). For growth on H$_2$ + CO$_2$, the gas atmosphere of the culture tubes is exchanged before autoclaving. After autoclaving, the pH of the bicarbonate-CO$_2$ buffered medium is 6.8–7.0.

The media may be stored for several months in the anaerobic glove box or for about one week on the laboratory bench prior to the addition of the sulfide solution. One hour before inoculation, one part of the sterile sulfide solution is added anaerobically to 50 parts of sterile medium. Some species appear to be especially sensitive to the products of sulfide oxidation. When small inocula are used, more reproducible results are obtained by adding the sulfide solution 24 hours before the inoculation.

For growth of fastidious methanogens, the media may be amended (per liter) with 100–300 ml of clarified rumen fluid, 10 ml of a volatile fatty acid mixture, or 2 ml of boiled cell extract (BCE) of a methanogen (Mah and Smith, 1981; Tanner and Wolfe, 1988). To prepare rumen fluid, samples of the rumen contents are obtained from a fistulated animal, a mouth tube, or a slaughtered animal. The fluid is separated from the large solids by filtration through 8 layers of cheese cloth into an Erlenmeyer flask. The fluid is then centrifuged under anoxic conditions at $10,000 \times g$ for 20 minutes to remove microbial cells and small particulate matter. The supernatant is decanted and sparged slowly with CO_2 overnight. Then it is dispensed into serum bottles, flushed with CO_2, and autoclaved. The volatile fatty acid mixture is modified from Bryant et al. 1971 and is composed as follows: 46 ml of butyric, 46 ml of isobutyric, 55 ml of isovaleric, 55 ml of DL-2-methylbutyric, 37 ml of propionic, and 54 ml of valeric acids are added to 500 ml of water in a fume hood, neutralized (by litmus) with 2 M NaOH, and brought to a final volume of 1 liter.

Some methanogens require growth factors that are only produced by other methanogenic bacteria. These factors may be obtained from rumen fluid or BCE (Tanner and Wolfe, 1988). To prepare BCE, 10 g wet weight of methanogen cells are resuspended in 20 ml of anaerobic 20 mM potassium phosphate buffer, pH 7.0, and incubated under a stream of N_2 gas in a boiling water bath for one hour. After cooling, the suspension is centrifuged anaerobically at $20,000 \times g$ for 30 min at 4°C. The resulting supernatant, BCE, is stored anaerobically at −20°C.

Mineral media may be prepared by omitting the organic supplements: yeast extract, trypticase, acetate, and cysteine. However, cysteine also functions as a reducing agent. Therefore, the medium will remain oxidized until the sulfide solution is added. In this case, the sulfide solution should be added 24 h before inoculation.

For growth of the alkalophilic methanogens, the medium is prepared under a N_2 atmosphere. The $NaHCO_3$ concentration is increased to 10 g/l (pH 8.5) or replaced with 3 g/l of $NaHCO_3$ plus 2 g/l of Na_2CO_3 (pH 9.3; Mathrani et al., 1988; Worakit et al., 1986). If H_2 is the substrate, the culture is initially pressurized with 50 kPa H_2. CO_2 is not added initially to avoid acidification of the medium. The gas is then replenished with $H_2 + CO_2$ (75 : 25 vol/vol) during growth, stoichiometrically replacing the CO_2 reduced to methane. For mass culture, gas additions are controlled by a pH auxostat (Boone et al., 1987).

Solid medium is prepared by the addition of agar or Gelrite during the medium preparation. For roll tubes or bottles, the medium plus sulfide with 2% agar or 1% Gelrite is allowed to coat the walls to provide a surface for inoculation (Harris, 1985; Hermann et al., 1986; Hungate, 1969). The agar surface may be streaked or the cells may be added as a soft agar overlay. Alternatively, molten agar may be cooled to 45°C and inoculated prior to solidification. Petri plates containing 0.8–2% agar or Gelrite (Kelco Div., Merck & Co., Inc., San Diego, Calif.) may be prepared in an anaerobic glove box (Balch et al., 1979). In this case, the plates are prepared without sulfide (Jones et al., 1983c). After inoculation, the plates are pressurized with $H_2 + CO_2$ or $N_2 + CO_2$ in a cannister and removed from the chamber. An open tube containing 0.75 ml of 20% $Na_2S \cdot 9H_2O$ per liter of cannister volume is added just prior to pressurization. Because the CO_2 neutralizes the sulfide solution, volatile H_2S (about 1.5%) is generated inside the cannister. Because H_2S is toxic, the cannister must either be opened in a fume hood or be flushed with $N_2 + CO_2$ gas before opening in the anaerobic glove box.

Measurement of Growth

In many cases, the growth of methanogens can be measured by normal microbiological techniques. However, some organisms grow as aggregates or have very low cell yields, and it is difficult to measure growth turbidimetrically or by other common techniques. An alternative is to measure CH_4 accumulation in the headspace by gas chromatography. However, whenever product formation is used to estimate growth rate, it is important to insure that cell growth is balanced, that is, product formation is proportional to the increase in cell mass. Thus, control experiments should demonstrate a correlation between CH_4 formation and cell mass. This method is additionally complicated because upon inoculation, only cells and not the initial product are transferred (Powell, 1983). Therefore, plots of the logarithm of CH_4 accumulated versus time are nonlinear for about four generations, and the growth curves appear to be convex. This artifact results in an overestimate of the specific growth rate. To correct for this effect, the logarithm of the rate of methane formation versus time should be plotted (Powell, 1983). This replot is linear, and the specific growth rate is equal to the slope.

Maintenance of Stock Cultures

Many of the same techniques used for storage of stock cultures of other anaerobes are suitable for at least some of the methanogens (Hippe, 1984). While freezing in liquid nitrogen is a very reliable method, less-expensive alternatives may be employed for laboratory collections. Lyophilization in horse serum plus 7.5% glucose and 3 mg/ml of ferrous sulfide followed by storage at either 8° or −70°C is suitable for many species of Methanobacteriaceae and *Methanosarcina* (Hippe, 1984). However, this method is less effective for *Methanococcus* and *Methanospirillum*. Many methanogens can also be stored on agar slants at 4°C for one year if oxygen is excluded (Winter, 1983). Likewise, storage at −18°C in medium containing 50% glycerol in sealed glass ampules is effective for 20 months. *Methanococcus* cultures can be preserved in medium containing 25% glycerol in screw-top vials at –70°C for 30 months without special anaerobic precautions (Whitman et al., 1986). *Methanobrevibacter* species and some other methanogens can be stored in biphasic cultures at –76°C for 6–12 months (Miller, 1989). Culture tubes are prepared with a slant of double-strength medium containing 3% agar and twice the usual concentration of reducing agents. One-third volume of reduced, single-strength broth is added. The tube is inoculated and allowed to grow to a heavy density. The culture is then cooled for 1 hour at 4°C prior to freezing at –76°C.

Cultures of the type strains of most species of methanogens are available through the Deutsche Sammlung von Mikroorganismen (Braunschweig, FRG) and the OGC Collection of Methanogenic Archaeobacteria (Oregon Graduate Institute, Beaverton, Oregon, USA).

Identification

Three major difficulties are encountered in the identification of isolates of methanogenic bacteria: 1) Most of the species descriptions are based upon the examination of only a few strains. Therefore, it is frequently not known if the phenotypic characterization is representative of the species. 2) Probably only a fraction of the methanogens in nature have been described. Therefore, a new isolate may represent an undescribed species or subspecies. 3) Because of their restricted catabolism coupled with extreme genetic diversity, phenotypic characters alone are often insufficient to identify methanogens. For these reasons, phylogenetic methods such as DNA-DNA hybridization and ribosomal RNA sequence analysis are frequently necessary for a definitive assignment (Boone and Whitman, 1988). Other techniques that have proven useful in the identification of methanogens include analysis of their antigenic relationships, polyamine content, molecular weights of the methylreductase subunits, and molecular weights of the polar lipids. Cross-reactivity of specific immunoglobulins has proven useful for identifying methanogens at the species and genus level (Conway de Macario et al., 1981). For distinguishing among members of different families, data on the relative distribution of putrescine, spermidine, spermine, and homospermidine, as well as the molecular weight of the small subunit of component C of the methylreductase, are useful (Kneifel et al., 1986; Rouvière and Wolfe, 1987). Likewise, the distribution of polar lipids distinguishes the families of methanogens (Koga et al., 1987; Morii et al., 1988).

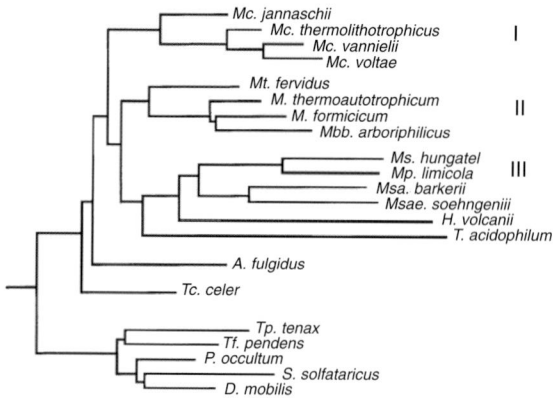

Fig. 1. Phylogenetic tree for the methanogenic bacteria and other archaebacteria based upon 16S rRNA sequences. *Mc. Methanococcus; Mt., Methanothermus; M., Methanobacterium; Mbb., Methanobrevibacter; Ms., Methanospirillum; Mp., Methanoplanus; Msa., Methanosarcina; Msae., Methanosaeta; H., Halobacterium; T., Thermoplasma; A., Archaeoglobus; Tc., Thermococcus; Tp., Thermoproteus; Tf., Thermofilum; P., Pyrodictium; S., Sulfolobus;* and *D., Desulfurococcus.* (P. E. Rouvière and C. R. Woese, personal communication.)

Nineteen genera and more than 50 species of methanogenic bacteria have been described. The current taxonomy reflects the phylogeny described by the oligonucleotide catalogs of the 16S rRNAs (Balch et al., 1979). More recent information, which includes nearly complete sequences of the 16S rRNA and many new species, has confirmed many of these earlier conclusions. The current phylogenetic tree supports the classification of the methanogens into three major groups that correspond to the orders proposed by Balch et al., 1979 (Fig. 1). The orders have been further divided into six families, and the creation of additional families may be warranted to include other deep branches of the major lineages. However, the proposal to place *Methanoplanus* in a separate family (Wildgruber et al., 1982) is not supported by the most recent information.

At present, three orders and six families are recognized within the methanogens (Tables 4 and 5). As indicated in the footnotes to Table 4, most genera have not been formally assigned to an order and family. While the present scheme is consistent with the current phylogenetic and phenotypic data, future changes are to be expected.

Two families, the Methanobacteriaceae and the Methanothermaceae, are closely related. These methanogens possess cell walls composed in part of pseudomurein (Kandler and König, 1985). The cell walls of the Methanothermaceae also contain an additional surface layer composed of protein. The family Methanothermaceae contains one genus, *Methanothermus,*

Table 4. Major taxonomic groups of methanogenic bacteria.

Order	Family	Genus
Methanobacteriales	Methanobacteriaceae (type)[a]	*Methanobacterium* (type)[a]
	"	*Methanobrevibacter*
	"	*Methanosphaera*[b]
	Methanothermaceae	*Methanothermus* (type)
Methanococcales	Methanococcaeae (type)	*Methanococcus* (type)
Methanomicrobiales	Methanosarcinaceae	*Halomethanococcus*[b]
	"	*Methanococcoides*
	"	*Methanohalobium*[b]
	"	*Methanohalophilus*[b]
	"	*Methanolobus*
	"	*Methanosarcina* (type)
	"	*Methanosaeta* ("*Methanothrix*")
	Methanomicrobiacaea (type)	*Methanoculleus*[c]
	"	*Methanogenium*
	"	*Methanolacinia*[c]
	"	*Methanomicrobium* (type)
	"	*Methanoplanus*[b]
	"	*Methanospirillum*
	Methanocorpusculaceae	*Methanocorpusculum* (type)

[a]Type family of the order or type genus of the family.
[b]Not formally placed within the family.
[c]While not formally placed within the family, it includes species originally placed within a genus of the family.

Table 5. Some characteristics of the methanogen families.

Family	Characteristics
Methanobacteriaceae[a]	Long or short rods, which use $H_2 + CO_2$ and sometimes formate or alcohols as substrates for methanogenesis; cocci, which utilize only $H_2 +$ methanol; mostly Gram-positive; contain pseudomurein; nonmotile; GC content, 23–61 mol%.
Methanothermaceae	Rods; substrate for methanogenesis is $H_2 + CO_2$; Gram-positive; contain pseudomurein; nonmotile; extreme thermophiles; GC content, 33–34 mol%.
Methanococcaceae	Irregular cocci; substrates for methanogenesis are $H_2 + CO_2$ and formate; Gram-negative; motile; GC content, 29–34 mol%.
Methanomicrobiaceae	Rods, spirals, plates, or irregular cocci; substrates for methanogenesis are $H_2 + CO_2$, frequently formate, and sometimes alcohols; Gram-negative; motile or nonmotile; GC content, 39–61 mol%.
Methanocorpusculaceae	Small, irregular cocci; substrates for methanogenesis are $H_2 + CO_2$, formate, and sometimes alcohols; Gram-negative; motile or nonmotile; GC content, 48–52 mol%.
Methanosarcinaceae	Pseudosarcina, irregular cocci, sheathed rods; substrates for methanogenesis are sometimes $H_2 + CO_2$, acetate, and methyl compounds; formate is never used; Gram-positive or negative; frequently nonmotile; GC content, 36–52 mol%.

[a]Including *Methanosphaera*.

both species of which are extremely thermophilic bacilli with temperature optima of 83–88°C. Like in many of the Methanobacteriaceae, the only substrate for methanogenesis is $H_2 + CO_2$. The family Methanobacteriaceae contains two genera, composed of thermophilic as well as mesophilic species. These genera, *Methanobacterium* and *Methanobrevibacter*, are bacilli that utilize either $H_2 + CO_2$ alone or $H_2 + CO_2$ and formate as substrates for methanogenesis. An additional genus, *Methanosphaera*, is closely related to the *Methanobacteriaceae*, although it has not yet been formally placed within this family. It includes two coccoid species that utilize $H_2 +$ methanol as substrates. These organisms are unique in that they are incapable of

both the reduction of CO_2 to methane and the oxidation of methanol to CO_2 (Miller and Wolin, 1983).

The family Methanococcaceae contains one genus of irregular cocci, *Methanococcus*, most species of which utilize both $H_2 + CO_2$ and formate as substrates. The six known species of *Methanococcus* contain protein cell walls and are either mesophilic or extremely thermophilic. Most species are also autotrophic. By the criterion of the 16S rRNA sequences, this genus is very diverse (Fig. 1). To maintain genera of equal phylogenetic depth, the two thermophilic species should be placed in two new genera.

The family Methanosarcinaceae contains all the methanogens capable of utilizing acetate as a substrate for methanogenesis or oxidizing the methyl groups of C-1 compounds. Formate is not utilized by any species in this family and most species are also unable to grow on $H_2 + CO_2$. Seven genera have been described. The genus *Methanosarcina* contains six species of cocci, which may occur singly, in packets, or in large pseudoparenchyma (Zhilina, 1976). The aggregates are held together by methanochondroitin, a polymer of N-acetyl-D-galactosamine and D-glucuronic acid similar to animal chondroitin (Kreisl and Kandler, 1986). Pseudomurein is absent. Pseudoparenchyma formation is species and strain specific and varies even in strains that do form aggregates, depending upon the culture conditions and growth stage. Therefore, the absence of aggregation is not sufficient evidence to exclude an isolate from this genus. Another typical property of this genus is the ability to utilize both acetate and C-1 compounds as substrates for methanogenesis. In addition, some species also utilize $H_2 + CO_2$. Other genera in this family that contain cocci are distinctive in their ability to utilize only C-1 compounds as substrates for methanogenesis. These genera include *Methanolobus, Methanococcoides, Methanohalobium, Halomethanococcus,* and *Methanohalophilus.* The last three genera contain the moderately and extremely halophilic species of methanogens, but have not been formally placed within this family. An additional species, *Methanococcus halophilus,* should probably also be classified with these organisms on the basis of its 5S rRNA sequence (Lysenko and Zhilina, 1985). The last genus, *Methanosaeta* ("*Methanothrix*"), includes large sheathed rods that can only utilize acetate as a substrate for methanogenesis.

The family Methanomicrobiaceae contains six genera of different morphological types with similar physiology. With one exception, all species utilize both $H_2 + CO_2$ and formate as substrates. Two genera, *Methanogenium* and *Methanoculleus,* are cocci that can be distinguished from the methanococci by their slower growth rates under optimal conditions, more complex nutritional requirements, higher GC content, and lower NaCl requirements. The genus *Methanospirillum* has a distinctive spiral morphology, and the genus *Methanoplanus* contains plate-like or irregular disk-shaped species. Lastly, the genera *Methanomicrobium* and *Methanolacinia* contain rod-shaped species. These species may be distinguished from *Methanobrevibacter* because they stain Gram-negative due to the absence of pseudomurein and they have a higher GC content.

The family Methanocorpusculaceae is related to the family Methanomicrobiaceae. It contains one genus, *Methanocorpusculum,* of very small cocci. The complex nutritional requirements of the species of this family distinguishes them from species of *Methanococcus.* The absence of a requirement for NaCl for good growth distinguishes them from species of *Methanogenium* and *Methanoculleus.*

Characteristics of Methanogenic Bacteria

Some of the most important and most distinctive features of methanogenic species will be described in this section. More detailed descriptions of the various taxa of bacteria may be found in the original citations, which are included in the footnotes to Tables 6 to 16. Whenever possible, the descriptions summarize the properties of all the described strains as well

Table 6. Some characteristic properties of species of the genus.

Species	Cell width (μm)	Cell length (μm)	Temperature optimum (°C)	pH optimum	Required organic growth factors[a]	GC content (mol%)
alcaliphilum	0.5–0.6	2–25	37	8.1–9.1	Peptone	57
bryantii	0.5–1.0	10–15	37–39	6.9–7.2	None	33–38
espanolae	0.8	6	35	5.6–6.2	B-vit, (Ac)	34
formicicum	0.4–0.8	2–15	37–45	6.6–7.8	None	38–42
ivanovii	0.5–0.8	1.2	37–45	7.0–7.4	None	37
"*palustre*"	0.5	2.5–5	37	7.0	None	34
thermoaggregans	0.4	4–8	65	7.0–7.5	None	42
thermoalcaliphilum	0.3	3–4	58–62	7.5–8.5	YE	39
thermoautotrophicum	0.4–0.6	3–120	65–70	7.2–7.6	None	48–52
thermoformicicum	0.3–0.6	2–120	45–60	7.0–8.0	None	43
uliginosum	0.2–0.6	2–4	37–40	6.0–8.5	None	29–34
wolfei	0.4	2.5	55–65	7.0–7.7	None	61

[a]None, autotrophic growth with CO_2 as sole carbon source; YE, yeast extract; B-vit, B-vitamins are stimulatory; (Ac), acetate added to medium but a growth requirement was not reported.

References: Belyaev et al., 1983, 1986; Blotevogel and Fischer, 1985; Blotevogel et al., 1985; Boone et al., 1986; Bryant and Boone, 1987b; Bryant et al., 1971; Bryant et al., 1967; Jain et al., 1987; König, 1984; Patel et al., 1990; Winter et al., 1984; Worakit et al., 1986; Zeikus and Wolfe, 1972; Zellner et al., 1989a; Zhao et al., 1986; Zhilina and Ilarionov, 1984.

Table 7. Some characteristic properties of species of the genus *Methanobrevibacter*.

Species	Cell width (μm)	Cell length (μm)	Temperature optimum (°C)	Catabolizes formate[a]	Required or stimulatory growth factors[b]	GC content (mol%)
ruminantium	0.7	0.8–1.7	37–39	+	Ac, MB, CoM, peptone	31
arboriphilicus	0.5	1–3	30–37	–	None	28–32
smithii	0.5–0.7	1–1.5	37–39	+	Ac, Peptone, YE, B-vit	28–31

[a]+, growth; –, no growth—with formate as an electron donor for methanogenesis.
[b]None, autotrophic growth with CO_2 as a sole carbon source; Ac, acetate; B-vit, B-vitamins; CoM, coenzyme M; MB, 2-methylbutyrate; YE, yeast extract.
References: Bryant et al., 1971; Lovley et al., 1984; Miller et al., 1982, 1986; Smith and Hungate, 1958; Zeikus and Hennig 1975.

Table 8. Some characteristic properties of species of the genus *Methanosphaera*.

Species	Cell diameter (μm)	Temperature optimum (°C)	pH optimum	Stimulatory or required growth factors[a]	GC content (mol%)
stadtmaniae	1.0	37	6.5–6.9	Ac, CO_2, ile, leu, thiamin, biotin	26
"cuniculi"	0.6–1.2	35–40	6.8	Ac[b]	23

[a]Ac, acetate; ile, isoleucine; leu, leucine.
[b]The requirement for other growth factors was not tested. The medium contained trypticase and yeast extract.
References: Biavati et al., 1988; Miller and Wolin, 1985.

Table 9. Some characteristic properties of species of the genus *Methanothermus*.

Species	Cell width (μm)	Cell length (μm)	Temperature optimum (°C)	pH optimum	Required organic growth factors[a]	GC content (mol%)
fervidus	0.3–0.4	1–3	83	6.5	YE	33
sociabilis	0.3–0.4	3–5	88	6.5	None	33

[a]None, autotrophic growth with CO_2 as the sole carbon source. YE, yeast extract.
References: Lauerer et al., 1986; Stetter et al., 1981.

Table 10. Some characteristic properties of species of the genus *Methanococcus*.

Species	Cell diameter (μm)	Temperature optimum (°C)	pH optimum	Optimal salinity (M or NaCl)	Required organic growth factors[a]	GC content (mol%)
vannielii	1.3	35–40	7.0–9.0	0.1–0.4	None	33
voltae	1.5	35–40	6.5–8.0	0.2–0.4	Ac, ile, leu	29–32
maripaludis	1.0	35–40	6.5–8.0	0.2–0.4	None	33–34
thermolithotrophicus	1.0	65	7.0	0.3–0.7	None	34
jannaschii	1.0	85	6.0–7.0	0.4–0.7	None	31–33
"aeolicus"	1.7	ND[b]	ND	0.2–0.4	None	30

[a]None, autotrophic growth with CO_2 as the sole source of carbon; Ac, acetate; ile, isoleucine; leu, leucine.
[b]ND, not determined.
References: Huber et al., 1982; Jones et al., 1983a, 1983b, 1989; Stadtman and Barker, 1951; Whitman et al., 1982, 1986; Zhao et al., 1988; W. B. Whitman, unpublished observations.

as the type strain. Because many of the growth descriptions depend greatly upon the experimental conditions, some caution must be exercised in the evaluation of results from different laboratories. In particular, variation in the growth optima may occur depending on whether growth is measured by turbidity, methane formation, growth rate, or growth yield.

The Family Methanobacteriaceae

The presence of a peptidoglycan chemically different from murein (called pseudomurein) is ubiquitous among species of the family Methanobacteriaceae. This peptidoglycan, like its counterpart in eubacteria, confers shape to these rods and coccobacilli, which have a wall structure typical of Gram-positive bacteria (Kandler and König, 1985). While the cells frequently stain Gram-positive, they are not formally considered Gram-positive because the cell wall is not composed of true murein. Many of the rod-shaped methanogens are in this family, and the morphologies of representative species are shown in Fig. 2. The exceptions are a few other species that have protein cell walls and are sensitive to

Table 11. Some characteristic properties of species of the genera *Methanogenium* and *Methanoculleus*.

Species	Diameter (μm)	Flagella	Temperature optimum (°C)	Optimal salinity (M of NaCl)	Required organic growth factors[a]	GC content (mol%)
Methanogenium						
cariaci	2.6	Peritrichous	40–45	<1.0	Ac, YE	52
organophilum	0.5–1.5	None[b]	30–35	0.3	Ac, PABA, biotin, B$_{12}$	47
tationis	3	Peritrichous	37–40	0.1	Ac	54
"frittonii"	1.0–2.5	None	57	0.1	None	49
"liminatans"	1.5	Present	40	<0.6	Ac	60
Methanoculleus						
marisnigri	1.3	Peritrichous	40–45	0.2	Peptones	61
thermophilicum	0.7–1.8	Single	55–60	0.2	Ac	56–60
bourgense	1–2	None	35–40	0.2	Ac	59
olentangyi	1.0–1.5	None	37	0.2	Ac	54

[a]Ac, acetate; YE, yeast extract; PABA, 4-aminobenzoate.
[b]Flagella not observed.
References: Corder et al., 1983; Harris et al., 1984; Maestrojuan et al., 1990; Ollivier et al., 1986; Rivard and Smith, 1982; Romesser et al., 1979; Widdel, 1986; Widdel et al., 1988; Zabel et al., 1984, 1985; Zellner et al., 1990.

Table 12. Some characteristic properties of species of the genera *Methanolacinia*, *Methanomicrobium*, *Methanoplanus*, and *Methanospirillum*.

Species	Cellular morphology (dimensions, μm)	Flagella	Catabolizes formate[a]	Required organic growth factors[b]	GC content (mol%)
Methanomicrobium					
mobile	Rod (0.7 × 1.5)	Simple, polar	+	Complex	49
Methanoplanus					
limicola	Plate (1.5 × 1.6–2.8)	Polar tuft	+	Ac	48
endosymbiosus	Disc (1.6–3.4)	Peritrichous[d]	+	ND	39
Methanolacinia					
paynteri	Pleomorphic (0.6 × 1.5–2.5)	Flagellated[c]	–	Ac	44–45
Methanospirillum					
hungatei	Spiral (0.5 × 7.4)	Polar tuft	+	(Ac)	47–50

[a]+, Growth; –, no growth with formate as an electron donor for methanogenesis.
[b]Ac, acetate; (Ac), required or stimulatory depending on the strain; complex, includes acetate, isobutyrate, isovalerate, 2-methylbutyrate, tryptophan (or indole), pyridoxine, thiamine, biotin, vitamin B$_{12}$, 4-aminobenzoate, and an unidentified growth factor from methanogen cell extracts; ND, not determined.
[c]Type not described.
[d]Peritrichous flagella or pili.
References: Ferry et al., 1974; Ferry and Wolfe, 1977; Patel et al., 1976; Paynter and Hungate, 1968; Rivard et al., 1983; Tanner and Wolfe, 1988; van Bruggen et al., 1986; Widdel et al., 1988; Wildgruber et al., 1982; Zellner et al., 1989b.

Table 13. Some characteristic properties of species of the genus *Methanocorpusculum*.

Species	Cellular morphology	Flagella	Temperature optimum (°C)	Optimal salinity (M of NaCl)	Required organic growth factors[a]	GC content (mol%)
Methanocorpusculum						
parvum	Coccus	Single	37	<0.80	Ac + YE	48.5
labreanum	Coccus	None	37	<0.25	Peptone	50
aggregans	Coccus	None	35	<0.10	Ac + peptone	52
sinense	Coccus	Flagellated[b]	30	NR[c]	RF + YE	52
bavaricum	Coccus	Flagellated[b]	37	NR[c]	RF + YE	48

[a]Ac, acetate; YE, yeast extract; RF, rumen fluid.
[b]Arrangement of flagella was not specified.
[c]NaCl is not required for growth.
References: Ollivier et al., 1985; Xun et al., 1989; Zellner et al., 1987a; Zhao et al., 1989; Zellner et al., 1989c.

Table 14. Some characteristic properties of species of the genus *Methanosarcina*.

Species	Morphology	Gram reaction[a]	Catabolic substrates[b] H₂	Me	Ac	Temperature optimum (°C)	GC content (mol%)
barkeri	Pseudoparenchyma	+	+	+	+	30–40	41–43
mazei	Coccus, macrocyst	–/V	+/–	+	+/–	30–40	42
thermophila	Irregular aggregate	+	+/–	+	+	50	42
acetivorans	Coccus, macrocyst	–	–	+	+	35–40	41
vacuolata	Small packet	+	+	+	+	37–40	36
frisia	Coccus	–	+	+	–	36	38

[a]+, Gram-positive; –, Gram-negative; V, Gram-variable.
[b]Me, methylated C-1 compounds and methanol; Ac, acetate; +, growth; –, no growth; +/–, very slow growth with the indicated substrates for methanogenesis.
References: Blotevogel and Fischer, 1989; Blotevogel et al., 1986; Bryant and Boone, 1987a; Liu et al., 1985; Mah, 1980; Murray and Zinder, 1985; Sowers et al., 1984a, 1984b; Touzel and Albagnac, 1983; Zhilina and Zavarzin, 1979a, 1987a; Zinder and Mah, 1979; and Zinder et al., 1985.

Table 15. Some characteristic properties of the obligately methylotrophic cocci.

Genus	Species	Cell diameter (μm)	Optimal salinity (M of NaCl)	Range of salinity (M of NaCl)	pH optimum	Required organic growth factor[a]	GC content (mol%)
Methanococcoides	*methylutens*	1	0.4	0.2–1.0	7.0–7.5	Biotin	42
Methanolobus	*tindarius*	0.8–1.3	0.5	0.05–1.3	6.5	None	40
Methanohalophilus	*oregonense*	1.0–1.5	0.5	0.1–1.6	8.6	Thiamine	41
Methanohalophilus	*zhilinae*	0.8–1.5	0.7	0.2–2.1	9.2	None	38
"*Methanococcoides*	*euhalobius*"	1.0–2.5	1.0	0.2–2.4	6.8–7.3	YE	43
Methanococcus	*halophilus*	0.5–2.0	1.2–1.5	0.3–2.6	6.5–7.4	None	41
Methanohalophilus	*mahii*	0.8–1.8	1.0–2.5	0.5–3.5	7.5	ND[b]	49
Halomethanococcus	*doii*	0.3–1.5	3.0	>1.8	6.8	Ac, RF	43
Methanohalobium	*evestigatum*	0.2–2.0	4.3	2.6–5.1	7.0–7.5	B-vitamins	37

[a]YE, yeast extract; Ac, acetate; RF, rumen fluid.
[b]ND, not determined.
References: König and Stetter, 1982; Liu et al., 1990; Lysenko and Zhilina, 1985; Mathrani and Boone, 1985; Mathrani et al., 1988; Obraztsova et al., 1987; Paterek and Smith, 1985, 1988; Sowers and Ferry, 1983, 1985; Stetter, 1989; Yu and Kawamura, 1987; Zhilina, 1983; Zhilina and Svetlichnaya, 1989; Zhilina and Zavarzin, 1987b.

Table 16. Some characteristic properties of species of the genus *Methanosaeta* ("*Methanothrix*").

Species	Culture purity	Cell width (μm)	Cell length (μm)	Temperature optimum (°C)	Required organic growth factors	GC content (mol%)
"*soehngenii*"	–	0.8	2	37	None[a]	52
concilii	+	0.8–3.5	2–7	35–40	Vitamins	49–50
thermoacetophila	+	1.0–1.3	2–6	65	None	57
strain CALS-1	+	1.0–1.2	5	60	Biotin	ND

[a]Growth in medium containing vitamins, but a requirement has not been demonstrated.
References: Huser et al., 1982; Nozhevnikova and Chudina, 1984; Nozhevnikova and Yagodina, 1982; Patel, 1984; Patel and Sprott, 1990; Touzel et al., 1988; Zehnder et al., 1980; Zinder et al., 1987.

lysis by detergents (*Methanomicrobium*) or are enclosed within a sheath that appears to confer the shape (*Methanosaeta*). The family Methanobacteriaceae contains three genera.

METHANOBACTERIUM. The species of *Methanobacterium* vary widely in length, and filaments are common (Table 6). One species, *M. thermoaggregans,* forms large multicellular aggregates. The rod-shaped cells are often irregularly crooked. All species grow with H₂ + CO₂ as a substrate for methanogenesis. In addition, formate is used by *M. formicicum,* "*M. palustre,*" and *M. thermoformicicum.* The secondary alcohols 2-propanol and 2-butanol are also utilized by *M. bryantii, M. formicicum,* and "*M. palustre*" (Widdel et al., 1988; Zellner and Winter, 1987a). In addition, 2-propanol and 2-butanol support low levels of methane synthesis but not growth of the moderately acidophilic species *M. espanolae* (Patel et al., 1990). Most species of *Methanobacterium* are capable of autotrophic growth

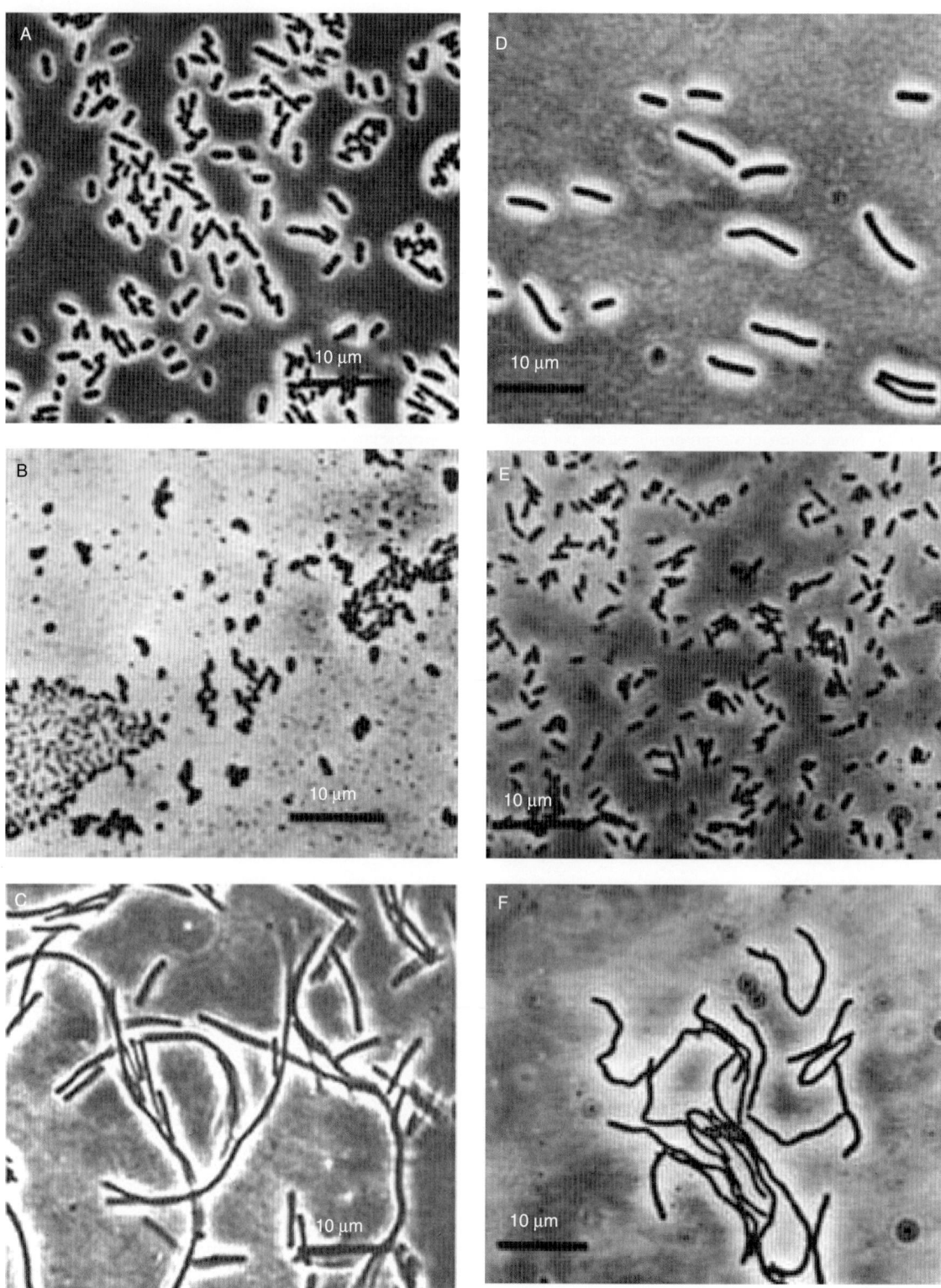

Fig. 2. Phase contrast photomicrographs of some rod-shaped methanogenic bacteria of the genera *Methanobrevibacter,* *Methanobacterium,* and *Methanomicrobium.* (A) *Methanobrevibacter ruminantium;* (B) *Methanomicrobium mobile* (stained preparation); (C) *Methanobacterium formicicum;* (D) *Methanobacterium bryantii;* (E) *Methanobrevibacter arboriphilicus;* (F) *Methanobacterium thermoautotrophicum.* (From Mah and Smith, 1981.)

with CO_2 as a sole carbon source. However, acetate, cysteine, and yeast extract are frequently stimulatory. *M. bryantii* is also stimulated by a mixture of B-vitamins. The alkaliphilic species, *M. alcaliphilum* and *M. thermoalcaliphilum*, both require yeast extract for growth, and their major carbon sources are not known.

While the nitrogen and sulfur sources have not been investigated systematically, all species of *Methanobacterium* that have been tested use ammonium, sulfide, and elemental sulfur. In addition, *M. ivanovii* uses glutamine as a sole nitrogen source (Bhatnagar et al., 1984) and *M. thermoautotrophicum* uses glutamine and urea. Note that at the high temperature at which the thermophile is grown glutamine decomposes rapidly in the medium (Friedmann and Thauer, 1987). *M. ivanovii*, *M. bryantii*, and *M. thermoautotrophicum* use cysteine as a sole sulfur source and *M. ivanovii* also uses methionine. *M. thermoautotrophicum* also uses sulfite and thiosulfate as sole sulfur sources (Daniels et al., 1986).

The species of *Methanobacterium* have been isolated from anaerobic digestors, sewage sludge, manure, groundwater, and formation water of oil-bearing rocks. In general, these habitats contain low NaCl concentrations, which are optimal for growth. Concentrations above 0.2 M are frequently inhibitory. *M. wolfei* also has a growth requirement for 8 μM tungstate, which is an unusually high concentration.

METHANOBREVIBACTER. Species of the genus *Methanobrevibacter* are oblong cocci or very short rods (Table 7). Flagella or fimbriae are not observed. All species utilize $H_2 + CO_2$ as substrates for methanogenesis. Formate is utilized by some species. Secondary alcohols were not utilized by the species tested, *M. arboriphilicus* and *M. smithii* (Widdel et al., 1988; Zellner and Winter, 1987a). The optimum pH for growth is near 7, except for *M. arboriphilicus*, which has an optimum near 8.0. *M. ruminantium* was isolated from the rumen and has very complex nutritional requirements. These requirements may be met by rumen fluid, acetate, 2-methylbutyrate, coenzyme M, or a mixture of amino acids. In addition, several similar strains have been isolated from the rumen that do not require coenzyme M (Lovley et al., 1984; Miller et al., 1986). At present, it is not known whether these strains represent new species or atypical strains of *M. ruminantium*. *M. smithii* appears to be a common methanogen in the human gastrointestinal tract (Miller and Wolin, 1982; Weaver et al., 1986). Acetate, trypticase, yeast extract, and B-vitamins are required or stimulatory for growth. In contrast to *M. ruminantium* and *M. arboriphilicus*, growth of *M. smithii* is not inhibited by bile salts.

Methanobrevibacter species have also been isolated from the feces of the horse, pig, goose, and rat (Miller and Wolin, 1986). The strain from the rat may be a new and undescribed species. On the basis of serological evidence, it has also been proposed that an additional species of *Methanobrevibacter* may predominate in some human feces (Misawa et al., 1986).

METHANOSPHAERA. *Methanosphaera* is distinct from other Methanobacteriaceae because it is coccoid and cannot reduce CO_2 to CH_4. Cells are capable of growth only by using H_2 to reduce methanol. However, the cell walls contain pseudomurein, and sequence analysis of the 16S rRNA suggests that it is closely related to the Methanobacteriaceae. Cells are spherical and occur singly or in pairs, tetrads, or small clusters (Table 8). So far, they are found only in the mammalian colon, and they are resistant to bile salts. Cells are chemoorganotrophic, and they require CO_2, acetate, some vitamins, and amino acids for growth. In addition to the amino acids, *M. stadtmaniae* requires NH_3 as a nitrogen source. Although the two species are very similar phenotypically, they can be easily distinguished by DNA-DNA hybridization or serology.

The Family Methanothermaceae

This family contains one genus, *Methanothermus*, which is extremely thermophilic and rod-shaped (Table 9). The temperature optimum is between 80° and 90°C and no growth occurs below 60°C or above 97°C. Its habitat is geothermal (85° to 101°C) waters and muds in Iceland, and it has not been found in similar habitats in other geographical locations. $H_2 + CO_2$ is the only substrate for methanogenesis. Colonies do not grow on agar plates; therefore, polysilicate plates have been used. Unlike those of the Methanobacteriaceae, the cell wall of *Methanothermus* has an S-layer external to the pseudomurein layer. Two species of *Methanothermus* have been described: *M. fervidus* grows as dispersed cells and never forms filaments whereas *M. sociabilis* forms aggregates as large as 3 mm in diameter and contains pili-like appendages. These species also differ in their lipid and pseudomurein composition (Lauerer et al., 1986). Growth of each species is inhibited in vessels made of borosilicate glass, so soda-lime glass vessels are used.

The Family Methanococcaceae

Methanococcaceae contains one genus, *Methanococcus*, which is composed of six species of mesophilic and thermophilic organisms (Table 10). An additional species has been proposed, *M. deltae*, but it is probably a subspecies of *M. mari-*

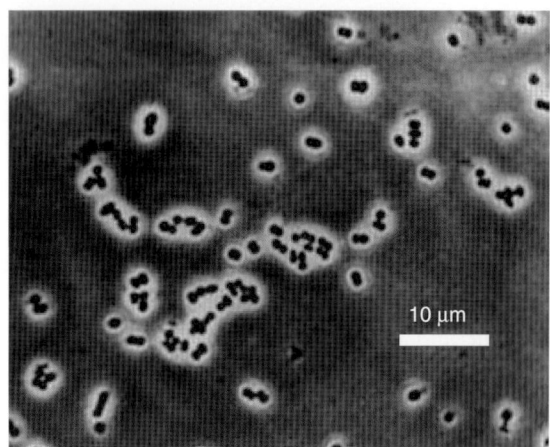

Fig. 3. Phase contrast photomicrograph of *Methanococcus voltae*. (From Mah and Smith, 1981.)

paludis (Corder et al., 1983; Whitman, 1989). The methanococci are all irregular cocci, 1–2 μm in diameter (Fig. 3). Thus, they are not easily distinguished morphologically from each other or other coccal methanogens. The cell wall is composed of an S-layer containing hexagonally arranged protein subunits (Koval and Jarrell, 1987; Nusser and König, 1987). The molecular weights of all the major wall proteins are less than 90,000, which distinguishes them from the S-layer proteins of the members of the family Methanomicrobiaceae. In addition, the S-layer proteins of the methanococci are not glycosylated. Cells lyse rapidly in 0.01% sodium dodecyl sulfate and in distilled water and they disintegrate upon heat fixation to microscope slides. The methanococci are motile, with one or two polar tufts of flagella. Isolated flagella of *M. voltae* are composed of two flagellins with molecular weights of 31,000 and 33,000 (Kalmokoff et al., 1988). The basal body is knob-like and lacks the complex ultrastructure typical of the eubacterial basal body.

The methanococci utilize $H_2 + CO_2$ as a substrate for methanogenesis. Formate is also an electron donor for all the methanococci except for two of the three described strains of *M. jannaschii*. None of the methanococci utilize acetate, methyl compounds, or alcohols as substrates for methanogenesis. Except for *M. voltae*, the methanococci are capable of rapid autotrophic growth with CO_2 as the sole carbon source. *M. voltae* requires acetate, isoleucine (or propionate or 2-methylbutyrate), and leucine (or isovalerate) as carbon sources (for a review see Jarrell and Koval, 1989). *M. voltae* grows mixotrophically, and its cellular carbon is derived from both the assimilation of exogenous acetate and autotrophic acetyl-CoA biosynthesis (Shieh et al., 1988). *M. maripaludis* also assimilates acetate, glycine, and a variety of nonpolar and basic amino acids (Whitman et al., 1987). In contrast, *M. vannielii* and "*M. aeolicus*" do not assimilate amino acids when they are provided at low concentrations, and acetate is not stimulatory to growth. Pantoyl lactone stimulates the growth of *M. voltae*, but water-soluble vitamins do not affect the growth rates of the other mesophilic methanococci.

The methanococci utilize ammonia as a sole nitrogen source. *M. maripaludis* and *M. thermolithotrophicus* can also use N_2; *M. maripaludis* can use alanine; and *M. vannielii* can use purines as sole nitrogen sources (DeMoll and Tsai, 1986). The methanococci utilize H_2S and elemental sulfur as sole sulfur sources. In addition to these, *M. thermolithotrophicus* utilizes thiosulfate, sulfite, and sulfate as sole sulfur sources (Daniels et al., 1986). The methanococci are not known to use organic sulfur sources, although *M. voltae* readily assimilates coenzyme M (Santoro and Konisky, 1987).

The methanococci appear to be restricted to marine environments. *M. vannielii* was isolated from a formate enrichment of black mud from San Francisco Bay, CA, USA. Only a single strain has been described. A number of strains of *M. voltae* and *M. maripaludis* were isolated from estuarine sediments and *Spartina alterniflora* marshes. The source of "*M. aeolicus*" has not been reported. The single described strain of *M. thermolithotrophicus* was isolated from geothermally heated sediments at Stufe di Nerone near Naples, Italy. Three strains of *M. jannaschii* were isolated from hydrothermal vents in the East Pacific Rise and the Guaymas Basin.

The Family Methanomicrobiaceae

The family Methanomicrobiaceae contains H_2-utilizing cocci (*Methanogenium* and *Methanoculleus*), as well as disc-shaped (*Methanoplanus*), rod-shaped (*Methanomicrobium* and *Methanolacinia*), and spiral-shaped (*Methanospirillum*) methanogens. The phylogentic tree is shown in Fig. 4. The coccoid Methanomicrobiaceae have protein S-layers, which cause them to stain Gram-negative and to be osmotically fragile. Cells lyse when exposed to dilute detergents or hypotonic shock. They all grow by reducing CO_2, using H_2 and formate as electron donors. Acetate is generally required as a growth factor and peptones are often required or stimulatory as well; many strains require tungstate and nickel. Motility is rare, although electron micrographs indicate that several strains have flagella.

METHANOGENIUM AND *METHANOCULLEUS*. Nine species have been described within the genera *Methanogenium* and *Methanoculleus* (Table 11).

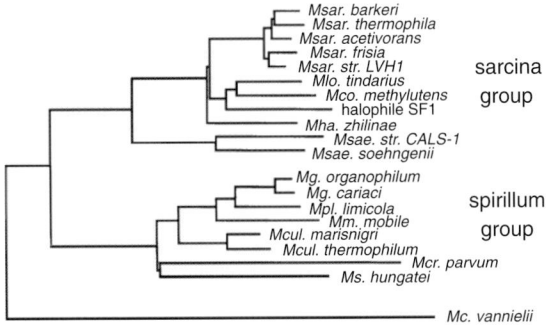

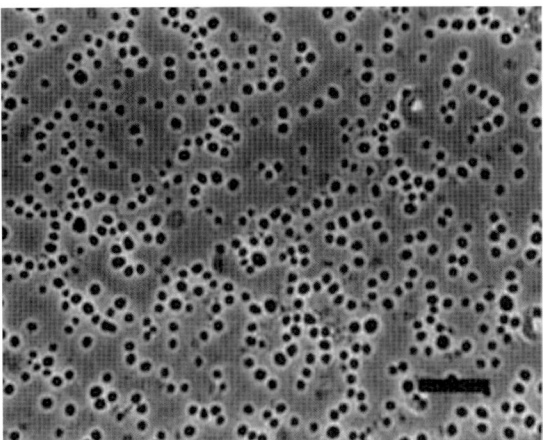

Fig. 4. Phylogenetic tree for the families *Methanomicrobiaceae* (spirillum group) and *Methanosarcinaceae* (sarcina group) based upon 16S rRNA sequences. *Msar., Methanosarcina; Mlo., Methanolobus; Mco., Methanococcoides;* SF1 is a strain closely related to *Methanohalophilus mahii; Mha., Methanohalophilus; Msae., Methanosaeta; Mg., Methanogenium; Mpl., Methanoplanus; Mm., Methanomicrobium; Mcul., Methanoculleus; Mcr., Methanocorpusculum; Ms., Methanospirillum;* and *Mc., Methanococcus.* (P. E. Rouvière and C. R. Woese, personal communication.)

Fig. 5. *Methanogenium cariaci.* Bar = 10 μm. (Courtesy of D. Boone.)

In addition, *Methanogenium aggregans,* which is not listed, has recently been reclassified as *Methanocorpusculum* (Xun et al., 1989). Since the time when the genus *Methanogenium* was originally defined (Romesser et al., 1979), the circumscription of the genus has gradually and informally been broadened to include most coccoid, H₂- and formate-utilizing methanogens that did not fit into the genus *Methanococcus.* Recently, these organisms were shown to fall into three phylogenetic groups (Xun et al., 1989): The first group, comprised of *Methanogenium cariaci* (the type species) (Fig. 5) and *M. organophilum,* includes the two species with a requirement for concentrations of NaCl close to that found in seawater. The second group, which is the most diverse physiologically, has been reclassified as a new genus called *Methanoculleus* (Maestrojuán et al., 1990). This group includes both mesophiles and thermophiles that require low concentrations of NaCl for optimal growth but may also be halotolerant. The species in this group are *Methanoculleus marisnigri, M. bourgense, M. olentangyi,* and *M. thermophilicum. Methanogenium tationis* constitutes the third group, and, on the basis of DNA-DNA hybridization, it is not closely related to other species of *Methanogenium* or *Methanoculleus.* However, at this time, it has not been reclassified. Likewise, the phylogenetic positions of "*M. frittonii*" and "*M. liminatans*" have not been examined in detail at this time.

The substrates for methanogenesis for all nine species in this group include H₂ + CO₂ and formate. Four species, *M. marisnigri, M. thermophilicum* (one out of three strains), *M. organo-*

philum, and "*M. liminatans,*" also utilize secondary alcohols. In addition, *M. organophilum* utilizes ethanol and 1-propanol. With the exception of "*M. frittonii,*" all species require organic carbon supplements. Growth of *M. marisnigri* requires trypticase, and other peptones and yeast extract do not substitute. Peptones or yeast extract also stimulate the growth of *M. organophilum, M. thermophilicum, M. bourgense, M. tationis,* and "*M. frittonii.*" For *M. tationis,* peptones can be partially replaced by a heavy metal solution. Certain strains of *M. thermophilicum* also have growth requirements for 4-aminobenzoate, biotin, nickel, molybdate, or tungstate.

Methanogenium and *Methanoculleus* species have been isolated from a variety of habitats. *M. cariaci, M. marisnigri, M. organophilum,* and *M. thermophilicum* were isolated from marine sediments. Additional strains of *M. thermophilicum* were also isolated from apple, potato, and kelp fed bioreactors. *M. bourgense* was isolated from a bioreactor fed tannery byproducts. *M. olentangyi* and "*M. frittonii*" were isolated from freshwater sediments. *M. tationis* was isolated from a moderately thermophilic solfataric pool in Chile. "*M. liminatans*" was isolated from treated industrial wastewater. Because the *Methanogenium* species grow more slowly than many other H₂- and formate-utilizing methanogens, they may be missed during viable-cell enumerations. In broth culture, they may be overgrown by *Methanococcus* species or other organisms. In roll tubes, colonies of *Methanogenium* may take several weeks to become visible. The addition of 0.05–0.20 M NaCl may improve their recovery, however.

METHANOPLANUS. The cellular morphology of *Methanoplanus* is very distinctive (Table 12): by phase contrast microscopy, *M. limicola* forms rectangular plates, and *M. endosymbiosus* is disc shaped. As the cells rotate in solution, they can be viewed on edge and appear as very thin rods. The dimensions of *M. limicola* are 0.07–0.30 μm thick, 1.6–2.8 μm long, and 1.5 μm wide. *M. endosymbiosus* is more disc shaped, with a diameter of 1.6–3.4 μm. Both species utilize H_2 and formate as substrates for methanogenesis. Best growth is obtained at neutral pH, mesophilic temperatures, and low NaCl (0.20–0.25 M). Growth of *M. limicola* is stimulated by yeast extract or peptones plus vitamins. The nutritional requirements of *M. endosymbiosus* have not been fully investigated. The cell envelope of each species contains an S-layer composed of hexagonally arranged subunits. The major proteins stain with the periodate-Schiff reagent and have apparent molecular weights of 143,000 and 110,000 for *M. limicola* and *M. endosymbiosus*, respectively. *M. limicola* was isolated from a small swamp formed from drilling wastes. *M. endosymbiosis* was isolated from a marine sapropelic ciliate.

METHANOMICROBIUM AND *METHANOLACINIA*. *Methanomicrobium* cells are short rods, but unlike the Methanobacteriaceae, they contain no pseudomurein (Table 12). The physical structure of the cell is apparently conferred by a protein cell wall. Exposure to low concentrations of sodium dodecyl sulfate and other detergents causes rapid cell lysis. Cells stain Gram-negative (Fig. 2). H_2 and formate are substrates for methanogenesis. Growth is most rapid at temperatures near 40°C and between pH 6.1 to 6.9. *M. mobile* has exceptionally complex nutritional requirements and was isolated from the rumen of a fistulated heifer fed alfalfa hay, where it was very abundant.

Methanolacinia cells are pleomorphic, and they have been described both as short rods and as highly irregular and lobed cocci. Although flagella have been observed, motility has not. Until recently, the single species, *M. paynteri*, was classified in the genus *Methanomicrobium*. The single isolate utilizes H_2 and secondary alcohols as substrates for methanogenesis (Zellner and Winter, 1987a). Growth is stimulated by yeast extract, trypticase, and 0.15 M NaCl. The temperature and pH optima are 40°C and pH 7, respectively. *M. paynteri* may readily be distinguished from *Methanomicrobium mobile* by differences in its polyamine and lipid content (Zellner et al., 1989b). *Methanolacinia paynteri* was isolated from marine sediment.

Methanospirillum

Methanospirillum is a spiral-shaped cell (0.4 μm wide by 7.4–10 μm long), twisted into a gentle

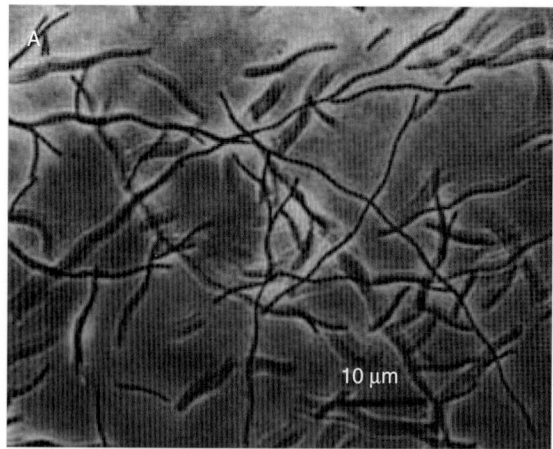

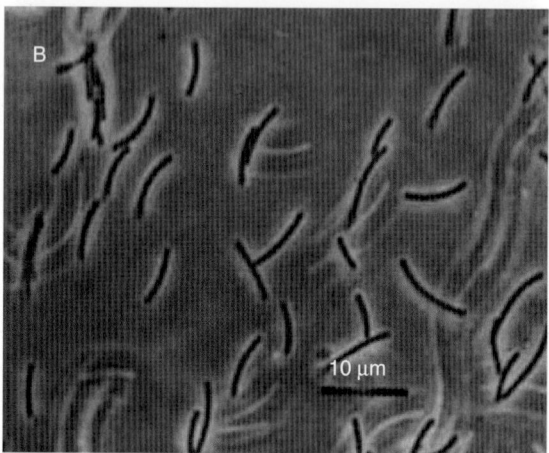

Fig. 6. *Methanospirillum* (A) Long, wavy filaments. (B) Individual cells. (From Mah and Smith, 1981.)

helix (Table 12). The helical shape may be conferred by a sheath that surrounds the cell (Patel et al., 1986). Cells may form either long chains, or filaments, encased by a continuous sheath or be single (Fig. 6; Patel et al., 1979). *M. hungatei* also lyses slowly in 1% sodium dodecyl sulfate. Colonies in solidified medium have a wavy appearance, probably conferred by the helical shape of the cells. *M. hungatei* is the only species in this genus that has been described. $H_2 + CO_2$ and formate are substrates for methanogenesis, and some strains use 2-propanol and 2-butanol as well. The temperature optimum for growth is 35–40°C. Acetate is required or stimulatory. In addition, one strain is stimulated by peptones and B-vitamins. *M. hungatei* has been isolated from sewage sludge and a pear-waste digestor.

The Family Methanocorpusculaceae

This family contains only one genus, *Methanocorpusculum*, and is more closely related to the Methanomicrobiaceae than to any other families of methanogens. The five species of the genus are very small, irregular cocci that utilize H_2 and formate as substrates for methanogenesis (Table 13). *M. parvum* and *M. bavaricum* also utilize secondary alcohols (Zellner and Winter, 1987a; Zellner et al., 1989c). The diameter of the cells varies somewhat with the growth conditions. For *M. parvum*, the cell diameter is less than or equal to 1 μm. The cell diameters of *M. labreanum* and *M. aggregans* vary from about 0.5–2.0 μm. *M. aggregans* also forms large, multicellular clumps. All species have complex nutritional requirements for peptones, rumen fluid, or yeast extract. In addition, *M. labreanum* is stimulated by acetate, and *M. parvum* requires 1 μM tungstate. *M. parvum* was isolated from a whey digestor and is unusually halotolerant. *M. aggregans* was isolated from a sewage digestor. *M. labreanum* was isolated from lake sediment near the La Brea Tar Pits, California. *M. sinense* was isolated from a biogas plant in Chengdu, China. *M. bavaricum* was isolated from a wastewater treatment pond in Germany.

The Family Methanosarcinaceae

All of the aceticlastic methanogens belong to this family, as well as all methanogens that disproportionate methanol, methylamines, or other methyl-containing compounds. The only other species that catabolize methanol are *Methanosphaera* species, which require H_2 to reduce it to methane. The family Methanosarcinaceae can be divided into three physiological and morphological types: 1) the genus *Methanosarcina*, which contains coccoid and pseudosarcinal cells and can disproportionate methanol and catabolize acetate and $H_2 + CO_2$; 2) the genus *Methanosaeta*, which grows only by the aceticlastic reaction; 3) the halophilic, methylotrophic organisms of the genera *Methanolobus*, *Methanococcoides*, and *Methanohalophilus*. Phylogenetically, only the first of these groups is composed of closely related species (Fig. 4).

METHANOSARCINA. *Methanosarcina* species disproportionate methanol and methylamines, forming CH_4 and CO_2 in a ratio of approximately 3:1 (Table 14). Most strains also use acetate, and most also use H_2 to reduce either CO_2 or methanol to CH_4. Formate is never used. Cells appear to have a protein cell wall adjacent to the cytoplasmic membrane. External to the protein wall, many strains have a layer of methanochondroitin, referred to as a matrix. Cells without the matrix are individual, irregular cocci, but when the matrix is present, packets or pseudoparenchyma are formed (Fig. 7). Within the packets, the cells are irregular in shape, the division

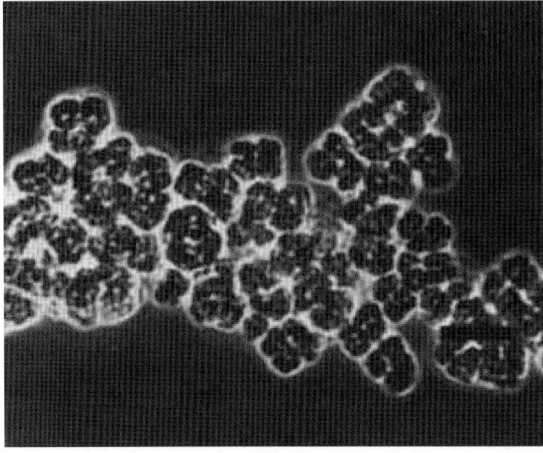

Fig. 7. *Methanosarcina barkeri* strain MS. Bar = 10 μm. (Courtesy of D. Boone.)

planes are not always perpendicular, and the volume of the daughter cells is not always equal. Three morphotypes of *Methanosarcina* have been described. Morphotype 1 includes some strains of *M. barkeri* that form aggregates of packets of cocci termed pseudoparenchyma, which may be visible to the unaided eye (Zhilina, 1976). Morphotype 2 includes *M. vacuolata*, which forms small packets and tends to remain dispersed and settles slowly in liquid medium. Morphotype 3 includes *M. mazei*, which forms single cocci, packets of pseudococci, and large macrocysts (Zhilina and Zavarzin, 1979b; Robinson, 1986).

While morphology and substrate range are important characteristics of *Methanosarcina*, they are not always reliable indicators of species differences. These characteristics frequently vary when closely related strains are tested, or the same strain may give varied results under different conditions. For example, a single enzyme may be responsible for changing large pseudoparenchymal aggregates into individual coccoid cells, and this enzyme may be produced only under certain growth conditions (Harris, 1987; Sowers and Gunsalus, 1988; Xun et al., 1988). The ability to use substrates often depends on the conditions of the inoculum (Boone et al., 1987), and the ability to form gas vacuoles is not a constant characteristic of strains (R. A. Mah, personal communication). Thus, variations in these characteristics within species may be commonplace, and they may not always be reliable for the placement of strains into species.

The diameter of the cocci or cells within the aggregates is generally 1–2 μm, although both smaller and larger cells are observed. The cells of *Methanosarcina* is nonmotile, and in those species that stain Gram negative, the cells lyse in the presence of sodium dodecyl sulfate. Cells of

M. vacuolata contain numerous gas vacuoles tightly clustered in the cytoplasm. The *Methanosarcina* do not require additional carbon sources, and species that use $H_2 + CO_2$ as a substrate for methanogenesis can grow autotrophically. However, peptones or yeast extract are stimulatory for *M. mazei, M. vacuolata,* and *M. frisia,* and riboflavin is stimulatory or required for at least one strain of *M. barkeri* (Scherer and Sahm, 1981a). The chemicals 4-aminobenzoate and calcium chloride (0.7 mM) are required for growth of *M. thermophila. M. barkeri* grows with dinitrogen as its sole nitrogen source (Bomar et al., 1985; Murray and Zinder, 1984). Methionine, cysteine, thiosulfate, and elemental sulfur are also sulfur sources for some strains of *M. barkeri* (Mazumder et al., 1986; Scherer and Sahm, 1981b).

M. barkeri, M. mazei, M. thermophila, and *M. vacuolata* were isolated from anaerobic digestors. *M. barkeri* is also found in freshwater and marine sediments, rumens of ungulates, and animal-waste lagoons. *M. mazei* has been found in garden soil, sewage sludge, and various other sources. *M. vacuolata* is common in freshwater sediments, marshes and wetlands. *M. acetivorans* and *M. frisia* were isolated from marine sediments and for optimal growth require 0.2 and 0.3–0.4 M NaCl, respectively. *M. acetivorans* also requires 0.05–0.10 M magnesium for optimal growth. In addition to the type strain of *M. thermophila,* two strains of thermophilic *Methanosarcina* have been described (Ollivier et al., 1984; Touzel et al., 1985), whose nutritional characteristics resemble those of *M. thermophila.* However, both the pseudoparenchymal and coccoid morphologies are present, and the taxonomic status of these latter strains is uncertain.

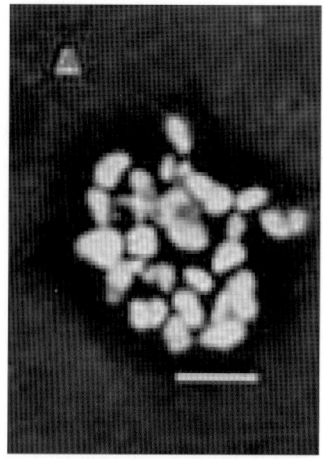

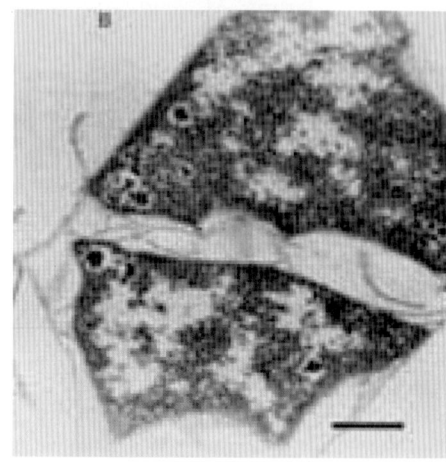

Fig. 8. *Methanohalobium evestigatum.* (A) Cells and aggregates, anoptral contrast Reichert. Bar = 5 μm. (B) Ultrathin section of late exponential phase cells. Layers of the envelope slip from the cell surface. Nuclear region and star-shaped granules (possibly glycogen) are present. Bar = 0.5 μm. (Courtesy of T. Zhilina.)

OBLIGATELY METHYLOTROPHIC COCCI. Eight species and five genera of obligately methylotrophic, irregular cocci have been described (Table 15). In addition, two species of *Methanolobus, M. siciliae* and *M. vulcani,* have been proposed, but a complete description is unavailable (Stetter, 1989). At present, the taxonomy of these organisms is uncertain. Therefore, they have been arranged in the table in the order of increasing optimal salinity for growth, which varies from marine to extremely halophilic. Phylogenetic studies by RNA sequencing, RNA-DNA hybridization and DNA-DNA hybridization suggest that *Methanococcoides methylutens, Methanolobus tindarius, Methanohalophilus zhilinae, M. mahii,* and *Methanohalobium evestigatum* probably represent five different genera rather than only four (Sowers et al., 1984b; Chumakov et al., 1987; also see Fig. 4). The placement of the three remaining species listed in Table 15 is not known.

These species grow by disproportionation of methanol, methylamines, and other C-1 compounds. H_2, formate, and acetate are not substrates. Cells are small irregular cocci, and some species form small aggregates. Cells of *M. evestigatum* are flat and polygonal, and they often occur in sheets (Fig. 8). *Methanococcus halophilus* produces slime. All the species are nonmotile, except for *Methanolobus tindarius,* which has monotrichous flagellation. With exception of *Methanohalobium evestigatum,* they stain Gram negative and lyse rapidly in dilute sodium dodecyl sulfate (or Sarkosyl for *Methanococcus halophilus*). These species are mesophilic, and their temperature optima are 30–35°C with three exceptions: the temperature optimum for *Methanolobus tindarius* is 25°C; the temperature optimum for *Methanohalophilus zhilinae* is 45°C; the temperature optimum for *Methanohalobium evestigatum* is 50°C. *Methanohalophilus zhilinae* and *M. oregonense* are also alkalophilic.

In addition to NaCl, *M. euhalobius* and *M. mahii* require high concentrations of magnesium and calcium for good growth. *M. methylutens* and *M. oregonense* require high concentrations of magnesium. Although the organic growth factors of the type strain of *M. mahii* are not known, strain SF-1, which appears to be closely related, requires yeast extract. Vitamins stimulate the growth of *M. tindarius* and peptones that of *M. zhilinae*. In contrast, greater than 0.05% yeast extract or peptones inhibit the growth of *M. halophilus* and *M. evestigatum*.

M. tindarius and *M. methylutens* were isolated from marine sediments. *M. mahii*, *Halomethanococcus doii*, *M. zhilinae*, and *M. oregonense* were isolated from sediments of saline or saline and alkaline lakes. *M. halophilus* was isolated from a sample containing the cyanobacterial mat and associated mud of a stromatolite in Shark Bay, Australia. *M. evestigatum* was isolated from a cyanobacterial mat in a hypersaline lagoon close to Arabat, Sivash. *M. euhalobius* was isolated from the stratal liquid of an exploratory oil well.

METHANOSAETA. In contrast to the other genera of the family Methanosarcinaceae, which are coccoid, the species of *Methanosaeta* (formerly called *Methanothrix*) are sheathed rods. These organisms grow only by the aceticlastic reaction (Table 16). Acetate is the only substrate for methanogenesis, and H_2, formate, and methyl compounds are not utilized. Because of their very long generation time, 1 to 3 days under optimal conditions, pure cultures have been difficult to obtain. For instance, the original culture of "*M. soehngenii*" is not pure, and vancomycin is required to suppress the growth of contaminants (Patel and Sprott, 1990; Touzel et al., 1988). Likewise, the description of *M. thermoacetophila* is based upon a monoculture, although a pure culture has recently been obtained. Two other pure cultures have also been described. *M. concilii* appears to be closely related to the methanogen in "*M. soehngenii*" cultures, and strain CALS-1 appears to be similar to *M. thermoacetophila*, but insufficient cell mass has been obtained to complete the species description.

Methanosaeta is a short, fat, nonmotile rod with flat ends. Cells stain Gram negative and are sheathed. Except for CALS-1, stationary cultures may form long filaments with a contiguous sheath (Fig. 9). The filaments may form large bundles, mats, or flocs. In shaken cultures, the filaments tend to fragment into short filaments or individual cells. *M. thermoacetophila* and CALS-1 contain gas vacuoles and *M. concilii* produces a capsule.

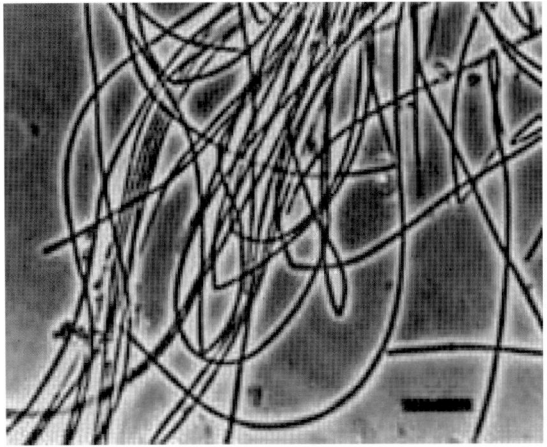

Fig. 9. *Methanosaeta concilii*. Bar = 10 μm. (Courtesy of D. Boone.)

Marine or halophilic species have not been described, and NaCl does not need to be added to the media above trace amounts. The pH optima of "*M. soehngenii*," *M. concilii*, and CALS-1 are 7.4–7.8, 7.1–7.5, and 6.5, respectively. Yeast extract inhibits the growth of *M. concilii*, and penicillin inhibits the growth of "*M. soehngenii*" and *M. thermoacetophila*. CALS-1 is very sensitive to the reducing agents in the medium, and 1 mM sulfide and 1 mM 2-mercaptoethanesulfonate appear to be optimal.

Methanosaeta is widely distributed in nature. Its distinctive morphology has allowed it to be identified in sewage sludge, digestors, animal wastes, and sanitary landfills. Monocultures of the thermophiles have been obtained from different types of thermophilic digestors and thermal sediments. "*M. soehngenii*" and CALS-1 were obtained from sludge digestors. *M. concilii* was obtained from a pear waste digestor. *M. thermoacetophila* was obtained from the sediments of a thermal lake.

Methanogens of Uncertain Affiliation

A thermophile, TAM, that morphologically resembles *Methanosaeta* has been isolated from a sewage sludge digestor (Ahring and Westermann, 1984, 1985). Cells are 0.8–1 μm wide and 4–5 μm in length, and they normally form short filaments. A sheath is present. Cells stain Gram positive to variable. In contrast to *Methanosaeta*, $H_2 + CO_2$ and formate are substrates for methanogenesis in addition to acetate. Growth factors are not required during growth on acetate. The pH optimum is 7.3–7.5, and the temperature optimum is 60°C. Penicillin and other typical eubacterial antibiotics are inhibitory at high concentrations.

The isolation of a wall-less methanogen, "*Methanoplasma elizabethii*," from an anaerobic chemostat fed glucose was reported but not confirmed (Rose and Pirt, 1981). "*M. elizabethii*" uses $H_2 + CO_2$ and formate as substrates for methanogenesis.

Biochemical and Physiological Properties

Coenzymes of Methanogenesis

Methanogens contain many novel coenzymes that are associated with the biochemistry of methane synthesis. Thus, what should be a chemically simple reduction of C-1 compounds to methane is biochemically complex. Because the structure and synthesis of many of the coenzymes have recently been reviewed (Jones et al., 1987), only the most recent citations are included below.

Methanofuran (MFR) (Fig. 10A) functions as a formyl carrier in methanogenesis. It is required for the initial activation of CO_2 to formyl-MFR, the first stable product of CO_2 reduction. MFR has been found in all methanogens examined at levels ranging from 0.5 to 2.5 mg per kg dry weight of cells. At least five different forms of methanofuran exist among the methanogens

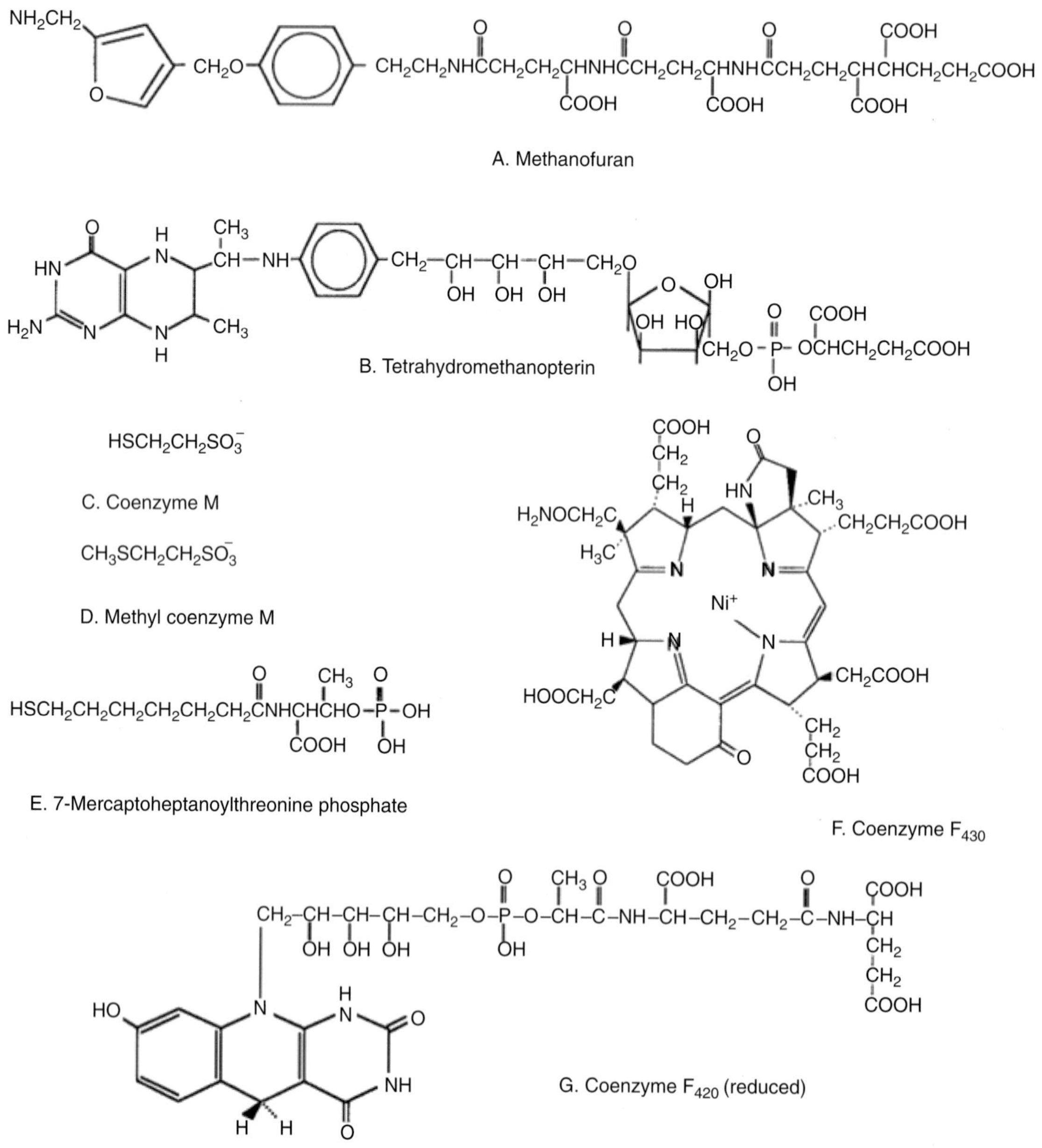

A. Methanofuran

B. Tetrahydromethanopterin

$HSCH_2CH_2SO_3^-$

C. Coenzyme M

$CH_3SCH_2CH_2SO_3^-$

D. Methyl coenzyme M

E. 7-Mercaptoheptanoylthreonine phosphate

F. Coenzyme F_{430}

G. Coenzyme F_{420} (reduced)

Fig. 10. The seven coenzymes of methanogenesis.

(White, 1988). This coenzyme has not been detected in either nonmethanogenic archaebacteria, with the exception of *Archaeoglobus fulgidus*, or eubacteria. The 2,4-disubstituted furan moiety and the 4,5-dicarboxy octanoate moiety are unique structures in nature.

Tetrahydromethanopterin (H$_4$MPT) is a one-carbon carrier similar to folate in structure and function (Fig. 10B). Likewise, biosynthesis of H$_4$MPT is also similar to tetrahydrofolate biosynthesis in eubacteria. The coenzyme was first noticed as a yellow fluorescent compound (YFC) that was rapidly labeled by $^{14}CO_2$ in cell extracts (Daniels and Zeikus, 1978). H$_4$MPT is also required for several biosynthetic reactions that are folate-dependent in eubacteria, such as serine hydroxymethyltransferase.

A third one-carbon carrier unique to methanogens is coenzyme M (HS-CoM) (Fig. 10C), which chemically is 2-mercaptoethanesulfonic acid. HS-CoM serves as the terminal carbon carrier in methanogenesis where methyl coenzyme M (CH$_3$-S-CoM) (Fig. 10D) is reduced to CH$_4$. Since HS-CoM has been found in all methanogens examined, it may serve as a sensitive biomarker for the qualitative identification of methanogens in various ecological niches.

7-Mercaptoheptanoylthreonine phosphate (HS-HTP) (Fig. 10E) is a colorless coenzyme that is required for the final step of methanogenesis. This coenzyme is the electron donor for CH$_3$-S-CoM reduction, and it is proposed to participate in the energy-conserving step of methanogenesis (Ellermann et al., 1988). It is also required for the activation of CO$_2$ reduction (Bobik and Wolfe, 1988). HS-HTP is probably biosynthesized from α-ketoglutarate by repeated α-keto-acid chain elongation in a process similar to the aminoadipate pathway of lysine biosynthesis in eubacteria (White, 1989).

Coenzyme F$_{430}$ (Fig. 10F) is a nickel-containing coenzyme required for the final step of methanogenesis, and it is tightly associated with component C of the methylreductase system. The coenzyme was first noticed in cell extracts by J. LeGall and reported by Gunsalus and Wolfe (1978). It is interesting to note that biosynthesis of coenzyme F$_{430}$ proceeds from 5-aminolevulenate via uroporphyrinogen III, which indicates that nickel tetrapyrroles share a common biosynthetic pathway with all other porphinoid compounds (Pfaltz et al, 1987).

Autofluorescence of methanogenic bacteria is largely due to the presence of high levels of coenzyme F$_{420}$ (Fig. 10G). Coenzyme F$_{420}$ is ubiquitous among methanogens, and it is found at low levels in some other organisms. It is a deazaflavin and participates in two-electron transfer reactions. In this regard, it functions in a manner analogous to NADH, but its redox potential of −340 to −350 mV is lower. The coenzyme is an electron donor in methanogenesis. In addition, coenzyme F$_{420}$ is coupled to hydrogenase, formate dehydrogenase, carbon monoxide dehydrogenase, NADP+ reductase, pyruvate synthase, and α-ketoglutarate synthase (Keltjens and van der Drift, 1986).

In addition to the novel coenzymes described above, methanogens contain a number of common vitamins (Noll and Barber, 1988). These include thiamine, riboflavin, pyridoxine, corrinoids, biotin, niacin, and pantothenate. Flavins are known to function as electron carriers for hydrogenase, NADH reductase, formate dehydrogenase, and the methylreductase system in *Methanobacterium*. Many methanogens contain abundant amounts of unusual corrinoids, with factor III or pseudovitamin B$_{12}$ being the most predominant (Stupperich and Kräutler, 1988). Corrinoids have also been implicated in methane synthesis, where they function as methyl carriers. In contrast, folates are absent or present in very small amounts in methanogens (Worrell et al., 1988). Presumably, methanopterin substitutes for folate as a methyl carrier. Methanogens have also been found to contain ferredoxin, thioredoxin, and cytochromes *b* and *c*. Ferredoxin is abundant in the acticlastic methanogens, where it is coupled with the catabolic carbon monoxide dehydrogenase, and has been isolated from *Methanococcus thermolithotrophicus* (Hatchikian et al., 1989; Terlesky and Ferry, 1988a). Thioredoxin was isolated from *Methanobacterium thermoautotrophicum*, but its function is unknown (Schlicht et al., 1985). Cytochromes function as electron carriers during methanogenesis from methanol, methylamines, and acetate. They are implicated in the oxidation of these substrates during their disproportionation (Terlesky and Ferry, 1988b). In support of this hypothesis, cytochromes are absent from *Methanosphaera*, which must use H$_2$ as a reductant during growth on methanol.

The Pathway of Methanogenesis

The pathway of methanogenesis has been extensively reviewed (Jones et al., 1987; Rouvière et al., 1988; Rouvière and Wolfe, 1988). The reduction of CO$_2$ to CH$_4$ involves seven steps. The source of electrons may be either H$_2$ via hydrogenase or formate via formate dehydrogenase. Formate itself does not appear to be an intermediate. Initially, CO$_2$ is activated to form formylmethanofuran. Next, the formyl group is transferred to H$_4$MPT, where it is reduced to the methylene and methyl levels. Last, the methyl group is transferred to coenzyme M and reduced to methane by the methylreductase system.

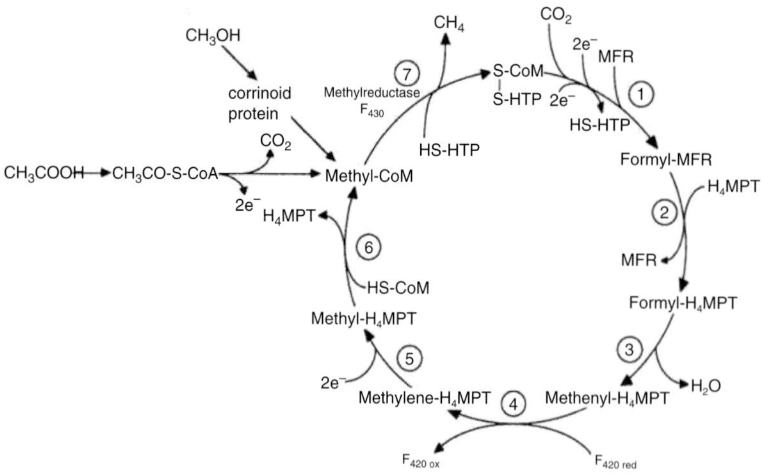

Fig. 11. The pathway of methane formation from acetate, methanol, and CO_2. Methyl-CoM is the key intermediate for the reduction of CO_2, acetate, and methanol to methane. The numbers refer to the seven steps of the cycle. (Adapted from Rouvière et al., 1988.)

The first step of CO_2 activation is not clearly understood (Fig. 11, step 1). It requires ATP, but not in stoichiometric amounts. The energy for the reaction is derived from the final step of methane formation (Fig. 11, step 7). This conclusion is supported by the observation that CO_2 activation in vitro requires small amounts of CH_3-S-CoM or a strong reductant such as titanium citrate (Bobik and Wolfe, 1989). The formyl group is transferred from MFR to H_4MPT by formyl methanofuran:tetra-hydromethanopterin formyltransferase (Fig. 11, step 2). The formyl group then undergoes sequential reduction beginning with its conversion to 5,10-methenyl-H_4MPT by the enzyme 5,10-methenyltetrahydromethanopterin cyclohydrolase (Fig. 11, step 3). The methenyl group is reduced by coenzyme F_{420} in a reaction catalyzed by methylenetetrahydromethanopterin:coenzyme F_{420} oxidoreductase to yield 5,10-methylene-H_4MPT (Fig. 11, step 4). The enzymes for steps 5 and 6 have not been purified to homogeneity. Step 5 is presumably catalyzed by methylene-H_4MPT reductase to yield methyl-H_4MPT. Conversion of methyl-H_4MPT to CH_3-S-CoM involves a corrinoid-containing methyltransferase (Kengen et al., 1988; Fig. 11, step 6).

The terminal reaction, catalyzed by the methylreductase system, represents the completion of the cycle with the release of CH_4 and the activation of CO_2 (Fig. 11, step 7). The methylreductase system catalyzes the conversion of CH_3-S-CoM and HS-HTP to CH_4 and CoM-S-S-HTP. Although the exact process is not known, methane formation is also coupled to ATP synthesis via a chemiosmotic mechanism (Blaut and Gottschalk, 1985). However, this coupling may occur during the reduction of the disulfide bond of CoM-S-S-HTP rather than during the reduction of CH_3-S-CoM (Ellermann et al., 1988). CoM-S-S-HTP is also required for the coupling of CH_3-S-CoM reduction to CO_2 activation (Bobik and Wolfe, 1988).

The process of CH_3-S-CoM reduction is complex and requires four protein components and at least three unique coenzymes. Components A1, A2, A3, and C are proteinaceous. A1 and A3 are oxygen-labile and have not been highly purified. A1 has coenzyme F_{420}-dependent hydrogenase activity and is probably required for the reduction of CoM-S-S-HTP (Rouvière et al., 1988). A3 has been resolved into two fractions: A3a and A3b. A3a is a large iron-sulfur protein, which may be involved in providing electrons for component C. A3b has methyl viologen-dependent hydrogenase activity (Rouvière and Wolfe, 1989). A2 is an oxygen-stable component and has been purified to homogeneity. A2 and A3 are probably required for the ATP-dependent reductive activation of component C (Rouvière et al., 1988). Component C is an oxygen-stable protein, which is the methyl-S-CoM methylreductase proper. It contains 2 moles of F_{430} and 2 moles of HS-CoM per mole of enzyme, and it has three types of subunits with a stoichiometry of $\alpha_2\beta_2\gamma_2$. The enzyme represents about 10% of the total cellular protein. The genes for component C are organized into a five-gene cluster, mcrBDCGA. The primary sequences of these genes are highly conserved in all methanogens examined. Subunits α, β, and γ are coded for by mcrA, mcrB, and mcrG, respectively. The functions of mcrC and mcrD are unknown (Allmansberger et al., 1989; Weil et al., 1989).

The methylreductase system is also required for methanogenesis from substrates other than CO_2 (Fig. 11). During the catabolism of acetate, the methyl and carboxyl groups are converted to CH_4 and CO_2, respectively, via an aceticlastic reaction. This reaction involves the initial activation of acetate as acetyl-CoA (Terlesky et al.,

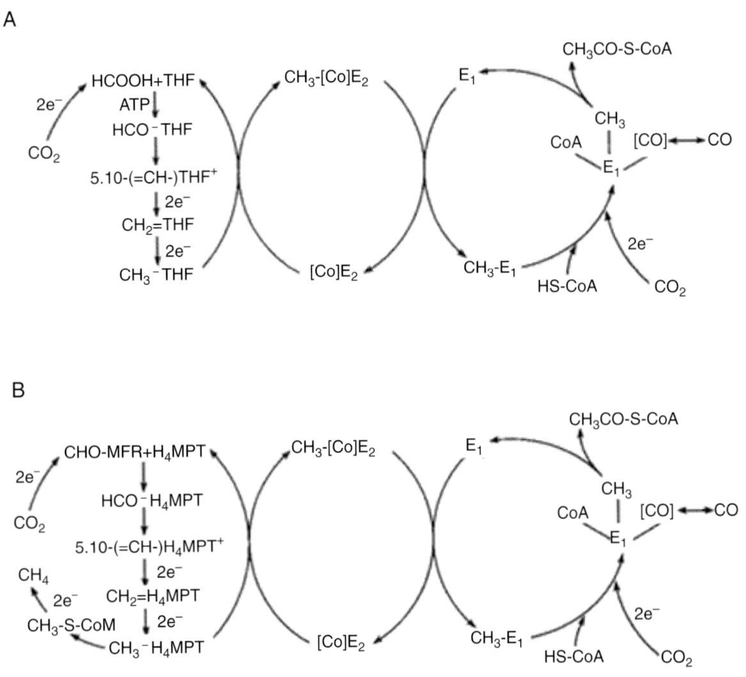

Fig. 12. The two acetogenic pathways for the autotrophic fixation of CO_2 to acetyl-CoA in (A) *Clostridium thermoaceticum* and (B) *Methanobacterium thermoautrotrophicum*. THF, tetrahydrofolate; E1, carbon monoxide dehydrogenase; E2, the corrinoid enzyme involved in methyl transfer. (Adapted from Jones et al., 1987.)

1987). The carbon-carbon bond is then cleaved by the carbon monoxide dehydrogenase system, producing HS-CoA, an enzyme-bound CO, and a methyl group. The methyl group is transferred to HS-CoM via H_4MPT with the aid of an unidentified corrinoid enzyme (Fischer and Thauer, 1989; van de Wijngaard et al., 1988). Oxidation of the enzyme-bound CO to H_2 and CO_2 provides electrons for the reduction of CH_3-S-CoM to CH_4 (Terlesky and Ferry, 1988b; Bott and Thauer, 1989).

Methane is also produced from the catabolism of methanol and methylamines. Methanogenesis from methanol can occur in two different ways depending on the electron donor: 1) if it is H_2, the methyl group of methanol is transferred to HS-CoM via two methyltransferases (MT1 and MT2). The former contains an oxygen-sensitive cobamide moiety as a methyl carrier. CH_3-S-CoM is then reduced to CH_4 by H_2. 2) When H_2 is absent, some of the methanol is oxidized in a disproportionation reaction to serve as reductant for the methylreductase system. Likewise, for methylamines the methyl group is either reduced to CH_4 or oxidized to CO_2. After growth in the presence of trimethylamine, *Methanosarcina barkeri* contains a trimethylamine:HS-coenzyme M methyltransferase similar to the methanol-induced methyltransferases.

Carbon Metabolism

Methanogens use only simple organic or inorganic carbon compounds as energy sources. About one-half of methanogens are also capable of autotrophic growth, and they obtain all their organic carbon from the assimilation of CO_2. The proposed methanogenic autotrophic pathway, the Ljungdahl-Wood pathway, is unlike classic autrotrophic CO_2 fixation (the ribulose bisphosphate pathway) since the primary intermediate is acetyl-CoA. The Ljungdahl-Wood pathway was discovered in the acetigenic clostridia and has been extensively studied in *Clostridium thermoaceticum* (Fig. 12A; Ljungdahl, 1986; Wood et al., 1986). The key enzyme is a nickel-containing carbon monoxide dehydrogenase system (CODH). In methanogens, the one-carbon carrier of the clostridial pathway, folate, is replaced by tetrahydromethanopterin (H_4MPT) (Fig. 12B; Fuchs, 1986). The pathways of autotrophy and methanogenesis are closely linked, and they share common intermediates. Thus, many of the methanogenic coenzymes serve both anabolic and catabolic functions. CO_2 is first reduced to formyl-MFR. The formyl group is then transferred to H_4MPT and reduced to a methyl group. At this point the anabolic and catabolic pathways diverge. The subsequent transfer of this methyl group to HS-CoA is believed to involve a cobamide-containing protein. The carboxyl moiety is derived from CO_2 via CODH and a carbonylation reaction. At this step the carboxyl group freely exchanges with CO in the environment. Although acetogenesis in methanogens appears similar to the pathway in eubacteria, it is distinctive due to the absence of folates and the lack of an ATP requirement for the activation of CO_2.

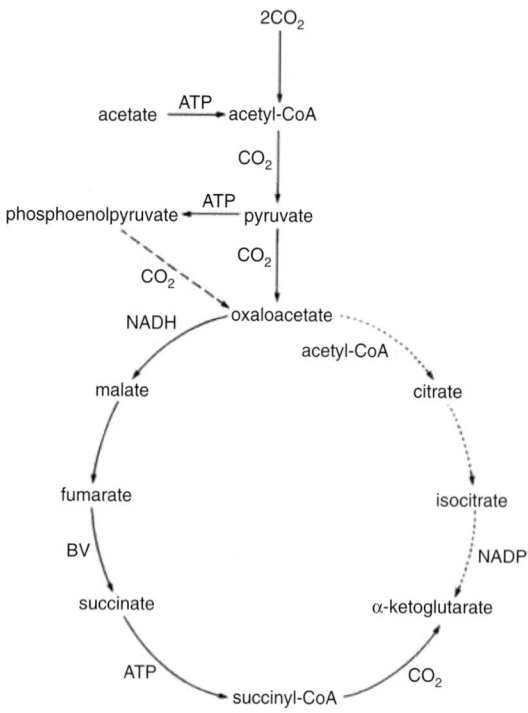

Fig. 13. The pathways of carbon assimilation in methanogens. Solid lines represent the incomplete reductive TCA cycle found in *Methanobacterium thermoautotrophicum* and *Methanococcus maripaludis*. Broken arrows represent the incomplete oxidative TCA cycle found in *Methanosarcina barkeri*. (Shieh and Whitman, 1987.)

Methanogens utilize the anabolic tricarboxylic acid (TCA) cycle to synthesize cell carbon from acetyl-CoA (Jones et al., 1987; Fig. 13). However, the exact nature of the cycle varies among different groups of methanogens. In *Methanosarcina barkeri,* α-ketoglutarate is formed by an oxidative branch. *Methanospirillum, Methanococcus,* and *Methanobacterium* utilize the reductive or reverse pathway (Fuchs and Stupperich, 1986). Hexoses are derived from acetate by gluconeogenesis. The major storage product for carbohydrates is glycogen. Very little information is available on how this polysaccharide reserve is used, since no known methanogens can catabolize exogenous sugars. *Methanosarcina thermophila* is capable of mobilizing its reserves and actually producing small amounts of acetate when starved for acetate (Murray and Zinder, 1987). Likewise, *Methanolobus* produces small amounts of methane upon mobilization of its glycogen reserves (König et al., 1985).

Heterotrophic methanogens rely on exogenous acetate for growth. An incomplete, reductive TCA cycle has been found in *Methanococcus voltae,* indicating the presence of a mechanism of anabolic carbon metabolism that is similar to the one found in autotrophic methanogens (Shieh and Whitman, 1987).

Nitrogen and Sulfur Metabolism

Ammonia serves as the sole nitrogen source during autotrophic growth, although some methanogens are capable of utilizing a variety of other nitrogenous compounds. The presence of glutamate synthase in *Methanobacterium thermoautotrophicum* enables it to use glutamate as a sole nitrogen source (Kenealy et al., 1982). Some methanogens have also been shown to utilize urea, purines, and dinitrogen (Bhatnagar et al., 1984; DeMoll and Tsai, 1986; Lobo and Zinder, 1988). Nitrogen fixation (diazotrophy) in *Methanosarcina barkeri* requires molybdenum and is inhibited by tungsten, like in the eubacteria (Lobo and Zinder, 1988). For at least one strain, vanadium will substitute for molybdenum (Scherer, 1989). The rate of biosynthesis of the nitrogenase enzyme decreases in the presence of ammonia. In addition, the decrease in growth rate suggests diazotrophy is an energetically expensive process, as it is in eubacteria. Although the enzyme has not been purified, genetic evidence suggests sequence homology between methanogen and eubacterial nitrogenase genes (Sibold et al., 1985).

Sulfide is commonly included as a reductant and a sulfur source in the cultivation of methanogens. However, several species have the ability to use other inorganic sulfur sources, including elemental sulfur, sulfate, sulfite, and thiosulfate (Daniels et al., 1986). The observed reduction of sulfate and sulfite is strictly assimilatory in methanogens. Therefore, it is unlikely that these compounds can serve as alternate electron acceptors. Organic sulfur sources include methionine and cysteine (Mazumder et al., 1986; Scherer and Sahm, 1981b).

Cell Envelope Structure

The diversity of the methanogenic bacteria is demonstrated by the large differences in cell envelope structure. These topics are reviewed in detail by König (1988). The three known types of cell walls are composed of pseudomurein, protein, and heteropolysaccharide.

The cell wall of the Methanobacteriaceae (Fig. 14A and B) is composed of pseudomurein, a peptidoglycan analogous to eubacterial murein (Fig. 15). Pseudomurein contains L-talosaminuronic acid instead of muramic acid, β(1–3) linkages instead of β(1–4) linkages in the polysaccharide backbone, and different sequences of crosslinking amino acids. When these differences are considered, it is obvious

Fig. 14. (A) *Methanobacterium thermoautotrophicum* prepared by freeze-substitution in osmium-acetone. The wall (W) averages 15–16 nm in thinkness and the membrane (PM) 8 nm. (B) *M. thermoautotrophicum* prepared by chemical fixation: glutaraldehyde-formaldehyde mixture followed by osmium tetroxide. Appearance is very similar to freeze-substituted cell (A and B from Aldrich et al., 1987). (C) Chemically fixed cells of *Methanosarcina mazei*. Wall (CW) in this preparation appears expanded and fluffy. Wall thickness is 40–50 nm. Small arrows indicate plasma membrane. (D) Freeze-fractured cell of *Methanosarcina barkeri*, showing large expanse of plasma membrane. Arrows delimit areas in which 12–13-nm-diameter intramembrane particles are arranged in regular arrays. (E) Cell of *M. barkeri* freeze-substituted in osmium-acetone. Wall (CW) appears dense and compact and measures about 20 nm in thickness. Trilaminar plasma membrane (PM) measures 7 nm. (F) Freeze-substituted cell of *M. mazei*. Wall and membrane are similar in size and appearance to those shown in (E). Bars = 100 nm. (Figs. 10C–F are courtesy of H.C. Aldrich.)

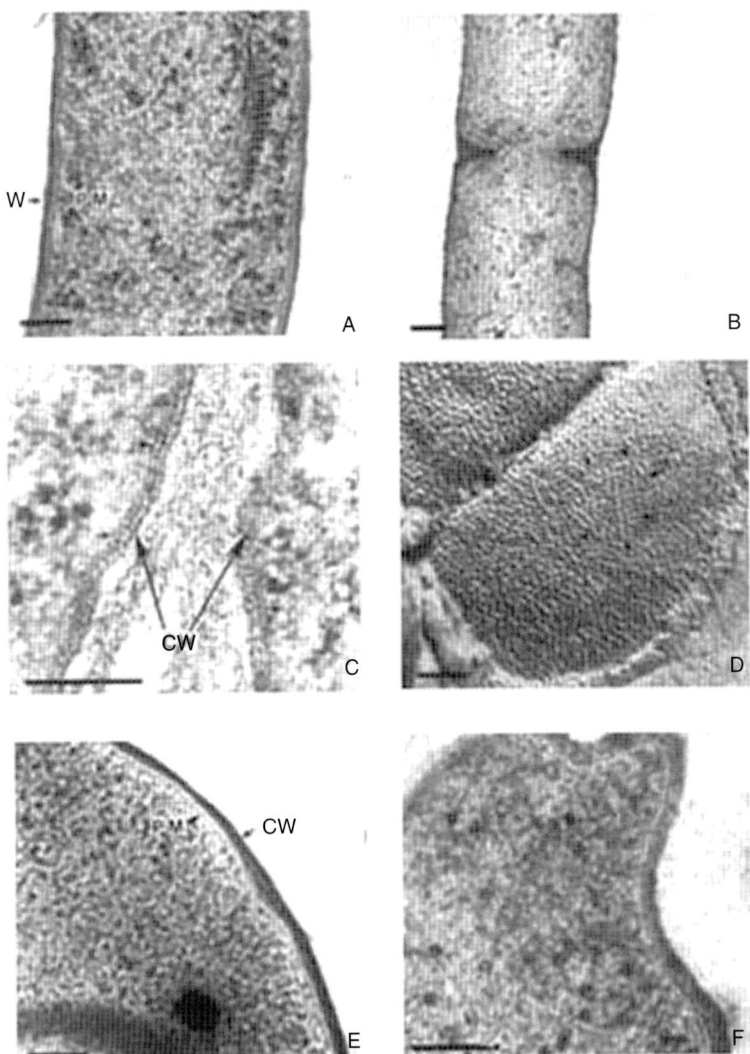

why *Methanobacterium* is resistant to the β-lactam antibiotics that inhibit eubacterial cell wall biosynthesis. Variation of pseudomurein structure within the Methanobacteriaceae occurs by alteration of the amino acid residues.

In contrast is the simple protein cell wall of the Methanococcaceae and most of the Methanomicrobiaceae. It consists of a crystalline arrangement of proteins or glycoproteins called an S-layer (Fig. 16). Unlike the rigid pseudomurein, the S-layer provides only limited support; therefore, the cells are osmotically fragile. The composition of the S-layer varies between genera based on the presence of glycosylation and between species by molecular weight of the subunits and antigenicity.

Members of the Methanosarcinaceae have a third type of cell wall, referred to as a heteropolysaccharide. The cell wall of *Methanosarcina* is composed of a polysaccharide matrix (Fig. 14 C–F), which is similar in structure to eukaryotic chondroitin found in connective tissue. This polysaccharide stabilizes the cell aggregates formed by *Methanosarcina* species. The work of Aldrich et al. (1986) and Kreisl and Kandler (1986) has elucidated the chemical structure and arrangement of these polysaccharides on the cell surface as a loose matrix.

Methanosaeta ("*Methanothrix*") and *Methanospirillum*, which are characteristically filamentous, have an outer protein sheath (Figs. 17 and 18). The cells are surrounded by an electron-dense inner wall of unknown composition and are separated by septa or plugs. The crystalline structure of the sheath has been studied in considerable detail by Beveridge et al. (1986) and Patel et al. (1986).

The cell membranes of methanogens are characterized by several unusual lipids (see reviews by De Rosa and Gambacorta, 1988; Langworthy, 1985; and Jones et al., 1987). In the majority of methanogens, the lipids are polar, either diphyt-

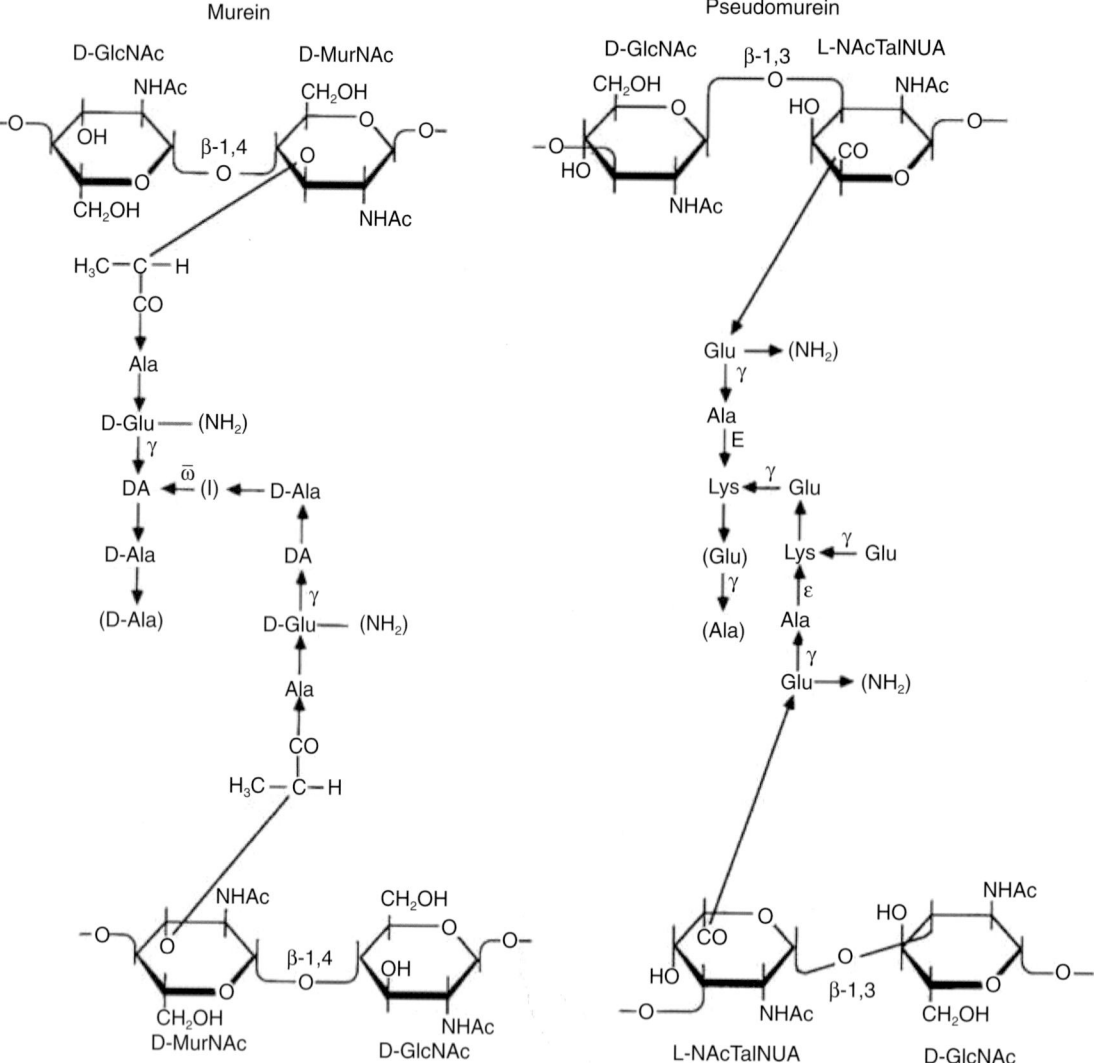

Fig. 15. Comparison of murein and pseudomurein structure. D-GlcNAc, D-N-acetylglucosamine; D-MurNAc, D-N-acetylmuramic acid; L-NAcTalNOA, L-N-acetyltalosaminuronic acid; L-DA, L-diamino acid; (I), interpeptide bridge. (Adapted from Kandler and König, 1985.)

anyl-glycerol-diethers or dibiphytanyl-diglyc-erol-tetraethers (Fig. 19). Thus, the membrane structure in many methanogens differs from the lipid bilayer typical of eubacteria. The tetra-ethers appear to span the membrane, with the polar head groups at opposite sides. Therefore, the membranes are probably arranged in a monolayer with bilayer regions resulting from interspersed diethers. As in eubacteria and eukaryotes, these lipids are often present as phospho- or glyco-derivatives. The major nonpo-lar lipids consist of isoprenoid and hydroiso-prenoid hydrocarbons.

Molecular Biology of the Methanogens

Because the methanogens and other archaebac-teria represent a very ancient line of descent,

there has been a great deal of interest in their molecular biology. Thus, the molecular features that the methanogens share with the eubacteria and eukaryotes may have been inherited from the common ancestor of all three kingdoms. Only a brief overview of this topic is presented here, but a number of detailed reviews on different aspects are available (Brown et al., 1989; Dennis, 1986; Fox, 1985; Jones et al., 1987; Kandler and Zillig, 1986; Matheson, 1985; Thomm et al., 1986; Woese and Wolfe, 1985; Zillig et al., 1988). Furthermore, an issue of the *Canadian Journal of Microbiology* (1989; vol. 35, no. 1) is devoted to this topic.

Methanogens have some genetic characteris-tics of both eubacteria and eukaryotes. Sequence analysis of methanogen genes suggests that they are arranged in operons similar to those in

Fig. 16. (A) Thin section of *Methano-coccus voltae* showing the S-layer (S) and plasma membrane (PM). (From Jarrell and Koval, 1989.) (B) Freeze-etch replica of *Methanococcus voltae* showing the hexagonal arrangement of the S-layer protein (arrows). (From Koval and Jarrell, 1987.) Bars = 100 nm.

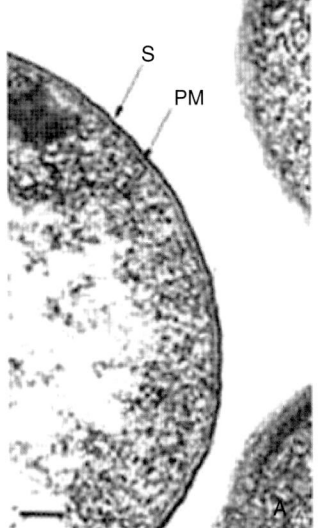
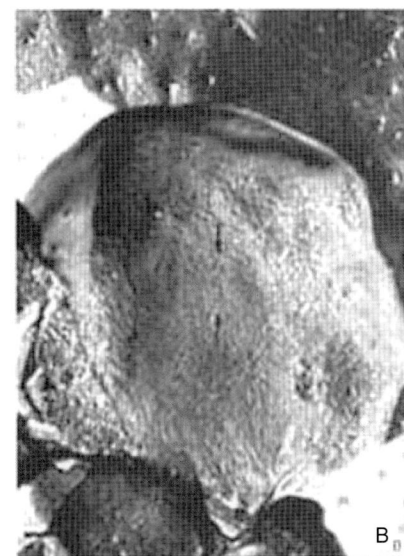

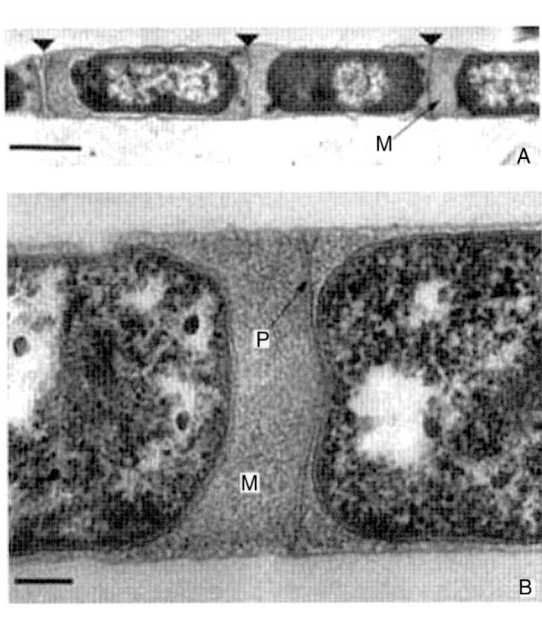
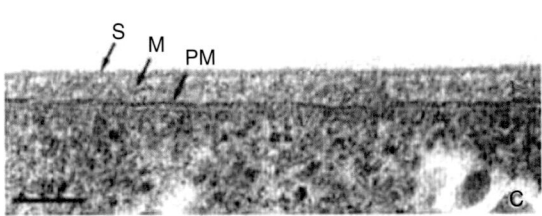

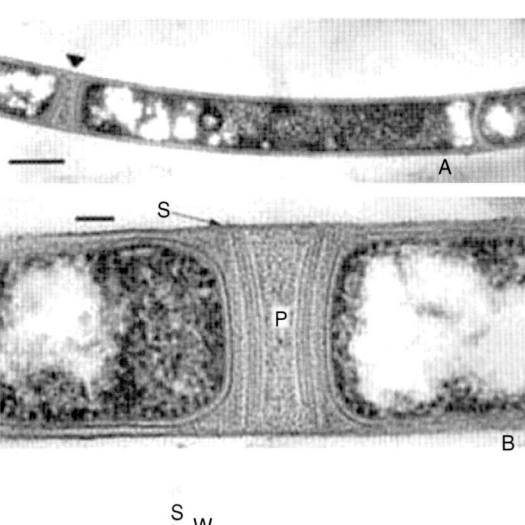
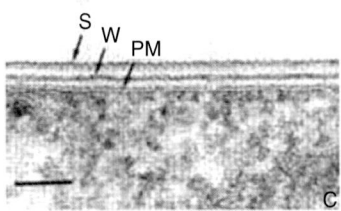

Fig. 17. (A) Thin section of a *Methanosaeta consilii* filament showing spacer plugs (arrow heads) separating the cells and the amorphous matrix (M) surrounding each cell. Bar = 1 μm. (B) Thin section showing the spacer plug (P) and the matrix (M) between two cells. Bar = 100 nm. (C) Thin section show-ing the sheath (S), the matrix (M), and the plasma membrane (PM). Bar = 100 nm. (From Beveridge et al., 1986.)

Fig. 18. (A) Thin section of a *Methanospirillum hungatei* fil-ament showing a spacer plug (arrow head) separating two cells. Bar = 500 nm. (B) Thin section showing spacer plug (P) and sheath (S). Bar = 100 nm. (C) Thin section showing sheath (S), cell wall (W), and plasma membrane (PM). Bar = 100 nm. (From Beveridge et al., 1987.)

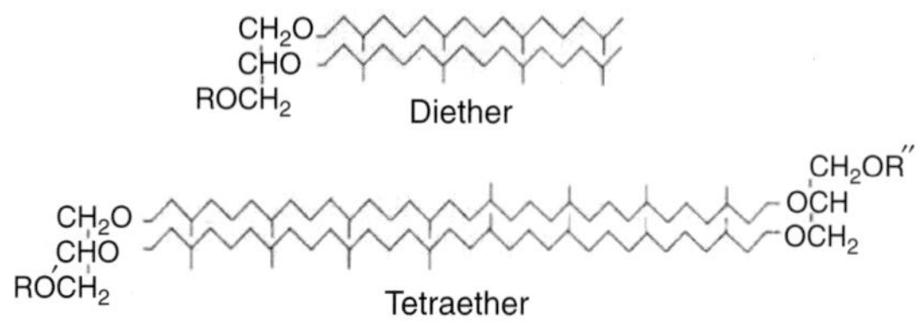

Fig. 19. Structures of a characteristic diether lipid and tetraether lipid found in the membranes of methanogenic bacteria. (Adapted from Jones et al., 1987.)

eubacteria. However, the process of transcription initiation and the termination sequences of certain genes are more similar to those found in eukaryotes. Moreover, the methanogen DNA polymerases that have been purified resemble both eubacterial and eukaryotic enzymes.

The DNA-dependent RNA polymerase of methanogens is genetically more closely related to the enzyme of eukaryotes than of eubacteria (Bernghofer et al., 1988). As a consequence, it might be expected that the methanogenic promoter sequences should also resemble eukaryotic sequences. Analysis of these sequences in numerous methanogen operons has revealed two conserved sequences. One has the consensus sequence of AAANNTTTATATA, where N represents any of the four nucleotides (Allmansberger et al., 1988). This sequence is found 25 bp upstream of the start site of transcription and resembles the "TATA" box found in eukaryotes. Binding of purified RNA polymerase to this sequence has been demonstrated in the methylreductase operon, *hisA*, and in the ribosomal RNA genes of *Methanococcus vannielii* (Thomm et al., 1989). The second conserved region, TGCAAGT, has been found at the start site of transcription of the ribosomal RNA genes. Termination of transcription has not been studied in detail. However, the gene *mcrA*, which codes for the α subunit of methylreductase, has a sequence analogous to the rho-independent terminator of eubacteria. In contrast, analysis of rRNA genes suggests a termination mechanism similar to that of eukaryotes, which use AT-rich sequences.

DNA polymerases in methanogens are heterogeneous. The enzyme purified from *Methanococcus vannielii* resembles the type A enzyme of eukaryotes in its sensitivity to the inhibitor aphidicolin (Zabel et al., 1987). In contrast, the purified enzyme of *Methanobacterium thermoautotrophicum* is resistant to aphidicolin and is similar to DNA polymerase I of *Escherichia coli* (Klimczak et al., 1986). The enzymes of *M. vannielii* and *M. thermoautotrophicum* also differ from each other in molecular weight.

The methanogen ribosome has a "typical" prokaryotic composition. The ribosome is 70S in size and dissociates into 30S and 50S subunits. The 30S subunit contains 16S rRNA, and the 50S subunit contains 5S and 23S rRNAs. A 5.8S rRNA, as found in eukaryotes, is not present. However, the structure of the methanogen ribosome differs from the eubacterial ribosome in that the 30S subunit has the characteristic archaebacterial "bill," a small protrusion for which the function is unknown. A similar "bill" is also present on the eukaryotic 40S subunit (Lake et al., 1982). Ribosomal proteins of methanogens and other archaebacteria are unusually acidic, in contrast to the basic ribosomal proteins of eubacteria and eukaryotes. However, the number of acidic ribosomal proteins correlates well with the internal K^+ concentration in the strains studied. The ribosomal proteins of *M. vannielii* are encoded by a transcriptional unit homologous in organization to the "spectinomycin operon" of *E. coli*. However, the deduced amino acid sequences of the methanogen proteins more closely resemble those of their eukaryotic homologs (Auer et al., 1989; Köpke and Wittmann-Liebold, 1989).

Sequencing of methanogen rRNAs has been instrumental in establishing their phylogenetic relationship with eubacteria and eukaryotes. The secondary structure of the methanogen 5S rRNA has characteristics of both the eubacterial and eukaryotic molecule. However, the methanogen rRNA molecule has characteristics that are unique to archaebacteria. Sequence data for the 16S rRNA support the hypothesis of the extreme diversity of the archaebacteria. The secondary structure of the methanogen 16S rRNA is more similar to that of the eubacterial structure than the eukaryotic secondary structure. Less information is available concerning the secondary structure of the 23S rRNA molecule. However, it does have characteristics of both the eubacterial and eukaryotic molecules. The genes for rRNA are arranged in a typical operon similar to that of eubacteria. The 16S, 23S, and 5S genes are closely linked and interspersed with several tRNA genes. *Methanococcus vannielii* and *M. voltae* also have unlinked 5S rRNA genes that are associated with tRNA-encoding genes.

Antibiotic Sensitivity and Resistance

The lack of sensitivity of methanogens and other archaebacteria to most antibiotics is a reflection of their unique biochemical nature and has been reviewed previously (Böck and Kandler, 1985). Methanogens are resistant to many commonly used antibiotics simply because the specific target is not present. Therefore, comparison of the antibiotic sensitivity spectrum of methanogens

Table 17. Comparison of the antibiotic sensitivity of several species of methanogens, eubacteria and eukaryotes.

Target	Antibiotic	Size of zone of inhibition (mm)									
		Methanogen[a]						Eubacterium[b]		Eukaryote[c]	
		A	B	C	D	E	F	G	H	I	J
Cell wall	Fosfomycin	—[d]	—	—	—	—	—	10	12	—	—
	D-Cycloserine	—	—	—	9	—	—	13	16	—	—
	Vancomycin	—	—	—	—	—	—	—	10	—	—
	Penicillin G	—	—	—	—	—	—	7	26	—	—
	Cephalosporin C	—	—	—	—	—	—	2	22	—	—
	Nocardicin A	—	—	—	—	—	—	2	—	—	—
	Bacitracin	40	27	40	17	—	—	—	8	—	—
	Gardimycin	30	16	6	—	—	—	—	3	—	—
	Nisin	11	8	10	3	—	—	—	3	—	—
	Enduracidin	20	16	10	5	—	8	—	10	—	—
	Flavomycin	—	—	—	—	—	—	3	16	—	—
	Subtilin	4	7	5	10	—	—	—	6	—	—
RNA polymerase	α-Amanitin	—	—	—	ND[e]	—	ND	ND	ND	ND	ND
	Rifampicin	2	—	—	2	—	—	9	22	—	—
Protein biosynthesis	Cycloheximide	—	—	—	—	—	—	—	—	23	22
	Chloramphenicol	25	13	22	28	4	40	11	10	—	—
	Virginiamycin	5	14	—	15	—	—	5	19	—	—
	Gentamycin	—	25	12	—	—	—	9	13	—	—
	Tetracycline	—	—	8	8	—	—	13	18	—	—
	Oleandomycin	—	—	—	—	—	—	—	13	—	—
	Erythromycin	—	—	—	3	—	—	3	17	—	—
	Kanamycin	—	—	—	—	—	—	9	13	—	—
Membrane	Garmicidin S	7	7	15	10	4	16	—	4	3	4
	Garmicidin D	—	—	11	—	—	—	—	4	—	—
	Polymyxin	—	—	8	—	—	—	5	—	4	3
	Amphotericin B	—	—	—	—	—	—	—	—	6	5
	Valinomycin	—	—	—	—	—	—	2	2	—	—
	Nonactin	—	—	—	—	—	—	2	5	—	—
	Monensin	—	5	25	12	15	25	—	6	—	—
	Lasalocid	22	25	42	21	40	40	—	15	—	—

[a]Species identification: A, *Methanobrevibacter smithii*; B, *Methanobacterium bryantii*; C, *M. thermoautotrophicum*; D, *Methanococcus vannielii*; E, *Methanospirillum hungatei*; F, *Methanosarcina barkeri*.
[b]G, *Escherichia coli*; H, *Staphylococcus aureus*.
[c]I, *Saccharomyces cerevisiae*; J, *Hansenula* sp.
[d]—, No inhibition.
[e]ND, not determined.
Adapted from Hilpert et al. (1981).

with those of eubacteria and eukaryotes can provide insight into their biochemical similarities and differences.

The differences in chemical structure of methanogen cell envelopes to those of eubacteria have been well established. These differences are supported by the lack of sensitivity of methanogens to many cell-wall inhibitors (Table 17). For example, *Methanobacterium thermoautotrophicum* is resistant to penicillin G, since the crosslinking of pseudomurein, which replaces peptidoglycan in the cell wall, is not inhibited by the antibiotic. However, antibiotics such as bacitracin, which inhibit the formation of lipid-bound precursors of murein in eubacteria, may inhibit formation of similar lipid-bound precursors involved in biosynthesis of lipids and glycoproteins in methanogens.

In addition to genetic homology, the antibiotic sensitivity of DNA-dependent RNA polymerase reflects the similarity of the methanogen enzyme to the eukaryotic enzyme (Table 17). Both enzymes are resistant to rifampicin, an antibiotic which inhibits RNA polymerase of eubacteria. The methanogen enzyme is also resistant to α-amanitin, which inhibits the type B and C polymerases of eukaryotes but not the type A enzyme.

Antibiotics that inhibit protein synthesis may block aminoacyl-tRNA synthesis, elongation, or translation. Pseudomonic acid is a potent inhibitor of *Methanococcus voltae* and inhibits isoleucyl-tRNA synthetase activity (Possot et al., 1988). Translation in methanogens is resistant to many of the antibiotics that inhibit translation by either the eubacterial 70S or eukaryotic 80S ribosomes (Böck and Kandler, 1985). While growth of the methanogens is sensitive to chloramphenicol, this inhibition is not due to an effect on protein synthesis. Instead, inhibition is probably due to the sensitivity of the particulate hydrogenase of *Methanobacterium* to this antibiotic and to other halogenated compounds. However, the elongation factor aEF-2 is sensitive to diphtheria toxin, which is also characteristic of the eukaryotic protein (Lechner et al., 1988). Sequence analysis of the genes encoding aEF-2 indicates that the methanogen enzyme is more closely related to its eukaryotic than its eubacterial homolog. In addition, all archaebacterial aEF-2 genes examined possess a histidine residue that is posttranslationally modified to diphthamide, a phenomenon analogous to that of the eukaryotic EF-2. Translation in the methanogens is also resistant to most antibiotics in the streptomycin class of aminoglycosides, as well as to tetracycline and the macrolides. However, methanogens are sensitive to deoxystreptamine aminoglycosides, such as gentamycin and neomycin, as well as to virginiamycin. Therefore, the translation process in methanogens is different from that of the eubacterial process with respect to antibiotic sensitivity.

Antibiotics that act on membrane integrity also have varying effects on methanogens, as they do with eubacteria. Ionophores such as monensin, lasalocid, and gramicidin S are very effective inhibitors of methanogens. This inhibition is presumably the result of interference with Na^+ transport and thus with methanogenesis (Perski et al., 1982). Monensin and lasalocid inhibit both growth and methanogenesis (Dellinger and Ferry, 1984). However, this inhibition depends on the culture conditions, since they both stimulate methanogenesis at low Na^+ concentrations in vivo (Schönheit and Beimborn, 1986).

Motility

Motility in methanogens has been associated with the presence of flagella. However, little is known about their structure, attachment to the cell envelope, and the energetics of motility. Kalmokoff et al. (1988) found that the flagella of *Methanococcus voltae*, consisting of two protein subunits with molecular weights of 33,000 and 31,000, have a filament diameter of 13 nm, unlike those of eubacteria which generally consist of only one monomer. The basal structure is knoblike in appearance, as opposed to the multiple ring structure found in eubacteria. This may be due to the simpler structure of the methanococcal S-layer. It will be interesting to compare the basal structures of methanogens that have more complex cell envelopes. The arrangement of flagella varies greatly among methanogen genera, as it does in eubacteria. *Methanococcus jannaschii* has lophotrichous flagella with two bundles of flagella at one pole arranged in a "corkscrew" fashion (Jones et al., 1983a). In contrast, *Methanogenium* species have peritrichous flagella (Romesser et al., 1979).

Chemotaxis in methanogens has been described in *Methanospirillum hungatei*, for which acetate is a chemoattractant (Migas et al., 1989). In addition, *Methanococcus voltae* responds to acetate, isoleucine, and leucine as attractants but not to histidine (Sment and Konisky, 1989). *M. voltae* has an absolute requirement for all these nutrients except histidine. Therefore, it is evident that some methanogens respond positively to higher nutrient concentrations. The advantage of this in certain habitats, such as marine marshes where substrate concentrations may be low, is self evident.

Applications

Methanogenic bacteria have numerous applications, especially in the degradation of wastes rich in organic matter to methane and CO_2. Methane is the major component of natural gas, a clean fuel which is already in wide use. This topic has been reviewed extensively (Daniels, 1984; Hobson et al., 1981; van den Berg, 1984; Wodzinski et al., 1987), and only a few major points will be discussed here.

Because of the narrow substrate specificity of methanogens, pure cultures have limited usefulness in waste treatment. For most applications, consortia containing both methanogenic bacteria and heterotrophic bacteria capable of converting the complex organic materials common in waste materials to methanogenic substrates are used. The consortia are usually obtained from an inoculum of manure, sewage sludge, freshwater sediment, or some other natural source, and they are formed upon enrichment with the waste material of interest under methanogenic conditions. In this manner, consortia have been obtained for the degradation of a wide variety of wastes and other materials. Moreover, thermophilic consortia have been obtained by performing the enrichments at high temperatures.

For example, methanogenic consortia that degrade whey, a major by-product in milk and cheese production, have been described (Chartrain and Zeikus, 1986a, 1986b; Wildenauer and Winter, 1985; Zellner et al., 1987b; Zellner and Winter, 1987b). These consortia were composed of sugar- and protein-fermenting strains of *Lactobacillus*, *Eubacterium*, *Clostridium*, *Klebsiella*, and *Leuconostoc*. Their fermentation products were then metabolized by acetogenic strains of *Desulfovibrio*, *Eubacterium*, and *Clostridium*. The acetate, H_2, and formate formed were substrates in turn for the methanogens present, *Methanobacterium*, *Methanobrevibacter*, *Methanospirillum*, *Methanocorpusculum*, *Methanosarcina*, and *Methanosaeta*. These complex consortia were capable of the nearly complete conversion of whey to methane and CO_2 within 5 days.

Another common application of methanogens is in the treatment of domestic sewage. Whether it be a simple cesspool or an advanced anaerobic digestor, the methane produced from sewage can be collected and used to offset the cost of an otherwise obligatory process. In countries such as India and China, where the economic resources required for such digestors are not available, animal and human wastes are still used as a source of energy. Instead of advanced digestors, these countries have developed inexpensive small scale digestors for production of methane. In addition, the spent slurry can be used as a natural fertilizer. Obvious drawbacks of these digestors are their limited size and requirement for manual loading and emptying. The digestors are also limited to temperate climates since they must rely on the ambient temperature to eliminate the additional heating costs required to maintain an active biomass.

While consortia can be obtained for the conversion of many types of organic wastes to methane and CO_2, the growth of these consortia is usually slow when compared to growth of those composed of aerobic microorganisms. This can cause problems since high waste input can result in loss of cell biomass, and thus waste decomposition is limited by the growth rate of the bacteria. Therefore, to increase efficiency, specialized bioreactors have been designed to retain the microbial population in the bioreactor while allowing for a high input of waste. In this way, the rate of waste decomposition is not limited by the growth rate of the bacteria. The five types of bioreactors commonly used have been reviewed by van den Berg (1984) and are described below:

1. The anaerobic contact reactor uses a sludge, which settles to the bottom and contacts the raw waste. Thus, the settling ability and mixing of the sludge with the waste are important in the efficiency of bioconversion. This bioreactor type works well for particulate wastes, which settle easily and are completely biodegradable.
2. The anaerobic filter reactor retains suspended bacteria and waste in a packing material or solid support where degradation occurs. It is advantageous for treatment of dilute soluble wastes. However, it is easily plugged by suspended particulate waste.
3. The upflow anaerobic sludge bed reactor avoids plugging by reducing the volume of the packing material and using a gas collector to encourage settling of the sludge. High concentrations of microbial biomass are needed in addition to proper development of a granular sludge to promote settling.
4. The anaerobic fluidized or expanded bed reactors allow growth of biomass on inert particles, which settle and are suspended by a rapid flow of waste.
5. The downflow stationary fixed-film reactor avoids plugging of the anaerobic filter by forming an active biomass film. This reactor can accommodate a wide variety of wastes.

Another area of potential applications of methanogenic consortia is in degradation of xenobiotics (for reviews see Berry et al., 1987; Young, 1984). Especially important is the degradation of halogenated aromatic and aliphatic compounds, which are common pollutants. For instance, tetrachloroethylene is a common pollutant in ground waters due to improper disposal in landfills (Fatherpure and Boyd, 1988). In the presence of sewage sludge or pure cultures of methanogens, it is reductively dehalogenated to trichloroethylene. In contrast, tetrachloroethylene is not rapidly degraded under aerobic conditions. Likewise, low concentrations of chloroform, carbon tetrachloride, and 1,2-dichloroethane are almost completely oxidized to CO_2 by methanogenic consortia (Bouwer and McCarty, 1983). Haloaromatic compounds are also reductively dehalogenated by methanogenic consortia (Suflita et al., 1983). Thus, chlorinated phenols and chlorinated benzoates are converted to phenol and benzoate, respectively. Phenol may then be carboxylated to form benzoate (Knoll and Winter, 1989). Benzoate is then converted to methane and CO_2, presumably by a reductive route via cyclohexane carboxylate.

Methanogens have also been implicated in anaerobic biocorrosion, which is a significant economic problem. By the process of cathodic depolarization, elemental iron is oxidized and protons in water are reduced to H_2 (Daniels et al., 1987). Methanogens accelerate this process by maintaining a very low partial pressure of H_2. The mechanism is similar to that proposed for sulfate-reducing bacteria, which also consume H_2. Protons from water combine with electrons from iron to form hydrogen, which is utilized by the hydrogenase enzyme.

Acknowledgements. The authors wish to thank Drs. Terry Miller, Josef Winter, and Tatjana

N. Zhilina for their critical reading of this manuscript.

Literature Cited

Ahring, B. K., P. Westermann. 1984. Isolation and characterization of a thermophilic, acetate-utilizing methanogenic bacterium. FEMS Microbiol. Lett. 25:47–52.

Ahring, B. K., P. Westermann. 1985. Methanogenesis from acetate: physiology of a thermophilic acetate-utilizing methanogenic bacterium. FEMS Microbiol. Lett. 28:15–19.

Aldrich, H. C., D. B. Beimborn, P. Schönheit. 1987. Creation of artifactual internal membranes during fixation of *Methanobacterium thermoautotrophicum*. Can. J. Microbiol. 33:844–849.

Aldrich, H. C., R. W. Robinson, D. S. Williams. 1986. Ultrastructure of *Methanosarcina mazei*. Syst. Appl. Microbiol. 7:314–319.

Allmansberger, R., M. Bokranz, L. Kröckel, J. Schallenberg, A. Klein. 1989. Conserved gene structures and expression signals in methanogenic archaebacteria. Can. J. Microbiol. 35:52–57.

Allmansberger, R., S. Knaub, A. Klein. 1988. Conserved elements in the transcription initiation regions preceding highly expressed structural genes of methanogenic archaebacteria. Nucl. Acids Res. 16:7419–7436.

Auer, J., G. Spicker, A. Böck. 1989. Organization and structure of the *Methanococcus* transcriptional unit homologous to the *Escherichia coli* "spectinomycin operon": implications for the evolutionary relationships of 70 S and 80 S ribosomes. J. Mol. Biol. 209:21–36.

Balch, W. E., G. E. Fox, L. J. Magrum, C. R. Woese, R. S. Wolfe. 1979. Methanogens: reevaluation of a unique biological group. Microbiol. Rev. 43:260–296.

Balch, W. E., R. S. Wolfe. 1976. New approach to the cultivation of methanogenic bacteria: 2-mercaptoethanesulfonic acid (HS-CoM)-dependent growth of *Methanobacterium ruminantium* in a pressurized atmosphere. Appl. Environ. Microbiol. 32:781–791.

Belay, N., R. Johnson, B. S. Rajagopal, E. Conway de Macario, L. Daniels. 1988. Methanogenic bacteria from human dental plaque. Appl. Environ. Microbiol. 54:600–603.

Belyaev, S. S., A. Ya. Obraztsova, K. S. Laurinavichus, L. V. Bezrukova. 1986. Characteristics of rod-shaped methane-producing bacteria from an oil pool and description of *Methanobacterium ivanovii* sp. nov. Microbiology (Engl. Trans. Mikrobiologiya) 55:821–826.

Belyaev, S. S., R. Wolkin, W. R. Kenealy, M. J. DeNiro, S. Epstein, J. G. Zeikus. 1983. Methanogenic bacteria from the Bondyuzhskoe oil field: general characterization and analysis of stable-carbon isotopic fractionation. Appl. Environ. Microbiol. 45:691–697.

Berghöfer, B., L. Kröckel, C. Körtner, M. Truss, J. Schallenberg, A. Klein. 1988. Relatedness of archaebacterial RNA polymerase core subunits to their eubacterial and eukaryotic equivalents. Nucl. Acids Res. 16:8113–8128.

Berry, D. F., A. J. Francis, J.-M. Bollag. 1987. Microbial metabolism of homocyclic and heterocyclic aromatic compounds under anaerobic conditions. Microbiol. Rev. 51:43–59.

Beveridge, T. J., B. J. Harris, G. D. Sprott. 1987. Septation and filament splitting in *Methanospirillum hungatei*. Can. J. Microbiol. 33:725–732.

Beveridge, T. J., G. B. Patel, B. J. Harris, G. D. Sprott. 1986. The ultrastructure of *Methanothrix concilii*, a mesophilic aceticlastic methanogen. Can. J. Microbiol. 32:703–710.

Bhatnagar, L., M. K. Jain, J.-P. Aubert, J. G. Zeikus. 1984. Comparison of assimilatory organic nitrogen, sulfur, and carbon sources for growth of *Methanobacterium* species. Appl. Environ. Microbiol. 48:785–790.

Biavati, B., M. Vasta, J. G. Ferry. 1988. Isolation and characterization of "*Methanosphaera cuniculi*" sp. nov. Appl. Environ. Microbiol. 54:768–771.

Blaut, M., G. Gottschalk. 1985. Evidence for a chemiosmotic mechanism of ATP synthesis in methanogenic bacteria. Trends Biochem. Sci. 10:486–489.

Bleicher, K., G. Zellner, J. Winter. 1989. Growth of methanogens on cyclopentanol/CO_2 and specificity of alcohol dehydrogenase. FEMS Microbiol. Lett. 59:307–312.

Blotevogel, K.-H., U. Fischer. 1985. Isolation and characterization of a new thermophilic and autotrophic methane producing bacterium: *Methanobacterium thermoaggregans* spec. nov. Arch. Microbiol. 142:218–222.

Blotevogel, K.-H., U. Fischer. 1989. Transfer of *Methanococcus frisius* to the genus *Methanosarcina* as *Methanosarcina frisia* comb. nov. Int. J. Syst. Bacteriol. 39:91–92.

Blotevogel, K. H., U. Fischer, K. H. Lüpkes. 1986. *Methanococcus frisius* sp. nov., a new methylotrophic marine methanogen. Can. J. Microbiol. 32:127–131.

Blotevogel, K.-H., U. Fisher, M. Mocha, S. Jannsen. 1985. *Methanobacterium thermoalcaliphilum* spec. nov., a new moderately alkaliphilic and thermophilic autotropic methanogen. Arch. Microbiol. 142:211–217.

Bobik, T. A., R. S. Wolfe. 1988. Physiological importance of the heterodisulfide of coenzyme M and 7-mercaptoheptanoylthreonine phosphate in the reduction of carbon dioxide to methane in *Methanobacterium*. Proc. Natl. Acad. Sci. USA 85:60–63.

Bobik, T. A., R. S. Wolfe. 1989. Activation of formylmethanofuran synthesis in cell extracts of *Methanobacterium thermoautotrophicum*. J. Bacteriol. 171:1423–1427.

Böck, A., O. Kandler. 1985. Antibiotic sensitivity of archaebacteria. 525–544. C. R. Woese and R. S. Wolfe (ed.) The bacteria, vol. 8. Academic Press. New York.

Bomar, M., K. Knoll, F. Widdel. 1985. Fixation of molecular nitrogen by *Methanosarcina barkeri*. FEMS Microbiol. Ecology 31:47–55.

Boone, D. R. 1987. Replacement of the type strain of *Methanobacterium formicicum* and reinstatement of *Methanobacterium bryantii* sp. nov. nom. rev. (ex Balch and Wolfe, 1981) with M. O. H. (DSM 863) as the type strain. Int. J. Syst. Bacteriol. 37:172–173.

Boone, D. R., R. L. Johnson, Y. Liu. 1989. Diffusion of the interspecies electron carriers H_2 and formate in methanogenic ecosystems and implications in the measurement of K^m for H_2 or formate uptake. Appl. Environ. Microbiol. 55:1735–1741.

Boone, D. R., R. Mah. 1988. Transitional bacteria. 35–47. W. H. Smith and J. R. Frank (ed.) Methane from biomass: a systems approach. Elsevier Applied Science Publications. New York.

Boone, D. R., J. A. G. F. Menaia, J. E. Boone, R. A. Mah. 1987. Effects of hydrogen pressure during growth and effects of pregrowth with hydrogen on acetate degradation by *Methanosarcina* species. Appl. Environ. Microbiol. 53:83–87.

Boone, D. R., W. B. Whitman. 1988. Proposal of minimal standards for describing new taxa of methanogenic bacteria. Int. J. Syst. Bacteriol. 38:212–219.

Boone, D. R., S. Worakit, I. M. Mathrani, R. A. Mah. 1986. Alkaliphilic methanogens from high-pH lake sediments. Syst. Appl. Microbiol. 7:230–234.

Bott, M., R. K. Thauer. 1989. Proton translocation coupled to the oxidation of carbon monoxide to CO_2 and H_2 in *Methanosarcina barkeri*. Eur. J. Biochem. 179:469–472.

Bouwer, E. J., P. L. McCarty. 1983. Transformations of 1- and 2-carbon halogenated aliphatic organic compounds under methanogenic conditions. Appl. Environ. Microbiol. 45:1286–1294.

Brown, J. W., C. J. Daniels, J. N. Reeve. 1989. Gene structure, organization, and expression in archaebacteria. CRC Crit. Rev. Microbiol. 16:287–338.

Brusa, T., R. Conca, A. Ferrara, A. Ferrari, A. Pecchioni. 1987. Presence of methanobacteria in human subgingival plaque. J. Clin. Periodontol. 14:470–471.

Bryant, M. P., D. R. Boone. 1987a. Emended description of strain MS^T (DSM 800^T), the type strain of *Methanosarcina barkeri*. Int. J. Syst. Bacteriol. 37:169–170.

Bryant, M. P., D. R. Boone. 1987b. Isolation and characterization of *Methanobacterium formicicum* MF. Int. J. Syst. Bacteriol. 37:171.

Bryant, M. P., S. F. Tzeng, I. M. Robinson, A. E. Joyner, Jr. 1971. Nutrient requirements of methanogenic bacteria. Adv. Chem. Ser. 105:23–40.

Bryant, M. P., E. A. Wolin, M. J. Wolin, R. S. Wolfe. 1967. *Methanobacillus omelianskii*, a symbiotic association of two species of bacteria. Arch. Mikrobiol. 59:20–31.

Chartrain, M., J. G. Zeikus. 1986a. Microbial ecophysiology of whey biomethanation: intermediary metabolism of lactose degradation in continuous culture. Appl. Environ. Microbiol. 51:180–187.

Chartrain, M., J. G. Zeikus. 1986b. Microbial ecophysiology of whey biomethanation: characterization of bacterial trophic populations and prevalent species in continuous culture. Appl. Environ. Microbiol. 51:188–196.

Chumakov, K. M., T. N. Zhilina, I. S. Zvyaginntseva, A. L. Tarasov, G. A. Zavarzin. 1987. 5S RNA in archaebacteria. Zhurnal Obstchey Biologi (Russ.) 48:167–181.

Cicerone, R. J., R. S. Oremland. 1988. Biogeochemical aspects of atmospheric methane. Global Biogeochem. Cycles 2:299–327.

Conway de Macario, E., M. J. Wolin, A. J. L. Macario. 1981. Immunology of achaebacteria that produce methane gas. Science 214:74–75.

Corder, R. E., L. A. Hook, J. M. Larkin, J. I. Frea. 1983. Isolation and characterization of two new methane-producing cocci: *Methanogenium olentangyi*, sp. nov., and *Methanococcus deltae*, sp. nov. Arch. Microbiol. 134:28–32.

Daniels, L. 1984. Biological methanogenesis: physiological and practical aspects. Trends in Biotechnol. 2:91–98.

Daniels, L, J. G. Zeikus. 1978. One-carbon metabolism in methanogenic bacteria: analysis of short-term fixation products of $^{14}CO_2$ and $^{14}CH_3OH$ incorporated into whole cells. J. Bacteriol. 136:75–84.

Daniels, L., N. Belay, B. S. Rajagopal. 1986. Assimilatory reduction of sulfate and sulfite by methanogenic bacteria. Appl. Environ. Microbiol. 51:703–709.

Daniels, L., N. Belay, B. S. Rajagopal, P. S. Weimer. 1987. Bacterial methanogenesis and growth from CO_2 with elemental iron as the sole source of electrons. Science 237:509–511.

Daniels, L., G. Fuchs, R. K. Thauer, J. G. Zeikus. 1977. Carbon monoxide oxidation by methanogenic bacteria. J. Bacteriol. 132:118–126.

Dellinger, C. A., J. G. Ferry. 1984. Effect of monensin on growth and methanogenesis of *Methanobacterium formicicum*. Appl. Environ. Microbiol. 48:680–682.

DeMoll, E., L. Tsai. 1986. Utilization of purines or pyrimidines as the sole nitrogen source by *Methanococcus vannielii*. J. Bacteriol. 167:681–684.

Dennis, P. P. 1986. Molecular biology of archaebacteria. J. Bacteriol. 168:471–478.

De Rosa, M., A. Gambacorta. 1988. The lipids of archaebacteria. Prog. Lipid Res. 27:153–175.

Doddema, H. J., G. D. Vogels. 1978. Improved identification of methanogenic bacteria by fluorescence microscopy. Appl. Environ. Microbiol. 36:752–754.

Ellermann, J., R. Hedderich, R. Böcher, R. K. Thauer. 1988. The final step in methane formation. Investigations with highly purified methyl-CoM reductase (component C) from *Methanobacterium thermoautotrophicum* (strain Marburg). Eur. J. Biochem. 172:669–677.

Fatherpure, B. Z., S. A. Boyd. 1988. Reductive dechlorination of perchloroethylene and the role of methanogens. FEMS Microbiol. Lett. 49:149–156.

Ferry, J. G., P. H. Smith, R. S. Wolfe. 1974. *Methanospirillum*, a new genus of methanogenic bacteria, and characterization of *Methanospirillum hungatii* sp. nov. Int. J. Syst. Bacteriol. 24:465–469.

Ferry, J. G., R. S. Wolfe. 1977. Nutritional and biochemical characterization of *Methanospirillum hungatii*. Appl. Environ. Microbiol. 34:371–376.

Fischer, R., R. K. Thauer. 1989. Methyltetrahydromethanopterin as an intermediate in methanogenesis from acetate in *Methanosarcina barkeri*. Arch. Microbiol. 151:459–465.

Fox, G. E. 1985. The structure and evolution of archaebacterial ribosomal RNA. 257–310. C. R. Woese and R. S. Wolfe (ed.) The bacteria, vol. 8. Academic Press. New York.

Friedmann, H. C., R. K. Thauer. 1987. Non-enzymatic ammonia formation from glutamine under growth conditions for *Methanobacterium thermoautotrophicum*. FEMS Microbiol. Lett. 40:179–182.

Fuchs, G. 1986. CO_2 fixation in acetogenic bacteria: variations on a theme. FEMS Microbiol. Rev. 39:181–213.

Fuchs, G., E. Stupperich. 1986. Carbon assimilation pathways in archaebacteria. Syst. Appl. Microbiol. 7:364–369.

Godsy, E. M. 1980. Isolation of *Methanobacterium bryantii* from a deep aquifer by using a novel broth-antibiotic disk method. Appl. Environ. Microbiol. 39:1074–1075.

Harris, J. E. 1985. GELRITE as an agar substitute for the cultivation of mesophilic *Methanobacterium* and *Methanobrevibacter* species. Appl. Environ. Microbiol. 50:1107–1109.

Harris, J. E. 1987. Spontaneous disaggregation of *Methanosarcina mazei* S-6 and its use in the development of genetic techniques for *Methanosarcina* spp. Appl. Environ. Microbiol. 53:2500–2504.

Harris, J. E., P. A. Pinn, R. P. Davis. 1984. Isolation and characterization of a novel thermophilic, freshwater methanogen. Appl. Environ. Microbiol. 48:1123–1128.

Hatchikian, E. C., M. L. Fardeau, M. Bruschi, J. P. Belaich, A. Chapman, R. Cammack. 1989. Isolation, characterization, and biological activity of the *Methanococcus*

LIVERPOOL JOHN MOORES UNIVERSITY
LEARNING & INFORMATION SERVICES

thermolithotrophicus ferredoxin. J. Bacteriol. 171:2384–2390.

Hermann, M., K. M. Noll, R. S. Wolfe. 1986. Improved agar bottle plate for isolation of methanogens or other anaerobes in a defined gas atmosphere. Appl. Environ. Microbiol. 51:1124–1126.

Hilpert, R., J. Winter, W. Hammes, O. Kandler. 1981. The sensitivity of archaebacteria to antibiotics. Zentrabl. Bakteriol. Mikrobiol. Hyg., Abt. 1, Orig. C2:11–20.

Hippe, H. 1984. Maintenance of methanogenic bacteria. 69–81. B. E. Kirsop, and J. J. S. Snell (ed.) Maintenance of microorganisms. Academic Press. London.

Hobson, P. N., S. Bousfield, R. Summers. 1981. Methane production from agricultural and domestic wastes. Applied Science Publishers, Ltd. London.

Huber, H., M. Thomm, H. König, G. Thies, K. O. Stetter. 1982. *Methanococcus thermolithotrophicus*, a novel thermophilic lithotrophic methanogen. Arch. Microbiol. 132:47–50.

Hungate, R. E. 1969. A roll tube method for cultivation of strict anaerobes. 117–132. J. R. Norris and D. W. Ribbons (ed.) Methods in microbiology, vol. 3B. Academic Press. New York.

Huser, B. A., K. Wuhrmann, A. J. B. Zehnder. 1982. *Methanothrix soehngenii* gen. nov. sp. nov., a new acetotrophic non-hydrogen-oxidizing methane bacterium. Arch. Microbiol. 132:1–9.

Jain, M. K., T. E. Thompson, E. Conway de Macario, J. G. Zeikus. 1987. Speciation of *Methanobacterium* strain Ivanov as *Methanobacterium ivanovii*, sp. nov. Syst. Appl. Microbiol. 9:77–82.

Jarrell, K. F., S. F. Koval. 1989. Ultrastructure and biochemistry of *Methanococcus voltae*. CRC Crit. Rev. Microbiol. 17:53–87.

Jones, W. J., J. A. Leigh, F. Mayer, C. R. Woese, R. S. Wolfe. 1983a. *Methanococcus jannaschii* sp. nov., an extremely thermophilic methanogen from a submarine hydrothermal vent. Arch. Microbiol. 136:254–261.

Jones, W. J., D. P. Nagel, Jr., W. B. Whitman. 1987. Methanogens and the diversity of archaebacteria. Microbiol. Rev. 51:135–177.

Jones, W. J., M. J. B. Paynter, R. Gupta. 1983b. Characterization of *Methanococcus maripaludis* sp. nov., a new methanogen isolated from salt marsh sediment. Arch. Microbiol. 135:91–97.

Jones, W. J., C. E. Stugard, H. W. Jannasch. 1989. Comparison of thermophilic methanogens from submarine hydrothermal vents. Arch. Microbiol. 151:314–319.

Jones, W. J., W. B. Whitman, R. D. Fields, R. S. Wolfe. 1983c. Growth and plating efficiency of methanococci on agar media. Appl. Environ. Microbiol. 46:220–226.

Kalmokoff, M. L., K. F. Jarrell, S. F. Koval. 1988. Isolation of flagella from the archaebacterium *Methanococcus voltae* by phase separation with Triton X-114. J. Bacteriol. 170:1752–1758.

Kandler, O., H. König. 1985. Cell envelopes of archaebacteria. 413–457. C. R. Woese and R. S. Wolfe (ed.) The bacteria, vol. 8. Academic Press. New York.

Kandler, O., W. Zillig (ed.). 1986. Archaebacteria '85. Gustav Fischer Verlag. Stuttgart.

Keltjens, J. T., C. van der Drift. 1986. Electron transfer reactions in methanogens. FEMS Microbiol. Rev. 39:259–303.

Kenealy, W. R., T. E. Thompson, K. R. Schubert, J. G. Zeikus. 1982. Ammonia assimilation and synthesis of alanine, aspartate, and glutamate in *Methanosarcina barkeri* and *Methanobacterium thermoautotrophicum*. J. Bacteriol. 150:1357–1365.

Kengen, S. W. M., J. J. Mosterd, R. L. H. Nelissen, J. T. Keltjens, C. van der Drift, G. D. Vogels. 1988. Reductive activation of the methyl-tetrahydromethanopterin: coenzyme M methyltransferase from *Methanobacterium thermoautotrophicum* strain ΔH. Arch. Microbiol. 150:405–412.

Kiene, R. P., R. S. Oremland, A. Catena, L. G. Miller, D. G. Capone. 1986. Metabolism of reduced methylated sulfur compounds in anaerobic sediments and by a pure culture of an estuarine methanogen. Appl. Environ. Microbiol. 52:1037–1045.

Kiener, A., T. Leisinger. 1983. Oxygen sensitivity of methanogenic bacteria. Syst. Appl. Microbiol. 4:305–312.

Klimczak, L. J., F. Grummt, K. J. Burger. 1986. Purification and characterization of DNA polymerase from the archaebacterium *Methanobacterium thermoautotrophicum*. Biochemistry 25:4850–4855.

Kneifel, H., K. O. Stetter, J. R. Andreesen, J. Wiegel, H. König, S. M. Schoberth. 1986. Distribution of polyamines in representative species of archaebacteria. Syst. Appl. Microbiol. 7:241–245.

Knoll, G., J. Winter. 1989. Degradation of phenol via carboxylation to benzoate by a defined, obligate syntrophic consortium of anaerobic bacteria. Appl. Microbiol. Biotechnol. 30:318–324.

Koga, Y., M. Ohga, M. Nishihara, H. Morii. 1987. Distribution of a diphytanyl ether analog of phosphatidylserine and an ethanolamine-containing tetraether lipid in methanogenic bacteria. Syst. Appl. Microbiol. 9:176–182.

König, H. 1984. Isolation and characterization of *Methanobacterium uliginosum* sp. nov. from a marshy soil. Can. J. Microbiol. 30:1477–1481.

König, H. 1988. Archaebacterial cell envelopes. Can. J. Microbiol. 34:395–406.

König, H., E. Nusser, K. O. Stetter. 1985. Glycogen in *Methanolobus* and *Methanococcus*. FEMS Microbiol. Lett. 28:265–269.

König, H., K. O. Stetter. 1982. Isolation and characterization of *Methanolobus tindarius* sp. nov., a coccoid methanogen growing only on methanol and methylamines. Zentralbl. Bakteriol. Parasitenkd. Infektionskr. Hyg. Abt. 1 Orig. Reihe C 3:478–490.

Köpke, A. K. E., B. Wittmann-Liebold. 1989. Comparative studies of ribosomal proteins and their genes from *Methanococcus vannielii* and other organisms. Can. J. Microbiol. 35:11–20.

Koval, S. F., K. F. Jarrell. 1987. Ultrastructure and biochemistry of the cell wall of *Methanococcus voltae*. J. Bacteriol. 169:1298–1306.

Kreisl, P., O. Kandler. 1986. Chemical structure of the cell wall polymer of *Methanosarcina*. System. Appl. Microbiol. 7:293–299.

Kristjansson, J. K., P. Schönheit, R. K. Thauer. 1982. Different K, values for hydrogen of methanogenic bacteria and sulfate reducing bacteria. Arch. Microbiol. 131:278–282.

Lake, J. A., E. Henderson, M. W. Clark, A. T. Matheson. 1982. Mapping evolution with ribosome structure: intralineage constancy and interlineage variation. Proc. Natl. Acad. Sci. USA 79:5948–5952.

Langworthy, T. A. 1985. Lipids of archaebacteria. 459–497. C. R. Woese, and R. S. Wolfe (ed.) The bacteria, vol. 8. Academic Press. New York.

Lauerer, G., J. K. Kristjansson, T. A. Langworthy, H. König, K. O. Stetter. 1986. *Methanothermus sociabilis* sp. nov., a second species within the *Methanothermaceae* growing at 97°C. Syst. Appl. Microbiol. 8:100–105.

Lechner, K., G. Heller, A. Böck. 1988. Gene for the diphtheria toxin-susceptible elongation factor 2 from *Methanococcus vannielii*. Nucl. Acids Res. 16:7817–7826.

Lee, M. J., P. J. Schreurs, A. C. Messer, S. H. Zinder. 1987. Association of methanogenic bacteria with flagellated protozoa from a termite hindgut. Curr. Microbiol. 15:337–341.

Liu, Y., D. R. Boone, C. Choy. 1990. *Methanohalophilus oregonense* sp. nov., a methylotrophic methanogen from an alkaline, saline aquifer. Int. J. Syst. Bacteriol. 40:111–116.

Liu, Y., D. R. Boone, R. Sleat, R. A. Mah. 1985. *Methanosarcina mazei* LYC, a new methanogenic isolate which produces a disaggregating enzyme. Appl. Environ. Microbiol. 49:608–613.

Ljungdahl, L. G. 1986. The autotrophic pathway of acetate synthesis in acetogenic bacteria. Ann. Rev. Microbiol. 40:415–450.

Ljungdahl, L. G., J. Wiegel. 1986. Working with anaerobic bacteria. 84–96. A. L. DeMain and N. A. Solomon (ed.) Manual of industrial microbiology and biotechnology. American Society for Microbiology. Washington, D. C.

Lobo, A. L., S. H. Zinder. 1988. Diazotrophy and nitrogenase activity in the archaebacterium *Methanosarcina barkeri* 227. Appl. Environ. Microbiol. 54:1656–1661.

Lovley, D. R. 1985. Minimum threshold for hydrogen metabolism in methanogenic bacteria. Appl. Environ. Microbiol. 49:1530–1531.

Lovley, D. R., R. C. Greening, J. G. Ferry. 1984. Rapidly growing rumen methanogenic organism that synthesizes coenzyme M and has a high affinity for formate. Appl. Environ. Microbiol. 48:81–87.

Lysenko, A. M., T. N. Zhilina. 1985. Taxonomic position of *Methanosarcina vacuolata* and *Methanococcus halophilus* determined by the technique of DNA-DNA hybridization. (Russ.). Microbiology 54:501–502.

Macy, J. M., J. E. Snellen, R. E. Hungate. 1972. Use of syringe methods for anaerobiosis. Am. J. Clin. Nutrit. 25:1318–1323.

Maestrojuán, G. M., D. R. Boone, L. Xun, R. A. Mah, L. Zhang. 1990. Transfer of *Methanogenium bourgense*, *Methanogenium marisnigri*, *Methanogenium olentangyi*, and *Methanoculleus* gen. nov., emendation of *Methanoculleus marisnigri* and *Methanogenium*, and description of new strains of *Methanoculleus bourgense* and *Methanoculleus marisnigri*. Inter. J. System. Bacteriol. 40:117–122.

Mah, R. A. 1980. Isolation and characterization of *Methanococcus mazei*. Curr. Microbiol. 3:321–326.

Mah, R. A., M. R. Smith. 1981. The methanogenic bacteria. 948–977. A. Balows, H. G. Trüper, M. Dworkin, W. Hander, and K. H. Schleifer (ed.) The prokaryotes. Springer-Verlag. New York.

Matheson, A. T. 1985. Ribosomes of archaebacteria. 345–377. C. R. Woese and R. S. Wolfe (ed.) The bacteria, vol. 8. Academic Press. New York.

Mathrani, I. M., D. R. Boone. 1985. Isolation and characterization of a moderately halophilic methanogen from a solar saltern. Appl. Environ. Microbiol. 50:140–143.

Mathrani, I. M., D. R. Boone, R. A. Mah, G. E. Fox, P. P. Lau. 1988. *Methanohalophilus zhilinae* sp. nov., an alka-

liphilic, halophilic, methylotrophic methanogen. Int. J. Syst. Bacteriol. 38:139–142.

Mazumder, T. K., N. Nishio, S. Fukuzaki, S. Nagai. 1986. Effect of sulfur-containing compounds on growth of *Methanosarcina barkeri* in defined medium. Appl. Environ. Microbiol. 52:617–622.

Migas, J., K. L. Anderson, D. L. Cruden, A. J. Markovetz. 1989. Chemotaxis in *Methanospirillum hungatei*. Appl. Environ. Microbiol. 55:264–265.

Miller, T. L. 1989. Genus II. *Methanobrevibacter*. 2178–2183. J. T. Staley (ed.) Bergey's manual of systematic bacteriology, vol. 3. Williams and Wilkins. Baltimore.

Miller, T. L., M. J. Wolin. 1982. Enumeration of *Methanobrevibacter smithii* in human feces. Arch. Microbiol. 131:14–18.

Miller, T. L., M. J. Wolin. 1983. Oxidation of hydrogen and reduction of methanol to methane is the sole energy source for a methanogen isolated from human feces. J. Bacteriol. 153:1051–1055.

Miller, T. L., M. J. Wolin. 1985. *Methanosphaera stadtmaniae* gen. nov., sp. nov.: a species that forms methane by reducing methanol with hydrogen. Arch. Microbiol. 141:116–122.

Miller, T. L., M. J. Wolin. 1986. Methanogens in human and animal intestinal tracts. Syst. Appl. Microbiol. 7:223–229.

Miller, T. L., M. J. Wolin, E. Conway de Macario, A. J. L. Macario. 1982. Isolation of *Methanobrevibacter smithii* from human feces. Appl. Environ. Microbiol. 43:227–232.

Miller, T. L., M. J. Wolin, Z. Hongxue, M. P. Bryant. 1986. Characteristics of methanogens isolated from bovine rumen. Appl. Environ. Microbiol. 51:201–202.

Misawa, H., T. Hoshi, F. Kitame, M. Homma, K. Nakamura. 1986. Isolation of an antigenically unique methanogen from human feces. Appl. Environ. Microbiol. 51:429–431.

Morii, H., M. Nishihara, Y. Koga. 1988. Composition of polar lipids of *Methanobrevibacter arboriphilicus* and structure determination of the signature phosphoglycolipid of *Methanobacteriaceae*. Agric. Biol. Chem. 52:3149–3156.

Murray, P. A., S. H. Zinder. 1984. Nitrogen fixation by a methanogenic archaebacterium. Nature 312:284–286.

Murray, P. A., S. H. Zinder. 1985. Nutritional requirements of *Methanosarcina* sp. strain TM-1. Appl. Environ. Microbiol. 50:49–55.

Murray, P. A., S. H. Zinder. 1987. Polysaccharide reserve material in the acetotrophic methanogen, *Methanosarcina thermophila* strain TM-1: accumulation and mobilization. Arch. Microbiol. 147:109–116.

Nagle, D. P., R. S. Wolfe. 1985. Methanogenesis. 425–438. A. T. Bull and H. Dalton (ed.) Comprehensive biotechnology, vol. 1. Pergamon Press. Oxford.

Noll, K. M., T. S. Barber. 1988. Vitamin contents of archaebacteria. J. Bacteriol. 170:4315–4321.

Nozhevnikova, A. N., V. I. Chudina. 1984. Morphology of the thermophilic acetate methane bacterium *Methanothrix thermoacetophila* sp. nov. Arch. Mikrobiol. Zh. 53:756–760.

Nozhevnikova, A. N., T. G. Yagodina. 1982. A thermophilic acetate methane-producing bacterium. Microbiology (Engl. Transl.) 51:534–541.

Nusser, E., H. König. 1987. S layer studies on three species of *Methanococcus* living at different temperatures. Can. J. Microbiol. 33:256–261.

Obraztsova, A. Ya., O. V. Shipin, L. V. Bezrukova, S. S. Belyaev. 1987. Properties of the coccoid methylotrophic methanogen, *Methanococcoides euhalobius* sp. nov. Microbiology (Engl. Transl.) 56:523–527.

Ollivier, B., A. Lombardo, J. L. Garcia. 1984. Isolation and characterization of a new thermophilic *Methanosarcina* strain (strain MP). Ann. Microbiol. (Inst. Pasteur) 135B:187–198.

Ollivier, B. M., R. A. Mah, J. L. Garcia, D. R. Boone. 1986. Isolation and characterization of *Methanogenium bourgense* sp. nov. Int. J. Syst. Bacteriol. 36:297–301.

Ollivier, B. M., R. A. Mah, J. L. Garcia, R. Robinson. 1985. Isolation and characterization of *Methanogenium aggregans* sp. nov. Int. J. Syst. Bacteriol. 35:127–130.

Patel, G. B. 1984. Characterization and nutritional properties of *Methanothrix concilii* sp. nov., a mesophilic, aceticlastic methanogen. Can. J. Microbiol. 30:1383–1396.

Patel, G. B., L. A. Roth, G. D. Sprott. 1979. Factors influencing filament length of *Methanospirillum hungatei.* J. Gen. Microbiol. 112:411–415.

Patel, G. B., L. A. Roth, L. van den Berg, D. S. Clark. 1976. Characterization of a strain of *Methanospirillum hungatii.* Can. J. Microbiol. 22:1404–1410.

Patel, G. B., G. D. Sprott. 1990. *Methanosaeta concilii* gen. nov., sp. nov. ("*Methanothrix concilii*") and *Methanosaeta thermoacetophila* nom. rev., comb. nov. Inter. J. System. Bacteriol. 40:79–82.

Patel, G. B., G. D. Sprott, R. W. Humphrey, T. J. Beveridge. 1986. Comparative analyses of the sheath structures of *Methanothrix concilii* GP6 and *Methanospirillum hungatei* strains GP1 and JF1. Can. J. Microbiol. 32:623–631.

Patel, G. B., G. D. Sprott, J. E. Fein. 1990. Isolation and characterization of *Methanobacterium espanolae* sp. nov., a mesophilic, moderately acidiphilic methanogen. Inter. J. System. Bacteriol. 40:12–18.

Paterek, J. R., P. H. Smith. 1985. Isolation and characterization of a halophilic methanogen from Great Salt Lake. Appl. Environ. Microbiol. 50:877–881.

Paterek, J. R., P. H. Smith. 1988. *Methanohalophilus mahii* gen. nov., sp. nov., a methylotrophic halophilic methanogen. Int. J. Syst. Bacteriol. 38:122–123.

Paynter, M. J. B., R. E. Hungate. 1968. Characterization of *Methanobacterium mobilis,* sp. n., isolated from the bovine rumen. J. Bacteriol. 95:1943–1951.

Perski, H. J., P. Schönheit, R. K. Thauer. 1982. Sodium dependence of methane formation in methanogenic bacteria. FEBS Lett. 143:323–326.

Pfaltz, A., A. Kobelt, R. Hüster, R. K. Thauer. 1987. Biosynthesis of coenzyme F430 in Methanogenic bacteria identification of $15,17^3$-seco-F430-17^3-acid as an intermediate. Eur. J. Biochem. 170:459–467.

Possot, O., P. Gernhardt, A. Klien, L. Sibold. 1988. Analysis of drug resistance in the archaebacterium *Methanococcus voltae* with respect to potential use in genetic engineering. Appl. Environ. Microbiol. 54:734–740.

Powell, G. E. 1983. Interpreting gas kinetics of batch cultures. Biotechnol. Lett. 5:437–440.

Rimbault, A., P. Niel, H. Virelizier, J. C. Darbord, G. Leluan. 1988. L-Methionine, a precursor of trace methane in some proteolytic clostridia. Appl. Environ. Microbiol. 54:1581–1586.

Rivard, C. J., J. M. Henson, M. V. Thomas, P. H. Smith. 1983. Isolation and characterization of *Methanomicrobium paynteri* sp. nov., a mesophilic methanogen isolated from marine sediments. Appl. Environ. Microbiol. 46:484–490.

Rivard, C. J., P. H. Smith. 1982. Isolation and characterization of a thermophilic marine methanogenic bacterium, *Methanogenium thermophilicum* sp. nov. Int. J. Syst. Bacteriol. 32:430–436.

Robinson, R. W. 1986. Life cycles of the methanogenic archaebacterium *Methanosarcina mazei.* Appl. Environ. Microbiol. 52:17–27.

Romesser, J. A., R. S. Wolfe, F. Mayer, E. Speiss, A. Walther-Mauruschat. 1979. *Methanogenium,* a new genus of marine methanogenic bacteria, and characterization of *Methanogenium cariaci* sp. nov. and *Methanogenium marisnigri* sp. nov. Arch. Microbiol. 121:147–153.

Rose, C. S., S. J. Pirt. 1981. Conversion of glucose to fatty acids and methane: roles of two mycoplasmal agents. J. Bacteriol. 147:248–254.

Rouvière, P. E., T. A. Bobik, R. S. Wolfe. 1988. Reductive activation of the methyl coenzyme M methylreductase system of *Methanobacterium thermoautotrophicum* ΔH. J. Bacteriol. 170:3946–3952.

Rouvière, P. E., R. S. Wolfe. 1987. Use of subunits of the methylreductase protein for taxonomy of methanogenic bacteria. Arch. Microbiol. 148:253–259.

Rouvière, P. E., R. S. Wolfe. 1988. Novel biochemistry of methanogenesis. J. Biol. Chem. 263:7913–7916.

Rouvière, P. E., R. S. Wolfe. 1989. Component A3 of the methylcoenzyme M methylreductase system of *Methanobacterium thermoautotrophicum* ΔH: resolution into two components. J. Bacteriol. 171:4556–4562.

Santoro, N., J. Konisky. 1987. Characterization of bromoethanesulfonate-resistant mutants of *Methanococcus voltae:* evidence of a coenzyme M transport system. J. Bacteriol. 169:660–665.

Schauer, N. L., W. B. Whitman. 1989. Formate growth and pH control by volatile formic and acetic acids in batch cultures of methanococci. J. Microbiol. Methods 10:1–7.

Scherer, P. A. 1989. Vanadium and molybdenum requirement for the fixation of molecular nitrogen by two *Methanosarcina* strains. Arch. Microbiol. 151:44–48.

Scherer, P. A., H. Sahm. 1981a. Effect of trace elements and vitamins on the growth of *Methanosarcina barkeri.* Acta Biotechnol. 1:57–65.

Scherer, P. A., H. Sahm. 1981b. Influence of sulphur-containing compounds on the growth of *Methanosarcina barkeri* in defined medium. Eur. J. Appl. Microbiol. Biotechnol. 12:28–35.

Schlicht, F., G. Schimpff-Weiland, H. Follmann. 1985. Methanogenic bacteria contain thioredoxin. Naturwissenschaften 72:328–330.

Schönheit, P., D. B. Beimborn. 1986. Monensin and gramicidin stimulate CH_4 formation from H_2 and CO_2 in *Methanobacterium thermoautotrophicum* at low external Na^+ concentration. Arch. Microbiol. 146:181–185.

Shieh, J., M. Mesbah, W. B. Whitman. 1988. Pseudoauxotrophy of *Methanococcus voltae* for acetate, leucine, and isoleucine. J. Bacteriol. 170:4091–4096.

Shieh, J., W. B. Whitman. 1987. Pathway of acetate assimilation in autotrophic and heterotrophic methanococci. J. Bacteriol. 169:5327–5329.

Sibold, L., D. Pariot, L. Bhatnagar, M. Henriquet, J.-P. Aubert. 1985. Hybridization of DNA from methanogenic bacteria with nitrogenase structural genes *(nif-HDK).* Mol. Gen. Genet. 200:40–46.

Sment, K. A., J. Konisky. 1989. Chemotaxis in the archaebacterium *Methanococcus voltae*. J. Bacteriol. 171:2870–2872.

Smith, P. H., R. E. Hungate. 1958. Isolation and characterization of *Methanobacterium ruminantium* n. sp. J. Bacteriol. 75:713–718.

Sowers, K. R., S. F. Baron, J. G. Ferry. 1984a. *Methanosarcina acetivorans* sp. nov., an acetotrophic methane-producing bacterium isolated from marine sediments. Appl. Environ. Microbiol. 47:971–978.

Sowers, K. R., J. G. Ferry. 1983. Isolation and characterization of a methylotrophic marine methanogen, *Methanococcoides methylutens* gen. nov., sp. nov. Appl. Environ. Microbiol. 45:684–690.

Sowers, K. R., J. G. Ferry. 1985. Trace metal and vitamin requirements of *Methanococcoides methylutens* grown with trimethylamine. Arch. Microbiol. 142:148–151.

Sowers, K. R., R. P. Gunsalus. 1988. Adaptation for growth at various saline concentrations by the archaebacterium *Methanosarcina thermophila*. J. Bacteriol. 170:998–1002.

Sowers, K. R., J. L. Johnson, J. G. Ferry. 1984b. Phylogenetic relationships among the methylotrophic methane-producing bacteria and emendation of the family *Methanosarcinaceae*. Int. J. Syst. Bacteriol. 34:444–450.

Stadtman, T. C., H. A. Barker. 1951. Studies on the methane fermentation. X. A new formate-decomposing bacterium, *Methanococcus vannielii*. J. Bacteriol. 62:269–280.

Stetter, K. O. 1989. Genus II. *Methanolobus*. 2205–2207. J. T. Staley (ed.) Bergey's manual of systematic bacteriology, vol. 3. Williams and Wilkins. Baltimore.

Stetter, K. O., M. Thomm, J. Winter, G. Wildgruber, H. Huber, W. Zillig, D. Janécovic, H. König, P. Palm, S. Wunderl. 1981. *Methanothermus fervidus*, sp. nov., a novel extremely thermophilic methanogen isolated from an Icelandic hot spring. Zentralbl. Bakteriol. Parasitenkd. Infektionskr. Hyg. Abt. 1 Orig. Reihe C 2:166–178.

Stupperich, E., B. Kräutler. 1988. Pseudo vitamin B_{12} or 5-hydroxybenzimidazolyl-cobamide are the corrinoids found in methanogenic bacteria. Arch. Microbiol. 149:268–271.

Suflita, J. M., J. A. Robinson, J. M. Tiedje. 1983. Kinetics of microbial dehalogenation of haloaromatic substrates in methanogenic environments. Appl. Environ. Microbiol. 45:1466–1473.

Tanner, R. S., R. S. Wolfe. 1988. Nutrient requirements of *Methanomicrobium mobile*. Appl. Environ. Microbiol. 54:625–628.

Terlesky, K. C., M. J. Barber, D. J. Aceti, J. G. Ferry. 1987. EPR properties of the Ni-Fe-C center in an enzyme complex with carbon monoxide dehydrogenase activity from acetate-grown *Methanosarcina thermophila*. Evidence that acetyl-CoA is a physiological substrate. J. Biol. Chem. 262:15392–15395.

Terlesky, K. C., J. G. Ferry. 1988a. Purification and characterization of a ferredoxin from acetate-grown *Methanosarcina thermophila*. J. Biol. Chem. 263:4080–4082.

Terlesky, K. C., J. G. Ferry. 1988b. Ferredoxin requirement for electron transport from the carbon monoxide dehydrogenase complex to a membrane-bound hydrogenase in acetate-grown *Methanosarcina thermophila*. J. Biol. Chem. 263:4075–4079.

Thiele, J. H., J. G. Zeikus. 1988. Control of interspecies electron flow during anaerobic digestion: significance of formate transfer versus hydrogen transfer during syntrophic methanogenesis in flocs. Appl. Environ. Microbiol. 54:20–29.

Thomm, M., J. Madon, K. O. Stetter. 1986. DNA-dependent RNA polymerases of the three orders of methanogens. Biol. Chem. Hoppe-Seyler 367:473–481.

Thomm, M., G. Wich, J. W. Brown, G. Frey, B. A. Sherf, G. S. Beckler. 1989. An archaebacterial promoter sequence assigned by RNA polymerase binding experiments. Can. J. Microbiol. 35:30–35.

Touzel, J. P., G. Albagnac. 1983. Isolation and characterization of *Methanococcus mazei* strain MC₃. FEMS Microbiol. Lett. 16:241–245.

Touzel, J. P., D. Petroff, G. Albagnac. 1985. Isolation and characterization of a new thermophilic *Methanosarcina*, the strain CHTI 55. Syst. Appl. Microbiol. 6:66–71.

Touzel, J. P., G. Prensier, J. L. Roustan, I. Thomas, H. C. Dubourguier, G. Albagnac. 1988. Description of a new strain of *Methanothrix soehngenii* and rejection of *Methanothrix concilii* as a synonym of *Methanothrix soehngenii*. Int. J. Syst. Bacteriol. 38:30–36.

van Bruggen, J. J. A., C. K. Stumm, K. B. Zwart, G. D. Vogels. 1985. Endosymbiotic methanogenic bacteria of the sapropelic amoeba *Mastigella*. FEMS Microbiol. Ecol. 31:187–192.

van Bruggen, J. J. A., K. B. Zwart, J. G. F. Hermans, E. M. van Hove, C. K. Stumm, G. D. Vogels. 1986. Isolation and characterization of *Methanoplanus endosymbiosus* sp. nov., an endosymbiont of the marine sapropelic ciliate *Metopus contortus* Quennerstedt. Arch. Microbiol. 144:367–374.

van Bruggen, J. J. A., K. B. Zwart, R. M. van Assema, C. K. Stumm, G. D. Vogels. 1984. *Methanobacterium formicicum*, an endosymbiont of the anaerobic ciliate *Metopus striatus* McMurrich. Arch. Microbiol. 139:1–7.

van de Wijngaard, W. M. H., C. van der Drift, G. D. Vogels. 1988. Involvement of a corrinoid enzyme in methanogenesis from acetate in *Methanosarcina barkeri*. FEMS Microbiol. Lett. 52:165–172.

van den Berg, L. 1984. Developments in methanogenesis from industrial waste water. Can. J. Microbiol. 30:975–990.

Weaver, G. A., J. A. Krause, T. L. Miller, M. J. Wolin. 1986. Incidence of methanogenic bacteria in a sigmoidoscopy population: an association of methanogenic bacteria and diverticulosis. Gut 27:698–704.

Weil, C. F., B. A. Sherf, J. N. Reeve. 1989. A comparison of the methyl reductase genes and gene products. Can. J. Microbiol. 35:101–108.

White, R. H. 1988. Structural diversity among methanofurans from different methanogenic bacteria. J. Bacteriol. 170:4594–4597.

White, R. H. 1989. Biosynthesis of the 7-mercaptoheptanoic acid subunit of component B [(7-mercaptoheptanoyl) threonine phosphate] of methanogenic bacteria. Biochemistry 28:860–865.

Whitman, W. B. 1989. *Methanococcales*. 2185–2190. J. T. Staley, M. P. Bryant, and N. Pfennig (ed.) Bergey's manual of systematic bacteriology, vol. 3. Williams & Wilkens. Baltimore.

Whitman, W. B., E. Ankwanda, R. S. Wolfe. 1982. Nutrition and carbon metabolism of *Methanococcus voltae*. J. Bacteriol. 149:852–863.

Whitman, W. B., J. Shieh, S. Sohn, D. S. Caras, U. Premachandran. 1986. Isolation and characterization of 22 mesophilic methanococci. Syst. Appl. Microbiol. 7:235–240.

Whitman, W. B., S. H. Sohn, S. U. Kuk, R. Y. Xing. 1987. Role of amino acids and vitamins in nutrition of mesophilic *Methanococcus* sp. Appl. Environ. Microbiol. 53:2373–2378.

Widdel, F. 1986. Growth of methanogenic bacteria in pure culture with 2-propanol and other alcohols as hydrogen donors. Appl. Environ. Microbiol. 51:1056–1062.

Widdel, F., P. E. Rouvière, R. S. Wolfe. 1988. Classification of secondary alcohol-utilizing methanogens including a new thermophilic isolate. Arch. Microbiol. 150:477–481.

Wildenauer, F. X., J. Winter. 1985. Anaerobic digestion of high-strength acidic whey in a pH-controlled up-flow fixed film loop reactor. Appl. Microbiol. Biotechnol. 22:367–372.

Wildgruber, G., M. Thomm, H. König, K. Ober, T. Ricchiuto, K. O. Stetter. 1982. *Methanoplanus limicola*, a plate-shaped methanogen representing a novel family, the *Methanoplanaceae*. Arch. Microbiol. 132:31–36.

Winfrey, M. R., J. G. Zeikus. 1979. Microbial methanogenesis and acetate metabolism in a meromictic lake. Appl. Environ. Microbiol. 37:213–221.

Winter, J. 1983. Maintenance of stock cultures of methanogens in the laboratory. Syst. Appl. Microbiol. 4:558–563.

Winter, J., C. Lerp, H.-P. Zabel, F. X. Wildenauer, H. König, F. Schindler. 1984. *Methanobacterium wolfei*, sp. nov., a new tungsten-requiring, thermophilic, autotrophic methanogen. Syst. Appl. Microbiol. 5:457–466.

Wodzinski, R. L., R. N. Gennaro, M. H. Scholla. 1987. Economics of the bioconversion of biomass to methane and other vendable products. 37–88. A. I. Laskin (ed.) Advances in applied microbiology, vol. 32. Academic Press. New York.

Woese, C. R. 1987. Bacterial evolution. Microbiol. Rev. 51:221–271.

Woese, C. R., R. S. Wolfe (ed.). 1985. The bacteria. Academic Press. New York. Archaebacteria 8.

Wolin, M. J., T. L. Miller. 1987. Bioconversion of organic carbon to CH_4 and CO_2. Geomicrobial J. 5:239–259.

Wolin, E. A., M. J. Wolin, R. S. Wolfe. 1963. Formation of methane by bacterial extracts. J. Biol. Chem. 238:2882–2886.

Wood, H. G., S. W. Ragsdale, E. Pezacka. 1986. A new pathway of autotrophic growth utilizing carbon monoxide or carbon dioxide and hydrogen. Biochem. Int. 12:421–440.

Worakit, S., D. R. Boone, R. A. Mah, M.-E. Abdel-Samie, M. M. El-Halwagi. 1986. *Methanobacterium alcaliphilum* sp. nov., an H_2-utilizing methanogen that grows at high pH values. Int. J. Syst. Bacteriol. 36:380–382.

Worrell, V. E., D. P. Nagle, Jr., D. McCarthy, A. Eisenbraun. 1988. Genetic transformation system in the archaebacterium *Methanobacterium thermoautotrophicum* Marburg. J. Bacteriol. 170:653–656.

Xing, R. Y., W. B. Whitman. 1987. Sulfometuron methyl-sensitive and -resistant acetolactate synthases of the archaebacteria *Methanococcus* spp. J. Bacteriol. 169:4486–4492.

Xun, L., D. R. Boone, R. A. Mah. 1988. Control of the life cycle of *Methanosarcina mazei* S-6 by manipulation of growth conditions. Appl. Environ. Microbiol. 54:2064–2068.

Xun, L., D. R. Boone, R. A. Mah. 1989. Deoxyribonucliec acid hybridization study of *Methanogenium* and *Methanocorpusculum* species, emendation of the genus *Methanocorpusculum*, and transfer of *Methanogenium aggregans* to the genus *Methanocorpusculum* as *Metha-*

nocorpusculum aggregans comb. nov. Int. J. Syst. Bacteriol. 39:109–111.

Young, L. Y. 1984. Anaerobic degradation of aromatic compounds. 487–523. D. T. Gibson (ed.) Microbial degradation of organic compounds. Marcel Dekker, Inc. New York.

Yu, I. K., F. Kawamura. 1987. *Halomethanococcus doii* gen. nov., sp. nov.: an obligately halophilic methanogenic bacterium from solar salt ponds. J. Gen. Appl. Microbiol. 33:303–310.

Zabel, H.-P., E. Holler, J. Winter. 1987. Mode of inhibition of the DNA polymerase a *Methanococcus vannielii* by aphidicolin. Eur. J. Biochem. 165:171–175.

Zabel, H. P., H. König, J. Winter. 1984. Isolation and characterization of a new coccoid methanogen, *Methanogenium tatii* spec. nov. from a solfataric field on Mount Tatio. Arch. Microbiol. 137:308–315.

Zabel, H.-P., H. König, J. Winter. 1985. Emended description of *Methanogenium thermophilicum*, Rivard and Smith, and assignment of new isolates to this species. Syst. Appl. Microbiol. 6:72–78.

Zehnder, A. J. B., B. A. Huser, T. D. Brock, K. Wuhrmann. 1980. Characterization of an acetate-decarboxylating, non-hydrogen-oxidizing methane bacterium. Arch. Microbiol. 124:1–11.

Zeikus, J. G., A. Ben-Bassat, P. W. Hegge. 1980. Microbiology of methanogenesis in thermal, volcanic environments. J. Bacteriol. 143:432–440.

Zeikus, J. G., D. L. Henning. 1975. *Methanobacterium arboriphilum* sp. nov. An obligate anaerobe isolated from wetwood of living trees. Antonie van Leeuwenhoek 41:543–552.

Zeikus, J. G., R. S. Wolfe. 1972. *Methanobacterium thermoautotrophicus* sp. nov., an anaerobic, autotrophic, extreme thermophile. J. Bacteriol. 109:707–713.

Zellner, G., J. Winter. 1987a. Secondary alcohols as hydrogen donors for CO_2-reduction by methanogens. FEMS Microbiol. Lett. 44:323–328.

Zellner, G., J. Winter. 1987b. Analysis of a highly efficient methanogenic consortium producing biogas from whey. System. Appl. Microbiol. 9:284–292.

Zellner, G., C. Alten, E. Stackebrandt, E. Conway de Macario, J. Winter. 1987a. Isolation and characterization of *Methanocorpusculum parvum*, gen. nov., spec. nov., a new tungsten requiring, coccoid methanogen. Arch. Microbiol. 147:13–20.

Zellner, G., P. Vogel, H. Kneifel, J. Winter. 1987b. Anaerobic digestion of whey and whey permeate with suspended and immobilized complex and defined consortia. Appl. Microbiol. Biotechnol. 27:306–314.

Zellner, G., K. Bleicher, E. Braun, H. Kneifel, B. J. Tindall, E. Conway de Macario, J. Winter. 1989a. Characterization of a new mesophilic, secondary alcohol-utilizing methanogen, *Methanobacterium palustre* spec. nov. from a peat bog. Arch. Microbiol. 151:1–9.

Zellner, G., P. Messner, H. Kneifel, B. J. Tindall, J. Winter, E. Stackebrandt. 1989b. *Methanolacinia* gen. nov., incorporating *Methanomicrobium paynteri* as *Methanolacinia paynteri* comb. nov. J. Gen. Appl. Microbiol. 35:185–202.

Zellner, G., U. G. Sleytr, P. Messner, H. Kneifel, J. Winter. 1990. *Methanogenium liminatans* spec. nov., a new coccoid, mesophilic methanogen able to oxidize secondary alcohols. Arch. Microbiol. 153:287–293.

Zellner, G., E. Stackebrandt, P. Messner, B. J. Tindall, E. Conway de Macario, H. Knelfel, U. B. Sleytr. 1989c.

Methanocorpusculaceae fam. nov., represented by *Methanocorpusculum parvum, Methanocorpusculum sinense* spec. nov. and *Methanocorpusculum bavaricum* spec. nov. Arch. Microbiol. 151:381–390.

Zhao, Y., D. R. Boone, R. A. Mah, J. E. Boone, L. Xun. 1989. Isolation and characterization of *Methanocorpusculum labreanum* sp. nov. from the LaBrea tar pits. Int. J. Syst. Bacteriol. 39:10–13.

Zhao, Y., H. Zhang, D. R. Boone, R. A. Mah. 1986. Isolation and characterization of a fast-growing, thermophilic *Methanobacterium* species. Appl. Environ. Microbiol. 52:1227–1229.

Zhao, H., A. G. Wood, F. Widdel, M. P. Bryant. 1988. An extremely thermophilic *Methanococcus* from a deep sea hydrothermal vent and its plasmid. Arch. Microbiol. 150:178–183.

Zhilina, T. N. 1972. Death of *Methanosarcina* in the air. Microbiology (Engl. Transl.) 41:980–981.

Zhilina, T. N. 1976. Biotypes of methanosarcina. Microbiology (Russ.) 45:481–489.

Zhilina, T. N. 1983. New obligate halophilic methane-producing bacterium. Microbiology (Engl. Transl.) 52:290–297.

Zhilina, T. N. 1986. Methanogenic bacteria from hypersaline environments. Syst. Appl. Microbiol. 7:216–222.

Zhilina, T. N., S. A. Ilarionov. 1984. Characteristics of formate-assimilating methane bacteria and description of *Methanobacterium thermoformicicum* sp. nov. Microbiology (Engl. Trans.) 53:647–651.

Zhilina, T. N., T. P. Svetlichnaya. 1989. The ultrafine structure of *Methanohalobium evestigatum* an extreme halophilic bacterium producing methane. Microbiology (Russ.) 58:312–318.

Zhilina, T. N., G. A. Zavarzin. 1979a. Comparative cytology of methanosarcinae and description of *Methanosarcina vacuolata* sp. nova. Microbiology (Engl. Transl.) 48:223–228.

Zhilina, T. N., G. A. Zavarzin. 1979b. Cyst formation by methanosarcina. Microbiology (Russ) 48:451–456.

Zhilina, T. N., G. A. Zavarzin. 1987a. *Methanosarcina vacuolata* sp. nov., a vacuolated species of methanosarcina. Int. J. Syst. Bacteriol. 37:281–283.

Zhilina, T. N., G. A. Zavarzin. 1987b. *Methanohalobium evestigatus*, nov. gen., nov. sp. The extremely halophylic methanogenic *Archaebacterium*. Dokl. Akad. Nauk SSSR 293:464–468.

Zillig, W., P. Palm, W.-D. Reiter, F. Gropp, G. Pühler, H.-P. Klenk. 1988. Comparative evaluation of gene expression in archaebacteria. Eur. J. Biochem. 173:473–482.

Zinder, S. H., T. Anguish, A. L. Lobo. 1987. Isolation and characterization of a thermophilic acetotrophic strain of *Methanothrix*. Arch. Microbiol. 146:315–322.

Zinder, S. H., R. A. Mah. 1979. Isolation and characterization of a thermophilic strain of *Methanosarcina* unable to use H_2-CO_2 for methanogenesis. Appl. Environ. Microbiol. 38:996–1008.

Zinder, S. H., K. R. Sowers, J. G. Ferry. 1985. *Methanosarcina thermophila* sp. nov., a thermophilic, acetotrophic, methane-producing bacterium. Int. J. Syst. Bacteriol. 35:522–523.

Prokaryotes (2006) 3:208–230
DOI: 10.1007/0-387-30743-5_10

CHAPTER 10

The Order Methanomicrobiales

JEAN-LOUIS GARCIA, BERNARD OLLIVIER AND WILLIAM B. WHITMAN

Characteristics of Methanomicrobiales

Although their morphology is diverse, Methanomicrobiales can be distinguished from other methanogens by growth properties, cell wall and lipid composition, and rDNA sequence. All Methanomicrobiales can use $H_2 + CO_2$ as a substrate for methanogenesis, most species can utilize formate, and many species utilize alcohols (Table 1). They cannot use acetate and methylated C-1 compounds such as methanol, methylamines and methyl sulfides as substrates for methanogenesis, and this property distinguishes them from the *Methanosarcinales*. Notably, even though the Methanomicrobiales cannot use acetate as a substrate for methanogenesis, many species require acetate as a carbon source. The absence of hydroxyarchaeol in their lipids further distinguishes them from the *Methanosarcinales* (Koga et al., 1998). In addition, all of the Methanomicrobiales so far examined contain aminopentanetetrols, galactose and glycerol in their lipids, and aminopentanetetrols are unique to this taxon. The absence of pseudomurein further distinguishes the Methanomicrobiales from both the *Methanobacteriales* and *Methanopyrales*. Additional distinctive features are listed in Table 1. Based upon the phylogeny of the 16S rRNA genes as well as phenotypic and genotypic characteristics, the order Methanomicrobiales has been divided into three families and nine genera of hydrogenotrophic methanogens (Boone et al., 1993; Rouvière et al., 1992; Garcia et al., 2000; Tables 2 and 3). Twenty-four species have been described so far within this order (Tables 3 and 4). Because some species epithets are not in agreement with the International Nomenclature Code for Taxonomy, proposed changes for the original spelling are shown in brackets after the current name (Euzéby, 1997; Chong and Boone, 2004; Ferry and Boone, 2004).

Habitats of Methanomicrobiales

Like methanogens belonging to other orders, Methanomicrobiales inhabit diverse anaerobic habitats comprising marine and fresh water sediments, anaerobic digesters and the rumen. Although based upon classical microbiological characterizations, this conclusion has been well supported recently by a wide variety of molecular and immunological studies (Kudo et al., 1997; Munson et al., 1997; Grossköpf et al., 1998; Sekiguchi et al., 1999). Moreover, a hybridization probe specific to the three families belonging to this order has been designed and characterized for environmental and determinative microbiological studies (Raskin et al., 1994).

Eight of the 24 validly published species within the order Methanomicrobiales including species of the genera *Methanoculleus*, *Methanofollis*, and *Methanocorpusculum* have been isolated from anaerobic digestors or sewage sludge (Whitehead and Cotta, 1999). *Methanoculleus bourgense* was isolated from a tannery-byproducts enrichment-culture inoculated with sewage sludge (Ollivier et al., 1986), whereas *Methanoculleus palmolei* was obtained from a digester treating wastewater from a palm oil mill in Indonesia (Zellner et al., 1998). *Methanoculleus* sp. was shown to be involved in the syntrophic oxidation of acetate in digesters containing high concentrations of ammonium salts and volatile fatty acids (Schnurer et al., 1999). *Methanofollis liminatans* was isolated from the effluent of an anaerobic reactor treating industrial wastewater in Germany (Zellner et al., 1990). *Methanocorpusculum aggregans* originated from a municipal sewage sludge digester in France (Ollivier et al., 1985). Subsequently, a similar organism, *Methanocorpusculum parvum*, was enriched from an anaerobic sour whey digester inoculated with sewage sludge (Zellner et al., 1987). Thereafter, *Methanocorpusculum bavaricum* and *Methanocorpusculum sinense* were isolated from a wastewater pond of a sugar factory in Germany and from a pilot plant for treatment of distillery wastewater in China, respectively (Zellner et al., 1989b). Ecological

Table 1. Distinguishing features of the Methanomicrobiales.

Feature	Methanomicrobiales	Methanococcales	Methanobacteriales	Methanosarcinales	Methanopyrales
Morphology	Small rod, plate, irregular coccus, curved rod	Irregular coccus	Rod, short rod, coccus	Irregular coccus sometimes in aggregates, rod, flat polygon	Rod
Motility	+/−	+	−	−	+
Substrates[a]	H_2 (for, alc)	H_2 (for)	H_2 (for, alc)	(me, H_2, ace)	H_2
Temperature range, °C	0–60	18–94	20–97	2–70	84–110
Cell wall	Glycoprotein (sheath)	Protein	Pseudomurein	Glycoprotein (methanochondrotin, sheath)	Pseudomurein
Distinctive lipid components	Galactose, aminopentanetetrols, glycerol	Serine	myo-Inositol, serine	Hydroxyarchaeol, myo-inositol, ethanolamine, glycerol	Galactose, mannose
Mol% G+C	38–61	30–33	26–61	36–54	60

Symbols: +, present in most or all taxa; −, present in a few or no taxa; and +/−, present in many but not all taxa.
Abbreviations: for, formate; alc, alcohols; me, methylamines; and ace, acetate.
[a]Substrates (in parentheses) are utilized by many but not all taxa.

Table 2. Some characteristics of the Methanomicrobiales families.

Family	Characteristics
Methanomicrobiaceae	Small rod, plate, or irregular coccus; substrates for methanogenesis are H_2 + CO_2, frequently formate, and sometimes alcohols; Gram negative; motile or nonmotile; G+C content, 39–50mol%.
Methanocorpusculaceae	Small, irregular coccus, substrates for methanogenesis are H_2 + CO_2, formate, and sometimes alcohols; Gram negative; motile or nonmotile; G+C content, 48–52mol%.
Methanospirillaceae	Curved rods; substrates for methanogenesis are H_2 + CO_2, formate, and sometimes alcohols; Gram negative; motile; G+C content, 45–49mol%.

Table 3. Taxonomic groups of order Methanomicrobiales.

Family	Genus	Species
Methanomicrobiaceae[T]	Methanomicrobium[T]	M. mobile[T]
	Methanolacinia	M. paynteri[T]
	Methanogenium	M. cariaci[T] (cariacoense)[a]
		M. frigidum
		M. frittonii
		M. organophilum (organiphilum)[a]
	Methanoculleus	M. bourgense[T] (bourgensis)[b]
		M. olentangyi (olentangyense)[a]
		M. marisnigri (marinigri)[a]
		M. oldenburgensis
		M. palmolei
		M. thermophilicum (thermophilicus)[b]
	Methanoplanus	M. endosymbiosus
		M. limicola[T]
		M. petrolearius
	Methanofollis	M. liminatans
		M. tationis[T] (tatioense)[a]
Methanocorpusculaceae	Methanocorpusculum[T]	M. bavaricum
		M. labreanum
		M. parvum[T]
		M. aggregans
		M. sinense
Methanospirillaceae	Methanospirillum[T]	M. hungatei[T] (hungateii)[c]
Genus incertae sedis	Methanocalculus	M. halotolerans[T]
		M. pumilus

Abbreviation: [T], type family of the order, type genus of the family or type species of the genus.
[a]Proposed change for the original spelling (Euzéby, 1997).
[b]Proposed change for the original spelling (S. C. Chong and D. R. Boone, in press).
[c]Proposed change for the original spelling (J. G. Ferry and D. R. Boone, in press).

Table 4. Some characteristics of methanogenic archaea of the order Methanomicrobiales.

Organism	Dimensions (μm)	Flagella	pH	Temp. (°C)	NaCl (M)	Required organic growth factors[b]	G+C content mol%
Family *Methanomicrobiaceae*							
Methanomicrobium							
mobile	0.7 × 1.5–2	One polar	6.1–6.9	40	nd	Complex	49
Methanolacinia							
paynteri	0.6 × 1.5–2.5	Flagellated[a]	7	40	0.15	ac	38
Methanogerium							
cariaci	φ 1.3–2.6	Peritrichous	6.8–7.3	20–25	0.5	ac, YE	52
frigidum	φ 1.2–2.5	None	7.5–7.9	15	0.3–0.6	ac	52
frittonii	φ 1–2.5	None	7–7.5	57	0	None	49
organophilum	φ 0.5–1.5	None	6.4–7.3	30–35	0.3	ac, PABA, biotin, B_{12}, tung	47
Methanoculleus							
bourgense	φ 1–2	None	6.7	35–40	0.18	ac	59
= olentangyi	φ 1–1.5	None	nd	37	0.2	ac	54
marisnigri	φ <1.3	Peritrichous	6.2–6.6	20–25	0.1	Peptones	61
oldenburgensis	φ 0.5–1.5	None	7.5–8	45	<0.25	ac	49
palmolei	φ 1.25–2	Flagellated[a]	6.9–7.5	40	nd	ac	59–60
thermophilicum	φ 0.7–1.8	Single	6.5–7.2	55–60	0.3	ac, peptones, vit	56–60
Methanoplanus							
endosymbiosus	0.5–1 × 1.6–3	Peritrichous	6.8–7.3	32	0.25	p-Cresol, tung	39
limicola	0.1–0.3 × 1.5–2.8	Polar tuft	7	40	0.17	ac	48
petrolearius	φ 1–3	None	7	37	0.1–0.5	ac	50
Methanofollis							
liminatans	φ 1.5	Flagellated[a]	7	40	0–0.6	ac	60
tationis	φ 3	Peritrichous	7	37–40	0.1–0.2	ac, YE, peptones, tung	54
Family *Methanocorpusculaceae*							
Methanocorpusculum							
bavaricum	φ <1	Flagellated[a]	7	37	nd	RF	51
labreanum	φ 0.4–2	None	7	37	0–0.2	YE, peptones	50
parvum	φ 0.5–1	Single	6.8–7.5	37	0	ac, YE, tung	49
= aggregans	φ 0.5–2	None	6.6	35	<0.18	ac, YE/peptones	52
sinense	φ <1	Flagellated[a]	7	30	0	RF	50
Family *Methanospirillaceae*							
Methanospirillum							
hungatei	0.5 × 7.4–10	Polar tuft	6.6–7.4	30–45	nd	(ac)	47–50
Genus insertae sedis							
Methanocalculus							
halotolerans	φ 0.8–1	Peritrichous	7.6	38	0.8	ac	55
pumilus	φ 0.8–1	None	7	35	0.17	ac	52

Abbreviations: nd, not determined; =, indicates synonyms; φ, diameter; ac, acetate; (ac), acetate required or stimulatory depending on the strain; Complex, a combination of acetate, isobutyrate, isovalerate, 2-methylbutyrate, tryptophan (or indole), pyridoxine, thiamine, biotin, vitamin B_{12}, 4-aminobenzoate and 7-mercaptoheptanoylthreonine; PABA, 4-aminobenzoate; tung, tungsten; YE, yeast extract; vit, vitamins; and RF, rumen fluid.
[a]Type not described.

References: Blotevogel et al., 1991; Corder et al., 1983; Franzmann et al., 1997; Ferry et al., 1974; Harris et al., 1984; Maestrojuan et al., 1990; Mori et al., 2000; Ollivier et al., 1985, 1986, 1997, 1998; Paynter and Hungate, 1968; Rivard and Smith, 1982; Rivard et al., 1983; Romesser et al., 1979; Tanner and Wolfe, 1988; van Bruggen et al., 1986; Widdel et al., 1988; Wildgruber et al., 1982; Xun et al., 1989; Zabel et al., 1984, 1985; Zellner et al., 1987, 1989a,b, 1990, 1998, 1999; and Zhao et al., 1989.

studies of a number of bioreactors have found varying contributions of the Methanomicrobiales to the methanogenic microflora. In a wood-fermenting bioreactor, *Methanomicrobium mobile* was the predominant methanogen, representing nearly 90% of all methanogens detectable by the immunological methods used (Macario et al., 1991). Such predominance of a single methanogen subpopulation has not been reported in other bioreactors so far and might be specific to wood-fermenting bioreactors. Nevertheless, in the 21 anaerobic sewage sludge digesters studied by Raskin et al. (1995), Methanomicrobiales together with *Methanosarcinales* represented the majority of the methanogens detected with oligonucleotide probes. In contrast, the Methanomicrobiales represented only a small fraction of the methanogens found in the granular consortia formed in an upflow anaerobic sludge blanket reactor (Visser et al., 1991).

Among the order Methanomicrobiales, six species belonging to genera *Methanolacinia*, *Methanogenium* and *Methanoculleus* were recovered from marine sediments and water. *Methanolacinia paynteri* was isolated from the edges of the mosquito control canals within mangrove swamps located in the Cayman Islands, British West Indies (Rivard et al., 1983). *Methanogenium cariaci*, *M. marisnigri* and *M. organophilum* were found in sediments from Cariaco Trench, the Black Sea, and near Venice (Italy), respectively (Romesser et al., 1979; Widdel, 1986; Widdel et al., 1988). The psychrophilic *Methanogenium frigidum* was isolated from the perennially cold, anoxic hypolimnion of a seawater-derived Antarctic meromictic lake (Franzmann et al., 1997), and the thermophilic *Methanoculleus thermophilicum* originated from sediments underlying the high temperature effluent channel of a nuclear power plant in Florida (Rivard and Smith, 1982). Moreover, examination of 16S rDNA libraries created with archaea-specific primers found sequences closely related to those of *Methanoculleus* and *Methanogenium* spp. from coastal salt-marsh sediment samples (Munson et al., 1997).

Members of the order Methanomicrobiales also were recovered from fresh water sediments. *Methanogenium frittonii* inhabits a freshwater lake in England (Harris et al., 1984). *Methanoculleus olentangyi* (Corder et al., 1983) and *Methanoculleus oldenburgensis* (Blotevogel et al., 1991) were isolated from river sediments in the United States and Germany, respectively. Finally, *Methanocorpusculum labreanum* was isolated from the surface sediments of Tar Pit Lake in Los Angeles (in the United States; Zhao et al., 1989). Likewise, small numbers of methanogens immunologically crossreactive with *Methanoculleus*

marisnigri and *Methanogenium cariaci* have been detected in other river and lake sediments (Cairó et al., 1991).

Members of the order Methanomicrobiales have also been found (using in situ hybridization with a domain-specific oligonucleotide probe) on the surfaces of rice roots and in the surrounding paddy soil (Grossköpf et al., 1998). An irregular coccus phylogenetically related to *Methanoculleus marisnigri* was also isolated from a French ricefield soil (Joulian et al., 1998). Using molecular techniques, *Methanogenium*- and *Methanoculleus*-like organisms were shown to be also present in diverse ricefield soils in Japan (Kudo et al., 1997).

Subterrestrial ecosystems such as oil reservoirs and groundwaters are suitable habitats for Methanomicrobiales. *Methanoplanus petrolearius* and *Methanocalculus halotolerans* were isolated from an African offshore oil field (Ollivier et al., 1997) and from a French oil field (Ollivier et al., 1998), respectively. Isolates affiliated to *Methanoculleus* species have been recently obtained from Californian oil reservoirs (Orphan et al., 2000). Members of the genus *Methanomicrobium* have been identified in granitic groundwater (Pedersen, 1997). Other habitats for Methanomicrobiales include a swamp of drilling waste in Italy (*Methanoplanus limicola*; Wildgruber et al., 1982) and the mud of a small, moderately thermophilic solfatara pool at an altitude of 4,750 m in Chile (*Methanofollis tationis*; Zabel et al., 1984). Indirect immunofluorescence has also detected cells related to *Methanoculleus thermophilicum* in an alkaline Icelandic hot spring (Sonnehansen and Ahring, 1997).

Large quantities of methane are produced during the rumen fermentation. Species of the genus *Methanobrevibacter*, order *Methanobacteriales*, are the predominant methanogens in bovine and caprine rumina, whereas Methanomicrobiales are the predominant methanogens in the ovine rumen (Lin et al., 1997). *Methanomicrobium mobile* has also been found in high numbers in the bovine rumen and is probably a significant contributor to the total amount of methanogenesis (Paynter and Hungate, 1968; Jarvis et al., 2000). This conclusion is supported by molecular ecological studies that indicate Methanomicrobiales are the second most abundant methanogens in the bovine rumen and in model rumen systems, where they account for 12% of the total small subunit (SSU) rRNA in rumen fluid (Sharp et al., 1998). In addition, there is one report that *Methanogenium* spp. are the predominant methanogens in the gastrointestinal tracts of chickens and turkeys (Miller and Wolin, 1986).

Termites harbor symbiotic methanogens in their gut. Nucleotide sequences of PCR-amplified 16S ribosomal RNA genes from four termite species demonstrated that Methanomicrobiales species distinct from known species were present and constitute a unique phylogenetic group (Ohkuma et al., 1999).

Methanoarchaea ascribable to the genus *Methanogenium* also were recovered in low numbers from vegetables, meat, fish and cheese where they are thought to be chance contaminants (Brusa et al., 1998).

Although the Methanomicrobiales are generally free-living, some species are endosymbionts of anaerobic protozoa. Within the protozoa, they are associated with the hydrogenosome, suggesting that the symbiosis is based upon interspecies H_2 transfer from the protozoan to the methanogen (Berger and Lynn, 1992; Embley et al., 1992). For instance, *Methanoplanus endosymbiosus* is an endosymbiont of the sapropelic marine ciliate *Metopus contortus* (van Bruggen et al., 1986). Endosymbionts closely related to *Methanocorpusculum* and *Methanoplanus* spp. also have been identified within the anaerobic protozoa *Metopus*, *Trimyema* and *Pelomyxa*. These endosymbionts appear to be different from their free-living phylogenetic relatives, suggesting that some species may have adapted specifically to this lifestyle (Embley and Finlay, 1993; Finlay et al., 1993).

Isolation

ENRICHMENT MEDIA FOR GROWTH OF METHANOMICROBIALES Numerous types of media are appropriate for the growth of Methanomicrobiales. Representative media for isolating and culturing these microorganisms will be described in this section. All Methanomicrobiales are hydrogen-oxidizing organisms; they can grow on a mixture of molecular hydrogen (H_2) and carbon dioxide (CO_2; or bicarbonate [HCO_3^-]) according the following reaction:

$$4H_2 + HCO_3^- + H^+ \rightarrow CH_4 + 3H_2O$$

$$\Delta G^{\circ\prime} = -135 kJ/mol \text{ of } CH_4$$

All except *Methanolacinia paynteri* can also use formate to produce methane (CH_4):

$$4HCO_2^- + H^+ + H_2O \rightarrow CH_4 + 3HCO_3^-$$

$$\Delta G^{\circ\prime} = -145 kJ/mol \text{ of } CH_4$$

Some species can also use secondary alcohols and CO_2 to produce methane and the organic acid or ketone:

$$2CH_3CH_2OH + HCO_3^- \rightarrow 2CH_3COO^- + H^+$$

$$+ CH_4 + H_2O \quad \Delta G^{\circ\prime} = -116 kJ/mol \text{ of } CH_4$$

The composition of growth medium will depend on the type of methanogen sought. Although most Methanomicrobiales will grow in one of these media described below, optimal growth may require 1) modification in the concentration of the mineral or organic components or 2) additional nutrients.

Standard Growth Medium 1

Values are per liter of medium.

K_2HPO_4	0.3 g
KH_2PO_4	0.3 g
$(NH_4)_2SO_4$	0.3 g
NaCl	0.6 g
$MgSO_4 \cdot 7H_2O$	0.13 g
$CaCl_2 \cdot 2H_2O$	8 mg
$FeSO_4 \cdot 7H_2O$	2 mg
Sodium acetate	0.5 g
Yeast extract	0.5 g
Trypticase	0.5 g
Cysteine-HCl	0.5 g
Trace minerals solution (see below)	10 ml
Resazurin	1 mg

This medium, adapted from that described by Balch et al. (1979), is recommended for isolation of species from freshwater sediments, sewage sludge or digesters.

The trace minerals solution (Balch et al., 1979) is composed of (per liter):

Nitrilotriacetic acid	1.5 g
$MgSO_4 \cdot 7H_2O$	3.0 g
$MnSO_4 \cdot 2H_2O$	0.5 g
NaCl	1.0 g
$FeSO_4 \cdot 7H_2O$	0.1 g
$CoSO_4$ or $CoCl_2$	0.1 g
$CaCl_2 \cdot 2H_2O$	0.1 g
$ZnSO_4$	0.1 g
$CuSO_4 \cdot 5H_2O$	0.01 g
$AlK(SO_4)_2$	0.01 g
H_3BO_3	0.01 g
$Na_2MoO_4 \cdot 2H_2O$	0.01 g

To prepare the trace minerals solution, dissolve the nitrilotriacetic acid in 800 ml of water and adjust the pH to 6.5 with KOH. Then dissolve the minerals in order, adjust the pH to 7.0, and bring the volume to 1 liter.

The vitamin solution (Balch et al., 1979) is composed of (per liter):

Biotin	2 mg
Folic acid	2 mg
Pyridoxine hydrochloride	10 mg
Thiamine hydrochloride	5 mg
Riboflavin	5 mg
Nicotinic acid	5 mg
DL-Calcium pantothenate	5 mg
Vitamin B_{12}	0.1 mg
p-Aminobenzoic acid	5 mg
Lipoic acid	5 mg

Standard Growth Medium 2

Values are per liter of medium:

K_2HPO_4	0.14 g
NH_4Cl	0.25 g
KCl	0.34 g

MgSO$_4$ · 7H$_2$O	0.35 g
CaCl$_2$ · 2H$_2$O	0.14 g
MgCl$_2$ · 2H$_2$O	2.75 g
NaCl	18 g
Fe(NH$_4$)$_2$(SO$_4$)$_2$ · 7H$_2$O	2 mg
Sodium acetate	1.0 g
Yeast extract	0.5 g
Trypticase	0.5 g
Cysteine-HCl	0.5 g
Trace minerals solution (see above)	10 ml
Vitamin solution (see above)	10 ml
Resazurin	1 mg

This medium, also adapted from that of Balch et al. (1979), is recommended for growth of marine species.

Medium 3 is for moderately halophilic Methanomicrobiales (from saline to hypersaline ecosystems). It contains the same components as Medium 1, but the saline (NaCl) concentration will vary depending on the (slight, moderate or extreme) salinity of the ecosystem studied. Among the species described so far, the upper NaCl limit for growth of any hydrogenotrophic methanogen belonging to the order Methanomicrobiales is 125 g/liter (Ollivier et al., 1998).

Medium 4 is for growth of species from the rumen. For enriching and isolating *Methanomicrobium* species from the rumen, clarified rumen fluid (rumen contents centrifuged at 25,000 × g for 15 min) is added to Medium 1. For *Methanomicrobium mobile*, the requirement for rumen fluid can be replaced by a complex mixture containing acetate, volatile fatty acids, amino acids, vitamins, and 7-mercaptoheptanoylthreonine phosphate (Kuhner et al., 1991; Tanner and Wolfe, 1988). Growth of *Methanoplanus endosymbiosus* is also dependent on rumen fluid, which can be replaced by the eluate of a rumen-derived anaerobic digester (Poirot et al., 1991). Para-cresol was identified as a growth-stimulatory component, and 50 nM of it supported the half-maximal growth.

All four media may be prepared anaerobically as follows. The pH of the medium is adjusted to 7.0 with 10 mM KOH. The medium is boiled under a stream of oxygen (O$_2$)-free nitrogen (N$_2$) gas and cooled to room temperature. It is then dispensed into Hungate tubes, serum bottles or flasks under a stream of N$_2$ + CO$_2$ (80:20, v/v) at atmospheric pressure. Vessels are autoclaved, and prior to culture inoculations, Na$_2$S · 9H$_2$O and NaHCO$_3$ are added from sterile anaerobic solutions to final concentrations of 0.04% and 0.2% (v/v), respectively. Sodium formate, added from a sterile anaerobic solution to a final concentration of 40 mM, or H$_2$ + CO$_2$ (80/20%; 2 bars) serves as energy source. Growth on alcohol (2-propanol, 1-propanol, ethanol, 2-butanol, 1-butanol or cyclopentanol) is tested at a final concentration of 20–30 mM (Widdel, 1986). Because gas is produced from formate and the alcohols,

care must be taken to avoid exceeding the pressure limit of the culture vessels during growth. Moreover, during growth on H$_2$ + CO$_2$, methanogens consume 5 moles of gas for every mole of CH$_4$ produced, and the vessel will quickly develop a negative pressure. A low internal pressure increases the possibility of contamination by air. In addition, the decrease in partial pressure of CO$_2$ renders the culture medium strongly alkaline, thus causing growth inhibition and cell lysis. To minimize these problems, the volume of the headspace should be at least five times the volume of the medium, and the culture vessels should be pressurized periodically throughout the growth period. The headspace also may be repressurized with H$_2$ + CO$_2$ to fully replenish the CO$_2$ consumed. For the thermophilic species, attention must also be given to the expansion of gas at high temperature.

Solid medium is prepared by the addition of 1.5–2% agar during the medium preparation. For the roll tube technique (Hungate, 1969), Hungate tubes are placed into a 45°C water bath after sterilization to prevent solidification of the molten agar. After inoculation, the tubes are rolled and rapidly cooled with an ice cube before incubation at the appropriate temperature. Petri plates may also be prepared in an anaerobic glove box and incubated in pressure cylinders (Balch et al., 1979) before removal from the chamber.

To measure the growth of methanogens, Hungate tubes with liquid cultures are directly inserted into a spectrophotometer, and the optical density at 580 nm is recorded.

ISOLATION AND CHARACTERIZATION For isolating Methanomicrobiales, enrichment cultures are initiated by inoculating the sample (10%) into serum bottles containing basal medium and H$_2$ + CO$_2$ or sodium formate as growth substrate. The inoculated serum bottles are incubated without shaking, except when H$_2$ is used as the energy source. Pure cultures are obtained by the repeated use of the Hungate roll tube technique (Hungate, 1969) using basal growth medium solidified with 1.5% (w/v) Noble agar (Difco). For thermophilic enrichments, the agar concentration is increased up to 2%. Characterization of the axenic strain should follow the guidelines of the International Committee on Systematic Bacteriology (ICSB) Subcommittee for the Taxonomy of Methanogens (Boone and Whitman, 1988). Growth rates at various pH values, temperatures and salt concentrations are determined in Hungate tubes of basal growth medium containing 40 mM sodium formate with a N$_2$ + CO$_2$ (80:20, v/v) atmosphere or in the absence of sodium formate with a H$_2$ + CO$_2$ (80:20, v/v) atmosphere. The pH is adjusted to the desired value by injecting appropriate volumes of anaerobic sterile 10% (v/v) $NaHCO_3$ or Na_2CO_3 stock solutions. Both the initial and final pH must be recorded because the pH can change rapidly during growth. To determine the salt requirement for growth, NaCl is weighed directly in Hungate tubes, and the medium is subsequently dispensed as described above. The strain is subcultured at least once at each salt concentration prior to measurement of the growth rate.

To test substrate utilization, substrates are added from sterile stock solutions to the basal medium at a final concentration of 20 mM (ethanol, 1-propanol, 2-propanol, 1-butanol or isobutanol) or 40 mM (formate). The gas phase is N_2 + CO_2 (80:20, 200 kPa). Hydrogen oxidation is tested using H_2 + CO_2 (80:20, 200 kPa) in the gas phase. All experiments must be performed in duplicate. Also, susceptibility to lysis by detergents and hypotonic solutions, Gram staining, and motility should be determined. Phase contrast, fluorescence and electron microscopy are also used for phenotype characterization. Methane is quantified by gas chromatography to verify that it is the major catabolic product. If possible, it is also extremely desirable to determine the nutritional requirements for isolates obtained from complex medium.

The G+C content of DNA is determined by using high-performance liquid chromatography (HPLC), as described by Mesbah et al. (1989), after isolation and purification of the DNA by chromatography on hydroxyapatite. Nonmethylated lambda DNA (Sigma) is used as the standard. Sequencing of the gene for the 16S rRNA is also very desirable and may be necessary for final placement of an isolate with the Methanomicrobiales. A primer pair, designated "FARCH-9" (5′-CTGGTTGATCCTGCCAG-3′) and "Rd1" (5′-AAGGAGGTGATCCAG CC-3′), is used to amplify the 16S rRNA gene from genomic DNA of Methanomicrobiales (Ollivier et al., 1998). The amplified product is purified (Andrews and Patel, 1996) and the sequence determined with an ABI automated DNA sequencer in conjunction with a Prism dideoxy terminator cycle sequencing kit and the protocol recommended by the manufacturer (Applied Biosystems Inc.). The primers used for sequencing are F2 (5′-CAGGATTAGATACC CTGGTAG-3′), R2 (5′-GTATTACCGCGGCT GCTG-3′), R4 (5′-CCGTCAATTCCTTTGAG TTT-3′) and the two amplification primers FARCH9 and Rd1 described above. The 16S rRNA gene sequence is manually aligned with reference sequences of various members of the domain *Archaea* by using the alignment editor "ae2" (Maidak et al., 2000). Reference sequences are obtained from the Ribosomal Database Project (Maidak et al., 2000). Positions

of sequence and alignment uncertainty are omitted from the analysis. Pairwise evolutionary distances based on a number of unambiguous nucleotides are computed using the method of Jukes and Cantor (1969) and dendrograms are constructed from these distances using the neighbor-joining method. Both programs form part of the PHYLIP package (Felsenstein, 1993).

THE   FAMILY   METHANOMICROBIACEAE
Methanomicrobiaceae comprises seven genera with morphologies ranging from rods to highly irregular cocci or plane-shaped cells. The cell walls are proteinaceous and the lipids include both C_{20} and C_{40} isopranyl glycerol ethers. The sugars in glycolipids include glucose and galactose. The polar head groups contain aminopentanetetrols and glycerol. The mol% G+C for members of this family ranges from 38 to 62. Almost all strains can use formate. The use of secondary alcohols may be observed.

Methanomicrobium Genus *Methanomicrobium* is represented by a single mesophilic species, *Methanomicrobium mobile*.

Methanomicrobium mobile. These Gram-negative, slightly curved, short rods (0.7 × 1.5–2 µm; Fig. 1) are sluggishly motile (with one polar flagellum) and subject to frequent lysis. Colonies are round, 1 µm in diameter, convex, smooth, translucent, with entire edges. Methane is produced from H_2 + CO_2 or from formate. Growth requires acetate and rumen fluid or a mixture of volatile fatty acids (VFA: isobutyrate, isovalerate,

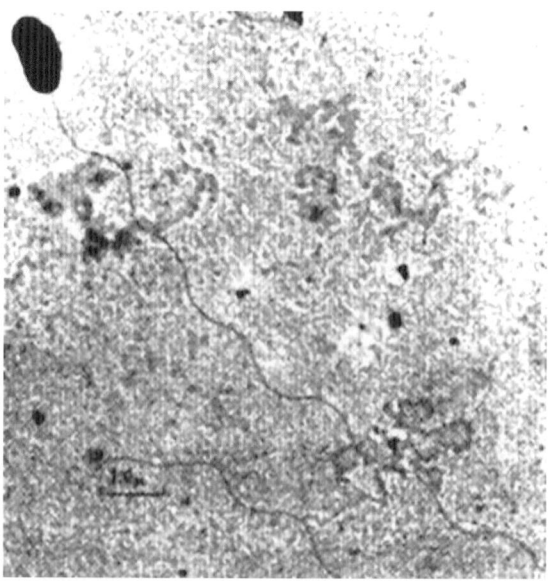

Fig. 1. Electron micrograph of *Methanomicrobium mobile* showing the single polar flagellum. Bar = 1 µm. (From Paynter and Hungate, 1968.)

and 2-methylbutyrate), amino acids (tryptophan or indole) and vitamins (pyridoxine, thiamine, biotin, vitamin B_{12}, and p-aminobenzoïc acid), and 7-mercaptoheptanoylthreonine phosphate (component B). This latter compound can be replaced by 7-mercaptoheptanoate if the medium is also supplemented with an unidentified growth factor found in rumen fluid and extracts of *Methanobacterium thermoautotrophicum*. Optimum temperature and pH for growth are 40°C (range 30–45°C) and 6.1–6.9 (range 5.9–7.7), respectively. This organism was isolated from rumen fluid. Its DNA G+C content is 49 mol%. Its Gen-Bank 16S rRNA sequence accession number is M59142. It is the type (and only) species of this genus; the type strain is DSM 1539 (≡ ATCC 35094). Refer to Paynter and Hungate (1968); Tanner and Wolfe (1988); and Kuhner et al. (1991).

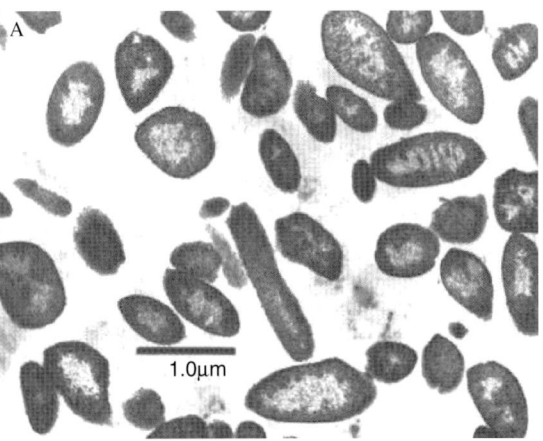

Fig. 2. Electron photomicrograph of a group of typical cells of *Methanolacinia paynteri*. (From Rivard et al., 1983.)

Methanolacinia

Genus *Methanolacinia* has only one representative, *Methanolacinia paynteri* (formerly *Methanomicrobium paynteri*).

Methanolacinia paynteri. These Gram-negative, pleomorphic, short and highly irregular coccoid to lobe-shaped cells (about 1.5–2.0 µm in diameter; Fig. 2) usually occur singly and are flagellated but only weakly motile or nonmotile. Cells lyse in detergents. The cell envelope consists of a hexagonally arranged S-layer with a glycoprotein of Mr 155,000 and a lattice constant of 15.3 nm (Fig. 3). Colonies are circular, 1–2 µm in diameter, off-white, with entire edges. Methane is produced from $H_2 + CO_2$, 2-propanol + CO_2, 2-butanol + CO_2, or cyclopentanol + CO_2. No growth or methane production is detected on formate, acetate, methylamines, ethanol, 1-propanol, 1-butanol, and cyclohexanol. Growth requires acetate. Polar lipids consist of di- and tetraether lipids and include phosphatidyl glycerol diether (PG), phosphatidyltrimethylaminopentanetetrol diether (PPTAD), diglycosyldiether, presumptive phosphatidylaminopentanetetrol diether (PPAD) and the corresponding phosphoglycolipid ethers. Putrescine and spermidine are the only polyamines. Optimum growth temperature and pH are 40°C (range 20–45°C) and 7.0 (range 6.6–7.3), respectively. Growth is optimal in 0.15 M NaCl (range 0–0.8 M). Minimum generation time is 4.8 h. The G+C content of DNA is 44 mol% (by buoyant density; BD) and 38 mol% (by thermal denaturation; Tm). This organism was isolated from marine sediments. *Methanolacinia paynteri* is the type species of the genus; its type strain is G-2000 (≡ DSM 2545 or ATCC 33997). Refer to Rivard et al. (1983) and Zellner et al. (1989a).

Fig. 3. Electron microphotograph of a freeze-etched preparation of *Methanolacinia paynteri*; bar = 0.1 µm. (From Zellner et al., 1989a.)

Methanogenium

Genus *Methanogenium* comprises four species. These highly irregular cocci are Gram negative. Although motility has not been observed, flagella are sometimes present. The cell wall is composed of regular glycoprotein subunits. Cells require growth factors, use formate, and are readily lysed in dilute detergents. The G+C content varies from 47 to 52 mol%. Two species were shown to use both CO_2 and secondary alcohols to form methane. *Methanogenium frittonii* is a thermophilic species (optimum temperature for growth, 57°C), whereas *M. frigidum* is psychrophilic (optimum temperature for growth, 15°C).

Methanogenium cariaci (cariacoense). Though motility has never been observed, these highly irregular cocci (1.3–2.6 µm in diameter) have peritrichous thin flagella up to 18 µm long and long thin pili about 4 nm in diameter. Though the cell is Gram negative, its cell wall lacks an outer membrane and has a periodic surface pattern consisting of structural units about 14 nm in diameter; these particles seem to be composed of several subunits (Fig. 4). The cells are sensitive and 1% SDS. Colonies are circular, umbonate, greenish yellow with entire edges, shiny, 0.5 μm in diameter after two weeks of incubation, and 4 μm after 14 weeks. Methane is produced from $H_2 + CO_2$ or from formate. Growth requires yeast extract and acetate. Optimum temperature and pH for growth are 20–25°C (range 15–35°C) and 6.8–7.3 (range 6–7.6), respectively. Isolated from marine sediments, these organisms grow optimally in 0.54 M NaCl (range 0.2–0.8 M). The G+C content of DNA is 52 mol% (BD). The GenBank 16S rRNA sequence accession number is M59130. *Methanogenium cariaci* is the type species of its genus; the type strain is JR1 (≡ DSM 1497, ATCC 35093 or OCM 155). Refer to Romesser et al. (1979).

Methanogenium frigidum. These Gram-negative, irregular cocci (1.2–2.5 μm in diameter) are nonmotile and sensitive to 0.1% of SDS; cells occur singly without flagella or pili. The cell wall has an S-layer exterior to the plasma membrane and a fibrous coat exterior to the S-layer. Colonies are circular, smooth, green-yellow with entire edges. Methane is produced from $H_2 + CO_2$ or from formate. No growth or methane production is detected on acetate, trimethylamine or methanol. Growth requires acetate and is stimulated by yeast extract and peptones. Sulfide can serve as a sole sulfur source, and ammonia serves as a sole nitrogen source. An obligate psychrophile, this species grows optimally at 15°C (range 0–17°C). Optimal growth occurs at pH 7.5–7.9 (range 6.3–8.0) and in 0.35–0.6 M NaCl (range >0.1–<0.85 M). The minimum generation time is 2.9 days. The G+C content of DNA is 52 mol%; the GenBank 16S rRNA sequence accession number is AF009219. This

species was isolated from the perennially cold, anoxic hypolimnion of Ace Lake in Antarctica. The type strain is Ace-2T (≡ SMCC 459W or OCM 469). Refer to Franzmann et al. (1997).

Methanogenium frittonii. These irregular cocci (1–2.5 μm in diameter) are nonmotile, fimbriated, and occur singly or in pairs. Cells are surrounded by a protein envelope (20 nm thick) and are disrupted by pronase and trypsin digestion, glass-distilled water and 1% SDS. Colonies (1–2 μm in diameter) are circular, convex, shiny, dark yellow with entire edges. Methane is produced from $H_2 + CO_2$ or from formate. Growth is stimulated by yeast extract, tryptone, and casamino acids. Optimum temperature for growth is 57°C (range 26–62°C); optimum pH is 7.0–7.5 (range 6.0–8.25). NaCl is not required and is inhibitory above 2%. The G+C content of DNA is 49.2 mol% (BD). This organism was isolated from lake sediment. The type strain is FR-4 (≡ DSM 2832 or OCM 200). Refer to Harris et al. (1984).

Methanogenium organophilum (*organiphilum*). These irregular cocci (0.5–1.5 μm in diameter) are nonmotile and subject to detergent lysis, e.g., by 0.01% or less SDS. Colonies are smooth and yellowish. Methane is produced from $H_2 + CO_2$, formate, 2-propanol + CO_2, 2-butanol + CO_2, ethanol + CO_2 and 1-propanol + CO_2. No growth or methane production is detected on acetate or methanol. Secondary alcohols are oxidized to ketones, and primary alcohols are oxidized to monocarboxylic acids. Acetate has to be added as a carbon source for growth on substrates other than ethanol. 4-Aminobenzoate, biotin, vitamin B_{12}, and tungstate are required for growth. Optimum temperature for growth is 30–35°C (maximum 39°C); optimum pH is 6.4–7.3; growth is optimal with 20 g/liter of NaCl and 3 g/liter of $MgCl_2$. Minimum generation time is 6 h on H_2 and 11 h on 2-propanol + CO_2. The G+C content of DNA is 46.7 mol% (Tm); the GenBank 16S rRNA sequence accession number is M59131. This species was isolated from marine mud. The type strain is CV (≡ DSM 3596 or OCM 72). Refer to Widdel et al. (1988).

Methanoculleus

The genus *Methanoculleus* (Maestrojuan et al., 1990) consists of five mesophilic species (including one subject synonym) and one thermophilic species of Gram-negative, highly irregular cocci. Motility has not been observed, although flagella are present in some species. Formate is used by five species. The G+C content range is between 49 and 62 mol%. The type species is *Methanoculleus bourgense*. However, DNA hybridization and rRNA sequence similarity suggest that this

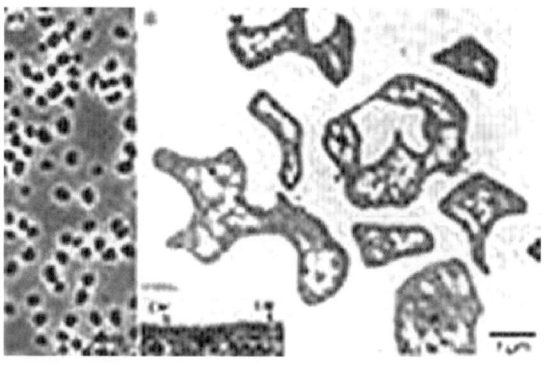

Fig. 4. Electron microphotograph of a freeze-etched preparation of (A) *Methanogenium cariaci* and (B) *Methanogenium marisnigri* showing particulate nature of the cell walls. (From Romesser et al., 1979.)

species may be a subjective synonym of *Methanoculleus olentangyi* (Rouvière et al., 1992; Boone et al., 1993). If this proposal is borne out by further investigations, strains of *M. bourgense* will be reclassified as *M. olentangyi* because it is the senior synonym, and the genus *Methanoculleus* will become invalid without action by the ICSB.

Methanoculleus bourgense (*bourgensis*). These Gram-negative, irregular cocci (1–2 μm in diameter) are nonmotile and subject to lysis by 0.02% SDS. Colonies are circular, convex and white to yellowish. Methane is produced from H_2 + CO_2 or from formate. Acetate is required for growth. Growth is stimulated by yeast extract and trypticase. Optimum temperature for growth is 35–40°C (range 30–50°C); optimum pH is 6.7 (range 5.5–8.0); and growth is optimal with 10 g/liter of NaCl (range 0–20 g/liter). The G+C content of DNA is 59 mol% (BD). The GenBank 16S rRNA sequence accession number is AF095269. The organism was isolated from a tannery-byproduct enrichment-culture inoculated with sewage sludge. It is the type species of genus *Methanoculleus*; the type strain is MS2 (≡ DSM 3045, ATCC 43281 or OCM 15). Refer to Ollivier et al. (1986) and Maestrojuan et al. (1990).

Methanoculleus marisnigri (*marinigri*). These Gram-negative, irregular cocci (1.3 μm in diameter) have peritrichous flagella (13 μm long). The cell wall has a periodic surface pattern of 14 nm in diameter. Cells are sensitive to lysis by 1% SDS. Colonies are circular, convex, yellow, shiny and have entire edges. Methane is produced from H_2 + CO_2, formate, 2-propanol + CO_2, or 2-butanol + CO_2. No growth or methane production is detected on acetate or methanol. Growth requires trypticase. Optimum temperature for growth is 20–25°C (range 15–45°C); optimum pH is 6.2–6.6 (range 5.7–7.6); growth is optimal in 0.1 M NaCl (range 0–0.7 M). The G+C content of DNA is 61.2 mol% (BD); the GenBank 16S rRNA sequence accession number is M59134. This organism was isolated from marine sediments. Its type strain is JR1 (≡ DSM 1498 or ATCC 35101). Refer to Romesser et al. (1979) and Maestrojuan et al. (1990).

Methanoculleus oldenburgensis. These irregular cocci (0.5–1.5 μm in diameter) are nonmotile, fimbriated single cells, seldom in pairs, and highly sensitive to detergents, e.g., 0.01% Triton X-100 or SDS. The cell envelope consists of a protein S-layer. The cells contain putrescine, sym-homospermidine and spermine as typical polyamines. Methane is produced from H_2 + CO_2 or from formate. Growth requires acetate. Optimum temperature for growth is 45°C (range 25–50°C); optimum pH is 7.5–8.0 (range 6.5–8.5); NaCl is not required and inhibitory at

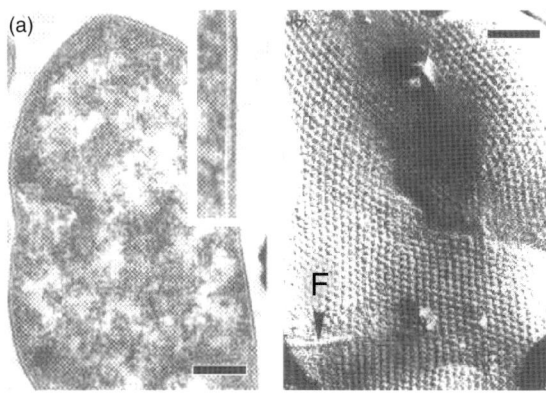

Fig. 5. (a) Electron micrographs of a thin section of a cell of *Methanocelleus palmolei*; bar = 0.1 μm. (b) Electron micrograph of a freeze-etched preparation of *Methanocelleus palmolei* showing the hexagonal arrangement of the S-layer glycoproteins and a piece of a flagellum (marked F); bar = 0.1 μm. (From Zellner et al., 1998.)

levels exceeding 1.5%. The G+C content of DNA is 48.6 ± 1 mol% (Tm). This organism was isolated from river sediments. Its type strain is CB-1 (≡ DSM 6216). Refer to Blotevogel et al. (1991).

Methanoculleus olentangyi (*olentangyense*). These irregular cocci (1–4.5 μm in diameter) are nonmotile and subject to lysis by 0.001% SDS and distilled water. Colonies are circular, convex, mucoid, shiny and yellow. Methane is produced from H_2 + CO_2 or from formate. No growth or methane production is detected on methanol and methylamines. Growth requires acetate. Optimum temperature for growth is 37°C (range 30–45°C); growth is optimal in 1% NaCl (range 0–3%). The G+C content of DNA is 54.4 mol% (BD). The GenBank 16S rRNA sequence accession number is AF095270. This organism was isolated from river sediments. Its type strain is RC/ER (= DSM 2772, ATCC 35293 or OCM 52). It is a subjective synonym of *M. bourgense*. Refer to Corder et al. (1983) and Maestrojuan et al. (1990).

Methanoculleus palmolei. The motility of these Gram-negative, highly irregular cocci (1.25–2 μm in diameter), though flagellated, has not been observed. Cell walls have S-layers of hexagonally arranged glycoprotein subunits (Mr 120,000; Fig. 5). The cells contain putrescine, sym-homospermidine and spermine as typical polyamines. Methane is produced from H_2 + CO_2, formate, 2-propanol + CO_2, 2-butanol + CO_2, or cyclopentanol + CO_2. No growth or methane production is detected on acetate, methanol, ethanol, 1-propanol, 2-pentanol + CO_2, 2,3-butanediol, dimethylamine and lactate. Growth requires acetate and is stimulated by potassium and tungstate ions. Optimum temperature for growth is 40°C

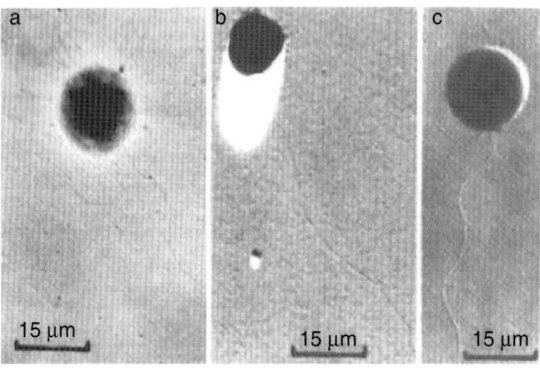

Fig. 6. Platinum-iridium shadowed cells of *Methanoculleus thermophilicus* (a) strain Ratisboa, (b) DSM 2373 and (c) strain Los Angeles. (From Zabel et al., 1985.)

(range 22–50°C); optimum pH is 6.9–7.5 (range 6.5–8.0); minimum generation time is 13.5 h. The G+C content of DNA is 59–59.5 mol% (Tm, HPLC); the GenBank 16S rRNA sequence accession number is Y16382. This organism was isolated from a digester treating wastewater of a palm oil mill in Indonesia. The type strain is INSLUZ T (≡ DSM 4273T). Refer to Zellner et al. (1998).

Methanoculleus thermophilicum (*thermophilicus*). These Gram-negative, irregular cocci to coccobacilli (0.7–1.8 μm in diameter; Fig. 6) occur singly or in pairs and have occasional internal membrane components. Cell envelope is composed of protein subunits and contains a glycoprotein (130 kD). One flagellum and pili are present; motility is not observed or weak. Colonies are circular, beige, with entire edges. Methane is produced from $H_2 + CO_2$ or from formate. No growth or methane production is detected on acetate, methanol, ethanol, propionate, pyruvate, dimethylamine and trimethylamine. Growth requires acetate, a trace vitamin solution and trypticase (or peptone or yeast extract). Optimum temperature for growth is 55–60°C; optimum pH is 6.7–7.2; optimum NaCl concentration is 0–0.3 M. The G+C content of DNA is 56–60 mol% (Tm); the GenBank 16S rRNA sequence accession number is M59129. This organism was isolated from digesters and marine sediments. The type strain is CR-1 (≡ DSM 2373, ATCC 33837 or OCM 174). Refer to Rivard and Smith (1982), Zabel et al. (1985), and Maestrojuan et al. (1990).

Methanoplanus

The genus *Methanoplanus* comprises three species of plane-shaped organisms. The cell walls contain at least one major glycoprotein. Formate is used for methanogenesis. One species is an endosymbiont of marine ciliates and is found in close association with microbodies that are thought to provide H_2 to the methanogen. The methanogen functions as an electron sink in the oxidation steps of the carbon flow in the ciliates. This symbiotic relationship is thought to be responsible for a total conversion of metabolites to CO_2 and CH_4 in marine sediments. The G+C range of the genus is 39 to 50 mol%.

Methanoplanus endosymbiosus. These Gram-negative, irregular discs with a diameter of 1.6–3.4 μm occur singly and are nonmotile despite the presence of peritrichous flagella or pili. The cell envelope shows a regular hexagonal surface pattern and consists of proteins. Cells are lysed by 0.001% SDS or 0.01% Triton X-100 in 2% NaCl. Colonies are circular with entire margins, shiny, convex and whitish yellow. Methane is produced from $H_2 + CO_2$ or from formate. No growth or methane production is detected on acetate, methanol or methylamine. Tungsten (0.1 mM) is required for growth, which is stimulated by yeast extract, tryptone or rumen fluid. *p*-Cresol replaces the requirement for rumen fluid. Optimum temperature for growth is 32°C (range 16–36°C); optimum pH is 6.8–7.3 (range 6.3–7.8); and optimum NaCl concentration is 0.25 M (range 0–0.75M). Minimum generation time is 7–12 h. The G+C content of DNA is 38.7 mol% (Tm). The GenBank 16S rRNA sequence accession number is Z29435. This organism was found as an endosymbiont in the marine ciliate *Metopus contortus*. The type strain is DSM 3599. Refer to Van Bruggen et al. (1986) and Poirot et al. (1991).

Methanoplanus limicola. These Gram-negative, angular, crystal-like plates with sharp edges (0.07–0.3 μm thick, 1.6–2.8 μm long, 1.5 μm wide) occur singly (Fig. 7). The cells are sometimes branched, without septa, and contain electron-dense round inclusions. The cell envelope shows a hexagonal surface pattern and contains a dominant glycoprotein. A polar tuft of flagella can be seen, and the cells are weakly motile. Cells are lysed by 2% SDS. Colonies on polysilicate are round, smooth, bright, ochre-colored, and flat (about 2 μm in diameter). Cells are resistant to vancomycin, penicillin, kanamycin and tetracycline. Methane is produced from $H_2 + CO_2$ or from formate. No growth or methane production is detected on acetate, methanol or methylamines. Acetate (0.1%) is required for growth, which is stimulated by yeast extract (or peptones and vitamins). Optimum temperature for growth is 40°C (range 17–41°C); optimum pH is 7 (range 6.5–7.5); and optimum NaCl concentration is 1% (range 0.4–5.4%). The G+C content of DNA is 47.5 mol% (Tm). This organism was isolated from a swamp of drilling waste in Italy. Its GenBank 16S rRNA sequence

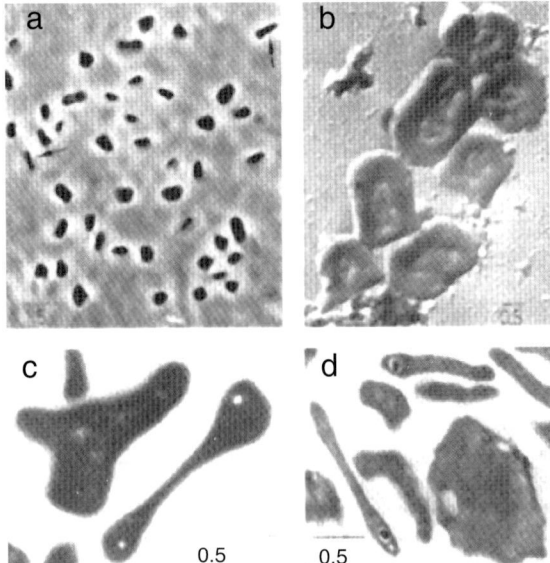

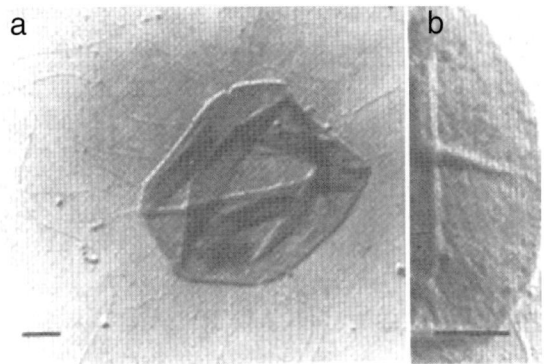

Fig. 8. Electron photomicrographs of platinum-iridium shadowed cells of *Methanofollis tationis*: (a) whole cells with peritrichous flagellation; (b) cell surface showing the hexagonal subunit structure; both bars = 0.5 μm. (From Zabel et al., 1984.)

Fig. 7. Light and electron micrographs of *Methanoplanus limicola*. (a) Phase contrast of exponentially growing cells; arrows indicate bacteria appearing in profile; (b) platinum shadowed; (c, d) thin sections. The bars are of indicated size in μm. (From Wildgruber et al., 1982.)

accession number is M59143. *Methanoplanus limicola* is the type species of its genus. The type strain is M3 (≡ DSM 2279,ATCC 35062 or OCM 101). Refer to Wildgruber et al. (1982).

Methanoplanus petrolearius. These irregular disc-shaped cells (1–3 μm in diameter) occur singly or in pairs and are nonmotile. Colonies are round (1–2 μm in diameter). Methane is produced from $H_2 + CO_2$, formate or 2-propanol + CO_2. No growth or methane production is detected on acetate, methanol, trimethylamine, lactate, glucose, 1-propanol, 1-butanol, or isobutanol. Acetate is required for growth, which is stimulated by yeast extract. Optimum temperature for growth is 37°C (range >25–<45°C); optimum pH is 7 (range 5.3–8.4); optimum NaCl concentration is 1–3% (range 0–5%). Minimum generation time is 10 h. The G+C content of DNA is 50 mol% (HPLC). This organism was isolated from an oil-producing well. The GenBank 16S rRNA sequence accession number is U76631. The type strain is SEBR 4847T (≡ DSM 11571 or OCM 486). Refer to Ollivier et al. (1997).

Methanofollis

This genus has been proposed to reclassify *Methanogenium tationis* and *M. liminatans*. These two species use formate and have a G+C content of 54–61 mol%.

Methanofollis liminatans. These irregular cocci (1.5 μm in diameter) are flagellated and motile. The cell envelope consists of hexagonally arranged glycoproteins. Cells grow in a synthetic, acetate-containing low-salt medium with substrates $H_2 + CO_2$, formate, 2-propanol + CO_2, 2-butanol + CO_2, or cyclopentanol + CO_2. No growth or methane production occurs with acetate, methanol, ethanol or dimethylamine. Tungstate (1–2 mM) stimulates growth. Optimum temperature for growth is 40°C (range >25–<45°C); optimum pH is 7; optimum NaCl concentration is 0–0.6 M (range 0–<0.8 M). Minimum generation time is 7.5–8.5 h. The G+C content of DNA is 59.3 mol% (Tm) and 60.5 mol% (HPLC). This organism was isolated from effluent of a digester for the treatment of industrial waste water. The GenBank 16S rRNA sequence accession number is AF095271. The type strain is GKZPZ (≡ DSM 4140). Refer to Zellner et al. (1990) and Zellner et al. (1999).

Methanofollis tationis (*tatioense*). These Gram-negative, regular or highly irregular cells, depending on the salt concentration in the medium, approximate 3 μm in diameter (Fig. 8), have peritrichous flagella, and lyse in 1% SDS. The cell envelope is composed of hexagonal protein subunits, which seem to contain a glycoprotein. The cytoplasm has numerous polyphosphate inclusions. Methane is produced from $H_2 + CO_2$ or from formate. No growth or methane production is detected on acetate, methanol, ethanol, ethyl acetate, dimethylamine or trimethylamine. Yeast extract, peptone and acetate are required for optimal growth. Yeast extract and peptone can partially be replaced by a heavy metal solution. Optimum temperature for growth is 37–40°C (range 25–45°C); optimum pH is 7 (range 6.3–8.8); optimum NaCl concentration is 0.8–1.2% (range 0–7%). Minimum generation time is 12 h at 40°C. The G+C content of DNA is 54 mol% (Tm). This organism was isolated from solfataric pools at an altitude of 4,750 m. The GenBank 16S rRNA sequence accession number is AF095272. *Methanofollis tationis* is the type species of this genus. The type strain is DSM 2702. Refer to Zabel et al. (1984) and Zellner et al. (1999).

The Family Methanocorpusculaceae

Methanocorpusculum

The family *Methanocorpusculaceae* (Zellner et al., 1989b) comprises one genus, *Methanocorpusculum*, and five mesophilic species (including one subjective synonym) of irregular cocci. They all use $H_2 + CO_2$ and formate, and some species can use 2-propanol + CO_2. The type species is *Methanocorpusculum parvum* (Zellner et al., 1987). However, *Methanocorpusculum aggregans* (formerly *Methanogenium aggregans*) was isolated prior to *M. parvum* and has been shown to be closely related, with about 70% DNA reassociation (Ollivier et al., 1985; Xun et al., 1989). The high level of DNA reassociation suggests that *M. parvum* and *M. aggregans* may be subjective synonyms (Boone et al., 1993). If further investigations support this conclusion, *M. aggregans* would have precedence over *M. parvum*, which would then have to be reclassified as *M. aggregans* subspecies *parvum*. This reclassification would then invalidate the genus *Methanocorpusculum* as well as the family *Methanocorpusculaceae*. *Methanocorpusculum parvum* is of special interest because it requires high levels of tungstate for growth (Zellner et al., 1987). It is also the first hydrogenotrophic methanogen found to possess cytochromes. Presumably, these cytochromes are involved in the oxidation of 2-propanol. The mol% G+C of this genus ranges from 48 to 52.

Methanocorpusculum aggregans. These Gram-negative, irregular cocci (0.5–2 μm in diameter) occur singly or in aggregates (Fig. 9) and are nonmotile and sensitive to 0.03% SDS. Colonies (3–4 μm in diameter) are round, convex, and white to yellow. Methane is produced from $H_2 + CO_2$ or from formate. No growth or methane production is detected on acetate, methanol or methylamines. Acetate and yeast extract or trypticase are required for growth. Optimum temperature for growth is 35°C (range 27–38°C); optimum pH is 6.6 (range 6.2–7.5); optimum NaCl concentration is 0% (range 0–2%). Mini-

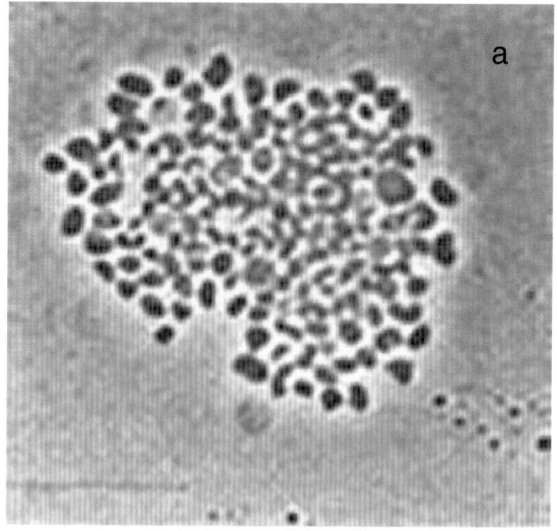

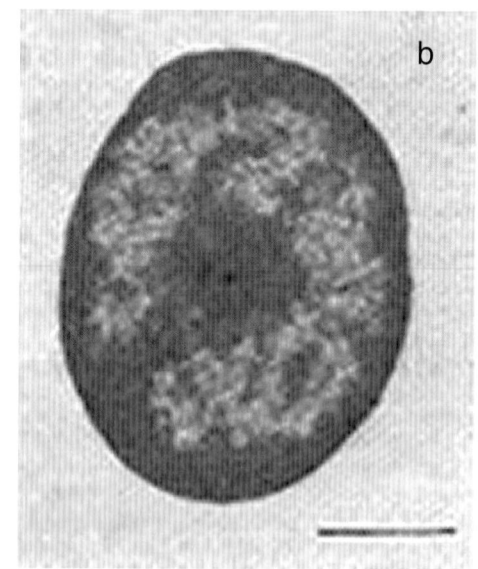

Fig. 9. (a) Phase-contrast photomicrograph of aggregates of *Methanocorpusculum aggregans*; bar = 10 μm; (b) thin-section electron photomicrograph of an individual cell of *Methanocorpusculum aggregans*; bar = 0.5 μm. (From Ollivier et al., 1985.)

mum generation time is 8 h. The G+C content of DNA is 52 mol% (BD). This organism was isolated from a sewage sludge digestor. The type strain is Mst (=DSM 3027 or OCM 21). It is possibly a subjective synonym of *M. parvum*. Refer to Ollivier et al. (1985) and Xun et al. (1989).

Methanocorpusculum bavaricum. These Gram-negative, small, irregular cocci (<1 μm in diameter) occur singly and are flagellated and weakly motile. The cell envelope consists of a hexagonally arranged S-layer with a center-to-center spacing of the glycoprotein subunits of 16 nm (Fig. 10). Methane is produced from $H_2 +$

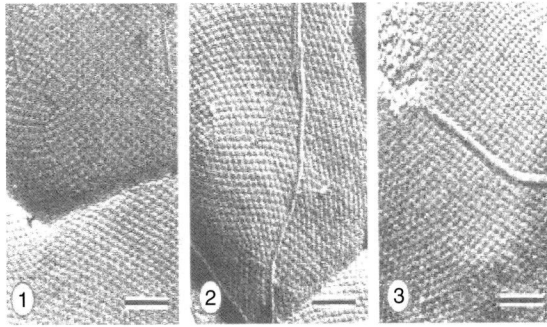

Fig. 10. Electron micrograph of a freeze-etched preparation of (1) *Methanocorpusculum parvum*; (2) *M. sinense*; (3) *M. bavaricum*; bar = 0.1 μm. (From Zellner et al., 1989b.)

CO$_2$, formate, 2-propanol or 2-butanol. No growth or methane production is detected on acetate, methanol, methylamines, ethanol, 1-propanol, 1-butanol or cyclohexanol. Rumen fluid is required for growth, which is stimulated by tungstate (1 mM). Optimum temperature for growth is 37°C (range 15–45°C); optimum pH is 7. The G+C content of DNA is 47.7 mol% (Tm), 51% (HPLC). This organism was isolated from anaerobic sediment of a sugar factory's wastewater treatment pond. Its GenBank 16S rRNA sequence accession number is AF042197. The type strain is SZSXXZ (≡ DSM 4179 or OCM 127). Refer to Zellner et al. (1989b).

Methanocorpusculum labreanum. These Gram-negative, irregular cocci are 0.4–2 μm in diameter. The degree of irregularity is dependent on the physiological state of the cells and on the ionic strength of the medium. The cells are non-motile, sensitive to 0.02% SDS, and have a protein cell wall. Colonies are circular (0.5 μm in diameter), clear, and convex with entire edges. Methane is produced from H$_2$ + CO$_2$ or formate. No growth or methane production is detected on acetate, propionate, methanol, trimethylamine or ethanol. Trypticase, peptone or yeast extract is required for growth, which is stimulated by acetate. Optimum temperature for growth is 37°C (range 25–40°C); optimum pH is 7 (range 6.5–7.5); optimum NaCl concentration is 0–1.5% (range 0–<3%). The G+C content of DNA is 50 mol% (BD). This organism was isolated from surface sediments of Tar Pit Lake in Los Angeles, California. The GenBank 16S rRNA sequence accession number is AF095267. The type strain is ZT (≡ DSM 4855, ATCC 43576, or OCM 1). Refer to Zhao et al. (1989).

Methanocorpusculum parvum. These Gram-negative, irregular cocci (= 1 μm in diameter) occur singly or in pairs. Each cell has one flagellum and is weakly motile. The cell envelope consists of hexagonally arranged protein sub-structures (Fig. 10). Complete lysis occurs with 1% SDS. Methane is produced from H$_2$ + CO$_2$, formate or 2-propanol + CO$_2$. No growth or methane production is detected on acetate, propionate, methanol, methylamines, ethanol, pyruvate, lactate, butyrate, 1-butanol, 2-pentanol, formaldehyde, L-alanine, L-lysine or L-leucine. Yeast extract, acetate and tungsten (1 mM) are required for growth, which is stimulated by clarified rumen fluid. Optimum temperature for growth is 37°C (range 20–40°C); optimum pH is 6.8–7.5; optimum NaCl concentration is 0% (range 0–5%). Minimum generation time is 8 h. The G+C content of DNA is 48.5 mol% (Tm).

This organism was isolated from an anaerobic sour whey digester inoculated with sewage sludge. The GenBank 16S rRNA sequence accession number is M59147.

Methanocorpusculum parvum is the type species of this genus and possibly a subjective synonym of *M. aggregans*. The type strain is XII (≡ DSM 3823, ATCC 43721 or OCM 63). Refer to Zellner et al. (1987).

Methanocorpusculum sinense. These Gram-negative organisms occur singly as flagellated, weakly motile, small, irregular cocci (<1 μm in diameter). The cell envelope consists of a hexagonally arranged S-layer with a center-to-center 15.8-nm spacing of the glycoprotein subunits (Fig. 10). Colonies are circular, convex with entire edges. Methane is produced from H$_2$ + CO$_2$ or from formate. No growth or methane production is detected on acetate, methanol, methylamines, ethanol, 1-propanol, 2-propanol, 1-butanol, 2-butanol or cyclohexanol. Clarified rumen fluid is required for growth. Optimum temperature for growth is 30°C (range 15–45°C); optimum pH is 7. The G+C content of DNA is 52 mol% (Tm) and 50 mol% (HPLC). This organism was isolated from an anaerobic biogas plant treating distillery wastewater. The GenBank 16S rRNA sequence accession number is AF095268. The type strain is CHINAZ (≡ DSM 4274 or OCM 128). Refer to Zellner et al. (1989b).

The Family Methanospirillaceae

Methanospirillum

Based upon its unique morphology and low 16S rRNA sequence similarity to the other Methanomicrobiales, the family *Methanospirillaceae* was created to include the single genus *Methanospirillum* (Boone et al., 1993). Members of the genus are mesophilic and have been reported from a wide range of habitats. However, only one species, *Methanospirillum hungatei*, has been described so far.

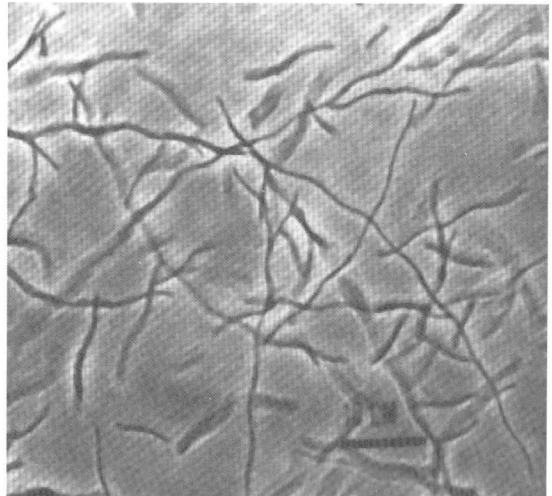

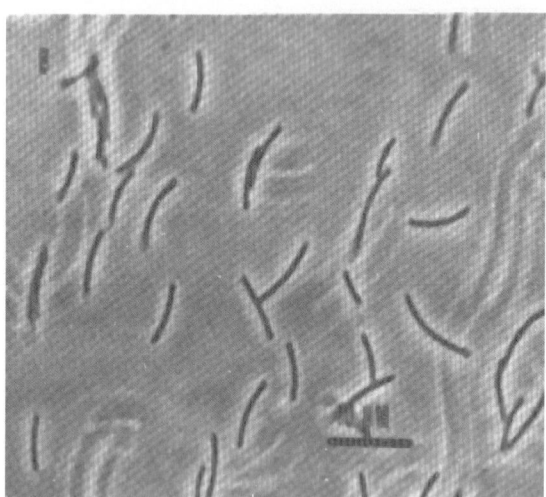

Fig. 11. Phase-contrast photomicrograph of (a) long, wavy filament and (b) individual cells of *Methanospirillum hungatei*. Both bars = 10 μm. (From Mah and Smith, 1981.)

Methanospirillum hungatei (*hungateii*). These Gram-negative, curved rods (7.4 μm × 0.5 μm) form filaments from 15 μm often to several hundred μm in length (Fig. 11). Possessing polar, tufted flagella and sheaths, cells exhibit weak motility and progressive movement. The cell envelope and sheath have been extensively studied (see below). Cells are striated on the surface and resistant to lysozyme. Colonies are circular, <3 μm in diameter, light yellow, convex with lobate margins, uniquely striated when observed under low-power magnification. The type species forms methane from $H_2 + CO_2$ or from formate, and some strains (GP1 and SK) are able to use 2-propanol + CO_2 or 2-butanol + CO_2 (Widdel, 1986; Widdel et al., 1988). No growth or methane production is detected on acetate, methanol, ethanol, pyruvate or benzoate.

Methanospirillum hungatei has a positive chemotactic response to acetate (Migas et al., 1989). Yeast extract and trypticase are stimulatory for growth. Optimum temperature for growth is 30–37°C; optimum pH is 6.6–7.4. The G+C content is 45–49 mol% (BD). This hydrogenotrophic methanogen shows a high affinity for hydrogen and is often utilized for isolation of syntrophic bacteria in place of sulfate-reducing bacteria when sulfate is not available in the culture medium. This organism was isolated from various anaerobic environments. Its GenBank 16S rRNA sequence accession number is M60880. It is the type species of genus *Methanospirillum*. Its type strain is JF1 (≡ DSM 864, ATCC 27890 or OCM 16). Refer to Ferry et al. (1974), Patel et al. (1976), Ferry and Wolfe (1977), and Widdel et al. (1988).

Genus *insertae sedis*

Methanocalculus

It is a newly described genus that encompasses two irregular cocci, *Methanocalculus halotolerans*, an isolate from an offshore oil well, and *M. pumilus*, isolated from a waste-disposal site. *Methanocalculus halotolerans* is a hydrogenotrophic halotolerant methanogen. Further investigation may lead to the reclassification of this genus within the family *Methanocorpusculaceae*.

Methanocalculus halotolerans. These irregular cocci (0.8–1 μm in diameter; Fig. 12) occur singly or in pairs and possess 2–3 peritrichous flagella. Colonies are round, reaching 1 μm in diameter after 10 weeks incubation at 37°C. Methane is produced from $H_2 + CO_2$ or from formate. No growth or methane production is detected on acetate, methanol, trimethylamine, lactate, glucose, 1-propanol, 2-propanol, 1-butanol or 2-butanol. Acetate is required for growth stimulated by yeast extract. For growth, the optimum temperature is 38°C (range >24–<50°C); optimum pH is 7.6 (range 7–8.4); optimum NaCl concentration is 5% (range 0–12.5%; the widest reported range to date for any hydrogenotrophic methanogen, including members of the orders *Methanobacteriales*, *Methanococcales* and *Methanomicrobiales*). Minimum generation time is 12 h. The G+C content of DNA is 55 mol% (HPLC). This organism was isolated from an oil-producing well. Its GenBank 16S rRNA sequence accession number is AF033672. It is the type species of genus:

Methanocalculus. The type strain is SEBR 4845^T (OCM 470 ^{T}). Refer to Ollivier et al. (1998).

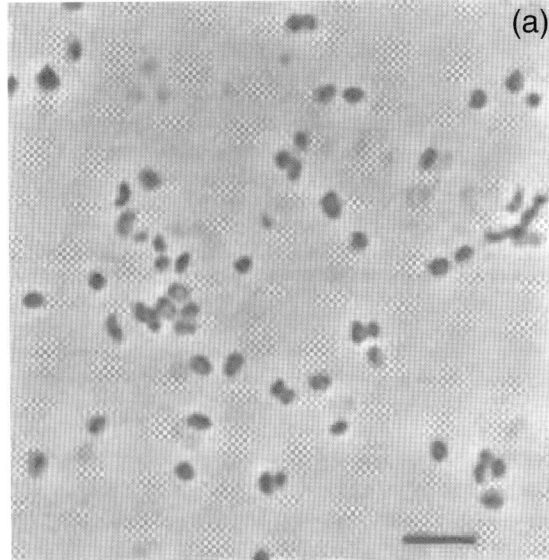

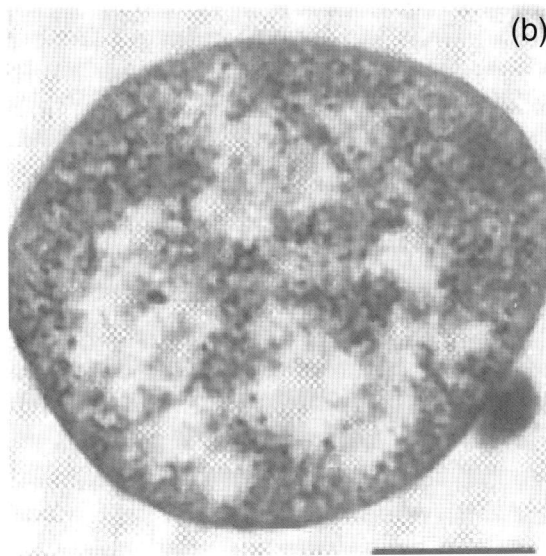

Fig. 12. (a) Phase-contrast micrograph of *Methanocalculus halotolerans* showing irregular coccoid cells; bar = 5 μm; (b) electron micrograph of an ultrathin section showing the cell wall structure; bar = 0.2 μm. (From Ollivier et al., 1998.)

Methanocalculus pumilus. These irregular cocci (0.8–1 μm in diameter) occur singly or in pairs and are nonmotile. They lyse in 0.01% SDS and under hypotonic conditions. Methane is produced from $H_2 + CO_2$ or from formate. No growth or methane production is detected on acetate, methanol, trimethylamine, lactate, glucose, 1-propanol, 2-propanol, 1-butanol or 2-butanol. Acetate is required for growth stimulated by yeast extract. Cells are tolerant to heavy metals. For growth, optimum temperature is 35°C (range 24–45°C); optimum pH is 7 (range 5.5–9); and optimum NaCl concentration is 1% (range 0–7%). The minimum generation time is

12 h. The G+C content of DNA is 51.9 mol% (HPLC). This organism was isolated from a waste-disposal site. Its GenBank 16S rRNA sequence accession number is AB008853. Its type strain is MHT-1^T ($\equiv$ DSM 12632^T or JCM 10627^T). Refer to Mori et al. (2000).

Biochemical and Physiological Properties

Coenzymes and Enzymes Within the Methanomicrobiales

In those Methanomicrobiales that have been tested, the activities of methane biosynthesis enzymes are comparable to those found in other methanogens (Schwörer and Thauer, 1991; Raemakers-Franken et al., 1990; Berk and Thauer, 1997). Specifically, high levels of formyl-methanofuran dehydrogenase, coenzyme F_{420}-dependent methylenetetrahydro-methanopterin dehydrogenase, methylenetetrahydromethanopterin reductase, and heterodisulfide reductase have been found. The proton-reducing methylenetetrahydro-methanopterin dehydrogenase was not found. The Methanomicrobiales also contain high levels of the methyl coenzyme M reductase (Rouvière and Wolfe, 1987). Methanopterin and sarcinapterin are common C-1 carriers in most methanogens. However, the Methanomicrobiales contain three unique pterins in place of methanopterin (Gorris and van der Drift, 1994). Two novel pterins, called "tatiopterin-O" and "tatiopterin-I," have been isolated and characterized from *Methanofollis tationis* (Raemakers-Franken et al., 1991b). Tatiopterin-I (a methanopterin-like compound) lacks the characteristic methyl group on the 7-position of the pterin and has additional aspartyl and glutamyl residues on the side chain. Tatiopterin-O is similar to tatiopterin-I, except the glutamyl residue is missing from the side chain. In addition, *Methanoculleus thermophilicum* contains a pterin similar to tatiopterin-O, except the aniline moiety contains two hydroxyl residues (Raemakers-Franken et al., 1991a). Interestingly, extracts of *M. hungatei* contain serine hydroxymethyltransferase activity that is dependent on tetrahydrofolate, but not on tetrahydromethanopterin and tetrahydrosarcinopterin (Lin and Sparling, 1998). This observation may indicate that folates are also present in these methanogens. Alternatively, a tatiopterin-dependent enzyme may be present whose substrate specificity includes folate.

The Methanomicrobiales also contain high levels of coenzyme F_{420} and vitamin B_{12} (Gorris and van der Drift, 1994). The predominant coenzymes F_{420-2} and F_{420-3} have two and three glutamyl residues on their side chains, respectively. Coenzyme F_{420-4} and coenzyme F_{420-5}, which are abundant in the *Methanosarcinales*, are either present at much lower concentrations or absent. Like most methanogens, the Methanomicrobiales contain high levels of the vitamin B_{12} compound, 5-hydroxybenzimidazoylcobamide. The ability to utilize alcohols as electron donors for methanogenesis is widely distributed among the Methanomicrobiales (Widdel, 1986; Zellner and Winter, 1987; Widdel et al., 1988). The secondary alcohol dehydrogenases responsible for these activities are either coenzyme F_{420}- or NADP⁺-dependent (Bleicher et al., 1989; Frimmer and Widdel, 1989). The coenzyme F_{420}-dependent enzymes have been purified from *M. thermophilicum* and *M. liminatans* (Widdel and Wolfe, 1989; Bleicher and Winter, 1991). In both cases, the enzymes were composed of subunits with an Mr of about 39,000 and were unstable during storage. The enzyme from *M. liminatans* had a broad substrate specificity that included 2-propanol, R(-)-2-butanol, S(+)-2-butanol, 2 pentanol, cyclopentanol, and the corresponding ketones. Like other coenzyme F_{420}-dependent enzymes, the *M. thermophilicum* enzyme possesses Si-face stereospecificity with respect to the C-5 position of coenzyme F_{420} (Klein et al., 1996). The ethanol dehydrogenase activity in extracts of *M. organophilum* has also been characterized (Frimmer and Widdel, 1989). This activity is NADP⁺-dependent and has a pH optimum of 10. It also has a Re-face stereospecificity at the C-4 position of NADP⁺ (Berk et al., 1996). Organisms which utilize the NADP⁺-dependent alcohol dehydrogenase also possess high levels of coenzyme F_{420}-dependent NADP⁺ reductase (Berk

and Thauer, 1997). This enzyme is necessary to couple alcohol oxidation with the coenzyme F_{420}-dependent steps in methanogenesis. The enzyme from *M. organophilum* has been purified and found to be similar in molecular weight and other properties to the enzymes from other methanogens (Berk and Thauer, 1997).

The bioenergetics of the Methanomicrobiales have not been studied in detail. In *Methanospirillum hungatei*, low affinity potassium transport and Na⁺/H⁺ antiport systems have been described (Sprott et al., 1985; Rusnák et al., 1992).

Cell Envelope Structure

With the exception of *Methanospirillum hungatei*, the cell envelopes of the Methanomicrobiales are composed of a regularly structured protein layer (S-layer) and a cytoplasmic membrane (Sleytr et al., 1986; Sprott and Beveridge, 1993). The S-layer forms a paracrystalline array, usually with hexagonal (p6) symmetry and center-to-center spacing of 14–16 nm (Table 5). The molecular weight and antigenicity of the S-layer proteins vary between species and genera. Except for *Methanogenium cariaci* and *Methanospirillum hungatei*, the S-layer proteins are generally glycosylated. Unlike cells containing pseudomurein in their envelopes, cells where the envelope is composed entirely of a S-layer are frequently sensitive to detergents. The S-layers of *Methanoculleus marisnigri* and *Methanoplanus limicola* have been analyzed in some detail. In *M. marisnigri*, the S-layer-membrane complex forms a tight but noncovalent association that is deformable and not rigid (Bayley and Koval,

Table 5. Envelopes of Methanomicrobiales.

| Envelope type | Species | S-layer | | | | |
| | | Monomer | | | | |
		kDa	Glycosylation	Lattice	Spacing (nm)	Refs
1. RS-layer + CM	*Methanocorpusculum parvum*	90	G	H	14.3	1
	Methanocorpusculum sinense	92	G	H	15.8	1
	Methanocorpusculum bavaricum	94	G	H	16.0	1
	Methanoplanus limicola	135	G	H	14.7	2
	Methanoculleus thermophilicum	130	G	H	—	3
	Methanoculleus marisnigri	138	G	H	14	3,4
	Methanoculleus cariaci	117	NG	H	14	3,4
	Methanoculleus palmolei	120	G	H	14	6
	Methanofollis liminatans	118	G	H	15.4	6
	Methanofollis tationis	120	G	H	—	7
	Methanolacinia paynteri	155	G	H	15.3	8
2. Sheath + S-layer + CM	*Methanospirillum hungatei* GP1	110	NG	H	15	9

Abbreviations: RS, regularly structured; CM, cytoplasmic membrane; G, glycosylated; NG, nonglycosylated; and H, hoop.
References: 1, Zellner et al., 1989b; 2, Cheong et al., 1991; 3, Zabel et al., 1985; 4, Romesser et al., 1979; 5, Zellner et al., 1998; 6, Zellner et al., 1990; 7, Zabel et al., 1984; 8, Zellner et al., 1989a; and 9, Firtel et al., 1993.
Data from Sprott and Beveridge, 1993.

1994). In *M. limicola*, the molecular architecture of the S-layer was reconstructed from a tilt series of negatively stained preparations and surface relief reconstructions of metal-shadowed preparations as well as from scanning tunneling microscopy (Cheong et al., 1991; Cheong et al., 1993). The surface layer has hexagonal or p6 symmetry, a lattice constant of 14.7 nm, and thickness of approximately 6.5 nm. The S-layer protein has an apparent molecular weight of 135 kDa, shifting to 115 kDa after removal of the carbohydrate components with anhydrous trifluoromethanesulfonic acid. In addition, quantitative estimations revealed a total neutral sugar content of 240 mg/g of polypeptide. To form an irregular coccus (as is common in the Methanomicrobiales), the S-layer must incorporate local faults or edge dislocations in the paracrystalline structure (Pum et al., 1991). Likewise, edge dislocations are required for invagination of the envelope during cell division.

The cell envelope of *Methanospirillum hungatei* has an unique ultrastructure. Cells contain an S-layer similar to that found in other Methanomicrobiales (Firtel et al., 1993; Table 5). Individual cells or chains of cells are then further enclosed by a sheath. Within the chains, cells are separated by a spacer region composed of multiple lamellae or spacer plugs. The sheath is extremely resilient to denaturants, salts, proteases, and other enzymes (Beveridge et al., 1985), but it can be disassociated into hoop structures and free polypeptides by treatment with sulfhydryl reagents (Sprott et al., 1986; Southam and Beveridge, 1991). In addition, about 20% of the mass of the sheath is made up of phenol-soluble proteins that confer rigidity (Southam and Beveridge, 1992). The intact sheath is a paracrystalline array of 2.8-nm particles and is composed of stacked hoops (Stewart et al., 1985; Beveridge et al., 1990; Blackford et al., 1994), and the tight packing of the sheath particles produces a barrier of low porosity that limits the movement of even small molecules (Beveridge et al., 1991). Measurements of the elasticity of the sheath by atomic force microscopy indicate that it can withstand pressures in the range of 300–400 atm (Xu et al., 1996). It is hypothesized that the accumulation of intracellular CH_4 to high pressures causes expansion of the sheath, which opens pores in the sheath and allows the release of CH_4 and uptake of H_2 and CO_2. In this way, the sheath is envisioned to act as a pressure regulator (Xu et al., 1996). In contrast to the sheath, the spacer plugs at the ends of the cell are highly permeable (Beveridge et al., 1991). These plugs are proteinaceous disks that span the sheath and consist of two types of paracrystalline layers sandwiched between amorphous layers of unknown composition (Firtel et al.,

1994). One type of paracrystalline layer is composed of 14-nm particles with 18-nm interparticle spacing to produce a highly porous structure. The second layer is net-like with large, 12.5-nm pores. These layers are attached to each other and the cell by intervening amorphous layers.

Cell Envelope Lipids

Similarly to other methanogens, the cell membranes of Methanomicrobiales are characterized by many unusual lipids. These lipids are polar, either archaeol (diphytanyl-glycerol-diethers) or caldarchaeol (dibiphytanyl-diglycerol-tetraethers). Caldarchaeol appears to span the membrane, with the polar head groups at opposite sides. Therefore, the membranes are probably arranged in a monolayer with bilayer regions resulting from interspersed diethers (Whitman et al., 1992). These lipids are then present as phospho- or glyco-derivatives. Among the Methanomicrobiales, the major sugars in the glycolipids are glucose (glc) and galactose (gal; Koga et al., 1998). The major polar head groups in the phospholipids are aminopentanetetrols and glycerol. In *Methanospirillum hungatei*, four of the polar lipids had as one head group either α-glc(p)-(1-2)-β-gal(f)- or β-gal(f)-(1-6)-β-gal(f)- in glycosidic linkage to the first glycerol of the lipid backbone and either a *N,N*-dimethylaminopentanetetrol or a *N,N*-trimethylaminopentanetetrol moiety in phosphodiester linkage to the second glycerol of the backbone (Sprott et al., 1994). A fifth lipid was a tetraether structure with carbohydrate moieties at both head group positions, namely α-glc(p)-(1-2)-gal(f) and β-gal(f)-. Two other lipids, a diether and a tetraether, had a single head group consisting of α-glc(p)-(1-2)-β-gal(f)- modified by O-acetylation of the gal(f) residue at C-6. Lastly, the diether and tetraether analogs of phosphatidylglycerol were found.

Motility

Flagella are observed in species of the order Methanomicrobiales. In *Methanospirillum hungatei* mono- or bipolar flagella are inserted through the end plugs of the filaments (Cruden et al., 1989; Southam et al., 1990). Filaments, with a tuft of flagella at each of the terminal cells on opposite ends of the filament, are able to coordinate flagellar rotation so that propulsion is achieved (Sprott and Beveridge, 1993). Isolated flagellar filaments from *Methanospirillum hungatei* can be dissociated by low concentrations (0.5% [v/v]) of Triton X-100 (Faguy et al., 1992). The filaments are composed of multiple glycosylated flagellins, where the flagellins represent distinct gene products rather than differentially glycosylated forms of the same protein. Similarly, the purified flagellar filaments isolated from *Methanoculleus marisnigri* were shown to be composed of two flagellins (Kalmokoff et al., 1992). Lastly, the flagellar filaments have a simple knob basal structure with no apparent ring or hook structures such as found in bacteria (Faguy et al., 1994). When cultivated at 37°C, *Methanospirillum hungatei* grows as single cells or short chains of cells (typically 10–30 μm long), and both forms are motile. When grown in low Ca^{2+} concentrations or with the divalent cation chelator EDTA, nonflagellated filaments (up to 900 μm long) are produced. Likewise, at suboptimal growth temperatures, the cells form short filaments that do not possess flagella. The amount of flagellin present appears to be equal in both nonflagellated and flagellated cultures (Faguy et al., 1993).

Compatible Solutes

Methanogens respond to osmotic stress by accumulating a series of organic molecules which function as compatible solutes. Four key organic solutes were observed in *Methanogenium cariaci*. They include L-α-glutamate, β-glutamate, N(e)-acetyl-β-lysine and betaine. Though L-α-glutamate, β-glutamate and N(e)-acetyl-β-lysine are synthesized de novo, betaine is preferentially assimilated from the medium (Robertson et al., 1992). In the absence of betaine, N(e)-acetyl-β-lysine is the dominant osmolyte.

Miscellaneous

Methanomicrobium mobile is light sensitive. Growth is inhibited by light in the blue end of the visible spectrum (370 to 430 nm; Olson et al., 1991).

Literature Cited

Andrews, K. T., and B. K. C. Patel. 1996. Fervidobacterium gondwanense sp. nov., a new thermophilic anaerobic bacterium isolated from nonvolcanically heated geothermal waters of the Great Artesian Basin of Australia. Int. J. Syst. Bacteriol. 46:265–269.

Balch, W. E., G. E. Fox, L. J. Magrum, C. R. Woese, and R. S. Wolfe. 1979. Methanogens: Reevaluation of a unique biological group. Microbiol. Rev. 43:260–296.

Bayley, D. P., and S. F. Koval. 1994. Membrane association and isolation of the S-layer protein of Methanoculleus marisnigri. Can. J. Microbiol. 40:237–241.

Berger, J., and D. H. Lynn. 1992. Hydrogenosome-methanogen assemblages in the echinoid endocommensal plagiopylid ciliates, Lechriopyla mystax Lynch, 1930 and Plagiopyla minuta Powers, 1933. J. Protozool. 39:4–8.

Berk, H., W. Buckel, R. K. Thauer, and P. A. Frey. 1996. Reface stereospecificity at C4 of NAD(P) for alcohol dehydrogenase from Methanogenium organophilum and for (R)-2-hydroxyglutarate dehydrogenase from Acidaminococcus fermentans as determined by 1H-NMR spectroscopy. FEBS Lett. 399:92–94.

Berk, H., and R. K. Thauer. 1997. Function of coenzyme F_{420}-dependent NADP reductase in methanogenic archaea containing an NADP-dependent alcohol dehydrogenase. Arch. Microbiol. 168:396–402.

Beveridge, T. J., M. Stewart, R. J. Doyle, and G. D. Sprott. 1985. Unusual stability of the Methanospirillum hungatei sheath. J. Bacteriol. 162:728–737.

Beveridge, T. J., G. Southam, M. H. Jericho, and B. L. Blackford. 1990. High-resolution topography of the S-layer sheath of the archaebacterium Methanospirillum hungatei provided by scanning tunneling microscopy. J. Bacteriol. 172:6589–6595.

Beveridge, T. J., G. D. Sprott, and P. Whippey. 1991. Ultrastructure, inferred porosity, and Gram-staining character of Methanospirillum hungatei filament termini describe a unique cell permeability for this archaeobacterium. J. Bacteriol. 173:130–140.

Blackford, B. L., W. Xu, M. H. Jericho, P. J. Mulhern, M. Firtel, and T. J. Beveridge. 1994. Direct observation by scanning tunneling microscopy of the two-dimensional lattice structure of the S-layer sheath of the archaeobacterium Methanospirillum hungatei GP1. Scanning Microscopy 8:507–512.

Bleicher, K., G. Zellner, and J. Winter. 1989. Growth of methanogens on cyclopentanol/CO₂ and specificity of alcohol dehydrogenase. FEMS Microbiol. Lett. 59:307–312.

Bleicher, K., and J. Winter. 1991. Purification and properties of F_{420}- and NADP+-dependent alcohol dehydrogenases of Methanogenium liminatans and Methanobacterium palustre, specific for secondary alcohols. Eur. J. Biochem. 200:43–51.

Blotevogel, K. H., R. Gahl-Janben, S. Jannsen, U. Fisher, F. Pilz, G. Auling, A. J. L. Macario, and B. J. Tindall. 1991. Isolation and characterisation of a novel mesophilic, fresh-water methanogen from river sediment Methanoculleus oldenburgensis sp. nov. Arch. Microbiol. 157:54–59.

Boone, D. R., and W. B. Whitman. 1988. Proposal of minimal standards for describing new taxa of methanogenic bacteria. Int. J. Syst. Bacteriol. 38:212–219.

Boone, D. R., W. B. Whitman, and P. Rouvière. 1993. Diversity and taxonomy of methanogens. *In:* J. G. Ferry (Ed.) Methanogenesis. Chapman & Hall. New York, NY. London, 35–80.

Brusa, T., F. Ferrari, and E. Canzi. 1998. Methanogenic bacteria—presence in foodstuffs. J. Basic Microbiol. 38:79–84.

Cairó, J., A. J. L. Macario, M. Bardulet, E. Conway de Macario, and J. M. París. 1991. Psychrophilic ecosystems of interest for wastewater treatment: Microbiologic and immunologic elucidation of their methanogenic flora. Syst. Appl. Microbiol. 14:85–92.

Cheong, G. W., Z. Cejka, J. Peters, K. O. Stetter, and W. Baumeister. 1991. The surface protein layer of Methanoplanus limicola-3-dimensional structure and chemical characterization. Syst. Appl. Microbiol. 14:358–363.

Cheong, G. W., R. Guckenberger, K.-H. Fuchs, H. Gross, and W. Baumeister. 1993. The structure of the surface layer of Methanoplanus limicola obtained by a combined electron microscopy and scanning tunneling microscopy approach. J. Structural Biol. 111:125–134.

Chong, S. C., and D. R. Boone. 2004. Methanoculleus. *In:* D. R. Boone and R. W. Castenholz (Eds.) Bergey's Manual of Systematic Bacteriology, 2nd ed. Springer-Verlag. New York, NY. 1.

Corder, R. E., L. A. Hook, J. M. Larkin, and J. I. Frea. 1983. Isolation and characterization of two new methane-producing cocci: Methanogenium olentangyi, sp. nov., and Methanococcus deltae, sp. nov. Arch. Microbiol. 134:28–32.

Cruden, D., R. Sparling, and A. J. Markovetz. 1989. Isolation and ultrastructure of the flagella of Methanococcus thermolithotrophicus and Methanospirillum hungatei. Appl. Environ. Microbiol. 55:1414–1419.

Embley, T. M., B. J. Finlay, and S. Brown. 1992. RNA sequence analysis shows that the symbionts in the ciliate Metopus contortus are polymorphs of a single methanogen species. FEMS Microbiol. Lett. 97:57–62.

Embley, T. M., and B. J. Finlay. 1993. Systematic and morphological diversity of endosymbiotic methanogens in anaerobic ciliates. Anton. Leeuwenhoek 64:261–271.

Euzéby, J. P. 1997. List of bacterial names with standing in nomenclature: A folder available on the Internet. Int. J. Syst. Bacteriol. 47:590–592.

Faguy, D. M., S. F. Koval, and K. F. Jarrell. 1992. Correlation between glycosylation of flagellin proteins and sensitivity of flagellar filaments to triton X-100 in methanogens. FEMS Microbiol. Lett. 90:129–134.

Faguy, D. M., S. F. Koval, and K. F. Jarrell. 1993. Effect of changes in mineral composition and growth temperature on filament length and flagellation in the archaeon Methanospirillum hungatei. Arch. Microbiol. 159:512–520.

Faguy, D. M., S. F. Koval, and K. F. Jarrell. 1994. Physical characterization of the flagella and flagellins from Methanospirillum hungatei. J. Bacteriol. 176:7491–7498.

Felsenstein, J. 1993. PHYLIP (Phylogenetic Inference Package) version 3.51c (Distributed by the author). Department of Genetics, University of Washington. Seattle, WA.

Ferry, J. G., and D. R. Boone. 2004. Methanospirillum. *In:* D. R. Boone and R. W. Castenholz (Eds.) Bergey's Manual of Systematic Bacteriology, 2nd ed. Springer-Verlag. New York, NY. 1.

Ferry, J. G., P. H. Smith, and R. S. Wolfe. 1974. Methanospirillum, a new genus of methanogenic bacteria, and characterization of Methanospirillum hungatii sp. nov. Int. J. Syst. Bacteriol. 24:465–469.

Ferry, J. G., and R. S. Wolfe. 1977. Nutritional and biochemical characterization of Methanospirillum hungatii. Appl. Environ. Microbiol. 34:371–376.

Finlay, B. J., T. M. Embley, and T. Fenchel. 1993. A new polymorphic methanogen, closely related to Methanocorpusculum parvum, living in stable symbiosis within the anaerobic ciliate Trimyema sp. J. Gen. Microbiol. 139:371–378.

Firtel, M., G. Southam, G. Harauz, and T. J. Beveridge. 1993. Characterization of the cell-wall of the sheathed methanogen Methanospirillum hungatei Gp1 as an S-layer. J. Bacteriol. 175:7550–7560.

Firtel, M., G. Southam, G. Harauz, and T. J. Beveridge. 1994. The organization of the paracrystalline multilayered spacer-plugs of Methanospirillum hungatei. J. Struct. Biol. 112:160–171.

Franzmann, P. D., Y. Liu, D. L. Balkwill, H. C. Aldrich, E. Conway de Macario, and D. R. Boone. 1997. Methanogenium frigidum sp. nov., a psychrophilic, H₂-using methanogen from Ace Lake, Antartica. Int. J. Syst. Bacteriol. 47:1068–1072.

Frimmer, U., and F. Widdel. 1989. Oxidation of ethanol by methanogenic bacteria: Growth experiments and enzymatic studies. Arch. Microbiol. 152:479–483.

Garcia, J.-L., B. K. C. Patel, and B. Ollivier. 2000. Taxonomic, phylogenetic, and ecological diversity of methanogenic Archaea. Anaerobe 6:205–226.

Gorris, L. G. M., and C. van der Drift. 1994. Cofactor contents of methanogenic bacteria reviewed. BioFactors 4:139–145.

Grosskopf, R., S. Stubner, and W. Liesack. 1998. Novel euryarchaeotal lineages detected on rice roots and in the anoxic bulk soil of flooded rice microcosms. Appl. Environ. Microbiol. 12:4983–4989.

Harris, J. E., P. A. Pinn, and R. P. Davis. 1984. Isolation and characterization of a novel thermophilic, freshwater methanogen. Appl. Environ. Microbiol. 48:1123–1128.

Hungate, R. E. 1969. A roll tube method for cultivation of strict anaerobes. *In:* J. R. Norris and D. W. Ribbons (Eds.) Methods in Microbiology. Academic Press. New York, NY. 3B:117–132.

Jarvis, G. N., C. Strompl, D. M. Burgess, L. C. Skillman, E. R. B. Moore, and K. N. Joblin. 2000. Isolation and identification of ruminal methanogens from grazing cattle. Curr. Microbiol. 40:327–332.

Joulian, C., B. Ollivier, B. K. C. Patel, and P. A. Roger. 1998. Phenotypic and phylogenetic characterization of dominant culturable methanogens isolated from ricefield soils. FEMS Microbiol. Ecol. 25:135–145.

Jukes, T. H., and C. R. Cantor. 1969. Evolution of protein molecules. *In:* H. N. Munro (Ed.) Mammalian Protein Metabolism. Academic Press. New York, NY. 211–232.

Kalmokoff, M. L., S. F. Koval, and K. F. Jarrell. 1992. Relatedness of the flagellins from methanogens. Arch. Microbiol. 157:481–487.

Klein, A. R., H. Berk, E. Purwantini, L. Daniels, and R. K. Thauer. 1996. Si-face stereospecificity at C5 of coenzyme F₄₂₀ for F₄₂₀-dependent glucose-6-phosphate dehydrogenase from Mycobacterium smegmatis and F₄₂₀-dependent alcohol dehydrogenase from Methanoculleus thermophilicus. Eur. J. Biochem. 239:93–97.

Koga, Y., H. Morii, M. Akagawa-Matsushita, and M. Ohga. 1998. Correlation of polar lipid composition with 16S rRNA phylogeny in methanogens: Further analysis of lipid component parts. Biosci. Biotechnol. Biochem. 62:230–236.

Kudo, Y., T. Nakajima, T. Miyaki, and H. Oyaizu. 1997. Methanogen flora of paddy soils in Japan. FEMS Microbiol. Ecol. 22:39–48.

Kuhner, C. H., S. S. Smith, K. M. Noll, R. S. Tanner, and R. S. Wolfe. 1991. 7-mercaptoheptanoylthreonine phosphate substitutes for heat-stable factor (mobile factor) for growth of Methanomicrobium mobile. Appl. Environ. Microbiol. 57:2891–2895.

Lin, C. Z., L. Raskin, and D. A. Stahl. 1997. Microbial community structure in gastrointestinal tracts of domestic animals—comparative analyses using ribosomal-RNA-targeted oligonucleotide probes. FEMS Microbiol. Ecol. 22:281–294.

Lin, Z., and R. Sparling. 1998. Investigations of serine hydroxymethyltransferase in methanogens. Can. J. Microbiol. 44:652–656.

Macario, A. J. L., M. W. Peck, E. Conway de Macario, and D. P. Chynoweth. 1991. Unusual methanogenic flora of a wood-fermenting anaerobic bioreactor. J. Appl. Bacteriol. 71:31–37.

Maestrojuan, G. M., D. R. Boone, L. Xun, R. A. Mah, and L. Zhang. 1990. Transfer of Methanogenium bourgense, Methanogenium marisnigri, Methanogenium olentangyi, and Methanogenium thermophilicum to the genus Methanoculleus gen. nov., emendation of Methanoculleus marisnigri and Methanogenium, and description of new strains of Methanoculleus bourgense and Methanoculleus marisnigri. Int. J. Syst. Bacteriol. 40:117–122.

Mah, R. A., and M. R. Smith. 1981. The methanogenic bacteria. In: M. P. Starr, H. Stolp, H. G. Trüper, A. Balows and H. G. Schlegel (Eds.) The Prokaryotes. Springer-Verlag. Berlin, 1:948–977.

Maidak, B. L., J. R. Cole, T. G. Lilburn, C. T. Parker Jr., P. R. Saxman, J. M. Stredwick, G. M. Garrity, B. Li, G. J. Olsen, S. Pramanik, T. M. Schmidt, and J. M. Tiedje. 2000. The RDP (Ribosomal Database Project) continues. Nucleic Acids Res. 28:173–174.

Mesbah, M., U. Premchandran, and W. B. Whitman. 1989. Precise measurement of the G+C content of deoxyribonucleotic acid by high performance liquid chromatography. Int. J. Syst. Bacteriol. 39:159–167.

Migas, J., K. L. Anderson, D. L. Cruden, and A. J. Markovetz. 1989. Chemotaxis in Methanospirillum hungatei. Appl. Environ. Microbiol. 55:264–265.

Miller, T. L., and M. J. Wolin. 1986. Methanogens in human and animal intestinal tracts. Syst. Appl. Microbiol. 7:223–229.

Mori, K., H. Yamamoto, Y. Kamagata, M. Hatsu, and K. Takamizawa. 2000. Methanocalculus pumilus sp. nov., a heavy-metal-tolerant methanogen isolated from a waste-disposal site. Int. J. Syst. Evol. Microbiol. 50:1723–1729.

Munson, M. A., D. B. Nedwell, and T. M. Embley. 1997. Phylogenetic diversity of Archaea in sediment samples from a coastal salt-marsh. Appl. Environ. Microbiol. 63:4729–4733.

Ohkuma, M., S. Noda, and T. Kudo. 1999. Phylogenetic-relationships of symbiotic methanogens in diverse termites. FEMS Microbiol. Lett. 171:147–153.

Ollivier, B. M., R. A. Mah, J.-L. Garcia, and R. Robinson. 1985. Isolation and characterization of Methanogenium aggregans sp. nov. Int. J. Syst. Bacteriol. 35:127–130.

Ollivier, B., R. A. Mah, J.-L. Garcia, and D. R. Boone.1986. Isolation and characterization of Methanogenium bourgense sp. nov. Int. J. Syst. Bacteriol. 36:297–301.

Ollivier, B., J.-L. Cayol, B. K. C. Patel, M. Magot, M.-L. Fardeau, and J.-L. Garcia. 1997. Methanoplanus petrolearius sp. nov., a novel methanogenic bacterium from an oil-producing well. FEMS Microbiol. Lett. 147:51–56.

Ollivier, B., M.-L. Fardeau, J.-L. Cayol, M. Magot, B. K. C. Patel, G. Prensier, and J.-L. Garcia. 1998. Methanocalculus halotolerans gen. nov., sp. nov., isolated from an oil-producing well. Int. J. Syst. Bacteriol. 48:821–828.

Olson, K. D., C. W. McMahon, and R. S. Wolfe. 1991. Light sensitivity of methanogenic archaebacteria. Appl. Environ. Microbiol. 57:2683–2686.

Orphan, V. J., L. T. Taylor, D. Hafenbradl, and E. F. Delong. 2000. Culture-dependent and culture-independent characterization of microbial assemblages associated with high-temperature petroleum reservoirs. Appl. Environ. Microbiol. 66:700–711.

Patel, G. B., L. A. Roth, L. van den Berg, and D. S. Clark. 1976. Characterization of a strain of Methanospirillum hungatii. Can. J. Microbiol. 22:1404–1410.

Paynter, M. J. B., and R. E. Hungate. 1968. Characterization of Methanobacterium mobilis, sp. n., isolated from the bovine rumen. J. Bacteriol. 95:1943–1951.

Pedersen, K. 1997. Microbial life in deep granitic rock. FEMS Microbiol. Rev. 20:399–414.

Poirot, C. C. M., G. J. W. M. Vanalebeek, J. T. Keltjens, and G. D. Vogels. 1991. Identification of para-cresol as a growth-factor for Methanoplanus endosymbiosus. Appl. Environ. Microbiol. 57:976–980.

Pum, D., P. Messner, and U. B. Sleytr. 1991. Role of the S layer in morphogenesis and cell division of the archaebacterium Methanocorpusculum sinense. J. Bacteriol. 173:6865–6873.

Raemakers-Franken, P. C., A. J. Kortstee, C. van der Drift, and G. D. Vogels. 1990. Methanogenesis involving a novel carrier of C1 compounds in Methanogenium tationis. J. Bacteriol. 172:1157–1159.

Raemakers-Franken, P. C., R. Bongaerts, R. Fokkens, C. Vanderdrift, and G. D. Vogels. 1991a. Characterization of 2 pterin derivatives isolated from Methanoculleus thermophilicum. Eur. J. Biochem. 200:783–787.

Raemakers-Franken, P. C., C. H. M. Vanelderen, C. Vanderdrift, and G. D. Vogels. 1991b. Identification of a novel tatiopterin derivative in Methanogenium tationis. Biofactors 3:127–130.

Raskin, L., J. M. Stromley, B. E. Rittman, and D. A. Stahl. 1994. Group-specific 16S ribosomal-RNA hybridization probes to describe natural communities of methanogens. Appl. Environ. Microbiol. 60:1232–1240.

Raskin, L., D. D. Zheng, M. E. Griffin, P. G. Stroot, and P. Misra. 1995. Characterization of microbial communities in anaerobic bioreactors using molecular probes. Anton. Leeuwenhoek 68:297–308.

Rivard, C. J., and P. H. Smith. 1982. Isolation and characterization of a thermophilic marine methanogenic bacterium, Methanogenium thermophilicum sp. nov. Int. J. Syst. Bacteriol. 32:430–436.

Rivard, C. J., J. M. Henson, M. V. Thomas, and P. H. Smith. 1983. Isolation and characterization of Methanomicrobium paynteri sp. nov., a mesophilic methanogen isolated from marine sediments. Appl. Environ. Microbiol. 46:484–490.

Robertson, D. E., D. Noll, and M. F. Roberts. 1992. Free amino-acid dynamics in marine methanogens—Beta-amino acids as compatible solutes. J. Biol. Chem. 267:4893–4901.

Romesser, J. A., R. S. Wolfe, F. Mayer, E. Spiess, and A. Walther-Mauruschat. 1979. Methanogenium, a new genus of marine methanogenic bacteria, and characterization of Methanogenium cariaci sp. nov. and Methanogenium marisnigri sp. nov. Arch. Microbiol. 121:147–153.

Rouvière, P., and R. S. Wolfe. 1987. Use of subunits of the methylreductase protein for taxonomy of methanogenic bacteria. Arch. Microbiol. 148:253–259.

Rouvière, P., L. Mandelco, S. Winker, and C. R. Woese. 1992. A detailed phylogeny for the Methanomicrobiales. Syst. Appl. Microbiol. 15:363–371.

Rusnák, P., P. Smigán, and M. Greksák. 1992. Evidence for Na+/H+ antiport in Methanospirillum hungatei. Folia Microbiol. 37:12–16.

Schnurer, A., G. Zellner, and B. H. Svensson. 1999. Mesophilic syntrophic acetate oxidation during methane formation in biogas reactors. FEMS Microbiol. Ecol. 29:249–261.

Schwörer, B., and R. K. Thauer. 1991. Activities of formylmethanofuran dehydrogenase, methylenetetrahydromethanopterin dehydrogenase, methylenetetrahydromethanopterin reductase, and heterodisulfide reductase in methanogenic bacteria. Arch. Microbiol. 155:459–465.

Sekiguchi, Y., Y. Kamagata, K. Nakamura, A. Ohashi, and H. Harada. 1999. Fluorescence in-situ hybridization using 16S ribosomal-RNA-targeted oligonucleotides reveals localization of methanogens and selected uncultured bacteria in mesophilic and thermophilic sludge granules. Appl. Environ. Microbiol. 65:1280–1288.

Sharp, R., C. J. Ziemer, M. D. Stern, and D. A. Stahl. 1998. Taxon-specific associations between protozoal and methanogen populations in the rumen and a model rumen system. FEMS Microbiol. Ecol. 26:71–78.

Sleytr, U. B., P. Messner, M. Sara, and D. Pum. 1986. Crystalline envelope layers in archaebacteria. Syst. Appl. Microbiol. 7:310–313.

Sonnehansen, J., and B. K. Ahring. 1997. Anaerobic microbiology of an alkaline Icelandic hot-spring. FEMS Microbiol. Ecol. 23:31–38.

Southam, G., M. L. Kalmokoff, K. F. Jarrell, S. F. Koval, and T. J. Beveridge. 1990. Isolation, characterization, and cellular insertion of the flagella from two strains of the archaebacterium Methanospirillum hungatei. J. Bacteriol. 172:3221–3228.

Southam, G., and T. J. Beveridge. 1991. Dissolution and immunochemical analysis of the sheath of the archaeobacterium Methanospirillum hungatei GP1. J. Bacteriol. 173:6213–6222.

Southam, G., and T. J. Beveridge. 1992. Characterization of novel, phenol-soluble polypeptides which confer rigidity to the sheath of Methanospirillum hungatei GP1. J. Bacteriol. 174:935–946.

Sprott, G. D., K. M. Shaw, and K. F. Jarrell. 1985. Methanogenesis and the K+ transport system are activated by divalent cations in ammonia-treated cells of Methanospirillum hungatei. J. Biol. Chem. 260:9244–9250.

Sprott, G. D., T. J. Beveridge, G. B. Patel, and G. Ferrante. 1986. Sheath disassembly in Methanospirillum hungatei strain GP1. Can. J. Microbiol. 32:847–854.

Sprott, G. D., and T. J. Beveridge. 1993. Microscopy. In: J. G. Ferry (Ed.) Methanogenesis. Chapman & Hall. New York, NY. London, 81–127.

Sprott, G. D., G. Ferrante, and I. Ekiel. 1994. Tetraether lipids of Methanospirillum hungatei with head groups consisting of phospho-N,N-dimethylaminopentanetetrol, phospho-N,N-trimethylaminopentanetetrol and carbohydrates. Biochim. Biophys. Acta—Lipids Lipid Metabol. 1214:234–242.

Stewart, M., T. J. Beveridge, and G. D. Sprott. 1985. Crystalline order to high resolution in the sheath of Methanospirillum hungatei: A cross-beta structure. J. Molec. Biol. 183:509–515.

Tanner, R. S., and R. S. Wolfe. 1988. Nutritional requirements of Methanobacterium mobile. Appl. Environ. Microbiol. 54:625–628.

Van Bruggen, J. J. A., K. B. Zwart, J. G. F. Hermans, E. M. van Hove, C. K. Stumm, and G. D. Vogels. 1986. Isolation and characterization of Methanoplanus endosymbiosus sp. nov., an endosymbiont of the marine sapropelic ciliate Metopus contortus Quennerstedt. Arch. Microbiol. 144:367–374.

Visser, F. A., J. B. van Lier, A. J. L. Macario, and E. Conway de Macario. 1991. Diversity and population dynamics of methanogenic bacteria in a granular consortium. Appl. Environ. Microbiol. 57:1728–1734.

Whitehead, T. R., and M. A. Cotta. 1999. Phylogenetic diversity of methanogenic archaea in swine waste storage pits. FEMS Microbiol. Lett. 179:223–226.

Whitman, W. B., T. L. Bowen, and D. R. Boone. 1992. The methanogenic bacteria. In: A. Balows, H. G. Trüper, M. Dworkin, W. Harder, and K.-H. Schleifer (Eds.) The Prokaryotes, 2nd ed. Springer-Verlag. Berlin, 1:719–767.

Widdel, F. 1986. Growth of methanogenic bacteria in pure culture with 2-propanol and other alcohols as hydrogen donors. Appl. Environ. Microbiol. 51:1056–1062.

Widdel, F., P. E. Rouvière, and R. S. Wolfe. 1988. Classification of secondary alcohol-utilizing methanogens including a new thermophilic isolate. Arch. Microbiol. 150:477–481.

Widdel, F., and R. S. Wolfe. 1989. Expression of secondary alcohol dehydrogenase in methanogenic bacteria and purification of the F_{420}-specific enzyme from Methanogenium thermophilum strain TCI. Arch. Microbiol. 152:322–328.

Wildgruber, G., M. Thomm, H. König, K. Ober, T. Ricchiuto, and K. O. Stetter. 1982. Methanoplanus limicola, a plate-shaped methanogen representing a novel family, the Methanoplanaceae. Arch. Microbiol. 132:31–36.

Xu, W., P. J. Mulhern, B. L. Blackford, M. H. Jericho, M. Firtel, and T. J. Beveridge. 1996. Modeling and measuring the elastic properties of an archaeal surface, the sheath of Methanospirillum hungatei, and the implication for methane production. J. Bacteriol. 178:3106–3112.

Xun, L., D. R. Boone, and R. A. Mah. 1989. Deoxyribonucleic acid hybridization study of Methanogenium and Methanocorpusculum species, emendation of the genus Methanocorpusculum, and transfer of Methanogenium aggregans to the genus Methanocorpusculum as Methanocorpusculum aggregans comb. nov. Int. J. Syst. Bacteriol. 39:109–111.

Zabel, H. P., H. König, and J. Winter. 1984. Isolation and characterization of a new coccoid methanogen, Methanogenium tatii spec. nov. from a solfataric field on Mount Tatio. Arch. Microbiol. 137:308–315.

Zabel, H. P., H. König, and J. Winter. 1985. Emended description of Methanogenium thermophilicum, Rivard and Smith, and assignment of new isolates to this species. Syst. Appl. Microbiol. 6:72–78.

Zellner, G., and J. Winter. 1987. Secondary alcohols as hydrogen donors for CO_2-reduction by methanogens. FEMS Microbiol. Lett. 44:323–328.

Zellner, G., C. Alten, E. Stackebrandt, E. Conway de Macario, and J. Winter. 1987. Isolation and characterization of Methanocorpusculum parvum, gen. nov., spec. nov., a new tungsten requiring, coccoid methanogen. Arch. Microbiol. 147:13–20.

Zellner, G., P. Messner, H. Kneifel, B. J. Tindall, J. Winter, and E. Stackebrandt. 1989a. Methanolacinia gen. nov., incorporating Methanomicrobium paynteri as Methanolacinia paynteri comb. nov. J. Gen. Appl. Microbiol. 35:185–202.

Zellner, G., E. Stackebrandt, P. Messner, B. J. Tindall, E. Conway de Macario, H. Knelfel, and U. B. Sleytr. 1989b. Methanocorpusculaceae fam. nov., represented by Methanocorpusculum parvum, Methanocorpusculum sinense spec. nov. and Methanocorpusculum bavaricum spec. nov. Arch. Microbiol. 151:381–390.

Zellner, G., U. B. Sleytr, P. Messner, H. Kneifel, and J. Winter. 1990. Methanogenium liminatans spec. nov., a new coccoid, mesophilic, methanogen able to oxidize secondary alcohols. Arch. Microbiol. 153:287–293.

Zellner, G., P. Messner, J. Winter, and E. Stackebrandt. 1998. Methanoculleus palmolei sp. nov., an irregularly coccoid methanogen from an anaerobic digestor treating wastewater of a palm oil plant in North Sumatra, Indonesia. Int. J. Syst. Bacteriol. 48:1111–1117.

Zellner, G., D. R. Boone, J. Keswani, W. B. Whitman, C. R. Woese, A. Hagelstein, B. J. Tindall, and E. Stackebrandt. 1999. Reclassification of Methanogenium tationis and Methanogenium liminatans as Methanofollis tationis gen. nov., comb. nov. and Methanofollis liminatans comb. nov. and description of a new strain of Methanofollis liminatans. Int. J. Syst. Bacteriol. 49:247–255.

Zhao, Y., D. R. Boone, R. A. Mah, and L. Xun. 1989. Isolation and characterization of Methanocorpusculum labreanum sp. nov. from the Labrea Tar pits. Int. J. Syst. Bacteriol. 39:10–13.

Prokaryotes (2006) 3:231–243
DOI: 10.1007/0-387-30743-5_11

CHAPTER 11

The Order Methanobacteriales

ADAM S. BONIN AND DAVID R. BOONE

Characteristics of Methanobacteriales

Members of the order Methanobacteriales are distinguished from other methanogens by their limited range of catabolic substrates, their morphology, lipid composition, and rDNA sequence. The Methanobacteriales are generally hydrogenotrophic, using H_2 to reduce CO_2 to CH_4. Some members of this order can use formate, CO, or secondary alcohols as electron donors for CO_2 reduction. However, members of one genus within this order, *Methanosphaera*, use H_2 to reduce methanol to methane. The predominant cell wall polymer of *Methanobacteriales* is pseudomurein, which distinguishes this order from the Methanomicrobiales. Lipids composing the cell membranes include caldarchaeol and *myo*-inositol. Cells usually stain Gram positive and are generally rod-shaped, often forming chains or long filaments up to 40 μm in length. The order Methanobacteriales is divided into two families, the Methanobacteriaceae and Methanothermaceae, on the basis of phylogenetic analysis of 16S rRNA gene sequences as well as phenotypic characteristics (Boone, 2001).

Habitats of Methanobacteriales

Similarly to other methanogenic orders, members of the Methanobacteriales are very strict anaerobes inhabiting a variety of anoxic habitats including freshwater and marine sediments, groundwater, rice paddy fields, terrestrial subsurface environments, anaerobic sewage digestors, environments containing accumulated geothermal H_2, and the gastrointestinal tracts of animals. Specific habitats known to harbor specific strains are indicated below in the sections describing each species.

Isolation and Characterization

Enrichment Media for Growth of Methanobacteriales

MS Mineral Medium and MS Medium are appropriate for the growth of Methanobacteriales. The recipes for these media are given below. MS Mineral Medium with H_2 added as a catabolic substrate is an effective medium for the enrichment of autotrophic members of this order. MS Medium, which contains yeast extract and trypticase peptones (2 g each per liter), is recommended for the enrichment of species that have organic nutrient requirements. Reducing the concentrations of yeast extract and trypticase peptones to 0.5 g of each per liter (MS Enrichment Medium) can reduce the numbers of heterotrophic bacteria growing in enrichment cultures. The use of the higher concentrations of organic compounds in MS medium has some advantages in the routine cultivation of pure cultures: it is easy to detect the growth of heterotrophic contaminants, when they occur, in this rich medium.

The Methanobacteriales generally grow and produce methane by reducing CO_2 with H_2, although some cells can use formate, CO, secondary alcohols, or methanol as electron donor. Species of *Methanospaera* grow only by reducing methanol to methane, with H_2 as the electron donor.

MS Medium (Boone et al., 1989)

NaHCO₃	0.1 mol
Yeast extract	2.0 g
Trypticase peptones	2.0 g
Mercaptoethanesulfonic acid	0.5 g
Na₂S · 9H₂O	0.25 g
NH₄Cl	1.0 g
K₂HPO₄ · 3H₂O	0.4 g
MgCl₂ · 6H₂O	1.0 g
CaCl₂ · 2H₂O	0.4 g
Resazurin	1.0 mg
Trace mineral solution (see below)	10 ml

Instead of $NaHCO_3$, an equimolar amount of NaOH equilibrated with CO_2 may be added. Dissolve all ingredients and dilute up to 1 liter with distilled water.

Table 1. Distinguishing features of Methanobacteriales.

Feature	Methanobacteriales	Methanococcales	Methanomicrobiales	Methanosarcinales	Methanopyrales
Morphology	Rod, coccus, and rarely coccobacillus	Irregular coccus	Small rod, plate, irregular coccus, and curved rod	Irregular coccus sometimes in aggregates, rod, and flat polygon	Rod
Substrates[a]	H_2 (for, alc, methanol)	H_2 (for)	H_2 (for, alc)	(me, H_2, ac)	H_2
Motility	+/−	+	+/−	−	+
Temperature range, °C	15–97	18–94	0–60	2–70	84–110
Cell wall	Pseudomurein	Protein	Glycoprotein (sheath)	Glycoprotein (methanochondroitin, sheath)	Pseudomurein
Distinctive lipid components	Caldarchaeol and *myo*-inositol	Serine	Galactose, aminopentanetetrols, and glycerol	Hydroxyarchaeol, *myo*-inositol, ethanolamine, and glycerol	Galactose and mannose
Mol% G + C	23–61	30–33	38–61	36–54	60

Symbols: +, present in most or all taxa; −, present in a few or no taxa; and +/−, present in many but not all taxa.
Abbreviations: for, formate; alc, alcohols; me, methylamines; and ac, acetate.
[a]Substrates (in parentheses) are utilized by many but not all taxa.
Adapted from Garcia et al. (2003).

Table 2. Some characteristics of the Methanobacteriales families.

Family	Characteristics
Methanobacteriaceae	Short lancet-shaped cocci, short-to-long rods, or filamentous rods; substrates for methanogenesis are $H_2 + CO_2$, formate and CO (which may also be oxidized), methanol (which may be reduced), and secondary alcohols + CO_2; cells typically stain Gram positive; nonmotile; little or no growth above 70°C; and G+C content, 23–61 mol%.
Methanothermaceae	Single rods and short rod chains; substrates for methanogenesis are $H_2 + CO_2$; stain Gram positive; motile; no growth below 60°C and optimal growth generally above 70°C; and G+C content, 33 mol%.

Trace Mineral Solution (adapted from Ferguson and Mah, 1983)

NaEDTA · $2H_2O$	500 mg
$CoCl_2$ · $6H_2O$	150 mg
$MnCl_2$ · $4H_2O$	100 mg
$FeSO_4$ · $7H_2O$	100 mg
$ZnCl_2$	100 mg
$AlCl_3$ · $6H_2O$	40 mg
Na_2WO_4 · $2H_2O$	30 mg
CuCl	20 mg
Ni_2SO_4 · $6H_2O$	20 mg
H_2SeO_3	10 mg
H_3BO_3	10 mg
Na_2MoO_4 · $2H_2O$	10 mg

Dissolve in 1 liter of distilled water and adjust pH to 3.0 with HCl.

MS Mineral Medium

This is MS medium without any mercaptoethanesulfonate, trypticase peptones, or yeast extract. Sulfide is increased to 0.5 g per liter.

Prepare medium by first dissolving the sodium bicarbonate (or NaOH) in oxygen-free water equilibrated with CO_2 gas. For MSH Medium, the water should also contain 87.75 g (1.5 mol) of NaCl, 5 g of $MgCl_2$ · $6H_2O$, and 1.5 g of KCl per liter. Then add yeast extract and trypticase peptones to a flask flushed with CO_2 followed by CO_2-saturated $NaHCO_3$ solution, and bubble CO_2 or a gas mixture containing N_2 and CO_2 through the liquid while the other constituents of the medium are added. For convenience, use stock solutions of minerals. Per liter of medium, add 10 ml of solution A (100X, which is per liter: NH_4Cl, 100 g; $MgCl_2$ · $6H_2O$, 100g; $CaCl_2$ · $2H_2O$, 40 g; adjust to pH 4 with HCl), 2 ml of solution B (500X, which is per liter: K_2HPO_4 · $3H_2O$, 200 g), 2 ml of solution C (500X, which is per liter: resazurin, 0.5 g), and 10 ml of the trace mineral solution (described above). Store these solutions as anoxic, nonsterile solutions at room temperature.

Add any soluble substrates, such as sodium formate or methanol, prior to dispensing the medium into individual vessels, or add these later as sterile anoxic stock solutions. Add mercapto-ethanesulfonic acid, a reducing agent, prior to dispensing the medium to individual vessels. Add Na_2S only after dispensing the medium and sealing the vessels because H_2S is volatile. In general

Table 3. Taxonomic groups of order Methanobacteriales.

Family	Genus	Species
Methanobacteriaceae[T]	*Methanobacterium*[T]	*M. formicicum*[T]
		M. alcaliphilum
		M. bryantii
		M. congolense
		M. espanolense
		M. ivanovii
		M. oryzae
		M. palustre
		M. subterraneum
		M. uliginosum
	Methanobrevibacter	*M. ruminantium*[T]
		M. acididurans
		M. arboriphilus
		M. curvatus
		M. cuticularis
		M. filiformis
		M. gottschalkii
		M. oralis
		M. smithii
		M. thaueri
		M. woesei
		M. wolinii
	Methanosphaera	*M. stadtmanae*[T]
		M. cuniculi
	Methanothermobacter	*M. themautotrophicus*[T]
		M. defluvii
		M. marburgensis
		M. thermoflexus
		M. thermophilus
		M. wolfeii
Methanothermaceae	*Methanothermus*[T]	*M. fervidus*[T]
		M. sociabilis

Abbreviation: [T], type family of the order, type genus of the family, or type species of the genus.

add as a sterile, anoxic stock solution (usually 2.5%) just prior to using the medium. Use CO_2 or a mixture containing N_2 and CO_2 to flush individual vessels. To control the pH (which is ca. 6.7 with 100% CO_2 and 7.1 with 30% CO_2) adjust the proportion of CO_2 in the gas phase.

Add insoluble constituents such as agar or cellulose to the medium and mix with a magnetic stirrer while dispensing. Alternately, weigh such materials in individual vessels and then suspend them in the appropriate volume of medium.

Modifications of Media

Medium can be modified for experiments by direct additions to individual vessels. Substrates can also be added from sterile, anoxic stock solutions. When many vessels with the same substrate are desired, the substrate can be added to the medium before it is dispensed. For gaseous substrates such as H_2, pure gas may be added as an overpressure after inoculation (it is difficult to inoculate tubes which have been pressurized). H_2 mixtures with CO_2 should not be used during this initial pressurization because this would change the pH of the medium. To restore H_2 levels depleted during growth of H_2-using methanogens, it should be added in a 3 : 1 mixture of H_2 and CO_2 with shaking. Adding this gas while maintaining the same pressure used for the initial pressurization re-establishes the proper pH and CO_2 content of the gas.

Solutions for addition to individual vessels of medium should normally be prepared at a concentration about 50-fold greater than that desired in the medium and at a neutral pH (e.g., organic acids should normally be added as sodium salts). By using 50X solutions, a convenient amount (0.1 ml per 5-ml tube) can be added to serum tubes or serum bottles or even to medium during preparation. However, stock solutions of higher concentrations may be more convenient for larger vessels (e.g., for 50 ml of medium, 1 ml of a 50X solution would be required; 0.2 ml of a 250X solution may be more convenient). Solutions may be prepared in a manner similar to that used for medium preparation, by dissolving components in anoxic water, sealing, and autoclaving.

Table 4. Some characteristics of methanogenic archaea of the order Methanobacteriales.

Organism	Dimensions (μm) width	Dimensions (μm) length	Flagella	pH	Temp. (°C)	NaCl (M)	Requires organic growth factors	G+C content mol%
Family *Methanobacteriaceae*								
Methanobacterium								
formicicum	0.4–0.8	2–15	—	7–7.5	30–45	nd	None	38–42
alcaliphilum	0.5–0.6	2–25	—	8–9	37	0	YE, peptones	57
bryantii	0.5–1.0	10–15	—	6.5–7.5	30–45	nd	None	33–38
congolense	0.4–0.5	2–10	—	7.2	37–42	nd	None	39.5
espanolense	0.8	3–9	—	5–6.5	30–45	nd	nd	34
ivanovii	0.5–0.8	1–15	—	7.0–7.5	30–45	nd	None	37
oryzae	0.3–0.4	3–10	—	7	40	0–0.4	None	31
palustre	0.5	2.5–5	—	6.5–7.5	30–45	0.2	None	34
subterraneum	0.1–0.15	0.6–1.2	—	7.8–8.8	20–40	0.2	None	54.5
uliginosum	0.2–0.6	1.9–3.8	—	5–7.5	30–45	nd	None	30–34
Methanobrevibacter								
ruminantium	0.7	0.8–1.7	—	6.3–6.8	37–39	nd	ac, B-vit, CoM, 2-MBA, AAs	30.6
acididurans	0.3–0.5	0.3–0.5	—	6	35	nd	RF, ac, AAs	nd
arboriphilus	0.5	1.2–1.4	1 polar[a]	7.5–8	30–37	nd	B-vit	25.8
curvatus	0.34	1.6	—	7.1–7.2	30	nd	Complex	nd
cuticularis	0.4	1.2	—	7.7	37	nd	None	nd
filiformis	0.23–0.28	4	—	7–7.2	30	nd	YE	nd
gottschalkii	0.7	0.9	—	7	37	nd	ac and/or YE and/or TP	29
oralis	0.4–0.5	0.7–1.2	—	6.9–7.4	35–38	0.01–1	Fecal extract	28
smithii	0.6–0.7	~1	1 polar[b]	6.9–7.4	nd	nd	ac, B-vit	30–31
thaueri	0.5	0.6–1.2	—	7	37	nd	ac and/or YE and/or peptones	38
woesei	0.6	1	—	7	37	nd	ac and/or YE and/or peptones	31
wolinii	0.6	1.0–1.4	—	7	37	nd	ac and/or YE and/or peptones	33
Methanosphaera								
stadtmanae	~1	~1	—	6.5–6.9	30–40	nd	Thiamine, ac	25.8
cuniculi	0.6–1.2	0.6–1.2	—	6.8	35–40	nd	ac	23
Methanothermobacter								
themautotrophicus	0.35–0.6	3–7	—	7.2–7.6	65–70	nd	None	49
defluvii	0.4	3–6	—	nd	nd	nd	CoM	62.2
marburgensis	0.4–0.6	3–6	—	nd	nd	nd	None	47.6
thermoflexus	0.4	7–20	—	nd	nd	nd	CoM	55
thermophilus	0.36	1.4–6.5	—	nd	nd	nd	CoM, nd	44.7
wolfeii	0.4–0.6	2.5–6	—	nd	nd	nd	nd	61
Family *Methanothermaceae*								
Methanothermus								
fervidus	0.3–0.4	1–3	Perit	6.5	80–85	nd	None	33
sociabilis	0.3–0.4	1–3	Perit	6.5	88	nd	None	33

Abbreviations: nd, not determined; AAs, mixture of amino acids; ac, acetate; B-vit, B vitamins; CoM, 2-mercaptoethane-sulfonic acid (coenzyme M); Complex, complex nutritional requirements including rumen fluid and nutrient broth; 2-MBA, 2-methylbutyric acid; perit, peritrichous; RF, rumen fluid; TP, trypticase peptones; and YE, yeast extract.
[a]Refers to strain AZ.
[b]Refers to strain PS.
Data in this table are from references cited in sections of this chapter describing individual species.

Removal of liquid from vials of stock solution results in decreased gas pressure. To prevent the development of negative pressure (which could result in oxygen entering the vial), these vials may be pressurized to about 35 kPa (5 psig [pounds per square inch gauge]) overpressure of N_2.

In some cases with large vessels it may be convenient to add directly very concentrated substrates from solutions or liquids without first flushing to remove oxygen. For instance, to add 100 mM methanol to a 50-ml serum bottle, one could add 0.202 ml of pure methanol. Pure methanol is sterile (when removed from the bottle with sterile syringe) and 0.2 ml contains an insignificant amount of O_2. Acids also may be added this way and neutralized in the medium by addition of sterile, oxygen-free sodium hydroxide solution.

Sulfide solution is generally added to medium about 1 to 24 h before use. For agar slants to be used within a week, the sulfide can be added prior to autoclaving. However, sulfide should always be added to individual vessels and not to the medium in the flask. At neutral pH values, dissolved sulfide is in equilibrium with gaseous H_2S, so it gradually would be lost while the flask was flushed with N_2 and CO_2. If slants are to be stored for more than one week before inoculation, the sulfide should be added the day before inoculation; it may be added without re-melting the slants. For roll tubes to be used on the day of preparation, sulfide may be added to individual tubes before autoclaving. If the roll-tube media will be stored, the sulfide may be added later, either before or after melting the agar.

Sulfide solution should be prepared in oxygen-free water. Removal of oxygen from water may be accomplished by boiling or by flushing with oxygen-free (and CO_2-free) gas.

To prepare sulfide solution in boiled water, add 100 ml of distilled water to a 250-ml flask, add boiling chips, and mark the level in the flask. Add about 10 ml of water and boil to the mark. When the level approaches the mark, begin flushing with oxygen-free N_2 gas. (It is important not to use mixtures containing CO_2! This would acidify the sulfide solution, converting sulfide to H_2S, which is volatile.) Place a stopper into the neck of the flask and cool the water while flushing the flask.

To prepare sulfide in water deoxygenated by flushing with gas, bubble with N_2 for 30 min.

Weigh out 2.5 g of $Na_2S \cdot 9H_2O$, choosing large, clear crystals. Clean small, wet, or off-color crystals by immersing them in distilled water for a short time, followed by drying with tissue or a paper towel. Add the sulfide to the oxygen-free water and swirl to dissolve. Dispense, seal, and autoclave. Never boil the sulfide solution. Use or discard the solution within 4 weeks.

Enrichment, Isolation and Characterization

Members of the *Methanobacteriales* may be enriched and subsequently isolated from the environment by using serum-tube modifications of the Hungate technique developed for the cultivation of strictly anaerobic bacteria (Hungate, 1969; Miller and Wolin, 1974; Balch and Wolfe,

1976). All genera within this order can grow with H_2/CO_2, with the exception of species within the genus *Methanosphaera*, which require H_2 and methanol. This characteristic distinguishes this genus from all other methanogens. Therefore, H_2 added to the headspace of serum tubes containing MS Medium (which also contains CO_2) provides effective conditions for the enrichment of most members of this order. However, for the enrichment of *Methanospaera*, a 10% inoculum from an environmental sample into MS Medium with 0.6% methanol (Boone, 2001) and added H_2 in the headspace is effective. MS Medium and MS Mineral Medium can be employed in separate culturing vessels during enrichment so that autotrophic and heterotrophic methanogens can be concomitantly cultivated from the environment. MS Mineral Medium will allow for the enrichment of autotrophs while preventing heterotrophs from potentially outcompeting or overwhelming the culture. For more successful enrichments and isolations, it is recommended that the medium be prepared to simulate the chemical conditions of the environment sampled.

Acetogenic bacteria can grow faster and use H_2 plus CO_2 (H_2/CO_2) faster than can methanogens. Sequential transfer of enrichment cultures containing both acetogens and methanogens can result in eventual outcompetition by the acetogens. Often the initial number of methanogens is greater, so outcompetition of acetogens can be avoided by an initial dilution series so that higher dilutions contain methanogens but not acetogens. When acetogens are initially present in equal or greater numbers, the recovery of methanogens may require the use of antibiotics to suppress growth of acetogens.

Isolations are accomplished by the Hungate roll tube technique for strict anaerobes (Hungate, 1969). However, some *Methanobacteriales*, such as *Methanothermus fervidus*, cannot grow on agar (Stetter et al., 1981). *Methanothermus fervidus* can be isolated by picking individual colonies from medium 1 (Balch et al., 1979), which uses polysilicate rather than agar for solidification. To ensure purity, cultures obtained by picking colonies from roll tube media should be subjected to the procedure a second time.

Guidelines for the characterization of new taxa of methanogens have been proposed (Boone and Whitman, 1988). These minimal standards include the employment of a pure culture to determine the cellular morphology, colony morphology on or within a solid surface, susceptibility to lysis, Gram stain reaction, substrate range, product formation, and nutrient requirements and the measurement of growth rate under various growth conditions supporting growth and methanogenesis. The range and optimum values of pH, temperature, and salinity (NaCl) should be determined. Elements of the physiological characterization (not required but highly recommended) include the determination of G+C content of the DNA, electron microscopy, antigenic fingerprinting, lipid analysis, protein analysis, and nucleic acid hybridization and sequencing. However, it should be noted that a phylogenetic analysis of the 16S rRNA gene sequence is paramount when making taxonomic determinations. In this analysis, 16S rRNA gene sequences are aligned and compared with those within GenBank® or other databases.

The Family Methanobacteriaceae

The family Methanobacteriaceae is represented by four genera including *Methanobacterium*, *Methanobrevibacter*, *Methanosphaera* and *Methanothermobacter*. Members of this family are very strict anaerobes and all grow by oxidizing H_2, although some species may also oxidize formate and/or CO. The terminal electron acceptor is CO_2 for all Methanobacteriaceae except members of the genus *Methanosphaera*, which reduce methanol. The microbes of this order are widely distributed in nature and rarely exist in systems that exceed 70°C.

Methanobacterium

The genus *Methanobacterium* is represented by nine species described below. Members of this genus are rod-shaped mesophiles that grow by reducing CO_2 with H_2 usually; however, some strains can use formate, secondary alcohols, and CO as electron donors. A tenth species, *Methanobacterium thermoaggregans*, a thermophilic autotrophic methanogen recovered from a cattle pasture mud sample, was reported in 1985 (Blotevogel and Fischer, 1985). However, this organism was obtained via a highest dilution strategy and was not isolated from a single cell or colony. Therefore, *M. thermoaggregans* was not included as a species within this genus in the second edition of *Bergey's Manual for Systematic Bacteriology* (Boone, 2001) or in this chapter of *The Prokaryotes*.

METHANOBACTERIUM FORMICICUM. The type strain of this species was isolated from a liquid sewage sludge sample collected from a domestic anaerobic digestor in Urbana, Illinois in 1966 (Bryant and Boone, 1987). This organism may be found at high densities in anaerobic digestors and freshwater sediments. This species has also been reported to be present at low densities in cattle rumen (van Bruggen et al., 1984) and within the anaerobic ciliate, *Metopus striatus*, where it survives as an endosymbiont

(Magingo and Stumm, 1991). Cells of this organism are long crooked rods, occurring singly and often forming rod chains and filaments. CO_2 may be reduced with H_2 or formate to support growth and methanogenesis. Some strains may be autotrophic, and acetate and cysteine are often stimulatory. The type strain of this species is strain MF (DSM 1535).

The literature regarding the type strain of this species is somewhat confusing. The species was originally described by Schnellen (1947), but the type strain was lost. Another strain was isolated and characterized by Mylorie and Hungate (1954), but that strain was also lost. Later, strain MF was isolated by Marvin Bryant (Bryant and Boone, 1987), and this strain was widely regarded as the type strain of *Methanobacterium formicicum*, although this was never formalized. Later, the *Approved Lists* (Skerman et al., 1989) incorrectly designated strain M.o.H. as the type strain of *Methanobacterium formicicum*. Subsequently, an Opinion of the Judicial Commission corrected the errors by making strain MF the neotype strain of *Methanobacterium formicicum* and assigning strain M.o.H. as the type strain of *Methanobacterium bryantii* (Boone, 1987; Judicial Commission, 1992).

METHANOBACTERIUM ALCALIPHILUM. Four alkaliphilic strains of this species were isolated from the sediments of four high-pH lakes in the Wadi el Natrum of Egypt (Worakit et al., 1986). Prior to the discovery of this species, the majority of methanogens identified had been shown to grow at pH values near neutrality. *Methanobacterium alcaliphilum* not only grows at high pH values, but prefers such conditions, growing fastest at pH values between 8.1 and 9.1. Cells of this organism tend to be long rods, occurring singly or in pairs and rarely in chains or filaments. H_2/CO_2 are the sole substrates supporting growth and methanogenesis. Components of yeast extract and/or peptones are required nutrients. The type strain of this species is WeN4 (DSM 3387).

METHANOBACTERIUM BRYANTII. This methanogen was separated and subsequently isolated from a syntrophic coculture containing *Methanobacillus omelianskii*, which was originally enriched from an anaerobic digestor (Bryant et al., 1967). This strain may also be found in freshwater sediments. Cells are generally rods with blunt, rounded ends often forming irregular or crooked chains and filaments. H_2/CO_2 are the sole substrates supporting growth and methanogenesis. Components of yeast extract and/or trypticase peptones are required for growth. Strain M.o.H. (DSM 863) is the type strain of *Methanobacterium bryantii*, but the *Approved Lists* (Skerman et al., 1989) incorrectly designated this strain as the type strain of

Methanobacterium formicicum. This error made *M. bryantii* a junior objective synonym of *M. formicicum*. The problem was subsequently resolved by an Opinion of the Judicial Commission that made strain MF the neotype strain of *Methanobacterium formicicum* and reinstated *M. bryantii* with strain M.o.H. as the type strain (Boone, 1987; Judicial Commission, 1992).

METHANOBACTERIUM CONGO-LENSE. This organism was isolated from an anaerobic digestor that was fed raw cassava peel in the Congo of Central Africa in 1992 (Cuzin et al., 2001). Cells are rod-shaped, usually occurring individually or in pairs though infrequently forming chains. H_2/CO_2 are the sole substrates supporting autotrophic growth; however, methane may be produced from 2-propanol/CO_2, 2-butanol/CO_2, and cyclopentanol/CO_2 without supporting growth. Strain C (DSM 7095) is the type strain of *M. congolense*.

METHANOBACTERIUM ESPANO-LENSE. This moderately acidiphilic methanogen (optimum pH is between 5.6 and 6.2, though no growth occurs below pH 4.7) was isolated from an anaerobic digestor degrading solids from a bleach-craft mill in Espanola, Ontario, Canada (Patel et al., 1990). Cells are rod-shaped, often forming chains and filaments. H_2/CO_2 are the sole substrates supporting growth and methanogenesis. Vitamins may be stimulatory though it appears that this organism is autotrophic, which has yet to be confirmed. Further, *M. espanolense* was shown to grow in minimal medium containing acetate (Boone, 2001). The type strain of this species is strain GP9 (NRC 5912 = DSM 5982).

METHANOBACTERIUM IVANOVII. The interest in methanogenesis in deep subsurface environments led to the discovery of this species within an oil-bearing sedimentary rock core from 1650 m deep in the Bondyuzshkoe oil field in the former Soviet Union. This location represents the deepest natural habitat providing a methanogenic isolate. This strain is similar to *Methanobacterium bryantii*, but is sufficiently distinct (both phenotypically and phylogenetically) to be considered as a separate species (Jain et al., 1987). Cells are rod-shaped and tend to grow singly or in groups of 2–3, not forming chains or filaments. H_2/CO_2 are the sole substrates supporting autotrophic growth and methanogenesis. Growth is stimulated with acetate. The type strain of this species is Ivanov (DSM 2611).

METHANOBACTERIUM ORYZAE. This species was isolated from rice field soil in the Philippines (Joulian et al., 2000), a habitat often exploited by the strictly-anaerobic methanogens. Rod-shaped cells occur individually and often form long chains up to 40 μm in length. *Metha-*

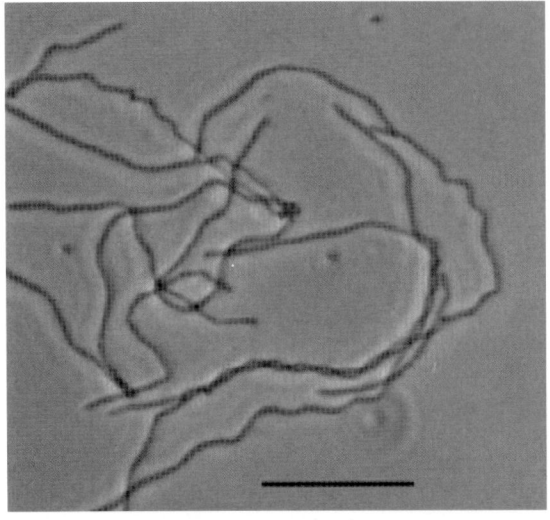

Fig. 1. Phase-contrast micrograph of *Methanobacterium oryzae*, strain FPi[t], showing long cell chains. Bar = 10 μm. Courtesy of Joulian et al. (2000).

nobacterium oryzae can grow by using H_2/CO_2 or formate. *Methanobacterium oryzae* is an autotroph, though its growth is stimulated by yeast extract. The type strain of this species is FPi (DSM 11106).

METHANOBACTERIUM PALUSTRE. This organism was isolated from a peat bog marshland of the Sippenauer Moor, Germany, and described in 1989 (Zellner et al., 1989). Substrates supporting growth include H_2/CO_2 and formate. In addition, 2-propanol/CO_2 and 2-butanol/CO_2 can be utilized as catabolic substrates; however, the use of these alcohols does not lead to measurable growth (Zellner et al., 1989). *Methanobacterium palustre* does not require growth factors. Cells are rod-shaped, sometimes forming filaments up to 65 μm long. Strain F (DSM 3108) is the type strain of this species.

METHANOBACTERIUM SUBTERRA-NEUM. Three strains of this species were isolated from deep granitic rock groundwater under the island of Aspo within the hard rock laboratory tunnel in southeastern Sweden following sample collections from 1995 to 1996 (Kotelnikova et al., 1998). Cells are small, thin rods, often growing in aggregates (Fig. 2) but not in chains or filaments. Catabolic substrates include H_2/CO_2 and formate in minimal medium, therefore supporting autotrophic growth. The type strain of this species is A8p (DSM 11074).

METHANOBACTERIUM ULIGINOSUM. This organism was isolated from marshy soil (König, 1984). H_2/CO_2 is the sole catabolic substrate regime supporting its autotrophic growth. Cells are rod-shaped with an irregular cell surface, and spherical cells are often produced at the ends of some rods, which are either retained or

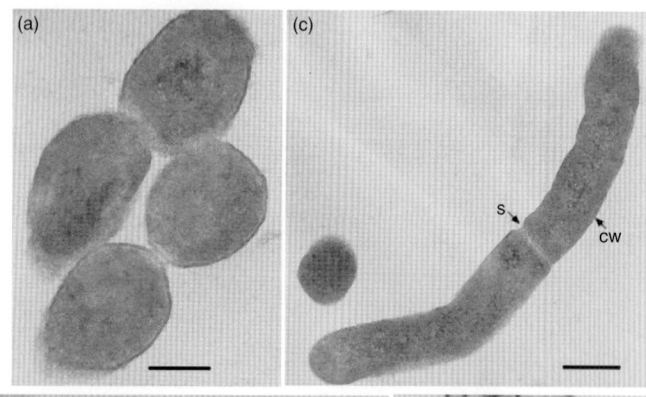

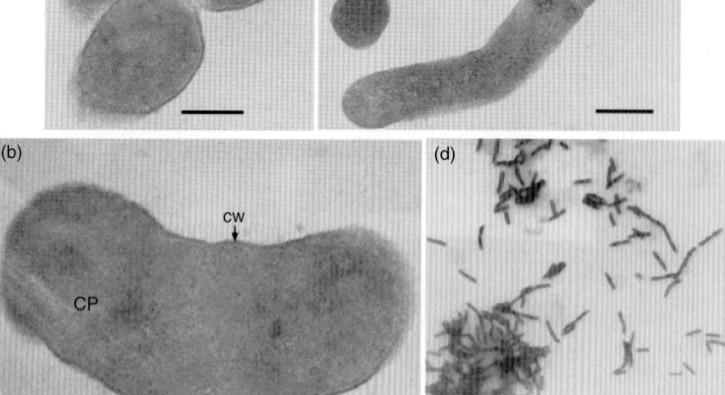

Fig. 2. Electron micrographs of cell thin sections (a–c) and phase-contrast micrograph showing cell aggregation in culture of *Methanobacterium subterraneum*. Bar = 50 nm in (a) and (b); bar = 100 nm in (c); and bar = 1 μm in (d). Abbreviations include M, cytoplasmic membrane; CW, cell wall; CP, cytoplasm; IN, invagination of cytoplasmic membrane prior to binary fission; and S, septum. Courtesy of Kotelnikova et al. (1998).

released in culture. The type strain of this species is P2St (DSM 2956).

Methanobrevibacter

The genus *Methanobrevibacter* includes the twelve species described below.

METHANOBREVIBACTER RUMINANTUM. This species, the type species of the genus *Methanobrevibacter*, was isolated from a bovine rumen and is the first methanogen to be isolated using H_2 as a catabolic substrate for isolation (Smith and Hungate, 1958). Previous isolations relied on soluble substrates contained in deep agar, making the exclusion of O_2 less difficult. Substrates supporting methanogenesis include H_2/CO_2 and formate. However, this species also requires acetate as a major cell carbon source (Bryant et al., 1971) and has complex nutritional requirements including amino acids, at least one B vitamin, 2-mercaptoethanesulfonic acid (coenzyme M), and 2-methylbutyric acid (Bryant, 1965; Bryant et al., 1971; Taylor et al., 1974). The type strain of this species is M1 (DSM 1093).

METHANOBREVIBACTER ACIDIDURANS. This methanogen was isolated from a distillery waste anaerobic digestor, which was initially seeded with cattle dung, and developed in the laboratory (Savant et al., 2002). Strain ATM (DSM 15163) is the type strain and the only strain of *M. acididurans*. It was enriched from a sour acidogenic digestor (pH 5) and is acid-tolerant, growing fastest at pH 6. H_2/CO_2 are the only catabolic substrates supporting growth, which occurs only in the presence of acetate, rumen fluid, and a mixture of amino acids. The rumen fluid appears to supply an unknown growth factor. Cells are coccus-shaped, occurring singly or in pairs and/or chains (Fig. 3).

METHANOBREVIBACTER ARBORIPHILUS. Five strains of *Methanobrevibacter arboriphilus* have been isolated and characterized from a variety of anoxic habitats. The type strain, strain DH1 (DSM 1125), was isolated from an enrichment of decaying wetwood tissue of a cottonwood tree (Zeikus and Henning, 1975). The four other strains were isolated from anaerobic sewage sludge (strains AZ and A2), a laboratory-scale anaerobic digestor (strain DC), and a rice paddy soil (strain SA). Cells are short rods with rounded ends, occurring singly or in pairs. Nutritional requirements include at least one B vitamin for growth; trypticase peptones, yeast extract, and rumen fluid are stimulatory. Two strains, A2 and SA, can catabolize formate as well as H_2/CO_2 (Asakawa et al., 1993).

METHANOBREVIBACTER CURVATUS. One strain of *Methanobrevibacter curvatus*, type strain RFM-2 (DSM 2462), was isolated from the termite hindgut of *Reticulitermes flavipes* (Leadbetter and Breznak, 1996). Cells are curved rods with tapered ends, occuring singly or in pairs. This species can utilize H_2/CO_2 only as catabolic substrates. Nutritional requirements are complex, including rumen fluid extract and nutrient broth.

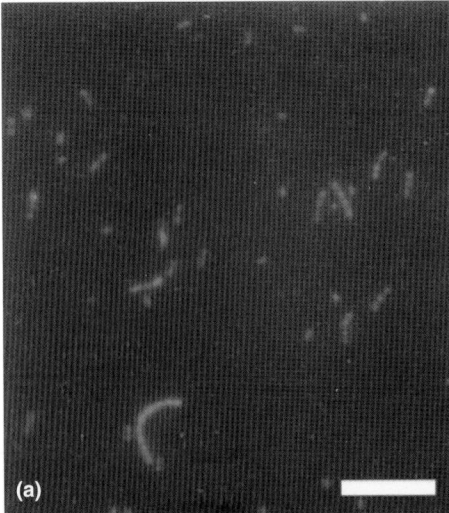

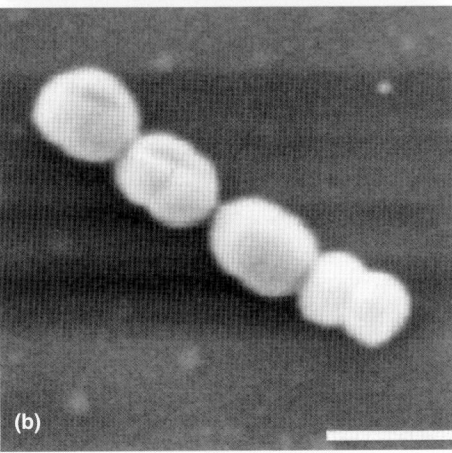

Fig. 3. (a) Fluorescence micrograph of *Methanobrevibacter acididurans* showing cellular morphology and cell chains. Bar = 25 μm. (b) Electron micrograph of cell chain. Bar = 1 μm. Courtesy of Savant et al. (2002).

METHANOBREVIBACTER CUTICU-LARIS. One strain of this species, type strain RFM-1 (DSM 11139), was isolated from the termite hindgut along with *Methanobrevibacter curvatus*, strain RFM-2 (Leadbetter and Breznak, 1996). Cells are short rods with tapered ends, occurring singly, in pairs, or in chains. Unlike *M. curvatus*, this species can use formate as well as H_2/CO_2 as catabolic substrate; however, growth is poor on formate. This species does not require any growth factors, which further distinguishes it from *M. curvatus*. In addition, its optimal temperature is 37°C, in contrast to the 30°C optimal temperature for *M. curvatus*.

METHANOBREVIBACTER FILIFORMIS. This is a third species of *Methanobrevibacter* whose type strain, strain RFM-3 (DSM 11501), was isolated from the subterraneum termite, *Reticulitermes flavipes* (Leadbetter et al., 1998). The cellular morphology of this strain differed from those of the other two species from the

termite. This species forms long filaments >50 μm in length. H_2/CO_2 are the sole substrates used for methanogenesis. Yeast extract is required for growth. This species was determined to be the most dominant methanogen within the particular sample of termite hindgut examined from Woods Hole, Massachusetts. However, samples analyzed from other geographic locations do not consistently yield cells or molecular evidence of *M. filiformis*.

METHANOBREVIBACTER GOTTSC-HALKII. Lin and Miller (1998) reported the phylogenetic analyses and comparisons of several species of *Methanobrevibacter* isolated from animal feces. *Methanobrevibacter gottschalkii* as well as three other species within this genus (*Methanobrevibacter thaueri, M. woesei* and *M. wolinii*) is a unique species (Miller and Lin, 2002). The type strain HO (DSM 11977) was isolated from enrichments of horse feces, and another strain of this species, strain PG, was isolated from the feces of a pig. Cells are coccobacilli, occurring in pairs or short chains. Growth and methanogenesis are supported by H_2/CO_2 only (Miller et al., 1986). Nutritional requirements include acetate and/or components of yeast extract or trypticase peptones. Growth is effectively maintained in salt concentrations approaching that of seawater.

METHANOBREVIBACTER ORALIS. Ferrari et al. (1994) isolated the type strain of this species, strain ZR (DSM 7256), from human subgingival plaque. Cells are short rods with tapered ends often occurring in pairs and/or short rod chains. *Methanobrevibacter oralis* is slightly smaller than other species of this genus, and its cell wall is tristratified with deep invaginations, similar to those of *M. ruminatium*. In contrast, a monostratified wall has been reported in other *Methanobrevibacter* species including *M. smithii* and *M. arboriphilicus*. H_2/CO_2 are the sole substrates supporting growth and methanogenesis. Fecal extract is a required nutrient, and a volatile fatty acid mixture enhances growth.

METHANOBREVIBACTER SMITHII. The type strain of this species, strain PS (DSM 861), was isolated from an anaerobic sewage-sludge digestor (Miller et al., 1982). However, additional strains have been isolated from human feces. Cells are short oval rods occurring most often in pairs or chains of 4–6 rods. H_2/CO_2 and formate are utilized as substrates, although the latter does not support measurable growth. Acetate is required as a major source of cell carbon and at least one B vitamin is stimulatory for growth.

METHANOBREVIBACTER THAUERI. The type strain, CW (DSM 11995), was isolated from an enrichment of cow feces (Miller and Lin, 2002). Cells are coccobacilli, occurring in pairs and short chains. H_2/CO_2 support growth and methanogenesis (Miller et al., 1986). Nutritional requirements include acetate for cell carbon and at least one component from either yeast extract or trypticase peptones. Cells grow in salt concentration comparable to that of seawater.

METHANOBREVIBACTER WOESEI. This methangen was isolated from an enrichment of goose feces (Miller et al., 1986). The type strain, strain GS (DSM 11979), is a coccobacillus that occurs in pairs or in short chains. *Methanobrevibacter woesei* grows and produces methane with H_2/CO_2 only and grows well at salt concentration similar to that of seawater. Acetate and at least one component of yeast extract or trypticase peptones are required for growth.

METHANOBREVIBACTER WOLINII. This organism was enriched and isolated from sheep feces (Lin and Miller, 1998). Similar to *M. thaueri*, the type strain, SH (DSM 11976), does not grow in medium with salt concentrations approaching those of seawater. Cells are coccobacilli with tapered or rounded ends, occurring in pairs or short chains. Similarly to *M. gottschalkii*, *M. thaueri* and *M. woesei*, this organism requires acetate and/or one or more components of yeast extract or trypticase peptones. However, nutritional requirements may also include coenzyme M and branched-chain fatty acids.

Methanosphaera

Genus *Methanosphaera* comprises two species, *Methanosphaera stadtmanae* and *Methanosphaera cuniculi*. Members of this genus are spherical chemorganotrophic methanogens requiring methanol and H_2 to support growth and methanogenesis. In contrast, the majority of methanogens isolated from the rumen and gastrointestinal tract of animals utilize H_2/CO_2 for growth and methanogenesis. CO_2 and acetate are required as cell carbon sources.

METHANOSPHAERA STADTMANAE. This species, the type species of the genus *Methanosphaera*, was isolated from human feces and therefore most likely inhabits the human large intestine (Miller and Wolin, 1985). Cells are spherical and occur in pairs, tetrads, and aggregates. A cleavage furrow in rapidly dividing cells is visible via electron microscopy as are inclusion bodies (Fig. 4). Methanol is reduced to methane with H_2. Thiamine, acetate, and CO_2 are required as nutrients; biotin is highly stimulatory. The type strain of this species is MCB3 (DSM 3091).

METHANOSPHAERA CUNICULI. The type strain of this organism, 1R7 (DSM 4103), was isolated from the intestinal tract of a rabbit (Biavati et al., 1988). Cells are round and frequently occur in pairs. Methanol and H_2 are the only substrates that support growth and metha-

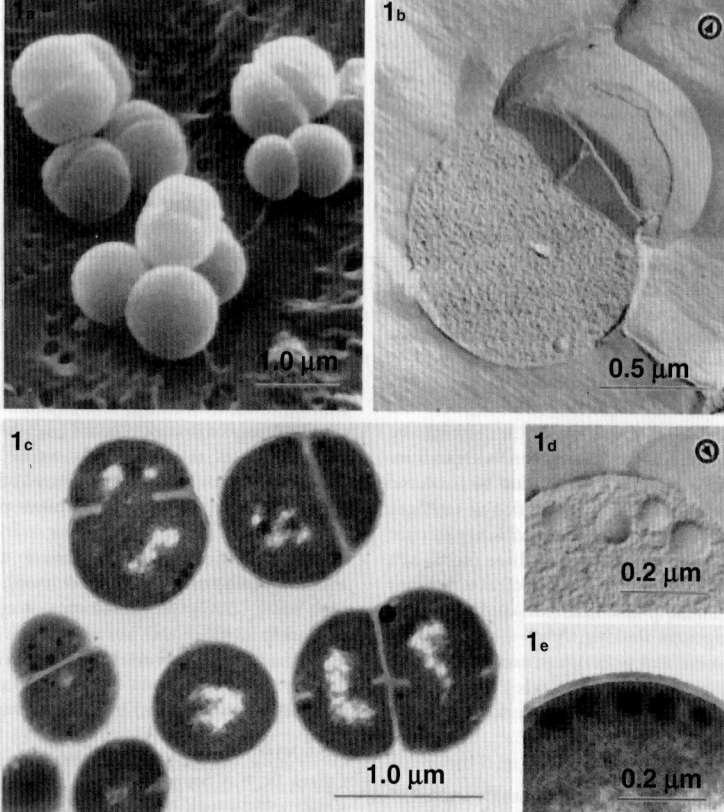

Fig. 4. Electron micrographs of *Methanosphaera stadtmanae*. (a) Scanning electron micrograph showing cellular morphology and the cleavage furrow present during cell division. (b) Freeze-fracture through actively dividing cells. (c) Thin section of dividing cells showing inclusion bodies and septa. (d) Freeze-fracture through cells showing inclusion bodies. Encircled arrows in (b) and (d) indicate the direction of the shadow in the two boxes. (e) Thin section through inclusion bodies adjacent to cell wall. Courtesy of Miller and Wolin (1985).

nogenesis consistent with this genus. Acetate is required for growth; however, other potential nutritional requirements contained in complex rumen fluid medium, which is routinely employed for the cultivation of this organism, have not yet been evaluated.

Methanothermobacter

The genus *Methanothermobacter* is represented by the six species described below. Members of this genus are thermophilic and generally grow optimally between 55°C and 65°C. H_2/CO_2 are used as substrates for methanogenesis and growth. In addition, some strains can use formate to reduce CO_2 to form methane.

METHANOTHERMOBACTER THERMAUTOTROPHICUS. This species, originally included in the genus *Methanobacterium*, was isolated from sewage sludge and reported and described by Zeikus and Wolfe (1972). Cells are irregularly-shaped crooked rods, often forming long filaments up to 120 μm long. As the species name indicates, this organism is autotrophic and thermophilic; therefore it does not require organic growth factors and inhabits environments at elevated temperatures, such as thermophilic anaerobic sewage-sludge digestors. H_2/CO_2 are primarily utilized for methanogenesis and growth; however, some strains can also use formate to reduce CO_2. Strain ΔH (DSM 1053) is the type strain of this species, which is the type species of the genus *Methanothermobacter*.

METHANOTHERMOBACTER DEFLUVII. This organism was isolated from an anaerobic sewage sludge digestor and may often be found in such environments (Kotelnikova et al., 1993). Cells are curved or crooked rods. The type strain is ADZ (DSM 7466).

METHANOTHERMOBACTER MARBURGENSIS. Originally, strain Marburg (the type strain of *Methanothermobacter marburgensis* [DSM 2133]) was included as a strain of *Methanothermobacter thermautotrophicum* (previously *Methanobacterium thermoautotrophicum*). However, Wasserfallen et al. (2000) used phylogenetic analyses to reclassify this organism as a novel species, *M. marburgensis*, with strain Marburg as the type strain. Strain Marburg was isolated from mesophilic sewage sludge in Marburg, Germany, although other strains of this species may also be found in hot springs. Cells are slender, cylindrical rods, generally occurring in pairs or filaments up to 20 μm in length. Growth is autotrophic with H_2/CO_2 although when acetate is present it may be assimilated as cellular carbon (Wasserfallen et al., 2000).

METHANOTHERMOBACTER THERMOFLEXUS. This species was isolated from an anaerobic sewage sludge digestor and reported and described in 1993 (Kotelnikova et al., 1993). Cells are crooked rods frequently forming chains. The type strain is IDZ (DSM 7268).

METHANOTHERMOBACTER THERMOPHILUS. This organism was isolated from a thermophilic, anaerobic sewage sludge digestor (Laurinavichus et al., 1987). Cells are slender, irregularly crooked rods frequently forming filaments up to 30 μm long. Unlike the other species of this genus, *M. thermophilus* requires coenzyme M for growth and is therefore not autotrophic. The type strain of this species is M (DSM 6529).

METHANOTHERMOBACTER WOLFEII. This species, originally described by Winter et al. (1984) as *Methanobacter wolfei*, is now recognized as a unique species of *Methanothermobacter* based upon phylogenetic data supporting its reclassification (Wasserfallen et al., 2000). Cells are generally rod-shaped although cultures of this organism may occur as cocci. This organism is autotrophic. It requires tungstate. The type strain is DSM 2970.

The Family Methanothermaceae

The family Methanothermaceae is represented by one genus, *Methanothermus*. Members of this family are very strict anaerobes and all grow by oxidizing H_2 to reduce CO_2 as the terminal electron acceptor. The species within *Methanothermus* have only been isolated from thermal springs with temperatures approaching 85°C at pH 6.5. Thus far it does not appear that these hyperthermophiles are widely distributed in thermal, anoxic habitats.

Methanothermus

The genus *Methanothermus* is represented by two species, *Methanothermus fervidus* and *Methanothermus sociabilis*.

METHANOTHERMUS FERVIDUS. This species, the type species of the genus *Methanothermus*, was isolated from an Icelandic hot spring and reported by Stetter et al. (1981). Subsequently, this species has only been detected in this particular system of thermal springs. Cells are slightly curved rods occurring singly and in short rod chains (Fig. 5). Unlike the majority of species within the order Methanobacteriales, this organism (as well as *M. sociabilis*) is motile via bipolar peritrichous flagella. This hyperthermophilic methanogen operates at an optimum temperature of 80–88°C by using H_2/CO_2 exclusively for growth and methanogenesis. The type strain is V24S (DSM 2088).

METHANOTHERMUS SOCIABILIS. This species, also isolated from an Icelandic solfatara,

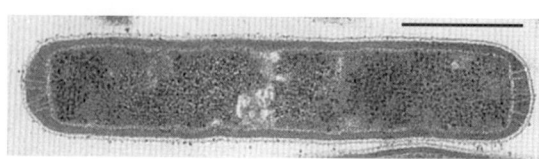

Fig. 5. Electron micrograph of an ultrathin section of *Methanothermus fervidus*. Bar = 0.5 μm. Courtesy of K.O. Stetter.

is the second species that has been found in the family Methanothermaceae (Lauerer et al., 1986). This species has not been detected outside of this collection of solfatara fields. The cells are rod-shaped and tend to grow in large clusters, which differentiates this species from *M. fervidus*. The optimal temperature supporting growth and methanogenesis on H_2/CO_2 is 88°C, and it can grow at temperatures up to ~97°C, as occurs for the other species of this genus. The lipid pattern within the cell wall is dissimilar to that of *M. fervidus*, further distinguishing the phenotypes of the two species. The type strain of this species is Kfl-F1 (DSM 3496).

Literature Cited

Asakawa, S., H. Morii, M. Akagawa Matsushita, Y. Koga, and K. Hayano. 1993. Characterization of Methanobrevibacter arboriphilicus SA isolated from a paddy field soil and DNA-DNA hybridization among M. arboriphilicus strains. Int. J. Syst. Bacteriol. 43:683–686.

Balch, W. E., and R. S. Wolfe. 1976. New approach to the cultivation of methanogenic bacteria: 2-mercaptoethanesulfonic acid (HS-CoM)-dependent growth of Methanobacterium ruminatium in a pressurized atmosphere. Appl. Environ. Microbiol. 32:781–791.

Balch, W. E., G. E. Fox, L. J. Magrum, C. R. Woese, and R. S. Wolfe. 1979. Methanogens: Reevaluation of a unique biological group. Microbiol. Rev. 43:260–296.

Biavati, B., M. Vasta, and J. G. Ferry. 1988. Isolation and characterization of "Methanosphaera cuniculi" sp. nov. Appl. Environ. Microbiol. 54:768–771.

Blotevogel, K. H., and U. Fischer. 1985. Isolation and characterization of a new thermophilic and autotrophic methane producing bacterium: Methanobacterium thermoaggregans spec. nov. Arch. Microbiol. 142:218–222.

Blotevogel, K. H., U. Fischer, M. Mocha, and S. Jannsen. 1985. Methanobacterium thermoalcaliphilum spec. nov., a new moderately alkaliphilic and thermophilic autotrophic methanogen. Arch. Microbiol. 142:211–217.

Boone, D. R. 1987. Replacement of the type strain of Methanobacterium formicicum and reinstatement of Methanobacterium bryantii sp. nov. nom. rev. (ex Balch and Wolfe, 1981) with M.o.H. (DSM 863) as the type strain. Int. J. Syst. Bacteriol. 37:172–173.

Boone, D. R., and W. B. Whitman. 1988. Proposal of minimal standards for describing new taxa of methanogenic bacteria. Int. J. Syst. Bacteriol. 38:212–219.

Boone, D. R., R. L. Johnson, and Y. Liu. 1989. Diffusion of the interspecies electron carriers H2 and formate in methanogenic ecosystems and its implications in the measurement of K_m for H_2 or formate uptake. Appl. Environ. Microbiol. 55:1735–1741.

Boone, D. R. 2001. Class I: Methanobacteria class. nov. *In:* D. R. Boone and R. W. Catenholz (Eds.) Bergey's Manual of Systematic Bacteriology, 2nd ed. Springer-Verlag. New York, NY. 1:213–235.

Bryant, M. P. 1965. Rumen methanogenic bacteria. *In:* R. W. Dougherty, R. S. Allen, W. Burroughs, N. L. Jacobson, and A. D. McGilliard (Eds.) Physiology of Digestion in the Ruminant. Butterworths. Washington, DC. 411–418.

Bryant, M. P., E. A. Wolin, M. J. Wolin, and R. S. Wolfe. 1967. Methanobacillus omelianskii, a symbiotic association of two species of bacteria. Arch. Mikrobiol. 59:20–31.

Bryant, M. P., S. F. Tzeng, I. M. Robinson, and A. E. Joyner. 1971. Nutrient requirements of methanogenic bacteria. *In:* R. F. Gould (Ed.) Anaerobic Biological Treatment Processes. American Chemical Society. Washington, DC. Advances in Chemistry Series 105:23–40.

Bryant, M. P., and D. R. Boone. 1987. Isolation and characterization of Methanobacterium formicicum MF. Int. J. Syst. Bacteriol. 37:171.

Cuzin, N., A. S. Ouattara, M. Labat, and J. L. Garcia. 2001. Methanobacterium congolense sp. nov., from a methanogenic fermentation of cassava peel. Int. J. Syst. Bacteriol. 51:489–493.

Ferguson, T. J., and R. A. Mah. 1983. Effect of H_2-CO_2 on methanogenesis from acetate and methanol in Methanosarcina spp. Appl. Environ. Microbiol. 46:348–355.

Ferrari, A., T. Brusa, A. Rutilik, E. Canzi, and B. Biavati. 1994. Isolation and characterization of Methanobrevibacter oralis sp. nov. Curr. Microbiol. 29:7–12.

Garcia, J. L., B. Ollivier, and W. B. Whitman. 2003. The Order Methanomicrobiales. *In:* M. Dworkin, S. Falkow, E. Rosenberg, K.-H. Schleifer, and E. Stackebrandt (Eds.) The Prokaryotes, 3rd ed. Springer-Verlag. New York, NY. 2.

Hungate, R. E. 1969. A roll tube method for cultivation of strict anaerobes. *In:* J. B. Norris and D. W. Ribbons (Eds.) Methods in Microbiology. Academic Press. London, UK. 3B:117–132.

Jain, M. K., T. E. Thompson, E. Conway de Macario, and J. G. Zeikus. 1987. Speciation of Methanobacterium strain Ivanov as Methanobacterium ivanovii, sp. nov. Syst. Appl. Microbiol. 9:77–82.

Judicial Commission. 1992. Designation of strain MF (DSM 1535) in place of strain MoH (DSM 863) as the type strain of Methanobacterium formicicum Schnellen 1947, and designation of strain MoH (DSM 863) as the type strain of Methanobacterium bryantii (Balch and Wolfe in Balch, Fox, Magrum, Woese, and Wolfe 1979, 284) Boone 1987, 173. Int. J. Syst. Bacteriol. 42:654.

Joulian, C., B. K. C. Patel, B. Ollivier, J. L. Garcia, and P. A. Roger. 2000. Methanobacterium oryzae sp. nov., a novel methanogenic rod isolated from a Philippines ricefield. Int. J. Syst. Bacteriol. 50:525–528.

König, H. 1984. Isolation and characterization of Methanobacterium uliginosum, new species from a marshy soil. Can. J. Microbiol. 30:1477–1481.

Kotelnikova, S. V., A. Y. Obraztsova, K. H. Blotevogel, and I. N. Popov. 1993. Methanobacterium thermoflexum sp. nov. and Methanobacterium defluvii sp. nov.: Thermophilic rod-shaped methanogens isolated from anaerobic digestor sludge. Syst. Appl. Microbiol. 16:427–435.

Kotelnikova, S. V., A. J. L. Macario, and K. Pederson. 1998. Methanobacterium subterraneum sp. nov., a new alkaliphilic, eurythermic and halotolerant methanogen isolated from deep granitic groundwater. Int. J. Syst. Bacteriol. 48:357–367.

Lauerer, G., J. K. Kristjansson, T. A. Langworthy, H. König, and K. O. Stetter. 1986. Methanothermus sociabilis sp. nov., a second species within the Methanothermaceae growing at 97°C. Syst. Appl. Microbiol. 8:100–105.

Laurinavichus, K. S., S. V. Kotelnikova, and A. Y. Obraztsova. 1987. Methanobacterium thermophilum, a new species of thermophilic methane-forming bacterium. Mikrobiol. 57:1035–1041.

Leadbetter, J. R., and J. A. Breznak. 1996. Physiological ecology of Methanobrevibacter cuticularis sp. nov. and Methanobrevibacter curvatus sp. nov., isolated from the hindgut of the termite Reticulitermes flavipes. Appl. Environ. Microbiol. 62:3620–3631.

Leadbetter, J. R., L. D. Crosby, and J. A. Breznak. 1998. Methanobrevibacter filiformis sp. nov., a filamentous methanogen from termite hindguts. Arch. Microbiol. 169:287–292.

Lin, C., and T. L. Miller. 1998. Phylogenetic analysis of Methanobrevibacter isolated from feces of humans and other animals. Arch. Microbiol. 169:397–403.

Magingo, F. S. S., and C. K. Stumm. 1991. Nitrogen fixation by Methanobacterium formicicum. FEMS Microbiol. Lett. 81:273–278.

Miller, T. L., and M. J. Wolin. 1974. A serum bottle modification of the Hungate technique for cultivating obligate anaerobes. Appl. Microbiol. 27:985–987.

Miller, T. L., M. J. Wolin, E. Conway de Macario, and A. J. L. Macario. 1982. Isolation of Methanobrevibacter smithii from human feces. Appl. Environ. Microbiol. 43:227–232.

Miller, T. L., and M. J. Wolin. 1985. Methanosphaera stadtmaniae, gen. nov., sp. nov.: A species that forms methane by reducing methanol with hydrogen. Arch. Microbiol. 141:116–122.

Miller, T. L., M. J. Wolin, and E. A. Kusel. 1986. Isolation and characterization of methanogens from animal feces. Syst. Appl. Microbiol. 8:234–238.

Miller, T. L., and C. Lin. 2002. Description of Methanobrevibacter gottschalkii sp. nov., Methanobrevibacter thaueri sp. nov., Methanobrevibacter woesi sp. nov. and Methanobrevibacter wolinii sp. nov. Int. J. Syst. Bacteriol. 52:819–822.

Mylorie, R. L., and R. E. Hungate. 1954. Experiments on the methane bacteria of sludge. Can. J. Microbiol. 1:55–64.

Patel, G. B., G. D. Sprott, and J. E. Fein. 1990. Isolation and characterization of Methanobacterium espanolae sp. nov., a mesophilic, moderately acidiphilic methanogen. Int. J. Syst. Bacteriol. 40:12–18.

Savant, D. V., Y. S. Shouche, S. Prakash, and D. R. Ranade. 2002. Methanobrevibacter acididurans sp. nov., a novel methanogen from a sour anaerobic digester. Int. J. Syst. Bacteriol. 52:1081–1087.

Schnellen, C. G. T. P. 1947. Onderzoekingen over de Methaangisting [thesis]. Delft, The Netherlands. 1–137.

Skerman, V. B. D., V. McGowan, and P. H. A. Sneath. 1989. Approved Lists of Bacterial Names: Amended Edition. American Society for Microbiology. Washington, DC. 87.

Smith, P. H., and R. E. Hungate. 1958. Isolation and characterization of Methanobacterium ruminatium n. sp. J. Bacteriol. 75:713–718.

Stetter, K. O., M. Thomm, J. Winter, G. Wildgruber, H. Huber, W. Zillig, D. Jane-Covic, H. König, P. Palm, and S. Wunderl. 1981. Methanothermus fervidus, sp. nov., a novel extremely thermophilic methanogen isolated from an Icelandic hot spring. Zbl. Bacteriol. Mikrobiol. Hyg. C 2:166–178.

Taylor, C. D., B. C. McBride, R. S. Wolfe, and M. P. Bryant. 1974. Coenzyme M, essential for growth of a rumen strain of Methanbacterium ruminatium. J. Bacteriol. 120:974–975.

van Bruggen, J. J. A., K. B. Zwart, R. M. van Assema, C. K. Stumm, and G. D. Vogels. 1984. Methanobacterium formicicum, an endosymbiont of the anaerobic ciliate Metopus striatus McMurrich. Arch. Microbiol. 139:1–7.

Wasserfallen, A., J. Nolling, P. Pfister, J. Reeve, and E. Conway de Macario. 2000. Phylogenetic analysis of 18 thermophilic Methanobacterium isolates supports the proposals to create a new genus, Methanothermobacter gen. nov., and to reclassify several isolates in three species, Methanothermobacter thermoautotrophicus comb. nov., Methanothermobacter wolfeii comb. nov., and Methanothermobacter marburgensis sp. nov. Int. J. Syst. Evol. Microbiol. 50:43–53.

Winter, J., C. Lerp, H. P. Zabel, F. X. Wildenauer, H. König, and F. Schindler. 1984. Methanobacterium wolfeii sp. nov., a new tungsten-requiring, thermophilic, autotrophic methanogen. Syst. Appl. Microbiol. 5:457–466.

Worakit, S., D. R. Boone, R. A. Mah, M. E. Abdel-Samie, and M. M. El-Halwagi. 1986. Methanobacterium alcaliphilum sp. nov., an H_2-utilizing methanogen that grows at high pH values. Int. J. Syst. Bacteriol. 36:380–382.

Zeikus, J. G., and R. S. Wolfe. 1972. Methanobacterium thermoautotrophicus sp. n., an anaerobic, autotrophic, extreme thermophile. J. Bacteriol. 109:707–713.

Zeikus, J. G., and D. L. Henning. 1975. Methanobacterium arbophilicum sp. nov. an obligate anaerobe isolated from wetwood of living trees. Ant. v. Leeuwenhoek 41:543–552.

Zellner, G., K. Bleicher, E. Braun, H. Kneifel, B. J. Tindall, E. Conway de Macario, and J. Winter. 1989. Characterization of a new mesophilic, secondary alcohol-utilizing methanogen, Methanobacterium palustre spec. nov. from a peat bog. Arch. Microbiol. 151:1–9.

Prokaryotes (2006) 3:244–256
DOI: 10.1007/0-387-30743-5_12

Chapter 12

The Order Methanosarcinales

MELISSA M. KENDALL AND DAVID R. BOONE

Introduction

Methanogens (methane-producing microbes) belong to the domain Archaea, and 16S rRNA gene sequences indicate that they form a phylogenetically distinct clade within the kingdom Euryarchaeota. The order Methanosarcinales is one of the five orders of methanogens. These microbes catalyze the terminal step in the degradation of organic matter in anoxic environments where light and terminal electron-acceptors other than CO_2 are limiting. Representatives of the Methanosarcinales are widespread and cosmopolitan in anaerobic environments. They are found in freshwater, marine environments, and extremely halophilic sediments as well as in anaerobic sludge digestors and the gastrointestinal tracts of animals. Use of molecular techniques has also identified Methanosarcinales in the deep terrestrial subsurface environments such as in the gold mines in South Africa.

Members of the order Methanosarcinales are coccoids, pseudosarcinae, or sheathed rods. Most cells have a protein cell wall, and some are surrounded by a sheath or acidic heteropolysaccharide; no peptidoglycan or pseudomurein is present. Hydroxyarchaeol is usually present in the lipids, which contain *myo*-inositol, ethanolamine and glycerol as polar headgroups. Cells are very strictly anaerobic and obtain energy with the concomitant production of methane. Methanosarcinales have the widest substrate range of methanogens: many can grow by reducing CO_2 with H_2, by dismutating methyl compounds, or by the splitting of acetate. Some species can use only one of those catabolic schemes, but others can use all three.

The order Methanosarcinales comprises two families, Methanosarcinaceae and Methanosaetaceae. Some of their distinctive characteristics are listed in Table 1.

Habitats and Metabolism of Methanosarcinales

Similarly to other methanogens, Methanosarcinales are widespread in diverse anaerobic habitats, including freshwater and marine mud and sediments, rumens of ungulates, animal waste lagoons, sludge from anaerobic sewage sludge digestors, and animal feces. Genetic sequences of 16S rRNA that are apparently from organisms that belong to Methanosarcinales have been detected in DNA from fissure water of deep South African gold mines (Takai et al., 2001), gas industry pipelines (Zhu et al., 2003), and a human periodontal pocket (Robichauz et al., 2003). In most natural environments where Methanosarcinales occur, they grow by splitting acetate to methane (CH_4) and carbon dioxide (CO_2), or they catabolize methyl compounds (methanol, methyl amines, methyl sulfides, etc.). In anoxic environments that lack light and alternate electron acceptors other than CO_2, methanogenesis is the major fate of organic matter that is degraded, and acetate splitting is the major source of methane. There are no other organisms than Methanosarcinales that are capable of catabolizing acetate in this manner. It is accomplished by members of one genus of the family Methanosarcinaceae (*Methanosarcina*) and by the only genus of the family Methanosetaceae (*Methanosaeta*). Acetate splitting is the sole catabolic pathway for this latter group.

The other important catabolic regime for Methanosarcinales is methanogenic catabolism of methyl groups. This catabolic pathway is accomplished by every known species of Methanosarcinaceae and is virtually exclusive to this family. (Members of the genus *Methanosphaera*, of the order Methanobacteriales, grow exclusively by reducing methanol with H_2.) Members of the genus *Methanosarcina* can dismutate methyl compounds such as methanol and methyl amines (sometimes methyl sulfides as well), producing one CO_2 and three CH_4 for each methyl compound dismutated. Most of these strains can also use H_2 to reduce methyl compounds when H_2 is also available. Other genera of Methanosarcinaceae lack the ability to use H_2. These genera occur in saline environments and probably acquire methyl compounds as degradation products of compatible solutes such as trimethylamine oxide, glycine betaine, or dimethylsulfoniopropionate.

Table 1. Some characteristics of the Methanobacteriales families.

Family	Characteristics
Methanobacteriaceae	Short lancet-shaped cocci, short-to-long rods, or filamentous rods; substrates for methanogenesis are $H_2 + CO_2$, formate and CO (which may also be oxidized), methanol (which may be reduced), and secondary alcohols $+ CO_2$; cells typically stain Gram positive; nonmotile; little or no growth above 70°C; and G+C content, 23–61 mol%.
Methanothermaceae	Single rods and short rod chains; substrates for methanogenesis are $H_2 + CO_2$; stain Gram positive; motile; no growth below 60°C and optimal growth generally above 70°C; and G+C content, 33 mol%.

The Methanosarcinales occur in anoxic sediments throughout the entire range of salinities, from anaerobic digestors and freshwater sediments to saturated brines. In low-salt environments, acetate and H_2 are the important methanogenic substrates. Although many Methanosarcinales can use H_2 to reduce CO_2, they appear to be outcompeted for these substrates by other methanogens. But Methanosarcinales are entirely responsible for methanogenesis from acetate. All members of the genus *Methanosaeta* and most strains of *Methanosarcina* can grow on acetate. Strains of *Methanosaeta* have a lower K_m for acetate than do strains of *Methanosarcina*, so *Methanosaeta* often outcompete *Methanosarcina* in environments in which turnover is slow and acetate concentrations are low. *Methanosarcina* strains often predominate at lower pH values and when acetate concentrations are higher.

Members of the genus *Methanosaeta* are found in anaerobic digestors and sediments. The type strain of *Methanosaeta concilii* as well as several other strains of this species was isolated from anaerobic digestors (Patel, 1984; Ohtsubo et al., 1991), and *Methanosaeta thermophila* was isolated from mud in the thermal Khlorid Lake, in Kamchatka (Kamagata et al., 1992).

The genus *Methanosarcina* includes slightly halotolerant and slightly halophilic species. They are particularly common in anaerobic digestors. The type species of *Methanosarcina barkeri* as well as *Methanosarcina mazeii*, *Methanosarcina vacuolata* and *Methanosarcina thermophila* was originally isolated from anaerobic sewage sludge digestors (Barker, 1936; Mah and Kuhn, 1984; Zinder et al., 1985; Bryant and Boone, 1987; Zhilina and Zavarzin, 1987b). Other species of this genus have been isolated from various sediment environments, including marine and estuarine sediments. Some of these marine species of *Methanosarcina* are methylotrophic (i.e., they grow by dismutating methyl compounds). *Methanosarcina acetivorans* was obtained from marine sediments (Sowers et al., 1984); *Methanosarcina baltica* was isolated from brackish sediments in the Gotland Deep of the Black Sea

(von Klein et al., 2002a); and *Methanosarcina semesiae* was isolated from mangrove sediment in Tanzania (Lyimo et al., 2000). *Methanosarcina lacustris* was isolated from freshwater anoxic lake sediments (Switzerland; Simankova et al., 2001), and *Methanosarcina siciliae* was obtained from sediments at the Lakes of Marinello, Italy (Ni and Boone, 1991; Ni et al., 1994). Additionally, another strain of *Methanosarcina siciliae*, strain C2J, was isolated from submarine canyon sediments near southern California (Elberson and Sowers, 1997). *Methanosarcina barkeri* strain CM1 has also been isolated from sheep rumen (Jarvis et al., 2000), although *Methanosarcina* do not occur commonly in rumen microflora.

Several genera of Methanosarcinaceae are slight halophiles, growing in environments with salinity near that of the ocean. These genera, including *Methanolobus*, *Methanococcoides* and *Methanosalsum*, are exclusively methylotrophic.

The major habitat of the genus *Methanolobus* is aquatic sediments. All current species were originally obtained from saline aquatic sediments. *Methanolobus tindarius* was isolated from seawater ponds near Tindari, Italy (König and Stetter, 1982). Two species have been isolated from marine sediments: *Methanolobus bombayensis* was isolated from Arabian Sea sediments near Bombay, India (Kadam et al., 1994), and *Methanolobus vulcanii* was isolated from submarine fumarole sediments near Vulcano Island, Italy (Stetter, 1989). *Methanolobus taylorii* was obtained from estuarine sediments from San Francisco Bay (Oremland and Boone, 1994). *Methanolobus oregonensis* was isolated from anoxic, subsurface sediments of a saline, alkaline aquifer near Alkali Lake, an alkaline, desert lake in south central Oregon (Liu et al., 1990).

The genus *Methanococcoides* contains two species, both from environments with salinity near that of seawater. *Methanococcoides methylutens* was isolated from submarine canyon sediments that contained large deposits of organic material off the coast of southern California (Sowers and Ferry, 1983), and *Methanococcoides burtonii* was obtained from the anoxic hypolimnion of Ace Lake, Anarctica (Franzmann et al., 1992).

The habitat of the genus *Methanosalsum* appears to be restricted to hypersaline soda lakes. *Methanosalsum zhilinae* was isolated from alkaline and saline sediments of a lake of the Wadi el Natrun in Egypt, and a second strain (strain Z-7936) was isolated from Lake Magadi, another alkaline soda lake (Mathrani et al., 1988).

Other Methanosarcinaceae are moderate halophiles, growing in hypersaline environments with Na^+ concentration between about 1.0 M and 3.0 M. These, along with the extreme halophiles of the genus *Methanohalobium*, are also exclusively methylotrophic.

Members of the genus *Methanohalophilus* have been isolated from diverse anoxic environments, in all cases with a salinity greater than that of seawater. *Methanohalophilus mahii* was obtained from the sediments of the Great Salt Lake (Paterek and Smith, 1985). *Methanohalophilus halophilus* was isolated from the ooze under a cyanobacterial mat and bottom deposits at Hamelin Pool, Shark Bay in Northwestern Australia (Zhilina, 1983; Zhilina and Kevrin, 1985; Wilharm et al., 1991), and *Methanohalophilus portucalensis* was retrieved from sediments of a solar salter in Figeira da Foz, Portugal (Boone et al., 1993). Saline subsurface water in an oil field yielded the culture "*Methanohalophilus euhalobius*" (Charakhchyan et al., 1989; Davidova et al., 1997).

Methanohalobium are generally found in extremely hypersaline environments. The type species, *Methanohalobium evestigatum*, was isolated from NaCl-saturated sediment of saline lagoons near Arabata, Sivash (Zhilina and Zavarzin, 1987a).

Cultivation

Methanosarcinales, like other methanogens, are strict anaerobes, and they must be cultivated by anaerobic techniques as described by Hungate (Hungate, 1969) or modifications of those techniques (Sowers and Noll, 1995). The use of serum tubes with butyl-rubber stoppers as culture vessels eliminates the difficulties in using pipets and mouth tubes, which require culture vessels to be opened anoxically for transfer of cultures. Modern techniques that use serum-stoppered vessels allow syringes to be used for inoculation. Culture fluid is transferred with syringes previously flushed with O_2-free gas.

Methanosarcinales are strict anaerobes, so culture medium must be prepared anoxically. Several types of media with various substrates will support the growth of Methanosarcinales.

A wide range of culture media are suitable for enriching, isolating, and cultivating these microbes. In this section, we describe culture media used by the Oregon Collection of Methanogens (Portland, Oregon, USA), which can be used to grow all of the methanogens described in this section.

Culture Media

The growth of very strictly anaerobic bacteria such as methanogens requires culture media that are completely free of O_2. The first description of methods to produce O_2-free culture media was provided by Hungate (1969), and several subsequent modifications have simplified the procedures for making medium (Sowers and Noll, 1995). One of the time-consuming procedures for preparing media for methanogens is boiling the water to remove O_2. In some laboratories, culture medium is successfully prepared without boiling the water or making any effort to remove O_2. We have found medium is more dependable when most of the O_2 is removed by flushing the liquid with O_2-free gas. We prepare bicarbonate buffer as the basal fluid for our medium by dissolving 0.1M NaOH or $NaHCO_3$ in water and equilibrating it with CO_2. When NaOH is added rather than $NaHCO_3$ it is important to equilibrate with CO_2 before adding minerals; otherwise precipitates may form. Alternately, the water can be boiled to remove oxygen and cooled under a gas phase containing CO_2. Then 100 mM sodium bicarbonate is added to establish the bicarbonate-CO_2 buffer.

The pH of this buffer when equilibrated with a gas phase of N_2 and CO_2 (7 : 3) is about 7.2 at 37°C, about 7.4 at 60°C, and about 7.15 at 4°C. Other pH values between about 6.6 and 8 can be achieved by adjusting the percentage of CO_2 in the gas phase. A gas phase of pure CO_2 (partial pressure of 1 atm) establishes the pH at about 6.6–6.7. Lower pH values can be achieved by adjusting with HCl before the vessels are sealed. After equilibration of the medium with a gas phase of N_2 and CO_2 (7 : 3), its pH can be adjusted to values higher than 7.2 by subsequently decreasing the CO_2 content of the gas. This should be done after autoclaving to prevent precipitation. Flushing culture medium with N_2 increases the pH to about 8. Higher pH values can be attained by adjusting with NaOH solution. Media are not strongly buffered by this system at pH values of about 8–8.6. At high pH values, the media described below may develop precipitates, such as apatites. These can be avoided by decreasing the concentration of phosphate or calcium. Also, at pH values above 8.5, ammonia toxicity is enhanced; it may be necessary to reduce ammonia concentration to avoid this.

MS Medium (Boone et al., 1989)

NaOH	4.0 g
Yeast extract	2.0 g
Trypticase peptones	2.0 g
Mercaptoethanesulfonic acid	0.5 g
$Na_2S \cdot 9H_2O$	0.25 g
NH_4Cl	1.0 g
$K_2HPO_4 \cdot 3H_2O$	0.4 g
$MgCl_2 \cdot 6H_2O$	1.0 g
$CaCl_2 \cdot 2H_2O$	0.4 g
Resazurin	1.0 mg
Sodium EDTA dihydrate	5.0 mg
$CoCl_2 \cdot 6H_2O$	1.5 mg
$MnCl_2 \cdot 4H_2O$	1.0 mg
$FeSO_4 \cdot 7H_2O$	1.0 mg
ZnCl	1.0 mg
$AlCl_3 \cdot 6H_2O$	0.4 mg
$Na_2WO_4 \cdot 2H_2O$	0.3 mg
$CuCl_2 \cdot 2H_2O$	0.2 mg
$NiSO_4 \cdot 6H_2O$	0.2 mg
H_2SeO_3	0.1 mg
H_3BO_3	0.1 mg
$Na_2MoO_4 \cdot 2H_2O$	0.1 mg

Dissolve the NaOH and equilibrate NaOH solution with CO_2 gas before its addition to the other ingredients. Dissolve all ingredients in gassed, distilled water and dilute up to 1 liter.

MH Medium (Boone et al., 1989)

MS medium	1.0 liter
NaCl	87.75 g
$MgCl_2 \cdot 6H_2O$	5.0 g
KCl	1.5 g

MSH Medium (Ni and Boone, 1991)

MS medium	2 parts
MH medium	1 part

MS, MSH or MH Mineral and Enrichment Media

Prepare MS mineral medium by adding all ingredients except the organic substrates such as yeast extract or trypticase peptone. Additionally, prepare an enrichment medium containing reduced concentrations of yeast extract and trypticase peptone (i.e., 0.5 g of each per liter) and increased concentration of sulfide (i.e., 0.5 g per liter; Ni and Boone, 1991). In these enrichment media, eliminate the mercaptoethane sulfonate to prevent the enrichment of bacteria that decompose this compound, and reduce yeast extract and peptone concentrations to limit the growth of heterotrophs while still providing the nutrients needed by methanogens for their growth. Prepare MS, MSH, or MH mineral and enrichment media by following the changes described below.

Enrichment cultures of *Methanosarcina* may be initiated in one of the enrichment media described above, with salinity and pH adjusted to match the environment of the sample. A catabolic substrate such as acetate (50–100 mM), methanol (30 mM), or methylamines (mono-, di- or trimethylamine; 30 mM) should be added. Organic growth factors that are required by some strains are provided by the yeast extract and peptones of the enrichment medium. For rapid enrichment of *Methanosarcina*, trimethylamine is the preferred substrate. Acetate may be the most selective substrate for the isolation of *Methanosarcina*, especially when the pH of the medium is lower than 7.0, but growth is slower than on trimethylamine, and not all species of this genus can catabolize acetate.

Methanolobus may be enriched in MSH medium to which trimethylamine or methanol has been added as catabolic substrate. For the growth of *Methanolobus bombayensis*, 1.7 g of KCl should be added per liter of medium.

Cultures of *Methanococcoides* can also be enriched with MSH medium to which trimethylamine has been added as catabolic substrate. When grown in a defined mineral medium, *Methanococcoides methylutens* requires biotin (Sowers and Ferry, 1985a). *Methanococcoides burtonii* can be enriched with MSH enrichment medium with trimethylamine by incubating at 15°C (Franzmann, 1992).

All *Methanohalophilus* may be enriched in MH medium with 20 mM trimethylamine as catabolic substrate.

Methanosalsum may be cultivated by using MSH culture medium (pH 8–9.5) containing methylamines or methanol as catabolic substrate.

The growth of *Methanohalobium* requires higher salt concentrations than those provided in MH medium. *Methanohalobium* can be grown by increasing the NaCl concentration from 1.5 M to 3 M. No organic substrates are required for growth. Yeast extract is inhibitory at 500 mg/liter, but at concentrations of 50–100 mg/liter, yeast extract can replace vitamins in the medium.

Species of *Methanosaeta* can be enriched in MS mineral medium containing acetate as the main organic source of carbon and energy for growth. *Methanosaeta thermophila* can be grown in MS medium with a final concentration of 10% MSH medium added. Vitamins should be added to the culture medium, as some strains of *Methanosaeta* require and others may be stimulated by vitamins. Yeast extract may be inhibitory to these organisms, so it should be omitted from the medium. *Methanosaeta* species are generally outcompeted by faster growing, acetate-utilizing *Methanosarcina* species, which can cause problems in their isolation. Janssen has suggested starting with batch enrichment cultures in which acetone and isopropanol are used as growth substrates for the enrichment step. These substrates are slowly fermented by certain bacteria to acetate, which then allows the *Methanosaeta* species to maintain acetate concentration at levels below the threshold required for growth of *Methanosarcina* species. This can result in dense populations of *Methanosaeta* which can subsequently be separated from contaminating microbes to yield pure cultures (Janssen, 2003).

Maintenance

Cultures of *Methanosarcina*, *Methanohalophilus* and *Methanosalsum* may be maintained by regular subculturing as described by Hippe (1984). Strains of *Methanococcoides* are maintained by transfer every 3 months on agar slants or in liquid medium stored at room temperature (or at lower temperatures for *Methanococcoides burtonii*) in the dark. For longer term storage, cultures can be maintained by freezing in liquid growth medium and glycerol (3:1 [v/v]; Tumbula et al., 1995). Cultures stored in this fashion have remained viable for 10 years. Cultures of *Methanohalophilus* can survive weeks or months after growth by storing cultures at room temperature in the dark, as long as the culture remains anoxic. Working cultures of *Methanosaeta* can be maintained by transfer into appropriate broth media at regular intervals of 1–4 weeks, depending on the strain. Often large inocula (10–50%) are required to initiate growth.

All species of *Methanosarcinales* can be stored as frozen stocks; 5% glycerol is added as a cryoprotectant before cell suspensions are anoxically distributed into vials, slowly frozen at a rate of 1°C/min to –40°C, and then stored at liquid nitrogen temperatures (Boone, 1995).

The Family Methanosarcinaceae

Methanosarcinaceae includes six genera of coccoidal or pseudosarcinal bacteria. Most cells have a protein cell wall. Other cells are surrounded by an acidic heteropolysaccharide. No cells contain peptidolglycan or pseudomurein. The lipids contain *myo*-inositol, ethanolamine and glycerol as polar head groups. All strains can gain energy by dismutating methyl compounds. The use of acetate or CO_2 reduction with H_2 may be observed; however, no strains catabolize formate.

Methanosarcina

The genus *Methanosarcina* comprises nine species: *Methanosarcina barkeri* (Fig. 1), *Methanosarcina acetivorans*, *Methanosarcina siciliae*

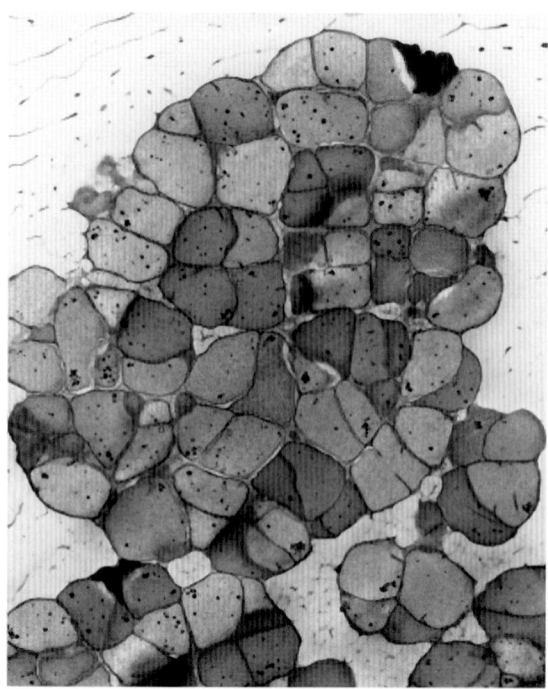

Fig. 1. Thin section of aggregates of *Methanosarcina barkerii*. Courtesy of Henry Aldrich, Department of Microbiology and Cell Science, University of Florida, Gainesville, FL.

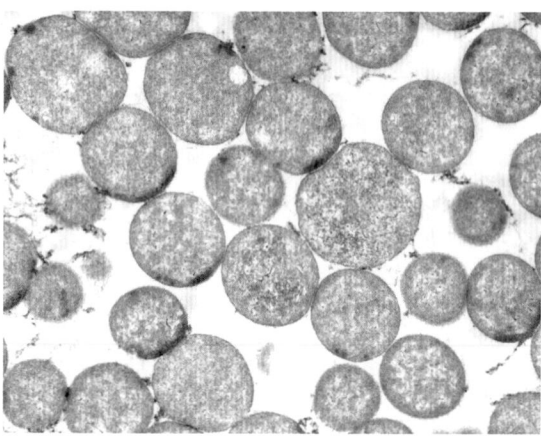

Fig. 2. Electron micrograph of *Methanosarcina siciliae*. Courtesy of Henry Aldrich, Department of Microbiology and Cell Science, University of Florida, Gainesville, FL.

(Fig. 2), *Methanosarcina thermophila* (Fig. 3), *Methanosarcina mazeii* (Fig. 4), *Methanosarcina vacuolata*, *Methanosarcina baltica*, *Methanosarcina lacustris* and *Methanosarcina semesiae*. All strains are nonmotile and can catabolize acetate, methanol, methylamines and CO. Some strains are able to gain energy by the reduction of CO_2 with H_2. Refer to Tables 2–5 for further information about these species' phenotypic and physiological characteristics.

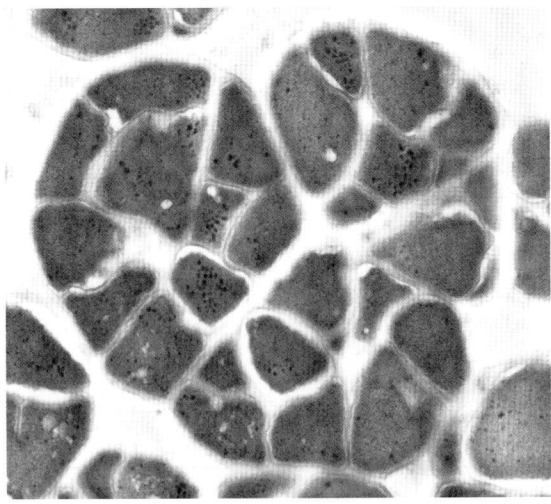

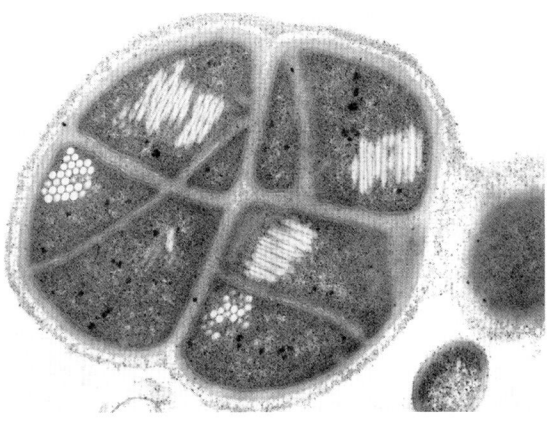

Fig. 4. Thin section of gas vesicles of *Methanosarcina mazeii*. Courtesy of Donna Williams and Henry Aldrich, Department of Microbiology and Cell Science, University of Florida, Gainesville, FL.

Fig. 3. Electron micrograph of *Methanosarcina thermophila*. Courtesy of Henry Aldrich, Department of Microbiology and Cell Science, University of Florida, Gainesville, FL.

Table 2. List of described groups in the order Methanosarcinales.

Species[a]	Culture collection	Sequence accession number[b]	Effective publication	Validation
Methanosarcina barkeri[T] MS	DSMZ 800, OCM38	AJ012094	Schnellen, 1947 Bryant and Boone, 1987 (emended description)	Bryant and Boone, 1987
Methanosarcina acetivorans C2A	ATCC 35395, DSMZ Z2834, OCM 95	M59137	Sowers et al., 1984	Sowers et al., 1986
Methanosarcina mazeii S6	DSMZ 2053, OCM 26	AJ01209, U20151	Barker, 1936 Mah and Kuhn, 1984	Mah and Kuhn, 1984
Methanosarcina siciliae T4/M	DSMZ 3028, OCM 156	U20153	Stetter, 1989	Ni et al., 1994
Methanosarcina thermophila TM-1	DSMZ 1825, OCM 12	M59140	Zinder and Mah, 1979	Zinder et al., 1985
Methanosarcina vacuolata Z-761	ATCC 35090, DSMZ 1232, OCM 85	U20150	Zhilina and Zavarzin, 1979	Zhilina and Zavarzin, 1987b
Methanosarcina baltica GS1-A	DSM 14042, JCM 11281	AJ238648	von Klein et al., 2002a	von Klein et al., 2002b
Methanosarcina lacustris ZS	DSM 13486, VKM B-2268	AF432127	Simankova et al., 2001	Simankova et al., 2002
Methanosarcina semesiae MD1	DSM 12914	AJ012742	Lyimo et al., 2000	Lyimo et al., 2000
Methanolobus tindarius[T] T3	DSMZ 2278, ATCC 35996, OCM 150	M59135	König and Stetter, 1982	König and Stetter, 1983
Methanolobus bombayensis B-1	OCM 438, DSMZ 7082	U20148	Kadam et al., 1994	Kadam et al., 1994
Methanolobus oregonensis WAL-1	DSMZ 5435, OCM 99	U20152	Liu et al., 1990	Boone and Baker, 2002
Methanolobus taylorii GS-16	DSMZ 9005, OCM 58	U20154	Oremland et al., 1989	Oremland and Boone, 1994
Methanolobus vulcani PL-12/M	DSMZ 3029, OCM 157	U20155	Stetter, 1989; Kadam, 1995	Stetter, 1989
Methanococcoides methylutens[T] TMA-10	ATCC 33938, DSMZ 2657, OCM 158	M59127	Sowers and Ferry, 1983	Sowers and Ferry, 1985b
Methanococcoides burtonii	DSMZ 6242, OCM 468	X65537	Franzmann et al., 1992	Franzmann et al., 1993

Table 2. *Continued*

Species[a]	Culture collection	Sequence accession number[b]	Effective publication	Validation
Methanohalobium evestigatum[T] Z-7303	DSMZ 3721, OCM 161	U20149	Zhilina and Zavarzin, 1987a	Zhilina and Zavarzin, 1988
Methanohalophilus mahii[T] SLP	ATCC 35705, DSMZ 5219, OCM 68	M59133	Paterek and Smith, 1985	Paterek and Smith, 1988
Methanohalophilus halophilus Z-7982	DSMZ 3094, OCM 160	U22259 (MCR I) X6860 (5S rDNA)	Zhilina, 1983	Wilharm et al., 1991
Methanohalophilus portucalensis FDF-1	OCM 59, DSMZ 7471	U22239	Boone et al., 1993	Boone et al., 1993
"*Methanohalophilus euhalobius*"	DSMZ 10369	X98192	Obraztsova et al., 1987 Davidova et al., 1997	Not validated
Methanosalsum zhilinae[T] WeN5	DSMZ 4017, OCM 62	U22252 (MCR I)	Mathrani et al., 1988	Boone and Baker, 2002
Methanosaeta concilii[T] GP6	ATCC 35969, DSMZ 3671, NRC 2989, OCM 69	35969	Patel, 1984	Patel and Sprott, 1990
Methanosaeta thermophila PT	DSMZ 6194, OCM 780	AB071701	Kamagata et al., 1992	Boone and Kamagata, 1998

Abbreviations: [T], type species of the genus; DSMZ, Deutsche Sammlung von Mikroorganismen und Zellkulturen, Braunschweig, Germany; ATCC, American Type Culture Collection, Manassas, VA., United States; OCM, Oregon Collection of Methanogens, Portland, OR., United States; and NRC, National Research Council of Canada.
[a]Strains listed are the type strain of each species.
[b]Accession numbers are for 16S rDNA sequences unless otherwise indicated.

Table 3. Phenotypic properties of the described species within the Methanosarcinales.

Species	Shape	Forms aggregates	Forms filaments	Cysts	Size (μm)	Motile	Flagella	G + C content (mol%)	Gram stain	Gas vacuoles	Plasmids
Methanosarcina barkeri	Coccoid	+	–	–	1.5–2.0	–	–	39–44	+	+	nd
Methanosarcina acetivorans	Coccoid	+	–	+	1.7–2.1	–	–	41	–	–	nd
Methanosarcina mazeii	Coccoid	+	–	+	1.0–3.0	–	–	42	+	–	nd
Methanosarcina siciliae	Coccoid	+	–	–	nd	nd	nd	41–43	–	–	nd
Methanosarcina thermophila	Coccoid	+	–	–	100	–	–	42	+	–	nd
Methanosarcina vacuolata	Coccoid	+	–	–	0.5–2.0	–	–	36	+	+	nd
Methanosarcina baltica	Irregular coccoid	+	–	nd	1.5–3.0	nd	Monotrichous	nd	nd	nd	nd
Methanosarcina lacustris	Coccoid	+	–	nd	1.5–3.5	–	–	43.4	+	nd	nd
Methanosarcina semesiae	Irregular coccoid	–	–	nd	0.8–2.1	nd	–	nd	+	nd	nd
Methanolobus tindarius	Coccoid	+	–	nd	0.8–1.25	+	Monotrichous	45.9	–	nd	–
Methanolobus bombayensis	Irregular coccoid	–	–	nd	1.0–1.5	–	–	39	–	nd	nd
Methanolobus oregonensis	Irregular Coccoid	+	–	nd	1.0–1.5	–	–	40.9	–	nd	nd

Table 3. *Continued*

	Shape	Forms aggregates	Forms filaments	Cysts	Size (µm)	Motile	Flagella	G + C content (mol%)	Gram stain	Gas vacuoles	Plasmids
Methanolobus taylorii	Coccoid	+	–	nd	0.5–1.0	–	–	41	–	nd	nd
Methanolobus vulcani	Irregular coccoid	+	–	nd	0.8–1.25	–	–	39	–	nd	+
Methanococcoides methylutens	Irregular coccoid	–	–	nd	0.8–1.8	–	–	42	–	nd	nd
Methnococcoides burtonii	Irregular coccoid	+	–	nd	0.8–1.8	+	Monotrichous	39.6	V	nd	nd
Methanohalobium evestigatum	Flat, polygonal, irregular spheroids	+	–	nd	0.2–2 (single) 5–10 (aggregate)	–	–	37	nd	nd	nd
Methanohalophilus mahii	Irregular coccoid	+	–	nd	0.8–1.8	–	–	41	–	nd	–
Methanohalophilus halophilus	Irregular coccoid	+	–	+	0.5–2.0	–	–	39	–	nd	nd
Methanohalophilus portucalensis	Irregular coccoid	+	–	nd	0.6–2.0	–	–	41	–	nd	nd
Methanosalsum zhilinae	Irregular coccoid	+	–	nd	0.75–1.5	+	Mono/ditrichous	38–39.5	–	nd	nd
Methanosaeta concilii	Straight rods	–	+	nd	0.8–1.3, wide 2.0–7.0, long	–	–	49 ± 1.25	–	–	nd
Methanosaeta thermophila	Straight rods	–	+	nd	0.8–1.3, wide 2.0–7.0, long	–	–	52.7–54.2	–	+	nd

Abbreviations: nd, not determined; V, variable.

Table 4. Physiological properties of the described species within the Methanosarcinales.

Species	G+C content (mol%)	Temperature optimum (°C)	Temperature range (°C)	pH optimum	pH range	NaCl optimum (M)	NaCl range (%)
Methanosarcina barkeri	39–44	45	25–50	7.0	6.5–7.5	<0.2	0.1–0.7
Methanosarcina acetivorans	41	35–40	15–48	6.5–7.0	5.4–8.5	0.2	0.1–1.0
Methanosarcina mazeii	42	40–42	25–45	6.8–7.2	5.8–8.0	0.2–0.4	0.1–0.7
Methanosarcina siciliae	41–43	40	15–42	6.5–6.8	5.0–7.8	0.4–0.6	0.2–0.6
Methanosarcina thermophila	42	50	<35–55	6.0	5.5–8.0	0.6	0.0–1.2
Methanosarcina vacuolata	36	40	20–45	7.5	6.0–8.0	0.1	0.1–0.5
Methanosarcina baltica	nd	25	4–27	6.5–7.5	4–8.5	0.3–0.4	nd
Methanosarcina lacustris	43.4	25	1–35	7.0	4.5–8.5	nd	nd
Methanosarcina semesiae	nd	30–35	18–39	6.5–7.5	6.2–8.3	0.2–0.6	0–1.4
Methanolobus tindarius	40	25	10–45	6.5	5.5–8.0	0.49	0.06–1.27
Methanolobus bombayensis	39	37	20–42	7.2	6.2–8.2	0.5	0.2–2.2
Methanolobus oregonensis	40.9	35	25–42	8.6	8.2–9.2	0.48	0.1–1.6
Methanolobus taylorii	40.8	37	5–42	8	5.5–9.2	0.5	0.1–1.5
Methanolobus vulcani	39	40	13–45	7.0	5.8–7.8	0.5	0.1–1.4
Methanococcoides methylutens	42	30–35	15–35	7.0–7.5	6.0–8.0	0.24–0.64	0.1–1.0
Methanococcoides burtonii	39.6	23.4	1.7–29.5	7.7	6.8–8.2	0.2	0.2–0.5
Methanohalobium evestigatum	37	50	25–60	7.0–7.5	6.0–8.3	4.3	1.7–5.1
Methanohalophilus mahii	48.5	35	10–45	7.5	6.8–8.2	2.0	0.4–3.5
Methanohalophilus halophilus	39	26–36	18–42	6.5–7.4	6.3–7.4	1.2	0.7–2.6
Methanohalophilus portucalensis	41	nd	nd	nd	nd	nd	nd
Methanosalsum zhilinae	38	45	20–50	9.2	8.0–10	0.7	0.2–2.1
Methanosaeta concilii	49 ± 1.25	35–40	>10–≤45	7.1–7.4	≤6.6–>7.8	nd	nd
Methanosaeta thermophila	52.7–54.2	55–60	>30–≤70	6.5–6.7	>5.5–≤8.4	nd	nd

Abbreviation: nd, not determined.

Table 5. Substrate specificity and growth requirements.

Species	H$_2$ + CO$_2$	Acetate	Formate	Methanol	Monomethylamine	Dimethylamine	Trimethylamine	Methyl sulfides	Chemoautotrophic	Compounds stimulating growth
Methanosarcina barkeri	+	+	−	+	+	+	+	nd	+	V
Methanosarcina acetivorans	−	+	−	+	+	+	+	nd	+	AA, YE, CA
Methanosarcina mazeii	+	+	−	+	+	+	+	nd	+	P, YE
Methanosarcina siciliae	−	−	−	+	+	+	+	+	+	YE
Methanosarcina thermophila	−	+	−	+	+	+	+	nd	−	PABA
Methanosarcina vacuolata	+	+	−	+	+	+	+	nd	+	P, YE
Methanosarcina baltica	−	+	−	+	+	+	+	−	nd	V, YE, P
Methanosarcina lacustris	+	−	−	+	+	+	+	−	−	YE
Methanosarcina semesiae	−	−	−	+	+	+	+	+	nd	V
Methanolobus tindarius	−	−	−	+	+	+	+	−	+	V
Methanolobus bombayensis	−	−	−	+	+	+	+	+	+	YE, P
Methanolobus oregonensis	−	−	−	+	+	+	+	+	−	V, P, YE
Methanolobus taylorii	−	−	−	+	+	+	+	+	−	B
Methanolobus vulcani	−	−	−	+	+	+	+	−	−	B, YE, P
Methanococcoides methylutens	−	−	−	+	+	+	+	nd	−	V, P, YE, RF
Methnococcoides burtonii	−	−	−	+	+	+	+	nd	+	YE, P
Methanohalobium evestigatum	−	−	−	−	+	+	+	nd	−	V, YE
Methanohalophilus mahii	−	−	−	+	+	+	+	nd	−	B, Th
Methanohalophilus halophilus	−	−	−	+	+	+	+	nd	+	nd
Methanohalophilus portucalensis	−	−	−	+	+	+	+	nd	−	B
Methanosalsum zhilinae	−	−	−	+	+	+	+	+	+	P, YE, RF
Methanosaeta concilii	−	+	−	−	−	−	−	−	nd	B, PABA, Th, SF
Methanosaeta thermophila	−	+	−	−	−	−	−	−	nd	nd
Methanosarcina vacuolata	−	+	−	−	−	−	−	−	nd	nd

Symbols and abbreviations: +, growth; −, no growth; V, vitamins; AA, amino acids; YE, yeast extract; P, peptones; PABA, *p*-aminobenzoate; B, biotin; Th, thiamine; RF, rumen fluid; SF, sludge fluid; CA, casamino acids; and nd, not determined.

The original type strain of *Methanosarcina barkeri*, isolated by Schnellen (1947), has been lost. In 1966, Bryant isolated strain MS (Bryant and Boone, 1987), which has been adopted as the neotype strain, and this strain is one of the most extensively studied methanogens. Additional information on *Methanosarcina barkerii* can be found in Scherer and Bochem (1983) and Maestrojuán and Boone (1991).

Certain strains may dechlorinate chloroform. For example, *Methanosarcina* sp. strain DCM and *Methanosarcina mazeii* S-6 catalyze this reaction; however, $^{14}CO_2$ is formed rather than $^{14}CH_4$. This suggests that the mechanism is not reductive (Mikesell and Boyd, 1990), but other data suggest that dechlorination is reductive (Holliger et al., 1992).

Methanolobus

The genus *Methanolobus* is represented by five species with coccoidal morphology, *Methanolobus tindarus*, *Methanolobus bombayensis*, *Methanolobus oregonensis* (Fig. 5), *Methanolobus taylorii* and *Methanolobus vulcani*. All cells are surrounded by a unit membrane and a protein S-layer. Cells may grow on methanol, methylamines, and sometimes on methyl sulfides; no strains can grow on H$_2$/CO$_2$, formate, acetate or alcohols (other than methanol). For further descriptive information about each strain, see Tables 2–5.

Methanococcoides

The genus *Methanococcoides* comprises two species, *Methanococcoides methylutens* (Figs. 6 and 7) and *Methanococcoides burtonii*. Cells are irregular cocci, and the cell wall consists of a very thin protein monolayer approximately 10-nm thick. Cells can dismutate methylamines and methanol for growth but cannot catabolize acetate, dimethylsulfide, H$_2$/CO$_2$ or formate. Refer to Tables 2–5 for further descriptive information.

Methanohalobium

The genus *Methanohalobium* contains a single species, *Methanohalobium evestigatum*. This

Fig. 5. A) Electron micrograph of *Methanohalophilus oregonense* cells. B) Single cell of *Methanohalophilus oregonense*. Courtesy of Henry Aldrich, Department of Microbiology and Cell Science, University of Florida, Gainesville, FL.

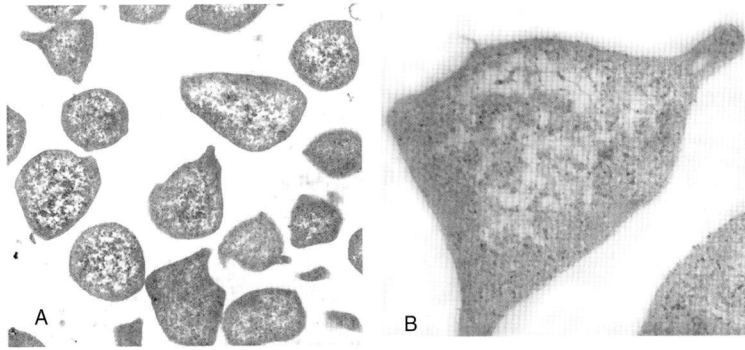

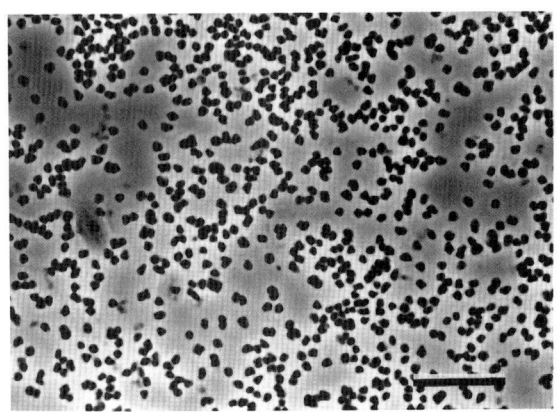

Fig. 6. Phase-contrast photomicrograph of *Methanococcoides methylutens*. Bar = 10 µm. Courtesy of K. Sowers, Center for Marine Biotechnology, University of Maryland Biotechnology Institute, Baltimore, MD.

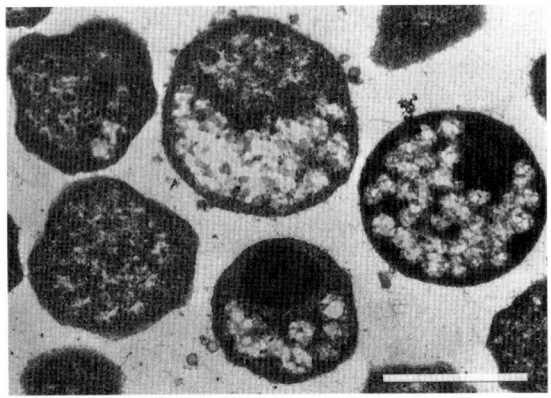

Fig. 7. Thin-section transmission electron photomicrograph of *Methanococcoides methylutens* stained with uranyl acetate. Bar = 1 µm. Courtesy of K. Sowers, Center for Marine Biotechnology, University of Maryland Biotechnology Institute, Baltimore, MD.

species is extremely halophilic and moderately thermophilic. Substrates for growth include methylamines but not acetate, formate, or H_2/CO_2. Methanol may be catabolized at very low concentrations (<5 mM) but not at concentrations of 20 mM or greater (Zhilina and Zavarzin, 1987a; Boone et al., 1993). The type strain of this species requires B-vitamins or yeast extract for growth. Tables 2–5 provide more descriptive information.

Methanohalophilus

This genus comprises three species, *Methanohalophilus mahii*, *Methanohalophilus halophilus* and *Methanohalophilus portucalensis*. Cells multiply by constriction of cell into two daughter cells rather than by fission. Cells are nonmotile. Cells grow on methylamines and methanol; however, methanol is toxic at concentrations of 40 mM or above. Cells do not grow on secondary alcohols, acetate, formate or H_2/CO_2. The addition of yeast extract and trypticase peptones to culture media at a concentration of 0.05% inhibits growth. Tables 2–5 describe further phe-

notypic and physiological characteristics of these organisms.

Methanosalsum

Methanosalsum zhilinae is the sole species in this genus. The cell wall is an S-layer. Energy and methane production are possible when this strain is grown on methylamines, methanol or dimethylsulfides but not on acetate, formate or H_2/CO_2 (Boone et al., 1986; Mathrani et al., 1988; Kadam and Boone, 1996; Kevbrin et al., 1997). Dimethylsulfide at 20 mM or greater concentration inhibits methane production (Mathrani et al., 1988). Refer to Tables 2–5 for more information about this organism.

The Family Methanosaetaceae

The family Methanosaetaceae includes one genus, *Methanosaeta*. Lipids are composed of *myo*-inositol and ethanolamine, with galactose in the polar head groups. All cells utilize acetate as the sole energy source.

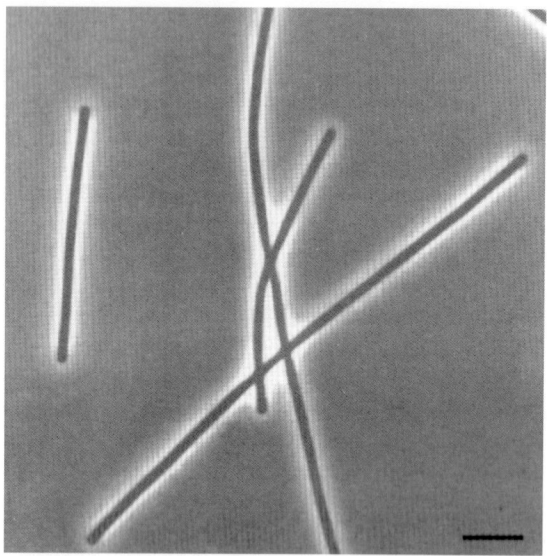

Fig. 8. *Methanosaeta concilli* culture observed under phase contrast microscopy. Bar = 5 μm. Courtesy of Girish Patel, National Research Council of Canada, Institute for Biological Sciences, Ottawa, Canada.

The genus *Methanosaeta* is represented by two species, *Methanosaeta concilii* (Fig. 8) and *Methanosaeta thermophila*. Both organisms split acetate into methane and CO_2, and acetate is the sole energy substrate. Tables 2–5 further describe phenotypic and physiological characteristics of these organisms.

Originally, aceticlastic, rod-shaped methanogens were described within the genus *Methanothrix*, with *Methanothrix soehngenii* strain Opfikon as the type species and strain, respectively. However, the first description of this genus and its type species was based on a culture that probably was not pure, thus the genus *Methanothrix* was later considered illegitimate. A pure culture of aceticlastic, mesophilic, rod-shaped methanogens was described by Patel (Patel, 1984; Patel, 1985) as *Methanothrix concilii*, with strain GP6 as the type strain for this species. Patel later transferred the species *Methanothrix concilii* to a new genus, *Methanosaeta*, with *Methanosaeta concilii* as the type species and strain GP6 as the type strain for this species. Comparisons between strain GP6 and strain Opfikon revealed that these two organisms most likely were subjective synonyms. Responding to a Request for Opinion (Boone, 1991), the Judicial Commission of the International Committee on Systematic Bacteriology denied the request to assign strain GP6 as a neotype for *Methanothrix soehngenii* (Wayne, 1994), which reaffirmed strain GP6 as the type strain of *Methanosaeta concilii*. *Methanothrix soehngenii* is now widely considered to be illegitimate because its description was not based on a pure culture.

Additionally, the first descriptions of thermophilic rod-shaped aceticlastic methanogens, *Methanothrix thermophila*, most likely were based on impure cultures (Nozhevnikova and Chudina, 1985; Zinder et al., 1987). It has recently been proposed that *Methanothrix thermophila* be transferred to the genus *Methanosaeta* as *Methanosaeta thermophila* comb. nov. Boone and Kamagata (1998) have submitted a Request for Opinion to reject the species *Methanothrix* as *nomina confusa* and to place these two names on the list of rejected names. On the basis of this proposal, thermophilic strains originally validly described as *Methanothrix thermophila* (Kamagata et al., 1992) are presented in this chapter as *Methanosaeta thermophila*. Only the descriptions of aceticlastic, rod-shaped, sheathed methanogens that have been based on pure cultures have been included in this chapter. Thus, to date, *Methansaeta* includes two species, *Methanosaeta conciili* and *Methanosaeta thermophila*.

Literature Cited

Barker, H. A. 1936. Studies upon the methane-producing bacteria. Arch. Microbiol. 7:420–438.

Boone, D. R., S. Worakit, I. M. Mathrani, and R. A. Mah. 1986. Alkaliphilic methanogens from high-pH lake sediments. Syst. Appl. Microbiol. 7:230–234.

Boone, D. R., R. L. Johnson, and Y. Liu. 1989. Diffusion of the interspecies electron carriers H_2 and formate in methanogenic ecosystems and its implications in the measurement of Km for H_2 or formate uptake. Appl. Environ. Microbiol. 55:1735–1741.

Boone, D. R. 1991. Strain GP6 is proposed as the neotype strain of Methanothrix soehngenii[VP] pro synon. Methanothrix concilii[VP] and Methanosaeta concilii[VP]: Request for an opinion. Int. J. Syst. Bacteriol. 41:588–589.

Boone, D. R., I. M. Mathrani, Y. Liu, J. A. G. F. Menaia, R. A. Mah, and J. E. Boone. 1993. Isolation and characterization of Methanohalophilus portucalensis sp. nov. and DNA reassociation study of the genus Methanohalophilus. Int. J. Syst. Bacteriol. 43:430–437.

Boone, D. R. 1995. Short- and long-term maintenance of methanogenic stock cultures. In: K. R. Sowers and H. J. Schreier (Eds.) Archaea: A Laboratory Manual—Methanogens. Cold Spring Harbor Laboratory Press. Plainview, NY. 79–83.

Boone, D. R., and Y. Kamagata. 1998. Rejection of the species Methanothrix soehngenii[VP] as nomina confusa, and transfer of Methanothrix thermophila[VP] to the genus Methanosaeta[VP] as Methanosaeta thermophila sp. nov.: Request for an opinion. Int. J. Syst. Bacteriol. 48:1079–1080.

Boone, D. R., and C. C. Baker. 2002. Validation of publication of new names and new combinations previously effectively published outside the IJSEM. List No. 85. Int. J. Syst. Evol. Microbiol. 52:685–690.

Bryant, M. P., and D. R. Boone. 1987. Emended description of strain MS[T] (DSM800T), the type strain of Methanosarcina barkeri. Int. J. Syst. Bacteriol. 37:169–170.

Charakhchyan, I. A., L. L. Mitushina, V. G. Kuznetsova, and S. S. Belyaev. 1989. The morphology and ultrastructure of Methanococcoides euhalobius and the role of calcium in its morphogenesis. Microbiology 58:667–672.

Davidova, I. A., H. J. M. Harmsen, A. J. M. Stams, S. S. Belyaev, and A. J. B. Zehnder. 1997. Taxonomic description of Methanococcoides euhalobius and its transfer to Methanohalophilus genus. Ant. v. Leeuwenhoek 71:313–318.

Elberson, M. A., and K. R. Sowers. 1997. Isolation of an aceticlastic strain of Methanosarcina siciliae from marine canyon sediments and emendation of the species description for Methanosarcina siciliae. Int. J. Syst. Bacteriol. 47:1258–1261.

Franzmann, P. D., N. Springer, W. Ludwig, E. Conway de Macario, and M. Rohde. 1992. A methanogenic archaeon from Ace Lake, Antarctica: Methanococcoides burtonii, sp. nov. Syst. Appl. Microbiol. 15:573–581.

Franzmann, P. D., N. Springer, W. Ludwig, E. Conway de Macario, and M. Rohde. 1993. Validation of the publication of new names and new combinations previously effectively published outside the IJSB. List No. 45. Int. J. Syst. Bacteriol. 43:398–399.

Hippe, H. 1984. Maintenance of methanogenic bacteria. In: B. E. Kirsop and J. J. S. Snell (Eds.) Maintenance of Microorganisms: A Manual of Laboratory Methods. Academic Press. London, UK. 69–81.

Holliger, C., G. Schraa, E. Stupperich, A. J. M. Stams, and A. J. B. Zehnder. 1992. Evidence for the involvement of corrinoids and factor F_{430} in the reductive dechlorination of 1,2-dichloroethane by Methanosarcina barkeri. J. Bacteriol. 174:4427–4434.

Hungate, R. E. 1969. A roll tube method for cultivation of strict anaerobes. In: J. B. Norris and D. W. Ribbons (Eds.) Methods in Microbiology. Academic Press. New York, NY. 117–132.

Janssen, P. H. 2003. Selective enrichment and purification of cultures of Methanosaeta spp. J. Microbiol. Meth. 52:239–244.

Jarvis, G. N., C. Strömpl, D. M. Burgess, L. C. Skillman, E. R. B. Moore, and K. N. Joblin. 2000. Isolation and identification of ruminal methanogens from grazing cattle. Curr. Microbiol. 40:327–332.

Kadam, P. C., D. R. Ranade, L. Mandelco, and D. R. Boone. 1994. Isolation and characterization of Methanolobus bombayensis sp. nov., a methylotrophic methanogen that requires high concentratons of divalent cations. Int. J. Syst. Bacteriol. 44:603–607.

Kadam, P. C., and D. R. Boone. 1995. Physiological characterization and emended description of Methanolobus vulcani. Int. J. Syst. Bacteriol. 45:400–402.

Kadam, P. C., and D. R. Boone. 1996. Influence of pH on ammonia accumulation and toxicity in halophilic, methylotrophic methanogens. Appl. Environ. Microbiol. 62:4486–4492.

Kamagata, Y., H. Kawasaki, H. Oyaizu, K. Nakamura, E. Mikami, G. Endo, Y. Koga, and K. Uamasato. 1992. Characterization of three thermophilic strains of Methanothrix ("Methanosaeta") thermophila sp. nov., and rejection of Methanothrix ("Methanosaeta") thermoacetophila. Int. J. Syst. Bacteriol. 42:463–468.

Kevbrin, V. V., A. M. Lysenko, and T. N. Zhilina. 1997. Physiology of the alkaliphilic methanogen Z-7936, a new strain of Methanosalsus zhilinaeae isolated from Lake Magadi. Microbiology 66:261–266.

König, H., and K. O. and K. O. Stetter. 1982. Isolation and Characterization of Methanolobus tindarius, sp. nov., a coccoid methanogen growing only on methanol and methylamines. Zbl. Bakteriol. Microbiol. Hyg. 1 Abt. Orig. C 3:478–490.

König, H., and K. O. Stetter. 1983. Validation of the publication of new names and new combinations previously effectively published outside the IJSB. List No. 10. Int. J. Syst. Bacteriol. 33:438–440.

Liu, Y., D. R. Boone, and C. Choy. 1990. Methanohalophilus oregonensis sp. nov., a methylotrophic methanogen from an alkaline, saline aquifer. Int. J. Syst. Bacteriol. 40:111–116.

Lyimo, T. J., A. Pol, H. J. M. OpdenCamp, H. R. Harhangi, and G. D. Vogels. 2000. Methanosarcina semesiae sp. nov., a dimethylsulfide-utilizing methanogen from mangrove sediment. Int. J. Syst. Evol. Microbiol. 50:171–178.

Maestrojuán, G. M., and D. R. Boone. 1991. Characterization of Methanosarcina barkeri MS^T and 227, Methanosarcina mazei $S-6^T$, and Methanosarcinal vacuolata $Z-761^T$. Int. J. Syst. Bacteriol. 41:267–264.

Mah, R. A., and D. A. Kuhn. 1984. Transfer of the type species of the genus Methanococcus to the genus Methanosarcina, naming it Methanosarcina mazei (Barker 1936) comb. nov. et emend. and conservation of the genus Methanococcus (Approved Lists 1980) with Methanococcus vannielii (Approved Lists 1980) as the type species: Request for an opinion. Int. J. Syst. Bacteriol. 34:263–265.

Mathrani, I. M., D. R. Boone, R. A. Mah, G. E. Fox, and P. P. Lau. 1988. Methanohalophilus zhilinae, sp. nov., an alkaliphilic, halophilic, methylotrophic methanogen. Int. J. Syst. Bacteriol. 38:139–142.

Mikesell, M. D., and S. A. Boyd. 1990. Dechlorination of chloroform by Methanosarcina strains. Appl. Environ. Microbiol. 56:1198–1201.

Ni, S. and D. R. Boone. 1991. Isolation and characterization of a dimethyl sulfide-degrading methanogen, Methanolobus siciliae HI350, from an oil well, characterization of M. siciliae $T4/M^T$, and emendation of M. siciliae. Int. J. Syst. Bacteriol. 41:410–416.

Ni, S., C. R. Woese, H. C. Aldrich, and D. R. Boone. 1994. Transfer of Methanolobus siciliae to the genus Methanosarcina, naming it Methanosarcina siciliae, and emendation of the genus Methanosarcina. Int. J. Syst. Bacteriol. 44:357–359.

Nozhevnikova, A. N., and V. I. Chudina. 1985. Morphology of the thermophilic acetate methane bacterium Methanothrix thermoacetophila sp. nov. Microbiology 53:618–624.

Obraztsova, A. Y., O. V. Shipin, L. B. Bezrukova, and S. S. Belyaev. 1987. Properties of the coccoid methylotrophic methanogen, Methanococcoides euhalobius sp. nov. Microbiology 56:523–527.

Ohtsubo, S., H. Miyahara, K. Demizu, S. Kohno, and I. Miura. 1991. Isolation and characterization of new Methanothrix strains. Int. J. Syst. Bacteriol. 41:358–362.

Oremland, R. S., R. P. Kiene, I. Mathrani, and M. J. Whiticar. 1989. Description of an estuarine methylotrophic methanogen which grows on dimethyl sulfide. Appl. Environ. Microbiol. 55:994–1002.

Oremland, R. S., and D. R. Boone. 1994. Methanolobus taylorii sp. nov., a new methylotrophic, estuarine methanogen. Int. J. Syst. Bacteriol. 44:573–575.

Patel, G. B. 1984. Characterization and nutritional properties of Methanothrix concilii sp. nov., a mesophilic, aceticlastic methanogen. Can. J. Microbiol. 30:1383–1396.

Patel, G. B. 1985. Validation of the publication of new names and new combinations previously effectively published outside the IJSB. List No. 17. Int. J. Syst. Bacteriol. 35:223–225.

Patel, G. B., and G. D. Sprott. 1990. Methanosaeta concilii gen. nov., sp. nov. ("Methanothrix concilii") and Methanosaeta thermoacetophila nom. rev., comb. nov. Int. J. Syst. Bacteriol. 40:79–82.

Paterek, J. R., and P. H. Smith. 1985. Isolation and characterization of a halophilic methanogen from Great Salt Lake. Appl. Environ. Microbiol. 50:877–881.

Paterek, J. R., and P. H. Smith. 1988. Methanohalophilus mahii gen. nov., sp. nov., a methylotrophic halophilic methanogen. Int. J. Syst. Bacteriol. 38:122–123.

Robichauz, M., M. Howell, and R. Boopathy. 2003. Methanogenic activity in human periodontal pocket. Curr. Microbiol. 46:53–58.

Scherer, P. A., and H.-P. Bochem. 1983. Ultrastructural investigation of 12 Methanosarcinae and related species grown on methanol for occurrence of polyphosphatelike inclusions. Can. J. Microbiol. 29:1190–1199.

Schnellen, C. G. T. P. 1947. Onderzoekingen over de methaangisting [PhD thesis]. University of Delft. Delft, The Netherlands.

Simankova, M. V., S. N. Parshina, T. P. Tourova, T. V. Kolganova, A. J. B. Zehnder, and A. N. Nozhevnikova. 2001. Methanosarcina lacustris sp. nov., a new psychrotolerant methanogenic archaeon from anoxic lake sediements. Syst. Appl. Microbiol. 24:362–367.

Simankova, M. V., S. N. Parshina, T. P. Tourova, T. V. Kolganova, A. J. B. Zehnder, and A. N. Nozhevnikova. 2002. Validation of publication of new names and new combinations previously effectively published outside the IJSEM. List No. 85. Int. J. Syst. Evol. Microbiol. 52:685–690.

Sowers, K. R., and J. G. Ferry. 1983. Isolation and characterization of a methylotrophic marine methanogen, Methanococcoides methylutens, gen. nov., sp. nov. Appl. Environ. Microbiol. 45:684–690.

Sowers, K. R., S. F. Baron, and J. G. Ferry. 1984. Methanosarcina acetivorans sp. nov., an acetotrophic methane-producing bacterium isolated from marine sediments. Appl. Environ. Microbiol. 47:971–978.

Sowers, K. R., and J. G. Ferry. 1985a. Trace metal and vitamin requirements of Methanococcoides methylutens grown with trimethylamine. Arch. Microbiol. 142:148–151.

Sowers, K. R., and J. G. Ferry. 1985b. Validation of the publication of new names and new combinations previously effectively published outside the IJSB. List No. 17. Int. J. Bacteriol. 35:223–225.

Sowers, K. R., S. F. Baron, and J. G. Ferry. 1986. Validation of the publication of new names and new combinations previously effectively published outside the IJSB. List No. 20. Int. J. Syst. Bacteriol. 36:354–356.

Sowers, K. R., and K. M. Noll. 1995. Techniques for anaerobic growth. In: F. T. Robb, K. R. Sowers, H. J. Schreier, S. DasSarma and E. M. Fleischmann (Eds.) Archaea: A Laboratory Manual. Cold Spring Harbor Laboratory Press. Plainview, NY. 15–47.

Stetter, K. O. 1989. Genus II: Methanolobus. In: J. T. Staley, M. P. Bryant, N. Pfennig, and J. G. Holt (Eds.) Bergey's Manual of Systematic Bacteriology. Williams and Wilkins. Baltimore, MD. 2205–2207.

Takai, K., D. P. Moser, M. DeFlaun, T. C. Onstott, and J. K. Fredrickson. 2001. Archaeal diversity in waters from waters from deep South African gold mines. Appl. Environ. Microbiol. 67:5750–5760.

Tumbula, D. L., J. Keswani, J. Shieh, and W. B. Whitman. 1995. Long-term maintenance of methanogenic stock cultures in glycerol. In: F. T. Robb, K. R. Sowers, H. J. Schreier, S. DasSarma and E. M. Fleischmann (Eds.) Archaea: A Laboratory Manual. Cold Spring Harbor Laboratory Press. Plainview, NY. 85–87.

von Klein, D., H. Arab, H. Volker, and M. Thomm. 2002a. Methanosarcina baltica, sp. nov., a novel methanogen isolated from the Gotland Deep of the Baltic Sea. Extremophiles 6:103–110.

von Klein, D., H. Arab, H. Volker, and M. Thomm. 2002b. Validation of publication of new names and new combinations previously effectively published outside the IJSEM. List No. 85. Int. J. Syst. Evol. Microbiol. 52:685–690.

Wayne, L. G. 1994. Actions of the Judicial Committee on Systematic Bacteriology on Requests for Opinions published between January 1985 and July 1993. Int. J. Syst. Bacteriol. 44:177–178.

Wilharm, T., T. N. Zhilina, and P. Hummel. 1991. DNA-DNA hybridization of methylotrophic halophilic methanogenic bacteria and transfer of Methanococcus halophilus[VP] to the genus Methanohalophilus as Methanohalophilus halophilus. Int. J. Syst. Bacteriol. 41:558–562.

Zhilina, T. N., and G. A. Zavarzin. 1979. Comparative cytology of Methanosarcinae and description of Methanosarcina vacuolata sp. nov. Microbiology 48:279–285.

Zhilina, T. N. 1983. A new obligate halophilic methane-producing bacterium [in Russian]. Mikrobiologiya 52:375–382.

Zhilina, T. N., and V. V. Kevbrin. 1985. Culturing a halophilic methane-forming coccus on monomethylamine. Microbiolgy 54:93–99.

Zhilina, T. N., and G. A. Zavarzin. 1987a. Methanohalobium evestigatus, gen. nov. sp. nov., the extremely halophilic methanogenic archaebacterium. Dokl. Akad. Nauk. SSSR 293:464–468.

Zhilina, T. N., and G. A. Zavarzin. 1987b. Methanosarcina vacuolata sp. nov., a vacuolated methanosarcina. Int. J. Syst. Bacteriol. 37:281–283.

Zhilina, T. N., and G. A. Zavarzin. 1988. Validation of the publication of new names and new combinations previously effectively published outside the IJSB. List No. 24. Int. J. Syst. Bacteriol. 38:136–137.

Zhu, X. Y., J. Lubeck, and J. J. Kilbanell. 2003. Characterization of microbial communities in gas industry pipelines. Appl. Environ. Microbiol. 69:5354–5363.

Zinder, S. H., and R. A. Mah. 1979. Isolation and characterization of a thermophilic strain of Methanosarcina unable to use H_2-CO_2 for methanogenesis. Appl. Environ. Microbiol. 38:996–1008.

Zinder, S. H., K. R. Sowers, and J. G. Ferry. 1985. Methanosarcina thermophila sp. nov., a thermophilic, acetotrophic, methane-producing bacterium. Int. J. Syst. Bacteriol. 35:522–523.

Zinder, S. H., T. Anguish, and A. L. Lobo. 1987. Isolation and characterization of a thermophilic acetotrophic strain of Methanothrix. Arch. Microbiol. 146:315–322.

Prokaryotes (2006) 3:257–273
DOI: 10.1007/0-387-30743-5_13

CHAPTER 13

Methanococcales

WILLIAM B. WHITMAN AND CHRISTIAN JEANTHON

Phylogeny

The sequencing of representative genes of the methane-producing archaea in the order Methanococcales suggests that this lineage is ancient and possesses a high degree of genetic diversity. For example, the mesophile *Methanococcus maripaludis* and hyperthermophile *Methanocaldococcus jannaschii* represent the range of diversity within this group. The sequence similarity of their 16S rRNA genes is 88%. Likewise, homologous open reading frames in these two organisms typically possess 60–80% amino acid sequence identity (W. B. Whitman, unpublished observation). For comparison, genes within *Escherichia* and *Yersinia* possess a comparable level of similarity.

In spite of this apparent genetic diversity, the phenotypes of members of the Methanococcales are similar. They have all been isolated from marine habitats and require sea salts for optimal growth. They are all obligately anaerobic methane-producers and use carbon dioxide as the electron acceptor. Hydrogen and sometimes formate are electron donors. Acetate, C-1 compounds such as methanol and methylamines, and alcohols such as isopropanol and ethanol are not utilized as substrates for methanogenesis. They are all irregular cocci with a diameter of 1–3 μm. The cell wall is composed of a protein S-layer, and glycoproteins and cell wall carbohydrates have not been detected. Cells are usually motile by means of flagellar tufts or bundles. However, there is tremendous diversity in the temperature range for growth, which varies from mesophilic to hyperthermophilic. Within each of these temperature ranges, these organisms are among the fastest growing methanogens known, with generation times of 2 hours at 37°C and less than 30 minutes at 85°C. Lastly, many of the Methanococcales require selenium for optimal growth.

Currently, the Methanococcales have been divided into two families and four genera on the basis in part of the temperature optima for growth (Table 1). For the most part this taxonomy reflects the apparent phylogeny of the 16S rRNA gene, but there are some points of ambiguity (Keswani et al., 1996). The two hyperthermophilic genera *Methanocaldococcus* and *Methanotorris* are placed in the family Methanocaldococcaceae, and this grouping appears robust (Fig. 1). It has strong bootstrap support, and it is found by more than one algorithm including the Fitch-Margoliash, neighbor-joining, and maximum likelihood analyses. However, the differences in mol% G+C content of the rRNAs of the hyperthermophiles and mesophiles within this group can bias these computations (Burggraf et al., 1990). When the phylogenetic analyses are performed with only transversions, which ameliorates this bias, the mesophiles appear as a sister group to the hyperthermophile *Methanotorris*, and the Methanocaldococcaceae are no longer a phylogenetic group. To resolve this question, the phylogeny of additional genes should be examined.

Similarly, within the Methanococcaceae, the mesophiles are assigned to the genus *Methanococcus* and the thermophiles to the genus *Methanothermococcus* (Whitman et al., 2001). Within the *Methanococcus*, the deepest phylogenetic group is represented by the patent strain "*Mc. aeolicus*" (Fig. 1). However, upon phylogenetic analyses of the rRNA gene, this strain forms a clade with the thermophile *Methanothermococcus okinawensis* (Takai et al., 2002; Fig. 1). This clade is also supported by transversion analysis. Even though bootstrap support is modest, the common alternative topologies found during the bootstrap analysis do not group the mesophiles to the exclusion of the thermophiles. Although phylogenetic analyses of other molecules are needed to confirm this result, it implies that the genera *Methanococcus* and *Methanothermococcus* are so deeply branched that they include distantly related and phenotypically dissimilar organisms. A taxonomic treatment more consistent with the phylogeny would probably place "*Mc. aeolicus*" and *Mtc. okinawensis* into two novel genera.

Table 1. Taxa in the order Methanococcales[a]

Family	Genus	Species
Methanococcaceae	*Methanococcus*[T]	"aeolicus"
		maripaludis
		voltae
		vannielii[T]
	Methanothermococcus	okinawensis
		thermolithotrophicus[T]
Methanocaldococcaceae	*Methanocaldococcus*[T]	fervens
		infernus
		jannaschii[T]
		vulcanius
	Methanotorris	igneus[T]

[a]Nomenclatural types are indicated with a superscript "T." Species that have not been validated are in quotes.

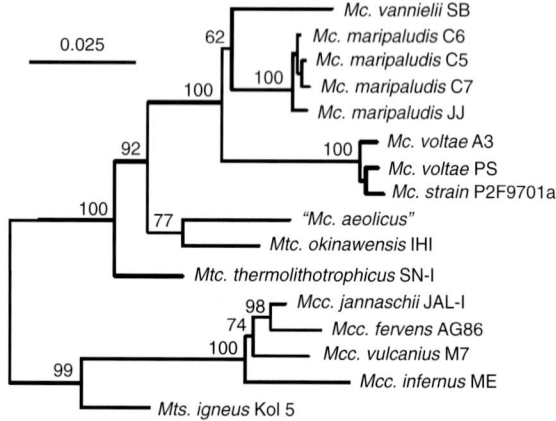

Fig. 1. Phylogeny of the 16S rRNA gene of the Methanococcales. The alignment was manually edited to include 1227 positions of unambiguous alignment. Evolutionary distances were calculated by the Kimura two-parameter model, and the tree was constructed by the Fitch-Margoliash algorithm in PHYLIP (phylogenetic analysis package). Bootstrap analysis was performed with 100 replicates, and values greater than 50 are reported on the nodes. Mc. = *Methanococcus*, Mtc. = *Methanothermococcus*, Mts. = *Methanotorris*, and Mcc. = *Methanocaldococcus*. The scale bar represents evolutionary distance.

Taxonomy

An overview of the current taxonomy of the Methanococcales is given in Table 1. This taxonomy groups organisms with similar temperature optima. All the mesophiles and thermophiles are found within the family Methanococcaceae. The mesophiles are assigned to the genus *Methanococcus* and the thermophiles to the genus *Methanothermococcus*. For the hyperthermophiles, two genera, *Methanocaldococcus* and *Methanotorris* are recognized within the family Methanocaldococcaceae. Although this taxonomy is in general agreement with the phylogeny of the rRNA, some ambiguity remains. To resolve these issues, additional strains and the phylogenies of

other molecules will probably be required (see above).

Historically, the genus *Methanococcus* included methane-producing cocci that did not form regular packets (i.e., *Methanosarcina*) or chains (i.e., some species of *Methanobacterium*, now *Methanobrevibacter*; Bryant, 1974). At that time, the type species, *Methanococcus mazei*, was not available in pure culture, and only one other species, *M. vannielii*, was known. Upon analysis of the partial sequence of the 16S rRNAs of methanogens (Balch et al., 1979) and isolation of an archaeon with the phenotype of *M. mazei* (Mah, 1980), it became apparent that *M. mazei* was related to the Methanosarcinaceae and that *M. vannielii* was related to a new species, *Methanococcus voltae*. Thus, it was proposed that *M. vannielii* become the new type species for the genus and *M. mazei* be reclassified as *Methanosarcina mazei* (Balch et al., 1979; Mah and Kuhn, 1984; Judicial Commission, 1986). Likewise, species more closely resembling *M. mazei* were placed in the Methanosarcinaceae. Thus, *Methanococcus halophilus*, which utilizes methylamines for methane synthesis, was not classified with the Methanococcaceae (Zhilina, 1983). Similarly, *Methanococcus frisius* resembles *Methanosarcina mazei* by nutritional and morphological criteria (Blotevogel et al., 1986) and was also classified with the Methanosarcinaceae.

Although the remaining methanococcal species were more closely related to each other than other methane-producing bacteria, they were not a closely knit group (Whitman, 1989). In particular, the sequence of the 16S rRNA of *M. jannaschii* was different enough from the other methanococci to justify creation of a new genus (Jones et al., 1983a). The eventual isolation of additional hyperthermophilic methanococci and the sequencing of the 16S rRNAs of most of the methanococci led to the recommendation that the genus be further subdivided into four genera to reduce the genetic diversity (Boone et al.,

1993). This proposal left the mesophilic species in the genus *Methanococcus* and placed the thermophiles and hyperthermophiles in novel genera. The 16S rRNA sequence similarities (>91%) and DNA hybridizations (>3%) among the remaining, mesophilic species were still somewhat lower than those found in many other bacterial and archaeal genera, suggesting that this group was still rather diverse (Keswani et al., 1996). However, in the absence of additional species, there seemed to be little benefit in subdividing this genus at this time.

The more moderate thermophiles, with 16S rRNA sequence similarities of <91% to the mesophiles as well as the hyperthermophiles, were placed in the new genus *Methanothermococcus*. The hyperthermophiles, which possessed <86% rRNA sequence similarity to the mesophiles, were placed in two new genera, *Methanocaldococcus* and *Methanotorris* (originally called "*Methanoignis*," see below). Members of these new genera possessed low 16S rRNA sequence similarity (<93%) to each other as well as differences in nutritional and other phenotypic properties. Within the genus *Methanocaldococcus*, the 16S rRNA sequence similarity was greater than 95%. The genus name "*Methanotorris*" was proposed to replace the original name "*Methanoignis*," which was incorrectly formed (Whitman et al., 2001). The correct latinization was "*Methanignis*." Therefore, *Methanotorris* was proposed to preserve the prefix "*methano-*" in the genus name.

The present descriptions of some of the methanococci are based on the description of a single isolate. Therefore, the known phenotypic characteristics may not be truly representative of the species. For instance, the growth responses of strains of *Methanococcus maripaludis* to NaCl and MgCl$_2$ are variable, and care must be taken when making taxonomic distinctions on this basis (Whitman et al., 1986). More reliable and rapid comparison of strain collections can be made by sodium dodecyl sulfate-polyacrylamide gel electrophoresis (SDS-PAGE) of cellular proteins and restriction fragment length polymorphism (RFLP) analyses of the 16S rRNA (Keswani et al., 1996; Jeanthon et al., 1999a).

Habitat

To date, methanococci have only been isolated from marine environments. *Methanococcus vannielii* was isolated from the shore of the San Francisco Bay (Stadtman and Barker, 1951). The type strain of *M. voltae* was isolated from sediments from the mouth of the Waccasassa River estuary in Florida (Ward, 1970). The type strain of *M. maripaludis* was isolated from salt-marsh

sediments near Pawley's Island, South Carolina (Jones et al., 1983b). Additional strains of *M. voltae* and *M. maripaludis* have been isolated from salt-marsh sediments in Georgia and Florida and an estuary in Taiwan (Whitman et al., 1986; Keswani et al., 1996; Lai and Shih, 2001). An unnamed methanococcal isolate has also been obtained from the biofilm of a ship hull (Boopathy and Daniels, 1992).

Methanothermococci have been isolated from coastal geothermally heated marine sediments at Stufe di Nerone near Naples (Italy; Huber et al., 1982) and reservoir water from a North Sea oil field (Nilsen and Torsvik, 1996). A new *Methanothermococcus* species (*M. okinawaensis*) has been recently isolated from a deep-sea hydrothermal vent of the Okinawa Trough (Pacific Ocean; Takai et al., 2002). Environmental clone sequences closely related to this genus have also been obtained from continental high-temperature oil reservoirs (Orphan et al., 2000).

Methanocaldococci are widespread in deep submarine hydrothermal systems. *Methanocaldococcus* species have so far been isolated from chimney material of deep-sea hydrothermal vents at the East Pacific Rise (13°N and 21°N; Jones et al., 1983a; Jeanthon et al., 1999a) and Mid-Atlantic Ridge (14°45'N and 23°N; Jeanthon et al., 1998; Jeanthon et al., 1999a) and hydrothermally heated sediment from Guaymas Basin (Gulf of California; Zhao et al., 1988; Jones et al., 1989; Jeanthon et al., 1999a). Environmental clone sequences closely related to this genus have also been retrieved from hot formation waters of a continental oil reservoir in western Siberia (Nercessian et al., 2000).

Methanotorris was originally isolated from sediments and venting water of a shallow submarine hydrothermal vent at Kolbeinsey Ridge located on the Mid-Atlantic Ridge (north of Iceland; Burggraf et al., 1990). Fifteen additional strains have been isolated from sediments from a Guaymas Basin hydrothermal site (Gulf of California) and chimney samples from the Mid-Atlantic Ridge (23°N; Jeanthon et al., 1999a).

Isolation

The Methanococcales, like other methane-producing archaea, are strict anaerobes that require specialized techniques for their cultivation. This methodology has been reviewed recently by Sowers and Noll (1995) and will not be discussed here. The mesophilic methanococci may be easily isolated after enrichment under H$_2$ + CO$_2$ (80:20) in pressurized tubes or bottles (Miller and Wolin, 1974; Balch and Wolfe, 1976). Because of their rapid growth, methanococci frequently outgrow other H$_2$-utilizing methanogens in marine sediments. Therefore, this enrichment is somewhat specific, and enrichment cultures that take longer than 5 days to develop seldom contain methanococci. The enrichments are transferred to medium containing antibiotics (penicillin G [0.2 mg/ml], erythromycin [0.2 mg/ml] and streptomycin sulfate [0.2 mg/ml]) before plating on agar plates or roll tubes (Jones et al., 1983b; Jones et al., 1983c; Tumbula et al., 1995a). In some cases, it is necessary to include antibiotics in the solid medium to prevent growth of spreading bacteria over colonies of methanococci. Isolated colonies are picked with a syringe needle and transferred to liquid medium. Purity may be demonstrated by microscopic examination, restreaking on agar medium and absence of growth in mineral medium supplemented with 1% casamino acids and glucose under $N_2 + CO_2$ (80:20).

A useful medium for isolation and rapid growth of methanococci consists of:

Glass distilled water	500 ml
General salts solution	500 ml
K_2HPO_4 solution	10 ml
Trace mineral solution	10 ml
Iron stock solution	5 ml
Resazurin solution	1 ml
NaCl solution	75 ml
$NaHCO_3$	5 g

The concentrations of stock K_2HPO_4, NaCl, and resazurin solutions are 14 g/liter, 293 g/liter and 0.1 g/100 ml of glass distilled water, respectively. The iron stock solution is prepared by adding 0.2 g of $Fe(NH_4)_2(SO_4)_2 \cdot 6H_2O$ to a small bottle, adding 2 drops of concentrated HCl followed by 100 ml of glass distilled water. This solution is prepared fresh monthly. The oxygen indicator resazurin is optional. These components are combined, and the medium is brought to a boil under a stream of $N_2 + CO_2$ (80:20). After boiling, cysteine or 2-mercaptoethanesulfonate, 0.5 g/liter, is added to reduce the medium. When hot, the medium will form a precipitate that goes back into solution upon cooling. After dispensing the medium anaerobically into crimp seal tubes or serum bottles, the gas is exchanged for $H_2 + CO_2$ (80:20, 100 kPa), and the medium sterilized by autoclaving. The medium may then be stored for several months in an anaerobic chamber (Coy Laboratories, Ann Arbor, Michigan). Within one day of inoculation, one part of sterile 2.5% $Na_2S \cdot 9H_2O$ (w/v) is added to 50 parts of medium. After inoculation, the tubes are pressurized to 200 kPa with $H_2 + CO_2$. Tubes are repressurized periodically throughout growth. During the period of rapid growth immediately following the lag phase, cultures need to be repressurized 3–4 times per day. If the CO_2 in the headspace is allowed to become too low, the pH of the medium will rise. Under alkaline conditions, the cells lyse (Schauer and Whitman, 1989). For growth in vessels that will not maintain a pressure >100 kPa, the $NaHCO_3$ concentration is reduced to 2 g/liter.

For *M. voltae*, the mineral medium must be supplemented with either yeast extract, 2 g/liter, or sodium acetate $\cdot 3H_2O$, 0.14 g/liter, L-isoleucine, 0.5 g/liter, L-leucine, 0.5 g/liter, and pantoyllactone, 1.3 mg/liter (Whitman et al., 1986). Similarly, the addition of casamino acids (2 g/liter) and sodium acetate $\cdot 3$ H_2O (0.14 g/liter) is stimulatory for *M. maripaludis*.

General Salts Solution (modified from Romesser et al., 1979)

KCl	0.67 g
$MgSO_4 \cdot 7H_2O$	6.9 g
$MgCl_2 \cdot 6H_2O$	5.5 g
NH_4Cl	1.0 g
$CaCl_2 \cdot 2H_2O$	0.28 g
Glass distilled water	1000 ml

Trace Metal Solution (modified from Wolfe's minerals; Balch et al., 1979)

Nitrilotriacetic acid, neutralized with KOH	1.5 g
$MnSO_4 \cdot 2H_2O$	0.1 g
$Fe(NH_4)_2(SO_4)_2 \cdot 6H_2O$	0.2 g
$CoCl_2 \cdot 6H_2O$	0.1 g
$ZnSO_4 \cdot 7H_2O$	0.1 g
$CuSO_4 \cdot 5H_2O$	10 mg
$NiCl_2 \cdot 6H_2O$	25 mg
Na_2SeO_3	0.2 g
$Na_2MoO_4 \cdot 2H_2O$	0.1 g
$Na_2WO_4 \cdot 2H_2O$	0.1 g

Sodium Sulfide Solution

A pellet of NaOH is added to 100 ml of distilled water, and the water is brought to a boil under a stream of N_2 gas and allowed to cool under N_2 in the fume hood. Then a large crystal of about 2.5 g of $Na_2S \cdot 9H_2O$ is briefly washed in a 25-ml beaker of distilled water. The crystal is blotted dry on a paper towel, reweighed and added to the NaOH solution. Because sulfide is toxic, chemically impermeable gloves are worn, and this procedure is performed in the fume hood. The solution is stoppered, brought into the anaerobic chamber, and dispensed into tubes and stoppered. The tubes are pressurized to 100 kPa with N_2, autoclaved, and stored in the anaerobic chamber for up to one month.

Methanothermococci and methanocaldococci may be isolated after enrichment under H_2/CO_2 (80:20, 200 kPa) in 50- or 100-ml serum bottles (Jeanthon et al., 1998).

Methanothermococci-Methanocaldococci Culture Medium

Sea salts (Sigma)	30 g
NH_4Cl	1 g
KH_2PO_4	0.35 g
PIPES	3.46 g
$NaHCO_3$	1 g
Difco yeast extract	2 g
Cysteine $\cdot$ HCl	0.5 g
Na_2SeO_4	0.5 mg
$Na_2WO_4 \cdot 2H_2O$	2 mg
Vitamin B_{12}	0.05 mg
Trace element mixture (see below)	1 ml
Vitamin mixture (see below)	1 ml
Thiamine solution (see below)	1 ml
Growth-stimulating factors (see below)	1 ml
Rezasurin	1 mg
Distilled water	1000 ml

Trace Element Mixture (Widdel and Bak, 1992)

HCl (25% = 7.7 M)	12.5 ml
FeSO$_4$ · 7H$_2$O	2.1 g
H$_3$BO$_3$	30 mg
MnCl$_2$ · 4H$_2$O	100 mg
CoCl$_2$ · 6H$_2$O	190 mg
NiCl$_2$ · 6H$_2$O	24 mg
CuCl$_2$ · 2H$_2$O	2 mg
ZnSO$_4$ · 7H$_2$O	144 mg
Na$_2$MoO$_4$ · 2H$_2$O	36 mg
Distilled water	987 ml

The solution is autoclaved under a N$_2$ (100%) atmosphere in serum bottles closed with tightly fitting rubber stoppers and fixed with aluminum seals.

Vitamin Mixture (Widdel and Bak, 1992)

Sodium phosphate buffer (10 mM; pH 7.1)	100 ml
4-Aminobenzoic acid	4 mg
D(+)-Biotin	1 mg
Nicotinic acid	10 mg
Calcium D(+)-pantothenate	5 mg
Pyridoxine dihydrochloride	15 mg

The solution is filter-sterilized (pore size, 0.2 µm) and kept in the dark at 4°C under a N$_2$ atmosphere (100 kPa).

Thiamine Solution (Widdel and Bak, 1992)

Ten mg of thiamine chloride dihydrochloride is dissolved in 100 ml of 25 mM sodium phosphate buffer, pH 3.4. The solution is filter-sterilized and kept at 4°C under a N$_2$ atmosphere (100 kPa).

Growth-stimulating Factors Solution (Pfennig et al., 1981)

Isobutyric acid	0.5 g
Valeric acid	0.5 g
2-Methyl-butyric acid	0.5 g
3-Methyl-butyric acid	0.5 g
Caproic acid	0.2 g
Succinic acid	0.6 g
Distilled water	100 ml

The solution, adjusted to pH 9 with NaOH, is autoclaved under a N$_2$ (100%) atmosphere in serum bottles closed with tightly fitting rubber stoppers and fixed with aluminum seals.

The pH of the Methanothermococci-Methanocaldococci culture medium is adjusted to 6.5 using 1 M HCl before autoclaving, and the medium is reduced by adding sodium sulfide Na$_2$S · 9H$_2$O to a final concentration of 0.05% (w/v). Solid medium is prepared by the addition of 0.7% (w/v) Phytagel (Sigma Chemical, Co., St. Louis, MO) and reduced with a titanium (III) citrate solution (Zehnder and Wuhrman, 1976). For methanothermococci, agar (2%, w/v) can be used as gelling agent. Medium is dispensed into Petri dishes in an anaerobic glove box. Plates are incubated at 60°C (for methanothermococci) or at 75°C (for methanocaldococci) in anaerobic jars for 3–5 days under a H$_2$/CO$_2$ atmosphere (80:20; 250 kPa).

Identification

To identify the mesophilic *Methanococcus* spp., isolates must be first distinguished from a num-

ber of morphologically and nutritionally similar species of *Methanogenium* and *Methanoculleus* that are also present in marine environments (Romesser et al., 1979; Rivard and Smith, 1982; Ferguson and Mah, 1983; Rivard et al., 1983). Methanococci may be distinguished from these other species by their faster growth rate, requirement for higher concentrations of NaCl for optimal growth, lack of organic growth requirements (except *M. voltae*) and the lower mol% G+C of their DNA. Unlike some members of these other genera, the methanococci have not been found to use secondary alcohols or ethanol as electron donors for methanogenesis. However, identification of isolates based solely on morphological and growth characteristics is equivocal, and use of salt or mineral requirements has been particularly deceptive (see below). Thus, antigenic crossreactivity (Conway de Macario et al., 1981) and 16S rRNA sequencing (Keswani et al., 1996) are helpful for final identification of new isolates.

Strains of methanococci may also be rapidly screened by one-dimensional SDS-PAGE of cellular proteins (Whitman, 1989; Keswani et al., 1996). Cultures are grown to an absorbance of 1.0 cm^{-1} at 600 nm, and 5-ml cultures are harvested by centrifugation. The cells are resuspended in 0.1 ml of mineral medium or a salt solution prepared without reducing agents. This cell suspension may be stored at –20°C prior to electrophoresis. After thawing, the suspension is vortexed to form an even suspension, and 15 µl are added to 60 µl of sample buffer containing sodium dodecyl sulfate and 2-mercaptoethanol. A portion, 35 µl, of this mixture is subjected to electrophoresis on a 12% polyacrylamide gel. The protein profile on SDS-PAGE is sufficiently distinctive to distinguish species of mesophilic methanococci from each other or from other methanogenic bacteria like *Methanogenium* species (data not shown).

Methanothermococcus is distinguished from the mesophile *Methanococcus* and the hyperthermophiles *Methanocaldococcus* and *Methanotorris* by its moderate thermophily. It differs from most other moderately thermophilic methanogens (such as *Methanothermobacter*) by its coccoid morphology. So far, *Methanoculleus thermophilicum* is the only other moderately thermophilic coccus described (Rivard and Smith, 1982; Ferguson and Mah, 1983). *Methanoculleus thermophilicum* requires acetate for growth, grows in low concentrations of NaCl, has a mol% G+C of 56–60, and is easily distinguished by these properties. Antigenic crossreactivity (Conway de Macario et al., 1981; Bryniok and Trosch, 1989; Nilsen and Torsvik, 1996), 16S rRNA sequencing (Keswani et al., 1996), and SDS-PAGE of cellular proteins (Nilsen and

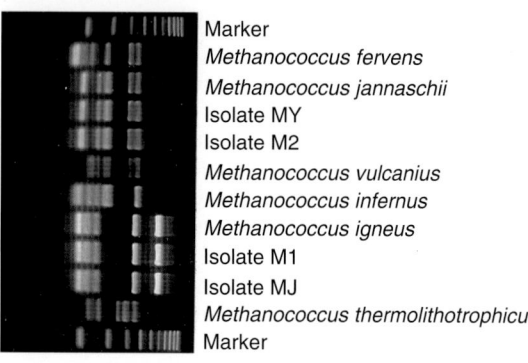

Marker
Methanococcus fervens
Methanococcus jannaschii
Isolate MY
Isolate M2
Methanococcus vulcanius
Methanococcus infernus
Methanococcus igneus
Isolate M1
Isolate MJ
Methanococcus thermolithotrophicus
Marker

Fig. 2. Restriction patterns of 16S rRNA genes of the type strains of hyperthermophilic methanococci and isolates from deep-sea hydrothermal vents selectively amplified and digested with BstUI. The marker is a 100-bp ladder. From Jeanthon et al. (1999a), and used with permission of the publisher.

Torsvik, 1996) are also helpful for final identification of new isolates.

Methanocaldococcus and Methanotorris are distinguished from the mesophile Methanococcus and the thermophile Methanothermococcus by their hyperthermophily. The hyperthermophiles may be distinguished from each other because Methanocaldococcus requires selenium for optimal growth and possesses flagellar tufts. Restriction fragment length polymorphism (RFLP) of the gene encoding the 16S rRNA has proven useful for distinguishing species of Methanocaldococcus from each other as well as from strains of Methanotorris and Methanothermococcus (Jeanthon et al., 1999a). Distinctive RFLP patterns are obtained after restriction of the amplified 16S rRNAs of the species of the three genera with HhaI, MspI and BstUI endonucleases (Fig. 2). The profiles obtained with HaeIII show distinctive patterns for all the type strains except for M. jannaschii and M. fervens. It is difficult to distinguish species of Methanocaldococcus on phenotypic properties alone. Of the four species currently described, only M. fervens is resistant to rifampicin, and only M. jannaschii is not stimulated by yeast extract.

A number of compounds have also proven to be useful chemotaxonomic markers. The core lipids of Methanococcus are composed of archaeol and hydroxyarchaeol (Koga et al., 1998). In Methanothermococcus, caldarchaeol is also present. In Methanocaldococcus and Methanotorris, hydroxyarchaeol is absent and a cyclic derivative of archaeol is found. In all four genera, the polar head groups are composed of glucose, N-acetylglucosamine, serine and ethanolamine (in some species). The most abundant polyamine is spermidine in Methanococcus and Methanothermococcus (Kneifel et al., 1986; Hamana et al., 1998). Spermine is the most

abundant polyamine in Methanocaldococcus. Unusual compatible solutes have not been detected in Methanococcus spp. In Methanothermococcus, β-glutamate and N-acetyl-β-lysine are abundant, depending on the growth condition (Robertson et al., 1990; Robertson et al., 1992). In Methanocaldococcus, β-glutamate is the most abundant compatible solute. Methanotorris contains di-myo-inositol-1,1'-phosphate in addition to β-glutamate (Ciulla et al., 1994).

Preservation

Because the methanococci lyse shortly after the cessation of growth in liquid media, stock cultures are grown below the temperature optimum (30°C) and stored at room temperature for up to 3 weeks. Strains of all the mesophilic methanococci have been stored with little loss in viability for up to 3 years in 25% glycerol at –70°C (Whitman et al., 1986; Tumbula et al., 1995b). Cultures 10 years or older have been routinely revived. Tube cultures are first concentrated by centrifugation and resuspended in a one-fifth volume of medium containing yeast extract and 25% glycerol (v/v). Portions of the cell suspension are transferred to sterile 1-ml screw-top glass vials in an anaerobic chamber. The vials are then stored at –70°C without anaerobic precautions. To revive the cultures, 0.2 ml of the cell suspension are allowed to thaw in an anaerobic chamber and transferred to fresh medium. Methanococci have also been stored by freeze-drying (Hippe, 1984) and freezing in glycerol (Winter, 1983).

For long-term preservation of the methanothermococci, strains are grown in suspension overnight, repressurized, and stored at 4°C, after which these suspensions can serve as an inoculum for at least 6 months. Using the same procedure, Methanothermococcus thermolithotrophicus strain ST22, stored at room temperature, was viable for at least 2 years (Nilsen and Torsvik, 1996). Cultures can also be stored at –80°C in fresh culture medium containing 20% (w/v) glycerol.

Characteristics of Methanococcales

Methanococcus

The genus Methanococcus is represented by four species whose properties are summarized in Table 2. Growth occurs at mesophilic temperatures, with the optima near 35–40°C. The pH optima for growth are between 6 and 8. During balanced growth, cells are slightly irregular and uniform in size, between 1 and 2 μm in diameter (Fig. 3). Pairs of cells are common. In stationary

Table 2. Descriptive characteristics of the species of the genus *Methanococcus*.

Characteristic	*vannielii*	*voltae*	*maripaludis*	*"aeolicus"*
Irregular coccus	+	+	+	+
Cell diameter, μm	1.3	1.3–1.7	0.9–1.3	1.7
Motile	+	+	+	+
Substrates for methane synthesis	$H_2 + CO_2$, and formate	$H_2 + CO_2$, and formate	$H_2 + CO_2$, and formate	$H_2 + CO_2$, and formate
Autotrophic growth	+	−	+	+
Growth requirement	None	Ac, Ile, Leu, and Ca^{2+}	None	None
Growth stimulatory	Se	Se, and pantoyllactone	Se, acetate, and amino acids	Se
Sulfur sources	S^{-2}, and S^0	S^{-2}, and S^0	S^{-2}, S^0, and $(S_2O_3^{-2})$	S^{-2}, and S^0
Nitrogen sources	NH_3, and purines	NH_3	NH_3, N_2, and alanine	NH_3, and N_2
Temperature range, °C	<20–45	<20–45	<20–45	<20–45
pH range	6.5–8	6.5–8	6.5–8	6.5–8
NaCl optimum, %	0.6–2	1–2	0.6–2	1–2
NaCl range, %	0.3–5	0.6–6	0.3–5	1->5
Mol% G+C[a]	31 (BD) and 33 (LC)	30 (LC) and 31 (BD)	33 (BD and LC)	32 (LC)
Type strain	SB	PS	JJ	PL-15/H[b]
Culture collection	DSM 1224 OCM 148	DSM 1537 OCM 70	DSM 2067 OCM 175	

Symbols: +, property of the species; and −, not a property of the species.
Abbreviations: BD, buoyant density method; LC, liquid chromatography; DSM, Deutsche Sammlung von Mikroorganismen und Zellkulturen; and OCM, Oregon Collection of Methanogens.
[a]Of the type strain.
[b]Because this species has not been formally described, this strain is not a nomenclatural type.

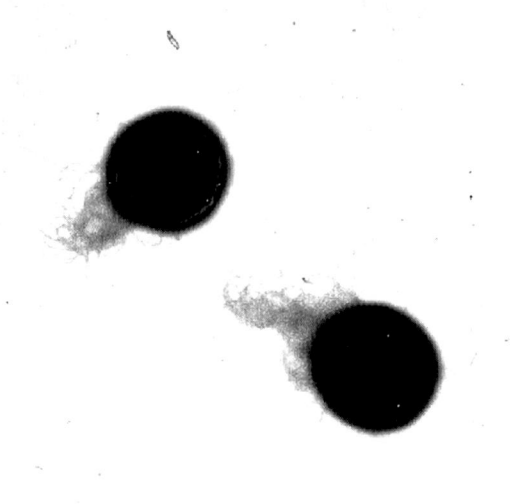

Fig. 3. *Methanococcus voltae* showing the polar flagellar tuft. The cells were negatively stained with 1% phosphotungstate (pH 7) and viewed with a JEOL 1200EX transmission electron microscope at 80 kV. (Electron micrograph courtesy of Shin-Ichi Aizawa and Ken Jarrell.)

cultures, colonies or enrichment cultures, cell shape is very irregular, and large cells up to 10 μm in diameter are observed (Jones et al., 1977b). In wet mounts, a few cells in a preparation may slowly swell and burst (Ward, 1970). Cells on the edge of a slide where drying may occur are much larger, less irregular, and more transparent than cells from the center of the slide. Cells from older cultures are mechanically fragile and rupture during vigorous stirring or upon harvesting by some continuous centrifugation devices. Cells are also osmotically fragile, and they lyse rapidly in distilled water. Cell integrity is maintained in 2% NaCl (w/v). Cells lyse rapidly in 0.01% sodium dodecyl sulfate (SDS) and contain a protein cell wall or S-layer. In *M. vannielii* and *M. voltae*, the outer cell surface is composed of hexagonally ordered structures (Jarrell and Koval, 1989). Cells are motile, and *M. vannielii* contains two tufts or bundles of flagella at the same pole.

Methanococcus spp. are obligate anaerobes and obligate methanogens. Molecular hydrogen (H_2) and formate serve as electron donors. Acetate, methanol and methylamines are not substrates for methanogenesis. All the strains tested are unable to utilize alcohols such as ethanol, isopropanol, isobutanol and cyclohexanol as electron donors for CO_2 reduction (Zellner and Winter, 1987). While an unnamed isolate of *Methanococcus* has also been reported to utilize methylfurfural compounds as a substrate for methanogenesis, the only strain of *Methanococcus maripaludis* tested did not utilize methylfurfural (Boopathy, 1996).

Except for *M. voltae*, the methanococci will grow in mineral medium with sulfide as the sole reducing agent and carbon dioxide as the sole carbon source (Whitman et al., 1986). Autotrophic CO_2 fixation is by the modified Ljungdahl-Wood pathway of acetyl-CoA biosynthesis (Shieh and Whitman, 1987; Shieh and Whitman, 1988; Ladapo and Whitman, 1990).

Methanococcus maripaludis is a facultative autotroph, and acetate and amino acids are stimulatory to growth. The amino acids are incorporated into cellular protein with a high efficiency but not further metabolized (Whitman et al., 1987). In contrast, the growth of *M. vannielii* and "*M. aeolicus*" is not affected by acetate or amino acids. *Methanococcus voltae* requires acetate, isoleucine and leucine for growth (Whitman et al., 1982). Isovalerate and 2-methylbutyrate can substitute for leucine and isoleucine, respectively. Pantoyllactone and pantoic acid, which are formed from pantothenate during autoclaving, can also stimulate growth. These requirements are the same for the type strain as well as other strains isolated from the coastal regions in the southeastern United States (Whitman et al., 1986). However, a strain isolated from Taiwan does not possess an absolute requirement for acetate or amino acids, although acetate and tryptone are stimulatory (Lai and Shih, 2001). The reason for this discrepancy is not known. Since DNA hybridization was not performed, it is possible that this strain may represent a novel but closely related species. Alternatively, the difference in growth responses may be due to different growth conditions employed in different laboratories, or the growth requirements may not be a characteristic of all members of the species. Glycogen has been identified as a storage product in most methanococci (König et al., 1985; Yu et al., 1994).

Nitrogen sources for the methanococci include ammonium, N₂ gas and alanine. Ammonium is sufficient as a nitrogen source for all methanococci and is required by *M. voltae* even during growth with amino acids (Whitman et al., 1982). Molecular nitrogen (N₂ gas) and alanine are additional nitrogen sources for *M. maripaludis*. *Methanococcus vannielii* cannot utilize N₂ gas or amino acids as nitrogen sources, but it will utilize purines (DeMoll and Tsai, 1986; Whitman, 1989).

Sulfide is sufficient as a sulfur source for all methanococci. Elemental sulfur is also reduced to sulfide (Stetter and Gaag, 1983; W. B. Whitman, unpublished data). Cysteine, dithiothreitol and sulfate do not substitute for sulfide (Whitman et al., 1982; Whitman et al., 1987). Some strains of *M. maripaludis* utilize thiosulfate as a sulfur source (Rajagopal and Daniels, 1986).

In addition to NaCl, high concentrations of magnesium salts are stimulatory or required by the methanococci (Whitman et al., 1982; Whitman et al., 1986; Corder et al., 1983; Jones et al., 1983b). Calcium is required by *M. voltae* (Whitman et al., 1982). Selenium is stimulatory to all species tested (Jones and Stadtman, 1977a; Whitman et al., 1982; Jones et al., 1983b). Iron, nickel and cobalt are required or stimulatory for

M. voltae (Whitman et al., 1982), and tungsten and nickel are required or stimulatory for *M. vannielii* (Jones and Stadtman, 1977a; Diekert et al., 1981).

Like other archaea, methanococci are generally resistant to low concentrations of many common antibiotics (Jones et al., 1977b). Some antibiotics that are inhibitory at low concentrations are: adriamycin, chloramphenicol, efrapeptin, leucinostatin, metonidazole, monensin, pleuromutilin, pyrrolnitrin and virginiamycin (Elhardt and Böck, 1982; Böck and Kandler, 1985). Methanococci are also sensitive to low concentrations of organic tin-containing compounds such as: phenyltin, tripropyltin and triethyltin (Boopathy and Daniels, 1991).

METHANOCOCCUS AEOLICUS This irregular coccus has an average diameter of 1.7 μm and occurs singly or in pairs. For growth, the temperature range is ≥20–45°C, the pH range is 6.5–8.0, the range of NaCl concentration is 1→5%, and the optimum is 1–2%. Ammonium and N₂ can serve as sole nitrogen sources, and selenium but not acetate or amino acids stimulates growth. Cells are susceptible to lysis by 0.01% sodium dodecyl sulfate (SDS) and hypotonic solutions. The mol% G+C is 32.0 (liquid chromatography method [LC]). The type strain has not been designated, and this taxon is currently represented solely by a patent strain (Schmid et al., 1984; Keswani et al., 1996; Whitman, 2001).

METHANOCOCCUS DELTAE Although validly published, this species is a subjective synonym of *M. maripaludis*. Hence, it is not further described here (Corder et al., 1983; Keswani et al., 1996).

METHANOCOCCUS MARIPALUDIS This irregular coccus has an average diameter of 0.9–1.3 μm and occurs singly or in pairs. For growth, the temperature range is ≥20–45°C, the pH range is 6.5–8.0, the range of NaCl concentration is 0.3–5%, and the optimum is 0.6–2%. Ammonium, alanine and N₂ can serve as sole nitrogen sources. Selenium, acetate and amino acids are stimulatory for growth. Cells are susceptible to lysis by 0.01% SDS and hypotonic solutions. The mol% G+C is 33–35 (LC). The type strain is JJ (= DSMZ 2067 = OCM 175; Jones et al., 1983b; Keswani et al., 1996; Whitman, 2001).

METHANOCOCCUS VANNIELII Cells are irregular cocci, with an average diameter of 1.3 μm, and occur singly or in pairs. For growth, the temperature range is ≥20–45°C, the pH range is 6.5–8.0, the range of NaCl is 0.3–5%, and the optimum is 0.6–2%. Ammonium and purines, but not alanine and N₂, can serve as sole nitrogen sources. Selenium stimulates growth, whereas acetate and amino acids do not. Cells are susceptible to lysis by 0.01% SDS and hypotonic solutions. The mol% G+C is 32.5 (LC). *Methanococcus vannielii* is the type species of the genus. The type strain is SB (= ATCC 35089 = DSMZ 1224 = OCM 148; Stadtman and Barker, 1951; Whitman, 2001).

METHANOCOCCUS VOLTAE Cells are irregular cocci, with an average diameter of 1.3–1.7 μm, and occur singly or in pairs. For growth, the temperature range is ≥20–45°C, the pH range is 6.5–8.0, the range of NaCl is 0.3–5%, and the optimum is 1–2%. Ammonium but not alanine and N₂ can serve as sole nitrogen source. Selenium is stimulatory for growth. Acetate and amino acids leucine and isoleucine are required or greatly stimulate growth. Cells are susceptible to lysis by 0.01% SDS and hypotonic solutions. The mol% G+C is 29–32 (LC). The type strain is PS (= ATCC 33273 = DSMZ 1537 = OCM 70; Ward, 1970; Whitman, 2001).

Methanothermococcus

The genus *Methanothermococcus* is represented by two species, *Methanothermococcus thermolithotrophicus* and *M. okinawensis*, whose properties are summarized in Table 3. In both cases, cells are Gram-negative irregular cocci and motile by means of a polar tuft of flagella. They lyse immediately in distilled water and in the presence of dilute solutions of SDS. They are thermophilic, with a temperature optimum of 60–70°C. NaCl is required for growth. They grow autotrophically in defined mineral medium with

H₂ and CO₂. H₂ and formate are used as electron donors for methanogenesis. Acetate, methanol, and methylamines are not substrates for methane production. Organic carbon sources are not stimulatory for growth. The mol% G+C is 31–32. The type species is *M. thermolithotrophicus*.

METHANOTHERMOCOCCUS OKINAWENSIS These irregular cocci have an average diameter of 1.0–1.5 μm and occur singly or in pairs. They are vigorously motile by means of a polar bundle of flagella. For growth, the temperature range is 40–70°C, the optimum is 60–65°C, the pH range is 4.5–8.5, optimum is 6–7, the concentration of sea salts is 1.2–9.6%, and the optimum is 2.0–5.0%. Ammonium is the nitrogen source, and selenium and magnetite (Fe₃O₄) are stimulatory for growth. Cells are susceptible to lysis by 0.1% SDS and hypotonic solutions. This species was first isolated from a deep-sea hydrothermal vent chimney at the Iheya Ridge, in the Okinawa Trough. The type strain of the species is IH 1 (= JCM11175 = DSM 14208; Takai et al., 2002).

METHANOTHERMOCOCCUS THERMOLITHOTROPHICUS Cells are regular to irregular cocci (diameter, 1.5 μm) that occur singly or in pairs. On agar and in Gelrite, round yellowish colonies around 1 mm in diameter are formed. About 20 flagella are inserted at a distinct area on the cell surface. The cells lyse immediately in the presence of 2% SDS; the cell envelope consists of protein subunits. Their optimum temperature for growth is 60–65°C (range, 17–70°C); optimum pH, 6–7.5 (range, 4.9–9.8); and optimum NaCl concentration, 1.8–4% (range, 0.6–9.4%). Methane is

Table 3. Descriptive characteristics of the species of the genus *Methanothermococcus*.

Characteristic	*thermolithotrophicus*	*okinawensis*
Irregular coccus	+	+
Cell diameter, μm	1.5	1.0–1.5
Motility	+	+
Substrates for methane synthesis	H₂ + CO₂, and formate	H₂ + CO₂, and formate
Autotrophic growth	+	+
Sulfur sources	S⁻², S⁰, S₂O₃⁻², SO₃⁻², and SO₄⁻²	S⁻²
Nitrogen sources	NH₃, NO₃⁻, and N₂	NH₃
Temperature optimum, °C	60–65	60–65
Temperature range, °C	17–70	40–75
pH optimum	5.1–7.5	6–7
pH range	4.9–9.8	4.5–8.5
NaCl optimum, %	2–4	2.5–5.0[a]
NaCl range, %	0.6–9.4	1.2–9.6[a]
Mol% G+C	31 (TM), and 34 (LC)	33.5 (LC)
Type strain	SN1	IH1
Culture collections	DSM 2095, and OCM 138	DSM 14208, and JCM 11175

Symbols: +, property of the species; and –, not a property of the species.
Abbreviations: TM, thermal denaturation; LC, liquid chromatography; DSM, Deutsche Sammlung von Mikroorganismen und Zellkulturen; and JCM, Japan Collection of Microorganisms.
[a]Sea salts.

formed from H_2/CO_2 and from formate. Organic material does not stimulate growth. Both N_2 and nitrate can serve as sole nitrogen sources (Belay et al., 1984; Belay et al., 1990). Sulfur is reduced to hydrogen sulfide with inhibition of methanogenesis. The DNA base composition is 31–32 mol% G+C. It is the type species of the genus *Methanothermococcus*. Type strain of the species is strain SN1[T] (= DSM2095 = ATCC35097 = JCM10549 = OCM 138). The species was first isolated from the sandy geothermally heated sediment of a beach at Stufe di Nerone close to Naples (Italy; Huber et al., 1982; Nilsen and Torsvik, 1996).

Methanocaldococcus

Genus *Methanocaldococcus* is composed of four species (Table 4). Cells are irregular cocci motile by means of polar tufts of flagella and readily lyse in distilled water and in the presence of dilute solutions of SDS. The genus is hyperthermophilic (optimum temperature for growth, 80–85°C). NaCl is required for growth. Cells grow autotrophically in defined mineral medium with H_2 and CO_2. Formate, acetate, methanol, and methylamines are not substrates for methane production. Selenium and tungsten are stimulatory for growth. The G+C content is 31–33 mol%. It is the type genus of the family Methanocaldococcaceae. The type species of the genus is *M. jannaschii*.

Methanocaldococcus fervens These regular to irregular cocci (diameter, 1–2 μm) occur singly and in pairs, and form whitish, translucent, and round colonies about 0.5 mm in diameter on Gelrite plates. Their optimum temperature for growth is 85°C (range, 48–92°C); optimum pH, 6.5 (range, 5.5–7.6); and optimum NaCl concentration, 3% (range, 0.5–5%). Yeast extract, casamino acids, and trypticase are stimulatory for growth. DNA base composition is 33 mol% G+C (thermal denaturation method). The type strain is AG86[T] (= DSM4213). The species was first isolated from a deep-sea hydrothermal vent core sample from Guaymas Basin, Gulf of California, at a depth of 2,003 m (Jeanthon et al., 1999b).

Methanocaldococcus infernus These irregular cocci (diameter, 1–3 μm) occur singly and in pairs. Cells exhibit a tumbling motility by means of at least three tufts of flagella. The cell envelope consists of a hexagonally arranged S-layer with a lattice constant of 12.2 nm (Fig. 4). Pale yellow colonies about 1 mm in diameter form on "Phytagel" plates. Optimum temperature for growth is 85°C (range, 55–91°C); optimum pH, 6.5 (range, 5.25–7.0); and optimum NaCl concentration, 2% (range, 0.8–3.5). Yeast extract is stimulatory for growth. Elemental sulfur is reduced to hydrogen sulfide in the presence of CO_2 and H_2. DNA base composition is 33 mol% G+C (thermal denaturation method). The type strain

Table 4. Descriptive characteristics of the species of the hyperthermophilic genera *Methanocaldococcus* and *Methanotorris*.[a]

| Characteristic | *Methanocaldococcus* | | | | *Methanotorris* |
	jannaschii	*infernus*	*vulcanius*	*fervens*	*igneus*
Irregular coccus	+	+	+	+	+
Cell diameter (μm)	1.5	1–3	1–3	1–2	1–2
Flagella	2 tufts	3 tufts	3 tufts	Not described	±[a]
Substrates for methane synthesis	$H_2 + CO_2$	$H_2 + CO_2$	$H_2 + CO_2$	$H_2 + CO_2$	$H_2 + CO_2$
Autotrophic growth	+	+	+	+	+
Yeast extract stimulates growth	–	+	+	+	–
Selenium stimulates growth	+	+	+	+	–
Sulfur sources	S^{-2} and S^0	S^{-2} and S^0	S^{-2} and S^0	S^{-2} and S^0	S^{-2} and S^0
Nitrogen sources	NH_3	NH_3 and NO_3^-	NH_3 and NO_3^-	NH_3 and NO_3^-	NH_3
Temperature optimum, °C	85	85	80	85	88
Temperature range, °C	50–91	55–91	49–89	48–92	45–91
pH optimum	6.0	6.5	6.5	6.5	5.7
pH range	5.2–7.0	5.25–7.0	5.2–7.0	5.5–7.6	5.0–7.5
NaCl optimum, %	3.0	2.0	2.5	3.0	1.8
NaCl range, %	1.0–5.0	0.8–3.5	0.6–5.6	0.5–5.0	0.45–7.2
Mol% G+C	31 (BD)	33 (TD)	31 (TD)	33 (TD)	31 (TD)
Type strain	JAL-1	ME	M7	AG86	Kol 5
Culture collections	DSM 2661, and ATCC 43067	DSM 11812	DSM 12094, and ATCC 700851	DSM 4213	DSM 5666

Symbols: +, property of the species; –, not a property of the species; and ( ), not relevant or not tested.
Abbreviations: BD, buoyant density method; TD, thermal denaturation; DSM, Deutsche Sammlung von Mikroorganismen und Zellkulturen; and ATCC, American Type Culture Collection.
[a]Although nonmotile, a few flagella-like structures are observed by electron microscopy.

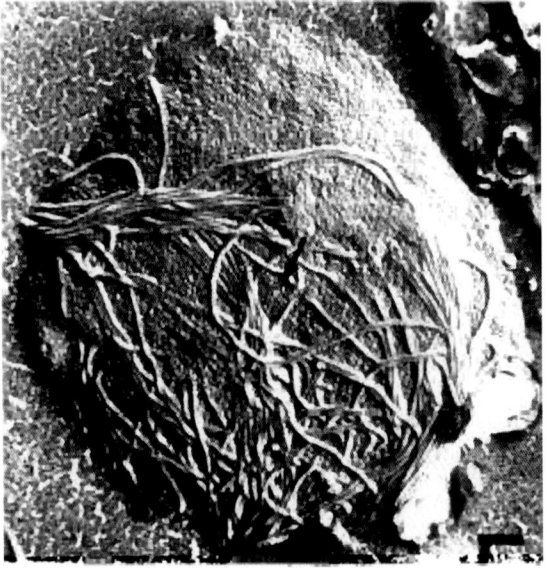

Fig. 4. Freeze-etched and shadowed preparation of cells of *M. infernus* showing the flagella and the S-layer (bar = 100 nm). From Jeanthon et al. (1998), and used with permission of the publisher.

Fig. 5. Freeze-etched and shadowed preparation of cells of *M. vulcanius* showing the three tufts of flagella and the S-layer (bar = 100 nm). Arrows indicate the presence of the S-layer lattice. From Jeanthon et al. (1999b), and used with permission of the publisher.

is strain ME^T (= DSM11812). The species was first isolated from a deep-sea hydrothermal vent chimney in the Mid-Atlantic Ridge (14°45′N; Jeanthon et al., 1998).

METHANOCALDOCOCCUS JANNASCHII These irregular cocci measure 1.5 µm in diameter and occur singly or in pairs. The cell surface is composed of hexagonally ordered substructures; two bundles of flagella are inserted close to the same cell pole. Colonies are convex, circular, yellowish in color with a smooth shiny surface. Optimum temperature for growth is 85°C (range, 50–91°C); optimum pH, 6.0 (5.2–7.0); and optimum NaCl concentration, 3%. Acetate, formate, yeast extract, trypticase, and vitamins are not stimulatory for growth. Sulfide is required. DNA base composition is 31 mol% G+C (buoyant density method). The species was first isolated from sedimentary material at the base of a white smoker submarine hydrothermal vent (21°N at 2,600-m depth on the East Pacific Rise). It is the type species of the genus *Methanocaldococcus*. The type strain is strain JAL-1^T (= DSM 2661 = ATCC 43067 = JCM 10045 = OCM 168; Jones et al., 1983a; Jeanthon et al., 1999a).

METHANOCALDOCOCCUS VULCANIUS These irregular cocci (diameter, 1–3 µm) occur singly and in pairs. Cells exhibit a tumbling motility by means of three tufts of flagella (Fig. 5). Pale yellow round colonies about 1 mm in diameter form on

"Phytagel" plates. Optimum temperature for growth is 80°C (range, 49–89°C); optimum pH, 6.5 (range, 5.25–7); and optimum NaCl concentration, 2.5% (range, 0.6–5.6%). Yeast extract is stimulatory for growth. Sulfur is reduced to hydrogen sulfide in the presence of CO_2 and H_2. DNA base composition is 31 mol% G+C (thermal denaturation method). The type strain is M7^T (= DSM 12094 = ATCC700851). The species was first isolated from a deep-sea hydrothermal vent chimney on the East Pacific Rise (13°N; Jeanthon et al., 1999b).

Methanotorris

Genus *Methanotorris* is represented by a single hyperthermophilic species, *Methanotorris igneus*.

METHANOTORRIS IGNEUS This irregular coccus (diameter, 1.3–1.8 µm) occurs singly and in pairs. Cells are nonmotile, and no tufts of flagella are visible; a few flagella-like filaments are visible on some cells. Pale yellow round smooth colonies are formed on agar. Optimum temperature for growth is 88°C (range, 45–91°C); optimum pH, 5.7 (range, 5–7.5); optimum NaCl, 1.8% (range, 0.45–7.2%). Cells form methane from H_2 and CO_2. Selenium is not required for growth and selenium, yeast extract and peptone are not stimulatory for growth. Formate, acetate, methanol, and methylamines are not substrates for methane production. Sulfur is reduced to hydrogen sulfide in the presence of CO_2 and H_2. DNA base composition is 31 mol% G+C (thermal denaturation method). It is the type species of the genus *Methanotorris*. The type strain is Kol5[T] (= DSM5666). The species was first isolated from a shallow submarine hydrothermal vent system at the Kolbeinsey Ridge (Mid-Atlantic Ridge north of Iceland, at a depth of 106 m; Burggraf et al., 1990; Jeanthon et al., 1999a).

Biochemical and Physiological Properties

Genomic Sequence

Methanocaldococcus jannashii was the first archeaon and hyperthermophile whose genome was sequenced, and it has served as model for other hyperthermophiles and archaea (Bult et al., 1996). While a review of the genome exceeds the scope of this review, an updated annotation that summarizes many of the recent advances is available (Graham et al., 2001). The genomic sequence has also been used to reconstruct the metabolism of this autotroph (Selkov et al., 1997).

Flagellum and Motility

Although the appearance is similar, the flagella of the methanococci and other archaea are very different from those of bacteria (for recent reviews, see Faguy and Jarrell, 1999; Jarrell et al., 1996, and Thomas et al., 2001a). The flagellum itself is somewhat narrower, about 10–14 nm, than bacterial flagella, which are >20 nm. The conspicuous hook and basal structure common on bacterial flagella is not observed; instead, a hook without the basal structure is usually seen (Fig. 6). In some studies, a simple basal body has also been visualized. The archaeal flagellins have sequence similarity to the bacterial type IV pilins and not the bacterial flagellins. They are also biosynthesized with a leader peptide, which implies different mechanisms of assembly and transport across the membrane than are found with the bacterial flagella. Thus, the archaeal and bacterial structures do not appear to be homologous. In support of this conclusion, homologs of genes encoding the bacterial flagellins, basal body and motility apparatus are not found in archaeal genomes.

Effects of Pressure on Thermophilic Methanogens

Methanocaldococcus jannaschii was the first hyperthermophilic organism isolated from deep-sea hydrothermal vents. Owing to the depth of

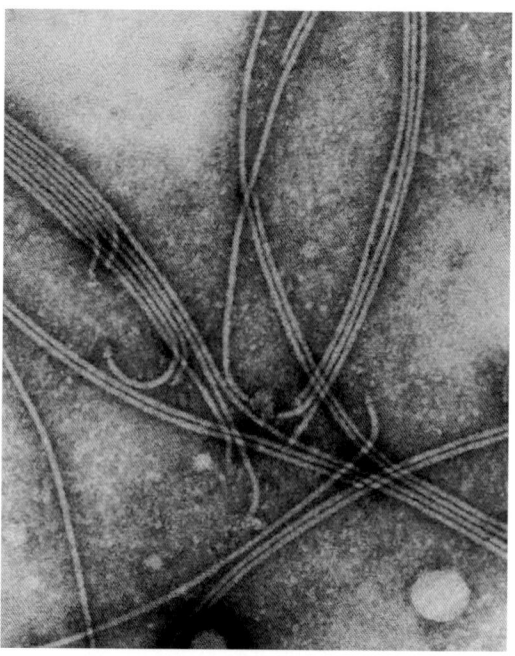

Fig. 6. Flagella of *Methanococcus voltae*. A hook region similar to that found on bacterial flagella is present on the flagella from methanococci. While a basal body has not been observed in these preparations, it is possible that the bodies are present in the cells but lost upon purification. Whole cells were concentrated and solubilized with 1% of the detergent OP-10 in the presence of DNase and RNase. Following removal of the membranes by centrifugation, the flagella were precipitated using 2% polyethylene glycol and 100 mM NaCl. The flagella were then further purified on a KBr gradient. (Electron micrograph courtesy of Shin-Ichi Aizawa and Ken Jarrell.)

these vents, the in situ pressure is generally between 20 and 30 MPa (hydrostatic pressure increases by about 10 MPa or 100 atm for every kilometer of seawater). *Methanocaldococcus jannaschii* has therefore served as a model for the effects of pressure on the growth and metabolism of deep-sea thermophiles and for comparison with organisms isolated from the surface at low pressure (i.e., *M. thermolithotrophicus*). High hyperbaric pressures of helium (up to 75 MPa) increase the growth rate of *M. jannaschii* at 86 and 90°C but do not to extend its upper temperature limit for growth (Miller et al., 1988). However, increased pressures extend the upper temperature limit for methanogenesis from less than 94°C at 0.78 MPa to 98°C at 25 MPa. In contrast, following growth at 50 MPa, cells of *M. thermolithotrophicus* are anomalously large and elongated, and the cellular amino acid composition and protein pattern are dramatically changed (Jaenicke et al., 1988). Its growth rate is also enhanced without extending its temperature range (Bernhardt et al., 1988).

Understanding the mechanisms that stabilize proteins, especially at extreme temperatures, is a challenging problem in both biochemistry and biotechnology. The initial findings raised the question of how pressure could extend the thermal stability of thermophilic enzymes. Hei and Clark (1994) compared the pressure effect on the thermal stability of a partially purified hydrogenase from *M. jannaschii* to that from *M. igneus*, *M. thermolithotrophicus* and *M. maripaludis*, all isolated from shallow marine sites. Application of 50 MPa increased the thermal half-life of hydrogenase from *M. jannaschii* 4.8-fold at 90°C. Since hydrogenase from the other hyperthermophile, *M. igneus*, was also substantially stabilized by pressure whereas the hydrogenases from *M. thermolithotrophicus* and *M. maripaludis* were destabilized by pressure, it was suggested that pressure stabilization of enzymes was related to their thermophilicity. In further studies, the thermostability of a protease from *M. jannaschii* also increased at high pressure (Michels and Clark, 1997). However, the destabilization by pressure of the adenylate kinase from *M. jannaschii* demonstrated that hyperbaric stabilization was not an intrinsic property of all enzymes from deep-sea thermophiles (Konisky et al., 1995). In studying the pressure effects on lipids from *M. jannaschii*, Kaneshiro and Clark (1995) showed that pressure had a lipid-ordering effect over the full range of growth temperatures for this organism. The shift toward macrocyclic archaeol from archaeol and caldarchaeol was shown to decrease membrane fluidity.

Genetics

Genetic systems have been utilized extensively in the mesophiles *Methanococcus voltae* and *M. maripaludis* (for a review, see Tumbula and Whitman, 1999). Methods that have become available include efficient transformation systems; antibiotic resistance markers for puromycin and neomycin; reporter genes such as those for β-galactosidase, β-glucuronidase and trehalase; and shuttle and expression vectors. Evidence for a phage transduction system has also been described for *M. voltae* strain PS (Bertani, 1999). This system depends upon VTA (or *voltae* transfer agent) and allows for low frequencies of genetic exchange with a wide variety of markers. The VTA appears to be a small polyhedral phage that carries 4.4-kb fragments of circular DNA that are derived from the host genome (Eiserling et al., 1999). This phage-like particle is similar to the gene transfer agent described for *Rhodobacter capsulata* but different from another

phage discovered in *M. voltae* strain A3 (Wood et al., 1985).

Genetic methods have been utilized to examine the expression of the hydrogenase and motility genes of *M. voltae* and the nitrogen fixation genes of *M. maripaludis*. *Methanococcus voltae* contains two sets of hydrogenase genes. One set encodes the selenium-dependent enzymes that contain selenocysteinyl residues at the active site. The second set encodes the selenium-free enzymes. Both sets include both the coenzyme F_{420}-reducing and coenzyme F_{420}-independent hydrogenases (reviewed in Sorgenfrei et al., 1997). The genes encoding the selenium-free enzymes are divergently transcribed from a 453-bp intergenic region that contains the *cis* elements for transcriptional regulation. Expression is only observed in the absence of selenium in the growth medium and appears to be both negatively and positively regulated. Negative regulation results from a silencer region in the middle of the intergenic region (Noll et al., 1999). Positive regulation results from binding of a transcriptional activator immediately upstream of both promoters (Müller and Klein, 2001).

The flagellar biosynthetic genes have been extensively studied in *M. voltae* (reviewed in Thomas et al., 2001a). These genes are arranged into two transcriptional units. The first unit encodes only *flaA*, one of the four flagellin genes present. The second unit encodes the remaining three flagellin genes as well as a number of other genes whose functions are not fully known. Homologs to these other genes are found in the motility gene clusters of other archaea, so they are probably required for flagellar assembly or motility. Although the functions of these genes are not known, mutants with an insertional inactivation of at least one of these genes are no longer flagellated (Thomas et al., 2001b). This experiment provides direct evidence for a role in motility.

Methanococcus maripaludis fixes nitrogen, and the regulation of the *nif* operon, which encodes the nitrogenase genes as well as other genes important for nitrogen fixation, has been studied in detail (for a review, see Leigh, 1999). Expression of the *nif* operon is under control of a repressor (Cohen-Kupiec et al., 1997). In addition, cellular nitrogenase activity is further regulated by two GlnB homologs. In various bacteria, GlnB homologs posttranslationally regulate the activity of a number of key enzymes in nitrogen assimilation, including nitrogenase and glutamine synthetase. Although the specific mechanism is not known, the methanococcal GlnB homologs are responsible for the rapid and reversible "switching off" of nitrogenase activity (Kessler et al., 2001).

Literature Cited

Balch, W. E. and R. S. Wolfe. 1976. New approach to the cultivation of methanogenic bacteria: 2-mercaptoethane-sulfonic acid (HS-CoM)-dependent growth of Methanobacterium ruminantium in a pressurized atmosphere. Appl. Environ. Microbiol. 32:781–791.

Balch, W. E., C. E. Fox, L. J. Magrum, C. R. Woese, and R. S. Wolfe. 1979. Methanogens: Reevaluation of a unique biological group. Microbiol. Rev. 43:260–296.

Belay, N., R. Sparling, and L. Daniels. 1984. Dinitrogen fixation by a thermophilic methanogenic bacterium. Nature 312:286–288.

Belay, N., K.-Y. Jung, B. S. Rajagopal, J. D. Kremer, and L. Daniels. 1990. Nitrate as a sole nitrogen source for Methanococcus thermolithotrophicus and its effect on growth of several methanogenic bacteria. Curr. Microbiol. 21:193–198.

Bernhardt, G., R. Jaenicke, H.-D. Ludemann, H. Konig, and K. O. Stetter. 1988. High pressure enhances the growth rate of the thermophilic archaebacterium Methanococcus thermolithotrophicus without extending its temperature range. Appl. Environ. Microbiol. 54:1258–1261.

Bertani, G. 1999. Transduction-like gene transfer in the methanogen Methanococcus voltae. J. Bacteriol. 181:2992–3002.

Blotevogel, K.-H., U. Fischer, and K. H. Lupkes. 1986. Methanococcus frisius sp. nov., a new methylotrophic marine methanogen. Can. J. Microbiol. 32:127–131.

Böck, A., and O. Kandler. 1985. Antibiotic sensitivity of archaebacteria. In: C. R. Woese and R. S. Wolfe (Eds.) The Bacteria. Academic Press. New York, NY. 8:525–544.

Boone, D. R., W. B. Whitman, and P. Rouviere. 1993. Diversity and taxonomy of methanogens. In: J. G. Ferry (Ed.) Methanogenesis: Ecology, Physiology, Biochemistry and Genetics. Chapman & Hall. New York, NY. 35–80.

Boopathy, R., and L. Daniels. 1991. Pattern of organotin inhibition of methanogenic bacteria. Appl. Environ. Microbiol. 57:1189–1193.

Boopathy, R., and L. Daniels. 1992. Isolation and characterization of a marine methanogenic bacterium from the biofilm of a shiphull in Los Angeles Harbor. Curr. Microbiol. 25:157–164.

Boopathy, R. 1996. Methanogenic transformation of methylfurfural compounds to furfural. Appl. Environ. Microbiol. 62:3483–3485.

Bryant, M. P. 1974. Methane-producing bacteria. In: R. E. Buchanan and N. E. Gibbons (Eds.) Bergey's Manual of Determinative Bacteriology, 8th ed. Williams and Wilkins. Baltimore, MD. 472–477.

Bryniok, D., and W. Trosch. 1989. Taxonomy of methanogens by ELISA techniques. Appl. Microbiol. Biotechnol. 32:243–247.

Bult, C. J., and 39 others. 1996. Complete genome sequence of the methanogenic archaeon, Methanococcus jannaschii. Science 273:1058–1073.

Burggraf, S., H. Fricke, A. Neuner, J. Kristjansson, P. Rouvier, L. Mandelco, C. R. Woese, and K. O. Stetter. 1990. Methanococcus igneus sp. nov., a novel hyperthermophilic methanogen from a shallow submarine hydrothermal system. Syst. Appl. Microbiol. 13:263–269.

Ciulla, R. A., S. Burggraf, K. O. Stetter, and M. F. Roberts. 1994. Occurrence and role of di-myo-inositol-1,1'-

phosphate in Methanococcus igneus. Appl. Environ. Microbiol. 60:3660–3664.

Cohen-Kupiec, R., C. Blank, and J. A. Leigh. 1997. Transcriptional regulation in Archaea: In vivo demonstration of a repressor binding site in a methanogen. Proc. Natl. Acad. Sci. USA 94:1316–1320.

Conway de Macario, E., M. J. Wolin, and A. J. L. Macario. 1981. Immunology of archaebacteria that produce methane gas. Science 214:74–75.

Corder, R. E., L. A. Hook, J. M. Larkin, and J. I. Frea. 1983. Isolation and characterization of two new methane-producing cocci: Methanogenium olentangyi, sp. nov., and Methanococcus deltae, sp. nov. Arch. Microbiol. 134:28–32.

DeMoll, E., and L. Tsai. 1986. Utilization of purines or pyrimidines as the sole nitrogen source by Methanococcus vannielii. J. Bacteriol. 167:681–684.

Diekert, G., U. Konheiser, K. Piechulla, and R. K. Thauer. 1981. Nickel requirement and factor F_{430} content of methanogenic bacteria. J. Bacteriol. 148:459–464.

Eiserling, F., A. Pushkin, M. Gingery, and G. Bertani. 1999. Bacteriophage-like particles associated with the gene transfer agent of Methanococcus voltae PS. J. Gen. Virol. 80:3305–3308.

Elhardt, D., and A. Böck. 1982. An in vitro polypeptide synthesizing system from methanogenic bacteria: Sensitivity to antibiotics. Molec. Gen. Genet. 188:128–134.

Faguy, D. M., and K. F. Jarrell. 1999. A twisted tale: The origin and evolution of motility and chemotaxis in prokaryotes. Microbiology 145:279–281.

Ferguson, T. J. and R. A. Mah. 1983. Isolation and characterization of an H_2-oxidizing thermophilic methanogen. Appl. Environ. Microbiol. 45:265–274.

Graham, D. E., N. Kyrpides, I. J. Anderson, R. Overbeek, and W. B. Whitman. 2001. Genome of Methanocaldococcus (Methanococcus) jannaschii. Meth. Enzymol. 330:40–123.

Hamana, K., M. Niitsu, K. Samejima, T. Itoh, H. Hamana, and T. Shinozawa. 1998. Polyamines of the thermophilic eubacteria belonging to the genera Thermotoga, Thermodesulfovibrio, Thermoleophilum, Thermus, Rhodothermus and Meiothermus, and the thermophilic archaebacteria belonging to the genera Aeropyrum, Picrophilus, Methanobacterium and Methanococcus. Microbios 94:7–21.

Hei, D. J. and D. S. Clark. 1994. Pressure stabilization of proteins from extreme thermophiles. Appl. Environ. Microbiol. 60:932–939.

Hippe, H. 1984. Maintenance of methanogenic bacteria. In: B. E. Kirsop and J. J. S. Snell (Eds.) Maintenance of Microorganisms: A Manual of Laboratory Methods. Academic Press. London, 69–81.

Huber, H., M. Thomm, H. Konig, G. Thies, and K. O. Stetter. 1982. Methanococcus thermolithotrophicus, a novel thermophilic lithotrophic methanogen. Arch. Microbiol. 132:47–50.

Jaenicke, R., G. Bernhardt, H.-D. Lüdemann, and K. O. Stetter. 1988. Pressure-induced alterations in the protein pattern of the thermophilic archaebacterium Methanococcus thermolithotrophicus. Appl. Environ. Microbiol. 54:2375–2380.

Jarrell, K. F., and S. F. Koval. 1989. Ultrastructure and biochemistry of Methanococcus voltae. Crit. Rev. Microbiol. 17:53–87.

Jarrell, K. F., D. P. Bayley, and A. S. Kostyukova. 1996. The archaeal flagellum: A unique motility structure. J. Bacteriol. 178:5057–5064.

Judical Commission. 1986. Opinion 62: Transfer of the type species of the genus Methanococcus to the genus Methanosarcina as Methanosarcina mazei (Barker 1936) comb. nov. et emend. Mah and Kuhn 1984 and conservation of the genus Methanococcus (Approved Lists, 1980) emend. Mah and Kuhn 1984 with Methanococcus vannielii (Approved Lists, 1980) as the type species. Int. J. Syst. Bacteriol. 36:491.

Jeanthon, C., S. L'Haridon, A. L. Reysenbach, M. Vernet, P. Messner, U. B. Sleytr, and D. Prieur. 1998. Methanococcus infernus sp. nov., a novel hyperthermophilic lithotrophic methanogen isolated from a deep-sea hydrothermal vent. Int. J. Syst. Bacteriol. 48:913–919.

Jeanthon, C., S. L'Haridon, N. Pradel, and D. Prieur. 1999a. Rapid identification of hyperthermophilic methanococci isolated from deep-sea hydrothermal vents. Int. J. Syst. Bacteriol. 49:591–594.

Jeanthon, C., S. L'Haridon, A. L. Reysenbach, E. Corre, M. Vernet, P. Messner, U. B. Sleytr, and D. Prieur. 1999b. Methanococcus vulcanius sp. nov., a novel hyperthermophilic methanogen isolated from East Pacific Rise and identification of Methanococcus spp. DSM 4213 as Methanococcus fervens sp. nov. Int. J. Syst. Bacteriol. 49:583–589.

Jones, J. B., and T. C. Stadtman. 1977a. Methanococcus vannielii: Culture and effects of selenium and tungsten on growth. J. Bacteriol. 130:1404–1406.

Jones, J. B., B. Bowers, and T. C. Stadtman. 1977b. Methanococcus vannielii: Ultrastructure and sensitivity to detergents and antibiotics. J. Bacteriol. 130:1357–1363.

Jones, W. J., J. A. Leigh, F. Mayer, C. R. Woese, and R. S. Wolfe. 1983a. Methanococcus jannaschii sp. nov., an extremely thermophilic methanogen from a submarine hydrothermal vent. Arch. Microbiol. 136:254–261.

Jones, W. J., M. J. B. Paynter, and R. Gupta. 1983b. Characterization of Methanococcus maripaludis sp. nov., a new methanogen isolated from salt marsh sediment. Arch. Microbiol. 135:91–97.

Jones, W. J., W. B. Whitman, R. D. Fields, and R. S. Wolfe. 1983c. Growth and plating efficiency of methanococci on agar media. Appl. Environ. Microbiol. 46:220–226.

Jones, W. J., C. E. Stugard, and H. W. Jannasch. 1989. Comparison of thermophilic methanogens from submarine hydrothermal vents. Arch. Microbiol. 151:314–318.

Kaneshiro, S. M., and D. S. Clark. 1995. Pressure effects on the composition and thermal behavior of lipids from the deep-sea thermophile Methanococcus jannaschii. J. Bacteriol. 177:3668–3672.

Kessler, P. S., C. Daniel, and J. A. Leigh. 2001. Ammonia switch-off of nitrogen fixation in the methanogenic archaeon Methanococcus maripaludis: Mechanistic features and requirement for the novel GlnB homologues, NifI$_1$ and NifI$_2$. J. Bacteriol. 183:882–889.

Keswani, J., S. Orkand, U. Premachandran, L. Mandelco, M. J. Franklin, and W. B. Whitman. 1996. Phylogeny and taxonomy of mesophilic Methanococcus spp. and comparison of rRNA, DNA hybridization, and phenotypic methods. Int. J. Syst. Bacteriol. 46:727–735.

Kneifel, H., K. O. Stetter, J. R. Andreesen, J. Wiegel, H. Konig, and S. M. Schoberth. 1986. Distribution of polyamines in representative species of archaebacteria. Syst. Appl. Microbiol. 7:241–245.

Koga, Y., H. Morii, M. Akagawa-Matsushita, and M. Ohga. 1998. Correlation of polar lipid composition with 16S rRNA phylogeny in methanogens: Further analysis of lipid component parts. Biosci. Biotechnol. Biochem. 62:230–236.

Konig, H., E. Nusser, and K. O. Stetter. 1985. Glycogen in Methanolobus and Methanococcus. FEMS Microbiol. Lett. 28:265–269.

Konisky, J., P. C. Michels, and D. S. Clark. 1995. Pressure stabilization is not a general property of thermophilic enzymes: The adenylate kinases of Methanococcus voltae, Methanococcus maripaludis, Methanococcus thermolithotrophicus, and Methanococcus jannaschii. Appl. Environ. Microbiol. 61:2795–2764.

Ladapo, J., and W. B. Whitman. 1990. Method for isolation of auxotrophs in the methanogenic archaebacteria: Role of the acetyl-CoA pathway of autotrophic CO$_2$ fixation in Methanococcus maripaludis. Proc. Natl. Acad. Sci. USA 87:5598–5602.

Lai, M.-C., and C.-J. Shih. 2001. Characterization of Methanococcus voltaei P2F9701a: A new methanogen isolated from estuarine environment. Curr. Microbiol. 42:432–437.

Leigh, J. A. 1999. Transcriptional regulation in Archaea. Curr. Opin. Microbiol. 2:131–134.

Mah, R. A. 1980. Isolation and characterization of Methanococcus mazei. Curr. Microbiol. 3:321–326.

Mah, R. A., and D. A. Kuhn. 1984. Transfer of the type species of the genus Methanococcus to the genus Methanosarcina, naming it Methanosarcina mazei (Barker 1936) comb. nov. et emend. and conservation of the genus Methanococcus (Approved Lists 1980) with Methanococcus vannielii (Approved Lists 1980) as the type species. Int. J. Syst. Bacteriol. 34:263–265.

Michels, P. C., and D. S. Clark. 1997. Pressure-enhanced activity and stability of a hyperthermophilic protease from a deep-sea methanogen. Appl. Environ. Microbiol. 63:3985–3991.

Miller, T. L., and M. J. Wolin. 1974. A serum bottle modification of the Hungate technique for cultivating obligate anaerobes. Appl. Microbiol. 27:985–987.

Miller, J. F., N. N. Shah, C. M. Nelson, J. M. Ludlow, and D. S. Clark. 1988. Pressure and temperature effects on growth and methane production of the extreme thermophile Methanococcus jannaschii. Appl. Environ. Microbiol. 54:3039–3042.

Müller, S., and A. Klein. 2001. Coordinate positive regulation of genes encoding [NiFe] hydrogenases in Methanococcus voltae. Molec. Genet. Genomics 265:1069–1075.

Nercessian, O., E. Corre, and C. Jeanthon. 2000. Phylogenetic analysis of the microbial communities from a deep hot oil reservoir in Western Siberia. In: G. Antranikian, R. Grote & K. Sahn (Eds.) Third International Congress on Thermophiles, Hamburg, Germany, September 3–7, 2000. Technical University Hamburg-Harburg, Institute of Technical Microbiology. Hamburg, Germany. Abstract P10:75.

Nilsen, R. K., and T. Torsvik. 1996. Methanococcus thermolithotrophicus isolated from North Sea oil field reservoir water. Appl. Environ. Microbiol. 62:728–731.

Noll, I., S. Müller, and A. Klein. 1999. Transcriptional regulation of genes encoding the selenium-free [NiFe]-hydrogenases in the archaeon Methanococcus voltae involves positive and negative control elements. Genetics 152:1335–1341.

Orphan, V. J., L. T. Taylor, D. Hafenbradl, and E. F. DeLong. 2000. Culture-dependent and culture-independent characterization of microbial assemblages associated with high-temperature petroleum reservoirs. Appl. Environ. Microbiol. 66:700–711.

Pfennig, N., F. Widdel, and H. G. Trüper. 1981. The dissimilatory sulfate-reducing bacteria. In: M. Starr, H. Stolp, H. G. Trüper, A. Balows, and H. G. Schlegel (Eds.) The Prokaryotes, 2nd ed. Springer-Verlag. New York, NY. 926–940.

Rajagopal, B. S., and L. Daniels. 1986. Investigation of mercaptans, organic sulfides, and inorganic sulfur compounds as sulfur sources for the growth of methanogenic bacteria. Curr. Microbiol. 14:137–144.

Rivard, C. J., and P. H. Smith. 1982. Isolation and characterization of a thermophilic marine methanogenic bacterium, Methanogenium thermophilicum sp. nov. Int. J. Syst. Bacteriol. 32:430–436.

Rivard, C. J., J. M. Henson, M. V. Thomas, and P. H. Smith. 1983. Isolation and characterization of Methanomicrobium paynteri sp. nov., a mesophilic methanogen isolated from marine sediments. Appl. Environ. Microbiol. 46:484–490.

Robertson, D. E., M. F. Roberts, N. Belay, K. O. Stetter, and D. R. Boone. 1990. Occurrence of β-glutamate, a novel osmolyte, in marine methanogenic bacteria. Appl. Environ. Microbiol. 56:1504–1508.

Robertson, D. E., D. Noll, and M. F. Roberts. 1992. Free amino acid dynamics in marine methanogens: β-amino acid amino acids as compatible solutes. J. Biol. Chem. 267:14893–14901.

Romesser, J. A., R. S. Wolfe, F. Mayer, E. Spiess, and A. Walther-Mauruschat. 1979. Methanogenium, a new genus of marine methanogenic bacteria, and characterization of Methanogenium cariaci sp. nov. and Methanogenium marisnigri sp. nov. Arch. Microbiol. 121:147–153.

Schauer, N. L., and W. B. Whitman. 1989. Formate growth and pH control by volatile formic and acetic acids in batch cultures of methanococci. J. Microbiol. Meth. 10:1–7.

Schmid, K., M. Thomm, A. Laminet, F. G. Laue, C. Kessler, K. O. Stetter, and R. Schmitt. 1984. Three new restriction endonucleases MaeI, MaeII, and MaeIII from Methanococcus aeolicus. Nucleic Acids Res. 12:2619–2628.

Selkov, E., N. Maltsev, G. J. Olsen, R. Overbeek, and W. B. Whitman. 1997. A reconstruction of the metabolism of Methanococcus jannaschii from sequence data. Gene 197:GC10–25.

Shieh, J. S., and W. B. Whitman. 1987. Pathway of acetate assimilation in autotrophic and heterotrophic methanococci. J. Bacteriol. 169:5327–5329.

Shieh, J. S., and W. B. Whitman. 1988. Acetyl coenzyme A biosynthesis in Methanococcus maripaludis. J. Bacteriol. 170:3072–3079.

Sorgenfrei, O., S. Müller, M. Pfeiffer, I. Sneizko, and A. Klein. 1997. The [NiFe] hydrogenases of Methanococcus voltae: Genes, enzymes and regulation. Arch. Microbiol. 167:189–195.

Sowers, K. R., and K. M. Noll. 1995. Techniques for anaerobic growth. In: K. R. Sowers and H. J. Schreier (Eds.) Archaea: A Laboratory Manual, Methanogens. Cold Spring Harbor Laboratory Press. Cold Spring Harbor, NY. 15–47.

Stadtman, T. C., and H. A. Barker. 1951. Studies on the methane fermentation. X: A new formate-decomposing

bacterium, Methanococcus vannielii. J. Bacteriol. 62:269–280.

Stetter, K. O., and G. Gaag. 1983. Reduction of molecular sulphur by methanogenic bacteria. Nature 305:309–311.

Takai, K., A. Inoue, and K. Horikoshi. 2002. Methanothermococcus okinawensis sp. nov., a thermophilic, methane-producing archaeon isolated from a Western Pacific deep-sea hydrothermal vent system. Int. J. Syst. Evol. Microbiol. 52(Pt 4):1089–1095.

Thomas, N. A., S. L. Bardy, and K. F. Jarrell. 2001a. The archaeal flagellum: A different kind of prokaryotic motility structure. FEMS Microbiol. Rev. 25:147–174.

Thomas, N. A., C. T. Pawson, and K. F. Jarrell. 2001b. Insertional inactivation of the flaH gene in the archaeon Methanococcus voltae results in non-flagellated cells. Molec. Genet. Genomics 265:596–603.

Tumbula, D. L., T. L. Bowen, and W. B. Whitman. 1995a. Growth of methanogens on solidified medium. In: K. R. Sowers and H. J. Schreier (Eds.) Archaea: A Laboratory Manual, Methanogens. Cold Spring Harbor Laboratory Press. Cold Spring Harbor, NY. 49–55.

Tumbula, D. L., J. Keswani, J. Shieh, and W. B. Whitman. 1995b. Long-term maintenance of methanogen stock cultures in glycerol. In: K. R. Sowers and H. J. Schreier (Eds.) Archaea: A Laboratory Manual, Methanogens. Cold Spring Harbor Laboratory Press. Cold Spring Harbor, NY. 85–87.

Tumbula, D. L, and W. B. Whitman. 1999. Genetics of Methanococcus: Possibilities for functional genomics in Archaea. Molec. Microbiol. 33:1–7.

Ward, J. M. 1970. The Microbial Ecology of Estuarine Methanogenesis (Masters thesis). University of Florida. Gainesville, FL.

Whitman, W. B., E. Ankwanda, and R. S. Wolfe. 1982. Nutrition and carbon metabolism of Methanococcus voltae. J. Bacteriol. 149:852–863.

Whitman, W. B., J. Shieh, S. Sohn, D. S. Caras, and U. Premachandran. 1986. Isolation and characterization of 22 mesophilic methanococci. Syst. Appl. Microbiol. 7:235–240.

Whitman, W. B., S. Sohn, S. Kuk, and R. Xing. 1987. Role of amino acids and vitamins in nutrition of mesophilic Methanococcus spp. Appl. Environ. Microbiol. 53:2373–2378.

Whitman, W. B. 1989. Order II: Methanococcales. In: J. T. Staley, M. P. Bryant, N. Pfennig, and J. G. Holt (Eds.) Bergey's Manual of Systematic Bacteriology, 1st ed. Williams & Wilkins. Baltimore, MD. 3:2185–2190.

Whitman, W. B., D. R. Boone, and Y. Koga. 2001. Methanococcales. In: D. R. Boone, R. W. Castenholtz, and G. M. Garrity (Eds.) Bergey's Manual of Systematic Bacteriology, 2nd ed. Springer-Verlag. New York, NY. 1:236.

Widdel, F., and F. Bak. 1992. Gram-negative mesophilic sulfate-reducing bacteria. In: A. Balows, H. G. Trüper, M. Dworkin, W. Harder, and K.-H. Schleifer (Eds.) The Prokaryotes, 2nd ed. Springer-Verlag. New York, NY. 3352–3378.

Winter, J. 1983. Maintenance of stock cultures of methanogens in the laboratory. Syst. Appl. Microbiol. 4:558–563.

Wood, A. G., W. B. Whitman, and J. Konisky. 1985. A newly-isolated marine methanogen harbors a small cryptic plasmid. Arch. Microbiol. 142:259–261.

Yu, J.-P., J. Ladapo, and W. B. Whitman. 1994. Pathway of glycogen metabolism in Methanococcus maripaludis. J. Bacteriol. 176:325–332.

Zehnder, A. J. B., and K. Wuhrman. 1976. Titanium(III) citrate as a non-toxic, oxidation-reduction buffering system
for the culture of obligate anaerobes. Science 194:1165–
1166.
Zellner, G., and J. Winter. 1987. Secondary alcohols as hydrogen donors for CO$_2$-reduction by methanogens. FEMS
Microbiol. Lett. 44:323–328.

Zhao, H., A. G. Wood, F. Widdel, and M. P. Bryant. 1988. An
extremely thermophilic Methanococcus from a deep sea
hydrothermal vent and its plasmid. Arch. Microbiol.
150:178–183.
Zhilina, T. N. 1983. New obligate halophilic methane-producing bacterium. Microbiology 52:290–297.

Prokaryotes (2006) 3:274–280
DOI: 10.1007/0-387-30743-5_14

CHAPTER 14

Nanoarchaeota

HARALD HUBER, MICHAEL J. HOHN, REINHARD RACHEL AND KARL O. STETTER

Introduction

All cultivated *Archaea* known so far belong to one of two phyla, *Crenarchaeota* or *Euryarchaeota* (Woese et al., 1990). The *Crenarchaeota* are composed of three orders, Desulfurococcales (Huber et al., 2001), Sulfolobales and Thermoproteales (Burggraf et al., 1997; Garrity and Holt, 2001a). The *Euryarchaeota* harbor different orders of methanogenic and halophilic microorganisms, sulfate reducers (Archaeoglobales), Thermoplasmatales, and Thermococcales (Garrity and Holt, 2001b). In addition, several years ago, a third archaeal phylum was proposed, the so-called "*Korarchaeota*" (Barns et al., 1996). However, this group is only indicated by 16S rDNA sequences obtained from environmental DNAs and still nothing is known of the corresponding organisms. In contrast, a novel phylum named "*Nanoarchaeota*" was described recently by Huber et al. (2002), which is represented by a nano-sized hyperthermophilic symbiont that grows attached to the surface of a new *Ignicoccus* species. "*Nanoarchaeum equitans*," the first representative of this phylum, was obtained from a sample of submarine hot rocks taken at the Kolbeinsey Ridge, north of Iceland, from a water depth of 106 m (Huber et al., 2002). It has a cell diameter of only 400 nm and grows under strictly anaerobic conditions at temperatures between 75 and 98°C. Molecular investigations revealed that "*Nanoarchaeum equitans*" harbors a highly divergent 16S rDNA sequence which exhibits several base exchanges even in previously "universal" sequence signatures (= primer sequences). In addition, several environmental 16S rDNA sequences from high temperature biotopes have been obtained which also belong to representatives of this phylum (Hohn et al., 2002).

Phylogeny

On the basis of 16S rDNA sequence comparisons, the phylum "*Nanoarchaeota*" represents an isolated deep lineage within the Archaea (Fig. 1). It forms a separate cluster and is distinct from the three other phyla, the *Crenarchaeota*, *Euryarchaeota* and the "*Korarchaeota*." "*Nanoarchaeum equitans*" and three other 16S rRNA sequences representing this phylum (LPC33, OP9, and CU1) exhibit phylogenetic similarities between 83 and 100% (Hohn et al., 2002). The sequence from the East Pacific Rise (LPC33) was identical to the sequence of "*Nanoarchaeum equitans*," while the others had previously unknown primary structures. In agreement with this, the three sequences consistently grouped together in phylogenetic analyses with high bootstrap support (98–100%), regardless of the dataset or phylogenetic algorithm used (distance matrix: neighbor joining; Fitch-Margoliash algorithm; maximum parsimony; or maximum-likelihood (fastDNAml) methods; Ludwig and Strunk, 2001). In comparison to all other archaeal species, the phylogenetic similarities were between 67 and 80% (Hohn et al., 2002). These values indicate that the phylum is well defined by 16S rRNA gene sequence comparisons, as expressed in the corresponding phylogenetic trees (Fig. 1). In all phylogenetic analyses, the "*Nanoarchaeota*" branch off very deeply within the archaeal domain. The placement of the "*Nanoarchaeota*" branch within the Archaea, however, is problematic. Their position varies significantly, depending on the analytical method used and the domain-specific filter sets applied. Also, in all calculations, very low (and therefore insignificant) bootstrap values are obtained for the branching point of the "*Nanoarchaeota*" lineage. Therefore, the definite branching position for the "*Nanoarchaeota*" is unclear at the moment. Possibly, additional organisms and/or sequences of this novel lineage may stabilize the 16S rRNA-based trees.

Phylogenetic investigations based on analyses of the 23S rRNA sequence of "*Nanoarchaeum equitans*" resulted in similar tree topologies and were consistent with the 16S rRNA trees (M. J. Hohn and H. Huber, unpublished observation).

Taxonomy

The phylum "*Nanoarchaeota*" was recently described by Huber et al. (2002). At present, the "*Nanoarchaeota*" are comprised of one genus

Fig. 1. Phylogenetic trees based on 16S rRNA sequences. The trees were calculated using NJ, neighbor joining; FM, Fitch Margoliash; PA, maximum parsimony; and FML, maximum likelihood calculation program. Scale bar: 10 estimated exchanges within 100 nucleotides.

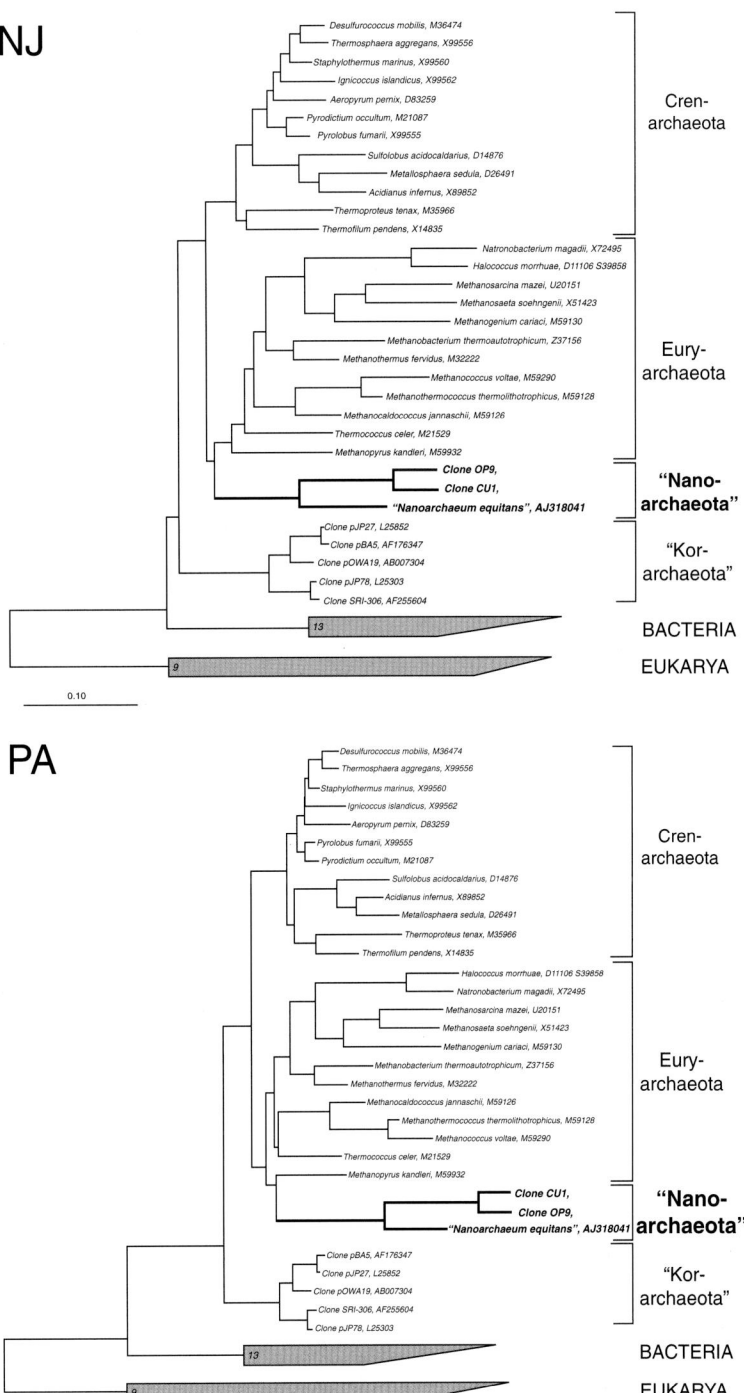

with one species: "*Nanoarchaeum equitans.*" However, this species is not yet formally described, and therefore the name has to be written in quotation marks. Besides this organism, three further "nanoarchaeotal" 16S rDNA sequences from environmental DNAs are known which were obtained from different high temperature biotopes (sequence LPC33 is identical to "*Nanoarchaeum equitans*"; Hohn et al., 2002). So far, it is unclear whether these new "*Nanoarchaeota*" sequences represent small symbiontic organisms similar to "*Nanoarchaeum equitans.*"

Habitats

"*Nanoarchaeum equitans*" was obtained from a sample of hot rocks taken at the Kolbeinsey

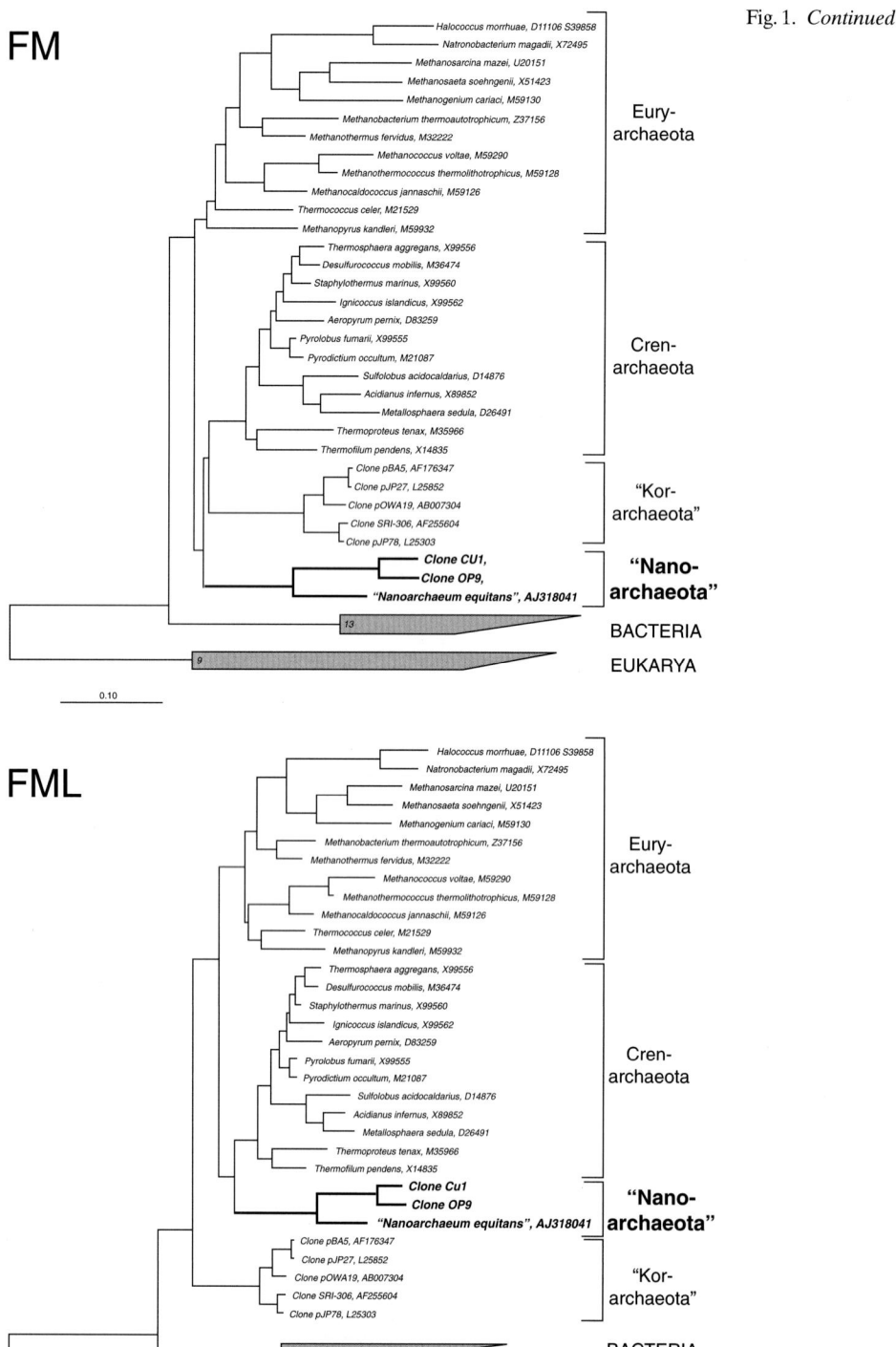

Fig. 1. *Continued*

Ridge, north of Iceland (Huber et al., 2002). This hydrothermal system is located at the subpolar Mid-Atlantic Ridge at a depth of around 106 m (Fricke et al., 1989). For the detection of "nanoarchaeotal" 16S rRNA genes, environmental DNAs obtained from a sample collection from high temperature submarine sandy sediments, venting water, and material from black smokers were investigated. Furthermore, muds, sediments, and spring waters from continental solfataric areas from the same collection were screened. Positive polymerase chain reaction (PCR) amplifications of "nanoarchaeotal" 16S rRNA genes were obtained from a sample of the

East Pacific Rise, designated "LPC33" (black smoker fragment; pH 6.5), and from two continental samples: one found in Obsidian Pool, Yellowstone National Park, United States, designated "OP9" (original temperature 80°C; pH 6.0) and the other in Uzon Caldera, Kamchatka, Russia, designated "CU1" (original temperature 85°C; pH 5.5; Hohn et al., 2002). These results document a broad distribution of members of the "Nanoarchaeota" in high temperature biotopes in the deep sea, in shallow marine areas, and in solfataric fields on land. Moreover, the habitats are located on different continents, suggesting a worldwide distibution for "Nanoarchaeota."

Isolation

Enrichment

"Nanoarchaeum equitans" has been enriched in 1/2 SME medium, prepared by the use of the anaerobic techniques of Balch and Wolfe (Balch and Wolfe, 1976; see the section Media for Cultivation in this Chapter) by anaerobic incubation at 90°C in the presence of S^0 and H_2/CO_2 (80:20, v/v, 300 kPa) without organic components. Under these conditions a new autotrophic sulfur-reducing species of the archaeal genus Ignicoccus could be enriched. In contrast to the Ignicoccus species known so far, several Ignicoccus cells were covered by very tiny cocci which turned out to represent "Nanoarchaeum equitans." In addition, some "Nanoarchaeum equitans" cells existed in the free state.

Isolation Procedures

Anaerobic culture medium is inoculated with 0.5–1 ml (or grams) of the environmental sample. The culture attempts should be checked by phase-contrast microscopy over a period of up to several weeks for growth. "Nanoarchaeum equitans" can be physically isolated either by using "optical tweezers" (Huber et al., 1995; see also the chapter Thermoproteales in this Volume) or by ultrafiltration (pore width: 0.45 µm). However, all attempts to grow "Nanoarchaeum equitans" in pure culture have failed so far. In contrast, the isolation of a combination of a "Nanoarchaeum equitans" cell attached to an Ignicoccus cell using "optical tweezers" resulted in a defined coculture.

Identification

By electron microscopy, "Nanoarchaeum equitans" cells exhibit a rather constant cell diameter of about 400 nm. Their surface is directly attached to the surface of the Ignicoccus cells, an outer membrane (Huber et al., 2000; Rachel et al., 2002; Fig. 2a, b and c). No specific attachment structures have been detected so far at the sites of contact, suggesting a loose connection between "Nanoarchaeum equitans" cells and Ignicoccus cells. In line with this assumption, they can be removed by mild sonication (30 W). "Nanoarchaeum equitans" cells do not divide by septa formation and spores are not formed. The organisms stain Gram-negative. Freeze-fracturing, freeze-etching, and ultrathin sections show that the cytoplasm of "Nanoarchaeum equitans" cells is surrounded by a cytoplasmic membrane, a 20-nm wide periplasmic space, and an S-layer (regular surface layer) with six-fold symmetry and a lattice constant of 15 nm as the only cell wall component (Huber et al., 2002; R. Rachel, unpublished observation; Figs. 2b and c). The cells can be stained by DNA-specific fluorescence microscopy (DAPI staining). The G+C-content of the genomic DNA is 32 mol%.

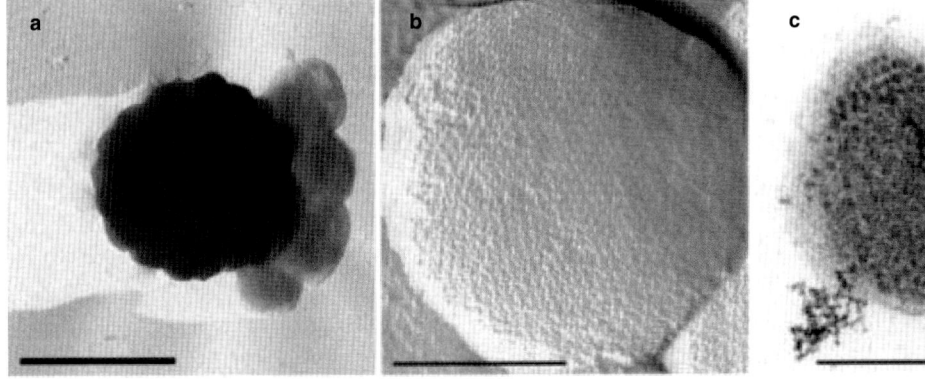

Fig. 2. Electron micrographs of cells of "Nanoarchaeum equitans" attached to Ignicoccus sp. Kin4/I: a) platinum-shadowed; b) freeze-etched; and c) thin section. Bars: 1 µm (a) and 0.2 µm (b and c).

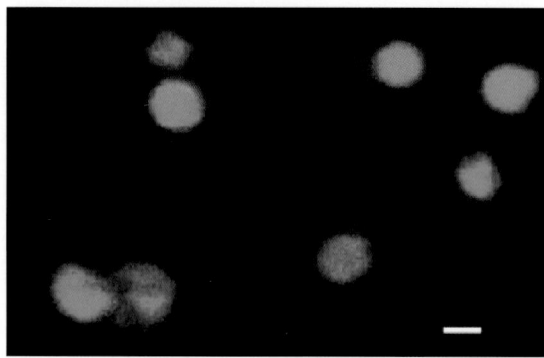

Fig. 3. Confocal laser scanning micrograph of "*Nanoarchaeum equitans*" attached to *Ignicoccus* sp. Kin4/I, after hybridization with two oligonucleotide probes specifically designed for the rRNA sequence of each microorganism. "*Nanoarchaeum*": red (CY3) and *Ignicoccus*: green (rhodamine green). Bar: 5 μm.

KH_2PO_4	0.5 g
NH_4Cl	0.5 g
$CaCl_2 \cdot 2H_2O$	0.38 g
$NaHCO_3$	0.16 g
KCl	0.33 g
NaBr	50 mg
H_3BO_3	15 mg
$SrCl_2$	7.5 mg
KI	0.05 mg
Elemental sulfur	10 g
Resazurin	1 mg
$Na_2S \cdot 9H_2O$	0.5 g
Distilled water	ad 1 liter

Adjust pH to 6.0, and incubate in a gas phase of H_2/CO_2 (80 : 20) under pressure (300 kPa).

This medium is based on the originally described "SME" medium (Stetter et al., 1983), modified by Pley et al. (1991), and described as "1/2 SME-medium" by Blöchl et al. (1997) and Huber et al. (2000).

Cultivation in Liquid Media

Ignicoccus sp. strain Kin4/I and "*Nanoarchaeum equitans*" are cultivated using the anaerobic technique of Balch and Wolfe (1976).

For small volumes, the organisms are grown in stoppered and pressurized 120-ml serum bottles (Bormioli, Italy) made of "type III" glass. One liter of the corresponding medium is flushed with N_2/CO_2 for 20 min in a 1-liter stoppered glass bottle. Then dissolved oxygen is removed by the addition of 0.5 g of $Na_2S \cdot 9H_2O$, and the pH is adjusted with H_2SO_4. The medium is dispensed in an anaerobic chamber and finally the culture vessels are pressurized with H_2/CO_2 (80 : 20; 300 kPa) by gas exchange at a gas station. Growth on solidified media is not described for *Ignicoccus* and the "*Nanoarchaeota*."

As a major characteristic, "*Nanoarchaeum equitans*" only grows in coculture with *Ignicoccus* strain KIN4/I, indicating a symbiontic mode of life. The final cell density of "*Nanoarchaeum equitans*" in serum bottles is about 3×10^7 cells $\cdot$ ml^{-1}, which is in the same range as the maximal cell density of *Ignicoccus* strain KIN4/I. In such a culture, about half of the *Ignicoccus* cells harbor "*Nanoarchaeum equitans*" cells. The number of attached "*Nanoarchaeum equitans*" cells can vary between one to more than ten cells per *Ignicoccus* cell.

The sequence of the 16S rRNA gene of "*Nanoarchaeum equitans*" is so far unique. It harbors many base exchanges, even in so-called "highly conserved regions" usually employed as primer targets for PCR (Huber et al., 2002). Owing to this great divergence, cells of "*Nanoarchaeum equitans*" do not stain by fluorescence *in situ* hybridization (FISH) using 16S rRNA-targeted oligonucleotide probes directed against *Crenarchaeota* and *Euryarchaeota* (e.g., EURY498R and CREN499R; Burggraf et al., 1994). However, after redesigning these oligonucleotide probes on the basis of the "*Nanoarchaeum equitans*" sequence, they exhibit a bright fluorescence (Fig. 3), indicating that "*Nanoarchaeum equitans*" possesses ribosomal RNA harboring the target sequence.

Preservation

Ignicoccus sp. strain Kin4/I and "*Nanoarchaeum equitans*" can be stored under anaerobic conditions at 4°C or at room temperatures for several months. The best results for long-term storage are achieved in culture media containing 5% dimethyl sulfoxide and storage in the gas phase over liquid nitrogen (around −140°C). No loss of cell viability is observed after storage of two years.

Cultivation

Medium for Cultivation

Culture Medium for "*Nanoarchaeum equitans*"— *Ignicoccus* coculture (1/2 SME-medium, modified)

NaCl	13.85 g
$MgSO_4 \cdot 7H_2O$	3.50 g
$MgCl_2 \cdot 6H_2O$	2.75 g

Physiology

Since growth of "*Nanoarchaeum equitans*" depends on propagating *Ignicoccus* cells (strain KIN4/I), very similar growth parameters are observed for both organisms. Growth of the coculture is only obtained under strictly anaerobic conditions between 70 and 98°C. Optimal growth occurs at 90°C, pH 6, and a salt concentration of 2% NaCl. The final cell density of "*Nanoarchaeum equitans*" in serum bottles (about 3×10^7 cells · ml^{-1}) can be increased about 10-fold in an enamel protected fermentor (300-liter) by adjusting the gassing rate to 30 liters · min^{-1} (H$_2$/ CO$_2$ = 80 : 20). This procedure improves hydrogen supply and efficiently removes H$_2$S, the main metabolic end product of *Ignicoccus*. Under these conditions, the cell density of *Ignicoccus* remains unchanged. During the late exponential growth phase of such a culture, about 80% of the "*Nanoarchaeum*" cells detach from their host cells and occur freely in suspension.

The host of "*Nanoarchaeum equitans*," *Ignicoccus* strain KIN4/I, is an obligate chemolithoautotroph using molecular hydrogen as the sole electron donor and elemental sulfur as the electron acceptor (S^0-H$_2$ autotrophy). It produces high amounts of H$_2$S as a metabolic end product. Similar to those described for *Ignicoccus* species, the final cell density and shortest doubling time are not significantly influenced by the addition of organic compounds to the culture medium. In contrast, for "*Nanoarchaeum equitans*" only a few metabolic properties are known so far. Isolated cells do not grow on cell homogenates of *Ignicoccus*; they require an actively growing *Ignicoccus* culture. A wide variety of single and complex organic compounds have been tested but none significantly increased the final cell density of "*Nanoarchaeum equitans*." "*Nanoarchaeum*" cells locally separated by dialysis bags from *Ignicoccus* cells are unable to propagate. Therefore, a direct cell-cell contact with the host appears to be a prerequisite for growth.

Genetics

The genome of "*Nanoarchaeum equitans*" has been recently sequenced by DIVERSA Corp. (San Diego, CA, United States). It has a length of about 490 kb and is therefore the smallest archaeal genome known so far. Its analysis may provide more insights into the physiology and phylogenetic position of "*Nanoarchaeum equitans*."

Ecology

"*Nanoarchaeum equitans*" and its host *Ignicoccus* strain KIN4/I have been isolated from a marine hydrothermal system. Other high temperature marine and terrestrial environments harbor further "nanoarchaeotal" sequences. Interestingly, these include continental biotopes with low ionic strength, which are so far unknown biotopes for *Ignicoccus* species. This

gives rise to the questions: is it characteristic for the "*Nanoarchaeota*" to live as symbionts or even as parasites and what organisms can serve as hosts? The existence of "*Nanoarchaeum equitans*" demonstrates that close relationships (symbiosis or parasitism) occur not only between members of different domains, but even within the Archaea.

Acknowledgments. The authors thank Peter Hummel, Bärbel Schwarz and Daniela Näther for help with electron microscopy and Brian Hedlund for critically reading the manuscript. We wish to thank Verena C. Wimmer for providing equipment and advisory help in confocal laser scanning microscopy.

This work was supported by a grant of the Deutsche Forschungsgemeinschaft to HH (Förderkennzeichen HU 703/1-1) and the Fonds der chemischen Industrie to KOS.

Literature Cited

Balch, W. E., and R. S. Wolfe. 1976. New approach to the cultivation of methanogenic bacteria: 2-mercaptoethanesulfonic acid (HS-CoM)-dependent growth of Methanobacterium ruminantium in a pressurized atmosphere. Appl. Environ. Microbiol. 32:781–791.

Barns, S. M., C. F. Delwiche, J. D. Jeffrey, and N. R. Pace. 1996. Perspectives on archaeal diversity, thermophily and monophyly from environmental rRNA sequences. Proc. Natl. Acad. Sci. USA 93:9188–9193.

Blöchl, E., R. Rachel, S. Burggraf, D. Hafenbradl, H. W. Jannasch, and K. O. Stetter. 1997. Pyrolobus fumarii, gen. and sp. nov., represents a novel group of Archaea, extending the upper temperature limit for life to 113°C. Extremophiles 1:14–21.

Burggraf, S., T. Mayer, R. Amann, S. Schadhauser, C. R. Woese, and K. O. Stetter. 1994. Identifying members of the domain Archaea with rRNA-targeted oligonucleotide probes. Appl. Environ. Microbiol. 60:3112–3119.

Burggraf, S., H. Huber, and K. O. Stetter. 1997. Reclassification of the crenarchaeal orders and families in accordance with 16S rRNA sequence data. Int. J. Syst. Bact. 47:657–660.

Fricke, H., O. Giere, K. O. Stetter, G. A. Alfredsson, and J. K. Kristjansson. 1989. Hydrothermal vent communities at the shallow subpolar Mid-Atlantic Ridge. Mar. Biol. 50:425–429.

Garrity, G. M., and J. G. Holt. 2001a. Phylum AI: Crenarchaeota phy. nov. In: G. Garrity (Ed.)Bergey's Manual of Systematic Bacteriology, 2nd ed. Springer-Verlag. New York, NY. 1:169.

Garrity, G. M., and J. G. Holt. 2001b. Phylum AII: Euryarchaeota phy. nov. In: G. Garrity (Ed.)Bergey's Manual of Systematic Bacteriology, 2nd ed. Springer-Verlag. New York, NY. 1:211.

Hohn, M. J., B. P. Hedlund, and H. Huber. 2002. Detection of 16S rDNA sequences representing the novel phylum "*Nanoarchaeota*": indication for a broad distribution in high temperature. Syst. Appl. Microbiol. 25:551–554.

Huber, R., S. Burggraf, T. Mayer, S. M. Barns, P. Rossnagel, and K. O. Stetter. 1995. Isolation of a hyperthermophilic archaeum predicted by in situ RNA analysis. Nature 367:57–58.

Huber, H., S. Burggraf, T. Mayer, I. Wyschkony, R. Rachel, and K. O. Stetter. 2000. Ignicoccus gen. nov., a novel genus of hyperthermophilic, chemolithoautotrophic Archaea, represented by two new species, Ignicoccus islandicus sp. nov. and Ignicoccus pacificus. sp. nov. Int. J. Syst. Evol. Microbiol. 50:2093–2100.

Huber, H., and K. O. Stetter. 2001. Order II: Desulfurococcales. *In:* G. Garrity (Ed.) Bergey's Manual of Systematic Bacteriology, 2nd ed. Springer-Verlag. New York, NY. 1:179–180.

Huber, H., M. J. Hohn, R. Rachel, T. Fuchs, V. C. Wimmer, and K. O. Stetter. 2002. A new phylum of Archaea represented by a nanosized hyperthermophilic symbiont. Nature 417:63–67.

Ludwig, W. and O. Strunk. 2001. ARB: A software environment for sequence data.

Pley, U., J. Schipka, A. Gambacorta, H. W. Jannasch, H. Fricke, R. Rachel, and K. O. Stetter. 1991. Pyrodictium abyssi sp. nov. represents a novel heterotrophic marine archaeal hyperthermophile growing at 110°C. Syst. Appl. Microbiol. 14:245–253.

Rachel, R., I. Wyschkony, S. Riehl, and H. Huber. 2002. The ultrastructure of Ignicoccus: Evidence for a novel outer membrane and for intracellular vesicle budding in an archaeon. Archaea 1:9–18.

Stetter, K. O., H. König, and E. Stackebrandt. 1983. Pyrodictium gen. nov., a new genus of submarine disc-shaped sulphur reducing archaebacteria growing optimally at 105°C. Syst. Appl. Microbiol. 4:535–551.

Woese, C. R., O. Kandler, and M. L. Wheelis. 1990. Towards a natural system of organisms: Proposal for the domains Archaea, Bacteria and Eukarya. Proc. Natl. Acad. Sci. USA 87:4576–4579.

Prokaryotes (2006) 3:281–289
DOI: 10.1007/0-387-30743-5_15

CHAPTER 15

Phylogenetic and Ecological Perspectives on Uncultured Crenarchaeota and Korarchaeota

SCOTT C. DAWSON, EDWARD F. DELONG AND NORMAN R. PACE

Introduction

Molecular phylogenetic analyses divide the domain Archaea into two lineages: Euryarchaeota and Crenarchaeota. Potentially a third main line, Korarchaeota (Woese et al., 1990; Barns et al., 1996), is indicated by only a few environmental sequences and so remains to be further documented. In the late 1970s, at the time of the initial recognition of Archaea as a third domain of life separate from Bacteria or Eucarya, the clade Archaea was thought to consist mostly of "extremeophiles," organisms inhabiting hostile or unusual settings such as geothermal or hypersaline environments. This perception, which proved to be highly distorted, was founded in the fact that most of the cultivated representatives of Archaea derived from such environments. The diverse physiologies of the cultured examples of Archaea seemed to reflect their settings. Cultured types of Euryarchaeota are physiologically varied, including methanogens, halophiles, and thermophiles. Cultivated members of the Crenarchaeota, on the other hand, have seemed more phenotypically homogeneous. Essentially all cultured Crenarchaeota are sulfur-dependent thermophiles and hyperthermophiles. More recently it has become evident that the cultured organisms represent a minority of crenarchaeal diversity, only one of several phylogenetic groups of Crenarchaeota. The uncultured kinds of organisms have only been encountered through environmental sequence studies. Considering the environmental settings of the uncultured organisms, the majority of types of Crenarchaeota are not thermophiles or "extremophiles" but rather must be viewed as mesophiles or psychrophiles of uncertain physiology.

Our new understanding of the diversity of Crenarchaeota is due largely to the use of molecular techniques that survey microbial diversity in natural environments without the traditional requirement to pure-culture microorganisms in order to identify them. These molecular techniques mainly use cloning and sequencing rRNA genes from environmental DNA in order to identify the phylogenetic types of organisms, the "phylotypes," that comprise the ecosystem analyzed. Since 1992 (DeLong, 1992), environmental crenarchaeal rDNAs have been encountered in such "mundane" temperate settings as the open ocean, terrestrial soils, and freshwater and marine sediments. In addition to the cosmopolitan, non-thermophilic representatives of Crenarchaeota, rRNA sequences have revealed novel thermophilic lineages in geothermal environments around the world, some related specifically to the novel mesophiles. The applications of molecular methods have expanded spectacularly our view of crenarchaeal diversity. The abundance and wide distribution of Crenarchaeota in the environment attest that they play important, still unknown roles in the global ecosystem.

Phylogeny of Crenarchaeota

Our perception of the phylogeny of Crenarchaeota is limited currently to only about 300 small-subunit rRNA sequences, mostly derived from environmental samples. A diagrammatic phylogenetic tree based on these sequences is shown in Figure 1. Although the deeper branchings among the Crenarchaeota are not well resolved, it is nonetheless clear that this primary archaeal clade is comprised of several major sub-clades. There is no formal nomenclature for the uncultured phylogenetic groups thus far established; names of the groups are anecdotal. The phylogeny and taxonomy of specific cultivated representatives of the Crenarchaeota have been summarized (Burggraf et al. 1997a) and are discussed elsewhere in this volume.

Figure 1 is a phylogenetic tree of Crenarchaeota showing positions of uncultivated lineages. This diagrammatic tree is based on multiple bootstrapped analyses using three different phylogenetic methods (maximum likelihood, parsimony, and minimum evolution) to infer the tree topologies. The bootstrap values, determined as percentages of 100 trees inferred by each type of analysis, are given for branches having greater than 50% support (parsimony and minimum

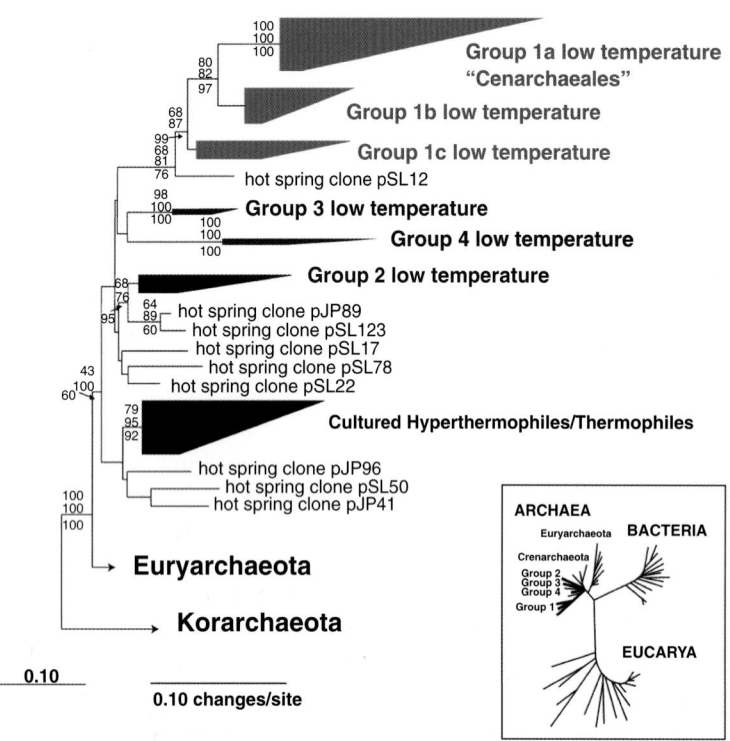

Fig. 1. Phylogenetic analysis of uncultivated Crenarchaeota.

evolution values, above lines respectively, and maximum likelihood values below lines). The scale bar indicates 0.10 changes/site. 56 representative rRNA sequences incorporating 856 unambiguously homologous nucleotide positions were used to create the tree. The size of the wedges is roughly proportional to the number of rRNA sequences in that particular phylogenetic group.

Cultured crenarchaeotes, all thermophiles, are seen to fall into only one of several main relatedness groups, distinct from several other, still uncultured thermophilic and low-temperature clades of equivalent branch-depth in the tree. These Crenarchaeota are only distantly related to the other major clades with no cultivars, about the same in rRNA sequence-divergence as between the bacterial divisions. Thus, by far the majority of known crenarchaeal diversity has no cultured representation. Since only a small number of environmental settings has been examined for representatives of Crenarchaeota, it is unquestionable that our understanding of their evolutionary history and distribution will expand substantially with further study.

rRNA-based phylogenetic trees suggest that the low-temperature lineages generally arose from thermophilic lines. This is based on the fact that low-temperature sequences generally group with thermophilic ones in phylogenetic trees and

the assumption that modern organisms that represent the short branches are most similar in nature to the ancestral kinds of organisms. The relatively long line-segments of low-temperature sequences compared to the thermophilic ones thus would indicate that low-temperature lines were derived from thermophilic ancestry. According to the topology of the emerging crenarchaeal tree (Fig. 1), this happened multiple times.

Korarchaeota

The clade "Korarchaeota" is represented by only a handful of small-subunit rRNA gene sequences; there are no cultivated representatives. The group is significant because of its deeply branching property in phylogenetic trees (Fig. 1). Analyses with small-subunit sequences indicate that the group branches more deeply than even the Crenarchaeota/Euryarchaeota separation and thus would merit the status of a third "kingdom" of Archaea. This result remains to be confirmed, however. The deeply divergent nature of Korarchaeota in phylogenetic analyses is dependent technically on the inclusion of representative eukaryotic rRNA sequences in the analysis. Analyses that include only bacterial and archaeal sequences associate Korarchaeota with Crenarchaeota (Barns et al., 1996). The status of this group needs to be tested and consolidated by further study of other gene sequences and descriptions of organismal properties.

First encountered in a survey of archaeal rRNA genes in Yellowstone's Obsidian Pool (formerly Jim's Black Pool), sequences representing Korarchaeota have been detected in other Yellowstone hot springs (unpublished). Considering their relatively high abundance where detected, korarchaeotes are likely to be abundant in similar settings globally. Obsidian Pool contains high concentrations of reduced iron, sulfur compounds and probably hydrogen. The biomass there is largely associated with paving stone-like stromatolitic structures. The relatively high abundance of korarchaeotes in that setting, a few percent of rRNA genes in bulk biomass, may indicate that they are engaged in primary production and so are likely to be chemolithotrophic.

Although not yet pure-cultured, the Obsidian Pool organisms detected as rDNA clones JP27 and JP78 have been maintained in consortium and visualized using fluorescent rRNA probes (Burggraf, Heyder et al., 1997b). Both are rod-shaped and occur only attached to surfaces, both common traits of organisms in the environment.

Low-Temperature Crenarchaeota

Low-temperature crenarchaeotes were first discovered during rRNA-based surveys of marine microbiota (DeLong, 1992). Nonthermophilic Crenarchaeota subsequently have been detected in freshwater and marine sediments, terrestrial soils, anaerobic digesters, deep subsurface paleosols, and in association with metazoan species (Table 1). Molecular phylogenetic analyses of the evolutionary relationships among the Crenarchaeaota based on small subunit rRNA genes suggest at least four main lineages of nonthermophilic Crenarchaeota (Fig. 1).

Figure 2 gives a summary of rRNA-based molecular phylogenetic surveys of microbial assemblages in which various groups of uncultivated low-temperature representatives of the Crenarchaeota have been identified. Phylogenetic groups of low-temperature Crenarchaeota are shown in Figure 2.

Densities of Group 1 Archaea and Bacteria sampled at the indicated depths were determined by fluorescence hybridization as described (DeLong et al., 1999). The sampling site was approximately 259 km offshore of Moss Landing, CA.

Two of the low-temperature lineages (Groups 1 and 2) are specifically associated phylogeneti-cally with crenarchaeal sequences identified from Obsidian Pool, a hydrothermal spring in Yellowstone National Park (Barns et al., 1994). The marine crenarchaeal types (Group 1a) first identified by DeLong in the open ocean (DeLong, 1992) are related specifically to rDNA phylotypes that are widely distributed in freshwater aquatic sediments and terrestrial soils (Group 1b). Groups 1a, 1b, and 1c Crenarchaeota are most closely affiliated with the hot spring clone pSL12 and form the most commonly identified clade of non-thermophilic Crenarchaeota in environmental molecular surveys. The order name Cenarchaeales (DeLong and Preston, order nov.) and the family name Cenarchaeaceae (DeLong and Preston, fam. nov.) have been proposed for the clade that contains a metazoan symbiont (below), planktonic and abyssal marine Crenarchaeota (Group 1a) (Preston et al., 1996).

Group 2 low-temperature Crenarchaeota have been encountered in both freshwater and marine anoxic sediments (Hershberger et al., 1996; Schleper et al., 1997a; Chandler et al., 1998) and are most closely related to the hot spring clones pJP89 and pSL123. Apart from Group 1 Crenarchaeota, this clade is the most ecologically widespread. Group 3 and Group 4 Crenarchaeota are represented by only a few rDNA sequences at this time. They are likely sister groups to one another and are not specifically related to a hot spring clone as are the other non-thermophilic lineages (Fig. 1). Sequences representing both Group 3 and Group 4 clades seem to be less abundant types (Table 1). Low bootstrap values associated with positions of Group 3 and Group 4 in phylogenetic trees indicate that their specific branch points are not certain. This could be resolved in the future with the identification of additional members of these groups and their inclusion in tree analyses.

The crenarchaeal evolutionary tree has so far been resolved using small subunit rRNA gene phylogenies. Genomic sequence scans of fosmid clones (below) of Group 1a low-temperature Crenarchaeota, however, have detected other genes that are useful for phylogenetic analyses, including the translational elongation factor EF2 and DNA-dependent RNA polymerase II. Protein phylogenies using either of these genes support the results based on small subunit rRNA phylogenies (Stein et al., 1996; Schleper, 1997b).

Ecological Distribution and Abundance

Estimations of crenarchaeal abundance in the environment can provide clues to the ecology of that environment and provide hints to the phys-

Table 1. Summary of small subunit rDNA surveys of uncultivated low-temperature Crenarchaeotal phylogenetic diversity.

Habitat Type	Group 1a	Group 1b	Group 1c	Group 2	Group 3	Group 4	References
Marine plankton	+	–	–	–	–	–	(DeLong 1992; Fuhrman 1993; DeLong 1994; Murray, Preston et al. 1996; Massana, Murray et al. 1997; McInerney, Mullarkey et al. 1997; Takai and Horikoshi 1999)
Marine fish	+	–	–	–	–	–	(van Der Maarel, Artz et al. 1998; van der Maarel, Sprenger et al. 1999)
Marine sediments	+	–	–	–	+	–	(Vetriani, Reysenbach et al. 1998; Bidle, Kastner et al. 1999)
Marine invertebrates	+	–	–	–	–	–	(Preston, Wu et al. 1996)
Freshwater plankton	–	+	–	–	–	–	(Ovreas, Forney et al. 1997)
Freshwater sediment	–	+	–	+	+	–	(Hershberger, Barns et al. 1996; Macgregor, Moser et al. 1997; Schleper, Holben et al. 1997)
Deep paleosol	–	+	–	+	–	–	(Chandler, Brockman et al. 1998)
Forest soil	–	–	+	–	–	–	(Jurgens, Lindstrom et al. 1997; Jurgens and Saano 1999)
Agricultural soil	–	+	–	–	–	–	(Bintrim, Donohue et al. 1997; Buckley, Graber et al. 1998)
Coastal salt marsh	–	+	–	+	–	+	(Munson, Nedwell et al. 1997; Dawson 2000)
Contaminated soils	–	+	–	–	–	–	(Kudo, Shibata et al. 1997; Dojka, Hugenholtz et al. 1998; Sandaa, Enger et al. 1999)
Rice paddy soil	–	+	–	+	–	–	(Ueda, Suga et al. 1995; Kudo, Shibata et al. 1997)
Anaerobic digestor	–	–	–	+	–	–	(Godon, Zumstein et al. 1997)

iological roles of these organisms. In the case of uncultured Crenarchaeota it has been necessary so far to rely on rRNA-based molecular approaches to quantify and visualize them. There are, so far, too few studies to generalize regarding representatives of the uncultured high-temperature clades. Limited surveys in Yellowstone thermal springs indicate that crenarchaeal rRNA genes generally constitute a few percent of total rRNA genes in these settings. (Dominant forms are members of the hydrogen-metabolizing bacterial divisions Aquificales and Thermotogales.) The places uncultured crenarchaeotes occur, hot springs rich in hydrogen, Fe(II) and sulfur at many oxidation states, perhaps suggest that their metabolic themes are similar to those of the cultured crenarchaeotes, i.e. hydrogen-sulfur metabolism.

Based on the relative frequencies of different crenarchaeal rDNAs in several globally distributed ecosystems, the Group 1 clade seems to be the most abundant and widespread of all the low-temperature Crenarchaeota (DeLong, 1998). In terms of numbers, the uncultivated low-temperature Crenarchaeota can represent a significant portion of biomass in some environments. For instance, molecular hybridization methods that quantify bulk archaeal or bacterial rRNA in environmental DNA indicate a high abundance of crenarchaeal rRNA in Antarctic waters, ranging from 5–30% of total planktonic rRNA. Further, in surveys of open ocean waters, Group 1a Crenarchaeota appear to reach maximal abundance (20–30% of planktonic rRNA) below 100 m depth (Massana et al., 1997; Murray et al., 1998). An example of the vertical distribu-

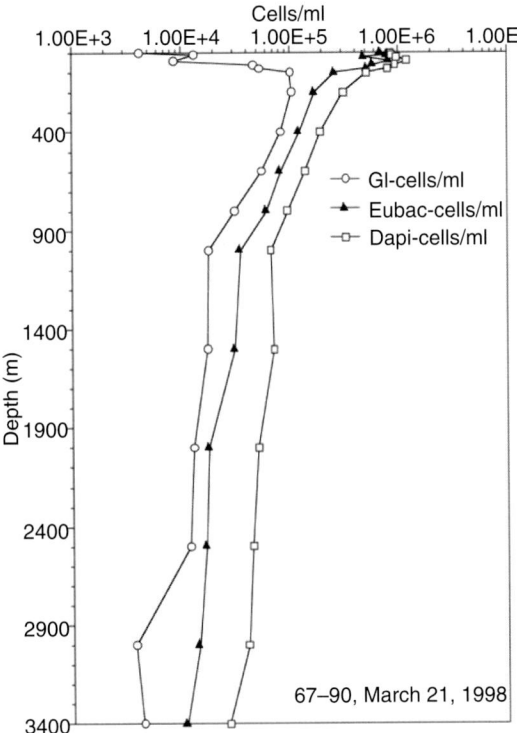

Fig. 2. Vertical distribution of Group 1a Crenarchaeota.

tion of Group 1a crenarchaeotes in the ocean is shown in Figure 2 (DeLong et al., 1999). The results indicate that these kinds of organisms constitute a significant component, about 30 percent, of the microbes in the deep sea.

Cloning and rRNA-targeted probing experiments indicate that low temperature Crenarchaeota are more rare in environments that are less oligotrophic than the marine setting, for instance freshwater and soil environments, in which they also are commonly found. In a molecular survey of Archaea in Lake Michigan sediment, members of the Group 1b type of nonthermophilic Crenarchaeota were reported to account for roughly 10% of bulk archaeal rRNA (Macgregor et al., 1997). In addition, non-thermophilic Crenarchaeota in this freshwater environment were found to be of greater abundance in the oxic region, although their range extended into the anoxic zone (Macgregor et al., 1997). A relatively low abundance (roughly 1% of total community rRNA) of Group 1b Crenarchaeota also has recently been reported in agricultural soils (Buckley et al., 1998).

Quantification of bulk archaeal lipids (archaeols) in order to estimate organism abundance is an alternative and complementary approach to rDNA analysis. In contrast to the ester-linked lipids of Bacteria and Eucarya, archaeal lipids (archaeols) are ether-linked. Analysis of such lipids has been used to detect Archaea indirectly in hypersaline and anoxic environments. The characteristic cyclic and acyclic diphytanyl glycerol tetraether lipids have been used as markers for the presence of Archaea in marine planktonic environments. Cyclic ether-linked lipids are not found in cultured Archaea among the thermophilic Crenarchaeota, but they co-occur with low-temperature Crenarchaeota and so likely occur in those kinds of organisms as well. Based on the abundance of these lipids, crenarchaeotes would seem to be the major biological source of tetraether lipids in the marine planktonic environment. This is additional confirmation of the ecological significance of low-temperature Crenarchaeota (Delong et al., 1998).

General Biology and Potential Physiologies of Uncultured Crenarchaeota

The physiological properties of uncultured Crenarchaeota are unknown. They are sufficiently different from cultured organisms that such inferences based on known organisms cannot be made reliably. In terms of energy metabolism, cultured representatives of the Archaea as a whole tend toward chemolithoautotrophy, including methanogenesis, and the reaction of sulfur compounds and hydrogen. Aerobic and anaerobic heterotrophic oxidation of carbon compounds are less common. Phototrophic metabolism (probably not resulting in primary production) is limited to the halophiles and, based on culture studies, the distribution of nitrogen fixation is limited. The thermophilic clade of the Crenarchaeota with cultivars is predominantly composed of anaerobic hydrogen and sulfur metabolizers. Uncultured thermophilic crenarchaeotes are abundant in geothermal areas, so they may also be chemolithotrophic. These generalities may hint that low-temperature Crenarchaeota, as well, carry out hydrogen and sulfur metabolism (Hershberger et al., 1996).

Deduction of the physiology or metabolism of the uncultured Crenarchaeota on the basis of properties of their cultured relatives is complicated by several factors. First, the uncultured clades of Crenarchaeota are only distantly related to any cultured and described organisms. Secondly, physiologic traits are not necessarily always coherent within phylogenetic groups. In some cases, phylogenetic information indeed is useful in inferring physiological properties. For instance, all known methanogens are members of the Archaea, so it is reasonable to presume that methanogenesis is carried out by an uncultivated microorganism that happens to be closely related to a known methanogen. Such phylogenetic and physiological coherence does not necessarily hold, however. Closely related microbes commonly possess disparate physiologies. *Nitrobacter winogradski* and *Rhodopseudomonas palustris*, for instance, are closely related phylogenetically, but one is a non-phototrophic chemolithotroph and the other is an anaerobic phototroph. These overtly distinct metabolisms, however, belie the fundamental biochemical similarity that is indicated by the close relationships of the organisms. The remote relatedness of the multiple and distinct phylogenetic lineages of thermophilic and low-temperature Crenarchaeota make it possible that each lineage may possess strikingly different physiologies. Perhaps low-temperature crenarchaeal metabolism will prove to be as varied as that of the bacterial divisions.

In the absence of cultures, alternate approaches are required to gain insight into the physiology and metabolism of uncultured crenarchaeotes. Other experimental lines of evidence that do not depend upon cultivation can provide information on the physiology of uncultured Crenarchaeota or any other organism. These include both ecological characterizations and genomic data.

The specific abiotic and biotic properties of the ecological niches of representatives of particular uncultured crenarchaeal lineages offer clues to physiology. For instance, Groups 1a, 1b, and 1c low-temperature Crenarchaeota are common in oxic zones such as agricultural soils and the open ocean and thus likely are aerobic (DeLong, 1994; Hershberger et al., 1996; Bintrim et al., 1997; Jurgens et al., 1997; Macgregor et al., 1997; Buckley et al., 1998). On the other hand, Group 2 low-temperature Crenarchaeota occur in the anaerobic environments of sediments, marshes (Hershberger et al., 1996; Schleper et al., 1997b), and an anaerobic digester (Godon et al., 1997) and so presumably conduct anaerobic metabolism. Group 3 and Group 4 Crenarchaeota also are likely to be anaerobes, based on the occurrence of their rDNA sequences in anoxic marine and freshwater sulfide-rich sediments (Hershberger et al., 1996; Vetriani et al., 1998; Dawson, 2000). Groups 2, 3 and 4 may engage in some form of sulfur metabolism, as similar mechanisms are used by their thermophilic relatives. Both marine and freshwater sediments tend to be rich in sulfur derivatives due to microbial sulfate respiration. Hydrogen metabolism, a common theme among the thermophilic cultivars, is another possibility for some low-temperature crenarchaeotes. Anaerobic environments, particularly, can be rich in molecular hydrogen due to metabolism or to inorganic reactions such as between Fe (II) and water.

Genomic sequences of uncultivated microorganisms offer opportunities to gain insight into their physiological potential in the absence of cultivation studies. Bulk environmental high molecular weight DNA from the environment can be shotgun cloned into vectors capable of carrying large inserts, to create community genomic libraries. Using rRNA sequences as specific tags for a particular microorganism, more genes from particular organisms can be identified by sequencing upstream or downstream from the small subunit rRNA operons. Appropriate vectors can contain up to several percent of a typical archaeal genome, and so, in principle, an entire genome can be assembled from contiguous segments. Two types of non-thermophilic Crenarchaeota have been characterized in this way to some extent: a planktonic (200 m depth) marine crenarchaeote and a sponge-associated organism, *Cenarchaeum symbiosum*.

Several genes in the marine crenarchaeote, including the translational elongation factor EF2, were identified and subsequently used to provide an independent assessment and confirmation of rRNA-based phylogenies. Two other genes, an RNA helicase and GSAT (glutamate-1-semialdehyde aminotransferase), had not previously been found among the Archaea (Stein et al., 1996). GSAT is a component of the pathway for the synthesis of 5-aminolevulinic acid. 5-aminolevulinic acid can be synthesized via two pathways: a C4 pathway (present in animals, fungi, and some bacteria) or a C5 pathway (present in some plants). The crenarchaeal GSAT shares higher similarity to that of the plant GSAT proteins. This is intriguing because the C5 pathway leads to the biosynthesis of chlorophyll.

Cenarchaeum symbiosum—A Model for Low-Temperature Crenarchaeota

An unexpected niche of low-temperature Crenarchaeota is in the specific association of a single crenarchaeal phylotype, *Cenarchaeum symbiosium*, with the sponge *Axinella mexicana* (Preston et al., 1996) (Fig. 3a, b). This archaeal-metazoan association represents the first described symbiosis involving Crenarchaeota. The sponge-crenarchaeote association is specific, with a single crenarchaeal phylotype inhabiting a particular sponge host species. This natural "enrichment" of nonthermophilic Crenarchaeota offers a laboratory source of cells for experimental analysis.

The archaeon *C. symbiosum* inhabits the sponge *Axinella mexicana* (Panel A) and can be cultured in the sponge tissues in appropriate

Fig. 3. *Cenarchaeum symbiosum*, a crenarchaeal sponge symbiont.

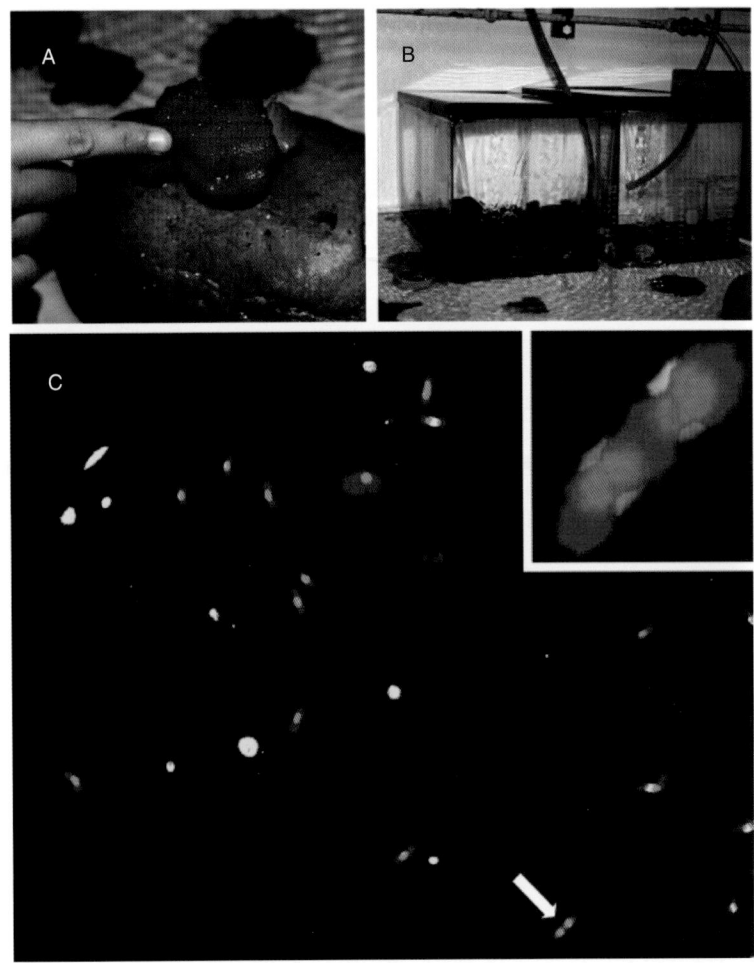

aquaria (Panel B). Differential fluorescent staining of *C. symbiosum* reveals that cellular DNA (stained with diamidinophenylindole [DAPI], yellow) is segregated from ribosomes (stained with an rRNA probe, red) (Preston et al., 1996).

Whole cell *in situ* hybridization studies using crenarchaeal-specific oligonucleotide probes have enabled the visualization and quantification of *Cenarchaeum symbiosium* in the tissues of the sponge. This crenarchaeote is a member of the Group 1a marine low-temperature clade and forms small rod-shaped cells of approximately 0.8 μm in length by 0.5 μm in width (Fig. 3c, d). Based on rDNA content, *C. symbiosum* accounts for 65% of microorganisms associated with the sponge (Preston et al., 1996). Actively dividing *C. symbiosum* have been visualized using whole cell hybridization, providing for the first time conclusive evidence of crenarchaeal growth at roughly 10°C (Fig. 3c). Moreover, several genes other than rRNA have been identified in genomic libraries made from purified cell fractions of *C. symbiosium*. One, a DNA polymerase, was studied in some detail. This polymerase showed the highest sequence-similarity to other crenarchaeal DNA polymerases, many of which are thermostable over 80°C for more than an hour. The sponge-symbiont DNA polymerase was expressed, purified, and studied with regard to some of its properties. The enzyme showed a half-life of ten minutes at 46°C and was inactivated at temperatures greater than 40°C. The thermally sensitive nature of this enzyme is consistent with the non-thermophilic phenotype of the sponge-symbiont, which has been maintained in association with the sponge in aquaria at temperatures of 10°C for several years.

Cenarchaeum symbiosum could prove a useful model for understanding cell division in the Archaea—an area in which little is known and the opportunities are many. Genomic analysis of some instances of Euryarchaeota suggests a unique mechanism of cell division in comparison to that of Bacteria or Eucarya (Bernander, 1998). Additionally, in situ hybridization experiments show that ribosomes are excluded from the nuclear region of *C. symbiosium*, a morphology more akin to the nucleus of eucaryotes than to the dispersed nucleoid of Bacteria (Preston et al., 1996; Fig. 3c). Recent microscopic studies with Sulfolobus (Poplowski, 1997) and genomic analysis of methanogens suggest a mechanism for chromosome segregation in the Archaea that may be more comparable to that of the Eucarya than that of the Bacteria. Perhaps the archaeal nuclear zone is a rudimentary version of our own nuclear matrix.

Conclusions

Molecular phylogenetic surveys of representatives of Crenarchaeota in natural environments show that temperature optimum of growth is not a defining characteristic of Crenarchaeota. Indeed, the growth-temperature optima of Crenarchaeota span the full range of life: from −1.5°C in Antarctic waters to over 100°C in sulfuric hotsprings. Molecular phylogenies of the Crenarchaeota suggest that non-thermophilic Crenarchaeota derive from relatives adapted to living at higher temperatures. The cosmopolitan distribution of multiple lineages of non-thermophilic Crenarchaeota implies that they play important roles in the terrestrial biosphere. Phylogenetically directed enrichment strategies, combined with ecological and genomic studies, should prove fruitful in determining the physiological and biochemical properties of all these yet-uncultivated microorganisms.

The deeply divergent clade Korarchaeota is so far indicated only by a few rRNA gene sequences, all from high-temperature ecosystems. This phylogenetic group remains to be substantiated by additional sequence or other information.

Literature Cited

Barns, S. M., R. E. Fundyga, et al. 1994. Remarkable archaeal diversity detected in a Yellowstone National Park hot spring environment. Proceedings of the National Academy of Sciences of the United States of America 91:1609–1613.
Barns, S. M., C. F. Delwiche, et al. 1996. Perspectives on archeal diversity, thermophily and monophyly from environmental rRNA sequences. Proceedings of the National Academy of Sciences of the United States of America 93:9188–9193.
Bernander, R. 1998. Archaea and the cell cycle. Mol. Microbiol. 29:955–961.
Bidle, K. A., M. Kastner, et al. 1999. A phylogenetic analysis of microbial communities associated with methane hydrate containing marine fluids and sediments in the Cascadia margin ODP site 892B. FEMS Microbiology Letters 177:101–108.
Bintrim, S. B., T. J. Donohue, et al. 1997. Molecular phylogeny of Archaea from soil. Proceedings of the National

Academy of Sciences of the United States of America 94:277–282.
Buckley, D. H., J. R. Graber, et al. 1998. Phylogenetic analysis of nonthermophilic members of the kingdom Crenarchaeota and their diversity and abundance in soils. Applied and Environmental Microbiology 64:4333–4339.
Burggraf, S., P. Heyder, et al. 1997. A pivotal Archaea group. Nature 385:780.
Burggraf, S., H. Huber, et al. 1997. Reclassification of the crenarchaeal orders and families in accordance with 16S rRNA sequence data. International Journal of Systematic Bacteriology 47:657–660.
Chandler, D. P., F. J. Brockman, et al. 1998. Phylogenetic diversity of archaea and bacteria in a deep subsurface paleosol. Microbial Ecology 36:37–50.
Dawson, S. C. 2000. Evolution of the Eucarya and Archaea: perspectives from natural microbial assemblages. Molecular and Cellular Biology University of California-Berkeley. Berkeley, CA. 118.
DeLong, E. F. 1992. Archaea in coastal marine environments. Proc. Natl. Acad. Sci. USA. 89:5685–5689.
DeLong, E. F., K. Y. Wu, R. V. Prézelin, M. Jovine. 1994. High abundance of Archaea in Antarctic marine picoplankton. Nature 371:695–697.
DeLong, E. F. 1998. Archael means and extremes. Science 280:542–3.
Delong, E. F., L. L. King, et al. 1998. Dibiphytanyl ether lipids in nonthermophilic crenarchaeotes. Applied and Environmental Microbiology 64:1133–1138.
DeLong, E. F., L. T. Taylor, et al. 1999. Visualization and enumeration of marine planktonic archaea and bacteria by using polyribonucleotide probes and fluorescent in situ hybridization. Appl Environ Microbiol 65:5554–5563.
Dojka, M. A., P. Hugenholtz, et al. 1998. Microbial diversity in a hydrocarbon- and chlorinated-solvent-contaminated aquifer undergoing intrinsic bioremediation. Applied and Environmental Microbiology 64:3869–3877.
Fuhrman, J. A., McCallum, K., Davis, A. A. 1993. Phylogenetic diversity of subsurface marine microbial communities from the Atlantic an Pacific oceans. Applied and Environmental Microbiology 59:1294–1302.
Godon, J.-J., E. Zumstein, et al. 1997. Molecular microbial diversity of an anaerobic digestor as determined by small-subunit rDNA sequence analysis. Applied and Environmental Microbiology 63:2802–2813.
Hershberger, K. L., S. M. Barns, et al. 1996. Wide diversity of Crenarchaeota [letter]. Nature 384:420.
Jurgens, G., K. Lindstrom, et al. 1997. Novel group within the kingdom Crenarchaeota from boreal forest soil. Applied and Environmental Microbiology 63:803–805.
Jurgens, G., and A. Saano. 1999. Diversity of soil Archaea in boreal forest before, and after clear-cutting and prescribed burning. FEMS Microbiology Ecology 29:205–213.
Kudo, Y., S. Shibata, et al. 1997. Peculiar archaea found in Japanese paddy soils. Bioscience Biotechnology and Biochemistry 61:917–920.
Macgregor, B. J., D. P. Moser, et al. 1997. Crenarchaeota in Lake Michigan sediment. Applied and Environmental Microbiology 63:1178–1181.
Massana, R., A. E. Murray, et al. 1997. Vertical distribution and phylogenetic characterization of marine planktonic Archaea in the Santa Barbara channel. Applied and Environmental Microbiology 63:50–56.

McInerney, J. O., M. Mullarkey, et al. 1997. Phylogenetic analysis of group I marine archaeal rRNA sequences emphasizes the hidden diversity within the primary group Archaea. Proceedings of the Royal Society of London Series B Biological Sciences 264:1663–1669.

Munson, M. A., D. B. Nedwell, et al. 1997. Phylogenetic diversity of Archaea in sediment samples from a coastal salt marsh. Applied and Environmental Microbiology 63:4729–4733.

Murray, A. E., C. M. Preston, et al. 1996. Molecular analyses of marine archaea and bacteria of the Antarctic Peninsula. Abstracts of the General Meeting of the American Society for Microbiology 96:347.

Murray, A. E., C. M. Preston, et al. 1998. Seasonal and spatial variability of bacterial and archaeal assemblages in the coastal waters near anvers island, Antarctica. Applied and Environmental Microbiology 64:2585–2595.

Ovreas, L., L. Forney, et al. 1997. Distribution of bacterioplankton in meromictic Lake Saelenvannet, as determined by denaturing gradient gel electrophoresis of PCR-amplified gene fragments coding for 16SrRNA. Applied and Environmental Microbiology 63:3367–3373.

Poplowski, A., Bernander, R. 1997. Nucleoid structure and distribution in thermophilic Archaea. Journal of Bacteriology 179:7625–7630.

Preston, C. M., K. Y. Wu, et al. 1996. A psychrophilic crenarchaeon inhabits a marine sponge: Cenarchaeum symbiosum gen. nov., sp. nov. Proceedings of the National Academy of Sciences of the United States of America 93:6241–6246.

Sandaa, R.-A., O. Enger, et al. 1999. Abundance and diversity of Archaea in heavy-metal-contaminated soils. Applied and Environmental Microbiology 65:3293–3297.

Schleper, C., Swanson, R. V., Mathur, E. J., DeLong, E. F. 1997. Characterization of a DNA polymerase from the uncultivated psychrophilic archaeon Cenarchaeum symbiosum. Journal of Bacteriology 179:7803–7811.

Schleper, C., W. Holben, et al. 1997. Recovery of crenarchaeotal ribosomal DNA sequences from freshwater-lake sediments. Applied and Environmental Microbiology 63:321–323.

Stein, J. L., T. L. Marsh, et al. 1996. Characterization of uncultivated prokaryotes: Isolation and analysis of a 40-kilobase-pair genome fragment from a planktonic marine archaeon. Journal of Bacteriology 178:591–599.

Takai, K., and K. Horikoshi. 1999. Genetic diversity of archaea in deep-sea hydrothermal vent environments. Genetics 152:1285–97.

Ueda, T., Y. Suga, et al. 1995. Molecular phylogenetic analysis of a soil microbial community in a soybean field. European Journal of Soil Science 46:415–421.

van der Maarel, M. J. E. C., R. R. E. Artz, et al. 1998. Association of marine Archaea with the digestive tracts of two marine fish species. Applied and Environmental Microbiology 64:2894–2898.

van der Maarel, M. J. E. C., W. Sprenger, et al. 1999. Detection of methanogenic archaea in seawater particles and the digestive tract of a marine fish species. FEMS Microbiology Letters 173:189–194.

Vetriani, C., A.-L. Reysenbach, et al. 1998. Recovery and phylogenetic analysis of archaeal rRNA sequences from continental shelf sediments. FEMS Microbiology Letters 161:83–88.

Woese, C. R., O. Kandler, et al. 1990. Towards a natural system of organisms: proposal for the domains Archaea, Bacteria, and Eucarya. Proc Natl Acad Sci U S A. 87:4576–4579.

Bacteria

Firmicutes (Gram-Positive Bacteria)

Firmicutes with High GC Content of DNA

Prokaryotes (2006) 3:297–321
DOI: 10.1007/0-387-30743-5_16

CHAPTER 1.1.1

Introduction to the Taxonomy of Actinobacteria

ERKO STACKEBRANDT AND PETER SCHUMANN

Introduction

During the past 25 years, comparative analysis of sequences of homologous and genetically stable semantides has demonstrated that several classification systems based on morphology and physiology do not reflect the natural relationships among actinomycetes and related organisms. Though today's taxonomists can better recognize the phylogenetic branching pattern of lineages than previous generations of systematists could, this ability obscures the view that previous classification systems also were established in the belief that they reflected conceptual and methodological progress. In this respect, no difference exists in the history of the actinomycetes classification and that of other groups of bacteria. In hindsight, it is their tempo and mode of evolution that make actinomycetes and their relatives difficult to classify outside the context of phylogeny: though the majority of taxa evolved rather late in Earth's history, individual characteristics (traditionally used in early classification) are highly diverse and only rarely reflect phylogenetic relationships.

This chapter highlights the progress in the elucidation of the natural relationships among actinomycetes leading to the inclusion of phylogenetic data into a gene sequence-based higher classification system. Although this taxon, with the introduction of the *Approved Lists of Bacterial Names* (Skerman et al., 1980) and guidelines for the circumscription of species (Wayne et al., 1987), was demonstrated to be a rather coherent entity, significant changes occurred at higher taxonomic ranks. This is true not only for the genus level, for which traditional descriptors play the same key role as in the past, but also more obviously for the ranks between genera and class.

The biology of most actinomycete genera has been covered in the previous edition of *The Prokaryotes* (Balows et al., 1992); some of them have been reclassified or dissected since then. Many additional genera were added in the last decade to the actinobacteria proper, increasing the phylogenetic and epigenetic diversity of the class, even though the medical, ecological and/or biotechnological importance of these new taxa have not yet been evaluated thoroughly. Table 1 compiles the validly and some invalidly described genera and organizes them into families until January 2000, following the scheme of Stackebrandt et al. (1997). The chapter also covers the taxonomy of new, mainly monogeneric families, as well as several genera for which, as compared to the 2nd edition of *The Prokaryotes* (Balows et al., 1992), no substantial increase in information has been recorded. These taxa may subsequently receive renewed attention that would justify their recognition in future updates. For the time being, only the most important taxonomic features of these taxa will be summarized.

A Note on Early Classification Systems of Actinomycetes

Probably no other group of organisms has been given more taxonomic names than the ray fungi ("Strahlenpilze") or mold bacteria. In some cases, the names were assigned to true fungi and true bacteria. Waksman (1950) lists 31 synonyms of generic names of actinomycetes, among which *Actinomyces*, *Nocardia*, *Micromonospora* and *Streptomyces* are today recognized as valid taxa. Names such as *Asteroides*, *Streptothrix*, *Cladothrix*, *Discomyces*, *Oospora*, *Actinocladothrix* and *Cohnistreptothrix* remain interesting from an historical point of view only.

Between 1877 and the early 1920s, questions about the priority of the name Actinomyces were a major issue (Flügge, 1896; Lieske, 1921; Buchanan, 1925). The earliest actinomycete genus described was *Strepthrotrix* (Cohn, 1875). However, *Streptothrix* was originally assigned to a true fungus (Corda, 1839), and for reasons of priority, the bacterial genus *Streptothrix* was declared invalid.

The terms "actinomycete" and "actinomyces" originate from the generic name *Actinomyces*, described for the causative agent of actinomycosis, the thread bacterium *A. bovis* (Corda, 1839). As in the case for *Streptothrix*, the validity of the

Table 1. Families and genera of the class Actinobacteria.

Family	Genus	Family	Genus
Acidimicrobiaceae	*Acidimicrobium*	*Actinosynnemaceae*	*Actinosynnema*
Rubrobacteraceae	*Rubrobacter*		*Lenzea*
Sphaerobacteraceae	*Sphaerobacter*		*Kutzneria*
Coriobacteraceae	*Coriobacterium*		*Streptoalloteichus*
	Atopobium		*Saccharothrix*
	Slackia	*Streptomycetaceae*	*Streptomyces*
	Egerthella		*Kitasatospora*
Actinomycetaceae	*Actinomyces*		"*Trichotomospora*"
	Actinobaculum	*Streptosporangiaceae*	*Streptosporangium*
	Arcanobacterium		*Microbispora*
	Mobiluncus		*Planomonospora*
Micrococcaceae	*Micrococcus*		*Planotetraspora*
	Arthrobacter		*Herbidospora*
	Renibacterium		*Planobispora*
	Kocuria		"*Sebekia*"
	Stomatococcus		"*Cathayosporangium*"
	Nesterenkonia	*Nocardiopsaceae*	*Nocardiopsis*
	Rothia		
Cellulomonadaceae	*Cellulomonas*		*Thermobifida*
	Oerskovia		*Prauserella*
Promicromonosporaceae	*Promicromonospora*	*Thermomonosporaceae*	*Thermomonospora*
Dermatophilaceae	*Dermatophilus*		*Spirillospora*
Dermacoccaceae	*Dermacoccus*		*Actinomadura*
	Kytococcus		
	Demetria		"*Parvopolysora*"
			"*Streptomycoides*"
Brevibacteriaceae	*Brevibacterium*	*Actinocoralliaceae*	*Actinocorallia*
Dermabacteraceae	*Dermabacter*		"*Sarraceniospora*"
	Brachybacterium	*Micromonosporaceae*	*Micromonospora*
Intrasporangiaceae	*Intrasporangium*		*Actinoplanes*
	Janibacter		*Catenuloplanes*
	Terrabacter		*Couchioplanes*
	Terracoccus		*Catellatospora*
Bogoriellaceae	*Bogoriella*		*Dactylosporangium*
Sanguibacteraceae	*Sanguibacter*		*Pilimelia*
Rarobacteraceae	*Rarobacter*		*Spirilliplanes*
Jonesiaceae	*Jonesia*		*Verrucosispora*
Microbacteriaceae	*Microbacterium*	Frankineae	*Frankia*
	Agrococcus	*Geodermatophilaceae*	*Geodermatophilus*
	Agromyces		*Blastococcus*
	Clavibacter		*Modestobacterium*
	Curtobacterium	*Microsphaeraceae*	*Microsphaera*
	Leucobacter	*Sporichtyaceae*	*Sporichtya*
	Rathayibacter	*Acidothermaceae*	*Acidothermus*
Corynebacteriaceae	*Corynebacterium*	*Cryptosporangiaceae*	*Cryptosporangium*
Mycobacteriaceae	*Mycobacterium*	*Propionibacteraceae*	*Propionibacterium*
Nocardiaceae	*Nocardia*		*Friedmaniella*
	Rhodococcus		*Luteococcus*
Gordoniaceae	*Gordonia*		*Microlunatus*
	Skermania		*Propioniferax*
Tsukamurellaceae	*Tsukamurella*		*Tessaracoccus*
Dietziaceae	*Dietzia*	*Nocardioideaceae*	*Nocardioides*
Pseudonocardiaceae	*Pseudonocardia*		*Aeromicrobium*
	Actinobispora		"*Hongia*"
	"*Actinoalloteichus*"	*Glycomycetaceae*	*Glycomyces*
	Actinopolyspora	*Bifidobacteraceae*	*Bifidobacterium*
	Amycolatopsis		*Gardnerella*
	Saccharomonospora		
	Kibdelosporangium	*Kineococcaceae*[a]	*Kineococcus*
	Saccharopolyspora		*Kineosporia*
	Thermocrispum		

[a]Tentative.

name "Actinomyces" was challenged initially. Several taxonomists pointed out that the existence of the fungus species *Actinomyce horkelii* (later recognized as an algae; Lieske, 1921) precludes the validity of *Actinomyces bovis* Harz. Nevertheless, for two reasons the name *Actinomyces* for a bacterial genus was accepted: firstly, the name was spelled differently, and secondly, the name "Actinomyce" was apparently only used by the original author Meyen (1827) and "universally regarded as invalid" (Buchanan, 1925). To credit the contribution of Cohn (1872; 1875), the generic name *Streptomyces* combined the names first given to actinomycetes, "Streptothrix" and "Actinomyces" (Waksman and Henrici, 1943).

Classification

The nomenclatural problems also highlight the confusing situation in classification. This period of about 90 years lasted until stringent taxonomic rules were introduced and chemical and molecular methods, helpful in assessment of chemotaxonomic coherency and monophyletic structure of genera, were applied. The early problems, by no means restricted to the actinomycetes, were symptomatic for the decades following the introduction of methods for obtaining pure cultures.

Botanists, by comparing spore formation, budding, and the formation of branches and conidia in bacteria to superficially similar processes in fungi, considered bacteria the lower life forms of Hyphomycetes and Ascomycetes. Consequently, emphasis in taxonomy was placed on morphological criteria (Cohn, 1872; Migula, 1894). Changes in concepts occurred when new properties were determined in bacteria, such as formation of end products, cultural characteristics, color and motility. Over decades, several classification systems were presented simultaneously but the taxonomic units were diffuse, often not described properly and not justified satisfactorily. Guidelines did not exist and it was left to the taxonomist to decide upon which alternative scheme to follow. The eight editions of *Bergey's Manual of Determinative Bacteriology* and the *Index Bergeyana* (Buchanan et al., 1966) are probably the richest historic source compiling changes in concepts in bacterial taxonomy.

Most of the early taxonomic literature states the morphological similarities between actinomycetes (Steptotrichae) and fungi. As summarized by Waksman (1950), the actinomycetes were often considered not only to occupy a position between the true fungi or the Hyphomycetes and the bacteria, but also to represent the ancestral prototype from which these two groups of organisms evolved. Some workers, like Kruse (1896), stated very clearly that the similarities between the *Streptotricheae* and fungi are superficial only, inasmuch as they are closely related to bacteria, especially to causative agents of diphtheria and tuberculosis. These two groups of bacteria were described as *Corynebacterium* and *Mycobacterium* (Lehmann and Neumann, 1896; Lehmann and Neumann, 1896) and were later, together with the actinomycete genera *Streptothrix* and *Oospora*, united in the family *Mycobacteriaceae* (Chester, 1897). Winslow et al. (1917), stressing the artificial nature of this classification system, included in *Mycobacteriaceae* not only *Actinomyces*, *Mycobacterium* and *Corynebacterium*, but also *Nocardia*, *Leptotrichia* and *Fusiformis*. Shortly thereafter, Buchanan (1917; 1918) proposed the order Actinomycetales, containing the families *Actinomycetaceae* and *Mycobacteriaceae*, a proposal that was generally accepted and published with slight taxonomic revisions in textbooks. Waksman (1950) has given a thorough compilation of classification systems of mycelium-forming actinomycetes from 1904 to 1943.

Because of the massive description of new sporoactinomycete and coryneform species in the 1950s and 1960s, the limitations of the unsatisfying classification schemes became obvious. Küster (1967), referring to the actinomycetes, stated "our conception and system of classification is entirely arbitrary" and Waksman (1967) concluded that it is the lack of an international agreement in establishing criteria dealing with the principles of nomenclature and classification that has been partly, if not largely, responsible for this confusion. The evaluation of chemotaxonomic properties in the 1970s finally convincingly demonstrated the phenotypic heterogeneity of almost all actinomycete genera. At the same time, the establishment of the *Approved Lists of Bacterial Names* (Skerman et al., 1980) marked a turning point in bacterial taxonomy. Only those taxa recognized as being validly described by an international committee of microbiologists were included on this list, and taxa not included lost their status in nomenclature. However, this important achievement did not prevent the presentation of higher taxa that in hindsight were chemotaxonomically and phylogenetically heterogeneous. Major taxonomic rearrangements at the genus and family level were initiated shortly after 1980, first by results of chemotaxonomic analyses alone, and later in concert with results of 16S rRNA sequencing analyses when correcting taxonomic conclusions according to the genomic homogeneity of taxa became feasible (Table 2).

Table 2. Example of classification changes that occurred as the value of taxonomic properties was reassessed.

Description year[a]	Genus	Description year[a]	Genus
1830	Monas	1857	Nosema
1843	Mycothamnion	1869	Tillatia
1845	Erebonemia	1870	Zoogloea
1872 Micrococcus (Cohn)			
1872	Bacterium	1894	Planococcus
1879	Sarcomyces	1900	Pseudomonas
1880	Chromobacterium	1900	Acetobacter
1882	Botromyces	1906	Rhodococcus
1883	Leucocystis	1906	Caphococcus
1884	Discomyces	1906	Albococcus
1885	Bacteriopsis	1908	Aurococcus
1885	Gaffkya	1912	Gyrococcus
1886	Lampropedia	1913	Actinococcus
1886	Palmella	1919	Nigrococcus
1886	Sphaerococcus	1920	Brucella
1886	Neisseria	1920	Erwinia
1886	Ascococcus	1923	Serratia
1887	Arthrobacterium	1925	Nitrosococcus
1887	Coccus	1933	Veilonella
1887	Pasteurella	1936	Methanococcus
1888	Jodococcus	1941	Proactinomyces
1888	Pediococcus	1942	Flavobacterium
1888	Botryococcus	1948	Streptomyces
1889	Bacillus	1953	Brevibacterium
1889	Staphylococcus	1957	Peptostreptococcus
1889	Streptococcus	1978	Stomatococcus
1889	Klebsiella	1995	Arthrobacter
1890	Photobacterium	1995	Nesterenkonia
1891	Mycobacterium	1995	Dermacoccus
1891	Diplococcus	1995	Kocuria
1892	Coccobacterium	1995	Kytococcus
1892	Actinomyces		
1894	Planosarcina		

[a]The year that genera with prior affiliation to the *Micrococcus* (Cohn, 1872) were established. Species of these genera were *Micrococcus* spp. at least once in the history of microbial taxonomy.

The Phylogenetic Aspect

The best known phylogenetic trees of the non-molecular area were first proposed by Kluyver and Van Niel (1936) and, based thereon, by Stanier and Van Niel (1941). According to their schemes, morphologically complex forms (such as sarcinas, spirilla and actinomycetes) evolved independently from morphologically simple spherical forms (Fig. 1). Krassil'nikov (1959), on the other hand, proposed the polyphyletic origin of four individual bacterial lines, one of which was the class Actinomycetes. Gram-positive cocci, through the stage of mycococci, were thought to be degenerated forms of the mycobacteria. Propionic acid bacteria were the bridge between mycobacteria and the lactic acid bacteria. If one excludes the organisms with low DNA G+C content and regards the invalid Pseudobacterium to represent the core of *Bifidobacterium*,

Krassil'nikov's proposal for the composition of this class matches closely that of the 16S rDNA-based actinomyces line of descent (Fig. 2). More recently, the use of morphology to depict evolutionary trends was outlined in a scheme (Prauser, 1975) in which the nocardioform bacteria (Prauser, 1976) were believed to represent the bridging stage between the simple coryneform organisms and the morphologically complex sporeforming actinomycetes. This idea is not supported by the placement of nocardioform organisms (other than sporeforming species) in individual lines of descent.

THE PRESENT STATUS OF ACTINOBACTERIAL CLASSIFICATION Following the recognition that mycelium-forming and non-mycelium-forming Gram-positive organisms with a DNA G+C content of <55 mol% are members of a major line

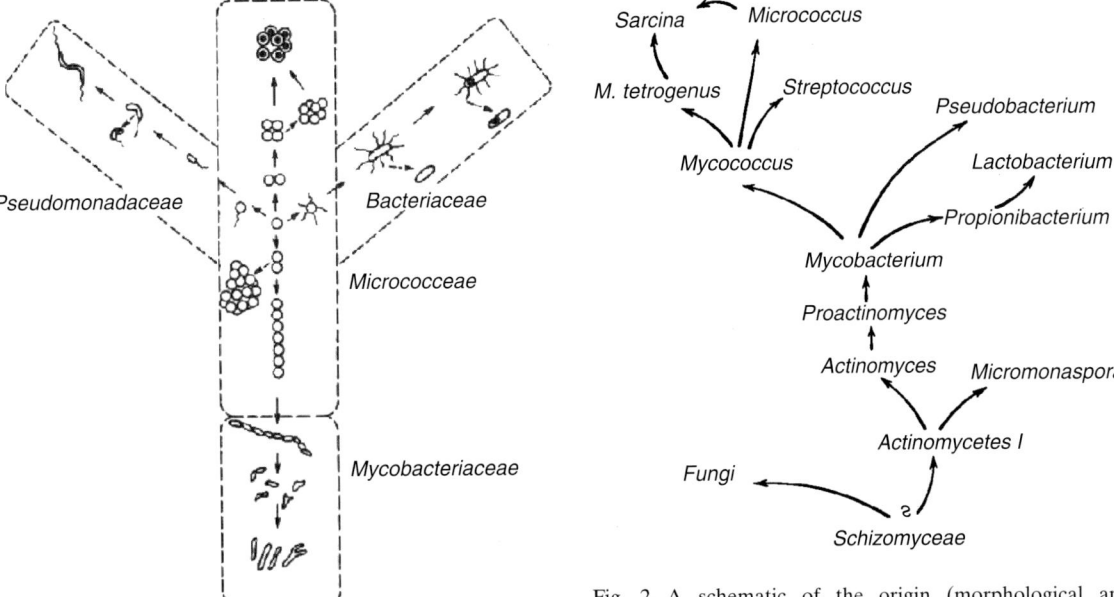

Fig. 1. Illustration of a possible evolutionary development of bacteria from a single coccus (after Kluyver and Van Niel, 1936; from Clifton, 1950).

Fig. 2. A schematic of the origin (morphological and physiological properties) of actinomycetes and relative taxa (Krassil'nikov, 1959).

of descent (Stackebrandt and Woese, 1979; Stackebrandt and Woese, 1981a; Stackebrandt et al., 1983b), numerous studies were carried out to determine the phylogenetic depth and breadth of this lineage (Embley and Stackebrandt, 1994; Koch et al., 1994; Stackebrandt et al., 1980; Stackebrandt et al., 1981b; Stackebrandt et al., 1983a) and to affiliate organisms to novel taxa in order to match classification with genomic relatedness (Collins et al., 1989; Stackebrandt et al., 1995). Another approach that specifically concentrates on elucidating natural relationships among the actinobacteria is illustrated by the search for possible correlations between the phylogenetic position of taxa and production of pharmacophores by these taxa. Phylogenetic research on more than 110 currently recognized genera of the actinobacteria has resulted in the description of a hierarchic classification system in which a rich taxonomic structure between class and genera was proposed (Stackebrandt et al., 1997). Although the basis for a revised classification scheme is based on the 16S rDNA, this proposal would not have been made without support from the analysis of other molecular markers. Whereas some markers (such as 23S rDNA, ATPase, elongation factor Tu [Schleifer and Ludwig, 1989], and the chaperonine Hsp60 [or GroEL]; Viale et al., 1994) are equivalent to 16S rDNA in covering a broad range of prokaryotic taxa, others such as the Hsp65 (Swanson et al., 1997), the RNAse RNA gene (Cho et al.,

1998), the *gyrB* gene (Kasai et al., 2000) and the ribosomal protein AT-L30 sequences (Ochi, 1992; Ochi, 1993) are suitable to cover a narrower range of taxa.

THE HIGHER TAXA The hierarchic system of the actinobacteria (schematically shown in Fig. 3) comprises taxa which have been delineated exclusively on the basis of their phylogenetic coherence and the presence of a pattern of differentiating signature nucleotides (Stackebrandt et al., 1997). This figure includes novel higher taxa which have been described recently or for which novel families are indicated from the phylogenetic position of recently described genera. Phenotypic properties traditionally used to circumscribe higher taxa may coincide with the phylogenetic groupings, but these properties do not influence the decision for the clustering of genera into higher taxa. It is not the presence or absence of a defined nucleotide but the uniqueness of the signature nucleotide pattern that circumscribes a higher taxon (Table 3). The rationale for selecting phenetic (instead of phenotypic) characters of the genome to circumscribe higher taxa lies in the availability of genetically stable markers. However, as in other classification systems, the addition of novel taxa may lead to conflicts with the data basis of the present system: some nucleotides may fit into the set of designated properties, whereas others may differ. Another problem is the shift of the phylo-

FAMILIES   SUBORDERS   ORDERS

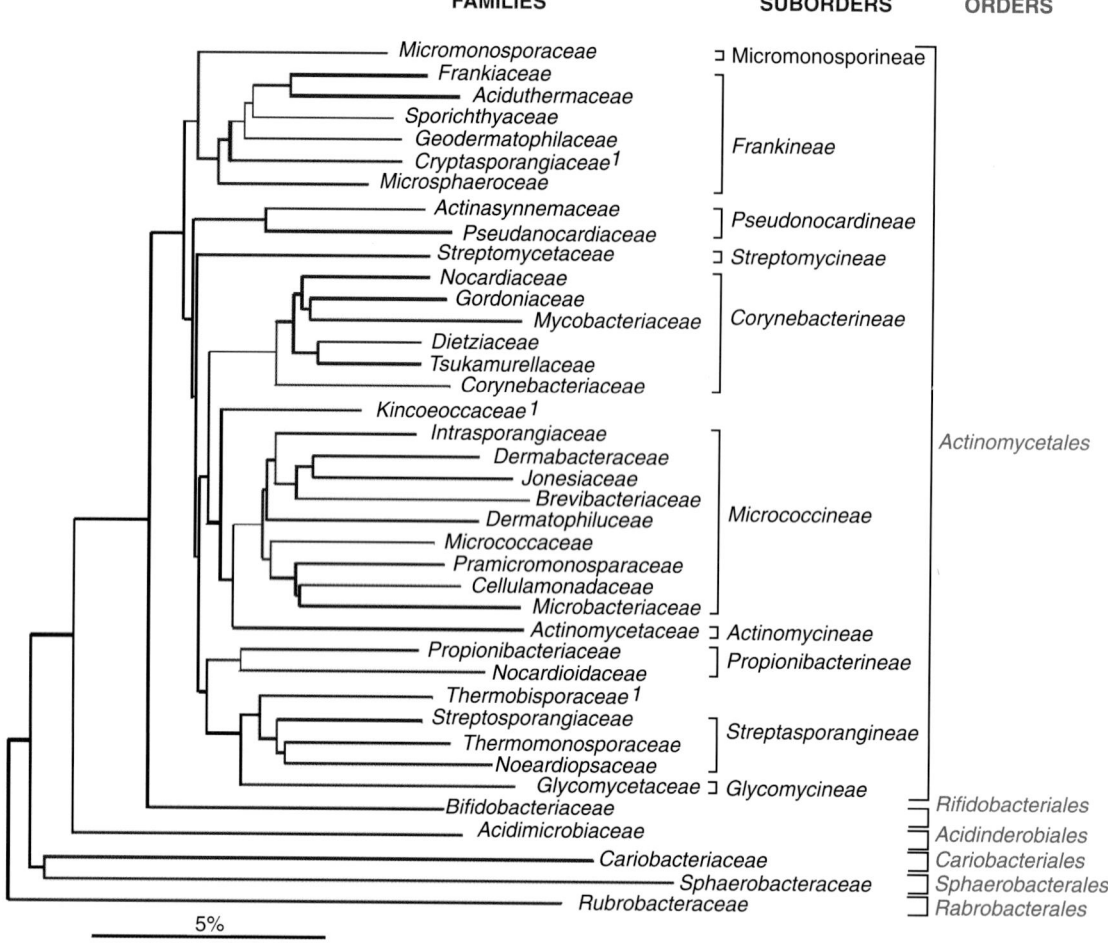

Fig. 3. Intraclass relatedness of Actinobacteria, based upon 16S rDNA/rRNA sequence comparison. The scale bar represents 5 nucleotide substitutions per 100 nucleotides. (Modified from Stackebrandt et al., 1997, with permission.)

genetic place of a taxon or the split of a hitherto homogeneous taxon by the introduction of a genus (Ludwig, 1999; Stackebrandt, 1999), which requires a redefinition of the circumscribing properties. The branching points of suprageneric lineages within the class Actinobacteria are not supported by high statistical significance and shifts in the order of lineages may occur, however, without changing the species composition of genera. For example, the introduction of several novel species into established genera and of novel genera into the suborder *Micrococcineae* has changed the branching order of some lineages, which resulted in the removal of the genera *Rarobacter* and *Sanguibacter* from the families *Cellulomonadaceae* and *Intrasporangiaceae*, respectively, and has led to the dissection of *Dermatophilaceae* (Stackebrandt and Schumann, 2000). Taxonomists are aware that changes are an intrinsic part of any classification

strategy, but newcomers to the field may be confused by the constant but unavoidable corrections of the phylogeny-based system to keep classification in step with insights into natural relationships.

The system as outlined by Stackebrandt et al. (1997) for the ranks between the ranks of genus and class is based on the most thoroughly studied gene(s) available at the time of publication. It cannot be excluded that emphasis will shift to other or to additional homologous genes that may indicate more relatedness or offer more power to discriminate than the 16S rDNA does. One example (the gyraseB gene sequence) may offer valuable nucleotide information for spreading distances at the intrageneric level.

SPECIES AND GENERA Though information on the describing properties of ranks above the genera has little importance in the identification

Table 3. Patterns of selected 16S rDNA signature nucleotides defining families of the suborder *Micrococcineae*.

Position	Microbacteriaceae	Brevibacteriaceae	Micrococcaceae	Cellulomonadaceae	Promicromonosporaceae	Dermacoccaceae	Dermatophilaceae	Dermabacteraceae	Jonesiaceae	Bogoriellaceae	Rarobacteraceae	Sanguibacteraceae	Intrasporangiaceae
41–401	G-C	U-A	G-C	G-C	G-C	G-C	G-C	G-C	G-C	G-C	G-C	G-C	G-C
45–396	U-A	U-G	U-G	U-G	U-G	U-G	U-G	U-G	U-G	U-G	U-G	U-G	U-G
69–99	R-U	C-U	A-U	G-U	G-U	R-U	A-U	A-U	A-U	A-U	A-U	G-U	G-U
144–178	C-G	U-G	C-G	C-G	U-G	C-G	U-A^a, C-G^b	U-G	U-G	U-G	U-G	C-G	C-G
140–223	G-Y	G-C	C-G	C-G	C-G	C-G, G-C	C-G	C-G	U-U	G-U	C-G	G-C	G-Y
142–221	U-A	U-A	C-G	C-G	C-G	C-G	C-G	C-G	C-G	C-G	C-G	C-G	C-G
157–164	variable	G-C	U-G	U-G	U-G	variable	C-G, G-C	G-C	U-G	U-G	C-G	G-C	G-C, U-G
248–276	C-G	C-C	C-G	C-G	C-G	C-G	C-G	U-G	C-G	C-G	C-G	C-G	C-G
258–268	A-U	G-C	A-U	G-C	G-C	G-C	A-U	A-U	G-C	G-C	G-C	A-U	A-U
293–304	G-U	G-C	G-U	G-C	G-C	G-Y	G-U, G-C	G-C	G-C	G-U	G-C	G-C	G-C
379–384	C-G	G-C	C-G	C-G	C-G	C-G	C-G	C-G	G-U	C-G	C-G	C-G	C-G
407–435	A-U	C-G	A-U	C-G	A-U	A-U, G-C	A-U	G-Y	G-C	A-U	A-U	A-U	A-U
502–543	R-Y	A-U	R-Y	G-C	G-C	A-U	A-U	R-Y	G-C	G-C	G-C	G-C	A-U
586–755	C-G	U-A	C-G	C-G	C-G	C-G	C-G	U-A	C-G	C-G	U-G	C-G	Y-R
589–650	U-A	U-A	C-G	U-A	U-A	U-A	U-G	C-G	U-G	C-G	U-A	U-A	U-A
591–648	U-A	G-U	U-A	U-A	U-A	U-A	U-A	U-A	U-A	U-A	U-A	U-A	U-A
610	variable	A	G	A	U	R	A	A	U	A	A	U	A
602–636	C-G	C-G	C-G	C-G	G-U	C-G	C-G	C-G	U-G	C-G	G-U	G-U	C-G
612–628	C-G	G-C	C-G	C-G	C-G	Y-G	C-G, UG	Y-G	C-G	C-G	U-A	U-A	C-G
615–625	A-U	A-U	G-C	A-U	U-A	G-C	G-C	A-U	A-U	G-C	C-G	A-U	R-Y
616–624	G-C	C-G	G-Y	G-C	G-U	G-Y	G-C	G-C	U-G	G-C	G-C	G-C	G-C
660–745	G-C	A-U	variable	G-C	U-A	G-C	G-C	G-C	G-C	G-C	G-C	G-C	G-C
668–738	A-U	A-U	A-U	U-A	A-U	A-U	A-U	A-U	U-A	A-U	A-U	U-A	A-U
670–736	A-U	U-A	A-U	A-U	A-U	A-U	A-U	A-U	A-U	A-U	A-U	A-U	A-U
839–847	G-U	A-U	A-U	C-G	C-G	U-A	C-G, U-A	R-U	U-A	C-G	U-A	U-A	U-A
863	U	U	U	A	U	U	U	U	U	U	A	U	U
1133–1141	A-U	A-U	A-U	G-C	G-C	A-U	A-U	A-U	A-U	A-U	G-C	A-U	A-U
1134–1140	C-G	C-G	C-G	G-C	C-G	C-G	C-G	C-G	C-G	C-G	G-C	C-G	C-G
1244–1293	C-G	U-A	C-G	C-G	C-G	C-G	C-G	C-G	C-G	C-G	C-G	C-G	C-G
1254–1283	G-C	A-C	G-C	G-C	G-C	G-C	U-A, G-C	G-C	G-C	G-C	G-C	G-C	G-C
1263–1272	A-U	C-G	A-U	A-U	A-U	A-U	A-U	A-U	A-U	A-U	A-U	A-U	A-U
1310–1327	A-U	U-A	R-Y	G-C	G-C	G-C	G-C	G-C	G-C	G-C	G-C	G-C	G-C
1414–1486	U-A	C-G	C-G	U-G	C-G	C-G	C-G	C-G	C-G	C-G	C-G	C-G	C-G

Abbreviations: G, guanine; C, cytosine; A, adenine; U, uracil; R, purine; Y, pyrimidine.
a *Dermatophilus congolensis.*
b *Dermatophilus chelonae.*
Adapted from Stackebrandt et al., 1997.

process, the availability of a rich spectrum of information describing species is indeed crucial. The polyphasic approach to classification provides the microbiologists with sufficient data to decide unambiguously whether a new isolate represents a new or an established species of a new or established genus (Stackebrandt, 1988; Stackebrandt, 1992c). At the generic level, the high correlation between phylogenetic position of the 16S rDNA and chemotaxonomic properties is striking. The finding of a unique pattern of chemotaxonomic characteristics in a novel isolate will in almost all cases be confirmed by the emergence of a novel 16S rDNA lineage falling outside the radiation of a known genus, and vice versa. In a few instances only, despite the presence of differentiating chemical properties, sequence data have revealed that these lineages fall within the realm of a genus; examples are the presence of ornithine and lysine in the peptidoglycan of some bifidobacterial species, of meso-diaminopimelic acid (meso-A$_2$pm) and L, L-diaminopimelic acid (LL-A$_2$pm) in members of *Streptomyces*, *Kitasotospora* and *Kineosporangium*, and the markedly different-peptidoglycan compositions found in closely related species of *Micrococcus luteus* and *M. lylae*. As explained for the importance of working with signature nucleotides of 16S rDNA in the definition of higher taxa, it is not the presence or absence of an individual chemotaxonomic marker but the profile of different markers that is valuable in recognizing and classifying genera. Tables 4 to 11 provide some examples from the suborder *Micrococcineae* which demonstrate the superiority of chemical data in the delineation of genera within Actinobacteria.

It has been argued that care is needed to distinguish morphologically similar organisms, such as members of *Dermacoccus*, *Kocuria*, *Micrococcus*, *Nesterenkonia* and *Stomatococcus*, from each other, from phylogenetically related taxa (Fig. 4) and from other Gram-positive and catalase-positive cocci (Goodfellow, 1998). Other examples are likewise true for the delineation of mycelium-producing organisms and rod-shaped forms. This word of warning is justified if addressed to taxonomists of the prechemotaxonomic and pre-molecular era. Classification and subsequently identification have developed into a demanding biological discipline in which taxonomic judgment has lost much of its subjective basis. The affiliation of an organism to its nearest phylogenetic neighbor by determination and comparative analysis of appropriate nucleotide sequences has become easier and faster than determination of a wide range of chemotaxonomic properties, fulfilling the same purpose.

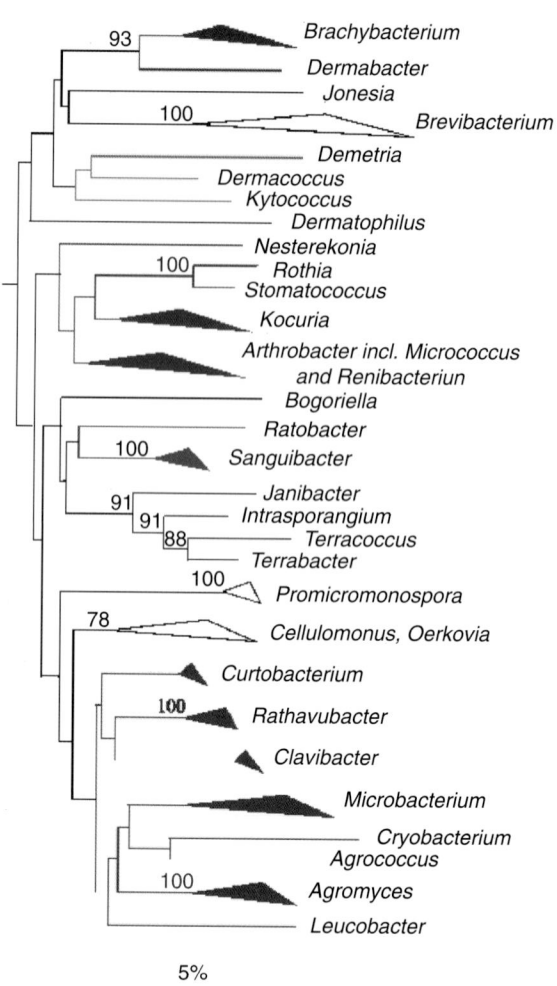

Fig. 4. Phylogenetic relatedness among genera of the suborder *Micrococcineae*, order *Actinomycetales*, class *Actinobacteria*, based upon 16S rDNA/rRNA sequence comparison. Numbers within the dendrogram indicate the percentages of occurrence of the branching order in 500 bootstrapped trees (only values of 70 and above are shown). Sequences of species of other actinobacterial families served as a root. The scale bar represents 5 nucleotide substitutions per 100 nucleotides.

CLASSIFICATION OF TAXA NOT COVERED IN DETAIL The following subchapters are short summaries of taxonomic information on either genera that have been compiled in the 2nd edition of *The Prokaryotes* (Balows et al., 1992) but are not included in the present edition or genera that have been described since 1991. The current information about the latter taxa is often not sufficient to be covered in individual chapters. Once these organisms receive the attention they deserve on the basis of their unique phylogenetic position, this situation may change. Some of the genera dealt with in *The Prokaryotes*, 2nd edition, are covered only briefly here because not much information other than the phylogenetic position has changed during the past decade. Still representing the state-of-the-art knowledge on these organisms, these chapters on *Micrococcus* (Kocur et al., 1992), *Renibacterium* (Embley, 1992), *Stomatococcus* (Stackebrandt, 1992c), *Dermatophilaceae* (Stackebrandt, 1992a), and *Arthrobacter* (Jones and Keddie, 1992) include thorough information on isolation, preservation and identification of strains and species and provide descriptions of habitats, biochemical and physiological properties and the biotechnological potential of these strains and species. Classification into risk groups, i.e., according to their pathogenic potential, follows the guidelines of the European Commission 90/679/EWG from November 26th, 1990.

The Deeply Branching Actinobacterial Families

Rubrobacteraceae The monogeneric family *Rubrobacteraceae* has been described as the most deeply branching lineage of the Actinobacteria proper. The genus *Rubrobacter* contains two species of Gram-positive, thermophilic bacteria (48–60°C), namely *R. radiotolerans* (the former *Arthrobacter radiotolerans*; Yoshinaka et al., 1973) and *R. xylanophilus* (Carreto et al., 1996). The type strains are chemotaxonomically characterized by peptidoglycan of type A3α (L-Lys-L-Ala), major respiratory quinone—menaquinone MK-8, fatty acids—12-methylhexadecanoic and 14-methyloctadecanoic acids, and several phospholipids. *Rubrobacter radiotolerans* has been isolated from a hot spring in Japan after gamma irradiation of water samples, and it has an inactivation dose of 10,000 Gy, which is 2,000 times greater than that of human cells (Saito et al., 1994). Neither the natural habitat of *R. xylanophilus* (isolated from a thermally polluted industrial runoff; Laura et al., 1996) nor the inactivation dose of gamma irradiation of its strains is known.

Several strains that phylogenetically cluster with *Rubrobacter* strains were isolated from soil from an Atacama Desert site in Chile. Analysis of a 16S rDNA clone library, generated from DNA extracted from the same environment, also revealed *Rubrobacter*-type sequences that differed from those of described species and the isolates (F. A. Rainey, personal communication). This highly arid soil is constantly bombarded with high doses of ultraviolet (UV) light, and soil thus may constitute one natural niche of these organisms. The DNAs recovered from a microbial mat associated with the Great Artesian Basin, Australia (Byers et al., 1998), from a subtropical Australian forested soil (Liesack and Stackebrandt, 1992), from a peat bog in North-

ern Germany (Rheims et al., 1996), and from various other environments have provided 16S rDNA clone sequences that appear to remotely link these uncultured organisms to the *Rubrobacter* lineage. However, their branching points are not supported by high statistical significance (Rheims et al., 1999). The degree of phenotypic similarity between *Rubrobacter* strains and the organisms from which DNA has been analyzed must await their cultivation.

Sphaerobacteraceae This family contains the monospecific genus *Sphaerobacter* with the type species *S. thermophilus*. The organism is aerobic, Gram-positive, pleomorphic and nonsporeforming and constitutes part of the dominating flora of aerobe-thermophilically treated sewage sludge (Demharter et al., 1989). Like rubrobacteria, *S. thermophilus* is thermophilic (optimum 55°C), the principal menaquinone is MK-8 and the G+C content of DNA 65 mol%; L-ornithine is the diamino acid at position 3 of the peptide subunit and the interpeptide bridge consists of β-alanine (A3β-type). Information on polar lipids and fatty acids is absent. Reports (other than the original description) on the presence and distribution, the ecological significance, and additional genomic or phenetic properties also are unavailable.

Acidimicrobiaceae This family embraces the monospecific genus *Acidimicrobium*. Rod-shaped, *A. ferrooxidans* is Gram-positive, moderately thermophilic (45–50°C), ferrous-iron-oxidizing and acidophilic (optimum around pH 2.0), with a DNA G+C content of 69 mol% (Clark and Norris, 1996). Autotrophic growth occurs on ferrous iron. The strains recognized so far were isolated from a test copper leaching system and a hot spring in the United States (strain TH3; Brierley, 1978) and a pyrite enrichment from an Icelandic geothermal site (type strain DSM 10331). The two strains differ in their morphology in that strain TH3 grows in filaments. Growing mixotrophically, strains of *Acidimicrobium* and *Sulfobacillus* oxidize ferrous iron under air more extensively than either strain does in pure culture (Clark and Norris, 1996).

Acidimicrobium strains are the only actinobacteria reported to thrive in warm, acidic, iron-, sulfur- or mineral-sulfide-rich environments, thus expanding the physiological capabilities of members of this class to lithotrophy.

Acidimicrobium thermooxidans is remotely related to "Microthrix parvicella" (87% 16S rDNA sequence similarity; Blackall et al., 1994). "M. parvicella" is a Gram-positive straight filamentous bacterium (300 to 500 μm long) that is the major causative agent of foaming and bulking in activated-sludge sewage-treatment plants (Christensson et al., 1998). Its poor growth on artificial media did not allow physiological testing, and the organisms could not be stored successfully. Similar to strains of *Acidimicrobium*, "M. parvicellum" requires reduced forms of sulfur to grow on artificial media (Blackall et al., 1994).

One of the most surprising results of molecular environmental studies was the emergence of several novel 16S rDNA sequence clusters with probable world-wide distribution in different terrestrial environments. Phylogenetic analyses revealed remote relationships of hitherto uncultured organisms to the family *Acidimicrobiaceae*, where they formed several deeply rooting lineages (Rheims et al., 1999). Members of these clusters originate in subtropical forest soil from Mount Coo-tha, Australia (Stackebrandt et al., 1993), peat bogs of Northern Germany (Rheims et al., 1996), grass soil of The Netherlands (Felske et al., 1997), paddy fields of Japan (Ueda et al., 1995), geothermically heated soil of New Zealand (Rainey et al., 1993), and even in the marine environments (Fuhrman et al., 1993).

CORIOBACTERACEAE Two genera have originally been included into this family, *Coriobacterium* and *Atopobium* (Rainey and Stackebrandt, 1994), whereas two new genera, *Slackia* and *Eggerthella*, have recently been added to the family. The anaerobic species *Coriobacterium glomerans* has been isolated from the intestine of the red soldier bug *Pyrrhocoris apterus* (Haas and König, 1988), and *Atopobium* species were described for some misclassified *Lactococcus* and *Streptococcus* species (Collins and Wallbanks, 1992) isolated from human gingival crevices or from blood, wounds and abscesses (Olsen et al., 1991). Species of these genera share some morphological and chemotaxonomic properties. Cells of both taxa are Gram-positive, nonsporeforming rods to cocci with cells containing central swellings or large spherical intercellular involution forms, respectively. Peptidoglycan is of the A6-type, either the lysine- aspartic acid (Lys-Asp) crosslinkage (*A. parvulum* and *C. glomerans*) or the lysine-serine-glutamic acid (Lys-Ser-Glu) crosslinkage (*A. minutum*). End products of glucose fermentation include lactic and acetic acids. Differences between members of the two genera are found in the composition of additional end products of glucose fermentation, but above all in the base composition of DNA, which is 76 mol% G+C for *C. glomerans* and 35–46 mol% G+C for *Atopobium* species. The latter values are low considering that the vast majority of actinobacteria have mol% G+C values above 55%.

The novel genera were added to the family following 16S rDNA analyses. *Eubacterium exiguum* and *Peptostreptoccus heliothrinreducens* were reclassified as members of *Slackia*. "Denitrobacterium detoxificans," a ruminal bacterium that respires on nitrocompounds, branches distantly to members of this genus (16S rDNA sequence cited as "unpublished" in *European Molecular Biology Laboratory* [EMBL]; Anderson et al., 1997). *Eubacterium lentum* was reclassified as *Eggerthella lenta* (Wade et al., 1999), whereas *Eubacterium fossor* branches within the radiation of the genus *Atopobium*, representing a new species of this genus.

GLYCOMYCINEAE This family has been described for the genus *Glycomyces* that forms a deeply branching individual 16S rDNA line of descent within the order Actinomycetales. The type strains of the three species of the genus are aerobic, produce nonfragmenting vegetative hyphae and form chains of conidia on aerial sporophores (Labeda et al., 1985; Evtushenko et al., 1991). Chemotaxonomically, the amycolate strains are defined by peptidoglycan containing *meso*-A_2pm and glycine, xylose and arabinose as whole cell sugars, phosphatidylinositol, diphosphatidyglycerol and numerous acylated phosphatidylinositol mannosides, menaquinones of type MK-9(H_6), MK-10(H_2), MK-10(H_6), and MK-11(H_6), and a DNA G+C content ranging between 71 and 73 mol%. (Note that MK-9, -10 and -11 are unsaturated menaquinones with nine, ten and eleven isoprene units, respectively; MK-X (H_2) and (H_6) are partially saturated menaquinones and one and three of X isoprene units are hydrogenated, respectively.) Fatty acids determined for the type strain of *G. tenuis* (Evtushenko et al., 1991) are mainly branched *iso*- and *anteiso* acids. The validity of the genus was subsequently confirmed by 16S rDNA analysis which remotely related this genus to the families *Streptosporangiaceae*, *Nocardiopsaceae*, *Thermomonosporaceae* and to the genus *Thermobispora*. The species *G. harbinensis* produces the antibiotic azaserine.

KINEOCOCCACEAE Based upon 16S rDNA sequence analysis, members of the two morphologically very different genera (Table 4) are distantly related (92% similarity) and they were shown to form a separate line of descent within the order Actinomycetales (Miyadoh, 1977; Kudo et al., 1998). The statistical significance of the branching point of this lineage, however, although reported to be as high as 98% by Kudo et al. (1998), is not stable and changes with the selection of reference organisms. If, for example, emphasis is placed on members of the suborder *Micrococcineae* or *Frankineae*, the *Kineococcus-*

Table 4. Morphological and chemotaxonomic characteristics of the genera *Kineococcus* and *Kineosporia*.

Genus	Morphology	Diamino acid	Glycine present	Major menaquinone[a]	Polar lipids	Predominant cellular fatty acid(s)	Whole cell sugars	Mol% G+C
Kineococcus	Motile spherical cells in clusters	*meso*-A$_2$pm	–	MK-9(H$_2$)	PG, DPG, PL	ai-C$_{15:0}$	Ara, Gal	74
Kineosporia	Vegetative hyphae, "spore domes," motile spores	*meso*-A$_2$pm and/or LL-A$_2$pm	+	MK-9(H$_4$)	PC	C$_{16:0}$, C$_{18:1}$, 10-Me-C$_{18:0}$	Gal, Glc, Man, Rib, Rha, 3-O-Me-Rha	69–71

Abbreviations: A$_2$pm, diaminopimelic acid; PG, phosphatidylglycerol; ai-C$_{15:0}$, 12-methyltetradecanoic acid; DPG, diphosphatidylglycerol; PL, unidentified phospholipid; Ara, arabinose; Gal, galactose; LL-A$_2$pm, L,L-diaminopimelic acid; PC, phosphatidylcholine; C$_{16:0}$, hexadecanoic acid; C$_{18:1}$, octadecenoic acid; 10-Me-C$_{18:0}$, 10-methyloctadecanoic acid (tuberculostearic acid); Glc, glucose; Man, mannose; Rha, rhamnose; and Rib, ribose; 3-O-Me-Rha, 3-O-methylrhamnose.
[a]MK-9(H$_2$) and MK-9(H$_4$) are partially saturated menaquinones with two and four of nine isoprene units hydrogenated, respectively.
Data are from Yokota et al., 1993; Kudo et al., 1988.

Kineosporia lineage is not branching outside but inside the radiation of the respective suborders. In fact, binary similarity values between individual species of *Kineococcus* and *Kineosporia* and members of these suborders are as high as those found for some species of *Kineococcus* and *Kineosporia*. Also, almost all of the 16S rDNA signature nucleotides defined for taxa of *Micrococcineae* are present in the sequences of the latter genera. Hence, the taxonomic affiliation of the two genera is not yet settled and sequences of more representatives of these genera are needed to obtain a more stable view of their relatedness. Only then can the question be answered of whether the two genera are monophyletic and members of the same family. It is noteworthy that the spores (2 µm in diameter) of *Kineosporia* (Pagani and Parenti, 1978) and the cells (1–1.5 µm in diameter) of *Kineococcus* are motile due to a polar tuft of flagella. This finding could lead to the suggestion that *Kineococcus* strains may represent the locked stage of the *Kineosporia* spore stage.

Kineosporia strains were isolated from plant samples found in Japan, and *Kineococcus aurantiacus* was isolated from a soil sample in India. Information other than taxonomic descriptions is not available for these organisms (Table 4).

MEMBERS OF THE SUBORDER MICROCOCCINEAE This suborder contains a rich family structure embracing organisms with very diverse morphology and chemotaxonomic characteristics; one example is *Microbacteriaceae*, differing markedly from the other organisms in the peculiar crosslinkage of peptidoglycan. The addition of novel organisms has made rearrangement of the taxonomic structure to some extent necessary (Stackebrandt and Schumann, 2000) to maintain the correlation of taxonomic ranks with their phylogenetic position (Stackebrandt et al., 1997).

BOGORIELLACEAE The genus *Bogoriella* with the type species *Bogoriella caseilytica* was described to accommodate a single alkaliphilic strain with unusual taxonomic properties (Groth et al., 1997). The strain was isolated from soda soil (pH 10) near Lake Bogoria in the Kenyan-Tanzanian Rift Valley, Africa. Cells of the organism are more or less aggregated irregular rods or cocci with spiky structures distributed over the surface of each cell. In addition to the morphological features unique among coryneform bacteria, the organism displays uncommon chemotaxonomic characteristics. The peptidoglycan type A4α based on L-lysine (Schleifer and Kandler, 1972) in combination with the occurrence of MK-8(H$_4$) as predominating menaquinone has not been reported earlier and was only reported recently in the single strain of *Demetria terragena* (Groth et al., 1997a). *Bogoriella caseilytica* differs from *D. terragena*, e.g., in its peptidoglycan interpeptide bridge consisting of two residues of L-alanine and one of L-glutamic acid (Table 5), which represents a novel variation of the peptidoglycan type A4α. The 16S rDNA sequence comparison revealed that *B. caseilytica* represents a distinct lineage within the order Actinomycetales and exhibits less than 94% similarity to the next phylogenetic neighbors (Groth et al., 1997). The family *Bogoriellaceae* has recently been proposed to accommodate the monogeneric taxon *Bogoriella* (Stackebrandt and Schumann, 2000).

Table 5. Chemotaxonomic characteristics of the genus *Bogoriella* representing the family *Bogoriellaceae*.

Genus	Diamino acid	Interpeptide bridge	Major menaquinone	Polar lipids	Predominant cellular fatty acids	Mol% G+C
Bogoriella	L-Lys	L-Ala-L-Ala-L-Glu	MK-8(H₄)	PI, PG, DPG, PL	ai-$C_{15:0}$, i-$C_{15:1}$, i-$C_{16:0}$	70

Abbreviations: Lys, lysine; Glu, glutamic acid; MK-8(H$_4$), partially saturated menaquinone with two of eight isoprene units hydrogenated; PI, phosphatidylinositol; Ala, alanine; i-$C_{15:1}$, 13-methyltetradecenoic acid; and $C_{16:0}$, hexadecanoic acid. Refer to the footnotes in Table 4 for definitions of all other abbreviations.
Data are from Groth et al., 1997.

Table 6. Chemotaxonomic characteristics of genera of the family *Dermabacteraceae*.

Genus	Diamino acid[b]	Interpeptide bridge[b]	Major menaquinones[c]	Polar lipids[d]	Predominant cellular fatty acids[e]	Mol% G+C	Acyl type[b]
Dermabacter	meso-A$_2$pm	None	MK-9, MK-8, MK-7	PG, DPG, PL, GL	ai-$C_{17:0}$, ai-$C_{15:0}$, i-$C_{16:0}$	62	n.d.
Brachybacterium	meso-A$_2$pm	D-Glu-D-Glu[a] or D-Asp-D-Glu[a]	MK-7, MK-8	PG, DPG, GL	ai-$C_{15:0}$, ai-$C_{17:0}$	68–72	Glycolyl

Abbreviations: MK-7, -8, -9, unsaturated menaquinones with seven, eight and nine isoprene units, respectively; ai-$C_{17:0}$, 14-methylhexadecanoic acid; n.d., not determined; GL, unidentified glycolipid; and Asp, aspartic acid. Refer to the footnotes in Tables 4 and 5 for definitions of all other abbreviations.
[a]Glycine amide is bound to glutamic acid at position 2 of the peptide subunit.
Data are from Bogdanovsky et al., 1971; Collins et al., 1988; Jones and Collins, 1988; Gvozdyak et al., 1992; Takeuchi et al., 1995; Schubert et al., 1996.

DERMABACTERACEAE The family *Dermabacteraceae* (Stackebrandt et al., 1997) was established on the basis of 16S rDNA sequence data and circumscribed by a set of specific signature nucleotides. It contains the genera *Dermabacter* (Jones and Collins, 1988) and *Brachybacterium* (Collins et al., 1988). The genus *Dermabacter* (Jones and Collins, 1988) is monospecific and contains the type species *Dermabacter hominis*. Strains of this human cutaneous coryneform bacterium are facultatively anaerobic and characterized by a directly crosslinked peptidoglycan based on meso-A$_2$pm (variation A1γ; Schleifer and Kandler, 1972). Mycolic acids are absent. *D. hominis* can be distinguished from morphologically similar species of the genus *Brevibacterium*, exhibiting the same combination of the latter two characteristics, by its totally unsaturated menaquinones (Table 6), physiological properties and its separate phylogenetic position as revealed by 16S rDNA sequence analysis (Cai and Collins, 1994). A normal inhabitant of healthy human skin, *D. hominis* is considered non-pathogenic.

Brachybacterium faecium was described to accommodate three strains isolated from poultry deep litter (Collins et al., 1988). Two of these strains were later reclassified as members of two new *Brachybacterium* species, strain NCIB 9859 (together with nine strains of "Micrococcus conglomeratus") as members of *B. conglomeratum* and strain NCIB 9861 as the type strain of *B.*

paraconglomeratum (Takeuchi et al., 1995). Takeuchi et al. (1995) also proposed the strain isolated from corn steep liquor, *B. rhamnosum*, so-called because it contains large amounts of rhamnose in its cell wall. *Brachybacterium nesterenkovii* strains were isolated from dairy products (Gvozdyak et al., 1992), whereas *B. alimentarium* and *B. tyrofermentans* were isolated from the surface of French cheeses (Schubert et al., 1996). All representatives of the genus *Brachybacterium* exhibit a rod-coccus growth cycle and are characterized by the peptidoglycan variation A4γ based on meso-A$_2$pm (Schleifer and Kandler, 1972), muramic acid in the glycolylated form, absence of mycolic acids, MK-7 as principal menaquinone, and predominance of anteiso-branched cellular fatty acids (Table 6). Whereas the peptidoglycan of the A4γ variation of *B. alimentarium* and *B. tyrofermentans* is cross-linked via an interpeptide bridge consisting of D-aspartic acid and D-glutamic acid, the interpeptide bridges of all other *Brachybacterium* species contain two D-glutamic acid residues. Except for *B. nesterenkovii*, all *Brachybacterium* species have glycine amide bound to the α-carboxyl group of D-glutamic acid at position 2 of the peptide subunit. Members of the genera are considered nonpathogenic.

DERMATOPHILACEAE AND DERMACOCCACEAE The family *Dermatophilaceae* (Austwick, 1958) was initially described to include the genera

Table 7. Chemotaxonomic characteristics of genera of the families *Dermatophilaceae* and *Dermacoccaceae*.

Genus	Diamino acid	Interpeptide bridge	Major menaquinone(s)	Polar lipids	Predominant cellular fatty acids	Mol% G+C
Dermatophilus	*meso*-A$_2$pm	None	MK-8(H$_4$)	PG, DPG, PI	C$_{16:0}$, C$_{15:0}$, C$_{14:0}$	57–59
Dermacoccus	L-Lys	L-Ser$_{1-2}$-D-Glu or L-Ser$_{1-2}$-L-Ala-D-Glu	MK-8(H$_2$)	PI, PG, DPG	i-C$_{17:0}$, ai-C$_{17:0}$, i-C$_{17:1}$	66–71
Kytococcus	L-Lys	D-Glu$_2$	MK-8, MK-9, MK-10	PI, PG, DPG	ai-C$_{17:0}$, C$_{17:0}$, i-C$_{17:1}$	68–69
Demetria	L-Lys	L-Ser-D-Asp	MK-8(H$_4$)	PI, PE, PG, DPG, PL	C$_{18:1}$, C$_{18:0}$, C$_{17:0}$, ai-C$_{17:0}$	66

Abbreviations: C$_{15:0}$, pentadecanoic acid; C$_{14:0}$, tetradecanoic acid; MK-8(H$_2$) and MK-8(H$_4$), partially saturated menaquinones with one and two of eight isoprene units hydrogenated, respectively; i-C$_{17:0}$, 15-methylhexadecanoic acid; i-C$_{17:1}$, 15-methylhexadecenoic acid; MK-8, -9, and -10, unsaturated menaquinones with eight, nine and ten isoprene units, respectively; Ser, serine; PE, phosphatidylethanolamine; and C$_{18:0}$, octadecanoic acid. Refer to the footnotes in Tables 4, 5 and 6 for definitions of all other abbreviations.
Data are from references Gordon, 1989; Groth et al., 1997a; Kocur, 1986; Stackebrandt et al., 1983; Kroppenstedt (Ph.D. thesis), 1977; and Stackebrandt et al., 1995.

Dermatophilus and *Geodermatophilus* which have in common the same type of cell division, i.e., in both transverse and longitudinal planes. Because members of both genera have markedly different growth characteristics, chemotaxonomic properties and 16S rRNA cataloguing data (Stackebrandt et al., 1983c), it was proposed that the family *Dermatophilaceae* be restricted to one genus *Dermatophilus* (Stackebrandt, et al., 1992). Based on 16S rDNA sequence data and presence of specific signature nucleotides, the family *Dermatophilaceae* was emended (Stackebrandt et al., 1997) to contain the type genus *Dermatophilus* (Van Saceghem, 1915; Gordon, 1964) as well as the genera *Dermacoccus* (Stackebrandt et al., 1995) and *Kytococcus* (Stackebrandt et al., 1995).

With the addition of novel species into *Micrococcineae*, the incoherence of the family *Dermatophilaceae* became obvious in that the type genus of *Dermatophilus* clustered separately from those of the remaining two genera (Fig. 4). Consequently, the family has been split into *Dermatophilaceae* and Dermacoccacea, with the latter family to embrace the genera *Dermacoccus*, *Demetria* and *Kytococcus* (Stackebrandt and Schumann, 2000). The genus *Demetria* (Groth et al., 1997a) should be included because of its phylogenetic position and the presence of all 16S rDNA signature nucleotides as defined for the *Dermacoccaceae*.

Dermatophilaceae The genus *Dermatophilus* contains the type species *D. congolensis* (Van Saceghem, 1915; Gordon, 1964), which was isolated from tissue specimens of infected mammals and lizards and identified as the cause of the skin disease streptotrichosis, and *D. chelonae* (Masters et al., 1995), which was isolated from infected chelonids (turtles and a tortoise). Both *Dermatophilus* species produce branching filaments that divide by transverse and longitudinal septa to form zoospores. The zoospores form germ tubes, elongate into filaments and repeat the cycle. *D. chelonae* differs from *D. congolensis* in the formation of thin capsules around mature filaments (Masters et al., 1995). Phylogenetic analysis of the two species reveals only a moderate degree of relatedness (94.7%), and phylogenetic analysis including a varying selection of reference organisms does not consistently group them as phylogenetic neighbors.

Members of the genus show a directly cross-linked peptidoglycan with *meso*-A$_2$pm as diagnostic diamino acid (variation A1γ; Schleifer and Kandler, 1972). Cellular fatty acids of *D. congolensis* are of the straight-chain saturated type (R. M. Kroppenstedt, 1977; Table 7). Mycolic acids are absent. Because they were identified as causative organisms of streptotrichosis in mammals and reptiles and were also isolated from humans, both *Dermatophilus* species were affiliated to risk group 2.

Dermacoccaceae The description of the genera *Dermacoccus* and *Kytococcus* resulted from the taxonomic dissection of the genus *Micrococcus* (Cohn, 1872; Stackebrandt et al., 1995). These two genera, as well as *Demetria*, are characterized by the peptidoglycan variation A4α based on L-lysine (Schleifer and Kandler, 1972). No pathogenic potential has been reported for strains of these genera.

Dermococcus. The type species *D. nishinomiyaensis* (Oda, 1935; Skerman et al., 1980) of the genus *Dermacoccus* (Stackebrandt et al., 1995) was isolated from mammalian skin. Strains produce spherical cells occurring in pairs, tetrads

Table 8. Chemotaxonomic characteristics of the *Rarobacter*.

Genus	Diamino acid[b]	Interpeptide bridge[b]	Major menaquinone[c]	Predominant cellular fatty acids[d]	Mol% G+C
Rarobacter	L-Orn	Ser-Glu or Glu	MK-9	ai-$C_{15:0}$, i-$C_{16:0}$, $C_{14:0}$ (aerobic)	65–66

Abbreviations: Orn, ornithine; refer to the footnotes in Tables 4 through 7 for definitions of all other abbreviations.
Data are from reference Goto-Yamamoto et al., 1993.

Table 9. Chemotaxonomic characteristics of the *Sanguibacter*.

Genus	Diamino acid	Interpeptide bridge	Major menaquinone	Predominant cellular fatty acid(s)	Mol% G+C
Sanguibacter	L-Lys	Ser-D-Glu	MK-9(H_4)	$C_{16:0}$, ai-$C_{15:0}$, $C_{18:0}$	69–70

Refer to the footnotes in Tables 4 through 7 for definitions of all abbreviations.
Data are from references Fernández-Garayzábal et al., 1995 and Pascual et al., 1996.

or irregular clusters of tetrads. The peptidoglycan is cross-linked via interpeptide bridges consisting of one or two residues of L-serine and one of either D-glutamic acid alone or D-glutamic acid plus L-alanine (Table 7).

KYTOCOCCUS. The type species *K. sedentarius* (ZoBell and Upham, 1944; Skerman et al., 1980) of the genus *Kytococcus* (Stackebrandt et al., 1995) was isolated from human skin. Cells are spherical and usually occur in tetrads, which can be arranged in cubical packets. The genus is characterized by the unique combination of only nonhydrogenated menaquinones and a peptidoglycan interpeptide bridge consisting of two D-glutamic acid residues (Table 7).

DEMETRIA. The genus *Demetria* and its type species *D. terragena* were described for a single strain isolated from compost soil (Groth et al., 1997a). Cells are irregular coccoid-to-rod shaped and occur singly, in pairs, short chains or irregular clusters. The combination of the peptidoglycan type A4α (Schleifer and Kandler, 1972) and MK-8(H_4) as predominating menaquinone is rare and has so far only been reported for *Bogoriella caseilytica* (Groth et al., 1997).

RAROBACTERACEAE The genus *Rarobacter* (Yamamoto et al., 1988), containing the type species *R. faecitabidus* (Yamamoto et al., 1988) and *R. incanus* (Goto-Yamamoto et al., 1993), was assigned to the emended family *Cellulomonadaceae* (Stackebrandt et al., 1997) following phylogenetic analysis of 16S rDNA sequences. Members of the genus are yeast-lysing facultative anaerobic bacteria which require hemin or hemoproteins as essential factors for aerobic growth and carbon dioxide for anaerobic growth. *Rarobacter* cells adhere to and agglutinate yeast cells and subsequently lyse them to utilize the digested cells as nutrients. Several lytic enzymes have been identified, one of which recognizes mannose chains of mannoproteins (Shimoi et al., 1995). Strains of *R. faecitabidus* were isolated

from wastewater treatment systems of alcoholic beverage factories (Yamamoto et al., 1988). Species are considered nonpathogenic and they are characterized by the peptidoglycan variation A4β with L-ornithine as diagnostic diamino acid (Schleifer and Kandler, 1972), unsaturated menaquinones with nine isoprenoid units and 12-methyltetradecanoic acid as the predominating cellular fatty acid. *Rarobacter faecitabidus* shows a peptidoglycan interpeptide bridge consisting of serine and glutamic acid, whereas serine is lacking in the cell wall of *R. incanus* (Goto-Yamamoto et al., 1993; Table 8). The latter species is also described to lack catalase activity. The cells are motile irregular pleomorphic rods. Recent 16S rDNA analyses indicated that *Rarobacter* should not be included in *Cellulomonadaceae* inasmuch as it forms an individual line of descent worthy of family status within the suborder *Micrococcineae* (Stackebrandt and Schumann, 2000).

SANGUIBACTERACEAE The genus *Sanguibacter* has been affiliated to the family *Intrasporangiaceae* on the basis of 16S rDNA sequence comparison and shared unique 16S rDNA nucleotides (Stackebrandt et al., 1997). *Sanguibacter* embraces the type species *S. keddieii* (Fernández-Garayzábal et al., 1995), *S. suarezii* (Fernández-Garayzábal et al., 1995) and *S. inulinus* (Pascual et al., 1996). Strains show the peptidoglycan variation A4 α-based on L-lysine (Schleifer and Kandler, 1972) with an interpeptide bridge consisting of serine and D-glutamic acid and a tetrahydrogenated menaquinone with nine isoprenoid units (Table 9). Cells are short irregular motile rods. All strains were isolated from bovine blood samples and were therefore allocated to risk group 2, indicating a moderate pathogenic potential. *Sanguibacter* species were differentiated by physiological tests and DNA-DNA hybridization (Pascual et al., 1996). The inclusion of new genera into the family *Intraspo-*

rangiaceae resulted in the exclusion of the genus *Sanguibacter* (Fig. 4) and a consequent redefinition of signature nucleotides for the emended family *Intrasporangiaceae* (Stackebrandt and Schumann, 2000).

JONESIACEAE This family has been described for the monospecific genus *Jonesia*. The species *J. denitrificans*, originally described as *Listeria denitrificans* (Prevot, 1961), clustered with members of *Cellulomonas* and *Promicromonospora* in early 16S rDNA-based phylogenetic studies (Rocourt et al., 1987), but the branching points of the *Jonesia* and *Promicromonospora* lineages were not stable in subsequent studies based on almost complete 16S rDNA molecules (Stackebrandt and Prauser, 1992b). Following extensive phylogenetic studies, *Jonesia* was excluded from the family *Cellulomonadaceae* (Rainey et al., 1995) and described as a family of *Micrococcineae* (Stackebrandt et al., 1997) on the basis of its well separated phylogenetic position. Thus, the rationale for keeping *Jonesia* within *Cellulomonadaceae* based on morphological and biochemical properties (Stackebrandt and Prauser, 1992b) became less significant once its isolated position within the actinomycete 16S rDNA tree (Fig. 4) was ascertained.

Jonesia denitrificans, isolated from boiled ox blood (Sohier et al., 1948), is characterized by typical coryneform morphology, i.e., straight to irregularly bent rods, forming hyphal or even mycelium-like structures that undergo fragmentation. Cells are motile by means of peritrichous flagella. Chemotaxonomically, cells are defined by a DNA G+C content of 57% (Seeliger and Jones, 1986), peptidoglycan of the Lys-Ser-D-Glu type (variation A4α; Fiedler and Seger, 1983), menaquinones of the MK-9 type (Collins et al., 1979), and the presence of an *N*-acetyl galactoseamine-substituted poly(ribitol phosphate) teichoic acid (Fiedler et al., 1984). *Jonesia denitrificans* is classified as a pathogenic organism (risk group 2).

PROMICROMONOSPORACEAE The genus *Promicromonospora* contains three species, one of which, *P. enterophila* (Jáger et al., 1983), is a member of the Cellulomonas/Oerskovia cluster, sharing 99.6% 16S rDNA similarity with *C. turbata* (Rainey et al., 1995). The other two species, the type *P. citrea* (Rainey et al., 1995) and *P. sukumoe* (Takahashi et al., 1987), are closely related (98.5% 16S rDNA sequence similarity; F. A. Rainey, E. Stackebrandt, P. Schumann and H. Prauser, unpublished observations) and form a separate lineage of family rank adjacent to members of *Cellulomonadaceae* and *Microbacteriaceae*.

These soil and compost inhabitants display a nocardioform life cycle that most closely resembles those of Nocardia and Nocardioides. The hyphae of an extended substrate mycelium fragment into nonmotile elements of different shape and size, which subsequently may give rise to new mycelia. Aerial mycelium may be formed on some synthetic media (Shirling and Gottlieb, 1966) and also undergoes fragmentation.

Chemotaxonomically, authentic *Promicromonospora* species are characterized by a Lys-Ala-Glu (variation A4α) peptidoglycan type (only *P. citrea* has been investigated; Evtushenko et al., 1984), menaquinone mainly of the MK-9(H_4) type (Collins and Jones, 1981), and major fatty acids 12-methyltetradecanoic acid (ai-$C_{15:0}$) and 13-methyl tetradecanoic acid (i-$C_{15:0}$; Andreyev et al., 1983). The G+C content of DNA ranges from 70 to 75 mol% (Evtushenko et al., 1984). Members of the genus can be rapidly identified by phage typing using a set of two phages that do not attack members of phylogenetically neighboring genera (Stackebrandt and Prauser, 1992b).

The species are classified as nonpathogens. The type strain of *P. sukumoe* produces an antibiotic called 7-hydro-8-methyl-pteroyl glutamyl-glutamic acid (Murata et al., 1987).

INTRASPORANGIACEAE The family *Intrasporangiaceae* (Stackebrandt et al., 1997), established on phylogenetic grounds, was initially defined to include the type genus *Intrasporangium* as well as Terrabacter (Stackebrandt et al., 1997) and the peripherally related genus *Sanguibacter* (Fernández-Garayzábal et al., 1995). The genera *Terracoccus* (Prauser et al., 1997) and *Janibacter* (Martin et al., 1997), described just after the establishment of the family *Intrasporangiaceae*, have to be included into this family because of correspondence in their signature nucleotides. The addition of new taxa into the family led to the exclusion of the genus *Sanguibacter* that, together with *Rarobacter*, formed a separate subline of descent within *Micrococcineae*. Consequently the set of signature nucleotides specific for the emended family Intrasporangiaceae was redefined (Table 4; Stackebrandt and Schumann, 2000).

All genera of the family are monospecific and form a rather heterogeneous group with respect to their phenotypic characteristics. Except for *Janibacter limosus*, all members of the family have in common the peptidoglycan type A3γ (Schleifer and Kandler, 1972), variation A41.2 (*Deutsche Sammlung von Mikroorganismen und Zellculturen (DSMZ)–Catalogue of Strains*, DSMZ–German Collection of Microorganisms, Braunschweig, Germany), which is based on LL-A2pm, a triglycine interpeptide bridge and a

Table 10. Chemotaxonomic characteristics of genera of the family *Intrasporangiaceae*.

Genus	Diamino acid	Interpeptide bridge	Menaquinone	Polar lipids	Predominant cellular fatty acid(s)	Mol% G+C
Intrasporangium	LL-A$_2$pm	Gly$_3$[a]	MK-8	PI, PIM, PG, DPG	i-C$_{15:0}$, ai-C$_{15:0}$, i-C$_{16:0}$	68
Janibacter	meso-A$_2$pm	None	MK-8(H$_4$)	PI, PG, DPG	C$_{17:1}$, C$_{17:0}$, i-C$_{16:0}$	70
Terrabacter	LL-A$_2$pm	Gly$_3$[a]	MK-8(H$_4$)	PE, PI, DPG, PL	i-C$_{15:0}$, i-C$_{14:0}$, i-C$_{16:0}$	70–73
Terracoccus	LL-A$_2$pm	Gly$_3$[a]	MK-8(H$_4$)	PE, PI, PG, DPG	i-C$_{15:0}$, ai-C$_{15:0}$, C$_{16:0}$	73

Abbreviations: Gly, glycine; PIM, phospatidylinositol mannosides; i-C$_{15:0}$, 13-methyltetradecanoic acid; and C$_{17:1}$, heptadecenoic acid. Refer to the footnotes in Tables 4 through 7 for definitions of all other abbreviations.
[a]Glycine bound to glutamic acid at position 2 of peptide subunit.
Data are from Schleifer and Kandler, 1972; Kalakoutskii, 1989; Collins et al., 1989; Martin et al., 1997; Prauser et al., 1997; and Schumann et al., 1997.

glycine residue bound to the α-carboxyl group of D-glutamic acid. The occurrence of the peptidoglycan type A3γ variation A41.2 is unique within the bacterial kingdom. All members are considered nonpathogenic to humans and other vertebrates.

INTRASPORANGIUM. *Intrasporangium calvum*, the type species of the genus *Intrasporangium*, was isolated as an airborne organism from a school dining room (Kalakoutskii et al., 1967). It forms a branching mycelium, which tends to break into irregular fragments, i.e., typically nocardioform. An aerial mycelium does not occur. Strains possess totally unsaturated menaquinone MK-8 instead of MK-8(H$_4$) which is the characteristic menaquinone of all other representatives of the family (Table 10).

TERRABACTER. *Terrabacter tumescens*, type species of the genus, was isolated from soil and forms irregular rods with tendency to extensive primary branching. Aerial mycelia are not produced. A typical coryneform rod-coccus growth cycle occurs during cultivation on complex media. Molecular analysis of the microbial composition of sludge revealed the occurrence of a high fraction of organisms closely related to *T. tumescens* (Christensson et al., 1998), and these organisms may be involved in the removal of phosphorous from sludge.

The taxonomic position of this organism has been controversial for a long time. First described as *Corynebacterium tumescens* (Jensen, 1934), it was then affiliated to the genus *Arthrobacter* as *Arthrobacter tumescens* (Conn and Dimmick, 1947) and later reclassified on chemotaxonomic grounds as *Pimelobacter tumescens* (Suzuki and Komagata, 1983). A 16S rRNA sequence comparison revealed that the organism occupies a separate line of descent warranting the generic status as type species of the genus *Terrabacter* (Collins et al., 1989).

TERRACOCCUS. *Terracoccus luteus*, type species of the genus *Terracoccus*, was isolated from the water-soil interface of a duck pond (Prauser et al., 1997). Cells of the organism are invariably coccoid and occur singly, in pairs or in small irregular clusters. *Terracoccus luteus* differs from *Intrasporangium calvum* and *Terrabacter tumescens* in its fatty acid profile with lower amounts of 14-methyl pentadecanoic acid (i-C$_{16:0}$) and higher amounts of ai-C$_{15:0}$ (Prauser et al., 1997).

JANIBACTER. The type species *J. limosus* has been isolated from 1-year-old sludge of a waste-water treatment plant (Martin et al., 1997). Whereas the type strain (DSM 11140) exhibits a rod-coccus growth cycle, cells of a second strain (DSM 11141) are invariably coccoid. Strains are characterized by the directly cross-linked peptidoglycan type A1γ (Schleifer and Kandler, 1972), based on meso-A$_2$pm. The cellular fatty acid profile of *J. limosus* differs from that of other genera of the family in containing high amounts of heptadecanoic (C$_{17:0}$) and heptadecenoic (C$_{17:1}$) acids (Table 10).

MICROCOCCACEAE The family *Micrococcaceae* (Pribham, 1929) was emended on the basis of 16S rDNA/rRNA sequence-based phylogenetic clustering and circumscription by a set of family-specific 16S rDNA signature nucleotides (Stackebrandt et al., 1997). The emendation of this family was a consequence of preceding phylogenetic analyses (Koch et al., 1994; Koch et al., 1995; Stackebrandt et al., 1995) of members of the genera *Micrococcus* and *Arthrobacter*. The emended family *Micrococcaceae* contains the type genus *Micrococcus*, as well as the genera *Arthrobacter* (Koch et al., 1995), *Kocuria* (Stackebrandt et al., 1995), *Nesterenkonia* (Stackebrandt et al., 1995), *Renibacterium* (Sanders and Fryer, 1980), *Rothia* (Georg and Brown, 1967; Skerman et al., 1980), and *Stomatococcus* (Bergan and Kocur, 1982).

All genera of the family *Micrococcaceae* are characterized by the occurrence of L-lysine as diagnostic diamino acid of the peptidoglycan and predominance of *iso-* and *anteiso*-branched cellular fatty acids but differ markedly in the structures of their interpeptide bridges and in the

Table 11. Chemotaxonomic characteristics of genera of the family *Micrococcaceae*.

Genus	Diamino acid	Interpeptide bridge	Major menaquinone	Polar lipids	Predominant cellular fatty acid(s)	Mol% G+C
Micrococcus	L-Lys	Peptide subunit or D-Asp	MK-8, MK-8(H$_2$)	DPG, PI, PG, PL, GL	ai-C$_{15:0}$, i-C$_{15:0}$	69–76
Arthrobacter, globiformis group	L-Lys	MCA$_{var}$	MK-9(H$_2$)	DPG, PG, PI, DMDG	ai-C$_{15:0}$, i-C$_{15:0}$, i-C$_{16:0}$	61–66
Arthrobacter, nicotianae group	L-Lys	DCA$_{var}$	MK-8, MK-9	DPG, PG, PI, DMDG	ai-C$_{15:0}$	60–65
Kocuria	L-Lys	L-Ala$_{3-4}$	MK-7(H$_2$), MK-8(H$_2$)	DPG, PG, (PI, PL, GL)	ai-C$_{15:0}$, ai-C$_{17:0}$, i-C$_{16:0}$	66–75
Nesterenkonia	L-Lys	Gly-L-Glu	MK-8, MK-9	DPG, PG, PI, PL, GL	ai-C$_{15:0}$, ai-C$_{17:0}$, i-C$_{16:0}$	70–72
Renibacterium	L-Lys	L-Ala-Gly^f	MK-9, MK-10	DPG, GL	ai-C$_{15:0}$, ai-C$_{17:0}$	52–54
Rothia	L-Lys	L-Ala$_3$	MK-7	DPG, PG	ai-C$_{15:0}$, ai-C$_{17:0}$, C$_{16:0}$	49–53
Stomatococcus	L-Lys	L-Ala, L-Ser or Gly	MK-7	DPG, PG	ai-C$_{15:0}$, i-C$_{16:0}$, C$_{16:0}$	56–60

Abbreviations: MCA, monocarboxylic amino acid; DCA, dicarboxylic amino acid; var, variable amino acid composition in the interpeptide bridge; DMDG, dimannosyldiacylglycerol. Refer to the footnotes in Tables 4, 5, 6, 7 and 10 for definitions of all other abbreviations.

^{a}D-Alanine amide is bound to glutamic acid at position 2 of the peptide subunit.

Data are from Schleifer and Kandler, 1972; Collins, 1982; Embley et al., 1983; Jones and Collins, 1986; Gerencser and Bowden, 1986; Stackebrandt et al., 1983; Schleifer, 1986; Fiedler and Draxl, 1986; and Stackebrandt et al., 1995.

composition of their menaquinone and polar lipid patterns (Table 11).

MICROCOCCUS. The genus *Micrococcus* (Cohn, 1872) was emended after taxonomic dissection resulting from phylogenetic and chemotaxonomic analyses (Stackebrandt et al., 1995). It now contains two species: *M. luteus* as type species and *M. lylae*. Because of its high DNA G+C content of about 74 mol%, *M. luteus* has been included as a Gram-positive representative in comparative evolutionary studies, e.g., on the sequences and structures of RNases P RNAs (Haas et al., 1996), the organization and codon usage of the streptomycin operon (Ohama et al., 1987), the spectinomycin operon (Ohama et al., 1989), and anticodon composition of tRNAs (Kano et al., 1991). The UV endonuclease (DNA glycosilase/abasic lyase) of *M. luteus* has been cloned and characterized (Piersen et al., 1995; Shiota and Nakayama, 1997), as has been the structural genes encoding essential components of the enzyme hexaprenyl diphosphate synthase (Shimizu et al., 1998), which is involved in the generation of hexaprenyl diphosphate (the precursor of the prenyl side chain of menaquinone).

Both *Micrococcus* species have coccoid morphology and can be differentiated by their different interpeptide bridges and menaquinone patterns: *M. luteus* has an interpeptide bridge (peptidoglycan variation A2) and predominantly menaquinone MK-8, whereas *M. lylae* contains a D-aspartic acid interpeptide bridge (peptidoglycan variation A4α) and MK-8(H$_2$) as major menaquinone (Schleifer and Kandler, 1972;

Stackebrandt et al., 1995). The primary habitat is mammalian skin. Although micrococci are considered nonpathogenic, the increasing number of reports on micrococci isolated from clinical specimens necessitates further studies on their role as opportunistic pathogens (Goodfellow, 1998). This applies also to the members of the genera *Kocuria*, *Dermacoccus*, *Kytococcus* and *Nesterenkonia*.

ARTHROBACTER. Strains belonging to *Arthrobacter* (Conn and Dimmick, 1947), later emended by Koch et al. (1995), show a rod-coccus growth cycle and fall at least into two groups as revealed early by chemotaxonomic studies (Minnikin et al., 1978; Collins and Jones, 1981; Collins and Kroppenstedt, 1983; Stackebrandt et al., 1983). The type species *A. globiformis* (Conn and Dimmick, 1947) represents the core of the first, the so-called "globiformis" group, while the second, the "nicotianae" group, is centered around the species *A. nicotianae*. The globiformis group is characterized by the presence of a peptidoglycan of the A3α variation (i.e., an interpeptide bridge made up of one to four residues of L-amino acids like L-alanine, L-threonine or L-serine; Schleifer and Kandler, 1972) and of a dihydrogenated menaquinone with nine isoprenoid units MK-9(H$_2$). The "nicotianae" group comprises organisms that have peptidoglycan variation A4α (i.e., an interpeptide bridge made up of a dicarboxylic amino acid like glutamic or aspartic; Schleifer and Kandler, 1972) and menaquinones that are completely unsaturated with eight to ten isoprenoid units.

Arthrobacter species were subdivided into the following groups based on the amino acid sequence of their peptidoglycan interpeptide bridges (Komagata and Suzuki, 1987):

I. Ser-Thr-Ala, *A. oxydans* (Sguros, 1954; Skerman et al., 1980), *A. polychromogenes* (Schippers-Lammertse et al., 1963; Skerman et al., 1980).

II. Ala-Thr-Ala, *A. aurescens* (Phillips, 1953; Skerman et al., 1980), *A. ilicis* (Mandel et al., 1961; Collins et al., 1981a), *A. ureafaciens* (Krebs and Eggleston, 1939; Clark, 1955; Skerman et al., 1980), *A. histidinolovorans* (Adams, 1954; Skerman et al., 1980), *A. nicotinovorans* (Kodama et al., 1992).

III. Ala1-4, *A. globiformis*, *A. pascens* (Lochhead and Burton, 1953; Skerman et al., 1980), *A. ramosus* (Jensen, 1960; Skerman et al., 1980), *A. crystallopoietes* (Ensign and Rittenberg, 1963; Skerman et al., 1980).

IV. Ser-Ala2-3, *A. atrocyaneus* (Kuhn and Starr, 1960; Skerman et al., 1980).

V. Thr-Ala2, *A. citreus* (Sacks, 1954; Skerman et al., 1980).

VI. Ala-Glu, *A. nicotianae*, *A. creatinolyticus* (Hou et al., 1998), *A. uratoxydans* (Stackebrandt et al., 1983), *A. protophormiae* (Lysenko, 1959; Stackebrandt et al., 1983).

VII. Glu, *A. sulfureus* (Stackebrandt et al., 1983).

With two exceptions these groups are in good agreement with 16S rRNA sequence-based phylogenetic clustering: *A. crystallopoietes* represents a separate line of descent outside of group III and *A. sulfureus* clusters together with organisms of group VI despite its interpeptide bridge consisting solely of glutamic acid (Funke et al., 1996; Hou et al., 1998; Fig. 5). The species *A. agilis* (interpeptide bridge of Thr-Ala; Ali-Cohen, 1889), its comb. nov. (Koch et al., 1995a; Koch et al., 1995b), *A. woluwensis* (interpeptide bridge of D-Asp; Funke et al., 1996) and *A. cumminsii* (interpeptide bridge of Ser-Glu; Funke et al., 1996) can not be affiliated to any of these groups, and each of them represents an individual line of descent (Funke et al., 1996; Hou et al., 1998).

The following species cannot be affiliated because 16S rDNA sequence data are lacking: *A. mysorens* (Nand and Rao, 1972; Skerman et al., 1980) belonging to group VI on the basis of a Ala-Glu interpeptide bridge (Komagata and Suzuki, 1987) and *A. duodecadis* (Lochhead, 1958; Skerman et al., 1980).

Arthrobacter globiformis has been subjected to some molecular studies: the overexpression in *Escherichia coli* of an esterase which stereospecifically hydrolyzes the (+)-*trans* (1R,3R) stereoisomer of ethyl chrysanthemate, which is

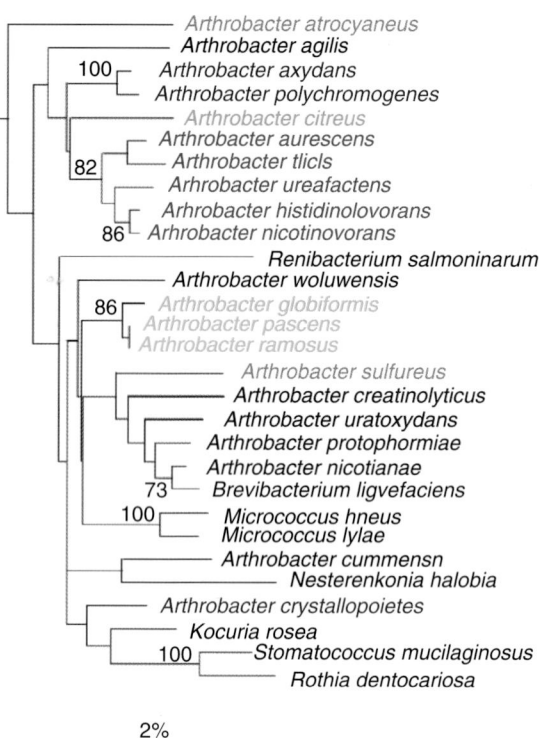

Fig. 5. Phylogenetic relatedness among authentic species of the genus *Arthrobacter*, which have been subgrouped based on results of 16S rDNA analysis. Organisms, the names of which are displayed in the same color, exhibit the same peptidoglycan structure. Numbers within the dendrogram indicate the percentages of occurrence of the branching order in 500 bootstrapped trees (only values of 70 and above are shown). Sequences of species of other genera of the family *Micrococcaceae* served as a root. The scale bar represents 2 nucleotide substitutions per 100 nucleotides.

the acidic part of pyrethrin, a naturally occurring insecticidal compound (Nishizawa et al., 1995); the manganese-dependent dioxygenase, involved in the degradation of aromatic compounds (Boldt et al., 1995); and the cloning and sequencing of the inulin fructotransferase genes (Haraguchi et al., 1995) and genes for glucodextranase and endodextranase (T. Oguma et al., unpublished observation).

The following two species cannot be considered authentic members of the genus *Arthrobacter* because 16S rDNA analysis clusters them separately from members of this genus (B. J. Tindall, E. Stackebrandt and J. Swiderski, unpublished observation); the type strain of *A. viscosus* (Gasdorf et al., 1965; Skerman et al., 1980) DSM 7303 and strain DSM 7287 are identical and cluster with *Rhizobium* species. *Arthrobacter viscosus* DSM 20159, on the other hand, is closely related to *Bacillus simplex* and *B. maroccanus*. The type strain of *A. siderocapsulatus* (i.e., National Collection of Industrial and Marine Biology (NCIMB) 11286; Dubinina and

Zhdanov, 1975; Skerman et al., 1980) and strain NCIMB 11287 (= DSM 7286) are highly related and cluster with *Pseudomonas putida* and related species.

The following *Arthrobacter* species, originally affiliated to the genus on morphological grounds, have been reclassified following chemotaxonomic and phylogenetic analyses: *A. flavescens* (Lochhead, 1958; Skerman et al., 1980) as *Microbacterium flavescens* (Collins et al., 1983; Takeuchi and Hatano, 1998), *A. picolinophilus* (Tate and Ensign, 1974; Skerman et al., 1980) as *Rhodococcus erythropolis* (Koch et al., 1995a), *A. radiotolerans* (Yoshinaka et al., 1973; Skerman et al., 1980) as *Rubrobacter radiotolerans* (Suzuki et al., 1988), *A. simplex* (Jensen, 1934; Lochhead, 1957; Skerman et al., 1980) as *Nocardioides simplex* (O'Donnell et al., 1982), *A. terregens* (Lochhead and Burton, 1953; Skerman et al., 1980) as *Microbacterium terregens* (Collins et al., 1983; Takeuchi and Hatano, 1998), *A. tumescens* (Jensen, 1934; Conn and Dimmick, 1947; Skerman et al., 1980) as *Terrabacter tumescens* (Collins et al., 1989), and *A. variabilis* (Müller, 1961; Skerman et al., 1980) as *Corynebacterium variabile* (Collins, 1987).

The species *A. cumminsii* and *A. woluwensis* are the only species of the genus isolated from clinical specimens (Funke et al., 1996) and consequently affiliated to risk group 2. *Arthrobacter creatinolyticus* was isolated from human urine (Hou et al., 1998), but this as well as all other species of the genus is regarded as nonpathogenic.

NESTERENKONIA. The genera *Nesterenkonia* and *Kocuria* were established simultaneously with the taxonomic dissection of the genus *Micrococcus* (Stackebrandt et al., 1995). *Nesterenkonia* (Stackebrandt et al., 1995) is monospecific, and its coccoid type species is *N. halobia* (Onishi and Kamekura, 1972; Skerman et al., 1980). The moderately halophilic organism was isolated from unrefined solar salt (Onishi and Kamekura, 1972) and is characterized by a Gly-L-Glu interpeptide bridge of the peptidoglycan and by completely unsaturated menaquinones with eight or nine isoprenoid units (Table 11).

KOCURIA. The genus *Kocuria* (Stackebrandt et al., 1995) comprises five species: the type species *K. rosea* (Fluegge, 1896; Skerman et al., 1980), isolated from soil and water; *K. varians* (Migula, 1900; Skerman et al., 1980), isolated from mammalian skin, soil and water; *K. kristinae* (Kloos et al., 1974; Skerman et al., 1980), isolated from healthy human skin; *K. palustris* (Kovács et al., 1999) and *K. rhizophila* (Kovács et al., 1999), both isolated from the rhizosphere of *Typha angustifolia*. The species *K. erythromyxa* (Brooks and Murray, 1981; Rainey et al., 1997) has recently been transferred to *K. rosea*

(Schumann et al., 1999). All species of the genus *Kocuria* have in common a peptidoglycan interpeptide bridge, which consists of three to four alanine residues (Table 11).

STOMATOCOCCUS. This genus is monospecific, and the type species *S. mucilaginosus* (Bergan and Kocur, 1982) was isolated from the human oral cavity and the upper respiratory tract. The genus accommodates facultatively anaerobic cocci, which are characterized by a single L-alanine, L-serine or glycine residue as interpeptide bridge of their peptidoglycan and by a completely unsaturated menaquinone with seven isoprenoid units (Table 11). *Stomatococcus mucilaginosus* has been isolated in several cases from blood cultures taken mainly from immunocompromised patients (Goodfellow, 1998) and has been classified as a risk group 2 organism.

ROTHIA. The genus *Rothia* (Georg and Brown, 1967; Skerman et al., 1980) is monospecific, and the type species *R. dentocariosa* (Onishi, 1949; Skerman et al., 1980) is a normal inhabitant of the human oral cavity, most frequently isolated from supragingival dental plaque. *Rothia dentocariosa* is facultatively anaerobic, and the culture usually consists of cocci, irregular rods, and filaments. Strains possess an interpeptide bridge consisting of three alanine residues resembling that of *Kocuria* species, but differ from these species in their low G+C content of only 49–53 mol% and in the occurrence of the completely unsaturated menaquinone with seven isoprenoid units (Table 11). Although classified as a nonpathogenic organism, *R. dentocariosa* has been identified as the causative agent of pneumonia and septicemia in immunocompromised patients and it has been isolated from various inflammatory processes (Schaal, 1992).

RENIBACTERIUM. The type species of the monospecific genus, *R. salmoninarum* (Sanders and Fryer, 1980), is an obligate pathogen of salmonid fish. It usually causes a chronic, systemic, granulomatous infection of the kidney, which can be fatal for the fish under appropriate conditions (Evelyn, 1993). Based upon multilocus enzyme electrophoresis and analysis of the intergenic spacer region between 16 and 23S rRNA genes, recent epidemiological studies on *R. salmoninarum* isolates from a broad variety of fish from different locations indicated a low level of genetic diversity (Starliper, 1996; Grayson et al., 1999), although random amplified polymorphic DNA (RAPD) analysis allowed reproducible differentiation between isolates obtained from different hosts and different geographic regions. Possible virulence factors of *R. salmoninarum* have been sought by screening a genomic library for expression of virulence factors in *E. coli*. A single hemolytic clone was identified and the gene encoded resembled that of bacterial zinc-metalloproteases (Grayson et al., 1995). Additional information is available for the gene coding for the 57-kDa major soluble antigen (Chien et al., 1992).

The slowly growing cells occur as short regular rods often in pairs and sometimes in short chains. *Renibacterium salmoninarum* is characterized by a peptidoglycan with an L-alanine-glycine interpeptide bridge and D-alanine amide bound to D-glutamic acid at position 2 of the peptide subunit, by completely unsaturated menaquinones with nine and ten isoprenoid units and by a low G+C value of 52–54 mol% (Table 11). Because of its potential pathogenicity for salmonid fish, *R. salmoninarum* is allocated to risk group 2.

Literature Cited

Anderson, R. C., M. A. Rasmussen, A. A. DiSpirito, and M. J. Allison. 1997. Characteristics of a nitropropanol-metabolizing bacterium isolated from the rumen. Can. J. Microbiol. 43:617–624.

Andreyev, L. V., L. I. Evtushenko, and N. S. Agre. 1983. Fatty acid composition of Promicromonospora citrea. Mikrobiologiya 52:58–63.

Austwick, P. K. C. 1958. Curaneous streptotrichosis, mycotic dermatitis and strawberry foot root and the genus Dermatophilus Van Saceghem. Vet. Rev. Annot. 4:33–38.

Balows, A., Trüper, H. G., Dworkin, M. Harder, W., and K.-H. Schleifer. 1992. The prokaryotes: A handbook on the biology of bacteria: Ecophysiology, isolation, identification, applications. Springer-Verlag. New York, NY.

Byers, H. K., Stackebrandt, E., Hayward, C., and L. L. Blackall. 1998. Molecular investigation of a microbial mat associated with the Great Artesian basin. FEMS Micro. Ecol. 25:391–403.

Bergan, T., and M. Kocur. 1982. Stomatococcus mucilaginosus gen. nov. ep. rev., a member of the family Micrococcaceae. Int. J. Syst. Bacteriol. 32:374–377.

Blackall, L. L., Seviour, E. M., Cunningham, M. A., Seviour, R. J., and P. Hugenholtz. 1994. "Microthrix parvicella" is a novel deep branching member of the actinomycetes subphylum. System. Appl. Microbiol. 17:513–518.

Bogdanovsky, D., E. Interschick-Niebler, K. H. Schleifer, F. Fiedler, and O. Kandler. 1971. γ-Glutamyl-glutamic acid, an interpeptide bridge in murein of some micrococci and Arthrobacter sp. Eur. J. Biochem. 22:173–178.

Boldt, Y. R., M. J. Sadowsky, L. B. M. Ellis, L. Que Jr., and L. P. Wackett. 1995. A manganese-dependent dioxygenase from Arthrobacter globiformis CM-2 belongs to the major extradiol dioxygenase family. J. Bacteriol. 177:1225–1232.

Brierley, J. A. 1978. Thermophilic iron-oxydizing bacteria found in copper leaching dumps. Appl. Environ. Microbiol. 36:523–525.

Brooks, B. W., and R. G. E. Murray. 1981. Nomenclature for "Micrococcus radiodurans" and other radiation-resistant cocci: Deinococcaceae fam. nov. and Deinococcus gen. nov., including five species. Int. J. Syst. Bacteriol. 31:353–360.

Buchanan, R. E. 1917. Studies in the nomenclature and classification of the bacteria. II. The primary subdivisions of the Schizomycetes. J. Bacteriol. 2:155–164.

Buchanan, R. E. 1918. Studies in the classification and nomenclature of the bacteria. VIII. The subgroups and genera of the Actinomycetales. J. Bacteriol. 3:403–406.

Buchanan, R. E. 1925. General Systematic Bacteriology. The William and Wilkins Co. Baltimore, MD.

Buchanan, R. E., J. G. Holt, and E. F. Lessel Jr. 1966. Index Bergeyana. Williams and Wilkins Co. Baltimore, 3–1472.

Cai, J., and M. D. Collins. 1994. Phylogenetic analysis of species of the meso-diaminopimelic acid-containing genera Brevibacterium and Dermabacter. Int. J. Syst. Bacteriol. 44:583–585.

Carreto, L., E. Moore, M. F. Nobre, R. Wait, P. W. Riley, R. J. Sharp, and M. S. da Costa. 1996. Rubrobacter xylanophilus sp. nov., a new thermophilic species isolated from a thermally polluted effluent. Int. J. Syst. Bacteriol. 46:460–465.

Chester, F. D. 1897. Report of the mycologist: bacteriological work. Ann. Rep. Del. agric. Exp. Sta. 10:47–137.

Chien, M.-S., T. L. Gilbert, C. Huang, M. L. Landolt, P. J. O'Hara, and J. R. Winton. 1992. Molecular cloning and sequence analysis of the gene coding for the 57-kDs major soluble antigen of the salmonid fish pathogen Renibacterium salmoninarum. FEMS Microbiol. Lett. 96:259–266.

Cho, M., J.-H. Moon, S.-B. Kim, and Y. H. Park. 1998. Application of the ribonuclease P (RNase P) RNA gene sequence for phylogenetic analysis of the genus Saccharomonospora. Int. J. Syst. Bacteriol. 48:1223–1230.

Christensson, M., L. L. Blackall, and T. Welander. 1998. Metabolic transformations and characterisation of the sludge community in an enhanced biological phosphorous removal system. Appl. Micro. Biotech. 49:226–234.

Clark, D. A., and P. R. Norris. 1996. Acidimicrobium ferrooxydans gen. nov., sp. nov.: mixed-culture ferrous iron oxidation with Sulfobacillus species. Microbiology (Reading) 142:785–790.

Clifton, C. E. 1950. Introduction to the bacteria. McGraw-Hill. New York.

Cohn, F. 1872. Untersuchungen über Bakterien. Beitr. Biol. Pfl. Heft 2 1:127–224.

Cohn, F. 1975. Untersuchungen über Bakterien. II. Beitr. Biol. Pfl. Heft 3 1:141–207.

Collins, M. D. 1987. Transfer of Arthrobacter variabilis (Müller) to the genus Corynebacterium, as Corynebacterium variabilis comb. nov. Int. J. Syst. Bacteriol. 37:287–288.

Collins, M. D. 1982. Lipid composition of Renibacterium salmoninarum (Sanders and Fryer). FEMS Microbiol. Lett. 13:295–297.

Collins, M. D., and R. M. Kroppenstedt. 1983. Lipid composition as a guide to the classification of some coryneform bacteria containing an A4 type peptidoglycan (Schleifer and Kandler). System. Appl. Microbiol. 4:95–104.

Collins, M. D., and D. Jones. 1981. Distribution of isoprenoid quinone structural types in bacteria and their taxonomic implications. Microbiol. Rev. 45:316–354.

Collins, M. D., and S. Wallbanks. 1992. Comparative sequence analysis of the 16S rRNA genes of Lactobacillus minutus, Lactobacillus rimae and Streptococcus parvulus: proposal for the creation of a new genus Atopobium. FEMS Microbiol. Lett. 95:235–240.

Collins, M. D., J. Brown, and D. Jones. 1988. Brachybacterium faecium gen. nov., sp. nov., a coryneform bacterium from poultry deep litter. Int. J. Syst. Bacteriol. 38:45–48.

Collins, M. D., M. Dorsch, and E. Stackebrandt. 1989. Transfer of Pimelobacter tumescens to Terrabacter gen. nov. as Terrabacter tumescens comb. nov. and of Pimelobacter jensenii to Nocardioides as Nocardioides jensenii comb. nov. Int. J. Syst. Bacteriol. 39:1–6.

Collins, M. D., D. Jones, and R. M. Kroppenstedt. 1981a. Reclassification of Corynebacterium ilicis (Mandel, Guba and Litsky) in the genus Arthrobacter, as Arthrobacter ilicis comb. nov. Zbl. Bakt. Hyg., I. Abt. Orig. C 2:318–323.

Collins, M. D., D. Jones, R. M. Keddie, R. M. Kroppenstedt, and K. H. Schleifer. 1983. Classification of some coryneform bacteria in a new genus Aureobacterium. Syst. Appl. Microbiol. 4:236–252.

Collins, M. D., M. Goodfellow, and D. E. Minnikin. 1979. Isoprenoid quinones in the classification of coryneform and related bacteria. J. Gen. Microbiol. 110:127–136.

Conn, H. J., and I. Dimmick. 1947. Soil bacteria similar in morphology to Mycobacterium and Corynebacterium. J. Bacteriol. 54:291–303.

Corda, A. C. J. 1839. Pracht-Flora Europaeischer Schimmelbildungen. Gerhard Fleischer. Leipzig.

Demharter, W., Hensel, R. Smida, J., and E. Stackebrandt. 1989. Sphaerobacter thermophilus gen. nov., spec. nov. a deeply rooting member of the actinomycetes subdivision isolated from thermophilically treated sewage plant. System. Appl. Microbiol. 11:261–266.

Embley, T. M. 1992. The genus Renibacterium. A. Balows, H. G. Trüper, M. Dworkin, W. Horder, and K.-H. Schleifer. The prokaryotes: A handbook on the biology of bacteria: Ecophysiology, isolation, identification, applications. Springer-Verlag. New York, 1312–1319.

Embley, T. M., M. Goodfellow, D. E. Minnikin, and B. Austin. 1983. Fatty acid, isoprenoid quinone and polar lipid composition of Renibacterium salmoninarum. J. Appl. Bacteriol. 55:31–37.

Embley, T. M., and Stackebrandt, E. 1994. The molecular phylogeny and systematics of the Actinomycetes. Annu. Rev. Microbiol. 48:257–289.

Evelyn, T. P. T. 1993. Bacterial kidney disease-BKD. Ingli, V., R. J. Roberts, and N. R. Bromage. Bacterial diseases of fish. Blackwell. Oxford, UK. 177–195.

Evtushenko, L. I., G. F. Levanova, and N. S. Agre. 1984. Nucleotide composition of DNA and amino acid composition of A4 peptidoglycan in Promicromonospora citrea. Mikrobiologiya 53:519–520.

Evtushenko, L. I., S. D. Taptykova, V. N. Akimov, S. A. Semyonova, and L. V. Kalakoutskii. 1991. Glycomyces tenuis sp. nov. Int. J. Syst. Bacteriol. 41:154–157.

Felske, A., Rheims, H., Wolterink, A., Stackebrandt, E., and Akkermans, A. D. L. 1997. Ribosome analysis reveals prominent activity of an uncultured member of the class Actinobacteria in grassland soils. Microbiology (Reading) 143:2983–2989.

Fernández-Garayzábal, J. F., L. Dominguez, C. Pascual, D. Jones, and M. D. Collins. 1995. Phenotypic and phylogenetic characterization of some unknown coryneform bacteria isolated from bovine blood and milk: description of Sanguibacter gen. nov. Lett. Appl. Microbiol. 20:69–75.

Fiedler, F., and R. Draxl. 1986. Biochemical and immunochemical properties of the cell surface of Renibacterium salmoninarum. J. Bacteriol. 168:799–804.

Fiedler, F., and J. Seger. 1983. The murein types of Listeria grayi, Listeria murrayi, and Listeria denitrificans. System. Appl. Microbiol. 4:440–450.

Fiedler, F., J. Seger, A. Schrettenbrunner, and H. P. R. Seeliger. 1984. The biochemistry of murein and cell wallteichoic acids in the genus Listeria. Syst. Appl. Microbiol. 5:52–96.

Flügge, C. 1896. Die Mikroorganismen. F. C. W. Vogel. Leipzig.

Fuhrman, J. A., McCallum, K., and Davis, A. A. 1993. Phylogenetic diversity of subsurface marine microbial communities from the Atlantic and Pacific oceans. Appl. Environ. Microbiol. 59:1294–1302.

Funke, G., R. A. Hutson, K. A. Bernard, G. E. Pfyffer, G. Wauters, and M. D. Collins. 1996. Isolation of Arthrobacter spp. from clinical specimens and description of Arthrobacter cumminsii sp. nov. and Arthrobacter woluwensis sp. nov. J. Clin. Microbiol. 34:2356–2363.

Gerencser, M. A., and G. H. Bowden. 1986. Genus Rothia Georg and Brown 1967. P. H. A. Sneath, N. S. Mair, M. E. Sharpe, and J. G. Holt. Bergey's manual of systematic bacteriology. The Williams and Wilkins Co. Baltimore, 2:1342–1346.

Goodfellow, M. 1998. The Actinomycetes: Micrococcus and related genera. L. Collier, A. Balows, and M. Sussman. Topley & Wilson's Microbiology and Microbial Infections. Arnold. London, 491–506.

Gordon, M. A. 1964. The genus Dermatophilus. J. Bact. 88:509–522.

Gordon, M. A. 1989. Genus Dermatophilus Van Saceghem 1915. S. T. Wiliams, M. E. Sharpe, and J. G. Holt Bergey's manual of systematic bacteriology. The Williams and Wilkins Co. Baltimore, MD. 4:2409–2410.

Goto-Yamamoto, N., S. Sato, H. Miki, Y. K. Park, and M. Tadenuma. 1993. Taxonomic studies on yeast-lysing bacteria, and a new species Rarobacter incanus. J. Gen. Appl. Microbiol. 39:261–272.

Grayson, T. H., L. F. Cooper, F. A. Atienzar, M. R. Knowles, and M. L. Gilpin. 1999. Molecular Differentiation of Renibacterium salmoninarum isolated from worldwide locations. Appl. Environ. Microbiol. 65:961–968.

Grayson, T. H., A. J. Evenden, M. L. Gilpin, K. L. Martin, and C. B. Munn. 1995. A gene from Renibacterium salmoninarum encoding a product which shows homology to zinc-matalloproteases. Microbiology (Reading) 141:1331–1341.

Groth, I., P. Schumann, F. A. Rainey, K. Martin, B. Schuetze, and K. Augsten. 1997a. Demetria terragena gen. nov., sp. nov., a new genus of actinomycetes isolated from compost soil. Int. J. Syst. Bacteriol. 47:1129–1133.

Groth, I., P. Schumann, F. A. Rainey, K. Martin, B. Schuetze, and K. Augsten. 1997b. Bogoriella caseilytica gen. nov., sp. nov., a new alkaliphilic actinomycete from a soda lake in Africa. Int. J. Syst. Bacteriol. 47:788–794.

Gvozdyak, O. R., T. M. Nogina, and P. Schumann. 1992. Taxonomic study of the genus Brachybacterium: Brachybacterium nesterenkovii sp. nov. Int. J. Syst. Bacteriol. 42:74–78.

Haas, E. S., A. B. Banta, K. J. Harris, N. R. Pace, and J. W. Brown. 1996. Structure and evolution of ribonuclease P RNA in Gram-positive bacteria. Nucl. Acids Res. 24:4775–4782.

Haas, F., and H. König. 1988. Coriobacterium glomerans gen. nov., sp. nov. from the intestinal tract of the red soldier bug. Int. J. Syst. Bacteriol. 38:382–384.

Haraguchi, K., K. Seki, M. Kishimoto, T. Nagata, T. Kasumi, K. Kainuma, and S. Kobayashi. 1995. Cloning and nucleotide sequence of the inulin fructotransferase (DFA I-producing) gene of Arthrobacter globiformis S14-3. Biosci. Biotech. Biochem. 59:1809–1812.

Harz, C. O. 1877. Actinomyces bovis, ein neuer Schimmel in den Geweben des Rindes. Jahresber. K. Cent. Thierärzt. Schule München (1877–1878) 125–140.

Hou, X.-G., Y. Kawamura, F. Sultana, S. Shu, K. Hirose, K. Goto, and T. Ezaki. 1998. Description of Arthrobacter creatinolyticus sp. nov., isolated from human urine. Int. J. Syst. Bacteriol. 48:423–429.

Jáger, K., K. Márialigeti, M. Hauck, and G. Barabás. 1983. Promicromonospora enterophila sp. nov., a new species of monospore actinomycetes. Int. J. Syst. Bacteriol. 33:525–531.

Jensen, H. L. 1934. Studies on saprophytic mycobacteria and corynebacteria. Proc. Linn. Soc. NSW 59:19–61.

Jones, D., and M. D. Collins. 1986. Irregular, non-sporing Gram-positive rods. P. H. A. Sneath, N. S. Mair, M. E. Sharpe, and J. G. Holt. Bergey's manual of systematic bacteriology. The Williams and Wilkins Co. Baltimore, MD. 2:1261–1266.

Jones, D., and M. D. Collins. 1988. Taxonomic studies on some human cutaneous coryneform bacteria: Description of Dermabacter hominis gen. nov., sp. nov. FEMS Microbiol. Lett. 51:51–56.

Jones, D., and R. M. Keddie. 1992. The genus Arthrobacter. Balows, A., H. G. Trüper, M. Dworkin., W. Harder and K.-H. Schleifer. The Prokaryotes, 2nd ed. Springer-Verlag. New York, 1283–1299.

Kalakoutskii, L. V. 1989. Genus Intrasporangium Kalakoutskii, Kirillova and Krassil'nikov. S. T. Wiliams, M. E. Sharpe, and J. G. Holt. Bergey's manual of systematic bacteriology. The Williams and Wilkins Co. Baltimore, MD. 4:2395–2397.

Kalakoutskii, L. V., I. P. Kirillova, and N. A. Krassil'nikov. 1967. A new genus of the Actinomycetales-Intrasporangium gen. nov. J. Gen. Microbiol. 48:79–85.

Kano, A., Y. Andachi, T. Ohama, and S. Osawa. 1991. Novel anticodon composition of transfer RNAs in Micrococcus luteus, a bacterium with a high genomic G+C content. J. Mol. Biol. 221:387–401.

Kluyver, A. J., and C. B. Van Niel. 1936. Prospects for a natural system of classification of bacteria. Zbl. Bakt. Abt. 2 94:369–403.

Kocur, M., W. E. Kloos, and K.-H. Schleifer. 1992. The genus Micrococcus. Balows, A., H. G. Trüper, M. Dworkin., W. Harder and K.-H. Schleifer. The Prokaryotes, 2nd ed. Springer-Verlag. New York, NY. 1300–1311.

Koch, C., S. Klatte, P. Schumann, J. Burghardt, R. M. Kroppenstedt, and E. Stackebrandt. 1995a. Transfer of Arthrobacter picolinophilus Tate and Ensign 1974 to Rhodococcus erythropolis. Int. J. Syst. Bacteriol. 45:576–577.

Koch, C., F. A. Rainey, and E. Stackebrandt. 1994. 16S rDNA studies on members of Arthrobacter and Micrococcus, an aid for their future taxonomic restructuring. FEMS Microbiol. Lett. 123:167–172.

Koch, C., P. Schumann, and E. Stackebrandt. 1995. Reclassification of Micrococcus agilis Ali-Cohen 1889 to the genus Arthrobacter as Arthrobacter agilis comb. nov. and emendation of the genus Arthrobacter. Int. J. Syst. Bacteriol. 45:837–839.

Kocur, M. 1986. Genus Micrococcus Cohn 1872. P. H. A. Sneath, N. S. Mair, M. E. Sharpe, and J. G. Holt. Bergey's

manual of systematic bacteriology. The Williams and Wilkins Co. Baltimore, MD. 2:1004–1008.

Kodama, Y., H. Yamamoto, N. Amano, and T. Amachi. 1992. Reclassification of two strains of Arthrobacter oxydans and proposal of Arthrobacter nicotinovorans sp. nov. Int. J. Syst. Bacteriol. 42:234–239.

Komagata, K., and K. Suzuki. 1987. Lipid and cell-wall analysis in bacterial systematics. R. R. Colwell and R. Grigorova. Methods in microbiology, vol. 19. Academic Press Ltd. London, 161–207.

Kovács, G., J. Burghardt, S. Pradella, P. Schumann, E. Stackebrandt, and K. Márialigeti. 1999. Kocuria palustris sp. nov. and Kocuria rhizophila sp. nov., isolated from the rhizoplane of the narrow-leaved cattail (Typha angustifolia). Int. J. Syst. Bacteriol. 49:167–173.

Krassil'nikov, N. A. 1959. Diagnostik der Bakterien und Actinomyceten. VEB Gustav Fischer Verlag. Jena.

Krassil'nikov, N. A., L. V. Kalakoutskii, and N. E. Kirillova. 1961. A new genus of ray fungi-Promicromonospora. gen. nov. (in Russian). Izv. Akad. Nauk SSSR (Ser Biol.) 1:107–112.

Kruse, W. 1896. Systematik der Streptotricheen und Bakterien. Flügge, C. Die Mikroorganismen, 3nd. edition. 185–526.

Kudo, T., Matsushima, K., Itoh, T., Sasaki, J., and K. Suzuki. 1998. Description of four new species of the genus Kineospora: Kineospora succinea sp. nov., Kineospora rhizophila sp. nov., Kineospora mikuniensis sp. nov. and Kineospora rhamnosa sp. nov., isolatd from plant samples, and amended description of the genus Kineospora. Int. J. Syst. Bacteriol. 48:1245–1255.

Küster, E. 1967. Taxonomy of soil Actinomycetales and related organisms. T. R. G. Gray and D. Parkinson. The ecology of the soil bacteria: An international symposium. Liverpool University Press. Liverpool, 322–336.

Labeda, D. P., R. T. Testa, M. P. Lechevalier, and H. Lechevalier. 1985. Glycomyces, a new genus of the Actinomycetales. Int. J. Syst. Bacteriol. 35:417–421.

Laura, C. More, E., Nobre, M. F. Wait, R. Riley, P. W., Sharp, R. J., and M. S. DaCosta. 1996. Rubrobacter xylanophilus sp. nov., a new thermohilic species isolated from a thermally polluted effluent. Int. J. Syst. Bacteriol. 46:460–465.

Lehmann, K. B., and Neumann, R. 1896. Atlas und Grundriss der Bakteriologie und Lehrbuch der speciellen bacteriologischen Diagnostik. München.

Liesack, W., and E. Stackebrandt. 1992. Occurrence of novel groups of the domain Bacteria as revealed by analysis of genetic material isolated from an Australian terrestrial environment. J. Bacteriol. 174:5072–5078.

Lieske, R. 1921. Morphologie und Biologie der Strahlenpilze (Actinomyceten). Borntraeger. Leipzig.

Martin, K., P. Schumann, F. A. Rainey, B. Schuetze, and I. Groth. 1997. Janibacter limosus gen. nov., sp. nov., a new actinomycete with meso-diaminopimelic acid in the cell wall. Int. J. Syst. Bacteriol. 47:529–534.

Masters, A. M., T. M. Ellis, J. M. Carson, S. S. Sutherland, and A. R. Gregory. 1995. Dermatophilus chelonae sp. nov., isolated from chelonids in Australia. Int. J. Syst. Bacteriol. 45:50–56.

Meyen, F. J. F. 1827. Actinomyce, Strahlenpilz. Eine neue Pilz-Gattung. Linnaea 2:433–444.

Migula, W. 1894. Ueber ein neues System der Bakterien. Arb. bact. Inst. Karlsruhe 1:235–238.

Minnikin, D. E., M. Goodfellow, M. D. Collins. 1978. Lipid composition in the classification and identification of coryneform and related taxa. I. J. Bousfield and A. G. Callely. Coryneform bacteria. Academic Press. London, 85–160.

Miyadoh, S., M. Hamada, K. Hotta, T. Kudo, A. Seino, G. Vobis, and A. Yokota. 1977. Atlas of Actinomycetes. Asakura Publishing Co. Tokyo.

Murata, M., H. Tanaka, and S. Omura. 1987. 7-Hydro-8-methyl-pteroyl glutamylglutamic acid, a new antifolate. fermentation, isolation, structure, and biological activity. J. Antibiot. 40:251–257.

Nishizawa, M., M. Shimizu, H. Ohkawa, and M. Kanaoka. 1995. Stereoselective production of ($^+$)-trans-chrysanthemic acid by a microbial esterase: cloning, nucleotide sequence, and overexpression of the esterase gene of Arthrobacter globiformis in Escherichia coli. App. Envir. Microbiol. 61:3208–3215.

Ochi, K. 1992. Electrophoretic heterogeneity of ribosomal protein AT L-30 among actinomycete genera. Int. J. Syst. Bacteriol. 42:144–150.

Ochi, K. 1993. A taxonomic review of the genus Microbispora by analysis of ribosomal protein AT-L30. Int. J. Syst. Bacteriol. 43:58–62.

O'Donnell, A. G., M. Goodfellow, and D. E. Minnikin. 1982. Lipids in the classification of Nocardioides: reclassification of Arthrobacter simplex (Jensen) Lochhead in the genus Nocardioides (Prauser) emend. O'Donnell et al. as Nocardioides simplex comb. nov. Arch. Microbiol. 133:323–329.

Ohama, T., A. Muto, and S. Osawa. 1989. Spectinomycin operon of Micrococcus luteus: evolutionary implications of organization and novel codon usage. J. Mol. Evol. 29:381–395.

Ohama, T., F. Yamao, A. Muto, and S. Osawa. 1987. Organization and codon usage of the streptomycine operon in Micrococcus luteus, a bacterium with a high genomic G+C content. J. Bacteriol. 169:4770–4777.

Olsen, I., J. L. Johnson, L. V. H. Moore, and W. E. C. Moore. 1991. Lactobacillus uli sp. nov. and Lactobacillus rimae, sp. nov., from the human gingival crevice and emended descriptions of Lactobacillus minutus and Streptococcus parvulus. Int. J. Syst. Bacteriol. 41:261–266.

Prévot, S. 1961. Traité de bactériologie systématique. Dunod. Paris.

Onishi, M. 1949. Studies on the actinomyces isolated from the deeper layer of carious dentine. J. Dent. Res. 6:273–282.

Onishi, H., and M. Kamekura. 1972. Micrococcus halobius sp. n. Int. J. Syst. Bacteriol. 22:233–236.

Pagani, H., and F. Parenti. 1978. Kineosporia, a new genus of the order Actinomycetales. Int. J. Syst. Bacteriol. 28:401–406.

Pascual, C., M. D. Collins, P. A. D. Grimont, and J. F. Fernández-Garayzábal. 1996. Sanguibacter inulinus sp. nov. Int. J. Syst. Bacteriol 46:811–813.

Piersen, C. E., M. A. Prince, M. L. Augustine, M. L. Dodson, and R. S. Lloyd. 1995. Purification and cloning of Micrococcus luteus ultraviolet endonuclease, an N-glycosylase/abasic lyase that proceeds via an imino enzyme-DNA intermediate. J. Biol. Chem. 270:23475–23484.

Prauser, H. 1975. The Actinomycetales—an order? T. Hasegawa. Proc. 1st Intersect. Congr. IAMS. Science council of Japan. Tokyo, 19–33.

Prauser, H. 1976. New nocardioform organisms and their relationship. T. AraiActinomycetes: The boundary microorganisms. Toppan Co., Ltd. Tokyo, 193–207.

Prauser, H., P. Schumann, F. A. Rainey, R. M. Kroppenstedt, and E. Stackebrandt. 1997. Terracoccus luteus gen. nov., sp. nov., an LL-diaminopimelic acid-containing coccoid actinomycete from soil. Int. J. Syst. Bacteriol. 47:1218–1224.

Pribham, E. 1929. A contribution to the classification of microorganisms. J. Bacteriol. 18:361–394.

Rainey, F. A., and E. Stackebrandt. 1994. Coriobacterium and Atopobium are phylogenetic neighbors within the actinomycetes line of descent. System. Appl. Microbiol. 17:202–205.

Rainey, F. A., Ward, N. L., and Stackebrandt, E. 1993. Molecular ecology study of a New Zealand acidothermal soil. Thermophiles '93, Hamilton. University of Waikato. Hamilton, New Zealand. Abstract A7.

Rainey, F. A., N. Weiss, and E. Stackebrandt. 1995. Phylogenetic analysis of the genera Cellulomonas, Promicromonospora, and Jonesia and proposal to exclude the genus Jonesia from the family Cellulomonadaceae. Int. J. Syst. Bacteriol. 45:649–652.

Rainey, F. A., M. F. Nobre, P. Schumann, E. Stackebrandt, and M. S. Da Costa. 1997. Phylogenetic diversity of the deinococci as determined by 16S ribosomal DNA sequence comparison. Int. J. Syst. Bacteriol. 47:510–514.

Rheims, H., C. Sproer, F. A. Rainey, and E. Stackebrandt. 1996. Molecular biological evidence for the occurrence on uncultured members of the actinomycete line of descent in different environments and geographical locations. Microbiol. (UK) 142:2863–2870.

Rheims, H., Felske, A., Seufert, S., and E. Stackebrandt. 1999. Molecular monitoring of an uncultured group of the class Actinobacteria in two terrestrial environments. J. Microbial. Meth. 36:65–75.

Rocourt, J., U. Wehmeyer, and E. Stackebrandt. 1987. Transfer of Listeria denitrificans into a new genus Jonesia gen. nov. as Jonesia denitrificans comb. nov. Int. J. Syst. Bacteriol. 37:266–270.

Saito, T., H. Terato, and O. Yamamoto. 1994. Pigments of Rubrobacter radiotolerans. Arch. Microbiol. 162:412–421.

Shirling, E. B., and D. Gottlieb. 1966. Methods for characterization of Streptomyces species. Int. J. Syst. Bacteriol. 16:313–340.

Sanders, J. E., and J. L. Fryer. 1980. Renibacterium salmoninarum gen. nov., sp. nov., the causative agent of bacterial kidney disease in salmonid fishes. Int. J. Syst. Bacteriol. 30:496–502.

Schaal, K. P. 1992. The genera Actinomyces, Arcanobacterium, and Rothia. A. Balows, H. G. Trüper, M. Dworkin, W. Horder, and K.-H. Schleifer. The prokaryotes: A handbook on the biology of bacteria: Ecophysiology, isolation, identification, applications. Springer-Verlag. New York, 850–905.

Schleifer, K. H. 1986. Gram-positive cocci. P. H. A. Sneath, N. S. Mir, M. E. Sharpe, and J. G. Holt. Bergey's manual of systematic bacteriology. The Williams and Wilkins Co. Baltimore, MD. 999–1003.

Schleifer, K. H., and O. Kandler. 1972. Peptidoglycan types of bacterial cell walls and their taxonomic implications. Bacteriol. Rev. 36:407–477.

Schleifer, K. H., and W. Ludwig. 1989. Phylogenetic relationships among bacteria. Fernholm, B., Bremer, K., and Jörnwall, H. The hierarchy of life. Elsevier Science Publishers. Amsterdam, 103–117.

Schubert, K., W. Ludwig, N. Springer, R. M. Kroppenstedt, J.-P. Accolas, and F. Fiedler. 1996. Two coryneform bacteria isolated from the surface of French Gruyère and Beaufort cheeses are new species of the genus Brachybacterium: Brachybacterium alimentarium sp. nov. and Brachybacterium tyrofermentans sp. nov. Int. J. Syst. Bacteriol. 46:81–87.

Schumann, P., H. Prauser, F. A. Rainey, E. Stackebrandt, and P. Hirsch. 1997. Friedmanniella antarctica gen. nov., sp. nov., an LL-diaminopimelic acid-containing actinomycete from antarctic sandstone. Int. J. Syst. Bacteriol. 47:278–283.

Schumann, P., C. Spröer, J. Burghardt, G. Kovács, and E. Stackebrandt. 1999. Reclassification of the species Kocuria erythromyxa (Brooks and Murray, 1981) as Kocuria rosea (Flügge, 1886). Int. J. Syst. Bacteriol. 49:393–396.

Seeliger, H. P. R., and D. Jones. 1986. Genus listeria pirie 1940. Sneath, P. H. A., N. S. Mair, N. E. Sharpe and J. G. Holt. Bergey's manual of determinative bacteriology, 8th ed. Williams & Wilkins. Baltimore, MD. 1235–1245.

Shimizu, N., Koyama, T., and K. Ogura. 1998. Molecular cloning, expression, and characterization of the genes encoding the two essential protein components of Micrococcus luteus B-P 26 hexaprenyl diphosphate synthase. J. Bacteriol. 180:1578–1581.

Shimoi, H., Y. Iimura, and T. Obata. 1995. Molecular cloning of CWP1: a gene encoding a Saccharomyces cerevisiae cell wall protein solubilized with Rarobacter faecitabidus protease I. J. Biochem. 118:302–311.

Shiota, S., and H. Nakayama. 1997. UV endonuclease of Micrococcus luteus, a cyclobutane pyrimidine dimer-DNA glycolsylase/abasic lyase: cloning and characterization of the gene. Proc. Natl. Acad. Sci. USA 94:593–598.

Skerman, V. B. D., V. McGowan, and P. H. A. Sneath. 1980. Approved lists of bacterial names. Int. J. Syst. Bacteriol. 30:225–420.

Sohier, R., F. Benazet, and M. Piechaud. 1948. Sur un germe du genre Listeria apparemment non pathogene. Ann. Inst. Pasteur. 74:54–57.

Stackebrandt, E., and C. R. Woese. 1979. A phylogenetic dissection of the family Micrococcaceae. Curr. Microbiol. 2:317–322.

Stackebrandt, E., B. J. Lewis, and C. R. Woese. 1980. The phylogenetic structure of the coryneform group of bacteria. Zbl. Bakt. I. Abt. Orig. C2:137–149.

Stackebrandt, E., and C. R. Woese. 1981a. Towards a phylogeny of actinomycetes and related organisms. Curr. Microbiol. 5:131–136.

Stackebrandt, E., B. Wunner-Füssl, V. J. Fowler, and K. H. Schleifer. 1981b. Molecular genetic relatedness of sporeforming members of the order Actinomycetales. I. Deoxyribonucleic acid homologies and ribosomal ribonucleic acid similarities. Int. J. Syst. Bacteriol 31:420–431.

Stackebrandt, E., V. J. Fowler, F. Fiedler, and H. Seiler. 1983. Taxonomic studies on Arthrobacter nicotianae and related taxa: description of Arthrobacter uratoxydans sp. nov. and Arthrobacter sulfureus sp. nov. and reclassification of Brevibacterium protophormiae as Arthrobacter protophormiae comb. nov. System. Appl. Microbiol. 4:470–486.

Stackebrandt, E., C. Scheuerlein, and K.-H. Schleifer. 1983a. Phylogenetic and biochemical studies on Stomatococcus mucilaginosus. System. Appl. Microbiol. 4:207–217.

Stackebrandt, E., W. Ludwig, E. Seewaldt, and K. H. Schleifer. 1983b. Phylogeny of sporeforming members of the order Actinomycetales. Int. J. Syst. Bacteriol 33:173–180.

Stackebrandt, E., R. M. Kroppenstedt, and V. J. Fowler. 1983c. A phylogenetic analysis of the family Dermatophilaceae. J. Gen. Microbiol. 129:1831–1838.

Stackebrandt, E. 1988. Phylogenetic relationships vs. phenotypic diversity: how to achieve a phylogenetic classification system of the eubacteria. Can. J. Microbiol. 34:552–556.

Stackebrandt, E. 1992a. The family Dermatophilaceae. A. Balows, H. G. Trüper, M. Dworkin, W. Horder, and K.-H. Schleifer The prokaryotes: A handbook on the biology of bacteria: Ecophysiology, isolation, identification, applications. Springer-Verlag. New York, 1346–1350.

Stackebrandt, E., and H. Prauser. 1992b. The family Cellulomonadaceae. A. Balows, H. G. Trüper, M. Dworkin, W. Harder and K. H. Schleifer. The Prokaryotes, 2nd ed. Springer-Verlag. New York, NY. 1323–1345.

Stackebrandt, E. 1992c. The genus Stomatococcus. Balows, H. G. Trüper, M. Dworkin, W. Harder and K. H. Schleifer. The Prokaryotes, 2nd. ed. Springer-Verlag. New York, NY. 1320–1322.

Stackebrandt, E., W. Liesack, and B. M. Goebel. 1993. Bacterial diversity ina soil sample from a subtropical Australian environment as determined by 16S rDNA analysis. FASEB 7:232–236.

Stackebrandt, E., C. Koch, O. Gvozdiak, and P. Schumann. 1995. Taxonomic dissection of the genus Micrococcus: Kocuria gen. nov., Nesterenkonia gen. nov., Kytococcus gen. nov., Dermacoccus gen. nov., and Micrococcus Cohn 1872 gen. emend. Int. J. Syst. Bacteriol. 45:682–692.

Stackebrandt, E., F. A. Rainey, and N. L. Ward-Rainey. 1997. Proposal for a new hierarchic classification system, Actinobacteria classis nov. Int. J. Syst. Bacteriol. 47:479–491.

Stackebrandt, E. 1999. Defining taxonomic ranks. Dworkin, M., S. Falkow, E. Rosenberg, K.-H. Schleifer and E. Stackebrandt. The Prokaryotes. Springer-Verlag. New York, NY.

Staliper, C. E. 1996. Genetic diversity of North American isolates of Renibacterium salmoninarum. Dis. Aquat. Org. 27:207–213.

Stanier, R. Y., and C. B. Van Niel. 1941. The main outlines of bacterial classification. J. Bacteriol. 42:437–466.

Suzuki, K.-I., and K. Komagata. 1983. Pimelobacter gen. nov., a new genus of coryneform bacteria with LL-diaminopimelic acid in the cell wall. J. Gen. Microbiol. 29:59–71.

Suzuki, K., M. D. Collins, E. Iijima, and K. Komagata. 1988. Chemotaxonomic characterization of a radiotolerant bacterium Arthrobacter radiotolerans: description of Rubrobacter radiotolerans gen. nov., comb. nov. FEMS Microbiol. Lett. 52:33–40.

Swanson, D. S., V. Kapur, K. Stockbauer, X. Pan, R. Frothingham, and J. M. Musser. 1997. Subspecific differentiation of Mycobacterium avium complex strains by automated sequencing of a region of the gene (hsp65) encoding a 65-kilodalton heat shock protein. Int. J. Syst. Bacteriol. 47:414–419.

Takahashi, Y., Y. Tanaka, Y. Iwai, and S. Omura. 1987. Promicromonospora sukumoe sp. nov., a new species of the Actinomycetales. J. Gen. Appl. Microbiol. 33:507–519.

Takeuchi, M., and K. Hatano. 1998. Union of the genera Microbacterium Orla-Jensen and Aureobacterium Collins et al. in a redefined genus Microbacterium. Int. J. Syst. Bacteriol. 48:739–747.

Takeuchi, M., C.-X. Fang, and A. Yokota. 1995. Taxonomic study of the genus Brachybacterium: Proposal of Brachybacterium conglomeratum sp. nov., nom. rev., Brachybacterium paraconglomeratum sp. nov., and Brachybacterium rhamnosum sp. nov. Int. J. Syst. Bacteriol. 45:160–168.

Ueda, T., Suga, Y., and Matsuguchi, T. 1995. Molecular phylogenetic analysis of a soil microbial community in a soybean field. Eur. J. Soil Sci. 46:415–421.

Van Saceghem, R. 1915. Dermatose contagieuse (impetigo contagieux). Bull. Soc. Path. Exot. 8:354–359.

Viale, A. M., A. K. Arakaki, F. C. Soncini, and R. G. Ferreyra. 1994. Evolutionary relationships among eubacterial groups as inferred from GroEL (chaperonin) sequence comparisons. Int. J. Syst. Bacteriol. 44:527–533.

Wade, W. G., J. Downes, D. Dymock, S. H. Hiom. A. J. Weightman, F. E. Dewhirst, B. J. Paster, N. Tzellas, and B. Coleman. 1999. The family Coriobacteriaceae: reclassification of Eubacterium exiguum (Poco et al. 1996) and Peptostreptococcus heliotrinreducens (Lanigan 1976) as Slackia exigua gen. nov., comb. nov. and Slackia heliotrinireducens gen. nov., comb. nov., and Eubacterium lentum (Prevot 1938) as Eggerthella lenta gen. nov., comb. nov. Int. J. Syst. Bacteriol. 49:595–600.

Waksman, S. A. 1950. The Actinomycetes. Their nature, occurrence, activities, and importance. Chronica Botanica Company. Waltham, MA.

Waksman, S. A. 1967. The Actinomycetes. A summary of current knowledge. Ronald Press Co. New York.

Waksman, S. A., and Henrici, A. T. 1943. The nomenclature and classification of the actinomycetes. J. Bacteriol. 46:337–341.

Wayne, L., D. J. Brenner, R. R. Colwell, P. A. D. Grimont, O. Kandler, M. I. Krichevsky, L. H. Moore, W. E. C. Moore, R. G. E. Murray, E. Stackebrandt, M. P. Starr, H. G. Trüper. 1987. International Committee on Systematic Bacteriology: Report of the ad hoc committee on reconciliation of approaches to bacterial systematics. Int. J. Syst. Bacteriol. 37:463–464.

Winslow, C.-E., J. Broadhurst, R. E. Buchanan, C. Krumwiede Jr., L. A. Rogers, and G. H. Smith. 1917. The families and genera of the bacteria. Preliminary report of the Committee of the Society of American Bacteriologists on characterization and classification of bacterial types. J. Bacteriol. 2:505–566.

Yamamoto, N., S. Sato, K. Saito, T. Hasuo, M. Tadenuma, K. Suzuki, J. Tamaoka, and K. Komagata. 1988. Rarobacter faecitabidus gen. nov., sp. nov., a yeast-lysing coryneform bacterium. Int. J. Syst. Bacteriol. 38:7–11.

Yokota, A., T. Tamura, T. Nishii, and T. Hasegawa. 1993. Kineococcus aurantiacus gen. nov., sp. nov., a new aerobic Gram-positive, motile coccus with meso-diaminopimelic acid and arabinogalactane in the cell wall. Int. J. Syst. Bacteriol. 43:52–57.

Yoshinaka, T., Yano, K., and H. Yamagushi. 1973. Isolation of highly radioresistant bacterium Arthrobacter radiotolerans nov. sp. Agric. Biol. Chem. 37:2269–2275.

Prokaryotes (2006) 3:322–382
DOI: 10.1007/0-387-30743-5_17

CHAPTER 1.1.2

The Family Bifidobacteriaceae

BRUNO BIAVATI AND PAOLA MATTARELLI

Family Bifidobacteriaceae

Family Bifidobacteriaceae fam. nov., Stacke-brandt, Rainey and Ward-Rainey. Bifidobacteri-aceae (Bi.fi.do.bac.teri.a'ce.ae. ending to denote a family; M.L. fem. pl. n. Bifidobacteriaceae, the Bifidobacterium family). The pattern of 16S rDNA signatures consists of nucleotides at positions 122-239 (G-U), 128-233 (C-G), 450-483 (C-G), 602-636 (C-G), 681-709 (C-G), 688-699 (A-U), 823-877 (A-U), 1118-1155 (C-G) and 1311-1326 (A-U; Stackebrandt et al., 1997).

The family contains the type genus *Bifidobacterium* (Orla-Jensen, 1924) with 32 species (type species *Bifidobacterium bifidum*, Tissier; Orla-Jensen, 1924) as well as *Gardnerella* (Greenwood and Pickett, 1980) with *Gardnerella vaginalis* as the only species. A phylogenetic structure of the genera *Bifidobacterium* and *Gardnerella* has been published (Maidak et al., 1994; Leblond-Bourget et al., 1996).

Characteristics of the Family

The family consists of pleomorphic rods that occur singly or in many-celled chains or clumps. Cells have no capsule and they are nonspore-forming, nonmotile, and nonfilamentous. They are Gram positive except for *G. vaginalis*, which is Gram variable. They are anaerobic (some *Bifidobacterium* species can tolerate O_2 only in presence of CO_2) or facultatively anaerobic (*Gardnerella*). They are negative for the following: indole, gelatin hydrolysis, catalase (except for *B. indicum* and *B. asteroides* when grown in presence of air), and oxidase. Optimum growth temperature is 35–39°C. They possess fructose-6-phosphoketolase (EC 4.1.2.2), which cleaves fructose-6-phosphate into acetylphosphate and erythrose-4-phosphate. The G+C content varies from 42–67 mol%. They are chemoorganotrophs having a fermentative type of metabolism. They produce acid but no gas from a variety of carbohydrates. They occur in animal and human habitats. They are nonpathogenic except for

bifidobacteria isolated from dental caries (and probably involved in caries pathology) and for *G. vaginalis* that often assumes pathogenic character (such as in bacterial vaginosis and in urogenital tract infections of both sexes).

Genus *Bifidobacterium*

Phylogeny

Tissier discovered bifidobacteria in infant feces and called them "Bacillus bifidus" (Tissier, 1899; Tissier, 1900). In 1924, Orla-Jensen recognized the existence of the genus *Bifidobacterium* as a separate taxon, but given their similarities to the genus *Lactobacillus*, bifidobacteria were included in the genus *Lactobacillus* as listed in the seventh edition of *Bergey's Manual of Determinative Bacteriology* (Breed et al., 1957).

In the eighth edition of *Bergey's Manual of Determinative Bacteriology* (Buchanan et al., 1974) bifidobacteria were classified in the genus *Bifidobacterium*, the name initially adopted by Orla-Jensen. The genus comprised eight species and was included in the family Actinomyceta-ceae of the order Actinomycetales of the Actino-mycetes and related organisms, although with regard to the murein structure of the cell wall, bifidobacteria are more similar to Lactobacil-laceae than to Actinomycetaceae (Kandler and Lauer, 1974).

The introduction of rRNA gene sequence analysis in studies on the phylogeny of bacteria has confirmed the assignment of bifidobacteria to an "Actinomyces" group (Fox et al., 1980).

In the first edition of *Bergey's Manual Systematic Bacteriology* (Sneath et al., 1986), as well as the ninth edition of *Bergey's Manual of Determinative Bacteriology* (Holt et al., 1993), organisms of the genus *Bifidobacterium* were grouped and defined as "irregular, nonsporing, gram-positive rods." The rRNA sequence analysis was mostly used to determine relationships among genera, families and other ranks. Molecular phylogeny, derived from high correlation of several molecules (e.g., 16S RNA genes coding for 16S rRNA or 16S rDNA; Woese, 1987), 23S rDNA (Collins and Wallbanks, 1992), elongation factors involved in translation, and the β-subunit of ATPase (Ludwig et al., 1993), led to the description of domains for the three highest taxa: Archea, Bacteria and Eukarya (Woese et al., 1990). Stackebrandt et al. (1997), through 16S rDNA analysis, proposed a novel hierarchical structure for the phylogenetic group of "Actinomycetes, bacteria and relatives." This proposal does not change the current description of species and genera based upon morphological, chemotaxonomic and physiological characters but "provides the descriptions of taxa above the genus level incorporating the characteristics of the individual genera contained therein." The genus *Bifidobacterium* together with the genus *Gardnerella* has been collected into the single family of Bifidobacteriaceae in the order of Bifidobacteriales in the subclass of Actinobacteridae in the class of Actinobacteria in the lineage of Firmicutes and in the domain of Bacteria (Stackebrandt et al., 1997; Fig. 1).

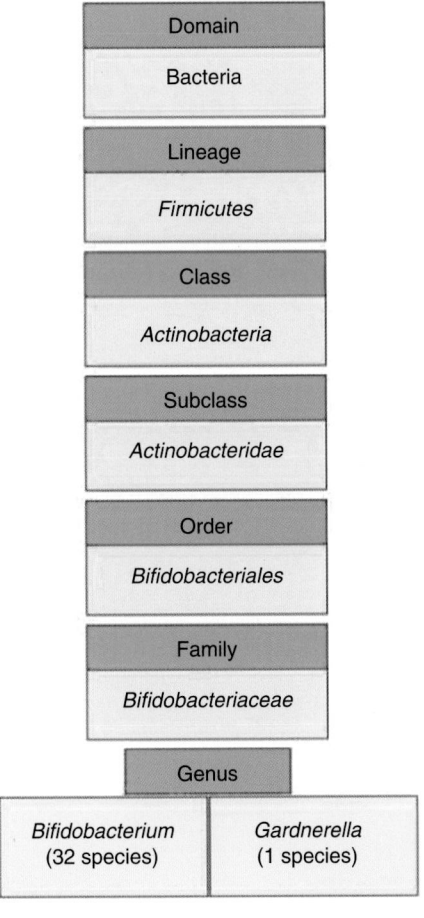

Fig. 1. Phylogeny of the genera *Bifidobacterium* and *Gardnerella*.

Taxonomy

Bifidobacteria were first discovered by Tissier, who observed and isolated a bacterium with a strange and characteristic Y shape in the feces of breast-fed infants. This bacterium was anaerobic, Gram-positive and did not produce gas during its growth (Tissier, 1899). Because of this morphology, he termed this organism "*Bacillus bifidus*" (Tissier, 1900). Winslow et al. (1917) proposed the family Lactobacillaceae, and three years later, Holland (1920) named the Tissier strain *Lactobacillus bifidus*.

During what Poupard et al. (1973) call the "first period" in the history of these bacteria, namely from 1900 to 1957, few advances were made in the knowledge of their biochemistry and taxonomy. In the seventh edition of *Bergey's Manual of Determinative Bacteriology* (Breed et al., 1957), only Lactobacillus bifidus was reported, although in 1924 Orla-Jensen had already recognized the existence of the genus *Bifidobacterium* as a separate taxon (Orla-Jensen, 1924).

In 1957, Dehnert first recognized the existence of multiple biotypes of *Bifidobacterium* and proposed a scheme for the differentiation of five groups of these bacteria based on their carbohydrate fermentation patterns (Dehnert, 1957). Reuter (1963) recognized and named seven species of *Bifidobacterium*, in addition to the known *B. bifidum*, on the basis of fermentative and serological characters and presented a scheme for their identification. The species of Reuter's scheme were: *B. bifidum* var. *a* and *b*; *B. infantis*; *B. parvulorum* var. *a* and *b*; *B. breve* var. *a* and *b*; *B. liberorum*; *B. lactentis*; *B. adolescentis* var. *a*, *b*, *c* and *d*; and *B. longum* var. *a* and *b*. The next major classification scheme was presented by Mitsuoka (1969). In this scheme, new fermentative biotypes added to *B. longum* species (*B. longum* subsp. animalis *a* and *b*) and two new species, *B. thermophilum* and *B. pseudolongum*, found in the feces of a variety of animals (e.g., pigs, chickens, calves and rats) were included. In the same year, Scardovi et al. (1969b) isolated *B. ruminale* (synonym of *B. thermophilum*) and *B. globosum* from cattle rumen. Furthermore, Scardovi and Trovatelli (1969a) found *B. asteroides*, *B. indicum* and *B. coryneforme* in the intestine of honeybees, three new species morphologically quite different from the previously known bifidobacteria (see Fig. 2).

Twenty-four species of the genus *Bifidobacterium* are listed in the first edition of *Bergey's Manual of Systematic Bacteriology* (Scardovi, 1986). This marked expansion of the genus was based on significant contributions of Biavati et al. (1982), Dehnert (1957), Matteuzzi et al. (1971), Mitsuoka (1969), Reuter (1963),

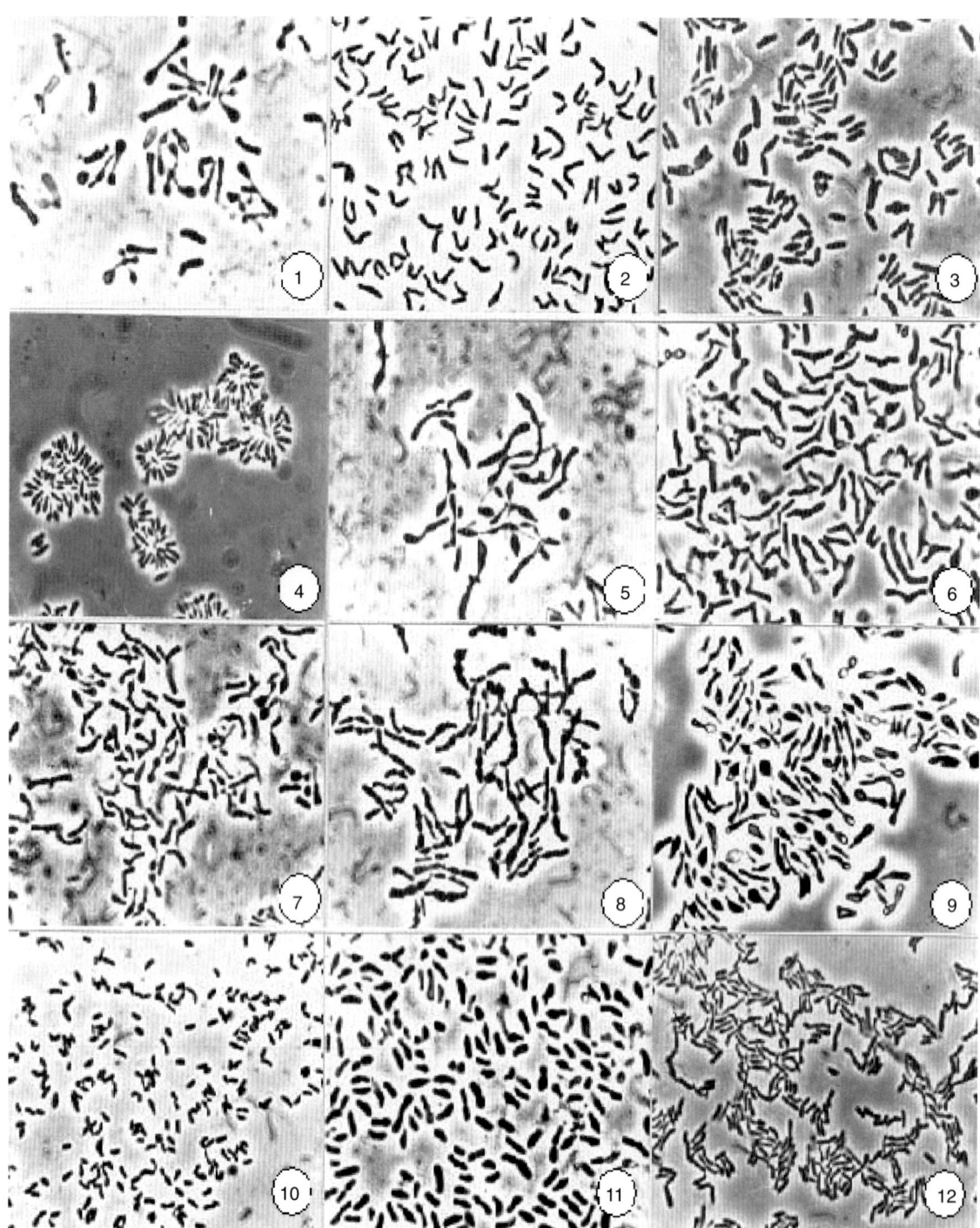

Fig. 2. Panels 1–12. The varied cellular morphology of the genus *Bifidobacterium*. Cells of the type strains were grown in trypticase-phytone-yeast extract (TPY) medium stabs. Phase-contrast photomicrographs show (1) *B. adolescentis*, (2) *B. angulatum*, (3) *B. animalis*, (4) *B. asteroides*, (5) *B. bifidum*, (6) *B. boum*, (7) *B. breve*, (8) *B. catenulatum*, (9) *B. choerinum*, (10) *B. coryneforme*, (11) *B. cuniculi*, and (12) *B. denticolens*.

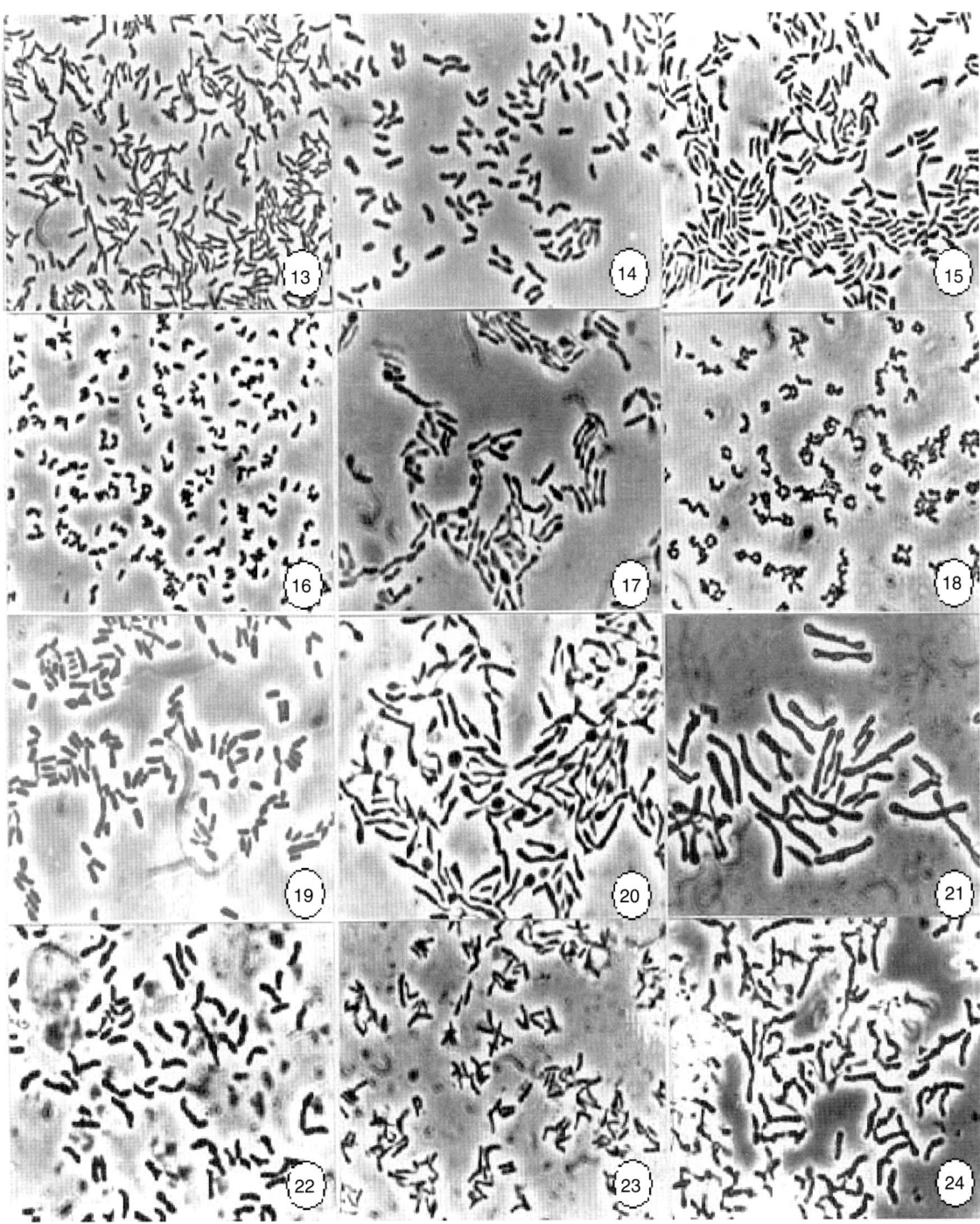

Fig. 2. Panels 13–24 show (13) *B. dentium*, (14) *B. infantis*, (15) *B. gallicum*, (16) *B. gallinarum*, (17) *B. indicum*, (18) *B. inopinatum*, (19) *B. lactis*, (20) *B. longum*, (21) *B. magnum*, (22) *B. merycicum*, (23) *B. minimum*, and (24) *B. pseudocatenulatum*.

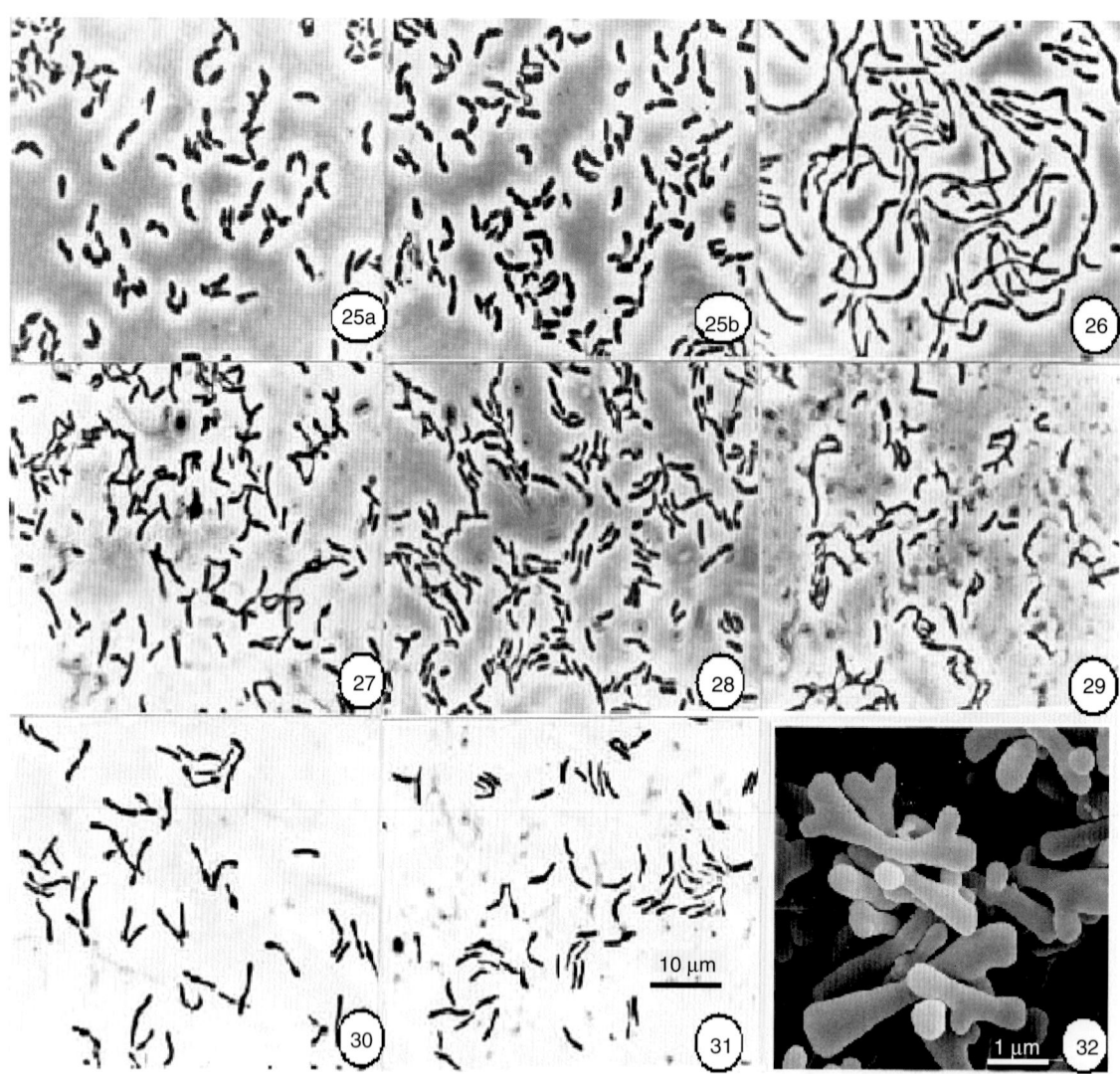

Fig. 2. Panels 25–32 show (25a) *B. pseudolongum* subsp. *pseudolongum*, (25b) *B. pseudolongum* subsp. *globosum*, (26) *B. pullorum*, (27) *B. ruminantium*, (28) *B. saeculare*, (29) *B. subtile*, (30) *B. suis*, (31) *B. thermophilum*, and (32) a scanning electron micrograph of *Bifidobacterium* spp. (Reproduced with permission from V. Bottazzi, Universit Cattolica del Sacro Cuore, Piacenza, Italy.)

Scardovi and Crociani (1974a), Scardovi and Trovatelli (Scardovi and Trovatelli, 1969a; Scardovi and Trovatelli, 1974b), Scardovi and Zani (1974c), Scardovi et al. (Scardovi et al., 1971b; Scardovi et al., 1979b) and Trovatelli et al. (1974). Afterwards, the description of eight additional species and the description of *B. globosum* as subspecies of *B. pseudolongum* (Yaeshima et al., 1992) were introduced: *B. gallicum* isolated from human feces (Lauer, 1990), *B. merycicum* and *B. ruminantium* isolated from rumen of cattle (Biavati and Mattarelli, 1991), *B. saeculare* isolated from rabbit feces (Biavati et al., 1991a), *B. denticolens* and *B. inopinatum* both isolated from dental caries (Crociani et al., 1996b), *B. lactis* isolated from yogurt (Meile et al., 1997), and *B. thermacidophilum* isolated from anaerobic digesters (Dong et al., 2000). The habitat and type strain of each species are listed in Table 1.

The pathway of hexose fermentation in bifidobacteria has, in the meantime, been elucidated by radioactive-carbon distribution and assay of enzymes in cellular extracts by Scardovi and Trovatelli (1965) and De Vries et al. (1967). The key enzyme (fructose-6-phosphate phosphoketolase [F-6-PPK]) of this pathway (the fructose-6-phosphate shunt; Fig. 3), purified later by Sgorbati et al. (1976), splits hexose phosphate to erythrose-4-phosphate and acetyl phosphate (Schramm et al., 1958). From tetrose and hexose phosphates, through the successive action of

Table 1. Type strains and habitats of species of the genus *Bifidobacterium*.

Species	Original label	ATCC	DSMZ	Habitat	Reference
	Type strain[a]				
B. adolescentis	E 194a	15703	20083	Feces of human adult; bovine rumen; and sewage	Reuter (1963)
B. angulatum	B 677	27535	20098	Sewage; and feces of human adult	Scardovi and Crociani (1974)
B. animalis	R 101-8	25527	20104	Feces of rat, chicken, rabbit, calf and guinea pig; and sewage	Scardovi and Trovatelli (1974)
B. asteroides	C 51	25910	20089	Intestine of *Apis mellifera* (subsp. *mellifera*, *ligustica*, and *caucasica*)	Scardovi and Trovatelli (1969)
B. bifidum	Ti	29521	20456	Feces of human adult, infant and suckling calf; and human vagina	Orla-Jensen (1924)
B. boum	RU 917	27917	20432	Bovine rumen; and feces of piglet	Scardovi et al. (1979b)
B. breve	S 1	15700	20213	Feces of infant and suckling calf; human vagina; and sewage	Reuter (1963)
B. catenulatum	B 669	27539	20103	Feces of infant and human adult; human vagina; and sewage	Scardovi and Crociani (1974)
B. choerinum	SU 806	27686	20434	Feces of piglet; and sewage	Scardovi et al. (1979b)
B. coryneforme	C 215	25911	20216	Intestine of *Apis mellifera* subsp. *mellifera*	ex Scardovi and Trovatelli (1969) Biavati et al. (1982)
B. cuniculi	RA 93	27916	20435	Feces of rabbit	Scardovi et al. (1979b)
B. denticolens	B 3028		10105	Human dental caries	Crociani et al. (1996b)
B. dentium	B 764	27534	20436	Human dental caries and oral cavity; feces of human adult; human vagina; abscesses and appendix	Scardovi and Crociani (1974)
B. gallicum	P 6		20093	Human feces	Lauer (1990)
B. gallinarum	Ch 206-5	33777	20670	Chicken cecum	Watabe et al. (1983)
B. indicum	C 410	25912	20214	Intestine of *Apis cerana* and *A. dorsata*	Scardovi and Trovatelli (1969)
B. infantis	S 12	15697	20088	Feces of infant and suckling calf; and human vagina	Reuter (1963)
B. inopinatum	B 3109		10107	Human dental caries	Crociani et al. (1996b)
B. lactis	UR 1		10140	Fermented milk	Meile et al. (1997)
B. longum	E 194b	15707	20219	Feces of human adult, infant and suckling calf; human vagina; and sewage	Reuter (1963)
B. magnum	RA 3	27540	20222	Feces of rabbit	Scardovi and Zani (1974)
B. merycicum	RU 915B	49391	6492	Bovine rumen	Biavati and Mattarelli (1991)
B. minimum	F 392	27538	20102	Sewage	Biavati et al. (1982)
B. pseudocatenulatum	B 1279	27919	20438	Feces of infant and suckling calf; and sewage	Scardovi et al. (1979b)
B. pseudolongum subsp. *pseudolongum*	PNC-2-9G	25526	20099	Feces of pig, chicken, bull, calf, rat and guinea pig	Mitsuoka (1969)
subsp. *globosum*	RU 224	25865	20092	Feces of piglet, suckling calf, rat, rabbit and lamb; sewage; and bovine rumen	ex Scardovi et al. (1969) Biavati et al. (1982)
B. pullorum	P 145	27685	20433	Feces of chicken	Trovatelli et al. (1974)
B. ruminantium	RU 687	49390	6489	Bovine rumen	Biavati and Mattarelli (1991)
B. saeculare	RA 161	49392	6531	Feces of rabbit	Biavati et al. (1991a)
B. subtile	F 395	27537	20096	Sewage	Biavati et al. (1982)
B. suis	SU 859	27533	20211	Feces of piglet	Matteuzzi et al. (1971)
B. thermacidophilum	36[b]			Anaerobic digester	Dong et al. (2000)
B. thermophilum	P 2-91	21525	20210	Feces of pig, chicken and suckling calf; bovine rumen; and sewage	Mitsuoka (1969)

[a]Culture collection strain numbers for American Type Culture Collection (ATCC) and Deutsche Sammlung von Mikroorganismen und Zellkulturen GmbH (DSMZ; German collection of microorganisms and cell cultures).
[b]The type strain has been deposited in China General Microbiological Culture Collection Center (CGMCC) with the number AS 1.

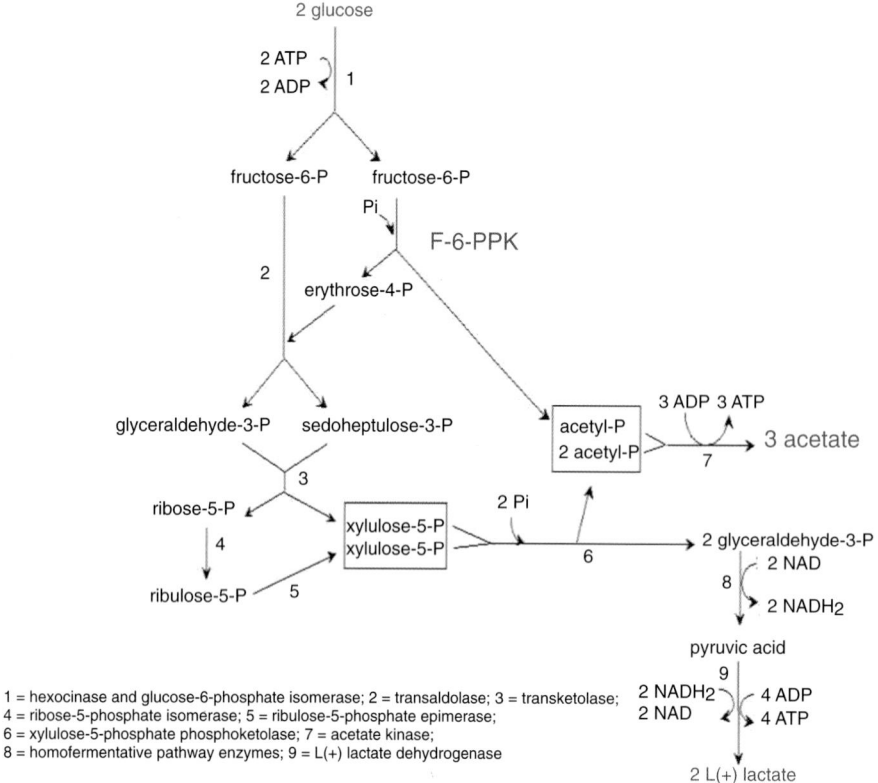

Fig. 3. Hexose metabolic pathways in *Bifidobacterium*.

transaldolase and transketolase, pentose phosphates are formed that, via the usual 2–3 cleavage, give rise to lactic acid and acetic acid so that lactic and acetic acids are formed in the theoretical ratio 1.0 : 1.5.

Phosphoroclastic cleavage of some pyruvate to formic and acetic acids and reduction of acetate to ethanol can often alter the fermentation balance (De Vries and Stouthamer, 1968) to a highly variable extent (Lauer and Kandler, 1976). During the metabolism of hexoses, there is no production of carbon dioxide, except in the degradation of gluconate.

Recently the presence of F-6-PPK has been found in *Gardnerella vaginalis* (Gavini et al., 1996). The genera *Gardnerella* and *Bifidobacterium* are phylogenetically very similar (Embley and Stackebrandt, 1994) but can be easily distinguished on a phenotypic basis. The guanine and cytosine (G+C) content of DNA is quite high (55–67 mol%), with differences among species (Scardovi, 1986; Table 2). The only exception is *B. inopinatum*, one of the three species found in dental caries, with a G+C mol% of 45 (Crociani et al., 1996b).

In 1970, Scardovi and colleagues started to extensively apply the DNA-DNA filter hybridization procedure to assess the validity of the bifidobacterial species previously described

(Scardovi et al., 1970; Scardovi et al., 1971b) and to recognize new DNA homology groups among the strains they were isolating in large numbers from diverse ecological niches, such as the feces of humans and animals, the rumen of cattle, sewage, the human vagina, dental caries and honeybee intestine (see Table 1). Some emendations of Reuter and Mitsuoka's species identifications were suggested (Scardovi et al., 1971b) and accepted (Rogosa, 1974). *Bifidobacterium liberorum* and *B. lactentis*, for example, were proposed as synonyms of *B. infantis*, and *B. parvulorum* was proposed as a synonym of *B. breve*, *B. longum* subsp. *animalis*. Mitsuoka was elevated to the species rank as *B. animalis* (Scardovi and Trovatelli, 1974b). Four additional species isolated from animal feces were described and proposed in 1979: *B. pseudocatenulatum*, *B. boum*, *B. choerinum*, and *B. cuniculi* (Scardovi et al., 1979b).

In 1982, the use of polyacrylamide gel electrophoresis (PAGE) of soluble cellular proteins was first introduced to distinguish species of bifidobacteria (Biavati et al., 1982). More than 1,000 strains of bifidobacteria were characterized using this technique. An excellent correlation with the data obtained through DNA-DNA homology was found. One of the results of this work was the proposal for the reinstatement of *B. coryneforme* (Scardovi and Trovatelli, 1969a) and *B.*

Table 2. Murein type, mol% G+C values and electrophoretic patterns of 6-phosphogluconate dehydrogenase and transaldolase enzymes.

| Species | Murein type[a] | % G+C[b] | Electrophoretic patterns of enzymes[c] | |
			6-Phosphogluconate dehydrogenase	Transaldolase
B. infantis		60.5	(3)-**4**-(5)	5-(6)-(**8**)
B. longum	L-Orn-L-Ser-L-Ala-L-Thr-L-Ala	61	5-(**6**)	(5)-6-**8**
B. suis		62	5-8	6
B. animalis		60	8-**9**	5
B. choerinum		66	4	3
B. cuniculi	L-Lys(L-Orn)-L-Ser-(L-Ala)-L-Ala$_2$	64	4	1
B. lactis		62	nt	nt
B. ruminantium		57	Present	Present
B. bifidum	L-Orn-D-Ser-D-Asp	61	7-(8)	7
B. catenulatum		54	**6**-8	5
B. magnum	L-Lys(L-Orn)-L-Ala$_2$-L-Ser	60	7	5
B. pseudocatenulatum		57.5	**1**-3	**4**-(5)
B. gallicum	L-Lys-L-Ala-L-Ser	61	nt	Present
B. pseudolongum				
subsp. globosum	L-Orn(L-Lys)-L-Ala$_{2-3}$	59.5	7	2
subsp. pseudolongum		64	(3$_a$)-(4)-(5)-(**6**)-(7)	2
B. adolescentis		59	5	8
B. denticolens			nt	nt
B. dentium	L-Lys(L-Orn)-D-Asp	61	(2)	4
B. merycicum		59	Present	Present
B. saeculare		63	Absent	Present
B. thermophilum	L-Orn(L-Lys)-D-Glu	60	7-8-**9**-(9$_a$)	(7)-**8**
B. boum	L-Lys-D-Ser-D-Glu	60	**8**-9-9$_a$	6
B. minimum	L-Lys-L-Ser	61.5	6	10
B. angulatum		59	5	5
B. coryneforme		nt	6	6
B. gallinarum	L-Lys-D-Asp	66	Absent	Present
B. indicum		60	6-6$_a$-(7)-8-(**9**)-(9$_a$)-(9$_b$)	(6)-7-8-**9**
B. pullorum		67.5	Not detectable	2
B. subtile		61.5	2	3
B. breve	L-Lys-Gly	58	(5)-6-6$_a$-**7**	6
B. asteroides		59	(9)-(9$_a$)(9$_b$)-(10)-**10**$_a$ 10$_b$-(11)-(12)-(13)	(6)-(7)-(7$_a$)-**8** (8a)-(8$_b$)-(9)-(9$_a$)
B. inopinatum	nt	45	nt	nt
B. thermacidophilum	nt	58	nt	nt

Symbol: nt, not tested.

[a]Amino acid in parenthese occasionally replaces the preceding amino acid; data from Deutsche Sammlung von Mikroorganismen und Zellkulturen GmbH (DSMZ; German collection of microorganisms and cell cultures).

[b]Average values; data from Scardovi (1986); for the values of the recently described species see under single species description.

[c]Numbers 1–10 and 1–13 were given to isozymes of transaldolase and 6-phospho-D-gluconate dehydrogenase: NAD(P)oxidoreductase (6PGD), respectively, in the order of decreasing mobility (subscripts indicate additional isozymes). Boldface numbers are the isozymes of the type strains. Numbers in parentheses are isozymes found in less than 10% of the strains studied. The 6PGD is undetectable by spot staining in most strains of B. dentium and in all strains of B. pullorum. "Present" and "Absent" indicate presence and absence as determined by other methods. Data from Scardovi et al. (1979a).

globosum (Scardovi et al., 1969b), which had not appeared in the Approved Lists of Bacterial Names (Skerman et al., 1980). Moreover, B. minimum and B. subtile homology groups (Scardovi and Trovatelli, 1974b) were elevated to the species level.

A study of DNA-DNA homology and several phenotypic characteristics of the type strains of Bifidobacterium spp. was done by Lauer and Kandler (1983). On the basis of their data, they suggested that the three species B. longum, B. suis and B. infantis be merged into only one species, namely B. longum; that B. pseudocatenulatum and B. catenulatum be recognized as two subspecies rather than two species; and that B. pseudolongum be merged into B. globosum, and B. indicum into B. coryneforme. The fact that some species of the genus Bifidobacterium are closely related is well established. However, when all the data available from the large

Table 3. Key to the identification of 25 species of the genus *Bifidobacterium*.

| | Acid produced from | | | | | | | | | | Electrophoretic pattern[a] of | | |
Sorbitol	Arabinose	Raffinose	Ribose	Starch	Lactose	Inulin	Cellobiose	Melezitose	Gluconate	6-Phosphogluconate dehydrogenase	Transaldolase	Suggested species
+	+							+				*B. adolescentis*
		+		+			+	−				*B. pseudocatenulatum*
							−					*B. angulatum*
			−	−			−					*B. catenulatum*
	−			+	+							*B. breve*
−					−							*B. subtile*
	+	+		+				+	+		**4**	*B. dentium*
		+	+								**8**	*B. adolescentis*
									−			*B. animalis*
								−			**4**	*B. pseudocatenulatum*
										7	**2**	*B. pseudolongum* subsp. *pseudolongum*
										(3$_a$)-(4)-(5)-**6**-(7)	**2**	*B. pseudolongum* subsp. *globosum*
			−	−							**5**	*B. angulatum*
						+		+				*B. gallinarum*
					+	−		−				*B. longum*
										(3)-**4**-(5)	**5**	*B. infantis*
										5-(**6**)	**8**	*B. longum*
										7		*B. magnum*
					−		+			6		*B. coryneforme*
							−			9-13		*B. asteroides*
												B. pullorum
		−										*B. suis*
	−		−	+								*B. cuniculi*
		+	+	−	+							*B. breve*
					+							*B. infantis*
			−		−							*B. indicum*
											3	*B. choerinum*
											6	*B. boum*
											8	*B. thermophilum*
					+							*B. bifidum*
		−										*B. minimum*

Symbols: +, 90% or more of strains are positive; −, 90% or more of strains are negative.

[a]Electrophoretic patterns of enzymes taken from Scardovi et al. (1979a). Numbers were given to isozymes of phosphogluconate dehydrogenase and transaldolase in order of decreasing anodic mobility (numbers with subscript letters indicate additional isozymes). Numbers in parentheses are the isozymes of the type strains. Numbers in boldface are isozymes found in less than 10% of strains. The enzyme phosphogluconate dehydrogenase is undetectable by spot staining in most strains of *B. dentium* and in all strains of *B. pullorum*. The lines indicate which reactions are useful as characteristic markers.

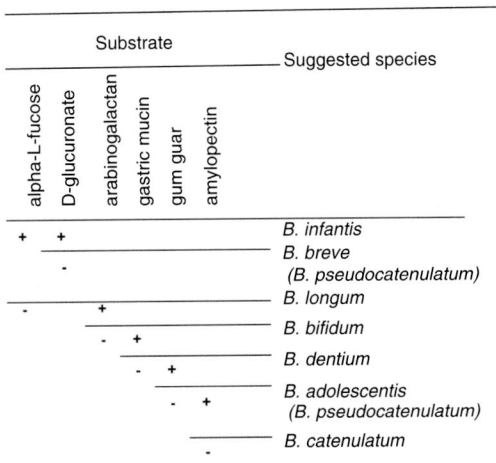

Species in brackets is less likely.
B. angulatum and *B. gallicum* are not included because of the low number of strains available.

Fig. 4. Key for differentiation of *Bifidobacterium* species of human origin.

1900 - Tissier - *Bacillus bifidus*
1917 - Winslow *et al.* - *Lactobacillaceae*
1920 - Holland - *Lactobacillus bifidus*
1957 - Bergey's manual VIIth edition -
     Lactobacillus bifidus
1965 - Scardovi - Fructose-6-phosphate-phosphoketolase (F-6-PPK)
1974 - Bergey's manual VIIIth edition -
     Family: *Actinomycetaceae*;
     Genus: Bifidobacterium (11 spp.)
1986 - bergey's Manual Systematic Bacteriology I[th] edition
     - (24 species)
1997 - Stackebrandt *et al.* - New hierarchic classification;
     Order: *Bifidobacteriaies*; Family: *Bifidobacteriaceae*;
     genera *Bifidobacterium* and *Gardnerella*
2000 - 32 species

Fig. 5. Milestones in the history of *Bifidobacterium* taxonomy.

number of strains studied are considered, the presence of different characters emerges, justifying the present separation scheme (Scardovi, 1986). To this day, only *B. globosum* has been merged with *B. pseudolongum* and has been described as a subspecies of *B. pseudolongum* (Yaeshima et al., 1992). A key for the identification of the species based on isoenzyme patterns and fermentation tests (so far described only in the second edition of *The Prokaryotes*; Biavati et al., 1991b) is presented in Table 3. A new identification key for human bifidobacteria based on complex carbohydrates and mucin fermentation has been introduced by Crociani et al. (Crociani et al., 1994; Fig. 4).

A study on rRNA sequences at the end of the 1970s provided a key to prokaryote phylogeny (Fox et al., 1980). The introduction of 16S rRNA/DNA sequence analysis, better than any other taxonomic method, placed an organism in the framework of phylogenetic relationships. This brought about the "molecular revolution" that led to the reorganization of the taxonomy above the genus level. Because of this, the genera *Gardnerella* and *Bifidobacterium* have been included in one family, Bifidobacteriaceae (Stackebrandt et al., 1997).

Milestones in the *Bifidobacterium* taxonomy are reported in Fig. 5.

Habitat

The presence of bifidobacteria in the alimentary tracts of human adults and infants has stimulated much interest among bacteriologists and nutritionists. Many factors control the number and the composition of microbial populations in dif-ferent regions of the gastrointestinal tract. The ways of restoring or maintaining a proper microbial balance have been extensively investigated (see "Applications"). In the intestinal tract of animals and humans, bifidobacteria coexist with a large variety of bacteria, most of which are obligate anaerobes; components of this microflora are different in the different areas of the tract (Moore et al., 1969). The obligate anaerobes of this complex microflora are in part unknown at present, although significant advances have been made in recent times with the appropriate techniques of identification such as molecular probes. Furthermore, if one considers that particular ecotypes (or biovars) probably have more ecological significance in these habitats than do the species to which they belong (Benno and Mitsuoka, 1986; Mitsuoka, 1972; Mitsuoka, 1984; Sears and Brownlee, 1952; Sears et al., 1950), all factors that pertain to and influence natural genetic variation of single bacterial populations should be studied (Milkman, 1975).

In studies on the ecology of bifidobacteria performed at the Institute of Agricultural Microbiology at Bologna University, Italy, a large number of strains (at present more than 7,000) has been isolated from many different habitats (see Table 1). Studies on the distribution of bifidobacteria in feces of both infants (Biavati et al., 1984) and adults (Biavati et al., 1986), the human vagina (Crociani et al., 1973), or "pathological conditions" such as dental caries (Crociani et al., 1996b) and hypochlorhydria (Brandi et al., 1996) point out that the adaptations of the species of human origin to the five habitats (Fig. 6) are different. *Bifidobacterium bifidum* and *B. longum*, for example, are the most represented species in the intestinal tract of infants and adults, whereas the group "*B. denticolens*, *B. inopinatum* and *B. dentium*" is typical of dental caries and hypochlorhydric stomach. *Bifidobacterium ado-*

lescentis, B. bifidum, B. longum and *B. breve* are typically found in the vagina.

Furthermore, bifidobacteria have been isolated from many animals and, among insects, from honeybees. In general, *Bifidobacterium* species are specific either for humans or for animals; the same *Bifidobacterium* species found in the intestinal microflora of suckling calves and breast-fed infants is the exception. Some species,

12 out of 16, are host-specific, and they are typical of a given animal habitat (Fig. 7).

Eleven species of *Bifidobacterium* have been found in sewage (Fig. 8): 5 are from humans, 4 are from animals, and *B. minimum* and *B. subtile* are found only in sewage. This raises the exciting question of the possible development of bifidobacteria in extraenteral ecological niches. *Bifidobacterium lactis*, a new species recently described by Meile et al. (1997), was isolated from yogurt, but this cannot be considered an *extra*-body habitat because these bifidobacteria are added to yogurt because of their probiotic properties. *Bifidobacterium thermacidophilum* is the most recently described species, described by Dong et al. (2000), and has been isolated from an

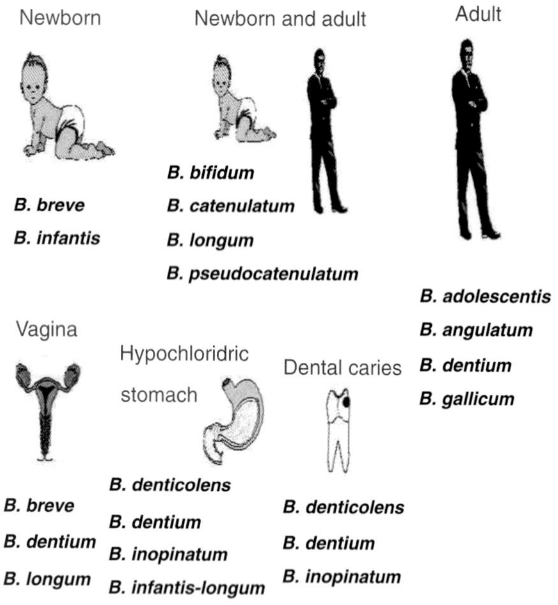

Fig. 6

Newborn
- B. breve
- B. infantis

Newborn and adult
- B. bifidum
- B. catenulatum
- B. longum
- B. pseudocatenulatum

Adult

Vagina
- B. breve
- B. dentium
- B. longum

Hypochloridric stomach
- B. denticolens
- B. dentium
- B. inopinatum
- B. infantis-longum

Dental caries
- B. denticolens
- B. dentium
- B. inopinatum

- B. adolescentis
- B. angulatum
- B. dentium
- B. gallicum

Fig. 6. Habitats of the genus *Bifidobacterium* in humans.

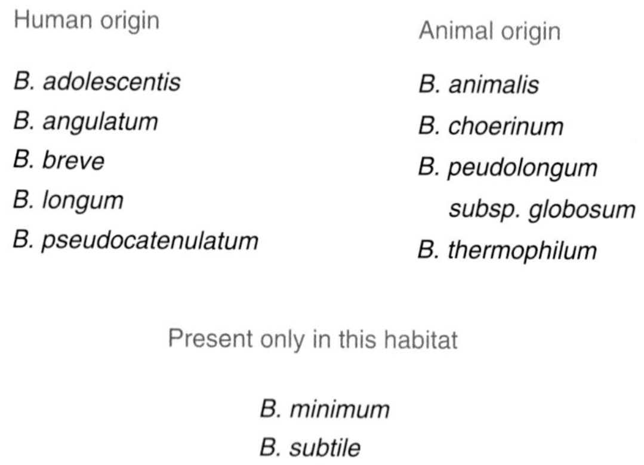

Fig. 8

Human origin
- B. adolescentis
- B. angulatum
- B. breve
- B. longum
- B. pseudocatenulatum

Animal origin
- B. animalis
- B. choerinum
- B. peudolongum
  subsp. globosum
- B. thermophilum

Present only in this habitat

- B. minimum
- B. subtile

Fig. 8. Bifidobacteria found in sewage.

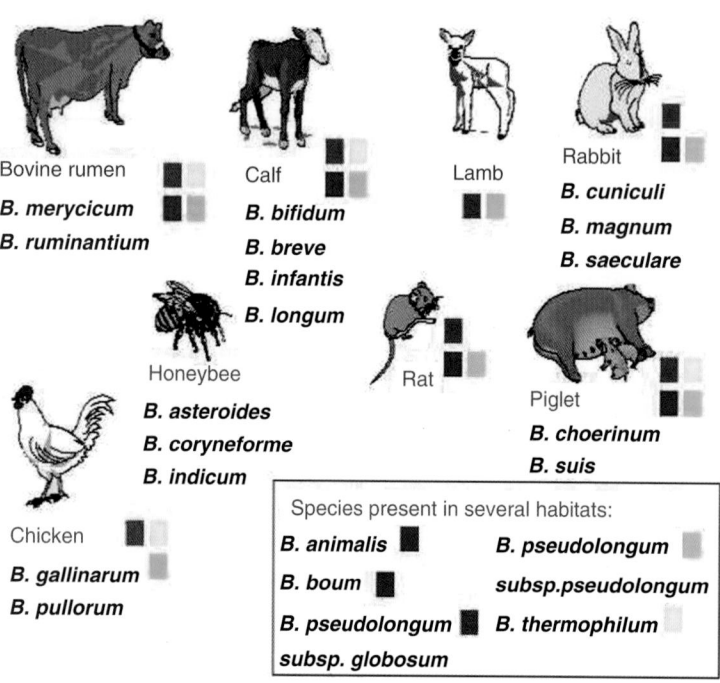

Bovine rumen
- B. merycicum
- B. ruminantium

Calf
- B. bifidum
- B. breve
- B. infantis
- B. longum

Lamb

Rabbit
- B. cuniculi
- B. magnum
- B. saeculare

Honeybee
- B. asteroides
- B. coryneforme
- B. indicum

Rat

Piglet
- B. choerinum
- B. suis

Chicken
- B. gallinarum
- B. pullorum

Species present in several habitats:
- B. animalis
- B. boum
- B. pseudolongum subsp. globosum
- B. pseudolongum subsp.pseudolongum
- B. thermophilum

Fig. 7. *Bifidobacterium* species found in the animals.

Fig. 9. *Bifidobacterium* species found in different habitats.

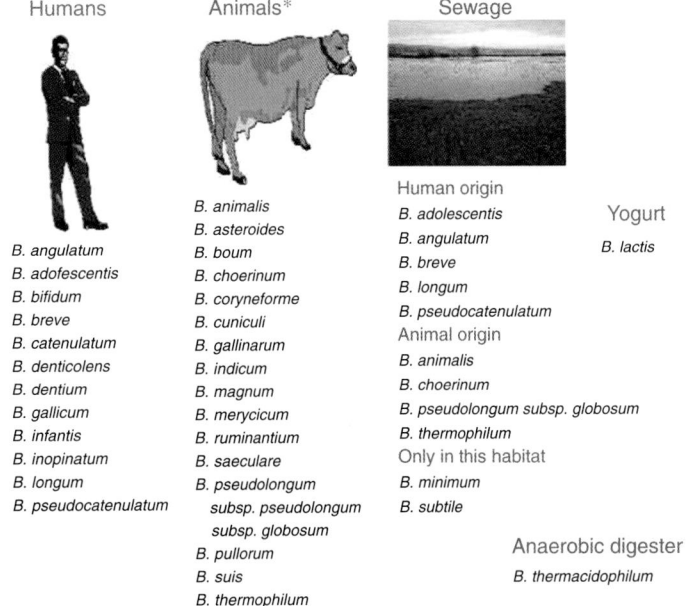

Humans

B. angulatum
B. adofescentis
B. bifidum
B. breve
B. catenulatum
B. denticolens
B. dentium
B. gallicum
B. infantis
B. inopinatum
B. longum
B. pseudocatenulatum

Animals*

B. animalis
B. asteroides
B. boum
B. choerinum
B. coryneforme
B. cuniculi
B. gallinarum
B. indicum
B. magnum
B. merycicum
B. ruminantium
B. saeculare
B. pseudolongum
  subsp. pseudolongum
  subsp. globosum
B. pullorum
B. suis
B. thermophilum

Sewage

Human origin
B. adolescentis
B. angulatum
B. breve
B. longum
B. pseudocatenulatum
Animal origin
B. animalis
B. choerinum
B. pseudolongum subsp. globosum
B. thermophilum
Only in this habitat
B. minimum
B. subtile

Yogurt

B. lactis

Anaerobic digester

B. thermacidophilum

*Only exception is caves which harbour human species; see text.

anaerobic digester used to treat wastewater from a bean-curd farm. This species differed from other species of *Bifidobacterium* by its growth at temperature of 49.5°C and by its ability to grow in medium with the starting pH of 4.0.

The habitats from which bifidobacteria were isolated are shown in Fig. 9.

Isolation

Many media have been devised for isolating or enumerating the bifidobacteria. They can be divided into nonselective and selective media.

NONSELECTIVE MEDIA   The most frequently used nonselective media are described in Table 4. Preliminary trials made with bifidobacteria from human feces, bovine rumen, and honeybee intestine showed that the ingredients ensuring their maintenance in stab cultures were trypticase and phytone (BBL). Finally, the following medium was adopted, which allowed for the isolation and cultivation of strains belonging to widely different genetic species of *Bifidobacterium* from all the habitats so far investigated.

Trypticase-Phytone-Yeast Extract (TPY) Medium

The medium contains per liter:

Trypticase (BBL)	10 g
Phytone (BBL)	5.0 g
Glucose	15 g
Yeast extract (Difco)	2.5 g
Tween 80	1 ml
Cysteine hydrochloride	0.5 g
K₂HPO₄	2.0 g
MgCl₂ · 6H₂O	0.5 g

Final pH is about 6.5 after autoclaving at 110°C for 30 min. Dilution can be made in the same medium. Vented Petri dishes are incubated at 37°C anaerobically. After 3–4 days of incubation, colonies are transferred into stab cultures or TPY medium with 0.5 % agar. After growth has occurred on stabs, these are stored at 3–4°C in an anaerobic jar. Transfer should be made every 2–3 weeks. The same liquid medium is used for the cultivation of bifidobacteria. Any size of tube or flask can be used as long as anaerobic conditions are respected and temperature controlled.

Because the medium is not selective, streptococci, lactobacilli, and other forms may grow profusely. Because bifidobacterial colonies cannot be differentiated from nonbifid ones, all morphological types grown on plates should be scored and tested as indicated in the section on identification.

SELECTIVE MEDIA   To improve selectivity, antibiotics and other factors inhibiting the growth of contaminant microorganisms are used (lithium chloride, sodium propionate or propionic acid, low pH [such as 5 or 5.5], iodoacetic acid, etc.), as well as factor components which improve the growth of bifidobacteria (riboflavin, nitrogenous bases and pyruvic acid, simple or complex carbohydrates such as raffinose, lactose, lactulose, oligosaccharides, etc.).

The selective substrates mainly used are listed in Table 4. In some cases, the use of selective media can be very helpful, but owing to great intraspecific variation in the resistance of bifidobacteria to antibiotics (Matteuzzi et al., 1983) or in their utilization of elective substances (Cro-

Table 4. The most nonselective and selective media for isolation and cultivation of bifidobacteria.

Nonselective media	References
de Man, Rogosa and Sharpe (MRS)	de Man et al., 1960
Tomato juice agar (TJA)	
Reinforced clostridial agar (RCA)	Willis et al., 1973
Briggs liver (BL)	Mitsuoka, 1984
Columbia blood agar	
Tryptone-phytone-yeast extract (TPY)	Scardovi, 1986
Selective media	
RCA (pH 5.0)	Willis et al., 1973
Paromomycin, neomycin, sodium propionate, and lithium chloride (BS1)	Mitsuoka et al., 1965
Neomycin, paromomycin, lithium chloride and nalidixic acid (NPLN)	Teraguchi et al., 1978
Nalidixic acid, riboflavin, nitrogenous bases and pyruvic acid	Tanaka and Mutai, 1980
Nalidixic acid, neomycin, and bromocreosol green (YN-6)	Resnick and Levin, 1981
Sorbitol (YN-A)	Mara and Oragui, 1983
Transgalacto-oligosaccharides agar (TOS-agar)	Sonoike et al., 1986
Nalidixic acid, polymyxin, kanamycin, iodoacetic acid and 2,3,5-triphenyltetrazolium chloride (BIM-25)	Muñoa and Pares, 1988
Propionic acid and pH 5.0	Beerens, 1990
Dicloxacillin	Sozzi et al., 1990
Lithium chloride and sodium propionate (LP)	Lapierre et al., 1992
Modified BIM-25 with lithium chloride and sodium propionate	Arroyo et al., 1994
RCA with Prussian blue (RCPB)	Ghoddusi and Robinson, 1996
Transgalacto-oligosaccharides and sodium propionate	Ji et al., 1994
Blood-glucose-liver agar with oxgall and gentamycin	Lim et al., 1995
Raffinose, lithium chloride and sodium propionate	Hartemink et al., 1996
Fructo-oligosaccharides	Sghir et al., 1998
Lactulose	Nebra and Blanch, 1999
MRS with oxgall blue, cysteine, and dicloxacillin (MRS-BCD)	Ingham, 1999

ciani et al., 1994), the media proposed neither guarantee complete selectivity nor allow the growth of all the species of bifidobacteria. Rather, these media permit satisfactory growth of the largest number of bifidobacterial types presently known. There is no general recipe; therefore, the choice of substrate and level of selectivity are suggested by the type of sample: for the media used to determine bifidobacteria counts in dairy products, for example, the presence of substances inhibiting dairy bacteria (such as lactobacilli and streptococci) is important.

In our experience the slightly modified medium proposed by Beerens (1990), which includes TPY instead of Columbia medium, can be very useful for isolation from both feces and dairy product samples.

Identification

Identification of a bacterial strain as *Bifidobacterium* is difficult. The morphology is of little help in identifying an isolate as a *Bifidobacterium*, but of great help in the recognition of a bifidobacterial species. Some morphological traits, such as disposition and number of branches, cell contours, dimensions, and arrangement in groups, are typical of many known *Bifidobacterium* species grown in solid TPY medium. Therefore, the morphologies illustrated in Fig. 2 (panels 1–31)

are useful for species differentiation. Cultural and physiological characters are grossly shared by many other genera, e.g., *Actinomyces*, *Corynebacterium* and *Lactobacillus*. One of the more practical approaches to the primary differentiation of bifidobacteria from related groups is the one proposed by Holdeman et al. (1977), which is based on identification by gas chromatography of the fermentation products, among which acetic acid generally predominates over lactic acid as the main final product. Side reactions, however, can lead to formation of substantial and variable amounts of ethanol and of formic and succinic acids, so that the pattern can be difficult to interpret, especially for the inexperienced worker. The most direct and reliable assignment of a bacterial strain to the genus *Bifidobacterium* is the one based upon the demonstration (in cellular extracts) of fructose-6-phosphate phosphoketolase, the key enzyme of bifidobacterial hexose metabolism (Fig. 3). Its validity is not only dictated by the results of previous investigation (De Vries et al., 1967; Scardovi and Trovatelli, 1965), but is confirmed with organisms isolated from the most diverse habitats.

Although DNA-DNA hybridization remains one of the more valuable tests in the identification of *Bifidobacterium* species, it cannot be routinely carried out in most laboratories. Identification of *Bifidobacterium* species, however, can be achieved with the use of the fermentation tests (particularly complex carbohydrate fermentations are useful for differentiation of human species; see Fig. 4) and/or cell-wall structure type and/or electrophoretic tests of cellular proteins.

In the last few years, certain molecular techniques based on sequence comparisons of DNA or RNA have been developed that provide the genus or species or strain characterization as well as a classification scheme that predicts natural evolutionary relations. Techniques which allow phylogenetic and typing characterization of intestinal microflora include 16S rRNA sequence analysis, rec A and internal transcribed spacer (ITS) sequence analysis; techniques for monitoring the distribution and prevalence of specific microbes include colony hybridization with nucleic acid probes, pulsed-field gel electrophoresis (PFGE), ribotyping, restriction fragment length polymorphism (RFLP) of the 16S rRNA gene and random amplified polymorphic DNA (RAPD). The use of such molecular techniques to analyze microflora is an additional aid to advancement of intestinal microflora complexity studies. In the future, the impact of such approaches on the study of this intestinal niche will be fundamental, and such studies may provide a scientific basis for application of potential probiotic bacteria to the maintenance of intestinal health. Such molecular techniques are not yet universally applicable to all species of *Bifidobacterium* genus; it is therefore important to develop further these applications. The DNA-DNA homology technique remains the most reliable method to identify the *Bifidobacterium* species.

The most common techniques for the identification of *Bifidobacterium* genus, species or strain are shown in Fig. 10. Procedures routinely performed in our laboratories are described in this chapter. For more detailed information please contact the authors.

PHENOTYPIC IDENTIFICATION

Genus Identification Fructose-6-phosphate phosphoketolase (F-6-PPK) is present in all bifidobacterial species and to this day the F-6-PPK assay (Scardovi and Trovatelli, 1969a) is the test that allows a strain to be identified reliably as a *Bifidobacterium* sp. This enzyme is present in *Gardnerella vaginalis* as well, but the genera *Gardnerella* and *Bifidobacterium* can be easily distinguished phenotypically. Fructose-6-phosphate phosphoketolase is involved in the breakdown of fructose-6-phosphate to erythrose-4-phosphate and acetyl phosphate in the presence of inorganic phosphate (Fig. 3); sodium pyrophosphate, ATP and ADP are mod-

Genus identification
- Fructose-6-phosphate phosphoketolase test
- Gas lipuid cromatography of fermentation products
- Use of the genus-specific PCR primers

Species identification
- Cell morphology
- Fermentation tests
- Electrophoretic patterns of enzymes
- Electrophoretic patterns of cellular proteins
- Cell-wall structure
- DNA-DNA reassociation
Molecular thecniques such as:
- Internal Transcribed Spacer (ITS) sequence analysis
- Use of the species-specific PCR primers
- Amplified Ribosomal DNA Restriction Analysis (ARDRA)
- Random Amplified Polymorphic DNA (RAPD) or Arbitrary Primed-PCR (AP-PCR)
- Fluorescent In Situ Hybridization (FISH)

Strain identification
- Restriction Fragment Lenght Polymorphism (RFLP) of 16S rRNA gene
- Pulsed Field Gel Electrophoresis (PFGE)
- Ribotyping

Fig. 10. Techniques used to identify genus, species and strain of bifidobacteria.

ulators of its activity, whereas copper (Cu^{++}) ions, *p*-chloromercuribenzoic acid (*p*CMB) and mercuric acetate greatly inhibit activity (Grill et al., 1995). The comparison of F-6-PPK electrophoretic mobility revealed three different bands of which one is present only in human species while the others are in animals (Scardovi et al., 1971a). Considering that such an enzyme is so useful for identification of *Bifidobacterium* at the genus level, it may be interesting to develop a specific probe based on the nucleotide sequence of such an enzyme.

The F-6-PPK Procedure
  Reagents:

1) 0.05 M phosphate buffer (pH 6.5) containing cysteine (500 mg/liter)
2) NaF (6 mg/ml) and potassium (or sodium) iodoacetate (10 mg/ml)
3) Fructose-6-phosphate (sodium salt; 70% purity; 80 mg/ml of water)
4) Hydroxylamine·HCl (13.9 g/100 ml) freshly neutralized (to pH 6.5) with NaOH
5) Trichloroacetic acid (TCA; 15% [w/v]) solution in water
6) 4 M HCl
7) $FeCl_3 \cdot 6H_2O$ (5% [w/v] in 0.1 M HCl)

Cells harvested from 100 ml of TPY liquid medium are washed twice with reagent 1 and resuspended in the same buffer (1 g of cells/ml of buffer). The cells are disrupted by sonication in the cold. Reagents 2 (0.25 ml) and 3 (0.25ml) are added to 1.0 ml of cellular extract. After a 30-min incubation at 37°C, the reaction is stopped with 1.5ml of reagent 4. After 10 min at room temperature, 1.0ml each of reagents 5 and 6 are added. The mixture may be stored at room temperature prior to the addition of 1.0 ml of the color-developing reagent 7. The tubes are mixed by inversion. Any reddish-violet color (absorption maximum 505 nm) that develops immediately is taken as a positive result. A tube without fructose-6-phosphate can serve as a blank for visual comparison (Pechère and Capony, 1968). The color is more visually evident after standing to allow debris and protein to settle. The formation of acetyl phosphate from fructose-6-phosphate is detected by the reddish-violet color of the ferric chelate of its hydroxamate (Lipmann and Tuttle, 1945).

Warning: Avoid heating during sonication, as the enzyme is heat sensitive.

In the bifidus pathway (Fig. 3) from tetrose-phosphate and hexose phosphate, through the subsequent action of transaldolase and transketolase, pentose phosphates are formed and then split in acetic and lactic acids. The latter are produced in the theoretical ratio of 1.0 : 1.5. However, the balance of the reaction can be altered by the phosphoroclastic cleavage of pyruvate into formic and acetic acids and by the reduction of acetate into ethanol (De Vries and Stouthamer, 1968; Lauer and Kandler, 1976).

Gas Chromatography Analysis Extraction of Volatile Acids. The extraction is performed in ether. To 1 ml of culture the following are added in succession: 0.2 ml of 50% H_2SO_4, 0.4 g of NaCl and 1 ml of ethylether. After agitation and centrifugation, the ether is collected and a pinch of 4–20 mesh anhydrous $CaCl_2$ is added to remove traces of water.

Extraction of Nonvolatile Acids. The extraction is performed in chloroform. To 1 ml of culture, 2 ml of methanol and 0.4 ml of 50% H_2SO_4 are added in succession. After incubation at 60°C for 30 min, 1 ml of H_2O and 0.5 ml of chloroform are added. Following agitation and centrifugation, chloroform is collected at the bottom of the test tube.

The Chromatographic Analysis. The sample (10 μl) is injected. The analysis is performed as previously described (Holdeman et al., 1977).

Rapid and preliminary identification of *Bifidobacterium* isolated from dairy products may be made using α-galactosidase as a marker, inasmuch as this enzyme is not detected in *Lactobacillus* strains (Chevalier et al., 1990). This enzymatic test is included in the APY-ZYM system (bioMerieux, Marcy L'Etoile, France). Identification must be confirmed by other more reliable methods, such as the F-6-PPK test.

Species Identification This is based on morphological features, fermentation testing, the electrophoretic patterns of enzymes and other cell proteins, and determination of cell wall structures.

Morphological Features. Generally, cells of the genus *Bifidobacterium* have a rod-shaped form and can be characterized as Gram-positive, nonmotile and nonsporeforming anaerobes. The rods (2–5 μm) often bulge at one end, in the form of a cudgel or spatula, and sometimes branch. The Y- and V-forms are frequent, but more regular and coccoid forms can also be observed, depending upon the species to which the cells belong. Some species possess a characteristic cell shape that can help in the identification of the species itself. However, different cellular forms can be found within a given species, depending on the culture conditions. Tissier already noted that his bacteria took on bulge forms in response to acidity, extreme temperatures or inappropriate nitrogen sources in the substrate (Tissier, 1899). The deficit of certain factors involved in the synthesis of the cellular wall results in an increase of branches. Among these factors, there are aminosugars such as *N*-acetyl-glucosamine (Glick et al., 1960) and aminoacids such as alanine, aspartic acid, glutaminic acid and serine (Husain et al., 1972). The concentrations of sodium salts and calcium ions also influence the formation of branches (Kojima et al., 1968; Kojima et al., 1970). Therefore, pleomorphism seems to be affected by a defective synthesis of the cell wall, rather than degenerative processes, as previously suggested. Biavati et al. (1992a) observed the phenomenon called "phase variation" (Trüper and Krämer, 1981) in *B. animalis*. The transition of colony phenotype accompanied by a dramatic change in cell morphology was described. The cells forming transparent (T) colonies, which were minute and mostly spherical, were seen to transition into cells forming opaque (O) colonies, which had species-specific shapes and dimensions (Fig. 11).

Fermentation Tests. One of the early identification schemes of the species of *Bifidobacterium* genus was based on a simple carbohydrate fermentation pattern (Mitsuoka, 1969). This method is still in use, but data obtained from fermentation tests cannot be considered final but need to be confirmed as by DNA-DNA hybridization or electrophoresis of cellular proteins. The simple carbohydrate fermentation characteristics of species of *Bifidobacterium* are reported in Table 5. Encouraging data were obtained from studies of the fermentation of complex polysaccharides and of mucin (Tables 6a and b). The largest number of species could ferment D-galactosamine, D-glucosamine, amylose and amylopectin. Many of the species isolated from animal habitats showed reduced fermentation activity. Some species of human origin selectively ferment certain components, and this allows the development of a key for the differentiation of human *Bifidobacterium* species (Fig. 4). *Bifidobacterium bifidum* was the only species to ferment porcine gastric mucin, *B. infantis* was the only species to ferment D-glucuronic acid; strains of *B. longum* fermented

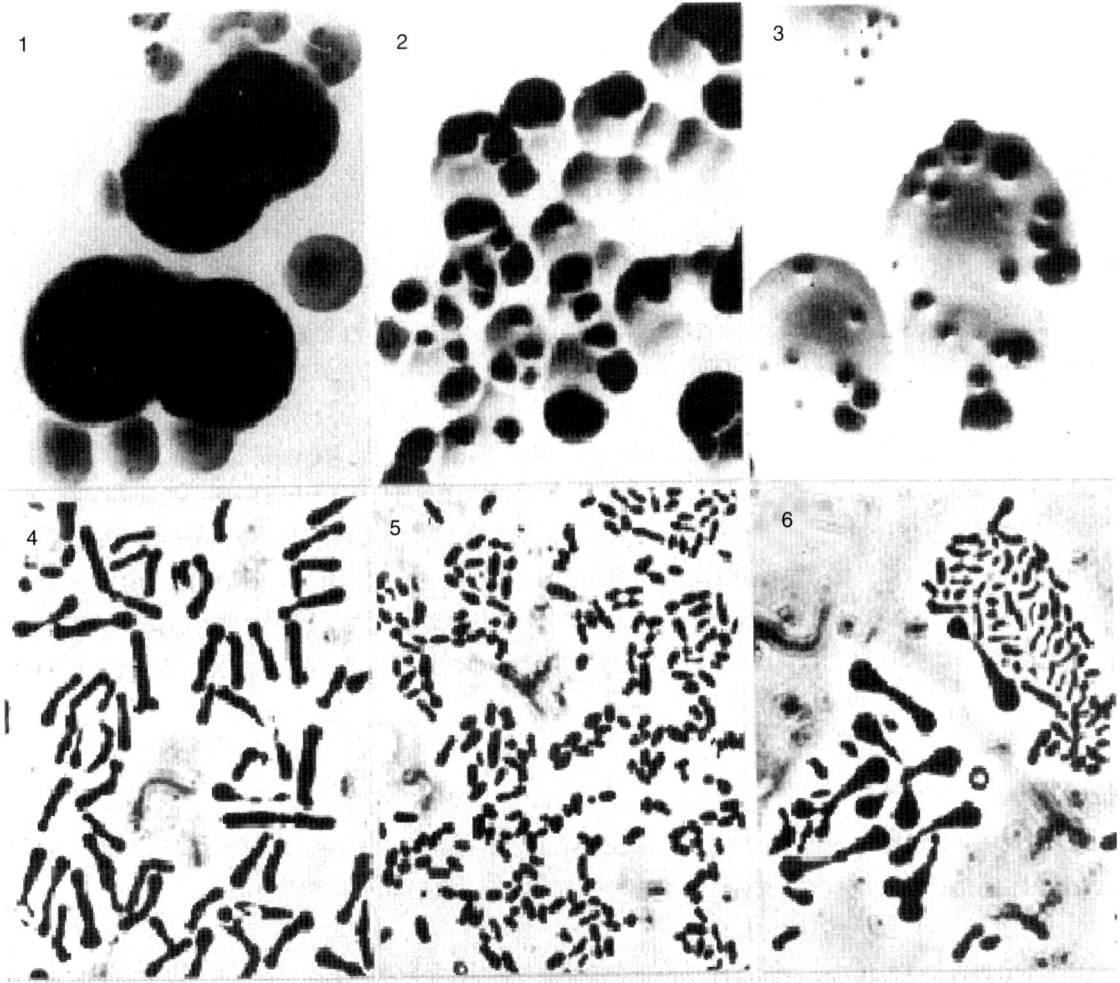

Fig. 11. Colonies on trypticase-phytone-yeast extract (TPY) stabs: (1) strain P 23 (ATCC 27536), (2) T and O, and (3) T colonies with O papillae; and cells from (4) O and (5) T colonies and from (6) the mixed type colonies.

arabinogalactan and the gums arabic, ghatti and tragacanth; α-L-fucose was fermented by strains of *B. breve*, *B. infantis* and *B. pseudocatenulatum*; strains of *B. dentium* fermented guar gum and locust bean (Crociani et al., 1994). Only *B. inopinatum* and *B. denticolens* strains fermented dextran (Crociani et al., 1996b).

Fermentation Test Procedures and Carbohydrates Conventionally Used. For the tests of carbohydrate fermentation, the modified TPY basal medium with half quantities of Trypticase, Phytone and Yeast extract without glucose (TPYM, pH 7.5) is used. As a pH indicator, bromocresol purple is added (30 mg/liter). About 9 ml of TPYM is dispensed into tubes that are then sterilized at 110°C for 30 min. Solutions (5%) of the substrates to be tested are sterilized by Tyndallization (except for arabinose and xylose which are sterilized by filtration) and added (1 ml) to the tubes containing the TPYM. Thus a final concentration of each carbohydrate is 0.5%. The following carbohydrates are tested and routinely used to characterize the species: arabinose, cellobiose, dextrin, fructose, galactose, glycerol, gluconate, inulin, lactate, lactose, maltose, mannitol, mannose, melibiose, raffinose, rhamnose, ribose, saccharose, salicin, sorbitol, starch, trehalose and xylose. Glucose is included as a positive test control. The cell suspension used to inoculate the assay media is from 10 ml-TPYM cultures containing 0.15% glucose. The cells from these cultures are centrifuged and resuspended in 5 ml of TPYM. Three drops of cell suspension are added to each assay tube. Strain positivity is determined from growth, color changes and pH measurements (at least one unit of pH decrease in respect to test without glucose is considered for strain positivity) after 5–7 days of incubation at 37°C in anaerobiosis. Strains fermenting gluconate produce CO_2 from this compound and less acid than that produced

Table 5. Fermentative characteristics distinguishing the species of the genus *Bifidobacterium*.[a]

Species	Xylose	D-Mannose	D-Fructose	D-Galactose	Sucrose	Maltose	Trehalose	Melibiose	Mannitol	Salicin	Sorbitol	L-Arabinose	Raffinose	D-Ribose	Starch	Lactose	Inulin	Cellobiose	Melezitose	Gluconate
1. *B. bifidum*	−	−	+[b]	+	v[c]	−[d]	−	v	−	−	−	−	−	−	−	+	−	−	−	−
2. *B. longum*	v	v	+	+	+	+	−	+	−	−	−	+	+	+	−	+	−	−	+	−
3. *B. infantis*	v	v	+	+	+	+	−	+	−	−	−	−	+	+	−	+	v	v	v	−
4. *B. breve*	−	+	+	+	+	+	v	+	v	+	v	−	+	+	−	+	v	−	−	−
5. *B. adolescentis*	+	v	+	+	+	+	v	+	v	+	v	+	+	+	+	+	v	+	+	+
6. *B. angulatum*	+	−	+	+	+	+	−	+	−	+	v	+	+	+	+	+	+	−	−	v
7. *B. catenulatum*	+	−	+	+	+	+	v	+	v	+	+	+	+	+	−	+	v	+	−	v
8. *B. pseudocatenulatum*	+	+	+	+	+	+	v	+	−	+	v	+	+	+	+	+	−	v	−	v
9. *B. gallicum*	+	−	+	+	+	+	−	−	−	+[g]	−	+	−	+	+	−	−	−	−	−
10. *B. dentium*	+	+	+	+	+	+	+	+	+	+	−	+	+	+	+	+	−	+	+	+
11. *B. denticolens*	−	−	+	+	+	+	v	+	−	+	nd	v	+	+	+	+	+	+	−	−
12. *B. inopinatum*	+	−	+	−	+	+	−	+[b]	−	v	nd	−	+[b]	+	+	v	v	−	v	−
13. *B. pseudolongum*																				
subsp. *pseudolongum*	+	+[g]	+	+	+	+	−	+	−	−	−	+	+	+	+	v	−	v	v	−
subsp. *globosum*	v	−	+[e]	+	+	+	−	+	−	−[f]	−	v	+	+	+	+	−	−	−	−
14. *B. cuniculi*	+	−	−	+	+	+	−	+	−	−	−	+	−	−	+	−	−	−	−	−
15. *B. choerinum*	−	−	−	+	+	+	−	+	−	−	−	−	+	−	+	+	−	−	−	−
16. *B. animalis*	+	v	+	+	+	+	v	+	−	+	−	+	+	+	+	+	−	v	v	−
17. *B. lactis*	+	−	−	−	+	+	−	+	−	−	−	+	+	+	−	+	−	−	−	−
18. *B. thermophilum*	−	−[h]	+	+	+	+	v	+	−	v	−	−	+	−	+	v	v	v	v	−
19. *B. thermacidophilum*	−[d]	−	+	+	+	nd	−	+	−	−[d]	+	+[i]	+	+[i]	v	−[d]	−	nd	v	−[i]
20. *B. boum*	−	−	+	+[i]	+	+	−	+	−	−	−	−	+	−	+	v	+	−	−	−
21. *B. merycicum*	+	−	+[i]	+	+	+	−	+	−	+[i]	−	+	+	+	+	+	−	v	−	−
22. *B. ruminantium*	−	−	+	+	+	+	−	+	+	+[i]	−	−	+	+	+	+	−	−	−	−
23. *B. magnum*	+	−	+	+	+	+	−	+	−	−	−	+	+	+	−	+	−	−	−	−
24. *B. pullorum*	+	+	+	+[g]	+	+	+	−	+	−	+	+	+	−	−	+	−	+	−	−
25. *B. gallinarum*	+	v	+	+	+	+	+	+	−	+	−	+	+	+	−	+	+	v	v	nd
26. *B. saeculare*	+	+[b]	+	+[g]	+	+[b]	+	+	−	v	−[d]	+	+	+	−	+[g]	+	−	+	−
27. *B. suis*	+	v[j]	v[j]	+	+	+	−	+	−	−	−	+	+	−	+	+	−	−	−	−
28. *B. minimum*	−	−	+	−	+	+	−	−	−	−	−	−	−	+	−	−	−	−	−	−
29. *B. subtile*	−	−	+	+	+	+	v	+	−	v	+	−	+	+	+	−	v	−	+	+
30. *B. coryneforme*	+	−	+	nd	+	+	−	+	−	+	−	−	+	+	−	−	nd	+	−	+
31. *B. asteroides*[k]	+[l]	−[d]	+	v	+	v	−	+	−	+	−	−	+	+	−	−	−	+	−	+
32. *B. indicum*[k]	−	v	+	v	+	v	−	+	−	+	−	−	+	+	−	−	−	+	−	+

Symbols: +, 90% or more of strains are positive; −, 90% or more of strains are negative; v, 11–89% of strains are positive; and nd, not determined.
[a] All the strains tested ferment glucose, but do not ferment glycerol, lactate and rhamnose.
[b] A few strains do not ferment this sugar.
[c] When positive, it is fermented slowly.
[d] Some strains ferment this sugar.
[e] Some strains are negative, especially from rat and rabbit feces.
[f] Some strains can ferment it weakly.
[g] Generally delayed or slight fermentation.
[h] Some strains from sewage ferment this sugar.
[i] Some strains are weak fermenters.
[j] Reported as "sometimes not fermented" (Matteuzzi et al., 1971).
[k] Sugars indicated "v" give mostly erratic results.
[l] A few strains do not ferment pentoses.
Data for *B. gallicum* is from Lauer (1990); for *B. lactis* from Meile et al. (1997); for *B. thermacidophilum* from Dong et al. (2000); and for *B. gallinarum* from Watabe et al. (1983).

from glucose or other sugars, and cells are generally very minute (Sgorbati et al., 1970); the indicator turns yellow more slowly and acid development should be carefully compared with that obtained in the absence of added sugars.

Complex Carbohydrates, Polyalcohols, Mucins and Gums. The fermentative tests are carried out with some variations. The compounds which can be used are *N*-acetyl-glucosamine, alginate, D-amygdalin, amylopectin, amylose,

Table 6a. Additional fermentation patterns of strains belonging to human bifidobacterial species.

| | Substrates | | | | | | | | | | | | | | | |
| | Monosaccharides | | | | Polysaccharides | | | | | | | Gums | | | | |
Species	α-L-Fucose	D-Galactosamine	D-Glucosamine	D-Glucuronate	Amylopectin	Amylose	Arabinogalactan	Pectin	Xylan	Dextran	Porcine gastric mucin	Arabic	Ghatti	Guar	Locust bean	Tragacanth
B. adolescentis	−[a]	−	−	−	+[b]	v	−	−	−	−	−	v	−	−	−	−
B. angulatum	−	v	v	−	+	v	−	−	−	−	+	−	−	−	−	−
B. bifidum	−	−	v	−	v	v	−	−[a]	−[a]	−	+	−	−	−	−	−
B. breve	+	v	v	−	v	v	−	v	v	−	−	−	−	−	−	−
B. catenulatum	−	v	v	−	−[a]	v	−	v	v	−	−	−	−	−	−	−
B. denticolens	−	−[a]	+[b]	−	−	−	−	−	−	+[b]	−	−	−[a]	−	+	−
B. dentium	−	v	v	−	+	+	−	v	v	−	−	−	−	+	+	−
B. gallicum	−	−	+	−	+	+	−	−	−	−	−	−	−	−	−	−
B. infantis	+[b]	−	v	+	−	−	−	−	−	−	−	−	−	−	−	−
B. inopinatum	−	−	−	−[a]	−	−	+[b]	−[a]	−	+[b]	−	−	−	−	−[a]	v
B. longum	−	v	v	−	−	−	−	−[a]	−	−	−	v	v	−	−[a]	−
B. pseudocatenulatum	v	−	−[a]	−	+[b]	+[b]	−	−	v	−	−	−	v	−	−	−

Symbols: +, 90% or more of strains are positive; −, 90% or more of strains are negative; and v, 11–89% of strains are positive.

[a]A few strains ferment this sugar.

[b]A few strains do not ferment this sugar. Substrates not fermented by any of the strains tested: alginate, bovine submaxillary mucin, chondroitin sulfate, alpha-D-fucose, D-galacturonate, gum karaya, heparin, hyaluronate, laminarin, ovomucoid and polygalacturonate.

Table 6b. Additional fermentation patterns of strains belonging to animal and extra-body environmental bifidobacterial species.

Species	Monosaccharides		Polysaccharides					Gums		
	D-Galactosamine	D-Glucosamine	Amylopectin	Amylose	Arabinogalactan	Pectin	Xylan	Guar	Locust bean	Tragacanth
B. animalis	+[b]	v	v	−	−	v	−[a]	−[a]	v	−
B. asteroides	v	−[a]	−[a]	−[a]	−[a]	−[a]	−[a]	−	−	−
B. boum	+	+	+	+[b]	−[a]	−	−	−[a]	−	−
B. choerinum	+[b]	−[a]	+	+	−	−	−	v	v	−
B. coryneforme	+	+[b]	−[a]	−[a]	−[a]	−[a]	−	−	−	−
B. cuniculi	v	+[b]	+	+[b]	−	−	−	−	−	−
B. gallinarum	−	−	−	−	−	−	−	−	−	−
B. indicum	+	+	−	−[a]	v	v	−	−	−	v
B. lactis	nt	nt	nt	nt	nt	nt	nt	nt	nt	nt
B. magnum	v	+	−[a]	−[a]	−	−	−	−	−	−
B. merycicum	+[b]	−	v	v	−	−	−	−	v	−
B. minimum	−	−	+	+	−	−	−	−	−	−
B. pseudolongum sub. pseudolongum	−	v	+	+	−	−	−	−	−	−[a]
sub. globosum	v	v	+	v	−[a]	−	−[a]	−	−	−
B. pullorum	+	+	−	−	−	+	+	−	−	−
B. ruminantium	+	+[b]	−[a]	−[a]	−	−	+	−	−	−
B. saeculare	+	+	−	+[b]	−	v	+	−	−	−
B. subtile	+	+	+[b]	+[b]	−	v	−	−	−	−
B. suis	−[a]	−[a]	v	v	v	v	−	−	−	−
B. thermacidophilum	nt	nt	nt	nt	nt	nt	nt	nt	nt	nt
B. thermophilum	+[b]	v	+[b]	+[b]	−[a]	v	v	−[a]	−[a]	−

Symbols: +, 90% or more of strains are positive; −, 90% or more of strains are negative; v, 11–89% of strains are positive; and nt, not tested.
[a]A few strains ferment this sugar.
[b]A few strains do not ferment this sugar. Substrates not fermented: alginate, bovine submaxillary mucin, chondroitin sulfate, dextran, α-D-fucose, α-L-fucose, D-galacturonate, D-glucuronate, heparin, hyaluronate, laminarin, ovomucoid, polygalacturonate, porcine gastric mucin, gums: arabic, ghatti and karaya.

arabinogalactan, bovine submaxillary mucin, chondroitin sulfate A, dextran, heparin, alpha-D-fucose, alpha-L-fucose, D-galactosamine, D-galacturonate, D-glucosamine, D-glucuronate, gum arabic, gum ghatti, gum guar, gum karaya, gum locust bean, gum tragacanth, hyaluronate, laminarin, ovomucoid, pectin, polygalacturonate, porcine gastric mucin, and xylan. They were all purchased from Sigma Chemical Co. (St. Louis, MO). Glucose is added to one assay tube as a positive control to check growth under the assay conditions. Basal medium TPYM (adjusted to pH 7.5) is used throughout the procedure and dispensed (2 ml) into test tubes, which are autoclaved twice at 100°C separated by a 24-h interval. Substrates are added up to a final concentration of 0.5% (and 1% for bovine submaxillary mucin, porcine gastric mucin and ovomucoid). D-Galactosamine, D-glucosamine, heparin, ovomucoid and pectin are added to autoclaved medium as filter-sterilized solutions because the pH of these solutions in the basal medium varies markedly when autoclaved; the pHs of the other autoclaved substrates are 7.2–7.5. Inoculum (0.025 ml) consists of cells derived from 10 ml of TPYM (containing 0.15% glucose), harvested by centrifugation and resuspended in 5 ml of TPYM to avoid introduction of glucose into the test medium. Bromocresol purple (30 mg/ml) is added to the test medium as an indicator. Strain positivity is determined from growth, color changes and pH measurements (at least one unit of pH decrease in respect to test without glucose is considered for strain positivity) after 5–7 days of incubation at 37°C in anaerobic jars.

Electrophoretic Pattern of Enzymes As shown in Table 3, 6-phosphogluconate dehydrogenase (6-PGD) and transaldolase isozymes should be used as additional characters to distinguish bifidobacterial species. The electrophoretic pattern of 6-PGD and transaldolase of some selected species is shown in Fig. 12.

Fourteen isozymes of transaldolase and 19 of 6-PGD, isolated from 1,206 strains of 24 bifidobacterial species, were identified and numbered (Table 2). Sixty percent of the strains were identifiable on the basis of these isozymes (Scardovi et al., 1979a). Antisera against eight purified transaldolases further established natural relationships among the species (Sgorbati, 1979a; Sgorbati and London, 1982a; Sgorbati and Scardovi, 1979b). The horizontal starch-gel electrophoresis system of Smithies (1955) is the preferred method for enzyme detection.

The Electrophoresis Procedure for Enzyme Detection
Transaldolase. TRIS (hydroxymethyl) aminomethane (TRIS; 16.3 g/liter) and citric acid monohydrate (9.0 g/liter) are used as bridge

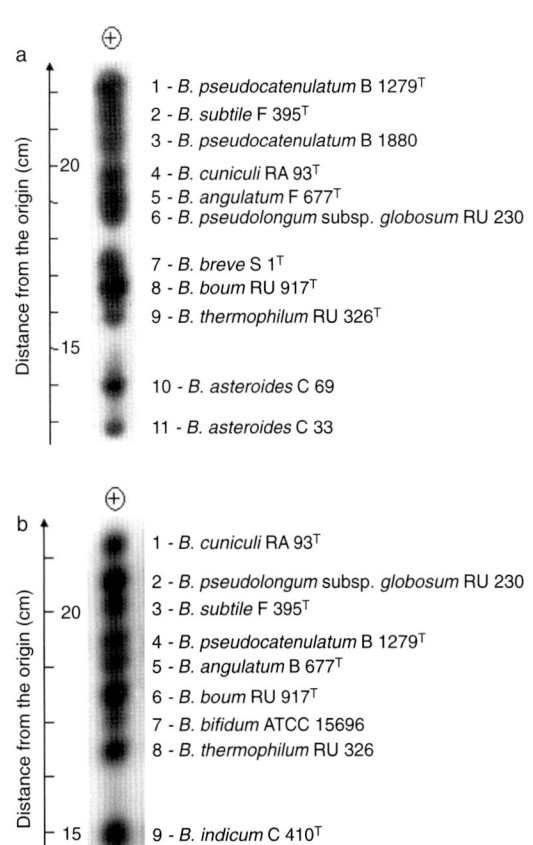

Fig. 12. Coelectrophoretic run of pooled cell-free extracts from 11 (*a*) and 10 (*b*) *Bifidobacterium* reference strains after staining for 6-phosphogluconate dehydrogenase (*a*) and transaldolase (*b*).

buffer (pH 7.0; dilute this 1:15 for use as gel buffer). Hydrolyzed starch (90 g; Sigma, cat. number S-4501) is added per liter of buffer. The mixture is agitated while boiling for 5 min, and gas is removed under reduced pressure. The liquid is poured into a plastic three-frame mold (the dimensions 12.0 × 37.0 × 0.9 cm are suitable to accommodate 12 samples at a time).

Samples (5–10 μl on 0.5 × 0.5 cm Whatman 3-mm paper cuts) of bacterial extracts (cells suspended in 0.05 M phosphate buffer, pH 7.0, sonicated and centrifuged) are generally run for 15–20 h with a current of 15–20 mA. The middle slab is preferably used for staining by a flooding technique. The developing solution contains (per 100 ml of distilled water): fructose-6-phosphate (Na salt, 98% purity, Sigma), 400 mg; sodium arsenate, 370 mg; glycine, 240 mg; NAD, 13 mg; D-erythrose-4-phosphate (60–75% purity, Sigma), 16 mg; phenazine methosulfate, 2 mg; nitroblue tetrazolium (NBT, Sigma), 20 mg; and glyceraldehyde-3-phosphate dehydrogenase, about 130–150 IU.

6-Phosphogluconate Dehydrogenase (6-PGD). Use trisodium citrate·2H$_2$O (120 g/liter; adjusted to pH 7.0 with citric acid) as a bridge buffer and histidine (0.75 g/liter) plus NaCl (1.5 g/liter; pH 7.0) as a gel buffer. Prepare the gel in the same way as the gel used for detecting transaldolase was prepared. The developing solution is made as follows: 0.5 M TRIS-HCl buffer (pH 7.0), 10 ml; 6-phosphogluconate (trisodium salt, Sigma), 250 mg; NADP, 20 mg; NBT, 20 mg; phenazine methosulfate, 2 mg; and distilled water, 90 ml.

Electrophoretic Patterns of Cellular Proteins The electrophoretic patterns of cellular proteins were used to confirm the taxonomic identification of 1,094 bifidobacterial strains belonging to the known species of the genus: excellent correlation was found between data obtained by electrophoresis and by DNA-DNA hybridization. In addition, a sort of "genus band" is clearly visible in all gel electrophoresis patterns; this band migrates to the same position for all strains, with the exception of *B. boum*, where it is somewhat less anodic. The presence of this "genus band" is a good indication that an unknown organism belongs to the *Bifidobacterium* genus (Biavati et al., 1982). Within species, strains that have 80% or greater DNA homology have identical or nearly identical profiles. *Bifidobacterium animalis* is the only exception: in such species, the PAGE pattern of strains isolated from rats appears to differ from that of strains isolated from sewage, rabbits and chickens (Biavati et al., 1982). The intestinal origin of strains isolated from sewage and fermented milk products was determined by comparing electrophoretograms of cellular soluble proteins (Mattarelli et al., 1992). This procedure was confirmed as one of the most useful in rapidly identifying unknown bifidobacterial isolates. Provided that highly standardized conditions are used throughout the procedure of cultivation and electrophoresis, computer-assisted numerical comparisons of protein patterns are feasible, and a database can be created for identification purposes. This allows large numbers of strains to be compared and grouped in clusters of closely related homology.

The Electrophoresis Procedure for Detecting Cellular Proteins Stock solution of the resolving gel contained 24.4 g of acrylamide, 0.64 g of bis-acrylamide and 250 ml of 0.4 M TRIS-HCl, pH 8.8. After filtration the solution is stored in the dark at 4°C. For each gel (16 × 20 cm vertical gels, 1.5-mm thick) 45 ml of stock solution is drawn into a 50-ml syringe. About 16 µl of TEMED and 0.35 ml of freshly prepared aqueous 10% ammonium persulfate are injected through the tip of the 50-ml syringe. An 18-gauge needle is affixed and the gels are cast in previously assembled plates. Stock solution for the stacking gel contains 3.46 g of acrylamide, 0.09 g of bis-acrylamide and 70 ml of 0.2 M TRIS-HCl, pH 7. After filtration, this solution is stored in the dark at 4°C. At the moment of the use, 10 µl of TEMED and 0.4 ml of tracking dye (0.25% bromophenol blue in water [w/v]) are added to 10 ml of solution and the mix is drawn into a 50-ml syringe. Freshly prepared aqueous 10% ammonium persulfate (0.25 ml) is added, and the solution is injected over the resolving gel while a 20-tooth comb is inserted. Samples are prepared from 10 ml of TPY cultures. Bacterial cells are harvested by centrifugation and washed once in 0.2 M TRIS-HCl, pH 7. Cells suspended in 0.15 ml of the same buffer in a tube containing approximately 0.15 g of 75–110 µm diameter glass beads are disrupted using a mechanical agitator (for total of 30 min, applying 2-min periods of agitation separated by 1-min periods of rest). Cell walls and residual cells are harvested by centrifugation. About 20 mg of saccharose was added to the supernatant to increase its density; 35 µl of samples are layered under the buffer solution for each well. Electrophoresis is run at 15°C at constant voltage of 200 V and about 33 mA initial current. The running buffer contains 3 g of TRIS-HCl and 15 g of glycine per liter. Electrophoresis is stopped when the tracking dye reaches the bottom of the gel. The gel is placed in 12% trichloroacetic acid for 30 min before staining. The trichloroacetic acid solution is replaced with Coomassie blue solution (2 g of Coomassie Brillant blue, 450 ml of methanol, 200 ml of acetic acid, and 800 ml of H$_2$O) for 30 min. The gel is destained with 25% ethanol plus 8% acetic acid (v/v) for about 12 h. After 1 or 2 h at room temperature, the depth of the color is adjusted by adding fresh destaining solution. The gels are then photographed and kept in small plastic bags, after the destaining solution is discarded.

Cell Wall Structure The most extensive study of murein (peptidoglycan) types in bifidobacteria cell walls has been conducted by Kandler and Lauer (1974) and Lauer and Kandler (1983). A considerable variety of murein types was found in the genus *Bifidobacterium*, and many species are clearly distinguishable on this basis. The same authors report that bifidobacteria cell walls contain significant amounts of polysaccharides, which usually consist of glucose and galactose often accompanied by rhamnose. The structure and composition of the cell wall polysaccharides have also been investigated (Habu et al., 1987; Nagaoka et al., 1988). Traces of lipoteichoic acid associated with the fraction of the cell wall that remains insoluble after lysozyme treatment were found in *Bifidobacterium bifidum* (Veerkamp et al., 1983). Its chemical composition and structure have been studied (Fischer et al., 1987; Op Den Camp et al., 1984). Immunochemical studies indicated that lipoteichoic acid is a common antigen within the genus *Bifidobacterium* (Op Den Camp et al., 1985b). Furthermore, proteins and lipoteichoic acid are responsible, either independently or in dynamic complexes, for the hydrophobic character of the bifidobacterial surface (Op Den Camp et al., 1985a).

Mukai et al. (1997) suggest that the cell surface proteinaceous components are involved in the adhesion of *B. adolescentis*. The study of cellular structures (proteins, lipoteichoic acids, etc.) as well as the interaction of cells with substrate (in vivo and in vitro adhesion) is particularly important for understanding the mechanisms that regulate bacterial adhesion and therefore colonization. Protein ligands present on the cell surfaces and/or in the culture medium have been identified by Bernet et al. (1993) in some strains of bifidobacteria of human origin (*B. breve, B. longum, B. bifidum* and *B. infantis*). These studies have shown that different strains belonging to the same species adhere differently to intestinal cells of the Caco-2 and HT29MTX lines. More recently Pérez et al. (1998) have studied the adhesive capacity of some strains of human origin, showing that in many cases adhesion was linked to the self-agglutination capacity of such strains: strains with a good self-agglutination capacity tested positive by hemagglutination and showed a high surface hydrophobicity.

The study of the cell wall-associated proteins in *B. pseudolongum* subsp. *globosum* is part of a study of the plasmid functions of bifidobacteria: nearly 40% of 150 strains of different animal origin were shown to have 3–8 cell wall-associated proteins whose distribution was ascertained with PAGE and with immunoblotting techniques (Mattarelli et al., 1993). Using plasmid-free clones obtained by "curing," the involvement of certain plasmids in determining the surface protein patterns was ascertained. Sensitivity to the lytic and bactericidal activities of lysozyme and cationic polypeptides was found to be related to the presence of a plasmid-dependent surface protein of high molecular weight (Mattarelli et al., 1997). The effect of the growth temperature on the biosynthesis of cell wall proteins has been studied in 70 strains of *B. pseudolongum* subsp. *globosum*. The production of cell-wall proteins is temperature-dependent. There is a positive correlation between the increase in growth temperature and the expression of the majority of bifidobacterial outer proteins (BIFOPs), which appears to be a common feature of *B. pseudolongum* subsp. *globosum* species and present in all the strains tested. Furthermore, at high-growth temperatures, the presence of a new common protein, which could be regarded as a temperature inducible protein, was detected in all the strains studied. The relationship of BIFOP, growth temperature and hydrophobicity has been investigated for some strains. At medium-growth temperatures, the cell hydrophobicity was strictly correlated with the BIFOP expression, whereas at low- and high-growth temperature, the presence of BIFOP only partially influenced the hydrophobic features (Mattarelli et al., 1999c).

A deeper understanding of the ecological variability of the surface structure in bifidobacteria and their functions in the multifaceted relationships of host and bacterium is thought to be a prerequisite for a less empirical use of the bifidobacteria in foods and pharmaceutical preparations for humans and animals.

GENOTYPIC IDENTIFICATION PHYLOGENETIC APPROACH. The two main tools used for comparative phylogenetic analysis of *Bifidobacterium* are described below.

Rec A Analysis. A fragment of the Rec A gene, which is usually present in bacteria and highly conserved, is amplified by polymerase chain reaction (PCR) and sequenced. The phylogenetic relationship obtained correlates well with the analysis of the complete rRNA gene and, given its speed and ease, is a valuable tool for comparative phylogenetic analysis of human intestinal isolates. This concept was applied to the genus *Bifidobacterium* by Kullen et al. (1997).

Sequence Analysis Using 16S rRNA. Using primers directed at universally conserved regions at both ends of the gene, the 16S rRNA directly from colonies is PCR-amplified. The amplicon (1.5 kb) is then sequenced and compared to the rRNA database. This allows an accurate typing of unknown isolates and determination of their phylogenetic position. This technique has greatly helped the understanding of phylogenetic relationships among the major microbial genera in the human intestine. Leblond-Bourget et al. (1996) comparing 16S rRNAs concluded that the evolutionary distances exhibited by most bifidobacteria are such that these organisms belong to the same genus. The 16S rRNA analysis was useful in discriminating between genera and confirmed the separation between genus *Gardnerella* and *Bifidobacterium*. Despite considerable phenotypic diversity, all of the bifidobacteria whose sequences have been determined are closely related (levels of 16S rRNA similarity >93%). Furthermore Miyake et al. (1998) showed that, based on a 16S rRNA analysis, 21 *Bifidobacterium* species belonged to a cluster phylogenetically distinct from the other genera; *Gardnerella vaginalis* ATCC 14018 is included in this cluster. The cluster was divided into two subclusters: subcluster 1 composed of most species of *Bifidobacterium* and *G. vaginalis* and subcluster 2 consisting of two species, *B. denticolens* and *B. inopinatum*. Langendijk et al. (1995) have developed 16S rRNA gene probes specific for the genus *Bifidobacterium* which enable the exclusive detection of bifidobacteria in fecal samples through in situ hybridization.

SPECIES IDENTIFICATION CLASSICAL TECHNIQUES. These include determination of G+C content and DNA-DNA homology.

Percent G+C Content The DNA of bifidobacteria has quite a high content of guanine and cytosine (G+C; Table 2). The G+C value varies from 55–67 mol%, with differences among species (Scardovi, 1986). The only exception, *B. inopinatum* (one of the three species found in dental caries), has a G+C content of 45% (Crociani et al., 1996b).

The Procedure for Determining G+C Content DNA base composition was calculated from melting temperatures determined with a response series UV-visible spectrophotometer equipped with a Response II thermal programmer. *Escherichia coli* K-12 DNA was used as a standard. The mol% of G+C is calculated by the formula of Marmur and Doty (1962).

DNA-DNA Homology DNA-DNA homology is the gold standard method to establish relationships within and between the species. A species is defined as an entity which includes "... strains with approximately 70% or greater DNA-DNA relatedness and with 5°C or less δTm." (Wayne et al., 1987). Both values must be considered and phenotypic characteristics must be similar in every species. DNA-DNA homology is not widely applied because it is time-consuming and laborious even if it is still the most reliable method for species identification.

Procedure for Determining DNA-DNA Homology The percentage of homology between DNAs is assessed through the direct-binding membrane method (Johnson, 1981). The DNA is from cells grown in 1 liter of TPY medium. Cells harvested by centrifugation are washed twice in saline-EDTA buffer (0.15 M NaCl and 0.1 M EDTA, pH 8.0) and suspended in 40 ml of saline-EDTA; lysozyme is then added up to a final concentration of 3 mg/ml. Subsequently, DNA is extracted using the Marmur (1961) technique. The reference DNA, diluted in 0.1 X SSC (1 X SSC: NaCl 0.15M, trisodium citrate 0.015 M, pH 7) at a concentration of

50 μg/ml, is denatured with NaOH (final concentration 0.25 M) for 15 min at 60°C, then cooled and quickly brought to pH 7 with HCl. The solution is diluted 1:10 with cool 6X SSC, then slowly filtered twice at low temperature through a 15-cm nitrocellulose membrane (Sartorius Membrane filter SM 11306) presoaked in 6X SSC. The filter is dried in air for a few minutes, then in the oven at 80°C for two hours. The filter is cut with a paper punch into small filters 10.5 mm in diameter and then stored at 4°C. The amount of DNA fixed is about 25 μg/cm². The labeled DNA is obtained by nick translation using [1',2',5-³H] dCTP (Amersham-Pharmacia-Biotech, cat. number TRK 625-9.25 MBq) following the protocol of the Nick Translation Kit (Amersham-Pharmacia-Biotech, cat. number N 5000). This DNA is sheared by passage through a syringe needle 3–4 times and denatured by heating in a boiled water bath for 10 min prior to use. Hybridization consists in the comparison between the labeled DNA with the DNA fixed on the filters. The filters with immobilized DNA (reference DNA) are preincubated at 60°C in 0.8 ml of the pre-incubation mixture (2X SSC with 0.02% bovine serum albumin, 0.02% polyvinylpyrrolidone, 0.02% Ficoll 400) for 1 h at 60°C. The filters are removed and the excess water blotted off. The filters were then placed individually in 4-ml screw-cap vials containing 0.15 ml of sterile H₂O, 0.04 ml of 11.1X SSC, and 10 μl of labeled DNA. The vials were gently shaken with a wrist-action shaker for 12 h at 69°C, 25°C below the melting temperature (Tm) as determined in 1X SSC. After incubation, the filters are washed twice with 0.8 ml of 2X SSC at 69°C, dried under an infrared lamp, and counted with a liquid-scintillation counter. Homology values were calculated in percent of binding with respect to 100% homology of the reference strain.

Recently Described Molecular Techniques These include internal transcribed spacer (ITS) sequence analysis; colony hybridization with nucleic acid probe techniques; pulse-field gel electrophoresis (PFGE); ribotyping; restriction fragment length polymorphism (RFLP) of the 16S rRNA gene; amplified ribosomal DNA restriction analysis (ARDRA); random amplified polymorphic DNA (RAPD), or arbitrary primed-PCR (AP-PCR); fluorescent in situ hybridization (FISH); and *ldh* gene sequence comparison.

Internal Transcribed Spacer Sequence Analysis. The analysis of the internal transcribed spacer, a region of DNA between the 16S and 23S rRNA gene universally present in bacteria, has been used to complement other rRNA sequence analysis. This molecule exhibits very low sequence conservation and for this reason cannot be used as an accurate phylogenetic marker. In addition, the ITS regions within the same bacterial strains can be heterogeneous. However, this analysis is feasible because PCR can be used to amplify the entire molecule from colonies. The ITS has been used for a more detailed analysis of bifidobacteria. The work by Leblond-Bourget et al. (1996) showed that 16S rRNA and ITS analysis are useful tools to dissect the bifidobacteria taxonomy. The ITS analysis was able to discriminate intra- and inter-specific differences that were in good correlation with 16 S rRNA data. Furthermore the ITS approach allowed the authors to propose *B. infantis*, *B. suis* and *B. longum* as belonging to the same species.

Colony Hybridization with Nucleic Acid Probes. Colonies can be directly examined for the presence of a specific DNA sequence recognized by a labeled probe. Short probes (20 bases) can be directed to rRNA regions known to be genus- or species-specific. In the study of bifidobacteria, this approach has been used to obtain species-specific probes (Ito et al., 1992; Mangin et al., 1995; Yamamoto et al., 1992) and genus-specific probes (Kaufmann et al., 1997; Langendijk et al., 1995).

Pulsed-Field Gel Electrophoresis. The PFGE method allows very large fragments of DNA to migrate through an agarose gel under a pulsed electric field. Genomic DNA is digested with a rare restriction enzyme. The unique pattern of DNA fragment distribution in the gel is the fingerprint of the strain. Owing to the large size of the DNA fragments, all the technical operations need to be carried out on cells embedded in an agarose gel (inserts). It is perhaps the most discriminatory technique at the strain level, even if it has no value in taxonomic identification. Bifidobacteria strains of different species were characterized by PFGE by different authors (Bourget et al., 1993; Roy et al., 1996; O'Riordan and Fitzgerald, 1997). The usefulness of this technique has been demonstrated by McCartney et al. (1996) and Kimura et al. (1997), who analyzed *Bifidobacterium* and *Lactobacillus* populations in fecal samples obtained from human subjects in a one-year period. It was demonstrated that each subject harbored numerically predominant strains that were characteristic of the particular human host.

Ribotyping. The total DNA is isolated and completely digested with a frequently cutting restriction enzyme. The restricted fragments are then separated by agarose gel electrophoresis and hybridized with a probe specific for the 16S (most commonly used), 23S or 5S rRNA genes. The visualized pattern represents a characteristic fingerprint. The method is well suited for the analysis of bacteria possessing multiple copies of rRNA genes in their genome, but it is not useful for those bacteria possessing only a single copy of rRNA genes. This technique is very reproducible. Its effectiveness has been demonstrated for the analysis of human intestinal bifidobacteria (Kimura et al., 1997; McCartney et al., 1996). This fingerprinting technique is limited by the fact that it is not as discriminative as PFGE, requires culturing of bacteria, and is labor intensive.

Restriction Fragment Length Polymorphism of the 16S rRNA Gene. The gene for 16S rRNA is amplified by PCR with primers recognizing universally conserved regions in the gene. The amplicon is then restricted and the resulting fragments are analyzed by electrophoresis, giving a characteristic pattern of distribution. Kullen et al. (1997) used this technique to differentiate an ingested bifidobacteria from the indigenous bifidobacteria in human subjects. Kok et al. (1996) developed three 16S rRNA-targeted primers specific to probiotic strain LW 420, probably *B. animalis*, to monitor the presence of this strain in the human intestine. As this is a PCR-based technique, it can be carried out on very few cells without culturing.

Amplified Ribosomal DNA Restriction Analysis. ARDRA is a potent and discriminating tool for organisms at the species level. In this technique, the DNA sequence of a 16S rRNA region is amplified by PCR using genus-specific primers, then the products of amplification are digested with restriction enzymes, and the pattern fragment obtained is used to distinguish different species. Roy and Sirois (2000), by utilizing this technique, are able to discriminate among *B. infantis*, *B. longum* and *B. animalis*.

Random Amplified Polymorphic DNA or Arbitrary Primed-PCR. A PCR-based technique, RAPD or AP-PCR uses only a single start primer (10–12 bases) with an arbitrarily chosen sequence. Usually the stringency of the reaction is reduced, allowing the primer to bind to regions where it exhibits greatest homology. The amplified products are separated by agarose gel electrophoresis. The pattern obtained from different samples can be compared. This is a rapid technique, very discriminative and can be applied to organisms for which no sequence information is available; however, it is not very reproducible because a subtle change in reaction conditions can dramatically change the binding patterns (Williams et al., 1990; Welsh and McClelland, 1990). In addition, RAPD analysis has been applied to the characterization of genetic variability of bifidobacteria isolated from rats fed particular diets (Fanedl et al., 1998). In this study, results indicated the presence of a large degree of genetic variability among bifidobacteria in the rat gut and demonstrated the usefulness of RAPD analysis for the study of microbial diversity in complex ecosystems. Vincent et al. (1998), using five primers, distinguished three different species of *Bifidobacterium* (*B. breve*, *B. bifidum* and *B. adolescentis*) based on similarity of the RAPD profiles to known reference strains. This technique is also useful for placement of industrial strains into specific clusters (either *B. longum/infantis* or *B. animalis/lactis*).

Fluorescent In Situ Hybridization. Oligonucleotide probes labeled with fluorescent molecules are allowed to hybridize directly onto cells fixed on glass slides. Subsequently the probes can be visualized using fluorescence microscopy. The technique can be applied to the determination of specific mRNA within cells. Langendijk et al. (1995) and Harmsen et al. (Harmsen et al., 1999; Harmsen et al., 2000) utilized this technique to detect bifidobacteria in fecal samples, providing quantitative data on gut flora.

ldh Gene Sequence Comparison. Roy and Sirois (2000) utilized *ldh* gene sequence comparison to distinguish between *B. longum* and *B. infantis* and between *B. animalis* and *B. lactis*. This technique permits differentiation of *B. longum* from *B. infantis*, but not *B. animalis* from *B. lactis*. Owing to the low number of species tested, the validity of this method is difficult to judge.

Preservation

SHORT-TERM METHODS A chopped meat medium (Holdeman et al., 1977) is probably the best one for preservation of strains of bifidobacteria in liquid media. To avoid dehydration and diffusion of O_2, sterilized chopped meat (pre-reduced anaerobically) in tubes sealed with butyl stoppers (Holdeman et al., 1977) is inoculated, and after growth, the cultures can be stored at room temperature or at 5°C for several months.

LONG-TERM METHODS Longer maintenance can be achieved both by freezing below 130°C (deep freezer) or in liquid nitrogen and by freeze-drying. The following cryoprotective agents may be used: skim milk solution (skim milk 10%, lactose 3%, yeast extract 0.3%; the solution after shaking for 20 min is autoclaved at 110°C for 30 min) or saccharose solution (saccharose 12%, yeast extract 0.3%, peptone 0.5%; the solution is autoclaved at 110°C for 30 min). Both cryoprotective agents allow for sufficient survival of the strains in the survival process: the level of survival is a characteristic linked to strains rather than species (B. Biavati and P. Mattarelli, unpublished data). In our experience lyophilized cultures kept at room temperature remain alive after 30 years. For details and other methods, see Gherna (1994).

Physiology

SENSITIVITY TO OXYGEN Bifidobacteria are anaerobic microorganisms. However, the sensitivity to oxygen changes according to the species and the different strains of each species. The mechanisms which are at the basis of their tolerance to oxygen are controversial (De Vries and Stouthamer, 1968). The reasons why bifidobacteria are anaerobic may vary: some require a high redox potential or enzymes that are sensitive to H_2O_2. Phosphoketolase is indeed sensitive to H_2O_2, but direct toxicity of O_2 has not yet been demonstrated. Bifidobacteria are all catalase-negative, yet among the species tolerant to air, *B. indicum* and *B. asteroides* become catalase-positive if they are grown in the presence of air, respectively with or without addition of hemin (Scardovi and Trovatelli, 1969a). Some strains of *B. bifidum* seem to be able to synthesize at a low extent some catalase when exposed to air (Bezkorovainy, 1989).

OPTIMUM TEMPERATURE Optimum temperature for growth is 37–41°C, whereas no growth occurs below 20°C and above 46°C with the exception of *B. thermacidophilum*, which is able to grow at 49.5°C (Dong et al., 2000). Some changes occur depending upon the habitat of origin. Growth at 45°C seems to discriminate between animal and human strains, since most of the animal but not the human strains are able to grow at this temperature (Gavini et al., 1991).

OPTIMUM pH The optimum pH at the beginning of the growth is between 6.5 and 7.0. No growth is recorded at pH lower than 4.5 or higher than 8.5 with the exception of *B. thermacidophilum*, which is able to grow at pH 4.0 (Dong et al., 2000). Bifidobacteria are acid-tolerant but are not acidophilic microorganisms.

HEXOSE METABOLISM Hexose metabolism in bifidobacteria occurs through a characteristic sequence called the "fructose-6-phosphate shunt" or "bifidus shunt" (Scardovi and Trovatelli, 1965; De Vries et al., 1967; Fig. 3). The key enzyme of this pathway is fructose-6-phosphate-phosphoketolase (EC 4.1.2.22), which cleaves hexose phosphate to erythrose-4-phosphate and acetyl-phosphate. Pentose phosphates are formed from tetrose phosphate and hexose phosphate, through the subsequent action of transaldolase and transketolase, and then split into acetic and lactic acids. The latter are produced in the theoretical ratio of 1.0:1.5. However, the balance of the reaction can be altered by the phosphoroclastic cleavage of pyruvate into formic and acetic acids and by the reduction of acetate into ethanol (De Vries and Stouthamer, 1968; Lauer and Kandler, 1976). During the metabolism of hexose, there is no production of carbon dioxide, except in the degradation of gluconate. Different bifidobacterial species produce different amounts of acetate, lactate, ethanol and formate under the same conditions. Furthermore, variations of growth conditions, such as quality and quantity of the carbon source, may result in the production of varying amounts of fermentation products.

Occasionally, claims are made that other pathways may also exist. It has been stated that the glycolytic enzyme aldolase (EC 4.1.2.13) and the hexose monophosphate shunt enzymes glucose-6-phosphate dehydrogenase (EC 1.1.1.49) and 6-phosphogluconate dehydrogenase (EC 1.1.1.44) were absent from many human bifidobacteria but not necessarily from animal species. *Bifidobacterium asteroides* and *B. indicum* (isolated from honeybee), in fact, both have hexose monophosphate shunt enzymes, but not aldolase (Scardovi and Trovatelli, 1969a). Both aldolase and the two hexose monophosphate shunt enzymes are present in the rumen bacteria *B. pseudolongum* subsp. *globosum* and *B. thermophilum* (Scardovi et al., 1969b). *Bifidobacterium catenulatum*, *B. longum* and *B. angulatum* among the human species contain both aldolase and 6-phosphogluconate dehydrogenase. The physiological role of these enzymes in bifidobacteria is unclear.

The enzymes of the Leloir pathway of galactose metabolism, i.e., galactokinase (EC 2.7.1.6), hexose-1-phosphate uridyltransferase (EC 2.7.7.12) and UDP galactose 4-epimerase (EC 5.1.3.2), are constitutive in glucose-grown cells of bifidobacteria, whereas in other microorganisms these enzymes are induced by galactose or fucose (Lee et al., 1980). The enzymatic carboxylation of phosphoenolpyruvate to oxaloacetate in some bifidobacteria from human feces and from honeybees has been compared to the corresponding activity in strains of *Actinomyces bovis* and *Actinomyces israelii*. In bifidobacteria, this activity is independent of the phosphate acceptor and is irreversible, whereas in *Actinomyces* it is inosine- or guanosine-diphosphate-dependent (Chiappini, 1966).

ULTRASTRUCTURE The ultrastructure of bifidobacteria has received little attention. Overman and Pine (1963) first reported electron micrographs of *B. bifidum* subsp. *pennsylvanicum*. Zani and Severi (1982) made a more extensive investigation of the ultrastructure of *B. bifidum* strain S28a of Reuter (ATCC 15696). Comparative studies of the ultrastructure of *Bifidobacterium* species comprising this genus also have been made (Novik et al., 1994; Novik, 1998).

ENZYMES *Bifidobacteria* utilize a great variety of mono- and disaccharides as carbon sources and are able to metabolize also complex carbohydrates that are normally not digested in the small intestine. This latter feature should give an ecological advantage to colonizers of the intestinal environment where complex carbohydrates, such as mucin, are present either because of production by the host epithelium or introduction through the diet. The selective stimulation of the growth of bifidobacteria by simple or complex carbohydrates is the basis of the prebiotic concept. The study of enzymes involved in the bifidobacterial carbohydrate degradation has therefore received great attention. Carbohydrates that are considered bifidogenic are: oligosaccharides (fructo-oligosaccharides, *trans*-β-D-galactosyl-oligosaccharides, α-D-galactose-oligosaccharides such as raffinose and stachyose, the latter found in legumes or beans, and endogenous carbohydrates such as mucus and glycoproteins) and polysaccharides such as inulin.

Bifidobacteria possess an array of enzymes that allow them to utilize different types of carbohydrates to adapt and compete in an environment with changing nutritional conditions; these enzymes are inducible in the presence of specific substances. Amongst these enzymes α- and β-glycosidases, neuraminidases (EC 3.2.1.18), α-glucosidases (EC 3.2.1.20) and β-D-glucosidases (EC 3.2.1.21), α-D-galactosidase (EC 3.2.1.22) and β-D-galactosidase (EC 3.2.1.23) are included. Extracellular α- and β-glycosidases that degrade intestinal mucin oligosaccharides and glycosphingolipids of the lacto-series type 1 chain were found in human fecal bifidobacteria (Falk et al., 1990). A β-glucosidase that also has β-D-fucosidase activity was found in *B. breve* (Nunoura et al., 1996). A novel β-glucosidase, which is inducible and capable of catalyzing the hydrolysis of sennosides, was purified from a strain of *Bifidobacterium* sp. (Yang et al., 1996). α-D-Galactosidase, which exhibited a high substrate specificity for α-galactoside linkages, seems to be highly specific for bifidobacteria and *Actinomycetes* of human origin (Desjardins and Roy, 1990; Hartemink et al., 1996); this enzyme is a specific test in the APY identification system for bifidobacteria. Transglycosidase activity of *B. adolescentis* α-galactosidase has been recently studied to utilize new types of α-galactosides as prebiotic and synbiotic substrates (Van Laere et al., 1999). Several isoenzymes of β-galactosidase were detected among strains of bifidobacteria by means of electrophoresis (Roy et al., 1994). Interestingly in this study, the authors assess that dairy-related bifidobacteria (*B. bifidum*, *B. breve*, *B. infantis* and *B. longum*) as well as *B. animalis* could be better differentiated from other bifidobacteria by comparison of their β-galactosidase electrophoretic patterns.

In *B. adolescentis* cells grown on transgalactosylated oligosaccharides (TOS), in addition to the lactose-degrading b-galactosidase, another b-galactosidase named "β-galactosidase II" (β-gal II) was detected, showing activity towards TOS but not towards lactose. β-Gal II activity was at least 20-fold higher when cells were grown on TOS than when cells were grown on galactose, glucose and lactose (Van Laere et al., 2000).

β-1,3-Galactosyl-N-acetylhexosamine phosphorylase, a new enzyme from *B. bifidum*, which catalyzed the reversible phosphorolytic cleavage of β-1,3-galacto-oligosaccharides, has been characterized (Derensy-Dron et al., 1999).

The presence of β-fructofuranosidase, which degrades fructo-oligosaccharides, from some strains of Bifidobacterium was recently investigated. This enzyme differs (in terms of relative activity towards different substrates) from yeast invertases (Muramatsu et al., 1992).

A new β-fructofuranosidase has been isolated from *B. infantis* (Imamura et al., 1994). This enzyme catalyzes the hydrolysis of sucrose, 1-ketose, nystose, inulin and raffinose, but does not catalyze the hydrolysis of maltose or cellobiose, indicating that this fructo-oligosaccharides enzyme is a novel type of β-fructofuranosidase.

Some enzymes are species-specific. Experimental studies carried out by Crociani et al. (1994) have shown that only *B. bifidum* may degrade intestinal gastric mucin. In the same study, it also has been observed that two out of three species found in the oral cavity, *B. denticolens* and *B. inopinatum*, are able to degrade dextran. From *Bifidobacterium* spp. isolated from dental caries, an extracellular dextranase (EC 3.2.1.11) was partially characterized. This extracellular enzyme, which is able to degrade α-D-(1→6) linkages of dextrans, uses an endohydrolytic mode of cleavage as it is able to liberate saccharides larger than 1 glucose unit (Kaster and Brown, 1983).

Knowledge of the enzymatic network of bifidobacteria, the study of which is expanding, may help to better understand the mechanisms by which probiotic organisms may exert health effects.

NUTRITION Most species of the genus *Bifidobacterium* are able to utilize ammonium salts as their sole source of nitrogen (Hassinen et al., 1951), but *B. suis*, *B. magnum*, *B. choerinum*, *B. cuniculi* and *B. pseudolongum* subsp. *globosum* do not grow without organic nitrogen (Matteuzzi et al., 1978; Mattarelli and Biavati, 1999a). In an organic nitrogen-free medium, considerable amounts of various amino acids are excreted by strains of *B. thermophilum*, *B. adolescentis*, *B.*

dentium, *B. animalis* and *B. infantis*. Generally the amino acids produced in the largest amounts are alanine, valine, and aspartic acid. Exceptionally, *B. bifidum* can produce up to 150 mg/liter of threonine (Matteuzzi et al., 1978).

Analogue-resistant mutants obtained from *B. thermophilum* (*B. ruminale*) showed increased production of isoleucine and valine (Crociani et al., 1977; Matteuzzi et al., 1976). Homoserine dehydrogenase and threonine deaminase activities as well as some aspects of their regulation have been studied in many species of the genus (Selli et al., 1986).

Studies on nutrition of plasmid-negative and -positive clones of *B. globosum* (now named "*B. pseudolongum* subsp. *globosum*") have revealed the auxotrophy for L-methionine for both clones (Mattarelli and Biavati, 1999a). The auxotrophy for L-leucine, shown only for plasmid-positive clones, pointed out the involvement of plasmids in this requirement. The lack of β-isopropylmalate dehydrogenase enzyme activity is considered responsible for the L-leucine auxotrophy in the plasmid-positive clones (Mattarelli and Biavati, 1999b). Minimal amino acid requirements were investigated. Only six or seven amino acids were required for growth, regardless of the ones chosen, including L-leucine and/or L-methionine for plasmid-positive and -negative clones, respectively.

A survey of vitamin and growth factor requirements was made in 10 species of bifidobacteria. The different species showed a great deal of heterogeneity (Trovatelli and Biavati, 1978). Comparative studies on the synthesis of water-soluble vitamins (thiamine, folic acid, nicotinic acid, pyridoxine, cyanocobalamin and riboflavin) among 24 strains of five species isolated from human feces have shown that many strains can synthesize all of the vitamins tested, with the exception of riboflavin. The concentration of the vitamins accumulated varied widely among different species or strains (Deguchi et al., 1985). Noda et al. (1994) have also demonstrated the ability of bifidobacteria to synthetize biotin.

Some factors called "bifidus factor" (BF) able to stimulate the growth of bifidobacteria have been studied in *B. bifidum*. The BF discovered by Petuely and Kristen (1953) is nothing but lactulose that, if added to milk, produces an increase in intestinal bifidobacteria. Actually, we cannot speak of a specific BF because Terada et al. (1992) found that also the number of *Lactobacillus* and *Streptococcus* increases. It should be noted that this molecule is active only in vivo.

The BF1 factor is a complex of oligosaccharides present in the milk of women and also, in minor portions, in the milk of several other mammals (Gyorgy et al., 1954a; Gyorgy et al., 1954b).

These oligosaccharides have been classified into two groups, neutral and acidic. The "bifidogenic" activity is proportional to the number of *N*-acetylglucosamine residues contained in the oligosaccharide (Von Nicolai and Zilliken, 1972). The BF2 factor is a proteolysate specifically obtained from bovine casein hydrolysis by trypsin, pancreatic or papain (Gyorgy and Rose, 1955; Rose and Gyorgy, 1963). A total acid hydrolysis totally inactivates casein hydrolysate (Seka Assy, 1982). Proteins and oligosaccharides also are classified as "bifidogenic factor." In the former case, the glycoproteins contained in human colostrum (Hirano et al., 1968) and in bovine colostrum (Nichols et al., 1974) help the growth of bifidobacteria. Even lactoferrin, an iron-binding protein, promotes the growth of bifidobacteria. The study by Petschow et al. (1999) suggests that bovine lactoferrin from mature milk increases the growth of *B. infantis* and *B. breve* in a dose-dependent way in vitro, whereas human lactoferrin promotes the growth of *B. bifidum* but not that of *B. infantis* and *B. breve*. Among the numerous oligosaccharides, soybean oligosaccharides, which consist mainly of the trisaccharide raffinose and the tetrasaccharide stachyose (Benno et al., 1987; Hayakawa et al., 1990; Yazawa et al., 1978), stimulate the growth of bifidobacteria. More recently, other oligosaccharides have been identified and termed "bifidogenic factors," such as fructo-oligosaccharides (FOS; Gibson and Wang, 1994; Roberfroid et al., 1998b), synthetic oligosaccharides, and transgalactosylated oligosaccharides (TOSs; Bouhnik et al., 1997; Ito et al., 1993). Gluco-oligosaccharides (GOSs) also stimulate the growth of bifidobacteria although less intensely than FOSs and TOSs do (Djouzi and Andrieux, 1997; Jaskari et al., 1998). The polysaccharide inulin also increases fecal bifidobacteria, having a similar prebiotic effect of oligofructose (Gibson and Roberfroid, 1995). In conclusion, the "bifidogenic factors" can be classified in two large groups: oligosaccharides and protein hydrolysates. In recent years, the need of bifidus factors for growth has been questioned because it is possible to obtain bifidobacteria multiplication in the absence of these factors. Therefore, they cannot be defined as growth factors.

The metalloelement requirements in *B. bifidum* have been extensively studied by Bezkorovainy and colleagues (Bezkorovainy and Topouzian, 1983; Bezkorovainy et al., 1986; Bezkorovainy et al., 1987; Topouzian et al., 1984). They examined the aspects of iron metabolism and the mechanisms of iron uptake that do not depend on siderophore-type carriers but on a proton-motive-force-associated electrogenic pump or pumps. There is hope that these investigations, which are important for bifidobacterial ecology, will be extended to other species of the genus.

UREASE ACTIVITY In a study on the urease activity of more than 400 strains representing 21 species of bifidobacteria, the strongest ureolytic activity was found for most *B. suis* strains and only a few *B. breve*, *B. magnum* and *B. subtile* strains. High levels of urease activity were present in cells grown in the absence of urea, suggesting that this enzyme is not inducible (Crociani and Matteuzzi, 1982).

Hatanaka et al. (1987) have shown that glutamate dehydrogenase and glutamine synthetase, only found in *B. adolescentis*, *B. bifidum*, *B. breve*, *B. infantis*, *B. longum*, *B. pseudocatenulatum* and *B. thermophilum*, may be involved in ammonia intake.

NITRATE REDUCTION *Bifidobacteria* usually do not reduce nitrates. However, cells grown in the presence of lysed cells may be capable of nitrate reduction. Cytochromes *b* and *d* are synthesized under these growth conditions (Van der Wiel-Korstanje and De Vries, 1973).

RESISTANCE TO ANTIBIOTICS AND OTHER BACTERICIDAL SUBSTANCES The antimicrobial susceptibility of 459 strains of bifidobacteria representing 15 species was determined by the broth dilution method. Penicillin G, erythromycin, clindamycin, vancomycin, bacitracin, chloramphenicol and lincomycin were the most active compounds. Resistance to polymyxin B, nalidixic acid, kanamycin, gentamicin, and metronidazole was a common feature among the fifteen species studied. Neomycin, streptomycin and tetracycline had the most variable activity (Matteuzzi et al., 1983). In a study by Charteris et al. (1998) on 15 human bifidobacterial strains and one strain of *B. animalis*, the same results were obtained except for vancomycin susceptibility. In this study, vancomycin resistance was observed as a general characteristic, opposite to what Matteuzzi et al. (1983) had found. This difference may be due to divergent assay methodologies inasmuch as vancomycin is reported to diffuse poorly in agar media (Thomson et al., 1995). The *b*-lactamase activity of four microorganism-drug combinations was evaluated. Its absence suggested that cell-wall impermeability was responsible for cephalosporin resistance among bifidobacteria (Charteris et al., 1998).

Mangin et al. (1999), using rRNA gene restriction analysis, carried out an in vivo study on the influence of antibiotics on bifidobacterial microflora. The predominant bifidobacterial strains of intestinal flora in four human subjects were monitored before and after eight days of treatment with oral amoxicillin-clavulanic acid (Augmentin). These antimicrobial agents are known to strongly inhibit bifidobacteria (Dubreuil et al., 1996). The combination of amoxicillin and clavulanic acid did not involve significant long-term changes in predominant bifidobacterial populations of intestinal flora, as was the case when rifampicin and streptomycin were administered (Mangin et al., 1999). Members of the normal bifidobacterial flora cannot be eradicated by these antimicrobial treatments. The antimicrobial effect is probably dose-related and removal of the antibiotic allows prompt regrowth of their population.

Encouraging results also have been obtained through the use of essential oils as antimicrobial agents against some species of bifidobacteria (Biavati et al., 1997; Crociani et al., 1997).

BACTERIOCIN PRODUCTION Bacteriocins have the potential to inhibit growth of pathogens and to prevent microbial food spoilage. The bactericidal action of bifidobacteria is ascribed to the production of some kind of antimicrobial agent in addition to their pH-reducing effect (Fujiwara et al., 1997; Gibson and Wang, 1994; Meghrous et al., 1990). Bifidocin B, for example, is a bacteriocin produced by *B. bifidum* NCTB 1454; it has a broad antimicrobial spectrum, inhibiting the growth of species of *Listeria*, *Enterococcus*, *Bacillus*, *Lactobacillus*, *Leuconostoc* and *Pediococcus* (Yildirim and Johnson, 1999). The finding could be very important in the exploitation of bifidobacteria as biopreservatives, i.e., preventing growth of pathogens or contaminating bacteria present in food.

Genetics

The genetic analysis and manipulation of bifidobacteria are very important to better understand their phylogeny and probiotic roles. However, the genetic work on bifidobacteria is quite new. Initially, the work by Sgorbati et al. (1982b) reported the presence of plasmids in members of the genus *Bifidobacterium*. Plasmids were found in *B. longum*, the predominant species in the human intestine, in *B. pseudolongum* subsp. *globosum*, the most common *Bifidobacterium* species in animals, and in *B. asteroides* and *B. indicum*, species found exclusively in the Western and Asiatic honeybee hindguts, respectively. *Bifidobacterium infantis*, the species most closely related to *B. longum*, did not carry plasmids, although strains of both species, generally isolated from the same specimens, were studied; the same was true for the 45 strains studied as *B. infantis-B. longum* "intermediates" found in calf feces. Other "human" bifidobacteria species, i.e., *B. dentium*, *B. breve*, *B. bifidum*, *B. pseudocatenulatum* and *B. catenulatum*, were found to be plas-

mid-free. In *B. breve*, however, the presence of plasmids with 5 different profiles has been shown by Iwata and Morishita (1989). Strains of *B. longum* may harbor defective lysogenic phages, but no apparent correlation exists between the presence of plasmids and the production of phages induced by UV or mitomycin C (Sgorbati et al., 1982b; Sgorbati et al., 1983). Preliminary data on structural relatedness among plasmids were obtained by means of blot hybridization using many selected, unrestricted plasmid probes. Seven different plasmids were found in *B. longum* (123 strains), 13 in *B. asteroides* (70 strains), three in *B. pseudolongum* subsp. *globosum* (28 strains), and three in *B. indicum* (73 strains). The frequency and distribution of such plasmids among samples and geographical areas (*B. asteroides* and *B. indicum*) were reported (Sgorbati et al., 1986a; Sgorbati et al., 1986b). In *B. pseudolongum* subsp. *globosum*, auxotrophy for L-leucine has been detected in plasmid-positive but not -negative clones. The conclusion of this study is that the presence of the plasmid confers the negative feature of auxotrophy for L-leucine.

From human intestinal bifidobacteria, only *B. longum* and *B. breve* have been shown to possess plasmids (Sgorbati et al., 1982b; Iwata and Morishita, 1989). All the plasmids so far investigated are cryptic. Few plasmids (pVS809 [Mattarelli et al., 1994], pMB1 [Rossi et al., 1996], pCIBb1 [O'Riordan and Fitzgerald, 1999] and pKJ50 [Park et al., 1999b]) have been characterized. Since these first observations, other studies have been carried out. Vectors capable of transfecting many bifidobacteria and with distinct replication origin and markers for *Escherichia coli* have been described (Missich et al., 1994; Park et al., 1997; Rossi et al., 1996; Rossi et al., 1998) as being derived from the small cryptic plasmid pMB1 or others (Matsumura et al., 1997). Initially, the transformation efficiency of many bifidobacteria with these plasmids was very low; recent developments (Argnani et al., 1996) have considerably increased transformation efficiency for all strains tested so far.

Numerous sequences of bifidobacteria belonging to a large part of the described species are deposited at international data banks (European Molecular Biology Laboratory [EMBL] and GenBank).

Ecology

BIFIDOBACTERIA IN HUMANS GASTROINTESTINAL TRACT. The microbial succession in the human and animal alimentary tract in the first weeks of life is similar even if newborn animals are exposed to a greater number of microorganisms compared to human newborns. In the first few days of life, microbiota are characterized by coliforms and streptococci. Soon clostridia, *Bacteroides* and lactobacilli appear. Microbial colonization proceeds via a constant succession of species and the occurrence of substantial changes in the appearance of the microbiota after weaning. Biota succession was studied in several animals (Moughan et al., 1992; Smith, 1965). In most of these studies, fecal samples were examined. Rarely were samples taken from several areas of the gastrointestinal tract. Autochthonous microorganisms were almost never distinguished from allochthonous ones, even though this distinction is very important to avoid being led astray in the interpretation of the presence of the several microbial types. The study of microbial successions is difficult because of the extreme variability contributed by great variation in rearing and feeding practices and by the limitations in the sampling and analysis of the intestinal microbial communities. Microbial colonization is regulated by: 1) external factors such as microbial change of the surrounding environment, food and eating habits, composition of maternal microbiota, diet and heat stress; 2) host-related factors such as endogenous nutrients, the physiology of the host and microbiota; and 3) factors inherent to microorganisms such as adhesive features and resistance to an acid environment (e.g., the stomach). The mechanism by which the intestine selects a limited number of species from the vast array of microbes with which it comes in contact is still unknown. The ability to degrade complex carbohydrates generated by the host epithelium, such as mucin, is restricted to a few species of the intestinal microflora. Therefore, bacteria such as bifidobacteria and *Ruminoccus*, which are able to hydrolyze oligosaccharide chains, are positively favored in the colonization.

The succession of microbial communities in the human gastrointestinal tract is amongst the most widely studied. Within the first 24 hours,

E. coli and *Streptococci*, which reach a level of 10^8–10^{10} per g of feces, colonize the gastrointestinal tract of all newborns. These bacteria create a reduced environment favorable to the settlement of anaerobic species such as *Bacteroides*, *Bifidobacterium*, *Clostridium* and *Lactobacillus*. In breast-fed newborns, the number of *E. coli*, streptococci, *Bacteroides* and *Clostridium* decreases, whereas the group of bifidobacteria becomes dominant. The distribution of the various microbial groups in the intestinal microbiota of breast-fed or bottle-fed infants is shown in Fig. 13. Once dietary supplementation begins, the microbiota of breast-fed infants become similar to those of bottle-fed infants in which bifidobacteria are not dominant. After solid food introduction and weaning, breast-fed and formula-fed infants have similar fecal microbiota, which resemble those of an adult by about the second year of life (Conway, 1997). The distribution of the major microbial groups in adult microbiota is shown in Fig. 14. Generally, fecal microbiota

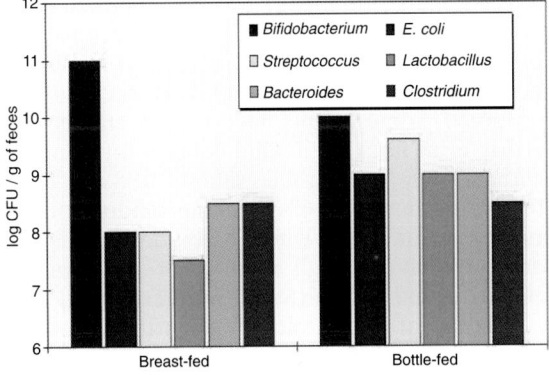

Fig. 13. Comparison of fecal flora in breast-fed and bottle-fed infants. Data from Mitsuoka and Kaneuchi (1977); Kleessen et al. (1995); Harmsen et al. (2000).

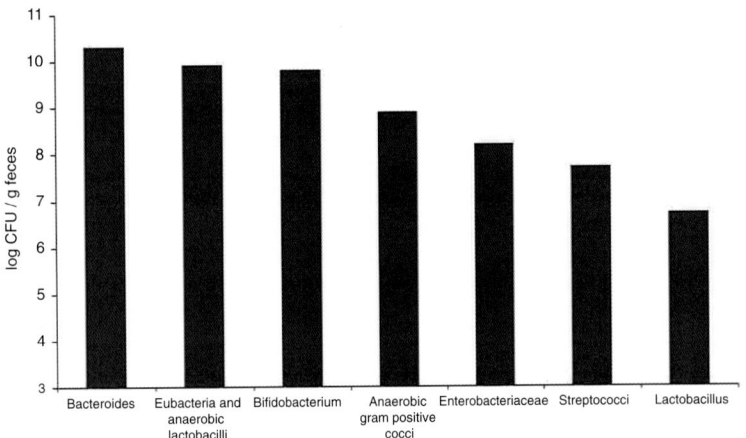

Fig. 14. Distribution of the major microbial groups in the fecal flora of human adults. Data from Mitsuoka and Kaneuchi, 1977; Finegold et al., 1983; Mata et al., 1969.

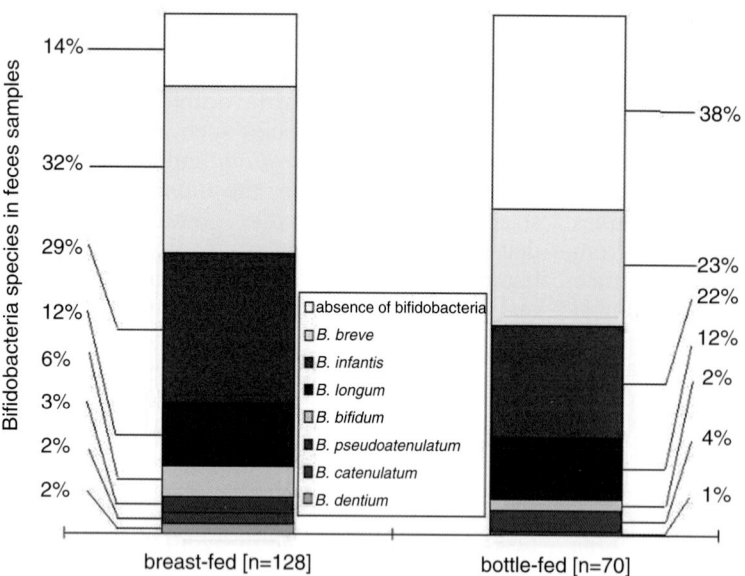

Fig. 15. Distribution of bifidobacterial species in the feces of breast-fed and bottle-fed infants. Data from Biavati et al. (1986).

fluctuations are greater in children with larger day-to-day variations in diet than in adults. Relatively small amounts of formula supplementation of breast-fed infants shift the pattern from a breast-fed to a formula-fed one. A great variation in the distribution of bifidobacteria has been observed. Some infants have no detectable bifidobacterial flora until solid food is introduced, with more formula-fed infants (38%) than breast-fed infants (16%) lacking bifidobacteria. In general, infants in whom no bifidobacteria are detected have a high number of *Bacteroides*, clostridia and *E. coli* (Stark and Lee, 1982).

Succession of the microbiota in the intestinal tract at the group level as determined by selective enumeration gives no indication about the distribution of species and strains.

Concerning species distribution, previous studies showed that *B. breve* and *B. infantis* are species typical of breast-fed or formula-fed infants whereas *B. adolescentis* was typical of children, adults and elderly people (Benno et al., 1984; Biavati et al., 1984; Biavati et al., 1986; Mitsuoka, 1984). *Bifdobacterum breve* was the dominant species in breast-fed infants, whereas *B. longum* and *B. adolescentis* were more frequent in formula-fed infants (Mevissen-Verhage et al., 1987; Fig. 15). *Bifidobacterum pseudocatenulatum*, *B. longum* and *B. adolescentis* are dominant in adults (Biavati et al., 1986; Fig. 16). This is consistent with the concept that supplementation of infant formula initiates the development of a microbiota more closely resembling the adult profile (Conway, 1997). In agreement with previous data (Biavati et al., 1986), a recent study of bifidobacterial species distribution by Matsuki et al. (1999) using

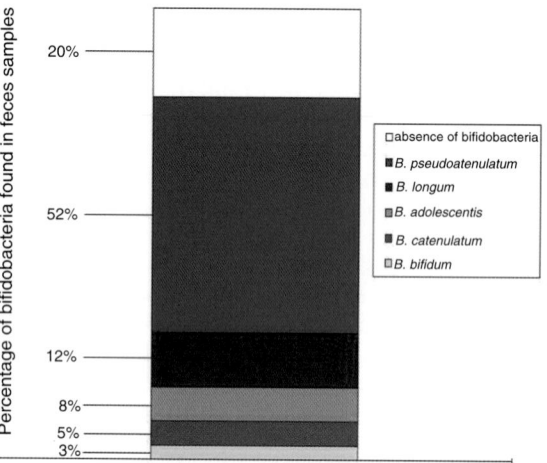

Fig. 16. Distribution of bifidobacterial species in the feces of human adults. Data from Biavati et al. (1984).

species-specific PCR primers showed that the *B. catenulatum* was the most common of the adult intestinal bifidobacterial flora (detected in 44 of 48 samples [92%]), followed by *B. longum* and *B. adolescentis*, and that *B. breve*, *B. infantis* and *B. longum* were frequently found in the intestinal tracts of infants.

Regarding the presence of simple or complex bifidobacterial microflora at the species level, the study of Biavati et al. (1986) reveals that 5 of 11 subjects examined had one species of bifidobacteria, whereas the remaining 6 subjects had two, three or four species.

The fine-tuning of molecular techniques (molecular probes, rybotyping, PFGE, etc.) gave great impulse to the study of microbial ecology, allowing monitoring of single strains of intestinal microflora.

In a recent study of fecal samples collected from 10 human subjects, analysis by pulsed-field gel electrophoresis and ribotyping revealed that the collection of bifidobacterial strains in feces of each subject was unique. Only one strain was detected in two different subjects. A "simple" bifidobacterial microflora, arbitrarily defined as a collection of four or fewer bifidobacterial strains, was found in five of the ten subjects studied, whereas the other subjects had a "complex" (more than four bifidobacterial strains) microflora. Complexity or simplicity of the bifidobacterial flora was not related to gender. In this study, the strains were characterized by pulsed-field gel electrophoresis and ribotyping but not classified (Kimura et al., 1997). Studies based on Southern analysis using ribosomal 23S DNA probes confirmed that bifidobacterial microflora is typical of each individual, with the possible coexistence of several species in the same flora (as already observed in previous studies) and even several distinct strains of the same species (up to three different *B. longum* or *B. adolescentis* for one subject; Mangin et al., 1999). Although the fecal microflora of adult humans shows considerable stability in composition at the level of bacterial population both over time and in response to variation of diet, dominant species and strains may be subjected to variations (Stark and Lee, 1982). Studies regarding the fluctuations within populations of bifidobacteria as well as other bacterial groups are scarce. A monitoring of the bifidobacterial fecal population of two humans over 12 months showed that one subject had a remarkably stable microfloral population with one strain predominating, whereas the other had a more dynamic microflo-ral population with varying strain composition (McCartney et al., 1996). The proportion of the human population harboring a stable simple bifidobacterial flora compared to that harboring a complex, dynamic microflora at the strain level is not known. It is possible that subjects with a complex and dynamic microflora are more easily colonized by exogenous bifidobacteria; however, studies in this field are limited and a complete knowledge of these complex phenomena is lacking. Such knowledge might determine a more rational use of probiotics and prebiotics. The studies carried out so far on the introduction of exogenous bifidobacteria (probiotics) through diet confirm the stability of intestinal microbiota; in fact they remain only during the period of ingestion to gradually decrease until they disappear a few days after treatment. Even with antibiotic treatment, such as the combination of amoxicillin and clavulanic acid, which substantially decreases the bifidobacterial microflora, the intestinal microflora regrows to normal levels with the same members about two months after the antibiotic treatment ends (Mangin, 1999). In general, we may conclude that each individual, apart from the dramatic biological succession that occurs in the digestive tract following birth and terminating before childhood, has a microflora stable in time as far as the anaerobic genera are concerned. Microflora in some humans may be subject to fluctuations as far as species or most important strains are concerned. Even in advanced age, a substantial deviation (Fig. 17) from the adult microfloral composition may occur, probably as a result of physiological and alimentary changes of the host. Studies on microbial ecology are therefore still in their initial stages and several questions are awaiting a clearer answer. A more in-depth knowledge of microbial ecology of the digestive

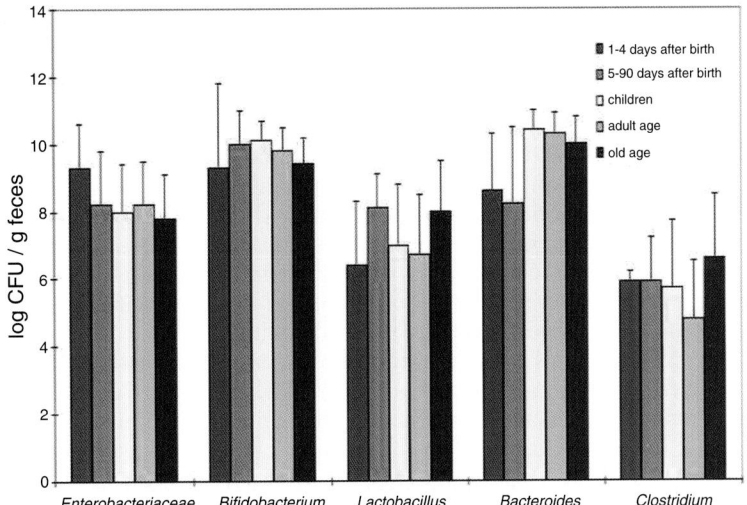

Fig. 17. Changes in distribution of some groups of the intestinal microbiota in humans from birth to old age. Data from Mitsuoka and Kaneuchi (1977); Finegold et al. (1983); Kleessen et al. (1995); Harmsen et al. (1999).

tract will improve the selection of probiotics and prebiotics.

ORAL CAVITY. More than 300 species have been isolated in the oral cavity, although in healthy subjects this number is 50–100. Bifidobacteria are very early colonizers of the gastrointestinal tract and are regularly present in the oral cavity of children. Later they can become normal inhabitants of the adult dental plaque. However, these microorganisms have been isolated only a few times in the oral cavity, probably due to the use of inappropriate methods of isolation. In one study, they were isolated in the saliva of only one of 10 subjects and none were found to host bifidobacteria in their gingival spaces (Sanyal and Russell, 1978). Other researchers were able to show that 7 out of 62 isolates in gingival spaces were bifidobacteria. They fermented esculin and were identified as *B. adolescentis* (Maeda, 1980). Bifidobacteria were found also in the dental plaque, accounting for 6% of all the isolates (Kolenbrander and Williams, 1981). In his work on oral microflora, Sutter has stressed the difficulty in having careful bacterial profiles from the different parts of the oral cavity (Sutter, 1984). In this habitat the presence of bifidobacteria is understimated owing to the use of inappropriate media and methods of identification. Scardovi and Crociani (1974a) and Crociani et al. (1996b) specifically looked at oral bifidobacterial microflora and found their presence was almost constant.

The species *B. dentium* was isolated from dental caries in 1974 (Scardovi and Crociani, 1974a). In 1996, a new study about dental caries led to the isolation and identification of two new species from this habitat: *B. denticolens* and *B. inopinatum* (Crociani et al., 1996b). *Bifidobacterium inopinatum* displays a very unusual morphology and a very low G+C mol% value compared to the other species of bifidobacteria. The samples studied highlighted the presence of one or more out of the three species.

VAGINA. Few studies are present in the literature dealing with the presence of bifidobacteria in vagina. From available data, bifidobacteria are present in 22–26% of healthy women (Werner and Seeliger, 1963; Crociani et al., 1973), but no epidemiological data are available on vaginal bifidobacteria. *Bifidobacterium breve* and *B. adolescentis* have been the most frequently isolated species, whereas *B. longum* and *B. bifidum* were present to a smaller extent (Werner and Seeliger, 1963; Crociani et al., 1973; Korschunov et al., 1999). In this habitat, bifidobacteria could play a role in maintaining vaginal homeostasis by producing organic acids and bacteriocins antagonistic towards pathogens.

Korschunov et al. (1999) proposed the use of bifidobacterial strains as probiotics in the correc-

tion of the microflora of the urogenital tract in females.

HYPOCHLORHYDRIC STOMACH. Interesting data have emerged from studies of the hypochlorhydric stomach by autoimmune atrophic gastritis or by omeprazole treatment; *B. denticolens*, *B. inopinatum*, *B. dentium* (the same species found in the oral cavity), but also "*B. infantis-longum*," colonize this habitat, pointing out that the hypochlorhydric stomach represents a new ecological niche different from all the others of the body (Brandi et al., 1996).

BIFIDOBACTERIA IN ANIMALS Studies of intestinal microflora were performed mainly on domestic animals and revealed a complex microflora: *Bacteroides*, eubacteria, anaerobic lactobacilli, anaerobic Gram-positive cocci, spirillaceae and often bifidobacteria. Bifidobacteria are present in almost all chickens, dogs, pigs, rats and hamsters even if less abundantly than lactobacilli, whereas in monkeys and guinea pigs, the amount of bifidobacteria is greater than the amount of lactobacilli. Mice, rabbits and horses rarely display bifidobacteria, and cats and minks never have them (Mitsuoka and Kaneuchi, 1977). The composition of bifidobacteria microflora in animals varies with the age, species and diet of the host.

Some species apparently are host-specific: *B. magnum* and *B. cuniculi* have only been found in rabbit fecal samples, *B. pullorum* (Trovatelli et al., 1974) and *B. gallinarum* (Watabe et al., 1983) only in the intestine of chickens and *B. suis* only in piglet feces (Matteuzzi et al., 1971; Fig. 7).

Feces of suckling calves and breast-fed infants harbor the same bifidobacterial species (the only case where species typical of humans also are found in animals), except for a group of 180 strains isolated from the feces of suckling calves that could not be identified as members of a species because of their high degree (over 80%) of DNA homology with both *B. infantis* and *B. longum*. These strains are referred to as "intermediate" between the two species (Scardovi et al., 1979a) pending a definition of their taxonomic status.

Bifidobacterum asteroides is the only species found in *Apis mellifera* intestine, irrespective of the geographical area of origin (Scardovi and Trovatelli, 1969a), whereas *A. cerana* and *A. dorsata* (from the Philippines and Malaysia) harbor specific biovars of the species *B. indicum* that have different transaldolase isozymes but are indistinguishable by DNA-DNA hybridization (Scardovi et al., 1979a). Such subtle dependence on host phylogenetic position was heretofore unsuspected. Furthermore, *B. asteroides* and *B. indicum* in the loci studied so far, using the zymogram technique, are much more variable than any other species of *Bifidobacterium* (Scardovi et al., 1979a). The exogenous and/or endogenous factors controlling this variability should be further studied.

Among the large biodiversity of animals, only a few species have been considered for studies of the intestinal microflora. The isolation of bifidobacteria from honeybees, whose significance and origin in the gut are at present unknown, could form an interesting chapter of insect microbiology if they could be found elsewhere in this class of animals. In conclusion, the study of bifidobacteria ecology is at its starting point, and interesting aspects of probiotics use in animals could stimulate its investigation.

BIFIDOBACTERIA IN *EXTRA*-BODY ENVIRONMENTS
Bifidobacteria are found in *extra*-body environments only as a consequence of fecal contamination. They can be considered to be useful indicators of human and animal fecal pollution when they are correctly identified (Mara and Oragui, 1983). A study of 75 samples from a highly contaminated stream near Bologna (Italy) reveals that *B. longum*, *B. adolescentis*, *B. pseudocatenulatum*, *B. catenulatum* and *B. thermophilum* are the most representative species, whereas *B. globosum*, *B. angulatum*, *B. breve*, *B. animalis*, *B. choerinum*, *B. subtile* and *B. minimum* occur only in low numbers (Crociani et al., 1996a). Field data from Rhodes and Kator (1999) indicate a better likelihood of bifidobacteria recovery from cool (10°C) than from warm water (23–30°C).

Bifidobacteria can be considered also as indicators of fecal contamination in meat and meat products. The conventional indicator, *E. coli*, does not provide any information on the origin of fecal contamination; this can be obtained by identifying the species of bifidobacteria that are present. The study of Gavini and Beerens (1999) showed the presence of bifidobacteria of animal origin in meat and meat products, except for 4% of strains of human origin, is due to artificial contamination of meat by manual handling.

Disease

Bifidobacteria have been occasionally isolated from infective material (Biavati et al., 1982), and although implicated in some opportunistic infections (Miller-Catchpole, 1989), they are generally regarded as nonpathogenic. Predisposing factors such as ongoing infections (Prévot et al., 1967) or contact with contaminated material (Ha et al., 1999) may favor finding bifidobacteria in association with infections. Some strains of bifidobacteria also have been reported to participate in soft-tissue infections initiated by mixed populations. These include some strains of bifi-

dobacteria associated with peritonitis, dental pyorrhea and various abscesses (Prévot et al., 1967; Biavati et al., 1982). Only species isolated from the dental caries and plaques, *B. dentium*, *B. denticolens* and *B. inopinatum*, appear to have the most pathogenic potential in cariogenic processes. The role played by bifidobacteria in this pathology is still unclear. *Bifidobacterium* species of the oral cavity are in fact capable of degrading dextran, an important component of plaque, which plays a crucial role in the colonization of a potential cariogenic organism such as *Streptococcus mutans*. Therefore in this light, bifidobacteria may be antagonists of the cariogenic process, impairing plaque formation. On the other hand, bifidobacteria produce acids, which may damage the enamel and cause cariogenic phenomena.

Applications

PROBIOTIC The "probiotic" concept, referring to consumption of live microbes, mainly lactic acid bacteria and bifidobacteria, to improve the health and well-being of the host, dates back to Metchnikoff (1907). The term was used for the first time by Parker (1974) to describe "every substance or microorganism that contributes to gastrointestinal balance." Over time, the meaning of the term was slightly changed. In fact Fuller (1989) describes it as "a live microbial feed supplement which beneficially affects the host animal by improving its microbial balance." This definition stresses the importance of viability and avoids the use of the too broad term, "substances," which could even include antibiotics. Such a term has been used by Haavenar and Huis in't Veld (1992) to describe a "mono- or mixed culture of live microorganisms which, when applied to animal or man, beneficially affect the host by improving the properties of the indigenous gastrointestinal microflora." In recent years, there has been a tendency to widen the concept of probiotics to apply not just to the gastrointestinal tract but to other mucous surfaces such as the urogenital tract and the upper respiratory tract and to use such a term as a synonym for "biotherapy," which means a therapy of microorganisms with antagonistic activity towards pathogens (Elmer et al., 1996).

The composition of the intestinal microflora is important for the well-being of the host although its precise functions are very difficult to study. Probiotic microorganisms, mainly lactobacilli and bifidobacteria, which influence the composition of bacterial microflora, may play an important role in the prophylaxis and therapy of gastrointestinal diseases. The main effects of probiotics on the host with particular reference to bifidobacteria (Fig. 18) are as follows.

- **Regulation effect of the gut microflora**

- **Production of antimicrobial substances**

- **Alleviation of lactose intolerance**

- **Enhancement of immune system**

- **Anticarcinogenic activity**

- **Production of vitamins**

Fig. 18. Benefits contributed by bifidobacteria to their host.

ALLEVIATION OF LACTOSE INTOLER-
ANCE. Bifidobacteria may alleviate lactose
intolerance (Rambaud, 1990), both directly by
providing β-galactosidase and indirectly by lowering the pH of the environment, thereby contributing to growth of the intestinal microflora
that produce β-galactosidase (Jiang et al., 1996;
Jiang and Savaiano, 1997; Vesa et al., 1996).
The decrease of bifidobacteria appeared to be
at least in part responsible for the lower β-
galactosidase activity in patients with active or
quiescent Crohn's disease. An alternative therapy for Crohn's disease is to increase the bifidobacterial counts in the human colon (Favier
et al., 1997).

PREVENTION OF DIARRHEA. In the
human large intestine, the microflora coexist in
constant balance (Macfarland and Cumming,
1991). Occasionally homeostasis may be upset by
an invasion of microorganisms which are not
part of the host flora. This alteration may induce
diarrhea (a serious problem in the world), which
is often a manifestation of acute inflammation.
The microorganisms present in the microflora
protect the host from pathogens thanks to the
formation of a front-line mucosal defense. In
fact autochthonous intestinal bacteria prevent
intestinal colonization by pathogenic bacteria
by directly competing with them for essential
nutrients or for epithelial attachment sites.
Moreover, thanks to the production of antimicrobial compounds and volatile fatty acids, bifidobacteria create unfavorable environmental
conditions for the growth of many enteric pathogens. Gibson and Wang (1994) have shown that
the antagonistic effect of a low pH environment
on the growth of E. coli and Clostridium perfringens may not be the sole mechanism of inhibition. Bifidobacteria in fact were able to excrete
some inhibitory factors having bactericidal or
bacteriostatic action towards a range of both
Gram-positive and Gram-negative potentially
pathogenic bacteria. These inhibitory factors,
also called "bacteriocins," are being studied
(Yildirim and Johnson, 1999).

Recent studies have shown that some lactobacilli and bifidobacteria preparations may have
potential in controlling diarrhea and intestinal
side effects of radiation (Korschunov et al.,
1996). Bifidobacterium longum decreased the
level of fecal clostridial species and reduced the
frequency of erythromycin-associated diarrhea
(Colombel et al., 1987). Orrhage et al. (1994a)
have shown that supplementation with a combination of B. longum and Lactobacillus acidophilus reduced the ecological changes in the
intestinal microbiota induced by administration
of clindamycin. Some strains of Bifidobacterium
spp. have been shown to inhibit the multiplication of Clostridium difficile in vivo (Rolfe et al.,
1981). The feeding of a formula supplemented
with B. bifidum and Streptococcus thermophilus
substantially reduces the incidence of acute diarrheal disease in infants (Hoyos, 1999) and the
rate of rotaviral shedding (Saavendra et al.,
1994). Moreover, Yasui et al. (1999) described
one strain of B. breve that activates the humoral
immune system and can protect against rotavirus
infection and influenza infection in infants. These
findings are in agreement with animal-model
studies. Mice infected with rotavirus have shown
a passive protection against the virus when fed
with B. breve (Yasui et al., 1995) or B. bifidum
(Duffy et al., 1993). Bifidobacterium animalis
also provided protection of immunodeficient
mice against mucosal and systemic candidias
(Wagner et al., 1997); it also plays a protective
action against damage to intestinal villi induced
by low zinc diets (Mengheri et al., 1999). Yasui
et al. (1999) suggest the use of B. breve YIT4064
as a live antigens-delivery-vehicle for oral vaccines to various pathogens.

ENHANCEMENT OF IMMUNE SYSTEM.
Bifidobacteria are able to enhance several
immune functions: activation of lymphocytes
and macrophages (Hatcher and Lambrecht,
1993; Sekine et al., 1994) and stimulation of
antibody production as well as the mitogenic
response in Peyer's patches (Kado-oka et al.,
1991; Lee et al., 1993; Takahashi et al., 1993;
Yasui and Ohwaki, 1991). This stimulation of
the immune response by bifidobacteria seems to
be mediated by the ability of bifidobacteria to
induce the production of cytokines (Marin et al.,
1996). Cytokines are protein mediators involved
in all aspects of immunoregulation. Bifidobacteria are able to stimulate macrophages and
helper-T cells to produce tumor necrosis factor-
α and interleukins, respectively, when tested in
vitro (Marin et al., 1996; Park et al., 1999a). The
ability to induce cytokine production seems to be
strain-dependent (Park et al., 1999a). Bifidobacteria also can induce production of large quantities of IgA in vitro using a Peyer's patch cell
culture method (Yasui et al., 1992). If the IgA

produced in vitro is also important in vivo, this could justify the use of bifidobacteria as a probiotic therapy able to build and stimulate an immunological barrier to foreign material, particularly pathogenic microorganisms, allergenic food proteins and carcinogens (Majamaa and Isolauri, 1997).

CANCER PREVENTION. Colorectal cancer is one of the leading causes of cancer morbidity and mortality in Western countries. Bifidobacteria have been shown to possess antimutagenic and anticarcinogenic properties. While there is no direct evidence for cancer suppression in humans, a wealth of indirect evidence is present in the literature, based on laboratory studies (Ishibashi and Shimamura, 1993; Kulkarni and Reddy, 1994; Mitsuoka, 1990; Rao et al., 1986; Singh et al., 1997).

In animal models of colon carcinogenesis, administration of *B. longum* as a lyophilized preparation was able to reduce azoxymethane-induced colon cancer as measured by the reduction in the number of aberrant crypt foci (Challa et al., 1997; Kulkarni and Reddy, 1994; Reddy, 1999; Singh et al., 1997) and tumor markers such as ornithine decarboxylase and in the expression of ras-p21 oncoprotein.

The precise mechanism by which bifidobacteria exert their antitumorigenic activity is not clear. A number of studies (Okawa et al., 1993; Orrhage et al., 1994b; Sekine et al., 1995; Zhang and Ohta, 1991; Zhang and Ohta, 1993) pointed out that the antimutagenic activity could reside in the cell wall.

The mechanism leading to tumorigenic suppression could comprise one or more of the following proposed events (Hirayama and Rafter, 1999): 1) enhancement of host immune responses; 2) direct binding and degradation of potential carcinogens; 3) alteration of intestinal microflora which produces putative carcinogens; 4) alteration of the physicochemical environment in the colon; and 5) production of antitumorigenic substances.

FEATURES OF BIFIDOBACTERIA FOR HUMAN CONSUMPTION. There is general agreement on the characteristics that probiotic microorganisms should possess to be useful for the host (Fig. 19):

1) They should survive in sufficient numbers to colonize or to persist in, albeit for short periods, the gastrointestinal tract and withstand the acidic gastric environment and bile acids as well as possess adhesion mechanisms to the intestinal mucosa. Regarding this issue, studies have been conducted on survival in an acidic environment (Biavati et al., 1992c) and bacterial adhesion (Bernet et al., 1993; Crociani et al., 1995; Dunne et al., 1999), which are important both to learn

- Microbial cells should be viable
- The strains used should belong to a species present in the specific habitat
- The strains used should colonize the specific habitat and have the ability to positively influence the host
- They should exhibit resistance to technological processes (i.e. lyophilization, etc.)
- Microbial cells should be administered together with substrates that promote their growth
- They should be nonpathogenic and nontoxic
- At least $15\text{-}20 \times 10^9$ microorganisms/day should be administered *

*This value represents approximately the content of a commercial yogurt preparation (125 ml)

Fig. 19. Important characteristics of "probiotic preparations" of bifidobacteria.

about the mechanisms that regulate such processes and to identify potentially probiotic strains.

2) They should be of human or animal origin depending on the intended use.

3) They should have the ability to positively influence the host and to colonize the specific habitat. Clinical trials need to be conducted to assess probiotic effects in normal humans and animals towards intestinal disorders and pathogenic conditions.

4) They should exhibit resistance to technological processes such as lyophilization and ease of obtaining a large microbial mass.

5) Microbial cells should be administered together with substrates that promote their growth.

6) They should be nonpathogenic and nontoxic.

7) At least $15\text{--}20 \times 10^9$ microorganisms/day should be administered (Zoppi et al., 1982).

Studies on potentially probiotic bifidobacteria carried both in vitro and in vivo highlight guidelines both for features of the strains and for the interactions with the host bacteria, which should lead to the best strains, thus creating a collection of strains adequate for probiotic use (Figs. 20 and 21).

BIFIDOBACTERIA AND THE HUMAN ENVIRONMENT. Several studies were conducted on fecal recovery of viable ingested probiotic bifidobacteria. The level of exogenous bifidobacteria was maintained as long as the probiotic preparation was consumed. After ingestion was stopped, the exogenous bifidobacteria gradually decreased and/or were no longer detectable eight days after cessation (Bouhnik et al., 1992; Biavati et al., 1995; Fig. 22).

In the development of alimentary and pharmaceutical probiotic products, bifidobacteria have been most commonly used in these last few years. Often, however, the bifidobacteria currently used in products for human consumption are not of human origin. Previous studies of dairy products such as yogurt, fermented milk, etc. showed the presence of *B. animalis*, which is known to live in animal habitats, in six different products sold in Europe (Biavati et al., 1992b) and in eleven out of sixteen products sold in Europe, Japan and the United States (Yaeshima et al., 1996). None of the products under study declared the presence of *B. animalis*. The product label instead provided the generic term *Bifidobacterium* spp. or a wrong specification such as *B. longum* or *B. bifidum*. In addition, in a recent study of an Italian pharmaceutical product, *B. animalis* was detected without being specified

- Origin
- Resistance to environmental stress
  (acidity, temperature, oxygen, etc.)
- Fermentative capability
- Adhesion characteristics
- Presence of plasmids
- Productivity
- Genotypic and phenotypic stability

Fig. 20. Guidelines used to select *in vitro* probiotic bifidobacteria.

- Study of probiotic strain distribution
- Proving probiotic effect in the health subject
- Clinical trials to assess probiotic effects
  in human and animal normal versus pathogenic
  conditions

Fig. 21. Guidelines used to test *in vivo* probiotic bifidobacteria.

(unpublished data). The difficulties of distinguishing *B. animalis* on the basis of common phenotypes together with the fact that *B. animalis* was described as a subspecies of *B. longum* are probably the main reasons for the presence of *B. animalis* as *B. longum* in many commercial fermented milk products. To avoid possible future mistakes in the use of bifidobacteria, strict identification methods should be applied (Biavati et al., 1992b).

Despite the generally acknowledged positive effects of bifidobacteria on human health, the basic mechanisms for this effect of probiotic microorganisms are still largely unknown. Food integrators, dairy products and pharmaceutical products that use probiotics will develop their full potential when a better knowledge of probiotic interaction with human microflora and the human gastrointestinal tract is gained. The development of in vitro and in vivo techniques to test the presence and survival of probiotics in the host intestine, the study of host immune system-probiotic interactions, the characterization of antimicrobial properties of probiotics and development of molecular biology techniques able to characterize both the probiotic and the resident intestinal microflora are crucial steps towards understanding probiotic physiology and ecology.

PREBIOTIC There are two approaches for increasing the number of health-promoting bacteria in the gastrointestinal tract: oral administration of live beneficial organisms (probiotics) and oral adminstration of dietary supplements (prebiotics) to selectively modify the composition of the microflora. Prebiotics have been defined (Gibson and Roberfroid, 1995) as "non-digestible food ingredients that beneficially affect the host by selectively stimulating the growth and/or activity of one or a limited number of bacteria in the colon, that can improve the host's health."

The primary requirement of a prebiotic is that it has to be at least partially undigested and

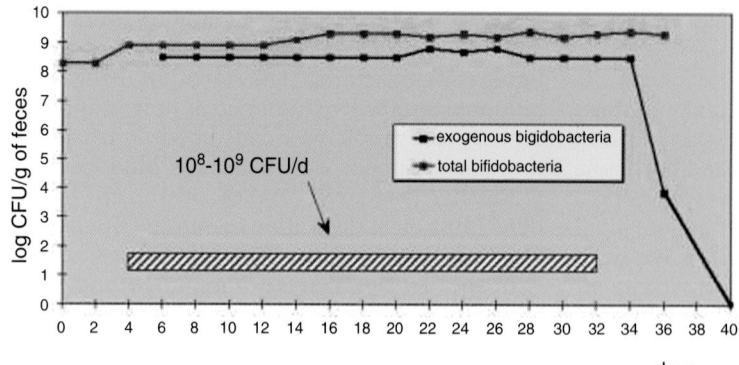

Fig. 22. Human fecal concentration of exogenous and total bifidobacteria during and after ingestion of probiotic bifidobacteria. Data from Biavati et al. (1995).

unabsorbed in the small intestine to provide a fermentable substrate for the colon microflora. It also should stimulate the growth of health-promoting bacteria (mainly bifidobacteria and lactobacilli) and not that of the pathogenic bacteria (Gibson, 1998).

Prebiotics, acting differently from probiotics, can stimulate the growth of the host resident flora, allowing the proliferation of bacterial strains specific to the host instead of replacing them with new strains. Prebiotics not only increase the number of beneficial bacteria, but can also increase their metabolic activity, by supplying fermentable substrates (Roberfroid, 1998a).

The list of prebiotics employed up to now comprises di- and oligosaccharides, fructo- and galacto-oligosaccharides, soybean-oligosaccharides, inulin, resistant starch and other non-digestible oligosaccharides.

Some of the effects on host health of prebiotic administration are: 1) to increase colonization resistance (i.e., the ability to protect the host from exogenous intestinal pathogens; Oli et al., 1998); 2) to reduce risk factors associated with colon cancer (Reddy et al., 1997; Rowland et al., 1998); 3) to decrease serum lipid concentrations (Fiordaliso et al., 1995; Davidson et al., 1998); and 4) to enhance the uptake of minerals (Fiordaliso et al., 1995).

Gibson and Roberfroid (1995) have shown that a small alteration in diet by substituting sucrose (15 g/day) with oligofructose (15 g/day) or inulin may lead to significant increases in bifidobacteria. In particular, dietary supplementation of oligofructose has been shown to decrease a number of potential pathogens. Similar results were obtained by Bouhnik et al. (Bouhnik et al., 1996; Bouhnik et al., 1997), Sghir et al. (1998) and Kruse et al. (1999). It is important to remark that the administration of inulin can cause a moderate increase in gastrointestinal symptoms such as flatulence and bloatedness, whereas blood lipids and short-chain fatty acids remain unaffected (Kruse et al., 1999). In vitro investigation showed that the growth of bifidobacteria was stimulated by oligofructose and the related carbohydrate inulin. In addition, soybean oligosaccharide and transgalacto-oligosaccharides contribute to an increased fecal bifidobacteria number (Hayakawa et al., 1990).

SYNBIOTIC The modulation of host intestinal microflora may be regulated by the simultaneous administration of prebiotics (specific substrates) and probiotics (live microbial preparations). This combination goes under the name "synbiotics" (Collins and Gibson, 1999).

The beneficial effects generated by the presence of the probiotic organism and its immedi-

ately available specific substrate can be more than simply additive. In rat colon carcinogenesis, a single administration of bifidobacteria or oligofructose did not modify the formation of chemically induced aberrant crypts; protection was evident, though, when the synbiotics (bifidobacteria + oligofructose) were coadministered (Gallaher and Jinmo, 1999).

Genus *Gardnerella*

Phylogeny

The first data dealing with the phylogenetic relationships of the genus *Gardnerella* derive from Maidak et al. (1994). These researchers, using 16S rRNA sequencing, demonstrated the close relationship between *Gardnerella vaginalis* and the genus *Bifidobacterium* and grouped in one single taxon or division the type strain of the species *G. vaginalis* and the type strains of 19 species of the genus *Bifidobacterium*. Van Esbroeck et al. (1996) agree with this proposal and concluded that *Gardnerella* and *Bifidobacterium* represent distinct genera belonging to a single phylogenetic lineage. Phylogenetic grouping of these two genera is supported by the common feature of the presence of the enzyme fructose-6-phosphoketolase (Gavini et al., 1996). In 1997, Stackebrandt et al. described the new order Bifidobacteriales, which comprises the family Bifidobacteriaceae with the two genera *Gardnerella* and *Bifidobacterium* (Fig. 23).

Taxonomy

The taxonomy of genus *Gardnerella* was the object of many discussions during the past years, having been assigned to different genera such as *Haemophilus* and *Corynebacterium*.

Gardnerella vaginalis takes its name from H. L. Gardner, who was the first to describe the vaginitis associated with this organism; it was isolated for the first time by Leopold (1953) and it was described in the genus *Haemophilus vagina-*

1953 - First recognized by Leopold
1955 - Named *Haemophilus vaginalis* by Gardner and Dukes
1963 - Proposal of reclassification such as *Corynebacterium vaginale* by Zinnemann and Turner *Butyribaecterium* or *Propionibacterium* by Reyn et al.
1980 - Description of the new genus *Gardnerella* with the species *G. vaginalis* by Greenwood and Pickett and by Piot et al.
1997 - Stackebrandt et al. - New hierarchic classification; Order: *Bifidobacteriales*; Family: *Bifidobacteriaceae*; genera *Bifidobacterium* and *Gardnerella*

Fig. 23. Milestones of the taxonomy of *Gardnerella vaginalis*.

lis by Gardner and Dukes (1955) because it was a Gram-negative or Gram-variable, rod-shaped bacterium isolated successfully on blood agar medium but not on other agar media. Furthermore, it was considered the etiologic agent of bacterial vaginosis, even if its role in this pathology was controversial, because no appropriate methods of detection and identification for this fastidious organism were available. Today it has been recognized that bacterial vaginosis is also associated with other microorganisms such as *Bacteroides* spp., *Mobiluncus* spp., *Mycoplasma hominis*, *Peptostreptococcus* spp., etc. Zinnemann and Turner (1963) suggested the removal of *Haemophilus vaginalis* from the genus *Haemophilus* and its reclassification in the genus *Corynebacterium* as *Corynebacterium vaginalis*; this statement was based on the similarity of Gram stain reaction and cellular morphology to the member of this genus and on the absence of requirements for X and V factors (hemin and NAD, respectively) which are needed for the growth of *Haemophilus* species. In the first study of *Haemophilus vaginalis* by electron microscopy, Reyn et al. (1966) showed that the fine structure of its cell wall resembled that of a Gram-positive bacterium, and it could be placed in the genus *Corynebacterium* or *Butyribacterium*. These authors also indicated that this bacterium may be closely related to the genus *Propionibacterium*. Differently from Reyn et al. (1966), Criswell et al. (Crisweel et al., 1971; Crisweel et al., 1972) in their study concluded that the fine structure examined by electron microscopy and the biochemical composition of the cell walls of *Haemophilus vaginalis* were typical of a Gram-negative bacterium. Moss and Dunkelberg (1969), based on a study on the volatile acid production, showed that *Haemophilus vaginalis* should be placed neither in the genus *Corynebacterium* or *Butyribacterium* nor in the genus *Lactobacillus* (as was suggested by Buchanan and Gibbons, 1974), because the main fermentation product is acetic acid rather than lactic, butyric or propionic acid.

In 1980, Greenwood and Pickett described the new genus *Gardnerella* with the species *Gardnerella vaginalis* in the absence of existing genera with genetic features compatible with those of *Haemophilus vaginalis* (Greenwood and Pickett, 1980). Piot et al. (1980) in the same year came to the same conclusion. In the 1986 edition of *Bergey's Manual of Systematic Bacteriology* (Sneath et al., 1986), *G. vaginalis* has been included in the group of facultatively anaerobic Gram-negative rods. In agreement with Sadhu et al. (1989), a recent study of cell-wall ultrastructure of *G. vaginalis* type strain confirms the Gram-positive nature of its cell wall (Muli et al., 1999); this report stated that any Gram variability staining can be ascribed to variations in the thickness of the cell wall, as was pointed out before.

The phylogenetic study based on 16S RNA homology clarified the taxonomic position of genus *Gardnerella* which, together with genus *Bifidobacterium*, has been included in the family of Bifidobacteriaceae (Stackebrandt et al., 1997). *Gardnerella vaginalis*, besides the phylogenetic affinity, shares the presence of the key enzyme fructose-6-phosphate phosphoketolase in common with bifidobacteria (Gavini et al., 1996).

Milestones in the *Gardnerella* taxonomy are reported in Fig. 23.

Habitat

Gardnerella vaginalis is commonly isolated from female genital tract and can occur as an endogenous organism of the vaginal flora in a population of healthy women. Overgrowth of *G. vaginalis*, which is present normally in low number, is involved in bacterial vaginosis. *Gardnerella vaginalis* transiently colonizes the genital tracts of healthy males as a result of continued passive acquisition from their female partners (Holst, 1990) or this microorganism may have been part of bacterial flora propria of these subjects, as suggested by Smith et al. (1992). In males, *G. vaginalis* can be involved in genital and urinary tract infections (Smith et al., 1992).

Gardnerella vaginalis can be found also in the anorectal flora of both sexes as well as children (Holst, 1990). In addition, *Gardnerella vaginalis* has been isolated from the reproductive tract of mares; probably they are normal inhabitants of the equine genital tract (Salmon et al., 1991).

Isolation

SPECIMEN COLLECTION, TRANSPORT AND STORAGE Material for the culture of *G. vaginalis* is obtained by swabbing the vaginal introitus with a cotton-tipped swab. It is preferable to take two swabs: one for immediate examination and the other for culturing, such as is necessary for epidemiological studies. If the culture media cannot be directly inoculated, then the swab should be placed in a transport medium and culture should be done within 24 h (Funke and Bernard, 1999). It is noteworthy that *G. vaginalis* is susceptible to sodium polyanetholesulfonate (SPS), so an SPS-free medium (or an SPS medium supplemented with gelatin, which overcomes the inhibitory effect of SPS) should be used to achieve optimal recovery of *G. vaginalis* from a blood culture system (Reimer and Reller, 1985a).

NONSELECTIVE MEDIA Nonselective or differential media for the detection of small numbers of G. vaginalis organisms, among other commensal microorganisms frequently found in the genital tract, support fairly well the growth of G. vaginalis. None of these media inhibit lactobacilli and other common vaginal Gram-positive bacteria such as coryneforms and streptococci. Casman agar (Leopold, 1953), chocolate agar (Pheifer et al., 1978), GC agar (Difco), and peptone-starch-dextrose agar (Dunkelberg et al., 1970) have been frequently used to grow G. vaginalis.

Chocolate Agar This medium contains GC agar base (BBL), 1% ISOVITALEX enrichment (BBL) and heated "chocolatized" 5% sheep blood. *Gardnerella vaginalis* is identified as pinpoint colonies appearing after 48 or 72 h; they produce no surrounding green discoloration of the agar reference medium (Pheifer et al., 1978).

Peptone-Starch-Dextrose Agar (PSD) This medium contains 2% proteose peptone No. 3 (Difco), 1% soluble starch, 0.2% dextrose, 0.1% $Na_2HPO_4·H_2O$ and 1.5% agar (Dunkelberg et al., 1970).

Dunkelberg and McVeigh (1969) have observed that only two commercial peptones (Difco proteose peptone no. 3 and possibly BBL myosate) of all the peptones tested contain the required growth factors for *G. vaginalis*. Smith (1975) modified PSD by adding a pH indicator to the medium, allowing the detection of *G. vaginalis* by the production of acid from starch around the colonies. The addition of 0.2% (final concentration) Tween 80 to PSD medium enhances the growth of the majority of strains of *G. vaginalis*, perhaps because it may compensate for some nutritional deficiency in the basal peptone used in the pre-reduced medium (Malone et al., 1975).

The use of nonselective media permits the relative quantification of *G. vaginalis* growth, whereas media designed to yield a higher isolation rate of *G. vaginalis* fail to indicate the relative numbers of this organism in mixed genital flora. With nonselective media, a higher degree of correlation has been observed between clinical symptoms and specimens yielding a pure growth of *G. vaginalis* than between clinical symptoms and specimens yielding predominant, mixed, or light growths (Ratnam and Fitzgerald, 1983).

Selective Media

Media for selective growth of *G. vaginalis* have been developed. β-Hemolysis observed in human or rabbit (not sheep) blood-containing media allows β-hemolytic *G. vaginalis* to be distinguished from among numerous nonhemolytic colonies. *Gardnerella vaginalis* is resistant to several antibiotics including amphotericin, colistin, nalidixic acid and gentamicin, which can be incorporated into selective media (Taylor-Robinson, 1984). Anaerobic or facultatively anaerobic Gram-negative vaginal flora are inhibited by colistin and nalidixic acid, and yeasts are inhibited by amphotericin B (Piot and Van Dick, 1983).

The organism grows best at 35–37°C in a 5–7% CO_2 atmosphere. Plates may be checked for the growth of diffuse β-hemolytic colonies of <0.5 mm in diameter after 24 h, but very often *G. vaginalis* is better observed after 48 or 72 h. Gram-staining of the suspected colonies confirms the diagnosis of *G. vaginalis*. The performance of confirmatory tests usually requires subculturing and incubation for another 24–48 h.

HUMAN BLOOD BILAYER TWEEN (HBT) AGAR The ability to detect *G. vaginalis* has improved with the development in 1982 of an HBT medium by Totten et al. (1982). It was composed of a basal layer of 7 ml of CNA agar (Columbia agar base containing colistin and nalidixic acid, BBL Microbiology Systems), 1% proteose peptone no. 3 (Difco), and 0.0075% Tween 80 with amphotericin B (2 µg/ml) and of a 14-ml overlayer of the same composition plus 5% human blood. Tween 80, amphotericin B and human blood were added after autoclaving. Tween 80 improves hemolysis and enhances growth. Colonies of *G. vaginalis* are identified as small white colonies that produce β-hemolysis after 48–72 h of incubation in 5% CO_2 (Totten et al., 1982). Because zones are viewed through a shallower layer of blood-containing medium, hemolysis is easier to detect on a double-layer as opposed to a single-layer HBT medium. Significantly, *G. vaginalis* was isolated from vaginal fluids more often on HBT than on Vaginalis agar (P=0.007; Totten et al., 1982). Holst (1990) used HBT with good results to isolate *G. vaginalis* from anorectal samples. This medium is the most satisfactory selective medium at present, even if it fails to inhibit a variety of Gram-positive bacteria such as lactobacilli, coryneforms and streptococci of the endogenous and bacterial vaginosis-associated flora (Piot and Van Dick, 1983; Spiegel, 1991). Isolates of Gram-positive or Gram-variable pleomorphic rods that showed β-hemolysis in this medium were identified as *G. vaginalis* if they produced acid from starch and maltose and were negative for catalase and oxidase (Spiegel et al., 1980). *Gardnerella vaginalis* when grown anaerobically on HBT medium with some other anaerobic constituents is also β-hemolytic. The addition to the HBT medium of gentamicin (4 mg/liter) instead of colistin

LIVERPOOL
JOHN MOORES UNIVERSITY

increases the inhibition towards non-*Gardnerella* species (Ison et al., 1982).

COLISTIN AND NALIDIXIC ACID MEDIUM (CNA) This is a semiselective medium containing colistin (10 µg/ml) and nalidixic acid (15 µg/ml) incorporated in a Columbia agar base (Goldberg and Washington, 1976). The addition of 1% cornstarch allows the discrimination of *G. vaginalis* colonies by their hydrolytic clearing of the opaque agar (Mickelsen et al., 1977).

VAGINALIS (V) AGAR This medium is an enriched and selective medium. It contains GC agar; the addition of 2% hemoglobin and 5% fetal bovine serum enhances the recovery of the microorganism. Vancomycin at 3 µg/ml is added to the medium to inhibit contaminating bacteria (Chapin and Murray, 1999).

COLISTIN-OXOLINIC ACID BLOOD AGAR (COBA-10) Columbia agar base contains 5% defibrinated sheep blood, 10 mg of sodium colistimethate per liter and 10 mg of oxolinic acid per liter (Thompson, 1985). This selective medium is useful for isolating *G. vaginalis* from vaginal and cervical specimens. However, COBA-10 grew lactobacilli, yeast, hemolytic streptococci, and viridans group streptococci equally well, whereas other vaginal diphtheroids and some enterococci were inhibited by COBA-10. Incubation occurs at a temperature of 37°C for 48 h in a humidified atmosphere of air plus 5–10% CO_2.

Commercial blood culture media, routinely with the anticoagulant SPS added, are inhibitory for *G. vaginalis* (Reimer and Reller, 1984). The adverse effect of SPS is overcome by the addition of gelatin to SPS-containing media (Reimer and Reller, 1985a). Failure to detect bacteremia where *G. vaginalis* is present may be due to SPS inhibition (Gibbs et al., 1987; Lee et al., 1987).

SORBAROD BIOFILM MEDIA Sorbarod biofilm media to investigate the growth characteristics of *G. vaginalis* and *Lactobacillus acidophilus* separately or together have been developed. This system is labor intensive but it appears suitable for studying the interaction of bacteria in biofilms (Muli and Struthers, 1998).

Identification

Clinical Diagnosis of Bacterial Vaginosis

The isolation of *G. vaginalis*, although not recommended for routine laboratory procedures, can support the diagnosis of bacterial vaginosis. In fact massive growth of *G. vaginalis* (about 10^8 CFUs/ml of vaginal fluid) in the absence or with a low number of lactobacilli is a better predictor of bacterial vaginosis than predominant, mixed or light growth. In clinical diagnosis, the "gold standard" for identification of bacterial vaginosis is the direct examination of vaginal secretions based on these parameters:

1) an abnormal vaginal discharge;
2) the presence of clue cells (vaginal epithelial cells whose peripheral borders are obscured because they are covered with adherent bacteria) on a wet mount of a smear examined microscopically; *G. vaginalis* has been shown to be the dominant organism adherent to exfoliated clue cells (Cook et al., 1989). The Gram-staining of the same microscopy preparation can give another indication of the presence of short Gram-negative or Gram-variable bacilli corresponding to *G. vaginalis* (Ison et al., 1982); normally *G. vaginalis*, besides having a Gram-positive type of organization because of the unusual thinness of the cell walls, tends to display Gram-negative or Gram-varible reaction (Sadhu et al., 1989);
3) a positive amine test performed by mixing a drop of 10% KOH solution with the discharge (milky homogenous vaginal discharge). The release of an amine (fishy) odor is a positive test. In bacterial vaginosis, the combination of *G. vaginalis* and other anaerobes actually produces organic acids (other than lactic acid) as well as several amines (Chen et al., 1979). These amines, in the presence of an elevated vaginal pH, readily volatilize to produce putrescine, cadaverine, methylamine, etc., which are the substances primarily responsible for the fishy odor in bacterial vaginosis. *Gardnerella vaginalis* itself is not thought to produce the polyamine but may elaborate catabolic products that facilitate the growth of anaerobes (such as *Mobilunculus*, which for example produces trimethylamine in bacterial vaginosis; Cruden and Galask, 1988).

A new diamine test for a rapid diagnosis of bacterial vaginosis was described by O'Dowd et al. (1996). Additional information can be obtained by examining the vaginal pH, which is greater than 4.5 in bacterial vaginosis. These diagnostic criteria cannot be readily applied to prepubertal children because the immature vagina is an alkaline environment. Also, the absence of *Trichomonas vaginalis* is an additional predictor for bacterial vaginosis diagnosis.

Additional tests currently commercially available use the detection of bacterial by-products (e.g., PIP, proline endopeptidase, an enzyme frequently found in vaginal secretions of women with bacterial vaginosis), elevated pH and amine, or oligonucleotide probes for *Gardnerella* (Schwebke, 1999; Sheiness et al., 1992). Furthermore, PCR 16S RNA gene methods have been set up for the identification of *G. vaginalis* (Nath et al., 2000).

Tests for the Identification of *G. vaginalis*

The predictive value of the growth of *G. vaginalis* in the diagnosis of bacterial vaginosis was only 49%, but a negative culture for *G. vaginalis* practically excluded bacterial vaginosis (Piot and Van Dick, 1983). It is unclear whether women who carry *G. vaginalis* in the vagina ultimately develop bacterial vaginosis and why some women carrying *G. vaginalis* have bacterial vaginosis and others do not.

The presence of positive results using the diagnostic criteria might prove to be particularly suitable in the first-contact clinic and could rule out the need for identification of *G. vaginalis*. The incidental finding of *G. vaginalis* from a routine vaginal culture should not be used to diagnose bacterial vaginosis because many women with *G. vaginalis* organisms show no signs of bacterial vaginosis. The identification of *G. vaginalis* is important also in the study of urinary tract infections in men because urinary tract infections caused by this microorganism have also been described (Smith et al., 1992).

The major problem in the diagnostic laboratory is the differentiation of *G. vaginalis* from other catalase-negative coryneform or diphtheroid coccobacilli, which occur commonly in vaginal specimens. For the identification of *G. vaginalis*, it is possible to use some confirmatory tests which, on the other hand, may not always be required by experienced microbiology technologists, considering they can usually identify the organism by colony morphology and Gram stain appearance (Reimer and Reller, 1985a).

Bailey et al. (1979) explored the possibility of using API 20A (Analylab Products Inc., Plainview, N.Y.), Minitek, and the rapid buffered-substrate fermentation (RTF) systems (Brown, 1974) and compared them to conventional tests. Results confirmed by Greenwood et al. (1977) showed API and Minitek gave variable and unreliable results, although RTF gave results comparable with conventional tests.

The following protocols are more frequently applied for identification:

CELLULAR MORPHOLOGY AND GRAM-STAINING *Gardnerella vaginalis* cells (Fig. 24) are generally small, pleomorphic rods that are nonmotile, nonsporeforming and do not possess flagella, endospores or typical capsules (Greenwood and Pickett, 1980). Cellular morphology varies with the medium on which the organism was grown. Bacilli from chocolate agar are slightly larger and more pleomorphic than those from Schlaedler agar (Oxoid) or Bacto-peptone agar.

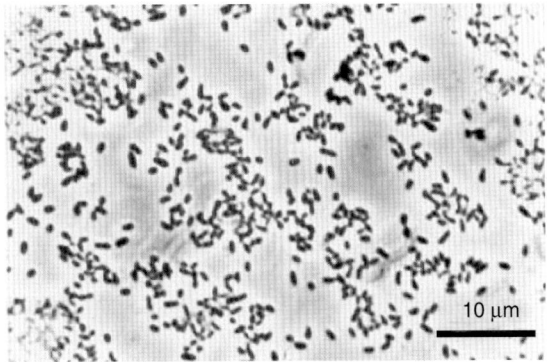

Fig. 24. Cellular morphology of *G. vaginalis*. ATCC 14018ᵀ.

In these latter media, the morphology resembles the control organism (NCTC 10287) in being predominantly coccobacillary (Jolly, 1983). The Gram stain varies considerably depending on the medium and the age of the culture. Consequently cellular morphology and size are more helpful than the Gram reaction itself. Structurally, it is a Gram-positive microorganism, but it has an unusual cell-wall structure that does not retain Gram's stain and generally appears as Gram-variable or Gram-negative (Sadhu et al., 1989).

RAPID IDENTIFICATION SYSTEM This system is known as the rapid-strip bacterial identification system, the API 20A (Analylab Products Inc., Plainview, N.Y.), Rapid ID32 Strep, and the API Coryne galleries (bioMerieux, La Balme les Grottes Montalieu-Vercieu, France). According to Taylor and Phillips (1983), many of the performance results differed quite considerably from those of Piot et al. (1980) and so this system of enzyme detection should not be used for the definitive identification of *G. vaginalis*.

PRESENCE OF CATALASE AND OXIDASE Growth from a 48-h chocolate agar plate is placed on a glass slide using a wooden applicator stick, overlaid with a drop of 3% H_2O_2 and then observed for evolution of bubbles. For the oxidase test, growth is smeared with a platinum loop on a filter paper strip saturated with a solution of 1% (w/v) tetramethyl-*p*-phenylenediamine dihydrochloride in 0.2% (w/v) ascorbic acid (Greenwood and Pickett, 1979). The results of both tests are negative for *Gardnerella vaginalis*.

HYDROLYSIS OF HIPPURATE *N*-Benzoylglycine (hippurate) hydrolysis is performed according to Hwang and Ederer (1975). A stock solution of 1% sodium hippurate is dispensed in 0.4-ml aliquots into test tubes. A loopful of the test organism from the overnight growth plate is emulsified in the hippurate solution and incubated for 2 h at 37°C. Then 0.2 ml of ninhydrin solution is added and re-incubated for 10 min. A positive reaction is indicated by the development of a purple color. The optimal pH of the substrate solution giving no false-negative or false-positive reactions is 6.4. Shaking of the test solution occasionally produces false-positive reactions (Piot et al., 1982). This test was less specific in respect to an SPS-disk test or an α-hemolytic (α-strep) test. Reimer and Reller (1985a) obtained a positive reaction in 50 out of 62 strains of G. vaginalis tested; also Piot et al. (1982) observed that 58% of unclassified coryneforms hydrolyzed hippurate. Although Reimer and Reller (1985a) do not consider this to be an important screening test, Jolly (1983) considers it fundamental to an accurate identification of G. vaginalis. As this test is not positive in 100% of the tested strains, it is only partially useful for the identification of G. vaginalis.

Susceptibility to Metronidazole and Sodium Polyanetholesulfonate A useful characteristic for the identification of G. vaginalis is the presence of a zone of inhibition around the disks containing 50 µg of metronidazole (Totten et al., 1982) or 1 mg of sodium polyanetholesulfonate (SPS; Reimer and Reller, 1985b). According to Jones (1985), the SPS test helps to exclude catalase-positive vaginal coryneforms and lactobacilli, but does not differentiate true G. vaginalis from the atypical, catalase-negative "G. vaginalis-like" organisms from the vagina; this inability remains the major problem in vaginal bacteriology.

HEMOLYSIS, COLONIAL AND CELL MORPHOLOGY ON HBT MEDIUM A presumptive identification in the clinical laboratory can be carried out using the culture of vaginal swabs on HBT medium and by the observation of the colonial morphology, clear β-hemolysis with a diffuse edge on HBT agar, negative catalase production, and typical cellular morphology in the Gram stain. All G. vaginalis strains as well as about 50% of catalase-negative coryneforms are β-hemolytic. However, hemolysis by catalase-negative coryneforms strains is nearly always weaker and the zone is of smaller diameter than that of Gardnerella-type hemolysis. Catalase-negative coryneforms generally yield smaller colonies and are more often pleomorphic, fairly thick and predominantly Gram-positive rods. Gardnerella vaginalis cells are rather thin, Gram-negative or Gram-variable, short rods. The quality of the hemolysis is, however, dependent on the age of the blood, which should just have time-expired (Ison et al., 1982). The diffuse β-hemolysis produced by G. vaginalis on human (but not on horse) blood

agar has proven very useful in differentiating it from other vaginal organisms and is not affected by the antibiotic added to the substrate (Ison et al., 1982).

LIPASE ACTIVITY Lipase activity was shown by the oily, iridescent sheen on an enriched egg yolk medium after anaerobic incubation. The plates were examined each day for 5 days to detect lipase (Holdeman et al., 1977). A more recent lipase test method uses an oleate substrate (Briselden and Hillier, 1990). The lipase reaction when present is helpful in detecting G. vaginalis, particularly in mixed cultures. Unfortunately, not all strains uniformly give this reaction (Greenwood and Pickett, 1979). This test is used in the determination of G. vaginalis biotypes (Briselden and Hillier, 1990).

CARBOHYDRATE FERMENTATION Fermentation of maltose, dextrose and fructose, in addition to starch, has been considered in many identification procedures (Shaw et al., 1981; Piot et al., 1982; Taylor and Phillips, 1983). Dunkelberg et al. (1970), however, have suggested that it is superfluous to test for dextrose and maltose fermentation because an organism which can acidify starch can likely acidify dextrose and maltose also.

INHIBITION BY α-HEMOLYTIC STREPTOCOCCUS The α-hemolytic streptococcus (a-strep) test is performed by inoculating a 0.5 McFarland standard suspension on a chocolate agar. A streak with Streptococcus sanguis ATCC 35557 is then made across the center of the plate, and the plate is incubated for 24–48 h at 37°C in 5% CO_2. A positive test requires inhibition of growth along the streptococcal streak. Preliminary tests showed that the test was less reliable with media other than chocolate, when incubated under anaerobic conditions, or using other strains of streptococci (Reimer and Reller, 1985b).

FATTY ACID COMPOSITION The fatty acid composition is characteristic and relatively simple. The differences among isolates are insignificant. Gardnerella vaginalis possesses polar lipids and fatty acids (i.e., linear chain saturated and unsaturated non-hydroxylated fatty acids: hexadecanoic acid [16:0], octadecenoic acid [18:1], and octadecanoic acid [18:1]). Polar lipids are mostly characterized by the presence of diphosphatidylglycerol, phosphatidylglycerol or phosphatidylinositol (Csango et al., 1982; Van Esbroeck et al., 1996).

FRUCTOSE-6-PHOSPHATE PHOSPHOKETOLASE TEST Fructose-6-phosphate phosphoketolase (F-6-

PPK) is present in all bifidobacterial species (Biavati et al., 1991) and in *G. vaginalis* as well (Gavini et al., 1996), but the two genera can be easily distinguished on another basis. For the procedure of F-6-PPK test, see identification section of *Bifidobacterium* genus.

The above tests are useful for the identification of *G. vaginalis* but usually confirmatory tests have to be obtained by a combination of tests: a limited number of biochemical reactions could fail to give a definitive identification. The most valuable characteristics are the presence of α-glucosidase, the absence of β-glucosidase, the hydrolysis of starch and hippurate, and the hemolysis of human but not sheep blood (Catlin, 1992; Reimer and Reller, 1985b).

Preservation of Cultures

Short-term Methods

A Casmaan's medium base (Difco cat. number 0290) with 5% rabbit blood is used for preservation of strains of *G. vaginalis* in liquid media. The cultures are transferred every week and are incubated in an atmosphere containing a 5% CO_2-enriched atmosphere.

Long-term Methods

Longer maintenance can be achieved by freezing below 130°C (deep freezer), of liquid nitrogen or by freeze-drying. The following cryoprotective agent may be used: skim milk solution (10% skim milk, 3% lactose and 3% yeast extract; the solution, after shaking for 20 min, is autoclaved at 110°C for 30 min). For details and other methods, see Gherna (1994).

Physiology

The most important physiological characteristics of *G. vaginalis* are shown in Table 7.

Temperature Requirements

Optimum growth occurs at temperatures between 35 and 37°C; however, growth can occur between 25 and 45°C (Greenwood and Pickett, 1979).

pH Requirements

Optimum growth occurs at a pH between 6 and 6.5; no growth occurs at pH 4 and slight growth occurs at pH 4.5 and pH 8.0 (Greenwood and Pickett, 1979; Buchanan and Gibbons, 1974).

Table 7. Features of *G. vaginalis* strains.

Oxidase	−
Catalase	−
Urease	−
Indole production	−
Methyl red	+
Voges-Proskauer	−
Lipase	v
Hydrolysis of	
Hippurate	+
Casein	−
Gelatin	−
Hemolysis of	
Human blood	+
Rabbit blood	+
Sheep blood	−
Gas from glucose	−
Litmus milk	v
Growth at	
25°C	+
30°C	+
42°C	v
pH 4	−
pH 8	+
NO_3 from NO_2	−
α-Glucosidase	+
β-Glucosidase	−
α-Galactosidase	−
β-Galactosidase	v

Symbols: +, 90% or more of strains are positive; and −, 90% or more of strains are negative.

Anaerobiosis

Gardnerella vaginalis has a fermentative metabolism with acetic acid as the major end product (Moss and Dunkelberg, 1969). It is facultatively anaerobic and uses carbohydrates as its major energy source (Greenwood and Pickett, 1980). It grows well in microaerophilic conditions in a 5–7% CO_2 atmosphere. Malone et al. (1975) described six strains that were strictly anaerobic on original isolation. One strain, however, when re-tested after lyophilization grew on the surface of plates incubated in a 5–7% CO_2 atmosphere. Piot and Van Dick (1983) also isolated four of 145 vaginal isolates after incubation in anaerobic conditions, of which two failed to grow in microaerophilic conditions after 10 subcultures. Anaerobic strains of *G. vaginalis* seem to be rare, and routine anaerobic incubation is not recommended (Piot and Van Dick, 1983). There were no differences among the characteristics of anaerobic and facultatively anaerobic strains, except for the occurrence of mannose fermentation (very weak in anaerobic strains). Today, it is more practical to use a commercially available kit to grow *G. vaginalis* in microaerophilic conditions.

Enzymes

Gardnerella vaginalis has a variable biochemical lipase and β-galactosidase profile. It possesses esterase (C4), esterase lipase, and α-glucosidase, but does not possess β-glucosidase (Greenwood and Pickett, 1979; Piot et al., 1982).

Carbohydrate Fermentation

Carbohydrate reactions were very variable owing to differences in the methodology used. Dextrose fermentation, for example, was reported to be positive for 100% of 63 strains tested with buffered single substrates (Greenwood et al., 1977), whereas only two of 16 were positive for cystine-trypticase agar (Smith, 1975). Similar variability has been reported for sucrose (Dunkelberg et al., 1970). The results of fermentation of other carbohydrates are reported in Table 8.

Table 8. Fermentative characteristics of *G. vaginalis*.[a]

Substrate	
L-Arabinose	v
Raffinose	−
Sorbitol	−
D-Ribose	+[b]
Starch	+
Lactose	v
Inulin	v
Cellobiose	−
Melezitose	−
Gluconate	+
Xylose	v
Mannose	v
Fructose	+[b]
Galactose	v
Sucrose	v
Maltose	+[b]
Trehalose	−[c]
Melibiose	−
Mannitol	−
Salicin	−
Rhamnose	−
Maltose	+[b]
Inositol	−
Dextrin	+
Lactose	−[c]
Glycerol	−
Dextrose	+[b]
Arbutin	
Glucose	+

[a]Data from Greenwood and Pickett, 1979; Piot et al., 1980, 1982; Taylor and Phillips, 1983.
Symbols: +, 90% or more of strains are positive; −, 90% or more of strains are negative; v, 11–89% of strains are positive. [b]A few strains do not ferment this sugar. [c]A few strains ferment this sugar.

Toxin Production

A cytolytic extracellular toxin released by *G. vaginalis* during growth in a supplemented peptone broth containing starch and Tween 80 was described by Rottini et al. (1990). This hemolysin was specific for human erythrocytes and had no activity on horse, rabbit, sheep or guinea pig erythrocytes. This toxin can produce lysis of human polymorphonuclear leukocytes and human endothelial cells. This cytolytic toxin forms a voltage-dependent cationic channel when incorporated into lipid membranes (Moran et al., 1992).

Phospholipase A$_2$ Activity

Gardnerella vaginalis was identified as a member of the vaginal flora having phospholipase A$_2$ activity, which initiates labor (Bejar et al., 1981).

Nutrition

Gardnerella vaginalis is a fastidious microorganism, but does not require either hemin or NAD (X and V factors, respectively) for growth. A semi-defined liquid medium that permits the growth of two *G. vaginalis* strains, 594^T (ATCC 14018) and 317 (ATCC 14019), was developed (Dunkelberg and McVeigh, 1969). Interestingly, the strain 594 required adenine for growth, whereas the strain 317 did not. Neither strain grew in the absence of purine and pyrimidine bases. The omission of any one of the six bases (except adenine) from the medium did not reduce growth.

Iron plays an important role in the virulence potential of many bacterial pathogens (Griffits, 1991). The ability of different strains of *G. vaginalis* to acquire iron from the medium was studied by Jarosik et al. (1998). The authors found that all strains examined were able to acquire iron from ferric and ferrous inorganic substrates as well as from hemin, catalase, hemoglobin, lactoferrin but not transferrin. In *G. vaginalis*, several proteins were recognized to be iron-regulated.

It would be useful to study the nutritional requirements of *G. vaginalis* strains because of the scarcity of such data in the literature.

Antimicrobial Susceptibility

Ninety percent of the strains of *G. vaginalis* are susceptible to penicillin (0.07 mg/liter), ampicillin (0.125 mg/liter), vancomycin (0.5 mg/liter), clindamycin (<0.06 mg/liter), gentamycin (4 mg/liter) and amphotericin B (2 mg/liter). *Gardnerella vaginalis* is susceptible to metronidazole (4mg/liter) and its hydroxy metabolite (1 mg/

liter), which is one of the main metabolites of metronidazole produced in vivo (McCarthy et al., 1979; Bannatyne et al., 1987). It is resistant to cephalexin (64 mg/liter), tetracycline (64 mg/liter), nalidixic acid (>128 mg/liter), colistin (>128 mg/liter), sulfadiazine (>128 mg/liter), and quinolones (>4 mg/liter; Jones et al., 1982; Tjiam et al., 1986; Ison et al., 1982). The minimal inhibitory concentration (MIC) of tetracycline showed a bimodal distribution (Kharsany et al., 1993; McCarthy et al., 1979); high-level resistance (>64 mg/liter) has been attributed to the Tet(M) conjugative transposon located on the chromosome (Roberts and Hillier, 1990; Huang et al., 1997).

It is insensitive to sulfonamides and resistant to trimethoprim. Metronidazole is the most efficacious drug for the treatment of bacterial vaginosis related to the presence of G. vaginalis. Evidence that bacterial vaginosis is susceptible to treatment with metronidazole (a drug active against anaerobic bacteria) supports the hypothesis that bacterial vaginosis is due to a set of anaerobic bacteria that are sustained by the presence of G. vaginalis. Ampicillin can be used just for the treatment of a G. vaginalis infection, such as a urinary tract infection or a postpartum sepsis. Nevertheless, ampicillin treatment failed to eradicate bacterial vaginosis where G. vaginalis was present with species of anaerobes. These other anaerobes presumably inactivate the drug via β-lactamase inhibition. Augmentin, in which clavulanic acid protects the β-lactam ring of amoxicillin from enzymatic cleavage, was used successfully for the treatment of bacterial vaginosis (Symonds and Biswas, 1986).

Ultrastructure

Electron microscopy studies have produced two opinions, one confirming that G. vaginalis is Gram positive (Reyn et al., 1966; Sadhu et al., 1989) and another stating that G. vaginalis is Gram negative, the latter based on both electron microscopic and biochemical evidence (Crisweel et al., 1971; Boustouller et al., 1987). The recent electron microscopy study of Muli et al. (1999) on the cell wall of G. vaginalis grown on conventional medium and in biofilm systems confirms the Gram-positive nature of the cell wall. The Gram variability can be ascribed to variations in the thickness of the cell wall as observed by Sadhu et al. (1989).

Furthermore, G. vaginalis cell walls do not contain the classical lipopolysaccharides found in Gram-negative organisms and contain major amounts of alanine, glutamic acid, glycine and lysine; glucose, galactose and 6-deoxytalose, a diagnostic sugar, are also found (Harper and Davis, 1982).

Exopolysaccharides and Pili

The microcapsular material that is visible in electron microscope preparations has been demonstrated to be a flocculent or fibrillar external polysaccharidic layer named "exopolysaccharide" (Scott et al., 1989; Greenwood and Pickett, 1980; Sadhu et al., 1989; Crisweel et al., 1972). It possesses a role in the adhesive nature of G. vaginalis and accounts in part for the clustering of cells in culture.

Gardnerella vaginalis possesses pili with a diameter ranging from 3 to 7.5 nm. The presence of pili is more frequent in clinically isolated than laboratory cultured strains (Boustouller et al., 1987; Johnson and Davies, 1984).

Genetics

Few studies on G. vaginalis are present. Nath et al. (1992) demonstrated that G. vaginalis in bacterial vaginosis represents a genetically mixed population; this conclusion was drawn using restriction endonuclease analysis (REA) and restriction fragment length polymorphism (RFLP) as well as by biochemical studies (biotyping based on β-galactosidase, lipase and hippurate hydrolysis; Briselden and Hillier, 1990). Nath et al. (1991) used a nonradioactive probe derived from a 5.7-kb DNA fragment from G. vaginalis ATCC 14018 to perform RFLP on 12 G. vaginalis strains and demonstrated the usefulness of this detection method for the study of this organism.

Similarly, Pao et al. (1990) used a DNA cloned from digested genomic potentially G. vaginalis DNA to serve as a probe and described this probe as having no crossreactions with a number of non-Gardnerella microorganisms.

Van Belkum et al. (1995) developed a species-specific PCR assay for the determination of G. vaginalis using the internal transcribed spacer (ITS) region of G. vaginalis DNA (between the 16S and 23S RNA genes) and the 23S rRNA gene. This PCR analysis in 10% of women, irrespective of their clinical status, detected G. vaginalis, whereas 10 out of the 11 patients with bacterial vaginosis were carrying G. vaginalis. Nath and Galdi (1995) refined the procedure to isolate G. vaginalis DNA from cultures by reducing the total extraction time and using an NaCl salting-in/out procedure to remove proteinaceous material.

Wu et al. (1996) analyzed 20 biotype "1" G. vaginalis strains, 10 from patients with bacterial vaginosis and 10 from patients without bacterial vaginosis. None had the same DNA fingerprint pattern. However, irrespective of the clinical status, 18 of the 20 samples were positive for a 1.18-kb Hind III fragment using RFLP.

Ecology

The composition of the endogenous vaginal flora is influenced by interactions (although not yet completely understood) among the host, the bacteria and exogenous factors. For example, estrogen produced by adult females increases glycogen in the vaginal epithelium, which is fermented by lactobacilli. Through this fermentation, lactobacilli produce acid and fulfill a central role in vaginal homeostasis of protecting the urogenital tract from pathogenic infections (Eschenbach, 1993). Before puberty, the female vagina is basic and does not produce glycogen. After menopause, glycogen disappears and the pH and the flora resemble their prepubertal status. In this stage of life, lactobacilli are absent and the vaginal flora consist mostly of staphylococci, diphtheroides and *E. coli*. In clinically healthy women also, some investigators have described the presence of *Bifidobacterium* species, such as *B. breve*, *B. adolescentis*, *B. longum* and *B. bifidum* (Crociani et al., 1973; Korschunov et al., 1999). The role played by bifidobacteria has not been elucidated; it is possible, however, bifidobacteria similar to lactobacilli are able to produce inhibitory substances such as organic acids and bacteriocins that inhibit different pathogens. The factors that initiate the shift in the ecology of the vagina, in which there is overgrowth of organisms already present or newly introduced, are incompletely understood. These changes in microbial flora with a prevalence of microorganisms such as *G. vaginalis*, *Mycoplasma* spp., anaerobic Gram-negative rods (*Prevotella* spp.), *Peptostreptococcus* spp., *Ureaplasma urealyticum*, and often *Mobilunculus* spp. (Hill, 1993) result in nonspecific vaginitis also named "bacterial vaginosis." Much remains to be learned about the interactions among members of the vaginal flora. Skarin and Sylwan (1986) found that vaginal microflora of *Lactobacillus acidophilus* and *L. casei* can inhibit growth of *G. vaginalis* and species of *Mobiluncus*, *Bacteroides* and anaerobic cocci on solid media by producing acid. Another possible antibacterial agent is hydrogen peroxide. Interestingly H_2O_2-producing lactobacilli are predominant in healthy women. *Lactobacillus* species which do not produce hydrogen peroxide are present in 36% of women with bacterial vaginosis (Eschenbach et al., 1989).

Gardnerella vaginalis may occasionally ascend to the upper genital tract of females and it has been isolated from cases of endometritis and salpingitis (Watts et al., 1989). *Gardnerella vaginalis* also has been isolated from wound infections and blood specimens (Sturm et al., 1983; Reimer and Reller, 1984; Holst, 1990). The reservoir of *G. vaginalis* and other microorganisms such as *Mobilunculus* and *Mycoplasma hominis* involved in bacterial vaginosis is also in the intestinal tract. They may be isolated from the rectums of women, men and children. The occurrence of these microorganisms is higher in the rectums of women who develop bacterial vaginosis compared to that found in the rectums of healthy subjects. The data regarding microbial composition of urethral samples from male sexual partners of women with bacterial vaginosis do not support the notion of a sexual transmission of bacteria associated with this pathology, even though men can acquire *G. vaginalis* through sexual contact (Holst, 1990; Bump and Buesching, 1988).

Gardnerella vaginalis has been isolated from the reproductive tract of mares as well. It is not known whether these microorganisms contributed to the reproductive problems of mares, inasmuch as *G. vaginalis* has been isolated in very small numbers, usually along with other microorganisms. Their role as pathogens is unclear; probably they are common inhabitants of the equine genital tract (Salmon et al., 1991).

Epidemiology

Bacterial vaginosis is associated with serious upper genital tract infections and an adverse pregnancy outcome (Sweet, 1995). In particular, the presence of bacterial vaginosis in pregnant women increases the risk of preterm delivery and there is now compelling evidence that bacterial vaginosis is a cause of preterm delivery (Hillier et al., 1995). In addition, the composition of the vaginal microflora has a significant impact on the overall health of women (Hill et al., 1984), and the abnormal and often foul-smelling discharge that accompanies bacterial vaginosis is of great concern to many women. These considerations combined with the high frequency of occurrence prompt the necessity for a greater understanding of the pathogenesis of bacterial vaginosis.

Gardnerella vaginalis was found to be present in 30–50% of normal women without evidence of bacterial vaginosis. Differences in the literature on the percentage of healthy women infected by *G. vaginalis* arise mainly from the different methodologies applied. However, using selective media, a high prevalence of *G. vaginalis* was documented in patients with bacterial vaginosis (98–100%).

Nevertheless, epidemiological studies of bacterial vaginosis suffer from a lack of clinical microbiological precision. Data from several studies propose that the prevalence varies from 15% in private gynecological patients (Bump et al., 1984) to 10–30% in pregnant women

(Gardner et al., 1957; Hill et al., 1983; Minkoff et al., 1984), whereas it was found to be 5–25% in college students (Spiegel et al., 1980; Eschenbach et al., 1988). Its prevalence range in other populations is extremely broad.

The definition of risk factors for bacterial vaginosis varies significantly, but two factors seem to be important: non-white race and the use of intrauterine devices. The effects of previous pregnancy or sexual activity are inconsistent from study to study, but it is clear that bacterial vaginosis is not exclusively a sexually transmitted disease. *Gardnerella vaginalis* has been isolated from urine samples in men with a clinical incidence of 0.1%; *G. vaginalis* may play a role in causing urinary tract infection in at least 67% of males patients from whom it was isolated (Smith et al., 1992).

Biotyping schemes based on the presence of lipase, hippurate hydrolysis and β-galactosidase have identified specific biotypes associated with bacterial vaginosis. The urethra of men and their partners who had bacterial vaginosis were colonized with the same biotype of *G. vaginalis* when the partners are cultured within 24 h of each other. The distribution of biotypes was the same in women with and without bacterial vaginosis. The treatment regimen used, and not the presence of a new sexual partner, was associated with the acquisition of a new biotype (Briselden and Hillier, 1990). Restriction endonuclease analysis of isolates from women with bacterial vaginosis has shown that a genetically mixed population is present in the vagina (Nath et al., 1992). Amplified ribosomal DNA restriction analysis (ARDRA) recognized different *G. vaginalis* genotypes. At least three to four genotypes were identified using different restriction enzymes, some of which showed a particular prevalence of distribution in certain centers from which they were collected. Although no correlation was found between bacterial vaginosis and any of the genotypes identified, the ARDRA method could prove to be a useful tool for studying the etiopathology and epidemiology of *G. vaginalis* (Ingianni et al., 1997).

Disease

In 1955, Gardner and Dukes described the newly discovered microorganism, *Haemophilus vaginalis*, which caused a newly defined specific infection called "*Haemophilus vaginalis* vaginitis" previously classified as "nonspecific vaginitis" (Gardner and Dukes, 1955). They also suggested some criteria for the identification of this pathology that still are the basis of the diagnosis today, as described by Amsel et al. (1983). These criteria consider the presence of a homogenous (not

clumpy) vaginal discharge, an elevated pH (5.5–6.0), a positive "whiff" test and the presence of clue cells in the vaginal fluid wet mount preparation. The discoveries of Gardner and Dukes were important in defining the clinical disease and the association of the microorganism *H. vaginalis* with the syndrome. The taxonomy of *Haemophilus vaginalis* underwent changes, that is, from *Haemophilus vaginalis* to *Corynebacterium vaginale* and to its present designation *Gardnerella vaginalis* (Greenwood and Pickett, 1980). Following the introduction of the new techniques for the isolation of anaerobic microorganisms, it appeared evident that "*Haemophilus vaginalis* vaginitis" is not only caused by *G. vaginalis* but also by a polymicrobial complex formed by a variety of anaerobic microorganisms such as *Mobiluncus*, *Bacteroides*, *Prevotella* and *Mycoplasma* species. *Haemophilus vaginalis* vaginitis, also called "nonspecific vaginitis" and "*G. vaginalis* vaginitis," today might be more appropriately termed "bacterial vaginosis," which better reflects the microbiological complexity of this clinical disease. The term "vaginosis" is therefore preferred to "vaginitis" because of the absence of clinical signs of inflammation; in addition, inflammatory cells are typically absent in vaginal discharge in bacterial vaginosis (Gardner and Dukes, 1955; Sobel, 1989). The clinical criteria, even in the presence of similar symptomatologies (and as abnormal vaginal discharge), can discriminate between vaginitis due to gonococci, yeasts or trichomonads where an increased level of leukocytes is observed and bacterial vaginosis where an increased concentration of anaerobes, a fishy or amine odor and a pH > 4.5 is observed. However, around half the women with this condition are asymptomatic. Diagnosis is best made by microscopic examination of a Gram-stained smear of vaginal secretions. Bacterial vaginosis, determined by the presence of microorganisms which are opportunistic pathogens, is an important vaginal infection because of its potential cause of upper genital tract infection in specific circumstances (Eschenbach, 1993). Recent work has linked bacterial vaginosis to numerous upper genital tract complications such as preterm labor and preterm delivery, preterm premature rupture of the membranes, chorioamnionitis, and postpartum endometritis (Watts et al., 1989; Holst et al., 1994; Kimberlin and Andrews, 1998). The findings from recent prospective randomized trials suggest that treatment of bacterial vaginosis in certain women who are at a high risk of preterm delivery decreases the rate of preterm birth (Hay et al., 1994; Kimberlin and Andrews, 1998). *Gardnerella vaginalis* and the anaerobes associated with bacterial vaginosis are often associated with an increased rate of postgynecological infection after a hysterectomy

(Kristiansen et al., 1990). The cause of salpingitis is suspected in part to be also by bacterial vaginosis. The consequences of bacterial vaginosis need to be confirmed by more studies, but it has been demonstrated that high concentrations of microorganisms (found in patients with bacterial vaginosis) can infect adjacent tissues.

Most therapeutic trials evaluate treatment outcome at the completion of treatment day 7, or at best, 28–35 days following the termination of therapy, and longer follow-up is rare. Moreover, Hillier and Holmes (1990) reported that 80% of patients with acute bacterial vaginosis can be expected to have a recurrence within 9 months after metronidazole therapy.

In males, *G. vaginalis* has been reported to be the cause of urethritis, prostatitis or urinary tract infections (Smith et al., 1992; Abercrombie et al., 1978; Woolfrey et al., 1986).

A correlation between *G. vaginalis* and activation of human immunodeficiency virus (HIV) type 1 has been found. *Gardnerella vaginalis* lysates were found to significantly stimulate HIV expression in monocytoid cells and in certain T cell lines. The activation of HIV production by *G. vaginalis* suggests that genital infection with *G. vaginalis* increases the risk of HIV transmission by increasing HIV expression in genital tract. This may explain, at least in part, the increased rate of HIV transmission in women with bacterial vaginosis (Hashemi et al., 1999).

Concerning therapy of bacterial vaginosis, treatment is with metronidazole or clavulanic acid-amoxicillin (Augmentin). The indications for treatment of asymptomatic bacterial vaginosis are not clear, but women should probably be treated before any invasive gynecological procedure, including intrauterine contraceptive device insertion.

Literature Cited

Abercrombie, G. F., J. Allen, and R. Maskell. 1978. Corynebacterium vaginale urinary-tract infection in a man. Lancet 1:766.

Amsel, R., P. A. Totten, C. A. Spiegel, K. C. Chen, D. A. Eschenbach, and K. K. Holmes. 1983. Nonspecific vaginitis: Diagnostic criteria and microbial and epidemiologic associations. Am. J. Med. 74:14–22.

Argnani, A., R. J. Leer, N. Van Luijk, and P. H. Pouwels. 1996. A convenient and reproducible method to genetically transform bacteria of the genus Bifidobacterium. Microbiology 142:109–114.

Arroyo, L., L. N. Cotton, and J. H. Martin. 1994. AMC agar—a composite medium for selective enumeration of Bifidobacterium longum. Cult. Dairy Prod. J. 30:12–15.

Bailey, R. K., J. L. Voss, and R. F. Smith. 1979. Factors affecting isolation and identification of Haemophilus vaginalis (Corynebacterium vaginale). J. Clin. Microbiol. 9:65–71.

Bannatyne, R. M., J. Jackowski, R. Cheung, and K. Biers. 1987. Susceptibility of Gardnerella vaginalis to metronidazole, its bioactive metabolites and tinidazole. Am. J. Clin. Pathol. 87:640–641.

Beerens, H. 1990. An elective and selective isolation medium for Bifidobacterium spp. Lett. Appl. Microbiol. 11:155–157.

Bejar, R., V. Curbelo, C. Davis, and L. Gluck. 1981. Premature labor II: Bacterial sources of phospholipase. Obstet. Gynecol. 57:479–482.

Benno, Y., K. Sawada, and T. Mitsuoka. 1984. The intestinal microflora of infants: omposition of fecal flora in breast-fed and bottle-fed infants. Microbiol. Immunol. 28:975–986.

Benno, Y., and T. Mitsuoka. 1986. Development of intestinal microflora in humans and animals. Bifidobacteria and Microflora 5:13–25.

Benno, Y., K. Endo, N. Shirsagami, K. Sayama, and T. Mitsuoka. 1987. Effects of raffinose intake on human fecal microflora. Bifidobacteria and Microflora 6:59–63.

Bernet, M.-F., D. Brassart, J.-R. Neeser, and A. L. Servin. 1993. Adhesion of human bifidobacterial strains to cultured human intestinal epithelial cells and inhibition of enteropathogen-cell interactions. Appl. Environ. Microbiol. 59:4121–4128.

Bezkorovainy, A., and N. Topouzian. 1983. Aspects of iron metabolism in Bifidobacterium bifidum var. pennsylvanicus. Int. J. Biochem. 15:316–366.

Bezkorovainy, A., R. Miller-Catchpole, M. Poch, and L. Solberg. 1986. The mechanism of ferrous iron binding by suspensions of Bifidobacterium bifidum var. pennsylvanicus. Biochim. Biophys. Acta 884:60–66.

Bezkorovainy, A., L. Solberg, M. Poch, and R. Miller-Catchpole. 1987. Ferrous iron uptake by Bifidobacterium bifidum var. pennsylvanicus: The effect of metals and metabolic inhibitors. Int. J. Biochem. 19:517–522.

Bezkorovainy, A. 1989. Nutrition and metabolism of bifidobacteria. In: A. Bezkorovainy and R. Miller-Catchpole (Eds.) Biochemistry and Physiology of Bifidobacteria. CRC Press. Boca Raton, FL. 93–129.

Biavati, B., V. Scardovi, and W. E. C. Moore. 1982. Electrophoretic patterns of proteins in the genus Bifidobacterium and proposal of four new species. Int. J. Syst. Bacteriol. 32:358–373.

Biavati, B., P. Castagnoli, F. Crociani, and L. D. Trovatelli. 1984. Species of the Bifidobacterium in the feces of infants. Microbiologica 7:341–345.

Biavati, B., P. Castagnoli, and L. D. Trovatelli. 1986. Species of the genus Bifidobacterium in the feces of human adults. Microbiologica 9:39–45.

Biavati, B., and P. Mattarelli. 1991. Bifidobacterium ruminantium sp.nov. and Bifidobacterium merycicum sp.nov. from the rumens of cattle. Int. J. Syst. Bacteriol. 41:163–168.

Biavati, B., P. Mattarelli, and F. Crociani. 1991a. Bifidobacterium saeculare, a new species isolated from feces of rabbit. Syst. Appl. Microbiol. 14:389–392.

Biavati, B., B. Sgorbati, and V. Scardovi. 1991b. The genus Bifidobacterium. In: A. Balows, H. G. Trüper, M. Dworkin, W. Harder, and K.-H. Schleifer (Eds.) The Prokaryotes, 2nd ed. Springer-Verlag. New York, NY. 1:816–833.

Biavati, B., F. Crociani, P. Mattarelli, and V. Scardovi. 1992a. Phase variations in Bifidobacterium animalis. Curr. Microbiol. 25:51–55.

Biavati, B., P. Mattarelli, and F. Crociani. 1992b. Identification of bifidobacteria from fermented milk products. Microbiologica 15:7–14.

Biavati, B., T. Sozzi, P. Mattarelli, and L. D. Trovatelli. 1992c. Survival of bifidobacteria from human habitat in acidified milk. Microbiologica 15:197–200.

Biavati, B., P. Mattarelli, A. Alessandrini, F. Crociani, and M. Guerrini. 1995. Survival in fermented milk products of Bifidobacterium animalis and its recovery in human feces. Mikroökologie und Therapie 25:231–235.

Biavati, B., S. Franzoni, H. Ghazvinizadeh, and R. Piccaglia. 1997. Antimicrobial and antioxidant properties of plant essential oils. In: C. Franz, Á. Máthé, and G. Buchbauer (Eds.) Essential Oils: Basic and Applied Research. Proceedings of 27th International Symposium on Essential Oils. Allured Publishing Corporation. Carol Stream, IL. 326–331.

Bouhnik, Y., P. Pochart, P. Marteu, G. Arlet, I. Goderel, and J. C. Rambaud. 1992. Fecal recovery in humans of viable Bifidobacterium sp. ingested in fermented milk. Gastroenterology 102:875–878.

Bouhnik, Y., B. Flourié, M. Riottot, N. Bisetti, M.-F. Gailing, A. Guibert, F. Bornet, and J. C. Rambaud. 1996. Effects of fructo-oligosaccharides ingestion on fecal bifidobacteria and selected metabolic indexes of colon carcinogenesis in healthy humans. Nutr. Cancer 26:21–29.

Bouhnik, Y., B. Flourié, L. D'Agay-Abensour, P. Pochart, G. Gramet, M. Durand, and J.-C. Rambaud. 1997. Administration of trangalactoligosaccharides increases fecal bifidobacteria and modifies colonic fermentation metabolism in healthy humans. J. Nutr. 127:444–448.

Bourget, N., J. M. Simonet, and B. Decaris. 1993. Analysis of the genome of the five Bifidobacterium breve strains: Plasmid content, pulsed-field gel electrophoresis genome size estimation and rrn loci number. FEMS Microbiol. Lett. 110:11–20.

Boustouller, Y. L., A. P. Johnson, and D. Taylor-Robinson. 1987. Pili on Gardnerella vaginalis studied by electron-microscopy. J. Med. Microbiol. 23:327–329.

Brandi, G., S. Sarchielli, P. Mordenti, S. Tambieri, C. Calabrese, P. Mattarelli, B. Biavati, and G. Biasco. 1996. Colonization of human achloridric stomach by bifidobacteria. Gut 39, Suppl. 3:A229.

Breed, R. S., E. G. D. Murray, and N. R. Smith (Eds.) 1957. Bergey's Manual of Determinative Bacteriology, 7th ed. Williams and Wilkins. Baltimore, MD.

Briselden, A. M., and S. L. Hillier. 1990. Longitudinal study of the biotypes of Gardnerella vaginalis. J. Clin. Microbiol. 28:2761–2764.

Brown, W. J. 1974. Modification of the rapid fermentation test for Neisseria gonorrhoeae. Appl. Microbiol. 27:1027–1030.

Buchanan, R. E., and N. E. Gibbons (Eds.) 1974. Bergey's Manual of Determinative Bacteriology, 8th ed. Williams and Wilkins. Baltimore, MD.

Bump, R. C., F. P. Zuspan, W. J. Buesching, 3rd, L. W. Ayers, and T. J. Stephens. 1984. The prevalence, six-month persistence, and predictive values of laboratory indicators of bacterial vaginosis (non specific vaginitis) in asymptomatic women. Am. J. Obstet. Gynecol. 150:917–924.

Bump, R. C., and W. J. Buesching, 3rd. 1988. Bacterial vaginosis in virginal and sexually active adolescent females: Evidence against exclusive sexual transmission. Am. J. Obstet. Gynecol. 158:935–939.

Catlin, B. W. 1992. Gardnerella vaginalis: Characteristics, clinical considerations, and controversies. Clin. Microbiol. Rev. 5:213–237.

Challa, A., D. R. Rao, C. B. Chawan, and L. Shackelford. 1997. Bifidobacterium longum and lactulose suppress azoxymethane-induced colonic aberrant crypt foci in rats. Carcinogenesis 18:517–521.

Chapin, K. C., and P. R. Murray. 1999. Media. In: P. R. Murray, E. Y. Baron, M. A. Pfaller, F. C. Tenover, and R. H. Yolken (Eds.) Manual of Clinical Microbiology, 7th ed. ASM Press. Washington DC, 1687–1707.

Charteris, W. P., P. M. Kelly, L. Morelli, and J. K. Collins. 1998. Antibiotic susceptibility of potentially probiotic Bifidobacterium isolates from the human gastrointestinal tract. Lett. Appl. Microbiol. 26:333–337.

Chen, K. C., P. S. Forsyth, T. M. Buchanan, and K. K. Holmes. 1979. Amine content of vaginal fluid from untreated and treated patients with nonspecific vaginitis. J. Clin. Invest. 63:828–835.

Chevalier, P., D. Roy, and P. Ward. 1990. Detection of Bifidobacterium species by enzymatic methods. J. Appl. Bacteriol. 68:619–624.

Chiappini, M. G. 1966. Carbon dioxide fixation in some strains of the species Bifidobacterium bifidum, Bifidobacterium constellatum, Actinomyces bovis and Actinomyces israelii. Annali di Microbiologia ed Enzimologia 16:25–32.

Collins, M. D., and S. Wallbanks. 1992. Comparative sequence analysis of the 16S rRNA genes of Lactobacillus minutus, Lactobacillus rimae and Streptococcus parvulus: Proposal for a creation of a new genus Atopobium. FEMS Microbiol. Lett. 95:235–240.

Collins, M. D., and G. R. Gibson. 1999. Probiotics, prebiotics and synbiotics: Approaches for modulating the microbial ecology of the gut. Am. J. Clin. Nutr. 69:S1052–S1057.

Colombel, J. F., A. Cortot, C. Neut, and C. Romond. 1987. Yoghurt with Bifidobacterium longum reduces erytromicin-induced gastrointestinal effects. Lancet 2:43.

Conway, P. 1997. Development of intestinal microbiota. In: R. I. Mackie, B. A. White, and R. E. Isacson (Eds.) Gastrointestinal Microbiology. Chapman and Hall. New York, NY. 2:3–38.

Cook, R. L., G. Reid, D. G. Pond, C. A. Schmith, and J. D. Sobel. 1989. Clue cells in bacterial vaginosis immunofluorescent identification of the adherent Gram-negative bacteria as Gardnerella vaginalis. J. Infect. Dis. 160:490–496.

Crisweel, B. S., J. H. Marston, W. A. Stenback, S. H. Black, and H. L. Gardner. 1971. Haemophilus vaginalis 594, a Gram negative organism? Can. J. Microbiol. 17:865–869.

Crisweel, B. S., W. A. Stenback, S. H. Black, and H. L. Gardner. 1972. Fine structure of Haemophilus vaginalis. J. Bacteriol. 109:930–932.

Crociani, F., D. Matteuzzi, and H. Ghazvinizadeh. 1973. Species of the genus Bifidobacterium found in human vagina. Zentralbl. Bakteriol. Parasitenkd. Infektionskr. Hyg. Abt. 1, Orig. Reihe A223 298–302.

Crociani, F., O. Emaldi, and D. Matteuzzi. 1977. Increase in isoleucine accumulation by alpha-aminobutyric acid-resistant mutants of Bifidobacterium ruminale. Eur. J. Appl. Microbiol. 4:177–179.

Crociani, F., and D. Matteuzzi. 1982. Urease activity in the genus Bifidobacterium. Ann. Microbiol. (Paris, France) 133A:417–423.

Crociani, F., A. Alessandrini, M. M. Mucci, and B. Biavati. 1994. Degradation of complex carbohydrates by Bifidobacterium spp. Int. J. Food Microbiol. 24:199–210.

Crociani, J., J.-P. Grill, M. Hupper, and J. Ballongue. 1995. Adhesion of different bifidobacteria strains to human enterocyte like Caco-2 cells and comparison with in vivo study. Lett. Appl. Microbiol. 21:146–148.

Crociani, F., A. Alessandrini, B. Biavati, and P. Mattarelli. 1996a. Presence of bifidobacteria in sewage. SOMED 21st International Congress on Microbial Ecology and Disease, October 28–30. Institute Pasteur. Paris, France. 74.

Crociani, F., B. Biavati, A. Alessandrini, C. Chiarini, and V. Scardovi. 1996b. Bifidobacterium inopinatum sp. nov. and Bifidobacterium denticolens sp. nov., two new species isolated from human dental caries. Int. J. Syst. Bacteriol. 46:564–571.

Crociani, F., B. Biavati, A. Alessandrini, and G. Zani. 1997. Growth inhibition activity of essential oils and other antimicrobial agents towards bifidobacteria from dental caries. In: C. Franz, Á. Máthé, and G. Buchbauer (Eds.) Essential Oils: Basic and Applied Research. Proceedings of 27th International Symposium on Essential Oils. Allured Publishing Corporation. Carol Stream, IL. 40–44.

Cruden, D. L., and R. P. Galask. 1988. Reduction of trimethylamine oxide to trimethylamine by Mobiluncus strains isolated from patients with bacterial vaginosis. Microb. Ecol. Health Dis. 1:95–100.

Csango, P. A., N. Hagen, and G. Jagars. 1982. Method for isolation of Gardnerella vaginalis (Haemophilus vaginalis). Acta Pathol. Microbiol. Immunol. Scand. 90:89–93.

Davidson, M. H., K. C. Maki, C. Synecki, S. A. Torri, and K. B. Drennan. 1998. Effects of dietary inulin on serum lipids in men and women with hypercholesterolemia. Nutr. Res. 18:503–517.

Deguchi, Y., T. Morishita, and M. Mutai. 1985. Comparative studies on the synthesis of water soluble vitamins among human species of bifidobacteria. Agric. Biol. Chem. 49:13.

Dehnert, J. 1957. Untersuchungen über die Gram positive Stuhlflora des Brustmilchkinder. Zentrabl. Bakteriol. Parasitenkd. Infektionskr. Hyg. Abt. I Orig. Reihe A169 66–79.

DeMan, J. C., M. Rogosa, and M. E. Sharpe. 1960. A medium for the cultivation of lactobacilli. J. Appl. Microbiol. 8:95–98.

Derensy-Dron, D., F. Krzewinski, C. Brassart, and S. Bouquelet. 1999. Beta-1,3-galactosyl-N-acetylhexosamine phosphorylase from Bifidobacterium bifidum DSM 20082: Characterization, partial purification and relation to mucin degradation. Biotechnol. Appl. Biochem. 29:3–10.

Desjardins, M.-L., and D. Roy. 1990. Growth of bifidobacteria and their enzyme profiles. J. Dairy Sci. 73:299–307.

De Vries, W., S. J. Gerbrandy, and A. H. Stouthamer. 1967. Carbohydrate metabolism in Bifidobacterium bifidum. Biochim. Biophys. Acta 136:415–425.

De Vries, W., and A. H. Stouthamer. 1968. Fermentation of glucose, lactose, galactose, mannitol, and xylose by bifidobacteria. J. Bacteriol. 96:472–478.

Djouzi, Z., and C. Andrieux. 1997. Compared effects of three oligosaccharides on metabolism of intestinal microflora in rats inoculated with a human faecal flora. Br. J. Nutr. 78:313–324.

Dong, X., Y. Xin, W. Jian, X. Liu, and D. Ling. 2000. Bifidobacterium thermacidophilum sp. nov., isolated from an anaerobic digester. Int. J. Syst. Evol. Microbiol. 50:119–125.

Dubreuil, L., I. Houcke, Y. Mouton, and J. F. Rossignol. 1996. In vitro evaluation of activities of nitazoxanide and tizoxanide against anaerobes and aerobic organisms. Antimicrob. Agents Chemother. 40:2266–2270.

Duffy, L. C., M. S. Zielezny, and M. Riepenhoff-Talty. 1993. Effectiveness of Bifidobacterium bifidum in mediating the clinical course of murine rotavirus diarrhea. Pediatr. Res. 33:983–986.

Dunkelberg Jr., W. E., and I. McVeigh. 1969. Growth requirements of Haemophilus vaginalis. Ant. v. Leeuwenhoek 35:129–145.

Dunkelberg Jr., W. E., R. Skaggs, and D. S. Kellogg. 1970. A study and a new description of Corynebacterium vaginale (Haemophilus vaginalis). Am. J. Clin. Pathol. 53:370–377.

Dunne, C., S. O'Halloran, L. O'Mahony, M. Feeney, G. O'Sullivan, and J. K. Collins. 1999. Epithelial adhesion of probiotic microrganism in vitro and in vivo. Gastroenterology 116:G3058.

Elmer, G. W., C. M. Surawicz, and L. V. McFarland. 1996. Biotherapeutic agents: A neglected modality for the treatment and prevention of selected intestinal and vaginal infections. JAMA 275:870–876.

Embley, T. M., and E. Stackebrandt. 1994. The molecular phylogeny and systematics of the Actinomycetes. Ann. Rev. Microbiol. 48:257–289.

Eschenbach, D. A., S. Hillier, C. Critchlow, C. Stevens, T. DeRouen, and K. K. Holmes. 1988. Diagnosis and clinical manifestations of bacterial vaginosis. Am. J. Obstet. Gynecol. 158:819–828.

Eschenbach, D. A., P. R. Davick, B. L. Williams, S. J. Klebanoff, K. Young-Smith, C. M. Critchlow, and K. K. Holmes. 1989. Prevalence of hydrogen peroxide-producing Lactobacillus species in normal women and women with bacterial vaginosis. J. Clin. Microbiol. 27:251–256.

Eschenbach, D. A. 1993. History and review of bacterial vaginosis. Am. J. Obstet. Gynecol. 169:441–445.

Falk, P., L. C. Hoskins, and G. Larson. 1990. Bacteria of the human intestinal microbiota produce glycosidases specific for lacto-series glycosphingolipids. J. Biochem. (Tokyo) 108:466–474.

Fanedl, L., F. V. Nekrep, and G. Avgustin. 1998. Random amplified polymorphic DNA analysis and demonstration of genetic variability among bifidobacteria isolated from rats fed with raw kidney beans. Can. J. Microbiol. 44:1094–1101.

Favier, C., C. Neut, C. Mizon, A. Cortot, J. F. Colombel, and J. Mizon. 1997. Fecal beta-D-galactosidase production and bifidobacteria are decreased in Crohn's disease. Digest. Dis. Sci. 42:817–822.

Finegold, S. M., V. L. Sutter, and G. E. Mathisen. 1983. Normal indigenous microflora. In: D. J. Hentges (Ed.) Human Intestinal Microflora in Health and Disease. Academic Press. London, 65–284.

Fiordaliso, M., N. Kok, J. P. Desager, F. Goethals, D. Deboyser, M. Roberfroid, and N. Delznne. 1995. Dietary oligofructose lowers triglycerides, phospholipids and cholesterol in serum and very low density lipoproteins in rats. Lipids 30:163–167.

Fischer, W., W. Bauer, and M. Feigel. 1987. Analysis of the lipoteichoic-acid-like macroamphiphile from Bifidobacterium bifidum subsp. pennsylvanicum by one- and two-dimensional, H- and C-NMR spectroscopy. Eur. J. Biochem. 165:647–652.

Fox, G. E., E. Stackebrandt, R. B. Hespell, J. Gibson, J. Maniloff, T. A. Dyer, R. S. Wolfe, W. E. Balch, R. S. Tanner, L. J. Magrum, L. B. Zablen, R. Blakemore, R. Gupta, L. Bonen, B. J. Lewis, D. A. Stahl, K. R. Luehrsen, K. N. Chen, and C. R. Woese. 1980. The phylogeny of prokaryotes. Science 209:457–463.

Fujiwara, S., H. Hashiba, T. Hirota, and J. F. Forstner. 1997. Proteinaceous factor(s) in culture supernatant fluids of bifidobacteria which prevents the binding of enterotoxigenic Escherichia coli to gangliotetraosylceramide. Appl. Environ. Microbiol. 63:506–512.

Fuller, R. 1989. Probiotics in man and animals. J. Appl. Bacteriol. 66:365–378.

Funke, G., and K. A. Bernard. 1999. Coryneform Gram-positive rods. In: P. R. Murray, E. Y. Baron, M. A. Pfaller, F. C. Tenover, and R. H. Yolken (Eds.) Manual of Clinical Microbiology, 7th ed. ASM Press. Washington DC, 319–346.

Gallaher, D. D., and K. Jinmo. 1999. The effect of synbiotics on colon carcinogenesis in rats. J. Nutr. 129:S1483–S1487.

Gardner, H. L., and C. H. Dukes. 1955. Haemophilus vaginalis vaginitis: A newly defined specific infection previously classified as non specific vaginitis. Am. J. Obstet. Gynecol. 69:962–976.

Gardner, H. L., T. K. Dampeer, and C. D. Dukes. 1957. The prevalence of vaginitis. Am. J. Obstet. Gynecol. 73:1080–1087.

Gavini, F., A.-M. Pourcher, C. Neut, D. Monget, C. Romond, C. Oger, and D. Izard. 1991. Phenotypic differentiation of bifidobacteria of human and animal origin. Int. J. Syst. Bacteriol. 41:548–557.

Gavini, F., M. Van Esbroeck, J. P. Touzel, A. Fourment, and H. Goossens. 1996. Detection of fructose-6-phosphate phosphoketolase (F6PPK), a key enzyme of the bifid-shunt in Gardnerella vaginalis. Anaerobe 2:191–193.

Gavini, F., and H. Beerens. 1999. Origin and identification of bifidobacteria strains isolated from meat and meat products. Int. J. Food Microbiol. 46:81–85.

Gherna, R. L. 1994. Culture preservation. In: P. Gerhard, R. G. E. Murray, W. A. Wood, and N. R. Krieg (Eds.) Methods for General and Molecular Bacteriology. American Society for Microbiology. Washington DC, 278–292.

Ghoddusi, H. B., and R. K. Robinson. 1996. Enumeration of starter cultures in fermented milks. J. Dairy Res. 63:753–761.

Gibbs, R. S., M. H. Weiner, K. Walmer, and P. J. St. Clair. 1987. Microbiologic and serologic studies of Gardnerella vaginalis in intra-amniotic infection. Obstet. Gynecol. 70:187–190.

Gibson, G. R., and X. Wang. 1994. Regulatory effects of bifidobacteria on the growth of other colonic bacteria. J. Appl. Bacteriol. 77:412–420.

Gibson, G. R., and M. B. Roberfroid. 1995. Dietary modulation of the human colonic microbiota: Introducing the concept of prebiotics. J. Nutr. 125:1401–1412.

Gibson, G. R. 1998. Dietary modulation of the human gut microflora using prebiotics. Br. J. Nutr. 80:S209–S212.

Glick, M. C., T. Sall, F. Zilliken, and S. Mudd. 1960. Morphologycal changes in Lactobacillus bifidus var. pennsylvanicus produced by a cell wall precursor. Biochem. Biophys. Acta 37:361–368.

Goldberg, R. L., and J. A. Washington, 2nd. 1976. Comparison of isolation of Haemophilus vaginalis (Corynebacterium vaginale) from peptone-starch-dextrose agar and columbia colistin-nalidixic acid agar. J. Clin. Microbiol. 4:245–247.

Greenwood, J. R., M. J. Pickett, W. J. Martin, and E. G. Mack. 1977. Haemophilus vaginalis (Corynebacterium vaginale): Method for isolation and biochemical rapid identification. Health Lab. Sci. 14:102–106.

Greenwood, J. R., and M. J. Pickett. 1979. Salient features of Haemophilus vaginalis. J. Clin. Microbiol. 9:200–204.

Greenwood, J. R., and M. J. Pickett. 1980. Transfer of Haemophilus vaginalis Gardner and Dukes to a new genus, Gardnerella: G. vaginalis (Gardner and Dukes) comb. nov. Int. J. Syst. Bacteriol. 30:170–178.

Griffits, E. 1991. Iron and bacterial virulence—a brief overview. Biol. Metals 4:7–13.

Grill, J. P., J. Crociani, and J. Ballongue. 1995. Characterization of fructose-6-phosphate phosphoketolases purified from Bifidobacterium species. Curr. Microbiol. 31:49–54.

Gyorgy, P., R. Kuhn, C. S. Rose, and F. Zilliken. 1954a. Bifidus factor II: Its occurrence in milk from different species and in other natural products. Arch. Biochem. Biophys. 48:202–209.

Gyorgy, P., R. F. Norris, and C. S. Rose. 1954b. Bifidus factor I: A variant of Lactobacillus bifidus requiring a special growth factor. Arch. Biochem. Biophys. 48:193.

Gyorgy, P., and C. S. Rose. 1955. Further observations on the metabolic requirements of Lactobacillus bifidus var. pensylvanicus. J. Bacteriol. 69:483–490.

Ha, G. Y., C. H. Yang, H. Kim, and Y. Chong. 1999. Case of sepsis caused by Bifidobacterium longum. J. Clin. Microbiol. 37:1227–1228.

Haavenar, R., J. H. J. Huis in't Veld. 1992. Probiotics, a general view. In: B. B. Wood (Ed.) The Lactic Acid Bacteria in Health and Disease. Elsevier. London, 1:209–224.

Habu, Y., M. Nagaoka, T. Yokokura, and I. Azuma. 1987. Structural studies of cell wall polysaccharides from Bifidobacterium breve YIT 4010 and related Bifidobacterium species. J. Biochem. 102:1423–1432.

Harmsen, H. J. M., G. R. Gibson, P. Elfferich, G. C. Raangs, A. C. M. Wildeboer-Veloo, A. Argaiz, M. B. Roberfroid, G. W. Welling. 1999. Comparison of viable cell counts and fluorescence in situ hybridization using specific rRNA-based probes for the quantification of human fecal bacteria. FEMS Microbiol. Lett. 183:125–129.

Harmsen, H. J. M., Wildeboer-Veloo, G. C. Raangs, A. A. Wagendorp, N. Klijn, J. B. Bindels, and G. W. Welling. 2000. Analysis of intestinal flora development in breast-fed and formula-fed infants by using molecular identification methods. J. Pediatr. Gastroent. Nutr. 30:61–67.

Harper, J. J., and G. H. G. Davis. 1982. Cell wall analysis of Gardnerella vaginalis (Haemophilus vaginalis). Int. J. Syst. Bacteriol. 32:48–50.

Hartemink, R., B. J. Kok, G. H. Weenk, and F. M. Rombouts. 1996. Raffinose-Bifidobacterium (RB) agar, a new selective medium for bifidobacteria. J. Microbiol. Meth. 27:33–34.

Hashemi, B. F., M. Ghassemi, K. A. Roebuck, and G. T. Spear. 1999. Activation of human immunodeficiency virus type 1 expression by Gardnerella vaginalis. J. Infect. Dis. 179:924–930.

Hassinen, J. B., G. T. Durbin, R. M. Tomarelli, and F. W. Bernhart. 1951. The minimal nutritional requirements of Lactobacillus bifidus. J. Bacteriol. 62:771–777.

Hatanaka, M., T. Tachiki, H. Kumagai, and T. Tochikura. 1987. Distribution and some properties of glutamine syntetase and glutamine dehydrogenase in bifidobacteria. Agric. Biol. Chem. 51:251–257.

Hatcher, G. E., and R. S. Lambrecht. 1993. Augmentation of macrophage phagocitic activity by cell-free extracts of selected lactic acid producing bacteria. J. Dairy Sci. 76:2485–2492.

Hay, P. E., D. J. Morgan, C. A. Ison, S. A. Bhide, M. Romney, P. McKenzie, J. Pearson, R. F. Lamont, and D. Taylor-Robinson. 1994. A longitudinal study of bacterial vaginosis during pregnancy. Br. J. Obstet. Gynaecol. 101:1048–1053.

Hayakawa, K., J. Mizutani, K. Wada, T. Masai, I. Yoshihara, and T. Mitsuoka. 1990. Effect of soybean oligosaccharides on human fecal microflora. Microb. Ecol. Health Dis. 3:293–303.

Hill, L. H., H. Ruparelia, and J. A. Embil. 1983. Nonspecific vaginitis and other genital infections in three clinic populations. Sex. Transm. Dis. 10:114–118.

Hill, G. B., D. A. Eschenbach, and K. K. Holmes. 1984. Bacteriology of the vagina. Scand. J. Urol. Nephrol. Suppl. 86:23–39.

Hill, G. B. 1993. The microbiology of bacterial vaginosis. Am. J. Obstet. Gynecol. 169:450–454.

Hillier, S. L., and K. K. Holmes. 1990. Bacterial vaginosis. In: K. K. Holmes, P. A. Mardh, P. F. Sparling, and P. J. Wiesner (Eds.) Sexually Transmitted Disease, 2nd ed. McGraw-Hill. New York, NY. 547–559.

Hillier, S. L., R. P. Nugent, D. A. Eschenbach, M. A. Krohn, R. S. Gibbs, D. H. Martin, M. F. Cotch, R. Edelman, J. G. Pastorek, 2nd, A. V. Rao, D. McNellis, J. A. Regan, J. C. Carey, and M. A. Klebanoff. 1995. Association between bacterial vaginosis and preterm delivery of a low birth-weight infant: The vaginal infections and prematurity study group. N. Engl. J. Med. 333:1737–1742.

Hirano, S., H. Hayashi, T. Terabayashi, K. Onodera, and S. Iseki. 1968. Biologically active glycopeptides in human colostrum. J. Biochem. 64:563–568.

Hirayama, K., and J. Rafter. 1999. The role of lactic acid in colon cancer prevention: Mechanistic considerations. Ant. v. Leeuwenhoek 76:391–394.

Holdeman, L. V., E. P. Cato, and W. E. C. Moore (Eds.) 1977. Anaerobe Laboratory Manual, 4th ed. Virginia Polytechnic Institute and State University. Blacksburg, VI.

Holland, D. F. 1920. Generic index of the commoner forms of bacteria. J. Bacteriol. 5:191–229.

Holst, E. 1990. Reservoir of four organisms associated with bacterial vaginosis suggests lack of sexual transmission. J. Clin. Microbiol. 28:2035–2039.

Holst, E., A. R. Goffeng, and B. Andersch. 1994. Bacterial vaginosis and vaginal microorganisms in idiopathic premature labor and association with pregnancy outcome. J. Clin. Microbiol. 32:176–186.

Holt, J. G., N. R. Krieg, P. H. A. Sneath, T. Staley, and S. T. Williams (Ed.) 1993. Bergey's Manual of Determinative Bacteriology, 9th ed. Williams and Wilkins. Baltimore, MD.

Hoyos, A. B. 1999. Reduced incidence of necrotizing enterocolitis associated with enteral administration of Lactobacillus acidophilus and Bifidobacterium infantis to neonates in an intensive care unit. Int. J. Infect. Dis. 3:197–202.

Huang, R., D. M. Gascoyne-Binzi, P. M. Hawkey, M. Yu, J. Heritage, and A. Eley. 1997. Molecular evolution of the tet(M) gene in Gardnerella vaginalis. J. Antimicrob. Chemother. 40:561–565.

Husain, I., J. A. Poupard, R. F. Norris. 1972. Influence of nutrition on the morphology of a strain of Bifidobacterium bifidum. J. Bacteriol. 111:841–844.

Hwang, M.-N., and G. M. Ederer. 1975. Rapid hippurate hydrolysis method for presumptive identification of group B streptococci. J. Clin. Microbiol. 1:114–115.

Imamura, L., K. Hisamitsu, and K. Kobashi. 1994. Purification and characterization of beta-fructofuranosidase from Bifidobacterium infantis. Biol. Pharm. Bull. 17:596–602.

Ingham, S. C. 1999. Use of modified Lactobacillus selective medium and Bifidobacterium iodoacetate medium for differential enumeration of Lactobacillus acidophilus and Bifidobacterium spp. in powdered nutritional products. J. Food Prot. 62:77–80.

Ingianni, A., S. Petruzzelli, G. Morandotti, and R. Pompei. 1997. Genotypic differentiation of Gardnerella vaginalis by amplified ribosomal DNA restriction analysis (ARDRA). FEMS Immunol. Med. Microbiol. 18:61–66.

Ishibashi, N., and S. Shimamura. 1993. Bifidobacteria: Research and development in Japan. Food Technol. 47:126–135.

Ison, C. A., S. G. Dawson, J. Hilton, G. W. Csonka, and C. S. Easmon. 1982. Comparison of culture and microscopy in the diagnosis of Gardnerella vaginalis infection. J. Clin. Pathol. 35:550–554.

Ito, M., T. Ohno, and R. Tanaka. 1992. A specific DNA probe for identification of Bifidobacterium breve. Microbiol. Ecol. Health Dis. 5:185–192.

Ito, M., Y. Deguchi, K. Matsumoto, M. Kimura, T. Kan, A. Miyamori-Watabe, and T. Yajima. 1993. Effects of trangalactosylated disaccharides on the human intestinal microflora and their metabolism. J. Nutr. Sci. Vitaminol. (Tokyo) 39:279–288.

Iwata, M., and T. Morishita. 1989. The presence of plasmids in Bifidobacterium breve. Lett. Appl. Microbiol. 9:165–168.

Jarosik, G. P., C. B. Land, P. Duhon, R. Chandler Jr., and T. Mercer. 1998. Acquisition of iron by Gardnerella vaginalis. Infect. Immun. 66:5041–5047.

Jaskari, J., P. Kontula, A. Siitonen, H. Jousimies-Somer, T. Mattila-Sandholm, and K. Poutanen. 1998. Oat-beta-glucan and xylan hydrolysates as selective substrates for Bifidobacterium and Lactobacillus strains. Appl. Microbiol. Biotechnol. 49:175–181.

Ji, G. E., S. K. Lee, and I. H. Kim. 1994. Improved selective medium for isolation and enumeration of Bifidobacterium sp. Korean J. Food Sci. Technol. 26:526–531.

Jiang, T., A. Mustapha, and D. A. Savaiano. 1996. Improvement of lactose digestion in humans by ingestion of unfermented milk containing Bifidobacterium longum. J. Dairy Sci. 79:750–757.

Jiang, T., and D. A. Savaiano. 1997. Modification of colonic fermentation by bifidobacteria and pH in vitro. Digest. Dis. Sci. 42:2370–2377.

Johnson, J. L. 1981. Genetic characterization. In: P. Gerhard, R. G. E. Murray, R. N. Costilow, E. W. Nester, W. A. Wood, N. R. Krieg, and G. B. Phillips (Eds.) Manual of Methods for General Microbiology. American Society for Microbiology. Washington DC, 450–472.

Johnson, A. P., and H. A. Davies. 1984. Demonstration by electron microscopy of pili of Gardnerella vaginalis. Br. J. Vener. Dis. 60:396–397.

Jolly, J. L. 1983. Minimal criteria for the identification of Gardnerella vaginalis isolated from the vagina. J. Clin. Pathol. 36:476–478.

Jones, B. M., G. R. Kinghorn, and I. Geary. 1982. In vitro susceptibility of Gardnerella vaginalis and Bacteroides organisms, associated with nonspecific vaginitis, to sulfonamide preparations. Antimicrob. Agents Chemother. 21:870–872.

Jones, B. M. 1985. Sodium polyanetholesulfonate in the identification of Gardnerella vaginalis. J. Clin. Microbiol. 22:324–325.

Kado-oka, Y., S. Fujiwara, and T. Hirota. 1991. Effects of bifidobacteria cells on mitogenic response of splenocytes and several functions of phagocytes. Milchwissenshaft 46:626–630.

Kandler, O., and E. Lauer. 1974. Neuere Vorstellungen zur Taxonomic der Bifidobacterien. Zentralbl. Bakteriol. Parasitenkd. Infektionskr. Hyg. Abt. 1, Orig. Reihe A228 29–45.

Kaster, A. G., and Brown, L. R. 1983. Extracellular dextranase activity produced by human oral strain of the genus Bifidobacterium. Infect. Immun. 42:716–720.

Kaufmann, P., A. Pfefferkorn, M. Teuber, and L. Meile. 1997. Identification and quantification of Bifidobacterium species isolated from food with genus-specific 16s rRNA-targeted probes by colony hybridization and PCR. Appl. Environ. Microbiol. 62:1268–1273.

Kharsany, A. B., A. A. Hoosen, and J. Van den Ende. 1993. Antimicrobial susceptilities of Gardnerella vaginalis. Antimicrob. Agents Chemother. 37:2733–2735.

Kimberlin, D. F., and W. W. Andrews. 1998. Bacterial vaginosis: Association with adverse pregnancy outcome. Semin. Perinatol. 22:242–250.

Kimura, K., A. L. McCartney, M. A. McConnell, and G. W. Tannock. 1997. Analysis of fecal populations of bifidobacteria and lactobacilli and investigation of the immunological responses of their human hosts to the predominant strains. Appl. Environ. Microbiol. 63:3394–3398.

Kleessen, B., H. Bunke, K. Tovar, J. Noack, and G. Sawatzki. 1995. Influence of two infant formulas and human milk on the development of the faecal flora in newborn infants. Acta Paediatr. 84:1347–1356.

Kojima, M., S. Suda, S. Hotta, and K. Hamada. 1968. Induction of pleomorphism in Lactobacillus bifidus. J. Bacteriol. 102:217–220.

Kojima, M., S. Suda, S. Hotta, and A. Suganuma. 1970. Necessity of calcium for cell division in Lactobacillus bifidus. J. Bacteriol. 104:1010–1013.

Kok, R. G., A. de Waal, F. Schut, G. W. Welling, G. Weenk, and K. J. Hellingwerf. 1996. Specific detection and analysis of a probiotic Bifidobacterium strain in infant feces. Appl. Environ. Microbiol. 62:3668–3672.

Kolenbrander, P. E., and B. L. Williams. 1981. Lactose-reversible coaggregation between oral actinomycetes and Streptococcus sanguis. Infect. Immun. 33:95–102.

Korschunov, V. M., V. V. Smeyanov, B. A. Efimov, N. P. Tarabrina, A. A. Ivanov, and A. E. Baranov. 1996. Therapeutic use of an antibiotic-resistant Bifidobacterium preparation in men exposed to high-dose gamma irradiation. J. Med. Microbiol. 44:70–74.

Korschunov, V. M., Z. A. Gudieva, B. A. Efimov, A. P. Pikina, V. V. Smeianov, G. Reid, O. V. Korshunova, V. L. Tiutiunnik, and I. I. Stepin. 1999. The vaginal Bifidobacterium flora in women of reproductive age. Zh. Mikrobiol. Epidemiol. Immunobiol. 4:74–78.

Kristiansen, F. V., L. Frost, B. Korsager, and B. R. Moller. 1990. Gardnerella vaginalis in posthysterectomy infection. Eur. J. Obstet. Gynecol. Reprod. Biol. 35:69–73.

Kruse, H.-P., B. Kleessen, and M. Blaut. 1999. Effects of inulin on faecal bifidobacteria in human subjects. Br. J. Nutr. 82:375–382.

Kulkarni, N., and B. S. Reddy. 1994. Inhibitory effect of Bifidobacterium longum cultures on the azoxymethane-induced aberrant crypt foci formation and fecal bacterial beta-glucuronidase. Proc. Soc. Exp. Biol. Med. 207:278–283.

Kullen, M. J., L. J. Brady, and D. J. O'Sullivan. 1997. Evaluation of using a short region of the Rec A gene for rapid and sensitive speciation of dominant bifidobacteria in the human large intestine. FEMS Microbiol. Lett. 154:377–383.

Langendijk, P. S., F. Schut, G. J. Jansen, G. C. Raangs, G. R. Kamphuis, M. H. Wilkinson, and G. W. Welling. 1995. Quantitative fluorescence in situ hybridization of Bifidobacterium spp. with genus-specific 16S rRNA-targeted probes and its application in fecal samples. Appl. Environ. Microbiol. 61:3069–3075.

Lapierre, L., P. Undeland, and L. J. Cox. 1992. Lithium cloride-sodium propionate agar for the enumeration of bifidobacteria in fermented dairy products. J. Dairy Sci. 75:1192–1196.

Lauer, E., and O. Kandler. 1976. Mechanism of the variation of the acetate/lactate/ratio during glucose fermentation by bifidobacteria. Arch. Microbiol. 110:271–277.

Lauer, E., and O. Kandler. 1983. DNA-DNA homology, murein types and enzyme patterns in the type strains of the genus Bifidobacterium. Syst. Appl. Microbiol. 4:42–64.

Lauer, E. 1990. Bifidobacterium gallicum sp.nov. isolated from human feces. Int. J. Syst. Bacteriol. 40:100–102.

Leblond-Bourget, N., H. Philippe, I. Mangin, and B. Decaris. 1996. 16S rRNA and 16S to 23S internal transcribed spacer sequence analyses reveal inter- and intraspecific Bifidobacterium phylogeny. Int. J. Syst. Bacteriol. 46:102–111.

Lee, L. J., S. Kinosahita, H. Kumagai, and T. Tochikura. 1980. Galactokinase metabolism in Bifidobacterium bifidum. Agric. Biol. Chem. 44:2961–2966.

Lee, W., L. E. Phillips, R. J. Carpenter, M. G. Martens, and S. Faro. 1987. Gardnerella vaginalis chorioamnionitis: A report of two cases and a review of the pathogenic role of G. vaginalis in obstetrics. Diagn. Microbiol. Infect. Dis. 8:107–111.

Lee, J., A. Ametani, A. Enomoto, Y. Sato, H. Motoshima, F. Ike, and S. Kaminogawa. 1993. Screening for the immunopotentiating activity of food microrganisms and enhancement of the immune response by Bifidobacterium adolescentis M101-4. Biosci. Biotechnol. Biochem. 57:2127.

Leopold, S. 1953. Heretofore undescribed organism isolated from the genitourinary system. US Armed Forces Med. J. 4:263–266.

Lim, K. S., C. S. Huh, Y. J. Baek, and H. U. Kim. 1995. A selective enumeration medium for bifidobacteria in fermented dairy products. J. Dairy Sci. 78:2108–2112.

Lipmann, F., and L. C. Tuttle. 1945. A specific micromethod for determination of acyl-phosphates. J. Biol. Chem. 159:21–28.

Ludwig, W., J. Neumaier, N. Klugbauer, E. Brockmann, C. Roller, S. Jilg, K. Reetz, I. Schachtner, A. Ludvigsen, M. Bachleitner, U. Fisher, and K.-H. Schleifer. 1993. Phylogenetic relationships of bacteria based on comparative sequence analysis of elongation factor Tu and ATP-syntase b-subunit genes. Ant. v. Leeuwenhoek 64:285–305.

Macfarland, G. T., and J. H. Cummings. 1991. The colonic flora, fermentation and large bowel digestive function. *In:* S. F. Phillips, J. H. Pemberton, and R. G. Shorter (Eds.) The Large Intestine: Physiology, Pathophysiology and Disease. Raven Press. New York, NY. 51–92.

Maeda, N. 1980. Anaerobic, Gram-positive, pleomorphic rods in human gingival crevice. Bull. Tokyo Med. Dent. Univ. 27:63–70.

Maidak, B. L., N. Larsen, M. J. McCaughey, R. Overbeek, G. J. Olsen, K. Fogel, J. Blandy, and C. R. Woese. 1994. The Ribosomal Database Project. Nucl. Acids Res. 22:3485–3487.

Majamaa, H., and E. Isolauri. 1997. Probiotics: A novel approach in the management of food allergy. J. Allergy Clin. Immunol. 99:179–185.

Malone, B. H., Schreiber M., Schneider, N. J., and L. V. Holdeman. 1975. Obligately anaerobic strains of Corynebacterium vaginale (Haemophilus vaginalis). J. Clin. Microbiol. 2:272–275.

Mangin, I., N. Bourget, J. M. Simonet, and B. Decaris. 1995. Selection of species-specific DNA probes which detect strain restriction polymorphism in four Bifidobacterium species. Res. Microbiol. 146:59–71.

Mangin, I., Y. Bouhnik, N. Bisetti, and B. Decaris. 1999. Molecular monitoring of human intestinal Bifidobacterium strain diversity. Res. Microbiol. 150:343–350.

Mara, D. D., and J. I. Oragui. 1983. Sorbitol-fermenting bifidobacteria as specific indicators of human faecal pollution. J. Appl. Bacteriol. 55:349–357.

Marin, M. L., J. H. Lee, J. Murtha, Z. Ustunol, and J. J. Pestka. 1996. Differential cytokine production in clonal macrophage and T-cell lines cultured with bifidobacteria. J. Dairy Sci. 80:2713–2720.

Marmur, J. 1961. A procedure for the isolation of deoxyribonucleic acids. J. Molec. Biol. 3:208–218.

Marmur, J., and P. Doty. 1962. Determination of the base composition of deoxyribonucleic acid from its thermal danaturation temperature. J. Molec. Biol. 5:109–118.

Mata, L. J., C. Carrillo, and E. Villatoro. 1969. Fecal microflora in healthy persons in a preindustrial region. Appl. Microbiol. 17:596–602.

Matsuki, T., K. Watanabe, R. Tanaka, M. Fukuda, and H. Oyaizu. 1999. Distribution of bifidobacterial species in human intestinal microflora examined with 16S rRNA-gene-targeted species-specific primers. Appl. Environ. Microbiol. 65:4506–4512.

Matsumura, H., A. Takeuchi, and Y. Kano. 1997. Construction of Escherichia coli-Bifidobacterium longum shuttle vector transforming B. longum 105-A and 108-A. Biosci. Biotechnol. Biochem. 61:1111–1212.

Mattarelli, P., F. Crociani, M. M. Mucci, and B. Biavati. 1992. Different electrophoretic patterns of cellular soluble proteins in Bifidobacterium animalis. Microbiologica 15:71–74.

Mattarelli, P., B. Biavati, F. Crociani, V. Scardovi, and G. Prati. 1993. Bifidobacterial cell-wall proteins (BIFOP) in Bifidobacterium globosum. Res. Microbiol. 144:581–590.

Mattarelli, P., B. Biavati, A. Alessandrini, F. Crociani, and V. Scardovi. 1994. Characterization of the plasmid pVS809 from Bifidobacterium globosum. Microbiologica 17:327–331.

Mattarelli, P., F. Crociani, and B. Biavati. 1997. Bactericidal activity of poly-D-lysine and lysozyme in Bifidobacterium globosum strains. Annali di Microbiologia ed Enzimologia 47:185–191.

Mattarelli, P., and B. Biavati. 1999a. Influence of aminoacid requirement on the growth of Bifidobacterium globosum strains. Microbiologica 22:69–72.

Mattarelli, P., and B. Biavati. 1999b. L-leucine auxotrophy in Bifidobacterium globosum. Microbiologica 22:73–76.

Mattarelli, P., B. Biavati, M. Pesenti, and F. Crociani. 1999c. Effect of temperature stress on cell-wall protein expression in Bifidobacterium globosum. Res. Microbiol. 150:117–127.

Matteuzzi, D., F. Crociani, G. Zani, and L. D. Trovatelli. 1971. Bifidobacterium suis n. sp.: A new species of the genus Bifidobacterium isolated from pig feces. Allg. Mikrobiol. 11:387–395.

Matteuzzi, D., F. Crociani, O. Emaldi, A. Selli, and R. Viviani. 1976. Isoleucine production in bifidobacteria. Eur. J. Appl. Microbiol. 2:185–194.

Matteuzzi, D., F. Crociani, and O. Emaldi. 1978. Amino acids produced by bifidobacteria and some clostridia. Ann. Microbiol. Inst. Pasteur (Paris, France) 129B:175–181.

Matteuzzi, D., F. Crociani, and P. Brigidi. 1983. Antimicrobial susceptibility of Bifidobacterium. Ann. Inst. Pasteur Microbiol. (Paris, France) 134A:339–349.

McCarthy, L. R., P. A. Mickelsen, and E. G. Smith. 1979. Antibiotic susceptibility of Haemophilus vaginalis (Corynebacterium vaginale) to 21 antibiotics. Antimicrob. Agents Chemother. 16:186–189.

McCartney, A. L., W. Wennzhi, and G. W. Tannock. 1996. Molecular analysis of the composition of the bifidobacterial and Lactobacillus microflora of humans. Appl. Environ. Microbiol. 62:4608–4613.

Meghrous, J., P. Euloge, A. M. Junelles, J. Ballongue, and H. Petitdemange. 1990. Screening of Bifidobacterium strains for bacteriocin production. Biotechnol. Lett. 12:575–580.

Meile, L., W. Ludwig, U. Rueger, C. Gut, P. Kaufmann, G. Dasen, S. Wenger, and M. Teuber. 1997. Bifidobacterium lactis sp.nov., a moderately oxygen tolerant species isolated from fermented milk. Syst. Appl. Microbiol. 20:57–64.

Mengheri, E., F. Nobili, F. Vignolini, M. Pesenti, G. Brandi, and B. Biavati. 1999. Bifidobacterium animalis protects intestine from damage induced by zinc deficiency in rats. J. Nutr. 129:2251–2257.

Metchnikoff, E. 1907. The prolongation of life. William Heinemann. London.

Mevissen-Verhage, E. A. E., J. H. Marcelis, and N. M. de Vos, W. C. Harmsen-VanAmerongen, J. Verhoef. 1987. Bifidobacterium, Bacteroides and Clostridium, spp. in fecal samples from breast-fed and bottle-fed infants with and without iron supplement. J. Clin. Microbiol. 25:285–289.

Mickelsen, P. A., L. R. McCarthy, and M. E. Mangum. 1977. New differential medium for the isolation of Corynebacterium vaginale. J. Clin. Microbiol. 5:488–489.

Milkman, R. 1975. Allozyme variation in Escherichia coli of diverse natural origins. *In:* C. L. Markert (Ed.) Isozymes IV: Genetics and Evolution. Academic Press. New York, NY. 273–285.

Miller-Catchpole, R. 1989. Bifidobacteria in clinical microbiology and medicine. *In:* A. Bezkorovainy and R. Miller-Catchpole (Eds.) Biochemistry and Physiology of Bifidobacteria. CRC Press. Boca Raton, FL. 177–200.

Minkoff, H., A. N. Grunebaum, R. H. Schwarz, J. Feldman, M. Cummings, W. Crombleholme, L. Clark, G. Pringle, and W. M. McCormack. 1984. Risk factors for prematurity and premature rupture of membranes: Prospective study of the vaginal flora in pregnancy. Am. J. Obstet. Gynecol. 150:965–972.

Missich, R., B. Sgorbati, and D. J. LeBlanc. 1994. Transformation of Bifidobacterium longum with pRM2, a constructed Escherichia coli-B. longum shuttle vector. Plasmid 32:208–211.

Mitsuoka, T., T. Sega, and S. Yamamoto. 1965. Eine verbesserte methodik der qualitativen und quantitative Analyse der Darmflora von Menschen und Tieren. Zentralbl. Bakteriol. Parasitenkd. Infektionskr. Hyg. Abt. 1, Orig. Reihe A195 455–469.

Mitsuoka, T. 1969. Comparative studies on bifidobacteria isolated from the alimentary tract of man and animals. Zentralbl. Bakteriol. Parasitenkd. Infektionskr. Hyg. Abt. 1, Orig. Reihe A210 52–64.

Mitsuoka, T. 1972. Bacteriology of fermented milk with special reference to the implantation of lactobacilli in the intestine. In: Proceedings of the VIth International Symposium on Conversion and Manufacture of Foodstuff by Microorganisms. Saikon Publishing. Tokyo, 169–179.

Mitsuoka, T., and C. Kaneuchi. 1977. Ecology of bifidobacteria. Am. J. Clin. Nutr. 30:1799–1810.

Mitsuoka, T. 1984. Taxonomy and ecology of bifidobacteria. Bifidobacteria and Microflora 3:11–28.

Mitsuoka, T. 1990. Bifidobacteria and their role in human health. J. Indust. Microbiol. 6:2633–2638.

Miyake, T., K. Watanabe, T. Watanabe, and H. Oyaizu. 1998. Phylogenetic analysis of the genus Bifidobacterium and related genera based on 16S rDNA sequences. Microbiol. Immunol. 42:661–667.

Moore, W. E. C., E. P. Cato, and L. V. Holdeman. 1969. Anaerobic bacteria of the gastrointestinal flora and their occurrence in clinical infections. J. Infect. Dis. 119:641–649.

Moran, O., O. Zegarra-Moran, C. Virginio, L. Gusmani, and G. D. Rottini. 1992. Physical characterization of the pore forming cytolysine from Gardnerella vaginalis. FEMS Microbiol. Immunol. 5:63–69.

Moss, C. W., and W. E. Dunkelberg Jr. 1969. Volatile and cellular fatty acids of Haemophilus vaginalis. J. Bacteriol. 100:544–546.

Moughan, P. J., M. J. Birtles, P. D. Cranwell, W. C. Smith, and M. Pedraza. 1992. The piglet as a model animal for studying aspects of digestion and absorption in milk-fed human infants. In: A. P. Simopoulos, A. P. (Ed.) Nutritional Triggers for Health and Disease. Karger. Basle, Switzerland. 40–113.

Mukai, T., T. Taba, and H. Ohori. 1997. Collagen binding of Bifidobacterium adolescentis. Curr. Microbiol. 34:326–331.

Muli, F. W., and J. K. Struthers. 1998. The growth of Gardnerella vaginalis and Lactobacillus acidophilus in Sorbarod biofilms. J. Med. Microbiol. 47:401–405.

Muli, F. W., J. K. Struthers, and P. A. Tarpey. 1999. Electron microscopy studies on Gardnerella vaginalis grown in conventional and biofilm systems. J. Med. Microbiol. 48:211–213.

Muñoa, F. J., and R. Pares. 1988. Selective medium for isolation and enumeration of Bifidobacterium spp. Appl. Environ. Microbiol. 54:1715–1718.

Muramatsu, K., S. Onodera, M. Kikuchi, and N. Shiomi. 1992. The production of beta-fructofuranosidase from Bifidobacterium spp. Biosci. Biotechnol. Biochem. 56:1451–1454.

Nagaoka, M., M. Muto, T. Yokokura, and M. Mutai. 1988. Structure of 6-deoxytalose-containing polysaccharide from the cell wall of Bifidobacterium adolescentis. J. Biochem. 103:618–621.

Nath, K., D.-J. Choi, and D. Devlin. 1991. The characterization of Gardnerella vaginalis DNA using non-radioactive probes. Res. Microbiol. 142:573–583.

Nath, K., D. Devlin, and A. M. Beddoe. 1992. Heterogeneity patterns of Gardnerella vaginalis isolated from individuals with bacterial vaginosis. Res. Microbiol. 143:199–209.

Nath, K., and J. Galdi. 1995. Rapid salt-based mini-scale Gardnerella vaginalis DNA isolation procedure. BioTechniques 19:738–740.

Nath, K., J. W. Sarosy, and S. P. Stylianou. 2000. Suitability of a unique 16S rRNA gene PCR product as an indicator of Gardnerella vaginalis. BioTechniques 28:222–226.

Nebra, Y., and A. R. Blanch. 1999. A new selective medium for Bifidobacterium spp. Appl. Environ. Microbiol. 65:5173–5176.

Nichols, J. H., A. Bezkorovainy, and W. Landau. 1974. Human colostral whey M-1 glycoproteins and their Lactobacillus bifidus var. pennsylvanicus growth promoting activities. Life Science 14:967–976.

Noda, H., N. Akasaka, and M. Ohsugi. 1994. Biotin production of bifidobacteria. J. Nutr. Sci. Vitaminol. 40:181–188.

Novik, G. I., V. V. Vysotskii, and Zh. N. Bogdanovskaia. 1994. Cellular ultrastructure of various species of the genus Bifidobacterium. Mikrobiologiia 63:515–522.

Novik, G. I. 1998. Structure-functional organization of bifidobacteria. Mikrobiologiia 67:376–383.

Nunoura, N., K. Ohdan, Y. T. ano, K. Yamamoto, and H. Kumagai. 1996. Purification and characterization of beta-D-glucosidase (beta-D-fucosidase) from Bifidobacterium breve clb acclimated to cellobiose. Biosci. Biotechnol. Biochem. 60:188–193.

O'Dowd, T. C., R. R. West, P. J. Winterburn, and J. Ewlins. 1996. Evaluation of a rapid diagnostic test for bacterial vaginosis. Br. J. Obstetric. Gynecol. 103:366–370.

Okawa, T., H. Niibe, T. Arai, K. Sekiba, K. Noda, S. Takeuki, S. Hashimoto, and N. Ogawa. 1993. Effect of LC9018 combined with radiation therapy on carcinoma of the uterine cervix. Cancer 72:1949–1954.

Oli, M. W., B. W. Petschow, and R. K. Buddington. 1998. Evaluation of fructooligosaccharide supplementation of oral electcrolite solution for treatment of diarrhea. Digest. Dis. Sci. 43:138–147.

Op den Camp, H. J. M., H. J. Veerkamp, A. Oosterhof, and H. Van Halbeek. 1984. Structure of the lipoteichoic acids from Bifidobacterium bifidum subsp. pennsylvanicum. Biochem. Biophys. Acta 795:301–313.

Op den Camp, H. J. M., A. Oosterhof, and J. H. Veerkamp. 1985a. Cell surface hydrophobicity of Bifidobacterium bifidum subsp. pennsylvanicum. Ant. v. Leeuwenhoek 51:303–312.

Op den Camp, H. J. M., P. A. M. Peeters, A. Oosterhof, and J. H. Veerkamp. 1985b. Immunochemical studies on the lipoteichoic acids of Bifidobacterium bifidum subsp. pennsylvanicum. J. Gen. Microbiol. 131:616–668.

O'Riordan, K., and G. F. Fitzgerald. 1997. Determination of genetic diversity within the genus Bifidobacterium and estimation of chromosomal size. FEMS Microbiol. Lett. 156:259–264.

O'Riordan, K., and G. F. Fitzgerald. 1999. Molecular characterisation of a 5.75-kb cryptic plasmid from Bifidobacterium breve NCFB 2258 and determination of mode of replication. FEMS Microbiol. Lett. 174:285–294.

Orla-Jensen, S. 1924. La classification des bactéries lactiques. Lait 4:468–474.

Orrhage, K., B. Brismar, and C. E. Nord. 1994a. Effects of supplements with Bifidobacterium longum and Lactobacillus acidophilus on the intestinal microbiota during administration of clindamycin. Microb. Ecol. Health Dis. 7:17–25.

Orrhage, K., E. Sillerstrom, J. A. Gustafsson, C. E. Nord, and J. Rafter. 1994b. Binding of mutagenic heterocyclic amines by intestinal and lactic acid bacteria. Mutation Res. 311:239–248.

Overman, J. R., and L. Pine. 1963. Electron microscopy of cytoplasmic structures in facultative and anaerobic Actinomyces. J. Bacteriol. 86:656–665.

Pao, C. C., S.-S. Lin, and T.-T. Hsienh. 1990. The detection of Gardnerella vaginalis DNA sequences in uncultured clinical specimens with cloned G. vaginalis DNA as probes. Molec. Cell. Probes 4:367–373.

Park, M. S., K. H. Lee, and G. E. Ji. 1997. Isolation and characterization of two plasmids from Bifidobacterium longum. Lett. Appl. Microbiol. 25:5–7.

Park, S. Y., G. E. Ji, Y. T. Ko, H. K. Jung, Z. Ustunol, and J. J. Pestka. 1999a. Potentiation of hydrogen peroxide, nitric oxide, and cytokine production in RAW 264.7 macrophage cells exposed to human and commercial isolates of Bifidobacterium. Int. J. Food Microbiol. 46:231–241.

Park, M. S., D. W. Shin, K. H. Lee, and G. E. Ji. 1999b. Sequence analysis of plasmid pKJ50 from Bifidobacterium longum. Microbiology 145:585–592.

Parker, R. B. 1974. Probiotics, the other half of the antibiotic story. Anim. Nutr. Health. 29:4–8.

Pechère, J. F., and J. P. Capony. 1968. On the colorimetric determination of acyl phosphates. Analyt. Biochem. 22:536–539.

Pérez, P. F., Y. Minnard, E. A. Disalvo, and G. L. De Antoni. 1998. Surface properties of bifidobacterial strains of human origin. Appl. Environ. Microbiol. 64:21–26.

Petschow, B. W., R. D. Talbott, and R. P. Batema. 1999. Ability of lactoferrin to promote the growth of Bifidobacterium spp. in vitro is independent of receptor binding capacity and iron saturation level. J. Med. Microbiol. 48:541–549.

Petuely, F., and G. Kristen. 1953. Investigation of the bifidus factor. III: The bifidus factor, an essential active material for the infant. Osterr. Z. Kinderheilk. Kinderfursorge 6:173–190.

Pheifer, T. A., P. S. Forsyth, M. A. Durfee, H. M. Pollock, and K. K. Holmes. 1978. Nonspecific vaginitis: Role of Haemophilus vaginalis and treatment with metronidazole. N. Engl. J. Med. 298:1429–1434.

Piot, P., E. Van Dick, M. Goodfellow, and S. Falkow. 1980. A taxonomic study of Gardnerella vaginalis (Haemophilus vaginalis) Gardner and Dukes 1955. J. Gen. Microbiol. 119:373–396.

Piot, P., E. Van Dick, P. A. Totten, and K. K. Holmes. 1982. Identification of Gardnerella (Haemophilus) vaginalis. J. Clin. Microbiol. 15:19–24.

Piot, P., and E. Van Dick. 1983. Isolation and identification of Gardnerella vaginalis. Scand. J. Infect. Dis. 40:S15–S18.

Poupard, J. A., I. Husain, and R. F. Norris. 1973. Biology of the bifidobacteria. Bacteriol. Rev. 37:136–165.

Prévot, A. R., A. Turpin, and P. Kaiser. 1967. Evolution de la systématique des anaèrobies. Les bactéries anaérobies. Dunod. Paris, France. 1840–1878.

Rambaud, C. 1990. Effect of microbial lactase activity in yogurt on the intestinal absorption of lactose. Br. J. Nutr. 64:71–79.

Rao, D. R., S. R. Pulasani, and C. B. Chawan. 1986. Natural inhibitors of carcinogenesis: Fermented milk product. In: B. S. Reddy and L. A. Cohen (Eds.) Diet, Nutrition, and Cancer: A Critical Evaluation. CRC Press. Boca Raton, FL. 2:63–75.

Ratnam, S., and B. L. Fitzgerald. 1983. Semiquantitative culture of Gardnerella vaginalis in laboratory determination in nonspecific vaginitis. J. Clin. Microbiol. 18:344–347.

Reddy, B. S., R. Hamid, and C. V. Rao. 1997. Effect of dietary oligofructose and inulin on colonic preneoplastic aberrant crypt foci inhibition. Carcinogenesis 18:1371–1374.

Reddy, B. S. 1999. Possible mechanisms by which pro- and prebiotics influence colon carcinogenesis and tumor growth. J. Nutr. 129:S1478–S1482.

Reimer, L. G., and L. B. Reller. 1984. Gardnerella vaginalis bacteremia: A review of thirty cases. Obstet. Gynnecol. 64:170–172.

Reimer, L. G., and L. B. Reller. 1985a. Effect of sodium polyanetholesulfonate and gelatin on the recovery of Gardnerella vaginalis from blood culture media. J. Clin. Microbiol. 21:686–688.

Reimer, L. G., and L. B. Reller. 1985b. Use of a sodium polyanetholesulfonate disk for the identification of Gardnerella vaginalis. J. Clin. Microbiol. 21:146–149.

Resnick, I. G., and M. A. Levin. 1981. Quantitative procedure for enumeration of bifidobacteria. Appl. Environ. Microbiol. 42:427–432.

Reuter, G. 1963. Vergleichende Untersuchunge über die Bifidus-Flora im Säuglings- und Erwachsenenstuhl. Zentralbl. Bakteriol. Parasitenkd. Infektionskr. Hyg. Abt. 1, Orig. Reihe A191 486–507.

Reyn, A., A. Birch-Andersen, and S. P. Lapage. 1966. An electron microscope study of thin sections of Haemophilus vaginalis (Gardner and Dukes) and some possibly related species. Can. J. Microbiol. 12:1125–1136.

Rhodes, M. W., and H. Kator. 1999. Sorbitol-fermenting bifidobacteria as indicators of diffuse human faecal pollution in estuarine watersheds. J. Appl. Microbiol. 87:528–535.

Roberfroid, M. B. 1998a. Prebiotics and synbiotics: Concepts and nutritional properties. Br. J. Nutr. 80:S197–S202.

Roberfroid, M. B., J. A. Van Loo, and G. R. Gibson. 1998b. The bifidogenic nature of chicory inulin and its hydrolysis products. J. Nutr. 128:11–19.

Roberts, G., and S. L. Hillier. 1990. Genetic basis of tetracycline resistance in urogenital bacteria. Antimicrob. Agents Chemother. 34:261–264.

Rogosa, M. 1974. Genus III, Bifidobacterium Orla-Jensen. In: Buchanan, R. E., and N. E. Gibbons (Eds.) Bergey's Manual of Determinative Bacteriology, 8th ed. Williams and Wilkins. Baltimore, MD. 669–676.

Rolfe, R., S. Helebian, and S. Finegold. 1981. Bacterial interference between Clostridium difficile and normal microflora. J. Infect. Dis. 143:470–475.

Rose, C. S., and P. Gyorgy 1963. Bifidus factor 2 for growth of Lactobacillus bifidus. Proc. Soc. Exp. Biol. NY 112:923–926.

Rossi, M., P. Brigidi, Y. Gonzalez Vara, A. Rodriguez, and D. Matteuzzi. 1996. Characterization of the plasmid pMB1 from Bifidobacterium longum and its use for shuttle vector construction. Res. Microbiol. 147:133–143.

Rossi, M., P. Brigidi, and D. Matteuzzi. 1998. Improved cloning vectors for Bifidobacterium spp. Lett. Appl. Microbiol. 26:101–104.

Rottini, G., A. Dobrina, O. Forgiarini, E. Nardon, G. A. Amirante, and P. Patriarca. 1990. Identification and partial characterization of a cytolytic toxin produced by Gardnerella vaginalis. Infect. Immun 58:3751–3758.

Rowland, I. R., C. J. Rumney, J. T. Coutts, and L. C. Lievense 1998. Effect of Bifidobacterium longum and inulin on gut bacterial metabolism and carcinogen-induced aberrant crypt foci in rats. Carcinogenesis 19:281–285.

Roy, D., J. L. Berger, and G. Reuter. 1994. Characterization of dairy-related Bifidobacterium spp. based on their beta-galactosidase electrophoretic patterns. Int. J. Food Microbiol. 23:55–70.

Roy, D., P. Ward, and G. Champagne. 1996. Differentiation of bifidobacteria by use of pulsed-field gel electrophoresis and polymerase chain reaction. Int. J. Food Microbiol. 129:11–29.

Roy, D., and S. Sirois. 2000. Molecular differentiation of Bifidobacterium species with amplified ribosomal DNA restriction analysis and alignment of short regions of the ldh gene. FEMS Microbiol. Lett. 191:17–24.

Saavendra, J. M., N. A. Bauman, I. Oung, J. A. Perman, and R. H. Yolken. 1994. Feeding of Bifidobacterium bifidum and Streptococcus thermophilus to infants in hospital for prevention of diarrhoea and shedding of rotavirus. Lancet 344:1046–1049.

Sadhu, K., P. A. G. Dominigue, A. W. Chow, J. Nellingan, N. Cheng, and J. W. Costerton. 1989. Gardnerella vaginalis has a Gram-positive cell-wall ultrastructure and lacks classical cell-wall lipopolysaccharide. J. Med. Microbiol. 29:229–235.

Salmon, A. S., R. D. Walker, C. L. Carleton, S. Shah, and B. E. Robinson. 1991. Characterization of Gardnerella vaginalis and Gardnerella vaginalis-like organisms from the reproductive tract of the mare. J. Clin. Microbiol. 29:1157–1161.

Sanyal, B., and C. Russell. 1978. Nonsporing, anaerobic, Gram-positive rods in saliva and the gingival crevice of humans. Appl. Environ. Microbiol. 35:678–678.

Scardovi, V., and L. D. Trovatelli. 1965. The fructose-6-phosphate shunt as peculiar pattern of hexose degradation in the genus Bifidobacterium. Annali di Microbiologia ed Enzimologia 15:19–29.

Scardovi, V., and L. D. Trovatelli. 1969a. New species of bifidobacteria from Apis mellifica L. and Apis indica F: A contribution to the taxonomy and biochemistry of the genus Bifidobacterium. Zentralbl. Bakteriol. Parasitenkd. Infektionskr. Hyg. Abt. 2, Reihe 123 64–88.

Scardovi, V., L. D. Trovatelli, F. Crociani, and B. Sgorbati. 1969b. Bifidobacteria in bovine rumen. New species of the genus Bifidobacterium; B. globosum sp. nov. and B. ruminale sp. nov. Archiv. Mikrobiol. 68:278–294.

Scardovi, V., G. Zani, and L. D. Trovatelli. 1970. Deoxyribonucleic acid homology among the species of the genus Bifidobacterium isolated from animals. Arch. Mikrobiol. 72:318–325.

Scardovi, V. B. Sgorbati, and G. Zani. 1971a. Starch gel electrophoresis of fructose-6-phosphate phosphoketolase in the genus Bifidobacterium. J. Bacteriol. 106:1036–1039.

Scardovi, V., L. D. Trovatelli, G. Zani, F. Crociani, and D. Matteuzzi. 1971b. Deoxyribonucleic acid homology relationships among species of the genus Bifidobacterium. Int. J. Syst. Bacteriol. 21:276–294.

Scardovi, V., and F. Crociani. 1974a. Bifidobacterium catenulatum, Bifidobacterium dentium, and Bifidobacterium angulatum: Three new species and their deoxyribonucleic acid homology relationships. Int. J. Syst. Bacteriol. 24:6–20.

Scardovi, V., and L. D. Trovatelli. 1974b. Bifidobacterium animalis (Mitsuoka) comb. nov. and the "minimum" and "subtile" groups of new bifidobacteria found in sewage. Int. J. Syst. Bacteriol. 24:21–28.

Scardovi, V., and G. Zani. 1974c. Bifidobacterium magnum sp. nov., a large, acidophilic Bifidobacterium isolated from rabbit feces. Int. J. Syst. Bacteriol. 24:29–34.

Scardovi, V., F. Casalicchio, and N. Vincenzi. 1979a. Multiple electrophoretic forms of transaldolase and 6-phosphogluconate dehydrogenase and their relationships to the taxonomy and ecology of bifidobacteria. Int. J. Syst. Bacteriol. 29:312–327.

Scardovi, V., L. D. Trovatelli, B. Biavati, and G. Zani. 1979b. Bifidobacterium cuniculi, Bifidobacterium choerinum, Bifidobacterium boum, and Bifidobacterium pseudocatenulatum: Four new species and their deoxyribonucleic acid homology relationships. Int. J. Syst. Bacteriol. 29:291–311.

Scardovi, V. 1986. Genus Bifidobacterium Orla-Jensen 1924, 472[al]. In: P. H. A. Sneath, N. S. Mair, M. E. Sharpe, and J. G. Holt (Eds.) Bergey's Manual of Systematic Bacteriology, 1st ed. Williams and Wilkins. Baltimore, MD. 2:1418–1434.

Schramm, M., V. Klybas, and E. Racker. 1958. Phosphorolytic cleavage of fructose-6-phosphate by fructose-6-phosphate phosphoketolase from Acetobacter xylinum. J. Biol. Chem. 233:1283–1288.

Schwebke, J. R. 1999. Diagnostic methods for bacterial vaginosis. Int. J. Gynecol. Obstet. 67:S21–S23.

Scott, T. G., B. Curran, and C. J. Smyth. 1989. Electron microscopy of adhesive interaction between Gardnerella vaginalis and vaginal epithelial cells, McCoy cells and human red blood cells. J. Gen. Microbiol. 135:475–480.

Sears, H. J., I. Brownlee, and J. K. Uchiyama. 1950. Persistence of individual strains of Escherichia coli in the intestinal tract of man. J. Bacteriol. 59:293–301.

Sears, H. J., and I. Brownlee. 1952. Further observations on the persistence of individual strains of Escherichia coli in the intestinal tract of man. J. Bacteriol. 63:47–57.

Seka Assy, N. 1982. Contribution à l'étude des facteurs bifidogènes présents dans le lait maternel (thesis). Université des Sciences et Technique de Lille I. Lille, France.

Sekine, K., E. Watanabe-Sekine, T. Toida, T. Kawashima, T. Kataoka, and Y. Hashimoto. 1994. Adjiuvant activity of the cell wall of Bifidobacterium infantis for in vivo immune responses in mice. Immunopharmacol. Immunotoxicol. 16:589–609.

Sekine, K., J. Ohta, M. Onishi, T. Tatsuki, Y. Shimokawa, T. Toida, T. Kawashima, and Y. Hashimoto. 1995. Analysis of antitumor properties of effector cell stimulated with a cell-wall preparation (WPG) of Bifidobacterium infantis. Biol. Pharm. Bull. 18:148–153.

Selli, A., F. Crociani, D. Matteuzzi, and G. Crisetig. 1986. Feedback inhibition of homoserine dehydrogenase and threonine deaminase in the genus Bifidobacterium. Curr. Microbiol. 13:33–38.

Sghir, A., J. M. Chow, and R. I. Mackie. 1998. Continuous culture selection of bifidobacteria and lactobacilli from human faecal samples using fructooligosaccharide as selective substrate. J. Appl. Microbiol. 85:769–777.

Sgorbati, B., G. Zani, L. D. Trovatelli, and V. Scardovi. 1970. Gluconate dissimilation by the bifidobacteria of the honey bee. Annali di Microbiologia ed Enzimologia 20:57–64.

Sgorbati, B., G. Lenaz, and F. Casalicchio. 1976. Purification and properties of two fructose-6-phosphate phosphoketolases in Bifidobacterium. Ant. v. Leeuwenhoek 42:49–57.

Sgorbati, B. 1979a. Preliminary quantification of immunological relationships among the transaldolase of the genus Bifidobacterium. Ant. v. Leeuwenhoek 45:557–564.

Sgorbati, B., and V. Scardovi. 1979b. Immunological relationships among transaldolases in the genus Bifidobacterium. Ant. v. Leeuwenhoek 45:129–140.

Sgorbati, B., and J. London. 1982a. Demonstration of phylogenetic relatedness among members of the genus Bifidobacterium using the enzyme transaldolase as an evolutionary marker. Int. J. Syst. Bacteriol. 32:37–42.

Sgorbati, B., V. Scardovi, and D. J. LeBlanc. 1982b. Plasmids in the genus Bifidobacterium. J. Gen. Microbiol. 128:2121–2131.

Sgorbati, B., M. B. Smiley, and T. Sozzi. 1983. Plasmids and phages in Bifidobacterium longum. Microbiologica 6:169–173.

Sgorbati, B., V. Scardovi, and D. J. LeBlanc. 1986a. Related structures in the plasmid profiles of Bifidobacterium asteroides, B. indicum and B. globosum. Microbiologica 9:443–456.

Sgorbati, B., V. Scardovi, and D. J. LeBlanc. 1986b. Related structures in the plasmid profiles of Bifidobacterium longum. Microbiologica 9:415–422.

Shaw, C. E., M. E. Forsyth, W. R. Bowie, and W. A. Black. 1981. Rapid presumptive identification of Gardnerella vaginalis (Haemophilus vaginalis) from human blood agar media. J. Clin. Microbiol. 14:108–110.

Sheiness, D., K. Dix, S. Watanabe, and S. L. Hillier. 1992. High levels of Gardnerella vaginalis detected with an oligonucleotide probe combined with elevated pH as a diagnostic indicator of bacterial vaginosis. J. Clin. Microbiol. 30:642–648.

Singh, J., A. Rivenson, M. Tomita, S. Shimamura, N. Ishibashi, and B. S. Reddy. 1997. Bifidobacterium longum, a lactic acid-producing intestinal bacterium inhibit colon cancer and modulates the intermediate biomarkers of colon carcinogenesis. Carcinogenesis 18:833–841.

Skarin, A., and J. Sylwan. 1986. Vaginal lactobacilli inhibiting growth of Gardnerella vaginalis, Mobiluncus and other bacterial species cultured from vaginal content of women with bacterial vaginosis. Acta Pathol. Microbiol. Immunol. Scand., Sect. B 94:399–403.

Skerman, V. B. D., V. McGowan, and P. H. A. Sneath. 1980. Approved lists of bacterial names. Int. J. Syst. Bacteriol. 30:225–420.

Smith, H. W. 1965. Development of the flora of the alimentary tract in young animals. J. Pathol. Bacteriol. 90:495–513.

Smith, R. F. 1975. New medium for isolation of Corynebacterium vaginale from genital specimens. Health Lab. Sci. 12:219–224.

Smith, S. M., T. Ogbara, and R. H. Eng. 1992. Involvement of Gardnerella vaginalis in urinary tracts infections in men. J. Clin. Microbiol. 30:1575–1577.

Smithies, O. 1955. Zone electrophoresis in starch gels: Group variations in the serum proteins of normal human adults. Biochem. J. 61:629–641.

Sneath, P. H. A., N. S. Mair, M. E. Sharpe, and J. G. Holt (Eds.) 1986. Bergey's Manual of Systematic Bacteriology. Williams and Wilkins. Baltimore, MD. 2:

Sobel, J. D. 1989. Bacterial vaginosis—an ecologic mystery. Ann. Int. Med. 111:551–553.

Sonoike, K., M. Mada, and M. Mutai. 1986. Selective agar medium for counting viable cells of bifidobacteria in fermented milk. J. Food Hyg. Soc. Japan 27:238–244.

Sozzi, T., P. Brigidi, O. Mignot, and D. Matteuzzi. 1990. Use of dicloxacillin for the isolation and counting of bifidobacteria from dairy products. Lait 70:357–361.

Spiegel, C. A., R. Amsel, D. Eschenbach, F. Schoenknecht, and K. K. Holmes. 1980. Anaerobic bacteria in nonspecific vaginitis. N. Engl. J. Med. 303:601–607.

Spiegel, C. A. 1991. Bacterial vaginosis. Clin. Microbiol. Rev. 4:485–502.

Stackebrandt, E., F. A. Rainey, and N. L. Ward-Rainey. 1997. Proposal for a new hierarchic classification system, Actinobacteria classis nov. Int. J. Syst. Bacteriol. 47:479–491.

Stark, P. L., and A. Lee. 1982. The microbial ecology of the large bowel of brest-fed and formula-fed infants during the first year of life. J. Med. Microbiol. 15:189–203.

Sturm, A. W., J. H. de Leeuw, and N. T. de Pree. 1983. Postoperative wound infection with Gardnerella vaginalis. J. Infect. 7:264–266.

Sutter, V. L. 1984. Anaerobes as normal oral flora. Rev. Infect. Dis. 6:S62–S66.

Sweet, R. L. 1995. Role of bacterial vaginosis in pelvic inflammatory disease. Clin. Infect. Dis. 20:S271–S275.

Symonds, J., and A. K. Biswas. 1986. Amoxycillin, augmentin, and metronidazole in bacterial vaginosis associated with Gardnerella vaginalis. Genitourin. Med. 62:136.

Takahashi, T., T. Oka, H. Iwana, T. Kuwata, and Y. Yamamoto. 1993. Immune response of mice to orally administered lactic acid bacteria. Biosci. Biotechnol. Biochem. 57:1557–1559.

Tanaka, R., and M. Mutai. 1980. Improved medium for selective isolation and enumeration of Bifidobacterium. Appl. Env. Microbiol. 40:866–869.

Taylor, E., and I. Phillips. 1983. The identification of Gardnerella vaginalis. J. Med. Microbiol. 16:83–92.

Taylor-Robinson, D. 1984. The bacteriology of Gardnerella vaginalis. Scand. J. Urol. Nephrol. Suppl. 86:41–55.

Terada, A., H. Hara, M. Katoaka, and T. Mitsuoka. 1992. Effect of lactulose on the composition and metabolic activity of the human faecal flora. Microbiol. Ecol. Health Dis. 5:43–50.

Teraguchi, S., M. Uehara, K. Ogasa, and T. Mitsuoka. 1978. Enumeration of bifidobacteria in dairy products. Nippon Saikingaku Zasshi 33:753–761.

Thompson, J. S. 1985. Colistin-oxolinic acid blood agar: A selective medium for the isolation of Gardnerella vaginalis. J. Clin. Microbiol. 21:843.

Thomson, K. S., J. S. Bakken, and C. C. Sanders. 1995. Antimicrobial susceptibility testing within the clinic. In: M. R. W. Brownand P. Gilbert (Eds.) Microbiological Quality Assurance. CRC Press. London, 275–288.

Tissier, M. H. 1899. La réaction chromophile d'Escherich et Bacterium Coli. C. R. Soc. Biol. 51:943–945.

Tissier, M. H. 1900. Réchérches sur la flore intestinale normale et pathologique du nourisson (thesis). University of Paris. Paris, France. 1–253.

Tjiam, K. H., J. H. Wagenvoort, B. van Klingeren, P. Piot, E. Stolz, and M. F. Michel. 1986. In vitro activity of the two new 4-quinolones A56619 and A56620 against Neisseria gonorrhoeae, Chlamydia trachomatis, Mycoplasma hominis, Ureaplasma urealyticum and Gardnerella vaginalis. Eur. J. Clin. Microbiol. 5:498–501.

Topouzian, N., B. J. Joseph, and A. Bezkorovainy. 1984. Effects of various metals and calcium metabolism inhibitors on the growth of Bifidobacterium var. pennsylvanicus. J. Pediatr. Gastroent. Nutr. 3:137–142.

Totten, P. A., R. Amsel, J. Hale, P. Piot, and K. K. Holmes. 1982. Selective differential human blood bilayer media for isolation of Gardnerella (Haemophilus) vaginalis. J. Clin. Microbiol. 15:141–147.

Trovatelli, L. D., F. Crociani, M. Pedinotti, and V. Scardovi. 1974. Bifidobacterium pullorum sp. nov. A new species isolated from chicken feces and a related group of bifidobacteria isolated from rabbit feces. Arch. Mikrobiol. 98:187–198.

Trovatelli, L. D., and B. Biavati. 1978. Esigenze nutrizionali di alcune specie del genere Bifidobacterium. In: Lombardo Atti XVIII Congresso Nazionale della Società Italiana di Microbiologia. Fiuggi, giugno. Rome, 330–333.

Trüper, H. G., and J. Krämer. 1981. Principles of characterization and identification of prokaryotes. In: M. P. Starr, H. Stolp, H. G. Trüper, A. Balows, and H. G. Schlegel (Eds.) The Prokaryotes. Springer-Verlag. Berlin, 1:176–193.

Van Belkum, A., A. Koeken, P. Vandamme, M. van Esbroeck, H. Goossens, J. Koopmans, J. Kuijpers, E. Falsen, and W. Quint. 1995. Development of a species-specific polymerase chain reaction assay for Gardnerella vaginalis. Molec. Cell. Probes 9:167–174.

Van der Wiel-Korstanje, J. A. A., and W. De Vries. 1973. Cytocrome synthesis by Bifidobacterium during growth in media supplemented with blood. J. Gen. Microbiol. 75:417–419.

Van Esbroeck, M., P. Vandamme, E. Falsen, M. Vancanneyt, E. Moore, B. Pot, F. Gavini, K. Kersters, and H. Goossens. 1996. Polyphasic approach to the classification and identification of Gardnerella vaginalis and unidentified Gardnerella vaginalis-like coryneforms present in bacterial vaginosis. Int. J. Syst. Bacteriol. 46:675–682.

Van Laere, K. M. J., R. Hartemink, G. Beldman, S. Pitson, C. Dijkema, H. A. Schols, and A. G. Voragen. 1999. Transglycosidase activity of Bifidobacterium adolescentis DSM 20083 alpha-galactosidase. Appl. Microbiol. Biotechnol. 52:681–688.

Van Laere, K. M., T. Abee, H. A. Schols, G. Beldman, and A. G. Voragen. 2000. Characterization of a novel beta-galactosidase from Bifidobacterium adolescentis DSM 20083 active towards transgalactooligosaccharides. Appl. Environ. Microbiol. 66:1379–1384.

Veerkamp, J. H., G. E. J. M. Hoelen, and H. J. M. Op den Camp. 1983. The structure of a mannitol teichoic acid from Bifidobacterium bifidum spp. pennsylvanicum. Biochim. Biophys. Acta 755:439–451.

Vesa, T. H., P. Marteau, S. Zidi, F. Briet, P. Pochart, and J. C. Rambaud. 1996. Digestion and tolerance of lactose from yoghurt and different semi-solid fermented dairy products containing Lactobacillus acidophilus and bifidobacteria in lactose maldigesters: Is bacterial lactase important? Eur. J. Clin. Nutr. 50:730–733.

Vincent, D., D. Roy, F. Mondou, and C. Dery. 1998. Characterization of bifidobacteria by random DNA amplification. Int. J. Food Microbiol. 43:185–193.

Von Nicolai, H., and F. Zilliken. 1972. Neuraminidase aus Lactobacillus bifidus var. pennsylvanicus Hoppe-Seylers Z. Physiol. Chem. 353:1015–1016.

Wagner, R. D., C. Pierson, T. Warner, M. Dohnalek, J. Farmer, L. Roberts, M. Hilty, and E. Balish. 1997. Biotherapeutic effects of probiotic bacteria on candidiasis in immunodeficient mice. Infect. Immun. 65:4165–4172.

Watabe, J., Y. Benno, and T. Mitsuoka. 1983. Bifidobacterium gallinarum sp. nov.: A new species isolated from the ceca of chickens. Int. J. Syst. Bacteriol. 33:127–132.

Watts, D. H., D. A. Eschenbach, and G. E. Kenny. 1989. Early postpartum endometritis: The role of bacteria, genital mycoplasmas, and Chlamydia trachomatis. Obstet. Gynecol. 73:52–60.

Wayne, L. G., D. J. Brenner, R. R. Colwell, P. A. D. Grimont, O. Kandler, M. I. Krichyevsky, L. H. Moore, W. E. C. Moore, R. G. E. Murray, E. Stackebrandt, M. P. Starr, and H. G. Trüper. 1987. Report of the ad hoc committee on the reconciliation of approaches to bacterial systematic. Int. J. Syst. Bacteriol. 37:463–464.

Welsh, J., and M. McClelland. 1990. Fingerprinting genomes using PCR with arbitrary primers. Nucl. Acids Res. 18:7213–7218.

Werner, H., and H. P. R. Seeliger. 1963. Kulturelle untersuchungen uber die vaginalflora unter besonderer berucksichtigung der bifidusbakterien. Path. Microbiol. 26:53–73.

Williams, J. G., A. R. Kubelik, K. J. Livak, J. A. Rafalsky, and S. V. Tingey. 1990. DNA polymorphisms amplified by arbit rary primers are useful as genetic markers. Nucl. Acids Res. 18:6531–6535.

Willis, A. T., C. L. Bullen, K. Williams, C. G. Fagg, A. Bourne, and M. Vignon. 1973. Breast milk substitute: A bacteriological study. Br. Med. J. 4:67–72.

Winslow, C. E. A., J. Broadhurst, R. E. Buchanan, C. Krumwiede, L. A. Rogers, and G. H. Smith. 1917. The families and genera of the bacteria: reliminary report of the Committee of the Society of American Bacteriologists on Characterization and Classification of Bacterial Types. J. Bacteriol. 2:505–566.

Woese, C. R. 1987. Bacterial evolution. Microbiol. Rev. 51:221–271.

Woese, C. R., O. Kandler, and M. L. Wheelis. 1990. Towards a natural system of organisms: Proposal for the domains archaea, bacteria, and eukarya. Proc. Natl. Acad. Sci. USA 87:4576–4579.

Woolfrey, B. F., G. K. Ireland, and R. T. Lally. 1986. Significance of Gardnerella vaginalis in urine cultures. Am. J. Clin. Pathol. 86:324–329.

Wu, S.-R., S. L. Hillier, and K. Nath. 1996. Genomic DNA fingerprint analysis of biotype 1 Gardnerella vaginalis from patients with and without bacterial vaginosis. J. Clin. Microbiol. 34:192–195.

Yaeshima, T., T. Fujisawa, and T. Mitsuoka. 1992. Bifidobacterium globosum, subjective synonym of Bifidobacterium pseudolongum, and description of Bifidobacterium pseudolongum subsp. pseudolongum comb. nov. and Bifidobacterium pseudolongum subsp. globosum comb. nov. Syst. Appl. Microbiol. 15:380–385.

Yaeshima, T., S. Takahashi, N. Ishibashi, and S. Shimamura. 1996. Identification of bifidobacteria from dairy products and evaluation of a microplate hybridization method. Int. J. Food Microbiol. 30:303–313.

Yamamoto, T., M. Morotomi, and R. Tanaka. 1992. Species-specific oligonucleotide probes for five Bifidobacterium species detected in human intestinal microflora. Appl. Environ. Microbiol. 58:4076–4079.

Yang, L., T. Akao, K. Kobashi, and M. Hattori. 1996. Purification and characterization of a novel sennoside-hydrolyzing beta-glucosidase from Bifidobacterium sp. strain SEN, a human intestinal anaerobe. Biol. Pharm. Bull. 19:705–709.

Yasui, H., and M. Ohwaki. 1991. Enhancement of immune response in Peyer's patch cells cultured with Bifidobacterium breve. J. Dairy Sci. 74:1187–1195.

Yasui, H., N. Nagaoka, A. Mike, K. Hayakawa, and M. Ohwaki. 1992. Detection of Bifidobacterium strains that induce large quantities of IgA. Microbial Ecol. Health Dis. 5:155–162.

Yasui, H., J. Kiyoshima, and H. Ushijima. 1995. Passive protection against rotavirus-induced diarrhea of mouse pups born to and nursed by dams fed Bifidobacterium breve YIT4064. J. Infect. Dis. 172:403–409.

Yasui, H., K. Shida, T. Matsuzaki, and T. Yokokura. 1999. Immunomodulatory function of lactic acid bacteria. Ant. v. Leeuwenhoek 76:383–389.

Yazawa, K., K. Imai, and Z. Tamura. 1978. Oligosaccharides and polysaccharides specifically utilizable by bifidobacteria. Chem. Pharm. Bull. 26:3306–3311.

Yildirim, Z., and M. G. Johnson. 1999. Characterization and antimicrobial spectrum of bifidocin B, a bacteriocin produced by Bifidobacterium bifidum NCFB 1454. J. Food Prot. 61:47–51.

Zani, G., and A. Severi. 1982. Cellular ultrastructure and morphology in Bifidobacterium bifidum. Microbiologica 5:225–267.

Zhang, X. B., and Y. Ohta. 1991. Binding of mutagens by fractions of cell wall skeleton of lactic acid bacteria on mutagens. J. Dairy Sci. 74:1477–1481.

Zhang, X. B., and Y. Ohta. 1993. Microorganisms in the gastrointestinal tract of the rat prevent absorption of the mutagen-carcinogen 3-amino-1,4-dimethyl-5H-pyrido [4,3-b] indole. Can. J. Microbiol. 39:841–845.

Zinnemann, K., and G. C. Turner. 1963. The taxonomic position of Haemophilus vaginalis (Corynebactreium vaginale). J. Pathol. Bacteriol. 85:213–219.

Zoppi, G., A. Deganello, G. Benoni, and F. Saccomani. 1982. Oral bacteriotherapy in clinical practice. Eur. J. Pediatr. 139:18–21.

Prokaryotes (2006) 3:383–399
DOI: 10.1007/0-387-30743-5_18

CHAPTER 1.1.3

The Family Propionibacteriaceae: The Genera *Friedmanniella*, *Luteococcus*, *Microlunatus*, *Micropruina*, *Propioniferax*, *Propionimicrobium* and *Tessarococcus*

ERKO STACKEBRANDT AND KLAUS P. SCHAAL

Introduction

A recent systematic-phylogenetic reevaluation of Gram-positive bacteria with a base composition (G+C) of DNA higher than 50 mol% has led to the description of the class Actinobacteria (Stackebrandt et al., 1997) that embraces six orders. The order Actinomycetales (Buchanan 1917, emend Stackebrandt, Rainey and Ward-Rainey 1997) has been defined to include 10 suborders, one of which is Propionibacterineae, containing the family Propionibacteriaceae (Delwiche 1957, emend Rainey, Ward-Rainey and Stackebrandt, 1997) with the genera *Propionibacterium* (Orla-Jensen, 1909), *Luteococcus* (Tamura et al., 1994), *Microlunatus* (Nakamura et al., 1995) and *Propioniferax* (Yokota et al., 1994). Recently, four new genera have been added to the family, i.e., *Friedmanniella* (Schumann et al., 1997), *Tessaracoccus* (Maszenan et al., 1999b), *Micropruina* (Shintani et al., 2000) and *Propionimicrobium* (Stackebrandt et al., 2002; Table 1). As the majority of the recently described species are defined by a single strain only, the metabolic properties of the description may change with the inclusion of new strains. Thus, it cannot be decided whether properties decribed for the type strain will allow affiliation of novel isolates to species.

Membership to the family is based on distinct phylogenetic position and an exclusive set of some 16S rDNA signature nucleotides not found in other families (Stackebrandt et al., 1997). As is also true for nonmolecular diagnostic features, these signatures should be reevaluated with new members added to the family.

Members of the Propionibacteriaceae thrive in diverse habitats, covering human epidermal surfaces, dairy products, silage, soil, water, Antarctic sandstone and sewage treatment plants. They are either aerobic or facultatively anaerobic, have different morphologies, and exhibit different peptidoglycan types and variations, and the base composition of DNA ranges between 53 and 73 mol%. However, with respect to chemotaxo-nomic properties (such as the combination of patterns of polyamines [Busse and Schumann, 1999], major menaquinones and fatty acids), members of the Propionibacteriaceae appear rather homogeneous. As judged from phylogenetic analyses (Fig. 1) and the occurrence of certain chemotaxonomic markers, the genus *Propionibacterium* appears to be diverse: the intrageneric 16S rDNA differences separating the most unrelated species of the *Propionibacterium* (90–94% similarity) are as low as those separating the other genera of the family. Within *Propionibacterium*, three clearly separated clusters emerge which do not follow the classical separation into dairy propionibacteria and the cutaneous propionibacteria commonly found on skin.

Applying different treeing algorithms to the dataset of 28 almost complete 16S rDNA sequences, including those of 21 validly described type strains (Table 1), the family contains two subgroups, one embracing members of *Propionibacterium*, the other embracing representatives of the other genera. While the phylogenetic coherence of multispecies genera is maintained, the relative order may change at which genera emerge. Only the position of certain species, i.e., *Propionibacterium lymphophilum*, *Propionibacterium propionicum* and *Tessaracoccus bendigoensis*, is less stable, and their position changes with the algorithm applied (Fig. 1). The distinct phylogenetic position of *P. lymphophilum* (branching with *Luteococcus japonicus*) has also been depicted by Dasen et al. (1998). The presence of distinct chemotaxonomic properties in *P. lymphophilum* and the remote phylogenetic position led to the reclassification of this species as *Propionimicrobium lymphophilum* (Stackebrandt et al., 2002).

The type strains of *Friedmanniella antarctica*, *Microlunatus phosphovorus*, *Luteococcus japonicus* and *Propioniferax innocua* were included in a sequencing study on ribonuclease P RNA (Yoon and Park, 2000). Surprisingly, *Luteococcus japonicus* possessed a nucleotide

Table 1. Validly published species in the family Propionibacteriaceae.

Genus	Species	Type strain	16S rDNA accession number
Friedmanniella	antarctica	DSM 11053	Z78206
	capsulata	ACM 5120	AF084529
	lacustris	DSM 11465	AJ132943
	spumicola	ACM 5121	AF062535
Luteococcus	japonicus	IFO 12422	Z78208
	peritonei	CCUG 38120	AJ132334
Microlunatus	phosphovorus	JCM 9379	D26169
Micropruina	glycogenica	JCM 10248	Ab012607
Propioniferax	innocua	NCTC 11082	AF227165
Propionimicrobium	lymphophilum	ATCC 27520	AJ003056
Tessaracoccus	bendigoensis	ACM 5119	AF038504
Propionibacterium	acidipropionici	ATCC 25562	X53221
	acnes	ATCC 6919	X53218
	avidum	ATCC 25577	AJ003055
	cyclohexanicum	IAM 14535	D82046
	freudenreichii	ATCC 6207	X53217
	granulosum	ATCC 25564	AJ003057
	jensenii	ATCC 4868	X53219
	propionicum	DSM 43307	X53216
	thoenii	ATCC 4874	X53220

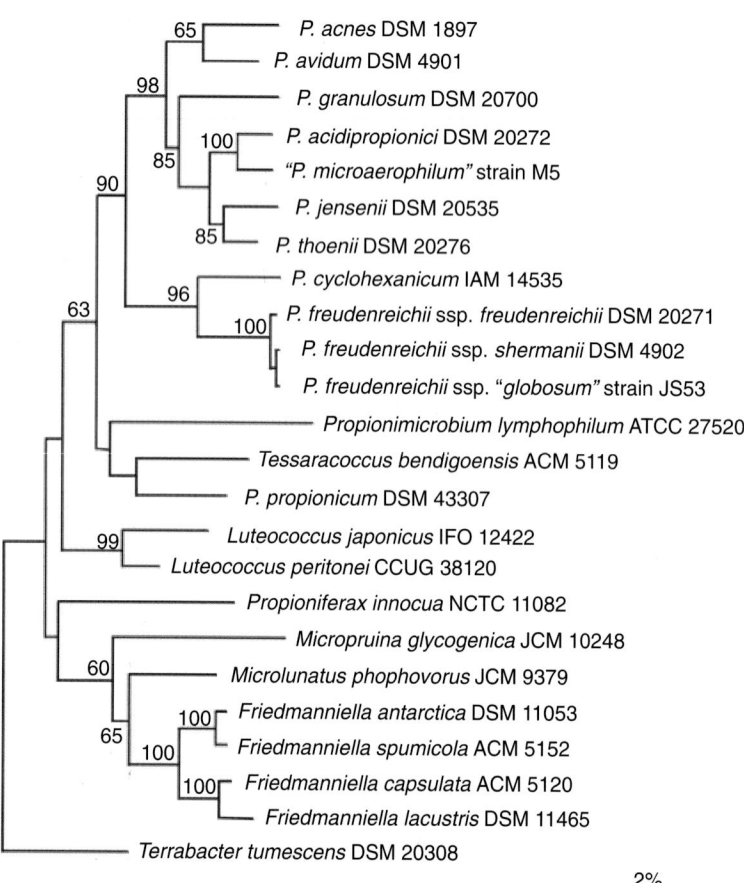

Fig. 1. Distance matrix analysis (De Soete, 1983) of almost complete 16S rDNA sequences of species of the family Propionibacteriaceae. Numbers refer to bootstrap values (500 resamplings).

sequence that was nearly identical to that of *Nocardioides jensenii* (>99% similarity), a member of the sister clade of Propionibacteriaceae. The sequences of the type strains of the other three genera were less closely related (81.0–83.8% similarity) but clustered together to the exclusion of other LL-diaminopimelic acid (A_2pm)-containing actinomycetes.

This chapter will be restricted to genera other than *Propionibacterium*. The authors feel that the clinical and biotechnological significance of members of this genus merit a separate chapter for the genus *Propionibacterium*.

The Genus *Friedmanniella*

The genus *Friedmanniella* contains four Gram-positive, nonmotile, coccoid, packet-forming species, two of which, viz., *F. antarctica* (Schumann et al., 1997) and *F. lacustris* (Lawson et al., 2000), originate from Antarctic environments, while the other two species, viz., *F. spumicola* and *F. capsulata* (Maszenan et al., 1999a), were isolated from sewage treatment plants in Australia. These organisms were described as members of the same genus mainly on phylogenetic, chemotaxonomic and morphological evidence. They exhibit the same rare peptidoglycan type A3γ, in which one glycine residue is defining in the interpeptide bridge while a second one is found at position 1 of the peptide subunit. The same peptidoglycan type defines the genera *Microlunatus* and *Tessaracoccus*, as well as *Propionibacterium propionicum*. The strains accumulate polyphosphate (*F. lacustris* has not been investigated). The phylogenetic neighbors of *Friedmanniella* species are members of *Microlunatus* and *Micropruina*. Interestingly, the two Antarctic species do not cluster together, but each of them is more closely related to species isolated from Australian activated sludge. *Friedmanniella antarctica* has 98.8% 16S rDNA similarity with *F. spumicola*, and the type strains share 50% DNA-DNA reassociation, as determined by the renaturation method (Maszenan et al., 1999a). *Friedmanniella capsulata* shares 99.0% 16S rDNA similarity with *F. lacustris*; on DNA-DNA hybridization are not available. Reassociation values obtained for *F. capsulata* with *F. antarctica* and *F. spumicola* are 27 and 29%, respectively. Thus, the overall DNA relatedness is in accord with the phylogenetic clustering as unravelled by gene sequence analysis.

Isolation and Maintenance

Owing to their isolation from different habitats (sandstone, hypersaline lake and activated sludge) and by different reseach groups, the isolation procedures are quite different.

Friedmanniella antarctica AA-1043[T] (DSM 11053[T]) has been isolated by Peter Hirsch (Kiel, Germany) from a sandstone sample containing a cryptoendolithic microbial community at 1,600 m above ocean level from the Linnaeus Terrace, McMurdo Dry Valleys, Asgard Range,

Transarctic Mountains, Antarctica. The pH of the sampling site was 4.8. Loosened material was sprinkled onto the surface of PYGV (pH 6.9) agar plates (see below). The plates were incubated for 5 months at 9°C in dim light. Individual colonies were picked and purified on PYGV agar after 5 months incubation at 4–6°C. Cells can also be cultivated on Rich medium (DSMZ medium 736; Yamada and Komagata, 1972), trypticase soy broth (TSB) medium (Difco), enriched nutrient broth (Serva), and enriched nutrient agar (pH 7.2; Serva). Biomass for chemotaxonomic, physiological and most of the biochemical studies was produced on R-agar slants at 22°C. For long-term storage, freeze drying, followed by storage at 8°C, is recommended.

Medium PYGV

(DSM 621; Staley, 1968)

Mineral salt sol. ("Hutner/Cohen-Bazire")	20.00 ml
Peptone (Bacto)	0.25 g
Yeast extract (Bacto)	0.25 g
Agar (Bacto)	15.00 g
Distilled water	965.00 ml

Sterilize 20 min.at 121°C. After cooling to 60°C, add the following to the medium:

Glucose sol. (2.5%, sterile-filtered)	10.00 ml
Vitamin sol. (double conc.)	5.00 ml

Adjust pH to 7.5 (the medium is only weakly buffered; one needs approx. 10 drops of 6N KOH per liter of medium).

Mineral Salt Solution

Nitrilotriacetic acid (NTA)	10.00 g
$MgSO_4 \cdot 7H_2O$	29.70 g
$CaCl_2 \cdot 2H_2O$	3.34 g
$Na_2MoO_4 \cdot 2H_2O$	12.67 mg
$FeSO_4 \cdot 7H_2O$	99.00 mg
Metallic salt solution	50.00 ml
Distilled water	900.00 ml

Dissolve NTA first by neutralizing with KOH, then add other salts. Adjust pH to 7.2 with KOH or H_2SO_4. Adjust volume to 1000.0 ml with distilled water.

Metallic salt solution

Na-EDTA	250.000 mg
$ZnSO_4 \times 7 H_2O$	1095.000 mg
$FeSO_4 \times 7 H_2O$	500.000 mg
$MnSO_4 \times H_2O$	154.000 mg
$CuSO_4 \times 5 H_2O$	39.200 mg
$Co(NO_3)_2 \times 6 H_2O$	24.800 mg
$Na_2B_4O_7 \times 10 H_2O$	17.700 mg
Distilled water	100.000 ml

Dissolve the EDTA and add a few drops of concentrated H_2SO_4 to retard precipitation of the heavy metal ions.

Vitamin Solution (double conc.)

Biotin	4.00 mg
Folic acid	4.00 mg
Pyridoxine-HCl	20.00 mg
Riboflavin	10.00 mg
Thiamine-HCl $\cdot$ 2H$_2$O	10.00 mg

Nicotinamide	10.00 mg
D-Ca-pantothenate	10.00 mg
Vitamin B$_{12}$	0.20 mg
p-Aminobenzoic acid	10.00 mg
Distilled water	1.0 liter

Store in the dark and cold (5°C).

Friedmanniella spumicola strain Ben 107^T (ACM 5121^T) was isolated by micromanipulation (Skerman, 1968) from the stable surface foam of an activated sludge plant aerobic reactor treating mainly wastewater from an orange-juice-processing plant in Mildura, Victoria, Australia. *Friedmanniella capsulata* strain Ben 108^T (ACM 5120^T; DSM 12936^T) was micromanipulated from an activated sludge biomass sample from Haman Island, Queensland, Australia. Both strains grew on standard methods agar (SMA [Difco]) supplemented with 1% sterile horse serum at 25°C, and colonies developed after 7–10 days. Subsequent purification was done on R2A agar (see below). Cells were stored at –80°C in R2A medium in 20% glycerol. At the DSMZ, for short-term storage, strains are suspended in trypticase soy yeast extract medium, and for long-term storage, cultures are freeze dried.

Medium R2A (DSMZ Medium 830)

Yeast extract	0.50 g
Proteose peptone (Difco no. 3)	0.50 g
Casamino acids	0.50 g
Glucose	0.50 g
Soluble starch	0.50 g
Na-pyruvate	0.30 g
K$_2$HPO$_4$	0.30 g
MgSO$_4$ · 7H$_2$O	0.05 g
Agar	15.00 g
Distilled water	1.0 liter

Adjust to pH 7.2 with crystalline K$_2$HPO$_4$ or KH$_2$PO$_4$ before adding agar. Add agar, heat medium to boiling to dissolve agar, and autoclave for 15 min at 121°C.

The species *F. lacustris* EL-17A^T (DSM 11465^T) was isolated from an Ekho Lake water sample taken at 1 m depth, collected in January 1990. The hypersaline and meromictic lake is located in the ice-free area of the Vestfold Hills in East Antarctica. The sample had a salinity of 9.5%, a temperature of 2.8°C, and a pH of 8.04. Enrichment was done in medium PYGV (pH 8.0; Staley, 1968), prepared with filtered water from the sampling site. Growth occurred within 12 days at 15°C, and single colonies were subcultured on the same medium. Freeze drying, followed by storage at 8 °C, was the long-term storage method.

Morphological and Cultural Characteristics

On PYGV medium, single cells of strain *F. antarctica* DSM 11053^T are spherical or nearly spherical and range from 1.2 to 1.5 μm in diam-

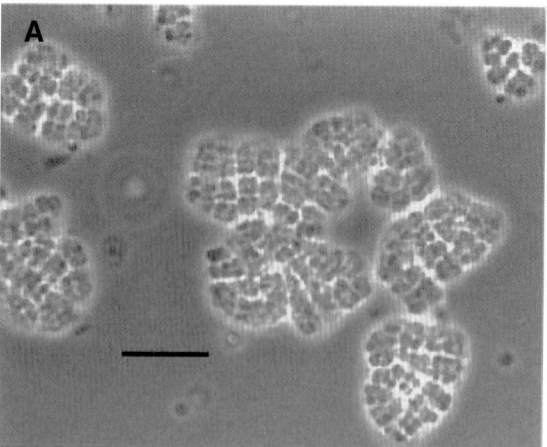
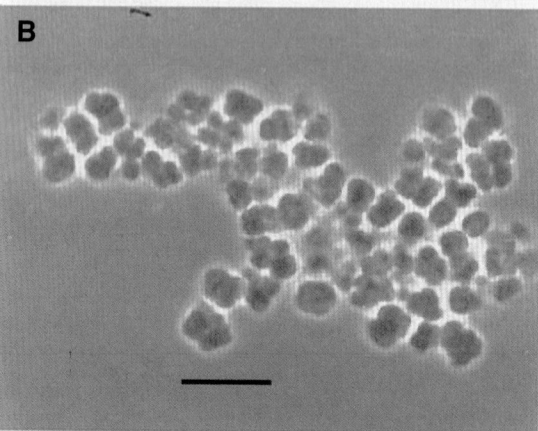
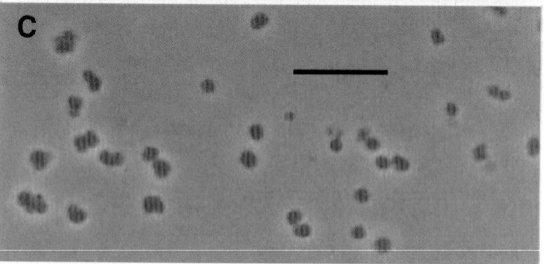

Fig. 2. Phase contrast photomicrographs of *Friedmanniella antarctica* DSM 11053^T. Bar = 10 μm. (A) Clusters grown on PYGV broth for 4 weeks at 20°C, (B) clusters grown on R medium for 5 days at 22°C, and (C) single cells, partly dividing. Bars = 10 μm. (Reproduced with permission from P. Schumann; Schumann et al., 1997.)

eter (Fig. 2C). Occasionally, larger cells are observed that are 2.0–2.2 μm in diameter. On R agar, the cells vary from 0.5 to 2.2 μm in diameter and are ellipsoidal. Generally, the cells occur in packets (Figs. 2A and 2B) which result from cell division in three perpendicular planes. The packets adhere to one another, forming clusters. On R agar, colony color is bright orange but less intense when the organism is grown in the dark; in aging cultures, the color changes from orange to yellow and finally to faint yellow.

On PYGV medium prepared with saline lake water, colonies of *Friedmanniella lacustris* DSM

Fig. 3. Electron micrograph, C/PT shadowed, of *Friedmanniella lacustris* DSM 11465[T]. Bar = 1 μm. (Reproduced with permission from P. Hirsch; from Lawson et al., 2000.)

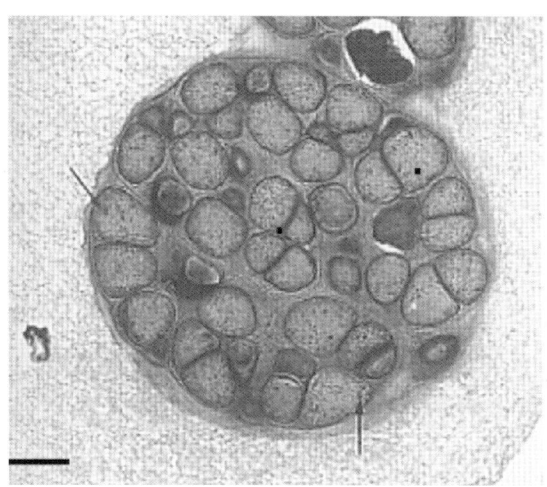

Fig. 5. TEM micrograph of a thinly sectioned cryosubstituted cell cluster of strain *Friedmanniella capsulata* ACM 5120[T], displaying large amounts of the dense extracellular capsular material that surrounds the cluster and divides it internally into smaller packages of cells. Small electron-dense granules (arrows) are present in the cytoplasm. Bar = 1 μm. (Reproduced with permission from R. J. Seviour; from Maszenan et al., 1999.)

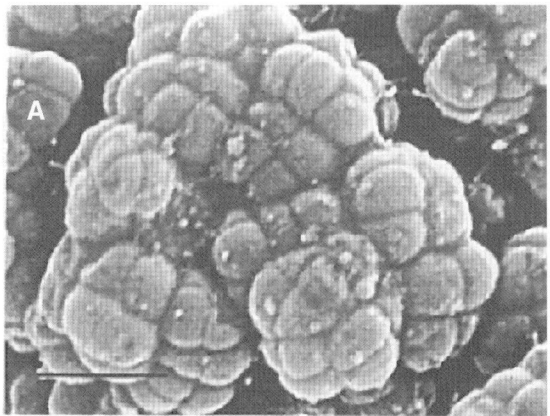

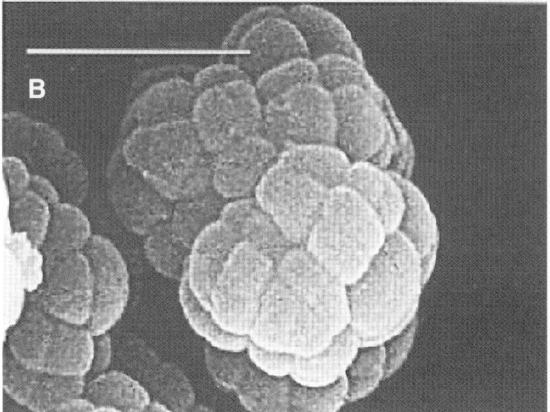

Fig. 4. Scanning electron micrograph of cell aggregates of (A) *Friedmanniella spumicola* ACM 5121[T] and (B) *F. capsulata* ACM 5120[T]. Bar = 2 μm. (Reproduced with permission from R. J. Seviour; from Maszenan et al., 1999.)

11465[T] are orange. Cells are often arranged in packets, but short rods may also be formed, 0.9–1.3 × 0.9–1.5 μm (Fig. 3). Packets are often surrounded by extracellular polymer.

Friedmanniella spumicola ACM 5121[T] and *F. capsulata* ACM 5120[T] grow as yellow to pale orange colonies on different media. Both organisms grew as clusters of coccoid cells (Figs. 4a and

4b, respectively). Spherical cells have a diameter of 0.5–1.4 and 0.6–1.2 μm for strains ACM 5121[T] and ACM 5120[T], respectively. Cells remain closely associated with each other after cell division as a result of production of extracellular capsular material (Fig. 5).

Physiological, Biochemical and Chemotaxonomic Properties

Friedmanniella antarctica, F. capsulata and *F. spumicola* were characterized by the API ZYM (bioMerieux) and Microbact (Oxoid) systems. Carbon utilization test of the latter two species was performed with the BIOLOG GN and GP (Special Diagnostics) systems and compared to the methods of Goodfellow (1971). The results obtained by these two different methods agreed in most cases (Maszenan et al., 1999a), though certain differences were noticed in the negative response of the BIOLOG system to the utilization of glycerol, maltose, melizitose, sucrose and trehalose. *Friedmanniella lacustris* was characterized by using the methods of Labrenz et al. (1999) developed for the characterization of a new Gram-negative species from Ekho Lake, i.e., *Roseovarius tolerans*. Other tests followed Smibert and Krieg (1994), Bradford (1976) and the BIOLOG GP system. Chemotaxonomic properties (Table 3) were determined using methods as listed in Table 2.

A detailed listing of properties is indicated in the species descriptions (Schumann et al., 1997;

Table 2. Methods used in the determination of chemotaxonomic properties.

Genus	Peptidoglycan	Fatty acids	Polyamines	Polar lipids	Menaquinones	Mol% G+C of DNA
Friedmanniella	Schleifer and Kandler, 1972; Groth et al., 1996; Uchida and Aida, 1984	Groth et al., 1996[a,b]	Altenburger et al., 1977	Groth et al., 1996	Groth et al., 1996	Meshba et al., 1989[d]
Luteococcus	Schleifer and Kandler, 1972; Harper and Davis, 1979; Uchida and Aida, 1984	Suzuki and Komagata, 1983[b]; Collins et al., 2000[c]	Altenburger et al., 1977	Yokota et al., 1993	Yokota et al., 1993; Collins et al., 1985	Meshba et al., 1989[d]
Propioniferax	Schleifer and Kandler, 1972	Kroppenstedt, 1985[b,c]	Altenburger et al., 1977	n.d.	Kroppenstedt, 1985	Owen and Pitcher, 1985
Propionimicrobium	Schleifer and Kandler, 1972	Groth et al., 1996[a,b]	n.d.	n.d.	Groth et al., 1996	Meshba et al., 1989[d]
Microlunatus	Komagata and Suzuki, 1987	Komagata and Suzuki, 1987[a,b]	Altenburger et al., 1977	n.d.	Hiraishi et al., 1992	Kamagata and Mikami, 1991[d]
Micropruina	Komagata and Suzuki, 1987	Komagata and Suzuki, 1987[a,b]	n.d.	n.d.	Tamaoka et al., 1983	Shintani et al., 2000[d]
Tessaracoccus	Schleifer and Kandler, 1972	Maszenan et al., 1999a[a,b]	n.d.	Groth et al., 1996	Groth et al., 1996	Porteous, 1994[e]; Owen and Lapage, 1976[e]

Abbreviations: GC-MS, gas chromatography-mass spectrometry; and n.d., not determined.
[a]GC-MS—gas chromatography-mass spectrometry.
[b]GC—gas chromatography.
[c]MIS—MIDI Sherlock Microbial Identification System manufactured by Microbial Identification, Inc. (also known as "MIDI").
[d]HPLC—high performance liquid chromatography.
[e]Thermal denaturation.

CHAPTER 1.1.3 The Family Propionibacteriaceae 389

Table 3. Morphological and chemotaxonomic characteristics of genera of the family Propionibacteriaceae.

Genus	Morphology	Major menaquinones	Diamino acid of peptidoglycan	Murein type[a,b]	Major polyamines[b]	Polar lipid	Major fatty acids	G+C (mol%)
Friedmanniella	Cocci, arranged in packets	MK-9(H$_4$) or MK-9(H$_4$), MK-9(H$_2$), and MK-7(H$_2$) or MK-9(H$_4$) and MK-9(H$_2$)	LL-A$_2$pm	A3-γ	SPD, SPM	PI, PG, DPG, PL	*iso* C$_{15:0}$, *anteiso* C$_{15:0}$	69–74
Luteococcus	Cocci, arranged in pairs and in tetrads or pleomorphic rods	MK-9(H$_4$)	LL-A$_2$pm	A3-γ	SPD, SPM	PI, PG, DPG, GL	C$_{16:1}$, C$_{17:1}$	65–68
Microlunatus	Cocci, arranged singly and in pairs	MK-9(H$_4$)	LL-A$_2$pm	A3-γ	SPD, SPM	PI, PG, DPG, PL	*iso* C$_{15:0}$, *anteiso* C$_{15:0}$	68
Micropruina	Cocci, arranged in pairs, packets or clusters	MK-9(H$_4$)	*meso*-A$_2$pm	A1γ	n.d.	n.d.	*anteiso* C$_{15:0}$	71
Propioniferax	Pleomorphic rods	MK-9(H$_4$)	LL-A$_2$pm	A3-γ	SPD, SPM	PE, PG, PL, GL	*iso* C$_{15:0}$, *anteiso* C$_{15:0}$	62–63
Propionimicrobium	Pleomorphic rods, coccoid, V and Y-shaped	MK-9(H$_4$)	Lys-Asp	A4α	n.d.	n.d.		53–56
Tessaracoccus	Cocci, arranged in tetrads	MK-9(H$_4$) and MK-7(H$_4$)	LL-A$_2$pm	A3-γ	n.d.	PI, PG, DPG, PL	*anteiso* C$_{15:0}$	74
Propionibacterium	Pleomorphic rods, coccoid	MK-9(H$_4$)[b]	*meso*-A$_2$pm; LL-A$_2$pm; *meso*- and LL-A$_2$pm	A1γ; A3γ, A3γ; A1γ and A3γ	n.d.	n.d.	Branched, or straight, or ω-cyclohexane	57–68[b]

Abbreviations: A$_2$pm, diaminopimelic acid; SPD, spermidine; SPM, spermine; DPG, diphosphatidylglycerol; PI, phosphatidylinositol; PL, unknown phospholipid; PG, phosphatidylglyc-erol; PE, phosphatidylethanolamine; GL, unknown glycolipid; PE, phosphatidylethanolamine; PG, phosphatidylglyc-erol; PI, phosphatidylinositol; PL, unknown phospholipid; and n.d., not determined.
[a]According to Schleifer and Kandler (1972).
[b]Kusano et al. (1997).

Maszenan et al., 1999a; Lawson et al., 2000). Table 4 only lists those properties which have also been determined for the majority of phylogenetically related taxa with spherical morphology. The original descriptions list many more reactions which are helpful in the identification of novel isolates. The metabolic properties clearly differentiate between the four species, though the intraspecies diversity of strains has yet to be elucidated with more strains affiliated. The metabolically most versatile organism is *Friedmanniella capsulata* that utilizes most of the substrates provided by the BIOLOG GN and GP Microplate panels, followed by *Friedmanniella lacustris*. According to 16S rDNA analysis, these two species are phylogenetic neighbors. *Friedmanniella spumicola* utilizes the BIOLOG and GP Microplate substrates poorly only, while *F. antarctica* shows almost no positive reaction (except for formate, xylose, ribose, arabinose, erythritol, D-saccharic acid, formic acid, adonitol, 3-methyl glucose, bromosuccinic acid and L-proline). Except for *F. antarctica*, each of the species exhibits certain characters which clearly distinguished it from the other species of the genus. They can also be distinguished from each other by the primary structure of their 16S rDNA sequence. *Friedmanniella lacustris* has a vitamin requirement of biotin, thiamine and nicotinic acid, and it hydrolyzes gelatin and tolerates up to 6% (w/v) NaCl; *Friedmanniella spumicola* has a complex pattern of menaquinones (Table 3) and a low DNA base composition of 69 mol%. Growth of both *F. spumicola* and *F. capsulata* is inhibited by NaCl, and the latter utilizes a significant portion of carbohydrates provided by commercial kits.

The Genus *Luteococcus*

This genus comprises two species, which are morphologically distinct but share chemotaxonomic and phylogenetic similarities. The type species *Luteococcus japonicus* (Tamura et al., 1994), containing two strains (IFO 12422^T and IFO15385), has been isolated from soil and water and is spherical. *Friedmanniella peritonei* (Collins et al., 2000) has been isolated from a human specimen and the type (and only) strain CCUG 38120^T exhibits a pleomorphic rod morphology. The 16S rDNA similarity for these two type strains is 96% (at least 3% higher than similarities separating these two species from other members of the family Propionibacteriaceae), which justified the affiliation of these morphologically different species to the same genus. The most salient feature of the genus is the presence of monounsaturated long-chain fatty acids (>80% of total), i.e., $C_{15:1}$, $C_{16:1}$, $C_{17:1}$, and $C_{18:1}$.

While $C_{16:1}$ predominates in *L. japonicus* (36–60%), $C_{17:1}$ (38%) is the major fatty acid in *L. peritonei*.

Isolation and Maintenance

Strain IFO 12422^T was isolated from soil on Tokara Island, Japan, while strain IFO15385 was originally obtained from water for brewing "miyamizu" in Hyogo prefecture, Japan (Oda et al., 1935). Miyamizu is short for "Nishinomiya no mizu," the water of Nishomiya, used for the production of Nada sake. Both strains were routinely maintained on nutrient agar (Difco) or grown in shake cultures in nutrient broth (Difco). Strain CCUG 38120^T was isolated from human peritoneum during a fetal autopsy and obtained from the culture collection of the University of Göteborg, Sweden. For long-term storage, cultures were lyophilized and stored at 8°C. Cells were grown on chocolate agar at 30°C for fatty acid analysis.

Morphological and Cultural Characteristics

Colonies of *Friedmanniella japonicus* is cream colored to yellow. Cells are 0.7–1.0 μm in diameter, occurring singly, in pairs or in tetrads (Figs. 6 and 7). Cells of *Friedmanniella peritonei*, on the other hand, are nonmotile pleomorphic rods (dimensions and color of colonies not indicated in the original description).

Physiological, Biochemical and Chemotaxonomic Properties

A detailed listing of properties is indicated in the species description (Tamura et al., 1994; Collins et al., 2000). Chemotaxonomic properties (Table 3) were determined using the methods listed in Table 2. Table 4 only lists those properties which have also been determined for the majority of phylogenetically related taxa.

The Genus *Microlunatus*

This genus has been described to accomodate the Gram-positive, nonmotile, coccoid and aerobic organism *Microlunatus phosphovorus* (Nakamura et al., 1995). The description is based on the properties of two strains NM-1 and NM-2; the latter strains differ from the type strain NM-1^T (JCM 9379^T) only by flocculent growth. These strains, isolated from activated sludge in Japan, are chemoorganotrophic organisms that store polyphosphate. In contrast to other spherical members of Propionibacteriaceae, which predominantly form tetrads or packets, *M. phosphovorus* occurs singly or in pairs. The original

Table 4. Comparative phenotypic properties of type strains of species of Propionibacteriaceae, characterized by spherical morphology.

Phenotypic properties	Friedmanniella antarctica	Friedmanniella spumicola	Friedmanniella capsulata	Friedmanniella lacustris	Tessaracoccus bendigoensis	Microlunatus phosphovorus[a,b]	Micropruina glycogenica	Luteococcus japonicus	Luteococcus peritonei
Color of colonies	Beige-orange	Yellow-orange	Bright orange	Brownish (young), orange (old)	Beige	Cream color (young); yellowish (old)	White	Yellow	Nonpigmented
O₂ requirement	Aerobe	Aerobe	Aerobe	Aerobe	Facultative anaerobe	Aerobe	Aerobe	Facultative anaerobe	Facultative anaerobe
Optimum growth temperature (°C) (range)	22 (9–25)	25 (15–37)	20–25 (>15<30)	26 (<3–>43.5)	25 (20–37)	25–30 (5–35)	30 (20–35)	26–28 (12–38)	n.d.
Optimum growth pH (range)	6–7.2 (5.1–8.7)	7–7.5 (5.5–8.0)	6.5–7 (5.5–7.5)	7–8 (<5.5–>9.5)	7.5 (5.5–9.3)	7 (5–9)	7 (6–8)	n.d.	n.d.
Storage products	Polyphosphate[a]	Polyphosphate[c]	Polyphosphate[c]	n.d.	Polyphosphate	Polyphosphate	Glycogen	n.d.	n.d.
Oxidase	–	–	–	+ (weak)	–	+ (weak)	+	+	–
Catalase	+	+	+	n.d.	+	+	+	+	+
Nitrate reduction	–	–	–	–	–	+	+	–	+
Urease	+	+	+	n.d.	–	+	n.d.	–	–
Indole production	–	–	–	–	–	+	n.d.	–	–
Carbohydrates utilized									
Lactose	–	–	–	+	+	–	+	n.d.	+
Raffinose	–	–	–	+	+	±	n.d.	+	–
Mannitol	+	+	–	+	+	+	+	+	+
Glycerol	–	–	–	v[d]	+	–	n.d.	+	n.d.
Arabinose	+ (weak)	+	+	+	+	–	+	+ (D-anomer) / – (L-anomer)	– (L-anomer)
Inositol	–	–	–	–	–	–	+	+	n.d.
Sucrose	–	–	+	+	+	–	+	+	+
Fructose	+	+	+	+	+	–	n.d.	+	n.d.
Glucose	–	–	–	+	+	–	+	+	+
Mannose	–	–	–	+	+	–	+	+	n.d.
Galactose	+	+	+	+	+	–	–	+	n.d.
Trehalose	–	–	–	+	–	–	+	–	–
Maltose	–	–	+	+	+	+	+	+	n.d.
Ribose	+	+	+	+	+	–	+	+	–
Xylose	+ (weak)	+	+	+	–	–	n.d.	–	–
Rhamnose	–	–	+	+	–	–	n.d.	–	n.d.
Adonitol	–	–	–	+	–	–	n.d.	n.d.	n.d.
Cellobiose	n.d.	n.d.	n.d.	n.d.	–	+	+	n.d.	n.d.

Symbols and abbreviations: +, positive in all strains; –, negative in all strains; ±, variable; and n.d.—not determined.
[a]Maszenan et al. (1999a).
[b]Nakamura et al. (1995).
[c]Maszenan et al. (1999b).
[d]v, differing reactions between Biolog and API 50 CH tests.

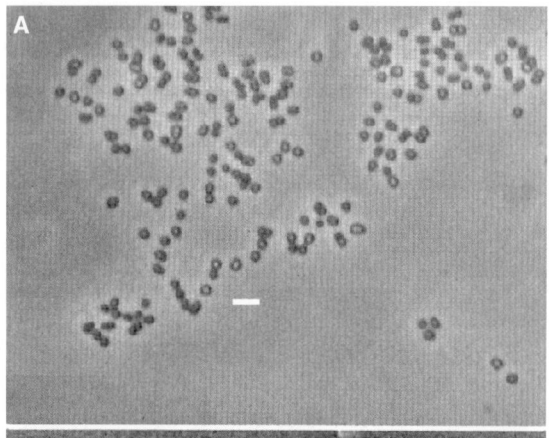

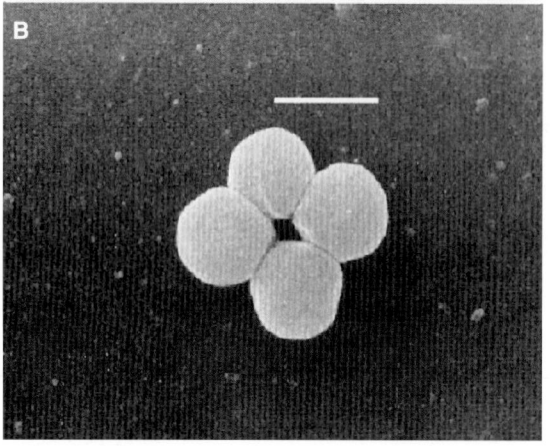

Fig. 6. Phase contrast A) and scanning electron micrograph B) of *L. japonicus* IFO 12422^T grown on yeast extract-malt extract agar. Bars = 3 and 1 μm, respectively. (Reproduced with permission from T. Tamura; from Tamura et al., 1994.)

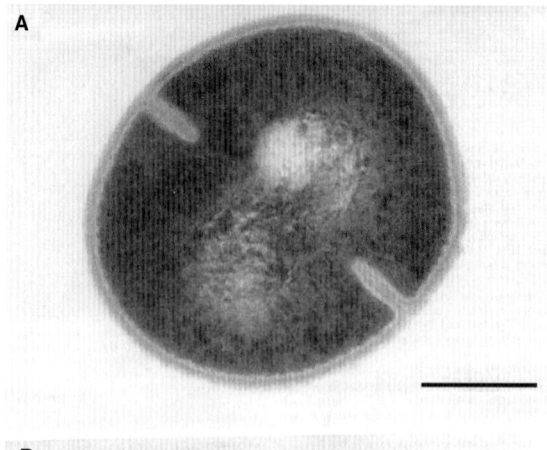

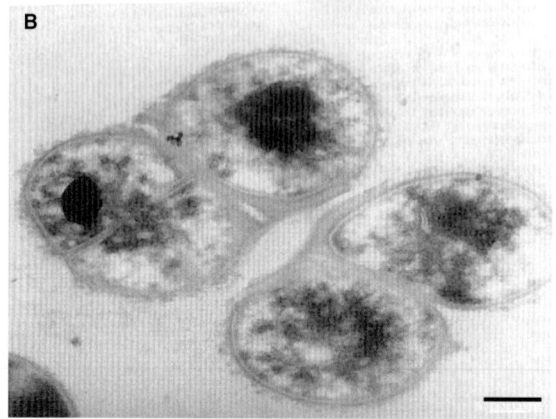

Fig. 7. Thin section electron micrographs of *L. japonicus* IFO 12422^T. Cell division occurs alternately in two planes: the first division (A) and the second division (B). Bar = 0.2 μm. (Reproduced with permission from T. Tamura; from Tamura et al., 1994.)

description did not link *M. phosphovorus* to the genera *Luteococcus*, *Propioniferax* and *Propionibacterium*, but it was more closely affiliated to *Aeromicrobium* and *Nocardioides*, later included in the family Nocardioidaceae (Stackebrandt et al., 1997). Only with the inclusion of novel members of Propionibacteriaceae did the membership of *Microlunatus* to this family become obvious. Two other strains have recently been investigated with respect to the 16S rDNA analysis. Strain UMT-1 (partial sequence; 16S rDNA accession number D32041) is reported to be a strain of *M. phosphovorus*. Strain Y-73 (16S rDNA accession number D50065), proposed as "*Microlunatus confertus*" (but not yet validly published), shows 98% 16S rDNA sequence similarity to *M. phosphovorus*.

Isolation and Maintenance

The strains were first mentioned as tentative members of *Micrococcus* or *Arthrobacter* in a study which reported on polyphosphate-accumulating bacteria and their ecological characteristics in the activated sludge process (Nakamura et al., 1989). However, these organisms differed from members of these Gram-positive genera in chemotaxonomic properties and from *Acinetobacter* strains by their ability to release polyphosphate after the addition of glucose. The isolation of strains NM-1^T and NM-2 was described by Nakamura et al. (1991).

The strains were cultured aerobically at 25°C in the following medium.

Microlunatus Medium (DSM Medium 776; Nakamura et al., 1995)

Glucose	0.5 g
Peptone	0.5 g
Yeast extract	0.5 g
Na-glutamate	0.5 g
KH$_2$PO$_4$	0.5 g
(NH$_4$)$_2$SO$_4$	0.1 g
MgSO$_4$ · 7H$_2$O	0.1 g
Distilled water	1.0 liter

Adjust pH to 7.0.

Cells are freeze dried and stored at 8°C.

Fig. 8. A) Phase contrast photomicrograph of *Microlunatus phosphovorus* JCM 9379^T. B) Scanning electron micrographs of a single cell and C) of a pair of cells of JCM 9379^T. Bars = 1 μm. (Reproduced with permission from K. Nakamura; Nakamura et al., 1995.)

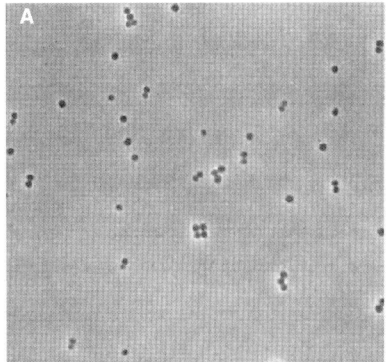
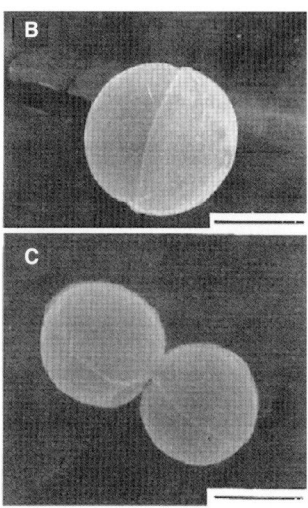

Morphological and Cultural Characteristics

The type strain JCM 9379^T is characterized by coccoid cells (0.8–2.0 μm in diameter) that occur singly or in pairs (Fig. 8). Sometimes, cells are arranged in small irregular clusters. Cell size depends on the growth stage; cells are larger at the early stage of a culture than at the late stationary phase. The type of cell division is unusual; thin-section electron microscopy revealed a cell wall-like structure in the middle of the cell. The type strain exhibits low growth rates even under optimal growth conditions. The doubling time is about 13 h in a liquid medium at pH 7.0 and 25°C. Visible colonies appear on agar plates after 5 days of incubation. The colonies are circular (diameters, 0.5–1 mm), smooth, convex, and cream colored at the early stage of growth. Colonies are yellowish and 1–2 mm in diameter after 10–14 days of incubation.

Physiological, Biochemical and Chemotaxonomic Properties

Utilization of substrates was determined by adding 1.8 g of each substrate to a basal medium that is a modification of DSM medium 776.

Basal Substrate Utilization Medium (Modified DSM Medium 776)

Peptone	0.1 g
Yeast extract	0.1 g
KH₂PO₄	0.44 g
(NH₄)₂SO₄	0.1 g
MgSO₄ · 7H₂O	0.4 g
Distilled water	1.0 liter
Adjust pH to 7.0.	

Chemotaxonomic properties (Table 3) were determined using methods listed in Table 2. A detailed listing of properties is indicated in the species description (Nakamura et al., 1995); for comparative reasons, some of the carbon utiliza-

tion data have been generated by Maszenan et al. (1999). Table 4 only lists those differentiating properties which have also been determined for the majority of phylogenetically and morphologically related taxa.

The Genus *Micropruina*

The genus *Micropruina* with *M. glycogenica* as the only species has also been isolated from an anaerobic-aerobic sequential batch biofilter reactor exhibiting enhanced biological phosphorous removal (Shintani et al., 2000). This Gram-positive, nonmotile spherical species is represented by a single strain Lg2^T (JCM 10248^T) which lies phylogenetically between *Friedmanniella antarctica* and *Microlunatus phosphovorus*. The position remained stable in the maximum-likelihood and consensus tree (Felsenstein, 1993) analyses when sequences of all type strains of Propionibacteriaceae are included (Fig. 1). As also observed for *Tessaracoccus bendigoensis*, *M. glycogenica* is a slow-growing organism. It accumulates glycogen intracellularly. In contrast to most other members of the family, *meso-* and not LL-diaminopimelic acid (A₂pm) is the diagnostic amino acid in the peptidoglycan.

Recently, the 16S rDNA sequences of three novel strains have been published, which are phylogenetically related to *M. glycogenica*. Strain VeSm15 originated from anoxigenic soil containing rice plants (Chin et al., 1999), in which it appeared to be abundant. Strain VeSm15 grew fermentatively and produced acetate and propionate from carbohydrates. The 16S rDNA analysis (accession number AJ229243) revealed 96% similarity with that of *M. glycogenica*. Strain VeSm15 is the closest relative to the uncultured bacterium SJA-1 (98% similarity).

The sequence of clone SJA-1 (accession number AJ009505), recovered from DNA isolated from a fluidized bed reactor inoculated with an anaerobic consortium transforming trichlorobenzene (von Wintzingerode et al., 1999), was 96% similar to the *M. glycogenica* sequence. Due to the fact that at the time of publication of SJA-1, *Micropruina* had not been published, the 16S rDNA sequence was shown to be related to that of *Microlunatus phosphovorus*. The third member is an unidentified bacterium sbr-gs28 (C. Jeon and J. Park, unpublished observation; accession number AF281118) displaying between 94 and 95% sequence similarity to the other members of the *Micropruina* lineage.

Isolation and Maintenance

The type strain has first been isolated by Liu et al. (1997) on *Microlunatus* (NM-1; DSM medium 776) agar (Nakamura et al., 1995). This medium, omitting agar, was subsequently used in the determination of morphological and metabolic properties. Long-term conservation is achieved by freeze-drying, followed by storage at 8°C.

Morphological and Cultural Characteristics

Strain JCM 1028^T grows slowly on NM-1 medium, developing white and punctated colonies after 2–3 weeks. Growth is very slow in the liquid NM-1 medium under aerobic conditions. The strain could grow in the presence of NaCl up to 3.00% (w/v). Spherical cells (0.5–2.2 μm in diameter) occur in pairs, packets or clusters and are surrounded by a capsule. When observed by electron microscopy, a septum is often seen in the middle of each growing cell (Fig. 9).

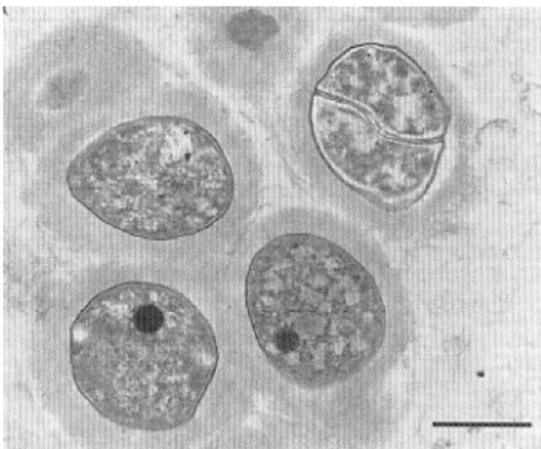

Fig. 9. Transmission electron micrograph of *Micropruina glycogenica* JCM 10248^T showing capsules outside the cells. Cells were grown in NM-1 medium at 30°C under aerobic conditions. Bar = 1 μm. (Reproduced with permission from K. Nakamura; from Shintani et al., 2000.)

Physiological, Biochemical and Chemotaxonomic Properties

A detailed listing of properties is indicated in the species description (Shintani et al., 2000). Chemotaxonomic properties (Table 3) were determined using methods as listed in Table 2. Table 4 only lists those properties which have also been determined for the majority of phylogenetically related taxa. Glycogen has been identified as a storage product; the glycogen content increased significantly in the exponential phase of growth and gradually decreased in the stationary phase. Up to 8.4% of cellular glycogen was determined, based on dry cell weight.

The Genus *Propioniferax*

Propionibacterium innocuum, a Gram-positive, non-acid fast, nonmotile, nonsporeforming organism, was isolated from human epidermal surface by Pitcher and Collins (1991). It was affiliated to *Propionibacterium* because of the presence of genus-specific characteristics such as the presence of LL-A$_2$pm in the peptidoglycan, coryneform morphology, base composition of DNA, and the formation of propionic acid as the main endproduct of glucose fermentation. 16S rRNA analysis placed *P. innocuum* as a remotely related member of the genus *Propionibacterium*. However, outside reference organisms that would have shown the position of *P. innocuum* in relation to other coryneform and actinomycete species were not included in the phylogenetic analysis. With the description of *Luteococcus japonicus*, *P. innocuum* emerged as the nearest phylogenetic neighbor (94.5% 16S rDNA similarity), and both species branched as a sister clade to *Propionibacterium* species. In the light of this information, certain properties of *P. innocuum* were reevaluated: in contrast to the descriptions of most *Propionibacterium* species (Cummins and Johnson, 1986; Charfreitag et al., 1988; Kusano et al., 1997), strains of *P. innocuum* showed aerobic and facultatively anaerobic growth, contained arabinose in cell wall hydrolysates, and did not require blood, serum, or Tween 80 for growth. As it differed from *Luteococcus* in morphology and from other LL-A$_2$pm-containing taxa, such as members of *Aeromicrobium* and *Nocardioides*, and by the formation of propionic acid, *P. innocuum* was reclassified as the type species of a new genus *Propioniferax* and renamed "*Propioniferax innocua*" (Yokota et al., 1994).

Isolation and Maintenance

Strains of *P. innocua* originated from human skin, isolated from laboratories in Philadelphia (United States), Leiden (The Netherlands), and London (United Kingdom). Strains were routinely cultured aerobically on nutrient agar (Oxoid) at 37°C. Anaerobic growth on nutrient agar at an atmosphere of 95% H_2 and 5% CO_2 was much reduced. Cells also grow well in CPY broth (casein peptone, 10 g; yeast extract, 5 g; glucose, 5 g; NaCl, 5 g; and H_2O, 1 liter; pH 7.2) and tryptone soy broth (Difco), supplemented with 1% yeast extract.

Morphological, Cultural, Physiological and Chemotaxonomic Characteristics

Cells are pleomorphic rods appearing in clusters and V forms. Colonies are white, shining, and convex to domed. The optimum growth temperature is approximately 37°C. Growth also occurs at 10–40°C. The species is facultatively anaerobic, but substantial growth occurs aerobically. On nutrient or horse blood agar, colonies are approximately 0.5–3 mm in diameter. The principal carboxylic acid produced from glucose is propionic acid. Methods used in chemotaxonomic analyses are indicated in Table 2, and data are presented in Table 3.

The Genus *Propionimicrobium*

Originally described as "*Bacillus lymphophilus*" (Torrey, 1916), "*Corynebacterium lymphophilum*" (Torrey, 1916) Eberson 1918 and "*Mycobacterium lymphophilum*" (Torrey, 1916) Krasil'nikov 1949, strain VIP 0202 has been included in a taxonomic study by Johnson and Cummins (1972) on coryneforms and propionibacteria. The type strain ATCC 27520[T] was tentatively classified as *Propionibacterium lymphophilum*, because of low DNA reassociation with members of *Corynebacterium*, anaerobic growth, and the formation of propionic acid. A combination of chemotaxonomic distinctness and phylogenetic position, i.e., sequence analysis of 16S rDNA, has been the main basis for the description of novel actinobacterial genera; in the past decade, *Propionibacterium lymphophilum* was transferred to a new genus *Propionimicrobium*. *Propionimicrobium lymphophilum* DSM 4309[T] shares lower than 91.8% sequence similarity with the other *Propionibacterium* species (Dasen et al., 1998), which themselves show higher than 93% similarity among each other. Depending upon the algorithm used, analysis of the 16S rDNA indicates that *P. lymphophilum* either forms the deepest branch of the genus (as shown by neighbor-joining, Felsenstein, 1993; Koussémon et al., 2001) or branches even among the other genera of Propionibacteriaceae as shown by maximum-likelihood (Felsenstein, 1993) or distance matrix analyses (De Soete, 1983; Fig. 1).

Isolation and Maintenance

The organism was first isolated by Torrey (1916) from lymph glands in a Hodgkin's disease patient. Later, strains were also isolated from urinary tract infections and the mesenteric ganglion of a monkey inoculated with "*Actinobacterium*." Strains of the species have also been reported to thrive in soil of rice paddy fields (Haashi and Furusaka, 1980) and in preparations of green olives (Cancho et al., 1980), but as pointed by Cummins and Johnson (1986), the identification of these strains did not include the verification of the distinct cell wall composition and base composition of DNA.

The organism is anaerobic, producing no growth on agar surface incubated aerobically, but growth develops in deep broth incubated aerobically. Cells are maintained on modified Peptone-Yeast-Glucose agar (DSMZ medium 104, see below) or Columbia agar containing 5% defibrinated sheep blood (Becton Dickinson and Company). Plates are sealed in Anerocult A (Merck) bags. For storage, cultures are frozen in 20% glycerol and maintained at –80°C or lyophilized.

Modified Peptone-Yeast-Glucose Medium (DSMZ Medium 104)

Trypticase peptone	5.00 g
Peptone	5.00 g
Yeast extract	10.00 g
Bf extract	5.00 g
Glucose	5.00 g
K_2HPO_4	2.00 g
Tween	801.00 ml
Resazurin	1.00 mg
Salt solution (see below)	40.00 ml
Distilled water	950.00 ml
Hemin solution (see below)	10.00 ml
Vitamin K1 solution (see below)	0.20 ml
Cysteine-HCl · H_2O	0.50 g

The vitamin K1, hemin solution, and cysteine are added after the medium has been boiled and cooled under CO_2. Adjust pH to 7.2 using 8 N NaOH. Distribute under N_2 and autoclave.

Salt Solution

$CaCl_2 \cdot 2H_2O$	0.25 g
$MgSO_4 \cdot 7H_2O$	0.50 g
K_2HPO_4	1.00 g
KH_2PO_4	1.00 g
$NaHCO_3$	10.00 g
NaCl	2.00 g
Distilled water	1.0 liter

Hemin Solution

Dissolve 50 mg of hemin in 1 ml of 1 N NaOH; make up to 100 ml with distilled water. Store refrigerated.

Vitamin K1 Solution

Dissolve 0.1 ml of vitamin K1 in 20 ml of 95% ethanol and filter sterilize. Store refrigerated in a brown bottle.

Morphological and Cultural Characteristics

After a 4-d incubation at 37°C, colonies (≤1 mm in diameter) on Columbia agar are punctiform and on horse blood (≤0.5 mm), punctiform, circular, entire, convex to pulvinate, glistening and smooth. Cells are pleomorphic rods (0.5–0.8 μm in width × 1–2.5 μm in length), often diphtheroid or club-shaped. Cells may be coccoid. PYG broth cultures (24 h) are turbid, becoming clear, with a ropy sediment and terminal pH of 5.4–5.7.

Cells may occur singly, in pairs or short chains, in V or Y configurations, or in clumps. A drawing of the morphology is given by Holdeman et al. (1977).

Physiological, Biochemical and Chemotaxonomic Properties

Main endproducts of glucose fermentation are propionic acid, acetic acid, succinic acid and *iso*-valeric acid. Formic acid is produced in smaller amounts. Physiologically this species can be differentiated from members of the genus *Propionibacterium* by the lack of esculin hydrolysis and indole production as well as by a combination of acid production from adonitol, erythritol, maltose and ribose (Holdeman et al., 1977). Johnson and Cummins (1972) found lysine instead of diaminopimelic acid as the dicarboxylic amino acid of the peptidoglycan. Using the methods of Schleifer and Kandler (Schleifer and Kandler, 1972; Table 2), we confirmed the presence of lysine and demonstrated the peptidoglycan type to be A4α (Lys-Asp) in which aspartic acid forms the interpeptide bridge (Table 3). Using the method of Meshbah et al. (1989), it was shown that the DSMZ culture has a DNA base composition of 56 mol% G+C, which is slightly higher than 53–54 mol% reported for VIP 202 (Johnson and Cummins, 1972). Investigation of the principal isoprenoid quinone (Stackebrandt et al., 2002) revealed MK-9(H$_4$), which is present in all members of the family Propionibacteriaceae. Major fatty acids, determined according to Miller and Berger (1984), are C$_{18:1\omega9c}$ (30%), *anteiso*-C$_{15:0}$ (29%) and C$_{16:0}$ (15%); smaller amounts (>2–<5%) are C$_{14:0}$, i-C$_{15:0}$, ai-C$_{17:0}$, and C$_{18:1\omega7c}$. This composition differs significantly from those reported for *Propionibacterium* strains, in which either branched fatty acids (Moss et al., 1969) or ω-cyclohexane (Kusano et al., 1997) dominate.

The Genus *Tessaracoccus*

The genus *Tessaracoccus* with the single species *T. bendigoensis* has been described for the non-motile, coccoid, facultatively anaerobic Gram-positive strain Ben 106 (ACM 5119^T, DSM 12906^T; Maszenan et al., 1999b), isolated from activated sludge biomass. The sludge, originating from the biological nutrient removal plant in Bendigo, Australia, had been processed in a laboratory-scale sequencing batch reactor. The morphology of this strain, i.e., spherical or clusters of cocci arranged in tetrads, resembles that of the Gram-negative so-called "G-bacteria" (Cech and Hartman, 1993; Carucci et al., 1994), commonly detected in activated sludge samples.

The G-bacteria constitute a phylogenetically diverse group of organisms belonging to the actinobacteria and Proteobacteria (Seviour et al., 2000). The description of *Tessarococcus bendigoensis* extends the range of spherical species decribed to thrive in similar habitats, e.g., four Gram-negative species of *Amaricoccus* (Maszenan et al., 1997) from wastewater treatment plants in Italy, Czech Republic, Macau and Australia and the Gram-positive species *Microlunatus phosphovorus* (Nakamura et al., 1995) from activated sludge and *Microsphaera multipartita* (Yoshimi et al., 1996) from sugar-containing synthetic wastewater, both from Japanese plants. While *Amaricoccus* is a member of the α-subclass of Proteobacteria, *Microsphaera* is a member of the actinobacterial suborder Frankineae, and *Microlunatus* belongs to Propionibacteriaceae. The role of these organisms in this habitat is not yet settled, but they are all defined by depositing storage polymers intracellularly under aerobic conditions. *Tessarococcus bendigoensis* stores polyphosphate and may thus participate in phosphate removal (Seviour et al., 2000).

Isolation and Maintenance

The difficulties encountered in the isolation of strain ACM 5119^T required the use of a micro-manipulator (Skerman, 1968). Pure cultures were obtained by streaking out colonies arising from manipulated cells several times onto fresh GS medium (Williams and Unz, 1985), incubated at 25°C. Cultivation is at 28°C. For chemotaxonomic analyses, cells are scraped from R-agar plates (Yamada and Komagata, 1972) after growth at 22°C for 2–3 weeks.

GS (Glucose Sulfide) Medium (DSMZ Medium 851)

Glucose	0.15 g
Yeast extract	1.00 g
(NH$_4$)$_2$SO$_4$	0.50 g
CaCO$_3$	0.10 g
Ca(NO$_3$)$_2$	0.10 g
KCl	0.05 g
K$_2$PO$_4$	0.05 g
MgSO$_4$ · 7H$_2$O	0.05 g
Na$_2$S · 9H$_2$O	0.20 g
Vitamin solution (Medium 131)	10.00 ml
Distilled water	990.00 ml

Adjust pH to 7.3.

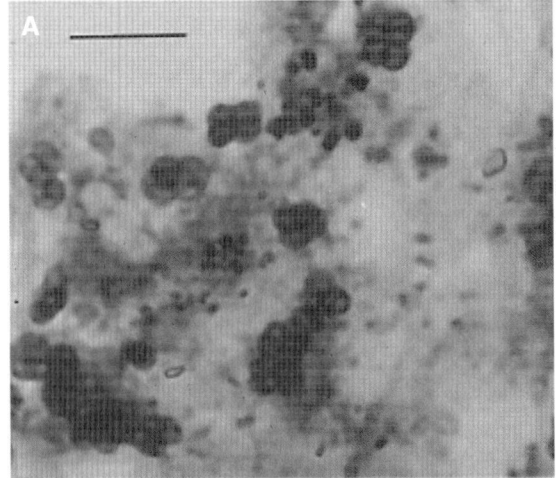

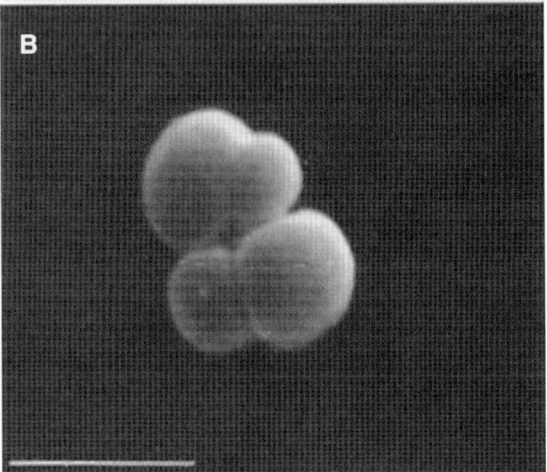

Fig. 10. Morphology of *Tessaracoccus bendigoensis* ACM 5119[T]. A) Cells as observed in activated sludge biomass. Bar = 10 μm. B) Scanning electron micrograph showing the characteristic tetrad morphology. Bar = 2 μm. (Reproduced with permission from R. J. Seviour; from Maszenan et al., 1999.)

Vitamin Solution (Medium 131)

Biotin	2.000 mg
Folic acid	2.000 mg
Pyridoxine-HCl	10.000 mg
Thiamine-HCl · 2H$_2$O	5.000 mg
Riboflavin	5.000 mg
Nicotinic acid	5.000 mg
Ca-pantothenate	5.000 mg
p-Aminobenzoic acid	1.000 mg
Vitamin B$_{12}$	0.010 mg
Distilled water	1 liter

R (rich) Medium (DSMZ Medium 736)

Bacto peptone (Difco)	10.00 g
Yeast extract	5.00 g
Casamino acids (Difco)	5.00 g
Meat extract (Difco)	2.00 g
Malt extract (Difco)	5.00 g
Glycerol	2.00 g
MgSO$_4$ · 7H$_2$O	1.00 g
Tween 80	0.05 g

Agar	20.00 g
Distilled water	1 liter

Adjust pH to 7.2.

Storage is at –80°C in GS medium containing 20% glycerol (Maszenan et al., 1999b). Lyophilized cultures can be stored long term at 8°C.

Morphological and Cultural Characteristics

Colonies are beige, visible after 2–3 d incubation at 25°C. Cells are arranged in tetrads, as seen in large numbers in the activated sludge biomass (Fig. 10A). The diameter of the individual spherical cells ranges from 0.5 to 1.1 μm (Fig. 10B). The type strain grows between 20 and 37°C (optimally at 25°C) and at pH 6–9 (optimally at pH 7.5).

Physiological, Biochemical and Chemotaxonomic Properties

Cells were grown on GS agar at 25°C for 5 days, suspended in physiological saline to the turbidity standard recommended by the manufacturer, and plates were incubated at 25°C for 24 h. Chemotaxonomic properties (Table 3) were determined using methods as listed in Table 2. A detailed listing of properties is indicated in the species description (Maszenan et al., 1999b). Table 4 only lists those properties which have also been determined for the majority of phylogenetically related taxa.

Literature Cited

Altenburger, P., P. Kämpfer, V. N. Akimov, W. Lubitz, and H. Busse. 1997. Polyamine distribution in actinomycetes with group B peptidoglycan and species of the genera Brevibacterium, Corynebacterium, and Tsukamurella. Int. J. Syst. Bacteriol. 47:270–277.

Bradford, M. M. 1976. A rapid and sensitive method for the quantification of microgram quantities of proteins utilizing the principle of protein-dye binding. Annals Biochem. 72:248–254.

Busse, H. J., and P. Schumann. 1999. Polyamine profiles within genera of the class Actinobacteria with LL-diaminopimelic acid in the peptidoglycan. Int. J. Syst. Bacteriol. 49:179–184.

Cancho, F. G., L. R. Navarro, and R. de la Borbolla, Y. Alcala. 1980. La formacion de acido propionico durante la conversacion ds las aceitunas verdes de mesa. III: Microorganismos responsables. Ghrasas Aceites 31:245–250.

Carucci, A., M. Majone, R. Ramadori, and S. Rossetti. 1994. Dynamics of phosphorus and organic substrates in anaerobic and aerobic phases of a sequencing batch reactor. Water Sci. Technol. 30:237–246.

Cech, J. S., and P. Hartman. 1993. Competition between polyphosphate and polysaccharide accumulating bacteria in enhanced biological phosphate removal systems. Water Res. 27:1219–1225.

Charfreitag, O., M. D. Collins, and E. Stackebrandt. 1988. Reclassification of Arachnia propionica as Propionibacterium propionicum. Int. J. Syst. Bacteriol. 38:354–357.

Chin, K.-J., D. Hahn, U. Henstmann, W. Liesack, and P. H. Janssen. 1999. Characterization and identification of numerically abundant culturable bacteria from the anoxic bulk soil of rice paddy microcosms. Appl. Environ. Microbiol. 65:5042–5049.

Collins, M. D. 1985. Analysis of isoprenoid quinones. Meth. Microbiol. 18:329–366.

Collins, M. D., P. A. Lawson, N. Nikolaitchouk, and E. Falsen. 2000. Luteococcus peritonei sp. nov., isolated from the human peritoneum. Int. J. Syst. Evol. Microbiol. 50:179–181.

Cummins, C. S., and J. L. Johnson. 1986. Genus 1: Propionibacterium Orla-Jensen 1909, 337AL. In: P. H. A. Sneath, N. S. Mair, M. E. Sharpe, and J. G. Holt (Eds.) Bergey's Manual of Systematic Bacteriology. Williams and Wilkins. Baltimore, MD. 2:1346–1353.

Cummins, C. S., and J. L. Johnson. 1991. The genus Propionibacterium. In: A. Balows, H. G. Trüper, M. Dworkin, W. Harder, and K. H. Schleifer (Eds.) The Prokaryotes, 2nd ed. Springer-Verlag. New York, NY. 835–849.

Dasen, G., J. Smutny, M. Teuber, and L. Meile. 1998. Classification and identification of propionibacteria based on ribosomal RNA genes and PCR. Syst. Appl. Microbiol. 21:251–259.

De Soete, G. 1983. A least square algorithm for fitting additive trees to proximity data. Psychometrika 48:621–626.

DSMZ. 2001. German Collection of Microorganisms and Cell Cultures. DSMZ catalogue of strains 2001, 7th edition. GmbH. Braunschweig, Germany.

Felsenstein, J. 1993. PHYLIP (Phylogenetic Inference Package) version 3.5.1 (Distributed by the author). Department of Genetics, University of Washington. Seattle, WA.

Goodfellow, M. 1971. Numerical taxonomy of some nocardioform bacteria. J. Gen. Microbiol. 69:33–80.

Groth, I., P. Schumann, N. Weiss, K. Martin, and F. Rainey. 1996. Agrococcus jenensis gen. nov., sp. nov., a new genus of actinomycetes with diaminobutyric acid in the cell wall. Int. J. Syst. Bacteriol. 46:234–239.

Harper, J. J., and G. H. G. Davis. 1979. Two-dimensional thin-laer chromatography for amino acid analysis of bacterial cell walls. Int. J. Syst. Bacteriol. 29:56–58.

Haashi, S., and C. Furusaka. 1980. Enrichment of Propionibacterium in paddy soils by addition of various organic substances. Ant. van Leeuwenhoek 46:313–332.

Hiraishi, A., Y. K. Shin, and J. Sugiyama. 1992. Rapid profiling of bacterial quinones by two-dimensional thin-laer chromatography. Lett. Appl. Microbiol. 14:170–173.

Holdeman, L. V., E. P. Cato, and W. E. C. Moore. 1977. Anaerobe Laboratory Manual, 4th ed. Virginia Polytechnic Institute and State University. Blacksburg, VA.

Johnson, J. L., and C. S. Cummins. 1972. Cell wall composition and deoxyribonucleic acid similarities among the anaerobic coryneforms, classical propionibacteria and strains of Arachnia propionica. J. Bacteriol. 109:1047–1066.

Kamagata, Y., and E. Mikami. 1991. Isolation and characterization of a novel Methanosaeta strain. Int. J. Syst. Bacteriol. 41:191–196.

Komagata, K., and K. Suzuki. 1987. Lipid and cell wall analysis in bacterial systematics. Methods Microbiol. 19:161–207.

Koussémon, M., Y. Combet-Blanc, B. K. C. Patel, J.-L. Caol, P. Thomas, J.-L. Garcia, and B. Ollivier. 2001. Propionibacterium microaerophilum sp. nov., a microaerophilic bacterium isolated from olive mill wastewater. Int. J. Syst. Evol. Microbiol. 51:1373–1382.

Kroppenstedt, R. M. 1985. Fatty acid and menaquinone analysis of actinomycetes and related organisms. In: M. Goodfellow and D. E. Minnikin (Eds.) Chemical Methods in Bacterial Systematics. Academic Press. London, 173–199.

Kusano, K., H. Yamada, M. Niwa, and K. Yamasoto. 1997. Propionibacterium cyclohexanicum sp. nov., a new acid-tolerant–cyclohexyl fatty acid containing propionibacterium isolated from spoiled orange juice. Int. J. Syst. Bacteriol. 47:825–831.

Labrenz, M., M. D. Collins, P. A. Lawson, B. J. Tindall, P. Schumann, and P. Hirsch. 1999. Roseovarius tolerans gen. nov., sp. nov., a budding bacterium with variable chlorophyll a production from hypersaline Ekho Lake. Int. J. Syst. Bacteriol. 49:137–147.

Lawson, P. A., M. D. Collins, P. Schumann, B. J. Tindall, P. Hirsch, and M. Labrenz. 2000. New LL-diaminopimelic acid-containing actinomycetes from hypersaline, heliothermal and meromictic Antarctic Ekho Lake: Nocardioides aquaticus sp. nov., and Friedmanniella lacustris. Syst. Appl. Microbiol. 23:219–229.

Liu, W.-T., K. Nakamura, T. Matsuo, and T. Mino. 1997. Internal energy-based competition between polyphophate- and glycogen accumulating bacteria in biological phosphorous removal reactor- effect of the P/C feeding ratio. Water. Res. 31:1430–1438.

Maszenan, A. M., R. J. Seviour, B. K. C. Patel, G. N. Rees, and B. M. McDougall. 1997. Amaricoccus gen. nov., a Gram-negative coccus occurring in regular packages or tetrads isolated from activated sludge biomass, and description of Amaricoccus veronensis sp. nov., Amaricoccus tamworthensis sp. nov., Amaricoccus macauensis sp. nov., and Amaricoccus kaplicensis sp. nov. Int. J. Syst. Bacteriol. 47:727–734.

Maszenan, A. M., R. J. Seviour, B. K. C. Patel, P. Schumann, J. Burghardt, R. I. Webb, J. A. Soddell, and G. N. Rees. 1999a. Friedmanniella spumicola sp. nov and Friedmanniella capsulata sp. nov. from activated sludge foam: Gram-positive cocci that grow in aggregates of repeating groups of cocci. Int. J. Syst. Bacteriol. 49:1667–1680.

Maszenan, A. M., R. J. Seviour, B. K. C. Patel, P. Schumann, and G. N. Rees. 1999b. Tessarococcus bendigoensis gen. nov., sp. nov., a Gram-positive coccus occurring in regular packages or tetrads, isolated from activated sludge. Int. J. Syst. Bacteriol. 49:459–468.

Meshbah, M., U. Premachandran, and W. B. Whitman. 1989. Precise measurement of the G+C content of deoyribonucleic acid by high-performance chromatography. Int. J. Syst. Bacteriol. 39:159–167.

Miller, L., and T. Berger. 1984. Bacterial identification by gas chromatography and whole cell fatty acids. Gas chromatography application note 228-41:Hewlett-Packard, Co.. Avondale, PA.

Moss, C. W., V. R. Dowell Jr., D. Farshtchi, L. J. Raines, and W. B. Cherry. 1969. Cultural characteristics and fatty acid composition of propionibacteria. J. Bacteriol. 97:561–570.

Nakamura, K., K. Masuda, and E. Mikami. 1989. Polyphosphate accumulating bacteria and their ecological characteristics in activated sludge process. In: T. Hattori, Y. Ishida, Y. Maruyama, R. Y. Morita, and A. Uchida (Eds.)

Recent Advances in Microbial Ecology. Japan Scientific Societies Press. Tokyo, Japan. 427–431.

Nakamura, K., K. Masuda, and E. Mikami. 1991. Isolation of a new type of polyphosphate accumulating bacterium and its phosphate removal characteristics. J. Ferment. Technol. 4:258–263.

Nakamura, K., A. Hiraishi, Y. Yoshimi, M. Kawaharasaki, K. Masuda, and Y. Kamagata. 1995. Microlunatus phosphovorus gen. nov., sp. nov., a new Gram-positive polyphosphate-accumulating bacterium isolated from activated sludge. Int. J. Syst. Bacteriol. 45:17–22.

Oda, M. 1935. Bacteriological studies on water used for brewing sake (part 6). I: Bacteriological studies on "miyamizu". Micrococcus and Actinomyces isolated from "miyamizu" [in Japanese]. Jozogaku Zasshi 13:1202–1228.

Orla-Jensen, S. 1909. Die Hauptlinien des naturlichen Bakteriensystems. Zentralblatt für Bakteriologie usw. Abt. 2 22:305–346.

Owen, R. J., and S. P. Lapage. 1976. The thermal denaturation of partly purified bacterial deoxiribonucleic acid and its taxonomic applications. J. Appl. Bacteriol. 41:335–340.

Owen, R. J., and D. G. Pitcher. 1985. *In:* M. Goodfellow and D. E. Minnikin (Eds.) Chemical Methods in Bacterial Systematics. Academic Press. London, 67–93.

Pitcher, D. G., and M. D. Collins. 1991. Phylogenetic analysis of some LL-diamino acid-containing coryneform bacteria from human skin: Description of Propionibacterium innocuum sp. nov. FEMS Microbiol. Lett. 84:295–300.

Porteous, L. A., J. L. Armstrong, R. J. Seidler, and L. S. Waltrud. 1994. An effective method to extract DNA from environment samples for polymerase chain reaction amplification and DNA fingerprint analysis. Curr. Microbiol. 29:301–307.

Schleifer, K. H., and O. Kandler. 1972. Peptidoglycan types of bacterial cell walls and their taxonomic implications. Bacteriol. Rev. 36:407–477.

Schumann, P., H. Prauser, F. A. Rainey, E. Stackebrandt, and P. Hirsch. 1997. Friedmanniella antarctica gen. nov., spec. nov., an LL-diaminopimelic acid-containing actinomycete from Antarctic sandstone. Int. J. Syst. Bacteriol. 47:278–283.

Seviour, R. J., A. M. Maszenan, J. A. Soddell, V. Tandoi, B. K. C. Patel, Y. Kong, and P. Schumann. 2000. Microbiology of the "G-bacteria" in activated sludge. Environ. Microbiol. 2:581–593.

Shintani, T., W.-T. Liu, S. Hanada, Y. Kamagata, S. Miyaoka, T. Suzuki, and K. Nakamura. 2000. Micropruina glycogenica gen. nov., sp. nov., a new Gram-positive glycogen-accumulating bacterium isolated from activated sludge. Int. J. Syst. Evol. Microbiol. 50:201–207.

Skerman, V. B. D. 1968. A new type of micromanipulator and microforge. J. Gen. Microbiol. 54:287–197.

Smibert, R. M., and N. R. Krieg. 1994. Phenotypic characterization. *In:* P. Gerhardt, R. G. E. Murra, W. A. Wood, and N. R. Krieg (Eds.) Methods for General and Molecular Bacteriology. American Society for Microbiology. Washington, DC, 607–654.

Stackebrandt, E., F. A. Rainey, and N. Ward-Rainey. 1997. Proposal for a new hierarchic classification system, Actinobacteria classis nov. Int. J. Syst. Bacteriol. 47:479–491.

Stackebrandt, E., P. Schumann, K. P. Schaal, and N. Weiss. 2002. Propionimicrobium gen. nov., a new genus to accommodate Propionibacterium lymphophilum (Torrey 1916) Johnson and Cummins 1972, 1057[AL] as Propionimicrobium lymphophilum comb. nov. Int. J. Syst. Bacteriol. 52:1925–1927.

Staley, J. T. 1968. Prosthecomicrobium and Ancalomicrobium: New prosthecate freshwater bacteria. J. Bacteriol. 95:1921–1942.

Suzuki, K., and K. Komagata. 1983. Taxonomic significance of cellular fatty acid composition in some coryneform bacteria. Int. J. Syst. Bacteriol. 33:188–193.

Tamaoka, J., Y. Kataama-Fujimura, and H. Kuraishi. 1983. Analysis of bacterial menaquinone mixtures by high performance liquid chromatography. J. Appl. Bacteriol. 54:31–36.

Tamura, T., M. Takeuchi, and A. Yokota. 1994. Luteococcus japonicus gen. nov., sp. nov., a new Gram-positive coccus with LL-diaminopimelic acid in the cell wall. Int. J. Syst. Bacteriol. 44:348–356.

Torrey, J. C. 1916. Bacteria associated with certain types of abnormal lymph glands. J. Med. Res. 34:65–80.

Uchida, K., and K. Aida. 1984. An improved method for the glycolate test for simple identification of the acyl type of bacterial wall. J. Gen. Appl. Microbiol. 30:131–134.

Von Wintzingerode, F., B. Selent, W. Hegemann, and U. Göbel. 1999. Phylogenetic analysis of an anaerobi, trochlorobenzene-transforming microbial consortium. Appl. Environ. Microbiol. 65:283–286.

Williams, T. M., and R. F. Unz. 1985. Isolation and characterization of filamentous bacteria present in bulking activated sludge. Appl. Microbiol. Biotechnol. 22:272–282.

Yamada, K., and K. Komagata. 1972. Taxonomic studies on coryneform bacteria. IV: Morphological, cultural, biochemical, and physiological characteristics. J. Gen. Microbiol. 18:399–416.

Yokota, A., T. Tamura, T. Nishii, and T. Hasegawa. 1993. Kineococcus aurantiacus gen. nov., sp. nov., a new aerobic, Gram-positive, motile coccus with meso-diaminopimelic acid and arabinogalactane in the cell wall. Int. J. Syst. Bacteriol. 43:52–57.

Yokota, A., T. Tamura, N. Weiss, and E. Stackebrandt. 1994. Transfer of Propionibacterium innocuum Pitcher and Collins 1992 to Propioniferax gen.nov. as Propioniferax innocuum comb. nov. Int. J. Syst. Bacteriol. 44:579–582.

Yoon, J.-H., and Y.-H. Park. 2000. Cooperative sequence analyses of the ribonuclease P (RNase P) RNA genes from LL-2,6-diaminopimelic acid-containing actinomycetes. Int. J. Syst. Evol. Microbiol. 50:221–2029.

Yoshimi, Y., A. Hiraishi, and K. Nakamura. 1996. Isolation and characterization of Multisphaera multipartita gen. nov., sp. nov., a polysaccharide accumulating Gram-positive bacterium from activated sludge. Int. J. Syst. Bacteriol. 46:519–525.

Prokaryotes (2006) 3:400–418
DOI: 10.1007/0-387-30743-5_19

CHAPTER 1.1.4

Family Propionibacteriaceae: The Genus *Propionibacterium*

ERKO STACKEBRANDT, CECIL S. CUMMINS AND JOHN L. JOHNSON

The suborder Propionibacterineae is one of the ten suborders of the class Actinobacteria (Stackebrandt et al., 1997), containing the family Propionibacteriaceae Delwiche 1957, emend Rainey, Ward-Rainey and Stackebrandt 1997, with the genera *Propionibacterium* (Orla-Jensen, 1909), *Luteococcus* (Tamura et al., 1994), *Microlunatus* (Nakamura et al., 1995) and *Propioniferax* (Yokota et al., 1994). Recently, four new genera have been added to the family, i.e., *Friedmanniella* (Schumann et al., 1997), *Tessaracoccus* (Maszenan et al., 1999), *Micropruina* (Shintani et al., 2000) and *Propionimicrobium* (Stackebrandt et al., 2002; Table 1). Members of the Propionibacteriaceae genera are either aerobic or facultative anaerobic, exhibit different morphologies, peptidoglycan types and variations, and have a wide range of DNA G+C content (53–73 mol%); however, with respect to chemotaxonomic properties (such as the pattern of polyamines [Busse and Schumann, 1999], major menaquinones, and fatty acids), they appear rather homogeneous (see the chapter The Family Propionibacteriaceae: The Genera Friedmaniella, Luteococcus, Microlunatus, Micropruina, Propioniferax, Propionimicrobium and Tessarococcus in this Volume). The family contains two phylogenetic clades, one embracing members of *Propionibacterium*, the other embracing representatives of the other family members (Stackebrandt et al., 2002).

The intrageneric 16S rDNA differences, which separate the most unrelated species of *Propionibacterium* (90–94% similarity), are as low as those separating the other genera of the family. The phylogenetic and chemotaxonomic distinctness of *P. lymphophilum* (Dasen et al., 1998) led to its reclassification as the type species of a new genus *Propionimicrobium* (Stackebrandt et al., 2002). This species was originally described as "*Bacillus lymphophilus*" (Torrey 1916), "*Corynebacterium lymphophilum*" (Torrey 1916) Eberson 1918 and "*Mycobacterium lymphophilum*" (Torrey 1916) Krasil'nikov 1949, and the type strain ATCC 27520^T was tentatively classified as *Propionibacterium lymphophilum*. A second *Propionibacterium* species that has

recently been excluded from this genus is *Propionibacterium innocuum* (Pitcher and Collins, 1991), now *Propioniferax innocua* (Yokota et al., 1994). The reclassification was based on phylogenetc position of the type strain which, in contrast to the descriptions of most *Propionibacterium* species (Cummins and Johnson, 1986; Charfreitag and Stackebrandt, 1988; Kusano et al., 1997), shows aerobic and facultatively anaerobic growth, contains arabinose in cell wall hydrolysates, and does not require blood, serum or Tween 80 for growth.

Three new species have recently been added to the genus. *Propionibacterium cyclohexanicum* has been isolated from spoiled fruit juice (Kusano et al., 1997), *P. australiense* from granulomatous bovine lesions (Bernard et al., 2002), and *P. microaerophilum* from olive mill wastewater (Koussémon et al., 2001).

The first systematic investigations into the organisms responsible for the formation of "eyes" in cheese were made by Von Freudenreich and Orla-Jensen (1906), although the earlier work of Fitz (Fitz, 1878; Fitz, 1879) had already shown that organisms from cheese would ferment lactate to propionic and acetic acids and liberate carbon dioxide in the process. The name "*Propionibacterium*" was suggested by Orla-Jensen (1909) for these organisms because they were characterized by the production of large amounts of propionic acid during growth. The economic importance of these organisms primarily derives from their role in the cheese industry. However, many of them also produce commercially valuable amounts of vitamin B_{12} under suitable conditions, and they have also been used for the commercial production of propionic acid.

The strains originally classified in the genus all came from cheese or dairy products. Traditionally, the term "classical propionibacteria" has been used for these organisms. Later it has been shown that some of the so called "anaerobic coryneform organisms" that form a major part of the skin flora of humans share so many properties with the strains of dairy origin that it seemed justified to regard them also as members of *Propionibacterium*. Douglas and Gunter

Table 1. Validly published species in the family Propionibacteriaceae, the 16S rRNA gene accession number of the type strain, and the diagnostic amino acid of their peptidoglycan.

Genus	Species	Type strain	16S rDNA accession number	Diagnostic amino acid in peptidoglycan
Propionibacterium	*Acidipropionici*	ATCC 25562	X53221	LL-A$_2$pm
	Acnes	ATCC 6919	X53218	Mostly LL-A$_2$pm
	Australiense	CCUG 46075	AF225962	*Meso*-A$_2$pm
	Avidum	ATCC 25577	AJ003055	LL- and *meso*-A$_2$pm
	Cyclohexanicum	IAM 14535	D82046	*Meso*-A$_2$pm
	Freudenreichii	ATCC 6207	X53217	*Meso*-A$_2$pm
	Granulosum	ATCC 25564	AJ003057	LL-A$_2$pm
	Jensenii	ATCC 4868	X53219	LL-A$_2$pm
	Microaerophilum	DSM 13435	AF234623	nd
	Propionicum	DSM 43307	X53216	LL-A$_2$pm
	Thoenii	ATCC 4874	X53220	*Meso*-A$_2$pm
Propionimicrobium	*Lymphophilum*	ATCC 27520	AJ003056	Lys-Asp
Friedmanniella	*Antarctica*	DSM 11053	Z78206	LL-A$_2$pm
	Capsulata	ACM 5120	AF084529	LL-A$_2$pm
	Lacustris	DSM 11465	AJ132943	LL-A$_2$pm
	Spumicola	ACM 5121	AF062535	LL-A$_2$pm
Tessaracoccus	*Bendigoensis*	ACM 5119	AF038504	LL-A$_2$pm
Propioniferax	*Innocua*	NCTC 11082	AF227165	LL-A$_2$pm
Luteococcus	*Japonicus*	IFO 12422	Z78208	LL-A$_2$pm
	Peritonei	CCUG 38120	AJ132334	LL-A$_2$pm
Microlunatus	*Phosphovorus*	JCM 9379	D26169	LL-A$_2$pm
Micropruina	*Glycogenica*	JCM 10248	AB012607	*Meso*-A$_2$pm

Abbreviations: A$_2$pm, diaminopimelic acid; ATCC, American Type Culture Collection, Manassas, VA, United States; CCUG, Culture Collection, University of Göteborg, Dept. of Clinical Bacteriology, Göteborg, Sweden; IAM, Institute of Applied Microbiology, University of Tokyo, Institute of Molecular and Cellular Bioscience, Tokyo, Japan; DSM, Deutsche Sammlung von Mikroorganismen und Zellkulturen Gm bH, Braunschweig, Germany; ACM, Australian Collection of Microorganisms, Department of Microbiology, University of Queensland; NCTC, National Collection of Type Cultures, Central Public Health Laboratory, London, United Kingdom; IFO, Institute For Fermentation, Osaka, Japan; JCM, Japan Collection Of Microorganisms, The Institute of Physical and Chemical Research, Hirosawa, Wako-shi, Japan; and nd, not determined.

(1946) and Moore and Cato (1963) showed that propionic acid was a major end product of their metabolism. Subsequent work has shown that the anaerobic coryneforms from skin differ in several fundamental ways from the classical corynebacteria (e.g., *Corynebacterium diphtheriae*). For example, they do not produce mycolic acids, the diamino-acid of the cell wall peptidoglycan is not *meso*-diaminopimelic acid (A2pm) but LL-A2pm in almost all cases, and their cell wall polysaccharide does not contain arabinose (Johnson and Cummins, 1972; Rogosa et al., 1974). The term "cutaneous propionibacteria" refers to the organisms from skin, but only a few species are clinically significant. Recent reviews on these organisms were published by Brook and Frazier (1991), Eady and Ingham (1994), and Funke et al. (1997a).

The organisms in both groups are Gram-positive, diphtheroid, rod-shaped bacteria that may bifurcate or even branch; they are nonsporeforming, nonmotile and anaerobic, though not strictly anaerobic but microaerophilic. The cutaneous propionibacteria, however, are generally slender and often quite irregular and curved, while the classical propionibacteria are usually rather short and thick. Despite being predominantly anaerobic, propionibacteria are generally catalase positive, and in many cases strongly so.

A Historic View of Classical and Cutaneous Propionibacteria

Classical Propionibacteria

A number of investigators (Troili-Petersson, 1904; Von Freudenreich and Orla-Jensen, 1906; Thöni and Allemann, 1910; Sherman, 1921; Shaw and Sherman, 1923; Sherman and Shaw, 1923) described the isolation of different kinds of propionic acid bacteria from cheese, but Van Niel (1928) was the first to deal with the classification of these organisms in any systematic way. He reviewed the earlier work, described his own investigations, and recognized eight species, *Propionibacterium freudenreichii*, *Propionibacterium jensenii*, "*Propionibacterium peterssonii*," *Propionibacterium shermanii*, "*Propionibacterium pentosaceum*," "*Propionibacterium rubrum*," *Propionibacterium thoenii* and "*Propionibacterium technicum*." He also recognized

Propionibacterium jensenii var. *raffinosaceum* as a variety of *Propionibacterium jensenii*. Since van Niel's classic paper, a number of taxonomic studies have been done. Werkman and Kendall (1931) have redesignated the variety of *Propionibacterium jensenii* as a species, "*Propionibacterium raffinosaceum*," and Hitchner (1932) named two more species, "*Propionibacterium zeae*" and "*Propionibacterium arabinosum*." Sakaguchi et al. (1941) have proposed five additional species: *Propionibacterium globosum*, "*Propionibacterium amylaceum*," "*Propionibacterium japonicum*," "*Propionibacterium orientum*" and "*Propionibacterium coloratum*," with one variety of "*Propionibacterium amylaceum*," "*Propionibacterium amylaceum* var. *aurantricum*." In addition, Janoschek (1944) named three more species, "*Propionibacterium casei*," "*Propionibacterium pituitosum*" and "*Propionibacterium sanguineum*."

Several authors (e.g., Van Niel, 1928; Werkman and Brown, 1933; Janoschek, 1944; Holdeman et al., 1977) have constructed keys for the differentiation of the species primarily on the basis of pigment production and fermentation of various carbohydrates. However, the total number of strains examined is rather small for many species, and the results of fermentation tests by different investigators do not always agree. Malik et al. (1968), using 38 morphological and physiological features to examine 56 strains, grouped the strains into four clusters by numerical taxonomy, and Johnson and Cummins (1972), using cell wall and DNA hybridization analysis on 29 strains, also found four groups that agree in many respects with those proposed by Malik and his colleagues. Table 2 gives the DNA similarity data of strains arranged in the DNA similarity groups described by Johnson and Cummins (1972). The principal difference between this scheme and that of Malik et al. (1968) is in the disposition of strains of "*Propionibacterium rubrum*," which Malik found showed a rather low level of similarity with strains of *Propionibacterium thoenii*, whereas by homology they appear to be closely related. Baer (1987) has also found that four groups can be distinguished by electrophoresis of soluble proteins, and the groups correspond exactly with those found by DNA studies (Table 1).

The species described by Janoschek (1944) and Sakaguchi et al. (1941) appear to have been lost. Of Janoschek's three species, it seems likely from the original descriptions that "*Propionibacterium casei*" is closely related to *Propionibacterium freudenreichii*, while "*Propionibacterium pituitosum*" and "*Propionibacterium sanguineum*" fall into the *Propionibacterium thoenii* group. Of the five species described by Sakaguchi et al. (1941), "*Propionibacterium orientum*," "*Propionibacterium globosum*, and "*Propionibacterium coloratum*" are probably related to *Propionibacterium freudenreichii*, while "*Propionibacterium amylaceum*" and "*Propionibacterium japonicum*," which are nitrate negative but ferment sucrose and maltose, would probably fall into the *Propionibacterium jensenii* group.

In summary, all investigators seem to agree that strains of *Propionibacterium freudenreichii* and "*Propionibacterium shermanii*" and "*Propionibacterium globosum*" are closely related and should be included in a single species, *Propionibacterium freudenreichii*. Today, this move has been done by treating "*Propionibacterium shermanii*" and "*Propionibacterium globosum*" as subspecies of *Propionibacterium freudenreichii*.

Table 2. DNA-DNA a and DNA-rDNA b relatedness between *Propionibacterium* species (% DNA similarity).

	Propionibacterium freudenreichii		*Propionibacterium acidipropionici*		*Propionibacterium thoenii*	
	Method 1[a]	Method 2[b]	Method 1[a]	Method 2[b]	Method 1[a]	Method 2[b]
P. freudenreichii (*P. shermanii*)[c]	90–100	100	ND		ND	
P. acidipropionici (*P. arabinosum*) (*P. pentosaceum*)	8–28	1–5	87–100		ND	
P. thoenii (*P. rubrum*)	12–20	7–8	30–35	16–17	96–100	
P. jensenii (*P. zeae*) (*P. technicum*) (*P. raffinosum*)	17–26	5–9	30–38	13–16	51–53	32–40

Abbreviation: ND, no data.

[a]Filter hybridization method; see Johnson and Cummins (1972).

[b]Membrane filter hybridization assay using acetylaminofluorene-labelled rRNA as a probe; see de Carvalho et al. (1994).

[c]Names in parentheses refer to invalid species. *Propionibacterium shermanii* is now a subspecies of *Propionibacterium freudenreichii*.

These strains also differ from the other classical propionibacteria in fermenting a more restricted range of carbohydrates, in being heat resistant (Malik et al., 1968), and in having a rather distinctive pattern of cell wall components (Table 1). Strains of "*Propionibacterium arabinosum*" and "*Propionibacterium pentosaceum*" are also closely related to each other, being the only strains in which catalase production is weak ("*Propionibacterium pentosaceum*") or absent ("*Propionibacterium arabinosum*"). They have therefore been placed into a single species, *Propionibacterium acidipropionici*. The two groups *Propionibacterium thoenii* and *Propionibacterium jensenii*, which show the highest degree of DNA similarity (51–53%; Table 2), also show a considerable range of phenotypic variation. The original species designations within these groups were based on various fermentation tests (Breed et al., 1957).

Fermentation patterns have been the basis of distinction between named strains in almost all identification schemes, and most authors (e.g., Van Niel, 1928; Werkman and Brown, 1933; Janoschek, 1944) have devised identification keys based on such tests. In the eighth edition of *Bergey's Manual* (Moore and Holdeman, 1974), the classical propionibacteria were classified in four species, *Propionibacterium freudenreichii* (with "*Propionibacterium shermanii*"), *Propionibacterium thoenii* (including "*Propionibacterium rubrum*"), *Propionibacterium jensenii* (including "*Propionibacterium zeae*," "*Propionibacterium technicum*," "*Propionibacterium raffinosum*" and "*Propionibacterium petersonii*"), and *Propionibacterium acidipropionici* (including "*Propionibacterium arabinosum*" and "*Propionibacterium pentosaceum*"), primarily on the basis of homology groupings, and the same scheme has been followed in the first edition of *Bergey's Manual of Systematic Bacteriology* (Cummins and Johnson, 1986).

A species of doubtful taxonomic status is the invalid "*Propionibacterium coccoides*." As described by Vorob'eva et al. (1983), these are Gram-positive, nonmotile, nonsporeforming organisms isolated from Russian cheese. The organisms are spherical at all stages of growth, are facultative anaerobes, growing well at 22–30°C (but also down to 8–10°C), and are halotolerant up to 6.5% NaCl. The G+C content is 63.4 mol% and they are reported to show 48% DNA sequence similarity to a strain of *Propionibacterium jensenii*. They are catalase positive, produce vitamin B_{12} and ferment lactate with the production of propionate, acetate and CO_2 as principal end products, apparently by the succinate-methylmalonylCoA pathway. These organisms are reported to ferment a wide variety of carbohydrates, and this, along with their ability to grow at temperatures as low as 8–10°C, would appear to distinguish them from *Propionibacterium freudenreichii*, the cells of which also can often appear coccal.

Cutaneous Propionibacteria

Organisms in this group were regarded as belonging to the genus *Corynebacterium* until Douglas and Gunter (1946) and Moore and Cato (1963) showed that propionic acid was a major end product of their metabolism. Subsequent work has shown that the anaerobic coryneforms from skin differ in several fundamental ways from the classical corynebacteria (e.g., *Corynebacterium diphtheriae*). For example, they do not produce mycolic acids, the diamino-acid of the cell wall peptidoglycan is LL-A2pm in almost all cases, and their cell wall polysaccharide does not contain arabinose (Johnson and Cummins, 1972; Rogosa et al., 1974). At one time, as many as 12 species of anaerobic coryneforms were described (Prévot and Fredette, 1966). However, an examination of 80 strains by cell wall analysis and by DNA hybridization analysis showed only three major groups that seemed sufficiently distinct to be regarded as species (Johnson and Cummins, 1972). In *Propionibacterium acnes* and *Propionibacterium avidum*, two serotypes are found; these can be distinguished by precipitin tests using trichloroacetic acid extracts (Cummins, 1975). The three groups can also be recognized by the gel electrophoresis pattern of soluble proteins (Gross et al., 1978; Nordstrom, 1985). The organism generally referred to as *Corynebacterium parvum*, which causes an unusual degree of reticulostimulation and macrophage-activation in animals, was found to be indistinguishable from *Propionibacterium acnes* (Cummins and Johnson, 1974).

Most strains of *Propionibacterium acnes* and *Propionibacterium granulosum* grow poorly, if at all, under aerobic conditions, although they are not sensitive to oxygen and organisms will remain viable for several hours or longer if plate cultures are left exposed to air. Strains of *Propionibacterium avidum*, however, will grow quite well under aerobic conditions. The requirement to grow the organisms anaerobically has led to the phrase "anaerobic coryneforms" or "anaerobic corynebacteria." However, McGinley et al. (1978) have used the phrase "cutaneous propionibacteria," and this seems in many ways more appropriate because the skin does appear to be the main reservoir of these organisms.

In most cases, the major problem in identifying these organisms is one of distinguishing them from morphologically similar organisms also isolated from skin swabs or clinical specimens. All members of the group produce major amounts of propionic and acetic acids as end products of hexose metabolism (generally about 2–3 moles of propionate for 1 mole of acetate), and this pattern can readily be established by gas chromatography of culture supernatants (Holdeman et al., 1977). Also, almost all strains contain the L-isomer of diaminopimelic acid (A$_2$pm) as the diamino-acid of peptidoglycan (Johnson and Cummins, 1972). Most morphologically similar groups, e.g., *Actinomyces israelii*, *Actinomyces naeslundii*, and aerobic skin diphtheroids resembling *Corynebacterium xerosis*, either do not have A$_2$pm (for example, *Actinomyces israelii*) or have the *meso* isomer (aerobic skin diphtheroids).

Strains of *Propionibacterium acnes* and *Propionibacterium granulosum* grow rather slowly on solid media, and, especially with *Propionibacterium acnes*, colonies may still be under 1 mm in diameter at 4 days. Colonies of *Propionibacterium granulosum* are generally a little larger and more creamy and opaque than those of *Propionibacterium acnes*. Strains of *Propionibacterium avidum* grow considerably faster than either of the other species and show good growth in 48 h.

On blood agar, many strains of *Propionibacterium acnes* and most strains of *Propionibacterium avidum* are β-hemolytic on human, rabbit, or horse blood (Hoeffler, 1977). However, except for *Propionibacterium avidum*, there is no hemolysis on sheep blood agar. Cummins and Johnson (1991) have found strains of *Propionibacterium granulosum* to be nonhemolytic, but Hoeffler (1977) reported them to be β-hemolytic on rabbit blood agar.

Habitat

The classical propionibacteria have been traditionally isolated from dairy products, especially cheese, and it appears that no systematic search has been made for them in other habitats. However, Van Niel (1957) reported that strains of "*Propionibacterium peterssonii*" had been isolated from soil and that "*Propionibacterium zeae*" had been isolated from silage, but gave no further details. Prévot and Fredette (1966) said that "*Propionibacterium pentosaceum*" had been isolated from soil but gave no reference. Propionibacteria have also been reported from fermenting olives (Plastourgos and Vaughn, 1957; Cancho et al., 1970; Cancho et al., 1980) and from the soil of rice paddies (Hayashi and Furusaka, 1980). The non-dairy strains have been isolated from different habitats, i.e., *Propionibacterium cyclohexanicum* from pasteurized spoiled off-flavor orange juice in Japan (Kusano et al., 1997) and *P. microaerophilum* from a decantation reservoir of olive mill wastewater

of an olive oil factory in Southern France (Koussémon et al., 2001).

Propionibacteria are important in the development of flavor during the ripening process in the manufacture of Swiss cheese, especially the Emmental variety. The growth of propionibacteria during the "warm-room" period of cheese ripening (72–78°F) produces propionic acid and CO$_2$ from the lactate left by the earlier action of lactobacilli. The CO$_2$ is responsible for the "eyes" or gas vacuoles characteristic of such cheeses, while the propionic acid, and probably also proline and other amino acids, is important in the flavor (Langsrud and Reinbold, 1973a; Langsrud and Reinbold, 1973b; Langsrud and Reinbold, 1973c; Langsrud and Reinbold, 1973d). The starter cultures of propionibacteria used in cheese manufacture are usually described as *Propionibacterium freudenreichii* subsp. *freudenreichii* or subsp. *shermanii*, and these strains appear to be the most important for flavor. Strains of other species are present in relatively small numbers.

In Emmental-type cheese made in Finland, Merilaeinen and Antila (1976) found that over 60% of the strains isolated could be described as *Propionibacterium freudenreichii* subsp. *freudenreichii* or subsp. *shermanii*. They isolated no strains that they could identify as *Propionibacterium thoenii*, and apparently these highly pigmented strains only occur as undesirable contaminants in cheese making (Langsrud and Reinbold, 1973d).

Cutaneous propionibacteria, then classified as coryneform organisms, were first observed in material from *acne comedones* (the lesions of the disease *acne vulgaris*) by Unna (1893) and were cultured from the lesions by Sabouraud (1897). It was assumed that these organisms were the cause of acne until 1911, when Lovejoy and Hastings (1911) isolated an apparently identical organism from the skin of persons without acne; since then, the relationship between the acne bacillus and the lesions of acne has been a matter of continued debate. The distribution of the various species of cutaneous propionibacteria on the human body has been investigated by McGinley et al. (1978). *Propionibacterium acnes* is the predominant organism, especially in areas rich in sebaceous glands such as the forehead and the naso-labial folds. *Propionibacterium granulosum* is the next most common and has a similar distribution to *Propionibacterium acnes*, except that it is especially common in the alae nasi region (wing of the nose). *Propionibacterium avidum* has a more restricted distribution and is found mostly in moist rather than oily regions, such as the axilla, perineum, or anterior nares. Organisms of this group have also been isolated from the mouth, the female genital tract, and from feces. They are also common contaminants in anaerobic cultures, probably coming from the skin of the operator during subculture.

Phylogeny

The first phylogenetic study on 16S rDNA catalogues showed *Propionibacterium freudenreichii* and *Propionibacterium acnes* to form a separate line of descent within the order Actinomycetales (Stackebrandt and Woese, 1981). Comparison of almost complete 16S rDNA sequences of these two species, *Actinomyces* species, and *Arachnia propionica* led to the reclassification of the latter species as *Propionibacterium [propionicus] propionicum* (Charfreitag and Stackebrandt, 1988). More comprehensive studies on the genus followed (Charfreitag and Stackebrandt, 1989; Dasen et al., 1998), and the 16S rDNA of all type strains has by now been completed (Kusano et al., 1997; Koussémon et al., 2001; Bernard et al. 2002). Strains isolated from dairy sources are *Propionibacterium freudenreichii*, *Propionibacterium acidipropionici*, *Propionibacterium microaerophilum*, *Propionibacterium jensenii* (rare humans), and *Propionibacterium thoenii*. The latter four species cluster apart from strains isolated from human skin, viz., *Propionibacterium acnes*, *Propionibacterium avidum* and *Propionibacterium granulosum*. *Propionibacterium freudenreichii*, on the other hand, forms a loose phylogenetic cluster with two non-dairy organisms, i.e., *Propionibacterium cyclohexanicum* (from spoiled orange juice) and *Propionibacterium australiense* (from bovine lesions). Thus, the traditional separation of *Propionibacterium* into dairy propionibacteria and the cutaneous propionibacteria is not fully matched by their phylogenetic relationships.

Different studies agree on the topology of trees with the exception of the position of *Propionibacterium propionicum*: in the study of Charfreitag and Stackebrandt (1989), the type strain DSM 43307^T (accession number X53216) clusters distantly from *Propionibacterium acnes*, but as depicted in Fig. 1, clusters outside the main *Propionibacterium* cluster, next to *Tessarococcus bendigoensis*. The isolated position of *Propionibacterium propionicum* was also confirmed by Kusano et al. (1997), and the sequence was also obtained from a *Propionibacterium propionicus* strain VA1535 isolated from a patient with canaliculitis (unpublished observation of W. Geissdoerfer et al.; accession number AF285117). In contrast, a new sequence of strain DSM 43307^T obtained by Dasen et al. (1998) showed this species to be closely related to *Propionibacterium acnes* and *Propionibacterium avidum*. Similarly, Koussémon et al. (2001) and Bernard et al. (2002) showed *Propionibacterium propionicum* (with the accession numbers X53216 and AJ003058, respectively) to be related to *Propionibacterium avidum* and *Propionibacterium acnes*. To solve this problem, a new strain of DSM 43307^T was sequenced for this communication and the result of Charfreitag and Stackebrandt (1989) was confirmed. Obviously, more than a single, genetically defined type strain is in use in different laboratories and this problem should be solved before *Propionibacterium propionicum* can be excluded from the genus *Propionibacterium*.

Nucleic acid reassociation studies have been performed to verify the species status (Johnson and Cummins, 1972), to affiliate strains to

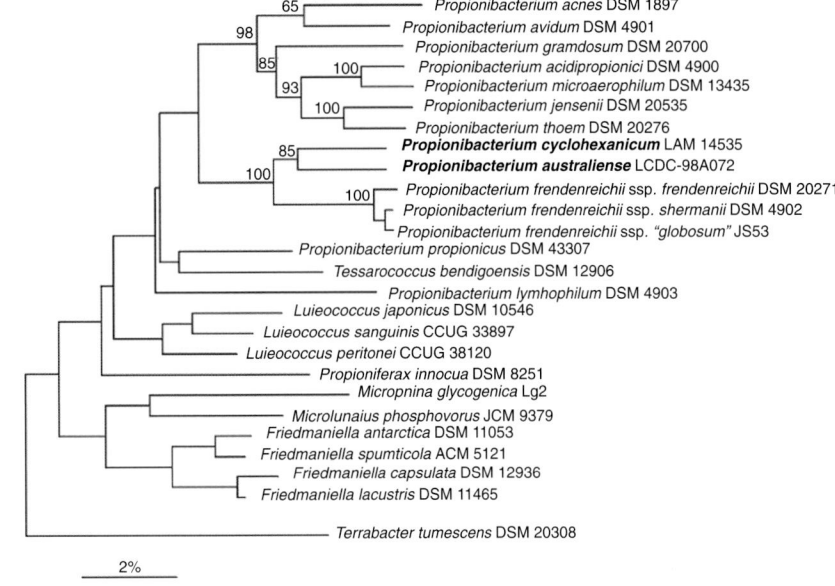

Fig. 1. Distance matrix analysis (De Soete, 1983) of almost complete 16S rDNA sequences of species of the family Propionibacteriaceae. Numbers refer to bootstrap values (500 resamplings). Names in red: cutaneous propionibacteria; names in blue: classical propionibacteria.

Table 3. DNA-DNA relationships within and between cutaneous species (% DNA similarity).

Species	*Propionibacterium acnes*, type I	*Propionibacterium avidum*, type I	*Propionibacterium granulosum*
P. acnes, type I	97	51	16
P. acnes, type II	ND	ND	ND
P. avidum, type I	50	90	17
P. avidum, type II	ND	ND	ND
P. granulosum	12	15	95

Abbreviation: ND, no data.

From Johnson and Cummins (1992) and Cummins and Johnson (1974).

described species (De Carvalho et al., 1994; Table 3), and to monitor the existence of propionibacteria in natural samples (Sunde et al., 2000). The studies on described species revealed the genomic distinctness of *Propionibacterium freudenreichii*, *Propionibacterium acidipropionici*, *Propionibacterium thoenii* and *Propionibacterium jensenii* (Table 2), of *Propionibacterium acnes*, *Propionibacterium avidum* and *Propionibacterium granulosum* (Table 3), as well as that of *Propionibacterium acidiurici* and *P. microaerophilum* (56% DNA reassociation; Koussémon et al., 2001). A DNA array-like "checkerboard" DNA-DNA hybridization technique has been performed to identify *Propionibacterium acnes* in periapical endodontic lesions of asymptomatic teeth. The occurrence of members of these species was verified in marginal and submarginal incision performed to expose the lesion, though the frequency of bacteria and their concentration of *P. acnes* varied significantly in these two types of incisions (Sunde et al., 2000).

Molecular Identification

To reliably identify prokaryotic organisms, molecular identification systems have been developed during the past decade. Derived from the suitability of ribosomal RNA genes to unravel phylogenetic relationships, methods were applied to rapidly identify *Propionibacterium* strains. These techniques were used to affiliate strains to named species and to identify strains directly in environmental samples. Approaches are based on restriction fragment analysis of total DNA (Gauthier et al., 1996), restriction analysis of amplified 16S rDNA (Riedel et al., 1998; Hall et al., 2001), or on polymerase chain reaction (PCR) assays targeting 16S rDNA (Dasen et al., 1998). One or the earliest studies used manual rRNA gene restriction patterns (riboprinting; Grimont and Grimont, 1986; De Carvalho et al., 1994; Riedel et al., 1994). These patterns are obtained by cleavage of total DNA with restriction endonucleases,

one-dimensional separation of DNA fragments, and hybridization of fragments with labeled 16+23S rRNA from *Escherichia coli*. Today this method is provided by an automated system (Dupont, Wilmington, DE, United States). In the study of De Carvalho et al. (1994), the six species analyzed with *Cla*I, i.e., *Propionibacterium freudenreichii*, *Propionibacterium jensenii*, *Propionibacterium thoenii*, *Propionibacterium acnes*, *Propionibacterium propionicum* and *Propionibacterium acidipropionici*, gave different restriction patterns, consisting of two to four bands; the method was sensitive enough to even discriminate between the two subspecies of *Propionibacterium freudenreichii*, i.e., subsp. *freudenreichii* and subsp. *shermanii*. Culture collection strains affiliated to certain species were often found to be misidentified, belonging to different described species, or even to represent novel taxonomic entities. Another study applied randomly amplified polymorphic DNA (RAPD) PCR (Rossi et al., 1998; Langsrud and Reinbold, 1973a).

Alternatively, ribosomal RNA genes are amplified, using a combination of conserved and *Propionibacterium*-specific oligonucleotide primers (e.g., primer gd1 in a multiplex-PCR; Dasen et al., 1998). The use of the latter probe resulted in the formation of a 900-bp long fragment, identifying members of *Propionibacterium*, whereas members of other genera, lacking the *Propionibacterium*-specific target site of the gd1 primer, yielded a 1500-bp amplificate. The assay can be performed within 1 day and the detection limit is about 10^3 colony forming units (cfu) of propionibacteria. The study of Rossi et al. (1999) used different primer sets for *Propionibacterium acnes* and *Propionibacterium* species occurring most frequently in raw milk. The target sites for these primers were located between positions 435 and 478 (*E. coli* nomenclature) of the 16S rDNA. The test allowed the detection of less than 10 cells per PCR assay from milk and cheese and 10^2 cells per PCR assay from forage and soil. Greisen et al. (1994) used a different target stretch (pos. 1376–1400 of the 16S rDNA); all of the four *Propionibacterium* species but none of the Gram-positive and Gram-negative reference strains of clinical significance tested positive. Nested 16S rDNA primers were used to identify *Propionibacterium propionicum* in patients with primary and persistent endodontic infections (Siqueira and Rocas, 2003).

Strains of several *Propionibacterium* species were included as references in a study by Kaufmann et al. (1997) in which a 16S rDNA-targeted probe directed against species of *Bifidobacterium* was evaluated by colony hybridization. Of all the *Propionibacterium* references tested, *Propionibacterium freudenreichii* subsp. *shermanii* gave a slight signal. This is surprising as the target stretch is identical in all propionibacterial 16S rDNAs analyzed.

Random amplification of polymorphic DNA (RAPD) was used in epidemiological studies of *P. acnes*, in the discrimination of *Propionibacterium acnes* strains from *Propionibacterium granulosum* and *Propionibacterium avidum* (Perry et al., 2003), and in the identification of dairy propionibacteria (Rossi et al., 1999). The discrimination power of the arbitrary primed PCR (AP-PCR) approach was lower than that of the pulsed-field gel electrophoresis approach (Jenkins et al., 2002), and the authors concluded that *Propionibacterium freudenreichii* strains used for Swiss cheese production in the United States were genetically diverse.

Pulsed–field gel electrophoresis has been used for strain typing of staphylococci and propionibacteria to the discriminate primary pathogens causing inflammatory reactions following cardiac surgery (Tammelin et al., 2002). Molecular identification of as yet uncultured and cultured microorganisms including propionibacteria followed the widely applied cloning and sequence analysis of PCR amplified 16S rDNA genes.

Clinical samples investigated included endodontic infections (Rolph et al., 2001) and noma lesions (Paster et al., 2002), demonstrating the presence of *P. acnes*. The two types of *P. acnes* (I and II) could be discriminated on the basis of the two dimensional separation patterns of ribosomal proteins, though the differences were small as compared to those between *Propionibacterium acnes* and *Propionibacterium granulosum* (5 versus 50 proteins; Dekio et al., 1989).

Phenotypic Properties

A compilation of tests differentiating species in the genus *Propionibacterium* is given in Tables 4 and 5. While the information is based on the analyses of many strains originating from skin, human flora or dairy sources, only six strains are available for *Propionibacterium australiense*, while the species *Propionibacterium microaerophilum* and *Propionibacterium cyclohexanicum* are represented only by their type strain.

Several studies on the evaluation of commercial identification kits have included strains of *Propionibacterium* species, but the number of strains and species was generally low. Comparing the results obtained by methods described by the Virginia Polytechnic Institute and State University with results obtained by the RapID-ANAII system (Innovative Diagnostic Systems, Inc., Atlanta, GA, United States) the latter system performed well in that 91% of the 23 strains of *P. acnes* were correctly identified (Celig and Schreckenberger, 1991). Similarly, the ATB 32A system (API System SA, Montalieu-Vercieu, France) correctly identified 100% of the 10 strains of *P. acnes* when compared to conventional identification methods (Looney et al., 1990). Also satisfying were the results obtained

Table 4. Characters useful in identification of *Propionibacterium acnes*, *Propionibacterium avidum* and *Propionibacterium granulosum*.

Character	Propionibacterium acnes		Propionibacterium avidum	Propionibacterium granulosum
	type I	type II		
Fermentation of				
Glucose	+	+	+	+
Sucrose	–	–	+	+
Maltose	–	–	+	+
Sorbitol	+/–	–	–	–
Esculin hydrolysis	–	–	+	–
Indole production	+	+	–	–
Reduction of nitrate	+	+	–	–
Gelatin liquefaction	+	+	+	–
Casein digestion	+	+	++	–
Colonies at 4 days	Small, semi-opaque, grayish; less than 1mm, reddish color may develop later		Large opaque, creamy, 1–2mm	Intermediate, opaque white to cream, ca. 1mm

Symbols: +, present; –, absent; +/–, variable; and ++, strongly present.

Table 5. Tests differentiating *Propionibacterium* species.

Test	1	2	3	4	5	6	7	8	9	10	11
Catalase	+	−	−	v	+	v	+	−	v	+	−
Indole	−	−	−	−	−	−	−	nd	+	−	−
Nitrate	v	−	+	−	−	+	−	+	+	−	+
Esculin hydrolysis	+	+	−	+	+	+	−	−	−	+	−
Gelatin hydrolysis	−	−	−	−	−	−	v	nd	+	+	v
Starch hydrolysis	−	−	−	−	v	−	−	nd	−	−	v
Urea hydrolysis	−	+	−	nd	nd	nd	nd	nd	nd	nd	−
Acid from sucrose	−	+	−	+	+	+	+	nd	−	+	+
Acid from maltose	−	+	−	+	+	+	+	nd	−	+	+
Growth in 20% bile	+	nd	−	+	−	+	−	nd	+	+	−
Mol% G+C of DNA	64–67	67	nd	65–68	66–67	66–68	61–63	68	57–60	62–63	63–65

Symbols: +, >90% of strains are positive; v, 11–89% of strains are positive; −, 0–10% of strains positive; and nd, not determined.
[a]1, *Propionibacterium freudenreichii*; 2, *Propionibacterium cyclohexanicum*; 3, *Propionibacterium australiens*; 4, *Propionibacterium jensenii*; 5, *Propionibacterium thoenii*; 6, *Propionibacterium acidipropionici*; 7, *Propionibacterium granulosum*; 8, *Propionibacterium microaerophilum*; 9, *Propionibacterium acnes*; 10, *Propionibacterium avidum;* and 11, *Propionibacterium propionicum*.
From Kusano et al. (1997), Kossemon et al. (2001), and Bernard et al. (2002).

with the BBL Crystal Anaerobe (ANR, Becton Dickinson) identification system in that 26 strains of *Propionibacterium acnes*, *Propionibacterium avidum* and *Propionibacterium granulosum* were correctly identified (Cavallaro et al., 1997). In contrast, identification problems were reported to occur for *Propionibacterium acnes* and *Propionibacterium avidum*, using the API (RAPID) Coryne System with database 2.0 (bioMerieux, La-Balme-les-Grottes, France; Funke et al., 1997b). *Propioniferax innocua*, not included in the API Coryne database 2.0, was identified as *Brevibacterium epidermidis* or *casei*.

Fluorogenic 4-methylumbelliferyl-linked substrate tests have been applied to rapid characterization of periodontal bacterial isolates, tested against reference strains. The type strains of *Propionibacterium avidum*, *Propionibacterium granulosum*, *Propionibacterium propionicum* and *Propionibacterium acnes* gave characteristic profiles but the patterns obtained from fresh isolates differed to some extent from those of the type strains (Maiden et al., 1996).

An alternative rapid identification method not based on physiological properties is pyrolysis mass spectrometry (PyMS). The spectra obtained from reference *Propionibacterium* strains and isolates were used to train artificial neural networks (Goodacre et al., 1994), which were able to recognize strains from dogs as human wild type *P. acnes*. In a later study (Goodacre et al., 1996), strains of *P. acnes* isolated from the forehead of healthy humans could be differentiated. Biochemical characteristics in combination with PyMS revealed the significant differences among the strains, some of which occurred simultaneously in the same habitat.

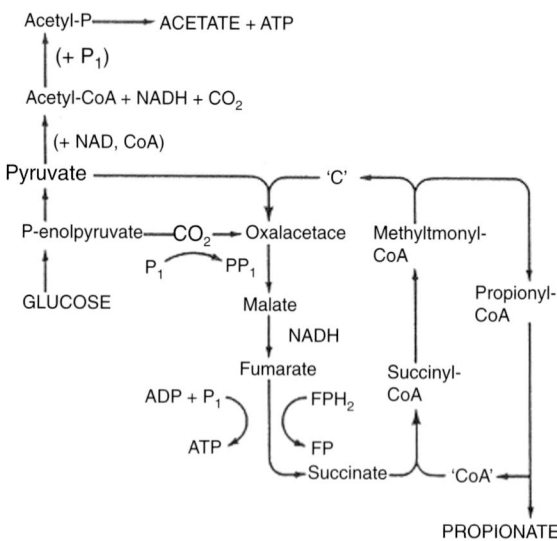

Fig. 2. Propionic acid fermentation and the formation of acetate, CO_2, propionate and ATP. FP, flavoprotein; FPH_2, reduced flavoprotein.

General Properties

Metabolism and Nutritional Requirements

The production of large amounts of propionic acid is characteristic of the propionibacteria. Hexoses are converted to pyruvate by the Embden-Meyerhof pathway, and propionate and acetate are formed by the reactions shown in Fig. 2. This diagram of propionic acid fermentation is slightly modified from that given by Allen et al. (1964) and is based on the extensive work of H. G. Wood and his collaborators (for references, see Allen et al., 1964). The background to propionic acid fermentations is also critically discussed in the reviews by Hettinga and Reinbold (Hettinga and Reinbold, 1972a; Hettinga and Reinbold, 1972b; Hettinga and Reinbold, 1972c). The fact that propionic and acetic acids are the main products of hexose fermentation can be readily shown by gas chromatography of culture supernatants (for methods, see Holdeman et al., 1977). The ratio of propionic to acetic acid is generally about 2 : 1 but may vary widely and be as high as 5 : 1 or more. Lactic acid is produced in addition to propionic acid and acetic acid by strains of *Propionibacterium cyclohexanicum* and *Propionibacterium propionicum*. The latter species, as well as *Propionibacterium australiense*, form also succinic acid (Kusano et al., 1997; Bernhard et al., 2002).

The nutritional requirements for all propionibacteria are basically very similar, which suggests a close resemblance between the overall metabolisms of the two (the classical and the cutaneous) groups. All strains require the vitamins pantothenate and biotin (Delwiche, 1949); some need thiamine and *p*-aminobenzoic acid as well. A number of other unknown factors in potato and yeast extract are stimulatory for growth. It appears from the investigations of Wood et al. (1938) that many strains of propionibacteria will grow in a basal medium without the addition of amino acids. However, growth is much improved when amino acids are added: a digest of casein (e.g., casamino acids [Difco, Detroit, MI, United States]) will supply the requirements of all strains. Strains of *Propionibacterium acnes* require pantothenate, biotin, thiamine and nicotinamide. Strains of *Propionibacterium avidum* and *Propionibacterium granulosum* require pantothenate, biotin and thiamine only. Most strains need a full complement of 18 amino acids for good growth, and growth is further improved by the addition of 0.1–0.2% lactate, pyruvate and ketoglutarate and of guanine and adenine (Ferguson and Cummins, 1978).

Antigens

The polysaccharides of the walls of the classical propionibacteria can be extracted with trichloroacetic acid and give good immunoprecipitation reactions with antisera prepared against suspensions of whole cells (C. S. Cummins and P. Hall, unpublished observations). All polysaccharides except those from *Propionibacterium freudenreichii* contain 2, 3-diaminohexuronic acid (Cummins and White, 1983; Cummins, 1985).

A number of strains of cutaneous propionibacteria give suspensions that are unstable in saline and so are unsuitable for agglutination tests using suspensions of whole cells. However, antisera are commercially available (Difco, Detroit, MI, United States) that can be used for the identification of *Propionibacterium acnes* by the slide agglutination test. A clearer separation of the three species *Propionibacterium acnes*, *Propionibacterium granulosum* and *Propionibacterium avidum* and of their serotypes can be obtained by using cell wall polysaccharide antigens extracted from the cells by 10% trichloroacetic acid at 56°C (Cummins, 1975). A number of strains of *Propionibacterium avidum* are heavily capsulated (C. S. Cummins et al., unpublished observations), and it has been found that better antisera to cell wall antigens are produced if noncapsulated strains are used for immunization.

Sensitivity to Muralytic Enzymes

The intact cells and the isolated cell walls of the propionibacteria are lysozyme resistant (unless acetylated), except for the isolated cell walls of *Propionibacterium freudenreichii*, which are sensitive (C. S. Cummins, unpublished observation). However, all propionibacteria are moderately sensitive to some other muralytic enzymes, such as mutanolysin or achromopeptidase.

Bacteriophages

The bacteriophages of classical propionibacteria do not appear to have been investigated, although Hettinga and Reinbold (1972c) reported failure to isolate bacteriophage from their strains. They ascribed their failure to interference by slime layers.

A number of bacteriophage types have been described for *Propionibacterium acnes* (e.g., Prévot and Thouvenot, 1961; Jong et al., 1975; Webster and Cummins, 1978). Bacteriophages for *Propionibacterium avidum* and *Propionibacterium granulosum* have not been investigated; however, strains of these two species are not lysed by any of the *Propionibacterium acnes* phages. Some phage strains, e.g., strain 174 of Zierdt et al. (1968), will lyse most stains of *Propionibacterium acnes* and can be used for rapid presumptive species identification. Bacteriophages active against *Propionibacterium acnes* can readily be detected in filtrates of skin washings (Marples, 1974).

Propionicins

Few bacteriocins have been described among dairy propionibacteria active against Gram-positive and Gram-negative bacteria, yeasts and molds (Lyon and Glatz, 1993; Faye et al., 2000; Faye et al., 2002; Ben-Shushan et al., 2003; Gollop et al., 2003). These propionicins may include protease-activated antibacterial peptides. The propionicin SM1 from *Propionibacterium jensenii* DF1 shows strong bacteriocidal action against

Propionibacterium jensenii DSM 20274. The gene is located on a plasmid and the propionicin amino acid sequence shows significant homologies to a protein excreted from *Lactobacillus lactis* (Miescher et al., 2000). Horizontal gene transfer could explain the presence in one but not in a second strain of the same species.

A mixture of *Propionibacterium jensenii* SM11 and different strains of *Lactobacillus paracasei* subsp. *paracasei* showed inhibitory activities against spoilage yeasts in dairy products at refrigerator temperature (Schwenninger and Meile, 2004).

Probiotics and Growth Stimulators

Growth stimulators for bifidobacteria were found in culture broth of *Propionibacterium freudenreichii* (Isawa et al., 2002): Isawa et al. (2002) identified one stimulator as 1,4-hydroxy-2-naphthoic acid, while Mori et al. (2000) identified another factor as 2-amino-3-carboxy-1,4-naphthoquinone. Also, short- chain fatty acids, such as propionate, stimulated the growth of bifidobacteria while being highly inhibitory to the growth of Gram-negative facultative and obligatory anaerobes (Kaneko et al., 1994).

Chemotaxonomic Properties

In contrast to most other genera of the class Actinobacteria, members of *Propionibacterium* exhibit different peptidoglycan types. The diamino acid of peptidoglycan is *meso*-A_2pm (type A1γ), LL-A_2pm (type A3γ) or a combination of both amino acids (Schleifer and Kandler, 1972; Kusano et al., 1997; Table 1). Other distinctive wall components are sugars composed of mainly glucose, mannose and rhamnose. Galactose may be present. Kusano et al. (1997) display chemotaxonomic data of all *Propionibacterium* species described until 1997. The major menaquinone is MK-9(H_4) (a quinone with 4 hydrogen atoms on the side chain containing 9 isoprene units) and the mol% G+C of DNA is 57–68. The fatty acid profiles are dominated by branched acids ($C_{i15:0}$ or $C_{a15:0}$ or both) while straight chain fatty acids ($C_{15:0}$, $C_{16:0}$, and $C_{17:0}$) occur in significantly lower amounts (Moss et al., 1969; Cummins and Moss, 1990; Bernard et al., 1991; Bernard et al., 2002). ω–Cyclohexyl undecanoic acid is the major fatty acid compound in *Propionibacterium cyclohexanicum* (57%; Kusano et al., 1997). The most comprehensive compilation of fatty acid data is presented by Bernard et al. (2002), which however omits data of *Propionibacterium cyclohexanicum*. Mannose-containing phospholipids have been reported in *Propionibacterium freudenreichii* subsp. *shermanii* (Brennan and Ballou,

1968; Prottey and Ballou, 1968). Complex lipids that have chemoattractant properties for phagocytes have been extracted from strains of *Propionibacterium acnes* (Russel et al., 1976).

Hettinga and Reinbold (1972c), quoting unpublished work by Skogen (1970), reported that a strain of *Propionibacterium jensenii* (*"Propionibacterium zeae"*) produced slime and capsular material composed of glucose and galactose. Cummins and Johnson (1991) report that a number of strains of *Propionibacterium thoenii* and *Propionibacterium jensenii* are capsulated (C. S. Cummins and P. Hall, unpublished observations), as have Skogen et al. (1974), and a number of species produce extracellular slime that makes cultures in liquid medium quite viscous. The capsular material is polysaccharide.

Sensitivity to Antimicrobial Agents

No very consistent or unusual pattern has been reported, except that all strains are highly resistant to sulfonamides and appear to be more resistant to semisynthetic penicillins (such as oxacillin) than to penicillin G (Reddy et al., 1973a). When disk sensitivity tests are used, some strains will grow in the presence of 1000 μg/ml sulfadiazine. Reddy et al. (1973b) have shown that some subspecies of *Propionibacterium freudenreichii* and *Propionibacterium thoenii* and some strains of *Propionibacterium acidipropionici* can synthesize folic acid, while other strains of *Propionibacterium acidipropionici*, *Propionibacterium thoenii* and *Propionibacterium jensenii* cannot. However, the latter strains are still resistant to sulfonamides, and Reddy et al. (1973a) concluded that this resistance may be due to the failure of the drugs to enter the cell. Among other antimicrobial substances, nisin (from streptococci) has an inhibitory effect on the growth of propionibacteria in Emmental cheese (Galesloot, 1957; Winkler and Fröhlich, 1957).

Strains of *Propionibacterium acnes* were found to be sensitive to penicillin, erythromycin, tetracyclines, chloramphenicol and novobiocin and resistant to streptomycin and sulfonamides (Pochi and Strauss, 1961). The strains were particularly resistant to sulfonamides and would grow in the presence of concentrations of more than 500 μg/ml.

Isolation and Maintenance

Classical Strains

Most investigators from Van Niel (1928) onward have relied primarily on yeast extract-sodium lactate media, with or without the addition of peptone. A typical formula is that of Malik et al. (1968).

Yeast Extract-Sodium Lactate (YEL) Medium for Isolation and Maintenance of Propionibacteria (Malik et al., 1968)

Trypticase (BBL)	1%
Yeast extract (Difco)	1%
Sodium lactate	1%
KH_2PO_4	0.25%
$MnSO_4$	0.0005%
Agar (Difco)	1.5%

Dilute and dissolve in distilled water up to 1 liter and adjust the pH to 7.0.

Hettinga et al. (1968) have devised a method whereby 2% lactate agar is placed in a pouch made of a plastic film of low gaseous diffusibility to maintain sufficiently anaerobic conditions.

Media of this type have been used both for isolation and for the maintenance of stock cultures. Sufficiently anaerobic conditions were maintained by agar overlay (Malik et al., 1968), by the addition of 0.5% sodium sulfite coupled with an overlay of paraffin oil (Demeter and Janoschek, 1941), or by growth in candle oats jars (Vedamuthu and Reinbold, 1967). Probably the easiest method of isolation is to supplement the medium of Malik et al. (1968) with 0.05% cysteine and 0.05% Tween 80 and to incubate the plates in a Brewer-type anaerobe jar containing 10–20% CO_2. Chopped-meat medium in stoppered tubes under CO_2 (Holdeman et al., 1977) is excellent for preserving stock cultures. Cultures remain viable for many months at room temperature, but at refrigerator temperatures (e.g., 4°C), cultures may die out rather rapidly. For stock cultures, it is better to omit glucose from the medium.

Larger cultures for biochemical or cellular analysis may conveniently be grown in Erlenmeyer flasks by the method described by Cummins and Johnson (1971) and explained in detail by Cummins and Johnson (1991).

A medium supporting good growth of all the species of propionibacteria is as follows:

Trypticase-Yeast Extract-Glucose Medium for Growth of Propionibacteria (Johnson and Cummins, 1972)

Trypticase (BBL)	1%
Yeast extract (Difco)	0.5%
Glucose	1%
$CaCl_2$	0.002%
$MgSO_4$	0.002%
NaCl	0.002%
Potassium phosphate buffer	0.05 M
Tween 80	0.05%
$NaHSO_2 \cdot CH_2O \cdot 2H_2O$ (Eastman Organic Chemicals)	0.05%
$NaHCO_3$	0.1%

Dissolve and dilute ingredients in 0.5 M potassium phosphate (equal molar mono- and dibasic), add $NaHCO_3$ as a sterile solution at the time of inoculation, and adjust pH to 7.0.

Completely synthetic media for the growth of propionibacteria have been devised by Kurmann (1960) and Reddy et al. (1973a).

Cutaneous Strains

Anaerobic coryneforms from the skin or other epithelial surfaces are easily obtained by swabbing suitable areas. Other methods of sampling are scraping with a sterile scalpel blade (Evans et al., 1950) or with the edge of a Teflon stirrer. The technique of Williamson and Kligman (1965), although originally designed for aerobes, is also very satisfactory for isolating *Propionibacterium acnes* or similar organisms from the skin surface, provided that anaerobic conditions are employed for cultivation. (The method is given in its original form below.)

Isolation of Anaerobic Coryneforms from the Skin (Williamson and Kligman, 1965)

1. Scrub the area (3.8 cm²), which is delineated by a sterile glass cylinder held firmly to the skin by two attached handles.
2. Pipet 1 ml of wash solution—0.1% Triton X-100 in 0.075 M phosphate buffer, pH 7.9—and scrub the area with moderate pressure for 1 min using a sterile Teflon stirrer.
3. Aspirate the wash fluid, replace it with a fresh 1-ml aliquot, and scrub again.
4. Pool the two washes and dilute an aliquot in 10-fold steps using 0.05% Triton X-100 in 0.0375 M phosphate buffer as diluent to prevent any reaggregation of organisms.
5. Plate the appropriate dilutions (usually 10^0, 10^{-1}, 10^{-2} for normal skin; 10^{-3} and 10^{-4} for areas of high bacterial density) in 15–20 ml of tryptic soy agar per plate.
6. After 48 h incubation at 37°C, count colonies and calculate viable cells in the original sample by standard methods.

Using this method, suitable media for anaerobic coryneforms are blood agar and peptone-yeast extract-glucose agar, pH 6.5, containing 0.1% Tween 80 (see also Kishishita et al., 1980). The plates need to be incubated anaerobically (e.g., GasPak jars containing H_2 and CO_2) for up to 7 days. Remember that strains of *Propionibacterium avidum* will frequently grow aerobically, although more slowly than on anaerobic plates.

This technique calls for special glass cylinders that can be held against the skin to contain fluid. However, satisfactory results can be obtained using swabs dipped in the detergent-buffer mixture and then squeezed to expel excess fluid. After sampling, the liquid in the swab is squeezed out into a measured volume of fluid to wash out organisms that have been picked up. Considerable variation in bacterial numbers is found from person to person (Evans et al., 1950), and it is important to plate out several dilutions, as described in the preceding procedure of Williamson and Kligman (1965).

A detailed comparison of the results from scraping versus those from swabbing is given in Evans and Stevens (1976). Scraping is more likely to yield organisms from the pilosebaceous glands, while swabbing picks up surface organisms only.

The method for bulk culturing described for the classical propionibacteria is also applicable for growing the cutaneous propionibacteria.

Pathogenic Cutaneous Propionibacteria

As summarized by Funke et al. (1997a), the predisposing conditions for *Propionibacterium acnes*, *Propionibacterium avidum* and *Propionibacterium granulosum* are the presence of foreign bodies (prosthetic valves or prosthetic joints), immunosuppression, preceding surgery trauma, diabetes, and obstruction of sinus ostia. These species have also been identified to cause endophthalmitis (Hykin et al., 1994), brain abscesses, meningitis, arthritis, osteomyelitis, endocarditis and infections of the central nervous system (literature summarized by Funke et al., 1997a). Incubation times may last from a few days up to 12 months, even 18 months (Barazi et al., 1995). Using the decrease in mitochondrial dehydrogenase activity of *Propionibacterium acnes*-infected viable HeLa and fibroblastic cell cultures as a measure of cytotoxicity, the bacteria under anaerobic conditions could be shown to produce cytotoxic effects for as long as 8 months (Csukas et al., 2004). Funke et al. (1997a) observed that the clinical significance is inversely proportional to the time of appearance in culture, unless the patient has been pretreated with antibiotics. Propionibacteria are rarely killed but are susceptible to a broad range of antibiotics. No differences were detected in the effects of 11 antibiotics between *Propionibacterium acnes* serotypes I and II (Kishishita et al., 1980).

Retinaldehyde has been used to significantly decrease the counts of viable *Propionibacterium acnes* suggesting that a daily total application of 0.05% retinaldehyde exerts antibacterial activity (Pechere et al., 1999). Likewise, concomitant application of 5% (w/w) benzoyl peroxide together with 3% (w/w) erythromycin was shown to bring about significant reductions in acne grade and lesion counts caused by erythromycin-resistant propionibacteria (Eady et al., 1996).

Propionibacterium acnes

Because it is widely distributed on the skin of man, *Propionibacterium acnes* is a common contaminant in clinical specimens sent for bacteriological examination, and strains of this species were isolated eight times more frequently than other *Propionibacterium* species (Funke et al., 1997a). However, occasionally it may be isolated in circumstances where it appears clearly to be a primary pathogen, for example in septic arthritis (Yocum et al., 1982) and in endocarditis (Felner and Dowell, 1970; Wilson et al., 1972). It has also been claimed that variant strains of *Propionibacterium acnes*, transmitted by house dust mites, are implicated in the causation of Kawasaki disease, an acute febrile illness in children, which may be accompanied by coronary arteritis (Kato et al., 1983).

The relationship of *Propionibacterium acnes* to the disease acne vulgaris is obscure. The essential lesion in acne is plugging of the orifice of the sebaceous glands, and overgrowth of *Propionibacterium acnes* (and *Propionibacterium granulosum*) in the obstructed gland may produce sufficient acid to irritate the tissues, or soluble antigens leaking out of the gland may cause an inflammatory reaction. Severe inflammation and scarring in acne are almost always due to an associated staphylococcal infection. Especially biotype 3 (B3) strains, showing higher lipase activity than those of B1, B2 and B4, were isolated from severe skin rashes (Higaki et al., 2000). The lipase activity of *P. acnes* B3 strains was found to be higher than that of *Propionibacterium granulosum* (Higaki et al., 2001).

RETICULOSTIMULATORY PROPERTIES OF *PROPIONIBACTERIUM ACNES* In 1966 it was shown that killed suspensions of an organism called "*Corynebacterium parvum*" could prevent the development of tumors from inocula of malignant cells in syngeneic mice (Halpern et al., 1966; Woodruff and Boak, 1966). Suspensions of this and other similar strains were known to increase the rate of clearance of carbon particles from the blood stream and cause considerable hepatosplenomegaly in mice and other animals. The majority of strains of "*Corynebacterium parvum*" were later identified as *Propionibacterium acnes*, and almost all the remainder were *Propionibacterium avidum*.

Vaccines from strains of *Propionibacterium acnes* and *Propionibacterium avidum* generally produce considerable hepatosplenomegaly (although some strains are inactive), while strains of *Propionibacterium granulosum* do not. However, suspensions of *Propionibacterium granulosum* may be active in preventing tumor development (see Cummins, 1984). The exact basis for the reticulostimulatory activity of these strains has not yet been established.

No statistically significant effect was found in the reduction of *Staphylococcus aureus* in quarter milk samples from untreated *Staphylococcus aureus*-infected lactating cows, *Staphylococcus aureus*-infected lactating cows treated with killed preparations of the putative immunostimulant *Propionibacterium acnes*, and untreated control (uninfected) cows (Dinsmore et al., 1995).

Propionibacterium avidum

This organism may be found in chronic infected sinuses, ulcers, abscesses, etc., but it usually is in combination with other organisms. Other than its occurrence in acne pustules, *Propionibacterium granulosum* has not been reported in pathological conditions.

Prophylactic application of *Propionibacterium avidum* KP-40 has been investigated in several studies for its immunostimulating effect in mice and swine. After application of this strain in combination with heparin to mice, the number of syngeneic sarcoma L-1 lung and liver tumor nodules decreased significantly (Beuth et al., 1987). Strain KP-40 is a potent stimulator of the macrophage-monocyte system and inducer of interferon as shown in comparative studies on vaccinated swine infected with classical swine fewer virus and bacterial infectious agents such as *Haemophilus pleuropneumoniae* or *Erysipelothrix rhusiopathiae* (Markowska-Daniel et al., 1992; Markowska-Daniel et al., 1993a; Markowska-Daniel et al., 1993b).

Propionibacterium propionicum

The organism appears to be a normal inhabitant of the human mouth and can be isolated from dental plaque but may also cause canaliculitis and dacryocystitis. It has also been demonstrated in cervico-vaginal smears by fluorescent antibody techniques. Like *Actinomyces israelii*, it may be found in the lacrimal duct as a cause of lacrimal canaliculitis, and in typical actinomycosis involving the cervico-facial area and occasionally elsewhere (Brock et al., 1973; Edminston, 1991). *Propionibacterium propionicum* was differentiated from *Propionibacterium acnes* by analyses of thin layer chromatography profiles of glycolipids (Mordarska and Pasciak, 1994).

MORPHOLOGY AND CULTURAL CHARACTERISTICS The organism is Gram-positive, nonmotile and non-acid-fast, but otherwise its appearance in stained smears is very variable. Usually some combination of short diphtheroidal elements and longer branched filaments is seen, depending on the medium and the age of the culture. Another characteristic feature is the presence of swollen coccoid forms. Filamentous forms are commoner in young cultures (24–48 h) and in clinical material (e.g., lacrimal sac infections).

Propionibacterium propionicum will generally grow well in standard complex media such as brain heart infusion or thioglycolate broth, especially if supplemented with 0.2% sterile rabbit serum. Schaal and Pulverer (1981) especially recommend the CC medium originally devised by Howell and Pine (1956). Good growth is obtained on the surface of the same media solidified with agar and supplemented with 4–5% rabbit or horse blood.

In liquid media the growth is generally floccular or granular: colonies on solid media may vary from smooth, regular, convex types to rough "breadcrumb" or "molar tooth" types reminiscent of *Actinomyces israelii*.

OXYGEN REQUIREMENTS AND BIOCHEMICAL REACTIONS *Propionibacterium propionicum* will grow both aerobically and anaerobically, but good growth is obtained earlier and with smaller inocula under anaerobic conditions. All strains are uniformly catalase, indole, and Voges-Proskauer negative; they all reduce nitrate and hydrolyze starch (Slack and Gerencser, 1975). A variety of sugars is fermented, and most strains were positive for adonitol, sorbitol, mannitol, fructose, sucrose, lactose, trehalose, glucose, galactose, mannose and raffinose (Slack and Gerencser, 1975; Schofield and Schaal, 1981).

ANTIBIOTIC SENSITIVITY *Propionibacterium propionicum* is normally sensitive to β-lactam antibiotics, tetracyclines, chloramphenicols, macrolides such as erythromycin, and a number of other antibiotics such as vancomycin. The organism is reported to be highly resistant to aminoglycosides such as gentamycin, to nitroimidazole compounds (such as metronidazole), and to peptide antibiotics (such as colistin; e.g., Schaal and Pape, 1980; Niederau et al., 1982).

SUBTYPES OF *PROPIONIBACTERIUM PROPIONICUM* Two serovars have been described using fluorescent antibody techniques and gel diffusion tests (Gerencser and Slack, 1967; Holmberg and Forsum, 1973; Slack and Gerencser, 1975; Schaal, 1986). The two serovars are represented by the type strain for *Propionibacterium propionicum* ATCC 14157 (serovar 1), and by ATCC 29326 (serovar 2: WVU 346, F. Lentze strain "Fleischmann"). Johnson and Cummins (1972) found no crossreaction between these strains by cell wall agglutination tests, and the serovar 2 strain

(ATCC 29326, VPI 5067) showed very low DNA homology (1%) to ATCC 14157. The two serovars were found to form distinct subclusters in numerical phenetic analyses (Schofield and Schaal, 1981). Therefore, the two serovars are in fact likely to be distinguished at the species level.

Isolation of *Propionibacterium propionicum* in Cases of Actinomycosis Despite its characteristics of cell wall structure, fermentation end products, and other properties, which have led to the conclusion that taxonomically this organism should be placed in *Propionibacterium*, it is important to remember that clinically the type of disease caused cannot be distinguished from that due to *Actinomyces israelii*. The reader is therefore referred to the chapter on The Family Actinomycetaceae: The Genera Actinomyces, Actinobaculum, Arcanobacterium, Varibaculum and Mobiluncus in this Volume for a discussion of procedures for isolation from clinical material, the details of which are essentially the same for both organisms.

Literature Cited

Allen, S. H. G., R. W. Kellermeyer, R. L. Stjernholm, and H. G. Wood. 1964. Purification and properties of enzymes involved in the propionic acid fermentation. J. Bacteriol. 87:171–187.

Allen, S. H. G., and B. A. Linehan. 1977. Presence of transcarboxylase in Arachnia propionica. Int. J. Syst. Bacteriol. 27:291–292.

Baer, A. 1987. Identification and differentiation of propionibacteria by electrophoresis of their proteins. Milchwissenschaft 41:431–433.

Barazi, S. A., K. K. Gnanalingham, I. Chopra, and J. R. van Dellen. 1995. Delayed postoperative intracerebral abscee caused by Propionibacterium acnes: Case report and review of the literature. Br. J. Neurosurg. 17:336–339.

Ben-Shushan, G., V. Zakin, and N. Gollop. 2003. Two different propiocins produced by Propionibacterium thoenii P-127. Peptides 24:1733–1740.

Bernard, K. A., M. Bellefeuille, and E. P. Ewan. 1991. Cellular fatty acid composition as an adjunct to the identification of asporogenous, aerobic Gram-positive rods. J. Clin. Microbiol. 29:83–89.

Bernard, K. A., L. Shuttleworth, C. Munro, J. C. Forbes-Faulkner, D. Pitt, J. H. Norton, and A. D. Thomas. 2002. Propionibacterium australiense sp. nov. derived from granulomatous bovine lesions. Anaerobe 8:41–47.

Beuth, J., H. L. Ko, G. Uhlenbruck, and G. Pulverer. 1987. Combined immunostimulation (Propionibacterium avidum KP 40) and anticoagulation(heparin) prevents metastatic lung and liver colonization in mice. J. Cancer Res. Clin. Oncol. 113:359–362.

Breed, R. S., E. G. D. Murray, and N. R. Smith. 1957. Bergey's Manual of Determinative Bacteriology, 7th ed. Williams & Wilkins. Baltimore, MD. 569–576.

Brennan, P., and C. E. Ballou. 1968. Phosphatidylmyoinositol monomannoside in Propionibacterium shermanii. Biochem. Biophys. Res. Comm. 30:69–75.

Brock, D. W., L. K. George, J. M. Brown, and M. D. Hicklin. 1973. Actinomycosis caused by Arachnia propionica. Am. J. Clin. Pathol. 59:66–77.

Brook, I., and E. Frazier. 1991. Infections caused by Propionibacterium species. Rev. Infect. Dis. 13:819–822.

Busse, H. J., and P. Schumann. 1999. Polyamine profiles within genera of the class Actinobacteria with LL-diaminopimelic acid in the peptidoglycan. Int. J. Syst. Bacteriol. 49:179–184.

Cancho, F. G., M. Nosti Vega, M. Fernandez Diaz, and N. J. Y. Buzcu. 1970. Especies de Propionibacterium relacionades con la zapateria: Factores que influyen en su desarrollo. Microbiol. Esp. 23:233–252.

Cancho, F. G., L. R. Navarro, and R. de la Borbolla, Y. Alcala. 1980. La formacion de acido propionico durante la concervacion de las aceitunas verdes de mesa III. Microorganismos responsables. Graces Aceites 31:245–250.

Cavallaro, J. J., L. S. Wiggs, and J. M. Miller. 1997. Evaluation of the BBL Crystal Anaerobic identification system. J. Clin. Microbiol. 35:3186–3191.

Celig, D. M., and P. C. Schreckenberger. 1991. Clinical evaluation of the PapID-AnaII panel for identification of anaerobic bacteria. J. Clin. Microbiol. 29:457–462.

Charfreitag, O., M. D. Collins, and E. Stackebrandt. 1988. Reclassification of Arachnia propionica as Propionibacterium propionicus. Int. J. Syst. Bacteriol. 38:354–357.

Charfreitag, O., and E. Stackebrandt. 1989. Inter- and intrageneric relationships of the genus Propionibacterium as determined by 16S rRNA sequences. J. Gen. Microbiol. 135:2065–2070.

Csukas, Z., B. Banizs, and F. Rozgonyi. 2004. Studies on the cytotoxic effects of Propionibacterium acnes strains isolated from cornea. Microb. Pathog. 36:171–174.

Cummins, C. S., and J. L. Johnson. 1971. Taxonomy of the clostridia: wall composition and DNA homologies in Clostridium butyricum and other butyric acid-producing clostridia. J. Gen. Microbiol. 67:33–46.

Cummins, C. S., and J. L. Johnson. 1974. Corynebacterium parvum: a synonym for Propionibacterium acnes. J. Gen. Microbiol. 80:433–442.

Cummins, C. S. 1975. Identification of Propionibacterium acnes and related organisms by precipitation tests with trichloroacetic acid extracts. J. Clin. Microbiol. 2:104–110.

Cummins, C. S., and R. H. White. 1983. Isolation, identification and synthesis of 2,3-diamino-2,3-dideoxyglucuronic acid: a component of Propionibacterium acnes cell wall polysaccharide. J. Bacteriol. 153:1388–1393.

Cummins, C. S. 1984. Corynebacterium parvum and its fractions. In: R. L. Fenichel and M. A. Chirigos (Eds.) Immune Modulation Agents and their Mechanisms. Marcel Dekker. New York, NY. 163–190.

Cummins, C. S. 1985. Distribution of 2,3-diaminohexuronic acid in strains of Propionibacterium and other bacteria. Int. J. Syst. Bacteriol. 35:411–416.

Cummins, C. S., and J. L. Johnson. 1986. Genus 1. Propionibacterium Orla-Jensen 1909, 337[AL]. In: P. H. A. Sneath, N. S. Mair, M. E. Sharpe, and J. G. Holt (Eds.) Bergey's Manual of Systematic Bacteriology. Williams & Wilkins. Baltimore, MD. 2:1346–1353.

Cummins, C. S., and C. W. Moss. 1990. Fatty acid composition of Propionibacterium propionicum (Arachnia propionicus). Int. J. Syst. Bacteriol. 30:307–308.

Cummins, C. S., and J. L. Johnson. 1991. The genus Propionibacterium. In: A. Balows, H. G. Trüper, M. Dworkin, W.

Harder, and K.-H. Schleifer (Eds.) The Prokaryotes. Springer-Verlag. New York, NY. 834–849.

Dasen, G., J. Smutny, M. Teuber, and L. Meile. 1998. Classification and identification of propionibacteria based on ribosomal RNA genes and PCR. Syst. Appl. Microbiol. 21:251–259.

De Carvalho, A. F., M. Gautier, and F. Grimont. 1994. Identification of diary Propionibacterium species by rRNA gene restriction patterns. Res. Microbiol. 145:667–676.

Dekio, S., K. Hashimoto, and M. Makino. 1989. Two-dimensional gel electrophoresis of ribosomal proteins from Propionibacterium acnes and granulosum. Zbl. Bakteriol. 271:442–445.

Delwiche, E. A. 1949. Vitamin requirements of the genus Propionibacterium. J. Bacteriol. 58:395–398.

Demeter, K. J., and A. Janoschek. 1941. Vorkommen und Entwicklung der Propionsaurebakterien in verschiedenen Kasearten. Zbl Bakteriol., Abt. 2 103:257–271.

De Soete, G. 1983. A least square algorithm for fitting additive trees to proximity data. Psychometrika 48:621–626.

Dinsmore, R. P., M. B. Cattell, R. D. Stevens, C. S. Gabel, M. D. Salman, and J. K. Collins. 1995. Efficacy of a Propionibacterium acnes immunostimulant for treatment of chronic Staphylococcus aureus mastitis. J. Dairy Sci. 78:1932–1936.

Douglas, H. C., and S. E. Gunter. 1946. The taxonomic position of Corynebacterium acnes. J. Bacteriol. 52:15–23.

Eady, E. A., and E. Ingham. 1994. Propionibacterium acnes: friend or foe? Rev. Med. Microbiol. 5:163–173.

Eady, E. A., R. A. Bojar, C. E. Jones, J. H. Cove, K. T. Holland, and W. J. Cunliffe. 1996. The effects of acne treatment with a combination of benzoyl peroxide and erythromycin on skin carriage of erythromycin-resistant propionibacteria. Br. J. Dermatol. 134:107–113.

Edminston, C. E. 1991. Arachnia and Propionibacterium: Casual commensals of opportunistic diphtheroids. Clin. Microbiol. Newsl. 12:57–59.

Evans, C. A., W. M. Smith, E. A. Johnson, and E. R. Giblett. 1950. Bacterial flora of the normal human skin. J. Inv. Dermatol. 15:305–323.

Evans, C. A., and R. J. Stevens. 1976. Differential quantitation of surface and subsurface bacteria of normal skin by the combined use of the cotton swab and scrub methods. J. Clin. Microbiol. 3:576–581.

Faye, T., T. Langsrud, I. F. Nes, and H. Holo. 2000. Biochemical and genetic characterization of propionicin T1, a new bacteriocin from Propionibacterium thoenii. Appl. Environ. Microbiol. 66:4230–4236.

Faye, T., D. A. Brede, T. Langsrud, I. F. Nes, and H. Holo. 2002. An antimicrobial peptide is produced by extracellular processing of a protein from Propionibacterium jensenii. J. Bacteriol. 184:3649–3656.

Felner, J. M., and V. R. Dowell. 1970. Anaerobic bacterial endocarditis. N. Engl. J. Med. 283:1188–1192.

Ferguson, D. A., and C. S. Cummins. 1978. Nutritional requirements of anaerobic coryneforms. J. Bacteriol. 135:858–867.

Fitz, A. 1878. Über Spaltpilzgährungen. Ber. Deutsch. Chemisch. Ges. 11:1890–1899.

Fitz, A. 1879. Über Spaltpilzgährungen. Ber. Deutsch. Chemisch. Ges. 12:474–481.

Funke, G., A. von Graevenitz, J. E. Clarridge, 3rd., and K. A. Bernard. 1997a. Clinical microbiology of coryneform bacteria. Clin. Microbiol. Rev. 10:125–159.

Funke, G., F. N. Renaud, J. Freney, and P. Roegel. 1997b. Multicenter evaluation of the updated and extended

API (Rapid) Coryne Database 2.0. J. Clin. Microbiol. 35:3122–3126.

Galesloot, T. E. 1957. Involved van Nisine op die Bacterien Welka Betrokken Zinj of Kunnen Zijn bij bacteriologische Processen in Kaas en Smeltkaas. Netherl. Milk Dairy J. 11:58–73.

Gauthier, M., A. de Carvalho, and A. Rouault. 1996. DNA fingerprinting of diary propionibacteria strains by pulsed field electrophoresis. Curr. Microbiol. 32:17–24.

Gerencser, M. A., and J. M. Slack. 1967. Isolation and identification of Actinomyces propionicus. J. Bacteriol. 94:109–115.

Gollop, N., D. Toubia, G. B. Shushan, and V. Zakin. 2003. High production system of the antibacterial peptide PLG-1. Biotechnol. Progr. 19:436–439.

Goodacre, R., M. J. Neal, L. W. Kell, W. C. Nobel, and R. G. Harvey. 1994. Rapid identification using pyrolysis mass spectrometry and artificial neural networks of Propionibacterium acnes isolated from dogs. J. Appl. Bacteriol. 76:124–134.

Goodacre, R., S. A. Howell, W. C. Noble, and M. J. Neal. 1996. Sub-species discrimination, using pyrolysis mass spectrometry and self-organising neural networks of Propionibacterium acnes isolated from normal human skin. Zbl. Bakteriol. 284:501–515.

Greisen, K., M. Loeffelholz, A. Purohit, and D. Leong. 1994. PCR primers and probes for the 16S rRNA gene of most species of pathogenic bacteria, including bacteria found in cerebrospinal fluid. J. Clin. Microbiol. 32:335–351.

Grimont, F., and P. A. D. Grimont. 1986. Ribosomal ribonucleic acid gene restriction patterns as potential taxonomic tools. Ann. Inst. Pasteur/Microbiol. 137B:165–175.

Gross, C. S., D. A. Ferguson, and C. S. Cummins. 1978. Electrophoretic protein patterns and enzyme mobilities in anaerobic coryneforms. Appl. Environ. Microbiol. 35:1102–1108.

Hall, V., T. Lewis-Evans, and B. I. Duerden. 2001. Identification of Actinomyces, propionibacteria, lactobacilli and bifidobacteria by amplified 16S DNA restriction analysis. Anaerobe 7:55–57.

Halpern, B. N., G. Biozzi, C. Stiffel, and D. Mouton. 1966. Inhibition of tumor growth by administration of killed Corynebacterium parvum. Nature 212:853–854.

Hayashi, S., and C. Furusaka. 1980. Enrichment of Propionibacterium in paddy soils by addition of various organic substances. Ant. v. Leeuwenhoek 46:313–320.

Hettinga, D. H., E. Vedamuthu, and G. W. Reinbold. 1968. Pouch method for isolating and enumerating propionibacteria. J. Dairy Sci. 51:1707–1709.

Hettinga, D. H., and G. W. Reinbold. 1972a. The propionic acid bacteriaa review. I: Growth. J. Milk Food Technol. 35:295–301.

Hettinga, D. H., and G. W. Reinbold. 1972b. The propionic acid bacteria—a review. II: Metabolism. J. Milk Food Technol. 35:358–372.

Hettinga, D. H., and G. W. Reinbold. 1972c. The propionic acid bacteria—a review. III: Miscellaneous metabolic activities. J. Milk Food Technol. 35:436–447.

Higaki, S., T. Kitagawa, M. Kagoura, M. Morohashi, and T. Yamagishi. 2000. Correlation between Propionibacterium acnes biotypes, lipase activity and rash degree in acne patients. J. Dermatol. 27:519–522.

Higaki, S., M. Nakamura, T. Kitagawa, M. Morohashi, and T. Yamagishi. 2001. Effect of lipase activities of Propionibacterum granulosum and Propionibacterium acnes. Drugs Exp. Clin. Res. 27:161–164.

Hitchner, E. R. 1932. A cultural study of the propionic acid bacteria. J. Bacteriol. 23:40–41.

Hoeffler, V. 1977. Enzymatic and hemolytic properties of Propionibacterium acnes and related bacteria. J. Clin. Microbiol. 6:555–558.

Holdeman, L. V., E. P. Cato, and W. E. C. Moore. 1977. Anaerobe Laboratory Manual, 4th ed. Virginia Polytechnic Institute and State University. Blacksburg, VA.

Holmberg, K., and V. Forsum. 1973. Identification of Actinomyces, Arachnia, Bacterionema, Rothia, and Propionibacterium species by defined immunofluorescence. Appl. Microbiol. 25:834–843.

Howell, A., and L. Pine. 1956. Studies on the growth of species of Actinomyces. I: Cultivation in a synthetic medium with starch. J. Bacteriol. 71:47–53.

Hykin, P. G., K. Tobal, G. McIntyre, M. M. Matheson, H. M. A. Towler, and S. L. Lightman. 1994. The diagnosis of delayed post-operative endophthalmitis by polymerase chain reaction of bacterial DNA in vitreous samples. J. Med. Microbiol. 40:408–415.

Isawa, K., K. Hojo, N. Yoda, T. Kamiyama, S. Makino, M. Saito, H. Sugano, C. Mizoguchi, S. Kurama, M. Shibasaji, and Y. Sato. 2002. Isolation and identification of a new bifidogenic growth stimulator produced by Propionibacterium freudenreichii ET-3. Biosci. Biotechnol. Biochem. 66:679–681.

Janoschek, A. 1944. Zur Systematik der Propionsäurebakterien. Zbl Bakteriol., Abt. 2 106:321–337.

Jenkins, J. K., W. J. Harper, and P. D. Courtney. 2002. Genetic diversity in Swiss cheese starter cultures assessed by pulsed field gel electrophoresis and arbitrary primed PCR. Lett. App. Microbiol. 35:423–427.

Johnson, J. L., and C. S. Cummins. 1972. Cell wall composition and deoxyribonucleic acid similarities among the anaerobic coryneforms, classifical propionibacteria and strais of Arachnia propionica. J. Bacteriol. 109:1047–1066.

Jong, E. C., H. L. Ko, and G. Pulverer. 1975. Studies on bacteriophages of Propionibacterium acnes. Med. Microbiol. Immunol. 161:263–271.

Kaneko, T., H. Mori, M. Iwata, and S. Meguro. 1994. Growth stimulator for bifidobacteria produced by Propionibacterium freudenreichii and several intestinal bacteria. J. Dairy Sci. 77:393–404.

Kato, H., O. Inoue, Y. Koga, M. Shingu, T. Fujimoto, M. Kondo, S. Yamomoto, K. Tominga, and Y. Sasaguri. 1983. Variant strain of Propionibacterium acnes: A clue to the aetiology of Kawasaki disease. Lancet ii:1383–1388.

Kaufmann, P., A. Pfefferkorn, M. Teuber, and L. Meile. 1997. Identification and quantification of Bifidobacterium species isolated from food with genus-specific 16S rRNA-targeted probes by colony hybridization and PCR. Appl. Environ Microbiol. 63:1268–1273.

Kishishita, M., T. Ushuijima, Y. Ozaki, and Y. Ito. 1980. New medium for isolating propionibacteria and its application to assay of normal flora of human facial skin. Appl. Environ. Microbiol. 40:1100–1105.

Koussémon, M., Y. Combet-Blanc, B. K. C. Patel, J.-L. Cayol, P. Thomas, J.-L. Garcia, and B. Ollivier. 2001. Propionibacterium microaerophilum sp. nov., a microaerophilic bacterium isolated from olive mill wastewater. Curr. Microbiol. 46:141–145.

Kurmann, J. 1960. Ein vollsynthetischer Nährboden fur Propionsäurebakterien. Pathologia et Microbiologia 23:700–711.

Kusano, K., H. Yamada, M. Niwa, and K. Yamasoto. 1997. Propionibacterium cyclohexanicum sp. nov., a new acid-tolerant ω-cyclohexyl fatty acid.containing Propionibacterium isolated from spoiled orange juice. Int. J. Syst. Bacteriol. 47:825–831.

Langsrud, T., and G. W. Reinbold. 1973a. Flavor development and microbiology of swiss cheese. I: Milk quality and treatments. J. Milk Food Technol. 36:487–490.

Langsrud, T., and G. W. Reinbold. 1973b. Flavor development and microbiology of swiss cheese. II: Starters, manufacturing processes and procedures. J. Milk Food Technol. 36:531–542.

Langsrud, T., and G. W. Reinbold. 1973c. Flavor development and microbiology of swiss cheese. III: Ripening and flavor production. J. Milk Food Technol. 36:593–609.

Langsrud, T., and G. W. Reinbold. 1973d. Flavor development and microbiology of swiss cheese. IV: Defects. J. Milk Food Technol. 37:26–40.

Looney, W. J., A. J. Galusser, and H. K. Modde. 1990. Evaluation of the ATB 32 A system for identification of anaerobic bacteria isolated from clinical specimen. J. Clin. Microbiol. 28:1519–1524.

Lovejoy, E. D., and T. W. Hastings. 1911. The isolation and growth of the acne bacillus. J. Cutan. Dis. 29:80–82.

Lyon, W. J., and B. A. Glatz. 1993. Isolation and purification of propionicin PLG-1, a bacteriocin produced by a strain of Propionibacterium thoenii. Appl. Environ. Microbiol. 59:83–88.

Maiden, M. F. J., A. Tanner, and P. J. Macuch. 1996. Rapid characterization of periodontal bacterial isolates by using fluorogenic substrate tests. J. Clin. Microbiol. 34:376–384.

Malik, A. C., G. W. Reinbold, and E. R. Vedamuthu. 1968. An evaluation of the taxonomy of Propionibacterium. Can. J. Microbiol. 14:1185–1191.

Markowska-Daniel, I., Z. Pejsak, S. Szmigielski, G. Sokolska, J. Jeljaszewicz, and G. Pulverer. 1992. Adjuvant properties of Propionibacterium avidum KP-40 in vaccination against endemic viral and bacterial infections. II: Swine immunized with inactivated Haemophilus pleuropneumoniae vaccine and experimentally infected with different virulent serotypes of H. pleuropneumoniae. Zbl Bakteriol. 277:538–546.

Markowska-Daniel, I., Z. Pejsak, S. Szmigielski, J. Jeljaszewicz, and G. Pulverer. 1993a. Prophylactic application of Propionibacterium avidum KP-40 in swine with acute experimental infections. I: Viral infections—Aujeszky's disease and classical swine fever. Deutsch. Tierärztl. Wochenschr. 100:149–151.

Markowska-Daniel, I., Z. Pejsak, S. Szmigielski, J. Jeljaszewicz, and G. Pulverer. 1993b. Prophylactic application of Propionibacterium avidum KP-40 in swine with acute experimental infections. II: Bacterial infections: pleuropneumonia and swine erysipelas. Deutsch. Tierärztl. Wochenschr. 100:185–188.

Marples, R. R. 1974. The microflora of the face and acne lesions. J. Inv. Dermatol. 62:326–331.

Maszenan, A. M., R. J. Seviour, B. K. C. Patel, P. Schumann, and G. N. Rees. 1999. Tessarococcus bendigoensis gen. nov., sp. nov., a Gram-positive coccus occurring in regular packages or tetrads, isolated from activated sludge. Int. J. Syst. Bacteriol. 49:459–468.

McGinley, K. J., G. F. Webster, and J. J. Leyden. 1978. Regional variation of cutaneous propionibacteria. Appl. Environ. Microbiol. 35:62–66.

Merilaeinen, V., and M. Antila. 1976. The propionic acid bacteria in Finnish Emmenthal cheese. Meijeritieellinen Aikakauskirja, Helsinki 34:107–116.

Miescher, S., M. P. Stierli, M. Teuber, and L. Meile. 2000. Propiocin SM1, a bacteriocin from Prpionibacterium jensenii DF1: Isolation and characterisation of the protein and its gene. Syst. Appl. Microbiol. 23:174–184.

Moore, W. E. C., and E. P. Cato. 1963. Validity of Propionibacterium acnes (Gilchrist) Douglas and Gunter comb. nov. J. Bacteriol. 85:870–874.

Moore, W. E. C., and L. V. Holdeman. 1974. Propionibacterium. In: R. E. Buchanan and N. E. Gibbons (Eds.) Bergey's Manual of Determinative Bacteriology, 8th ed. Williams & Wilkins. Baltimore, MD. 633–644.

Mordarska, H., and M. Pasciak. 1994. A simple method for differentiation of Propionibacterium acnes and Propionibacterium propionicum. FEMS Microbiol. Lett. 123:325–329.

Mori, H., Y. Sato, N. Takemoto, T. Kamiyama, Y. Yoshiyama, S. Meguro, H. Sato, and T. Kaneko. 2000. Isolation and structural identification of bifidogenic growth stimulator produced by Propionibacterium freudenreichii. J. Dairy Sci. 80:1959–1964.

Moss, C. W., V. R. Dowell Jr., D. Farshtchi, L. J. Raines, and W. B. Cherry. 1969. Cultural characteristics and fatty acid composition of propionibacteria. J. Bacteriol. 97:561–570.

Nakamura, K., A. Hiraishi, Y. Yoshimi, M. Kawaharasaki, K. Masuda, and Y. Kamagata. 1995. Microlunatus phosphovorus gen. nov., sp. nov., a new Gram-positive polyphosphate-accumulating bacterium isolated from activated sludge. Int. J. Syst. Bacteriol. 45:17–22.

Niederau, W., W. Pape, K. P. Schaal, V. Hoffler, and G. Pulverer. 1982. Zur Antibiotikehandlung der menschlichen Aktinomykosen. Deutsch. Med. Wochenschr. 107:1279–1283.

Nordstrom, K. M. 1985. Polyacrylamide gel electrophoresis (PAGE) of whole cell proteins of cutaneous Propionibacterium species. J. Med. Microbiol. 19:9–14.

Orla-Jensen, S. 1909. Die Hauptlinien des natürlichen Bakteriensystems. Zb. Bakteriol., Abt. 2 22:305–346.

Paster, B. J., W. A. Falkner Jr., C. O. Enwonwu, E. O. Ideigbe, K. O. Savage, V. A. Levanos, M. A. Tamer, R. L. Ericson, C. N. Lau, and F. E. Dewhirst. 2002. Prevalent bacterial species and novel phylotypes in advanced noma lesions. J. Clin. Microbiol. 40:2187–2191.

Pechere, M., J. C. Pechere, G. Siegenthaler, L. Germanier, and J. H. Saurat. 1999. Antibacterial activity of retinaldehyde against Propionibacterium acnes. Dermatology 199:29–31.

Perry, A. L., T. Worthington, A. C. Hilton, P. A. Lambert, A. J. Stirling, and T. S. J. Elliott. 2003. Analysis of clinical isolates of Propionibacterium acnes by optimised RAPD. FEMS Microbiol Lett. 228:51–55.

Pitcher, D. G., and M. D. Collins. 1991. Phylogenetic analysis of some LL-diamino acid-containing coryneform bacteria from human skin: Description of Propionibacterium innocuum sp. nov. FEMS Microbiol. Lett. 84:295–300.

Plastourgos, S., and R. H. Vaughn. 1957. Species of Propionibacterium associated with Zapateria spoilage of olives. Appl. Microbiol. 5:262–271.

Pochi, P. E., and J. S. Strauss. 1961. Antibiotic sensitivity of Corynebacterium acnes (Propionibacterium acnes). J. Inv. Dermatol. 36:423–429.

Prévot, A.-R., and H. Thouvenot. 1961. Essai de lysotypie des Corynebacterium anaerobies. Ann. Inst. Pasteur 101:966–970.

Prevot, A.-R., and V. Fredette. 1966. Manual for the Classification and Determination of the Anaerobic Bacteria. Lea & Febiger. Philadelphia, PA. 345–355.

Prottey, C., and C. E. Ballou. 1968. Diacyl myoinositol monomannoside from Propionibacterium shermanii. J. Biol. Chem. 243:6196–6201.

Reddy, M. S., G. W. Reinbold, and F. D. Williams. 1973a. Inhibition of propionibacteria by antibiotic and antimicrobial agents. J. Milk Food Technol. 36:564–569.

Reddy, M. S., F. D. Williams, and G. W. Reinbold. 1973b. Sulfonamide resistance of propionibacteria: Nutrition and transport. Antimicrob. Agents Chemother. 4:254–258.

Riedel, K. H. J., B. D. Wingfield, and T. J. Britz. 1994. Justification of the "classical" Propionibacterium species concept by restriction analysis of the 16S ribosomal RN genes. Syst. Appl. Microbiol. 17:536–542.

Riedel, K. H., B. D. Wingfield, and T. J. Britz. 1998. Identification of classical Propionibacterium species using 16S rDNA-restriction fragment length polymorphisms. Syst. Appl. Microbiol. 21:419–428.

Rogosa, M., C. S. Cummins, R. A. Lelliott, and R. M. Keddie. 1974. The coryneform group of bacteria. In: R. E. Buchanan and N. E. Gibbons (Eds.) Bergey's Manual of Determinative Bacteriology, 8th ed. Williams & Wilkins. Baltimore, MD. 599–617.

Rolph, H. J., A. Lennon, M. P. Riggio, W. P. Saunders, D. MacKenzie, L. Coldero, and J. Bagg. 2001. Molecular identification of microorganisms from endodontic infections. J. Clin. Microbiol. 29:3282–3289.

Rossi, F., S. Torriani, and F. Dellaglio. 1998. Identification and clustering of dairy propionibacteria by RAPD-PCR and CGE-REA methods. J. Appl. Microbiol. 85:956–964.

Rossi, F., S. Torriani, and F. Dellaglio. 1999. Genus- and species-specific PCR-based detection of dairy propionibacteria in environmental samples by using primers targeted to the genes encoding 16S rRNA. Appl. Environ. Microbiol. 65:4241–4244.

Russel, R. J., R. J. McInroy, P. C. Wilkison, and R. G. White. 1976. A lipid chemotactic factor from anaerobic coryneform bacteria and monocytes. Immunology 30:935–949.

Sabouraud, R. 1897. La seborrhée grasse et la pélade. Ann. Inst. Pasteur 11:134–159.

Sakaguchi, K., M. Iwaski, and S. Yamado. 1941. Studies on the propionic acid fermentation. J. Agric. Chem. Soc. Japan 17:127–138.

Schaal, K. P., and W. Pape. 1980. Special methodological problems in antibiotic testing of fermentative actinomyces. Infection 8 (Suppl. 2):176–182.

Schaal, K. P., and G. Pulverer. 1981. The genera Actinomyces, Agromyces, Arachnia, Bacterionema, and Rothia. In: M. P. Starr, H. Stolp, H. G. Trüper, A. Balows, and H. G. Schlegel (Eds.) The Prokaryotes. Springer-Verlag. Berlin, Germany. 1923–1950.

Schaal, K. P. 1986. The genus Arachnia. In: P. H. A. Sneath, N. S. Mair, M. E. Sharp, and J. G. Holt (Eds.) Bergey's Manual of Systematic Bacteriology, 1st ed. Williams & Wilkins. Baltimore, MD. 1332–1342.

Schleifer, K.-H., and O. Kandler. 1972. Peptidoglycan types of bacterial cell walls and their taxonomic implications. Bacteriol. Rev. 36:407–477.

Schofield, G. M., and K. P. Schaal. 1981. A numerical taxonomic study of members of the Actinomycetaceae and related taxa. J. Gen. Microbiol. 127:237–259.

Schumann, P., H. Prauser, F. A. Rainey, E. Stackebrandt, and P. Hirsch. 1997. Friedmanniella antarctica gen. nov., spec. nov., an LL-diaminopimelic acid-containing actinomycete from Antarctic sandstone. Int. J. Syst. Bacteriol. 47:278–283.

Schwenninger, S. M., and L. Meile. 2004. A mixed culture of Propionibacterium jensenii and Lactobacillus paracasei subsp. paracasei inhibits food spoilage yeasts. Syst. Appl. Microbiol. 27:229–237.

Shaw, R. H., and J. M. Sherman. 1923. The production of volatile fatty acids and carbon dioxide by propionic acid bacteria with special reference to their action in cheese. J. Dairy Sci. 6:303–309.

Sherman, J. M. 1921. The cause of eyes and characteristic flavor in Emmental or Swiss cheese. J. Bacteriol. 6:379–392.

Sherman, J. M., and R. H. Shaw. 1923. The propionic acid fermentation of lactose. J. Biol. Chem. 56:695–700.

Shintani, T., W.-T. Liu, S. Hanada, Y. Kamagata, S. Miyaoka, T. Suzuki, and K. Nakamura. 2000. Micropruina glycogenica gen. nov., sp. nov., a new Gram-positive glycogen-accumulating bacterium isolated from activated sludge. Int. J. Syst. Evol. Mocrobiol. 50:201–207.

Siqueira, J. F., and I. N. Rocas. 2003. Polymerase chain reaction detection of Propionibacterium propionicum and Actinomyces radicidentis in primary and persistent endodontic infections. Oral Surg. Oral Med. Pathol. Radiol. Endod. 96:215–222.

Skogen, L. O. 1970. [MS thesis]. Iowa State University. Ames, IA.

Skogen, L. O., G. W. Reinbold, and E. R. Vedamuthu. 1974. Capsulation of Propionibacterium. J. Milk Food Technol. 37:314–321.

Slack, J. M., and M. A. Gerencser. 1975. Actinomyces, Filamentous Bacteria: Biology and Pathogenicity. Burgess Publishing. Minneapolis, MN. 65–69.

Stackebrandt, E., and C. R. Woese. 1981. Towards a phylogeny of actinomycetes and related organisms. Curr. Microbiol. 5:131–136.

Stackebrandt, E., F. A. Rainey, and N. Ward-Rainey. 1997. Proposal for a new hierarchic classification system, Actinobacteria classis nov. Int. J. Syst. Bacteriol. 47:479–491.

Stackebrandt, E., P. Schumann, K. P. Schaal, and N. Weiss. 2002. Propionimicrobium gen. nov., a new genus to accommodate Propionibacterium lymphophilum (Torrey 1916) Johnson and Cummins 1972, 1057[AL] as Propionimicrobium lymphophilum comb. nov. Int. J. Syst. Bacteriol. 52:1609–1614.

Sunde, P. T., L. Tronstad, E. R. Eribe, P. O. Lind, and I. Olsen. 2000. Assessment of periradicular microbiota by DNA-DNA hybridization. Endod. Dent. Traumatol. 16:191–196.

Tammelin, A., A. Hambraeus, and E. Stahle. 2002. Mediastinitis after cardia surgery: improvement of bacteriological diagnosis by use of multiple tissue sample and strain typing. J. Clin. Microbiol. 40:2936–2941.

Tamura, T., M. Takeuchi, and A. Yokota. 1994. Luteococcus japonicus gen. nov., sp. nov., a new Gram-positive coccus with LL-diaminopimelic acid in the cell wall. Int. J. Syst. Bacteriol. 44:348–356.

Thöni, J., and O. Allemann. 1910. Über das vorkommen von gefarbten, makroskopischen Bakterienkolonien in Emmentalerkäsen. Zb. Bakteriol., Abt. 2 25:8–30.

Torrey, J. C. 1916. Bacteria associated with certain types of abnormal lymph glands. J. Med. Res. 34:65–80.

Troili-Petersson, G. 1904. Studien über die Mikroorganismen des schwedischen Güterkäses. Zbl. Bakteriol., Abt. 2 11:120–143.

Unna, P. J. 1893. Die Histopathologie der Hautkrankheiten. A. Hirschwald. Berlin, Germany.

Van Niel, C. B. 1928. The Propionic Acid Bacteria. J. W. Boissevain. Haarlem, The Netherlands.

Van Niel, C. B. 1957. The genus Propionibacterium. In: R. S. Breed, E. G. D. Murray, and N. R. Smith (Eds.) Bergey's Manual of Determinative Bacteriology, 7th ed. Williams & Wilkins. Baltimore, MD, pp. 569–576.

Vedamuthu, E. R., and G. W. Reinbold. 1967. The use of candle oats jar incubation for the enumeration, characterization and taxonomic study of propionibacteria. Milchwissenschaft 22:428–431.

Von Freudenreich, E., and S. Orla-Jensen. 1906. Über die im Emmentalerkase statfindene Propionsauregarung. Zbl. Bakteriol., Abt. 2 17:529–546.

Vorob'eva, L. I., T. P. Turova, N. I. Kraeva, and M. A. Alekseeva. 1983. Propionic acid cocci and their systematic position (English translation). Mikrobiologia 52:368–373.

Webster, G. F., and C. S. Cummins. 1978. Use of bacteriophage typing to distinguish Propionibacterium acnes types 1 & 2. J. Clin. Microbiol. 7:84–90.

Werkman, C. H., and S. E. Kendall. 1931. The propionic acid bacteria. I: Classification and nomenclature. Iowa State Journal of Science 6:17–32.

Werkman, C. H., and R. W. Brown. 1933. The propionic acid bacteria. II: Classification. J. Bacteriol. 26:393–417.

Williamson, P., and A. M. Kligman. 1965. A new method for the quantitative investigation of cutaneous bacteria. J. Inv. Dermatol. 45:498–503.

Wilson, W. R., W. J. Martin, C. J. Wilkowske, and J. A. Washington, 2nd. 1972. Anaerobic bacteremia. Mayo Clin. Proc. 47:639–646.

Winkler, S., and M. Fröhlich. 1957. Prüfung des Einflusses von Nisin auf die wichtigsten Reifungserreger beim Emmentalerkäse. Milchwiss. Ber. 7:125–137.

Wood, H. G., A. A. Anderson, and C. H. Werkman. 1938. Nutrition of the propionic acid bacteria. J. Bacteriol. 36:201–214.

Woodruff, M. F. A., and J. L. Boak. 1966. Inhibitory effect of injection of Corynebacterium parvum on the growth of tumor transplants in isogenic hosts. Br. J. Cancer. 20:345–355.

Yocum, R. C., J. McArthur, B. G. Petty, A. M. Diehl, and T. R. Moench. 1982. Septic arthritis caused by Propionibacterium acnes. J. Am. Med. Assoc. 248:1740–1741.

Yokota, A., T. Tamura, N., Weiss, and E. Stackebrandt. 1994. Transfer of Propionibacterium innocuum Pitcher and Collins 1992 to Propioniferax gen. nov. as Propioniferax innocuum comb. nov. Int. J. Syst. Bacteriol. 44:579–582.

Zierdt, C. H., C. Webster, and W. S. Rude. 1968. Study of the anaerobic corynebacteria. Int. J. Syst. Bacteriol. 18:33–47.

Prokaryotes (2006) 3:419–429
DOI: 10.1007/0-387-30743-5_20

CHAPTER 1.1.5

The Family Succinivibrionaceae

ERKO STACKEBRANDT AND ROBERT B. HESPELL

The rumen is a strictly anaerobic ecosystem inhabited mainly by bacteria and ciliated protozoa, plus smaller numbers of fungi. Although several hundred or more species of bacteria can be found in this ecosystem, only about thirty species are usually found at high enough levels to be considered of ecological significance. In the article written for the second edition of *The Prokaryotes* (Hespell, 1992), the species *Succinivibrio dextrinosolvens* (Bryant and Small, 1956) and *Succinimonas amylolytica* (Bryant et al., 1958) were covered in the same chapter because these species are major rumen species that have not been extensively studied. In addition, the species *Anaerobiospirillum succiniproducens* (Davis et al., 1976) was tentatively included in the chapter by Hespell (1992) since the phylogenetic relationships of *Succinivibrio dextrinosolvens*, *Succinimonas amylolytica* and *Anaerobiospirillum succiniproducens* and to other higher taxa were not known. *Succinivibrio dextrinosolvens* and *Succinimonas amylolytica* are usually found in animals with some grain in their diets. These two organisms are mainly isolated from ruminants but have also been isolated from humans suffering from bacteremia and other conditions. *Anaerobiospirillum succiniproducens* has been isolated from colonic contents of dogs but not from the rumen. *Anaerobiospirillum thomasii* (Malnick, 1997) was isolated from the throat and the colon of dogs and from human diarrheal feces. Phylogenetic analyses also linked *Ruminobacter amylophilus* (Hamlin and Hungate, 1956; Stackebrandt and Hippe, 1987) to this group. *Ruminobacter amylophilus* occurs sporadically in the rumen contents of cattle. When present, *Ruminobacter amylophilus* may be the predominant starch digester and may constitute as much as 10% of the bacterial population of the rumen.

The family Succinivibrionaceae (Hippe et al., 1999) are Gram-negative short or oval to long, or curved to helical, rods. Cells are motile (*Succinivibrio*, *Anaerobiospirillum* and *Succinimonas*) or nonmotile (*Ruminobacter*), do not form endospores, and are chemo-organotrophic and strictly anaerobic. Glucose and other carbohydrates are fermented with the production of succinate and acetate; low amounts of formate and lactate may be produced. CO_2 is absorbed, and gas is not produced. Cells are catalase-negative. Nitrate is not reduced. Quinones are not detected (*Anaerobiospirillum* not tested), and cyto-chromes are not found in *Ruminobacter amylophilus*, the only species tested (G. Gottschalk, personal communication). Major fatty acids are saturated (35–66%) and unsaturated (19–59%), straight-chained, even-numbered fatty acids, and 16:0 and 18:0 3-OH fatty acids (4–11%). *Iso-* and *anteiso-*branched fatty acids, cyclopropane fatty acids, and odd-numbered fatty acids are absent (Moore et al., 1994). Strains are isolated from the rumen of sheep and cattle and from feces of humans, cats and dogs and from the colon of dogs (*Anaerobiospirillum*). The DNA G+C content is 39–44 mol% (data available for *Anaerobiospirillum* and *Ruminobacter*). Phylogenetically, they are members of the Gammaproteobacteria. The type genus is *Succinivibrio* (Bryant and Small, 1956).

Habitats

Succinivibrio dextrinosolvens and *Succinimonas amylolytica* are inhabitants of the rumen of cattle and sheep. The numbers of these bacteria vary depending on the diet. *Succinovibrio amylolytica* is found in the rumen of animals fed mixed diets of forages and at least some grain (Bryant et al., 1958), whereas *S. dextrinosolvens* occurs in high numbers in animals fed high-starch or other diets containing large amounts of rapidly fermentable carbohydrates (Bryant and Small, 1956; Wozny et al., 1977). Both species are involved in the digestion of starch and its breakdown products in the rumen. The occurrence of both species in ecosystems other than the rumen has not been well documented. However, the presence of *Succinivibrio* species in the blood of humans suffering from bacteremia has been reported (Southern, 1975; Porschen and Chan, 1977). In both cases, the patients had gastrointestinal tract disorders and thus loss of gastrointestinal tract integrity might have permitted the organisms to enter the blood. If so, this suggests that *Succinivibrio* species might normally be present in low numbers in the colon of humans. *Anaerobiospirillum succiniproducens* was described on the basis of a study of three strains isolated from the throat and cecum from different beagle dogs. Spiral-shaped bacteria were also detected (but not identified) in seven of nine dogs by phase contrast microscopy of fecal homogenates or of cecal and colonic tissue sections and in four of nine dogs by scanning electron microscopy (SEM) of cecal and colon tissues. Two of the nine dogs consistently showed no spiral-shaped organisms in throat swabs. The second species, *A. thomasii*, contains animal and human strains, the latter of which were isolated during routine diagnostic examination of blood and fecal samples. The clinical strains and the methods used to isolate animal and human strains have been described previously (Malnick et al., 1990).

Ruminobacter amylophilus is one of the predominant anaerobic bacteria obtained in pure culture from the rumen contents of a Holstein cow. The bacterium thrives on basal medium supplemented with maltose and ammonia. A casein hydrolysate plus volatile fatty acids is only slightly stimulatory, and glucose and cellobiose cannot substitute for maltose (Bryant and Robinson, 1961). *Ruminobacter amylophilus* occurs sporadically in the rumen contents of cattle and, when present, may be the predominant starch digester and constitute as much as 10% of the bacterial population of the rumen. *Ruminobacter amylophilus* also occurs also in the ovine rumen (Blackburn and Hobson, 1962; Bryant and Robinson, 1961).

Isolation

Enrichment and Isolation

Currently, no procedures have been published regarding selective enrichment or isolation of either member of the family. All strains of *Ruminobacter amylophilus*, *Succinivibrio dextrinosolvens* or *Succinimonas amylolytica* have been isolated on nonspecific, habitat-simulating media used to enumerate total culturable bacteria from ruminal contents. As with other ruminal bacteria, both species are strict anaerobes. However, three strains of *S. dextrinosolvens* isolated under diverse conditions have been shown to grow on a defined medium (Gomez-Alarcon et al., 1982). This medium lacks the volatile fatty acids, peptides, and a number of other components often used in media to cultivate many species of ruminal bacteria. Thus, this medium

might be used to selectively enrich or isolate *S. dextrinosolvens*.

Anaerobiospirillum cells were originally isolated from tissue samples from the stomachs, ilea, ceca and colons of male beagle dogs. Samples were obtained by clamping the tissue with hemostats, cutting out about 1 cm from the clamped-off portion of the intestine, and quickly placing the excised tissue in pre-reduced transport broth (Aranki et al., 1969). Used for culture and microscopy, ileal and colonic tissues were removed from a region about 20 cm proximal and distal to the cecum, and stomach tissue was removed from the cardiac region. The tissues, after they were placed in a glove box, were homogenized in a blender; the homogenates were diluted and plated on a pre-re-duced medium in the glove box for 48 to 72 h prior to use. Tubed transport-broth dilution medium was prepared, sterilized in screw-capped tubes, and, to reduce its oxygen concentration, put into an anaerobic glove box (containing 80% N_2, 10% H_2, and 10% CO_2) with screw caps loose for at least 48 h prior to use (Davis et al., 1976). Using procedures to isolate human strains, *A. thomasii* strains were isolated during routine diagnostic examination of blood and fecal samples. The clinical strains and methods used to isolate animal and human strains have been described previously (Malnick et al., 1990).

Cultivation Media

Strains of *Succinivibrio* and *Succinimonas* species can be grown on a variety of complex, rumen fluid-containing media (Bryant and Burkey, 1953; Bryant and Robinson, 1961) or on more defined media in which the rumen fluid has been replaced by trypticase, yeast extract, hemin, and volatile fatty acids (Caldwell and Bryant, 1966; Hespell and Bryant, 1981). A typical complex medium for both species and a chemically defined medium for *S. dextrinosolvens* (adapted from Gomez-Alarcon et al., 1982) are shown in Table 1. Since both species produce succinate as a fermentation product, the use of a bicarbonate-carbonic acid system is needed to provide carbon dioxide and a means of buffering. Ammonium ions serve as a major nitrogen source for both species. *Succinimonas amylolytica* has a limited range of fermentable substrates (i.e., glucose, maltose, dextrin and starch). *Succinivibrio dextrinosolvens* also ferments these substrates plus galactose and xylose, but starch is only partially used and is incompletely hydrolyzed. Neither species ferments amino acids or peptides.

Anaerobiospirillum succiniproducens is routinely grown on Wikins-Chalgren Anaerobe Broth (Oxoid CM 643), supplemented with 1 g/l of $NaHCO_3$, 1 mg/l resazurin and 0.3 g/l L-

Table 1. Media for *S. amylolytica* and *S. dextrinosolvens*.[a]

| | Amount per 100ml of medium | |
Ingredient	Medium 1	Medium 2
Carbohydrate (5%)[b]	5.0 ml	5.0 ml
Trypticase	0.5 g	—
Serine	—	30 mg
Leucine	—	40 mg
Methionine	30 mg	30 mg
Yeast extract	0.2 g	—
p-Aminobenzoic acid	—	10 µg
Ammonium chloride (5.3%)	1.0 ml	1.0 ml
Resazurin (0.1%)	0.1 ml	0.1 ml
IVI VFA solution[c]	3.0 ml	—
L-Cysteine·HCl (2.5%)[b]	1.0 ml	1.0 ml
Mineral A[d]	4.0 ml	4.0 ml
Mineral B[e]	4.0 ml	4.0 ml
Ferrous sulfate	0.3 mg	0.3 mg
Trace minerals[f]	0.5 ml	0.5 ml
Hemin plus 1,4-napthoquinone[g]	1.0 ml	1.0 ml
Sodium carbonate (8%)[h]	5.0 ml	5.0 ml
Distilled water	76.0 ml	79.0 ml

Medium 1 is for both species and medium 2 for *S. dextrinosolvens*.

[a]Both media are prepared under carbon dioxide; final pH 6.8. The defined medium is for *S. dextrinosolvens* only.

[b]Autoclaved separately; prepared and stored under a nitrogen gasphase.

[c]Prepared by adding 7ml acetic acid, 3ml propionate, 2ml butyrate, and 0.6ml each of isobutyrate, 2-methylbutyrate, and isovalerate to 700ml of 0.2M NaOH. Adjust to pH 7.0 with NaOH and to a final volume of 1 liter.

[d]Mineral A = 0.5% K_2HPO_4.

[e]Mineral B = 1.0% KH_2PO_4, 1,2 NaCl, 0.58% $NaSO_4$, 0.16% $CaCl_2 \times 2H_2O$, and 0.25% $MgSO_4 \times 7H_2O$.

[f]trace minerals contained 10mg each of $ZnSO_4 \times 7H_2O$, H_3BO_3, $NaSeO_3$, $Na_2MoO_4 \times 2 H_2O$, $NiCl_2 \times 6 H_2O$; 5mg $CuSO_4 \times 5 H_2O$, and 2mg of $Al(SO_4)_2 \times 12 H_2O$ made to 100ml with distilled water.

[g]solution made by dissolving 25mg of 1,4-naphtoquinone in 2ml of 95% ethanol and combining with 10mg of hemin dissolved in 50ml of 0.01M NaOH and 38ml distilled water.

[h]autoclaved separately; prepared, equilibrated, and stored under a carbon dioxide gas phase.

cysteine. *A. succiniproducens* has also been cultivated on fastidious anaerobe agar (Lab M, Topley House, Bury, England) (Malnick, et al., 1990) and in the Bactec 9240 anaerobic system (Wecke and Horbach, 1999).

A selective medium has been described by Malnick et al. (1990) for *Anaerobiospirillum* spp, using Skirrow campylobacter medium (Skirrow, 1977) supplemented with polymyxin (2,500 IU/liter), vancomycin (20 mg/liter), sulfamethoxalole (100 mg/liter), Victoria blue B, (250 mg/liter, dissolved in 4 ml of ethanol) and horse blood (lysed with saponin, 40 ml/liter). After 48 h colonies of *Anaerobiospirillum* spp were 1–2 mm in diameter, uniform dark blue and low convex.

A. thomasii strains grow on fastidious anaerobe agar (Lab M, Topley House, Bury, England) with added horse blood (50 ml/liter) incubated at 37°C in an anaerobic hood. Alternatively, Wilkins Chalgrene Anaerobe Broth (Oxoid CM

643), prepared anaerobically under 80% N_2 + 20%, can be used.

Preservation of Cultures

All species can be maintained for long times by storage of cultures in liquid nitrogen or ultracold freezers (Hespell and Canale-Parola, 1970). All species can be maintained as freeze-dried cultures. Short-term storage is possible by placing glycerol cultures in normal (–20°C) freezers (Teather, 1982). *Ruminobacter amylophilus* is routinely maintained in the DSMZ-German Collection of Microorganisms and Cell Cultures in medium consisting per 1000 of distilled water 0.45 g (each) KH_2PO_4 and K_2HPO_4, 0.9g $(NH_4)_2SO_4$, 0.9g NaCl, 0.18g $MgSO_4 \times 7H_2O$, 5.0g soluble starch, 10.0g casitone (Difco), 1.0 mg resazurin, 0.5 g cysteine hydrochloride, and 6.0 g

NaHCO₃, pH 7.0, at a gas atmosphere of 100 CO_2.

Phylogeny

On the basis of 16S rRNA gene sequence similarities, members of the family Succinivibrion-aceae are members of the Gammaproteobacteria (Hippe et al., 1999). The phylogenetic distance between type species of the family is only remote (97.7–92.7% similarity) and even the type strains of the two *Anaerobiospirillum* species, *A. succiniproducens* and *A. thomasii*, share only 92.8% similarity. The family is a phylogenetic neighbor of the family Aeromonadaceae and of a cluster consisting of the genera *Oceanisphaera*, *Oceanimonas* and *Ferrimonas*. Sequence similarities between members of Succinivibrionaceae and type species of neighboring genera range between 84.6 and 88.5%. All species described so far can unambiguously be identified by their 16S rRNA gene sequence (Tee et al., 1998; Misawa et al., 2002). As the type strains share about 93% sequence similarity, differences of seven percent sequence dissimilarity account for about 100 nucleotides which are mainly located in the variable regions of the sequence (i.e., positions 68–87, 126–218, 446–473, and 991–1037). This distribution allows identification of at least those strains that are highly related to the type strains on the basis of a continuous stretch located between positions 50 and 500. This stretch can be amplified and sequenced using primers located at positions 18 forward and 530 reverse (Lane, 1991).

16S rRNA gene sequence-specific polymerase chain reaction (PCR) primers were used to detect Proteobacteria in rumen fluid taken from a ruminally fistulated Holstein cow. Though originally designed to detect type-1 methanotrophs (Wise et al., 1999) placed in the Gammaproteobacteria, the study identified five sequences moderately related to *Succinivibrio dextrinisolvens* (Mitsumori et al., 2002). Species-specific 16S rRNA gene primers were designed and used in a real-time PCR experiment to determine the population shift when the diet of cows was changed from hay to grain. Though the *Ruminococcus amylophilus* and *Succinivibrio dextrinisolvens* population was still high at day three after the shift, cell numbers of both species declined significantly after 28 days and were hardly detectable (Tajima et al., 2001). The primers for *Ruminococcus amylophilus* ATCC 29744 were CAACCAGTCGCATTCAGA (forward) and CACTACTCATGGCAACAT (reverse) and for *Succinivibrio dextrinisolvens* ATCC 19716ᵀ TGGGAAGCTACCTGATA GAG (forward) and CCTTCAGAGAGGTTCTCACT (reverse).

A phylogenetic dendrogram of the ATP/ADP-dependent phosphoenolpyruvate carboxykinase amino acid sequence of *Anaerobiospirillum succiniproducens* indicated that this species clusters with those of other members of Gammaproteobacteria, confirming the 16S rRNA gene sequence-based phylogenetic position (Laivenieks et al., 1997).

Taxonomy

Succinivibrio

Succinivibrio dextrinosolvens forms light tan, entire, translucent surface colonies that are generally only 1–2 mm in diameter. Occasionally, colonies may be irregular, raised, and yellowish. Subsurface colonies are lenticular and equally small. However, in media having low agar concentrations (0.2–0.6%), subsurface colonies take on a diffuse, white, furry appearance. The cells appear singly or in pairs of small, curved rods (usually 0.3–0.5 × 2.0–4.0 μm) with pointed ends. With newly isolated strains, helical or twisted filaments of 2–4 coils composed of two or more cells are common. These filaments often show rotational motility. This helical characteristic varies with strains and can be lost during successive subcultures in the laboratory, resulting in single or pairs of slightly curved rods. Under adverse cultural conditions many strains can form pleomorphic shapes—usually large, swollen rods or spherical bodies. Cells invariably stain Gram negatively and are motile. The motility is translational with a vibrating movement. The flagellation arrangement is polar and monotrichous.

A large production of acetic and succinic acids with an uptake of carbon dioxide is the most common fermentation pattern for *Succinivibrio* strains fermenting glucose (Bryant and Small, 1956; Scardovi, 1963). Small, variable amounts of formate and lactate can be made. Hydrogen gas, propionate, and butyrate are not made. Good growth occurs at incubation temperatures of 30–40°C, but no growth occurs at 22°C or 45°C. Biochemical tests for nitrate reduction, catalase, and production of indole, acetoin, or hydrogen sulfide are all negative. Presently, all ruminal strains are considered to constitute a single species, *Succinivibrio dextrinosolvens* (Bryant, 1984b). The succinovibrio-like strains isolated from blood samples (Southern, 1975; Porschen and Chan, 1977) appeared to have lophotrichous flagellation and are probably not true isolates of this species, but further studies are needed to verify this.

Succinimonas

Succinimonas strains form surface colonies that are light tan in color, translucent, smooth and convex. Even after prolonged incubation, the colonies are only 0.6–1.5 mm in diameter. Colonies within agar media are lenticular in shape. Typically, the cell morphology is that of a short, oval rod or coccobacillus (1–1.5 µm × 1.2–3.0 µm). The cells appear singly or in pairs but can form clumps in older cultures. The cells always stain Gram negatively. In wet-mount preparations, most cells show some degree of translational motility, which ceases when oxygen is present. Often cells which have become attached to the glass surface display a rotational movement about one cell end. The flagellar arrangement is polar and monotrichous. Some cells may show granules near the cell periphery. The nature of these granules is unknown, but they may be composed of glycogen, which is found in a number of species of ruminal bacteria. In liquid media, growth is even and cells are dispersed, but generally dense turbidities do not occur.

The major products formed from glucose are acetic and succinic acids, and the formation of these products requires a large uptake of carbon dioxide. No hydrogen gas, formate, lactate or ethanol is produced, but trace amounts of acetoin and/or propionate may be formed. No strains are known to hydrolyze gelatin and production of hydrogen sulfide, indole or catalase has not been observed. Growth occurs at 30–39°C but not at 22°C or 45°C. Strain differences are rather minimal and all strains are considered to be isolates of a single species, *Succinimonas amylolytica* (Bryant, 1984a).

Anaerobiospirillum

The genus is defined by Gram-negative helical rods with rounded ends (usually 0.6–0.8 µm × 3–15 µm), although some cells are up to 32 µm long. Cells are motile by means of bipolar tufts of flagella. Cells usually occur singly. Endospores are not formed. Cells are strictly anaerobic, catalase and oxidase negative, do not hydrolyze esculin, hippurate or urea, and do not reduce nitrate. Carbohydrate metabolism is fermentative. The major products from glucose metabolism are succinic and acetic acids; traces of lactic and formic acids may also be formed. The growth temperature is 33–43°C with an optimum of 37°C.

Anaerobiospirillum succiniproducens is a Gram-negative helical rod (0.6–8.0 µm × 3.0 µm) with rounded ends; some cells reach 20 µm in length. The helix diameter is 0. 9–1.1 µm and the wavelength is about 1.3–1.7 µm. *Anaerobiospirillum thomasii* cells are helical rods (0.5–0.7 × 5–32 µm). Flat colonies with even, sometimes

spreading edges are 1–2 mm in diameter after 24–48 h on horse blood agar. Hemolysis does not occur. The enzymes detected with an API ZYM kit are leucine arylamidase and *N*-acetyl-β-glucosaminidase. Enzymes not detected with an API ZYM kit include acid and alkaline phosphatases, esterase (C-4), esterase-lipase (C-8), lipase (C-14), valine arylamidase, cystine arylamidase, trypsin, chymotrypsin, phosphohydrolase, DL- and D-galactosidase, β-D-galactosidase, β-glucuronidase, β-glucosidase, mannosidase, and L-fucosidase.

Ruminococcus

Without phylogenetic information this species could be misclassified as *Prevotella ruminicola* (Shah and Collins, 1990) or *Fibrobacter succinogenes* (Montgomery et al., 1988), organisms previously classified as members of *Bacteroides*. *Ruminobacter amylophilus* differs considerably from these organisms, as well as *Fusobacterium*, *Leptotrichia* and members of genera previously grouped in the family Bacteroideaceae as defined by Holdeman et al. (1984) using a combination of properties, including morphology, the pattern and end products of carbohydrates utilized, fermentation products, mol% G+C of their DNA, and fatty acid composition. On the basis of metabolic properties, *Ruminobacter* can also be distinguished from species of several other genera defined by Gram-negative staining, but phylogenetically Gram-positive obligate-anaerobes, viz., *Acidaminobacter* (Stams and Hansen, 1984), *Ilyobacter* (Stieb and Schink, 1985), *Pelobacter* (Schink and Pfennig, 1982a), *Propiogenium* (Schink and Pfenning, 1982b) and *Propionispira* (Schink et al., 1982).

Following the loss of the first culture of *Bacteroides amylophilus* (Hamlin and Hungate, 1956), organisms conforming to the original description of *B. amylophilus* have been isolated frequently from the rumen (Blackburn and Hobson, 1962; Bryant and Robinson, 1961; Caldwell et al., 1969; Holdeman et al., 1977). A neotype strain of *B. amylophilus* was proposed (Cato et al., 1978) for strain H 18 isolated by Blackburn and Hobson (1962). As this species differed from authentic *Bacteroides* species in fatty acid composition (Miyagawa et al., 1979) and the lack of sphingophospholipids (Kunsman and Caldwell, 1974; Miyagawa et al., 1978; Miyagawa et al., 1979; and Shah and Collins, 1983), the removal of *B. amylophilus* from *Bacteroides* sensu stricto has been suggested. These conclusions were supported by the finding that *B. amylophilus* and authentic members of *Bacteroides* belong to different main phylogenetic lines of descent (Woese et al., 1985; Martens et al., 1987). *Bacteroides amylophilus* grouped certain members of the so called "γ-subclass of the class Proteobacteria" (today named "Gammaproteobacteria"), e.g., enterobacteria, vibrios, oceanospirilla, altero-monads, while authentic *Bacteroides* formed a coherent cluster within the *Bacteroides–Cytophaga–Flavobacterium* line of descent (Paster et al., 1985). Consequently, *Bacteroides amylophilus* was reclassified as *Ruminobacter amylophilus* (Stackebrandt and Hippe, 1986). The generic name "*Ruminobacter*," originally used to describe Gram-negative, nonmotile, anaerobic chemoheterotrophic cellulose-fermenting bacteria (Kaars Sijpesteijn, 1949), was considered genus incertae sedis (Prévot, 1966). But at the same time, *Bacteroides* species "*B. amylogenes*," *B. ruminicola*, *B. succinogenes* and the noncellulolytic species *B. amylophilus* were transferred into this genus. As the taxon *Ruminobacter* was not included in the Approved Lists of Bacterial Names (Skerman et al., 1980), it had no standing in nomenclature and could be used to accommodate the former *B. amylophilus*.

Ruminobacter amylophilus forms small, tan colonies on the surface of agar media. However, the cell morphology is distinctly different from that of the other members of the family (Table 2). The cells are larger ($0.9–1.2 \times 1–3$ μm long), tend to form pleomorphic, swollen shapes, and most importantly, are nonmotile (Hamlin and Hungate, 1956). In addition, *R. amylophilus* does not grow on glucose. *Ruminobacter amylophilus* produces acetate, formate and succinate as major fermentation products, and trace amounts of lactate and ethanol may also be formed.

Cells are Gram-negative, oval to long rods with tapered or round ends, and some cells are swollen forms and irregularly curved ($0.9–1.2$ μm $\times 1.1–8.0$ μm; Cato et al., 1978). Surface colonies on rumen fluid-glucose-cellobiose agar roll tubes are 1 mm in diameter, circular, entire, slightly convex, translucent, smooth, glistening, and white to tan. Colonies in laked-blood roll streak tubes are pinpoint to 2 mm in diameter, circular, entire to slightly erose, convex to slightly erose, convex to umbonate, translucent, colorless, shiny and smooth after incubation for 4 days. Growth is sparse on freshly prepared blood agar plates incubated anaerobically. Surface colonies on rumen fluid-glucose-cellobiose agar roll tubes are 1 mm in diameter, circular, entire, slightly convex, translucent, smooth, glistening, and white to tan. Colonies in deep agar are 0.8–1 mm in diameter, lenticular, entire or irregular, white, soft and butyrous. Growth is not stimulated by amino acids, volatile fatty acids (Miura et al., 1980), vitamins or hemin (Macy and Probst, 1979). No colonies appear on egg yolk agar plates and on the surface of plates incubated in an aerobic jar. Rumen fluid is not required

(Bryant and Robinson, 1961) for growth, but ammonia is essential as nitrogen. Carbon dioxide is fixed and ammonia is assimilated.

Biochemical and Physiological Properties

The carbohydrate fermentation pathways in *Succinivibrio dextrinosolvens* have been studied by Scardovi (1963) and O'Herrin and Kenealy (1993). Enzymatic activities of the Embden-Meyerhof pathway were present in cell-free extracts, but enzymes of the hexose monophosphate pathway were absent. Phosphoenolpyruvate was found to undergo carboxylation to form oxaloacetate as the product, with ADP acting as the phosphate acceptor. Presumably, the oxaloacetate mainly undergoes reduction to form succinate and is partially used in biosynthetic transamination reactions to form amino acids. The pathways for assimilation of ammonia by *S. dextrinosolvens* have been elucidated with strain C18 (Patterson and Hespell, 1985). This strain was shown to possess glutamine synthetase, urease, glutamate dehydrogenase, glutamate-oxaloacetate transaminase and aspartate synthase. When grown in continuous culture under ammonia limitation, cells contained high levels of glutamine synthetase and urease, but glutamate dehydrogenase activity was low. The opposite pattern is observed with growth in the presence of ample ammonia. The glutamate dehydrogenase was reduced nicotinamide adenine dinucleotide phosphate (NADPH)-linked. The glutamine synthetase was regulated by an adenylation-deadenylation mechanism, allowing for rapid enzyme inactivation when high levels of ammonia are present.

Zeikus and coworkers have performed detailed analysis of the phosphoenolpyruvate carboxykinase (PEP) catalyzing the reversible formation of oxalacetate and ATP from PEP, ADP and CO_2 in *Anaerobiospirillum succiniproducens*, e.g., purification and characterization (Podkovyrov et al., 1993; Jabalquinto et al., 1999), site-directed mutagenesis (Jabalquinto et al., 2002a; Jabalquinto et al., 2002b; Jabalquinto et al., 2003; Jabalquinto et al., 2004), as well as its cloning, sequencing and overexpression in *E. coli* (Laivenieks et al., 1997).

In the presence of CO_2, *Ruminococcus amylophilus* ferments maltose to succinate: 2 maltose + 2 CO_2 → 2 succinate + 2 acetate + 2 formate (Kühn, 1979). This method of succinic acid production has been patented (United States patent number 5,143,833). The organism is able to grow well in basal medium supplemented with starch. Starch broth cultures are turbid and have a final

Table 2. Morphological and metabolic properties differentiating members of the family Succinivibrionaceae.

Properties	Succinivibrio	Succinimonas amylolytica	Anaerobiospirillum		Ruminococcus
			succiniproducens	thomasii	
Morphology	Curved rods	Spherical	Helical rods, sometimes spheres	Helical rods	Oval to long rods with tapered or round ends
Motility	+	+	+	+	−
Flagella	Monotrichous, polar flagella	Monotrichous, polar flagella	Bipolar, tuft of 12–16 flagella[a]	Bipolar tufts	−
Major fermentation end products	Succinate, acetate	Succinate, acetate	Succinate, acetate	Succinate, and acetate	Succinate, acetate, and formate
Fermentation of carbohydrates	ND	ND	Fructose, glucose lactose, maltose, sucrose, raffinose, inulin, β-D-galactoside, and α-D-glucoside	Adonitol, galactose, glucose, and maltose	Dextrin, glycogen, maltose and starch
Not fermenting	ND	ND	Adonitol, L-arabinose, cellobiose, dulcitol, esculin, galactose, mannitol, rhamnose, salicin, sorbitol, and D-xylose	Fructose, cellobiose, inulin, lactose, mannitol, raffinose, salicin, sorbitol, sucrose, and trehalose	Amygdalin, arabinose, cellobiose, esculin, fructose, D-galactose, D-glucose, glycerol, i-inositol, inulin, lactose, D-mannitol, raffinose, L-rhamnose, ribose, salicin, sucrose, trehalose, D-xylose Cellulose milk and meat digestion
Negative for	Catalase nitrate reduction, and production of indole, acetoin, or hydrogen sulfide	ND	Catalase, oxidase nitrate reduction fermentation of gelatin liquefication, indole production, lipase, meat digestion, hydrolyze esculin, hippurate, or urea	Catalase, nitrate reduction oxidase activity, and hydrolysis of esculin, hippurate, and urea	
Mol% G+C	ND	ND	44	39–42	40–42
Habitat, isolated from	Rumen of cattle and sheep, septicemia	Rumen of cattle and sheep	Colon and throat of dogs	Feces of cats and dogs and from human diarrheal feces	Rumen of cattle and sheep
Type strain	None	None	NCTC 11536[T]	NCTC 12467[T]	ATCC 29744[T]

[a]Detailed ultrastructure analyses are provided by Wecke and Horbach (1999).
Symbols and abbreviations: +, present; −, absent; and ND, no data.
From Bryant and Small (1956), Bryant et al. (1958), Davis et al. (1976), Malnick (1979), and Stackebrandt and Hippe (1986).

LIVERPOOL JOHN MOORES UNIVERSITY
LEARNING & INFORMATION SERVICES

pH of 5.3–5.5. Amylase production is stimulated by Tween 80 (McWethy and Hartman, 1977).

Clinical Significance

Reports on *Succinivibrio dextrinosolvens* bacteremia (Southern, 1975; Porschen and Chan, 1977) and *Succinimonas* spp. bacteremia (Johnson and Finegold, 1987) are rare. On the other hand, there are several reports of *Anaerobiospirillum* infection in humans. The infections have been diagnosed via evidence of bacteremia or septicemia, but the full extent of the organisms's pathogenicity remains to be determined. The study by McNeil et al. (1987) indicated that most of the patients from which members of the genus had been isolated had an underlying disorder like arteriosclerosis, malignancy, liver cirrhosis, diabetes mellitus, and dental infections, and most suffered from gastrointestinal symptoms (Malnick et al., 1983; Park et al., 1986; Marcus et al., 1996; Goddard et al., 1998; Pienaar et al., 2003). Some of the patients owned pet dogs and cats (Goddard et al., 1998). Two *Anaerobiospirillum* strains (identified by 16S rRNA gene sequencing) with different morphologies were isolated from a puppy with bloody diarrhea (Misawa et al., 2002). Recently, strains of *A. succiniproducens* were isolated from blood of AIDS patients in Australia (Tee et al., 1998) and Europe (Wecke and Horbach, 1999). Strains of this species have also been isolated from the blood of a young child with bacteremia and no underlying disease; this led the authors Rudensky et al. (2002) to alert microbiologists and physicians to these organisms, which are resistant to antibiotics prescribed for infections caused by the morphologically similar *Campylobacter* spp.

Application

Anaerobiospirillum succiniproducens has been used in anaerobic fermentation processes for the production of succinate-based animal feed additive using whey as a carbohydrate source. Between 60 and 90% succinate yield was achieved depending on the production process (60% for continuous culture, 80% in batch cultivation, and 90% in a variable volume fed process). The end products were determined by the amount of CO_2 added to the process. Under conditions of excess CO_2, more than 90% of the whey lactose was consumed and the product ratio of succinate to acetate was 4:1. Lactate was the main end product when no excess CO_2 was provided (Samuelov et al., 1999). Even higher yields of succinate were obtained (93–

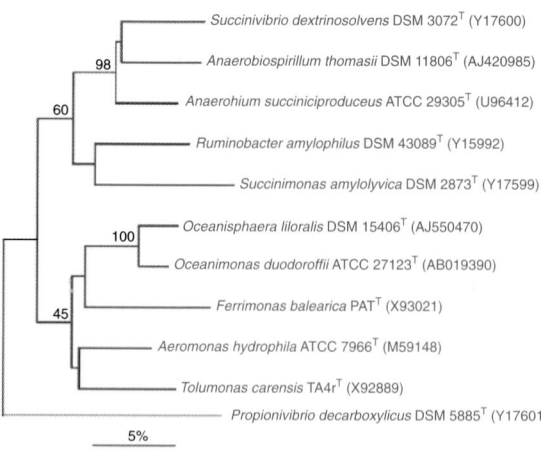

Fig. 1. Dendrogram showing the phylogenetic position of representatives of the family Succinivibrionaceae (Hippe et al.,1999) next to its phylogenetic neighbors within the Gammaproteobacteria. The tree was constructed by the neighbor-joining method (Saitou and Nei, 1987), using corrected distance values (Jukes and Cantor, 1969). Sequences of members of the Enterobacteriaceae were used to root the dendrogram. Bootstrap values (expressed as percentages of 500 replications) of 40% or more, are indicated at the branch points (Felsenstein, 1993). Bar = 5% sequence divergence.

95%) when yeast extract, polypeptones and glucose were added to whey and the ratio of succinic acid to acetic acid was increased to 5.1:1 (Lee et al., 2000). With glycerol as a carbon source, the yield of succinic acid reached 133% when the ratio of succinic acid to acetic acid was increased to 25.8:1. In the presence of yeast extract, the yield even increased to 160% (succinic acid:acetic acid = 31.7:1; Lee et al., 2001). When grown on minimal salts medium, wood hydrolysates (equivalent to 27 g of glucose per liter), and corn steep liquor (10 g), the succinate yield was 24 g per liter (88% [w/w] glucose; Lee et al., 2003).

Literature Cited

Aranki, A., Syed, S. A., Kenny, E. B., and R. Freter. 1969. Isolation of anaerobic bacteria from human gingiva and mouse cecum by means of a simplified glove box procedure. Appl. Micro. 17:568–576.

Aranki, A., and R. Freter. 1972. Use of anaerobic glove boxes for the cultivation of strictly anaerobic bacteria. Am. J. Clin. Nutr. 25:1329–34.

Blackburn, T. H., and P. N. Hobson. 1962. Further studies on the isolation of proteolytic bacteria from the sheep rumen. J. Gen. Microbiol. 29:69–81.

Bryant, M. P., and L. A. Burkey. 1953. Cultural methods and some characteristics of some of the more numerous groups of bacteria in the bovine rumen. J. Dairy Sci. 36:205–217.

Bryant, M. P., and N. Small. 1956. Characteristics of two genera of anaerobic curved rod-shaped bacteria of the rumen. J. Bacteriol. 72:22–26.

Bryant, M. P., N. Small, C. Bouma, and H. Chu. 1958. Species of succinic acid-producing anaerobic bacteria of the bovine rumen. J. Bacteriol. 76:15–23.

Bryant, M. P., and I. M. Robinson. 1961. An improved non-selective culture medium for ruminal bacteria and its use in determining diurnal variations in numbers of bacteria in the rumen. J. Dairy Sci. 44:1446–1456.

Bryant, M. P. 1984a. Genus V: Succinimonas Bryant, Small, Bouma, Chu 1958, 21. In: N. R. Krieg and J. G. Holt (Eds.) Bergey's Manual of Systematic Bacteriology. Williams & Wilkins. Baltimore, MD. 1:643–644.

Bryant, M. P. 1984b. Genus VI: Succinivibrio Bryant and Small 1956, 22. In: N. R. Krieg and J. G. Holt (Eds.) Bergey's Manual of Systematic Bacteriology. Williams & Wilkins. Baltimore, MD. 1:644–645.

Bryant, M. P., I. M. Robinson. 1961. An improved nonselective culture medium for ruminal bacteria and its use in determining diurnal variations in numbers of bacteria in the rumen. J. Dairy Sci. 44:1446–1456.

Caldwell, D. R., and M. P. Bryant. 1966. Medium without rumen fluid for nonselective enumeration and isolation of rumen bacteria. Appl. Microbiol. 14:794–801.

Caldwell, D. R., M. Keenly, and P. J. van Sorest. 1969. Effects of carbon dioxide on growth and maltose fermentation by Bacteroides amylophilus. J. Bacteriol. 98:668–676.

Cato, E. P., W. E. C. Moore, and M. P. Bryant. 1978. Designation of neotype strains for Bacteroides amylophilus Hamlin and Hungate 1956 and Bacteroides succinogenes Hungate 1950. Int. J. Syst. Bacteriol. 28:491–495.

Davis, C. P., D. Cleven, J. Brown, and E. Balish. 1976. Anaerobiospirillum, a new genus of spiral-shaped bacteria. Int. J. Syst. Bacteriol. 26:498–505.

Felsenstein, J. 1993. PHYLIP (Phylogeny Inference Package) Version 3.5.1. Department of Genetics, University of Washington. Seattle, WA.

Goddard, W. W., S. A. Bennett, and C. Parkinson. 1998. Anaerobiospirillum succiniciproducens septicaemia: Important aspects of diagnosis and management. J. Infect. 37:68–70.

Gomez-Alarcon, R. A., C. Dowd, J. A. Z. Leedle, and M. P. Bryant. 1982. 1,4-Napthoquinone and other nutrient requirements of Succinivibrio dextrinosolvens. Appl. Environ. Microbiol. 44:346–350.

Hamlin, L. J., and R. E. Hungate. 1956. Culture and physiology of a starch-digesting bacterium (Bacteroides amylophilus n. sp.) from the bovine rumen. J. Bacteriol. 72:548–554.

Hespell, R. B., and E. Canale-Parola. 1970. Spirochaeta litoralis sp. n., a strictly anaerobic marine spirochete. Arch. Mikrobiol. 74:1–18.

Hespell, R. B., and M. P. Bryant. 1981. The genera Butyrivibrio, Succinivibrio, Succinimonas, Lachnospira, and Selenomonas. In: M. P. Starr, H. Stolp, G. Trüper, A. Balows, and H. G. Schlegel (Eds.) The Prokaryotes. Springer-Verlag. Berlin, Germany. 1479–1494.

Hespell, R. B. 1992. The genera Succinivibrio and Succinimonas. In: A. Balows, H. G. Trüper, M. Dworkin, W. Harder, and K.-H. Schleifer (Eds). The Prokaryotes, 2nd ed. Springer-Verlag. New York, NY. 3979–3982.

Hippe, H., A. Hagelstein, I. Kramer, J. Swiderski, and E. Stackebrandt. 1999. Phylogenetic analysis of Formivibrio citricus, Propionivibrio dicarboxylicus, Anaerobiospirillum thomasii, Succinimonas amylolytica and Succinivibrio dextrinosolvens and proposal of Succinivibrionaceae fam. nov. Int. J. Syst. Bacteriol. 49:779–782.

Holdeman, L. V., E. P. Cato, and W. E. C. Moore. 1977. Anaerobe Laboratory Manual, 4th ed. Virginia Polytechnic Institute and State University. Blacksburg, VA.

Holdeman, L. V., R. W. Kelley, and W. E. C. Moore. 1984. Bacteroidaceae. In: N. R. Krieg and J. G. Holt (Eds.) Bergey's Manual of Systematic Bacteriology. Williams & Wilkins. Baltimore, MD. 1:602–603.

Jabalquinto, A. M., M. Laivenieks, J. G. Zeikus, and E. Cardemil. 1999. Characterization of the oxaloacetate decarboxylase and pyruvate kinase-like activities of Saccharomyces cerevisiae and Anaerobiospirillum succiniciproducens phosphoenolpyruvate carboxykinases. J. Prot. Chem. 18:659–664.

Jabalquinto, A. M., M. Laivenieks, M. Cabezas, J. G. Zeikus, and E. Cardemil. 2002a. The effect of active site mutations in the oxaloacetate decarboxylase and pyruvate kinase-like activities of Anaerobiospirillum succiniciproducens phosphoenolpyruvate carboxykinase. J. Prot. Chem. 21:443–445.

Jabalquinto, A. M., M. Laivenieks, F. D. Gonzalez-Nilo, A. Yevenes, M. V. Encinas, J. G. Zeikus, and E. Cardemil. 2002b. Evaluation by site-directed mutagenesis of active site amino acid residues of Anaerobiospirillum succiniciproducens phosphoenolpyruvate carboxykinase. J. Prot. Chem. 21:393–400.

Jabalquinto, A. M., M. Laivenieks, F. D. Gonzalez-Nilo, M. V. Encinas, J. G. Zeikus, and E. Cardemil. 2003. Anaerobiospirillum succiniciproducens phosphoenolpyruvate carboxykinase: Mutagenesis at metal site 2. J. Prot. Chem. 22:515–519.

Jabalquinto, A. M., F. D. Gonzalez-Nilo, M. Laivenieks, M. Cabezas, J. G. Zeikus, and E. Cardemil. 2004. Anaerobiospirillum succiniciproducens phosphoenolpyruvate carboxykinase: Mutagenesis at metal site 1. Biochimie 86:47–51.

Johnson, C. C., and S. M. Finegold. 1987. Uncommonly encountered, motile, anaerobic Gram-negative bacilli associated with infection. Rev. Infect. Dis. 9:1150–1162.

Jukes, T. H., and C. R. Cantor. 1969. Evolution of protein molecules. In: H. N. Munro (Ed.) Mammalian Protein Metabolism. Academic Press. New York, NY. 3:21–132.

Kaars Sijpesteijn, A. 1949. Cellulose-decomposing bacteria from the rumen of cattle. J. Microbiol. Serol. 15:49–52.

Kühn, W. 1979. Untersuchungen zur Vergärung von Maltose durch Bacteroides amylophilus und zu der daran beteiligten Fumarat Reduktase [diploma thesis]. Georg-August-University Göttingen. Göttingen, Germany.

Kunsman, J. E., and D. R. Caldwell. 1974. Comparison of the sphingolipid content of rumen Bacteroides species. Appl. Microbiol. 28:1088–1089.

Laivenieks, M., C. Vieille, and J. G. Zeikus. 1997. Cloning, sequencing, and overexpression of the Anaerobiospirillum succiniciproducens phosphoenolpyruvate carboxykinase (pckA) gene. Appl. Environ. Microbiol. 63:2273–2280.

Lane, D. J. 1991. 16S/23S rRNA sequencing. In: E. Stackebrandt and M. Goodfellow (Eds.) Nucleic Acid Techniques in Bacterial Systematics. Wiley. Chichester, UK. 125–175.

Lee, P. C., W. G. Lee, S. Kwon, S. Y. Lee, and H. N. Chang. 2000. Batch and continuous cultivation of Anaerobiospirillum succiniciproducens for the production of succinic acid from whey. Appl. Microbiol. Biotechnol. 54:23–27.

Lee, P. C., W. G. Lee, S. Y. Lee, and H. N. Chang. 2001. Succinic acid production with reduced by-product formation in the fermentation of Anaerobiospirillum succiniciproducens using glycerol as a carbon source. Biotechnol. Bioengin. 72:41–48.

Lee, P. C., S. Y. Lee, S. H. Hong, H. N. Chang, and S. C. Park. 2003. Biological conversion of wood hydrolysate to succinic acid by Anaerobiospirillum succiniciproducens. Biotechnol Lett. 25:111–114.

Macy, J. M., and I. Probst. 1979. The biology of gastrointestinal bacteroides. Annu. Rev. Microbiol. 33:561–594.

Malnick, H., M. E. Thomas, H. Lotay, and M. Robbins. 1983. Anaerobiospirillum species isolated from humans with diarrhoea. J. Clin. Path. 36:1097–1101.

Malnick, H., K. Williams, J. Phil-Ebosie, and A. S. Levy. 1990. Description of a medium for isolating Anaerobiospirillum spp., a possible cause of zoonotic disease, from diarrheal feces and blood of humans and use of the medium in a survey of human, canine, and feline feces. J. Clin. Microbiol. 28:1380–1384.

Malnick, H. 1997. Anaerobiospirillum thomasii sp. nov., an anaerobic spiral bacterium isolated from the feces of cats and dogs and from diarrheal feces of humans, and emendation of the genus Anaerobiospirillum. Int. J. Syst. Bacteriol. 47:381–384.

Marcus, L., E. W. Gove, M. L. van der Walt, H. J. Koornhof, and H. Malnick. 1996. First reported African case of Anaerobiospirillum succiniciproducens septicemia. Eur. J. Clin. Microbiol. Infect. Dis. 15:741–744.

Martens, B., H. Spiegl, and E. Stackebrandt. 1987. Sequence of a 16S ribosomal RNA gene of Ruminobacter amylophilus: The relation between homology values and similarity coefficients. Syst. Appl. Microbiol. 9:224–230.

McNeil, M. M., W. J. Martone, and V. R. Dowell Jr. 1987. Bacteremia with Anaerobiospirillum succiniciproducens. Rev. Infect. Dis. 9:737–742.

McWethy, S. J., and P. A. Hartman. 1977. Purification and some properties of an extracellular alpha amylase from Bacteroides amylophilus. J. Bacteriol. 129:1537–1544.

Misawa, N., K. Kawashima, F. Kondo, E. Kushima, K. Kushima, and P. Vandamme. 2002. Isolation and characterization of Campylobacter, Helicobacter, and Anaerobiospirillum strains from a puppy with bloody diarrhea. Vet. Microbiol. 22:353–64.

Mitsumori, M, N. Ajisaka, K. Tajima, H. Kajikawa, and M. Kurihara. 2002. Detection of Proteobacteria from the rumen by PCR using methanotroph-specific primers. Lett. Appl. Microbiol. 35:251–255.

Miyagawa, E., R. Azuma, and T. Suto. 1978. Distribution of sphingolipids in Bacteroides species. J. Gen. Appl. Microbiol. 2:341–348.

Miyagawa, E., R. Azuma, and T. Suto. 1979. Cellular fatty acid composition in Gram-negative obligately anaerobic rods. J. Gen. Appl. Microbiol. 25:41–51.

Montgomery, L., B. Flesher, and D. Stahl. 1988. Transfer of Bacteroides succinogenes (Hungate) to Fibrobacter gen. nov. as Fibrobacter succinogenes comb. nov. and description of Fibrobacter intestinalis sp. nov. Int. J. Syst. Bacteriol. 38:430–435.

Moore, L. V. H., D. M. Bourne, and W. E. C. Moore. 1994. Comparative distribution and taxonomic value of cellular fatty acids in thirty-three genera of anaerobic Gram-negative bacilli. Int J Syst Bacteriol. 44:338–347.

O'Herrin, S. M., and W. R. Kenealy. 1993. Glucose and carbon dioxide metabolism by Succinivibrio dextrinosolvens. Appl. Environ. Microbiol. 59:748–755.

Park, C. H., D. L. Hixon, J. F. Endlich, P. O'Connell, F. T. Bradd, and M. Mount. 1986. Anaerobiospirillum suc-

ciniproducens: Two case reports. Am. J. Clin. Pathol. 85:73–76.

Paster, B. J., W. Ludwig, W. Weisburg, E. Stackebrandt, R. B. Hespell, C. M. Hahn, H. Reichenbach, K. O. Stetter, and C. R. Woese. 1985. A phylogenetic grouping of the Bacteroides, cytophagas, and certain flavobacteria. Syst. Appl. Microbiol. 6:34–42.

Patterson, J. A., and R. B. Hespell. 1985. Glutamine synthetase activity in the ruminal bacterium Succinivibrio dextrinosolvens. Appl. Environ. Microbiol. 50:1014–1020.

Pienaar, C., A. J. Kruger, E. C. Venter, and J. D. Pitout. 2003. Anaerobiospirillum succiniciproducens bacteraemia. J. Clin. Pathol. 56:316–318.

Podkovyrov, S. M., and J. G. Zeikus. 1993. Purification and characterization of phosphoenolpyruvate carboxikinase, a catabolic CO_2-fixing enzyme from Anaerobiospirillum succiniproducens. J. Gen. Microbiol. 139:223–228.

Porschen, R. K., and P. Chan. 1977. Anaerobic vibrio-like organisms cultured from blood: Desulfovibrio desulfuricans and Succinivibrio species. J. Clin. Microbiol. 5:444–447.

Prévot, A. R. 1966. Manual for the classification and determination of the anaerobic bacteria. 1st. Amer. Ed., transl by V. Fredette. Lea and Febiger. Philadelphia, PA.

Rudensky, B., D. Wachtel, A. M. Yinnon, D. Raveh, and Y. Schlesinger. 2002. Anaerobiospirillum succiniciproducens bacteremia in a young child. Pediatr. Infect. Dis. J. 21:575–576.

Saitou, N., and M. Nei. 1987. The neighbor-joining method: A new method for reconstructing phylogenetic trees. Molec. Biol. Evol. 4:406–425.

Samuelov, N. S., R. Datta, M. K. Jain, and J. G. Zeikus. 1999. Whey fermentation by anaerobiospirillum succiniciproducens for production of a succinate-based animal feed additive. Appl Environ Microbiol. 65:2260–2263.

Scardovi, V. 1963. Studies in rumen microbiology. I. A succinic acid producing vibrio: Main physiological characters and enzymology of its succinate acid forming system. Annali di Microbiol. 13:171–187.

Schink, B., and N. Pfennig. 1982a. Fermentation of trihydrobenzenes by Pelobacter acidigallici gen. nov. sp. nov., a new strictly anaerobic non-sporeforming bacterium. Arch. Microbiol. 133:195–201.

Schink, B., and N. Pfennig. 1982b. Propiogenium modestum gen. nov. sp. nov. a new strictly anaerobic, nonsporing bacterium growing on succinate. Arch. Microbiol. 133:209–216.

Schink, B., T. E. Thompson, and J. G. Zeikus. 1982. Characterisation of Propionispira arboris gen. nov. sp. nov., a nitrogen-fixing anaerobe common to wetwoods of living trees. J. Gen. Microbiol. 128:2771–2780.

Shah, H. N., and M. D. Collins. 1983. A review. Genus Bacteroides: A chemotaxonomic perspective. J. Appl. Bacteriol. 55:403–416.

Shah, H. N., and M. D. Collins. 1990. Prevotella, a new genus to include Bacteroides melaninogenicus and related species formerly classified into the genus Bacteroides. Int. J. Syst. Bacteriol. 40:205–208.

Skerman, V. B. D., V. McGowan, and P. H. A. Sneath. 1980. Approved Lists of Bacterial Names. Int. J. Syst. Bacteriol. 30:225–420.

Skirrow, M. B. 1977. Campylobacter enteritis: A "new" disease. Br. Med. J. 2:9–11.

Southern, P. M. 1975. Bacteremia due to Succinivibrio dextrinosolvens. Am. J. Clin. Pathol. 64:540–543.

Stackebrandt, E., and H. Hippe. 1986. Transfer of Bacteri-
odes amylophilus to a new genus Ruminobacter gen.
nov., nom. rev. as Ruminobacter amylophilus comb. nov.
Syst. Appl. Microbiol. 8:204–207.

Stams, A. J. M., and T. A. Hansen. 1984. Fermentation of
glutamate and other compounds by Acidaminobacter
hydrogenoformans gen. nov. sp. nov. an obligate anaerobe isolated from black mud: Studies with pure cultures
and mixed cultures with sulfate-reducing and methanogenic bacteria. Arch. Microbiol. 137:329–337.

Stieb, M., and B. Schink. 1985. A new 3-hydroxybutyrate
fermenting anaerobe, Ilyobacter polytrophus, gen. nov.
sp. nov., possessing various fermentation pathways.
Arch. Microbiol. 140:139–146.

Tajima, K., R. I. Aminov, T. Nagamine, H. Matsui, M.
Nakamura, and Y. Benno. 2001. Diet-dependent shifts
in the bacterial population of the rumen revealed with
real-time PCR. Appl. Environ. Microbiol. 67:2766–
2774.

Teather, R. M. 1982. Maintenance of laboratory strains of
obligatedly anaerobic rumen bacteria. Appl. Environ.
Microbiol. 44:499–501.

Tee, W., T. M. Korman, M. J. Waters, A. Macphee, A. Jenney,
L. Joyce, and M. L. Dyall-Smith. 1998. Three cases of
Anaerobiospirillum succiniciproducens bacteremia confirmed by 16S rRNA gene sequencing. J. Clin. Microbiol.
36:1209–1213.

Wecke, J., and I. Horbach. 1999. Ultrastructural characterization of Anaerobiospirillum succiniciproducens and its
differentiation from Campylobacter species. FEMS
Microbiol. Lett. 170:83–88.

Wise, M. G., J. V. McArthur, and L. J. Shimkets. 1999. Meth-
anotroph diversity in landfill soil: isolation of novel type
I and type II methanotrophs whose presence was suggested by culture-independent 16S ribosomal DNA
analysis. Appl. Environ. Microbiol. 65:4887–4897.

Woese, C. R., W. G. Weisburg, C. M. Hahn, B. Paster, L. B.
Zablen, B. J. Lewis, T. J. Macke, W. Ludwig, and E.
Stackebrandt. 1985. The phylogeny of purple bacteria:
The gamma subdivision. Syst. Appl. Microbiol. 6:25–33.

Wozny, M. A., M. P. Bryant, L. V. Holdeman, and W. E. C.
Moore. 1977. Urease assay and urease-producing species
of anaerobes in the bovine rumen and human feces.
Appl. Environ. Microbiol. 33:1097–1104.

Prokaryotes (2006) 3:430–537
DOI: 10.1007/0-387-30743-5_21

CHAPTER 1.1.6

The Family Actinomycetaceae: The Genera *Actinomyces*, *Actinobaculum*, *Arcanobacterium*, *Varibaculum*, and *Mobiluncus*

KLAUS P. SCHAAL, ATTEYET F. YASSIN AND ERKO STACKEBRANDT

Introduction

The family Actinomycetaceae was created by Buchanan in 1918 and was originally used to accommodate many diverse organisms such as members of the genera *Actinobacillus*, *Leptotrichia*, *Actinomyces* and *Nocardia*. After several revisions, membership of the family was restricted to bacterial species that appeared to be linked taxonomically by the following phenotypic characteristics: ability to produce Grampositive, branching and, later on, fragmenting filaments without aerial hyphae and spores; comparatively exacting nutritional requirements; facultatively anaerobic (capnophilic) to anaerobic growth; and fermentative carbohydrate metabolism (Slack, 1974; Slack and Gerencser, 1975).

Taking into account these common characters, the family Actinomycetaceae was thought to include the genera *Actinomyces*, *Arachnia*, *Bifidobacterium*, *Bacterionema* and *Rothia* (Slack, 1974). However, the validity of this family concept was increasingly questioned after modern and more relevant taxonomic techniques such as chemotaxonomic, numerical phenetic, and molecular genetic procedures had been applied to the respective organisms.

As currently defined, the family Actinomycetaceae (Buchanan 1918) with the type genus *Actinomyces* (Harz 1877) is a member of the suborder Actinomycineae (Stackebrandt et al., 1997), order Actinomycetales (Buchanan 1918; Skerman et al., 1980), emend. Stackebrandt et al., 1997. This order has been included in the class Actinobacteria by Stackebrandt et al. (1997). The type species is *Actinomyces bovis* (Harz, 1877). Besides the genus *Actinomyces*, the family includes the genera *Arcanobacterium* (Collins et al., 1982c) with the type species *Arcanobacterium haemolyticum*, *Mobiluncus* (Spiegel and Roberts, 1984a), type species *Mobiluncus curtisii*, *Actinobaculum* (Lawson et al., 1997), type species *Actinobaculum suis*, and *Varibaculum* (Hall et al., 2003e), type species *Varibaculum cambriense*.

The species that are currently considered valid members of the family and the sources of their isolation have been compiled in Table 1. *Actinomyces* species that were reclassified are listed in Table 2, and synonyms and basonyms of members of the Actinomycetaceae have been compiled in Table 3. Many new species were described after the chapter on the genera *Actinomyces*. *Arcanobacterium* and *Rothia* had been published in the second edition of *The Prokaryotes* (Schaal, 1992b), and the phylogeny of the order Actinomycetales has been studied extensively. In contrast, little progress has been made with respect to physiology and chemotaxonomy, and the ecological significance of many of the novel species is not fully understood. This is also exemplified by the almost complete lack of information on as yet uncultured actinomycetes in public databases, which indicates that the natural environment of these organisms has hardly been covered by molecular surveys (exceptions are Oaster et al. [2001] and K. E. Kempsell et al. [unpublished observation], accession numbers AJ404525, AJ405526 and AJ405527). Furthermore, very little information is available on gene sequences other than 16S rRNA genes, coding for, e.g., fimbrial structural subunits and fimbria-associated proteins (Yeung et al., 1998), urease (Morou-Bermudez and Burne, 1999), sialidase (Henningsen et al., 1991), neuraminidase (Jost et al., 2001), phospholipase (Cuevas and Songer, 1993), and genes involved in carbohydrate metabolism (L. J. Bergeron, unpublished observation; AF228582). Thus, the emphasis of the biology of the Actinomycetaceae still focuses on their epidemiology and identification in the clinical field.

Phylogeny

Actinomyces bovis belonged to one of the first species subjected to 16S rRNA analysis (Stackebrandt and Woese, 1981). By now, almost complete 16S rRNA gene sequences are available for all of the type strains of the species

Table 1. Alphabetical list of species of the family Actinomycetaceae.

Described as	Type strains deposited in selected collections	Sources of isolation	Original description
Actinomyces			
Actinomyces bovis	ATCC 13683 = CCUG 31996 = DSM 43014	Bovine actinomycosis—lumpy jaw, cattle	Harz, 1877
Actinomyces bowdenii	CCUG 37421	Abscesses, dogs, and pleural fluid, cat	Pascual et al., 1999
Actinomyces canis	CCUG 41706	Vagina and purulent lesions, dogs	Hoyles et al., 2000
Actinomyces cardiffensis	CCUG 44997	Clinical sources including pleural fluid, brain, jaw, pericolic and ear abscesses, and intrauterine contraceptive devices, humans	Hall et al., 2002
Actinomyces catuli	CCUG 41709	Mediastinal tissue and blood clot, dogs	Hoyles et al., 2001a
Actinomyces coleocanis	CCUG 41708	Vagina, dog	Hoyles et al., 2002
Actinomyces denticolens	ATCC 43322 = NCTC 11490 = CCUG 32758 = DSM 20671	Supragingival dental plaque, dairy cattle	Dent and Williams, 1984a
Actinomyces europaeus	ATCC 700353 = CCUG 32789 A	Femur tissue, various abscesses, atheroma cyst, and decubital ulcer, humans	Funke et al., 1997
Actinomyces funkei	CCUG 42773 = CIP 106713	Blood cultures (endocarditis) and localized lesions, humans	Lawson et al., 2001
Actinomyces georgiae	ATCC 49285 = CCUG 32935 = DSM 6843	Gingival crevice with and without periodontitis, humans	Johnson et al., 1990
Actinomyces gerencseriae	ATCC 23860 = CCUG 32936 = DSM 6844	Parotid abscess and various other manifestations of actinomycosis, humans	Johnson et al., 1990
Actinomyces graevenitzii	ATCC 27294 = CIP 105737	Bronchial secretions, sputum, and osteitis of the jaw, humans	Pascual Ramos et al., 1997
Actinomyces hongkongensis	HKU8 = DSM 15629 = LMG 21939	Pus from pelvic actinomycosis, human	Woo et al., 2003
Actinomyces hordeovulneris	ATCC 35275 = CCUG 32937 = DSM 20732	Pleuritis, pericarditis, peritonitis, visceral abscesses, septic arthritis and recurrent localized infections, dogs	Buchanan et al., 1984
Actinomyces howellii	ATCC 43323 = CCUG 32757	Dental plaque, dairy cattle	Dent and Williams, 1984b
Actinomyces hyovaginalis	ATCC 51367 = CCUG 35604 = DSM 10695	Purulent vaginal discharge, pigs	Collins et al., 1993
Actinomyces israelii	ATCC 12102 = CCUG 18307 = DSM 43320	Various actinomycotic lesions including cervicofacial abscesses, pulmonary lesions, abdominal and pelvic abscesses, brain abscesses, and lacrimal canaliculitis, humans	Kruse, 1896
Actinomyces marimammalium	CCUG 41710 = CIP 106509	Various internal organs and small intestine, dead seals and a porpoise	Hoyles et al., 2001b
Actinomyces meyeri	ATCC 35568 = CCUG 21024 = DSM 20733	Periodontal sulcus, brain and other abscesses, pleural fluid, human	Cato et al., 1984
Actinomyces naeslundii	ATCC 12104 = (NCTC 10301, CCUG 2238) = DSM 43013	Dental plaque, cervicovaginal secretions, abscesses, empyemas, eye infections, human	Thompson and Lovestedt, 1951
Actinomyces nasicola	R2014 = CCUG 46092 = CIP 107668	Pus from *antrum nasi*, human	Hall et al., 2003a
Actinomyces neuii subsp. *anitratus*	ATCC 51849 = CCUG 32253 = DSM 8577	Abscesses, blood cultures, human	Funke et al., 1994
Actinomyces neuii subsp. *neuii*	ATCC 51847 = CCUG 32252 = DSM 8576	Abscesses, blood cultures, infected mammary hematoma, human	Funke et al., 1994
Actinomyces odontolyticus	ATCC 17929 = (NCTC 9935, CCUG 20536) = DSM 43760	Dental plaque, carious lesions, abscesses, eye infections, human	Batty, 1958
Actinomyces oricola	R5292 = CCUG 46090 = CIP 107639	Dental abscess, human	Hall et al., 2003b
Actinomyces radicidentis	CCUG 36733 = CIP 106352	Infected dental root canals, human	Collins et al., 2000

Table 1. *Continued*

Described as	Type strains deposited in selected collections	Sources of isolation	Original description
Actinomyces radingae	ATCC 51856 = CCUG 32394 = DSM 9169	Mixed wound infections, human	Wüst et al., 1995, emend.
Actinomyces slackii	ATCC 49928 = (NCTC 11923, CCUG 32792)	Supragingival plaque, dairy cattle	Vandamme et al., 1998 Dent and Williams, 1986
Actinomyces suimastitidis	CCUG 39276 = CIP 106779	Chronic granulomatous mastitis, pig	Hoyles et al., 2001c
Actinomyces turicensis	ATCC 51857 = CCUG 32401 = DSM 9168	Mixed wound infections, human	Wüst et al., 1995, emend. Vandamme et al., 1998
Actinomyces urogenitalis	CCUG 38702 = CIP 106421	Vaginal secretions, urine, genital infections, human	Nikolaitchouk et al., 2000
Actinomyces vaccimaxillae	CCUG 46091 = CIP 107423	Jaw lesion, cow	Hall et al., 2003
Actinomyces viscosus	ATCC 15987 = CCUG 14476 = (DSM 43327 [serovar 1], ATCC 19246 [serovar 2])	Subgingival plaque and periodontal disease, hamster; cervical plaque, rat; dental plaque, human; various suppurative processes in man and animals	Howell et al., 1965 Georg et al., 1969
Actinobaculum			
Actinobaculum massiliae	CIP 107404	Urine, human	Greub and Raoult, 2002
Actinobaculum schaalii	CCUG 27420 = CIP 105739	Blood and urine, human	Lawson et al. 1997
Actinobaculum suis	ATCC 33144 = CCUG 19206 = DSM 20639	Cystitis, pyelonephritis, metritis, urine, and semen, pigs	Wegienek and Reddy, 1982a Lawson et al., 1997
Actinobaculum urinale	CCUG 46093 = CIP 107424	Urine, human	Hall et al., 2003
Arcanobacterium			
Arcanobacterium bernardiae	ATCC 51727 = CCUG 33419 = DSM 9152	Blood, abscesses, and various other clinical sources, human	Funke et al. 1995
Arcanobacterium haemolyticum	ATCC 9345 = CCUG 17215 = DSM 20595	Pharyngitis, wound infections, soft tissue infections, endocarditis, septicemia, human	Pascual Ramos et al., 1997 ex Mac Lean et al., 1946 Collins et al., 1983
Arcanobacterium hippocoleae	CCUG 44697 = CIP 106850	Vaginal discharge, horse	Hoyles et al., 2002
Arcanobacterium phocae	CIP 105740 = DSM 10002	Various tissues and body fluids, gray seal and common seal	Pascual Ramos et al., 1997
Arcanobacterium pluranimalium	CCUG 42575 = DSM 13483	Dead harbor porpoise, dead sallow deer	Lawson et al., 2001
Arcanobacterium pyogenes	ATCC 19411 = (NCTC 5224, CCUG 13230) = DSM 20630	Various pyogenic infections, blood, abscesses, respiratory specimens, domestic animals, humans	Glage, 1903 Pascual Ramos et al., 1997
Mobiluncus			
Mobiluncus curtisii subsp. *curtisii*	ATCC 35241 = CCUG 21018	Vaginal secretions from patients with bacterial vaginosis	Spiegel and Roberts, 1984
Mobiluncus curtisii subsp. *holmesii*	ATCC 35242 = CCUG 17762	Vaginal secretions from patients with bacterial vaginosis	Spiegel and Roberts, 1984
Mobiluncus mulieris	ATCC 35243 = CCUG 20071	Vaginal secretions, human	Spiegel and Roberts, 1984
Varibaculum			
Varibaculum cambriense	CCUG 44998 = CIP 107344	Clinical sources including breast, brain, cheek, submandibular, postauricular, and ischiorectal abscesses and intrauterine contraceptive devices, human	Hall et al., 2003

Abbreviations: ATCC, American Type Culture Collection, Manassas, Virginia, USA; CCUG, Culture Collection, University of Göteborg, Dept. of Clinical Bacteriology, Göteborg, Sweden; DSM, Deutsche Sammlung von Mikroorganismen und Zellkulturen GmbH, Braunschweig, Germany; NCTC, National Collection of Type Cultures, Central Public Health Laboratory, London, United Kingdom; CIP, Institut Pasteur, Paris, France; HKU, Hong Kong University Culture Collection; and LMG, Universiteit Gent, Laboratorium voor Mikrobiologie, Gent, Belgium.

Table 2. Reclassified Actinomyces species.

Original description	Reclassified as	References
Actinomyces propionicus Buchanan and Pine 1962	*Arachnia propionica*	Pine and Georg, 1969
	Propionibacterium propionicum	Charfreitag et al., 1988
Actinomyces bernardiae Funke et al., 1995	*Arcanobacterium bernardiae*	Pascual Ramos et al., 1997
Actinomyces pyogenes (Glage 1903) Reddy et al., 1982	*Arcanobacterium pyogenes*	Pascual Ramos et al., 1997
Actinomyces suis (Wegienek and Reddy 1982a) Ludwig et al., 1992	*Actinobaculum suis*	Lawson et al., 1997
Actinomyces humiferus Gledhill and Casida 1969	*Cellulomonas humilata*	Collins and Pascual, 2000

of the family Actinomycetaceae. Phylogenetic trees have been included in various publications covering descriptions of novel species and phylogenetic analyses. As does the size of databases, selection of organisms of the family, outgroup organisms, and treeing algorithms differ in these surveys, and the phylogenetic trees also differ from each other in the intrafamily structure and position of the family within the suborder Actinomycineae (Stackebrandt et al., 1997). The interpretation of relatedness at the intrafamily level is based on the maximum-likelihood algorithm (ML; Fig. 1), included in the Phylip V version of Felsenstein (1993). As compared to the neighbor-joining dendrogram (Felsenstein, 1993) and the distance matrix tree of De Soete (1983), this tree is similar, though not identical, in the order of branching points of main lineages and in details of branching patterns within individual lineages (not shown). The consensus tree (Felsenstein, 1993) differs significantly from any of these trees. The term "cluster" will be used for lineages which appear to be distantly related to each other in Fig. 1, though the definition of cluster is objective.

Position at the Suborder Level

On the basis of both the phylogenetic position of type strains of species available in 1997 and on signature nucleotides of the 16S rRNA gene sequences, Stackebrandt et al. (1997) described the suborder Actinomycineae for the family Actinomycetaceae that branched between the suborders Micrococcineae and Corynebacterineae. According to the phylogenetic dendrogram generated by the ARB program, the family branches adjacent to the Bifidobacteriaceae, while the Ribosomal Database Project (http://rdp.cme.msu.edu/download/SSU_rRNA/SSU_Prok.phylo) groups the Actinomycetaceae with the Micrococcineae and the Bifidobacteriaceae (subdivision 2.30.1.9.2). In the description of *Arcanobacterium* (Pascual Ramos et al., 1997b), the Actinomycetaceae branched next to the

Micrococcineae (only two sequences included) and Propionibacterineae.

Intrafamily Relatedness

Similarity values of 16S rRNA gene sequence analyses for members of the family range from 88% to 99%, indicating the presence of several phylogenetically defined clusters (Fig. 1), which are separated by intra-cluster similarity values ranging from 88–90% to 89–92%. The idea that the genus *Actinomyces* may not represent a monophyletic taxon has been expressed before by Pascual Ramos et al. (1997b). Because of the lack of phenotypic and chemotaxonomic properties that support a taxonomic dissection of the family, the family serves as a phylogenetic dumping ground. Chemotaxonomic properties, demonstrated to correlate well with phylogenetic structure in other actinobacterial genera, are either not available or not exclusive in phylogenic clusters. Although tempting, a purely phylogeny based dissection is not favored, as novel sequences may blur the present topology of the family.

Three clusters correspond to *Mobiluncus*, *Arcanobacterium* and *Actinobaculum*, respectively. The latter two genera are phylogenetic neighbors, while *Mobiluncus* species branch adjacent to *Actinomyces neuii*. From a phylogenetic point of view, members of *Mobiluncus*, *Arcanobacterium* and *Actinobaculum* appear to have originated as descendants of an *Actinomyces* ancestor, but even this situation may change with inclusion of more sequences.

Members of *Actinomyces* form several phylogenetic clusters, the number of which depends on the cut-off branching points. As the branching is rarely supported by high bootstrap values, the level at which these clusters are considered taxonomically coherent entities should be decided on the basis of results of taxonomic investigations, e.g., chemotaxonomy, physiology and morphology. Even within individual clusters, the majority of species are remotely related, only showing more than 3% dissimilarity between their 16S rRNA gene sequences. Following the

Cluster

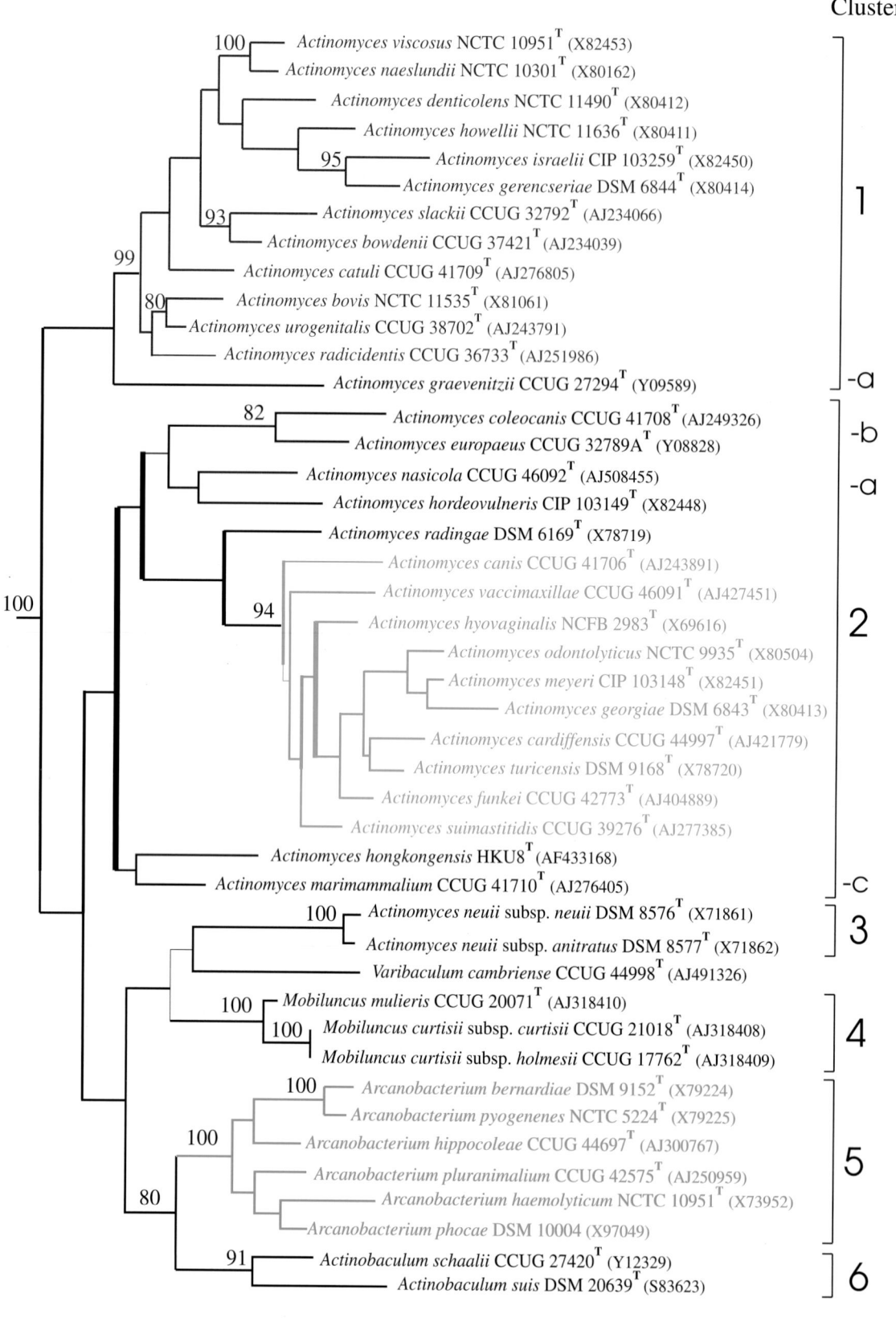

10%

suggestion of Stackebrandt and Goebel (1994), that at this low level of relatedness DNA-DNA hybridization would result in values significantly below the species delineation threshold value of around 70% (Wayne et al., 1987), most authors refrained from confirming the species status of the more recently described species by applying this technique to the type strains of the novel species.

On the basis of similar cut off points, *Actinomyces* species (excluding *A. neuii*) form two major clusters. Cluster 1 contains either 11 or 13 species, depending on whether *A. graevenitzii* and *A. radicidentis* are included or whether they represent individual lineages. On the basis of 16S rRNA gene signature nucleotides (Table 4), *A. radicidentis* should be considered a member of cluster 1, whereas *A. graevenitzii* shows deviations from the set of signature nucleotides and should be treated as a separate phylogenetic unit. The remaining *Actinomyces* species (except for *A. neuii*) form cluster 2. While the majority of species of cluster 2 are moderately related, four distantly related subclusters emerge, containing *A. radingae, A. marimammalium, A. hongkongensis, A. europaeus* and *A. coleocanis,* as well as *A. nasicola* and *A. hordeovulneris,* respectively. The individual lineages are defined by the signature nucleotides shown in Table 4.

Description of the Phylogenetic Clusters and Subclusters

CLUSTER 1, CORE CLUSTER 1 Members of core cluster 1, sharing 94–99% sequence similarity, can be considered authentic *Actinomyces* species as it contains the type species *A. bovis.* This species is closely related to *A. urogenitalis* (97.9% similarity; 93% bootstrap value) as originally pointed out in the description of this species (Nikolaitchouk et al., 2000). Additional pairs of phylogenetic neighbors are *A. israelii* and *A. gerencseriae* (96.8%, 95% bootstrap value), *A. slackii* and *A. bowdenii* (97.3%, 93% bootstrap value), and *A. viscosus* and *A. naeslundii* (99.2%, 100% bootstrap value). All other species form individual lines of descent. Results of DNA-DNA reassociation experiments, determined by the S1 nuclease method (Johnson et al., 1990) and the membrane filter hybridization method (Dent and Williams, 1984b), indicated that

A. israelii, A. gerencseriae, A. denticolens, A. naeslundii and *A. viscosus* constituted genomically well-separated species. The relatedness among the latter two species was higher than that found for the other species. *Actinomyces gerencseriae* (ATCC 23860[T]), formerly treated as a strain of *A. israelii* serovar 2, was identified as a taxon separate from *A. israelii* serovar 1 strains by oligonucleotide probing (Stackebrandt and Charfreitag, 1990) and low DNA reassociation values (Johnson et al., 1990).

CLUSTER 1A The distinct phylogenetic position of *A. graevenitzii,* sharing less than 94% 16S rRNA gene sequence similarity with other species of the genus (Pascual Ramos et al., 1997a) has been verified in subsequent descriptions of new *Actinomyces* species (Pascual et al., 1999; Nikolaitchouk et al., 2000; Hoyles et al., 2001a; Hoyles et al., 2001b; Hoyles et al., 2001c; Hoyles et al., 2002a). In these dendrograms, *A. graevenitzii* constituted the deepest branch of core cluster 2, or it branched off even deeper, giving the appearance of an isolated phylogenetic lineage within *Actinomyces.* The branching point of the *A. graevenitzii* lineage is supported by a 99% bootstrap value. The 16S rRNA gene sequence of the type strain CCUG 27294[T] (www.ncbi.nlm. nih.gov/entrez/query.fcgi?db=nucleotide=search =27294) contained several undetermined nucleotides, erroneous stretches and deletions. The corrected sequence (A. F. Yassin, unpublished data) changed neither the similarity values nor its branching point but supported the distinct signature of the organism.

CLUSTER 2, CORE CLUSTER 2 Except for *A. odontolyticus, A. meyeri* and *A. georgiae* described in 1958, 1984 and 1990, respectively, this cluster, supported by high statistical significance (94% bootstrap value), contains species that have been described within the past 10 years. Some species pairs are moderately related, e.g., *A. suimastiditis* and *A. hyovaginalis* (97.3% similarity), *A. cardiffensis* and *A. turicensis,* as well as *A. meyeri* and *A. odontolyticus.* Together with their phylogenetic neighbor *A. georgiae* (97.7%), the latter two species have been subjected to DNA-DNA reassociation studies (Johnson et al., 1990); low hybridization values confirmed their separate species status.

Fig. 1. Neighbor-joining tree (Felsenstein, 1993) of 16S rRNA gene sequences displaying the relationship of members of the family Actinomycetaceae. Numbers at branch points refer to bootstrap values (500 resamplings). This dendrogram differs in details from previously published dendrograms of novel species of individual actinomycete genera to the same extent as the latter dendrograms differ from each other. This can be explained by differences in the number and identity of the reference organisms included, which may significantly influence the topology of branching patterns. Scale bar: 10% sequence divergence. The designation of clusters on the right margin was solely based on phylogenetic grounds.

Table 3. List of synonyms and basonyms of species of Actinomycetaceae.

Species	Synonym	Basonym
Actinomyces bovis	"*Discomyces bovis*" (Harz 1877) Rivolta 1878	
	"*Sarcomyces bovis*" (Harz 1877) Rivolta 1879	
	"*Oospora bovis*" (Harz 1877) Sauvageau and Radais 1892	
	"*Actinocladothrix bovis*" (Harz 1877) Gasperini 1892	
	"*Nocardia bovis*" (Harz 1877) Blanchard 1896	
	"*Streptothrix bovis*" (Harz 1877) Chester 1901	
	"*Cladothrix bovis*" (Harz 1877) Macé 1901	
	"*Sphaerotilus bovis*" (Harz 1877) Engler 1907	
	Proactinomyces bovis" (Harz 1877) Henrici 1939	
Actinomyces israelii	"*Streptothrix israeli*" (sic) Kruse 1896	
	"*Actinobacterium israeli*" (Kruse 1896) Sampietro 1908	
	"*Cohnistreptothrix israeli*" (Kruse 1896) Pinoy 1913	
	"*Nocardia israeli*" (Kruse 1896) Castellani and Chalmers 1913	
	"*Oospora israeli*" (Kruse 1896) Sartory 1920	
	"*Brevistreptothrix israeli*" (Kruse 1896) Lignières 1924	
	"*Corynebacterium israeli*" (Kruse 1896) Haupt and Zeki 1933	
	"*Proactinomyces israeli*" (Kruse 1896) Negroni 1934	
	"*Discomyces israeli*" (Kruse 1896) Gedoelst 1902	
Actinomyces meyeri	"*Actinobacterium meyeri*" Prévot 1938	
Actinomyces viscosus	"*Odontomyces viscosus*" Howell et al., 1965	
Actinomyces pyogenes	"*Bacillus pyogenes*" Glage 1903	
Actinobaculum suis	"*Corynebacterium suis*" Soltys and Spratling 1957	*Corynebacterium pyogenes* (Glage 1903) Eberson 1918 *Eubacterium suis* (ex Soltys and Spratling 1957) Wegienek and Reddy 1982a, *Actinomyces suis* (Wegienek and Reddy 1982a) Ludwig et al., 1992.
Arcanobacterium haemolyticum	"*Corynebacterium haemolyticum*" Mac Lean et al., 1946	
Arcanobacterium pyogenes	"*Bacillus pyogenes*" Glage 1903	*Corynebacterium pyogenes* (Glage 1903) Eberson 1918, *Actinomyces pyogenes* (Glage 1903) Reddy et al., 1982.

Table 4. Signature a nucleotides of 16S rRNA gene sequence, defining individual subclusters of *Actinomyces, Arcanobacterium, Actinobaculum* and *Mobiluncus.*

	Cluster 1		Cluster 2					Cluster 6	Cluster 5	Cluster 4	Cluster 3
Position	Core cluster 1[b]	A. graevenitzii cluster 1a	Core cluster 2[c]	A. radingae cluster	A. hordeovulneris cluster 2a	A. europaeus A. coleocanis cluster 2b	A. marimam-malium cluster 2c	Actinoba-culum	Arcano-bacterium	Mobiluncus	A. neuii
125–236	U-G	U-G	U-G	U-G	U-G	U-G	U-G	U-G	U-A	U-G	U-G
146–176	U-G	U-G	G-Y	G-U	G-C	U-A	G-U	G-Y	G-Y	G-U	G
153–168	C-G	C-G	C-G	C-G	C-G	C-G	G-U	G-U	G-U	C-G	G-U
154–167	Y-R	U-A	G-U	G-U	G-U	Y-G	C-G	C-G	C-G	G-U	C-G
316–337	C-G	C-G	U-G	U-G	C-G	C-G	C-G	U-G	C-G	C-G	C-G
378–385	G-C	G-C	R-Y	G-C	G-C	G-C	G-C	G-C	A-U	A-U	A-U
407–435	A-U	A-U	A-U	A-U	A-U	A-U	A-U	G-U	G-C	G-C	G-C
408–434	G-C	G-C	G-C	G-C	G-C	G-C	G-C	G-Y	G-U	A-U	G-U
419–424	C-G	C-G	C-G	C-G	C-G	C-G	C-G	U-G	U-G	C-G	C-G
441	G	G	G	G	A	G	A	G	A	U	G
450–483	G-C	C-G	C-G	C-G	G-C	R-Y	G-C	C-G	Y-R	A-U	A-U
501–544	C-G	C-G	C-G	C-G	C-G	U-A	C-G	C-G	C-G	C-G	C-G
502–543	G-C	G-C	G-C	G-C	G-C	A-U	G-C	G-C	G-C	G-C	G-C
586–755	C-G	C-G	C-G	C-G	C-G	U-A	C-G	C-G	C-G	U-R	U-G
590–649	U-G	U-G	U-G	U-G	U-G	U-G	G-U	U-G	U-G	U-G	U-G
591–648	G-Y	G-Y	G, U-V	U-A	U-A	U-A	U-A	U-A	U-A	C-G	U-A
601–637	G-U	G-U	G-U	G-U	G-U	G-U	G-U	G-U	G-U	G-U	A-U
602–636	Y-G	Y-G	Y-G	U-G	U-G	U-G	U-G	G-Y	C-G	C-G	G-U
603–635	C-G	C-G	C-G	C-G	C-G	C-G	C-G	U-A	U-A	C-G	C-G
614–626	C-G	C-G	C-G	C-G	A-U	R-U	A-U	G-C	G-C	G-C	C-C
615–625	U-G	C-G	U-G	U-G	C-G	V-V	U-G	G-U	G-Y	U-R	G-C
613–627	Y-G	U-G	U-G	U-G	C-G	Y-G	A-U	C-G	C-G	A-U	U-G
835–851	G-C	A-U	G-C	G-C	G-C	G-C	G-C	G-C	G-C	G-C	G-C
668–738	A-U	U-A	A-U	A-U	A-U	A-U	A-U	U-A	U-A	U-A	U-A
669–737	G-C	A-U	G-C	G-C	G-C	G-C	G-C	G-C	A-U	G-C	A-U
722–733	G-G	G-G	G-G	G-G	G-G	G-G	G-G	A-A	A-A	A-A	G-G
838–848	G-Y	C-G	G-U	N-U	A-U	G-U	G-U	V-V	C-G	A-U	C-G
839–847	V-V	U-G	C-G	C-G	U-G	C-G	C-G	C-G	U-R	U-G	U-U
840–846	Y-G	C-G	C-G	C-G	U-G	Y-G	U-G	U-A	U-A	U-A	U-G
997–1044	U-A	U-A	U-G	U-G	U-A	U-G	U-G	U-A	U-A	C-G	U-A
1118–1155	U-A	U-A	C-G	C-G	U-A	U-A	U-A	U-A	U-A	A-U	A-U
1123–1151	G-C	G-C	A-U	A-U	A-U	G-C	G-C	G-C	G-C	U-G	U-G
1123–1150	U-G	U-G	U-G	U-G	U-G	U-U	U-G	U-A	U-A	C-G	C-G
1244–1293	C-R	U-G	U-R	A-U	U-G	Y-G	C-G	C-G	Y-R	C-G	G-C
1243–1294	C-G	C-G	C-G	C-G	C-G	C-G	C-G	C-G	C-G	A-U	U-A
1245–1292	G-C	G-U	G-Y	G-U	A-U	G-Y	G-U	G-Y	G-Y	A-U	U-G
1246–1291	G-C	G-C	G-C	G-C	G-C	G-C	G-C	G-C	G-C	G-C	C-G
1308–1329	C-G	C-G	Y-G	G-U	U-A	C-G	C-G	G-U	U-A	U-A	C-G
1308–1327	G-C	G-C	R-Y	G-U	G-C	U-A	A-U	V-V	G-C	G-C	C-C
1311–1326	U-A	U-A	R-Y	G-C	G-C	A-U	U-A	G-C	G-C	G-C	A-U
1308–1329	C-G	U-G	Y-R	C-G	U-A	C-G	C-G	C-G	U-A	U-A	C-G

Abbreviations: V, variable nucleotide composition; Y, pyrimidine; and R, purine.

a A signature is defined as the presence of a nucleotide unique in members (>90%) of a single cluster or more clusters but not in all clusters of the family.

b Contains *A. radicidentis.*

c Contains *A. canis.*

The 16S rRNA gene sequence similarity values found between *A. canis* and members of the core cluster 2 are 93.1–96.6%. Its separate phylogenetic position as a deep branching lineage of core cluster 2 organisms is in accord with the original description of Hoyles et al. (2000). The major deviation from the set of signatures defining core cluster 2 sequences is a stretch of 3 base pairs (positions 614–626, 615–625, and 616–624).

Actinomyces radingae is another remote relative of cluster 2, but the branching point of this lineage is not supported by a high bootstrap value. This is in contrast to the dendrogram shown by Pascual Ramos et al. (1997b), but the number of species included is smaller than that shown in Fig. 1. As the set of signature nucleotides (Table 4) is similar to that of other members of the cluster, *A. radingae* will be considered a tentative member of cluster 2.

CLUSTER 2A—*A. HORDEOVULNERIS* AND *A. NASICOLA* The phylogenetic position of *A. hordeovulneris* was not investigated in the original description of this species (Buchanan et al., 1984b) but included in the broad phylogenic survey of the genus by Pascual Ramos et al. (1997b). The isolated position of this species has been seen since then in many subsequent publications of *Actinomyces* species. Several deviations in the set of 16S rRNA gene sequence signatures of cluster 2 support the branching point of *A. hordeovulneris*. The species *A. nasicola* (Hall et al., 2003b), isolated from the nose of a human, was recently added to this lineage.

CLUSTER 2B—*A. EUROPAEUS* AND *A. COLEOCANIS* These two species share a remote relationship of 94.2% sequence similarity. The branching point of this lineage is not settled, as seen by low bootstrap values and different affiliations of either one or both species in dendrograms published in recent species descriptions (Pascual et al., 1999; Collins et al., 2000b; Hoyles et al., 2000; Hoyles et al., 2001c; Hoyles et al., 2002a; Nikolaitchouk et al., 2000). The presence of unique 16S rRNA gene nucleotide signatures (Table 4), shared specifically with members of other clusters, highlights the separate phylogenetic position of these two organisms. Affiliation to cluster 2 should be regarded as tentative.

CLUSTER 2C—*A. MARIMAMMALIUM* AND *A. HONGKONGENSIS* *Actinomyces marimammalium* and the recently described *A. hongkongensis* (Woo et al., 2003) branch deeply within the tree of Actinomycetaceae species, showing almost equidistant relationship to members of other clusters. The bootstrap value for the two type strains of this cluster is low. The signature nucleotides (Table 4) of *A. marimammalium* support

the isolated phylogenetic position. Affiliation to cluster 2 should be regarded as tentative.

CLUSTER 3—*ACTINOMYCES NEUII* AND *VARIBACULUM CAMBRIENSE* *Actinomyces neuii* (Funke et al., 1994) was described at a time when the phylogenetic diversity of the genus *Actinomyces* was not apparent. Since then, the separate position of the species, apart from other members of the genus, has been noted, but lack of distinct and differentiating phenotypic properties has hindered the formal description of a new genus for the former CDC group 1 (subspecies *neuii*) and CDC group 1-like coryneforms (subspecies *anitratus*). The two groups were considered subspecies because of phenotypic differences observed for the phylogenetically highly related type strains (99.7% sequence similarity). Nonquantitative DNA-DNA hybridization (Funke et al., 1993), however, indicated a low degree of overall genomic relatedness between members of the two subspecies. The branching of *Varibaculum cambriense* to the *A. neuii* lineage is without statistical significance.

CLUSTER 4—*MOBILUNCUS* Formerly considered a member of the Bacteroidaceae (Spiegel and Roberts, 1984a), the genus was transferred to the Actinomycetales by Lassnig et al. (1989) on the basis of 16S rRNA gene sequence analyses. The phylogenetic closeness between *Mobiluncus* and *Actinomyces* species became apparent, but only a few species of core cluster 1 were available for comparison. Similarities were found in the end products of carbohydrate fermentation and a requirement for CO_2, but one of the salient properties excluding reclassification of *Mobiluncus* species as species of *Actinomyces* was the low DNA mol% G+C content of only 49–52. In the past years, improved sequences have been published (Tiveljung et al., 1995) or submitted to public databases (L. Hoyles, unpublished data), but the branching of *Mobiluncus* within the radiation of the family Actinomycetaceae did not change. The phylogenetic coherence of *Mobiluncus* species and the significance of the topology of the branching pattern are supported by 16S rRNA gene sequence signatures. Though also depicted in the dendrograms published by Funke et al. (1997a), Nikolaitchouk et al. (2000), Collins et al. (1993), and Hoyles et al. (2001a), the branching of the genus *Mobiluncus* next to *Actinomyces neuii* is barely of statistical significance.

CLUSTER 5—*ARCANOBACTERIUM* This genus was established for the reclassified species "*Corynebacterium haemolyticum*" (Collins et al., 1982c). Later, *A. pyogenes*, *A. bernardiae*, *A. phocae*, *A.*

pluranimalium and *A. hippocoleae* (Pascual Ramos et al., 1997b; Lawson et al., 2001a; Hoyles et al., 2002b) were added to the genus. The phylogenetic topology (Fig. 1) agrees with that shown by Hoyles et al. (2002b). The phylogenetic coherency of the genus is supported by a set of distinct signature nucleotides. The position of *Arcanobacterium* species, adjacent to members of *Actinobaculum*, supports earlier reports in the more recent descriptions of *Actinomyces, Arcanobacterium* and *Actinobaculum* species.

CLUSTER 6—*ACTINOBACULUM* The genus was described on the basis of the distinct phylogenetic position and chemotaxonomic properties of its members (Lawson et al., 1997), which separated them from most representatives of the genus *Actinomyces*. *Actinobaculum* contains the distantly related species *Actinobaculum suis* (basonym *Actinomyces suis*) and *Actinobaculum schaalii* (94.2% 16S rRNA gene sequence similarity). The distinct position of these species, supported by a high bootstrap value of 91%, has also been described in the publications of Pascual Ramos et al. (1997b), Nikolaitchouk et al. (2000) and the more recent species descriptions. The set of signature nucleotides (Table 4) clearly identifies members of this genus.

The Genus *Actinomyces*

The first member of the genus *Actinomyces* was discovered in 1877 when the German veterinarian Otto Bollinger found that lesions of a chronically destructive disease of the jaw and tongue of cattle, previously mistaken for a sarcomatous malignant tumor, contained small particles resembling clusters of crystals ("drusen" or "vugs," later termed "sulfur granules" in the Angloamerican literature) which chiefly consisted of filamentous, fungus-like structures. The botanist C. O. Harz confirmed the microbial nature of these elements (Harz, 1877) and, because of the radial microscopic configuration of the particles, named the suspected microorganism *Actinomyces* ("ray fungus") *bovis*, although its culture was not possible until 1890 (Mosselman and Lienaux, 1890). A similar disease in humans was first described by James Israel (Israel, 1878), but attempts to cultivate the pathogen were again fruitless until Bujwid (1889) first succeeded in isolating the organism. A more comprehensive bacteriological description of these pathogens was published by Wolff and Israel (1891), and one of their current designations, *A. israelii*, was introduced by Lachner-Sandoval (1898).

Taxonomy

Breed and Conn (1919) first recommended that *Actinomyces* be considered a *genus conservandum* with *A. bovis* as the type species, and this was accepted by the Winslow Committee in 1920 (Winslow et al., 1920), thereby confirming that *Actinomyce* (*sic*) Meyen 1827, a name proposed for a fungus (*Tremella meteorica*), was not valid.

Originally, the description of the genus was solely based upon the morphology of the filamentous elements produced by *A. bovis* in the granules of bovine infections. The discovery of similar but not identical granules in human infections and the successful cultivation of both *A. bovis* and *A. israelii* added little to this description. Cultural, physiological, and chemical characters only slowly began to influence the definition of the taxon. Pine and Georg (1969) used cellular and colonial morphology, cell wall composition, fermentation end products, and certain physiological characters to define the genus *Actinomyces* in light of more modern approaches to taxonomy. Thus, *Actinomyces propionicus* was removed from the genus and reclassified as *Arachnia propionica* (now: *Propionibacterium propionicum*), as it produced propionic acid as a major fermentation end product and had LL-diaminopimelic acid in its cell wall (Pine and Georg, 1969; Table 2).

Because of persisting taxonomic and nomenclatural uncertainties, as well as the lack of suitable differential characters, it was essentially impossible for many years to separate *A. bovis* reliably from *A. israelii*, and both names were used interchangeably until the reports of Erikson (1940), Thompson (1950), and Pine et al. (1960) finally proved that the causative agents of human and bovine actinomycosis were two taxonomically distinct species. At the same time, the improved taxonomic techniques also allowed the delineation of a first group of additional *Actinomyces* species. Thompson and Lovestedt (1951) proposed the name *A. naeslundii* for a filamentous bacterium which had first been isolated from the human oral cavity by Carl Naeslund in 1925 (Table 1). *Actinomyces viscosus*, which had primarily been described under the genus designation "*Odontomyces*" (Howell et al., 1965), was included in the genus *Actinomyces* after its definition had been changed to accommodate both catalase-positive and catalase-negative bacteria (Georg et al., 1969; Table 3).

However, it soon became apparent that *A. naeslundii* and *A. viscosus* were closely related in terms of physiological characters and antigenic structure (Holmberg and Forsum, 1973a; Fillery et al., 1978; Schofield and Schaal, 1981; Schaal and Gatzer, 1985c). It was even suggested (Gerencser, 1979) that *A. naeslundii* and *A. vis-*

cosus were varieties of a single species. Nevertheless, detailed numerical phenetic analyses (Schaal and Schofield, 1981c; Schaal and Schofield, 1981d; Schofield and Schaal, 1981) revealed that strains labeled *A. naeslundii* and *A. viscosus* formed a set of comparatively stable subclusters, two of which could be equated with typical *A. naeslundii* and *A. viscosus* isolates, respectively. Very similar results were obtained in a recent evaluation of a new phenotypic differentiation system (Taxa Profile®, Merlin, Germany) in which four distinct numerical phenetic clusters could be delineated within the *A. naeslundii* and *A. viscosus* complex (Dahlen, 2004). At least in part, these results have been confirmed using molecular genetic analyses (Coykendall and Munzenmaier, 1979). In particular, the study of Johnson et al. (1990) showed that strains belonging to the *A. naeslundii* and *A. viscosus* complex may be assigned to four genospecies, which were provisionally termed "*A. naeslundii*, genospecies 1," "*A. naeslundii*, genospecies 2," "*Actinomyces* serotype WVA 963," and "*A. viscosus*, serotype I." *Actinomyces naeslundii*, genospecies 1, contains strains formerly designated "*A. naeslundii*, serotype I." The *A. naeslundii* genospecies 2 comprises strains of *A. naeslundii* serotype II, *A. naeslundii* serotype III, *A. viscosus* serotype II, and strains that react with both *A. naeslundii* and *A. viscosus* serotype II antisera (so-called "serotype NV strains"). *Actinomyces* serotype WVA 963 had formerly been designated "*A. naeslundii* serotype IV" but was found to be more distantly related to the other organisms of this complex and also occupied a separate position in numerical phenetic studies (Schaal and Schofield, 1981c; Schaal and Schofield, 1981d; Schofield and Schaal, 1981). The hamster isolates of *A. viscosus* (serotype I) appeared to form another separate genospecies.

Gram-positive, predominantly rod-shaped bacteria that had first been isolated from advanced human carious lesions were recognized as a further *Actinomyces* species by Batty (1958) and named "*A. odontolyticus*" (Table 1). The classical *Actinomyces* species mentioned above form two different phylogenetic clusters with *A. bovis*, *A. naeslundii*, *A. viscosus* and *A. israelii* belonging to cluster 1 and *A. odontolyticus* belonging to cluster 2 (Fig. 1).

After further improvement of the chemotaxonomic and physiological tests (and especially after introduction of DNA-DNA pairing and 16S rRNA gene sequencing technologies into the field of actinomycete taxonomy), numerous additional taxa were added to the list of *Actinomyces* species (Table 1). Dent and Williams described three groups of Gram-positive bacteria isolated from the dental plaque of dairy cattle which showed some phenotypic similarity to *A.*

naeslundii or *A. viscosus* but could be differentiated from these species on the basis of cell wall and DNA base composition, DNA-DNA similarity data, polypeptide molecular weight distribution, and a few physiological reactions (Dent and Williams, 1984a; Dent and Williams, 1984b; Dent and Williams, 1984c; Dent and Williams, 1986). The new species designations "*A. denticolens*," "*A. howellii*" and "*A. slackii*," respectively, were proposed for these isolates, which all belong to the phylogenetic cluster 1. In contrast, bacteria that were isolated from canine infections and named "*A. hordeovulneris*" by Buchanan et al. (1984b) appear to represent a separate phylogenetic line of descent (cluster 2a; Fig. 1).

Actinomyces meyeri, which was first described by Kurt Meyer (Meyer, 1911) and included in the genus "*Actinobacterium*" by Prévot (1938), was reclassified as *Actinomyces meyeri* by Cato et al. (Cato et al., 1984; Table 3). The genus designation "*Actinobacterium*" lost its standing in nomenclature after the type species of this genus, "*Actinobacterium*" *israelii*, had been transferred to the genus *Actinomyces* by Breed and Conn (1919), so that "*Actinobacterium*" *meyeri* remained "a species in search of a genus" for many years. In addition, 16S rDNA sequence data confirm the close affinity of this species to other members of the genus *Actinomyces*, namely in the phylogenetic cluster 2.

The dramatically increasing application of 16S rRNA or rDNA sequencing to Gram-positive rod-shaped bacteria in recent years has led to a host of new or reclassified *Actinomyces* species (Tables 1–3): As far as human isolates are concerned, Stackebrandt and Charfreitag (1990), as well as Johnson et al. (1990), found that strains formerly known as *A. israelii* serovar 2 showed a surprisingly low level of DNA relatedness to *A. israelii* serovar 1 isolates, which led the latter authors to propose the new species *A. gerencseriae* for the serovar 2 strains. In the same study, Johnson et al. (1990) described the species *A. georgiae*, whose members were recovered from gingival crevices of humans with and without periodontitis, but while *A. gerencseriae* belongs to the phylogenetic cluster 1, *A. georgiae* is recovered in cluster 2. Another group of organisms which had primarily been termed CDC group 1 coryneform bacteria and CDC group 1-like coryneform bacteria was assigned to the genus *Actinomyces* as *A. neuii* subsp. *neuii* and *A. neuii* subsp. *anitratus*, respectively, by Funke et al. (1994), although current phylogenetic data indicate that these bacteria, which were chiefly isolated from human abscesses in association with mixed anaerobic flora or blood cultures, represent a separate line of descent (see Phylogeny). *Actinomyces radingae* and *A. turicensis* (Wüst et al., 1995) were both mainly isolated from human mixed infections (including skin, genital, and urinary tract infections) and belong to the phylogenetic cluster 2.

Actinomyces-like bacteria that had chiefly been recovered from human respiratory tract secretions were termed "*A. graevenitzii*" by Pascual Ramos et al. (1997a). However, several dendrograms derived from 16S rDNA sequences show unequivocally that *A. graevenitzii* strains appear to constitute an isolated phylogenetic lineage within the *Actinomycetaceae*. Similarly, *A. europaeus*, which was described by Funke et al. (1997a) for isolates mainly derived from human abscesses but also from urinary tract infections (Sabbe et al., 1999), is only remotely related to the classical *Actinomyces* species when 16S rDNA sequence similarity is considered (Fig. 1). In contrast, *A. urogenitalis*, which was described by Nikolaitchouk et al. (2000) for human urogenital isolates, represents a member of the phylogenetic core cluster 1 of *Actinomyces* (see Phylogeny), while *A. funkei*, which was proposed by Lawson et al. (2001) for three human strains of different sources including blood cultures, is a member of the phylogenetic core cluster 2.

The most recent members of the genus *Actinomyces* from human sources apparently belong to different phylogenetic clusters. *Actinomyces cardiffensis*, which was described by Hall et al. in 2002, had been isolated from various clinical specimens such as pleural fluid, brain, jaw, pericolic and ear abscesses, as well as intrauterine contraceptive devices, and is also a member of the phylogenetic core cluster 2. In contrast, actinomycetes isolated from infected human dental root canals were described by Collins et al. (2000b) as *A. radicidentis* and belong to or are associated with the phylogenetic cluster 1. The single strain described as *A. nasicola* by Hall et al. (2003b) had been recovered from the purulent discharge found in the nasal antrum of an 81-year-old man and shows some 16S rRNA gene sequence relatedness to *A. hordeovulneris*, and to a lesser extent, also to *A. coleocanis* and *A. europaeus*. A second new *Actinomyces* species of human origin, *A. oricola* (Hall et al., 2003d), was isolated from a dental abscess and appears to be phylogenetically related to *A. israelii* and *A. gerencseriae*. The third new member of the genus *Actinomyces* described in 2003 from a human clinical specimen (pus of a patient with pelvic actinomycosis) was termed "*A. hongkongensis*" (Woo et al., 2003) and also showed some 16S rRNA gene relatedness to *A. marimammalium*. The definition of *A. houstonensis* which was proposed by Clarridge and Zhang in 2002 for three human isolates derived from serious subcutaneous abscesses has as yet chiefly been based upon 16S rDNA sequence data and its sufficient phenotypic characterization remains to be per-

formed. This species will therefore not be further considered in this chapter.

Additional new *Actinomyces* species were derived from various animal sources (Table 1): The species designation "*A. hyovaginalis*" was proposed by Collins et al. (1993) for a group of *Actinomyces*-like strains that had been isolated from purulent discharge and aborted fetuses of pigs. These organisms are recovered in the phylogenetic core cluster 2. Fermentative Gram-positive bacteria also derived from pigs and designated "*Actinomyces suis*" pose several taxonomic and nomenclatural problems (Table 2). This species epithet was first used by Gasperini (1892) but had not been validly published and was therefore declared a *nomen dubium* (Slack, 1974). Grässer (1957) isolated bacteria from mastitis of swine, which were again named "*A. suis*." In this case, the name was validly published, but the description was inadequate, and cultures are not available. More recently, Franke (1973) once more described actinomycetes isolated from the udder of swine, which he also called "*A. suis*." The description of Franke's isolates appears to differ from that of Grässer's strains but is strikingly similar to that of *A. hyovaginalis* (Collins et al., 1993), at least as far as physiological characteristics are concerned. Another single bacterial isolate from pig mastitis was found to represent a further new *Actinomyces* species (Hoyles et al., 2001a) in core cluster 2 and was named *A. suimastitidis* (Fig. 1).

Several further new *Actinomyces* species originated from canine and feline lesions. *Actinomyces bowdenii* (Pascual et al., 1999) was isolated from abscesses, pleural fluid, or a pyogranuloma of dogs and a cat, predominantly in mixed culture with additional microbes, and is a member of the phylogenetic cluster 1. *Actinomyces canis* (Hoyles et al., 2000), which was cultured from vaginal secretions or pus of dogs, belongs to cluster 2. A third new canine species, *A. catuli* (Hoyles et al., 2001b), originated from a polymicrobial infection of the lungs and pleural space. The single strain of *A. coleocanis* (Hoyles et al., 2002a) was isolated from the vagina of a cocker spaniel dog. While *A. catuli* is a member of the phylogenetic cluster 1, *A. coleocanis*, together with *A. europaeus*, occupies a separate phylogenetic position.

Actinomyces marimammalium (Hoyles et al., 2001c), which was isolated from seals and a porpoise, was found to branch very deep within the tree of the Actinomycetaceae, so that it also appears to represent a separate line of descent (Fig. 1). The as yet most recent animal *Actinomyces* species, *A. vaccimaxillae* (Hall et al., 2003a), was cultured from pus of a jaw lesion of an adult cow and was described to be phylogenetically related to *A. canis*, *A. hyovag-*

inalis, A. suimastitidis, A. funkei and *A. turicensis*, respectively.

The species *Actinomyces humiferus* was validly published by Gledhill and Casida in 1969. However, organisms conforming to this description differ considerably from typical members of the genus *Actinomyces* in that they grow at 30°C, have a high DNA G+C content (73 mol%), are sensitive to lysozyme, and occur in high numbers in organically rich soils (Gledhill and Casida, 1969). Also, they were found not to be related phylogenetically to other members of the genus *Actinomyces* and were thus reclassified as *Cellulomonas humilata* (Collins and Pascual, 2000a). Further reclassifications appeared to be appropriate for *Actinomyces bernardiae* (Funke et al., 1995b) and *Actinomyces pyogenes* (Reddy et al., 1982), both of which were transferred to the genus *Arcanobacterium* as *Arcanobacterium bernardiae* and *Arcanobacterium pyogenes*, respectively (Pascual Ramos et al., 1997b; Table 2). *Actinomyces suis* (Ludwig et al., 1992), which had primarily been described as *Eubacterium suis* (Wegienek and Reddy, 1982b) and which is apparently not identical with the swine isolates mentioned as "*A. suis*" above, was reclassified as *Actinobaculum suis* (Lawson et al., 1997; Table 2).

MORPHOLOGICAL CHARACTERISTICS

Cellular Morphology Cells of *Actinomyces* species have been described as being straight or slightly curved rods with or without clubbed ends, variable in shape and size (0.2–1.0 μm in diameter and 1.5–5.0 μm in length), and slender filaments (1 μm in width and 10–50 μm or more in length; Figs. 2 and 3), showing various degrees of true branching and a more or less wavy appearance (Slack, 1974; Slack and Gerencser,

1975; Schaal, 1986a). However, short rods (coccobacillary) or coccoid forms may also be seen and may occur singly, in pairs, or in diphtheroidal arrangements (Y, V, T forms or palisades). All of the *Actinomyces* species are Gram positive, although irregular staining may give rise to a beaded or barred appearance, and they are nonmotile and do not form endospores.

After the inclusion of many additional species the cellular morphology of the genus *Actinomyces* covers a wide range of forms so that morphological criteria have largely lost their relevance as taxonomic markers of the genus (Table 5). Thus, these organisms may resemble classical actinomycetes as well as bifidobacteria, corynebacteria, other rod-shaped bacterial species, and even streptococci. Furthermore, the cellular morphology may vary from strain to strain within one species and may be influenced additionally by the composition of the growth medium, its pH, cultural conditions, or the age of cultures.

Filamentous cells are chiefly produced by *A. israelii, A. gerencseriae, A. hordeovulneris,* and *A. oricola,* although these species may also appear as longer or shorter, possibly diphtheroidal rods with and without branching (Table 5). The majority of species predominantly present themselves as straight or curved rods that may show branching, and some species are primarily characterized by coccobacillary or even coccoid forms (Table 5).

Colony Morphology Microcolonies of filamentous *Actinomyces* species as well as those of *Propionibacterium propionicum* originate from rod-like propagules, usually by apical growth at one or both ends of the rod (Fig. 4), resulting in spider- or cobweb-like structures (Fig. 5) so that it is hardly possible to differentiate these *Actino-*

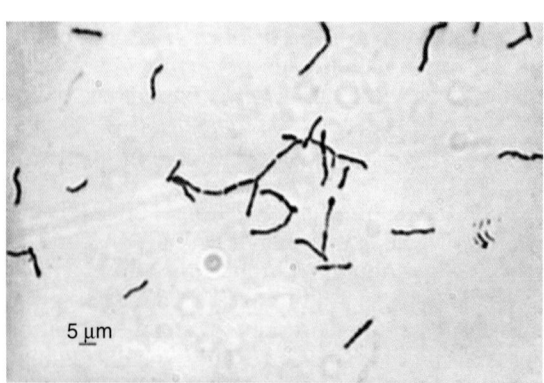

Fig. 2. *Actinomyces israelii*; Gram-stained smear from a mature colony (CC medium, 7 days at 36 ± 1°C, and Fortner's method).

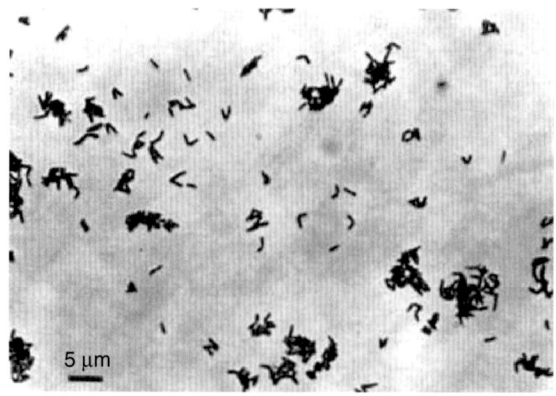

Fig. 3. *Actinomyces naeslundii*; Gram-stained smear from a mature colony (brain heart infusion agar, 7 days at 36 ± 1°C, and Fortner's method).

Table 5. Morphological characteristics of *Actinomyces* species.

Species	Cell morphology				Colony morphology				
	Cc	Cb	Br	Cf	Mf	R	S	M	Pig
A. bovis	–	d	d	d	d	d	d	–	W
A. bowdenii	–	–	d	–	ND	–	+	–	W
A. canis	–	–	d	–	ND	ND	ND	ND	ND
A. cardiffensis	–	–	d	–	ND	–	+	–	C/P
A. catuli	–	–	–	–	ND	d	d	–	W
A. coleocanis	–	–	d	–	ND	ND	ND	ND	ND
A. denticolens	–	–	d	d	d	–	+	–	P
A. europaeus	–	d	–	–	ND	–	+	–	W
A. funkei	–	–	d	–	ND	–	+	–	W
A. georgiae	–	–	(d)	–	ND	–	+	–	W
A. gerencseriae	–	–	+	+	+	d	d	–	W
A. graevenitzii	–	–	d	–	ND	–	+	–	DB
A. hongkongensis	–	–	–	–	ND	ND	ND	ND	ND
A. hordeovulneris	–	–	+	+	+	+	–	–	W
A. howellii	–	–	ND	d	ND	–	+	–	W
A. hyovaginalis	d	d	–	–	–	–	+	–	W
A. israelii	–	–	+	+	+	+	d	–	W
A. marimammalium	–	–	d	–	–	–	+	–	W
A. meyeri	–	d	–	(d)	–	–	+	–	W
A. naeslundii	–	–	+	+/–	d	d	d	(d)	W
A. nasicola	d	d	d	–	–	–	+	–	W
A. neuii	–	d	–	–	–	–	+	–	W
A. odontolyticus	–	d	d	d	(d)	(d)	+	–	B/R
A. oricola	–	–	+	d	ND	+	–	–	W
A. radicidentis	+	–	–	–	–	–	+	–	(B)
A. radingae	–	–	d	–	–	–	+	–	W
A. slackii	–	+	–	–	–	–	+	–	W
A. suimastitidis	–	–	–	–	–	ND	ND	ND	ND
A. turicensis	–	–	d	–	–	–	+	–	W
A. urogenitalis	–	–	–	–	ND	ND	ND	ND	(Rd)
A. vaccimaxillae	–	+	–	–	–	–	+	–	W
A. viscosus	–	–	+	d	d	d	d	d	W

Symbols: +, positive or present; –, negative or absent; +/–, predominantly positive or present; –/+, predominantly negative; and (d), rarely positive or present.

Abbreviations: Cc, cells coccoid; Cb, cells coccobacillary; Br, branching rods; Cf, cells filamentous; Mf, microcolonies filamentous; R, colonies rough; S, colonies smooth; M, colonies mucoid; Pig, colony pigmentation on blood agar: W, white, gray-white, creamy white; C/P, cream to pinkish; P, pink; DB, dark brown; B/R, brown to red; (B), may be brown; (Rd), may be reddish; d, strain differences; and ND, no data available.

Data from Dent and Williams (1986c), Schaal (1986b), Johnson et al. (1990), Collins et al. (1993, 2000), Funke et al. (1994, 1997), Wüst et al. (1995), Pascual Ramos et al. (1997), Pascual et al. (1999), Hoyles et al. (2000, 2001a, b, c, 2002), Nikolaitchouk et al. (2000), Lawson et al. (2001), Sarkonen et al. (2001), and Hall et al. (2002).

myces and *Propionibacterium* species on the basis of microcolony morphology. However, after 4–10 h of incubation, the filamentous growth produced by *Propionibacterium propionicum* often appears to be due to apical growth at only one end of the original rod, which places the latter in the periphery of the young microcolony, giving rise to asymmetrical, twig-like structures, whereas the microcolonies of *A. israelii, A. gerencseriae* and other *Actinomyces* species predominantly develop with radial symmetry, with the original propagule being located in the center of the "spider" microcolony (Fig. 6). As growth proceeds, a network of interwoven, branching filaments forms whose central parts may gradually undergo various degrees of frag-

mentation so that at the end of this life cycle the centers of the mature colonies may consist predominantly or exclusively of bacillary elements (Fig. 7) that lead to more or less soft and smooth colonies. As degree of primary filament formation (Fig. 8) and rate of fragmentation and cell separation vary considerably both between and within species, any intergrade between highly filamentous colonies with a hard or crumbly texture and a rough appearance and completely bacillary colonies with a soft texture and a smooth surface may be seen.

The filamentous microcolonies of *A. naeslundii* and *A. viscosus* usually grow considerably faster than those of *A. israelii*, and they show early fragmentation of the central parts of the

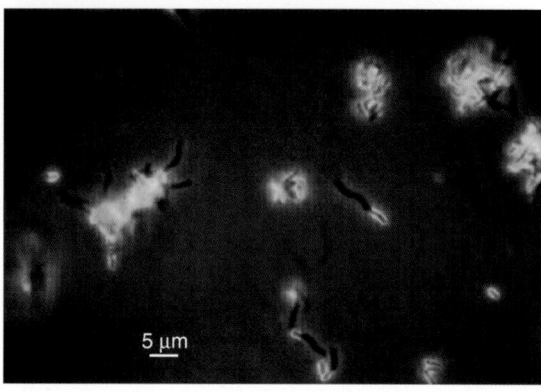

Fig. 4. *Actinomyces israelii*; simultaneous phase-contrast and incident-light fluorescence micrograph in situ after indirect FA staining of the inoculum (brain heart infusion agar slide culture, 8 h at 36 ± 1°C, and GasPak jar). Note the (dark) apical filamentous growth originating from one or both ends of the brightly FITC-stained original propagule.

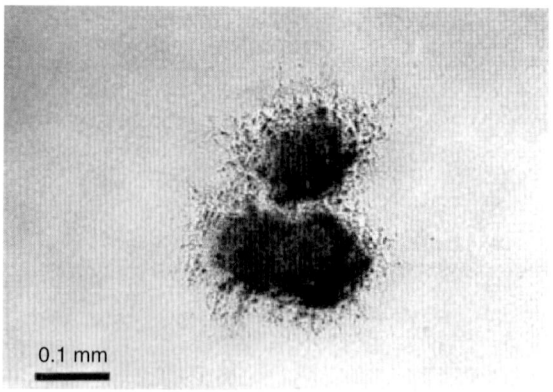

Fig. 7. *Actinomyces naeslundii*; small mature colonies with dense, granular centers and filamentous edges (CC medium, 7 days at 36 ± 1°C, and Fortner's method), transmitted-light in situ micrograph.

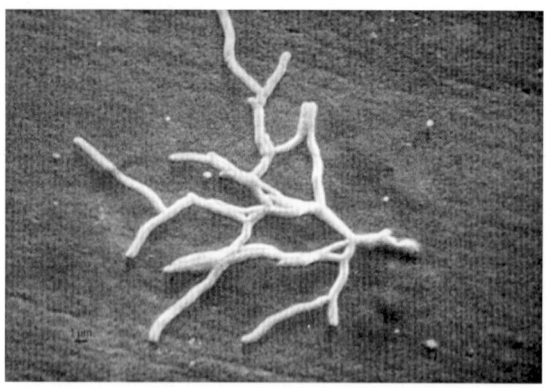

Fig. 5. *Actinomyces israelii*; scanning electron micrograph (SEM) of a microcolony in situ (40 h at 36 ± 1°C, and GasPak jar).

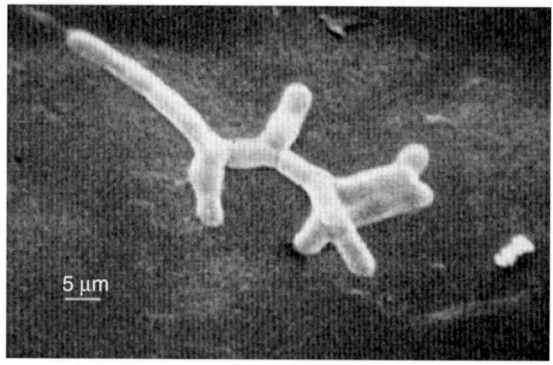

Fig. 8. *Actinomyces bovis*; scanning electron micrograph of a microcolony in situ (40 h at 36 ± 1°C, and GasPak jar).

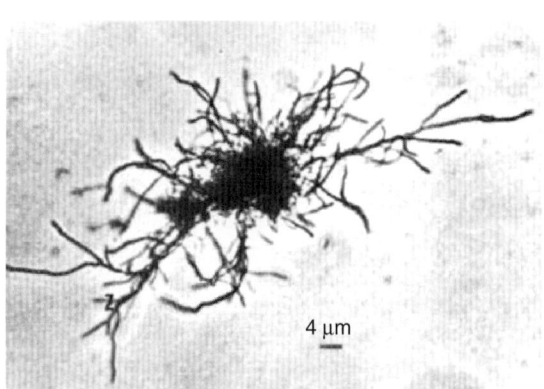

Fig. 6. *Actinomyces israelii*; microcolony (brain heart infusion agar slide culture, 48 h at 36 ± 1°C, and GasPak jar), phase-contrast micrograph in situ in lactophenol-cotton blue mounting fluid.

colony, with comparatively short and slightly thicker radially arranged peripheral hyphae. Furthermore, the filaments of these species develop predominantly on the agar surface, whereas those of *A. israelii* and *A. gerencseriae* tend to grow into the agar below its surface, which reduces their optical refractivity. Although *Actinomyces* species were described as not forming an aerial mycelium, an occasional strain of *A. israelii* or *A. gerencseriae* may produce short aerial hyphae that, because of their higher optical refractivity, are easily distinguished from the substrate filaments (Schaal, 1986a).

Actinomyces colonies mature between 2–5 and 7–17 days of incubation at 36 ± 1°C. Size, texture, and color of the colonies may vary considerably depending on the respective species affiliation as well as on the composition of the growth medium. Strains producing filamentous microcolonies predominantly form rough mature colonies (Table 5) that may retain much of their mycelial appearance, especially in the periphery. Most characteristic among these are the so-

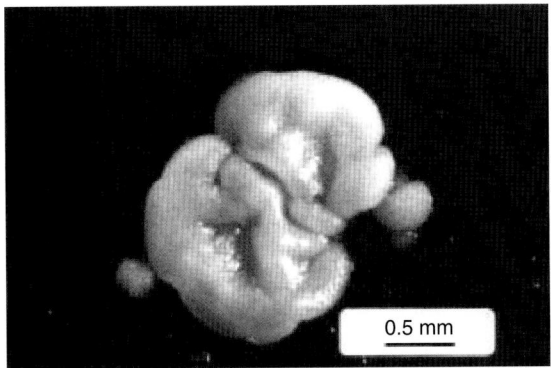

Fig. 9. *Actinomyces israelii;* mature colony ("molar tooth" type), micrograph in obliquely incident light (CC medium, 14 days at 36 ± 1°C, and Fortner's method).

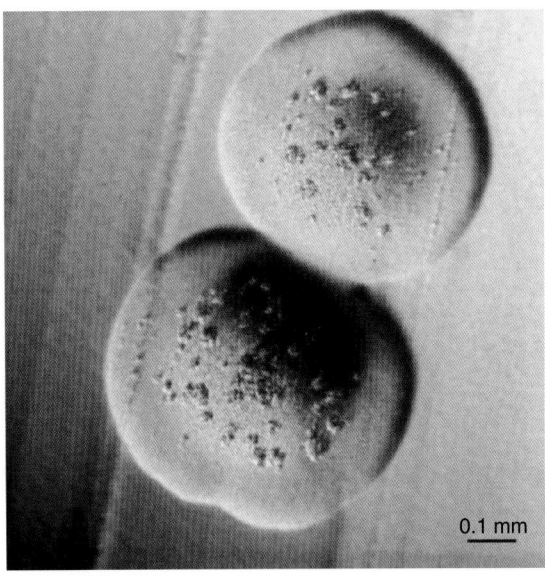

Fig. 10. *Actinomyces odontolyticus;* mature colony—transparent with some optically dense inclusions, slightly elevated, and centrally domed; micrograph in obliquely transmitted light (CC medium, 7 days at 36 ± 1°C, and Fortner's method).

called "molar tooth" colonies (Fig. 9), which may be seen in cultures of *A. israelii* and *A. gerencseriae* but also in those of *A. hordeovulneris.* The "bread-crumb" colony type is slightly smoother and only occasionally exhibits a filamentous fringe. The latter type may characteristically be found not only in *A. israelii, A. gerencseriae, A. hordeovulneris* and *A. oricola* cultures but also among rough variants of other taxa such as *A. bovis, A. odontolyticus, A. naeslundii, A. viscosus* and possibly also *A. catuli.* The two rough colony forms, especially the molar-tooth variant, may be very hard and either adhere firmly to the medium or come off *in toto* when an attempt is made to pick them.

Smooth mature colonies may occur in the species mentioned before but are more characteristic of the other *Actinomyces* species. They may be flat or convex with entire, circular or irregular edges and a soft texture. Their surface may be shiny or even mucoid but also granular or matte (Table 5). Usually, the actinomycete colonies are densely opaque although translucent forms (Fig. 10) are not uncommon in species that predominantly or exclusively produce flat smooth colonies (e.g., *A. meyeri* and *A. odontolyticus*).

Most of the *Actinomyces* colonies are white, gray-white, or creamy-white (Table 5). However, pigment production may be observed in a few species, especially when they are grown on blood agar. Thus, *A. cardiffensis* forms creamy to pinkish mature colonies. Those of *A. denticolens* may show a pink coloration. Dark brown growth may be seen in *A. graevenitzii* cultures. *Actinomyces odontolyticus* characteristically grows in brown to deep red colonies in particular after prolonged incubation. *Actinomyces radicidentis* and *A. urogenitalis* were reported to produce brownish or reddish colonies, respectively (Table 5).

In liquid media, *Actinomyces* species may grow as discrete cotton pad-like colonies or compact masses of variable size adhering to the

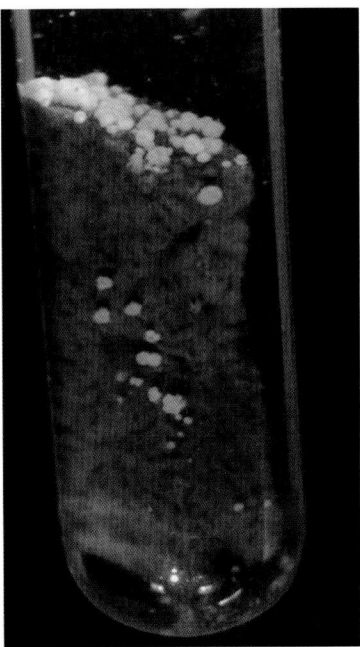

Fig. 11. *Actinomyces israelii;* 14 day-old culture in Tarozzi broth. Note the cotton pad- or snowball-like Actinomyces colonies on the piece of liver tissue.

inner glass surface of the tube or the tissue surface of Tarozzi broth (Fig. 11) or forming a granular or pellicular sediment so that the broth remains clear. Or they may produce diffuse turbidity with varying amounts of granular, flaky,

Table 6. Cell wall constituents and DNA G+C content of selected *Actinomyces* species and related bacteria.

Species	Orn	Lys	Asp	Gly	A₂pm[a]	Glu	Gal[a]	Rha[a]	Tal[a]	Ara[a]	Man	MA[a]	Mol% G+C[b]
A. bovis	–	+	+	–	–	(+)[c]	–	+	+	–	(+)	–	57–63
A. israelii	+	+	–	–	–	–	+	–	–	–	–	–	63–70
A. gerencseriae	+	+	–	–	–	–	+	–	–	–	–	–	70–71
A. odontolyticus	+	+	–	–	–	+	+	+/–[f]	+/–	–	+	–	61–62
A. naeslundii	+	+	–	–	–	+	–	+	+/–	–	(+)	–	63–67
A. viscosus	+	+	–	–	–	+/–	+/–	+	+	–	–	–	61–66
A. denticolens	+	+	–	–	–	–	–	+	–	–	–	–	66–68
A. slackii	+	+	–	–	–	+	+/–	+	–	–	–	–	64–67
A. howellii	+	+	–	–	–	+	–	+	–	–	–	–	66–67
A. neuii[e]	–	+	–	–	ND	ND	ND	ND	ND	ND	ND	ND	55
A. europaeus	–	+	–	–	ND	ND	ND	ND	ND	ND	ND	ND	62
A. hordeovulneris	+	+	–	–	–	+	+	–	–	–	ND	–	67–68
Arcanobacterium haemolyticum[e]	–	+	–	–	ND	ND	ND	ND	ND	ND	ND	ND	48–52
A. bernardiae	–	+	–	–	ND	ND	ND	ND	ND	ND	ND	ND	65
A. pyogenes	–	+	–	–	–	+	–	+	–	–	(+)	–	56–58
A. suis	–	+	–	–	–	–	–	+	–	–	+	–	55
P. propionicum	–	–	–	+	LL[f]	+	+	–	–	–	(+)	–	63–65
C. matruchotii	–	–	–	+	DL[g]	+/–	+	–	–	+	(+)	+	55–58

Symbols: –/+, predominantly negative; –, negative or absent; +, positive or present; (+), weakly positive; and +/–, predominantly positive or present.

Abbreviations: Orn, ornithine; Lys, lysine; Asp, aspartic acid, Gly, glycine; A₂pm, 2,6-diaminopimelic acid; Glu, glucose; Gal, galactose; Rha, rhamnose; Tal, 6-deoxytalose; Ara, arabinose; Man, mannose; MA, mycolic acids; ND, no data available.

[a]Whole-cell extracts may be used instead of cell wall preparations.

[b]Guanine plus cytosine content of the DNA.

[c]Present in smaller amounts.

[d]Different types.

[e]N. Weiss, unpublished.

[f]LL-A₂pm present.

[g]*meso*-A₂pm present.

From Slack and Gerencser (1975), Schaal (1986a, b, 1992), and Funke et al. (1995, 1997).

pellicular, or even slimy (predominantly *A. viscosus*) sediments.

CHEMOTAXONOMIC PROPERTIES Many of the 32 species that are currently thought to belong to the genus *Actinomyces* have been described inadequately and their inclusion in the genus was chiefly based upon data derived from 16S rRNA gene sequence analyses. However, comparative analysis of these 16S rRNA gene sequences revealed considerable phylogenetic diversity within the genus *Actinomyces*, with some species forming distinct subgroups while others appear to form individual lines of descent that may also represent the nuclei of additional genera (Pascual Ramos et al., 1997b). Nevertheless, apart from the phylogenetic evidence, little further information is available that could provide additional criteria for the delineation of such new genera. This is especially true for stable taxonomic characters such as chemotaxonomic properties.

Not surprisingly, therefore, the taxonomic affiliation of certain *Actinomyces* species should be considered tentative. Although the species currently classified in the genus apparently dis-

play a broad range of morphological, biochemical and physiological characteristics (Slack, 1974), the set of phenotypic tests used for delineating especially the recently described new taxa often lacks discriminatory power so that it is essentially impossible to distinguish them from each other and from classical species of the genus. Moreover, present members of the genus *Actinomyces* appear to be very heterogeneous chemotaxonomically, in particular with regard to their amino acid composition of the cell-wall peptidoglycan, DNA base ratio values, cell-wall sugars (Table 6), and menaquinone composition, as well as their polar lipid and cellular fatty acid composition.

The mode of cross-linkage and the amino acid composition of the tetrapeptide bridge of the peptidoglycan layer of members of the *Actinomycetaceae* are complex. Following the methods of Schleifer and Kandler (1972), modified by the use of cellulose sheets (Merck, Darmstadt, Germany) for ascending thin-layer chromatography (N. Weiss, unpublished observation), the crosslinking between the amino group of the diamino acid in position 3 of the peptide side chain and the carboxyl group of D-alanine in position 4 of the adjacent tetrapeptide varies between members of the Actinomycetaceae. The type abbreviated A5 (Schleifer and Seidl, 1985) is *Actinomyces europaeus*, and *Actinobaculum schaalii* and *Arcanobacterium pluranimalium* possess subtype A5α (L-Lys-L-Lys-D-Glu; Funke et al., 1997; Lawson et al., 1997, 2001). *Actinomyces radingae* also exhibits this subtype with L-lysine in position 3 of the tetrapeptide being partially replaced by L-ornithine (L-Lys[L-Orn]-L-Lys-D-Glu; Wüst et al., 1995). *Actinomyces turicensis* contains peptidoglycan subtype A5β with L-ornithine as the diamino acid in position 3 of the tetrapeptide subunit (L-Orn-L-Lys-D-Glu), which is identical to that described for *A. israelii* (Schleifer and Seidl, 1985; Wüst et al., 1995). In contrast, *A. bovis* was shown to belong to peptidoglycan type A4, which is characterized by L-lysine in position 3 of the tetrapeptide subunit and a D-β-asparagine residue forming the interpeptide bridge (L-Lys-D-Asp; Schleifer and Kandler, 1972).

Though the majority of species have not yet been analyzed with respect to peptidoglycan composition, the detailed structures do not correlate with the affiliation of members to individual genera as currently defined. For example, members of *Actinomyces* exhibit the types L-Orn-L-Lys-D-Glu (*A. georgiae, A. gerencseriae, A. hyovaginalis,* and *A. turicensis*), L-Lys(L-Orn)-L-Lys-D-Glu (*A. radingae*), or L-Lys-L-Lys-D-Glu (*A. europaeus* and *A. neuii*). *Actinobaculum suis* is defined by the L-Lys-L-Ala-L-Lys-D-Glu type, while members of *Arcanobacterium* exhibit either the latter type (*A. bernardiae* and *A. pyogenes*) or the L-Lys-L-Lys-D-Glu type (*A. haemolyticum* and *A. phocae*). Thus, the distribution of peptidoglycan types, matching by and large the phylogenetic clustering of members of other actinobacterial genera, cannot be used as a sole marker suitable for supporting a dissection of the genus *Actinomyces* on phylogenetic grounds.

Apart from *A. bovis, A. neuii* and *A. europaeus* (Table 6), all of the other species of the genus *Actinomyces* contain ornithine in the cell-wall peptidoglycan, and the reported base ratios of their DNA are 54–70 mol% G+C (Slack and Gerencser, 1975; Schaal, 1986a). The wall of *A. bovis* contains glucose, mannose, rhamnose and 6-deoxytalose. *Actinomyces israelii* contains only galactose, whereas *A. viscosus* contains galactose, rhamnose and 6-deoxytalose (Table 6 and 7). Fucose is found, in addition to other sugars, in the cell wall of *A. meyeri, A. turicensis* and *A. hyovaginalis*.

The principal respiratory quinones among the *Actinomyces* species are menaquinones with eight, nine, ten and eleven isoprene units. *Actinomyces canis, A. bovis, A. graevenitzii, A. hordeovulneris* and *A. nasicola* thus far appear to be the only *Actinomyces* species which contain fully unsaturated menaquinones, whereas the other species possess menaquinones with varying degrees of saturation.

Concerning the distribution of polar lipids among the *Actinomyces* species, there are generally two phospholipid types, namely phospholipid type I (with no diagnostic phospholipid) and phospholipid type III (with phosphatidylcholine as diagnostic phospholipid) *sensu* (Lechevalier et al., 1977; Table 7).

The principal cellular fatty acids in members of the genus are straight-chain saturated, straight-chain monounsaturated, and branched-chain fatty acids of the *iso* and *anteiso* type as well as fatty acids which contain a cyclopropane ring. These are variably distributed among the different species.

Comparative chemotaxonomic studies on members of the genus *Actinomyces* indicate that this genus is worthy of dissection into several

Table 7. Chemotaxonomic characteristics of selected members of the genus Actinomyces.

Organisms	Wall sugars	Acyl type	Phospholipid types	Menaquinones	Fatty acids
Actinomyces bovis	Glucose + mannose + rhamnose + 6-deoxytalose	Acetyl	PIII DPG, PC, PI, PIM	MK-8 + MK-9[a] + MK-10 + *methyl*-MK-10	S, U
Actinomyces israelii	Galactose	Acetyl	PIII DPG, PG, PC, PI PIM	MK-10(H$_4$)[a]	S, U
Actinomyces naeslundii	Galactose + mannose + rhamnose + 6-deoxytalose	Acetyl	PIII DPG, PG, PC, PI, PIM	MK-9(H$_4$) + MK-10(H$_4$)[a]	S, U

Abbreviations: DPG, diphosphatidylglycerol; PG, phosphatidylglycerol; PC, phosphatidylcholine; PI, phosphatidylinositol; PIM, phosphatidylinositol mannosides; MK-n(Hx) represents a partially hydrogenated menaquinone with x hydrogen atoms on the side chain containing n isoprene units; S, straight-chain saturated; and U, monounsaturated.
[a]Major component.

genera, which are characterized by different chemotaxonomic patterns. Details of three such patterns are shown in Table 7. One such pattern is represented by *A. bovis*, the type species of the genus, the second is found in *A. israelii*, and the third is characteristic of *A. viscosus*.

A detailed description of these three different chemotaxonomic patterns found in members of the genus *Actinomyces* follows.

Chemotaxonomic Pattern of A. bovis This pattern consists of glucose, mannose, rhamnose and 6-deoxytalose as diagnostic cell-wall sugars. The respiratory quinones are composed of fully unsaturated menaquinones with eight (MK-8), nine (MK-9), and ten (MK-10) isoprene units, with MK-9 as the major menaquinone. In addition to these menaquinones, *methyl*-MK-10 (which is characterized by two fragments in the mass spectrum—one occurs in the lower mass region at mz/239 and the second is a strong fragment in the high mass region at mz/866) is also detected. Polar lipids consist of phosphatidylcholine as the characteristic phospholipid, i.e., the organism possesses phospholipid type PIII *sensu* Lechevalier et al. (1977), in addition to diphosphatidylglycerol, phosphatidylinositol and phosphatidylinositol dimannosides. The fatty acid profile comprises straight-chain saturated and monounsaturated fatty acids. The muramic acid residue of the peptidoglycan is N-acetylated.

Chemotaxonomic Pattern of A. israelii In this pattern, galactose is the only cell-wall sugar. The respiratory quinones only consist of tetrahydrogenated menaquinones with ten isoprene units [MK-10(H$_4$)] as the major isoprenologue. Polar lipids comprise phosphatidylcholine as diagnostic phospholipids, i.e., phospholipid type III *sensu* (Lechevalier et al., 1977), in addition to diphosphatidylglycerol, phosphatidylglycerol, phosphatidylinositol and phosphatidylinositol dimannosides. The fatty acid profile consists of straight-chain saturated and monounsaturated fatty acids. However, the fatty acid profile of *Actinomyces gerencseriae* contains branched-chain fatty acids of the *iso* and *anteiso* type in addition. The muramic acid residue of the peptidoglycan is N-acetylated.

Chemotaxonomic Pattern of A. viscosus Diagnostic cell-wall sugars are galactose, mannose, rhamnose and 6-deoxytalose. The respiratory quinones are tetrahydrogenated menaquinones with nine [MK-9(H$_4$)] and ten [MK-10(H$_4$)] isoprene units, with MK-10(H$_4$) as the major isoprenologue. The polar lipids are diphosphatidylglycerol, phosphatidylcholine, phosphatidylinositol and phosphatidylinositol dimannosides, i.e., phospholipid type PIII *sensu* (Lechevalier

et al., 1977). The fatty acid profile consists of straight-chain saturated and monounsaturated as well as branched-chain fatty acids of the *iso* and *anteiso* type. The latter fatty acids were not detected in *A. naeslundii*. The muramic acid residue of the peptidoglycan is N-acetylated.

PHYSIOLOGICAL PROPERTIES Members of the genus *Actinomyces* display considerable diversity concerning their physiological characteristics. This is especially true for their optimum gaseous growth conditions in which they vary considerably from species to species and even from strain to strain within one species. Some species, especially many of the recently described ones, show good to moderate growth when incubated in air on the surface of suitable agar media. This applies in particular to *A. viscosus* (Howell, 1963a), *A. denticolens* (Dent and Williams, 1984a), *A. hordeovulneris* (Buchanan et al., 1984b), *A. slackii* (Dent and Williams, 1986), *A. hyovaginalis* (Collins et al., 1993), *A. neuii* (Funke et al., 1994), *A. radingae* and *A. turicensis* (Wüst et al., 1995), *A. europaeus* (Funke et al., 1997a), *A. graevenitzii* (Pascual Ramos et al., 1997a), *A. bowdenii* (Pascual et al., 1999), *A. canis* (Hoyles et al., 2000), *A. urogenitalis* (Nikolaitchouk et al., 2000), *A. radicidentis* (Collins et al., 2000b), *A. funkei* (Lawson et al., 2001), *A. marimammalium* (Hoyles et al., 2001c), *A. catuli* (Hoyles et al., 2001b), *A. suimastitidis* (Hoyles et al., 2001a), *A. coleocanis* (Hoyles et al., 2002), *A. nasicola* (Hall et al., 2003b), and many strains of *A. naeslundii* and *A. odontolyticus*. However, aerobic growth is usually greatly enhanced when the incubation atmosphere contains at least 5% carbon dioxide or when HCO$_3^-$ is added to the medium. In liquid media, aerobic growth is usually better, provided that a large inoculum is used.

In contrast, most strains of *A. israelii*, *A. gerencseriae*, *A. bovis*, *A. meyeri*, *A. hongkongensis*, *A. oricola* and possibly also *A. georgiae* (Schaal, 1986a; Johnson et al., 1990; Hall et al., 2003d; Woo et al., 2003) prefer anaerobic growth conditions (with added CO$_2$ or HCO$_3^-$), although a marked strain variation in aerotolerance has been observed. Anaerobically without CO$_2$, most strains will not grow at all or will produce only very scanty growth. The same is true for many aerobically grown *Actinomyces* strains when the gaseous atmosphere does not contain additional CO$_2$.

Because all of the *Actinomyces* species recognized so far are commensals or pathogens of warm-blooded vertebrates, their optimum growth temperature is 30–37°C (Schaal, 1986a). Detailed studies on the temperature requirements of *Actinomyces* species showed that growth may occur well below or above the 30–

37°C range. Most strains of *A. israelii*, *A. naeslundii*, and *A. viscosus* were found to also produce visible growth in the range of 28–32°C (Thompson and Lovestedt, 1951; Howell, 1963a). The hamster strains of *A. viscosus* (serovar 1) grew well at 23°C when the organisms were incubated in air with added CO_2 or in an atmosphere of CO_2 and N_2 on brain heart infusion agar (BHIA; Howell, 1963). This species and *A. naeslundii* may also grow at 45°C (Holmberg and Nord, 1975a). *Actinomyces meyeri* strains were reported to multiply equally well at 30°C and 37°C and nearly as well at 25°C. No growth was obtained at 45°C (Cato et al., 1984).

The minimal nutritional requirements of *Actinomyces* species are not well understood. However, good growth has only been obtained in or on complex media containing either rich biological substrates such as brain heart infusion, meat extract, yeast extract, or serum or a defined and complex mixture of a large variety of organic and inorganic compounds (Schaal, 1986a). Studies using synthetic media were reported (Howell and Pine, 1956; Christie and Porteus, 1962b; Christie and Porteus, 1962c; Christie and Porteus, 1962a; Keir and Porteus, 1962) but were understandably mostly restricted to a few strains. However, modifications or simplifications of the synthetic medium of Howell and Pine (1956) have been developed (Pine and Watson, 1959b; Georg et al., 1964; Heinrich and Korth, 1967) that were tested with a by far greater number of strains. Attempts to further simplify these media have so far always resulted in insufficient growth rates or loss of cultures during subculturing. Thus, the anabolic capacity of *Actinomyces* species is limited (Slack and Gerencser, 1975). These organisms apparently require organic nitrogen (e.g., peptides and amino acids), a fermentable carbohydrate and possibly also vitamins and other growth factors for optimum growth. For instance, *A. meyeri* was found to have an absolute requirement for vitamin K_1, and its multiplication is greatly stimulated by 0.02% Tween 80 and by a fermentable carbohydrate (Cato et al., 1984). Similarly, growth of *A. hordeovulneris* is considerably enhanced when the medium is supplemented with 10–20% (v/v) fetal bovine serum (Buchanan et al., 1984b). On the other hand, purines and pyrimidines were found to be either stimulatory or inhibitory to growth of certain *Actinomyces* strains (Christie and Porteus, 1962a; Keir and Porteus, 1962).

Fermentable carbohydrates are the preferred sources of carbon and energy and, when utilized, are fermented with production of acids but no gas (Lentze, 1938a; Howell et al., 1959; Slack, 1974; Slack and Gerencser, 1975; Dent and Williams, 1984a). The spectrum of carbohydrates and similar carbon sources that may be metabolized fermentatively by *Actinomyces* species is broad (Tables 8 and 9), with considerable species differences so that carbohydrate fermentation patterns form the traditional basis for phenotypic differentiation of these actinomycetes and related organisms. Glucose is the only sugar that is universally utilized by nearly all of the species currently included in the genus *Actinomyces* (Tables 8 and 9). Only *A. hongkongensis* appears to represent an exception to this rule (Woo et al., 2003).

Anaerobically with added CO_2 or HCO_3^-, the major acidic end products of the fermentative carbohydrate metabolism of *Actinomyces* species are succinic and lactic acids with smaller amounts of acetic and formic acids (Schaal, 1986a). The amount of succinic acid formed depends on the concentration of CO_2 and HCO_3^- available in the gaseous environment or medium, respectively. Anaerobically without added CO_2, the fermentation of "*A. israelii*" ATCC 10049, which is now considered to belong to the species *A. gerencseriae*, was found to be homolactic (Buchanan and Pine, 1965). In the presence of CO_2, the fermentation is heterolactic with formate, acetate, lactate and succinate as end products. In air with added CO_2, acetate and carbon dioxide are the main end products. Carbon dioxide when present is utilized (Howell and Pine, 1956).

In a detailed study on the carbohydrate metabolism of *A. viscosus* in continuous culture (Hamilton and Ellwood, 1983) under glucose or nitrogen limitation, it was shown that the molar growth yields were lower under nitrogen limitation than under glucose limitation, indicating that the supply of utilizable nitrogen sources may be as important for the growth of this organism in the oral cavity as the presence of suitable carbohydrates.

Energy is produced by substrate phosphorylation, but the amount of ATP formed depends on the cultural conditions. According to the results of Buchanan and Pine (1965), four mols of ATP were formed per mol of glucose when "*A. israelii*" (= *A. gerencseriae*) was grown in air with added CO_2. Only two mols of ATP were produced anaerobically without CO_2.

A few actinomycete enzymes involved in the dissimilation of exogenous carbohydrates or the energy yielding glycolytic pathways were characterized in some detail: The β-galactosidase of *A. viscosus* has a molecular weight of 4.2×10^5, a K_m for lactose of about 6 mM and a pH optimum of 6.0–6.5 (Kiel et al., 1977b). Invertase activity has been demonstrated both in *A. naeslundii* (Miller, 1974) and *A. viscosus* (Palenik and Miller, 1975; Kiel et al., 1977b). The characteristics of the *A.*

Table 8. Physiological properties of *Actinomyces* species from human sources.[a]

Property	1.	2.	3.	4.	5.	6.	7.	8.	9.	10.	11.	12.	13.	14.	15.	16.	17.	18.	19.	20.	21.	22.	23.	24.
Acid production from																								
N-Acetyl-β-glucosamine	-	nd	nd	nd	nd	+	nd	nd	nd	nd	nd	d	d	nd	nd	nd	nd	nd	nd	nd	nd	nd	nd	nd
Adonitol	nd	d	nd	d	nd	nd	nd	nd	d	-/+	nd	d	+	d	(+)	nd	nd	nd	nd	-	nd	nd	nd	nd
Amygdalin	-	-	nd	-	+/-	-	nd	d	d	-	-	nd	nd	nd	nd	nd	-	-	nd	-	-	-	-	-
D-Arabinose	nd	nd	nd	-	nd	-	nd	d	d	d	nd	-	-	nd	-	-	d	d	d	-	nd	nd	-	nd
L-Arabinose	-	-	-	-/+	-	-	nd	+/-	d	-	-	d	-	d	-	-	d	d	+	-	nd	-	-	-
D-Arabitol	nd	nd	nd	nd	nd	-	nd	nd	nd	nd	nd	+	+	nd	-	-	nd	nd	nd	nd	nd	d	nd	+
L-Arabitol	nd	nd	nd	nd	nd	-	nd	nd	nd	nd	-	d	d	nd	-	-	nd	nd	-	nd	nd	nd	nd	nd
Arbutin	-	-	-	-	+/-	-	nd	+	-	nd	-	d	d	d	-	-	d	d	nd	d	d	d	d	-
Cyclodextrin	-	nd	nd	nd	nd	nd	nd	+	+	+	+	+	nd	-	+	nd	nd	nd	nd	-	nd	nd	nd	nd
Cellobiose	-	nd	nd	-	+	-	nd	-	-	-	+	nd	nd	-	+	-	-	-	-	d	nd	d	d	nd
Dextrin	nd	nd	nd	nd	+	nd	nd	nd	+	nd	nd	nd	nd	nd	nd	nd	nd	nd	nd	d	nd	nd	nd	nd
Dulcitol	nd	nd	nd	nd	-	-	nd	+	-	nd	nd	nd	nd	nd	nd	nd	nd	nd	nd	-	nd	nd	nd	nd
iso-Erythritol	nd	nd	nd	+/-	+	+	nd	+	+	+	+	+	+	+	+	+	+	d	+	+	+	+	+	+
D-Fructose	nd	+	nd	nd	nd	+	nd	+	+	+	+	+	+	+	+	+	d	d	+	d	+	+	+	+
D-Fucose	nd	nd	nd	nd	nd	-	nd	nd	nd	nd	nd	d	d	d	nd	nd	d	nd	nd	nd	nd	nd	nd	nd
L-Fucose	nd	nd	nd	nd	nd	+	nd	nd	nd	+	nd	+	+	-	nd	nd	d	d	nd	d	+	+	+	nd
D-Galactose	nd	+	nd	nd	+	-	nd	+	+	+	nd	+	+	d	nd	nd	d	d	+	+	+	+	+	+
β-Gentiobiose	nd	-	nd	+	nd	+	-	-/+	d	nd	nd	d	d	d	nd	nd	d	nd	nd	d	nd	nd	nd	nd
D-Glucose	+	+	+	+	+	+	+	+	+	+	+	+	+	+	+	+	+	+	+	+	+	+	+	+
Glycerol	nd	nd	nd	d	nd	+	nd	d	d	d	d	d	d	+	d	+	+	+	-	d	-	+	+/-	+
Glycogen	-	d	-	-	-/+	+	nd	d	-	-/+	-	d	d	d	-	-	+	-	-	d	+/-	d	d	-
myo-Inositol	nd	nd	nd	-/+	+/-	+	nd	+/-	+	+	nd	+	+	-	-	-	d	d	nd	d	d	+/-	+	+
Inulin	nd	-	nd	nd	nd	-	nd	nd	-	d	nd	-	-	-	nd	-	-	d	nd	d	nd	nd	+	nd
2-Ketogluconate	nd	-	nd	nd	nd	+	nd	nd	d	nd	nd	+	+	nd	nd	nd	d	d	nd	d	nd	d	nd	nd
5-Ketogluconate	nd	nd	d	nd	nd	+	nd	+	nd	nd	-	+	+	nd	nd	+	+	+	+	+	+	+	+	+
Lactose	-	-	d	+/-	+	+	nd	+	d	d	-	+	+	d	-	+	d	d	+	d	d	d	d	(+)
D-Lyxose	nd	nd	d	nd	nd	+	nd	+/-	nd	+	-	+	+	nd	-	+	+	d	+	+	d	+/-	d	nd
Maltose	d	+	d	+	+	+	nd	+	d	+	-	+	+	d	d	+	+	+	+	+	+	+	+/-	+
D-Mannitol	-	-	-	-/+	+/-	-	nd	d	-	-	-	d	-	-	d	+	+	d	d	-	-	-/+	d	-
D-Mannose	d	d	nd	-/+	+/-	+	nd	+	d	+	-	+	+	-	nd	+	d	d	+	d	-/+	+/-	d	+
Melezitose	-	d	-	-/+	+/-	-	-	nd	d	-	-	-	-	nd	nd	nd	d	d	d	d	-/+	d	d	-
Melibiose	-	d	-	-	+/-	-	nd	nd	-	+	-	d	-	-	+	+	+	+	+	-	+	+	+	+
α-Methyl-D-glucoside	nd	d	nd	nd	nd	-	nd	+	-	-	nd	+	+	nd	nd	nd	+	nd	nd	d	nd	d	nd	nd
α-Methyl-D-mannoside	nd	d	nd	nd	nd	-	nd	-	d	d	nd	d	-	nd	-	-	+	nd	d	-	nd	d	nd	nd
Methyl-β-D-glucopyranoside	-	nd	-	nd	nd	nd	-	nd	nd	nd	-	+	nd	nd	-	-	nd	nd	nd	nd	nd	nd	nd	nd
Pullulan	-	nd	-	nd	nd	+	nd	+	+	+	nd	+	+	-	+	+	+	+	+	+	+	+	+	+
D-Raffinose	d	-	-	-/+	+/-	-	nd	-	-	-	-	-	-	d	d	-	d	nd	d	d	+/-	-/+	-	-
L-Rhamnose	nd	-	nd	+	-/+	-	nd	-	d	-	nd	-	-	d	nd	nd	-	-	-	d	+/-	-/+	+	-
D-Ribose	d	d	d	+	+/-	+	nd	+/-	+	d	-	+	+	d	-	+	+	+	d	d	d	d	-/+	-

Characteristic	1	2	3	4	5	6	7	8	9	10	11	12	13	14	15	16	17	18	19
Salicin	−	d	d	d	d	nd	d	d	nd	+	d	d	d	−	−	−	+/−	−	nd
D-Sorbitol	−	d	d	−/+	−	−	−	−	−	−	−	−	−	−	−	−	−/+	+	−
L-Sorbose	nd	nd	nd	nd	nd	nd	nd	nd	nd	nd	nd	nd	nd	nd	nd	nd	nd	nd	nd
Starch	−	d	+	+	d	+	+	+	+	d	nd	nd	+	+	d	+	+	+	+
Sucrose	+	+/−	+	+	+	+	+	+	d	−	+	+	−	+	+	+	+	−	+
D-Tagatose	nd	d	nd	nd	+	nd	nd	nd	+	d	nd	d	d	−	nd	nd	nd	−	nd
Trehalose	−	d	d	d	d	+	d	d	+	+	+	+	+	+	+	+	+	−	+
D-Turanose	nd	nd	nd	nd	nd	nd	nd	nd	nd	nd	−	d	−	−	nd	nd	nd	−	nd
Xylitol	nd	nd	−/+	−/+	nd	+	+	+	−	nd	nd	nd	−	nd	nd	nd	nd	nd	nd
D-Xylose	−	−/+	−/+	−/+	−	+	+	+	d	nd	−/+	+	+	+/−	+/−	−/+	−/+	+	+
L-Xylose	nd	nd	nd	nd	nd	nd	nd	nd	nd	nd	nd	nd	nd	nd	nd	nd	nd	nd	nd
Hydrolysis of																			
Casein	nd	+	+	nd	d	−	nd	nd	−	−	−	−	+	−	−	−	−	+	−
Esculin	+	+	+	d	d	d	d	+	+	d	d	+	−	−	−	+	+	−	+
Gelatin	−	−	−	−	−	−	−	−	−	−	−	−	−	nd	−	−	−	+	−
Hippurate	−	−	−	−	−	−	−	−	−	−	−	−	−	−	−	nd	+	−	+
Starch	+	−	−	d	−	d	d	−	−	−	−	−/+	−	−	−	−	−	−	−
Enzyme activities																			
N-Acetyl-β-glucosaminidase	−	nd	+	+	−	nd	+	nd	nd	−	+	−	nd	nd	nd	nd	−	+	−
Acid phosphatase	nd	+	+	+	+	+	+	+	+	+	+	+	+	−	+	+	+	+	−
Alanine arylamidase	+	nd	nd	nd	nd	nd	nd	nd	nd	−	+	+	+	nd	nd	nd	+	+	−
Alanine phenylalanine proline arylamidase	+	nd	nd	nd	nd	+	+	+	+	+	+	+	+	nd	nd	nd	+	+	−
Alkaline phosphatase	−	nd	−	−	−	−	d	−	−	−	−	−	−	nd	nd	nd	+	−	−
α-Arabinosidase	nd	−	nd	nd	nd	−	nd	nd	d	nd	nd	nd	nd	nd	nd	nd	−	−	−
Arginine arylamidase	+	nd	+	+	+	+	+	+	−	+	nd	+	+	nd	nd	nd	+	+	−
Arginine dihydrolase	−	nd	−	−	−	−	−	−	−	−	−	−	−	nd	nd	nd	−	−	−
Benzoyl-arginine arylamidase	nd	nd	nd	nd	nd	d	nd	nd	d	nd	+	nd	d	nd	nd	nd	d	d	−
Chymotrypsin	nd	−	nd	nd	nd	nd	nd	nd	nd	−	nd	−	nd	nd	nd	nd	d	d	d
Cysteine arylamidase	nd	−	nd	nd	nd	nd	nd	nd	nd	−	−	−	−	−	−	nd	d	d	d
DNAase	nd	nd	nd	nd	nd	+	+	+	nd	nd	nd	−	+	nd	nd	nd	d	d	d
Esterase C4	nd	+	+	+	+	+/−	+/−	+	+	nd	nd	−	−	−	nd	nd	−	−	−
Ester lipase C8	nd	+	+	+	+	+/−	+	+	+	nd	+	−	−	nd	nd	nd	d	−	−
α-Fucosidase	−	nd	−	−	−	nd	nd	nd	−	−	−	−	−	−	nd	nd	d	−	d
β-Fucosidase	nd	−	d	−/+	d	+	+	+	d	−	−	−	−	−	nd	nd	+	−	d
α-Galactosidase	+	nd	+	+	+	+	+	+	+	+	+	+	+	+	nd	nd	+	d	d
β-Galactosidase	+	−	+	+	+	+	+	+	d	d	+	+	+	−	nd	nd	+	+	d
α-Glucosidase	+	nd	+	+	+	+	+	+	+	+	+	+	+	−	nd	nd	+	+	d
β-Glucosidase	−	+	−	−	−	+	+	+	+	+	−	−	−	−	nd	nd	−	+	d
β-Glucuronidase	−	−	−	−	−	−	−	−	−	−	−	−	−	−	nd	nd	−	−	−

Table 8. *Continued*

Property	1.	2.	3.	4.	5.	6.	7.	8.	9.	10.	11.	12.	13.	14.	15.	16.	17.	18.	19.	20.	21.	22.	23.	24.
Glutamic acid decarboxylase	nd	nd	nd	nd	nd	nd	−	nd	nd	nd	−	nd	nd	nd	−	nd	nd	nd	nd	nd	nd	nd	nd	nd
Glutamyl glutamic acid arylamidase	nd	nd	nd	nd	nd	nd	−	nd	nd	nd	d	nd	nd	nd	−	nd	nd	nd	nd	nd	nd	nd	nd	nd
Glycine arylamidase	+	nd	nd	nd	nd	nd	+	nd	nd	nd	+	nd	nd	nd	+	nd	nd	nd	nd	nd	nd	nd	nd	nd
Glycyl tryptophan arylamidase	nd	nd	nd	nd	nd	nd	+	nd	nd	nd	−	nd	nd	nd	−	nd	nd	nd	nd	nd	nd	nd	nd	nd
histidine arylamidase	+	nd	nd	nd	nd	nd	+	nd	nd	nd	+	nd	nd	nd	−	nd	nd	nd	nd	nd	nd	nd	nd	nd
β-Lactosidase	nd	nd	nd	nd	nd	nd	−	nd	−	nd	+	nd	nd	nd	−	nd	nd	−	−	−	nd	nd	nd	nd
Lecithinase	nd	nd	nd	nd	nd	nd	nd	nd	nd	nd	−	nd	nd	nd	−	nd	nd	nd	nd	nd	nd	nd	nd	nd
Lipase C14	nd	−	−	nd	−	−	nd	−	−	nd	−	nd	nd	−	−	nd	−	−	−	−	nd	nd	nd	nd
Leucine arylamidase	+	+	nd	nd	nd	+	+	+	+	+	+	+	+	d	+	+	+	+/−	+	+	nd	nd	nd	nd
Leucyl glycine arylamidase	+	nd	nd	nd	nd	nd	+	nd	nd	+	+	+	nd	nd	+	nd	+	nd	nd	+	nd	nd	nd	nd
Lysine arylamidase	nd	nd	nd	nd	−	nd	+	nd	nd	nd	nd	nd	nd	nd	nd	nd	nd	nd	nd	+	nd	nd	nd	nd
α-Mannosidase	nd	−	−	nd	−	−	−	−	−	−	−	+	+	−	nd	−	−	−	d	−	nd	nd	nd	nd
β-Mannosidase	−	nd	−	nd	nd	nd	nd	nd	nd	nd	−	nd	nd	nd	−	−	nd	nd	−	nd	nd	nd	nd	nd
Proline arylamidase	+	nd	nd	nd	nd	nd	+	nd	nd	+	+	nd	nd	nd	+	nd	nd	nd	nd	nd	nd	nd	nd	nd
Phenylalanine arylamidase	+	nd	nd	nd	nd	nd	+	nd	nd	+	+	nd	nd	nd	+	nd	nd	nd	+	nd	nd	nd	nd	nd
Phosphoamidase	nd	nd	nd	nd	nd	nd	nd	nd	nd	nd	nd	+	nd	nd	+	+	+	+/−	nd	+	nd	nd	nd	nd
Pyrazinamidase	−	−	d	nd	nd	nd	nd	nd	nd	nd	nd	+	+	nd	+	+	+	+/−	−	−	nd	nd	nd	nd
Pyroglutamic acid arylamidase	nd	nd	nd	nd	nd	nd	−	nd	nd	nd	d	nd	nd	nd	nd	nd	nd	nd	nd	nd	nd	nd	nd	nd
Pyrrolidonyl arylamidase	−	−	−	nd	nd	−	nd	nd	nd	nd	−	nd	nd	nd	nd	nd	nd	+	+	nd	nd	nd	nd	nd
Serine arylamidase	+	nd	nd	nd	nd	nd	+	nd	nd	+	+	nd	nd	nd	−	nd	nd	−	nd	nd	nd	nd	nd	nd
Trypsin	nd	−	−	nd	−	−	nd	−	−	−	−	−	−	nd	nd	nd	−	−	−	−	nd	nd	nd	nd
Tyrosine arylamidase	+	nd	nd	nd	nd	nd	+	nd	nd	nd	+	nd	nd	nd	+	nd	nd	nd	+	+	nd	nd	nd	nd
Urease	−	−	−	−	−	−	−	−	d	d	+	−	−	−	−	d	−	−	−	d	−	d	−	−
Valine arylamidase	nd	+	nd	nd	+	nd	+	+	+	d	+	+	+	+	nd	−	−	−	+	+	+	d	+/−	+
β-Xylosidase	nd	nd	nd	nd	+	nd	−	−	−	−	−	nd	nd	−	nd	−	−	nd	d	−	nd	nd	nd	nd
Catalase production	−	−	−	−	−	−	−	−	−	−	−	+	+	+	−	+	−	−	−	+	−	d	−	+
Nitrate reduction	d	−	d	−/+	−/+	−	−	+/−	+	d	−	+	−	+	+	d	−	−	+	d	+	d	+/−	−
Nitrite reduction	nd	nd	nd	nd	nd	nd	nd	nd	d	d	nd	nd	nd	−	nd	nd	nd	−	nd	−	nd	nd	nd	nd
Acetoin production	−	nd	−	nd	−	nd	nd	−	−	−	−	nd	nd	−	−	(+)	nd	nd	d	−	nd	nd	nd	nd
Indole production	−	nd	nd	nd	nd	nd	nd	−	−	−	−	nd	nd	−	−	nd	nd	nd	nd	nd	nd	nd	nd	nd

Symbols: +, positive/present; −, negative/absent; d, positive or negative, possibly depending on the system used; +/−, usually positive; −/+, usually negative; (), weak reaction; and nd, no data available.

[a]1. *A. cardiffensis*; 2. *A. europaeus*; 3. *A. funkei*; 4. *A. georgiae*; 5. *A. gerencseriae*; 6. *A. graevenitzii*; 7. *A. hongkongensis*; 8. *A. israelii*; 9. *A. meyeri*; 10. *A. naeslundii* (classical definition); 11. *A. nasicola*; 12. *A. neuii* subsp. *neuii*; 13. *A. neuii* subsp. *anitratus*; 14. *A. odontolyticus*; 15. *A. oricola*; 16. *A. radicidentis*; 17. *A. radingae*; 18. *A. turicensis*; 19. *A. urogenitalis*; 20. *A. viscosus* (classical definition); 21. *A. naeslundii* genospecies 1; 22. *A. naeslundii*, genospecies 2; 23. *Actinomyces* serotype WVA 963; 24. *A. viscosus*, serovar 1 (reaction of 1 hamster strain). Data from Schofield and Schaal (1981), Cato et al. (1984), Schaal (1986), Johnson et al. (1990), Funke et al. (1994, 1997), Wüst et al. (1995), Pascual Ramos et al. (1997a), Collins et al. (2000), Nikolaitchouk et al. (2000), Lawson et al. (2001b), Sarkonen et al. (2001), Hall et al. (2002, 2003a, b), and Woo et al. (2003).

Table 9. Physiological properties of *Actinomyces* species from animal sources.[a]

Property	1.	2.	3.	4.	5.	6.	7.	8.	9.	10.	11.	12.	13.	14.
Acid production from														
N-Acetyl-β-glucosamine	nd	nd	nd	nd	–	nd	nd	nd	+	+	nd	nd	nd	nd
Adonitol	–	nd	nd	nd	nd	nd	nd	nd	+	nd	nd	nd	nd	nd
Amygdalin	–	nd	nd	nd	nd	–	nd	–	nd	nd	–	nd	nd	–
D-Arabinose	nd	nd	nd	nd	nd	nd	nd	nd	+	nd	nd	nd	nd	nd
L-Arabinose	–	–	+	d	–	–	–	d	+	–	–	(+)	+	–
D-Arabitol	nd	–	–	–	–	nd	nd	nd	–	–	nd	–	–	nd
L-Arabitol	nd	nd	nd	nd	nd	nd	nd	nd	+	nd	nd	nd	nd	nd
Cyclodextrin	nd	–	–	–	–	nd	nd	nd	nd	–	nd	–	–	nd
Cellobiose	–	nd	nd	nd	nd	–	+	–	d	nd	–	nd	nd	–
Dextrin	d	nd	nd	nd	nd	nd	nd	nd	nd	nd	nd	nd	nd	nd
Dulcitol	–	nd	nd	nd	nd	nd	nd	–	nd	nd	nd	nd	nd	nd
iso-Erythritol	–	nd	nd	nd	nd	nd	nd	nd	nd	nd	nd	nd	nd	nd
D-Fructose	+	nd	nd	nd	nd	nd	nd	nd	+	nd	nd	nd	nd	+
D-Galactose	+	nd	nd	nd	nd	nd	nd	nd	+	nd	nd	nd	nd	nd
β-Gentiobiose	nd	nd	nd	nd	nd	nd	nd	nd	–	nd	nd	nd	nd	nd
D-Glucose	+	+	+	+	+	+	+	+	+	+	+	+	+	+
Glycerol	–	nd	nd	nd	nd	–	–	–	nd	nd	–	nd	nd	nd
Glycogen	d	nd	+	–	+	nd	nd	nd	nd	d	nd	nd	d	–
myo-Inositol	d	nd	nd	nd	nd	d	–	–	d	nd	–/+	nd	nd	–
Inulin	–	nd	nd	nd	nd	d	nd	–	–	nd	–/+	nd	nd	nd
Lactose	d	+	+	+	+	+	+	d	d	+	d	–	d	w
D-Lyxose	nd	nd	nd	nd	nd	nd	nd	nd	–	nd	nd	nd	nd	nd
Maltose	+	+	+	+	+	+	+	+	+	+	nd	d	–	+
D-Mannitol	–	–	–	–	–	d	–	–	nd	–	–	–	d	–
D-Mannose	d	nd	nd	nd	nd	d	(+)	d	+	nd	d	nd	–	+
Melezitose	–	+	–	–	–	–	nd	nd	–	–	–	–	–	–
Melibiose	d	+	–	d	–	nd	(+)	d	–	–	nd	+	–	+
α-Methyl-D-glucoside	–	nd	nd	nd	nd	nd	nd	nd	–	nd	nd	nd	nd	nd
α-Methyl-D-mannoside	–	nd	nd	nd	nd	nd	nd	nd	nd	nd	nd	nd	nd	nd
Methyl-β-D-glucopyranoside	nd	+	–	d	–	nd	nd	nd	nd	–	nd	–	–	nd
Pullulan	nd	–	+	–	+	nd	nd	nd	nd	–	nd	+	–	nd
D-Raffinose	–	+	d	+	–	+	(+)	+	–	–	+	+	–	+
L-Rhamnose	–	nd	nd	nd	nd	–	–	–	–	nd	–	nd	nd	–
D-Ribose	d	+	+	+	–	d	–	–	nd	–	–/+	(+)	+	–
Salicin	–	nd	nd	nd	nd	+	nd	–	+	nd	+/(+)	nd	nd	–
D-Sorbitol	–	–	–	–	–	–	nd	nd	–	–	–	–	–	–
L-Sorbose	–	nd	nd	nd	nd	nd	nd	nd	nd	nd	nd	nd	nd	nd
Starch	+/–	nd	nd	nd	nd	nd	nd	nd	nd	nd	nd	nd	nd	–
Sucrose	+	+	d	+	–	+	nd	+	+	–	d	+	d	+
D-Tagatose	nd	d	–	–	–	nd	nd	nd	–	–	nd	–	–	nd
Trehalose	–	+	–	+	–	–	+	d	–	–	d	–	+	–
D-Turanose	nd	nd	nd	nd	nd	nd	nd	nd	+	nd	nd	nd	nd	nd
Xylitol	nd	nd	nd	nd	nd	nd	nd	nd	–	nd	nd	nd	nd	nd
D-Xylose	–	–	+	+	–	–	+	d	+	–	–	+	+	–
L-Xylose	nd	nd	nd	nd	nd	nd	nd	nd	–	nd	nd	nd	nd	nd
Hydrolysis of														
Casein	–	nd	nd	nd	nd	nd	nd	nd	nd	nd	nd	nd	nd	nd
Esculin	d	+	–	+	–	+	+	nd	+	–	–	+	+	+
Gelatin	–	–	–	–	–	nd	nd	nd	–	–	nd	–	–	–
Hippurate	–	–	–	–	–	nd	nd	nd	+	–	nd	–	–	nd
Starch	+/–	nd	nd	nd	nd	nd	nd	nd	nd	nd	nd	nd	–	–
Enzyme activities														
N-Acetyl-β-glucosaminidase	+	–	+	d	nd	–	nd	nd	nd	nd	nd	d	–	nd
Acid phosphatase	–	+	–	+	nd	d	nd	nd	nd	nd	nd	nd	(+)	nd
Alanine arylamidase	nd	nd	nd	nd	nd	nd	nd	nd	nd	nd	nd	nd	+	nd
Alanine phenylalanine prolinearylamidase	nd	+	+	–	+	nd	nd	nd	nd	+	nd	+	+	nd
Alkaline phosphatase	–	d	–	–	–	–	nd	nd	+	d	nd	d	d	nd
α-Arabinosidase	nd	nd	nd	nd	nd	nd	nd	nd	nd	nd	nd	nd	–	nd
Arginine arylamidase	nd	nd	nd	nd	nd	nd	nd	nd	nd	nd	nd	nd	+	nd
Arginine dihydrolase	–	–	–	d	–	nd	nd	nd	–	–	nd	–	–	nd
Chymotrypsin	–	–	–	–	nd	–	nd	nd	nd	nd	nd	nd	–	nd
Cystine arylamidase	–	–	+	–	nd	–	nd	nd	nd	nd	nd	nd	–	nd

Table 9. *Continued*

Property	1.	2.	3.	4.	5.	6.	7.	8.	9.	10.	11.	12.	13.	14.
DNAase	+	nd	nd	nd	nd	nd	nd	nd	nd	nd	nd	nd	nd	nd
Esterase C4	–	–	–	d	nd	–	nd	nd	nd	nd	nd	nd	–	nd
Ester lipase C8	–	–	+	+	nd	–	nd	nd	nd	nd	nd	nd	–	nd
α-Fucosidase	–	–	+	–	nd	–	nd	nd	nd	nd	nd	nd	+	nd
α-Galactosidase	–	+	+	d	–	d	nd	nd	+	–	nd	+	–	nd
β-Galactosidase	–	+	+	+	+	+	nd	nd	+	+	nd	+	–	nd
α-Glucosidase	–	d	+	+	+	+	nd	nd	nd	–	nd	+	d	nd
β-Glucosidase	–	+	–	+	–	d	nd	nd	nd	–	nd	+	d	nd
β-Glucuronidase	–	–	–	+	–	–	nd	nd	–	–	nd	–	–	nd
Glycine arylamidase	nd	nd	nd	nd	nd	nd	nd	nd	nd	nd	nd	nd	+	nd
Histidine arylamidase	nd	nd	nd	nd	nd	nd	nd	nd	nd	nd	nd	nd	+	nd
Lipase C14	–	–	–	–	nd	–	nd	nd	nd	nd	nd	nd	–	nd
Leucine arylamidase	+	+	+	+	nd	+	nd	nd	+	+	nd	nd	+	nd
Leucylglycine arylamidase	nd	nd	nd	nd	nd	nd	nd	nd	nd	nd	nd	nd	+	nd
α-Mannosidase	–	–	–	–	nd	–	nd	nd	nd	nd	nd	nd	–	nd
β-Mannosidase	nd	–	–	–	–	nd	nd	nd	–	nd	–	–	–	nd
Proline arylamidase	nd	nd	nd	nd	nd	nd	nd	nd	nd	nd	nd	nd	+	nd
Phenylalanine arylamidase	nd	nd	nd	nd	nd	nd	nd	nd	nd	nd	nd	nd	+	nd
Pyrazinamidase	nd	+/–	+	+	+	nd	nd	nd	–	–	nd	(+)	+	nd
Pyrrolidonyl arylamidase	nd	nd	–	+	–	nd	nd	nd	–	–	nd	–	–	nd
Serine arylamidase	nd	nd	nd	nd	nd	nd	nd	nd	nd	nd	nd	nd	+	nd
Trypsin	–	–	–	–	nd	–	nd	nd	nd	nd	nd	nd	–	nd
Tyrosine arylamidase	nd	nd	nd	nd	nd	nd	nd	nd	nd	nd	nd	nd	+	nd
Urease	–	–	–	–	–	nd	–	nd	–	–	nd	–	–	nd
Valine arylamidase	–	–	–	–	nd	–	nd	nd	nd	nd	nd	nd	–	nd
β-Xylosidase	–	nd	nd	nd	nd	nd	nd	nd	nd	nd	nd	nd	nd	nd
Catalase production	–	+	+	d	–	–	(+)	+	–	–	+	–	–	+
Nitrate reduction	–	+	–	+	–	+	–	nd	+	–	nd	–	–	–
Nitrite reduction	–	nd	nd	nd	nd	–	·nd	nd	nd	nd	nd	nd	nd	nd
Acetoin production	–	–	–	–	–	nd	–	nd	nd	–	nd	(+)	–	nd
Indole production	–	nd	nd	nd	nd	nd	nd	nd	nd	nd	nd	nd	–	nd

Symbols: +, positive/present; –, negative/absent; d, positive or negative; +/–, usually positive; –/+, usually negative; (), weak reaction; and nd, no data available.

[a]1. *A. bovis*; 2. *A. bowdenii*; 3. *A. canis*; 4. *A. catuli*; 5. *A. coleocanis*; 6. *A. denticolens*; 7. *A. hordeovulneris*; 8. *A. howellii*; 9. *A. hyovaginalis*; 10. *A. marimammalium*; 11. *A. slackii*; 12. *A. suimastitidis*; 13. *A. vaccimaxillae*; 14. *A. viscosus*.

From Schofield and Schaal (1981), Buchanan et al. (1984), Dent and Williams (1984, 1986), Schaal (1986b), Johnson et al. (1990), Collins et al. (1993), Pascual et al. (1999), Hoyles et al. (2000, 2001a, b, c, 2002b), and Hall et al. (2003b).

viscosus invertase were as follows: molecular weight 8.6×10^4; K_m for sucrose ~71 mM; pH optimum between 5.8 and 6.3. This enzyme was noncompetitively inhibited by fructose-6-phosphate and fructose-1,6-diphosphate (Kiel and Tanzer, 1977a). Fructose-1,6-diphosphate aldolase and phosphate acetyltransferase activities were demonstrated in cell-free extracts of *A. israelii* (Buchanan and Pine, 1967). A nicotinamide adenine dinucleotide-dependent lactate dehydrogenase of *A. viscosus* (Brown et al., 1975) was found to have a molecular weight of 1×10^5 and a pH optimum of 5.5–6.2 and was under negative control by adenosine 5×-triphosphate and inorganic phosphate. Malate and glutamate dehydrogenases were detected in extracts of *A. viscosus*, *A. naeslundii*, *A. israelii* and *A. gerencseriae* but differed in their electrophoretic mobilities (Fillery et al., 1978). 6-Phosphogluconate dehydrogenase was only demonstrated in *A. naeslundii* and *A. viscosus* but not in *A. israelii* or *A. gerencseriae* (Fillery et al., 1978).

The formation of extracellular or cell-associated polymers such as levan, dextran or glycogen, which enable certain *Actinomyces* species to attach directly to the tooth surface, has been studied in detail for *A. viscosus* (Howell and Jordan, 1967; Warner and Miller, 1978; Hamilton and Ellwood, 1983; Imai and Kuramitsu, 1983; Komiyama et al., 1988). The levan produced by this organism formed a capsule tenaciously adhering to the cells under certain conditions (Warner and Miller, 1978). The enzyme levansucrase, which is responsible for the production of this high-molecular-weight substance, was found to occur both in the growth medium and affixed to the cell wall (Pabst, 1977; Pabst et al., 1979). In addition, *A. viscosus* was also shown to exert levan-hydrolyzing activities (Miller and Somers, 1978a). Similar synthetic and degrading capacities for such polymers were also observed in *A. naeslundii* (Miller and Somers, 1978a; Komiyama et al., 1988).

Actinomyces species do not show pronounced proteolytic activities (Slack and Gerencser, 1975; Schofield and Schaal, 1981). Similarly, the ability to deaminate or decarboxylate amino acids is uncommon in the genus (Schaal, 1986a). However, various other enzymatic activities, which may be useful for classification and identification purposes, have been demonstrated using the API Zym test kit or fluorogenic substrates (Tables 8 and 9).

True β-hemolysis is also not common in the genus *Actinomyces*, although this property has been reported for certain strains of *A. bovis* and *A. odontolyticus*, depending on the animal source of the red blood cells.

Lack of catalase activity was once thought to be one of the major phenotypic properties differentiating *Actinomyces* species from morphologically or physiologically related organisms. However, phylogenetic data have shown that the genus *Actinomyces* contains both catalase-positive and catalase-negative members. *Actinomyces neuii, A. radicidentis, A. viscosus, A. bowdenii, A. canis*, some strains of *A. catuli, A. hordeovulneris, A. howellii* and *A. slackii* are catalase-positive.

Habitat

Members of the genus *Actinomyces* have exclusively been found as commensals or pathogens of humans and other warm-blooded animals but never as free-living saprophytes in the environment.

ACTINOMYCES SPECIES AS COMMENSALS OR PATHOGENS OF HUMANS Although the first descriptions of *Actinomyces bovis* and *A. israelii* seemed to indicate that they were obligate pathogens (Bollinger, 1877; Israel, 1878), it soon became apparent that these and possibly all of the other species of the genus *Actinomyces* were primarily natural inhabitants of the body surfaces of healthy humans and other animals. Bergey (1907) was probably the first to describe filamentous bacteria adhering to the teeth and being involved in the formation of dental plaque. Since then, numerous reports have demonstrated the presence of such microbes in the oral cavity of man, showing that the classical pathogenic species, as well as less virulent or non-pathogenic taxa, form a major component of the oral microflora.

Naeslund (1925) was the first to culture facultatively anaerobic, filamentous organisms from the human mouth that differed clearly from the pathogen *Actinomyces israelii*. Identical strains were isolated by Thompson and Lovestedt (1951) and validly designated *Actinomyces naeslundii*. The first indications that potentially pathogenic actinomycetes are generally present in the oral cavity of healthy humans were obtained by Lord (1910) who was able to produce actinomycosis-like lesions in guinea pigs that had been inoculated with the contents of carious teeth. Emmons (1938) examined an unselected series of 200 pairs of tonsils from routine tonsillectomies and observed organisms indistinguishable from *Actinomyces bovis* (at that time a synonym of *A. israelii*) in 37% of the tonsil pairs studied. From 11% of the specimens, he was even able to culture such microbes. However, no clinical or pathological signs of invasive actinomycete infection were found in these cases. Since then, many other workers (Slack, 1942; Howell et al., 1959; Howell et al., 1962; Slack et al., 1971; Sutter, 1984), using different techniques of demonstrating and identifying the bacteria, have proved that *A. israelii* is a facultatively pathogenic, primarily commensal, resident member of the oral microflora of man. This also applies to *Actinomyces gerencseriae*, which was formerly known as serovar 2 of *A. israelii* (Johnson et al., 1990; Sarkonen et al., 2000).

Similar results were obtained for several other members of the genus *Actinomyces*: *Actinomyces naeslundii, A. viscosus, A. odontolyticus, A. georgiae* and *A. graevenitzii* (Batty, 1958; Socransky, 1970; Slack et al., 1971; Collins et al., 1973; Hill et al., 1977; Sarkonen et al., 2000) were recovered regularly from the oral cavity of more-or-less healthy adults and even of young children. The principal natural habitat of *A. meyeri* appears to be the human periodontal sulcus (Cato et al., 1984).

In addition to *A. israelii*, also *A. naeslundii* and possibly other *Actinomyces* species have been demonstrated in human tonsils (Blank and Georg, 1968; Grüner, 1969; Hotchi and Schwarz, 1972; Garcia Ramos et al., 1984). Furthermore, *A. viscosus* and other fermentative actinomycetes may occasionally be isolated from the uninfected conjunctiva and cornea (Jones and Robinson, 1977; Schaal and Beaman, 1984b).

The occurrence of *Actinomyces* species as normal inhabitants of the intestinal tract has not been documented as clearly. Indirect evidence for the view that these organisms may form a small but significant component of the intestinal flora can be derived from cases of abdominal actinomycoses that developed endogenously after bowel surgery, appendix perforation, or injuries of the intestine. Similarly, the recovery of *A. israelii* from large proportions of diseased appendices (60.3% of the samples examined) suggests the presence of these bacteria in the lower digestive tract (Minsker and Moskovskaya, 1979).

Direct demonstration of fermentative actinomycetes in feces and in contents of the upper intestinal tract has also been possible, although number and conclusiveness of these studies are still limited. Some authors (Slack and Gerencser, 1975) regarded occasional isolations of *A. israelii* from feces (Sutter and Finegold, 1972) as accidental in the sense of a transient rather than a permanent colonization of the intestine. On the other hand, strong selective measures (Fritsche, 1964) apparently allow the isolation of *A. israelii* at least in low numbers from most stool specimens of healthy individuals (D. Fritsche, unpublished results). In addition, Noack-Loebel and co-workers, when comparing the composition of the fecal flora in two groups of primary school age children reported that the numbers of fermentative actinomycetes were significantly influenced by the individual dietary habits (Noack-Loebel et al., 1983). Children with a lacto-ovo-vegetarian diet showed an increase in the fecal numbers of fermentative actinomycetes along with increased Bifidobacterium and Enterobacteriaceae counts as compared to school children with normal, *ad libitum* diet. Furthermore, studies on the composition of the microflora in the duodenal and jejunal fluids suggest that *Actinomyces* species frequently represent an important component of the cultivable flora in the upper intestinal tract (Bernhardt and Knoke, 1984; Justesen et al., 1984).

Similar uncertainties still exist with regard to the actinomycete colonization of the female genital tract, although this problem has recently attracted much more attention than the intestinal occurrence. Previously, the vagina, the cervical canal, and the *cavum uteri* were not considered natural habitats of fermentative actinomycetes (Hanf and Hanf, 1955). Also, in more recent studies, these organisms often could not be demonstrated in healthy women who did not use intrauterine contraceptive devices (IUDs; Jones et al., 1983). However, some reports indicate that at least *A. israelii* and *Propionibacterium propionicum* may occur in the female genital tract independently of the use of IUDs, although these organisms were usually recovered in smaller numbers when no foreign body was present in the *cavum uteri* or cervical canal, respectively. Under the latter circumstances, the percentage of positive specimens ranged from 3% for *A. israelii* (Persson et al., 1983) and 19% for *Actinomyces* species in general (Grice and Hafiz, 1983) to more than 70% for *A. israelii* alone (Persson and Holmberg, 1984). In women wearing IUDs or vaginal pessaries, demonstration of these organisms is easier either in cervicovaginal secretions or on the IUD and in particular on its thread reaching from the *cavum uteri* to the vagina.

Nevertheless, the isolation or microscopic demonstration rates of actinomycetes in material from the female genital tract varied widely between 0% (Schiffer et al., 1978) to more than 80% (Gupta et al., 1976, 1978; Hager et al., 1979; Pine et al., 1981; Traynor et al., 1981; Duguid et al., 1982; Valicenti et al., 1982; Grice and Hafiz, 1983; Persson et al., 1983; Schaal and Pulverer, 1984c; Jarvis, 1985; Eibach et al., 1989). These differing results indicate that sensitivity and specificity of the detection methods used are not comparable and that there could also exist regional, ethnic, social, or even religious differences. Furthermore, it remains to be determined whether the colonization with fermentative actinomycetes in IUD-wearing women possesses any predictive value of existing or imminent gynecological infections. An additional question is posed by the observation that, as in the oral cavity and under pathological conditions, actinomycetes may also occur in association with amoebae in the female genital tract (Arroyo and Quinn, 1989).

Apart from *A. israelii* and *Propionibacterium propionicum*, additional *Actinomyces* species such as *A. gerencseriae*, *A. naeslundii*, *A. viscosus*, *A. odontolyticus*, *A. meyeri*, and *A. cardiffensis* were recovered from specimens of the female genital tract (Mitchell and Crow, 1984; Schaal and Pulverer, 1984c; Eibach et al., 1989; Schaal and Lee, 1992c; Hall et al., 2002). Occasionally, two different *Actinomyces* species may be isolated from a single genital specimen. In particular, associations of *A. israelii* and *A. viscosus* or *A. israelii* and *A. odontolyticus* were observed in low frequencies (0.6% of the specimens examined; K. P. Schaal, unpublished observations).

Essentially nothing is known about the natural habitat of the remaining new members of the genus *Actinomyces*. *Actinomyces europaeus*, *A. funkei*, *A. neuii*, *A. radingae*, *A. turicensis*, *A. radicidentis*, *A. urogenitalis*, *A. nasicola*, *A. oricola* and *A. hongkongensis* were up to now isolated only from pathological lesions, so that their natural reservoirs remain to be identified.

Apart from *A. georgiae*, essentially all of the human *Actinomyces* species mentioned in this section have been isolated from pathological lesions although in varying frequencies. These cover a broad spectrum of diseases and impairments, among which actinomycoses are the most characteristic disease entities. Clinical and pathological details of these conditions are given in the section on Pathogenicity.

ACTINOMYCES SPECIES AS COMMENSALS OR PATHOGENS OF ANIMALS The extent to which members of the genus *Actinomyces* form a part of the indigenous microflora of healthy domestic and wild animals is not known in detail. By analogy to the situation in humans, animal actinomycoses might also develop endogenously and their causative agents might therefore also belong to the normal microflora of animal mucous membranes (Slack and Gerencser, 1975; Schaal, 1986a). However, Dent and Williams (1984b), in their study on the microflora of the dental plaque in dairy cattle, were not able to identify as *A. bovis*, any filamentous or diphtheroidal, Gram-positive isolates from healthy cattle; the normal habitat of this species thus awaits definite clarification.

Other members of the genus *Actinomyces* have been demonstrated in healthy animals: Beighton (1985) studied the microbial colonization of the mouth of neonatal monkeys (*Macaca fascicularis*) and found that *A. naeslundii* and, to a lesser extent, *A. viscosus* are already established in the dental plaque of these animals during breast-feeding period, suggesting that these organisms belong to the basic plaque flora which is altered by the diet as the animals mature. *Actinomyces viscosus* (serovar 1) has been isolated from subgingival plaque of hamsters and from cervical plaque of rats (Howell, 1963a; Howell and Jordan, 1963b; Jordan and Keyes, 1964; Jordan and Keyes, 1965a; Bellack and Jordan, 1972). Similarly, *A. denticolens*, *A. howellii* and *A. slackii* were recovered from the dental plaque of cattle (Dent and Williams, 1984a; Dent and Williams, 1984b; Dent and Williams, 1986), as were *A. viscosus*, *A. hordeovulneris* and *A. denticolens* from the gingival margin of healthy cats (Love et al., 1990). The natural habitat of *A. hyovaginalis* was assumed to be the porcine genital tract (Collins et al., 1993), but the organism was also recovered from a large number of tonsils of piglets before and after weaning (Baele et al., 2001).

All of the other recently described *Actinomyces* species from animals were isolated from various clinical conditions in dogs (*A. canis*, *A. bowdenii*, *A. catuli* and *A. coleocanis*), cats (*A. bowdenii*), pigs (*A. suimastitidis*), seals or a porpoise (*A. marimammalium*), and a cow (*A. vaccimaxillae*), respectively, but their natural habitat has not been established. Some details on the clinical pictures of the infections caused by these and other animal *Actinomyces* species are given in the section on Pathogenicity. The sources of a few *A. israelii* infections reported in animals (Tyrrell et al., 2002) remain to be clarified.

Detection, Isolation and Cultivation

COLLECTION AND TRANSPORT OF CLINICAL SPECIMENS Materials suitable for diagnosing human and animal infections due to *Actinomyces* species are pus, sinus discharge, bronchial secretions, granulation tissue, and biopsy specimens.

During sampling, precautions must be taken against contamination with the indigenous mucosal flora, which physiologically contains fermentative actinomycetes. Whenever possible, pus or tissue should be obtained by transcutaneous puncture or needle biopsy through the carefully disinfected skin (Schaal, 1998). For the diagnosis of thoracic actinomycoses, bronchial secretions should be obtained by transtracheal aspiration; sputum is by no means appropriate. Transthoracic or transabdominal needle-aspiration biopsies are often the only ways of obtaining satisfactory specimens from pulmonary or abdominal lesions (Pollock et al., 1978; Schaal, 1998).

The transport of specimens to the bacteriological laboratory should be expeditious unless a reducing transport medium such as Stuart's transport medium is used, although *Actinomyces* species are less susceptible to oxidative damage than are certain strict anaerobes of the concomitant microflora. Adequate transport media are commercially available (e.g., Port-A-Cul [Becton-Dickinson] and Portagerm [bioMérieux]) and should be used according to the instructions of the manufacturers.

MACROSCOPIC AND MICROSCOPIC EXAMINATION OF CLINICAL SPECIMENS Pus from actinomycotic abscesses or the purulent discharge from sinus tracts occasionally contains large amounts of sulfur granules which give these exudates the macroscopic appearance of a semolina soup (Lentze, 1969; Schaal, 1984a). More often, however, sulfur granules are not immediately apparent and have to be specifically looked for. For this purpose, the purulent material is spread out in a sterile Petri dish and examined for the presence of comparatively hard, yellowish to brownish to reddish particles of up to 1 mm in diameter. These particles are picked with an inoculating needle or loop and used for further examinations.

As pus specimens may contain particles that resemble sulfur granules in terms of size, color or texture without being a typical sulfur granule (e.g., granules of mycetomas and actinomycetomas, microcolonies of other filamentous microorganisms, and particles of necrotic tissue), the nature of suspected granules has to be confirmed by subsequent microscopic examinations as follows.

Microscopy of Suspected Sulfur Granules (Schaal, 1992c)

Microscopy at Low Magnification

Place the granule in the middle of a glass slide. Add one drop of 1% methylene blue solution, place a coverslip on top, and press down gently. View under a microscope in transmitted light at about 100× magnification.

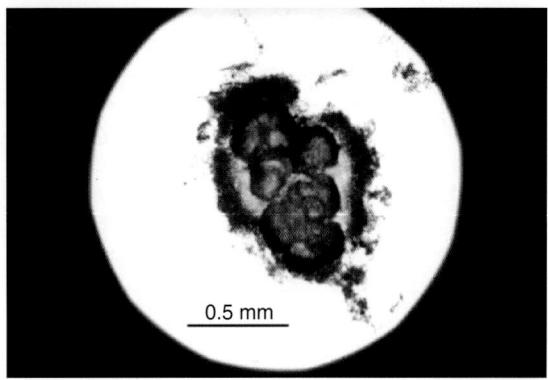

Fig. 12. Actinomycotic sulfur granule, partially disinte-grated—note the spherical segment-like structures repre-senting filamentous Actinomyces colonies and the surrounding blue-colored material which chiefly consists of leukocytes; micrograph after immersing in 1% methylene blue solution and gently pressing on the cover slip.

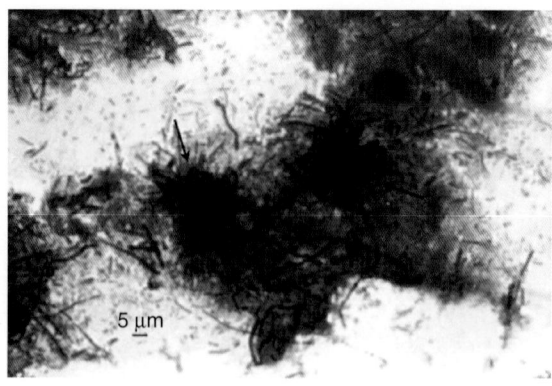

Fig. 13. Gram stain prepared from a smear of a crushed actinomycotic sulfur granule. Note the host of various Gram-positive and Gram-negative bacterial forms in addition to the nests ($\rightarrow$) of Gram-positive, interwoven and branching fila-ments which represent the causative Actinomyces species.

Under these conditions, the cauliflower-like appearance of the sulfur granules is easily demonstrated (Fig. 12). Because of its reducing capacity, the microbial center of the granule remains unstained, whereas the surrounding leukocytes, other tissue reaction material, and peripheral hyphae take up the blue color of the stain. After slightly harder pressing of the coverslip, sulfur granules usually disintegrate into spherical or partially spherical elements, which represent actinomycete colonies formed in vivo.

Microscopy at High Magnification

> Crush a particle between two slides and stain the smears thus obtained with Gram stain. Observe at high magnifi-cation (ca. 1000×) using an oil-immersion objective.

Actinomycotic sulfur granules typically show nests of Gram-positive, branched, unevenly curved and irregularly stained filaments and rods. In contrast to granules derived from acti-nomycetomas, the Gram stain also shows, together with leukocytes, the presence of a vari-ety of additional Gram-positive and Gram-negative bacteria, which represent the obligate concomitant flora (Schaal, 1984a; Fig. 13).

METHODS FOR OBTAINING CAPNOPHILIC OR ANAEROBIC GROWTH CONDITIONS Essentially all of the *Actinomyces* species are capnophilic, requiring increased CO_2 concentrations for opti-mal growth. Some species prefer also reduction of oxygen tension. In contrast to isolation of strict anaerobes, however, isolation of fermenta-tive actinomycetes usually does not require very elaborate techniques for obtaining an oxygen-free atmosphere. Strains from several species may even be reduced in growth rate when no oxygen at all is present.

Thus, the use of pre-reduced media, the roll-tube method, glove boxes, or similar complicated and expensive equipment often employed for growing strict anaerobes is usually not needed for culturing fermentative actinomycetes. How-ever, such procedures may be suitable for special research purposes that require constant and con-trolled growth conditions. When these methods and devices are used, the detailed descriptions and instructions given in the VPI Anaerobe Laboratory Manual (Holdeman et al., 1977) should be followed. Nevertheless, essentially all of the fermentative actinomycetes can be suc-cessfully isolated and subcultured by much simpler techniques.

Although it may seem old-fashioned, the eas-iest, least expensive, and most widely applicable technique for obtaining growth conditions with moderately reduced oxygen tension and increased CO_2 concentration is Fortner's method (Fortner, 1928). This method is based upon the observation that certain *Enterobacteriaceae*, in particular *Serratia marcescens*, are able to con-sume oxygen and to produce carbon dioxide when grown in a Petri dish which has been care-fully sealed before incubation to make it air-tight. Depending on the degree of anaerobiosis and CO_2 enrichment needed for an individual strain, growth of the organism to be cultivated in coculture with *Serratia* starts with some delay but usually within the first 4–8 h of incubation.

Fortner's Method for Obtaining Semi-anaerobic Growth Conditions (Fortner, 1928)

> Use only agar media, preferably transparent ones, in glass Petri dishes. Plastic dishes will inevitably cause problems! Inoculate one-half to two-thirds of the agar surface with the material to be examined or with the strain to be subcultured using a platinum loop. Inoculate the remain-ing agar surface heavily with Serratia using a spatula.

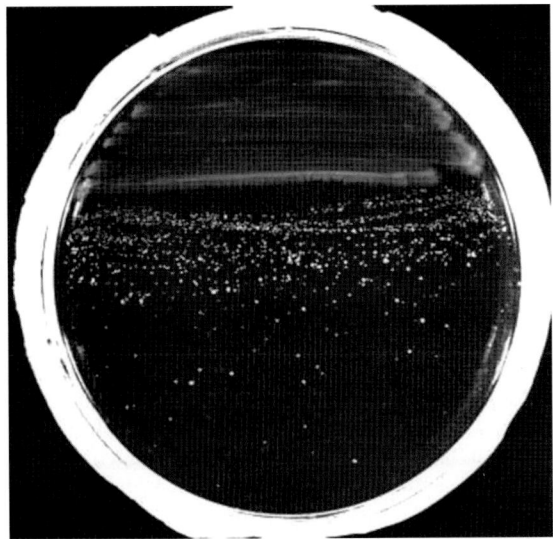

Fig. 14. Fortner's method. Top third of the Petri dish: confluent growth of *Serratia marcescens*; lower two thirds: growth of an *Actinomyces israelii* strain. Note its enhanced growth near *Serratia*.

Place the dish upside down upon a glass sheet of appropriate size.

Fix and seal the dish with plasticine to make the system air-tight (Fig. 14). Incubate at $36 \pm 1°C$ for up to 14 days and examine for microbial growth and filamentous colonies every two days using either a hand lens or a microscope with a long-distance 10× objective in transmitted light.

Leakage in the plasticine seal, which may occur when the plasticine is too brittle or the base plate is damp, can easily be detected because *Serratia marcescens* forms nonpigmented growth under shortage of oxygen but becomes red when the system is not completely air-tight.

According to the authors' experience, essentially all of the organisms treated in this chapter can easily be cultured by this method. Even the more aerophilic species such as *Actinomyces viscosus* or *Actinomyces neuii* will grow, although their mature colonies may remain slightly smaller than when incubated under full oxygen tension with added CO_2. Additional plates to be incubated under CO_2-enriched conditions are usually not necessary. Thus, Fortner's technique may be considered a universal means of culturing *Actinomyces* species. This method is particularly useful in the clinical microbiological laboratory where measures should be as simple and inexpensive as possible but still optimally efficient in order not to lose important clinical isolates. In this respect, it is an additional advantage that Fortner plates can be checked individually for bacterial growth without disturbing the semi-anaerobic atmosphere.

An alternative way of obtaining anaerobic and/or CO_2-enriched growth conditions is the use of anaerobic jars or cabinets. The Torbal anaerobic jar with a gas mixture of 80% N_2:10% H_2:10% CO_2 or the GasPak jar with a H_2-CO_2 or only CO_2 generating envelope has been recommended (Slack and Gerencser, 1975). For the more anaerobic strains, jars without catalysts are usually less satisfactory. The catalase-positive *Actinomyces* species, however, will grow only poorly if at all in an oxygen-free gaseous atmosphere. Therefore, when jars with such gas mixtures are used for primary isolation, additional plates have to be incubated aerobically with added CO_2 (5–10%, v/v). A candle jar will be most satisfactory for this purpose, but a GasPak jar with a CO_2-generating unit as mentioned above also can be employed. Recently, the Oxyrase OxyPlate anaerobe incubation system (Oxyrase, Inc., Mansfield, Ohio, USA) was evaluated (Wiggs et al., 2000) for its ability to support growth of various clinically significant anaerobes including *Actinomyces* species. While some of the strict anaerobes (29.9% of the total number of strains examined) grew better in the anaerobic chamber, nearly all of the *Actinomyces* strains tested showed good growth with the Oxyrase OxyPlate system. A glove box is not needed for growing fermentative actinomycetes.

If the oxygen requirements of an individual isolate are not known and cannot be predicted, tests for oxygen requirements must be performed in order not to lose the strain during subculturing.

Tests for Oxygen Requirements (Slack and Gerencser, 1975)

Agar Slant Method

Prepare the inoculum in the following way from about 3-day-old cultures, either in broth or on plates: Suspend cells obtained by centrifugation of the broth culture or by scraping off from the agar surface in 0.85% saline, and adjust the suspension to a density matching a MacFarland 3 standard. Inoculate eight suitable agar medium slants in cotton-plugged tubes with the suspension of the test organism using a capillary pipette.

Incubate the slants, in duplicate, under the following conditions: 1) For aerobic conditions, place two slants with the original cotton plugs directly into the incubator. 2) For aerobic conditions with added CO_2, clip off the cotton plugs of two tubes and push the remaining parts into the tube to just above the slant. Place small pledgets of absorbent cotton on top of the plugs, add 5 drops of 10% Na_2CO_3 and 5 drops of 1M KH_2PO_4 to each tube, and close it immediately with a rubber stopper. 3) For anaerobic conditions with added CO_2, prepare tubes as under 2. In this case, however, add 5 drops of 10% Na_2CO_3 and 5 drops of pyrogallol solution (100 g of pyrogallic acid in 150 ml of distilled water) to the absorbent cotton. 4) For anaerobic conditions without CO_2, add 5 drops of 10% KOH and 5 drops of pyrogallol solution to the properly prepared tubes (see 2). Record results after

3 and 7 days of incubation, usually at $36 \pm 1°C$. If the growth in two corresponding tubes does not seem equal, the test has to be repeated.

Agar Deep Method

Inoculate melted and cooled agar medium, while still liquid, with a suspension of the test organism using a capillary pipette. Push the tip of the pipette to the bottom of the tube and then slowly withdraw it while expelling a drop of inoculum. Gently mix the agar by rotating the tube and allow it to solidify in an upright position. Read the results after 3 and 7 days of incubation by measuring the distance in millimeters between the surface and the zone of maximum growth.

Anaerobiosis for fluid cultures may be obtained by placing tubes with suitable broth media into jars. However, another classical technique, based upon the oxygen-binding effect of sterilized animal tissue particles, can give equal or even better results, especially on primary isolation from clinical specimens. The three most common media that utilize this principle are the Tarozzi medium, cooked-meat medium, and Rosenow broth. In Tarozzi medium, when one solid piece of liver is used, fermentative actinomycetes grow especially well on the surface of the tissue and often produce macroscopically visible, "snowball"- or "cotton-pad"-like colonies (Fig. 11) that can be selectively aspirated with a capillary pipette for further microscopic and cultural examination.

MEDIA AND PROCEDURES FOR ISOLATION AND CULTIVATION FROM CLINICAL SPECIMENS Clinical microbiology laboratories which are involved in the diagnosis of anaerobic infections need reliable, simple and inexpensive isolation procedures which allow rapid and characteristic growth of all of the pathogens present to establish a definite diagnosis as early as possible. This does not necessarily mean that the growth conditions must be optimal and that maximum growth yields must be obtained.

General-Purpose Culture Media Various high-quality, general-purpose culture media have been recommended for primary isolation and subsequent *Actinomyces* isolates. These include fluid thioglycolate broth (THIO), possibly supplemented with 0.1–0.2% sterile rabbit serum; brain heart infusion broth (BHIB); trypticase soy broth (TSB); brain heart infusion agar (BHIA); trypticase soy agar (TSA); heart infusion agar or Columbia agar base supplemented with 5–6% defibrinated rabbit, sheep or horse blood and possibly also (e.g., for *A. meyeri*) with 0.02% Tween 80 and vitamin K_1 (10 μg/ml); and Schaedler broth or agar (Slack and Gerencser, 1975; Schaal and Pulverer, 1981b). Most of these media are commercially available and therefore

should be prepared and used according to the manufacturers' instructions.

Special Nonselective Culture Media Nevertheless, none of these commercial media is equally satisfactory under all clinical circumstances and for all of the agents to be expected. Especially in THIO and BHIB or on BHIA, which are widely employed in the bacteriological diagnosis of actinomycotic infections, growth of the actinomycetes is sometimes poor or cannot be recognized because of the rapid and abundant multiplication of concomitant bacteria. Therefore, in the German Consulting Laboratory for Pathogenic Actinomycetes, an additional medium (CC medium) has been used, which was developed more than 30 years ago at the Institute of Hygiene of the University of Cologne (Heinrich and Korth, 1967) as a modification of the synthetic medium of Howell and Pine (1956). This medium has proved to be extremely useful since it promotes growth of pathogenic fermentative actinomycetes in a way that characteristic filamentous micro- and macrocolonies are formed (Fig. 15), and it also allows long-term maintenance of actinomycete cultures over several years by only requiring subculture of the organisms every six to ten weeks.

CC Medium for Isolating and Maintaining Fermentative Actinomycetes (Heinrich and Korth, 1967).

Mineral and Trace Element Solution

$MgSO_4 \cdot 7H_2O$	20.0 g
$CaCl_2 \cdot 2H_2O$	2.0 g
$FeSO_4 \cdot 7H_2O$	0.4 g
$MnSO_4 \cdot 2H_2O$	15.0 mg
$Na_2MoO_4 \cdot 2H_2O$	15.0 mg
$ZnSO_4$	4.0 mg
$CuSO_4 \cdot 5H_2O$	0.4 mg
$CoCl_2 \cdot 4H_2O$	0.4 mg
Boric acid	20.0 mg
Potassium iodide	10.0 mg

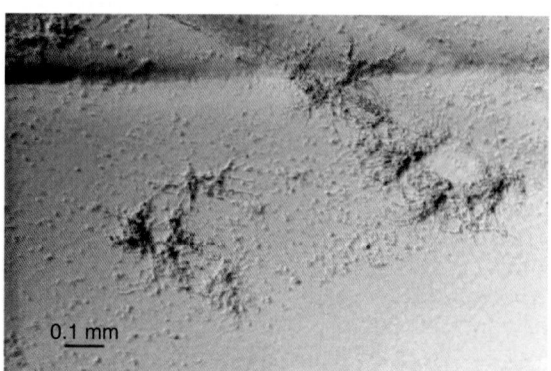

Fig. 15. *Actinomyces israelii*; filamentous, spider- or cobweb-like microcolonies, micrograph in situ (CC medium, 48 h at $36 \pm 1°C$, and Fortner's method).

Dissolve in 1 liter of distilled water and acidify with 10 ml of 10% HCl.

Vitamin Solution

Thiamine · HCl	20 mg
Pyridoxine · HCl	20 mg
Biotin	1 mg
Folic acid	5 mg
Vitamin B_{12}	(1 mg/100 ml) 1 ml
p-Aminobenzoic acid	20 mg
myo-Inositol	20 mg
Nicotinamide	10 mg
Nicotinic acid	10 mg
Ca-Pantothenate	20 mg

Dissolve in 100 ml of distilled water.

Amino Acid and Vitamin Solution

Casein hydrolysate	12 g
Yeast extract	12 g
L-Cysteine · HCl	500 mg
L-Asparagine	30 mg
DL-Tryptophan	20 mg
Vitamin solution (see above)	12 ml

Dilute with distilled water up to a total amount of 100 ml. Filter-sterilize.

The Final Medium

Dissolve 4 g of KH_2PO_4 in 250 ml of distilled water and adjust to pH 7.6 with NaOH. Add 10 ml of mineral and trace element solution, 500 mg of potato starch dissolved in 70 ml of boiling distilled water, about 20 g of agar (depending on quality), and distilled water to give a final volume of 900 ml. Sterilize by autoclaving at 121°C for 15 min. After cooling to about 50°C, add the entire amino acid and vitamin solution under aseptic conditions. Adjust the final pH to 7.3. Pour the medium into glass Petri dishes.

For cultural analyses of clinical specimens, three to four different media, including a fluid medium, preferably Tarozzi broth, should be inoculated. When Tarozzi broth is used, the test tube containing the medium is heated in a boiling water bath for about 10 min immediately before use, then quickly cooled down under running tap water, inoculated, and sealed with about 2 ml of sterilized melted petrolatum or paraffin wax. Use of paraffin oil is not recommended.

Modified Tarozzi Broth for the Isolation and Subcultivation of Fermentative Actinomycetes (Schaal, 1986a)

Basal Nutrient Broth	
Meat extract	10 g
Peptone (e.g., peptone P; Oxoid)	12 g
NaCl	3 g
K_2HPO_4	2 g

Dissolve the ingredients in a total of 1 liter of distilled water while heating at 80°C and boil subsequently for 20 min. Adjust the pH to 7.5 with NaOH solution.

Preparation of Liver Pieces

Cut fresh guinea pig or beef liver into pieces measuring approximately 2 × 1 × 1 cm and wash them thoroughly in several changes of fresh saline.

Preparation of the Medium

Place the liver pieces into test tubes (1 piece per tube). Add 8 ml of basal nutrient broth supplemented with 0.1% sodium thioglycolate (w/v) to each tube. Stopper the tubes with a cotton plug and sterilize by autoclaving at 121°C for 15 min. After inoculation, seal the tube by adding about 1 ml of sterilized melted petrolatum (petroleum jelly, Vaseline) or paraffin wax.

When using Fortner's method for agar cultures and Tarozzi broth for fluid cultures, the media can be checked daily for growth of typical mycelial colonies without disturbing the semi-anaerobic atmosphere. When anaerobic jars or incubators or glove boxes are employed, a duplicate set of media should be inoculated and examined after 3 and 7–10 days of incubation at 36 ± 1°C. In addition, a third set of media has to be incubated in air with added CO_2 (candle jar or GasPak jar with a CO_2-generating unit).

For detecting typical mycelial microcolonies of the actinomycetes including those of *Propionibacterium propionicum* in early growth stages, it is best to examine surface cultures on transparent media under a microscope at 80–100× magnification (Fig. 15). If a long-distance objective is available, cultures can be examined through the medium without opening Fortner plates (Lentze, 1938a).

Removal of concomitant bacteria to obtain pure cultures of the actinomycetes for further examination (Fig. 16) is occasionally difficult, even when the organisms have been subcultured several times or when dilution techniques have been applied. Subculture media supplemented with suitable antibiotics may help in these cases but may inhibit growth of the actinomycetes as well. Commercial disks produced for antibiotic susceptibility testing can be used alternatively.

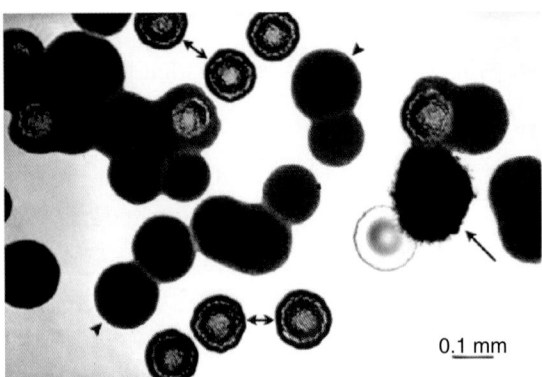

Fig. 16. Synergistic actinomycotic flora; (→) single colony of *A. israelii*; (↔) colonies of *Actinobacillus actinomycetemcomitans*; (◄) additional concomitant bacteria; micrograph *in situ* in transmitted light (brain heart infusion agar, 14 days at 36 ± 1°C, and Fortner's method).

These disks are placed upon the agar surface after the strain to be purified has been inoculated. Depending on the contaminating organisms, disks that contain metronidazole, colistin or nalidixic acid in usual concentrations can be employed. From their inhibition zones, actinomycete colonies can often be picked easily and transferred to another nonselective plate.

ISOLATION FROM THE ORAL CAVITY OR OTHER NATURAL HABITATS IN HUMANS AND ANIMALS Ecological studies on the prevalence of fermentative actinomycetes in various secretions and concretions of mucous membranes usually require the qualitative demonstration of the various species, their differential quantification, and the determination of their numerical relationships to other members of the respective microflora. Therefore, any losses of viable cells or uncontrolled multiplication must be avoided during sampling and transportation of the specimens. Furthermore, since samples from typical habitats of fermentative actinomycetes (especially gingival crevice, sulcus fluid, plaque material, feces, or cervicovaginal secretions) harbor very high numbers of many different additional indigenous microorganisms, selective culture media are necessary to facilitate the detection and enumeration of specific organisms.

Collection and Transport of Specimens from Natural Habitats For ecological studies or for bacteriological analyses in periodontitis or caries, plaque material, dental calculus, and sulcus fluid represent suitable materials for isolating or demonstrating actinomycetes by molecular methods. When quantitative results are to be obtained, the sampling technique must allow measurements of the original amount of material collected.

Different types of transport media have been applied to oral microbes, including the Stuart medium, a general-purpose transport medium widely used in medical microbiology (Loesche et al., 1972; Syed and Loesche, 1972). The reduced transport fluid (RTF) of Syed and Loesche (Syed and Loesche, 1972) was specifically designed for specimens from the oral cavity.

Reduced Transport Fluid (RTF; Syed and Loesche, 1972)

Stock Mineral Salt Solution No. 1	
K₂HPO₄	0.6%
Stock Mineral Salt Solution No. 2	
NaCl	1.2%
(NH₄)₂SO₄	1.2%
KH₂PO₄	0.6%
MgSO₄	0.25%

Final Transport Medium

Stock solution no. 1	75 ml
Stock solution no. 2	75 ml
Ethylenediaminetetraacetate (0.1 M)	10 ml
Na₂CO₃ (8%, w/v)	5 ml
Dithiothreitol (1%; freshly prepared)	20 ml
Resazurin (0.1%; optional)	1 ml
Distilled water	814 ml

Sterilize by membrane filtration (pore size, 0.22 μm). Dispense into 16 × 125-mm screw-cap tubes (dilution tubes) and 18 × 150-mm test tubes (sample-collection tubes).

The pH should be 8 ± 0.2 without adjustment and decrease to 7 in 48 h in the anaerobic glove box atmosphere (85% N_2, 10% H_2, and 5% CO_2).

Microscopic Examination of Specimens from Natural Habitats Gram-stained smears from materials such as dental plaque, cervicovaginal secretions or the microbial growth on IUDs and their threads often contain large amounts of Gram-positive, branched rods and filaments suggestive of actinomycetes. Using fluorescent antibody techniques these structures may be identified immediately as belonging to one of the traditional *Actinomyces* species. However, as specific antisera to all of the actinomycete species and serovars potentially present are usually not available, and as filamentous cells may be produced by various other bacteria, microscopy alone does mostly not suffice to demonstrate actinomycetes with adequate certainty.

Media and Procedures for Isolation and Cultivation from Natural Habitats Usually, plaque samples or similar specimens have to be dispersed before dilution and plating. Dispersal can be achieved by sonic or mechanical (Potter-type homogenizer or Vortex shaker) treatment. The homogeneous suspensions thus obtained can be diluted as usual using prereduced fluids. To grow fastidious and oxygen-sensitive members of the human indigenous microflora, all procedures should be carried out in an anaerobic chamber (glove box). If only actinomycetes are being considered, exposure to air for a shorter period of time does not significantly decrease the number of colony-forming units.

For cultivating fermentative actinomycetes, the various nonselective media recommended for examining clinical specimens before can also be used for ecological purposes. However, selective principles understandably facilitate their isolation and identification. On the other hand, selective media including those devised for fermentative actinomycetes so far may also inhibit certain strains of the species that are looked for, so that the qualitative and quantitative results obtained from such selective media may not necessarily reflect the natural interrelations *in vivo* with sufficient accuracy.

The recipes for four different media devised for the selective isolation and enumeration of oral actinomycetes are given below. These media differ to some extent in their selectivity against other oral microbes and also in their inhibitory activity to individual *Actinomyces* species so that they should be selected according to the special aims of a particular study.

Partially Selective Medium for *A. viscosus* and *A. naeslundii* (CNAC-20 Medium; Ellen and Balcerzak-Raczkowski, 1975)

Add $3CdSO_4 \cdot 8H_2O$ to Columbia CNA agar base (Difco) prepared according to the manufacturer's instructions to give a final concentration of 20.0 µg/ml. Autoclave. Incubate plates containing CNAC-20 medium at $36 \pm 1°C$ in 90% air and 10% CO_2 to encourage growth of *A. viscosus* and *A. naeslundii* while inhibiting the more anaerobic actinomycetes and other Gram-positive anaerobes.

Selective Medium for Oral Actinomycetes (Beighton and Colman, 1976)

Basal Culture Medium (BYS medium)	
Brain heart infusion broth	3.7 g
Yeast extract powder	0.5 g
Polyvinylpyrrolidone	1.0 g
Cysteine · HCl	0.1 g
Agar	1.5 g

Add these ingredients to 100 ml of distilled water and autoclave for 15 min at 121°C. Cool to 45°C and supplement with 5 ml of sterile horse serum.

Selective Enrichment Medium (FC medium)

NaF solution (25 mg/ml)	1.0 ml
Colistin sulfate solution (1 mg/ml)	0.5 ml

Sterilize the solutions separately by autoclaving at 121°C for 15 min before adding them to 100 ml of BYS medium.

Selective GMC Medium for *A. viscosus* and *A. naeslundii* (Kornman and Loesche, 1978)

Basal Medium

The enriched gelatin agar of Syed (1976) is recommended, but other complex media may yield similar results.

Selective Medium

Add Cadmium sulfate ($3CdSO_4 \cdot 8H_2O$) and Metronidazoleto to 1 liter of the basal medium to give final concentrations of 20 µg/ml and 10 µg/ml, respectively.

Autoclave cadmium sulfate together with the basal medium. Cool to 45–50°C and add a filter-sterilized solution of metronidazole aseptically.

The latter medium was reported to allow 98% recovery of *A. viscosus* and 73% recovery of *A. naeslundii*, while suppressing 76% of the total count of other oral organisms. *Propionibacterium propionicum* strains also grew quite well, whereas *A. israelii* and *A. odontolyticus* were inhibited (Kornman and Loesche, 1978).

Selective Medium for Detection and Enumeration of *A. viscosus* and *A. naeslundii* (CFAT Medium; Zylber and Jordan, 1982)

Trypticase soy broth (BBL)	30.0 g
Glucose	5.0 g
Agar	15.0 g
Cadmium sulfate ($3CdSO_4 \cdot 8H_2O$)	13.0 mg
Sodium fluoride (NaF)	80.0 mg
Neutral acriflavin	1.20 mg
Potassium tellurite	2.50 mg
Basic fuchsin	0.25 mg
Defibrinated sheep blood	50.0 ml

Dissolve ingredients in distilled water and dilute up to 1 liter. Adjust the final pH to 7.3.

This medium was reported to give considerably higher counts for *A. viscosus* and *A. naeslundii* than those from the FC and CNAC-20 media, respectively. The major interfering organisms on this medium were *Aerococcus* and *Rothia*, whereas FC and CNAC-20 media were less selective and supported growth of significant numbers of other oral organisms in the study of Zylber and Jordan (1982).

A different means of selectively isolating fermentative actinomycetes, especially *A. israelii* and *A. gerencseriae*, was described by Fritsche (1964).

Selective Isolation of *A. israelii* (Fritsche, 1964)

Suspend heavily contaminated materials such as feces in a suitable reduced transport or culture fluid (e.g., reduced transport fluid of Syed and Loesche [1972] or thioglycolate medium). Add 1 ml of the suspension thus obtained to 1 ml of toluene in a screw-cap tube. Agitate the tube on a mechanical shaker (high speed) for 20–25 min. Remove the watery suspension at the bottom of the tube carefully with a capillary pipette and add it to 10 ml of the same transport or broth medium. Remove remaining droplets of toluene on the surface of the medium with a Bunsen flame. Centrifuge so that the Actinomyces cells can settle, and streak the sediment onto nonselective agar media.

On first isolation, the colony morphology of the actinomycetes may be altered by this treatment but will become typical again upon subculture. *Actinomyces israelii* and possibly *A. gerencseriae* have been proven to survive this treatment in all instances. However, whether and to what extent the colony counts decrease after this procedure have not been investigated.

In a study of Lewis et al. (1995), the combination of mupirocin and metronidazole was used as selective principle for isolating *Actinomyces* species from clinical specimens. These authors reported that significantly more *Actinomyces* species were found on this selective medium than on corresponding nonselective media from both dental specimens and intrauterine contraceptive devices.

CULTIVATION Cell mass of *Actinomyces* species may be required in larger quantities for chemotaxonomic, molecular or physiologic analyses. Broth cultures usually applied for these purposes only give satisfactory results with species or strains which do not form persisting mycelial growth and which are comparatively rapid growers. Highly filamentous strains usually produce only scanty growth in fluid cultures unless the medium is continuously agitated or stirred during incubation. Furthermore, broth cultures are easily contaminated with propionibacteria from the skin of laboratory personnel. Growth of these contaminants may be facilitated by the slow and granular growth of certain actinomycetes and may be difficult to detect because the microscopic morphology of *Actinomyces* and *Propionibacterium* species is often very similar.

Mass cultures of filamentous *Actinomyces* species are therefore obtained most easily and economically when the membrane culture technique described in the section on Chemotaxonomy is used. In contrast, the more coryneform members of the genus *Actinomyces* can well be cultivated in fluid broth media.

The agar and broth media most suitable for growing fermentative actinomycetes together with certain supplements, which may be necessary, are described in detail in the sections on Isolation from Clinical Specimens and Natural Habitats.

PRESERVATION *Actinomyces* species may cause problems on continuous subcultivation, in particular when the growth medium contains fermentable carbohydrates that rapidly lead to toxically low pH values. Furthermore, cultures of these organisms tend to be readily contaminated, especially with propionibacteria, as mentioned above. Therefore, procedures that avoid both contamination and loss of viability should be used especially for long-term storage.

Cultures in routine use may be maintained by weekly transfer in thioglycolate broth containing 0.2% (v/v) rabbit serum or by monthly transfer on Fortner plates containing CC medium with 0.2% (v/v) rabbit or fetal bovine serum but no carbohydrates. Because of the risk of contamination of broth cultures, these have to be checked carefully for purity upon each transfer. On Fortner plates, accidental contaminants are usually detected more easily, but contamination problems may be caused by the *Serratia* strain simultaneously grown on the plate especially when the agar surface is too wet upon inoculation.

For longer preservation (up to 1 year) and long-term storage (several years to 10 years or longer), freezing or lyophilization is recommended (Slack and Gerencser, 1975; Schaal, 1986a; Schaal, 1986b). For freezing, 3–5 ml of brain heart infusion broth in screw-capped tubes are inoculated with 0.1 ml of a fresh broth culture or a dense suspension of the organism in brain heart infusion broth prepared from a 3–7 day-old agar culture, and the inoculated broth is incubated at $36 \pm 1°C$ for 3–5 (to 7) days under appropriate gaseous conditions (in air, in air with added CO_2, or in an anaerobic jar) with the caps loosened. As soon as good growth is obtained, the caps are tightened and the tubes are placed in a freezer at –70°C.

For freeze-drying, the sediment of broth cultures or—preferably—biomass scraped from agar plates or dialysis membranes are suspended in sterilized skim milk or lyophilization medium and lyophilized by standard techniques (Slack and Gerencser, 1975; Schaal, 1986a; Schaal, 1986b).

Lyophilization Medium for Long-Term Storage of *Actinomyces* Species (Schaal, 1986b)

Brain heart infusion broth	1 part
Horse or fetal bovine serum	1 part

Mix and add sucrose to give a final concentration of 7% (w/v) in the whole mixture.

Using lyophilized cultures, Slack and Gerencser (1975) reported viability after 10 years of storage at room temperature. We were even able to revive most of the lyophilisates that had been stored for 20–30 years at room temperature. Frozen broth cultures will survive for at least 1 year.

Identification

Although *Actinomyces* species possess several differential characteristics potentially useful for their identification, they have been known for many years to be notoriously difficult to identify in medical or veterinary microbiology laboratories because morphology and traditional biochemical and physiological tests are often misleading and may be responsible for inadequate diagnoses or treatment failures. With the exception of serological methods, there is essentially no simple traditional test that will reliably differentiate *Actinomyces* species from morphologically or physiologically similar organisms. The same is true when individual isolates have to be identified to species level. It is for this reason that modern taxonomic techniques such as chemotaxonomy, fermentation end product analysis or molecular genetic methods have contributed so much to the classification and identification of these organisms. For the same reason, serological tests such as immunofluorescence procedures have widely been employed for identification purposes because they represent rapid, simple, and supposedly reliable diagnostic tools. However, more detailed serological as well as molecular studies (Schaal and Gatzer, 1985c;

Johnson et al., 1990; Stackebrandt and Charfreitag, 1990) have indicated that the antigenic and taxonomic structure even of those species that have been recognized for many years is more heterogeneous than previously thought. Therefore, identification results should not be based solely upon serology when a complete and reliable bacteriological diagnosis is to be achieved.

Morphological Characteristics

Morphological criteria were previously considered especially important for the identification of various actinomycetes because branching filamentous cells and mycelial growth appeared to be stable and reliable diagnostic characters. However, detailed morphological observations as well as modern taxonomic developments have shown that morphology may considerably vary within species and even between different cultures of the same strain, that morphological similarities do not necessarily reflect phylogenetic relationships, and that there is considerable overlapping between the microscopic and cultural characteristics of different species, genera and even families. Thus, microscopic or macroscopic examinations of clinical specimens or cultures usually do not allow reliable identification of the disease or the organisms present.

Nevertheless, cellular and colonial features may provide presumptive evidence for the systematic position of an unknown isolate and may therefore facilitate the selection of proper tests for rapidly and economically achieving a definite identification. When Gram-stained smears from clinical specimens show fragmenting filaments that are branching, irregularly curved and unevenly stained (Figs. 2 and 13), this is highly suggestive of an actinomycete infection or colonization, although not specific for a certain *Actinomyces* species because *Propionibacterium propionicum* and obligately aerobic actinomycetes (such as *Nocardia* or *Actinomadura* species) may be morphologically similar. On the other hand, fermentative actinomycetes often appear as diphtheroid rods or even coccoid forms resembling corynebacteria (Figs. 3 and 17), cutaneous propionibacteria, or streptococci, especially *Streptococcus mutans*, which tends to form elongated cells. Furthermore, *Actinomyces hordeovulneris* has been shown to produce L-phase variants spontaneously with coincident calcium deposition (Buchanan and Scott, 1984a) that are difficult to recognize as actinomycete elements.

Ørskov (Erikson, 1940) introduced the method of microscopically examining growing colonies of actinomycetes *in situ* on the agar surface. This method proved to be extremely useful for the presumptive recognition of these organisms, especially when early growth stages (after

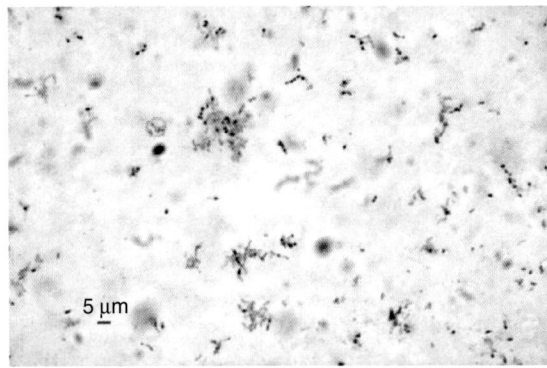

Fig. 17. *Actinomyces meyeri*; Gram-stained smear from a mature colony (brain heart infusion agar, 14 days 36 ± 1°C, and Fortner's method). Note the uneven uptake of the dye.

24–48 h of incubation) are examined. Lentze (1938a), using Fortner's method for cultivating actinomycetes, modified Ørskov's technique by using a long-distance, low-magnification objective (10×) to examine the plates from the bottom in transmitted light without opening them and thus without disturbing their semi-anaerobic microenvironment (Figs. 15 and 16). For applying this modification under routine conditions, the organisms are incubated anaerobically for 18–48 h on a suitable transparent agar medium (e.g., BHIA or CC medium) in glass Petri dishes with even-bottom surfaces. These are then searched microscopically for growth of filamentous colonies *in situ*. For more detailed morphological studies with higher magnifications, the following modification of the slide culture technique introduced by Schaal (Schaal, 1986a; Schaal, 1992b) is recommended.

Slide Culture Technique for the Detailed Microscopic Observation of Filamentous Actinomycete Microcolonies (Schaal, 1986a)

Preparation of Slide Cultures

Cover sterile glass slides with a thin layer of BHIA by briefly immersing one surface of the slide in melted BHIA of about 50°C. For solidification of the agar, transfer the slide to a sterile *moist* chamber at ambient temperature to prevent rapid drying of the thin agar layer. These chambers can be prepared from Petri dishes containing a bottom layer of sterilized filter paper, which has been moistened with sterile saline or distilled water. Perform all of the necessary manipulations under strict aseptic conditions, preferably in a clean bench.

Culture

Inoculate the agar-coated slides with the strain to be examined using a very thin glass filament. This can be prepared easily from a Pasteur pipette by softening the glass in a Bunsen flame and rapidly pulling both ends apart. The resulting filament is sterile and can be used immediately as an inoculating loop or needle. Depending on the organism to be examined, incubate the slide cultures inside their moistened dishes in an anaerobic or CO_2-enriched jar for 5–20 h at 36 ± 1°C.

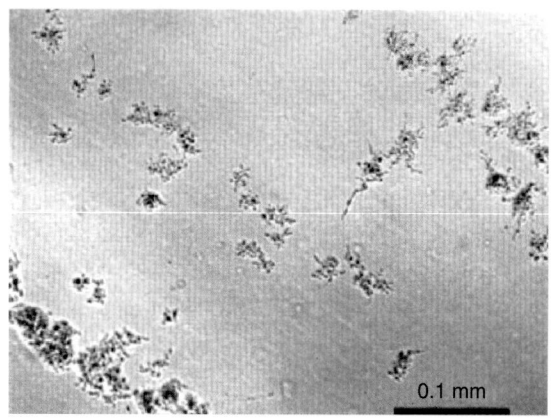

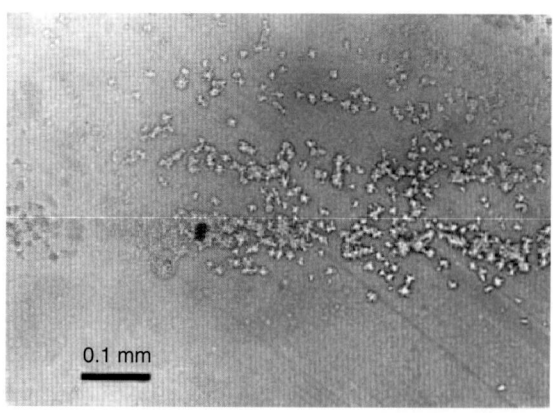

Fig. 18. *Actinomyces israelii*; microcolonies with longer and shorter filaments and visible fragmentation; micrograph in situ in transmitted light (CC medium, 24 h at 36 ± 1°C, and Fortner's method).

Fig. 19. *Actinomyces bovis*; small, irregular, but not filamentous microcolonies; micrograph in situ in obliquely transmitted light (CC medium, 24 h 36 ± 1°C, and Fortner's method).

Microscopy

After incubation, remove the slides from the moist chambers and allow them to dry in air for 10–15 min. Thereafter, place 1 drop of lactophenol-cotton blue mounting fluid in the center of the agar layer and cover it with a clean cover slip. View under a microscope using phase-contrast equipment with objectives between 50× and 100× magnification (Fig. 6).

Lactophenol-cotton Blue Mounting Solution (Hendrickson, 1985)

Phenol crystals	20 g
Lactic acid	20 ml
Glycerol	40 ml
Distilled water	20 ml

Dissolve the ingredients by heating the flask in a hot-water bath. Add 0.05 g of cotton blue (Poirier blue).

Under such conditions, the microcolonies of fermentative actinomycetes may be highly filamentous without signs of fragmentation ("spider colonies"; Fig. 15), but the filaments may also be shorter although easy to identify (Fig. 18), or fragmentation may occur in very early growth stages.

Although there is some relation between species affiliation and the morphology of the microcolonies formed, this character is considerably variable and depends on growth conditions and medium composition as well as on the individual strain under examination. Filamentous microcolonies are usually not found in *A. odontolyticus*, *A. meyeri*, *A. georgiae*, *A. graevenitzii*, *A. radingae*, *A. turicensis*, *A. europaeus*, *A. neuii*, *A. funkei*, *A. radicidentis*, *A. urogenitalis*, *A. cardiffensis*, *A. nasicola* and *A. hongkongensis*. The same is apparently true for the animal species *Actinomyces bovis* (Fig. 19), *A. denticolens*, *A. howellii*, *A. slackii*, *A. hyovaginalis*, *A. bowdenii*, *A. canis*, *A. catuli*, *A. suimastitidis*, *A.*

marimammalium and *A. coleocanis*, although the description of these species was primarily based upon 16S rDNA sequence data and a few physiological characters, whereas a detailed morphological characterization is as yet missing. Thus, among the members of the genus *Actinomyces* as currently defined, only *A. israelii*, *A. gerencseriae*, *A. naeslundii*, *A. viscosus*, *A. hordeovulneris* and possibly *A. oricola* usually or occasionally develop characteristic filamentous microcolonies.

Mature colonies may be observed with the naked eye or with a hand lens, but microscopy may provide more detailed information. Depending on experience and preference of the investigator, the following microscopic techniques may be applied: 1) standard microscopy in transmitted light using transparent media and a long-distance, low-magnification objective; 2) reflected light microscopy using a standard laboratory microscope, a long-distance objective, and an additional, external source of light (the path of which is adjusted to hit the area below the objective at an acute angle); and 3) reflected light microscopy using a dissecting, stereoscopic microscope. Irregularly raised, opaque colonies are best observed under reflected light. The examination of flat, smooth, and translucent colonies is easier, however, when the usual transmitted-light microscopy is applied.

Mature *Actinomyces* colonies may retain much of their mycelial appearance, especially in the periphery (Fig. 7). The typical "molar tooth" colonies of *A. israelii* (Fig. 9) and *A. gerencseriae* usually also show discernible peripheral filaments that radiate from the dense center of the colony. But in many strains of the primarily filamentous *Actinomyces* species, the actinomorphous edge may gradually get lost with further

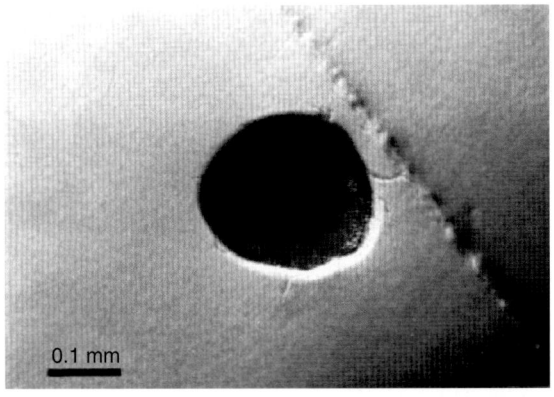

Fig. 20. *Actinomyces israelii*; nearly completely fragmented macrocolony with a few single filaments left in the periphery; micrograph in situ in obliquely transmitted light (CC medium, 7 days at 36 ± 1°C, and Fortner's method).

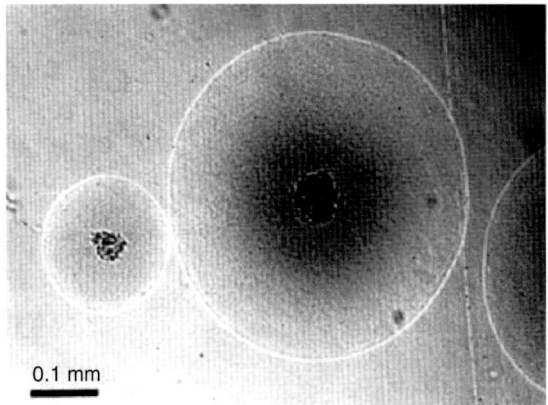

Fig. 22. *Actinomyces meyeri*; larger and smaller translucent macrocolonies with optically dark centers; micrograph in transmitted light (CC medium, 14 days at 36 ± 1°C, and Fortner's method).

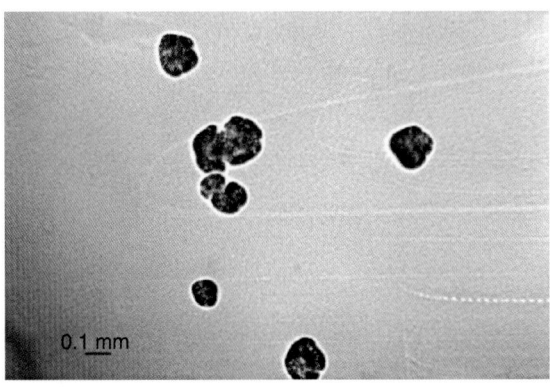

Fig. 21. *Actinomyces bovis*; small, comparatively dense, nonfilamentous colonies; micrograph in obliquely transmitted light (CC medium, 14 days at 36 ± 1°C, and Fortner's method).

development of the colony (Fig. 20). Most *Actinomyces* species, especially the majority of the recently described new ones, produce entirely smooth nonfilamentous colonies. These colonies may still be dense and opaque such as in *A. bovis* or *A. neuii* (Fig. 21), but they may also be flat and translucent such as in *A. odontolyticus* (Fig. 10) and *A. meyeri* (Fig. 22).

Acid End Product Analysis

As with strict anaerobes, the identification of end products from the fermentative metabolism of actinomycetes and related organisms provides additional criteria for allocation of an unknown isolate to the genus *Actinomyces*. Usually, gas-liquid chromatography (GLC) as described in the VPI Anaerobe Laboratory Manual (Holdeman et al., 1977) is used for this purpose, but more modern techniques such as high per-

formance liquid chromatography (HPLC) may give similar or even better results.

Determination of Fermentation End Products (Holdeman et al., 1977; Schaal, 1985b).

Culture Conditions

Inoculate peptone-yeast extract-glucose medium (PYG) or chopped meat-carbohydrate medium (CMC; Holdeman et al., 1977) heavily from an actively growing (3–7 days old, depending on the generation time of the test strain) agar culture (e.g., BHIA) by scraping off the growth using an inoculating needle or a sterile scalpel or by employing the dialysis membrane culture technique described in the section Chemotaxonomic Characters. Disperse the inoculum in the broth medium as homogeneously as possible by aspirating and expelling the colonial material several times with a Pasteur pipette. Incubate the inoculated tubes at 36 ± 1°C for 3–7 days under anaerobic conditions with added CO_2 (e.g., in an anaerobic jar or chamber), or until visible growth is obtained.

Preparation of Extracts for Analysis of Volatile Fatty Acids

Pipette 1 ml of the broth culture into a glass tube. Add 0.2 ml of 50% (v/v) H_2SO_4, 0.4 g of NaCl and 1 ml of diethyl ether, and close the tube tightly with a screw cap lined with an inert polytetrafluoroethylene seal. Mix the components by inverting the tube gently about 20 times and allow the ether layer to separate from the aqueous phase either by placing the tube for a short time on the bench or by centrifugation. Pipette the upper ether layer, which contains the extracted volatile fatty acids, off the aqueous layer using a microliter syringe. Use this ether extract directly for gas chromatographic analysis, when the instrument is equipped with a flame ionization detector. When a thermal conductivity detector is used instead, remove the ether layer carefully from the tube and transfer it to another glass tube to which anhydrous 4–20 mesh $CaCl_2$ is added to equal about one-fourth of the volume of the ether in the tube. Close the tube and let it stand for about 5 min to remove most of the residual water.

Preparation of Extracts for Analysis of Nonvolatile Acids

Pipette 1 ml samples of the original culture into glass tubes as before, add 2 ml of methanol and 0.4 ml of 50% H_2SO_4, and seal the tubes tightly. Heat the mixture at 60°C in a heating block or water bath for 30 min so that methyl esters can form. Add distilled water (1 ml) and chloroform (0.5 ml), close the tube, and mix the contents by gentle inversion about 20 times. Any emulsion that may form can easily be broken by brief centrifugation. Aspirate the lower chloroform layer, which contains the methylated derivatives, directly from the bottom of the tube for gas chromatographic analysis using a microliter syringe.

Gas-liquid Chromatography

Comparatively simple and inexpensive instruments with either thermal conductivity or flame ionization detectors can be used. A temperature programming unit is not necessary but may facilitate the identification of the acids. Various types of column packings in glass or steel columns as well as capillary columns have been used successfully. FFAP or LAC-1-R 296 (WGA Düsseldorf, no. 6035) on Chromosorb-G as stationary phases has proved especially useful. The operating conditions routinely followed in the authors' laboratories are as follows: injection block, 160°C; detector block, 180°C; column oven programmed to rise from 80 to 160°C at a rate of 2°C/min; attenuation and carrier-gas flow rate depends on the types of columns and detectors used; and recorder speed may be 0.5 or 1.0 cm/min. Volatile and methylated products are run on the same column using the same operating conditions.

Procedure

Draw the ether or chloroform extracts (10 μl in case of use of a thermal conductivity detector and 1 μl in case of use of a flame ionization detector) into an appropriate microliter syringe, wipe off the outside of the injection needle, and inject the extract onto the column. Compare the elution times of the various compounds to be expected with those of a standard mixture (Holdeman et al., 1977) treated in the same way. Usually, qualitative interpretation of the results with visual comparison of the peak heights in the same chromatogram is sufficient for identification purposes. Alternatively, a more detailed semiquantitative analysis using a standard curve of peak heights can also be performed. After testing approximately 20 methylated samples, recondition columns by injecting about 10 μl of methanol.

Actinomyces species characteristically produce comparatively large amounts of succinic acid as compared to lactic and acetic acids (Table 10) when grown under anaerobic conditions with added CO_2. Morphologically similar strains of *Propionibacterium propionicum* or other *Propionibacterium* species can easily be differentiated by their production of major amounts of propionic acid, which is not produced by any member of the genus *Actinomyces*. Arcanobacteria produce considerable amounts of acetic and lactic acid, but small amounts of succinic acid, which may be difficult to detect. For comparison, some other morphologically similar genera are included in Table 10 to stress the differentiating

power of GLC end product analysis when applied to nonsporeforming Gram-positive, fermentative bacteria.

Chemotaxonomic Identification Procedures

Comparatively simple modifications of several chemotaxonomic procedures have been developed that are easily applicable for identification purposes under routine conditions (Becker et al., 1964; Alshamaony et al., 1977; Schaal, 1985b). This is especially true for the demonstration of diaminopimelic acid (A_2pm) or certain sugars in whole-cell hydrolysates, for the analysis of long-chain fatty acids in whole-cell extracts, and for the determination of cytoplasmic polypeptide patterns using sodium dodecylsulfate (SDS) polyacrylamide gel electrophoresis (PAGE). For clinical isolates that exhibit typical actinomycete morphology and which possess the basic physiological properties of fermentative actinomycetes, the application of chemotaxonomic methods is not necessarily required when reliable identification results are to be obtained. However, in case of morphologically and physiologically less characteristic isolates or in case of insufficient experience of the microbiologist with these organisms, the application of chemotaxonomic tests is strongly recommended unless suitable molecular methods are used to keep the number of misidentifications low.

Cell wall analysis of actinomycetes is undoubtedly a very valuable tool for the classification and identification of these organisms (Cummins and Harris, 1956; Cummins and Harris, 1958). However, even simplified techniques for the preparation of cell walls (Bousfield et al., 1985) still make the whole procedure so elaborate and time-consuming that most laboratories will not be able to perform such tests routinely. The use of whole cells instead of cell walls clearly saves time and labor, although it also reduces the amount of information that may be derived from these techniques. Thus, several simplified chemotaxonomic techniques have been used in the authors' laboratories for many years and have been found to be suitable for application in the routine diagnostic laboratory.

PREPARATION OF BIOMASS FOR CHEMOTAXONOMIC ANALYSES For most chemical analyses, comparatively large amounts of biomass are needed that should be free of contaminating material from the culture media. For many microbes, this requirement is readily fulfilled by using carefully washed cell mass from fluid cultures. Broth cultures of fermentative actinomycetes, however, are especially prone to microbial contamination, which is more easily controlled when the organisms are grown on solid media. On the other

Table 10. Morphological, physiological, and chemotaxonomic characters useful for differentiating the genera of the family Actinomycetaceae from each other and from phenotypically related taxa under routine diagnostic conditions.

Genus	Morphological and chemotaxonomic characters				Physiological characters				
	FM	A$_2$pm	MA	CFA	AG	Cat	Nit	Mot	Fermentation end products[a]
Actinomyces	D[b]	–	–	S, U, (C), (I), (A)	D	D	D	–	A/a, L/l, S/s
Actinobaculum	–	–	–	S, U	D	–	–	–	A, (S)
Arcanobacterium	–	–	–	S, U	+	–(D)[c]	D	–	A/a, L, S/s
Mobiluncus	–	–	–	S, U	–	–	D	+	a, (l), S
Varibaculum	–	–	–	ND	–	–	D	–	L, S
Propionibacterium	D	LL/DL[d]	–	S, I, A, C	D	D	D	–	a, P, (iv), l, (s)
Corynebacterium	D	DL[e]	+	S, U, (T)	+	+	D	–	A, (p), L, (s)
Erysipelothrix	–	–	–	S, I, A, U	+	–	–	–	A, L, s
Listeria	–	DL	–	S, I, A	+	+	D	+	(a), L, (s)
Bifidobacterium	–	–	–	S, U	–	–/D	–	–	A, L, (s)
Eubacterium	–	DL/–[f]	–	ND	–	–	D	D	a, (p), (B), (ib), (iv), (C), l, (s)
Lactobacillus	–	–	–	S, U, (C)	D	–	–	–/D	(a), L, (s)
Gardnerella	–	–	–	S, U	D	–	–	–	A, (l), (s)
Rothia	D	–	–	S, I, A	+	D	+	–	(a), L, (s)

Symbols: –, not present; +, present; (), may or may not be present; and ND, no data.
Abbreviations: FM, filamentous microcolonies; A$_2$pm, 2,6-diaminopimelic acid; MA, mycolic acids; CFA, cellular fatty acids: S, straight-chain saturated; U, monounsaturated; A, *anteiso*-methyl-branched; I, *iso*-methyl-branched; C, cyclopropane ring; T, 10-methyl branched (tuberculostearic acid); AG, aerobic growth; Cat, catalase activity; Nit, nitrate reduction; Mot, motility; A/a, acetic acid; P/p, propionic acid; B, *n*-butyric acid; ib, *iso*-butyric acid; iv, *iso*-valeric acid; C, *n*-caproic acid; L/l, lactic acid; S/s, succinic acid.
[a]Capital letters, present in major amounts; lower-case letters, present in smaller amounts; capital/small letters, present in major or smaller amounts.
[b]Species differences.
[c]Usually absent, but species variation may occur.
[d]LL- and/or *meso*-A$_2$pm present.
[e]*meso*-A$_2$pm present.
[f]*meso*- or no A$_2$pm present.
Modified from Schaal (1992).

hand, the substrate mycelium of *Actinomyces* species may be difficult to remove from an agar medium. To overcome these difficulties, the following simple and convenient procedure for obtaining noncontaminated cell mass of fermentative actinomycetes is recommended.

Preparation of Biomass of Fermentative Actinomycetes for Chemotaxonomic Tests (Schaal, 1985b)

Media

Brain heart infusion agar (BHIA) is satisfactory for essentially all of the organisms treated in this chapter. However, other media that provide good growth of the test organisms may also be used.

Membrane Culture Technique

Place autoclaved sheets of dialysis tubing on top of the agar surface of BHIA (in a Petri dish) and inoculate them with a heavy suspension of the test strain. Depending on its growth conditions, incubate the plates aerobically, under increased CO$_2$ tension, or anaerobically (anaerobic jar) at $36 \pm 1°C$ until sufficient growth is obtained.

Cell Mass Preparation

After sufficient incubation, remove the dialysis membranes with the microbial growth on top from the agar surface with a sterile forceps and transfer them to Erlenmeyer flasks containing either distilled water or saline and a few glass beads (~5 mm in diameter). Inactivate the organisms either by adding formaldehyde or by heating the flasks in a water bath at 75°C for 45 min. Wash growth off the membranes by shaking the flasks in a rotary shaker at 200–250 rpm and collect the cell mass by centrifugation. Wash the sediment in two changes of distilled water. Dry the cells by lyophilization (amino acid and sugar analyses may also be carried out on cells dried overnight in absolute ethanol).

AMINO ACID ANALYSIS One of the chemotaxonomic tests that are comparatively easy to perform is the demonstration of 2,6-diaminopimelic acid (A$_2$pm) in whole-cell hydrolysates (Schaal, 1985b). Although this test does not aid in the differentiation of *Actinomyces, Actinobaculum* or *Arcanobacterium* species from each other, it allows the reliable differentiation of *A. israelii* and *A. gerencseriae* from *Propionibacterium propionicum* (which are very similar morphologically) and of many of the new *Actinomyces* species from corynebacteria, propionibacteria, or listeriae (Table 10).

Demonstration of 2,6-Diaminopimelic Acid in Whole-Cell Hydrolysates (Schaal, 1985b)

Preparation of Extracts

Hydrolyze dried cells (1–5 mg) with 1 ml of 6N HCl in a sealed Pyrex tube at 100°C for 18 h. Cool and filter the hydrolysate through paper and wash the residue with 1 ml of distilled water. To remove most of the HCl, place the filtrate in a sand bath and evaporate the excessive fluid without burning the residue. Repeat this procedure 2 or 3 times after further addition of distilled water, or use a rotary evaporator to dry the filtrate under reduced pressure at 40°C.

Thin-layer Chromatography

After final washing, take up the extract in 0.3 ml of distilled water and spot 5 μl onto cellulose-coated thin-layer chromatography (TLC) plastic sheets (Merck 5577). For separating the amino acids, use ascending, single-dimensional TLC in a solvent system containing methanol-pyridine-10N HCl-water (80:10:2.5:17.5, v/v). A second development in the same direction and solvent mixture with intermediate air-drying can improve separation and identification of A_2pm.

Identification of the A_2pm Isomers

Visualize amino acid spots by spraying chromatograms with ninhydrin in acetone (0.1%, w/v) followed by air-drying and heating at 100–110°C for about 2 min. The isomers of A_2pm produce slowly moving, olive-green spots that fade to yellow; hydroxy-A_2pm has the lowest R_f value and is followed by meso-A_2pm and LL-A_2pm, respectively. Most of the other amino acids present in whole-cell hydrolysates give purple or blue spots. A standard solution containing meso-A_2pm (1%, w/v), which is usually contaminated with traces of LL-A_2pm, should always be included. DD-A_2pm is not separated from meso-A_2pm by this method.

SUGAR ANALYSIS   The sugar patterns derived from TLC analyses of whole-cell hydrolysates are more difficult to evaluate than the simple A_2pm patterns. The various diagnostic sugars may originate from the cell wall, wall-associated structures, the cytoplasm, or from the culture medium when the cell mass has not been washed sufficiently before hydrolysis. Even trace amounts of contaminating sugars from the growth medium may become apparent when highly concentrated extracts are used. Nevertheless, TLC sugar patterns are especially useful for differentiating *Actinomyces* species.

Determination of Sugars in Whole-Cell Hydrolysates of Fermentative Actinomycetes (Schaal, 1985b)

Preparation of Extracts

Hydrolyze dried cells (50 mg) with 1 ml of 2N H_2SO_4 in a sealed Pyrex tube held at 100°C for 2 h. Neutralize the hydrolysate (pH 5.0–5.5 with methyl red as an internal indicator) by adding barium hydroxide (0.5 g) and shaking the mixture several times. Deposit the resulting precipitate and the remaining $Ba(OH)_2$ by centrifugation at 6000 rpm. Remove the supernatant carefully and evaporate to dryness either by lyophilization or on a rotary

evaporator under reduced pressure at 40°C. Take up the residue in 0.4 ml of distilled water and add pyridine (~1 ml). Mix and separate the pyridine phase from the aqueous phase by centrifugation at 6000 rpm. Transfer the upper pyridine layer, which contains the partially purified sugar extract, to a small glass vial and repeat the extraction procedure. Combine the pyridine extracts and concentrate the mixture in a stream of nitrogen at room temperature to give a final volume of approximately 0.3 ml.

Thin-layer Chromatography

Spot 5–10 μl of the concentrated pyridine extract onto cellulose-coated TLC aluminum sheets (Merck 5552) and separate the sugars by ascending, single-dimensional TLC using the following two solvent mixtures applied consecutively. Solvent 1: n-butanol-water-acetic acid (60:20:20, v/v), in which the chromatograms are run until the solvent front has reached about 6 cm. Remove the sheets from the tank and dry in air. Solvent 2: ethyl acetate-pyridine-water-acetic acid (100:35:25:5, v/v), in which the chromatograms are run three times, with intermediate air-drying, to an optimal distance of about 12 cm.

Identification of Sugars

Visualize the sugar spots by spraying chromatograms with an aniline-phthalate reagent (aniline, 2 ml; phthalic acid, 3.3 g; and water-saturated n-butanol, 100 ml), air-drying, and subsequent heating at 100°C for 2–5 min, whereupon hexoses give brown, pentoses red, and 6-deoxy sugars gray-green spots. The R_f values increase in the following order: galactose, glucose, mannose, arabinose, xylose, fucose, ribose, rhamnose and 6-deoxytalose. A mixture of these sugars (1%, w/v) in pyridine should be used as a standard.

In contrast to pathogenic aerobic actinomycetes, *Actinomyces* as well as *Arcanobacterium*, *Actinobaculum* or *Varibaculum* species do not contain detectable amounts of arabinose, xylose or madurose, and the same is true for propionibacteria, bifidobacteria, eubacteria and lactobacilli. The only filamentous, *Actinomyces*-like, fermentative organism that possesses arabinose in whole-cell hydrolysates is *Corynebacterium matruchotii*. The characteristic sugar patterns for *Actinomyces* species are summarized in Table 6 together with additional chemotaxonomic markers.

MYCOLIC ACID ANALYSIS AND OTHER CHEMOTAXONOMIC TESTS   Mycolic acid analysis, which is very useful for differentiating aerobic actinomycetes and mycobacteria, only allows the separation of *Corynebacterium matruchotii* and other corynebacteria from members of the family Actinomycetaceae. Therefore, the simplified technique for analyzing mycolic acids is not described in detail in this chapter but has been published in the literature (Schaal, 1984a; Schaal, 1985b).

Additional useful criteria for identification may be derived from the analysis of long-chain fatty acids and isoprenoid quinones (Collins and Jones, 1981; Kroppenstedt, 1985) and from electrophoretic whole-cell protein patterns (Jackman, 1985). However, as fatty acid analysis requires some experience and as the results may be influenced by media composition and growth conditions, the methods for fatty acid analyses are also not considered in detail in this chapter. For the identification of human isolates of *Actinomyces, Arcanobacterium* and *Actinobaculum* species, determination of whole-cell protein patterns is usually not necessary. Nevertheless, this technique may facilitate the reliable identification of certain animal isolates or some of the recently described new species of human origin.

Whole-cell protein patterns have been used widely to indicate intraspecies relatedness of newly described species (e.g., Pascual Ramos et al., 1997a; Collins et al., 2000b; Hoyles et al., 2000), thus avoiding more laborious DNA-DNA hybridization experiments. While easy to perform by an experienced laboratory, the disadvantage of this method is the lack of a universal cumulative database, thus preventing interlaboratory comparison of patterns.

Determination of Electrophoretic Polypeptide Patterns (Mauff et al., 1981; Dent and Williams, 1985)

Preparation of Protein Extracts

Wash about 1 g (wet weight) of the biomass (obtained as described) 3 times in distilled water with intermediate centrifugation at about 6000 *g*. Use one of the following extraction procedures. Procedure 1: Resuspend cells in distilled water and disintegrate the suspension in a Mickle shaker (Mickle Engineering Ltd., Gomshall, Surrey, UK) with glass beads (grade 12) for 1.5 h. Centrifuge at 17,000 g for 15 min and then at 150,000 *g* for 1 h to remove all traces of membranes. Procedure 2: Alternatively, resuspend the cells in 35 ml of 6.5 mM Tris · HCl (pH 8.4) containing 0.35 mg of DNAse. Pass this suspension through a "French pressure cell" at 2200 psi. Centrifuge the crude extract thus obtained at 20,000 *g* for 15 min and subsequently at 48,000 *g* for 1 h. Concentrate the cell-free extract about 50 times using an Amicon B15 cell.

Electrophoresis

Electrophorese the cell-free extracts in 10% polyacrylamide-SDS gels as described by Swindlehurst et al. (1977) or using gradient polyacrylamide gel electrophoresis (G-PAGE; 4–30% gradient [Pharmacia]), 40 mM Tris-borate buffer (pH 8.4), 200 V, approximately 70 mA for 3 h at 4°C in an ORTEC flat bed electrophoresis apparatus (EGG Instruments, Munich, Germany; Mauff et al., 1981). After electrophoresis, stain the gels with 1% Coomassie brilliant blue (SERVA, Heidelberg, Germany).

Evaluation of Protein Patterns: Two Methods

Method 1. Use visual recognition and comparison of polypeptide patterns as a surveying technique in examinations also employing other taxonomic techniques. This simple and convenient procedure allows the identification of taxonomically identical or very similar strains in comparison to well-defined reference organisms, and it also allows the differentiation of isolates that give similar physiological test results but differ taxonomically.

Method 2. When the electrophoretic polypeptide patterns are produced under highly standardized conditions, scan in a microdensitometer and use the data to perform cluster analysis or computer identification (Kersters and De Ley, 1975; Kersters and De Ley, 1980; Mauff et al., 1981). More recent densitometric analysis, normalization and interpretation of protein patterns are often done with the Gelcompare software package (Applied Maths, Kortrijk, Belgium). Levels of correlation of patterns are usually expressed as percentages of similarity.

Apart from preparing cell-free, more or less purified soluble protein fractions, whole-cell samples produced by incubating a 6.4 M urea solution with an equal volume of packed *Actinomyces* cells may directly be used for PAGE (McCormick et al., 1985). The patterns thus obtained contain many more bands than those derived from cell-free extracts, but they appear to be reproducible and well suited for classification, identification or typing.

Methods for Testing Physiological Properties

Physiological markers, such as growth on different media, oxygen tolerance, CO_2 requirement, fermentation of carbohydrates, catalase activity, nitrate and nitrite reduction and others, have been used for many years to identify *Actinomyces* species but have recently been neglected in favor of DNA sequence data. As far as the genus level of these organisms is concerned, there is no doubt (Bowden and Hardie, 1973; Slack, 1974; Slack and Gerencser, 1975; Collins and Cummins, 1986; Schaal, 1986b) that physiological criteria need support from chemotaxonomic techniques, end product analyses or molecular methods to make them really effective (Tables 6, Table 7 and Table 10). On the species level, however, physiological tests still provide reliable and cost-effective means of identification when the procedures have been adjusted carefully to this special purpose and when a comparatively large battery of tests is used (Slack and Gerencser, 1975; Schofield and Schaal, 1980a; Schofield and Schaal, 1980b; Schofield and Schaal, 1981; Schaal and Schofield, 1984d; Schaal, 1986a; Schaal, 1986b).

A compilation of the traditional procedures for testing physiological markers of fermentative actinomycetes and related organisms was given by Slack and Gerencser (Slack and Gerencser, 1975). These authors proposed the use of thioglycolate fermentation base as mediator of fermentation tests.

Fermentation Tests of *Actinomyces* Species (Slack and Gerencser, 1975)

Basal Medium
Fluid thioglycolate medium without glucose   1 liter
or indicator

| Yeast extract (glucose-free) | 2 g |
| Bromocresol purple (1% aqueous solution) | 2 ml |

To this medium, add low molecular-weight carbohydrates (mono- and disaccharides) aseptically to give a final concentration of 1%; use other compounds at 0.5%.

It was shown that these and similar media, when used in amounts of 1 ml or more in regular test tubes, did not always yield satisfactory and reproducible results, especially when highly filamentous strains were to be identified. This became especially apparent in several numerical taxonomic studies in which analogous techniques were used (Melville, 1965; Holmberg and Hallander, 1973b; Holmberg and Nord, 1975a). These problems are obviously due to the inadequacy of the inoculum size and growth rate of the organisms compared to the volume of the substrate-containing medium. In contrast, the results obtained from miniaturized test procedures (Schofield and Schaal, 1979a; Schofield and Schaal, 1979b; Schofield and Schaal, 1980a; Schofield and Schaal, 1980b; Schofield and Schaal, 1981; Buchanan et al., 1984b) indicate that these modifications of the traditional physiological tests appear to be much more suitable for testing of fermentative actinomycetes because in this case the inoculum-substrate relation is more favorable for slowly and inhomogeneously growing organisms. Among the various commercial miniaturized identification systems, only the Minitek differentiation system (BD) was shown to be applicable to the identification of all members of the genus *Actinomyces* (Schofield and Schaal, 1979a). However, since this system is no longer available on the market, the following modification can be used.

Modified Miniaturized System for Testing Acid Production from Carbohydrates by Fermentative Actinomycetes and Related Organisms (Schofield and Schaal, 1979a; Schofield and Schaal, 1980a; Schofield and Schaal, 1981)

Basal Medium

Thioglycolate broth without glucose or indicator, supplemented with 0.2% (w/v) yeast extract.

Preparation of Carbohydrate-containing Filter Paper Disks

Soak sterile blank filter paper disks (Oxoid) in 3% (w/v) solutions of the various test carbohydrates (filter-sterilized under aseptic conditions) and dry by lyophilization.

Test Procedure

Prepare the inoculum from 4–7 day-old cultures at 36 ± 1°C on BHIA or CC medium by removing colony material carefully without adhering agar particles, and suspend it in basal medium (thioglycolate) using a platinum loop or a sterile spatula. For strains whose colonies adhere firmly to the agar, use the dialysis membrane culture technique described for chemotaxonomic analyses. Homogenize the suspensions using a Vortex mixer or a

similar mixing device after addition of a few sterile glass beads to the tube. Add 2–3 drops of a densely turbid suspension thus obtained to each well of a microtiter tray with flat bottoms already containing one of the sugar-impregnated disks.

Cover the trays with a plastic tape or a plastic cover (when using a tape, perforate it over each well with a sterile injection needle to allow gaseous exchange), and incubate for 2–3 days in an anaerobic jar at 36 ± 1°C. After incubation, add 1 drop of a 0.2% (w/v) solution of bromothymol blue to each well. Acid production is indicated by a yellow color of the indicator; when the carbohydrate has not been fermented, the color of the indicator is green. A blue-green or blue color is usually due to contamination of the inoculum or a test strain not belonging to the genus *Actinomyces*.

Determination of Nitrate and Nitrite Reduction and Urease and Arginine Dihydrolase Activities by Fermentative Actinomycetes (Schofield and Schaal, 1980b; Schofield and Schaal, 1981)

Basal Medium

Bacto peptone (Difco)	5.0 g
Yeast extract (Difco)	3.0 g
Glucose	0.5 g
K_2HPO_4	2.0 g
Agar	0.5 g

Dissolve in and dilute up to 1 liter with distilled water. Adjust the pH to 6.8 and sterilize the medium by autoclaving at 121°C for 15 min.

Test Procedure

Add the test substrates to basal medium from filtered solutions to obtain the following final concentrations:

Sodium nitrate	0.1%
Sodium nitrite	0.005%
Urea	0.5%
Arginine	0.5%

Pipette the various solutions of the test substances in basal medium into the wells of microtiter trays in volumes of 0.15 ml. Inoculate with suspensions of the test organisms in basal medium containing no agar, seal the wells with a plastic tape, which is perforated as described above, or use a plastic cover. After an incubation of 3–4 days, assess the formation (nitrate reduction) or disappearance (nitrite reduction) of nitrite by adding one drop of the usual nitrite reagent (8.0 g of sulfanilic acid + 1000 ml of 5N acetic acid and 5.0 g of α-naphthylamine + 1000 ml of 5N acetic acid, mixed in equal amounts immediately before use). The development of a brick-red color immediately after addition of the reagent indicates the presence of nitrite. If no color develops in the nitrate reduction test, add a small amount of fine granular zinc (which would reduce nitrate to nitrite when the former is still present in the well) to differentiate between negative nitrate reduction and reduction of both nitrate and nitrite. Assess urease activity and deamination of arginine by means of the ammonia released into the medium. To detect the presence of ammonia, add 1–2 drops of Nessler's reagent to the wells whereupon a heavy, dark yellow-to-ochre precipitate forms immediately. Slowly developing precipitates may be derived from other ingredients of the test medium and must be considered as negative test results.

The test for esculin hydrolysis may be performed using a standard esculin hydrolysis test medium dispensed into microtiter wells as described for the nitrate reduction test. Further characters useful for the identification of fermentative actinomycetes and related organisms can be derived from carbon source utilization tests. However, in organisms such as the fermentative actinomycetes, which require complex media for good growth, these tests are comparatively difficult to perform and only provide reproducible results when the basal medium is carefully adjusted (Schofield and Schaal, 1979b). Therefore, results from the latter tests were not included in Tables 11 and 12, which summarize the physiological characters most useful for the identification of *Actinomyces* species.

Tests for enzymatic activities detectable by means of "chromogenic" or "fluorogenic" substrates were also shown to possess differential power when applied to the classification and identification of fermentative actinomycetes and related organisms (Schofield and Schaal, 1981; Casin et al., 1984; Kilian, 1987; Lämmler and Blobel, 1988; Brander and Jousimies-Somer, 1992; Maiden et al., 1996; Sarkonen et al., 2001), but they were again found to be difficult to standardize because many of the enzymes which might be detected in this way are inducible, so that the test result highly depends on the nature of the growth medium used for preparation of the inoculum. Furthermore, enzyme reactions are not equally suited for the identification of all of the *Actinomyces* species described so far (Brander and Jousimies-Somer, 1992).

Various commercial test systems, which are based upon such enzyme tests completely or in part, have been used for the characterization and identification of *Actinomyces* species, especially of the recently described new taxa. These systems include the API ZYM, the RapID ANA II, the API 20 anaerobes, the API 20 Strep, the API rapid ID 32 Strep, the API CORYNE, the API (RAPID) CORYNE (bioMérieux), and the Rosco diagnostic tablets (Rosco, Taastrup, Denmark) systems. Furthermore, a test based on 4-methylumbelliferyl (4-MU) derivatives of various substrates has also been applied successfully to the phenotypic identification of these organisms (Sarkonen et al., 2001).

4-MU Fluorogenic Substrate Tests (Whiley et al., 1990; Maiden et al., 1996)

4-MU substrates such as (4-MU)-α-L-arabinoside, 2′-(4-MU)-α-D-N-acetylneuraminic acid and many others for the detection of, e.g., α-L-arabinosidase or sialidase may be purchased from Sigma Chemical Co. Ltd., UK.

Preparation of Substrate Solutions

Dissolve the various substrates in a minimum volume of dimethyl sulfoxide and dilute in 50 mM *N*-tris (hydroxymethyl)methyl-2-aminoethanesulfonic acid buffer (pH 7.5) (TES buffer, Sigma) to a final concentration of 100 µg/ml.

Inoculum Preparation

Grow the bacteria for 2 days, usually in an anaerobic jar, at $36 \pm 1°C$ on BHIA, remove colonies from the medium using sterile swabs and place the growth in TES buffer; adjust the suspension to an optical density at 620 nm of 0.1 (approximately 10^8 organisms per ml).

Tests for Enzyme Activity

Mix 20 µl of substrate solution with 50 µl of bacterial suspension in a flat-bottomed clear microdilution plate well and incubate at $36 \pm 1°C$ for 3 h. Visualize substrate degradation (release of 4-methylumbelliferone) by viewing the trays on an ultraviolet (UV) transilluminator or with a hand-held UV lamp under long-wave UV rays (366 nm). Positive reactions are indicated by a bright blue-white fluorescence. Use positive (e.g., *Prevotella oris* ATCC 33573) and negative (e.g., *Veillonella parvula* ATCC 10790) and noninoculated controls.

Miniaturized multitest systems such as the API 20 anaerobes (Casin et al., 1984), the API 20 Strep (Morrison and Tillotson, 1988), the API rapid ID 32 Strep (Nikolaitchouk et al., 2000), the API CORYNE (Hoyles et al., 2001), and the API (RAPID) CORYNE (Funke et al., 1997b) systems contain carbohydrate fermentation reactions in addition to the enzyme tests, or they consist exclusively of fermentation reactions such as the API 50 CH system (bioMérieux). When applied to the identification of fermentative actinomycetes, reliability and reproducibility of the identification results highly depend upon the growth characteristics of the individual test strain, especially as far as acid production from carbohydrates is concerned. Species and strains that form predominantly smooth, nonfilamentous growth, so that homogeneous cell suspensions can easily be prepared, may give satisfactory results (Funke et al., 1997b). This is particularly true for most of the recently described new *Actinomyces, Actinobaculum* and *Arcanobacterium* species. On the other hand, highly filamentous, slowly fragmenting strains such as members of the traditional species *A. israelii, A. gerencseriae* or *A. hordeovulneris* are often misidentified or cannot be identified at all when using these systems because key reactions among the carbohydrate fermentation tests tend to be falsely negative. Furthermore, it has been shown (Hall et al., 2003a) that the patterns of physiological characters produced by different test systems may differ considerably for a given actinomycete strain. This is true not only when comparing conventional tests with commercial test systems but also when comparing different commercial

Table 11. Physiological characteristics useful for differentiating *Actinomyces* species from human sources.

Species	Production of									Acid production from												
	Cat	Nit	Arg	Ure	Esc	α-Fuc	α-Glu	β-NAG	β-Gal	Ara	Cel	Ino	Lac	Man	Mlz	Raf	Rha	Rib	Sor	Suc	Tre	Xyl
A. cardiffensis	-	d	-	-	-	-	+	-	-	-	-	ND	-	-	ND	d	ND	d	-	+/-	-	-
A. europaeus	-	-/+	ND	-	d	-	+	-	+	-	ND	ND	d	-	d	-	-	d	-	d	d	-/+
A. funkei	-	d	-	-	-	-/+	+	d	d	-	ND	ND	-	-	-	-	ND	d	-	+	-	+
A. georgiae	-	d	ND	-	+	-	+	-	+	-	-	d	+	-/+	-/+	-	+	+	-	+	+	+
A. gerencseriae	-	d	d	-	+	-	+	-	+	-	d	d	+	+/-	d	+/-	-/+	+/-	-/+	+	+	+/-
A. graevenitzii	-	-	ND	-	-	-	-/+	+	+	-	-	+	+	-	-	-/+	-	+	-	+	-/+	(+)
A. hongkongensis	-	-	+	-	-	-	-	+	-	+/-	ND	ND	ND	ND	ND	-	ND	ND	ND	ND	-	-
A. israelii	-	+/-	+	-	+	-	+	+	-	+/-	+	+/-	+	+/-	d	+	-/+	+/-	d	+	-	+
A. meyeri	-	-	+	-/+	-	-	+	+	+	d	d	-	d	-	-	-	-	-	-	+	d	+
A. naeslundii	-	+	d	d	+	-	d	-	+/-	-	d	+	d	-/+	-/+	+	-	d	-	+	+	-
A. nasicola	-	-	d	-	-	-	+	+	+	-	+	ND	+	-	+	-	ND	-	-	-	-	-
A. neuii ssp. *neuii*	+	+	ND	-	-	-	+	-	+	d	d	+	+	+	+	+	-	+	-	+	+	+/-
A. neuii ssp. *anitratus*	+	-	ND	-	-/+	-	+	-	+	-	d	+	+	+	+	+	-	+	-	+	+	+
A. odontolyticus	-	+	-	-	d	d	-/+	-	+/(+)	d	-	-	d	-	-	-	d	-/+	-	d	-/+	+/-
A. oricola	-	-	-	-	+	-	+	-	d	-	+	ND	-	-	-	d	ND	-	-	d	d	-
A. radicidentis	+	d	-	d	(+)	-	+	d	+	-	ND	ND	+	+	+	+	-	+	-	+	+	-
A. radingae	-	-	ND	-	(+)/-	+	+	-	+	+	d	d	d	-/(+)	d	-/(+)	-	+	-	+	d	+/-
A. turicensis	-	-	ND	-	-/+	d	+	d	-	-/+	d	d	d	-/+	-/+	-/+	-	+	-	+	d	+
A. urogenitalis	+	+	+	-	+	-	+	+	+	d	ND	ND	+	d	+	+/-	(+)	d	-	+	(+)	+
A. viscosus	+	+/-	+	d	-/+	-	+	+	+/-	-	d	d	d	-	-	-	-	d	-	+	d	-

Symbols: –, negative; +, positive; d, strain differences, possibly depending on the system used; +/–, predominantly positive; –/+, predominantly negative; (+), weak reaction; +/(+), positive or weak reaction; (+)/–, weak or negative reaction; –/(+), negative or weak reaction; and ND, no data available.

Abbreviations: Cat, catalase activity; Nit, nitrate reduction; Arg, arginine dihydrolase activity; Ure, urease activity; Esc, esculin hydrolysis; α-Fuc, fucosidase; α-Glu, α-glucosidase; β-NAG, N-acetyl-β-glucosaminidase; β-Gal, β-galactosidase; Ara, L-arabinose; Cel, cellobiose; Ino, *myo*-inositol; Lac, lactose; Man, mannitol; Mlz, melezitose; Raf, raffinose; Rha, rhamnose; Rib, D-ribose; Sor, sorbitol; Suc, sucrose; Tre, trehalose; and Xyl, D-xylose.

Data taken from Schaal (1986b), Johnson et al. (1990), Funke et al. (1994, 1997), Wüst et al. (1995), Pascual Ramos et al. (1997), Collins et al. (2000), Nikolaitchouk et al. (2000), Lawson et al. (2001), Sarkonen et al. (2001), and Hall et al. (2002, 2003a, b).

Table 12. Physiological characteristics useful for differentiating *Actinomyces* species from animal sources.

Species	Production of									Acid production from													
	Cat	Nit	Ure	Esc	Arg	α-Fuc	α-Glu	β-NAG	β-Gal	Ara	Cel	Ino	Lac	Man	Mlz	Raf	Rha	Rib	Sor	Suc	Tre	Xyl
A. bovis	–	–	–	d	–	–	–	+	–	–	–	d	+	–	–	–	–	–	–	+	–	–
A. bowdenii	+	+	–	+	–	–	d	–	+	–	ND¹	ND	+	–	+	–	ND	+	–	–	+	–
A. canis	+	–	–	–	–	+	+	+	+	+	ND	ND	+	–	+	d	ND	+	–	d	–	+
A. catuli	d	+	–	+	d	–	+	d	+	d	ND	ND	+	–	–	+	ND	+	–	+	+	+
A. coleocanis	–	–	–	–	–	ND	+	ND	+	–	ND	ND	+	–	–	–	ND	–	–	+	–	–
A. denticolens	–	+	ND	+	ND	–	+	ND	+	–	(d)	+/–	+	d	–	+	–/(+)	d	–	+	–	–
A. hordeovulneris	(+)	–	–	+	ND	ND	ND	ND	ND	–	+	–	+	–	ND	(+)	–	–	ND	ND	+	d
A. howellii	+	ND	ND	ND	ND	ND	ND	ND	ND	d	–	d	d	–	ND	+	–	–	ND	+	d	d
A. hyovaginalis	–	+	–	+	–	ND	ND	ND	+	+	d	d	d	ND	–	–	–	ND	–	+	–	+
A. marimammalium	–	–	+	+	–	ND	–	ND	+	–	ND	ND	+	–	–	–	ND	–	–	–	–	–
A. slackii	+	ND	ND	–	ND	ND	ND	ND	ND	–	–	–/+	d	–	–	+	–	d	–	d	d	–
A. suimastitidis	–	ND	–	+	–	ND	+	d	+	(+)	ND	ND	–	–	–	+	ND	(+)	–	+	–	+
A. vaccimaxillae	–	–	–	+	–	+	–	–	–	+	ND	ND	–	–	–	–	ND	(+)	–	+	–	+

Symbols and abbreviations: please refer to footnotes to Table 11.
From Dent and Williams (1986c), Schaal (1986b), Collins et al. (1993), Pascual et al. (1999), and Hoyles et al. (2000, 2001a, b, c, 2002).

systems, even from the same manufacturer. In a recent comparative study of four commercial systems (RapID ANA II, Rapid ID 32 A, RapID CB Plus, and BBL Crystal) for identification of *Actinomyces* species and some closely related bacteria (Santala et al., 2004), a strikingly poor applicability of these systems to the fermentative actinomycetes was found that relates not only to the new but also to the "classical" species of this bacterial group. Updating these kits to improve their performance as proposed by Santala et al. (2004) clearly requires both extension of the respective databases and inclusion of additional characters with more appropriate differential power.

In view of the problems discussed above, the physiological characteristics listed in the diagnostic Tables 11 and 12 and in Tables 8 and 9 should be used with care because only some of the data were derived from identical or comparable techniques. The latter is particularly true for all of the *Actinomyces* species included in volume 2 of *Bergey's Manual of Systematic Bacteriology* (Schaal, 1986a) and for *A. hordeovulneris*, which were tested using modifications of the Minitek® system (Schofield and Schaal, 1979a; Schofield and Schaal, 1979b; Schofield and Schaal, 1980a; Schofield and Schaal, 1980b; Schofield and Schaal, 1981; Buchanan et al., 1984b). The data for all of the other *Actinomyces* species were taken from the respective original descriptions in which the various API systems were preferentially applied. In Tables 8 and 9, which summarize many of the physiological characters available so far for *Actinomyces* species, this methodological aspect could not be taken into account, so that this table is merely a compilation of the rough literature data rather than an exact physiological description of the respective species.

Because of difficulties in standardizing the various physiological tests, it appears by no means appropriate to construct a flowchart for identifying fermentative actinomycetes and related organisms that is solely based upon physiological reactions (Sarkonen et al., 2001). When branching points such as esculin hydrolysis, urease activity, β-*N*-acetylglucosaminidase (NAG) activity, or α-fucosidase activity are used in these flowcharts, standardization problems will inevitably lead to misidentifications.

For the same reason, biotyping has gained no practical importance for differentiating *Actinomyces* species below the species level. In general, little information is available on the typing possibilities of these organisms. A bacteriophage typing system has not been developed. Lytic phages have only been isolated for *A. viscosus*, both from domestic sewage and human dental plaque (Delisle et al., 1978; Delisle et al., 1988;

Tylenda et al., 1985). These phages were found to exhibit high host-cell specificity and productively infected only certain *A. viscosus* strains. However, the suitability of such phages for probing surface components of human oral actinomycetes has been discussed (Tylenda et al., 1985; Delisle et al., 1988). Thus, only serological and molecular typing methods appear as yet suitable for being applied to fermentative actinomycetes.

Serological Identification and Typing

The early papers on the serology of fermentative actinomycetes (see Slack et al., 1951 and Slack et al., 1955) already indicated that the differentiation of these organisms is possible by means of serological methods. However, the traditional serological procedures such as agglutination or complement fixation tests presented methodological problems. Thus, serology only became widely applicable for diagnostic and scientific identification and typing purposes after the introduction of fluorescent antibody techniques into the field of actinomycete classification and identification (Slack et al., 1961).

Since then, other serological methods, including gel precipitation, enzyme immunoassay and immunoelectrophoresis, have also been used, and some of the antigens of *Actinomyces* species have been isolated and characterized in some detail (Ayakawa et al., 1983; Schaal, 1986a; Happonen et al., 1987; Firtel and Fillery, 1988). The latter investigations showed that antigens common to several different species are generally present, but they also showed that species-specific, partial antigens are usually strong enough to allow reliable serological identification when absorbed or properly diluted antisera are used (Slack and Gerencser, 1975; Schaal and Gatzer, 1985c). Firtel and Fillery (1988), investigating the antigenic structure of *A. viscosus* and *A. naeslundii* using a set of 18 monoclonal antibodies against whole cells of these species, found 11 different antigenic determinants which were arranged in a complicated mosaic. Although some of the monoclonal antibodies appeared to be useful for identification, not all of the antigenic groups delineated within the *A. viscosus* and *A. naeslundii* complex could be identified with comparable reliability. Nevertheless, the antigenic profiles defined by these 18 monoclonal antibodies corresponded at least to some extent to the distinctions made within this group of actinomycetes by DNA hybridization experiments (Johnson et al., 1990).

For routine application, a simple fluorescent antibody (FA) procedure has been devised by Slack and Gerencser (1975).

Direct Fluorescent Antibody Test (Slack and Gerencser, 1975)

Preparation of Smears

For clinical material, make two smears of the material on a clean glass slide containing two marked circles. Fix air-dried smears by flooding with methanol for 1 min. Pour off the alcohol and allow the slide to air-dry. For cultures (from a suspension of the organism obtained by centrifugation of a broth culture or from an agar culture), make smears on glass slides, air-dry, and gently heat-fix.

Staining Procedure

Place one drop of conjugated specific antiserum on each smear and incubate the slide in a moist chamber for 30 min at room temperature. Pour off excess conjugate and wash the slide in two changes of pH 7.2 buffer (FTA hemagglutination buffer, BD) for 5 min each. Counterstain in 0.5% Evans blue for 5 min. Remove excess Evans blue by dipping the slide briefly into distilled water. Then, wash the slide in two changes of pH 9.0 buffer for 1 min each. Allow smears to air-dry. Place one drop of buffered glycerol mounting fluid (9 parts chemically pure glycerol, 1 part pH 9.0 buffer) on each smear and cover it with a coverslip. For examination with a microscope equipped for FA work, preferably use 54× oil-immersion or 50× water-immersion objectives.

For routine use of this technique, it is disadvantageous that fluorescein isothiocyanate (FITC)-labeled specific antisera to *Actinomyces* species are not available except for antisera to *A. israelii* and *A. naeslundii* (Biological Reagent Section, Centers for Disease Control, Atlanta, Georgia, United States). Thus, most antisera have to be prepared and conjugated by the user. To avoid part of this elaborate and time-consuming work, which is required for the application of the direct fluorescent antibody (FA) technique, Schaal and Pulverer (1973) have described an indirect FA staining procedure that provides similarly specific identification results and is even more sensitive than the direct FA modification. In this method, commercially available, FITC-conjugated goat antisera to rabbit globulins are used. Nevertheless, the specific antisera to the various *Actinomyces* species still have to be produced in the diagnostic laboratory by immunization of rabbits. The following immunization scheme has proved especially effective in the authors' laboratory.

Immunization Scheme for Preparing Antisera to Whole-Cell Antigens of Fermentative Actinomycetes (Schaal and Pulverer, 1973; Schaal and Gatzer, 1985c)

Preparation of Antigens

Biomass of the reference strains is produced using the dialysis tubing membrane technique on BHIA as described for chemotaxonomic tests or BHI broth cultures when nonfilamentous species are to be examined. Incubate plates anaerobically (anaerobic jar) or aerobically with added CO_2 (5–10%) depending on the species examined at $36 \pm 1°C$ for 7 days. Remove the dialysis membranes from the agar surface with sterile forceps. Transfer them

to Erlenmeyer flasks containing sterile saline with added phenol (1%, v/v) and wash the cell mass off the membranes by shaking with glass beads. Centrifuge broth cultures and resuspend the sediment in sterile saline with added phenol. Inactivate the cell suspensions thus obtained by heating at 60°C for 20 min in a water bath. Homogenize them in a Potter tissue homogenizer, and adjust to a concentration of 20 mg of wet weight per ml.

Immunization Scheme

Most types of ordinary laboratory rabbits can be used for immunization. Cross-bred animals often give best immune responses. Use the following schedule: First week—day 1, 1.5 ml, i.m.; day 3, 0.5 ml, i.v.; day 4, 1.0 ml, i.v.; day 5, 1.5 ml, i.v.; and day 7, 1.5 ml, i.m. Second week—day 10, 1.5 ml, i.v.; and day 13, 1.5 ml, i.v. Fourth week—day 22, test for antibody production. Possibly use cardiac puncture for obtaining the antiserum. If the test for antibody production (indirect FA) shows titers below 1 : 1000, make 2–4 (to 6) booster injections at half-weekly intervals. Lyophilize the antisera (obtained in the usual way, by cardiac puncture in ether anesthesia and subsequent separation from fibrin and blood cells) in 1-ml aliquots until use.

For antiserum production, at least the type and reference strains (see Schaal, 1986a and Schaal, 1986b) of the classical *Actinomyces* species should be used. However, these strains apparently do not cover completely the antigenic variation present in these taxa, especially in the species *A. israelii* (Ayakawa et al., 1983; Schaal and Gatzer, 1985c), and they do not include the recently described new species whose antigenic structure and relationships are largely unknown.

Because of the higher sensitivity of the indirect immunofluorescence technique, use of absorbed antisera is usually necessary for convenient and reliable identification to the species level. Crossreacting activities can easily be removed by traditional sorption procedures using the respective cellular antigens at concentrations of 60 mg wet weight per ml of serum and a final dilution of the absorbed antiserum of 1 : 40. Alternatively, highly diluted (depending upon the homologous antibody titer) antisera or comparative end point titrations can be used.

Indirect Fluorescent-antibody Procedure for Identifying *Actinomyces* species (Schaal and Pulverer, 1973; Schaal and Gatzer, 1985c)

Preparation of Smears

Prepare smears as described for the direct modification.

Staining Procedure

Place 1 drop of absorbed specific rabbit antiserum on each smear and incubate the slide in a moist chamber for 30 min at ambient temperature. Pour off excess antiserum and wash the slide in two changes of pH 7.2 buffer (FTA hemagglutination buffer, BD) for 5 min each. Allow the slide to briefly air-dry. Place 1 drop of FITC-conjugated goat antiserum to rabbit immunoglobulins on the same smear and incubate again in a moist chamber for about 30 min at room temperature. Pour off excess conjugate

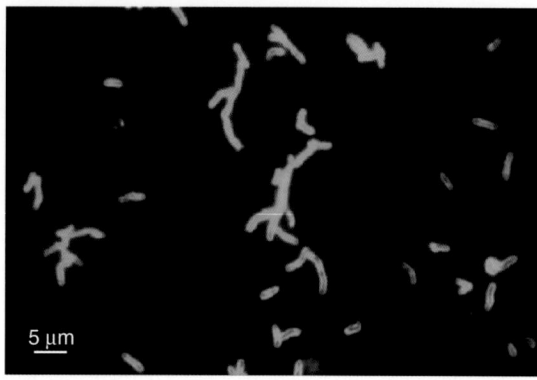

Fig. 23. *Actinomyces israelii*; indirect immunofluorescence with an absorbed rabbit antiserum to *A. israelii* and FITC-labeled goat anti-rabbit immunoglobulin; incident-light fluorescence micrograph.

and wash the slide in two changes of pH 7.2 buffer. Counterstain the smear in 0.5% Evans blue for 5 min. Remove excess Evans blue by dipping the slide briefly into distilled water and wash the slide in two changes of pH 9.0 buffer for 1 min each. Allow the smear to air-dry. Place 1 drop of buffered glycerol mounting fluid on each smear and cover it with a coverslip. For examination with a microscope equipped for FA work, use a 50× water-immersion or an analogous oil-immersion objective.

Provided that staining procedure and reagents have been sufficiently standardized (Holmberg and Forsum, 1973a; Schaal and Gatzer, 1985c), immunofluorescence techniques allow rapid and comparatively reliable identification of at least most of the classical *Actinomyces* species (Fig. 23). They also allow the recognition of the serovars of those classical species for which serovars were defined. However, these techniques are especially valuable for clinical microbiological laboratories because the methods are directly applicable to clinical specimens as well as to cultures (Slack and Gerencser, 1975). Furthermore, the identification and enumeration of filamentous organisms in dental plaque, gingival crevice, or cervicovaginal secretions can be achieved by these methods. It should be kept in mind, however, that antigenically aberrant strains and hitherto unknown or insufficiently characterized serotypes or species may be lost or misidentified if immunofluorescence procedures are applied as the sole means of demonstration and identification (Ayakawa et al., 1983; Schaal and Gatzer, 1985c).

Antifibril antisera have been used to identify the subgroups delineated within the species *A. viscosus* and *A. naeslundii* by numerical phenetic methods, using either immunofluorescence or whole bacterial cell agglutination (Masuda et al., 1983; Ellen and Grove, 1985a).

Serological Diagnosis of Actinomycete Infections

Because of difficulties in obtaining suitable clinical specimens and isolating and identifying pathogenic actinomycetes, there has always been a demand for diagnosing actinomycotic infections by means of serological techniques. Although several different serological methods have been tested for their suitability for this purpose, the results have usually not been satisfactory.

The only detailed serological studies that appear to indicate at least a certain diagnostic value of serology in actinomycete infections were performed by Holmberg and co-workers (Holmberg et al., 1975b; Holmberg, 1981; Persson and Holmberg, 1984). Using counterimmunoelectrophoresis or crossed immunoelectrophoresis, these authors reported sensitivities of 83–86% (for their test system when comparing cultural and histological with serological results) and comparatively high specificities, although a certain number of patients suffering from nocardiosis, tuberculosis, candidiasis or aspergillosis also gave positive reactions (Holmberg, 1981). According to the authors' experience with immunofluorescence and immunoassay procedures for more than 20 years, these techniques are only able to detect long-lasting and generalizing actinomycotic processes with sufficient reliability, but they are nearly useless for diagnosing localized infections. Therefore, these serological tests cannot be used for excluding or confirming the diagnosis actinomycosis when suitable specimens for cultural examinations are difficult or impossible to obtain, such as in cases of deep-seated actinomycotic abscesses of the internal organs including the female genital tract. The nature and importance of precipitation reactions against *A. israelii* antigens in uterine secretions still remain to be definitely clarified (Persson and Holmberg, 1985).

Natural human antibody response to *Actinomyces* antigens has also been studied in relation to caries and periodontal disease (Levine et al., 1984; Levine and Movafagh, 1986; Morinushi et al., 1989). From these studies, it was concluded that such natural antibodies might exert a protective effect against caries and subgingival plaque formation. Animal studies (Haber and Grinnell, 1989) appear to support these conclusions. In addition, these antibodies, especially to *A. viscosus* or *A. naeslundii* antigens, are obviously very common in otherwise healthy individuals and they tend to interfere with the serological tests for actinomycosis because *Actinomyces* species usually have one or more antigens in common, so that cross-reactions are frequently observed.

Identification and Typing Using DNA Technologies

DNA:DNA hybridization studies may be performed according to the procedures devised by Johnson et al. (1990) and Dent and Williams (1985). Barsotti et al. (1987) described a rapid technique for isolating DNA from fermentative actinomycetes. This method may facilitate DNA work with these organisms, which are difficult to lyse gently and effectively.

While these classical quantitative methods are mainly used to determine the threshold value for species differentiation, DNA probe hybridization methods (i.e., dot blot, subtraction, and checkerboard hybridizations) have been applied to identify *Actinomyces* species. Among others, these approaches have been applied to identify oral actinomycetes thriving in supra- and subgingival habitats (Ximenez-Fyvie et al., 1999) and in periapical endodontic lesions of asymptomatic teeth (Gatti et al., 2000; Sunde et al., 2000), *A. hyovaginalis* in necrotic samples of pigs (Storms et al., 2002), to differentiate *A. israelii* and *A. gerencseriae* strains (Jauh-Hsun et al., 1999), or to detect and identify *Mobiluncus* species (Påhlson et al., 1992).

Other molecular methods used in identification are based on characterization of the rRNA gene, such as direct sequence analysis of this gene (Clarridge and Zhang, 2002; Storms et al., 2002; Woo et al., 2002), sequence analysis of the rRNA spacer region (Storms et al., 2002), and restriction analysis of amplified genes (amplified rDNA restriction analysis [ARDRA]; Hall et al., 1999). The advantage of using sequences for reliable identification was demonstrated by the study of Woo et al. (2002). While an unidentified isolate from pelvic actinomycosis was identified as *Propionibacterium granulosum* by the VITEK (ANI) system (99% confidence), the API system (20A) affiliated this strain with 78% confidence to the *A. meyeri/odontolyticus* cluster; 16S rRNA gene sequence analysis, however, indicated absolute match with the sequence of *A. odontolyticus*. Kiyama et al. (1996) developed a system for detecting *Actinomyces* species based upon nonradioactive riboprobes coupled with polymerase chain reaction (PCR). Using a portion of the sialidase gene of *A. naeslundii* as a target for a digoxigenin-labeled probe, the system was reported to be three orders of magnitude more sensitive in detecting specific chromosomal DNA than a conventional riboprobe system.

Brailsford et al. (1999) used repetitive extragenic palindromic PCR (REP-PCR) for typing 56 *A. israelii* and 46 *A. gerencseriae* strains derived from nine active root caries lesions. Their results indicated that the genotypes of both *A. israelii* and *A. gerencseriae* populations were heterogeneous (i.e., not sharing the same REP-PCR patterns) within individual lesions.

Ecology

Knowledge of the biologic role of members of the genus *Actinomyces* in their natural habitats under healthy conditions is still fragmentary. The same is true for their interactions with other members of the respective indigenous microflora. Analysis of the biologic functions of fermentative actinomycetes in the lower intestine is difficult because their numbers are strikingly lower than those of other members of the normal bowel flora. In the upper intestinal tract, which until recently was considered to harbor no or only small numbers of cultivable microorganisms, the significance of actinomycetes and other aerobic and anaerobic bacteria remains to be defined in general.

The occurrence of *Actinomyces* species in the female genital tract raises several questions. The first is whether these microbes are able to colonize the internal female genital organs without presence of a foreign body such as an IUD. Although several authors have demonstrated *Actinomyces* species—and *Propionibacterium propionicum*—in genital secretions of women who did not wear IUDs or vaginal pessaries, two observations suggest that their presence may reflect change in the physiological balance in the female genital tract: 1) these organisms directly adhere to the device and its thread (Fig. 24), often buried in a thick biofilm, and 2) the actinomycetes disappear or become difficult to isolate after removal of the IUD. This change might have medical implications, although a considerable percentage of those women do not show any

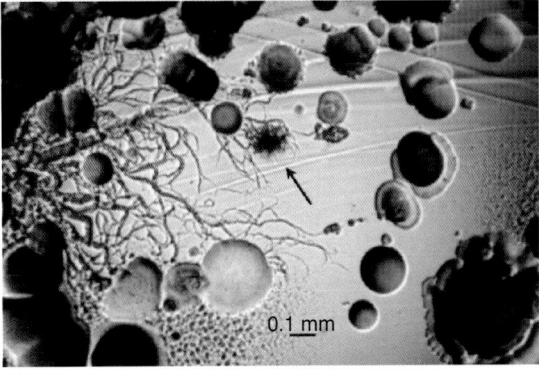

Fig. 24. Mixed actinomycotic flora cultured from the thread of an intrauterine contraceptive device. Small actinomycete colony (arrow) among a variety of other bacterial colony forms; micrograph in situ in obliquely transmitted light (brain heart infusion agar, 14 days at 36 ± 1°C, and Fortner's method).

subjective or objective symptoms of disease (Mao and Guillebaud, 1984; Pine et al., 1985; K. P. Schaal, unpublished observations).

The occasional recovery of *Actinomyces* species from the uninfected eye is probably accidental and does not imply that these organisms constitute normal inhabitants of the conjunctiva or the tear organs given the physiological connection between conjunctiva and upper respiratory tract via the lacrimal canaliculi.

Much more information has accumulated in recent years concerning the role of actinomycetes in the oral cavity. Although most of these studies were performed to clarify the pathogenesis of caries, periodontal disease, or gingivitis, some of the possible disease-contributing factors may also be present without signs of impairment and thereby represent at least one aspect of the biologic functions of these organisms in their natural environment.

Studies of the ability of fermentative actinomycetes to colonize the oral cavity, to resist natural and artificial cleaning mechanisms, and to utilize organic compounds present in the oral cavity as nutritional sources, as well as the complex interactions between various members of the oral microflora and between them and the mucosal and dental surfaces, have been extensive. A total of 1.9×10^7 fermentative actinomycete cells were estimated per gram wet weight of human dental calculus (Collins et al., 1973). Subgingival plaque samples of different groups of periodontal patients who had been successfully treated and maintained were populated by higher proportions of Gram-positive bacteria including *Actinomyces viscosus* and *A. odontolyticus* (Asikainen et al., 1987), whereas in patients with active periodontitis, spirochetes, *Porphyromonas* or *Bacteroides* species predominated, indicating that the actinomycetes were commensals rather than pathogens in these particular conditions (Loesche et al., 1985).

Recent studies on the oral colonization with *Actinomyces* species in 2–24 month healthy infants have provided a more detailed insight into the composition of the actinomycete oral flora during this early period (Sarkonen et al., 2000). Within the first two years, the frequency of the total actinomycete flora increased from 31% to 97%. *Actinomyces odontolyticus* appeared to be the most prominent colonizer at 2, 6, 12, 18, and 24 months of age. *Actinomyces naeslundii* (a species regarded as one of the pioneer colonizers of tooth surfaces) was also commonly encountered but not before the age of one year (Kolenbrander et al., 1999). The recently described species *A. graevenitzii* was also rather regularly identified in the mouths of children (Sarkonen et al., 2000), and *A. viscosus, A. gerencseriae, A. israelii* and *A. georgiae* were found

in the oral cavity of children and in gingival crevices of periodontally healthy adults (Cato et al., 1984; Johnson et al., 1990).

A special ecological situation arises when natural teeth are replaced by artificial implants, particularly by complete dentures. The subgingival microflora under these conditions may resemble qualitatively the plaque flora of healthy individuals but also that of progressive periodontitis (Wahl and Schaal, 1989). When comparing the microbial colonization of two types of connection parts of an IMZ-implant-supported Dolder bar, Wahl et al. (1992) found significant differences between the flora of elastic and titanium devices. The number of aerobically growing bacterial species, as well as that of oral capnophils and anaerobes, was higher on the elastic than on the titanium elements. However, essentially no difference in actinomycete colonization between the connection elements was seen: *A. viscosus* clearly predominated while *A. israelii, A. gerencseriae* and *Propionibacterium propionicum* were rarely identified in both locations.

Probably the most important feature that allows actinomycetes to colonize the oral cavity and even the smooth surfaces of teeth (and that prevents them from being washed off by salivary flow) is their ability to attach directly to the tooth surfaces, to adhere to mucosal cells and to co-aggregate with other microbes present in the same place. The simplest of these attachment mechanisms is the production of sticky extracellular or cell-associated polymers (glycocalyx), especially of levan, dextran, glycogen, and *N*-acetylglucosamine-rich slime polysaccharides, which facilitate attachment to enamel, artificial denture materials and even glass. *Actinomyces viscosus* (Howell and Jordan, 1967; Pabst, 1977; Miller et al., 1978b; Hamilton and Ellwood, 1983; Imai and Kuramitsu, 1983; Ooshima and Kuramitsu, 1985; Komiyama et al., 1988) and *A. naeslundii* (Slack and Gerencser, 1975; Komiyama et al., 1988) were shown to produce these polymers.

Important cell-associated structures of oral actinomycetes (including several *Actinomyces* species), which mediate both interbacterial aggregation and adhesion processes, are the so-called "surface fibrils" or "fimbriae." These fimbriae were found to consist of high molecular-weight proteins with some carbohydrate and up to 14.3% nitrogen (Cisar and Vatter, 1979; Masuda et al., 1983). That attachment and coaggregation might be due to such fimbrial proteins was first suggested by the finding that proteolytic enzyme or heat treatment impaired coaggregation between *A. naeslundii* or *A. viscosus* on the one hand and *Streptococcus sanguinis* or *Streptococcus mitis* on the other hand (Ellen and Balcerzak-Raczkowski, 1977; McIntire et al., 1978).

Soon it became apparent that a lectin-like mechanism was involved in some of these interbacterial coaggregations. McIntire et al. (1978) reported that the coaggregation between *A. viscosus* T14V and *Streptococcus sanguinis* 34 required a protein or glycoprotein on *A. viscosus* and a carbohydrate on *S. sanguinis*. (This coaggregation, as well as the binding of *A. viscosus* to glycoprotein-coated latex beads and its hemagglutinating properties, was specifically inhibited by lactose, β-metyl-D-galactoside and D-galactose, but not (or only weakly) by α-methyl-D-galactoside, melibiose, maltose, cellobiose, sucrose and a number of monosaccharides. (Costello et al. 1979; Heeb et al. 1985) found that *A. naeslundii* and *A. viscosus* were able to agglutinate human AB and horse erythrocytes.) This agglutinating property was enhanced by pretreatment of the red blood cells with neuraminidase and specifically inhibited by the compounds mentioned before. Furthermore, it was suggested that the actinomycetes were able to prime erythrocytes and human buccal epithelial cells for hemagglutination or attachment by removing sialic acid to expose more penultimate β-galactosides on the surface of the cells. Thus, neuraminidase removal of terminal sialic acid and lectin-like binding to exposed β-galactoside-associated sites on the erythrocytes appeared to be responsible for the hemagglutinating and adherence capabilities of *A. naeslundii* and *A. viscosus* (Saunders and Miller, 1983).

Other studies have shown that *A. viscosus* possesses two different types of fimbriae: type 1 fimbriae are involved in the attachment of this organism to saliva-coated hydroxyapatite, whereas type 2 fimbriae mediate the lactose-inhibitable, lectin-dependent coaggregation with *Streptococcus sanguinis* (Cisar et al., 1984a; Cisar et al., 1984b; Cisar et al., 1988; Clark et al., 1986; Mergenhagen et al., 1987). Using specific polyclonal and monoclonal antibodies, Ellen et al. (1989) demonstrated that anti-type 2 antibodies bound to antigens on long fibrils, whereas anti-type 1 antibodies were mostly localized close to the cell body or on shorter appendages.

A. naeslundii has type 2 fimbriae, which contain a lectin with a specificity similar to but not identical with that of *A. viscosus* (McIntire et al., 1983; Cisar et al., 1984a; Cisar et al., 1984b). These β-galactoside-reactive fimbrial lectins have specificities similar to those of certain plant lectins (Brennan et al., 1984) and bind to glycosphingolipids (gangliosides and globosides) and a 160-kDa surface glycoprotein of mammalian cells, particularly epithelial cells (Brennan et al., 1986; Brennan et al., 1987). Cloning, sequencing, and expression of type 1 and type 2 fimbrial antigens of *A. viscosus* and of type 2 fimbriae of *A. naeslundii* in *Escherichia coli* have been successful (Yeung et al., 1978; Donkersloot et al., 1985; Yeung and Cisar, 1988), so that a more detailed insight into the structure of type 1 and type 2 fimbriae could be obtained.

The intergeneric bacterial coaggregations and the adhesion to oral tissues are mediated not only by fimbrial lectins but clearly also by several other mechanisms (Mizuno et al., 1983; Gibbons, 1989). Furthermore, coaggregation of bacteria in the oral cavity is a complex phenomenon that involves not only direct cell-to-cell interactions of coaggregating cell types, but also the formation of bridges (especially by *Bacteroides/Prevotella/Porphyromonas* species and fusobacteria) that mediate coaggregations between two primarily noncoaggregating cell types (Kolenbrander et al., 1985; Weiss et al., 1987; Kolenbrander, 2000). Bacteria that serve as coaggregation bridges may use different adhesion mechanisms for binding to the different partners (Weiss et al., 1987), and binding-site competition does occur between organisms with similar coaggregation mechanisms (Kolenbrander et al., 1985). In addition, plaque formation is influenced by inhibitory activities exerted by some oral bacteria against other organisms (e.g., *Prevotella melaninogenica* against *Actinomyces* species; Takazoe et al., 1984; Tompkins and Tagg, 1986).

Saliva influences the adherence to teeth and interbacterial coaggregation and may have antibacterial properties; it may also serve as a source of nutrients for oral microorganisms (Kolenbrander and Phucas, 1984; De Jong et al., 1986; Koop et al., 1989). A unique family of salivary components, the so-called "acidic proline-rich proteins" (PRPs), is especially involved in non-lectin attachment mechanisms. These compounds adsorb to hydroxyapatite surfaces, thereby undergoing conformational changes that enable oral bacteria such as *A. viscosus* to bind to the PRPs (Gibbons et al., 1988; Gibbons, 1989). In addition, zeta potential, surface energy, and hydrophobic and electrostatic interactions may influence bacterial adhesion to uncoated and saliva-coated human enamel and dentine (Gibbons and Etherden, 1983; Wheeler et al., 1983; Clark et al., 1985; Weerkamp et al., 1988).

All of these factors finally result in an extremely complex ecological system (i.e., the oral biofilm or the oral microbial-plaque community; Kolenbrander, 2000). This system is characterized by special mutualistic relationships (Palmer et al., 2001) that not only can determine whether the partners grow preferentially in the planktonic or in the biofilm state, but also may be responsible for the special spatial arrangement and associative behavior of the partners (Guggenheim et al., 2001) and for resistance to detrimental factors such as oxygen contact. Bradshaw et al. (1998) showed convincingly that coaggregation-mediated interactions between *Fusobacterium nucleatum* and other species facilitated the survival of strict anaerobes in an aerated environment.

Thus, apart from physical interactions (such as coaggregation and coadhesion), physiological and metabolic interactions also facilitate communication between members of the genetically diverse biofilm flora (Kolenbrander, 2000). One of the most obvious consequences of the latter is the production of various organic acids. They evolve from the fermentative metabolic processes that occur *in vivo* as well as in vitro and may be strong enough to lower the pH values locally (Ellen and Onose, 1978). Acid production may lead not only to enamel demineralization and cell alteration, but also to influences on growth and metabolism of other oral microorganisms (McDermid et al., 1986). These influences may be attributed to different sensitivities of the various oral microbes to low pH values (Phan et al., 2000), to the ability of certain bacteria such as veillonellae to utilize the end products of the actinomycete metabolism for their nutrition in the sense of a food chain (Distler et al., 1980), or to the simultaneous utilization of several carbon and energy sources (Van der Hoeven et al., 1984). In *A. viscosus* and *Lactobacillus casei*, differences in acid tolerance can be related to the activity of proton-translocating membrane ATPases (Bender and Marquis, 1987).

A variety of enzymes involved in the initial dissimilation of exogenous carbohydrates or the energy yielding glycolytic pathway has been identified in *Actinomyces* species; thus, α- and β-galactosidases, β-glucuronidases, α- and β-glucosidases, β-xylosidases and α-fucosidases have been detected in several *Actinomyces* species (Fiehn and Moe, 1984; Tables 8 and 9). In addition, invertase activity has been demonstrated in both *A. naeslundii* (Miller, 1974) and *A. viscosus* (Palenik and Miller, 1975; Kiel et al., 1977b). And *A. viscosus* and *A. naeslundii* were shown to possess extracellular and cell-associated enzymes able to degrade levans of microbial origin (Miller and Somers, 1978a; Igarashi et al., 1987), indicating that additional types of food chains do exist which make the nutrition of some actinomycetes partially independent of exogenous carbohydrate sources from the host's diet.

The relative numbers of actinomycetes in the oral environment may be influenced by inhibitory factors released by other bacteria (Rogers et al., 1978), by fluorides or other measures for preventing caries (Beighton and McDougall, 1977; Johansen et al., 1997; Phan et al., 2000), or by eating habits of the host (Pfister et al., 1984; Minah et al., 1985). Furthermore, an active role of actinomycete bacteriophages with high host

cell specificity in the oral microbial ecology has been discussed (Tylenda et al., 1985). In any case, the qualitative and quantitative composition of the oral microflora differs between the saliva (in the planktonic state) and the mucosal and tooth surfaces (both in the biofilm state) and even varies from tooth to tooth and at different sites on a single tooth. It has been reported that Gram-positive, facultative and anaerobic bacteria constitute 15–20% of the cultivable flora in saliva and on the tongue, 40% of that in plaque, and 35% of that in gingival crevice (Socransky and Manganiello, 1971). Actinomycetes amounted to about 40% of the cultivable organisms in plaque (Loesche et al., 1972; Ellen et al., 1985b; Brown et al., 1986; Nyvad and Fejerskov, 1989). The incidence of individual species has been given above, although these figures (or the relative incidences) vary considerably between different authors and different groups of test persons.

Finally, the oral microflora changes with age and life circumstances of the host. Anaerobic bacteria are generally found in large numbers during the period of life when natural teeth are present (Socransky and Manganiello, 1971; Russell and Melville, 1978). Before dentition and after loss of all teeth in later life, the counts for anaerobes and so for actinomycetes are lower (Kostecka, 1924; Rosenthal and Gootzeit, 1942; Berger et al., 1959; Loesche, 1960). The age distribution of actinomycotic infections as well as studies on the oral colonization with *Actinomyces* species in early childhood (Sarkonen et al., 2000) indicates, however, that these organisms are present during the whole lifetime of their hosts.

Epidemiology

For the region of Cologne, Germany, annual morbidity of human cervicofacial actinomycoses averaged 1 : 40,000 (Schaal, 1979a) or 1 : 71,000 (Lentze, 1970). In contrast, Durie (1958) estimated 10 cases of the disease occurred annually in the population of Sydney/Australia, and Hemmes (1963) reported one case per year occurred in 119,000 inhabitants of Holland. Thus, differences in the frequency with which cervicofacial processes are diagnosed appear to be the chief reason for differences in the anatomic distribution of this disease. This, in turn, might reflect a difference in the approach to the bacteriological diagnosis of cervicofacial infections in Germany versus other parts of the world. On the other hand, there are indications that the cervicofacial form has also been decreasing recently in Germany.

Because of their endogenous origin, human actinomycoses are world-wide in distribution.

Nevertheless, the sex distribution of the disease is skewed toward males. Most tabulations of sex ratios (see Slack and Gerencser, 1975) agree with our data (Pulverer, 1974; Pulverer and Schaal, 1978; Pulverer and Schaal, 1984a; Pulverer and Schaal, 1984b; Schaal, 1981a; Schaal and Lee, 1992c; Pulverer et al., 2003), which indicate that the ratio is about three to four males per one female. However, this proportion only applies to sexually mature patients suffering from cervicofacial infections; practically no sex distribution differences can be demonstrated before puberty and in advanced age (Pulverer and Schaal, 1978; Pulverer and Schaal, 1984a; Pulverer and Schaal, 1984b; Schaal and Lee, 1992c; Pulverer et al., 2003), so that the proportions vary with age from 0.9 : 1 to 4.0 : 1. Nevertheless, actinomycoses may be acquired during the whole lifetime (Slack and Gerencser, 1975; Pulverer and Schaal, 1978; Pulverer and Schaal, 1984a; Pulverer and Schaal, 1984b), as the age of patients with bacteriologically confirmed actinomycoses was found to range from 28 days to 82 years, with the highest incidence observed in males 20–40 years old and females 10–30 years old (Pulverer and Schaal, 1978; Pulverer and Schaal, 1984a; Pulverer and Schaal, 1984b; Pulverer et al., 2003).

As far as caries and periodontitis are concerned, some epidemiological hints are given in the sections on Ecology and Pathogenicity. Nearly nothing is known about the epidemiology of infections due to species other than *A. israelii, A. gerencseriae, A. naeslundii* or *A. viscosus.* The same is true for animal *Actinomyces* infections, although the classical bovine actinomycosis appears to occur sporadically world-wide.

Pathogenicity

Since the description of the first member of the genus *Actinomyces* (Bollinger, 1877), fermentative actinomycetes have been known to be able to produce characteristic and potentially severe inflammatory diseases in man and animals. These pathogenic species belong to the normal indigenous microflora of their hosts, so that they must be considered facultative pathogens, which only invade tissue under certain facilitating conditions and are usually not transmissible (Lentze, 1938a; Lentze, 1938b; Lentze, 1969; Slack and Gerencser, 1975; Pulverer and Schaal, 1978; Pulverer and Schaal, 1984a; Pulverer and Schaal, 1984b; Schaal, 1979a; Schaal, 1986a; Schaal, 1996; Schaal, 1998; Schaal and Beaman, 1984b; Schaal and Pulverer, 1984c; Schaal and Lee, 1992c). Therefore, in a sense, these organisms may be called opportunistic pathogens, although they differ from classical opportunists in that they generally do not require an immunocompromised host for invasiveness but only proper local starting conditions.

Human Diseases due to *Actinomyces* Species

Human Actinomycoses The most typical disease entity caused by fermentative actinomycetes is called "actinomycosis," a name that dates back to the first extensive description of the disease in animals by Bollinger (1877) and in humans by Israel (1878). As currently recognized, actinomycosis is a subacute-to-chronic, granulomatous inflammatory syndrome that gives rise to suppuration, abscess formation, and draining sinus tracts during its course (Slack and Gerencser, 1975; Pulverer and Schaal, 1978; Pulverer and Schaal, 1984a; Pulverer and Schaal, 1984b; Schaal, 1979a; Schaal, 1996; Schaal, 1998; Schaal and Beaman, 1984b; Fig. 25). In the beginning, actinomycotic lesions may appear as slowly progressing infiltrations or as acute abscesses or phlegmons. Advanced processes show a typical and dangerous preference to penetrate the tissue irrespective of natural organ borders (Lentze, 1969; Lentze, 1970) and a tendency to remit and exacerbate with and without antibiotic therapy (Lentze, 1969; Slack and Gerencser, 1975; Schaal, 1979a; Schaal, 1996; Schaal, 1998; Bennhoff, 1984; Schaal and Beaman, 1984b).

Sulfur Granules In 10–25% of the cases, depending on the course of the disease (Pulverer

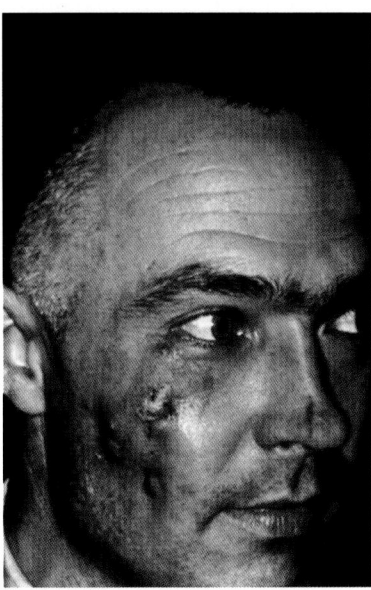

Fig. 25. Chronic cervicofacial actinomycosis in a 45-year-old man presenting with multiple fresh and older abscesses, non-healing incision wounds, and sinus tract as well as central scar formation.

and Schaal, 1978; Pulverer and Schaal, 1984a; Pulverer and Schaal, 1984b; Schaal, 1979a; Schaal, 1981a; Schaal, 1996; Schaal, 1998; Schaal and Beaman, 1984b), the purulent discharge from actinomycotic processes contains macroscopically visible (up to 1 mm in diameter), yellowish to reddish to brownish particles. These particles represent a conglomerate of filamentous actinomycete microcolonies formed *in vivo* and surrounded by tissue-reaction material, especially polymorphonuclear granulocytes (Fig. 12). These diagnostically typical structures, which, when derived from human infections, exhibit on microscopic examination a cauliflower-like appearance at low magnification and clusters of interwoven branching filaments with radially arranged peripheral hyphae at higher magnification (Lentze, 1969; Slack and Gerencser, 1975; Schaal, 1979a; Schaal, 1981a; Schaal, 1984a; Fig. 13), were originally designated "Drusen" in Harz's first description of *Actinomyces bovis* (Harz, 1877), and this name was subsequently used also for corresponding particles in human infections although the morphology of human and animal Drusen is not completely identical. In the English literature, these particles are usually referred to as "sulfur granules," a term which solely relates to their yellow color but not to a strikingly high sulfur content. Such granules, when encountered in tissue sections, and less frequently also in purulent discharge, often show a club-shaped layer of hyaline material on the tips of peripheral filaments which can aid in the differentiation of *Actinomyces* granules from macroscopically similar particles of various other microbial and nonmicrobial origins.

Clinical Pictures The source of typical human actinomycoses is nearly always endogenous (i.e., the causative agents are derived from the patient's own indigenous mucosal microflora). Therefore, the disease primarily develops in tissue adjacent to the mucosal surfaces that harbor the microorganisms. Clinically, these predilection sites are mostly referred to as cervicofacial, thoracic, abdominal or pelvic. In the extensive material that has been collected at the Hygiene Institute of the University of Cologne by Lentze, Pulverer, and Schaal and in the German National Consulting Laboratory for Actinomycetes at the Institute for Medical Microbiology and Immunology of the University of Bonn by Schaal over a 50-year period, cervicofacial actinomycoses have been observed most frequently (in 96% of more than 4000 cases examined; Pulverer et al., 2003). In the same series of cases, thoracic infections only amounted to about 2% and abdominal and pelvic manifestations to about 1% (Pulverer and Schaal, 1984a;

Pulverer and Schaal, 1984b; Schaal and Pulverer, 1984c; Schaal, 1988).

Although the above-mentioned data reflect the situation for Germany as a whole rather than that for the Cologne and Bonn region only, the anatomical distribution observed by these German authors differs considerably from that reported for other countries, especially for the United States (Slack and Gerencser, 1975), where 40–50% of human actinomycotic infections were found to be located in thoracic and abdominal sites. These differences are still difficult to explain but may relate in part to the higher overall incidence of the cervicofacial form of the disease in Germany (see Epidemiology).

Cervicofacial actinomycoses predominantly develop in the soft tissue around the mandible, but they may also be found adjacent to the maxilla or may extend to the orbita, sinuses, ear or neck (Pape et al., 1984; Olson et al., 1989). Even primary actinomycosis of the thyroid gland has been observed (Dan et al., 1984). In contrast to typical animal actinomycoses, which usually affect the skeleton, osseous involvement is rarely seen in humans. Nevertheless, actinomycotic osteitis or osteomyelitis of the mandible, the maxillae, and even the skull and the cervical vertebrae may be observed sporadically (Yenson et al., 1983; Gupta et al., 1986; Vannier et al., 1986).

Actinomycotic infections of the central nervous system (CNS) are also rare and may either develop hematogenously from distant sites (lung, abdomen and pelvis) or by direct extension from contiguous foci such as ear, orbital cavity, sinus or cervicofacial lesions (Smego, 1987; Nithyanandam et al., 2001). Such CNS infections predominantly present as brain abscesses but may also become manifest as meningitis, meningoencephalitis, subdural empyema, or epidural abscess (Peacock et al., 1984; Prager et al., 1984; Tvede et al., 1985; Smego, 1987; Tsai et al., 2001).

Thoracic actinomycoses usually have a history of preceding cervicofacial lesions, which descend along the cervical soft tissue to the mediastinum, or of aspiration, but they may also develop hematogenously. Diagnosis is mostly difficult to establish and many cases reported in the literature as well as the authors' observations indicate that the disease is primarily often misdiagnosed as bronchial carcinoma until abscess or sinus tract formation and the detection of sulfur granules suggest the proper diagnosis (Suzuki and Delisle, 1984; Masters et al., 1985; Massart et al., 1986; Philipsen et al., 1988; Schaal, 1996; Schaal, 1998; Fig. 26). Early diagnosis is usually only possible when aggressive methods of obtaining suitable specimens for microbiological examination (such as transtracheal aspiration of bronchial secretions and transthoracic needle biopsy or

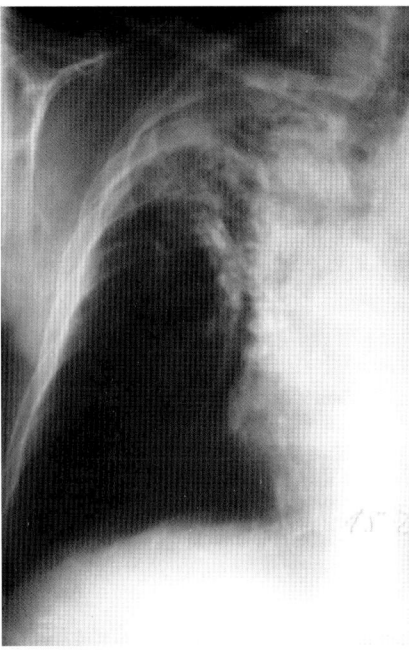

Fig. 26. Thoracic actinomycosis in a 64-year-old man presenting as massive infiltration of the right upper lung accompanied by a huge subcutaneous abscess over the right shoulder; X-ray photograph.

abscess puncture) are applied (Orloff et al., 1988). The nature and clinical relevance of a disease that some authors have termed "endobronchial actinomycosis" remain to be definitely clarified (Jin et al., 2000).

Like pulmonary and thoracic infections, actinomycoses of the abdomen and small pelvis tend to be misdiagnosed as carcinoma. This is especially true for infections of the rectum and for pelvic actinomycoses following the use of IUDs. Abdominal actinomycoses often follow appendicitis, abdominal trauma, or bowel surgery, may be located retro- or intraperitoneally (Adachi et al., 1985; Wohlgemuth and Gaddy, 1986; Levine and Doyle, 1988), and may involve a variety of organs such as the colon (Heer et al., 1986), rectum (Stein and Schaal, 1987), perianal tissue (Shimada et al., 1986; Gayraud et al., 2000), gallbladder (Van Steensel and Kwan, 1988), and liver (Mongiardo et al., 1986; Forgan-Smith et al., 1989; Logan et al., 1989).

The incidence of pelvic actinomycoses appears to be increasing in recent years, and this increase is probably related to the colonization of the cervical canal, the *cavum uteri,* and contraceptive devices by fermentative actinomycetes. Such infections may remain restricted to tissue in the small pelvis (uterus, Fallopian tubes, and ovaries) but may also spread continuously or hematogenously to the liver and other intraabdominal organs (Lininger and Frable, 1984; Persson and

Brihmer, 1986; Brihmer et al., 1987; Shurbaji et al., 1987; O'Connor et al., 1989; Eibach et al., 1992). Even involvement of the placenta and of the urinary bladder presenting with sulfur granules in the urine has been observed (Wajszczuk et al., 1984; Zakut et al., 1987).

Primary actinomycoses of the skin and the extremities are rare and usually have a history of trauma resulting from human bites or fist fights ("punch actinomycosis"), thus representing the only exogenously acquired forms of the disease. Hematogenous dissemination to the soft tissue, the bones, or joints of the extremities has also been observed (Mesgarzadeh et al., 1986; Reiner et al., 1987; Blinkhorn et al., 1988), and even a total hip arthroplasty was found to be infected (Strazzeri and Anzel, 1986).

Generalization of actinomycotic processes with the formation of multiple abscesses in various organ systems is rare nowadays and usually only occurs when diagnosis and proper treatment have been delayed (Schaal and Pulverer, 1984c; Schaal et al., 1984e). The same is true for the dissemination of the disease via the lymphatic vessels with involvement of the regional lymph nodes (Amrikachi et al., 2000).

Opportunistic actinomycotic infections in immunocompromised patients are not common. However, actinomycoses following osteoradionecrosis of the jaw (Happonen et al., 1983), radiotherapy of a tonsillar carcinoma (Spapen et al., 1989), renal transplantation (Kammoun et al., 2001), as well as infections in AIDS patients (Gresser et al., 1988) indicate that immunocompromising conditions or treatment methods may also be able to initiate actinomycotic processes, particularly when they result in tissue damage adjacent to the mucous membranes.

Etiopathology The etiology of human actinomycoses has been difficult to elucidate, especially because diseases are almost never caused by classical facultatively pathogenic actinomycetes demonstrable by isolation in pure culture. Nearly all of the typical human actinomycotic lesions appear as polymicrobial infections in which the pathogenic *Actinomyces* acts as the "guiding organism" (Lentze, 1948; Lentze, 1969; Lentze, 1970) that is responsible for the typical clinical picture and course of the disease. The concomitant bacteria, which are always present but may vary in number and species composition, are obviously necessary to strengthen the comparatively low invasive power of the actinomycetes (Lentze, 1948; Lentze, 1953; Lentze, 1969; Lentze, 1970; Holm, 1950; Holm, 1951; Pulverer and Schaal, 1978; Pulverer and Schaal, 1984a; Pulverer and Schaal, 1984b; Schaal, 1979a; Schaal, 1981a; Schaal, 1985a), either by providing low oxygen tension in the tissue or by supplementing the guiding organism with toxic, necrotizing or otherwise aggressive extracellular products.

In about 50% of the cases examined in the laboratories in Cologne and Bonn, the mixed actinomycotic flora was composed of facultative to strict anaerobes together with aerobically growing microorganisms. In the other half of the cases, solely anaerobes were cultured from the purulent exudate (Lentze, 1969; Pulverer, 1974; Pulverer and Schaal, 1978; Pulverer and Schaal, 1984a; Pulverer and Schaal, 1984b; Schaal, 1979a; Schaal, 1985a; Schaal, 1988; Schaal and Lee, 1992c; Pulverer et al., 2003). Among the aerobically growing concomitant bacteria, *Staphylococcus epidermidis*, *Staphylococcus aureus* and α- and β-hemolytic streptococci were the most prevalent. The carboxyphilic and anaerobic component of the concomitant actinomycotic flora consisted of carboxyphilic ("microaerophilic") streptococci (*S. milleri* group), anaerobic streptococci such as *Peptococcus*, *Peptostreptococcus*, *Anaerococcus*, *Peptoniphilus*, *Finegoldia*. *Micromonas* or *Atopobium* species, black-pigmented Bacteroidaceae, other *Bacteroides* and *Prevotella* species, fusobacteria, cutaneous propionibacteria, *Leptotrichia buccalis*, *Eikenella corrodens*, *Capnocytophaga* species, and *Actinobacillus actinomycetemcomitans*. The latter organism was encountered in about 25% of the cases examined and, when present, usually led to a particularly chronic and serious course of the disease.

Although the concomitant bacteria constitute a necessary factor in the pathogenesis of human actinomycosis in the sense of synergistic companions, the development of typical chronically progressive actinomycotic lesions is impossible without the presence of the pathogenic actinomycete species. However, as the concomitant bacteria also belong to the resident or transient indigenous microflora of human mucous membranes, localized anaerobic infections other than actinomycosis do occur at the same predilection sites and may be difficult to distinguish clinically from true actinomycoses, especially in their early manifestations. In addition, clinical specimens may be contaminated with actinomycetes from the mucosal surface or may contain two different actinomycete species, and this may cause confusion concerning their etiological role.

These considerations illustrate the problems that exist when posing the question: which oral actinomycetes produce typical human actinomycoses and which do not? Numerous reports in the world literature leave no doubt that *Actinomyces israelii* is the most characteristic and most common cause of human actinomycosis. Out of 1042 strains of fermentative actinomycetes isolated in the authors' laboratories from typical clinical conditions over a 27-year period and identified to species level, 588 (56.4%) were found to belong to the species *A. israelii* (Schaal and Lee, 1992c). The number of isolates identified as *A. gerencseriae*, which had formerly been subsumed under the designation *A. israelii*, serovar 2, amounted to 259 (24.9%) in this study, so that these two classical human pathogens were found to be responsible for 81.3% of the actinomycosis cases examined. Several other *Actinomyces* species such as *Actinomyces naeslundii*, *A. viscosus*, *A. odontolyticus* and *A. meyeri* have also been recovered from actinomycotic processes (Schaal and Lee, 1992c), but their pathogenic potential under these clinical circumstances has not been documented with similar certainty and has chiefly been deduced from comparatively small numbers of isolations or case reports. However, there is no doubt that another fermentative filamentous bacterial species, *Propionibacterium propionicum* (formerly *Arachnia propionica*), is also able to produce typical human actinomycoses (Gerencser and Slack, 1967; Brock et al., 1973; Slack and Gerencser, 1975; Conrad et al., 1978; Schaal and Lee, 1992c). According to the authors' observations, this species was encountered in 3–4% of the infections suspected to be of actinomycotic nature (Pulverer and Schaal, 1978; Pulverer and Schaal, 1984a; Pulverer and Schaal, 1984b; Schaal, 1979a; Schaal, 1985a; Schaal, 1988; Schaal and Lee, 1992c; Pulverer et al., 2003).

LACRIMAL CANALICULITIS AND OTHER EYE INFECTIONS Lacrimal canaliculitis is another, less severe and usually noninvasive infection that may be caused by fermentative actinomycetes. Branching, filamentous organisms in lacrimal concretions were already described by Ferdinand Cohn (1875). More recent examinations showed that these filamentous microbes can be identified as *Actinomyces israelii*, *A. gerencseriae*, *A. naeslundii*, *A. viscosus*, *A. odontolyticus* or *Propionibacterium propionicum* (Pine and Hardin, 1959a; Pine et al., 1960; Ellis et al., 1961; Buchanan and Pine, 1962; Slack and Gerencser, 1975; Jones and Robinson, 1977; Blanksma and Slijper, 1978; Schaal and Pulverer, 1984c; Schaal, 1988; Schaal and Lee, 1992c; Schütt-Gerowitt et al., 1999; Takemura et al., 2002). According to the authors' experience, the incidence of *Propionibacterium propionicum* is much higher in lacrimal canaliculitis than in cervicofacial and other actinomycotic infections, in that this organism was isolated from lacrimal concretions approximately as frequently as was *A. israelii* (Schaal and Lee, 1992c; Schütt-Gerowitt et al., 1999). Other inflammatory processes of the eye such as conjunctivitis, dacryocystitis, keratitis, hordeolum and granuloma, or even intraocular infections, may also be due to *Actinomyces* species (Schaal, 1988; Schaal and Lee, 1992c; Schütt-Gerowitt et al., 1999). However, it appears that in these conditions species other than *A. israelii*, *A. gerencseriae* and *Propionibacterium propionicum* are the predominant etiologic agents. Furthermore, two different *Actinomyces* species such as *A. israelii* plus *A. naeslundii* may occasionally be recovered simultaneously from the same eye specimen (Schaal and Pulverer, 1984c).

OTHER INFECTIONS DUE TO *ACTINOMYCES* SPECIES There is growing evidence that *Actinomyces* species are most likely to play an etiological role in a variety of other human inflammatory processes (*nonspecific infections*) that do not present as clinically characteristic actinomycoses. This is especially true for most of the recently described new species, but also for some of the classical ones. Information about the pathogenicity of *Actinomyces naeslundii* and *A. viscosus* as far as invasive infections are concerned is still comparatively meager, although these organisms may be recovered from clinical specimens rather frequently (Coleman and Georg, 1969; Gerencser and Slack, 1969; Karetzky and Garvey, 1974; Schaal and Lee, 1992c). In material from the authors' laboratories, the isolation rates of *A. naeslundii* and *A. viscosus* in infections suspected of being of an actinomycotic nature amounted to 7.6% and 4.1%, respectively (Schaal and Lee, 1992c). However, in a comparative study of such infections in which the ways of obtaining specimens (intraoral incision/puncture vs. extraoral incision/puncture) were compared, it could be demonstrated that the isolation of these *Actinomyces* species is often related to contamination of the specimen with oral secretions (Schaal and Pulverer, 1984c). Therefore, involvement of these organisms in cervicofacial or pulmonary infections remains to be definitely clarified. On the other hand, *A. naeslundii* has been reported convincingly as the causative agent of gallbladder infections (Freland et al., 1987), pelvic infections (Bonnez et al., 1985), a precostal abscess resulting from mediastinal actinomycosis (Chapoy et al., 1984), and a disseminating infection in a mentally retarded child (Dobson and Edwards, 1987). Furthermore, a case was observed of recurring empyema of the knee joint in which *A. naeslundii* was isolated on several occasions as the only pathogen (Schaal and Pulverer, 1984c), and similar infections were reported for a total knee arthroplasty (Ruhe et al., 2001) and a hip prosthesis (Wüst et al., 2000).

Similarly, *A. viscosus* has been isolated from thoracic infections, chest wall abscesses, an abscess following a dog bite (del Carmen Pinilla and Ciniglio, 1983; Spiegel and Telford, 1984b), and even from blood samples of patients with endocarditis (Mardis and Many, 2001).

Reports on *Actinomyces odontolyticus* as a human pathogen are still scarce although their number has apparently increased recently. In material from the authors' laboratories, the relative incidence of this species was 1.8% of all actinomycete isolates encountered (Schaal and Lee, 1992c). Both the clinical picture and course of infections due to *A. odontolyticus* are usually noncharacteristic, so whether such infections should be termed "actinomycosis" is doubtful (Peloux et al., 1983; Klaaborg et al., 1985). Nevertheless, various types of inflammatory processes have been attributed to *A. odontolyticus* as the causative agent: These were an ulcer of the oral mucosa (Alamillos-Granados et al., 2000), peritonsillar abscesses (Civen et al., 1993) from which *A. odontolyticus* was isolated in 23% of the specimens examined (Jousimies-Somer et al., 1993), enterocutaneous fistulation (Klaaborg et al., 1985; K. P. Schaal, unpublished observation), multiple liver abscesses (Ruutu et al., 1982), thoracic infections including lung infections with or without extension to the pleural space, mediastinum, or pericardium (Baron et al., 1979; Hooi et al., 1992; Ibanez-Nolla et al., 1993; Verrot et al., 1993; Dontfraid and Ramphal, 1994; Bassiri et al., 1996; Perez-Castrillon et al., 1997; Litwin et al., 1999), brain abscess (Simpson et al., 1996), and even septicemia (in association with *Fusobacterium necrophorum*; Raoult et al., 1982).

Actinomyces meyeri has been isolated frequently from brain abscesses and pleural fluid (Cato et al., 1984) and less often from cervicofacial abscesses (Pordy, 1988), lung and breast abscesses (Rose et al., 1982; Rippon and Kathuria, 1984; Ferrier et al., 1986; Allen, 1987), pneumonitis (Allworth et al., 1986), and abscesses with or without osteomyelitis of hip, leg and foot (Cato et al., 1984; Ferrier et al., 1986; Pang and Abdalla, 1987; Machet et al., 1993) as well as from spleen, liver or perianal infections and infected bite wounds (Cato et al., 1984; Garcia-Corbeira and Esteban-Moreno, 1994; Garduno et al., 2000; Gayraud et al., 2000; Harsch et al., 2001). Furthermore, *A. meyeri* infections show a striking tendency to disseminate so that various organs including heart valves or the brain may become involved (Lentino et al., 1985; Ferrier et al., 1986; Marty and Wüst, 1989; Kuijper et al., 1992; Apotheloz and Regamey, 1996; Chaumentin et al., 1997; Van Mook et al., 1997; Huang et al., 1998). Even an infection of the umbilical cord has been described (Wright et al., 1994).

Both *Actinomyces radingae* and *A. turicensis* were primarily isolated from various polymicrobial infections (Wüst et al., 1995) but occasionally also as pure cultures (Clarridge and Zhang, 2002). In recent studies by Sabbe et al. (1999) and Clarridge and Zhang (2002), *A. radingae* was found in patients with skin-related and soft tissue conditions. The main sources of *A. turicensis* were genital infections, followed by skin (Lepe et al., 1998) and urinary tract infections (Clarridge and Zhang, 2002). Involvement of the appendix, gallbladder, ear, nose and throat was also noted, as was bacteremia.

The two subspecies of *Actinomyces neuii* were isolated mainly from abscesses in association with mixed anaerobic flora and from human blood cultures (Funke et al., 1994) but not from typical actinomycotic lesions. Further conditions in which this coryneform species may be etiologically involved are infected atheromas (Funke and Von Graevenitz, 1995a), infected mammary prostheses (Brunner et al., 2000), chronic osteomyelitis (Van Bosterhaut et al., 2002), postoperative endophthalmitis (Garelick et al., 2002), and neonatal sepsis following chorioamnionitis (Mann et al., 2002).

Actinomyces europaeus was found in abscesses of various locations (breast abscess, suprapubic abscess, perianal abscess, and labial abscess) but also in femur tissue, secretions from a decubital ulcer, and an atheroma cyst (Funke et al., 1997a; Clarridge and Zhang, 2002). It was also detected in patients suffering from urinary tract infections (Sabbe et al., 1999).

The few isolates of *Actinomyces graevenitzii* encountered so far were cultured from bronchial secretions and osteitis material (Pascual Ramos et al., 1997a), *A. radicidentis* was derived from infected root canals of human teeth (Collins et al., 2000b; Kalfas et al., 2001), *A. urogenitalis* was detected in urine, vaginal or urethral secretions (Nikolaitchouk et al., 2000) and scrapings from a penis ulcer (K. P. Schaal, unpublished observation), and the three presently described strains of *A. funkei* were from a case of endocarditis, a human sternum wound, and an abdominal incision (Lawson et al., 2001b; Westling et al., 2002). The eight strains of *A. cardiffensis* described so far originated from intrauterine contraceptive devices without reported signs of inflammation (three strains), from multiple abscesses in a patient four weeks after mastoidectomy together with a complex concomitant bacterial flora, from pleural fluid of a patient suffering from shortness of breath and wheezing, from pus of an actinomycotic jaw abscess, from a pericolic abscess, and from a right antral washout of a patient with sinusitis (Hall et al., 2002). The single isolates described as *A. nasicola* (Hall et al., 2003b), *A. oricola* (Hall et al., 2003d), and *A. hongkongensis* (Woo et al., 2003) originated from pus from the nasal antrum, a dental abscess, and a patient with suspected pelvic actinomycosis, respectively.

Actinomyces georgiae is the only human *Actinomyces* species described so far that has not been associated with human disease (Johnson et al., 1990).

CARIES, PERIODONTITIS AND GINGIVITIS From an economical and social point of view, caries, periodontal disease (periodontitis), and gingivitis are probably the most common and most important impairments in which *Actinomyces* species are etiologically involved. Although these conditions are complex in their pathogenesis, certain bacteria have been incriminated as contributors to their development (Slack and Gerencser, 1975; Newman, 1984; Page, 1986). In recent years, innumerable reports have been published which tried to elucidate the role these bacteria (including the fermentative actinomycetes) may play in the etiology of caries, periodontal disease, and gingivitis. This overwhelming quantity of information cannot be discussed in any detail in this chapter. Thus, only a brief summary of the major findings will be given.

Because of the adherence and coaggregation mechanisms which *Actinomyces* species manifest (see the subsection on Ecology in this Chapter), these organisms play an important role in the formation of dental plaque (see Slack and Gerencser, 1975), which is thought to be a prerequisite for the development of caries, periodontal disease, and gingivitis (Winford and Haberman, 1966; Socransky, 1970; Jordan and Hammond, 1972; Jordan and Sumney, 1973), although it must not in itself be considered pathological. Filamentous bacteria were found to contribute markedly to the volume of plaque and to supply an enormous additional surface area for the attachment of other bacteria (Boyd and Williams, 1971; Boue et al., 1987; Fure et al., 1987).

One of these filamentous bacteria, *Corynebacterium matruchotii*, was shown to form intracellular deposits of calcium phosphate that are indistinguishable from bone and tooth apatite (Ennever, 1960; Takazoe, 1961; Ennever et al., 1971; Ennever et al., 1978; Boyan-Salyers et al., 1978). This interesting ability contributes to the calcification of plaque, which leads to increased gingival irritation and subsequent inflammation (Slack and Gerencser, 1975).

Because of these and other striking features, fermentative actinomycetes including *Actinomyces* species have attracted the attention of many workers who are interested in the pathophysiology, cure, and prevention of oral and dental diseases and impairments (Winford and Haberman, 1966; Socransky, 1970; Jordan and Hammond, 1972; Jordan and Sumney, 1973; Moore et al., 1982; Loesche et al., 1983; Hoshino, 1985; Björndal and Larsen, 2000; Sunde et al., 2000;

Ximenez-Fyvie et al., 2000a; Ximenez-Fyvie et al., 2000b), although many other oral microorganisms, especially the cariogenic streptococci, have an even more important etiological role to play.

The capacity of fermentative saccharolytic bacteria including the actinomycetes to produce various organic acids thereby lowering the pH values in dental plaque has been mentioned in the subsection on Physiology in this Chapter; pH values of 5 or lower initiate the process of demineralization of the enamel, which is the first step in the development of caries and which can also be demonstrated *in vitro* (Clarkson et al., 1987; Kaufman et al., 1988). The carbohydrate sources for this acid production may be either of exogenous origin from the host's diet or of endogenous origin from the synthetic action of plaque bacteria (e.g., glycogen, levan, and dextran; Zero et al., 1986). In contrast, ureolytic activities of microbes of the dental plaque may be responsible for alkali production that causes fluoridated apatite to precipitate in the matrix (Gallagher et al., 1984).

Many additional factors that might contribute to the pathogenic effects of actinomycetes have been detected. Oral microorganisms including actinomycetes were found to possess fibrinogenolytic and fibrinolytic activities (Wikstrom et al., 1983). *Actinomyces israelii* and *A. odontolyticus* elaborate factors that adversely affect fibroblasts (Duguid, 1985). Specific surface components of *A. viscosus* exhibit mitogenicity for splenocytes including the B-cell subpopulation (Engel et al., 1984; Baker, 1985; Halfpap et al., 1985) and induce *in vitro* proliferation of T lymphocytes (Burckhardt, 1978). In addition, similar components were found to activate the alternate complement pathway (Baker and Billy, 1983a). The same species exerts chemotactic effects, is able to mark fibroblasts for immune-mediated damage, and possesses amphipathic antigens (Engel et al., 1976; Engel et al., 1978; Wicken et al., 1978). Furthermore, the type 2 fimbrial lectin of *A. viscosus* was demonstrated to stimulate superoxide and lactoferrin release from polymorphonuclear leukocytes, and it also mediates phagocytosis and subsequent killing of the bacteria (Taichman et al., 1978; Sandberg et al., 1986). On the other hand, sialic acid localized on the surface of *A. viscosus* has been implicated in the inhibition of the alternate complement pathway activation and subsequent opsonophagocytosis (Jones et al., 1986).

In the pulps and periapical tissue of diseased teeth, actinomycetes were identified histologically (Levine et al. 1985), immunocytochemically (Happonen et al., 1985), and by molecular methods (Sunde et al., 2000). *Actinomyces radicidentis* was exclusively found in infected root canals,

especially after conventional root canal treatment (Kalfas et al., 2001). In children with nursing caries, *A. israelii* was found to constitute 18.2% of the flora of the carious lesions, but this species was not identified in corresponding plaque samples in which *A. naeslundii* and *A. odontolyticus* predominated (Marchant et al., 2001). Penetrating *A. viscosus* and *A. naeslundii* cells and antigens were found to be always associated with antibodies and frequently also with complement (Pekovic et al., 1987). The involvement of plasma cells and T cells was also demonstrated and cytotoxic and Arthus type immunopathological reactions occurred (Pekovic and Fillery, 1984). In patients with hereditary fructose intolerance, the common gingival bleeding was related to the more frequent occurrence of *Actinomyces odontolyticus, Veillonella parvula* and *Campylobacter rectus* (Saxen et al., 1989).

Indirect evidence for the pathogenic potential of the microflora of dental plaque has been obtained from studies in which various antimicrobial substances were used to prevent or reduce plaque formation. Disinfectants in mouthrinse formulations such as bisguanides, phenols, quaternary ammonium compounds, oxygenating compounds, plant extracts, and fluorides, as well as antibiotics such as tetracyclines and antimicrobial combinations, were all found to inhibit growth and polymer production of oral microbes *in vitro* and in vivo, thereby reducing the intensity of plaque formation (Baker et al., 1983b; Slee and O'Connor, 1983; Meiers and Schachtele, 1984; Pallanza et al., 1984; Wegman et al., 1984; Gaffar et al., 1985; Hall et al., 1987; Schaeken and De Haan, 1989; Modesto et al., 2000; Phan et al., 2000; Feres et al., 2001). Furthermore, it could be demonstrated that the topical and, in case of the antibiotics, systemic application of the substances mentioned above was able to inhibit the development of caries, periodontal disease, and gingivitis (Paolino and Kashket, 1985; Gusberti et al., 1988; Walker, 1988).

TREATMENT OF ACTINOMYCES INFECTIONS AND ANTIMICROBIAL SUSCEPTIBILITY OF *ACTINOMYCES* SPECIES In addition to typical actinomycoses, the vast majority of the other diseases that may be caused by *Actinomyces* species also have a polymicrobial etiology. Therefore, successful treatment of these conditions depends not only on the type of disease and its localization, but also, often more importantly, on the composition of the synergistic concomitant flora. Caries and periodontitis can usually be treated successfully using the various local measures of conservative and operative dentistry, although administration of suitable antibiotics may facilitate therapy of periodontitis and especially of gingivitis. Lacrimal canaliculitis and other, noninvasive, eye infections usually also respond immediately to local measures such as removal of lacrimal concretions and the local application of β-lactam antibiotics or other drugs effective against the causal microflora. β-Lactam antibiotics are also the drugs of choice for treating monoinfections due to one of the recently described *Actinomyces* species.

In contrast, typical actinomycotic lesions have been known for many years to be notoriously difficult to treat. Penicillin G, which had been considered the drug of choice for many years (Slack and Gerencser, 1975; Lerner, 1988) because of its high *in vitro* activity against the pathogenic actinomycetes, is often not able to completely cure these infections even when the drug is administered in high doses for weeks or even months (Akgun et al., 1985; McNeil and Schaal, 1998). According to statistics based upon more than 4000 cases, failure of penicillin G treatment (including relapses) can be estimated at about 15–20% (Schaal and Pape, 1980). These treatment failures and relapses within a period of weeks or months after discontinuing antibiotic therapy can be attributed to the presence of penicillin-resistant concomitant organisms, especially *Actinobacillus actinomycetemcomitans* and *Bacteroides* species or other β-lactamase-producing organisms. In cases in which the penicillin resistance of the concomitant bacteria (such as in *Actinobacillus actinomycetemcomitans*) is not related to β-lactamase production, administration of penicillin G often results in elimination of the actinomycetes, but the inflammatory process continues because of the pathogenic potential of the remaining penicillin-resistant members of the synergistic flora. When β-lactamase producing organisms are present, they usually protect the actinomycetes from the action of the drug. On the other hand, clindamycin administration (Martin, 1985; Nielsen and Novak, 1987; Schaal, 1996; Schaal, 1998) may result in the selection of *Actinobacillus actinomycetemcomitans* without cessation of the clinical symptoms.

Alternative therapeutic measures for treating human actinomycoses, besides their good efficacy against fermentative actinomycetes, should also be sufficiently active against *Actinobacillus actinomycetemcomitans* and the common *Bacteroides*, *Prevotella* and *Porphyromonas* species (Schaal and Pape, 1980; Niederau et al., 1982; McNeil and Schaal, 1998). The former is achieved by using aminopenicillins instead of penicillin G (Schaal and Pape, 1980; Martin, 1984), the latter by combining an aminopenicillin with either a β-lactamase inhibitor or an agent effective against Gram-negative anaerobes (Niederau et al., 1982; Nord and Kager, 1983;

Narikawa, 1986; Collignon et al., 1988; Schaal, 1996; Schaal, 1998). Thus, in the past 20 years, by far the best results have been observed from combinations of ampicillin or amoxicillin with clavulanic acid or with metronidazole or clindamycin. Since amoxicillin + clavulanic acid in high doses was introduced as the treatment of choice, the authors have not observed any further treatment failure, even in chronic cases resistant to treatment with penicillin G and other antibiotics for months or years. Further alternatives that may be required if the patient is allergic to aminopenicillins are cefoxitin (Schaal et al., 1979b), imipenem (Dubreuil et al., 1987; McNeil and Schaal, 1998), or minocycline (Schaal, 1979a; Martin, 1985).

Because of the comparatively slow and anaerobic growth of *Actinomyces* species, special conditions are required for antimicrobial susceptibility testing of these organisms (Schaal and Pape, 1980). Mueller-Hinton broth or agar (commonly used for susceptibility testing of bacterial pathogens) does not promote the growth of fermentative actinomycetes adequately so that minimal inhibitory concentrations (MICs) found with these media are usually by far too favorable. The same is true for partially defined media such as Iso Sensitest Agar (Oxoid). However, reproducible results that correlate well with clinical efficacy can be obtained when DST agar (Oxoid, CM 261) is employed (Schaal and Pape, 1980).

The slow and often filamentous growth of fermentative actinomycetes also does not allow sufficient standardization of the agar dilution and broth dilution tests, respectively. Thus, essentially the only technique found to give reproducible and clinically suitable results of susceptibility testing of the bacteria is the agar dilution test using DST agar as test medium and microscopic assessment of results after only 24 h to a maximum of 48 h of incubation (Schaal and Pape, 1980). Using this procedure, it has been shown that nearly all of the penicillins, cephalosporins, and carbapenems currently in use are highly active against the classical *Actinomyces* species, aminopenicillins being particularly effective (Schaal et al., 1979b; Schaal and Pape, 1980; Niederau et al., 1982). The same is true for lincomycins, macrolides, rifamycins and vancomycin. In contrast, no or only very weak activity was observed for aminoglycosides, imidazoles and colistin.

Animal Diseases Caused by *Actinomyces* Species

ANIMAL ACTINOMYCOSES Since the first description of bovine actinomycosis (Bollinger, 1877), cattle have been known to suffer frequently from this infection. Bovine actinomycosis, which is also called "lumpy jaw," is a chronic suppurative

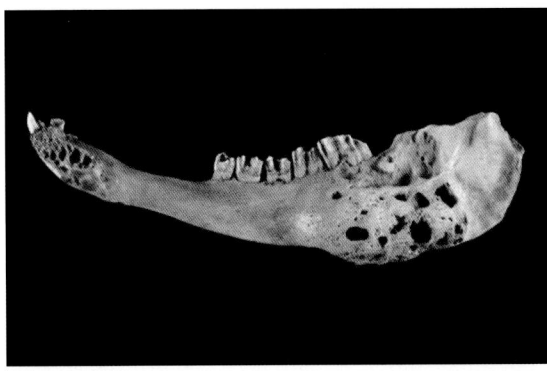

Fig. 27. Bovine mandibula showing a sponge-like osseous deformation caused by actinomycotic simultaneous osteoclastic and osteoblastic processes.

disease that, as in humans, tends to form multiple abscesses and draining sinus tracts. In contrast to human cases, however, bones are usually involved, mainly the mandible and occasionally the maxilla. The affected bones are slowly deformed by a destructive and, at the same time, proliferative osteitis (Fig. 27) (Slack and Gerencser, 1975). Primary lung infections in cattle have been reported (Biever et al., 1969).

The source of infection in cattle is not definitely established, but in analogy to humans, has been assumed to be endogenous although the principal etiological agent of bovine actinomycosis, *Actinomyces bovis* (Bollinger, 1877; Slack and Gerencser, 1975), has not been demonstrated in the oral cavity of healthy cattle even when modern isolation and identification methods were applied (Dent and Williams, 1984b). Bovine infections due to *A. israelii* were also reported (King and Meyer, 1957; Cummins and Harris, 1959; Pine et al., 1960) but remain to be confirmed using modern reliable identification methods. The same is true for the isolation of *A. israelii* from different organs and muscles of dead, disabled, diseased, dying and normal carcasses of sheep (Goda et al., 1986) and for the role of *A. viscosus* as an etiological agent of abortion in cattle (Okewole et al., 1989).

Clinically similar, but not identical diseases in which fermentative actinomycetes were incriminated as causative agents were reported from swine (Magnusson, 1928; Franke, 1973). These infections appear to involve the udder, lungs or other internal organs more frequently than the neck or bones. "*Actinomyces suis*" (Grässer, 1957; Franke, 1973), *A. israelii* (Magnusson, 1928), and *A. viscosus* (Georg et al., 1972) were reported from these infections. "*Actinomyces suis*" Franke 1973 was listed as species *incertae sedis* in *Bergey's Manual of Systematic Bacteriology* (Schaal, 1986a), but an immunohistochemical study of Murakami et al. (1998) appears to indicate that this species might be separable

from other actinomycetes found in swine. One of these other hog species that had been isolated from purulent vaginal discharge, resembling "*A. suis*" phenotypically, was recently described as *Actinomyces hyovaginalis* (Collins et al., 1993). A further new species from pig mastitis was named "*Actinomyces suimastitidis*" (Hoyles et al., 2001). To make the situation even more complicated, an organism formerly known as "*Eubacterium suis*" (Wegienek and Reddy, 1982b) was first transferred to the genus *Actinomyces* as *Actinomyces suis* (Ludwig et al., 1992) and then to the new genus *Actinobaculum* as *Actinobaculum suis* (Lawson et al., 1997). The latter organism has been isolated from pigs with cystitis and pyelonephritis as well as pregnant sows with metritis (Soltys and Spratling, 1957; Soltys, 1961), so that it remains to be clarified which of these species are the most important swine pathogens.

Comparatively frequently, *Actinomyces* infections were also reported from dogs and cats (McGaughey et al., 1951; Georg et al., 1972; Davenport et al., 1974; Davenport et al., 1975; Moens and Verstraeten, 1980; Hardie and Barsanti, 1982). These included infections of the soft tissue of the jaw as well as thoracic and abdominal manifestations. Tail, scrotum, lymph nodes, epidural space, and vertebrae or other bones may also be affected (Bestetti et al., 1977; Johnson et al., 1984; Murakami et al., 1997). *Actinomyces viscosus* was found to cause most of the infections encountered in dogs or cats (Georg et al., 1972; Davenport et al., 1974; Davenport et al., 1975; Bestetti et al., 1977; Moens and Verstraeten, 1980; Hardie and Barsanti, 1982; Johnson et al., 1984; Murakami et al., 1997). In one dog with a subvertebral mass and quadriplegia (Edwards et al., 1988), *Actinomyces odontolyticus* was identified as the causative agent.

Further *Actinomyces* species that are or may be specific pathogens of dogs and cats have been described (Buchanan et al., 1984b; Pascual et al., 1999; Hoyles et al., 2000; Hoyles et al., 2001). *Actinomyces hordeovulneris* (Buchanan et al., 1984b) was first isolated from pleuritis, peritonitis, visceral abscesses, septic arthritis, and recurrent localized infections of dogs in California and was found to be frequently associated with injuries caused by awns of the grass genus *Hordeum*. These awns, which easily penetrate skin or mucous membranes, are propelled forward through the tissue with any adjacent muscle contraction, leaving trails of inflammation and necrosis (Brennan and Ihrke, 1983). Infections due to *A. hordeovulneris* in dogs have also been reported in Hungary (Pelle et al., 2000) and were occasionally seen in Germany (K. P. Schaal, unpublished observation). The ability of this pathogen to produce L-phase variants spontaneously with coincident calcium deposition

(Buchanan and Scott, 1984a) was believed to be related to sulfur granule formation *in vivo*. As *A. hordeovulneris* could be recovered from the gingival margin of cats (Love et al., 1990), it appears sensible to assume that the organism is introduced from the oral cavity or intestinal tract when mucous membranes are penetrated by the awn or the animal licks or bites a wound. In addition or alternatively, the pathogen may have spread hematogenously to the necrotic focus resulting from an awn injury.

OTHER ANIMAL INFECTIONS *Actinomyces bowdenii* (Pascual et al., 1999) is another new species that was isolated from canine and feline clinical specimens, in particular from an abscess under the mandible of a dog, from feline pleural fluid, from a canine neck abscess, and from a canine pyogranuloma. All of the isolates were found to be associated with aerobically or anaerobically growing concomitant organisms such as *Pasteurella multocida* or *Bacteroides*, *Prevotella* or *Fusobacterium* species.

Actinomyces canis (Hoyles et al., 2000) was isolated in mixed culture with other bacteria from the vagina and pus specimens of dogs. The two strains identified so far as *Actinomyces catuli* (Hoyles et al., 2001) were of similar origin, as was the single strain of *Actinomyces coleocanis* (Hoyles et al., 2002), which was detected in vaginal secretions of a cocker spaniel dog together with "*Corynebacterium genitalium*."

Actinomycosis-like lesions have been reported from sheep, goats, horses, deer, moose, antelope, mountain sheep, gazelles and even an arctic fox (Slack and Gerencser, 1975; Raju et al., 1986; Snyder et al., 1987). However, little detailed information on the *Actinomyces* species involved is available. In one case of pyogenic granulomas found in the abdomen of a mandrill, *A. israelii* was identified as the etiologic agent by immunofluorescence (Altman and Small, 1973). *Actinomyces marimammalium* (Hoyles et al., 2001) was isolated from two dead seals and a dead porpoise, but its pathogenic potential has not been determined yet. *Actinomyces vaccimaxillae* (Hall et al., 2003) was cultured from pus of a jaw lesion in an adult cow, thus resembling in this respect *A. bovis*. Finally, *A. viscosus*, serovar 1, and possibly other *Actinomyces* species have been isolated from dental plaque of hamsters and beagle dogs with naturally occurring periodontitis (Howell, 1963a; Jordan and Keyes, 1964; Jordan and Keyes, 1965a; Syed et al., 1981).

EXPERIMENTAL ANIMAL INFECTIONS Various domestic and laboratory animals including horses, cattle, sheep, goats, pigs, dogs, rabbits, guinea pigs, rats, hamsters and mice have been used to produce experimental infections with *Actinomyces* species (Slack and Gerencser,

1975). After intraperitoneal, intravenous or subcutaneous injection, *Actinomyces bovis*, *A. israelii*, *A. naeslundii*, *A. odontolyticus* and *A. viscosus* were all found to cause abscess formation; some of these abscesses resembled naturally occurring actinomycotic lesions (Schaal, 1986a). However, progressive infections rarely developed from these abscesses, and the infected animals usually survived. Furthermore, detailed histopathological examinations showed that the actinomycotic lesions produced by *A. israelii* in mice differed considerably from those following challenge with *A. viscosus* or *A. naeslundii* (Behbehani et al., 1983a). These and other findings indicate that the virulence of *Actinomyces* species apparently varies both between species and between individual strains, and the susceptibility of the test animals may also vary considerably (Wolff and Israel, 1891; Pine et al., 1960; Coleman and Georg, 1969; Georg and Coleman, 1970; Georg et al., 1972; Beaman et al., 1979). In addition, it has been demonstrated that the simultaneous challenge with *A. israelii* and *Eikenella corrodens* decreased the minimal infecting dose of *A. israelii* considerably, stressing the role of mixed infections in the naturally occurring disease (Jordan et al., 1984). Similar results were obtained when *A. israelii* cells were packed in alginate gel particles, which led to the formation of structures comparable with sulfur granules (Sumita et al., 1998). Using an ischemic mouse thigh model for evaluation of pathogenicity, Chatterjee and Chakraborti (1989) found that *A. naeslundii*, under these experimental conditions, exhibited an especially high degree of virulence, causing death in all test animals within 24 h.

The only animal *Actinomyces* species that appear to be nonpathogenic in general are *Actinomyces denticolens*, *A. howellii* and *A. slackii*, although detailed pathogenicity studies have not been performed with these organisms.

Numerous animal experiments were performed on the ability of actinomycetes to colonize teeth (De Jong et al., 1983; Crawford and Clark, 1986) and on their role in the development of caries and periodontitis. After oral inoculation, *A. viscosus* and *A. naeslundii* were both shown to produce periodontal disease with alveolar bone loss and root surface caries in hamsters (Jordan and Keyes, 1964; Hernandez, 1985; Kametaka et al., 1989) as well as in conventional and gnotobiotic rats (Jordan et al., 1965b; Socransky et al., 1970; Llory et al., 1971; Brecher et al., 1978; Crawford et al., 1978; Brecher and Van Houte, 1979; Burckhardt et al., 1981; Shakespeare et al., 1985; Firestone et al., 1987; Firestone et al., 1988). Similar observations were made for *Actinomyces israelii* orally implanted into gnotobiotic rats, which produced extensive plaque formation accompanied by root surface caries and bacterial invasion of the pulp (Behbehani et al., 1983b).

The factors of pathogenicity and virulence, which enable *Actinomyces* species to invade the tissue and to cause necrosis and abscess formation, remain to be elucidated in detail. Some of the virulence factors that are involved in the development of caries, gingivitis and periodontal disease have been mentioned in the sections on the respective human diseases.

The Genus *Actinobactulum*

The genus *Actinobaculum* was introduced by Lawson et al. (1997) to accommodate bacterial strains that had previously been assigned to various other genera and that were named "*Actinobaculum suis,*" together with recent *Actinomyces*-like isolates from human clinical specimens for which the designation "*A. schaalii*" was proposed. The taxonomic and nomenclatural history of *A. suis* is unusually varied: In 1957, Soltys and Spratling proposed the name "*Corynebacterium suis*" for an organism which was associated with cases of cystitis and pyelonephritis in pigs. This generic assignment was apparently based almost exclusively on the diphtheroid morphology of the organism, a common practice at that time. Neither "*Corynebacterium suis*" Soltys and Spratling nor "*Corynebacterium suis*" Hauduroy et al. (1937) was included in the Approved Lists of Bacterial Names (Skerman et al., 1980), and therefore neither has nomenclatural standing.

In a taxonomic study of "*Corynebacterium suis,*" Wegienek and Reddy (1982b) found that "*Corynebacterium suis*" strain Soltys 50052 is anaerobic, has rhamnose and lysine as major cell components, and produces acetate, formate and ethanol as major end products of carbohydrate metabolism. In contrast, representatives of the genus *Corynebacterium* are aerobic or facultatively anaerobic organisms characterized by cell walls containing arabinose and galactose as the major sugar components and *meso*-diaminopimelic acid as the major diamino acid. Furthermore, they produce major amounts of acetate, propionate and formate and variable amounts of other acids as products of carbohydrate metabolism (Reddy and Kao, 1978). Thus, Wegienek and Reddy (1982b) concluded that this strain did not belong to the genus *Corynebacterium* and they proposed, despite differences in cell wall composition and DNA base ratios, that the organism should be included in the genus *Eubacterium.* Their proposal was based on a rather limited number of characteristics such as anaerobiosis, morphology, as well as absence

of propionate, lactate and succinate among the end products of carbohydrate metabolism.

Comparative sequence analysis of the 16S rRNA gene of "*Eubacterium suis*" showed that this organism is a close relative (93.8% sequence similarity) of "*Actinomyces pyogenes*" (Ludwig et al., 1992). These two organisms share a common cell wall composition (lysine is the diamino acid; rhamnose and traces of mannose are the cell wall sugars), they both contain type *c* cytochromes, and they have similar DNA G+C contents (55 and 56–58 mol%, respectively; Soltys and Spratling, 1957; Schaal and Pulverer, 1981b; Wegienek and Reddy, 1982b; Schaal, 1986a). Differences in the biochemical characteristics of "*Eubacterium suis*" Soltys and Spratling and the species *incertae sedis* "*Actinomyces suis*" Franke 1973 indicate that these two organisms are not identical. Since no type strain of "*Actinomyces suis*" Franke is extant, Ludwig et al. (1992) proposed that "*Eubacterium suis*" should be transferred to the genus *Actinomyces* as "*Actinomyces suis.*" The reclassification in the newly created genus *Actinobaculum* as *Actinobaculum suis* by Lawson et al. (1997) was then a consequence of the results of a comparative 16S rRNA gene sequence analysis.

Taxonomy

The genus *Actinobaculum* currently comprises four validly described species (Table 1). The taxonomic history of *Actinobaculum suis* has been discussed above. The species *Actinobaculum schaalii* was proposed by Lawson et al. (1997) for human bacterial isolates that originated from urine or blood. Two other isolates from human urine were described as *Actinobaculum urinale* (Hall et al., 2003c) and *Actinobaculum massiliae* (Greub and Raoult, 2002), respectively.

CHEMOTAXONOMIC PROPERTIES Apart from phylogenetic evidence, the inclusion of these organisms in the genus *Actinobaculum* has not been substantiated by other taxonomically relevant properties such as chemotaxonomic features which usually provide useful additional criteria for the delineation of genera. Thus, despite differences in the cell wall murein types of *Actinobaculum suis* (L-Lys-L-Ala-L-Lys-D-Glu) and *Actinobaculum schaalii* (L-Lys-L-Lys-D-Glu), the assignment of both species to the same genus was considered to be justified because of the close affinity suggested by their 16S rRNA gene sequence similarity (approximately 94%) and a statistically significant phylogenetic association with one another (bootstrap value of 100%).

In an extensive chemotaxonomic study of *Actinobaculum suis*, the type species of the genus *Actinobaculum*, this species was found to be

Table 13. Comparative chemotaxonomic characteristics of members of three validly described genera of the family Actinomycetaceae.

Organisms	Wall sugars	Acyl type	Phospholipid type	Menaquinones	Fatty acids
Actinomyces bovis	Glucose + mannose + rhamnose + 6-deoxytalose	Acetyl	PIII DPG, PC, PI, PIM	MK-8 + MK-9[a] + MK-10 + *methyl*-MK-10	S, U
Actinobaculum suis	Galactose + mannose + rhamnose	Acetyl	PI DPG, PG, PI, PIM	MK-10(H$_4$)[a]	S, U
Arcanobacterium haemolyticum	Rhamnose + glucose	Acetyl	PIII DPG, PG, PC, PI, PIM	MK-8(H$_4$) + MK-9(H$_4$)[a]	S, U

Abbreviations: please refer to footnote in Table 7.
[a]Major component.

characterized by a cell-wall sugar pattern composed of galactose, mannose and rhamnose and by a Phospholipid Type PI *sensu* Lechevalier et al. (1977) in that it contains the phospholipids diphosphatidylglycerol, phosphatidylglycerol, phosphatidylinositol, and phosphatidylinositol dimannosides (Table 13). Its respiratory quinones consist of tetrahydrogenated menaquinones with ten isoprene units (MK-10[H$_4$]) as the only isoprenoid quinones; the fatty acid profile comprises straight-chain saturated and monounsaturated fatty acids; its muramic acid residue is N-acetylated (Table 13). However, chemotaxonomic investigations on cell wall components of the type strain of *Actinobaculum schaalii* showed that the latter species possesses a completely different chemotaxonomic pattern (A. F. Yassin, unpublished observation; data not shown) as compared to that of *Actinobaculum suis*. Thus, these findings revealed considerable chemotaxonomic heterogeneity within the genus *Actinobaculum*, so that the genus *Actinobaculum* as currently defined apparently needs taxonomic revision.

Detailed long-chain cellular fatty acid profiles were given for *A. schaalii*, *A. massiliae* and *A. urinale* and were found to consist of straight-chain saturated and monounsaturated fatty acids.

The DNA base composition of members of the genus was reported to be 55–57 mol% G+C.

MORPHOLOGICAL CHARACTERISTICS Additional characteristics that were used to circumscribe the genus *Actinobaculum* consist of a rather limited number of common features. Among them, the morphology of members of the genus has been summarized as follows (Fig. 28): "On blood agar, cells are straight to slightly curved rods, some of which exhibit branching. Cells are Grampositive, not acid fast, and non-motile and do not form spores" (Lawson et al., 1997). In the descriptions of *A. suis* and *A. urinale* (Hall et al., 2003), branching cells were not mentioned.

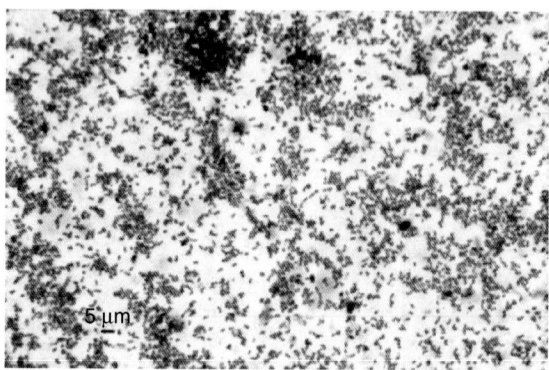

Fig. 28. *Actinobaculum schaalii*; Gram-stained smear from a mature colony (brain heart infusion agar, 14 days at 36 ± 1°C, and Fortner's method).

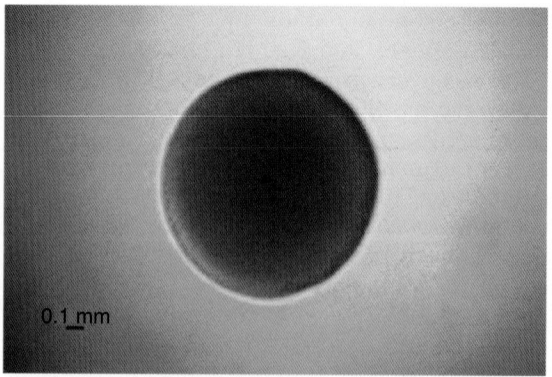

Fig. 29. *Actinobaculum schaalii*; mature, slightly raised, translucent colony without internal structures; micrograph in transmitted light (brain heart infusion agar, 14 days at 36 ± 1°C, and Fortner's method).

Colonies are predominantly smooth with entire to slightly irregular margins. They are gray to white and, only in the case of *A. urinale*, weakly β-hemolytic (Fig. 29).

PHYSIOLOGICAL PROPERTIES In the description of the genus (Lawson et al., 1997), production of acetate as major end product of glucose and maltose metabolism was given as one of its main diagnostic characteristics. However, *A. urinale* (Hall et al., 2003) appears to differ from this profile in that this species was reported to produce lactic acid as major end product of glucose fermentation, together with minor amounts of acetic acid.

Actinobaculum species were described as anaerobic or facultatively anaerobic, *A. suis* being the only species that appears to prefer anaerobic growth conditions.

As do *Actinomyces* species, members of the genus *Actinobaculum* apparently require complex media for good growth, indicating that their anabolic capacity is limited.

The catabolic activities of *Actinobaculum* species have not been studied comprehensively. The data so far available are summarized in Table 14 together with the respective information for *Arcanobacterium* and *Varibaculum* species. The following physiological properties are common to all four members of the genus: acid production from maltose but not from glucose (which is not fermented by *A. suis*), no esculin and gelatin hydrolysis, lack of arginine dihydrolase, no

Table 14. Physiological properties of *Actinobaculum, Arcanobacterium, Varibaculum* and *Mobiluncus* species.

Property	1.	2.	3.	4.	5.	6.	7.	8.	9.	10.	11.	12.	13.	14.
Acid production from														
N-Acetyl-β-glucosamine	nd	–	nd	nd	nd	nd	nd	+	nd	nd	nd	nd	nd	nd
Adonitol	nd	nd	–	nd	+	nd	nd	–	nd	d[b]	nd	nd	nd	nd
Amygdalin	nd	nd	–	nd	–	–	nd	–	nd	–	–	nd	nd	nd
D-Arabinose	nd	nd	nd	nd	nd	nd	nd	–	nd	nd	nd	nd	nd	nd
L-Arabinose	nd	d	–	–	nd	–	–	–	–	d	–	–	–	–/(+)
D-Arabitol	nd	–	nd	–	d	–	–	–	–	nd	–	nd	nd	nd
L-Arabitol	nd	nd	nd	nd	d	nd	nd	–	nd	nd	nd	nd	nd	nd
Cyclodextrin	nd	–	nd	–	nd	nd	–	nd	–	nd	–	nd	nd	nd
Cellobiose	nd	nd	–	nd	nd	–	nd	–	nd	d	–	nd	nd	nd
Dextrin	nd	nd	nd	nd	nd	+	nd	nd	nd	d	nd	nd	nd	nd
Dulcitol	nd	nd	–	nd	nd	–	nd	–	nd	–	nd	nd	nd	nd
iso-Erythritol	nd	nd	–	nd	d	–	nd	–	nd	d	nd	nd	nd	nd
D-Fructose	nd	nd	–	nd	+	+	nd	+	nd	+	d	(d)	(d)	d
D-Galactose	nd	nd	–	nd	nd	+	nd	+	nd	+	nd	–/+	(d)	((d))
β-Gentiobiose	nd	nd	nd	nd	nd	nd	nd	–	nd	nd	nd	nd	nd	nd
D-Glucose	+	+	–	+	+	+	+	+	+	+	+	(d)	–/(+)	d
Glycerol	nd	nd	–	nd	+	–	nd	+	nd	–/+	nd	nd	nd	nd
Glycogen	+	–	+	–	d	–	–	+	d	+	–	((+))	((+))	+
myo-Inositol	nd	nd	–	nd	nd	–	nd	+/–	nd	d	nd	–	–	+/–
Inulin	nd	nd	–	nd	nd	nd	nd	–	nd	–	nd	nd	nd	nd
Lactose	nd	–	–	–	–	+	+	+	d	+	–	–/(+)	–	–/(+)
D-Lyxose	nd	nd	nd	nd	nd	nd	nd	–	nd	nd	nd	nd	nd	nd
Maltose	+	+	+	d	+	+	d	+	d	+	–/+	(+)	((+))	+
D-Mannitol	–	–	–	–	–	–	–	d	–	–/(+)	d	nd	nd	nd
D-Mannose	–	d	–	–	nd	nd	nd	+/–	nd	+	–	–/(+)	–	((d))
Melezitose	nd	–	–	–	nd	–	–	d	–	+	–	nd	nd	nd
Melibiose	nd	–	–	–	nd	–	–	–	–	–	–	+/–	–/+	–
α-Methyl-D-glucoside	nd	nd	nd	nd	nd	–	nd	–	nd	d	nd	nd	nd	nd
α-Methyl-D-mannoside	nd	nd	nd	nd	–	nd	nd	–	nd	–	nd	nd	nd	nd
Methyl-β–D-glucopyranoside	nd	–	nd	–	nd	nd	–	nd	nd	nd	–	nd	nd	nd
Pullulan	nd	–	nd	–	+	–	–	–	–	nd	–	nd	nd	nd
D-Raffinose	(+)	–	–	–	nd	–	–	–	–	–	–	–/(+)	–	–
L-Rhamnose	nd	nd	–	nd	nd	–	nd	–	nd	–	nd	–	–	–/(+)
D-Ribose	+	+	nd	+	d	d	–	d	d	+	d	–	–	(d)
Salicin	nd	nd	–	nd	nd	nd	nd	–	nd	–	–	nd	nd	nd
D-Sorbitol	–	–	–	–	nd	–	–	–	–	d	–	nd	nd	nd
L-Sorbose	nd	nd	nd	nd	nd	nd	nd	–	nd	nd	nd	nd	nd	nd
Starch	nd	d	+	nd	+	nd	nd	+	nd	+	nd	–	((d))	d
Sucrose	nd	d	–	d	–	d	–	+	d	–/+	+/–	–	–	d
D-Tagatose	nd	–	nd	–	nd	nd	–	d	–	nd	–	nd	nd	nd
Trehalose	+	d	–	–	nd	d	–	+/–	–	+	–/+	–	–	d
D-Turanose	nd	nd	nd	nd	nd	nd	nd	+	nd	nd	nd	nd	nd	nd
Xylitol	nd	nd	nd	nd	+	–	nd	–	nd	+	nd	nd	nd	nd

Table 14. *Continued*

Property	1.	2.	3.	4.	5.	6.	7.	8.	9.	10.	11.	12.	13.	14.
D-Xylose	+	+	nd	–	–	–	–	d	d	+	d	(+)	((+))	+
L-Xylose	nd	nd	nd	nd	nd	nd	nd	–	nd	nd	nd	nd	nd	nd
Hydrolysis of														
Casein	nd	nd	nd	nd	nd	–	nd	nd	nd	+	nd	nd	nd	nd
Esculin	–	–	–	–	–	–	(+)	–	(+)	–	–	–	–	–
Gelatin	–	–	–	–	d	–	–	–	+	+	–	–	–	–
Hippurate	+	+	nd	+	–	–	+	–	d	+/–	+/–	+	+	–
Starch	nd	nd	+	nd	nd	d	nd	nd	nd	d	–	nd	nd	nd
Enzyme activities														
N-Acetyl-β-glucosaminidase	–	nd	nd	–	nd	+	d	–	–	+	–	nd	nd	nd
Acid phosphatase	nd	nd	nd	–	–	+	–	+	–	–	(+)/–	nd	nd	nd
Alanine arylamidase	nd	+	+?	–	nd	nd	nd	nd	nd	nd	nd	nd	nd	nd
Alanine phenylalanine proline arylamidase	nd	+	+?	–	+	+	–	+	+	nd	d	nd	nd	nd
Alkaline phosphatase	–	–	+?	–	–	nd	d	+	–	–	–	nd	nd	nd
α-Arabinosidase	nd	nd	nd	–	nd	nd	nd	nd	nd	nd	nd	nd	nd	nd
Arginine arylamidase	nd	nd	nd	+?	nd	nd	nd	nd	nd	nd	nd	nd	nd	nd
Arginine dihydrolase	–	–	–	–	nd	–	–	nd	–	–	–	+	+	–
Chymotrypsin	nd	nd	nd	–	–	nd	–	–	nd	–	–	nd	nd	nd
Cysteine arylamidase	nd	nd	nd	–	–	nd	nd	+	nd	–	nd	nd	nd	nd
DNAase	nd	nd	nd	nd	–	+	nd	nd	nd	+	nd	nd	nd	nd
Esterase C4	nd	nd	nd	–	nd	nd	–	+	nd	–	(+)/–	nd	nd	nd
Esterase-lipase C8	nd	nd	nd	–	nd	nd	–	+	nd	–	(+)/–	+	+	+
α-Fucosidase	nd	nd	nd	–	–	–	–	–	nd	–	–	nd	nd	nd
α-Galactosidase	–	–	nd	–	–	–	–	(+)	–	d	–	+	+	–
β-Galactosidase	–	–	nd	–	–	d	+	+	–	+	–/(+)	+	+	–
α-Glucosidase	+	+	+?	–	+	nd	+	+	–	–	+	nd	nd	nd
β-Glucosidase	nd	nd	nd	–	+	nd	–	–	–	–	–	nd	nd	nd
β-Glucuronidase	–	–	+?	+	–/(+)	+	–	+	+	–	–	nd	nd	nd
Glycine arylamidase	nd	+	+?	–	nd	nd	nd	nd	nd	nd	nd	nd	nd	nd
Histidine arylamidase	nd	nd	nd	–	nd	nd	nd	nd	nd	nd	nd	nd	nd	nd
Lipase C14	nd	nd	nd	–	nd	nd	–	–	nd	–	–	–	–	–
Leucine arylamidase	–	+	+?	–	+	nd	+	+	nd	d	+	+	+	+
Leucylglycine arylamidase	nd	nd	nd	–	nd	nd	nd	nd	nd	nd	nd	nd	nd	nd
α-Mannosidase	nd	nd	nd	–	–	nd	–	–	nd	–	–	nd	nd	nd
β-Mannosidase	nd	–	nd	–	–	–	–	nd	–	nd	–	nd	nd	nd
Proline arylamidase	nd	nd	nd	+	nd	nd	nd	nd	nd	nd	nd	nd	nd	nd
Phenylalanine arylamidase	nd	nd	nd	–	nd	nd	nd	nd	nd	nd	nd	nd	nd	nd
Pyrazinamidase	+	d	–	–	+	+	–	+	d	nd	–	nd	nd	nd
Pyrrolidonyl arylamidase	–	d	nd	nd	d	d	–	d	+	nd	–	nd	nd	nd
Serine arylamidase	nd	nd	nd	–	nd	nd	nd	nd	nd	nd	nd	nd	nd	nd
Trypsin	nd	nd	nd	–	–	nd	–	d	nd	–	–	nd	nd	nd
Tyrosine arylamidase	nd	+	+?	–	nd	nd	nd	nd	nd	nd	nd	nd	nd	nd
Urease	–	–	++	d	–	–	–	–	nd	–	–	–	–	–
Valine arylamidase	nd	nd	nd	–	nd	nd	–	–	nd	–	–	nd	nd	nd
β-Xylosidase	nd	nd	nd	nd	nd	nd	nd	nd	nd	nd	nd	nd	nd	nd
Catalase production	–	–	–	–	–	–/(+)	–	d	+	–/(+)	–	–	–	–
Nitrate reduction	–	–	–	–	–	+/–	–	–	–	–	+/–	–	+	–/+
Nitrite reduction	nd	nd	nd	nd	nd	–	nd	nd	nd	–	nd	nd	nd	nd
H₂S production	nd	nd	nd	nd	nd	nd	nd	nd	nd	nd	nd	–	–	–
Acetoin production	nd	–	nd	–	nd	–	–	nd	–	+	–	nd	nd	nd
Indole production	nd	nd	–	–	nd	–	nd	nd	nd	–	–	–	–	–

Symbols: +, positive/present; –, negative/absent; +/–, usually positive; –/+, usually negative; (+), weakly positive; ((+)), very weakly positive; +?, discrepancies in literature; ++, strongly positive; d, strain variation; (d), strain variation, always weak reaction; and nd, no data available.

[a]1. *Actinobaculum massiliae*; 2. *Actinobaculum schaalii*; 3. *Actinobaculum suis*; 4. *Actinobaculum urinale*; 5. *Arcanobacterium bernardiae*; 6. *Arcanobacterium haemolyticum*; 7. *Arcanobacterium hippocoleae*; 8. *Arcanobacterium phocae*; 9. *Arcanobacterium pluranimalium*; 10. *Arcanobacterium pyogenes*; 11. *Varibaculum cambriense*; 12. *Mobiluncus curtisii* subsp. *curtisii*; 13. *Mobiluncus curtisii* subsp. *holmesii*; 14. *Mobiluncus mulieris*.

From Reddy et al. (1982), Spiegel and Roberts (1984), Collins and Cummins (1986), Moore and Holdeman Moore (1986), Schaal (1986b), Ludwig et al. (1992), Spiegel (1992), Funke et al. (1995), Lawson et al. (1997), Pascual Ramos et al. (1997b), Lawson et al. (2001a), Hoyles et al. (2002a), Greub and Raoult (2002), and Hall et al. (2003a, c).

catalase production, and no reduction of nitrate. Other physiological features vary from species to species and may therefore be used as differential characteristics.

Habitat and Ecology

Actinobaculum suis was originally isolated from cases of cystitis, pyelonephritis and metritis in pregnant sows (Soltys and Spratling, 1957; Soltys, 1961; Walker and MacLachlan, 1989; Carr and Walton, 1993; Liebhold et al., 1995; Biksi et al., 1997). Although not cultured from healthy sows, the organism was frequently recovered from urine and semen of apparently healthy boars. It was also isolated from the preputium or preputial diverticulum of boars (Sobestiansky et al., 1993; Biksi et al., 1997). The habitat of *A. schaalii* was assumed to be the human genital or urinary tract. This species was isolated from human blood and urine. Similarly, *A. urinale* was cultured under anaerobic conditions from urine of a human female with pyuria, and *A. massiliae*, from urine of an elderly woman with recurrent cystitis.

Isolation and Cultivation

A more detailed study on the nutritional and metabolic features was only performed for *A. suis* (Wegienek and Reddy, 1982a). Excellent growth of this organism was obtained using a peptone-yeast extract-starch (PYS) medium containing trypticase (BD), yeast extract, starch, minerals, cysteine, and sodium carbonate, together with anaerobic conditions. Growth was found to be considerably less when the starch in this medium was replaced by maltose. Deletions of starch or maltose resulted in negligible growth, indicating that a fermentable carbohydrate is required for growth. Furthermore, CO_2 or Na_2CO_3, yeast extract, and trypticase were also required. Yeast extract could be replaced in PYS by a defined mixture of purine and pyrimidine bases, vitamins, and amino acids, but deletion of trypticase resulted in no detectable growth, suggesting a possible peptide requirement of *A. suis*.

The nutritional requirements of the other *Actinobaculum* species are not known in detail. *Actinobaculum schaalii* was successfully cultured on 5% horse blood agar (Columbia base, BD) in a 5% CO_2 atmosphere. *Actinobaculum urinale* grew well on the same medium under anaerobic conditions. *Actinobaculum massiliae* was successfully cultured on Columbia sheep blood agar in a 5% CO_2 atmosphere.

Identification

As the genus *Actinobaculum* is heterogeneous chemotaxonomically as well as physiologically, there is no single test or a narrow spectrum of physiological markers that would allow the identification of these organisms on the genus or species level. Nevertheless, Gram-positive straight or slightly curved rod-shaped bacteria with or without branching that grow under increased CO_2 tension or anaerobically, produce predominantly acetate or lactate as end products of their carbohydrate metabolism, show acid production from maltose, and are catalase-, nitrate reductase-, esculin hydrolysis-, and gelatinase-negative (Table 14) may well belong to the genus *Actinobaculum*.

Differentiation of the four species of the genus may be achieved by a combination of a number of physiological properties or by molecular methods. A strong urease activity together with acid production from starch and maltose but not from glucose is characteristic of *A. suis* (Table 14). Weak acid production from raffinose, fermentation of trehalose and glycogen, and pyrazinamidase activity may be indicative of *A. massiliae*. *Actinobaculum schaalii* and *A. urinale* may be differentiated from one another and from the two other *Actinobaculum* species by sucrose and xylose fermentation, alanine arylamidase-, alanine phenylalanine proline arylamidase-, α-glucosidase-, leucine arylamidase-, glycine arylamidase-, and tyrosine arylamidase-activities (Table 14).

The methods useful for identifying these bacteria are by and large the same as those described for *Actinomyces* species. However, the evaluation of four commercial identification systems for anaerobes showed (Walker et al., 1990) that two of these systems could identify *A. suis* when colony morphology and Gram reaction were considered in addition. The remaining two systems were of limited or no use for identifying this species.

An indirect immunofluorescence procedure was successfully used both for identifying *A. suis* directly in swabs from the preputial diverticulum of boars (Langfeldt et al., 1990; Sobestiansky et al., 1993) and for detecting antibodies in infected pigs (Wendt and Amtsberg, 1995). For the latter test, the authors found a sensitivity of 78.9% when the specificity was 100% for animals actually infected with *A. suis*.

Epidemiology

Essentially nothing is known about the epidemiology of infections due to the recently described human *Actinobaculum* species because only a few cases have been reported.

Actinobaculum suis is an important pathogen in the urogenital tract of pigs. It is a common cause of cystitis and pyelonephritis in breeding sows in Canada, Europe and Australia. Disease isolates have also been reported from Norway, Denmark, Holland, Hong Kong, Switzerland, Malaysia, Germany, Brazil and the United States (Jones, 1992). However, disease is apparently only produced in sows, whereas boars are only colonized and transmit the organism sexually.

Pathogenicity

Actinobaculum suis is able to produce cystitis, pyelonephritis, metritis and abortion in sows (Soltys and Spratling, 1957; Soltys, 1961; Yamini and Slocombe, 1988). Significant bacteriuria often appears to be the only clinical sign of beginning cystitis (Liebhold et al., 1995). However, more advanced cases frequently show macrohematuria and urinary pH values above 8.0, which are apparently due to the heavy production of ammonia from urea. Primarily, cystitis is the result of colonization of the bladder by *A. suis*, which is facilitated by its ability to adhere to bladder epithelial cells by numerous fimbriae (Larsen et al., 1986). Using scanning electron microscopy, Wendt et al. (1994) were able to demonstrate that the hemorrhagic cystitis, thus induced by *A. suis*, is accompanied by a total loss of superficial cells of the infected bladder mucosa. Nevertheless, these severe alterations appeared to be only possible when previous infections had debilitated the urothelium, thereby supporting infection of the bladder with *A. suis*. This is in accord with the observation that sows can be infected artificially by intrarenal injection of live organisms only when 5% saponin is added to the suspension (Soltys, 1961). Exotoxin activities could be demonstrated in this pathogen (Wegienek and Reddy, 1982b).

The five human isolates of *A. schaalii* described by Lawson et al. (1997) originated from blood (two strains) or urine (three strains). Thus, this organism appears to be a pathogen of the human urinary tract causing cystitis and pyelonephritis possibly associated with bacteremia or urosepsis. The single isolate of *A. urinale* described so far was also cultured from the urine of a human female patient suffering from pyuria (Hall et al., 2003). Similarly, *A. massiliae* was isolated from the urine of a female patient with catheter-associated cystitis (Greub and Raoult, 2002).

Comprehensive studies on the treatment of *Actinobaculum* infections have not been performed. Only one report (Wendt and Sobestiansky, 1995b) deals with the therapy of urinary tract infections in sows due to *A. suis*. Enrofloxacin and ampicillin were used in this study and proved to be efficient even in cases of hemorrhagic cystitis.

The Genus *Arcanobacterium*

The genus *Arcanobacterium* was described by Collins et al. (1982c) to accommodate bacterial strains originally isolated from infected American soldiers and previously named "*Corynebacterium haemolyticum*" (MacLean et al., 1946). However, the species exhibits little similarity to typical corynebacteria and its placement in the genus *Corynebacterium* was questioned by several workers (Cummins and Harris, 1956; Barksdale et al., 1957; Barksdale, 1970; Jones, 1975; Minnikin et al., 1978; Schofield and Schaal, 1981; Collins et al., 1982b). Furthermore, the relationship of "*C. haemolyticum*" to the species "*C. pyogenes*" (Glage) remained unclear. In a study of cell-wall compositions in some Gram-positive bacteria, Cummins and Harris (1956) noted that the cell-wall compositions of "*C. pyogenes*" and "*C. haemolyticum*" were obviously similar to one another but differed both in sugar and amino acid composition from these of the other corynebacteria (i.e., neither species contains arabinose or galactose, lysine is the diamino acid of the peptidoglycan, and A_2pm is absent). On the other hand, Cummins and Harris (1956) found that rhamnose was present in both organisms, and this, together with the fact that alanine, glutamic acid and lysine were the major amino acid components, led these authors to suggest that the two organisms were related to the streptococci. This view was upheld by Barksdale et al. (1957), who suggested not only that "*C. haemolyticum*" and "*C. pyogenes*" should be reclassified in the genus *Streptococcus*, but also that "*C. haemolyticum*" was a mutant form of "*C. pyogenes*."

In the eighth edition of *Bergey's Manual of Determinative Bacteriology*, both taxa were listed in an addendum to the genus *Corynebacterium* (Cummins et al., 1974), and "*C. haemolyticum*" does not appear in the Approved Lists of Bacterial Names (Skerman et al., 1980).

Later, numerical phenetic (Schofield and Schaal, 1981) and chemical (Collins et al., 1982b) studies showed that "*C. haemolyticum*" and "*C. pyogenes*" are two distinct taxa. The discovery of tetrahydrogenated menaquinones in "*C. pyogenes*" and "*C. haemolyticum*" (Collins et al., 1982b) was not in accord with the inclusion of these taxa in the genus *Streptococcus*. The majority of streptococci completely lack respiratory quinones, although some unsaturated naphthoquinones have been detected in a few group D and group N streptococci (Collins and Jones, 1979a; Collins and Jones, 1979b). In addition, fatty acid data (Collins et al., 1982b) did not support the view of Barksdale et al. (1957) that "*C. haemolyticum*" should be reclassified in the genus *Streptococ-*

cus. "*Corynebacterium haemolyticum*" possesses major amounts of monounsaturated fatty acids of the oleic acid series (synthesized via an aerobic pathway; Collins et al., 1982b). In contrast, members of the genus *Streptococcus* possess monounsaturated fatty acids of the *cis*-vaccenic acid series (synthesized via an anaerobic pathway; Kroppenstedt and Kutzner, 1978). Furthermore, the menaquinone patterns of "*C. pyogenes*" and "*C. haemolyticum*" are also incompatible with the retention of these species in the genus *Corynebacterium*. True corynebacteria generally possess dihydrogenated menaquinones with eight and nine isoprene units (Yamada et al., 1976; Collins et al., 1977). Tetrahydrogenated menaquinones with ten and nine isoprene units have, however, been reported in the genera *Actinomyces* (Collins et al., 1977) and *Propionibacterium* (Schwartz, 1973; Sone, 1974), respectively. Therefore, Reddy et al. (1982) as well as Collins and Jones (1982a) proposed that "*C. pyogenes*" should be reclassified in the genus *Actinomyces*, as "*Actinomyces pyogenes*." The presence of a peptidoglycan based upon L-lysine in "*C. pyogenes*" supported this view since this amino acid is present in the peptidoglycan of *A. bovis* (Schleifer and Kandler, 1972). "*Corynebacterium pyogenes*" is physiologically similar to *A. bovis*, although it differs from *A. bovis* in being actively proteolytic (Cummins et al., 1974; Slack and Gerencser, 1975). However, the DNA base ratio of "*C. pyogenes*" (58 mol% G+C) is not incompatible with its inclusion in the genus of *A. bovis*, which has 63 mol% G+C (Johnson and Cummins, 1972).

The taxonomic position of "*C. haemolyticum*" remained equivocal. The results of lipid analyses did not support the view of Barksdale et al. (1957) that "*C. haemolyticum*" is a mutant of "*C. pyogenes*." Phenotypically, "*C. haemolyticum*" is very similar to *A. bovis* and also contains lysine in the cell wall peptidoglycan. The menaquinone composition of "*C. haemolyticum*" is distinct from that of *A. bovis* (Collins et al., 1977) and resembles that of the propionibacteria (Schwartz, 1973; Sone, 1974), but the results of cell wall and fatty acid analyses do not support this latter relationship. On the basis of cell wall and lipid composition, the taxon "*C. haemolyticum*" appeared quite distinct from all other coryneform and actinomycete taxa, thus warranting a new genus (Collins et al., 1982b). Therefore, on the basis of phenetic, peptidoglycan, fatty acid, menaquinone and DNA data (Schleifer and Kandler, 1972; Schofield and Schaal, 1981; Collins et al., 1982b), this taxon was reclassified by Collins et al. (1982c) in the new genus *Arcanobacterium* as *Arcanobacterium haemolyticum*.

Taxonomy

The early description of the genus was mainly based upon morphological characteristics together with some chemotaxonomic information such as peptidoglycan composition, long-chain fatty acid content and respiratory quinone composition (Collins et al., 1982c).

The genus *Arcanobacterium* currently comprises six validly described species, namely *Arcanobacterium bernardiae*, *Arcanobacterium haemolyticum*, *Arcanobacterium hippocoleae*, *Arcanobacterium phocae*, *Arcanobacterium pluranimalium* and *Arcanobacterium pyogenes* (Table 1). This current taxonomic structure has involved the inclusion of species that were transferred to *Arcanobacterium* from other genera. Thus, *Arcanobacterium pyogenes* and *Arcanobacterium bernardiae* were transferred to this genus by Pascual Ramos et al. (1997b) after they had meanwhile been included in the genus *Actinomyces* as "*Actinomyces pyogenes*" and "*Actinomyces bernardiae*," respectively.

CHEMOTAXONOMIC PROPERTIES The salient chemotaxonomic features of the type strains of *Arcanobacterium haemolyticum*, the type species of the genus, *Arcanobacterium pyogenes*, *Arcanobacterium phocae* and *Arcanobacterium bernardiae* include cell wall chemotype V (Lechevalier and Lechevalier, 1970) with lysine as wall diamino acid, rhamnose as diagnostic whole-cell sugar, a type PIII phospholipid pattern (Lechevalier et al., 1977) with phosphatidylcholine as key diagnostic phospholipid, and tetrahydrogenated menaquinones with eight (MK-8[H$_4$]) and nine (MK-9[H$_4$]) isoprene units, the predominant menaquinone being MK-9(H$_4$); the long-chain fatty acids are straight-chain saturated and monounsaturated (oleic acid series; Table 15).

Other chemotaxonomic properties include lack of mycolic acids, a muramic acid residue of the peptidoglycan that is N-acetylated, and DNA G+C contents between 50 and 58 mol%. This pattern of chemotaxonomic characteristics is unique among the genera of the family Actinomycetaceae (Tables 13 and 15).

MORPHOLOGICAL CHARACTERISTICS When Collins et al. (1982) introduced the genus *Arcanobacterium* for bacteria formerly named "*Corynebacterium haemolyticum*," the cellular morphology of the genus was described as follows: "On blood agar plates, slender, irregular, bacillary forms predominate during the first 18 h; many cells are arranged at an angle to give V-formations (Fig. 30). As growth proceeds, organisms become granular and segmented so that they resemble small and irregular cocci. Both rods and coccoid

Table 15. Chemotaxonomic characteristics of members of the genus *Arcanobacterium*.

Organisms	Wall sugars	Acyl type	Phospholipid types	Menaquinones	Fatty acids
Arcanobacterium haemolyticum	Rhamnose + glucose	Acetyl	PIIIDPG, PG, PC, PI, PIM	MK-8(H$_4$) + MK-9(H$_4$)[a]	S, U
Arcanobacterium bernardiae	Rhamnose + glucose + mannose	Acetyl	PIIIDPG, PG, PC, PI, PIM	MK-8(H$_4$) + MK-9(H$_4$)[a]	S, U
Arcanobacterium pyogenes	Rhamnose + glucose	Acetyl	PIIIDPG, PG, PC, PI, PIM	MK-8(H$_4$) + MK-9(H$_4$)[a]	S, U
Arcanobacterium phocae	Rhamnose	Acetyl	PIIIDPG, PG, PC, PI	MK-8(H$_4$) + MK-9(H$_4$)[a]	S, U, *iso*, *anteiso*

Abbreviations: please refer to the footnote in Table 7.
[a]Major component.

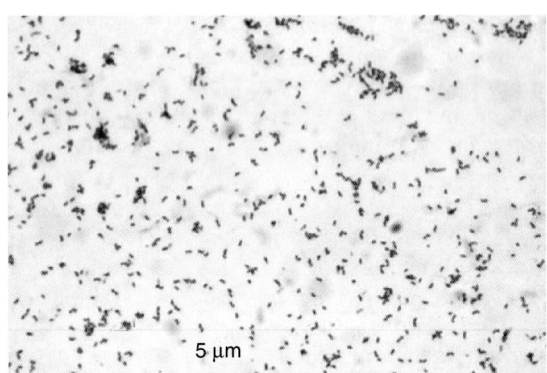

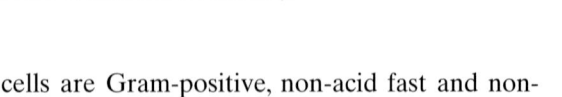

Fig. 30. *Arcanobacterium bernardiae*; Gram-stained smear from a mature colony (brain heart infusion agar, 14 days at 36 ± 1°C, and Fortner's method).

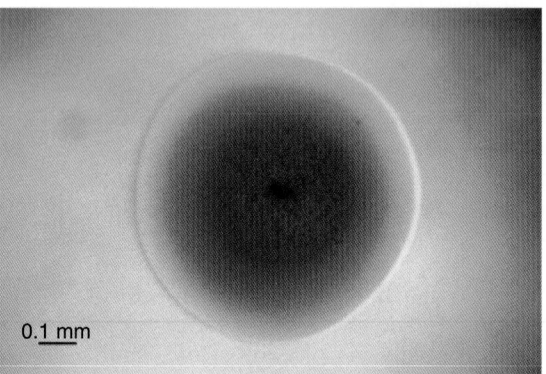

Fig. 31. *Arcanobacterium bernardiae*; mature, flat, translucent colony with a small, optically darker center; micrograph in transmitted light (brain heart infusion agar, 14 days at 36 ± 1°C, and Fortner's method).

cells are Gram-positive, non-acid fast and non-motile; endospores are not formed."

The morphology of *A. pyogenes* by and large conforms to this description (Reddy et al., 1982; Schaal, 1986b). The same is true for *A. phocae* and *A. bernardiae*, whereas *A. hippocoleae* and *A. pluranimalium* appear to be less coccoid.

Surface colonies of *A. haemolyticum* on blood agar were described to be small after 24 h of incubation at 36 ± 1°C, becoming larger upon extended incubation (Collins and Cummins, 1986). They are circular, discoid, slightly raised, opaque and nonpigmented with a butyrous consistency, and they show β-hemolysis (Fig. 31). Colonies of *A. pyogenes* are very similar, while *A. bernardiae* varies concerning β-hemolysis. *Arcanobacterium phocae* also produces a potent hemolysin, whereas *A. hippocoleae* is weakly β-hemolytic, and *A. pluranimalium* was reported to be α-hemolytic.

PHYSIOLOGICAL PROPERTIES Arcanobacteria are facultatively anaerobic, having a fermentative type of carbohydrate metabolism. Fermentation end products are acetic and lactic acids; the amount of succinic acid produced may vary from species to species and may even be difficult to detect. Catalase production is negative in *A. bernardiae* and *A. hippocoleae*; it is positive in *A. pluranimalium* and variable in *A. phocae*. Catalase activity in *A. haemolyticum* is usually negative, but some strains may show weak catalase production.

Acid production from various carbohydrates, enzyme activities of *Arcanobacterium* species, and several other physiologic properties are compiled in Table 14. Nearly all members of the genus ferment glucose and maltose, and none produces urease. All of the other physiological characters vary among the *Arcanobacterium* species and may therefore be used for differentiating or identifying them (Table 14).

Habitat and Ecology

Information on the natural habitat of arcanobacteria is scarce, although it has been assumed that these organisms occur as commensals of the mucous membranes of warm-blooded animals, in particular domestic animals such as cattle, sheep, swine and goats (Reddy et al., 1982; Billington et al., 2002a). Thus, Baele et al. (2001), in a study on the Gram-positive tonsillar and nasal flora of piglets before and after weaning, found that *A. pyogenes* was one of the common colonizers of the tonsils of pigs before and after weaning but not of the nasal conchae. Furthermore, this organism was also isolated from semen of beef breed bulls (Sprecher et al., 1999) without a significant relationship between the cultural results and possible sperm abnormalities. This is in accordance with the findings of Olson et al. (2002) that *A. pyogenes*, as certain *Actinomyces* species, is able to produce biofilms experimentally.

Another interesting finding is that *A. haemolyticum* could be identified as one of the major secondary colonizers of leprosy skin ulcers (Sturm et al., 1996) and may contribute to the inflammatory reaction of these lesions. The vast majority of *Arcanobacterium* isolates originated, however, from various pathological conditions that were thought to be etiologically related to these microbes and that will be discussed in the section on Pathogenicity.

Isolation and Cultivation

For primary isolation and maintenance, organically complex media such as BHI broth (or agar) or trypticase soy broth (or agar) have been recommended and are usually satisfactory. Blood agar, especially Columbia agar with addition of 5% horse, sheep or human blood, in most cases also gives satisfactory results. The blood agar may be layered upon a nutrient or digest agar base (Collins and Cummins, 1986). Good growth of *A. haemolyticum* is more difficult to obtain in liquid media. However, addition of serum may enhance growth considerably. Thus, a suitable liquid medium for this organism consists of brain heart infusion broth supplemented with 5% horse serum (Collins and Cummins, 1986).

Recognition of *A. haemolyticum* in cultures from clinical specimens is often hampered by its delayed β-hemolysis and the presence of additional pathogens (e.g., streptococci) or microbes from the indigenous flora of the affected mucous membranes. Thus, media for selective isolation or improved recognition of this organism were developed: Coman et al. (1996) used a medium containing 5% sheep blood agar and 3.5% NaCl for selective isolation of *A. haemolyticum* from pharyngeal swabs of children. Jurankova and Votava (2001) applied sheep blood agar with a streak of *Staphylococcus aureus* to recognize *A. haemolyticum* on the basis of the reverse CAMP (Christie, Atkins and Munch-Petersen) phenomenon. The medium of Votava et al. (2000) con-

taining Columbia blood agar base and 5% washed sheep erythrocytes sensitized with equi factor (EF) of *Rhodococcus equi* is based upon the same principle.

All of the other *Arcanobacterium* species apparently grow well on Columbia blood agar supplemented with 5% horse or sheep blood and incubated under increased CO_2 tension (~5%) at $36 \pm 1°C$.

Identification

In view of their unique chemotaxonomic profile, arcanobacteria are most reliably identified to genus level by chemotaxonomic methods. The methodological details of the respective techniques were described in the section on *Actinomyces*.

Using physiological markers, identification to species level may be achieved by a set of carbohydrate fermentation and enzyme tests that can be selected from Table 14. This set may include acid production from α-methyl-D-glucoside, glycogen, lactose, melezitose, pullulan, sucrose, trehalose, xylitol and xylose, hydrolysis of gelatin and hippurate, the activities of acid phosphatase, α- and β-galactosidases, α-glucosidase, and β-glucuronidase as well as catalase production, acetoin production, and nitrate reduction (Table 14). Characteristically, essentially all of the *Arcanobacterium* species ferment glucose and maltose and are negative for acid production from melibiose and urease activity. Whether *A. haemolyticum* possesses α-mannosidase activity has been a matter of debate: Carlson and Kontiainen (1994a) reported that α-mannosidase activity was a very good character for identifying this species and differentiating it from *A. pyogenes*. However, Von Graevenitz (1994) commented that this enzyme activity could be influenced by various accidental factors, thus making the value of this differential feature questionable. In Table 14, mainly the results of the API system are given as far as enzyme activities are concerned, and with this system, essentially all of the arcanobacteria are α-mannosidase-negative. Additional characters useful for differentiating *A. pyogenes* and *A. haemolyticum* in clinical microbiology laboratories are xylitol and α-methyl-D-glucoside fermentation, acetoin production (Voges-Proskauer test), and α-galactosidase activity (Carlson et al., 1995a).

As with the *Actinomyces* species, various commercial identification systems were used for the primary description of the new *Arcanobacterium* species and for identifying members of the genus (Guerin-Faublee et al., 1992). However, Brander and Jousimies-Somer (1992) found that the RapID ANA II system misidentified all *A.*

haemolyticum strains as *A. pyogenes*, although all of the *Actinomyces odontolyticus* and 65% of the *Actinomyces israelii* isolates were correctly identified. On the other hand, the API ZYM profiles were useful for differentiating *A. haemolyticum* and *A. pyogenes*. In contrast, the performance of the Rapid CORYNE system appeared to be considerably better (Gavin et al., 1992), although Ding and Lämmler (1992) and Guerin-Faublee et al. (1992) reported that the "classical" API CORYNE system failed to identify 6 out of 42 *A. pyogenes* and one of five *A. haemolyticum* strains, or only 58 of 103 *A. pyogenes* cultures, respectively. As in *Actinomyces*, 4-methylumbelliferyl-linked substrates may provide additional characters for identifying *A. haemolyticum* (Kämpfer, 1992).

Tentative serological identification of *Actinomyces pyogenes* may be achieved with antisera against group G streptococci using either immunodiffusion or latex agglutination test procedures (Lämmler and Blobel, 1986). Furthermore, Lämmler (1990) reported typing of this organism by its production of and susceptibility to bacteriocin-like inhibitors. A set of morphological and physiological characteristics such as colony morphology, β-hemolysis on horse blood agar, β-glucuronidase activity, and ability to produce acid from sucrose and trehalose was used by Carlson et al. (1994d) to define two biovars of *A. haemolyticum*. A bacteriophage typing system for these organisms has not been developed.

Serological Diagnosis of *Arcanobacterium* Infections

Serology may have some practical importance in diagnosing *Arcanobacterium* infections (unlike *Actinomyces* infections). Votava et al. (2001) used a neutralization test carried out on nonnutrient blood agar prepared from sheep erythrocytes sensitized with equi factor of *Rhodococcus equi* for the detection of antibodies to phospholipase D (PLD) of *A. haemolyticum*. When testing 433 sera from 404 patients with signs of acute tonsillitis, these authors found antibodies to PLD in 28 patients (6.9%). Among 116 sera from individuals without signs of pharyngitis, antibodies to PLD were detected in only one case (0.9%).

Using a similar assay, Skalka et al. (1998) investigated sera of humans with spontaneous *A. haemolyticum* infections and mice experimentally infected with this organism and found that the titers of neutralizing antibodies to the homologous PLD were always higher than those to the heterologous PLD produced by *Corynebacterium pseudotuberculosis*.

Using SDS-PAGE and Western blot analyses, specific antibodies to *A. haemolyticum* could be demonstrated in sera from patients with acute infections or from convalescents (Nyman et al., 1997). These antibodies reacted primarily with four cell wall-associated proteins with estimated molecular weights of 80, 60, 50, and 30 kDa, respectively.

Epidemiology

In contrast to essentially all other members of the family Actinomycetaceae, *Arcanobacterium* species may cause inflammatory diseases that are not exclusively endogenous in origin but may well be transmitted from animal to animal, from animals to humans, or even between humans. This is especially true for *A. pyogenes*, which is able to induce mastitis in cows and may, therefore, appear in the milk. In cows with undiagnosed mastitis, the organism may then easily be transmitted to other animals or humans (Mazura, 1989).

Bovine mastitis occurs throughout the year, but infections due to *A. pyogenes* were encountered most often in late autumn, winter and spring (Jones and Ward, 1989; Waage et al., 1999). At parturition, the causative agents of mastitis such as *Escherichia coli*, *A. pyogenes*, streptococci and staphylococci were isolated in about equal frequencies, but in the late lactation period *E. coli* predominated (Jones and Ward, 1989). Mastitis in cattle has often been associated with insufficient milking hygiene or drug instillations into the teats, but *A. pyogenes*, in contrast to certain other mastitis agents, may also cause epizootic outbreaks of "summer mastitis" in heifers and even in young calves that are too young for breeding (Lean et al., 1987). In these latter cases, a large number of flies, intersucking of calves, or preceding viral disease was assumed to influence the outbreak. Furthermore, various forms of teat lesions were found to increase the risk of mastitis due to *A. pyogenes* and other organisms (Mulei, 1999).

Arcanobacterium pyogenes also is a typical and common agent of peritonitis and pleuritis in swine and of pneumonia in calves (Glage, 1903; Roberts, 1968; Reddy et al., 1982; Seno and Azuma, 1983; Vogel et al., 2001). Reliable epizootic conclusions cannot be drawn on the basis of occasional reports on *A. pyogenes* infections of other body sites or other domestic or feral animals.

The same is true for human infections due to this organism or *A. haemolyticum* and for infections caused by the other *Arcanobacterium* species. Nevertheless, Kotrajaras and Tagami (1987), when investigating the role of *A. pyogenes* in the etiology of endemic leg ulcers in Thailand, found that skin injuries followed by *A. pyogenes* infection were important for the development of this human disease. In addition, Oriental-eye flies were assumed to carry the pathogen to the traumatized skin. As far as human bacterial pharyngitis is concerned, Linder (1997) reported that 0.5–2.5% of this infection was due to *A. haemolyticum.* In a more detailed study of a pediatric and adolescent population, Mackenzie et al. (1995) found that the incidence of infection caused by this organism was highest in the 15–18 year-old age group (2.5%). No single isolate of *A. haemolyticum* was cultured from healthy control persons (2241 healthy students). Comparable results were obtained in a study on army conscripts (Carlson et al., 1995b), with an overall incidence of 1.4% in individuals with sore throat, while 23% of this study group yielded growth of β-hemolytic streptococci.

Pathogenicity

ANIMAL INFECTIONS *Arcanobacterium pyogenes* is primarily an animal pathogen. The pyogenic pathological conditions caused by this organism in domestic animals most typically include mastitis in cows, often associated with *Peptoniphilus* (*Peptostreptococcus indolicus*), and peritonitis and pleuritis in swine (Glage, 1903; Roberts, 1968; Reddy et al., 1982; Seno and Azuma, 1983). *Arcanobacterium pyogenes* was also found to be etiologically involved in pneumonia in calves (Vogel et al., 2001), in hematogenous osteomyelitis in cattle (Firth et al., 1987), in purulent osteomyelitis in fattening pigs (Burgi et al., 2001), in a pyothorax in a one-month-old female kitten (Gulbahar and Gurtuk, 2002), in chronic otitis externa in a cat and urinary tract infection in a dog (Billington et al., 2002a), in hypertrophic osteopathy in a dog, in the retention of fetal membranes in ewes (Tzora et al., 2002), in facial and mandibular abscesses in blue duiker, often associated with *Fusobacterium necrophorum* (Roeder et al., 1989), in intracranial abscesses in white-tailed deer (Baumann et al., 2001) and a steer (Strain et al., 1987), and in lymphadenitis in camels (Moustafa, 1994). The development of renal medullary amyloidosis in Dorcas gazelles also appeared to be related to chronic or recurring *A. pyogenes* infections (Rideout et al., 1989).

Experimental mastitis induced by intramammary challenge with *A. pyogenes* was found to be more severe in dry than in lactating glands and was very difficult to eliminate in the dry period even by antibiotic treatment, whereas lactating glands recovered more easily either spontaneously or after administration of antibiotics. Combined infections of *A. pyogenes* with *Peptoniphilus indolicus* took an especially severe clinical course and showed a higher frequency of systemic involvement (Hillerton and Bramley, 1989).

Arcanobacterium haemolyticum has been known as an occasional cause of infections in farm animals (Collins and Cummins, 1986). Recently, the organism was identified as one of the pathogens responsible for mandibular and maxillary abscesses in pet rabbits (Tyrrell et al., 2002).

Of the remaining *Arcanobacterium* species, *A. bernardiae* has as yet only been isolated from human sources. *Arcanobacterium phocae* has been recovered in mixed culture from various tissues and fluids of common seals (*Phoca vitulina*) and gray seals (*Halichoerus grypus*), but its pathological significance remains unclear. *Arcanobacterium pluranimalium* has been isolated from the spleen of a dead harbor porpoise and a lung abscess of a dead sallow deer (Lawson et al., 2001a), and *A. hippocoleae* has been, cultured from a female Arab horse with vaginal discharge (Hoyles et al., 2002b).

HUMAN INFECTIONS Human infections due to *A. pyogenes* usually present as acute pharyngitis, urethritis or as cutaneous or subcutaneous suppurative processes (Barksdale et al., 1957; Collins and Jones, 1982a; Reddy et al., 1982; Gahrn-Hansen and Frederiksen, 1992), but bacteremia has also been reported (Barnham, 1988). Additional human *A. pyogenes* infections include septic arthritis in a diabetic farmer (Lynch et al., 1998) and endemic leg ulcers in Thailand (Kotrajaras and Tagami, 1987).

The clinical symptoms of *A. haemolyticum* infections are similar, this organism being predominantly isolated from pharyngeal infections, skin lesions and septicemia (Collins and Cummins, 1986; Dethy et al., 1986; Moreno-Montesinos et al., 1989; Carlson et al., 1994e; Gaston and Zurowski, 1996; White and Foshee, 2000; Jurankova and Votava, 2001) with and without metastatic abscess formation (Dieleman et al., 1989). Primary deep-seated abscesses (Barnham and Bradwell, 1992), cellulitis including a fulminant tubo-ovarian soft tissue infection (in coculture with fusobacteria; Batisse-Milton et al., 1995) and an orbital cellulitis (Ford et al., 1995), surgical wound infections, an infected foot wound, and endocarditis due to this organism have also been reported (Ritter et al., 1993; Esteban et al., 1994; Alos et al., 1995; Skov et al., 1998; Dobinsky et al., 1999).

Arcanobacterium bernardiae has been isolated from various clinical sources including blood, abscesses, and eye infections (Funke et al., 1995b). More recently, urinary tract infections and septic arthritis due to this organism were also described (Adderson et al., 1998; Lepargneur et al., 1998). The remaining *Arcanobacterium* species appear to occur only in animals.

TREATMENT OF *ARCANOBACTERIUM* INFECTIONS AND ANTIMICROBIAL SUSCEPTIBILITY Studies on the antibiotic susceptibility of *A. pyogenes* were mainly performed on strains of animal origin. Yoshimura et al. (2000), investigating the resistance patterns of 49 bovine and porcine isolates from Japan, reported that all of the isolates were highly susceptible to benzyl penicillin and ampicillin and also to other penicillins and cephems. In contrast, varying numbers of strains were resistant to dihydrostreptomycin, gentamicin and oxytetracycline. Resistance to dihydrostreptomycin appeared to be more pronounced among porcine (85.7%) than among bovine (52.4%) isolates; resistance to gentamicin occurred in only 7.1% of the bovine strains; oxytetracycline resistance was also more frequently encountered among porcine (85.7%) than among bovine (57.1%) isolates.

All bovine isolates were susceptible to erythromycin and lincomycin, but two porcine strains were resistant to both antibiotics. Chloramphenicol as well as florfenicol and thiamphenicol were all active against most of the animal *A. pyogenes* isolates. In analogy to other Actinomycetaceae, the fluoroquinolones enrofloxacin and ofloxacin were not as active against this organism as penicillins and macrolides were.

In a study on the resistance of *A. pyogenes* isolates from the United States to macrolides, lincosamides and tetracyclines, Trinh et al. (2002), using a broth microdilution test, found that 25% were resistant to the former two groups and 41.7% to the latter class of antibiotics. Tetracycline resistance could be related to the presence of a gene which was highly similar at the DNA level (92% identity) to the *tet(W)* gene from the rumen bacterium *Butyrivibrio fibrisolvens* encoding a ribosomal protection tetracycline resistance protein (Billington et al., 2002b).

Arcanobacterium haemolyticum was reported to be resistant to oxytetracycline but susceptible to trimethoprim-sulfamethoxazole and amikacin (Schofield and Schaal, 1981; Collins and Cummins, 1986). Susceptibility to β-lactam compounds was reported by Carlson et al. (1994c) and also by Arikan et al. (1997) from Turkey, together with susceptibility to vancomycin, erythromycin, azithromycin, clindamycin, doxycycline and ciprofloxacin. However, according to these authors *A. haemolyticum* was resistant to trimethoprim-sulfamethoxazole.

The antibiotic susceptibility pattern of *A. bernardiae* as reported in the original description of seven strains of this species (Funke et al., 1995b) was as follows: All strains were susceptible to clindamycin, erythromycin, penicillin G, rifampin, tetracycline and vancomycin. Most of the strains were susceptible to gentamicin but resistant to ciprofloxacin.

Detailed studies on the antibiotic susceptibility of the remaining *Arcanobacterium* species have not been reported.

VIRULENCE FACTORS OF *ARCANOBACTERIUM* SPECIES *Arcanobacterium pyogenes* has been shown to produce soluble toxic and hemolytic activities which can be neutralized by antitoxin and which are fatal to mice and rabbits after intravenous injection (Lovell, 1944; Reddy et al., 1982). This hemolytic exotoxin, called "pyolysin" (PLO), belongs to the thiol-activated, cholesterol-binding pore-forming family of toxins (Billington et al., 2001) and has a molecular mass of 56 kDa (Ikegami et al., 2000). Inactivation of the *plo* gene results in a significant reduction in virulence of *A. pyogenes* (Jost et al., 1999). Furthermore, a recombinant PLO-based subunit vaccine protected mice from experimental *A. pyogenes* infection. Analysis of the functional domains of this toxin revealed that the N-terminal region of PLO was important for the hemolytic activity (Imaizumi et al., 2001).

Additional virulence factors of *A. pyogenes* are two different neuraminidases that obviously play a role in adhesion of this organism to host epithelial cells (Jost et al., 2001; Jost et al., 2002), as do neuraminidases of other bacterial pathogens. Acyl-neuraminate pyruvate lyase is also produced (Müller, 1973). In cattle with necrotizing bronchopneumonia, *A. pyogenes* induces the expression of nitric oxide synthase (iNOS) (Fligger et al., 1999), the expressing cells being largely restricted to the cellular zone surrounding necrotic areas. The ability of *A. pyogenes* to bind to human α_2-macroglobulin and haptoglobin may further contribute to its pathogenicity (Lämmler et al., 1985).

Arcanobacterium haemolyticum produces a phospholipase D (PLD) as one of its soluble toxins (Cuevas and Songer, 1993). This toxin is a protein of approximately 31.5 kDa with a pI of about 9.4 and is related to, but not identical with, the phospholipases D of *Corynebacterium pseudotuberculosis* and *C. ulcerans*. Sequence comparison of PLD coding regions revealed 65% DNA-DNA similarity of the *pld* genes of these three species. Comparison of the amino acid sequences of *C. pseudotuberculosis* bv. *equi*, *C. pseudotuberculosis* bv. *ovis*, *C. ulcerans* and *A. haemolyticum* PLDs showed that these four enzymes share 64–97% similarity (McNamara et al., 1995).

Like *A. pyogenes*, *A. haemolyticum* produces neuraminidase and acylneuraminate pyruvate lyase (Müller, 1973), and both species also possess DNAse activity, which may contribute to their pathogenicity as well. In addition, the findings of Osterlund (1995) suggest that *A.*

haemolyticum may be able to invade host cells and to survive intracellularly.

Virulence factors of the other *Arcanobacterium* species have not been studied.

Genetics

Electroporation was used to transform *A. pyogenes* (Jost et al., 1997). The transformants were able to express resistance to chloramphenicol, erythromycin, kanamycin and streptomycin encoded by the introduced plasmid DNA.

The Genus *Varibaculum*

The genus *Varibaculum* with the single species *Varibaculum cambriense* was proposed by Hall et al. (2003e) for anaerobic, catalase-negative, Gram-positive diphtheroidal bacteria isolated from human sources on the basis of phenotypic and molecular methods. For the sake of grammatical correctness, the species had to be renamed *Varibaculum cambriense* (see Phylogeny).

Phylogeny and Taxonomy

16S rRNA gene sequence analysis of five strains of *V. cambriense* revealed that the almost complete sequences of these strains were nearly identical (99.1–100% sequence similarity). The phylogenetic placement confirmed their association with members of the family Actinomycetaceae, supporting phenotypic, in particular morphological, resemblance. The new genus contributes further to the phylogenetic diversity of the family (Hall et al., 2003e), as *Varibaculum* forms a relatively long and distinct line of descent, about equidistantly related to *Actinomyces neuii* and *Mobiluncus* species (Hoyles et al., 2003).

MORPHOLOGICAL CHARACTERISTICS Cells of *Varibaculum cambriense* are short, straight or curved, diphtheroid rods which stain Gram-positive and are non-acid-fast and nonmotile (Fig. 32). Colonies after 48 h of anaerobic incubation on horse blood-containing complex media are pinpoint, convex, glistening, translucent white or gray with entire edges (Fig. 33). No hemolysis is observed.

CHEMOTAXONOMIC PROPERTIES Whole-cell protein profiling using PAGE demonstrated a considerable phenotypic coherence of members of the species *V. cambriense* and their clear distinctness from *Mobiluncus* species (Hall et al., 2003e). The G+C content of the DNA is 51.7 mol%.

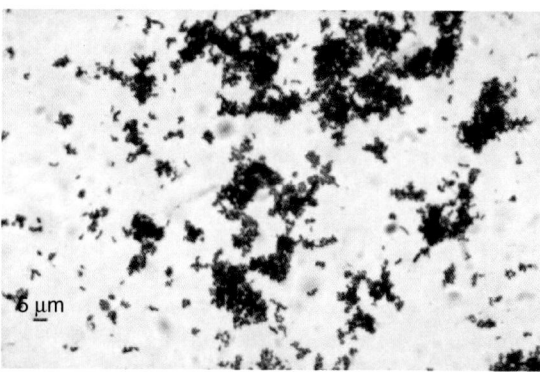

Fig. 32. *Varibaculum cambriense*; Gram-stained smear from a mature colony (brain heart infusion agar, 14 days at 36 ± 1°C, and Fortner's method).

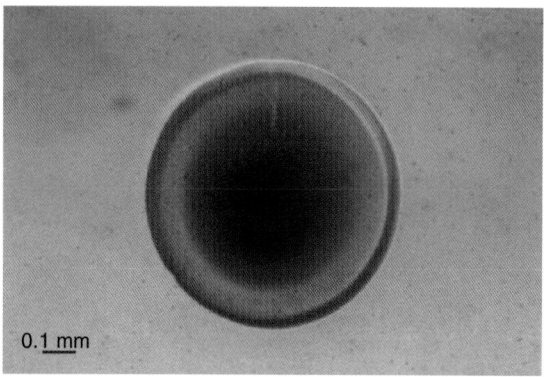

Fig. 33. *Varibaculum cambriense*; mature, slightly raised, translucent colony without internal structures; micrograph in transmitted light (brain heart infusion agar, 14 days at 36 ± 1°C, and Fortner's method).

PHYSIOLOGICAL PROPERTIES *Varibaculum cambriense* is anaerobic and catalase-negative. Major end products of its fermentative glucose metabolism are lactic and succinic acids.

Other physiological properties are summarized in Table 14. Markers characteristic of the species are acid production from glucose and only a few additional carbohydrates, hydrolysis of hippurate (but not of esculin, gelatin and starch), and α-glucosidase (but no urease) activity. Nitrate is reduced by the majority of strains (Table 14).

Isolation and Cultivation

Varibaculum cambriense appears to grow well on Columbia agar supplemented with 5% horse serum at 36 ± 1°C under anaerobic conditions. Alternatively, Fastidious Anaerobe Agar (LabM, Bury, UK) may be used. Anaerobic incubation requires 48–72 h until colonies are easily visible. Strains may grow well or poorly in air plus 5%

CO_2, but they grow poorly or not at all in air, thus behaving like typical members of the Actinomycetaceae.

Identification

Apart from 16S rRNA gene sequencing, identification of *V. cambriense* may be achieved by a set of morphological and physiological characters. Cellular and colonial morphology, growth conditions together with lack of catalase production, positive nitrate reduction, and lactic and succinic acids as major end products of glucose fermentation may form the basis of recognizing *V. cambriense*. Additional differential characters may be selected from Table 14.

Pathogenicity

Given the fact that only 15 strains of *V. cambriense* have been encountered so far, it is not surprising that essentially nothing is known about the habitat, ecology and epidemiology of these organisms.

The strains were isolated from human abscesses (1 brain abscess following ear and mastoid problems together with a complex anaerobic flora; 1 postauricular abscess; 1 large ischiorectal abscess; 1 submandibular abscess; 3 breast abscesses; 1 cheek abscess; and 2 from a non-characterized abscess), intrauterine contraceptive devices or vaginal swabs (3 strains), and hidradenitis and a fistula (1 strain each). Several isolates were cultured together with other potentially pathogenic bacteria, and others were obtained without clear indications of an inflammatory process. Therefore, the pathogenicity of *V. cambriense* to humans remains to be elucidated.

The Genus *Mobiluncus*

Unlike any other member of the family *Actinomycetaceae*, *Mobiluncus* species are curved, non-branching, motile, anaerobic bacteria that stain Gram-variable or Gram-negative. Such *Vibrio*-like organisms were already seen in vaginal secretions by Krönig as early as 1895 and were first isolated in pure culture from vaginal and uterine specimens of a woman suffering from postpartum endometritis by Curtis in 1913. Prévot (1940), in his manual on classification and identification of anaerobic bacteria, used the designation "*Vibrio mulieris*" for these bacteria of Curtis. However, no further description of curved rod-shaped bacteria from the female genital tract was published until Moore's characterization of "anaerobic vaginal vibrios" (Moore, 1954). Similar or identical "vibrions succinoproducteurs" were separated into two groups by

Durieux and Dublanchet (1980) on the basis of cellular morphology, fermentation of glucose, and reduction of nitrate.

In 1984, Spiegel and Roberts as well as Hammann et al., referring to "*Vibrio mulieris*" of Prévot (1940), proposed the genus designations "*Falcivibrio*" and "*Mobiluncus*," respectively, for two different species of anaerobic curved rods from the human vagina. Although the genus designation *Falcivibrio* appeared first in the literature, it was not validated first, so that *Mobiluncus* has priority. Recently, Hoyles et al. (2003), in a polyphasic taxonomic study, formally proposed to transfer the members of the genus *Falcivibrio* to the genus *Mobiluncus*. These organisms are etiologically involved in a syndrome called "bacterial vaginosis," "nonspecific vaginitis," "*Haemophilus vaginalis* vaginitis" (Gardner and Dukes, 1955), anaerobic vaginosis (Blackwell et al., 1983), or "amine kolpitis."

Taxonomy

Although members of the genus *Mobiluncus* appeared to be clearly distinct from the *Actinomycetaceae*, the *Bacteroidaceae*, and the *Propionibacteriaceae phenotypically*, Spiegel and Roberts (1984a) tentatively placed them in the family Bacteroidaceae. They noted, however, that *Mobiluncus* bacteria, despite the variable to negative reaction in the Gram stain, revealed a multilayered Gram-positive cell wall under the electron microscope and did not possess an outer membrane characteristic of Gram-negative bacteria. A more detailed chemical analysis of *Mobiluncus* cell walls (Carlone et al., 1986) confirmed that their composition was not Gram-negative. However, in view of the morphological characteristics, it was still surprising that the results of partial reverse transcriptase sequencing of 16S rRNA suggested that the genus *Mobiluncus* was not a member of the family Bacteroidaceae, but apparently belonged to the order Actinomycetales (Lassnig et al., 1989). As outlined in the Phylogeny section, this affiliation has definitely been confirmed since then.

Thus far, the genus *Mobiluncus* only comprises the two species described by Spiegel and Roberts (1984a) under the designations *Mobiluncus curtisii* and *M. mulieris*. These correspond to the organisms described by Hammann et al. (1984) as "*Falcivibrio vaginalis*" (*M. curtisii.*) and "*Falcivibrio grandis*" (*M. mulieris*; Hoyles et al., 2003).

The separation of two subspecies (*M. curtisii* subsp. *curtisii* and *M. curtisii* subsp. *holmesii*) within *M. curtisii* was chiefly based upon a few phenotypic differences but was not supported by DNA-DNA similarity studies (Spiegel, 1992) or comparative partial 16S rRNA gene analysis

(Tiveljung et al., 1996). However, the analysis of surface antigens of *Mobiluncus* using monoclonal antibodies in an enzyme-linked immunosorbent assay and indirect immunofluorescence revealed four monoclonal antibodies (MAbs), and thus surface components, that appeared to be subspecies-specific for *M. curtisii* subsp. *curtisii* (Fohn et al., 1988). One MAb reacted with an epitope present on *M. curtisii* subsp. *holmesii* and *M. mulieris* but not on *M. curtisii* subsp. *curtisii*. A more detailed analysis of the 16S rRNA genes of *Mobiluncus* species showed that so-called "atypical *Mobiluncus* strains" could be separated from both *M. curtisii* and *M. mulieris* (Tiveljung et al., 1995), and Hoyles et al. (2003) stated that *M. curtisii* should be considered a single but complex species which cannot be differentiated into distinct subspecies on the basis of physiological markers alone.

MORPHOLOGICAL CHARACTERISTICS *Mobiluncus* cells are Gram-variable to Gram-negative, curved rods with tapered ends, do not form endospores, and occur singly or in pairs producing a gull wing appearance. They are motile by means of multiple, either subpolar or more centrally located, flagella (Spiegel and Roberts, 1984a; Spiegel, 1992). *Mobiluncus curtisii* has one to six flagella per cell with a common origin, while *M. mulieris* possesses one to eight flagella with multiple origins (Hammann et al., 1984; Spiegel, 1992). *Mobiluncus curtisii* subsp. *curtisii* was reported to migrate through soft (0.25%) agar, whereas *M. curtisii* subsp. *holmesii* was not; however, migration through soft agar is also variable in *M. mulieris* (Spiegel and Roberts, 1984a), while this feature was recently shown not to exactly correspond to *M. curtisii* subsp. *curtisii* (Hoyles et al., 2003).

Electron micrographs revealed that *Mobiluncus* species possess a multilayered Gram-positive cell wall lacking an outer membrane. The thinness of these walls was considered to be the reason for the tendency of mobilunci to stain Gram-negative (Spiegel and Roberts, 1984a). Pili (fimbriae) were not observed (Hammann et al., 1984).

Cells of the two *Mobiluncus* species differ considerably in size: While *M. curtisii* cells are <0.5 μm wide and on average 1.7 μm long rods (Fig. 34), *M. mulieris* cells are significantly longer, measuring <0.5 μm in width but 2.9 μm in mean length (Fig. 35).

Colonies of both species are white to gray or slightly yellowish, translucent, smooth, convex, and entire, and reach a maximum diameter of 1–4 mm after 5 days of incubation (Fig. 36).

CHEMOTAXONOMIC PROPERTIES The G+C content of the DNA of the *Mobiluncus* type strains is 49–

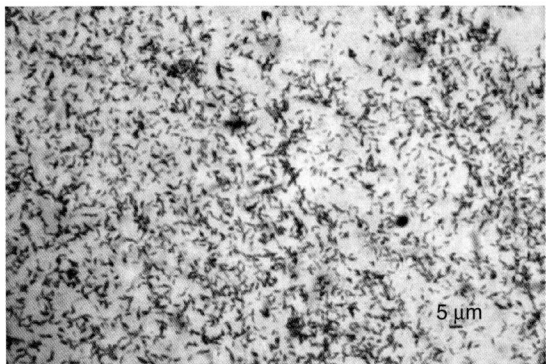

Fig. 34. *Mobiluncus curtisii* subsp. *holmesii*; Gram-stained smear from a mature colony (Columbia sheep blood agar, 14 days at 36 ± 1°C, and GasPak jar).

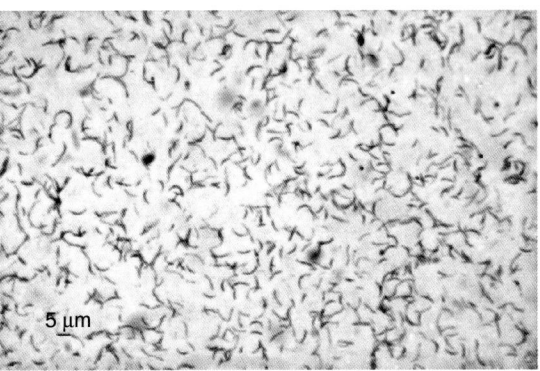

Fig. 35. *Mobiluncus mulieris*; Gram-stained smear from a mature colony (Columbia sheep blood agar, 14 days at 36 ± 1°C, and GasPak jar).

Fig. 36. *Mobiluncus mulieris*; mature, slightly raised, translucent colony with finely granular internal structures; micrograph in transmitted light (Columbia sheep blood agar, 14 days 36 ± 1°C, and GasPak jar).

52 mol% (Spiegel and Roberts, 1984a), although values as high as 53.5 ± 0.6 for *M. mulieris* and 55.2 ± 0.6 for *M. curtisii* have been reported (Hammann et al., 1984).

Lipopolysaccharide (LPS) could not be demonstrated, although four *Mobiluncus* strains were observed in the *Limulus* amoebocyte lysate (LAL) test to have activities intermediate between those of "*Corynebacterium genitalium*" (no detectable LAL reactivity) and those of *Pseudomonas aeruginosa* (Carlone et al., 1986).

Detailed analysis of the carbohydrate content of hydrolyzed whole cells of *Mobiluncus* strains revealed the absence of 2-keto-3-deoxyoctonate (KDO) and heptose but the presence of rhamnose, galactose, glucosamine, and muramic acid in comparable amounts.

Mobiluncus mulieris extracts contained approximately three times as much ribose as the *M. curtisii* strains, while the latter had two times as much glucose as the *M. mulieris* isolates. Mannosamine was detected in both *M. curtisii* subsp. *curtisii* and *M. curtisii* subsp. *holmesii* but was not found in the *M. mulieris* strains. Galactosamine was present in both species, but in the *M. mulieris* strains, only in amounts of less than 0.5%.

The fatty acid profile of *Mobiluncus* strains consisted of tetradecanoic (14:0), hexadecenoic (16:1), hexadecanoic (16:0), heptadecanoic (17:0), octadecadienoic (18:2), octadecenoic (18:1), and octadecanoic (18:0) acids (16:0, 18:2, and 18:1 being the major components in all strains tested). Fatty acid aldehydes were not detected by Carlone et al. (1986) but were reported by Skarin et al. (1982b); hydroxylated fatty acids were also not found.

Electrophoretic patterns of bacterial cellular proteins were often found to possess differential power when traditional phenotypic markers did not provide clear-cut species-specific profiles. However, determination of PAGE protein patterns requires highly standardized conditions to provide reproducible results. Thus, several authors obtained conflicting results (Vetere et al., 1987; Spiegel, 1992) concerning number and specificity of protein bands in *Mobiluncus* species. More recently, however, Drouet et al. (1991), studying 35 *Mobiluncus* strains (including three reference strains), reported "species-related" protein patterns as well as intraspecies variation possibly useful for epidemiological purposes. Combination of SDS-PAGE and immunoblotting (Schwebke et al., 1990) demonstrated a strikingly high degree of protein heterogeneity within the genus *Mobiluncus*, suggesting that further taxonomic divisions might be appropriate. Similar conclusions were drawn from the results of a combined morphological, biochemical and serological study performed by Garlind et al. (1989).

PHYSIOLOGICAL PROPERTIES At least for primary isolation, mobilunci require anaerobic growth conditions (Spiegel, 1992). However, after multiple transfers, growth has also been reported in 10% CO_2 or 5% oxygen (Hammann et al., 1984). The optimum temperature for multiplication is 35–37°C (Holst et al., 1982; Hammann et al., 1984; Spiegel and Roberts, 1984a), and there is poor or no growth at 20, 43, or 45°C (Durieux and Dublanchet, 1980; Peloux and Thomas, 1981; Holst et al., 1982). Heating at 55°C for 15 min kills *Mobiluncus* cells (Curtis, 1913). Reports on growth at 22, 27, 30, or 42°C are contradictory (Spiegel, 1992).

When growing in peptone yeast extract glucose medium (PYG), succinic, acetic, and lactic acids are the major fermentation end products (Holst et al., 1982; Hammann et al., 1984; Spiegel and Roberts, 1984a; Vetere et al., 1987). Addition of rabbit serum and glycogen to peptone yeast extract broth (PY) enhances the production of acetic and succinic acids as compared to the same medium supplemented with rabbit serum and glucose or rabbit serum alone (Spiegel and Roberts, 1984a). When fumarate is added to the medium, malic acid is produced (Hammann et al., 1984; Spiegel and Roberts, 1984a).

Most of the metabolic activities of *Mobiluncus* species have been included in Table 14. Note, however, that the determination of such "biochemical" characters is highly method-depending, so that there is still some controversy about some of the physiological properties (Spiegel, 1992; Hoyles et al., 2003). This may be especially true for the detection of weak or very weak acid production from carbohydrates, which may be influenced by size and viability of the inoculum, methods used for the fermentation tests, or techniques used for obtaining anaerobic growth conditions.

Physiological properties that are characteristic of the genus *Mobiluncus* and that may aid in the differentiation of the species of this genus are summarized in Table 16.

Habitat and Ecology

Numerous studies using a variety of methods such as microscopy of wet mounts or immunofluorescence-labeled smears, culture, or DNA probes were performed to assess the occurrence of *Mobiluncus* species or simply motile, curved rods in human vaginal secretions. Such curved rods were detected in up to 7.2% of unselected women undergoing gynecological examinations (Popp, 1977; Dropsy, 1978). However, these organisms were mainly present in women complaining of foul-smelling discharge who also had increased numbers of epithelial cells in their vaginal fluids without an elevated level of leukocytes. These are key symptoms of a condition called "bacterial vaginosis" (Gardner and Dukes, 1955; Amsel

Table 16. Characteristics useful for identifying and differentiating *Mobiluncus* species.

Characteristic	*Mobiluncus curtisii* subsp. curtisii	*Mobiluncus curtisii* subsp. holmesii	*Mobiluncus mulieris*
Stimulated by serum	+	+	+
Catalase production	−	−	−
Esculin hydrolysis	−	−	−
Leucine arylamidase	+	+	+
Esterase-lipase	+	+	+
Lipase	−	−	−
Fermentation end products	Su, Ac, La	Su, Ac, La	Su, Ac, La
Acid production from			
Glycogen	−	−	+
Melibiose	+/−	−/+	−
Trehalose	−	−	d
Arginine dihydrolase	+	+	−
Hippurate hydrolysis	+	+	−
α-Galactosidase	+	+	−
β-Galactosidase	+	+	−
Nitrate reduction	−	+	−/+
Stimulated by arginine	+	+	−
Migration through soft agar	+	−	d
Length of cells (μm)	1.7	1.7	2.9

Symbols and abbreviations: +, positive/present; −, negative/absent; d, positive or negative; +/−, usually positive; −/+, usually negative; Su, succinic acid; Ac, acetic acid; and La, lactic acid.
From Spiegel and Roberts (1984).

et al., 1983; see the Pathogenicity section in this Chapter), in which detection rates of mobilunci between 10.9% and 77% were reported (Durieux and Dublanchet, 1980; Durieux and Dublanchet, 1981; Spiegel et al., 1983b; Holst et al., 1984a; Holst et al., 1984b; Holst et al., 1984c; Roberts et al., 1984; Sprott et al., 1984; Thomason et al., 1984; Hallén et al., 1987; Hallén et al., 1988; Teo et al., 1987; Schnadig et al., 1989; Hillier et al., 1991; Kaneko et al., 1992). Using PCR, the percentage of women with bacterial vaginosis even reached 84.5% (Schwebke and Lawing, 2001). In contrast, very little is known about the occurrence of mobilunci in healthy women and possible other natural reservoirs.

With no signs of bacterial vaginosis, *Mobiluncus* species were detected in human vaginal secretions either not at all or only in low numbers (and in a small proportion, i.e, 6%; Skarin and Mårdh, 1982; Spiegel et al., 1983b; Sprott et al., 1983; Hallén et al., 1987). These findings were confirmed, to some extent, in recent studies (Hillier et al., 1991; Kaneko et al., 1992; Schwebke and Lawing, 2001). However, while Hillier et al. (1991) and Kaneko et al. (1992), using conventional bacteriological techniques, found the prevalence of *Mobiluncus* species in healthy women was 0.7–4.0%, Schwebke and Lawing (2001), using a PCR-based methodology, found that 38% of women without infection harbored these organisms. *Mobiluncus curtisii* was rarely detected in this latter group. Notably, in the study of Schwebke and Lawing (2001), sen-

sitivity and specificity of Gram stain compared with PCR were 46.9% and 100%, respectively. Thus, *Mobiluncus* species appear to be more common in the vagina of healthy women than assumed before, with *M. mulieris* predominating.

Hallén et al. (1988) first studied simultaneous colonization of vagina and rectum in women attending a sexually transmitted diseases department. Of 43 women showing symptoms of bacterial vaginosis (BV), 32 had *Mobiluncus* species in the vaginas and 23 in the rectum. In 20 individuals, the same *Mobiluncus* species was found in both sites, and two only showed rectal colonization. In a comparative study of the occurrence of the four organisms associated with BV (namely, *Mobiluncus mulieris, M. curtisii, Mycoplasma hominis* and *Gardnerella vaginalis*) in 374 persons (women with and without BV, their male sexual partners, four homosexual men, and children), Holst (1990) reported that all four species were isolated from the recta of 45–62% of women with BV and of 10–14% of women without BV. These organisms also occurred in the recta of males and children. Vaginal occurrence, in this study, was restricted to women with BV (97%).

The prevalence of *Mobiluncus* species in males has been investigated by Hillier et al. (1990), Holst (1990), and Vetere et al. (1987). Hillier et al. (1990) found that semen samples from 37 men attending a special infertility clinic contained nearly always aerobic or anaerobic bacteria or both with a mean isolation rate of 5.2 strains per seminal specimen. Predominant organisms were coagulase-negative staphylococci (89%), α-hemolytic streptococci (65%), coryneforms (86%), anaerobic streptococci (62%), Bacteroidaceae (27%) and *Gardnerella vaginalis* (19%). *Mobiluncus* species were recovered from fewer than 10% of the samples. Already in 1987, Vetere et al. had also reported one *Mobiluncus* isolate from seminal fluid from an infertile man. In the study of Holst (1990), mobilunci were infrequently isolated from genital samples of 135 males. However, these organisms were cultured from the urethrae and coronal sulci of 10 out of 44 male consorts of women showing symptoms of BV. After two weeks of condom use during sexual intercourse, only *Mycoplasma hominis* remained in the urethra of one man, but the *Mobiluncus* species had completely disappeared.

Other extragenital occurrence of *Mobiluncus* species in men has always been related to disease and will therefore be discussed in the Pathogenicity section. Environmental sources of the organisms have not been reported; however, Doyle et al. (1991) stated that *Mobiluncus curtisii* subsp. *curtisii* belonged to the most common bacterial isolates from the vaginas of 37 healthy rhesus macaques (*Macaca mulatta*).

Isolation, Cultivation and Detection

As in many other members of the family Actinomycetaceae, *Mobiluncus* species require organically complex media and a suitable gaseous environment for good growth. As they grow slowly, they are easily overgrown by more rapidly multiplying organisms simultaneously present. Selective media or principles therefore facilitate the isolation and recognition of mobilunci.

NONSELECTIVE SOLID AND LIQUID CULTURE MEDIA Agar media usually yield better growth of *Mobiluncus* species than do liquid culture media. Several general purpose solid media were reported to support growth of mobilunci including Columbia agar, brain heart infusion agar, DST agar (Oxoid), peptone yeast extract glucose agar, and Schaedler agar (Spiegel, 1992). Growth on these media is stimulated by the addition of 4–15% horse (Moore, 1954; Sprott et al., 1982), human (Curtis, 1913; Moore, 1954; Sprott et al., 1983; Hammann et al., 1984), sheep (Curtis, 1913; Moore, 1954; Hammann et al., 1984; Smith and Moore, 1988), guinea pig (Moore, 1954), goat (Curtis, 1913), or rabbit blood (Smith and Moore, 1988). Addition of serum is also stimulatory (Spiegel, 1992). Supplementation with vitamin K and hemin is not necessary (Sprott et al., 1983).

Growth of *Mobiluncus* species in liquid media is usually poor or at least poorer than on solid media (Vetere et al., 1987; Taylor-Robinson and Taylor-Robinson, 2002). Addition of several putative growth factors does not necessarily improve multiplication of the organisms (Taylor-Robinson and Taylor-Robinson, 2002), although supplementation with serum from horse (Moore, 1954; Durieux and Dublanchet, 1980; Durieux and Dublanchet, 1981; Taylor-Robinson and Taylor-Robinson, 2002), rabbit (Spiegel et al., 1983b; Spiegel and Roberts, 1984a), or cattle (Taylor-Robinson and Taylor-Robinson, 2002) was found to be beneficial.

More important for acceptable growth yields in liquid culture is, however, the selection of an appropriate commercial basal medium. From nine media evaluated by Taylor-Robinson and Taylor-Robinson (2002), only two appeared to be satisfactory. These were Columbia blood broth (Oxoid) and peptone starch dextrose broth (Oxoid) supplemented with 10% horse serum. Provided that a strain is chosen that multiplies easily, these media allow the generation of sufficient amounts of cell mass for, e.g., antigen production or the analysis of virulence factors.

PRETREATMENT PROCEDURES In view of the fastidiousness and the slow growth of *Mobiluncus* species, isolation of the organisms from their natural habitats is hampered by their occurrence with a variety of other, often rapidly multiplying microbes that tend to overgrow them. Thus, pretreatment techniques and selective media are required to obtain optimum detection rates.

Pretreatment measures have been used in conjunction with selective and nonselective culture media. Alkaline pretreatment (Påhlson and Forsum, 1985; Påhlson et al., 1986b) was found to allow survival of more than 90% of *Mobiluncus* cells exposed to pH 12 buffer for 30 and 5 min, respectively. In contrast, less than 80% of other vaginosis-associated organisms survived when clinical specimens were treated in the same way. After exposure to alkaline conditions, the samples were inoculated onto a nonselective medium consisting of brain heart infusion agar supplemented with vitamin K, hemin, and 5% sheep blood. This procedure led to a *Mobiluncus* recovery rate of 58% as compared with 17% without alkaline pretreatment (Påhlson et al., 1986b).

An alternative pretreatment method is the cold-enrichment technique introduced by Smith and Moore (1988) in combination with the application of selective culture media. When specimens are stored at 4–5°C for 1–21 h prior to cultivation immediately after they had been obtained, the recovery rate of mobilunci increased from 18% to 77% compared with samples primarily held at room temperature. These findings were confirmed by Menolascina et al. (1999).

SELECTIVE CULTURE MEDIA A variety of antibiotic or other inhibitory additives has been tried to make agar media selective for *Mobiluncus* species. These include colistin plus nalidixic acid (Durieux and Dublanchet, 1980; Durieux and Dublanchet, 1981), trimethoprim-sulfamethoxazole plus polymyxin B, tinidazole plus colistin or nalidixic acid (Holst et al. [1984a], cited by Spiegel [1992]), and colistin, nalidixic acid, tinidazole, and Nile blue A (Spiegel, 1992). When using tinidazole it has to be noted that concentrations >4 mg/liter of this drug may be inhibitory to some strains of *M. mulieris* (Spiegel, 1987).

Details on the composition and preparation of the selective media used in conjunction with the cold-enrichment technique (Smith and Moore, 1988) are given below.

Rlk Medium (Smith and Moore, 1988)

Columbia CNA agar (BD)	42.5 g
Yeast extract	6.0 g
Peptone	20.0 g

Dissolve in and dilute up to 1 liter with distilled water and autoclave. After cooling, add laked rabbit or sheep blood (50.0 ml), tinidazole (48.0 mg), and nalidixic acid (20.0 mg).

Air-dry 1–2 days and incubate under reducing conditions 1–7 days before inoculation.

SA Medium (Smith and Moore, 1988)

Columbia CNA agar (BD)	42.5 g
Laked rabbit or sheep blood	16.0 ml
Rabbit serum	20.0 ml

Dissolve in and dilute up to 1 liter with distilled water and autoclave. After cooling, add tinidazole (48.0 mg) and nalidixic acid (20.0 mg). Air-dry 1–2 days and incubate under reducing conditions 1–7 days before inoculation.

Further selective media were reported by Spiegel and Krueger (1986) and Holst et al. (1984a).

Mobi Agar (Spiegel and Krueger [1986], cited by Spiegel [1992])

Columbia broth (Difco/BD)	35.0 g
Bacto agar (Difco/BD)	15.0 g
Cysteine · HCl	0.4 g
Soluble starch (Difco/BD)	10.0 g
Resazurin solution (11 mg/44 ml)	4.0 ml
Colistin methane sulfonate	10.0 mg
Nalidixic acid	15.0 mg

Dilute up to 1 liter with distilled water and boil until solids are dissolved, autoclave, and cool to 45–50°C. Aseptically add rabbit serum (20.0 ml), tinidazole (0.6 mg), and 1.2 ml of Nile blue (Eastman Kodak Company, Rochester, NY; 75% dye content, 0.5 g/10 ml).

Selective Agar Medium (Holst et al. [1984a], cited by Spiegel [1992])

Columbia agar (BD)	42.5 g
Tinidazole	1.0 mg
Colistin	10.0 mg
Nalidixic acid	15.0 mg

Dilute up to 1 liter with distilled water and autoclave.

METHODS FOR DIRECT DETECTION OF *MOBILUNCUS* SPECIES IN CLINICAL SPECIMENS Gram-stained smears of vaginal discharge or other clinical materials and wet mounts of vaginal fluids may allow the microscopic detection of Gram-labile to Gram-negative curved rods or of motile curved rods, respectively (Teo et al., 1987). However, recognition of these structures highly depends on their concentration, the amount of other microbes present, and the epithelial or blood cell content of the specimen. Furthermore, the presence of Gram-negative and motile curved rods does not suffice to identify them as members of the genus *Mobiluncus*.

High specificity in detecting mobilunci microscopically may be achieved, however, when immunofluorescence procedures are applied. Påhlson et al. (1986a) used species-specific monoclonal antibodies and an indirect immunofluorescence test to detect and identify *Mobiluncus* species in vaginal specimens with considerable success, although they noted that two atypical strains of *M. curtisii* were not detected with their set of antibodies. Similar results were reported by Fohn et al. (1988), who employed species-, subspecies-, and genus-specific monoclonal antibodies for direct detection of *Mobiluncus* species in clinical specimens by indirect immunofluorescence. Comparably specific monoclonal antibodies were also produced by Ison et al. (1989). Polyclonal antibodies exhibited varying specificity depending on the type of antigen (untreated whole cells versus heated whole cells) used for immunization (Moi and Danielsson [1984], cited by Spiegel [1992]).

Nucleic acid probes and PCR-based techniques were also successfully used for identifying and directly detecting *Mobiluncus* species. Roberts et al. (1984), applying ^{32}P-labeled or biotinylated DNA and a filter blot method, were able to identify *Mobiluncus* strains to species but not subspecies level. The same probes detected mobilunci in 60–70% of specimens that were culture-positive. In a more detailed comparative study (Roberts et al., 1985b) of vaginal secretions analyzed by culture and DNA-probing, cultivation was found to be more sensitive (77%) than DNA-probe hybridization, although the sensitivity of culture appeared to be especially high in this study. However, detection by probe hybridization was considerably faster than by cultivation, being comparable with Gram stain, which of course does not allow speciation.

A hybridization assay using a ^{32}P-labeled synthetic oligonucleotide probe complementary to a nucleotide sequence in the variable region V8 of the *Mobiluncus* 16S rRNA (Påhlson et al., 1992) was reported to be sensitive and to react with 62 of 68 tested typical or atypical *Mobiluncus* strains. The specificity was high so that reactions with related organisms such as *Actinomyces* species were not observed. An advantage of this test appeared to be that rRNA was used as target molecule, which may be present in copy numbers as high as 10^4. A particularly high sensitivity was found for a PCR-based technique (Schwebke and Lawing, 2001) applied to vaginal fluids. Using this technique, 84.5% of women with BV and 38% of women without this condition gave positive results, with apparent species differences between these two groups.

Identification

Apart from the nucleic acid methodologies and serological methods briefly mentioned above, identification of *Mobiluncus* species is traditionally based upon phenotypic characteristics.

MORPHOLOGICAL AND CHEMOTAXONOMIC MARKERS In view of the unique cellular morphology of these organisms, simple microscopic techniques often suffice for detection and preliminary identification. In Gram-stained smears, the curved rods representing mobilunci may appear Gram-positive, Gram-variable, or Gram-negative depending on the cell age (Hammann et al., 1984), the staining method (Sprott et al., 1982), or the medium from which cells were derived (Moore, 1954), but motility and anaerobic growth are stable additional properties indicative of a *Mobiluncus* strain. The active motility of *Mobiluncus* isolates may be assessed microscopically in addition, *M. curtisii* subsp. *curtisii* was reported to be able to migrate through soft agar (unlike *M. curtisii* subsp. *holmesii*), so that it thereby can be differentiated from *M. curtisii* subsp. *holmesii* (Spiegel and Roberts, 1984a; Table 16).

The cell size is an additional morphological character that differentiates between the two proposed subspecies of *M. curtisii* on the one hand and *M. mulieris* on the other hand (Curtis, 1913; Moore, 1954; Durieux and Dublanchet, 1980; Durieux and Dublanchet, 1981; Påhlson et al., 1983; Spiegel et al., 1983b; Hammann et al., 1984; Spiegel and Roberts, 1984a). The smaller form measures 0.8–3 μm in length with a mean length of 1.7 μm, is only slightly bent, may appear coryneform, is more likely to stain Gram-variable to Gram-positive, and usually represents *M. curtisii* (Fig. 34). The larger form (1.9–6.0 μm in length; mean length 2.9 μm) appears moon-like and may even form a half circle, almost invariably appears Gram-negative, and generally represents *M. mulieris* (Table 16; Fig. 35). Colonies are colorless, translucent, smooth, slightly convex, and entire; they reach a maximum diameter of 2–3 mm after 5 days of incubation (Fig. 36).

The pattern of cellular fatty acids (see the section on Chemotaxonomy in this Chapter) may also aid in the recognition of the genus.

PHYSIOLOGICAL CHARACTERISTICS The determination of the metabolic end products from growth of mobilunci in peptone yeast extract glucose medium (PYG) helps to separate *Mobiluncus* species from other curved bacteria and to relate them to the family Actinomycetaceae, the major end products being succinic, acetic and lactic acids (Hjelm et al., 1982; Holst et al., 1982; Hammann et al., 1984; Spiegel and Roberts, 1984a; Vetere et al., 1987; Table 16).

Various methods and commercial systems have been used to assess the fermentative degradation of carbohydrates and the enzymatic activity of *Mobiluncus* strains. These include the Minitek® system (Becton-Dickinson; Hjelm et al., 1982; Påhlson et al., 1983), prereduced anaerobically sterilized media with and without added serum (Holst et al., 1982; Spiegel and Roberts, 1984a), the Rapid ID 32A system (Pattyn et al., 1993), and the Crystal anaerobe identification system (Becton-Dickinson; Cavallaro et al., 1997). The results obtained by these methods and systems differed to a considerable extent, and some commercial systems were not able to distinguish between subspecies and even species. These uncertainties also apply to the results of enzyme activity tests as obtained by the API ZYM (bioMérieux) or combined systems of this manufacturer that include enzyme tests (Holst et al., 1982; Påhlson et al., 1983; Spiegel and Roberts, 1984a).

Nevertheless, it appears appropriate to select certain carbohydrate fermentation and enzyme tests (Table 16) that are obviously suitable for identifying and differentiating *Mobiluncus* species and subspecies. When supplemented with morphological and growth characteristics, they clearly provide an acceptable level of diagnostic reliability.

SEROLOGICAL IDENTIFICATION AND SEROLOGICAL DIAGNOSIS OF *MOBILUNCUS* INFECTIONS As *M. curtisii* and *M. mulieris* were shown to differ clearly in their respective antigenic compositions (Roberts et al., 1985a), serological methods may be used not only for direct detection in clinical specimens but also for identification from culture. Monoclonal antibodies have especially been used for this purpose, immunofluorescence being a simple and reliable but somewhat expensive technique because of personnel costs (Fohn et al., 1988). Furthermore, it should be kept in mind that antigenically atypical variants occur so that especially highly specific monoclonal antibodies may misidentify some isolates (Påhlson et al., 1986a). Two of these atypical groups of isolates were characterized by Schwebke et al. (1991) biochemically, antigenetically, morphologically as well as by DNA hybridization, and one of them was found to belong to *M. mulieris* and the other one to *M. curtisii*. Gatti et al. (1997), in a study of 30 vaginal *Mobiluncus* isolates (22 *M. curtisii* and 8 *M. mulieris*), identified two antigenic profiles within *M. curtisii* and some antigenic variability within *M. mulieris* using immunoblotting but were still able to identify mobilunci at the species level.

Schwebke et al. (1996), also using immunofluorescence techniques, studied the prevalence of antibodies to *M. curtisii* in pregnant women, pediatric patients, and sexually inexperienced women. In these three groups, the immunoglobulin G antibody prevalence was 75, 6, and 0%, respectively. This was interpreted as an indication that serum antibodies could possibly be used as a diagnostic marker for BV.

Epidemiology

The etiological role of *Mobiluncus* species in the pathological lesions from which they have been recovered remains to be elucidated in detail. However, numerous reports leave no doubt that these organisms show a striking although not obligate association with a condition commonly termed bacterial vaginosis (BV). This condition is defined as a characteristic change in the vaginal flora from lactobacilli to strict and facultative anaerobes, among which *Gardnerella vaginalis*, *Mycoplasma hominis*, certain Bacteroidaceae and *Mobiluncus* spp. are most prominent (Mårdh, 1993).

Studies on the epidemiology of BV from the United States, Serbia, Spain, Italy, Sweden, Thailand and Japan showed some differences in the incidence of this condition in relation to the clinical background of the women examined, sexual activity, and race and ethnicity. Georgijevic et al. (2000) reported an average incidence of 10–35% in women visiting gynecological wards, 10–30% in patients visiting obstetric wards, and 20–60% in women visiting facilities that serve those affected by sexually transmitted diseases. More than 50% of these women were asymptomatic. Risk factors for acquiring BV in this study were reported to be multiple sexual partners, exposure to semen, previous trichomoniasis, use of intrauterine contraceptive devices, smoking, indigence, and frequent use of scented soap (Georgijevic et al., 2000).

In a pregnant population of Barcelona, Spain, the incidence of BV was 7.5% (Martinez de Tejada et al., 1998). No correlation between race, parity, education, marital status, smoking, or drug use and BV could be demonstrated in this study. However, a significant association of BV was observed with non-use of birth control methods, presence of sexually transmitted diseases during pregnancy, and HIV seropositivity. Comparing Japanese and Thai pregnant women, Puapermpoonsiri et al. (1996) found a prevalence of BV of 13.6% and 15.9%, respectively. In an Italian population of asymptomatic pregnant women, the incidence of BV was reported to be 4.9% (Cristiano et al., 1996). Holst et al. (1994) were able to correlate the presence of BV with obstetrical problems: BV was diagnosed in 41% of women who had both preterm labor and preterm delivery, while the incidence of the condition was 11% in women who had preterm labor but term delivery and also in women who had both term labor and delivery.

Royce et al. (1999), studying the relationship of BV during pregnancy and black race and ethnicity, demonstrated significant differences between black and white women (22.3% vs. 8.5%). Adjustment for sociodemographics, sexual activity, sexually transmitted diseases, health behavior, and sexual hygiene did not explain these differences. In an attempt to relate these differences to known risk factors, Ness et al. (2003) enrolled 900 black and 235 white women in a study from five United States sites. The results of this study confirmed that black women were more likely than white women to have BV. Furthermore, black women were also more likely to be older, have lower educational attainment and family incomes, and have a history of sexually transmitted diseases. However, after adjustment for demographic and lifestyle factors, blacks remained at elevated risk for BV, so that the observed racial disparity in the occurrence of BV could not be explained by risk factor differences.

Although there is a significant correlation between BV and the presence of *Mobiluncus* species in vaginal secretions, these organisms are not always present when BV is diagnosed (84.5% according to Schwebke and Lawing, 2001), and they may be recovered in up to 38% of women without signs of BV (Schwebke and Lawing, 2001). Also Kaneko et al. (1992), investigating the occurrence of *Mobiluncus* species in Japanese women, found a significant difference between patients with BV and clinically healthy pregnant and nonpregnant women (27.3% vs. 0.7–3.2%). In women attending a sexually transmitted disease clinic, the detection rate of mobilunci was 53% of those with and 4% of those without BV (Hillier et al., 1991).

As far as extragenital *Mobiluncus* infections are concerned, epidemiological conclusions cannot be drawn owing to the rare and sporadic occurrence of these conditions.

Pathogenicity

BACTERIAL VAGINOSIS Despite the unambiguous association of BV and the presence of *Mobiluncus* species in vaginal fluids, the etiological role of these organisms in the pathogenesis of this condition is difficult to assess definitively. The reasons for this difficulty are 1) mobilunci have never been recovered in pure culture from vaginal discharge but rather in combination with various other microbes, and 2) mobilunci may be encountered without clinical or bacteriological signs of disease or impairment. However, this is a problem characteristic of all of the indigenous, facultatively pathogenic microbes etiologically involved in human or animal disease and particularly applies to facultative or strict anaerobes including most members of the family Actinomycetaceae. The observation that differences in the severity of BV between women with or without *Mobiluncus* species or in their response to treatment (Spiegel et al., 1983; Jones et al., 1985) cannot be demonstrated does not exclude the pathogenic potential of the organisms, as this is also true for several other anaerobic species that predominantly or exclusively occur in synergistic mixed infections.

Bacteria Potentially Involved in the Etiology of Bacterial Vaginosis The typical bacterial flora associated with clinical signs of BV consists of *Prevotella bivia*, *Prevotella disiens*, *Porphyromonas* spp., *Mobiluncus* spp., anaerobic streptococci, and *Mycoplasma hominis* (Spiegel, 1991; Kimberlin and Andrews, 1998). Both the exact species composition and their relative concentrations vary among women having this condition, indicating that etiology and pathogenesis of BV are comparable to that of other polymicrobial infections, especially to those of endogenous origin.

More detailed analyses of the prevalence of the two *Mobiluncus* species in patients with BV revealed that *M. curtisii* appears to be significantly more often associated with this condition than *M. mulieris*. Schwebke and Lawing (2001), using a PCR-based procedure, reported that *M. curtisii* was rarely detected in women without signs of BV; in contrast, this species was identified in 65.3% of the patients suffering from BV. Moi et al. (1991), using an indirect immunofluorescence assay, demonstrated significantly higher titers of IgG antibodies to *M. curtisii* (p = <0.01) than to *M. mulieris* in women with BV. Furthermore, these titers were higher in patients with BV than in women without this condition and in healthy males. No secretory IgA antibodies to *Mobiluncus* species were found in vaginal washings.

Potential Virulence Factors of Mobiluncus *spp.* Using luminol-enhanced chemiluminescence, *M.*

curtisii was shown to induce a significantly less pronounced oxidative metabolism of polymorphonuclear leukocytes than that of *M. mulieris*, indicating that the former species is apparently able to escape phagocytosis more easily. In light of these findings, the earlier report of DeBoer and Plantema (1988), who saw curved rods inside phagocytic vacuoles of vaginal polymorphonuclear leukocytes, could mean that the intraphagocytic organisms were *M. mulieris* rather than *M. curtisii*. The same authors, investigating the *in situ* adherence of *Mobiluncus* to detached vaginal epithelial cells by electron microscopy, found that the curved rods attached both directly to the epithelial cell surface and at various distances from it. From this finding, they concluded that after initial adherence the motile bacteria were able to grow on the epithelial surface in sessile microcolonies. This ability appeared to be due to the inclusion in a matrix of exopolysaccharides, and the production of such a glycocalyx could be demonstrated *in vitro* for both *M. curtisii* subsp. *curtisii* and *M. mulieris*.

Both *Mobiluncus* species were shown to produce a cytotoxin active against Vero cells and four other cell lines (Taylor-Robinson et al., 1993). The marked cytopathic effect led to destruction of the cells within 72 h. In bovine oviduct organ cultures, eight *M. curtisii* and two *M. mulieris* strains caused loss of ciliary activity, which was complete after 60 h. Histologic changes were also observed. The toxin was identified to be extracellular and relatively heat-stable; at pH 9 and pH 3, it lost its activity.

Ecological Aspects of Alterations of the Vaginal Flora There is no doubt that these potential virulence factors do not suffice to explain the development of BV, especially because *Mobiluncus* species cannot be detected in all of the patients exhibiting clinical symptoms of the condition. The most obvious finding characteristic of all cases of BV is the typical alteration of the vaginal flora, which can be demonstrated microscopically, culturally, or by molecular techniques. This alteration of the vaginal flora consists of a dramatic decrease in numbers of indigenous lactobacilli and replacement of these bacteria by *Prevotella* spp., *Porphyromonas* spp., *Gardnerella vaginalis*, *Mycoplasma hominis* and *Mobiluncus* spp. (Goldacre et al., 1979; Spiegel et al., 1980; Taylor et al., 1982; Blackwell et al., 1983; Rabe et al., 1988).

As a result of these changes in the vaginal ecosystem, various alterations of the vaginal milieu occur that can be related to activities of the above-specified mixed facultative and anaerobic flora. The absence of *Lactobacillus* species has two major consequences: Lactic acid concentration due to these bacteria is decreased in the vaginal fluid, facilitating elevated pH values, and the antagonistic effect of hydrogen peroxide on nonphysiological microbes is missing (Skarin and Sylwan, 1986; Spiegel, 1991). In a detailed study, Nagy et al. (1991) found that 79% (n = 47) of *Lactobacillus* strains derived from women without BV produced H_2O_2, while only 23% (n = 39) of the isolates from patients with BV did so. Five of 20 H_2O_2-producing and two of 26 non-producing strains exhibited inhibitory activities against four of 12 strains of anaerobic streptococci and two of 10 *Mobiluncus* isolates. None of another 41 different anaerobic and facultatively anaerobic bacterial strains was inhibited by any of the *Lactobacillus* isolates tested. Increasing the iron content of the medium by adding $FeCl_3$ (0.01–1 mM) decreased or completely abolished the antibiosis.

Several authors have demonstrated that metabolic or enzymatic activities of the altered vaginal flora of BV patients directly contribute to the development of the condition. Gonzales Pedraza Aviles et al. (1999) related the production of aminopeptidases and decarboxylases by anaerobes of the BV flora to the conversion of protein and amino acids to amines that are responsible for raising the vaginal pH and producing the typical odor of the vaginal discharge characteristic of the syndrome. The biological activity of the fermentation end products of vaginal anaerobes also has attracted attention. Sturm (1989a) as well as Al-Mushrif et al. (2000) studied the role of succinic acid on the chemotactic response of white blood cells. Sturm (1989) observed that certain succinate-producing anaerobes including mobilunci inhibited the chemotactic response of granulocytes while "*Peptostreptococcus productus*" (*Ruminococcus productus*) and *Escherichia coli* did not. Al-Mushrif et al. (2000) exposed a monocytic cell line (MonoMac 6) to varying concentrations of pure succinic, acetic and lactic acids. In this model, succinic acid exhibited high inhibition of chemotaxis; lower inhibition and no inhibition were observed with acetic and lactic acids, respectively. Correspondingly, succinic and acetic acids were detected in high concentrations in the vaginal fluids of patients with BV, and these fluids but not those of healthy women had the same inhibitory effect on the monocytic cell line.

Complications of Bacterial Vaginosis The importance of BV as a disease rather than a mere imbalance of the vaginal microflora is underlined by numerous observations of various types of complications associated with BV. Vaginal bleeding has been reported in women with BV although *Mobiluncus* spp. were not recovered from all of them (Larsson and Bergman, 1986). Especially well documented is an association of

BV late in pregnancy with an increased risk of developing chorioamnionitis, premature labor, and preterm delivery (Hillier et al., 1988; Martius et al., 1988; McGregor et al., 1994). Mucinases and sialidases produced by BV bacteria were thought to contribute to the development of these adverse pregnancy outcomes. McGregor et al. (1994) not only could confirm that presence of BV in pregnant women at intake was associated with increased risk of preterm birth, premature rupture of membranes, and preterm premature rupture of membranes, but also found that mucinase and sialidase activities were more commonly encountered in these women and that these enzymes occurred in higher concentrations. Furthermore, a treatment study with 2% clindamycin vaginal cream revealed that persistence of sialidase-producing organisms in numbers sufficient to increase vaginal fluid sialidase activity obviously was a risk factor for possibly preventable subclinical intrauterine infection and preterm birth. Sialidase activity was found to be associated with *Gardnerella vaginalis, Mobiluncus* spp., *Mycoplasma hominis* as well as *Chlamydia trachomatis* and yeasts; mucinase activity was only detected with BV-linked organisms (McGregor et al., 1994).

Additional complications and sequelae reported in conjunction with BV are cervicitis, salpingitis, endometritis, pelvic infections, postoperative infections, and even urinary tract infections (Georgijevic et al., 2000), although Pattman reported that signs of ascending infections or urinary tract symptoms were not more common in BV patients with and without *Mobiluncus* species (Pattman [1984], cited in Spiegel [1992]).

EXTRAGENITAL   MOBILUNCUS   INFECTIONS
Extragenital infections due to *Mobiluncus* species have increasingly been reported. In particular, these were nonpuerperal breast abscesses in females (Glupczynski et al., 1984; Sturm and Sikkenk, 1984; Weinbren et al., 1986; Bennett et al., 1989; Edmiston et al., 1989; Sturm, 1989b; Sota et al., 1993) and in a male (Sturm, 1989b). Sturm (1989b) analyzed 109 specimens from nonpuerperal breast masses and found anaerobes or mixed infections in 43% of them. Seven (16%) of 44 samples containing anaerobes also yielded growth of *Mobiluncus* spp., of which six were identified as *M. curtisii* and one as *M. mulieris*. The author concluded that the development of these abscesses might be related to sexual activity. The one nonpuerperal breast abscess reported by Edmiston et al. (1989) was attributed to *M. curtisii* subsp. *holmesii*.

Other extragenital localized lesions assigned to mobilunci were an abdominal abscess (Mayer et al., 1994), a mixed-infected suppuration of the naval, and an exudative postoperative wound infection after radical mastectomy (Glupczynski et al., 1984). Furthermore, bacteremia due to *M. curtisii* following septic abortion (Gomez-Garces et al., 1994) and a severe sepsis due to *M. curtisii* subsp. *curtisii* (Hill et al., 1998) were observed. In the latter case, the generalizing infection became life threatening by causing septic shock, renal failure, disseminated intravascular coagulation, adult respiratory distress syndrome, and spontaneous splenic rupture. Despite these serious complications, the 54-year-old female patient survived with full intensive care support and intravenous application of ceftriaxone.

Treatment of *Mobiluncus* Infections and Antimicrobial Susceptibility Treatment of bacterial vaginosis has been attempted using local application of clindamycin phosphate cream or oral metronidazole, erythromycin, or ofloxacin (Nygaard et al., 1991; Nagayam et al., 1992; Wathne et al., 1993; McGregor et al., 1994; Mikamo et al., 1996; Porozhanova and Bozhinova, 2000). Good therapeutic results were reported for local clindamycin cream as well as for oral metronidazole, the former being possibly slightly more effective (Mikamo et al., 1996). Erythromycin appeared to be clearly less effective than metronidazole (Wathne et al., 1993); the same is true for ofloxacin although some efficacy of this drug was observed (Nagayam et al., 1997).

Susceptibility tests of *Mobiluncus* spp. revealed good in vitro activities of clindamycin, imipenem, cefmetazole, amoxicillin, and amoxicillin plus clavulanic acid (Puapermpoonsiri et al., 1997). However, metronidazole was found to be ineffective, which had already been reported by Spiegel (1992) for *M. curtisii* and some strains of *M. mulieris*. Rifaximin, a new topical rifamycin derivative, also showed very good in vitro activity against mobilunci (Hoover et al., 1993). Among many other human bacterial isolates, also *M. curtisii* carried the tetracycline resistance determinant *tetQ*, which was found to be integrated into the chromosome (Leng et al., 1997). In contrast to gene transfer from other anaerobes possessing *tetQ*, transfer of this gene from one *M. curtisii* strain into an *Enterococcus faecalis* recipient failed.

In view of the difficulties encountered when cultivating mobilunci, Croco et al. (1994) compared the E-test with standard procedures for susceptibility testing and found that the results for *Mobiluncus* isolates correlated well with those of the standard technique used.

Literature Cited

Adachi, A., G. J. Kleiner, G. H. Bezahler, W. M. Greston, and G. H. Friedland. 1985. Abdominal wall actinomycosis associated with an IUD: A case report. J. Reprod. Med. 30:145–148.

Adderson, E. E., A. Croft, R. Leonard, and K. Carroll. 1998. Septic arthritis due to Arcanobacterium bernardiae in an immunocompromised patient. Clin. Infect. Dis. 27:211–212.

Akgun, Y., E. Ustunel, and E. Tarlak. 1985. A case of cervico-facial actinomycosis. Mikrobiol. Bull. 19:104–108.

Alamillos-Granados, F. J., A. Dean-Ferrer, A. Garcia-Lopez, and F. Lopez-Rubio. 2000. Actinomycotic ulcer of the oral mucosa: An unusual presentation of oral actinomycosis. Br. J. Oral Maxillofac. Surg. 28:121–123.

Allen, J. N. 1987. Actinomyces meyeri breast abscess. Am. J. Med. 83:186–187.

Allworth, A. M., H. K. Ghosh, and N. Saltos. 1986. A case of Actinomyces meyeri pneumonia in a child. Med. J. Australia 145:33.

Alos, J. I., C. Barros, and J. L. Gomez-Garces. 1995. Endocarditis caused by Arcanobacterium haemolyticum. Eur. J. Clin. Microbiol. Infect. Dis. 14:1085–1088.

Alshamaony, L., M. Goodfellow, D. E. Minnikin, G. H. Bowden, and J. M. Hardie. 1977. Fatty and mycolic acid composition of Bacterionema matruchotii and related organisms. J. Gen. Microbiol. 98:205–213.

Altman, N. H., and J. D. Small. 1973. Actinomycosis in a primate confirmed by fluorescent antibody techniques in formalin-fixed tissue. Lab. Animal Sci. 23:696–700.

Amrikachi, M., B. Krishnan, C. J. Finch, and I. Shabab. 2000. Actinomyces and Actinobacillus actinomycetemcomitans- and Actinomyces-associated lymphadenopathy mimicking lymphoma. Arch. Pathol. Lab. Med. 124:1502–1505.

Amsel, R., P. A. Totten, C. A. Spiegel, K. C. S. Chen, D. Eschenbach, and K. K. Holmes. 1983. Nonspecific vaginitis: Diagnostic criteria and microbial and epidemiologic associations. Am. J. Med. 74:14–22.

Apotheloz, C., and C. Regamey. 1996. Disseminated infection due to Actinomyces meyeri: Case report and review. Clin. Infect. Dis. 22(4):621–625.

Arikan, S., S. Erguven, and A. Gunalp. 1997. Isolation, in vitro antimicrobial susceptibility and penicillin tolerance of Arcanobacterium haemolyticum in a Turkish university hospital. Zbl. Bakteriol. 286:487–493.

Arroyo, G., and J. A. Quinn Jr. 1989. Association of amoebae and actinomycetes in an intrauterine contraceptive device user. Acta Cytol. 33:298–300.

Asikainen, S., H. Jousimies-Somer, A. Kanervo, and P. Summanen. 1987. Certain bacterial species and morphotypes in localized juvenile periodontitis and in matched controls. J. Periodontol. 58:224–230.

Ayakawa, G. Y., B. L. Williams, and G. E. Kenny. 1983. Identification and preliminary characterization of a major heat-stable surface antigen of Actinomyces israelii by two-dimensional (crossed) immunoelectrophoresis. Infect. Immun. 41:11–18.

Baele, M., K. Chiers, L. A. Devriese, H. E. Smith, H. J. Wisselink, M. Vaneechoutte, and F. Haesebrouck. 2001. The Gram-positive tonsillar and nasal flora of piglets before and after weaning. J. Appl. Microbiol. 91:997–1003.

Baker, J. J., and S. A. Billy. 1983a. Activation of the alternate complement pathway by peptidoglycan of Actinomyces viscosus, a potentially pathogenic oral bacterium. Arch. Oral Biol. 28:1073–1075.

Baker, P. J., R. T. Evans, R. A. Coburn, and R. J. Genco. 1983b. Tetracycline and its derivatives strongly bind to and are released from the tooth surface in active form. J. Periodontol. 54:580–585.

Baker, J. J. 1985. Peptidoglycan from the potentially pathogenic oral bacterium Actinomyces viscosus is a B-cell mitogen. Arch. Oral Biol. 30:291–294.

Barksdale, W. L., K. Li, C. S. Cummins, and H. Harris. 1957. The mutation of Corynebacterium pyogenes to Corynebacterium haemolyticum. J. Gen. Microbiol. 16:749–758.

Barksdale, W. L. 1970. Corynebacterium diphtheriae and its relatives. Bacteriol. Rev. 34:378–422.

Barnham, M. 1988. Actinomyces pyogenes bacteremia in a patient with carcinoma of the colon. J. Infect. 17:231–234.

Barnham, M., and R. A. Bradwell. 1992. Acute peritonsillar abscess caused by Arcanobacterium haemolyticum. J. Laryngol. Otol. 106:1000–1001.

Baron, E. J., J. M. Angevine, and W. Sundström. 1979. Actinomycotic pulmonary abscess in an immunosuppressed patient. Am. J. Clin. Pathol. 72:637–639.

Barsotti, O., F. Renaud, J. Freney, G. Benay, D. Decoret, and J. Dumont. 1987. Rapid isolation of DNA from Actinomyces. Ann. Inst. Pasteur Microbiol. 138:529–536.

Bassiri, A. G., R. E. Girgis, and J. Theodore. 1996. Actinomyces odontolyticus thoracopulmonary infections: Two cases in lung and heart-lung transplant recipients and a review of the literature. Chest 109:1109–1111.

Batisse-Milton, S. E., R. M. Gander, and D. D. Colvin. 1995. Tubo-ovarian and peritoneal effusion caused by Arcanobacterium haemolyticum. Clin. Microbiol. Newsl. 17:118–120.

Batty, I. 1958. Actinomyces odontolyticus, a new species of actinomycete regularly isolated from deep carious dentine. J. Pathol. Bacteriol. 75:455–459.

Baumann, C. D., W. R. Davidson, D. E. Roscoe, and K. Beheler-Amass. 2001. Intracranial abscessation in white-tailed deer of North America. J. Wildl. Dis. 37:661–670.

Beaman, B. L., M. E. Gershwin, and S. Maslan. 1979. Infectious agents in immunodeficient murine models: Pathogenicity of Actinomyces israelii, serotype 1, in congenitally athymic (nude) mice. Infect. Immun. 24: 583–585.

Becker, B., M. P. Lechevalier, R. E. Gordon, and H. Lechevalier. 1964. Rapid differentiation between Nocardia and Streptomyces by paper chromatography of whole-cell hydrolysates. Appl. Microbiol. 12:421–423.

Becker, M. R., B. J. Paster, E. J. Leys, M. L. Moeschberger, S. G. Kenyon, J. L. Galvin, S. K. Boches, F. E. Dewhirst, and A. L. Griffen. 2002. Molecular analysis of bacterial species associated with childhood caries. J. Clin. Microbiol. 40:1001–1009.

Behbehani, M. J., J. D. Heeley, and H. V. Jordan. 1983a. Comparative histopathology of lesions produced by Actinomyces israelii, Actinomyces naeslundii, and Actinomyces viscosus in mice. Am. J. Pathol. 110:267–274.

Behbehani, M. J., H. V. Jordan, and J. D. Heeley. 1983b. Oral colonization and pathogenicity of Actinomyces israelii in gnotobiotic rats. J. Dent. Res. 62:69–74.

Beighton, D., and G. Colman. 1976. A medium for the isolation and enumeration oral Actinomycetaceae from dental plaque. J. Dent. Res. 55:875–878.

Beighton, D., and W. A. McDougall. 1977. The effects of fluoride on the percentage bacterial composition of dental plaque, on caries incidence, and on the in vitro growth of Streptococcus mutans, Actinomyces viscosus, and Actinobacillus sp. J. Dent. Res. 56:1185–1191.

Beighton, D. 1985. Establishment and distribution of the bacteria Actinomyces viscosus and Actinomyces naeslundii in the mouth of monkeys (Macaca fascicularis). Arch. Oral. Biol. 30:403–407.

Bellack, S., and H. V. Jordan. 1972. Serological identification of rodent strains of Actinomyces viscosus and their relationship to Actinomyces of human origin. Arch. Oral Biol. 17:175–182.

Bender, G. R., and R. E. Marquis. 1987. Membrane ATPases and acid tolerance of Actinomyces viscosus and Lactobacillus casei. Appl. Environ. Microbiol. 53:2124–2128.

Bennhoff, D. F. 1984. Actinomycosis: Diagnostic and therapeutic considerations and a review of 32 cases. Laryngoscope 94:1198–1217.

Berger, U., M. Kapovits, and G. Pfeifer. 1959. Zur Besiedlung der kindlichen Mundhöhle mit anaeroben Mikroorganismen. Zeitschr. Hyg. Infektionskrankh. 145:564–573.

Bergey, D. H. 1907. Actinomyces der Mundhöhle. Zbl. Bakteriol. Parasitenkd. Infektionskr., Abteilung I 40:361.

Bernhardt, H., and M. Knoke. 1984. Growth of anaerobes of the upper small intestine using the glove box technique. Nahrung 28:723–726.

Bestetti, G., V. Bühlmann, J. Nicolet, and R. Frankhauser. 1977. Paraplegia due to Actinomyces viscosus infection in a cat. Acta Neuropathol. 39:231–235.

Biever, L. J., G. W. Robertstad, K. van Steenbergh, E. E. Scheez, and G. F. Kennedy. 1969. Actinomycosis in a bovine lung. Am. J. Vet. Res. 30:1063–1066.

Biksi, I., L. Fodor, O. Szenci, and F. Vetesi. 1997. The first isolation of Eubacterium suis in Hungary. Zentralbl. Veterinärmed. 44:547–550.

Billington, S. J., G. Songer, and B. H. Jost. 2001. Molecular characterization of the pore-forming toxin, pyolysin, a major virulence determinant of Arcanobacterium pyogenes. Vet. Microbiol. 82:261–274.

Billington, S. J., K. W. Post, and B. H. Jost. 2002a. Isolation of Arcanobacterium (Actinomyces) pyogenes from cases of feline otitis externa and canine cystitis. J. Vet. Diagn. Invest. 14:159–162.

Billington, S. J., J. G. Songer, and B. H. Jost. 2002b. Widespread distribution of the tet W determinant among tetracycline-resistant isolates of the animal pathogen Arcanobacterium pyogenes. Antimicrob. Agents Chemother. 46:1281–1287.

Björndal, L., and T. Larsen. 2000. Changes in the cultivable flora in deep carious lesions following a stepwise excavation procedure. Caries Res. 34:502–508.

Blackwell, A. L., A. R. Fox, I. Phillips, and D. Barlow. 1983. Anaerobic vaginosis (non-specific vaginitis): Clinical, microbiological, and therapeutic findings. Lancet ii:1379–1382.

Blank, C. H., and L. K. Georg. 1968. The use of fluorescent antibody methods for the detection and identification of Actinomyces species in clinical material. J. Lab. Clin. Med. 71:283–293.

Blanksma, L. J., and J. Slijper. 1978. Actinomycotic dacryocystitis. Ophthalmologica 176:145–149.

Blinkhorn Jr., R. J., V. Strimbu, D. Effron, and P. J. Spagnuolo. 1988. "Punch" actinomycosis causing osteomyelitis of the hand. Arch. Intern. Med. 148:2668–2670.

Bollinger, O. 1877. Über eine neue Pilzkrankheit beim Rinde. Zbl. Med. Wissensch. 15:481–485.

Bonnez, W., G. Lattimer, N. A. Mohanraj, and T. H. Johnson. 1985. Actinomyces naeslundii as an agent of pelvic actinomycosis in the presence of an intrauterine device. J. Clin. Microbiol. 21:273–275.

Boue, D., E. Armau, and G. Tiraby. 1987. A bacteriological study of rampant caries in children. J. Dent. Res. 60:23–28.

Bousfield, I. J., R. M. Keddy, T. R. Dando, and S. Shaw. 1985. Simple rapid methods of cell wall analysis as an aid in the identification of aerobic coryneform bacteria. In: M. Goodfellow and D. E. Minnikin (Eds.) Actinomycetales: Characteristics and Practical Importance. Academic Press. London, UK. 221–236.

Bowden, G. H., and J. M. Hardie. 1973. Commensal and pathogenic Actinomyces species in man. In: G. Sykes and F. A. Skinner (Eds.) Actinomycetales: Characteristics and Practical Importance. Academic Press. London, UK. 277–295.

Boyan-Salyers, B. D., J. J. Vogel, and J. Ennever. 1978. Basic biological sciences; preapatitic mineral deposition in Bacterionema matruchotii. J. Dent. Res. 57:291–295.

Boyd, A., and R. A. D. Williams. 1971. Estimation of the volumes of bacterial cells by scanning microscopy. Arch. Oral Biol. 16:259–267.

Bradshaw, D. J., P. D. Marsh, G. K. Watson, and C. Allison. 1998. Role of Fusobacterium nucleatum and coaggregation in anaerobe survival in planktonic and biofilm oral microbial communities during aeration. Infect. Immun. 66:4729–4732.

Brailsford, S. R., R. B. Tregaskis, H. S. Leftwich, and D. Beighton. 1999. The predominant Actinomyces spp. isolated from infected dentin of active root caries lesions. J. Dent. Res. 78:1525–1543.

Brander, M., and H. Jousimies-Somer. 1992. Evaluation of the RapID ANA II and API ZYM systems for identification of Actinomyces species from clinical specimens. J. Clin. Microbiol. 30:3112–3116.

Brecher, S. M., J. van Houte, and W. F. Hammond. 1978. Role of colonization in the virulence of Actinomyces viscosus strains T14-Vi and T14-Av. Infect. Immun. 22:603–614.

Brecher, S. M., and J. van Houte. 1979. Relationship between host age and susceptibility to oral colonization by Actinomyces viscosus in Sprague-Dawley rats. Infect. Immun. 26:1137–1145.

Breed, R. S., and H. J. Conn. 1919. The nomenclature of the Actinomycetaceae. J. Bacteriol. 4:585–602.

Brennan, K. E., and P. J. Ihrke. 1983. Grass awn migration on dogs and cats: A retrospective study of 182 cases. J. Am. Vet. Med. Assoc. 182:1201–1204.

Brennan, M. J., J. O. Cisar, A. E. Vatter, and A. L. Sandberg. 1984. Lectin-dependent attachment of Actinomyces naeslundii to receptors on epithelial cells. Infect. Immun. 46:459–464.

Brennan, M. J., J. O. Cisar, and A. L. Sandberg. 1986. A 160-kilodalton epithelial cell surface glycoprotein recognized by plant lectins that inhibit the adherence of Actinomyces naeslundii. Infect. Immun. 52:840–845.

Brennan, M. J., R. A. Joralmon, and J. O. Cisar. 1987. Binding of Actinomyces naeslundii to glycosphingolipids. Infect. Immun. 55:487–489.

Brihmer, C., I. Callings, C.-E. Nord, and J. Brundin. 1987. Salpingitis; aspects of diagnosis and etiology: A 4-year study from a Swedish capital hospital. Eur. J. Obstet. Gynecol. Reprod. Biol. 24:211–220.

Brock, D. W., L. K. Georg, J. M. Brown, and M. D. Hicklin. 1973. Actinomycosis caused by Arachnia propionica. Am. J. Clin. Pathol. 59:66–77.

Brown, A. T., C. P. Christian, and R. L. Eifert. 1975. Purification, characterization, and regulation of a nicotinamide adenine dinucleotide-dependent lactate dehydrogenase from Actinomyces viscosus. J. Bacteriol. 122:1126–1135.

Brown, L. R., R. J. Billings, and A. G. Kaster. 1986. Quantitative comparison of potentially cariogenic microorganisms cultured from non-caries and caries root and coronal tooth surfaces. Infect. Immun. 51:765–770.

Brunner, S., S. Graf, P. Riegel, and M. Altwegg. 2000. Catalase-negative Actinomyces neuii subsp. neuii from an infected mammary prosthesis. Int. J. Med. Microbiol. 290:285–287.

Buchanan, R. E. 1917. Studies in the nomenclature and classification of bacteria. II: The primary subdivisions of the Schizomycetes. J. Bacteriol. 2:155–164.

Buchanan, R. E. 1918. Studies in the classification and nomenclature of bacteria. VIII: The subgroups and genera of the Actinomycetales. J. Bacteriol. 3:403–406.

Buchanan, B. B., and L. Pine. 1962. Characterization of a propionic acid producing actinomycete, Actinomyces propionicus, sp. nov. J. Gen. Microbiol. 28:305–323.

Buchanan, B. B., and L. Pine. 1965. Relationship of carbon dioxide to aspartic acid and glutamic acid in Actinomyces naeslundii. J. Bacteriol. 89:729–733.

Buchanan, B. B., and L. Pine. 1967. Path of glucose breakdown and cell yields of a facultative anaerobe Actinomyces naeslundii. J. Gen. Microbiol. 46:225–236.

Buchanan, A. M., and J. L. Scott. 1984a. Actinomyces hordeovulneris, a canine pathogen that produces L-phase variants spontaneously with coincident calcium deposition. Am. J. Vet. Res. 45:2552–2560.

Buchanan, A. M., J. L. Scott, M. A. Gerencser, B. L. Beaman, S. Jang, and E. L. Biberstein. 1984b. Actinomyces hordeovulneris sp. nov., an agent of canine actinomycosis. Int. J. Syst. Bacteriol. 34:439–443.

Bujwid, O. 1889. Über die Reinkultur des Actinomyces. Zbl. Bakteriol. Hyg. 6:630–633.

Burgi, E., T. Sydler, S. Ohlerth, L. Corboz, and G. Nietlispach. 2001. Purulent osteomyelitis in fattening pigs. Schweiz. Arch. Tierheilkd. 143:93–98.

Burckhardt, J. J. 1978. Rat memory t lymphocytes: in vitro proliferation induced by antigens of Actinomyces viscosus. Scand. J. Immunol. 7:167–172.

Burckhardt, J. J., R. Gaegauf-Zollinger, R. Schmid, and B. Guggenheim. 1981. Alveolar bone loss in rats after immunization with Actinomyces viscosus. Infect. Immun. 31:971–977.

Carlone, G. M., M. L. Thomas, R. J. Arko, G. O. Guerrant, C. Wayne Moss, J. M. Swenson, and S. A. Morse. 1986. Cell wall characteristics of Mobiluncus species. Int. J. Syst. Bacteriol. 36:288–296.

Carlson, P., and S. Kontiainen. 1994a. Alpha-mannosidase: A rapid test for identification of Arcanobacterium haemolyticum. J. Clin. Microbiol. 32:854–855.

Carlson, P., and S. Kontiainen. 1994b. Evaluation of a commercial kit in the identification of Arcanobacterium haemolyticum and Arcanobacterium pyogenes. Eur. J. Clin. Microbiol. Infect. Dis. 13:507–509.

Carlson, P., S. Kontainen, and O. V. Renkonen. 1994c. Antimicrobial susceptibility of Arcanobacterium haemolyticum. Antimicrob. Agents Chemother. 38:142–143.

Carlson, P., K. Lounatmaa, and S. Kontiainen. 1994d. Biotypes of Arcanobacterium haemolyticum. J. Clin. Microbiol. 32:1654–1657.

Carlson, P., O. V. Renkonen, and S. Kontiainen. 1994e. Arcanobacterium haemolyticum and streptococcal pharyngitis. Scand. J. Infect. Dis. 26:283–287.

Carlson, P., E. Eerola, and S. Kontiainen. 1995a. Additional tests to differentiate Arcanobacterium haemolyticum and Arcanobacterium pyogenes. Zbl. Bakteriol. 282:232–236.

Carlson, P., S. Kontiainen, O. V. Renkonen, A. Sivonen, and R. Visakorpi. 1995b. Arcanobacterium haemolyticum and streptococcal pharyngitis in army conscripts. Scand. J. Infect. Dis. 27:17–18.

Carr, J., and J. R. Walton. 1993. Bacterial flora of the urinary tract of pigs associated with cystitis and pyelonephritis. Vet. Rec. 132:575–577.

Casin, I., M. Ortenberg, M. J. Sanson Le Pors, J. J. Denis, C. Denis, and Y. Perol. 1984. Actinomyces israelii. Methods for isolation and identification: Apropos of ten cases. Pathol. Biol. Paris 32:153–159.

Cato, E. P., W. E. C. Moore, G. Nygaard, and L. V. Holdemann. 1984. Actinomyces meyeri sp. nov., specific epithet rev. Int. J. Syst. Bacteriol. 34:487–489.

Cavallaro, J. J., L. S. Wiggs, and J. M. Miller. 1997. Evaluation of the BBL Crystal Anaerobe identification system. J. Clin. Microbiol. 35:3186–3191.

Chapoy, P., J. Roux, J. P. Giraud, and D. Sephieng. 1984. Thoracic actinobacteriosis due to Actinomyces naeslundii. Arch. Franc. Ped. 41:701–703.

Chatterjee, B. D., and C. K. Chakraborty. 1989. Ischaemic mouse thigh model for evaluation of pathogenicity of non-clostridial anaerobes. Indian J. Med. Res. 89:36–39.

Chaumentin, G., C. Pariset, T. Stouls, A. Boibieux, M. E. Reverdy, J. Baulieux, P. Spitalier, F. Biron, and D. Peyramond. 1997. Actinomyces meyeri disseminated actinomycosis disclosing pulmonary carcinoma. Rev. Med. Interne. 18:563–565.

Christie, A. O., and J. W. Porteus. 1962a. Growth of several strains of Actinomyces israelii in chemically defined media. Nature 195:408–409.

Christie, A. O., and J. W. Porteus. 1962b. The cultivation of a single strain of Actinomyces israelii in a simplified and chemically defined medium. J. Gen. Microbiol. 28:443–454.

Christie, A. O., and J. W. Porteus. 1962c. The growth factor requirements of the Wills strain of Actinomyces israelii growing in a chemically defined medium. J. Gen. Microbiol. 28:455–460.

Cisar, J. O., and A. E. Vatter. 1979. Surface fibrils (fimbriae) if Actinomyces viscosus T15V. Infect. Immun. 24:523–531.

Cisar, J. O., V. A. David, S. H. Curl, and A. E. Vatter. 1984a. Exclusive presence of lactose-sensitive fimbriae on a typical strain (BVU 45) of Actinomyces naeslundii. Infect. Immun. 46:453–458.

Cisar, J. O., A. L. Sandberg, and S. E. Mergenhagen. 1984b. The function and distribution of different fimbriae on strains of Actinomyces viscosus and Actinomyces naeslundii. J. Dent. Res. 63:393–396.

Cisar, J. O., A. E. Vatter, W. B. Clark, S. H. Curl, S. Hurst Calderone, and A. L. Sandberg. 1988. Mutants of Actinomyces viscosus T14V lacking type 1, type 2, or both types of fimbriae. Infect. Immun. 56:2984–2989.

Civen, R., M. L. Vaisanen, and S. M. Finegold. 1993. Peritonsillar abscess, retropharyngeal abscess, mediastinitis, and nonclostridial anaerobic myonecrosis: A case report. Clin. Infect. Dis. 16 (Suppl. 4):299–303.

Clark, W. B., M. D. Lane, J. E. Beem, S. L. Bragg, and T. T. Wheeler. 1985. Relative hydrophobicities of Actinomyces viscosus and Actinomyces naeslundii strains and their adsorption to saliva-treated hydroxyapatite. Infect. Immun. 47:730–736.

Clark, W. B., T. T. Wheeler, M. D. Lane, and J. O. Cisar. 1986. Actinomyces adsorption mediated by type-1 fimbriae. J. Dent. Res. 65:1166–1168.

Clarkson, B. H., D. Krell, J. S. Wefel, J. Crall, and F. F. Feagin. 1987. In vitro caries-like lesions production by Streptococcus mutans and Actinomyces viscosus using sucrose and starch. J. Dent. Res. 66:795–798.

Clarridge, 3rd, J. E., and Q. Zhang. 2002. Genotypic diversity of clinical Actinomyces species: phenotype, source, and disease correlation among genospecies. J. Clin. Microbiol. 40:3442–3448.

Cohn, F. 1875. Untersuchungen über Bakterien II. Beiträge zur Biologie der Pflanzen III:141–207.

Coleman, R. M., and L. K. Georg. 1969. Comparative pathogenicity of Actinomyces naeslundii and Actinomyces israelii. Appl. Microbiol. 18:427–432.

Collignon, P. J., R. Munro, and G. Morris. 1988. Susceptibility of anaerobic bacteria to antimicrobial agents. Pathology 20:48–52.

Collins, P. A., M. A. Gerencser, and J. M. Slack. 1973. Enumeration and identification of Actinomycetaceae in human dental calculus using the fluorescent antibody technique. Arch. Oral Biol. 18:145–153.

Collins, M. D., T. Pirouz, M. Goodfellow, and D. E. Minnikin. 1977. Distribution of menaquinones in actinomycetes and corynebacteria. J. Gen. Microbiol. 100:221–230.

Collins, M. D., and D. Jones. 1979a. The distribution of isoprenoid quinones in streptococci of serological group D. In: M. T. Parker (Ed.) Pathogenic Streptococci: Proceedings of the VII Symposium on Streptococci and Staphylococcal Diseases. Reedbooks. Chertsey, UK. 249–250.

Collins, M. D., and D. Jones. 1979b. The distribution of isoprenoid quinones in streptococci of serological groups D and N. J. Gen. Microbiol. 114:27–33.

Collins, M. D., and D. Jones. 1981. Distribution of isoprenoid quinone structural types in bacteria and their taxonomic implications. Microbiol. Rev. 45:316–354.

Collins, M. D., and D. Jones. 1982a. Reclassification of Corynebacterium pyogenes (Glage) in the genus Actinomyces, as Actinomyces pyogenes comb. nov. J. Gen. Microbiol. 128:901–903.

Collins, M. D., D. Jones, R. M. Kroppenstedt, and K. H. Schleifer. 1982b. Chemical studies as a guide to the classification of Corynebacterium pyogenes and "Corynebacterium haemolyticum." J. Gen. Microbiol. 128:335–341.

Collins, M. D., D. Jones, and G. M. Schofield. 1982c. Reclassification of "Corynebacterium haemolyticum" (Mac Lean, Liebow & Rosenberg) in the genus Arcanobacterium gen. nov. as Arcanobacterium haemolyticum nom. rev., comb. nov. J. Gen. Microbiol. 128:1279–1281.

Collins, M. D., and C. S. Cummins. 1986. Genus Arcanobacterium. In: P. H. A. Sneath, N. S. Mair, M. E. Sharpe, and J. G. Holt (Eds.) Bergey's Manual of Systematic Bacteriology. Williams & Wilkins. Baltimore, MD. 2:1287–1288.

Collins, M. D., S. Stubbs, J. Hommez, and L. A. Devriese. 1993. Molecular taxonomic studies of Actinomyces-like bacteria isolated from purulent lesions in pigs and description of Actinomyces hyovaginalis sp. nov. Int. J. Syst. Bacteriol. 43:471–473.

Collins, M. D., and C. Pascual. 2000a. Reclassification of Actinomyces humiferus (Gledhill and Casida) as Cellulomonas humilata nom. corrig., comb. nov. Int. J. Syst. Evol. Microbiol. 50:661–663.

Collins, M. D., L. Hoyles, S. Kalfas, G. Sundquist, T. Monsen, N. Nikolaitchouk, and E. Falsen. 2000b. Characterization of Actinomyces isolates from infected root canals of teeth: description of Actinomyces radicidentis sp. nov. J. Clin. Microbiol. 38:3399–3403.

Coman, G., C. Panzaru, and C. Dahorea. 1996. The isolation of Arcanobacterium haemolyticum from the pharyngeal exudate of children. Bacteriol. Virusol. Parazitol. Epidemiol. 41:141–144.

Conrad, S. E., D. City, J. Breivis, and M. A. Fried. 1978. Vertebral osteomyelitis, caused by Arachnia propionica and resembling actinomycosis. J. Bone Joint Surg. 60-A:549–553.

Costello, A. H., J. O. Cisar, P. E. Kolenbrander, and O. Gabriel. 1979. Neuraminidase-dependent haemagglutination of human erythrocytes by human strains of Actinomyces viscosus and Actinomyces naeslundii. Infect. Immun. 26:563–572.

Coykendall, A. L., and A. J. Munzenmaier. 1979. Deoxyribonucleic acid hybridization among strains of Actinomyces viscosus and Actinomyces naeslundii. Int. J. Syst. Bacteriol. 29:234–240.

Crawford, J. M., M. A. Taubman, and D. J. Smith. 1978. The natural history of periodontal bone loss in germ-free and gnotobiotic rats infected with periodontopathic microorganisms. J. Periodont. Res. 13:316–325.

Crawford, P. C., and W. B. Clark. 1986. Actinomyces viscosus colonization of mouse teeth. J. Dent. Res. 65:105–108.

Cristiano, L., S. Rampello, C. Noris, and V. Valota. 1996. Bacterial vaginosis: Prevalence in an Italian population of asymptomatic pregnant women and diagnostic aspects. Eur. J. Epidemiol. 12:383–390.

Cuevas, W. A., and J. G. L. Songer. 1993. Arcanobacterium haemolyticum phospholipase D is genetically and functionally similar to Corynebacterium pseudotuberculosis phospholipase. Infect. Immun. 61:4310–4316.

Cummins, C. S., and H. Harris. 1956. The chemical composition of the cell wall in some Gram-positive bacteria and its possible value as a taxonomic character. J. Gen. Microbiol. 14:583–600.

Cummins, C. S., and H. Harris. 1958. Studies on the cell-wall composition and taxonomy of Actinomycetales and related groups. J. Gen. Microbiol. 18:173–189.

Cummins, C. S., and H. Harris. 1959. Cell-wall composition in strains of Actinomyces isolated from human and bovine lesions. J. Gen. Microbiol. 21:ii.

Cummins, C. S., R. A. Lelliott, and M. Rogosa. 1974. Genus Corynebacterium. In: R. E. Buchanan and N. E. Gibbons (Eds.) Bergey's Manual of Determinative Bacteriology, 8th ed. Williams & Wilkins. Baltimore, MD. 602–617.

Curtis, A. H. 1913. A motile curved anaerobic bacillus in uterine discharges. J. Infect. Dis. 12:165–169.

Dahlen, N. A. 2004. [MD thesis]. Bonn, Germany.

Dan, M., A. Garcia, and C. von Westarp. 1984. Primary actinomycosis of the thyroid mimicking carcinoma. J. Otolaryngol. 13:109–112.

Davenport, A. A., G. R. Carter, and R. G. Schirmer. 1974. Canine actinomycosis due to Actinomyces viscosus: Report of six cases. Vet. Med. Small Anim. 69:1442–1447.

Davenport, A. A., G. R. Carter, and M. J. Patterson. 1975. Identification of Actinomyces viscosus from canine infections. J. Clin. Microbiol. 1:75–78.

De Jong, M. H., M. J. Schaeken, C. W. van den Kieboom, and J. S. van der Hoeven. 1983. Colonization of the teeth of rats by human and rodent oral strains of the bacterium Actinomyces viscosus. Arch. Oral Biol. 28:247–252.

De Jong, M. H., J. S. van der Hoeven, and J. H. van Os. 1986. Growth of microorganisms from supragingival dental plaque on saliva agar. J. Dent. Res. 65:85–88.

del Carmen Pinilla, L., and M. Ciniglio. 1983. Isolation of Actinomyces viscosus from an abscess caused by a dog bite. Rev. Latinoam. Microbiol. 25:155–156.

Delisle, A. L., R. K. Nauman, and G. E. Minah. 1978. Isolation of a bacteriophage for Actinomyces viscosus. Infect. Immun. 20:303–306.

Delisle, A. L., J. A. Donkersloot, P. E. Kolenbrander, and C. A. Tylenda. 1988. Use of a lytic bacteriophage for Actinomyces viscosus T14V as a probe for cell surface components mediating intergeneric coaggregation. Infect. Immun. 56:54–59.

Dent, V. E., and R. A. D. Williams. 1984a. Actinomyces denticolens Dent & Williams sp. nov.: a new species from the dental plaque of cattle. J. Appl. Bacteriol. 56:183–192.

Dent, V. E., and R. A. D. Williams. 1984b. Actinomyces howellii, a new species from the dental plaque of dairy cattle. Int. J. Syst. Bacteriol. 34:316–320.

Dent, V. E., and R. A. D. Williams. 1984c. Deoxyribonucleic acid reassociation between Actinomyces denticolens and other Actinomyces species from dental plaque. Int. J. Syst. Bacteriol. 34:501–502.

Dent, V. E., and R. A. D. Williams. 1985. A combined biochemical approach to the taxonomy of Gram-positive rods. In: M. Goodfellow and D. E. Minnikin (Eds.) Chemical Methods in Bacterial Systematics. Academic Press. London, UK. 341–357.

Dent, V. E., and R. A. D. Williams. 1986. Actinomyces slackii sp. nov. from dental plaque of dairy cattle. Int. J. Syst. Bacteriol. 36:392–395.

De Soete, G. 1983. On the construction of "optimal" phylogenetic trees. Z. Naturforsch. 38:156–158.

Dethy, M., P. Hantson, B. van Bosterhaut, C. Swine, and A. Sassine. 1986. Septicemia caused by Arcanobacterium haemolyticum (Corynebacterium haemolyticum) and Streptococcus milleri. Acta Clin. Belg. 41:115–118.

Dieleman, L. A., S. de Marie, R. P. Mouton, J. L. Bloem, W. E. Peters, A. J. Bos, and K. P. Schaal. 1989. Paravertebral abscess due to nondiphtheria coryneform bacteria as a complication of ingrown toenails. Infection 17:26–27.

Ding, H., and C. Lämmler. 1992. Evaluation of the API CORYNE test system for identification of Actinomyces pyogenes. Zentralbl. Veterinärmed. 39:273–276.

Distler, W., K. Ott, and A. Kröncke. 1980. Wechselwirkungen von Streptococcus mutans, Actinomyces und Veillonella in vitro—ein vereinfachtes Modell für den Kohlenhydratmetabolismus in der Plaque. Dtsch. Zahnärztl. Zeitschr. 35:548–553.

Dobinsky, S., T. Noesselt, A. Rucker, J. Maerker, and D. Mack. 1999. Three cases of Arcanobacterium haemolyticum associated with abscess formation and cellulitis. Eur. J. Clin. Microbiol. Infect. Dis. 18:804–806.

Dobson, S. R., and M. S. Edwards. 1987. Extensive Actinomyces naeslundii infection in a child. J. Clin. Microbiol. 25:1327–1329.

Donkersloot, J. A., J. O. Cisar, M. E. Wax, R. J. Harr, and B. M. Chassy. 1985. Expression of Actinomyces viscosus antigens in Escherichia coli: Cloning of structural gene (fimA) for type 2 fimbriae. J. Bacteriol. 162:1075–1078.

Dontfraid, F., and R. Ramphal. 1994. Bilateral pulmonary infiltrates in association with disseminated actinomycosis. Clin. Infect. Dis. 19:143–145.

Doyle, L., C. L. Young, S. S. Jang, and S. L. Hillier. 1991. Normal vaginal aerobic and anaerobic bacterial flora of the rhesus macaque (Macaca mulatta. J. Med. Primatol. 20:409–413.

Dropsy, G. 1978. A propos de l'article: Les vaginites à Campylobacter. Med. Mal. Infect. 8:163–165.

Drouet, E. B., M. Boude, and G. A. Denoyel. 1991. Diversity of Mobiluncus strains as demonstrated by their electrophoretic protein patterns. Zbl. Bakteriol. 276:9–15.

Dubreuil, L., J. Devos, and C. Romond. 1987. In vitro activity of imipenem against Gram-positive anaerobic bacteria. Int. J. Clin. Pharmacol. Res. 7:39–43.

Duguid, H. L. D., D. Parratt, R. Traynor, D. Taylor, I. D. Duncan, and J. Elias-Jones. 1982. Studies on uterine tract infections and the IUCD with special reference to actinomycetes. Br. J. Obstet. Gynecol. 89:32–40.

Duguid, R. 1985. Inhibition of [3 H]-thymidine uptake in human gingival fibroblasts by extracts from human dental plaque, oral bacteria of the Streptococcus and Actinomyces species. Arch. Oral Biol. 30:89–91.

Durie, E. B. 1958. A critical survey of mycological research and literature for the years 1946–1956 in Australia. Mycopathologia et Mycologia Applicata 9:80–96.

Durieux, R., and A. Dublanchet. 1980. Les "vibrions" anaérobies des leucorrhées. I: Technique d'isolement et sensibilité aux antibiotiques. Med. Mal. Infect. 10:109–115.

Durieux, R., and A. Dublanchet. 1981. Isolement de "vibrions" anaérobies stricts de la flore vaginal. Rev. Inst. Pasteur (Lyon) 14:157–162.

Edwards, D. F., T. G. Nyland, and J. P. Weigel. 1988. Thoracic, abdominal, and vertebral actinomycosis: Diagnosis and long-term therapy in 3 dogs. J. Vet. Int. Med. 2:184–191.

Eibach, H. W., A. Bolte, G. Pulverer, K. P. Schaal, and G. Küpper. 1989. Klinische Relevanz und pathognomonische Bedeutung der Aktinomyzetenbesiedlung von Intrauterinpessaren. Geburtsh. Frauenheilkd. 49:972–976.

Eibach, H. W., W. Neuhaus, W. Günther, A. Bolte, G. Pulverer, and K. P. Schaal. 1992. Clinical relevance and pathognomonic significance of actinomycotic colonization of intrauterine pessaries. Int. J. Feto-Maternal Med. 5:40–42.

Ellen, R. P., and I. B., Balcerzak-Raczkowski. 1975. Differential medium for detecting dental plaque bacteria resembling Actinomyces viscosus and Actinomyces naeslundii. J. Clin. Microbiol. 2:305–310.

Ellen, R. P., and I. B., Balcerzak-Raczkowski. 1977. Interbacterial aggregation of Actinomyces naeslundii and dental plaque streptococci. J. Periodontol. Res. 12:11–20.

Ellen, R. P., and H. Onose. 1978. pH measurements of Actinomyces viscosus colonies grown on media containing dietary carbohydrates. Arch. Oral Biol. 23:105–111.

Ellen, R. P., E. D. Fillery, K. H. Chan, and D. A. Grove. 1980. Sialidase-enhanced lectin-like mechanism for Actinomyces viscosus and Actinomyces naeslundii haemagglutination. Infect. Immun. 27:335–343.

Ellen, R. P., and D. A. Grove. 1985a. Assignment of Actinomyces viscosus and Actinomyces naeslundii strains to numerical taxonomy clusters by immunofluorescence based on antifibril antisera. J. Clin. Microbiol. 21:850–853.

Ellen, R. P., D. W. Banting, and E. D. Fillery. 1985b. Longitudinal microbiological investigation of a hospitalized population of older adults with a high root surface caries risk. J. Dent. Res. 64:1377–1381.

Ellen, R. P., I. A. Buivids, and J. R. Simardone. 1989. Actinomyces viscosus fibril antigens detected by immuno gold electron microscopy. Infect. Immun. 57:1327–1331.

Ellis, P. P., S. C. Bausor, and J. M. Fulmer. 1961. Streptothrix canaliculitis. Am. J. Ophthalmol. 52:36–43.

Emmons, C. W. 1938. The isolation of Actinomyces bovis from tonsillar granules. Publ. Health Rep. 53:1967.

Engel, D., D. van Epps, and J. Clagett. 1976. In vivo and in vitro studies on possible pathogenic mechanisms of Actinomyces viscosus. Infect. Immun. 14:548–554.

Engel, D., H. E. Schroeder, and R. C. Page. 1978. Morphological features and functional properties of human fibroblasts exposed to Actinomyces viscosus substances. Infect. Immun. 19:287–295.

Engel, D., S. Monzingo, P. Rabinovitch, J. Clagett, and R. Stone. 1984. Mitogen-induced hyperproliferation response of peripheral blood mononuclear cells from patients with severe generalized periodontitis: Lack of correlation with proportions of T cells and T-cell subsets. Clin. Immunol. Immunopathol. 30:374–386.

Ennever, J. 1960. Intracellular calcification by oral filamentous microorganisms. J. Periodontol. 31:304–307.

Ennever, J., J. J. Vogel, and J. L. Streckfuss. 1971. Synthetic medium for calcification of Bacterionema matruchotii. J. Dent. Res. 50:1327–1330.

Ennever, J., L. J. Riggan, J. J. Vogel, and B. Boyan-Salyers. 1978. Characterization of Bacterionema matruchotii calcification nucleator. J. Dent. Res. 57:637–642.

Erikson, D. 1940. Pathogenic anaerobic organisms of the Actinomyces group. Medical Research Council, Special Report Series 240:5–63.

Esteban, J., J. Zapardiel, and F. Soriano. 1994. Two cases of soft tissue infection caused by Arcanobacterium haemolyticum. Clin. Infect. Dis. 18:835–836.

Felsenstein, J. 1993. PHYLIP (Phylogenetic Inference Package) Version 3.5.1. Department of Genetics, University of Washington. Seattle, WA.

Feres, M., A. D. Haffajee, K. Allard, S. Som, and S. S. Socransky. 2001. Change in subgingival microbial profiles in adult periodontitis subjects receiving either systemically-administered amoxicillin or metronidazole. J. Clin. Periodontol. 28:597–609.

Ferrier, M. C., A. Janin Mercier, A. Meyer, J. Beytout, M. Cambon, J. Sirot, and P. Souteyrand. 1986. Actinomyces meyeri actinomycosis: A case with thoracic and tibial localization. Ann. Med. Int. Paris 137:649–651.

Fiehn, N. E., and D. Moe. 1984. Sucrase and maltase activities in supragingival dental plaque in humans of streptococcal, Actinomyces, and Lactobacillus species. Scand. J. Dent. Res. 92:97–108.

Fillery, E. D., G. H. Bowden, and J. M. Hardie. 1978. A comparison of strains of bacteria designated Actinomyces viscosus and Actinomyces naeslundii. Caries Res. 12:299–312.

Firestone, A. R., C. Graves, P. W. Caufield, and F. F. Feagin. 1987. Root surface caries subsequent to gingivectomy in rats inoculated with Streptococcus sobrinus (mutans) and Actinomyces viscosus. J. Dent. Res. 66:1583–1586.

Firestone, A. R., C. N. Graves, and F. F. Feagin. 1988. The effects of different levels of dietary sucrose on root caries subsequent to gingivectomy in conventional rats infected with Actinomyces viscosus M-100. J. Dent. Res. 67:1342–1345.

Firtel, M., and E. D. Fillery. 1988. Distribution of antigenic determinants between Actinomyces viscosus and Actinomyces naeslundii. J. Dent. Res. 67:15–20.

Firth, E. C., A. W. Kersjes, K. J. Dik, and F. M. Hagens. 1987. Haematogenous osteomyelitis in cattle. Vet. Rec. 120: 148–152.

Fligger, J. M., A. S. Waldvogel, H. Pfister, and T. W. Jungi. 1999. Expression of inducible nitric oxide synthase in spontaneous bovine bronchopneumonia. Vet. Pathol. 36:397–405.

Fohn, M. J., S. A. Lukehart, and S. L. Hillier. 1988. Production and characterization of monoclonal antibodies to Mobiluncus species. J. Clin. Microbiol. 26:2598–2603.

Ford, J. G., R. P. Yeatts, and L. B. Givner. 1995. Orbital cellulitis, subperiostal abscess, sinusitis, and septicemia caused by Arcanobacterium haemolyticum. Am. J. Ophthalmol. 120:261–262.

Forgan-Smith, J. R., P. Mowat, and G. Strutton. 1989. A report of actinomycosis involving the lung and liver. Med. J. Australia 150:153–155.

Fortner, J. 1928. Ein einfaches Plattenverfahren zur Züchtung strenger Anaerobier. Zentralbl. Bakteriol. Parasitenkd. Infektionskr., I. Abt. Orig. 108:155–159.

Franke, F. 1973. Untersuchungen zur Ätiologie der Gesäugeaktinomykose des Schweins. Zbl. Bakteriol. Parasitenkd. Infektionskr. Hyg. I. Abt. Orig. A 223:111–124.

Freland, C., W. Massoubre, J. M. Horeau, J. Caillon, and H. W. Drugeon. 1987. Actinomycosis of the gall bladder due to Actinomyces naeslundii. J. Infect. 15:251–257.

Fritsche, D. 1964. Die Benzol- und Toluolresistenz des Actinomyces israelii, ein Hilfsmittel für die Strahlenpilzdiagnostik. Zbl. Bakteriol. Parasitenkd. Infektionskr. Hyg., I. Abt. Orig. 194:241–244.

Funke, G., G. Martinetti Lucchini, G. E. Pfyffer, M. Marchiani, and A. von Graevenitz. 1993. Characteristics of CDC Group 1 and Group 1-like coryneform bacteria isolated from clinical specimens. J. Clin. Microbiol. 31:2907–2912.

Funke, G., S. Stubbs, A. von Graevenitz, and M. D. Collins. 1994. Assignment of human-derived CDC Group 1 coryneform bacteria and CDC Group 1-like coryneform bacteria to the genus Actinomyces as Actinomyces neuii subsp. neuii sp. nov., subsp. nov., and Actinomyces neuii subsp. anitratus subsp. nov. Int. J. Syst. Bacteriol. 44:167–171.

Funke, G., and A. von Graevenitz. 1995a. Infections due to Actinomyces neuii (former "CDC Coryneform Group 1" bacteria. Infection 23:73–75.

Funke, G., C. Pascual Ramos, J. F. Fernandez-Garayzabal, N. Weiss, and M. D. Collinsb. 1995. Description of human-derived Centers for Disease Control Coryneform Group 2 bacteria as Actinomyces bernardiae sp. nov. Int. J. Syst. Bacteriol. 45:57–60.

Funke G., N. Alvarez, C. Pascual Ramos, E. Falsen, E. Akervall, L. Sabbe, L. Scjouls, N. Weiss, and M. D. Collins. 1997a. Actinomyces europaeus sp. nov., isolated from human clinical specimens. Int. J. Syst. Bacteriol. 47:687–692.

Funke, G., F. N. R. Renaud, J. Freney, and P. Riegel. 1997b. Multicenter evaluation of the updated and extended API (RAPID) Coryne Database 2.0. J. Clin. Microbiol. 35:3122–3126.

Fure, S., M. Romaniec, C. G. Emilson, and B. Krasse. 1987. Proportions of Streptococcus mutans, lactobacilli and Actinomyces spp. in root surface plaque. Scand. J. Dent. Res. 95:119–123.

Gaffar, A., E. J. Coleman, A. Esposito, H. Niles, and R. J. Gibbons. 1985. Non-bactericidal approach to reduce colonization on plaque microflora on teeth in vitro and in vivo. J. Pharmacol. Sci. 74:1228–1232.

Gahrn-Hansen, B., and W. Frederiksen. 1992. Human infections with Actinomyces pyogenes (Corynebacterium pyogenes. Diagn. Microbiol. Infect. Dis. 15:349–354.

Gallagher, I. H., E. I. Pearce, and E. M. Hancock. 1984. The ureolytic microflora of immature dental plaque before and after rinsing with a urea-based mineralizing solution. J. Dent. Res. 63:1037–1039.

Garcia-Corbeira, P., and J. Esteban-Moreno. 1994. Liver abscess due to Actinomyces meyeri. Clin. Infect. Dis. 18:491–492.

Garcia Ramos, E., N. Kichick Tello, J. Rozco, R. Caballero, and P. Cardona Carillo. 1984. Isolation of Actinomyces species and other microorganisms from 140 hypertrophic tonsils in children. Rev. Latinoam. Microbiol. 26:251–260.

Gardner, H. L., and C. D. Dukes. 1955. Haemophilus vaginalis vaginitis. Am. J. Obstet. Gynecol. 69:962–976.

Garduno, E., M. Rebollo, M. A. Asencio, J. Carro, J. M. Pascasio, and J. Blanco. 2000. Splenic abscesses caused by Actinomyces meyeri in a patient with autoimmune hepatitis. Diagn. Microbiol. Infect. Dis. 37:213–214.

Garelick, J. M., A. J. Khodabakhsh, and R. G. Josephberg. 2002. Acute postoperative endophthalmitis caused by Actinomyces neuii. Am. J. Ophthalmol. 133:145–147.

Garlind, A., C. Påhlson, and U. Forsum. 1989. Phenotypic complexity in Mobiluncus. Acta Pathol. Microbiol. Immunol. Scand. 97:38–42.

Gasperini, G. 1892. Ricerche morfologiche e biologiche sul genere Actinomyces Harz come contributo allo studio delle relative micosi. Ann. Ist. d'Igiene, Universitá Roma 2:167–231.

Gaston, D. A., and S. M. Zurowski. 1996. Arcanobacterium haemolyticum pharyngitis and exanthem: Three case reports and literature review. Arch. Dermatol. 132:61–64.

Gatti, M., R. Aschbacher, C. Cimmino, and R. Valentini. 1997. Antigenic profiles for the differentiation of Mobiluncus curtisii and Mobiluncus mulieris by immunoblotting technique. New Microbiol. 20:247–252.

Gatti, J. J., J. M. Dobeck, C. Smith, R. R. White, S. S. Socransky, and Z. Skobe. 2000. Bacteria of asymptomatic periradicular endodontic lesions identified by DNA-DNA hybridization. Endod. Dent. Traumatol. 16:197–204.

Gavin, S. E., R. B. Leonard, A. M. Briselden, and M. B. Coyle. 1992. Evaluation of the rapid CORYNE identification system for Corynebacterium species and other coryneforms. J. Clin. Microbiol. 30:1692–1695.

Gayraud, A., C. Grosieux-Dauger, A. Durlach, V. Salmon-Ehr, A. Elia, E. Grosshans, and P. Bernard. 2000. Cutaneous actinomycosis in the perianal area and buttocks. Ann. Dermatol. Venereol. 127:393–396.

Georg, L. K., G. W. Robertstad, and S. A. Brinkman. 1964. Identification of species of Actinomyces. J. Bacteriol. 88:477–490.

Georg, L. K., L. Pine, and M. A. Gerencser. 1969. Actinomyces viscosus comb. nov.: A catalase positive, facultative member of the genus Actinomyces. Int. J. Syst. Bacteriol. 19:291–293.

Georg, L. K., and R. M. Coleman. 1970. Comparative pathogenicity of various Actinomyces species. *In:* H. Prauser (Ed.) The Actinomycetales: The Jena International Symposium on Taxonomy. Jena Germany. VEB Gustav Fischer Verlag, 35–45.

Georg, L. K., J. M. Brown, H. J. Baker, and G. H. Cassell. 1972. Actinomyces viscosus as an agent of actinomycosis in the dog. Am. J. Vet. Res. 33:1457–1470.

Georgijevic, A., S. Cjukic-Ivancevic, and M. Bujko. 2000. Bacterial vaginosis: Epidemiology and risk factors. Srp. Arh. Celok. Lek. 128:29–33.

Gerencser, M. A., and J. M. Slack. 1967. Isolation and characterization of Actinomyces propionicus. J. Bacteriol. 94:109–115.

Gerencser, M. A., and J. M. Slack. 1969. Identification of human strains of Actinomyces viscosus. Appl. Microbiol. 18:80–87.

Gerencser, M. A. 1979. The application of fluorescent antibody techniques to the identification of Actinomyces and Arachnia. *In:* T. Bergan and J. R. Norris (Eds.) Methods in Microbiology. Academic Press. London, UK. 13:287–321.

Gibbons, R. J., and I. Etherden. 1983. Comparative hydrophobicities of oral bacteria and their adherence to salivary pellicles. Infect. Immun. 41:1190–1196.

Gibbons, R. J., D. I. Hay, J. O. Cisar, and W. B. Clark. 1988. Adsorbed salivary proline-rich protein 1 and statherin: Receptors for type 1 fimbriae of Actinomyces viscosus T14V-J1 on apatitic surfaces. Infect. Immun. 56:2990–2993.

Gibbons, R. J. 1989. Bacterial adhesion to oral tissues: A model for infectious diseases. J. Dent. Res. 68:750–760.

Glage, F. 1903. Über den Bazillus pyogenes suis Grips, den Bazillus pyogenes bovis Künnemann und den bakteriologischen Befund bei den chronischen, abszedierenden Euterentzündungen der Milchkühe. Zeitschr. Fleisch Milch Hyg. 13:166–175.

Gledhill, W. E., and L. E. Casida. 1969. Predominant catalase-negative soil bacteria. II: Occurrence and characterization of Actinomyces humiferus, sp. n. Appl. Microbiol. 18:114–121.

Goda, F. F., N. A. Wassef, A. A. Ibrahim, and S. Roushdy. 1986. Studies on microorganisms secured from different organs of slaughtered sheep with special reference to the microbial load in certain muscles. Beitr. trop. Landwirtsch. Veterinärmed. 24:85–95.

Grässer, R. 1957. Vergleichende Untersuchungen an Actinomyceten von Mensch, Rind und Schwein [thesis]. Leipzig, Germany.

Gresser, U., V. Preac-Mursic, H. Dörfler, F. A. Spengel, and N. Zöllner. 1988. Aktinomykose der Haut bei HIV-Infektion. Klin. Wochenschr. 66:651–653.

Greub, G., and D. Raoult. 2002. "Actinobaculum massiliae," a new species causing chronic urinary tract infection. J. Clin. Microbiol. 40:3938–3941.

Grice, G. C., and S. Hafiz. 1983. Actinomyces in the female genital tract: A preliminary report. Br. J. Ven. Dis. 59: 317–319.

Grüner, O. P. N. 1969. Actinomyces in tonsillar tissue: A histological study of tonsillectomy material. Acta Pathol. Microbiol. Scand. 76:239–244.

Guerin-Faublee, V., S. Karray, B. Tilly, and Y. Richard. 1992. Actinomyces pyogenes: Conventional and API system

bacteriologic study of 103 strains isolated from ruminants. Ann. Rech. Vet. 23:151–160.

Guggenheim, H., S. Shapiro, R. Gmür, and B. Guggenheim. 2001. Spatial arrangements and associative behavior of species in an in vitro oral biofilm model. Appl. Environ. Microbiol. 67:1343–1350.

Gulbahar, M. Y., and K. Gurtuk. 2002. Pyothorax associated with a Mycoplasma sp. and Arcanobacterium pyogenes in a kitten. Australian Vet. J. 80:344–345.

Gupta, P. K., D. H. Hollander, and J. K. Frost. 1976. Actinomycetes in cervicovaginal smears: An association with IUD usage. Acta Cytol. 20:295–297.

Gupta, P. K., Y. S. Erozan, and J. K. Frost. 1978. Actinomycetes and the IUD: An update. Acta Cytol. 22:281–282.

Gupta, D. S., M. K. Gupta, and N. G. Naidu. 1986. Mandibular osteomyelitis caused by Actinomyces israelii. Report of a case. J. Maxillofac. Surg. 14:291–293.

Gusberti, F., A., P. Sampathkumar, D. E. Siegrist, and N. P. Lang. 1988. Microbiological and clinical effects of chlorhexidine digluconate and hydrogen peroxide mouth rinses on developing plaque and gingivitis. J. Clin. Periodontol. 15:60–67.

Haber, J., and C. Grinnell. 1989. Analysis of the serum antibody response to type 1 and type 2 fimbriae in mice immunized with Actinomyces viscosus T14V. J. Periodont. Res. 24:81–87.

Hager, W. D., B. Douglas, B. Majmudar, Z. M. Naib, O. J. Williams, C. Ramsey, and J. Thomas. 1979. Pelvic colonization with Actinomyces in women using intrauterine contraceptive devices. Am. J. Obstet. Gynecol. 133:60–64.

Halfpap, L. M., D. A. Brown, J. A. Clagett, and D. C. Birdsell. 1985. The mitogenicity for murine splenocytes of specific surface components of the oral periodontopathic bacterium, Actinomyces viscosus. Arch. Oral Biol. 30:661–666.

Hall, R. E., G. Bender, and R. E. Marquis. 1987. Inhibitory and cidal antimicrobial actions of electrically generated silver ions. J. Oral Maxillofacial Surg. 45:779–784.

Hall, V., G. L. O'Neill, J. T. Magee, and B. I. Duerden. 1999. Development of amplified 16S ribosomal DNA restriction analysis for identification of Actinomyces species and comparison with pyrolysis-mass spectrometry and conventional biochemical tests. J. Clin. Microbiol. 37:2255–2261.

Hall, V., M. D. Collins, R. Hutson, E. Falsen, and B. I. Duerden. 2002. Actinomyces cardiffensis sp. nov. from human clinical sources. J. Clin. Microbiol. 40:3427–3431.

Hall, V., M. D. Collins, R. Hutson, E. Inganäs, E. Falsen, and B. I. Duerden. 2003a. Actinomyces vaccimaxillae sp. nov., from the jaw of a cow. Int. J. Syst. Evol. Microbiol. 53:603–606.

Hall, V., M. D. Collins, P. A. Lawson, E. Falsen, and B. I. Duerden. 2003b. Actinomyces nasicola sp. nov., isolated from a human nose. Int. J. Syst. Evol. Microbiol. 53: 1445–1448.

Hall, V., M. D. Collins, R. A. Hutson, E. Falsen, E. Inganäs, and B. I. Duerden. 2003c. Actinobaculum urinale sp. nov., from human urine. Int. J. Syst. Evol. Microbiol. 53:679–682.

Hall, V., M. D. Collins, R. A. Hutson, E. Inganäs, E. Falsen, and B. I. Duerden. 2003d. Actinomyces oricola sp. nov., from a human dental abscess. Int. J. Syst. Evol. Microbiol. 53:1515–1518.

Hall, V., M. D. Collins, P. A. Lawson, R. A. Hutson, E. Falsen, E. Inganäs, and B. Duerden. 2003e. Characterization of some Actinomyces-like isolates from human clinical sources: Description of Varibaculum cambriensis gen. nov., sp. nov. J. Clin. Microbiol. 41:640–644.

Hallén, A., C. Påhlson, and U. Forsum. 1987. Bacterial vaginosis in women attending STD clinic: Diagnostic criteria and prevalence of Mobiluncus spp. Genitourin. Med. 63:386–389.

Hallén, A., C. Påhlson, and U. Forsum. 1988. Rectal occurrence of Mobiluncus species. Genitourin. Med. 64:273–275.

Hamilton, I. R., and D. C. Ellwood. 1983. Carbohydrate metabolism by Actinomyces viscosus growing in continuous culture. Infect. Immun. 42:19–26.

Hammann, R., A. Kronibus, A. Viebahn, and H. Brandis. 1984. Falcivibrio grandis gen. nov., sp. nov., and Falcivibrio vaginalis gen. nov., sp. nov., a new genus and species to accommodate anaerobic motile curved rods formerly described as "Vibrio mulieris" (Prévot 1940) Breed et al. 1948. Syst. Appl. Microbiol. 5:81–96.

Hanf, U., and G. Hanf. 1955. Ein Beitrag zum Infektionsmodus der weiblichen Genitalaktinomykose. Geburtsh. Frauenheilkd. 15:366–373.

Happonen, R. P., M. Viander, L. Pelliniemi, and K. Aitasalo. 1983. Actinomyces israelii in osteoradionecrosis of the jaw: Histopathologic and immunochemical study of five cases. Oral Surg. Oral Med. Oral Pathol. 55:580–588.

Happonen, R. P., E. Soderling, M. Viander, L. Linko Kettunen, and L. J. Pelliniemi. 1985. Immunocytochemical demonstration of Actinomyces species and Arachnia propionica in periapical infections. J. Oral Pathol. 14:405–413.

Happonen, R. P., P. Arstila, M. Viander, E. Soderling, and M. Viljanen. 1987. Comparison of polyclonal and monoclonal antibodies to Actinomyces and Arachnia species. Scand. J. Dent. Res. 95:136–143.

Hardie, E. M., and J. A. Barsanti. 1982. Treatment of canine actinomycosis. J. Am. Vet. Med. Assoc. 180:537–541.

Harrington, B. J. 1966. A numerical taxonomical study of some corynebacteria and related organisms. J. Gen. Microbiol. 45:31–40.

Harsch, I. A., J. Benninger, G. Niedobitek, G. Schindler, H. T. Schneider, E. G. Hahn, and G. Nusko. 2001. Abdominal actinomycosis: Complication of endoscopic stenting in chronic pancreatitis? Endoscopy 33:1065–1069.

Harz, C. O. 1877. Actinomyces bovis ein neuer Schimmel in den Geweben des Rindes. Deutsche Zeitschrift für Thiermedizin 1877–1878 5:125–140.

Hauduroy, P., G. Ehringer, A. Urbain, G. Guillot, and J. Magrou. 1937. Dictionaire des bactéries pathogènes. Masson et Co. Paris, France.

Heeb, M. J., A. M. Marini, and O. Gabriel. 1985. Factors affecting binding of galacto ligands to Actinomyces viscosus lectin. Infect. Immun. 45:61–67.

Heer, M., A. Hany, R. Rauch, and H. Sulser. 1986. Aktinomykose des Colons: Klinische, endoskopische, serologische und therapeutische Gesichtspunkte. Schweiz. Med. Wochenschr. 116:514–518.

Heinrich, S., and H. Korth. 1967. Zur Nährbodenfrage in der Routinediagnostik der Aktinomykose: Ersatz unsicherer biologischer Substrate durch ein standardisiertes Medium. In: H.-J. Heite (Ed.) Krankheiten durch Aktinomyceten und verwandte Erreger. Springer-Verlag. Berlin, Germany. 16–20.

Hemmes, G. D. 1963. Einige bevindingen over actinomycose. Nederl. Tijdschr. Geneeskde. 107:193.

Henningsen, M., P. Roggentin, and R. Schauer. 1991. Cloning, sequencing and expression of the sialidase gene from Actinomyces viscosus DSM 43798. Biol. Chem. Hoppe Seyler 372:1065–1072.

Hernandez, M. D. 1985. Experimental induction of periodontal disease in hamsters using Actinomyces viscosus isolated from humans and a diet enriched with brown sugar. Acta Clin. Odontol. 8:35–55.

Hesselink, J. W., and J. G. van den Tweel. 1990. Hypertrophic osteopathy in a dog with a chronic lung abscess. J. Am. Vet. Med. Assoc. 196:760–762.

Hill, P. E., K. W. Knox, R. G. Schamschula, and M. Tabua. 1977. The identification and enumeration of Actinomyces from plaque of New Guinea indigenes. Caries Res. 11:327–335.

Hillerton, J. E., and A. J. Bramley. 1989. Infection following challenge of the lactating and dry udder of dairy cows with Actinomyces pyogenes and Peptostreptococcus indolicus. Br. Vet. J. 145:148–158.

Hillier, S. L., L. K. Rabe, C. H. Muller, P. Zarutskie, F. B. Kuzan, and M. A. Stenchever. 1990. Relationship of bacteriologic characteristics to semen indices in men attending an infertility clinic. Obstet. Gynecol. 75:800–804.

Hillier, S. L., C. W. Critchlow, C. E. Stevens, M. C. Roberts, P. Wolner-Hanssen, D. A. Eschenbach, and K. K. Holmes. 1991. Microbiological, epidemiological and clinical correlates of vaginal colonisation by Mobiluncus species. Genitourin. Med. 67:26–31.

Hjelm, E., A. Hallén, U. Forsum, and J. Wallin. 1982. Motile anaerobic curved rods in non-specific vaginitis. Eur. J. Sex. Trans. Dis. 1:9–14.

Holdeman, L. V., E. P. Cato, and W. E. C. Moore. 1977. V.P.I. Anaerobe Laboratory Manual, 4th ed. Southern Printing Co. Blacksburg, VI.

Holm, P. 1950. Studies on the aetiology of human actinomycosis. I: The "other microbes" of actinomycosis and their importance. Acta Pathol. Microbiol. Scand. 27:736–751.

Holm, P. 1951. Studies on the aetiology of human actinomycosis. II: Do the "other microbes" of actinomycosis possess virulence? Acta Pathol. Microbiol. Scand. 28:391–406.

Holmberg, K., and U. Forsum. 1973a. Identification of Actinomyces, Arachnia, Bacterionema, Rothia, and Propionibacterium species by defined immunofluorescence. Appl. Microbiol. 25:834–843.

Holmberg, K., and H. O. Hallander. 1973b. Numerical taxonomy and laboratory identification of Bacterionema matruchotii, Rothia dentocariosa, Actinomyces naeslundii, Actinomyces viscosus, and some related bacteria. J. Gen. Microbiol. 76:43–63.

Holmberg, K., and C.-E. Nord. 1975a. Numerical taxonomy and laboratory identification of Actinomyces and Arachnia and some related bacteria. J. Gen. Microbiol. 91:17–44.

Holmberg, K., C.-E. Nord, and T. Wadström. 1975b. Serological studies of Actinomyces israelii by crossed immunoelectrophoresis: Standard antigen-antibody system for A. israelii. Infect. Immun. 12:387–397.

Holmberg, K. 1981. Immunodiagnosis of human actinomycosis. In: K. P. Schaal and G. Pulverer (Eds.) Actinomycetes: Proceedings of the Fourth International Symposium on Actinomycete Biology, Cologne 1979. Gustav Fischer Verlag. Stuttgart, Germany. 259–261.

Holst, E., A. Skarin, and P.-A. Mårdh. 1982. Characteristics of anaerobic comma-shaped bacteria recovered from the female genital tract. Eur. J. Clin. Microbiol. 1:310–316.

Holst, E., H. Hofmann, and P.-A. Mårdh. 1984a. Anaerobic curved rods in genital samples of women: Performance of different selective media, comparison of detection by microscopy and culture studies, and recovery from different sampling sites. *In:* P.-A. Mårdh and D. Taylor-Robinson (Eds.) Bacterial Vaginosis. Almqvist and Wiksell International. Stockholm, The Netherlands. 117–124.

Holst, E., P.-A. Mårdh, and I. Thelin. 1984b. Recovery of anaerobic curved rods and Gardnerella vaginalis from the urethra of men, including male heterosexual consorts of female carriers. *In:* P.-A. Mårdh and D. Taylor-Robinson (Eds.) Bacterial Vaginosis. Almqvist and Wiksell International. Stockholm, The Netherlands. 173–178.

Holst, E., L. Svensson, A. Skarin, L. Westrom, and P.-A. Mårdh. 1984c. Vaginal colonization with Gardnerella vaginalis and anaerobic curved rods. *In:* P.-A. Mårdh and D. Taylor-Robinson (Eds.) Bacterial Vaginosis. Almqvist and Wiksell International. Stockholm, Sweden. 147–152.

Holst, E. 1990. Reservoir of four organisms associated with bacterial vaginosis suggests lack of sexual transmission. J. Clin. Microbiol. 28:2035–2039.

Holst, E., A. R. Goffeng, and B. Andersch. 1994. Bacterial vaginosis and vaginal microorganisms in idiopathic premature labor and association with pregnancy outcome. J. Clin. Microbiol. 32:176–186.

Hooi, L. N., B. S. Na, and K. S. Sin. 1992. A case of empyema thoracis caused by actinomycosis. Med. J. Malaysia 47:311–315.

Hoshino, E. 1985. Predominant obligate anaerobes in human caries dentin. J. Dent. Res. 64:1195–1198.

Hotchi, M., and J. Schwarz. 1972. Characterization of actinomycotic granules by architecture and staining methods. Arch. Pathol. 93:392–400.

Howell Jr., A., and L. Pine. 1956. Studies on the growth of species of Actinomyces. I: Cultivation in a synthetic medium with starch. J. Bacteriol. 71:47–53.

Howell Jr., A., W. C. Murphy, 3rd, F. Paul, and R. M. Stephan. 1959. Oral strains of Actinomyces. J. Bacteriol. 78:82–95.

Howell Jr., A., R. M. Stephan, and F. Paul. 1962. Prevalence of Actinomyces israelii, A. naeslundii, Bacterionema matruchotii, and Candida albicans in selected areas of the oral cavity and saliva. J. Dent. Res. 41:1050–1059.

Howell Jr., A. 1963a. A filamentous microorganism isolated from periodontal plaque in hamsters. I: Isolation, morphology, and general cultural characteristics. Sabouraudia 3:81–92.

Howell Jr., A., and H. V. Jordan. 1963b. A filamentous microorganism isolated from periodontal plaque in hamsters. II: Physiological and biochemical characteristics. Sabouraudia 3:93–105.

Howell, A., H. V. Jordan, L. K. Georg, and L. Pine. 1965. Odontomyces viscosus gen. nov. spec. nov.: A filamentous microorganism isolated from periodontal plaque in hamsters. Sabouraudia 4:65–67.

Howell, A., and H. V. Jordan. 1967. Production of an extracellular levan by Odontomyces viscosus. Arch. Oral. Biol. 12:571–573.

Hoyles, L., E. Falsen, G. Foster, C. Pascual Ramos, C. Greko, and M. D. Collins. 2000. Actinomyces canis sp. nov., isolated from dogs. Int. J. Syst. Evol. Microbiol. 50:1547–1551.

Hoyles, L., E. Falsen, G. Holmström, A. Persson, B. Sjöden, and M. D. Collins. 2001a. Actinomyces suimastitidis sp. nov., isolated from pig mastitis. Int. J. Syst. Evol. Microbiol. 51:1323–1326.

Hoyles, L., E. Falsen, C. Pascual Ramos, B. Sjöden, G. Foster, D. Henserson, and M. D. Collins. 2001b. Actinomyces catuli sp. nov., from dogs. Int. J. Syst. Evol. Microbiol. 51:679–682.

Hoyles, L., C. Pascual Ramos, E. Falsen, G. Foster, J. M. Grainger, and M. D. Collins. 2001c. Actinomyces marimammalium sp. nov., from marine mammals. Int. J. Syst. Evol. Microbiol. 51:151–156.

Hoyles, L., E. Falsen, G. Foster, and M. D. Collins. 2002a. Actinomyces coleocanis sp. nov., from the vagina of a dog. Int. J. Syst. Evol. Microbiol. 52:1201–1203.

Hoyles, L., E. Falsen, G. Foster, F. Rogerson, and M. D. Collins. 2002b. Arcanobacterium hippocoleae sp. nov., from the vagina of a horse. Int. J. Syst. Evol. Microbiol. 52:617–619.

Huang, K. L., S. M. Beutler, and C. Wang. 1998. Endocarditis due to Actinomyces meyeri. Clin. Infect. Dis. 27:909–910.

Ibanez-Nolla, J., J. Carratala, J. Cucurull, A. Corbella, V. Curull, J. Linares, and F. Gudiol. 1993. Thoracic actinomycosis. Enferm. Infect. Microbiol. 11:433–436.

Igarashi, T., M. Takahashi, A. Yamamoto, Y. Etoh, and K. Takamori. 1987. Purification and characterization of levanase from Actinomyces viscosus ATCC 19246. Infect. Immun. 55:3001–3005.

Ikegami, M., N. Hashimoto, T. Kaidoh, T. Sekizaki, and S. Takeuchi. 2000. Genetic and biochemical properties of a hemolysin (pyolysin) produced by a swine isolate of Arcanobacterium (Actinomyces) pyogenes. Microbiol. Immunol. 44:1–7.

Imai, S., and H. Kuramitsu. 1983. Chemical characterization of extracellular polysaccharides produced by Actinomyces viscosus. Infect. Immun. 39:1059–1066.

Imaizumi, K., A. Serizawa, N. Hashimoto, T. Kaidoh, and S. Takeuchi. 2001. Analysis of the functional domains of Arcanobacterium pyogenes pyolysin using monoclonal antibodies. Vet. Microbiol. 81:235–242.

Ison, C. A., B. Kolator, J. H. Reid, E. Dermott, J. Clark, and C. S. Easmon. 1989. Characterisation of monoclonal antibodies for detection of Mobiluncus spp. in genital specimens. J. Med. Microbiol. 30:129–136.

Israel, J. 1878. Neue Beobachtungen auf dem Gebiete der Mykosen des Menschen. Archiv für Pathologische Anatomie und Physiologie und für Klinische Medicin 74:15–53.

Jackman, P. J. H. 1985. Bacterial taxonomy based on electrophoretic whole-cell protein patterns. *In:* M. Goodfellow and D. E. Minnikin (Eds.) Chemical Methods in Bacterial Systematics. Academic Press. London, UK. 115–129.

Jarvis, D. 1985. Isolation and identification of actinomycetes from women using intrauterine contraceptive devices. J. Infect. 10:121–125.

Jauh-Hsun, C., T. Vinh, J. K. Davies, and D. Figdor. 1999. Molecular approach to the differentiation of Actinomycetes species. Oral Microbiol. Immunol. 14:250–256.

Jin, S. L., H. P. Lee, J. I. Kim, J. Y. Chin, S. J. Choi, M. Joo, and H. K. Yum. 2000. A case of endobronchial actinomycosis. Korean J. Intern. Med. 15:240–244.

Johansen, C., P. Falholt, and L. Gram. 1997. Enzymatic removal and disinfection of bacterial biofilms. Appl. Environ. Microbiol. 63:3724–3728.

Johnson, J. L., and C. S. Cummins. 1972. Cell wall composition and deoxyribonucleic acid similarities among the anaerobic coryneforms, classical propionibacteria, and strains of Arachnia propionica. J. Bacteriol. 109:1047–1066.

Johnson, K. A., G. R. Lomas, and A. K. Wood. 1984. Osteomyelitis in dogs and cats caused by anaerobic bacteria. Australian Vet. J. 61:57–61.

Johnson, J. L., L. V. H. Moore, B. Kaneko, and W. E. C. Moore. 1990. Actinomyces georgiae sp. nov., Actinomyces gerencseriae sp. nov., designation of two genospecies of Actinomyces naeslundii, and inclusion of A. naeslundii serotypes II and III and Actinomyces viscosus serotype II in A. naeslundii genospecies 2. Int. J. Syst. Bacteriol. 40:273–286.

Jones, D. 1975. A numerical taxonomic study of coryneform and related bacteria. J. Gen. Microbiol. 87:52–96.

Jones, D. B., and N. M. Robinson. 1977. Anaerobic ocular infections. Treat. Am. Acad. Ophthalmol. Otolaryngol. 83:309–331.

Jones, J. B., W. Kaplan, J. M. Brown, and W. White. 1983. Studies of cervicovaginal smears for the presence of actinomycetes. Mycopathology 83:53–55.

Jones, B. M., I. Geary, A. B. Alawattiegama, G. R. Kinghorn, and B. I. Duerden. 1985. In-vitro and in-vivo activity of metronidazole against Gardnerella vaginalis, Bacteroides spp. and Mobiluncus spp. in bacterial vaginosis. J. Antimicrob. Chemother. 16:189–197.

Jones, G. F., and G. E. Ward. 1989. Cause, occurrence, and clinical signs of mastitis and anorexia in cows in a Wisconsin study. J. Am. Vet. Med. Assoc. 195:1108–1113.

Jones, J. E. T. 1992. Eubacterium (Corynebacterium) suis. In: A. D. Leman, E. Straw (Eds.) Diseases of Swine, 7th ed. Iowa State University Press. Ames, IA.

Jordan, H. V., and P. H. Keyes. 1964. Aerobic, Gram-positive, filamentous bacteria as etiological agents of experimental periodontal disease in hamsters. Arch. Oral Biol. 9:401–414.

Jordan, H. V., and P. H. Keyes. 1965a. Studies on the bacteriology of hamster periodontal disease. Am. J. Pathol. 46:843–857.

Jordan, H. V., R. J. Fitzgerald, and H. R. Stanley. 1965b. Plaque formation and periodontal pathology in gnotobiotic rats infected with an oral actinomycete. Am. J. Pathol. 47:1157–1167.

Jordan, H. V., and B. F. Hammond. 1972. Filamentous bacteria isolated from root surface caries. Arch. Oral Biol. 17:1–12.

Jordan, H. V., and D. L. Sumney. 1973. Root surface caries: Review of the literature and significance of the problem. J. Periodontol. 44:158–163.

Jordan, H. V., D. M. Kelly, and J. D. Heeley. 1984. Enhancement of experimental actinomycosis in mice by Eikenella corrodens. Infect. Immun. 46:367–371.

Jost, B. H., S. J. Billington, and J. G. Songer. 1997. Electroporation-mediated transformation of Arcanobacterium (Actinomyces) pyogenes. Plasmid 38:135–140.

Jost, B. H., J. G. Songer, and S. J. Billington. 1999. An Arcanobacterium (Actinomyces) pyogenes mutant deficient in production of the pore-forming cytolysin pyolysin has reduced virulence. Infect. Immun. 67:1723–1728.

Jost, B. H., J. G. Songer, and S. J. Billington. 2001. Cloning, expression, and characterization of a neuraminidase gene from Arcanobacterium haemolyticum. Infect. Immun. 69:4430–4437.

Jost, B. H., J. G. Songer, and S. J. Billington. 2002. Identification of a second Arcanobacterium pyogenes neuraminidase and involvement of neuraminidase activity in host cell adhesion. Infect. Immun. 70:1106–1112.

Jousimies-Somer, H., S. Savolainen, A. Makitie, and J. Ylikoski. 1993. Bacteriologic findings in peritonsillar abscesses in young adults. Clin. Infect. Dis. 16 (Suppl. 4): 292–298.

Jurankova, J., and M. Votava. 2001. Detection of Arcanobacterium haemolyticum in primoculture using the reverse CAMP test. Epidemiol. Mikrobiol. Immunol. 50:71–73.

Justesen, T., O. H. Nielsen, I. E. Jacobsen, J. Lave, and S. N. Rasmussen. 1984. The normal cultivable microflora in upper jejunal fluid in healthy adults. Scand. J. Gastroenterol. 19:279–282.

Kalfas, S., D. Figdor, and G. Sundqvist. 2001. A new bacterial species associated with failed endodontic treatment: Identification and description of Actinomyces radicidentis. Oral Surg. Oral Med. Oral Pathol. Oral Radiol. Endod. 92:208–214.

Kametaka, S., T. Miyazaki, Y. Inoue, S. Hayashi, A. Takamori, Y. Miyake, and H. Suginaka. 1989. The effect of ofloxacin on experimental periodontitis in hamsters infected with Actinomyces viscosus ATCC 15987. J. Periodontol. 60:285–291.

Kammoun, K., V. Garrigue, C. Bouloux, G. Chong, P. Baldet, and S. Mourad. 2001. Actinomycosis after renal transplantation: Apropos of 1 case and review of the literature. Nephrologie 22:21–23.

Kämpfer, P. 1992. Differentiation of Corynebacterium spp., Listeria spp., and related organisms by using fluorogenic substrates. J. Clin. Microbiol. 30:1067–1071.

Kaneko, T., T. Kubota, M. Takada, and T. Oguri. 1992. Colonization rates of Mobiluncus spp. in female lower genital tract and its relationship with bacterial vaginosis. Kansenshogaku Zasshi 66:382–389.

Karetzky, M. S., and J. W. Garvey. 1974. Empyema due to Actinomyces naeslundii. Chest 65:229–230.

Kaufman, H. W., J. J. Pollock, and A. J. Gwinnett. 1988. Microbial caries induction in the root of human teeth in vitro. Arch. Oral Biol. 33:499–503.

Keir, H. A., and J. W. Porteus. 1962. The amino acid requirements of a single strain of Actinomyces israelii growing in a chemically defined medium. J. Gen. Microbiol. 28:193–201.

Kersters, K., and J. De Ley. 1975. Identification and grouping of bacteria by numerical analysis of their electrophoretic protein patterns. J. Gen. Microbiol. 87:333–342.

Kersters, K., and J. De Ley. 1980. Classification and identification of bacteria by electrophoresis of their proteins. In: M. Goodfellow and R. G. Board (Eds.) Microbial Classification and Identification. Academic Press. London, UK. 273–297.

Kiel, R. A., and J. M. Tanzer. 1977a. Regulation of invertase of Actinomyces viscosus. Infect. Immun. 17:510–512.

Kiel, R. A., J. M. Tanzer, and F. N. Woodiel. 1977b. Identification, separation, and preliminary characterization of invertase and β-galactosidase in Actinomyces viscosus. Infect. Immun. 16:81–87.

Kilian, M. 1987. Rapid identification of Actinomycetaceae and related bacteria. J. Clin. Microbiol. 74:234–238.

Kimberlin, D. F., and W. W. Andrews. 1998. Bacterial vaginosis: Association with adverse pregnancy outcome. Semin. Perinatol. 22:242–250.

King, S., and E. Meyer. 1957. Metabolic and serological differentiation of Actinomyces bovis and anaerobic diphtheroids. J. Bacteriol. 74:234–238.

Kiyama, M., K. Hiratsuka, S. Saito, T. Shiroza, H. Takiguchi, and Y. Abiko. 1996. Detection of Actinomyces species using nonradioactive riboprobes coupled with polymerase chain reaction. Biochem. Molec. Med. 58:151–155.

Klaaborg, K. E., O. Kronborg, and H. Olsen. 1985. Enterocutaneous fistulization due to Actinomyces odontolyticus: Report of a case. Dis. Colon Rectum 28:526–527.

Kolenbrander, P. E., and C. S. Phucas. 1984. Effect of saliva on coaggregation of oral Actinomyces and Streptococcus species. Infect. Immun. 44:228–233.

Kolenbrander, P. E., R. N. Andersen, and L. V. Holdeman. 1985. Coaggregation of moral Bacteroides species with other bacteria: Central role on coaggregation bridges and competitions. Infect. Immun. 48:741–746.

Kolenbrander, P. E., and J. London. 1993. Adhere today, here tomorrow: oral bacterial adherence. J. Bacteriol. 175:3247–3252.

Kolenbrander, P. E., R. N. Andersen, D. L. Clemans, C. J. Whittaker, and C. M. Klier. 1999. Potential role of functionally similar coaggregation mediators in bacterial succession. In: H. N. Newman and M. Wilson (Eds.) Dental Plaque Revisited: Oral Biofilms in Health and Disease. Bioline. Cardiff, UK. 171–186.

Kolenbrander, P. E. 2000. Oral microbial communities: biofilms, interactions, and genetic systems. Ann. Rev. Microbiol. 54:413–437.

Komiyama, K., R. L. Khandelwal, and S. E. Heinrich. 1988. Glycogen synthetic and degradative activities by Actinomyces viscosus and Actinomyces naeslundii of root surface caries and noncaries sites. Caries Res. 22:217–225.

Krönig, I. 1895. Über die Natur der Scheidenkeime, speciell über das Vorkommen anaerober Streptokokken in Scheidensekret Schwangerer. Zentralbl. Gynaecol. 19:409–412.

Koop, H. M., M. Valentijn Benz, A. V. Nieuw Amerongen, P. A. Roukaema, and J. De Graaff. 1989. Aggregation of 27 oral bacteria by human whole saliva. Influence of culture medium, calcium, and bacterial cell concentration, and interference by autoaggregation. Ant. v. Leeuwenhoek 55:277–290.

Kornman, K. S., and W. J. Loesche. 1978. New medium for isolation of Actinomyces viscosus and Actinomyces naeslundii from dental plaque. J. Clin. Microbiol. 7:514–518.

Kostecka, F. 1924. Relation of the teeth to the normal development of microbial flora in the oral cavity. Dental Cosmos, 66:927–935.

Kotrajaras, R., and H. Tagami. 1987. Corynebacterium pyogenes. Its pathogenic mechanism in endemic leg ulcers in Thailand. Int. J. Dermatol. 26:45–50.

Kroppenstedt, R. M., and H. J. Kutzner. 1978. Biochemical taxonomy of some problem actinomycetes. Zbl. Bakteriol. Hyg., I. Abt., Suppl. 6:125–133.

Kroppenstedt, R. M. 1985. Fatty acid and menaquinone analysis of actinomycetes and related organisms. In: M. Goodfellow and D. E. Minnikin (Eds.) Chemical Methods in Bacterial Systematics. Academic Press. London, UK. 173–199.

Kruse, W. 1896. Systematik der Streptothricheen und Bakterien. In: C. Flügge (Ed.) Die Mikroorganismen, 3rd ed. Vogel. Leipzig, Germany. 2:48–96.

Kuijper, E. J., H. O. Wiggerts, G. J. Jonker, K. P. Schaal, and J. de Gans. 1992. Disseminated actinomycosis due to Actinomyces meyeri and Actinobacillus actinomycetemcomitans. Scand. J. Infect. Dis. 24:667–672.

Lachner-Sandoval, V. 1898. Über Strahlenpilze [Inaugural Dissertation, Strassburg]. Universitäts Buchdruckerei von Carl Georgi. Bonn, Germany.

Lämmler, C., G. S. Chhatwal, and H. Blobel. 1985. Binding of α2-macroglobulin and haptoglobin to Actinomyces pyogenes. Can. J. Microbiol. 31:657–659.

Lämmler, C., and H. Blobel. 1986. Tentative identification of Actinomyces pyogenes with antisera against group G streptococci. Zbl. Bakteriol. Hyg. A 262:357–360.

Lämmler, C., and H. Blobel. 1988. Comparative studies on Actinomyces pyogenes and Arcanobacterium haemolyticum. Med. Microbiol. Immunol. 177:109–114.

Lämmler, C. 1990. Typing of Actinomyces pyogenes by its production and susceptibility to bacteriocin-like inhibitors. Zbl. Bakteriol. 273:173–178.

Langfeldt, N., M. Wendt, and G. Amtsberg. 1990. Comparative studies of the detection of Corynebacterium suis infections in swine by indirect immunofluorescence and culture. Berl. Münch. Tierärztl. Wochenschr. 103:273–276.

Larsen, J. L., P. Hogh, and K. Hovind-Hougen. 1986. Hemagglutinating and hydrophobic properties of Corynebacterium (Eubacterium) suis. Acta Vet. Scand. 27:520–530.

Lassnig, C., M. Dorsch, E. Schaper, H. Stöffler, J. Wolters, and E. Stackebrandt. 1989. Phylogenetic evidence for a relationship of Mobiluncus to the genus Actinomyces. FEMS Microbiol. Lett. 65:17–22.

Lawson, P. A., E. Falsen, E. Åkervall, P. Vandamme, and M. D. Collins. 1997. Characterization of some Actinomyces-like isolates from human clinical specimens: reclassification of Actinomyces suis (Soltys and Spratling) as Actinobaculum suis comb. nov. and description of Actinobaculum schaalii sp. nov. Int. J. Syst. Bacteriol. 47:899–903.

Lawson, P. A., E. Falsen, G. Foster, E. Eriksson, N. Weiss, and M. D. Collins. 2001a. Arcanobacterium pluranimalium sp. nov., isolated from porpoise and deer. Int. J. Syst. Evol. Microbiol. 51:55–59.

Lawson P. A., N. Nikolaitchouk, E. Falsen, K. Westling, and M. D. Collins. 2001b. Actinomyces funkei sp. nov., isolated from human clinical specimens. Int. J. Syst. Evol. Microbiol. 51:853–855.

Lean, J. J., A. J. Edmondson, G. Smith, and M. Villanueva. 1987. Corynebacterium pyogenes mastitis outbreak in inbred heifers in a California dairy. Cornell Vet. 77:367–373.

Lechevalier, H. A., and M. P. Lechevalier. 1970. A critical evaluation of the genera of aerobic actinomycetes. In: H. Prauser (Ed.) The Actinomycetales. Gustav Fischer. Jena, Germany. 393–405.

Lechevalier, M. P., C. de Bievre, and H. A. Lechevalier. 1977. Chemotaxonomy of aerobic actinomycetes: phospholipid composition. Biochem. Syst. Ecol. 5:249–260.

Lentino, J. R., J. E. Allen, and M. Stachowski. 1985. Haematogenous dissemination of thoracic actinomycosis due to Actinomyces meyeri. Ped. Infect. Dis. 4:698–699.

Lentze, F. A. 1938a. Die mikrobiologische Diagnostik der Aktinomykose. Münch. Med. Wochenschr. 47:1826–1836.

Lentze, F. A. 1938b. Zur Bakteriologie und Vakzinetherapie der Aktinomykose. Zbl. Bakteriol. Parasitenkd. Infektionskr. Hyg. I. Abt. Orig. 141:21–36.

Lentze, F. A. 1948. Die Aetiologie der Aktinomykose des Menschen. Dtsch. Zahnärztl. Zeitschr. 3:913–919.

Lentze, F. A. 1953. Zur Aetiologie und mikrobiologischen Diagnostik der Aktinomykose. Estratto dagli Atti del

VI Congresso Internazionale di Microbiologia, Roma 5(14):145–148.

Lentze, F. A. 1969. Die Aktinomykose und die Nocardiosen. *In:* A. Grumbach and O. Bonin (Eds.) Die Infektionskrankheiten des Menschen und ihre Erreger, 2nd ed. Georg Thieme Verlag. Stuttgart, Germany.

Lentze, F. A. 1970. Klinik, Diagnostik und Therapie der Aktinomykosen. *In:* Diagnostik und Therapie der Pilzkrankheiten und neuere Erkenntnisse in der Biochemie der pathogenen Pilze (Kongressreferate, 6. Tagung der Deutschsprachigen Mykologischen Gesellschaft am 15. Juli 1966). Grosse Verlag. Berlin, Germany.

Lepargneur, J. P., R. Heller, R. Soulie, and P. Riegel. 1998. Urinary tract infection due to Arcanobacterium bernardiae in a patient with urinary tract diversion. Eur. J. Clin. Microbiol. Infect. Dis. 17:399–401.

Lepe, J. A., J. de Leon, A. de la Iglesia, and M. de la Iglesia. 1998. The first description of infection by Actinomyces radingae. Enferm. Infect. Microbiol. 16:75–78.

Lerner, P. I. 1988. The lumpy jaw: Cervicofacial actinomycosis. Infect. Dis. Clin. North Am. 2:203–220.

Levine, M., D. E. Parker, and J. A. Stober. 1984. Human serum precipitins to moral bacteria related to dental caries. Arch. Oral Biol. 29:191–194.

Levine, S., S. Friedman, and A. J. Nevins. 1985. Mycotic infection in necrotic pulp tissue: A histologic report. Oral Med. Oral Pathol. 59:414–417.

Levine, M., and B. F. Movafagh. 1986. Analysis of the specificity of natural human antibody to Actinomyces. Molec. Immunol. 23:255–261.

Levine, L. A., and C. J. Doyle. 1988. Retroperitoneal actinomycosis: A case report and review of the literature. J. Urol. 140:367–369.

Lewis, R., D. McKenzie, J. Bagg, and A. Dickie. 1995. Experience with a novel selective medium for isolation of Actinomyces spp. from medical and dental specimens. J. Clin. Microbiol. 33:1613–1616.

Liebhold, M., M. Wendt, F. J. Kaup, and W. Drommer. 1995. Clinical, and light and electron microscopical findings in sows with cystitis. Vet. Rec. 137:141–144.

Linder, R. 1997. Rhodococcus equi and Arcanobacterium haemolyticum: Two "coryneform" bacteria increasingly recognized as agents of human infection. Emerg. Infect. Dis. 3:145–153.

Lininger, J. R., and W. J. Frable. 1984. Diagnosis of pelvic actinomycosis by fine needle aspiration: A case report. Acta Cytol. 28:601–604.

Lipton, M., and G. Sonnenfeld. 1980. Actinomyces meyeri osteomyelitis: An unusual cause of chronic infection of the tibia. Clin. Orthoped. 148:169–171.

Litwin, K. A., K. Jadbabaie, and M. Villanueva. 1999. Case of pleuropericardial disease caused by Actinomyces odontolyticus that resulted in cardiac tamponade. Clin. Infect. Dis. 29:219–220.

Llory, H., W. Guillo, and R. M. Frank. 1971. A cariogenic Actinomyces viscosus: A bacteriological and gnotobiotic study. Helvet. Odontol. Acta 15:134–138.

Loesche, W. J. 1960. Importance of nutrition in gingival crevice microbial ecology. Periodontics 6:245–249.

Loesche, W. J., R. N. Hochett, and S. A. Syed. 1972. The predominant cultivated flora of tooth surface plaque removed from institutionalized subjects. Arch. Oral. Biol. 17:1311–1325.

Loesche, W. J., F. Gusberti, G. Mettraux, T. Higgins, and S. A. Syed. 1983. Relationship between oxygen tension and subgingival bacterial flora in untreated human periodontal pockets. Infect. Immun. 42:659–667.

Loesche, W. J., S. A. Syed, E. Schmidt, and E. C. Morrison. 1985. Bacterial profiles of subgingival plaque in periodontitis. J. Periodontol. 56:447–456.

Logan, M. N., P. J. Stanley, A. Exley, C. Gagg, and I. D. Farrell. 1989. Actinomycetes in pyogenic liver abscess. Eur. J. Clin. Microbiol. Infect. Dis. 8:394–396.

Lord, F. T. 1910. The etiology of actinomycosis. J. Am. Med. Assoc. 55:1261–1263.

Love, D. N., R. Vekselstein, and S. Collings. 1990. The obligate and facultatively anaerobic bacterial flora of the normal feline gingival margin. Vet. Microbiol. 22:267–275.

Lovell, R. 1944. Further studies on the toxin of Corynebacterium pyogenes. J. Pathol. Bacteriol. 56:525–529.

Ludwig, W., G. Kirchhof, M. Weizenegger, and N. Weiss. 1992. Phylogenetic evidence for the transfer of Eubacterium suis to the genus Actinomyces as Actinomyces suis comb. nov. Int. J. Syst. Bacteriol. 42:161–165.

Lynch, M., J. O'Leary, D. Murnaghan, and B. Cryan. 1998. Actinomyces pyogenes septic arthritis in a diabetic farmer. J. Infect. 37:71–73.

Machet, L., M. C. Machet, E. Esteve, J. M. Delarbre, C. Pelucio-Lopes, F. Pruvost, and G. Lorette. 1993. Actinomyces meyeri cutaneous actinomycosis with pulmonary localization. Ann. Dermatol. Venereol. 120:896–899.

Mackenzie, A., L. A. Fuite, F. T. Chan, J. King, U. Allen, N. MacDonald, and F. Diaz-Mitoma. 1995. Incidence and pathogenicity of Arcanobacterium haemolyticum during a 2-year study in Ottawa. Clin. Infect. Dis. 21:177–181.

MacLean, P. D., A. A. Liebow, and A. A. Rosenberg. 1946. A haemolytic corynebacterium resembling Corynebacterium ovis and Corynebacterium pyogenes in man. J. Infect. Dis. 79:69–90.

Magnusson, H. 1928. The commonest forms of an actinomycosis in domestic animals and their etiology. Acta Pathol. Microbiol. Scand. 5:170–245.

Maiden, M., A. Tanner, and P. Macuch. 1996. Rapid characterization of periodontal bacteria isolates using fluorogenic substrate tests. J. Clin. Microbiol. 34:376–384.

Mann, C., S. Dertinger, G. Hartmann, R. Schurz, and B. Simma. 2002. Actinomyces neuii and neonatal sepsis. Infection 30:178–180.

Mao, K., and J. Guillebaud. 1984. Influence of removal of intrauterine contraceptive devices on colonization of the cervix by Actinomyces-like organisms. Contraception 30:535–544.

Marchant, S., S. R. Brailsford, A. C. Twomey, G. J. Roberts, and D. Beighton. 2001. The predominant microflora of nursing caries lesions. Caries Res. 35:397–406.

Mårdh, P.-A. 1993. The definition and epidemiology of bacterial vaginosis. Rev. Fr. Gynecol. Obstet. 88:195–197.

Mardis, J. S., and W. J. Many Jr. 2001. Endocarditis due to Actinomyces viscosus. South. Med. J. 94:240–243.

Martin, M. V. 1984. The use of oral amoxycillin for the treatment of actinomycosis: A clinical and in vitro study. Br. Dent. J. 156:252–254.

Martin, M. V. 1985. Antibiotic treatment of cervicofacial actinomycosis for patients allergic to penicillin: A clinical and in vitro study. Br. J. Oral Maxillofacial Surg. 23:428–434.

Martinez de Tejada, B., O. Coll, M. de Flores, S. L. Hillier, and D. V. Landers. 1998. Prevalence of bacterial vaginosis in an obstetric population of Barcelona. Med. Clin. (Barcelona) 110:201–204.

Marty, H. U., and J. Wüst. 1989. Disseminated actinomycosis caused by Actinomyces meyeri. Infection 17:154–155.

Massart, V., J. Soots, E. C. Fournier, A. Mallart Voisin, B. Gosselin, J. Remy, and A. B. Tonnel. 1986. Thoracic actinomycosis: Apropos of five cases. Rev. Pneumol. Clin. 42:219–225.

Masters, B., P. D. Phelan, and A. B. Auldist. 1985. Pulmonary actinomycosis in a child. Australian Ped. J. 21:129–130.

Masuda, N., R. P. Ellen, E. D. Fillery, and D. A. Grove. 1983. Chemical and immunological comparison of surface fibrils of strains representing 6 taxonomic groups of Actinomyces viscosus and Actinomyces naeslundii. Infect. Immun. 39:1325–1333.

Mauff, G., M. Herrmann, and K. P. Schaal. 1981. Electrophoretic protein patterns of nocardiae and their possible taxonomic relevance. Zbl. Bakteriol. Hyg., I. Abt. Suppl. 11:33–38.

Mazura, F. 1989. Bacteriological findings in samples of milk from cows studied for the causative agent of mastitis. Vet. Med. Praha 34:149–156.

McCormick, S. S., H. F. Mengoli, and M. A. Gerencser. 1985. Polyacrylamide gel electrophoresis of whole-cell preparations of Actinomyces spp. Int. J. Syst. Bacteriol. 35:429–433.

McDermid, A. S., A. S. McKee, D. C. Ellwood, and P. D. Marsh. 1986. The effect of lowering the pH on the composition and metabolism of a community of 9 oral bacteria grown in a chemostate. J. Gen. Microbiol. 32:1205–1214.

McGaughey, C. A., J. K. Bateman, and P. Z. McKenzie. 1951. Actinomycosis in the dog. Br. Vet. J. 107:428–430.

McIntire, F. C., A. E. Vatter, J. B. Baros, and J. Arnold. 1978. Mechanism of coaggregation between Actinomyces viscosus T14V and Streptococcus sanguis 34. Infect. Immun. 21:978–988.

McIntire, F. C., L. K. Crosby, J. J. Barlow, and K. L. Matta. 1983. Structural preferences of beta-galactoside reactive lectins on Actinomyces viscosus T14V and Actinomyces naeslundii WVU45. Infect. Immun. 41:848–850.

McNamara, P. J., W. A. Cuevas, and J. G. Songer. 1995. Toxic phospholipases D of Corynebacterium pseudotuberculosis, C. ulcerans and Arcanobacterium haemolyticum: Cloning and sequence homology. Gene 156:113–118.

McNeil, M. M., and K. P. Schaal. 1998. Actinomycoses. In: V. L. Yu, T. C. Merigan Jr., and S. L. Barriere (Eds.) Antimicrobial Therapy and Vaccines. Williams & Wilkins. Baltimore, MD. 14–22.

Meiers, J. C., and C. F. Schachtele. 1984. The effect of an antibacterial solution on the microflora of human incipient fissure caries. J. Dent. Res. 63:47–51.

Melville, T. H. 1965. A study of the overall similarity of certain actinomycetes mainly of oral origin. J. Gen. Microbiol. 40:309–315.

Menolascina, A., B. Nieves, E. Velazco, N. Rivero, and Z. Calderas. 1999. Clinical, epidemiological and microbiological aspects of Mobiluncus spp. in bacterial vaginosis. Enferm. Infect. Microbiol. Clin. 17:219–222.

Mergenhagen, S. E., A. L. Sandberg, B. M. Chassy, M. J. Brennan, M. K. Yeung, J. A. Donkersloot, and J. O. Cisar. 1987. Molecular basis of bacterial adhesion in the oral cavity. Rev. Infect. Dis. 9 (Suppl. 5):467–474.

Mesgarzadeh, M., A. Bonakdarpour, and P. D. Redecki. 1986. Case report 395: Hematogenous Actinomyces osteomyelitis (calcaneus). Skeletal Radiol. 15:584–588.

Meyer, K. 1911. Über eine anaerobe Streptothrix-Art. Zbl. Bakteriol. Hyg., I. Abt. Orig. 60:75–78.

Miller, C. H. 1974. Degradation of sucrose by whole cells and plaque of Actinomyces naeslundii. Infect. Immun. 10:1280–1291.

Miller, C. H., and P. J. B. Somers. 1978a. Degradation of levan by Actinomyces viscosus. Infect. Immun. 22:266–274.

Miller, C. H., C. J. Palenik, and K. E. Stamper. 1978b. Factors affecting the aggregation of Actinomyces naeslundii during growth and in washed cell suspensions. Infect. Immun. 21:1003–1009.

Minah, G. E., E. S. Solomon, and K. Chu. 1985. The association between dietary sucrose consumption and microbial population shifts at oral sites in man. Arch. Oral Biol. 30:397–401.

Minnikin, D. E., M. Goodfellow, and M. D. Collins. 1978. Lipid composition in the classification and identification of coryneform and related taxa. In: I. J. Bousfield and A. G. Callely (Eds.) Coryneform Bacteria. Academic Press. London, UK. 85–160.

Minsker, O. W., and M. A. Moskovskaya. 1979. Abdominal actinomycosis: Some aspects of pathogenesis, clinical manifestation, and treatment. Mykosen 22:393–408.

Mitchell, J., J. O. Cisar, A. E. Vatter, P. V. Fennessey, and F. C. McIntire. 1983. Actinomyces odontolyticus isolated from the female genital tract. J. Clin. Pathol. 37:1379–1383.

Mizuno, J., J. O. Cisar, A. E. Vatter, P. V. Fennessey, and F. C. McIntire. 1983. A factor from Actinomyces viscosus T14V that specifically aggregates Streptococcus sanguis H1. Infect. Immun. 40:1204–1213.

Modesto, A., K. C. Lima, and M. de Uzeda. 2000. Effects of three different infant dentifrices on biofilms and on oral microorganisms. J. Clin. Pediatr. Dent. 24:237–243.

Moens, Y., and W. Verstraeten. 1980. Actinomycosis due to Actinomyces viscosus in a young dog. Vet. Rec. 106:344–345.

Moi, H., and D. Danielsson. 1984. Studies on rabbit hyperimmune, patient and blood donor serum with regards to bactericidal activity and serum antibodies against anaerobic curved rods from patients with bacterial vaginosis. In: P.-A. Mårdh and D. Taylor-Robinson (Eds.) Bacterial Vaginosis. Almqvist and Wiksell International. Stockholm, The Netherlands. 89–92.

Moi, H., H. Fredlund, E. Tornqvist, and D. Danielsson. 1991. Mobiluncus species in bacterial vaginosis: Aspects of pathogenesis. Acta Pathol. Microbiol. Immunol. Scand. 99:1049–1054.

Mongardio, N., B. De Rienzo, G. Zanchetta, G. Lami, F. Pellegrino, and F. Squadrini. 1986. Primary hepatic actinomycosis. J. Infect. 12:65–69.

Moore, B. 1954. Observations on a group of anaerobic vaginal vibrios. J. Pathol. Bacteriol. 67:461–473.

Moore, W. E., L. V. Holdeman, R. M. Smibert, I. J. Good, J. A. Burmeister, K. G. Palcanis, and R. R. Ranney. 1982. Bacteriology of experimental gingivitis in young adult humans. Infect. Immun. 38:651–667.

Moore, W. E. C., and L. V. Holdeman Moore. 1986. Genus Eubacterium Prévot 1938. In: P. H. A. Sneath, N. S. Mair, M. E. Sharpe, and J. G. Holt (Eds.) Bergey's Manual of Systematic Bacteriology. Williams & Wilkins. Baltimore, MD. 2:1353–1373.

Moreno-Montesinos, M., A. Valle-Vallencia, and L. Aguilar-Alfaro. 1989. Pharyngitis caused by Arcanobacterium haemolyticum. Ann. Espagn. Ped. 30:209–210.

Morinushi, T., D. E. Lopatin, S. A. Syed, G. Bacon, C. J. Cowalski, and W. J. Loesche. 1989. Humoral immune response to selected subgingival plaque microorganisms in insulin-dependent diabetic children. J. Periodontol. 60:199–204.

Morou-Bermudez, E., and R. A. Burne. 1999. Genetic and physiologic characterization of urease of Actinomyces naeslundii. Infect. Immun. 67:504–512.

Morrison, J. R., and G. S. Tillotson. 1988. Identification of Actinomyces (Corynebacterium) pyogenes with the API 20 Strep system. J. Clin. Microbiol. 26:1865–1866.

Mosselman, G., and E. Lienaux. 1890. L'actinomycose et son agent infecteur. Ann. Med. Vet. 39:409–426.

Moustafa, A. M. 1994. First observation of camel (Camelus dromedarius) lymphadenitis in Libya: A case report. Rev. Elev. Med. Vet. Pays Trop. 47:313–314.

Mulei, C. M. 1999. Teat lesions and their relationship to intramammary infections on small-scale dairy farms in Kiambu district in Kenya. J. S. Afr. Vet. Assoc. 70:156–157.

Müller, H. E. 1973. Neuraminidase und Acylneuraminat-Pyruvat-Lyase bei Corynebacterium haemolyticum und Corynebacterium pyogenes. Zbl. Bakteriol. Parasitenkd. Infektionskr. Hyg., Abt. I Orig. A 225:59–65.

Murakami, S., M. W. Yamanishi, and R. Azuma. 1997. Lymph node abscess due to Actinomyces viscosus in a cat. J. Vet. Med. Sci. 59:1079–1080.

Murakami, S., R. Azuma, T. Koeda, H. Oomi, T. Watanabe, and H. Fujiwara. 1998. Immunohistochemical detection of Actinomyces sp. in swine tonsillar abscess and granulomatous mastitis. Mycopathologia 141:15–19.

Naeslund, C. 1925. Studies of Actinomyces from the oral cavity. Acta Pathol. Microbiol. Scand. 2:110–140.

Narikawa, S. 1986. Distribution of metronidazole susceptibility factors in obligate anaerobes. J. Antimicrob. Chemother. 18:565–574.

Ness, R. B., S. Hillier, H. E. Richter, D. E. Soper, C. Stamm, D. C. Bass, R. L. Sweet, and P. Rice. 2003. Can known risk factors explain racial differences in the occurrence of bacterial vaginosis. J. Natl. Med. Assoc. 95:201–212.

Newman, M. G. 1984. Anaerobic oral and dental infection. Rev. Infect. Dis. 6 (Suppl. 1):101–114.

Niederau, W., W. Pape, K. P. Schaal, U. Höffler, and G. Pulverer. 1982. Zur Antibiotikabehandlung der menschlichen Aktinomykosen. Dtsch. Med. Wochenschr. 107:1279–1283.

Nielsen, P. M., and A. Novak. 1987. Acute cervicofacial actinomycosis. Int. J. Oral Maxillofac. Surg. 16:440–444.

Nikolaitchouk, N., L. Hoyles, E. Falsen, J. M. Grainger, and M. D. Collins. 2000. Characterization of Actinomyces isolates from samples from the human urogenital tract: description of Actinomyces urogenitalis sp. nov. Int. J. Syst. Evol. Microbiol. 50:1649–1654.

Nithyanandam, S., O. D'Souza, S. S. Rao, R. R. Battu, and S. George. 2001. Rhinoorbitocerebral actinomycosis. Ophthalmol. Plast. Reconstr. Surg. 17:134–136.

Noack-Loebel, C., E. Küster, V. Rusch, and K. Zimmermann. 1983. Influence of different dietary regimens upon the composition of the human fecal flora. Progress Food Nutr. Sci. 7:127–131.

Nord, C.-E., and L. Kager. 1983. Tinidazole—microbiology, pharmacology and efficacy in anaerobic infections. Infection 11:54–60.

Nyman, M., K. R. Alugupalli, S. Stromberg, and A. Forsgren. 1997. Antibody response to Arcanobacterium haemolyticum infection in humans. J. Infect. Dis. 175:1515–1518.

Nyvad, B., and O. Fejerskov. 1989. Structure of dental plaque and the plaque-enamel interface in human experimental caries. Caries Res. 23:151–158.

O'Connor, K. F., M. N. Bagg, M. R. Croley, and S. I. Schabel. 1989. Pelvic actinomycosis associated with intrauterine devices. Radiology 170:559–560.

Okewole, P. A., P. S. Odeyemi, R. A. Ocholi, E. A. Irokanulo, E. S. Haruna, and I. L. Oyetunde. 1989. Actinomyces viscosus isolated from a case of abortion in a Friesian heifer. Vet. Rec. 124:464.

Olson, T. S., A. B. Seid, and S. M. Pransky. 1989. Actinomycosis of the middle ear. Int. J. Ped. Otorhinolaryngol. 17:51–55.

Olson, M. E., H. Ceri, D. W. Morck, A. G. Buret, and R. R. Read. 2002. Biofilm bacteria: Formation and comparative susceptibility to antibiotics. Can. J. Vet. Res. 66:86–92.

Ooshima, T., and H. K. Kuramitsu. 1985. Actinomyces viscosus cell-free synthesis of extracellular slime polysaccharide. Microbiol. Immunol. 29:479–485.

Orloff, J. J., M. J. Fine, and J. D. Rihs. 1988. Acute cardiac tamponade due to cardiac actinomycosis. Chest 93:661–663.

Osterlund, A. 1995. Are penicillin treatment failures in Arcanobacterium haemolyticum pharyngotonsillitis caused by intracellularly residing bacteria? Scand. J. Infect. Dis. 27:131–134.

Pabst, M. J. 1977. Levan and levan sucrase of Actinomyces viscosus. Infect. Immun. 15:518–526.

Pabst, M. J., J. O. Cisar, and C. L. Trummel. 1979. The cell wall-associated levansucrase of Actinomyces viscosus. Biochem. Biophys. Acta 566:274–282.

Page, R. C. 1986. Gingivitis. J. Clin. Periodontol. 13:345–359.

Påhlson, C., U. Forsum, A. Hallén, E. Hjelm, and J. Wallin. 1983. Characterization of motile anaerobic curved rods isolated from women with lower genital tract infection in three different countries. Eur. J. Sex. Trans. Dis. 1:73–75.

Påhlson, C., and U. Forsum. 1985. Rapid detection of Mobiluncus species. Lancet i:927.

Påhlson, C., A. Hallén, and U. Forsum. 1986a. Curved rods related to Mobiluncus: Phenotypes as defined by monoclonal antibodies. Acta Pathol. Microbiol. Immunol. Scand. Sect. B 94:113–116.

Påhlson, C., A. Hallén, and U. Forsum. 1986b. Improved yield of Mobiluncus species from clinical specimens after alkaline treatment. Acta Pathol. Microbiol. Immunol. Scand. Sect. B 94:117–125.

Påhlson, C., J. G. Mattson, P. G. Larsson, H. Gersdorf, U. B. Göbel, U. Forsum, and K. E. Johansson. 1992. Detection and identification of Mobiluncus species by direct filter hybridization with an oligonucleotide probe complementary to rRNA. Acta Pathol. Microbiol. Immunol. Scand. 100:655–662.

Palenik, C. J., and C. H. Miller. 1975. Extracellular invertase activity from Actinomyces viscosus. J. Dent. Res. 54:186.

Palmer Jr., R. J., K. Kazmerzak, M. C. Hansen, and P. E. Kolenbrander. 2001. Mutualism versus independence: strategies of mixed-species oral biofilms in vitro using saliva as the sole nutrient source. Infect. Immun. 69:5794–5804.

Pang, D. K., and M. Abdalla. 1987. Osteomyelitis of the foot due to Actinomyces meyeri: A case report. Foot Ankle 8:169–171.

Paolino, V. J., and S. Kashket. 1985. Inhibition by coco extracts of biosynthesis of extracellular polysaccharide by human oral bacteria. Arch. Oral Biol. 30:359–363.

Pape, H.-D., K. P. Schaal, and J. Braun. 1984. Erreger- und Resistenzspektrum bei odontogenen Infektionen. *In:* G. Pfeifer and N. Schwenzer (Eds.) Fortschritte der Kiefer- und Gesichtschirurgie, Band XXIX: Septische Mund-Kiefer-Gesichtschirurgie. Georg Thieme Verlag. New York, NY. 86–88.

Pascual, C., G. Foster, E. Falsen, K. Bergström, C. Greko, and M. D. Collins. 1999. Actinomyces bowdenii sp. nov., isolated from canine and feline clinical specimens. Int. J. Syst. Bacteriol. 49:1873–1877.

Pascual Ramos, C., E. Falsen, N. Alvarez, E. Åkervall, B. Sjöden, and M. D. Collins. 1997a. Actinomyces graevenitzii sp. nov., isolated from human clinical specimens. Int. J. Syst. Bacteriol. 47:885–888.

Pascual Ramos, C., G. Foster, and M. D. Collins. 1997b. Phylogenetic analysis of the genus Actinomyces based on 16S rRNA gene sequences: Description of Arcanobacterium phocae sp. nov., Arcanobacterium bernardiae comb. nov., and Arcanobacterium pyogenes comb. nov. Int. J. Syst. Bacteriol. 47:46–53.

Paster, B. J., S. K. Boches, J. L. Galvin, R. E. Ericson, C. Lau, V. A. Levanos, A. Sahasrabudhe, and F. E. Dewhirst. 2001. Bacterial diversity in human subgingival plaque. J. Bacteriol. 183:3770–3783.

Pattyn, S. R., M. Ieven, and L. Buffet. 1993. Comparative evaluation of the Rapid ID 32A kit system, miniaturized standard procedure and a rapid fermentation procedure for the identification of anaerobic bacteria. Acta Clin. Belg. 48:81–85.

Peacock Jr., J. E., M. R. McGinnis, and M. S. Cohen. 1984. Persistent neutrophilic meningitis: Report of 4 cases and review of the literature. Med. Baltimore 63:379–395.

Pekovic, D. D., and E. D. Fillery. 1984. Identification of bacteria in immunopathologic mechanisms of human dental pulp. Oral Surg. Oral Med. Oral Pathol. 57:652–661.

Pekovic, D. D., V. W. Adamkiewicz, A. Shapiro, and M. Gornitsky. 1987. Identification of bacteria in association with immune components in human carious dentine. J. Oral Pathol. 16:223–233.

Pelle, G., L. Makrai, L. Fodor, and M. Dobos-Kovacs. 2000. Actinomycosis of dogs caused by Actinomyces hordeovulneris. J. Comp. Pathol. 123:72–76.

Peloux, Y., and P. Thomas. 1981. A propos de quelques bactéries mobiles anaérobies Gram négatives. Rev. Inst. Pasteur (Lyon) 14:103–111.

Peloux, Y., H. Chardon, E. Lagier, P. Jauffret, G. Latil, and J. P. de Cuttoli. 1983. Pathogenic role of Actinomyces apart from actinomycosis: Apropos of two cases of acute suppurations with Actinomyces odontolyticus. Semaine Hopital Paris 59:3063–3064.

Perez-Castrillon, J. L., C. Gonzales-Castaneda, F. del Campo-Matias, J. Bellido-Casado, and G. Diaz. 1997. Empyema necessitatis due to Actinomyces odontolyticus. Chest 111:1144.

Persson, E., K. Holmberg, S. Dahlgren, and L. Nilsson. 1983. Actinomyces israelii in the genital tract of women with and without intrauterine contraceptive devices. Acta Obstet. Gynaecol. Scand. 62:563-568.

Persson, E., and K. Holmberg. 1984. Clinical evaluation of precipitin tests for genital actinomycosis. J. Clin. Microbiol. 20:917–922.

Persson, E., and K. Holmberg. 1985. Study of precipitation reactions to Actinomyces israelii antigens in uterine secretions. J. Clin. Pathol. 38:99–102.

Persson, E., and C. Brihmer. 1986. Actinomyces israelii-associated salpingitis. Eur. J. Obstet. Gynecol. Reprod. Biol. 21:173–175.

Pfister, W., M. Sprossig, P. Gangler, and M. Mirgorod. 1984. Bacteriological characterization of gingivitis-inducing plaque depending on different sugar levels of the diet. Zbl. Bakteriol. Mikrobiol. Hyg., Reihe A 257:364–371.

Phan, T.-N., J. S. Reidmiller, and R. E. Marquis. 2000. Sensitization of Actinomyces naeslundii and Streptococcus sanguis in biofilms and suspensions to acid damage by fluoride and other weak acids. Arch. Microbiol. 174:248–255.

Philipsen, E. K., S. Larsen, and K. D. Jensen. 1988. Subcutaneous abscess and pulmonary infiltrate due to Actinomyces infection: Case report. Acta Chirur. Scand. 154:675–677.

Pine, L., and H. Hardin. 1959a. A. israelii, a cause of lacrimal canaliculitis in man. J. Bacteriol. 78:164–170.

Pine, L., and S. J. Watson. 1959b. Evaluation of an isolation and maintenance medium for Actinomyces species and related organisms. J. Lab. Clin. Med. 54:107–114.

Pine, L., A. Howell, and S. J. Watson. 1960. Studies of the morphological, physiological and biochemical characters of Actinomyces bovis. J. Gen. Microbiol. 23:403–424.

Pine, L., and L. K. Georg. 1969. Reclassification of Actinomyces propionicus. Int. J. Syst. Bacteriol. 19:267–272.

Pine, L., G. Bradley Malcolm, E. M. Curtis, and J. M. Brown. 1981. Demonstration of Actinomyces and Arachnia species in cervicovaginal smears by direct staining with species-specific fluorescent-antibody conjugate. J. Clin. Microbiol. 13:15–21.

Pine, L., E. M. Curtis, and J. M. Brown. 1985. Actinomyces and the intrauterine contraceptive device: Aspects of the fluorescent antibody stain. Am. J. Obstet. Gynecol. 152:287–290.

Pollock, P. G., F. P. Koontz, T. F. Viner, C. J. Krause, D. S. Meyers, and J. F. Valicenti Jr. 1978. Cervicofacial actinomycosis. Rapid diagnosis by thin-needle aspiration. Arch. Otolaryngol. 104:491–494.

Popp, W. 1977. The diagnosis and treatment of mixed anaerobic vaginal discharges. Geburtsh. Frauenheilkd. 37:432–437.

Pordy, R. C. 1988. Lumpy jaw due to Actinomyces meyeri: Report of the first case and review of the literature. Mt. Sinai J. Med. NY 55:190–193.

Prager, J., B. S. Zaret, R. Davidson, and T. W. Smith. 1984. Gliosarcoma at the site of a surgically treated Actinomyces cerebral abscess. Neurosurg. 15:868–872.

Prévot, A. R. 1938. Etude de systèmatique bactérienne. III: Invalidité du genre Bacteroides Castellani et Chalmers, demembrement et réclassification. Ann. Inst. Pasteur 60:285–307.

Prévot, A. R. 1940. Manuel de classification et de détermination des bactéries anaérobies. Masson et Cie. Paris, France.

Puapermpoonsiri, S., N. Kato, K. Watanabe, K. Ueno, C. Chongsomchai, and P. Lumbiganon. 1996. Vaginal microflora associated with bacterial vaginosis in Japanese and Thai pregnant women. Clin. Infect. Dis. 23:748–752.

Pulverer, G. 1974. Problems of human actinomycosis. Postepy Hiegieny i Meycyny Doswwiadczalnej 28:253–260.

Pulverer, G., and K. P. Schaal. 1978. Pathogenicity and medical importance of aerobic and anaerobic actinomycetes.

In: M. Mordarski, W. Kurylowicz, and J. Jeljaszewicz (Eds.) Nocardia and Streptomyces. Gustav Fischer Verlag. Stuttgart, Germany. 417–427.

Pulverer, G., and K. P. Schaal. 1984a. Human actinomycoses. Drugs: Experimental and Clinical Research X(3):187–196.

Pulverer, G., and K. P. Schaal. 1984b. Medical and microbiological problems in human actinomycoses. *In:* L. Ortiz-Ortiz, L. F. Bojalil, and V. Yakoleff (Eds.) Biological, Biochemical, and Biomedical Aspects of Actinomycetes. Academic Press. Orlando, FL. 161–170.

Pulverer, G., H. Schütt-Gerowitt, and K. P. Schaal. 2003. Human cervicofacial actinomycoses: Microbiological data for 1,997 cases. Clin. Infect. Dis. 37:490–497.

Raju, N. R., R. F. Langham, C. Kispert, and A. Koestner. 1986. Suppurative spinal meningitis caused by an Actinomyces sp. in an arctic fox. J. Am. Vet. Med. Assoc. 189:1194–1195.

Raoult, D., J. L. Kohler, H. Gallais, E. Estrangin, Y. Peloux, and P. Casanova. 1982. Fusobacterium necrophorum associated with Actinomyces odontolyticus septicemia. Pathol. Biol. (Paris) 30:576–580.

Reddy, C. A., and M. Kao. 1978. Value of metabolic products in identification of certain corynebacteria. J. Clin. Microbiol. 7:428–433.

Reddy, C. A., C. P. Cornell, and A. M. Fraga. 1982. Transfer of Corynebacterium pyogenes (Glage) Eberson to the genus Actinomyces as Actinomyces pyogenes (Glage) comb. nov. Int. J. Syst. Bacteriol. 32:419–429.

Reiner, S. L., J. M. Harrelson, S. E. Miller, G. B. Hill, and H. A. Gallis. 1987. Primary actinomycosis of an extremity: A case report and review. Rev. Infect. Dis. 9:581–589.

Rideaut, B. A., R. J. Montali, R. S. Wallace, M. Bush, L. G. Phillips Jr., T. T. Antonovych, and S. G. Sabnis. 1989. Renal medullary amyloidosis in Dorcas gazelles. Vet. Pathol. 26:129–135.

Rippon, J. W., and S. K. Kathuria. 1984. Actinomyces meyeri presenting as an asymptomatic lung mass. Mycopathology 84:187–192.

Ritter, E., A. Kaschner, C. Becker, E. Becker-Boost, C. H. Wirsing von König, and H. Finger. 1993. Isolation of Arcanobacterium haemolyticum from an infected foot wound. Eur. J. Clin. Microbiol. Infect. Dis. 12:473–474.

Roberts, R. J. 1968. Biochemical reactions of Corynebacterium pyogenes. J. Pathol. Bacteriol. 95:127–130.

Rogers, A. H., J. S. van der Hoeven, and F. H. M. Mikx. 1978. Inhibition of Actinomyces viscosus by bacteriocin-producing strains of Streptococcus mutans in the dental plaque of gnotobiotic rats. Arch. Oral Biol. 23:477–485.

Roberts, M. C., S. L. Hillier, F. D. Schoenknecht, and K. K. Holmes. 1984. Nitrocellulose filter blots for species identification of Mobiluncus curtisii and Mobiluncus mulieris. J. Clin. Microbiol. 20:826–827.

Roberts, M. C., E. J. Baron, S. M. Finegold, and G. E. Kenny. 1985a. Antigenic distinctiveness of Mobiluncus curtisii and Mobiluncus mulieris. J. Clin. Microbiol. 21:891–893.

Roberts, M. C., S. L. Hillier, F. D. Schoenknecht, and K. K. Holmes. 1985b. Comparison of Gram stain, DNA probe, and culture for the identification of species of Mobiluncus in female genital specimens. J. Infect. Dis. 152:74–77.

Roeder, B. L., M. M. Chengappa, K. F. Lechtenberg, T. G. Nagaraja, and G. A. Varga. 1989. Fusobacterium necrophorum and Actinomyces pyogenes associated facial and mandibular abscesses in blue duiker. J. Wildl. Dis. 25:370–377.

Rose, H. D., B. Varkey, and C. P. Kutty. 1982. Thoracic actinomycosis caused by Actinomyces meyeri. Am. Rev. Respir. Dis. 125:251–254.

Rosenthal, T., and E. H. Gootzeit. 1942. The incidence of B. fusiformis and spirochaetes in the edentulous mouth. J. Dent. Res. 21:373–374.

Royce, R. A., T. P. Jackson, J. M. Thorp Jr., S. L. Hillier, L. K. Rabe, L. M. Pastore, and D. A. Savitz. 1999. Race/ethnicity, vaginal flora patterns, and pH during pregnancy. Sex. Transm. Dis. 26:96–102.

Ruhe, J., K. Holding, and D. Mushatt. 2001. Infected total knee arthroplasty due to Actinomyces naeslundii. Scand. J. Infect. Dis. 33:230–231.

Russell, C., and T. H. Melville. 1978. A review: Bacteria in the human mouth. J. Appl. Bacteriol. 44:163–181.

Ruutu, P., P. J. Pentikainen, U. Larinkari, and M. Lempinen. 1982. Hepatic actinomycosis presenting as repeated cholestatic reactions. Scand. J. Infect. Dis. 14:235–238.

Sabbe, L. J., D. van de Merve, L. Schouls, A. Bergmans, M. Vaneechoutte, and P. Vandamme. 1999. Clinical spectrum of infections due to the newly described Actinomyces species A. turicensis, A. radingae, and A. europaeus. J. Clin. Microbiol. 37:8–13.

Sandberg, A. L., L. L. Mudrick, J. O. Cisar, M. J. Brennan, S. E. Mergenhagen, and A. E. Vatter. 1986. Type 2 fimbrial lectin-mediated phagocytosis of oral Actinomyces spp. by polymorphonuclear leucocytes. Infect. Immun. 54:472–476.

Santala, A.-M., N. Sarkonen, V. Hall, P. Carlson, H. Jousimies-Somer, and E. Könönen. 2004. Evaluation of four commercial test systems for identification of Actinomyces and some closely related species. J. Clin. Microbiol. 42:418–420.

Sarkonen, N., E. Könönen, P. Summanen, A. Kanervo, A. Takala, and H. Jousimies-Somers. 2000. Oral colonization with Actinomyces species in infants by two years of age. J. Dent. Res. 79:864–867.

Sarkonen, N., E. Könönen, P. Summanen, M. Könönen, and H. Jousimies-Somer. 2001. Phenotypic identification of Actinomyces and related species isolated from human sources. J. Clin. Microbiol. 39:3955–3961.

Saunders, J. N., and C. H. Miller. 1983. Neuraminidase-activated attachment of Actinomyces naeslundii ATCC 12104 to human buccal epithelial cells. J. Dent. Res. 62:1038–1040.

Saxen, L., H. Jousimies-Somer, A. Kaisla, A. Kanervo, P. Summanen, and I. Sipila. 1989. Subgingival microflora, dental and periodontal conditions in patients with hereditary fructose intolerance. Scand. J. Dent. Res. 97:150–158.

Schaal, K. P., and G. Pulverer. 1973. Fluoreszenzserologische Differenzierung von fakultativ anaeroben Aktinomyzeten. Zbl. Bakteriol. Hyg., I. Abt. Orig. A 225:424–430.

Schaal, K. P. 1979a. Die Aktinomykosen des Menschen: Diagnose und Therapie. Dtsch. Ärztebl. 31:1997–2006.

Schaal, K. P., H. Schütt-Gerowitt, and W. Pape. 1979b. Cefoxitin-Empfindlichkeit pathogener aerober und anaerober Aktinomyzeten. Infection 7 (Suppl. 1):47–51.

Schaal, K. P., and W. Pape. 1980. Special methodological problems in antibiotic susceptibility testing of fermentative actinomycetes. Infection 8 (Suppl. 2):176–182.

Schaal, K. P. 1981a. Actinomycoses. Rev. Inst. Pasteur (Lyon) 14:279–288.

Schaal, K. P., and G. Pulverer. 1981b. The genera Actinomyces, Agromyces, Arachnia, Bacterionema, and Rothia.

In: M. P. Starr, H. Stolp, H. G. Trüper, A. Balows, and H. G. Schlegel (Eds.) The Prokaryotes. Springer-Verlag. Berlin, Germany. 2:1923–1950.

Schaal, K. P., and G. Schofield. 1981c. Current ideas on the taxonomic status of the Actinomycetaceae. Zbl. Bakteriol. Hyg., I. Abt. Suppl. 11:67–78.

Schaal, K. P., and G. Schofield. 1981d. Taxonomy of Actinomycetaceae. Rev. Inst. Pasteur (Lyon) 14:27–39.

Schaal, K. P. 1984a. Laboratory diagnosis of actinomycete diseases. *In:* M. Goodfellow, M. Mordarski, and S. T. Williams (Eds.) The Biology of the Actinomycetes. Academic Press. London, UK. 425–456.

Schaal, K. P., and B. L. Beaman. 1984b. Clinical significance of actinomycetes. *In:* M. Goodfellow, M. Mordarski, and S. T. Williams (Eds.) The Biology of the Actinomycetes. Academic Press. London, UK. 389–424.

Schaal, K. P., and G. Pulverer. 1984c. Epidemiologic, etiologic, diagnostic, and therapeutic aspects of endogenous actinomycete infections. *In:* L. Ortiz-Ortiz, L. F. Bojalil, and V. Yakoleff (Eds.) Biological, Biochemical, and Biomedical Aspects of Actinomycetes. Academic Press. New York, NY. 13–32.

Schaal, K. P., and G. Schofield. 1984d. Classification and identification of clinically significant Actinomycetaceae. *In:* L. Ortiz-Ortiz, L. F. Bojalil, and V. Yakoleff (Eds.) Biological, Biochemical, and Biomedical Aspects of Actinomycetes. Academic Press. Orlando, FL. 505–520.

Schaal, K. P., M. Herzog, H.-D. Pape, G. Pulverer, and S. Herzog. 1984e. Kölner Therapiekonzepte zur Behandlung der menschlichen Aktinomykosen von 1952–1982. *In:* G. Pfeifer and N. Schwenzer (Eds.) Fortschritte der Kiefer- und Gesichtschirurgie, Band XXIX: Septische Mund-Kiefer-Gesichtschirurgie. Georg Thieme Verlag. Stuttgart, Germany. 151–156.

Schaal, K. P. 1985a. Die Aktinomykosen des Menschen. Int. Welt 8:32–38.

Schaal, K. P. 1985b. Identification of clinically significant actinomycetes and related bacteria using chemical techniques. *In:* M. Goodfellow and D. E. Minnikin (Eds.) Chemical Methods in Bacterial Systematics. Academic Press. London, UK. 359–381.

Schaal, K. P., and R. Gatzer. 1985c. Serological and numerical phenetic classification of clinically significant fermentative actinomycetes. *In:* T. Arai, K. Terao, M. Yamazaki, M. Miyaji, and T. Unemoto (Eds.) Filamentous Microorganisms: Biomedical Aspects. Japan Scientific Societies Press. Tokyo, Japan. 85–109.

Schaal, K. P. 1986a. Genus Actinomyces Harz 1877. *In:* P. H. A. Sneath, N. S. Mair, M. E. Sharpe, and J. G. Holt (Eds.) Bergey's Manual of Systematic Bacteriology. Williams & Wilkins. Baltimore, MD. 2:1383–1418.

Schaal, K. P. 1986b. Genus Arachnia Pine and Georg 1969. *In:* P. H. A. Sneath, N. S. Mair, M. E. Sharpe, and J. G. Holt (Eds.) Bergey's Manual of Systematic Bacteriology. Williams & Wilkins. Baltimore, MD. 2:1332–1342.

Schaal, K. P. 1988. Actinomycetes as human pathogens. *In:* Y. Okami, T. Beppu, and H. Ogawara (Eds.) Biology of Actinomycetes '88: Proceedings of the 7th International Symposium on Biology of Actinomycetes. Japan Scientific Societies Press. Tokyo, Japan. 277–282.

Schaal, K. P. 1992a. Fakultativ bis obligat anaerobe, Grampositive, sporenlose Stäbchenbakterien. *In:* F. Burkhardt (Ed.) Mikrobiologische Diagnostik. Thieme. New York, NY. 209–223.

Schaal, K. P. 1992b. The genera Actinomyces, Arcanobacterium, and Rothia. *In:* A. Balows, H. G. Trüper, M.

Dworkin, W. Harder, and K.-H. Schleifer (Eds.) The Prokaryotes, 2nd ed. Springer-Verlag. New York, NY.

Schaal, K. P., and H. J. Lee. 1992c. Actinomycete infections in humans: A review. Gene 115:201–211.

Schaal, K. P. 1996. Actinomycoses. *In:* D. J. Weatherall, J. G. G. Ledingham, and D. A. Warrell (Eds.) Oxford Textbook of Medicine, 3rd ed. Oxford University Press. Oxford, UK. 1:680–686.

Schaal, K. P. 1998. Actinomycoses, actinobacillosis, and related diseases. *In:* L. Collier, A. Balows, and M. Sussman (Eds.) Topley and Wilson's Microbiology and Microbial Infections, 9th ed., Volume 3: Bacterial Infections. Arnold. London, UK. 777–798.

Schaal, K. P., A. Crecelius, G. Schumacher, and A. A. Yassin. 1999. Towards a new taxonomic structure of the genus Actinomyces and related bacteria. Nov. Acta Leopoldina 80:83–91.

Schaeken, M. J., and P. de Haan. 1989. Effects of the sustained-release chlorhexidine acetate on the human dental plaque flora. J. Dent. Res. 68:119–123.

Schiffer, M. A., A. Elguezabal, and M. C. Allen. 1978. Actinomycosis infections associated with intrauterine contraceptive devices and a vaginal pessary. Adv. Planned Parenthood 12:183–192.

Schleifer, K.-H., and O. Kandler. 1972. Peptidoglycan types of bacterial cell walls and their taxonomic implications. Bacteriol. Rev. 36:407–477.

Schleifer, K.-H., and P. H. Seidl. 1985. Chemical composition and structure of murein. *In:* M. Goodfellow and D. E. Minnikin (Eds.) Chemical Methods in Bacterial Systematics. Academic Press. London, UK. 201–219.

Schnadig, V. J., K. D. Davie, S. K. Shafer, R. B. Yandell, M. Z. Islam, and E. V. Hannigan. 1989. The cytologist and bacterioses of the vaginal-ectocervical area: Clues, commas and confusion. Acta Cytol. 33:287–297.

Schofield, G., and K. P. Schaal. 1979a. A simple basal medium for carbon source utilization tests with the anaerobic actinomycetes. FEMS Microbiol. Lett. 5:309–310.

Schofield, G., and K. P. Schaal. 1979b. Application of the Minitek differentiation system in the classification and identification of Actinomycetaceae. FEMS Microbiol. Lett. 5:311–313.

Schofield, G., and K. P. Schaal. 1980a. Carbohydrate fermentation patterns of facultatively anaerobic actinomycetes using micromethods. FEMS Microbiol. Lett. 8:67–69.

Schofield, G., and K. P. Schaal. 1980b. Rapid micromethods for detecting deamination and decarboxylation of amino acids, indole production, and reduction of nitrate and nitrite by facultatively anaerobic actinomycetes. Zbl. Bakteriol. Hyg., I. Abt. Orig. A 247:383–391.

Schofield, G., and K. P. Schaal. 1981. A numerical taxonomic study of members of the Actinomycetaceae and related taxa. J. Gen. Microbiol. 127:237–259.

Schütt-Gerowitt, H., K. P. Schaal, and G. Pulverer. 1999. The role of actinomycetes in the etiology of lacrimal canaliculitis and other eye infections. Nova Acta Leopoldina 80:227–233.

Schwartz, A. C. 1973. Terpenoid quinones of the anaerobic Propionibacterium shermanii. I. (II,3)-Tetrahydromenaquinone-9. Arch. Mikrobiol. 91:273–279.

Schwebke, J. R., S. L. Hillier, M. J. Fohn, and S. A. Lukehart. 1990. Demonstration of heterogeneity among the antigenic proteins of Mobiluncus species. J. Clin. Microbiol. 28:463–468.

Schwebke, J. R., S. A. Lukehart, M. C. Roberts, and S. L. Hillier. 1991. Identification of two new antigenic subgroups within the genus Mobiluncus. J. Clin. Microbiol. 29:2204–2208.

Schwebke, J. R., S. C. Morgan, and S. L. Hillier. 1996. Humoral antibody to Mobiluncus curtisii, a potential serological marker for bacterial vaginosis. Clin. Diagn. Lab. Immunol. 3:567–569.

Schwebke, J. R., and L. F. Lawing. 2001. Prevalence of Mobiluncus spp. among women with and without bacterial vaginosis as detected by polymerase chain reaction. Sex. Transm. Dis. 28:195–199.

Seno, N., and R. Azuma. 1983. A Study on Zin Japan and its Causative Microorganisms. National Institute of Animal Health. Tokyo, Japan. 23:82–91.

Shakespeare, A. P., D. B. Drucker, and R. M. Green. 1985. The comparative cariogenicity and plaque-forming ability in vivo of four species of the bacterium Actinomyces in gnotobiotic rats. Arch. Oral Biol. 30:855–858.

Shimada, M., T. Kotani, S. Ohtaki, S. Tateno, H. Tanigawa, and T. Katsuki. 1986. Primary perianal actinomycosis over a thirty year period. Jpn. J. Surg. 16:302–304.

Shurbaji, M. S., P. K. Gupta, and M. M. Newman. 1987. Hepatic actinomycosis diagnosed by fine-needle aspiration. Acta Cytol. 31:751–755.

Simpson, A. J., S. S. Das, and I. J. Mitchelmore. 1996. Polymicrobial brain abscess involving Haemophilus paraphrophilus and Actinomyces odontolyticus. Postgrad. Med. J. 72:297–298.

Skalka, B., I. Literak, P. Chalupa, and M. Votava. 1998. Phospholipase D-neutralization in serodiagnosis of Arcanobacterium haemolyticum and Corynebacterium pseudotuberculosis infections. Zbl. Bakteriol. 288:463–470.

Skarin, A., and P.-A. Mårdh. 1982a. Comma-shaped bacteria associated with vaginitis. Lancet i:342–343.

Skarin, A., L. Larsson, E. Holst, and P. A. Mårdh. 1982b. Gas chromatographic study of cellular fatty acids of comma-shaped bacteria isolated from the vagina. Eur. J. Clin. Microbiol. 1:307–309.

Skerman, V. B. D., V. McGowan, and P. H. A. Sneath. 1980. Approved Lists of Bacterial Names. Int. J. Syst. Bacteriol. 30:225–420.

Skov, R. L., A. K. Sanden, V. H. Danchell, K. Robertsen, and T. Ejlertsen. 1998. Systemic and deep-seated infections caused by Arcanobacterium haemolyticum. Eur. J. Clin. Microbiol. Infect. Dis. 17:578–582.

Slack, J. M. 1942. The source of infection in actinomycosis. J. Bacteriol. 43:193–209.

Slack, J. M., E. H. Ludwig, H. H. Bird, C. M. Canawy. 1951. Studies with microaerophilic actinomycetes. I: The agglutination reaction. J. Bacteriol. 61:721–735.

Slack, J. M., R. G. Spears, W. G. Snodgrass, and R. J. Kuchler. 1955. Studies with microaerophilic actinomycetes. II: Serological groups as determined by the reciprocal agglutinin adsorption technique. J. Bacteriol. 70:400–404.

Slack, J. M., A. Winger, and D. W. Moore Jr. 1961. Serological grouping of Actinomyces by means of fluorescent antibodies. J. Bacteriol. 82:54–65.

Slack, J. M., S. Landfried, and M. A. Gerecser. 1971. Identification of Actinomyces and related bacteria in dental calculus by the fluorescent antibody technique. J. Dent. Res. 50:78–82.

Slack, J. M. 1974. Family Actinomycetaceae Buchanan 1918 and genus Actinomyces Harz 1877. In: R. E. Buchanan and N. E. Gibbons (Eds.) Bergey's Manual of Determinative Bacteriology, 8th ed. Williams & Wilkins. Baltimore, MD. 659–667.

Slack, J. M., and M. A. Gerencser. 1975. Actinomyces, Filamentous Bacteria. Burgess Publishing. Minneapolis, MN.

Slee, A. M., and J. R. O'Connor. 1983. In vitro antiplaque activity of octenidine dehydrochloride (WIN 41464-2) against preformed plaques of selected oral plaque-forming microorganisms. Antimicrob. Agents Chemother. 23:379–384.

Smego Jr., R. A. 1987. Actinomycosis of the central nervous system. Rev. Infect. Dis. 9:855–865.

Smith, H., and H. B. Moore. 1988. Isolation of Mobiluncus species from clinical specimens using cold enrichment and selective media. J. Clin. Microbiol. 26:1134–1137.

Snyder, J. R., J. R. Pascoe, and D. C. Hirsh. 1987. Antimicrobial susceptibility of microorganisms isolated from equine orthopedic patients. Vet. Surg. 16:197–201.

Sobestiansky, J., M. Wendt, R. Perestrelo, and A. Ambrogi. 1993. Studies on the prevalence of Eubacterium suis in boars on farms in Brazil, Portugal and Argentina by indirect immunofluorescence technique. Dtsch. Tierärztl. Wochenschr. 100:463–464.

Socransky, S. S. 1970. Relationship of bacteria of the etiology of periodontal disease. J. Dent. Res. 49:203–222.

Socransky, S. S., and S. D. Manganiello. 1971. The oral microbiota of man from birth to senility. J. Periodontol. 42:485–496.

Soltys, M. A., and F. R. Spratling. 1957. Infectious cystitis and pyelonephritis in pigs: A preliminary communication. Vet. Rec. 69:500–504.

Soltys, M. A. 1961. Corynebacterium suis associated with specific cystitis and pyelonephritis in pigs. J. Pathol. Bacteriol. 81:441–446.

Sone, N. 1974. Isolation of a novel menaquinone with a partially hydrogenated side chain from Propionibacterium arabinosum. J. Biochem. 76:133–136.

Spapen, H. D., P. de Quint, F. de Geeter, R. Sacre, and S. J. Belle. 1989. Cervico-facial actinomycosis in a patient treated for tonsillar carcinoma. Eur. J. Surg. Oncol. 15:383–385.

Spiegel, C. A., P. Davick, P. A. Totten, K. C. S. Chen, D. A. Eschenbach, R. Amsel, and K. K. Holmes. 1983a. G. vaginalis and anaerobic bacteria in the etiology of bacterial (nonspecific) vaginosis. Scand. J. Infect. Dis. Suppl. 40:41–46.

Spiegel, C. A., D. A. Eschenbach, R. Amsel, and K. K. Holmes. 1983b. Curved anaerobic bacteria in bacterial (nonspecific) vaginosis and their response to antimicrobial therapy. J. Infect. Dis. 148:817–822.

Spiegel, C. A., and M. Roberts. 1984a. Mobiluncus gen. nov., Mobiluncus curtisii subsp. curtisii sp. nov., Mobiluncus curtisii subsp. holmesii subsp. nov., and Mobiluncus mulieris sp. nov., curved rods from the human vagina. Int. J. Syst. Bacteriol. 34:177–184.

Spiegel, C. A., and G. Telford. 1984b. Isolation of Wolinella recta and Actinomyces viscosus from an actinomycotic chest wall mass. J. Clin. Microbiol. 20:1187–1189.

Spiegel, C. A. 1987. Susceptibility of Mobiluncus species to 23 antimicrobial agents and 15 other compounds. Antimicrob. Agents Chemother. 31:249–252.

Spiegel, C. A. 1991. Bacterial vaginosis. Clin. Microbiol. Rev. 4:485–502.

Spiegel, C. A. 1992. The Genus Mobiluncus. In: A. Balows, H. G. Trüper, M. Dworkin, W. Harder, and K.-H. Schleifer (Eds.) The Prokaryotes, 2nd ed. Springer-Verlag. New York, NY. 906–917.

Sprecher, D. J., P. H. Coe, and R. D. Walker. 1999. Relationships among seminal culture, seminal white blood cells, and the percentage of primary sperm abnormalities in bulls evaluated prior to the breeding season. Theriogenology 51:1197–1206.

Sprott, M. S., R. S. Pattman, H. R. Ingham, G. R. Short, N. K. Narang, and J. B. Selkon. 1982. Anaerobic curved rods in vaginitis. Lancet i:54.

Sprott, M. S., H. R. Ingham, R. S. Pattman, R. L. Eisenstadt, G. R. Short, H. K. Narang, P. R. Sisson, and J. B. Selkon. 1983. Characteristics of motile curved rods in vaginal secretions. J. Med. Microbiol. 16:175–182.

Sprott, M. S., H. R. Ingham, R. S. Pattman, L. M. Clarkson, A. A. Codd, and H. K. Narang. 1984. Motile curved bacilli: Isolation and investigation. *In:* P.-A. Mårdh and D. Taylor-Robinson (Eds.) Bacterial Vaginosis. Almqvist and Wiksell International. Stockholm, Sweden. 107–112.

Stackebrandt, E., and C. R. Woese. 1981. Towards a phylogeny of actinomycetes and related organisms. Curr. Microbiol. 5:131–136.

Stackebrandt, E., and O. Charfreitag. 1990. Partial 16S rRNA primary structure of five Actinomyces species: Phylogenetic implications and development of an Actinomyces israelii-specific oligonucleotide probe. J. Gen. Microbiol. 136:37–43.

Stackebrandt, E., and B. M. Goebel. 1994. A place for DNA-DNA reassociation and 16S rRNA sequence analysis in the present species definition in bacteriology. Int. J. Syst. Bacteriol. 44:846–849.

Stackebrandt, E., F. A. Rainey, and N. L. Ward-Rainey. 1997. Proposal for a new hierarchic classification system, Actinobacteria classis nov. Int. J. Syst. Bacteriol. 47:479–491.

Stein, E., and K. P. Schaal. 1987. Die menschlichen Aktinomykosen aus heutiger Sicht. Colo-proct. 9:37–42.

Storms, V., J. Hommez, L. A. Devriese, M. Vaneechoutte, T. De Baere, M. Baele, R. Coopman, G. Verschraegen, M. Gillis, and F. Haesebrouck. 2002. Identification of a new biotype of Actinomyces hyovaginalis in tissues of pigs during diagnostic bacterial examination. Vet. Microbiol. 84:93–102.

Strain, G. M., M. S. Claxton, S. E. Turnquist, and J. M. Kreeger. 1987. Evoked potential and electroencephalographic assessment of central blindness due to brain abscesses in a steer. Cornell Vet. 77:374–382.

Strazzeri, J. C., and S. Anzel. 1986. Infected total hip arthroplasty due to Actinomyces israelii after dental extraction: A case report. Clin. Orthop. 210:128–131.

Sturm, A. W., B. Jamil, K. P. McAdam, K. Z. Khan, S. Parveen, T. Chiang, and R. Hussain. 1996. Microbial colonizers in leprosy skin ulcers and intensity of inflammation. Int. J. Lepr. Other Mycobact. Dis. 64:274–281.

Sumita, M., E. Hoshino, and M. Iwaku. 1998. Experimental actinomycosis in mice induced by alginate gel particles containing Actinomyces israelii. Endod. Dent. Traumatol. 14:137–143.

Sunde, P. T., L. Tronstad, E. R. Eribe, P. O. Lind, and I. Olsen. 2000. Assessment of periradicular microbiota by DNA-DNA hybridization. Endod. Dent. Traumatol. 16:191–196.

Sutter, V. L., and S. M. Finegold. 1972. Anaerobic Bacteriology Manual. UCLA. Los Angeles, CA.

Sutter, V. L. 1984. Anaerobes as normal oral flora. Rev. Infect. Dis. Suppl. 1:62–66.

Suzuki, J. B., and A. L. Delisle. 1984. Pulmonary actinomycosis of periodontal origin. J. Periodontol. 55:581–584.

Swindlehurst, C. A., H. N. Shah, C. W. Parr, and R. A. D. Williams. 1977. Sodium dodecyl sulphate polyacrylamide gel electrophoresis of polypeptides from Bacteroides melaninogenicus. J. Appl. Bacteriol. 43:319–324.

Syed, S. A., and W. J. Loesche. 1972. Survival of human dental plaque flora in various transport media. Appl. Microbiol. 24:638–644.

Syed, S. A., M. Syanberg, and G. Svanberg. 1981. The predominant cultivable dental plaque flora of beagle dogs with periodontitis. J. Clin. Periodontol. 8:45–56.

Taichman, N. S., B. F. Hammond, C.-C. Tsai, P. C. Baehni, and W. P. McArthur. 1978. Interaction of inflammatory cells and oral microorganisms. VII: In vitro polymorphonuclear response to viable bacteria and to subcellular components of avirulent and virulent strains of Actinomyces viscosus. Infect. Immun. 21:594–604.

Takazoe, I. 1961. Study on the intracellular calcification of oral aerobic leptotrichia. Shika Gakuho 61:394–401.

Takazoe, I., T. Nakamura, and K. Okuda. 1984. Colonization of the subgingival area by Bacteroides gingivalis. J. Dent. Res. 63:422–426.

Takemura, M., N. Yokoi, Y. Nakamura, A. Komuro, J. Sugita, and S. Kinoshita. 2002. Canaliculitis caused by Actinomyces in a case of dry eye with punctal plug occlusion. Nippon Ganka Gakkai Zasshi 106:416–419.

Taylor-Robinson, A. W., and D. Taylor-Robinson. 2002. Evaluation of liquid culture media to support growth of Mobiluncus species. J. Med. Microbiol. 51:491–494.

Teo, C., L. Kwong, and R. Benn. 1987. Incidence of motile, curved anaerobic rods (Mobiluncus species) in vaginal secretions. Pathology 19:193–196.

Thomason, J. L., P. C. Schreckenberger, L. J. LeBeau, L. M. Wilcoski, and W. N. Spellacy. 1984. A selective and differential agar for anaerobic comma-shaped bacteria recovered from patients having motile rods and nonspecific vaginosis. *In:* P.-A. Mårdh and D. Taylor-Robinson (Eds.) Bacterial Vaginosis. Almqvist and Wiksell International. Stockholm, Sweden. 125–128.

Thompson, L. 1950. Isolation and comparison of Actinomyces from human and bovine infections. Proc. Staff Meet. Mayo Clin. 25:81–86.

Thompson, L., and S. A. Lovestedt. 1951. An Actinomyces-like organism obtained from the human mouth. Proc. Staff Meet. Mayo Clin. 6:169–175.

Tiveljung, A., J. Backstrom, U. Forsum, and H. J. Monstein. 1995. Broad-range PCR amplification and DNA sequence analysis reveals variable motifs in 16S rRNA genes of Mobiluncus species. Acta Pathol. Microbiol. Immunol. Scand. 103:755–763.

Tiveljung, A., U. Forsum, and H. J. Monstein. 1996. Classification of the genus Mobiluncus based on comparative partial 16S rRNA gene analysis. Int. J. Syst. Bacteriol. 46:332–336.

Tompkins, G. R., and J. R. Tagg. 1986. Incidence and characterization of antimicrobial effects produced by Actinomyces viscosus and Actinomyces naeslundii. J. Dent. Res. 65:109–112.

Traynor, R. M., D. Parratt, H. L. D. Duguid, and I. D. Duncan. 1981. Isolation of actinomycetes from cervical specimens. J. Clin. Pathol. 34:914–916.

Trinh, H. T., S. J. Billington, A. C. Field, J. G. Songer, and B. H. Jost. 2002. Susceptibility of Arcanobacterium pyogenes from different sources to tetracycline, macrolide and lincosamide antimicrobial agents. Vet. Microbiol. 85:353–359.

Tsai, M. S., J. J. Tarn, K. S. Liu, Y. L. Chou, and C. L. Shen. 2001. Multiple Actinomyces brain abscesses: Case report. J. Clin. Neurosci. 8:183–186.

Tvede, M., J. Bodenhoff, and B. Bruun. 1985. Actinomycotic infections of the central nervous system: Two case reports. Acta Pathol. Microbiol. Immunol. Scand. B 93:327–330.

Tylenda, C. A., C. Calvert, P. E. Kolenbrander, and A. Tylenda. 1985. Isolation of Actinomyces bacteriophage from human dental plaque. Infect. Immun. 49:1–6.

Tyrrell, K. L., D. M. Citron, J. R. Jenkins, and E. J. Goldstein. 2002. Periodontal bacteria in rabbit mandibular and maxillary abscesses. J. Clin. Microbiol. 40:1044–1047.

Tzora, A., L. S. Leontides, G. S. Amiridis, G. Manos, and G. C. Fthenakis. 2002. Bacteriological and epidemiological findings during examination of the uterine content of ewes with retention of fetal membranes. Theriogenology 57:1809–1817.

Valicenti, J. F., A. A. Pappas, C. D. Graber, H. O. Williamson, and N. F. Willis. 1982. Detection and prevalence of IUD-associated Actinomyces colonization and related morbidity. J. Am. Med. Assoc. 247:1149–1152.

Van Bosterhaut, B., P. Boucquey, M. Janssen, G. Wauters, and M. Delmee. 2002. Chronic osteomyelitis due to Actinomyces neuii subspecies neuii and Dermabacter hominis. Eur. J. Clin. Microbiol. Infect. Dis. 21:486–487.

Vandamme, P., E. Falsen, M. Vancanneyt, M. Van Esbroeck, D. Van de Merve, A. Bergmans, L. Schouls, and L. Sabbe. 1998. Characterization of Actinomyces turicensis and Actinomyces radingae strains from human clinical samples. Int. J. Syst. Bacteriol. 48:503–510.

Van der Hoeven, J. S., M. H. de Jong, A. H. Rogers, and P. J. Kamp. 1984. A conceptual model for the co-existence of Streptococcus spp. and Actinomyces spp. in dental plaque. J. Dent. Res. 63:389–392.

Van Mook, W. N., F. S. Simonis, P. M. Schneeberger, and J. L. van Opstal. 1997. A rare case of disseminated actinomycosis caused by Actinomyces meyeri. Netherl. J. Med. 51:39–45.

Vannier, J. P., G. Schaison, B. George, and I. Casin. 1986. Actinomycotic osteomyelitis of the skull and atlas with late dissemination: A case of transient neurosurgical syndrome. Eur. J. Ped. 145:316–318.

Van Steensel, C. J., and T. S. Kwan. 1988. Actinomycosis of the gallbladder. Netherl. J. Surg. 40:23–25.

Verrot, D., P. Disdier, J. R. Harle, Y. Peloux, L. Garbes, A. Arnaud, and P. J. Weiller. 1993. Pulmonary actinomycosis: Caused by Actinomyces odontolyticus? Rev. Med. Intern. 14:179–181.

Vetere, A., S. P. Borriello, E. Fontaine, P. J. Reed, and D. Taylor-Robinson. 1987. Characterisation of anaerobic curved rods (Mobiluncus spp.) isolated from the urogenital tract. J. Med. Microbiol. 23:279–288.

Vogel, G., J. Nicolet, J. Martig, P. Tschudi, and M. Meylan. 2001. Pneumonia in calves: characterization of the bacterial spectrum and the resistance pattern to antimicrobial drugs. Schweiz. Arch. Tierheilkd. 143:341–350.

Von Graevenitz, A. 1994. Alpha-mannosidase in Arcanobacterium haemolyticum. J. Clin. Microbiol. 32:854–855.

Votava, M., B. Skalka, P. Ondrovcik, F. Ruzicka, J. Svoboda, and V. Woznicova. 2000. A diagnostic medium for Arcanobacterium haemolyticum and other bacterial species reacting with hemolytic synergism to the equifactor of Rhodococcus equi. Epidemiol. Mikrobiol. Immunol. 49:123–129.

Votava, M., B. Skalka, V. Woznicova, F. Ruzicka, O. Zahradnicek, P. Ondrovcik, and L. Klapacova. 2001. Detection of Arcanobacterium haemolyticum phospholipase D neutralizing antibodies in patients with acute tonsillitis. Epidemiol. Mikrobiol. Immunol. 50:111–116.

Waage, S., T. Mork, A. Roros, D. Aasland, A. Hunshamar, and S. A. Odegaard. 1999. Bacteria associated with clinical mastitis in dairy heifers. J. Dairy Sci. 82:712–719.

Wahl, G., and K. P. Schaal. 1989. Mikroben in subgingivalen Implantatspalträumen. Zbl. Zahnärztl. Implantol. V:287–291.

Wahl, G., F. Müller, and K. P. Schaal. 1992. Die mikrobielle Besiedlung von Implantatelementen aus Kunststoff und Titan. Schweiz. Monatsschr. Zahnmed. 102:1321–1326.

Wajszczuk, C. P., T. F. Logan, A. W. Pasculle, and M. Ho. 1984. Intraabdominal actinomycosis presenting with sulphur granules in the urine. Am. J. Med. 77:1126–1128.

Walker, C. B. 1988. Microbial effects of mouthrinses containing antimicrobials. J. Clin. Periodontol. 15:499–505.

Walker, R. L., and N. J. MacLachlan. 1989. Isolation of Eubacterium suis from sows with cystitis. J. Am. Vet. Med. Assoc. 195:1104–1107.

Walker, R. L., R. T. Greene, and T. M. Gerig. 1990. Evaluation of four commercial systems for identification of Eubacterium suis. J. Vet. Diagn. Invest. 2:318–322.

Warner, T. N., and C. H. Miller. 1978. Cell-associated levan of Actinomyces viscosus. Infect. Immun. 19:711–719.

Wayne, L. G., D. J. Brenner, R. R. Colwell, P. A. D. Grimont, P. Kandler, M. I. Krichevsky, L. H. Moore, W. E. C. Moore, R. G. E. Murray, E. Stackebrandt, M. P. Starr, and H. G. Trüper. 1987. Report of the ad hoc committee on reconciliation of approaches to bacterial systematics. Int. J. Syst. Bacteriol. 37:463–464.

Weerkamp, A. H., H. N. Uyen, and H. J. Busscher. 1988. Effect of zeta potential and surface energy on bacterial adhesion to uncoated and saliva-coated human enamel and dentin. J. Dent. Res. 67:1483–1487.

Wegienek, J., and C. A. Reddy. 1982a. Nutritional and metabolic features of Eubacterium suis. J. Clin. Microbiol. 15:895–901.

Wegienek, J., and C. A. Reddy. 1982b. Taxonomic study of "Corynebacterium suis" Soltys and Spratling: Proposal of Eubacterium suis (nov. rev.) comb. nov. Int. J. Syst. Bacteriol. 32:218–228.

Wegman, M. R., A. D. Eisenberg, M. E. Curzon, and S. L. Handelman. 1984. Effects of fluoride, lithium, and strontium on intracellular polysaccharide accumulation in S. mutans and A. viscosus. J. Dent. Res. 63:1126–1129.

Weiss, E. I., P. E. Kolenbrander, J. London, A. R. Hand, and R. N. Anderson. 1987. Fimbria-associated proteins of Bacteroides loeschei PK 1295 mediate intergeneric coaggregation. J. Bacteriol. 169:4215–4222.

Wendt, M., M. Liebhold, and W. Drommer. 1994. Scanning electron microscopic studies of the urinary bladder of sows with special reference to a Eubacterium suis infection. Zentralbl. Veterinärmed. 41:126–138.

Wendt, M., and G. Amtsberg. 1995a. Serologic examination for the detection of antibodies against Eubacterium suis in swine. Schweiz. Arch. Tierheilkd. 137:129–136.

Wendt, M., and J. Sobestiansky. 1995b. The therapy of urinary tract infections in sows. Dtsch. Tierärztl. Wochenschr. 102:21–27.

Westling, K., C. Lidman, and A. Thalme. 2002. Tricuspid valve endocarditis caused by a new species of actinomyces: Actinomyces funkei. Scand. J. Infect. Dis. 34:206–207.

Wheeler, T. T., W. B. Clark, M. D. Lane, and T. E. Grow. 1983. Influence of physicochemical parameters on adsorption of Actinomyces viscosus to hydroxyapatite surfaces. Infect. Immun. 39:1095–1101.

Whiley, R. A., H. Fraser, J. M. Hardie, and D. Beighton. 1990. Phenotypic differentiation of Streptococcus intermedius, Streptococcus constellatus, and Streptococcus anginosus strains within the "Streptococcus milleri group. J. Clin. Microbiol. 28:1497–1501.

White, C. B., and W. S. Foshee. 2000. Upper respiratory tract infections in adolescents. Adolesc. Med. 11:225–249.

Wicken, A. J., K. W. Broady, J. D. Evans, and K. W. Knox. 1978. New cellular and extracellular amphipathic antigens from Actinomyces viscosus NY1. Infect. Immun. 22:615–616.

Wiggs, L. S., J. J. Cavallaro, and J. M. Miller. 2000. Evaluation of the Oxyrase OxyPlate anaerobe incubation system. J. Clin. Microbiol. 38:499–507.

Wikstrom, M. B., G. Dahlen, and A. Lind. 1983. Fibrinogenolytic and fibrinolytic activity in oral microorganisms. J. Clin. Microbiol. 17:759–760.

Winford, T. E., and S. Haberman. 1966. Isolation of aerobic Gram-positive filamentous rods from diseased gingivae. J. Dent. Res. 45:1159–1167.

Winslow, C. E. A., J. Broadhurst, R. E. Buchanan, C. Krumwiede, L. A. Rogers, and G. H. Smith. 1920. The families and genera of bacteria. J. Bacteriol. 5:191–229.

Wohlgemuth, S. D., and M. C. Gaddy. 1986. Surgical implications of actinomycosis. South. Med. J. 79:1574–1578.

Wolff, M., and J. Israel. 1891. Über Reincultur des Actinomyces und seine Übertragbarkeit auf Thiere. Archiv der Pathologischen Anatomie, Physiologie und Klinischen Medicin 126:11–59.

Woo, P. C., A. M. Fung, S. K. Lau, E. Hon, and K. Y. Yuen. 2002. Diagnosis of pelvic actinomycosis by 16S ribosomal RNA gene sequencing and its clinical significance. Diagn. Microbiol. Infect. Dis. 43:113–118.

Woo, P. C. Y., A. M. Y. Fung, S. K. P. Lau, J. L. L. Teng, B. H. L. Wong, M. K. M. Wong, E. Hon, G. W. K. Tang, and K.-Y. Yuen. 2003. Actinomyces hongkongensis sp. nov.: A novel Actinomyces species isolated from a patient with pelvic actinomycosis. System. Appl. Microbiol. 26:518–522.

Wright Jr., J. R., D. Stinson, A. Wade, D. Haldane, and S. A. Heifetz. 1994. Necrotizing funisitis associated with Actinomyces meyeri infection: A case report. Pediatr. Pathol. 14:927–934.

Wüst, J., S. Stubbs, N. Weiss, G. Funke, and M. D. Collins. 1995. Assignment of Actinomyces pyogenes-like (CDC Coryneform Group E) bacteria to the genus Actinomyces as Actinomyces radingae sp. nov. and Actinomyces turicensis sp. nov. Lett. Appl. Microbiol. 20:76–81.

Wüst, J., U. Steiger, H. Vuong, and R. Zbinden. 2000. Infection of a hip prosthesis by Actinomyces naeslundii. J. Clin. Microbiol. 38:929–930.

Ximenez-Fyvie, L. A., A. D. Haffajee, L. Martin, A. Tanner, P. Macuch, and S. S. Socransky. 1999. Identification of oral Actinomyces species using DNA probes. Oral Microbiol. Immunol. 14:257–265.

Ximenez-Fyvie, L. A., A. D. Haffajee, and S. S. Socransky. 2000a. Comparison of the microbiota of supra- and subgingival plaque in health and periodontitis. J. Clin. Periodontol. 27:648–657.

Ximenez-Fyvie, L. A., A. D. Haffajee, and S. S. Socransky. 2000b. Microbial composition of supra- and subgingival plaque in subjects with adult periodontitis. J. Clin. Periodontol. 27:722–732.

Yamada, Y., G. Inouye, Y. Tahara, and K. Kondo. 1976. The menaquinone system in the classification of coryneform and nocardioform bacteria and related organisms. J. Gen. Appl. Microbiol. 22:203–214.

Yamini, B., and R. F. Slocombe. 1988. Porcine abortion caused by Actinomyces suis. Vet. Pathol. 25:323–324.

Yenson, A., H. O. deFries, and Z. E. Deeb. 1983. Actinomycotic osteomyelitis of the facial bones and mandibula. Otolaryngol. Head Neck Surg. 91:173–176.

Yeung, M. K., B. M. Chassy, and J. O. Cisar. 1978. Cloning and expression of a type 1 fimbrial subunit of Actinomyces viscosus T14V. J. Bacteriol. 169:1678–1683.

Yeung, M. K., and J. O. Cisar. 1988. Cloning and nucleotide sequence of a gene for Actinomyces naeslundii WVU45 type 2 fimbriae. J. Bacteriol. 170:3803–3809.

Yeung, M. K., J. A. Donkersloot, J. O. Cisar, and P. A. Ragsdale. 1998. Identification of a gene involved in assembly of Actinomyces naeslundii T14V type 2 fimbriae. Infect. Immun. 66:1482–1491.

Yoshimura, H., A. Kojima, and M. Ishimaru. 2000. Antimicrobial susceptibility of Arcanobacterium pyogenes isolated from cattle and pigs. J. Vet. Med. Bur. Infect. Dis. Vet. Public Health 47:139–143.

Zakut, H., R. Achiron, O. Treschan, and E. Kutin. 1987. Actinomyces invasion of placenta as a possible cause of preterm delivery. Clin. Exp. Obstet. Gynecol. 14:89–91.

Zero, D. T., J. van Houte, and J. Russo. 1986. Enamel demineralization by acid produced from endogenous substrate in oral streptococci. Arch. Oral Biol. 31:229–234.

Zylber, L. J., and H. V. Jordan. 1982. Development of a selective medium for detection and enumeration of Actinomyces viscosus and Actinomyces naeslundii in dental plaque. J. Clin. Microbiol. 15:253–259.

Prokaryotes (2006) 3:538–604
DOI: 10.1007/0-387-30743-5_22

CHAPTER 1.1.7

The Family Streptomycetaceae, Part I: Taxonomy

PETER KÄMPFER

Phylogeny and Taxonomy

The family Streptomycetaceae was created by Waksman and Henrici (1943). Originally this family harbored only the type genus *Streptomyces*. Zhang et al. (1997) proposed that the genus *Kitasatospora* be included, and recently, a third genus, *Streptacidiphilus*, was added (Kim et al., 2003).

Description of the family Streptomycetaceae Waksman and Henrici 1943 emend, Kim et al. (2003) (Strep.to.my.ce.ta'ce.ae. ending to denote a family; M.L. masc. n. *Streptomyces*, type genus of the family) is based on data taken from Williams et al. (1989), Zhang et al. (1997) and Kim et al. (2003). These aerobic, Gram-positive, non-acid-alcohol fast actinomycetes form an extensively branched substrate mycelium that rarely fragments. The aerial mycelium forms chains of three to many spores. Members of a few species bear short chains of spores on the substrate mycelium. The organisms produce a wide range of pigments responsible for the color of the substrate and aerial mycelium. The organisms grow within different pH ranges, namely 5.5–9 (*Kitasatospora*), 5–11.5 (*Streptomyces*), and 3.5–6.0 (*Streptacidiphilus*). They are chemoorganotrophic with an oxidative type of metabolism. The substrate mycelium contains either LL-(*Streptacidiphilus* and *Streptomyces*) or *meso*-(*Kitasatospora*) diaminopimelic acid as the predominant diamino acid; aerial or submerged spores contain LL-diaminopimelic acid. In whole-organism sugar profiles, either major amounts of galactose or galactose and rhamnose (*Kitasatospora* and *Streptacidiphilus*) can be detected. Lipid profiles typically contain hexa- and octa-hydrogenated menaquinones with nine isoprene units as the predominant isoprenologues. The polar lipid profiles are composed of diphosphatidylglycerol, phosphatidylethanolamine, phosphatidylinositol, and phosphatidylinositol mannosides. Fatty acids are complex mixtures of saturated, *iso*- and *anteiso*-fatty acids. Mycolic acids are not present. The mol% G + C of the DNA ranges generally between 66 and 74%. Members of all three taxa are widely distributed in terrestrial habitats, especially soil. Very few species are pathogens for animals (including man) and plants.

A phylogenetic tree showing selected representatives of all three genera (all species of *Streptacidiphilus* and *Kitasatospora* and selected *Streptomyces* "species") representing the clusters of the numerical taxonomic study of Williams et al. (1983a) is shown in Fig. 1. The genera are difficult to differentiate on the basis of phenotypic features (including chemotaxonomic markers). Some characteristic features are shown in Table 1.

History

Early investigations of actinomycetes, including streptomycetes, were dominated by a strong emphasis of morphology and the high degree of morphological diversity was subsequently considered to be sufficient for their assignment to genera and families (Waksman, 1961; Cross and Goodfellow, 1973). A short summary of early classification systems of actinomycetes is given in Introduction to the Classification of the Actinomyces in this Volume. Streptomycetes are the producers of more than 5000 known bioactive compounds (Anderson and Wellington, 2001), and estimates of the total number of antimicrobial compounds produced by representatives of *Streptomyces* screened for new antibiotics are of the order of 100,000 (Watve et al., 2001). In addition, not only has the overall versatility of these compounds been studied in great detail, but also a high proportion of them have known biological effects, which is unparalleled in the living world (Kieser et al., 2000).

The family Streptomycetaceae was originally proposed by Waksman and Henrici (1943) and contained at that time only two genera: the genus *Streptomyces* and the genus *Micromonospora*. *Streptomyces* was described as "Streptomycetaceae," forming spores in chains on aerial hyphae. Spores are apparantly endogeneous in origin, formed by a segregation of protoplasm within the hyphae into a series of round oval or cylindrical bodies. Chains of spores are often spirally coiled. Sporophores may be simple or branched (Waksman and Henrici, 1943). Figure 2 shows the morphology of the aerial mycelium of

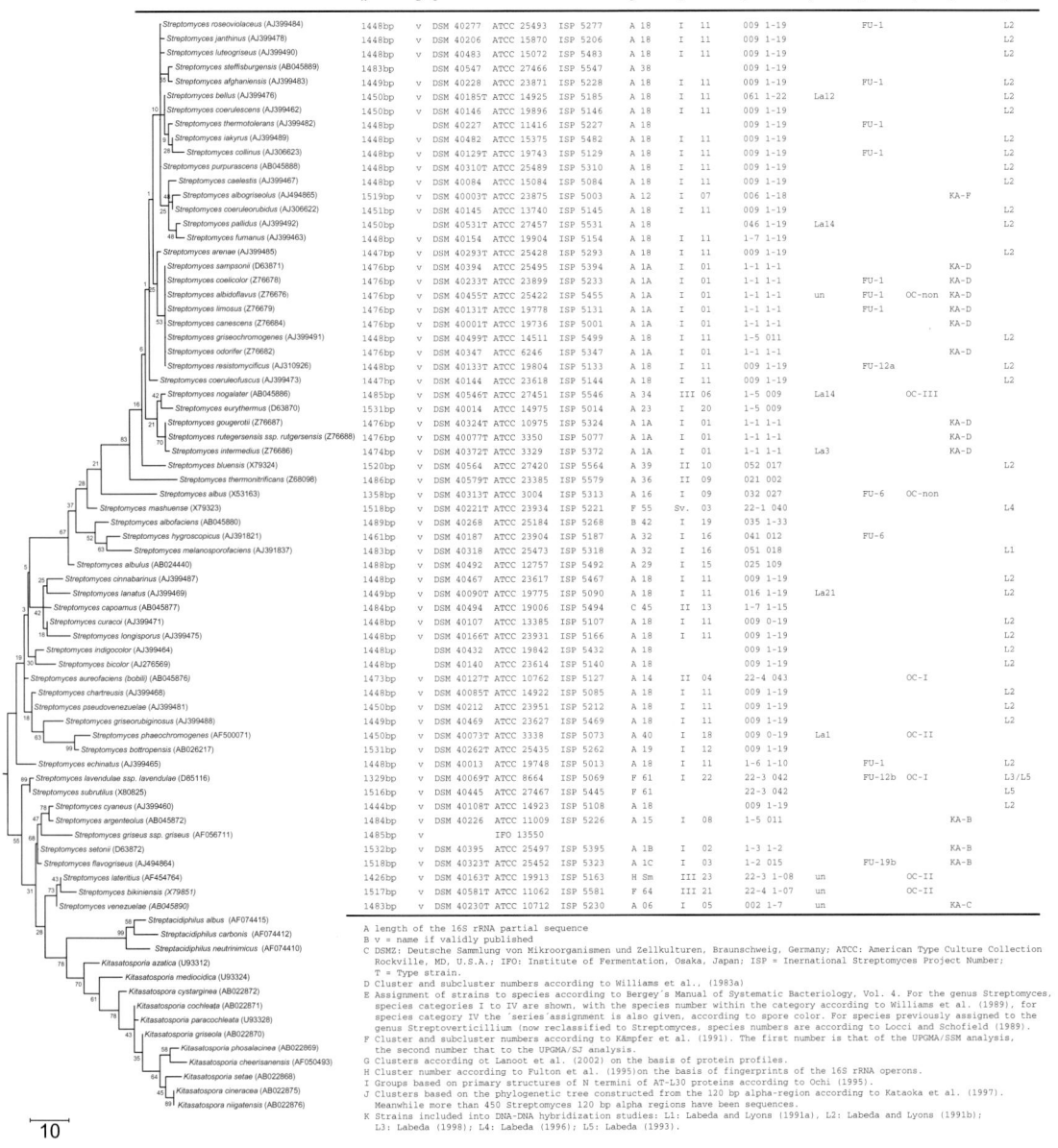

Fig. 1. Phylogenetic analysis based on 16S rRNA gene sequences available from the European Molecular Biology Laboratory data library (accession numbers are given in brackets) constructed after multiple alignment of data. Calculations of distances (distance options according to the Kimura-2 model) and clustering with maximum parsimony method were performed using the software package Mega (Molecular Evolutionary Genetics Analysis) version 2.1. Bootstrap values based on 1000 replications are listed as percentages at the branching points.

three streptomycetes. Although they are important (see Table 2 of The Family Nocardiopsaceae in this Volume), morphological differences between members of the actinomycete genera do not represent the extensive diversity of genera and species of the sporoactinomycetes.

Beginning in the 1950s, new developments in the application of numerical phenetics, numerical taxonomy, chemosystematics, and finally molecular systematics revolutionized the classification of actinomycetes. These developments have been excellently reviewed for actinomycete systematics by Goodfellow et al. (1999).

The determination of major cell wall sugars and peptidoglycan composition (Lechevalier and Lechevalier, 1970; Schleifer and Kandler, 1972) led to a classification system of well-characterized chemotypes and peptidoglycan types. The pioneering work of Lechevalier and coworkers (Becker et al., 1964; Lechevalier and Lechevalier, 1970) clearly showed that *Streptomyces* and the other genera of the family Strep-

Table 1. Chemotaxonomic, morphological and physiological characteristics of *Kitasatospora*, *Streptacidiphilus* and *Streptomyces* strains.

Characteristics	*Streptomyces*	*Kitasatospora*	*Streptacidiphilus*
Long chains of spores formed on aerial hyphae	+	+	+
Major menaquiones	MK-9(H$_6$, H$_8$)	MK-9(H$_6$, H$_8$)	MK-9(H$_6$, H$_8$)
Predominant phospholipids	DPG, PE, PI, and PIMs	DPG, PE, PI, and PIMs	DPG, PE, PI, and PIMs
Diagnostic sugars in whole-organism hydrolysates	None	Galactose[a]	Galactose and rhamnose
Fatty acid pattern[d]	2c	2c	2c
G+C content of DNA (mol %)	66–73	70–74	70–72
Isomer(s) of diaminopimelic acids in whole-organism hydrolysates	LL-A$_{2pm}$	LL-/*meso*A$_{2pm}$[b]	LL-A$_{2pm}$
Optimal pH range	6.5–8.0[c]	ND	4.5–5.5
pH range for growth	5.0–11.5	5.5–9.0	3.5–6.0

Symbols and abbreviations: +, present; 2c, fatty acid group *sensu* Kroppenstedt (1985); DPG, diphosphatidylglycerol; PE, phosphatidylethanol; PI, phosphatidylinositol; PIMs, phosphatidylinositol mannosides; MK-9(H$_6$, H$_8$), hexa-and octa-hydrogenated menaquinones with nine isoprene units; and ND, not determined.
[a]Rhamnose was detected in whole-organism hydolysates of *Kitasatospora mediocidica* (Labeda, 1988).
[b]Aerial and submerged spores contain LL-A$_{2pm}$ and vegetative mycelia *meso*-A$_{2pm}$.
[c]Alkalophilic strains, which grow between pH 8.0 and 11.5, have an optimum at pH 9–9.5 (Mikami et al., 1982).
From Kim et al. (2003) and the previous studies of Shirling and Gottlieb (1976), Omura et al. (1989), Lonsdale (1985), Williams et al. (1989) and Nakagaito et al. (1992).

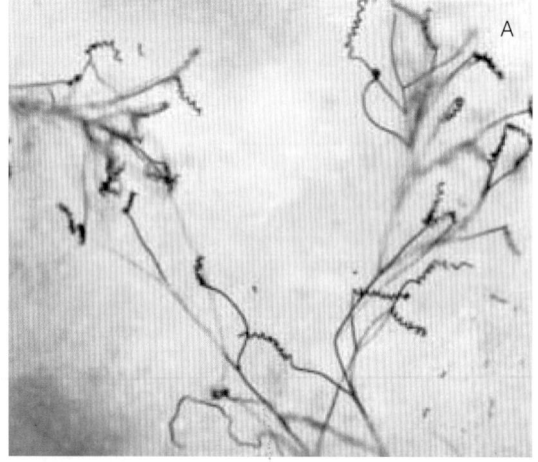

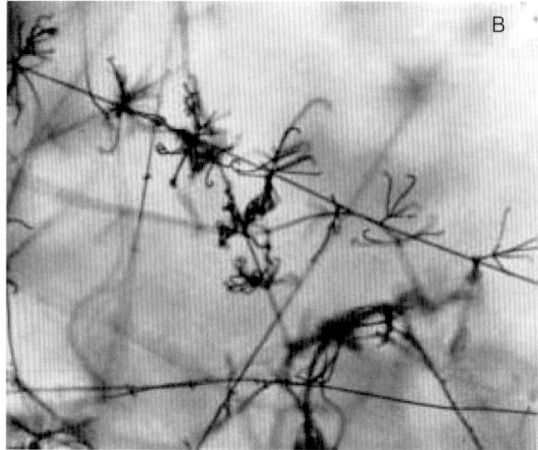

Fig. 2. Morphology of the aerial mycelium of three streptomycetes. (A) A *Streptomyces* species: sympodially branched aerial hyphae; spore chains form spirals with up to 10 turns. (B) A *Streptoverticillium* species: spore chains arranged in typical verticils along straight, long aerial hyphae; the end of the spore chain is sometimes hook-like or forms one to two turns. (C) "*Streptomyces pallidus*": despite the verticil-like arrangement of spore chains, this organism was described as *Streptomyces* by Shirling and Gottlieb (1972). All photos: ×250.

Table 2. Key biochemical markers of Streptomycetaceae genera and some other selected actinomycete genera (belonging to different families) producing an aerial mycelium.[a]

Family and genus	Diaminopimelic acid (A_{2pm})[b]	Glycine in IPB[b]	Peptidoglycan type[b]	Sugar type[b]	Phospholipid type[c]	Mycolic acids[b]	Fatty acid pattern[d]	Menaquinones[e]	G+C content (mol%)
Streptomycetaceae									
Kitasatospora	LL/*meso*	+	A3γ	C/E	PII	–	2c	9(H_6)/(H_8)	66–73
Streptomyces	LL	+	A3γ	–	PII	–	2c	9(H_6)/(H_8)	69–78
Streptacidophilus	LL	+	A3γ	E	PII	–	2c	9(H_6)/(H_8)	70–72
Pseudonocardiaceae									
Amycolatopsis	*meso*	–	A1γ	A	PII	–	3f	9(H_4)/(H_2)	66–69
Kibdelosporangium[f]	*meso*	–	A1γ	A	PII	–	3f	ND	66
Pseudonocardia	*meso*	–	A1γ	A	PIII	–	2f	8(H_4)	79
Saccharopolyspora	*meso*	–	A1γ	A	PIII	–	2c/3e	9(H_4)/10(H_4)/9(H_2)	70–72
Saccharomonospora	*meso*	–	A1γ	A	PII	–	2a	9(H_4)/8(H_4)	69–74
Actinopolyspora	*meso*	–	A1γ	A	PIII	–	2c	9(H_6)/9(H_4)	64
Genera belonging to different families of the Actinobacteria[g]									
Sporichthya	LL	+	A3γ	–	ND	–	3a	9(H_6)/9(H_8)	ND
Kineosporia[h]	LL/*meso*	(+)	ND	C	PIII	–	1	9(H_4)	69
Nocardioides	LL	+	A3γ	–	PI	–	3c	8(H_4)	66–73
Actinomadura[i]	*meso*	–	A1γ	B	PI	–	3a	9(H_6)/(H_4)/(H_8)	66–72
Microtetraspora[i]	*meso*	–	A1γ	B	PIV	–	3c	9(H_4)/(H_2)/(H_0)	66–69
Glycomyces	*meso*	+	ND	D	PI	–	2c	9(H_4)/10(H_4)	71–73
Saccharothrix	*meso*	–	ND	C/E	PII	–	3f	9(H_4)/10(H_4)	70–76
Nocardi[j]	*meso*	–	A1γ	A	PII	+	1b	cyclo 8(H_4)/9(H_2)	64–72
Nocardiopsis	*meso*	–	ND	C	PIII	–	3d	10(H_2)/(H_4)/(H_6)	64–69
Streptoalloteichus	*meso*	–	ND	C	PII	–	ND	9(H_6)/10(H_6)	ND

Symbols and abbreviations: +, present; –, not applicable for LL-A_{2pm}; IPB; interpeptide bridge; A, arabinose and xylose; E, rhamnose and galactose; C, no diagnostic sugars; B, madurose; D, arabinose and galactose; PI, phosphatidylglycerol (variable); PII, only phosphatidylethanolamine, PIII, phosphatidylcholine (with phosphatidylethanolamine, phosphatidylmethylethanolamine and phosphatidylglycerol variable and no phospholipids containing glucosamine); PIV, phospholipids containing glucosamine (with phosphatidylethanolamine and phosphatidylmethylethanolamine variable); Menaquinones: number indicates number of isoprene units, H_x indicates presence of x hydrogenated menaquinones; and ND, not determined.

[a]The genus *Kineosporia* does not produce aerial mycelium.
[b]Data from Goodfellow (1989).
[c]Data from Lechevalier et al. (1977, 1981).
[d]Data from Kroppenstedt (1985).
[e]Data from Kroppenstedt (1987) and R. Kroppenstedt (personal communication).
[f]Data from Bowen et al. (1989).
[g]For details, see Introduction to the Classification of the Actinomyces.
[h]Data from Itoh et al. (1989).
[i]Data from R. Kroppenstedt (personal communication).
[j]Data for menaquinones from Howarth et al. (1986).
Modified according to Korn-Wendisch and Kutzner (1992).

tomycetaceae, proposed by then, contained LL-diaminopimelic acid (LL-A$_2$pm) in its peptidoglycan (cell wall type I), whereas *meso*-A$_2$pm was found in most of the other actinomycetes described at that time. The genera containing LL-A$_2$pm in their peptidoglycan contained an interpeptide bridge composed of a glycine residue (type A3γ of Schleifer and Kandler, 1972). In addition to these chemical traits, which had the advantage of higher genetic stability in comparison with morphological features, the pattern of sugars in whole-cell hydrolysates (Lechevalier and Lechevalier, 1970), phospholipids (Lechevalier et al., 1977a), fatty acids (Kroppenstedt, 1985), menaquinones (Alderson et al., 1985; Kroppenstedt, 1985), and acetylated muramic acid residues (Uchida and Aida, 1977) were shown to be of essential importance for the classification of actinomycetes. Major chemotaxonomic markers of the genus *Streptomyces* and other actinomycete genera are given in Table 2. In combination with other phenotypic properties, like physiological and biochemical characteristics, these traits were also helpful in defining genera within the family Streptomycetaceae. This resulted in the reclassification of six additional genera (*Actinopycnidium, Actinosporangium, Chainia, Elytrosporangium, Kitasatoa* and *Microellobosporia*), described mainly on the basis of morphological features, to the genus *Streptomyces* (Williams et al., 1983a; Goodfellow et al., 1986b; Goodfellow et al., 1986c; Goodfellow et al., 1986d; Goodfellow et al., 1986e).

The application of 16S rRNA oligonucleotide cataloguing (Stackebrandt and Woese, 1981) and subsequently the sequencing of the 16S rRNA genes provided a basis for studies of the natural relationships among actinomycetes and related organisms (for details, see Stackebrandt et al., [1997] and Introduction to the Classification of the Actinomyces in this Volume). On the basis of these data, the description of the family Streptomycetaceae was emended by Wellington et al. (1992) and Witt and Stackebrandt (1990), who proposed the unification of the genera *Kitasatosporia* and *Streptoverticillium* with the genus *Streptomyces*, and more recently by Stackebrandt et al. (1997), who excluded the genus *Sporichthya*. Zhang et al. (1997) demonstrated, however, that the genus *Kitasatosporia* formed a stable subbranch in *Streptomyces*, when sequences from the almost complete 16S rRNA genes were compared. In addition, members of the genus *Kitasatosporia* can be distinguished from *Streptomyces* by the ratio of *meso*-DAP to LL-DAP and the presence of galactose in whole-cell hydrolysates (Zhang et al., 1997; Table 1). The genera *Kineosporia* and *Sporichthya*, both sharing chemotaxonomic similarities with members of the genus *Streptomyces* and considered

to be members of this genus (Logan, 1994), have been shown by 16S rRNA sequencing to be independent genera: *Sporichthya* is a member of the family Sporichthyaceae of the suborder Frankineae (Stackebrandt et al., 1997), and the genus *Kineosporia* is grouped together with *Kineococcus* (Kudo et al., 1998) into the tentative family "Kineococcaceae" (see Introduction to the Classification of the Actinomyces in this Volume). Recently the genus *Streptacidiphilus* has been proposed by Kim et al. (2003) to accommodate acidophilic actinomycetes forming a distinct clade within the family Streptomycetaceae.

Note that although 16S rRNA sequence analyses have provided a framework for prokaryotic classification, the current classification system based on this molecule has not yet solved the taxonomic problems within the genera (especially within the genus *Streptomyces*). Several studies have attempted to use sequence data from variable regions of 16S rRNA to establish taxonomic structure within the genus, but the variation is too limited to resolve problems of species differentiation (see Witt and Stackebrandt [1990], Stackebrandt et al., [1991], Stackebrandt et al., [1992], Anderson and Wellington [2001], and the references therein).

The discovery of antibiotics produced by streptomycetes in the 1940s, which led to extensive screening for novel bioactive compounds, and the subsequent need for patenting, which led to an extreme overclassification of the genus, complicated the situation. Producers of novel natural products were described as new species and patented. Species described within the genus *Streptomyces* increased from approximately 40 to over 3000 (Trejo, 1970). The current status of streptomycete taxonomy including phylogeny has been summarized by Anderson and Wellington (2001) and will be treated briefly in the next sections. Of the 539 species and subspecies listed under the List of Bacterial Names with Standing in Nomenclature as of December 9, 2003, 376 are on the Approved lists (Tables 3 and 4).

Genus *Kitasatospora*

Zhang et al. (1997) revived the genus *Kitasatospora* to accommodate actinomycete strains forming a stable, separate subbranch on the basis of phylogenetic analyses within the family Streptomycetaceae and containing major amounts of *meso*-DAP in their whole-cell hydrolysates. Phylogenetic trees were also constructed by using 16S-23S rRNA gene spacers, leading to groupings similar to those based on 16S rRNA sequence data (Zhang et al., 1997).

The substrate mycelium of members of *Kitasatospora* is as well developed as the *Streptomy*-

Table 3. Studies on numerical classification of streptomycetes.

Nature of material	Number of strains	Number of characters (features)	Number of clusters	Number of unclustered strains	References
"Species"	159	105	24	16	Silvestri et al., 1962
Isolates	18	46	5		Williams et al., 1969
ISP "species"	448	31	14/21	168/37	Kurylowicz et al., 1975[a]
ISP "species"	618	24	15	218	Gyllenberg, 1976
Streptomyces with verticils and pseudoverticils, formerly *Streptoverticillium*	111	185	24		Locci et al., 1981
394 ISP species plus others	475	139	73	28	Williams et al., 1983a
394 ISP species plus others	821	329	15 (major) 34 (minor)	40	Kämpfer et al., 1991

Abbreviation: ISP, International *Streptomyces* Project.
[a]Fourteen and 168 were obtained by the Wroclaw taxonomy method (dendrite method); 21 and 37 were obtained by the centrifugal correlation method. Modified according to Korn-Wendisch and Kutzner (1992).

ces substrate mycelium. The aerial mycelium bears long spore chains with more than 20 spores. Galactose is present in whole-cell hydrolysates of *Kitasatospora*. Specific nucleotide signatures in the sequences of both 16S rRNA and 16S-23S rRNA gene spacers can differentiate *Kitasatospora* from *Streptomyces* (for details, see Zhang et al., 1997); however, phenotypic differences between *Streptomyces* and *Kitasatospora* are not pronounced so that the separate genus status of *Kitasatospora* may be questioned.

To date, eleven species of the genus *Kitasatospora* have been recognized: *Kitasatospora setae* (Omura et al. 1982), *Kitasatospora phosalacinea* (Takahashi et al., 1984a), *Kitasatospora griseola* (Takahashi et al., 1984a), *Kitasatospora mediocidica* (Labeda, 1988), *Kitasatospora cystarginea* (Kusakabe and Isono, 1988), *Kitasatospora cochleata* (Nakagaito et al., 1992b; Zhang et al., 1997), *Kitasalospora paracochleata* (Nakagaito et al., 1992b; Zhang et al., 1997), *Kitasatospora azatica* (Nakagaito et al., 1992b; Zhang et al., 1997), *Kitasalospora cheerisanensis* (Chung et al., 1999), *Kitasatospora cineracea* (Tajima et al., 2001) and *Kitasatospora niigatensis* (Tajima et al., 2001).

The designations of some species to this genus are open to discussion. *Kitasatospora cystarginea*, *Kitasatospora griseola*, *Kitasatospora mediocidica*, *Kitasatospora phosalacinea* and *Kitasatospora setae* are synonyms of *Streptomyces cystargineus*, *Streptomyces griseolosporeus*, *Streptomyces mediocidicus*, *Streptomyces phosalacineus* and *Streptomyces setae*, respectively. For these species, and according to scientific opinion, an author may use *Kitasatospora* or *Streptomyces*. See the List of Bacterial Names with Standing in Nomenclature for detailed comments.

Genus *Streptacidophilus*

The genus *Streptacidophilus* was proposed by Kim et al. (2003) to accommodate acidophilic actinomycetes isolated from acidic soils and litter. On the basis of 16S rRNA sequence analysis, it could be shown that the 11 isolates formed a stable clade within the family Streptomycetaceae. These organisms showed a distinctive pH profile, showed a unique 16S rDNA signature, and contained major amounts of LL-diaminopimelic acid, galactose and rhamnose in whole-cell hydrolysates (Kim et al., 2003). The members of the genus form an extensively branched, nonfragmenting mycelium carrying long chains of spores in aerial mycelia at maturity (Kim et al., 2003). To date, three species have been recognized: *Streptacidophilus albus*, *Streptacidophilus neutrinimicus* and *Streptacidophilus carbonis*.

Similar to *Kitasatospora*, *Streptomyces* and *Streptacidophilus* have no pronounced phenotypic differences, so that a separate genus status also of *Streptacidophilus* may be questioned.

Genus *Streptomyces*

The genus *Streptomyces* Waksman and Henrici (1943) is the type genus of the family. Most of the general characteristics described below also apply to members of the genera *Kitasatospora* and *Streptacidiphilus*, unless stated otherwise.

GENERAL CHARACTERISTICS Streptomycetes are Gram-positive aerobic members of the order Actinomycetales within the class Actinobacteria (Stackebrandt et al., 1997) and have a DNA G+C content of 69 ± 78 mol%. The vegetative hyphae (0.5–2.0 μm in diameter) produce an extensively branched mycelium that rarely fragments. The aerial mycelium at maturity forms chains of three to many spores. Some species may bear

Table 4. *Streptomyces* species listed in alphabetical order included in comprehensive taxonomic studies since 1980.

	Species name	Strain number(s)	1	2	3	4	5	6	7	8	bp	rRNA data strain no. if different from the first column	Acc. no. (EMBL)[b]	9
			Cluster numbers of numerical taxonomic studies and Bergey categories and species no.[a]			Groups according to different studies[a]								Strains included in DNA/DNA hybridisation studies[c]
>	S. abikoensis	DSM 40831T NRRL B-2113T				Ha1					1358bp		X53168	L4
>	S. aburaviensis	DSM 40033T ATCC 23869 ISP 5033	A 02	II 01	22-3 043		un		OC-I	KA-B	120bp	JCM 4613	D44265	
>	S. achromogenes	DSM 40028 ATCC 12767 ISP 5028	A 19	I 12	1-1 009			FU-1			120bp	JCM 4561	D44232	
>	S. achromogenes ssp. achromogenes													
>	S. achromogenes ssp. rubradiris	DSM 40789 NRRL 3061		I 12	028 009						1530bp		D63865	
>	S. acidiscabies	DSM 41668 ATCC 49003			22-3 042							JCM 7913	D85106	
	S. acidomyceticus	DSM 40798 ATCC 11611		1-3	1-2									
	S. acidoresistans	DSM 40540 ATCC 27413 ISP 5540			1-3 010									
>	S. acrimycini	DSM 40135T ATCC 19885 ISP 5135		IV 04 (green series)	1-7 1-19						120bp	JCM 4339	D44060	
>	S. actuosus	DSM 40337 ATCC 25421 ISP 5337												L2
>	S. aculeolatus	DSM 41644												
>	S. afghaniensis	DSM 40228 ATCC 23871 ISP 5228	A 18	I 11	009 1-19			FU-1			1449bp		AJ399483	
>	S. alanosinicus	DSM 40606T ATCC 15710 ISP 5606		IV 01 (gray series)	009 1-19						120bp	JCM 4714	D44304	L2
>	S. albaduncus	DSM 40478T ATCC 14698 ISP 5478		IV 02 (gray series)	006 1-10						120bp	JCM 4715		
>	S. albiaxialis	DSM 41799												
>	S. albidochromogenes	DSM 41800												
>	S. albidoflavus	DSM 40455T ATCC 25422 ISP 5455	A 1A	I 01	1-1 1-1		un	FU-1	OC-non	KA-D	1476bp		Z76676	
>	S. albidochromogenes	DSM 40880		I 01	020 032								Z76683	
	S. albidus	DSM 40320 ATCC 25423			1-3 1-2									
	S. albidus	DSM 40793 NRRL B-1672 ISP 5320	A 1B	I	1-3 1-2									
	S. albidus	DSM 40869			1-1 1-1									
	S. albiflaviniger	NRRL B-1356T									1466bp		AJ391812	
>	S. albireticuli	DSM 40051T ATCC 19721 ISP 5051	F SM	v. 11	076 069	Ha5					120bp	JCM 4116	D44009	
	S. albocinerescens	DSM 40794 NRRL 3419			002 1-7									
	S. albocyaneus	DSM 40197 ATCC 15845 ISP 5197	A Sm		007 003									
>	S. albofaciens	DSM 40268 ATCC 25184 ISP 5268	B 42	I 19	035 1-33						1489bp	JCM 4342	AB045880	
>	S. alboflavus	DSM 40045T ATCC 12626 ISP 5045	E 54	III 20	033 1-33				OC-IV		120bp	JCM 4615	D44266	
>	S. alboflavus	DSM 40761 NCIB 9453		III 20	035 1-33									
>	S. albogriseolus	DSM 40003T ATCC 23875 ISP 5003	A 12	I 07	006 1-18		un			KA-F	1519bp		AJ494865	
	S. albohelvatus	DSM 40410 ATCC 19820 ISP 5410			22-3 1-08									
>	S. albolongus	DSM 40570 ATCC 27414 ISP 5570	F 63	II 15	22-4 043									
>	S. alboniger	DSM 40043T ATCC 12461 ISP 5043	A 1B	I 02	1-6 1-31						120bp	JCM 4716	D44306	L4
	S. alborubidus	DSM 40465 ATCC 23612 ISP 5465	A 12		066 034									
>	S. albospinus	DSM 41422		IV 03 (gray series)	013 1-19									
>	S. albosporeus ssp. albospo	DSM 40795T ATCC 15394		IV 01 (red series)	063 049		La1				120bp	JCM 4135	D44013	
>	S. albosporeus ssp. labilom	DSM 41672									120bp	JCM 4135	D44013	
>	S. alboverticillatus	DSM 41678T				Ha6								
>	S. albovinaceus	DSM 40136T ATCC 15823 ISP 5136	A 1B	I 02	1-3 008					KA-B	120bp	JCM 4343	D44063	
>	S. alboviridis	DSM 40326 ATCC 25425 ISP 5326	A 1B	I 02	1-3 1-2					KA-B	120bp	JCM 4449	D44146	

	Species	DSM	Other strain	ISP	Morph	Group	Code	Ha	La	FU	OC	KA	bp	JCM	GenBank	L
>	*S. albulus*	DSM 40492	ATCC 12757	ISP 5492	A 29	I 15	025 109						1488bp		AB024440	
>	*S. albus*	DSM 40313T	ATCC 3004	ISP 5313	A 16	I 09	032 027						1358bp		X53163	
>	*S. albus*	DSM 40652				I 09	030 1-34									
>	*S. albus*	DSM 40763	IMRU 3888			I 09	030 1-34									
>	*S. albus*	DSM 40785				I 09	066 034									
>	*S. albus*	DSM 40832	NRRL 2490			I 09	014 021						1476bp		Z76689	
>	*S. albus*	DSM 40890				I 09	1-1 1-1									
>	*S. albus*	DSM 40946				I 09	030 1-34									
>	*S. albus*	DSM 40947				I 09	030 1-34									
>	*S. albus*	DSM 40948				I 09	030 1-34									
>	*S. albus*	DSM 40949				I 09	030 1-34									
>	*S. albus*	DSM 40950				I 09	22-5 039									
>	*S. albus*	DSM 40951				I 09	030 1-34									
>	*S. albus*	DSM 40963				I 09	030 1-34									
>	*S. albus*	DSM 40964				I 09	030 1-34									
>	*S. albus*	DSM 40965				I 09	030 1-34									
>	*S. albus* ssp. *albus*															
>	*S. albus* ssp. *pathocidicus*					I 09	030 1-34			FU-6	OC-non		1485bp	JCM 10204	AB045884	
>	*S. almquistii*	DSM 40447	ATCC 618	ISP 5447	A	I 09	030 1-34						121bp	JCM 4451	D44148	
	S. alni	DSM 40557	ATCC 27415	ISP 5557	A 16	1-1	1-1 1-1									
>	*S. albhioticus*	DSM 40092	ATCC 19724	ISP 5092	B 12	I 07	006 1-18		un		OC-I		121bp	JCM 4344	AB018205	L2
>	*S. amakusaensis*	DSM 40219T	ATCC 23876	ISP 5219	B 12	III 12	079 063						120bp	JCM 4617	D44268	
>	*S. ambofaciens*	DSM 40053T	ATCC 23877	ISP 5053	A Sm	I 20	006 1-18		La5				121bp	JCM 4618	D44269	
>	*S. aminophilus*	DSM 40186T	ATCC 14961	ISP 5186	A 16	I 09	031 1-34						121bp	JCM 4275	D44040	
>	*S. anandii*	DSM 40535	ATCC 19388	ISP 5535	B 42	I 19	021 1-05			FU-6			121bp	JCM 5058	D44427	
>	*S. anthocyanicus*	DSM 41422T			A 31	I 21	1-7 1-15						121bp	JCM 4620	D44270	
>	*S. antibioticus*	DSM 40234T	ATCC 8663	ISP 5234	A 1B	IV 05 (gray series)	051 018				OC-IV		121bp	JCM 4228	D44034	
>	*S. antimycoticus*	DSM 40284T	ATCC 23880	ISP 5284	A 12	I 02	047 1-35						120bp	JCM 4721	D44309	
>	*S. anulatus*	DSM 40361T	ATCC 27416	ISP 5361	A 1B	I 07	006 1-18		La22		OC-I	KA-B	121bp	JCM 4622	D44271	
>	*S. arabicus*	DSM 40252	ATCC 23881	ISP 5225	A 12	Sv. 03	22-1 040	Ha2					120bp	JCM 4543	D44223	L4
>	*S. ardus*	DSM 40527	ATCC 27417	ISP 5527	A 18	I 11	009 1-19						1447bp		AJ399485	L2
>	*S. arenae*	DSM 40525		ISP 5525	A 15	I 08	1-5 011									
>	*S. argenteolus*	DSM 40293T	ATCC 25428	ISP 5293	A 18		009 1-19					KA-B	1484bp	JCM 4623	AB045872	L2
>	*S. armeniacus*	DSM 40226	ATCC 11009	ISP 5226	A 15		1-5 011						1532bp	JCM 3070	AB018092	L2
	S. aschabadicus	DSM 43125T	NRRL B-5643					Ha13								
>	*S. asiaticus*	DSM 41761T / SM	ATCC 14808	ISP 5565	F 59	Sv. 08	22-1 040						1483bp	A14P1	AJ391830	
>	*S. aspergilloides*	DSM 41673T														
>	*S. asterosporus*															
>	*S. atratus*															
>	*S. atroaurantiacus*	DSM 40475	ATCC 27418	ISP 5475	A 33	II 20	009 1-09		La23		OC-I		120bp	JCM 3386	D43986	
>	*S. atrofaciens*	DSM 40137T	ATCC 19725	ISP 5137	A 03		006 0-10									
>	*S. atroolivaceus*	DSM 40412T	ATCC 19822	ISP 5412	C 45	II 13	012 019									
>	*S. atrovirens*	DSM 40138T	ATCC 19887	ISP 5138	A Sm	III 09	1-5 011									
	S. aurantiacus	DSM 40386	ATCC 19823	ISP 5386	A 03	II 20	033 1-33		La1		OC-non		120bp	JCM 4453	D44150	
	S. aurantiogriseus	DSM 40127T	ATCC 10762	ISP 5127	A 14	II 04	22-4 043		La16		OC-IV		120bp	JCM 4346	D44065	
	S. aureocirculatus	DSM 40414	ATCC 19824	ISP 5414	E 54		033 1-33						120bp	JCM 4454	D44151	
	S. aureofaciens	DSM 40416	ATCC 19825	ISP 5416			011 1-20				OC-I		1473bp	JCM 4624 81	AB045876	
	S. aureofasciculus	DSM 40914														
	S. aureomonopodiales															
	S. aureomonopodiales															

(Continued)

Table 4. *Continued*

	Species name	Strain number(s)	1	2	3	4	5	6	7	8	bp	rRNA data strain no. if different from the first column	Acc. no. (EMBL)[b]	Strains included in DNA/DNA hybridisation studies[c] (9)
			Cluster numbers of numerical taxonomic studies and Bergey categories and species no.[a]			Groups according to different studies[a]								
✓	S. aureorectus	DSM 4087T ATCC 15853 ISP 5387		Sv. 05	22-1 040	Ha7					121bp	JCM 4457	D44154	L4
✓	S. aureoversilis	DSM 40080 ATCC 15854 ISP 5080	A 10	I 06	033 1-33						121bp	JCM 4347	D44066	L3
	S. aureoverticillatus										1448bp	B7319T	AY094368	
	S. aureus										1411bp	B-CR4	AY094369	
	S. aureus	DSM 40862 ATCC 3309			1-3 1-20									
	S. aureus	DSM 40867 ATCC 15437			1-7 1-19									
✓	S. avellaneus	DSM 40554 ATCC 23730 ISP 5554		II 17	002 1-7						120bp	JCM 4725	D44312	
✓	S. avermectinius										1485bp	MA-4680	AB078897	
✓	S. avermitilis										1517bp	MA-4680T	AF145223	
✓	S. avidinii	DSM 40526 ATCC 27419 ISP 5526	F 56		023 004						120bp	JCM 4726	D44313	
✓	S. azaticus (K. azaticus)										1481bp		U93312	
✓	S. azureus	DSM 40106 ATCC 14921 ISP 5106	A 18	I 11	009 1-19						120bp	JCM 4349	AJ399470	L2
✓	S. baarnensis	DSM 40232 ATCC 23885 ISP 5232	A 1B	I 02	006 1-2			FU-1		KA-B	120bp	JCM 4727	D44067	
✓	S. bacillaris	DSM 40598 ATCC 15855 ISP 5598	A 1B	I 02	1-3 1-2					KA-B	120bp		D44314	
✓	S. badius	DSM 40139T ATCC 19888 ISP 5139	C Sm	III 15	1-1 1-1		un		OC-I		1349bp	JCM 4350	D44069	
✓	S. baldaccii	DSM 40845T ATCC 23654	A Sm	Sv. 01	22-1 040	Ha7					120bp		X53164	L4
✓	S. bambergiensis	DSM 40590T ATCC 13879 ISP 5590		III 10	075 1-25		La20	FU-12b	OC-non		1499bp	JCM 4728	D44315	
✓	S. beijiangensis										1450bp	YIM6	AF385681	
✓	S. bellus	DSM 40185T ATCC 14925 ISP 5185	A 18	I 11	061 1-22		La12				1448bp		AJ399476	L2
✓	S. bicolor	DSM 40140 ATCC 23614 ISP 5140	A 18		009 1-19						1517bp		AJ276569	L2
✓	S. bikiniensis	DSM 40581T ATCC 11062 ISP 5581	F 64	III 21	22-4 1-07		un		OC-II		120bp		X79851	L4
✓	S. biverticillatus	DSM 40272 ATCC 23615 ISP 5272	F 58	Sv. 01	22-1 040	Ha7					120bp	JCM 4431	D44139	
✓	S. blastmyceticus	DSM 40029T ATCC 19731 ISP 5029	A 39	Sv. 02	22-1 040	Ha3					1520bp	JCM 4184	D44025	L4
✓	S. bluensis	DSM 40564 ATCC 27420 ISP 5564		II 10	052 017						120bp		X79324	
✓	S. bobili	DSM 40056T ATCC 3310 ISP 5056		IV 02 (white series)	1-7 1-15						120bp	JCM 4627	D44274	L2
✓	S. bottropensis	DSM 40262T ATCC 25435 ISP 5262	A 19	I 12	009 1-19						1531bp		AB026217	
✓	S. brasiliensis	DSM 43159T									1356bp		X53162	
✓	S. brunneus (K. brunnea)	IFO 14627T									1475bp		U93314	
✓	S. bungoensis													
✓	S. cacaoi sp. cacaoi	DSM 40057T ATCC 3082 ISP 5057	A 16	I 09	031 1-34		La5				121bp	JCM 4352	D44070	
✓	S. cacoissp.asoensis													
	S. caelestis	DSM 40084 ATCC 15084 ISP 5084	A 18	I 11	009 1-19						1448bp		AJ399467	L2
	S. caelicus	DSM 40835 NRRL 2957			039 1-28		La21				1448bp			
✓	S. caeruleus	DSM 40292 ISP 4292			058 050		La19							
✓	S. caeruleus	DSM 40103T ATCC 19828 ISP 5103		IV 07 (gray series)	006 1-18		La19				120bp	JCM 4670	D44236	
	S. caesius	DSM 40419 ATCC 27421 ISP 5419	A 21											
	S. caespitosum	DSM 40603 ATCC 27442 ISP 5603		Sv.	22-1 040									
✓	S. californicus	DSM 40058T ATCC 3312 ISP 5058	A 09	II 02	1-3 030		La22	FU-6	OC-I		120bp	JCM 4567	D44236	L2
	S. californicus	DSM 40801 ATCC 15436		II 02	1-3 1-2									
✓	S. calvus	DSM 40010 ATCC 13382			006 1-18						121bp	JCM 4628	D44275	
✓	S. canadiensis	DSM 40837 ATCC 17776 ISP 5010	A 12	I 07	020 1-20									

Species		DSM	Strain	ISP								bp	JCM	Accession	
S. canaries	v	DSM 40528T	ATCC 27423	ISP 5528	A 20	I 13	009 1-19					120bp	JCM 4629	D44276	
S. candidus	v	DSM 40141T	ATCC 19891	ISP 5141	A 03		002 1-7					1476bp		Z76684	
S. canescens	v	DSM 40001T	ATCC 19736	ISP 5001	A 1A	I 01	1-1 1-1					1482bp		AJ391831	
S. cangkringensis	v	DSM 41769T									KA-D				
S. caniferus	v												D13P3		
S. canus	v	DSM 40017T	ATCC 12237	ISP 5017	A 25	III 02	009 1-19	La21	OC-IV			120bp	JCM 4569	D44238	
S. capillispiralis	v	DSM 41695T										121bp	JCM 5075	D44439	
S. capoamus	v	DSM 40494	ATCC 19006	ISP 5494	C 45	II 13	1-7 1-15					1484bp	JCM 4734	AB045877	
S. capuensis		DSM 40402	ATCC 25436	ISP 5402	B 42		035 1-33								
S. carnosus		DSM 40294	ATCC 25437	ISP 5294	A 15		006 1-18								
S. carpaticus	v														
S. carpinensis	v	DSM 43835T										120bp	JCM 3301	D43982	
S. catenulae	v	DSM 40258	ATCC 12476	ISP 5258	C 43	II 11	035 041					121bp	JCM 4353	D44071	
S. cattleya	v		ATCC 51928									1484bp	JCM 4925	AB045871	
S. caviscabies	v											1523bp		AF112160	
S. cavourensis ssp. *cavourensis*	v	DSM 40300	ATCC 14889	ISP 5300	A 1B	I 02	1-3 1-2			FU-6	KA-A	120bp	JCM 4555	D44228	
S. cavourensis ssp. *washingtonensis*	v														
S. cellostaticus	v	DSM 40189	ATCC 23894	ISP 5189	A 06	I 05	007 003					120bp	JCM 4631	D44277	
S. celluloflavus	v	DSM 40839T	ATCC 29806			IV 01 (yellow series)	020 032					121bp	JCM 4126	D44011	
S. cellulolyticus	v														
S. cellulosae	v	DSM 40362T	ATCC 25439	ISP 5362	A 13	II 03	006 1-18	La15	OC-non			1476bp		Z76690	
S. champavatii	v	DSM 40802	ATCC 3313				1-1 1-1					120bp	JCM 5066	D44435	
S. chartreusis	v	DSM 40841T NRRL B-5682	ATCC 14922	ISP 5085	A 18	I 11	009 1-19					1448bp		AJ399468	L2
S. chattanoogensis	v	DSM 40002T	ATCC 19739	ISP 5002	A 24	II 05	009 1-19	un	OC-non			121bp	JCM 4299	AL44047	
S. chibaensis	v	DSM 40220	ATCC 23895	ISP 5220	B 42		035 1-33					120bp	JCM 4632	D44278	
S. chrestomyceticus	v	DSM 40545	ATCC 14947	ISP 5545	A 15	I 19	006 1-18					121bp	JCM 4735	D44319	
S. chromofuscus	v	DSM 40273T	ATCC 23896	ISP 5273		I 08	020 032	La6	OC-III			120bp	JCM 4354	D44072	
S. chromogenes		DSM 40765													
S. chryseus	v	DSM 40420	ATCC 19829	ISP 5420	A 17	I 10	22-3 1-08			FU-22	KA-B	120bp	JCM 4737	D44321	L3
S. chrysomallus	v	DSM 40128	ATCC 11523	ISP 5128	A 1B		1-3 1-2								
S. chrysomallus ssp. *chrysomallus*	v	DSM 40870					020 032					120bp	JCM 4296	D44046	
S. chrysomallus ssp. *fumigatus*	v	DSM 40685					082 077								
S. cinereorectus	v	DSM 40012	ATCC 19740	ISP 5012	A 05	I 04	002 038					120bp	JCM 4572	D44240	
S. cinereoruber ssp. *cinereoruber*	v	DSM 40692	NRRL 2588			I 04	006 1-18			FU-6					
S. cinereoruber ssp. *fructofermentans*															
S. cinereospinus	v	DSM 43033T										120bp	JCM 3040	D43974	
S. cinereus	v	DSM 41651T										120bp	JCM 3385	D43985	
S. cinerochromogenes	v														
S. cinnabarinus	v	DSM 40467 ATCC 23617 / DSM 40897 ATCC 25186		ISP 5467	A 18	I 11 Sv. 02	009 1-19 / 22-1 040					1448bp		AJ399487	L2
S. cinnamoneus ssp. *albosporus*															
S. cinnamonensis	v	DSM 40803T	ATCC 12308			IV 02 (red series)	22-3 042					120bp	JCM 4019	D43988	L4
S. cinnamonensis	v	DSM 40804	ATCC 15413				033 1-33								

(Continued)

Table 4. *Continued*

Species name		Strain number(s)	Cluster numbers of numerical taxonomic studies and Bergey categories and species no.[a]					Groups according to different studies[a]			bp	rRNA data strain no. if different from the first column	Acc. no. (EMBL)[b]	Strains included in DNA/DNA hybridisation studies[c]
			1	2	3	4	5	6	7	8				9
S. cinnamoneus	>	DSM 40005T ATCC 11874 ISP 5005	F 55	Sv. 02	22-1 040	Ha4					1345bp		X53171	L4
S. cinnamoneus ssp. azacolutus	>	DSM 40646 ATCC 12686		Sv. 03	22-1 040									L4
S. cinnamoneus ssp. lanosus		DSM 40898 ATCC 25187		Sv. 02	22-1 040									
S. cinnamoneus ssp. sparsus		DSM 40899 ATCC 25185		Sv. 02	22-1 040									
S. cinnerocrocatus		DSM 40876			076 069									
S. cirratus	>	DSM 40479 ATCC 14699 ISP 5479	F 62	II 14	22-3 042						120bp	JCM 4738	D44322	
S. ciscaucasicus		DSM 40275T ISP 5275									120bp	JCM 4384	D44099	
S. citreofluorescens	>	DSM 40265T ATCC 15858 ISP 5265	A 1B	I 02	1-3 1-2			FU-19b		KA-B	120bp	JCM 4356	D44074	
S. citreus		DSM 40364 ATCC 25441 ISP 5364	A 1A		1-1 1-1									
S. clavifer	>	DSM 40843T									120bp	JCM 5059	D44428	
S. clavuligerus	>	DSM 40751T ATCC 27064		IV 10 (gray series)	22-5 036						1486bp	JCM 4710	AB045869	
S. cochleatus	>													
S. coelescens	>	DSM 40421 ATCC 19830 ISP 5421	A 21	I 14	006 1-18						121bp	JCM 4739	D44323	
S. coeliatus		DSM 40422 ATCC 19833 ISP 5422			009 1-19									
S. coelicoflavus	>													
S. coelicolor	>	DSM 40233T ATCC 23899 ISP 5233	A 1A	I 01	1-1 1-1			FU-1		KA-D	1476bp		Z76678	
S. coerulatus		DSM 40424 ATCC 19834 ISP 5424			009 1-19									
S. coeruleofuscus	>													
S. coeruleoflavus	>	DSM 40144 ATCC 23618 ISP 5144	A 18	I 11	009 1-19						1447bp		AJ399473	L2
S. coeruleoprunus	>													
S. coeruleoroseus	>	NRRL B-5642												L2
S. coeruleorubidus	>	NRRL 12372												L2
S. coeruleorubidus	>	NRRL 3045												L2
S. coeruleorubidus	>	DSM 40145 ATCC 13740 ISP 5145	A 18	I 11	009 1-19						1451bp		AJ306622	L2
S. coerulescens	>	DSM 40146 ATCC 19896 ISP 5146	A 18	I 11	009 1-19						1450bp		AJ399462	L2
S. collinus	>	DSM 40129T ATCC 19743 ISP 5129	A 18	I 11	009 1-19			FU-1			1448bp		AJ306623	L2
S. colombiensis	>	DSM 40558 ATCC 27425 ISP 5558	F 61	I 22	22-3 042			FU-12b			120bp	JCM 4740	D44324	L5
S. coralus		DSM 40256 ATCC 23901 ISP 5256	A 19	I 13	009 1-19									
S. corchorusii	>	DSM 40340 ATCC 25444 ISP 5340	A 20	I 13	009 1-19						120bp	JCM 4467	D44162	
S. coriofaciens		DSM 40485 ATCC 14155 ISP 5485	A 1A		1-1 1-1									
S. costaricanus	>													
S. cremeus	>	DSM 40147 ATCC 19897 ISP 5147	A 1B	I 02	002 1-7			FU-21			120bp	JCM 4362	D44079	
S. cretaceus		DSM 40561 ATCC 3005	A 03		1-3 1-2					KA-B				
S. crystallinus		DSM 40945		IV 03 (red series)	009 1-09									
S. curacoi	>	DSM 40107 ATCC 13385 ISP 5107	A 18	I 11	009 0-19						1448bp	JCM 5067	D44436	L2
S. cuspidosporus	>	DSM 41425		IV 11 (gray series)	22-4 1-06						1448bp		AJ399471	
S. cyaneofuscus	>	DSM 40148 ATCC 23619 ISP 5148	A 1B	I 02	1-3 1-2						120bp	JCM 4316	D44052	
S. cyaneus	>	DSM 40108T ATCC 14923 ISP 5108	A 18	I 02	009 1-19						120bp	JCM 4364	D44081	
S. cyanoalbus	>	DSM 40198T ATCC 15859 ISP 5198	A 37	I 17	007 003		La17				1444bp		AJ399460	
S. cyanocolor	>	DSM 40425 ATCC 19835 ISP 5425	A 21		006 1-18						120bp	JCM 4363	D44080	L2
S. cyanogenus		DSM 40426 ATCC 19836 ISP 5426			006 1-18									

Species		DSM	Other collection	ISP										bp	Culture coll.	GenBank	
S. cyanoglomerus ssp. *cellulose*		DSM 40427	ATCC 19837	ISP 5427				013 1-19									
S. cyanogriseus	v	DSM 40534T			H Sm			019 023		La20				1482bp	JCM 7356	U93318	L3
S. cystargineus (*K. cystarginea*)	v																L3
S. daghestanicus	v	DSM 40149	ATCC 23620 / NRRL B-2710	ISP 5149	A 17			006 010						120bp	JCM 4365	D44082	
S. daghestonicus	v																L4
S. diastaticus	v	DSM 40496T ATCC 3315		ISP 5496	A 19	I 12		1-1 1-1		un	FU-1	OC-non		1354bp		X53161	
S. diastaticus ssp. *ardesiacus*	v																
S. diastatochromogenes	v	DSM 40449T ATCC 12309		ISP 5449	A 19	I 12		009 1-19						1531bp		AB026218	
S. diastatochromogenes	v	DSM 40700 ATCC 12309				I 12		1-7 1-15	Hal4							D63867	
S. distallicus	v	DSM 40846 NCIB 8936				Sv. 01		22-1 040									
S. djakartensis	v	DSM 40743T ATCC 13441				IV 12 (gray series)		035 1-33						121bp	JCM 4957	D44420	
S. durhamensis	v	DSM 40539 ATCC 23194		ISP 5539	A 30	II 06		009 1-19						120bp	JCM 4747	D44331	
S. eburosporeus	v	DSM 40944						064 1-30									
S. echinatus	v	DSM 40013 ATCC 19748		ISP 5013	A 18	I 11		1-6 1-10			FU-1			1448bp		AJ399465	L2
S. echinatus	v	DSM 40730						1-7 1-15									L2
S. echinoruber	v	DSM 41696T						013 1-19						120bp	JCM 5016	D44426	
S. ederensis	v	DSM 40741T ATCC 15304				IV 14 (gray series)		22-1 040						120bp	JCM 4958	AB018209	
S. ehimensis	v	DSM 40253T ATCC 23903		ISP 5253		Sv. 09		009 1-19	Hal					120bp	JCM 4162	D44021	
S. endus	v	DSM 40187 NRRL 2339								La8							L1
S. enissocaesilis	v	DSM 40941T ATCC 23266				IV 15 (gray series)		035 1-33						120bp	JCM 5060	D44429	
S. erumpens	v																
S. erythraeus	v	DSM 40116T ATCC 27427 / NRRL 5729		ISP 5116		IV 04 (red series)		074 1-27		La15				1518bp		X80826	
S. erythrogriseus	v																
S. espinosus	v																
S. eurocidicus	v	DSM 40604T ATCC 27428		ISP 5604	F 56	Sv. 02		22-1 040	Ha5	un				120bp	JCM 4029	D43989	
S. europaeiscabiei	v													1492bp	CFBP 4497	AJ007423	L4
S. eurythermus	v	DSM 40014 ATCC 14975		ISP 5014	A 23	I 20		1-5 009						1531bp		D63870	
S. exfoliatus	v	DSM 40060T ATCC 12627		ISP 5060	A 05	I 04		002 1-7		un		OC-II	KA-C	120bp	JCM 4366	D44083	
S. fasiculatus	v	DSM 40054 ATCC 19751		ISP 5054	A 29			025 005									
S. felleus	v	DSM 40130T ATCC 19752		ISP 5130	A 1A	I 01		1-1 1-1					KA-D	1476bp		Z76681	
S. felleus	v	DSM 40647 NRRL 2251				I 01		22-3 1-07									
S. felleus	v	DSM 40976				I 01		002 1-7									
S. fervens ssp. *fervens*	v	DSM 40086 ATCC 27429		ISP 5086	A 05	Sv. 01		22-1 040	Ha7								
S. fervens ssp. *melrosporus*	v	DSM 40905T NRRL 3117			A 30	Sv. 01		22-1 040									L4
S. filamentosus	v	DSM 40022 ATCC 19753		ISP 5022	A 05	I 04		002 1-7		La10				120bp	JCM 4576	D44244	
S. filipinensis	v	DSM 40112T ATCC 23905		ISP 5112	A 30	II 06		009 1-19				OC-III		120bp	JCM 4369	D44086	
S. fimbriatus	v	DSM 40942T ATCC 15051			A 1B	I 02		006 1-18						1484bp	JCM 4910	AB045868	
S. fimicarius	v	DSM 40322 ATCC 25449		ISP 5322	I Sm	IV 16 (gray series)		1-3 1-2			FU-9			120bp	JCM 4472	D44167	
S. finlayi	v	DSM 40218T ATCC 23340		ISP 5218	A 24	III 24		22-4 043		un		OC-I		120bp	JCM 4637	D44279	
S. flaveolus	v	DSM 40061T ATCC 3319		ISP 5061		II 05		1-6 1-13		La12		OC-III		121bp	JCM 4577	D44245	
S. flavescens	v	DSM 40428 ATCC 19838		ISP 5428				22-3 1-08									
S. flaveus	v	DSM 43153T								La21				120bp	JCM 3035	D43971	
S. flavidofuscus	v	DSM 40150T ATCC 19900		ISP 5150		IV 03 (yellow series)		026 033		La22				120bp	JCM 4474	D44169	
S. flavidovirens	v	DSM 40270				I 08		017 007						121bp	JCM 4751	D44333	
S. flaviscleroticus	v	DSM 40541 ATCC 14841		ISP 5541	A 05			002 1-7			FU-NC		KA-G				
S. flavochromogenes		DSM 40651						009 1-19									
S. flavochromogenes																	
S. flavofungini	v	DSM 40366 ATCC 27430		ISP 5366	B 42			033 1-33						120bp	JCM 4753	D44335	

(Continued)

Table 4. *Continued*

Species name		Strain number(s)	Cluster numbers of numerical taxonomic studies and Bergey categories and species no.[a]			Groups according to different studies[a]					bp	rRNA data strain no. if different from the first column	Acc. no. (EMBL)[b]	Strains included in DNA/DNA hybridisation studies[c]
			1	2	3	4	5	6	7	8				9
S. flavofuscus	✓	DSM 40323T ATCC 25452 ISP 5323	A 1C	I 03	1-2 015					KA-B	1518bp		AJ494864	
S. flavogriseus	✓	DSM 40990 ATCC 3331		I 03	1-4 1-4			FU-19b						
S. flavogriseus														
S. flavopersicus	✓	DSM 40093T ATCC 19756 ISP 5093	F 56		22-1 040	Ha14					120bp	JCM 4307	D44049	L4
S. flavotricini	✓	DSM 40152 ATCC 23621 ISP 5152	F 61	I 22	22-3 042			FU-1			120bp	JCM 4371	D44087	L3/L5
S. flavovariabilis	✓													
S. flavovirens		DSM 40062 ATCC 3320 ISP 5062	A 1C		1-2 015						268bp		U72171	
S. flavoviridis		DSM 40153 ATCC 19903 ISP 5153	A 28		006 1-10						121bp	JCM 4372	D44088	
S. flocculus		DSM 40327T ATCC 25453 ISP 5327	A 16		030 1-34						121bp	JCM 4476	D44171	
S. floridae		DSM 40938 NCIB 9345		IV 04 (yellow series)	1-3 1-2						120bp	JCM 5068	D44437	
S. fluorescens		DSM 40203 ATCC 15860 ISP 5203	A 1B	I 02	1-3 1-2					KA-B	121bp	JCM 4373	D44089	
S. fradiae		DSM 40063T ATCC 10745 ISP 5063	G 68	II 18	22-5 039		un		OC-I		120bp	JCM 4579	D44246	
S. fragilis		DSM 40044T ATCC 23908 ISP 5044	G SM	III 22	078 058				OC-III		120bp	JCM 4638	D44280	
S. griseoplanus		DSM 40009 ATCC 19766 ISP 5009	A 29		078 060						120bp	JCM 4300	D44048	
S. fulvissimus		DSM 40593T ATCC 27431 ISP 5593	A 10	I 15	034 1-33		un		OC-IV		120bp	JCM 4754	D44336	L3
S. fulvissimus	✓	DSM 40767			1-3 1-2									
S. fulvorobeus	✓	DSM 40210 ATCC 15863 ISP 5210	A 03		1-3 1-20									
S. fulvoviridis														
S. fumanus	✓	DSM 40154 ATCC 19904 ISP 5154	A 18		1-7 1-19						1448bp		AJ399463	
S. fumigatiscleroticus	✓	DSM 43154T		I 11							121bp	JCM 3101	D43979	
S. fungicidicus		DSM 40020 ATCC 27432 ISP 5020	A 16		035 1-33									
S. fungicidicus		DSM 40811 ATCC 13853			024 005									
S. galbus	✓	DSM 40089 ATCC 23910 ISP 5089	A 15	I 08	006 1-10						1517bp		X79852	
S. galbus		DSM 40480 ATCC 14077 ISP 5480		I 08	1-5 011						1517bp		X79325	
S. galilaeus	✓	DSM 40481 ATCC 14969 ISP 5481	A 19	I 12	1-7 1-15						1484bp		AB045878	
S. gancidicus	✓	DSM 40935T NRRL B-1872		IV 17 (gray series)	006 1-18						121bp	JCM 4171	D44022	
S. gardneri	✓	DSM 40064 ATCC 9604 ISP 5064	A 04		002 1-07			FU-23		KA-C	120bp	JCM 4375	D44091	
S. gelaticus	✓	DSM 40065T ATCC 3323 ISP 5065	A Sm	III 11	003 1-3		un							
S. geldanomyceticus		NRRL 3602T									1129bp		AJ391824	
S. geysiriensis		DSM 40742T ATCC 15303		IV 18 (gray series)	006 1-18						121bp	JCM 4962	D44421	
S. ghanaensis		DSM 40746T ATCC 14672		IV 05 (green series)	1-7 1-21						121bp	JCM 4963	D44422	
S. gibsonii		DSM 40959		IV 05 (white series)	030 1-34						121bp	JCM 5061	D44430	
S. glaucescens	✓	DSM 40716												L2
S. glaucescens	✓	DSM 40155T ATCC 23622 NRRL 12514 ISP 5155	A 28	III 05	006 1-10		La16		OC-III		1519bp		X79322	L2
S. glaucogriseus	✓													
S. glaucosporus	✓													
S. glaucus	✓													
S. globisporus ssp. caucasicus		DSM 40814 ATCC 19907		I 02	1-1 1-1									
S. globisporus ssp. flavofuscus	✓													
S. globisporus ssp. globisporus		DSM 40199T ATCC 15864 ISP 5199	A 1B	I 021	1-3 1-2					KA-B	120bp	JCM 4378	D44093	
S. globosus	✓	DSM 40815T ATCC 14979		IV 19 (gray series)	22-3 042						120bp	JCM 4225	D44032	

Table (continued). *Streptomyces* species — strain designations, morphological/cluster codes, and sequence data.

Species	>	DSM	ATCC / NRRL / IFO	ISP	Morph	Grp	Cluster	Ha	La	FU	OC	KA	bp	JCM	Accession	L3/L5
S. glomeratus	>	DSM 40429		ISP 5429									120bp	JCM 4761	D44340	
S. glomeroaurantiacus	>	DSM 41701T											120bp	JCM 5062	D44431	
S. gobitricini	>	DSM 40190	ATCC 23914	ISP 5190									120bp	JCM 4294	D44044	
S. goshikiensis	>	DSM 40324T	ATCC 10975	ISP 5324	F 61	I 22	22-3 042					KA-D	1476bp		Z76687	
S. gougerotii	>				A 1A	I 01	1-1 1-1						120bp	JCM 4762	D44341	
S. graminearus	>	DSM 40559T	ATCC 12705 / NRRL B-1865T	ISP 5559	A 26	III 03	004 1-23				OC-I		1495bp		AJ391818	
S. graminofaciens	>	DSM 40047	ATCC 23915	ISP 5047	A 1B	I 02	1-3 1-2						120bp	JCM 4379	D44094	
S. griseiniger	>	DSM 40430	ATCC 19840	ISP 5430	A 12	I 07	1-7 1-15			FU-6		KA-B	120bp	JCM 4763	D44342	
S. griseinus	>	DSM 40066	ATCC 19762	ISP 5066	A 1B	I 02	1-3 1-2			FU-6		KA-A	120bp	JCM 4380	D44095	
S. griseoaurantiacus	>	DSM 40915			A 12	I 02	020 1-08									
S. griseobrunneus	>	DSM 40004T	ATCC 12628	ISP 5004	F 55	Sv. 03	22-1 040									
S. griseobrunneus	>	DSM 40499T	ATCC 14511	ISP 5499	A 18	I 11	1-5 011									
S. griseocarneus		DSM 40816	ATCC 13180		A 37		002 1-7	Ha6	un	FU-12b			1515bp		X99943	L4
S. griseochromogenes	>	DSM 40456T	ATCC 25456	ISP 5456	A 12	I 17	006 1-18		La4		OC-non		1448bp		AJ399491	L2
S. griseoflavus		DSM 40698	NRRL 2717		A 37	I 17	1-7 1-14						121bp			
S. griseoflavus		DSM 40191	ATCC 23916	ISP 5191	A 12	I 07	1-6 1-16									
S. griseofuscus	>	DSM 40274T	ATCC 23623	ISP 5274	A 13	II 03	006 1-18			FU-6			121bp	JCM 4479	D44174	
S. griseoincarnatus		DSM 40385	ATCC 25457	ISP 5385			22-3 042		La15			KA-G	121bp	JCM 4381	D44096	
S. griseolavendus	>	DSM 40468T	ATCC 23624	ISP 5468		IV 05 (yellow series)	)017 007						121bp	JCM 4480	D44175	L5
S. griseoalbus	>	DSM 40067	ATCC 3325	ISP 5067	A 1C	I 03	1-2 015									
S. griseolosporeus	>	DSM 40854	ATCC 11796			I 03	023 004			FU-24		KA-B	120bp	JCM 4043	D43990	
S. griseolus	>	DSM 40392T	ATCC 12768	ISP 5392	C 43	II 11	1-5 1-16		La24		OC-III		120bp	JCM 4765	D44344	
S. griseolus	>	DSM 40159	ATCC 23625	ISP 5159	A 12	I 07	006 1-10						121bp	JCM 4382	D44097	
S. griseoluteus	>	DSM 40768	ATCC 12125		A 12	I 07	006 010									
S. griseomycini	>	DSM 40160	ATCC 19909	ISP 5160	A 12	I 07	006 1-18					KA-F	121bp	JCM 4383	D44098	
S. griseoroseus	>	DSM 40281T	ATCC 23919	ISP 5281	A 21	I 14	018 023						1436bp		AY094585	
S. griseorubens	>	DSM 40469	ATCC 23627	ISP 5469	A 18	I 11	009 1-19		un		OC-I		1449bp		AJ399488	L2
S. griseoruber	>	DSM 40562	ATCC 27435	ISP 5562	A 23	I 20	1-7 1-19						120bp	JCM 4766	D44345	
S. griseorubiginosus	>	DSM 40161T	ATCC 23628	ISP 5161	F 60	IV 06 (green series)	006 1-10						121bp	JCM 4385	D44100	
S. griseosporeus	>	DSM 40507T		ISP 5507	F 58	I 10	22-1 040						120bp	JCM 4202	D44028	
S. griseostramineus	>	DSM 40229T	ATCC 23920	ISP 5229	A 17	I 02	006 010	Ha4	La6		OC-III		120bp	JCM 4643	D44283	L3
S. griseoverticillatus	>	DSM 40855	ATCC 10137			I 03	1-3 1-2						1478bp		Y15501	
S. griseoviridis	>	DSM 40937	NRRL B-2249			I 02	1-3 1-2									
S. griseus	>	DSM 40561T		ISP 5561		I 02	1-3 1-2						120bp	JCM 4742	D44326	
S. griseus ssp. *alpha*	>	DSM 40932														
S. griseus ssp. *cretosus*	>		IFO 13550										1485bp		AB045866	
S. griseus ssp. *farinosus*	>	DSM 40933	NRRL B-1561		I 02		1-1 1-1									
S. griseus ssp. *griseus*	>	DSM 40236T	ATCC 23345	ISP 5236	A 1B	I 021	1-3 1-2		La22	FU-19b		KA-B	1537bp	JCM 4331	AF056711	L4
S. griseus ssp. *solvifaciens*	>	DSM 40114T	ATCC 19769	ISP 5114	F 55	Sv. 04	22-1 040	Ha4		FU-NC		KA-B	120bp	JCM 4052	D44054	
S. griseus ssp. *griseus*	>	DSM 40068T	ATCC 10897	ISP 5068	A 1C	I 03	1-2 015			FU-24	OC-I		120bp		D43991	
S. hachijoensis	>	DSM 40863	ATCC 13449		A 18	I 03	1-3 1-2						1448bp		AJ399466	L2
S. halstedii	>	DSM 40042	ATCC 12236	ISP 5042		I 11	009 1-19									
S. halstedii	>	DSM 40328	ATCC 25460	ISP 5328		I 14	1-6 1-14									
S. hawaiiensis	>	DSM 40431	ATCC 19841	ISP 5431	F 62	II 14	22-3 043						120bp	JCM 4768	D44346	
S. heimi		DSM 40123	ATCC 23922	ISP 5123	A 02	II 01	22-4 043						120bp	JCM 4138	D44014	
S. heliomycini	>	DSM 40037	ATCC 19772	ISP 5037	F 57	Sv. 01	22-1 040	Ha7		FU-NC			120bp	JCM 4098	D44005	
S. helvaticus	>	DSM 40095T	ATCC 3008	ISP 5095			1-1 1-1						120bp	JCM 4587	D44249	L4
S. herbaricolor	>	DSM 40770														
S. hirsutus	>	DSM 40263	ATCC 12760	ISP 5263	A 19	I 12	009 1-19						269bp		U72169	

(Continued)

Table 4. *Continued*

Column group headers: **Cluster numbers of numerical taxonomic studies and Bergey categories and species no.[a]** (columns 1–3); **Groups according to different studies[a]** (columns 4–8); **Strains included in DNA/DNA hybridisation studies[c]** (column 9).

Species name		Strain number(s)	1	2	3	4	5	6	7	8	bp	rRNA data strain no. if different from the first column	Acc. no. (EMBL)[b]	9
S. humifer		DSM 40602 ATCC 13748 ISP 5602	A 1C		035 1-33									
S. humiferus	>	DSM 43030T												
S. hydrogenans	>	DSM 40586 ATCC 19631 ISP 5586	A 05	I 04	002 1-7						121bp	JCM 3037	D43972	
S. hygroscopicus	>	ATCC 21431 NRRL 1477									1517bp		X79853	
S. hygroscopicus	>	NRRL 2387T									1483bp		AJ391819	
S. hygroscopicus	>										1465bp		AJ391820	
S. hygroscopicus ssp. ascomyceticus		DSM 40822 ATCC 14891			056 016		La7							
S. hygroscopicus ssp. decoyicus	>													L1
S. hygroscopicus ssp. geldanus		NRRL 3602												
S. hygroscopicus ssp. glebosus	>	IFO 13598									1485bp		AB045864	
S. hygroscopicus ssp. hygroscopicus	>	DSM 40187 ATCC 23904 ISP 5187	A 32	I 16	041 012			FU-6			1461bp		AJ391821	
S. hygroscopicus ssp. hygroscopicus	>	DSM 40578T ATCC 27438 ISP 5578	A 32	I 16	085 012		La8	FU-6						L1
S. hygroscopicus ssp. hygroscopicus	>			I 16	009 1-19									
S. hygroscopicus ssp. ossamyceticus	>	DSM 40824 ATCC 15420			009 1-19									
S. iakyrus	>	DSM 40482 ATCC 15375 ISP 5482	A 18	I 11	009 1-19						1448bp		AJ399489	L2
S. indiaensis	>													
S. indigocolor	>	DSM 40432 ATCC 19842 ISP 5432	A 18	I 11	009 1-19						1448bp	JCM 4646	AJ399464	L2
S. indigoferus	>	DSM 40124T ISP 5124									120bp	A4 R2	D44285	
S. indoniensis	>	DSM 41759T									1481bp		AJ391835	
S. intermedius	>	DSM 40372T ATCC 3329 ISP 5372	A 1A	I 01	1-1 1-1		La3				1474bp		Z76686	
S. insulatus	>	DSM 41441T								KA-D	121bp	JCM 4988	D44424	L2
S. ipomoeae	>	DSM 40383T ATCC 25462 ISP 5383		IV 02 (blue series)	077 074									L2
S. ipomoeae	>	DSM 40818 ATCC 11747		IV 02 (blue series)	009 1-19									
S. janthinus	>	DSM 40206 ATCC 15870 ISP 5206	A 18	I 11	009 1-19						1448bp	B22 P3	AJ399478	L2
S. javensis	>	DSM 41764T									1471bp		AJ391833	
S. kanamyceticus	>	DSM 40500T ISP 5500	C 44		22-4 035		La11				120bp	JCM 4775	D44352	
S. karnatakensis		DSM 40345 ATCC 25463 ISP 5345												
S. kashmirensis	>	DSM 40336T ISP 5336				Ha8					120bp	JCM 4776	D44353	
S. kasugaensis	>	DSM 40819T ISP 5819									1488bp	M338-M1	AB024441	
S. katrae	>	DSM 40550 ATCC 27440 ISP 5550	F 61	I 22	22-3 042						120bp	JCM 4777	D44354	L5
S. kentuckensis	>	DSM 40052 ATCC 12691 ISP 5052	F SM	Sv. 11	22-1 040	Ha14					120bp	JCM 4153	D44019	L4
S. kifunensis	>										1481bp	JCM 9081	U93322	
S. kishiwadensis	>	DSM 40397T ATCC 25464 ISP 5397		Sv. 15	22-1 040	Ha11					121bp	JCM 4486	D44180	
S. krainskii	>	DSM 40321 ATCC 25465 ISP 5321	A 1A		1-1 1-1									
S. krestomyceticus		DSM 40820			035 1-33									

Species		DSM	Other collection	ISP	A	Series	Code	Group	FU/OC	KA	bp	JCM/strain	Accession	L
S. kunmingensis	>	DSM 41681T			F 60	IV 20 (gray series)	025 1-15				120bp	JCM 7473	D44441	
S. karssanovii	>	DSM 40162T	ATCC 15824	ISP 5162							120bp	JCM 4388	D44103	
S. labedae	>										1127bp	C762	AY094365	
S. laceyi											1389bp	C765	AY094366	
S. laceyi											1446bp	C7654T	AY094367	
S. laceyi								Ha12			1357bp		X53167	L4
S. ladakanum	>	DSM 40587T	NRRL 3191T	ISP 5090	A 18	I 11	016 1-19	La21			1449bp		AJ399469	L2
S. lanatus	>	DSM 40090T	ATCC 19775								120bp	JCM 5063	D44432	
S. larentii	>	DSM 41684T						un	OC-II		1426bp	JCM 4389	AF454764	
S. lateritius	>	DSM 40163T	ATCC 19913	ISP 5163	H Sm	III 23	22-3 1-08				120bp	JCM 4391	D44106	
S. lavendofoliae	>	DSM 40217T	ATCC 15872	ISP 5217		IV 07 (red series)	22-3 1-08				1514bp		D85114	
S. lavendulae	>		IFO 14028											
S. lavendulae	>		NRRL B-2243											L3/L5
S. lavendulae	>		NRRL B-2402											L3/L5
S. lavendulae	>		NRRL B-3080											L3/L5
S. lavendulae	>	DSM 40748	ATCC 14159			I 22	22-3 042						D85109	
S. lavendulae	>	DSM 41570	ATCC 11924			I 22	22-3 042						D85110	
S. lavendulae	>	DSM 41571	ATCC 13664			I 22	1-3 1-2						D85111	
S. lavendulae	>	DSM 41573	ATCC 14158			I 22	22-3 042						D85112	
S. lavendulae	>	DSM 41576	ATCC 14162			I 22	22-3 042				120bp	JCM 4056	D43992	
S. lavendulae ssp. *grasserius*	>	DSM 40385T												
S. lavendulae ssp. *avirens*			NRRL B-16576											L5
S. lavendulae ssp. *brasilicus*			NRRL B-2937T T						FU-12b					L3/L5
S. lavendulae ssp. *inositophilus*			NRRL B-3904T T											L3/L5
S. lavendulae ssp. *lavendulae*	>	DSM 40069T	ATCC 8664	ISP 5069	F 61	I 22	22-3 042		FU-12b OC-I		1329bp		D85116	L3/L5
S. lavenduligriseus	>	DSM 40487T	ATCC 13306	ISP 5487	A 34	Sv. 02	1-5 009				120bp	JCM 4545	AJ399487	L4
S. lavendulocolor	>	DSM 40216	ATCC 15871	ISP 5216	F 61	I 22	22-3 1-08				120bp	JCM 4390	D44105	L5
S. levis	>	DSM 40202	ATCC 15876	ISP 5202	A 29	I 15	1-1 1-1							
S. levoris	>	DSM 40555	ATCC 23732	ISP 5555			025 005				121bp	JCM 4781	D44357	
S. libani	>	DSM 40254T	ATCC 23930	ISP 5254	A 1A	Sv. 16	22-1 040	Ha8			121bp	JCM 4188	D44027	
S. libani ssp. *rufus*	>	DSM 40131T	ATCC 19778	ISP 5131	A 19	I 01	1-1 1-1		FU-1	KA-D				
S. lilacinus	>	DSM 40355	ATCC 25466	ISP 5355		I 12	009 1-19				1476bp		Z76679	
S. limosus	>	DSM 40070	ATCC 3331	ISP 5070	A 1B	I 02	1-3 1-2		FU-9	KA-B	1519bp		X79854	
S. lincolnensis		DSM 40752	ATCC 27357			I 02	079 061				1484bp	JCM 4711	AB045961	
S. lincomycini		DSM 40297				Sv.	067 034							
S. lipmanii	>	DSM 40164	ATCC 19914	ISP 5164	A 05	I 04	002 1-7		FU-6	KA-C	120bp	JCM 4394	D44109	
S. lipmanii		DSM 40434	ATCC 19844	ISP 5434	A 21		006 1-18						AB03756	
S. listeri		DSM 40825	ATCC 11415				029 1-32							
S. litmocidini		DSM 41428				IV 03 (blue series)	009 1-19							
S. lividans	>	DSM 40165T	ATCC 19915	ISP 5165	A 39	II 10	005 010				121bp	JCM 4866	D44415	
S. loidensis	>	DSM 40599	ATCC 27443	ISP 5599	A 10	I 06	033 1-33				120bp	JCM 4396	D44111	
S. lomondensis	>	DSM 40749	ATCC 13931			I 06	033 1-33						D44359	
S. longisporoflavus	>	DSM 40166T	ATCC 23931	ISP 5166	A 18	I 19	009 1-19	un	OC-non	KA-F	121bp	JCM 4784	AJ399475	L3
S. longispororuber	>	DSM 40435	ATCC 19850	ISP 5435		I 11	033 1-33							L2
S. longispororuber	>	DSM 40826	ATCC 14562		E 54		064 1-30				1448bp			
S. longisporus	>													
S. longissimus	>													
S. longissimus														

(*Continued*)

Table 4. *Continued*

Species name		Strain number(s)	Cluster numbers of numerical taxonomic studies and Bergey categories and species no.[a]			Groups according to different studies[a]					bp	rRNA data strain no. if different from the first column[b]	Acc. no. (EMBL)[b]	Strains included in DNA/DNA hybridisation studies[c]
			1	2	3	4	5	6	7	8				9
S. longwoodensis	✓	DSM 41677T	A 31	I 21	1-5 1-16						120bp	JCM 4976	D44423	
S. lucensis	✓	DSM 40317 ATCC 17804	F 62	II 14	22-3 1-08						120bp	JCM 4490	D44183	
S. luridus	✓	DSM 40081T ATCC 19782 ISP 5081			006 1-18		La17		OC-II		120bp	JCM 4591	D44252	
S. lusitanus	✓	DSM 40568 ATCC 15842 ISP 5568	C 44	II 12	038 025						121bp	JCM 4785	D44360	
S. luteofluorescens		ATCC 15469 ISP 5398	A 18	I 11	009 1-19						1448bp		AJ399490	L2
S. luteogriseus	✓	DSM 40483 ATCC 15072 ISP 5483	A 1B		1-3 1-20									
S. luteolutescens		DSM 40600 ATCC 27445 ISP 5600		Sv.	1-8 1-17	Ha9					1354bp		X53172	L4
S. luteoreticuli		DSM 40509 ATCC 27446 ISP 5509				Ha10								
S. luteosporeus	✓	DSM 40833T												
S. luteoverticillatus	✓	DSM 40038T ATCC 23933 ISP 5038	F 55	Sv. 03	22-1 040	Ha1		FU-12b			120bp	JCM 4099		
S. lydicus	✓	DSM 40461T ATCC 25470 ISP 5461	A 29	I 15	025 005		La9	FU-21	OC-non		1481bp		Y15507	
S. macrosporus	✓	DSM 40096 ATCC 19783 ISP 5096	A 38		006 010									
S. macrosporus	✓	DSM 41449T									1484bp		Z68099	
S. macrosporus	✓	DSM 41476T									1490bp		Z68100	
S. majorciensis		NRRL 15167 DSM 40167 ISP 5167	A 12		060 051		La14							L5
S. malachiticus		DSM 40332 ATCC 25471 ISP 5332			006 1-18									
S. malachitofuscus	✓	DSM 40333 ATCC 25472 ISP 5333			006 1-18						120bp	JCM 4493	D44185	
S. malachitorectus	✓													
S. malachitospinus	✓													
S. malaysiensis	✓	DSM 41697T	F 55		22-1 040	Ha11					1475bp	ATB-11	AF117304	L4
S. mashuensis	✓	DSM 40221T ATCC 23934 ISP 5221		Sv. 03	22-1 040						1518bp		X79323	
S. mashuensis		DSM 40896		Sv. 03	22-1 040									
S. massasporeus	✓	DSM 40035T ATCC 19785 ISP 5035	D SM	III 19	015 1-19		La12		OC-III		121bp	JCM 4593	D44253	
S. matensis	✓	DSM 40188 ATCC 23935 ISP 5188	A 12	I 07	006 1-18			FU-1			122bp	JCM 4651	D44286	
S. mauvecolor	✓	DSM 41702T									120bp	JCM 5002	D44425	
S. mediocidicus	✓	DSM 40021 ATCC 23936 ISP 5021	F Sm		22-1 040	Ha3					120bp	JCM 4060	D43994	
S. mediocidicus	✓	DSM 40864 ATCC 13278			1-3 1-2									
S. mediocidicus	✓	DSM 40865 ATCC 13279			1-3 1-2									
S. mediolani	✓	DSM 41058T									120bp	JCM 5076	D44440	
S. mediterranei	✓	DSM 40773 DSM 41476T			064 1-30						1490bp		Z68100	
S. megasporus	✓	DSM 40192T ATCC 23937 ISP 5192	A 33	II 07	009 1-09						1490bp			
S. melanogenes	✓		A 32	I 16	051 018						120bp	JCM 4398	D44113	
S. melanosporofaciens	✓	DSM 40318 ATCC 25473 ISP 5318									1433bp		AJ391837	L1
S. melanosporofaciens	✓	DSM 40318T NRRL B-12234									1433bp		AJ271887	
S. mexicanus	✓	DSM 40015 ATCC 14970 ISP 5015	A 06	I 05	005 029						1450bp	CH-M-1035T	AF441168	
S. michiganensis	✓	DSM 40031T ATCC 13231 ISP 5331	A 23	I 20	1-3 1-2		La22		OC-I		120bp	JCM 4594	D44254	
S. microflavus	✓	DSM 40031 ATCC 19787 ISP 5031	A 19								120bp	JCM 4496	D44188	
S. minoensis		DSM 40301 ATCC 17757 ISP 5301	A 15	I 19	009 1-19									
S. minutiscleroticus	✓	DSM 40553 ATCC 27447 ISP 5553	A 19	I 08	006 1-18					KA-G	121bp	JCM 4790	D44365	
S. mirabilis	✓	DSM 40222T ATCC 23938 ISP 5222	F 66		1-7 1-19						1466bp		AF112180	
S. misakiensis	✓	DSM 40306 ATCC 14991 ISP 5306	A 31	II 16	22-4 043		La18		OC-non		120bp	JCM 4653	D44287	
S. misionensis	✓	DSM 40847T ATCC 29032		I 21	1-6 1-16						120bp	JCM 4497	D44189	
S. mobaraensis	✓	DSM 40587 ISP 5587			22-1 040	Ha12		FU-12b			120bp	JCM 4778	D44355	L4
S. moderatus	✓	DSM 40529 ATCC 23443 ISP 5529												
S. monomycini	✓		A Sm	Sv. 07	033 1-33									

Species		DSM no.	ATCC no.	ISP no.					Ha	La	FU	OC	KA	Size	JCM no.	Accession	
S. morookaensis	>	DSM 40503T	ATCC 19166	ISP 5503	F 59	Sv. 08	22-1	040	Ha13					120bp	JCM 4333	D44056	L3
S. murinus	>	DSM 40091	ATCC 19788	ISP 5091	A 17	I 10	1-6	1-16					KA-E	121bp	JCM 4400	D44115	
S. mutabilis	>	DSM 40169	ATCC 19919	ISP 5169	A 12	I 07	006	1-18									
S. mutomycini	>																
S. naganishii	>	DSM 40282	ATCC 23939	ISP 5282	A 31	I 21	1-6	1-15						120bp	JCM 4596	D44255	
S. naraensis	>	DSM 40508	ATCC 13788	ISP 5508	A 1C		003	1-3									
S. narbonensis	>	DSM 40016	ATCC 19790	ISP 5016	A 04	I 04	002	1-7					KA-C	120bp	JCM 4498	D44190	
S. nashvillensis	>	DSM 40314	ATCC 25476	ISP 5314	A 05	I 04	002	1-7									
S. netropsis	>	DSM 40259T	ATCC 23940	ISP 5259	F 56	Sv. 01	22-1	040	Ha14		FU-21			120bp	JCM 4063	D43995	L4
S. neyagawaensis	>	DSM 40588	ATCC 27449	ISP 5588	A 18	I 11	009	1-19			FU-24			1449bp		AB026219	L2
S. niger	>	DSM 40302			A 40	I 18	069	1-26						121bp	JCM 3158	D43980	
S. nigrescens	>	DSM 40276T	ATCC 23941	ISP 5276	A 29	I 15	025	005		La2				121bp	JCM 4401	D44116	
S. nigrifaciens	>	DSM 40071	ATCC 19791	ISP 5071	A 1C	I 03	1-2	015					KA-B	120bp	JCM 4223	D44031	
S. nitrosporeus	>	DSM 40023T		ISP 5023										120bp		D43996	
S. niveoruber	>	DSM 40638T	ATCC 14971		A 1B	IV 08 (red series)	013	1-19		La1				120bp	JCM 4234	D44035	
S. niveus	>	DSM 40088T	ATCC 19793	ISP 5088	A 10	I 02	043	013		La19				120bp	JCM 4599	D44256	
S. nobilis	>	DSM 40441	ATCC 19251	ISP 5441	A 33	II 07	033	1-33				OC-I		120bp	JCM 4557	D44229	
S. noboritoensis	>	DSM 40223T	ATCC 25477	ISP 5223	A 20	I 13	009	1-09		La19		OC-III		121bp	JCM 4499	D44191	
S. olivaceoviridis	>	DSM 40334T	ATCC 23630	ISP 5334	A 35	II 08	009	1-19		La21		OC-III		120bp	JCM 4656	AF114034	
S. nodosus	>	DSM 40109	ATCC 14899	ISP 5109	A 34	III 06	006	1-11				OC-III		1485bp		AB045886	
S. nogalater	>	DSM 40546T	ATCC 27451	ISP 5546			1-5	009		La14				120bp	JCM 4799	D43984	
S. nojiriensis	>	DSM 41655T					025	1-09						121bp	JCM 3382	D44419	
S. noursei	>	DSM 40635T	ATCC 11455		J Sm	IV 23 (gray series)	004	006				OC-IV		120bp	JCM 4922	D44371	
S. novaecaesareae	>	DSM 40358T	ATCC 27452	ISP 5358		III 25	069	1-26				OC-non		121bp	JCM 4800	D44372	
S. ochraceiscleroticus	>	DSM 40594T				III 08	1-1	1-1						1476bp	JCM 4801	Z76682	
S. ochroleucus	>	DSM 40591	ATCC 3006	ISP 5591													
S. odorifer	>	DSM 40347	ATCC 6246	ISP 5347	A 1A	I 01	1-1	1-1					KA-D				
S. oligocarbophilus	>	DSM 40589	ATCC 27453	ISP 5589	A 1B		1-3	1-2									
S. olivaceiscleroticus	>	DSM 40595T	ATCC 15722	ISP 5595	A 1C	IV 24 (gray series)	069	1-26		La23							
S. olivaceus	>	DSM 40072T	ATCC 3335	ISP 5072	A 19	I 03	042	014			FU-1			121bp	JCM 4805	D44375	
S. olivochromogenes	>	DSM 41538	ATCC 21379			I 03	1-4	1-4						709bp		AF318046	
S. olivochromogenes	>	DSM 40451	ATCC 3336	ISP 5451		I 12	009	1-19									
S. olivochromogenes ssp. cytovirinus	>	DSM 40828	ATCC 12791			I 12	009	1-19			FU-1			1448bp		AY094370	
S. olivomycini	>	DSM 40105T		ISP 5105					Ha1					120bp	JCM 4176	D44023	
S. olivoreticuli ssp. olivoreticuli	>																
S. olivoverticillatus	>	DSM 40196		ISP 5196					Ha18								
S. olivoverticillatus	>	DSM 40250T NRRL B-1994T							Ha15					120bp	JCM 4400	D44007	L4
S. olivoviridis	>	DSM 40211	ATCC 15882	ISP 5211	A 03	II 20	1-3	010						120bp	JCM 4432	D44140	
S. omiyaensis	>	DSM 40552	ATCC 27454	ISP 5552	A 05	I 04	002	1-7						120bp	JCM 4806	D44376	
S. orinoci	>	DSM 40571T	ATCC 23202	ISP 5571	F 58	Sv. 17	22-1	040					KA-C	120bp	JCM 4546	D44225	
S. ornatus	>	DSM 40307	ATCC 23265	ISP 5307	A 1B		1-3	1-2						1518bp		X79326	
S. ostreogriseus	>	DSM 40511	ATCC 27455	ISP 5511	A 25		025	1-23									
S. pactum	>	DSM 40530T	ATCC 27456	ISP 5530	C 44	II 12	22-4	035		La11		OC-II		121bp	JCM 4809	D44377	
S. pallidus	>	DSM 40531T	ATCC 27457	ISP 5531	A 18		046	1-19		La14				1450bp		AJ399492	L2
S. paracochleatus	>	DSM 43350T															
S. paradoxus	>	DSM 40567	ATCC 27458	ISP 5567			033	1-33						121bp	JCM 3052	D43975	
S. paraguayensis	>	DSM 40473T	ATCC 12568	ISP 5473	C 46	Sv. 02	22-1	040	Ha1					120bp	JCM 4694	D44302	
S. parvisporogenes	>																
S. parvullus	>	DSM 40722					048	026									
S. parvullus	>	DSM 40728					006	1-18									
S. parvullus	>	DSM 40912				I 07	011	1-20									

(Continued)

Table 4. *Continued*

Species name		Strain number(s)	Cluster numbers of numerical taxonomic studies and Bergey categories and species no.[a]			Groups according to different studies[a]					bp	rRNA data strain no. if different from the first column	Acc. no. (EMBL)[b]	Strains included in DNA/DNA hybridisation studies[c]
			1	2	3	4	5	6	7	8			9	
S. parvulus	v	DSM 40048T ATCC 12434 ISP 5048	A 12	I 02	006 1-18		La24			KA-B				
S. parvus	v	DSM 40348 ATCC 12433 ISP 5348	A 1B		1-3 1-2			FU-6			120bp	JCM 4069	D43998	
S. parvus	v	DSM 40829 ATCC 12320		I 02	005 029									
S. paucisporogenum		DSM 40315 ATCC 12596 ISP 5315	F 55	Sv. 02	22-1 040									
S. pentaticum ssp. jenense		DSM 40848		Sv. 03	22-1 040									L4
S. peruviensis		DSM 40592 ATCC 27459 ISP 5592	A 18	IV 09 (red series)	009 1-19								AJ399494	
S. peucetius	v	DSM 40754 NCIB 10972	A 40	I 18	035 1-33						1486bp	JCM 9920	AB045887	
S. phaeochromogenes	v	DSM 40073T ATCC 3338 ISP 5073		I 18	009 0-19		La1		OC-II		1450bp		AF500071	
S. phaeochromogenes	v	DSM 40788 NRRL B-1266			076 071									
S. phaeofaciens	v	DSM 40367T ISP 5367									120bp	JCM 4814	D44381	
S. phaeoluteigriseus		NRRL 5182									1479bp		AJ391815	
S. phaeopurpureus	v	DSM 40125 ATCC 23946 ISP 5125	A 09	II 02	009 1-19						120bp	JCM 4660	D44290	
S. phaeoviridis	v	DSM 40285 ATCC 23947 ISP 5285	A 19	I 12	009 1-19						120bp	JCM 4661	D44291	
S. phosalacineus	v										1475bp	JCM 3340	U93330	
S. pilosus	v	DSM 40097T ATCC 19797 ISP 5097	A 37	I 17	006 1-10						121bp	JCM 4403	D44118	
S. platensis	v	DSM 40041 ATCC 13865 ISP 5041	A 29	I 15	025 005			FU-21			1488bp	JCM 4662	AB045882	
S. plicatus	v	DSM 40319 ATCC 25483 ISP 5319	A 12	I 07	006 1-18					KA-E	121bp	JCM 4504	D44194	
S. pluricolorescens	v	DSM 40019 ATCC 25798 ISP 5019	A 1B	I 02	1-3 1-2					KA-B	120bp	JCM 4602	D44258	
S. polychromogenes	v	DSM 40316 ATCC 12595 ISP 5316	F 61	I 22	22-3 042						120bp	JCM 4505	D44195	L3/L5
S. poonensis	v	DSM 40596T ATCC 15723 ISP 5596	A 22	II 19	071 1-19		La4		OC-III		121bp	JCM 4815	D44382	
S. praecox	v	DSM 40393T ATCC 3374 ISP 5393		IV 08 (yellow series)	)1-3 1-2						120bp	JCM 4506	D44196	
S. prasinopilosus	v	DSM 40098T ATCC 19799 ISP 5098	A 37	I 17	007 003		La20				119bp	JCM 4404	D44119	
S. prasinosporus	v	DSM 40506T ATCC 17918 ISP 5506	A 38	III 07	22-2 1-15		L10		OC-III		119bp	JCM 4816	D44383	
S. prasinus	v	DSM 40099T ATCC 19800 ISP 5099	A 37	I 17	007 003						120bp	JCM 4603	D44259	
S. pristinaespiralis	v	DSM 40338 ATCC 25486 ISP 5338	A 26	III 01	004 1-23									
S. prunicolor	v	DSM 40335T ATCC 25487 ISP 5335	A 11	III 01	1-1 1-1				OC-II		120bp	JCM 4508	D44198	
S. psammoticus	v	DSM 40341T ATCC 25488 ISP 5341	F 67	II 17	011 1-21		un		OC-I		120bp	JCM 4434	D44141	
S. pseudoechinosporeus	v	NRRL 3985									1516bp		X80827	
S. pseudogriseolus	v	DSM 40026 ATCC 12770 ISP 5026	A 12	I 07	006 1-18					KA-G	121bp	JCM 4071	D43999	
S. pseudogriseolus	v	DSM 40212 ATCC 23951 ISP 5212	A 18	I 11	009 1-19									
S. pseudovenezuelae	v	DSM 40213		I 11	22-3 042						1450bp		AJ399481	L2
S. pseudovenezuelae	v	DSM 40566 ATCC 13849 ISP 5566	A 12	I 18	006 1-18									L3
S. pulcher		DSM 41657T												
S. pulveraceus	v	DSM 40083 ATCC 19801 ISP 5083	A 09	II 02	005 029						120bp	JCM 7545	D44442	
S. puniceus	v	DSM 40283T ATCC 23952 ISP 5283			22-3 043						120bp	JCM 4406	D44121	
S. purpeofuscus	v	DSM 40310T ATCC 25489 ISP 5310	A 18	IV 26 (gray series)	009 1-19						121bp	JCM 4665	D44294	
S. purpurascens	v	DSM 43360T		I 11	009 1-19						1448bp		AB045888	L2
S. purpureus	v	DSM 43362T									1350bp		X53170	
S. purpureus	v	DSM 40271	A 40	I 23	22-3 1-05		La18		OC-I					
S. purpurogeneiscleroticus	v	DSM 40024 ATCC 23953 ISP 5024			069 1-26						121bp	JCM 4818	D44385	
S. pyridomyceticus					22-4 062									
S. racemochromogenes	v	DSM 40194 ATCC 23954 ISP 5194	F 61	I 22	22-3 042						120bp	JCM 4407	D44122	L5

Table (continued) — *Streptomyces* species strain data

Species		DSM	Collection	ISP	C Sm								bp	Strain/JCM	Accession	
S. rameus	v	DSM 41685T														
S. ramulosus	v	DSM 40100T	ATCC 19802	ISP 5100		III 16	035 041		un			OC-non	120bp	JCM 5064	D44433	
S. rangoonensis	v	DSM 40452T	ATCC 6860	ISP 5452		IV 07 (white series)	030 1-34						121bp	JCM 4604	D44260	
S. recifensis	v	DSM 40115	ATCC 19803	ISP 5115	A 23	I 20	1-5 059						120bp	JCM 4510	D44200	
S. rectiverticillatus	v	DSM 40436T	ATCC 19845	ISP 5436	F 57	Sv. 18	22-1 040						120bp	JCM 4408	D44123	
S. rectiviolaceus	v							Ha7								
S. regalis	v	DSM 40532	ATCC 27460	ISP 5532	A 20	I 13	009 1-19						120bp	JCM 4820	D44387	
S. regensis	v	DSM 40551	ATCC 27461	ISP 5551		I 11	009 1-19						1448bp		AJ399472	L2
S. resistomycificus	v	DSM 40133T	ATCC 19804	ISP 5133	A 18	I 11	009 1-19			FU-12a						
S. reticuli	v	DSM 40776	ATCC 23384				1-3 1-2									
S. reticuliscabiei													1418bp	CFBP4531	AJ007428	L4
S. reticulum		DSM 40893	ATCC 25607			Sv.	22-1 040									
S. reticulum ssp. *protomycicum*		DSM 40849T				Sv. 19	22-1 040									
S. rhizosphaericus	v	DSM 41760T											1480bp	A10P1	AJ391834	
S. rimosus	v												6067bp	R6-554T	X62884	
S. rimosus ssp. *paromomycinus*	v															
S. rimosus ssp.*rimosus*	v	DSM 40260T	ATCC 10970	ISP 5260	B 42	I 19	035 1-33		La9	FU-12a		OC-non	1485bp	JCM 4667	AB045883	
S. rishiriensis	v	DSM 40489	ATCC 14812	ISP 5489	A 19	I 12	1-7 1-15						120bp	JCM 4821	D44388	
S. rochei	v	DSM 40231T	ATCC 10739	ISP 5231	A 12	I 07	006 1-18		La13		KA-E	OC-III	869bp		AJ291995	
S. rosa		DSM 40533	ATCC 27462	ISP 5533	A 17		006 010									
S. roseiscleroticus	v	DSM 40303	ATCC 17755	ISP 5303		II 19	049 022						121bp	JCM 4823	AB018206	
S. roseochromogenes		DSM 40463	ATCC 13400	ISP 5463			1-3 004						120bp	JCM 4295	D44045	
S. roseochromogenes		DSM 40856	ATCC 3347				1-3 1-2						1412bp	JCM 4154	AF369704	
S. roseochromogenes		DSM 40879	IFO 3363				22-3 042						1523bp		AF290616	
S. roseodiastaticus	v	DSM 41703T														
S. roseoflavus	v		ATCC 19920		A 14	IV 10 (red series)	22-5 105									
S. roseoflavus	v	DSM 40536	ATCC 13167	ISP 5536		II 04	002 1-7						120bp	JCM 4605	D44261	
S. roseofulvus	v	DSM 40172	ATCC 19921	ISP 5172			009 1-19									
S. roseogriseus		DSM 40488	ATCC 12414	ISP 5488		II 18	22-5 039									
S. roseolilacinus	v	DSM 40173	ATCC 19922	ISP 5173	G 68	I 04	002 1-7						121bp	JCM 4335	D44057	
S. roseolus	v	DSM 40174	ATCC 23210	ISP 5174	A 05		1-6 1-16						120bp	JCM 4411	D44126	
S. roseoluteus		DSM 40240	ATCC 23975	ISP 5240	A 17	I 04	002 1-7									
S. roseosporus	v	DSM 40122	ATCC 23958	ISP 5122	A 05	I 04	22-1 040						120bp	JCM 4412	D44127	
S. roseoverticillatus	v	DSM 40039T	ATCC 19807			Sv. 01	22-1 040									L4
S. roseoverticillatus ssp. *albosporus*	v	DSM 40900	ATCC 25189			Sv. 01	22-1 040	Ha7								
S. roseoviolaceus	v	DSM 40277	ATCC 25493 / IFO 12845	ISP 5277	A 18	I 11	009 1-19			FU-1			1197bp		AB072837	L2
S. roseoviridis	v	DSM 40175	ATCC 23959	ISP 5175	A 05	I 04	22-2 037						1448bp		AJ399484	
S. roseus		DSM 40076	ATCC 19808	ISP 5076	A 07		002 1-7						120bp	JCM 4414	D44128	
S. ruber	v	DSM 40304				IV 11 (red series)	049 022						121bp	JCM 3131	AB018203	
S. rubescens		DSM 40777	NRRL B-1519				076 071									
S. rubiginosohelvolus	v	DSM 40176T	ATCC 19926	ISP 5176		IV 12 (red series)	006 1-2						120bp	JCM 4415	D44129	
S. rubiginosus	v	DSM 40177	ATCC 19927	ISP 5177	A 12	I 07	006 1-18						121bp	JCM 4416	D44130	
S. rubrochlorinus		DSM 40850	NRRL B-12558			Sv. 01	22-1 040	Ha7								L4
S. rubroverticillatus		DSM 40851						Ha1								L4
S. rubroverticillatus		DSM 41489	NRRL B-16433													
S. rugersensis ssp. *castelarensis*	v	DSM 40830	ATCC 15191			I 01	055 018									

(Continued)

Table 4. *Continued*

Species name		Strain number(s)	Cluster numbers of numerical taxonomic studies and Bergey categories and species no.[a]			Groups according to different studies[a]					bp	rRNA data strain no. if different from the first column	Acc. no. (EMBL)[b]	Strains included in DNA/DNA hybridisation studies[c]
			1	2	3	4	5	6	7	8				9
S. rugersensis ssp. rugersensis	✓	DSM 40077T ATCC 3350 ISP 5077	A 1A	I 01	1-1 1-1					KA-D	1476bp		Z76688	
S. salmonis	✓										1340bp	DPUD 0098T	X53169	L4
S. salmonis	✓	DSM 40895T NRRL B-1472		Sv. 05	22-1 040	Ha7								
S. sampsonii	✓	DSM 40394 ATCC 25495 ISP 5394	A 1A	I 01	1-1 1-1					KA-D	1476bp		D63871	
S. sannanensis	✓	DSM 41675T				Ha4								
S. sapporonensis	✓	DSM 40537 ATCC 3351 ISP 5537	A 1A		1-1 1-1									
S. saprophyticus	✓	DSM 40241 ATCC 25496 ISP 5241 ATCC 49173	A 29		050 019									
S. saraceticus	✓										1530bp		D63862	
S. scabiei (scabies)	✓	DSM 40078 ATCC 23962 ISP 5078	A 03		1-3 010									
S. scabiei	✓	DSM 40611 ATCC 3352			1-4 1-4									
S. scabiei	✓	DSM 40859 IFO 3111			1-2 015									
S. scabiei	✓	DSM 40960			006 1-18									
S. scabiei	✓	DSM 40961			045 1-24									
S. scabiei	✓	DSM 40962			014 021									
S. scabiei	✓	DSM 40994			013 1-19									
S. scabiei	✓	DSM 40995			008 019									
S. scabiei	✓	DSM 40996			002 1-7									
S. scabiei	✓	DSM 40997			009 1-19									
S. scabiei	✓	DSM 40998			009 1-19									
S. scabiei	✓	DSM 40999			008 019									
S. scabiei	✓	DSM 41000			009 1-19									
S. scabiei	✓	DSM 40269												
S. sclerotialus	✓			I 18	069 1-26						121bp	JCM 3039	D43973	
S. scopiformis	✓										1402bp	A25T	AF184081	
S. seoulensis	✓										1479bp	IMSNU 2126	Z71365	
S. septatus	✓	DSM 40577T ATCC 27464 ISP 5577	F 55	Sv. 02	22-1 040	Ha6								
S. setae	✓	DSM 43861									1464bp		M55220	
S. setonii	✓	DSM 40395 ATCC 25497 ISP 5395	A 1B	I 02	1-3 1-2					KA-B	1532bp	JCM 4830	D63872	
S. showdoensis	✓	DSM 40504 ATCC 15105 ISP 5504	A 06	I 05	22-2 037					KA-B	120bp	JCM 4669	D44393	
S. sindenensis	✓	DSM 40255T ATCC 23963 ISP 5255	A 1B		1-3 1-2						120bp	JCM 4418	D44297	
S. sioyaensis	✓	DSM 40032 ATCC 13989 ISP 5032	A 29	I 15	025 005						121bp		D44131	
S. somaliensis	✓	DSM 40760 ATCC 14817	A 34		035 1-33						1483bp		AJ007399	
S. spadicis	✓	DSM 40476 ATCC 19017 ISP 5476	A 32		1-5 009									
S. sparsogenes	✓	DSM 40356T ATCC 25498 ISP 5356		I 16	010 1-19		La7				1493bp		AJ391817	L1
S. spectabilis	✓	DSM 40512 NRRL 2792T ISP 5512									121bp	JCM 4832	D44395	L3
S. speibonae	✓	DSM 41797T									1490bp		AF452714	
S. speleomycini	✓	DSM 40292 ATCC 23965 ISP 5292	A 1B	I 02	040 048		La19							
S. spheroides	✓										120bp	JCM 4670	D44298	
S. spinoverrucosus	✓													
S. spiralis	✓	DSM 43836									121bp	JCM 3302	D43983	
S. spiroverticillatus	✓	DSM 40036 ATCC 19811 ISP 5036	A 06	I 05	002 1-7	Ha7				KA-A	120bp	JCM 4609	D44263	
S. spitsbergensis	✓													

Species		Strain(s)	ISP	Code	Group	No.	Ha	Misc	OC	bp	Other strain	GenBank	L
S. sporiferum		DSM 40901 ATCC 25188											
S. sporocinereus	>												
S. sporoclivatus	>												
S. sporocochleatus (K. paracochleatus)	>	IFO 14769T			Sv. 02	22-1 040				1475bp		U93328	
S. spororaveus	>												
S. sporoverrucosus	>	DSM 40547 ATCC 27466	ISP 5547	A 38	III 17	009 1-19				1483bp	JCM 4833	AB045889	
S. steffisburgensis		DSM 41683T					Ha16			1478bp	CFBP 4521	AJ007429	L5
S. stelliscabiei	>												
S. stramineus		DSM 40200 ATCC 15886	ISP 5200			1-3 1-2				1516bp		X80825	
S. streptomycini	>	DSM 40445 ATCC 27467 ATCC 31892	ISP 5445	F 61		22-3 042				620bp		AF318042	
S. subrutilus	>												
S. sulfonofaciens	>	DSM 40104T ATCC 27468	ISP 5104	C Sm	III 17	068 002	Ha14	un	OC-non	120bp	JCM 4835	D44397	
S. sulphureus	>	DSM 41480											L4
S. syringium		DSM 41499 NRRL B-16435			Sv. 09	22-1 040	Ha1			120bp		D44299	L4
S. taitoensis		DSM 40576 ATCC 27469	ISP 5576		IV 30 (gray series)	002 1-7				1492bp	JCM 4671	AB045879	
S. takataensis	>	DSM 40195T ATCC 23967	ISP 5195	A 19	I 07	012 019		La14		1530bp	JCM 4837	D63873	
S. tanashiensis	>	DSM 40560 ATCC 27470	ISP 5560	A 12		006 1-18				120bp	JCM 4518	D44206	
S. tauricus	>	DSM 40101T ATCC 19812	ISP 5101	A 05	I 04	026 027		KA-E		1286bp	TA 56T	AJ000284	
S. tendae	>	DSM 40477 ATCC 17920	ISP 5477	A 1A		22-2 037				1502bp		U94487	
S. tenebrarius		DSM 40329 ATCC 25499	ISP 5329			1-1 1-1				1502bp		U94489	
S. termitum	>	DSM 40585 ATCC 27471	ISP 5585							1500bp		U94490	
S. tetanusemus		DSM 41741T								1499bp	B19	AJ007402	
S. thermoalcalitolerans	>	DSM 44294								1483bp		AB018096	
S. thermoautotrophicus	>	DSM 44296T								1540bp	CCTCCAA 97	AF056712	
S. thermocarboxydovorans	>	DSM 44293								1540bp	CCTCCAA 97	AF056714	
S. thermocarboxydovorans		DSM 41700T								1481bp		Z68097	
S. thermocarboxydus	>	DSM 40573T ATCC 27472	ISP 5573	A 1C	I 03	006 1-18				1486bp		Z68098	
S. thermocoprophilus	>	DSM 40574 ATCC 27473	ISP 5574	A 36		1-7 1-15				1522bp	AT 10	AF333113	
S. thermodiastaticus	>	DSM 41451T						FU-1		1448bp		AJ399482	
S. thermoflavus	>	DSM 40579T ATCC 23385	ISP 5579	A 36	II 09	021 002				1350bp	AB106	AY029353	
S. thermogriseus	>	DSM 40365 ATCC 19282	ISP 5365	A 15		006 1-18				1211bp	JCM 4312	D44050	
S. thermogriseus		DSM 40227 ATCC 11416	ISP 5227	A 18		009 1-19				1483bp		Z68096	
S. thermolineatus	>	DSM 41392T						La13		1486bp		Z68094	
S. thermonitrificans (thermovulgaris)		DSM 40443T ATCC 19283	ISP 5443	C 45	II 13	004 006				120bp	JCM 4087	D44001	
S. thermophilus		DSM 40444T ATCC 19284	ISP 5444	A 36	II 09	021 002	Ha17	un	OC-non	1211bp	JCM 4872	D44416	
S. thermospinosisporus	>	DSM 40027T ATCC 12310	ISP 5027	F Sm	Sv. 21	22-1 040				120bp	JCM 4421	D44133	L3/L5
S. thermotolerans		DSM 40894T NRRL B-3889			IV 31 (gray series)	009 1-19				1211bp		D44434	
S. thermoviolaceus ssp.	>	DSM 40178 ATCC 19813	ISP 5178	F 61		22-3 042				121bp	JCM 5065	D44230	L4
S. thermoviolaceus ssp. apingens	>	DSM 40030 ATCC 19814	ISP 5030	A 29		013 1-19	Ha6	La2	OC-non	121bp	JCM 4558	AF503490	
S. thermoviolaceus ssp. termoviolaceus		DSM 42704								1455bp	JCM 4846		
S. thioluteus		DSM 40520 ATCC 17963	ISP 5520	F 55	Sv. 03	22-1 040							
S. thiovulgaris	>	DSM 40261T ATCC 25502	ISP 5261	C 47	III 14	025 005							
S. torulosus	>	DSM 40505		A 21	I 14	006 1-18							
S. toxytricini													
S. toyocaensis													
S. tricolor	>												
S. tropicalensis	>												
S. tubercidicus													
S. tuirus	>												

(Continued)

Table 4. *Continued*

Species name	Strain number(s)	1	2	3	4	5	6	7	8	bp	rRNA data strain no. if different from the first column	Acc. no. (EMBL)[b]	Strains included in DNA/DNA hybridisation studies[c]
S. turgidiscabies	ATCC 702348T T			1-6 1-16						1528bp	S27	AF361782	
S. turgidiscabies				1-6 1-12						1865bp		AB026221	
S. umbrinus	DSM 40278 ATCC 19929 ISP 5278	A 05	I 04							1200bp	JCM 4521	D44209	
S. umbrosus	DSM 40242 ATCC 25504 ISP 5242	A 33											
S. valinus	NRRL B-5644												L2
S. variabilis	DSM 40179 ATCC 19930 ISP 5179	A 12	I 07	006 1-18					KA-F	121bp	JCM 4422	D44134	
S. variegatus													
S. varsoviensis	DSM 40346T ATCC 25505 ISP 5346	C 46	III 13	037 028		La12		OC-II		121bp	JCM 4523	D44211	
S. vastus	NRRL 8037												
S. vellosus										1520bp		X99942	
S. vendargensis	DSM 40379 ATCC 25507 ISP 5379			035 1-33									
S. venezuelae	DSM 40230T ATCC 10712 ISP 5230	A 06	I 05	002 1-7		un			KA-C	1483bp	JCM 4526	AB045890	
S. verne	DSM 40079 ATCC 3353 ISP 5079	A 40		069 1-26									
S. versipellis	DSM 40491 ATCC 27475 ISP 5491	A Sm		1-7 1-15									
S. verticillium	DSM 40903 ATCC 15003		Sv. 06	22-1 040									
S. vinaceus	DSM 40257 ATCC 11861 ISP 5257	A 06	I 05	1-3 1-2						120bp	JCM 4849	D44405	
S. vinaceus	DSM 40515 ATCC 27476 ISP 5515	A 06	I 05	22-3 042					KA-A	120bp	JCM 4425	D85123	
S. virginiae	DSM 40094 ATCC 19817 ISP 5094	F 61	I 22	22-3 042			FU-12b			121bp	JCM 4529	D44214	
S. vinaceusdrappus	DSM 40470 ATCC 25511 ISP 5470	A 12	I 07	006 1-18					KA-E	121bp	JCM 4530	D44215	L3/L5
S. violaceochromogenes	DSM 40181T ATCC 19932 ISP 5181		IV 33 (gray series)	009 1-19						121bp	JCM 4531	D44216	
S. violaceolatus	DSM 40438 ATCC 19847 ISP 5438	A 21	I 14	006 1-18						121bp	JCM 4532	D44217	
S. violaceorectus	DSM 40279 ATCC 25514 ISP 5279	A 05	I 04	002 1-7						120bp	JCM 4423	D44135	
S. violaceoruber (coelicolor)	DSM 40049T ATCC 14980 ISP 5049	A 05	IV 34 (gray series)	069 1-26						2300bp	A(3)2	X60514	
S. violaceus	DSM 40082T ATCC 15888 ISP 5082	A 06	I 05	009 0-19				OC-III		121bp	JCM 4533	D44218	
S. violaceusniger	DSM 40182		I 16	009 020									
S. violaceusniger	DSM 40563T ATCC 27477 ISP 5563	A 32	I 16	051 018		La7		OC-I		1480bp		AJ391823	L1
S. violaceusniger	DSM 40699		I 16	041 012									L1
S. violaceusniger	DSM 41598 NRRL B-1356		I 16	054 018									L1
S. violaceusniger	DSM 41599 NRRL B-1477		I 16	053 018									L1
S. violaceusniger	DSM 41600 NRRL B-1478		I 16	053 018									L1
S. violaceusniger	DSM 41602 NRRL B-16257		I 16	054 018									L1
S. violaceusniger	NRRL 8097												
S. violaceusniger	DSM 40710 NRRL 2834		I 16	1-6 1-16						1497bp		AJ391816	L1
S. violaceusniger	DSM 41601 NRRL B-5799		I 16	1-5 1-16						1341bp		AJ391813	L1
S. violaceorubidus						La12				1484bp		AJ391814	L1
S. violarus	DSM 40205 ATCC 15891 ISP 5205	A 18	I 11	009 1-19						1447bp		AJ399477	L2
S. violascens	DSM 40183 ATCC 23968 ISP 5183	A 06	I 05	002 1-7						120bp	JCM 4424	D44136	
S. violatus	DSM 40209T ATCC 15892 ISP 5209	A 18	I 11	050 019						1448bp		AJ399480	
S. violens	DSM 40597 ATCC 15898 ISP 5597	A 40	I 18	069 1-26						121bp	JCM 3072	D43977	L2
S. violochromogenes	DSM 40207 ATCC 15893 ISP 5207	A 18		009 1-19									
S. virens													

Species	v	DSM	ATCC/IFO	ISP	Code	Category	Cluster no.	bp	KA	JCM	OC	La	Accession	Cluster
S. virginiae	v		IFO 3729T		F 66		22-4 043	1514bp					D85119	
S. viridifaciens	v	DSM 40239	ATCC 11989	ISP 5239	A 27		009 1-19							
S. viridiflavus		DSM 40381	ATCC 15732	ISP 5381			076 069							
S. viridis		DSM 40637			A 27									
S. viridis														
S. viridiviolaceus	v	DSM 40280T	ATCC 27478	ISP 5280		IV 35 (gray series)	006 1-18	120bp		JCM 4855			D44408	
S. viridobrunneus	v													
S. viridochromogenes	v	DSM 40110T	ATCC 14920	ISP 5110	A 27	III 04	009 1-19	1494bp		JCM 5013	OC-III		AF045858	L2
S. viridochromogenes	v	DSM 40249T	ATCC 25518	ISP 5249			006 1-18	121bp		JCM 4856			D44409	
S. viridodiastaticus		DSM 40237	ATCC 12631	ISP 5237	J 70	IV 36 (gray series)	22-1 040	121bp		JCM 4536			D44221	
S. viridoflavum		DSM 40454	ATCC 3372		A 03	Sv. 24	006 1-4							
S. viridogenes				ISP 5454										
S. viridosporus	v	DSM 40243	ATCC 27479	ISP 5243	A 15	I 08	006 1-18	121bp		JCM 4859			D44410	
S. vitaminophilus	v													
S. vulgaris		DSM 40201	ATCC 15895	ISP 5201			1-3 1-2							L4
S. waksmanii		DSM 40464				Sv. 09	22-1 040							
S. wedmorensis	v													
S. werraensis	v	DSM 40486	ATCC 14424	ISP 5486	A 12	I 07	006 1-18	121bp	KA-G	JCM 4860			D44411	
S. willmorei	v	DSM 40459T	ATCC 6867	ISP 5459	A 1B	I 02	1-3 1-2	120bp	KA-B	JCM 4861			D44412	
S. xanthochromogenes	v	DSM 40111T	ATCC 19818	ISP 5111	F 63	II 15	005 029	120bp		JCM 4612		La23	D44264	
S. xanthocidicus	v	DSM 40575T	ATCC 27480	ISP 5575	F 66	II 16	22-4 043	120bp		JCM 4862	OC-I	La18	AB018208	
S. xantholiticus	v	DSM 40244T	ATCC 27481	ISP 5244	C 24	II 05	062 024	120bp		JCM 4863		La21	D44413	
S. xanthophaeus	v	DSM 40134	ATCC 19819	ISP 5134	F 61	I 22	084 067	120bp		JCM 4426			D44138	L5
S. yatensis		DSM 41771				III 18	080 066	1493bp		SFOCin 76	OC-I		AF336800	
S. yerevanensis	v	DSM 43167T						120bp		JCM 3065			D43976	
S. yerevanensis		DSM 41766T												
S. yogyakartensis	v							1481bp		C4R3(S3)			AJ391827	
S. yokosukanensis	v	DSM 40224	ATCC 25520	ISP 5224	A 30	II 06	009 1-19	120bp		JCM 4559			D44231	
S. yunnanensis	v							1521bp		YIM 41004T			AF346818	
S. zaomyceticus	v	DSM 40196	ATCC 27482	ISP 5196	A 05	I 04	002 1-7	120bp	KA-C	JCM 4864			D44414	

Abbreviations: EMBL, European Molecular Biology Laboratory; v, valid name; T, type strain; DSM, Deutsche Sammlung von Mikroorganismen und Zellkulturen; ATCC, American Type Culture Collection; ISP, International *Streptomyces* Project; NRRL, ARS Culture Collection, Northern Regional Research Laboratory, U.S. Department of Agriculture, Peoria, Illinois, USA; JCM, Japan Collection Of Microorganisms; IMRU, Institute of Microbiology, Rutgers State University; IMSNU, Institute of Microbiology, Seoul National University, Seoul, Korea; and CFBP, Collection Francaise des Bacteries Phytopathogenes.

a)1) Cluster and subcluster numbers according to Williams et al. (1983a). 2) Assignment of strains to species according to *Bergey's Manual of Systematic Bacteriology, Vol. 4*. For the genus *Streptomyces*, species categories I to IV are shown, with the species number within the category according to Williams et al. (1989); for species category IV, the 'series' assignment is also given, according to spore color. For species previously assigned to the genus *Streptoverticillium* (now reclassified to *Streptomyces*), species numbers are according to Locci and Schofield (1989). 3) Cluster and subcluster numbers are according to Kämpfer et al. (1991). The first number is that of the UPGMA/S$_{SM}$ analysis, the second number is that of the UPGMA/S analysis. 4) Groups are according to the study of Hatano et al. (2003): Ha1 (synonyms of *S. abikoensis*), Ha2 (synonyms of *S. ardus*), Ha3 (synonyms of *S. blastmyceticus*), Ha4 (synonyms of *S. cinnamoneus*), Ha5 (synonyms of *S. eurocidicus*), Ha6 (synonyms of *S. griseocarneus*), Ha7 (synonyms of *S. hiroshimensis*), Ha8 (synonyms of *S. lilacinus*), Ha9 ("*S. luteoreticuli*"), Ha10 (*S. luteosporeus*), Ha11 (synonyms of *S. luteoreticuli*"), Ha12 (synonyms of *S. mashuensis*), Ha13 (synonyms of *S. morookaense*), Ha14 (synonyms of *S. netropsis*), Ha15 (*S. orinoki*), Ha16 (*S. stramineus*), Ha17 (*S. thioluteus*), and Ha18 (synonyms of *S. viridflavus*). 5) Clusters are according to Lanoot et al. (2002) on the basis of protein profiles. 6) Cluster numbers are according to Fulton et al. (1995) on the basis of fingerprints of the 16S rRNA operons. 7) Groups based on primary structures of N termini of AT-L30 proteins are according to Ochi (1995). 8) Clusters based on the phylogenetic tree constructed from the 120-bp alpha-region are according to Kataoka et al. (1997). More than 450 *Streptomyces* 120-bp alpha regions have been sequenced since those reports.

b)The strain numbers and the accession numbers of the databases are given in the adjacent columns. Accession numbers (EMBL) of complete 16S rRNA sequences are also given in those columns in addition to the number of bases in the sequence.

9)Strains included in the DNA-DNA hybridization studies L1: Labeda and Lyons (1991a): on the basis of 70% DNA-DNA similarity, *S. hygroscopicus* NRRL 2387T and *S. endus* are synonymous (together with several strains of *S. violaceusniger*; the remaining strains represent single species); L2: Labeda and Lyons (1991b): on the basis of 70% DNA-DNA similarity, *S. bellus*, *S. curacoi*, and *S. coeruleorubidus* are synonymous; *S. afghaniensis*, *S. janthinus*, *S. roseoviolaceus*, *S. violatus*, and *S. purpurascens* are synonymous; the remaining strains represent single species; L3: Labeda (1998): on the basis of 80% DNA-DNA similarity, *S. griseoviridis* and *S. daghestonicus* are synonymous; *S. chryseus* and *S. longispororuber* are synonymous; the remaining strains represent single species; L4: Labeda (1996): on the basis of 70% DNA-DNA similarity, *S. abikoensis*, *S. waksmanii*, and *S. takataensis* are synonymous; *S. biverticillatus*, *S. fervens*, *S. baldaccii*, *S. roseoverticillatus* are synonymous; *S. netropsis*, *S. kentuckensis*, *S. flavopersicus* are synonymous; *S. cinnamoneus* subsp. *azacoluta*, *S. hachijoensis* are synonymous; the remaining strains represent single species; L5: Labeda (1993): on the basis of 80% DNA-DNA similarity, *S. lavendulae* subsp. *avirens*, *S. lavendulae* subsp. *grasserius*, *S. columbiensis* are synonymous; and the remaining strains represent single species.

short chains of spores on the substrate mycelium. Sclerotia, pycnidial-, sporangia-, and synnemata-like structures may be formed by some species. The spores are nonmotile. On complex agar media, discrete and lichenoid, leathery or butyrous colonies are formed. Colonies are initially relatively smooth surfaced, but later they develop an aerial mycelium that may appear floccose, granular, powdery or velvety.

Members of the genus *Streptomyces* undergo a complex life cyle, which has been studied most intensively for strain "*S. coelicolor*" A2(3). *Streptomyces* colonies are multicellular, differentiated organisms exhibiting temporal and spatial control of gene expression, morphogenesis, metabolism and the flux of metabolites (see chapter 2 of Kieser et al., [2000] for more details).

Strains belonging to the genus *Streptomyces* may produce a wide variety of pigments responsible for the color of the vegetative and aerial mycelia (Figs. 3 and 4). In addition, colored diffusible pigments may also be formed. Note that the production of pigments largely depends on the medium composition and cultivation conditions (Figs. 3 and 4). Many strains produce one or more antibiotics (more details are given in The Family Streptomycetaceae, Part II: Moleular Biology in this Volume). The metabolism is oxidative and chemoorganotrophic. The catalase reaction is positive, and generally, nitrates are reduced to nitrites. Most representatives can degrade polymeric substrates like casein, gelatin, hypoxanthine, starch and also cellulose. In addition, a wide range of organic compounds is used as sole sources of carbon for energy and growth (Williams et al., 1983a; Kämpfer et al., 1991b; Korn-Wendisch and Kutzner, 1992a). The optimum temperature for most species is 25–35°C; however, several thermophilic and psychrophilic species are known. The optimum pH range for growth is 6.5–8.0.

The Embden-Meyerhof-Parnas (glycolysis) pathway of glucose catabolism has been found in many streptomycetes (Cochrane, 1961), but also the hexose monophosphate shunt (Salas et al., 1984) was detected in *S. antobioticus*. Several streptomycetes are able to switch from glycolysis to the hexose monophosphate shunt during secondary metabolism (Kieser et al., 2000). At present, no streptomycete is known to use the Entner-Doudoroff pathway. Sugar transport is mediated in connection with phosphorylation by specific kinases (Sabater et al., 1972; Ikeda et al., 1984). The phosphoenolpyruvate:fructose phosphotransferase system (PTS) for the transport and phosphorylation of fructose has recently been detected in *S. coelicolor*, *S. lividans* and *S. griseofuscus* (Titgemeier et al., 1995). More details about specific metabolic pathways, including nitrogen metabolism and the regula-

tion processes involved, are given in chapter 1 of Kieser et al. (2000) and the references therein.

On the basis of 16S rRNA/DNA sequence comparisons, members of the genus *Streptomyces* form a separate line of descent, and Stackebrandt et al. (1997) proposed the emendation of the family Streptomycetaceae in the suborder Streptomycinae and the order Actinomycetales. The intrageneric phylogenetic relationships of many of the 346 recognized species in *Bergey's Manual of Systematic Bacteriology* (Williams et al., 1989) inferred from the 350 complete 16S rRNA sequences, however, are clearly restricted by the limited resolving power of the method to discriminate between related species and are often in contrast with a morphologically and physiologically based classification. Though about 350 almost complete 16S rRNA sequences are available to date, the high degree of conservation within 16S rRNA genes causes problems for resolving phylogenetic relationships at the intergeneric level.

Notably, the different methods used for grouping of the *Streptomyces* species often lead to contradictory results. In Table 4, all 376 *Streptomyces* species and subspecies with valid names (as of December 9, 2003; taken from the List of Bacterial Names with Standing in Nomenclature); and some additional species with names not validly published (but included in taxonomic studies) are given with their grouping according to different studies.

The chemotaxonomic features for the identification of strains at the genus level (for details see below) are of high value and can be summarized as follows: The cell wall peptidoglycan contains major amounts of LL-diaminopimelic acid (LL-A$_2$pm). Genus members lack mycolic acids, contain major amounts of saturated, *iso-* and *anteiso*-fatty acids, possess either hexa- or octahydrogenated menaquinones with nine isoprene units as the predominant isoprenolog, and have complex polar lipid patterns that typically contain diphosphatidylglycerol, phosphatidylethanolamine, phosphatidylinositol, and phosphatidylinositol mannosides (Table 2; Figs. 5 and 6). In addition to these traits, the acyl type of the muramyl residues in the cell-wall peptidoglycans is acetyl (Uchida and Seino, 1997). Strains are widely distributed and abundant in soil, including composts (see detailed description below). A few species are pathogenic for animals and man, and others are phytopathogens. The type species is *Streptomyces albus* (Rossi-Doria 1891) Waksman and Henrici (1943).

CELL WALL COMPOSITION

Peptidoglycan The cell walls of streptomycetes show the typical ultrastructure and chemical

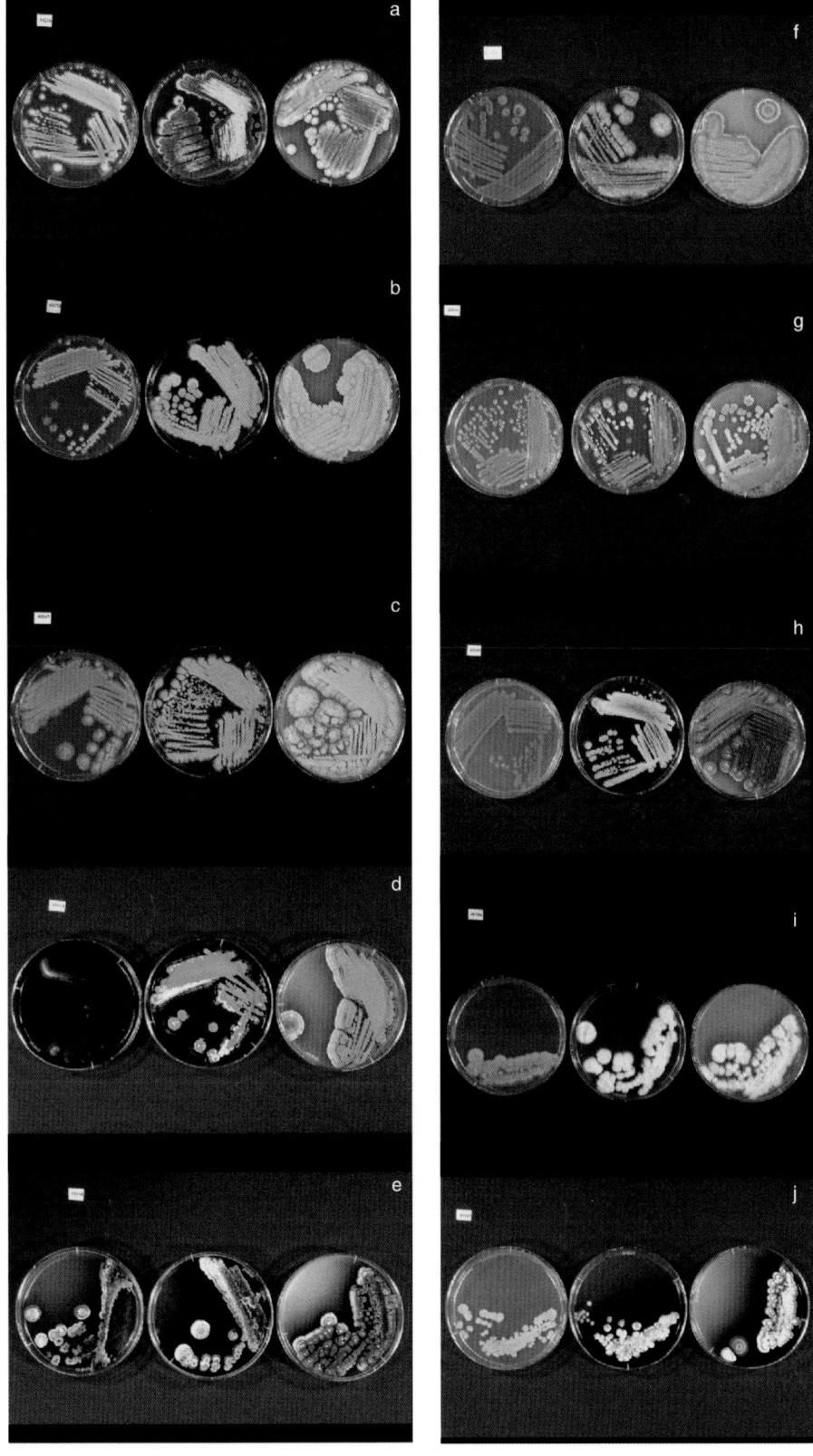

Fig. 3. A–J: Color of the aerial mycelium of *Streptomyces* strains grown on different agar media after 3 weeks of incubation at 28°C. Left: starch-casein-nitrate agar; middle: GYM agar; right: oatmeal agar (for compositions, see Tables 10 and 12). Species names and strain numbers are given in Table 6.

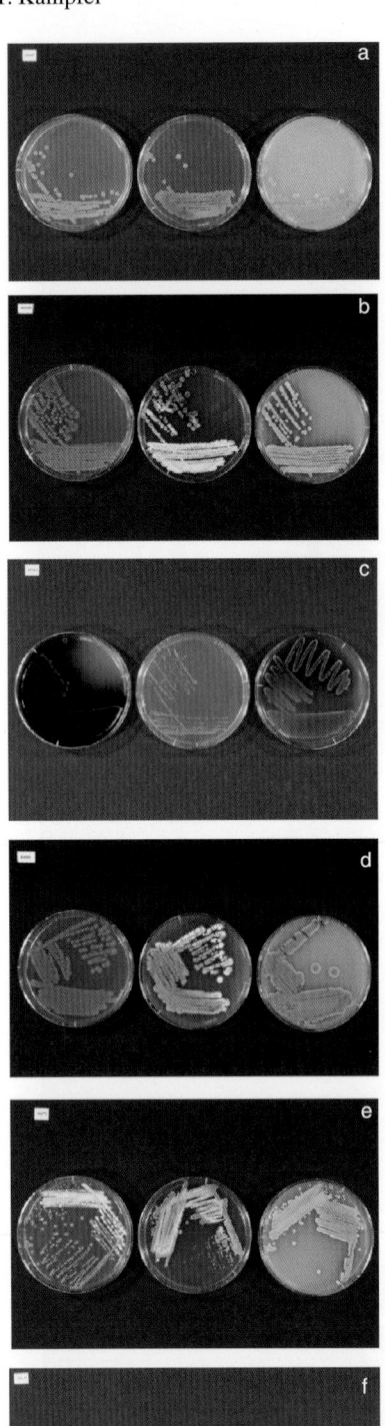

Fig. 4. A–F: Color of the substrate mycelium and soluble pigments of *Streptomyces* strains grown on different agar media after 7 days of incubation at 28°C. Left: starch-casein-nitrate agar; middle: GYM agar; right: oatmeal agar (for compositions, see Tables 10 and 12). Species names and strain numbers are given in Table 7.

composition of Gram-positive bacteria (Schleifer and Kandler, 1972). They appear under the electron microscope as homogeneous less electron dense layers of about 16–35 nm. The cell walls have a multilayered structure of peptidoglycan strands. The peptidoglycan is a heteropolymer consisting of heteropolysaccharide chains cross-linked through short peptide units. The so-called "sugar back bone" of the peptidoglycan is constructed of alternating β-1,4-linked units of *N*-acetylglucosamine and *N*-acetylmuramic acid. The carboxyl group of muramic acid is substituted by an oligopeptide of alternating D- and L-amino acids (Schleifer and Kandler, 1972). *Streptomyces* is characterized by the tetrapeptide L-Ala–D-Glu–LL-A2pm–D-Ala. This tetrapeptide is crosslinked by a pentaglycine bridge which extends from the C-terminal D-alanine of the peptide unit to the amino group located on the D carbon of LL-A$_2$pm, resulting in the macromolecule structure forming the cell envelope. This LL-A$_2$pm-Gly$_5$, or A3γ peptidoglycan type (Schleifer and Kandler, 1972), is diagnostic for streptomycetes and some other combined-wall chemotype I actinomycetes (Lechevalier and Lechevalier, 1970).

The aerobic actinomycetes were grouped (using specific amino acids in purified cell walls) into four so-called "wall chemotypes" by Lechevalier and coworkers. Cell walls with *meso*-DAP and LL-DAP were detected early. In another study, Takahashi et al. (1984b) reported that strains belonging to this group change cell wall composition during sporulation. They found in submerged mycelium LL-DAP and glycine (wall chemotype I) whereas in spores, only *meso*-DAP could be detected (wall chemotype III according to Lechevalier and Lechevalier, 1970). In 11 streptomycetes, the cell wall compositions of aerial, substrate and submerged mycelium differed in the quantitative distribution of cell wall amino acids and cell wall sugars. *N*-Acetylmuramic acid is found in the glycolyl type of cell wall of *Streptomyces*, as in all other actinomycetes (Uchida and Aida, 1977). Muramic acid phosphate residues are the attachment points to teichoic acids, which are of diagnostic value for Gram-positive bacteria. The cell wall teichoic acids (polymeric substances containing repeating phosphodiester groups) consist of polyols (i.e., the sugar alcohols glycerol and ribitol) or *N*-acetylamino sugars or both. The teichoic acids of streptomyces are of the same structure as those of other Gram-positive bacteria, containing either ribitol phosphate or glycerol phosphate polymers; significantly, the teichoic acid of actinomycetes does not contain ester-bound D-alanine but does have ester-linked acetic acid and sometimes succinic acid residues (Naumova et al., 1980).

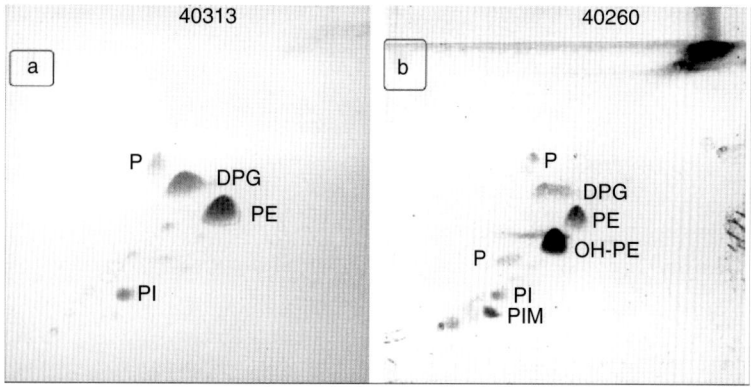

Fig. 5. Two-dimensional thin layer chromatograms of polar lipids of A) *Streptomyces albus* (DSM 40313) and B) *Streptomyces rimosus* (DSM 40260). Abbreviations: DPG, diphosphatidyglycerol; PI, phosphatidylinositol; PIM, phosphatidylinositol mannosides; PE, phosphatidylethanolamine; OH-PE, hydroxy-phosphatidylethanolamine; and P, phospholipids of unknown structure. (Courtesy of R. M. Kroppenstedt.)

For streptomycetes, the synthesis of either ribitol phosphate (*S. streptomycinii* and *S. violaceus*) or glycerol phosphate polymers (*S. thermovulgaris*, *S. levoris*, *S. rimosus* and *S. antibioticus*) has been reported (Naumova et al., 1980). In ribitol teichoic acids, positions 1 and 5 of ribitols are connected to the phosphates; in glycerol teichoic acids, position 1 is commonly connected to 3; and in other types, position 1 connected to 2 (as in *S. antibioticus*) is not common. The polyol phosphates can be substituted with various combinations of sugars or amino sugars or both, which are linked to glycerols or ribitols via glycosidic bonds. At present, few strains (species) have been investigated in detail and so the role of teichoic acid in the taxonomy of *Streptomyces* is not clear (Naumova et al., 1980).

Cell Wall Polysaccharides These compounds seem to be of no diagnostic value in those strains where LL-DAP is found in whole cell hydrolysates (Lechevalier et al., 1971). Some of the diagnostic sugars found in other actinomycetes, like xylose, galactose and arabinose, were reported occasionally in streptomycetes. Hundreds of streptomyces were analyzed for the presence of diagnostic sugars (Kroppenstedt, 1977), and mainly ribose, mannose and glucose were usually found in small amounts.

Phospho- and Glycolipids The lipids of streptomycetes comprise mainly diphosphatidylglycerol (DPG), phosphatidylethanolamine (PE), phosphatidylinositol (PI), and phosphatidylinositolmannosides (PIMs). Lipid composition has been extensively investigated and summarized by Lechevalier et al. (1977b). Glycolipids do not occur consistently in streptomycetes, and their qualitative and quantitative lipid composition depends largely on culture conditions. Under phosphate limiting conditions, the amount of glycolipids increase, significantly.

The taxonomic significance of polar lipids in actinomycetes was demonstrated by Lechevalier et al. (1977b). From the phospholipid results of 97 actinomycete strains representing 20 genera, Lechevalier et al. (1977b) proposed a classification of five phospholipid types. These five groups are based on the presence or absence of certain nitrogenous phospholipids. The marker lipids of type II (PII) are phosphatidylethanolamine (PE), methyl-PE, hydroxy-PE, and lyso-PE. In addition to various other families, members of the family Streptomycetaceae contain the lipids of phospholipid type II. Additional lipids (e.g., phosphomonoester [PME] and OH-PE) and the presence or absence of PI and PG allow further differentiation (Fig. 5).

MENAQUINONES Streptomycetes contain only menaquinones (Collins and Jones, 1981), and like the majority of actinomycetes, they synthesize quinones that have a partly saturated isoprenoid side chain at position 3 of the naphthoquinone ring. Menaquinone composition is very useful for differentiation of actinomycetes because of the different numbers of isoprene units, the different degree of hydrogenation, and the position of hydrogenated isoprene units (Table 2). These three variations are useful for classification and identification. Streptomycetes synthesize menaquinones with a highly hydrogenated isoprenoid chain. Three to four (rarely five) isoprene units are saturated. The actinomycetes which belong to this type synthesize menaquinones with the same chain length but different degree of saturation (Fig. 6).

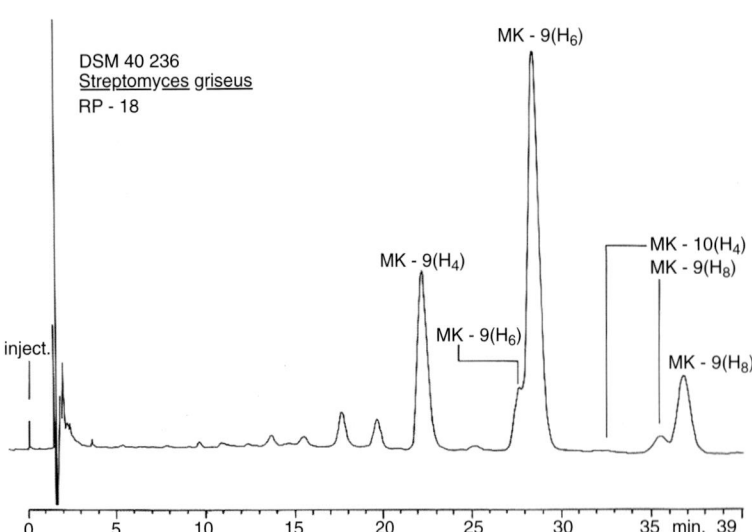

Fig. 6. Menaquinone profile of *Strep-tomyces griseus* (DSM 40236). The extent of hydrogenation of the iso-prene units is shown by the subscript of the abbreviation. For instance, MK-9 (H$_8$) is a menaquinone with four hydrogenated isoprene units. (Courtesy of R. M. Kroppenstedt.)

Phenotypic Methods for Classification within the Genus *Streptomyces*

Phenotypic methods comprise all those that are not directed towards DNA or RNA. They include also chemotaxonomic techniques. Between 1916 and 1943, most of the studies on streptomyces were published by soil microbiologists, who were mainly interested in ecological questions. Only few species were described at that time, mainly on the basis of morphological criteria, pigmentation and ecological requirements (Waksman and Curtis, 1916; Waksman, 1919; Jensen, 1930).

The discovery of actinomycin from *S. antibioticus* (Waksman and Woodruf, 1940) was the starting point of the investigations of antibiotics and other bioactive substances produced by streptomycetes in the 1940s and this led to extensive screening approaches for novel bioactive compounds in the following two decades.

The description of each producer of a novel natural product as a new species (often patented) led to an explosion of species descriptions and resulted in an overclassification of the genus. In the 1970s, the number of species increased to over 3000 (Trejo, 1970).

Reduction in the number of species names was first attempted in 1964 by the International *Streptomyces* Project (ISP), which introduced standard criteria for determining species (described in Shirling and Gottlieb, 1968a, Shirling and Gottlieb, 1968b, Shirling and Gottlieb, 1969, and Shirling and Gottlieb, 1972) to reduce the number of poorly described synonymous species. The major drawback of these descriptions was that they were based mainly on morphology (i.e., spore chain morphology, spore

surface ornamentation, color of spores, substrate mycelium, soluble pigments, and production of melanin pigment), in addition to a few physiological properties, which were mainly restricted to utilization tests of different carbon sources.

In these classical papers, more than 450 *Streptomyces* species were redescribed, and type strains were deposited in internationally recognized culture collections. Although intended, the efforts of the ISP did not result in an applicable identification scheme.

A first step in this direction was the development of numerical taxonomic methods in the 1960s including the methods of numerical identification. The first numerical taxonomic studies of streptomycetes by Silvestri et al. (1962) found considerable diversity within the genus but also groups that corresponded to the initial morphological descriptions. These studies did not result in nomenclatural changes, and despite the development of other small databases for identification of streptomycetes (Kurylowicz et al., 1975; Gyllenberg, 1976), these studies had no impact on streptomyces systematics (Table 3). Data from a large-scale numerical taxonomic study by Williams et al. (1983a) of 475 strains (including 394 *Streptomyces* type cultures from the ISP) for 139 unit characters were analyzed with simple matching, the Jaccard coefficient, and the average linkage algorithm. Consequently, the genus *Streptomyces* was subdivided into species groups. *Streptomyces* type strains (394) were clustered according to similarities obtained from the phenetic tests. At the $77 \pm 5\%$ simple matching coefficient (S_{SM}) level, 19 major, 40 minor and 18 single strain clusters were recovered. Many of the minor clusters consisted of less than five strains. Major clusters varied in size from 6 to 71 strains. Each cluster was addressed as a single "species" despite the high diversity observed within some clusters, and these therefore were addressed as "species groups." The largest species group is *Streptomyces albidoflavus* (cluster 1), containing 71 strains, including 44 type strains, 15 invalidly published species, and 12 unnamed strains. This cluster is further subdivided into three clusters: cluster 1a, *Streptomyces albidoflavus* subsp. *albidoflavus* (20 strains), cluster 1b, *Streptomyces albidoflavus* subsp. *anulatus* (38 strains), and cluster 1c, *Streptomyces albidoflavus* subsp. *halstedii* (13 strains; Williams et al., 1989).

The high phenotypic diversity of this cluster is obvious from the different test pattern. All strains produced yellow gray colonies, produced smooth spores in straight chains and no melanin, and exhibited resistance to a number of antibiotics including penicillin, lincomycin and cephaloridine. Many of the strains showed also antimicrobial activity; 39% produced compounds with antifungal activity, 32% produced compounds active against Gram-positive microorganisms and 10% against Gram-negative microorganisms (Williams et al., 1983b), showing the large diversity within one cluster and exemplifying clearly the problems with streptomycete systematics (Anderson and Wellington, 2001).

Nevertheless, the comprehensive survey of Williams et al. (1983a) resulted subsequently in a reduction of the number of described *Streptomyces* species; however, the problem of overspeciation remained. Numerous species and subspecies were described and many natural isolates did not match the reference strains used to construct the identification matrices (Goodfellow and Dickenson, 1985). Although probability matrices for identification purposes were published (Williams et al., 1983b; Langham et al., 1989), these matrices were not widely adopted by the scientific community.

This study was the basis of the taxonomic scheme for streptomyces presented in the 1989 edition of *Bergey's Manual of Systematic Bacteriology*, in which 142 species are listed (Williams et al., 1989), in contrast to 463 species described in the 1974 edition of *Bergey's Manual of Determinative Bacteriology* (Pridham and Tresner, 1974). A further numerical taxonomic analysis by Kämpfer et al. (1991b) included more strains and more than one strain of each species when available. A total of 821 strains were tested for 329 physiological properties, and the resulting cluster analysis was compared with the data published by Williams et al. (1983a) in addition to published genetic and chemotaxonomic data.

Many of the clusters defined by Williams et al. (1983a) were again recognized; for example the *S. albidoflavus, S. anulatus, S. griseus, S. halstedii* group appeared as cluster 1 in both studies, in which 28 of the *S. griseus* strains were grouped.

Interestingly, most of the strains sharing the same specific epithet were grouped together, indicating previous identification was reliable, but some exceptions were also observed. For example *S. hygroscopicus* strains were recovered in cluster 1 but also in several other clusters and subclusters.

On the basis of this study, a probability matrix was constructed (Kämpfer and Kroppenstedt, 1991a), but this matrix was likewise not widely used by other research groups.

Parallel with the numerical taxonomic studies, additional chemotaxonomic and also molecular methods were developed that are now often used together with (often) few physiological tests to study streptomycetes; however, a clear species concept is still pending. In Table 4, the cluster allocation of the species is given in comparison.

Other phenotypic methods include cell wall analysis (Lechevalier and Lechevalier, 1970), fatty acid profiling (Hofheinz and Grisbach, 1965; Lechevalier, 1977a; Saddler et al., 1986; Saddler et al., 1987; Kroppenstedt, 1992), rapid biochemical assay for utilization of 4-methyl-umbelliferone-linked substrates (Goodfellow et al., 1987c), serological assay (Ridell et al., 1986), phage typing (Wellington and Williams, 1981a; Korn-Wendisch and Schneider, 1992b), and protein profiling (Manchester et al., 1990; Goodfellow and O'Donnell, 1993; Lanoot et al., 2002), including comparison of ribosomal protein patterns (Ochi, 1989; Ochi, 1992; Ochi, 1995).

Fatty acids

The initial studies on actinomycete fatty acids were carried out by Hofheinz and Grisebach (1965) on *Saccharopolyspora erythraeus* (formerly "*Streptomyces erythraeus*") and *Streptomyces halstedii* to elucidate the biosynthetic pathway of branched fatty acids. It was shown that streptomyces synthesize terminally branched fatty acids. *Anteiso*-branched fatty acids are synthesized from 2-methylbutyrate, leading to *anteiso* fatty acids with an odd number of carbon atoms. In contrast, isovalerate and isobutyrate as starting compounds lead to the formation of *iso*-branched fatty acids with even and odd numbers of C-atoms, respectively. For this reason *iso*- and *anteiso*-branched fatty acids appear in pairs with odd numbers of C-atoms only.

In their early studies, Hofheinz and Grisebach (1965) separated the fatty acids as their methyl esters by gas chromatography on different stationary phases. Identification of the individual fatty acids was obtained by comparing the equivalent chain lengths of unknown fatty acids with those of standard mixtures. The results were confirmed by preparative gas-chromatography and by physical methods such as mass spectrometry and nuclear magnetic resonance (NMR) spectrometry. In both species, *iso-* and *anteiso*-branched fatty acids with chain lengths of 15 and 17 carbon atoms were detected. High amounts of 14-methyl pentadecanoic acid (*iso*-C16:0) were found in addition, while minor amounts of unbranched fatty acids, tuberculostearic acid and their homologues, could be detected in "*S. erythraeus*" (now *Saccharopolyspora erythraea*) but not in *Streptomyces halstedii*. These results are congruent with those of several other studies (Lechevalier et al., 1977a; Saddler et al., 1985; Saddler et al., 1987) in which 10-methyl branched fatty acids could not be detected among streptomycetes. Usually only small amounts of hydroxy fatty acids are synthesized by a limited number of streptomycetes under optimal oxygen supply. The hydroxy fatty acids are easily destroyed in a non-deactivated injection port of capillary gas chromatography system. Therefore hydroxy fatty acids in streptomycetes often go unnoticed. If streptomyces are grown under reproducible culture conditions, the hydroxy fatty acids they produce are highly diagnostic for some streptomyces species. Hydroxy fatty acids were detected in all strains of *S. coelicolor* (30), *S. rimosus* (14), and *S. violaceusniger* (18) and in 20 of 27 *S. hygroscopicus* strains but not in *S. violaceoruber* (16), *S. lavendulae* (18), *S. griseus* (22), *S. fradiae* (25), *S. viridochromogenes* (25), *S. glaucescens* (8) and *S. albus* (33; Kroppenstedt, 1992; R. M. Kroppenstedt, unpublished observation). Standardized growth and cultivation conditions are a general prerequisite for the use of fatty acid patterns below the genus level (Saddler et al., 1986). Saddler et al. (1987) used fatty acid profiles to investigate the taxonomy of *Streptomyces cyaneus* strains and soil isolates showing also blue spores. The *S. cyaneus* cluster harbors 13 of 19 blue-spored strains of streptomycetes (Hütter, 1962; Pridham and Tresner, 1974; Korn et al., 1978). In the study of Saddler et al. (1987), 8 of their 10 blue-spored isolates clustered, while 17 of the 34 *S. cyaneus* strains were assigned to a separate cluster. The conclusions of the fatty acid study (Saddler et al., 1987) and Williams et al. (1983a) agree that conventional features like spore chain morphology, color and ornamentation of spores may be helpful for presumptive identification but are not definitive for classification of streptomycetes. The same combination of features may be found in different clusters, yet one cluster may have members with different features. The study of Saddler et al. (1987), however, demonstrated also the heterogeneity of the *Streptomyces cyaneus* taxon as defined by Williams et al. (1983a). Fatty acid patterns in general cannot delimit *Streptomyces* species (Phillips, 1992; R. M. Kroppenstedt, unpublished observation), but using standardized conditions, they are still of high

value for the rapid characterization (independent of the taxonomic status) of large numbers of wild-type streptomycetes isolated from the environment (Saddler et al., 1987). By using the automated commercially available MIDI system consisting of a Hewlett-Packard model 5890 capillary gas chromatograph and a computer with specific software (Microbial ID, Inc., Newark, DE), the fatty acids are automatically identified and quantified by the computer using fatty acid standard mixtures for comparison. In their chapter on The Family Nocardiopsaceae in this Volume, Kroppenstedt and Evtushenko give a table of types and the fatty acids diagnostic for different genera of Actinomycetales, including *Streptomyces*. The comparison of different methods revealed that the use of numerical methods to determine taxonomy lumped too many strains together into some clusters (Williams et al., 1983a; Kämpfer et al., 1991b).

Curie-point Pyrolysis Mass Spectrometry (PyMS)

This method has also been applied to the classification and identification of actinomycetes (Sanglier et al., 1992). Similar to fatty acid profiling, highly standardized conditions are necessary. Whole cells are subject to high temperatures and subsequent nonoxidative thermal degradation. The resulting pyrolysate is then analyzed using mass spectrometry, resulting in a fingerprint for each organism.

Sanglier et al. (1992) applied this method to strains belonging to the largest *Streptomyces* species group, *Streptomyces albidoflavus*. Interestingly, *Streptomyces albidoflavus* and *Streptomyces anulatus* strains could be separated into distinct groups. Three of the six *Streptomyces halstedii* strains investigated also clustered into a distinct group, whereas the remaining strains clustered into two other groups. The study of Kämpfer et al. (1991b) also found that *Streptomyces albidoflavus* strains and *Streptomyces anulatus* strains grouped separately. Interestingly, it was confirmed that *Streptomyces anulatus* ISP 5361[T], the strain used to name the *Streptomyces anulatus* cluster, formed also a single-member cluster (Table 4).

Serology

Few results using serological methods have been published. Antisera against the mycelia from streptomycetes, streptoverticillia and *Nocardiopsis* species (Ridell et al., 1986) were used to confirm the high similarity between *Streptomyces lavendulae* and the streptoverticillia (Witt and Stackebrandt, 1990; Kämpfer et al., 1991b). The antisera of Kirby and Rybick (1986) raised against *Streptomyces griseus* (*Streptomyces anu-*

Table 5. Species-specific actinophages of the genus Streptomyces (modified according to Anderson and Wellington, 2001).

Phage	Host	Host species group	Host cluster no.[a]	Cluster no.[b]	References
98	S. coelicolor Müller ATCC 23899[T]	S. albidoflavus	1A	1-1	Wellington and Williams, 1981
14, 24, 233	S. coelicolor Müller ATCC 23899[T]	S. albidoflavus	1A	1-1	Korn-Wendisch and Schneider, 1992
89, DP 9	S. griseus ATCC 23345[T]	S. albidoflavus	1B	1-3	Wellington and Williams, 1981
90	S. griseinus ATCC 23915[T]	S. albidoflavus	1B	1-3	Wellington and Williams, 1981
33	'S. scabies' ATCC 23962	S. atroolivaceus	3	1-3	Wellington and Williams, 1981
SV1, SV2	S. venezuelae ATCC 10712[T]	S. violaceus	6	2	Stuttard, 1982
41	S. matensis ATCC 23935[T]	S. rochei	12	6	Wellington and Williams, 1981
S3	S. albus DSM 40313[T]	S. albus	16	32	Korn-Wendisch and Schneider, 1992
SAt1	S. azureus ATCC 14921[T]	S. cyaneus	18	9	Ogata et al., 1985
100	'S. caesius' ATCC 19828	S. griseoruber	21	6	Wellington and Williams, 1981
4, 5a, 5b, 49	S. violaceoruber DSM 40049[T]	S. violaceoruber	SMC	69	Korn-Wendisch and Schneider, 1992

Abbreviations: [T], type strain; SMC, single-member cluster; DSM, Deutsche Sammlung von Mikroorganismen und Zellkulturen; and ATCC, American Type Culture Collection.
[a]According to Williams et al. (1983); SMC according to Williams et al. (1989).
[b]According to Kämpfer et al. (1991).
Modified from Anderson and Wellington (2001).

latus, cluster 1B of Williams et al., 1983a) and "*Streptomyces cattleya*" (cluster 47) were shown to be genus-specific and to a certain degree also group-specific. The monoclonal antibody produced by Wipat et al. (1994) to "*Streptomyces lividans*" 1326 was shown to be specific for "*Streptomyces lividans*" strain 1326. Interestingly, antigenic reaction was observed also for strains grouped into cluster 21 of Williams et al. (1983a).

Phage Typing

Phage typing can be used for host identification at the genus and the species level (Welsch et al., 1957; Kutzner, 1961a; Kutzner, 1961b; Korn et al., 1978; Wellington and Williams, 1981a). Many actinophages (most of them virulent) for phage typing have been described. Streptomycete phages can be either polyvalent (e.g., C31; Chater et al., 1986b) or species-specific (Anderson and Wellington, 2001; Table 5). The specificity of actinophages at the genus level (e.g., Wellington and Williams, 1981a; Korn-Wendisch, 1982; Prauser, 1984) was an additional feature that justified the transfer of *Actinopycnidium*, *Actinosporangium*, *Chainia*, *Elytrosporangium*, *Microellobosporia*, *Kitasatoa* and *Streptoverticillium* to the genus *Streptomyces* (Goodfellow et al., 1986b; Goodfellow et al., 1986c; Goodfellow et al., 1986d; Goodfellow et al., 1986e; Witt and Stackebrandt, 1990). Other transfers justified by phage specificity include *Actinoplanes armeniacus* to the genus *Streptomyces* (Kroppenstedt et al., 1981; Wellington et al., 1981b) and "*S. erythraeus*" to the genus *Saccharopolyspora* (Labeda, 1987). Species or group identification of *Streptomyces* using phage typing has been less successful, but there are a few exceptions (Table 5).

Phages are also useful in industrial microbiology studies (Carvajal, 1953; Ogata, 1980) and in genetic studies (for review, see Chater [1986a] and chapter 12 of Kieser et al. [2000]). One of the best-investigated actinophages is C31, a temperate phage with a broad host range within the genus *Streptomyces* (Lomovskaya et al., 1980). This phage has become the subject of extensive studies and has been employed for many purposes (e.g., transfection, transduction, detection of transposon-like elements of host DNA, and cloning). Details are given in chapter 12 of Kieser et al. (2000).

Protein Profiling

Polyacrylamide gel electrophoresis (PAGE) of total protein extracts generate more or less complex banding patterns. These patterns can be used to differentiate species and subspecies within various bacterial genera. Protein patterns can be determined using one-dimensional (1-D) or two-dimensional (2-D) protein electrophoresis. The protein profiles of streptomycetes were first analyzed by Manchester et al. (1990), who investigated 37 *Streptomyces* strains (among them 5 streptoverticillia). Some taxonomic correlations were found between these profiles and the phenotypic groupings observed by Williams et al. (1983a) and Kämpfer et al. (1991b) in addition to some DNA hybridization groupings (Table 4). But Lanoot et al. (2002) confirmed only a few of these correlations. For *Streptomyces* isolates that are the causal agent of common potato scab, Paradis et al. (1994) used both PAGE and DNA-DNA hybridizations to elucidate the taxonomy of their strain. Isolates obtained from potato tubers were divided into two groups with a correlation coefficient of 0.75

using sodium dodecylsulfate (SDS)-PAGE analysis. The same two groups were resolved at approximately 44% similarity using DNA-DNA hybridization analysis. The fatty acid analysis results of the same study did not correlate with the SDS-PAGE and the DNA-DNA hybridization groupings, which can be explained by the influence of growth conditions on the profiles obtained (Saddler et al., 1986; Saddler et al., 1987). Protein profiling was not able to differentiate pathogenic from nonpathogenic strains.

Other more specific patterns are obtained with multilocus enzyme electrophoresis (MLEE) and depend on the relative mobilities of cellular enzymes in a gel matrix. Oh et al. (1996) studied 24 *Streptomyces* strains and demonstrated how MLEE could be used for both inter- and intraspecific characterization of streptomycetes, provided the appropriate enzymes were used. However, because only a restricted set of strains was studied, no general recommendations can be made for the usefulness of this method. In a more comprehensive study, 93 *Streptomyces* reference strains were investigated using SDS-PAGE of whole-cell proteins (Lanoot et al., 2002). Subsequent computer-assisted numerical analysis revealed 24 clusters encompassing strains with very similar protein profiles. Five of them included several type strains with visually identical patterns. DNA-DNA hybridizations revealed similarities higher than 70% among these type strains. On the basis of these results, consideration of *Streptomyces albosporeus* subsp. *albosporeus* LMG 19403^T as a subjective synonym of *Streptomyces aurantiacus* LMG 19358^T, *Streptomyces aminophilus* LMG 19319^T as a subjective synonym of *Streptomyces cacaoi* subsp. *cacaoi* LMG 19320^T, *Streptomyces niveus* LMG 19395^T and *Streptomyces spheroides* LMG 19392^T as subjective synonyms of *Streptomyces caeruleus* LMG 19399^T, and *Streptomyces violatus* LMG 19397^T as a subjective synonym of *Streptomyces violaceus* LMG 19360^T was proposed (Table 4).

Two-dimensional PAGE of the total cellular proteins allows a finer resolution of the individual gene products. This technique results in very complex patterns and seems to be too sensitive to investigate proteins with high rates of evolution (Hori and Osawa, 1987). Mikulik et al. (1982) and later Ochi (1989) were the first to apply 2-D PAGE to determine the variability of ribosomal proteins for use in streptomycete taxonomy. These studies were later extended by focusing on AT-L30 proteins, which give genus-specific profiles (Ochi, 1992). In an even more specific analysis, Ochi (1995) correlated the N termini sequences of the ribosomal AT-L30 protein of 81 streptomycete strains from different taxonomic groups to phylogenetic groupings

within the genus and pointed out that on this basis the genus *Streptomyces* seems to be well described. However, Ochi's groupings did not correlate with those of Williams et al. (1983a) and Kämpfer et al. (1991b). For details of these groupings, see Table 4.

More detailed taxonomic studies of *Streptomyces* have been performed by isolating and sequencing specific proteins. For example, Taguchi et al. (1996) used the *Streptomyces* subtilisin inhibitor protein (SSI), which plays unidentified role(s) in physiological or morphological regulation, to investigate the taxonomic status of the *Streptomyces coelicolor* strains. The amino acid sequence, of SSI from "*Streptomyces lividans*" 66, *Streptomyces coelicolor* Müller ISP 5233^T and "*Streptomyces coelicolor*" A3(2) were compared. The alignments supported ribosomal sequence comparisons, indicating that "*Streptomyces coelicolor*" A3(2) is more closely related to "*Streptomyces lividans*" 66 (cluster 21 of Williams et al., 1983a) than to the type strain, *Streptomyces coelicolor* Müller ISP 5233^T (cluster 1).

Genotypic Methods for Classification within the Genus *Streptomyces*

Genotypic methods comprise all those that are directed towards DNA or RNA molecules (Schleifer and Stackebrandt, 1983; Vandamme et al., 1996). The development of molecular methods to analyze bacterial genomes has provided a new basis for studying bacterial taxonomy and in some cases phylogenetic relationships of the prokaryotes at the genus, species and subspecies level. These methods are widely used in modern taxonomic studies. The general taxonomic values of different molecular techniques are given by Vandamme et al. (1996) and in Prokaryote Characterization and Identification in Volume 1. Their usefulness in taxonomic studies to delimit species within the genus *Streptomyces* is discussed briefly in the following chapters and has also been reviewed by Anderson and Wellington (2001).

Problems in assigning new strains to existing species on the basis of this molecule alone persist and doing this will not be possible in the future. As already pointed out by Stackebrandt and Schumann in Introduction to the Classification of the Actinomyces in this Volume, comparative analysis of sequences of homologous and genetically stable semantides has demonstrated that several classification systems based on morphology and physiology do not reflect the natural relationships among actinomycetes and related organisms. In this respect, rRNA sequence comparison is a powerful tool in modern taxonomy and has revolutionized our insight in phylogenetic lineages of major taxonomic groups. However, the resolving power of 16S rRNA sequences is not sufficient to delimit species. Nevertheless, rRNA sequence comparisons are important in the taxonomy of *Streptomyces* and the results have also been used to answer questions about horizontal transfer of genes within the genus (Huddleston et al., 1997). Genes for 16S RNAs are highly conserved within bacteria. Within the genus *Streptomyces*, three regions within the gene have enough sequence variation to be useful as genus-specific (*a* and *b* regions) and species-specific (*c* regions) probes (see Stackebrandt et al., 1991a; Stackebrandt et al., 1991b; Stackebrandt et al., 1992; Anderson and Wellington, 2001). Not only 16S rRNA genes, but also 23S rRNA and 5S rRNA genes (Mehling et al., 1995), 16S-23S rRNA internally transcribed spacer (ITS) sequences (Song et al., 2003), and ribosomal protein sequences have also been used to investigate species relationships within the genus *Streptomyces* (Liao and Dennis, 1994; Ochi, 1995). Wenner et al. (2002) studied the nucleotide composition of the ITS sequences of the six rDNA operons of two *Streptomyces ambofaciens* strains. Their findings suggested that recombination frequently occurs between the rDNA loci, leading to the exchange of nucleotide blocks, and confirmed that a high degree of ITS variability is a common characteristic among *Streptomyces* spp. Note that rRNA sequences cannot be used alone because of the intraspecific variation and intragenomic heterogeneity.

DNA-DNA Hybridizations

The percent DNA-DNA hybridization and the decrease in thermal stability of the hybrid are at present the "gold standard" methods of species delineation in bacteriology (Wayne et al., 1987). An ad hoc committee for the re-evaluation of the species definition in bacteriology (Stackebrandt et al., 2003) has encouraged investigators to verify the species concept with other methods; however, DNA-DNA similarity and change in melting temperature (ΔTm; Wayne et al., 1987) remain the acknowledged standard for definition of species.

DNA-DNA hybridizations of total chromosomal DNA have also been used within the genus *Streptomyces*. In an initial study, Mordarski et al. (1986) compared the numerical and phenetic groupings of strains of the *Streptomyces albidoflavus* cluster 1 of Williams et al. (1983a) with the results of DNA-DNA hybridizations (reassociation of labeled DNA on nitrocellulose filters). In congruence with numerical studies, DNA-DNA hybridization studies showed this cluster could be subdivided into three subclusters and confirmed the homogeneity of the *Streptomyces albidoflavus* subcluster *albidoflavus*. Two further subclusters obtained by DNA-DNA hybridizations were not congruent with the groupings of *S. anulatus* or *halstedii* by Williams et al. (1983a); however, some correlations were found with the groupings of Kämpfer et al. (1991b). Unification using DNA-DNA similarity (Witt and Stackebrandt, 1990) of *Streptoverticillium* strains with the genus *Streptomyces* (reassociation of labeled DNA on filters) was confirmed by phenotypic data in the numerical study of Kämpfer et al. (1991b).

The most extensive application of DNA-DNA hybridizations (thermal renaturation method) to the study of the major streptomycete phenetic groups of Williams et al. (1983a) was by Labeda and coworkers. In studies using *Streptomyces cyaneus* (Labeda and Lyons, 1991a), *Streptomyces violaceusniger* (Labeda and Lyons, 1991b), *Streptomyces lavendulae* (Labeda, 1993), the verticil-forming streptomycetes (formerly *Streptoverticillium* species; Labeda, 1996; Hatano et al., 2003), and *S. fulvissimus* and *S. griseoviridis* phenotypic clusters (Labeda, 1998), the degree of DNA similarities was often not congruent with the phenetic groupings of Williams et al. (1983a). But again, more correlations were found with the groupings of Kämpfer et al. (Kämpfer et al., 1991b; Table 4). The usefulness of the DNA-DNA hybridization technique to delineate species within the genus *Streptomyces* has been questioned. In support of the usefulness of this technique is the considerable genetic instability of certain regions within the *Streptomyces* chromosome (Redenbach et al., 1993). The two complete *Streptomyces* genome sequences available at this time (Omura et al., 2001; Ikeda et al., 2003) indicate that the central core region contains mostly the essential housekeeping genes, while the chromosome arms comprise laterally acquired contingency genes. Another issue is the presence of large plasmids in strains of *Streptomyces*, which can considerably influence the results of DNA-DNA hybridizations. General properties of *Streptomyces* plasmids and their use for gene cloning are given in chapter 11 of Kieser et al. (2000).

Fingerprinting Techniques

RANDOMLY AMPLIFIED POLYMORPHIC DNA (RAPD) POLYMERASE CHAIN REACTION (PCR) RAPD-PCR is used as a rapid screening method to detect similarity among streptomycete strains. Single primers with arbitrary nucleotide sequences to amplify DNA are used in addition to a low annealing temperature so that polymorphisms can be detected. Stringent standardization of the reaction parameters is required. These include primer sequence, annealing temperatures, buffer components, concentration and quality of template DNA. The resulting characteristic fingerprint of PCR products enables detection of chromosomal differences between individual isolates without having any prior knowledge of the chromosomal sequence (Williams et al., 1990). Applying this technique, Mehling et al. (1995) could not detect any characteristic banding patterns for closely related species unless a highly specific actinomycete primer was used. As a major disadvantage, the resulting fingerprints contained only few bands (one to four), reducing the effectiveness of this method. Huddleston et al. (1995) reported similar results. They evaluated the use of RAPD-PCR for the resolution of interspecific relationships among members of the *Streptomyces albidoflavus* cluster of Williams et al. (1983a). Anzai et al. (1994) demonstrated the variation in fingerprints obtained when a single base was substituted on the arbitrary primer; 11 primers were investigated and the number of bands ranged from zero to 20. The most significant differences were observed when the sequence at the 3′ end was altered. Anzai et al. (1994) also investigated the relationship of *Streptomyces virginiae* strains to *Streptomyces lavendulae* strains by RAPD-PCR. Williams et al. (1983a) found that *Streptomyces virginiae* was a synonym of *Streptomyces lavendulae*. These strains were also grouped into the same cluster in the study of Kämpfer et al. (1991b). In addition to RAPD-PCR, DNA-DNA hybridization, low-frequency restriction fragment analysis (LFRFA), and cultural and physiological tests were performed. Consistent results were obtained using all these methods after RAPD PCR optimization. It was however not possible to clarify the interspecific relationship of *Streptomyces lavendulae* and *Streptomyces virginiae*.

RESTRICTION DIGESTS OF TOTAL CHROMOSOMAL DNA Similar to the RAPD PCR techniques, low-frequency restriction fragment analysis (LFRFA) uses the entire bacterial chromosome, which is digested with restriction endonucleases that cut infrequently. Because streptomycetes belong to bacteria with a high DNA G + C content, rare AT cutters are used. The fragments obtained were separated by pulsed-field gel electrophoresis (PFGE). In a first study, Beyazova and Lechevalier (1993) included 59 strains from eight species groups and found the method useful for the clustering of related strains. However, some discrepancies were found, for example for strains grouped into the *Streptomyces cyaneus* cluster of Williams et al. (1983a). Again this method seems to be useful for the detection of

very closely related strains, but similar to RAPD-PCR, it cannot resolve interspecific relationships. In addition, Rauland et al. (1995) found that large chromosomal amplifications or deletions may also lead to misinterpretations of resulting banding patterns.

NUCLEIC ACID SEQUENCE COMPARISONS OF 16S rRNA AND OTHER GENES The application of 16S rRNA gene sequence analysis to the study of the taxonomy of streptomycetes is reviewed by Stackebrandt et al. (1992), who highlighted the importance of the region selected for comparison. Sequence analysis of rRNA genes has already been applied at the genus, species and strain level. The relationships obtained differed considerably depending on the variable region (*a*, *b* or *c*). Kataoka et al. (1997) subsequently investigated the *c* region from 89 streptomycete type strains representing several clusters of Williams et al. (1983a). Though these variable regions were useful for resolving inter- and intraspecies relationships within the streptomycetes, they were too variable for determining generic relationships. Among the 89 strains studied, 57 variants were detected and 42 strains were found to have unique sequences. After this publication, the *c* regions from 485 streptomycete strains were sequenced and deposited in GenBank by this group. At present, this is the largest publically available set of streptomycete 16S rDNA sequence data.

Anderson and Wellington (2001) published a phylogenetic tree based on comparison of the *c* regions from representatives of the major cluster groups defined by Williams et al. (1983a). The taxonomic status of the phenotypic groups was confirmed, although they did not cluster together. Only *Streptomyces olivaceoviridis* and *Streptomyces griseoruber* strains (which are found in clusters 20 and 21 of Williams et al. [1983a], but in a single cluster 9 of Kämpfer et al. [1991b]) had identical *c* regions. The *Streptomyces albidoflavus* group, which was previously divided into three species groups by Williams et al. (1983a) and contained more than 60 strains (Williams et al., 1989), was now divided into six groups using sequence comparisons of the *c* region (Kataoka et al., 1997). The three phenotypic subgroups of Williams et al. (1983a) were maintained but did not cluster together.

Hain et al. (1997) designed 16S rRNA oligonucleotide probes to determine intraspecific relationships within the *Streptomyces albidoflavus* group. As a result, sequence comparisons were found to be useful for species delimitation but of no value for strain differentiation.

Also the intergenic 16S-23S rRNA spacer regions were investigated in detail and they were obviously more suitable for delineation of the intraspecific relationships within that cluster. Genus specific probes based on 23S rRNA gene sequences (Mehling et al., 1995) and 5S rRNA gene sequences (Park et al., 1991) have been developed. The reclassification of the genera *Chainia*, *Elytrosporangium*, *Kitasatoa*, *Microellobosporia* and *Streptoverticillium* into the genus *Streptomyces* (Park et al., 1991) was confirmed using 5S rRNA gene sequence evaluation.

At present about 350 complete 16S rRNA sequences are available from public databases. Although 16S rRNA sequence analyses have provided a framework for prokaryotic classification, note that the current classification system based on this molecule has not yet solved the taxonomy within the genera (especially within the genus Streptomyces). Several studies have attempted to use sequence data from variable regions of 16S rRNA to establish taxonomic structure within the genus, but the variation is too limited to help resolve problems of species differentiation (Witt and Stackebrandt, 1990; Stackebrandt et al., 1991b; Stackebrandt et al., 1992; Anderson and Wellington, 2001).

The situation is complicated by the fact that *Streptomyces* species may harbor different 16S rRNA gene sets. For example, *S. coelicolor* A3(2), *S. lividans* and several *Streptomyces* species harbor six ribosomal rRNA gene sets. Each set comprises one gene copy for 16S, 26S, and 5S rRNA (Van Wezel et al., 1991) and lacks tRNA genes.

A comprehensive study of the phylogenetic relationships between 64 whorl-forming streptomycetes using partial *gyrB* sequences (the structural gene of the B subunit of the DNA gyrase) has been published by Hatano et al. (2003). The strains in this study consisted of 46 species and eight subspecies and in addition 13 species whose names have not been validly published (including 10 strains examined by the International *Streptomyces* Project [ISP]). Two major groups were found. The typical whorl-forming species (59 strains) were further divided into six major clusters of three or more species, seven minor clusters of two species, and five single-member clusters on the basis of the threshold value of 97% *gyrB* sequence similarity. Major clusters contained *Streptomyces abikoensis*, *Streptomyces cinnamoneus*, *Streptomyces distallicus*, *Streptomyces griseocarneus*, *Streptomyces hiroshimensis* and *Streptomyces netropsis*. Phenotypically, members of each cluster resembled each other closely except for those in the *S. distallicus* cluster (which was divided phenotypically into the *S. distallicus* and *Streptomyces stramineus* subclusters) and the *S. netropsis* cluster (which was divided into the *S. netropsis* and *Streptomyces eurocidicus* subclusters). Strains in each minor cluster closely resembled each other phenotypi-

cally. At the conclusion, 59 strains of typical whorl-forming *Streptomyces* species were placed into the following 18 species, including subjective synonym(s): *S. abikoensis*, *Streptomyces ardus*, *Streptomyces blastmyceticus*, *S. cinnamoneus*, *S. eurocidicus*, *S. griseocarneus*, *S. hiroshimensis*, *Streptomyces lilacinus*, "*Streptomyces luteoreticuli*," *Streptomyces luteosporeus*, *Streptomyces mashuensis*, *Streptomyces mobaraensis*, *Streptomyces morookaense*, *S. netropsis*, *Streptomyces orinoci*, *S. stramineus*, *Streptomyces thioluteus* and *Streptomyces viridiflavus* (Table 4). All strains showing 98.5–100% *gyrB* sequence similarity also had a high DNA-DNA similarity (70–100%), showing better resolution with *gyrB* sequences than with 16S rRNA sequences.

Other genes known to be conserved between species, such as housekeeping genes (e.g., elongation factors and ATPase subunits), may be useful as primary target genes for study (Ludwig and Schleifer, 1994). Huddleston et al. (1997) used tryptophan synthase genes in addition to 16S rRNA gene comparisons to determine the phylogeny of streptomycin-producing streptomycetes and provide evidence for the horizontal transfer of antibiotic resistance genes. Predictably, sequence information on other genes (i.e., other housekeeping genes) will lead to better insight into the intraspecific structure of the genus *Streptomyces* (Stackebrandt et al., 2003).

Rapid Methods for Gene Analysis in Streptomycete Taxonomy

Several alternative methods have been described that do not involve sequencing, using either restriction analysis (Clarke et al., 1993; Fulton et al., 1995) or specialized gel electrophoresis techniques to monitor the mobility of the product (Hain et al., 1997; Heuer et al., 1997). Restriction fragment length polymorphism (RFLP) patterns of purified rRNA were used by Clarke et al. (1993) with strains from the *Streptomyces albidoflavus* cluster (subgroups 1A and 1B of Williams et al., 1983a). The following combination of enzymes was used: *BglI*, *EcoRI*, *PstI* and *PvuII*. The resulting RFLP profiles varied considerably between species groups but were found to enable differentiation of phenotypically similar strains. Fulton et al. (1995) performed ribosomal restriction analysis using *MseI* fingerprints of rRNA operons (RiDiTS) to group 98 named streptomycete strains including members of cluster group A (comprising clusters 1–41) and F (comprising clusters 55–67) of Williams et al. (1983a) and other strains. The resulting RiDiTS belonged to 11 pattern types with varying degrees of similarity to the Williams subclusters. At low resolution (70% similarity), cluster groups A and F could be differentiated, but individual clusters could not.

Further studies are based on genotypic variation monitored using denaturing gradient gel electrophoresis (DGGE; Muyzer et al., 1993) with or without DNA-binding agents (Hain et al., 1997). Anderson and Wellington (2001) recommended DGGE in combination with other techniques. Using the variable 16S rRNA regions, this method enables delimitation of genus groups and species-groups. Huddleston et al. (1995) allocated isolates ASB33, ASB37 and ASSF22 to *Streptomyces albidoflavus*, *Streptomyces griseoruber* and *Streptomyces albidoflavus*, respectively, using a combination of techniques including numerical taxonomy, PFGE and sequence comparisons (Huddleston et al., 1997).

Identification of Streptomycetaceae at the Genus Level

Sequencing of 16S rRNA genes and comparison of these sequences after careful alignment with published sequences is currently the most reliable method for assigning unknown organisms to the different genera. The calculation of phylogenetic trees at the subgeneric level should be done very carefully and may lead to misinterpretations. Notably, phylogenetic analysis on the basis of 16S rRNA comparisons does not allow species delineation.

Colony morphology (color of the aerial mycelium, color of the substrate mycelium, and soluble pigment) is (especially in the case of the genus *Streptomyces*) very useful (Tables 6 and 7; Figs. 3 and 4). Here the traditional methods extensively described by Korn-Wendisch and Kutzner (1992a) are highly recommended.

A microscopic characterization (particularly the morphologies of the aerial mycelium, arthrospores and vegetative mycelium) is of high value (Figs. 7–11). See Korn-Wendisch and Kutzner (1992a) and chapter 3 of Kieser et al. (2000) for details of the methods on microscopy.

Furthermore, the detection of LL-A$_2$pm in cell wall or whole-cell hydrolysates, the lack of mycolic acids, the predominance of mainly *iso*- and *anteiso*-methyl branched fatty acids, and the 16S rRNA sequence are well suited for genus identification (Table 1).

Identification of Species

Kitasatospora Novel isolates can be readily identified as members of the genus by 16S rRNA gene and 16S-23S rRNA gene spacer analyses. The presence of *meso*-A$_2$pm in whole cell

hydrolysates is an important chemotaxonomic character for differentiation from *Streptomyces* (Table 1). Additional chemotaxonomic investigations (polar lipids, fatty acid patterns, and menaquinone type) are helpful for allocation of an unknown isolate to the genus. The *meso*-A$_2$pm content is 49–89% in *Kitasatospora* strains and 1–16% in *Streptomyces* strains (Zhang et al., 1997). Strains belonging to the genus *Streptacidophilus* contain (like *Streptomyces*) LL-diaminopimelic acid as predominant diamino acid (Kim et al., 2003). For further species identification, the characters shown in Table 8 are helpful.

Streptacidiphilus Novel isolates can be readily identified as members of the genus by 16S rRNA gene analysis. The presence of LL-A$_2$pm in whole cell hydrolysates and other chemotaxonomic characters (e.g., polar lipids, fatty acid patterns, and menaquinone type; see family description) are shared by members of the genus *Streptomyces*. The three species of the genus *Streptacidophilus* can be differentiated on the basis of some phenotypic properties (Table 9).

Streptomyces The identification of species poses severe problems. Because of the high number of validly published species (Table 4), most of which are based on a single strain description, it is at present not possible to recommend a single method or even a set of methods for identification at the species level. Because a clear species concept within the genus *Streptomyces* is still pending, investigators should be very careful with species allocations on the basis of the results of one or few methods. The ICSP Subcommittee on the Systematics of Streptomycetaceae (Kämpfer and Labeda, 2003) has recommended that before a species concept of the genus *Streptomyces* is formulated more genomic information should be evaluated, and it was agreed "that the proposal of new species should only be accepted on the basis of very careful studies done with sufficient practice and considering all other species." In recent years, only a few species have been proposed on the basis of 16S rRNA sequence analysis and phenotypic characterization (most often restricted to those species closely related by phylogenetic analysis [e.g., Bouchek-Mechine et al., 2000; Kim et al., 2000; Kim and Goodfellow, 2002a, b; Li et al., 2002a; Li et al., 2002b; Meyers et al., 2003; Petrosyan et al., 2003; Zhang et al., 2003]). The ICSP Subcommittee on the Systematics of Streptomycetaceae (Kämpfer and Labeda, 2003) recommended a careful look at synonymy as a first step to reduce the number of "species" within the genus.

At present two complete *Streptomyces* genomes are available (Omura et al., 2001; Ikeda

Table 6. Spore colors for the grouping of streptomycetes and representatives of each color (according to Korn-Wendisch and Kutzner 1992).

Representative species (DSM no.)[a]	Figure (strain no.)[b]	Color of aerial mycelium
S. griseus (40236); S. coelicolor (40233)	3a (40236)	Yellow-gray: "griseus"
S. fradiae (40063); S. toxytricini (40178)	3b (40178)	Pink/light violet
S. lavendulae (40069); S. flavotricini (40152)	3c (40069)	Gray-pink/lavender: "cinnamomeus"
S. eurythermus (40014); S. fragilis (40044)	3d (40014)	Brown (plus gray or red)
S. viridochromogenes (40110); S. cyaneus (40108)	3e (40108)	Blue: "azureus"
S. glaucescens (40155)	3f (40155)	Blue-green: "glaucus"
S. prasinus (40099); S. hirsutus (40095)	3g (40099)	Green: "prasinus"
S. violaceoruber (40049); S. echinatus (40013)	3h (40049)	Gray: "cinereus"
S. albus (40313); S. longisporus (40166)	3i (40166)	White: "niveus"
S. alboniger (40043); S. rimosus (40260)	3j (40043)	Not definable: white plus various light-colored shades

Abbreviations: DSM, Deutsche Sammlung von Mikroorganismen und Zellkulturen; and ISP, International *Streptomyces* Project.
[a]DSM no. 40XXX = ISP no. 5XXX; e.g., 40236 = ISP 5236.
[b]Figures 3a–j show the aerial mycelia of the strain after cultivation on three different media for 21 days (left: starch-casein-nitrate agar, middle: GYM agar, right: oatmeal agar; for compositions, see Tables 10 and 12).
From Korn-Wendisch and Kutzner (1992).

Table 7. Colors of substrate mycelium and soluble pigment occurring in streptomycetes.

Representative species (DSM no.)[a]	Figure (strain no.)[b]	Color
S. aurantiacus (40412); S. griseoruber (40275)	4a (40412)	Orange to dark red
S. longispororuber (40599); S. spectabilis (40512)		(mainly endopigment)
S. californicus (40058); S. cinereoruber (40012)	4b (40058)	Red to blue/violet
S. violaceus (40082); S. purpurascens (40310)		(mainly endopigment)
S. coelicolor (40233); S. cyaneus (40108)	4c (40163)	Red-violet to blue
S. violaceoruber (40049); S. lateritius (40163)		(endo- or exopigment or both)
S. atroolivaceus (40137); S. canarius (40528)	4d (40089)	Yellow-orange/greenish-yellow
S. galbus (40089); S. tendae (40101)		(endo- and exopigment)
S. flavoviridis (40210); S. olivoviridis (40211)	4e (40071)	Green to gray-olive
S. viridochromogenes (40110); S. nigrifaciens (40071)		(endo- and exopigment)
"S. malachiticus" (40167); "S. malachitorectus" (40333)		Green (endopigment)
S. badius (40139); S. eurythermus (40014)	4f (40100)	Red-brown to dark-brown
S. phaeochromogenes (40073); S. ramulosus (40100)		(endo- and exopigment)
S. alboniger (40043); S. hygroscopicus (40578)		Gray-brown to black
S. purpeofuscus (40283); S. mirabilis (40553)		(mainly endopigment)

Abbreviations: DSM, Deutsche Sammlung von Mikroorganismen und Zellkulturen; and ISP, International *Streptomyces* Project.
[a]DSM no. 40XXX = ISP no. 5XXX.
[b]Figures 4a–f show the substrate mycelia of the strain after cultivation on three different media for 7 days; left: starch-casein-nitrate agar, middle: GYM agar, right: oatmeal agar; for compositions, see Tables 10 and 12).
From Korn-Wendisch and Kutzner (1992).

et al., 2003). Bentley et al. (2002) reported the first complete genome sequence of *S. coelicolor* and described a model for the evolution of the large linear chromosome, where the central core region contains mostly the essential housekeeping genes and the "arms" contain laterally acquired contingency genes. A comparison of the *S. averilitils* and *S. coelicolor* genomes supports this model (Bentley et al., 2003). Most of the secondary metabolism gene clusters are located in the arms. The detailed study of genomes of different streptomycetes and housekeeping genes may provide a more reliable basis for an intrageneric subdivision (Stackebrandt et al., 2003).

Ecophysiology and Habitat

The following sections are largely based on the information summarized by Korn-Wendisch and Kutzner (1992a). Members of the family Streptomycetaceae are ubiquitous in nature. Members of the genus *Kitasatospora* have been predominantly isolated from soils (Zhang et al., 1997), and *Streptacidiphilus* species have been isolated from acidic soils and litter (Kim et al., 2003). Streptomycetes can be isolated in high numbers in soil, which is their natural habitat. Most streptomycetes can degrade complex and

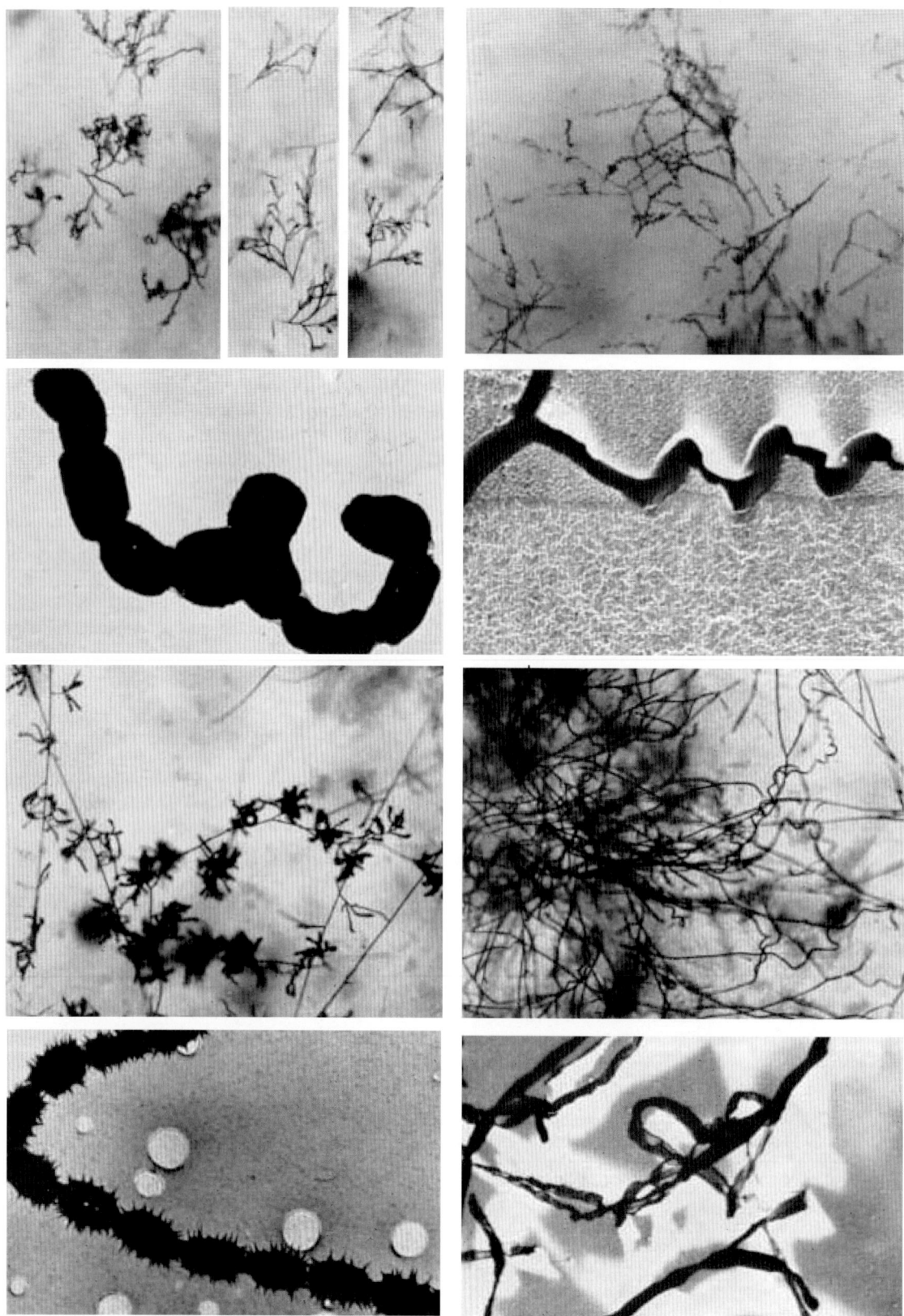

Fig. 7. Aerial mycelium of the fertile (left) and the sterile (right) strain of two streptomycetes. First and third lines, light microscopy (@ 250); second and fourth lines, electron microscopy (×15,000). (From Kutzner [1956], with permission.)

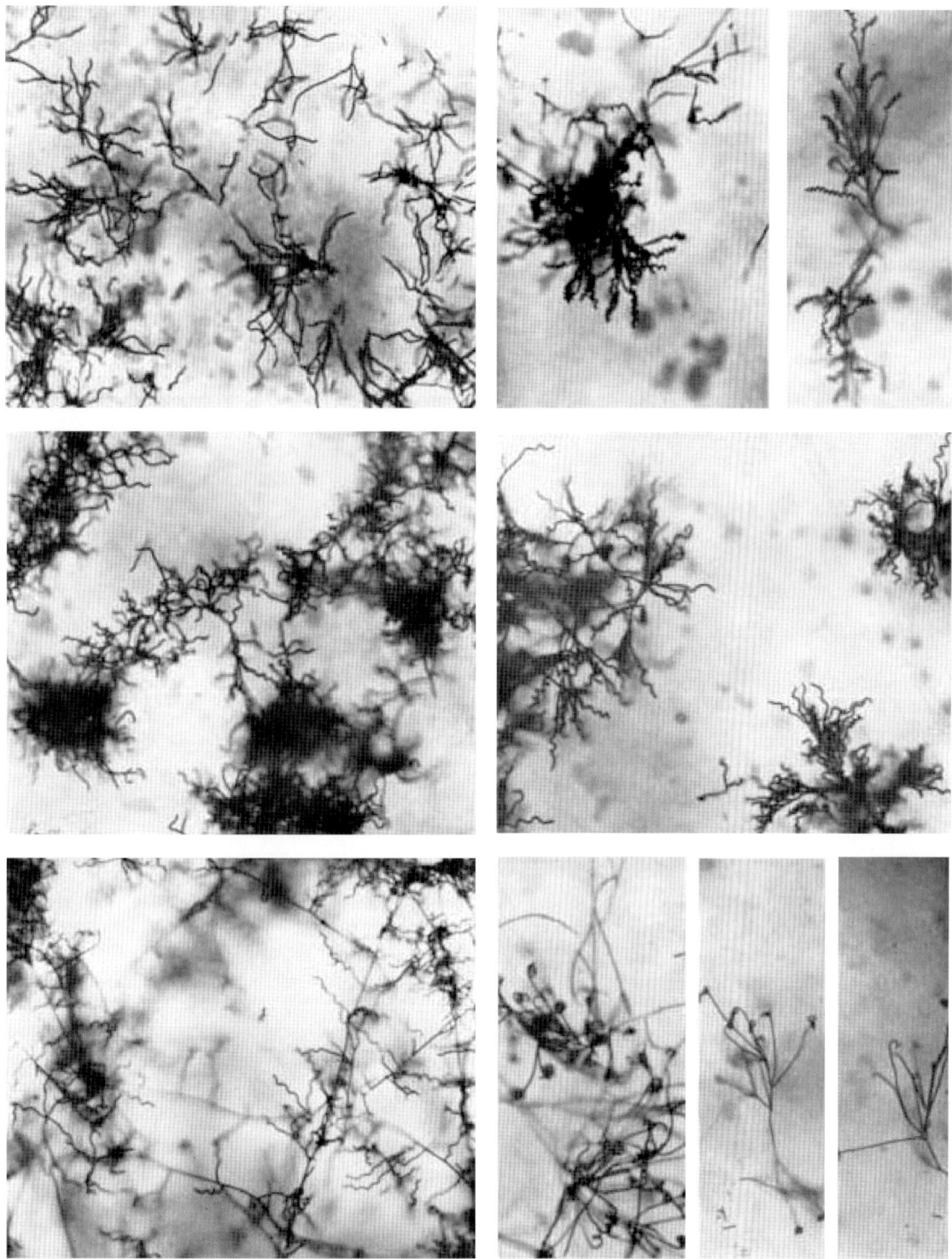

Fig. 8. Morphology of the aerial mycelium of some strains of *Streptomyces* (×250). (From Flaig and Kutzner [1960b], with permission.)

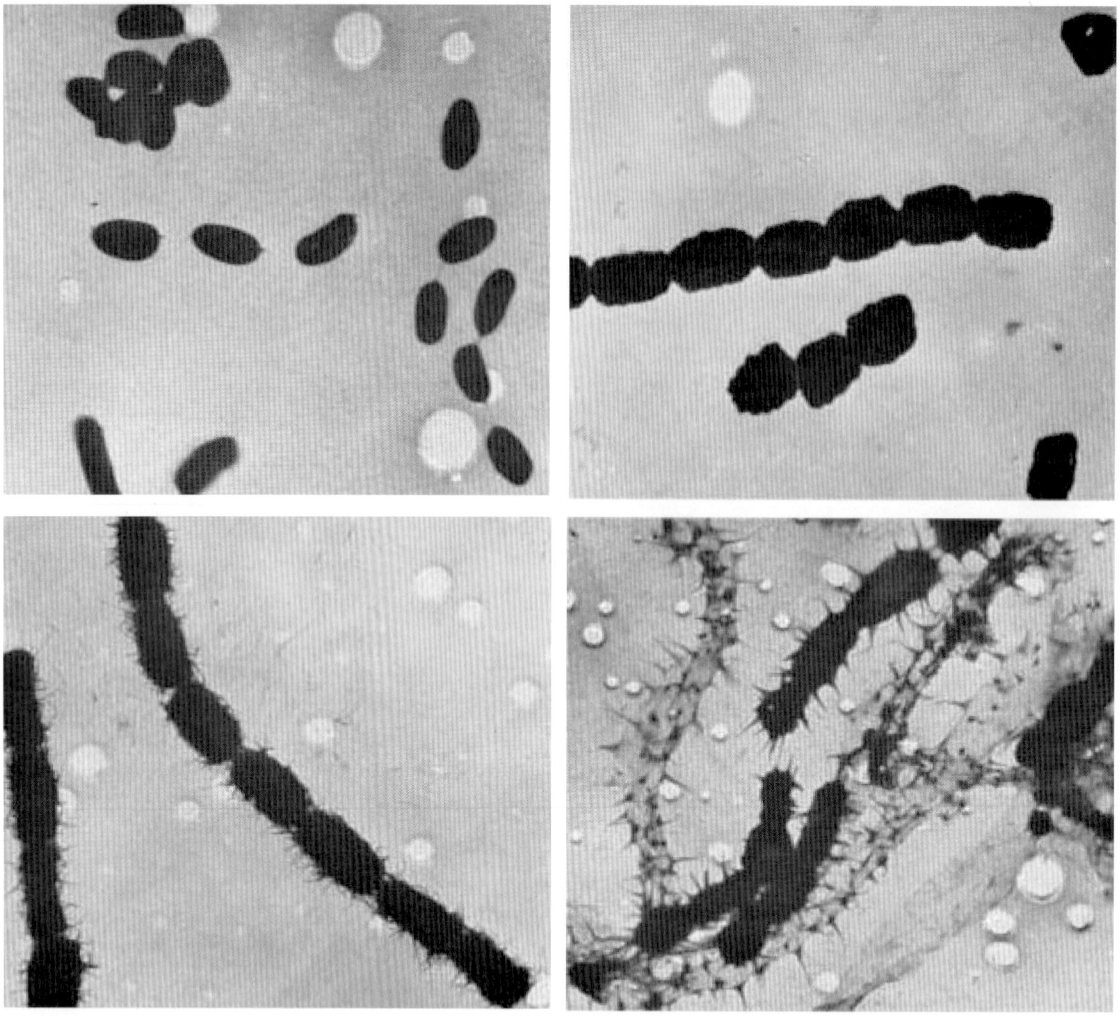

Fig. 9. Electron micrographs of four types of arthrospores of streptomycetes: smooth, warty, hairy and spiny. The spores are about 1 m long. (From Kutzner [1956], with permission.)

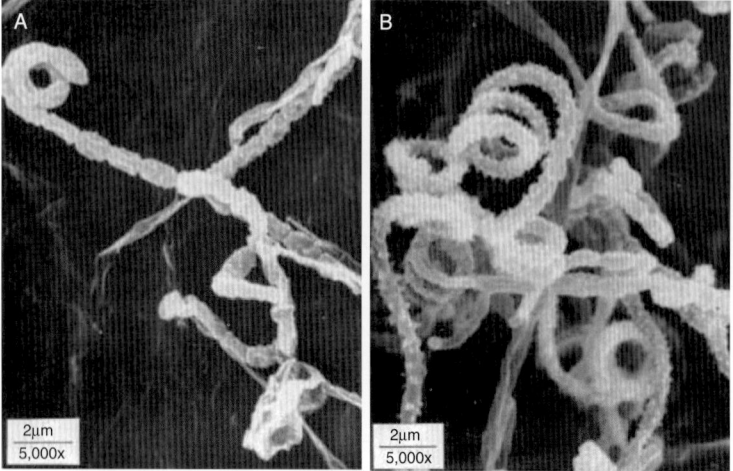

Fig. 10. Scanning electron microscopy of arthrospore chains. (A) *Streptomyces torulosus* (knobby); (B) *S. bluensis* (spiny); (C) *S. antimycoticus* (rugose); and (D) "*S. karnatakensis*" (hairy). (Courtesy of A. Dietz.)

Fig. 10. *Continued*

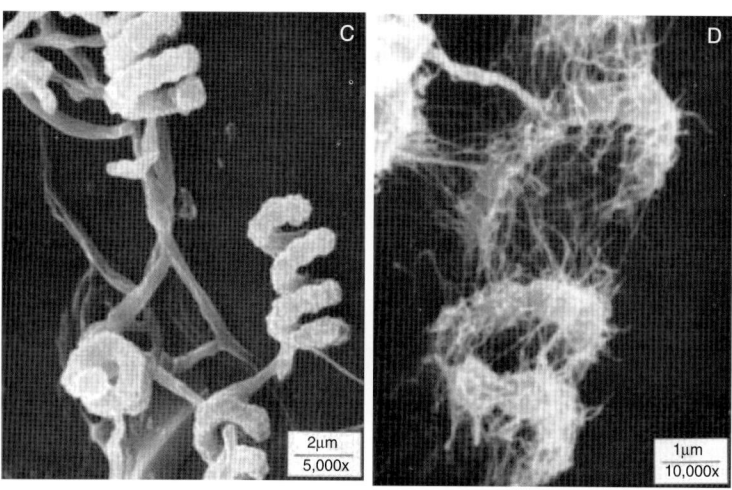

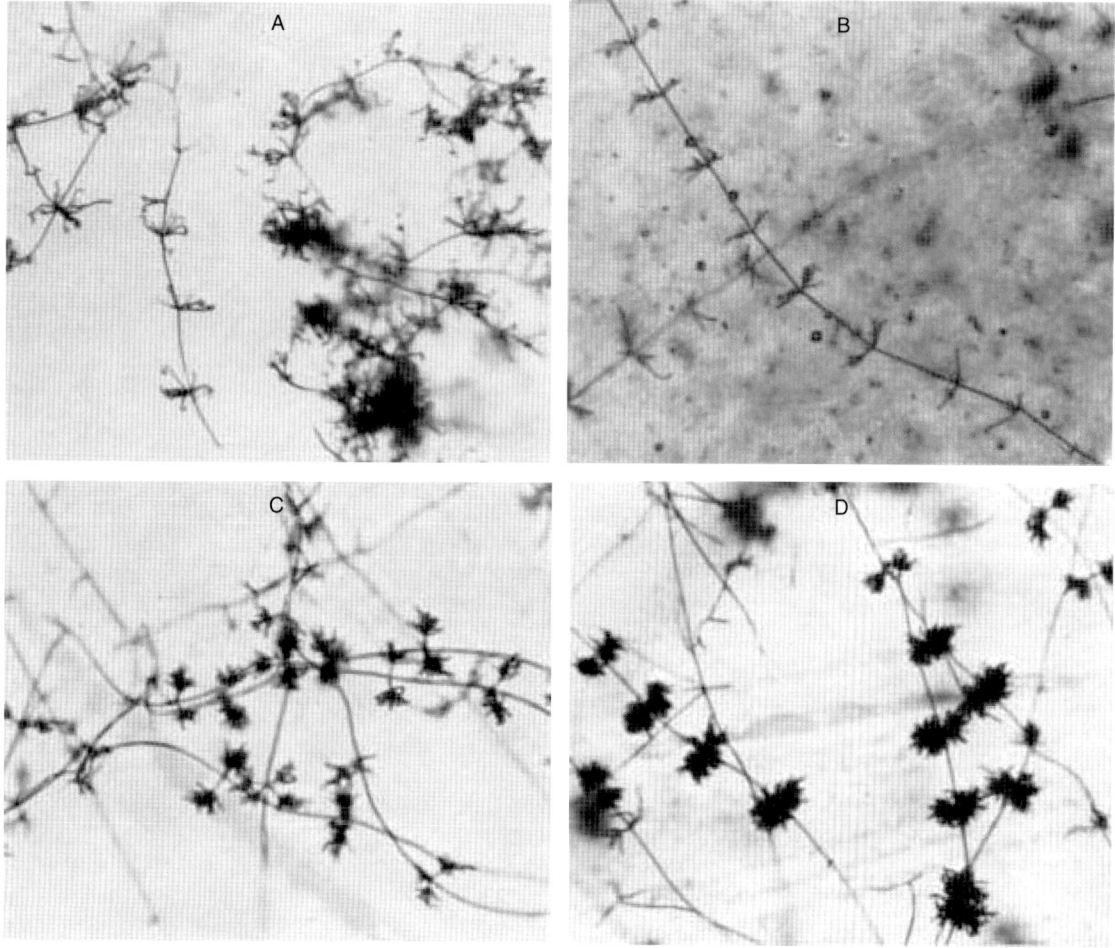

Fig. 11. Morphology of the aerial mycelium of some species of *Streptoverticillium*. (A)–(E) Light microscopy (×250). (Courtesy of C. Mütze.) (A) Sv. netropsis (DSM 40259). (B) "*Sv. reticulum*" (DSM 40893). (C) "*Sv. cinnamomeum* subsp. azacolutum" (DSM 40646). (D) *Sv. septatum* (DSM 40577). (E) Sv. mobaraense (DSM 40847). (F) Scanning electron microscopy: a *Streptoverticillium* species (×6,200).

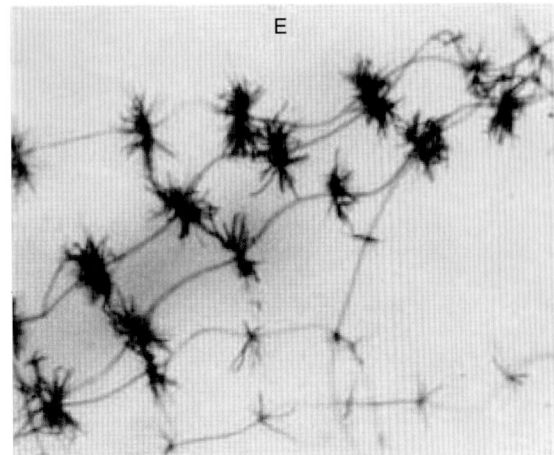

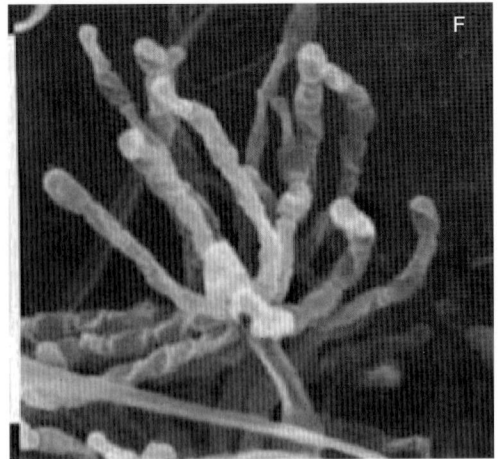

Fig. 11. *Continued*

Table 8. Differentiation of Kitasatospora species (according to Tajima et al., 2001).[a]

Characteristic	1	2	3	4	5	6	7	8
Fermentation of melanoid pigment	–	–	–	–	–	+	–	+
Reduction of nitrate	–	+	–	+	–	–	–	+
Coagulation of milk	–	–	+	–	+	–	–	–
Utilization of								
L-Arabinose	+	+	+	+	+	+	–	–
D-Xylose	+	+	+	+	+	–	–	–
Raffinose	+	–	–	+	+	–	–	–
Melibiose	+	–	–	–	–	ND	ND	ND
D-Fructose	–	–	–	+	–	–	–	+
D-Rhamnose	+	–	–	+	–	–	–	–
Sucrose	–	–	–	+	–	–	–	–
Cellulose	–	–	–	–	–	ND	ND	ND
Temperature for growth (°C)	15–37	15–41	15–37	15–42	15–37	13–38	17–40	11–38
DNA G+C content (mol %)	73.7	73.5	73.1	73.4	73.1	72.4	70.6	73.1

Symbols and abbreviations: +, positive; –, negative; ND, not determined; [T], type strain; and IFO, Institute for Fermentation, Osaka, Japan.

[a]Taxa are identified as: 1) strain SK-3255[T]; 2) strain SK-3406[T]; 3) *K. setae* KM-6054[T]; 4) *K. phosalacinea* KA-338[T]; 5) *K. griseola* AM-9660[T]; 6) *K. chochleata* IFO 14768[T]; 7) *K. cystarginea* IFO 14836[T]; and 8) *K. paracochleata* IFO 14769[T]. All strains are positive for peptonization of milk, hydrolysis of starch, and utilization of D-glucose. All strains are negative for liquefaction of gelatin and utilization of inositol and D-mannitol.

Data were from Tajima et al. (2001) for strains SK-3255[T] and SK-3406[T], Ōmura et al. (1982) for *K. setae* KM-6054[T], Takahashi et al. (1984b) for *K. phosalacinea* KA-338[T] and *K. griseola* AM-9660[T], Nakagaito et al. (1992a) for *K. cochleata* IFO 14768[T] and *K. paracochleata* IFO 14769[T] and Nakagaito et al. (1992b) for *K. cystarginea* IFO 14836[T]. From Tajima et al. (2001).

recalcitrant plant and animal materials, often polymeric residues including polysaccharides (e.g., starch, pectin, cellulose and chitin), proteins (e.g., keratin and elastin), lignocellulose, and aromatic compounds.

Members of the genus *Streptomyces* are involved in the biodegradation of various polymers abundant in soil owing to their ability to produce extracellular enzymes. The biodegradatve activities of actinomycetes in general were reviewed in the 1980s by Lechevalier (Lechevalier, 1981a; Lechevalier, 1988), Crawford (1988), and Peczynska-Czoch and Mordarski (1988). Streptomycetes are among the very few bacteria able to degrade lignin which occurs in nature closely associated with cellulose and xylan (hemicellulose), i.e., in the lignocellulose complex. Although fungi play the more important role in lignin decomposition (Crawford, 1981; Crawford, 1988; Janshekar and Fiechter, 1983; Kirk and Farell, 1987), evidence from experiments using ^{14}C-labeled lignin now shows that streptomycetes (Crawford, 1978; Antai and Crawford, 1981) and also several other genera of actinomycetes are involved in this process (McCarthy and Broda, 1984a; McCarthy et al., 1984b; McCarthy et al., 1986). As a constituent of the lignocellulose complex, cellulose can be degraded by the few ligninolytic streptomycetes. More details are given by Ramachandra et al. (1988), Wang et al. (1990),

Table 9. Differentiation of *Streptacidiphilus* species (according to Kim et al., 2003) using phenotypic properties.

Characteristic	No. of strains	*Streptacidiphilus albus* 7[a]	*Streptacidiphilus neutrinimicus* 5[a]	*Streptacidiphilus carbonis* 6[a]
Growth at pH 6.0		+	−	+
Growth on sole carbon sources (1%, w/v)				
D-Gluconic acid		−	−	+
D-Glucosamine hydrochloride		+	+	−
meso-Inositol		−	−	+
Inulin		−	−	+
L-Rhamnose		v	−	+
D-Ribose		+	−	−
Growth in presence of ($\mu g \cdot ml^{-1}$)				
Cadmium acetate (50)[b]		+	−	−
Lead acetate (100)		+	+	−
Cephaloridine hydrochloride (2)		+	+	−
Penicillin-G (16)		+	+	−
Streptomycin sulfate (16)		+	+	−

Symbols: +, positive; −, negative; and v, variable.
From Kim et al. (2003).
[a]The number of species.
[b]The numbers in parentheses are the concentrations of the compounds in μg per ml.

Crawford et al. (1993), Chamberlain and Crawford (2000), Kormanec et al. (2001), Gottschalk et al. (2003), and Kaneko et al. (2003).

In addition, multicomponent cellulases consisting of several endoglucanases and exoglucanases have been found in thermophilic and mesophilic streptomycetes (Enger and Sleeper, 1965; Crawford and McCoy, 1972; MacKenzie et al., 1984; Schrempf and Walter, 1995; Harchand and Singh, 1997; Marri et al., 1997; Ulrich and Wirth, 1999; Wirth and Ulrich, 2002). Also, xylanases involved in the decomposition of the lignocellulose complex have been found in streptomycetes (Kluepfel and Ishaque, 1982; Kluepfel et al., 1986; Deobald and Crawford, 1987; Godden et al., 1989; Schäfer et al., 1996; Morosoli et al., 1999). Again, xylanases seem to be more widespread among thermophilic actinomycetes (McCarthy et al., 1985). Pectinolytic streptomycetes have been reported (Sato and Kaji, 1975; Sato and Kaji, 1977; Sato and Kaji, 1980a; Sato and Kaji, 1980b) and chitinolytic complexes consisting of chitinase and chitobiase have been isolated from various streptomycetes: *Streptomyces griseus* (Berger and Reynolds, 1958), *Streptomyces antibioticus* (Jeuniaux, 1966), and othe streptomycetes (Beyer and Diekmann, 1985). For more details, see The Family Streptomycetaceae, Part II: Molecular Biology in this Volume).

The ability to decompose starch (the primary material for the textile, paper and food industry) is widespread among fungi and bacteria. The involved enzymes, amylases, have been detected in various streptomycetes (Mordarski et al., 1970; Suganuma et al., 1980; Fairbairn et al., 1986; McKillop et al., 1986).

In addition to degrading polymeric compounds (as described above), streptomycetes can play an important role in the destruction of other organic materials used by humans for diverse purposes, among them cotton and plant fibers (Khan et al., 1978; Lacey and Lacey, 1987), wool (Noval and Nickerson, 1959), hydrocarbons in jet fuel and emulsions (Genner and Hill, 1981), and rubber (Cundell and Mulcock, 1975; Hutchinson et al., 1975). The biodeterioration of natural and synthetic substances has been reviewed in detail by Lacey (1988) and Behal (2000). More details are given in The Family Streptomycetaceae, Part II: Molecular Biology in this Volume.

Most isolated streptomycetes are nonfastidious; they do not require organic nitrogen sources or vitamins and other growth factors. Soil as a habitat gives support to their mycelial growth. Furthermore, spore formation enables streptomycetes to adapt to various physical conditions in soil (such as shifts in aeration, moisture tension, and pH), periods of drought, frost, hydrostatic pressure, and anaerobic conditions which may change dramatically and quickly.

The spores can be regarded as a semi-dormant stage in the life cycle that facilitates survival in soil for long periods (Mayfield et al., 1972; Ensign, 1978). Morita (1985) reported viable cultures from 70-year-old soil samples. A relatively high number of streptomycetes in soil is almost always present as inactive spores. The very low germination efficiencies often obtained may be caused by competition with other indigenous microorganisms, but pre-germinated spores are found to grow for a short time and then re-sporulate (Lloyd, 1969a). Germination may depend on the presence of special signaling factors, and there is evidence that exogenous nutrients, water and Ca^{2+} are required (Ensign, 1978). Furthermore the nutrient status of the germination site influences the extent of hyphal growth and the time to differentiation into aerial hyphae. Many other "habitats" may come into contact with soil owing to human or other activities. As listed by Korn-Wendisch and Kutzner (1992a), these are 1) fodders and other organic material and 2) freshwater and marine habitats as well as potable water systems. Mesophilic and especially thermophilic streptomycetes are involved in the degradation of many natural substrates (e.g., hay, fodder, grain, and wood) and can degrade synthetic products (e.g., cotton textiles, fabric, paper, rubber, plastics and plasticizers). Drainage after heavy rainfalls causes creeks and rivers to become contaminated with soil streptomycetes that find their way into the sediments of freshwater lakes and even to marine biotopes. According to some reports, drinking water supplies may also become contaminated with streptomycetes; some of them produce odorous compounds leading to the spoilage of the water. Streptomycetes play only a minor role as plant pathogens, and although very few streptomycetes have been isolated from pathological material so far, their role as agents of infectious diseases cannot be ignored (more details are given below).

Soil as Habitat

A number of biotic and abiotic properties characterize any habitat and determine the current composition of the community and also the numbers of microorganisms. These are: 1) vegetation and content and kind of organic matter, 2) soil type, 3) season and climate, 4) temperature, 5) circulation of water and air, and 6) pH. As already pointed out by Korn-Wendisch and Kutzner (1992a), the reports published on the occurrence of streptomycetes in soil are numerous and too extensive to be treated quantitatively. The reader is referred to reviews by Lechevalier (Lechevalier, 1981a; Lechevalier, 1988), Williams (1982a), Goodfellow and Williams (1983), Williams et al. (1984b), Goodfellow and Simpson (1987a), Korn-Wendisch and Kutzner (1992a) and to a series of papers by Williams and coworkers dealing specifically with the ecology of actinomycetes in various soils and under specific conditions: Davies and Williams (1970), Williams and Mayfield (1971b), Williams et al. (Williams et al., 1971c; Williams et al., 1972; Williams et al., 1977), Mayfield et al., (1972), Ruddick and Williams (1972), Watson and Williams (1974), Khan and Williams (1975), Flowers and Williams (1977), Williams and Robinson (1981). By direct observation of the soil microflora, several authors have shown that streptomycetes perform their typical life cycle in this natural habitat. For details of this life cycle and its genetic control, the reader is referred

to Kieser et al. (2000). As summarized by Korn-Wendisch and Kutzner (1992a), in most soils, 10^4 to 10^7 colony forming units (CFU) per g can be expected, accounting for about 1–20% of the total viable count; in some soils however streptomycetes dominate. The number of streptomycetes and also the number of subgroups vary under different conditions. Details can be found in the studies of Flaig and Kutzner (1960a), Misiek (1955), Szabó and Marton (1964), and Küster (1976).

Note that the detection and localization of different *Streptomyces* species or types in their natural habitat are based mainly on cultivation dependent techniques. In addition, the difficulties in the intrageneric classification of the genus *Streptomyces* often do not allow a comparison of these ecophysiological studies. With respect to moisture, it has been shown by Williams et al. (1972) that streptomycetes resist desiccation because they form arthrospores. In addition they need a lower water tension for growth than other bacteria need, but they may be very sensitive to water-logged conditions.

Most streptomycetes prefer neutral to alkaline soils as a natural habitat (e.g., Flaig and Kutzner, 1960a). But several studies of acidic soils employing media adjusted to acid pH and supplemented with antifungal agents found numerous acidophilic as well as acido-tolerant streptomycetes. Khan and Williams (1975), Hagedorn (1976), and Williams et al. (1977) showed in detail that acidic forest soils and other acidic habitats contained different groups or species of streptomycetes. These streptomycetes also have unusual properties when compared with neutrophilic strains, e.g., production of specific amylases (Williams and Flowers, 1978b) and chitinases (Williams and Robinson, 1981). Only a few reports of alkalophilic, acid-sensitive actinomycetes have been published (Taber, 1959; Taber, 1960; Mikami et al., 1982; Mikami et al., 1985).

Streptomycetes, like other soil bacteria, may also be found in the intestinal tract of earthworms (Brüsewitz, 1959; Parle, 1963b; Parle, 1963a), the gut of arthropods (Szabó et al., 1967; Bignell et al., 1980; Bignell et al., 1981; Bignell, 1984), and the pellets produced by millipedes and woodlice (Márialigeti et al., 1984).

Studies on the occurence of streptomycetes in the rhizosphere have been published by various authors (for a review, see Goodfellow and Williams, 1983). The competitive advantage of antibiotic-producing organisms over nonproducing microbes in soil has been suggested since the time these compounds were first discovered. However, evidence for the in situ production of antibiotics in soil is still not clear (Williams, 1982a). This may be due to 1) the instability and low concentrations in soil (Brian, 1957; Williams,

1982a) and 2) possible adsorption to soil colloids in combination with inadequate and insensitive detection methods (Williams, 1982a) and 3) the ephemeral growth of producers in response to nutrient shortage (Williams and Khan, 1974; Williams, 1982a).

However, Rothrock and Gottlieb (1984) reported antibiotic production occurs in sterilized soil supplemented with nutrients and inoculated with a potent producer. Many soil microbiologists support the assumption that streptomycetes play an important role in the control of fungal root pathogens (Williams, 1978a; Williams, 1982a; Sing and Mehrotra, 1980; Rothrock and Gottlieb, 1981). In addition many streptomycetes are often successful in competition with other rhizosphere bacteria such as pseudomonads and bacilli, especially in relatively dry soil.

Thermophilic Streptomycetes

The genus *Streptomyces* contains mainly mesophilic species in addition to some thermotolerant (growing up to 45°C) and a few thermophilic species. A detailed taxonomic study of thermophilic streptomycetes has been published by Kim et al. (1999). Additional thermophilic species (*S. thermocoprophilus* and *S. thermospinisporus*) have been described by Kim et al. (2000) and Kim and Goodfellow (2002b). The thermophilic streptomycetes described so far grow at 28–55°C, and several grow at even higher temperatures.

Thermophilic streptomycetes go through an interesting cycle in nature in regard to their dispersal: active growth takes place at sites of high temperature such as in compost, manure, and self-heating hay or grain. The vegetative phase ends with the formation of a large number of spores. These are returned with the compost or manure to the fields and pastures where they infect plant material and hay directly or via soil dust (Korn-Wendisch and Kutzner, 1992a). Not surprisingly, the majority of actinomycetes isolated from bioaerosols in the surroundings of composting facilities belong to the genus *Streptomyces* (P. Kämpfer et al., unpublished observation). For this reason thermophilic actinomycetes are widespread and can be isolated from various sources like soils (Tendler and Burkholder, 1961; Craveri and Pagani, 1962), pig feces (Ohta and Ikeda, 1978), sewage-sludge compost (Millner, 1982), and freshwater habitats (Cross, 1981).

Freshwater Environments, Water Supplies and Marine Environments

Actinomycetes can easily be isolated from fresh water and especially from sediments of rivers and lakes. Cross (1981) stated, however, that most of these organisms may not be active at these sites. Nevertheless, these wash-in forms ("aliens") from surrounding terrestrial environments can survive as dormant spores in aquatic habitats for a long time (Al-Diwany and Cross, 1978), and especially rivers carry a load of various actinomycetes, among them also streptomycetes. In a study on the occurence of actinomycetes in the river Thames, Burman (1973) found 59–200 streptomycetes per ml and 10–20 micromonosporae per ml. These organisms were found to grow on decaying vegetation on riverbanks and mud flats at low water or on floating mats of decaying algae or other vegetation. Under these conditions odorous substances are produced, and subsequent increase in river levels washes them into the water, thus giving rise to "earthy taste" complaints. Of these odorous compounds, geosmin and methyl iso-borneol are most often detected (Gerber, 1979a; Gerber, 1979b). As summarized by Wood et al. (1983), the prevention of earthy tastes in reservoirs and water supply systems depends on locating the production sites and determining the patterns of distribution of these compounds (Silvey and Roach, 1975; Lechevalier et al., 1980).

Burman (1973) studied the fate of actinomycetes of river water in the course of production of drinking water. Filtration processes reduce the number of streptomycetes considerably. In the distribution system, a new type named "aquatic strains of *Streptomyces*" has appeared, which can be enriched (for details, see Burman, 1973).

Okazaki and Okami (1976), Cross (1981), Weyland (1981b), Weyland (1981a), Weyland and Helmke (1988), and Goodfellow and Haynes (1984) have reviewed the occurrence of streptomycetes in marine habitats including sediments. Two localities can be distinguished: 1) the littoral and inshore zone and 2) deep-sea sediments. From both localities streptomycetes can be isolated; however, similarly to streptomycetes in freshwater environments, most of these organisms are "survivors" rather than constituents of the autochthonous microflora. From the littoral zone, streptomycetes have been isolated both from sediments (Roach and Silvey, 1959) and from decaying seaweed (Siebert and Schwartz, 1956). These isolates were able to grow on polymeric substances, e.g., agar and chitin (Humm and Shepard, 1946), alginate and laminarin (Chesters et al., 1956), and cellulose (Chandramohan et al., 1972), characteristic for these microsites.

In sediments, both the depth and location of sample sites play important roles in determining the ratio of different taxa of actinomycetes (Weyland, 1981b; Weyland and Helmke, 1988). Sites in the open sea generally contain only low numbers of actinomycetes (viable counts were about 100 CFU per ml of wet sediment). The distribution (horizontal as well as vertical) is assumed to correlate with the physiological properties of the three taxa *Streptomycetes*, *Micromonosporae* and *Rhodococci barotolerance* (Helmke, 1981), halotolerance, and psychrophilism (Weyland, 1981a). Goodfellow and Haynes (1984), however, were not able to find a correlation between salinity, pH, or depth and the number of actinomycetes recovered from marine sediments. In this study numerous isolates are described in detail, and from a total of 732 organisms, 250 belonged to *Streptomyces*, 250 to *Micromonospora*, 140 to *Rhodococcus*, and 92 to *Thermoactinomyces*. A selected number of isolates belonging to the genus *Streptomyces* was subsequently identified using 41 diagnostic tests and applying the MATIDEN program and the *Streptomyces* probability matrix (Williams et al., 1983b). Around 50% of these streptomycetes were similar to those grouped into cluster 1 of Williams et al. (1983a).

Okami and Okazaki (1978) found streptomycetes (300–1270 colonies per cm^3) mainly in the sediments of shallow seas (70–520 m deep), whereas in samples 700–1600 m deep, *Micromonospora* dominated. No actinomycetes were obtained from depths of 2800 and 5000 m in the Pacific Ocean. A higher salt tolerance of marine streptomycetes than of their terrestrial counterparts was observed. However, already Tresner et al. (1968) found that salt tolerance among streptomycetes is widespread and this feature may be due to selection of the more tolerant organisms in marine habitats. Few streptomycetes isolated from marine environments were found to be obligate halophiles (Okazaki and Okami, 1976).

Notably, marine habitats are essential for screening programs in the search of new antimicrobial and anti-insecticidal compound producers. In early studies, Nissen (1963) found a high percentage of antibiotic-producing streptomycetes from decaying seaweed, and subsequent investigations confirmed these findings (Okami and Okazaki, 1972; Okami et al., 1976; Hotta et al., 1980).

Isolation

The procedures of isolation of streptomycetes (extensively summarized by Korn-Wendisch and Kutzner, 1992a) are partly summarized in the next paragraphs. Additional information about isolation for special purposes, growth of streptomycetes, and preservation of streptomycetes can be obtained from the excellent textbook *Practical Streptomyces Genetics* (Kieser et al., 2000).

Generally, all procedures for the isolation of microorganisms are influenced by the nature of the microorganism and the number of propagules relative to the number of other microbes within the habitat (Stolp and Starr, 1981). If the organism to be isolated is best adapted to the selected isolation conditions, then direct plating of a serial dilution on a nutrient agar medium can readily lead to a pure culture. This procedure does not work well for isolation of streptomycetes. These actinomycetes are isolated usually by enrichment or use of selective media and specific isolation conditions or both.

As already pointed out by Korn-Wendisch and Kutzner (1992a), Streptomycetaceae members can be isolated using general selective isolation procedures (Williams and Wellington, 1982b; Williams and Wellington, 1982c; Williams et al., 1984a). These procedures require 1) choice of the material containing the selected microorganisms, 2) pretreatment of the sample, and in some cases, enrichment of the chosen microbial groups, 3) use of selective media or selective incubation conditions or both, and 4) colony selection on the basis of colony morphology and purification.

Streptomycetes can be isolated from a wide variety of habitats, and most isolation procedures involve extraction from soil or another environmental sample followed by dilution of the cells (cell aggregates) to allow cultivation on solid media.

Isolation and Enrichment from Soil

Because the vegetative mycelium and spore chains are often closely associated with the mineral and organic particles of the soil, vigorous shaking of the sample with the diluent is often needed to suspend the spores of mycelial fragments. Use of glass beads and agitation on a shaker may aid suspension. In the literature, methods involving mechanical devices such as the Ultrasonics sonicator-disrupter, Ultra-Turrax homogenizer, Turmix blender, Waring blender, or a mortar and pestle are described, but a detailed comparison of the dilution efficiency of these pretreatments is still missing. Chemical disruption methods are also reported in literature. Gently shaking soil samples with an ion-exchange resin Chelex-100 (Biorad) followed by differential centrifugation and filtration was used to separate the mycelium from the spores (Herron and Wellington, 1990).

A subsequent treatment of samples (i.e., preparing dilutions and plating) differs little from general bacteriological practice. Most often the coarse particles of the soil suspension are allowed to settle before dilutions are made. But soil particles may also be used directly for incubation of "soil plates" (Warcup, 1950), which are also used to isolate fungi. The addition of lime to soil can enrich for streptomycetes (see chapter 2 of Kieser et al. [2000] and references therein). Isolation plates may be surface-inoculated with a sterile glass rod (or Drigalski spatula).

Spread of motile bacteria via water films can be avoided by drying the plates at 45°C before incubation or mixing the soil suspension with the molten agar, which is highly recommended (Korn-Wendisch and Kutzner, 1992a). The addition of CaCO$_3$ to air-dried soil samples (10:1 w/w) and subsequent incubation at 26°C for 7–9 days in a water-saturated atmosphere can lead to a 100-fold increase of streptomycete colonies on isolation plates (Tsao et al., 1960; El-Nakeeb and Lechevalier, 1963).

The enrichment of streptomycetes by soil amendment with keratin was first carried out by Jensen (1930). Later the addition of chitin to soil was found to stimulate growth of actinomycetes (Williams and Mayfield, 1971b). Williams and Robinson (1981) similarly obtained the enrichment of acidophilic and neutrophilic streptomycetes in acidic soil and litter containing pure and fungal chitin. Another isolation strategy using chitin in the form of insect wings has been used as a baiting method (Veldkamp, 1955; Jagnow, 1957; Okafor, 1966). Porter and Wilhelm (1961) studied isolation using various other organic materials, like salmon viscera meal, peanut meal, cottonseed meal, and dried blood flour (15 mg/g of soil). They found that incubation of the enrichment cultures under moist conditions led to an up to 1000-fold increase in the number of streptomycetes.

Besides the frequently used arginine glycerol agar (El-Nakeeb and Lechevalier, 1963), the following media (details below) are also often applied for selective isolation of streptomycetes: HV agar (Hayakawa and Nonamura, 1987a, 1987b), colloidal chitin agar (Hsu and Lockwood, 1975), and reduced arginine starch salts agar.

Several physical, chemical and biological methods have been studied to reduce or inhibit other microbes (for review, see Goodfellow and Williams, 1986a). Nüesch (1965) centrifuged soil suspensions for 20 min at 1600 × g to separate the spores of streptomycetes (in the supernatant) from other bacteria and spores of fungi (in the sediment); however, the method has not been very successful. Using a similar approach, El-Nakeeb and Lechevalier (1963) obtained a significantly smaller number of streptomycete colonies as compared with the control. Voelskow (1988/1989) described a simple sedimentation method in which 1 g of soil was suspended in 15 ml of salt solution, vigorously shaken, and then treated with ultrasonic vibrations. After 1, 2, and 4 h of sedimentation, samples were taken

from different levels of this solution, further diluted, and plated on agar surfaces.

Initial drying and heating procedures were applied because arthrospores have a relatively high resistance to low moisture tension. Drying of the sample or prolonged storage at ambient temperatures for mesophiles and at 50–60°C for thermophiles led to a relative increase in streptomycete concentrations. Williams et al. (1972) showed that heat treatment of soil (40–50°C, 2–16 h) leads to a significant reduction of the vegetative bacterial proportion without affecting the colony counts of streptomycetes.

Membrane filtration has been mainly used for the enrichment of streptomycetes from water samples (Burman et al., 1969) and from seawater and mud (Okami and Okazaki, 1972), but it has also been a first step in the isolation of streptomycetes from soil (Trolldenier, 1966). These authors used 1 ml of a series of 10-fold dilutions, which were membrane-filtered (0.3-μm pore size). This filter was then placed upside down on a suitable agar medium, which was supplemented with 10% compost soil. Colonies developed between the agar surface and the membrane filter, and the streptomycetes (but not other bacteria or fungi) were able to grow through the pores. Using this method, the selective effects of the physical barrier (the membrane) and the soil lead to a 3–5-fold increase in the number of streptomycete colonies in comparison with poured plates without soil.

Hirsch and Christensen (1983) introduced the use of cellulose ester membrane filters (pore size 0.01–3.0 μm), which were placed onto nutrient agar containing antifungal antibiotics (cycloheximide and candicidin). These plates were inoculated with different samples from soil, water and vegetable materials. After 4 days, the hyphae of actinomycetes penetrated the filter pores and grew on the underlying agar medium, whereas the growth of the other bacteria was restricted to the surface of the filter. To allow further development of actinomycete colonies, the membrane filter was removed and the plates were reincubated. Filters (0.22–0.45 μm) were also found to be suitable for the exclusive recovery of actinomycetes. Polsinelli and Mazza (1984) and Hanka et al. (1985) used this approach independently.

Several authors have added chemicals to improve the isolation efficiency. Phenol treatment of a dense soil suspension (1.4% for 10 min) was recommended to eliminate bacteria and fungi, but El-Nakeeb and Lechevalier (1963) obtained less favorable results with this method. Chloramine, ammonia, and sodium hypochlorite were mainly employed for the treatment of water samples, because it has been found that streptomycetes and other actinomycetes are slightly more resistant to these agents than other bacteria are (Burman et al., 1969).

Isolation of Airborne Spores

For the isolation of *Streptomyces* spores from self-heating material such as hay or compost, samples can be agitated in a wind channel (Lacey and Dutkiewicz, 1976b) or sedimentation chamber (see below; Lacey and Dutkiewicz, 1976a). Plates are then inoculated with the aerosol using an Andersen sampler (Goodfellow and Williams, 1986a). This method, widely employed for the isolation of thermophilic actinomycetes, may also be used for the isolation of mesophilic streptomycetes from soil.

In addition, other devices like filtration samplers (e.g., Sartorius MD 8) are suitable for the sampling of airborne streptomycetes.

Use of Selective Media and Incubation

Selective media always play an important role in the isolation of the desired microorganisms. A number of factors can be varied: 1) nutrient composition and concentration of the isolation medium, i.e., choice of carbon and nitrogen sources preferred by the organisms; 2) addition of chemical substances to inhibit selectively the accompanying flora of the natural habitat or those which are stimulating the desired organisms; 3) pH, for acidophilic, neutrophilic, and alkalophilic organisms; and 4) temperature, e.g., for the isolation of thermophiles or psychrophiles.

Many different media have been formulated empirically and proposed for the isolation of streptomycetes. Selected carbon and nitrogen compounds listed in Table 10 are especially suitable for the isolation of these organisms. The most frequently used media with their formulas are listed in Tables 11 and 12. Alternatively, streptomycetes can be grown on very poor media such as water agar.

DIFFERENT CARBON AND NITROGEN SOURCES FOR ENRICHMENT Because it was early recognized that streptomycetes can degrade chitin (Veldkamp, 1955; Jagnow, 1957), a chitin medium was devised by Lingappa and Lockwood (1962) for selective isolation. However, the authors and later also El-Nakeeb and Lechevalier (1963) found that their chitin agar was only a little better than water agar. Hsu and Lockwood (1975) added mineral salts to this medium (Table 11), which was shown to be useful for the isolation of actinomycetes (*Streptomyces*, *Nocardia* and *Micromonospora*) from water samples but had little effect when isolating actinomycetes from soil. Note that chitinolytic activity is not a genus specific feature for streptomyces. Williams et al. (1983a) found that only 25% of over 300 strains were strongly chitinolytic, so this widely used medium selects the chitinolytic streptomycete strains, which may not be the most abundant strains in soil. Starch is degraded by the vast majority of streptomycetes and can therefore be used as a selective carbon source. An early finding was that the combination of starch with nitrate, which is utilized by many streptomycetes (in contrast to other bacteria) as nitrogen source, is very useful for the selective isolation of streptomycetes (Flaig and Kutzner, 1960a). Küster and Williams (1964), who improved this medium, stated: "The three best media, allowing good development of streptomycetes while suppressing bacterial growth, were those containing starch or glycerol as the carbon source with casein, arginine or nitrate as the nitrogen source."

Table 10. Nutritional substances and selective agents for isolation of streptomycetes from soil (according to Korn-Wendisch and Kutzner, 1992).

Preferred C and N source	Selective agents in the medium Antibiotic(s)	Others	References
Starch and KNO$_3$	None	None	Flaig and Kutzner, 1960b
Starch, casein, and KNO$_3$	None	None	Küster and Williams, 1964
Chitin	None	None	Lingappa and Lockwood, 1962
Glycerol and arginine	None	None	El-Nakeeb and Lechevalier, 1963
Glycerol, casein, and KNO$_3$	None	None	Küster and Williams, 1964
Raffinose, histidine	None	None	Vickers et al., 1984
Starch, casein, and KNO$_3$	Rifampicin	None	Vickers et al., 1984
Starch, casein, and KNO$_3$	Cycloheximide, nystatin, penicillin, and polymyxin	None	Williams and Davies, 1965
Glycerol and arginine	Cycloheximide, pimaricin, and nystatin	None	Porter et al., 1960
Dextrose and asparagine	Cycloheximide	None	Corke and Chase, 1956
Asparagine	None	Propionate	Crook et al., 1950
Starch, casein, and KNO$_3$	Cycloheximide	Rose bengal	Ottow, 1972

From Korn-Wendisch and Kutzner (1992).

Table 11. Some media useful for the selective isolation of streptomycetes.

References[a] Ingredients (g/liter)	1 Starch-casein-KNO$_3$ agar	2 Glycerol-arginine agar	3 Actinomyces isolation agar	4 Chitin agar	5 Raffinose-histidine agar
Chitin (colloidal)	—	—	—	4.0	—
Starch	10.0[b]	—	—	—	—
Glycerol	—	12.5	5.0[c]	—	10.0
Raffinose	—	—	—	—	—
Sodium propionate	—	—	4.0	—	—
KNO$_3$	2.0	—	—	—	—
Casein	0.3	—	—	—	—
Sodium caseinate	—	—	2.0	—	—
Asparagine	—	—	0.1	—	—
Arginine	—	1.0	—	—	—
Histidine	—	—	—	—	1.0
NaCl	2.0	1.0	—	—	—
KH$_2$PO$_4$	—	1.0	0.5	0.3	1.0
K$_2$HPO$_4$	2.0	0.5	0.1	0.7	0.5
MgSO$_4 \cdot$ 7H$_2$O	0.05	—	—	0.5	—
CaCO$_3$	0.02	—	—	—	—
Fe$_2$(SO$_4$)$_3 \cdot$ 6H$_2$O	—	0.01	—	—	—
FeSO$_4 \cdot$ 7H$_2$O	0.01	—	0.001	0.01	0.01
CuSO$_4 \cdot$ 5H$_2$O	—	0.001	—	—	—
ZnSO$_4 \cdot$ 7H$_2$O	—	0.001	—	0.001	—
MnSO$_4 \cdot$ H$_2$O	—	0.001	—	—	—
MnCl$_2 \cdot$ 4H$_2$O	—	—	—	0.001	—
Agar[d]	18.0	15.0	15.0	20.0	12.0
pH	Adjusted to 7.0–7.5 or lower or higher depending on the flora to be isolated.	Adjusted to 7.0–7.5 or lower or higher depending on the flora to be isolated.	Adjusted to 7.0–7.5 or lower or higher depending on the flora to be isolated.	Adjusted to 7.0–7.5 or lower or higher depending on the flora to be isolated.	Adjusted to 7.0–7.5 or lower or higher depending on the flora to be isolated.

[a]References: Küster and Williams, 1964; El-Nakeeb and Lechevalier, 1963; Hsu and Lockwood, 1975; and Vickers et al., 1948; Difco Laboratories.
[b]Alternatively, glycerol at 10g/liter can be used.
[c]Not contained in the dehydrated medium; added at the time of preparation.
[d]The different amounts of the agar are due to the varying quality used by the individual authors.

Table 12. Composition of some media suitable for the cultivation of streptomycetes.[a]

Ingredients		Comments
1. Glucose-yeast extract-malt extract (GYM) agar		
Glucose	4.0g	Addition of CaCO₃ (2.0g/liter) is advantageous for the growth of many streptomycetes. Adjust pH to 7.2.
Yeast extract	4.0g	
Malt extract	10.0g	
Agar	12.0g	
Distilled water	1 liter	
2. Oatmeal agar		
Oatmeal	20.0g	Cook 20.0g of oatmeal in 1 liter of distilled water for 20min. Filter through cheesecloth. Add distilled water to restore the volume of the filtrate to 1 liter, then add trace salts solution and agar. Adjust pH to 7.2.
Agar	12.0g	
Trace salts solution (see no. 5)	1.0ml	
Distilled water	1 liter	
3. Inorganic salts-starch agar		
Starch (soluble)	10.0g	Make a paste of the starch with a small amount of cold distilled water and bring to a volume of 1 liter; then add the other ingredients. Adjust pH (if necessary) to 7.0–7.4.
$(NH_4)_2SO_4$	2.0g	
K_2HPO_4 (anhydrous basis)	1.0g	
$MgSO_4 \cdot 7H_2O$	1.0g	
NaCl	1.0g	
$CaCO_3$	2.0g	
Trace salts solution (see no. 5)	1.0ml	
Agar	12.0g	
Distilled water	1 liter	
4. Glycerol-asparagine agar		
Glycerol	10.0g	The pH should be about 7.0–7.4. Do not adjust if it is within this range.
L-Asparagine (anhydrous)	1.0g	
K_2HPO_4	1.0g	
Trace salts solution (see no. 5)	1.0ml	
Agar	12.0g	
Distilled water	1 liter	
5. Trace salts solution		
$FeSO_4 \cdot 7H_2O$	0.1g	
$MnCl_2 \cdot 4H_2O$	0.1g	
$ZnSO_4 \cdot 7H_2O$	0.1g	
Distilled water	100.0ml	
6. Trace elements solution SPV-4		
$CaCl_2 \cdot 2H_2O$	4.0g	SPV-4 is used as an alternative to (5). Five ml of this stock solution is added to 1 liter of medium.
Fe (III) citrate	1.0g	
$MnSO_4$	0.2g	
$ZnCl_2$	0.1g	
$CuSO_4 \cdot 5H_2O$	0.04g	
$CoCl_2$	0.022g	
$Na_2MoO_4 \cdot 2H_2O$	0.025g	
$Na_2B_4O_7 \cdot 10H_2O$	0.1g	
Distilled water	1 liter	

[a]Recipes 1–5 are from Shirling and Gottlieb (1966) and recipe 6 from Voelskow (1988/89).
From Korn-Wendisch and Kutzner (1992).

Benedict et al. (1955) were the first to report that the combination of glycerol and arginine favored streptomycete isolation. El-Nakeeb and Lechevalier (1963) found that this medium (Tables 11 and 12) was superior to nine other media, resulting in higher number and proportion of streptomycete colonies.

Other compounds (e.g., pectin [Wieringa, 1955], poly-ß-hydroxybutyrate [Delafield et al., 1965], rubber [Nette et al., 1959], cholesterol [Brown and Peterson, 1966], elemental sulfur [Wieringa, 1966], and natural and artificial humic acids [Hayakawa and Nonomura, 1987a, 1987b]) have been used successfully, and most of them strongly select certain organisms producing visible zones of clearing or other changes in the medium.

The use of compounds with antifungal activity (antibiotics) as supplements to isolation media has also been widely used to suppress fungal growth (Table 10). The most frequently used compound is cycloheximide (actidione, 50–100 µg/ml) by Williams and Davies (1965). Pimaricin and nystatin (each 10–50 µg/ml) were found to be even more effective (Williams and Davies, 1965).

The use of compounds with antibacterial activity is more restricted because actinomycetes are often also sensitive to them. Williams and Davies (1965) found polymyxin (5 µg/ml) and penicillin (1 µg/ml) suppressed bacterial flora; however, they also inhibited streptomycetes. Preobrazhenskaya et al. (1978) showed that the genera of Actinomycetales may differ significantly in their sensitivity to antibacterial antibiotics and that streptomycetes were the most sensitive group. Thus, the use of antibacterial compounds may be more helpful in isolating other genera of this order (Cross, 1982).

In contrast, some antibiotics may facilitate the isolation of certain species or groups of *Streptomyces*. For example, the selective isolation of members of the *Streptomyces diastaticus* cluster sensu Williams et al. (1983a) was achieved on starch casein medium containing rifampicin (50 µg/ml; Vickers et al., 1984). Wellington et al. (1987) found a similar effect with several media containing different C and N sources as well as with media supplemented with inhibitors.

Hanka et al. (1985) described a selective isolation medium for streptoverticil-producing *Streptomyces* species containing cycloheximide and nystatin (each 50 g/ml, to control fungal growth) and oxytetracycline (25 µg/ml, to suppress other actinomycete genera and other *Streptomyces* groups). Hanka and Schaadt (1988) enhanced selectivity of this medium by adding lysozyme (1000 µg/ml). Also, sodium propionate can suppress the competing fungi (Crook et al., 1950; Table 11), and Rose Bengal (35 mg/liter) in starch casein nitrate agar (Ottow, 1972) can suppress most of the bacteria and inhibit the spreading growth of fungi.

pH of the Isolation Medium and Incubation Temperature Most streptomycetes grow optimally at neutral pH values, i.e., are neutrophilic. Therefore, the pH of most isolation media is 7.0–7.5. However, for the isolation of acidophilic streptomycetes, the medium pH is 4.5 (Khan and Williams, 1975), and for alkalophilic strains, it is 10–11 (Mikami et al., 1982). Most streptomycetes isolated from soils are mesophilic, and therefore plates are most often incubated at 22–37°C (mostly at 28°C). Psychrophilic strains (e.g., from marine environments) grow at 15–20°C. In contrast, thermotolerant and thermophilic representatives can be isolated at higher temperatures (40, 45, 50, or 55°C). Note that thermophiles often form colonies after a short period of incubation (within 2–5 days), and mesophilic members produce visible colonies within 7–14 days. However, marine and other psychrophilic organisms often need several weeks (up to 10) for the formation of visible colonies.

Colonies of *Streptomyces* are in most cases readily recognized by their macroscopic and microscopic appearance. In most cases, *Streptomyces* are easily purified by picking colonies and transfer to a nonselective medium. According to Williams and Wellington (1982b), purification is "undoubtedly the most time-consuming and often the most frustrating stage of the isolation procedure." Acidiphilic members of the Streptomycetaceae can be isolated on acidified starch casein agar containing cycloheximide and nystatin (Kim et al., 2003).

Isolation of Antibiotic Producers and Strains for Genetic Studies

The isolation of antibiotic producing streptomycetes follows the same procedures as given above. Most often, the activity is tested after isolation of pure cultures, but this can also be combined with the isolation procedure. Thus, strains exhibiting antibiotic activity can be recognized even on the initial dilution plates if they are treated with an appropriate test organism, either by flooding or by spraying. Zones of inhibition can be detected after further incubation (Lindner and Wallhäusser, 1955; Wilde, 1964). As an alternative, a simple replication procedure allows the examination of the antibiotic activity of the colonies against selected sensitive organisms (Lechevalier and Corke, 1953). Selective techniques for isolation and screening of actinomycetes that produce antibiotics and other secondary metabolites of clinical relevance have been summarized by Nolan and Cross (1988). In addition, Kieser et al. (2000) have provided a set of protocols for selective isolation of streptomycetes, generating spore suspensions, and several other more sophisticated procedures.

Isolation of Thermophilic Streptomycetes

As already stated, thermophilic streptomycetes and other thermophilic actinomycetes are often isolated from samples derived from high temperature environments (e.g., compost materials, manure heaps, and fodders). The high temperatures (45–60°C) can serve as a selective condition that favors enrichment of the desired organisms (Festenstein et al., 1965).

As pointed out by Greiner-Mai et al. (1987), a very important requisite for the isolation of thermophilic streptomycetes is a humid atmosphere, which can be achieved by incubating plates in large jars with water in the bottom. Alternatively, the sealing of Petri dishes with masking tape is also effective.

Interestingly, the media recommended for the isolation of thermophilic actinomycetes, including streptomycetes, contain higher nutrient concentrations than those used for mesophilic strains. In addition, antifungal agents and antibacterial agents are sometimes added as supplements (Lacey and Dutkiewicz, 1976b; Goodfellow et al., 1987c). For some special isolation techniques, the reader is referred to the papers of Uridil and Tetrault (1959), Fergus (1964), Gregory and Lacey (1963), and Cross (1968). Additional information can be obtained from the publications of Kim et al. (Kim et al., 1996; Kim et al., 1998; Kim et al., 2000).

Isolation from Aquatic Habitats

For the isolation of streptomycetes from water, the media listed in Table 11 can be used. In a comparative study on the suitability of media, Hsu and Lockwood (1975) found that chitin-agar was superior to the other four (egg albumin, glycerol arginine, starch casein, and *Actinomyces* isolation agar; see also Table 11).

Water samples can be directly streaked onto the solid medium after dilution. When low numbers are expected, the samples can be concentrated by membrane filtration (for details, see Burman et al. [1969] or Trolldenier [1967]).

Streptomycetes from marine habitats were successfully isolated on media containing 25 or 75% seawater (Weyland, 1981a; Weyland, 1981b), artificial seawater (Goodfellow and Haynes, 1984), or deionized water supplemented with 3.0% NaCl (Okami and Okazaki, 1978). For further details, see also Weyland (1981b) and Goodfellow and Haynes (1984).

Isolation from Infected Plants

For the isolation of streptomycetes from diseased plant tissue (i.e., from scabby potato or beet surface layers), three general steps have been recommended (see also Korn-Wendisch and Kutzner, 1992a): 1) sterilization of the surfaces of tubers, beets, or roots; 2) maceration of the plant tissues; and 3) use of appropriate media for plating.

Methods for isolating *Streptomyces scabies* from potatoes have been described in detail by several authors (Taylor, 1936; KenKnight and Munzie, 1939; Menzies and Dade, 1959; Adams and Lapwood, 1978; Archuleta and Easton, 1981).

Cultivation

Nutritional Requirements and Media for Sporulation

The vast majority of streptomycetes are nonfastidious organisms, having a chemoorganotrophic metabolism. The nutritional requirements are (in most cases) restricted to an organic carbon source (e.g., starch, glucose, glycerol and lactate) and an inorganic nitrogen source (NH_4^+ or NO_3^-), in addition to the essential mineral salts for growth. The necessity to amend media with specific trace elements has not been studied in detail. Many of the early used media (even the "synthetic" media) were not supplemented with trace elements, although the positive effect of trace elements in soil on the growth of streptomycetes has been reported (Spicher, 1955). Other authors used quite different recipes (Tables 11 and 12), each containing only a select number of metal ions. A rather complete mixture (SPV-4; Table 12) has been found to be optimal for actinomycetes and other bacteria (Voelskow, 1988/1989).

Because no specific requirement for vitamins or organic growth factors has been described, "synthetic media" can be used for their cultivation. Notably, complex organic substrates (e.g., oatmeal, yeast extract, or malt extract) are well utilized and enhance growth rates and biomass production. A combination of a complex organic carbon source with a single amino acid as nitrogen source (e.g., glutamic acid, arginine or asparagine) is also suitable.

Several authors have proposed "general media" for streptomycetes that allow the completion of the streptomyces' life cycle, i.e. germination of spores, growth of substrate and aerial mycelium, and formation of spores (visible because of the typical color of the spores). Some of these media have been used in the International *Streptomyces* Project (ISP). Of the great number of useful media (Pridham et al., 1956/ 1957), four are of particular value and most often used (Table 12). Additional media are listed by Waksman (1961) and by Williams and Cross (1971a). $CaCO_3$ added to some media not only supplies Ca^{2+} for growth but also neutralizes acids produced by many streptomycetes. These media also allow good sporulation. Because macroscopically heavy aerial mycelia may contain very few spores and aerial mycelia hardly detectable by the naked eye may be a good source of spores, it is advisable to check these cultures microscopically. Specialized media, especially for genetic studies, are given by Kieser et al. (2000).

Media Containing Soil, Clay, Minerals and Calcium Humate

Addition of soil to isolation media increases the number of colonies as it promotes growth, sporulation and pigmentation (Trolldenier, 1966). Martin et al. (1976) observed the stimulation of both growth and metabolic activity of some actinomycetes when montmorillonite or Ca-humate was added to a liquid medium. A similar effect was observed for clay in dialysis tubes after a short lag period, which was explained by a possible adsorption of one or more inhibitory substances produced during growth. Martin et al. (1976) also reported a positive effect of these adsorbing materials on the genetic stability of other bacteria and on fungi.

Temperature and Oxygen

As pointed out above, most streptomycetes are mesophilic organisms; however, psychrophilic as well as thermotolerant and thermophilic species are also known. Note that in many instances, the optimum temperature for fast growth or maximal yield may not be the best choice for studying the production of secondary metabolites (e.g., antibiotics and pigments). Most streptomycetes have an obligately aerobic metabolism, but many streptomycetes are able to reduce nitrate to nitrite under strictly anaerobic conditions.

In semisolid agar with a nutrient medium, they grow at the surface of the agar column; however, in semisolid agars with a poor medium or non-utilizable carbon source, they grow microaerophilically. A stationary liquid culture grows as a pellicle at the surface, and the medium remains completely clear.

Cultivation and Preparation of Inoculum

For subcultivation and maintenance, and for most diagnostic tests, streptomycetes are preferably cultivated on solid media in dishes or slants. On solid media many strains produce aerial mycelia and spores when the entire surface is covered by confluent growth. Since *Streptomyces* colonies, in contrast to most molds, spread over a limited distance, a point inoculation will often not lead to a confluent growth.

Some strains show sporulation only when the plates are cross-hatched inoculated, i.e., by a method which leaves empty spaces between the streaks. Sporulation generally occurs better under dry conditions. For this reason, slants should be incubated horizontally for the first two days to allow the liquid to soak into the surface

of the agar (Hopwood et al., 1985). The starting material should be a suspension of inoculum in liquid (Kieser et al., 2000).

The propagation of cultures by successive rounds of mass culture should be avoided, but instead a single colony should be picked and streaked to start the next culture. Especially in genetic studies, this reduces the accumulation of revertants or the gradual loss of selected plasmids or both (Kieser et al., 2000).

Note that morphological heterogeneity is often observed when streptomycetes are cultivated on solid media (for more details, see Kieser et al., 2000).

Because it may be difficult to produce a homogeneous suspension from the grown colonies (a prerequisite for inoculation of some diagnostic tests; Kämpfer et al., 1991b), precultivation in liquid media is sometimes useful. This is the case for certain physiological studies (e.g., degradation tests), for the provision of cell material for biochemical analysis, and for the production of secondary metabolites (e.g., antibiotics) or enzymes. For these purposes, streptomycetes can be cultivated in liquid medium with agitation.

The use of liquid cultures started from an inoculum of spores is also recommended for many detailed studies, e.g., for preparation of protoplasts for fusion, transformation or transfection.

Notably, the multicellular lifestyle of streptomycetes complicates the study of metabolic properties, where all cells of the initial suspension should be in the same physiological condition. In general, streptomycetes grow by mycelial elongation and branching. But when central parts of the colony become nutrient limited, physiological homogeneity cannot be sustained. To overcome these problems, spore germlings are used for physiological studies, even though a large amount of spores is needed. Other solutions include liquid cultures supplemented with dispersants, like sucrose, polyethylene glycol, Junlon®, starch, agar and carboxymethylcellulose. Chapter 2 of Kieser et al. (2000) summarizes the advantages and disadvantages. Since streptomycetes are highly aerobic, the cultures need to be shaken during incubation. Use of Erlenmeyer flasks with indentations or stainless steel springs is recommended, but for small quantities (3–5 ml being enough for some physiological tests), tubes in a slanted position on a shaker or roller also allow excellent supply of oxygen. Note that some secondary metabolites (e.g., antibiotics and pigments, which are produced on solid media) may fail to be synthesized under these conditions.

Korn-Wendisch and Kutzner (1992a) recommend two media that have been widely used for the submerged cultivation of streptomycetes (g/liter): 1) GPYB broth (glucose, 10.0; peptone from casein, 5.0; yeast extract, 5.0; beef extract,

5.0; $CaCl_2 \cdot 2H_2O$, 0.74; pH 7.2) and 2) soybean meal-mannitol nutrient medium (soybean meal, 20.0; mannitol, 20.0; pH 7.2). Two kinds of noculation material can be employed for subculturing streptomycetes: 1) arthrospores and 2) vegetative mycelium, occasionally including "submerged spores" (Wilkin and Rhodes, 1955). Kieser et al. (2000) recommend similar procedures.

Spore suspensions can be used over a period of several weeks when stored at 4°C, but since the spores tend to settle and clump, the addition of a few glass beads to the screw-cap tube helps to resuspend the spores before use. Chapters 8 and 9 of Kieser et al. (2000) describe the preparation of mycelia for detailed DNA or RNA studies.

Preservation

Several different procedures have been employed for the short- and long-term preservation of microorganisms (Kirsop and Snell, 1984).

Three short-term preservation methods have been described by Korn-Wendisch and Kutzner (1992a). First, agar slope cultures may be stored at 4°C for few months. Second, spore suspensions can be mixed with soft water agar and kept at 4°C (Kutzner, 1972). And third, glycerol can be added to spore suspensions (final concentration, 10%, v/v) and stored at –20°C (Wellington and Williams, 1978); these cultures (after thawing) can serve as inoculum for most diagnostic tests except carbon utilization (Williams et al., 1983a). Kieser et al. (2000) recommended for long-term preservation the preparation of a spore suspension in 20% glycerol and freezing at –20°C. Another procedure is the growth of strains in complex media (like trypticase soy broth [TSB] agar), addition of 20% glycerol plus 10% lactose, and storage in the vapor phase of liquid nitrogen. In addition, drying on unglazed porcelain beads (Lange and Boyd, 1968), followed by soil culture (Pridham et al., 1973), and lyophilization (Hopwood and Ferguson, 1969) are used. For longer preservation (see The Family Nocardiopsaceae in this Volume), spore suspensions or homogenized mycelia are mixed with glycerol to a final concentration of 25% and kept at 25°C (Wellington and Williams, 1978). Alternatively, spores and mycelia suspended in 10% skim milk are lyophilized. A very simple, reliable, and time-saving method is liquid nitrogen cryopreservation of living cells in small polyvinyl chloride (PVC) tubes ("straws") at –196°C. The procedure has been tested for various actinomycetes. The strains are harvested from well-sporulated cultures grown on suitable agar media in Petri dishes. A 2×25-mm piece of sterile PVC tubing is pressed into the mycelial mat and agar and carefully raised to excise the agar plug. This is repeated until the tube is filled with agar. The filled tubes are placed in a sterile cryovial (the screw cap marked with the strain accession number). A 1.8-ml vial will hold up to 13 tubes. Two vials are prepared for each strain and then fixed to a metal clamp for freezing in the gas phase of a liquid nitrogen container. After 10–15 min, when temperature falls below –130°C, the clamp can be immersed in the liquid phase at –196°C. A container with a capacity of 250 liters will hold at least 8000 vials or 4000 strains. For viability testing, one tube is removed from the vial within the nitrogen gas atmosphere of the container and placed directly and thawed on a suitable agar medium. After a few days incubation, the mycelium will be visible. For those strains that do not produce an abundant mycelium, the plugs may be pushed out of the tubes by a sterile needle.

Detection of *Streptomyces* Using Cultivation-independent Methods

Microorganisms in the environment can be identified without cultivation by retrieving and sequencing macromolecules and using oligonucleotide probes (largely based on small subunit rRNA; Amann et al., 1995; Rappé and Giovannoni, 2003). Stackebrandt et al. (1991b) developed 16S rRNA-targeted oligonucleotide probes specific for certain *Streptomyces* species and subsequently studied bacterial diversity in a soil sample from a subtropical Australian environment (Stackebrandt et al., 1993). They found that most sequences were from alpha subclass Proteobacteria and only a few from streptomycetes. Hahn et al. (1992), using the in situ hybridization approach, were unable to analyze bacterial populations in soil without prior activation by adding nutrients. Growing cells, e.g., *Streptomyces scabies* hyphae, were easily detected. The use of specific primers in connection with environmental clone libraries is a powerful approach for study of the microbial diversity in soil (Felske et al., 1997; Rheims et al., 1999; Rintala et al., 2001; Courtois et al., 2003) and discovery of novel bioactive metabolites (Donadio et al., 2002). More details are given in The Family Streptomycetaceae, Part II: Molecular Biology in this Volume.

Acknowledgments. The basis of the sections on ecophysiology, isolation and habitats has been the excellent and comprehensive treatise of Korn-Wendisch and Kutzner (1992a) from the second edition of *The Prokaryotes*, which is still recommended for deeper study of classical approaches in *Streptomyces* biology.

Literature Cited

Adams, M. J., and D. H. Lapwood. 1978. Studies on the lenticel development, surface microflora and infection by common scab (Streptomyces scabies) of potato tubers growing in wet and dry soils. Ann. Appl. Biol. 90:335–343.

Alderson, G., M. Goodfellow, and D. E. Minnikin. 1985. Menaquinone composition in the classification of Streptomyces and other sporoactinomycetes. J. Gen. Microbiol. 131:1671–1679.

Al-Diwany, L. J., and T. Cross. 1978. Ecological studies on nocardioforms and other actinomycetes in aquatic habitats. In: M. Mordarski, W. Kurylowicz, and J. Jeljaszewicz (Eds.), Nocardia and Streptomyces: Proceedings of the International Symposium on Nocardia and Streptomyces, Warsaw, 1976. Gustav Fischer-Verlag. Stuttgart, Germany. 153–160.

Amann, R. I, W. Ludwig, and K.-H. Schleifer. 1995. Phylogenetic identification and in situ detection of individual microbial cells without cultivation. Microbiol. Rev. 59:143–169.

Anderson, A. S., and E. M. H. Wellington. 2001. The taxonomy of Streptomyces and related genera. Int. J. Syst. Evol. Microbiol. 51:797–814.

Antai, S. P., and D. L. Crawford. 1981. Degradation of softwood, hardwood, and grass lignocelluloses by two Streptomyces strains. Appl. Environ. Microbiol. 42:378–380.

Anzai, Y., T. Okuda, and J. Watanabe. 1994. Application of the random amplified polymorphic DNA using the polymerase chain reaction for accient elimination of duplicate strains in microbial screening. II: Actinomycetes. J. Antibiot. 47:183–193.

Archuleta, J. G., and G. D. Easton. 1981. The cause of deep-pitted scab of potatoes. Am. Potato J. 58:385–392.

Becker, B., M. P. Lechevalier, R. E. Gordon, and H. A. Lechevalier. 1964. Rapid differentiation between Nocardia and Streptomyces by paper chromatography of whole-cell hydrolysates. Appl. Microbiol. 12:421–423.

Behal, V. 2000. Bioactive products from Streptomyces. Adv. App. Microbiol. 47:113–156.

Benedict, R. G., T. G. Pridham, L. A. Lindenfelser, H. H. Hall, and R. W. Jackson. 1955. Further studies in the evaluation of carbohydrate utilization tests as aids in the differentiation of species of Streptomyces. Appl. Microbiol. 3:1–6.

Bentley, S. D, K. F. Chater, A. M. Cerdeno-Tarraga, G. L. Challis, J. K. D. Thomson, D. E. Harris, M. A. Quail, H. Kieser, D. Harper, A. Bateman, S. Brown, G. Chandra, C. W. Chen, M. Collins, A. Cronin, A. Fraser, A. Gob, J. Hidalgo, T. Hornsby, S. Howarth, C. H. Huang, T. Kieser, L. Larke, L. Murphy, K. Oliver, S. O'Neil, E. Rabbinowitsch, M. A. Rajandream, K. Rutherford, S. Rutter, K. Seeger, D. Saunders, S. Sharp, R. Squares, S. Squares, K. Taylor, T. Warren, A. Wietzorrek, J. Woodward, J. Barrell Parkhill, and D. A. Hopwood. 2002. Complete genome sequence of the model actinomycete Streptomyces coelicolor A3(2). Nature 417:141–147.

Bentley, S. D., N. R. Thomson, S. Mohammed, L. C. Crossman, and J. Parkhill. 2003. The devil is in the detail. Trends Microbiol. 11:256–258.

Berger, D. R., and D. M. Reynolds. 1958. The chitinase system of a strain Streptomyces griseus. Biochim. Biophys. Acta 29:522–534.

Beyazova, M., and M. P. Lechevalier. 1993. Taxonomic utility of restriction endonuclease fingerprinting of large DNA fragments from Streptomyces strains. Int. J. Syst. Bacteriol. 43:674–682.

Beyer, M., and D. Diekmann. 1985. The chitinase system of Streptomyces sp. ATCC 11238 and its significance for fungal cell wall degradation. Appl. Microbiol. Biotechnol. 23:140–146.

Bignell, D. E., H. Oskarsson, and J. M. Anderson. 1980. Colonization of the epithelial face of the peritrophic membrane and the ectoperitrophic space by actinomycetes in a soil-feeding termite. J. Invertebr. Pathol. 36:426–428.

Bignell, D. E., H. Oskarsson, and J. M. Anderson. 1981. Association of actinomycetes with soil-feeding termites: a novel symbiotic relationship? In: K. P. Schaal and G. Pulverer (Eds.), Actinomycetes: Proceedings of the 4th International Symposium on Actinomycete Biology, Cologne, 1979. Gustav Fischer-Verlag. Stuttgart, Germany. 201–206.

Bignell, D. E. 1984. The arthropod gut as an environment for microorganisms. In: J. M. Anderson, A. D. M. Rayner, and D. W. H. Walton (Eds.), Invertebrate-Microbial Interactions. Cambridge University Press. Cambridge, UK. 205–227.

Bouchek-Mechine, K., L. Gardan, P. Normand, and B. Jouan. 2000. DNA relatedness among strains of Streptomyces pahtogenic to potato in France: description of three new species, S. europaeiscabici sp. nov. and S. stelliscabiei sp. nov. associated with common scab, and S. reticuliscabiei sp. nov. associated with netted scab. Int. J. Syst. Evol. Microbiol. 50:91–99.

Bowen, T., E. Stackebrandt, M. Dorsch, and T. M. Embley. 1989. The phylogeny of Amycolata autotrophica, Kibdelosporangium aridum and Saccharothrix australiensis. J. Gen. Microbiol. 135:2529–2536.

Brian, P. W. 1957. The ecological significance of antibiotic production. In: R. E. O. Williams and C. C. Spicer (Eds.), Microbial Ecology. Cambridge University Press. Cambridge, UK. 168–188.

Brown, R. L., and G. E. Peterson. 1966. Cholesterol oxidation by soil actinomycetes. J. Gen. Microbiol. 45:441–450.

Brüsewitz, G. 1959. Untersuchungen über den Einfluss des Regenwurms auf Zahl, Art und Leistungen von Mikroorganismen im Boden. Arch. Microbiol. 33:52–82.

Burman, N. P., C. W. Oliver, and J. K. Stevens. 1969. Membrane filtration techniques for the isolation from water, of coli-aerogenes, Escherichia coli, faecal streptococci, Clostridium perfringens, actinomycetes and microfungi. In: D. A. Shapton and G. W. Gould (Eds.), Isolation Methods for Microbiologists. Academic Press. London, UK. Society for Applied Bacteriology Technical Series No. 3::127–134.

Burman, N. P. 1973. The occurrence and significance of actinomycetes in water supply. In: G. Sykes, and F. A. Skinner (Eds.), Actinomycetales: Characteristics and Practical Importance. Academic Press. London, UK. 219–230.

Carvajal, F. 1953. Phage problems in the streptomycin fermentation. Mycologia 45:209–234.

Chamberlain, K., and D. L. Crawford. 2000. Thatch biodregradation and antifungal activities of two lignocellulolytic Streptomyces strains in laboratory cultures and in golf green turfgrass. Can. J. Microbiol. 46:550–558.

Chandramohan, D., S. Ramu, and R. Natarajan. 1972. Cellulolytic activity of marine streptomycetes. Curr. Sci. 41:245–246.

Chater, K. F. 1986a. Streptomyces phages and their applications of Streptomyces genetics. *In:* St. W. Queener and L. E. Day (Eds.), The Bacteria, Volume 9: Antibiotic-producing Streptomyces. Academic Press. Orlando, FL. 119–158.

Chater, K. F., N. D. Lomovskaya, T. A. Voeykova, I. A. Sladkova, N. M. Mkrtumian, and G. L. Muravnik. 1986b. Streptomyces ÈC31-like phages: cloning vectors, genome changes and host range. *In:* G. Szabo, S. Biro, and M. Goodfellow (Eds.), Biological, Biochemical and Biomedical Aspects of Actinomycetes. Akademiai Kiado. Budapest, Hungary. 45–54.

Chesters, C. G. C., A. Apinis, and M. Turner. 1956. Studies of the decomposition of seaweeds and seaweed products by microorganisms. Proc. Linn. Soc. Lond. 166:87–97.

Chung, Y. R., K. C. Sung, H. K. Mo, D. Y. Son, J. S. Nam, J. Chun, and K. S. Bae. 1999. Kitasatospora cheerisanensis sp. nov., a new species of the genus Kitasatospora that produces an antifungal agent. Int. J. Syst. Bacteriol. 49:753–758.

Clarke, S. D., D. A. Ritchie, and S. T. Williams. 1993. Ribosomal DNA restriction fragment analysis of some closely related Streptomyces species. Syst. Appl. Microbiol. 16:256–260.

Cochrane, V. W. 1961. Physiology of Actinomycetes. Ann. Rev. Microbiol. 15:1–26.

Collins, M. D., and D. Jones. 1981. Distribution of isoprenoid quinone structural types in bacteria and their taxonomic implications. Microb. Rev. 45:316–345.

Corbaz, R., P. H. Gregory, and M. E. Lacey. 1963. Thermophilic and mesophilic actinomycetes in mouldy hay. J. Gen. Microbiol. 32:449–456.

Corke, C. T., and F. E. Chase. 1956. The selective enumeration of actinomycetes in the presence of large numbers of fungi. Can. J. Microbiol. 2:12–16.

Courtois, S., C. M. Cappellano, M. Ball, F.-X. Francou, P. Normand, G. Helynck, A. Martinez, S. J. Kolvek, J. Hopke, M. S. Osburne, P. R. August, R. Nalin, M. Guerineau, P. Jeannin, P. Simonet, and J. L. Pernodet. 2003. Recombinant environmental libraries provide access to microbial diversity for drug discovery from natural products. Appl. Environ. Microbiol. 69:49–55.

Craveri, R., and H. Pagani. 1962. Thermophilic microorganisms among actinomycetes in the soil. Annali di Microbiologia 12:115–130.

Crawford, D. L., and E. McCoy. 1972. Cellulases of Thermomonospora fusca and Streptomyces thermodiastaticus. Appl. Microbiol. 24:150–152.

Crawford, D. L. 1978. Lignocellulose decomposition by selected Streptomyces strains. Appl. Environ. Microbiol. 35:1041–1045.

Crawford, R. L. 1981. Lignin Biodegradation and Transformation. John Wiley. New York, NY.

Crawford, D. L. 1988. Biodegradation of agricultural and urban wastes. *In:* M. Goodfellow, S. T. Williams, and M. Mordarski (Eds.), Actinomycetes in Biotechnology. Academic Press. London, UK. 433–459.

Crawford, D. L., J. D. Doyle, Z. Wang, C. W. Hendricks, S. A. Bentjen Bolton Jr., H., J. K. Fredrickson, and B. H. Bleakley. 1993. Effects of lignin peroxidase-expressing recombinant, Streptomyces lividans TK23.1, on biogeochemical cycling and the numbers and activities of microorganisms in soil. Appl. Environ. Microbiol. 59:508–518.

Crook, P., C. C. Carpenter, and P. F. Klens. 1950. The use of sodium propionate in isolating actinomycetes from soils. Science 111:656.

Cross, T. 1968. Thermophilic actinomycetes. J. Appl. Bacteriol. 31:36–53.

Cross, T., and M. Goodfellow. 1973. Taxonomy and classification of the actinomycetes. *In:* G. Sykes and F. A. Skinner (Eds.), Actinomycetales: Characteristics and Practical Importance. Academic Press. London, UK. 11–112.

Cross, T. 1981. Aquatic actinomycetes: A critical survey of the occurrence, growth and role of actinomycetes in aquatic habitats. J. Appl. Bacteriol. 50:397–423.

Cross, T. 1982. Actinomycetes: A continuing source of new metabolites. Devel. Indust. Microbiol. 23:1–18.

Cundell, A. M., and A. P. Mulcock. 1975. The biodegradation of vulcanized rubber. Devel. Indust. Microbiol. 16:88–96.

Davies, F. L., and S. T. Williams. 1970. Studies on the ecology of actinomycetes in soil. I: The occurrence and distribution of actinomycetes in a pine forest soil. Soil Biol. Biochem. 2:227–238.

Delafield, F. P., M. Doudoroff, N. J. Palleroni, C. J. Lusty, and R. Contolpoulos. 1965. Decomposition of polyhydroxybutyrate by pseudomonads. J. Bacteriol. 90:1455–1466.

Deobald, L. A., and D. L. Crawford. 1987. Activities of cellulase and other extracellular enzymes during lignin solubilization by Streptomyces viridosporus. Appl. Microbiol. Biotechnol. 26:158–163.

Donadio, S., P. Monciardini, R. Alduina, P. Mazza, C. Chiocchini, L. Cavaletti, M. Sosio, and A. M. Puglia. 2002. Microbial technologies for the discovery of novel bioactive metabolites. J. Biotechnol. 99:187–198.

El-Nakeeb, M. A., and H. A. Lechevalier. 1963. Selective isolation of aerobic actinomycetes. Appl. Microbiol. 11:75–77.

Enger, M. D., and B. P. Sleeper. 1965. Multiple cellulase system from Streptomyces antibioticus. J. Bacteriol. 89:23–27.

Ensign, J. C. 1978. Formation, properties, and germination of actinomycete spores. Ann. Rev. Microbiol. 32:185–219.

Fairbairn, D. A., F. G. Priest, and J. R. Stark. 1986. Extracellular amylase synthesis by Streptomyces limosus. Enz. Microb. Technol. 8:89–92.

Felske, A., H. Rheims, A. Wolterink, E. Stackebrandt, and A. D. Akkermans. 1997. Ribosome analysis reveals prominent activity of an uncultured member of the class Actinobacteria in grassland soils. Microbiology 143:2983–2989.

Fergus, C. L. 1964. Thermophilic and thermotolerant molds and actinomycetes of mushroom compost during peak heating. Mycologia 56:267–284.

Festenstein, G. N., J. Lacey, F. A. Skinner, P. A. Jenkins, and J. Pepys. 1965. Self-heating of hay and grain in Dewar flasks and the development of farmer's lung antigen. J. Gen. Microbiol. 41:389–407.

Flaig, W., and H. J. Kutzner. 1960a. Beitrag zur Ökologie der Gattung Streptomyces Waksman et Henrici. Arch. Mikrobiol. 35:207–228.

Flaig, W., and H. J. Kutzner. 1960b. Beitrag zur Systematik der Gattung Streptomyces Waksman et Henrici. Arch. Mikrobiol. 35:105–138.

Flowers, T. H., and S. T. Williams. 1977. The influence of pH on the growth rate and viability of neutrophilic and acidophilic streptomycetes. Microbios 18:223–228.

Fulton, T. R., M. C. Losada, E. M. Fluder, and G. T. Chou. 1995. Ribosomal-RNA operon restriction derived taxa for streptomycetes (RIDITS). FEMS Microbiol. Lett. 125:149–158.

Genner, C., and E. C. Hill. 1981. Fuels and oils. *In:* A. H. Rose (Ed.), Economic Microbiology, Volume 6: Microbial Biodeterioration. Academic Press. London, UK. 259–306.

Gerber, N. N. 1979a. Odorous substances from actinomycetes. Devel. Indust. Microbiol. 20:225–238.

Gerber, N. N. 1979b. Volatile substances from actinomycetes: Their role in the odor pollution of water. Crit. Rev. Microbiol. 9:191–214.

Godden, B., T. Legon, P. Helvenstein, M. Penninckx. 1989. Regulation of the production of hemicellulotyic and cellulolytic enzymes by a Steptomyces sp. growing on lignocellulose. J. Gen. Microbiol. 135:285–292.

Goodfellow, M., and S. T. Williams. 1983. Ecology of actinomycetes. Ann. Rev. Microbiol. 37:189–216.

Goodfellow, M., and J. A. Haynes. 1984. Actinomycetes in marine sediments. *In:* L. Ortiz-Ortiz, L. F. Bojalil, and V. Yakoleff (Eds.), Biological, Biochemical and Biomedical Aspects of Actinomycetes: Proceedings of the 5th International. Symposium on Actinomycetes Biology, Oaxtepec, Mexico, 1982. Academic Press. Orlando, FL. 453–472.

Goodfellow, M., and C. H. Dickenson. 1985. Delineation and description of microbial populations using numerical methods. *In:* M. Goodfellow, D. Jones, and F. G. Priest (Eds.), Computer-assisted Bacterial Systematics. Academic Press. London, UK. 165–226.

Goodfellow, M., S. T. Williams, and G. Alderson. 1986a. Transfer of Elytrosporangium brasiliense Falcao de Morais et al., Elytrosporangium carpinense Falcao de Morais et al., Elytrosporangium spirale Falcao de Morais et al., Microellobosporia cinerea Cross et al., Microellobosporia flavea Cross et al., Microellobosporia grisea (Konev et al.) Pridham and Microellobosporia violacea (Tsyganov et al.) Pridham to the genus Streptomyces, with emended description of the species. Syst. Appl. Microbiol. 8:48–54.

Goodfellow, M., S. T. Williams, and G. Alderson. 1986b. Transfer of Actinosporangium violaceum Krasil'nikov and Yuan, Actinosporangium vitaminophilumI Shomura et al., and Actinopycnidium caeruleum Krasil'nikov to the genus Streptomyces, with amended descriptions of the species. Syst. Appl. Microbiol. 8:61–64.

Goodfellow, M., S. T. Williams, and G. Alderson. 1986c. Transfer of Chainia species to the genus Streptomyces with emended description of species. Syst. Appl. Microbiol. 8:55–60.

Goodfellow, M., S. T. Williams, and G. Alderson. 1986d. Transfer of Elytrosporangium brasiliense Falcao de Morais et al., Elytrosporangium carpinense Falcao de Morais et al., Elytrosporangium spirale Falcao de Morais et al., Microellobosporia cinerea Cross et al., Microellobosporia Øavea Cross et al., Microellobosporia grisea (Konev et al.) Pridham and Microellobosporia violacea (Tsyganov et al.) Pridham to the genus Streptomyces, with emended descriptions of the species. Syst. Appl. Microbiol. In: A. S. Anderson and E. M. H. Wellington (Ed.), International Journal of Systematic and Evolutionary Microbiology. vol. 51, pp. 811 8:48–54.

Goodfellow, M., S. T. Williams, and G. Alderson. 1986e. Transfer of Kitasatoa purpurea Matsumae and Hata to the genus Streptomyces as Streptomyces purpureus comb. nov. Syst. Appl. Microbiol. 8:65–66.

Goodfellow, M., and K. E. Simpson. 1987a. Ecology of streptomycetes. Front. Appl. Microbiol. 2:97–125.

Goodfellow, M., J. Lacey, and C. Todd. 1987b. Numerical classification of thermophilic streptomycetes. J. Gen. Microbiol. 133:3135–3149.

Goodfellow, M., C. Lonsdale, A. L. James, and O. C. MacNamara. 1987c. Rapid biochemical tests for the characterisation of streptomycetes. FEMS Microbiol. Lett. 43:39–44.

Goodfellow, M. 1989. The actinomycetes I: Suprageneric classification of actinomycetes. *In:* S. T. Williams, M. E. Sharpe, and J. G. Holt (Eds.), Bergey's Manual of Systematic Bacteriology. Williams and Wilkins. Baltimore, MD. 4:2333–2339.

Goodfellow, M., and A. G. O'Donnell. 1993. Roots of bacterial systematics. *In:* M. Goodfellow and A. G. O'Donnell (Eds.), Handbook of New Bacterial Systematics. Academic Press. London, UK. 3–56.

Goodfellow, M., K. Isik, and E. Yates. 1999. Actinomycete systematics: An unfinished synthesis. Nova Acta Leopoldina 80:47–82.

Gottschalk, L. M., R. Nobrega, and E. P. Bon. 2003. Effect of aeration on lignin peroxidase production by Streptomyces viridosporus T7A. Appl. Biochem. Biotechnol. 105:799–807.

Gregory, P. H., and M. E. Lacey. 1963. Mycological examination of dust from mouldy hay associated with farmer's lung disease. J. Gen. Microbiol. 30:75–88.

Greiner-Mai, E., R. M. Kroppenstedt, F. Korn-Wendisch, and H. J. Kutzner. 1987. Morphological and biochemical characterization and emended descriptions of thermophilic actinomycetes species. Syst. Appl. Microbiol. 9:97–106.

Gyllenberg, H. G. 1976. Application of automation to the identification of streptomycetes. *In:* T. Arai (Ed.), Actinomycetes: The Boundary Microorganisms. Toppan. Tokyo, Japan. 299–321.

Hagedorn, C. 1976. Influences of soil acidity on Streptomyces populations inhabiting forest soils. Appl. Environ. Microbiol. 32:368–375.

Hahn, D., R. I. Amann, W. Ludwig, A. D. Akkermanns, and K.-H. Schleifer. 1992. Detection of micro-organisms in soil after in situ hybridization with rRNA-targeted, fluorescently labelled oligonucleotides. Gen. Microbiol. 138:879–887.

Hain, T., N. Ward-Rainey, R. M. Kroppenstedt, E. Stackebrandt, and F. A. Rainey. 1997. Discrimination of Streptomyces albidoflavus strains based on the size and number of 16S-23S ribosomal DNA intergenic spacers. Int. J. Syst. Bacteriol. 47:202–206.

Hanka, L. J., P. W. Rueckert, and T. Cross. 1985. A method for isolating strains of the genus Streptoverticillium from soil. FEMS Microbiol. Lett. 30:365–368.

Hanka, L. J., and R. D. Schaadt. 1988. Methods for isolation of Streptoverticillia from soils. J. Antibiot. 41:576–578.

Harchand, R. K., and S. Singh. 1997. Extracellular cellulase system of a thermotolerant streptomycete: Streptomyces albaduncus. Acta Microiol. Immunol. Hung. 44:229–239.

Hatano, K., T. Nishii, and H. Kasai. 2003. Taxonomic re-evaluation of whorl-forming Streptomyces (formerly Steptoverticillium) species by using phenotypes, DNA-DNA hybridization and sequences of gyrB, and proposal of Streptomyces luteireticuli (ex Katoh and Arai 1957) corrig., sp. nov, nom. rev. Int. J. Syst. Evol. Microbiol. 53:1519–1529.

Hayakawa, M., and H. Nonomura. 1987a. Efficacy of artificial humic acid as a selective nutrient in HV agar used for the isolation of soil actinomycetes. J. Ferment. Technol. 65:609–616.

Hayakawa, M., and H. Nonomura. 1987b. Humic acid-vitamin agar, a new medium for the selective isolation of soil actinomycetes. J. Ferment. Technol. 65: 501–509.

Heitzer, R. 1981. Numerische Taxonomie der Actinomyceten-Gattungen Streptomyces und Streptoverticillium [PhD dissertation]. TH Darmstadt. Darmstadt, Germany.

Held, T. 1990. Regulation und Genetik des Tyrosinstoffwechsels von Streptomyces michiganensis DSM 40 015 [PhD dissertation]. TH Darmstadt. Darmstadt, Germany.

Helmke, E. 1981. Growth of actinomycetes from marine and terrestrial origin under increased hydrostatic pressure. In: K. P. Schaal and G. Pulverer (Eds.), Actinomycetes: Proceedings of the 4th International Symposium on Actinomycete Biology, Cologne, 1979. Gustav Fischer-Verlag. Stuttgart, Germany. 321–327.

Herron, P., and E. M. H. Wellington. 1990. New method for the extraction of streptomycete spores from soil and application to the study of lysogeny in sterile amended and nonsterile soil. Appl. Environ. Microbiol. 56:1406–1412.

Heuer, H., M. Krsek, P. Baker, K. Smalla, and E. M. H. Wellington. 1997. Analysis of actinomycete communities by specific amplifiation of genes encoding 16S rRNA and gel-electrophoretic separation in denaturing gradients. Appl. Environ. Microbiol. 63:3233–3241.

Hill, L. R., and L. G. Silvestri. 1962. Quantitative methods in the systematics of Actinomycetales. III. The taxonomic significance of physiological-biochemical characters and the construction of a diagnostic key. Giorn. Microbiol. 10:1–28.

Hirsch, C. F., and D. L. Christensen. 1983. Novel method for selective isolation of actinomycetes. Appl. Environ. Microbiol. 46:925–929.

Hofheinz, W., and H. Grisebach. 1965. Die Fettsäuren von Streptomyces erythreus and Streptomyces halstedii. Z. Naturforsch. 20B:43.

Hopwood, D. A., and H. M. Ferguson. 1969. A rapid method for lyophilizing Streptomyces cultures. J. Appl. Bacteriol. 32:434–436.

Hopwood, D. A., M. J. Bibb, K. F. Chater, T. Kieser, C. J. Bruton, H. M. Kieser, D. J. Lydiate, C. P. Smith, J. M. Ward, and H. Schrempf. 1985a. Genetic Manipulation of Streptomyces: A Laboratory Manual. The John Innes Foundation. Norwich, UK.

Hori, H., and S. Osawa. 1987. The rates of evolution in some ribosomal components. J. Molec. Evol. 9:191–201.

Hotta, K., N. Saito, and Y. Okami. 1980. Studies on new aminoglycoside antibiotics, istamycins, from an actinomycete isolated from a marine environment. I: The use of plasmid profiles in screening antibiotic-producing streptomycetes. J. Antibiot. (Tokyo) 33:1502–1509.

Howarth, O. W., E. Grund, R. M. Kroppenstedt, and M. D. Collins. 1986. Structural determination of a new naturally occurring cyclic vitamin K. Biochem. Biophys. Res. Commun. 140:916–923.

Hsu, S. C., and J. L. Lockwood. 1975. Powdered chitin agar as a selective medium for enumeration of actinomycetes in water and soil. Appl. Microbiol. 29:422–426.

Huddleston, A. S., J. L. Hinks, M. Beyazova, A. Horan, D. I. Thomas, S. Baumberg, and E. M. H. Wellington. 1995. Studies on the diversity of streptomycin-producing streptomycetes. Biotekhnologia 7+8:242–253.

Huddleston, A. S., N. Cresswell, M. C. P. Neves, J. E. Beringer, S. Baumberg, D. I. Thomas, and E. M. H. Wellington. 1997. Molecular detection of streptomycin-producing streptomycetes in Brazilian soils. Appl. Environ. Microbiol. 63:1288–1297.

Humm, J. H., and K. S. Shepard. 1946. Three new agar-digesting actinomycetes. Duke Univ. Mar. Sta. Bull. 3:76–80.

Hutchinson, M., J. W. Ridgway, and T. Cross. 1975. Biodeterioration of rubber in contact with water, sewage and soil. In: D. W. Lovelock and R. A. Gilbert (Eds.), Microbial Aspects of Deterioration of Materials. Academic Press. London, UK. 187–202.

Hütter, R. 1962. Zur Systematik der Actinomyceten. 8: Quirlbildende Streptomyceten. Arch. Microbiol. 43:365–391.

Ikeda, H., E. T. Seno, C. J. Bruton, and K. F. Charter. 1984. Genetic mapping, clonino, and physiological aspects of the glucose kinase gene of Streptomyces coelicolor. Molec. Gen. Genet. 196:501–507.

Ikeda, H, J. Ishikawa, A. Hanamoto, M. Shinose, H. Kikuchi, T. Shiba, Y. Sakaki, M. Hattori, and S. Omura. 2003. Complete genome sequence and comparative analysis of the industrial microorganism Streptomyces avermitilis. Nature Biotechnol. 21:526–531.

Itoh, T., T. Kudo, F. Parenti, and A. Seino. 1989. Amended description of the genus Kineosporia, based on chemotaxonomic and morphological studies. Int. J. Syst. Bacteriol. 39:168–173.

Jagnow, G. 1957. Beiträge zur Ökologie der Streptomyceten. Arch. Mikrobiol. 26:175–191.

Janshekar, H., and A. Fiechter. 1983. Lignin: Biosynthesis, application and biodegradation. Adv. Biochem. Engin. Biotechnol. 27:120–178.

Jensen, H. L. 1930. Decomposition of keratin by soil microorganisms. J. Agricult. Sci. 20:390–398.

Jeuniaux, C. 1966. Chitinases. Meth. Enzymol. 8:644–650.

Kämpfer, P., and R. M. Kroppenstedt. 1991a. Probabilistic identification of streptomycetes using miniaturized physiological tests. J. Gen. Microbiol. 137:1892–1902.

Kämpfer, P., R. M. Kroppenstedt, and W. Dott. 1991b. A numerical classification of the genera Streptomyces and Streptoverticillium using miniaturized physiological tests. J. Gen. Microbiol. 137:1831–1891.

Kämpfer, P., and D. P. Labeda. 2003. International Committee on Systematics of Prokaryotes, Subcommittee on the Taxonomy of Streptomycetaceae: Minutes of the Meeting, 30 July 2002, Paris, France. Int. J. Syst. Evol. Microbiol. 53:925.

Kaneko, M., Y. Ohnishi, and S. Horinouchi. 2003. Cinnamate: Coenzyme A ligase from the filamentous bacterium streptomyces coelicolor A3(2). J. Bacteriol. 185:20–27.

Kataoka, M., K. Ueda, T. Kudo, T. Seki, and T. Yoshida. 1997. Application of the variable region in 16S rDNA to create an index for rapid species identification in the genus Streptomyces. FEMS Microbiol. Lett. 151:249–255.

KenKnight, G., and J. H. Munzie. 1939. Isolation of phytopathogenic actinomycetes. Phytopathology 29:1000–1001.

Khan, M. R., and S. T. Williams. 1975. Studies on the ecology of actinomycetes in soil. 8: Distribution and characteristics of acidophilic actinomycetes. Soil Biol. Biochem. 7:345–348.

Khan, M. R., S. T. Williams, and M. L. Saha. 1978. Studies on the microbial degradation of jute. Bangladesh J. Jute Fibre Res. 3:45–52.

Kieser, T., M. J. Bibb, M. J. Buttner, K. F. Chater, and D. A. Hopwood. 2000. Practical Streptomyces Genetics. The John Innes Foundation. Norwich, UK.

Kim, D., Chun, J., Sahin, N., Hah, Y.-C. and Goodfellow, M. 1996. Analysis of thermophilic clades within the genus Streptomyces by 16S ribosomal DNA sequence comparisons. Int. J. Syst. Bacteriol. 46:581–587.

Kim, S. B., Falconer, C., Williams, E. and Goodfellow, M. 1998. Streptomyces thermocarboxydovorans sp. nov. and Streptomyces thermocarboxydus sp. nov., two moderately thermophilic carboxydotrophic species from soil. Int. J. Syst. Bacteriol. 48:56–68.

Kim, B., Sahin, N., Minnikin, D. E., Zakrzewska-Czerwinska, J., Mordarski, M., and M. Goodfellow. 1999. Classification of thermophilic streptomycetes, including the description of Streptomyces thermoalcalitolerans sp. nov. Int. J. Syst. Bacteriol. 49:7–17.

Kim, B., A. M. Al-Tai, S. B. Kim, P. Somasundaram, and M. Goodfellow. 2000. Streptomyces thermocoprophilus sp. nov., a cellulase-free endo-xylanase-producing streptomycete. Int. J. Syst. Evol. Microbiol. 50:505–509.

Kim, S. B., and M. Goodfellow. 2002a. Streptomyces avermitilis sp. nov. rev., a taxonomic home for the avermectin-producing streptomycetes. Int. J. Syst. Evol. Microbiol. 52:211–2014.

Kim, S. B., and M. Goodfellow. 2002b. Streptomyces thermospinisporus sp. nov., a moderately thermophilic carboxydotrophic streptomycete isolated from soil. Int. J. Syst. Evol. Microbiol. 52:1225–1228.

Kim, B. S., J. Lonsdale, C.-N. Seong, and M. Goodfellow. 2003. Streptacidiphilus gen. nov., acidophilic actinomycetes with wall chemotype I and emendation of the family Streptomycetaceae (Waksman and Henrici (1943)[AL]) emend. Rainey et al., 1997. Ant. v. Leeuwenhoek 83:107–116.

Kirby, R., and E. P. Rybicki. 1986. Enzyme-linked immunosorbent assay (ELISA) as a means of taxonomic analysis of Streptomyces and related organisms. J. Gen. Microbiol. 132:1891–1894.

Kirk, T. K., and R. L. Farell. 1987. Enzymatic combustion: The microbial degradation of lignin. Ann. Rev. Microbiol. 41:465–505.

Kirsop, B. E., and J. J. S. Snell (Eds.). 1984. Maintenance of Microorganisms. Academic Press. London, UK.

Kluepfel, D., and M. Ishaque. 1982. Xylan-induced cellulolytic enzymes in Streptomyces flavogriseus. Devel. Indust. Microbiol. 23:389–395.

Kluepfel, D., F. Shareck, F. Mondou, and R. Morosoli. 1986. Characterization of cellulase and xylanase activities of Streptomyces lividans. Appl. Microbiol. Biotechnol. 24:230–234.

Kormanec, J., R. Novakova, D. Hamerova, and B. Rezuchova. 2001. Streptomyces aureofaciens sporulation-specific sigma factor sigma (rpoZ) directs expression of a gene encoding protein similar to hydrolases involved in degradation of the lignin-related biphenyl compounds. Res. Microbiol. 152:883–888.

Korn, F., B. Weingärtner, and H. J. Kutzner. 1978. A study of twenty actinophages: Morphology, serological relationship and host range. In: E. Freerksen, I. Tarnok, and J. Thumin (Eds.), Genetics of the Actinomycetales. Gustav Fischer-Verlag. Stuttgart, Germany. 251–270.

Korn-Wendisch, F. 1982. Phagentypisierung und Lysogenie bei Actinomyceten [Ph. D. dissertation]. TH Darmstadt. Darmstadt, Germany.

Korn-Wendisch, F., and H. J. Kutzner. 1992a. The family Streptomycetaceae. In: A. Balows, H. G. Trüper, M. Dworkin, W. Harder, and K.-H. Schleifer (Eds.), The Prokaryotes. Springer-Verlag. New York, NY. 921–995.

Korn-Wendisch, F., and J. Schneider. 1992b. Phage typing: a useful tool in actinomycete systematics. Gene 115:243–247.

Kroppenstedt, R. M. 1977. Untersuchungen zur Chemotaxonomie der Ordnung Actinomycetales Buchanan 1917 [PhD thesis]. Universität Darmstadt. Darmstadt, Germany.

Kroppenstedt, R. M., F. Korn-Wendisch, V. J. Fowler, and E. Stackebrandt. 1981. Biochemical and molecular genetic evidence for a transfer of Actinoplanes armeniacus into the family Streptomycetaceae. Syst. Appl. Microbiol. 4:254–262.

Kroppenstedt, R. M. 1985. Fatty acid and menaquinone analysis of actinomycetes and related organisms. In: M. Goodfellow and D. E. Minnikin (Eds.), Chemical Methods in Bacterial Systematics. Academic Press. London, UK. SAB Technical Series 20:173–199.

Kroppenstedt, R. M. 1987. Chemische Untersuchungen an Actinomycetales und verwandte Taxa, Korrelation von Chemosystematik und Phylogenie [Habilitationsschrift]. TH Darmstadt. Germany.

Kroppenstedt, R. M. 1992. The genus Nocardiopsis. In: A. Balows, H. G. Trüper, M. Dworkin, W. Harder, and K.-H. Schleifer. The Prokaryotes. Springer-Verlag. New York, NY. 1139–1156.

Kudo, T., K. Matsushima, T. Itoh, J. Sasaki, and K. Suzuki. 1998. Description of four new species of the genus Kineosporia: Kineosporia succinea sp. nov., Kineosporia rhizophila sp. nov., Kineosporia mikuniensis sp. nov. and Kineosporia rhamnosa sp.nov., isolated from plant samples, and amended description of the genus Kineosporia. Int. J. Syst. Bacteriol. 48:1245–1255.

Kurylowicz, W., A. Paszkiewicz, W. Woznicka, W. Kurzatkowski, and T. Szulga. 1975. Classification of Streptomyces by different numerical methods. Postepy higieny i medycyny doswiadczalnej. 29:281–355.

Kurylowicz, W., A. Paszkiewicz, W. Woznicka, W. Kurzatkowski, and T. Szulga. 1976. Numerical taxonomy of streptomycetes (ISP strains). In: T. Arai (Ed.), Actinomycetes: The Boundary Microorganisms. Toppan. Tokyo, Japan. 323–340.

Kusakabe, H., and K. Isono. 1988. Taxonimic studies on Kitasatosporia cystarginea sp. nov., which produces a new antifungal antibiotic cystargin. J. Antibiot. 41:1758–1762.

Küster, E., and S. T. Williams. 1964. Selection of media for isolation of streptomycetes. Nature 202:928–929.

Küster, E. 1976. Ecology and predominance of soil streptomycetes. In: T. Arai (Ed.), Actinomycetes: The Boundary Microorganisms. Toppan. Tokyo, Japan. 109–121.

Kutzner, H. J. 1956. In: Beitrag zur Systematik und Ökologie der Gettung Streptomyces Waksman et Henrici [PhD dissertation]. Hohenheim University. Hohenheim, Germany.

Kutzner, H. J. 1961a. Effect of various factors on the efficiency of plating and plaque morphology of some Streptomyces phages. Pathol. Microbiol. 24:30–51.

Kutzner, H. J. 1961b. Specificity of actinophages within a selected group of Streptomyces. Pathol. Microbiol. 24:170–191.

Kutzner, H. J. 1972. Storage of streptomycetes in soft agar and other methods. Experientia 28:1395.

Labeda, D. P. 1987. Transfer of the type strain of Streptomyces erythraeus (Waksman 1923) Waksman and Henrici 1948 to the genus Saccharopolyspora Lacey and Goodfellow 1975 as Saccharopolyspora erythraea sp. nov. and designation of a neotype strain for Streptomyces erythraeus. Int. J. Syst. Bacteriol. 37:19–22.

Labeda, D. P. 1988. Kitasatospora mediocidica sp. nov. Int. J. Syst. Bacteriol. 38:287–290.

Labeda, D. P., and A. J. Lyons. 1991a. Deoxyribonucleic-acid relatedness among species of the Streptomyces cyaneus cluster. Syst. Appl. Microbiol. 14:158–164.

Labeda, D. P., and A. J. Lyons. 1991b. The Streptomyces violaceusniger cluster is heterogeneous in DNA relatedness among strains: emendation of the descriptions of S. violaceusniger and Streptomyces hygroscopicus. Int. J. Syst. Bacteriol. 41:398–401.

Labeda, D. P. 1993. DNA relatedness among strains of the Streptomyces lavendulae phenotypic cluster group. Int. J. Syst. Bacteriol. 43:822–825.

Labeda, D. P. 1996. DNA relatedness among verticil-forming Streptomyces species (formerly Streptoverticillium species). Int. J. Syst. Bacteriol. 46:699–703.

Labeda, D. P. 1998. DNA relatedness among the Streptomycesfilvissimus and Streptomyces griseoviridis phenotypic cluster group. Int. J. Syst. Bacteriol. 48:829–832.

Lacey, J., and J. Dutkiewicz. 1976a. Isolation of actinomycetes and fungi from mouldy hay using a sedimentation chamber. J. Appl. Bacteriol. 41:315–319.

Lacey, J., and J. Dutkiewicz. 1976b. Methods for examining the microflora of mouldy hay. J. Appl. Bacteriol. 41:13–27.

Lacey, J., and M. E. Lacey. 1987. Airborne spores in cotton mills. Ann. Occup. Hyg. 31:1–19.

Lacey, J. 1988. Actinomycetes as biodeteriogens and pollutants of the environment. In: M. Goodfellow, S. T. Williams, and M. Mordarski (Eds.), Actinomycetes in Biotechnology. Academic Press. San Diego, CA. 359–432.

Lange, B. J., and W. J. R. Boyd. 1968. Preservation of fungal spores by drying on porcelain bead. Phytopathology 58:1711–1712.

Langham, C. D., S. T. Williams, P. H. A. Sneath, and A. M. Mortimer. 1989. New probability matrices for identification of Streptomyces. J. Gen. Microbiol. 135:121–133.

Lanoot, B., M. Vancanneyt, I. Cleenwerck, L. Wang, W. Li, Z. Liu, and J. Swings. 2002. The search for synonyms among streptomycetes by using SDS-PAGE of whole-cell proteins. Emendation of the species Streptomyces aurantiacus, Streptomyces cacaoi subsp. cacaoi, Streptomyces caeruleus and Streptomyces violaceus. Int. J. Syst. Evol. Microbiol. 52:823–829.

Lawrence, C. H. 1956. A method of isolating actinomycetes from scabby potato tissue and soil with minimal contamination. Can. J. Bot. 34:44–47.

Lechevalier, H. A., and C. T. Corke. 1953. The replica plate method for screening antibiotic producing organism. Appl. Microbiol. 1:110–112.

Lechevalier, M. P., and H. A. Lechevalier. 1970. Chemical composition as a criterion in the classification of aerobic actinomycetes. Int. J. Syst. Bacteriol. 20:435–443.

Lechevalier, H. A., Lechevalier, M. P. and N. N. Gerber. 1971. Chemical composition as a criterion in the classification of actinomycetes. Adv. Appl. Microbiol. 14:47–72.

Lechevalier, M. P. 1977a. Lipids in bacterial taxonomy: a taxonomist's view. CRC Crit. Rev. Microbiol. 5:109–210.

Lechevalier, M. P., C. deBievre, and H. Lechevalier. 1977b. Chemotaxonomy of aerobic actinomycetes: Phospholipid composition. Biochem. Syst. Ecol. 5:249–260.

Lechevalier, M. P., R. J. Seidler, and T. M. Evans. 1980. Enumeration and characterization of standard plate count bacteria in chlorinated and raw water supplies. Appl. Environ. Microbiol. 40:922–930.

Lechevalier, M. P. 1981a. Ecological associations involving actinomycetes. In: K. P. Schaal, and G. Pulverer (Eds.), Actinomycetes: Proceedings of the 4th International Symposium on Actinomycete Biology, Cologne, 1979. Gustav Fischer-Verlag. Stuttgart, Germany. 159–166.

Lechevalier, M. P., A. E. Stern, and H. A. Lechevalier. 1981b. Phospholipids in the taxonomy of actinomycetes. In: K. P. Schaal and G. Pulverer (Eds.), Actinomycetes: Proceedings of the 4th Int. Symposium on Actinomycete Biology, Cologne, 1979. Gustav Fischer-Verlag. Stuttgart, Germany. 111–116.

Lechevalier, M. P. 1988. Actinomycetes in agriculture and forestry. In: M. Goodfellow, S. T. Williams, and M. Mordarski (Eds.), Actinomycetes in Biotechnology. Academic Press. San Diego, CA. 327–358.

Li, W., B. Lanoot, Y. Zhang, M. Vancanneyt, J. Swings, and Z. Liu. 2002a. Streptomyces scopiformis sp. nov., a novel streptomycete with fastigiate spore chains. Int. J. Syst. Evol. Microbiol. 52:1629–1633.

Li, W.-J., L.-P. Zhang, P. Xu, X.-L. Cui, Z.-T. Lu, L.-H. Xu, and C.-L. Jiang. 2002b. Streptomyces beijiangensis sp. nov., a psychotolerant actinomycete isolated from soil in China. Int. J. Syst. Evol. Microbiol. 52:1695–1699.

Liao, D., and P. P. Dennis. 1994. Molecular phylogenies based on ribosomal protein L11, L1, L10, and L12 sequences. J. Molec. Evol. 38:405–419.

Lindner, F., and K. H. Wallhäusser. 1955. Die Arbeitsmethoden der Forschung zur Auffindung neuer Antibiotica. Arch. Mikrobiol. 22:219–234.

Lingappa, Y., and J. L. Lockwood. 1962. Chitin media for selective isolation and culture of actinomycetes. Phytopathology 52:317–323.

Lloyd, A. B. 1969a. Behaviour of streptomycetes in soil. J. Gen. Microbiol. 56:156–170.

Lloyd, A. B. 1969b. Dispersal of streptomycetes in air. J. Gen. Microbiol. 57:35–40.

Locci, R., J. Rogers, P. Sardi, and G. M. Schofield. 1981. A preliminary numerical study of named species of the genus Streptoverticillium. Ann. Microbiol. Enzimol. 31:115–121.

Locci, R., and G. M. Schofield. 1989. Genus Streptoverticillium Baldacci 1958, 15, emend. Mut.char. Baldacci, Farina and Locci 168AL. In: S. T. Williams, M. E. Sharpe, and J. G. Holt (Eds.), Bergey's Manual of Determinative Bacteriology. Williams and Wilkins. Baltimore, MD. 4:2492–2504.

Logan, N. A. 1994. Bacterial Systematics. Blackwell Scientific Publications. Oxford, UK.

Lomovskaya, N. D., K. F. Chater, and N. M. Mkrtumian. 1980. Genetics and molecular biology of Streptomyces bacteriophages. Microbiol. Rev. 44:206–229.

Lonsdale, J. T. 1985. Aspects of the biology of acidophilic actinomycetes. PhD University of Newcastle, Newcastle.

Ludwig, W., and K.-H. Schleifer. 1994. Bacterial phylogeny based on 16S and 23S rRNA sequence analysis. FEMS Microbiol. Rev. 15:155–173.

MacKenzie, C. R., D. Bilous, and K. G. Johnson. 1984. Purification and characterization of an exoglucanase from Streptomyces flavogriseus. Can. J. Microbiol. 30:1171–1178.

Manchester, L., B. Pot, K. Kersters, and M. Goodfellow. 1990. Classification of Streptomyces and Streptoverticillium species by numerical analysis of electrophoretic protein patterns. Syst. Appl. Microbiol. 13:333–337.

Manfio, G. P., E. Atalan, J. Zakrzweska-Czerwinska, M. Modarski, C. Rodríguez, M. D. Collins, and M. Goodfellow. 2003. Classification of novel soil streptomycetes as Streptomyces aureus sp. nov.,Streptomyces laceyi sp. nov. and Streptomyces sanglieri sp. nov. Ant. v. Leeuwenhoek 83:245–255.

Márialigeti, K., K. Jáger, I. M. Szabó, M. Pobozsny, and A. Dzingov. 1984. The faecal Actinomycete flora of Protracheoniscus amoenus (woodlice; Isopoda). Acta Microbiologica Hungarica 31:339–344.

Marri, L., E. Barboni, T. Irdani, B. Perito, G. Mastromei. 1997. Restriction enzyme and DNA hybridization analysis of cellulolytic Streptomyces isolates of different origin. Can. J. Microbiol. 43:395–299.

Martin, J. P., Z. Filip, and K. Haider. 1976. Effect of montmorillonite and humate on growth and metabolic activity of some actinomycetes. Soil Biol. Biochem. 8:409–413.

Mayfield, C. I., S. T. Williams, S. M. Ruddick, and H. L. Hatfield. 1972. Studies on the ecology of actinomycetes in soil. IV: Observations on the form and growth of streptomycetes in soil. Soil Biol. Biochem. 4:79–91.

McCarthy, A. J., and P. Broda. 1984a. Screening for lignin-degrading actinomycetes and characterization of their activity against [14 C] lignin-labelled wheat lignocellulose. J. Gen. Microbiol. 130:2905–2913.

McCarthy, A. J., M. J. MacDonald, A. Paterson, and P. Broda. 1984b. Lignocellulose degradation by actinomycetes. J. Gen. Microbiol. 130:1023–1030.

McCarthy, A. J., E. Peace, and P. Broda. 1985. Studies on extracellular xylanase activities of some thermophilic actinomycetes. Appl. Microbiol. Biotechnol. 21:238–244.

McCarthy, A. J., A. Paterson, and P. Broda. 1986. Lignin solubilisation by Thermomonospora mesophila. Appl. Microbiol. Biotechnol. 24:347–352.

McKillop, C., P. Elvin, and J. Kenten. 1986. Cloning and expression of an extracellular-amylase gene from Streptomyces hygroscopicus in Streptomyces lividans 66. FEMS Microbiol. Lett. 36:3–7.

Mehling, A., U. F. Wehmeier, and W. Piepersberg. 1995. Application of random amplfied polymorphic DNA (RAPD) assays in identifying conserved regions of actinomycete genomes. FEMS Microbiol. Lett. 128:119–126.

Menzies, J. D., and C. E. Dade. 1959. A selective indicator medium for isolating Streptomyces scabies from potato tubers or soil. Phytopathology 49:457–458.

Meyers, P. R., D. S. Porter, C. Omorogie, J. M. Pule, and T. Kwetane. 2003. Streptomyces speibonae sp. nov., a novel streptomycete with blue substrate mycelium isolated from South African soil. Int. J. Syst. Evol. Microbiol. 53:801 and 805.

Mikami, Y., K. Miyashita, and T. Arai. 1982. Diaminopimelic acid profiles of alkalophilic and alkaline-resistant strains of actinomycetes. J. Gen. Microbiol. 128:1709–1712.

Mikami, Y., K. Miyashita, and T. Arai. 1985. Alkalophilic actinomycetes. The Actinomycetes 19(3):176–191.

Mikulik, K., I. Janda, J. Weiser, and A. Jiranova. 1982. Ribosomal proteins of Streptomyces aureofaciens producing tetracycline. Biochim. Biophys. Acta 699:203–210.

Millner, P. D. 1982. Thermophilic and thermotolerant actinomycetes in sewage-sludge compost. Devel. Indust. Microbiol. 23:61–78.

Misiek, M. 1955. Comparative studies of Streptomyces populations in soils [Ph D dissertation]. Syracue University. Syracuse, NY.

Mordarski, M., J. Wieczorek, and B. Jaworska. 1970. On the condition of amylase production by Actinomycetes. Archiwum Immunol. Ther. Experimentalis 18:375–381.

Mordarski, M., M. Goodfellow, S. T. Williams, and P. H. A. Sneath. 1986. Evaluation of species groups in the genus Streptomyces. In: G. Szabó, S. Bíró, and M. Goodfellow (Eds.), Biological, Biochemical, and Biomedical Aspects of Actinomycetes: Proceedings of the 6th International Symposium on Actinomycetes Biology, Debrecen, Hungary, 1985. Akademiai Kiado. Budapest, Hungary. 517–525.

Morita, R. Y. 1985. Starvation and miniaturization of heterotrophs, with special emphasis on maintenance of the starved viable state. In: M. Fletcher and G. D. Floodgate (Eds.), Bacteria in their Natural Environments. Academic Press. London, United Kingdom. 111–130.

Morosoli, R., S. Ostiguy, and C. Dupont. 1999. Effect of carbon source, growth and temperature on the expression of the sec genes of Streptomyces lividans 1326. Can. J. Microbiol. 45:1043–1049.

Muyzer, G., E. C. de Waal, and A. G. Uitterlinden. 1993. Profiling of complex microbial populations by denaturing gradient gel electrophoresis analysis of polymerase chain reaction-amplified genes coding for 16S rRNA. Appl. Environ. Microbiol. 59:695–700.

Nakagaito, Y., A. Shimazu, A. Yokata, and T. Hasegawa. 1992a. Proposal of Streptomyces atroaurantiacus sp. nov. and Streptomyces kifunensis sp. nov. and transferring Kitasatosporia cystarginea Kusakabe and Isono to the genus Streptomyces as Streptomyces cystargineus comb. nov. J. Gen. Appl. Microbiol. 38:627–633.

Nakagaito, Y., A. Yokogata, and T. Hasegawa. 1992b. Three new species of the genus Streptomyces: Streptomyces cochleatus sp. nov., Streptomyces paracochleatus sp. nov., and Streptomyces azaticus sp. nov. J. Gen. Appl. Microbiol. 38:105–120.

Naumova, I. B, V. D. Kuznetsov, K. S: Kudrina, and A. P. Bezzubenkova. 1980. The occurrence of teichoic acids in streptomycetes. Arch. Microbiol. 126:71–75.

Nette, I. T., N. J. Pomorzeva, and E. I. Koslova. 1959. Destruction of caoutchouc by microorganisms. Mikrobiologiya 28:881–886.

Nissen, T. V. 1963. Distribution of antibiotic-producing actinomycetes in Danish soil. Experientia 19:470–471.

Nolan, R. D., and T. Cross. 1988. Isolation and screening of actinomycetes. In: M. Goodfellow, S. T. Williams, and M. Mordarski (Eds.), Actinomycetes in Biotechnology. Academic Press. San Diego, CA. 1–32.

Noval, J. J., and W. J. Nickerson. 1959. Decomposition of native keratin by Streptomyces fradiae. J. Bacteriol. 77:251–263.

Nüesch, J. 1965. Isolierung und Selektionierung von Actinomyceten, in Symposium ("Anreicherungskultur und Mutantenauslese") Göttingen, April 1964. Zentralbl. Bakteriol. Parasitenkd. Infektionskr. Hyg., Abt. 1 Suppl. 1:234–252.

Ochi, K. 1989. Heterogeneity of ribosomal proteins among Streptomyces species and its application to identification. J. Gen. Microbiol. 135:2635–2642.

Ochi, K. 1992. Polyacrylamide gel electrophoresis analysis of ribosomal protein: a new approach for actinomycete taxonomy. Gene 115:261–265.

Ochi, K. 1995. A taxonomic study of the genus Streptomyces by analysis of ribosomal protein AT-L30. Int. J. Syst. Bacteriol. 45:507–514.

Ogata, S. 1980. Bacteriophage contamination in industrial processes. Biotechnol. Bioeng. 22 (Suppl. 1):177–193.

Ogata, S., H. Suenaga, and S. Hayashida. 1985. A temperate phage from Streptomyces azureus. Appl. Environ. Microbiol. 49:201–204.

Oh, C., M. Ahn, and J. Kim. 1996. Use of electrophoretic enzyme patterns for streptomycete systematics. FEMS Microbiol. Lett. 140:9–13.

Ohta, Y., and M. Ikeda. 1978. Deodorization of pig feces by actinomycetes. Appl. Environ. Microbiol. 36:487–491.

Okafor, N. 1966. The ecology of microorganisms on, and the decomposition of, insect wings in the soil. Plant Soil 25:211–237.

Okami, Y., and T. Okazaki. 1972. Studies on marine microorganisms. I: Actinomycetes in Sagami Bay and their antibiotic substances. J. Antibiot. 25:456–460.

Okami, Y., T. Okazaki, T. Kitahara, and H. Umezawa. 1976. Studies on marine microorganisms. V: A new antibiotic, aplasmomycin, produced by a streptomycete isolated from shallow sea mud. J. Antibiot., Ser. A 29:1019–1025.

Okami, Y., and T. Okazaki. 1978. Actinomycetes in marine environments. In: M. Mordarski, W. Kurylowicz, and J. Jeljaszewicz (Eds.), Nocardia and Streptomyces: Proceedings of the International Symposium on Nocardia and Streptomyces, Warsaw, 1976. Gustav Fischer-Verlag. Stuttgart, Germany. 145–151.

Okazaki, T., and Y. Okami. 1976. Studies on actinomycetes isolated from shallow sea and their antibiotic substances. In: T. Arai (Ed.), Actinomycetes: The Boundary Microorganisms. Toppan. Tokyo, Japan. 123–161.

Omura, S., Y. Iwai, Y. Takahashi, K. Kojima, K. Otoguru, and R. Oiwa. 1981. Type of diaminopimelic acid different in aerial and vegetative mycelia of setamycin-producing actinomycete KM-60054. J. Antibiot. 34:1633–1634.

Omura, S., Y. Takahashi, Y. Iwai, and H. Tanaka. 1982. Kitasatosporia, a new genus of the order Actinomycetales. J. Antibiot. 35:1013–1019.

Omura, S., Y. Takahashi, Y. Iwai, and H. Tanaka. 1985. Revised nomenclature of Kitasatosporia setalba. Int. J. Syst. Bacteriol. 35:221.

Omura, S., Y. Takahashi, and Y. Iwai. 1989. Genus Kitasatosporia^VP. In: Williams, S. T., Sharpe, M. E. and Holt, J. G. Bergey's Manual of Systematic Bacteriology. Williams & Wilkins. Baltimore, MD. 4:2594–2598.

Omura, S., H. Ikeda, J. Ishikawa, A. Hanamoto, C. Takahashi, M. Shinose, Y. Takahashi, H. Horikawa, H. Nakazawa, T. Osonoe, H. Kikuchi, Y. Shiba Sakaki, and M. Hattori. 2001. Genome sequence of an idustrial microoganism Streptomyces avermitilis: deducing the ability of producing secondary metabolites. Proc. Natl. Acad. Sci. USA. 21:12215–12220.

Ottow, J. C. G. 1972. Rose bengal as a selective aid in the isolation of fungi and actinomycetes from natural sources. Mycologia 64:304–315.

Paradis, E., C. Goyer, N. C. Hodge, R. Hogue, R. E. Stall, and C. Beaulieu. 1994. Fatty acid and protein profiles of Streptomyces scabies strains isolated in eastern Canada. Int. J. Syst. Bacteriol. 44:561–564.

Park, Y.-H., D.-G. Yim, E. Kim, Y.-H. Kho, T.-I. Mheen, J. Lonsdale, and M. Goodfellow. 1991. Classification of acidophilic, neutrotolerant and neutrophilic streptomycetes by nucleotide sequencing of 5S ribosomal RNA. J. Gen. Microbiol. 137:2265–2269.

Parle, J. N. 1963a. A microbiological study of earthworm casts. J. Gen. Microbiol. 31:13–22.

Parle, J. N. 1963b. Microorganisms in the intestines of earthworms. J. Gen. Microbiol. 31:1–11.

Peczynska-Czoch, W., and M. Mordarski. 1988. Actinomycete enzymes. In: M. Goodfellow, S. T. Williams, and M. Mordarski (Eds.), Actinomycetes in Biotechnology. Academic Press. San Diego, CA. 219–283.

Petrosyan, P., M. García-Varela, A. Luz-Madrigal, C. Huitrón, and M. E. Flores. 2003. Streptomyces mexicanus sp. nov., a xylanolytic micro-organism isolated from soil. Int. J. Syst. Evol. Microbiol. 53:269–273.

Phillips, L. 1992. The Distribution of Phenotypic and Genotypic Characters within Streptomycetes and their Relationship to Antibiotic Production [PhD thesis]. University of Warwick. Warwick, UK.

Polsinelli, M., and P. G. Mazza. 1984. Use of membrane filters for selective isolation of actinomycetes from soil. FEMS Microbiol. Lett. 22:79–83.

Pommer, E.-H, and G. Lorenz. 1986. The behaviour of polyester and polyether polyurethanes towards microorganisms. In: K. J. Seal (Ed.), Biodeterioration and Biodegradation of Plastics and Polymers. Biodeterioration Society Occasional Publication 1:77–86.

Porter, J. N., J. J. Wilhelm, and H. D. Tresner. 1960. Method for the preferential isolation of actinomycetes from soils. Appl. Microbiol. 8:174–178.

Porter, J. N., and J. J. Wilhelm. 1961. The effect on Streptomyces populations of adding various supplements to soil samples. Devel. Indust. Microbiol. 2:253–259.

Prauser, H. 1984. Phage host ranges in the classification and identification of Gram-positive branched and related bacteria. In: Ortiz-Ortiz, Bojalil and Yakoleff (Eds.), Biological, Biochemical and Biomedical Aspects of Actinomycetes. Academic Press. Orlando, FL. 617–633.

Preobrazhenskaya, T. P., M. A. Sveshnikova, L. P. Terekhova, and N. T. Chormonova. 1978. Selective isolation of soil actinomycetes. In: M. Mordarski, Kurylowicz and Jeljaszweicz (Eds.), Nocardia and Streptomyces. Gustav Fischer Verlag, Stuttgart. New York, NY. 119–123.

Pridham, T. G., P. Anderson, C. Foley, H. A. Lindenfelser, C. W. Hesseltine, and R. G. Benedict. 1956/1957. A selection of media for maintenance and taxonomic study of Streptomyces. Antibiot. Ann. 1956/57:947–953.

Pridham, T. G., A. J. Lyons, and B. Phronpatima. 1973. Viability of Actinomycetales stored in soil. Appl. Microbiol. 26:441–442.

Pridham, T. G., and H. G. Tresner. 1974. Genus I. Streptomyces Waksman and Henrici 1943, 339. In: R. E. Buchanan and N. E. Gibbons (Eds.), Bergey's Manual of Determinative Bacteriology, 8th ed. Williams and Wilkins. Baltimore, MD. 748–829.

Ramachandra, M., D. L. Crawford, and G. Hertel. 1988. Characterization of an extracellular lignin peroxidase of the lignocellulolytic actinomycete Streptomyces viridosporus. Appl. Environ. Microbiol. 54:3057–3063.

Rappé, M. S., and S. J. Giovannoni. 2003. The uncultured microbial majority. Ann. Rev. Microbiol. 57:369–394.

Rauland, U., I. Glocker, M. Redenbach, and J. Cullum. 1995. DNA amplifications and deletions in Streptomyces lividans 66 and the loss of one end of the linear chromosome. Molec. Gen. Genet. 246:37–44.

Redenbach, M., F. Flett, W. Piendl, I. Glocker, U. Rauland, O. Wafzig, R. Kliem, P. Leblond, and J. Cullum. 1993. The Streptomyces lividans 66 chromosome contains a 1 MB deletogenic region flanked by two amplifiable regions. Molec. Gen. Genet. 241:255–262.

Rheims, H., A. Felske, S. Seufert, and E. Stackebrandt. 1999. Molecular monitoring of an uncultured group of the class Actinobacteria in two terrestrial environments. J. Microbiol. Meth. 36:65–75.

Ridell, M., G. Wallerström, and S. T. Williams. 1986. Immunodiffusion analyses of phenetically defined strains of Streptomyces, Streptoverticillium and Nocardiopsis. Syst. Appl. Microbiol. 8:24–27.

Rintala, H., A. Nevalainen, E. Ronka, and M. Suutari. 2001. PCR primers targeting the 16S rRNA gene for the specific detection of streptomycetes. Molec. Cell. Probes 15:337–347.

Roach, A. W., and J. K. G. Silvey. 1959. The occurrence of marine actinomycetes in Texas gulf coast substrates. Am. Midland Naturalist 62:482–499.

Rossi-Doria, T. 1891. Su di alcune specie di "Streptotrix" trovate nell'aria studiate in rapporto a quelle giá note e specialment all' "Actinomyces." Ann Ist Igiene Sper Univ Roma 1:399–438.

Rothrock, C. S., and D. Gottlieb. 1981. Importance of antibiotic production in antagonism of selected Streptomyces species to two soil-borne plant pathogens. J. Antibiot. 34:830–835.

Rothrock, C. S., and D. Gottlieb. 1984. Roles of antibiosis in antagonism of Streptomyces hygroscopicus var. geldanus to Rhizoctonia solani in soil. Can. J. Microbiol. 30:1440–1447.

Ruddick, S. M., and S. T. Williams. 1972. Studies on the ecology of actinomycetes in soil. V: Some factors influencing the dispersal and adsorption of spores in soil. Soil Biol. Biochem. 4:93–103.

Sabater, B., J. Sebastian, and C. Asensio. 1972. Identification and properties of an inducible mannokinase from Streptomyces violaceoruber. Biochim. Biophys. Acta. 284:406–413.

Saddler, G. S., M. Goodfellow, D. E. Minnikin, and A. G. O'Donnell. 1986. Influence of the growth cycle on the fatty acid and menaquinone composition of Streptomyces cyaneus NCIB 9616. J. Appl. Bacteriol. 60:51–56.

Saddler, G. S., A. G. O'Donnell, M. Goodfellow, and D. E. Minnikin. 1987. SIMCA pattern recognition in the analysis of streptomycete fatty acids. J. Gen. Microbiol. 133:1137–1147.

Salas, J. A., L. M. Quiros, and C. Hardisson. 1984. Pathways of glucose catabolism during germination of Streptomyces spores. FEMS Microbiol. Lett. 25:229–233.

Sanglier, J. J., D. Whitehead, G. S. Saddler, E. V. Ferguson, and M. Goodfellow. 1992. Pyrolysis mass-spectrometry as a method for the classification, identification and selection of actinomycetes. Gene 115:235–242.

Sato, M., and A. Kaji. 1975. Purification and properties of pectate lyase produced by Streptomyces fradiae IFO 3439. Agric. Biol. Chem. 39:819–824.

Sato, M., and A. Kaji. 1977. Purification and properties of a pectate lyase produced by Streptomyces nitrosporeus. Agricult. Biol. Chem. 41:2193–2197.

Sato, M., and A. Kaji. 1980a. Another pectate lyase produced by Streptomyces nitrosporeus. Agricult. Biol. Chem. 44:1345–1349.

Sato, M., and A. Kaji. 1980b. Exopolygalacturonate lyase produced by Streptomyces massasporeus. Agricult. Biol. Chem. 44:717–721.

Schäfer, A., R. Konrad, T. Kuhnigk, P. Kämpfer, H. Hertel, and H. König. 1996. Hemicellulose-degrading bacteria and yeasts from the termite gut. J. Appl. Bacteriol. 80:471–478.

Schleifer, K.-H., and O. Kandler. 1972. Peptidoglycan types of bacterial cell walls and their taxonomic implications. Bacteriol. Rev. 36:407–477.

Schleifer, K.-H., and E. Stackebrandt. 1983. Molecular systematics of prokaryotes. Ann. Rev. Microbiol. 37:143–187.

Schrempf, H., and S. Walter. 1995. The cellulolytic system of Streptomyces reticuli. Int. J. Biol. Macromol. 17:353–355.

Sabater, B., J. Sebastian, and C. Asensio. 1972. Identification and properties of an inducibale mannokinase from Streptomyces violaceoruber. Biochim. Biophys. Acta 25:406–413.

Sembiring, L., A. C. Ward, and M. Goodfellow. 2000. Selective isolation and characterisation of members of the Streptomyces violaceusniger clade associated with the roots of Paraserianthes falcataria. Ant. v. Leeuwenhoek 78:353–366.

Shirling, E. B., and D. Gottlieb. 1966. Methods for characterization of Streptomyces species. Int. J. Syst. Bacteriol. 16:313–340.

Shirling, E. B., and D. Gottlieb. 1968a. Cooperative description of type cultures of Streptomyces. II: Species description from the first study. Int. J. Syst. Bacteriol. 18:69–189.

Shirling, E. B., and D. Gottlieb. 1968b. Cooperative description of type cultures of Streptomyces. III: Additional species description from first and second studies. Int. J. Syst. Bacteriol. 18:279–399.

Shirling, E. B., and D. Gottlieb. 1969. Cooperative descriptions of type cultures of Streptomyces. IV: Species descriptions from the second, third and forth studies. Int. J. Syst. Bacteriol. 19:391–512.

Shirling, E. B., and D. Gottlieb. 1972. Cooperative description of type cultures of Streptomyces. V: Additional descriptions. Int. J. Syst. Bacteriol. 22:265–394.

Siebert, G., and W. Schwartz. 1956. Untersuchungen über das Vorkommen von Mikroorganismen in entstehenden Sedimenten. Arch. Hydrobiol. 52:331–366.

Silvestri, L., M. Turri, L. E. Hill, and E. Gilardi. 1962. A quantitative approach to the systematics of actinomycetes based on overall similarity. In: G. C. Ainsworth and P. H. Sneath (Eds.), Microbial Classification: 12th Symposium of the Society for General Microbiology. Cambridge University Press. Cambridge, UK. 333–360.

Silvey, J. K. G., and A. W. Roach. 1975. The taste and odor producing aquatic actinomycetes. Crit. Rev. Environ. Control 5:233–273.

Sing, P. J., and R. S. Mehrotra. 1980. Biological control of Rhizoctonia bataticola on grain by coating seed with Bacillus and Streptomyces ssp. and their influence on plant growth. Plant Soil 56:475–483.

Song, J., S.-C. Lee, J.-W. Kang, H.-J. Baek, and J.-W. Suh. 2003. Phylogenetic analysis of Streptomyces spp. isolated from potato scab lesions in Korea on the basis of 16S rRNA gene and 16S-23S rDNA internally transcribed specer sequences. Int. J. Syst. Evol. Microbiol. 54:203–209.

Spicher, G. 1955. Untersuchungen über die Wirkung von Erdextrakt und Spurenelementen auf das Wachstum verschiedener Streptomyzeten. Zbl. Bakteriol. Parasitenkd. Infektionskrankh. Hyg. Abt. 2 108:577–587.

Stackebrandt, E., and C. R. Woese. 1981. Towards a phylogeny of the actinomycetes and related organisms. Curr. Microbiol. 5:197–202.

Stackebrandt, E., W. Liesack, R. Webb, and D. Witt. 1991a. Towards a molecular identification of Streptomyces species in pure culture and in environmental samples. Actinomycetologia 5:38–44.

Stackebrandt, E., D. Witt, C. Kemmerling, R. Kroppenstedt, and W. Liesack. 1991b. Designation of streptomycete 16S and 23S rRNA-based target regions for oligonucleotide probes. Appl. Environ. Microbiol. 57:1468–1477.

Stackebrandt, E., W. Liesack, and D. Witt. 1992. Ribosomal RNA and rDNA sequence analyses. Gene 115:255–260.

Stackebrandt, E., W. Liesack, and B. M. Goebel. 1993. Bacterial diversity in a soil sample from a subtropical Australian environment as determined by 16S rDNA analysis. FASEB J. 7:232–236.

Stackebrandt, E., F. A. Rainey, and N. L. Ward-Rainey. 1997. Proposal for a new hierarchic classification system, Actinobacteria classis nov. Int. J. Syst. Bacteriol. 47:479–491.

Stackebrandt, E., W. Frederiksen, G. M. Garrity, P. A. D. Grimont, P. Kämpfer, M. C. J. Maiden, X. Nasme, R. Roselló-Mora, J. Swings, H. G. Trüper, L. Vauterin, A. C. Ward, and W. B. Withman. 2003. Report of the ad hoc committee for the re-evaluation of the species definition in bacteriology. Int. J. Syst. Evol. Microbiol. 52:1043–1047.

Stolp, H., and M. P. Starr. 1981. Principles of isolation, cultivation, and conservation of bacteria. In: M. P. Starr, H. Stolp, H. G. Trüper, A. Balows, and H. G. Schlegel (Eds.), The Prokaryotes. Springer-Verlag. Berlin, Germany. 135–175.

Stuttard, C. 1982. Temperate phages of Streptomyces venezuelae: lysogeny and host speciÆcity shown by phages SV1 and SV2. J. Gen. Microbiol. 128:115–121.

Suganuma, T., T. Mizukami, K. Moari, M. Ohnishi, and K. Hiromi. 1980. Studies of the action pattern of an amylase from Streptomyces praecox NA-273. J. Biochem. 88:131–138.

Szabó, I., and M. Marton. 1964. Zur Frage der spezifischen Bodenmikrofloren. Ein Versuch zur systematischen Bestimmung der Strahlenpilzflora einer mullartigen (Wald-) Rendzina. Zbl. Bakteriol. Parasitenkd. Infektionskrankh. Hyg. Abt. 2 118:265–306.

Szabó, G., I. Bekesi, and S. Vitális. 1967. Mode of action of factor C, a substance of regulatory function in cytodifferentiation. Biochim. Biophys. Acta 145:159–165.

Taber, W. A. 1959. Identification of an alkaline-dependent Streptomyces as Streptomyces caeruleus Baldacci and characterization of the species under controlled conditions. Can. J. Microbiol. 5:335–344.

Taber, W. A. 1960. Evidence for the existence of acid-sensitive actinomycetes in soil. Can. J. Microbiol. 6:503–514.

Taguchi, S., S. Kojima, K. Miura, and H. Momose. 1996. Taxonomic characterisation of closely-related Streptomyces spp. based on the amino acid sequence analysis of protease inhibitor proteins. FEMS Microbiol. Lett. 135:169–173.

Tajima, K., Y. Takahashi, A. Seino, Y. Iwai, and S. Omura. 2001. Description of two novel species of the genus Kita-

satospora Omura et al., 1982, Kitasatospora cineracea sp. nov. and Kitasatospora niigatensis sp. nov. Int. J. Syst. Evol. Microbiol. 51:1765–1771.

Takahashi, Y., Y. Iwai, and S. Omura. 1984a. Two new species of the genus Kitasatosporia, Kitasatosporia phosalacinea sp. nov. and Kitsatosporia griseola sp. nov. J. Gen. Appl. Microbiol. 30:377–387.

Takahashi, Y., T. Kuwana, Y. Iwai, and S. Omura. 1984b. Some characteristics of aerial and submerged spores of Kitasatospora. J. Gen. Appl. Microbiol. 30:223–229.

Takahashi, Y., A. Matsumoto, A. Seino, J. Ueno, Y. Iwai, and S. Omura. 2002. Streptomyces avermectinius sp. nov., an avermectin-producing strain. Int. J. Syst. Evol. Microbiol. 52:2163–2168.

Taylor, C. F. 1936. A method for isolation of actinomycetes from scab lesions on potato tubers and beet roots. Phytopathology 26:287–288.

Tendler, M. D., and P. R. Burkholder. 1961. Studies on the thermophilic actinomycetes. I: Methods of cultivation. Appl. Microbiol. 9:394–399.

Titgemeier, F., J. Walkenhorst, J. Reizer, M. H. Stuiver, X. Ciu, and M. H. Saier. 1995. Identification and characterization of phosphoenolpyruvate:fructose phosphotransferase system in three Streptomyces species. Microbiology 141:51–58.

Trejo, W. H. 1970. An evaluation of some concepts and criteria used in the speciation of streptomycetes. Trans. NY Acad. Sci. Ser. II 32:989–997.

Tresner, H. D., J. A. Hayes, and E. J. Backus. 1968. Differential tolerance of streptomycetes to sodium chloride as a taxonomic aid. Appl. Microbiol. 16:1134–1136.

Trolldenier, G. 1966. Über die Eignung Erde enthaltender Nährsubstrate zur Zählung und Isolierung von Bodenmikroorganismen auf Membranfiltern. Zbl. Bakteriol. Parasitenkd. Infektionskrankh. Hyg. Abt. 2 120:496–508.

Trolldenier, G. 1967. Isolierung und Zählung von Bodenactinomyceten auf Erdplatten mit Membranfiltern. Plant Soil 27:285–288.

Trujillo, M. E., and M. Goodfellow. 2003. Numerical phenetic classification of clinically significant aerobic sporoactinomycetes and related organisms. Ant. v. Leeuwenhoek 84:39–68.

Tsao, P. H., C. Leben, and G. W. Keitt. 1960. An enrichment method for isolating actinomycetes that produce diffusible antifungal antibiotics. Phytopathology 50:88–89.

Uchida, K., and K. Aida. 1977. Acyl type of bacterial cell wall: Its simple identification by colorometric methods. J. Gen. Appl. Microbiol. 23:249–260.

Uchida, K., and A. Seino. 1997. Intra- and intergeneric relationships of various actinomycete strains based on the acyl types of the muramyl residue in cell wall peptidoglycans examined in a glycolate test. Int. J. Syst. Bacteriol. 47:182–190.

Ulrich, A., and S. Wirth. 1999. Phylogenetic diversity and population densities of cultural cellulolytic soil bacteria across an agricultural encatchment. Microb. Ecol. 37:238–247.

Uridil, J. E., and P. A. Tetrault. 1959. Isolation of thermophilic streptomycetes. J. Bacteriol. 78:243–246.

Vandamme, P., B. Pot, M. Gillis, P. De Vos, K. Kersters, and J. Swings. 1996. Polyphasic taxonomy, a consensus approach to bacterial systematics. Microbiol. Rev. 60:407–438.

Van Wezel, G. P., E. Vijgenboom, and L. Bosch. 1991. A comparative study of the ribosomal RNA operons of

Streptomyces coelicolor A3(2) and sequence analysis of rrnA. Nucl. Acids Res. 25:4399–4403.

Veldkamp, J. 1955. A study of the aerobic decomposition of chitin by microorganisms. Medelingen van de Landbouwhogeschool te Wageningen/Nederland 55:127–174.

Vickers, J. C., S. T. Williams, and G. W. Ross. 1984. A taxonomic approach to selective isolation of streptomycetes from soil. *In:* In: L. Ortiz-Ortiz, L. F. (Eds.), Biological, Biochemical and Biomedical Aspects of Actinomycetes. Academic Press. Orlando, FL. 553–561.

Voelskow, H. 1988/1989. Methoden der zielorientierten Stammisolierung. *In:* P. Präve, M. Schlingmann, W. Crueger, K. Esser, R. Thauer, and F. Wagner (Eds.), Jahrbuch Biotechnologie. Carl Hanser-Verlag. Munich, Germany. 2:343–361.

Waksman, S. A., and R. Curtis. 1916. The actinomyces of the soil. Soil Sci. 1:99–134.

Waksman, S. A. 1919. Cultural studies of species of Actinomyces. Soil Sci. 8:71–215.

Waksman, S. A., and H. B. Woodruff. 1940. The soil as a source of microorganisms antagonistic to disease-producing bacteria. J. Bacteriol. 40:581–600.

Waksman, S. A., and A. T. Henrici. 1943. The nomenclature and classification of the actinomycetes. J. Bacteriol. 46:337–341.

Waksman, S. A. 1961. The Actinomycetes, Volume 2: Classification, Identification and Descriptions of Genera and Species. Williams and Wilkins. Baltimore, MD.

Wang, Z. M., B. H. Bleakley, D. L. Crawford, G. Hertel, and F. Rafii. 1990. Cloning and expression of a lignin peroxidase gene from Streptomyces viridosporus in Streptomyces lividans. J. Biotechnol. 13:131–144.

Warcup, J. H. 1950. The soil-plate method for isolation of fungi from soil. Nature 166:117–118.

Watson, E. Z., and S. T. Williams. 1974. Studies on the ecology of actinomycetes in soil. 7. Actinomycetes in a coastal sand belt. Soil Biol. Biochem. 6:43–52.

Watve, M. G., Tickoo, R., Jog, M. M., and B. D. Bhole. 2001. How many antibiotics are produced by the genus Streptomyces? Arch. Microbiol. 176:386–390.

Wayne, L. G., D. J. Brenner, R. R. Colwell, et al. 1987. International Committee on Systematic Bacteriology: Report of the ad hoc committee on reconciliation of approaches to bacterial systematics. Int. J. Syst. Bacteriol. 37:463–464.

Wellington, E. M. H., and S. T. Williams. 1978. Preservation of actinomycete inoculum in frozen glycerol. Microbios Lett. 6:151–157.

Wellington, E. M. H., and S. T. Williams. 1981a. Host ranges of phages isolated to Streptomyces and other genera. *In:* K. P. Schaal and G. Pulverer (Eds.), Actinomycetes: Proceedings of the 4th International Symposium on Actinomycetes Biology, Cologne, 1979. Gustav Fischer-Verlag. Stuttgart, Germany. 93–98.

Wellington, E. M. H., and S. T. Williams. 1981b. Transfer of Actinoplanes armeniacus Kalakoutskii and Kuznetsov to Streptomyces: Streptomyces armeniacus (Kalakoutskii and Kuznetsov) comb. nov. Int. J. Syst. Bacteriol. 31:77–81.

Wellington, E. M. H., M. Al-Jawadi, and R. Bandoni. 1987. Selective isolation of Streptomyces species-groups from soil. Devel. Indust. Microbiol. 28:99–104.

Wellington, E. M. H., E. Stackebrandt, D. Sanders, J. Wolstrup, and N. O. G. Jorgensen. 1992. Taxonomic status of Kitasatosporia, and proposed unfication with Streptomyces on the basis of phenotypic and 16S rRNA analysis

and emendation of Streptomyces Waksman and Henrici 1943, 339AL. Int. J. Syst. Bacteriol. 42:156–160.

Welsch, M., R. Corbaz, and L. Ettlinger. 1957. Phage typing of streptomycetes. Schweiz. Z. Allgem. Path. Bakteriol. 20:454–458.

Wenner, T., V. Roth, B. Decaris, and P. Leblond. 2002. Intragenomic and intraspecific polymorhpism of the 16S-23 rDNA internally transcribed sequences of Streptomyces ambofaciens. Microbiology 148:633–642.

Weyland, H. 1981a. Characteristics of actinomycetes isolated from marine sediments. *In:* K. P. Schaal and G. Pulverer (Eds.), Actinomycetes: Proceedings of the 4th International Symposium on Actinomycete Biology, Cologne, 1979. Gustav Fischer-Verlag. Stuttgart, Germany. 309–314.

Weyland, H. 1981b. Distribution of actinomycetes on the sea floor. *In:* K. P. Schaal and G. Pulverer (Eds.), Actinomycetes: Proceedings of the 4th International Symposium on Actinomycete Biology, Cologne, 1979. Gustav Fischer-Verlag. Stuttgart, Germany. 185–193.

Weyland, H., and E. Helmke. 1988. Actinomycetes in the marine environment. *In:* Y. Okami, T. Beppu, and H. Ogawara (Eds.), Biology of Actinomycetes '88: Proceedings of the 7th International Symposium on Biology of Actinomycetes, Tokyo, 1988. Japan Scientific Societies Press. Tokyo, Japan. 294–299.

Wieringa, K. T. 1955. Der Abbau der Pektine; der erste Angriff der organischen Pflanzensubstanz. Z. Pflanzenernährung 69:150–155.

Wieringa, K. T. 1966. Solid media with elemental sulphur for detection of sulphur-oxidizing microbes. Ant. v. Leeuwenhoek 32:183–186.

Wilde, P. 1964. Gezielte Methoden zur Isolierung antibiotisch wirksamer Boden-Actinomyceten. Z. Pflanzenkrankh. 71:179–182.

Wilkin, G. D., and A. Rhodes. 1955. Observations on the morphology of Streptomyces griseus in submerged culture. J. Gen. Microbiol. 12:259–264.

Williams, S. T., and F. L. Davies. 1965. Use of antibiotics for selective isolation and enumeration of actinomycetes in soil. J. Gen. Microbiol. 38:251–261.

Williams, S. T., F. L. Davies, D. M. Hall. 1969. A practical approach to the taxonomy of actinomycetes isolated from soil. *In:* J. G. Sheals (Ed.), The Soil Ecosystem. The Systematics Association. London, UK. 8:107–117.

Williams, S. T., and T. Cross. 1971a. Actinomycetes. *In:* J. R. Norris and D. W. Ribbons (Eds.), Methods in Microbiology. Academic Press. London, UK. 4:295–334.

Williams, S. T., and C. I. Mayfield. 1971b. Studies on the ecology of actinomycetes in soil. III: The behavior of neutrophilic streptomycetes in acid soil. Soil Biol. Biochem. 3:197–208.

Williams, S. T., F. L. Davies, C. I. Mayfield, and M. R. Khan. 1971c. Studies on the ecology of actinomycetes in soil. II: The pH requirements of streptomycetes from two acid soils. Soil Biol. Biochem. 3:187–195.

Williams, S. T., M. Shameemullah, E. T. Watson, and C. I. Mayfield. 1972b. Studies on the ecology of actinomycetes in soil. VI: The influence of moisture tension on growth and survival. Soil Biol. Biochem. 4:215–225.

Williams, S. T., and M. R. Khan. 1974. Antibiotics—a soil microbiologist's viewpoint. Postepy Higieny I Medycyny Doswiadczalnej 28:395–408.

Williams, S. T., T. McNeilly, and E. M. H. Wellington. 1977. The decomposition of vegetation growing on metal mine waste. Soil Biol. Biochem. 9:271–275.

Williams, S. T. 1978a. Streptomycetes in the soil ecosystem. *In:* M. Mordarski, W. Kurylowicz, and J. Jeljaszewicz (Eds.), Nocardia and Streptomyces: Proceedings of the International Symposium on Nocardia and Streptomyces, Warsaw, 1976. Gustav Fischer-Verlag. Stuttgart, Germany. 137–144.

Williams, S. T., and T. H. Flowers. 1978b. The influence of pH on starch hydrolysis of neutrophilic and acidophilic streptomycetes. Microbios 20:99–106.

Williams, S. T., and C. S. Robinson. 1981. The role of streptomycetes in decomposition of chitin in acidic soils. J. Gen. Microbiol. 127:55–63.

Williams, S. T. 1982a. Are antibiotics produced in soil? Pedobiologia 23:427–435.

Williams, S. T., and E. M. H. Wellington. 1982b. Principles and problems of selective isolation of microbes. *In:* J. D. Bu'Lock, L. J. Nisbet, and D. J. Winstanly (Eds.), Bioactive Products: Search and Discovery. Academic Press. New York, NY. 9–26.

Williams, S. T., and E. M. H. Wellington. 1982c. Actinomycetes. *In:* A. L. Page, R. H. Miller, and D. R. Keeney (Eds.), Methods of Soil Analysis, Part 2: Chemical and Microbiological Properties. American Society of Agronomy and Soil Science Society of America. Madison, WI. 969–987.

Williams, S. T., M. Goodfellow, G. Alderson, E. M. H. Wellington, P. H. A. Sneath, and M. J. Sackin. 1983a. Numerical classification of Streptomyces and related genera. J. Gen. Microbiol. 129:1743–1813.

Williams, S. T., M. Goodfellow, E. M. H. Wellington, J. C. Vickers, G. Alderson, P. H. A. Sneath, M. J. Sackin, and A. M. Mortimer. 1983b. A probability matrix for identification of some streptomycetes. J. Gen. Microbiol. 129:1815–1830.

Williams, S. T., M. Goodfellow, and J. C. Vickers. 1984a. New microbes from old habitats? *In:* D. P. Kelley and N. G. Carr (Eds.), The Microbe 1984, Part 2: Prokaryotes and Eukaryotes. Society for General Microbiology Symposium 36. Cambridge University Press. Cambridge, UK. 219–256.

Williams, S. T., S. Lanning, and E. M. H. Wellington. 1984b. Ecology of actinomycetes. *In:* M. Goodfellow, M. Mordarski, and S. T. Williams (Eds.), The Biology of the Actinomycetes. Academic Press. London, UK. 481–528.

Williams, S. T., M. Goodfellow, and G. Alderson. 1989. Genus Streptomyces Waksman and Henrici 1943, 339AL. *In:* S. T. Williams, M. E. Sharpe, and J. G. Holt (Eds.), Bergey's Manual of Systematic Bacteriology. Williams and Willkins. Baltimore, MD. 4:2453–2492.

Williams, J. G., A. R. Kubelik, K. J. Livak, J. A. Rafalski, and S. V. Tingey. 1990. DNA polymorphisms amplfied by arbitrary primers are useful as genetic markers. Nucl. Acids Res. 18:6531–6535.

Wipat, A., E. M. H. Wellington, and V. A. Saunders. 1994. Monoclonal antibodies for Streptomyces lividans and their use for immunomagnetic capture of spores from soil. Microbiology 140:2067–2076.

Wirth, S., and A. Ulrich. 2002. Cellulose-degrading potentials and phylogenetic classification of carboxymethyl-cellulose decomposing bacteria isolated from soil. Syst. Appl. Microbiol. 25:584–591.

Witt, D., and E. Stackebrandt. 1990. Unification of the genera Streptoverticillium and Streptomyces, and amendation of Streptomyces Waksman and Henrici, 1943, 339AL. Syst. Appl. Microbiol. 13:361–371.

Wood, S., S. T. Williams, and W. R. White. 1983. Microbes as a source of earthy flavors in potable water—a review. Int. Biodeterior. Bull. 19:83–97.

Zhang, Z., Y. Wang, and J. Ruan. 1997. A proposal to revive the genus Kitasatospora (Omura, Takahashi, Iwai, and Tanaka 1982). Int. J. Syst. Bacteriol. 47:1048–1054.

Zhang, Q., W.-J. Li, Y.-L. Cui, M.-G. Li, L.-H. Xu, and C.-L. Jiang. 2003. Streptomyces yunnanensis sp.nov., a mesophile from soils in Yunnan, China. Int. J. Syst. Evol. Microbiol. 53:217–221.

Prokaryotes (2006) 3:605–622
DOI: 10.1007/0-387-30743-5_23

CHAPTER 1.1.8

The Family Streptomycetaceae, Part II: Molecular Biology

HILDGUND SCHREMPF

Introduction

Some members of the order of Actinomycetales were already identified about a century ago (for review, see Kutzner, 1981). The current phylogenetic tree includes Bifidobacteriaceae, Actinomycetaceae, Arthrobacteriaceae, Cellulomonadaceae, Microbacteriaceae, Dermathophylaceae, Propionobacteriaceae, Nocardioidaceae, Frankiaceae, Corynebacteriaceae, Mycobacteriaceae, Nocardiaceae, Actinoplanaceae, Pseudonocardiaceae, Streptomycetaceae and Streptosporangiaceae.

The family Streptomycetaceae includes the genus *Streptomyces*, which comprises strains formerly classified as *Chania*, *Elytrosporangium*, *Kitasatoa*, *Kitasosporia*, *Actinosporangium* and *Streptoverticillium*. All members of the Streptomycetaceae family have a complex life cycle, featuring an extended vegetative substrate mycelium and aerial hyphae, in which spores form upon depletion of nutrients. They contain specific menaquinones, incorporate LL-diaminopimelic acid (a diagnostic amino acid) into their peptidoglycan, but lack a diagnostic sugar (Kutzner, 1981). In addition they contain characteristic signatures within the genes for 23S and 16S rRNA (Roller et al., 1992; Embley and Stackebrandt, 1994). An enormous number (>800) of *Streptomyces* species has been described by numerical taxonomy (Goodfellow et al., 1990). Kämpfer presents the current methods for classification in an another chapter, The Family Streptomycetaceae, Part I: Taxonomy, in this Volume.

Sixty years ago, streptomycin was the first antibiotic which was classified from a *Streptomyces* strain. Since then, streptomycetes have proved to be the richest sources of thousands of low-molecular weight, chemically different compounds having antibacterial, antifungal, antiparasitic, agro-active, cytostatic or other biological activities. Additionally, they have provided many enzymes which are required for the natural turnover of many macromolecules in soil.

As outlined in this chapter, the tools of molecular biology in combination with physiological, biochemical, and genetic studies have considerably deepened understanding of the biology of streptomycetes. This knowledge is the current and future basis for exploiting the seemingly endless repertoire of metabolites and activities of the family of Streptomycetaceae.

The Developmental Growth Cycle

Growth Characteristics

Streptomycetes grow by tip extension as long, branching vegetative hyphae, which rarely have septae. The compartments within the substrate hyphae contain numerous copies of the chromosomal DNA. Aerial hyphae begin to form upon depletion of nutrients. The aerial structures contain several surface layers. One of them is the hydrophobic rodlet layer, which comprises the proteins RdlA and RdlB. The corresponding genes within *Streptomyces coelicolor* A3(2) and *S. lividans* are expressed within growing aerial hyphae but not within spores (Claessen et al., 2002). Partitioning of the chromosomes occurs within the aerial hyphae, in which a comparatively synchronous septation entails the formation of compartments, resulting in the formation of spores after maturation. Each of the spores (or the majority of them) contains a single chromosome. Using fusions of the gene of interest with the gene for green fluorescent protein (GFP), the distribution of the corresponding fusion protein can be visualized during development (Flardh, 2003).

Bald (bld) mutants lack aerial mycelia and thus have a shiny (bald) appearance. In addition to the block in differentiation, *bld* mutations lead to pleiotrophic effects, which often cause defects in carbon catabolite repression and cell-cell signaling and sometimes prevent antibiotic production. Whereas most *bld* genes within *S. coelicolor* A3(2) determine regulatory proteins, *bld*K can encode a transporter for an extracellular peptide. BldA specifies a tRNA for a rare codon and *bld*N determines a member of the extracytoplasmatic function subfamily of RNA polymerase sigma factors. During differentiation, the pro-sigma

BldN-form is processed to the mature Bld (Bibb and Buttner, 2003). As expected *S. coelicolor* A3(2) and *S. griseus* share several orthologous genes which are important for the development of the aerial hyphae (Chater and Horinouchi, 2003). The genes *ram*C and *ram*R are required for the production of the aerial hyphae, yet are not essential for vegetative growth. The production of RamC requires developmental regulatory genes (*bld*D, *crp*A and *ram*R, but not *bld*N and *bld*M; O'Connor et al., 2002).

The action of different *whi* (white) genes induces a curling of the aerial hyphae, their septation, and finally spore formation. Early regulatory *whi* genes (*A, B, G, H, I* and *J*) from *S. coelicolor* A3(2) are needed for septation during sporulation and the *whiA* gene plays a key part in switching the extension of aerial hyphae towards septation (Ainsa et al., 2000). The genome of *S. coelicolor* A3(2) encodes about 60 sigma (σ) factors. A subfamily of nine σ factors controls late sporulation, and σ^H is involved in the development of aerial hyphae (in addition to responses to various stresses), whereas spore maturation is governed by σ^F.

Bacterial cell division is initiated by the polymerization of the FtsZ protein on the inner surface of the cytoplasmatic membrane to form the Z-ring structure at the future division site (Lutkenhaus, 1997; Margolin, 2003). FtsZ is a homologue of tubulin and interacts with several proteins of auxiliary functions (i.e., Z-ring anchoring, stabilization and others). Transcription studies in *S. griseus* have suggested that *ftsZ* is expressed during both vegetative growth and sporulation (Dharmatilake and Kendrick, 1994). The *S. coelicolor* A3(2) FtsZ is required for septation within the vegetative substrate mycelium as well as for the synchronous formation of septae within the developing aerial hyphae prior to the detectable partitioning of nucleoids (Grantcharova et al., 2003).

The differentiation cycle generally occurs on solid media, although some strains (i.e., *S. griseus*; McCue et al., 1996) synchronously undergo sporulation in liquid culture. Recently differentiation of *S coelicolor* A3(2) has been demonstrated in standing liquid minimal media (Van Keulen et al., 2003). Interestingly, gas vesicles are present within hyphae at the air interface of standing liquid culture. *Streptomyces coelicolor* A2(3) has been shown to contain a gene cluster encoding proteins resembling gas vesicles of cyanobacteria and their homologues within halophilic archaea.

Several *Streptomyces* strains show a transient slow down during growth in liquid culture before entering the stationary phase. This transition phase is characterized by an increase in ppGpp (guanosine 3′, 5′-bispyrophosphate), by a decrease in GTP, as well as by the activation of genes required for secondary metabolism. The synthesis of two ribosomal proteins is drastically reduced when the culture approaches the stationary phase (Blanco et al., 1994).

Autoregulating Factors

Autoregulating factors form during the late stage of growth, presumably under the control of global regulatory systems. These autoregulating compounds (Khokhlov et al., 1967; Horinouchi and Beppu, 1992) have a 2,3-disubstituted σ-butyrolactone skeleton in common. They have been classified into three groups: the A-factor type (*S. griseus*), virginiae butanolide (VB) type (*S. virginiae*), and IM-2 type (*S. spec.* FRI-5). Although the structural differences among the butyrolactones are slight, each of them is effective at very low concentrations (10^{-9} M). Their mode of action resembles that of a highly specific hormone or morphogen for a given strain, controlling the formation of various secondary metabolites (such as antibiotics and pigments), as well as the formation of aerial mycelium. Presumably, a given σ-butyrolactone is also an effective chemical signal for communication between different parts of the mycelium within one strain (Horinouchi and Beppu, 1992). The receptor proteins ArpA, BarA and IM2 deduced from the corresponding genes (*arp*A in *S. griseus*, *bar*A in *S. virginiae*, and *far*A in *Streptomyces* spec. FRT-5) share an NH_2-terminally located helix-turn-helix (HTH) motif and each includes a binding domain with high specificity for the respective ligand. Recent studies suggest that ArpA and BarA represent transcriptional regulators recognizing a specific DNA motif, and it was suggested that they prevent the expression of certain key gene(s) during the early growth phase before a critical level of the corresponding ligand is reached. Upon binding to the ligands, the key gene(s) for secondary metabolites and morphogenesis are transcribed. Like ArpA, CprB from *S. coelicolor* A3(2) acts as a negative regulator and affects actinorhodin formation and sporulation. In contrast, the second *S. coelicolor* A3(2) homologue, CprA, is a positive regulator accelerating synthesis of secondary metabolite(s) and sporulation (Onaka et al., 1998).

Accumulating evidence reveals that a pair (or sometimes multiple pairs) of genes encoding a probable γ-butyrolactone biosynthetic enzyme (an AfsA-like protein and its specific receptor ArpA-like protein) is contained in a number of biosynthetic gene clusters for secondary metabolites in various *Streptomyces* species. It controls the biosynthesis of respective metabolites by activating pathway-specific regulatory genes (Horinouchi, 2002).

Geosmin, Storage Compounds, and Exopolysaccharides

Geosmin (1, 10-*trans*-dimethyl-*trans*-9-decalol) is the characteristic odor constituent of streptomycetes. Its synthesis is initiated by cyclization of farnesyldiphosphate to a germacradiene-type intermediate (Spiteller et al., 2002).

Under a range of conditions which allow sporulation, glycogen accumulates in sporogenic hyphae and is presumably converted into trehalose during the final period of spore maturation (Rueda et al., 2001). The operon required for glycogen-trehalose metabolism within *S. coelicolor* A3(2) has been characterized (Schneider et al., 2000).

Triacylglycerols (TAG) are fatty acid triesters of glycerol which are abundant in yeast, fungi, plants and animals and have also been found in bacteria. They do, however, seem to be widespread and vary in composition among actinomycetes including species of *Streptomyces*, *Rhodococcus*, *Nocardia* and *Mycobacterium*. Poly-3-hydroxybutyric acid (PHB) has been detected within different *Streptomyces* strains (Verma et al., 2002). A gene cluster encoding for exopolysaccharide formation has recently been identified within a *Streptomyces* strain (Wang et al., 2003).

Chromosomal DNA

Structure and Genomic Information

Most *Streptomyces* strains contain a large genome. On the basis of genetic studies, a circular genome was concluded to be present in *S. coelicolor* A3(2). Ordered cosmid libraries in combination with comparisons of physical maps from the wildtype and mutant strains of *S. coelicolor* A3(2) and *S. lividans* lead to the deduction that the chromosome occurs in a linear and in a circular form (Lin et al., 1993; Redenbach et al., 1996). Pulse-field gel electrophoresis (PFGE) studies revealed the existence of a linear chromosome in other streptomycetes, including *S. ambofaciens* (Leblond et al., 1996), *S. antibioticus*, *S. moderatus*, *S. lipmanii*, *S. parvulus*, *S. rochei* (Lin et al., 1993), *S. griseus* (Lezhava et al., 1995) and *S. hygroscopicus* (Pang et al., 2002). Linear topology of the chromosomal DNA has been demonstrated for a *Streptoverticillium* sp. belonging to the family of Streptomycetaceae and for representatives of other actinomycetes, including *Actinoplanes philipinensies*, *Nocardia asteroides*, *Saccharopolyspora erythrea* (Redenbach et al., 1998) and *Rhodococcus fascians* (Crespi et al., 1992). In this context, it is interesting that genomes of different *Myco-*

bacterium sp. were found to be circular (Philipp et al., 1996).

The *S. coelicolor* A3(2) linear chromosome comprises about 8.667 mega base pairs (Mbp) from which 7825 genes can be deduced. Twenty gene clusters encode known or predicted secondary metabolites (Bentley et al., 2002). Recently the *S. avermitilis* genome was determined to comprise about 9.025 Mbp (average G+C content 70.7 mol%) and to encode 7574 potential open reading frames, 35% of which constitute 721 paralogous families. Thirty gene clusters could encode secondary metabolites. One region of 6500 Mbp has been highly conserved with respect to gene order in the *S. avermitilis* and *S. coelicolor* A3(2) genomes and contains predicted essential genes. However the terminal regions are not conserved and preferentially contain nonessential genes (Ikeda et al., 2003). An ancient synteny (conservation of gene-order) has been revealed between the central core of the *S. coelicolor* A3(2) chromosome and the whole chromosome of *Mycobacterium tuberculosis* and *Corynebacterium diphtheriae* (Bentley et al., 2002).

Initiation of DNA Replication and Terminal Repeats

The structure of the chromosomal replication origin (*oriC*) region is highly conserved among the investigated *Streptomyces* strains *S. coelicolor* A3(2) (Calcutt and Schmidt, 1992), *S. lividans* (Zakrzewska-Czerwinska and Schrempf, 1992), *S. ambofaciens*, *S. antibioticus*, *S. chrysomallus* and *S. reticuli* (Jakimowicz et al., 1998). Contrary to the high overall G+C content (69–73 mol%) of *Streptomyces* DNA, the *oriC* region is comparatively rich in A+T (64 mol%) and contains 19 DnaA boxes (consensus sequence TTGTCCACA) arranged in two clusters (134 bps). Thus, interactions between DnaA molecules may bring two clusters of DnaA boxes separated by the spacer into functional contact by loop formation (Jakimowicz et al., 2000). The *oriC* of *S. lividans* and *S. coelicolor* A3(2) is located centrally, opposite the terminal ends located in the formerly designated "silent region" of the chromosome. EshA (a 52-kDa protein whose expression is developmentally controlled) forms during the late growth phase. It has been proposed that the *S. griseus* EshA protein positively affects (or regulates) the replication of DNA in wildtype cells during the late growth phase but leads to aberrant phenotypes in mutant cells owing to disturbed DNA replication (Saito et al., 2003).

The chromosomal ends of a few *Streptomyces* species contain terminal inverted repeats (TIRs) covalently bound to protein, presumably at their

5′-ends. Depending on the strains, the length of the TIRs varies between 20 kb and 550 kb. They have a considerable sequence-divergence except for the first, about 200 bps comprising palindromic sequences. Modeling predicts that the 3′-ends of these sequences can form extensive hairpin structures resembling those characterized within the 3′-ends of the single-stranded parvoviral DNA. Most of the putative hairpins contain a GCGCAGC-sequence able to form a single C-residue loop closed by a sheared G:A base pair. As the *Streptomyces* chromosome is replicated from the centrally located replication origin, 3′-single-stranded gaps arise which are assumed to be patched by an unknown mechanism involving the covalently bound terminal proteins (TP; Huang et al., 1998). The *tpgC* gene encodes the TP (TpgC) protein of the *S. coelicolor* A3(2) genome and is predicted to have a DNA-binding motif (which is found in reverse transcriptase) and an amphiphilic domain, possibly involved in protein-protein interactions or membrane-binding. Whereas one chromosomal *tpg* gene has been found in *S. coelicolor* A3(2) and *S. rochei*, three homologues are present within the *S. lividans* chromosome (Bao and Cohen, 2001; Yang et al., 2002). The TPs (about 185 amino acids) of *Streptomyces* form a conserved group that is divergent from those of linear genomes of bacteriophages (i.e., Ø29, PRD1) and viruses (i.e., adenoviruses, 600–660 amino acids) or eukaryotic linear plasmids. Recent results indicate that the Tap protein functions to recruit and position the Tpg protein at the telomers of replication intermediates (Bao and Cohen, 2003).

Deletions and Amplifications

Among colonies—or sectors of colonies—of *Streptomyces* species, a high spontaneous variability of pigmentation, sporulation, antibiotic biosynthesis, or A-factor production is noticeable. Several unstable genes encode various antibiotic resistances, A-factor formation, and synthesis of tyrosinase or arginosuccinate. Genetic instability can be stimulated by mutagens including mitomycin, ethidium bromide, and ultraviolet light and by gyrase- (topoisomerase II) inhibiting antibiotics. These variations are the result of large chromosomal deletions which occur preferentially at the telomeric and subtelomeric regions and may include up to 2 Mbp of DNA (Hütter and Eckhardt, 1988; Schrempf et al., 1989; Leblond and Decaris, 1994; Chen, 1995). Deletions may occur in high frequencies (up to 10^{-2}), in several cases sequentially at either one or both ends of the chromosome. Concomitantly, amplification(s) of amplifiable unit(s) of DNA (AUD) may occur.

In some cases, amplifications of certain genes correlate with enhanced levels of gene products. Accordingly, the 4.3-kb AUD (*S. lividans*) encodes a surface-located protein which is over-expressed in large quantities after amplification (Betzler et al., 1997). Upon selection for high resistance towards spectinomycin (Hornemann et al., 1986) or chloramphenicol (Dittrich et al., 1991), variants comprising amplifications of the corresponding gene have been encountered. Moreover, it has been suggested (Qin and Cohen, 2002) that spontaneous circularization of linear chromosomes may be the consequence of telomere damage that cannot be repaired by homologous recombination. An *S. lividans* mutant deficient in the *recR* gene that encodes one of the proteins of the RecF pathway (Pelaez et al., 2001) segregates a large portion of small colonies, the majority of which fail to produce aerial mycelium. Many of them are nonviable or have chromosomal deletion(s) including the chloramphenicol resistance gene. The RecA protein within different *Streptomyces* species is larger than the corresponding one of other bacteria as it contains an extra long C-terminus (Ahel et al., 2002).

The variability of the chromosomal DNA is increased by its interaction with linear and circular plasmids, phages, transposons, and insertion-elements (see below). As streptomycetes inhabit quickly changing environments (see below), those variants that are best adapted are expected to rapidly multiply. The high plasticity of the genome is likely an effective prerequisite for quick adaptation.

Repetitive DNA (mostly insertion-sequences or directly repeated GC-elements) is used for the efficient classification of mycobacterial strains (Kamerbeek et al., 1997). Comparative investigations of genomic sequences have permitted the deduction of additional diagnostic elements for mycobacteria (Philipp et al., 1996). Comparative strategies are expected to facilitate in future the classification *Streptomyces* strains.

Plasmids, Transposons, and Phages

Circular Plasmids and Their Vector Derivatives

In addition to the chromosomal DNA, *Streptomyces* strains contain various types of circular plasmids of different sizes. The low-copy number circular plasmid SCP2 was the first one to be isolated from a streptomycete (Schrempf et al., 1989). Despite its relatively large size, it is used as a cloning vector. The plasmid SCP2* (a variant of SCP2; Hopwood et al., 1985) has recently been sequenced and deduced to encode 34 proteins, most of them lacking similarity to known ones. The replication region contains the *rep*I and the *rep*H genes encoding small proteins. While the *tra*A gene is essential for DNA transfer and pock-formation, 10 additional genes are suspected to be involved in conjugation and DNA spreading. Among them are genes required for plasmid stability. The ParA gene product shows similiarity to a family of ATPases (Haug et al., 2003).

The natural plasmid pIJ101 (about 8.8 kb) is a high-copy number conjugative *Streptomyces* plasmid. Regions required for its replication, stability, transfer and distribution have been identified. Efficient conjugation among streptomycetes necessitates the plasmid-encoded *tra* gene and the *cis*-acting locus of transfer (clt; Ducote et al., 2000). A variety of nonconjugative derivatives carrying a selectable marker have been constructed (Hopwood et al., 1985). The derivative pIJ702 (5.8 kb), containing a tyrosinase and a thiostrepton-resistance gene, has been frequently used as a vector. After in vitro mutagenesis, a derivative of pIJ702 exhibiting a temperature-sensitive replication phenotype has been selected. Several *Escherichia coli*-*Streptomyces* bifunctional vectors have been constructed using derivatives of pIJ101. However, some of them are not particularly stable. Nonetheless, pWHM3 and pWHM4 (Vara et al., 1989) have been used successfully for gene cloning.

The plasmid pSG5 (initially found in *Streptomyces ghanaensis*) is 12.3 kb, is naturally temperature-sensitive, and cells contain approximately 50 copies per chromosome. A number of vectors, several of them bifunctional, have been constructed using pSG5. The replicon of pSG5 is naturally temperature-sensitive. All vectors based on this replicon can be eliminated easily at nonpermissive temperatures and can be used as suicide delivery systems (see below) for transposons (Baltz et al., 1997) as well as for gene disruption and gene replacement experiments (Muth et al., 1995). A pSG5-based cosmid vector has been shown to be transferable to a *Micromonospora aurantia* strain by intergenic conjugative transfer (Rose and Steinbuchel, 2002).

The multicopy plasmids pSN22 (Kataoka et al., 1991), pJV1 (Bailey et al., 1986) and pSMA2 (Pernodet et al., 1984) replicate via a rolling circle mechanism (Hagege et al. 1993; Servín-González, 1993). All Rep proteins encoded by these plasmids carry at their N-terminus a motif resembling zinc fingers found also in DNA topoisomerase I or other unwinding proteins. The deduced Rep protein (Muth et al., 1995) of pSG5 differs from the above-mentioned ones insofar as it is most closely related to one of the pC194 family in other Gram-positive bacteria.

The 11-kb pSMA2 (initially identified within *S. ambofaciens*) occurs in an autonomous form and is subject to rolling circle DNA replication. The pSMA2 plasmid is self-transmissible and mobilizes chromosomal markers. Specific recombination with the chromosomal attB attachment site requires the pSMA2-encoded integrase; the plasmid can stably integrate or can subsequently re-excise. Replicase, excisonase, and integrase genes constitute an operon that is positively regulated by the regulator gene *pra* (Sezonov et al., 1998). Recent studies have suggested that during conjugation, pSAM2 is transferred as double-stranded DNA, depending on the presence of the plasmid-located *trasA2* gene. Notably, TraSA (Possoz et al., 2001) shows similarities to other *Streptomyces* plasmid transfer proteins but also with SpoIIIE involved in chromosome partitioning into the prespore in *Bacillus subtilis* and the *E. coli* FtsK involved in movement of double-stranded DNA. Replicative and integrative vectors using pSMA2 were constructed.

Linear Plasmids

Although linear plasmids are abundant among streptomycetes, only some of their encoding functions have been identified. These include antibiotic production (Kinashi et al., 1991; Gravius et al., 1994) and mercury resistance (Ravel et al., 1998). Some linear plasmids are efficiently transferred during conjugation. The *tra* genes of the linear plasmid pBL1 have been characterized by insertional inactivation (Zotchev and Schrempf, 1994). Among the features common to all linear plasmids are terminal inverted repeats of varying length and the presence of a terminal protein, the latter being linked to the 5'-ends as shown for SCP1 (*S. coelicolor* A3(2); Kinashi et al., 1991) or pSLA2 (*S. rochei*; Hirochika et al., 1984). The linear plasmid pHZ6 (70 kb) residing within *S. hygroscopicus* is a temperature-sensitive replicon, representing a hitherto unique feature among known natural linear plasmids (Pang et al., 2002).

The replication mechanism is best understood for pSLA2. In contrast to the linear Ø29 *Bacillus* phage DNA, its replication is initiated bidirectionally near the center and proceeds towards its telomeric ends, generating 3' leading-strand overhangs. The basic replicon (1.9 kb) contains the information required for autonomous replication in the circular form (Hiratsu et al., 2000). The 21-kDa terminal protein (Tpg) of pSLA2 had been isolated. Its gene encodes a protein with a reverse transcriptase-like domain. Interestingly, homologues of the *tpgR* gene are located on other linear plasmids or chromosomes (see above). As shown for *S. lividans* (Chen, 1995) and *S. rimosus* (Gravius et al.,

1994), large linear plasmids can interact with the ends of the chromosome.

The SLP1 (int) element (17.2 kb) specifically integrates into the *S. coelicolor* A3(2) chromosome. During mating with *S. lividans*, SLP1 (int) can excise and form a family of autonomously replicating conjugative plasmids (Hagege et al., 1999). SLP2 (50 kb) is a linear *S. lividans* plasmid that contains short terminal repeats and covalently bound terminal proteins. The right most part (15.4 kb) of SLP2 is identical to the *S. lividans* chromosome. The plasmid encodes many genes predicted to be involved in replication, partitioning, conjugational transfer, and intramycelial spread. Though the two telomers differ in their sequence, they retain similar secondary structures (Huang et al., 2003).

Interestingly, several other members of the order of Actinomycetales harbor genes required for isopropylbenzene and trichlorethylene catabolism (*Rhodococcus erythropolis*; Kebeler et al., 1996), biphenyl degradation (*Rhodococcus erythropolis* and *Rhodococcus globerulus*; Kosono et al., 1997), hydrogen autotrophy (*Rhodococcus opacus*, formerly *Nocardia opaca*; Kalkus et al., 1993) and fasciation in plants (*Rhodococcus fascians*; Crespi et al., 1992) on linear plasmids. Linear plasmids of varying sizes have also been discovered in several *Mycobacterium* species (*M. xenopi*, *M. branderi*, *M. celaturum* and *M. arium*). Their termini are related to those of linear plasmids from *Streptomyces* and *Rhodococcus* species (Picardeau and Vincent, 1998). Exploring whether the linear and circular plasmids are exchanged during conjugation among various actinomycetes should be of interest.

Transposons

Compared to other bacteria, relatively little knowledge exists for *Streptomyces* transposons. A truncated copy of Tn*4811* of *S. lividans* is present at the terminal inverted repeats of the *S. coelicolor* A3(2) giant linear plasmid SCP1 (Spychaj and Redenbach, 2001). The transposon Tn*4556* was discovered in *Streptomyces fradiae* and used for transposition studies (Chung and Crose, 1989; Ikeda et al., 1993). Tn*4560* (8.6 kb) is a derivative of Tn*4556*, a Tn*3*-like element from *Streptomyces fradiae*. It contains a viomycin resistance gene and its usefulness as a mutagenesis tool within *Mycobacterium smegmatis* was demonstrated by transformation of this host via a delivery plasmid (Bhatt et al., 2002). On the basis of the *S. lividans* insertion element IS*493*, transposons containing an antibiotic resistance gene have been developed which are suitable for transposon mutagenesis of various *Streptomyces* strains (Baltz et al., 1997). The IS*6100* element

(originally identified in *Mycobacterium fortuitum*; Martin et al., 1990) belonging to the IS*6* family normally forms a cointegrate as the endproduct of transposition. The IS*6100*-based minitransposon Tn*1792* has been developed as a genetic tool. The resolution of Tn*1792* cointegrates within *S. avermitilis* has facilitated the isolation of insertion mutants and the subsequent rescue of Tn*1792*-tagged sequences. Subsequently, genes involved in the production of secondary metabolites have been identified (Pitman et al., 2002). An engineered derivative of the *E. coli* transposon Tn*5* also transposes randomly into the *S. lividans* genome (Volff and Altenbuchner, 1997).

Phages

As outlined previously, phages of broad or narrow host range can be gained easily from soil and several of them have been used for classifying strains (for review, see Kutzner, 1981).

The *Streptomyces* temperate phage *j*31 has a relatively broad host spectrum. It has been revealed that the *j*C31 receptor is glycosylated. The *S. coelicolor* A3(2) gene encoding the homologue to dolichol-phosphate-mannose synthase is essential for phage sensitivity (Cowlishaw and Smith, 2002). Also, *j*31 has a linear genome (41.5 kb) with 3′ overhanging cohesive ends (Lomovskaya et al., 1980). The integrase from *j*C31 is a member of the serine recombinase family of site-specific recombinases. The integrase recombines an attP site in the phage genome with a chromosomal attB site of its *Streptomyces* host (Combes et al., 2002). Thus, *j*31 has been used to develop cloning and integration vectors as well as cosmids (Hopwood et al., 1985).

The transducing phages (DAH2, DAH4, DAH5 and DAH6) have been shown to transduce multiple chromosomal markers at frequencies ranging from 10^5 to 10^9 per plaque-forming unit. The host range comprises distantly related species including *S. coelicolor* A3(2), *S. veticillus* and *S. avermitilis* (Burke et al., 2001).

Role of Secreted Proteins

Recognition and Hydrolysis of Polysaccharides

Streptomycetes play a very important role in the turnover of chitin, the second most abundant polysaccharide in nature. Contrary to other chitinolytic bacteria, almost every *Streptomyces* species uses chitin not only as a carbon but also as a nitrogen source (Blaak and Schrempf, 1995). The few analyzed *Streptomyces* strains produce several chitinases (Robbins et al., 1988;

Miyashita et al., 1991). Some of the chitinase genes were cloned, and the corresponding *Streptomyces* transformants secreted the corresponding enzymes. The efficient degradation of crystalline forms of chitin is dependent on the presence of a chitin-binding domain in a chitinase (Blaak and Schrempf, 1995). During growth on chitin, many streptomycetes secrete, in addition to chitinases, small proteins (Kolbe et al., 1998) that act as an adhesive, mediating a biofilm-like contact between chitin-containing substrates or organisms.

Enzymes catalyzing the hydrolysis of soluble forms of cellulose have been investigated for a few *Streptomyces* strains (including *S. lividans*) and cloning of several corresponding genes has been carried out (Crawford and McCoy, 1972; Kluepfel et al., 1986; Nakai et al., 1988). *Streptomyces reticuli* specifically adheres to crystalline cellulose via a membrane-anchored protein (Walter et al., 1998) and efficiently degrades crystalline forms of cellulose to forms of cellobiose (Schlochtermeier et al., 1992), which serve as nutrients.

The synthesis of amylases and their inhibitors is common among streptomycetes. Some of the genes have been cloned (Virolle and Bibb, 1988) and studies pertaining to a better understanding of the regulation of the glucose-repressed genes have been initiated (Nguyen et al., 1997). Sequence information suggests that the *S. coelicolor* A3(2) genome has many genes for glucosyltransferases. Those belonging to family 13 are predicted to encode starch-converting enzymes.

Xylan consists of a *b*-1-4-linked xylose polymer and commonly contains side branches of arabinosyl, glucoronosyl, acetyl, uronyl and mannosyl residues. Enzymes degrading this complex polymer have been used for different purposes, including biopulping and biobleaching in the paper industry and bioconversion. Xylanases and their genes were identified from *S. lividans* (Pagé et al., 1996), *S. halstedii* (Ruiz-Arribas et al., 1998), and the thermophilic *S. thermoviolaceus* (Tsujibo et al., 1997). Laccases including those produced by *S. cyameus* can be efficiently applied for biobleaching of kraft pulps (Arias et al., 2003).

Various *Streptomyces* strains produce enzyme inhibitors including allosamidine (chitinase inhibitor), isoflavone (Lee and Lee, 2001), gentisin (α-glucosidase inhibitor) and tendamistat (amylase inhibitor).

Proteases

Extracellular proteases are abundant among streptomycetes, and several corresponding genes have been characterized (Kim and Lee, 1995). Streptomycetes also contain a multitude of genes for protease inhibitors (Taguchi et al., 1995) including leupeptin and subtilisin (Hiraga et al., 2000). Keratinases are frequently encountered (for review, see Kutzner, 1981), as exemplified by the identification of keratinolytic proteinases in *S. pactum* catalyzing the extracellular reduction of disulfide bonds in the disintegration of chicken feathers (Böckle and Müller, 1997).

Lipases

The interest in bacterial lipases has increased considerably. However, only a few extracellular lipases and their genes have been characterized from *Streptomyces* strains (Servín-González et al., 1997; Sommer et al., 1997). A member of the family II of lipolytic enzymes secreted by *S. rimosus* has been characterized (Vujaklia et al., 2002).

Modification of Various Compounds

Streptomyces strains are dominant among isolates of latex rubber-degrading actinomycetes (Jendrossek et al., 1997). The average weight of synthetic rubber showed shifting to lower values after incubation with degrading *Streptomyces* strains. As diketone derivatives of oligo (*cis*-1,4) isoprene were identified, an oxidative pathway by β-oxidation has been suggested (Bode et al., 2001). Recently, genes involved in the rubber degradation have been identified from a *Micromonospora aurantiaca* strain (Rose and Steinbuchel, 2002).

Streptomycetes produce, in addition to enzymes that degrade macromolecules, a large repertoire of enzymes, including those for the modification of pharmacologically relevant compounds and xenobiotics (Peczynska-Czoch and Mordarski, 1988).

Secretion of Enzymes and Proteins

As in other bacteria, secretion of many proteins depends on the Sec pathway (Morosoli et al., 1999). Type I signal peptidases (SPases) are responsible for proteolytic cleavage of secreted proteins. The *S. lividans* chromosome has four adjacently located *sip* genes encoding functional SPases (Geukens et al., 2001). *Streptomyces lividans* harbors, as do several other bacteria, a functional twin-arginine translocation (Tat) pathway (Schaerlaekens et al., 2001) for folded proteins.

Streptomyces lividans has been shown to be a suitable host for the secretion of proteins, including enzymes (Geueke and Hummel, 2003) and homologous and heterologous proteins (Lammertyn and Anné, 1998).

General Metabolism

Transport Systems

Streptomycetes secrete a variety of enzymes (see above) that hydrolyze complex macromolecules, and the resulting low molecular weight compounds can frequently serve as carbon sources and in some cases (i.e., degradation-products of chitin, see above) as nitrogen sources. Some of the nutrients are taken up via a specific type of an ATP-binding cassette (ABC) transporter, which comprises a lipid-anchored, specific ligand-binding protein, interacting efficiently with either cellobiose and cellotriose, or chitobiose and N-acetylglucosamine, maltose, or others. In addition to specific membrane-bound permeases for each type of these transporters, the required ATP-binding protein can be shared by several transporters (Schlösser et al., 1997; Wang et al., 2002; Xiao et al., 2002). The genomic sequences of S. coelicolor A3(2) predict the presence of a large diversity of ABC transporters, most of which still require investigation with regard to ligand-specificity. Phosphoenolpyruvate (PEP) dependent phosphotransferase (PTS) uptake has been demonstrated for N-acetylglucosamine (Xiao et al., 2002) as well as for fructose. Interestingly, only very few PTS systems can be predicted from genomic sequences: An analysis of the S. coelicolor A3(2) genome has indicated the presence of genes encoding components of the PTS systems including permeases belonging to the mannitol-glucose and the glucose-sucrose family (Parche et al., 2000).

Homologous conventional ion transport systems can be deduced from genomic information, but they have not yet been investigated. However, a Streptomyces gene for a potassium ion channel (named KscA) has been identified. The functionally reconstituted KcsA protein carries the basic features of a eukaryotic potassium ion channel. Because of its small size, it has become a key model system for understanding the basic features of channel functioning (Schrempf et al., 1995; Meuser et al., 2001).

Streptomycetes have different types of transport systems for antibiotics, few of which have been analyzed with biochemical methods (Dittrich et al., 1991; Méndez and Salas, 1998; Gandlur et al., 2004), and genes for predicted transporters are frequently located within clusters for antibiotic biosynthesis (see below, the section Pharmacologically Active Substances).

Pathways of Primary Metabolites

Knowledge of primary metabolism within streptomycetes is relatively limited. Genes encoding key enzymes, which include fructose 1,6-bisphosphate aldolase and glucose-6-phosphate dehydrogenases, have been identified. Streptomyces lividans harbors two zwf genes determining isozymes of glucose-6-phosphate dehydrogenases (the first enzyme in the oxidative pentose phosphate pathway [PPP]) and one gene (devB) encoding 6-phosphogluconolactonase (Butler et al., 2002). The fluxes through the pentose phosphate pathway and the tricarboxylic cycle relative to glucose uptake have been evaluated in S. noursei using C^{13} labeling experiments (Jonsbu et al., 2001). The final goal will be to analyze metabolic fluxes (Wehmeier, 2001; Bentley et al., 2002).

Glutamine synthetase I (GSI) in S. coelicolor A3(2) is posttranslationally controlled by adenylyltransferase (as in enteric bacteria), but the regulation of GSI differs (Hesketh et al., 2002). A new class of glutamate dehydrogenases (GDH) has been identified within S. clavuligerus (Minambres et al., 2000). During balanced growth either mineral or organic nitrogen sources are readily used by S. pristinaespiralis. Glutamate and alanine serve as both nitrogen and carbon sources, whereas valine is only used as a nitrogen source, which consequently leads to excretion of 2-ketoisovalerate.

Upon phosphate limitation, glycerol originating from the breakdown of teichoic acids is released, leading to the recovery of phosphate from the cell wall (Voelker and Altaba, 2001). The biosynthesis of most secondary metabolites is strongly depressed by inorganic phosphate. The two-component system, PhoR (phosphate regulon sensor protein) and PhoP (DNA-binding response regulator protein), was shown to control pathways for primary and secondary metabolites (Sola-Landa et al., 2003).

A novel pathway utilizing acetate as the sole carbon source has been discovered within S. collinus (Han and Reynolds, 1997). Streptomyces coelicolor A3(2) can use fatty acids (C4-C18) as the sole carbon source, and enzymes for their in vivo β-oxidation have been identified (Banchio and Gramajo, 1997). The glyoxylate cycle comprising isocitrate lyase and malate synthase is an anaplerotic pathway and essential for growth with acetate as the sole carbon source. Corresponding genes have been identified within S. clavuligerus (Soh et al., 2001). Malonate is a three-carbon dicarboxylic acid and it is well known as a competitive inhibitor of succinate dehydrogenase. The genes matB and matC have been successfully used in generating industrial strains of Streptomyces employed in the production of antibiotics (Kim, 2002).

It will be necessary to investigate the pathways and genes required for the biosynthesis of primary compounds, examine the regulation (Rodríguez-García et al., 1997), and explore metabolic fluxes (Obanye et al., 1996), as the precursors of pharmacologically active compounds are derived from primary metabolites. Proteins involved in primary and secondary metabolism are currently under investigation using proteomic analysis (Hesketh et al., 2002). Expression patterns of many genes involved in metabolic pathways can also be investigated by using microarrays (Huang et al., 2001). Recent data suggest that global changes in gene expression during stages of diauxic growth, adaptation or starvation can be analyzed by gathering proteomic and metabolomic data (Novotna et al., 2003). Recently, the basis for transcriptome analysis has been established (Bucca et al., 2003).

Pharmacologically Active Substances

Diversity of Compounds

Streptomycetes synthesize an amazing variety of chemically distinct substances, many of which act as antibiotics, fungicides, cytostatics, or modulators of immune responses, and a huge diversity of ever increasing numbers of inhibitors for many different cellular processes. New *Streptomyces* isolates provide sources of novel natural compounds.

Strategies for the Identification of Gene Clusters

Genes encoding these pharmacologically active substances are located within DNA stretches of 20 kb to more than 100 kb. Cloning had been achieved by complementing mutants, by screening total genomic DNA or gene libraries with homologous or heterologous gene probes generated by cloning or with the help of polymerase chain reactions (PCR), and by transposon mutagenesis. The biosynthetic genes for antibiotics are frequently located near one or more genes mediating resistance to the corresponding antibiotic. Accordingly, the initial identification of such resistance genes has often facilitated the identification of a biosynthetic cluster. Protein-patterns change drastically during the production of secondary metabolites. Proteins deduced to be involved in the synthesis of secondary metabolites can be analyzed by various biochemical methods as well as by mass spectroscopy. NH_2-terminal and internal amino acid sequences of characteristic proteins can be determined to design oligonucleotides useful for screening gene libraries and thereby identifying the desired gene cluster.

The gene-cluster for the synthesis of the polyketide actinorhodin (Malpartida and Hopwood, 1984) was achieved by comple-

mentation of mutants. Subsequently, many other gene clusters for polyketides were cloned using a gene-probe for the predicted key step for polyketide synthesis. These polyketides include: daunorubicin (Stutzman-Engwall and Hutchinson, 1989), frenolicin (Bibb et al., 1994), granaticin (Sherman et al., 1989), griseusin B (Yu et al., 1994), jadomycin B (Han et al., 1994), mithramycin (Lombo et al., 1996), tetracyclines (Binnie et al., 1989), tetracenomycin C (Motamedi and Hutchinson, 1987), tetrangomycin (Hong et al., 1997), and urdamycin A (Decker and Haag, 1995). The utilization of *pks* (polyketide synthase) genes as probes has made possible the identification of a continuously increasing number of biosynthetic gene clusters. Analysis by PCR has led to the discovery of many *pks* genes in various microorganisms, including hitherto nonculturable ones.

Genes for several clusters of macrolides were identified, including the genes for carbomycin (Epp et al., 1987), oleandomycin (Swan et al., 1994), rapamycin (Schwecke et al., 1995), and tylosin (Fishman et al., 1987). Furthermore, the genes for peptide antibiotics (such as actinomycin [Stindl and Keller, 1994; Hsieh and Jones, 1995] and biolaphos [Murakami et al., 1986]) and cyclopentenoid antibiotics (such as methylenomycin; Chater and Bruton, 1985) were identified. Additionally genes were discovered for synthesis of nikkomycin (a nucleoside-peptide; Bormann et al., 1996), nosiheptide (a thiopeptide; Dosch et al., 1988), undecylprodigiosin (a pyrrole; Feitelson and Hopwood, 1983; Malpartida et al., 1990), ansamycins (such as rubradirin [Sohng et al., 1997] and rifamycin [August et al., 1998]), aminoglycosides (such as puromycin [Lacalle et al., 1992] and streptomycin [Ohnuki et al., 1985; Distler et al., 1987]), carbapenems (Nakata et al., 1989), cephamycin (Aharonowitz and Cohen, 1992; Paradkar et al., 1996) and cyclopilins (Pahl et al., 1997).

Following the above-cited pioneering work, cloning of genes for many biosynthetic pathways has been extended enormously (see reviews by Champness [1999], Meurer and Hutchinson [1999], and Challis and Hopwood [2003]). In addition to conventional studies of transcription (Kelemen et al., 1994), novel approaches apply the application of DNA microarrays to instigate growth phase responsive expression of genes for antibiotic biosynthetic pathways (Huang et al., 2001).

Towards Engineering of Pathways

Polyketides constitute a diverse class of substances, including antibiotics, anticancer and antiparasitic pigments, and immunosuppressants (see also above). Because gene clusters for polyketides are easy to detect, they are among the most studied. Consequently, a wealth of information has accumulated. The synthesis of polyketides requires polyketide synthases (PKSs) which are either multifunctional (type I; Cortes et al., 1990; Donadio et al., 1991; Weber et al., 2003) or monofunctional (type II; Bibb et al., 1989; Fernandez-Moreno et al., 1992). Type I PKSs are multifunctional enzymes consisting of modules, which control the sequential addition of acylthioester units to a growing polyketide chain and the subsequent modification of the condensation products. In contrast, Type II PKSs are composed of three to seven separate mono- or bifunctional proteins that are required repeatedly for the formation of the polyketide chain and some of its modifications. Both PKS types are biochemically similar to fatty acid synthases and catalyze decarboxylative condensations between a thioester-linked nascent carbon chain and short-chain fatty acid extender units. However, PKSs can use different starter molecules (acetate, propionate and butyrate) and chain extenders (malonate, methylmalonate and ethylmalonate) at various steps. Whereas condensations of fatty acids are typically followed by a trio of reactions (β-ketoreduction, dehydration and enoylreduction), a PKS may employ all, some or none of these reactions after each condensation step. Thus, the structural diversity of polyketides is due to the variable numbers and types of acyl units, reduction, dehydration, cyclization and aromatization reactions of the initial β-ketoacyl condensation products (Hopwood and Sherman, 1990; Donadio et al., 1991; Aparicio et al., 1994; Hutchinson and Fujii, 1995).

Multifunctional enzymes specify the choice of substrates and their sequence during a nonribosomal process for synthesis of oligopeptide antibiotics. Each amino acid (which may also be a nonprotein type) may undergo C_2-epimerization, C- or N-methylation, halogenation, and various types of oxidation reactions during or after assembly of the carbon backbone. Oligopeptides can be linear or circular and may occasionally comprise attached acyl groups or sugars (Lawen and Zocher, 1990; Von Döhren and Kleinkauf, 1997; Zuber and Marahiel, 1997).

Having established the key steps, the principal rules for the synthesis of polyketides, peptide antibiotics, β-lactams, aminoglycosides and other classes have emerged. Thorough analysis of this accumulated information meanwhile allows the design of a broader diversity of biosynthetic pathways. The modular arrangement of biosynthetic gene clusters facilitates the construction of hybrid gene clusters. Consequently, a tailoring of antibiotics will complement the screening for new antibiotics and thereby help overcome the problems linked to the broad repertoire of antibiotic resistances developed by pathogenic microorganisms.

Ecology

Habitats

Soil is the most important habitat of streptomycetes. Most soils contain 10^4 to 10^7 colony-forming units of streptomycetes per gram, representing 1–20% or even more of the total viable counts. Grass vegetation or soil rich in organic matter contain the highest numbers of streptomycetes. Most *Streptomyces* species prefer a neutral to mildly alkaline pH. However, several reports prove that some strains also grow under acidic (ca pH 3.5) or alkaliphilic (pH 8–11.5) conditions. As expected, streptomycetes settle in soil as microcolonies on particles of organic matter. In terrestrial habitats, streptomycetes constitute the most abundant actinomycete (90% or more). Mesophilic streptomycetes play an important part during the initial decomposition stages of organic material in compost (Kutzner, 1981). This process is achieved by the efficient enzyme-repertoire leading to the degradation of polysaccharides and other macromolecules (see above, the section Role of Secreted Proteins). Streptomycetes inhabiting aquatic and marine environments have also been investigated (Okami et al., 1976; Ravel et al., 1998). Thorough studies of lake sediments showed a dominance of *Micromonospora* species, whereas *Streptomyces* species were the secondmost abundant microorganisms (Jiang and Xu, 1996). Like strains from soils, aquatic streptomycetes produce compounds with antibacterial and antifungal activities (see the section Pharmacologically Active Substances in this Chapter) and have the capacity to hydrolyze a wide range of macromolecules including cellulose, chitin and proteins (see above). Furthermore, they have been suggested to play a major part in the turnover of toxic compounds.

Streptomycetes occur at higher concentrations in the gut of earthworms and arthropods. Some highly cellulolytic strains were isolated from the gut of termites (Kutzner, 1981; see the The Family Streptomycetaceae, Part I: Taxonomy in this Volume).

Actinomycetes, including various *Streptomyces* species, were identified within hydrocarbon-polluted soils and were found to utilize them as sole carbon sources. So far, the biochemical details have not been exploited. Streptomycetes were also found in soils contaminated with heavy metals. The molecular mechanisms of their resistances to heavy metals were until now barely known. Recently, nickel-resistance was attributed to a transporter (Amoroso et al., 2000).

Some *Streptomyces* strains utilize a wide range of structurally diverse phosphonates as the sole phosphorus source and it is very likely they have a hitherto unique pathway for breakage of C-P bonds (Obojska and Lejczak, 2003). *Streptomycetes thermoautotrophicus* is a thermophilic, CO- and H_2-oxidizing obligate chemolithoautotrophic organism (Gadkari et al., 1992) whose ability for N_2 fixation involves a molybdenum dinitrogenase and a manganese-superoxide oxidoreductase (Ribbe et al., 1997). A unique system leading to N_2 and nitrous oxide (N_2O) through denitrification and codenitrification has been discovered within *S. antibioticus* (Kumon et al., 2002).

Actinomycetes have been detected and localized by the dilution plate technique, either without enrichment or under special selective isolation conditions (for details, see The Family Streptomycetaceae, Part I: Taxonomy in this Volume). Recently however, hybridization techniques using fluorescent, genus-specific 16S rRNA probes have been developed for a broad range of bacteria (Roller et al., 1992). Such oligonucleotides can be used to detect uncultivable streptomycetes in situ. The increasing application of molecular tools should considerably deepen our knowledge of the ecological role of actinomycetes.

Interaction with Plants

The rhizosphere of plants comprises varying amounts of actinomycetes. Presumably, secretion of antibacterial and antifungal compounds plays an important role in natural habitats (Kutzner, 1981). Using molecular detection methods, streptomycin-producing *Streptomyces* strains have been found to colonize the rhizosphere of soybeans (Huddleston et al., 1997). Previous investigations suggested that some streptomycetes serve as potential biocontrol agents against fungi within the rhizosphere (Kutzner, 1981) and recent studies support this interpretation (Blaak and Schrempf, 1995; Yuan and Crawford, 1995; Kolbe et al., 1998).

Plant-pathogenic streptomycetes produce different types of potato scab (for review, see Kutzner, 1981). Studies of DNA-DNA hybridization and rRNA studies revealed that the pathogenic *Streptomyces* strains are phylogenetically diverse (see The Family Streptomycetaceae, Part I: Taxonomy in this Volume). The production of thaxtomins (a family of 4-nitroindol-3-yl-containing 2,5 dioxopiperazines) has been correlated with plant pathogenicity (King and Lawrence, 1996). Recent evidence shows that the peptide synthetase genes *txt*AB (*S. acidiscabies*)

are required for thaxtomin biosynthesis (Healy et al., 2000). The *nec1* gene encodes a putative virulence factor (Bukhalid and Loria, 1997). A gene encoding an extracellular esterase of an *S. scabies* strain was cloned and has so far only been identified in streptomycetes causing potato scab. Nonpathogenic, antibiotic-producing streptomycetes reduce potato scab in soil (Neeno-Eckwall et al., 2001). *Streptomyces ipomoeae* strains are sweet potato pathogens. The bacteriocin ipomicin has been found to lead to interstrain inhibition (Zhang et al., 2003).

Yokonolide B, a spiroketal-macrolide, inhibits the expression of auxin-inducible genes in *Arabidopsis*. As this compound blocks auxin-dependent cell division and auxin-regulated epinastic growth mediated by auxin-binding protein 1, it provides a new tool to dissect auxin signal transduction (Hayashi et al., 2003).

Perspectives

The development and application of many molecular tools have considerably deepened our understanding of the biology of streptomycetes. We can expect that the currently available as well as future knowledge will enormously extend our ability to tailor hybrid antibiotics, cytostatics, antifungals, and antiparasitics, as well as other biologically active substances. The usage of current as well as new tools will lead to a much better understanding of the many processes within the environment that are due to streptomycete activities including humus formation, breakdown of many different macromolecules, modification of xenobiotics, and detoxification of harmful compounds. An exciting avenue is the elucidation of the molecular basis for interactions between streptomycetes and other microbes, plants and animals. To resolve this complexity, the ever-increasing knowledge about genomic processes, proteomes, transcriptomes, global networks, and signal transduction pathways will need to be integrated with studies of physiology, biochemistry and chemistry.

Literature Cited

Aharonowitz, Y., and G. Cohen. 1992. Penicillin and cephalosporin biosynthesis genes: Structure, organization, regulation, and evolution. Ann. Rev. Microbiol. 46:461–495.

Ahel, I., D. Vujaklija, A. Mikoc, and V. Gamulin. 2002. Transcriptional analysis of the recA gene in Streptomyces rimosus: Identification of the new type of promoter. FEMS Microbiol. Lett. 209:133–137.

Ainsa, J. A., N. J. Ryding, N. Hartley, K. C. Findlay, C. J. Bruton, and K. F. Chater. 2000. WhiA, a protein of unknown function conserved among Gram-positive bacteria, is essential for sporulation in Streptomyces coelicolor A3(2). J. Bacteriol. 182:5470–5478.

Amoroso, M. J., D. Schubert, P. Mitscherlich, P. Schumann, and E. Kothe. 2000. Evidence for high affinity nickel transporter genes in heavy metal resistant Streptomyces spec. J. Basic Microbiol. 40:295–301.

Aparicio, J. F., P. Caffrey, A. F. A. Marsden, J. Stauton, and P. F. Leadlay. 1994. Limited proteolysis and active-site studies of the first multienzyme component of the erythromycin-producing polyketide synthase. J. Biol. Chem. 269:8524–8528.

Arias, M. E., M. Arenas, J. Rodriguez, J. Soliveri, A. S. Ball, and M. Hernandez. 2003. Kraft pulp biobleaching and mediated oxidation of a nonphenolic substrate by laccase from Streptomyces cyaneus CECT 3335. Appl. Environ. Microbiol. 69:1953–1958.

August, P. R., L. Tang, Y. J. Yoon, S. Ning, R. Muller, T. W. Yu, M. Taylor, D. Hoffmann, C. G. Kim, X. H. Zhang, C. R. Hutchinson, and H. G. Floss. 1998. Biosynthesis of the ansamycin antibiotic rifamycin—deductions from the molecular analysis of the rif biosynthetic gene cluster of Amycolatopsis mediterranei S699. Chem. Biol. 5:69–79.

Bailey, C. R., C. J. Bruton, M. J. Butler, K. F. Chater, J. E. Harris, and D. A. Hopwood. 1986. Properties of in vitro recombinant derivatives of pJV1, a multi-copy plasmid from Streptomyces phaeochromogenes. J. Gen. Microbiol. 132:2071–2078.

Baltz, R. H., M. A. McHenney, C. A. Cantwell, S. W. Queener, and P. J. Solenberg. 1997. Applications of transposition mutagenesis in antibiotic producing streptomycetes. Ant. v. Leeuwenhoek 71:179–187.

Banchio, C., and H. C. Gramajo. 1997. Medium- and long-chain fatty acid uptake and utilization by Streptomyces coelicolor A3(2): First characterization of a Gram-positive bacterial system. Microbiology 143:2439–2447.

Bao, K., and S. N. Cohen. 2001. Terminal proteins essential for the replication of linear plasmids and chromosomes in Streptomyces. Genes Dev. 15:1518–1527.

Bao, K., and S. N. Cohen. 2003. Recruitment of terminal protein to the ends of Streptomyces linear plasmids and chromosomes by a novel telomere-binding protein essential for linear DNA replication. Genes Dev. 17:774–785.

Bentley, S. D., K. F. Chater, A. M. Cerdeno-Tarraga, G. L. Challis, N. R. Thomson, K. D. James, D. E. Harris, M. A. Quail, H. Kieser, D. Harper, A. Bateman, S. Brown, G. Chandra, C. W. Chen, M. Collins, A. Cronin, A. Fraser, A. Goble, J. Hidalgo, T. Hornsby, S. Howarth, C. H. Huang, T. Kieser, L. Larke, L. Murphy, K. Oliver, S. O'Neil, E. Rabbinowitsch, M. A. Rajandream, K. Rutherford, S. Rutter, K. Seeger, D. Saunders, S. Sharp, R. Squares, S. Squares, K. Taylor, T. Warren, A. Wietzorrek, J. Woodward, B. G. Barrell, J. Parkhill, and D. A. Hopwood. 2002. Complete genome sequence of the model actinomycete Streptomyces coelicolor A3(2). Nature 9:141–147.

Betzler, M., I. Tlolka, and H. Schrempf. 1997. Amplification of a Streptomyces lividans 4.3 kb DNA element causes overproduction of a novel hypha- and vesicle-associated protein. Microbiology 143:1243–1252.

Bhatt, A., H. M. Kieser, R. E. Melton, and T. Kieser. 2002. Plasmid transfer from Streptomyces to Mycobacterium smegmatis by spontaneous transformation. Molec. Microbiol. 43:135–146.

Bibb, M. J., S. Biro, H. Motamedi, J. F. Collins, and C. R. Hutchinson. 1989. Analysis of the nucleotide sequence of the Streptomyces glaucescens tcmI genes provides key information about the enzymology of polyketide antibiotic biosynthesis. EMBO J. 8:2727–2736.

Bibb, M. J., D. H. Sherman, S. Omura, and D. A. Hopwood. 1994. Cloning, sequencing and deduced functions of a cluster of Streptomyces genes probably encoding biosynthesis of the polyketide antibiotic frenolicin. Gene 142:31–39.

Bibb, M. J., and M. J. Buttner. 2003. The Streptomyces coelicolor developmental transcription factor sigma(BldN) is synthesized as a proprotein. J. Bacteriol. 185:2338–2345.

Binnie, C., M. Warren, and M. J. Butler. 1989. Cloning and heterologous expression in Streptomyces lividans of Streptomyces rimosus genes involved in oxytetracycline biosynthesis. J. Bacteriol. 171:887–895.

Blaak, H., and H. Schrempf. 1995. Binding and substrate specificities of a Streptomyces olivaceoviridis chitinase in comparison with its proteolytically processed form. Eur. J. Biochem. 229:132–139.

Blanco, G., M. R. Rodicio, A. M. Puglia, C. Méndez, C. J. Thompson, and J. A. Salas. 1994. Synthesis of ribosomal proteins during growth of Streptomyces coelicolor. Molec. Microbiol. 12:375–385.

Böckle, B., and R. Müller. 1997. Reduction of disulfide bonds by Streptomyces pactum during growth on chicken feathers. Appl. Environ. Microbiol. 63:790–792.

Bode, H. B., K. Kerkhoff, and D. Jendrossek. 2001. Bacterial degradation of natural and synthetic rubber. Biomacromolecules 2:295–303.

Bormann, C., V. Möhrle, and C. Bruntner. 1996. Cloning and heterologous expression of the entire set of structural genes for nikkomycin synthesis from Streptomyces tendae Tü901 in Streptomyces lividans. J. Bacteriol. 178:1216–1218.

Bucca, G., A. M. Brassington, G. Hotchkiss, V. Mersinias, and C. P. Smith. 2003. Negative feedback regulation of dnaK, clpB and lon expression by the DnaK chaperone machine in Streptomyces coelicolor, identified by transcriptome and in vivo DnaK-depletion analysis. Molec. Microbiol. 50:153–166.

Bukhalid, R. A., and R. Loria. 1997. Cloning and expression of a gene from Streptomyces scabies encoding a putative pathogenicity factor. J. Bacteriol. 179:7776–7783.

Burke, J., D. Schneider, and J. Westpheling. 2001. Generalized transduction in Streptomyces coelicolor. Proc. Natl. Acad. Sci. USA 98:6289–6294.

Butler, M. J., P. Bruheim, S. Jovetic, F. Marinelli, P. W. Postma, and M. J. Bibb. 2002. Engineering of primary carbon metabolism for improved antibiotic production in Streptomyces lividans. Appl. Environ. Microbiol. 68:4731–4739.

Calcutt, M. J., and F. J. Schmidt. 1992. Conserved gene arrangement in the origin region of the Streptomyces coelicolor chromosome. J. Bacteriol. 174:3220–3226.

Challis, G. L., and D. A. Hopwood. 2003. Synergy and contingency as driving forces for the evolution of multiple secondary metabolite production by Streptomyces species. Proc. Natl. Acad. Sci. USA 100 (Suppl. 2):14555–14561. Epub;14555-14561.

Champness, W. 1999. Cloning and analysis of regulatory genes involved in streptomycete secondary metabolite biosynthesis. In: A. L. Demain and J. E. Davies (Eds.) Manual of Industrial Microbiology and Biotechnology, 2nd ed. American Society for Microbiology. Washington, DC. 725–739.

Chater, K. F., and C. J. Bruton. 1983. Mutational cloning in Streptomyces and the isolation of antibiotic production genes. Gene 26:67–78.

Chater, K. F., and C. J. Bruton. 1985. Resistance, regulatory and production genes for the antibiotic methylenomycin are clustered. EMBO J. 4:1893–1897.

Chater, K. F., and S. Horinouchi. 2003. Signalling early developmental events in two highly diverged Streptomyces species. Molec. Microbiol. 48:9–15.

Chen, C. W. 1995. The unstable ends of the Streptomyces linear chromosomes: A nuisance without cures? Trends Biotechnol. 13:157–160.

Chung, S.-T., and L. L. Crose. 1989. Streptomyces transposon Tn4556 and its applications. In: C. L. Hershberger, S. W. Queener, and G. Hegeman (Eds.) Genetics and Molecular Biology of Industrial Microorganisms. American Society for Microbiology. Washington, DC. 168–175.

Claessen, D., H. A. Wosten, G. van Keulen, O. G. Faber, A. M. Alves, W. G. Meijer, and L. Dijkhuizen. 2002. Two novel homologous proteins of Streptomyces coelicolor and Streptomyces lividans are involved in the formation of the rodlet layer and mediate attachment to a hydrophobic surface. Molec. Microbiol. 44:1483–1492.

Combes, P., R. Till, S. Bee, and M. C. Smith. 2002. The Streptomyces genome contains multiple pseudo-attB sites for the phiC31-encoded site-specific recombination system. J. Bacteriol. 184:5746–5752.

Cortes, J., S. F. Haydock, G. A. Roberts, D. J. Bevitt, and P. F. Leadlay. 1990. An unusually large multifunctional polypeptide in the erythromycin-producing polyketide synthase of Saccharopolyspora erythraea. Nature 348:176–178.

Cowlishaw, D. A., and M. C. Smith. 2002. A gene encoding a homologue of dolichol phosphate-beta-D-mannose synthase is required for infection of Streptomyces coelicolor A3(2) by phage (phi)C31. J. Bacteriol. 184:6081–6083.

Crawford, D. L., and E. McCoy. 1972. Cellulases of Thermomonospora fusca and Streptomyces thermodiastaticus. Appl. Microbiol. 24:150–152.

Crespi, M., E. Messens, A. B. Caplan, M. Vanmontagu, and J. Desomer. 1992. Fasciation induction by the phytopathogen Rhodococcus fascians depends upon a linear plasmid encoding a cytokinin synthase gene. EMBO J. 11:795–804.

Decker, H., and S. Haag. 1995. Cloning and characterization of a polyketide synthase gene from Streptomyces fradiae Tu2717, which carries the genes for biosynthesis of the angucycline antibiotic urdamycin A and a gene probably involved in its oxygenation. J. Bacteriol. 177:6126–6136.

Dharmatilake, A. J., and K. E. Kendrick. 1994. Expression of the division-controlling gene ftsZ during growth and sporulation of the filamentous bacterium Streptomyces griseus. Gene 147:21–28.

Distler, J., A. Ebert, K. Mansouri, K. Pissowotzki, M. Stockmann, and W. Piepersberg. 1987. Gene cluster for streptomycin biosynthesis in Streptomyces griseus: Nucleotide sequence of three genes and analysis of transcriptional activity. Nucleic Acids Res. 15:8041–8056.

Dittrich, W., M. Betzler, and H. Schrempf. 1991. An amplifiable and deletable chloramphenicol-resistance determinant of Streptomyces lividans 1326 encodes a putative transmembrane protein. Molec. Microbiol. 5:2789–2797.

Donadio, S., M. Staver, J. B. McAlpine, S. J. Swanson, and L. Katz. 1991. Modular organization of genes required for complex polyketide biosynthesis. Science 252:675–679.

Dosch, D. C., W. R. Strohl, and H. G. Floss. 1988. Molecular cloning of the nosiheptide resistance gene from Streptomyces actuosus ATCC 25421. Biochem. Biophys. Res. Commun. 156:517–523.

Ducote, M. J., S. Prakash, and G. S. Pettis. 2000. Minimal and contributing sequence determinants of the cis-acting locus of transfer (clt) of streptomycete plasmid pIJ101 occur within an intrinsically curved plasmid region. J. Bacteriol. 182:6834–6841.

Embley, T. M., and E. Stackebrandt. 1994. The molecular phylogeny and systematics of the Actinomycetes. Ann. Rev. Microbiol. 48:257–289.

Epp, J. K., S. G. Burgett, and B. E. Schoner. 1987. Cloning and nucleotide sequence of a carbomycin-resistance gene from Streptomyces thermotolerans. Gene 53:73–83.

Feitelson, J. S., and D. A. Hopwood. 1983. Cloning of a Streptomyces gene for an O-methyltransferase involved in antibiotic biosynthesis. Molec. Gen. Genet. 190:394–398.

Fernandez-Moreno, M. A., E. Martinez, L. Boto, D. A. Hopwood, and F. Malpartida. 1992. Nucleotide sequence and deduced functions of a set of cotranscribed genes of Streptomyces coelicolor A3(2) including the polyketide synthase for the antibiotic actinorhodin. J. Biol. Chem. 267:19278–19290.

Fishman, S. E., K. Cox, J. L. Larson, P. A. Reynolds, E. T. Seno, W. K. Yeh, R. van Frank, and C. L. Hershberger. 1987. Cloning genes for the biosynthesis of a macrolide antibiotic. Proc. Natl. Acad. Sci. USA 84:8248–8252.

Flardh, K. 2003. Essential role of DivIVA in polar growth and morphogenesis in Streptomyces coelicolor A3(2). Molec. Microbiol. 49:1523–1536.

Gadkari, D., G. Mörsdorf, and O. Meyer. 1992. Chemolithoautotrophic assimilation of dinitrogen by Streptomyces thermoautotrophicus UTB1: Identification of an unusual N_2-fixing system. J. Bacteriol. 174:6840–6843.

Gandlur, S. M., L. Wei, J. Levine, J. Russell, and P. Kaur. 2004. Membrane topology of the DrrB protein of the doxorubicin transporter of Streptomyces peucetius. J. Biol. Chem. 279:27799–27806.

Geueke, B., and W. Hummel. 2003. Heterologous expression of Rhodococcus opacus L-amino acid oxidase in Streptomyces lividans. Protein Expr. Purif. 28:303–309.

Geukens, N., V. Parro, L. A. Rivas, R. P. Mellado, and J. Ann. 2001. Functional analysis of the Streptomyces lividans type I signal peptidases. Arch. Microbiol. 176:377–380.

Goodfellow, M., L. J. Stanton, K. E. Simpson, and D. E. Minnikin. 1990. Numerical and chemical classification of Actinoplanes and related Actinomycetes. J. Gen. Microbiol. 136:19–36.

Grantcharova, N., W. Ubhayasekera, S. L. Mowbray, J. R. McCormick, and K. Flardh. 2003. A missense mutation in ftsZ differentially affects vegetative and developmentally controlled cell division in Streptomyces coelicolor A3(2). Molec. Microbiol. 47:645–656.

Gravius, B., D. Glocker, J. Pigac, K. Pandza, D. Hranueli, and J. Cullum. 1994. The 387 kb linear plasmid pPZG101 of Streptomyces rimosus and its interactions with the chromosome. Microbiology 140:2271–2277.

Hagege, J., J.-L. Pernodet, A. Friedmann, and M. Guérineau. 1993. Mode and origin of replication of pSAM2, a conjugative integrating element of Streptomyces ambofaciens. Molec. Microbiol. 10:799–812.

Hagege, J. M., M. A. Brasch, and S. N. Cohen. 1999. Regulation of transfer functions by the imp locus of the Streptomyces coelicolor plasmidogenic element SLP1. J. Bacteriol. 181:5976–5983.

Han, L., K. Yang, E. Ramalingam, R. H. Mosher, and L. C. Vining. 1994. Cloning and characterization of polyketide synthase genes for jadomycin B biosynthesis in Streptomyces venezuelae ISP5230. Microbiology 140:3379–3389.

Han, L., and K. A. Reynolds. 1997. A novel alternate anaplerotic pathway to the glyoxylate cycle in streptomycetes. J. Bacteriol. 179:5157–5164.

Haug, I., A. Weissenborn, D. Brolle, S. Bentley, T. Kieser, and J. Altenbuchner. 2003. Streptomyces coelicolor A3(2) plasmid SCP2*: Deductions from the complete sequence. Microbiology 149:505–513.

Hayashi, K. I., A. M. Jones, K. Ogino, A. Yamazoe, Y. Oono, M. Inoguchi, H. Kondo, and H. Nozaki. 2003. Yokonolide B, a novel inhibitor of auxin action, blocks degradation of Aux/IAA factors. J. Biol. Chem. 278:23797–23806.

Healy, F. G., M. Wach, S. B. Krasnoff, D. M. Gibson, and R. Loria. 2000. The txtAB genes of the plant pathogen Streptomyces acidiscabies encode a peptide synthetase required for phytotoxin thaxtomin A production and pathogenicity. Molec. Microbiol. 38:794–804.

Hesketh, A. R., G. Chandra, A. D. Shaw, J. J. Rowland, D. B. Kell, M. J. Bibb, and K. F. Chater. 2002a. Primary and secondary metabolism, and post-translational protein modifications, as portrayed by proteomic analysis of Streptomyces coelicolor. Molec. Microbiol. 46:917–932.

Hesketh, A., D. Fink, B. Gust, H. U. Rexer, B. Scheel, K. Chater, W. Wohlleben, and A. Engels. 2002b. The GlnD and GlnK homologues of Streptomyces coelicolor A3(2) are functionally dissimilar to their nitrogen regulatory system counterparts from enteric bacteria. Molec. Microbiol. 46:319–330.

Hiraga, K., T. Suzuki, and K. Oda. 2000. A novel double-headed proteinaceous inhibitor for metalloproteinase and serine proteinase. J. Biol. Chem. 275:25173–25179.

Hiratsu, K., S. Mochizuki, and H. Kinashi. 2000. Cloning and analysis of the replication origin and the telomeres of the large linear plasmid pSLA2-L in Streptomyces rochei. Molec. Gen. Genet. 263:1015–1021.

Hirochika, H., K. Nakamura, and K. Sakaguchi. 1984. A linear DNA plasmid from Streptomyces rochei with an inverted terminal repetition of 614 base pairs. EMBO J. 3:761–766.

Hong, S. T., J. R. Carney, and S. J. Gould. 1997. Cloning and heterologous expression of the entire gene clusters for PD 116740 from Streptomyces strain WP 4669 and tetrangulol and tetrangomycin from Streptomyces rimosus NRRL 3016. J. Bacteriol. 179:470–476.

Hopwood, D. A., M. J. Bibb, K. F. Chater, T. Kieser, C. J. Bruton, H. M. Kieser, D. J. Lydiate, C. P. Smith, J. M. Ward, and H. Schrempf. 1985. Genetic Manipulation of Streptomyces: A Laboratory Manual. John Innes Foundation. Norwich, UK.

Hopwood, D. A., and D. H. Sherman. 1990. Molecular genetics of polyketides and its comparison to fatty acid biosynthesis. Ann. Rev. Genet. 24:37–66.

Horinouchi, S., and T. Beppu. 1992. Autoregulatory factors and communication in Actinomycetes. Ann. Rev. Microbiol. 46:377–398.

Horinouchi, S. 2002. A microbial hormone, A-factor, as a master switch for morphological differentiation and secondary metabolism in Streptomyces griseus. Front. Biosci. 7:d2045–d2057.

Hornemann, U., C. J. Otto, and X. Y. Zhang. 1989. DNA amplification in Streptomyces achromogenes subsp. rubradiris is accompanied by a deletion, and the amplified sequences are conditionally stable and can be eliminated by two pathways. J. Bacteriol. 171:5817–5822.

Hsieh, C.-J., and G. H. Jones. 1995. Nucleotide sequence, transcriptional analysis, and glucose regulation of the phenoxazinone synthase gene (phsA) from Streptomyces antibioticus. J. Bacteriol. 177:5740–5747.

Huang, C.-H., Y.-S. Lin, Y.-L. Yang, S. Huang, and C. W. Chen. 1998. The telomeres of Streptomyces chromosomes contain conserved palindromic sequences with potential to form complex secondary structures. Molec. Microbiol. 28:905–916.

Huang, J., C. J. Lih, K. H. Pan, and S. N. Cohen. 2001. Global analysis of growth phase responsive gene expression and regulation of antibiotic biosynthetic pathways in Streptomyces coelicolor using DNA microarrays. Genes Dev. 15:3183–3192.

Huang, C.-H., C. Y. Chen, H. H. Tsai, C. Chen, Y.-S. Lin, and C. W. Chen. 2003. Linear plasmid SLP2 of Streptomyces lividans is a composite replicon. Molec. Microbiol. 47:1563–1576.

Huddleston, A. S., N. Cresswell, M. C. P. Neves, J. E. Beringer, S. Baumberg, D. I. Thomas, and E. M. H. Wellington. 1997. Molecular detection of streptomycin-producing streptomycetes in Brazilian soils. Appl. Environ. Microbiol. 63:1288–1297.

Hutchinson, C. R., and I. Fujii. 1995. Polyketide synthase gene manipulation: A structure-function approach in engineering novel antibiotics. Ann. Rev. Microbiol. 49:201–238.

Hütter, R., and T. Eckhardt. 1988. Genetic manipulation. In: M. Goodfellow, S. T. Williams, and M. Mordarski (Eds.) Actinomycetes in Biotechnology. Academic Press. London, UK. 89–184.

Ikeda, H., Y. Takada, C.-H. Pang, H. Tanaka, and S. Omura. 1993. Transposon mutagenesis by Tn4560 and applications with avermectin-producing Streptomyces avermitilis. J. Bacteriol. 175:2077–2082.

Ikeda, H., J. Ishikawa, A. Hanamoto, M. Shinose, H. Kikuchi, T. Shiba, Y. Sakaki, M. Hattori, and S. Omura. 2003. Complete genome sequence and comparative analysis of the industrial microorganism Streptomyces avermitilis. Nature Biotechnol. 21(5):526–531.

Jakimowicz, D., J. Majka, W. Messer, C. Speck, M. Fernandez, M. Cruz Martin, J. Sanchez, F. Schauwecker, U. Keller, H. Schrempf, and J. Zakrzewska-Czerwinska. 1998. Structural elements of the Streptomyces oriC region and their interactions with the DnaA protein. Microbiology 144:1281–1290.

Jakimowicz, D., J. Majka, G. Konopa, G. Wegrzyn, W. Messer, H. Schrempf, and J. Zakrzewska-Czerwinska. 2000. Architecture of the Streptomyces lividans DnaA protein-replication origin complexes. J. Molec. Biol. 298:351–364.

Jendrossek, D., G. Tomasi, and R. M. Kroppenstedt. 1997. Bacterial degradation of natural rubber: A privilege of Actinomycetes? FEMS Microbiol. Lett. 150:179–188.

Jiang, C.-L., and L.-H. Xu. 1996. Diversity of aquatic Actinomycetes in lakes of the Middle Plateau, Yunnan, China. Appl. Environ. Microbiol. 62:249–253.

Jonsbu, E., B. Christensen, and J. Nielsen. 2001. Changes of in vivo fluxes through central metabolic pathways during the production of nystatin by Streptomyces noursei in batch culture. Appl. Microbiol. Biotechnol, 56:93–100.

Kalkus, J., C. Dörrie, D. Fischer, M. Reh, and H. G. Schlegel. 1993. The giant linear plasmid pHG207 from Rhodococcus sp. encoding hydrogen autotrophy: Characterization of the plasmid and its termini. J. Gen. Microbiol. 139:2055–2065.

Kamerbeek, J., L. Schouls, A. Kolk, M. van Agterveld, D. van Soolingen, S. Kuijper, A. Bunschoten, H. Molhuizen, R. Shaw, M. Goyal, and J. D. A. van Embden. 1997. Simultaneous detection and strain differentiation of Mycobacterium tuberculosis for diagnosis and epidemiology. J. Clin. Microbiol. 35:907–914.

Kataoka, M., T. Seki, and T. Yoshida. 1991. Five genes involved in self-transmission of pSN22, a Streptomyces plasmid. J. Bacteriol. 173:4220–4228.

Kebeler, M., E. R. Dabbs, B. Averhoff, and G. Gottschalk. 1996. Studies on the isopropylbenzene 2,3-dioxygenase and the 3'-isopropylcatechol 2,3-dioxygenase genes encoded by the linear plasmid of Rhodococcus erythropolis BD2. Microbiology 142:3241–3251.

Kelemen, G. H., M. Zalacain, E. Culebras, E. T. Seno, and E. Cundliffe. 1994. Transcriptional attenuation control of the tylosin-resistance gene tlrA in Streptomyces fradiae. Molec. Microbiol. 14:833–842.

Khokhlov, A. S., I. I. Tovarova, L. N. Borisova, S. A. Pliner, L. A. Schevchenko, E. Y. Kornitskaya, N. S. Ivkina, and I. A. Rapoport. 1967. A-factor responsible for the biosynthesis of streptomycin by a mutant strain of Actinomyces streptomycini. Dokl. Akad. Nauk. SSSR 177:232–235.

Kim, I. S., and K. J. Lee. 1995. Physiological roles of leupeptin and extracellular proteases in mycelium development of Streptomyces exfoliatus SMF13. Microbiology 141:1017–1025.

Kim, Y. S. 2002. Malonate metabolism: Biochemistry, molecular biology, physiology, and industrial application. J. Biochem. Molec. Biol. 35:443–451.

Kinashi, H., M. Shimaji-Murayama, and T. Hanafusa. 1991. Nucleotide sequence analysis of the unusually long terminal inverted repeats of a giant linear plasmid, SCP1. Plasmid 26:123–130.

King, R. R., and C. H. Lawrence. 1996. Characterization of new thaxtomin A analogues generated in vitro by Streptomyces scabies. J. Agric. Food Chem. 44:1108–1110.

Kluepfel, D., F. Shareck, F. Mondou, and R. Morosoli. 1986. Characterisation of cellulase and xylanase activities of Streptomyces lividans. Appl. Microbiol. Biotechnol. 24:230–234.

Kolbe, S., S. Fischer, A. Becirevic, P. Hinz, and H. Schrempf. 1998. The Streptomyces reticuli α-chitin-binding protein CHB2 and its gene. Microbiology 144:1291–1297.

Kosono, S., M. Maeda, F. Fuji, H. Arai, and T. Kudo. 1997. Three of the seven bphC genes of Rhodococcus erythropolis TA421, isolated from a termite ecosystem, are located on an indigenous plasmid associated with biphenyl degradation. Appl. Environ. Microbiol. 63:3282–3285.

Kumon, Y., Y. Sasaki, I. Kato, N. Takaya, H. Shoun, and T. Beppu. 2002. Codenitrification and denitrification are dual metabolic pathways through which dinitrogen evolves from nitrate in Streptomyces antibioticus. J. Bacteriol. 184:2963–2968.

Kutzner, H. J. 1981. The family Streptomycetaceae. In: M. P. Starr, H. Stolp, H. G. Trüper, A. Balows, and H. Schlegel

(Eds.) The Prokaryotes. Springer-Verlag. Berlin, Germany. 2028–2090.

Lacalle, R. A., J. A. Tercero, and A. Jimenez. 1992. Cloning of the complete biosynthetic gene cluster for an aminonucleoside antibiotic, puromycin, and its regulated expression in heterologous hosts. EMBO J. 11:785–792.

Lammertyn, E., and J. Anné. 1998. Modifications of Streptomyces signal peptides and their effects on protein production and secretion. FEMS Microbiol. Lett. 160:1–10.

Lawen, A., and R. Zocher. 1990. Cyclosporin synthetase: The most complex peptide synthesizing multienzyme polypeptide so far described. J. Biol. Chem. 265:11355–11360.

Leblond, P., and B. Decaris. 1994. New insights into the genetic instability of Streptomyces. FEMS Microbiol. Lett. 123:225–232.

Leblond, P., G. Fischer, F. Francou, F. Berger, M. Guérineau, and B. Decaris. 1996. The unstable region of Streptomyces ambofaciens includes 210 kb terminal inverted repeats flanking the extremities of the linear chromosomal DNA. Molec. Microbiol. 19:261–271.

Lee, D. S., and S. H. Lee. 2001. Genistein, a soy isoflavone, is a potent alpha-glucosidase inhibitor. FEBS Lett. 501:84–86.

Lezhava, A., T. Mizukami, T. Kajitani, D. Kameoka, M. Redenbach, H. Shinkawa, O. Nimi, and H. Kinashi. 1995. Physical map of the linear chromosome of Streptomyces griseus. J. Bacteriol. 177:6492–6498.

Lin, Y. S., H. M. Kieser, D. A. Hopwood, and C. W. Chen. 1993. The chromosomal DNA of Streptomyces lividans 66 is linear. Molec. Microbiol. 10:923–933.

Lombo, F., G. Blanco, E. Fernandez, C. Mendez, and J. A. Salas. 1996. Characterization of Streptomyces argillaceus genes encoding a polyketide synthase involved in the biosynthesis of the antitumor antibiotic mithramycin. Gene 172:87–91.

Lomovskaya, N. D., K. F. Chater, and N. M. Mkrtumian. 1980. Genetics and molecular biology of Streptomyces bacteriophages. Microbiol. Rev. 44:206–229.

Lutkenhaus, J. 1997. Bacterial cytokinesis: Let the light shine in. Curr. Biol. 7:573–575.

Malpartida, F., and D. A. Hopwood. 1984. Molecular cloning of the whole biosynthetic pathway of a Streptomyces antibiotic and its expression in a heterologous host. Nature 309:462–464.

Malpartida, F., J. Niemi, R. Navarrete, and D. A. Hopwood. 1990. Cloning and expression in a heterologous host of the complete set of genes for biosynthesis of the Streptomyces coelicolor antibiotic undecylprodigiosin. Gene 93:91–99.

Margolin, W. 2003. Bacterial division: The fellowship of the ring. Curr. Biol. 13:16–18.

Martin, C., J. Timm, J. Rauzier, R. Gomez-Lus, J. Davies, and B. Gicquel. 1990. Transposition of an antibiotic resistance element in mycobacteria. Nature 345:739–743.

McCue, L. A., J. Kwak, J. Wang, and K. E. Kendrick. 1996. Analysis of a gene that suppresses the morphological defect of bald mutants of Streptomyces griseus. J. Bacteriol. 178:2867–2875.

Méndez, C., and J. A. Salas. 1998. ABC transporters in antibiotic-producing Actinomycetes. FEMS Microbiol. Lett. 158:1–8.

Meurer, G., and C. R. Hutchinson. 1999. Genes for the biosynthesis of microbial secondary metabolites. In: A. L. Demain et al. (Eds.) Industrial Microbiology and Biotechnology, 2nd ed. American Society for Microbiology. Washington, DC. 740–758.

Meuser, D., H. Splitt, R. Wagner, and H. Schrempf. 2001. Mutations stabilizing an open conformation within the external region of the permeation pathway of the potassium channel KcsA. Eur. Biophys. J. 30:385–391.

Minambres, B., E. R. Olivera, R. A. Jensen, and J. M. Luengo. 2000. A new class of glutamate dehydrogenases (GDH): Biochemical and genetic characterization of the first member, the AMP-requiring NAD-specific GDH of Streptomyces clavuligerus. J. Biol. Chem. 275:39529–39542.

Miyashita, K., T. Fujii, and Y. Sawada. 1991. Molecular cloning and characterization of chitinase genes from Streptomyces lividans 66. J. Gen. Microbiol. 137:2065–2072.

Morosoli, R., S. Ostiguy, and C. Dupont. 1999. Effect of carbon source, growth and temperature on the expression of the sec genes of Streptomyces lividans 1326. Can. J. Microbiol. 45:1043–1049.

Motamedi, H., and C. R. Hutchinson. 1987. Cloning and heterologous expression of a gene cluster for the biosynthesis of tetracenomycin C, the anthracycline antitumor antibiotic of Streptomyces glaucescens. Proc. Natl. Acad. Sci. USA 84:4445–4449.

Murakami, T., S. Anzai, S. Imai, A. Satoh, K. Nagaoka, and C. J. Thompson. 1986. The bialaphos biosynthetic genes of Streptomyces hygroscopicus: Molecular cloning and characterization of the gene cluster. Molec. Gen. Genet. 205:42–50.

Muth, G., M. Farr, V. Hartmann, and W. Wohlleben. 1995. Streptomyces ghanaensis plasmid pSG5: Nucleotide sequence analysis of the self-transmissible minimal replicon and characterization of the replication mode. Plasmid 33:113–126.

Nakai, R., S. Horinouchi, and T. Beppu. 1988. Cloning and nucleotide sequence of a cellulase gene, casA, from an alkalophilic Streptomyces strain. Gene 65:229–238.

Nakata, K., S. Horinouchi, and T. Beppu. 1989. Cloning and characterization of the carbapenem biosynthetic genes from Streptomyces fulvoviridis. FEMS Microbiol. Lett. 48:51–55.

Neeno-Eckwall, E. C., L. L. Kinkel, and J. L. Schottel. 2001. Competition and antibiosis in the biological control of potato scab. Can. J. Microbiol. 47:332–340.

Nguyen, J., F. Francou, M.-J. Virolle, and M. Guérineau. 1997. Amylase and chitinase genes in Streptomyces lividans are regulated by reg1, a pleiotropic regulatory gene. J. Bacteriol. 179:6383–6390.

Novotna, J. et al. 2003. Proteomic studies of diauxic lag in the differentiating prokaryote Streptomyces coelicolor reveal a regulatory network of stress-induced proteins and central metabolic enzymes. Molec. Microbiol. 48:1289–1303.

Obanye, A. I. C., G. Hobbs, D. C. J. Gardner, and S. G. Oliver. 1996. Correlation between carbon flux through the pentose phosphate pathway and production of the antibiotic methylenomycin in Streptomyces coelicolor A3(2). Microbiology 142:133–137.

Obojska, A., and B. Lejczak. 2003. Utilisation of structurally diverse organophosphonates by streptomycetes. Appl. Microbiol. Biotechnol. 62:567–563.

O'Connor, T. J., P. Kanellis, and J. R. Nodwell. 2002. The ramC gene is required for morphogenesis in Streptomyces coelicolor and expressed in a cell type-specific manner under the direct control of RamR. Molec. Microbiol. 45:45–57.

Ohnuki, T., T. Imanaka, and S. Aiba. 1985. Self-cloning in Streptomyces griseus of an str gene cluster for streptomycin biosynthesis and streptomycin resistance. J. Bacteriol. 164:85–94.

Okami, Y., T. Okazaki, T. Kitahara, and H. Umezawa. 1976. Studies on marine microorganisms. V: A new antibiotic, aplasmomycin, produced by a streptomycete isolated from shallow sea mud. J. Antibiot. Ser. A 29:1019–1025.

Onaka, H., T. Nakagawa, and S. Horinouchi. 1998. Involvement of two A-factor receptor homologues in Streptomyces coelicolor A3(2) in the regulation of secondary metabolism and morphogenesis. Molec. Microbiol. 28:743–753.

Pagé, N., D. Kluepfel, F. Shareck, and R. Morosoli. 1996. Effect of signal peptide alterations and replacement on export of xylanase A in Streptomyces lividans. Appl. Environ. Microbiol. 62:109–114.

Pahl, A., A. Gewies, and U. Keller. 1997. ScCypB is a novel second cytosolic cyclophilin from Streptomyces chrysomallus which is phylogenetically distant from ScCypA. Microbiology 143:117–126.

Pang, X., Y. Sun, J. Liu, X. Zhou, and Z. Deng. 2002a. A linear plasmid temperature-sensitive for replication in Streptomyces hygroscopicus 10-22. FEMS Microbiol. Lett. 19:208:25–28.

Pang, X., X. Zhou, Y. Sun, and Z. Deng. 2002b. Physical map of the linear chromosome of Streptomyces hygroscopicus 10-22 deduced by analysis of overlapping large chromosomal deletions. J. Bacteriol. 184:1958–1965.

Paradkar, A. S., K. A. Aidoo, A. Wong, and S. E. Jensen. 1996. Molecular analysis of a β-lactam resistance gene encoded within the cephamycin gene cluster of Streptomyces clavuligerus. J. Bacteriol. 178:6266–6274.

Parche, S., H. Nothaft, A. Kamionka, and F. Titgemeyer. 2000. Sugar uptake and utilisation in Streptomyces coelicolor: A PTS view to the genome. Ant. v. Leeuwenhoek 78:243–251.

Peczynska-Czoch, W., and M. Mordarski. 1988. Actinomycete enzymes. In: M. Goodfellow, S. T. Williams, and M. Mordarski (Eds.) Actinomycetes in Biotechnology. Academic Press. London, UK. 219–283.

Pelaez, A. I., R. M. Ribas-Aparicio, A. Gomez, and M. R. Rodicio. 2001. Structural and functional characterization of the recR gene of Streptomyces. Molec. Genet. Genom. 265:663–672.

Pernodet, J.-L., J.-M. Simonet, and M. Guérineau. 1984. Plasmids in different strains of Streptomyces ambofaciens: Free and integrated form of plasmid pSAM2. Molec. Gen. Genet. 198:35–41.

Philipp, W. J., S. Poulet, K. Eiglmeier, L. Pascopella, V. Balasubramanian, B. Heym, S. Bergh, B. R. Bloom, W. R. Jacobs Jr. and S. T. Cole. 1996. An integrated map of the genome of the tubercle bacillus, Mycobacterium tuberculosis H37Rv, and comparison with Mycobacterium leprae. Proc. Natl. Acad. Sci. USA 93:3132–3137.

Picardeau, M., and V. Vincent. 1998. Mycobacterial linear plasmids have an invertron-like structure related to other linear replicons in Actinomycetes. Microbiology 144:1981–1988.

Pitman, A., P. Herron, and P. Dyson. 2002. Cointegrate resolution following transposition of Tn1792 in Streptomyces avermitilis facilitates analysis of transposon-tagged genes. J. Microbiol. Meth. 49:89–96.

Possoz, C., C. Ribard, J. Gagnat, J. L. Pernodet, and M. Guérineau. 2001. The integrative element pSAM2 from

Streptomyces: Kinetics and mode of conjugal transfer. Molec. Microbiol. 42:159–166.

Qin, Z., and S. N. Cohen. 2002. Survival mechanisms for Streptomyces linear replicons after telomere damage. Molec. Microbiol. 45:785–794.

Ravel, J., H. Schrempf, and R. T. Hill. 1998. Mercury resistance is encoded by transferable giant linear plasmids in two Chesapeake Bay Streptomyces strains. Appl. Environ. Microbiol. 64:3383–3388.

Redenbach, M., H. M. Kieser, D. Denapaite, A. Eichner, J. Cullum, H. Kinashi, and D. A. Hopwood. 1996. A set of ordered cosmids and a detailed genetic and physical map for the 8 MB Streptomyces coelicolor A3(2) chromosome. Molec. Microbiol. 21:77–96.

Redenbach, M., J. Scheel, J. Cullum, and U. Schmidt. 1998. The chromosome of various Actinomycetes strains is linear (Abstract). In: G. Cohen and Y. Aharonowitz (Eds.) 8th International Symposium on the Genetics of Industrial Microorganisms, June 28–July 2, 1998, Jerusalem, Israel. 69–70.

Ribbe, M., D. Gadkari, and O. Meyer. 1997. N_2 fixation by Streptomyces thermoautotrophicus involves a molybdenum-dinitrogenase and a manganese-superoxide oxidoreductase that couple N_2 reduction to the oxidation of superoxide produced from O_2 by a molybdenum-CO dehydrogenase. J. Biol. Chem. 272:26627–26633.

Robbins, P. W., C. Albright, and B. Benfield. 1988. Cloning and expression of a Streptomyces plicatus chitinase (chitinase-63) in Escherichia coli. J. Biol. Chem. 263: 443–447.

Rodríguez-García, A., M. Ludovice, J. F. Martín, and P. Liras. 1997. Arginine boxes and the argR gene in Streptomyces clavuligerus: Evidence for a clear regulation of the arginine pathway. Molec. Microbiol. 25:219–228.

Roller, C., W. Ludwig, and K.-H. Schleifer. 1992. Gram-positive bacteria with a high DNA G+C content are characterized by a common insertion within their 23S rRNA genes. J. Gen. Microbiol. 138:167–175.

Rose, K., and A. Steinbuchel. 2002. Construction and intergeneric conjugative transfer of a pSG5-based cosmid vector from Escherichia coli to the polyisoprene rubber degrading strain Micromonospora aurantiaca W2b. FEMS Microbiol. Lett. 211:129–132.

Rueda, B., E. M. Miguelez, C. Hardisson, and M. B. Manzanal. 2001. Changes in glycogen and trehalose content of Streptomyces brasiliensis hyphae during growth in liquid cultures under sporulating and non-sporulating conditions. FEMS Microbiol. Lett. 194:181–185.

Ruiz-Arribas, A., G. G. Zhadan, V. P. Kutyshenko, R. I. Santamaría, M. Cortijo, E. Villar, J. M. Fernandez-Abalos, J. J. Calvete, and V. L. Shnyrov. 1998. Thermodynamic stability of two variants of xylanase (Xys1) from Streptomyces halstedii JM8. Eur. J. Biochem. 253:462–468.

Saito, N., K. Matsubara, M. Watanabe, F. Kato, and K. Ochi. 2003. Genetic and biochemical characterization of EshA, a protein that forms large multimers and affects developmental processes in Streptomyces griseus. J. Biol. Chem. 278:5902–5911.

Salas, J. A., C. Hernandez, C. Mendez, C. Olano, L. M. Quiros, A. M. Rodriguez, and C. Vilches. 1994. Intracellular glycosylation and active efflux as mechanisms for resistance to oleandomycin in Streptomyces antibioticus, the producer organism. Microbiologia 10:37–48.

Schaerlaekens, K., M. Schierova, E. Lammertyn, N. Geukens, J. Anné, and L. van Mellaert. 2001. Twin-arginine trans-location pathway in Streptomyces lividans. J. Bacteriol. 183:6727–6732.

Schlochtermeier, A., S. Walter, J. Schröder, M. Moormann, and H. Schrempf. 1992. The gene encoding the cellulase (Avicelase) Cel1 from Streptomyces reticuli and analysis of protein domains. Molec. Microbiol. 6:3611–3621.

Schlösser, A., and H. Schrempf. 1996. A lipid-anchored binding protein is a component of an ATP-dependent cellobiose/-triose transport system from the cellulose degrader Streptomyces reticuli. Eur. J. Biochem. 242: 332–338.

Schlösser, A., T. Kampers, and H. Schrempf. 1997. The Streptomyces ATP-binding component MsiK assists in cellobiose and maltose transport. J. Bacteriol. 179:2092–2095.

Schneider, D., C. J. Bruton, and K. F. Chater. 2000. Duplicated gene clusters suggest an interplay of glycogen and trehalose metabolism during sequential stages of aerial mycelium development in Streptomyces coelicolor A3(2). Molec. Gen. Genet. 263:543–553.

Schrempf, H., P. Dyson, W. Dittrich, M. Betzler, C. Habiger, B. Mahro, V. Brönneke, A. Kessler, and H. Düvel. 1989. Genetic instability in Streptomyces. In: Y. Okami, T. Beppu, and H. Ogawara (Eds.) Biology of Actinomycetes '88. Scientific Press. Tokyo, Japan. 145–150.

Schrempf, H. et al. 1995. A prokaryotic potassium ion channel with two predicted transmembrane segments from Streptomyces lividans. EMBO J. 14:5170–5178.

Schwecke, T., J. F. Aparicio, I. Molnar, A. Konig, L. E. Khaw, S. F. Haydock, M. Oliynyk, P. Caffrey, J. Cortes, J. B. Lester, G. A. Bohm, J. Staunton, and P. F. Leadlay. 1995. The biosynthesis gene cluster for the polyketide immunosuppressant rampamycin. Proc. Natl. Acad. Sci. USA 92:7839–7843.

Servín-González, L. 1993. Relationship between the replication functions of Streptomyces plasmids pJV1 and pIJ101. Plasmid 30:131–140.

Servín-González, L., C. Castro, C. Pérez, M. Rubio, and F. Valdez. 1997. bldA-dependent expression of the Streptomyces exfoliatus M11 lipase gene (lipA) is mediated by the product of a contiguous gene, lipR, encoding a putative transcriptional activator. J. Bacteriol. 179:7816–7826.

Sezonov, G., A.-M. Duchêne, A. Friedmann, M. Guérineau, and J.-L. Pernodet. 1998. Replicase, excisionase, and integrase genes of the Streptomyces element pSAM2 constitute an operon positively regulated by the pra gene. J. Bacteriol. 180:3056–3061.

Sherman, D. H., F. Malpartida, M. J. Bibb, H. M. Kieser, and D. A. Hopwood. 1989. Structure and deduced function of the granaticin-producing polyketide synthase gene cluster of Streptomyces violaceoruber TU22. EMBO J. 8:2717–2725.

Soh, B. S., P. Loke, and T. S. Sim. 2001. Cloning, heterologous expression and purification of an isocitrate lyase from Streptomyces clavuligerus NRRL 3585. Biochim. Biophys. Acta 1522:112–117.

Sohng, J. K., T. J. Oh, J. J. Lee, and C. G. Kim. 1997. Identification of a gene cluster of biosynthetic genes of rubradirin substructures in S. achromogenes var. rubradiris NRRL3061. Molec. Cells 7:674–681.

Sola-Landa, A., R. S. Moura, and J. F. Martin. 2003. The two-component PhoR-PhoP system controls both primary metabolism and secondary metabolite biosynthesis in Streptomyces lividans. Proc. Natl. Acad. Sci. USA 100:6133–6138.

Sommer, P., C. Bormann, and F. Götz. 1997. Genetic and biochemical characterization of a new extracellular lipase from Streptomyces cinnamomeus. Appl. Environ. Microbiol. 63:3553–3560.

Spiteller, D., A. Jux, J. Piel, and W. Boland. 2002. Feeding of [5,5-2H(2)]-1-desoxy-D-xylulose and [4,4,6,6,6-2H(5)]-mevalolactone to a geosmin-producing Streptomyces sp. and Fossombronia pusilla. Phytochemistry 61:827–834.

Spychaj, A., and M. Redenbach. 2001. The terminal inverted repeats of the linear plasmid SCP1 of Streptomyces coelicolor A3(2) possess a truncated copy of the transposon Tn4811 of Streptomyces lividans 66. Ant. v. Leeuwenhoek 79:49–52.

Stindl, A., and U. Keller. 1994. Epimerization of the D-valine portion in the biosynthesis of actinomycin D. Biochemistry 33:9358–9364.

Stutzman-Engwall, K. J., and C. R. Hutchinson. 1989. Multigene families for anthracycline antibiotic production in Streptomyces peucetius. Proc. Natl. Acad. Sci. USA 86:3135–3139.

Swan, D. G., A. M. Rodriguez, C. Vilches, C. Méndez, and J. A. Salas. 1994. Characterisation of a Streptomyces antibioticus gene encoding a type I polyketide synthase which has an unusual coding sequence. Molec. Gen. Genet. 242:358–362.

Taguchi, S., T. Endo, Y. Naoi, and H. Momose. 1995. Molecular cloning and sequence analysis of a gene encoding an extracellular serine protease from Streptomyces lividans 66. Biosci. Biotechnol. Biochem. 59:1386–1388.

Tsujibo, H., T. Ohtsuki, T. Iio, I. Yamazaki, K. Miyamoto, M. Sugiyama, and Y. Inamori. 1997. Cloning and sequence analysis of genes encoding xylanases and acetyl xylan esterase from Streptomyces thermoviolaceus OPC-520. Appl. Environ. Microbiol. 63:661–664.

Van Keulen, G., H. M. Jonkers, D. Claessen, L. Dijkhuizen, and H. A. Wosten. 2003. Differentiation and anaerobiosis in standing liquid cultures of Streptomyces coelicolor. J. Bacteriol. 185:1455–1458.

Vara, J., M. Lewandowska-Skarbek, Y. G. Wang, S. Donadio, and C. R. Hutchinson. 1989. Cloning of genes governing the deoxysugar portion of the erythromycin biosynthesis pathway in Saccharopolyspora erythraea (Streptomyces erythreus). J. Bacteriol. 171:5872–5881.

Verma, S., Y. Bhatia, S. P. Valappil, and I. Roy. 2002. A possible role of poly-3-hydroxybutyric acid in antibiotic production in Streptomyces. Arch. Microbiol. 179:66–69.

Virolle, M.-J., and M. J. Bibb. 1988. Cloning, characterization and regulation of an α-amylase gene from Streptomyces limosus. Molec. Microbiol. 2:197–208.

Voelker, F., and S. Altaba. 2001. Nitrogen source governs the patterns of growth and pristinamycin production in Streptomyces pristinaespiralis. Microbiology 147:2447–2459.

Volff, J. N., and J. Altenbuchner. 1997. High-frequency transposition of IS1373, the insertion sequence delimiting the amplifiable element AUD2 of Streptomyces lividans. J. Bacteriol. 179:5639–5642.

Von Döhren, H., and H. Kleinkauf. 1997. Enzymology of peptide synthetases. In: W. R. Strohl (Ed.) Biotechnol-

ogy of Antibiotics, 2nd ed. Marcel Dekker. New York, NY. 217–240.

Vujaklija, D. et al. 2002. A novel streptomycete lipase: Cloning, sequencing and high-level expression of the Streptomyces rimosus GDS(L)-lipase gene. Arch. Microbiol. 178:124–130.

Walter, S., E. Wellmann, and H. Schrempf. 1998. The cell wall-anchored Streptomyces reticuli Avicel-binding protein (AbpS) and its gene. J. Bacteriol. 180:1647–1654.

Wang, F., X. Xiao, A. Saito, and H. Schrempf. 2002. Streptomyces olivaceoviridis possesses a phosphotransferase system that mediates specific, phosphoenolpyruvate-dependent uptake of N-acetylglucosamine. Molec. Genet. Genom. 268:344–351.

Wang, L., S. Li, and Y. Li. 2003. Identification and characterization of a new exopolysaccharide biosynthesis gene cluster from Streptomyces. FEMS Microbiol. Lett. 220:21–27.

Weber, T., K. Welzel, S. Pelzer, A. Vente, and W. Wohlleben. 2003. Exploiting the genetic potential of polyketide producing streptomycetes. J. Biotechnol. 106:221–232.

Wehmeier, U. F. 2001. Molecular cloning, nucleotide sequence and structural analysis of the Streptomyces galbus DSM40480 fda gene: The S. galbus fructose-1,6-bisphosphate aldolase is a member of the class II aldolases. FEMS Microbiol. Lett. 197:53–58.

Xiao, X., F. Wang, A. Saito, J. Majka, A. Schlösser, and H. Schrempf. 2002. The novel Streptomyces olivaceoviridis ABC transporter Ngc mediates uptake of N-acetylglucosamine and N,N'-diacetylchitobiose. Molec. Genet. Genom. 267:429–439.

Yang, C. C., C. H. Huang, C. Y. Li, Y. G. Tsay, S. C. Lee, and C. W. Chen. 2002. The terminal proteins of linear Streptomyces chromosomes and plasmids: A novel class of replication priming proteins. Molec. Microbiol. 43:297–305.

Yu, T. W., M. J. Bibb, W. P. Revill, and D. A. Hopwood. 1994. Cloning, sequencing, and analysis of the griseusin polyketide synthase gene cluster from Streptomyces griseus. J. Bacteriol. 176:2627–2634.

Yuan, W. M., and D. L. Crawford. 1995. Characterization of Streptomyces lydicus WYEC108 as a potential biocontrol agent against fungal root and seed rots. Appl. Environ. Microbiol. 61:3119–3128.

Zakrzewska-Czerwinska, J., and H. Schrempf. 1992. Characterization of an autonomously replicating region from the Streptomyces lividans chromosome. J. Bacteriol. 174:2688–2693.

Zhang, X., C. A. Clark, and G. S. Pettis. 2003. Interstrain inhibition in the sweet potato pathogen Streptomyces ipomoeae: Purification and characterization of a highly specific bacteriocin and cloning of its structural gene. Appl. Environ. Microbiol. 69:2201–2208.

Zotchev, S. B., and H. Schrempf. 1994. The linear Streptomyces plasmid BL1: Analyses of transfer functions. Molec. Gen. Genet. 242:374–382.

Zuber, P., and M. A. Marahiel. 1997. Structure, function and regulation of genes encoding multidomain peptide synthetases. In: W. R. Strohl (Ed.) Biotechnology of Antibiotics, 2nd ed. Marcel Dekker. New York, NY. 187–216.

Prokaryotes (2006) 3:623–653
DOI: 10.1007/0-387-30743-5_24

CHAPTER 1.1.9

The Genus *Actinoplanes* and Related Genera

GERNOT VOBIS

Sporeforming actinomycetes related to the genus *Actinoplanes* are combined at present under the suprageneric group named "actinoplanetes" (Goodfellow, 1989). Originally, the term actinoplanetes was used ecologically to describe the strains of *Actinoplanes* and a small number of their relatives (Nonomura and Takagi, 1977). The members of the genus *Actinoplanes* all produce sporangia that release actively swimming spores. Later, the term actinoplanetes was expanded to encompass also the five genera *Amorphosporangium, Ampullariella, Pilimelia, Dactylosporangium*, and *Micromonospora*, which all have a common cell wall chemotype containing *meso-* and/or 3-hydroxy diaminopimelic acid and glycine in combination with xylose and arabinose as the characteristic sugars in the hydrolysates of whole organisms (Goodfellow and Cross, 1984).

The GC content of their DNA is in general 71–73 mol% (Vobis, 1989a), but recent studies have expanded the range to 67–76 mol% (Kothe, 1987). All genera of the actinoplanetes studied so far belong to one RNA homology cluster (Stackebrandt and Woese, 1981), excluding the sporangiate genera *Streptosporangium, Spirillospora, Planomonospora*, and *Planobispora*, which were classified for a long time within one family together with *Actinoplanes* (Bland and Couch, 1981). The genera excluded were placed temporarily into a suprageneric group called the "maduromycetes" (Goodfellow, 1989). According to Goodfellow et al. (1990) and Kothe (1987), the members of the actinoplanetes can now be harbored taxonomically in the newly defined family Micromonosporaceae, and those of the maduromycetes belong in the family Streptosporangiaceae (see this Volume).

The genus *Actinoplanes* was first described by Couch (1950), who called attention to its close similarities in colonial characteristics to the genus *Micromonospora* (Ørskov, 1923), which produces single nonmotile spores. Our knowledge of the sporangiate members was extended by subsequent descriptions of further genera: *Ampullariella* and *Amorphosporangium* (Couch, 1963), *Pilimelia* (Kane, 1966), and *Dactylosporangium* (Thiemann et al., 1967). In recent years, two new genera have been added: *Glycomyces* (Labeda et al., 1985) and *Catellatospora* (Asano and Kawamoto, 1986). Both genera form nonmotile spores in chains. Although their chemotaxonomic features suggest convincing arguments for placing them into the actinoplanetes group, their phylogenetic relationships still need to be investigated (E. Stackebrandt, personal communication).

This chapter was taken unchanged from the second edition.

Ecophysiology

Without applying selective procedures, it is difficult to isolate the members of *Actinoplanes* and related genera from soil or other natural substrates. Although they are mesophilic and aerobic organisms, the growth rate of their colonies is often very slow, and on routine isolation plates, the fast-growing streptomycetes can overrun them before they have developed conspicuous mycelia. Furthermore, they exhibit quite varied types of reproductive structures, which are connected with distinct life cycles: 1) the genera *Actinoplanes, Ampullariella, Pilimelia*, and *Dactylosporangium* form sporangia with flagellated spores and are adapted to a periodic wetting and drying of the habitat; 2) the hydrophilic, nonmotile spores of *Micromonospora* are passively disseminated in water and in soil, again distributed by wind or rain; and 3) the genera *Catellatospora* and *Glycomyces* produce nonmotile, hydrophobic spores in chains, presumably associated with other survival strategies.

Sporangiate Actinoplanetes

The life cycles of the sporangiate genera are based on an alternation between terrestrial and "aquatic" habitats. The growth of vegetative mycelium on plant or animal residues culminates in the differentiation into sporangia, which are produced in general on the surface of the substrate, directly in contact with the air (Figs. 1F, 2C, 4A, and 7B) (Bland and Couch, 1981). The sporangia can easily lose their connection to the degenerating mycelium and are disseminated as diaspores by the wind or by soil fauna such as mites, collembola, or arthropods. The sporangia can withstand prolonged desiccation and survive for many years (Makkar and Cross, 1982). The sporangial envelope is usually water repellent. But if sporangia become rehydrated by sufficient moisture, e.g., during periods of fog or rain, the spores inside the sporangia begin to swell, the sporangial envelope bursts, and the flagellated spores are released. Under laboratory conditions, this process takes 10 to 60 min (Higgins,

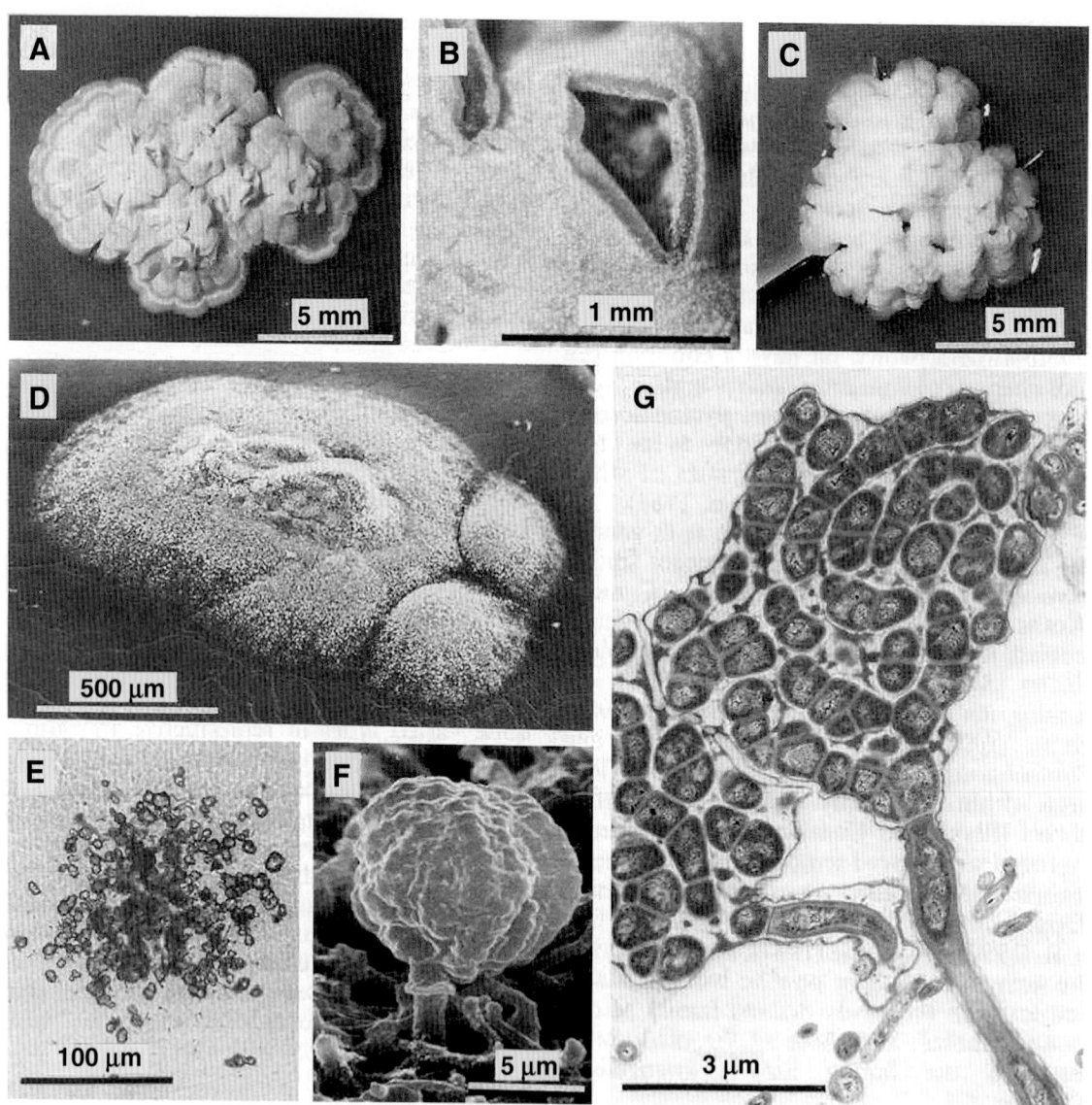

Fig. 1. Features of the genus *Actinoplanes*. (A) Colony with rough surface; marginal areas divided into radial and concentric sections, dissecting microscope (DM). (B) Burst substrate mycelium covered with a mass of sporangia (DM). (C) Elevated colony with squamules; smooth surface without sporangia (DM). (D) Flat colony with abundant sporangia visible on the substrate mycelium, scanning electron microscope (SEM). (E) Irregularly shaped sporangia on agar medium, light microscope (LM). (F) Globose sporangium at the tip of a palisade hypha (SEM). (G) Section of a sporangium with coiled chains of spores, transmission electron microscope (TEM). (G from Kothe, 1987; with permission.)

1967; Vobis, 1984, 1987). Spores of *Actinoplanes brasiliensis* retain their motility for more than one day in liquid mineral medium with glucose (Palleroni, 1983). Even without an exogenous source of energy, the zoospores of *Dactylosporangium* species are motile for two to three days before they germinate (Vobis, 1987). Reserve substances are included in their cytoplasm (Fig. 4C), possibly consisting of trehalose (J. C. Ensign, personal communication).

The zoospores of *Actinoplanes* exhibit chemotactic properties. In *A. brasiliensis,* Palleroni (1976) found bromide and chloride ions acting as attractants at a relatively high concentration (0.1 M). Addition of methionine stimulated this chemotactic effect, suggesting that protein methylation may be involved (Palleroni, 1983). Not all species of *Actinoplanes* are attracted by halides. Spores of *A. missouriensis* were attracted to fungal conidia, chlamydospores, and sclerotia and to exudates of them (Arora, 1986). Several sugars had the same function. An extract of cattle horn meal can be more attractive to the spores than chloride ions are (G. Vobis, unpublished observations). Phototactic effects could not be observed, but an apparent microaerophilic

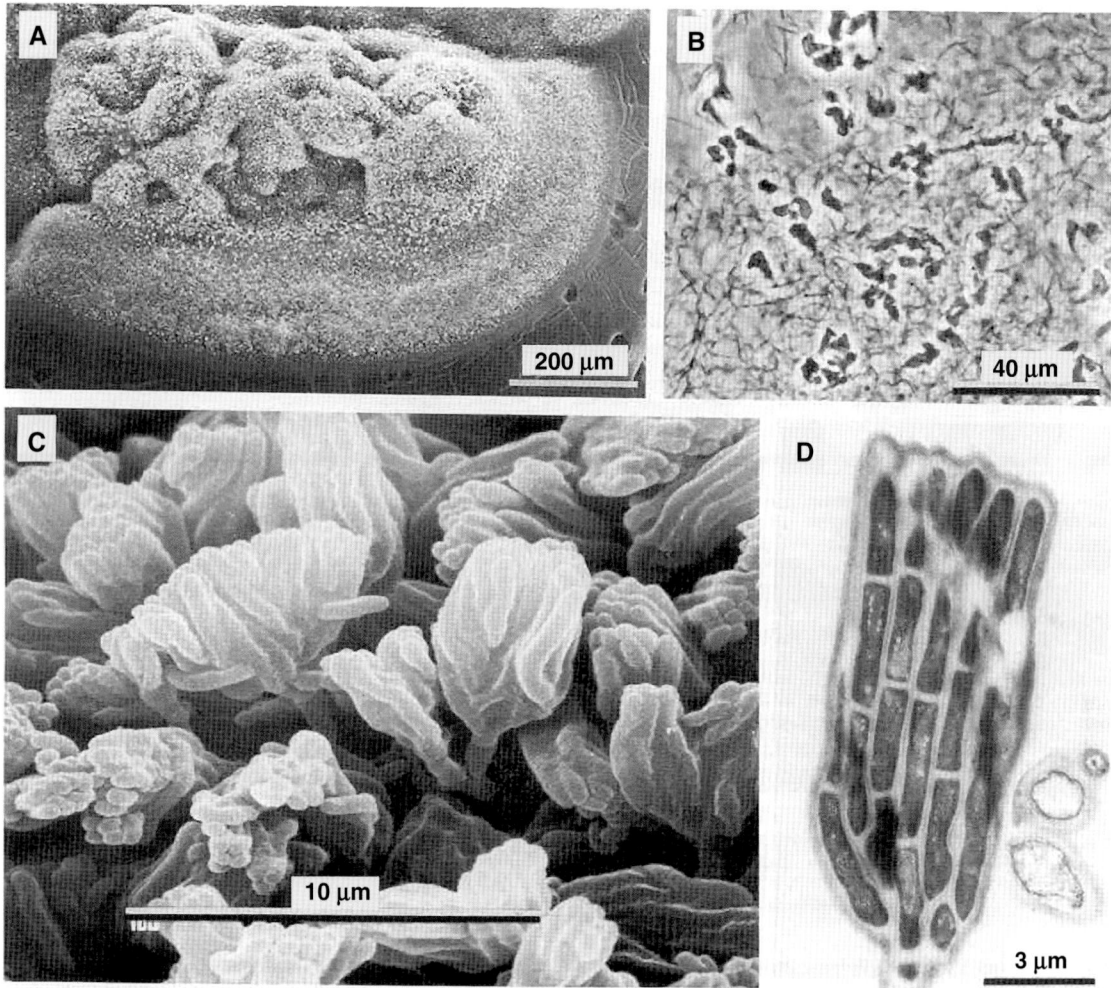

Fig. 2. Features of the genus *Ampullariella.* (A) Colony on agar medium, covered with a bloom-like layer of sporangia (SEM). (B) Sporangia on substrate mycelium, phase contrast (PHACO). (C) Campanulate sporangia; sporogenous hyphae are recognizable under the sporangial envelope (SEM). (D) Section of a cylindrical sporangium with parallel rows of bacilliform spores (TEM). (D from Vobis, 1987; with permission.)

behavior was seen in *A. brasiliensis* (Palleroni, 1976). In baiting experiments simulating an aquatic microhabitat, pollen or hair is exposed to the surface of water. The zoospores, once released from the submerged sporangia, are able to swim to the surface, fasten to the natural substrates, germinate, and colonize them within several days (Couch, 1963; Vobis, 1984). This may be a result of aerotactic and chemotactic behavior of the spores (Cross, 1986). Although the chemotactic response is used effectively in the isolation method of Palleroni (1980), the exact physiological explanation is not yet known.

The members of the genus *Dactylosporangium* have an additional variant in their life cycle. Beside the typical few-spored sporangia (Fig. 4A–C), they can develop "globose bodies" (Thiemann et al., 1967). These are spores that are borne singly on substrate hyphae (Fig. 4D–F).

Ensign (1978) demonstrated that these spores are able to germinate on 1.0% (w/v) yeast extract agar medium. The zoospores may function to insure a prompt dissemination in water, and the globose spores, which are constitutively dormant, may survive long periods of starvation or desiccation (Ensign, 1978). Likewise, treatment with dry heat (1 h at 120°C) seems to have no adverse effect on the dormant globose spores of *Dactylosporangium,* in contrast to the sporangiospores of other actinoplanetes (Shearer, 1987).

In general, the sporangiate actinoplanetes can be considered as normal inhabitants of the soil and leaf litter (Cross, 1981a), although they can also be isolated directly from lake or river water (Willoughby, 1969b, 1971). They colonize vegetable or animal debris preferably (Cross, 1981a). A frequent drying and wetting of the substrate

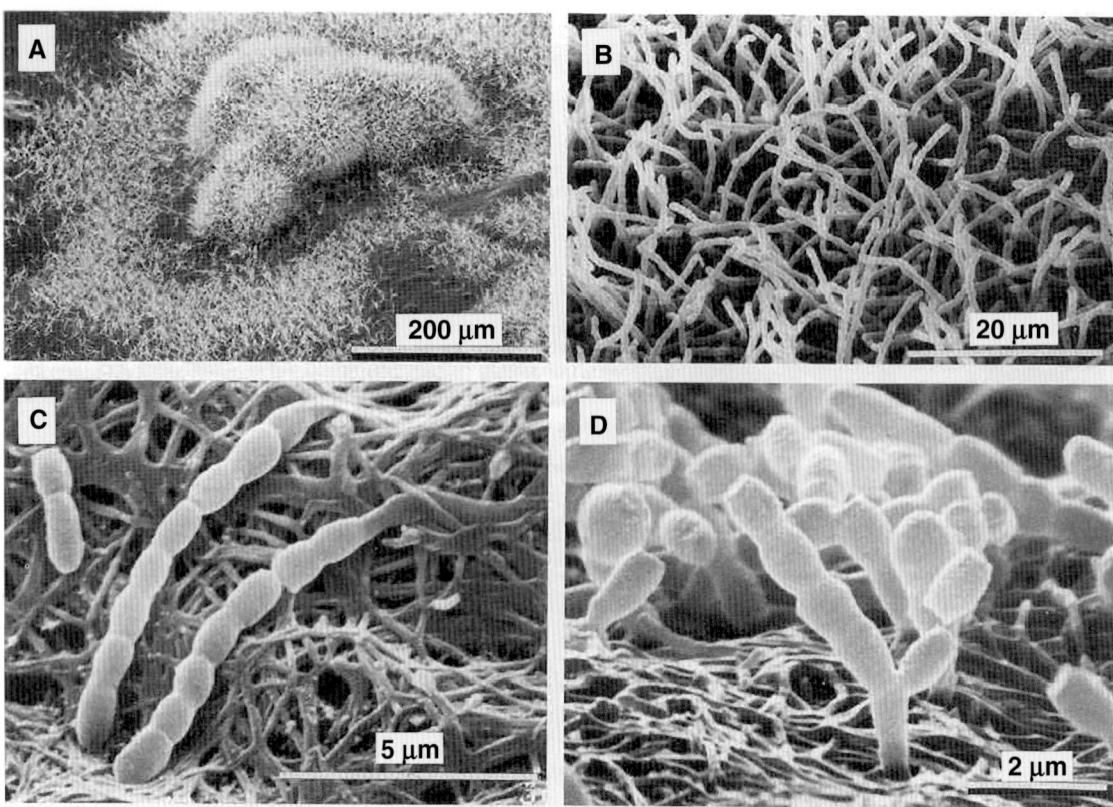

Fig. 3. Features of the genus *Catellatospora*. (All SEMs.) (A) A sporulating colony. (B) Sporeforming hyphae on the surface of agar medium. (C) Short spore chains emerging from substrate hyphae. (D) Branched spore chains on rudimentary sporophores containing cylindrical to ovoid spores.

increases their occurrence. Favored habitats are the edges of ponds, drainage ditches, and barnyards (Shearer, 1987). Sediments of rivers seem to be also a good source for the isolation of *Actinoplanes* species (Goodfellow et al., 1990). They can also be frequently isolated from twigs submerged in streams (Willoughby, 1971), muddy dead leaves that are caught and dried on branches of overhanging trees (Cross, 1981a), and from allochthonous leaf litter cast up on the shores of lakes (Willoughby, 1969a). Rarely, strains of *Dactylosporangium* can be encountered in lake sediments. Possibly survival depends on the globose spores (Johnston and Cross, 1976). Also, marine sediments are not very productive habitats for sporangiate actinoplanetes (Goodfellow and Haynes, 1984).

The genus *Actinoplanes* has a world-wide distribution (Couch, 1963; Gaertner, 1955; Vobis, 1987; Schäfer, 1973; Nonomura and Takagi, 1977). Strains of *Actinoplanes* occur in all types of soil, arid desert areas (Makkar and Cross, 1982), sand dune systems close to seashores (Palleroni, 1976), and subtropical and tropical regions. The most productive samples seem to originate from the latter (Shearer, 1987). In a

large-scale investigation of the distribution of actinoplanetes in soil in Japan, Nonomura and Takagi (1977) demonstrated a correlation among their abundance, the type of soil, its pH value, and the content of organic matter. Relatively few actinoplanetes occurred in soils with pH 4.0 to 5.0 and abundant organic matter content. Their number increased with lower humus content and a pH value between 6.4 to 7.2. Soils with a permanent high content of water (e.g., paddy rice fields) have no advantage compared with cultivated fields, which are dry for longer periods.

Strains of *Ampullariella* are widely distributed throughout the world (Couch, 1963). A little less than 10% of the isolated sporangiate actinoplanetes are represented by this genus (Schäfer, 1973; Vobis, 1987). *Ampullariella* could also be obtained from freshwater habitats (Willoughby, 1969b).

The function of sporangiate actinoplanetes in soil ecosystems is not really known. With the exception of a few strains, the *Actinoplanes* species cannot decompose cellulose (Parenti and Coronelli, 1979; Palleroni, 1989). Although they can be isolated on colloidal chitin agar medium,

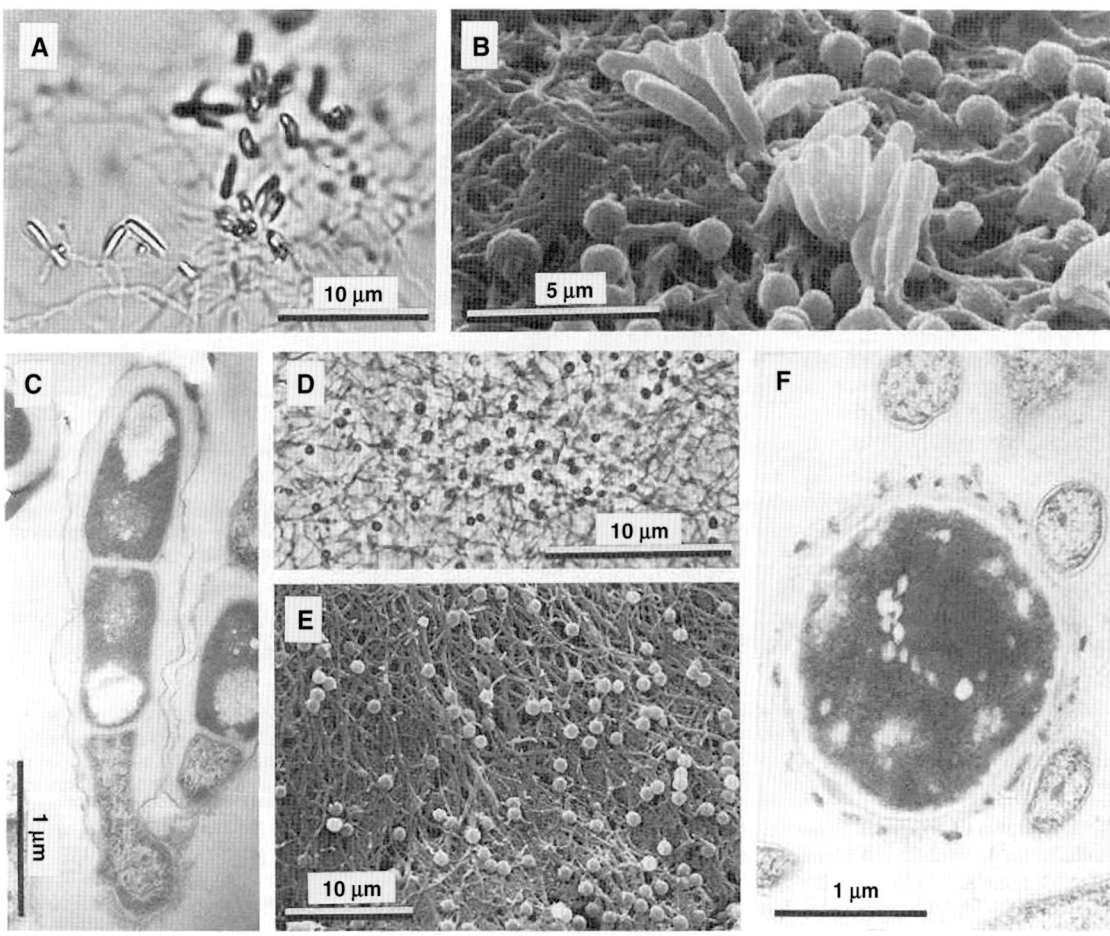

Fig. 4. Features of the genus *Dactylosporangium*. (A) Finger-like sporangia on agar medium (LM). (B) Bundles of sporangia and globose spores (SEM). (C) Section of a sporangium with two spores containing reserve material; sporangial envelope is thin and wavy (TEM). (D) Refractile globose spores dispersed in substrate mycelium (PHACO). (E) Globose spores on the surface of substrate mycelium (SEM). (F) Section of a globose spore with paracrystalline inclusion bodies and perispherical deposits (TEM). (C, D, and F from Vobis, 1987; with permission.)

degradation tests with chitin from insects and fungi gave negative results (Schäfer, 1973). Since they exhibit good growth on xylose and arabinose, it is conceivable that they play a role in decomposing pentosans of plant origin (Parenti and Coronelli, 1979). *Pilimelia* strains are able to colonize keratinic substrates like hair of mammalia or snake skin (Karling, 1954; Gaertner, 1955; Tribe and Abu El-Souod, 1979). They are distributed worldwide and occur statistically in about one of every five soil samples (Schäfer, 1973; Vobis et al., 1986). Although they can aggressively attack the scleroproteins of animals (Fig. 7C), they are not known as dermatophytes.

MICROMONOSPORA Members of the genus *Micromonospora* can be commonly isolated from neutral and alkaline soils (Jensen, 1930; 1932) and according to Shearer (1987), two or three strains may be expected from most soil samples. However their predominant incidence seems to be in aquatic ecosystems, including both freshwater and marine habitats (Cross, 1981a; Goodfellow and Haynes, 1984). Since they decompose chitin, cellulose, and lignin of lake sediments they might play an important role in lacustrine ecology (Erikson, 1941). The presence of *Micromonospora* in lake systems has been shown by investigations in many countries and was comprehensively reviewed by Cross (1981a, 1981b). Beside streptomycetes and nocardioforms, the micromonosporas were the predominant actinomycetes in the bottom sediments of Blelham Tarn UK, with numbers increasing from littoral to profundal mud samples (Willoughby, 1969b). This dominance was even more striking in deeper mud layers, as could be shown in studies of other lakes of the English Lake District (Johnston and Cross, 1976). Similar observations were made by Fernandez (1984) at a thermal lake, Lake Hévíz, in Hungary. Compared with the surface of the mud, the number of micromonosporas increases twofold in 20 cm of depth, whereas the number of streptomycetes decreases significantly in the same layer. Under those conditions, the spores of *Micromonospora* seem to be more resistant than the propagules of *Streptomyces* and nocardioform actinomycetes. This could be confirmed in investigations on the longevity of actinomycete spores in deep mud cores. Viable spores of *Micromonospora* were recorded from sediments deposited at least 100 years before (Cross and Attwell, 1974). Populations of *Micromonospora* species, accompanied by other actinomycetes, have also been frequently found in streams and rivers (Rowbotham and Cross, 1977; Al-Diwany and Cross, 1978). Their spores are hydrophilic and wettable (Fig. 6C) and can easily be removed from soil by the passage of water (Ruddick and Williams, 1972). The spores withstand ultrasonication, moist heat treatment (20 min at 60°C), and dry heat up to 75°C, and they are resistant to various chemical solutions. However, they are somewhat sensitive to acidic pH (Kawamoto, 1989). Thus, the conclusion can be drawn that the spores of *Micromonospora* are washed into the streams,

rivers, and lakes where they can survive as dormant propagules for many years (Cross, 1981a).

Micromonospora species have been isolated from many different marine habitats, ranging from coastal regions to deep-sea sediments. Hunter et al. (1981) found abundant micromonosporas in salt marsh ecosystems in New Jersey (USA), with seasonal fluctuations in quantity. In a study at the San Francisco Bay (USA) National Wildlife Refuge, Hunter et al. (1984) showed that micromonosporas occur more frequently in rhizospheric soils of seashore plants than in mud samples obtained from plant-free areas. Watson and Williams (1974) studied the actinomycetes in a coastal sand belt near Formby, Lancashire (UK). In sea water and beach strand, the *Micromonospora* strains predominated. They grew well on freshwater media and most of them tolerated seawater salinity. Okazaki and Okami (1972) and Okami and Okazaki (1978) isolated micromonosporas from littoral muds and from samples collected in shallow sea areas of the Pacific Ocean, occurring more frequently at the bottom than in the sea water. Weyland (1969, 1981) found that the micromonosporas predominated in the deep-sea sediments, and his results were confirmed by

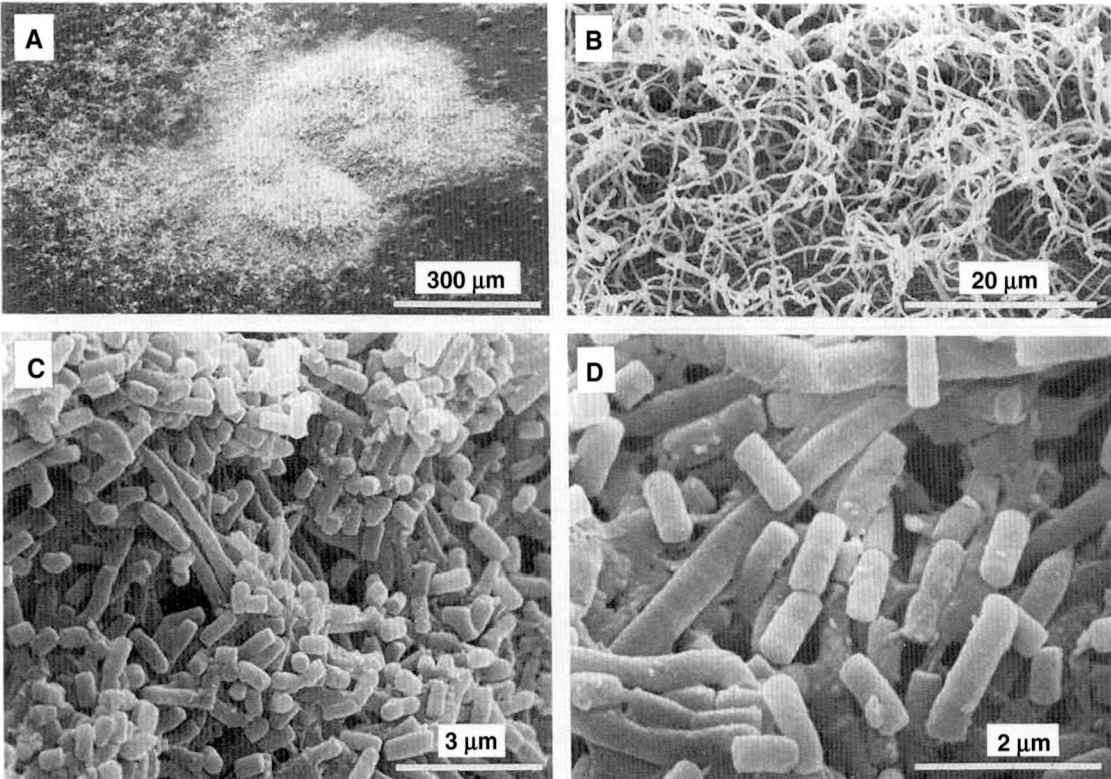

Fig. 5. Features of the genus *Glycomyces*. (All SEMs.) (A) A colony with aerial mycelium. (B) Irregularly oriented and branched aerial hyphae. (C) Mass of spores produced by fragmentation of aerial hyphae. (D) Square-ended spores of various lengths (0.5 × 0.7–1.8 μm).

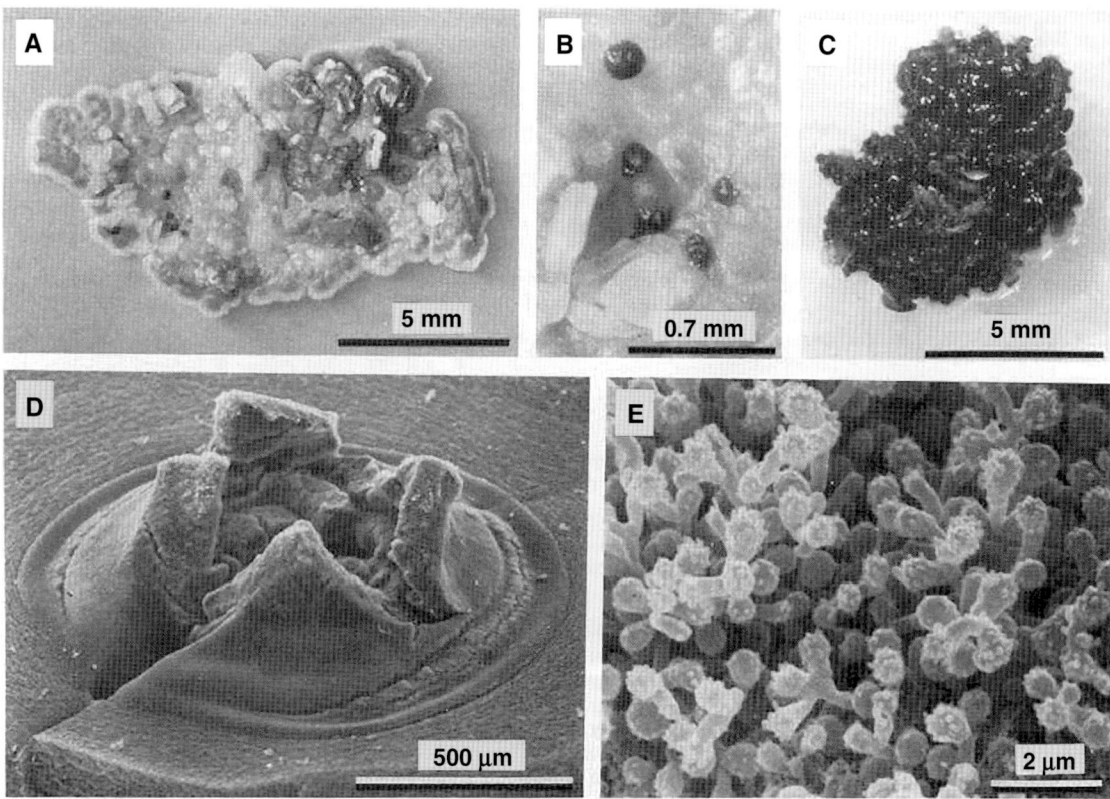

Fig. 6. Features of the genus *Micromonospora*. (A) Raised and folded colony with areas of different colors (DM). (B) Clusters of dark spore masses (DM). (C) Colony completely covered by a mucoid, black mass of spores (DM). (D) Crosswise-burst colony exposing the sporeforming substrate mycelium (SEM). (E) Cluster of spores formed on short side-branches of substrate hyphae; surface of the spores is covered with blunt spines (SEM).

Goodfellow and Haynes (1984). The ability of *Micromonospora* strains to tolerate reduced oxygen tensions (Watson and Williams, 1974) favored the view that after the spores sink into the sea, they settle into littoral or marine sediments where they can survive for long periods of time (Goodfellow and Williams, 1983). Other authors suggest that the actinomycetes are a part of the indigenous marine microflora, able to grow in seawater and its sediments (Okami and Okazaki, 1978; Weyland, 1981).

The presence of *Micromonospora* species could also be established in terrestrial habitats like straw and hay, in grain stores, or as pollutants of the aerial environment in homes. They also can be found in structural timbers, chiefly those found below the watertable in the foundation piles of buildings. They have also been found in the trunks of trees of the genera *Picea* and *Pinus* and in pulp from paper mills (Lacey, 1988).

Most *Micromonospora* species probably degrade biopolymers (Erikson, 1941), and they can even attack lignin complexes (McCarthy and Broda, 1984). Many of the salt marsh isolates of Hunter et al. (1981) were active in the decompo-

sition of chitin and cellulose. In particular, cellulose is frequently utilized as substrate (Jensen, 1930; Sandrak, 1977; Kawamoto, 1989). The cellulase studied from *Micromonospora melanosporea* was found to be more heat stable than those of the imperfect fungus *Trichoderma,* but less stable than the enzymes of thermophilic actinomycetes. The principle sugar released by *Micromonospora* cellulase from ball-milled bagasse and filter paper was cellobiose (Van Zyl, 1985).

It is possible that micromonosporas are able to grow under microaerophilic conditions, as would be found in wet soils (Goodfellow and Williams, 1983). However, the strictly anaerobic *Micromonospora* strains isolated from the intestinal tract of termites (Hungate, 1946) and from the rumen of sheep (Maluszyńska and Janota-Bassalik, 1974) urgently require taxonomical studies to clarify their affiliation to the genus *Micromonospora*.

CATELLATOSPORA AND GLYCOMYCES As Asano et al. (1989b) emphasized, all their isolates of *Catellatospora* species were recovered from woodland soils collected at various localities from the

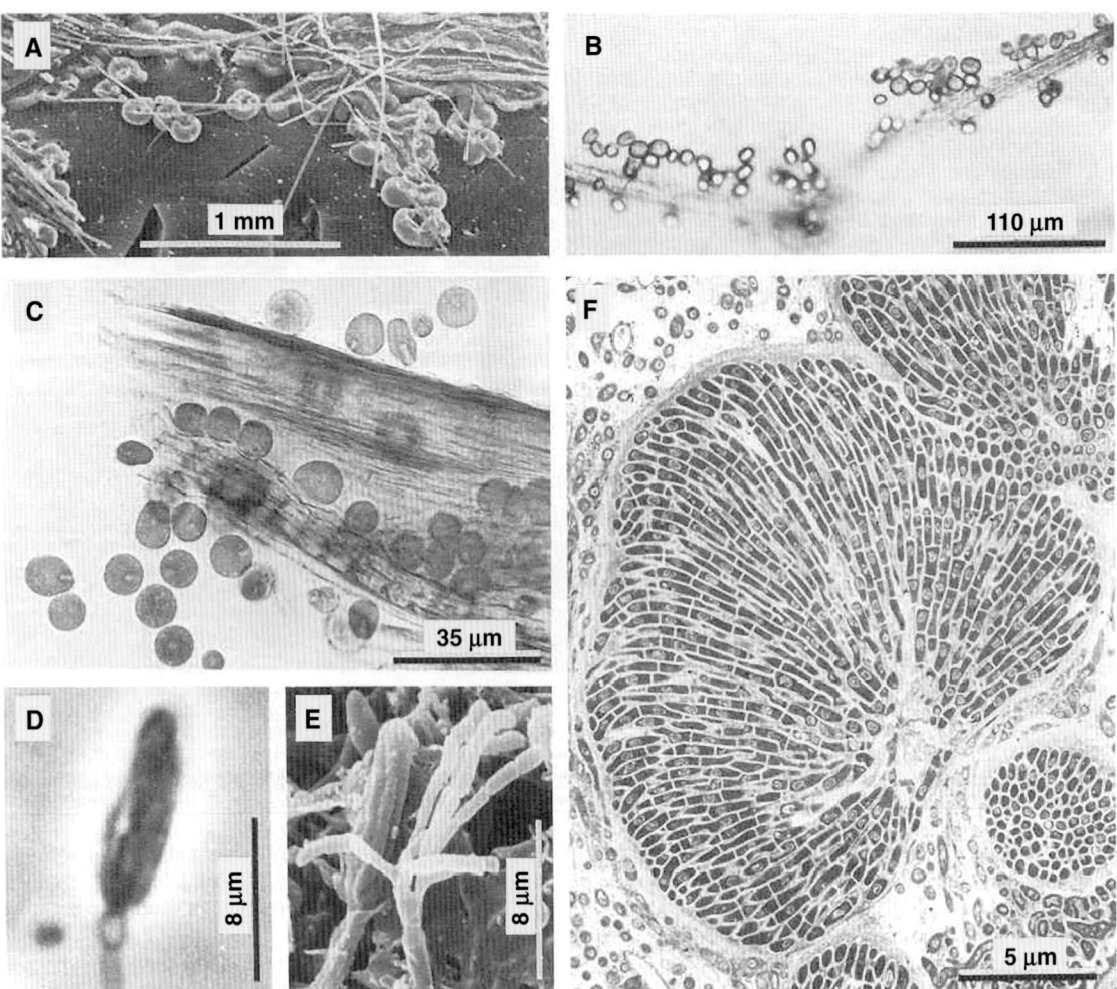

Fig. 7. Features of the genus *Pilimelia*. (A) Compact, small colonies on agar medium; hairs added as natural substrate (SEM). (B) Bundles of sporangia formed on hair (LM). (C) Globose to pyriform sporangia with internal columella; structure of the colonized part of the hair was destroyed (LM). (D) Cylindrical sporangium with an annulus at the base (PHACO). (E) Penicillate conidiophore with bacilliform conidia; the sporangium behind it has parallel-arranged sporogenous hyphae (SEM). (F) Section of a campanulate sporangium with branched spore chains (TEM). (C and D from Vobis et al., 1986; E from Vobis, 1987; F from Vobis, 1984; with permission.)

northern to the western part of Japan; no strains could be obtained from soils of agricultural fields or from sediments of lakes and rivers. The authors suggest that members of this genus are probably distributed widely in woodlands of temperate regions. *Catellatospora* has been isolated from a soil sample that originated from the pampas grassland south of Buenos Aires, Argentina (Vobis, 1987). Earlier descriptions of nonmotile strains related to *"Dactylosporangium"* (Lechevalier and Lechevalier, 1970a; Thiemann, 1970) and the recent report of Meyertons et al. (1987) suggest a global distribution of these rare organisms.

The role of the singly produced globose spores, which can occur in addition to the nonmotile spores in chains (Asano et al., 1989b), has not been studied yet, but they may have interesting ecophysiological aspects similar to those of the *Dactylosporangium* spores.

It is premature to give ecological data for *Glycomyces* species. The two type strains were isolated from soil samples from Harbin, China, and from a greenhouse in Trenton, New Jersey, USA, respectively (Labeda et al., 1985).

Isolation

In agreement with the wide distribution of actinoplanetes in nature, samples from various habitats have proved to be favorable sources of inoculum: soil, sediment, mud, water, and plant material. Microbial populations vary greatly, so that only about one in four samples may prove successful (Shearer, 1987). Special pretreatments of the freshly collected samples enhance the numbers of actinoplanetes and reduce the non-desirable concomitant microorganisms. Selective isolation media (see "Media for Isolation and Cultivation") favor their development and limit the growth of other bacteria and common soil actinomycetes like *Streptomyces* species. The addition of humic acid, for example, activates the germination of spores (Hayakawa and Nonomura, 1987), and media with low nutrient concentration favors the growth of *Micromonospora* (Rowbotham and Cross, 1977).

To avoid the growth of fungi on the isolation plates, cycloheximide (50 µg/ml) and/or nystatin (100 units/ml) can be added to the isolation medium (Shearer, 1987). The selective effect can be enhanced by the addition of other antibiotics like novobiocin, streptomycin, gentamicin, or tunicamycin, supplemented either solely or in combination (Table 1). *Actinoplanes, Ampullariella,* and *Micromonospora* strains could be isolated very selectively using the agar plate method with additions of cycloheximide, nystatin, or novobiocin (Torikata et al., 1978, 1983). A combination of tunicamycin and nalidixic acid permitted a selective growth of *Micromonospora* species (Nonomura and Hayakawa, 1988). Instead of antibiotic supplements, a special membrane filter technique can be used, since the branching hyphae of the actinoplanetes can penetrate the small filter pores, whereas bacterial growth is restricted to one side of the membrane (Hirsch and Christensen, 1983).

The differing behavior of the spores permits the use of two isolation methods: 1) direct isolation on selective agar media for the actinoplanetes having nonmotile spores and 2) techniques using spore motility to specifically enrich the sporangiate members. Obviously, strains belonging to one group can appear also on the isolation plates intended for the others.

The isolation plates and the enrichment cultures are usually incubated at 22 to 28°C. Because of the very slow growth rate of the actinoplanetes, the incubation time has to be extended up to several weeks in some cases. The use of a dissecting microscope is recommended to select the colonies grown on the isolation plates. The mycelia can be picked up and transferred with toothpicks or with a thin metal needle.

Isolation from Water

The freshly collected water samples should be stored at 4°C until processed. If necessary, the spores can be concentrated from a relatively large volume of water either by the membrane filtration technique described by Burman et al. (1969) or by centrifugation (Okami and Okazaki, 1972). To reduce the numbers of the concomitant vegetative bacterial cells, a pretreatment either with 1) mild heating or 2) chemical substances is recommended.

1) For the heat treatment procedure, 2 ml of the water sample is placed in a glass tube which is sealed and heated in a water bath. Various periods of incubation and temperatures have been used: 6 min at 55°C (Rowbotham and Cross, 1977), 10 min at 70°C (Cross, 1981b), or 60 min at 44°C (Burman et al., 1969).

2) An alternative pretreatment with chlorine was suggested by Burman et al. (1969) and Willoughby (1969b): the samples are first treated with 4 mg/l ammonia, followed by 2 mg/l chlorine (added as 1 ml of a hypochlorite solution containing 200 mg/l of available chlorine). Samples are allowed to stand for 10 to 30 min; then the chlorine is neutralized with sodium thiosulfate. The correct amount has to be calculated from titration of a blank sample.

After brief mixing of the pretreated samples (either heat or chlorine), spreading can be carried out immediately with 0.2 ml of the sample on each agar plate (Rowbotham and Cross, 1977). If necessary, dilutions can be made, either with sterile buffer (0.5 M KH_2PO_4 adjusted with NaOH to pH 7.2; Hsu and Lockwood, 1975) or with quarter-strength Ringer's solution containing gelatin (0.01% w/v; pH 7.0; Rowbotham and Cross, 1977). The inoculated plates are incubated at 28 or 30°C for 3 to 4 weeks.

As shown in Table 1, using selective media, *Micromonospora* strains are mainly obtained on the isolation plates from water sources. *Actinoplanes* may also be isolated from water samples (Willoughby, 1971; Torikata et al., 1978).

Isolation from Soil and Sediments from Freshwater and Marine Habitats

SOIL-DILUTION-PLATE TECHNIQUES Soil samples, marine sediments, or mud from lakes and rivers are air-dried at room temperature and then ground in a mortar (Shearer, 1987). About 1 g of the sample is added to 10 ml of saline in a 25-mm test tube. The suspension is mixed (vortex mixer) for 1 min and diluted in series with a sterile salt solution. The salt solution proposed by Wakisaka et al. (1982) contains 0.01% $MgSO_4 \cdot 7 H_2O$ and 0.002% Tween, from which air is eliminated by use of a vacuum desiccator for about 30 min. Instead of a salt solution, sterile water can also be used for suspension and serial dilutions (Hayakawa and Nonomura, 1987). Petridishes are prepared one day before plating and incubated at 37°C overnight to eliminate films of moisture on the agar surface (Shearer, 1987). An 0.1-ml inoculum of the proper dilution is placed on each plate and spread with a sterile glass rod. Plates are incubated at 28 to 30°C for 4 to 5 weeks.

Cellulose-decomposing micromonosporas can be isolated from soil adjacent to the roots of wheat and maize according to the method of Sandrak (1977). One ml of soil suspension is mixed with 0.67 g sterile cellulose powder (as used for thin-layer chromatography) and 2 ml of liquid Kadota's benzoate medium (Sandrak, 1977). The mixture is spread on plates with Kadota's benzoate agar. The cellulose layer is allowed to dry before the plates are incubated for 25–30 days at 28°C (Cross, 1981b).

For *Micromonospora* species from marine sediments, Goodfellow and Haynes (1984) incubated the isolation plates at 18°C for 10 weeks (duplicates at 4°C for 6 months).

Heat or chlorine treatment, as described for water samples, can also be used with soil and sediment dilutions (Cross, 1981b). An alkaline pretreatment method is suggested by Wakisaka et al. (1982): One ml of the diluent is mixed with 9 ml of 0.01 N NaOH. After standing for 5 to 10 min, the mixture is neutralized with 0.1 N HCl to pH 6 to 7 (with cooling) before serial dilution and plating. Nonomura and Hayakawa (1988) treated the soil-water suspension with 1.5% phenol at 30°C for 30 min.

The routine plating technique is preferable for the isolation of strains of the genera *Micromonospora*, *Dactylosporangium*, *Catellatospora*, and *Glycomyces*. Table 1 lists the results of the use of various selective agar media in combination with the additions of antibiotic agents.

DRY HEAT TECHNIQUE A procedure which involved dry heating of soil samples at extreme temperatures was originally developed by Nonomura and Ohara (1969) for the isolation of *Microbispora* and *Streptosporangium* species. Shearer (1987) demonstrated that this technique is also useful for the isolation of *Dactylosporangium* and *Micromonospora* strains (Table 1). The samples are first air-dried at room temperature and ground in a mortar. Then they are heated in a drying oven at 120°C for 60 min. One g of the heat-treated soil is added to 10 ml of saline solution and then processed as described for the routine dilution-plating technique. Inoculated plates are incubated at 28°C for 4 to 5 weeks. Arginine-vitamin agar and starch-casein-nitrate agar with B vitamins are used as selective media (Shearer, 1987).

STAMPING TECHNIQUE The stamp technique was used successfully in the study of actinomycete populations of salt marsh ecosystems (Hunter et al., 1984). Depending on the moisture content, the samples of soil or mud are air-dried in petri dishes for several days at room temperature. Two

methods of further pretreatment were suggested by Hunter-Cevera et al. (1986): 1) the dried samples are ground with a pestle in a mortar and heated for 2 hours at 60 to 65°C, and 2) dried samples are mixed with powdered chitin (1 : 1) and incubated for 2 to 3 weeks at 26°C.

The pretreated and ground samples are stamped onto the isolation plates using the following procedure: A small circular sponge (Dispo culture plug, 16 mm; Scientific Products) is pressed into the powdered sample and removed. The excess small crumbs are shaken off. A stack of a dozen plates with various different alternating selective media is then inoculated by successively "stamping" (lightly touching) the sponge to the agar surface 10 times in a circle around the perimeter and 3 times in the middle of each plate (Hunter et al., 1984). Continuously stamping with the same plug yields the desired dilution effect. Plates are incubated at 26–28°C for 2 weeks. Hunter-Cevera et al. (1986) recommend arginine-glycerol salts agar, starch-casein-nitrate agar, and thin pablum agar as the selective media for the isolation of *Micromonospora* strains (Table 1).

Isolation from Plant Material

A special wash technique was employed by Willoughby (1968, 1969a) for the investigation of actinomycetes populations on decomposing leaf litter. The leaves are collected at a fairly early stage of the decomposition. Small pieces of approximately 3 cm^2 are cut out, and each piece is transferred to a 100-ml conical flask containing 25 ml of sterile-filtered lake water. After 2 min of agitation on a rotary shaker, small aliquots of the leaf washing liquid are either incorporated into molten agar (0.5 ml/plate) or spread onto the surface of agar, 0.2 ml for each plate, using a right-angled glass rod. The plates are incubated at 25°C for 3 to 5 weeks (Willoughby, 1968). The most successful isolation medium for strains of *Actinoplanes*, *Ampullariella*, and *Micromonospora* was colloidal chitin agar with cycloheximide as the antifungal agent (Table 1).

Special Isolation Methods Using Motile Spores

The following very selective isolation methods are used when dormant sporangia are present in the substrate to be tested. The sporangia can release actively swimming spores when submerged in water. The individual spores must be motile for many hours and must show positive chemotaxis to specific chemical substances. Once fastened to a natural or cultural substrate, they must be able to germinate and form new mycelia

Table 1. Examples of the direct isolation of actinoplanetes using selective media.

Source	Isolation method	Isolation medium	Agent added	Genus selected	Reference
Water	Dilution plate	Water agar or colloidal chitin agar	Potassium tellurite (0.1% w/v)	*Actinoplanes*	Willoughby (1971)
Water, soil, and sediment	Dilution plate	Colloidal chitin agar	Cycloheximide (50 mg/l) and thiamine (4.0 mg/l)	*Micromonospora*	Hsu and Lockwood (1975)
	Dilution plate	M3 agar	—	*Micromonospora*	Rowbotham and Cross (1977)
Marine sediment	Dilution plate	Kadota's cellulose benzoate agar	Sodium benzoate (20 g/l)	*Micromonospora*	Sandrak (1977)
	Dilution plate	Cellulose asparagine agar (+ artificial seawater)	Cycloheximide (50 mg/l) and novobiocin (50 mg/l)	*Micromonospora*	Goodfellow and Haynes (1984)
Soil	Dilution plate	Humic acid-vitamin agar	Cycloheximide (50 mg/l)	*Micromonospora* and *Dactylosporangium*	Hayakawa and Nonomura (1987)
Soil	Dilution plate	Humic acid-vitamin agar	Tunicamycin (20 mg/l) and nalidixic acid (30 mg/l)	*Micromonospora*	Nonomura and Hayakawa (1988)
Soil	Dilution plate	Bennet's agar (modified)	Tunicamycin (25 mg/l) and cycloheximide (30 mg/l)	*Micromonospora*	Wakisaka et al. (1982)
Soil	Dilution plate	Gause no. 1 agar or Czapek sucrose agar	Novobiocin (25 mg/l) and streptomycin (15 mg/l) (+ antifungal agent)	*Glycomyces*	Labeda (1989); Nolan and Cross (1988)
Soil	Dilution plate	ND	ND	*Catellatospora*	Asano and Kawamoto (1986)
Soil	Dry-heat treatment and dilution plate	Starch-casein-nitrate agar (+ B-vitamins)	—	*Micromonospora* *Dactylosporangium*	Shearer (1987)
Soil and mud of salt-marsh	Stamping	Arginine glycerol salts agar	Gentamicin (5.0 mg/l)	*Micromonospora*	Hunter et al. (1984); Hunter-Cevera et al.(1986)
Plant material	Washing	Colloidal chitin agar	Cycloheximide (75 mg/l) and nystatin (75 mg/l) (+ artificial seawater)	*Actinoplanes, Ampullariella,* and *Micromonospora*	Willoughby (1968, 1969a)
			Cycloheximide (50 mg/l)		

ND, no data.

and, for use of the baiting technique, produce a new generation of sporangia.

BAITING TECHNIQUE The baiting technique is the classical isolation method for *Actinoplanes*, which made possible the first discovery of actinomycetes with motile spores (Couch, 1949). Although other powerful techniques are available, baiting is still the only way to isolate keratinophilic *Pilimelia* strains.

0.5 to 1.0 g of the sample is placed in a small, sterilized petri dish (3 or 4 cm in diameter) or in a chamber of a multi-well microtiter plate, which is then half flooded with sterile demineralized water. After cautiously stirring, the particles settle to the bottom. Natural baits are exposed singly or in combination on the surface of the water: pollen of *Pinus*, *Liquidambar*, or *Sparganium*, boiled *Paspalum* grass leaves, hair of mammalia (human, dog, deer, cattle, white mice, etc.), or bits of snake skin (Couch, 1949, 1954; Karling, 1954; Gaertner, 1955; Kane, 1966; Schäfer, 1973; Tribe and Abu El-Souod, 1979; Makkar and Cross, 1982). The baits must be presterilized, depending on their consistency, either chemically with ethanol or propylene dioxide or by autoclaving (Gaertner, 1955; Schäfer, 1973; Makkar and Cross, 1982). The baiting enrichment cultures are closed and stored undisturbed at room temperature for several weeks. The water level can be regulated by additions of sterile distilled water. The examination for actinoplanetes can begin after one week with a dissecting microscope using 100-× magnification and horizontal lighting (Bland and Couch, 1981). Further examination after 3 to 4 weeks is recommended for keratinophilic organisms (Schäfer, 1973).

Sporangia of actinoplanetes are recognizable as glistening beads on the air-exposed sides of the baits. Such baits are then removed carefully from the water and transferred to a 3% agar plate (Bland and Couch, 1981). Individual sporangia are separated from the bait and rolled several centimeters over the surface of agar, using a thin-pointed tungsten needle, which has a tip curved like a hockey stick. In this way, contaminants are removed from the sporangial surface. Cleaned sporangia can be transferred either directly or together with a small, cut-out agar block onto a petri dish with suitable agar medium. Media for isolation from pollen and grass leaves include Czapek sucrose agar, peptone-Czapek agar (Bland and Couch, 1981), half-concentrated casamino acids-peptone Czapek agar (Schäfer, 1973), or Emerson's yeast extract-starch agar. Sporangia from keratinic baits should be transferred to highly diluted skim milk-cattle horn-meal agar (Vobis, 1984).

The colonies originating from the individual sporangia are visible with the naked eye after 1

to 4 weeks of incubation and can partly be used as the inoculum for the new strain on slant cultures. The other part of the mycelium can be transferred onto sporulation agar for morphological identification.

DEHYDRATION-REHYDRATION TECHNIQUE This technique utilizes the ability of the sporangia to withstand desiccation and to release motile spores when they are subsequently in contact with water. Besides soil samples, it is also applicable to leaf litter, decaying plant material from aquatic habitats, organic debris, etc. (Makkar and Cross, 1982).

The samples are dried at 28 to 30°C for 7 days. For rehydration, 0.5 g of soil or corresponding substrate is mixed with 50 ml of sterile tap water in a 150-ml beaker or Erlenmeyer flask, which is covered with sterile aluminium foil (Shearer, 1987; Vettermann and Prauser, 1979). The suspension is incubated at 20 to 30°C for about 1 hour. During the first 30 minutes, the vessel can be shaken at irregular intervals. After that, the particles should be permitted to settle. From the supernatant, 0.5 to 1.0 ml are removed with a sterile Pasteur pipette and spread onto agar plates (Shearer, 1987). If it is necessary, dilutions can be prepared from the inoculation fluid (Makkar and Cross, 1982). For cultivation, the following media can be used: soil extract agar; colloidal chitin agar containing cycloheximide and nystatin (Makkar and Cross, 1982); oatmeal-soil extract agar; and starch-casein-sulfate agar (Shearer, 1987). Plates are incubated at 28°C for 2 to 4 weeks. Colonies of actinoplanetes can be selected under a dissecting microscope at 60-× magnification and used as inoculum for new strains.

CHEMOTACTIC METHOD The spores of *Actinoplanes* exhibit an apparently microaerophilic reaction and are attracted to chloride and bromide ions (Palleroni, 1976). Therefore, a chemotactic method can be used to isolate these strains. An essential part of this technique is a simple isolation chamber, a sterilizable plastic block (80 × 40 × 12 mm) with two circular holes (9 mm deep and 24 mm in diameter) whose centers are 32 mm apart. They are connected by a channel that is 2 mm wide and 3 mm deep (Palleroni, 1980). One gram of a soil sample is divided into two equal parts and then placed in each compartment. Sterile water is added nearly to the rim and stirred cautiously. After incubation for 1 hour at 30°C, the spores are released from the sporangia and move freely in the water. Using a sterilized tweezer, a sterile 1-μl glass capillary about 32-mm long is filled with 0.01 M phosphate buffer (pH 7.0) containing 0.01 M KCl and placed in the channel. The capillary must be submerged, connecting the two suspensions. After incubation at 30°C for 1 hour more, the attracted spores are concentrated in the lumen of the capillary, which is then removed and washed from the outside with a jet of sterile water. The contents of the capillary are blown into 1-ml sterile water or buffer. Portions of the dilution are taken with a sterile pipette and spread onto carefully dried agar plates. The plates are then incubated at 28°C. Starch-casein-sulfate agar is recommended as the isolation medium (Palleroni, 1980). Although colonies can be selected after 4 days, slowly growing actinomycetes may only be detectable after 3 weeks.

MOIST INCUBATION TECHNIQUE This method is suitable for the direct detection of actinoplanetes on natural substrate. Although the ability to produce motile spores obviously plays no role, *Actinoplanes* strains can be readily enriched (Willoughby, 1968). Portions of decaying leaves or other biological substrates, freshly collected from the field, are washed with sterile water to remove adhering detritus. They are placed in prepared petri dishes, the bottoms of which have been covered with very moist filter paper or layers of cellulose before autoclaving. The petri dishes, working as moist chambers, are sealed and incubated for about 4 weeks at 25°C. Examination with both dissecting and light microscopes is necessary to identify the sporangia of the actinoplanetes (Willoughby, 1969a).

Identification

As outlined by Labeda (1987) and Lechevalier (1989), a combined use of morphological and chemical criteria is still the best way to identify actinomycetes at the genus level. To start with simple and time-saving morphological studies can sometimes shorten large-scale chemotaxo-

nomical procedures. However, to verify the determination, biochemical analyses are still absolutely necessary.

Morphological Criteria

Because members of the actinoplanetes are highly differentiated morphologically, (Bland and Couch, 1981; Luedemann and Casmer, 1973; Vobis and Kothe, 1985), it is possible to use morphological criteria alone for identification at the genus level. Table 2 is a dichotomous key that can be used to determine the genus of the isolate being studied.

For morphological studies in general, the short guidelines compiled by Cross (1989) are useful. Tiny pieces of the mycelium from vigorously developing colonies are transferred to freshly prepared sporulation medium in small petri dishes. The temperature of incubation should be between 20 and 30°C, and observations are made after 2 days to 4 weeks of incubation (depending on the individual strain). Preliminary observations should be made with a dissecting microscope (Figs. 1A, B, and C; 6A, B, and C) and directly with a light microscope (Figs. 1E; 4A; 7B) before water mounts are prepared. The best magnification of undisturbed colonies is with 20 or 40× objectives so that appropriate structures can be accurately studied (Okami and Suzuki, 1958; Cross, 1981b). Spore chains and single spores can be detached by slight pressure on the colony with a cover slip. Sporangia should be cut out and mounted with water. Although observation with a bright-field microscope is generally sufficient, phase-contrast is often also helpful (Figs. 2B; 4D; 7D).

For observing spore release from the sporangia, water-mounted preparations should be observed continuously for 1 to 2 h; unchlorinated tap water or distilled water should be used. To avoid evaporation, slides should be stored in

Table 2. Determinative key for the actinoplanetes genera using morphological criteria.

1a Spores in chains	2
1b Spores not in chains	3
2a Spore chains long, on aerial hyphae	*Glycomyces*
2b Spore chains short, emerging from substrate hyphae	*Catellatospora*
3a Spores spherical, single on substrate hyphae, nonmotile	4
3b Spores enclosed in sporangia, motile by flagella	5
4a Diameter of spores: 0.7–1.5 μm, in clusters	*Micromonospora*
4b Diameter of spores: 1.7–2.8 μm, dispersed	*Dactylosporangium*
4c Diameter of spores: 0.4–0.6 μm, dispersed	*Catellatospora*
5a Sporangia oligosporous, with one row of spores	*Dactylosporangium*
5b Sporangia polysporous, with several rows of spores	6
6a Sporangiospores globose to subglobose	*Actinoplanes*
6b Sporangiospores bacilliform	7
7a Spores distinct rod-shaped, 2.0–4.0 μm long, tuft of flagella polar	*Ampullariella*
7b Spores rod-shaped to reniform, 0.7–1.5 μm long, tuft of flagella lateral	*Pilimelia*

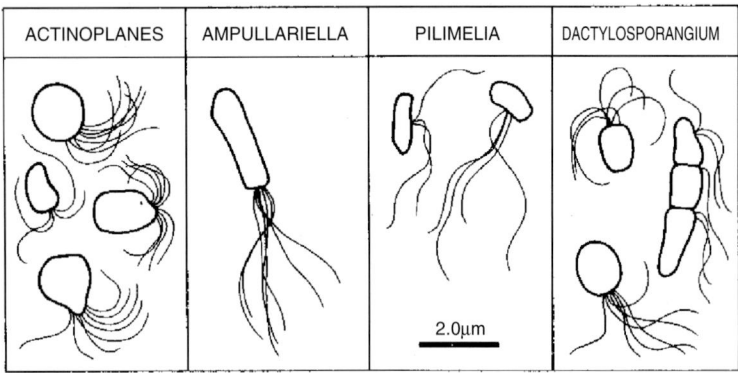

Fig. 8. Morphology of the motile sporangiospores of the genera of the
actinoplanetes. *Actinoplanes:* globose,
ellipsoidal, short bacilliform (0.8–2.0
μm), with a polar tuft of flagella. *Ampullariella:* bacilliform (0.5–1.0 × 2.0–
4.0 μm), with a polar tuft of flagella.
Pilimelia: bacilliform, reniform (0.3–
0.7 × 0.7–1.5 μm), with a lateral tuft of
flagella. *Dactylosporangium:* cylindrical, ovoid or pyriform (0.4–1.3 × 0.5–
1.8 μm), with a polar or subpolar tuft
of flagella. (From Vobis, 1987; with
permission.)

moist chambers. Actively swimming spores are easy to distinguish from Brownian molecular movement since the action of the flagellar tuft causes the spores to move and rotate rapidly. Based on the type of flagellation, the spores of *Ampullariella* rotate around the longitudinal axis, while those of *Pilimelia* rotate around the lateral axis (Fig. 8). To observe the flagella with the light microscope, staining methods must be used (Couch, 1950), but the exact location of flagellar insertion can only be determined by transmission electron microscopy.

Although many strains sporulate readily on isolation media or on rich media used for subculturing, special agar media have to be used to obtain good sporulation of certain strains: Czapek sucrose agar for *Glycomyces harbinensis* (Labeda et al., 1985), *Dactylosporangium vinaceum* (Shomura et al., 1983), and *Ampullariella* spp. (Nonomura et al., 1979); calcium malate agar for *Dactylosporangium* spp. (Thiemann et al., 1967), *Actinoplanes* spp. (Palleroni, 1989), and *Catellatospora* spp. (Asano and Kawamoto, 1986); inorganic salts-starch agar for *Dactylosporangium* spp. (Vobis, 1989c) and *Ampullariella* spp. (Nonomura et al., 1979); humic acid-vitamin agar for *Dactylosporangium* spp. and *Micromonospora* spp. (Hayakawa and Nonomura, 1987); humic acid-ion agar for *Actinoplanes* and *Ampullariella* spp. (Willoughby et al., 1968); soil extract agar for *Dactylosporangium* spp. (Thiemann et al., 1967) and *Actinoplanes* spp. (Parenti et al., 1975); diluted skim milk-hornmeal agar for *Pilimelia* spp. (Vobis, 1984); oatmeal agar for *Actinoplanes* spp. (Bland and Couch, 1981); and M3 agar, supplemented with 0.1% (w/v) fructose (Goodfellow et al., 1990). An excellent routine sporulation agar is artificial soil agar (Henssen and Schafer, 1971). The compositions of two of the above, inorganic salts-starch agar and oatmeal agar, are described by Shirling and Gottlieb (1966). The formulations of some of the other media are given below. For morphological stud-

ies, the same natural substrate can be used as that used for the isolation by baiting or moist incubation (Fig. 7B and C).

Individual strains of *Actinoplanes, Ampullariella,* and *Pilimelia* may produce incomplete sporangiate structures. The sporogenous hyphae are still developed at the tip of the supporting hypha, but the sporangial envelope appears to be absent. Sometimes, the division into spores is not complete, so that the hyphae look like brushes or little trees. If the sporogenous hyphae are fragmented, the resulting spores are always arranged in chains (Fig. 7E). It is remarkable that these free, developed spores never have flagella. Therefore, they have been called "conidia" and the supporting hyphae "conidiophores" (Couch, 1954, 1963; Kane, 1966; Willoughby, 1966).

Isolates of the *Actinoplanes* group exhibiting marked morphological characteristics can be identified to the genus level with aid of the simplified determinative key given in Table 2.

Chemotaxonomical Criteria

The taxonomy of the actinoplanetes can be established 1) by the composition of the amino acids of the cell walls; 2) by the sugar pattern of whole cell hydrolysates; and 3) by the absence of mycolic acids. Further chemotaxonomical markers, such as the composition of: 4) the phospholipids; 5) menaquinones; and 6) fatty acids, also have, in many cases, an important diagnostic value. Table 3 lists the most useful chemotaxonomical properties of the actinoplanetes. The chemical features are not only good indicators for identification, but they are also very important with regard to suprageneric groupings and phylogenetical relationships (Goodfellow and Cross, 1984; Stackebrandt and Schleifer, 1984).

CELL WALL CHEMOTYPE According to the classification scheme of Lechevalier and Lechevalier (1970b), the members of the *Actinoplanes* group

Table 3. Chemotaxonomical characters of the genus Actinoplanes and related genera.[a]

Genus	Cell wall chemotype[b]	Whole cell sugar pattern[c]	Phospholipid pattern[d]	MK-9 H_2	MK-9 H_4	MK-9 H_6	MK-9 H_8	MK-10 H_2	MK-10 H_4	MK-10 H_6	MK-10 H_8	MK-12 H_4	MK-12 H_6	MK-12 H_8	FA S	FA U	FA I	FA A	FA T
Actinoplanes	II	D	P II	V	+	V			+						+	+	+	+	−
Ampullariella	II	D	P II	V	+	V			+						+	+	+	+	−
Catellatospora	II	D	P II		+	+				+	+						ND		
Dactylosporangium	II	D	P II		+	+	+								V	V	+	+	−
Glycomyces	II	D	P I					+	+	+							ND		
Micromonospora	II	D	P II		+				+	+		+	+	+	+	+	+	+	+
Pilimelia	II	D	P II	+	+										+	+	+	V	−

(Major menaquinones[e]; Fatty acids)

[a]Compiled from: Asano et al. (1989b); Collins et al.(1984); Goodfellow and Cross (1984); Kroppenstedt (1985); Labeda (1989); Labeda et al. (1985); Lechevalier et al. (1981); Stackebrandt and Kroppenstedt (1987); Vobis (1989b). For further details and exceptions, see text.
[b]Cell wall chemotype II, glycine and *meso*- and/or 3-hydroxy diaminopimelic acid (A_2pm).
[c]Whole cell sugar pattern D, xylose and arabinose (Lechevalier and Lechevalier, 1970b).
[d]Phospholipid pattern: P I, nitrogenous phospholipid absent; P II, phosphatidylethanolamine present (Lechevalier et al. 1977).
[e]S, saturated; U, unsaturated; I, iso-; A, anteiso-; T, tuberculostearic acid (10-methyl); V, variable; +, present; −, absent; ND, no data.

are characterized by cell wall chemotype II. Glycine and *meso-* diaminopimelic acid (A$_2$pm) and/or 3-hydroxy-diaminopimelic acid are the amino acids characteristic of the peptidoglycan. The primary structure of the peptidoglycan was described by Kawamoto et al. (1981). Glycine, rather than L-alanine, is linked to muramic acid, and *meso*-diaminopimelic acid or its hydroxylated derivative is directly cross-linked to the D-alanine of an adjacent peptide subunit: Muramic acid is *N*-glycolated. The *N*-glycolyl muramic acid is a characteristic unique to the genera with cell wall type II and was found in many *Actinoplanes* and *Ampullariella* species (Stackebrandt and Kroppenstedt, 1987). It can be easily identified by a colorimetric method (Uchida and Aida, 1977).

Deviations from the typical wall chemotype II are found in some species of *Micromonospora*, which also contain LL-diaminopimelic acid (Kawamoto et al., 1981). A further exception concerns *Actinoplanes caeruleus,* which has lysine and serine as amino acids, instead of A$_2$pm (Horan and Brodsky, 1986a; Tille et al., 1982). The lack of A$_2$pm in a morphologically incontestable *Actinoplanes* species raises a lot of taxonomical problems (Stackebrandt and Kroppenstedt, 1987).

WHOLE CELL SUGAR PATTERN The sugars used for classification are, in particular, the components of the cell wall polysaccharides. They can be detected after hydrolyzation of whole cells. Based on the pattern of diagnostic sugars, the actinomycetes containing *meso*-A$_2$pm can be divided into further groups (Lechevalier and Lechevalier, 1970a). In general, the actinoplanetes have wall chemotype II and the sugar pattern D in common (Table 3). The pentoses xylose and arabinose are the characteristic sugars.

Some species of *Micromonospora* also contain other diagnostic sugars, such as galactose and madurose (3-O-methyl-D-galactose) (Meyertons et al., 1987). In *Catellatospora* strains, an unusual sugar detected was 3-O-methylrhamnose (Asano et al., 1989a).

The separation of whole cell sugars is usually carried out by thin-layer chromatography (Hasegawa et al., 1983; Meyertons et al., 1987).

MYCOLIC ACIDS The cell walls of the actinoplanetes have the *N*-glycolated muramic acid type in common with those of the *Mycobacterium-Nocardia-Rhodococcus-* group (Minnikin and O'Donnell, 1984). The mycelium of this group fragments into coccoid or bacillary elements. They can also be further characterized by the presence of mycolic acids, which are not present in the *Actinoplanes* group. For routine analysis, thin-layer chromatography is usually used (Minnikin et al., 1975).

PHOSPHOLIPIDS The composition of phospholipids is of high taxonomical value in actinomycetes in general, and five very distinct phospholipid types (P I to P V) have been observed (Lechevalier et al., 1977, 1981). Most of the genera of the actinoplanetes are of phospholipid type P II (phosphatidylethanolamine present, phosphatidylcholine and the unknown glucoseamine-containing phospholipid absent) (Table 3). However, *Glycomyces* is distinct since it has P I (phosphatidylcholine, phosphatidylethanol and the unknown glucoseamine-containing phospholipid absent) (Labeda et al., 1985). The phospholipid types are also useful taxonomic markers for separating other sporangia-forming genera with motile spores from the *Actinoplanes* group (Hasegawa et al., 1979). The unusual absence of phosphatidylethanolamine in *Ampullariella regularis* (Stackebrandt and Kroppenstedt, 1987) may indicate a peculiar variability in the phospholipid pattern in the genus *Ampullariella.*

Several methods for characterizing the phospholipids have been described (Lechevalier et al., 1977; Vitiello and Zanetta, 1978; O'Donnell et al., 1982).

MENAQUINONES Determination of the composition of menaquinones is helpful for differentiation at the genus level (Table 3). The genera *Actinoplanes, Ampullariella,* and many of the *Micromonospora* species contain menaquinones with 9 to 10 tetrahydrogenated isoprene side chains, MK-9 (H$_4$) and MK-10 (H$_4$), thus fitting into type 3b of the classification scheme of Kroppenstedt (1985). In *Actinoplanes auranticolor* and in *Micromonospora carbonaceae* subsp. *aurantiaca* only MK-9 (H$_4$) occurs (Stackebrandt and Kroppenstedt, 1987). Type 4 occurs in *Micromonospora echinospora* subsp. *pallida,* which has tetra-, hexa-, and octahydrogenated quinones with 12 isoprene units. Many other *Micromonospora* species also have type 4 menaquinones, with MK-10 (H$_4$) and MK-10 (H$_6$) (Collins et al., 1984; Kawamoto, 1989).

The menaquinone type 4b, with MK-9 (H$_4$), (H$_6$) and (H$_8$), is characteristic for *Dactylosporangium* species and type 4a with MK-9 (H$_2$) and (H$_4$) is characteristic for *Pilimelia* (Kroppenstedt, 1985). *Catellatospora* species have either type 4a with tetra- and hexahydrogenated menaquinones with nine isoprene units or type 4c with hexa- and octahydrogenated menaquinones with 10 units (Asano et al., 1989b). According to Labeda (1989), the genus *Glycomyces* has MK-10 (H$_2$) and MK-10 (H$_6$). In the species *Glycomyces rutgersensis,* MK-9 (H$_4$) could also be detected (R. M. Kroppenstedt, unpublished observations). This indicates that the common type 3b for *Actinoplanes* and *Micromonospora* species is also present in *Glycomyces*.

FATTY ACIDS The usefulness of fatty acids as chemotaxonomical markers was demonstrated by Kroppenstedt and Kutzner (1978) and Kroppenstedt (1979). In the classification scheme developed by Kroppenstedt (1985), three main types of fatty acid pattern were proposed. Among the actinoplanetes (Table 3), the genus *Micromonospora* has a separate position with type 3b, characterized by the occurrence of 10-methyl branched fatty acids (tuberculostearic acid). Recent studies indicate that some strains of *Dactylosporangium* belong to the same type (Stackebrandt and Kroppenstedt, 1987). In general, *Dactylosporangium* species have fatty acid type 2, as is also characteristic for *Ampullariella* (type 2d) and *Actinoplanes* (type 2c): iso/anteiso branched- and monosaturated *cis*-9,10-octadecanonic acids are the predominant fatty acids (Kroppenstedt, 1985; Stackebrandt and Kroppenstedt, 1987). A similar iso/anteiso fatty acid pattern is found in *Glycomyces rutgersensis* (R. M. Kroppenstedt, unpublished observation). Members of *Pilimelia* have fatty acid type 2b with a high content of iso-C_{15}, iso-$C_{15:1}$, and iso-$C_{17:1}$ fatty acids, lacking substantial amounts of C_{18} fatty acid (Stackebrandt and Kroppenstedt, 1987).

Cultural Characteristics of the Differentiation of the Actinoplanetes Genera and Species

For subculturing new isolates, several complex agar media should be tested, including the isolation medium, because the individual strains might have quite different nutrient requirements. In our laboratory, we usually start with Emerson's yeast extract-starch agar or casamino acids-peptone-Czapek agar. Vegetative growth is also supported by Bennett agar, Hickey-Tresner agar, Czapek sucrose agar, peptone-Czapek agar, and glucose-asparagine agar. For keratinic isolates, half-concentrated skim milk agar is used (Schäfer, 1973). For the compositions of the above-mentioned media, see "Media for Isolation and Cultivation" below.

Actinoplanes ((Couch)/Palleroni, 1989)

Isolates of *Actinoplanes* form compact colonies on solid agar media (Fig. 1A and C). Aerial mycelium is usually absent or only rudimentarily developed. Colonies can be covered with a whitish bloom, if abundant sporangia are produced on the surface of the substrate mycelium (Fig. 1B and D). The basic color of the mycelium is orange, presumably due to a carotenoid pigment (Szaniszlo, 1968), but a great variety of colors exists, depending on the individual strains: cream to yellow, brown, rusty brown, red, blue, violet, green, or black (Palleroni, 1989; Vobis, 1987).

In general, sporangia develop directly on the surface of a colony (Fig. 1D, E, and F). In *A. minutisporangius*, they are also submerged (Ruan et al., 1986). Frequently, the sporangia arise terminally from "palisade" hyphae, which are thicker in diameter and vertically oriented (Bland and Couch, 1981). Inside the sporangia, the spores are arranged in coils (Fig. 1G); but in *A. rectilineatus*, they run in parallel rows (Lechevalier and Lechevalier, 1975). If the sporangial envelopes are very thin and transient, individual sporangia may be attached (Fig. 1E). The spores of *Actinoplanes* are globose or subglobose to short bacilliform and possess a tuft of flagella (Fig. 8).

Species differentiation is based on a combination of morphological and physiological characters (Palleroni, 1989). The shape of the sporangia ranges from globose, subglobose, oval, umbelliform, cylindrical, or lobate to irregular. The average size of a sporangium is from 4 to 25 μm in diameter. Extreme dimensions of sporangia exist in *A. minutisporangius*, with 2 μm as a minimum (Ruan et al., 1986) and 47 μm as a maximum (G. Vobis, unpublished observations). The unusual presence of aerial hyphae is characteristic for *A. rectilineatus*, *A. ferrugineus*, *A. garbadinensis*, and *A. teichomyceticus* (Palleroni, 1989).

The color of the mycelium is another characteristic used in species differentiation: for instance, tan to blue in *A. caeruleus* (Horan and Brodsky, 1986a), violet in *A. ianthinogenes* (Coronelli et al., 1974), and rusty brown in *A. ferrugineus* (Palleroni, 1979). Beside yellowish or brownish soluble pigments in several species, a cherry red pigment is the distinctive mark of *A. italicus* (Beretta, 1973) and a soluble blue pigment is found in *A. cyaneus* (Terekhova et al., 1977). Physiological tests like degradation, hydrolysis, coagulation, peptonization, or liquefaction of various compounds or media can also be used (Palleroni, 1989). The assimilation tests with different carbon sources, routinely carried out with sugars, were extended to many more organic compounds (Palleroni, 1979). Palleroni (1989) listed a total of 15 species for the genus, including the two members of the genus formerly called *Amorphosporangium* (Couch, 1963). Using numerical and chemical classification methods, recently the numbers of *Actinoplanes* species could be enlarged by the description of five new species (Goodfellow et al., 1990).

Ampullariella (Couch, 1963)

The colonies of *Ampullariella* strains grown on solid agar media are often soft and can be pulled apart easily with a needle. They are elevated, with protuberances in the center of the colony. Their marginal areas can be ridged or flat (Fig. 2A). The basic orange color of the substrate mycelium is frequently overlayed by darker pigments. Depending on the individual strain, red, brown, olive or black variations can be observed. Different-colored segments may even occur in a single colony. Rudimentary aerial mycelium can be produced by several strains. Their colonies are consequently covered by a whitish gray layer.

The production of sporangia may occur in such extensive masses that the whole surface of the colony is completely covered (Fig. 2A and C). Other strains only develop a very few sporangia, which can be detected only by direct observation with the light microscope. Palisade hyphae supporting the sporangia have never been observed. The species of the genus *Ampullariella* have to be identified by the morphology of their spores and sporangia. The spores are rod-shaped and are usually four times longer than they are wide (Table 2; Fig. 2D). They have a polarly inserted tuft of flagella (Fig. 8). The spores are arranged within the sporangia in parallel rows. *A. regularis* has cylindrical or bottle-shaped sporangia, measuring $5–14 \times 8–30$ μm (Fig. 2D). Pyriform or bell-shaped sporangia, sometimes with irregularly rounded protuberances at the top, can be split one or several times along the longitudinal axis (Fig. 2B and C). Such sporangia with lobes occur in *A. campanulata* ($5–15 \times 6–12$ μm) and in *A. lobata* ($4–20 \times 12–32$ μm). The sporangia of *A. digitata* are smaller ($3–9 \times 6–12$ μm) and very deeply split (Couch, 1963; Vobis and Kothe, 1989). Beside the morphological features, a few physiological criteria can also be used for species differentiation (Schäfer, 1973; Vobis and Kothe, 1989). Melanoid pigments are produced in *A. digitata,* which has an olive- to-black substrate mycelium. Yellowish to greenish soluble pigments occur in several strains (Couch, 1963). "*A. violaceochromogenes*" can be identified by the presence of a violet-to-dark purple diffusible pigment (Nonomura et al., 1979).

Eight species and subspecies of *Ampullariella* are currently known (Vobis and Kothe, 1989) although half of them are still "species incertae sedis." Whether *Ampullariella* should be a separate genus from *Actinoplanes* is unclear. The molecular systematic data favor a unification of the two genera (Stackebrandt and Kroppenstedt, 1987; Kothe, 1987), whereas following traditional morphological concepts (Vobis, 1989a), two genera can be distinguished clearly by the shape and size of spores (Table 2; Figs. 2D and 8).

Catellatospora (Asano and Kawamoto, 1986)

Colonies of *Catellatospora* strains always produce substrate mycelium; true aerial mycelium is not developed. The color of the colonies varies from light wheat to yellow, mustard gold, orange, and reddish brown. Cultures grown on agar media that support poor growth can be crowded with short chains of spores (Fig. 3A), with 5 to 30 spores forming a chain which emerges directly from the substrate hypha (Fig. 3C). The spore chains are straight to flexuous, vertically oriented, and may be branched (Fig. 3B and D). The spores are nonmotile, cylindrical, barrel-shaped to ovoid, with smooth to slightly rough surfaces, measuring $0.6–0.8 \times 0.9–1.2$ μm (Asano and Kawamoto, 1986).

In addition to spore chains, a few strains of *C. tsunoense* and *C. ferruginea* produce single spores ("globose bodies") terminally on branches of substrate hyphae (Asano et al., 1989b). These spores are morphologically very similar to the globose spores of the genus *Dactylosporangium* (Fig. 4E). In *Catellatospora*, they are 0.4–0.6 μm in diameter, whereas in *Dactylosporangium*, they are more than double that size (Table 2).

The color of the substrate mycelium, the type of menaquinone, and the presence or absence of the specific cell wall sugar 3-O-methylrhamnose are used for species differentiation (Asano et al., 1989a, 1989b). *C. citrea* and *C. tsunoense* have major menaquinones with nine isoprene units (MK-9) and yellow-colored mycelium. The species *C. ferruginea* and *C. matsumotoense* have menaquinones with 10 isoprene units (MK-10) and are usually orange to reddish brown. Their cell walls also contain 3-O-methylrhamnose, and they are resistant to novobiocin (Asano et al., 1989b). Further species differentiation can be made by the requirement for thiamine in *C. tsunoense* and *C. ferrugineae* and the requirement for methionine in *C. citrea* subsp. *methionotrophica* (Asano and Kawamoto, 1988). The utilization pattern of carbon sources also supports the differentiation into the four species and subspecies mentioned above.

Dactylosporangium (Thiemann, Pagani and Beretta, 1967)

Dactylosporangium strains form compact colonies on agar media. The colonies are mostly flat, with a smooth to wrinkled surface, and look somewhat tough and leathery. Formation of coremia may occur but true aerial mycelium is not developed. The color of the substrate mycelium varies from pale orange to deep orange, amber, brownish with rose tinge, rose, and wine to brown or yellowish brown (Shomura et al., 1986; Vobis, 1989c).

Two completely different types of spores can be formed in *Dactylosporangium* strains: motile spores inside the sporangia and nonmotile spores, which are borne singly on substrate hyphae. The sporangia sit directly on the substrate, either singly or in bundles (Fig. 4A and B). They are finger-shaped or claviform, containing only one row of no more than four spores (Fig. 4C). The sporangiospores are variable in shape and are sometimes still connected while swimming with the aid of flagella (Fig. 8). The nonmotile, substrate spores are spherical (1.7–2.8 μm in diameter) and exhibit a typical phase-brightness (Fig. 4D). They arise terminally on short side-branches of substrate hyphae (Fig. 4E). Amorphous material can be deposited outside the spore wall (Fig. 7F). The cytoplasm includes crystalline proteins and structured bodies (Sharples and Williams, 1974). Morphologically similar globose spores may also occur in *Catellatospora* strains (Asano et al., 1989b), but they are smaller in diameter (Table 2). In *D. fulvum*, the sporangia and the globose spores may also be visible on coremia (Shomura et al., 1986).

Because of the minimal differences in morphology, chemotaxonomy, and physiology between the various *Dactylosporangium* strains, speciation is mainly based on the color of the substrate mycelium and the diffusible pigments (Vobis, 1989c). *D. aurantiacum* has orange-colored substrate mycelium without soluble pigments. The mycelium of *D. thailandense* is orange to brown, producing an amber to brown soluble pigment (Thiemann et al., 1967). *D. matsuzakiense* also has orange colonies, but is recognizable by a light brownish-pink diffusible pigment (Shomura et al., 1980). The substrate mycelium of *D. vinaceum* is red to brown, and the soluble pigments are red to deep red (Shomura et al., 1983). *D. roseum* is characterized by a rose-colored mycelium and *D. fulvum* by a yellowish-brown-colored one (Shomura et al., 1985, 1986).

At present, eight species of *Dactylosporangium* are described, but two of them are "species incertae sedis": "*D. variesporium*" and "*D. salmoneum*" (Vobis, 1989c).

Glycomyces (Labeda, Testa, Lechevalier, and Lechevalier, 1985)

Colonies of *Glycomyces* develop both substrate and aerial mycelium on agar media (Fig. 5A and B). The substrate mycelium is pale yellowish-white to tan. The aerial mycelium is white and is produced abundantly to sparsely, depending on

the medium used (Labeda, 1989). Slightly yellowish to brown colored soluble pigments can be produced (Labeda et al., 1985).

Long chains of spores are formed by fragmentation of aerial hyphae. The spores are non-motile, cylindrical with square ends, measuring 0.5 × 0.7–1.8 μm (Fig. 5C and D).

Only two species have been described. They can be differentiated by their utilization pattern of organic compounds (Labeda, 1989). Furthermore, *G. rutgersensis* produces acid from adonitol and melibiose, whereas *G. harbinensis* does not (Labeda et al., 1985).

The glycolipid-containing genus *Glycomyces* is tentatively classified within the *Actinoplanes* group because of the chemotaxonomic similarities in the composition of the cell wall, the menaquinones, and the fatty acids (R.M. Kroppenstedt, unpublished observations), but because it has phospholipid type P I instead of type P II, its position remains exceptional among the actinoplanetes (Table 3). Phospholipid type P I is characteristic of the genus *Actinomadura* and of some species of *Microtetraspora* and *Nocardioides*. This correlates with the production of true aerial mycelium and of nonmotile spores in chains, which *Glycomyces* has in common with the above-mentioned genera, but not with the actinoplanetes.

Micromonospora (Ørskov, 1923)

Colonies growing on solid media form only substrate mycelium, which is raised and folded (Fig. 6A). Some strains can develop short, sterile aerial hyphae, giving a pruinose surface. The color of young colonies is pale yellow to light orange, becoming deep orange, red, purple, brown, or blue-green with age, depending on the individual strain (Kawamoto, 1989). The upper mycelial layers may burst open (Fig. 6B and D) or may be completely covered by a mucous mass of spores, giving the colonies bright brown-black, green-black, or black-colored surfaces (Fig. 6C). Other strains produce the spores accumulated in distinct areas on the surface (Fig. 6B) or completely immersed in the substrate (Fig. 6D).

The spores of *Micromonospora* are formed singly on substrate mycelium (Fig. 6E). They are nonmotile, and spherical to oval in shape, with a diameter of 0.7–1.5 μm. The sporophores, often together in clusters, produce spores terminally on short hyphal branches (Luedeman and Casmer, 1973). The spore surfaces of almost all species have blunt-spiny projections (Kawamoto, 1989).

The description of many species of this genus began with the taxonomical studies of the gentamicin-producing *Micromonospora* strains (Luedeman and Brodsky, 1964). More than 12 species and subspecies could be listed recently (Kawamoto, 1989). The species concept is based on chemotaxonomical markers, pigment production, and physiological characteristics. The presence or absence of the 3-hydroxy-A$_2$pm isomer in the peptidoglycans of the cell walls divides the genus into two species groups (Kawamoto et al., 1981). The predominant menaquinones of *M. carbonacea* and *M. halophytica* have nine isoprene units (MK-9) and those of *M. echinospora* subsp. *pallida* have 12 units (MK-12). All the other remaining species have MK-10 menaquinones (Collins et al., 1984; Kawamoto, 1989).

Mycelial pigments are also of diagnostic value: *M. coerulea* has a blue-green colored mycelium. *M. echinospora* subsp. *echinospora* and subsp. *ferruginea* have a maroon-purple pigment in the substrate mycelium. The following species are characterized by soluble pigments of various colors: *M. halophytica* subsp. *halophytica* (red-brown), *M. chalcea* (yellow), *M. purpureochromogenes* and *M. olivasterospora* (olive-green), and *M. rosaria* (wine-red) (Horan and Brodsky, 1986b; Kawamoto, 1989). Further physiological parameters, such as growth on special media, carbon utilization profiles, glycosidase activity, nitrate reduction, and NaCl tolerance, support the identification of species and subspecies (Kawamoto, 1989).

Pilimelia (Kane, 1966)

Strains of *Pilimelia* only form small, compact colonies, which are about 5 mm in diameter after 4 weeks of incubation. Growth is supported by keratinic substances like hair (Fig. 7A) and cattle-horn meal (Vobis, 1984). Aerial mycelium is not developed. The color of the substrate mycelium ranges from pale to lemon yellow, yellow-gray, and golden yellow to orange (Vobis, 1989b).

The sporangia are developed directly on the surface of agar medium or on natural substrates (Fig. 7B). The shape of the sporangia is globose, ovoid, pyriform, campanulate, or cylindrical and approximately 10–15 µm in size. In some strains, each sporangium contains up to a thousand spores (Fig. 7F). The spores are rod like to reniform, equipped with a laterally inserted tuft of flagella (Fig. 8). Nonmotile spores ("conidia") may be also produced (Fig. 7E).

Species differentiation for *Pilimelia* is based on morphological and colonial characters. *P. terevasa* has spherical to campanulate sporangia with parallel rows of abundantly branched spore chains (Fig. 7B and F). The colonies have a soft consistency and are yellow to yellow-gray. *P. anulata* has cylindrical sporangia (Fig. 7E). The top segment of the sporangiophore is expanded to

form a small ring-like structure ("annulus") (Fig. 7D). The mycelium has a yellowish color and is soft and pasty. In contrast, the colonies of *P. columellifera* are very solid and spherical to pyriform sporangia are produced, with the spore chains inside arranged in swirls (Vobis, 1984). The sporangiophores are unseptate and reach into the lumen of the sporangium, where they are visible as small columns (Fig. 7C). The substrate mycelium of *P. columellifera* is either golden yellow to orange or colorless to pale brownish in the subspecies *pallida* (Vobis et al., 1986).

Pilimelia strains were considered for a long time to be keratinophilic members of *Ampullariella*, but recent studies on the chemotaxonomy, ultrastructure, and physiology of the three species (Vobis, 1989b) have supported recognition of them as a separate genus (Kroppenstedt, 1979; Schäfer, 1973; Vobis et al., 1986).

Media for Isolation and Cultivation

The recipes for the media used for isolation, sporulation, and vegetative growth are listed below. However, the media developed by the "International Streptomyces Project" (Shirling and Gottlieb, 1966) are not included. Routine sterilization is carried out at 121°C for 15 to 20 min by autoclaving. The vitamins and the antibiotic agents are separately sterilized by filtration and added to cooled (50°C) agar media. If necessary, after autoclaving, the pH should be adjusted with sterile acid or base.

Additions to the Agar Media

ANTIBIOTICS Many isolation media contain antifungal or antibacterial agents. The concentrations and/or combinations of these supplements are given in Table 1, in the text, or in the corresponding recipes. To dissolve cycloheximide, warm water (45°C) is recommended (Hunter-Cevera et al., 1986). Nystatin can be dissolved initially in dimethylsulfoxide and then diluted in 95% alcohol (Makkar and Cross, 1982). As nystatin is not totally soluble in water at pH 7.0 and is unstable at high pH, the pH of the aquatic solution can be increased with 1 M NaOH to 11.0, filter sterilized, and immediately lowered to 7.0 with HCl (Hunter-Cevera et al., 1986).

VITAMINS Some media require the addition of B vitamins (Nonomura and Ohara, 1969; Hayakawa and Nonomura, 1987; Shearer, 1987). The amounts indicated below are per liter of the corresponding medium.

B-Vitamins (Nonomura and Ohara, 1969)

Thiamine hydrochloride	0.5 mg
Riboflavin	0.5 mg
Niacin	0.5 mg
Pyridoxine hydrochloride	0.5 mg
Inositol	0.5 mg
Calcium pantothenate	0.5 mg
para-Aminobenzoic acid	0.5 mg
Biotin	0.25 mg

Recipes for Agar Media

The quantities below are all per liter of distilled water, unless otherwise stated.

Arginine Glycerol Salts Agar (El-Nakeep and Lechevalier, 1963)

Arginine monohydrochloride	1.0 g
Glycerol (specific gravity not less than 1.249 at 25°C)	12.5 g
K_2HPO_4	1.0 g
$MgSO_4 \cdot 7H_2O$	0.5 g
$Fe_2(SO_4)_3 \cdot 6H_2O$	0.01 g
$CuSO_4 \cdot 5H_2O$	0.001 g
$ZnSO_4 \cdot 7H_2O$	0.001 g
$MnSO_4 \cdot H_2O$	0.001 g
Agar	15.0 g

Adjust to pH 6.9 to 7.1.

Arginine Vitamin Agar (Nonomura and Ohara, 1969)

L-Arginine	0.3 g
Glucose	1.0 g
Glycerol	1.0 g
K_2HPO_4	0.3 g
$MgSO_4 \cdot 7H_2O$	0.2 g
NaCl	0.3 g
$Fe_2(SO_4)_3$	10.0 mg
$MnSO_4 \cdot 7H_2O$	1.0 mg
$CuSO_4 \cdot 5H_2O$	1.0 mg
$ZnSO_4 \cdot 7H_2O$	1.0 mg
Agar	15.0 g

Adjust to pH 6.4, and then add:

B-Vitamins	(see above)
Cycloheximide	50.0 mg
Nystatin	50.0 mg

Artificial Soil Agar (Henssen and Schäfer, 1971)

$CaSO_4 \cdot 2H_2O$	1.01 g
$Ca(NO_3)_2 \cdot 4H_2O$	0.49 g
$MgSO_4 \cdot 7H_2O$	0.70 g
K_2SO_4	0.025 g
K_2HPO_4	0.005 g
$NaHCO_3$	0.2 g
$FeCl_3$	trace
Yeast extract	0.1 g
Glucose	0.01 g
Soil extract	50 ml
Agar	15.0 g

Adjust to pH 6.6 to 6.8.

To prepare soil extract, equal volumes of leafy soil and tap water are boiled for 2 hours and then cleared by centrifugation.

Bennett's Agar (Waksman, 1961)

Glucose	10.0 g
Yeast extract	1.0 g
Beef extract	1.0 g
N-Z-Amine A (casein digest; Sheffield Farms Co.)	2.0 g
Agar	15.0 g

Adjust to pH 7.3.

Bennett's Agar (Modified) (Wakisaka et al., 1982)

Glucose	10.0 g
Casamino acids	2.0 g
Yeast extract	2.0 g
Beef extract	1.0 g
Agar	15.0 g

Adjust to pH 7.3, and then add:

Cycloheximide	30.0 mg

Casamino Acids-Peptone-Czapek Agar (Henssen and Schäfer, 1971)

Casamino acids	1.0 g
Peptone	2.0 g
K_2HPO_4	1.0 g
KCl	0.5 g
$MgSO_4 \cdot 7H_2O$	0.5 g
$FeSO_4 \cdot 7H_2O$	0.01 g
Sucrose	30.0 g
Agar	15.0 g

Adjust to pH 7.0.

Calcium Malate Agar (Waksman, 1961)

Glycerol	10.0 g
Calcium malate	10.0 g
NH_4Cl	0.5 g
K_2HPO_4	0.5 g
Agar	15.0 g

Adjust to pH 7.0.

Colloidal Chitin Agar (Composition of Minerals According to Willoughby, 1968)

Colloidal chitin (dry weight)	2.0 g
$CaCO_3$	0.02 g
$FeSO_4 \cdot 7H_2O$	0.01 g
KCl	1.71 g
$MgSO_4 \cdot 7H_2O$	0.05 g
$Na_2HPO_4 \cdot 12H_2O$	4.11 g
Agar	18.0 g

Adjust to pH 7.0.

Collodial Chitin Agar (Composition of Minerals According to Hsu and Lockwood, 1975)

Collodial chitin (dry weight)	2.0 g
K_2HPO_4	0.7 g
KH_2PO_4	0.3 g
$MgSO_4 \cdot 5H_2O$	0.5 g
$FeSO_4 \cdot 7H_2O$	0.01 g
$ZnSO_4$	0.001 g
$MnCl_2$	0.001 g
Agar	20.0 g

Adjust to pH 7.0.

Preparation of Colloidal Chitin (Makkar and Cross, 1982)

Crude chitin is washed alternately in 1 N NaOH and 1 N HCl for 24-h periods each, five times. Then, it is washed four times with 95% (v/v) ethanol. 15 g of the purified white chitin is dissolved with 100 ml of concentrated HCl and stirred in an ice bath for 20 min. The mixture is filtered through glass wool, and the solution is poured into cold distilled water to precipitate the chitin. The insoluble chitin on the glass wool is treated again with HCl, and the process is repeated until no more precipitate is obtained when the filtrate is added to cold water. The colloidal chitin is allowed to settle overnight and the supernatant is decanted. The remaining suspension is neutralized to pH 7.0 with NaOH. The precipitated chitin is centrifuged, washed, and stored as a paste at 4°C.

Various other procedures for preparing colloidal chitin can be found in Willoughby (1968), Hsu and Lockwood (1975), Johnston and Cross (1976), and Hunter-Cevera et al. (1986).

Czapek Sucrose Agar (Bland and Couch, 1981)

Sucrose	30.0 g
NaNO$_3$	3.0 g
K$_2$HPO$_4$	1.0 g
MgSO$_4$ · 7H$_2$O	0.5 g
KCl	0.5 g
FeSO$_4$	0.01 g
Agar	15.0 g

Adjust to pH 7.0 to 7.3.

Emerson's Yeast Extract-Starch Agar (Emerson, 1958)

Yeast extract	4.0 g
Soluble starch	15.0 g
K$_2$HPO$_4$	1.0 g
MgSO$_4$ · 7H$_2$O	0.5 g
Agar	20.0 g

Gause Number 1 Agar (Gause, 1958)

KNO$_3$	1.0 g
K$_2$HPO$_4$	0.5 g
MgSO$_4$	0.5 g
NaCl	0.5 g
FeSO$_4$	0.01 g
Starch	20.0 g
Agar	30.0 g
Tap water	1 liter

Glucose Asparagine Agar (Gordon and Smith, 1955)

Glucose	10.0 g
Asparagine	0.5 g
K$_2$HPO$_4$	0.5 g
Agar	15.0 g

Adjust to pH 6.8.

Hickey-Tresner Agar (Waksman, 1961)

Dextrin	10.0 g
Yeast extract	1.0 g
Beef extract	1.0 g
N-Z-Amine A (casein digest; Sheffield Farms Co.)	2.0 g
CoCl$_2$ · 7H$_2$O	0.02 g
Agar	20.0 g

Adjust to pH 7.3.

Humic Acid-Ion Agar (Willoughby et al., 1968)

5 g of humic acid in solid form (see below) is dissolved in 10 ml of 0.2 N NaOH, and the solution is made up to 1 liter with distilled water; pH 7.0 to 7.3. Then 20 g of agar (Oxoid Ionagar No. 2) is added.

Preparation of Humic Acid (Willoughby et al., 1968):

Air-dried (90°C for 24 hours) peat from a blanket bog or from compacted, decomposing tree leaves (oak and alder) is first extracted with acetone for 3 hours and then redried at 50°C overnight. The humic acids are extracted from the material (100 g) with 2 liters of 0.2 N NaOH by shaking occasionally, standing overnight at room temperature, and filtering off the insoluble residue on a no. 4 sintered glass funnel. The filtrate is acidified to pH 1.0 with concentrated HCl. The resulting precipitate is centrifuged off, washed twice with water, and left overnight at –20°C to freeze. After thawing, the now-granular particles are filtered off from the remaining liquors, washed, and air-dried at 50°C.

Humic Acid-Vitamin Agar (Hayakawa and Nonomura, 1987)

Humic acid (see below) (dissolved in 10 ml of 0.2 N NaC1)	1.0 g
Na$_2$HPO$_4$	0.5 g
KC1	1.71 g
MgSO$_4$ · 7H$_2$O	0.05 g
FeSO$_4$ · 7H$_2$O	0.01 g
CaCO$_3$	0.02 g
Agar	18.0 g

Adjust to pH 7.2, and then add:
B Vitamins (see above)

Cycloheximide	50.0 mg

Preparation of Humic Acid (Hayakawa and Nonomura, 1987)

500 g of soil sample (A-horizon of a forest) is suspended in 1 liter of 0.5% NaOH solution and left standing at room temperature for 24 hours, with occasional stirring. The precipitate of the suspension is removed by centrifugation (20 min at 7,000 rpm), and the supernatant is acidified to pH 1.0 with concentrated HCl. The resulting precipitate is centrifugated (20 min at 3,000 rpm), washed three times by centrifugation with 150 ml of water and suspended again in 150 ml of water. The suspension is frozen overnight at –20°C. After thawing, the granulated humic acid is filtered, washed, and air-dried.

Kadota's Cellulose Benzoate Agar (Sandrak, 1977)

1) Basal medium Sodium benzoate

NaNO$_3$	0.5 g
K$_2$HPO$_4$	1.0 g
MgSO$_4$ · 7H$_2$O	0.5 g
FeSO$_4$ · 7H$_2$O	0.01 g
Sodium benzoate	20.0 g
Agar	20.0 g

Adjust to pH 7.2

2) Cellulose powder (0.67g) is mixed with 1 ml soil suspension and 2 ml liquid basal medium before inoculation of the plates.

M3 Agar Medium (Rowbotham and Cross, 1977)

KH_2PO_4	0.466 g
Na_2HPO_4	0.732 g
KNO_3	0.10 g
NaCl	0.29 g
$MgSO_4 \cdot 7H_2O$	0.10 g
$CaCO_3$	0.02 g
Sodium propionate	0.20 g
$FeSO_4 \cdot 7H_2O$	200 µg
$ZnSO_4 \cdot 7H_2O$	180 µg
$MnSO_4 \cdot 4H_2O$	20 µg
Agar	18.00 g

Ajust to pH 7.0, and then add:

Cycloheximide	50.0 mg
Thiamine hydrochloride	4.0 mg

Oatmeal-Soil Extract Agar (Shearer, 1987)

Oatmeal agar (Difco)	5.5 g
Agar	16.0 g
Soil extract	500 ml
Deionized water	500 ml

Adjust to pH 7.2.

Peptone-Czapek Agar (Bland and Couch, 1981)

Sucrose	30.0 g
Peptone	5.0 g
K_2HPO_4	1.0 g
$MgSO_4 \cdot 7H_2O$	0.5 g
KCl	0.5 g
$FeSO_4 \cdot 7H_2O$	0.01 g
Agar	15.0 g

Adjust to pH 7.3.

Skim Milk Agar (Gordon and Smith, 1955)

Solution a:

Skim milk powder	50 g
Distilled water	500 ml

Solution b:

Agar	20 g
Distilled water	500 ml

Solutions a and b are autoclaved separately, cooled, and mixed at 45°C, and the pH is adjusted to 7.2.

Skim Milk-Cattle Hornmeal Agar (Vobis, 1984)

Solution a:

Cattle hornmeal (powdered)	10.0 g
$Ca(NO_3)_2 \cdot 4H_2O$	0.5 g
$MgSO_4 \cdot 7H_2O$	0.7 g
K_2HPO_4	0.005 g
$NaHCO_3$	0.2 g
$FeCl_3$	trace
Agar	20.0 g
Distilled water	900 ml

Solution b:

Skim milk powder	2.5 g
Distilled water	100 ml

Solutions a and b are autoclaved separately, cooled, and mixed at 48°C, and the pH is adjusted to 7.2.

Soil Extract Agar (Makkar and Cross, 1982)

First 150 g of garden soil is stirred in 600 ml of tap water, filtered immediately through Whatman No. 1 filter, and made up to 1 liter with tap water. To this is added:

Agar	18.0 g

Adjust to pH 7.2, and then add:

Cycloheximide	50 mg
Nystatin	50 mg

Starch-Casein-Nitrate Agar (Küuster and Williams, 1964)

Starch	10.0 g
Casein (Difco, vitamin-free)	0.3 g
KNO_3	2.0 g
NaCl	2.0 g
K_2HPO_4	2.0 g
$MgSO_4 \cdot 7H_2O$	0.05 g
$CaCO_3$	0.02 g
$FeSO_4 \cdot 7H_2O$	0.01 g
Agar	18.0 g

Adjust to pH 7.0 to 7.2.

Hunter-Cevera et al. (1986) added:

Cycloheximide	75 mg
Nystatin	75 mg

Shearer (1987) added:

B-Vitamins (see above)

Starch-Casein Sulfate Agar (Palleroni, 1980)

K_2HPO_4	0.5 g
$MgSO_4$ (anhydrous)	5.0 g
Soluble starch	10.0 g
Casein (dissolved in diluted NaOH)	1.0 g
Agar	15.0 g

Adjust to pH 7.0 to 7.5.

Thin Pablum Agar (Hunter-Cevera et al., 1986)

Pablum (boiled in cheesecloth for 20 to 30 min)	7.5 g
Agar	15.0 g
Tap water	1 liter

Adjust to pH 6.8 to 7.0, and then add:

Cycloheximide	75 mg
Nystatin	75 mg

Water Agar (Hunter-Cevera et al., 1986)

Tap water	1 liter
Crude agar flakes	17.5 g

Adjust to pH 6.8, and then add:

Cycloheximide	75 mg
Nystatin	75 mg

Applications

Many members of the actinoplanete genera produce useful enzymes and secondary metabolites, so they may have important applications in industry, biotechnology, and agriculture. For example, vitamin B_{12} can be produced by a

Micromonospora strain (Florent and Ninet, 1979), and strains of *Actinoplanes* and *Micromonospora* can be used for the bioconversion of complex compounds (Okami and Hotta, 1988; Boek et al., 1988). Applications which have been established already or might be used commercially in the future are described below.

Glucose Isomerase in the Food Industry

The enzyme glucose isomerase, which is primary a xylose isomerase, can be obtained from strains of *Actinoplanes*, *Ampullariella* and *Micromonospora*, in addition to strains of *Streptomyces*, *Bacillus*, and *Arthrobacter* (Crueger and Crueger, 1982; Peczyńsca-Czoch and Mordarski, 1988). The glucose isomerase converts D-glucose into D-fructose and is used commercially in the starch industry to obtain high-fructose corn syrup (Aunstrup et al., 1979). Starting from about 95% glucose syrup, a twice sweeter fructose syrup is produced which is usually composed of 53% of D-glucose, 42% of D-fructose, and 5% of oligosaccharides (Crueger and Crueger, 1982).

Actinoplanes missouriensis strain ATCC 14538 produces an intracellular, soluble glucose isomerase with a molecular weight of about 80,000 daltons. The optimal pH of the enzyme is 7.0 at temperatures between 60 and 65°C. A requirement for cobalt ions for optimal activity is eliminated if the proper amount of magnesium ions is used (Gong et al., 1980). The xylose isomerase of *Ampullariella* strain ATCC 31354 exhibits superior thermostability and activity over a wide range of conditions. However, the strain itself is difficult to use as a production organism, which makes it desirable to clone and express its enzyme in a more convenient microorganism (Saari et al., 1987).

Inhibitors of α-Glucosidase as Pharmaceutical Drugs

In the course of screening for inhibitors of amylases and other mammalian intestinal carbohydrate-splitting enzymes, strains of *Actinoplanes* exhibited higher amounts of activity than did those of *Streptomyces* and *Streptosporangium* (Frommer et al., 1979). Applied orally together with food carbohydrates like starch and other oligosaccharides, these glycoside hydrolase inhibitors slow down oligosaccharide decomposition and reduce or avoid postprandial hyperglycemia and hyperinsulinemia of type IV. Therefore, they may be useful to treat metabolic illnesses such as diabetes mellitus, adipositas, and hyperlipoproteinaemia (Truscheit et al.,

1981; Creutzfeldt, 1988). Pseudooligosaccharides with an essential core consisting of an unsaturated cyclitol and 4-amino-4,6-dideoxyglucose are the most important group of α-glucosidase inhibitors. In culture filtrates of the *Actinoplanes* strain SE 50, a very effective pseudotetrasaccharide with the generic name "acarbose" could be found. This low-molecular-weight compound is stabile to acid, alkali, and heat treatment and exhibits pronounced inhibition of sucrase, maltase, and amylase (Truscheit et al., 1981).

Antibiotics

The "rare" actinomycetes (i.e., nonstreptomycetes) have become increasingly interesting in the search for new active secondary metabolites (Nara et al., 1977; Okami and Hotta, 1988). In 1984, the number of known antibiotic compounds produced by all the actinomycetes together amounted to about 4,200 (Bérdy, 1984), and the proportion produced by actinoplanetes (*Micromonospora*, *Actinoplanes*, *Dactylosporangium*, and *Ampullariella*) increased from less than 1% in the year 1966 up to 10% in 1984 (Lechevalier and Lechevalier, 1967; Wagman and Weinstein, 1980; Bérdy, 1984). As can be seen in Table 4, each actinoplanetes genus covers only a part of all the chemical groups of antibiotics that are known from the genus *Streptomyces*. But they complement one another, so only the β-lactam antibiotics seem to be absent. Until now, members of *Catellatospora* and *Pilimelia* have not been included to any extent in antibiotic screening procedures.

ACTINOPLANES More than 120 antibiotics are known from *Actinoplanes* species. The most common are peptides/depsipeptides, polyene-type macrolides, and aromatic compounds, whereas the aminoglycosides, streptothricins, macrolides, and ansamycine seem to be absent (Table 4). About 50% of the antibiotics found in *Actinoplanes* are amino acid derivatives. The proline antimetabolite L-acetidine-2-carboxylic acid could be isolated from *A. ferrugineus*. This amino acid has not been found in any other prokaryotes and has only been found in eukaryotes (Palleroni, 1979). Additional amino acid derivatives are 5-azacytidin, azaserine, and the chromopeptide actinomycin, which are all properly typical *Streptomyces* antibiotics (Torikata et al., 1978). They are active as antitumor agents. The polypeptides generally exhibit activity against Gram-positive bacteria. The acidic peptide 41.012 is also an agent against mycobacteria (Celmer et al., 1977). The antibiotics A-10947, A-7413, taitomycin, and gardimicin are sulfur-

Table 4. Antibiotic groups produced by the genera of the actinoplanetes and by the streptomycetes.

Genus	Aminoglycoside	Macrolide	Ansamacrolide	β-Lactam	Peptide	Glycopeptide	Anthracycline	Tetracycline	Nucleoside	Polyene	Quinone
Actinoplanes					+	+			+	+	+
Ampullariella							+		+	+	
Catellatospora											
Dactylosporangium	+	+			+			+	+	+	
Glycomyces					+						
Micromonospora	+	+	+		+		+		+		+
Pilimelia											
Streptomyces	+	+	+	+	+	+	+	+	+	+	+

Data mainly based on Okami and Hotta (1988) and Nara et al. (1977); the additions made are discussed in the text.

containing polypeptides; the latter two are also active against anaerobic bacteria (Yaginuma et al., 1979; Parenti and Coronelli, 1979). The cyclic polypeptides A/287 and mycoplanecin show growth-promoting and antituberculosis effects, respectively (Hamill and Stark, 1974; Nakajima et al., 1983). The depsipeptide antibiotic plauracins A 17002 and A 2315 belong to the virginiamycin group, composed of a mixture of macrocyclic lactones and depsipeptides. They can be used for growth promotion in chicken, swine, and ruminants (Hamill and Stark, 1975; Parenti and Coronelli, 1979).

A well-studied example of the antifungal polyenic macrolides is the antibiotic 67–121 (Sch 16656). It is a complex of four polyene heptaenes produced by *A. caeruleus* (Horan and Brodsky, 1986a). It is also produced by *A. azureus*, a strain which also produces the plauracins (Parenti and Coronelli, 1979).

Some metabolites belonging to various other chemical groups are also found in strains of *Actinoplanes*. Purpuromycin is a naphthoquinone antibiotic of the rubromycin type, effective against bacteria and fungi (Coronelli et al., 1974). *A. teichomyceticus* produces the glycopetide teicoplanin (formerly called teichomycin A$_2$), which is composed of six factors. It belongs to the vancomycin family (Malabarba et al., 1984). The same strain also produces a phosphorus-containing glycolipid (teichomycin A$_1$). Both carbohydrate antibiotics are active against Gram-positive bacteria (Parenti et al., 1978). The recently described polycyclic xanthones actinoplanone A and B were found to be potent cytotoxins in in vitro assays with Hela cells (Kobayashi et al., 1988).

Some chemical novelties have been detected in *Actinoplanes* strains. For example, chuangxinmycin, clinically effective in cases of septicemia and urinary and biliary infections caused by *Escherichia coli*, is a completely new antibiotic composed of an unique bicyclic system formed of an indole nucleus fused to a thiopyran residue (Parenti and Coronelli, 1979). The antibiotic A/15104 Y is a chlorophenol derivative, active against bacteria and fungi. It represents the first example of a halogenated pyrrole from actinomycetes; the other biological sources have been sponges and pseudomonads (Cavalleri et al., 1978).

AMPULLARIELLA Only a dozen antibiotics are known to be produced by *Ampullariella* strains. Nevertheless, they exhibit quite a range of different chemical structures and antibiotic effects. Viriplanin is an anthracyclic antibiotic isolated from *A. regularis*, which shows activities against herpes simplex viruses (Hütter et al., 1986). The neplanecins are nucleosides produced by

another *A. regularis* strain. They are antitumor antibiotics with additional activities against phytopathogenic fungi (Yaginuma et al., 1981). Another antifungal antibiotic, candiplanecin, could be isolated from *A. regularis* subsp. *mannitophila* (Itoh et al., 1981).

DACTYLOSPORANGIUM At present, about 30 antibiotics are known from *Dactylosporangium* strains, belonging to several chemical divisions (Table 4). As with *Micromonospora*, the major antibiotics seem to be aminoglycosides. Dactimicin is produced by *D. matsuzakiense* and *D. vinaceum* (Shomura et al., 1980, 1983). The closely related aminoglycosides gentamicin, sisomicin, fortimicin, and antibiotic G-367 could also be isolated. Another carbohydrate antibiotic is known from *D. roseum*, namely the orthosomycin complex SF-2107 (Shomura et al., 1985). All the above-mentioned antibiotics are generally active against Gram-positive and Gram-negative bacteria. Capreomycin, a polypeptide compound previously known from *Streptomyces capreolus*, could be obtained also from *"D. variesporium."* It is of primary interest for its use as an antituberculosis agent (Tomita et al., 1977). *"D. salmoneum"* produces the polyether antibiotic compound 44,161, which is useful for the control of coccidiosis in poultry and improving feed efficiency in ruminants (Celmer et al., 1978).

The tetracycline antibiotic compound Sch 34164 (Patel et al., 1987) and the macrolide tiacumicin (Hochlowski et al., 1986) have been isolated from other *Dactylosporangium* species, suggesting that the capacity for producing antibiotic metabolites in this genus may be very large.

Glycomyces *Glycomyces harbinensis* produces the amino acid derivative azaserine and the antibiotic LL-D05139-beta (Labeda, 1989). No other antibiotics have been identified in any *Glycomyces* strains so far.

Micromonospora Among the antibiotics produced by actinoplanete genera, those of *Micromonospora* occupy the most important commercial position (Crueger and Crueger, 1982). An intensive screening of *Micromonospora* species as sources for new antibiotics began in 1963 with the discovery of gentamicin. Over 300 different antibiotics have been described (Bérdy, 1984), and the range of chemical structures is quite large (Table 4). Examples include the aminoglycosides, represented by gentamicins, sisomicin, and verdamicin; the antibiotics G-52, G-418, and JI-20; mannosidostreptomycin; kanamycin; neomycin B (antibiotic 460); sagamicin (gentamicin C$_{2b}$); paromamine;

fortimicins; and antibiotics 66–40 and SF 1854 (Nara et al., 1977; Wagman and Weinstein, 1980). The macrolides comprise megalomycins, rosaramicin, juvenimicins, the M-4365 complex, erythromycins, and antibiotic XK 41-B-2. Examples for ansa macrolides (ansamycins) are halomicins, rifamycins, and compound 32656. Bottromycin, microsporonin, the 70591 complex, and actinomycin should be cited as representatives of the peptide antibiotics. Other miscellaneous antibiotics isolated from *Micromonospora* species include the oligosaccharides everninomycin and antlermicin, the nucleosides PA-1322 and XK-101–2, and the quinone PA-2046 (Wagman and Weinstein, 1980).

Aminoglycosides of *Micromonospora* show antibiotic effects against both Gram-positive and Gram-negative bacteria and have been introduced into clinical practice. The gentamicins C_1, C_{1a}, and C_2 are produced by *M. purpurea* and *M. echinospora* and exhibit excellent activity against *Staphylococcus aureus* and species of *Pseudomonas* and *Proteus*. Because of their vestibular and nephrotoxicity, they are used in human therapy only for severe infections (Wagman and Weinstein, 1980; Crueger and Crueger, 1982; Lancini and Parenti, 1982). Sisomycin and fortimicins, derived from *M. inyonensis* and *M. olivoasterospora,* respectively, have a similar spectrum of effectivity as gentamicin and can be used against gentamicin-resistant organisms (Wagman and Weinstein, 1980).

Biological Control in Agriculture

Reports of hyperparasitism by actinoplanetes on parasitic *Peronosporales* and *Saprolegniales* demonstrate a possible biological control of serious diseases of economic plants (Lechevalier, 1988). The oospores of pink rot causing *Phytophthora megasperma* var. *sojae* or f. sp. *glycinea* can be parasitized by certain strains of *Actinoplanes*, *Ampullariella*, and *Micromonospora* (Sneh et al., 1977). The hyphae of *Actinoplanes missouriensis* penetrate the walls of the oogonia and the oospores without forming appressoria or haustoria or changing the morphological and internal structures of the oospores (Sutherland et al., 1984). In greenhouse experiments, the root rot of soybeans caused by *Phytophthora* could be reduced by *Actinoplanes missouriensis*, *A. utahensis*, and *Micromonospora* sp. (Filinow and Lockwood, 1985).

Acknowledgments. I wish to thank Dr. H.-W. Kothe (Burlington), Dr. R. M. Kroppenstedt (Braunschweig), Dr. H. Prauser (Jena), Dr. G. S. Saddler (Gerenzano), Prof. Dr. O. Salcher (Wuppertal), Dr. J.-J. Sanglier (Basel), and Dr. J. Wink (Frankfurt) for helpful cooperation.

Literature Cited

Al-Diwany, L. J., T. Cross. 1978. Ecological studies on nocardioforms and other actinomycetes in aquatic habitats. Zentralbl. Bacteriol. Parasitenkd. Infektionskr. Hyg. Abt. 1, Suppl. 6:153–160.

Arora, D. K. 1986. Chemotaxis of *Actinoplanes missouriensis* zoospores to fungal conidia, chlamydospores and sclerotia. J. Gen. Microbiol. 132:1657–1663.

Asano, K., I. Kawamoto. 1986. *Catellatospora,* a new genus of the *Actinomycetales.* Int. J. Syst. Bacteriol. 36:512–517.

Asano, K., I. Kawamoto. 1988. *Catellatospora citrea* subsp. *methionotrophica* subsp. nov., a methionine-deficient auxotroph of the Actinomycetales. Int. J. Syst. Bacteriol. 38:326–327.

Asano, K., H. Sano, I. Masunaga, I. Kawamoto. 1989a. 3-0-methylrhamnose: identification and distribution in *Catellatospora* species and related actinomycetes. Int. J. Syst. Bacteriol. 39:56–60.

Asano, K., I. Masunaga, I. Kawamoto. 1989b. *Catellatospora matsumotoense* sp. nov. and *C. tsunoense* sp. nov., actinomycetes found in woodland soils. Int. J. Syst. Bacteriol. 39:309–313.

Aunstrup, K., O. Andresen, E. A. Falch, T. K. Nielsen. 1979. Production of microbial enzymes. 281–309. H. J. Peppler and D. Perlman (ed.) Microbial technology, vol. 1. Academic Press. New York.

Bérdy, J. 1984. New ways to obtain new antibiotics. Chin. J. Antibiot. 7:348–360.

Beretta, G. 1973. *Actinoplanes italicus,* a new red-pigmented species. Int. J. Syst. Bacteriol. 23:37–42.

Bland, C. E., J. N. Couch. 1981. The family *Actinoplanaceae*. 2004–2010. M. P. Starr, H. Stolp, H. G. Trüper, A. Balows, and H. G. Schlegel (ed.) The prokaryotes: A handbook on habitats, isolation, and identification of bacteria. Springer-Verlag. Berlin.

Boeck, L. D., D. F. Fukuda, B. J. Abbott, M. Debono. 1988. Deacylation of A21978C, an acidic lipopeptide antibiotic complex, by *Actinoplanes utahensis.* J. Antibiot. 41:1085–1092.

Burman, N. P., C. W. Oliver, J. K. Stevens. 1969. Membrane filtration techniques for the isolation from water, of coli-aerogenes, *Escherichia coli*, faecal streptococci, *Clostridium perfringens*, actinomycetes and microfungi. 127–134. D. A. Shapton, and G. W. Gould (ed.) Isolation methods for microbiologists. Academic Press. London.

Cavalleri, B., G. Volpe, G. Tuan, M. Berti, F. Parenti. 1978. A chlorinated phenylpyrrole antibiotic from *Actinoplanes.* Curr. Microbiol. 1:319–324.

Celmer, W. D., W. P. Cullen, C. E. Moppett, J. B. Routien, M. T. Jefferson, R. Shibakawa, J. Tone. 1978. Polycyclic ether antibiotic produced by new species of *Dactylosporangium.* U.S. Patent 4,081,532, Mar. 28, 1978.

Celmer, W. D., C. E. Moppett, W. P. Cullen, J. B. Routien, M. T. Jefferson, R. Shibakawa, J. Tone. 1977. Antibiotic compound 41,012. U.S. Patent 4,001,397, Jan. 4, 1977.

Collins, M. D., M. Falkner, R. M. Keddie. 1984. Menaquinone composition of some sporeforming actinomycetes. System. Appl. Microbiol. 5:20–29.

Coronelli, C., H. Pagani, M. R. Bardone, G. C. Lancini. 1974. Purpuromycin, a new antibiotic isolated from *Actinoplanes ianthinogenes* n. sp. J. Antibiot. 27:161–168.

Couch, J. N. 1949. A new group of organisms related to *Actinomyces*. J. Elisha Mitchell Sci. Soc. 65:315–318.

Couch, J. N. 1950. *Actinoplanes,* a new genus of the Actinomycetales. J. Elisha Mitchell Sci. Soc. 66:87–92.

Couch, J. N. 1954. The genus *Actinoplanes* and its relatives. Trans. N.Y. Acad. Sci. 16:315–318.

Couch, J. N. 1963. Some new genera and species of the Actinoplanaceae. J. Elisha Mitchell Sci. Soc. 79:53–70.

Creutzfeldt, W. 1988. Acarbose for the treatment of diabetes mellitus. Springer-Verlag. Berlin.

Cross, T. 1981a. Aquatic actinomycetes: a critical survey of the occurrence, growth and role of actinomycetes in aquatic habitats. J. Appl. Bacteriol. 50:397–423.

Cross, T. 1981b. The monosporic actinomycetes. 2091–2102. M. P. Starr, H. Stolp, H. G. Trüper, A. Balows, and H. G. Schlegel (ed.) The prokaryotes: A handbook on habitats, isolation, and identification of bacteria. Springer-Verlag. Berlin.

Cross, T. 1986. The occurrence and role of actinoplanetes and motile actinomycetes in natural ecosystems. 265–270. F. Megusar, and M. Gantar (ed.) Perspectives in microbial ecology. Slovene Soc. Microbiol., Ljubljana.

Cross, T. 1989. Growth and examination of actinomycetes—some guidelines. 2340–2343. S. T. Williams (ed.) Bergey's manual of systematic bacteriology, vol. 4. Williams & Wilkins. Baltimore.

Cross, T., R. W. Attwell. 1974. Recovery of viable thermoactinomycete endospores from deep mud cores. 11–20. A. N. Barker, G. W. Gould, and J. Wolf (ed.) Spore research 1973. Academic Press. London.

Crueger, W., A. Crueger. 1982. Lehrbuch der Angewandten Mikrobiologie. Akademische Verlagsgesellschaft. Wiesbaden.

El-Nakeeb, M., H. A. Lechevalier. 1963. Selective isolation of aerobic actinomycetes. Appl. Microbiol. 11:75–77.

Emerson, R. 1958. Mycological organization. Mycologia 50:589–621.

Ensign, J. C. 1978. Formation, properties, and germination of actinomycete spores. Ann. Rev. Microbiol. 32:185–219.

Erikson, D. 1941. Studies on some lake-mud strains of *Micromonospora*. J. Bacteriol. 41:277–300.

Fernandez, C. 1984. Studies on the microflora of the curative bottom mud of the thermal lake Hévíz (W. Hungary). Acta Bot. Hung. 30:257–268.

Filinow, A. B., J. L. Lockwood. 1985. Evaluation of several actinomycetes and the fungus *Hyphochytrium catenoides* as biocontrol agents for phytophthora root rot of soybean. Plant disease 69:1033–1036.

Florent, J., L. Ninet. 1979. Vitamin B$_{12}$. 497–519. H. J. Peppler, and D. Perlman (ed.) Microbial technology, vol. 2. Academic Press. New York.

Frommer, W., B. Junge, L. Müller, D. Schmidt, E. Truscheit. 1979. Neue Enzyminhibitoren aus Mikroorganismen. Planta Medica 35:195–217.

Gaertner, A. 1955. Über zwei ungewöhnliche keratinophile Organismen aus Ackerböden. Arch. Mikrobiol. 23:28–37.

Gause, G. F. 1958. Zur Klassifizierung der Actinomyceten. VEB Gustav Fischer Verlag. Jena.

Goodfellow, M. 1989. Suprageneric classification of actinomycetes. 2333–2339. S. T. Williams (ed.) Bergey's manual of systematic bacteriology, vol. 4. Williams & Wilkins. Baltimore.

Goodfellow, M., T. Cross. 1984. Classification. 7–164. M. Goodfellow, M. Mordarski, and S. T. Williams (ed.) The biology of the actinomycetes. Academic Press. London.

Goodfellow, M., J. A. Haynes. 1984. Actinomycetes in marine sediments. 453–472. L. Ortiz-Ortiz, L. F. Bojalil, and V. Yakoleff (ed.) Biological, biochemical, and biomedical aspects of actinomycetes. Academic Press. Orlando.

Goodfellow, M., S. T. Williams. 1983. Ecology of actinomycetes. Ann. Rev. Microbiol. 37:189–216.

Goodfellow, M., L. J. Stanton, K. E. Simpson, D. E. Minnikin. 1990. Numerical and chemical classification of *Actinoplanes* and some related actinomycetes. J. Gen. Microbiol. 136:19–36.

Gong, C.-S., L. F. Chen, G. T. Tsao. 1980. Purification and properties of glucose isomerase of *Actinoplanes missouriensis*. Biotechnol. Bioeng. 22:833–845.

Gordon, R. E., M. M. Smith. 1955. Proposed group of characters for the separation of *Streptomyces* and *Nocardia*. J. Bacteriol. 69:147–150.

Hamill, R. L., W. M. Stark. 1974. Antibiotic A-287 and process for preparation thereof. U.S. Patent 3,824,305, July 16, 1974.

Hamill, R. L., W. M. Stark. 1975. Antibiotic A-2315 and process for preparation thereof. U.S. Patent 3,923,980, Dec. 2, 1975.

Hasegawa, T., M. P. Lechevalier, H. A. Lechevalier. 1979. Phospholipid composition of motile actinomycetes. J. Gen. Appl. Microbiol. 25:209–213.

Hasegawa, T., M. Takizawa, S. Tanida. 1983. A rapid analysis for chemical grouping of aerobic actinomycetes. J. Gen. Appl. Microbiol. 29:319–322.

Hayakawa, M., H. Nonomura. 1987. Humic acid-vitamin agar, a new medium for the selective isolation of soil actinomycetes. J. Ferment. Technol. 65:501–509.

Henssen, A., D. Schäfer. 1971. Emended description of the genus *Pseudonocardia* Henssen and description of a new species *Pseudonocardia spinosa* Schäfer. Int. J. Syst. Bacteriol. 21:29–34.

Higgins, M. L. 1967. Release of sporangiospores by a strain of *Actinoplanes*. J. Bacteriol. 94:495–498.

Hirsch, C. F., D. L. Christensen. 1983. Novel method for selective isolation of actinomycetes. Appl. Environ. Microbiol. 46:925–929.

Hochlowski, J. E., S. J. Swanson, D. N. Whittern, A. N. Buko, J. B. McAlpine. 1986. Tiacumicins, a novel series of 18-membered macrolide antibiotics. II. Isolation and elucidation of structures. 26th Interscience Congress on Antimicrobial Agents and Chemotherapy, Abstract 937.

Horan, A. C., B. Brodsky. 1986a. *Actinoplanes caeruleus* sp. nov., a bluepigmentd species of the genus *Actinoplanes*. Int. J. Syst. Bacteriol. 36:187–191.

Horan, A. C., B. C. Brodsky. 1986b. *Micromonospora rosaria* sp. nov., nom. rev., the rosaramicin producer. Int. J. Syst. Bacteriol. 36:478–480.

Hsu, S. C., J. L. Lockwood. 1975. Powdered chitin agar as a selective medium for enumeration of actinomycetes in water and soil. Appl. Microbiol. 29:422–426.

Hungate, R. E. 1946. Studies on cellulose fermentation. II. An anaerobic cellulose-decomposing actinomycete, *Micromonospora propionici* n. sp. J. Bacteriol. 51:51–56.

Hunter, J. C., D. E. Eveleigh, G. Casella. 1981. Actinomycetes of a salt marsh. Zentralbl. Bakteriol. Mikrobiol. Hyg. Abt. 1, Suppl. 11:195–200.

Hunter, J. C., M. Fonda, L. Sotos, B. Toso, A. Belt. 1984. Ecological approaches to isolation. Dev. Ind. Microbiol. 25:247–266.

Hunter-Cevera, J. C., M. E. Fonda, A. Belt. 1986. Isolation of cultures. 3–23. A. L. Demain, and N. A. Solomon (ed.) Manual of industrial microbiology and Biotechnology. Am. Soc. Microbiol., Washington, D.C.

Hütter, K., E. Baader, K. Frobel, A. Zeek, K. Bauer, W. Gau, J. Kurz, T. Schröder, C. Wünsche, W. Karl, D. Wendisch. 1986. Viriplanin, a new anthracycline antibiotic of the nogalamycin group. J. Antibiot. 39:1195–1204.

Itoh, Y., A. Torikata, C. Katayama, T. Haneishi, M. Arai. 1981. Candiplanecin, a new antibiotic from *Ampullariella regularis* subsp. *mannitophila* subsp. nov. II. Isolation, physico-chemical characterization and biological activities. J. Antibiot. 34:934–937.

Jensen, H. L. 1930. The genus *Micromonospora* Ørskov, a little known group of soil microorganisms. Proc. Linn. Soc. N.S. Wales 55:231–248.

Jensen, H. L. 1932. Contributions to our knowledge of the actinomycetales. III. Further observations on the genus *Micromonospora*. Proc. Linn. Soc. N.S. Wales 57:173–180.

Johnston, D. W., T. Cross. 1976. The occurrence and distribution of actinomycetes in lakes of the English Lake District. Freshw. Biol. 6:457–463.

Kane, W. D. 1966. A new genus of Actinoplanaceae, *Pilimelia,* with a description of two species, *Pilimelia terevasa* and *Pilimelia anulata.* J. Elisha Mitchel Sci. Soc. 82:220–230.

Karling, J. S. 1954. An unusual keratinophilic microorganism. Proc. Indiana Acad. Sci. 63:83–86.

Kawamoto, I. 1989. Genus *Micromonospora* Ørskov. 2442–2450. S. T. Williams (ed.) Bergey's manual of systematic bacteriology, vol. 4. Williams & Wilkins. Baltimore.

Kawamoto, I., T. Oka, T. Nara. 1981. Cell wall composition of *Micromonospora olivoasterospora, Micromonospora sagamiensis,* and related organisms. J. Bacteriol. 146:527–534.

Kobayashi, K., C. Nishino, J. Ohya, S. Sato, T. Mikawa, Y. Shiobara, M. Kodama. 1988. Actinoplanones A and B, new cytotoxic polycyclic xanthones from *Actinoplanes* sp. J. Antibiot. 41:502–511.

Kothe, H.-W. 1987. Die Gattung *Actinoplanes* und ihre Stellung innerhalb der Actinomycetales. Dissertation, Marburg.

Kroppenstedt, R. M. 1979. Chromatographische Identifizierung von Mikroorganismen, dargestellt am Beispiel der Actinomyceten. Kontakte (Merck) 2:12–21.

Kroppenstedt, R. M. 1985. Fatty acid and menaquinone analysis of actinomycetes and related organisms. 173–199. M. Goodfellow, and D. E. Minnikin (ed.) Chemical methods in bacterial systematics. Academic Press. New York.

Kroppenstedt, R. M., H. J. Kutzner. 1978. Biochemical taxonomy of some problem actinomycetes. Zentralbl. Bakteriol. Parasitenkd. Infektionskr. Hyg. Abt. 1, Suppl. 6:125–133.

Küster, E., S. T. Williams. 1964. Selection of media for isolation of streptomycetes. Nature 202:928–929.

Labeda, D. P. 1987. Actinomycete taxonomy: generic characterization. Dev. Ind. Microbiol. 28:115–121.

Labeda, D. P. 1989. Genus *Glycomyces* Labeda, Testa, Lechevalier and Lechevalier. 2586–2589. S. T. Williams (ed.) Bergey's manual of systematic bacteriology, vol. 4. Williams & Wilkins. Baltimore.

Labeda, D. P., R. T. Testa, M. P. Lechevalier, H. A. Lechevalier. 1985. *Glycomyces,* a new genus of the Actinomycetales. Int. J. Syst. Bacteriol. 35:417–421.

Lacey, J. 1988. Actinomycetes as biodeteriogens and pollutants of the environment. 359–432. M. Goodfellow, S. T. Williams, and M. Mordarski (ed.) Actinomycetes in biotechnology. Academic Press. London.

Lancini, G., F. Parenti. 1982. Antibiotics. An integrated view. Springer-Verlag. New York.

Lechevalier, H. A. 1989. A practical guide to generic identification of actinomycetes. 2344–2347. S. T. Williams (ed.) Bergey's manual of systematic bacteriology, vol. 4. Williams & Wilkins. Baltimore.

Lechevalier, H. A., M. P. Lechevalier. 1967. Biology of actinomycetes. Ann. Rev. Microbiol. 21:71–100.

Lechevalier, M. P. 1988. Actinomycetes in agriculture and forestry. 327–358. M. Goodfellow, S. T. Williams, and M. Mordarski (ed.) Actinomycetes in biotechnology. Academic Press. London.

Lechevalier, M. P., H. A. Lechevalier. 1970a. Composition of whole-cell hydrolysates as a criterion in the classification of aerobic actinomycetes. 311–316. H. Prauser (ed.) The Actinomycetales. VEB Gustav Fischer Verlag. Jena.

Lechevalier, M. P., H. Lechevalier. 1970b. Chemical composition as a criterion in the classification of aerobic actinomycetes. Int. J. Syst. Bacteriol. 20:435–443.

Lechevalier, M. P., H. A. Lechevalier. 1975. Actinoplanete with cylindrical sporangia, *Actinoplanes rectilineatus* sp. nov. Int. J. Syst. Bacteriol. 25:371–376.

Lechevalier, M. P., C. de Bievre, H. Lechevalier. 1977. Chemotaxonomy of aerobic actinomycetes: phospholipid composition. Biochem. Syst. Ecol. 5:249–260.

Lechevalier, M. P., A. E. Stern, H. A. Lechevalier. 1981. Phospholipids in the taxonomy of actinomycetes. Zentralbl. Bakteriol. Mikrobiol. Hyg. Abt. 1, Suppl. 11:111–116.

Luedemann, G. M., B. C. Brodsky. 1964. Taxonomy of gentamicin-producing *Micromonospora*. Antimicrob. Agents Chemother. 1963:116–124.

Luedemann, G. M., C. J. Casmer. 1973. Electron microscope study of whole mounts and thin sections of *Micromonospora chalcea* ATCC 12452. Int. J. Syst. Bacteriol. 23:243–255.

Makkar, N. S., T. Cross. 1982. Actinoplanetes in soil and on plant litter from freshwater habitats. J. Appl. Bacteriol. 52:209–218.

Malabarba, A., P. Strazzolini, A. Depaoli, M. Landi, M. Berti, B. Cavalleri. 1984. Teicoplanin, antibiotics from *Actinoplanes teichomyceticus* nov. sp. J. Antibiot. 37:988–999.

Maluszyńska, G. M., L. Janota-Bassalik. 1974. A cellulolytic rumen bacterium, *Micromonospora ruminantium* sp. nov. J. Gen. Microbiol. 82:57–65.

McCarthy, A. J., P. Broda. 1984. Screening for lignin-degrading actinomycetes and characterization of their activity against ^{14}C-lignin labelled wheat lignocellulose. J. Gen. Microbiol. 130:2905–2913.

Meyertons, J. L., D. P. Labeda, C. L. Cote, M. P. Lechevalier. 1987. A new thin-layer chromatographic method for whole-cell sugar analysis of *Micromonospora* species. The Actinomycetes 20:182–192.

Minnikin, D. E., L. Alshamaony, M. Goodfellow. 1975. Differentiation of *Mycobacterium, Nocardia,* and related taxa by thin-layer chromatographic analysis of whole-organism methanolysates. J. Gen. Microbiol. 88:200–204.

Minnikin, D. E., A. G. O'Donnell. 1984. Actinomycete envelope lipid and peptidoglycan composition. 337–388. M. Goodfellow, M. Mordarski, and S. T. Williams (ed.) The biology of the actinomycetes. Academic Press. London.

Nakajima, M., A. Torikata, Y. Ichikawa, T. Katayama, A. Shiraishi, T. Haneishi, M. Arai. 1983. Mycoplanecins, novel antimycobacterial antibiotics from *Actinoplanes awajinensis* subsp. *mycoplanecinus* subsp. nov. J. Antibiot. 36:961–964.

Nara, T., I. Kawamoto, R. Okachi, T. Oka. 1977. Source of antibiotics other than *Streptomyces*. Japanese J. Antibiot. Suppl. 30:174–189.

Nolan, R. D., T. Cross. 1988. Isolation and screening of actinomycetes. 1–32. M. Goodfellow, S. T. Williams, M. Mordarski (ed.) Actinomycetes in biotechnology. Academic Press. London.

Nonomura, H., M. Hayakawa. 1988. New methods for the selective isolation of soil actinomycetes. 288–293. Y. Okami, T. Beppu, and H. Ogawara (ed.) Biology of actinomycetes '88. Japan Sci. Soc. Press. Tokyo.

Nonomura, H., Y. Ohara. 1969. Distribution of actinomycetes in soil (VI) A culture method effective for both preferential isolation and enumeration of *Microbispora* and *Streptosporangium* strains in soil (part I). J. Ferment. Technol. 47:463–469.

Nonomura, H., S. Takagi. 1977. Distribution of actinoplanetes in soils of Japan. J. Ferment. Technol. 55:423–428.

Nonomura, H., S. Iino, M. Hayakawa. 1979. Classification of actinomycetes of genus *Ampullariella* from soils of Japan. Hakkokogaku 57:79–85.

O'Donnell, A. G., M. Goodfellow, D. E. Minnikin. 1982. Lipids in the classification of *Nocardioides*: reclassification of *Arthrobacter simplex* (Jensen) Lochhead in the genus *Nocardioides* (Prauser) emend. O'Donnell et al. as *Nocardioides simplex* comb. nov.Arch. Microbiol. 133:323–329.

Okami, Y., K. Hotta. 1988. Search and discovery of new antibiotics. 33–67. M. Goodfellow, S. T. Williams, and M. Mordarski (ed.) Actinomycetes in biotechnology. Academic Press. London.

Okami, Y., T. Okazaki. 1972. Studies on marine microorganisms. I. Isolation from the Japan sea. J. Antibiot. 25:456–460.

Okami, Y., T. Okazaki. 1978. Actinomycetes in marine environments. Zentralbl. Bakteriol. Parasitenkd. Infektionskr. Hyg. Abt. 1, Suppl. 6:145–151.

Okami, Y., M. Suzuki. 1958. A simple method for microscopical observation of streptomycetes and critique of *Streptomyces* grouping with reference to aerial structure. J. Antibiot. 11:250–253.

Okazaki, T., Y. Okami. 1972. Studies on marine microorganisms. II. Actinomycetes in Sagami Bay and their antibiotics substances. J. Antibiot. 25:461–466.

Ørskov, J. 1923. Investigations into the morphology of the ray fungi. Levin & Munksgaard Publishers. Copenhagen.

Palleroni, N. J. 1976. Chemotaxis in *Actinoplanes*. Arch. Microbiol. 110:13–18.

Palleroni, N. J. 1979. New species of the genus *Actinoplanes, Actinoplanes ferrugineus*. Int. J. Syst. Bacteriol. 29:51–55.

Palleroni, N. J. 1980. A chemotactic method for the isolation of Actinoplanaceae. Arch. Microbiol. 128:53–55.

Palleroni, N. J. 1983. Biology of *Actinoplanes*. The Actinomycetes 17:46–65.

Palleroni, N. J. 1989. Genus *Actinoplanes* Couch. 2419–2428.S. T. Williams (ed.) Bergey's manual of systematic bacteriology, vol. 4. Williams & Wilkins. Baltimore.

Parenti, F., G. Beretta, M. Berti, V. Arioli. 1978. Teichomycins, new antibiotics from *Actinoplanes teichomyceticus* nov. sp. I. Description of the producer strain, fermenta-

tion studies and biological properties. J. Antibiot. 31:276–283.

Parenti, F., C. Coronelli. 1979. Members of the genus *Actinoplanes* and their antibiotics. Ann. Rev. Microbiol. 33:389–411.

Parenti, F., H. Pagani, G. Beretta. 1975. Lipiarmycin, a new antibiotic from *Actinoplanes* I. Description of the producer strain and fermentation studies. J. Antibiot. 28:247–252.

Patel, M., V. P. Gullo, V. R. Hedge, A. C. Horan, J. A. Marquez, R. Vaughan, M. S. Puar, G. H. Miller. 1987. A new tetracyclone antibiotic from a *Dactylosporangium* species. J. Antibiot. 40:1414–1418.

Peczyńska-Czoch, W., M. Mordarski. 1988. Actinomycete enzymes. 219–283. M. Goodfellow, S. T. Williams, and M. Mordarski (ed.) Actinomycetes in biotechnology. Academic Press. London,

Rowbotham, T. J., T. Cross. 1977. Ecology of *Rhodococcus coprophilus* and associated actinomycetes in fresh water and agricultural habitats. J. Gen. Microbiol. 100:231–240.

Ruan, J., M. P. Lechevalier, C. Jiang, H. A. Lechevalier. 1986. A new species of the genus *Actinoplanes: Actinoplanes minutisporangius* n. sp. The Actinomycetes 19:163–175.

Ruddick, S. M., S. T. Williams. 1972. Studies on the ecology of actinomycetes in soil V. Some factors influencing the dispersal and adsorption of spores in soil. Soil Biol. Biochem. 4:93–103.

Saari, G. C., A. A. Kumar, G. H. Kawasaki, M. Y. Insley, P. J. O'Hara. 1987. Sequence of the *Ampullariella* sp. strain 3876 gene coding for xylose isomerase. J. Bacteriol. 169:612–618.

Sandrak, N. A. 1977. Degradation of cellulose by micromonosporas. Mikrobiologiya 46:478–481.

Schäfer, D. 1973. Beiträge zur Klassifizierung und Taxonomie der Actinoplanaceen. Dissertation, Marburg.

Sharples, G. P., S. T. Williams. 1974. Fine structure of the globose bodies of *Dactylosporangium thailandense (Actinomycetales)*. J. Gen. Microbiol. 84:219–222.

Shearer, M. C. 1987. Methods for the isolation of non-streptomycete actinomycetes. J. Indus. Microbiol. Suppl. 2:91–97.

Shirling, E. B., D. Gottlieb. 1966. Methods for characterization of *Streptomyces* species. Int. J. Syst. Bacteriol. 16:313–340.

Shomura, T., S. Amano, H. Tohyama, J. Yoshida, T. Ito, T. Niida. 1985. *Dactylosporangium roseum* sp. nov. Int. J. Syst. Bacteriol. 35:1–4.

Shomura, T., S. Amano, J. Yoshida, M. Kojima. 1986. *Dactylosporangium fulvum* sp. nov. Int. J. Syst. Bacteriol. 36:166–169.

Shomura, T., M. Kojima, J. Yoshida, M. ItóS. Amano, K. Totsugawa, T. Niwa, S. Inouye, T. ItóT. Niida. 1980. Studies on a new aminoglycoside antibiotic, dactimicin I. Producing organism and fermentation. J. Antibiot. 33:924–930.

Shomura, T., J. Yoshida, S. Miyadoh, T. Ito, T. Niida. 1983. *Dactylosporangium vinaceum* sp. nov. Int. J. Syst. Bacteriol. 33:309–313.

Sneh, B., S. J. Humble, J. L. Lockwood. 1977. Parasitism of oospores of *Phytophthora megasperma* var. *sojae*, *P. cactorum*, *Pythium* sp. and *Aphanomyces euteiches* in soil by oomycetes, chytridiomycetes, hyphomycetes, actinomycetes and bacteria. Phytopathology 67:622–628.

Stackebrandt, E., R. M. Kroppenstedt. 1987. Union of the genera *Actinoplanes* Couch, *Ampullariella* Couch, and

Amorphosporangium Couch in a redefined Genus *Actinoplanes*. Syst. Appl. Microbiol. 9:110–114.

Stackebrandt, E., K-H. Schleifer. 1984. Molecular systematics of actinomycetes and related organisms. 485–504. L. Ortiz-Ortiz, L. F. Bojalil, and V. Yakoleff (ed.) Biological, biochemical, and biomedical aspects of actinomycetes. Academic Press. Orlando.

Stackebrandt, E., C. R. Woese. 1981. The evolution of prokaryotes. 1–31. M. J. Carlile, J. F. Collins, and B. E. B. Moseley (ed.) Molecular and cellular aspects of microbial evolution. Cambridge University Press. Cambridge.

Sutherland, E. D., K. K. Baker, J. L. Lockwood. 1984. Ultrastructure of *Phytophthora megasperma* f. sp. *glycinea* oospores parasitized by *Actinoplanes missouriensis* and *Humicola fuscoatra*. Trans. Br. Mycol. Soc. 82:726–729.

Szaniszlo, P. J. 1968. The nature of the intramycelial pigmentation of some Actinoplanaceae. J. Elisha Mitchell Sci. Soc. 84:24–26.

Terekhova, L. P., O. A. Sadikova, T. P. Preobrazhenskaya. 1977. *Actinoplanes cyaneus* sp. nov. and its antagonistic properties. Antibiotiki 22:1059–1063.

Thiemann, J. E. 1970. Study of some new genera and species of the Actinoplanaceae. 245–257. H. Prauser (ed.) The Actinomycetales. VEB Gustav Fischer Verlag. Jena.

Thiemann, J. E., H. Pagani, G. Beretta. 1967. A new genus of the Actinoplanaceae: *Dactylosporangium,* gen. nov. Arch. Mikrobiol. 58:42–52.

Tille, D., R. Vettermann, H. Prauser. 1982. DNA-DNA reassociation between DNAs of actinoplanetes and related genera. 182. Fifth Int. Sym. Actino. Biol., Abstracts. Oaxtepec. México.

Tomita, K., S. Kobaru, M. Hanada, H. Tsukiura. 1977. Fermentation process. U.S. Patent 4,026,766, May 31, 1977.

Torikata, A., R. Enokita, H. Imai, Y. Itoh, M. Nakajima, T. Haneishi, M. Arai. 1978. Studies on the antibiotics from genus *Actinoplanes*. I. Taxonomy of the producers of three antibiotics and their isolation and identification with azaserine, 5-azacytidine and actinomycins. Ann. Rep. Sankyo Res. Lab. 30:84–97.

Torikata, A., R. Enokita, T. Okazaki, M. Nakajima, S. Iwado, T. Haneishi, M. Arai. 1983. Mycoplanecins, novel antimycobacterial antibiotics from *Actinoplanes awajinensis* subsp. *mycoplanecinus* subsp. nov. I. Taxonomy of producing organism and fermentation. J. Antibiot. 36:957–960.

Tribe, H. T., S. M. Abu El-Souod. 1979. Colonization of hair in soil-water cultures, with especial reference to the genera *Pilimelia* and *Spirillospora* (Actinomycetales). Nova Hedwigia 31:789–805.

Truscheit, E., W. Frommer, B. Junge, L. Müller, D. D. Schmidt, W. Wingender. 1981. Chemie und Biochemie mikrobieller α-Glucosidasen-Inhibitoren. Angew. Chem. 93:738–755.

Uchida, K., K. Aida. 1977. Acyl type of bacterial cell wall: its simple identification by colorimetric method. J. Gen. Appl. Microbiol. 23:249–260.

Van Zyl, W. H. 1985. A study of the cellulases produced by three mesophilic actinomycetes grown on bagasse as substrate. Biotechnol. Bioeng. 27:1367–1373.

Vettermann, R., H. Prauser. 1979. Comparative studies on the isolation of actinoplanetes. Fourth Int. Sym. Actino. Biol., Poster-presentation, Cologne.

Vitiello, F., J. P. Zanetta. 1978. Thin-layer chromatography of phospholipids. J. Chromatogr. 166:637–640.

Vobis, G. 1984. Sporogenesis in the *Pilimelia* species. 423–439. L. Ortiz-Ortiz, L. F. Bojalil, and V. Yakoleff (ed.) Biological, biochemical, and biomedical aspects of actinomycetes. Academic Press. Orlando.

Vobis, G. 1987. Sporangiate Actinoplaneten, Actinomycetales mit aero-aquatischem Lebenszyklus. Forum Mikrobiol. 11:416–424.

Vobis, G. 1989a. Actinoplanetes. 2418–2419. S. T. Williams (ed.) Bergey's manual of systematic bacteriology, vol. 4. Williams & Wilkins. Baltimore.

Vobis, G. 1989b. Genus *Pilimelia* Kane. 2433–2437. S. T. Williams (ed.) Bergey's manual of systematic bacteriology, vol. 4. Williams & Wilkins. Baltimore.

Vobis, G. 1989c. Genus *Dactylosporangium* Thiemann, Pagani and Beretta. 2437–2442. S. T. Williams (ed.) Bergey's manual of systematic bacteriology, vol. 4. Williams & Wilkins. Baltimore.

Vobis, G., H.-W. Kothe. 1985. Sporogenesis in sporangiate actinomycetes. 25–47.K. G. Mukerji, N. C. Pathak, and V. P. Singh (ed.) Frontiers in applied microbiology, vol. 1. Print House (India). Lucknow.

Vobis, G., H.-W. Kothe. 1989. Genus *Ampullariella* Couch. 2429–2433. S. T. Williams (ed.) Bergey's manual of systematic bacteriology, vol. 4. Williams & Wilkins. Baltimore.

Vobis, G., D. Schäfer, H.-W. Kothe, B. Renner. 1986. Descriptions of *Pilimelia columellifera* (ex Schäfer 1973) nom. rev. and *Pilimelia columellifera* subsp. *pallida* (ex Schäfer 1973) nom. rev. Syst. Appl. Microbiol. 8:67–74.

Wagman, G. H., M. J. Weinstein. 1980. Antibiotics from *Micromonospora*. Ann. Rev. Microbiol. 34:537–557.

Wakisaka, Y., Y. Kawamura, Y. Yasuda, K. Koizumi, Y. Nishimoto. 1982. A selective isolation procedure for *Micromonospora*. J. Antibiot. 35:822–836.

Waksman, S. A. 1961. The actinomycetes, vol.2. Williams & Wilkins. Baltimore.

Watson, E. T., S. T. Williams. 1974. Studies on the ecology of actinomycetes in soil—VII. Actinomycetes in a coastal sand belt. Soil Biol. Biochem. 6:43–52.

Weyland, H. 1969. Actinomycetes in North Sea and Atlantic Ocean sediments. Nature 223:858.

Weyland, H. 1981. Distribution of actinomycetes on the sea floor. Zentralbl. Bacteriol. Microbiol. Hyg. Abt. 1. Suppl. 11:185–193.

Willoughby, L. G. 1966. A conidial *Actinoplanes* isolate from Blelham Tarn. J. Gen. Microbiol. 44:69–72.

Willoughby, L. G. 1968. Aquatic Actinomycetales with particular reference to the Actinoplanaceae. Veroeff. Inst. Meeresforsch. Bremerh. Sonderband 3:19–26.

Willoughby, L. G. 1969a. A study on aquatic actinomycetes—the allochthonous leaf component. Nova Hedwigia 18:45–113.

Willoughby, L. G. 1969b. A study of the aquatic actinomycetes of Blelham Tarn. Hydrobiologia 34:465–483.

Willoughby, L. G. 1971. Observations on some aquatic actinomycetes of streams and rivers. Freshwater Biol. 1:23–27.

Willoughby, L. G., C. D. Baker, S. E. Foster. 1968. Sporangium formation in the Actinoplanaceae induced by humic acids. Experientia 24:730–731.

Yaginuma, S., N. Muto, M. Otani. 1979. A-10947, a new peptide antibiotic from *Actinoplanes*. J. Antibiot. 32:967–969.

Yaginuma, S., N. Muto, M. Tsujino, Y. Sudate, M. Hayashi, M. Otani. 1981. Studies on Neplanocin A, new antitumor antibiotic. I. Producing organism, isolation and characterization. J. Antibiot. 34:359–366.

Prokaryotes (2006) 3:654–668
DOI: 10.1007/0-387-30743-5_25

CHAPTER 1.1.10

The Family Actinosynnemataceae

DAVID P. LABEDA

Introduction and Phylogenetic Position

The family Actinosynnemataceae (Labeda and Kroppenstedt, 2000), as currently defined by phylogenetic studies based on the analysis of 16S rDNA sequences, contains the genera *Actinoki-neospora* (Hasegawa, 1988), *Actinosynnema* (Hasegawa et al., 1978), *Lechevalieria* (Labeda et al., 2001), *Lentzea* (Yassin et al., 1995; Labeda et al., 2001) and *Saccharothrix* (Labeda et al., 1984; Labeda and Lechevalier, 1989a). This family lies within the suborder Pseudonocardiniae in the class Actinobacteria (Stackebrandt et al., 1997). Earlier phylogenetic studies (Embley et al., 1988; Warwick et al., 1994) proposed that the genus *Saccharothrix* was associated with the family Pseudonocardiaceae, but this affiliation was not supported statistically in phylogenetic analyses, and diagnostic chemotaxonomic characteristics of *Saccharothrix* species were different from those of taxa whose placement in the Pseudonocardiaceae was well supported.

The genera included in the family Actinosynnemataceae share similar chemotaxonomic characteristics and exhibit a diagnostic whole cell sugar pattern (i.e., galactose, rhamnose and mannose are diagnostic sugars) distinct from the pattern observed for the taxa within the family Pseudonocardiaceae (i.e., arabinose and galactose). The family Actinosynnemataceae forms a coherent group (Fig. 1) within the actinobacterial phylogenetic tree calculated from ribosomal DNA sequences (Labeda and Kroppenstedt, 2000), and on the basis of current data, forms a lineage separate from, but phylogenetically closest to, the Pseudonocardiaceae. Moreover, a diagnostic nucleotide signature pattern of TA (823–975), GC (824–874) is observed in the 16S rDNA sequence for all taxa within the family Actinosynnemataceae.

Within the suborder, related genera (phylogenetically not quite within this family, but which share some properties with family members) include *Actinoalloteichus*, *Crossiella*, *Kutzneria* and *Streptoalloteichus*. These genera may well represent another family within the suborder Pseudonocardiniae, but insufficient taxa have been isolated and described to provide adequate data for resolution of their actual phylogenetic position.

The genera currently circumscribed by the family Actinosynnemataceae are all aerobic, catalase positive, non-acid-fast, and lysozyme resistant. Their common chemotaxonomic properties include type III cell wall composition (*meso*-diaminopimelic acid), whole cell sugar pattern consisting of galactose (in the absence of significant quantities of arabinose) along with varying quantities of rhamnose and mannose, a PII phospholipid profile typically displaying a significant amount of phosphatidylethanolamine, and menaquinones containing nine isoprenoid units (MK-9). The Actinosynnemataceae exhibit a range of morphological and physiological properties, including spore motility (expressed by several genera). Representatives of the family have been primarily isolated from soil or plant material, but their role in these environments is unclear. Some taxa have also been isolated from human or animal sources, but their capacity to cause infection or disease has not been confirmed.

Description of Genera

The genus *Actinokineospora* was described by Hasegawa (1988) to contain a single species, *Actinokineospora riparia*, which produces short aerial mycelia that fragment into motile arthrospores, but exhibit no sporangial structures. Subsequently, Tamura et al. (1995) isolated and described four additional species, *Actinokineospora diospyrosa*, *Actinokineospora globicatena*, *Actinokineospora inagensis* and *Actinokineospora terrae*, from soil and fallen leaves collected in the Yamanashi Prefecture of Japan; these species exhibit the same capacity to produce motile arthrospores from fragmentation of the aerial mycelia.

Hasegawa et al. (1978) described the genus *Actinosynnema* to contain actinomycetes that

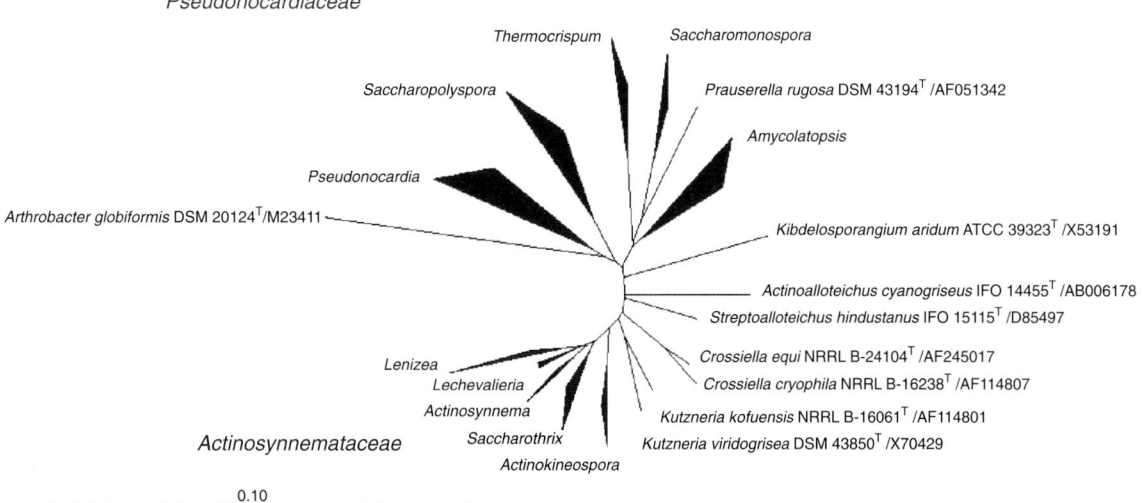

Fig. 1. Radial phylogenetic tree of the suborder Pseudonocardiniae calculated from 16S rDNA sequences using Kimura's evolutionary distance methods (Kimura et al., 1980) and the neighbor-joining method of Saitou and Nei (1987). This tree illustrates the relationship of the family Actinosynnemataceae to the family Pseudonocardiaceae.

were observed to have unique morphological properties. Strains of this genus were observed to produce synnemata or dome-like bodies on most media, and the aerial mycelia produced on the synnemata eventually fragment into peritrichously flagellated motile zoospores in aqueous environments. The genus currently contains *Actinosynnema mirum*, the type species, and *Actinosynnema pretiosum* and its subspecies *Actinosynnema pretiosum* subsp. *auranticum* (Hasegawa et al., 1983). All of the species of this genus formally described to date have been isolated from the surface of blades of grass.

The genus *Lechevalieria* was recently described by Labeda et al. (2001) and contains *Lechevalieria aerocolonigenes* and *Lechevalieria flava*; both species were transferred from *Saccharothrix aerocolonigenes* and *Saccharothrix flava*, respectively, when they were shown to be phylogenetically distinct from either *Saccharothrix* or *Lentzea* and exhibited chemotaxonomic differences from these genera.

The genus *Lentzea* was first described (Yassin et al., 1995) on the basis of a single strain obtained from a tissue specimen from the abdominal cavity of a patient with peritoneal carcinoma. Phylogenetic analysis based on 16S rDNA sequences indicated that this isolate was related to *Actinosynnema* and *Saccharothrix*, which have similar chemotaxonomic characteristics. The pathogenicity of this species was never confirmed, particularly since the description is based on only one isolate. Recently, Lee et al. (2000) proposed that the type species *Lentzea albidocapillata* be transferred to the genus *Saccharothrix* as *Saccharothrix albidocapillata*,

effectively abolishing this genus. A subsequent phylogenetic study (Labeda et al., 2001) reported phylogenetic and chemotaxonomic evidence that supported the revival of this genus and addition of the new species *Lentzea californiensis* and *Lentzea albida*, which had been proposed as the taxon "*Asiosporangium albidum*" (Itoh et al., 1987; Runmao et al., 1995; Tamura and Hatano, 1998). The transfer of *Saccharothrix violacea* and *Saccharothrix waywayandensis* to the genus *Lentzea* as *Lentzea violacea* and *Lentzea waywayandensis*, respectively, added two more species.

The genus *Saccharothrix* was described by Labeda et al. (1984) to contain nocardioform actinomycetes that morphologically resembled *Nocardiopsis* in the ability of their aerial and substrate mycelia to fragment into ovoid, nonmotile elements. *Saccharothrix* strains contain the *meso*-isomer of diaminopimelic acid in their cell walls, but are chemotaxonomically distinct from *Nocardiopsis* by containing rhamnose in addition to galactose as a diagnostic whole cell sugar. When first described, the genus consisted of only one species, *Saccharothrix australiensis*, but currently it consists of four species representing strains originally isolated from nature, i.e., *Saccharothrix australiensis*, *Saccharothrix espanaensis* (Labeda and Lechevalier, 1989), *Saccharothrix texasensis* (Labeda and Lechevalier, 1989b) and *Saccharothrix tangerinus* (Kinoshita et al., 1999), as well as 7 species and subspecies transferred from other genera, i.e., *Saccharothrix mutabilis* subsp. *capreolus* (Grund and Kroppentstedt, 1989), transferred from the genus *Nocardia*; and *Saccharothrix mutabilis* subsp.

mutabilis (Grund and Kroppentstedt, 1989; Labeda and Lechevalier, 1989), *Saccharothrix coeruleofusca* (Grund and Kroppentstedt, 1989), *Saccharothrix coeruleoviolacea* (Grund and Kroppentstedt, 1989), *Saccharothrix longispora* (Grund and Kroppentstedt, 1989) and *Saccharothrix syringiae* (Grund and Kroppentstedt, 1989), transferred from the genus *Nocardiopsis*.

Habitats of Each Genus

Actinokineospora

Strains of the species of *Actinokineospora* have been isolated from plant material or directly from soil. More specifically, *Actinokineospora riparia* and *Actinokineospora terrae* have been isolated from soil samples taken in the proximity of a pond (Hasegawa, 1988; Tamura et al., 1995), strains of *Actinokineospora inagensis* and *Actinokineospora diospyrosa* from fallen leaves (Tamura et al., 1995), and strains of *Actinokineospora globicatena* from both soil samples and fallen leaves collected adjacent to the shores of lakes and ponds (Tamura et al., 1995).

Actinosynnema

The strains of the described species of *Actinosynnema* have all been isolated directly from plant material, primarily the surfaces of blades of grass (Hasegawa, 1978) and sedge (*Carex* species; Hasegawa et al., 1983). Hayakawa et al. (2000) recently reported the use of a new technique that permits the isolation of *Actinosynnema* strains from several different soil samples.

Lechevalieria

The genus *Lechevalieria*, at present, consists of *Lechevalieria aerocolonigenes* and *Lechevalieria flava*, originally isolated from soil samples obtained in Japan and Russia, respectively.

Lentzea

The first described strain of *Lentzea*, the type species *Lentzea albidocapillata*, was isolated from a tissue sample from the peritoneal cavity of a cancer patient (Yassin et al., 1995), but species subsequently added to this genus are from soil samples obtained in China, Korea, and the United States, suggesting that the distribution of genus members in soils is probably global.

Saccharothrix

The original strain of *Saccharothrix australiensis*, the type species of the genus, was isolated from a soil sample from Australia. The genus appears to be ubiquitous in soils and has a worldwide distribution. Isolates described as members of this genus have come from soil samples collected in the United States, Japan, Panama, Africa, and Russia.

Isolation and Cultivation

Strains of *Actinokineospora* have been routinely isolated on humic acid-vitamin agar (HV agar; Hayakawa and Nonomura, 1987), using a modification of the procedure of Makkar and Cross (1982). Soil or leaf litter samples are air dried at 28°C for 7 days, 0.5 g of sample is mixed with 50 ml of sterile tap water, and then the mixture is incubated at 20°C for 55 minutes with occasional shaking. Aliquots (0.1 ml) of dilutions of the supernatant are spread on the surface of plates and incubated at 28°C for 2–3 weeks.

Hayakawa et al. (2000) recently described a new technique for the isolation of actinomycetes having motile zoospores. In this technique, a 0.5 g sample of air-dried soil or leaf litter is placed in a conical beaker (46 mm × 60 mm) and gently flooded with 50 ml of sterile 10 mM phosphate buffer (pH 7.0) containing 10% soil extract. The vessel is loosely covered and incubated statically at 30°C for 90 minutes to permit liberation of motile zoospores. An 8-ml aliquot of the supernatant is transferred to a 16.5 × 105 mm screw cap tube and centrifuged for 20 minutes at 1,500 g at room temperature. The tube contents are allowed to settle for 30 minutes, a portion of the supernatant is diluted serially with sterile tap water, and 0.2 ml aliquots are spread onto the surface of HV medium (containing cycloheximide [50 µg/ml]) with and without the addition of trimethoprim (20 µg/ml) and nalidixic acid (10 µg/ml). The plates are incubated for 2–3 weeks at 30°C and are then observed with a microscope fitted with a high-dry long working distance objective for the tentative identification of putative *Actinokineospora* based on morphological criteria.

Humic Acid-Vitamin (HV) Agar (Hayakawa and Nonomura, 1987)

Humic acid	1.0 g
Na_2HPO_4	0.5 g
KCl	1.7 g
$MgSO_4 \cdot 7H_2O$	50 mg
$FeSO_4 \cdot 7H_2O$	10 mg
$CaCO_3$	10 mg
Agar	15.0 g

Before it is added to the medium, the humic acid is dissolved in 10 ml of 0.2 N NaOH. The medium components are dissolved in 1 liter of distilled water, and the medium is adjusted to pH 7.2, then autoclaved, and finally amended with 1 ml of the filter-sterilized vitamin stock solution after tempering.

Vitamin Stock Solution

Thiamine hydrochloride	0.5 mg
Riboflavin	0.5 mg
Niacin	0.5 mg
Pyridoxine hydrochloride	0.5 mg
Inositol	0.5 mg
Calcium pantothenate	0.5 mg
p-Aminobenzoic acid	0.5 mg
Biotin	0.25 mg

These vitamins (per ml) are dissolved in distilled water and the solution is filter sterilized.

Strains of the genus *Actinosynnema* have been isolated from grass blades placed on yeast extract agar (0.02% yeast extract and 1.5% agar in distilled water). It has been reported that *Actinosynnema mirum* is not inhibited by the presence of nystatin at 100 μg/ml and candicidin at 50 μg/ml, so it might be possible to use these antibiotics to suppress the growth of fungal competitors. The plates containing the grass blades are incubated for 3 weeks at 28°C, after which the agar surface may be covered with various types of growth, but small synnemata can be observed on the grass blade itself using a stereoscopic microscope. These synnemata can be carefully picked off using a sterile loop and transferred to fresh media. The technique of Hayakawa et al. (2000) described above for *Actinokineospora* has also been reported to permit the isolation of *Actinosynnema* strains from soils and plant material.

Strains of *Lechevalieria*, *Lentzea* and *Saccharothrix* have been isolated from soil samples by spreading serial soil dilutions onto the surface of routine selective media (such as 1.5% crude agar and 0.4% casein hydrolysate in tap water) used for the general isolation of actinomycetes. The use of antibiotics to selectively isolate members of this group has also been reported (Shearer, 1987).

Typical actinomycete isolation media, such as AV (arginine-vitamin) agar or starch-casein agar, amended with a combination of penicillin G (5–10 μg/ml) and nalidixic acid (15 μg/ml), have been used to isolate *Saccharothrix* strains selectively (Shearer, 1987).

AV Agar (Nonomura and Ohara, 1969)

L-Arginine	0.3 g
Glucose	1.0 g
Glycerol	1.0 g
K_2HPO_4	0.3 g
$MgSO_4 \cdot 7H_2O$	0.2 g
NaCl	0.3 g
$Fe_2(SO_4)_3$	10 mg
$CuSO_4 \cdot 5H_2O$	1.0 mg
$MnSO_4 \cdot H_2O$	1.0 mg
$ZnSO_4 \cdot 7H_2O$	1.0 mg
Agar	15 g

The pH of the medium is adjusted to 6.4 prior to autoclaving. A vitamin stock solution (1 ml; for composition, see above) is added to each liter of sterilized, tempered medium just prior to dispensing.

Starch-Casein Agar (Kuster and Williams, 1964)

Soluble starch	10.0 g
KNO_3	2.0 g
Casein (vitamin-free)	0.3 g
K_2HPO_4	2.0 g
$MgSO_4 \cdot 7H_2O$	50 mg
NaCl	2.0 g
$FeSO_4 \cdot 7H_2O$	10 mg
$CaCO_3$	20 mg
Agar	18.0 g

The pH of this medium is adjusted to 7.0 to 7.2 prior to autoclaving.

More recently Lee et al. (2000) reported the use of tap water agar and oligotrophic medium (M5) for the isolation of *Lentzea* strains from serial dilutions of soils from a gold mine in Korea.

Oligotrophic Medium M5 (Lee et al., 2000)

Glucose	0.1 g
K_2HPO_4	0.5 g
NaH_2PO_4	0.7 g
KNO_3	0.1 g
$MgSO_4 \cdot 7H_2O$	0.1 g
NaCl	0.3 g
$CaCl_2 \cdot H_2O$	20 mg
$FeSO_4 \cdot 7H_2O$	200 mg
$CuSO_4 \cdot 5H_2O$	90 μg
$MnSO_4 \cdot 4H_2O$	20 μg
$ZnSO_4 \cdot 7H_2O$	180 μg
$CoSO_4 \cdot 7H_2O$	10 μg
$(NH_4)_6Mo_7O_{24} \cdot 4H_2O$	5 μg
H_3BO_3	200 μg
Agar	15 g

The ingredients are dissolved in one liter of tap water, adjusted to pH 7.2 and autoclaved.

Cultivation

Strains of the family Actinosynnemataceae are rather nonfastidious in their growth requirements and can be cultivated on a range of standard media typically used for actinomycetes, such as media described by Pridham et al. (1957) or used by the International *Streptomyces* Project (Shirling and Gottlieb, 1966): inorganic salts-starch agar, yeast extract-malt extract agar, or glycerol-asparagine agar. Two excellent media for the routine cultivation of strains are N-Z amine with soluble starch and glucose medium (ATCC Medium No. 172; Cote et al., 1984) and Bennett's Agar (ATCC Medium No. 174; Cote et al., 1984).

N-Z Amine with Soluble Starch and Glucose (ATCC Medium No. 172)

Glucose	10.0 g
Soluble starch	20.0 g
Yeast extract	5.0 g
N-Z amine type A	5.0 g
$CaCO_3$	1.0 g
Agar	15.0 g
Distilled water	1.0 liter

Adjust pH to 7.3 prior to autoclaving. A good substitute for N-Z amine type A is enzymatic hydrolyzate of casein (cat. no. C0626; Sigma Chemical Company, St. Louis, Missouri, USA).

Bennett's Medium (ATCC Medium No. 174)

Yeast extract	1.0 g
Beef extract	1.0 g
N-Z amine type A	2.0 g
Glucose	10.0 g
Agar	15.0 g
Distilled water	1.0 liter

Adjust pH to 7.3 prior to autoclaving. A good substitute for N-Z amine type A is enzymatic hydrolyzate of casein (cat. no. C0626; Sigma Chemical Company, St. Louis, Missouri, USA).

Preservation of Cultures

Members of the family Actinosynnemataceae can generally be preserved as mycelia/spore suspensions or colony plugs prepared from young plate cultures (7–14 days). Mycelia/spores harvested from liquid grown cultures or agar plugs are suspended in sterile 20% (vol/vol) aqueous glycerol and stored at –40°C or colder. The viability of strains frozen in this manner is quite good for considerable periods of time, but for long-term archival storage, these preparations should be stored in vapor phase liquid nitrogen or the strains lyophilized using standard procedures and stored at 4°C.

Identification Procedures

Morphology, as is typical for actinomycetes in general, plays an important role in the identification of the taxa classified within the family Actinosynnemataceae, particularly those forming distinctive sporulation structures, such as *Actinosynnema* and *Actinokineospora*. The micromorphological properties of the other genera within the family (i.e., *Lechevalieria*, *Lentzea* and *Saccharothrix*) are quite similar and thus are not as useful for generic differentiation. The morphology observed may vary with medium composition, and specific media may be required to encourage the formation of sporulation structures. Media used successfully to culture strains for this purpose are HV agar (Hayakawa and

Nonomura, 1987), tyrosine agar (Shirling and Gottlieb, 1966), and 1.5% agar in tap water. The micromorphology of colonies growing on agar plates can be observed using a long working distance objective (30–40X) and a standard binocular microscope. Agar blocks containing sporulating colonies can also be cut from agar plates, fixed with osmium tetroxide vapor, critical point dried, and mounted on a stub for coating and observation in a scanning electron microscope.

Chemotaxonomic markers have generally been found to be extremely stable and useful features for the classification of actinomycete genera and are likewise useful when applied to the identification of members of the family Actinosynnemataceae. All of the genera within the family share some common properties, such as *meso*-diaminopimelic as the diamino acid in the peptidoglycan of the cell wall, a whole cell sugar pattern containing galactose, a phospholipid pattern including a significant amount of phosphatidylethanolamine, and menaquinones having nine isoprenoid units. Other chemotaxonomic characteristics used to differentiate the genera of the family Actinosynnemataceae are shown in Table 1.

The chemotaxonomic properties of strains can be determined following the detailed procedures outlined by Lechevalier and Lechevalier (1970, 1980) and the thin-layer chromatography procedures suggested by Staneck and Roberts (1974) and Meyertons et al. (1988) for isomers of diaminopimelic acid and whole-cell sugars, respectively. Sugar content within hydrolyzates has also been determined by gas chromatography of alditol acetate derivatives (Englyst and Cummings, 1984; Saddler et al., 1991) and using high performance liquid chromatography (Tamura et al., 1994). Phospholipid content can be determined by the protocol outlined by Minnikin et al. (1984). Menaquinone analyses can be performed using the procedures outlined by Collins et al. (1977), Kroppenstedt (1982), and Tamaoka et al. (1983), and fatty acid profiles can be determined by the method of Kroppenstedt (1985).

Physiological characterization is extremely important in distinguishing between the species within each genus, and it is best evaluated using the methods described by Gordon et al. (1974), Kurup and Schmitt (1973), and Goodfellow (1971).

DNA-DNA hybridization still serves as the definitive means of determining species relatedness in bacteria. Several methods have been used to determine DNA relatedness in the family Actinosynnemataceae. DNA relatedness among species of *Lechevalieria*, *Lentzea* and *Saccharothrix* (as shown in Table 2) has been determined from $C_0t_{0.5}$ values (C_0t is the initial DNA concen-

Table 1. Chemotaxonomic characteristics of the family Actinosynnemataceae and related genera.[a]

Taxon	Whole-cell sugar pattern	Phospholipid[b] type	Phospholipids	Predominant menaquinones
Actinoalloteichus	Galactose, mannose, and ribose	PII	PE, DPG, PMI, PG, PIM	MK-9(H_4)
Actinokineospora	Galactose, mannose, and rhamnose	PII	PE, HO-PE	MK-9(H_4)
Actinosynnema	Galactose and mannose	PII	PE, HO-PE, PI, PIM, DPG	MK-9(H_4) MK-9(H_6)
Crossiella	Galactose, mannose, rhamnose, and ribose	PII	PE, DPG, PI, PIM, PME	MK-9(H_4)
Kutzneria	Galactose, trace rhamnose	PII	PE, HO-PE, PI, DPG	MK-9(H_4)
Lechevalieria	Galactose, mannose, and rhamnose	PII	PE	MK-9(H_4)
Lentzea	Galactose, mannose, and ribose	PII	PE, DPG, PG, PI	MK-9(H_4)
Saccharothrix	Galactose, rhamnose, and mannose (trace)	PII, PIV	PE, HO-PE, PI, PIM, DPG, PG (v)	MK-10(H_4) MK-9(H_4)
Streptoalloteichus	Galactose, mannose, and ribose	PII	PE	MK-10(H_4) MK-10(H_6)

[a]All genera have *meso*-diaminopimelic acid (DAP) as the cell wall diamino acid, are of cell wall chemotype III, and contain straight chain, monounsaturated *iso* and *anteiso* fatty acids.
[b]Phospholipid pattern type *sensu* Lechevalier et al., 1977.
Abbreviations: DPG, diphosphatidylglycerol; PE, phosphatidylethanolamine; PG, phosphatidylglycerol; PG (v), phosphatidylglycerol (variable presence); PI, phosphatidylinositol; PIM, phosphatidylinositol mannosides; PME, phosphatidylmethylethanolamine; HO-PE, phosphatidyl-ethanolamine containing hydroxylated fatty acids; and MK-9 and -10, menaquinones containing 9 and 10 isoprenoid units, respectively.

Table 2. DNA relatedness among species currently and formerly classified within the genus *Saccharothrix*.

Strain	% DNA relatedness to strain					
	NRRL 11239[Ta]	NRRL 15764[T]	NRRL B-16077[T]	NRRL B-16107[T]	NRRL B-16238[T]	NRRL B-3289[T]
Saccharothrix espanaensis NRRL 15764[T]	25					
Saccharothrix mutabilis subsp. *mutabilis* NRRL B-16077[T]	18	0				
Saccharothrix texasensis NRRL B-16107[T]	21	17	12			
Crossiella cryophila NRRL B-16238[T]	6	9	9	4		
Lechevalieria aerocolonigenes NRRL B-3298[T]	10	2	17	0	11	
Lentzea waywayandensis NRRL B-16159	3	12	4	7	4	44

[a]NRRL 11239[T] is the type strain of *Saccharothrix australiensis*.

tration [C_o] times time [t] and $C_ot_{1/2} = C_ot$ at which 1/2 of the DNA has reannealed) for renaturation in 5X standard saline citrate (SSC; 1X SSC is 0.15 M NaCl and 0.015 M trisodium citrate) and 20% dimethylsulfoxide at $T_m - 25°C$ (e.g., 66°C) using the method of Seidler and Mandel (1971) and Seidler et al. (1975) as modified by Kurtzman et al. (1980). More recent studies of DNA relatedness among the species of *Actinokineospora* have been performed fluorometrically using biotinylated DNA by the method of Ezaki et al. (1989).

Identification of Individual Genera

Actinokineospora

The cultural characteristics of all *Actinokineospora* species are rather similar: yellow to orange colonies with no visible aerial mycelia. Observation of mature cultures grown on an appropriate medium, such as HV agar or tyrosine agar, reveals the presence of short aerial hyphae that fragment into chains of spores (Figs. 2 and 3). These arthrospores are rod shaped and

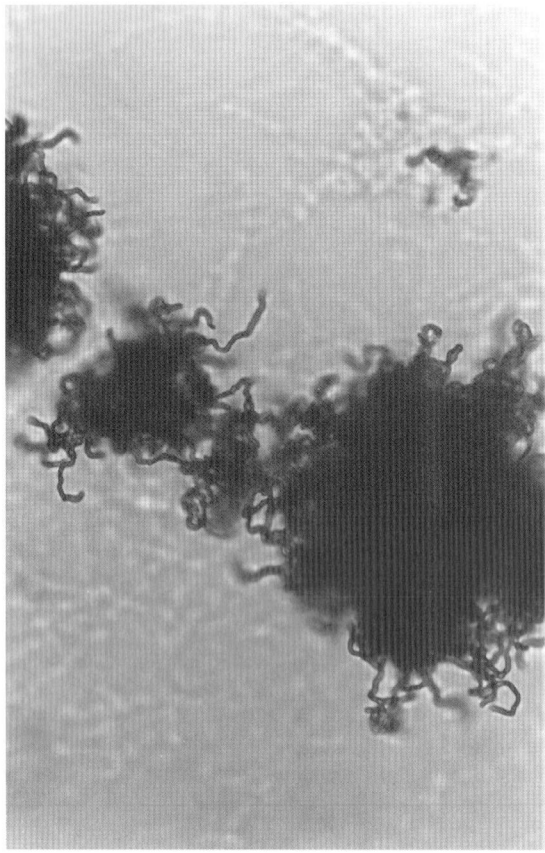

Fig. 2. Light micrograph of *Actinokineospora riparia*. Photograph graciously provided by Dr. Kazunori Hatano, Institute for Fermentation, Osaka. It is from the Atlas of Actinomycetes and is used with the permission of the editor, Shinji Miyadoh, on behalf of the Society for Actinomycetes, Japan.

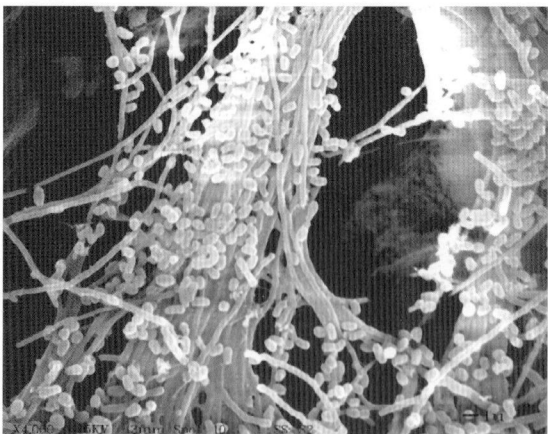

Fig. 3. Scanning electron micrograph of 21-day growth of *Actinokineospora riparia*.

become actively motile when suspended in sterile water.

Actinokineospora strains have galactose as the predominant sugar present in whole cell hydrolyzates, but smaller quantities of rhamnose, mannose and arabinose are also observed. Phosphatidylethanolamine and phosphatidylethanolamine containing 2-hydroxy-fatty acids were observed as the diagnostic phospholipids. The major menaquinone found to be present was MK-9(H_4).

The described species of *Actinokineospora* can be distinguished from each other on the basis of differential physiological properties as shown in Table 3.

Actinosynnema

The cultural characteristics of *Actinosynnema* strains differ between the described species and subspecies, with *A. mirum* producing a pale yellow to yellowish brown substrate mycelium on most media, with or without white to yellowish white sparse aerial mycelia. *Actinosynnema pretiosum* subsp. *pretiosum* produces yellow to pale orange-yellow substrate mycelia, while *A. pretiosum* subsp. *auranticum* produces distinctively yellow to orange substrate mycelia. A sparse white to yellowish-white aerial mycelium, where produced, is observed for all species. The most striking diagnostic characteristic of members of the genus *Actinosynnema* is the production of synnemata (also called "coremia"), or compacted groups of erect hyphae originating from the substrate mycelium which give rise to chains of conidia (Fig. 4). These synnemata can be large (50–200 μm × 200–1,500 μm), and the chains of conidia on these synnemata give rise to peritrichously flagellated rod-shaped to ellipsoidal spores which are motile for about 30 minutes after being introduced into a liquid medium (Fig. 5). Synnemata appear to form in greatest numbers on water agar, tyrosine agar, Bennett's agar, yeast dextrose agar, and TPC agar (Higgins et al., 1967).

Thin Potato-Carrot (TPC) Agar (Higgins et al., 1967)

| Potatoes | 30.0 g |
| Carrots | 2.5 g |

Boil potatoes and carrots in 1 liter of distilled water, filter, and make up to 1 liter with distilled water. Add 15 grams of agar per liter and autoclave.

Actinosynnema strains have galactose as the predominant whole cell sugar, but also contain mannose. The phospholipid content observed includes phosphatidylethanolamine, phosphatidylethanolamine containing 2-hydroxy-fatty acids, phosphatidylinositol, phosphatidylinositol mannosides, and diphosphatidylglycerol. The predominant menaquinones observed are MK-9(H_4) and MK-9(H_6).

Table 3. Differential physiological properties of *Actinokineospora* species.

Characteristic	*Actinokineospora diospyrosa* IFO 15665^T	*Actinokineospora globicatena* IFO 15664^T	*Actinokineospora inagensis* IFO 15663^T	*Actinokineospora riparia* IFO 14541^T	*Actinokineospora terrae* IFO 15668^T
Hydrolysis of:					
Gelatin	+	+	−	−	+
Starch	+	+	−	−	+
Peptonization of milk	+	−	−	−	+
Decomposition of					
Calcium malate	−	+	−	−	+
Utilization of					
D-Mannose	+	+	−	+	±
Sucrose	+	+	−	+	−
Reduction of nitrate	+	+	+	+	−
Growth in NaCl (%)	<3	<2	<4	<4	<4
Formation of clusters	−	+	−	−	−

Data from Tamura et al. (1995).

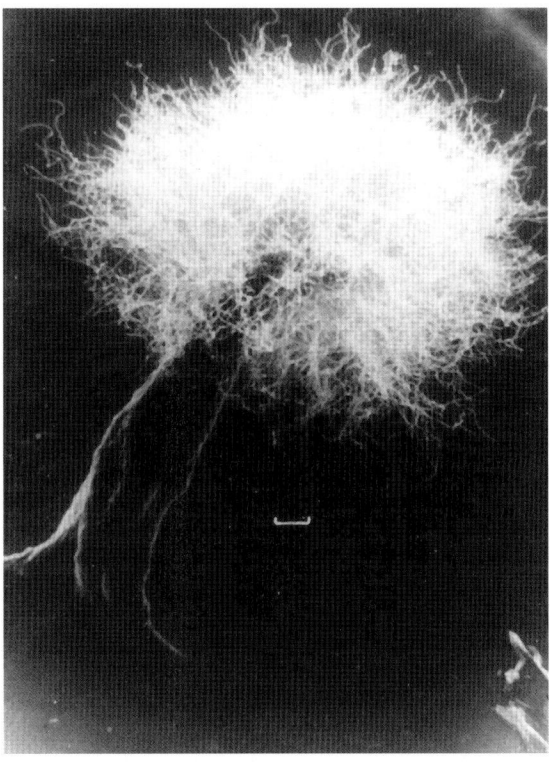

Fig. 4. Synnemata or coremia of *Actinosynnema mirum* growing on tyrosine agar (ISP-7). Photograph graciously provided by Dr. Kazunori Hatano, Institute for Fermentation, Osaka. It is from the Atlas of Actinomycetes and is used with the permission of the editor, Shinji Miyadoh, on behalf of the Society for Actinomycetes, Japan.

The described species of *Actinosynnema* can be distinguished from each other on the basis of differential physiological properties as shown in Table 4.

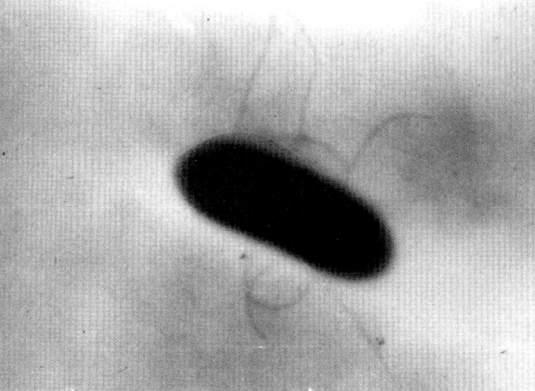

Fig. 5. Peritrichously flagellated motile zoospore from *Actinosynnema mirum*. Photograph graciously provided by Dr. Kazunori Hatano, Institute for Fermentation, Osaka. It is from the Atlas of Actinomycetes and is used with the permission of the editor, Shinji Miyadoh, on behalf of the Society for Actinomycetes, Japan.

Saccharothrix

All strains of *Saccharothrix* are quite similar in morphological appearance. Both substrate and aerial hyphae are approximately 0.5–0.7 μm in diameter and fragment into ovoid bacillary units typical of nocardioform actinomycetes (Fig. 6). The color of the substrate mycelium of *Saccharothrix* species ranges from yellow to yellowish-brown, and that of the aerial mycelium tends to be white to yellowish-white or gray, except for *S. coeruleofusca* and *S. longispora*, which produce blue aerial mycelia on glycerol-nitrate agar. *Saccharothrix mutabilis* is photochromogenic and tends to produce orange-yellow aerial mycelia on many media when cultivated in the light,

Table 4. Differential physiological properties of *Actinosynnema* species.

Characteristic	*Actinosynnema mirum* IFO 14064[T]	*Actinosynnema pretiosum* subsp. *auranticum* IFO 15620[T]	*Actinosynnema pretiosum* subsp. *pretiosum* IFO 15621[T]
Growth at			
10°C	+	−	−
38°C	−	+	+
Utilization of			
Melibiose	−	+	+
Raffinose	−	+	+
Fragmentation of substrate hyphae in liquid media	−	+	+

Data from Hasegawa et al. (1989).

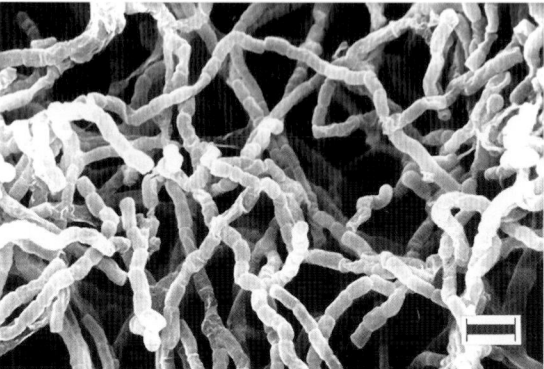

Fig. 7. Scanning electron micrograph of aerial hyphae of a 14-day culture of *Saccharothrix australiensis* on ATCC Medium No. 172. The aerial mycelium has fragmented into ovoid arthrospores and demonstrates typical "zig-zag" morphology. Bar = 1 μm.

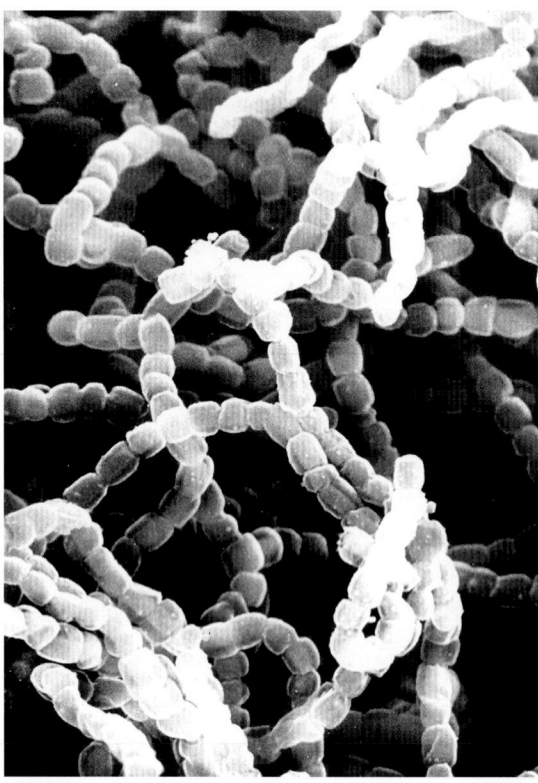

Fig. 6. Scanning electron micrograph of vegetative hyphae of a 28-day culture of *Saccharothrix australiensis* on ATCC Medium No. 172. Note fragmentation of the mycelium into coccoid elements.

while the aerial mycelia are white when cultured in the dark. Soluble pigments are produced by very few species (e.g., *S. australiensis* and *S. syringae*). Aerial hyphae often display the "zig-zag" morphology, which is also typical of the genus *Nocardiopsis* (Fig. 7), thus making the differentiation between these genera rather difficult based on morphological criteria. Aerial mycelium is best observed on minimal media such as 1.5% crude agar in tap water or Czapek's agar (Pridham and Lyons, 1980).

Saccharothrix strains have galactose as the predominant sugar in their whole cell sugar pattern, but also contain rhamnose and a trace of mannose. The predominant phospholipids are phosphatidylethanolamine, phosphatidylethanolamine containing 2-hydroxy fatty acids, phosphatidylinositol and phosphatidylinositol mannoside, and diphosphatidylglycerol. Glucosamine-containing phospholipids of unknown composition have been observed in *Saccharothrix mutabilis* subsp. *mutabilis* and *Saccharothrix espanaensis*. The major menaquinone observed in all species is MK-9(H$_4$), but MK-10(H$_4$) is observed in *Saccharothrix australiensis*, *Saccharothrix coeruleofusca* and *Saccharothrix tangerinus*, while MK-9(H$_2$) is observed in *Saccharothrix longispora*.

Diagnostic nucleotide signature patterns in the16S rDNA gene sequence, CACG (607-610), GTGG (617–620) and GTC (843–845), are extremely useful for differentiating *Saccharothrix* species from *Lentzea* and *Lechevalieria* (Fig. 8). *Saccharothrix tangerinus* is the most

	601	611	621	841	1001
Lechevalieria aerocoloni genes NRRL. B-3298[T]	AAACTTGGGG	CTTAACCCCG	AGCCTGCGGT	ACGTTCTCCG	GAAACCGGTA
Lechevalieria flava NRRL. B-16131[T]	....	....	....	....	....
Lentzea albida IFO 16102[T]	....	......T..A	...T......	...CC...T.	......A...
Lentzea albidocapillata DSM 44073[T]	....	......T..A	...T......	...CC...T.	....G.TC..
Lentzea californiensis NRRL. B-16137[T]	....	......T..A	...T...C..	...CC...T.	......A...
Lentzea violacea IMSNU 50388[T]	....	......T..A	...T......	...CC...T.	....G.TC..
Lentzea waywayandensis NRRL. B-16159[T]	....	......T..A	...T......	...CC...T.	......A...
Actinosynnema mirum DSM 43827[T]	....	....	....	....C.....	......A...
Actinosynnema pretiosum subsp. *pretiosum* NRRL B-16060[T]	....	....	...T......	....C.....	......A...
Saccharothrix australiensis NRRL. 11239[T]	......CAC.	......GTG.	....	....C.....	......TCC.
Saccharothrix coeruleofusca NRRL. B-16115[T]	......CAC.	......GTG.	....	....C.....	......T.C.
Saccharothrix espanaensis NRRL. 15764[T]	......CAC.	......GTG.	....	....C.....	......T.C.
Saccharothrix longispora NRRL. B-16116[T]	......CAC.	......GTG.	.......A..	....C.....	......A...
Saccharothrix mutabilis subsp. *capreolus* DSM 40225[T]	......CAC.	..N..GTG.	....	....C.....	......TCC.
Saccharothrix mutabilis subsp. *mutabilis* DSM 43853[T]	......CAC.	..N...GTG.	....	....C.....	......TCC.
Saccharothrix syringae NRRL. B-16468[T]	......CAC.	......GTG.	....	....C.....	......T.C.
Saccharothrix texasensis NRRL. B-16134[T]	......CAC.	......GTG.	.......A..	....C.....	......T.C.

Fig. 8. Signatures in 16S rDNA sequence for genera in the family Actinosynnemataceae.

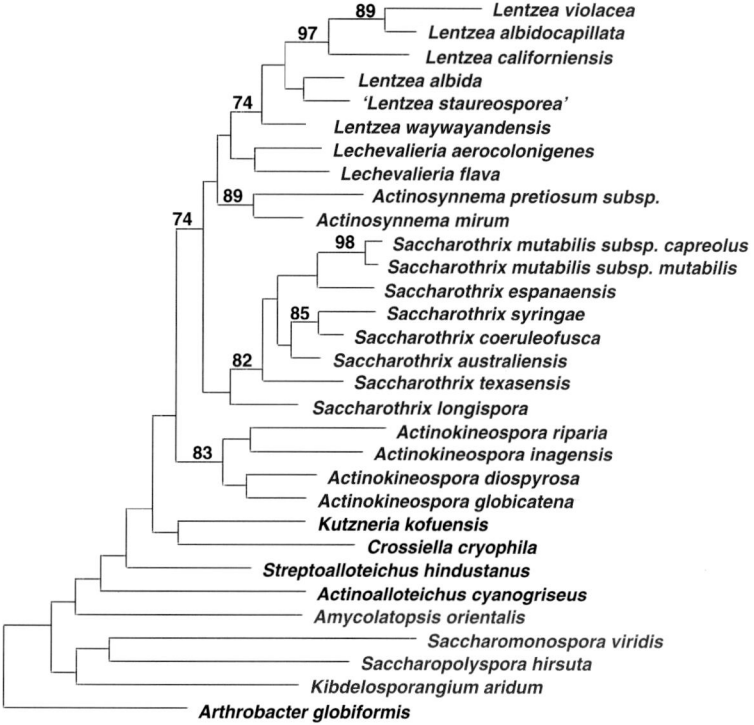

Fig. 9. Phylogenetic dendrogram reconstructed from evolutionary distances (Kimura, 1980) by the neighbor-joining method (Saitou and Nei, 1987) with stability of the groupings estimated by bootstrap analysis (Felsenstein, 1985), indicating the position of species within the family Actinosynnemataceae. Species in blue are within the family Actinosynnemataceae, and species in red are representatives of the family Pseudonocardiaceae. Scale bar represents 0.1 nucleotide substitutions per site.

recently described and validated species of the genus and has the appropriate chemotaxonomic properties. In the phylogenetic tree shown in Fig. 9, this species appears to be the most phylogenetically distinct member of the genus, and the signature patterns within the 16S rDNA gene are slightly different than those of "authentic" *Saccharothrix*. Additional sequence determination will be necessary to establish whether the published sequences are real or in error.

The described species of *Saccharothrix* can be best distinguished from each other on the basis of differential physiological properties as shown in Table 5.

Lentzea

Lentzea strains are quite similar in gross morphology to members of the genera *Saccharothrix* and *Lechevalieria*, producing aerial mycelium, sometimes exhibiting "zig-zag" morphology, and fragmenting into rod-shaped elements. The members of this genus are observed to lack phosphatidylethanolamine containing 2-hydroxy-fatty acids, although they contain phosphatidylinositol and diphosphatidylglycerol, which differentiates them from *Saccharothrix*. They also appear to have galactose, mannose and small quantities of ribose, but no rhamnose in

Table 5. Differential properties of *Saccharothrix* species.

Characteristic	S. australiensis NRRL 11239^T	S. coeruleofuscus DSM 43679^T	S. coeruleoviolacea DSM 43935^T	S. espanaensis NRRL 15764^T	S. longispora DSM 43749^T	S. mutabilis subsp. mutabilis NRRL B-16077^T	S. mutabilis subsp. capreolus DSM 40225^T	S. syringae DSM 43886^T	Sct. tangerinus JCM 10302^T	S. texasensis NRRL B-16134^T
Decomposition of										
Adenine	−	−	−	−	+	−	−	−	−	−
Hippurate	−	−	−	+	−	+	+	+	v	+
Hypoxanthine	−	−	−	+	−	+	+	−	+	+
Starch	−	+	−	−	+	+	+	+	+	+
Tyrosine	+	−	−	−	+	+	+	+	+	v
Urea	−	−	−	−	+	−	−	−	−	−
Production of										
Soluble pigments	+	−	+	−	−	−	−	+	−	−
Nitrate reductase	+	−	+	w	+	+	−	−	−	+
Assimilation of										
Citrate	−	−	+	v	+	+	−	−	+	−
Lactate	w	−	nd	+	+	+	−	−	nd	+
Malate	+	−	nd	+	+	+	+	+	nd	+
Acid from										
Arabinose	−	+	+	−	+	+	+	+	+	+
Dextrin	+	+	+	−	+	+	+	+	+	+
Inositol	−	−	−	−	−	+	+	−	+	+
Lactose	−	+	+	−	+	+	−	+	+	+
Melibiose	−	−	+	−	−	+	+	+	+	+
Raffinose	−	+	−	−	+	+	−	+	+	−
Rhamnose	−	+	+	−	w	−	−	+	w	+
Salicin	−	+	−	−	+	+	+	w	+	+
Sorbitol	+	+	−	−	+	−	−	−	+	−
Sucrose	−	+	−	+	+	+	+	+	+	+
Xylose	−	+	+	v	+	+	+	+	w	+
α-Methyl-D-glucoside	−	+	w	−	−	+	−	−	w	+
Growth in presence of										
4% NaCl	+	+	−	+	+	−	+	+	+	−
5% NaCl	−	+	−	−	+	−	+	+	w	−
Growth at										
37°C	−	+	+	+	+	+	+	+	−	+
45°C	+	+	+	−	−	+	+	+	−	−

Symbols: +, positive in all strains; −, negative in all strains; v, variable reaction; w, weak positive reaction; and nd, not determined.

Table 6. Differential physiological properties of *Lentzea* species.

Characteristic	*Lentzea albidocapillata* NRRL B-24057[T]	*Lentzea albida* NRRL B-24073[T]	*Lentzea californiensis* NRRL B-16137[T]	*Lentzea violacea* IMSNU 50388[Tb]	*Lentzea waywayandensis* NRRL B-16159[T]
Hydrolysis of					
Urea	w	w	+	+	+
Production of					
Nitrate reductase	–	–	+	–	+
Assimilation of					
Acetate	–	+	+	+	+
Citrate	–	+	+	–	+
Lactate	–	–	–	+	+
Malate	+	+	+	–	+
Acid from					
Adonitol	+	+	–	–	+
Cellobiose	+	+	+	–	+
Inositol	+	+	+	–	+
Maltose	+	+	+	–	+
Mannitol	+	+	+	–	+
Raffinose	w	–	+	+	+
Rhamnose	+	+	+	–	+
Sucrose	+	+	+	–	+
Trehalose	+	+	+	–	+
Xylose	+	+	+	–	+
Growth at					
10°C	+	–	+	+	+
37°C	+	+	+	+	w
42°C	–	+	–	–	–
45°C	–	+	–	–	–

Symbols: see footnote in Table 5.
Data from Lee et al. (2000).

their whole cell sugar profiles, which differentiates them from the genera *Saccharothrix* and *Lechevalieria*.

Lentzea species can also be differentiated from *Saccharothrix* and *Lechevalieria* species by virtue of the genus diagnostic nucleotide signature TCCA (617–620) and GCC (843–845) regions in their 16S rDNA gene sequence, as can be seen in Fig. 8.

The substrate mycelium is yellow to yellow-brown in *Lentzea albidocapillata* and *Lentzea californiensis*, yellowish-orange in *Lentzea albida*, violet in *Lentzea violacea*, and pale to dark yellow in *Lentzea waywayandensis*. White aerial mycelium is produced by all species. A reddish-brown soluble pigment is produced on some media by *Lentzea violacea*, while *Lentzea californiensis* produces orange soluble pigments, particularly on Czapek's agar.

The described species of *Lentzea* can be distinguished from each other on the basis of colonial morphology and the differential physiological characteristics shown in Table 6.

Lechevalieria

The species within the genus *Lechevalieria* have had a checkered taxonomic history, with both species having spent some time in at least four different genera, including *Saccharothrix*. *Lechevalieria aerocolonigenes* was originally described as *Streptomyces aerocolonigenes*, while *Lechevalieria flava* was originally described as *Actinomadura flava*. The members of this genus have morphology quite similar to that of *Saccharothrix* and *Lentzea*, i.e., branching vegetative mycelium and rudimentary aerial mycelium that fragments into coccoidal elements. The whole cell sugar pattern consists of galactose and mannose, with traces of rhamnose. The phospholipid pattern consists of significant quantities of phosphatidylethanolamine lacking hydroxylated fatty acids. The sequence of the 16rRNA gene contains genus-specific diagnostic nucleotide signature patterns of TT (844–845) and GGT (1107–1109; Fig. 8).

Both species of *Lechevalieria* exhibit similar gross morphology when growing in culture, with substrate mycelium of yellow shades and only traces of white aerial mycelia, and can be best distinguished through the use of the differential physiological characteristics shown in Table 7.

Applications

New strains of genera within the family Actinosynnemataceae isolated from nature in the

Table 7. Differential physiological properties of *Lechevalieria* species.

Characteristic	*Lechevalieria flava* NRRL B-3298[T]	*Lechevalieria flava* NRRL B-16131[T]
Growth in presence of		
4% NaCl	+	−
5% NaCl	+	−
Utilization of		
Lactate	+	−
Acid from		
Adonitol	+	w*a*
Salicin	+	−
Growth at		
45°C	−	+

Symbols: see footnote in Table 5.

Table 8. Antibiotics produced by species within the family Actinosynnemataceae.

Species	Strain no.	Antibiotic	Reference
Actinosynnema pretiosum subsp. *auranticum*	ATCC 31309	Ansamitocins	Higashide et al., 1977
Actinosynnema pretiosum subsp. *pretiosum*	ATCC 31281	Ansamitocins	Higashide et al., 1977
Lechevalieria aerocolonigenes	ATCC 39243	Rebeccamycin	Bush et al., 1987
Lechevalieria flava	INA 2171	Madumycin	Gauze et al., 1974
Saccharothrix australiensis	NRRL 11239	LL-BM782 complex	Tresner et al., 1980
Saccharothrix espanaensis	NRRL 15764	LL-C19004	Kirby et al., 1987
Saccharothrix mutabilis subsp. *capreolus*	NRRL 2773	Capreomycin	Stark et al., 1967
Saccharothrix mutabilis subsp. *mutabilis*	ATCC 31520	Polynitroxin	Jain et al., 1982
Saccharothrix syringae	INA 2240	Nocamycin	Gauze et al., 1977
Saccharothrix tangerinus	JCM 10302	Formamicin	Kinoshita et al., 1999

future should have significant biotechnological potential as sources of novel and useful secondary metabolites in natural products discovery research. The biosynthetic diversity of members of this family is illustrated by the list of antibiotics produced by *Actinokineospora*, *Actinosynnema*, *Lechevalieria*, *Lentzea* and *Saccharothrix* species (Table 8).

Literature Cited

Bush, J. A., B. H. Long, J. J. Catino, W. T. Bradner, and K. Tomita. 1987. Production and biological activity of rebeccamycin, a novel antitumor agent. J. Antibiot. 40:668–678.

Collins, M. D., T. Pirouz, M. Goodfellow, and D. E. Minnikin. 1977. Distribution of menaquinones in actinomycetes and corynebacteria. J. Gen. Microbiol. 100:221–230.

Cote, R., P.-M. Daggett, M. J. Gantt, R. Hay, S.-C. Jong, and P. Pienta. 1984. ATCC Media Handbook, 1st ed. American Type Culture Collection. Rockville, MD.

Embley, M. T., J. Smida, and E. Stackebrandt. 1988. The phylogeny of mycolate-less wall chemotype IV actinomycetes and description of Pseudonocardiaceae fam. nov. Syst. Appl. Microbiol. 11:44–52.

Englyst, H. N., and J. H. Cummings. 1984. Simplified method for the measurement of total non-starch polysaccharides by gas-liquid chromatography of constituent sugars as alditol acetates. Analyst 108:937–947.

Ezaki, T., Y. Hashimoto, and E. Yabuuchi. 1989. Fluorometric deoxyribonucleic acid-deoxyribonucleic acid hybridization in microdilution wells as an alternative to membrane filter hybridization in which radioisotopes are used to determine genetic relatedness among bacterial strains. Int. J. Syst. Bacteriol. 39:224–229.

Felsenstein, J. 1985. Confidence limits on phylogenies: An approach using the bootstrap. Evolution 39:783–791.

Gauze, G. F., T. S. Maksimova, O. L. Olkhovatova, M. A. Sveshnikova, G. V. Kochetkova, and G. B. Ilchenko. 1974. Production of madumycin, an antibacterial antibiotic, by Actinomadura flava sp. nov. Antibiotiki 9:771–775.

Gauze, G. F., M. A. Sveshnikova, R. S. Ukholina, G. N. Komorova, and V. S. Bashanov. 1977. Production of nocamycin, a new antibiotic, by Nocardiopsis syringae sp. nov. Antibiotiki 22:483–486.

Goodfellow, M. 1971. Numerical taxonomy of some nocardioform bacteria. J. Gen. Microbiol. 69:33–80.

Gordon, R. E., D. A. Barnett, J. E. Handerhan, and C. Pang. 1974. Nocardia coeliaca, Nocardia autotrophica, and the nocardin strain. Int. J. Syst. Bacteriol. 24:54–63.

Grund, E., and R. M. Kroppenstedt. 1989. Transfer of five Nocardiopsis species to the genus Saccharothrix. Syst. Appl. Microbiol. 12:267–274.

Hasegawa, T., M. P. Lechevalier, and H. A. Lechevalier. 1978. A new genus of the Actinomycetales, Actinosynnema gen. nov. Int. J. Syst. Bacteriol. 28:304–310.

Hasegawa, T., S. Tanida, K. Hatano, E. Higashide, and M. Yoneda. 1983. Motile actinomycetes: Actinosynnema pretiosum subsp. pretiosum sp. nov., subsp. nov., and Actinosynnema pretiosum subsp. auranticum subsp. nov. Int. J. Syst. Bacteriol. 33:314–320.

Hasegawa, T. 1988. Actinokineospora: a new genus of the Actinomycetales. Actinomycetologica 2:31–45.

Hasegawa, T., M. P. Lechevalier, and H. A. Lechevalier. 1989. Genus Actinosynnema Hasegawa, Lechevalier, and Lechevalier, 1978a, 204AL. In S. T. Williams, M. E. Sharpe, and J. G. Holt (Eds.) Bergey's Manual of Systematic Bacteriology 4:Williams & Wilkins. Baltimore, MD. 2560–2562.

Hayakawa, M., and H. Nonomura. 1987. Humic acid-vitamin agar, a new medium for the selective isolation of soil actinomycetes. J. Ferment. Technol. 65:501–509.

Hayakawa, M., M. Otoguro, T. Takeuchi, T. Yamazaki, and Y. Iimura. 2000. Application of a method incorporating differential centrifugation for selective isolation of motile actinomycetes in soil and plant litter. Ant. v. Leeuweenhoek 78:171–185.

Higashide, E., M. Asai, K. Ootsu, S. Tanida, Y. Kozai, T. Hasegawa, and T. Kishi. 1977. Ansamitocin, a group of novel maytansinoid antibiotics with antitumor properties from nocardia. Nature 220:721–722.

Higgins, M. L., M. P. Lechevalier, and H. A. Lechevalier. 1967. Flagellated actinomycetes. J. Bacteriol. 93:1446–1451.

Itoh, T., T. Kudo, and A. Seino. 1987. Chemotaxonomic studies on new genera of actinomycetes proposed in Chinese papers. Actinomycetol 1:43–59.

Jain, T. K. 1982. Polynitroxin antibiotics produced by Nocardiopsis mutabilis Shearer sp. nov. US Patent. 4,317,812.

Kimura, M. 1980. A simple method for estimating evolutionary rates of base substitutions through comparative studies of nucleotide sequences. J. Molec. Evol. 16:111–120.

Kinoshita, N., M. Igarashi, S. Ikeno, M. Hori, and M. Hamada. 1999. Saccharothrix tangerinus sp. nov., the producer of the new antibiotic formamicin: A taxonomic study. Actinomycetol 13:20–31.

Kirby, J. P., W. M. Maiese, R. T. Testa, and D. P. Labeda. 1987. Antibiotic LL- C19004. US Patent 4,699,790.

Kroppenstedt, R. M. 1982. Separation of bacterial menaquinones by HPLC using reverse phase (RP18) and a silver loaded ion exchanger as stationary phases. J. Liq. Chromatogr. 5:2357–2359.

Kroppenstedt, R. M. 1985. Fatty acid and menaquinone analysis of actinomycetes and related organisms. In: M. Goodfellow and D. E. Minnikin (Eds.) Chemical Methods in Bacterial Systematics. Academic Press. London, 173–199.

Kurtzman, C. P., M. J. Smiley, C. J. Johnson, L. J. Wickerham, and G. B. Fuson. 1980. Two new and closely related heterothallic species, Pichia amyophila and Pichia mississippiensis: Characterization by hybridization and deoxyribonucleic acid reassociation. Int. J. Syst. Bacteriol. 30:208–216.

Kurup, P. V., and J. A. Schmitt. 1973. Numerical taxonomy of Nocardia. Can. J. Microbiol. 19:1035–1048.

Kuster, E., and S. T. Williams. 1964. Selective media for the isolation of streptomycetes. Nature 202:928–929.

Labeda, D. P., R. T. Testa, M. P. Lechevalier, and H. A. Lechevalier. 1984. Saccharothrix: a new genus of the Actinomycetales related to Nocardiopsis. Int. J. Syst. Bacteriol. 34:426–431.

Labeda, D. P., and M. P. Lechevalier. 1989a. Amendment of the genus Saccharothrix and descriptions of Saccharothrix espanaensis sp. nov., Saccharothrix cryophilus sp. nov., and Saccharothrix mutabilis comb. nov. Int. J. Syst. Bacteriol. 39:419–423.

Labeda, D. P., and A. J. Lyons. 1989b. Saccharothrix texasensis sp. nov. and Saccharothrix waywayandensis sp. nov. Int. J. Syst. Bacteriol. 39:355–358.

Labeda, D. P., and R. M. Kroppenstedt. 2000. Phylogenetic analysis of Saccharothrix and related taxa: Proposal for Actinosynnemataceae fam. nov. Int. J. Syst. Evol. Microbiol. 50:331–336.

Labeda, D. P., K. Hatano, R. M. Kroppenstedt, and T. Tamura. 2001. Revival of the genus Lentzea and proposal for Lechevalieria gen. nov. Int. J. Syst. Evol. Microbiol. 51:1045–1050.

Lechevalier, M. P., and H. A. Lechevalier. 1970. Chemical composition as a criterion in the classification of aerobic actinomycetes. Int. J. Syst. Bacteriol. 34:435–444.

Lechevalier, M. P., C. D. Bievre, and H. A. Lechevalier. 1977. Chemotaxonomy of aerobic actinomycetes: phospholipid composition. Biochem. Syst. Ecol. 5:249–260.

Lechevalier, M. P., and H. A. Lechevalier. 1980. The chemotaxonomy of actinomycetes. In: A. Dietz and D. W. Thayer (Eds.) Actinomycete Taxonomy. Special Publication No. 6. Society for Industrial Microbiology. Arlington, VA. 227–291.

Lee, S. D., E. S. Kim, J.-H. Roe, J.-H. Kim, S.-O. Kang, and Y. C. Hah. 2000. Saccharothrix violacea sp. nov., isolated from a gold mine cave, and Saccharothrix albidocapillata comb. nov. Int. J. Syst. Evol. Microbiol. 50:1315–1323.

Makkar, N. S., and T. Cross. 1982. Actinoplanetes in soil and on plant litter from freshwater habitats. J. Appl. Bacteriol. 52:209–218.

Meyertons, J. L., D. P. Labeda, G. L. Cote, and M. P. Lechevalier. 1988. A new thin-layer chromatographic method for whole-cell sugar analysis of Micromonospora species. The Actinomycetes 20:182–192.

Minnikin, D. E., A. G. O'Donnell, M. Goodfellow, G. Alderson, M. Athalye, A. Schaal, J. H. Parlett. 1984. An integrated procedure for the extraction of isoprenoid quinones and polar lipids. J. Microbiol. Meth. 2:223–241.

Nonomura, H., and Y. Ohara. 1969. The distribution of actinomycetes in soil. VI: A selective plate-culture method for Microbispora and Streptosporangium strains. Part 1. J. Ferment. Technol. 47:463–469.

Pridham, T. G., P. Anderson, C. Foley, F. A. Lindenfelser, C. W. Hesseltine, and R. G. Benedict. 1957. A selection of media for maintenance and taxonomic study of Streptomyces. Antibiotic Ann. 1957/57:947–953.

Pridham, T. G., and A. J. Lyons. 1980. Methodologies for Actinomycetales with special reference to streptomycetes and streptoverticillia. In: A. Dietz and D. W. Thayer (Eds.) Actinomycete Taxonomy. Special Publication No. 6. Society for Industrial Microbiology. Arlington, VA. 153–224.

Runmao, H., L. Junying, and M. Lianjun. 1995. A new genus of Actinomycetales: Asiosporangium gen. nov. Actinomycetes 6:19–22.

Saddler, G. S., P. Tavecchia, S. Lociuro, M. Zanol, E. Colombo, and E. Selva. 1991. Analysis of madurose and other actinomycete whole cell sugars by gas chromatography. J. Microbiol. Meth. 14:185–191.

Saitou, N., and M. Nei. 1987. The neighbor-joining method: A new method for reconstructing phylogenetic trees. Molec. Biol. Evol. 4:406–425.

Seidler, R. J., and M. Mandel. 1971. Quantitative aspects of deoxyribonucleic acid renaturation: base composition, state of chromosome replication, and polynucleotide homologies. J. Bacteriol. 106:608–614.

Seidler, R. J., M. D. Knittel, and C. Brown. 1975. Potential pathogens in the environment: cultural reactions and nucleic acid studies on Klebsiella pneumonia from chemical and environmental sources. Appl. Microbiol. 29:819–825.

Shearer, M. C. 1987. Methods for the isolation of non-streptomycete actinomycetes. Devel. Ind. Microbiol. 28:91–97.

Shirling, E. B., and D. Gottlieb. 1966. Methods for characterization of Streptomyces species. Int. J. Syst. Bacteriol. 16:313-340.

Stackebrandt, E., F. A. Rainey, and N. L. Ward-Rainey. 1997. Proposal for a new heirarchic classification system, Actinobacteria classis nov. Int. J. Syst. Bacteriol. 47:479–491.

Staneck, J. L., and G. D. Roberts. 1974. Simplified approach to identification of aerobic actinomycetes by thin-layer chromatography. Appl. Microbiol. 28:226–231.

Stark, W. M., C. E. Higgens, R. N. Wolfe, M. M. Hoehn, and J. M. McGuire. 1965. Capreomycin, an new antimycobacterial agent produced by Streptomyces capreolus sp. nov. Antimicrob. Agents Chemother. 1962:596–606.

Tamaoka, J., Y. Katayama-Fujimura, and H. Kuraishi. 1983. Analysis of bacterial menaquinone mixtures by high performance liquid chromatography. J. Appl. Bacteriol. 54:31–36.

Tamura, T., Y. Nagagaito, T. Nishii, T. Hasegawa, E. Stackebrandt, and A. Yokota. 1994. A new genus of the order Actinomycetales, Couchiplanes gen. nov., with descriptions of Couchiplanes caeruleus (Horan and Brodsky 1986) comb. nov., and Couchiplanes caeruleus subsp. azureus subsp. nov. Int. J. Syst. Bacteriol. 44:193–203.

Tamura, T., M. Haykawa, H. Nonomura, A. Yokota, and K. Hatano. 1995. Four new species of the genus Actinokineospora: Actinokineospora inagensis sp. nov., Actinokineospora globicatena sp. nov., Actinokineospora terrae sp. nov., and Actinokineospora diospyrosa sp. nov. Int. J. Syst. Bacteriol. 45:371–378.

Tamura, T., and K. Hatano. 1998. Phylogenetic analyses on the strains belonging to invalidated genera of the order Actinomycetales. Actinomycetol. 12:15–28.

Tresner, H. D., A. A. Fantini, D. B. Borders, and W. J. McGahren. 1980. Antibacterial antibiotic BM782. US Patent. 4, 234, 717.

Warwick, S., T. Bowen, H. McVeigh, and T. M. Embley. 1994. A phylogenetic analysis of the family Pseudonocardiaceae and the genera Actinokineospora and Saccharothrix with 16S rRNA sequences and proposal to combine the genera Amycolata and Pseudonocardia in an emended genus Pseudonocardia. Int. J. Syst. Bacteriol. 44:293–302.

Yassin, A. F., F. A. Rainey, H. Brzenzinka, K.-D. Jahnke, H. Wessbrodt, H. Budzikiewicz, E. Stackebrandt, and K. P. Schaal. 1995. Lentzea gen. nov., a new genus of the order Actinomycetales. Int. J. Syst. Bacteriol. 45:357–363.

Prokaryotes (2006) 3:669–681
DOI: 10.1007/0-387-30743-5_26

CHAPTER 1.1.11

The Families Frankiaceae, Geodermatophilaceae, Acidothermaceae and Sporichthyaceae

PHILIPPE NORMAND

Introduction

The four families, Frankiaceae, Geodermatophilaceae, Acidothermaceae and Sporichthyaceae, embrace bacteria that have been poorly studied for several years, mostly owing to their slow growth rate and fastidious growth requirements. Apart from this common trait, the four families have little in common, which is why they had been considered unrelated until a new hierarchic classification system was proposed (Stackebrandt et al., 1997). These families, together with Microsphaeraceae, are today placed in the suborder Frankineae. The best known of these, the family Frankiaceae, comprises nitrogen-fixing bacteria that infect the roots of a number of woody dicotyledonous plants and induce the synthesis of nodules, studied for many years since their description 135 years ago (Woronin, 1866). The first confirmed isolation in pure culture of a *Frankia* strain was reported only in 1978 (Callaham et al., 1978) despite almost a century of unsuccessful efforts. The other bacteria were isolated from various environments such as soil (Geodermatophilaceae), thermal springs (Acidothermaceae), or compost (Sporichthyaceae). *Geodermatophilus* was grouped together with *Dermatophilus* in the family Dermatophilaceae until 16S rDNA studies revealed their phylogenetic unrelatedness (Stackebrandt et al., 1983). Only *Geodermatophilus* was considered morphologically related enough to be grouped together with *Frankia* in the family Frankiaceae. It was only later, with the advent of molecular techniques, that all these families were determined to be relatives of *Frankia*.

Phylogeny

The first phylogenetic study made on members of these four families was based partly on 16S reverse sequencing and mostly on 16S rRNA oligonucleotides sequencing (Hahn et al., 1989). The main conclusion of this work was that *Geodermatophilus* (Luedemann, 1968) and *Blastococcus* isolated from the Baltic Sea (Ahrens and Moll, 1970) were close relatives of *Frankia*, but the study also confirmed earlier work that showed the genus *Dermatophilus* formed a separate lineage, even though this genus shared a rare morphological feature (i.e., having hyphae dividing in more than one plane to produce multilocular sporangia).

From then on, several (more than 310 in April 2002) 16S rRNA genes have been deposited in sequence databanks and a comprehensive phylogenetic study was made (Normand et al., 1996), showing strains in genus *Frankia* could be grouped into four clusters (Fig. 1) that correspond roughly to host specificities: cluster 1 comprises strains infective on *Alnus* and also strains infective on *Casuarina*; cluster 2 comprises strains not cultivated in pure culture but present in the root nodules of members of plant families Rosaceae, Datiscaceae, and Coriariaceae; cluster 3 comprises strains infective on Eleagnaceae and *Gymnostoma*; and cluster 4 comprises a number of diverse strains isolated from different host plants but unable to fulfill Koch's postulates and also a few strains that infect *Alnus* but not efficiently (Fig. 1). The largest distance in the 16S rRNA gene between any two *Frankia* strains was around 4% (Normand et al., 1996).

As expected, given the unusual ecological niche and the particular morphological features, all *Frankia* strains formed a coherent cluster with *Geodermatophilus* and *Blastococcus*, which are close phylogenetic neighbors. Surprisingly, *Frankia* had even closer relatives sharing hardly any morphological features with it: *Sporichthya polymorpha* isolated from compost (Rainey et al., 1993b) and *Acidothermus cellulolyticus* isolated from hot springs (Rainey and Stackebrandt, 1993a). The distances between *Geodermatophilus* and the different *Frankia* 16S rRNA sequences are 5.3–7.2%, while the distances between *Acidothermus* and the *Frankia* sequences are 4.8–6.5% (Normand et al., 1996).

These findings were found sufficiently surprising to warrant a new study. The strains were

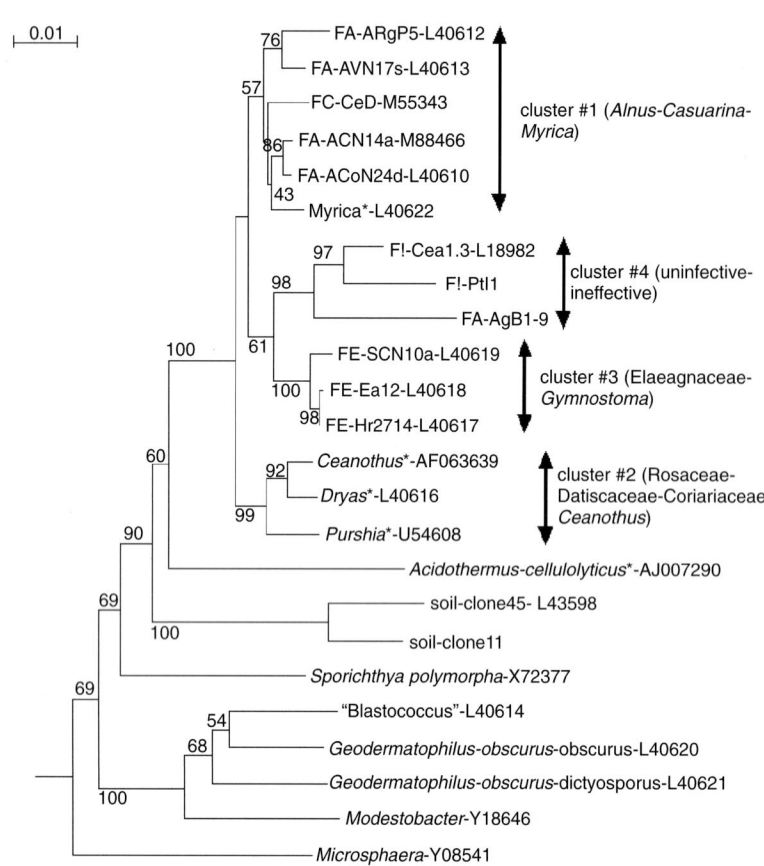

Fig. 1. Neighbor-joining (NJ) phylogenetic tree (Saitou and Nei, 1987) of the 16S rRNA gene of representative isolates belonging to the four families: Frankiaceae, Acidothermaceae, Geodermatophilaceae and Sporicthyaceae. In the *Frankia* isolates, FA refers to strains infective on Alnus, FE to those infective on *Eleagnus*, FC to those infective on *Casuarina*, F! to uninfective strains, and the asterisk to unisolated amplificates from plant tissues. *Kineococcus* (AB007420) was used as outgroup. The numbers beside the clusters are those previously given by Normand et al. (1996). Bootstrap replicates (Felsenstein, 1985) above 50% are indicated above the nodes. Bar represents 0.01 substitution/site.

reordered from the American Type Culture Collection (ATCC), their 16S rRNA sequenced, and their *recA* genes also studied. Several errors in the 16S rRNA gene were corrected that did not modify the phylogenetic topology, but *Acidothermus* was indeed found to be closer to *Frankia* than to *Geodermatophilus* (Maréchal et al., 2000). Besides, the *recA* gene was found to confirm the 16S rRNA gene-derived phylogeny showing close proximity of *Acidothermus* to *Frankia* (Fig. 2).

Other genes have been used to study the phylogenetic relations of *Frankia* strains, such as the *nif* genes coding for nitrogenase (Navarro et al., 1997), and the proximity of the *Gymnostoma*-infective strains to cluster 3 was thereby shown; however, these genes are absent in phyletic neighbors and also in several *Frankia* strains of cluster 4. This is also the case for *gln*II that codes for the assimilatory glutamine synthetase (Cournoyer and Lavire, 1999), but its use has remained modest despite its phylogenetic potential.

Some anomalies in phylogenetic groupings were noted when 16S rDNA trees based upon partial and complete genes were compared. These anomalies concerned mainly the strains infective on *Casuarina* (cluster 1). When partial

16S rRNA sequences are used (lacking 300 nucleotides from the 5′ end), these *Casuarina*-infective *Frankia* are separated from the *Alnus*-infective *Frankia* and group close to cluster 2 (Huguet et al., 2001). This topology with the *Casuarina*-infective *Frankia* close to the *Alnus*-infective *Frankia* appears to be consistent with conclusions drawn from studies of morphology and the sequence of the 23S rRNA insertion (Hönerlage et al., 1994).

Given that the bona fide *Frankia* strains form a coherent phylogenetic cluster relative to its neighbors, a *Frankia*-specific 16S rDNA primer was identified and tested. Only one region in the 5′ part of the gene was found appropriate and was used in combination with a universal primer at the 3′ part of the gene. All *Frankia* strains tested yielded positive amplicons with pure culture DNA, while DNA from non-*Frankia* strains was not amplified. Consequently, the primer pair was used to follow *Frankia* in soil, along a chronosequence of soil colonization by *Alnus* in the Alps. Several positive amplicons, two of which originated from the *Alnus* rhizosphere, were obtained and found to belong to undescribed close relatives of *Frankia* with about 5% distances in the 16S rRNA gene (Normand and Chapelon, 1997). This finding emphasized the

Fig. 2. Neighbor-joining (NJ) phylogenetic tree (Saitou and Nei, 1987) of the *rbcL* gene of representative actinorhizal species belonging to the eight families: Betulaceae, Casuarinaceae, Myricaceae, Rosaceae, Elaeagnaceae, Rhamnaceae, Coriariaceae and Datisaceae. *Pinus pinea* (AB019822) was used as outgroup. The numbers beside the family abbreviations refer to the cluster (Fig. 1) to which the majority (large characters) or a minority (small characters) of *Frankia* strains belong. Bootstrap replicates (Felsenstein, 1985) above 50% are indicated above the nodes. Bar represents 0.01 substitution/site.

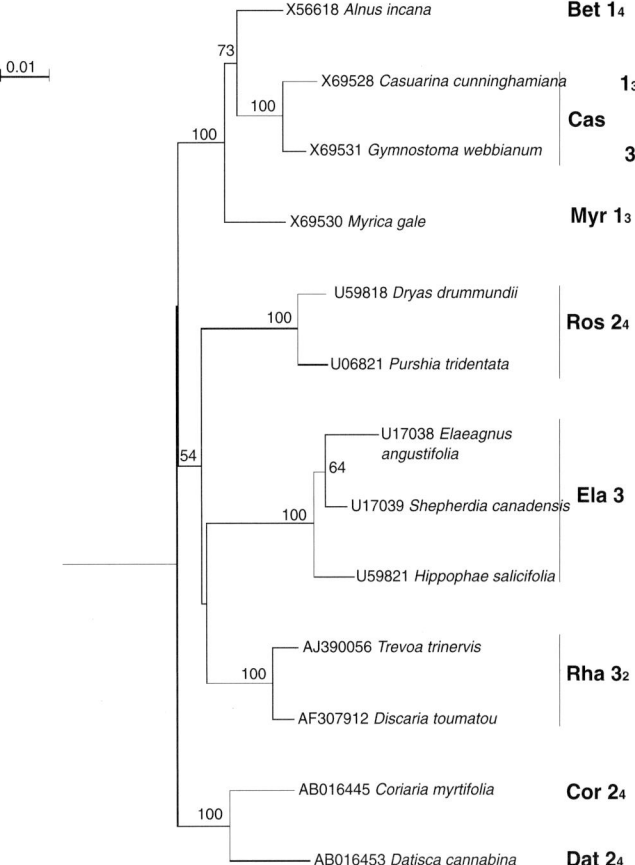

previous notion that the group of actinomycetes described here contains slow growers and may remain poorly defined until an appropriate screening strategy has been developed.

Some isolates are listed in Genbank as *Frankia* strains although a phylogenetic analysis shows that they do not belong to the cluster. This is true of strains G48 (accession number L11306) and L27 (M59075) described as *Frankia* strains isolated from *Podocarpus*, a plant genus belonging to the Coniferales for which nodulation by *Frankia* has been claimed (Benson et al., 1996). As this claim was later dropped, there is now a general agreement among scientists working on *Frankia* that *Podocarpus* is not in symbiosis with *Frankia*. The isolates that appear to belong to the actinomycetes genus *Micromonospora* are still wrongly labeled in Genbank.

One major phylogenetic question that arises from the study of the actinorhizal symbiosis is whether the phylogenetic trees obtained from the symbionts and the host organisms are sufficiently congruent to indicate coevolution. Indeed, the congruency between the 16S rRNA gene of the bacteria and the *rbcL* gene of the host plants is relatively convincing (Jeong et al., 1999). However, under closer scrutiny, several deviations (Fig. 2) from complete congruence

are noted: 1) among the Casuarinaceae, *Gymnostoma* is indeed close to *Casuarina* yet it is nodulated by strains of cluster 3; 2) Elaeagnaceae and Rhamnaceae both belong to the Rhamnales, yet the former is nodulated by strains of cluster 3, while genus *Ceanothus* of the Rhamnaceae (Clawson et al., 1998) is nodulated by strains of cluster 2; and 3) Rosaceae that are nodulated by strains of cluster 2 are closer to the Rhamnales nodulated mainly by strains of cluster 3 than to the Coriariaceae and Datiscaceae nodulated by strains of cluster 2. Finally, *Alnus* is nodulated by strains of clusters 4 and 1, while *Myrica* is nodulated by strains of clusters 3 and 1. All these inconsistencies indicate that the evolution of *Frankia* did not proceed in a linear fashion, dominated by gradual plant evolution, but rather that other factors (such as climate or soil formation) may have played a major role.

The genus *Geodermatophilus*, which was originally described for isolates obtained from desert soils, has new isolates regularly added to it. These are obtained from various environments (soil, stones, walls, etc.) and the current tally is 33 16S rDNA entries in Genbank. *Blastococcus*, which was isolated from the Baltic Sea (Ahrens and Moll, 1970) and which has very few phenotypic

Table 1. Discriminatory features of members of Frankineae.

	Frankia	Geodermatophilus	Acidothermus	Sporichthya
Hyphae	+	+	−	+/−
Spore motility	−	+	−	+
Multilocular sporangia	++	+	−	−
Growth temperature °C	25–30	24–28	55	28
N$_2$-fixation	Most	−	−	−
Nitrate reduction	−	+/−	?	?

Symbols: +, present; −, absent; and +/−, occasionally present.

features that distinguish it from *Geodermatophilus*, may eventually be reclassified as a member of *Geodermatophilus*, inasmuch as the single known strain was placed by 16S rRNA gene analysis in the middle of the *Geodermatophilus* cluster (Fig. 1). *Modestobacter multiseptatus* isolated from polar soil (Mevs et al., 2000) is positioned at the bottom of the Geodermatophilaceae cluster.

All these families (Frankiaceae, Geodermatophilaceae, Acidothermaceae and Sporichthyaceae) are listed on the National Center for Biotechnology Information (NCBI) taxonomy page as belonging to the Frankineae suborder as defined recently (Stackebrandt et al., 1997). Other families (not treated in this chapter) of this suborder are Kineosporiaceae (*Cryptosporangium*, *Kineococcus* and *Kineosporia*) and Microsphaeraceae (*Microsphaera*).

Taxonomy

The discriminatory phenotypic features of the families Frankiaceae, Geodermatophilaceae, Acidothermaceae and Sporichthyaceae are shown in Table 1.

Frankia strains are characterized as producing branched hyphae, multilocular sporangia and nonmotile spores. Most strains will have an optimal growth temperature of 25–30°C. Most strains will grow on nitrogen-free medium, except of course those belonging to cluster 4.

Species description of genus *Frankia* has taken a notably long period to achieve owing to the slow growth rate of many strains, and several actinorhizal plants have not yielded isolates even able to fulfill Koch's postulates (see below). The first DNA-DNA reassociations, made on a small number of isolates, resulted in the expected conclusion that isolates belonged to more than one species (An et al., 1985). Several isolates had high levels of similarity with an *Alnus*-infective strain, several with a *Casuarina*-infective strain, and none with an *Elaeagnus*-infective strain. A more comprehensive study (Fernandez et al., 1989) was done with 35 isolates. Within the group of isolates infective on *Alnus*, one genomic species dominated numerically and was named

"*Frankia alni*" by Becking (1970). Two other *Alnus*-infective genomic species emerged with four and one representatives. The group of strains infective on *Elaeagnus* comprised 5 genomic species, while the group of strains infective on *Casuarina* comprised only a single species. This is in line with the published diversity in the *nifH* restriction fragment length polymorphism (RFLP) patterns of these two groups of strains (Nazaret et al., 1989). Three other papers (Bloom et al., 1989; Akimov and Dobritsa, 1992; Lumini et al., 1996) have been published that show a profusion of genomic species defined by DNA-DNA reassociation values, but no species names have been proposed. Results of these studies agreed with the characterization of the *nif* region, done in parallel with hybridization studies.

Only one species, *Frankia alni*, was named because of the few available phenotypic characters that would circumscribe the genomic clusters identified by DNA-DNA hybridization data. This is true for both *Alnus*-infective strains and especially *Elaeagnus*-infective strains. Of these, the group of strains infective on *Casuarina* merit species status, and the strains sharing several phenotypic features (host spectrum and DNA sequence) should be named *Frankia casuarinae*. Discriminatory phenotypic features are listed below together with group-specific primers that target a hypervariable region located around position 990 in the 16S rRNA gene (Bosco et al., 1992; Table 2).

In the Geodermatophilaceae, species (and genera) have been named without DNA-DNA hybridization as recommended by the International Committee on Systematics of Prokaryotes. In genus *Geodermatophilus*, only one species has been named, *Geodermatophilus obscurus*, on the basis of phenotypic characterization, and it contains several subspecies (*amargosae*, *dictyosporus*, *obscurus*, and *utahensis*; Luedemann, 1968; Table 3).

Isolation

Even though nowadays several hundreds of *Frankia* isolates are available, isolation in pure

Table 2. Discriminant phenotypic features of groups of *Frankia* species.

	Species #1–3	Species #4–8	Species #9
Hyphae	Straight	Irregular	Straight, torulose
Pigment production	Rare	Frequent, orange	Rare
Sporangia	Large (5–50 μM)	Small (<5 μM)	Small (<5 μM)
Host	*Alnus*	*Elaeagnus*	*Casuarina*
16S rRNA signature	GGGGTCC<u>G</u>TA<u>A</u>GGGTC	GGGGTCC<u>T</u>TA<u>G</u>GGG<u>CT</u>	GGGGTCC<u>G</u>TA<u>A</u>GGGTC

Table 3. Discriminatory phenotypic features of the *Geodermatophilus obscurus* subspecies.

	amargosae	dictyosporus	obscurus	utahensis
Colony color	Black	Pink >dark brown	Black	Black
Spore motility	+/−	+	+	+
Gelatin	+/−	+	−	−
Inositol	+	−	+	+
β-Lactose	−	+	−	−
Nitrate reduction	−	+/−	+/−	+

Symbols: +, present; −, absent; and +/−, occasionally present.

culture of *Frankia* strains remains a difficult task. The isolation varies from one host plant to the other: the easiest to work with are the Elaeagnaceae and the most difficult are the Coriariaceae, Datiscaceae and Rosaceae. In these three plant families and in the genera *Allocasuarina* (Casuarinaceae) and *Ceanothus* (Rhamnaceae), no isolate is able to fulfill Koch's postulates. Though some of these isolates do belong to the genus *Frankia*, they do not fix nitrogen (Mirza et al., 1994).

Several procedures for isolating *Frankia* have been published and some, especially the historic ones, were complex with mechanical and enzymatic dissection steps (Callaham et al., 1978). Subsequent protocols were much simpler. Although time consuming, the dissection step is nevertheless considered crucial to avoid isolation of "atypical" isolates present in the outer cortical layers and unable to fulfill Koch's postulates (H. G. Diem, personal communication). The addition of activated charcoal or polyvinylpolypyrrolidone appears to increase the probability of obtaining bacterial growth because these agents remove inhibiting plant phenolic compounds (Lechevalier and Lechevalier, 1990). Different media, both liquid and with agar, have been used successfully to obtain isolates from nodule tissues of *Alnus, Elaeagnus, Casuarina,* etc., and in one instance, directly from the soil (Baker and O'Keefe, 1984). Most strains are microaerophilic, and thus agitation of liquid medium should be avoided or a soft agar overlay used in the case of solid medium. Most isolation attempts use complex media, for instance yeast dextrose agar (Baker and Torrey, 1979) or QmodB medium that contains lecithin (Lalonde and Calvert, 1979), although use of simpler media such as tap water agar has been advocated

(Lechevalier and Lechevalier, 1990) to reduce the growth of competing microbes. This notion raises a crucial point relating to the purity of the cultures used, especially when a poor medium has been used in the first steps of isolation. Most strains presently in use have not been derived from single cells, and thus it is not surprising to still find slow-growing contaminants, even after years of study. Optimal temperature for growth is 28°C, although tropical strains (e.g., from Casuarinaceae) have a higher optimum temperature. Incubation proceeds for 3–4 weeks before putative *Frankia* isolates are checked under the microscope and subcultured.

Geodermatophilus was first isolated from desert and forest soils (Luedemann, 1968) on a dilute medium containing yeast extract 0.1%, glucose 0.1%, soluble starch 0.1%, CaCO₃ 0.1% and 1.5% agar (to prevent overgrowth of fungi and bacteria following incubation at 28°C for 2–3 weeks). After microscopic observation, colonies were subcultured on richer medium (yeast extract, 0.5%; mixture of amines, 0.5%; glucose, 1%; soluble starch, 2%; CaCO₃, 0.1%; and 1.5% agar). A recently described technique (Hayakawa et al., 2000), based on the fact that *Geodermatophilus* has zoospores that are motile by means of terminal flagella, includes a gentle centrifugation step that sediments nonmotile actinomycetes prior to plating on a medium selective for actinomycetes. A recent paper (Mevs et al., 2000) describes the isolation from Antarctic soils on an oligotrophic medium of a bacterium designated "*Modestobacter multiseptatus,* sp. nov., gen. nov.," a close relative of *Geodermatophilus*. Isolates belonging to genus *Geodermatophilus* are now regularly described from dry soils and also from rock surfaces (Eppard et al., 1996; Urzi et al., 2001).

Acidothermus cellulolyticus is available as a single strain (ATCC 43068^{T}), although three strains were isolated initially during the screening program aimed at exploiting the natural microbial biodiversity in Yellowstone National Park (Mohagheghi et al., 1986). To obtain microbial isolates that could efficiently convert biomass into alcohol for fuel, samples of mud and decaying wood were inoculated in liquid medium containing cellulose as sole carbon source and incubated at 55°C and pH 5.2. No other isolate has been described, even though the procedure appears straightforward and the potential benefits are important.

Isolation of *Sporichthya polymorpha* was first reported in 1968 (Lechevalier et al., 1968) from soil samples using a dilute medium (tap water or one-tenth strength Czapek agar). This organism is considered a rare actinomycete as only five isolates have been obtained over 20 years. Recently, on the basis of the fact that *Sporichthya* also has motile zoospores, the same gentle centrifugation technique followed by plating on selective medium previously used for the isolation of *Geodermatophilus* was used and found successful for *Sporichthya* (Hayakawa et al., 2000). Suzuki (1999) used the same technique together with a highly selective medium (humic acid-vitamin agar [HVA] that contains soil humic acid as the sole source of carbon and nitrogen supplemented with gellan gum and calcium chloride to stimulate formation of spores and aerial mycelium) and reported numerous isolates of *Sporichthya*.

Preservation and Cultivation

Most *Frankia* strains are microaerophilic, grow slowly, and have an optimal growth temperature in the range of 25 to 33°C (Lechevalier and Lechevalier, 1990). This is why *Frankia* strains are routinely cultivated in liquid media. For particular purposes such as germination of spores to obtain single cell isolates (Prin et al., 1991), regeneration of protoplasts (Normand et al., 1987; Tisa and Ensign, 1987), or to assess resistance to metals (Richards et al., 2002), cells are in general embedded in a soft (0.8%) agar overlay maintained at 42°C before plating. For routine purposes, several media are used, including those already described for isolation, the defined propionate medium (DPM; Baker and O'Keefe, 1984) or Biotin-Ammonium-Propionate medium (BAP; Murry et al., 1984).

The most frequent preservation method of *Frankia* strains is maintenance in rich liquid medium with infrequent (once or twice per year) subcultures. Some comprehensive lyophilization programs were dropped when reports indicated

that reculture of such lyophilized material was often problematic. Skim milk suspension that appears to contain some inhibitory compounds should be avoided, and instead, bacteria in spent medium should be used (Lechevalier and Lechevalier, 1990). CpI1, the first strain that was deposited in the American Type Culture Collection (ATCC) and seemingly kept as lyophilized material, was found not to be viable and is now absent from the list of available strains. The only strain listed as available from the ATCC is AvcI1, a cluster 1 *Alnus*-infective strain (Baker et al., 1980), available freeze-dried.

Frankia spores are not able to withstand high temperatures. The highest reported temperature (55°C) is only slightly higher than the highest temperature (52°C) tolerated by hyphae (Lalonde and Calvert, 1979). Some strains isolated from *Casuarina* produce particular hyphae called "torulose hyphae," which have thickened walls, are produced in old cultures, and can be subcultured (Diem and Dommergues, 1985). Diazo-vesicles also have the potential to differentiate into hyphae (Schultz and Benson, 1989).

Geodermatophilus is relatively easy to grow at 28°C on solid rich medium (e.g., 0.5% yeast extract, 0.5% amine mixture, 1% glucose, 2% soluble starch, 0.1% CaCO$_3$, and 1.5% agar; Luedemann, 1968) or on TYB (tryptone, yeast extract, glucose and NaCl at pH 7; Ishiguro and Wolfe, 1970). Carbohydrate utilization in *Geodermatophilus* isolates is in general broad: L-arabinose, D-galactose, D-glucose, D-levulose, D-mannitol, sucrose and D-xylose (Luedemann, 1968).

Acidothermus is more tedious to work with mainly because of evaporation problems connected with its optimal growth above 55°C. Good growth occurs in low phosphate basal salts medium (Mohagheghi et al., 1986) containing yeast extract and cellulose. Different optimum temperatures have been reported but the average was 58°C, while the optimum pH was 5.0.

Sporichthya is grown at 26°C in standard yeast malt extract agar from Difco (Lechevalier et al., 1968).

Geodermatophilus, *Acidothermus* and *Sporichthya* are available from the ATCC in freeze-dried form.

Physiology

Nitrogen Fixation, Partial Pressure of Oxygen, and Vesicles

In contrast to most *Rhizobium* strains, nitrogen fixation has been found to occur not only in actinorhizal nodules but also in pure cultures of *Frankia* in the absence of fixed nitrogen. Growth without fixed nitrogen, however, is much slower and is associated with the synthesis of diazovesicles (specialized cells with thickened walls that provide a barrier to the diffusion of oxygen). It was even found that nitrogen fixation, which is a reductive process inhibited by molecular oxygen, could occur at normal oxygen pressure but also at hyperbaric oxygen concentration (Harris and Silvester, 1992). Increasing partial oxygen pressure results in increased thickness of the vesicles' cell wall and the compound involved was identified as hopanoid lipids (Berry et al., 1993). Hopanoid lipids, which behave like cholesterol as membrane hardeners, are present at such high concentrations in *Zymomonas mobilis* that cells are able to grow in up to 10% ethanol and in *Alicyclobacillus acidocaldarius* that cells are able to grow in acidic thermal springs (Poralla et al., 1980). Hopanoid lipids are also present at concentrations of 15% in *Acidothermus cellulolyticus* (Maréchal et al., 2000).

In *Frankia* cells, two types of glutamine synthetases (GSs) incorporate ammonium into glutamic acid to yield glutamine (Schultz and Benson, 1990). Both GSI (heat-stable, dodecameric, constitutive, and coded by *glnA*) and GSII (heat-labile, octameric, inducible by N-starvation, and coded by *glnB*) are present in vesicles at levels similar to those detected in vegetative hyphae from N_2-fixing cultures (Rochefort and Benson, 1990). However, glutamate synthase, glutamate dehydrogenase, and alanine dehydrogenase activities, all of which are involved in subsequent steps of ammonium assimilation, are restricted to the vegetative hyphae. Thus, diazovesicles apparently lack a complete pathway for assimilating ammonia beyond the glutamine stage.

Effect of Phenolic Compounds

In actinorhizal nodules, *Frankia* growth is restricted to the central part of the thickened cortex and it has been hypothesized that the plant controls the growth of its symbiont by preventing its access to some parts of the nodule. Given that root nodules contain high amounts of tannins (which can be described as phenolic), the effect of phenolic compounds on the growth and morphology was investigated and it was found that some compounds repressed sporulation (coumaric, ferulic, and *trans*-cinnamic acids), induced vesicle formation (benzoic and hydroxybenzoic acids), or repressed growth (coumaric acid; Perradin et al., 1983).

Superoxide Dismutase

Normal aerobic metabolism produces molecular oxygen (O_2), which is toxic to cell constituents in general and to the 4Fe-4S cluster containing enzymes in particular. Nitrogenase is one of the enzymes in the 4Fe-4S cluster and thus is particularly sensitive to oxygen radicals. Enzymes capable of metabolizing this compound, the superoxide dismutases (Sods), belong to two types: the manganese-iron Sods and the copper-zinc Sods. Given the sensitivity of nitrogenase to O_2, this enzyme has been looked for in numerous nitrogen-fixers and indeed detected in actinorhizal nodules (Alskog and Huss-Danell, 1997) and in pure cultures, where its level was among the highest reported in prokaryotes (Steele and Stowers, 1986). Using a two-dimensional (2D)-gel electrophoresis approach, the effect of plant exudates on *Frankia* in pure culture was investigated. Among the five spots overexpressed in response to *Alnus glutinosa* root exudates, one spot was found to be an iron Sod (Hammad et al., 2001).

Carbon/Nitrogen Exchange in Nodules

Frankia strains can grow on a variety of carbon sources, relatively well on small molecular weight compounds such as acetate or propionate, and also on malate, pyruvate and succinate, and particularly cluster 1 strains can grow on Tween 80. In contrast, growth on sugars is poor. Sugars are added to growth media to stimulate growth (and thus increase detection) of contaminants. ^{14}C-labeling experiments with cluster 1 strains, by far the most studied strains, showed that propionate metabolism proceeds via active transport followed by activation with coenzyme A, carboxylation to methyl-malonyl CoA and racemization to succinyl-CoA, which then can be processed in the tricarboxylic acid cycle (Stowers et al., 1986). Glycogen and trehalose have been identified as major storage compounds in *Frankia* (Lopez et al., 1984), with their level correlated negatively with the energy-demanding nitrogen fixation activity (Fontaine et al., 1984).

Hydrogenase

In contrast to leguminous nodules, actinorhizal nodules emit little or no hydrogen, a by-product of nitrogenase activity. The presence of an uptake hydrogenase, the enzyme involved in the recycling of the energy-rich gas, has been demonstrated in nodules (Benson et al., 1980) and in pure cultures (Lindblad and Sellstedt, 1989; Murry and Lopez, 1989). The enzyme's large subunit was located in the membrane by immunogold staining (Mattsson and Sellstedt, 2000) and its activity correlated with nitrogenase activity.

Genetics

The genetics of *Frankia* in pure culture has been limited by the lack of a transformation system. Several attempts at genetic transformation using naked DNA (Cournoyer and Normand, 1992) or conjugation have failed. However, a promising breakthrough has been the conjugation between *Enterococcus* and *Frankia* by means of a conjugative plasmid (Myers et al., 2001[Abstract]).

Organization of the *nif* Cluster

The *nif* genes that code for the nitrogenase enzyme complex are among the most conserved bacterial genes. The organization of the gene cluster in *Frankia* has been determined by hybridization with the *nifHDK* genes of *Klebsiella pneumoniae* (Normand et al., 1988). The gene order in the core system (HDKENX) is conserved as in most systems analyzed so far.

Glutamine Synthetase Genes

Frankia strain CpI1 has two genes coding for glutamine synthetase (Hosted et al., 1993), the protein responsible for adding ammonia to glutamate to yield glutamine. The arrangement of genes appears to result from an early (a 300 million-year-old) gene duplication and is conserved in several bacteria (Kumada et al., 1993). This region (especially the *glnB* gene) is highly variable (Cournoyer and Lavire, 1999) and thus suitable for phylogenetic analysis of closely related strains. The *Frankia glnB* gene was capable of complementing an *E. coli glnA* mutant but only when transcribed from the lac promoter (Rochefort and Benson, 1990).

Proteasome

The genes coding for the extracellular protease complex called the "proteasome" have been isolated and characterized from the cluster 1 *Frankia* strain ACN14a (Pouch et al., 2000). When compared to the homologous sequences in other actinobacteria (i.e., *Rhodococcus erythropolis*, *Mycobacterium tuberculosis* and *Streptomyces coelicolor*), the structure and gene order are conserved.

Sod

As explained above, the superoxide dismutase is especially important to nitrogen-fixing microorganisms. The Sod enzyme is present in two forms, the constitutive one containing Mn as ligand and the inducible one containing Fe as ligand; the enzyme is induced by switch to nitrogen-fixing conditions (Steele and Stowers, 1986) or exposure to plant root exudates (Hammad et al., 2001). The respective gene has been isolated and found to complement *E. coli* mutants. The expression level with its own promoter was much lower than that with an *E. coli* promoter (P. Normand, in preparation), suggesting that *Frankia* promoters are not well recognized by the *E. coli* expression machinery, as was previously reported for the *glnA* gene (Rochefort and Benson, 1990).

The *shc* Gene

This gene codes for a squalene hopene cyclase, an intermediary enzyme in the anabolic pathway from acetyl-CoA to hopanetetrol, and has been characterized from several *Frankia* strains by amplification with conserved primers (Dobritsa et al., 2001). This compound is involved in protecting nitrogenase from oxygen in thick-walled specialized diazovesicles (already described).

Plasmids

Several plasmids have been detected in various strains, essentially from the cluster 1 strains infective on *Alnus*, varying in size from 8 kb (Normand et al., 1983) to 190 kb, the largest carrying a copy of the *nif* genes (Simonet et al., 1986). Three of the smaller plasmids have been sequenced to understand why transformation in both *Frankia* and in heterologous hosts has failed repeatedly and to develop efficient vectors. The 8-kb plasmids were found to have genes resembling the classic *kor* (kill over-ride) system for repression of toxic *kil* function: *par* for partition, *rep* for replication of DNA, and a replication origin (Lavire et al., 2001). A larger 22-kb plasmid (pFQ12) present in the same strains was found to have similar functions plus some genes with similarity to transfer and mobilization determinants (John et al., 2001).

Codon Usage

Codon usage is highly skewed toward –G and —C triplets, except for the underrepresented glycine GGG. This bias, which is usual for actinomycetes, is a useful signature to confirm that *Frankia* is the source of a gene (Ligon and Nakas, 1988).

Cellulase Genes of Acidothermus cellulolyticus

Acidothermus cellulolyticus was isolated using the ability to grow at high temperature on cellulose as sole carbon source. The cellulase gene (U33212) was isolated and characterized as encoding a family 5 thermostable β-1,4-endoglucanase (Sakon et al., 1996). The G+C content of the region is lower (62.2 mol%) than that of *Frankia* genes known so far (66 mol% for *sodF*, 67 mol% for *recA* and *gln2*, 68 mol% for the *nifHDK* region, and 71 mol% for the proteasome region), and the highest similarity was found with a gene from *Paenibacillus polymyxa* (P23548, 63% similar), suggesting a recent lateral transfer. The gene was transformed within tobacco where it was transcribed, and the protein was targeted to the chloroplasts where it accumulated to 1.35% of soluble proteins (Dai et al., 2000).

Ecology

Frankia strains are ubiquitous in soils. In several ecological survey programs, *Alnus*, *Myrica* and *Elaeagnus* have consistently been found to be profusely nodulated (Bond, 1976), except in some anoxic situations. Some *Allocasuarina* species have also been found to be sparsely nodulated, but this has been interpreted as an evolutionary trend of the plant towards a narrower spectrum of strains (Maggia and Bousquet, 1994) and eventual elimination of the symbiont in the drier parts of Australia (Simonet et al., 1999). Even soils from sites with no known actinorhizal plants cause nodule formation on test plants, suggesting that *Frankia* can thrive in the soil in the absence of host plants (Normand and Lalonde, 1982), which was demonstrated experimentally (Rossi, 1964). It has been suggested that the ability to thrive as saprophytes varies markedly between strains. In particular, some strains that sporulate profusely in plants have been suggested to have different ecological strategies (van Dijk, 1978) and in particular to be less saprophytic.

pH Effect on Distribution of Strains

Several soil parameters are expected to play a role on the survival of spores or on the saprophytic life of hyphae. These are texture, organic matter content, pH, chemicals, pO_2, etc. However, few of these parameters have been tested. Acidity of the soil is known to be a critical factor, acting directly on the bacterial physiology and indirectly on nutrient availability. As expected, it was found that the diversity of *Frankia* strains infective on *Elaeagnus* was reduced in an acidic soil relative to a neutral soil (Jamann et al., 1992).

Effect of Soil and Host Plant on Distribution of Strains

Actinorhizal plants are known to be pioneer plants that invade recently deforested areas following volcanic eruptions, glacier melting, forest fires, landslides, or man-made processes like deposition of mine spoils, hydrodam construction, or sand dune formation. Such processes may result in unbalanced substrate for pH or chemicals acting as selective forces on strains. Thus, it is not surprising to have differences in 50% of lethal dose (LD_{50}) for a variety of heavy metals (Richards et al., 2002). In the case of *Frankia* strains infective on *Gymnostoma* spp., this was found to result in abundance, dependent on soils and the host plant species (Navarro et al., 1999). A similar conclusion was reached in the case of *Ceanothus*-infective strains, with some strains restricted to serpentine soils (Ritchie and Myrold, 1999).

Antibiotics

Differences in antibiotic resistance were noted between strains, even closely related ones (Dobritsa, 1998), even though most differences are linked to taxonomic position. Conversely, *Frankia* has been found to synthesize unusual compounds, such as a macrocycle containing imide and orthoamide functionalities (Klika et al., 2001) and benzo[a]naphthacene quinones (Rickards, 1989).

Applications

Actinorhizal plants are pioneer species that are dominant in re-vegetating areas, following natural events such as glacier retreat or volcanic eruptions. These plants can be self-reliant in nitrogen because of their symbiosis with *Frankia*, and they are also undemanding for other nutrients, in part because of the general occurrence of symbiosis with endomycorrhizal fungi (Diem, 1996). For all these reasons, actinorhizal plants have been used for several applications where it is necessary to rapidly establish a plant cover.

The Americas and Europe

Alnus viridis subsp. *crispa* has been planted on a commercial scale in Canada (Périnet et al., 1985) to stabilize dikes in the large hydroelectric facilities of Northern Quebec and for biomass production on abandoned farmland (Prégent and Camiré, 1985) and on reclaimed oil shale sites in Alberta (Gordon and Dawson, 1979).

Interplanting is another possibility, in that the valuable but demanding tree species such as walnut are grown together with slower growing, undemanding alder or Russian olive trees which form nitrogen-rich soil (Campbell and Dawson, 1989). Furthermore, small-scale production of seabuckthorn berries (*Hippophae rhamnoides*) is recommended on marginal lands or in mountainous areas (http://www.hort.purdue.edu/newcrop/) to yield alcohol, vitamin A and C-rich juice, and a variety of jams, candies and "natural" products (http://www.genres.de/bmlfao/natber3.htm). Finally, several undemanding ornamentals plants are actinorhizals, prominent among them *Alnus cordata*, *Hippophae rhamnoides*, *Elaeagnus* spp. and *Ceanothus* spp.

Africa, Asia and Oceania

Casuarina spp., which produce dense and hard wood and have few if any diseases (Dommergues et al., 1999), are now widely used in tropical climates for planting on sand dunes and other marginal sites (Diem and Dommergues, 1990). *Casuarina* spp., especially of *Casuarina equisetifolia*, planted in Africa, India and China's seacoasts as windbreak are now commonplace (Midgley et al., 1986). For instance, up to a million hectares have thus been planted in China (Turnbull, 1983). In Egypt, *Casuarina* spp. are planted in individual fields where not only does it protect against wind but also it provides fuel wood (El-Lakany, 1990). *Gymnostoma* spp. have been used for re-vegetation of nickel mine spoils in New Caledonia (McCoy et al., 1996). *Casuarina equisetifolia* may increase the soil's nitrogen from 80 kg·ha^{-1} to more than 300 kg·ha^{-1} (Dommergues et al., 1999). *Coriaria* spp. have been planted to stabilize hilly areas in Pakistan (Chaudhary and Mirza, 1987). *Alnus nepalensis*, the first plant that appears on landslides and abandoned farmlands in Northeastern India, is used in mixed plantations with cardamom and tea and to re-vegetate degraded lands (Dommergues et al., 1999).

Conclusion

Few groups of microbes are composed of such contrasting families with ecological niches and physiological characteristics as diverse as the one described here. It is only the advent of molecular techniques that has permitted the close proximity of the thermoacidophile Acidothermaceae, the soil Geodermatophilaceae and Sporichthyaceae, and the plant symbionts Frankiaceae to be recognized. The evolutionary history of this group remains to be established and the genome sequence of *Frankia* and its neighbors will certainly be revealing.

Literature Cited

Ahrens, R., and G. Moll. 1970. Ein neues knospendes Bakterium aus der Ostsee [A new budding bacterium from the Baltic Sea]. Arch. Mikrobiol. 70:243–265.

Akimov, V., and S. Dobritsa. 1992. Grouping of Frankia strains on the basis of DNA relatedness. Syst. Appl. Microbiol. 15:372–379.

Alskog, G., and K. Huss-Danell. 1997. SOD, catalase and N2ase activities of symbiotic Frankia (Alnus incana) in response to different oxygen tensions. Physiol. Plant. 99:286–292.

An, C., W. Riggsby, and B. Mullin. 1985. Relationships of Frankia isolates based on deoxyribonucleic acid homology studies. Int. J. Syst. Bacteriol. 35:140–146.

Baker, D., and J. Torrey. 1979. The isolation and cultivation of actinomycetous root nodule endophytes. *In:* J. C. Gordon, D. A. Perry, and O. R. Corvallis (Eds.) Symbiotic Nitrogen Fixation in the Management of Temperate Forests. Oregon State University, Forest Research Laboratory. Corvallis, OR. 38–56.

Baker, D., W. Newcomb, and J. Torrey. 1980. Characterization of an ineffective actinorhizal microsymbiont, Frankia sp. EuI1 (Actinomycetales). Can. J. Microbiol. 26:1072–1089.

Baker, D., and D. O'Keefe. 1984. A modified sucrose fractionation procedure for the isolation of frankiae from actinorhizal root nodules and soil samples. Plant Soil 78:23–28.

Becking, J. H. 1970. Frankiaceae fam. nov. (Actinomycetales) with one new combination and six new species of the genus Frankia Brunchorst 1886, 174. Int. J. Syst. Bacteriol. 20:201–220.

Benson, D., D. Arp, and R. Burris. 1980. Hydrogenase in actinorhizal root nodules and root nodule homogenates. J. Bacteriol. 142:138–144.

Benson, D., D. Stephens, M. Clawson, and W. Silvester. 1996. Amplification of 16s rRNA genes from Frankia strains in root nodules of Ceanothus griseus, Coriaria arborea, Coriaria plumosa, Discaria toumatou and Purshia tridentata. Appl. Environ. Microbiol. 62:2904–2909.

Berry, A., O. Harriott, R. Moreau, S. Osman, D. Benson, and A. Jones. 1993. Hopanoid lipids compose the Frankia vesicle envelope, presumptive barrier of oxygen diffusion to nitrogenase. Proc. Natl. Acad. Sci. USA 90:6091–6094.

Bloom, R., B. Mullin, and R. I. Tate. 1989. DNA restriction patterns and solution hybridization studies of Frankia isolates from Myrica pensylvanica (Bayberry). Appl. Environ. Microbiol. 55:2155–2160.

Bond, G. 1976. The results of the IBP survey of root nodule formation in non-leguminous angiosperms. *In:* P. Nutman (Ed.) Symbiotic Nitrogen Fixation in Plants. Cambridge University Press. London, UK. 443–474.

Bosco, M., M. Fernandez, P. Simonet, R. Materassi, and P. Normand. 1992. Evidence that some Frankia sp. strains are able to cross boundaries between Alnus and Elaeagnus host specificity groups. Appl. Environ. Microbiol. 58:1569–1576.

Callaham, D., P. Del Tredici, and J. Torrey. 1978. Isolation and cultivation in vitro of the actinomycete causing root nodulation in Comptonia. Science 199:899–902.

Campbell, G., and J. Dawson. 1989. Growth, yield and value projections for a black walnut interplanting with black alder and autumn olive. North J. Appl. Forestry 6:129–132.

Chaudhary, A., and M. Mirza. 1987. Isolation and characterization of Frankia from nodules of actinorhizal plants of Pakistan. Physiol. Plant. 70:255–258.

Clawson, M. L., M. Caru, and D. R. Benson. 1998. Diversity of Frankia strains in root nodules of plants from the families Elaeagnaceae and Rhamnaceae. Appl. Environ. Microbiol. 64:3539–3543.

Cournoyer, B., and P. Normand. 1992. Electropermeabilization of Frankia intact cells to plasmid DNA. Acta Oecologica 13:369–378.

Cournoyer, B., and C. Lavire. 1999. Analysis of Frankia evolutionary radiation using glnII sequences. FEMS Microbiol. Lett. 177:29–34.

Dai, Z., B. S. Hooker, D. B. Anderson, and S. R. Thomas. 2000. Expression of Acidothermus cellulolyticus endoglucanase E1 in transgenic tobacco: biochemical characteristics and physiological effects. Transgenic Res. 9:43–54.

Diem, H., and Y. Dommergues. 1985. In vitro production of specialized reproductive torulose hyphae by Frankia strain ORS 021001 isolated from Casuarina junghuhniana root nodules. Plant Soil 87:17–29.

Diem, H., and Y. Dommergues. 1990. Current and potential uses and management of Casuarinaceae in the tropics and subtropics. In: C. R. Schwintzer (Ed.) The Biology of Frankia and Actinorhizal Plants. Academic Press. New York, NY. 317–342.

Diem, H. G. 1996. Les mycorhizes des plantes actinorhiziennes. Acta Bot. Gallica 143:581–592.

Dobritsa, S. V. 1998. Grouping of Frankia strains on the basis of susceptibility to antibiotics, pigment production and host specificity. Int. J. Syst. Bacteriol. 48:1265–1275.

Dobritsa, S. V., D. Potter, T. E. Gookin, and A. M. Berry. 2001. Hopanoid lipids in Frankia: identification of squalene-hopene cyclase gene sequences. Can. J. Microbiol. 47:535–540.

Dommergues, Y., E. Duhoux, and G. Diem. 1999. Les arbres fixateurs d'azote: Caractéristiques fondamentales et rôle dans l'aménagement des écosystèmes méditérranéens et tropicaux. CIRAD-FAO-IRD. Paris, France. 1–499.

El-Lakany, M.1990. Provenance trials of Casuarina glauca and C. cunninghamiana in Egypt. In: M. H. El-Lakany and J. L. Brewbaker (Eds.) Advances in Casuarina Research and Utilization. Desert Development Center, American University. Cairo, Egypt. 12–22.

Eppard, M., W. E. Krumbein, C. Koch, E. Rhiel, J. T. Staley, and E. Stackebrandt. 1996. Morphological, physiological, and molecular characterization of actinomycetes isolated from dry soil, rocks, and monument surfaces. Arch. Microbiol. 166:12–22.

Felsenstein, J. 1985. Confidence limits on phylogenies: an approach using the bootstrap. Evolution 783–791.

Fernandez, M., H. Meugnier, P. Grimont, and R. Bardin. 1989. Deoxyribonucleic acid relatedness among members of the genus Frankia. Int. J. Syst. Bacteriol. 39:424–429.

Fontaine, M., S. Lancelle, and J. Torrey. 1984. Initiation and ontogeny of vesicles in cultured Frankia sp. strain HFPArI3. J. Bacteriol. 160:921–927.

Gordon, J., and J. Dawson. 1979. Potential uses of nitrogen-fixing trees and shrubs in commercial forestry. Bot. Gaz. 140 (Supplement):S88–S90.

Hahn, D., M. Lechevalier, A. Fischer, and E. Stackebrandt. 1989. Evidence for a close phylogenetic relationship between members of the genera Frankia, Geodermatophilus, and "Blastococcus" and emendation of the family Frankiaceae. Syst. Appl. Microbiol. 11:236–242.

Hammad, Y., J. Marechal, B. Cournoyer, P. Normand, and A. M. Domenach. 2001. Modification of the protein expression pattern induced in the nitrogen-fixing actinomycete Frankia sp. strain ACN14a-tsr by root exudates of its symbiotic host Alnus glutinosa and cloning of the sodF gene. Can. J. Microbiol. 47:541–547.

Harris, S., and W. Silvester. 1992. Oxygen controls the development of Frankia vesicles in continuous culture. New Phytol. 121:43–48.

Hayakawa, M., M. Otoguro, T. Takeuchi, T. Yamazaki, and Y. Iimura. 2000. Application of a method incorporating differential centrifugation for selective isolation of motile actinomycetes in soil and plant litter. Ant. v. Leeuwenhoek 78:171–185.

Hönerlage, W., D. Hahn, K. Zepp, J. Zeyer, and P. Normand. 1994. A hypervariable 23S rRNA region provides a discriminating target for specific characterization of uncultured and cultured Frankia. Syst. Appl. Microbiol. 17:433–443.

Hosted, T., D. Rochefort, and D. Benson. 1993. Close linkage of genes encoding glutamine synthetase I and II in Frankia alni CpI1. J. Bacteriol. 175:3679–3684.

Huguet, V., J. M. Batzli, J. F. Zimpfer, P. Normand, J. O. Dawson, and M. P. Fernandez. 2001. Diversity and specificity of Frankia strains in nodules of sympatric Myrica gale, Alnus incana, and Shepherdia canadensis determined by rrs gene polymorphism. Appl. Environ. Microbiol. 67:2116–2122.

Ishiguro, E. E., and R. S. Wolfe. 1970. Control of morphogenesis in Geodermatophilus: Ultrastructural studies. J. Bacteriol. 104:566–580.

Jamann, S., M. Fernandez, and A. Moiroud. 1992. Genetic diversity of Elaeagnaceae-infective Frankia strains isolated from various soils. Acta Oecolog. 13:395–405.

Jeong, S. C., N. J. Ritchie, and D. D. Myrold. 1999. Molecular phylogenies of plants and Frankia support multiple origins of actinorhizal symbioses. Molec. Phylogenet. Evol. 13:493–503.

John, T. R., J. M. Rice, and J. D. Johnson. 2001. Analysis of pFQ12, a 22.4-kb Frankia plasmid. Can. J. Microbiol. 47:608–617.

Klika, K. D., J. P. Haansuu, V. V. Ovcharenko, K. K. Haahtela, P. M. Vuorela, and K. Pihlaja. 2001. Frankiamide, a highly unusual macrocycle containing the imide and orthoamide functionalities from the symbiotic actinomycete Frankia. J. Org. Chem. 66:4065–4068.

Kumada, Y., D. Benson, D. Hillemann, T. Hosted, D. Rochefort, C. Thompson, W. Wohlleben, and Y. Tateno. 1993. Evolution of the glutamine synthetase gene, one of the oldest existing and functioning genes. Proc. Natl. Acad. Sci. USA 90:3009–3013.

Lalonde, M., and H. Calvert. 1979. Production of Frankia hyphae and spores as an infective inoculant for Alnus species. *In:* J. C. Gordon and D. A. Perry (Eds.) Symbi-otic Nitrogen Fixation in the Management of Temperate Forests. Oregon State University, Forest Research Lab-oratory. Corvallis, OR. 95–110.

Lavire, C., D. Louis, G. Perriere, J. Briolay, P. Normand, and B. Cournoyer. 2001. Analysis of pFQ31, a 8551-bp cryptic plasmid from the symbiotic nitrogen-fixing actinomycete Frankia. FEMS Microbiol. Lett. 197:111–116.

Lechevalier, M. P., H. A. Lechevalier, and P. E. Holbert. 1968. Sporichthya, un nouveau genre de Streptomycetaceae. Ann. Inst. Pasteur 114:277–286.

Lechevalier, M. P., and H. A. Lechevalier. 1990. Systematics, isolation and culture of Frankia. *In:* J. D. Schwintzer (Ed.) The Biology of Frankia and Actinorhizal Plants. Academic Press. San Diego, CA. 35–60.

Ligon, J. M., and J. P. Nakas. 1988. Nucleotide sequence of nifK and partial sequence of nifD from Frankia species strain FaC1 [published erratum appears in Nucleic Acids Res. Feb. 25 1990; 18(4): 1097]. Nucleic Acids Res. 16:11843.

Lindblad, P., and A. Sellstedt. 1989. Immunogold localization of hydrogenase in free-living Frankia CpI1. FEMS Microbiol. Lett. 60:311–316.

Lopez, M., M. Fontaine, and J. Torrey. 1984. Levels of treha-lose and glycogen in Frankia sp. HPFArI3 (Actinomyc-etales). Can. J. Microbiol. 30:746–752.

Luedemann, G. M. 1968. Geodermatophilus, a new genus of the Dermatophilaceae (Actinomycetales). J. Bacteriol. 96:1848–1858.

Lumini, E., M. Bosco, and M. P. Fernandez. 1996. PCR-RFLP and total DNA homology revealed three related genomic species among broad-host-range Frankia strains. FEMS Microbiol. Ecol. 21:303–311.

Maggia, L., and J. Bousquet. 1994. Molecular phylogeny of the actinorhizal Hamamelidae and relationships with host promiscuity towards Frankia. Molec. Ecol. 3:459–467.

Maréchal, J., B. Clement, R. Nalin, C. Gandon, S. Orso, J. H. Cvejic, M. Bruneteau, A. Berry, and P. Normand. 2000. A recA gene phylogenetic analysis confirms the close proximity of Frankia to Acidothermus. Int. J. Syst. Evol. Microbiol. 50:781–785.

Mattsson, U., and A. Sellstedt. 2000. Hydrogenase in Frankia KB5: expression of and relation to nitrogenase. Can. J. Microbiol. 46:1091–1095.

McCoy, S. G., J. Ash, and T. Jaffré. 1996. The effect of Gym-nostoma deplancheanum (Casuarinaceae) litter on seed-ling establishment of New Caledonian ultramafic maquis species. Second Australian Native Seed Biology for Revegetation Workshop. S.M. Bellairs and J.M. Osborne (eds.) Australian Centre for Minesite Rehabilitation Research. Kenmore, Australia. 127–135.

Mevs, U., E. Stackebrandt, P. Schumann, C. A. Gallikowski, and P. Hirsch. 2000. Modestobacter multiseptatus gen. nov., sp. nov., a budding actinomycete from soils of the Asgard Range (Transantarctic Mountains). Int. J. Syst. Evol. Microbiol. 50(1):337–346.

Midgley, S., J. Turnbull, and R. Johnston. 1986. Casuarina Ecology, Management and Utilization. Commonwealth Scientific and Industrial Research Organization. Mel-bourne, Australia.

Mirza, M., D. Hahn, S. Dobritsa, and A. Akkermans. 1994. Phylogenetic studies on uncultured Frankia populations in nodules of Datisca cannabina. Can. J. Microbiol. 40:313–318.

Mohagheghi, A., K. Grohmann, M. Himmel, L. Leighton, and D. M. Updegraff. 1986. Isolation and characterization of Acidothermus cellulolyticus gen. nov., sp. nov., a new genus of thermophilic, acido-philic, cellulolytic bacteria. Int. J. Syst. Bacteriol. 36:435–443.

Murry, M., M. Fontaine, and J. Torrey. 1984. Growth kinetics and nitrogenase induction in Frankia sp. HFPArI3 grown in batch culture. Plant Soil 78:61–78.

Murry, M., and M. Lopez. 1989. Interaction between hydro-genase, nitrogenase, and respiratory activities in a Frankia isolate from Alnus rubra. Can. J. Microbiol. 35:636–641.

Myers, A., T. Rawnsley, and L. S. Tisa. 2001. Developing genetic tools for Frankia, the bacterial partner of the actinorhizal symbiosis. P. Normand et al., eds. 12th Inter-national Meeting on Frankia and Actinorhizal Plants, June 2001. 12th International Meeting on Frankia and Actinorhizal Plants, June 2001. Carry-le-Rouet, France. Abstract 16:29.

Navarro, E., R. Nalin, D. Gauthier, and P. Normand. 1997. The nodular microsymbionts of Gymnostoma spp. are Elaeagnus-infective strains. Appl. Environ. Microbiol. 63:1610–1616.

Navarro, E., T. Jaffre, D. Gauthier, F. Gourbiere, G. Rinaudo, P. Simonet, and P. Normand. 1999. Distribution of Gym-nostoma spp. microsymbiotic strains in New Caledonia is related to soil type and to host-plant species. Molec. Ecol. 8:1781–1788.

Nazaret, S., P. Simonet, P. Normand, and R. Bardin. 1989. Genetic diversity among Frankia isolated from Casua-rina nodules. Plant Soil 118:241–247.

Normand, P., and M. Lalonde. 1982. Evaluation of Frankia strains isolated from provenances of two Alnus species. Can. J. Microbiol. 28:1133–1142.

Normand, P., P. Simonet, J. Butour, C. Rosenberg, A. Moiroud, and M. Lalonde. 1983. Plasmids in Frankia sp. J. Bacteriol. 155:32–35.

Normand, P., P. Simonet, Y. Prin, and A. Moiroud. 1987. Formation and regeneration of Frankia protoplasts. Physiol. Plant. 70:259–266.

Normand, P., P. Simonet, and R. Bardin. 1988. Conservation of nif sequences in Frankia. Molec. Gen. Genet. 213:238–246.

Normand, P., S. Orso, B. Cournoyer, P. Jeannin, C. Chapelon, J. Dawson, L. Evtushenko, and A. K. Misra. 1996. Molecular phylogeny of the genus Frankia and related genera and emendation of the family Frankiaceae. Int. J. Syst. Bacteriol. 46:1–9.

Normand, P., and C. Chapelon. 1997. Direct characterization of Frankia and of close phyletic neighbors from an Alnus viridis rhizosphere. Physiol. Plant. 99:722–731.

Périnet, P., J. Brouillette, J. Fortin, and M. Lalonde. 1985. Large scale inoculations of actinorhizal plants with Frankia. Plant Soil 87:175–183.

Perradin, Y., M. Mottet, and M. Lalonde. 1983. Influence of phenolics on in vitro growth of Frankia strains. Can. J. Bot. 61:2807–2814.

Poralla, K., E. Kannenberg, and A. Blume. 1980. A glycolipid containing hopane isolated from the acido-philic, thermophilic Bacillus acidocaldarius, has a cholesterol-like function in membranes. FEBS Lett. 113:107–110.

Pouch, M. N., B. Cournoyer, and W. Baumeister. 2000. Char-acterization of the 20S proteasome from the actino-mycete Frankia. Molec. Microbiol. 35:368–377.

Prégent, G., and C. Camiré. 1985. Biomass production by alders on four abandoned agricultural soils in Quebec. Plant Soil 87:185–193.

Prin, Y., L. Maggia, B. Picard, H. Diem, and P. Goullet. 1991. Electrophoretic comparison of enzymes from 22 single-spore cultures obtained from Frankia strain ORS140102. FEMS Microbiol. Lett. 77:223–228.

Rainey, F., and E. Stackebrandt. 1993a. Phylogenetic evidence for the classification of Acidothermus cellulolyticus into the subphylum of actinomycetes. FEMS Microbiol. Lett. 108:27–30.

Rainey, F., P. Schumann, H. Prauser, R. Toalster, and E. Stackebrandt. 1993b. Sporichthya polymorpha represents a novel line of descent within the order Actinomycetales. FEMS Microbiol. Lett. 109:263–268.

Richards, J. W., G. D. Krumholz, M. S. Chval, and L. S. Tisa. 2002. Heavy metal resistance patterns of Frankia strains. Appl. Environ. Microbiol. 68:923–927.

Rickards, R. W. 1989. Revision of the structures of the benzo[a]naphthacene quinone metabolites G-2N and G-2A from bacteria of the genus Frankia. J. Antibiot. 42:336–339.

Ritchie, N., and D. Myrold. 1999. Geographic distribution and genetic diversity of Ceanothus-infective Frankia strains. Appl. Environ. Microbiol. 65:1378–1383.

Rochefort, D. A., and D. R. Benson. 1990. Molecular cloning, sequencing, and expression of the glutamine synthetase II (glnII) gene from the actinomycete root nodule symbiont Frankia sp. strain CpI1. J. Bacteriol. 172:5335–5342.

Rossi, S. 1964. Propagation dans le sol de l'organisme causant les nodosites dans les racines d'Aune (Alnus glutinosa). Ann. Microbiol. Inst. Pasteur Ser. A 106:505–510.

Saitou, N., and M. Nei. 1987. The neighbor-joining method: a new method for reconstructing phylogenetic trees. Molec. Biol. Evol. 4:406–425.

Sakon, J., W. S. Adney, M. E. Himmel, S. R. Thomas, and P. A. Karplus. 1996. Crystal structure of thermostable family 5 endocellulase E1 from Acidothermus cellulolyticus in complex with cellotetraose. Biochem. 35:10648–10660.

Schultz, N. A., and D. R. Benson. 1989. Developmental potential of Frankia vesicles. J. Bacteriol. 171:6873–6877.

Schultz, N. A., and D. R. Benson. 1990. Enzymes of ammonia assimilation in hyphae and vesicles of Frankia sp. strain CpI1. J. Bacteriol. 172:1380–1384.

Simonet, P., J. Haurat, P. Normand, R. Bardin, and A. Moiroud. 1986. Localization of nif genes on a large plasmid in Frankia sp. strain ULQ0132105009. Molec. Gen. Genet. 204:492–495.

Simonet, P., E. Navarro, C. Rouvier, P. Reddell, J. Zimpfer, Y. Dommergues, R. Bardin, P. Combarro, J. Hamelin, A. M. Domenach, F. Gourbiere, Y. Prin, J. O. Dawson, and P. Normand. 1999. Co-evolution between Frankia populations and host plants in the family Casuarinaceae and consequent patterns of global dispersal. Environ. Microbiol. 1:525–533.

Stackebrandt, E., R. Kroppenstedt, and V. Fowler. 1983. A phylogenetic analysis of the family Dermatophylaceae. J. Gen. Microbiol. 129:1831–1838.

Stackebrandt, E., F. A. Rainey, and N. L. Ward-Rainey. 1997. Proposal for a new hierarchic classification system, actinobacteria classis nov. Int. J. Syst. Bacteriol. 47:479–491.

Steele, D., and M. Stowers. 1986. Superoxide dismutase and catalase in Frankia. Can. J. Microbiol. 32:409–413.

Stowers, M., R. Kulkarni, and D. Steele. 1986. Intermediary carbon metabolism in Frankia. Arch. Microbiol. 143:319–324.

Suzuki, S.-I., T. Okuda, and S. Komatsubara. 1999. Selective isolation and distribution of Sporichthya strains in soil. Appl. Environ. Microbiol. 65:1930–1935.

Tisa, L., and J. Ensign. 1987. Formation and regeneration of protoplasts of the actinorhizal nitrogen-fixing actinomycete Frankia. Appl. Environ. Microbiol. 53:53–56.

Turnbull, J. W. 1983. The use of Casuarina equisetifolia for protection forests in China. In: S. Midgley, J. W. Turnbull, and R. D. Johnston (Eds.) Casuarina ecology management and utilization. Commonwealth Scientific and Industrial Research Organization. Melbourne, Australia. 155–157.

Urzi, C., L. Brusetti, P. Salamone, C. Sorlini, E. Stackebrandt, and D. Daffonchio. 2001. Biodiversity of Geodermatophilaceae isolated from altered stones and monuments in the Mediterranean basin. Environ. Microbiol. 3:471–479.

van Dijk, C. 1978. Spore formation and endophyte diversity in root nodules of Alnus glutinosa (L.) Vill. New Phytol. 81:601–615.

Woronin, M. 1866. Über die bei der Schwarzerle (Alnus glutinosa) und der gewöhnlichen Garten-Lupine (Lupinus mutabilis) auftretenden Wurzelanschellungen. Mem. Acad. Sci. St. Petersburg 10:1–13.

Prokaryotes (2006) 3:682–724
DOI: 10.1007/0-387-30743-5_27

CHAPTER 1.1.12

The Family Thermomonosporaceae: *Actinocorallia, Actinomadura, Spirillospora* and *Thermomonospora*

REINER MICHAEL KROPPENSTEDT AND MICHAEL GOODFELLOW

Phylogeny and Taxonomy

The application of chemosystematic and molecular systematic methods has clarified the taxonomic position of the family Thermomonosporaceae. This taxon together with the families Nocardiopsaceae and Streptosporangiaceae is classified in the suborder Streptosporangineae, one of the ten suborders of the order Actinomycetales (Stackebrandt et al., 1997). The family Thermomonosporaceae (Stackebrandt et al., 1997) emended (Zhang et al., 2001) includes the genera *Actinomadura* (Lechevalier and Lechevalier, 1970a), *Actinocorallina* (Iinuma et al., 1994) emended (Zhang et al., 2001), *Spirillospora* (Couch, 1963), and the type genus *Thermomonospora* (Henssen, 1957) emended (Zhang et al., 2001). Members of these taxa form a distinct phyletic line in the 16S rRNA gene tree (Zhang et al., 1998; Zhang et al., 2001) and can be distinguished from one another using a combination of morphological, physiological and chemical features (Trujillo and Goodfellow, 1997; Zhang et al., 1998; Miyadoh et al., 2001; Zhang et al., 2001; Trujillo and Goodfellow, 2003).

Since the second edition of the *The Prokaryotes* many changes have occurred in the classification of the family Thermomonosporaceae: 1) the so called "*Actinomadura* (*Microtetraspora*) *pusilla* group" has been transferred to the new genus *Nonomuraea* (Zhang et al., 1998); 2) *Actinomadura aurantiaca, A. glomerata, A. libanotica* and *A. longicatena* have been transferred to the genus *Actinocorallia* (Zhang et al., 2001); 3) *Thermomonospora alba, T. fusca* and *T. mesouviformis* to the genus *Thermobifida* (Zhang et al., 1998); 4) *Thermomonospora formosensis* and *T. mesophila* to the genus *Actinomadura* (Zhang et al., 2001); and 5) *Spirillospora* from the family Streptosporangiaceae to the family Thermomonosporaceae (Stackebrandt et al., 1997).

The family Thermomonosporaceae was proposed for a morphologically diverse group of organisms classified in the genera *Actinomadura, Microbispora, Microtetraspora, Saccharomonospora* and *Thermomonospora* (Cross and Goodfellow, 1973). This avowedly heterogeneous assemblage was nevertheless seen to include genera that had more in common than the assortment of organisms previously classified in the family Nocardiaceae, an area long regarded as a dumping ground for aerobic actinomycetes (Lechevalier, 1976). The new family encompassed mesophilic and thermophilic actinomycetes that formed heat sensitive, motile and nonmotile spores carried singly, in pairs, or as short chains on aerial hyphae, on both aerial and substrate mycelia, or developed in spore vesicles (sporangia) as in *Spirillospora*. All members of this taxon have a cell wall containing *meso*-diaminopimelic acid (*meso*-A$_2$pm) but no other characteristic amino acids or sugars (wall chemotype III; Lechevalier and Lechevalier, 1970d; Lechevalier and Lechevalier, 1970b). The genus *Saccharomonospora*, which has *meso*-A$_2$pm as the diamino acid of the cell wall peptidoglycan and a polysaccharide fraction rich in arabinose and galactose (wall chemotype IV; Lechevalier and Lechevalier, 1970d; Lechevalier and Lechevalier, 1970c), was included in the family as it was considered to share key morphological properties with *Thermomonospora* species. Other wall chemotype III sporoactinomycetes subsequently thought to be related to the genus *Thermomonospora* included the genera *Actinobifida, Actinosynnema, Nocardiopsis, Saccharothrix* and *Streptoalloteichus* (Küster, 1974; Goodfellow and Cross, 1984; Goodfellow, 1989a; McCarthy, 1989).

The application of chemical and molecular systematic methods has provided invaluable data for the classification of actinomycetes above the genus level (Stackebrandt and Woese, 1981a; Goodfellow and Minnikin, 1985; Stackebrandt and Goodfellow, 1990; Stackebrandt et al., 1997; Zhang et al., 2001); such studies have shown that morphological features previously weighted for the circumscription of genera and families have little predictive value (Stackebrandt and Schleifer, 1984; Stackebrandt, 1986; Goodfellow, 1989b). It soon became evident that the family Thermomonosporaceae (Cross and Goodfellow, 1973) contained markedly diverse taxa and that the genus *Actinomadura* (Lechevalier and Lech-evalier, 1970a) included two groups of organisms with little in common (Fischer et al., 1983; Poschner et al., 1985). Ribosomal (r)RNA partial oligonucleotide sequencing (Fowler et al., 1985; Goodfellow et al., 1988) underlined this hetero-geneity and showed that *A. madurae*, the type species of the genus, showed a closer affinity to *Thermomonospora curvata* than to the *A. pusilla* group; members of the latter were seen to be related to *Streptosporangium roseum*.

The division of the genus *Actinomadura* into two disparate groups was formally recognized when Kroppenstedt et al. (1990) proposed that the genus be retained for *A. madurae* and related species and that the *A. pusilla* group be reclassi-fied in the genus *Microtetraspora*. On the basis of the results of comparative 16S rDNA sequenc-ing, all members of the former *A. pusilla* group were transferred from the genus *Microtetraspora* to the new genus *Nonomuraea* (Zhang et al., 1998), a proposition raised but never validated by Goodfellow (Goodfellow et al., 1988). Simi-larly, several genera associated with the genus *Thermomonospora* have been transferred to other suprageneric groups. The genera *Micro-bispora* and *Microtetraspora* have been assigned to the family Streptosporangiaceae (see The Family Streptosporangiaceae in this volume), and the genera *Saccharomonospora* and *Saccha-rothrix* to the family Pseudonocardiaceae (see The Family Pseudonocardiaceae in the second edition). The genus *Nocardiopsis* can be sharply separated from the genus *Thermomonospora* (Fowler et al., 1985; Kroppenstedt et al., 1990) and readily distinguished from the poorly stud-ied genera *Actinosynnema* and *Streptoallotei-chus*. *Nocardiopsis* is now the type genus of the family Nocardiopsaceae (Stackebrandt et al., 1997), which includes the genera *Thermobifida* (Zhang et al., 1998) and *Streptomonospora* (Cui et al., 2001; Li et al., 2003). These developments leave the genera *Actinomadura*, *Actinocorallina*, *Spirillospora* and *Thermomonospora* as a rela-tively homogeneous group with respect to their chemical markers and to the other actinomycete families (Stackebrandt et al., 1983a; Stacke-brandt et al., 1983b; Kroppenstedt et al., 1990; Zhang et al., 1998; Zhang et al., 2001).

It is becoming increasingly clear that the fam-ily Thermomonosporaceae should be retained for aerobic, Gram-positive, non-acid-alcohol-fast, chemo-organotrophic actinomycetes which form a branched substrate mycelium bearing aerial hyphae that differentiate into single or short chains of arthrospores (*Actinomadura*, *Actinocorallia* and *Thermomonospora*; Figs. 1 and 2) or produce spore vesicles that release motile zoospores (*Spirillospora*; Fig. 3). Constit-uent strains have *meso*-diaminopimelic acid in

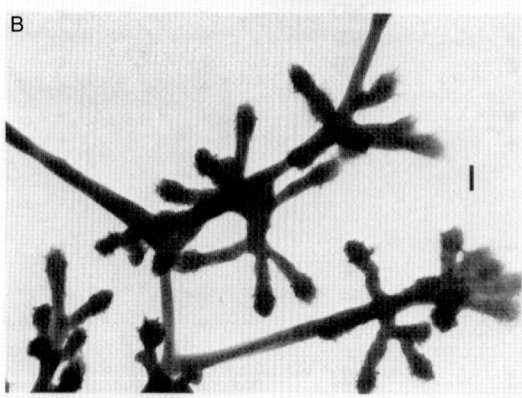

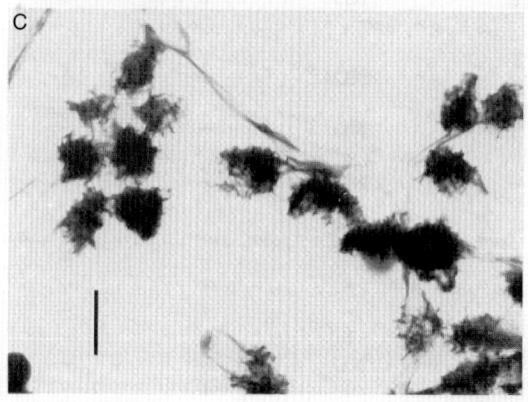

Fig. 1. Transmission electron micrographs showing spore for-mation in thermomonosporae. Bars = 1 μm. A) *Ther-momonospora curvata*, single spores on branched and unbranched sporophores. (From Greiner-Mai et al., 1987.) B) *Thermobifida alba*, single spores on relatively stiff sporo-phores. (Courtesy of F. Korn-Wendisch.) C) *Thermomono-spora chromogena*, single spores in clusters. (Courtesy of F. Korn-Wendisch.)

a wall peptidoglycan that lacks characteristic sugars (Lechevalier and Lechevalier, 1970d; Lechevalier and Lechevalier, 1970c), an A1γ peptidoglycan type (Schleifer and Kandler, 1972), N-acetylated muramic acid (Uchida and Aida, 1977), mixtures of straight and branched

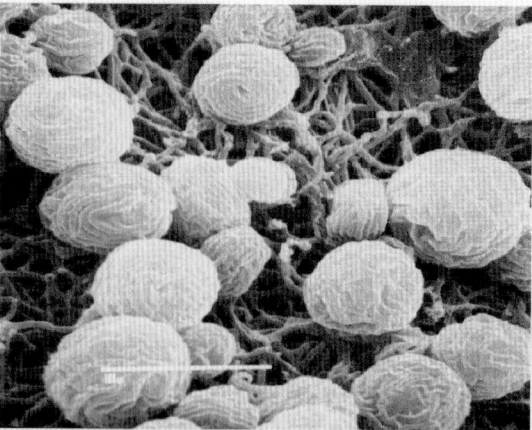

Fig. 3. *Spirillospora albida* ATCC 15331. A scanning electron microscopy image of spherical spore vesicles formed on aerial mycelium after 15 days of incubation. (From Vobis and Kothe (1989), with permission.)

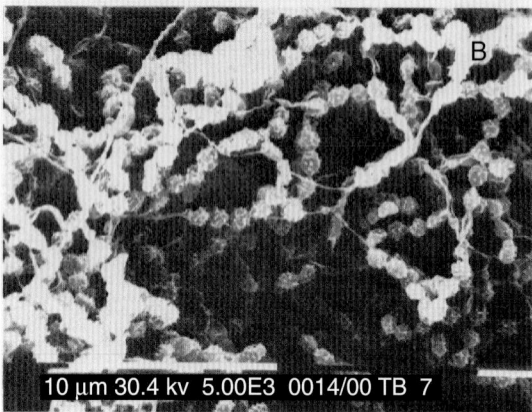

Fig. 2. Scanning electron micrographs showing spore formation. A) Spore clusters of *Thermobifida fusca*. B) Long arthrospore chains of *Actinomadura rubrobrunea*. Bars = 1 μm. (Both courtesy of A. Kempf.)

fatty acids, hydrogenated menaquinones with nine isoprenyl units, and large amounts of phosphatidylglycerol (PG), phosphatidylinositol (PI), and phosphatidylinositol mannosides (PIMs), with the occurrence of PG being variable (Kroppenstedt, 1985; Kroppenstedt, 1987; Kroppenstedt et al., 1990). In addition, whole-organism hydrolysates of thermomonosporae usually contain madurose (3-*O*-methyl-D-galactose; Lechevalier and Gerber, 1970b). There are two exceptions to this profile of properties: thermomonosporae and thermophilic *Actinomadura* strains either lack madurose or synthesize it in trace amounts (Kroppenstedt, 1987), whereas actinocorallinae contain the diagnostic phosphatidylethanolamine in addition to the nondiagnostic phospholipids present in the members of other genera of the family Thermomonosporaceae (Iinuma et al., 1994). The guanine plus cytosine (G+C) content of the DNA lies within the range 66–72 mol% (Fischer et al., 1983; Poschner et al., 1985; Hasegawa et al., 1986; Miyadoh et al., 1987; Table 2).

The revised genus *Actinomadura* (Zhang et al., 2001) currently accommodates 32 validly described species and a single subspecies (Table 1), the members of which characteristically form nonfragmenting, extensively branched substrate mycelia and aerial hyphae that carry up to 15 arthrospores. Spore chains may be straight, hooked (open loops), or irregular spirals (1–4 turns) and spore surfaces folded, irregular, smooth, spiny or warty. The temperature range for growth is 10–60°C. Actinomadurae contain major proportions of hexahydrogenated menaquinones with 9 isoprenyl units, saturated at sites II, III and VIII, and complex mixtures of fatty acids with hexadecanoic, 14-methylpentadecanoic (*iso*-16:0) and 10-methyloctadecanoic acid (tuberculostearic acid) predominating, i.e., fatty acid type 3a (Kroppenstedt, 1985). The thermophilic actinomadurae *A. rubrobrunea* and *A. viridilutea*, formerly *Excellospora* (Agre and Guzeva, 1975), can be separated from the mesophilic actinomadurae by differences in fatty acid composition. They contain relatively high proportions of *iso*-branched fatty acids (high melting point) and low amounts of 10-methyl branched acids (low melting point), but these differences can be attributed to the thermophilic nature of these strains. A similar effect can be seen in the sugar composition of these strains as whole-organism hydrolysates contain galactose, glucose, mannose, and ribose with madurose present in trace amounts if at all.

The generic name *Thermomonospora* was proposed by Henssen (1957) for thermophilic actinomycetes isolated from composted stable manure. The genus contained three species that formed single spores on the aerial mycelium. All of these organisms produced colorless to

Table 1. Chemotaxonomic markers of members of the family Thermomonosporaceae and those species formerly misclassified as members of the genera *Actinomadura* and *Thermomonospora* and phylogenetically and morphologically related sporoactinomycetes.[a]

	Wall chemotype[b]	Diagnostic sugars[c]	Sugar type	G+C (mol%)	Diagnostic phospho-lipids[d]	Phospho-lipid type[e]	Principal menaquinones[f] MK-	Mena-quinone type[g]	Fatty acid type[h]
Thermomonosporaceae									
Thermomonospora									
Tms. curvata	III	None	C	ND	PI, PG, DPG	I	$9(H_4), 9(H_6)[1], 9(H_8)$	4B2	3a
Tms. chromogena	III	None	C	ND	PI, PG, DPG	I	$9(H_4), 9(H_6)[1], 9(H_8)$	4B2	3a
Actinomadura sp.	III	Madurose	B	66–72	PI, PG, DPG	I	$9(H_4), 9(H_6)[1], 9(H_8)$	4B2	3a
Acm. (Microbispora) echinospora	III	Madurose	B	ND	PI, PG, DPG	I	$9(H_4), 9(H_6)[1], 9(H_8)$	4B2	3a
Acm. (Thermomonosp) formosensis	III	Madurose	B	72	PI, PG, DPG	I	$9(H_4), 9(H_6)[1], 9(H_8)$	4B2	3a
Acm. (Microtetraspora) viridis	III	Madurose	B	ND	PI, PG, DPG	I	$9(H_4), 9(H_6)[1], 9(H_8)$	4B2	3a
Acm. (Excellospora) viridilutea	III	Madurose	B	ND	PI, PG, DPG	I	$9(H_4), 9(H_6)[1], 9(H_8)$	4B2	3a
Actinocorallia									
Acr. herbida	III	Madurose	B	73	PE, PI, PIM, DPG	II	$9(H_4), 9(H_6)$	4B	3a
Acr. (Acm.) aurantiaca	III	Madurose	B	66–70	PE, PI, PIM, DPG	II	$9(H_4), 9(H_6)[1], 9(H_8)$	4B2	3a
Acr. (Acm.) glomerata	III	Madurose	B	66–70	PE, PI, PIM, DPG	II	$9(H_4), 9(H_6)[1], 9(H_8)$	4B2	3a
Acr. (Acm.) libanotica	III	Madurose	B	66–70	PE, PI, PIM, DPG	II	$9(H_4), 9(H_6)[1], 9(H_8)$	4B2	3a
Acr. (Acm.) longicatena	III	Madurose	B	66–70	PE; PI, PIM, DPG	II	$9(H_4), 9(H_6)[1], 9(H_8)$	4B2	3a
Spririllospora									
Spirillospora albida	III	Madurose	B	71–73	PI, PIM, DPG, (PE)	I/II	$9(H_4), 9(H_6)[1]$	4B2	3a
Spl. rubra	III	Madurose	B	ND	PI, PIM, DPG	I	$9(H_4), 9(H_6)[1]$	4B2	3a
Streptosporangiaceae									
Streptosporangium	III	Madurose	B	69–71	PME, OH-PE, NPG	IV	$9(H_2), 9(H_4)[2], 9(H_6)$	4A2	3c
Acrocarpospora	III	Madurose	B/C	68–69	PE, PME, NPG variable	II/IV	$9(H_2), 9(H_4), 9(H_6)$	4A	3e
Herbidospora	III	Madurose	B	71	PME, OH-PE, NPG	IV	$10(H_2), 10(H_4)[2], 10(H_6)$	4C	3c
Microbispora	III	Madurose	B	71–73	PME, OH-PE, NPG	IV	$9(H_2), 9(H_4)[2], 9(H_6)$	4A2	3c
Mbs. (Tms.) mesophila	III	None	C	ND	PME, OH-PE, NPG	IV	$9(H_2), 9(H_4)[2], 9(H_6)$	4A2	3c
Microtetraspora	III	Madurose	B	71–73	PME, OH-PE, NPG	IV	$9(H_2), 9(H_4)[2], 9(H_6)$	4A2	3c
Nonomuraea	III	Madurose	B	64–69	ME, OH-PE, NPG	IV	$9(H_2), 9(H_4)[2], 9(H_6)$	4A2	3c
Nmr. (Acm) roseoviolacea carminata	III	Madurose	B	ND	PME, OH-PE, NPG	IV	$9(H_2), 9(H_4)[2], 9(H_6)$	4A2	3c
Planobispora	III	Madurose	B	70–71	PME, OH-PE, NPG	IV	$9(H_2), 9(H_4)[2], 9(H_6)$	4A2	3c
Planomonospora	III	Madurose	B	72	PME, OH-PE, NPG	IV	$9(H_2), 9(H_4)[2], 9(H_6)$	4A2	3c
Planotetraspora	III	Madurose	D, A	70–71	PME, OH-PE, NPG	IV	$9, 9(H_2), 9(H_4)$	4A3c	–
Nocardiopsaceae									
Nocardiopis sp.	III	None	C	64–69	PC, PME, PI, GL	III	$10(H_0)$ to $10(H_6)$	4C	3d
Streptomonospora	III	None	C	73–74	PC, PME, PI, PG	III	$10(H_0)$ to $10(H_6)$	4C	2d

(Continued)

Table 1. Continued

	Wall chemotype[b]	Diagnostic sugars[c]	Sugar type	G+C (mol%)	Diagnostic phospho-lipids[d]	Phospho-lipid type[e]	Principal menaquinones[f] MK-	Mena-quinone type[g]	Fatty acid type[h]
Thermobifida									
Tbf. (Tms.) alba	III	None	C	ND	PE, PME, GL	II	$10(H_6), 10(H_8), 11(H_6)$	4D	3e
Tbf. (Tms mesouviformis) alba	III	None	C	ND	PE, PME, GL	II	$10(H_6), 10(H_8), 11(H_6)$	4D	3e
Tbf. (Tms.) fusca	III	None	C	ND	PE, PME, GL	II	$10(H_6), 10(H_8), 11(H_6)$	4D	3e
Other sporoactinomycetes									
Actinosynnema	III	None	C	71–73	PE, OH-PE, LPE	II	$9(H_4)^2$	2D	3f
Streptoalloteichus	III	None	C	ND	PE, PI DPG	II	$9(H_4), 10(H_4)$	3B	ND
Saccharothrix	III	Rhamnose	E	70–76	PE, PI, DPG	II	$9(H_6), 10(H_6)$	3C	3g
Faenia	IV	Arabinose	A	70–72	PC, PME, LPE	III	$9(H_2), 9(H_4), 10(H_4)$	3B	3e
Nocardia[i,j]	IV	Arabinose	A	64–72	PE, PI, PIM, DPG	II	$8(H_{4cyclic})^3$	2E	1b
Saccharomonospora	IV	Arabinose	A	69–74	PE, OH-PE	II	$8(H_4), 9(H_4)$	3A	2a
Saccharopolyspora	IV	Arabinose	A	77	PE, PME, LPE	II	$9(H_2), 9(H_4), 10(H_4)$	3B/4A1	3e
Streptomyces	I	Not characteristic		66–75	PE, OH-PE var.	II	$9(H_6)^4, 9(H_8)$	4B1	2c

Symbol and abbreviations: *Acm, Actinomadura; Acr, Actinocorrallina; Mbs, Microbispora; Nmr, Nonomuraea; Spl, Spirillospora; Tbf, Thermobifida; Tms, Thermomonospora.*

[a]All organisms lack mycolic acids and contain N-acetylated muramic acid (i) apart from the *Nocardia* strains which contain mycolic acids (i) and have N-glycolated (i) muramic in the wall peptidoglycan. ND, not determined; and ( ), genera names which changed since the second edition of "*The Prokaryotes.*"

[b]Major constituents in cell walls type: I, LL-diaminopimelic acid (LL-A_2pm); II, meso-A_2pm and glycine; III, meso-A_2pm direct; IV, meso-A_2pm direct and glycine (Schleifer and Kandler, 1972): I with A3 γ, and III and IV with A1γ.

[c]Whole-organisms sugar patterns of actinomycetes containing meso-A_2pm: A, arabinose and galactose; B, madurose (3-O-methyl-D-galactose); C, no diagnostic sugars; E, rhamnose; –, not applicable (Lechevalier and Lechevalier, 1970a; Lechevalier and Lechevalier, 1970c; Lechevalier and Lechevalier, 1970d; Lechevalier et al., 1971; Labeda et al., 1984).

[d]Abbreviations of phospholipids: PC, phosphatidylcholine; PE, phosphatidylethanolamine; OH-PE, hydroxy-phosphatidylethanolamine; l-PE, lyso-phosphatidylethanolamine; PME, phosphatidylmethylethanolamine. Other nondiagnostic phospholipids may also be found, i.e., DPG, diphosphatidylglycerol; PG, phosphatidylglycerol; PI, phosphatidylinositol, and PIM, phosphatidylinositol mannosides. The occurrence of OH-PE is species specific in *Streptomyces* (R. M. Kroppenstedt, unpublished results).

[e]Phospholipid types according to Lechevalier et al. (1977) and Lechevalier et al. (1981).

[f]Abbreviations exemplified by MK-9(H_6); menaquinones having three of the nine isoprene units saturated (Collins and Jones, 1981; Kroppenstedt et al., 1981; Collins et al., 1982; Collins et al., 1984; Collins et al., 1988; Athalye et al., 1984): $9(H_4)^1$ = MK-9(II, III, VIII H_6), saturated isoprene units in positions two, three and eight (Batrakov et al., 1976; Batrakov and Bergelson, 1978; Kroppenstedt, 1982; Yamada et al., 1982a); $9(H_4)^2$ = MK-9(III, VIII H_4), saturated isoprene units in positions three and eight (Collins et al., 1988); $8(H_{4cyclic})^3$ = MK-8(H_4-o-cycl.), i.e., II, III-tetrahydro-ω-(2,6,6-trimethylcyclohex-2-enylmethyl) menaquinone-6 (Collins et al., 1988; Howard et al., 1986; Kroppenstedt, 1982; $9(H_6)^4$ = MK-9(II, III, IX H_6) (Yamada et al., 1982b).

[g]Menaquinone types according to Kroppenstedt (1985) exemplified by: type one, no hydrogenation of isoprene chain, e.g., MK-9; type two, only one di- or tetrahydrogenated menaquinone present, e.g., MK-9(H_4), and type three, tetrahydrogenated multiprenyl menaquinones, e.g., MK-9(H_4) + MK-10(H_4), and type four, menaquinones with mainly the same isoprenoid chain length but increasing degrees of hydrogenation, e.g., MK-9(H_4), MK-9(H_6), and MK-9(H_8). The position of the double bonds in the isoprene chain can be diagnostic, e.g., MK-9(II, III-H_4) in *Saccharothrix* vis-à-vis MK-9(II, VIII-H_4) in *Microtetraspora*; 4A1 against 4A2; another example is MK-9(II, III, VIII-H_6) in *Actinomadura* vis-à-vis MK-9(II, III, IX-H_6) in *Streptomyces*, 4B1 against 4B2 (Kroppenstedt, 1985; Kroppenstedt, 1987).

[h]Fatty acid types according to Kroppenstedt (1985) and Kroppenstedt and Kutzner (1978); compositions of the fatty acid types are shown in Table 4 of The Family Nocardiopsaceae in this Volume.

[i]All strains contain N-glycolylglucoseamine in the cell wall detected using the simple glycolate test (Uchida and Aida, 1977).

[j]All strains synthesize mycolic acids, α-branched β-hydroxylated long-chain fatty acids (Minnikin et al., 1975).

Data taken from Kroppenstedt (1987) and R. M. Kroppenstedt (unpublished data), *Bergey's Manual of Systematic Bacteriology*, volume 4 (Williams et al., 1989), Kudo et al. (1993), Itoh et al. (1995), Zhang et al. (2001), and Li et al. (2003).

pale yellow colonies and a white aerial mycelium but were distinguished from one another by aerial mycelium morphology and the type of branching shown by the substrate hyphae. *Thermomonospora curvata*, the only species isolated and maintained in pure culture, was later named as the type species of the genus (Henssen and Schnepf, 1967). The description of the remaining two species, *Thermomonospora fusca* and *Thermomonospora lineata*, was based on morphological properties in contaminated preparations. Neither of these species was included in the Approved Lists of Bacterial Names (Skerman et al., 1980) even though *Thermomonospora fusca* had in the meantime been isolated in pure culture and described in detail (Crawford, 1975; Crawford and Gonda, 1977). The name "*Thermomonospora fusca*" was validated (Moore et al., 1985), but this taxon was later transferred to the genus *Thermobifida* on the basis of comparative 16S rDNA sequencing results (Zhang et al., 1998). A mesophilic monosporic actinomycete was assigned to the genus as *Thermomonospora mesophila* (Nonomura and Ohara, 1971c) but was later transferred to the genus *Microbispora* as *Microbispora mesophila* (Miyadoh et al., 1990). A second mesophilic species, *Thermomonospora mesouviformis* (Nonomura and Ohara, 1974), was assigned to the genus *Thermobifida* as a synonym of *Thermobifida alba* (Zhang et al., 1998). A third mesophilic species, *Thermomonospora formosensis*, was described by Hasegawa et al. (1986) but cited as a species incertae sedis in *Bergey's Manual of Systematic Bacteriology* by McCarthy (1989). This species was reclassified by Zhang and coworkers and transferred to the genus *Actinomadura* as *A. formosensis* (Zhang et al., 1998).

Krassil'nikov and Agre (1964a) proposed the genus *Actinobifida* for actinomycetes that formed single spores on dichotomously branched sporophores, but they failed to mention that dichotomous branching had previously been observed in the genus *Thermomonospora* (Henssen, 1957) and in species of *Micromonospora* (Jensen, 1930; Jensen, 1932; Krassil'nikov, 1941). The following year these workers introduced a second species, *Actinobifida chromogena*, and in doing so, suggested that all actinomycetes showing dichotomous branching be transferred to the genus *Actinobifida*. A third species, *Actinobifida alba*, was proposed by Locci et al. (1967). *Actinobifida dichotomica*, the type species of the genus, was later transferred to the genus *Thermoactinomyces* given its ability to produce endospores, and *Actinobifida alba* was transfered to the genus *Thermomonospora* as it formed heat-sensitive spores on substrate and aerial hyphae (Cross and Goodfellow, 1973).

A comprehensive numerical taxonomic survey of the genus *Thermomonospora* and related organisms (McCarthy and Cross, 1984b; McCarthy and Cross, 1984c) confirmed the status of *T. curvata* and provided strong evidence for the formal recognition of *T. fusca*. In contrast, *T. mesouviformis* was considered to be a synonym of *T. alba*. These taxa, termed the "white *Thermomonospora* group" because of their white aerial mycelium, were sharply distinguished from organisms which included the type strain of *Actinobifida chromogena* (Krassil'nikov and Agre, 1965), "*Thermomonospora falcata*" (Henssen, 1970), and similar actinomycetes from mushroom compost (McCarthy and Cross, 1981). The "*chromogena*" strains with reddish-brown colonies and a light brown aerial mycelium had been provisionally included in the genus *Thermomonospora* (Cross, 1981) because of their wall composition and morphology. In the meantime *Thermomonospora viridis* (Küster and Locci, 1963) had been transferred to the genus *Saccharomonospora* (Nonomura and Ohara, 1971c).

Members of the genus *Spirillospora*, which were originally assigned to the family Streptosporangiaceae, may be easily confused on morphological grounds with markers of some species of the genus *Streptosporangium*. All strains belonging to these taxa have multispored, usually spherical spore vesicles carried on aerial hyphae, but members of the genus *Spirillospora* produce zoospores whereas zoospores have never been detected in members of the genus *Streptosporangium*. Nevertheless, it is evident from nucleic reassociation studies, comparative 16S rDNA sequencing, and chemical data that members of the genus *Spirillospora* belong to the family Thermomonosporaceae (Stackebrandt et al., 1981b; Goodfellow, 1989b; Stackebrandt et al., 1997; Zhang et al., 2001). Like other members of this family, the type strain of *Spirillospora albida* has a type I polar lipid pattern (Lechevalier et al., 1981) and a type 3a fatty acid profile (Kroppenstedt, 1985) though previous workers reported a type II phospholipid pattern for *S. albida* ATCC 14541[T] (Hasegawa et al., 1978; Lechevalier et al., 1981).

Combining the descriptions of Henssen (1957), Couch (1963), McCarthy and Cross (1984b), McCarthy and Cross (1984c), Goodfellow et al. (1988), Vobis and Kothe (1989), Meyer (1989a), Meyer (1989b), Kroppenstedt et al. (1990), Goodfellow (1992), Kroppenstedt and Goodfellow (1992), Iinuma et al. (1994), Kudo (1997), Stackebrandt et al. (1997), Zhang et al. (1998), and Zhang et al. (2001)), the family Thermomonosporaceae can be described as follows: Aerobic, Gram-positive, non-acid-fast, chemo-organotrophic actinomycetes which produce a branched substrate mycelium which bears aerial hyphae that differentiate into either single or short chains of arthrospores or spore vesicles containing zoospores. The G+C content of the DNA lies in the range 66–72 mol%. The family Thermomonosporaceae contains four genera, namely *Thermomonospora*, *Actinocorallia*, *Actinomadura* and *Spirillospora*, which form a distinct clade in the suborder Streptosporangineae and are closely related to members of the families Streptosporangiaceae and Nocardiopsaceae. Members of the family Thermomonosporaceae share the same cell wall type (type III; *meso*-diaminopimelic acid), a similar menaquinone profile type 4B2 sensu Kroppenstedt (1985) [MK-9(H$_4$), MK-9(H$_6$) and MK-9(H$_8$)], in which MK-9(H$_6$) is predominant, and a fatty acid profile type 3a (Kroppenstedt, 1985), properties that differentiate them from members of the families Nocardiopsaceae and Streptosporangiaceae. The presence of madurose is variable, but this sugar can be found in members of most species of the family. The polar lipid profiles are characterized as phospholipid type PI sensu Lechevalier et al. (1977), which indicates the presence of PIM, PI, PG, and diphosphatidyl glycerol (DPG) in most species of *Thermomonospora*, *Actinomadura* and *Spirillospora*; *Actinocorallia* strains are characterized by a phospholipid type PII because of the presence of phosphatidylethanolamine (PE). Further studies need to be performed to clarify the taxonomic position of three *Actinomadura* species, *Spirillospora rubra*, and *Thermomonospora chromogena*. Comparative 16S rDNA sequencing of thermomonosporae (Zhang et al., 2001) revealed that *Thermomonospora curvata*, *Actinomadura echinospora* and *A. umbrina* form a fairly stable clade (bootstrap value 63%) which is separated from one containing the rest of the *Actinomadura* species, apart from *A. spadix*, which is distinct from all of the other members of the family. The separate position of *A. spadix*, is supported by numerical phenetic data (Athalye et al., 1985) and by an electrophoretic mobility study of ribosomal protein AT-L30 (Ochi et al., 1991). *Spirillospora albida*, the type species of *Spirillospora*, is closely related to many *Actinomadura* species. The 16S rRNA gene sequence similarity of *S. albida* to some *Actinomadura* strains is significantly higher than most similarity scores between representatives of *Actinomadura* species. In contrast, *S. rubra* is only distantly related to *S. albida* and to the other species of the family Thermomonosporaceae, which suggests that *Spirillospora rubra* may merit an independent genus status (Zhang et al., 2001).

The family Thermomonosporaceae, as defined above, encompasses the genera *Actinocorallia* and *Actinomadura* and the two additional genera that encompass *Spirillospora albida*, *S. rubra*, and *Thermomonospora curvata* and *T. chromogena* (Table 1; Fig. 4).

Habitat

Habitats of *Actinomadura* and *Actinocorallina*

Actinomadura and *Actinocorallina* species are widely distributed in soil (Nonomura and Ohara, 1971a; Nonomura and Ohara, 1971d; Preobrazhenskaya et al., 1978; Meyer, 1979; Athalye et al., 1981b; Labeda et al., 1985; Miyadoh et al., 1987; Iinuma et al., 1994; Trujillo and Goodfellow, 2003), where they probably have a role in the turnover of organic matter. *Actinomadura latina*, *A. madurae* and *A. pelletieri* are agents of actinomycete mycetoma. *Actinomadura latina* and *A. pelletieri* have only been isolated from clinical material, but *A. madurae* seems to be widespread in soil. Isolates of the latter from environmental samples lack the red endopigment of clinical isolates and sporulate more readily. The red pigments are characterized by a tripyrrole skeleton and have been identified as cyclononylprodiginine, nonylprodiginine and undecylprodiginine (Gerber, 1971; Gerber, 1973; Lechevalier et al., 1971).

Information on the occurrence and frequency of *Actinomadura* species in different soils has mainly been provided by Preobrazhenskaya and her coworkers (Preobrazhenskaya et al., 1978; Chormonova and Preobrazhenskaya, 1981; Galatenko et al., 1981; Terekhova et al., 1982), who found that the total number of *Actinomadura* strains was higher in cultivated than in uncultivated soil. The highest number of actinomadurae was isolated from chernozem soil (Kazakhstan) and the lowest from sierozem soil (Turkmenistan). Dark chestnut soil (cultivated) contained half the number of actinomadurae found in the chernozem soil, but the species diversity in each of the soil types was similar. *Actinomadura citrea* was the most frequent species in the uncultivated and cultivated soils, followed by *A. cremea* and *A. verrucosospora*. *Actinomadura viridilutea* was isolated from an arid soil from the Soviet Union, and *A. rubrobrunea* was isolated from Egyptian soils under maize and rice (Agre and Guzeva, 1975). There is evidence that both clinically and ecologically significant actinomadurae are underspeciated (Trujillo and Goodfellow, 2003).

Habitats of *Spirillospora*

Members of the genus *Spirillospora* occur in soil, albeit infrequently. They may be isolated from

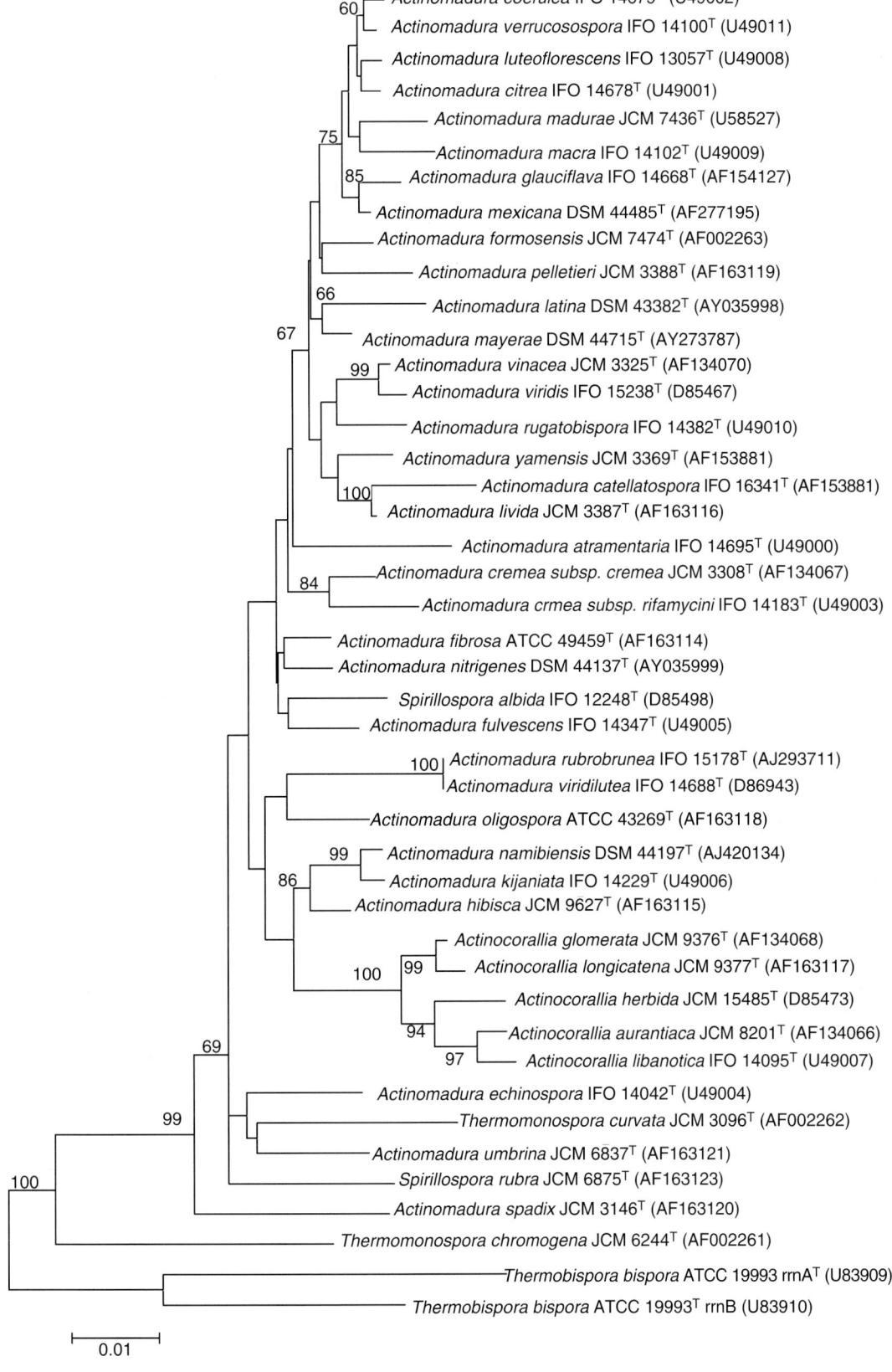

Fig. 4. Neighbor-joining tree (Saitou and Nei, 1987) based on nearly complete 16S rDNA sequences showing the relationship between members of the family Thermomonosporaceae. The numbers at the nodes indicate levels of bootstrap support based on a neighbor-joining analysis of 1000 resampled datasets; only values above 50% are given. Scale indicates 0.01 substitutions per nucleotide position. T, type strain.

this milieu by applying the bait technique (using pollen for the isolation of *S. alba* or hair for the recovery of *S. rubra*); these organisms account for less than 1% of the "sporangiate" actinomycetes isolated from soil using baits (Schäfer, 1973).

Habitats of *Thermomonospora*

The ability of thermomonosporae to secrete a variety of thermostable extracellular enzymes enables them to become established as dominant populations during high temperature composting of plant residues and other wastes (Fergus, 1964; Stutzenberger, 1971; Bernier et al., 1988b). In addition to the true *Thermomonospora* species (i.e., *T. curvata* and *T. chromogena*), other thermophilic monosporic actinomycetes need be considered here because it is possible that strains assigned to the species may belong to the genus *Thermobifida*, a genus of the family Nocardiopsaceae (Zhang et al., 1998).

Thermophilic thermomonosporae are common in overheated substrates such as bagasse, compost, fodders and manures. They are especially abundant in mushroom compost (Fergus, 1964; Lacey, 1974; Lacey, 1977; McCarthy and Cross, 1981; McCarthy and Cross, 1984b) and are highly cellulolytic (McCarthy, 1987; Ball and McCarthy, 1988) or hemicellulolytic or both (McCarthy et al., 1985; McCarthy et al., 1988). *Thermomonospora curvata* and *Thermobifida fusca* strains show endogluconase activity and attack cellulose. The production of endogluconase (e.g., cellobiohydrolase) by thermomonosporae is currently regarded as unsubstantiated. *Thermomonospora* strains also grow on rapeseed that has been subject to heat damage during storage (Mills and Bollen, 1976).

Thermomonosporae can extensively degrade cellulose and lignocellulose residues that make up the bulk of agricultural and urban wastes (McCarthy, 1987a; McCarthy, 1987b). *Thermomonospora curvata* is active in the decomposition of municipal waste compost (Stutzenberger et al., 1970; Stutzenberger, 1971; Stutzenberger, 1972a; Stutzenberger, 1972b); heavy metals in such composts have been shown to inhibit cellulase production (Stutzenberger and Sterpu, 1978). Similarly, thermomonosporae are very active against arabinoxylan and produce enzymes that have thermostability properties (McCarthy et al., 1985). Evidence that cellulase decomposition in mushroom compost involves a positive interaction between cellulolytic actinomycetes and certain bacteria (Staneck, 1972) underlines the importance of considering natural lignocellulose degradation as the product of an interactive heterogeneous microflora.

Isolation and Cultivation

Isolation and Cultivation of *Actinomadura*

Noncontaminated clinical samples, such as pus and biopsy material, thought to contain clinically significant actinomadurae, notably *A. madurae* and *A. pelletieri*, should be inoculated into media such as brain heart infusion (Oxoid CM261; Schaal, 1972), Sabouraud dextrose (Gordon, 1974) and yeast extract agars (Pridham et al., 1956–57). Media should be transparent to facilitate direct microscopic examination of colonies. Selective isolation media need to be used for the examination of clinical material that contains large numbers of other microbes. Actinomadurae form small colonies on media selective for nocardiae (see The Families Nocardiaceae Dietziaceae, Gordoniaceae, and Tsukamurellaceae in this Volume). All cultures should be inoculated aerobically at 2–27°C and at 36°C for up to three weeks and examined both macroscopically and microscopically for growth every two days (Schaal, 1984a). Actinomadurae can be recognized by their filamentous appearance, by leathery colonies, and by the production of red prodiginine pigments. Actinomycetoma granules should be washed in sterile tap water before they are crushed to obtain material for inoculation of culture media.

Media formulations supplemented with antifungal antibiotics such as actidione and cycloheximide can be used to isolate actinomadurae from environmental samples. Media that have proved to be effective include egg albumin (Lawson and Davey, 1972), glucose yeast extract (Athalye et al., 1981b), glycerol asparagine (ISP medium 5; Shirling and Gottlieb, 1966), inorganic salts and starch (ISP medium 4; Shirling and Gottlieb, 1966), oatmeal (ISP medium 3; Shirling and Gottlieb, 1966) and yeast extract agars (ISP medium 2; Shirling and Gottlieb, 1966). Strains may be isolated from agar plates after incubation for 14–21 days.

Isolation of actinomadurae from enriched soil can be achieved using pretreatment regimes and selective agents. Nonomura and Ohara (1971d) reduced the number of unwanted microorganisms by air-drying soil and applying dry heat at 100°C for 1 h before plating onto various media, notably arginine-vitamins (AV) and mineral-glucose-asparagine (MGA) agars, and incubating for several weeks at 28–30°C. Soviet microbiologists (Lavrova et al., 1972; Preobrazhenskaya et al., 1975b) increased the number of actinomadurae isolated by the addition of antibiotics to medium no. 2 of Gauze et al. (1957); the antibiotics were used to inhibit the growth of bacteria and the more frequently occurring streptomycetes, thereby providing more favorable conditions for the growth of the slow-growing actinomadurae. Bruneomycin (0.5, 1.0 or 2.0 μg/ml), rubomycin (5.0, 10.0 or 20 μg/ml), and streptomycin (0.5, 1.0 or 2.0 μg/ml) were the most successfully used antibiotics. Athalye et al. (1981b) combined drying and heat pre-treatment regimes with the use of rifampicin (5 μg/ml) as the selective agent. *Actinomadura rubrobrunea* was isolated by scattering crushed soil over the medium of Kosmachev (1960) and incubating at 55°C. Similarly, *A. formosensis* was obtained by plating soil dilutions onto starch casein agar (Waksman, 1961) supplemented with kabicidin (6.25 μg/ml) and rifampicin (12.5 μg/ml; Hasegawa et al., 1986).

Actinomadurae generally grow well on modified Bennett's agar (Jones, 1949), glucose yeast extract agar (Athalye et al., 1981b), and on formulations used for the cultivation of streptomycetes in the International *Streptomyces* Project (ISP; Shirling and Gottlieb, 1966). Most strains show abundant growth on oatmeal agar (ISP medium 3) at 30°C, but *A. kijaniata* and *A. macra* grow better on yeast extract malt extract agar (ISP medium 2). *Actinomadura spadix* requires vitamin B_{12} for good growth (Nonomura and Ohara, 1971d), and *A. vinacea* grows well on glucose peptone agar but poorly on the media mentioned above. *Actinomadura rubrobrunea* and *A. viridilutea* grow well on oatmeal and peptone maize agars at 50°C.

Isolation and Cultivation of *Actinocorallia*

Members of this genus are widely distributed in soil and can be isolated using the methods described for *Actinomadura* strains. A specific method was described by Iinuma et al. (1994) for *Actinocorallia herbida*, which was isolated from a soil sample in Bangkok, Thailand, by the dilution agar plating method, using colloidal chitin-vitamin agar. The isolate was subcultured on OMH agar containing 2.2% ISP medium 3, 0.1% yeast extract, 0.02% humic acid, and 1.5% agar or on yeast extract malt extract agar (ISP medium 2) at 28°C for 10 days.

Isolation and Cultivation of *Spirillospora*

Spirillospora strains have been recovered from soil by baiting with natural substrates (Couch, 1954; Bland and Couch, 1981). Using the baiting technique, Schäfer (1973) found that less than 1.0% of the isolates of "sporangiate" actinomycetes were represented by this genus. The following method was used: A small amount of soil, approximately one level teaspoonful, is placed in a sterile Petri dish and flooded with sterile water (distilled water or filtrated soil or charcoal water extracts may be used). Added pollen and hair float at the water surface; various types of pollen have been employed including that from members of the genera *Liquidamber, Pinus* and *Sparganium* (Schäfer, 1973). After 1–4 weeks, examination of the water surface with a dissecting microscope (100X) and strong horizontal lighting should reveal the presence of white glistening spore vesicles formed in the air at the surface of the water by "sporangiate" members of the families Micromonosporaceae, Streptosporangiaceae and the genus *Spirillospora* if they are present. The spore vesicles of *Spirillospora albida* develop on long aerial hyphae growing between the baits on the pollen grain whereas *Spirillospora rubra* could only be isolated from hair (Schäfer, 1973). Single spore vesicles can be picked up with a thin needle and placed on the surface of agar media in small Petri dishes. After 2–4 weeks, the young colonies can be transferred to slant cultures. Various complex media support the growth of *Spirillospora* strains. *Spirillospora albida* grows well on Czapek, peptone-Czapek, and oatmeal agars, whereas *S. rubra* prefers half concentrated corn-meal agar (Difco) supplemented with sterilized garden soil (50 g/liter) or half concentrated skim milk agar (Difco; Schäfer, 1973).

Isolation and Cultivation of *Thermomonospora*

Thermophilic *Thermomonospora* strains can be isolated from composts and overheated vegetable material by dilution plating on nonselective media, but recovery is poor owing to the rapid competing growth of *Bacillus* and *Thermoactinomyces* strains. The most effective isolation methods are those based on the use of a sedimentation chamber and an Andersen sampler (Andersen, 1958; Lacey and Dutkiewicz, 1976; McCarthy and Cross, 1981; McCarthy and Broda, 1984a). Dried environmental samples are shaken within the chamber to create an aerosol of propagules that, after 1–2 h of sedimentation, still contains many actinomycete spores but relatively few bacteria. Actinomycetes are isolated from this spore suspension using an Andersen sampler loaded with half-strength tryptone soy agar plates supplemented with cycloheximide (50 μg/ml) to prevent growth of fungi. The recovery of *Thermomonospora* strains can be enhanced by adjusting the isolation medium to a high pH. *Thermomonospora chromogena* strains can grow up to pH 10.0, and *T. curvata* show significant growth even at pH 11 (Kempf, 1995). Cellulolytic isolates may be detected by incorporating cellulose powder or ball-milled straw into the agar (Stutzenberger et al., 1970; McCarthy and Broda, 1984a).

Thermomonospora colonies can usually be recognized after 3–5 days incubation at 50°C. Care must be taken to distinguish these target colonies from colonies of other filamentous bacteria with white aerial mycelium that grow at increased temperatures, for example, *Streptomyces albus* and *Thermoactinomyces* strains. However, *Thermomonospora* colonies can be spotted by carefully searching isolation plates with a 40X long-working-distance objective and looking for the characteristic single spores on the aerial mycelium. The morphology of pure cultures can be confirmed by observing plate cultures or inclined cover slip preparations as described by Williams and Cross (1971).

Thermomonospora chromogena is readily isolated on selective media containing kanamycin (25 μg/ml; McCarthy and Cross, 1981) or rifampicin (5 μg/ml; Athalye et al., 1981b). Thermomonosporae have been isolated from soil samples using the selective procedure that involves heating environmental samples at 100°C prior to preparing suspensions for plating out (Nonomura and Ohara, 1971c; Nonomura and Ohara, 1974). Mixtures of organic matter and soil, inoculated in partially sealed polyethylene bags for a day, yield samples enriched in thermophilic actinomycetes, including thermomonosporae (McCarthy, 1989). Henssen (1957) used an anaerobic enrichment method to isolate her original *Thermomonospora* strains, but the procedure and apparatus involved were unwieldy and may not have produced anything like anaerobic conditions.

Thermomonospora chromogena and *T. curvata* strains grow well at pH 7.5 on any nutrient medium that contains some yeast extract. They are readily cultivated on Czapek peptone agar, glucose-yeast extract-malt extract agar (Greiner-Mai et al., 1987), Hickey and Tresner agar (Hickey and Tresner, 1952), nutrient and peptone-maize agar (Greiner-Mai et al., 1987), and ISP media (Shirling and Gottlieb, 1966) after 2–4 days incubation at their optimal temperature of 50°C (Greiner-Mai et al., 1987).

Media and Techniques for Isolation and Cultivation

The following are the formulae of the media mentioned in the preceding section that are not commercially available. Quantities are diluted to 1 liter with distilled water unless otherwise stated.

AV Agar with Vitamins (Nonomura and Ohara, 1969)

L-Arginine	0.3 g
Glucose	1.0 g
Glycerol	1.0 g
K_2HPO_4	0.3 g
$MgSO_4 \cdot 7H_2O$	0.2 g
NaCl	0.3 g
Agar	15 g

Adjust to pH 8.0 with NaOH.

Trace Salts Solutions

$CuSO_4 \cdot 5H_2O$	1.0 mg/ml
$Fe_2(SO_4)_3$	10 mg/ml
$MgSO_4 \cdot 7H_2O$	1.0 mg/ml
$ZnSO_4 \cdot 7H_2O$	1.0 mg/ml

Adjust pH of each to 8.0.

Vitamins and Antibiotics (dissolve in 10 ml)

p-Aminobenzoic acid	0.5 mg
Calcium pantothenate	0.5 mg
Inositol	0.5 mg
Niacin	0.5 mg
Pyridoxine · HCl	0.5 mg
Riboflavin	0.5 mg
Thiamine · HCl	0.5 mg
Biotin	0.25 mg
Cycloheximide	50 mg
Nystatin	50 mg
Nalidixic acid	0 or 20 mg
Penicillin G 0 or 0.8 mg Polymyxin	0 or 4 mg

Adjust final pH to 6.4.

Add to the sterilized basal medium aseptically, 10 ml of filter sterilized trace salt solution, and 10 ml of filter sterilized Vitamin and Antibiotics solution.

Bennett's Agar (Medium 548, DSMZ Catalogue; Jones, 1949)

Beef extract	1 g
Glucose	10 g
N-Z amine A (enzymatic digest of casein)	2 g
Yeast extract	1 g
Agar	15 g

Adjust to pH 7.3 with NaOH.

Modified Bennett's agar can be used where glucose is replaced by glycerol (10 g) and N-Z amine by Bacto-casein (2 g, Difco).

Chitin-Vitamins Agar (Iinuma et al., 1994)

Chitin	1.00 g
KCl	1.71 g
Na_2HPO_4	1.63 g
Agar	18 g
$CaCO_3$	20 mg
$FeSO_4 \cdot 7H_2O$	10 mg
Thiamine · HCl	1 mg
Riboflavin	1 mg
Niacin	1 mg
Pyridoxine · HCl	1 mg
Inositol	1 mg
Calcium pantothenate	1 mg

p-Aminobenzoic acid	1 mg
Biotin	0.5 mg
Cycloheximide	50 mg
Kabicidin	10 mg

Adjust to pH 7.2.

Czapek Agar (Medium 130, DSMZ Catalogue; Waksman, 1950)

Sucrose	30 g
$FeSO_4 \cdot 7H_2O$	0.01 g
KCl	0.5 g
K_2HPO_4	1.0 g
$MgSO_4 \cdot 7H_2O$	0.5 g
$NaNO_3$	3.0 g
Agar	15 g

Dissolve components, dispense the medium into flasks or tubes, and then autoclave. For peptone-Czapek agar, substitute 5 g of peptone for the 3 g of $NaNO_3$ (Bland and Couch, 1981).

GYM Agar (Medium 65, DSMZ Catalogue; Greiner-Mai et al., 1987)

Glucose	4 g
Yeast extract	410 g
Malt extract	10 g
$CaCO_3$	2 g
Agar	12 g

Adjust to pH 7.2.

Glucose Peptone Agar (Medium 85 DSMZ Catalogue)

Bacto peptone (Difco)	20 g
Glucose	10 g
Yeast extract	10 g
$CaCO_3$	10 g
Agar	15 g

Adjust to pH 7.3.

Glucose Yeast Extract Agar (Waksman, 1950)

Glucose	10 g
Yeast extract (50% solution)	20 ml
Agar	15 g

Adjust pH to 7.5.

Hickey and Tresner Agar (Medium 658, DSMZ Catalogue; Hickey and Tresner, 1952)

Dextrin	10 g
Beef extract (Difco)	1 g
NZ amine A or Tryptone (Difco)	2 g
Yeast extract (Difco)	1 g
$CoCl_2 \cdot 6H_2O$	2 mg
Agar	20 g

Adjust pH to 7.3.

The dextrin "Amidax" (Corn Products Refining Co., Argo, IL, United States) was used in the original formulation.

ISP Medium 2: Yeast Extract-Malt Extract Agar (Medium 987; DSMZ Catalogue; Pridham et al., 1956–57)

Dextrose (Difco)	4 g
Malt extract (Difco)	10 g
Yeast extract (Difco)	4 g

Adjust to pH 7.3, and then add 20 g of Bacto agar. Liquefy agar by steaming at 100°C for 15–20 min.

ISP Medium 3: Oatmeal Agar (Küster, 1959b)

Oatmeal	20.0 g
Agar	18.0 g

Cook or steam 20 g of oatmeal in 1 liter of distilled water for 20 min. Filter through cheesecloth. Add distilled water to restore volume of filtrate to 1000 ml. Add 1 ml trace salt solution ($FeSO_4 \cdot 7H_2O$, 0.1 g; $MnCl_2 \cdot H_2O$, 0.1 g; $ZnSO_4 \cdot 7H_2O$, 0.1 g; and distilled water, 100.0 ml). Adjust to pH 7.5 with NaOH. Add 18 g of agar; liquefy by steaming at 100°C for 15–20 min.

ISP Medium 4: Inorganic Salt-Starch Agar (Medium 547, DSMZ Catalogue; Küster, 1959a)

Solution I

Difco soluble starch	10.0 g

Make a paste of the starch with a small amount of cold distilled water and bring to a volume of 500 ml.

Solution II

$CaCO_3$	2 g
K_2HPO_4 (anhydrous)	1 g
$MgSO_4 \cdot 7H_2O$	1 g
NaCl	1 g
$(NH_4)_2SO_4$	2 g
Distilled water	500 ml
ISP medium 3 trace salt solution	1 ml

The pH should be 7.0–7.4. Do not adjust if it is within this range. Mix solutions I and II together. Add 20 g of agar. Liquefy agar by steaming at 100°C for 10–20 min.

ISP Medium 5: Glycerol-Asparagine Agar (Medium 993, DSMZ Catalogue; Pridham and Lyons, 1961)

L-Asparagine (anhydrous)	1.0 g
Glycerol	10.0 g
K_2HPO_4 (anhydrous)	1.0 g
ISP medium 3 trace salt solution	1.0 ml

Do not adjust if the pH of this solution is about 7.0–7.4. Add 20 g of agar. Liquefy agar by steaming at 100°C for 15–20 min.

ISP Medium 7: Tyrosine Agar (Shinobu, 1958)

L-Asparagine	1.0 g
Glycerol	15 g
L-Tyrosine	0.5 g
$FeSO_4 \cdot 7H_2O$	0.01 g
K_2HPO_4 (anhydrous)	0.5 g
$MgSO_4 \cdot 7H_2O$	0.5 g
NaCl	0.5 g
ISP medium 3 trace salt solution	1.0 ml

Adjust pH to 7.2–7.4. Add 20 g of agar. Liquefy agar by steaming at 100°C for 15–20 min.

Kosmachev's Medium (Kosmachev, 1960)

Yeast autolysate (30% [w/v])	15 ml
$CaCO_3$	4 g
$FeSO_4 \cdot 7H_2O$	0.01 g
KNO_3	1 g
$MgSO_4 \cdot 7H_2O$	0.5 g
Na_2HPO_4	1 g
$(NH_4)_2SO_4$	1 g
Agar	15 g

MGA Agar (Nonomura and Ohara, 1971b)

L-Asparagine	1 g
Glucose	2 g
K₂HPO₄	0.5 g
MgSO₄ · 7H₂O	0.5 g
Agar	20 g

Add the trace salt solutions as indicated for AV agar with vitamins. Add antibiotics: cycloheximide, 50 mg; nystatin, 50 mg; benzylpenicillin, 0.8 mg; and polymyxin B, 4 mg.

Medium No 2 (Gauze et al., 1957)

Glucose	10 g
Hottinger's broth	30 ml
Peptone	5 g
NaCl	10 g
Agar	15 g

Peptone Corn Meal Agar (Medium 85 831, DSMZ Catalogue; Greiner-Mai et al., 1987)

Corn meal	17 g
Peptone	5 g
Starch	10 g
CaCl₂ · 2H₂O	0.5 g
NaCl	5.0 g
Agar	15 g

Adjust pH to 7.2 before sterilization.

Modified Skim Milk-Mineral Agar (Schäfer, 1973)

Skim milk (powder)	2.5 g
Ca(NO₃)₂ · 4H₂O	0.5 g
MgSO₄ · 7H₂O	0.7 g
K₂HPO₄	0.005 g
NaHCO₃	0.2 g
FeCl₃	0.005 g
Agar	15 g

Autoclave the skim milk powder separately in 100 ml of H₂O and add it to the agar-medium at 60°C.

Starch Casein Agar (Waksman, 1961)

Casein	1.0 g
Soluble starch	10 g
K₂HPO₄	0.5 g
Agar	15 g

Adjust pH to 7.0–7.5.

Preservation

Heavily sporulated cultures are needed to maintain high viability irrespective of the preservation method. The highest survival rates for cultures that do not sporulate are achieved using cells from the logarithmic growth phase. The culture age of the organisms to be preserved is very important especially for the preservation of thermophilic strains that readily lyse at the stationary phase of growth.

Sporulated *Actinomadura* strains can be kept on oatmeal agar, or other suitable agar slants, at 4°C and transferred every four months for short-term preservation. The tubes should be sealed with silicone stoppers to prevent the agar drying out. The same procedure can be used for strains

of *Thermomonospora*, but in this case, cultures should be transferred every one to two months using Czapek Dox-yeast extract-casamino acids agar at pH 8.0 (McCarthy, 1989). Medium-term preservation for up to four years can be achieved by mixing spore suspensions or homogenized mycelia with glycerol (45%, v/v) and storing at –25°C (Zippel and Neigenfind, 1988). Lyophilization of spores and mycelia suspended in 10% skim milk is a convenient method for long-term storage. An alternative simple, reliable and quick method involves nitrogen cryopreservation of living cells in small polyvinyl chloride tubes ("straws") at –196°C (Hoffmann, 1989a; Hoffmann, 1989b). This procedure has been shown to be reliable for actinomycetes, including *Actinomadura*, *Actinocorallina*, *Spirillospora* and *Thermomonospora* (see The Family Nocardiopsaceae in this Volume).

Identification

Members of the genera *Actinomadura*, *Actinocorallina*, *Spirillospora* and *Thermomonospora* can be distinguished from all other actinomycetes using a combination of chemical and morphological features. Primary diagnostic information can be gained by the detection of marker amino acids, fatty acids, menaquinones, sugars, and polar lipids (Table 2). Simple procedures have been devised for the detection of these chemical features (Staneck and Roberts, 1974; Lechevalier and Lechevalier, 1980; Minnikin et al., 1984; Kroppenstedt, 1985).

One-dimensional thin-layer chromatography will determine if an organism contains diaminopimelic acid and whether the latter is in the LL- or *meso*-form. The presence of LL-A₂pm will lead to *Streptomyces* and related genera (see The Family Streptomycetaceae, Part I in this Volume and The Fanily Streptomycetaceae, Part II in this Volume). The detection of various combinations of *meso*-A₂pm, mixtures of straight and branched chain fatty acids, predominant amounts of hexahydrogenated menaquinones with nine isoprene units, and large amounts of DPG, PG and PI (phospholipid type I sensu Lechavalier et al. [1977]) can be used to separate strains of *Thermomonospora*, *Actinomadura* and *Spirillospora* from all other sporoactinomycetes, notably those classified in the families Actinoplanaceae (see The Family Actinomycetaceae in this Volume), Nocardiaceae (see The Family Nocardiaceae in this Volume), Pseudonocardiaceae (including *Saccharothrix*)(see The Family Pseudonocardiaceae in the second edition) and Streptosporangiaceae (see The Family Streptosporangiaceae in this Volume), and in the genus *Thermoactinomyces*. The reports on the phospholipid composition of *Spirillospora albida*

Table 2. Taxonomic history of actinomycetes designated or once assigned to the genera *Actinomadura*, *Actinocorallia*, *Spirillospora* and *Thermomonospora*.

Original assignment	DSMZ-accession numbers of type strains	Bergey's Manual of Systematic Bacteriology volume 4[a,c]	The Prokaryotes 1992[c]	Current designation Approved Lists of Bacterial Names[b,c]
Genus Actinocorallia				
Actinomadura aurantiaca (1)	DSM 43924	*Actinomadura aurantiaca*	*Actinomadura aurantiaca*	*Actinocorallia aurantiaca* (2, 3)
Actinomadura glomerata (4)	DSM 44360	—		*Actinocorallina glomerata* (3)
Actinocorallina herbida (5)	DSM 44254			*Actinocorallia herbida*
Actinomadura libanotica (20, 21)	DSM 43554	*Actinomadura libanotica*	*Actinomadura libanotica*	*Actinocorallia libanotica* (3)
Actinomadura longicatena (4)	DSM 44361			*Actinocorallina longicatena* (3)
Genus Actinomadura				
Actinomadura atramentaria (6)	DSM 43919	—	*Actinomadura atramentaria*	*Actinomadura atramentaria* (7)
Actinomadura catellatispora	DSM 44772	—		*Actinomadura catellatispora* (8)
Actinomadura citrea (9)	DSM 43461	*Actinomadura citrea*	*Actinomadura citrea*	*Actinomadura citrea*
Actinomadura coerulea (10)	DSM 43675	*Actinomadura coerulea*	*Actinomadura coerulea*	*Actinomadura coerulea*
Actinomadura cremea (10)	DSM 43676	*Actinomadura cremea*	*Actinomadura cremea* subsp. *cremea* (11)	*Actinomadura cremea* subsp. *cremea*
Actinomadura cremea subsp. *rifamicini* (11)	DSM 43936		*Actinomadura cremea* subsp. *rifamicini*	*Actinomad. cremea* subsp. *rifamicini*
Microbispora echinospora (12)	DSM 43163	*Microbispora echinospora*	*Actinomadura echinospora* (13, 14)	*Actinomadura echinospora*
Actinomadura fibrosa (15)	DSM 44224		*Actinomadura fibrosa*	*Actinomadura fibrosa*
Thermomonospora formosensis (16)	DSM 43997	*Thermomonospora formosensis*	*Thermomonospora formosensis*	*Actinomadura formosensis* (7)
Actinomadura fulvescens (17, 18)	DSM 43923		*Actinomadura fulvescens*	*Actinomadura fulvescens*
Actinomadura glauciflava	DSM 44770			*Actinomadura glauciflava* (8)
Actinomadura hibisca	DSM 44148			*Actinomadura hibisca*
Actinomadura kijaniata (19)	DSM 43764	*Actinomadura kijaniata*	*Actinomadura kijaniata*	*Actinomadura kijaniata*
Actinomadura latina (20)	DSM 43382	—		*Actinomadura latina* (20)
Actinomadura livida (1)	DSM 43677	*Actinomadura livida*	*Actinomadura livida*	*Actinomadura livida*
Streptomyces luteofluorescens (22)	DSM 40398	*Actinomadura luteofluorescens* (22a)	*Actinomadura luteofluorescens*	*Actinomadura luteofluorescens*
Actinomadura macra (23)	DSM 43862	*Actinomadura macra*	*Actinomadura macra*	*Actinomadura macra*
Actinomadura madurae (24)	DSM 43067	*Actinomadura madurae* (24)	*Actinomadura madurae*	*Actinomadura madurae*
Nocardia madurae (25)				
Streptomyces madurae (26)				
Streptothrix madurae (27)				
Actinomadura mexicana (28)	DSM 44485	—		*Actinomadura mexicana*
Actinomadura meyerae (28)	DSM 44715	—		*Actinomadura meyerae*
Actinomadura namibiensis (29)	DSM 44197	—		*Actinomadura namibiensis*
Actinomadura nitrigens (30)	DSM 44137			*Actinomadura nitrigens*
Actinomadura oligospora (31)	DSM 43930	*Actinomadura oligospora*	*Actinomadura oligospora*	*Actinomadura oligospora*
Actinomadura pelletieri (24)	DSM 43383	*Actinomadura pelletieri* (24)	*Actinomadura pelletieri*	*Actinomadura pelletieri*
Micrococcus pelletieri (32)				
Nocardia pelletieri (33)				

(Continued)

Table 2. Continued

Original assignment	DSMZ-accession numbers of type strains	Bergey's Manual of Systematic Bacteriology volume 4[a,c]	The Prokaryotes 1992[c]	Current designation Approved Lists of Bacterial Names[b,c]
Streptomyces pelletieri (26)				
Micropolyspora rubrobrunea (34)	DSM 43750	"Excellospora rubrobrunea"	Actinomadura rubrobrunea (36, 13)	Actinomadura rubrobrunea
Excellospora rubrobrunea (36)				
Micropolyspora viridinigra (34)	DSM 43751	"Excellospora viridinigra"	Actinomadura rubrobrunea (36)	Actinomadura rubrobrunea
Excellospora viridinigra (36)				
Microbispora viridis (37)	DSM 44130	Microbispora viridis	Actinomadura viridis	Atinomadura rugatobispora (38)
Actinomadura spadix (39)	DSM 43459	Actinomadura spadix	Actinomadura spadix	Actinomadura spadix
Actinomadura umbrina (40, 41)	DSM 43927		Actinomadura umbrina (13)	Actinomadura umbrina
Actinomadura verrucosospora (39)	DSM 43358	Actinomadura verrucosospora	Actinomadura verrucosospora	Actinomadura verrucosospora
Actinomadura vinacea (1)	DSM 43765	Actinomadura vinacea	Actinomadura vinacea	Actinomadura vinacea
Micropolyspora viridilutea (34)	DSM 44433	"Excellospora viridilutea"	"Excellospora viridilutea"	Actinomadura viridilutea (3)
Excellospora viridilutea (36)				
Microtetraspora viridis (42)	DSM 43175	Microtetraspora viridis	Actinomadura viridis (13, 14)	Actinomadura viridis
Actinomadura malachitica (9)	DSM 43462	Actinomadura malachitica	Actinomadura viridis (13, 14)	Actinomadura viridis
Actinomadura yumaensis (43)	DSM 43931	Actinomadura yumaensis	Actinomadura yumaensis	Actinomadura yumaensis
Genus Lechevalieria				
Actinomadura flava (44)	DSM 43885	Nocardiopsis flava (45)	Saccharothrix flava (46)	Lechevalieria flava (47)
Genus Microbispora				
Thermomonospora mesophila (48)	DSM 43048	Thermomonospora mesophila	Thermomonospora mesophila	Microbispora mesophila (38)
Genus Nocardiopsis				
Actinomadura dassonvillei (24)	DSM 43111	Nocardiopsis dassonvillei	Nocardiopsis dassonvillei (49)	Nocardiopsis dassonvillei subsp dassonvillei subsp. Nocardia dassonvillei (52) dassonvillei (50, 51) Streptothrix dassonvillei (53)
Genus Nonomuraea				
Actinomadura africana (44, 35)	DSM 43748	Nocardiopsis africana (45)	Microtetraspora africana (13)	Nonomuraea africana (7)
Actinomadura carminata (54)	DSM 44170	—	Actinomadura carminata	Nonomuraea roseoviolacea subsp. carminata (55)
Actinomadura fastidiosa (56)	DSM 43674	Actinomadura fastidiosa	Microtetraspora fastidiosa (13)	Nonomuraea fastidiosa (7)
Actinomadura ferruginea (20, 21)	DSM 43563	Actinomadura ferruginea	Microtetraspora ferruginea (13)	Nonomuraea ferruginea (7)
Thermopolyspora flexuosa (57)	DSM 43186	Actinomadura flexuosa (2)	Microtetraspora flexuosa (13)	Nonomuraea flexuosa (7)
Actinomadura helvata (39)	DSM 43142	Actinomadura helvata	Microtetraspora helvata (13)	Nonomuraea helvata (7)
Actinomadura polychroma (40, 41)	DSM 43925		Microtetraspora polychroma (13)	Nonomuraea polychroma (7)
Actinomadura pusilla (39)	DSM 43357	Actinomadura pusilla	Microtetraspora pusilla (13)	Nonomuraea pusilla (7)
Actinomadura recticatena (58, 59)	DSM 43937		Microtetraspora recticatena (13)	Nonomuraea recticatena (7)

Actinomadura roseola (1)	DSM 43767	Actinomadura roseola	Nonomurae roseola (7)
Actinomadura roseoviolacea (39)	DSM 43144	Actinomadura roseoviolacea	Nonomuraea roseoviolacea (7, 55)
Micromonospora rubra (60)	DSM 43768	Actinomadura rubra (61)	Nonomuraea rubra (7)
Actinomadura salmonea (10)	DSM 43678	Actinomadura salmonea	Nonomuraea salmonea (7)
Actinomadura spiralis (20, 21)	DSM 43555	Actinomadura spiralis	Nonomuraea spiralis (7)
Actinomadura turkmeniaca (17, 18)	DSM 43926	—	Nonomuraea turkmeniaca (7)
Genus Saccharothrix			
Actinomadura coeruleofusca (44)	DSM 43679	Nocardiospis coeruleofusca (45)	Saccharothrix coeruleofusca
Actinomadura coeruleoviolacea (62, 58)	DSM 43935	—	Saccharothrix coeruleoviolacea
Actinomadura longispora (44)	DSM 43749	Nocardiopsis longispora (45)	Saccharothrix longispora
Genus Spirillospora			
Spirillospora albida (63)	DSM 43034	Spirillospora albida	Spirillospora albida
Spirillospora rubra (64)	CBS571.75	Spirillospora rubra	Spirillospora rubra
Genus Thermobifida			
Actinobifida alba (65)	DSM 43795	Thermomonospora alba (66)	Thermobifida alba (7)
"Thermomonospora falcata"			
Thermomonospora fusca (68)	DSM 43792	Thermomonospora fusca (68)	Thermobifida fusca (7)
Thermomonospora mesouviformis (70)	DSM 43185	Thermomonospora alba (68)	Thermobifida alba (7)
Genus Thermomonospora			
Actinobifida chromogena (67)	DSM 43794	Thermomonospora chromogena (68, 69)	Thermomonospora chromogena
Thermomonospora curvata (71)	DSM 43183	Thermomonospora curvata	Thermomonospora curvata

Symbols and abbreviations: DSM, Deutsche Sammlung von Mikroorganismen und Zellkulturen (German Collection of Microorganisms and Cell Cultures); CBS, Centraalbureau voor Schimmelcultures; and −, name does not appear in this publication.

aWilliams et al. (1989).

bSkerman et al. (1980) plus subsequent listings as May 2004.

cThe numbers in parentheses indicate the references as follows: (1) Lavrova and Preobrazhenskaya (1975); (2) Iinuma et al. (1994); (3) Zhang et al. (2001); (4) Itoh et al. (1995); (5) Iinuma et al. (1994); (6) Miyadoh et al. (1987); (7) Zhang et al. (1998); (8) Lu et al. (2003); (9) Lavrova et al. (1972); (10) Preobrazhenskaya et al. (1975b); (11) Gauze et al. (1987); (12) Nonomura and Ohara (1971b); (13) Kroppenstedt et al. (1990); (14) Miyadoh et al. (1989); (15) Mertz and Yao (1990); (16) Hasegawa et al. (1986); (17) Terekhova et al. (1982); (18) Terekhova et al. (1987); (19) Horan and Brodsky (1982); (20) Trujillo and Goodfellow (1997); (21) Meyer (1989a); (22a) Preobrazhenskaya et al. (1975a); (23) Huang (1980); (24) Lechevalier and Lechevalier (1970a); (25) Blanchard (1896); (26) Waksman and Henrici (1948); (27) Vincent (1894); (28) Quintana et al. (2003); (29) Wink et al. (2003); (30) Lipski and Altendorf (1995); (31) Mertz and Yao (1986); (32) Laveran (1906); (33) Pinoy (1912); (34) Krassil'nikov et al. (1968); (35) Greiner-Mai et al. (1987); (36) Agre and Guzeva (1975); (37) Miyadoh et al. (1985); (38) Miyadoh et al. (1990); (39) Nonomura and Ohara (1971d); (40) Galatenko et al. (1981); (41) Galatenko et al. (1987); (42) Nonomura and Ohara (1971a); (43) Labeda et al. (1985); (44) Preobrazhenskaya and Sveshnikova (1974); (45) Preobrazhenskaya et al. (1982); (46) Grund and Kroppenstedt (1989); (47) Labeda et al. (2001); (48) Nonomura and Ohara (1971c); (49) Grund and Kroppenstedt (1990); (50) Miyashita et al. (1984); (51) Meyer (1976); (52) Liegard and Landrieu (1911); (53) Brocq-Rousseu (1904); (54) Gauze et al. (1973); (55) Gyobu and Miyadoh (2001); (56) Soina et al. (1975); (57) Krassil'nikov and Agre (1964b); (58) Preobrazhenskaya et al. (1987); (59) Gauze et al. (1984); (60) Sveshnikova et al. (1969); (61) Meyer and Sveshnikova (1974); (62) Preobrazhenskaya et al. (1976); (63) Couch (1963); (64) Vobis and Kothe (1989); (65) Locci et al. (1967); (66) Cross and Goodfellow (1973); (67) Krassil'nikov and Agre (1965); (68) McCarthy and Cross (1984b), McCarthy and Cross (1984c); (69) Moore et al. (1985); (70) Nonomura and Ohara (1974); and (71) Henssen (1957).

are confusing. One group detected DPG, PG and PI in *Spirillospora albida* (PL-type one; Lechevalier et al., 1977), whereas another group found phosphatidylethanolamine (PE) as an additional component (PL-type II; Hasegawa et al., 1979).

Actinocorallina, like the other members of the family Thermomonosporaceae, falls into madur-omycetes taxon. The name "maduromycetes" was proposed by Goodfellow (Goodfellow, 1989b; Goodfellow, 1989a) for sporoactino-mycetes able to synthesize madurose, a sugar which is always present in whole-cell hydroly-sates of members of the closely related families Thermomonosporaceae and Streptospo-rangiaceae. Madurose, a 3-*O*-methylgalactose (Lechevalier and Gerber, 1970b), got its name from *A. madurae* as it was found for the first time in this actinomycete (Lechevalier and Lecheva-lier, 1970c). *Actinocorallina* can be differentiated from the other maduromycetes by its pho-spholipid pattern type II: from members of the family Thermomonosporaceae, i.e., type II versus type I and from Streptosporangiaceae, type II versus type IV. All members of the family Thermomonosporaceae have the menaquinone type 4B2 in common, whereas in members of the family Streptosporangiaceae, the type 4A2 is found (Table 2). This is one of the examples of good correlation between chemotypes and genotypes in members of the order Actinomycetales.

Thermomonosporae in the broad sense are similar but not related to thermoactinomycetes with respect to morphology, cell wall chemotype, and natural habitat. The detection of endospores in the latter separates them from thermomono-sporae and all other monosporic actinomycetes. The genera *Actinosynnema* (Hasegawa et al., 1978) and *Streptoalloteichus* (Tomita et al., 1987) were proposed for actinomycetes considered to be morphologically unusual. The morphology of members of the two taxa is complex and differs fundamentally from *Actinomadura*, *Actinocoral-lia* and *Thermomonospora* strains as motile spores are produced. Members of the two genera can be differentiated from *Spirillospora* by the type of zoospore produced and the presence of spore vesicles. In *Actinosynnema*, peritrichously flagellated zoospores originate from aerial spore chains borne on synnemata, whereas in *Strep-toalloteichus*, spores propelled by single polar flagella are formed in sporangium like vessels carried on substrate hyphae, which show some similarities with those of *Spirillospora*. *Streptoal-loteichus* strains also produce chains of *Strepto-myces*-like arthrospores on aerial hyphae.

The genus *Actinomadura* can be distinguished from other sporoactinomycete taxa using a judi-cious selection of chemical and morphological

features (Meyer, 1989a; Meyer, 1989b; Kroppen-stedt et al., 1990; Wink, 2002; Tables 2 and 3). Quantitative analyses of fatty acid data allow the separation of mesophilic actinomadurae from corresponding members of the genera *Microtetraspora*, *Nocardiopsis*, *Saccharothrix* and *Thermomonospora* (Kroppenstedt, 1987). Actinomadurae and microtetrasporas can also be distinguished on the basis of menaquinone and polar lipid profiles (Figs. 5 and 6).

Members of *Actinomadura* and *Actinocorallia* species can be separated using biochemical, morphological, pigmentation and physiological properties (Tables 3–6). However, identification of most of these species is difficult, for in many instances, only one representative organism, usu-ally the type strain, has been examined. Even when several representatives of species have been studied, the results of biochemical and physiological tests have proved to be variable or inconsistent when data from the literature are compared. Even so, numerical taxonomic evi-dence shows that many of the validly described taxa do merit species status (Goodfellow et al., 1979; Goodfellow and Pirouz, 1982; Athalye et al., 1985; Trujillo and Goodfellow, 1997; Trujillo and Goodfellow, 2003). It is also encouraging that *Actinomadura* species have been shown to form distinct genomic species (Fischer et al., 1983; Poschner et al., 1985; Zhang et al. 1998; Zhang et al., 2001).

Clinically significant actinomadurae (such as *Actinomadura madurae* and *A. pelletieri*) can be separated from each other and from other patho-genic aerobic actinomycetes, notably as *Strepto-myces somaliensis*, *Rhodococcus equi*, *Nocardia* spp., *Gordonia* spp., *Tsukamurella* spp. and *Mycobacterium* spp., by DNA amplification and restriction endonuclease analysis. Steingrube et al. (1997) used the two primers TB11 and TB12 for amplification of the 439-bp segment (ampli-con) of the 65-kDa shock protein to differentiate between the type strains of representative species and clinically significant isolates. The amplicon was cleaved by five restriction endonu-cleases, namely *Bst*EII, *Hae*III, *Hin*fI, *Msp*I, and *Bsa*HI. This polymerase chain reaction-restriction fragment length polymorphism (PCR-RFLP) methodology distinguished between clinical isolates of aerobic actino-mycetes including actinomadurae with 96.8% accuracy.

Members of the redefined *Thermomonospora* share chemical markers in common with *Actino-madura* strains, i.e., wall chemotype III, phos-pholipid type I, menaquinone type 4B2, and fatty acid type 3a (Table 2; Fig. 6). The integrity of the group containing *T. curvata* and *T. chromogena* is also supported by fatty acid and menaquinone data. Variations in the quantitative fatty acid

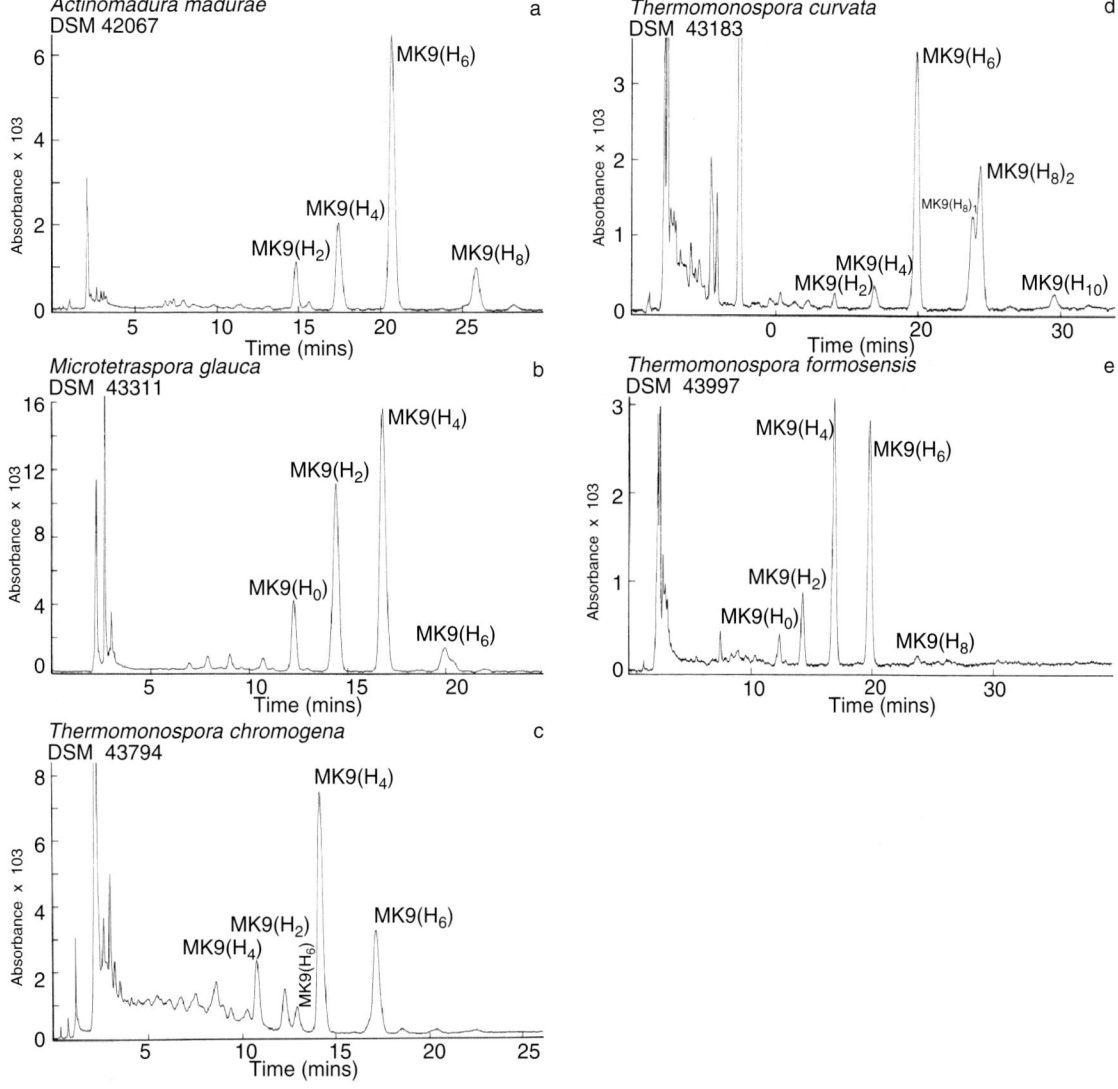

Fig. 5. Menaquinone profiles of: a) *Actinomadura madurae* DSM 43067[T], b) *Microtetraspora glauca* DSM 43311[T], c) *Thermomonospora chromogena* DSM 43794[T], d) *Thermomonospora curvata* DSM 43183[T] and e) *Actinomadura formosensis* DSM 43997[T].
Chromatographic conditions: column, 250 mm × 4 mm; stationary phase, RP 18, size 5 µm; mobile phase, acetonitrile-isopropanol (65:35, v/v); flow rate, 1 ml/min; temperature 40°C; injection volume, 5 µl of extract prepared after Minnikin et al. (1984) without further purification. Key to abbreviations: see table 2.

composition and in the menaquinone profiles of *T. curvata* and *T. chromogena* with respect to other members of this family may be attributed to differences in their temperature requirements for growth (Kroppenstedt et al., 1990). The lack of madurose (sugar type C) might be an effect of the elevated temperature for growth, because in members of thermophilic *Actinomadura* species (formerly *Excellospora*) this sugar is found only in traces. *Thermomonospora chromogena* and *Thermobifida fusca* can be distinguished from one another and from strains of other mono-sporic taxa previously assigned to the genus *Thermomonospora* (Table 7).

Members of the genus *Spirillospora* may easily be confused with some *Streptosporangium* species on morphological grounds, but they are easily differentiated from members of this taxon as they produce motile zoospores, and zoospores have never been detected in *Streptosporangium* strains (Table 8; see The Family Streptosporangiaceae in this Volume). All strains belonging to this taxon produce multispored, usually spherical spore vesicles carried on the aerial mycelium. Nevertheless, it is evident from comparative 16S rRNA gene sequencing and chemical data that spirillosporae belong to the family Thermomonosporaceae (Goodfellow 1989b;

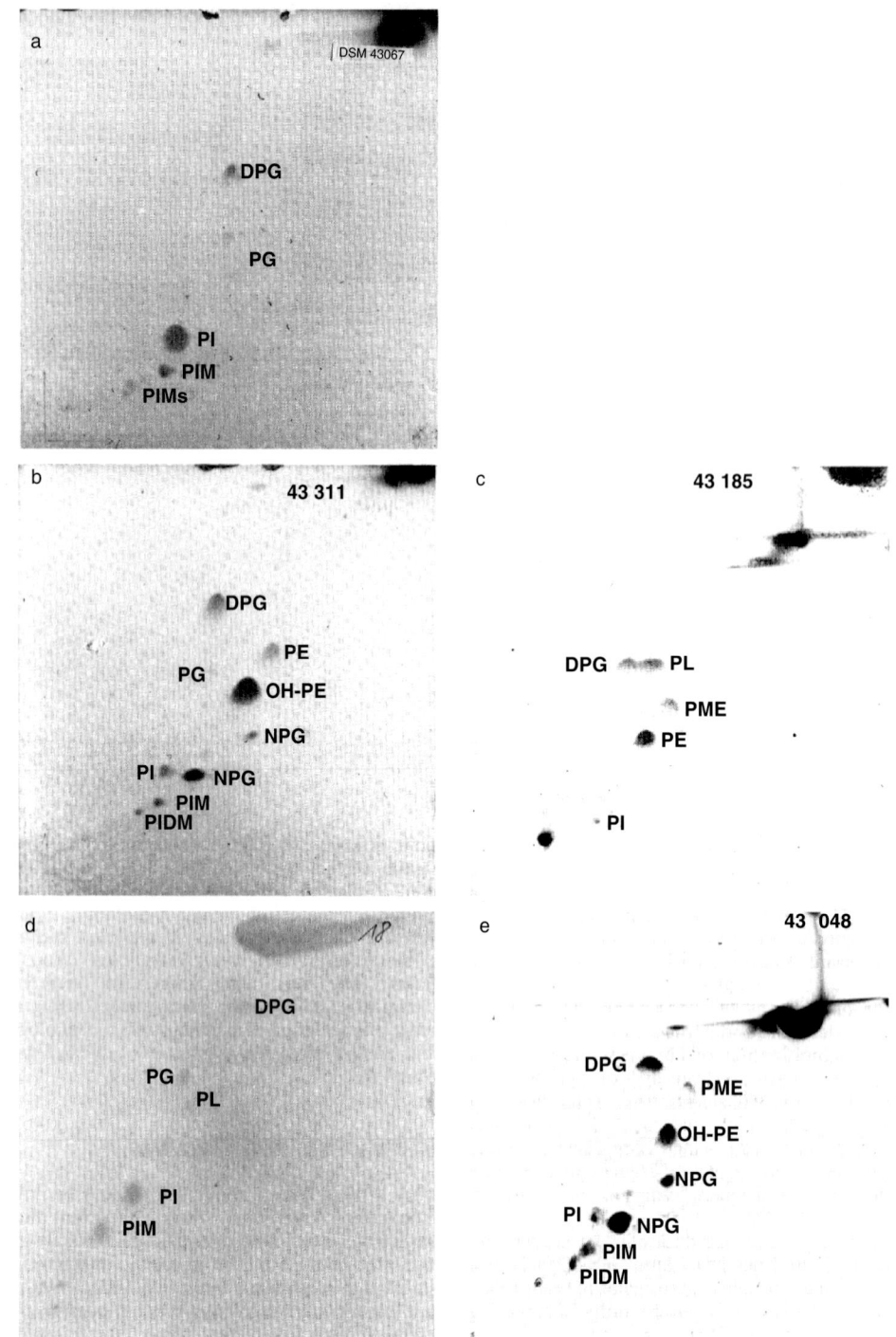

Fig. 6. Two dimensional thin-layer chromatograms of polar lipids: a) *Actinomadura madurae* DSM 43067[T] (PL-type I); b) *Microtetraspora glauca* DSM 43311[T] (PL-type IV); c) *Thermobifida alba* DSM 43185 (PL-type II); d) *Thermomonospora curvata* DSM 43183[T] (PL-type IV); and e) *Microbispora mesophila* DSM 43048[T] (PL-type IV). Example for PL-type III, see *Nocardiopsis dassonvillei* (The Family Nocardiopsaceae in this Volume).
Chromatographic conditions: 10 × 10 cm silica-gel-60 thin layer plates were spotted with 20 µl of the extracts from test strains using the procedures of Minnikin et al. (1984); solvent I: chloroform-methanol-water (65:25:4, v/v); solvent II: chloroform-acetic acid-methanol-water (80:15:12:4 v/v). The developed plates were sprayed with 50% sulfuric acid ammonium sulfate and charred at 180°C.
Abbreviations: PG, phosphatidylglycerol; DPG, diphosphatidylglycerol (cardiolipin); PE, phosphatidylethanolamine; OH-PE, hydroxy-PE; PME, phosphatidylmethylethanolamine; NPG, *n*-acetylglucosamine carrying phospholipid (GluNu; Lechevalier et al., 1981); PI, phosphatidylinositol; PIM, PIDM, phosphatidylinositolmannosides; PL, unknown phospholipids; PL, phospholipid types according to Lechevalier et al. (Lechevalier et al., 1977; Lechevalier et al., 1981).

Table 3. Macroscopic and microscopic characters separating species of *Actinocorallia*, *Actinomadura*, *Spirillospora* and *Thermomonospora*.

Species	Color on oatmeal agar		Diffusible pigments	Spore chain type	Spore surface type
	aerial mycelium	substrate mycelium			
Actinocorallia					
Acr. *aurantiaca*	Cream to pink	Yellow	None	Hooks to spirals	Warty
Acr. *glomerata*	White	Yellow to brown	None	Straight, pseudosporangia	Smooth
Acr. *herbida*	White to pale yellow	Ivory to pale yellow	None	Coralloid sporophores	Smooth
Acr. *libanotica*	White to pink	Yellow to brown	None	Hooks, curled	Folded
Acr. *longicatena*	White	Oak brown	None	Straight	Smooth
Actinomadura					
A. *atramentaria*	White	Colorless	Inky brown[a]	Cluster	Smooth
A. *catellatispora*	Yellow	Light yellow	None	Straight	Smooth
A. *citrea*	White to blue	Yellow	Yellow	Hooks or curled	Uneven
A. *coerulea*	Pink to blue	Colorless	None	Hooks to spirals	Warty
A. *cremea* subsp. *cremea*	White to yellow	Colorless	None	Hooks to spirals	Warty
A. *cremea* subsp. *rifamicini*	White to yellow	Colorless	None	Hooks to spirals	Warty
A. *echinospora*	Yellow to pink	Orange	Yellow	Cluster	Spiny
A. *fibrosa*	White	Brown	Brown	No spores	No spores
A. *formosensis*	White	Pink to light orange	None	Single	Warty
A. *fulvescens*	None	Red-brown	None	Spiral	Smooth
A. *glauciflava*	Bluish-green	Yellow to brown	Yellow	Hooks, spirals	Warty
A. *hibisca*	Light gray	Red violet	Carmine red	Straight	Smooth
A. *kijaniata*	Trace	Gray	None	Spirals	Smooth
A. *latina*	None	Cream to pink	None	None	None
A. *livida*	Trace	Gray to brown	Violet	Hooks to spirals	Uneven
A. *luteofluorescens*	Yellow to blue	Yellow to green	Yellow-green	Hooks, curled	Warty
A. *macra*	Cream to pink	Cream to pink	None	Hooks, curled	Smooth
A. *madurae*	Trace	Colorless	None	Hooks to spirals[b]	Warty

Table 3. *Continued*

Species	Color on oatmeal agar			Diffusible pigments	Spore chain type	Spore surface type
	aerial mycelium	substrate mycelium				
A. mexicana	None	Cream to yellow		None	Hooks	Warty
A. meyeri	Sparse	Cream to yellow		None	Hooks	Warty
A. namibiensis	White	Salmon pink		None	Spirals	Smooth
A. nitrigenes	Brown mahogany	Colorless		None	Hooks	Smooth
A. oligospora	None	Brown		Light brown	Straight	Uneven
A. pelletieri	None	Pink to brown		None	Hooks to spirals	Warty
A. rubrobrunea	Grayish-blue	Orange		None	Spirals[c]	Spiny
A. rugatobispora	Light to dusty green	Pastel yellow		None	Paired spores	Rugose with vertical ridges
A. spadix	Yellow/brown	Yellow-red-brown		None	Pseudosporangia	Smooth
A. umbrina	None	Colorless		None	Straight, hooks, spirals	Smooth
A. verrucosospora	Pink to blue	Orange to pink		None	Hooks to spirals	Warty
A. vinacea	None	Pink to red		None	Straight	Uneven
A. viridilutea	Blue green	Orange to yellow		None	Hooks to spirals	Spiny
A. viridis	Green	Yellow-brown to green		None	Straight	Smooth
A. yumaensis	Gray to yellow	Gray to yellow		None	Hooks	Smooth
Spirillospora						
Spl. albida	Whitish	Pale yellow to buff pink		Pale yellowish	Pseudosporangia	Rod shaped, motile
Spl. rubra	White	Red to reddish brown		None	Pseudosporangia	Rod shaped, motile
Thermomonospora						
Tms. chromogena	White	Yellow to orange		None	Single spores	Folded
Tms. curvata	White to brownish	Dark reddish brown		Dark brown	Single spores in cluster	Folded

[a]pigment found on tyrosine agar (ISP medium 7).
[b]spirals of 2 to 4 turns.
[c]coiled, 2 to 20 spores, some single.
Data from Athalye et al. (1985), Mertz and Yao (1990), Miyadoh et al. (1985), Meyer (1989a), Meyer (1989b), Quintana et al. (2003), and Wink et al. (2003).

Table 4. Physiological characteristics differentiating the species of *Actinomadura*.[a]

Species	Type strain DSM	Esculin hydrolysis	Arbutin hydrolysis	Casein	DNA	Elastin	Gelatin	Guanine	Hypo-xanthine	Nitrate reduction[b]	Starch	Testo-sterone	Tyrosine	Urease production	Xanthine
Actinomadura[c]															
A. atramentaria	43919	+	+	-	+	-	-	-	+	-	+	+	ND	-	-
A. catellatispora	44772	+	ND	-	ND	ND	ND	ND	-	+	-	ND	ND	-	ND
A. citrea	43461	+	+	+	-	+	+	d	+	+	+	+	+	-	-
A. coerulea	43675	-	+	+	v	+	+	+	+	+	-	+	-	+	-
A. cremea	43676	+	ND	+	ND	+	+	-	-	+	+	+	+	+	-
A. fibrosa	44224	+	+	+	d	ND	+	ND	d	ND	d	+	-	ND	-
A. fulvescens	43923	+	ND	+	d	+	+	ND	+	ND	+	+	+	ND	+
A. glauciflava	44770	+	ND	+	ND	ND	+	-	+	ND	+	-	-	+	+
A. kijaniata	43764	+	+	+	+	+	+	d	ND	ND	+	d	+	ND	+
A. latina	43383	+	+	+	d	d	+	+	-	+	+	+	+	-	+
A. livida	43677	+	d	+	+	-	+	+	+	d	-	+	+	-	-
A. luteofluorescens	40398	-	+	+	+	-	+	-	+	+	+	-	-	-	+
A. macra	43862	-	+	-	-	+	+	d	+	+	-	+	-	ND	-
A. madurae	43067	+	+	+	d	+	+	d	-	+	d	+	-	-	ND
A. mexicana	44485	+	+	+	ND	+	+	ND	+	+	-	ND	d	ND	-
A. meyerae	44715	+	ND	+	ND	-	+	ND	ND	+	+	ND	+	+	ND
A. namibiensis	44197	ND	ND	ND	ND	ND	-	-	ND	ND	ND	ND	-	-	ND
A. nitrigenes	44137	+	ND	+	ND	ND	ND	-	+	+[d]	ND	+	d	+	-
A. oligospora	43930	+	+	+	+	-	+	-	+	-	-	+	-	+	-
A. pelletieri	43383	-	-	+	-	-	+	-	d	+	-	-	-	-	-
A. spadix	43459	+	+	+	+	+	-	-	-	+	+	-	+	-	+
A. verrucosospora	43358	+	+	+	+	+	+	d	+	d	d	+	+	-	-
A. vinacea	43765	+	ND	+	-	ND	+	ND	-	ND	ND	-	-	ND	-
A. viridis	43751	ND	ND	ND	ND	ND	+	ND	ND	+	ND	ND	ND	ND	ND
A. yumaensis	43931	+	ND	+	ND	ND	+	-	d	+	+	ND	+	+	+

Symbols: +, 90% or more strains positive; -, 10% or less strains positive; d, 11–89% of strains positive; v, strain unstable; and ND, not determined.

[a]Based on data from Goodfellow et al. (1979), Athalye (1981a), Horan and Brodsky (1982), Athalye et al. (1985), Labeda et al. (1985), Mertz and Yao (1986, 1990), Lipski and Altendorf (1995), Trujillo and Goodfellow (1997, 2003), Lu et al. (2003), Quintana et al. (2003), and Wink et al. (2003).

[b]Nitrate reduction under aerobic conditions.

[c]No data available for *Actinomadura carminata, A. cremea* subsp. *rifamicini, A. echinospora, A. formosensis, A. hibisca, A. rubrobrunea, A. umbrina* and *A. viridilutea*.

Table 5. Enzymatic activities of *Actinomadura* and *Actinocorallia* type strains (API ZYM).[a]

	DSM no.	1	2	3	4	5	6	7	8	9	10	11	12	13	14	15	16	17	18	19
Actinomadura																				
A. atramentaria	43919	−	+	+	−	+	−	−	−	−	+	+	−	−	−	−	+	−	−	−
A. catellatispora	44772	+	(+)	(+)	−	+	+	−	−	−	+	+	−	−	−	−	−	−	−	−
A. citrea	43461	+	+	+	+	+	+	+	−	−	+	+	−	−	−	+	(+)	+	−	−
A. coerulea	43675	+	−	+	+	+	+	+	+	+	+	+	−	−	+	+	+	+	−	(+)
A. cremea	43676	+	+	+	+	+	+	+	−	+	+	(+)	−	−	+	+	+	+	−	−
A. echinospora	43163	+	+	+	+	+	+	+	+	+	+	(+)	−	−	−	+	+	+	−	−
A. fibrosa	44224	+	+	+	−	+	+	+	+	+	+	+	−	−	−	+	+	+	−	−
A. formosensis	43997	+	+	+	−	+	+	+	+	+	+	+	−	+	−	+	+	+	−	−
A. fulvescens	43923	+	+	+	−	+	+	(+)	+	+	+	+	−	+	−	+	+	+	−	−
A. glauciflava	44770	+	−	−	−	+	+	+	+	−	+	+	−	−	−	(+)	(+)	+	−	(+)
A. hibisca	44148	+	+	+	−	+	+	+	+	−	+	+	−	(+)	−	+	+	+	−	−
A. kijaniata	43764	+	+	+	+	+	−	(+)	+	−	+	+	−	+	−	(+)	+	+	−	−
A. latina	43382	+	+	+	−	+	+	+	+	+	+	+	−	−	−	+	+	+	−	−
A. livida	43677	+	+	+	+	+	−	+	−	+	+	+	−	−	−	+	+	+	−	−
A. luteofluorescens	40398	+	+	+	+	+	+	+	+	+	+	+	−	−	−	+	+	+	−	−
A. macra	43862	−	−	+	+	+	+	−	−	+	+	+	−	−	−	+	+	+	−	−
A. madurae	43067	+	+	+	+	+	+	+	+	+	+	+	−	+	−	−	+	+	−	−
A. mexicana	44485	+	−	+	+	+	+	+	+	−	+	+	−	+	−	+	−	+	−	−
A. meyerae	44715	+	−	+	+	+	−	+	+	+	+	+	−	+	−	+	+	+	+	−
A. namibiensis	44197	+	−	−	+	+	+	+	+	−	+	+	−	+	−	+	+	+	+	−
A. nitrigenes	44137	+	−	+	−	+	+	+	+	−	+	+	−	+	−	+	+	+	+	−
A. oligospora	43930	+	−	+	+	+	+	+	−	−	+	+	−	−	−	−	+	+	−	−
A. pelletieri	43383	+	−	−	−	+	+	(+)	(+)	+	+	+	−	−	−	−	−	+	−	−
A. rubrobrunea	43750	not determined (thermophilic)																		
A. rugatobispora	44130	+	−	+	−	+	+	+	+	+	+	+	−	+	−	+	−	+	(+)	−
A. spadix	43459	+	−	+	+	+	−	+	+	−	+	+	−	+	−	+	−	+	−	−
A. umbrina	43927	+	−	+	+	+	+	+	(+)	+	+	+	−	−	−	(+)	+	+	−	−
A. verrucosospora	43358	+	−	+	+	+	+	+	+	+	+	+	−	−	−	+	+	+	−	−
A. vinacea	43765	+	−	−	−	−	+	+	+	+	+	+	−	−	−	+	+	+	−	−
A. viridilutea	44433	+	+	+	−	(+)	(+)	(+)	(+)	(+)	+	+	−	+	−	+	+	+	−	−
A. viridis	43175	+	+	+	(+)	(+)	(+)	(+)	(+)	(+)	+	+	−	+	−	+	+	+	−	−
A. viridis (malachitica)	43462	+	−	+	−	+	+	−	−	−	+	+	−	+	−	+	−	+	−	−
A. yumaensis	43931	+	+	+	−	+	+	(+)	(+)	+	+	+	−	−	−	+	+	+	−	−
Actinocorallia																				
A. aurantiaca	43924	(+)	−	(+)	−	+	+	+	+	+	+	+	−	+	−	+	+	+	+	ND
A. glomerata	44360	+	−	+	+	+	+	+	+	−	+	+	−	+	−	+	+	+	+	−
A. herbida	44254	+	−	+	+	+	+	+	+	+	+	+	−	+	−	+	+	+	+	−
A. libanotica	43554	+	−	−	−	+	+	+	+	−	+	+	−	+	−	+	+	+	+	−
A. longicatena	44361	+	+	+	+	+	+	(+)	−	−	+	+	−	−	−	+	+	+	−	−

Symbols: +, present; −, absent; (+) and ND, not determined.
[a]1. Alkaline phosphatase, 2. esterase, 3. esterase (C4), 4. esterase (C8), 5. leucine arylamidase, 6. valine arylamidase, 7. cysteine arylamidase, 8. trypsin, 9. chymotrypsin, 10. phosphatase acid, 11. naphthol-AS-BI-phosphorylase, 12. α-galactosidase, 13. β-galactosidase, 14. α-glucuronidase, 15. α-galactosidase, 16. β-glucosidase, 17. N-acetyl-β-glucoseamidase, 18. α-mannosidase, and 19. α-fucosidase.
Data from Wink (2002), Wink et al. (2003) and R. M. Kroppenstedt (unpublished data).

Table 6. Characters differentiating species of *Actinocorallia*.[a]

Characters	Acr. aurantiaca	Acr. glomerata	Acr. herbida	Acr. libanotica	Acr. longicatena
Color of					
Aerial mycelium	Cream to pink	White	White to pale yellow	White to pink	White
Substrate mycelium	Yellow	Yellow to brown	Ivory to pale yellow	Yellow to brown	Oak brown
Excreted pigments	None	None	None	None	None
Melanoid pigments	None	None	Produced	None	None
Spore chain type	Hooks to spirals Spore chains	Straight chains and pseudosporangia	Coralloid sporophores	Hooks and curled spore chains	Straight chains
Spore surface type	Warty	Smooth	Smooth	Folded	Smooth
NaCl resistance (%)	ND	0	2.5	2.5	5–10
Assimilation of carbohydrates					
Glucose	+	+	+	+	+
Arabinose	–	+	+	+	+
Sucrose	–	+	+	+	+
Xylose	–	+	+	+	+
Inositol	–	+	+	+	+
Mannitol	–	+	+	+	–
Fructose	+	+	+	+	+
Rhamnose	+	+	+	+	+
Raffinose	–	+	+	+	+
Cellulose	–	+	+	+	+

Symbols and abbreviations: +, present; –, absent; and ND, not determined.
From Meyer (1979), Iinuma et al. (1994), Itoh et al. (1995) and Wink (2002).

Table 7. Characters differentiating between *Thermomonospora* and other monosporic species formerly classified in this taxon.[a]

	Thermomonospora		*Thermobifida*		*Microbispora*	*Actinomadura*
Characteristics	1	2	3	4	5	6
Color of substrate mycelium	Yellow/orange	Brown	Light-yellow	Light-yellow	Brown	Light-orange
Color of aerial mycelium	White	Light brown	Brown	White	White	White to pink
Spores on substrate mycelium	−		−	−	−	+
Spores on aerial mycelium						
C[b]	−	+	+	+	−	+
SB and SU S[c]	+	−	(+)	−	−	+
SS[d]	−	−	−	−	+	−
Spore ornamentation	Spiny	Spiny	Smooth	Smooth	Smooth	Warty
Biochemical tests						
Nitrate reduction	+	+	+	−	+	−
Oxidase	−	+	−	−	+	ND
Phosphatase	+	d	+	+	−	ND
Growth at (% of the strains)						
25°C	−	−	−	−	100	100
35°C	100	6	100	47	100	100
53°C	87	94	14	96	−	−
60°C	−	−	−	38	−	−
pH 6	−	56	14	13	100	ND
pH 10	100	89	100	100	−	ND
pH 11	100	−	100	98	−	+
Growth on sole carbon sources (1%, w/v)						
L-Arabinose	−	−	−	−	+	−
Galactose	−	+	d	+	+	+
Lactose	−	−	d	+	−	−
Mannitol	−	−	−	−	−	+
Ribose	+	d	−	−	−	ND
Sucrose	+	−	+	+	−	+

	1[a]	2	3	4	5	6
Degradation of						
Agar	+	−	+	+	+	ND
Cellulose powder (MN300)	+	−	+	+	−	ND
Carboxymethylcellulose	+	+	+	+	+	ND
DNA	−	−	−	−	−	ND
Elastin	+	+	−	−	−	ND
Keratin	+	+	v	+	+	ND
Chitin	−	−	−	−	−	ND
Pectin	−	d	+	+	+	ND
Xylan	+	+	+	+	+	ND
Starch	+	−	+	+	−	ND
Tweens 20 and 80	+	+	+	+	−	ND
T, H and X	−	+	−	−	+	ND
Growth in presence of:						
Crystal violet (0.2 µg/ml)	+	+	−	+	+	ND
Tetrazolium chloride (20 µg/ml)	d	d	−	+	−	ND
Thallous acetate (10 µg/ml)	d	d	−	+	+	ND
Kanamycin (25 µg/ml)	−	+	−	−	−	ND
Novobiocin (25 µg/ml)	−	−	−	−	+	ND

Symbols: +, 90% or more strains positive; −, 10% or less strains positive; (+), weakly positive; v, variable; ND, not determined; T, tyrosine; H, hypoxanthine; and X, xanthine.

[a] 1, *T. curvata*; 2, *T. chromogena*; 3, *T. alba*; 4, *T. fusca*; 5, *M. mesophila*; and 6, *A. formosensis*.

[b] In clusters.

[c] Single on branched and unbranched sporophores.

[d] Single sessile.

Table 8. Characters differentiating the species of *Spirillospora*.

Characteristics	S. alba	S. rubra
Color of aerial mycelium	White	White[a]
Color of substrate mycelium	White to pale yellow	Red to reddish brown
Shape of sporangium	Spherical to vermiform	Spherical
Flagella on zoospores	Subpolar, polytrichous	Subpolar, polytrichous
Degradation of		
Esculin	+	ND
Chitin	−	ND
Hypoxanthine	−	ND
Reduction of nitrate	−	−
Utilization of		
Adonitol	−	ND
Arabinose	(+)	ND
Cellulose	−	ND
Fructose	+	ND
Galactose	−	ND
Glucose	+	ND
Inositol	−	ND
Lactose	−	ND
Mannose	+	ND
Melezitose	+	ND
Raffinose	−	ND
Rhamnose	+	ND
Sucrose	+	ND
Trehalose	+	ND

Symbols: +, positive reaction; (+), weak growth; −, negative reaction; and ND, not determined.
[a]Aerial mycelium is usually absent, a sparse white aerial mycelium is produced on corn meal agar (Schäfer, 1973).
Based on data of Vobis and Kothe (1989) and Wink (2002).

Stackebrandt et al., 1997; and Zhang et al., 2001). There are some contradictory results on the phospholipid pattern of *Spirillospora albida*; Lechevalier and coworkers reported a phospholipid pattern I as expected (Lechevalier et al., 1981), whereas Hasegawa and his colleagues found PE in addition to DPG, PG and PI, i.e., a phospholipid pattern II (Hasegawa et al., 1979). The fatty acid profile of *Spirillospora* strains is type 3a sensu Kroppenstedt (Kroppenstedt, 1985).

Spirillospora albida and *S. rubra* produce a substrate mycelium that is usually white, pale yellow, pale buff pink, or red to reddish-brown; the aerial mycelium is white (Couch, 1963; Schäfer, 1973). Spherical to vermiform spore vesicles (5–24 μm in diameter) are formed on the aerial mycelium (Fig. 3). On initiation of "sporangial" development, the ends of aerial hyphae first coil and then wind in branches within a common sheath (Lechevalier et al., 1966; Vobis, 1985a). Small, two-layered cross-walls divide sporangeous hyphae into oblong segments that differentiate into spores. The latter, which are rod shaped and curved, are motile by means of one to seven subpolarly inserted flagella. The spore vesicles are considered to be resistant to desiccation. When flooded with water, zoospores are released from the spore vesicles through a rupture in the envelope or through a large irregular pore (Couch, 1963).

Pathogenicity

The only members of the Thermomonosporaceae known to be pathogenic for humans are *Actinomadura latina*, *A. madurae* and *A. pelletieri*. Members of these taxa cause mycetoma, a chronic suppurative granulomatous disease of subcutaneous tissue and bones that is characterized by swellings, abscesses, and discharge of pus. The pus contains granules; those found in pus infected with *A. madurae* strains are soft, white to yellow or reddish, usually 1–5 mm in diameter, and are spherical or angular with lobes, whereas those characteristic of *A. latina* and *A. pelletieri* infections are hard, deep red, 0.3–0.5 mm in diameter, and usually spherical to oval. Infection usually occurs through skin lesions contaminated with dust, soil or vegetable matter and remains localized with dissemination occurring by direct extension of the organisms through the tissues. As infection becomes increasingly chronic, the lesionsextend more deeply into the body with involvement of both muscle and bone. Animal models of infection with *Actinomadura* have not been developed though Rippon (1968) reported that virulent strains of *A. madurae* produce a collagenase that has a significant role in the pathogenicity of the organism.

Actinomadura pelletieri occurs mainly in Africa whereas *A. madurae* is widespread in all tropical and subtropical areas (Schaal and Beaman, 1984b; Gumaa et al., 1986; Develoux et al., 1988). Colonies resembling members of these species have been isolated from water (Lawson and Davey, 1972). There is evidence that climatic conditions may play a major role in the distribution of mycetomas. In Venezuela, most cases of actinomycetoma have been reported from semi-arid areas colonized by trees and bushes such as *Acacia tortuosa*, *Amarantus espinosus* and *Prosepis fuliflora* and cacti (e.g., *Opuntia caribea* and *Opuntia wentiana*; Serrano et al., 1986). Since the trees and cacti have spines and thorns, it is very common for agricultural workers to puncture their skin. Intensive studies on the occurrence, cause of mycetomas, and epidemiological considerations have been carried out in Yemen (Yu et al., 1993), Sudan (Boiron et al., 1998), and West Bengal, where 264 cases were diagnosed between 1981, and 2000 (Maiti et al., 2002).

Effective treatment of patients with mycetomas requires the identification of the causative agents, as the disease may be caused by members of several actinomycete taxa (e.g., *Nocardia asteroides*, *Nocardia brasiliensis* and *Streptomyces somaliensis* cause actinomycetomas) or by fungi (cause eumycetoma; McNeil and Brown, 1994; Goodfellow, 1997). All types of mycetoma infections should be treated with early surgical debridement and tissue culture. The clinical picture of an actinomycetoma is almost uniform irrespective of the causal agent. In addition to surgical management, these patients should be managed adjunctively with a prolonged course of chemotherapy. The choice of antimicrobial agents for therapy is currently based on non-comparable clinical observations (McNeil and Brown, 1994; Goodfellow, 1997). The most encouraging results have been obtained with sulfonamides. Alternative therapies mainly involve the use of minocycline, penicillin, and tetracycline. *Actinomadura madurae* infections have been successfully treated by resection and prolonged doxycycline chemotherapy (Davis et al., 1999).

The growth of thermophilic actinomycetes in high temperature environments leads to the release of spores that can cause allergic alveolitis. There are currently no grounds for implicating the spores of *Thermomonospora* species in such respiratory disorders (Lacey, 1988) though the cause of mushroom worker's disease, a form of allergic alveolitis, is still a matter for conjecture. Precipitins to *Faenia rectivirgula* and *Thermoactinomyces* species have been found (Moller et al., 1976) and *Thermoactinomyces dichotomicus* has been implicated (Molina, 1982), but these species have seldom been detected in compost (Lacey, 1973). Conversely, precipitins have not been demonstrated to any of the other actinomycetes tested, including *Thermomonospora chromogena* and *Thermobifida fusca* strains, which are found in high numbers in high temperature habitats. However, antigenic extracts may have been unsatisfactory, given the poor growth of these organisms in culture (Lacey, 1988). Both mesophilic and thermophilic monosporic actinomycetes once described as *Thermomonospora* have been isolated from soil (Krassil'nikov and Agre, 1965; Locci et al., 1967; Nonomura and Ohara, 1971c; Nonomura and Ohara, 1974).

Metabolism and Genetics

Knowledge of the genetics and metabolism of *Actinomadura* strains has increased during the last decade with the discovery that actinomadurae are a rich source of antibiotics (Table 9). Earlier work on such organisms is difficult to interpret owing to the uncertain taxonomic status of the organisms examined. Indeed, without further characterization it is not possible to know whether organisms simply labeled *Actinomadura* are correctly named or belong to the *Actinomadura pusilla* group now classified in the genus *Nonomuraea* (Goodfellow et al., 1988; Zhang et al., 1998).

Nevertheless it is clear that actinomadurae are chemo-organotrophic actinomycetes with an oxidative type of metabolism. Members of most species grow well in the range 25–40°C, but several are thermophilic, with an optimum growth temperature of 50–65°C. Actinomadurae can metabolize a wide range of sugars as sole carbon sources for energy and growth; proteolytic activity is shown by the capacity of most strains to attack casein and gelatin (Goodfellow et al., 1979; Goodfellow and Pirouz, 1982; Athalye et al., 1985; Trujillo and Goodfellow, 1997; Trujillo and Goodfellow, 2003; Quintana et al., 2003). In contrast, neither chitin nor xanthine or xylan is metabolized, but an uncharacterized *Actinomadura* strain has been implicated in the degradation of wheat lignocellulose (McCarthy and Broda, 1984a). Pathogenic strains of *A. madurae* produce collagenolytic enzymes (Rippon, 1968), and *Actinomadura* strain R39 synthesizes serine peptidases (Schindler et al., 1986).

All *Thermomonospora* strains grow within the temperature range 40–48°C and pH range 8.0–9.0. However, strains of *T. curvata* and *T.*

Table 9. Secondary metabolites produced by *Actinomadura*, *Actinocorallina*, *Spirillospora* and *Thermomonospora* strains.

Species	Name of antibiotic	Description	Class	Activity
"*Actinomadur aurea*"ᵃ	Catiomycin	Nakamura and Isono, 1983	Polyether	Gram-positive bacteria
		Ubukata et al., 1986		Anaerobic bacteria
Actinomadura atramentaria	SF 2197	Miyadoh et al., 1987		Gram-positive bacteria
"*Actinomadura azurea*"	Catiomycin	Delort et al., 1998	Carboxylic polyether	Gram-positive bacteria
"*Actinomadura brunnea*"	2′-N-methyl-8-methoxychlor-tetracycline	Patel et al., 1987	Tetracycline	Less active against Gram-negative
		Smith et al., 1986		bacteria
Actinomadura crema subsp. *rifamicinii*	Rifamycin O	Gauze et al., 1975	Ansamycin	Gram-positive bacteria
Actinomadura formosensis	Rifamycins O and S	Hasegawa et al., 1986		Gram-positive bacteria including
				mycobacteria
Actinomadura hibisca	Pradimicins	Oki et al., 1988	Anthracycline	Fungal
		Kakinuma et al., 1993		
		Dairi et al., 1997		
	Pradimicins A, B and C	Tomita et al., 1990		
		Oki et al. 1990		
	Pradimicins D and E	Sawada et al., 1990a		
	Pradimicins M, N, O	Sawada et al., 1990b		
Actinomadura kinjaniata	Kijanimicin	Waitz et al. 1981	Nitroaminoglycoside	Tumors
"*Actinomadura luzonensis*"	Antibiotic complex BBM-928	Ohkuma et al., 1980		
		Tomita et al. 1980		
Actinomadura macra	Antibiotic CP 47433,47434	Huang et al. 1980	Polyether	Gram-positive bacteria, anticoccidial activity
Actinomadura madurae	Prodigisins	Gerber, 1973	Pyrrole	
	Maduraferrin	Keller-Schierlein et al., 1988	Siderophore	
	Simaomicin	Maiese et al., 1990		Gram-positive bacteria, anticoccidial activity

Species	Compound	Chemical class	Reference	Activity
Actinomadura oligospora	Portmicin	Polyether	Mertz and Yao, 1986	Gram-positive bacteria, anticoccidial activity
Actinomadura pelletieri	Antibiotic MM 46115	Macrolide antibiotic	Ashton et al., 1990	Antiviral agent
	Prodigisins	Pyrrole antibiotics	Gerber, 1973	Gram-positive bacteria
"*Actinomadura pulveracea*"	WS 6049-A and WS 6049-B	Tumors	Iwami, 1985	
Actinomadura rugatobispora	Angucycline		Ohba et al., 1984	
Actinomadura sp. MK73-NF4	Decatromicins A and B		Momose et al., 1999a, b	Methicillin resistant *Staphylococcus aureus*
"*Actinomadura spinosa*"	Parimicin S	Anthracycline	Saitoh et al., 1993b	*Candida albicans*
Actinomadura spadix	Benanomicin	Anthracycline	Nonomura and Ohara, 1971d	
Actinomadura verrucosospora	Antibiotic CP 51532	Polyether antibiotic	Wink, 2002	
	Antibiotic SF 2487	Polyether antibiotic	Wink, 2002	Gram-positive bacteria, viruses, and anticoccidial activity
	Esperamicin X	Endien antibiotic	Wink, 2002	Tumors
	Pradamicin A	Anthracycline	Dairi et al., 1999	Fungal
	Pradimicins L and FL	Anthracycline	Saitoh et al., 1993	Fungal
	Verrucopeptin	Cyclic depsipeptide	Nishiyama et al., 1993	Tumors
Actinomadura yumaensis	Maduramycin	Monoglycoside polyether	Berger et al., 1988	Anticoccidial activity
Actinocorallina herbida	Azaserine like		Ashton et al., 1990	
Spirillospora albida	Spirillomycin		Domnas, 1968	Gram-positive bacteria
			McInnis and Domnas, 1970	
Spirillospora sp. strain HM 17	HM 17	Polyene, methylpentaene	Hacene et al., 1994	Fungal
Spirillospora sp. strain 719 H 107		Aminoglycoside	Hacene et al., 2000	Gram-negative bacteria, especially

[a]Names in quotation marks have not been validated, i.e., they have not been cited in the "Approved List of Bacterial Names."

chromogena differ in pH and temperature optimum (Table 7). The most commonly isolated thermomonosporae are thermophiles with a growth temperature range of 40–55°C and an optimum of 50°C. *Thermomonospora* strains have several enzymes in common, including catalase, carboxymethylcellulose, deaminase, β-glucosidase, β-galactosidase and β-xylanase (McCarthy and Cross, 1984b; McCarthy and Cross, 1984c; McCarthy, 1989). Esculin, arbutin, and testosterone are degraded, but chitinase activity is universally absent. All strains attack casein and gelatin, and usually agar, keratin, starch and the Tweens, and apart from *T. curvata*, pectin.

The capacity to degrade cellulosic substrates is an important property of many strains, and *Thermomonospora* cellulases (Crawford and McCoy, 1972; Stutzenberger, 1972b; Hägerdal et al., 1978; Bachmann and McCarthy, 1989; Ball and McCarthy, 1989a) and xylanases (McCarthy et al., 1985) have been partially characterized. Amylases from *Thermomonospora* species, including *T. curvata*, are extremely active and stable at 60–70°C and at slightly acid to neutral pH values (Kuo and Hartman, 1967; Stutzenberger and Carnell, 1977; Lupo and Stutzenberger, 1988). Separation of endoglucanase and exoglucanase activity in culture filtrates of *Thermomonospora curvata* has been described (Stutzenberger, 1972b; Stutzenberger and Lupo, 1986b; Lin and Stutzenberger, 1995).

In *T. curvata*, cellobiose has been reported as both a poor (Fennington et al., 1984) and good (Stutzenberger and Kahler, 1986a) cellulase inducer. One possible explanation for this apparent discrepancy is that the role of cellobiose as an inducer or repressor of cellulase production is concentration-dependent and that the threshold concentrations for induction and repression of cellulase production vary in different strains. In *T. curvata* it has been shown that a sufficient concentration of cellobiose to support rapid growth leads to repression of cellulase biosynthesis (Wood et al., 1984). These cultures also provided evidence implicating cyclic AMP as the mediator of catabolic repression of cellulase biosynthesis in *T. curvata*.

Optimization of cellulase production is usually achieved by empirical manipulation of growth conditions and is often species- or strain-specific (McCarthy, 1987). Cellulose degradation or cellulase production can be improved by altering the cellulose source and nitrogen component employed (Moreira et al., 1981; Fennington et al., 1984). Comparable levels of cellulase activity in *Thermomonospora* strains and the cellulolytic fungus *Trichoderma reesii* (commonly used as the reference standard) have been reported. However, different sources of cellulose can greatly affect the extent of hydrolysis realized

(Stutzenberger, 1979; Ferchak et al., 1980). *Thermomonospora curvata* releases sixteen times more β-glucosidases when grown on protein-extracted lucerne fiber compound compared with growth on cellulose or purified cellulose (Bernier and Stutzenberger, 1988a). Enhanced cellulase production has been achieved by mutation (Meyer and Humphrey, 1982), and in *T. curvata* (Fennington et al., 1984), the mutants have been identified as catabolite repressor-resistant.

Xylanase activity has received comparatively little attention even though it is more widespread than cellulose activity among actinomycetes. This imbalance is being redressed using *Thermomonospora* strains that have been extensively studied in relation to their cellulolytic activity (McCarthy et al., 1985; Ristroph and Humphrey, 1985a; Ristroph and Humphrey, 1985b; Ball and McCarthy, 1989a). The production of xylanases is correlated with the growth stage, and degradation is the result of endoxylanase activity. The pH and temperature relationships of *Thermomonospora* xylanases have much in common with those of the corresponding cellulases (McCarthy, 1987). Similarities include optimal activity within the pH range 5–8, increased thermostability of enzymes from thermophilic strains, and lower temperature optima for all-bound dimer hydrolases. *Thermomonospora* xylanases are sufficiently active to produce some direct saccharification of lignocellulose as shown by the generation of xylose from straw (McCarthy et al., 1985). Xylose inhibition of β-xylosidase has been demonstrated in a *Thermomonospora* strain (Ristroph and Humphrey, 1985b).

The genus *Thermomonospora* also contains lignin-degrading organisms (McCarthy and Broda, 1984a; McCarthy et al., 1986). The latter found that *Microbispora* (*Thermomonospora*) *mesophila* degraded lignocellulose to a water-soluble polymeric complex as the main product. The capacity to solubilize lignin is common among lignocellulose-degrading actinomycetes (Ball et al., 1989b).

Little information is available on the genetics of *Thermomonospora* species. Plasmids have been isolated from *Thermomonospora* isolates (Pidcock et al., 1985) and *T. chromogena* (McCarthy, 1989). Protoplast formation, regeneration and transformation of *Thermobifida fusca* (formerly *Thermomonospora*) with plasmid pIJ702 have also been reported (Pidcock et al., 1985). Plasmids such as pIJ702 (Katz et al., 1983) are now being used routinely for the cloning of antibiotic resistance, catabolic and biosynthetic genes of *Streptomyces*, with the most common gene recipient being *Streptomyces lividans* (Chater and Hopwood, 1984). A gene coding for one of the endoglucanases of *Thermobifida fusca* has been cloned into and expressed in *Bacillus subtilis, Escherichia coli* and *Streptomyces lividans* (Collmer and Wilson, 1983; Ghangas and Wilson, 1987) as a first step towards designing a vehicle for the production of high levels of thermostable cellulases. A number of bacteriophages that infect cellulolytic *Thermomonospora* strains (Lawrence et al., 1986) and members of *Faenia* and *Thermoactinomyces* species (Kurup and Heinzen, 1978) have been characterized.

Molecular genetic studies on *Thermomonospora chromogena* ATCC 43196ᵀ performed by Yap et al. (1999) revealed that the genome of this strain contained small interfering (si) rRNA operons (rrn), four of which were complete and two incomplete. Comparative analyses showed that the operon *rrnB* exhibits a high level of sequence variation compared to the other five, which have nearly identical sequences. The coding sequences for the 16S and 23S rRNA genes differ by approximately 6% and 10%, respectively, between the two types of operons. Normal functionality of the *rrnB* gene operon was concluded on the basis of the nonrandom distribution of nucleotide substitutions, the presence of compensation nucleotide covariations, the preservation of secondary and tertiary rRNA structures, and the detection of correctly processed rRNAs in the cell. Comparative sequence analysis also revealed a close evolutionary relationship between the *rrnB* operon of *Thermomonospora chromogena* and the *rrnA* operon of another thermophilic actinomycete, namely *Thermobispora bispora*. These results provide strong evidence that *Thermomonospora chromogena* acquired the *rrnB* operon from *Thermobispora bispora* or related organisms via horizontal gene transfer (Yap et al., 1999). Similar reports on 16S rRNA sequence diversity were reported by Reischl and coworkers for *Mycobacterium celatum* (Reischl et al., 1998) and for the archeon *Halobacterium marismortui* (Mylvaganam and Dennis, 1992). The taxonomic implications of the intraspecific variation (within and between strains) of prokaryotic rRNA sequences were discussed by Clayton and coworkers (Clayton et al., 1995).

Applications

Actinomadura and *Thermomonospora* strains are sources of new antibiotics, enzymes, and products with pharmacological activity. *Actinomadura carminata* produces carminomycin (Gauze et al., 1973), *A. citrea* rifamycin (Gauze et al., 1975), *A. kijaniata* kijanimicin (Waitz et al., 1981), *A. macra* polycyclic ether antibiotics (Huang, 1980), "*A. parvasota*" parvodicin (Christensen et al., 1987), and *A. oligospora*

and *A. yumaensis* polyether antibiotics (Labeda et al., 1985; Mertz and Yao, 1986). In addition, novel antitumor (Takahashi et al., 1988; Matson et al., 1989), glycolipid (Goldstein et al., 1987), nucleoside (Suhadolnik et al., 1989), tetracycline (Patel et al., 1987) and chlorine-containing polyether antibiotics (Cullen et al., 1987) have been isolated from poorly characterized strains labeled *Actinomadura*. An *Actinomadura viridilutea* strain produced a compound that inhibits reverse transcriptase of avian myeloblastosis virus (Sasaki et al., 1988), a mesophilic *Thermomonospora* strain, with a type II phospholipid pattern, produced a novel isochromane-quinone which inhibits thrombin-induced aggregation of human platelets (Patel et al., 1989), an organism labeled "*A. spiculosospora*" produced a specific and reversible inhibitor of angiotensin I converting enzyme (Koguchi et al., 1986), and unspecified thermophilic thermomonosporae produced a new naphthoquinone antibiotic (Hedge et al., 1986); further details are given in Table 9.

Some actinomycetes, including *Actinomadura* strain R39, excrete large amounts of DD-carboxypeptidases, enzymes that are generally membrane-bound. When isolated and characterized, DD-carboxypeptidases serve as model enzymes for elucidating the mechanism of β-lactam antibiotic action (Ghuysen et al., 1979). The inactivation of DD-carboxypeptidases from *Actinomadura* strain R39 by β-lactam antibiotics facilitated the development of a rapid and sensitive method for the qualitative determination of these antibiotics in biological fluids such as human serum and cow's milk (Frere et al., 1980; Schindler et al., 1986). *Actinomadura* strains are also a source of novel restriction endonucleases (Roberts, 1984; Kessler et al., 1985). Proteases from *Thermobifida fusca* have been used on an industrial scale (Desai and Dhala, 1969).

DNA fragments which contain the gene that encodes the extracellular lactamases of *Actinomadura* strain R39 have been introduced into *Streptomyces lividans* TK24 via the high-copy number, promotor-probe plasmid pIJ424 (Piron-Fraipont et al., 1989). Maximal levels of β-lactamase secretion were observed with *S. lividans* CM3 harboring the recombinant plasmid pDML150. The DNA segment in pDML 150 that encodes the β-lactamase has been sequenced (Houba et al., 1989). The β-lactamase precursor is a 304-amino acid polypeptide, the amino terminal region of which has the characteristic features of a signal peptide. The *Actinomadura* strain R39 β-lactamase is a marker of the class A β-lactamases; it shows a particularly high homology with the β-lactamases of *Bacillus licheniformis*.

Thermophilic monosporic actinomycetes have been evaluated for cellulose bioconversion processes designed to produce single cell protein (Bellamy, 1974; Bellamy, 1977; Humphrey et al., 1977; Wood, 1985; Crawford, 1988) or sugar syrups for fermentation to ethanol (Hägerdal et al., 1979; Lee and Humphrey, 1979; Ball and McCarthy, 1988). Several of these publications have referred to a cellulolytic *Thermoactinomyces* strain that has been labeled *T. cellulosae* (Vacca and Bellamy, 1976), but this organism is now known to be a member of the genus *Thermomonospora* (Hägerdal et al., 1980).

Thermobifida fusca and other thermomonosporae have been grown at 55°C on a pilot scale to convert high-cellulose low-lignin pulp mill waste into single cell protein (Crawford et al., 1973). The proteinaceous product had a high nutritional value as a feed supplement in the diet of chicks (Harkin et al., 1974). Thermostable cellulases are also attractive as catalysts in biomass conversion, since their temperature optima can reduce contamination problems in economically feasible nonaseptic applications (Wood, 1985).

A cellulase enhanced *Thermomonospora* mutant grown on chemically pretreated poplar wood was found to use 70% of the woody biomass after 62 h at 55°C but failed to grow well on wood that had not been chemically pretreated. Bellamy (1977) used the parental strain of the same organism, then known as "*Thermoactinomyces cellulosae*," for the bioconversion of chemically pretreated lignocellulosic feedlot waste into microbial biomass. The culture degraded up to 85% of the cellulosic and over 90% of the hemicellulosic fraction of the substrate. The process led to the production of a product with high protein content but was not economical because of the cost of the chemical pretreatment needed to obtain a suitable degradable substrate.

Some success has been achieved in the use of thermophilic actinomycetes in processes designed for the saccharification of purified celluloses. The cellulase complex of *Thermobifida* (*Thermomonospora*) *fusca* YX, the Bellamy organism, readily saccharifies finely ground, acid swollen Avicel, a purified chemically pretreated cellulose (Su and Paulavicius, 1975; Ferchak and Pye, 1980). In subsequent work (Ferchak and Pye, 1983), a process was developed for saccharification of cellulose using a cellulose and cellobiose mixture. In a 7-day, 50°C saccharification, 15–20% glucose syrups were generated from acid swollen Avicel. The availability of cellulase-overproducing strains of *Thermomonospora* species (Fennington et al., 1982; Meyer and Humphrey, 1982) should further improve the commercial potential of saccharification processes.

Ligninolytic *Thermomonospora* strains may have potential for the biological bleaching of pulps or in biological pulping processes given their capacity to solubilize lignin and separate it from the cellulose component of lignocellulose (McCarthy et al., 1986). It has also been suggested (Crawford, 1988) that the ability of some *Thermomonospora* strains to simultaneously attack all of the major components of lignocellulose (i.e., cellulose, lignin and hemicellulose) may be used to develop direct fermentations for the conversion of lignocellulosic wastes into specific high value chemicals and for single cell protein. The development of efficient industrial size bioreactors for solid substrate fermentations will help to foster commercially successful actinomycete-adapted bioconversion processes.

In a study on the degradation of natural rubber, 1220 bacteria from different culture collections were tested (Jendrossek et al., 1997). The results were unexpected because only actinomycetes were able to degrade natural rubber. Most of the positive strains belonged to the genera *Streptomyces* (31), *Actinoplanes* (3), *Dactylosporangium* (1), *Micromonospora* (5), and *Nocardia* (2), but only one strain, *Actinocorallia libanotica* DSM 44554[T], belonged to the family Thermomonosporaceae. It seems that the degradation of natural rubber is a hallmark of actinomycetes, as all of the rubber-degrading strains from the collections and those that have been isolated from different sources using natural rubber or rubber-containing baits were actinomycetes (Jendrossek et al., 1997; Linos et al., 1999; Linos et al., 2000, Linos et al., 2002; Arenskötter et al., 2000).

Literature Cited

Agre, N. S., and L. N. Guzeva. 1975. New genus of the actinomycetes: Excellospora gen. nov. Mikrobiologiya 44:518–522.

Andersen, A. A. 1958. A new sampler for the collection, sizing and enumeration of viable airborne particles. J. Bacteriol. 76:71–484.

Arenskötter, M., D. Baumeister, M. M. Berekaa, G. Pötter, R. M. Kroppenstedt, A. Linos, and A. Steinbüchel. 2001. Taxonomic characterization of two rubber degrading bacteria belonging to the species Gordonia polyisoprenivorans and analysis of hypervariable regions of 16S rDNA sequences. FEMS Microbiol. Lett. 205:277–282.

Ashton, R. J., M. D. Kenig, K. Luk, D. N. Planterose, and G. Scott-Wood. 1990. MM 46115, a new antiviral antibiotic from Actinomadura pelletieri: Characteristics of the producing cultures, fermentation, isolation, physicochemical and biological properties. J. Antibiot. 43:1387–1393.

Athalye, M. 1981a. Classification and Isolation of Actinomadurae [PhD thesis]. University of Newcastle upon Tyne. Newcastle upon Tyne, UK.

Athalye, M., J. Lacey, and M. Goodfellow. 1981b. Selective isolation and enumeration of actinomycetes using rifampicin. J. Appl. Bacteriol. 51:289–297.

Athalye, M., M. Goodfellow, and D. E. Minnikin. 1984. Menaquinone composition in the classification of Actinomadura and related taxa. J. Gen. Microbiol. 130:17–823.

Athalye, M., M. Goodfellow, J. Lacey, and R. P. White. 1985. Numerical classification of Actinomadura and Nocardiopsis. Int. J. Syst. Bacteriol. 35:86–98.

Bachmann, S. L., and A. J. McCarthy. 1989. Purification and characterization of a thermostable β-xylosidase from Thermomonospora fusca. J. Gen. Microbiol. 135:293–299.

Ball, A. S., and A. J. McCarthy. 1988. Saccharification of straw by actinomycete enzymes. J. Gen. Microbiol. 134:2139–2147.

Ball, A. S., and A. J. McCarthy. 1989a. Production and properties of xylanases from actinomycetes. J. Appl. Bacteriol. 66:439–444.

Ball, A. S., W. B. Betto, and A. J. McCarthy. 1989b. Degradation of lignin-related compounds by actinomycetes. Appl. Environ. Microbiol. 55:1642–1644.

Batrakov, S. G., A. G. Panosyan, B. V. Rozynov, I. V. Konova, and L. D. Bergelson. 1976. Menaquinones of Actinomyces olivaceus: the structure of MK-9(H$_6$), MK-9(H$_8$), MK-8(H$_6$), MK-8(H$_8$) [English translation]. Bioorganicheskaya Khimiya 2:1538–1546.

Batrakov, S. G., and L. D. Bergelson. 1978. Lipids of the streptomycetes. Structural investigation and biological interrelation. Chem. Phys. 21:1–29.

Bellamy, W. D. 1974. Single cell proteins from cellulosic wastes. Biotechnol. Bioengin. 16:869–880.

Bellamy, W. D. 1977. Cellulose and lignocellulose digestion by thermophilic actinomycetes for single cell protein production. Devel. Indust. Microb. 18:249–254.

Berger, H. D. L. Sharkey, and G. O. Gale. 1988. Evaluation of the efficacy of maduramicin ammonium in combination with roxarsone and avoparcin in caged broiler chicken. Br. Poultry Sci. 29:435–438.

Bernier, R., and F. Stutzenberger. 1988a. Extracellular and cell-associated forms of beta-glucosidase in Thermomonospora curvata. Lett. Appl. Microbiol. 7:103–107.

Bernier, R., M. Kopp, B. Trakas, and F. Stutzenberger. 1988b. Production of extracellular enzymes by Thermomonospora curvata during growth on protein-extracted lucerne fibers. J. Appl. Bacteriol. 65:411–418.

Blanchard, R. 1896. Parasites végétaux à l'exclusion des bactéries. In: Bouchard (Ed.) Traité de pathologie générale. G. Masson. Paris, France. II:811–932.

Bland, C. E., and J. N. Couch. 1981. The family Actinoplanaceae. In: M. P. Starr, H. Stolp, H. G. Trüper, A. Balows, and H. G. Schlegel (Eds.) The Prokaryotes. Springer-Verlag. Berlin, Germany. 2:2004–2010.

Boiron, P., R. Locci, M. Goodfellow, S. A. Gumaa, K. Isik, B. Kim, M. M. McNeil, M. C. Salinas-Carmona, and H. Shojaei. 1998. Nocardia, nocardiosis and mycetoma. Med. Mycol. 36 (Suppl. 1):26–37.

Brocq-Rousseu, D. 1904. Sur un Streptothrix cause de l'alteration des avoines moisies. Rev. Bot. 16:219–230.

Chater, K. F., and D. A. Hopwood. 1984. Streptomyces genetics. In: M. Goodfellow, M. Mordarski, and S. T. Williams (Eds.) The Biology of Actinomycetes. Academic Press. London, UK. 229–286.

Chormonova, N. T., and T. P. Preobrazhenskaya. 1981. Occurrence of Actinomadura in Kazakhstan soils. Antibiotiki 26:341–345.

Christensen, S. B., H. S. Allaudeen, M. R. Burke, S. A. Carr, S. K. Chung, P. DePhillips, J. J. Dingerdissen, M.

DiPaolo, A. J. Giovenella, S. L. Heald, L. B. Killmer, B. A. Mico, L. Mueller, C. H. Pan, B. L. Poehland, J. B. Rake, G. D. Roberts, M. C. Shearer, R. D. Sitrin, L. J. Nisbet, and P. W. Jeffs. 1987. Parvodicin, a novel glycopeptide from a new species Actinomadura parvosata: Discovery, taxonomy, activity and structure elucidation. J. Antibiot. 40:970–990.

Clayton, R. A., G. Sutton, P. S. Hinkle, Jr., C. Bult, and C. Fields. 1995. Intraspecific variation in small subunit rRNA sequences in GenBank: Why single sequences may not adequately represent prokaryotic taxa. Int. J. Syst. Bacteriol. 45:595–599.

Collins, M. D., and D. Jones. 1981. Distribution of isoprenoid quinone structural types in bacteria and their taxonomic implications. Microbiol. Rev. 45:316–354.

Collins, M. D., A. J. McCarthy, and T. Cross. 1982. New highly saturated members of the vitamin K2 series from Thermomonospora. Zbl. Bakt. Hyg. I. Abt. Orig. C3:358–363.

Collins, M. D., M. Faulkner, and R. M. Keddie. 1984. Menaquinone composition of some sporeforming actinomycetes. System. Appl. Microbiol. 5:20–29.

Collins, M. D., R. M. Kroppenstedt, J. Tamaoka, K. Komagata, and T. Kinoshita. 1988. Structures of the tetrahydrogenated menaquinones from Actinomadura angiospora, Faenia rectivirgula and Saccharothrix australiensis. Curr. Microbiol. 17:275–279.

Collins, B. S., C. T. Kelly, W. M. Fogarty, and E. M. Doyle. 1993. The high maltose-producing alpha-amylase of the thermophilic actinomycete, Thermomonospora curvata. Appl. Microbiol. Biotechnol. 39:31–35.

Collmer, A., and D. B. Wilson. 1983. Cloning and expression of a Thermomonospora YX endoglucanase in Escherichia coli. Biol. Technol. 1:494–501.

Couch, J. N. 1954. The genus Actinoplanes and its relatives. Trans. NY Acad. Sci. 16:315–318.

Couch, J. N. 1963. Some new genera and species of the Actinoplanaceae. J. Elisha Mitchell Sci. Soc. 79:53–70.

Crawford, D. L., and E. McCoy. 1972. Cellulases of Thermomonospora fusca and Streptomyces thermodiastaticus. Appl. Microbiol. 24:150–152.

Crawford, D. L., E. McCoy, J. M. Harkin, and P. Jones. 1973. Production of microbial protein from waste cellulose by Thermomonospora fusca, a thermophilic actinomycete. Biotechnol. Bioengin. 15:833–843.

Crawford, D. L. 1974. Growth of Thermomonospora fusca in lignocellulosic pulps of varying lignin content. Can. J. Microbiol. 20:1069–1072.

Crawford, D. L. 1975. Cultural, morphological and physiological characteristics of Thermomonospora fusca (strain 190Th). Can. J. Microbiol. 21:1842–1848.

Crawford, D. L., and M. A. Gonda. 1977. The sporulation process in Thermomonospora fusca as revealed by scanning electron microscopy and transmission electron microscopy. Can. J. Microbiol. 23:1088–1095.

Crawford, D. L. 1988. Biodegradation of agricultural and urban wastes. In: M. Goodfellow, S. T. Williams, and M. Mordarski (Eds.) Actinomycetes in Biotechnology. Academic Press. London, UK. 433–439.

Cross, T., and M. Goodfellow. 1973. Taxonomy and classification of actinomycetes. In: G. Sykes and F. A. Skinner (Eds.) Actinomycetales: Characteristics and Practical Importance. Academic Press. London, UK. 11–112.

Cross, T. 1981. The monosporic actinomycetes. In: M. P. Starr, H. Stolp, H. G. Trüper, A. Balows, and H. G. Schlegel (Eds.) The Prokaryotes. Springer-Verlag. Berlin, Germany. 2091–2100.

Cui, X.-L., P.-H. Mao, M. Zeng, W.-J. Li, L. P. Zhang, L.-H. Xu, and C.-L. Jiang. 2001. Streptomonospora salina gen. nov., sp. nov., a new member of the family Nocardiopsaceae. Int. J. Syst. Evol. Microbiol. 51:357–363.

Cullen, W. P., N. D. Celmer, L. R. Chappel, L. H. Huang, H. Maeda, S. Nishijama, R. Shibakawa, J. Sone, and P. C. Watts. 1987. CP-54-883 a novel chlorine-containing polyether antibiotic produced by a new species of Actinomadura: Taxonomy of the producing culture, fermentation, physio-chemical and biological properties of the antibiotic. J. Antibiot. 40:1490–14195.

Dairi, T., Y. Hamano, Y., Y. Igarashi, T. Furumai, and T. Oki. 1997. Cloning and nucleotide sequence of putative polyketide synthase genes for pradimicin biosynthesis from Actinomadura hibisca. Biosci. Biotechnol. Biochem. 61:1445–1453.

Dairi, T., Y. Hamano, Y., T. Furumai, and T. Oki. 1999. Development of a self-cloning system for Actinomadura verrucosospora and identification of polyketide synthase genes essential for production of the angucyclic antibiotic pradimicin. Appl. Environ. Microbiol. 65:2703–2709.

Davis, J. D., P. A. Stone, and J. J. McGarry. 1999. Recurrent mycetoma of the foot. J. Foot Ankle Surg. 38:55–60.

Delort, A. M., G. Jeminet, S. Sareth, and F. G. Riddle. 1998. Ionophore properties of catiomycin in large unilamellar vesicles studies by 23Na- and 39K-NMR. Chem. Pharm. Bull. 46:1680–1620.

Desai, A. J., and S. A. Dhala. 1969. Purification and properties of proteolytic enzymes from thermophilic actinomycetes. J. Bacteriol. 100:149–155.

Develoux, M., J. Audvin, T. Freguer, T. M. Vetter, A. Warter, and A. Cenac. 1988. Mycetoma in the Republic of Niger: Clinical features and epidemiology. Clin. J. Trop. Med. Hyg. 38:386–390.

Domnas, A. 1968. Pigments of the Actinoplanaceae. I: Pigment production by Spirillospora 1655. J. Elisha Mitchell Sci. Soc. 84:16–23.

Felsenstein, J. 1993. PHYLIP (Phylogenetic Inference Package), version 3.5 C. Department of Genetics, University of Washington. Seattle, WA.

Fennington, G., D. Lupo, and F. J. Stutzenberger. 1982. Enhanced cellulase production in mutants of Thermomonospora curvata. Biotechnol. Bioengin. 24:2487–2497.

Fennington, G., D. Neubauer, and F. J. Stutzenberger. 1984. Cellulase biosynthesis in a catabolite repression-resistant mutant of Thermomonospora curvata. Appl. Environ. Microbiol. 47:201–204.

Ferchak, J. D., and E. K. Pye. 1980. Saccharification of cellulose by the cellulolytic enzyme system of Thermomonospora species. I: Stability of cellulolytic activities with respect to time, temperature and pH. Biotechnol. Bioengin. 22:1515–1526.

Ferchak, J. D., and E. K. Pye. 1983. Effects of cellulose, glucose, ethanol and metal ions on the cellulose complex of Thermomonospora fusca. Biotechnol. Bioengin. 25:2862–2872.

Fergus, C. L. 1964. Thermophilic and thermotolerant molds and actinomycetes in mushroom compost during peak heating. Mycologia 56:267–284.

Fischer, A., R. M. Kroppenstedt, and E. Stackebrandt. 1983. Molecular-genetic and chemotaxonomic studies on Actinomadura and Nocardiopsis. J. Gen. Microbiol. 129:3433–3446.

Fitch, W. M., and E. Margoliash. 1967. Construction of phylogenetic trees: a method based on mutation distances as estimated from cytochrome C sequences is one of general applicability. Science 155:119–284.

Fowler, V. J., W. Ludwig, and E. Stackebrandt. 1985. Ribosomal ribonucleic acid cataloguing in bacterial systematics: The phylogeny of Actinomadura. In: M. Goodfellow and D. E. Minnikin (Eds.) Chemical Methods in Bacterial Systematics. Academic Press. London, UK. 17–40.

Frère, J. M., D. Klein, and J. M. Ghuysen. 1980. Enzymatic method for rapid and sensitive determination of β-lactam antibiotics. Antimicrob. Agents Chemother. 18:506–510.

Galatenko, O. A., L. P. Terekhova, and T. P. Preobrazhenskaya. 1981. New Actinomadura species isolated from Turkmen soil samples and their antagonistic properties. Antibiotiki 26:803–807.

Galatenko, O. A., L. P. Terekhova, and T. P. Preobrazhenskaya. 1987. Validation of the publications of new names and new combinations previously effectively published outside the IJSB. Int. J. Syst. Bacteriol. 37:179–180.

Gauze, G. F., T. P. Preobrazhenskaya, E. S. Kudrina, N. O. Blinov, I. D. Ryabova, and M. A. Sveshnikova 1957. Problems in the Classification of Antagonistic Actinomycetes. Medzig, Moscow State Publishing House for Medical Literature. Moscow, Russia.

Gauze, G. F., M. A. Sveshnikova, R. S. Ukholina, G. N. Gavrilina, V. A. Filicheva, and E. G. Gladkikh. 1973. Production of antitumor antibiotic carminomycin by Actinomadura carminata sp. nov. Antibiotiki 18:675–678.

Gauze, G. F., T. P. Preobrazhenskaya, N. V. Lavrova, R. S. Ukholina, G. V. Kochetkova, N. P. Nechaeva, N. V. Konstantinova, and I. V. Tolstykh. 1975. Actinomadura cremea var. rifamycini, a rifamicin-producing organism. Antibiotiki 20:963–966.

Gauze, G. F., L. P. Terekhova, O. A. Galatenko, T. P. Preobrazhenskaya, V. N. Borisova, and G. B. Fedorova. 1984. Actinomadura recticatena sp. nov., a new species and its antibiotic properties. Antibiotiki 29:3–7.

Gauze, G. F., L. P. Terekhova, O. A. Galatenko, T. P. Preobrazhenskaya, V. N. Borisova, and G. B. Federova. 1987. Validation of the publications of new names and new combinations previously effectively published outside the IJSB. List No. 23. Int. J. Syst. Bacteriol. 37:179–180.

Gerber, N. N. 1971. Prodigiosin-like pigments from Actinomadura (Nocardia) pelletieri. J. Antibiot. 24:636.

Gerber, N. N. 1973. Minor prodiginine pigments from Actinomadura madurae and Actinomadura pelletieri. J. Heterocycl. Chem. 10:925.

Ghangas, G. S., and D. B. Wilson. 1987. Expression of a Thermomonospora fusca cellulase gene in Streptomyces lividans and Bacillus subtilis. Appl. Environ. Microbiol. 53:1470–1475.

Ghuysen, J. M., J. M. Frère, M. Levy-Bouille, J. Coyette, J. Dusart, and M. Nguyen-Disteche. 1979. Use of model enzymes in the determination of the mode action of penicillins and delta³-cephalosporins. Chem. Rev. Biochem. 48:73–101.

Goldstein, B. P., E. Selva, L. Gastaldo, M. Berti, R. Pallanza, F. Ripamonti, P. Ferrara, M. Denaro, V. Arioli, and G. Cassani. 1987. A40926, a new glycolipid antibiotic with anti-Neisseria activity. Antimicrob. Agents Chemother. 31:1961–1965.

Goodfellow, M., G. Alderson, and J. Lacey. 1979. Numerical taxonomy of Actinomadura and related actinomycetes. J. Gen. Microbiol. 112:95–111.

Goodfellow, M., and T. Pirouz. 1982. Numerical classification of sporoactinomycetes containing meso-diaminopimelic acid in the cell wall. J. Gen. Microbiol. 128:503–507.

Goodfellow, M., and T. Cross. 1984. Classification. *In:* M. Goodfellow, M. Mordarski, and S. T. Williams (Eds.) The Biology of Actinomycetes. Academic Press. London, UK. 7–164.

Goodfellow, M., and D. E. Minnikin (Eds.). 1985. Chemical Methods in Bacterial Systematics. Academic Press. London, UK.

Goodfellow, M., E. Stackebrandt, and R. M. Kroppenstedt. 1988. Chemotaxonomy and actinomycete systematics. *In:* Y. Okami, T. Beppu and H. Ogawara (Eds.) Biology of Actinomycetes '88. Japan Scientific Societies Press. Tokyo, Japan. 233–238.

Goodfellow, M. 1989a. Maduromycetes. *In:* S. T. Williams, M. E. Sharpe and J. G. Holt (Eds.) Bergey's Manual of Systematic Bacteriology. Williams and Wilkins. Baltimore, MD. 4:2509–2510.

Goodfellow, M. 1989b. The actinomycetes. I: Supragenetic classification of actinomycetes. *In:* S. T. Williams, M. E. Sharpe and J. G. Holt (Eds.) Bergey's Manual of Systematic Bacteriology. Williams and Wilkins. Baltimore, MD. 4:2333–2339.

Goodfellow, M. 1992. The family Streptosporangiaceae. *In:* A. Balows, H. G. Trüper, M. Dworkin, W. Harder, and K.-H. Schleifer. The Prokaryotes, 2nd ed. Springer-Verlag. New York, NY. 1116–1138.

Goodfellow, M. 1997. Nocardia and related genera. *In:* A. Balows and B. I. Huerden (Eds.) Jopley and Wilson's Microbiology and Microbial Infections, 9th ed., Volume 2: Systematic Bacteriology. Edward Arnold. London, UK. 463–489.

Gordon, M. A. 1974. Aerobic pathogenic Actinomycetaceae. *In:* E. H. Lennette, E. H. Spaulding, and T. P. Truant (Eds.) Manual of Clinical Microbiology. American Society for Microbiology. Washington, DC. 175–188.

Greiner-Mai, E., R. M. Kroppenstedt, F. Korn?Wendisch, and H. J. Kutzner. 1987. Morphological and biochemical characterization and amended descriptions of thermophilic actinomycetes species. Syst. Appl. Microbiol. 9:97–109.

Grund, E., and R. M. Kroppenstedt. 1989. Transfer of five Nocardiopsis species to the genus Saccharothrix Labeda et al., 1984. J. Syst. Appl. Microbiol. 12:267–274.

Grund, E., and R. M. Kroppenstedt. 1990. Chemotaxonomy and numerical taxonomy of the genus Nocardiopsis Meyer 1976. Int. J. Syst. Bacteriol. 40:5–11.

Gumaa, S. A., E. S. Mahgoub, and M. A. El Sid. 1986. Mycetoma of the head and neck. Am. J. Trop. Med. Hyg. 35:594–600.

Gyobu, Y., and S. Miyadoh. 2001. Proposal to transfer Actinomadura carminata to a new subspecies of the genus Nonomuraea as Nonomuraea roseoviolacea subsp. carminata comb. nov. Int. J. Syst. Evol. Microbiol. 51:881–889.

Hacene, H., K. Kebir, D. S. Othmane, and G. Levebre. 1994. HM17, a new polyene antifungal antibiotic produced by a strain of Spirillospora. J. Appl. Bacteriol. 77:484–489.

Hacene, H., F. Daoudi-Hamdad, T. Bhatnagar, J. C. Baratti, and G. Lefebvre. 2000. H107, a new aminoglycoside anti-Pseudomonas antibiotic produced by a new strain of Spirillospora. Microbios 102:69–77.

Hägerdal, B. G. R., J. D. Ferchak, and E. K. Pye. 1978. Cellulolytic enzyme system of Thermoactinomyces sp. grown on microcrystalline cellulose. Appl. Environ. Microbiol. 36:606–612.

Hägerdal, B. G. R., H. Harris, and E. K. Pye. 1979. Association of β-glucosidase with intact cells of Thermoactinomyces. Biotechnol. Bioengin. 21:345–356.

Hägerdal, B. G. R., J. D. Ferchak, and E. K. Pye. 1980. Saccharification of cellulose by the cellulolytic enzyme system of Thermomonospora sp. I: Stability of cellulolytic activities with respect to time, temperature and pH. Biotechnol. Bioengin. 22:1515–1528.

Harkin, J. M., D. L. Crawford, and E. McCoy. 1974. Bacterial protein from pulps and papermill sludge. TAPPI 57:131–134.

Hasegawa, T., M. P. Lechevalier, and H. A. Lechevalier. 1978. A new genus of the Actinomycetales: Actinosynnema gen. nov. Int. J. Syst. Bacteriol. 28:304–310.

Hasegawa, T., M. P. Lechevalier, and H. A. Lechevalier. 1979. Phospholipid composition of motile actinomycetes. J. Gen. Microbiol. 25:209–213.

Hasegawa, T., S. Tanida, and H. Ono. 1986. Thermomonospora formosensis sp. nov. Int. J. Syst. Bacteriol. 36:20–23.

Hedge, V., T. Barrett, V. Gullo, A. Horan, A. T. McPhail, J., Marquez, M. Patel, and M. Puar. 1986. A novel naphthoquinone antibiotic Sch 3819 produced by a Thermomonospora. *In:* 28th Interscience Congress on Antimicrobial Agents and Chemotherapy. Abstract 908.

Henssen, A. 1957. Beiträge zur Morphologie und Systematik der thermophilen Actinomyceten. Arch. Microbiol. 26:373–414.

Henssen, A., and E. Schnepf. 1967. Zur Kenntnis thermophiler Actinomyceten. Arch. Mikrobiol. 57:214–231.

Henssen, A. 1970. Spore formation in thermophilic actinomycetes. *In:* H. Prauser (Ed.) The Actinomycetales. Gustav Fischer-Verlag. Jena, Germany. 205–210.

Hickey, R. T., and H. D. Tresner. 1952. A cobalt containing medium for sporulation of Streptomyces species. J. Bacteriol. 64:891–892.

Hoffmann, P. 1989a. Cryopreservation of basidiomycete cultures: Mushroom Science XII (Part I). Proceedings of the Twelfth International Congress on the Science and Cultivation of Edible Fungi (1987). Braunschweig, Germany.

Hoffmann, P. 1989b. Cryopreservation of Fungi. UNESCO/WFCC/Education Committee. Braunschweig, Germany. World Federation for Culture Collections Technical Information Sheet No. 5:

Horan, A. C., and B. C. Brodsky. 1982. A novel antibiotic-producing Actinomadura, Actinomadura kijaniata sp. nov. Int. J. Syst. Bacteriol. 32:195–200.

Houba, S., S. Willem, C. Dueng, C. Molitor, J. Dusart, J.-M. Frère, and J. M. Ghuysen. 1989. Cloning and amplified expression in Streptomyces lividans of the gene encoding the extracellular β-lactamase of Actinomadura R39. FEMS Microbiol. Lett. 65:241–246.

Howard, O. W., E. Grund, R. M. Kroppenstedt, and M. D. Collins. 1986. Structural determination of a novel naturally occurring cyclic vitamin K. Biochem. Biophys. Res. Commun. 140:916–923.

Huang, L. H. 1980. Actinomadura macra sp. nov., the producer of antibiotics CP-47,433 and CP-47434. Int. J. Syst. Bacteriol. 30:565–568.

Humphrey, A. E., A. Moreira, W. Armiger, and D. Zabriskie. 1977. Production of single cell protein from cellulose wastes. Biotechnol. Bioengin. Symp. 7:45–64.

Iinuma, S., A. Yokota, T. Hasegawas, and T. Kanamura. 1994. Actinocorallia gen. nov., a new genus of the order Actinomycetales. Int. J. Syst. Bacteriol. 44:235–245.

Itoh, T., T. Kudo, H. Oyaizu, and A. Seino. 1995. Two new species in the genus Actinomadura: A. glomerata sp. nov., and A. longicatena sp. nov. Actinomycetology 9:164–177.

Iwami, M., S. Kiyoto, M. Nishikawa, H. Terano, M. Koshaka, H. Aoki, and H. Imanaka. 1985. New antitumor antibiotics, FR-900405 and FR-900406. I: Taxonomy of the producing strain. J. Antibiot. 38:835–839.

Jendrossek, D., G. Tomasi, and R. M. Kroppenstedt. 1997. Bacterial degradation of natural rubber: A privilege of actinomycetes? FEMS Microbiol. Lett. 150:179–188.

Jensen, H. L. 1930. The genus Micromonospora Orskov, a little known group of soil microorganisms. Proc. Linn. Soc. NSW 55:231–248.

Jensen, H. L. 1932. Contribution to our knowledge of Actinomycetales. III: Further observations on the genus Micromonospora. Proc. Linn. Soc. NSW 57:173–180.

Jones, K. L. 1949. Fresh isolates of actinomycetes in which the presence of sporogenous aerial mycelium is a fluctuating characteristic. J. Bacteriol. 64:891–892.

Kakinuma, S., K. Suzuki, M. Hatori, K. Saitoh, T. Hasegawa, T. Furumai, and T. Oki. 1993. Biosynthesis of the pridimicin family of antibiotics. III: Biosynthesis pathway of both pradimicins and benanomicincs. J. Antibiot. 46:430–440.

Katz, E., C. J. Thompson, and D. A. Hopwood. 1983. Cloning expression of the tyrosinase gene from Streptomyces antibioticus in Streptomyces lividans. J. Gen. Microbiol. 129:2703–2714.

Keller-Schierlein, W., L. Hagmann, H. Zähner, and W. Huhn. 1988. Stoffwechselprodukte von Mikroorganismen. Maduraferrin, ein neuartiger Siderophor aus Actinomadura madurae. Helvetica Chemica Acta 71:1528–1540.

Kempf, A. 1995. Untersuchungen über thermophile Actinomyceten: Taxonomie, Ökologie, und Abbau von Biopolymeren [PhD dissertation]. University of Darmstadt. Darmstadt, Germany.

Kessler, C., P. S. Neumaier, and W. Wolf. 1985. Recognition sequences of restriction: Endonucleases and methylases, a review. Gene 33:1–102.

Kluge, A. G., and F. S. Ferris. 1969. Quantitative phyletics and the evolution of anurans. Syst. Zool. 18:1–32.

Koguchi, T., K. Yamada, M. Yamato, R. Okachi, K. Nakayama, and H. Kase. 1986. K-4, a novel inhibitor of angiotensin I converting enzyme produced by Actinomadura spiculospora. J. Antibiot. 39:364–371.

Kosmachev, A. K. 1964. A new thermophilic actinomycete Micropolyspora thermovirida sp. nov. Microbiology 33:235–237.

Krassil'nikov, N. A. 1941. Keys to Actinomycetales [English version: E. Rabinovitz (Ed. and Trans.), Israel Program for Scientific Translations, Jerusalem, Israel, 1966]. Izvest. Akad Nauk SSSR. Moscow, Russia.

Krassil'nikov, N. A., and N. S. Agre. 1964a. A new genus of the actinomycetes—Actinobifida: The yellow group Actinobifida dichotomica*. Mikrobiologiya 33:935–943.

Krassil'nikov, N. A., and N. S. Agre. 1964b. On two new species of Thermopolyspora. Hind. Antibiot. Bull. 6:97–107.

Krassil'nikov, N. A., and N. S. Agre. 1965. The brown group of Actinobifida chromogena sp. nov. Mikrobiologiya 34:284–291.

Krassil'nikov, N. A., N. S. Agre, and G. I. El-Registan. 1968. New thermophilic species of the genus Micropolyspora. Mikrobiologiya 37:1065–1072.

Kroppenstedt, R. M., and H. J. Kutzner. 1978. Biochemical taxonomy of some problem actinomycetes. Zbl. Bakteriol. Parasitenkd. Infektionskr. Hyg. Abt. 1 Orig. Reihe C Suppl. 6:125–133.

Kroppenstedt, R. M., F. Korn-Wendisch, V. J. Fowler, and E. Stackebrandt. 1981. Biochemical and molecular genetic evidence for a transfer of of Actinoplanes armeniacus into the family Streptomycetaceae. Zbl. Bakt. Hyg., I. Abt. Orig. C2:254–262.

Kroppenstedt, R. M. 1982. Separation of bacterial menaquinones by HPLC using reverse phase (RP-18) and a silver loaded ion exchanger. J. Liquid Chromatogr. 5:2359–2367.

Kroppenstedt, R. M. 1985. Fatty acid and menaquinone analysis of actinomycetes and related organisms. In: M. Goodfellow and D. E. Minnikin (Eds.) Chemical Methods in Bacterial Systematics. Academic Press. London, UK. 173–199.

Kroppenstedt, R. M. 1987. Chemische Untersuchungen an Actinomycetales und verwandten Taxa, Korrelation von Chemosystematik und Phylogenie [Habilitationsschrift]. Technische Hochschule Darmstadt. Darmstadt, Germany.

Kroppenstedt, R. M., E. Stackebrandt, and M. Goodfellow. 1990. Taxonomic revision of the actinomycete genera Actinomadura and Microtetraspora. Syst. Appl. Microbiol. 13:148–160.

Kroppenstedt, R. M., and M. Goodfellow, M. 1992. The family Thermomonosporacea. In: A. Balows, H. G. Trüper, M. Dworkin, W. Harder, and K.-H. Schleifer (Eds.) The Prokarytes, 2nd ed. Springer-Verlag. New York, NY. 1085–1114.

Kudo, T., T. Itoh, S. Miyado, T. Shomura, and A. Seino. 1993. Herbidospora gen. nov., a new genus of the family Streptosporangiaceae Goodfellow et al. 1990. Int. J. Syst. Bacteriol. 43:319–328.

Kudo, T. 1997. Family Thermomonosporaceae. In: S. Miyadoh (Ed.) Atlas of Actinomycetes. Asakura Publishing. Tokyo, Japan. 82–100.

Kuo, M. J., and P. A. Hartman. 1967. Purification and partial characterization of Thermomonospora vulgaris amylases. Can. J. Microbiol. 13:1157–1163.

Kurup, V. P., and R. J. Heinzen. 1978. Isolation and characterization of actinophages of Thermoactinomyces and Micropolyspora. Can. J. Microbiol. 24:794–797.

Küster, E. 1959a. Introductory remarks and welcome, "Round table conference on streptomycetes," 7th International Congress of Microbiology, Stockholm, August 4-5, 1958. Int. Bull. Bact. Nomencl. Taxon. 9:57–61.

Küster, E. 1959b. Outline of comparative study of criteria used in characterization of the actinomycetes. Int. Bull. Bact. Nomencl. Taxon. 9:98–104.

Küster, E., and R. Locci. 1963. Transfer of Thermoactinomyces viridis Schuurmans et al. 1956 to the genus Thermomonospora as Thermomonospora viridis comb. nov. Int. Bull. Bacteriol. Nomencl. Taxon. 13:214–216.

Küster, E. 1974. Family VIII: Micromonosporaceae Krasil'nikov 1938. *In:* R. E. Buchanan and N. E. Gibbon (Eds.) Bergey's Manual of Determinative Bacteriology, 8th ed. Williams and Wilkins. Baltimore, MD. 846.

Labeda, D. P., R. T. Testa, M. P. Lechevalier, and H. A. Lechevalier. 1984. Saccharothrix: A new genus of the Actinomycetales related to Nocardiopsis. Int. J. Syst. Bacteriol. 34:426–431.

Labeda, D. P., R. T. Testa, M. P. Lechevalier, and H. A. Lechevalier. 1985. Actinomadura yumaensis sp. nov. Int. J. Syst. Bacteriol. 35:333–336.

Labeda, D. P., K. Hatano, R. M. Kroppenstedt, and T. Tamura. 2001. Revival of the genus Lentzea and proposal for Lechevalieria gen. nov. Int. J. Syst. Evol. Microbiol. 51:1045–1050.

Lacey, J. 1973. Actinomycetes in soils, composts and fodders. *In:* F. A. Skinner and G. Sykes (Eds.) Actinomycetales: Characteristics and Practical Importance. Academic Press. London, UK. 231–251.

Lacey, J. 1974. Allergy in mushroom workers. Lancet 1:366.

Lacey, J., and J. Dutkiewicz. 1976. Isolation of actinomycetes and fungi using a sedimentation chamber. J. Appl. Bacteriol. 41:15–319.

Lacey, J. 1977. The ecology of actinomycetes in fodders and related substrates. Zbl. Bakteriol. Parasitenkd. Infektionskr. Hyg. Abt. 1, Suppl. 6:61–170.

Lacey, J. 1988. Actinomycetes as biodeteriogens and pollutants of the environment. *In:* M. Goodfellow, S. T. Williams, and M. Mordarski (Eds.) Actinomycetes in Biotechnology. Academic Press. London, UK. 359–432.

Laveran, M. 1906. Tumeur provoquée par un microcoque rose en zooglées. CR Hebd. Soc. Biol. 2:340–341.

Lavrova, N. V., T. P. Preobrazhenskaya, and M. A. Sveshnikova. 1972. Isolation of soil actinomycetes on selective media with rubomycin [in Russian]. Antibiotiki 11:965–970.

Lavrova, N. V., and T. P. Preobrazhenskaya. 1975. Isolation of new species of Actinomadura on selective media with rubromycin [in Russian]. Antibiotiki 20:483–488.

Lawrence, H. M., H. Merivuori, J. A. Sands, and K. A. Pidcock. 1986. Preliminary characterization of bacteriophages infecting the thermophilic actinomycete Thermomonospora. Appl. Environ. Microbiol. 52:631–636.

Lawson, E. N., and L. N. Davey. 1972. A waterborn actinomycete resembling strains causing mycetoma. J. Appl. Bacteriol. 35:389–394.

Lechevalier, H. A., M. P. Lechevalier, and P. E. Holbert. 1966. Electron microscopical observation of sporangial structure of strains of Actinoplanaceae. J. Bacteriol. 92:1228–1235.

Lechevalier, H. A., and M. P. Lechevalier. 1970a. A critical evaluation of the genera of aerobic actinomycetes. *In:* H. Prauser (Ed.) The Actinomycetales. Gustav Fischer-Verlag. Jena, Germany. 393–405.

Lechevalier, M. P., and N. N. Gerber. 1970b. The identity of 3-O-methyl-D-galactose with madurose. Carbohydr. Res. 13:51–454.

Lechevalier, M. P., and H. A. Lechevalier. 1970c. Chemical composition as a criterion in the classification of aerobic actinomycetes. Int. J. Syst. Bacteriol. 20:435–443.

Lechevalier, M. P., and H. A. Lechevalier. 1970d. Composition of whole-cell hydrolysates as a criterion in the classification of aerobic actinomycetes. *In:* H. Prauser (Ed.) The Actinomycetales. Gustav Fischer-Verlag. Jena, Germany. 311–316.

Lechevalier, H. A., M. P. Lechevalier, and N. N. Gerber. 1971. Chemical composition as a criterion in the classification of actinomycetes. Adv. Appl. Microbiol. 14:47–72.

Lechevalier, M. P. 1976. The taxonomy of the genus Nocardia: Some light at the end of the tunnel? *In:* M. Goodfellow, G. H. Brownell, and J. A. Serrano (Eds.) The Biology of the Nocardiae. Academic Press. London, UK. 1–38.

Lechevalier, M. P., C. de Biévre, and H. A. Lechevalier. 1977. Chemotaxonomy of aerobic actinomycetes: Phospholipid composition. Biochem. Ecol. Syst. 5:249–260.

Lechevalier, M. P., and H. A. Lechevalier. 1980. The chemotaxonomy of actinomycetes. *In:* A. Dietz and D. Thayer (Eds.) Actinomycete Taxonomy. Society for Industrial Microbiology. Arlington, VA. Special Publication 6:227–291.

Lechevalier, M. P., A. E. Stern, and H. A. Lechevalier. 1981. Phospholipids in the taxonomy of actinomycetes. *In:* K. P. Schaal and G. Pulverer (Eds.) Actinomycetes. Gustav Fischer-Verlag. Jena, Germany.

Lee, S. E., and A. E. Humphrey. 1979. Use of continuous culture techniques for determining the growth kinetics of cellulolytic Thermoactinomyces sp. Biotechnol. Bioengin. 21:1277–1288.

Li, W.-J., P. Xu, L.-P. Zhang, S.-K. Tang, X.-L. Cui, P.-H. Mao, L.-H. Xu, P. Schumann, E. Stackebrandt, and C.-L. Jiang. 2003. Streptomonospora alba spec. nov., a novel halophilic actinomycete, and emended description of the genus Streptomonospora Cui et al. 2001. Int. J. Syst. Evol. Microbiol. 53:1421–1425.

Liegard, H., and M. Landrieu. 1911. Un cas de mycose conjonctivale. Ann. Ocul. 146:418–426.

Lin, S. B., and F. J. Stutzenberger. 1995. Purification and characterization of the major beta-1,4-endoglucanase from Thermomonospora curvata. J. Appl. Bacteriol. 79:447–453.

Linos, A., A. Steinbüchel, C. Spröer, and R. M. Kroppenstedt. 1999. Gordonia poplyisoprenivorans sp. nov., a rubber-degrading actinomycete isolated from automobile tyre. Int. J. Syst. Bacteriol. 49:1785–1791.

Linos, A., M., M. Berekaa, R. Reichelt, U. Keller, J. Schmitt, H.-C. Flemming, R. M. Kroppenstedt, and A. Steinbüchel. 2002a. Biodegradation of cis-1,4-polyisoprene rubber by distinct actinomycetes: microbial strategies and detailed surface analyses. Appl. Environ. Microbiol. 66:1639–1645.

Linos, A., M. M. Barekaa, A. Steinbüchel, K. K. Kim, C. Spröer, and R. M. Kroppenstedt. 2002b. Gordonia westfalica sp. nov., a novel rubber-degrading actinomycete. Int. J. Syst. Evol. Microbiol. 52:1133–1139.

Lipski, A., and K. Altendorf. 1995. Actinomadura nitritigens sp. nov., isolated from experimental biolfilters. Int. J. Syst. Bacteriol. 45:717–723.

Locci, R., E. Baldacci, and B. Petrolini. 1967. Contribution to the study of oligosporic actinomycetes. I: Description of new species of Actinobifida: Actinobifida alba sp. nov. and revision of the genus. Gen. Microbiol. 15:79–91.

Lu, Z., L. Wang, Y. Zhang, Y. Shi, Z. Liu, E. T. Quintana, and M. Goodfellow. 2003. Actinomadura catellatispora sp. nov. and Actinomadura glauciflava sp. nov., from a sewage ditch and soil in southern China. Int. J. Syst. Evol. Microbiol. 53:137–142.

Lupo, D., and F. Stutzenberger. 1988. Changes in endoglucanase patterns during growth of Thermomonospora curvata on cellulose. Appl. Environ. Microbiol. 54:588–589.

Maiese, W. M., J. Korshally, J. Goodman, M. J. Torrey, S. Kantor, D. P. Labeda, and M. Greenstein. 1990. Simaomicin (LL-D42067), a novel antibiotic from Actinomadura. I: Taxonomy, fermentation and biological activity. J. Antibiot. 43:1059–1063.

Maiti, P. K., A. Ray, and S. Bandyopadhyay. 2002. Epidemiological aspects of mycetoma from a retrospective study of 264 cases in West Bengal. Trop. Int. Health. 7:788–792.

Matson, J. A., C. Claridge, J. A. Bush, J. Titus, W. T. Bradner, and T. W. Doyle. 1989. AT 2433-A1, AT 2433-A2, AT 2433-B1, and AT 2433-B2, novel antitumor antibiotic compounds produced by Actinomadura melliaura. Taxonomy, fermentation, isolation and biological properties. J. Antibiot. 42:1547–1555.

McCarthy, A. J., and T. Cross. 1981. A note on a selective isolation medium for the thermophilic actinomycete Thermomonospora chromogena. J. Appl. Bacteriol. 51:299–302.

McCarthy, A. J., and P. Broda. 1984a. Screening for lignin-degrading actinomycetes and characterisation of their activity against ^{14}C-lignin-labelled wheat lignocellulose. J. Gen. Microbiol. 130:2905–2913.

McCarthy, A. J., and T. Cross. 1984b. A taxonomic study of Thermomonospora and other monosporic actinomycetes. J. Gen. Microbiol. 130:5–25.

McCarthy, A. J., and T. Cross. 1984c. Taxonomy of Thermomonospora and related oligosporic actinomycetes. In: L. Ortiz-Ortiz, L. F. Bojalil, and V. Yakoleff (Eds.) Biological, Biochemical and Biomedical Aspects of Actinomycetes. Academic Press. San Diego, CA. 521–536.

McCarthy, A. T., E. Peace, and P. Broda. 1985. Studies on extracellular activities of some thermophilic actinomycetes. Appl. Microbiol. 21:238–244.

McCarthy, A. J., A. Paterson, and P. Broda. 1986. Lignin solubilization by Thermomonospora mesophila. Appl. Microbiol. 24:347–352.

McCarthy, A. J. 1987. Lignocellulose-degrading actinomycetes. FEMS Microbiol. Rev. 46:145–163.

McCarthy, A. T., A. S. Ball, and S. L. Bachmann. 1988. Ecological and biotechnological implication of lignocellulose degradation by actinomycetes. In: Y. Okami, T. Beppu, and H. Ogawara (Eds.) Biology of Actinomycetes '88. Japan Scientific Society Press. Tokyo, Japan. 283–287.

McCarthy, A. J. 1989. Genus Thermomonospora Henssen 1957. In: S. T. Williams, M. E. Sharpe and J. G. Holt (Eds.) Bergey's Manual of Systematic Bacteriology. Williams and Wilkins. Baltimore, MD. 4:2553–2559.

McInnis, T. M., and A. Domnas. 1970. Pigments in Actinoplanaceae III: A spirillomycin-type pigment from Spririllospora 1309-b. Z. Allgem. Mikrobiol. 10:129–136.

McNeil, M. M., and I. M. Brown. 1994. The medically important aerobic actinomycetes: epidemiology and microbiology. Clin. Microbiol. Rev. 7:357–417.

Mertz, F. P., and R. C. Yao. 1986. Actinomadura oligospora sp. nov., the producer of a new polyether antibiotic. Int. J. Syst. Bacteriol. 36:179–182.

Mertz, F. P., and R. C. Yao. 1990. Actinomadura fibrosa sp. nov. isolated from soil. Int. J. Syst. Bacteriol. 40:28–33.

Meyer, J., and M. Sveshnikova. 1974. Micromonospora rubra Sveshnikova et al. Actinomadura rubra comb. nov. Z. Allgem. Microbiol. 14:67–170.

Meyer, J. 1976. Nocardiopsis, a new genus of the order Actinomycetales. Int. J. Syst. Bacteriol. 26:487–493.

Meyer, J. 1979. New species of the genus Actinomadura. Z. Allgem. Microbiol. 19:37–44.

Meyer, H. P., and A. E. Humphrey. 1982. Cellulase production by a wild type and a new mutant strain of Thermomonospora sp. Biotechnol. Bioengin. 24:1901–1904.

Meyer, J. 1989a. Genus Actinomadura Lechevalier and Lechevalier 1970. In: S. T. Williams, M. E. Sharpe and J. G. Holt (Eds.) Bergey's Manual of Systematic Bacteriology. Williams and Wilkins. Baltimore, MD. 4:2511–2526.

Meyer, J. 1989b. Genus Nocardiopsis (Brocq-Rousseu) Meyer 1976. In: S. T. Williams, M. E. Sharpe and J. G. Holt (Ed.) Bergey's Manual of Systematic Bacteriology. Williams and Wilkins. Baltimore, MD. 4:2562–2569.

Mills, J. T., and G. J. Bollen. 1976. Microflora of heat damaged rapeseed. Can. J. Bot. 54:2893–2902.

Minnikin, D. E., L. Alshamaony, and M. Goodfellow. 1975. Differentiation of Mycobacterium, Nocardia and related taxa by thin-layer chromatographic analysis of whole-organism methanolysates. J. Gen. Microbiol. 88:200–204.

Minnikin, D. E., A. G. O'Donnell, M. Goodfellow, G. Alderson, M. Athalye, and J. H. Parlett. 1984. An integrated procedure for extraction of bacterial isoprenoid quinones and polar lipids. J. Microbiol. Meth. 2:33–241.

Miyadoh, S., H. Tohyama, S. Amano, T. Shomura, and T. Niida. 1985. Microbispora viridis, a new species of Actinomycetales. Int. J. Syst. Bacteriol. 35:281–284.

Miyadoh, S., S. Amano, H. Tohyama, and T. Shomura. 1987. Actinomadura atramentaria, a new species of the Actinomycetales. Int. J. Syst. Bacteriol. 37:342–346.

Miyadoh, S., H. Anzai, S. Amano, and T. Shomura. 1989. Actinomadura malachitica and Microtetraspora viridis are synonyms and should be transferred as Actinomadura viridis comb. nov. Int. J. Syst. Bacteriol. 39:152–158.

Miyadoh, S., S. Amano, H. Tohyama, and T. Shomura. 1990. A taxonomic review of the genus Microbispora and proposal to transfer two species to the genus Actinomadura and to combine ten species into Microbispora rosea. J. Gen. Microbiol. 136:1905–1913.

Miyadoh, S., and 121 coauthors. 2001. Identification Manual of Actinomycetes. Society of Actinomycetes Japan, Business Centre for Academic Societies. , Japan.

Miyashita, K., Y. Mikami, and T. Arai. 1984. Alkalophilic actinomycete, Nocardiopsis dassonvillei subsp. prasina subsp. nov., isolated from soil. Int. J. Syst. Bacteriol. 34:405–409.

Molina, C. 1982. Acquisition récentes sur les alvéolites allergiques professionelles. Schweiz. med. Wochenschr. 112:192–197.

Moller, B. B., P. Holberg, S. Gravesen, and B. Weeke. 1976. Precipitating antibodies against Micropolyspora faeni in sera from mushroom workers. Acta Allerg. 31:69–70.

Momose, I., S. Hirosawa, H. Nakamura, H. Naganawa, H. Iinuma, D. Ikeda, and T. Takeuchi. 1999a. Decatromicins A and B, new antibiotics produced by Actinomadura sp. MK73-NF4. II: Structure determination. J. Antibiot. 52:787–796.

Momose, I., H. Iinuma, N. Kinoshita, Y. Momose, S. Kunimoto, M. Hamada, and T. Takeuchi. 1999b. Decatromicins A and B, a new antibiotic produced by Actinomadura sp. MK73-NF4: Taxonomy, isolation, physico-chemical properties and biological activities. J. Antibiot. 52:781–786.

Moore, M. E. C., E. P. Cato, and L. V. H. Moore. 1985. Index of bacterial and yeast nomenclatural changes published in the International Journal of Systematic Bacteriology since the 1980 Approved Lists of Bacterial Names (1 January 1980 to 1 January 1985). Int. J. Syst. Bacteriol. 35:382–407.

Moreira, A. R., J. A. Phillips, and A. E. Humphrey. 1981. Production of cellulose by Thermomonospora spec. Biotechnol. Bioengin. 23:1339–1347.

Mylvaganam, S., and P. P. Dennis. 1992. Sequence heterogeneity between the two genes encoding 16S rRRNA from the halophilic archaebacterium Halobacterium marismortui. Genetics 130:399–410.

Nakamura, G., and K. Isono. 1983. A new species of Actinomadura producing a polyether antibiotic, cationomycin. J. Antibiot. 36:1468–1472.

Nishiyama, Y., K. Sugawara, K. Tomita, H. Yamamoto, H. Kamei, and T. Oki. 1993. Verucopeptin, a new antitumor antibiotic active against B16 melanoma. I: Taxonomy, production, isolation, physio-chemical properties and biological activity. J. Antibiot. 46:921–927.

Nonomura, H., and Y. Ohara. 1969. Distribution of actinomycetes in soil. VI: A culture method effective for both preferential isolation and enumeration of Microbispora and Microtetraspora strains in soil. J. Ferment. Technol. 47:463–469.

Nonomura, H., and Y. Ohara. 1971a. Distribution of actinomycetes in soil. VIII: Green-spore group of Microtetraspora, its preferential isolation and taxonomic characteristic. J. Ferment. Technol. 49:1–7.

Nonomura, H., and Y. Ohara. 1971b. Distribution of actinomycetes in soil. IX: New species of the genus Microbispora and Microtetraspora and their isolation methods. J. Ferment. Technol. 49:887–894.

Nonomura, H., and Y. Ohara. 1971c. Distribution of actinomycetes in soil. X: New genus and species of monosporic actinomycetes in soil. J. Ferment. Technol. 49:895–903.

Nonomura, H., and Y. Ohara. 1971d. Distribution of actinomycetes in soil. XI: Some new species of the genus Actinomadura Lechevalier et al. J. Ferment. Technol. 49:904–912.

Nonomura, H., and Y. Ohara. 1974. A new species of actinomycetes, Thermomonospora mesouviformis sp. nov. J. Ferment. Technol. 53:10–13.

Ochi, K., S. Miyadoh, and T. Tamura. 1991. Polyacrylamide gel electrophoresis analysis of ribosomal protein AT-L30 as a novel approach to actinomycetes taxonomy: application to genera Actinomadura and Microtetraspora. Int. J. Syst. Bacteriol. 41:234–239.

Ohba, K., S. Miyadoh, J. Yoshida, M. Ito, T. Shomura, M. Sezaki, and T. Itoh. 1984. Studies on a new peptide antibiotic SF-2240. Sci. Rept. Meijii Seika Kaisha 23:12–15.

Ohkuma, H., F. Sakai, Y. Nishiyama, M. Ohbayashi, H. Imanishi, M. Konishi, T. Miyaki, H. Koshiyama, and H. Kawaguchi. 1980. BBM-928, a new antitumor antibiotic complex. I: Production, isolation, characterization and antitumor activity. J. Antibiot. 33:1087–1097.

Oki, T., M. Konishi, K. Tomatsu, K. Tomita, K. Saitoh, M. Tsunakawa, M. Nishio, T. Miyaki, and H. Kawaguchi. 1988. Pradimicin, a novel class of potent antifungal antibiotics. J. Antibiot. 41:1701–1704.

Oki, T., O. Tenmyo, M. Hirano, K. Tomatsu, and H. Kamei. 1990. Pradimycins A, B and C: new antifungal antibiotics. II: In vitro and in vivo biological activities. J. Antibiot. 43:763–770.

Patel, M., V. P. Gullo, V. R. Hedge, A. C. Horan, F. Gentile, J. A. Marquez, G. H. Miller, M. S. Puar, and J. A. Waitz. 1987. A novel tetracycline from Actinomadura brunnea:

Fermentation, isolation and structure eluciation. J. Antibiot. 40:1408–1413.

Patel, M., V. Hedge, A. C. Horan, T. Barrett, R. Bishop, A. King, J. Marquez, R. Hare, and V. Gullo. 1989. Sch 38519, a novel platelet aggregation inhibitor produced by a Thermomonospora sp. Taxonomy, fermentation, isolation, physico-chemical properties, structure and biological properties. J. Antibiot. 42:1063–1069.

Pidcock, K. A., B. S. Montenecourt, and J. A. Sands. 1985. Genetic recombinations and transformations in protoplasts of Thermomonospora fusca. Appl. Environ. Microbiol. 50:693–695.

Pinoy, E. 1912. Isolement et culture d'une nouvelle oospora pathogene. Mycetome a grains rouges de la paroi thoracique. Bull. Soc. Path. Exot. 5:585–589.

Piron-Fraipont, C., C. Duez, A. Matagne, C. Molitor, J. Dusart, J.-M. Frère, and J. M. Ghuysen. 1989. Cloning and amplified expression in Streptomyces lividans of the gene encoding the extracellular β-lactamase of Actinomadura R39. Biochem. J. 262:849–854.

Poschner, J., R. M. Kroppenstedt, A. Fischer, and E. Stackebrandt. 1985. DNA-DNA-reassociation and chemotaxonomic studies on Actinomadura, Microbispora, Microtetraspora, Micropolyspora, and Nocardiopsis. Syst. Appl. Microbiol. 6:264–270.

Preobrazhenskaya, T. P., and M. A. Sveshnikova. 1974. New species of the genus Actinomadura. Mikrobiologiya 43:864–868.

Preobrazhenskaya, T. P., N. V. Lavrova, and N. O. Blinov. 1975a. Taxonomy of Streptomyces luteofluorescens. Mikrobiologiya 44:524–527.

Preobrazhenskaya, T. P., N. V. Lavrova, R. S. Ukholina, and N. P. Nechaeva. 1975b. Isolation of new species of Actinomadura on selective media with streptomycin and bruneomycin. Antibiotiki 20:404–409.

Preobrazhenskaya, T. P., L. P. Terekhova, A. V. Laiko, T. I. Selezneva, V. A. Zenkova, and N. O. Blinov. 1976. Actinomadura coeruleoviolacea sp. nov. and its antagonistic properties. Antibiotiki 21:779–784.

Preobrazhenskaya, T. P., M. A. Sveshnikova, and L. P. Terekhova. 1977. Key for the identification of the species of the genus Actinomadura. Biol. Actin. Rel. Org. 12:30–38.

Preobrazhenskaya, T. P., M. A. Sveshnikova, L. P. Terekhova, and N. T. Chormonova. 1978. Selective isolation of soil actinomycetes. *In:* M. Mordarski, W. Kurylowicz, and J. Jeljaszewicz (Eds.) Nocardia and Streptomyces. Gustav Fischer-Verlag. Stuttgart, Germany. 114–123.

Preobrazhenskaya, T. P., M. A. Sveshnikova, and G. F. Gauze. 1982. On the transfer of certain species of the genus Actinomadura Lechevalier et Lechevalier 1970 to the genus Nocardiopsis Meyer 1976. Mikrobiologiya 51:111–113.

Preobrazhenskaya, T. P., L. P. Terekhova, A. V. Laiko, T. I. Selezneva, V. A. Zenkova, and N. O. Blinov. 1987. Validation of the publication of new names and new combinations previously effectively published outside the IJSB. List No. 23. Int. J. Syst. Bacteriol. 37:179–180.

Pridham, T. G., P. Anderson, C. Foley, L. A. Lindenfelser, C. W. Hesseltine, and R. G. Benedict. 1956–57. A selection of media for maintenance and taxonomic study of streptomycetes. Antibiot. Ann. 1956/57:947–953.

Pridham, T. G., and A. T. Lyons Jr. 1961. Streptomyces albus (Rossi Doria) Waksman and Henrici: Taxonomic study of strains labeled Streptomyces albus. J. Bacteriol. 81:431–441.

Quintana, E. T., M. E. Trujillo, and M. Goodfellow. 2003. Actinomadura mexicana sp. nov. and Actinomadura meyerii sp. nov., two novel soil sporoactinomycetes. Syst. Appl. Microbiol. 26:511–517.

Reischl, U., K. Feldmann, L. Naumann, B. J. Gaugler, B. Ninet, and B. Hirschel. 1998. 16S rRNA sequence diversity in Mycobacterium celatum strains caused by presence of two different copies of 16S rRNA gene. J. Clin. Microbiol. 36:1761–1764.

Rippon, T. W. 1968. Extracellular collagenase produced by Streptomyces madurae. Biochim. Biophys. Acta 159:147–152.

Ristroph, D. L., and A. E. Humphrey. 1985a. Kinetic characterization of the extracellular xylanaes of Thermomonospora sp. Biotechnol. Bioengin. 27:832–836.

Ristroph, D. L., and A. E. Humphrey. 1985b. The β-xylosidase of Thermomonospora. Biotechnol. Bioengin. 27:909–913.

Roberts, R. J. 1984. Restriction and modification enzymes and their recognition sequences. Nucleic Acids Res. 12:167–204.

Saitoh, K., Y. Sawada, K. Tomita, T. Hatori, and T. Oki. 1993a. Pradimicins L and FL: new prdimicin congeners from Actinomadura verucosospora subsp. neohibisca. J. Antibiot. 46:387–397.

Saitoh, K., O. Tenmyo, S. Yamamoto, T. Furumai, and T. Oki. 1993b. Pradimicin S, a new pradimicin analog. I: Taxonomy, fermentation and biological activities. J. Antibiot. 46:580–588.

Saitou, N., and M. Nei. 1987. The neighbor-joining method: A new method for constructing phylogenetic trees. Molec. Biol. Evol. 4:406–425.

Sasaki, T., J. Yoshida, M. Itoh, S. Gomi, T. Shomura, and M. Senzaki. 1988. New antibiotic SF 2315A and B produced by an Excellospora sp. I: Taxonomy of the strain, isolation and characterization of antibiotic. J. Antibiot. 41:835–842.

Sawada, Y., M. Nishio, H. Yamamoto, M. Hatori, T. Miyaki, M. Konishi, and T. Oki. 1990a. New antifungal antibiotics, pradimicins D and E. Glycine analogs of pradimicins A and C. J. Antibiot. 43:771–777.

Sawada, Y., T. Tsuno, H. Yamamoto, M. Nishio, M. Konoshi, and T. Oki. 1990b. Pradiminins M. N. O. and P., new hidydrobenzo[a]naphthacenequinones produced by blocked mutants of Actinomadura hibisca P157-2. J. Antibiot. 43:1367–1374.

Schaal, K. P. 1972. Zur mikrobiologischen Diagnostik der Nocardiose. Zbl. Bakt. Hyg. I. Abt. Orig. A 220:242–246.

Schaal, K. P. 1984a. Laboratory diagnosis of actinomycete diseases. In: M. Goodfellow, M. Mordarski, and S. T. Williams (Eds.) The Biology of the Actinomycetes. Academic Press. London, UK. 425–456.

Schaal, K. P., and B. L. Beaman. 1984b. Clinical significance of actinomycetes. In: M. Goodfellow, M. Mordarski, and S. T. Williams (Eds.) The Biology of the Actinomycetes. Academic Press. London, UK. 389–424.

Schäfer, D. 1973. Beiträge zur Klassifizierung und Taxonomie der Actinoplanaceen [PhD dissertation]. University of Marburg/Lahn. Marburg, Germany.

Schindler, P. W., W. König, S. Chatterjee, and B. N. Ganguli. 1986. Improved screening for β-lactam antibiotics: A sensitive, high-throughput assay using DD-

carbopeptidase and a novel chromophore labelled substrate. J. Antibiot. 39:53–57.

Schleifer, K.-H., and O. Kandler. 1972. Peptidoglycan types of bacterial cell walls and their taxonomic implications. Bacteriol. Rev. 36:407–477.

Serrano, J. A., B. L. Beaman, T. E. Viloria, M. A. Mejia, and R. Zamora. 1986. Histological and ultrastructural studies on human actinomycetomas. In: G. Szabó, S. Biró and M. Goodfellow (Eds.) Biological, Biochemical and Biomedical Aspects of Actinomycetes. Akadémiai Kiadó. Budapest, Hungary. 647–662.

Shinobu, R. 1958. Physiological and cultural study for the identification of soil actinomycete species. Mem. Osaka Univ. Lib. Arts. Educ. Ser. B. Nat. Sci. 7:1–76.

Shinobu, R. 1962. A new Streptomyces species producing fluorescent-yellow soluble pigment. Mem. Osaka Univ. Lib. Arts Educ. Ser. B. Nat. Sci. 11:115–122.

Shirling, E. B., and D. Gottlieb. 1966. Methods for the characterization of Streptomyces species. Int. J. Syst. Bacteriol. 16:313–340.

Skerman, V. B. D., V. McGowan, and P. H. A. Sneath. 1980. Approved lists of bacterial names. Int. J. Syst. Bacteriol. 30:225–420.

Smith, E. B., H. K. Munayyer, M. J. Ryan, B. A. Mayles, V. R. Hedge, and G. H. Miller. 1986. Direct selection of a specifically blocked mutant of Actinomadura brunnea: Isolation of a third methoxy substituted chlortetracycline. J. Antibiot. 40:1419–1425.

Soina, V. S., A. A. Sokolov, and N. S. Agre. 1975. Ultrastructure of mycelium and spores of Actinomadura fastidiosa sp. nov. Mikrobiologiya 44:883–887.

Stackebrandt, E., and C. R. Woese. 1981a. The evolution of prokaryotes. In: M. T. Carlile, M. T. Collins, and B. E. B. Moseley (Eds.) Molecular and Cellular Aspects of Microbial Evolution. Cambridge University Press. Cambridge, UK. 1–31.

Stackebrandt, E., B. Wunner-Füssl, V. J. Fowler, and K. H. Schleifer. 1981. Deoxyribonucleic acid homologies and ribosomal ribonucleic acid similarities among spore-forming members of the order Actinomycetales. Int. J. Syst. Bacteriol. 31:420–431.

Stackebrandt, E., R. M. Kroppenstedt, and V. J. Fowler. 1983a. A phylogenetic analysis of the family Dermatophilaceae. J. Gen. Microbiol. 129:1831–1838.

Stackebrandt, E., W. Ludwig, E. Seewald, and K.-H. Schleifer. 1983b. Phylogeny of sporeforming members of the order Actinomycetales. Int. J. Syst. Bacteriol. 33:173–180.

Stackebrandt, E., and K.-H. Schleifer. 1984. Molecular systematics of actinomycetes and related organisms. In: L. Ortiz-Ortiz, L. F. Bojalil and V. Yakoleff (Eds.) Biological, Biochemical, and Biomedical Aspects of Actinomycetates. Academic Press. Orlando, FL. 485–504.

Stackebrandt, E. 1986. The significance of "Wall Types" in phylogenetically based taxonomic studies on actinomycetes. In: G. Szabó, S. Biró, and M. Goodfellow (Eds.) Biological, Biochemical and Biomedical Aspects of Actinomycetes. Akadémiai Kiadó. Budapest, Hungary. 497–506.

Stackebrandt, E., and M. Goodfellow. 1990. Nucleic Acid Techniques in Bacterial Systematics. John Wiley. Chichester, UK.

Stackebrandt, E., F. A. Rainey, and N. L. Ward-Rainey. 1997. Proposal for a new hierarchic classification system, Actinobacteria classis nov. Int. J. Syst. Bacteriol. 47:479–491.

Staneck, M. 1972. Microorganisms inhabiting mushroom compost during fermentation. Mushroom Sci. 8:797–811.

Staneck, J. L., and G. D. Roberts. 1974. Simplified approach to identification of aerobic actinomycetes by thin-layer chromatography. Appl. Microbiol. 28:226–231.

Steingrube, V. A., R. W. Wilson, B. A. Brown, K. C. Jost, J. R., Z. Blacklock, J. L. Gibson, and J. R. Wallace. 1997. Rapid identification of clinical significant species and taxa of aerobic actinomycetes, including Actinomadura, Gordona, Nocarida, Rhodococcus, Streptomyces and Tsukumarella isolates, by DNA amplification and restriction endonucleases analysis. J. Clin. Microbiol. 35:817–822.

Stutzenberger, F. J., A. J. Kaufman, and R. D. Lossin. 1970. Cellulolytic activity in municipal solid waste compost. Can. J. Microbiol. 16:553–560.

Stutzenberger, F. J. 1971. Cellulase production by Thermomonospora curvata isolated from municipal solid waste compost. Appl. Microbiol. 22:147–152.

Stutzenberger, F. J. 1972a. Cellulolytic activity of Thermomonospora curvata. I: Nutritional requirements for cellulase production. Appl. Microbiol. 24:77–82.

Stutzenberger, F. J. 1972b. Cellulolytic activity of Thermomonospora curvata. 2: Optimal conditions, partial purification and product of the cellulase. Appl. Microbiol. 24:83–90.

Stutzenberger, F. J., and R. Carnell. 1977. Amylase production by Thermomonospora curvata. Appl. Environ. Microbiol. 34:234–236.

Stutzenberger, F. J., and I. Sterpu. 1978. Effect of municipal refuse metals in cellulase production by Thermomonospora curvata. Appl. Environ. Microbiol. 36:201–204.

Stutzenberger, F. J. 1979. Degradation of cellulosic substances by Thermomonospora curvata. Biotechnol. Bioengin. 21:909–913.

Stutzenberger, F. J., and G. Kahler. 1986a. Cellulase biosynthesis during degradation of cellulose derivatives by Thermomonospora curvata. J. Appl. Bacteriol. 61:225–233.

Stutzenberger, F. J., and D. Lupo. 1986b. pH-dependent thermal activation of endo-1,4-β-glucanase in Thermomonospora curvata. Enz. Microb. Technol. 8:205–208.

Su, T. M., and D. Paulavicius. 1975. Enzymatic saccharification of cellulose by thermophilic actinomycetes. Appl. Polymer Symp. 28:221–236.

Suhadolnik, R. J., S. Pornbanlualap, D. C. Baker, K. N. Tiwari, and A. K. Hebbler. 1989. Stereospecific 2′-amination and 2′-chlorination of adenosine by Actinomadura in the biosynthesis of 2′-amino-2′-deoxyadenosine and 2′-chloro-2′-deoxycoformycin. Arch. Biochem. Biophys. 270:374–382.

Sveshnikova, M. A., T. S. Maksimova, and E. S. Kudrina. 1969. The species belonging to the genus Micromonospora Orskov, and their taxonomy. Mikrobiologiya 38:883–893.

Takahashi, I., K.-I. Takahashi, K. Asano, I. Kawamoto, T. Yasuzawa, T. Ashizawa, F. Tomita, and H. Nakano. 1988. DC92-B, a new antitumor antibiotic from Actinomadura. J. Antibiot. 41:1151–1153.

Terekhova, L. P., O. A. Galatenko, and T. P. Preobrazhenskaya. 1982. Actinomadura fulvescens sp. nov. and A. turkmeniaca sp. nov. and their antagonistic properties. Antibiotiki 27:87–92.

Terekhova, L. P., O. A. Galatenko, and T. P. Preobrazhenskaya. 1987. Validation of publication of new names and new combinations previously effectively published outside the IJSB. List No. 23. Int. J. Syst. Bacteriol. 37:179–180.

Tomita, K., Y. Hoshino, T. Sasahira, and H. Kawaguchi. 1980. BBM-928, a new antitumor complex. II: Taxonomic studies on the producing organism. J. Antibiot. 33:1098–1102.

Tomita, K., Y. Nakakita, Y. Hoshino, K. Numata, and H. Kawaguchi. 1987. New genus of the Actinomycetales: Streptoalloteichus hindustanicus gen. nov., nom. rev., sp. nov., nom. rev. Int. J. Syst. Bacteriol. 37:211–213.

Tomita, K., M. Nishio, K. Saitoh, H. Yamamoto, Y. Hoshino, H. Okhuma, M. Konishi, T. Miyaki, and T. Oki. 1990. Pradimicins A, B and C: New antifungal antibiotics. J. Antibiot. 43:755–762.

Trujillo, M. E., and M. Goodfellow. 1997. Polyphasic taxonomic study of clinically significant Actinomadura including the description of Actinomadura latina sp. nov. Zbl. Bakteriol. 285:212–233.

Trujillo, M. E., and M. Goodfellow. 2003. Numerical phenetic classification of clinically significant aerobic sporoactinomycetes and related organisms. Ant. v. Leeuwenhoek 84:39–68.

Ubukata, M., Y. Hamazaki, and K. Isono. 1986. Chemical modification of cationomycin and its structure-activity relationship. Agric. Biol. Chem. 50:1153–1160.

Uchida, K., and K. Aida. 1977. Acyl type of bacterial cell wall: Its simple identification by colorimetric method. J. Gen. Appl. Microbiol. 23:249–260.

Vacca, J. G., and W. D. Bellamy. 1976. Classification and morphological properties of a high temperature, cellulolytic actinomycete. In: Abstracts of the Annual Meeting of the American Society for Microbiology. 117.

Vincent, H. 1894. Étude sur le parasite du pied le madura. Ann. Inst. Pasteur 8:129–151.

Vobis, G. 1985a. Spore development in sporangia-forming actinomycetes. In: G. Szabó, S. Biró, and M. Goodfellow (Eds.) Biological, Biochemical and Biomedical Aspects of Actinomycetes. Akadémiai Kiadó. Budapest, Hungary. 443–452.

Vobis, G., and H.-W. Kothe. 1985b. Sporogenesis in sporangiate actinomycetes. Front. Appl. Microbiol. 1:25–47.

Vobis, G., and H.-W. Kothe. 1989. Genus Spirillospora Couch 1963. In: S. T. Williams, E. M. Sharpe, and J. G. Holt (Eds.) Bergey's Manual of Systematic Bacteriology. Williams and Wilkins. Baltimore, MD. 4:2543–2545.

Waitz, J. A., A. C. Horan, M. Kalyanpur, B. K. Lee, D. Loebenberg, J. A. Marquez, G. Miller, and M. G. Patel. 1981. Kijanimicin (Sch 25663), a novel antibiotic produced by Actinomadura kijaniata SCC 1256. Fermentation, isolation, characterization and biological properties. J. Antibiot. 34:1101–1106.

Waksman, S. A., and A. T. Henrici. 1948. Family Actinomycetaceae Buchanan and family Streptomycetaceae Waksman and Henrici. In: R. S. Breed, E. G. D. Murray, and A. P. Hitchens (Eds.) Bergey's Manual of Determinative Bacteriology, 6th ed. Williams and Wilkins. Baltimore, MD. 961.

Waksman, S. A. 1950. The Actinomycetes: Their nature, occurrence, activities and importance. Ann. Crypt. Phytopathol. 8:1–230.

Waksman, S. A. 1961. In: The Actinomycetes: Classification, Identification and Descriptions of Genera and Species. Williams and Wilkins. Baltimore, MD. 2:1–363.

Williams, S. T., and T. Cross. 1971. Actinomycetes. Meth. Microbiol. 4:295–234.

Williams, S. T., E. M. Sharpe, and J. G. Holt (Eds.) 1989. Bergey's Manual of Systematic Bacteriology. Williams and Wilkins. Baltimore, MD.

Wink, J. 2002. The Actinomycetales: An Electronic Manual. Aventis Pharma. Frankfurt, Germany.

Wink, J., R. M. Kroppenstedt, G. Seibert, and E. Stackebrandt. 2003. Actinomadura namibiensis sp. nov. Int. J. Syst. Evol. Microbiol. 53:721–724.

Wood, W. E., D. G. Neubauer, and F. J. Stutzenberger. 1984. Cyclic AMP levels during induction and repression of cellulase biosynthesis in Thermomonospora curvata. J. Bacteriol. 160:1047–1054.

Wood, W. A. 1985. Useful biodegradation of cellulose. Ann. Phythochem. Soc. Eur. 26:295–309.

Yamada, Y., K. Aoki, and Y. Tahara. 1982a. The structure of the hexahydrogenated isoprenoid side-chain menaquinone with nine isoprene units isolated from Actinomadura madurae. J. Gen. Microbiol. 28:321–429.

Yamada, Y., C. F. Hou, J. Sasaki, Y. Tahara, and H. Yoshioka. 1982b. The structure of the octahydrogenated iso-

prenoid side-chain menaquinone with nine isoprene units isolated from Streptomyces albus. J. Gen. Appl. Microbiol. 28:519–529.

Yap, W. H., Z. Zhang, and Y. Wang. 1999. Distinct types of rRNA operons exists in the genome of the actinomycete Thermomonospora chromogena and evidence for horizontal transfer of an entire rRNA operon. J. Bacteriol. 181:5201–5209.

Yu, A. M., S. Zhao, and L. Y. Nie. 1993. Mycetomas in northern Yemen: Identification of causative organisms and epidemiologic consideration. Am. J. Trop. Med. Hyg. 48:812–817.

Zhang, Z., Y. Wang, and J. Ruan. 1998. Reclassification of Thermomonospora and Microtetraspora. Int. J. Syst. Bacteriol. 48:411–422.

Zhang, Z., T. Kudo, Y. Nakajima, and Y. Wang. 2001. Clarification of the relationship between the members of the family Thermomonosporaceae on the basis of 16S rDNA, 16S-23S rRNA internal transcribed spacer and 23S rDNA sequences and chemotaxonomic analyses. Int. J. Syst. Evol. Microbiol. 51:373–383.

Zippel, M., and M. Neigenfind. 1988. Preservation of streptomycetes. J. Gen. Appl. Microbiol. 34:7–14.

Prokaryotes (2006) 3:725–753
DOI: 10.1007/0-387-30743-5_28

CHAPTER 1.1.13

The Family Streptosporangiaceae

MICHAEL GOODFELLOW AND ERIKA TERESA QUINTANA

Phylogeny

The founder members of the family Streptosporangiaceae Goodfellow et al. 1990 have a number of taxonomic pedigrees. Couch (1955a) classified the genus *Streptosporangium* in the family "Actinosporangiaceae" together with sporangiate actinomycetes belonging to the genus *Actinoplanes*. This family was renamed Actinoplanaceae by Couch (1955b). In addition to *Actinoplanes* (the type genus) and *Streptosporangium*, this taxon encompassed the genera *Amorphosporangium*, *Ampullariella*, *Dactylosporangium*, *Kitasatoa*, *Pilimelia*, *Planobispora*, *Planomonospora* and *Spirillospora* (Couch and Bland, 1974). Members of all of these genera were considered to form sporangia.

Representatives of the genera *Planobispora*, *Planomonospora*, *Spirillospora* and *Streptosporangium* were subsequently shown to form a DNA homology group that was readily separated from a second aggregate DNA relatedness group which encompassed the genera *Actinoplanes*, *Ampullariella* and *Dactylosporangium* (Farina and Bradley, 1970). Organisms in the first group contained *meso*-diaminopimelic acid (*meso*-A$_2$pm) and the sugar madurose (wall chemotype III/B sensu Lechevalier and Lechevalier, 1970b, Lechevalier and Lechevalier, 1970c) whereas those in the second taxon had a wall chemotype II, that is, they contained *meso*- or hydroxy-A$_2$pm or both and glycine (Lechevalier and Lechevalier, 1970b).

The genera *Actinoplanes*, *Dactylosporangium*, *Micromonospora* and *Pilimelia* are now known to have many properties in common and are classified in the family Micromonosporaceae (Krassil'nikov, 1938) emended Stackebrandt et al. 1997. The genus *Kitasatoa* has become a subjective synonym of the genus *Streptomyces* (Goodfellow et al., 1986), the genera *Amorphosporangium* and *Ampullariella* have been reduced to subjective synonyms of the genus *Actinoplanes* (Stackebrandt and Kroppenstedt, 1987), and the genus *Spirillospora* is now classified in the family Thermomonosporaceae (Zhang et al., 2001).

The oligosporic genera (*Actinomadura* Lechevalier and Lechevalier 1970b, *Microbispora* Nonomura and Ohara 1957, and *Microtetraspora* Thiemann et al. 1968b) and the sporangiate genera (*Planobispora* Thiemann and Beretta 1968a, *Planomonospora* Thiemann et al. 1967, *Spirillospora* Couch 1963, and *Streptosporangium* Couch 1955a) belong to an aggregate group, the maduromycetes (Goodfellow and Cross, 1984; Goodfellow, 1989a; Goodfellow, 1989b). Apart from representatives of the genus *Spirillospora*, these taxa formed a recognizable suprageneric group on the basis of 16S rRNA gene cataloguing and sequencing data (Stackebrandt, 1986).

The taxonomic status of the genera assigned to the maduromycetes was formalized by the proposal that *Streptosporangium* be recognized as the type genus of a new suprageneric taxon, the family Streptosporangiaceae Goodfellow et al. 1990. This family also served as a taxonomic niche for the genera *Microbispora*, *Microtetraspora* (including the *Actinomadura pusilla* group sensu Kroppenstedt et al. [1990]), *Planobispora*, *Planomonospora*, and initially for the genus *Spirillospora*. Additional genera have been added to the family, namely *Acrocarpospora* Tamura et al. 2000, *Herbidospora* Kudo et al. 1993, *Planotetraspora* Runmao et al. 1993, and *Nonomuraea* Zhang et al. 1998. In general, representatives of the nine genera classified in the family Streptosporangiaceae form distinct, but related, taxa on the basis of 16S rRNA gene sequence data (Fig. 1).

The family Streptosporangiaceae (Goodfellow et al. 1990) emended Ward-Rainey et al. 1997 encompasses aerobic, Gram-positive, non-acid-alcohol-fast, chemoorganotrophic actinomycetes that form a branched, nonfragmenting substrate mycelium. When formed, aerial hyphae differentiate into either short chains of arthrospores or spore vesicles containing one to many motile or nonmotile spores. In the absence of aerial hyphae, sporophores are borne on the substrate mycelium. Members of the family typically contain *meso*-A$_2$pm in a wall peptidoglycan that contains madurose as the diagnostic sugar, have an A1γ type peptidoglycan, *N*-acetylated muramic

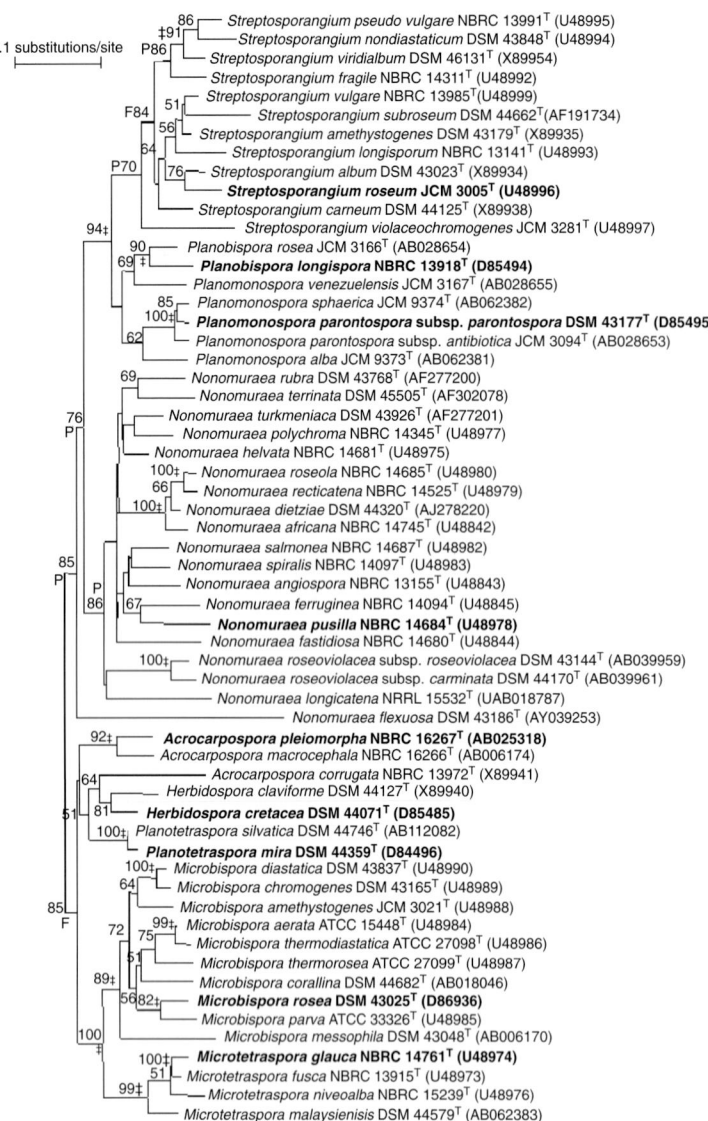

0.1 substitutions/site

Fig. 1. Neighbor-joining tree based on nearly complete 16S rRNA gene sequences showing positions of representative strains classified in the family Streptosporangiaceae. Asterisks indicate branches of the tree that were also found using the least-squares, maximum-likelihood and maximum-parsimony tree-making algorithms. The numbers at the nodes indicate the levels of bootstrap support (%) based on a neighbor-joining analysis of 1000 re-sampled datasets; only values above 50% are given. The bar indicates 0.1 substitutions per nucleotide position. [T], type strain. Type species are in bold. It has been proposed that the type strain of *Streptosporangium claviforme* be reduced to a synonym of *Herbidospora cretacea*, though this preposition needs to be checked.

acid, and major amounts of glucosamine-containing polar lipids. The cellular fatty acids include straight, *iso-* and *anteiso-*branched and 10-methyl acids. Most of the unsaturated, di- and tetrahydrogenated menaquinones have 9 or 10 isoprene units. The G+C content of the DNA lies within the range of 64–73 mol%. The type genus is *Streptosporangium* Couch 1955a[AL].

Members of the family Streptosporangiaceae are chemically homogeneous but morphologically diverse (Table 1). However, strains that bear spore vesicles (*Acrocarpospora, Planobispora, Planomonospora, Planotetraspora* and *Streptosporangium*) are closely related to organisms that carry one or more spores in spore chains (*Herbidospora, Microbispora, Microtetraspora* and *Nonomuraea*). Each spore vesicle contains a coiled chain of arthrospores formed by septation of an unbranched, spiral hypha

within an expanded sporangiophore sheath (Vobis and Kothe, 1985b). Spore formation is not endogenous, and hence the term "spore vesicle" has greater precision than does the original term "sporangium" (Cross, 1970; Sharples et al., 1974). Studies on spore maturation have shown that spores in spore vesicles and spore chains are formed in essentially the same way. In each case, spores are differentiated by fragmentation of a hypha within its sheath, and the latter either expands to form the vesicular envelope or remains around the spore chain (Lechevalier et al., 1966; Sharples et al., 1974; Vobis and Kothe, 1985b).

Streptosporangium Couch 1955a, 148[AL]

This taxon was proposed for sporangiate actinomycetes that formed nonmotile sporangiospores on abundant aerial hyphae. Initially, one species, *Streptosporangium roseum*, was rec-

Table 1. Morphological features and chemotaxonomic characteristics of members of the genera classified in the family Streptosporangiaceae.

Characteristics	Genus								
	Acrocarpospora	Herbidospora	Microbispora	Microtetraspora	Nonomuraea	Planobispora	Planomonospora	Planotetraspora	Streptosporangium
Vesicle formation	Club or globose spore vesicles on aerial hyphae	Spore chains on aerial hyphae	Spores in characteristic longitudinal pairs on aerial hyphae	Spore chains containing four or more spores on short aerial hyphae	Spore chains or pseudosporangia formed on aerial hyphae	Cylindrical to clavate spore vesicles containing longitudinal pairs of spores on aerial hyphae	Cylindrical to clavate spore vesicles containing single spores on aerial hyphae	Spore vesicles containing four spores on aerial hyphae	Globose spore vesicles on aerial hyphae
Motile spores	−	−	−	−	−	+	+	+	−
Cell-wall chemo-type[a]	III	III	III	III	III	III	III	III	III
Whole-organism sugar pattern[b]	B, C	B	B, C	B, C	B, C	B	B	D, A	B
Fatty-acid type[c]	3c	3c	3c	3c	3c	3c	3c	ND	3c
Major menaquinones (MK)[d]	−9[H$_2$, H$_4$, H$_6$]	−10[H$_4$, H$_6$, H$_8$]	−9[H$_0$, H$_2$, H$_4$]	−9[H$_2$, H$_4$, H$_6$]	−9[H$_0$, H$_2$, H$_4$]	−9[H$_2$, H$_4$]	−9[H$_2$]	ND	9[H$_4$]
Phospholipid type[e]	IV, II	IV	IV	IV	IV	IV	IV	ND	IV

Symbols: +, present; −, absent; and ND, not determined.
[a]Major constituents: alanine, glutamic acid, glucosamine and meso-A$_2$pm (Lechevalier and Lechevalier, 1970b).
[b]Whole-organism sugar patterns of actinomycetes containing meso-diaminopimelic acid: A, arabinose and galactose; B, madurose (3-O-methyl-D-galactose); C, no diagnostic sugar; D, arabinose and xylose (Lechevalier and Lechevalier, 1970a).
[c]Saturated fatty acids, unsaturated fatty acids, iso-fatty acids (variable) and methyl-branched fatty acids (Kroppenstedt, 1985).
[d]Herbidospora strains contain tetrahydrogenated menaquinone with ten isoprene units (Kudo et al., 1993), organisms in the remaining taxa contain tetrahydrogenated menaquinone with nine isoprene units (Kroppenstedt, 1982).
[e]Phospholipid patterns: PI, phosphatidylinositol; PI, phosphatidylglycerol (variable); PII, only phosphatidylethanolamine; PIV, phospholipids containing glucosamine (with phosphatidylmethylethanolamine variable; Lechevalier et al., 1977).
Data taken from Goodfellow (1989a, b), Goodfellow et al. (1990), Kroppenstedt et al. (1990), Kudo et al. (1993), and Tamura et al. (2000).

Table 2. Validly described species and subspecies of the genus *Streptosporangium*.

Taxon	Type strain	Source	References
Streptosporangium album	DSM 43023[T]	Japan, soil	Nonomura and Ohara, 1960
Streptosporangium amethystogenes subsp. *amethystogenes*	DSM 43179[T]	Japan, soil	Nonomura and Ohara, 1960
Streptosporangium amethystogenes subsp. *fukuiense*	DSM 44779[T]	River Tana, Nairobi, Kenya, soil	Iinuma et al., 1996
Streptosporangium carneum	NRRL 18437[T]	Soil	Mertz and Yao, 1990
Streptosporangium claviforme[a]	DSM 44127[T]	*Betula alba* leaf litter	Petrolini et al., 1992
Streptosporangium fragile	DSM 43847[T]	Anaikota, Sri Lanka, soil	Shearer et al., 1983
Streptosporangium longisporum	DSM 43180[T]	Turkey, soil	Schäfer, 1969
Streptosporangium nondiastaticum	DSM 43848[T]	Japan, soil	Nonomura and Ohara, 1969b
Streptosporangium pseudovulgare	DSM 43181[T]	Japan, soil	Nonomura and Ohara, 1969a
Streptosporangium roseum	DSM 43021[T]	Vegetable garden, soil	Couch, 1955a
Streptosporangium subroseum	DSM 44662[T]	China, soil	Zhang et al., 2003
Streptosporangium violaceochromogenes	DSM 43849[T]	Japan, soil	Kawamoto et al., 1975
Streptosporangium viridialbum	DSM 43801[T]	Yotei, Hokkaido, Japan, soil	Nonomura and Ohara, 1960
Streptosporangium vulgare	DSM 43802[T]	Anjo, Aichi Prefecture, Japan, soil	Nonomura and Ohara, 1960

Abbreviations: [T], type strain; DSM, Deutsche Sammlung von Mikroorganismen und Zellkulturen, Mascheroder Weg 1B, D-38124 Braunschweig, Germany; and NRRL, Northern Research and Development Division, United States Department of Agriculture, Peoria, Illinois, United States.
[a]Kudo et al. (1993) considered this organism to be a subjective synonym of *Herbidospora cretacea*.

ognized but others were soon added (Table 2). The genus was shown to be heterogeneous on the basis of spore and spore vesicular morphology (Nonomura, 1989c), electrophoretic mobility of ribosomal protein AT-L30 (Ochi and Miyadoh, 1992), 16S rRNA (Kemmerling et al., 1993) and 5S rRNA (Kudo et al., 1993) gene sequences, and discontinuous distribution of chemical markers (Stackebrandt et al., 1994).

Stackebrandt and his colleagues found that streptosporangiae had many chemical properties in common but could be assigned to two groups on the basis of chemical differences. Members of most species, including *S. roseum*, had a phospholipid pattern type IV and predominant proportions of MK-9 [H_2, H_4, H_6], whereas strains in the second group, which contained *S. albidum* and *S. viridogriseum*, had a MK-9 [H_4] as the predominant isoprenologue and a phospholipid pattern type II; these results were in excellent agreement with corresponding 16S rRNA gene sequence data (Kemmerling et al., 1993). Stackebrandt and his colleagues proposed that *Streptosporangium albidum* Fumurai et al. 1968, *Streptosporangium viridogriseum* subsp. *kofuense* Nonomura and Ohara 1969b, and *Streptosporangium viridogriseum* subsp. *viridogriseum* Okuda et al. 1966 be assigned to a new taxon, the genus *Kutzneria*, as *Kutzneria albida* comb. nov., *Kutzneria viridogrisea* comb. nov., and *Kutzneria kofuensis* comb. nov., respectively.

Streptosporangiae characteristically form aerial hyphae that carry, on either short or long sporophores, single or clustered spore vesicles that may be up to 40 μm in diameter (Figs. 2 and 3). They have a wall chemotype III (Lechevalier and Lechevalier, 1970b), that is, *meso*-A_2pm in

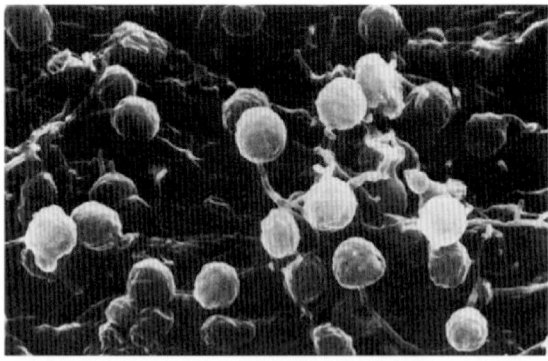

Fig. 2. *Streptosporangium album* CBS 426.61[T] on oatmeal agar. Scanning electron microscopy, gold splattered. Sporangiophores are short. From Nonomura (1989c), with permission.

a wall peptidoglycan that lacks characteristic sugars other than madurose (3-*O*-methyl-D-galactose; Lechevalier and Gerber, 1970a) and a peptidoglycan of the A1γ type (Schleifer and Kandler, 1972). Members of the taxon are rich in *iso*-, *anteiso*-, saturated, unsaturated, and methyl-branched fatty acids (pattern 3c; Kroppenstedt, 1985; Kudo et al., 1993; Whitham et al., 1993; Stackebrandt et al., 1994), contain di- and tetrahydrogenated menaquinones with nine isoprene units as predominant isoprenologues (Kroppenstedt, 1985; Kudo et al., 1993; Whitham et al., 1993; Stackebrandt et al., 1994), and have phospholipid patterns characterized by glucosamine-containing lipids with phosphatidylethanolamine, diphosphatidylglycerol and phosphatidylinositol (Lechevalier et al., 1977; Lechevalier et al., 1981b; Kudo et al., 1993;

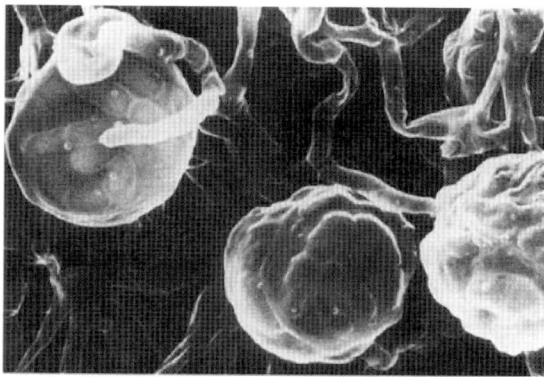

Fig. 3. *Streptosporangium album*. Scanning electron microscopy, gold splattered. Sporangial walls (membranes) are thin. From Nonomura (1989c), with permission.

Whitham et al., 1993; Stackebrandt et al., 1994). The G+C content of the DNA is 69–71 mol% (Tsyganov et al., 1966; Yamaguichi, 1967; Farina and Bradley, 1970; Stackebrandt et al., 1994). Type species: *Streptosporangium roseum* Couch 1955a, 151[AL]. Type strain: ATCC 12428[T]=DSM 43021[T].

Acrocarpospora Tamura et al. 2000, 1170[AL]

This aerobic, Gram-positive, non-acid-alcohol-fast, nonmotile actinomycete forms a stable, branched substrate mycelium. Spherical and club-shaped structures, which contain coiled chains of smooth-surfaced, oval or rod-like spores (0.6–08 0.7–1.0 μm in diameter), are carried on the tips of aerial hyphae. The organism grows well on oatmeal agar and at 20–30°C. White aerial hyphae and pale yellow substrate mycelium, but no diffusible pigments, are formed on most standard growth media. Cell walls contain alanine, glutamic acid and *meso*-A$_2$pm and *N*-acetylated muramic acid. The peptidoglycan is of the A1γ type. Strains contain di- and tetrahydrogenated menaquinones with nine isoprene units as predominant isoprenologues; glucose and madurose as major sugars; phosphatidylethanolamine as the diagnostic phospholipid; and *iso-* C$_{16:0}$, 10-methyl C$_{17:0}$, C$_{17:0}$ and C$_{17}$ as major fatty acids, but lack mycolic acids. The G+C content of the DNA is 68–69 mol%. Type species: *Acrocarpospora pleiomorpha* Tamura et al. 2000, 1170[AL]. Type strain: DSM 44706[T]=NBRC 16267[T].

In addition to the type species, there are two additional species, *A. corrugata* and *A. macrocephala*. The former was originally classified as *Streptosporangium corrugatum* (Williams and Sharples, 1976). From the 16S rRNA Streptosporangiaceae gene tree (Fig. 1), the representatives of the *Acrocarpospora* species are apparently closely related to one another and to the type strains of *Herbidospora cretacea*, *Planotetraspora mira* and *Planotetraspora silvatica*. However, the *Acrocarpospora* strains

can be distinguished from *Herbidospora cretacea* (including the type strain of *S. claviforme*) using morphological and menaquinone data and from the genus *Planotetraspora* using morphological properties and whole-organism sugar composition.

Herbidospora Kudo et al. 1993, 319[AL]

This aerobic mesophilic actinomycete forms a stable, branched substrate mycelium, but does not produce true aerial hyphae. Straight chains of nonmotile, smooth-surfaced spores (10–30 per chain) are borne at tips of sporophores branching in clusters from the vegetative mycelia. The substrate mycelia are yellow to brown on most media; distinctive exopigments are not formed. When sporulation occurs, the surface of the colony is white or brownish yellow. Thiamine is required for growth. The organism is susceptible to lysozyme. Cell walls contain *meso*-A$_2$pm and acetylated muramic acid but lack significant amounts of glycine. Whole-organism hydrolysates contain glucose, mannose, ribose, and a trace of madurose. Strains contain major amounts of *iso*-hexadecanoic, *n*-hexadecanoic, *n*-heptadecanoic, 10-methylheptadecanoic, and 2-hydroxy acids; phosphatidylethanolamine and glucosamine-containing phospholipids as diagnostic polar lipids; and major proportions of tetrahydrogenated menaquinones with ten isoprene units with hydrogenation at units III and IX (MK-10 [III, IX-H$_4$]), but lack mycolic acids. The G+C content of the DNA is 69–71 mol%. Type species: *Herbidospora cretacea* Kudo et al. 1993. Type strain: DSM 44071[T]=JCM 8553[T].

The type strains of *Herbidospora cretacea* Kudo et al. 1993 and *Streptosporangium claviforme* Petrolini et al. 1993 belong to the same genomic species and share key morphological features—similarities which led Kudo and his colleagues to propose that the latter be seen as a subjective synonym of the former. However, apparent from the 16S rRNA Streptosporangiaceae gene tree is that the two strains form distinct albeit related phyletic lines (Fig. 1).

Microbispora Nonomura and Ohara 1957, 307[AL]

This taxon was proposed for actinomycetes that form conspicuous aerial hyphae bearing longitudinal pairs of spores. It currently contains ten validly described species, excluding *Microbispora echinospora* Nonomura and Ohara 1971b and *Microbispora viridis* Miyadoh et al. 1985, which have been reclassified as *Actinomadura rugatobispora* (Miyadoh et al., 1985) Miyadoh et al., 1990 and *Actinomadura viridis* (Miyadoh et al., 1985) Miyadoh et al. 1990, respectively. *Microbispora bispora* (Lechevalier, 1965), which was originally described as *Thermopolyspora bispora* (Henssen, 1957), has been transferred to a new genus, *Thermobispora*, as

Thermobispora bispora (Henssen, 1957; Wang et al., 1996; and see The Family Thermomonosporaceae in this Volume). In contast, *Thermomonospora mesophilica* Nonomura and Ohara 1971b, which forms single spores, has been reclassified as *Microbispora mesophila* (Nonomura and Ohara, 1971b; Zhang et al., 1998).

Miyadoh et al. (1990) undertook a radical revision of the genus *Microbispora* in which they proposed that *M. amethystogenes*, *M. chromogenes*, *M. diastatica*, *M. indica*, *M. karnatakensis* and *M. rosea* be assigned to a single taxon as *M. rosea* subsp. *rosea* and that *M. aerata*, *M. thermodiastatica* and *M. thermorosea* be combined and recognized as *M. rosea* subsp. *aerata*. These proposals were based on DNA:DNA relatedness data, though it was acknowledged that most of the cut-off points used in the circumscription of the two taxa were below the 70% guideline recommended for the delineation of genomic species (Wayne et al., 1987). However, *M. indica* ATCC 35926^T shared 83% DNA relatedness with *M. rosea* JCM 3006^T; the corresponding figure between the type strains of *M. diastatica* and *M. karnatakensis* was 91%. *Microbispora* species form a relatively distinct monophyletic group in the 16S rRNA Streptosporangiaceae gene tree, with most of the type strains forming distinct phyletic lines (Fig. 1). From the 16S rRNA gene tree, the taxon is apparently closely related to the genus *Microtetraspora*, an association supported by all of the tree-making algorithms and 100% boostrap value in the neighbor-joining tree.

Microbisporae are aerobic, Gram-positive, nonmotile actinomycetes which typically form a conspicious aerial mycelium bearing longitudinal pairs of spores (Fig. 4) that may be closely arranged along the aerial hyphae, giving the appearance of catkins; spores are not usually formed on the substrate mycelium. In some cases, the spores are borne at longer intervals (Fig. 5). They first appear as club-shaped initials that later become transformed into the paired spores visible under the light microscope. Spores are either sessile or on short sporophores, spherical to oval (usually 1.2–1.6 μm in diameter) with smooth surfaces. Mature spores are easily detached from the sporophores and each other when placed in water. B vitamins, particularly thiamine, are essential for growth on synthetic media.

Mesophilic and thermophilic species have been described. Mesophilic strains generally produce a pale yellow to distinct pink aerial spore mass, and the reverse side of the colonies is yellowish-brown to orange. Thermophilic strains form a white or pale yellowish-brown to pale pinkish-brown aerial spore mass; the reverse side of colonies is either pale yellowish-brown or yellow brown. Cell walls contain *N*-acetylated muramic acid and major amounts of *meso*-A$_2$pm but no characteristic sugars. Madurose is present in whole-organism hydrolysates. The organism contains tetrahydrogenated menaquinone with nine isoprene units as the predominant isoprenologue, phosphatidylcholine, and unknown glucosamine- containing components (such as diagnostic polar lipids, tuberculostearic acid and its analogues), but not mycolic acids. The G+C content of the DNA is 71–73 mol%. Type species: *Microbispora rosea* Nonomura and Ohara 1957, 307^{AL}. Type strain: DSM 43839^T=JCM 3006^T.

Microtetraspora Thiemann et al. 1968b, 296^{AL}

This genus was proposed for actinomycetes that form short, sparsely branched aerial hyphae bearing chains of four spores (Fig. 6). This morphological trait was considered typical of the genus though chains of two or three spores, and more rarely five spores, have been reported. Initially, four species were recognized, *Micro-*

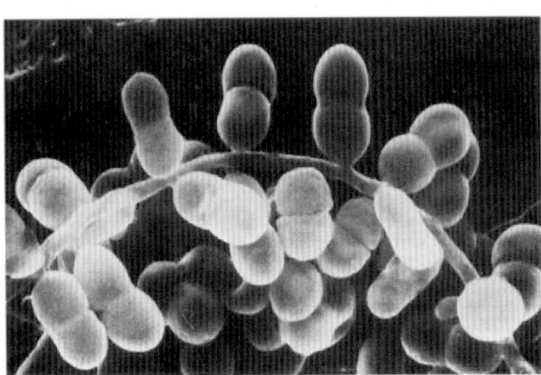

Fig. 4. *Microbispora rosea* ATCC 12950^T. Scanning electron microscopy, gold splattered. Paired spores on hyphae. From Nonomura (1989a), with permission.

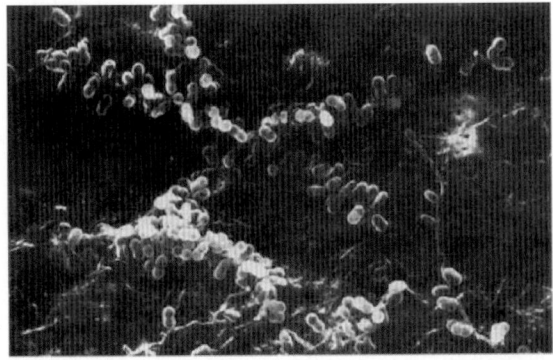

Fig. 5. Morphology of *Microbispora rosea* ATCC 12950^T on oatmeal agar. Scanning electron microscopy, gold splattered. Spores on entire mycelium. From Nonomura (1989a), with permission.

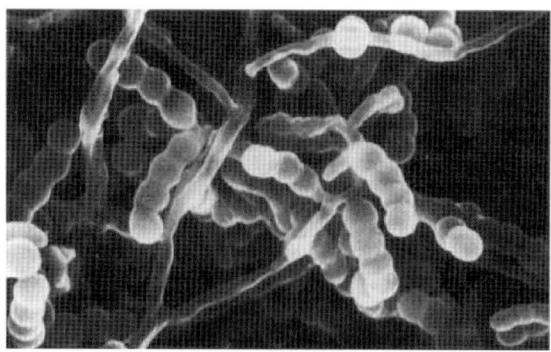

Fig. 6. Morphology of *Microtetraspora niveoalba* ATCC 27301[T] on inorganic salts starch agar. Scanning electron microscopy, gold splattered. From Nonomura (1989a), with permission.

tetraspora glauca, the type species, *M. fusca*, *M. niveoalba* and *M. viridis*; the latter has been reclassified as *Actinomadura viridis* (Nonomura and Ohara, 1971a; Miyadoh et al., 1989). The taxon provided a temporary refuge for the *Actinomadura pusilla* group (Fischer et al., 1983; Poschner et al., 1985; Goodfellow et al., 1988; Kroppenstedt et al., 1990) until it became clear that members of this taxon and the three bona fide *Microtetraspora* species mentioned above can be distinguished using numerical taxonomic (Athalye et al., 1985), electrophoretic mobility of ribosomal AT-L30 protein (Ochi et al., 1991; Ochi et al., 1993), and 16S rRNA gene sequence (Wang et al., 1998) data. The *A. pusilla* group was subsequently classified in a new taxon, the genus *Nonomuraea* Zhang et al. 1998. "*Microtetraspora tyrrkensis*" was proposed by Tomita et al. (1991) for an organism that formed hooked or spiral spore chains and other properties consistent with its assignment to the *A. pusilla* group. This organism probably belongs to the genus *Nonomuraea*, though the type strain is no longer available to test this proposition. An additional species, *M. malaysiensis*, has been described for strains isolated from a primary dipterocarp forest soil (Nakajima et al., 2003).

Microtetrasporae are aerobic, Gram-positive, non-acid-alcohol-fast, mesophilic, nonmotile actinomycetes which form stable, highly branched substrate and aerial mycelia. Spore chains, typically containing four spores, are borne exclusively on short aerial hyphae. Spores are spherical (1.2–1.5 µm in diameter) or oval to short cylindrical (1.0–1.4 to 1.2–1.7 µm in diameter) and have smooth surfaces. Some species require B vitamins for growth. The organism is chemoorganotrophic, having an oxidative type of metabolism. It grows well at 20–37°C. Cell walls contain *N*-acetylated muramic acid and major amounts of *meso*-A$_2$pm but no character-

istic sugars. Madurose is present in whole-organism hydrolysates. The organism contains major proportions of menaquinone with nine isoprene units with hydrogenation at units III and IV (MK-9 [III, IV-[H$_4$]]); phosphatidylcholine and unknown glucosamine-containing lipids are the major polar lipids. The G+C content of the DNA is 69–71 mol%. Type species: *Microtetraspora glauca* Thiemann et al. 1968b, 296[AL]. Type strain: ATCC 23057[T]=DSM 43311[T].

Nonomuraea Zhang et al. 1998, 419[AL]

This taxon was introduced to accommodate the species assigned to the *Microtetraspora pusilla* group (Fischer et al., 1983; Poschner et al., 1985; Goodfellow et al., 1988; Kroppenstedt et al., 1990). The genus encompasses aerobic, Gram-positive, non-acid-alcohol-fast strains that form extensively branched substrate and aerial hyphae. The latter bear chains of spores that may be hooked, spiral, straight, or enmeshed in pseudovesicles. Spore surfaces may be folded, irregular, smooth or warty. The growth temperature range is 20–45°C, and some strains grow up to 55°C. Cell walls contain *meso*-A$_2$pm, and madurose is present in whole-organism hydrolysates (cell wall type III/B sensu Lechevalier and Lechevalier, 1970b). The predominant menaquinones are MK-9 [H$_0$, H$_2$, H$_4$], the phospholipid pattern is characterized by glucosamine-containing lipids with phosphatidylethanolamine variable, phosphatidylmethylethanolamine, diphosphatidylglycerol and phosphatidylinositol (phospholipid type IV sensu Lechevalier et al., 1977), and the predominant fatty acids are 10-methyl-17- and *iso*-16-branched components (pattern 3c; Kroppenstedt, 1985). The G+C content of the DNA is 64–69 mol%. Type species: *Nonomuraea pusilla* Zhang et al. 1998, 419[AL]. Type strain: ATCC 27296[T]=DSM 43357[T].

The genus *Nonomuraea* contains eighteen validly described species, namely *N. africana* (Preobrazhenskaya and Sveshnikova, 1974) Zhang et al. 1998, *N. angiospora* (Zhukova et al., 1968) Zhang et al. 1998, *N. dietziae* Stackebrandt et al. 2001, *N. fastidiosa* (Soina et al., 1975) Zhang et al. 1998, *N. ferruginea* (Meyer, 1981) Zhang et al. 1998, *N. flexuosa* (Meyer, 1989) Zhang et al. 1998, *N. helvata* (Nonomura and Ohara, 1971c) Zhang et al. 1998, *N. longicatena* (Chiba et al., 1999) Zhang et al. 1998, *N. polychroma* (Galantenko et al., 1987) Zhang et al. 1998, *N. pusilla* (Nonomura and Ohara, 1971c) Zhang et al. 1998, *N. recticatena* (Gauze et al., 1984) Zhang et al. 1998, *N. roseola* (Lavrova and Preobrazhenskaya, 1975) Zhang et al. 1998, *N. roseoviolacea* subsp. *carminata* (Gauze et al., 1973) Gyobu and Miyadoh 2001, *N. roseoviolacea* subsp. *roseoviolacea* (Nonomura and Ohara, 1971c) Zhang et al. 1998, *N. rubra* (Sveshnikova et al., 1969) Zhang et al. 1998, *N. salmonea* (Preobrazhenskaya et al., 1975) Zhang et al. 1998, *N. spiralis* (Meyer, 1981) Zhang et al. 1998, *N. terrinata* Quintana et al. 2003, and *N. turkmeniaca* (Terekhova et al., 1987) Zhang et al. 1998. Representatives of each of these validly described species form a distinct phyletic line in the 16S rRNA Streptosporangiaceae gene tree (Fig. 1).

Planobispora Thiemann and Beretta 1968a, 157[AL]

This taxon encompasses aerobic, Gram-positive, non-acid-alcohol-fast, chemoorganotrophic actinomycetes that form irregular branched, occasionally septate, substrate hyphae (0.5–1.0 μm in diameter) and sparsely branched, rarely septate aerial hyphae (1 μm in diameter). Cylindrical to clavate spore vesicles (1.0–1.2 μm wide, 6.0–8.0 μm long), each containing a longitudinal pair of spores, are formed singly or in bundles on short ramifications of the aerial hyphae. Spores are straight or slightly curved with rounded ends (1.0–1.2 μm in length) and are motile by means of peritrichous flagella. The spores are pushed out of opposite ends of the spore vesicle, which is easily detached from the supporting hyphae; only a small percentage of spores show motility. They are only motile after being dispersed for some time and usually germinate with one or two polar germ tubes.

Still unclear is whether spores are formed endogenously (Williams and Wellington, 1980) or by simple transformation of sporangeous hyphae (Bland and Couch, 1981). A transverse septum or diaphragm connected to the vesicular envelope divides the two spores (Thiemann, 1970; Vobis and Kothe, 1985b; Fig. 7). The vesicular envelope is smooth and contains fibrillar elements (Vobis and Kothe, 1985b) that resemble those present in *Planomonospora* (Sharples et al., 1974). The type strains of *P. longispora* and *P. rosea* have many phenotypic properties in common; some of these distinguish them from other sporoactinomycetes with a wall chemotype III (Goodfellow and Pirouz, 1982).

The substrate mycelium of planobisporae is either without distinctive color or rose-colored. The aerial mycelium, which develops only on certain agar media, is white or has a light rose tinge. Good growth occurs at pH 6.0–9.0 and temperature 28–40°C but not 20°C or 45°C. Cell walls contain *meso*-A$_2$pm; madurose is the characteristic whole-organism sugar. Planobisporae contain diphosphatidylglycerol, phosphatidylethanolamine and unknown glucosamine-containing phospholipids as diagnostic polar lipids; major amounts of straight chain, unsaturated, *iso*- and 10-methyl branched fatty acids, and tetrahydrogenated menaquinone with nine isoprene units (MK-9 [III, IV-H$_4$]) as the predominant isoprenologue. The G+C content of the DNA is 70–71 mol%. Type species: *Plano-*

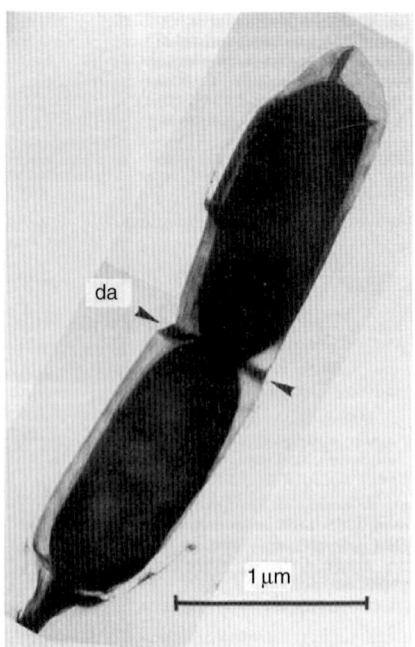

Fig. 7. Spore vesicle of *Planobispora rosea* strain MB-SE-893. Negative stained: transmission electron micrograph. A diaphragm (da) is visible between the two spores and the longitudinal fibrillar elements in the sporangial envelope. From Vobis (1989a), with permission.

bispora longispora Thiemann and Beretta 1968a, 157[AL]. Type strain: DSM 43041[T]=JCM 3092[T].

The genus contains an additional species, namely, *Planobispora rosea* Thiemann 1970.

Planomonospora Thiemann et al. 1967, 29[AL]

This taxon was proposed for actinomycetes that form cylindrical or clavate spore vesicles, each containing a single spore, on aerial hyphae. In *Planomonospora parontospora*, the type species, the spore vesicles are sessile and occur in double parallel rows on curved sporangiophores (Fig. 8). A single sporangiophore can bear up to 60 spore vesicles. In the other founder member of the taxon, *P. venezuelensis*, the spore vesicles are developed singly or in groups on short lateral branches (Fig. 9) forming a characteristic palm leaf pattern (Thiemann, 1970). The spores may be formed endogenously (Sharples et al., 1974), but in the *P. parontospora* spore vesicle, development begins with the growth of a sporangeous hypha inside a thin expanding sheath (Vobis, 1985a; Vobis and Kothe, 1985b). Through thickening, the sheath becomes a massive vesicular envelope. The spores, which are released through apical pores, become motile by peritrichous flagella about 30 min after being expelled. They remain motile for up to a day during which time spore germination may begin (Thiemann, 1970). The type strains of *P. parontospora* and *P. venezuelensis* were assigned to a distinct cluster in an

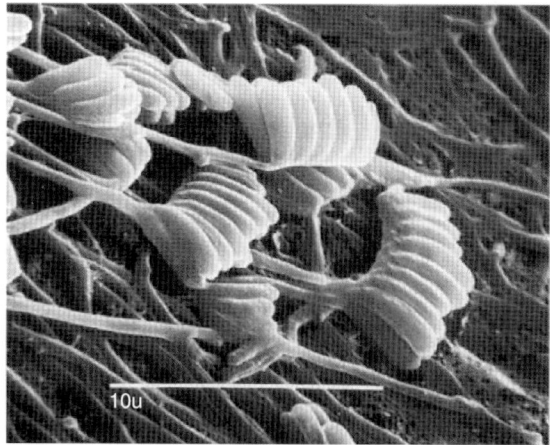

Fig. 8. *Planomonospora parontospora* ATCC 23863[T]. Scanning electron microscopy. Numerous monosporous spore vesicles in double parallel rows, arranged directly on bent aerial hyphae. The strain was cultivated for 10 days on soil agar. From Vobis (1989a), with permission.

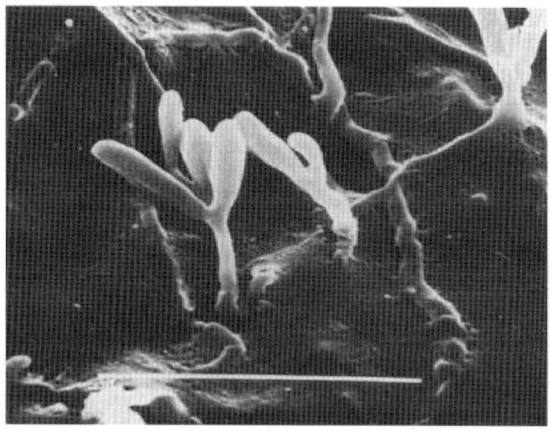

Fig. 9. *Planomonospora venezuelensis* ATCC 23865[T]. Scanning electron microscopy. Monosporous spore vesicles on aerial hyphae in young stages of formation of palm leaf pattern. Cultivation as in Fig. 8. From Vobis (1989b), with permission.

extensive numerical taxonomic analysis of sporoactinomycetes with a wall chemotype III (Goodfellow and Pirouz, 1982).

The genus *Planomonospora* contains five validly described taxa, namely, *Planomonospora alba* Mertz 1994, *Planomonospora parontospora* subsp. *antibiotica* Thiemann et al. 1967, *Planomonospora parontospora* subsp. *parontospora* Thiemann 1967, *Planomonospora sphaerica* Mertz 1994 and *Planomonospora venezuelensis* Thiemann 1970. The type strains of *P. alba* and *P. sphaerica* form spore vesicles in long parallel rows, which resemble rows of bananas (Mertz, 1994). *Planomonospora sphaerica* strains form large spherical bodies when grown on inorganic

salt starch agar (ISP medium 4; Shirling and Gottlieb, 1966).

Planomonosporae are aerobic, Gram-positive, non-acid-alcohol-fast, chemoorganotrophic actinomycetes that form branched, occasionally septate, nonfragmenting substrate hyphae (0.6–1.0 μm in diameter) and sparsely branched rarely septate aerial hyphae (0.5–1.0 μm in diameter). Cylindrical to clavate spore vesicles (1.0–1.5 μm wide, 3.5–5.5 μm long), each containing a single spore, are formed on the aerial mycelium. Spores are fusiform or cylindrical to clavate and motile by peritrichous flagella; they are 1.0–1.6 μm in diameter and 3.5–5.4 μm in length. Colonies are flat or elevated with smooth surfaces, occasionally wrinkled or slightly crustose. Substrate mycelia show a range of colors, including light orange, brown violet to light brown, and grayish yellow. Aerial mycelia are white with a rose tinge, grayish white or pink. Growth occurs at 20–50°C.

Cell walls contain *meso*-A_2pm. Variation is found in menaquinone, sugar and polar lipid composition. *Planomonospora alba* and *P. sphaerica* strains contain diphosphatidylglycerol, phosphatidylglycerol, phosphatidylinositol, phosphatidylethanolamine, hydroxyphosphatidylethanolamine and glucosamine-containing phospholipids and galactose, madurose and xylose as whole-organism sugars (Mertz, 1994). In contrast, *Planomonospora parontospora* and *P. venezuelensis* have madurose as the characteristic sugar (Kroppenstedt and Kutzner, 1978); the diagnostic phospholipids of *P. parontospora* consist of diphosphatidylglycerol, lysophosphatidylglycerol, phosphatidylethanolamine, phosphatidylinositol and unknown glucosamine-containing phospholipids (Hasegawa et al., 1979). Di- and tetrahydrogenated menaquinones with nine isoprene units are the major component in *P. parontospora* whereas tetrahydrogenated menaquinones with eight isoprene units predominate in *P. venezuelensis* (Collins et al., 1984); the major component of *P. alba* and *P. sphaerica* is dehydrogenated menaquinone with nine isoprene units (Mertz, 1994). Planomonosporae have complex mixtures of straight chains, unsaturated and branched fatty acids but lack mycolic acids. The G+C content of the DNA is 72 mol%. Type species: *Planomonospora parontospora* Thiemann et al. 1967. Type strain: ATCC 23863[T]=DSM 43177[T].

Planotetraspora Runmao et al. 1993, 468[VP]

The genus *Planotetraspora* was proposed for an isolate that formed long cylindrical spore vesicles that contained four spores in a single row at the ends of short sporangiophores on aerial hyphae. The organism, *Planotetraspora mira*, was reported to contain *meso*-A_2pm in the cell wall and arabinose, galactose, mannose, ribose and xylose in whole-organism hydrolysates. However, Kudo (2001) found that the strain contained madurose and rhamnose, but not arabinose or xylose in whole-organism hydrolysates. They also noted that it contained tetrahydrogenated menaquinone with nine isoprene units as the predominant isoprenologue and had a type IV phospholipids pattern. Stackebrandt et al. (1997) classified the genus in the family Streptosporangiaceae. A second species, *P. silvatica*, has been described for a strain isolated from a soil sample collected on Amami Island, Japan (Tamura and Sakane, 2004).

Planotetrasporae are aerobic, Gram-positive, non-acid-fast actinomycetes which form moderate, irregularly branching, stable substrate hyphae (0.3–6.0 μm width) and sparsely branched, rarely septate aerial hyphae (0.2–0.4 μm in diameter). Long cylindrical spore vesicles are formed at the ends of short sporophores on aerial hyphae, with each spore vesicle (size about 2.1–2.7 μm 0.6–0.9 μm) containing four spores in a single row. Spores are short, cylindrical, short rod-like or oval elements (0.4–1.4 μm, 0.8–1.5 μm) and may exhibit mobility. They are released and become motile by means of single polar flagella when spore vesicles are immersed in water; active movement of the spores begins 30 min after they are released. In general, the vegetative mycelia are pale yellow to white. Good growth occurs at 25–30°C. Cell walls contain alanine, glutamic acid and *meso*-A$_2$pm and *N*-acetylmuramic acid. Galactose, glucose, madurose, 3-*O*-methylmannose and rhamnose are found in whole-organism hydrolysates. The predominant isoprenologue is tetrahydrogenated menaquinone with nine isoprene units, phosphatidylethanolamine is the diagnostic phospholipid, and 10-methylated C$_{18:0}$ is the major cellular fatty acid. The G+C content of the DNA is 71 mol%.

Type species: *Planotetraspora mira* Runmao et al. 1993, 468^{AL}. Type strain: DSM 44359^T=JCM 9131^T.

Habitats

Members of the family Streptosporangiaceae are usually associated with soil, but little is known about their role within this milieu. However, improvements in selective isolation procedures are beginning to cast light on the occurrence, distribution, numbers and activity of actinomycete taxa in natural habitats (Suzuki et al., 2001a; Suzuki et al., 2001b). Members of the family Streptosporangiaceae are probably involved in the primary decomposition of plant material in soils.

Streptosporangiae were associated with leaf litter (Van Brummelen and Went, 1957;

Potekhina, 1965), as well as soil and dung (Nonomura and Ohara, 1969a) until the introduction of a selective isolation procedure (Nonomura and Ohara, 1969a) showed that these organisms were an integral part of the actinomycete community in soils. The number of streptosporangiae in various soils in Japan has been estimated at 10^4 to 10^6 colony forming units (cfu) per gram dry weight of soil (Nonomura and Ohara, 1969a; Nonomura, 1984). Slightly acid, humus-rich garden soils are a favorite habitat. They have also been isolated from lake sediments (Willoughby, 1969a; Johnstone and Cross, 1976), beach sand (Williams and Sharples, 1976), and pasture and woodland soils (Whitham et al., 1993), but organisms labeled *Streptosporangium* type I from stream water (Willoughby, 1969b), given their morphological properties and capacity to form motile spores, probably belong to the genus *Actinoplanes*. "*Streptosporangium bovinum*" was isolated from infected bovine hooves (Chaves Batista et al., 1963).

Few *Acrocarpospora* and *Herbidospora* strains have been isolated; hence little is known about their distribution in natural habitats. The single representatives of *A. corrugata*, *A. macrocephala* and *A. pleiomorpha* were isolated from beach sand (pH 7.8) at Freshfield, Lancashire, United Kingdom (Williams and Sharples, 1976) and from soil samples collected from Saitama Prefecture, Japan (Tamura et al., 2000), and in Louisiana, United States (Tamura et al., 2000), respectively. Similarly, *Herbidospora cretacea* strains have been isolated from soil and plant material collected from several locations in Japan (Kudo et al., 1993).

Microbispora strains are common in soils. Using selective isolation procedures, counts of between 10^4 and 10^6 cfu per gram dry weight of soil have been reported from various Japanese soils (Nonomura and Ohara, 1971b). Larger populations have been found in slightly acidic (pH 5–6), humus-rich garden soils (Nonomura and Ohara, 1969a; Nonomura and Hayakawa, 1988). Microbisporae have also been isolated from marine sediments (Weyland, 1969). *Microbispora corallina* strains were isolated from soil samples collected in a deciduous dipterocarp forest in Thailand (Nakajima et al., 1999). One species, *Microbispora rosea*, has been implicated in a case of pericarditis and pleuritis in a human (Louria and Gordon, 1960). Most of the species previously assigned to the *Actinomadura pusilla* group, and now part of the genus *Nonomuraea*, originated from soil (Nonomura and Ohara, 1971c; Meyer, 1979; Galatenko et al., 1981).

Microtetrasporae are common in soil, notably forest soils (Thiemann et al., 1968b; Nonomura and Ohara, 1971b; Hayakawa et al., 1988; Nakajima et al., 2003). Using humic acid agar, Nonomura and Hayakawa (1988) recorded average counts of 3.6 10⁴ cfu per gram dry weight soil for a number of forest soil samples collected in Japan. *Microtetraspora niveoalba* strains are particularly widely distributed albeit with counts of less than 10³ cfu per gram dry weight of soil (Nonomura and Ohara, 1971b). In contrast, *M. malaysiensis* strains have only been isolated from two locations, namely from soil collected from below the leaf litter of mainly *Shorea* spp. in a primary lowland dipterocarp forest at Pasok, Negere Sembilan, and from a step hill dipterocarp forest at the Virgin Jungle Reserve, Gombak, Selangor in Peninsular Malaysia (Nakajima et al., 2003). *Microtetraspora fusca* and *M. glauca* have been isolated from soil samples collected in Brazil, Italy and Thailand.

Planomonospora strains have a worldwide distribution in soils of arid, temperate and tropical regions. Thiemann (1970) isolated 37 strains of *P. parontospora* from 7 out of 454 soil samples (1.5%) collected from Argentina, Chile, India, Peru and Venezuela. He also isolated 7 strains of *P. venezuelensis* from 3 out of 454 soil samples (0.7%) originating from Venezuela. Similarly, Suzuki et al. (2001b) isolated 246 *Planomonospora* strains from 137 out of 1200 soil samples; 94% of these isolates were from neutral to slightly alkaline soils (pH 7.0–9.0). Strains assigned to the *P. parontospora* group were recovered from 131 of these soil samples, notably from ones collected in Ecuador, Greece and India. Strains classified in the *P. venezuelensis* group were isolated from 13 soil samples (1.1%) collected in Bolivia, Cyprus, Egypt, Greece, India, Japan, New Caledonia and Turkey. The single strains of *P. alba* and *P. sphaerica* were isolated from soil samples collected from The Sudan and India, respectively (Mertz, 1994). *Planomonospora* strains have also been isolated from soil samples collected in Africa, Europe and Central and North America (Vobis, 1989b) and from the arid northeastern region of the Republic of South Africa (Kizuka et al., 1997).

Until recently, planobisporae had rarely been isolated from soil. Several strains, including the type strains of *P. longispora* and *P. rosea*, were isolated from soil samples taken from a riverbank in Venezuela (Thiemann, 1970). A few additional strains were recovered from soil samples collected from near Windhoek, Namibia (Vobis, 1989a), and from arid regions of South Africa (Kizuka et al., 1997). Suzuki et al. (2001a) have shown that planobisporae are distributed over a much wider geographical area, as they isolated 119 strains from 51 soil samples (3.5% of the samples tested) collected in Ecuador, Egypt, French Guiana, India and Madagascar. Nearly 90% of these strains were isolated from soil samples with pH values ranging from 7.0 to

7.9, results which suggest that *Planobispora* strains prefer neutral to alkaline environments. Suzuki and his colleagues were unable to isolate planobisporae from temperate regions in Europe, North America and Oceania. Planotetrasporae, in contrast to planobisporae and planomonosporae, have only been isolated from two sources. The type strain of *Planotetraspora mira* was isolated from a soil sample collected in the village of Wolung, Sichuan, People's Republic of China (Runmao et al., 1993), and *P. silvatica* from a sample of forest soil originating from Amami Island, Kagoshima Prefecture, Japan (Tamura and Sakane, 2004).

Isolation and Cultivation

Acrocarposporae, Herbidosporae and Planotetrasporae

Information was not provided on the procedures used to isolate *A. macrocephala*, *A. pleiomorpha* or *P. mira* strains from soil (Runmao et al., 1993; Tamura et al., 2000). The type strain of *A. corrugata* (previously *Streptosporangium corrugatum*) was isolated on starch casein agar supplemented with antifungal antibiotics following inoculation with a suspension of beach sand (Williams and Sharples, 1976). Similarly, *H. cretacea* strains have been isolated by plating soil suspensions onto yeast extract-starch agar (Kudo et al., 1993) and humic acid-vitamin agar (Hayakawa and Nonomura, 1987b). Members of this species have been isolated on yeast extract agar supplemented with antifungal antibiotics and plant material which had been desiccated at 28°C for at least a week prior to being ground with a blender following the addition of sterile water; the resultant plates were incubated at 28°C for 2 weeks (Kudo et al., 1993). The type strain of *P. silvatica* (Tamura and Sakane, 2004) was isolated from a sample of forest soil on humic acid-vitamin agar (Hayakawa and Nonomura, 1987b) using the yeast extract-sodium dodecylsulfate method (Hayakawa and Nonomura, 1989).

Acrocarpospora, *Herbidospora* and *Planotetraspora* strains grow well on oatmeal agar (Shirling and Gottlieb, 1966). Similarly, acrocarposporae and herbidosporae show good growth on inorganic salts starch (Shirling and Gottlieb, 1966) and yeast extract-starch agars (Kudo et al., 1993), respectively. The type strain of *Planotetraspora mira* grows well and sporulates on calcium malate (Runmao et al., 1993) and humic acid-vitamin agars (Hayakawa and Nonomura, 1987b). The type strain of *P. silvatica* grows well on glycerol-asparagine, tyrosine, and yeast-extract-malt extract agars (Tamura and Sakane, 2004).

Microbisporae, Microtetrasporae, Nonomuraea and Streptosporangiae

Dry heat treatment of air-dried soil samples and dilution plate culture with selective synthetic media are used for the preferential isolation and enumeration of some members of the family Streptosporangiaceae. The procedures outlined below have been developed for the selective isolation of the genera *Microbispora* and *Streptosporangium* (Nonomura and Ohara, 1969a; Nonomura and Ohara, 1969b) and with modifications for the isolation of other actinomycete genera, notably *Microtetraspora* and *Nonomuraea* (Nonomura and Ohara, 1971b; Nonomura and Ohara, 1971c). There is evidence (Nonomura and Hayakawa, 1988) that pretreatment of soil suspensions with yeast extract (6%, w/v) and sodium dodecyl sulfate (0.05%, w/v) at 40°C for 20 min, followed by dilution with water, activates actinomycete spores but kills vegetative cells of other soil bacteria in the suspensions. This practice leads to an increase in the counts of actinomycetes on isolation plates.

After soil samples are dried slowly at room temperature, passed through a 2-mm sieve, ground slightly in a mortar, spread on filter paper, and heated in a hot air oven at 120°C for 1 h, the number of bacteria and streptomycetes is dramatically reduced, and the isolation frequency of *Microbispora*, *Microtetraspora* and *Streptosporangium* strains is enhanced. Heated soil is incorporated directly onto isolation media, or a suspension is used to make dilution plates. Initially, arginine-vitamin (AV) and mineral-glucose-asparagine plus soil extract (MGA-SE) agars were recommended for the selective isolation of microbisporae and streptosporangiae, but two additional formulations, chitin-V and humic-vitamin (HV) agars (Hayakawa and Nonomura, 1987a; Hayakawa and Nonomura, 1987b; Nonomura, 1989c), have been developed. These media are supplemented with antifungal antibiotic(s); sometimes penicillin and polymyxin B are also used. Inoculated plates are incubated for 4–6 weeks at 30°C (or 2–3 weeks at 50°C) and examined using a light microscope with a long-working-distance objective. The highest counts and cleanest plates are usually obtained with HV agar.

Hayakawa et al. (1991) introduced an improved procedure for the selective isolation of streptosporangiae from soil. The method is based on the ability of streptosporangial spores to withstand dry heat and treatment with benzethonium chloride (BC) and the capacity of streptosporangiae to grow in the presence of leucomycin and nalidixic acid. Initially an air-dried soil sample is ground in a mortar and heated in a hot-air oven for an hour. Half a ml of a 10^{-1} dilution in water

of the heated sample is transferred to 4.5 ml of sterile 5 mM phosphate buffer (pH 7.0) containing BC at a final concentration of 0.1% (w/v). The resultant preparation is maintained at 30°C for 30 min with occasional stirring, and a portion (1 ml) is then diluted with sterile tap water (1 : 10 or 1 : 15). Inocula of 0.1 ml or 0.2 ml of the dilution are then spread over the surface of plates of HV agar supplemented with leucomycin in ethanol (1 mg per liter) and nalidixic acid (20 mg per liter) and the plates are incubated at 30°C for 3–4 weeks. Actinomycetes which appear on the plates are examined by light microscopy (600X) and assigned to genera on the basis of characteristic morphological properties.

Microbispora strains can be preferentially isolated by treating suspensions of dry heat pretreated soil samples with 1.5% phenol at 30°C for 30 min, diluting in water, and plating onto HV agar supplemented with nalidixic acid (20 mg per liter). *Microbispora karnatakensis* (Rao et al., 1987) was isolated by plating a suspension of soil onto inorganic salts-starch agar (Küster, 1959).

Microtetraspora fusca, *M. glauca* and *M. malaysiensis* were isolated from soil samples using methods that have not been disclosed (Thiemann et al., 1968b; Nakajima et al., 2003). However, the pretreatment procedure described above has been used to isolate several *Nonomuraea* species, including *N. helvata*, *N. pusilla*, *N. roseoviolacea* and *N. spadix* (Nonomura and Ohara, 1971c). *Nonomuraea* spores appear to be particularly resistant to dry heat at 100–120°C thereby allowing the slow-growing nonomuraea to develop into recognizable colonies on dilution plates. Soil dilutions are plated onto various media, including AV and MGA-SE agars, and incubated for several weeks at 28–30°C (Nonomura and Ohara, 1971b). *Microtetraspora niveoalba* was isolated from dry-heated soil on MGA-SE agar incubated at 40°C for 1 month. Similarly, *M. glauca* strains have been isolated on plates of this medium incubated at 30°C.

Other *Nonomuraea* (previously *Microtetraspora*) species, such as *N. roseola* and *N. salmonea*, have been isolated from soil by Soviet investigators who supplemented media with antibiotics to improve their selectivity. Lavrova et al. (1972) added rubomycin (5, 10 or 20 μg·ml⁻¹) to medium no. 2 of Gauze et al. (1957); Preobrazhenskaya et al. (1975) added bruneomycin (0.5, 1 or 2 μg·ml⁻¹) or streptomycin (0.5, 1 or 2 μg·ml⁻¹). The use of these antibiotics led to the growth of more *Nonomuraea* colonies on isolation plates while reducing the number of streptomycetes. In contrast, *N. ferruginea* and *N. spiralis* were isolated by plating soil suspensions onto oatmeal agar or Gauze's no. 1 medium without addition of selective antibiotics (Meyer, 1979).

Microbisporae, microtetrasporae, nonomuraea and streptosporangiae grow well on rich media, including Bennett's (Jones, 1949), glucose-yeast extract (Waksman, 1950), oatmeal (ISP medium 3; Difco 0771), and yeast extract-malt extract agars (ISP medium 2 [Difco 0770]; Shirling and Gottlieb, 1966). Oatmeal-yeast extract agar is recommended for the growth of mesophilic microbisporae and glycerol agar for the corresponding thermophilic strains (Nonomura, 1989a). Streptosporangiae grow well and produce an abundant aerial spore mass on oatmeal-yeast extract agar (Nonomura and Ohara, 1960). Good growth of vegetative and sporing aerial mycelia was obtained for *Microtetraspora fusca* and *M. glauca* on Hickey and Tresner (1952) agar. *Microtetraspora malaysiensis* strains grow well on yeast-malt extract agar (Nakajima et al., 2003). *Microtetraspora niveoalba* requires B vitamins for growth on synthetic media (Nonomura and Ohara, 1971b).

Planobisporae and Planomonosporae

BAITING WITH NATURAL SUBSTRATES. The procedures used to isolate *Planobispora* and *Planomonospora* strains from soil were not revealed by Thiemann and his colleagues (Thiemann et al., 1967; Thiemann and Beretta, 1968a). However, members of these taxa have been isolated from soil by baiting with natural substrates (Couch, 1954; Bland and Couch, 1981) as follows: a small amount of soil, approximately one level teaspoonful, is placed in a sterile Petri dish and flooded with sterile water (distilled water or filtered soil or charcoal water extracts may be used). Added pollen and hair float at the water surface; various types of pollen have been employed including that from members of the genera *Liquidamber*, *Pinus* and *Sparganium* (Schäfer, 1973). After 1–4 weeks, examination of the water surface with a dissecting microscope (100X) and strong horizontal lighting should reveal white glistening spore vesicles formed in the air at the surface of the water by spore vesicle-forming members of the families Micromonosporaceae and Streptosporangiaceae. The characteristic two-spored vesicles of *Planobispora* develop on long aerial hyphae growing between the baits. Similarly, sporulating aerial hyphae of *Planomonospora* strains grow on pollen grains. Single spore vesicles or bundles of them can be picked up with a thin needle and placed on the surface of agar media in small Petri dishes. After 2–4 weeks, the young colonies can be transferred to slant cultures. Similarly, *Planomonospora alba* strain A82600^T and *Planomonospora sphaerica* strain A15460^T were isolated by immersing a soil sample in water enriched for growth of microorganisms with motile spores, with sterile grass floating on the water surface as bait (Mertz, 1994).

USE OF FLOODING SOLUTIONS, CENTRIFUGATION AND HUMIC ACID GELLAN GUM. A multistage procedure was developed by Suzuki et al. (2001a) for the selective isolation of planobisporae from soil. Air-dried soil samples (500 mg) are heated at 90°C for 60 min in a hot-air oven and then cooled to room temperature. Each heat treated sample is added to 2 ml of flooding solution (0.1% skim milk [neutralized], 0.01% Tween, 100 µg·ml^{-1} nalidixic acid in 5 mM *N*-cyclohexyl-2-amino-ethanesulfonic acid [CHES]; pH 9.0) and incubated at 35°C for 60 min with occasional stirring to stimulate zoospore motility. After centrifugation (1000 g) for 10 min at room temperature, 800 µl of supernatant is gently transferred to a sterile tube; 100-µl aliquots of this preparation are spread over humic acid-trace salts gellan gum medium (HSG) supplemented with cycloheximide (50 µg·ml^{-1}), enoxacin (20 µg·ml^{-1}), nalidixic acid (50 µg·ml^{-1}), nystatin (50 µg·ml^{-1}), sodium ampicillin (2 µg·ml^{-1}), streptomycin sulfate (1 µg·ml^{-1}) and trimethoprim (50 µg·ml^{-1}). Following incubation at 32°C for 14–21 days, planobisporae colonies growing on the HSG plates are recognized by their characteristic morphological features, as seen using a 40× long working district objective lens. Pure cultures are isolated by streaking onto HSG medium and tested for zoospore production using flooding solution containing 0.1% skim milk in 5 mM CHES (pH 9.0).

A similar multistage procedure is available for the selective isolation of planomonosporae from soil (Suzuki et al., 2001b). Air-dried soil samples (500 mg) are heated at 100°C for 60 min in a hot-air oven and cooled to room temperature. Each heat treated sample is added to 2 ml of sterile flooding solution (0.1% skim milk in 5 mM *N*-cyclohexyl-2-amino-ethanesulfonic acid [CHES]; pH 9.0) and incubated at 32°C for 90 min with occasional stirring to stimulate motility. The soil suspension is centrifuged (1000 g) for 10 min at room temperature, incubated at 32°C for 60 min, and 500 µl of supernatant gently transferred to a sterile tube and 100-µl aliquots spread over HSG medium supplemented with cycloheximide (50 µg·ml^{-1}), enoxacin (20 µg·ml^{-1}), nalidixic acid (20 µg·ml^{-1}), nystatin (50 µg·ml^{-1}), sodium ampicillin (2 µg·ml^{-1}) and trimethoprim (20 µg·ml^{-1}). Inoculated plates are incubated at 35°C for 14–21 days. Colonies of actinomycetes are observed directly under a phase-contrast microscope using a 40X long distance working objective lens.

Planomonospora colonies, identified using morphological features (clavate spore vesicles containing single spores), are purified by single colony isolation on HSG plates and incubated at

35°C for 14 days. Isolates are tested for motility with flooding solution containing 0.1% skim milk in 5 mM CHES (pH 9.0). Isolates with motile spores can be assigned to two groups on the basis of morphological features: the *P. parontospora* group (spore vesicles arranged in double parallel rows resembling bananas) and the *P. venezuelensis* group (spore vesicles arranged in palm leaf patterns).

Planobisporae and planomonosporae grow on standard media used for cultivating streptomycetes (Waksman, 1961); the first signs of visible growth appear after 3–4 days at 28–30°C. *Planobispora longispora* produces aerial hyphae and abundant spore vesicles on calcium malate, soil extract, and yeast extract-malt extract agars (Shirling and Gottlieb, 1966). Vesicular development in *P. rosea* is promoted by all media on which aerial mycelium is formed, notably soil extract and Hickey-Tresner agars. Spore vesicle development in *Planomonospora* strains is especially abundant on Bennett's, Hickey-Tresner, oatmeal and soil extract agars (Thiemann et al., 1967; Vobis, 1989b).

Media for Isolation and Cultivation

The following are the recipes for those media mentioned in the preceding section that are not commercially available. Quantities are for 1 liter of distilled water unless otherwise stated.

ISOLATION MEDIA Arginine-Vitamin (AV) Agar (Nonomura and Ohara, 1969a)

L-Arginine	0.3 g
Glucose	1.0 g
Glycerol	1.0 g
K$_2$HPO$_4$	0.3 g
MgSO$_4$ · 7H$_2$O	0.2 g
NaCl	0.3 g
Agar	15 g

Trace Salts Solution

CuSO$_4$ · 5H$_2$O	1.0 mg/ml
Fe$_2$(SO$_4$)$_3$	10.0 mg/ml
MnSO$_4$ · 7H$_2$O	1.0 mg/ml
ZnSO$_4$ · 7H$_2$O	1.0 mg/ml
Adjust to pH 8.0.	

Vitamins (final weight in medium): 0.5 mg each of p-aminobenzoic acid, calcium pantothenate, inositol, niacin, pyridoxine HCl, riboflavin, thiamine HCl; 0.25 mg of biotin.

Antibiotics: cycloheximide, 50 mg; nystatin, 50 mg; nalidixic acid, none or 20 mg; penicillin G, none or 0.8 mg; polymyxin, none or 4 mg. Adjust pH to 6.4.

Chitin-V Agar (Hayakawa and Nonomura, 1984)

Colloidal chitin	2 g (dry weight)
CaCO$_3$	0.02 g
FeSO$_4$ · 7H$_2$O	10 mg

K$_2$HPO$_4$	0.35 g
KH$_2$PO$_4$	0.15 g
MgSO$_4$ · 7H$_2$O	0.2 g
MnCl$_2$	1 mg
NaCl	0.3 g
ZnSO$_4$ · 7H$_2$O	1 mg
Agar	18 g

Add B vitamins, as for AV agar, and cycloheximide, 50 mg. Adjust pH to 7.2.

Humic Acid-Trace Salts Gellan Gum Medium (Suzuki et al., 2001a)

Nitrohumic acid	0.5 g
CaCl$_2$	0.33 g
FeSO$_4$ · 7H$_2$O	1 mg
MnCl$_2$ · 4H$_2$O	1 mg
NiSO$_4$ · 6H$_2$O	1 mg
ZnSO$_4$ · 7H$_2$O	1 mg
N-cyclohexyl-2-amino-ethanesulfonic acid	1 g
Gellan gum	7 g

Obtain the nitrohumic acid from Tokyo Chemical Industry, Tokyo, Japan and the gellan gum from Wako Pure Chemicals, Osaka, Japan. Adjust pH to 9.0. Autoclave at 121°C for 20 min.

Humic Acid-Trace Salts Gellan Gum Medium (Suzuki et al., 2001b)

Nitrohumic acid	0.5 g
CaCl$_2$	0.33 g
FeSO$_4$ · 7H$_2$O	0.001 g
MnCl$_2$ · 4H$_2$O	0.001 g
NiSO$_4$ · 6H$_2$O	0.001 g
ZnSO$_4$ · 7H$_2$O	0.001 g
N-cyclohexyl-2-amino-ethanesulfonic acid	1 g
Gellan gum	7 g

Obtain the nitrohumic acid from Tokyo Chemical Industry, Tokyo, Japan and the gellan gum from Wako Pure Chemicals, Osaka, Japan. Adjust pH to 9.0. Autoclave at 121°C for 20 min.

Humic-Vitamin (HV) Agar (Hayakawa and Nonomura, 1984; Nonomura, 1984)

Humic acid (see below)	1 g
CaCO$_3$	0.02 g
FeSO$_4$ · 7H$_2$O	0.01 g
KCl	1.7 g
MgSO$_4$ · 7H$_2$O	0.05 g
Na$_2$HPO$_4$	0.5 g
Agar	18 g

Humic acid (1 g) is used in the form of an alkaline solution. Artificial humic acid prepared from glycine and urea may be employed, as may natural humic acid from soil humus, but the pale brown humic acid designated as "Rp type" gives the best results. Add B vitamins, as for AV agar and cycloheximide, 50 mg; adjust to pH 7.2.

Inorganic Salts-Starch Agar (Küster, 1959)

Solution I: Difco soluble starch, 10 g. Make a paste of the starch with a small amount of cold distilled water and bring to a volume of 500 ml.

Solution II:

CaCO$_3$	2 g
K$_2$HPO$_4$ (anhydrous salt)	1 g
MgSO$_4$ · 7H$_2$O	1 g

NaCl	1 g
(NH₄)₂SO₄	2 g
Distilled water	500 ml
Trace salts solution	1 ml

Adjust to pH 7–7.4. Mix solutions I and II and add 20 g of agar. Trace salts include FeSO₄ · 7H₂O, 0.1 g; MnCl₂ · 4H₂O, 0.1 g; ZnSO₄ · 7H₂O, 0.1 g; and distilled water, 100 ml.

Medium No. 1 (Gauze et al., 1957)

Starch	20 g
FeSO₄ · 7H₂O	0.01 g
KNO₃	1 g
K₂HPO₄	0.5 g
MgSO₄	0.5 g
NaCl	0.5 g
Agar	30 g

Adjust to pH 7.2–7.4.

Mineral Glucose Asparagine-Soil Extract (MGA-SE) Agar (Nonomura and Ohara, 1971b)

L-Asparagine	1 g
Glucose	2 g
K₂HPO₄	0.5 g
Soil extract	200 ml
Distilled water	800 ml
Agar	20 g

Add antibiotics: cycloheximide, 50 mg; nystatin, 50 mg; benzylpenicillin, 0.8 mg; and polymyxin B, 4 mg. To prepare the soil extract, add 1000 g of soil to 1 liter water, autoclave for 30 min, then decant and filter. Adjust to a final pH of 8.0.

CULTIVATION MEDIA

Bennett's Agar (Jones, 1949)

Beef extract	1 g
Glucose	10 g
N-Z amine A (enzymatic digest of casein)	2 g
Yeast extract	1 g
Agar	15 g

Adjust to pH 7.3 with NaOH.

Modified Bennett's agar can be used, where glucose is replaced by glycerol (10 g) and N-Z amine A by Bacto-casitone (2 g; Difco).

Calcium Malate Agar (Waksman, 1961)

Calcium malate	10 g
K₂HPO₄	0.5 g
NH₄Cl	0.5 g
Agar	15 g

Czapek Agar (Waksman, 1950)

Sucrose	30 g
FeSO₄ · 7H₂O	0.01 g
KCl	0.5 g
K₂HPO₄	1 g
MgSO₄ · 7H₂O	0.5 g
NaNO₃	3 g
Agar	15 g

After components are dissolved, dispense the medium into flasks or tubes and then autoclave. For peptone-Czapek agar, substitute 5 g of peptone for 3 g of NaNO₃ (Bland and Couch, 1981).

Glucose-Yeast Extract Agar (Waksman, 1950)

Glucose	10 g
Yeast extract (50% solution)	20 ml
Agar	15 g

Adjust to pH 6.8.

Glycerol Agar (Nonomura and Ohara, 1969b)

Casamino acids	2 g
Glycerol	5 g
K₂HPO₄	0.3 g
MgSO₄ · 7H₂O	0.5 g
NaCl	0.3 g
Agar	20 g

Add trace salts and B vitamins as for AV agar; adjust final pH to 7.2.

Hickey-Tresner Agar (Hickey and Tresner, 1952)

Amidax or dextrin	10 g
Beef extract	1 g
N-Z amine A or tryptone	2 g
Yeast extract	1 g
CoCl₂ · 6H₂O	2 mg
Agar	20 g

Obtain Amidax from Corn Products Refining, Argo, Illinois, and beef extract, tryptone, and yeast extract from Difco. Adjust to pH 7.3.

Medium No. 2 (Gauze et al., 1957)

Glucose	10 g
Peptone	5 g
NaCl	10 g
Hottinger's broth	30 ml
Agar	15 g

Adjust to pH 7.0.

Oatmeal Agar (Bland and Couch, 1981)

| Baby oatmeal | 65 g |
| Agar | 15 g |

Autoclave for 30 min, leave for 24 h, then autoclave again before dispensing.

Soil Extract Agar (Thiemann et al., 1968b)

Air-dried garden soil	30 g
Agar	20 g
Tap water	1 liter

Adjust pH to 7.0. Autoclave for 20 min at 120°C.

Yeast Extract-Malt Extract Agar (Pridham et al., 1956–57)

Bacto-dextrose (Difco)	4 g
Bacto-malt extract (Difco)	10 g
Bacto-yeast extract (Difco)	4 g

Adjust to pH 7.3, then add 20 g of Bacto-agar. Liquify agar by steaming at 100°C for 15–20 min.

Preservation of Cultures

The most convenient method for short-term storage is by serial transfer from agar slants of appropriate media (see above) every two months (Meyer, 1989). The tubes should be tightly closed with cotton plugs dipped in melted paraffin wax. Sporulated spore cultures can be stored at 5°C and at room temperature. Lyophilization, storage in liquid nitrogen, and freezing in glycerol can be used for long-term preservation (Wellington and Williams, 1978; Meyer, 1989).

For lyophilization, the spore suspension or vegetative mycelium is suspended in a suitable fluid, such as serum plus 7.5% (w/v) glucose or skimmed milk plus 7.5% (w/v) glucose. For storage in liquid nitrogen, the microorganisms are inoculated into small test tubes containing the appropriate medium and incubated until satisfactory growth is visible. The tubes are then closed with cotton plugs dipped in melted paraffin wax and placed in a liquid nitrogen container. Glycerol suspensions are prepared by scraping aerial growth or substrate mycelium or both from heavily inoculated plates and making heavy suspensions in 3 ml of aqueous glycerol in small (e.g., bijoux) bottles, which are stored at –20°C. The frozen glycerol suspensions serve both as a practical means of long-term preservation and as convenient source of inoculum. Working inocula are obtained by thawing suspensions at room temperature prior to treating as for broth cultures. After use, glycerol suspensions are promptly frozen and stored again at –20°C.

Identification

Streptosporangiaceae strains can be distinguished from all other actinomycetes using a combination of chemotaxonomic and morphological features. Members of the family show a range of morphological properties but are relatively homogeneous from a chemotaxonomic perspective (Table 1). Simplified procedures are available for detecting chemical markers, notably, cell wall constituents (Staneck and Roberts, 1974; Uchida et al., 1977; Uchida et al., 1999; Hancock, 1994), fatty acids (Suzuki and Komagata, 1983; Kroppenstedt et al., 1990), menaquinones (Collins et al., 1977; Kroppenstedt, 1982; Suzuki et al., 1983; Minnikin et al., 1984) and polar lipids (Suzuki et al., 1983; Minnikin et al., 1984). Procedures such as these are being progressively complemented or replaced by molecular systematic methods, including the use of oligonucleotide primers (Monciardini et al., 2002) and 16S rRNA gene sequences (Stackebrandt et al., 1997; Tamura et al., 2000; Stach et al., 2003).

Primary diagnostic chemotaxonomic data can be gained by examination of whole-organism hydrolysates (Lechevalier and Lechevalier, 1980). Unidimensional thin layer chromatography will determine whether an organism contains diaminopimelic acid and whether the latter

is in the *LL*- or *meso*- form. Sporoactinomycetes rich in *LL*-A_2pm can be provisionally assigned to the genus *Streptomyces* (see The Family Streptomycetaceae, Part I: Taxonomy in this Volume and The Family Streptomycetaceae, Part II: Molecular Biology in this Volume). The detection of *meso*-A_2pm and madurose with the absence of characteristic sugars serves to separate strains of Streptosporangiaceae from those of *Actinoplanes* and related genera, *Nocardia* and related genera, *Pseudonocardia* and related genera, *Nocardiopsis* and related genera, and *Thermomonospora* and related genera, but not from the genera *Dermatophilus* and *Frankia*. The latter can be distinguished readily from *Streptosporangium* and allied taxa on morphological grounds. To date, the presence of madurose is associated with wall chemotype III actinomycetes, although there is an unconfirmed report of this sugar from a wall chemotype I actinomycete with a streptomycete morphology (Weyland et al., 1982). The discovery of 3-*O*-methylgalactosyl (madurosyl) units in the structure of teichoic acids of an *Nonomuraea roseoviolaceae* subsp. *carminata* strain (previously *Actinomadura carminata*; Naumova et al., 1986) is also noteworthy, as madurose is not perceived to be a cell wall constituent according to Lechevalier and Lechevalier (1981a).

Wall chemotype III actinomycetes which form spore vesicles can be identified to the genus level using morphological features (Table 1), though care must be taken to distinguish between *Spirillospora* and *Streptosporangium* strains even though only the former produces motile spores (Vobis and Kothe, 1989c). Note also that the sporogenic hyphae in the spore vesicles of spirillosporae are branched, whereas those in streptosporangiae are unbranched. Chemical analyses are required if the genera *Microbispora*, *Microtetraspora* and *Nonomuraea* are to be reliably separated from the genus *Actinomadura* (Kroppenstedt et al., 1990). Actinomadurae contain mostly hexahydrogenated menaquinones with nine isoprene units (MK-9 [H_6]) and diphosphatidylglycerol and phosphatidylinositol as predominant polar lipids (phospholipid type I sensu Lechevalier et al., 1977) whereas *Microbispora*, *Microtetraspora* and *Nonomuraea* strains have major amounts of MK-9 [H_4] saturated at positions III and IV, with major amounts of diphosphatidylglycerol, hydroxylated phosphatidylethanolamine, uncharacterized glycolipids, and a glucosamine-containing phospholipid (type IV phospholipid pattern sensu Lechevalier et al., 1977).

The continued separation of the genera *Microbispora*, *Microtetraspora* and *Nonomuraea* solely on morphological criteria is questionable. *Microbispora* is distinguished from related genera

(Table 1) primarily by the formation of paired spores on aerial hyphae. These are either sessile or borne on short sporophores. The latter in *M. rosea* were found to attach to the base of the spore by a ball and socket arrangement (Williams, 1970). *Microbispora* species are currently separated by using a range of phenotypic properties (Table 3).

Cultures of *M. aerata*, *M. amethystogenes*, and *M. parva* deposit crystals with a metallic sheen in the medium, particularly when grown on Pablum extract agar (Lechevalier and Lechevalier, 1957) for about 10 days (Gerber and Lechevalier, 1964). These crystals are composed of iodinin (1,6-phenazinediol-5,10-dioxide), a red, water-soluble pigment. In addition, *M. aerata* produces two brown-yellow pigments (2-aminophenoaxazine-3-one and 1,6-phenazinediol), a yellow pigment (2-acetamidophenoxazine-3-one; Gerber and Lechevalier, 1964), and an orange pigment (1,6-phenazinediol-5-oxide; Gerber and Lechevalier, 1965).

Microtetrasporae can be recognized by their ability to produce short spore chains which typically contain four spores. The identification of species within the genus is still based on phenotypic properties, notably on the color of the aerial and substrate mycelium (Table 4).

Nonomuraea strains are distinguished from members of related genera by their ability to form chains of spores or pseudovesicles on aerial hyphae (Figs. 10–12). The constituent species may be distinguished by means of spore chain morphology, spore wall ornamentation, color of mature sporulated aerial mycelium, and substrate mycelium pigmentation (Table 5). Nevertheless, identification of many of these species is difficult because in most instances only one (the type) strain or a few strains have been examined. Even when several strains have been studied, the

results of biochemical and physiological tests have proved to be variable or inconsistent when data from the literature are compared. However, numerical taxonomic evidence indicates that most of the validly described taxa merit species status (Goodfellow et al., 1979; Goodfellow and Pirouz, 1982).

Streptosporangium species can be distinguished from one another using a combination of phenotypic properties, notably morphological features (Table 6). They can be separated by spore vesicle size, sporangiophore length, spore shape, aerial spore mass color, and substrate mycelium pigmentation and subdivided according to the nature of the vesicular wall. At one extreme, the spore vesicular membrane of *S. fragile* is so thin that it cannot be detected by light microscopy (Shearer et al., 1983). This feature may lead to difficulty in differentiating

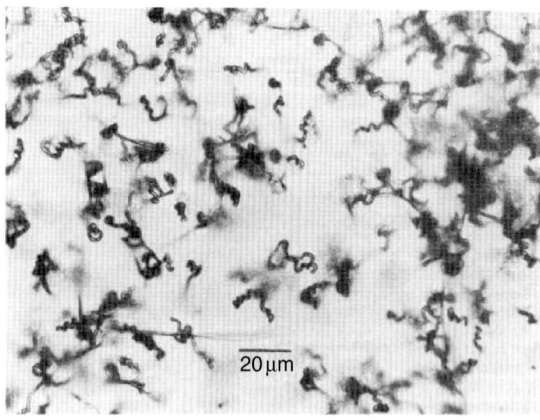

Fig. 11. *Nonomuraea spiralis*, IMET 9621[T]. Sporulating aerial mycelium. Oatmeal-nitrate agar, 28°C, 18 days. From Meyer (1989), with permission.

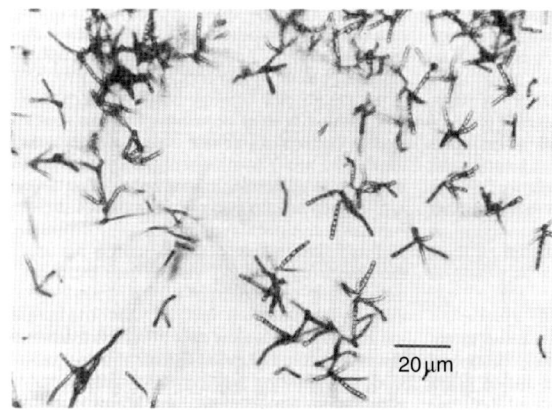

Fig. 10. *Nonomuraea roseola*, INA 1671[T]. Sporulating aerial mycelium. Cultivation on oatmeal-nitrate agar, 28°C, 12 days. From Meyer (1989), with permission.

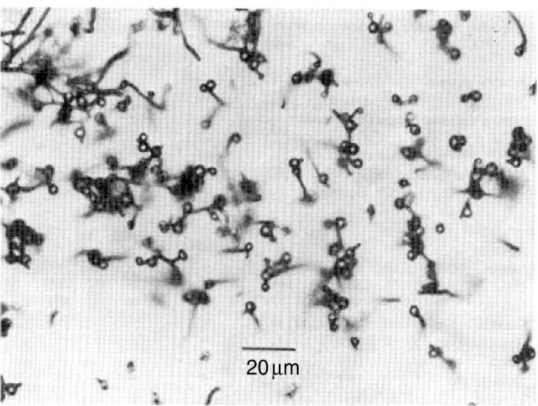

Fig. 12. *Nonomuraea pusilla*, ATCC 27296[T]. Sporulating aerial mycelium with "pseudosporangia." Cultivation on yeast extract-malt agar, 28°C, 9 days. From Meyer (1989), with permission.

Table 3. Characteristics differentiating validly described species of the genus *Microbispora*.

Characteristics	*M. aerata*	*M. amethystogenes*	*M. chromogenes*	*M. corallina*	*M. indica*	*M. karnatakensis*	*M. mesophilica*	*M. parva*	*M. rosea*	*M. thermodiastatica*	*M. thermorosea*
Morphology											
Aerial mycelium color	Pink	Pink	Pink	Pink	Pinkish-white	White	White	Pink	Pale pink	Pink	Pink
Substrate mycelium color	Yellowish-brown	Light brown	Orange	Coral pink to reddish	Violet-orange	Yellowish-pink	Brown	Light brown	Orange	Yellowish-brown	Yellowish-brown
Soluble pigments	None	−	Light yellow	−	Deep orange yellow	Deep orange yellow	−	−	−	−	−
Iodinin production	+	+	−	−	−	−	ND	+	−	−	−
Degradation of											
Hypoxanthine	−	−	+	v	+	+	+	−	+	−	+
Starch	+	−	+	ND	−	+	+	−	−	+	+
Testosterone	−	−	+	+	+	+	ND	+	+	+	+
Xanthine	−	−	−	−	−	−	+	−	−	−	−
Xylan	−	−	−	−	−	−	+	−	−	−	−
Nitrate reduction	+	−	+	−	+	+	+	−	? Data missing	−	−
Growth on sole carbon sources (1%, w/v)											
Arabinose	+	+	+	+	+	−	+	+	+	+	+
Glycerol	+	+	+	+	−	+	+	+	+	+	+
Inositol	−	+	+	+	−	+	−	−	−	−	−
Rhamnose	−	−	−	−	+	+	−	+	+	−	−
Growth at											
25°C	−	+	+	+	+	+	+	+	+	−	−
50°C	+	−	−	−	+	+	−	+	−	+	+
55°C	+	−	−	−	−	−	−	−	−	+	+
Requirement for											
Biotin	+	−	+	−	+	−	ND	−	+	+	+
Thiamine	+	+	+	+	+	+	ND	+	+	+	+

Symbols and abbreviations: +, positive; −, negative; ND, not determined; and v, variable.
From McCarthy and Cross (1984), Rao et al. (1987), Nonomura (1989b), and Nakajima et al. (1999).

Table 4. Characteristics differentiating the type strains of *Microtetraspora* species.

Characteristics	M. fusca DSM 43841[T]	M. glauca DSM 43311[T]	M. malaysiensis DSM 44579[T]	M. niveoalba DSM 43174[T]
Morphology				
Aerial spore mass color				
Blue-gray	−	+	−	−
Gray	+	+	−	−
White	−	−	+	+
Substrate mycelium color				
Cream-yellow	−	−	+	−
Greenish-blue	−	+	−	−
Purplish	+	−	−	−
Branched spore chains	−	−	−	+
Requirement for biotin	−	+	−	−
Biochemical tests				
Reduction of nitrate	−	+	ND	+
Urea hydrolysis	+	−	−	+
Degradation of				
Elastin	ND	−	−	+
Gelatin	−	+	ND	+
Hypoxanthine	−	+	−	+
Starch	−	+	ND	+
Testosterone	−	+	+	+
Xanthine	−	+	−	+
Xylan	−	+	−	−
Growth on sole carbon sources				
L(+) Arabinose	+	+	−	+
D(+) Fructose	−	+	+	+
D(+) Galactose	−	+	−	−
Glycerol	−	+	−	+
D(+) Mannitol	−	+	+	+
D(+) Mannose	+	+	−	+
meso-Inositol	−	+	−	+
L(+) Rhamnose	−	+	+	−
D(+) Trehalose	+	+	−	+
Xylitol	−	−	−	+
Citrate	+	+	+	−
Fumarate	−	+	−	+
Malate	−	+	−	+

Symbols and abbreviations: +, positive or present; −, negative or absent; ND, not determined; [T], type strain; and DSM, Deutsche Sammlung von Mikroorganismen und Zellkulturen, Mascheroder Weg 1B, D-38124 Braunschweig, Germany. From Nonomura (1989c) and Nakajima et al. (2003).

Streptosporangium from *Nonomuraea* strains, as members of some species of the latter produce pseudovesicles covered by a slimy substance (Nonomura and Ohara, 1971c). The remaining species of *Streptosporangium* form thin vesicular membranes that are readily disrupted in water. *Streptosporangium amethystogenes* produces violet crystals of iodine after a month's incubation at 30°C on oatmeal-yeast extract agar.

Improved phenotypic tests are needed for the identification of unknown streptosporangiae. Clearly such tests should be based on a representative set of strains. Whitham (1988) generated a probability matrix on the basis of 26 diagnostic properties for the identification of unknown streptosporangiae to established and novel *Streptosporangium* species. In a continuation of these studies, Kim (1993) assigned 65 out of 70

marker *Streptosporangium* strains and 12 out of 131 putative streptosporangiae isolated from soil to known species of *Streptosporangium*. A further 19 of the soil isolates were identified to known species when less stringent cut-off points were adopted for a positive identification.

The genera *Planobispora*, *Planomonospora* and *Planotetraspora* may be distinguished by the shape of their spore vesicles and by the number of encased spores. *Planobispora* and *Planomonospora* species can be separated using a judicious selection of phenotypic properties (Table 7). *Planobispora longispora* and *Planobispora rosea* have many properties in common (Goodfellow and Pirouz, 1982) but can be separated by using cultural characteristics (Thiemann, 1974a). *Planobispora longispora* produces a hyaline- to creamish-colored substrate mycelium and a

Table 5. Characteristics distinguishing the type strains of *Nonomuraea* species.

Characteristics	*N. africana*	*N. angiospora*	*N. dietziae*	*N. fastidiosa*	*N. ferruginea*	*N. flexuosa*	*N. helvata*	*N. longicatena*	*N. polychroma*	*N. pusilla*	*N. reticatena*	*N. roseola*	*N. roseoviolacea* subsp. *carminata*	*N. roseoviolacea* subsp. *roseoviolacea*	*N. rubra*	*N. salmonea*	*N. spiralis*	*N. terrinata*	*N. turkmeniaca*
Morphology																			
Spore chains	str	sp	str, sp	s, sp	h, s	h, s	h, psp	str	ND	psp	str	sp, str	sp, psp	psp	h, s, sp	h, s	sp	Irregular, psp	sp
Spore ornamentation	Smooth	Ridged	Cross-ridged, smooth- and rough	Irregular	Folded	Warty	Smooth	Smooth	ND	Smooth	Smooth	Folded	Smooth	Smooth	Smooth	Warty	Folded	Rugose	Smooth
No. of spores	4–10	4–15	Up to 30	4–10	4–10	4–10	4–10	10–30	ND	>10	4–20	4–20	ND	4–20	4–20	4–30	4–20	8–15	10–20
Growth on ISP medium 3																			
Aerial mycelium	Grayish/blue	White	Beige	White/pink	White/pink	White/yellow	White	White	Trace	White/cream	White/cream	Pink	Pink	Pink/violet	Trace	Pink	White/yellow	White/white	Trace
Substrate mycelium	Yellow	White/ochre	Beige	Colorless	Pink	Brown	Yellow/brown	Ochre	Colorless	Gray/brown	Dark yellow/brown	Brown/red	Old-wine	Violet	Orange red	Red	Yellow/brown	White/ochre	Violet/red
Soluble pigment	Yellowish/brown	None	Yellow	None	None	None	None	None	None	None	None	None	Wine-red	Violet	Red	None	None	None	Pink/violet
Biochemical tests																			
Esculin hydrolysis	+	+	ND	+	–	+	+	+	+	+	+	+	ND	+	–	+	+	+	+
Nitrate reductase	+	–	ND	+	+	+	+	–	–	+	+	+	–	+	+	+	+	–	+
Degradation tests																			
Casein	+	+	ND	+	+	+	–	+	–	–	–	–	ND	–	–	+	–	+	+
DNA	+	+	ND	+	+	+	–	–	–	+	–	–	ND	+	–	+	–	+	–
Elastin	–	+	ND	–	–	–	ND	+	+	–	+	–	ND	–	+	+	–	–	–
Gelatin	+	+	–	+	+	+	–	–	+	+	+	+	–	+	+	+	+	+	+
Hypoxanthine	+	+	ND	+	+	–	–	+	+	+	+	+	ND	+	+	+	–	+	+
Starch	+	–	ND	–	+	+	–	+	–	–	–	–	+	–	+	+	+	+	+
Tyrosine	+	+	ND	–	–	–	–	–	–	+	–	+	ND	–	+	+	+	–	–
Xanthine	–	–	ND	+	–	–	–	–	–	–	–	–	ND	–	–	–	–	–	–

Symbols and abbreviations: +, positive; –, negative; ND, not determined; NS, not seen; h, hooks, curled; psp, pseudovesicles; s, spirals of 1 to 2 turns; sp, spirals of 3 to 5 turns; and str, straight.
From Meyer (1989), Chiba et al. (1999), Gyobu and Miyadoh (2001), Stackebrandt et al. (2001), and Quintana et al. (2003).

Table 6. Characteristics distinguishing the type strains of *Streptosporangium* species.

	S. album DSM 43023[T]	S. amethystogenes DSM 43179[T]	S. carneum NRRL 18437[T]	S. fragile DSM 43847[T]	S. longisporum DSM 43180[T]	S. nondiastaticum DSM 43848[T]	S. pseudovulgare DSM 43181[T]	S. roseum DSM 43021[T]	S. subroseum DSM 44662[T]	S. violaceochromogenes DSM 43849[T]	S. viridialbum DSM 43801[T]	S. vulgare DSM 43802[T]
Morphology on ISP 3												
Color of substrate mycelium												
Brown-black	−	−	−	+	−	−	−	−	−	−	−	−
Red-orange	−	−	−	−	+	+	+	+	−	−	−	+
Yellowish-brown to brown	+	+	+	−	−	+	+	+	+	+	+	+
Color of aerial spore mass												
Greenish-gray	−	−	−	−	−	−	−	−	−	−	+	−
Pink	−	+	+	+	+	+	+	+	+	+	−	+
White	+	−	−	−	−	−	−	−	−	−	−	−
Spore vesicle size (μm)												
1–5	−	−	−	−	−	−	−	−	+	−	−	−
6–10	+	+	−	+	+	−	+	+	−	+	+	+
11–20	−	−	−	+	+	+	−	(+)	−	−	−	−
21–30	−	−	−	−	−	+	−	−	−	−	−	−
31–50	−	−	(+)	−	−	−	−	−	−	−	−	−
Sporangiophore size (μm)												
Short (10)	+	+	−	+	+	+	+	+	+	+	+	+
Long (50)	−	−	+	−	−	−	−	−	−	−	−	−
Spore shape												
Spherical to clavate	+	+	+	+	+	+	+	+	+	+	+	+
Rod-like	−	−	−	−	−	−	−	−	−	−	−	−
Soluble pigments												
Other than pale yellow-brown	−	−	−	+	−	−	−	+	−	+	−	−
B vitamins required	+	+	−	−	−	+	+	+	−	−	+	+
Growth at												
42°C	−	−	−	+	−	+	+	−	+	−	−	−
50°C	−	−	−	−	−	−	−	−	−	−	−	−
Biochemical test												
Nitrate reduction	−	+	−	+	(+)	+	+	+	+	+	d	−
Degradation tests												
Gelatin hydrolysis	+	−	−	−	ND	+	+	+	ND	(+)	d	d
Iodinin production	−	+	−	−	−	−	−	−	−	−	−	−
Starch hydrolysis	−	+	−	+	+	−	+	+	−	+	+	+
Sole carbon source utilization (1% w/v)												
Adonitol	+	+	−	−	+	+	+	+	ND	ND	−	+
L(+)-Arabinose	+	ND	−	+	+	+	+	+	+	ND	−	+
D(+)-Galactose	+	−	+	+	−	+	−	−	+	ND	−	+
Glycerol	−	ND	−	−	−	−	+	+	ND	ND	−	+
meso-Inositol	−	+	−	−	−	−	−	−	−	(+)	+	+
D(+)-Mannitol	+	ND	−	+	−	+	+	+	+	ND	+	−
L(+)-Rhamnose	−	+	−	+	−	−	−	−	+	(+)	+	+
D(+)-Turanose	+	ND	ND	+	−	+	+	+	ND	ND	+	+

Symbols and abbreviations: +, positive; (+), weak positive; −, negative; d, doubtful; ND, not determined; and [T], type strain; and DSM, Deutsche Sammlung von Mikroorganismen und Zellkulturen, Mascheroder Weg 1B, D-38124 Braunschweig, Germany.

From Nonomura (1989a), Mertz and Yao (1990), Whitham et al. (1993), and Zhang et al. (2002).

Table 7. Characteristics differentiating validly described taxa classified in the genera *Planobispora* and *Planomonospora*.

	Planobispora		Planomonospora				
Characteristics	*longispora*	*rosea*	*alba*	*parontospora* var. *antibiotica*	*parontospora* var. *parontospora*	*sphaerica*	*venezuelensis*
Morphology							
Color of aerial mycelium	White	Rose	White	Pink	Pink	Pink	Pink
Number of spores in spore vesicle	2	2	1	1	1	1	1
Biochemical tests							
Esculin hydrolysis	−	+	−	−	+	−	−
Nitrate reductase	ND	ND	−	+	+	+	+
Phosphatase	+	+	+	−	+	+	+
Decomposition of							
Gelatin	+	+	+	+	−	+	−
Hypoxanthine	+	−	−	−	−	−	+
Tyrosine	+	+	+	+	−	+	−
Utilization of							
L-Arabinose	+	+	+	+	−	+	−
Cellobiose	+	+	+	+	−	+	−
Citrate	ND	ND	+	+	+	+	−
Dextrin	ND	ND	+	+	−	+	−
Fructose	+	+	+	+	+	+	−
Galactose	−	+	+	+	+	+	−
Glycerol	−	−	+	−	−	−	−
Glycogen	+	+	−	+	−	+	−
Maltose	+	+	+	+	−	+	−
Mannitol	+	+	+	+	−	+	−
Mannose	−	−	+	+	−	+	−
Rhamnose	+	−	+	−	−	+	−
Salicin	−	+	−	+	−	−	−
Succinate	ND	ND	+	+	−	+	−
Starch	+	+	+	+	−	+	−
Sucrose	−	−	+	+	−	+	−
Trehalose	−	−	+	+	−	+	−
Xylose	+	+	−	+	−	+	−
Resistance to 5% NaCl	ND	ND	−	+	−	−	−
Growth at							
15°C	−	−	−	+	−	−	−
45°C	−	−	+	−	−	−	−

Symbols and abbreviation: +, positive; −, negative; and ND, not determined.
From Vobis (1989b) and Mertz (1994).

white aerial mycelium whereas *P. rosea* has a rose-colored substrate mycelium and an aerial mycelium with a light rose tinge.

The type strain of *P. mira*, unlike that of *P. silvatica*, produces acid from lactose, mannitol, mannose and rhamnose and uses glucose, mannitol and xylose as sole carbon sources. Conversely, the *P. silvatica* strain degrades xanthine and uses melibiose and raffinose as sole carbon sources. The two strains can also be distinguished using colonial characteristics.

Planomonospora parontospora and *P. venezuelensis* strains can be distinguished by the morphological arrangement of their spore vesicles (Figs. 8 and 9), different menaquinone profiles (Collins et al., 1984), and the characteristic color of the mycelium (Thiemann, 1974a). *Planomonospora sphaerica* can be distinguished from the other members of the genus by its ability to form large spherical bodies when grown on inorganic salts starch agar (Mertz, 1994). Members of the genera *Acrocarpospora* and *Herbidospora* can also be recognized on morphological grounds (Table 1). The identification of species of *Acrocarpospora* is based on the discontinuous distribution of a few phenotypic properties (Table 7).

Metabolism and Genetics

Little is known about the genetics and metabolism of members of the family Streptosporangiaceae despite their potential importance in exploitable biology, notably in the discovery of novel bioactive compounds. However, vectors capable of stably maintaining large segments of actinomycete DNA in *Escherichia coli* and of

Table 8. Characteristics distinguishing the type strains of *Acrocarpospora* species.

Characteristics	*A. corrugata* DSM 43316[T]	*A. macrocephala* DSM 44705[T]	*A. pleiomorpha* DSM 44706[T]
Nitrate reduction	–	–	+
Starch degradation	–	+	+
Utilization of			
L(+) Arabinose	+	–	–
D(+) Mannitol	–	+	+
D(+) Raffinose	–	+	+
α-L(–) Rhamnose	–	+	+
D(+) Xylose	+	–	–

Symbols and abbreviations: +, positive; –, negative; [T], type strain; and DSM, Deutsche Sammlung von Mikroorganismen und Zellkulturen, Mascheroder Weg 1B, D-38124 Braunschweig, Germany.
From Whitham et al. (1993) and Tamura et al. (2000).

integrating site specifically in the *Streptomyces* genome have been developed to facilitate the manipulation of uncommon actinomycete strains, including streptosporangiaceae and related taxa (Donadio et al., 2002). These vectors, designated "ESAC," an abbreviation for "*E. coli-Streptomyces* artificial chromosome," are suitable for the reconstruction of gene clusters from small segments of the cloned DNA, the preparation of large insert libraries from unusual actinomycete strains, and the construction of environmental libraries. Other examples of heterologous expression of entire gene clusters in model actinomycetes have been reported (Piel et al., 2000; Tang et al., 2000; Kwon et al., 2001).

The gene cluster coding for the biosynthesis of glycopeptide antibiotic A40926 in *Nonomuraea* strain ATCC 39727 has been isolated and characterized by Sosio et al. (2003). This glycopeptide, a member of the teichoplanin family of glycopeptides, is the precursor of dalbavancin, a second generation glycopeptide. Sosio and her colleagues also isolated the novel compound, dechloromannosyl-A40926 aglycone, following the construction of a *Nonomuraea* mutation by deleting *dbr* open reading frames 8–10. Prauser (1984) reported that attempts to isolate phage from Streptosporangiaceae strains had been unsuccessful.

Applications

Members of the family Streptosporangiaceae are expected to be an increasingly rich source of commercial products, notably antibiotics and enzymes. "*Microtetraspora tyrrkensis*" produces fluvirucins active against influenza A virus (Tomita et al., 1991); *Microbispora rosea*, deoxycephalomycin B (Okazaki and Naito, 1985); *Microbispora* strain SCC 1438, a novel fungal antibiotic (Patel et al., 1988); *N. roseoviolacea*,

carminomicins (Nakagawa et al., 1983; Nakagawa et al., 1989); *N. rubra*, maduromycin (Fleck et al., 1978), *N. pusilla*, actinotiocin (Tamura et al., 1973), and *N. spiralis*, pyralomicin (Naganawa et al., 2002). Similarly, *Streptosporangium albidum* produces aculeximycin (Murata et al., 1989); *S. pseudovulgare*, sporamycin (Komiyama et al., 1977); *S. roseum*, maytansin-type ansamacrolactam (Hacene et al., 1998), sporangirosamycin (Gazhal and Abl El-Aziz, 1993), and thiosporamycin (Celmer et al., 1978); *S. violaceochromogenes*, platomycins A and B (Takasawa et al., 1975) and victomycin (Kawamoto et al., 1975), and *S. vulgare*, sporacuaracins A and B (Atsushi et al., 1978). A novel anthracycline antibiotic has been isolated from *S. fragile* (Shearer et al., 1983), an antitumor antibiotic from an organism resembling *S. pseudovulgare* (Umezawa et al., 1976) and unspecified antimicrobial agents from *Microbispora indica* and *M. karnatakensis* (Rao et al., 1987).

Microbispora rosea is an excellent source of D-xylose (glucose) isomerase (Crueger and Crueger, 1982), which converts D-glucose into D-fructose. The enzyme is used to produce D-fructose on a commercial scale, and its biosynthesis, purification, and immobilization, as well as its application for the production of high fructose syrup, have been the subject of many reports and patents (Crueger and Crueger, 1984). Similarly, cystathionine γ-lyase has been detected in strains of *Streptosporangium* (Nagasawa et al., 1984). This enzyme has been shown to catalyse the α, γ-elimination reaction of L-cystathionine and also the γ-replacement of L-homoserine in the presence of various thiol compounds (Kanzaki et al., 1986a). An efficient method based on the reaction of γ-replacement has been developed (Kanzaki et al., 1986b) for the preparation of L-cystathionine, a product that may be useful because a deficiency of this compound has been observed in the brains of homocystinuric patients (Gerritsen and Waisman, 1964). The procedure allowed the total conversion of *O*-succinyl-L-homoserine and L-cysteine into L-cystathionine. *Microbispora rosea* also produces exoxylanases (Kusakabe et al., 1969).

Thermophilic microbisporae synthesize a wide range of enzymes that are involved in the degradation and modification of heteropolysaccharides, notably celluloses, lignocelloloses, and hemicelluloses (Henssen and Schnepf, 1967; McCarthy, 1987; Crawford, 1988). These enzyme systems have the potential for novel application in biotechnological processes, particularly for the enzymatic generation of fermentable sugar from agricultural residues (McCarthy et al., 1988; Zimmermann, 1989).

Literature Cited

Athalye, M., M. Goodfellow, J. Lacey, and R. P. White. 1985. Numerical classification of Actinomadura and Nocardiopsis. Int. J. Syst. Bacteriol. 35:86–98.

Atsushi, T., F. Rizuji, and K. Hirotada. 1975. Antibiotic Sporocuracin Production. Japan Patent, 75,125,094.

Bland, C. E., and J. N. Couch. 1981. The family Actinoplanaceae. *In:* M. P. Starr, H. Stolp, H. G. Trüper, A. Balows, and H. G. Schlegel (Eds.) The Prokaryotes. Springer-Verlag. Berlin, Germany. 2:2004–2010.

Celmer, W. D., W. P. Cullen, C. E. Moppett, J. B. Routien, P. C. Watts, R. Shibakawa, and J. Tone. 1978. Polypeptide Antibiotic Produced by New Subspecies of Streptosporangium. US Patent 4,083,963.

Chaves Batista, A., S. K. Shome, and J. Americo de Lima. 1963. Streptosporangium bovinum sp. nov. from cattle hoofs. Dermat. Trop. 2:49–54.

Chiba, S., M. Suzuki, and K. Ando. 1999. Taxonomic re-evaluation of "Nocardiopsis" sp. K-252^T (=NRRL 15532^T): A proposal to transfer this strain to the genus Nonomuraea as Nonomuraea longicatena sp. nov. Int. J. Syst. Bacteriol. 49:1623–1630.

Collins, M. D., T. Pirouz, M. Goodfellow, and D. E. Minnikin. 1977. Distribution of menaquinones in actinomycetes and corynebacteria. J. Gen. Microbiol. 100:221–230.

Collins, M. D., M. Faulkner, and R. M. Keddie. 1984. Menaquinone composition of some sporeforming actinomycetes. Syst. Appl. Microbiol. 5:20–29.

Couch, J. N. 1954. The genus Actinoplanes and its relatives. Trans. NY Acad. Sci. 16:315–318.

Couch, J. N. 1955a. A new genus and family of the Actinomycetales with a revision of the genus Actinoplanes. J. Elisha Mitchell Sci. Soc. 71:148–155.

Couch, J. N. 1955b. Actinosporangiaceae should be Actinoplanaceae. J. Elisha Mitchell Sci. Soc. 71:269.

Couch, J. N. 1963. Some new genera and species of the Actinoplanaceae. J. Elisha Mitchell Sci. Soc. 79:53–70.

Couch, J. N., and C. E. Bland. 1974. The Actinoplanaceae. *In:* R. E. Buchanan and N. E. Gibbons (Eds.) Bergey's Manual of Determinative Bacteriology, 8th ed. Williams and Wilkins. Baltimore, MD. 706–723.

Crawford, D. L. 1988. Biodegradation of agricultural and urban wastes. *In:* M. Goodfellow, S. T. Williams, and M. Mordarski (Eds.) Actinomycetes in Biotechnology. Academic Press. London, UK. 433–459.

Cross, T. 1970. The diversity of bacterial spores. J. Appl. Bacteriol. 33:95–102.

Crueger, W., and A. Crueger. 1982. Glukose isomerasen. Lehrbuch der Angewandten Mikrobiologie. Akademische Verlagsgesellschaft Weisbaden. Weisbaden, Germany. 166–184.

Crueger, A., and W. Crueger. 1984. Carbohydrates. *In:* H. J. Rehm and D. Reed (Eds.) Biotechnology. Verlag Chemie Weinheim. Weinheim, Germany. 6a:421–457.

Donadio, S., P. Monciardini, R. Alduina, P. Mazza, C. Chiocchini, L. Cavaletti, M. Sosio, and A. M. Puglia. 2002. Microbial technologies for the discovery of novel bioactive metabolites. J. Bacteriol. 99:187–198.

Farina, G., and S. G. Bradley. 1970. Reassociation of deoxyribonucleic acids from Actinoplanes and other actinomycetes. J. Bacteriol. 102:30–35.

Fischer, A., R. M. Kroppenstedt, and E. Stackebrandt. 1983. Molecular-genetic and chemotaxonomic studies on Actinomadura and Nocardiopsis. J. Gen. Microbiol. 129:3433–3446.

Fleck, W. F., D. G. Strauss, J. Meyer, and G. Porstendorfer. 1978. Fermentation, isolation, and biological activity of maduramycin: A new antibiotic from Actinomadura rubra. Z. Allg. Mikrobiol. 18:389–398.

Fumarai, T., H. Ogawa, and T. Okuda. 1968. Taxonomic study on Streptosporangium albidum sp. nov. J. Antibiot. (Tokyo) 21:179–181.

Galatenko, O. A., L. P. Terekhova, and T. P. Preobrazhenskaya. 1981. New Actinomadura species isolated from Turkmen soil samples and their antagonistic properties [in Russian]. Antibiotiki 26:803–807.

Gauze, G. F., T. P. Preobrazhenskaya, E. S. Kudrina, N. O. Blinov, I. D. Ryabova, and M. A. Sveshnikova. 1957. Problems in the Classification of Antagonistic Actinomycetes. Moscow State Publishing House for Medical Literature Medgiz. Moscow, Russia.

Gauze, G. F., M. A. Sveshnikova, R. S. Ukholina, D. V. Gaurilina, V. A. Filicheva, and K. G. Gladkikh. 1973. Production of antitumor antibiotic carminomycin by Actinomadura carminata sp. nov. Antibiotiki 18:675–678.

Gauze, G. F., L. P. Terekhova, O. A. Galatenko, T. P. Preobrazhenskaya, V. N. Borisova, and G. B. Federova. 1984. Actinomadura recticatena sp. nov., a new species and its antibiotic properties [in Russian]. Antibiotiki 29:3–7.

Gazal, S. A., and Z. K. Alb El-Aziz. 1993. Sporangiosomycin, a new chromopeptide antibiotic produced by Streptosporangium roseum subsp. antibioticus subsp. nova. Al-Azhar Bull. Sci. 4:265–274.

Gerber, N. N., and M. P. Lechevalier. 1964. Phenazones and phenoxazinones from Waksmania aerata sp. nov. and Pseudomonas iodina. Biochemistry 3:598–602.

Gerber, N. N., and M. P. Lechevalier. 1965. 1-6-phenazinediol-5-oxide from microorganisms. Biochemistry 4:176–180.

Gerritsen, T., and H. A. Waisman. 1964. Homocystonuria: Absence of cystathionine in the brain. Science 145:588.

Goodfellow, M., G. Alderson, and J. Lacey. 1979. Numerical taxonomy of Actinomadura and related actinomycetes. J. Gen. Microbiol. 112:95–111.

Goodfellow, M., and T. Pirouz. 1982. Numerical classification of sporoactinomycetes containing meso-diaminopimelic acid in the cell wall. J. Gen Microbiol. 128:503–527.

Goodfellow, M., and T. Cross. 1984. Classification. *In:* M. Goodfellow, M. Mordarski, and S. T. Williams (Eds.) The

Biology of the Actinomycetes. Academic Press. London, UK. 7–164.

Goodfellow, M., S. T. Williams, and G. Alderson. 1986. Transfer of Kitasatoa purpurea Matsumae and Hata to the genus Streptomyces as Streptomyces purpureus comb. nov. Syst. Appl. Microbiol. 8:65–66.

Goodfellow, M., E. Stackebrandt, and R. M. Kroppenstedt. 1988. Chemotaxonomy and actinomycete systematics. *In:* Y. Okami, T. Beppu, and H. Ogawara (Eds.) Biology of Actinomycetes. Japan Scientific Societies Press. Tokyo, Japan. 233–238.

Goodfellow, M. 1989a. Maduromycetes. *In:* S. T. Williams, M. E. Sharpe, and J. G. Holt (Eds.) Bergey's Manual of Systematic Bacteriology. Williams and Wilkins. Baltimore, MD. 4:2509–2510.

Goodfellow, M. 1989b. Suprageneric classification of actinomycetes. *In:* S. T. Williams, M. E. Sharpe, and J. G. Holt (Eds.) Bergey's Manual of Systematic Bacteriology. Williams and Wilkins. Baltimore, MD. 4:2333–2339.

Goodfellow, M., L. J. Stanton, K. E. Simpson, and D. E. Minnikin. 1990. Numerical and chemical classification of Actinoplanes and some related actinomycetes. J. Gen. Microbiol. 136:19–34.

Gyobu, Y., and S. Miyadoh. 2001. Proposal to transfer Actinomadura carminata to a new subspecies of the genus Nonomuraea subsp. carminata comb. nov. Int. J. Syst. Bacteriol. 51:881–889.

Hacene, H., F. Boudjellal, and G. Lefebvre. 1998. AH7, a non polyenic antifungal antibiotic produced by a new strain of Streptosporangium roseum. Microbios 96:103–109.

Hancock, I. C. 1994. Analysis of cell wall constituents of Gram-positive bacteria. *In:* M. Goodfellow and A. G. O'Donnell (Eds.) Chemical Methods in Prokaryotic Systematics. John Wiley. Chichester, UK. 63–84.

Hasegawa, T., M. P. Lechevalier, and H. A. Lechevalier. 1979. Phospholipid composition of motile actinomycetes. J. Gen. Appl. Microbiol. 25:209–213.

Hayakawa, M., and H. Nonomura. 1984. HV agar, a new selective medium for isolation of soil actinomycetes. *In:* Abstracts of Papers Presented at the Annual Meeting of the Actinomycetologists, Osaka, Japan. 6.

Hayakawa, M., and H. Nonomura. 1987a. Efficacy of artificial humic acid as a selective nutrient in HV agar used for the isolation of soil actinomycetes. J. Ferment. Technol. 65:609–616.

Hayakawa, M., and H. Nonomura. 1987b. Humic acid-vitamin agar, a new medium for the selective isolation of soil actinomycetes. J. Ferment. Technol. 65:501–509.

Hayakawa, M., K. Ishizawa, and H. Nonomura. 1988. Distribution of rare actinomycetes in Japanese soils. J. Ferment. Technol. 66:367–373.

Hayakawa, M., and H. Nonomura. 1989. A new method for the intensive selective isolation of actinomycetes from soil. Actinomycetologica 3:95–104.

Hayakawa, M., T. Kajiura, and H. Nonomura. 1991. New methods for the highly selective isolation of Streptosporangium and Dactylosporangium from soil. J. Ferment. Bioengin. 72:327–333.

Henssen, A. 1957. Beiträge zur Morphologie und systematik der thermophiler Actinomyceten. Arch. Microbiol. 26:374–414.

Henssen, A., and E. Schnepf. 1967. Zur Kenntnis thermophiler Actinomyceten. Arch. Mikrobiol. 57:214–231.

Henssen, A., and D. Schäfer. 1971. Emended description of the genus Pseudonocardia Henssen and description of a new species Pseudonocardia spinosa Schäfer. Int. J. Syst. Bacteriol. 21:29–34.

Hickey, R. J., and H. D. Tresner. 1952. A cobalt-containing medium for sporulation of Streptomyces species. J. Bacteriol. 64:891–892.

Johnstone, D. W., and T. Cross. 1976. The occurrence and distribution of actinomycetes in lakes of the English Lake District. Freshwater Biol. 6:457–463.

Jones, K. L. 1949. Fresh isolates of actinomycetes in which the presence of sporangous aerial mycelia is a fluctuating characteristic. J. Bacteriol. 57:141–145.

Kanzaki, H., M. Kobayashi, T. Nagasawa, and H. Yamada. 1986a. Synthesis of S-substituted L-homocysteine derivatives by cystathionine γ-lyase of Streptomyces phaeochromogenes. Agric. Biol. Chem. 50:391–397.

Kanzaki, H., T. Nagasawa, and H. Yamada. 1986b. Highly efficient production of L-cystathionine from O-succinyl-L-homoserine and L-cysteine by Streptomyces cystathionine γ-lyase. Appl. Microbiol. Biotechnol. 25:97–100.

Kawamoto, I., S. Takasawa, R. Okachi, M. Kohakura, I. Takahashi, and T. Nara. 1975. A new antibiotic victomycin (XK 49-1-B-2). I: Taxonomy and production of the producing organisms. J. Antibiot. 28:358–365.

Kemmerling, C., H. Gürtler, R. M. Kroppenstedt, R. Toalster, and E. Stackebrandt. 1993. Evidence for the phylogenetic heterogeneity of the genus Streptosporangium. Syst. Appl. Microbiol. 16:369–372.

Kim, B. 1999. Polyphasic Taxonomy of Thermophilic Actinomycetes [PhD thesis]. Department of Agricultural and Environmental Science, University of Newcastle. Newcastle upon Tyne, UK.

Kizuka, M., R. Enolata, K. Takahashi, and T. Okazaki. 1997. Distribution of the actinomycetes in the Republic of South Africa investigated using a newly developed isolation method. Actinomycetologica 11:54–58.

Komiyama, K., K. Sugimoto, H. Takeshima, and I. Umezawa. 1977. A new antitumour antibiotic, sporamycin. J. Antibiot. 30:202–208.

Krassil'nikov, N. A. 1938. Ray Fungi and Related Organisms, Actinomycetales. Izdatel'stvo Akademii Nauk SSSR Moscow. Moscow, Russia.

Kroppenstedt, R. M., and H. J. Kutzner. 1978. Biochemical taxonomy of some problem actinomycetes. Zbl. Bacteriol. Parasitenkd. Infektionskr. Hyg. Abt. 1., Suppl. 6: 125–133.

Kroppenstedt, R. M. 1982. Separation of bacterial menaquinones by HPLC using reverse phase (RP18) and a silver loaded ion exchanger as stationary phases. J. Liquid Chromat. 5:2359–2367.

Kroppenstedt, R. M. 1985. Fatty acid and menaquinone analysis of actinomycetes and related organisms. *In:* M. Goodfellow and D. E. Minnikin (Eds.) Chemical Methods in Bacterial Systematics. Academic Press. London, UK. 173–199.

Kroppenstedt, R. M., E. Stackebrandt, and M. Goodfellow. 1990. Taxonomic revision of the actinomycete genera Actinomadura and Microtetraspora. Syst. Appl. Microbiol. 13:148–160.

Kudo, T., T. Itoh, S. Miyadoh, T. Shomura, and A. Seino. 1993. Herbidospora gen. nov., a new genus of the family Streptosporangiaceae Goodfellow et al. 1990. Int. J. Syst. Bacteriol. 43:319–328.

Kudo, T. 2001. Family Streptosporangiaceae. *In:* S. Miyadoh, H. Hamada, K. Hotta, T. Kudo, A. Seino, G. Vobis, and A. Yakota (Eds.). Tokyo Business Center for Academic Societies Japan. Tokyo, Japan. 259–276.

Kusakabe, I., T. Yasui, and T. Kobayashi. 1969. Some properties of extracellular xylanase from Streptomyces. J. Agric. Chem. Soc. Japan. 43:145–153.

Küster, E. 1959. Outline of a comparative study of criteria used in characterization of the actinomycetes. Int. Bull. Bacteriol. Nomencl. Taxon. 9:98–104.

Kwon, H. J., W. C. Smith, L. Xiang, and B. Shan. 2001. Cloning and heterologous expression of the macrotetrolide biosynthetic gene cluster revealed a novel polyketide synthase that lacks an acyl carrier protein. J. Am. Chem. Soc. 123:3385–3386.

Lavrova, N. V., T. P. Preobrazhenskaya, and M. A. Sveshnikova. 1972. Isolation of soil actinomycetes on selective media with rubomycin [in Russian]. Antibiotiki 17:965–970.

Lavrova, N. V., and T. P. Preobrazhenskaya. 1975. Isolation of new species of the genus Actinomadura on selective media with rubromycin [in Russian]. Antibiotiki 20:438–448.

Lechevalier, M. P., and H. A. Lechevalier. 1957. A new genus of the Actinomycetales: Waksmania gen. nov. J. Gen. Microbiol. 17:104–111.

Lechevalier, H. A. 1965. Priority of the generic name Microbispora over Waksmania and Thermopolyspora. Int. Bull. Bacteriol. Nomencl. Taxon. 15:139–142.

Lechevalier, H. A., M. P. Lechevalier, and P. E. Holbert. 1966. Electron microscopic observation of the sporangial structure of strains of Actinoplanaceae. J. Bacteriol. 92:1228–1235.

Lechevalier, M. P., and N. N. Gerber. 1970a. The identity of 3-O-methyl-D-galactose with madurose. Carbohydr. Res. 13:451–454.

Lechevalier, H. A., and M. P. Lechevalier. 1970b. A critical evaluation of the genera of aerobic actinomycetes. In: H. Prauser (Ed.) The Actinomycetales. Gustav Fischer Verlag. Jena, Germany. 393–405.

Lechevalier, M. P., and H. A. Lechevalier. 1970c. Chemical composition as a criterion in the classification of aerobic actinomycetes. Int. J. Syst. Bacteriol. 20:435–443.

Lechevalier, M. P., C. De Biévre, and H. A. Lechevalier. 1977. Chemotaxonomy of aerobic actinomycetes: phospholipid composition. Biochem. Syst. Ecol. 5:249–260.

Lechevalier, M. P., and H. A. Lechevalier. 1980. The chemotaxonomy of actinomycetes. In: A. Dietz and D. Thayer (Eds.) Actinomycete Taxonomy. Society for Industrial Microbiology. Arlington, VA. Special Publication 6:227–291.

Lechevalier H. A., and M. P. Lechevalier. 1981a. Introduction to the order Actinomycetales. In: M. P. Starr, H. Stolp, H. G. Trüper, A. Balows, and H. G. Schlegel (Eds.) The Prokaryotes. Springer-Verlag. Berlin, Germany. 2:1915–1922.

Lechevalier, M. P., A. E. Stern, and H. A. Lechevalier. 1981b. Phospholipids in the taxonomy of actinomycetes. Zbl. Bakteriol. Suppl. 11:111–116.

Louria, D. B., and R. E. Gordon. 1960. Pericarditis and pleuritis caused by a recently discovered microorganism, Waksmania rosea. Am. Rev. Respir. Dis. 81:83–88.

McCarthy, A., and T. Cross. 1984. A taxonomic study of Thermomonospora and other monosporic actinomycetes. J. Gen. Microbiol. 130:5–25.

McCarthy, A. 1987. Lignocellulose-degrading actinomycetes. FEMS Microbiol. Rev. 46:145–163.

McCarthy, A. J., A. S. Ball, and S. L. Bachmann. 1988. Ecological and biotechnological implications of lignocellulose degradation by actinomycetes. In: Y. Okami, T.

Beppu, and H. Ogawara (Eds.). Japan Scientific Societies Press. Tokyo, Japan. 283–287.

Mertz, F. P., and R. C. Yao. 1990. Streptosporangium carneum sp. nov., isolated from soil. Int. J. Syst. Bacteriol. 40:247–253.

Mertz, F. P. 1994. Planomonospora alba sp. nov. and Planomonospora sphaerica sp. nov., two new species isolated from soil by baiting techniques. Int. J. Syst. Bacteriol. 44:274–281.

Meyer, J. 1979. New species of the genus Actinomadura. Z. Allgem. Mikrobiol. 19:37–44.

Meyer, J. 1981. Validation of the publication of new names and new combinations previously affectively published outside the IJSB List No. 6. Int J. Syst. Bacteriol. 31:215–218.

Meyer, J. 1989. Genus Actinomadura Lechevalier and Lechevalier 1970, 400[AL]. In: S. T. Williams, M. E. Sharpe, and J. G. Holt (Eds.) Bergey's Manual of Systematic Bacteriology. Williams and Wilkins. Baltimore, MD. 4:2511–2526.

Minnikin, D. E., A. G. O'Donnell, M. Goodfellow, G. Alderson, M. Athalye, A. Schaal, and J. H. Parlett. 1984. An integrated procedure for the extraction of bacterial isoprenoid quinones and polar lipids. J. Microbiol. Meth. 2:233–241.

Miyadoh S., H. Tohyama, S. Amano, T. Shomura, and T. Niida. 1985. Microbispora viridis, a new species of Actinomycetales. Int. J. Syst. Bacteriol. 57:342–346.

Miyadoh, S., H. Anzai, S. Amano, and T. Shomura. 1989. Actinomadura malachitica and Microtetraspora viridis are synonyms and should be transferred as Actinomadura viridis comb. nov. Int. J. Syst. Bacteriol. 39:152–158.

Miyadoh, S., S. Amano, H. Tohyama, and T. Shomura. 1990. A taxonomic review of the genus Microbispora and a proposal to transfer two species to the genus Actinomadura and to combine ten species into Microbispora rosea. J. Gen. Microbiol. 136:1905–1913.

Monciardini, P., M. Sosio, L. Cavaletti, C. Chiocchini, and S. Donadio. 2002. New PCR primers for the selective amplification of 16S rDNA from different groups of actinomycetes. FEMS Microbiol. Ecol. 42:419–429.

Murata, H., N. Kojima, K.-I. Harada, M. Suzuki, T. Ikemoto, T. Shibuya, T. Haneishi, and A. Torikata. 1989. Structural elucidation of aculescimycin. I: Further purification and glycosidic bond cleavage of aculescimycin. J. Antibiot. 42:691–700.

Naganawa, H., H. Hashizume, Y. Kubota, R. Sawa, Y. Takahashi, K. Arakawa, S. G. Bowers, and T. Mahmud. 2002. Biosynthesis of the aminocyclitol moeity ofpyralomicin 1a in Nonomuraea spiralis MI178-34F18. J. Antibiot. 55:578–584.

Nagasawa, T., H. Kanzaki, and N. Yamada. 1984. Cystathionine γ-lyase of Streptomyces phaeochromogenes—the occurrence of cystathionine γ-lyase in filamentous bacteria and its purification and characterisation. J. Biol. Chem. 259:10393–10403.

Nakagawa, M., Y. Hayakawa, H. Kawai, K. Imamura, H. Inoue, A. Shimazu, H. Seto, and N. Otake. 1983. A new anthracycline antibiotic N-formyl-13-dehydrocarminomycin. J. Antibiot. 36:457–458.

Nakagawa, M., Y. Hayakawa, K. Imamura, M. Seto, and N. Otake. 1989. Microbial conversion of anthracyclinones to carminomycins by a blocked mutant of Actinomadura roseoviolacea. J. Antibiot. 42:1698–1703.

Nakajima, Y., V. Kitpreechavanich, K.-I. Suzuki, and T. Kudo. 1999. Microbispora corallina sp. nov., a new species of the genus Microbispora isolated from Thai soil. Int. J. Syst. Bacteriol. 49:1761–1767.

Nakajima, Y., C. C. Ho, and T. Kudo. 2003. Microtetraspora malaysiensis sp. nov., isolated from Malaysian primary dipterocarp forest soil. J. Gen. Appl. Microbiol. 49:181–189.

Naumova, I. B., N. V. Potekhina, L. P. Terekhova, T. P. Preobrazhenskaya, and K. Digimbay. 1986. Wall polyol phosphate polymers of bacteria belonging to the genus Actinomadura. In: G. Szabó, S. Biró, and M. Goodfellow (Eds.) Biological, Biochemical and Biomedical Aspects of Actinomycetes. Akadémiai Kiadó Budapest,. Budapest, Hungary. 561–566.

Nonomura, H., and Y. Ohara. 1957. Distribution of actinomycetes in the soil. II: Microbispora, a new genus of the Streptomycetaceae. J. Ferment. Technol. 35:307–311.

Nonomura, H., and Y. Ohara. 1960. Distribution of actinomycetes in soil. IV: The isolation and classification of the genus Microbispora. J. Ferment. Technol. 38:401–405.

Nonomura, H., and Y. Ohara. 1969a. Distribution of actinomycetes in soil. VI: A culture method effective for both preferential isolation and enumeration of Microbispora and Streptosporangium strains in soil (Part 1). J. Ferment. Technol. 47:463–469.

Nonomura, H., and Y. Ohara. 1969b. Distribution of actinomycetes in soil. VII: A culture method effective for both preferential isolation and enumeration of Microbispora and Streptosporangium strains in soil (Part 2): Classification of isolates. J. Ferment. Technol. 47:701–709.

Nonomura, H., and Y. Ohara. 1971a. Distribution of actinomycetes in soil. VIII: Green-spore group of Microtetraspora, its preferential isolation and taxonomic characteristics. J. Ferment. Technol. 49:1–7.

Nonomura, H., and Y. Ohara. 1971b. Distribution of actinomycetes in soil. IX: New species of the genera Microbispora and Microtetraspora, and their isolation method. J. Ferment. Technol. 49:887–894.

Nonomura, H., and Y. Ohara. 1971c. Distribution of actinomycetes in soil. XI: Some new species of the genus Actinomadura Lechevalier et al. J. Ferment. Technol. 49:904–912.

Nonomura, H. 1984. Design of a new medium for isolation of soil actinomycetes. The Actinomycetes 18:206–209.

Nonomura, H., and M. Hayakawa. 1988. New methods for the selective isolation of soil actinomycetes. In: Y. Okami, T. Beppu, and H. Ogawara (Eds.) Biology of Actinomycetes. Japan Scientific Societies Press. Tokyo, Japan. 288–293.

Nonomura, H. 1989a. Genus Microbispora Nonomura and Ohara 1957, 307[AL]. In: S. T. Williams, M. E. Sharpe, and J. G. Holt (Eds.) Bergey's Manual of Systematic Bacteriology. Williams and Wilkins. Baltimore, MD. 4:2526–2531.

Nonomura, H. 1989b. Genus Microtetraspora Thiemann, Pagani and Beretta 1968, 296[AL]. In: S. T. Williams, M. E. Sharpe, and J. G. Holt (Eds.) Bergey's Manual of Systematic Bacteriology. Williams and Wilkins. Baltimore, MD. 4:2531–2536.

Nonomura, H. 1989c. Genus Streptosporangium Couch 1955, 148[AL]. In: S. T. Williams, M. E. Sharpe, and J. G. Holt (Eds.) Bergey's Manual of Systematic Bacteriology. Williams and Wilkins. Baltimore, MD. 4:2545–2551.

Ochi, K., S. Miyadoh, and T. Tamura. 1991. Polyacrylamide gel electrophoresis analysis of ribosomal protein AT-L30 as a novel approach to actinomycete taxonomy: Application to the genera Actinomadura and Microtetraspora. Int. J. Syst. Bacteriol. 41:234–239.

Ochi, K., and S. Miyadoh. 1992. Polyacrylamide gel electrophoresis analysis of ribosomal protein AT-L30 from an actinomycetes genus, Streptosporangium. Int. J. Syst. Bacteriol. 42:151–155.

Ochi, K., K. Haraguchi, and S. Miyadoh. 1993. A taxonomic review of the genus Microbispora by analysis of ribosomal protein AT-L30. Int. J. Syst. Bacteriol. 46:658–663.

Okazaki, T., and A. Naito. 1985. Studies in actinomycetes isolated from Australian soils. In: G. Szabó, S. Biró, and M. Goodfellow (Eds.) Biological, Biochemical and Biomedical Aspects of Actinomycetes. Akadémiai Kiadó. Budapest, Hungary. 739–741.

Okuda, T., Y. Itoh, T. Yamaguichi, T. Furumai, M. Suzuki, and M. Tsuruoka. 1966. Sporaviridin, a new antibiotic produced by Streptosporangium viridogriseum nov. sp. J. Antibiot., Ser A 19:85–87.

Patel, M., M. Conover, A. Horan, D. Loebenberg, J. Marquez, R. Mierzwa, M. S. Puar, R. Yarborough, and J. A. Waitz. 1988. Sch 31828, a novel antibiotic from a Microbispora sp.: taxonomy, fermentation, isolation and biological properties. J. Antibiot. 41:794–797.

Petrolini, B., S. Quaroni., P. Sardi, M. Saracchi, and N. Anterrollo. 1992. A sporangiate actinomycete with unusual morphological features: Streptosporangium claviforme sp. nov. Actinomycetes 3:45–50.

Piel, J., K. Hoong, and B. S. Moore. 2000. Metabolic diversity encoded by the enterocin biosynthesis gene cluster. J. Am. Chem. Soc. 122:5415–5416.

Poschner, J., R. M. Kroppenstedt, A. Fischer, and E. Stackebrandt. 1985. DNA : DNA reassociation and chemotaxonomic studies on Actinomadura, Microbispora, Microtetraspora, Micropolyspora and Nocardiopsis. Syst. Appl. Microbiol. 6:264–270.

Potekhina, L. L. 1965. Streptosporangium rubrum n. sp.—a new species of the Streptosporangium genus [in Russian]. Mikrobiologiya 34:292–299.

Prauser, H. 1984. Phage host ranges in the classification and identification of Gram-positive branched and related bacteria. In: L. Ortiz-Ortiz, L. F. Bojalil, and V. Yakoleff (Eds.) Biological, Biochemical and Biomedical Aspects of Actinomycetes. Academic Press. Orlando, FL. 617–633.

Preobrazhenskaya, T. P., and M. A. Sveshnikova. 1974. New species of the genus Actinomadura [in Russian]. Microbiologiya 43:864–868.

Preobrazhenskaya, T. P., N. V. Lavrova, R. S. Ukholina, and N. P. Nechaeva. 1975. Isolation of new species of Actinomadura on selective media with streptomycin and bruneomycin [in Russian]. Antibiotiki 20:404–409.

Pridham, T. G., P. Anderson, G. Foley, L. A. Lindenfelser, C. W. Hesseltine, and R. G. Benedict. 1956–57. A selection of media for maintenance and taxonomic study of streptomycetes. Antibiotics Ann. 1956/57:947–953.

Quintana, E., L. Maldonado, and M. Goodfellow. 2003. Nonomuraea terrinata sp. nov., a novel soil actinomycete. Ant. v. Leeuwenhoek 84:1–6.

Rao, V. A., K. K. Prabhu, B. P. Sridhar, A. Venkateswarlu, and P. Actor. 1987. Two new species of Microbispora from Indian soils: Microbispora karnatakensis sp. nov. and Microbispora indica sp. nov. Int. J. Syst. Bacteriol. 37:181–185.

Runmao, H., W. Guizhen, and L. Junying. 1993. A new genus of actinomycetes, Planotetraspora gen. nov. Int. J. Syst. Bacteriol. 43:468–470.

Schäfer, D. 1973. Beitrage zur Klassifizerung and Taxonomie der Actinoplanaceen [PhD dissertation]. University of Marburg/Lahn. Marburg, Germany.

Schleifer, K.-H., and O. Kandler. 1972. Peptidoglycan types of bacterial cell walls and their taxonomic implications. Bacteriol. Rev. 36:407–477.

Sharples, G. P., S. T. Williams, and R. M. Bradshaw. 1974. Spore formation in the Actinoplanaceae (Actinomycetales). Arch. Microbiol. 101:9–20.

Shearer, M. C., P. M. Colman, and C. H. Nash. 1983. Streptosporangium fragile sp. nov. Int. J. Syst. Bacteriol. 33:364–368.

Shirling, E. B., and D. Gottlieb. 1966. Methods for characterization of Streptomyces species. Int. J. Syst. Bacteriol. 16:313–340.

Soina, U. S., A. A. Sokolov, and N. S. Agre. 1975. Ultrastructure of mycelium and spores of Actinomadura fastidiosa sp. nov. [in Russian]. Microbiologiya 44:883–887.

Sosio, M., S. Stinchi, F. Beltrametti, A. Lazzarini, and S. Donadio. 2003. The gene cluster for the biosynthesis of the glycopeptide antibiotic A40926 by Nonomuraea species. Chem. Biol. 10:541–549.

Stach, J. E. M., L. A. Maldonado, A. C. Ward, M. Goodfellow, and A. T. Bull. 2003. New primers for the class Actinobacteria: application to marine and terrestrial environments. Environ. Microbiol. 5:828–841.

Stackebrandt, E. 1986. The significance of "wall types" in phylogenetically based taxonomic studies on actinomycetes. In: G. Szabó, S. Biró, and M. Goodfellow (Eds.) Biological, Biochemical and Biomedical Aspects of Actinomycetes. Akadémiai Kaidó. Budapest, Hungary. 497–506.

Stackebrandt, E., and R. M. Kroppenstedt. 1987. Union of the genera Actinoplanes Couch, Ampurallariella Couch, and Amorphosporangium Couch in a redefined genus Actinoplanes. Syst. Appl. Microbiol. 9:110–114.

Stackebrandt, E., R. M. Kroppenstedt, K. D. Jahnke, C. Kemmering, and H. Gürtler. 1994. Transfer of Streptosporangium viridogriseum (Okuda et al., 1966), Streptosporangium subsp. kofuense (Nonomura & Ohara, 1969), Streptosporangium albidum (Furumai et al., 1968) to Kutzneria gen. nov. as Kustneria viridogrisea comb. nov., Kutneria kofuensis com. nov., and Kutzneria albida comb. nov., and emendation of the genus Streptosporangium. Int. J. Syst. Bacteriol. 43:254–269.

Stackebrandt, E., F. A. Rainey, and N. L. Ward-Rainey. 1997. Proposal for a new hierarchic classification system, Actinobacteria classis nov. Int. J. Syst. Bacteriol. 47:479–491.

Stackebrandt, E., J. Wink, U. Steiner, and R. M. Kroppenstedt. 2001. Nonomuraea dietzii sp. nov. Int J. Syst. Evol. Microbiol. 51:1437–1441.

Staneck, J. L., and G. D. Roberts. 1974. Simplified approach to identification of aerobic actinomycetes by thin-layer chromatography. Appl. Microbiol. 28:226–231.

Suzuki, K., and K. Komagata. 1983. Taxonomic significance of cellular fatty acid composition in some coryneform bacteria. Int. J. Syst. Bacteriol. 33:188–200.

Suzuki, K., M. Goodfellow, and A. G. O'Donnell. 1993. Cell envelopes and classification. In: M. Goodfellow and A. G. O'Donnell (Eds.) Handbook of New Bacterial Systematics. Academic Press. London, UK. 195–250.

Suzuki, S.-I., T. Okuda, and S. Komatsubara. 2001a. Selective isolation and distribution of the genus Planomonospora in soils. Can. J. Microbiol. 47:253–263.

Suzuki, S.-I., T. Okuda, and S. Komatsubara. 2001b. Selective isolation and study of the global distribution of the genus Planobispora in soils. Can. J. Microbiol. 47:979–986.

Sveshnikova, M., T. Maxinova, and E. Kudrina. 1969. The species belonging to the genus Micromonospora Oerskov 1923, and their taxonomy [in Russian]. Microbiologiya 38:883–893.

Takasawa, S., I. Kawamoto, I. Takahashi, M. Kohakura, R. Okachi, S. Sata, M. Yamamoto, and T. Nara. 1975. Platomycins A and B. I: Taxonomy of the producing strain and production, isolation and biological properties of platomycins. J. Antibiot. 28:656–661.

Tamura, A., R. Furuta, S. Naruto, and H. Ishii. 1973. Actinotiocin, a new sulfur-containing peptide antibiotic from Actinomadura pusilla. J. Antibiot. 26:343–350.

Tamura, T., S. Suzuki, and H. Hatano. 2000. Acrocarpospora gen. nov., a new genus of the order Actinomycetales. Int. J. Syst. Evol. Microbiol. 50:1163–1171.

Tamura, T., and T. Sakane. 2004. Planotetraspora silvatica., an emended description of the genus Planotetraspora. Int. J. Syst. Evol. Microbiol. 54(Pt 6):2053–2056.

Tang, L., Shah, L. Chung, J. Carney, L. Katy, C. Khosfa, and B. Julien. 2000. Cloning and heterologous expression of the epothiline gene cluster. Science 287:640–642.

Terekhova, L. P., O. A. Galatenko, and T. P. Preobrazhenskaya. 1982. Actinomadura fulvescens sp. nov. and Actinomadura turkmeniaca sp. nov. and their antagonistic properties [in Russian]. Antibiotiki 27:87–92.

Terekhova, L. P., O. A. Galatenko, and T. P. Preobrazhenskaya. 1987. Validation of the publication of new names and combinations previously effectively published outside the IJSB List No. 23. Int. J. Syst. Bacteriol. 37:179–180.

Thiemann, J. E., H. Pagani, and G. Beretta. 1967. A new genus of the Actinoplanaceae: Planomonospora gen. nov. Giorn. Microbiol. 15:27–38.

Thiemann, J. E., and G. Beretta. 1968a. A new genus of the Actinoplanaceae: Planobispora gen. nov. Arch. Microbiol. 62:157–166.

Thiemann, J. E., H. Pagani, and G. Beretta. 1968b. A new genus of Actinomycetales: Microtetraspora gen. nov. J. Gen. Microbiol. 50:295–303.

Thiemann, J. E. 1970. Studies of some genera and species of the Actinoplanaceae. In: H. Prauser (Ed.) The Actinomycetales. Gustav Fischer Verlag. Jena, Germany. 245–257.

Thiemann, J. E. 1974a. Genus Planobispora Thiemann and Beretta. In: R. E. Buchanan and N. E. Gibbons (Eds.) Bergey's Manual of Determinative Bacteriology, 8th ed. Williams and Wilkins. Baltimore, MD. 720–721.

Thiemann, J. E. 1974b. Genus Planomonospora Thiemann, Pagani and Beretta. In: R. E. Buchanan and N. E. Gibbons (Eds.) Bergey's Manual of Determinative Bacteriology, 8th ed. Williams and Wilkins. Baltimore, MD. 719–720.

Tomita, K., N. Oda, Y. Hishino, N. Ohkusa, and H. Chikayawa. 1991. Fluviricins A1, A2, B1, B2, B3, B4 and B5, new antibiotics active against influenza A virus. IV: Taxonomy on the producing organism. J. Antibiot. 44:940–948.

Tsyganov, V. A., V. P. Namestinkova, and A. Krassykova. 1966. DNA composition in various genera of the Actinomyceteales [in Russian]. Microbiologiya 35:92–95.

Uchida, K., and K. Aida. 1977. Acyl type of bacterial cell wall: its simple identification by colorimetric method. J. Gen. Appl. Microbiol. 23:249–260.

Uchida, K., T. Kudo, K. Suzuki, and T. Nakase. 1999. A new rapid method of glycolate test by diethyl ether extraction, which is applicable to a small amount of bacterial cells of less than one milligram. J. Gen. Appl. Microbiol. 45:49–56.

Umezawa, I., K. Kamiyama, H. Takeshita, J. Awaya, and S. Omura. 1976. A new antitumour antibiotic, PO-357. J. Antibiot. 29:1249–1251.

Van Brummelen, J., and J. C. Went. 1957. Streptosporangium isolated from forest litter in the Netherlands. Ant. v. Leeuwenhoek 23:385–392.

Vobis, G. 1985a. Spore development in sporangia-forming actinomycetes. In: G. Szabó, S. Biró, and M. Goodfellow (Eds.) Biological, Biochemical and Biomedical Aspects of Actinomycetes. Akadémiai Kiadó. Budapest, Hungary. 443–452.

Vobis, G., and H. W. Kothe. 1985b. Sporogenesis in sporangiate actinomycetes. Front. Appl. Microbiol. 1: 25–47.

Vobis, G. 1989a. Genus Planobispora Thiemann and Beretta 1968, 157[AL]. In: S. T. Williams, M. E. Sharpe, and J. G. Holt (Eds.) Bergey's Manual of Systematic Bacteriology. Williams and Wilkins. Baltimore, MD. 4:2536–2539.

Vobis, G. 1989b. Genus Planomonospora Thiemann, Pagani and Beretta 1967, 29[AL]. In: S. T. Williams, M. E. Sharpe, and J. G. Holt (Eds.) Bergey's Manual of Systematic Bacteriology. Williams and Wilkins. Baltimore, MD. 4:2539–2543.

Vobis, G., and H. W. Kothe. 1989c. Genus Spirillospora 1963, 61[AL]. In: S. T. Williams, M. E. Sharpe, and J. G. Holt (Eds.) Bergey's Manual of Systematic Bacteriology. Williams and Wilkins. Baltimore, MD. 4:2543–2545.

Waksman, S. A. 1950. The actinomycetes: Their nature, occurrence, activities and importance. Ann. Crypt. Phytopath. 9:1–230.

Waksman, S. A. 1961. The Actinomycetes, Volume 2: Classification, Identification and Descriptions of Genera and Species. Bailliere, Tindall and Cox. London, UK.

Wang, Y., Z. Zhang, and J. Ruan. 1996a. A proposal to transfer Microbispora bispora (Lechevalier 1965) to a new genus, Thermobispora gen, nov., as Thermobispora bispora comb. nov. Int. J. Syst. Bacteriol. 46:933–938.

Wang, Y., Z. Zhang, and J. Ruan. 1996b. Phylogenetic analysis reveals new relationships among members of the genera Microtetraspora and Microbispora. Int. J. Syst. Bacteriol. 46:658–663.

Ward-Rainey, N. L., F. A. Rainey, and E. Stackebrandt. 1997. Proposal for a new hierarchic classification system Actinobacteria classis nov.: Family Streptosporangiaceae. Int. J. Syst. Bacteriol. 47:479–491.

Wayne, L. G., D. J. Brenner, R. R. Colwell, P. A. D. Grimont, O. Kandler, M. Krichevsky, L. H. Moore, W. E. C. Moore, R. G. E. Murray, E. Stackebrandt, M. P. Starr, and H. G. Trüper. 1987. Report of the Ad Hoc Committee on Reconciliation of Approaches to Bacterial Systematics. Int. J. Syst. Bacteriol. 37:463–464.

Wellington, E. M. H., and S. T. Williams. 1978. Preservation of actinomycete inoculum in frozen glycerol. Microbios Lett. 6:151–159.

Weyland, H. 1969. Actinomycetes in North Sea and Atlantic Ocean sediments. Nature (London) 223:858.

Weyland, H., E. Helmke, K. Weber, and T. Richter. 1982. Madurose in a LL-DAP containing actinomycete. In: Proceedings of the 5th International Symposium on Actinomycete Biology. Mexico.

Whitham, T. S. 1988. The Selective Isolation, Characterisation and Identification of Streptosporangia [PhD thesis]. Department of Microbiology, University of Newcastle. Newcastle upon Tyne, UK.

Whitham, T. S., M. Athalye, D. E. Minnikin, and M. Goodfellow. 1993. Numerical and chemical classification of Streptosporangium and related actinomycetes. Ant. v. Leeuwenhoek 64:357–386.

Williams, S. T. 1970. Further investigations of actinomycetes by scanning electron microscopy. J. Gen. Microbiol. 62:67–73.

Williams, S. T., and G. P. Sharples. 1976. Streptosporangium corrugatum sp. nov., an actinomycete with some unusual morphological features. Int. J. Syst. Bacteriol. 26:45–52.

Williams, S. T., and E. M. H. Wellington. 1980. Micromorphology and fine structure of actinomycetes. In: M. Goodfellow and R. G. Board (Eds.) Microbiological Classification and Identification. Academic Press. London, UK. 139–165.

Willoughby, L. G. 1969a. A study of aquatic actinomycetes: The allochthonous leaf component. Nova Hedwigia. 18:45–113.

Willoughby, L. G. 1969b. A study of the aquatic actinomycetes of Blenham Tarn. Hydrobiologiya 34:465–483.

Yamaguichi, T. 1967. Similarity in DNA of various morphologically distinct actinomycetes. J. Gen. Appl. Microbiol. 13:63–71.

Zhang, Z., Y. Wang, and J. Ruan. 1998. Reclassification of Thermomonospora and Microtetraspora. Int. J. Syst. Bacteriol. 48:411–422.

Zhang, Z., T. Kudo, Y. Nakajima, and Y. Wang. 2001. Classification of the relationships between members of the family Thermomonosporaceae on the basis of 16S rDNA 16S–23S rRNA internal transcribed spacer and 23S rDNA sequences and chemotaxonomic analyses. Int. J. Syst. Evol. Microbiol. 51:373–383.

Zhang, L.-P., C.-L. Jiang, and W.-X. Chen. 2002. Streptosporangium subroseum sp. nov., an actinomycete with an unusual phospholipid pattern. Int. J. Syst. Evol. Microbiol. 52:235–1238.

Zhukova, R. A., V. A. Tsyganov, and V. M. Morozov. 1968. A new species of Micropolyspora-Micropolyspora angiospora sp. nov. [in Russian]. Microbiologiya 97:724–728.

Zimmermann, W. 1989. Hemicellulolytic enzyme systems from actinomycetes. In: M. P. Coughlan (Ed.) Enzyme Systems for Lignocellulose Degradation. Elsevier Applied Science. London, UK. 167–181.

Prokaryotes (2006) 3:754–795
DOI: 10.1007/0-387-30743-5_29

CHAPTER 1.1.14

The Family Nocardiopsaceae

REINER MICHAEL KROPPENSTEDT AND LYUDMILA I. EVTUSHENKO

Phylogeny and Taxonomy

The family Nocardiopsaceae was created by Rainey and co-workers in 1996 (Rainey et al., 1996). One year later, this family was combined with Streptosporangiaceae and Thermomonosporaceae into Streptosporangineae, which is one of the ten suborders of the order Actinomycetales (Stackebrandt et al., 1997). Originally the family Nocardiopsaceae harbored only the type genus *Nocardiopsis*. Later, Zhang included *Thermobifida* into this family, a new genus which was the home for the two misclassified *Thermomonospora* species *Tm. alba* and *Tm. fusca* (Zhang et al., 1998). Recently, a third genus *Streptomonospora* was added to this family (Cui et al., 2001). The original spelling *Streptimonospora* has been corrected to *Streptomonospora* in the notification list of the same issue of the IJSEM.

Nocardiopsis

At present, the genus *Nocardiopsis* harbors 14 species and one subspecies:

Nocardiopsis alba Grund and Kroppenstedt, 1990

Nocardiopsis dassonvillei subsp. *albirubida* (Grund and Kroppenstedt, 1990) Evtushenko et al., 2000

Nocardiopsis dassonvillei subsp. *dassonvillei* (Brocq-Rousseau, 1904), Meyer, 1976

Nocardiopsis compostus Kämpfer et al., 2002

Nocardiopsis exhalans Peltola et al., 2001

Nocardiopsis halophila Al-Tai and Ruan, 1994

Nocardiopsis halotolerans Al-Zarban et al., 2002

Nocardiopsis kunsanensis Chun et al., 2000

Nocardiopsis listeri Grund and Kroppenstedt, 1990

Nocardiopsis lucentensis Yassin et al., 1993

Nocardiopsis prasina (Grund and Kroppenstedt, 1990) Yassin et al., 1997

Nocardiopsis synnemataformans Yassin et al., 1997

Nocardiopsis trehalosi (ex Dolak et al., 1980) Evtushenko et al., 2000

Nocardiopsis tropica Evtushenko et al., 2000

Nocardiopsis umdischolae Peltola et al., 2001.

The taxonomic positions and designations of some species and subspecies currently belonging to the genus *Nocardiopsis* or referred previously to this genus were changed from time to time (Table 1), reflecting the steps in the development of actinomycete classification. Originally, morphology and physiology were the only criteria used to differentiate *Nocardiopsis* from other mesophilic aerobic actinomycete genera such as *Streptomyces* and *Nocardia*, which also produce conidia- or arthrospores on aerial mycelia. Although morphology is a very valuable character for identification of aerobic spore forming actinomycetes, it has limited value in classification. The same morphological structures are found in different actinomycete genera and a high degree of overlap exists in morphology between phylogenetically unrelated taxa (Gordon and Horan, 1968; Goodfellow and Cross, 1984; Stackebrandt et al., 1983; Stackebrandt, 1986).

Table 2 lists the most important morphological characters of the major genera of sporoactinomycetes.

With the introduction of chemical markers, the taxonomy of Actinomycetales was on a solid basis. The first chemotaxonomic properties used in classification of actinomycetes were amino acids and sugars determined in purified cell walls (Cummins and Harris, 1956; 1958; Romano and Sohler, 1956; Hoare and Work, 1957; Cummins, 1962). This work was continued by Lechevalier and Lechevalier (Lechevalier and Lechevalier, 1970a; Lechevalier and Lechevalier, 1970b), who analyzed the cell walls and whole cells of more than 600 strains of actinomycetes. The results of this study led to the concept of "cell wall chemo types." The authors used specific differences in the major constituents of cell wall (amino acid composition and presence of arabinose and galactose) to differentiate aerobic spore-forming actinomycetes into four groups: I (LL-2,6-diaminopimelic acid-[A_2pm] and glycine as diagnostic components), II (glycine and *meso*-2,6-diaminopimelic acid combined sometimes with 2,6-diamino-3-hydroxypimelic acid), III (*meso*-A_2pm only), and IV (*meso*-A_2pm com-

Table 1. Key references for the history of *Nocardiopsis*, *Thermobifida* and *Streptomonospora species*.

DSM number	*Approved Lists of Bacterial Names* (2002)	*Bergey's Manual* Meyer (1989)	*Approved Lists of Bacterial Names* (1980)	Original assignment
43377[T]	N. alba (1,2)	—	—	"Noc. dassonvillei" (3)
40465[T]	N. dassonvillei subsp. albirubida (1,4)	—	—	"Act.(Stm). alborubidus" (5)
43111[T]	N. dass. subsp. dassonvillei (4,6)	N. dass. subsp. dass. (6,7)	N. dassonvillei (6)	"Stx. dassonvillei" (8)
43884	N. dass. subsp. dassonvillei (2)	N. antarcticus (9)	—	N. antarcticus (9)
44407[T]	N. exhalans (31)	—	—	N. exhalans (31)
44494[T]	N. halophila (10)	—	—	N. halophila (10)
44524[T]	N. kunsanensis (11)	—	—	N. kunsanensis (11)
40297[T]	N. listeri (1)	—	—	"Act.(Stm). listeri" (12)
44048[T]	N. lucentensis (13)	—	—	N. lucentensis (13)
43845[T]	N. prasina (1,2)	N. dass. subsp. prasina (7)	—	N. dass. subsp. prasina (7)
44143[T]	N. synnemataformans (2)	—	—	N. synnemataformans (2)
44380[T]	N. trehalosi (4)	"N. trehalosei" (14)	—	"N. trehalosei" (14)
44381[T]	N. tropica (4)	—	—	N. tropica (4)
44362[T]	N. umdischolae (31)	—	—	N. umdischolae (31)
43885[T]	L. flava (15)	N. flava (16)	A. flava (17)	A. flava (17)
43748[T]	Nm. africana (18)	N. africana (16, 19)	A. africana (20)	A. africana (20)
43679[T]	S. coeruleofusca (21)	N. coeruleofusca (16)	A. coeruleofusca (20)	A. coeruleofusca (20)
43749[T]	S. longispora (21)	N. longispora (16)	A. longispora (20)	A. longispora (20)
43853[T]	S. mutabilis subsp. mutabilis (22)	N. mutabilis (23)	—	N. mutabilis (23)
43886[T]	S. syringae (21)	N. syringae (24)	—	N. syringae (24)
43795[T]	Tb. alba (18)	Tm. alba (25)	Tm. alba (26)	Ab. alba (27)
43185	Tb. alba (18)	Tm. alba (25)	Tm. mesouviformis (28)	Tm. mesouviformis (28)
43792[T]	Tb. fusca (18)	Tm. fusca (25)	—	Tm. fusca (29)
44593[T]	Sm. salina (30)	—	—	Sm. salina (30)

Abbreviations: [T], type species; DSM, Deutsche Sammlung von Mikroorganismen und Zellkulturen, Braunschweig, Germany; *A*, *Actinomadura*; *Act*, *Actinomyces*; *Stm*, *Streptomyces*; *Ab*, *Actinobifida*; *L*, *Lechevalieria*; *M*, *Microtetraspora*; *N*, *Nocardiopsis*; *Noc*, *Nocardia*; *Nm*, *Nonomuraea*; *S*, *Saccharothrix*; *Sm*, *Streptomonospora*; *Stx*, *Streptothrix*; *Tb*, *Thermobifida*; and *Tm*, *Thermomonospora*.
Numbers in parentheses correspond to the following references: (1) Grund and Kroppenstedt (1990); (2) Yassin et al. (1997); (3) Goodfellow and Alderson (1979); (4) Evtushenko et al. (2000); (5) Kudrina (1957); (6) Meyer (1976); (7) Miyashita et al. (1984); (8) Brocq-Rousseau (1904); (9) Abyzov et al. (1983); (10) Al-Tai and Ruan (1994); (11) Chun et al. (2000); (12) Erikson (1935); (13) Yassin et al. (1993); (14) Dolak et al. (1981); (15) Labeda et al. (2001); (16) Preobrazhenskaya et al. (1982); (17) Gauze et al. (1974); (18) Zhang et al. (1998); (19) Greiner-Mai et al. (1987); (20) Preobrazhenskaya and Sveshnikova (1974); (21) Grund and Kroppenstedt (1989); (22) Labeda and Lechevalier (1989); (23) Shearer et al. (1983); (24) Gauze et al. (1977); (25) McCarthy and Cross (1984a); (26) Cross and Goodfellow (1973); (27) Locci et al. (1967); (28) Nonomura and Ohara (1974); (29) Crawford (1975); (30) Cui et al. (2001); and (31) Peltola et al. (2001).

bined with arabinose and galactose; Lechevalier and Lechevalier, 1970a).

On the basis of the chemical data and some specific morphological characters, Lechevalier and Lechevalier (Lechevalier and Lechevalier, 1970a; Lechevalier and Lechevalier, 1970b) created the new actinomycete genus *Actinomadura*. This genus harbored three species, *A. madurae*, *A. pelletieri*, and *A. dassonvillei*. The two unifying characters of these three aerobic actinomycete species were: 1) the ability to produce spore chains on aerial mycelia and 2) cell wall chemotype III, which corresponds to the peptidoglycan type A1γ (Schleifer and Kandler, 1972), i.e., *N*-acetyl-muramic acid, *N*-acetyl-glucosamine, alanine, glutamic acid, and *meso*-2, 6-diaminopimelic acid. Using cell wall chemistry, it was possible to separate these three *Actinomadura* species from morphologically related taxa showing a different cell wall composition, e.g. type I *Streptomyces*, IV *Nocardia*, *Pseudonocardia* and related genera. However, using only morphology and this limited set of chemical markers, the genus *Actinomadura* quickly became a dumping ground for many unrelated chemotype III isolates.

Progress in the classification of this taxon was achieved by the introduction of an additional chemotaxonomic marker, the sugar pattern of whole cell hydrolysates (Lechevalier and Lechevalier, 1970a; 1970b; Lechevalier et al., 1971). This made it possible to separate cell chemotype III organisms into those having the diagnostic sugar madurose (3-*O*-methyl-D-galactose), i.e.,

Table 2. Morphological differences between members of Nocardiopsaceae and some other sporoactinomycete genera.

Genus	Spore chains on special aerial hyphae	AM disintegrates into spores	Multiple cell division	Spore chains of 50 spores or more	Fragmentation of substrate mycelium	Sporangia or pseudosporangia	Motile spores
Nocardiopsis	–	+	–	+	+/–	–	–
Thermobifida	+	–	–	–	–	–	–
Streptomonospora	+	–	–	–	–	–	–
Saccharothrix	–	+	–	+	–	–	–
Glycomyces	–	+	–	+	–/+[a]	–	–
Actinomadura	+	–	–	–	–	–	–
Microtetraspora	+	–	–	–	–	–	–
Nonomuraea	+	–	–	–	–	+	–
Pseudonocardia	–	+	+	+	+	–[b]	–
Amycolatopsis	–	+/–	–	+	+	+	–
Kibdellosporangium	+	+	–	+	+/–	–	–
Actinopolyspora	+	–	+	–	–	–	–
Saccharopolyspora	–	+	–	+	+/–	–	–
Streptomyces	+	–	–	–/+[c]	–	–[b]	–
Kitasatospora	+	–	–	–	–/+	–	–
Nocardia	–	+	–	+/–	+	–	–

Symbols: +, present; –, absent; +/–, character is inconsistent among the species of this genus.

Abbreviations: AM, aerial mycelium.

[a]In *Glycomyces tenuis*, the aerial hyphae are not observed and the spore chains are formed on substrate mycelium.

[b]Sporangium-like structures or sclerotia may present in *Pseudonocardia* (the former *Amycolata*) and in *Streptomyces* (the former "*Actinosporangium*," "*Chainia*," etc.;see Williams et al., 1989).

[c]In streptomycetes, the spore chains may contain more than 50 spores.

Adapted from Meyer (1976); Williams and Wellington (1981); Preobrazhenskaya et al. (1982); Labeda et al. (1984; 1985); Lechevalier et al. (1986); Shearer et al. (1986); Goodfellow (1989); Kothe et al. (1989); Williams et al. (1989); Kroppenstedt et al. (1990); Zhang et al. (1997, 1998); and Chung et al. (1999).

Actinomadura, and those lacking this sugar, i.e., *Nocardiopsis*. On the basis of the characteristic development of spores, including the specific zig-zag formation of aerial hyphae before spore dispersal (Williams et al., 1974) and the lack of madurose (Lechevalier and Lechevalier, 1970b), Meyer (1976) created a new genus for *A. dassonvillei*. The name *Nocardiopsis* was selected because of the nocardia-like appearance of *N. dassonvillei*. Like *Nocardia*, *N. dassonvillei* does not produce special spore-forming hyphae but the whole aerial mycelium disintegrates completely into smooth long spores (Williams et al., 1974). Results of fatty acid and menaquinone composition analyses, as well as molecular genetic data, confirmed the separation of *Nocardiopsis* and *Actinomadura* into two genera (Agre et al., 1975; Kroppenstedt and Kutzner, 1976, 1978; Fischer et al., 1983; Athalye et al., 1984; Alderson et al., 1985). Preobrazhenskaya et al. (1982) transferred four additional *Actinomadura* species into the genus *Nocardiopsis*, i.e., *A. africana*, *A. coeruleofusca*, *A. flava* and *A. longispora*. However, subsequent chemotaxonomic studies and partial oligonucleotide sequencing of 16S ribosomal RNA from *Nocardiopsis* and *Actinomadura* revealed that these species did not belong to *Nocardiopsis* or *Actinomadura* and that both genera are heterogeneous (Poschner et al., 1985; Goodfellow et al., 1988; Streshinskaya et al., 1989; Grund and Kroppenstedt, 1990; Taptykova et al., 1990). The heterogeneity of these taxa was underlined by the results of numerical phenetic studies (Goodfellow and Alderson, 1979; Alderson et al., 1984; Athalye et al., 1984).

In the meantime, Labeda et al. (1984) described a nocardioform actinomycete isolated from a soil sample collected in Australia. Because of its unique combination of chemical markers, the authors created a new genus, *Saccharothrix*. Although its isolate resembles *Nocardiopsis dassonvillei* morphologically and in some chemotaxonomic characters, i.e., cell wall chemotype III/C, Labeda et al. (1984) did not classify this isolate as *Nocardiopsis*: it lacked phosphatidylcholine, always present in *Nocardiopsis*, and it synthesized the menaquinones MK9-(H_4) and MK-10(H_4). In addition, Labeda (1992) detected rhamnose in whole cell hydrolysates of all *Saccharothrix* strains. This sugar has diagnostic value because it was also present in *Streptoalloteichus* (Goodfellow, 1989), *Kutzneria* (Stackebrandt et al., 1994b), *Actinokineospora* (Tamura et al., 1995), and other genera (Labeda et al., 2001). These genera have now been classified into the Actinosynnemataceae, a new family of the suborder Pseudonocardineae (Labeda and Kroppenstedt, 2000). *Saccharothrix* and related taxa can clearly be separated from *Nocardiopsis*

because rhamnose has never been detected in species of the latter genus (Fischer et al., 1983; Grund and Kroppenstedt, 1989; Grund and Kroppenstedt, 1990; Naumova et al., 2001).

Results of 16S rRNA gene sequencing revealed that *Nocardiopsis* phylogenetically occupies a separate position, apart from all other actinomycetes (Goodfellow et al., 1988; Kroppenstedt et al., 1990). Thus, the genus *Nocardiopsis* was the only genus of the family Nocardiopsaceae (Rainey et al., 1996), a member of the suborder Streptosporangineae (Stackebrandt et al., 1997). The results of chemotaxonomic studies on nearly all type strains of *Nocardiopsis* and related genera showed that six of the eight *Nocardiopsis* species were misclassified (Grund and Kroppenstedt, 1989; Grund and Kroppenstedt, 1990). Five of them, *N. coeruleofusca*, *N. flava*, *N. longispora*, *N. mutabilis* and *N. syringae*, were reclassified in the genus *Saccharothrix*, and one, *N. africana*, was affiliated to the genus *Microtetraspora* (Poschner et al., 1985; Goodfellow et al., 1988; Kroppenstedt et al., 1990; Table 1). Later, *Saccharothrix flava* was transferred to the new genus *Lechevalieria* (Labeda et al., 2001), and *Microtetraspora africana* together with most other species of the genus *Microtetraspora* to *Nonomuraea* (Zhang et al., 1998).

Phenetic data indicated that the remaining single member species, *N. antarctica*, was a synonym of *N. dassonvillei* (Meyer, 1989). These data were confirmed by 16S rRNA gene sequences and DNA/DNA hybridization results by Yassin et al. (1997), who transferred *N. antarctica* to *N. dassonvillei* as a later subjective synonym. DNA-DNA hybridization data and numerical analyses using fatty acid results and phenetic data showed that all tested strains of *N. dassonvillei* should be divided into two species (Fischer et al., 1983; Grund, 1987). On the basis of these results and those of some additional studies, a new species, *N. alba*, was created (Grund and Kroppenstedt, 1990). This study showed in addition that *N. dassonvillei* subsp. *prasina* was actually more closely related to *N. alba* than to *N. dassonvillei*. This subspecies was therefore transferred to *N. alba* with the revision of its name to *N. alba* subsp. *prasina*. During a study on the new *Nocardiopsis* species *N. synnemataformans*, Yassin and co-workers described on the basis of 16S rRNA gene sequences and DNA/DNA similarity studies *N. alba* subsp. *prasina* as *Nocardiopsis prasina* (Yassin et al., 1997).

A study of the cell walls of about 450 International Streptomyces Project (ISP) streptomycetes (Shirling and Gottlieb, 1972) by Pridham and Lyons (1961) revealed that some of these species did not belong to the genus *Streptomyces* because they synthesized *meso*- A_2pm instead of LL-A$_2$pm as expected for members of the genus *Streptomyces*. The so-called "*meso*-DAP" streptomycetes were further characterized by Grund (1987). Two members of this group, "*Streptomyces alborubidus*" and "*Streptomyces listeri*," were reclassified into the genus *Nocardiopsis* as *N. alborubida* and *N. listeri* (Grund and Kroppenstedt, 1990). Later, Yassin and co-workers found high similarities between the 16S rRNA gene sequences of *N. alborubida* (DSM 40465^T) and *N. dassonvillei* (Yassin et al., 1997). As DNA/DNA similarity studies revealed 77 % similarity, *N. alborubida* was transferred to *N. dassonvillei* as a later subjective synonym of this species.

Strain *N. alborubida* VKM Ac-1882^T (= DSM 40465^T) and *Nocardiopsis dassonvillei* differed on the basis of physiological test results, in macroscopic appearance (Evtushenko et al., 2000), and in chemical composition (the cell wall teichoic acid representing up to 20% of the cell wall in these species; Shashkov et al., 1997; Tul'skaya et al., 1993). On the basis of a DNA/DNA similarity of 71% obtained between VKM Ac-1882^T and other members of *N. dassonvillei*, strain VKM Ac-1882 was reclassified as *Nocardiopsis dassonvillei* subsp. *albirubida* (Evtushenko et al., 2000).

Current progress in the development of intrageneric classification of the genus *Nocardiopsis* is linked to studies of cell wall teichoic acids and the finding that these polymers and their structural elements are chemotaxonomic markers for the *Nocardiopsis* species and subspecies (Shashkov, et al., 1997; Streshinskaya et al., 1996, 1998; Tul'skaya et al., 1993, 2000).

The cell wall teichoic acids are covalently linked by phosphodiester bridges to muramic acid residues in peptidoglycan (in contrast to lipoteichoic acids, which are associated with the plasma membrane) and placed between other cell wall layers and at the cell surface (Baddiley, 1972, 2001; Archibald, 1974). The polymers are widespread among Gram-positive bacteria and may compose up to 60% (v/v) of cell envelope in some actinomycetes (Shashkov and Naumova, 1997). The following structural types of cell wall teichoic acids are (Naumova et al., 2001): 1) polymers with chains of only polyol phosphate residues or 2) polymers with chains of only glycosylpolyol phosphate residues joined by phosphodiester bonds, and 3) mixed structures of glycosyl-1-phosphate in addition to polyol phosphate residues or 4) mixed structures of alternating poly(polyol phosphate) and poly(glycosylpolyol phosphate) units. Different polyol and sugar (amino sugar) residues in the main chain and in lateral branches (which are linked to the polyol residues via glycosyl bonds, various O-acyl residues, different kinds of phos-

phodiester bonds, and some other factors) contribute additionally to the structural diversity of cell wall teichoic acids (Naumova et al., 2001). The great structural diversity of cell wall teichoic acids expressed by the information coded in various genes, their involvement in vital cell functions, and their wide occurrence in Gram-positives, suppose a certain correlation of these polymers with other taxonomic properties and their taxonomic applicability to systematics. The fact that the species within a genus have structurally dissimilar teichoic acids was reported for the first time for staphylococci (Davison and Baddiley, 1963) and later published for some other bacteria, mainly actinomycetes.

Naumova et al. (2001) summarized the data on teichoic acids of 27 strains of the genus *Nocardiopsis* and demonstrated that the structures of these polymers or their combinations are indicative of each of the seven *Nocardiopsis* species and two subspecies which were analyzed. *Nocardiopsis dassonvillei* subsp. *dassonvillei* and *N. dassonvillei* subsp. *albirubida* are characterized by polymers whose cores are composed of alternating units of both glycerol phosphate and β-*N*-acetyl-galactosamine glycerol phosphate residues (Fig. 1). However, the phosphodiester bond in the teichoic acid of *N. dassonvillei* subsp. *dassonvillei* is positioned at the C-3 of the amino sugar (Fig. 1a), while the polymer of VKM Ac-1882 has the phosphodiester bond located at C-4, and the C-3 position is occupied by an *O*-succinyl substituent in most glycosyl residues (Fig. 1b). *Nocardiopsis trehalosi* contains 1,3-poly(glycerol phosphate) substituted with β-glucosyl residues (Fig. 1c). Teichoic acid of *N. tropica* is represented by an assortment of heterogeneous 1,5-poly(ribitol phosphate) chains with lateral glycerol phosphate oligomers (Fig. 1d). The species *N. alba*, *N. listeri*, *N. lucentensis* and *N. prasina* contain different combinations of the following three polymers: unsubstituted 3,5-poly(ribitol phosphates), i.e., teichoic acid type 1 (TA1); (Fig. 1e), 1,3-poly(glycerol phosphates) substituted with *N*-acetylglucosamine at C-2 of glycerol residues (TA2; Fig. 1f), and 1,5-poly(ribitol phosphates) with each ribitol phosphate unit carrying a 2,4-pyruvate ketal group (TA3; Fig. 1g). In the cell walls of the strains of *N. alba*, all these three teichoic acids are present in varying amounts, *N. prasina* and *N. listeri* contain two polymers each, TA1 and TA2 (*N. prasina*), TA2 and TA3 (*N. listeri*), while *N. lucentensis* has only TA2. As each of the species and subspecies discussed is definable by the composition of teichoic acids, the polymers can be considered a valuable chemotaxonomic marker for the intrageneric taxa of *Nocardiopsis*.

Fig. 1. Fragments of the primary structure of teichoic acids typical of the Nocardiopsis species and subspecies (Data from Naumova et al., 2001; Shashkov et al., 1997; Streshinskaya et al., 1989, 1996, 1998; Tul'skaya et al., 1993, 2000.)

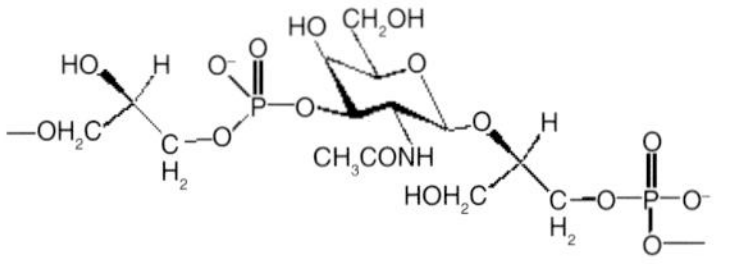

a) Poly(glycerol phosphate-β-N-acetylgalactosaminylglycerol phosphate) (*N. dassonvillei* subsp. *dassonvillei*)

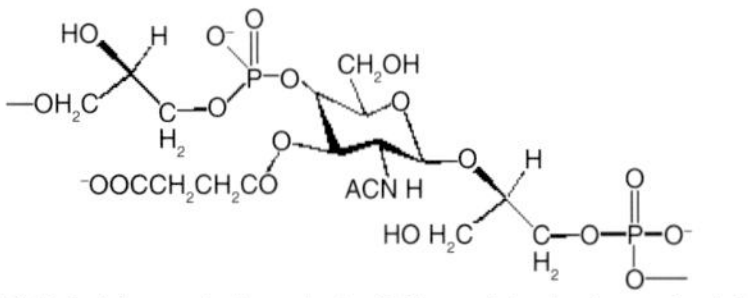

b) Poly(glycerol phosphate-β-N-acetylgalactosaminylglycerol phosphate) with O-succinyl residues (*N. dassonvillei* subsp. *albirubiba*)

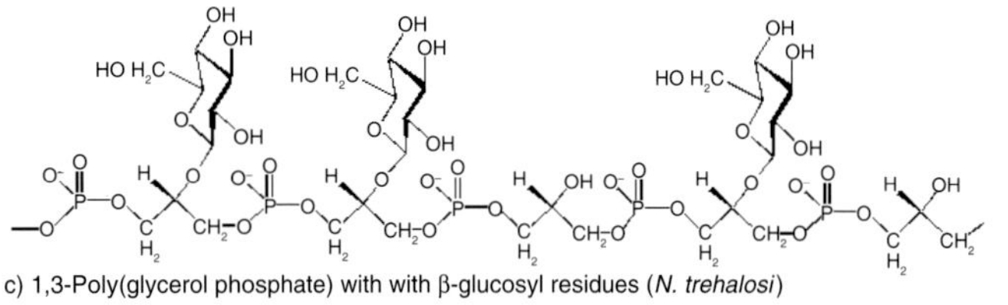

c) 1,3-Poly(glycerol phosphate) with with β-glucosyl residues (*N. trehalosi*)

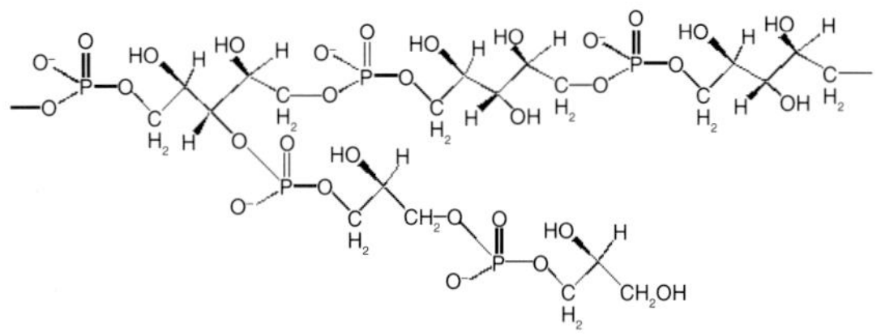

d) 1,5-Poly(ribitol phosphate) with side chains of glycerol phosphate oligomers (*N. tropica*)

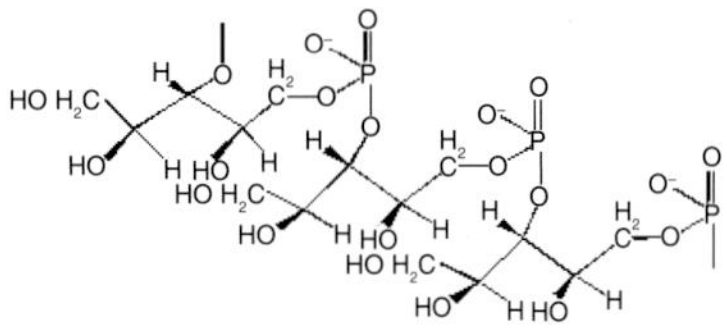

e) Unsubstituted 3,5-poly(ribitol phosphate), TA1 (*N. alba, N. prasina*)

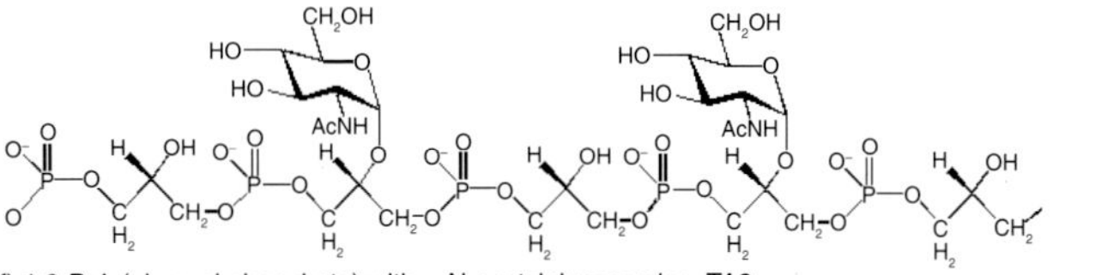

f) 1,3-Poly(glycerol phosphate) with α-N-acetylglucosamine, TA2 (*N. alba, N. listeri, N. lucentensis, N. prasina*)

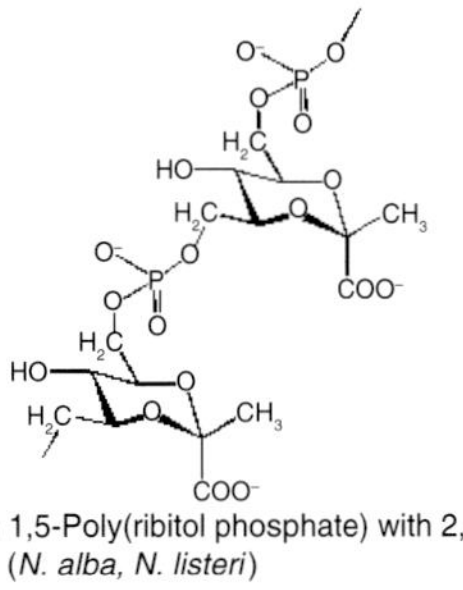

g) 1,5-Poly(ribitol phosphate) with 2,4-pyruvate ketal group, TA3 (*N. alba, N. listeri*)

Besides the fully identified structures of the polymers, the main products of acid degradation of the common fraction of the teichoic acids (e.g., glycerol, ribitol, glucose, glucosamine, galactosamine, pyruvate, and succinate in various combinations) can also serve as chemotaxonomic markers for the species and subspecies. Determination of some diagnostic products of alkaline hydrolysates of whole fractions of the polymers (e.g., glycerol, ribitol, their mono-, di- or triphosphates, and also certain phosphodiesters) gives additional information and increases taxonomic resolution of this approach. For instance, *N. alba* and *N. listeri*, which are characterized by similar profiles of acid degradation products, can be differentiated by the presence (absence) of mono- and diphosphates of ribitol in the alkaline hydrolysates of polymers. Certain components of teichoic acids determined in acid hydrolysates of the whole cell wall without the extraction of pure polymers are also of diagnostic value (see the section "Identification" for details in this Chapter). Finally, the ^{13}C-nuclear magnetic resonance (NMR) spectra of the polymers found in the *Nocardiopsis* species and subspecies are also highly indicative, and therefore the respective intrageneric taxa are well recognizable from NMR analysis of their polymers.

The genus *Nocardiopsis* may serve as a notable example for the usefulness of cell wall teichoic acids in the differentiation and circumscription of actinomycete species. Further comparative studies of teichoic acids and related polymers in other bacterial groups and the development of simple and efficient methods for the identification of the polymers, accompanied by revision and approval of bacterial species by extensive genomic and phenetic investigations, will contribute to further progress in bacterial taxonomy and the construction of a practicable and predictive classification scheme.

Thermobifida

The history of *Thermobifida* species is complex. The name "*Thermobifida fusca*" changed from *Micromonospora* to *Thermomonospora*, then to *Actinobifida*, then back to *Thermomonospora* and recently to *Thermobifida*. The earliest observation of organisms exhibiting a clustered spore morphology typical of the genus *Thermobifida* (*Micromonospora*) was reported by Waksman et al. (1939), who applied the species epithet "fusca" to their strains. The authors considered these isolates to be the species *Micromonospora fusca*. Later, similar strains were also described by Erikson (1953), who noticed that the strains showed a tendency to produce spore clusters on substrate mycelia. On the basis of a study of

isolates from composted horse manure, Henssen (1957) created the new genus *Thermomonospora* for thermophilic actinomycetes forming single spores on aerial mycelia. Two species, *Tm. fusca* and "*Tm. lineata*," were characterized by spore clusters similar to those described by Waksman et al. (1939). However, the respective strains isolated by these authors were lost, and the species name *Tm. fusca* appeared as nomen dubium in the eighth edition of *Bergey's Manual* (Buchanan and Gibson, 1974).

The first viable strain of the genus *Thermobifida*, *Tb. alba* DSM 43795^T (the type strain of the type species; Table 1) was isolated and described by Locci et al. (1967). The authors included their isolate with the genus "*Actinobifida*," which was created previously by Krasilnikov and Agre (1964). Later, Cross and Lacey (1970) showed the high similarity in morphological characters and physiology among species of "*Actinobifida*," *Thermomonospora* and *Thermoactinomyces* and questioned the nomenclatural validity of the genus "*Actinobifida*." On the basis of the heat resistance of their endospores, the "*Actinobifida*" species were transferred (i.e., "*Ab. alba*" to *Thermomonospora* and "*Ab. dichotomica*" to *Thermoactinomyces*; Cross and Goodfellow, 1973; Skerman et al., 1980).

Subsequently, Crawford (1975) identified Forbes strain 190Th (Forbes, 1969) as *Thermomonospora fusca* sensu Henssen and proposed it as the neotype strain of *Thermomonospora fusca* (Sneath, 1992). This proposal did not result in the validation of this name because it was not included in the *Approved Lists of Bacterial Names* (Skerman et al., 1980). On the basis of intensive study of thermophilic monosporic actinomycetes, McCarthy and Cross (McCarthy and Cross, 1984a; McCarthy and Cross, 1984b) emended the description of this species with the type strain 190Th and validated the species name *Tm. fusca* (McCarthy and Cross, 1984c).

Chemotaxonomic studies of the polar lipids, the menaquinones and the fatty acids of *Tm. alba*, *Tm. fusca* and *Tm. mesouviformis* show that these species synthesized an unique combination of chemical markers which were different from those of *Tm. curvata* (the type species of the genus *Thermomonospora*) and *Tm. chromogena* (Kroppenstedt and Goodfellow, 1992). All three species produce single spores borne mainly on aerial mycelia and sometimes on substrate mycelia, which is also in contrast to *Tm. curvata*. In addition, the spores of the *Thermobifida* species are formed in clusters, each singly on the tip of a short sporophore, with spore clusters originating from primary laterally branched aerial hyphae. Cross and Goodfellow (1973) described this structure as a repeatedly branched sporophore producing dense spore heads. These clusters have never been observed in other species of the genus *Thermomonospora* (Greiner-Mai et al., 1987). It was therefore proposed to exclude these species from the genus *Thermomonospora* (Kroppenstedt and Goodfellow, 1992). Zhang confirmed this proposal when he compared the 16S rRNA gene sequences of the *Thermomonospora* type strains with all other actinomycetes (Zhang et al., 1998). Because the sequences of *Tm. alba*, *Tm. fusca* and *Tm. mesouviformis* were not related to those of any known actinomycete taxa, a new genus *Thermobifida* was established to accommodate *Tb. alba* and *Tb. fusca*. The authors also showed that the sequences of *Thermomonospora alba* and *Tm. mesouviformis* were 98.7% identical and claimed that this similarity supports the reduction in status of *Thermomonospora mesouviformis* to a subjective synonym of *Tm. alba*, which was already proposed by McCarthy and Cross (1984a) on the basis of phenotypic study. However, McCarthy and Cross did not validate this proposal. In contrast, Kurup (1979) and Cross (1981) found similarities between *Tm. alba* and *Tm. fusca*, and they argued on their findings that these two species were synonymous but distinct from the mesophilic *Tm. mesouviformis* described by Nonomura and Ohara (1974). In addition, the results of these studies showed that the value of physiological properties for the classification of members of thermophilic actinomycetes including *Thermobifida* is limited. This is probably due to the fact that members of this genus are rather difficult to cultivate and in many cases yield poor inocula (Greiner et al., 1987) or because temperature change affects growth negatively during plating (R.M. Kroppenstedt and L. I. Evtushenko, personal observation). Taking into account these data and the limited usefulness of high 16S rRNA gene sequence similarity for differentiating species (Stackebrandt and Goebel, 1994a), the relationship of *Thermomonospora alba* and *Thermomonospora mesouviformis* species has to be proven by DNA/DNA hybridization.

Streptomonospora salina

The taxonomy of genera comprising the family Nocardiopsaceae is a good example of the dependence of the methods used in systematics in the past for classification of the actinomycetes. With the spread of molecular biological methods like 16S rRNA gene sequencing and the application of polyphasic taxonomy, classification became more and more stable and reliable. Recently the third genus of the family Nocardiopsaceae, *Streptomonospora*, was described. The classification was based on phylogenetic and phenotypic data (Cui et al., 2001). The only strain of this genus, *Streptomonospora salina* DSM 44593^T (CCTCC 99003^T), was isolated from a salt lake in China. *Streptomonospora salina* DSM 44593^T forms short spore chains on aerial mycelia and single spores on substrate mycelia. The substrate spores are produced singly on branched or simple sporophores, as can be observed in some *Thermobifida* strains. Results from analysis of *Streptomonospora salina* cell walls revealed an unusual combination of amino acids, i.e., *meso*-DAP, glycine, lysine, aspartate, and a trace of LL-DAP. Glucose, galactose, ribose, xylose, arabinose and mannose were detected in its whole cell hydrolysates. The phospholipid pattern was of the PII type and the major menaquinones were represented by MK-9(H$_6$), MK-10(H$_2$), and MK-10(H$_4$).

Habitat

Nocardiopsis

The first description of *Nocardiopsis dassonvillei* (*Streptothrix dassonvillei*) was based on strains isolated from mildewed grain and fodder (Brocq-Rousseau, 1904). Later studies showed that the natural habitat of *Nocardiopsis* is soil. In screening programs searching for new antimicrobial products or other metabolites, members of this genus are frequently isolated from different soils together with streptomycetes and some other actinomycetes (Dolak et al., 1980, 1981; Mikami et al., 1982; Mishra et al., 1987a; Al-Tai and Ruan, 1994; Jiang and Xu, 1998; Wang et al., 1999; Evtushenko et al., 2000; Xu et al., 2001).

Nocardiopsis species were also found in an Antarctic glacier (Abyzov et al., 1983), marine sediments (Dixit and Pant, 2000a; Evtushenko et al., 2000), actinoryzal plant rhizosphere (Boivin-Jahns et al., 1995), gut tract of animals (Vasanthi et al., 1992), as an endophyte of yam bean (Stamford et al., 2001), salterns (Chun et al., 2000), and active stalactites (Laiz et al., 2000). Lacey (1977) detected *N. dassonvillei* strains in cotton waste and occasionally in hay, whereas Mishra et al. (1988) isolated 22 *Nocardiopsis* strains from soil samples. In a study of actinomycetes from indoor environments, toxigenic strains of the two new species *N. exhalans* and *N. umdischolae* were isolated from the water-damaged building material of a children's day care centere (Peltola et al., 2001), *N. dassonvillei* from air of a cattle barn using an Anderson sampler (Andersson et al., 1998), and *Nocardiopsis compostus* from the atmosphere of a composting facility (Kämpfer et al., 2002). This strain could grow at temperatures up to 50°C but grew optimally at 30°C. Information on true thermophilic

LIVERPOOL JOHN MOORES UNIVERSITY
LEARNING & INFORMATION SERVICES

Nocardiopsis strains is rare. Nearly all *Nocardiopsis* species that have been validly described until now were found at sites with moderate temperatures. The thermotolerant strains (which may grow up to 42°C or even at 50°C) have their optimal growth temperature at 28°C (Evtushenko et al., 2000; Kämpfer et al., 2002) or, in the case of *Nocardiopsis kunsanensis*, at 37°C (Chun et al., 2000). There is only one report of a thermophilic *Nocardiopsis* isolate, KMD/8, that grows up to 65°C with an optimal growth temperature of 50°C (Kempf, 1995). This strain was among 176 thermophilic strains isolated (on glycerol-asparagine-propionate agar at 45°C) from municipal waste compost produced by a compost plant in Germany. Cryophilic *Nocardiopsis* strains were not isolated until now. Abyzov et al. (1983) found a single *Nocardiopsis* strain in an Antarctic glacier at a depth horizon of 85 m, which corresponded to an age of over 2000 years. He named this isolate "*N. antarcticus*," reflecting its origin. Although this culture was isolated from a cold region, it is a mesophilic species with an optimal growth temperature of 28°C and an inability to grow at 4°C. This species was later reclassified by Yassin et al. (1997), who could show that *N. antarcticus* was a subjective synonym of *N. dassonvillei* subsp. *dassonvillei*.

Members of this genus have been frequently isolated from areas with high salt concentrations, such as *Nocardiopsis lucentensis* from a salt marsh soil sample near Alicante, Spain (Yassin et al., 1993), *Nocardiopsis halotolerans* strain from the same environment at Al-Khiran, Kuwait (Al-Zarban et al., 2002), *Nocardiopsis halophila* from a saline soil sample in Iraq (Al-Tai and Ruan, 1994), and *Nocardiopsis kunsanensis* from a saltern in the Republic of Korea (Chun et al., 2000). Although these species are able to grow at high salt concentration (up to 10–20% NaCl) in medium, only *Nocardiopsis kunsanensis* was classified originally with moderately halophilic actinomycetes (Chun et al., 2000) in accordance with the scheme of Larsen (1986). The optimal growth of this strain was at 5–15% NaCl and it was unable to grow on medium without salt. The minimal NaCl concentration for growth was 3% and the maximum 20%, but sometimes growth could be obtained at 25% depending on culture conditions (Chun et al., 2000). Three other species may be classified as halo tolerant because the same growth can be obtained in the medium without salt. *Nocardiopsis lucentensis* and *Nocardiopsis halophila* were also cited as moderately halophilic species (Ventosa et al., 1998), in accordance to the most widely used classification of Kushner (1978). Their optimal NaCl concentration for growth was 5–10% and 15%, respectively. *Nocardiopsis halophila* did not grow at all in medium contain-

ing less than 5% NaCl (Al-Zarban et al., 2002). It should be pointed out that the salt requirement and tolerance of many bacterial species vary according to growth conditions such as temperature and medium composition (Ventosa et al., 1998).

In a screening program for alkaliphilic bacteria from soils, Mikami et al. (1982) isolated 20 actinomycetes that grew at pH 11.5. Nine of these isolates resembled *N. dassonvillei* morphologically and had the same III/C cell wall chemotype. These strains were later classified as *N. dassonvillei* subsp. *prasina* (Miyashita et al., 1984). At the same time, Mikami et al. (1985) screened the 420 International Streptomycetes Project (ISP) streptomycetes for alkaliphilic strains and found that only six of them were able to grow at a pH of 11.5 or above. Cell wall analyses of the alkaliphilic strains revealed that half of these strains were misclassified. One strain (ISP 5011[T], DSM 40011[T]) was later classified as *Amycolata autotrophica* (Lechevalier et al., 1986) and then transferred to the genus *Pseudonocardia* (Warwick et al., 1994). Another strain, ISP 5465[T], DSM 40465[T], was described as *Nocardiopsis albirubida* (Grund and Kroppenstedt, 1989), while *Streptomyces caeruleus* (ISP 5103[T], DSM 40103[T]) showed a close relationship to *Saccharothrix* based on chemical markers (Grund, 1987). Tsujibo isolated alkaliphilic strains from soil that grew at pH 10 and were identified as *N. dassonvillei* (Tsujibo et al., 1988).

Jiang and Xu screened for extremophilic actinomycetes at unusual habitats of Yunnan, China (Jiang and Xu, 1998). They isolated 49 actinomycetes from alkaline soils. Nine of them belonged to the genus *Nocardiopsis*, while the remainder were alkaliphilic or alkali tolerant streptomycetes. All *Nocardiopsis* strains could grow at extreme alkaline conditions (i.e., pH 12), whereas the alkaliphilic streptomycetes were not able to grow at this high pH. Six of the *Nocardiopsis* isolates were obligate alkaliphilic strains because they grew at pH 8–12 but not at pH 7, whereas the remaining three grew at pH 7 but not at pH 6. None of the *Nocardiopsis* isolates were acidophilic, thermophilic or psychrophilic.

On the basis of these findings, it can be concluded that soil is the natural habitat of *Nocardiopsis* species and that mildly alkaline conditions (pH 8) are best for optimal growth of most strains of this genus. For some *N. dassonvillei* strains (Tsujibo et al., 1988) and other species of this genus, i.e., *N. prasina* (Miyashita et al., 1984) and *N. dassonvillei* subsp. *albirubida* (Mikami et al., 1982; Mikami et al., 1985), growth is enhanced at even higher pH values (10). Inoculation of alkaliphilic strains on agar plates at the lowest possible pH for their growth results in slow growth, which increases as these bacteria change the pH of the plate until it reaches their optimal pH for growth (Horikoshi and Akiba, 1982; Mikami et al., 1985).

The clinical significance of *Nocardiopsis* strains is well documented. Liegard and Landrieu (1911) reported a case of ocular conjunctivitis from which they isolated a bacterium matching the description of Brocq-Rousseau (1904). They called this strain "*Nocardia dassonvillei*" because of its nocardia-like appearance. Later, Erikson (1935) isolated *Nocardiopsis* strains from other clinical material. The clinical importance of *Nocardiopsis* and related organisms has been discussed by Gordon and Horan (1968). They found that 15 of their 26 collection strains were clinical isolates. The authors speculated that similar cultures from other sources might usually be discarded in laboratories as contaminants because of their macroscopic resemblance to cultures of *Streptomyces griseus*, which is one of the most common actinomycetes in the environment. The clinical significance was later confirmed by Ajello et al. (1987), Beau et al. (1999), Sindhuphak et al. (1985), Singh et al. (1991), and Yassin et al. (Yassin et al., 1993; Yassin et al., 1997), who isolated *Nocardiopsis* strains from clinical material. Gugnani et al. (1998) and Mordarska et al. (1998) discuss the role of these isolates as the causal agents of disease.

Thermobifida

Thermobifida belongs to the autochthonic microflora of organic materials, which heat up during degradation (self heated), e.g., manure heaps, composts of different kinds, or damp stored hay (Lacey, 1973). Both species of the genus *Thermobifida* are thermophilic. The optimal growth temperature for *Tb. alba* is 40–45°C, whereas *Tb. fusca* grows better at 50–55°C. *Thermobifida* strains are able to secrete a variety of thermostable extracellular enzymes enabling them to become established as dominant populations during high temperature composting of plant residues and other wastes (Fergus, 1964). They are highly cellulolytic (McCarthy, 1987; Ball and McCarthy, 1988) and/or hemicellulolytic (McCarthy et al., 1985; McCarthy et al., 1988), and strains are abundant in mushroom compost (Fergus, 1964; Lacey, 1974; Lacey, 1977; McCarthy and Cross, 1984a, 1984b). Evidence that cellulose decomposition in mushroom compost involves a positive interaction between cellulolytic actinomycetes and certain bacteria (Stanek, 1972) underscores the possibility that natural lignocellulose degradation is the product of interaction between heterogeneous microflora.

Growth of thermophilic actinomycetes in high temperature environments leads to the release of spores that can cause an exogenic allergic alveolitis. This is well documented for *Saccharopolyspora rectivirgula* (Lacey and Crook, 1988b), *Saccharomonospora viridis* (Green et al., 1981), and some *Thermoactinomyces* species (Greiner-Mai, 1988). There is currently no evidence for implicating spores of *Thermobifida* species in such respiratory disorders (Lacey, 1988a), though the cause of mushroom worker's disease, a form of allergic alveolitis, is still a matter for conjecture. Nevertheless, Van den Borgard et al. (1993) and co-workers presumed *Tb. alba* and *Tb. fusca*, along with *Thermomonospora curvata* and misclassified actinomycete "*Excellospora flexuosa*," contribute to the occurrence of "mushroom worker's lung" (MWL) because 1) high numbers of spores of these organisms were present in air breathed by the affected mushroom workers and 2) serum titers of actinomycete spore antibodies were elevated in these workers. Vast numbers of spores of thermophilic actinomycetes, including spores of *Tb. alba* and *Tb. fusca*, were collected from the air in fermentation tunnels during the spawning of mushroom compost. Sera from ten mushroom growers affected by MWL were tested by a qualitative dot-enzyme-linked immunosorbent assay (ELISA) for antibodies against the spores of these micro-organismsmicroorganisms. All 10 sera were positive for one or more of the actinomycetes, *Nonomuraea flexuosa* ("Excellospora"), *Tb. alba*, *Tb. fusca* and *Tm. curvata*. In contrast, no antibodies were found against *Streptomyces thermovulgaris*, *Thermoactinomyces vulgaris* and *T. sacchari* or against fungi. Sera of 11 of 14 workers engaged in routine spawning of compost in tunnels reacted positively with one or more of the actinomycetes. Their serum titers increased with the duration of employment to an upper limit of 2.5. The sera of 19 non-exposed individuals were negative.

Streptomonospora

The description of the genus *Streptotomonospora* was based on a single isolate, *Streptoptomonospora salina* YIM 90002[T] (CCTCC 99003[T], DSM 44593[T]; Cui et al., 2001). The strain was isolated from a hypersaline habitat of soil samples collected from a salt lake in Xinjiang, a western province of China, during a taxonomic study on extremophilic actinomycetes. Owing to its optimal growth at 15% NaCl, this strain can be referred to as a moderately halophilic microorganism. Isolation and studies of additional strains of this genus will shed more light on the habitat and occurrence of members of this genus.

Isolation and Cultivation

Nocardiopsis

For enrichment and isolation of *Nocardiopsis*
strains, no specific procedures have been
described (Meyer, 1989), and the same methods described for the isolation of streptomycetes can generally be used (Korn-Wendisch
and Kutzner, 1992). Good results have been
obtained using the GAC-Agar medium of Nonomura and Ohara (Ohara, 1971a; Ohara,
1971b). The following special method for the
isolation of alkaliphilic *Nocardiopsis* strains has
been reported by Horikoshi (1971): a small
amount of soil is suspended in 1 ml of sterilized
water, 100 µl aliquots of the suspension are
spread on dry agar plates and the plates are
incubated for 7 to 14 days at 27°C. For isolation
of *N. halophila*, the modified complex medium
(SGA agar) containing 20% NaCl was used
(Al-Tai and Ruan, 1994), and *Nocardiopsis
kunsanensis* has been isolated on Bennett's
agar based on seawater (Chun et al., 2000).
Spores or mycelium of isolated colonies are
picked up with a needle and streaked on different cultivation media for color determination
and for macroscopic, microscopic and chemotaxonomic examination.

Solution A: GAC Agar (Nonomura and Ohara, 1971a;
1971b)

Glucose	1.0 g
L-Asparagine	1.0 g
K₂HPO₄	0.3 g
MgSO₄ · 7 H₂O	0.3 g
NaCl	0.3 g
Trace salts	1.0 ml
Distilled water	1.0 liter
Agar	15.0 g

Solution B:

Casamino acids	0.5 g
Adjust the pH to 8.0 with NaOH solution and autoclave	
Trace salt solution	
FeSO₄ · 7H₂O	10 mg
MnSO₄ · 7H₂O	1 mg
CuSO₄ · 5H₂O	1 mg
ZnSO₄ · 7H₂O	1 mg
Distilled water	1 ml
Adjust the pH to 8.0	

Dissolve solution A components in one liter of distilled water and solution B (casamino acids) in one
liter of distilled water and autoclave. Antibiotics
(cycloheximide, nystatin, each 50 µg/ml; polymyxin B,
4 µg/ml; and penicillin, 0.8 µg/ml) may be added to
autoclaved medium A to the final concentration. Four
ml of solution B is poured over 15 ml of hardened
agar plates.

Horikoshi Agar for Alkaliphilic *Nocardiopsis* Strains
(Horikoshi, 1971)

Glucose	10.0 g
Peptone	5.0 g
Yeast extract	5.0 g
K₂HPO₄	1.0 g
MgSO₄	0.2 g
NaCO₃	10.0 g
Distilled water	1.0 liter
Agar	20.0 g

Adjust the pH of the autoclaved medium with sterile
sodium carbonate to 10.

SGA Agar (Al-Tai and Ruan, 1994)

Casamino acids	7.5 g
Yeast extract	10.0 g
MgSO₄ · 7H₂O	20.0 g
Sodium citrate	3.0 g
KCl	1.0 g
NaCl	200 g
Trace salts	1 ml
Water	1.0 liter
Agar	15.0 g

Trace salt solution:	
FeSO₄ · 7H₂O	0.1 g
MnCl₂ · 4H₂O	0.1 g
ZnSO₄ · 7H₂O	0.1 g
Distilled water	100.0 ml

Dissolve all SGA agar components, autoclave and adjust
to pH 7.5.

Bennett's Agar Based on Seawater (Chun et al., 2000)

Glucose	2.0 g
Malt extract	1.0 g
Yeast extract	1.0 g
Peptone	2.0 g
Sea water	1 liter
Agar	20.0 g

Dissolve all ingrediens, autoclave and adjust to pH 7.2.

Most *Nocardiopsis* species grow very well at
28°C on agar media used in the ISP (Shirling and
Gottlieb, 1966). After 14–21 days of incubation
at optimal conditions, the colour of the white
aerial mycelia will change to yellow or gray, indicating the production of spore chains. For optimal growth of *N. alba* and some *N. dassonvillei*
strains, the pH of the ISP media should be
adjusted to pH 8.5 with sterile sodium carbonate
after autoclaving. Although *N. listeri* grows quite
well on the ISP media, no aerial mycelia are
produced. Hickey-Tresner agar (Hickey and
Tresner, 1952) is the only medium where sparse
white aerial mycelia are seen. The truly alkaliphilic species, i.e., *N. dassonvillei* subsp. *albirubida* and *N. prasina*, need an even higher pH of
10 to produce aerial mycelia and spores on yeast-
extract malt-extract agar. The pH has to be
adjusted to 10 or above with sodium hydroxide

after autoclaving. However, some of the strains can produce aerial mycelia and spores on ISP medium 3 (oatmeal agar) at pH 8.0. The halotolerant or moderately halophilic species, *Nocardiopsis halophila* and *Nocardiopsis kunsanensis*, grow optimally on most ISP media in the presence of 5–15% NaCl (Chun et al., 2000). Also, 5–10% NaCl should be added for optimal growth of *Nocardiopsis lucentensis* (Ventosa et al., 1998). The following media are used for cultivating most species of the genus *Nocardiopsis*:

ISP Medium 2: Yeast Extract Malt Extract Agar (Pridham et al., 1956–1957)

Yeast extract (Difco)	4.0 g
Malt extract (Difco)	10.0 g
Dextrose	4.0 g
Agar	20.0 g
Distilled water	1.0 liter

Dissolve and autoclave.

ISP Medium 3: Oatmeal Agar (Küster, 1959b)

Oatmeal	20.0 g
Agar	18.0 g
ISP-Ttrace salt solution:	
FeSO$_4$ · 7H$_2$O	0.1 g
MnCl$_2$ · 4H$_2$O	0.1 g
ZnSO$_4$ · 7H$_2$O	0.1 g
Distilled water	100.0 ml

Cook or steam 20g of oatmeal in 1 liter of distilled water for 20 minutes. Filter through cheesecloth. Add distilled water to restore volume of filtrate to 1 liter. Add salt solution. Adjust to pH 7.5 with NaOH. Add 18g of agar; liquefy by steaming at 100°C for 15–20 min.

ISP Medium 4: Inorganic Salts-starch Agar (Küster, 1959a)

Solution I:

Difco soluble starch	10.0 g

Make a paste of the starch with a small amount of cold distilled water and bring to a volume of 500 ml.

Solution II:

K$_2$HPO$_4$ (anhydrous)	1.0 g
MgSO$_4$ · 7H$_2$O	1.0 g
NaCl	1.0 g
(NH$_4$)$_2$SO$_4$	2.0 g
CaCO$_3$	2.0 g
Distilled water	500 ml
ISP-trace salt solution (see ISP medium 3) 1.0 ml	

Adjust to pH 7.5. Mix solutions I and II together. Add 20 g of agar. Liquefy agar by steaming at 100°C for 10–20 min.

ISP Medium 5: Glycerol-asparagine Agar (Pridham et al., 1956–1957)

L-asparagine (anhydrous)	1.0 g
Glycerol	10.0 g
K$_2$HPO$_4$ (anhydrous)	1.0 g
Distilled water	1.0 liter
ISP-trace salt solution	1.0 ml
Agar	20 g

Dissolve (and add trace salts [see ISP medium 3 above]) and adjust to pH 7.4.

ISP Medium 7: Tyrosine Agar Medium (Shinobu, 1958)

Glycerol	15.0 g
L-Tyrosine	0.5 g
L-Asparagine	1.0 g
K$_2$HPO$_4$ (anhydrous)	0.5 g
MgSO$_4$ · 7H$_2$O	0.5 g
NaCl	0.5 g
FeSO$_4$ · 7H$_2$O	0.01 g
Distilled water	1.0 liter
Trace salt solution	1.0 ml
Agar	20 g

Dissolve all ingredients (and add trace salt [see ISP medium 3 above]) and adjust to pH 7.5.

Hickey-Tresner Agar Medium (Hickey and Tresner, 1952)

Dextrin	10.0 g
Tryptone	2.0 g
Meat extract	1.0 g
Yeast extract	1.0 g
CoCl$_2$	2.0 g
Agar	15.0 g
Distilled water	1.0 liter

Dissolve and adjust to pH 7.5 and autoclave.

Thermobifida

Good sources for the isolation of *Thermobifida* strains are composted horse manure/straw (the so-called "mushroom compost"; Fergus, 1964), bio-composted paper (Kempf, 1995), municipal solid waste compost (Stutzenberger, 1971; Rész et al., 1977), or self-heated grass clippings (Kutzner, 2000). *Thermobifida* dominate the microflora in these habitats owing to their high cellulolytic activity. For the isolation of the actinomycetes, 1 g of air-dried sample is suspended in 100 ml of Ringer's solution and mixed with 10 g of sterile plastic beads. The sample is shaken at 180 rpm for 30 min to release the spores and mycelia from the compost particles. Subsequently, 0.1 ml of a decimal dilution series is plated on pre-dried starch-casein-KNO$_3$-agar plates. The Petri dishes should be pre-dried to reduce the swarming of *Bacillus* isolates. For reduction of the accompanying bacteria and fungi, Na-propionate (4 g/liter) is added (Lacey, 1974). However, the addition of Na-propionate results in the reduction of aerial mycelia of actinomycetes. Half-strength nutrient agar appears to be a good medium for *Thermobifida* strains.

Minimal salt-yeast extract nutrient medium (Ramachandra et al., 1988) supplemented with different carbon sources was shown to be good for stimulating the production of extracellular lignocellulose-degrading enzymes, e.g., endo-1,4-β-xylanase, endo-1,4-glucanase, and peroxidase, in *Thermobifida fusca*. The highest production of these enzymes was found at 50°C and pH 8 under high aeration and when the medium was supplemented with oat spelt xylan or ball-milled wheat straw. The highest activities of endoxylanase, endoglucanase, peroxidase, β-xylosidase, and α-L-arabinofuranosidase were detected on 0.6% or 0.8% (w/v) xylan, which corresponds to C:N ratios of 4:1 and 5.3:1, respectively. Straw 0.8% (w/v) was also a good growth substrate for the production of endoxylanase and endoglucanase, while peroxidase activities reached the highest level on 0.6% (w/v) straw. Maximum production of β-xylosidase and α-L-arabinofuranosidase was detected on 0.4% (w/v) straw (Tuncer et al.,1999).

Modified Starch-Casein-KNO$_3$-Agar (Küster and Williams, 1964, modified)

Starch (soluble)	10.0 g
Casein	0.3 g
KNO$_3$	2.0 g
KH$_2$PO$_4$	0.3 g
K$_2$HPO$_4$	0.7 g
MgSO$_4$ · 7H$_2$O	0.5 g
NaCl	1.0 g
Trace elements solution	5.0 ml
Distilled water	1 liter

Trace Elements Solution SPV-4 (Voelskow, 1988)

CaCl$_2$ · 2H$_2$O	4 g
Fe-III-citrate	1 g
MgSO$_4$	200 mg
ZnCl$_2$	100 mg
CuSO$_4$ · 5H$_2$O	40 mg
CoCl$_2$ · 6H$_2$O	22 mg
Na$_2$MO$_4$ · 2H$_2$O	25 mg
Na$_2$B$_4$O$_7$ · 10H$_2$O	100 mg
Water	1 liter

Adjust pH of the trace elements solution and modified starch-casein-KNO$_3$ to 7.2. Add 12g of agar.
Nutrient agar (1/2 conc.)

Peptone from casein	2.5 g
Meat extract	1.5 g
Distilled water	1 liter

Adjust to pH 7.2. Add 12g of agar.

Minimal Salt-Yeast Extract Nutrient Medium (Ramachandra et al., 1988)

Glucose	6.0 g
Yeast extract	0.1 g
(NH$_4$)$_2$SO$_4$	0.1 g
NaCl	0.3 g
MgSO$_4$	0.1 g
Ca$_2$CO$_3$	0.02 g
Trace elements solution SPV-4	1 ml
Water	1 liter
Trace elements solution	
Fe$_2$(SO$_4$)$_3$	1.0 g
ZnSO$_4$	0.9 g
MgSO$_4$	0.2 g
Water	1 liter

Adjust pH of the trace elements solution and minimal salt-yeast extract nutrient medium to 8.0 and autoclave.

Streptomonospora

Soils or other samples collected in areas with high salt concentrations may be sources for the isolation of halophilic or halotolerant strains of the genus *Streptomonospora* (Cui et al., 2001). Media supplemented with salt, e.g., the modified medium ISP 5, may be used for isolation of *Streptomonospora* (Cui et al., 2001).

Modified ISP Medium 5 for Isolation of *Streptomonospora salina* (Cui et al., 2001)

L-Asparagine (anhydrous)	1.0 g
Glycerol	10.0 g
K$_2$HPO$_4$ (anhydrous)	1.0 g
NaCl	150 g
Distilled water	1.0 liter
Trace salt solution	1.0 ml
Water	1 liter

Adjust to pH 7.0. Add 20 g of agar.

Streptomonospora salina grows very well at 28°C on most agar media used in the ISP (Shirling and Gottlieb, 1966). Only ISP medium 3 (oatmeal agar) does not support growth. Optimum growth occurs in media with salt at a concentration of 15% and pH 7.0 (Cui et al., 2001).

Identification

New micro-organisms can be identified as members of the family Nocardiopsaceae by both 16S rRNA gene analysis and phenotypic studies. The 16S rRNA gene sequences (full or partial sequences of about the first 500 nucleotides) of *Nocardiopsis*, *Thermobifida* and *Streptomonospora* are specific, and strains of these genera form closely aggregated clades (Rainey et al., 1996; Zhang et al., 1998; Cui et al., 2001). Many *Nocardiopsis* isolates or respective DNA samples may be also identified by the signature nucleotides from positions 1435 to 1466 of the *E. coli* numbering system of Brosius et al. (1978; Rainey et al., 1996). However, in order to affiliate new isolates to *Nocardiopsis*, *Thermobifida*, *Streptomonospora* or new genera, morphological and chemotaxonomic characteristics should be examined (Table 3). Identification of new strains at the species level requires additional phenotypic studies and in some cases determination of DNA/DNA relatedness. The phenotypic characteristics useful for identification of members of the family Nocardiopsaceae at different taxonomic levels and different approaches to identification based on these characteristics are given in the following sections.

Nocardiopsis

DIFFERENTIATION OF THE GENUS *NOCARDIOPSIS* FROM OTHER GENERA *Nocardiopsis* isolates can be differentiated from all other actinomycetes by a combination of phenotypic and chemotaxonomic markers. Although *Nocardiopsis* resem-

Table 3. Salient phenotypic characteristics that differentiate genera of the family Nocardiopsaceae.

Characteristic	*Nocardiopsis*	*Streptomonospora*	*Thermobifida*
Spore chains on AM	+	+	−
Long chains, AM disintegrated into spores	+	−	−
Short spore chains on sporophores	−	+	−
Single spores on short sporophores, in clusters	−	−	+
Fragmentation of SM	+[a]	−	−
Single spores on SM	−	+	−/+
Spore clusters on SM	−	−	−/+
Cell wall amino acids[b]	*meso*-DAP	*meso*-DAP, (LL-DAP), Gly, Lys, and Asp	*meso*-DAP, and (LL-DAP)
Cell wall type	III/C[c]	III/C[d]	III/C[c]
Phospholipid type	P III	P II	P II
Major menaquinones	10/0 to 10/8	9/6, 10/2, and 10/4	10/6, 10/8, and 11/6
Fatty acid type	3d	n.d.	3e
Growth temperature optimum, °C	28–37	28	40–50
Growth temperature range, °C	7–45	n.d.	35–60
Growth salinity optimum, NaCl%	0–15	15	<3
Growth salinity range, NaCl%	0–20	n.d.	<5

Symbols: see footnote in Table 2.

Abbreviations: AM, aerial mycelium; SM, substrate mycelium; Gly, glycine; Asp, aspartate; DAP, diaminopimelic acid; (LL-DAP), trace of LL-DAP; n.d., not determined; and see footnote in Table 3 for other definitions.

[a]Non-fragmenting hyphae may be present in some species.

[b]Alanine and glutamate are present in all genera.

[c]Teichoic acids are present, and glycerol or other main constituents of these polymers are determined in the cell wall (Naumova et al., 2001; see also Table 6).

[d]Glucose (39%), galactose (34%), ribose (15%), xylose (6%), arabinose (5%), and mannose (2%) were determined in whole cell hydrolysates.

Adapted from McCarthy and Cross (1984a); Grund and Kroppenstedt (1990); Yassin et al. (1993); Chun et al. (2000); Evtushenko et al. (2000); Cui et al. (2001); and Naumova et al. (2001).

bles many other actinomycetes morphologically (e.g., the presence of long sporulated aerial hyphae and fragmenting substrate mycelia), it shows quite specific chemical signatures (Tables 3 and 4). *Nocardiopsis* has a cell wall chemotype III/C, i.e., 2,6-*meso*-diaminopimelic acid, alanine, and glutamic acid in its peptidoglycan. The muramic acid of the peptidoglycan is acetylated (Kroppenstedt, 1987). In whole cell hydrolysates, glucose and galactose are also detected (type C). Diagnostic sugars, like arabinose, madurose, xylose or rhamnose, are absent (Lechevalier and Lechevalier, 1970b; Grund and Kroppenstedt, 1990; Yassin et al., 1993, 1997; Al-Tai and Ruan, 1994; Chun et al., 2000; Evtushenko et al., 2000; Naumova et al., 2001). Cell wall teichoic acids are present and composed of glycerol and other main constituents, like ribitol, glucosamine, galactosamine, succinic and propionic acids. Different *Nocardiopsis* species are determined by the cell wall teichoic acid composition (Naumova et al., 2001).

No mycolic acids are present in whole cell methanolysates (Mordarska et al., 1972; Minnikin et al., 1975). The phospholipid pattern is composed of phosphatidylcholine (cardiolipin), phosphatidylmethylethanolamine (PME), phosphatidylglycerol (PG), phosphatidyl inositol (PI), and small amounts of diphosphatidylglycerol (DPG). In addition, two spots of glycolipids identified as monomannosyl diglyceride and monoacetylated glucose (Mordarska et al., 1983) and 2–4 unknown phospholipids that run above the DPG spot are detected on two-dimensional thin layer chromatography (TLC)-plates (Minnikin et al., 1977; Grund and Kroppenstedt, 1990; Fig. 2). The occurrence of phosphatidylcholine in lipid extracts of *Nocardiopsis* corresponds with phospholipid type III, according to Lechevalier et al. (1977). *Nocardiopsis* synthesizes menaquinones with an isoprene side chain of 10 units. Depending on the species up to four of these isoprene units are saturated, i.e., MK-10(H$_0$) to MK-10(H$_8$). Small amounts of the MK-9 series are also found (Minnikin et al., 1978; Fischer et al., 1983; Grund and Kroppenstedt, 1990; Fig. 3).

Members of this genus are characterized by terminally branched and 10-methyl-branched fatty acids showing a chain length of 14 to 18 carbon atoms. Hydroxy fatty acids have never been detected in *Nocardiopsis* strains (Grund and Kroppenstedt, 1990). Among the terminally branched fatty acids, 14-methyl-heptadecanoic acid (*iso*-16:0; 21–35%) and 14-methyl-hexadecanoic acid (*anteiso*-17:0; 7–24%) are the main components. Substantial amounts (6–18%)

Table 4. Chemical markers useful to differentiate members of Nocardiopsaceae genera and other sporoactinomycetes with similar morphology and/or possessing cell wall type III and/or phospholipid type III.

Genus	Cell wall type[a]		Diagnostic sugars[b]	Sugar type[c]	Cell wall teichoic acids[d]	G+C content (mol%)	Diagnostic phospholipids	Phospholipid type[e]	Principal menaquinone[f]	Menaquinone type[g]	Fatty acid type[h]
Nocardiopsis	III	Acetyl	None	C	Present	64–71	PC, PME	III	10/0 to 10/8	4C2	3d
Thermobifida	III	Acetyl	None	C	Present	66–69	PE, PME, GL	II	10/6, 10/8, 11/6	4D	3e
Streptomonospora	III	n.d.	Gal, Xyl, Ara, Man	C	n.d.	n.d.	PE, PI, PG	II	9/6, 10/2, 10/4	3C?	n.d.
Actinopolyspora	IV	Acetyl	Ara, Gal	A	None	64	PC, PE	III	9/4, 9/6	4B2	2e
Saccharopolyspora	IV	Acetyl	Ara, Gal	A	None	67–77	PC, 1-PE, OH-PE, PME	III	9/4, 9/2, 10/4	3B/4A	3e
Pseudonocardia	IV	Acetyl	Ara, Gal	A	None	79	PC, 1-PE, OH-PE	III	8/4	2B	3f
Kineosporia[i,j]	I/III	Acetyl	Gal, Man (Rib)	C	n.d	69–71	PC	III	9/4	2D	1a/1c
Kribbella	I	Acetyl	None	n.c.	n.d.	68–70	PC	III	9/4	2D	3c
Hongia	I	Acetyl	None	n.c.	n.d.	71	PC, DPG, PG, PI	III	9/4	2D	3c
Catenuloplanes[i,k]	VI	Glycolyl	Xyl, Man (Rib)	n.c.	None	70–72	PC	III	10/4, 11/4	4C?	2c
Kitasatospora	III/I	Acetyl	Gal, Man	C	None	66–76	PE	II	9/4, 9/6, 9/8	4B1	2c
Actinosynnema[j]	III	Acetyl	Gal, Man	C	None	71	PE, PIM, PI, DPG	II	9/4, 9/6	3B	3f
Lentzea	III	Acetyl	Gal, Man, Rib	C	None	69–71	PE, DPG, PG, PI	II	9/4	2D	3f
Lechevalieria	III	Acetyl	Gal, Man, Rham	E	None	70	PE	II	9/4	2D	3f
Actinokineospora[j]	III	Acetyl	Gal, Man, Rham	E	None	69–70	PE, OH-PE	II	9/4	2D	2d
Saccharothrix	III	Acetyl	Gal, Rham, Man	E	None	72–76	PE, OH-PE, PI, PIMs, DPG, PG	II, IV	10/4, 9/4	2D/3B	3f
Streptoalloteichus[a,l]	III	Acetyl	Rham, Gal, Man	E	None	n.d.	PE, PI, DPG	II	9/6, 10/6	3C	3e
Actinoalloteichus	III	Acetyl	Rham, Gal, Man	E	n.d.	73	PE, PME, PI	II	9/4	2D	2d
Actinocorallia	III	Acetyl	Mad, Gal	B	Present	66	PE	II	9/4, 9/6, 9/8	4B2	3a
Actinomadura	III	Acetyl	Mad, Gal	B	Present	66–70	PI, PIM, DPG	I	9/4, 9/6, 9/8	4B2	3a
Thermomonospora	III	Acetyl	Mad, Gal	B	Present	66–70	PI, PIM, DPG	I	9/4, 9/6, 9/8	4B2	3a
Herbidospora	III	Acetyl	Mad, Gal	B	Present	69–71	PE, PGN	IV	10/4	4C	3c
Microtetraspora	III	Acetyl	Mad, Gal	B/C	Present	66	PGN, OH-PE	IV	9/4	2D	3c
Nonomuraea	III	Acetyl	Mad, Gal	B	Present	64–69	PE, DPG, PGN, OH-PE	IV	9/0, 9/2, 9/4	4A2	3c

Table 4. *Footnote continued*

Abbreviations: n. d., not determined; n. c., not characteristic; Rham, rhamnose; Ara, arabinose; Gal, galactose; Mad, madurose (3-*O*-methylgalactose); Xyl, xylose; PC, phosphatidylcholine; PME, phosphatidylmethylethanolamine; GL, glycolipid; PI, phosphatidylinositol; PG, phosphatidylglycerol; OH-PE, hydroxy-phosphatidylethanolamine; DPG, diphosphatidylglycerol; PIMs, phosphatidyl-*myo*-inositol mannosides; and PGN, phosphatidylglucoseamine.

[a]Cell wall types are given in H. A. Lechevalier and M. P. Lechevalier (1970a, b); cell wall types correspond to the following peptidoglycan types: I, A3 gamma; and II, III, and IV, A1 gamma (Schleifer and Kandler, 1972).

[b]Diagnostic sugars may be present, but have no taxonomic implication in cell wall type I[c]. Ribose (Rib) reported for some species is usually determined in whole cell hydrolysates of all actinomycetes.

[c]Whole cell sugar types given in M. P. Lechevalier and H. A. Lechevalier (1970a, b).

[d]Glycerol and/or ribitol present if cell wall contains teichoic acids.

[e]Phospholipid types according to Lechevalier et al. (1977).

[f]10/4, MK-10 (H₄), i.e., a menaquinone having 10 isoprene units, two of which are saturated.

The principal menaquinones with different degree of saturation are characteristic for different *Nocardiopsis* species (see Table 6). For *Streptoalloteichus, Actinopolyspora* and *Catenuloplanes*, the following major menaquinones were also reported: 10/4, 10/6; 9/4, 10/4; and 9/8, 10/8, respectively (Tamura et al. 2000b). In *Saccharopolyspora halophila*, menaquinones 9/6 and 9/8 were determined (Lu et al., 2001).

[g]Menaquinone types according to Kroppenstedt (1985) are exemplified as follows: type 1, no hydrogenation of the isoprenoid chain, e.g., MK-9; type 2, only one menaquinone present, di- or tetrahydrogenated, e.g., MK-9(H₄); type 3, a tetrahydrogenated multiprenyl menaquinone, e.g., MK-9(H₄) and MK-10(H₄); type 4, menaquinones with the same isoprenoid chain length but increasing degree of hydrogenation, e.g., MK-9(H₄), MK-9(H₆), and MK-9(H₈); positions of double bonds in isoprenoid chains are diagnostic, e.g., MK-9(II,III-H₄) vs. MK-9(III,VIII-H₄) 4A1 vs. 4A2; and MK-9(II,III,IX-H₆) vs. MK-9(II,III,VIII-H₆) 4B1 vs. 4B2. From Batrakov and Bergelson (1978); Yamada et al. (1978a, b); Howard, et al. (1986); Collins and Kroppenstedt (1987); and Collins et al. (1988).

[h]Fatty acid types according to Kroppenstedt (1985); for further information, see Table 11. Fatty acid type 2c was also reported for *Actinopolyspora* and *Saccharopolyspora* (Tamura et al., 2000b).

[i]Motile spores are formed in aqueous environment.

[j]No aerial mycelium was observed; rhamnose and 3-*O*-methylrhamnose may be present in cell wall of some species.

[k]Aerial mycelium is rudimentary, developed or absent; the spore chains are aggregated into clusters and may be developed by outer sheaths.

[l]Sporangiospores are found in addition.

Adapted from: Evtushenko et al. (1984, 2000); Kroppenstedt (1985); Naumova (1988); Streshinskaya et al. (1989); Grund and Kroppenstedt (1990); Labeda et al. (1992, 2001); Kudo et al. (1993, 1999); Tomita et al. (1993); Yassin et al. (1993, 1997); Al-Tai and Ruan (1994); Tamura et al. (1998, 2000a, b); Zhang et al. (1998, 2001); Chung et al. (1998, 2001); Park et al. (1999); Chun et al. (2000); Labeda and Kroppenstedt (2000); Lee et al. (2000); Lu et al. (2001); and Naumova et al. (2001).

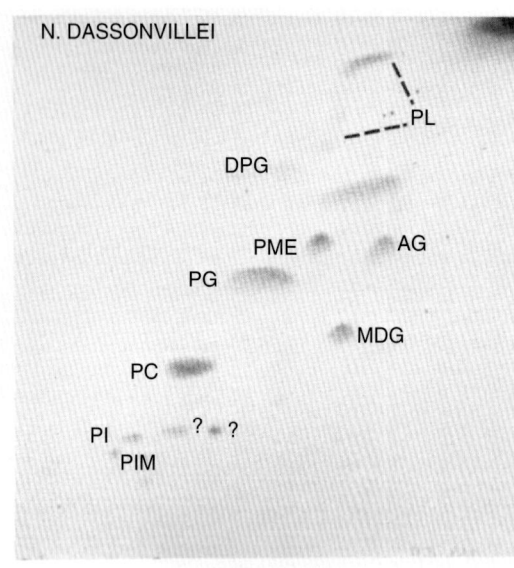
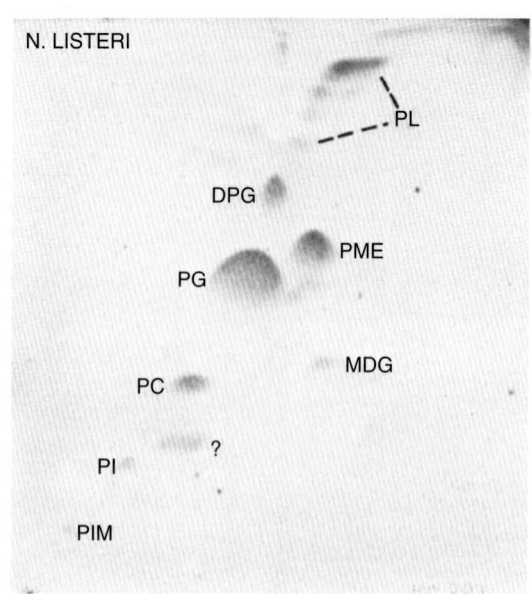

Fig. 2. Two-dimensional thin layer chromatograms of polar lipids of *Nocardiopsis dassonvillei* DSM 43111 and *N. listeri* DSM 40297. Abbreviations: DPG, diphosphatidyglycerol; PME, phosphatidylmethylethanolamine; PG, phosphatidyglycerol; PC, phosphatidylcholine; PI, phosphatidylinositol; PIM, phosphatidylinositolmannosides; AG, monoacetylated glucose; MDG, monomannosyl diglyceride; PL, unknown phospholipids specific for *Nocardiopsis*; β Lipids of unknown structure.

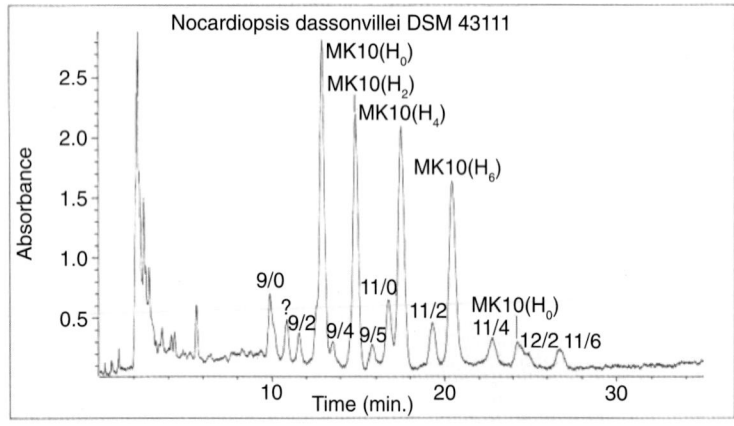

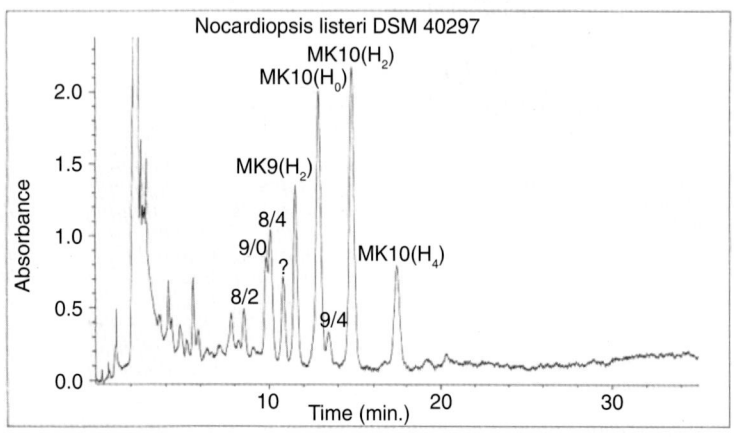

Fig. 3. Menaquinone profiles of *Nocardiopsis dassonvillei* (top) and *N. listeri* (bottom). The extent of hydrogenation of the 10 isoprene units is shown by the subscript of the abbreviation. For instance, MK-10(H$_8$) is a menaquinone with four hydrogenated isoprene units. Injection volume 5 µl; 5% of the extract prepared by the method of Minnikin et al (1984) without further purification.

of the 10-methyl branched tuberculostearic acid, i.e., 10-methyl-octadecanoic acid (10-methyl-18:0), and its precursor the unsaturated *cis* 9,10 octadecenoic acid (18:1 *cis*; 2–19%) have been found in addition (Grund and Kroppenstedt, 1990). According to Kroppenstedt (1985), this fatty acid pattern belongs to fatty acid type 3d. Table 5 lists the fatty acid types found among Actinomycetales. They can be combined into 3 main types and 14 subtypes (diagnostic fatty acids of the main types in bold). Type 1 is characterized by unbranched saturated and unsaturated fatty acids and those fatty acids for which unsaturated fatty acids are precursors, i.e., cyclopropane and 10-methyl-branched fatty acids. Type 2 is composed of *iso*- and *anteiso*-branched fatty acids. In type 3, both types, (the terminally branched [*iso/anteiso*] and the 10-methyl-branched fatty acids) are present. Subtypes of types 2 and 3 are mainly based on quantitative differences of the fatty acids and the occurrence of hydroxy fatty acids.

PRESUMPTIVE IDENTIFICATION OF *NOCARDIOPSIS* STRAINS BASED ON MORPHOLOGY Although *Nocardiopsis* strains resemble many other actinomycetes morphologically, their specific macroscopic and microscopic features may be used for presumptive identification. Members of this genus produce sparse-to-abundant white aerial mycelia that in most species become yellow-gray (griseus-colored) during spore formation. The aerial hyphae are long, moderately branched, straight to flexuous. They show a typical zig-zag formation before spore delimitation. The elongated smooth spores differ in length and can divide subsequently into smaller spores of irregular size. Spores are enclosed within a fibrillar sheath and have thickened polar walls.

Although many *Nocardiopsis* strains resemble *Streptomyces griseus* or other streptomycete strains macroscopically (Gordon and Horan, 1968; Shirling and Gottlieb, 1972), the two genera can be differentiated by their specific type of spore formation and spores (Table 2). In contrast to *Streptomyces* where spores of the same shape are delimited basipetally almost simultaneously, in *N. dassonvillei*, the cross walls are formed in a relatively uncoordinated manner resulting in spores of various lengths. *Actinomadura* (including the species currently belonging to *Nonomuraea*, *Saccharothrix* and *Lechevalieria*) has also been mixed up with *Nocardiopsis* (Table 1). However, *Nocardiopsis* differs from these genera in morphology and the specific mode of spore formation. In *Actinomadura* and *Nonomuraea*, special spore-forming hyphae develop on sterile aerial mycelia, either singly or in clusters, producing short chains of spores showing hooks or spirals with 1.5 to 2 turns, whereas in *Nocardio-*

psis, special spore-forming hyphae are absent. In this latter genus, the straight, spiralled aerial mycelium disintegrates completely into fragmentation spores (Table 2). However, it is quite difficult to differentiate *Nocardiopsis* from *Glycomyces harbinensis* and *Glycomyces rutgersensis* on morphological characters alone. Slight differences may be found in the formation of substrate and aerial mycelia between them. Labeda et al. (1985) reported stable substrate mycelia and scant aerial mycelia carrying short chains of square-ended conidia in *Glycomyces* species, whereas in many *Nocardiopsis* strains, the substrate mycelium fragments very early depending on strain and culture conditions. In addition, most strains of *Nocardiopsis* develop abundant aerial mycelia with long hyphae that fragment completely into chains of conidiospores (Meyer, 1989). Similarities also occur in morphology between *Nocardiopsis* and *Saccharothrix*, *Lentzea*, and *Lechevalieria* (Labeda et al., 1984, 2001; Labeda, 1992). At the macroscopic and microscopic levels, no clear dissimilarities may be observed between the *Nocardiopsis* strains and some representatives of the genera *Amycolatopsis* (Lechevalier et al., 1986; Mertz and Yao, 1993), *Kitasatospora* (Chung et al., 1999) or other actinomycetes exhibiting long spore chains and the fragmented vegetative myceliaum.

RAPID IDENTIFICATION OF THE GENUS *NOCARDIOPSIS* FROM OTHER GENERA BASED ON PHOSPHOLIPID PATTERN AND MAJOR CONSTITUENTS OF CELL WALL A rapid diagnosis for reliable differentiation of *Nocardiopsis* strains from all other actinomycete taxa is based on their specific phospholipid pattern type III, and in some cases, on the diagnostic diamino acid or sugars arabinose and galactose determined in whole cell hydrolysates (Table 3). Besides *Nocardiopsis*, only a few actinomycete genera, namely, *Kribella*, *Hongia*, *Kineosporia*, *Catenuloplanes*, *Actinopolyspora*, *Pseudonocardia* and *Saccharopolyspora*, show the same phospholipid type, with phosphatidylcholine as main diagnostic component. The *Nocardiopsis* strains can be easily separated from the first four genera by the presence of *meso*-DAP in their whole cell hydrolysates, and they are differentiated from *Pseudonocardia*, *Actinopolyspora*, and *Saccharopolyspora* by absence of arabinose and galactose in whole cell hydrolysates, presence of glycerol or ribitol in the cell wall, or by additional lipid markers.

ADDITIONAL LIPID MARKERS: INFORMATION FOR THE GENERIC LEVEL IDENTIFICATION OF THE GENUS *NOCARDIOPSIS* BY LIPIDS The *Nocardiopsis* strains synthesize phospho- and glycolipids

Table 5. Fatty acid types and the diagnostic fatty acids among different genera of Actinomycetales.[a]

Genus	Type	Iso-15:0	Ante-iso-15:0	Iso-16:0	16:0	10-Methyl 16:0	Iso-17:0	Ante-iso-17:0	17:0	17:1	10-Methyl 17:0	Iso-18:0	18:0	18:1	10-Methyl 18:0	2-OH	Cycl. 19:0
Corynebacterium[b]	1a	-	-	-	+++	-	-	-	-	-	-	-	+	+++	-	-	-
Nocardia[c]	1b	-	-	-	+++	++	-	-	-	-	-	-	+	++	+++	-	-
Actinomyces	1c	-	-	-	+++	-	-	-	-	-	-	-	+	++	-	-	++
Saccharomonospora	2a	+	-	+++	++	-	+	+	+	+	-	+	-	-	-	++	-
Thermoactinomyces	2b	++++	++	++	+	-	+++	++	-	-	-	-	-	-	-	-	-
Streptomyces[d]	2c	++	+++	+++	+	-	+	++	-	-	-	-	-	-	-	(v)	-
Actinoplanes	2d	+++	++	++++	++	-	++	-	+	+	-	+	+	++	-	-	-
Actinomadura	3a	-	-	++	+++	+	-	-	+	+	+	+	+	++	++	-	-
Micromonospora	3b	++++	+	++	-	-	++	++	+	++	+	+	-	+	-	-	-
Microtetraspora	3c	++	+	+++	+	+	+	+	++	+	++	+	+	-	-	++	-
Nocardiopsis	3d	+	+	+++	+	-	+	+++	+	+	++	+	+	++	++	-	-
Pseudonocardia	3e	++	-	++++	+	-	+	+	-	+	+	-	-	-	-	-	-
Amycolatopsis	3f	++	+	++++	+	-	+	+	++	++	+	-	-	-	-	+	-
Saccharothrix	3g	++	+	++++	+	-	+	++	+	+	+	+	-	-	-	+	-

Symbols: +, 1–5%; ++, 5–10%; +++, 15–20%; ++++, >25%; and (v), variable, usually less than 2% for one component.
[a]Diagnostic fatty acids of main types are emphasized by blue bold and enlarged crosses, those of subtypes by green bold crosses, i.e., type 1, straight chain saturated and unsaturated fatty acids; type 2, iso/anteiso-branched fatty acids; type 3, iso/anteiso-branched fatty acids plus 10-methyl-branched fatty acids.
[b]Except C. ammoniagenes, C. bovis, C. minutissimum, C. pilosum and C. viriablis which belong to type 1b.
[c]Including Dietzia, Gordonia, Mycobacterium, Rhodococcus, Skermania, Tsukamurella and Williamsia.
[d]Occurrence of hydroxy fatty acids is species specific among Streptomyces. Branched 2-hydroxy fatty acids could be found in all tested strains of S. coelicolor (30 strains), S. rimosus (14 strains), S. violaceusniger (18 strains) and in 20 of 27 strains of S. hygroscopicus. No hydroxy fatty acids could be detected in S. albus (33 strains), S. fradiae (25 strains), S. glaucescens (8 strains), S. griseus (22 strains), S. violaceoruber (16 strains), S. viridochromogenes (25 strains).
From Kawanami et al. (1969) and Y. Zhang and R. M. Kroppenstedt (unpublished observations).

that are not found in *Pseudonocardia*, *Actinopolyspora*, and *Saccharopolyspora*, which also contain *meso*-DAP in their cell walls. Separating the lipid extract on TLC plates using the solvent system of Minnikin et al. (1984), 3–4 additional unidentified phospholipids (PL) with high R_f-values above diphosphatidylglycerol (DPG) and two glycolipids (monomannosyl diglyceride and monoacetylated glucose) right below phosphatidyl methylethanolamine (PME) can be detected on TLC plates (Minnikin et al., 1977; Grund, 1987; Mordarska et al., 1998; Fig. 2).

In addition, these three genera with the above listed phospholipid type III may be further differentiated from *Nocardiopsis* by looking for specific phospholipids containing primary or secondary amino groups, i.e., phosphatidylethanolamine (PE), PME, hydroxy-PE and lyso-PE, (1) PME and lyso-PE in *Saccharopolyspora* species, (2) PE, lyso-PE, hydroxy-PE in *Pseudonocardia* and (3) PE in *Actinopolyspora*. In *Nocardiopsis*, only one ninhydrin-positive spot, PME, is present.

Nocardiopsis species show complex menaquinone patterns containing menaquinones from MK-10 to MK-10(H_8) (Fig. 3, Table 4). Small amounts of the MK-9 and/or MK-11 series are also found. Although the major menaquinones of MK-10 series are found in some other conidia- and arthrospore-forming actinomycetes, the latter can be distinguished because here MK-10(H_4), MK-10(H_2) or MK-10(H_6) usually occurs in different combinations, together with MK-9(H_4) as in *Saccharopolyspora* and *Saccharothrix*, MK-9(H_6) as in *Streptomonospora*, or high amounts of MK-10(H_8) and MK-11(H_6) as in *Thermobifida* strains or MK-11(H_4) typical of *Catenuloplanes* (Kudo et al., 1999). Only *Herbidospora* and *Spirilliplanes* were reported to contain the major menaquinone MK-10(H_4), along with other minor components of MK-10 series, e.g., MK-10(H_2) and MK-10(H_6) or MK-10(H_2) and MK-10(H_8) (Kudo et al., 1993; Tamura et al., 1997). However, these genera can be differentiated easily from *Nocardiopsis* spp. by morphology, polar lipids, and cell wall and fatty acid composition.

The fatty acid pattern is a significant taxonomic marker in *Nocardiopsis*. *Nocardiopsis* strains are easily separated from all other actinomycetes on the basis of their specific fatty acid patterns (Fischer et al., 1983; Poschner et al., 1985; Tables 3 and 4). For differentiation from other taxa, both qualitative and quantitative differences of the fatty acid patterns are useful (Grund and Kroppenstedt, 1989, 1990).

DIFFERENTIATION OF *NOCARDIOPSIS* SPECIES
Nocardiopsis species may be differentiated from each other by comparing the occurrence of aerial mycelia, morphology, color of substrate and aerial mycelia, pattern of carbon-source utilization and growth at different pH, salt concentrations and temperatures (Table 6). Most *Nocardiopsis* species produce white aerial mycelia, which become yellowish grey (griseus color) after 2–3 weeks. On specific media, the white aerial mycelia of some *Nocardiopsis* develop shades of different colors with age, e.g., *N. prasina* develops, on mineral starch agar and oat meal agar, a leek-green aerial mycelium; colonies of *N. dassonvillei* subsp. *albirubida* turn grey after 3–4 weeks. *Nocardiopsis listeri* fails to produce any aerial mycelia at all on nearly all media. The reverse sides of most *Nocardiopsis* colonies show no specific colors. Some show a reddish tone like *N. dassonvillei* subsp. *albirubida* (brownish-red), *N. halophila* (yellow-red), *N. trehalosi* (orange), and *N. synnemataformans* (pimento colored).

Growth on high salt concentration of up to 20% NaCl is of diagnostic value for *N. halophila* and *N. kunsanensis*, whereas *N. prasina* and *N. dassonvillei* subsp. *albirubida* can easily be differentiated from the other *Nocardiopsis* species by their alkaliphilic behavior. They are able to grow at pH up to 12 but not at pH 6. Their optimal pH for growth is pH 10. The other *Nocardiopsis* species prefer mild alkaline condition (pH 7.7–8), but they are able to grow at pH 6. *Nocardiopsis* are mesophilic actinomycetes, which grow best at about 28°C. Some species have an optimum at 37°C (*N. kunsanensis*) or grow up to 42°C, but only *N. trehalosi* is able to grow at 45°C (Table 6).

The menaquinone profiles can also be used for separation of *Nocardiopsis* strains at the species level. In *N. dassonvillei* subsp. *albirubida*, *N. listeri*, and *N. synnemataformans*, the highly saturated MK-10(H_6) and MK-10(H_8) are usually missing. In addition, *N. dassonvillei* subsp. *albirubida* and *N. listeri* synthesize a higher amount of MK-9(H_4) and can be separated from most other *Nocardiopsis* species by this difference (Fig. 3). Exceptions are found in some strains of *N. dassonvillei* subsp. *dassonvillei*. Most strains of this species, together with *N. alba* and *N. prasina*, have predominant menaquinone MK-10(H_4). The major menaquinone components of other species are highly saturated, up to four isoprene units.

Recently, Naumova and coworkers demonstrated that the structures and different combinations of cell wall teichoic acids are useful for differentiation of *Nocardiopsis* species (Naumova et al., 2001). The degradation products of whole cell walls derived from teichoic acids and the ^{13}C-NMR spectra of polymers were able to differentiate each of the eight species and subspecies studied (Table 7). These products (glucose, ribitol, glycerol, succinic and propionic

Table 6. Differentiating characters for *Nocardiopsis* species and subspecies.

Characteristics[1]	N. dassonvillei dassonvillei	N. dassonvillei albirubida	N. alba	N. halophila	N. kunsasensis	N. listeri	N. lucentensis	N. prasina	N. synnemata-formans	N. trehalosi	N. tropica
Color of AM	White to gray or to yellowish-gray	White	White	White	White	Lacking	White	Olive-greenish	White	Griseus	White
SM	Beige-brown	Brownish-red	Light yellow	Yellow-red	Yellow	Beige	Yellowish	Light olive	Pimento	Orange-yellow	Olive-yellow to ochra
Synnemata	–	–	–	–	–	–	–	–	+	–	–
Utilization of											
L-Arabinose	+	+	–	+	nd	+	–	+	–	+	+
Cellobiose	+	–v	+	w	–	+	+	–v	+	–	+
D-Galactose	(+)	+v	+	+	–	+v	–	–	+	+	+
Inositol	–	–	–	+	–	+v	+	–	–	–	–
D-Lactose	–	+v	–	nd	nd	–v	+	–	+	+	+
Maltose	(+)	+v	+	+v	–	–	+	–	+	+	+
Mannitol	+	+	–	–	nd	+v	+	–	+	+	+
Melibiose	–	–	–	nd	nd	+	–	–	nd	+	+
L-Rhamnose	+	+	–	+v	+	+	+	+	+	+	+
Sucrose	(+)	+	–	+	+	+v	+	–	–	–	+
Trehalose	+	+	+	+	–	–	+	–	–	+	+
D-Xylose	+	+	–	+	v	+	–	+	+	+	+
Acid from											
Adonitol	(–)	–	–	nd	nd	–v	–	–	+	–	–
L-Arabinose	(+)	+	–	+	nd	+	–	+	–	+	+
Galactose	(+)	+	–	–	nd	+	–	–	+	–	+
Inositol	(–)	–	–	+	nd	–	+	–	+	–	–
D-Lactose	–	–	–	nd	nd	+v	+	–	+	+	–
Mannitol	(+)	+	+v	+	nd	–	–	+	+	+	+
D-Mannose	(–)	+	–	–	nd	–	+	–	+	+	+
Melibiose	–	–	–	nd	nd	+v	–	–	+	–	+

L-Rhamnose	+	−	+	nd	+	+	+	−	+
Sucrose	(+)	+	−	nd	−	+	+	−	+
D-Xylose	(+)	−v	+	nd	+	−	nd	+	−
Decarboxylation of									
Lactate	(+)	+	nd	nd	+	+	−	+	+
Oxalate	(+)	−	nd	nd	−	−	nd	−	+
Propionate	−v	+	nd	nd	+	+	nd	+	+
Decomposition of									
Casein	+	+	+	+	+	+	+	+	−
Tyrosine	(+)	+	−	+	+	+	+	−	−
Tween 80	(−)	+	+	nd	−	−	nd	+	+
Tween 85	(+)	+	nd	nd	−	−	nd	−	+
Nitrate reductase	(−)	(−)	nd	−	+	+	+	−	−
Urease	(−)	(+)	+	+	−	−	+	+	+
Growth at									
pH (optimal)	10	9	7.5[3]	9	8	7.5	7.5[3]	7.5[3]	7.5[3]
0% NaCl	+	+	+v	−	+	+v	+	+	+
20% NaCl	−	−	+	+	−	−	−	−	−
10°C	(−)	−	−	nd	+	+	−	−	−v
42°C	(+)	(+)	−	nd	−	−	−v	−	−
45°C	−	−	−	nd	−	−	−	+	−

Symbols: +, all strains are positive; (+), most strains of the species are positive; −, all strains are negative; (−), most strains of the species are negative.

Abbreviations: AM, aerial mycelium; SM, substrate mycelium; nd, no data available; and v, variable between different publications and experiments.

[3]Optimal pH for growth was not tested

Data from A1-Tai and Ruan (1994); Chun et al. (2000); Evtushenko et al. (2000); Grund and Kroppenstedt (1990); Ventosa et al. (1998); Yassin et al. (1993, 1997).

Table 7. Differentiation of *Nocardiopsis* species and subspecies based on cell wall analysis (acid hydrolysis), menaquinone pattern and some growth and morphological properties.

Characteristics[a]	N. dassonvillei dassonvillei	N. dassonvillei alborubida	N. alba	N. halophila	N. kunsanensis	N. listeri	N. lucentensis	N. prasina	N. synnemata-formans	N. trehalosi	N. tropica
Products of acid degradation of cell wall											
Glycerol	+	+	+	nd	nd	+	+	+	nd	+	+
Ribitol	−	+	+	nd	nd	+	−	+	nd	−	+
Glucose	+	−	−	nd	nd	−	−	−	nd	+	−
Pyruvate	−	−	+	nd	nd	+	−	−	nd	−	−
Succinate	−	+	−	nd	nd	−	−	−	nd	−	−
Major menaquinones[b]	10/2 to 10/6	10/0	10/4, 10/6	10/6, 10/8	10/8	10/0, 10/2	10/6, 10/8	10/4	10/0, 10/2	10/4, 10/6	10/6, 10/8
Growth at:											
20% NaCl	−	−	−	+	+	−	−	−	−	−	−
45°C	−	−	−	−	−	−	−	−	−	+	−
pH (optimal)	8	10	9	7.5	9	8	7.5	10	7.5	7.5	7.5
Synnemata	−	−	−	−	−	−	−	−	+	−	−

[a]Data from: Chun et al. (2000); Grund and Kroppenstedt (1990); Evtushenko et al. (2000); Naumova et al. (2001); Yassin et al. (1993; 1997).
[b]Abbreviations for menaquinones are the same as in Table 3.

Table 8. Main characteristics which differentiate *Streptomonospora salina* from halophilic or moderately halophilic sporeforming actinomycetes containing meso-diaminopimelic acid in the cell wall.

Characteristic	Sm. salina	A. halophila	A. mortivallis	A. iraquiensis	N. lucentensis	N. halophila	N. kunsasensis
Spore chain on aerial mycelium	Short	Long (>20)	Short (<10)	Short (<15)	Long	Very long	Long
Shape of aerial spore	Oval/cylindrical	Oval/spheroidal	Cylindrical	Oval/cylindrical	Spheroidal	Cylindrical	Cylindrical
Spores on substrate mycelium	Single	Fragmentation	Fragmentation	Fragmentation	Fragmentation	Fragmentation	None
NaCl optimum, %	15	20	15–20	10–15	5–10	15	10
NaCl maximum, %	n.d.	30	30	20	n.d.	20	20
Growth temperature optimum, °C	28	37	45	28	28	28	37
Phospholipid pattern	P II	P III	P III	P III	P III	P III	III
Cell wall chemotype	III/C	IV	IV	IV	III/C	III/C	III/C
Major menaquinones	9/6, 10/2, 10/4	9/4	10/6, 10/8	9/4	10/8	10/6, 10/8	10/8

Data from Al-Tai and Ruan (1994); Chun et al. (2000); Cui et al. (2001); Gochnauer et al. (1975; 1989); Johnson et al. (1986); Ruan et al. (1994); Ventosa et al. (1998); Yassin et al. (1993); Yoshida et al. (1991).
Abbreviations for phospholipid pattern, cell wall chemotype and menaquinones as in table 3.

acids), in some cases combined with the major menaquinones, morphological features and the ability of strains to grow at different conditions, allow one to attribute new strains to the listed *Nocardiopsis* species.

Nocardiopsis strains isolated from saline or hypersaline soils or other habitats are easily identified using the criteria listed in Table 8.

Thermobifida

DIFFERENTIATION OF *THERMOBIFIDA* FROM OTHER GENERA Identification of *Thermobifida* strains at the generic level requires determination of several genus-specific markers, namely, the ability to grow at higher temperatures, formation of single spores on branched sporophores resulting in spore clusters on aerial mycelia and sometimes on non-fragmented substrate mycelia, cell wall chemotype III C, presence of cell wall teichoic acids, the highly saturated major menaquinones of the MK-10 series, phospholipid type P II, and fatty acid pattern type 3e (see Table 3).

Macroscopic and microscopic criteria as well as optimal growth temperature and medium are suitable for presumptive differentiation of *Thermobifida alba* and *Tb. fusca* from the other genera containing thermophilic species. Table 9 lists some thermophilic species of other genera with their conventional diagnostic markers. Chemotaxonomic markers support the separation of *Thermobifida* from the respective genera (Tables 3 and 4).

Menaquinones of *Thermobifida* with their highly saturated isoprenoid side chain are of diagnostic value. Up to five isoprene units may be saturated. In addition to the MK-10 series, minor amounts of menaquinones with 11 isoprene units, i.e., MK-11(H_{4-10}), are found. This pattern has been reported only for some mesophilic species of *Nocardiopsis* and *Micromonospora* (Fischer et al., 1983; Collins et al., 1984). Although the phospholipid type II is widely distributed among actinomycetes, it is useful for the differentiation of *Thermobifida* species from thermophilic species of the genera *Thermomonospora*, *Actinomadura*, *Microbispora*, *Nonomuraea*, *Saccharopolyspora*, or *Pseudonocardia* (Tables 4 and 8) or from mesophilic species of other genera which can grow at elevated temperatures, show similar morphological features, and/or synthesize hydrogenated major menaquinones of MK-10 or MK-11 series, e.g., *Nocardiopsis*, *Streptomonospora*, *Streptoalloteichus*, or *Catenuloplanes* (Table 4).

The analysis of fatty acids is quite useful to differentiate *Thermobifida* species from other actinomycete species with similar morphology or formerly referred to the genus *Thermomonospora* (Table 8). *Thermobifida* is characterized by

a high amount of *iso-* and *anteiso*-branched fatty acids, namely, 15-methyl-hexadecanoic acid (15–30%) and 14-methyl-hexadecanoic acid (15–30%), whereas *Thermomonospora* and *Actinomadura* species synthesize only traces of these fatty acids. In contrast, these latter taxa synthesize significant amounts of 10-methyl branched fatty acids, whereas this fatty acid is only a minor component in *Thermobifida* (1–5%). In *Thermobifida* species 2-hydroxy fatty acid (a diagnostic fatty acid in *Microbispora*) could not be detected. *Nocardiopsis* has a similar fatty acid type as *Thermobifida* (type 3e), but synthesizes significantly higher amounts of 10-methyl branched tuberculostearic acid (type 3d).

IDENTIFICATION OF *THERMOBIFIDA* SPECIES The two species of the genus *Thermobifida*, *Tb. alba* and *Tb. fusca*, can easily be differentiated from each other by the formation of spores on substrate mycelia, the ability to grow at different temperatures and pH ranges (Tables 9 and 10), resistance to specific inhibitors, and carbon utilization tests (McCarthy and Cross, 1984a). *Thermobifida alba* strains have the lower temperature optimum for growth and sporulation. McCarthy and Cross (1984a) reported that *Tb. alba* had an optimum of 40°C to 45°C as compared to 45°C to 50°C for *Tb. fusca* and that *Tb. alba* did not grow at 55°C on Czapek-Dox-yeast extract-casamino acid (CYC)-agar. Examination of temperature requirement of another set of strains on glucose-yeast extract-malt extract (GYM)-agar showed that *Tb. alba* strains grow best at 37–40°C whereas *Tb. fusca* strains prefer temperatures of 50 to 55°C. *Thermobifida fusca* prefers alkaline conditions for growth, while *Tb. alba* grows better at pH 7. In addition, the *Tb. fusca* strains were also reported to be more resistant to some growth inhibitors, e.g., crystal violet (0.00002%), sodium azide (0.01%), tetrazolium chloride (0.002%), and thallous acetate (0.001%; McCarthy and Cross, 1984a). Some strains of *Tb. alba* were shown to be unable to degrade elastin and keratin, in contrast to most other thermophilic monosporic actinomycetes, which are potent degraders of bio-polymers and usually demonstrate the almost identical degradation patterns (McCarthy and Cross, 1984a; Bös, 1994; Tables 10 and 11).

Strains of the two *Thermobifida* species show a different pattern of enzymes in cell free extracts, e.g., esterases and malate dehydrogenases in poly-acryl amide gel electrophoresis (Greiner-Mayi, 1988), which can be used to separate thermobifidas at the species level. In addition, the two *Thermobifida* species are well separated by quantitative differences in fatty acid pattern. *Thermobifida alba* synthesizes lower amounts of 15-methyl-hexadecanoic acid (13 to 17%) and higher amounts of tuberculostearic acid (5 to 8%), whereas 25–33% and

Table 9. Growth and morphological characteristics of *Thermobifida alba* and *Tb. fusca* and some other thermophilic actinomycetes.[a]

DSM number	Species	Aerial mycelium	Soluble pigment	Spores	Spore surface	Optimal medium	Optimal temp. °C
43795	*Thermobifida alba*	White	—	Singly in clusters	Smooth-ridged	GYM-agar	40
43792	*Thermobifida fusca*	Light	—	Singly in clusters	Smooth-ridged	GYM-agar	50
43750	*Actinomadura rubrobrunnea*	Green	—	Short, curved chains	Short spines	HT-agar	50
44461	*Actinopolyspora mortivallis*	Pale yellow	Pale brown	Long straight chains	Smooth	GYM-agar	45
44096	*Amycolatopsis methanolica*	White	—	Fragmenting hyphae	Smooth	GYEA	25–45[b]
44574	*Amycolatopsis thermoflava*	White	Yellowish	Sterile hyphae	—	GYEA	28–55[b]
43176	*Microbispora rosea* ssp. *aerata*	Pale pink	Brown	In pairs	Smooth	GYM-agar	50
43671	*Nonomuraea angiospora*	Green	Dark green	Short, straight chains	Smooth-warty	HT-agar	45
43186	*Nonomuraea flexuosa*	Pale blue	Brown	Short, curved chains	Warty	PC-agar	50
43027	*Pseudonocardia thermophila*	White	Orange	Fragmenting hyphae	Smooth	GYM-agar	45
43017	*Saccharomonospora viridis*	Green	Dark-green	Singly	Smooth-warty	PM-agar	50
43463	*Saccharopolyspora hirsuta*	White	—	Short, straight chains	Hairy	GYM-agar	45
43747	*Saccharopolyspora rectivirgula*	White	—	Short, curved chains	Smooth	TSA-NaCl	50
44575	*Saccharopolyspora thermophila*	White	—	Short, curved chains	Smooth	GYM-agar	45–55[b]
41741	*Streptomyces thermoalkalitolerans*	Gray	—	Spirals	Warty	Inorganic salt-starch agar	25–55[b]
41756	*Streptomyces thermogriseus*	Yellowish-gray	—	Curved chains	Smooth	Inorganic salt-starch agar	65–68[b]
40444	*Streptomyces thermovulgaris*	Gray	—	Short, spirals	Smooth-warty	GYM-agar	50
43038	*Thermobispora bispora*	White	—	In pairs	Smooth	PM-agar	50
44069	*Thermocrispum municipale*	White	—	Fragmenting hyphae	Smooth-ridged	CYC-agar	45–55[b]
43794	*Thermomonospora chromogena*	Pale brown	Brown	Singly in clusters	Smooth-ridged	PM-agar	50
43183	*Thermomonospora curvata*	White	—	Singly	Smooth-ridged	CYC-agar	50

Abbreviations: DSM, Deutsche Sammlung von Mikroorganismen und Zellkulturen, Braunschweig, Germany; GYM, glucose-yeast extract-malt extract; HT, Hickey-Tresner; GYEA, glycerol yeast extract agar; PC, plate count; PM, peptone maize; TSA, trypticase soy agar; and CYC, Czapek-Dox-yeast extract-casamino acid.

[a]The thermophilic genus *Thermoactinomyces* is not included because it does not belong to the order Actinomycetales. These organisms are easily differentiated from "true" actinomycetes by the thermoresistance of their endospores, which germinate after pasteurization.

[b]Temperature range as stated in various publications.

Table 10. Morphological and physiological criteria for the differentiation of *Thermobifida alba* and *Tb. fusca* from other monosporic, thermophilic actinomycetes formerly classified with the genus *Thermomonospora*.

| Characteristic | Thermobifida | | Thermomonospora | | Actinomadura formosensis | Microbispora mesophila |
	alba	fusca	curvata	chromogena		
Number of strains	7	55	15	18	1	1
Color of SM	Pale yellow	Pale yellow	Yellow orange	Brown	Pale orange	Brown
Color of AM	White	White	White	Pale brown	White to pink	White
Spores						
on SM	–	+	–	–	+	–
on AM						
in cluster	+	+	–	+	–	–
Single, sporophores	–	–	+	–	+	–
Single, sessile	–	–	–	–	–	+
Spore ornamentation	Smooth	Smooth	Spiny	Spiny	Warty	Smooth
Growth at (% of strains)						
25°C	–	–	–	–	100	100
35°C	100	47	100	6	100	100
53°C	14	96	87	94	–	–
60°C	–	38	–	–	–	–
pH6	14	13	–	56	nd	100
pH10	100	100	100	89	nd	–
pH11	100	98	100	–	+	–
Resistance to						
Kanamycin (25 µg/ml)	–	–	–	+	–	–
Novobiocin (25 µg/ml)	–	–	–	–	–	+
Lysozyme	v	v	–	–	nd	–
Degradation of						
Agar	+	+	+	–	nd	+
Starch	+	+	+	–	nd	+
Pectin	+	+	–	v	nd	+
Cellulose powder	+	+	+	v	nd	–
Carboxymethylcellulose	+	+	+	+	nd	+
Chitin	–	–	–	–	nd	–
Xylan	+	+	+	+	nd	+
Keratin	v	+	+	+	nd	+
Elastin	v	+	–	+	nd	+
Tween 20 and 80	+	+	+	+	nd	+

Symbols: see footnote in Table 5.
Abbreviations: SM, substrate mycelium; AM, aerial mycelium; and nd, not determined.
From McCarthy and Cross (1984a) and Kempf (1995).

1–2% of these acids, respectively, are found in *Tb. fusca*. However, these differences in fatty acid patterns might be a response to their different growth optima.

Since the value of phenotypic criteria for differentiation of the two current *Thermobifida* species is poor and the usefulness of 16S rDNA sequence similarity for identification of bacterial species at the molecular level is limited in some taxa, the new *Thermobifida* strains can be precisely attributed to *Tb. alba*, *Tb. fusca* or to new species only on the basis of polyphasic taxonomy, including examination of the DNA-DNA relatedness and other relevant studies (Vandamme et al., 1996; Stackebrandt et al., 2000; Rossello-Mora and Amann, 2001). New chemical compounds of cells and cell walls should also be evaluated for their taxonomic significance.

Streptomonospora salina

The genus *Streptomonospora* has a single species *S. salina* with only one isolate YIM 90002^T. It can be identified by unusual morphology, chemical markers, growth temperature, and salinity ranges (Table 3). This species produces white to pale yellow irregularly branched aerial mycelia where short chains of oval to rod-shaped spores (1.5–2 × 1 µm) with wrinkled surfaces are formed. The extensively branched substrate mycelium is nonfragmented. Single spores are oval to round (1.4–1.6 µm), borne on single or dichotomously branched sporophores developed on substrate hyphae. Spores of both types are non-motile. The cell wall of the type strain was reported to contain an unusual combination of amino acids (*meso*- and DD-diaminopimelic acids, glycine,

Table 11. Degradation of biopolymers by *Nocardiopsis* and *Thermobifida* species.

Species (Number of strains)	Growth Temperature °C	Agar	Car	Starch	Chitin	Xalan	Pectin	Gelatine	Substrates Casein	Elastin	Tributyrin	Tween 80	Lectine	PHB	DNA	Haem	Mic. lut.	Cand. alb.
Nocardiopsis dassonvillei (3)	28	100	100	(67)	100	100	100	100	100	100	100	100	0	67	(100)	100	100	0
N. dassonvill. alborubida (1)	28	100	100	(100)	100	100	100	100	100	100	100	100	0	0	(100)	100	100	0
Nocardiosis alba (4)	28	100	100	(100)	(100)	0	0	100	100	100	100	100	100	(75)	100	100	100	0
Nocardiopsis prasina (1)	28	100	100	100	(100)	0	0	100	100	100	100	100	100	0	100	100	100	100
Nocardiopsis listeri (1)	28	100	100	0	0	(100)	(100)	100	100	100	100	100	100	0	(100)	100	(100)	0
Nocardiopsis spec. DSM40936	37	100	100	100	100	0	100	0	100	100	100	100	100	100	100	100	100	100
Thermobifida alba (2)	40	100	n.d.	100	0	100	100	100	100	100	100	100	50	100	50	100	(100)	n.d.
Thermobifida fusca (8)	50	100	n.d.	100	0	100	100	100	100	100	100	100	75	100	38	100	(100)	n.d.

Abbreviations: Car, carrageenan; PHB, polyhydroxybutyric acid; Haem, haemolysis of erythrocytes; Mic. lut., lysis of *Micrococcus luteus* cells; Cand. alb., lysis of *Candida albicans* cells; (), weak reaction or clearing zone only beneath the colony.
Data from Kempf, 1995.

lysine and aspartate) but the structure of peptidoglycan has not been elucidated. Large amounts of glucose and galactose and smaller amounts of ribose, xylose, arabinose and mannose were present in whole cell hydrolysates. *Streptomonospora salina* is also characterized by the menaquinones MK-9(H$_6$) and with smaller amounts of MK-10(H$_6$) and MK-10(H$_4$) and by a phospholipid pattern of the P II type. Together with optima and ranges of temperatures and salinity for growth, *Streptomonospora salina* can be easily differentiated from *Thermobifida* and *Nocardiopsis*.

Menaquinone composition can be used to differentiate *Streptomonospora salina* from most other mesophilic spore-forming actinomycete genera containing hydrogenated menaquinones with relatively long isoprene side chains (Tables 3 and 4). A quite useful character for differentiation of *Streptomonospora salina* from most other spore-forming actinomycetes is its optimal growth at high salt concentration (15%). Besides *Streptomonospora salina*, only six halotolerant, moderately or extremely halophilic actinomycetes form long chains of spores and contain *meso*-diaminopimelic acid in the cell wall (Table 3). They belong to *Actinopolyspora* and *Nocardiopsis* (Gochnauer et al., 1975, 1989; Johnson et al., 1986; Yoshida et al., 1991; Yassin et al., 1993; Al-Tai and Ruan, 1994; Ruan et al., 1994; Ventosa et al., 1998; Chun et al., 2000). However, in contrast to *Streptomonospora salina*, these species have phospholipids of the P III type, their substrate hyphae are usually fragmented, and single spores (typical of *Streptomonospora*) are not formed on substrate hyphae. Compared with *Streptomonospora salina*, *Nocardiopsis* spp. produces much longer spore chains on aerial mycelia (aerial hyphae are almost fully disintegrated into spores), whereas actinopolysporas are characterized by IV cell wall chemotypes. Growth temperature, growth salinity range, and menaquinone composition may serve as additional characters for the differentiation of *Streptomonospora salina* from other taxa.

Current Taxonomic Position of *Nocardiopsis, Streptomonospora* and *Thermobifida*

After nearly 100 years of searching for a home for *Nocardiopsis dassonvillei* strains, the taxonomic position of this genus has now been elucidated. On the basis of chemotaxonomy, numerical taxonomy, DNA-DNA reassociation studies, and 16S rRNA gene sequence analyses (Fischer et al., 1983; Athalye et al., 1984, 1985; Kroppenstedt et al., 1985, 1990; Poschner et al., 1985; Stackebrandt, 1986; Goodfellow et al., 1988; Grund and Kroppenstedt, 1989, 1990;

Kroppenstedt et al., 1985, 1990; Meyer, 1989; Rainey et al., 1996; Stackebrandt et al., 1997; Zhang et al., 1998), many of the former *Nocardiopsis* species found their place within the genera *Lechevalieria* (Labeda et al., 2001), *Saccharothrix* (Grund and Kroppenstedt, 1989; Labeda and Lechevalier, 1989) and *Nonomuraea* (Zhang et al., 1998; Chiba et al., 1999). The current genus *Nocardiopsis* harbors twelve validly described species and one subspecies (Table 1).

The phylogenetic tree based on the comparison of the 16S rRNA gene sequences separates the *Nocardiopsis* species into four clusters (Fig. 4). The *N. dassonvillei* cluster includes the type species *N. dassonvillei* subsp. *dassonvillei* (strain DSM 43111) with its subspecies *N. dassonvillei* subsp *albidorubidus* (DSM 40465), *N. dassonvillei* subsp. *dassonvillei* (DSM 43884) formerly *N. antarctica*, *N. synnemataformans*, and "*N. halotolerans*" (DSM 44410). The second cluster of low bootstrap replication values includes *N. alba*, *N. exhalans* (not shown), *N. prasina*, *N. listeri*, *N. tropica*, *N. lucentensis*, *N. umdischolae* (not shown) and *N. kunsanensis*, while *N. trehalosi* and *N. halophila* are distant single member clusters.

The comparison of the 16S rRNA gene sequences of *Nocardiopsis* with those of other actinomycetes revealed that the *Nocardiopsis* species show a homogeneous cluster apart from the other actinomycetes. On the basis of the isolated phylogenetic position and the unique morpho- and chemotaxonomic properties of this genus, Rainey and co-workers created the family Nocardiopsaceae (Rainey et al., 1996). Subsequently, the three families Nocardiopsaceae, Streptosporangiaceae and Thermomonosporaceae were unified into the suborder Streptosporangineae, one of the 10 suborders of the order Actinomycetales, class Actinobacteria (Stackebrandt et al., 1997).

Kroppenstedt and Goodfellow (1992) could show from the differences in chemical markers that the genus *Thermomonospora* was heterogeneous. They found that *Tm. alba* including *Tm. mesouviformis* and *Tm. fusca* merit generic status and should be removed from the genus *Thermomonospora* (Kroppenstedt and Goodfellow, 1992). Zhang and co-workers showed that the two *Thermomonospora* species, *T. alba* and *T. fusca*, differed significantly in their 16S rRNA gene sequences from *Thermomonospora* sensu stricto and form a separate branch closely affiliated with the *Nocardiopsis* line of descent. For these two misclassified *Thermomonospora* species, the second genus *Thermobifida* was created (Zhang et al., 1998).

The recently proposed third genus of the family, *Streptomonospora*, is represented by one species, *Streptomonospora salina*, which is posi-

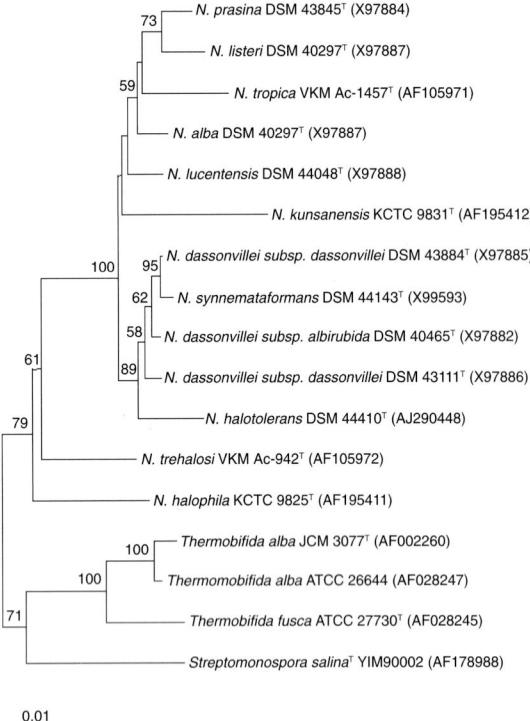

Fig. 4. Phylogenetic dendrogram obtained by distance matrix analysis showing the relationship among strains of the family *Nocardiopsaceae*. The sequence of *Actinomadura* served as outside reference. Scale bar inferred nucleotide substitution per 100 nucleotides. The numbers at the nodes are bootstrap values.

tioned within the Nocardiopsaceae phylogenetic clade and exhibits somewhat closer relatedness to thermobifidas than to most *Nocardiopsis* species (Cui et al., 2001).

All three genera of family Nocardiopsaceae (*Nocardiopsis*, *Thermobifida*, and *Streptomonospora*) are clearly separated from each other and from all other actinomycetes by their unique combination of chemical markers, morphologies, physiological properties (Cui et al., 2001; Kroppenstedt, 1985, 1987; Kroppenstedt and Goodfellow, 1992; Zhang et al., 1998; Table 3), and molecular characteristics (Stackebrandt et al., 1983, 1997; Stackebrandt, 1986; Goodfellow et al., 1988; Kroppenstedt et al., 1990; Rainey et al., 1996; Zhang et al., 1998).

Recent publications and the 16S rRNA gene sequence data available from GenBank show that the family Nocardiopsaceae harbors also many new organisms that are not yet validly described. Only those new actinomycete isolates should be included into the genera *Nocardiopsis*, *Thermobifida*, and *Streptomonospora* that show the phenotypic properties

characteristic of these genera and/or are proven to be in these genera by 16S rRNA gene sequence analyses and other relevant taxonomic studies.

Preservation

For short-term preservation, sporulated strains may be kept on agar slants at 4°C and transferred every 4 months. To prevent the agar from drying, the tubes are tightly sealed with silicone stoppers. For longer preservation, spore suspensions or homogenized mycelia are mixed with glycerol to a final concentration of 25% and kept at 25°C (Wellington and Williams, 1978). Zippel and Neigenfind (1988) tested the viability, stability of auxotrophic markers, and antibiotic production of strains kept at 25°C in 45% glycerol. No drastic changes of the markers were noted after one year. Good results were obtained for non-sporulating strains by this method because the high concentration of glycerol prevents freezing of the suspension at 25°C. For long-term preservation, lyophilization of spores and mycelia suspended in 10% skim milk is a convenient method.

A very simple, reliable, and time-saving method is liquid nitrogen (N₂) cryopreservation of living cells in small polyvinyl chloride (PVC) tubes ("straws") at –196°C. This method has been described for long-term preservation of basidiomycetes (Hoffmann, 1989a, 1989b). The procedure has been tested for actinomycetes, including *Nocardiopsis* and *Thermobifida*. The strains are harvested from well-sporulated cultures grown on suitable agar media in Petri dishes. A 2 × 25 mm piece of sterile PVC tube is pressed into the mycelial mat and the agar and carefully raised to excise the agar plug. This is repeated until the tube is filled with agar. The filled tubes are placed in a sterile cryo vial with the screw cap marked with strain accession number. A 1.8-ml vial will hold up to 13 tubes. Two vials are prepared for each strain and then fixed to a metal clamp for freezing in the gas phase of the liquid nitrogen container. After 10–15 min, when temperature falls below 130°C, the clamp can be positioned into the liquid phase at 196°C. A container with a capacity of 250 liters will hold at least 8,000 vials or 4,000 strains. For viability testing, one tube is removed from the vial within the gas atmosphere of the container and placed directly on a suitable agar medium for thawing. After a few days incubation, the mycelium will be visible. For those strains that do not produce an abundant mycelium, the plugs may be pushed out of the tubes by a sterile needle.

Physiology

The family harbors organisms with different temperature, salinity and alkalinity requirements for their growth and includes thermotolerant and thermophilic, halotolerant and moderately halophilic, as well as alkaliphilic or at least alkalitolerant species. Their enzymes, antibiotics and other biologically active metabolites have applications in biotechnology and medicine. Although most members of the family are strict saprophytes, there is evidence that members of the genus *Nocardiopsis* are pathogens because they are found in clinical material, and airborne spores of *Thermobifida* species can be associated with the respiratory allergies.

Nocardiopsis

A high percentage of *Nocardiopsis* strains are usually found enriched under alkaline conditions in soil sources. Miyashita et al. (1984) reported that under these conditions, half of the actinomycetes strains belong to the genus *Nocardiopsis*. Table 6 shows that this taxon harbours many alkaliphilic or at least alkali-tolerant species. As expected, the alkaliphilic strains synthesize many different alkaline enzymes (Horikoshi and Akiba, 1982). Mikami et al. (1985) examined alkaliphilic amylase from these strains and found that the pH optimum was about 10. In contrast, the neutrophilic *N. dassonvillei* strains, which grow better at pH 7.5, synthesize amylase with a pH optimum of pH 8. Recently, Stamford et al. (2001) reported on a thermostable α-amylase from *Nocardiopsis* sp., an endophyte of yam bean. Thermostable amylolytic enzymes have been currently investigated to improve industrial processes of starch degradation.

An alkaline protease was isolated from *N. dassonvillei* strain ATCC 21944 by Kim et al. (1993). This protease showed maximum activity at pH 8.5. Only a small decrease in protease activity was found at pH 11, indicating that the protease of this strain is strongly alkaliphilic. The optimal temperature of the alkaline protease was 55°C; at 65° C however, it retains 70% of the enzymatic activity (Kim et al., 1993). The relatively strong thermal stability of the enzyme is of particular interest considering that the optimum growth temperature for most *Nocardiopsis* species is around 25°C and usually no growth occurs above 40°C (Table 7).

In a screening program searching for microbial metabolites with interesting properties, Mishra et al. (1987a) isolated 942 microorganisms from soil. Of these, 832 were aerobic actinomycetes. While the largest number (302) were streptomycetes, the other 530 isolates were distributed among 17 different actinomycete genera, including 22 strains that were classified as *Nocardiopsis*. The *Nocardiopsis* strains had no effect on the tested algae or higher plants (Mishra et al., 1987a), but 6 of the 22 nocardiopsis isolates showed specific herbicidal properties. Four strains inhibited the germination of cress seeds and two affected barnyard grass seeds (Mishra et al., 1988). Bioassays for insecticidal properties from these strains were performed using mosquito larvae of *Aedesaegypti* and for nematicidical activities using the free-living nematode *Penagrellusrevidivus*. One *Nocardiopsis* strain showed insecticidal properties, but none showed nematicidal properties (Mishra et al., 1987b). The growth of the fungus *Fusariumoxysporum* f. sp. *albedensis*, a pathogen of date palm, was inhibited by a *Nocardiopsis* strain (Sabaou and Bounaga, 1987). The growth of some other soil fungi was not affected by *Nocardiopsis dassonvillei* extracts. Papeta and co-workers isolated from *Nocardiopsis* strain A6 a factor inhibiting the growth of some phytopathogenic fungi. This proteinaceous toxin was identified as a xylanase that showed 100% sequence homology with the xylanase from *Bacillus circulans* (Papeta et al., 1995).

Phenazine antibiotics were extracted from the mycelium of the alkaliphilic *Nocardiopsis* strain OPC-15 (Tsujibo et al., 1988). This strain produced different phenazine antibiotics under different culture conditions (incubation period, incubation temperature, and temperature shift). In addition, 1,6-dihydroxyphenazine (compound III) was obtained from the mycelium after incubation for 6–8 days at 27°C, while 1,6-dihydroxyphenazine 5,10 dioxide, known as iodinin (compound I), was isolated after incubation for 6 days at 27°C followed by further incubation for 2 days at 4°C. A new disaccharide antibiotic (3-trehalosamine) active against Gram-positive bacteria was reported to be synthesized by *Nocardiopsis trehalosi* (Dolak et al., 1980; Dolak et al., 1981). Members of the genus can produce an apoptolidin, a new apoptosis inducer in transformed cells (Kim et al., 1997).

The production of a new indole alkaloid, pendolmycin, by *Nocardiopsis* strain SA 1715, isolated from soil collected in a river near Shanghai, was reported (Yamashita et al., 1988). Pendolmycin is an inhibitor of phosphatidyl inositol turnover induced in human epidermoid carcinoma by epidermal growth factor. The structurally related metabolites, teleocidin B and lyngbyatoxin, have been isolated from a *Streptomyces* strain and from lipid extracts of a Hawaiian shallow-water variety of the *Lyngbya majuscula*, respectively. The production of pendolmycin (structurally similar to teleocidin A; Nishiwaki et al., 1991) and methylpendolmycin

(Sun et al., 1991) was attributed to other *Nocardiopsis* strains. A cytotoxic substance TS-1 was isolated from the alkaliiphilic *Nocardiopsis* strain OPC-553 regarded as a strain of *Nocardiopsis dassonvillei* subsp. *prasina* (Tsujibo et al., 1990c). It showed a marked inhibitory activity against L5178Y mouse leukemic cells in vitro. TS-1 was identified as the antifungal antibiotic, kalafungin, already isolated from the culture broth of *Streptomyces tanashiensis*. Thus, the same secondary metabolite may be synthesized in different taxa, indicating that the occurrence of a metabolite may not always have taxonomic relevance.

The antiflammatory and antiallergic effects of kinase C inhibitors are well known. Protein kinase C inhibitors of microbial origin are intensively studied because of their physiological and medical importance (Koizumi et al., 1988; Matsuda and Fukuda, 1988; Nakanishi et al., 1988; Ohmori et al., 1988; Iwabe et al., 1998; Kim and Kawai, 1998; Arvanov et al., 2000; Bishop et al., 2000; Takahashi et al., 2000). New protein kinase C inhibitor designated "K-252a" was found to be produced by an unclassified *Nocardiopsis* strain K-252 (Kase et al., 1986; Matsuda and Fukud, 1988). Three structurally related compounds, K-252b, c, and d, were isolated from a culture broth of a *Nocardiopsis* isolate K-290 (Nakanishi et al., 1986).

"*Nocardiopsis atra*" IFO 14198, a producer of nocardicin A, was described by Celmer et al. (1980). It should be mentioned that the latter two strains cited in this section as *Nocardiopsis* may have been misclassified because they were assigned to this genus solely on the basis of morphology and limited chemical markers such as cell wall composition. Subsequently, strain K-252 (=NRRL 15532) was reclassified as a new species of the genus *Nonomuraea, Nonomuraea longicatena* (Chiba et al., 1999). The other strain, "*Nocardiopsis atra*," possesses phospholipids of the P II type (Streshinskaya et al., 1989) and belongs to the Pseudonocardiaceae (L. I. Evtushenko, unpublished observation).

Members of the genus *Nocardiopsis* degrade biopolymers very actively (Kempf and Kutzner, 1988; Tsujibo et al., 1990a, 1990b, 1990c, 1990d, 1991; Kempf, 1995). Table 11 gives examples of the degradation capabilities of some selected *Nocardiopsis* strains. All of the 12 strains were able to degrade gelatin, casein, elastin, Tween 80, blood and *Micrococcus luteus* cells. Particularly total hemolysis is of special interest because this ability is not common among other sporoactinomycetes. Differences could be found in the degradation capabilities of the Nocardiopsis species in the degradation of starch, xylan, lecithin, and polyhydroxybutyrate (PHB). Starch was only slowly degraded by the mesophilic species,

whereas the thermotolerant and thermophilic strains had a very active amylase. Xylanase was only produced by *N. dassonvillei* strains, although a little xylanase activity could be detected in *N. listeri*. The data of Kempf (1995) support the results of Tsujibo et al. (1990d), (1991), who detected three xylanases in isolate OPC-18 assigned to *Nocardiopsis dassonvillei* subsp. *alba*. Protease activity was present in *N. prasina*. This confirms the results of Tsujibo et al. (1990a), (1990b), who found two serine proteases in this species. In contrast to the other *Nocardiopsis* species, *N. dassonvillei* was lecithinase negative. Two *N. dassonvillei* strains, two *N. alba* strains, and the thermotolerant *Nocardiopsis* strain DSM 40936 degraded PHB. There is also a report showing *Nocardiopsis* strain NCIMB 5124 from oil-contaminated marine environments can degrade hydrocarbons (crude oil) and utilize organic nitrogen by producing extracellular proteases (Dixit and Pant, 2000a; Dixit and Pant, 2000b). Both enzymes were identified as alkaline serine endopeptidases.

A screening program for metal leaching microorganisms from a siliceous alkaline slag dump led to the isolation of an actinomycete (Willscher and Bosecker, 2001). A polyphasic taxonomic study showed that this isolate belongs to a new *Nocardiopsis* species for which the tentative name "*Nocardiopsis metallicus*" was chosen, referring to its ability to separate metal from slag (Schippers et al., 2002).

Thermobifida

Biological and physiological features of *Thermobifida* reflect their adaptive potential to their natural habitat, which is mainly self-heated organic material. Thermobifidas share this habitat with other thermophilic and thermotolerant actinomycetes able to degrade plant material and other bio-polymers (McCarthy, 1987; Tables 10 and 11). From this environment *Tb. fusca* strains have been isolated that are able to mineralize plastic disposals and other anthropogenic xenobiotica (Kleeberg et al., 1998). Actinomycetes are well suited for this environment because they generally grow as branching hyphae and are well adapted to penetration and degradation of insoluble substrates such as lignocellulose. Within this taxon, thermobifidas are of particular interest because they produce multiple thermostable enzymes involved in the degradation of lignocellulose (McCarthy, 1987; Wong et al., 1988; Ball and McCarthy, 1989). Knowledge of lignocellulytic activity is important to the understanding of plant biomass recycling in nature and the use of enzymes in controlled lignocellulose conversation (Thomson, 1993). *Thermobifida* strains are also capable of producing ferments degrading other biopolymers. These enzymes were able to degrade all of the 16 tested polymers and substrates except chitin (Tables 9 and 11). A novel non-heme extracellular peroxidase was isolated from the thermophilic strain *Thermobifida* (*Thermomonospora*) *fusca* BD25 and its catalytic mechanism was characterized (Rob et al., 1995; Rob et al., 1996; Rob et al., 1997b; Rob et al., 1997a). Optimization of fermentation product yields under aerobic bioreactor conditions was reported (Tuncer et al., 1999).

Thermobifida and other thermophilic monosporic actinomycetes have been evaluated for cellulose bioconversion processes designed to produce single -cell protein (Bellamy, 1974, 1977; Wood, 1985; Crawford, 1988) or sugar syrups for fermentation to ethanol (Hägerdal et al., 1979; Lee and Humphrey, 1979; Ball and McCarthy, 1988). Several of these publications have referred to a cellulolytic *Thermoactinomyces* strain that has been labeled "*T. cellulosae*" (Vacca and Bellamy, 1976). This organism was later transferred to *Thermomonospora fusca* (Hägerdal et al., 1980), which was recently reclassified together with *Thermomonospora alba* to the new genus *Thermobifida* (Zhang et al., 1998).

Thermobifida fusca shows endogluconase activity and attacks cellulose. It is assumed that *Tb. fusca* is involved in the formation of humic substances in soil (Trigo and Ball, 1994). *Thermobifida* strains are capable of extensively degrading cellulose and lignocellulose residues that make up the bulk of agricultural and urban wastes (McCarthy, 1987). *Thermobifida fusca* strains are very active against arabinoxylan and produce thermostable enzymes (McCarthy et al., 1985).

Thermobifida fusca was grown at 55°C on a pilot scale to convert high cellulose, low lignin pulp mill waste into single-cell protein (Crawford et al., 1973). The proteinaceous product had a high nutritional value as a feed supplement in the diet of chicks (Harkin et al., 1974). Thermostable cellulases are also attractive as catalysts in biomass conversion, since their temperature optima can reduce contamination problems in economically feasible non-aseptic applications (Wood, 1985; Bachmann and McCarthy, 1991). A cellulase -enhanced *Thermobifida fusca* mutant grown on chemically pretreated poplar wood was found to use 70% of the woody biomass after 62 h at 55°C, but failed to grow well on wood that had not been chemically pretreated. Bellamy (1977) used the parental strain of the same organism, originally described as "*Thermoactinomyces cellulosae*," for the bioconversion of chemically pretreated lignocellulosic feedlot waste into microbial biomass. The culture degraded up to 85% of the cellulosic and over 90% of the hemicellulosic fraction of the substrate. The process led to the production of a high protein content product but was not economical, given the cost of the chemical pretreatment needed to obtain a suitable degradable substrate.

Some success has been achieved in the use of thermophilic actinomycetes in processes designed for the saccharification of purified celluloses. The cellulase complex of *Tb. fusca* YX readily saccharifies finely ground, acid swollen Avicel, a purified chemically pretreated cellulose (Su and Paulavicius, 1975; Ferchak and Pye, 1980). In subsequent work by Ferchak and Pye (1983), a process was developed for saccharification of cellulose using a cellulose/cellobiose mixture. In a 7-day, 50° C saccharification, 15 to 20% glucose syrups were generated from acid swollen Avicel. The availability of cellulase-overproducing strains of *Thermomonospora* and *Thermobifida* species (Fennington et al., 1982; Meyer and Humphrey, 1982) should further improve the commercial potential of saccharification processes. *Thermobifida fusca* was also reported to secrete thermostable β-mannanase (at a temperature optimum of 80°C), which hydrolyzes the O-glycosidic bonds in mannan, a hemicellulose constituent of plants, and has potential use in pulp and paper production. The three-dimensional structure of a mannan-degrading enzyme was reported for the first time and a possible mechanism of enzyme selectivity was described (Hilge et al., 1998; Hilge et al., 2001).

Ligninolytic thermomonosporas may have potential for the biological bleaching of pulps or in biological pulping processes, given their capacity to dissolve lignin and separate it from the cellulose component of lignocellulose (McCarthy et al., 1986). It has also been suggested (Crawford, 1988) that the ability of some *Thermobifida* (*Thermomonospora*) strains to simultaneously attack all of the major components of lignocellulose, (i.e., cellulose, lignin and hemicellulose) may be used to develop direct fermentations for the conversion of lignocellulosic wastes into specific high value chemicals and for single cell protein. The development of efficient industrial size bioreactors for solid substrate fermentations will help to foster commercially successful actinomycete-adapted bioconversion processes. The biochemistry and genetics involved in these processes have been intensively studied for *Tb. fusca* during the last years (Irwin et al., 1998; Irwin et al., 2000; Spiridonov and Wilson, 1998; Spiridonov and Wilson, 1999; Spiridonov and Wilson, 2001; Tuncer et al., 1999; Zhang et al., 2000a; 2000b).

During a screening program for micro organisms degrading plastic waste disposals like BTA (random aliphatic-aromatic copolyesters synthesized from 1,4-butanediol, adipic, and terephthalic acid), 20 BTA-degrading strains were isolated from the mature compost of green waste. Among these micro organisms, thermophilic actinomycetes obviously play an outstanding role and appear to dominate the initial degradation step of BTA. Two of the isolates, which exhibited about a 20-fold higher BTA degradation rate than usually observed in common compost tests, were identified as members of the species *Thermobifida fusca*. Further studies on strains received from culture collections showed that all tested *Thb. fusca* strains show the ability to degrade BTA (Kleeberg et al., 1998).

Members of the genus *Thermobifida* were also reported to be producers of substances of medical importance. A novel inhibitor of topoisomerases designated "topostatin" was isolated from the culture filtrate of *Thermomonospora alba* strain no. 1520 (Suzuki et al., 1998a). Topostatin inhibited the relaxation of supercoiled pBR322 DNA by calf thymus topoisomerase I and also inhibited the relaxation of supercoiled pBR322 DNA and decatenation of kinetoplast DNA by human placenta topoisomerase II. The inhibitor exhibited growth inhibition activity against tumour cells (SNB-75 and SNB-78) of the central nervous system, but did not exhibit any antimicrobial activity against Gram-positive and Gram-negative bacteria, yeasts and fungi (Suzuki et al., 1998a; Suzuki et al., 1998b). The inhibitor of topoisomerase I designated "isoaurostatin" was isolated, and its features and structure were studied. It was determined to be 6,4'-dihydroxyisoaurone, an inhibitor of the cleavable complex-nonforming type without DNA intercalation (Suzuki et al., 2001).

Pathogenicity for Humans

Nocardiopsis

Although *Nocardiopsis* strains were frequently isolated from patients (Liegard and Landrieu, 1911), doubt of their clinical significance was expressed for a long time. Sindhuphak and coworkers showed that actinomycetoma was caused by *Nocardiopsis dassonvillei* and repeatedly isolated *N. dassonvillei* strains from nodules and draining sinuses of the anterior aspect of the right leg below the knee of a 39-year-old man (Sindhuphak et al., 1985). This was the first reported case of mycetoma caused by *N. dassonvillei*. Recently a *Nocardiopsis dassonvillei* strain has been isolated from the blood of a patient suffering from cholangitis. The isolate was recovered from blood when the patient presented an acute onset of fever. The biliary and gastrointestinal tracts may be the entry routes, since *N.*

dassonvillei was recovered after retrograde cholangiography. Alternatively, an intravenous catheter inserted while the patient was hospitalized may have been the entry route. Since *N. dassonvillei* was isolated on the 10th day of hospitalization, it should be regarded as a potential nosocomial pathogen (Beau et al., 1999). Another *Nocardiopsis* species of clinical significance is *Nocardiopsis synnemataformans*. It was isolated from the sputum of a kidney transplant patient, but in this case, the pathogenicity of the isolate could not be verified (Yassin et al., 1997). The role of *Nocardiopsis dassonvillei* in broncho pulmonary disorders was clearly demonstrated by Gugnani et al. (1998). In a study of aerobic actinomycetes in human infections in Nigeria, they found two patients with *Nocardiopsis dassonvillei* among 42 patients with broncho pulmonary disorders. The patients were asthmatics and had long histories of corticosteroid therapy. In the same year, Mordarska et al. (1998) identified *N. dassonvillei* (by glycolipid markers) as the only etiological agent of a severe pulmonary infection. The identification by this marker was confirmed by DNA/DNA homology studies. These results demonstrated the value and generic specificity of glycolipid markers from *Nocardiopsis*. This was the first time that the immunologically active glycolipid markers G(1) and G(2) were used for rapid recognition of a clinical strain causing a nocardiosis.

Literature Cited

Abyzov, S. S., S. N. Philipova, and V. D. Kuznetsov. 1983. Nocardiopsis antarcticus, a new species of actinomycetes, isolated from the ice sheet of the central Antarctic glacier. Izv. Akad. Nauk SSSR Ser. Biol. 4:559–568.

Agre, N. S., T. P. Efirnova, and L. N. Guzeva. 1975. Heterogeneity of the genus Actinomadura Lechevalier and Lechevalier. Microbiology 44:200–233.

Ajello, L., J. Brown, E. Macdonald, and E. Head. 1987. Actinomycetoma caused by Nocardiopsis dassonvillei. Arch. Dermatol. 123:426.

Alderson, G., M. Athalye, and R. P. White. 1984. Numerical methods in the taxonomy of sporoactinomyetes. *In:* L. Ortiz-Ortiz, L. F. Bojalil, and V. Yakoleff (Eds.) Biological, Biochemical and Biomedical Aspects of Actinomycetes. Academic Press. London, UK. 597–615.

Alderson, G., M. Goodfellow, and D. E. Minnikin. 1985. Menaquinone composition in classification of Streptomyces and other sporoactinomycetes. J. Gen. Microbiol. 131:1671–1679.

Al-Tai, A. M., and J.-S. Ruan. 1994. Nocardiopsis halophila sp. nov., a new halophilic actinomycete isolated from soil. Int. J. Syst. Bacteriol. 44:474–478.

Al-Zarban, S. S., I. Abbas, A. A. Al-Musallam, U. Steiner, E. Stackebrandt, and R. M. Kroppenstedt. 2002. Nocardiopsis halotolerans sp. nov., isolated from salt march soil in Kuwait. Int. J. Syst. Evol. Microbiol. 52:525–529.

Andersson, M. A., R. Mikkola, R. M. Kroppenstedt, F. A. Rainey, J. Peltola, J. Helin, K. Sivonen, and M. S. Salkinoja-Salonen. 1998. The mitochondrial toxin produced by Streptomyces griseus strains isolated from indoor environment is valinomycin. Appl. Environ. Microbiol. 64:4767–4773.

Archibald, A. R. 1974. The structure, biosynthesis and function of teichoic acid. Adv. Microbiol Physiol. 11:53–95.

Arvanov, V. L., B. S. Seebach, and L. M. Mendell. 2000. NT-3 evokes an LTP-like facilitation of AMPA/kainate receptor-mediated synaptic transmission in the neonatal rat spinal cord. J. Neurophysiol. 84:752–758.

Athalye, M., M. Goodfellow, and D. E. Minnikin. 1984. Menaquinone composition in the classification of Actinomadura and related taxa. J. Gen. Microbiol. 130:817–823.

Athalye, M., M. Goodfellow, J. Lacey, and R. P. White. 1985. Numerical classification of Actinomadura and Nocardiopsis. Int. J. Syst. Bacteriol. 35:86–98.

Bachmann, S. L., and A. J. McCarthy. 1991. Purification and cooperative activity of enzymes constituting the xylan-degradation system of Thermomonospora fusca. Appl. Environ. Microbiol. 57:2121–2130.

Baddiley, J. 1972. Teichoic acids in cell walls and membranes of bacteria. Essays Biochem. 8:35–77.

Baddiley, J. 2001. Teichoic acids in bacterial coaggregation. Microbiology 146:1257–1258.

Ball, A. S., and A. J. McCarthy. 1988. Saccharification of straw by actinomycete enzymes. J. Gen. Microbiol. 134:2139–2147.

Ball, A. S., and A. J. McCarthy. 1989. Production and properties of xylanases from actinomycetes. J. Appl. Bacteriol. 66:439–444.

Batrakov, S. G., and L. D. Bergelson. 1978. Lipids of the streptomycetes. Structural investigation and biological interrelation. Chem. Phys. 21:1–29.

Beau, F., C. Bollet, T. Coton, E. Garnotel, and M. Drancourt. 1999. Molecular identification of a Nocardiopsis dassonvillei blood isolate. J. Clin. Microbiol. 37:3366–3368.

Bellamy, W. D. 1974. Single cell proteins from cellulosic wastes. Biotech. Bioeng. 16:869–880.

Bellamy, W. D. 1977. Cellulose and lignocellulose digestion by thermophilic actinomycetes for single cell protein production. Devel. Ind. Microb. 18:249–254.

Bishop, A. C., J. A. Ubersax, D. T., Petsch, D. P. Matheos, N. S. Gray, J. Blethrow, E. Shimizu, J. Z. Tsien, P. G. Schultz, M. D. Rose, J. L. Wood, D. O. Morgan, and K. M. Shokat. 2000. A chemical switch for inhibitor-sensitive alleles of any protein kinase. Nature 407 (6802):395–401.

Boivin-Jahns, V., A. Bianchi, R. Ruimy, J. Garcin, S. Daumas, and R. Christen. 1995. Comparison of phenotypical and molecular methods for the identification of bacterial strains isolated from a deep subsurface environment. Appl. Environ. Microbiol. 61:3400–3406.

Bös, D. 1994. Die Gattung Thermomonospora: Taxonomische Untersuchung und Charakterisierung von Thermomonospora-Phagen. Technische Hochschule Darmstadt. Darmstadt, Germany.

Brocq-Rousseau, D. 1904. Sur un Streptothrix. Références Générale Botanique 16:219–230.

Brosius, J., M. L. Palmer, P. J. Kennedy, and H. R. Noller. 1978. Complete nucleotide sequence of a 16S ribosomal RNA gene from Escherichia coli. Proc. Natl. Acad. Sci. USA 7:4801–4805.

Buchanan, R. E., and N. E. Gibson (Eds.). 1974. Bergey's Manual of Determinative Bacteriology, 8th ed. Williams and Wilkins. Baltimore, MD.

Celmer, W. D. et al. 1980. Process for producing nocardicin A. US Patent. 4,212,944.

Chiba, S., M. Suzuki, and K. Ando. 1999. Taxonomic re-evaluation of Nocardiopsis sp. K-252T (= NRRL 15532T): A proposal to transfer this strain to the genus Nonomuraea as Nonomuraea longicatena sp. nov. Int. J. Syst. Bacteriol. 49:1623–1630.

Chun, J., K. S. Bae, E. Y. Moon, S. O. Jung, H. K. Lee, and S. J. Kim. 2000. Nocardiopsis kunsanensis sp. nov., a moderately halophilic actinomycete isolated from a saltern. Int. J. Syst. Evol. Microbiol. 50:1909–1913.

Chung, Y. R., K. C. Sung, H. K. Mo, D. Y. Son, J. S. Nam, J. Chun, and K. S. Bae. 1999. Kitasatoa cheerisanensis sp. nov., a new species of the genus Kitasatospora that produces antifungal agent. Int. J. Syst. Bacteriol. 49:753–758.

Collins, M. D., M. Faulkner, and R. M. Keddie. 1984. Menaquinone composition of some spore forming actino-mycetes. Syst. Appl. Microbiol. 5:20–29.

Collins, M. D., and R. M. Kroppenstedt. 1987. Structures of the partially saturated menaquinones of Glycomyces rutgersensis. FEMS Microbiol. Lett. 44:215–219.

Collins, M. D., R. M. Kroppenstedt, J. Tamaoka, K. Komagata, and T. Kinoshita. 1988. Structures of the tetra hydrogenated menaquinones from Actinomadura angiospora, Faenia rectivirgula, and Saccharothrix australiensis. Curr. Microbiol. 17:275–279.

Crawford, D. L., E. McCoy, J. M. Harkin, and P. Jones. 1973. Production of microbial protein from waste cellulose by Thermomonospora fusca, a thermophilic actinomycete. Biotech. Bioeng. 15:833–843.

Crawford, D. L. 1975. Cultural, morphological and physiological characteristics of Thermomonospora fusca (strain 190Th). Can. J. Microbiol. 21:1842–1848.

Crawford, D. L. 1988. Biodegradation of agricultural and urban wastes. In: M. Goodfellow, S. T. Williams, and M. Mordarski (Eds.) Actinomycetes in Biotechnology. Academic Press. London, UK. 433–439.

Cross, T., and J. Lacey. 1970. Studies on the genus Thermomonospora. In: H. Prauser (Ed.) Actinomycetales VEB. Gustav Fischer Verlag. Jena, Germany. 211–219.

Cross, T., and M. Goodfellow. 1973. Taxonomy and classification of actinomycetes. In: G. Sykes and F. A. Skinner (Eds.) Actinomycetales: Characteristics and Practical Importance. Academic Press. London, UK. 11–112.

Cross, T. 1981. The monosporic actinomycetes. In: M. P. Starr, H. Stolp, H. G. Trüper, A. Balows and H. G. Schlegel (Eds.) The Prokaryotes. Springer-Verlag. New York, NY. 2091–2102.

Cui, X. L., P. H. Mao, M. Zeng, W. J. Li, L. P. Zhang, L. H. Xu, and C. L. Jiang. 2001. Streptomonospora salina gen. nov., sp. nov., a new member of the family Nocardiopsaceae. Int. J. Syst. Evol. Microbiol. 51:357–363.

Cummins, C. S., and H. Harris. 1956. A comparison of cell-wall composition in Nocardia, Actinomyces, Mycobacterium and Propionibacterium. J. Gen. Microbiol. 15:ix–x.

Cummins, C. S., and H. Harris. 1958. Studies on cell-wall composition and taxonomy of Actinomycetales and related groups. J. Gen. Microbiol. 18:173–189.

Cummins, C. S. 1962. La composition chimique des parois cellulares d'actinomycetes et son application taxonomique. Ann. Inst. Pasteur 103:67–73.

Davison, A. L., and J. Baddiley. 1963. The distribution of teichoic acids in staphylococci. J. Gen. Microbiol. 32:271–276.

Dixit, V. S., and A. Pant. 2000a. Hydrocarbon degradation and protease production by Nocardiopsis sp. NCIM 5124. Lett. Appl. Microbiol. 30:67–69.

Dixit, V. S., and A. Pant. 2000b. Comparative characterization of two serine endopeptidases from Nocardiopsis sp. NCIM 5124. Biochim. Biophys. Acta 1523:261–268.

Dolak, L. A., T. M. Castle, and A. L. Laborde. 1980. 3-Trehalosamine, a new disaccharide antibiotic. J. Antibiot. (Tokyo) 33:690–694.

Dolak, L. A., T. M. Castle, and L. A. Laborde. 1981. Biologically pure culture of Nocardiopsis trehalosei sp. nov. US Patent. 4,306,028.

Deutsche Sammlung von Mikroorganismen und Zellkulturen. 2001. Deutsche Sammlung von Mikroorganismen und Zellkulturen (DSMZ) Catalogue of Strains, 7th. ed. German Collection of Microorganisms and Cell Cultures. Braunschweig, Germany.

Erikson, D. 1935. The pathogenic aerobic organisms of the actinomyces group. Med. Res. Counc. (G.B.) Spec. Rep. Ser. 203:1–61.

Erikson, D. 1953. The reproductive pattern of Micromonospora vulgaris. J. Gen. Microbiol. 8:449–454.

Evtushenko, L. I., N. A. Ianushkene, G. M. Streshinskaya, I. B. Naumova, and N. S. Agre. 1984. Distribution of teichoic acids in representatives of the order Actinomycetales [in Russian]. Dokl. Akad. Nauk SSSR 278:237–239.

Evtushenko, L. I., V. V. Taran, V. N. Akimov, R. M. Kroppenstedt, J. M. Tiedje, and E. Stackebrandt. 2000. Nocardiopsis tropica sp. nov., nov. rev., Nocardiopsis trehalosi sp. nov., nom. rev. and Nocardiopsis dassonvillei subsp. albirubida subsp. nov., comb. nov. Int. J. Syst. Evol. Microbiol. 50:73–81.

Fennington, G., D. Lupo, and F. J. Stutzenberger. 1982. Enhanced cellulase production in mutants of Thermomonospora curvata. Biotech. Bioeng. 24:2487–2497.

Ferchak, J. D., and E. K. Pye. 1980. Saccharification of cellulose by the cellulolytic enzyme system of Thermomonospora species. I. Stability of cellulolytic activities with respect to time, temperature and pH. Biotech. Bioeng. 22:1515–1526.

Ferchak, J. D., and E. K. Pye. 1983. Effects of cellulose, glucose, ethanol, and metal ions on the cellulose complex of Thermomonospora fusca. Biotech. Bioeng. 25:2865–2872.

Fergus, C. L. 1964. Thermophilic and thermotolerant moulds and actinomycetes in mushroom compost during peak heating. Mycologia 56:267–284.

Fischer, A., R. M. Kroppenstedt, and E. Stackebrandt. 1983. Molecular-genetic and chemotaxonomic studies on Actinomadura and Nocardiopsis. J. Gen. Microbiol. 129:3433–3446.

Forbes, L. A. 1969. [MSc Thesis]. University of Wisconsin. Madison, Wisconsin.

Gauze, G. F., T. S. Maksimova, O. L. Olkhovotova, M. A. Sveshnikova, G. V. Kochetkova, and G. B. Ilchenko. 1974. Production of madumycin, an antibacterial antibiotic, by Actinomadura flava sp. nov. Antibiotiki 9:771–775.

Gauze, G. F., M. A. Sveshnikova, R. S. Ukholina, G. N. Komorova, V. S. Boshanov. 1977. Production of nocamycin, a new antibiotic, by Nocardiopsis syringae spec. nov. Antibiotiki 22:483–486.

Gochnauer, M. B., G. G. Leppard, P. Komaratat, M. Kates, T. Novitsky, and D. J. Kushner. 1975. Isolation and characterization of Actinopolyspora halophila, gen. et sp. nov., an extremely halophilic actinomycete. Can. J. Microbiol. 21:1500–1511.

Gochnauer, M., K. G. Jonson, and D. J. Kushner. 1989. Genus Actinopolyspora Gochnauer, Leppard, Komaratat, Kates, Novitsky and Kushner 1975, 1510[AL]. In: S. T. Williams, M. E. Sharpe, and J. G. Holt (Eds.) Bergey's Manual of Systematic Bacteriology. Williams and Wilkins. Baltimore, MD. 4:2398–2401.

Goodfellow, M., and G. Alderson. 1979. Numerical taxonomy of Actinomadura and related actinomycetes. J. Gen. Microbiol. 112:95–111.

Goodfellow, M., and T. Cross. 1984. Classification. In: M. Goodfellow, M. Modarski, and S. T. Wil-liams (Eds.) The Biology of Actinomycetes. Academic Press. London, UK. 7–164.

Goodfellow, M., E. Stackebrandt, and R. M. Kroppenstedt. 1988. Chemotaxonomy and actinomyete systematics. In: Y. Okami, T. Beppu, and H. Ogawara (Eds.) Biology of Actinomycetes 88. Japan Scientific Societies Press. Tokyo, Japan.

Goodfellow, M. 1989. The Actinomycetes 1: Supra genetic classification of actinomycetes. In: S. T. Williams, M. E. Sharpe, and J. G. Holt (Eds.) Bergey's Manual of Systematic Bacteriology. Williams and Wilkins. Baltimore, MD. 4:2333–2339.

Gordon, R., and A. Horan. 1968. Nocardia dassonvillei, a macroscopic replica of Streptomyces griseus. J. Gen. Microbiol. 50:235–240.

Green, J. G., M. W. Treuhaft, and R. M. Arusell. 1981. Hypersensivity pneumonitis due to Saccharomonospora viridis diagnosed by inhalation challenge. Ann. Allergy 47:449–452.

Greiner-Mai, E., R. M. Kroppenstedt, F. Korn-Wendisch, and H. J. Kutzner. 1987. Morphological and biochem-ical characterization and emended descriptions of thermophilic actinomycetes species. Syst. Appl. Microbiol. 9:97–109.

Greiner-Mai, E. 1988. Taxonomie thermophiler Actinomyceten im Hinblick auf ihre Bedeutung für allergische Erkrankung "Farmerlunge" [PhD Thesis]. Technische Hochschule Darmstadt. Darmstadt, Germany.

Grund, E. 1987. Untersuchungen zur Chernotaxonomie einiger Actinomyceten und coryneformer Bakterien [PhD Thesis]. Technische Hochschule Darmstadt. Darmstadt, Germany.

Grund, E., and R. M. Kroppenstedt. 1989. Transfer of five Nocardiopsis species to the genus Saccharothrix Labeda et al. 1984. J. Syst. Appl. Microbiol. 12:267–274.

Grund, E., and R. M. Kroppenstedt. 1990. Chemotaxonomy and numerical taxonomy of the genus Nocardiopsis Meyer 1976. Int. J. Syst. Bacteriol. 40:5–11.

Gugnani, H. C., I. C. Unaogu, F. Provost, and P. Boiron. 1998. Pulmonary infection due to Nocardiopsis dassonvillei, Gordona sputi, Rhodococcus rhodochrous, and Micromonospora sp. in Nigeria and literature review. J. Mycol. Medic. 8:21–25.

Hägerdal, B. G. R., H. Harris, and E. K. Pye. 1979. Association of β-glucosidase with intact cells of Thermoactinomyces. Biotech. Bioeng. 21:345–356.

Hägerdal, B. G. R., J. D. Ferchak, and E. K. Pye. 1980. Saccharification of cellulose by cellulolytic enzyme system of Thermomonospora sp. I: Stability of cellulytic activities with respect to time, temperature and pH. Biotech. Bioeng. 22:1515–1528.

Harkin, J. M., D. L. Crawford, and E. McCoy. 1974. Bacterial protein from pulps and papermill sludge. TAPPI 57:131–134.

Henssen, A. 1957. Beiträge zur Morphologie und Systematik der thermophilen Actinomyceten. Arch. Mikrobiol. 26:21–27.

Hickey, R. J., and H. D. Tresner. 1952. A cobalt containing medium for sporulation of Streptomyces species. J. Bacteriol. 64:891–892.

Hilge, M., S. M. Gloor, W. Rypniewski, O. Sauer, T. D. Heightman, W. Zimmermann, K. Winterhalter, and K. Piontek. 1998. High-resolution native and complex structures of thermostable beta-mannanase from Thermomonospora fusca: Substrate specificity in glycosyl hydrolase family 5. Structure 6:1433–1444.

Hilge, M., A. Perrakis, J. P. Abrahams, K. Winterhalter, K. Piontek, and S. M. Gloor. 2001. Structure elucidation of beta-mannanase: From the electron-density map to the DNA sequence. Acta Crystallographica, Section D—Biol. Crystallography 57:37–43.

Hoare, D. S., and E. Work. 1957. The stereoisomers of a,e diaminopimelic acid. Their distribution in the bacterial order Actinomycetales and certain Eubacteriales. Biochem. J. 65:441–447.

Hoffmann, P. 1989a. Cryopreservation of fungi. World Federation for Culture Collections UNESCO/WFCC/Education Committee. Braunschweig, Germany. Technical Information Sheet No. 5.

Hoffmann, P. 1989b. In: Cryopreservation of basidiomycete cultures: Mushroom Science XII, Part 1. Proceeding of the 12th International Congress on the Science and Cultivation of Edible Fungi, 1987. Braunschweig, Germany.

Horikoshi, K. 1971. Production of alkaline enzymes by alkaliphilic micro organisms. Part I: Alkaline protease produced by Bacillus no. 221. Agric. Biol. Chem. 35:1407–1414.

Horikoshi, K., and T. Akiba. 1982. Alkaliphilic Micro Organisms: A New World. Japan Scientific Press. Tokyo, Japan.

Howard, O. W., E. Grund, R. M. Kroppenstedt, and M. D. Collins. 1986. Structural determination of a novel naturally occurring cyclic vitamin. Biochem. Biophys. Res. Commun. 140:916–923.

Irwin, D., D. H. Shin, S. Zhang, B. K. Barr, J. Sakon, P. A. Karplus, and D. B. Wilson. 1998. Roles of catalytic domain and two cellulose binding domains of Thermomonospora fusca E4 in cellulose hydrolysis. J. Bacteriol. 180:1709–1714.

Irwin, D. C., S. Zhang, and D. B. Wilson. 2000. Cloning, expression and characterization of a Family 48 exocellulase, Cel48A, from Thermobifida fusca. Eur. J. Biochem. 267:4988–4997.

Iwabe, K., M. Teramura, K. Yoshinaga, S. Kobayashi, Y. Hoshikawa, T. Maeda, M. Hatakeyama, and H. Mizoguchi. 1998. K-252a-induced polyploidization and differentiation of a human megakaryocytic cell line, Meg-J: Transient elevation and subsequent suppression of cyclin B1 and cdc2 expression in the process of polyploidization. Br. J. Haematol. 102:812–919.

Jiang, C., and L. Xu. 1998. Actinomycete diversity in unusual habitats. In: C. Jiang and L. Xu (Eds.) Actinomycetes Research. Yunnan University Press. Yunnan, China. 259–270.

Johnson, K. G., P. H. Lanthier, and M. B. Gochnauer. 1986. Studies of two strains of Actinopolyspora halophila, an extremely halophilic actinomycete. Arch. Microbiol. 143:370–378.

Kämpfer, P., H.-J. Busse, and F. Rainey. 2002. Nocardiopsis compostus sp. nov., from the atmosphere of a composting facility. Int. J. Syst. Evol. Microbiol. 52:621–627.

Kase, H. K., K. Iwahashi, and Y. Matsuda. 1986. K-252a, a potent inhibitor of protein kinase c from microbial origin. J. Antibiot. 39:1059–1065.

Kawanami, J., A. Kimura, Y. Nakagawa, and H. Otsuka. 1969. Lipids of Streptomyces sioyaensis, V: On the 2-hydroxy-13-methyl-tetradecanoic acid from phosphatidylethanolamine. Chem. Phys. Lipids 3:29–38.

Kempf, A., and H.-J. Kutzner. 1988. Screening von biopolymerabbauenden Exoenzymen bei thermophilen Actinomyceten VDLUFA-Schriftenreihe 28. Kongressband Teil II. Oldenburg, Germany. 979–989.

Kempf, A. 1995. Untersuchungen über thermophile Actinomyceten: Taxonomie, Ökologie und Abbau von Biopolymeren [PhD Thesis]. Technische Hochschule Darmstadt. Darmstadt, Germany.

Kim, M.-J., H.-S. Chung, and S. J. Park. 1993. Properties of alkaline protease isolated from Nocardiopsis dassonvillei. Korean Biochem. J. 26:81–85.

Kim, J. W., H. Adachi, K. Shin-ya, Y. Hayakawa, and H. Seto. 1997. Apoptolidin, a new apoptosis inducer in transformed cells from Nocardiopsis sp. J. Antibiot. (Tokyo) 50:628–630.

Kim, Y. S., and A. Kawai. 1998. Studies on the antiviral mechanisms of protein kinase inhibitors K-252a and KT5926 against the replication of vesicular stomatitis virus. Biol. Pharm. Bull. 21:498–505.

Kleeberg, I., C. Hetz, R. M. Kroppenstedt, R.-J. Müller, and W.-D. Deckwer. 1998. Biodegradation of aliphatic-aromatic copolyester by Thermomonospora fusca and other thermophilic compost isolates. Appl. Environ. Microbiol. 64:1731–1735.

Koizumi, S., M. L. Contreras, Y. Matsuda, T. Hama, P. Lazarovici, and G. Guroff. 1988. K-252a: A specific inhibitor of the action of nerve growth factor on PC 12 cells. J. Neurosci. 8:715–721.

Korn-Wendisch, F., and H. J. Kutzner. 1992. The Family Streptomycetaceae. In: A. Balows, H. G. Trüper, M. Dworkin, W. Harder, and K.-H. Schleifer (Eds.) The Prokaryotes. Springer-Verlag. Berlin, Germany. 921–995.

Korn-Wendisch, F., F. Rainey, R. M. Kroppenstedt, A. Kempf, A. Majazza, H. J. Kutzner, and E. Stackebrandt. 1995. Thermocrispum gen. nov., a new genus of the order Actinomycetales, and description of Thermocrispum municipale sp. nov. and Thermocrispum agreste sp. nov. Int. J. Syst. Bacteriol. 45:67–77.

Kothe, H. W., G. Vobis, R. M. Kroppenstedt, and A. Henssen. 1989. A taxonomic study of mycolateless, wall chemotype IV actinomycetes. Syst. Appl. Microbiol. 12:61–69.

Krasilnikov, N. A., and N. S. Agre. 1964. A new genus of the actinomycetes-Actinobifida. The yellow group: Actinobifida dichotomica [in Russian]. Mikrobiologiya 33:935–943.

Kroppenstedt, R. M., and H. J. Kutzner. 1976. Biochemical markers in the taxonomy of Actinomycetales. Experientia 32:319–320.

Kroppenstedt, R. M., and H. J. Kutzner. 1978. Biochemical taxonomy of some problem actinomycetes. Zentralbl.

Bacteriol. Parasitenkd. Infektionskr. Hyg. Abt. 1 Orig. Reihe C. Suppl. 6:125–133.

Kroppenstedt, R. M. 1985. Fatty acid and menaquinone analysis of actinomycetes and related organisms. *In:* M. Goodfellow and D. E. Minnikin (Eds.) Chemical Methods in Bacterial Systematics No. 20. Academic Press. New York, NY. SAB Technical Series:173–199.

Kroppenstedt, R. M. 1987. Chemische Untersuchungen an Actinomycetales und verwandten Taxa, Korrelation von Chemosystematik und Phylogenie [Habilitationsschrift]. Technische Hochschule Darmstadt. Darmstadt, Germany.

Kroppenstedt, R. M., E. Stackebrandt, and M. Goodfellow. 1990. Taxonomic revision of the actinomycete genera Actinomadura and Microtetraspora. J. Syst. Appl. Bact. Microbiol. 13:148–160.

Kroppenstedt, R. M., and M. Goodfellow. 1992. The Family Thermomonosporaceae. *In:* A. Balows, H. G. Trüper, M. Dworkin, W. Harder, and K.-H. Schleifer (Eds.) The Prokaryotes. Springer-Verlag. Berlin, Germany. 1085–1114.

Kudo, T., T. Itoh, S. Miyadoh, T. Shomura, and A. Seino. 1993. Herbidospora gen. nov., a new genus of the family Streptosporangiaceae Goodfellow et al. 1990. Int. J. Syst. Bacteriol. 43:319–328.

Kudo, T., Y. Nakajima, and K. Suzuki. 1999. Catenuloplanes crispus (Petrolini et al. 1993) comb. nov.: Incorporation of the genus Planopolyspora Petrolini 1993 into the genus Catenuloplanes Yokota et al. 1993 with an amended description of the genus Catenuloplanes. Int. J. Syst. Bacteriol. 49:1853–1860.

Kudrina, E. S. 1957. Characteristics of actinomycetes-antagonists of the Albus series. *In:* G. F. Gauze, T. P. Preobrashinskaya, E. S. Kudrina, N. O. Blinoo, J. D. Ryabova, and M. K. Sveshnikova (Eds.) Problems of Classification of Actinomycetes Antagonists. Government Publishing House of Medical Literature. Moscow, USSR. 109–111.

Kurup, V. P. 1979. Characterization of some members of the genus Thermomonospora. Curr. Microbiol. 2:267–272.

Kushner, D. J. 1978. Life in high salt and solute concentrations: Halophilic bacteria. *In:* D. J. Kushner (Ed.) Microbial Life in Extreme Environments. Academic Press. London, UK. 317–368.

Küster, E. 1959a. Introductory remarks and welcome, "Round Table Conference on Streptomycetes," 7th International Congress of Microbiology, Stockholm, August 4–5, 1958. Int. Bull. Bact. Nomen. Taxon. 9:57–61.

Küster, E. 1959b. Outline of comparative study of criteria used in characterization of the actinomycetes. Int. Bull. Bact. Nomen. Taxon. 9:98–104.

Küster, E., and S. Williams. 1964. Selection of media for isolation of streptomycetes. Nature 202:928–929.

Kutzner, H.-J. 2000. Microbiology of composting. *In:* H.-J. Rehm and G. Reed (Eds.) Biotechnology. Wiley-VCH. Weinheim, Germany. 11c:35–100.

Labeda, D. P., R. T. Testa, M. P. Lechevalier, and H. A. Lechevalier. 1984. Saccharothrix: A new genus of the Actinomycetales related to Nocardiopsis. Int. J. Syst. Bacteriol. 34:426–431.

Labeda, D. R., R. T. Testa, M. P. Lechevalier, and H. A. Lechevalier. 1985. Glycomyces, a new genus of the Actinomycetales. Int. J. Syst. Bacteriol. 35:417–421.

Labeda, D. P., and M. P. Lechevalier. 1989. Amendment of the genus Saccharothrix Labeda et al. 1984 and descriptions of Saccharothrix espanensis sp. nov., Sac-

charothrix cryophilis sp. nov., and Saccharothrix mutabilis comb. nov. Int. J. Syst. Bacteriol. 39:429–423.

Labeda, D. P. 1992. The genus Saccharothrix. *In:* A. Balows, H. G. Trüper, M. Dworkin, W. Harder, and K.-H. Schleifer (Eds.) The Prokaryotes. Springer-Verlag. Berlin, Germany. 2:1061–1068.

Labeda, D. P., and R. M. Kroppenstedt. 2000. Phylogenetic analysis of Saccharothrix and related taxa: Proposal for Actinosynnemataceae fam. nov. Int. J. Syst. Ecol. Microbiol. 50:331–336.

Labeda, D. P., K. Hatano, R. M. Kroppenstedt, and T. Tamura. 2001. Revival of the genus Lentzea and proposal for Lechevalieria gen. nov. Int. J. Syst. Evol. Microbiol. 51:1045–1050.

Lacey, J. 1973. Actinomycetes in soils, composts and fodders. *In:* F. A. Skinner and G. Sykes (Eds.) Actinomycetales: Characteristics and Practical Importance. Academic Press. London, UK. 231–251.

Lacey, J. 1974. Allergy in mushroom workers. Lancet 1:366.

Lacey, J. 1977. The ecology of actinomycetes in fodders and related substrates: Proceeding of the Warsaw Symposium on Streptomyces and Nocardia. Zentralbl. Bakteriol. Parasitenkd. Infektionskr. Hyg. Abt. 1. Suppl. 6:161–170.

Lacey, J. 1988a. Actinomycetes as biodeteriogens and pollutants of the environment. *In:* M. Goodfellow, S. T. Williams, and M. Modarski (Eds.) Actinomycetes in Biotechnology. Academic Press. London, UK. 359–432.

Lacey, J., and B. Crook. 1988b. Fungal and actinomycete spores as pollutants of the workplace and occupational allergens. Ann. Occup. Hyg. 32:515–533.

Laiz, L., I. Groth, P. Schumann, F. Zezza, A. Felske, B. Hermosin, and C. Saiz-Jimenez. 2000. Microbiology of the stalactites from Grotta dei Cervi, Porto Badisco, Italy. Int. Microbiol. 3:25–30.

Larsen, H. 1986. Halophilic and halotolerant micro organisms: Overview and historical perspective. FEMS Microbiol. Rev. 39:3–7.

Lechevalier, M. P., and H. A. Lechevalier. 1970a. Chemical composition as a criterion in the classification of aero-bic actinomycetes. Int. J. Syst. Bacteriol. 20:435–443.

Lechevalier, M. P., and H. A. Lechevalier. 1970b. Composition of whole-cell hydrolysates as a criterion in the classification of aerobic actinomycetes. *In:* H. Prauser (Ed.) The Actinomycetales. VEB-Fischer. Jena, Germany. 311–316.

Lechevalier, H. A., M. P. Lechevalier, and N. N. Gerber. 1971. Chemical composition as a criterion in the clas-sification of actinomycetes. Adv. Appl. Microbiol. 14:47–72.

Lechevalier, M. P., C. de Bievre, and H. A. Lechevalier. 1977. Chemotaxonomy of aerobic actinomycetes: Phospholipid composition. Biochem. Ecol. Syst. 5:249–260.

Lechevalier, M. P., H. Prauser, D. P. Labeda, and J. S. Ruan. 1986. Two new genera of nocardioform actinomycetes: Amycolata gen. nov. and Amycolatopsis gen. nov. Int. J. Syst. Bacteriol. 36:29–37.

Lee, S. E., and A. E. Humphrey. 1979. Use of continuous culture techniques for determining the growth kinetics of cellulolytic Thermoactinomyces sp. Biotech. Bioeng. 21:1277–1288.

Lee, S. D., S. O. Kang, and Y. C. Hah. 2000. Hongia gen. nov., a new genus of the order Actinomycetales. Int. J. Syst. Evol. Microbiol. 50:191–199.

Liegard, H., and Landrieu. 1911. Un as de mycose con junctivale. Ann. Ocul. 146:418–426.

Locci, R., E. Baldacci, and B. Petrolini. 1967. Contribution to the study of oligosporic actinomycetes. I: Description of new species of Actinobifida: Actinobifida alba sp. nov. and revision of the genus. Gen. Microbiol. 15:79–91.

Lu, Z., Z. Liu, L. Wang, Y. Zhang, W. Qi, and M. Goodfellow. 2001. Saccharopolyspora flava sp. nov. and Saccharopolyspora thermophila sp. nov., novel actinomycetes from soil. Int. J. Syst. Evol. Microbiol. 51:319–325.

Matsuda, Y., and J. Fukuda. 1988. Inhibition by K-252, a new inhibitor of protein kinase, of nerve growth factor-induced neurite outgrowth of chick embryo dorsal root ganglion cells. Neurosci. Lett. 87:11–17.

McCarthy, A. J., and T. Cross. 1984a. A taxonomic study of Thermomonospora and other monosporic actinomycetes. J. Gen. Microbiol. 130:5–25.

McCarthy, A. J., and T. Cross. 1984b. Taxonomy of Thermomonospora and related oligosporic actinomycetes. In: L. Ortiz-Ortiz, L. F. Bojalil, and V. Yakoleff (Eds.) Biological, Biochemical and Biomedical Aspects of Actinomycetes. Academic Press. San Diego, CA. 521–536.

McCarthy, A. J., and T. Cross. 1984c. Validation of the publication of new names and new combinations previously effectively published outside the IJSB: List No. 15. Int. J. Syst Bacteriol. 34:355–357.

McCarthy, A. T., E. Peace, and P. Broda. 1985. Studies on extracellular activities of some thermophilic actinomycetes. Appl. Microbiol. 21:238–244.

McCarthy, A. J., A. Paterson, and P. Broda. 1986. Lignin solubilization by Thermomonospora mesophila. Appl. Microbiol. 24:347–352.

McCarthy, A. J. 1987. Lignocellulose-degrading actinomycetes. FEMS Microbiol. Rev. 46:145–163.

McCarthy, A. T., A. S. Ball, and S. L. Bachmann. 1988. Ecological and biotechnological implication of lignocellulose degradation by actinomycetes. In: Y. Okami, T. Beppu, and H. Ogawara (Eds.) Biology of Actinomycetes 88. Japan Scientific Society Press. Tokyo, Japan. 283–287.

Mertz, F. P., and R. C. Yao. 1993. Amycolatopsis alba sp. nov., isolated from soil. Int. J. Syst. Bacteriol. 43:715–720.

Meyer, J. 1976. Nocardiopsis, a new genus of the order Actinomycetales. Int. J. Syst. Bacteriol. 26:487–493.

Meyer, H. P., and A. E. Humphrey. 1982. Cellulase production by a wild type and a new mutant strain of Thermomonospora sp. Biotech. Bioeng. 24:1901–1904.

Meyer, J. 1989. Genus Nocardiopsis Meyer 1976, 487[AL]. In: S. T. Williams, M. E. Sharpe, and J. G. Holt (Eds.) Bergey's Manual of Systematic Bacteriology. Williams and Wilkins. Baltimore, MD. 4:2562–2569.

Mikami, Y., K. Miyashita, and T. Arai. 1982. Diaminopimelic acid profiles of alkaliphilic and alkali-resistant strains of actinomycetes. J. Gen. Microbiol. 128:1709–1712.

Mikami, Y., K. Miyashita, and T. Arai. 1985. Alkaliphilic actinomycetes. Actinomyetes 19:176–191.

Minnikin, D. E., L. Alshamaony, and M. Goodfellow. 1975. Differentiation of Mycobacterium, Nocardia, and related taxa by thin-layer chromatographic analysis of whole-organism methanolysates. J. Gen. Microbiol. 88:200–204.

Minnikin, D. E., T. Pirouz, and M. Goodfellow. 1977. Polar lipid composition in the classification of some Actinomadura species. Int. J. Syst. Bacteriol. 27:118–121.

Minnikin, D. E., M. D. Collins, and M. Goodfellow. 1978. Menaquinone patterns in the classification of nocardio-form and related taxa. Zentralbl. Bakteriol. Parasitenkd. Infektionskr. Hyg. Abt. 1, Orig. Reihe C. Suppl. 6:85–90.

Minnikin, D. E., A. G. O'Donnell, M. Goodfellow, G. Alderson, M. Athalye, K. P. Schaal, and J. H. Parlett. 1984. An integrated procedure for extraction of bacterial isoprenoid quinones and polar lipids. J. Microbiol. Meth. 2:233–241.

Mishra, S. K., J. E. Keller, J. R. Miller, R. M. Heisey, M. G. Nair, and A. R. Putnam. 1987a. Insecticidal and nematicidal properties of microbial metabolites. J. Indust. Microbiol. 2:267–276.

Mishra, S. K., W. H. Taft, A. R. Putnam, and S. K. Ries. 1987b. Plant growth regulatory metabolites from novel actinomycetes. J. Plant Growth Regul. 6:75–84.

Mishra, S. K., C. J. Whitenack, and A. R. Putnam. 1988. Herbicidal properties of metabolites from several genera of soil microorganisms. Weed Sci. 36:122–126.

Miyashita, M., Y. Mikami, and T. Arai. 1984. Alkaliphilic actinomycete, Nocardiopsis dassonvillei subsp. prasina subsp. nov. isolated from soil. Int. J. Syst. Bacteriol. 34:405–409.

Mordarska, H., M. Mordarski, and M. Goodfellow. 1972. Chemotaxonomic characteristics and classification of some nocardioform bacteria. J. Gen. Microbiol. 73:77–86.

Mordarska, H., A. Gamian, and J. Carrasco. 1983. Sugar-containing lipids in the classification of representative Actinomadura and Nocardiopsis species. Arch. Immunol. Exp. 31:135–143.

Mordarska, H., J. Zakrzewska-Czerwinska, M. Pasciak, M. Szponar, and S. Rowinski. 1998. Rare suppurative pulmonary infection caused by Nocardiopsis dassonvillei recognized by glycolipid markers. FEMS Immun. Med. Microbiol. 21:47–55.

Nakanishi, S., Y. Matsuda, K. Iwahashi, and H. Kase. 1986. K-252b, c and d, potent inhibitors of protein kinase c from microbial origin. J. Antibiot. 39:1066–1071.

Nakanishi, S., K. Yamada, H. Kase, S. Nakamura, and Y. Nonomura. 1988. K-252a, a novel microbial product, inhibits smooth muscle myosin light chain kinase. J. Biol. Chem. 263:6215–6219.

Naumova, I. B. 1988. The teichoic acids of actinomycetes. Microbiol. Sci. 5:275–279.

Naumova, I. B., A. S. Shashkov, E. M. Tul'skaya, G. M. Streshinskaya, Yu. I. Kozlova, N. V. Potekhina, L. I. Evtushenko, and E. Stackebrandt. 2001. Cell wall teichoic acids: Structural diversity, species-specificity in the genus Nocardiopsis and chemotaxonomic perspective. FEMS Microbiol. Rev. 25:269–284.

Nishiwaki, S., H. Fujiki, S. Yoshizawa, M. Suganuma, H. Furuya-Suguri, S. Okabe, M. Nakayasu, K. Okabe, H. Muratake, M. Natsume. 1991. Pendolmycin, a new tumor promoter of the teleocidin A class on skin of CD-1 mice. Jpn. J. Cancer Res. 82:779–783.

Nonomura, H., and Y. Ohara. 1974. A new species of actinomycetes, Thermomonospora mesouviformis sp. nov. J. Ferment. Technol. 53:10–13.

Nonomura, H., and Y. Ohara. 1971a. Distribution of actinomycetes in soil. X: New genus and species of monosporic actinomycetes in soil. J. Ferment. Technol. 49:895–903.

Nonomura, H., and Y. Ohara. 1971b. Distribution of actinomycetes in soil. J. Ferment. Technol. 49:904–912.

Ohmori, K., H. Ishii, H. Manabe, H. Satoh, T. Tamura, and H. Kase. 1988. Antiinflammatory and antiallergic effects of a novel metabolite of Nocardiopsis sp. as a potent protein kinase C inhibitor from microbial origin. Drug Res. 38:809–814.

Papeta, N. F., G. M., Tsilosani, N. T. Nikolaishvili, and G. I. Kvesitadze. 1995. Identification, purification, and characterization of xylanase from Nocardiopsis sp. inhibiting the growth of some phytopathogenic fungi. Biochem. 4:429–433.

Park, Y. H., J. H. Yoon, Y. K. Shin, K. Suzuki, T. Kudo, A. Seino, H. J. Kim, J. S. Lee, and S. T. Lee. 1999. Classification of 'Nocardioides fulvus' IFO 14399 and Nocardioides sp. ATCC 39419 in Kribbella gen. nov., as Kribbella flavida sp. nov. and Kribbella sandramycini sp. nov. Int. J. Syst. Bacteriol. 49:743–752.

Peltola, J. S. P., M. A. Andersson, P. Kämpfer, G. Auling, R. M. Kroppenstedt, H.-J. Busse, M. Salkinoja-Salonen, and F. A. Rainey. 2001. Isolation of toxigenic Nocardiopsis strains from indoor environments and description of two new Nocardiopsis species, N. exhalans sp. nov. and N. umdischolae sp. nov. Appl. Environ. Microbiol. 67:4293–4304.

Poschner, J., R. M. Kroppenstedt, A. Fischer, and E. Stackebrandt. 1985. DNA-DNA-reassociation and chemotaxonomic studies on Actinomadura, Microbispora, Microtetrapora, Micropolyspora, and Nocardiopsis. Syst. Appl. Microbiol. 6:264–270.

Preobrazhenskaya, T. P., and M. A. Sveshnikova. 1974. New species of the genus Actinomadura. Mikrobiolo-giya 43:864–868.

Preobrazhenskaya, T. P., M. A. Sveshnikova, and G. F. Gauze. 1982. On the transfer of certain species of the genus Actinomadura Lechevalier et Lechevalier 1970 to the genus Nocardiopsis Meyer 1976. Mikrobiologiya 51:111–113.

Pridham, T. G., P. Anderson, C. Foley, L. A. Lindenfelser, C. W. Hesseltine, and R. G. Benedict. 1956–1957. A selection of media for maintenance and taxonomic study of streptomycetes. Antibiot. Ann. 1956/1957:947–953.

Pridham, T. G., and A. T. Lyons Jr. 1961. Streptomyces albus (Rossi Doria) Waksman et Henrici: Taxonomic study of strains labelled Streptomyces albus. J. Bacteriol. 81:431–441.

Rainey, F. A., N. Ward-Rainey, R. M. Kroppenstedt, and E. Stackebrandt. 1996. The genus Nocardiopsis represents a phylogenetically coherent taxon and a distinct actinomycete lineage: Proposal of Nocardiopsaceae fam. nov. Int. J. Syst. Bacteriol. 46:1088–1092.

Ramachandra, M., D. L. Craeford, and G. Hertel. 1988. Characterisation of an extracellular lignin peroxidase of the lignocellulytic actinomycetes Streptomyces viridosporus. Appl. Environ Microbiol. 54:3057–3063.

Resz, A., J. Schwanbeck, and J. Knösel. 1977. Thermophile Actinomyceten aus Müllkompost: Temperaturansprüche und proteolytische Aktivität. Forum Städte-Hygiene 28:71–73.

Rob, A., A. S. Ball, M. Tuncer, and M. T. Wilson. 1995. Isolation and characterisation of a novel non-haem extracellular peroxidase produced by the thermophilic actinomycete Thermomonospora fusca BD25. Biochem. Soc. Trans. 23:507S.

Rob, A., A. S. Ball, M. Tuncer, G. D. Jones, P. D. Taylor, and M. T. Wilson. 1996. Redox reaction of the novel non-haem glycosylated peroxidases from thermophilic actinomycete Thermomonospora fusca BD25. Biochem. Soc. Trans. 24:455S.

Rob, A., A. S. Ball, M. Tuncer, and M. T. Wilson. 1997a. Catalytic mechanism of the novel non-haem iron con-

taining peroxidase produced by the thermophilic actinomycete Thermomonospora fusca BD25. Biochem. Soc. Trans. 25:64S.

Rob, A., A. S. Ball, M. Tuncer, and M. T. Wilson. 1997b. The detection and quantification of novel non-haem extracellular glycosylated peroxidases produced by the thermophilic actinomycete Thermomonospora fusca BD25 by means of PAGE-zymogram. Biochem. Soc. Trans. 25:37S.

Romano, A. H., and A. Sohler. 1956. Biochemistry of the Actinomycetales. II: A comparison of the cell wall composition of species of the genera Streptomyces and Nocardia. J. Bacteriol. 72:865–867.

Rossello-Mora, R., and R. Amann. 2001. The species concept for prokaryotes. FEMS Microbiol. Rev. 25:39–67.

Ruan, J.-S., A. M. Al-Tai, Z.-H. Zhou, and L.-H. Qu. 1994. Actinopolyspora iraquiensis sp. nov., a new halophilic actinomycete isolated from soil. Int. J. Syst. Bacteriol. 44:759–763.

Sabaou, N., and N. Bounaga. 1987. Actinomycetes parasite de champignons: Étude des espèces, specificit´ de l'action parasitaire au genre Fusarium et antagonist dans le sol envers Fusarium oxysporum f. sp. albedinis (Kilian et Maire) Gordon. Can. J. Microbiol. 33:445–451.

Schippers, A., K. Bosecker, S. Willscher, C. Spröer, P. Schumann, and R. M. Kroppenstedt. 2002. Nocardiopsis metallicus sp. nov., a metal leaching actinomycete isolated from an alkaline slag dump. Int. J. Syst Evol. Microbiol 52:2291–2295.

Schleifer, K. H., and O. Kandler. 1972. Peptidoglycan types of bacterial cell walls and their taxonomic implications. Bacteriol. Rev. 36:407–477.

Shashkov, A. S., G. M. Streshinskaya, Y. I. Kozlova, N. V. Potekhina, L. I. Evtushenko, V. V. Taran, and I. B. Naumova. 1997. Structure of teichoic acid from cell walls of Nocardiopsis alborubida. Biochemistry 64:1135–1139.

Shearer, M. C., P. M. Colman, and C. H. Nash, 3rd. 1983. Nocardiopsis mutabilis, a new species of nocardioform bacteria isolated from soil. Int. J. Syst. Bacteriol. 33:369–374.

Shearer, M. C., P. M. Colman, R. M. Ferrin, L. J. Nisbet, and C. H. Nash, 3rd. 1986. New genus of the Actinomycetales: Kibdelosporangium aridum gen. nov., sp. nov. Int. J. Syst. Bacteriol. 36:47–54.

Shinobu, R. 1958. Physiological and cultural study for the identification of soil actinomycete species. Mem. Osaka Univ. B. Nat. Sci. 7:1–76.

Shirling, E. B., and D. Gottlieb. 1966. Methods for the characterization of streptomycetes species. Int. J. Syst. Bacteriol. 16:313–340.

Shirling, E. B., and D. Gottlieb. 1972. Cooperative description of type cultures of Streptomyces. V: Additional species descriptions from first and second studies. Int. J. Syst. Bacteriol. 18:279–392.

Sindhuphak, W., E., Macdonald, and E. Head. 1985. Actinomycetoma caused by Nocardiopsis dassonvillei. Arch. Dermatol. 121:1332–1334.

Singh, S. M., J. Naidu, S. Mukerjee, and A. Malkani. 1991. Cutaneous infections due to Nocardiopsis dassonvillei (Brocq-Rousseau) Meyer 1976, endemic in members of a family up to fifth degree relatives, abstr. PS1.91. In: Program and Abstracts of the XI Congress of the International Society for Human and Animal Mycology. 85.

Skerman, V. B. D., V. McGowan, and P. H. A. Sneath. 1980. Approved lists of bacterial names. Int. J. Syst. Bacteriol. 30:225–420.

Sneath, P. H. A. 1992. International Code of Nomenclature of Bacteria. American Society for Microbiology. Washington, DC.

Spiridonov, N. A., and D. B. Wilson. 1998. Regulation of biosynthesis of individual cellulases in Thermomonospora fusca. J. Bacteriol. 180:3529–3532.

Spiridonov, N. A., and D. B. Wilson. 1999. Characterization and cloning of CelR, a transcriptional regulator of cellulase genes from Thermomonospora fusca. J. Biol. Chem. 274:13127–13132.

Spiridonov, N. A., and D. B. Wilson. 2001. Cloning and biochemical characterization of BglC, a beta-glucosidase from the cellulolytic actinomycete Thermobifida fusca. Curr. Microbiol. 42:295–301.

Stackebrandt, E., W. Ludwig, E. Seewald, and K.-H. Schleifer. 1983. Phylogeny of spore forming members of the order Actinomycetales. Int. J. Syst. Bacteriol. 33:173–180.

Stackebrandt, E. 1986. The significance of "Wall Types" in phylogenetically based taxonomic studies on actinomycetes. In: G. Szabo, S. Biro, and M. Goodfellow (Eds.) Biological, Biochemical and Biomed-ical Aspects of Actinomycetes. Akademiai Kiado. Bu-dapest, Hungary. 497–506.

Stackebrandt, E., and B. M. Goebel. 1994a. Taxonomic note: A place for DNA-DNA reassociation and 16S rRNA sequence analysis in the present species definition in bacteriology. Int. J. Syst. Bacteriol. 44:846–849.

Stackebrandt, E., R. M. Kroppenstedt, K.-D. Jahnke, C. Kemmerling, and H. Gürtler. 1994b. Transfer of Streptosporangium viridogriseum (Okuda et al. 1966) Streptosporangium viridogriseum subsp. kofuense (Nonomura and Ohara 1969), Streptosporangium albidum (Furumai et al. 1968) to Kutzneria gen. nov. as Kutzneria viridogrisea comb. nov., respectively, and emendation of the genus Streptosporangium. Int. J. Syst. Bacteriol. 44:265–269.

Stackebrandt, E., F. A. Rainey, and N. L. Ward-Rainey. 1997. Proposal for a new hierarchic classification system, Actinobacteria classis nov. Int. J. Syst. Bacteriol. 47:479–491.

Stackebrandt, E. 2000. Defining taxonomic ranks. In: M. Dworkin, S. Falkow, E. Rosenberg, K.-H. Schleifer, and E. Stackebrandt (Eds.) The Prokaryotes, 3rd ed. Springer-Verlag, New York. Release 3.4.

Stamford, T. L. M., N. P. Stamford, L. C. B. B. Coelho, and J. M. Araujo. 2001. Production and characterization of a thermostable alpha-amylase from Nocardiopsis sp. endophyte of yam bean. Bioresource Technol. 76:137–141.

Stanek, M. 1972. Microorganisms inhabiting mushroom compost during fermentation. Mushroom Sci. 8:797–811.

Streshinskaya, G. M., E. M. Tulskaya, L. P. Terekhova, O. A. Galatenko, I. B. Naumova, and T. P. Preobrazhenskaya. 1989. Some chemotaxonomic criteria of the genus Nocardiopsis [in Russian]. Dokl. Akad. Nauk SSSR. 309:477–480.

Streshinskaya, G. M., Yu. I. Kozlova, L. I. Evtushenko, V. V. Taran, A. S. Shashkov, and I. B. Naumova. 1996. Cell wall teichoic acid of Nocardiopsis subsp. VKM Ac-1457. Biochemistry (Moscow) 61:285–288.

Streshinskaya, G. M., E. M. Tul'skaya, A. S. Shashkov, L. I. Evtushenko, V. V. Taran, and I. B. Naumova. 1998. Teichoic acids of the cell wall of Nocardiopsis listeri, Nocardiopsis lucentensis, and Nocardiopsis tregalosei. Biochemistry (Moscow) 63:230–234.

Stutzenberger, F. J. 1971. Cellulase production by Thermomonospora curvata isolated from municipal solid waste compost. Appl. Microbiol. 22:147–152.

Su, T. M., and D. Paulavicius. 1975. Enzymatic saccharification of cellulose by thermophilic actinomycetes. Appl. Polymer Symp. 28:221–236.

Sun, H. H., C. B. White, J. Dedinas, R. Cooper, and D. M. Sedlock. 1991. Methylpendolmycin, an indolactam from a Nocardiopsis sp. J. Nat. Prod. 54:1440–1443.

Suzuki, K., K. Nagao, Y. Monnai, A. Yagi, and M. Uyeda. 1998a. Topostatin, a novel inhibitor of topoisomerases I and II produced by Thermomonospora alba strain No. 1520. I: Taxonomy, fermentation, isolation and biological activities. J. Antibiot. (Tokyo) 51:991–998.

Suzuki, K., S. Yahara, Y. Kido, K. Nagao, Y. Hatano, and M. Uyeda. 1998b. Topostatin, a novel inhibitor of Topoisomerases I and II produced by Thermomonospora alba strain No. 1520. II: Physico-chemical properties and structure elucidation. J. Antibiot. (Tokyo) 51:999–1003.

Suzuki, K., S. Yahara, K. Maehata, and M. Uyeda. 2001. Isoaurostatin, a novel topoisomerase inhibitor produced by Thermomonospora alba. J. Nat. Products 64:204–207.

Takahashi, T., A. Kamimura, A. Shirai, and Y. Yokoo. 2000. Several selective protein kinase C inhibitors including procyanidins promote hair growth. Skin Pharmacol. Appl. Skin Physiol. 13:133–142.

Tamura, T., M. Hayakawa, H. Nonomura, A. Yokota, and K. Hatano. 1995. Four new species of the genus Actinokineospora: Actinokineospora inagensis sp. nov., Actinokineospora globicatena sp. nov., Actinokineospora terrae sp. nov., and Actinokineospora diospyrosa sp. nov. Int. J. Syst. Bacteriol. 45:371–378.

Tamura, T., M. Hayakawa, and K. Hatano. 1997. A new genus of the order Actinomycetales, Spirilliplanes gen. nov., with description of Spirilliplanes yamanashiensis sp. nov. Int. J. Syst. Bacteriol. 47:97–102.

Tamura, T., M. Hayakawa, and K. Hatano. 1998. A new genus of the order Actinomycetales, Cryptosporangium gen. nov., with descriptions of Cryptosporangium arvum sp. nov. and Cryptosporangium japonicum sp. nov. Int. J. Syst. Bacteriol. 48:995–1005.

Tamura, T., S. Suzuki, and K. Hatano. 2000a. Acrocarpospora gen. nov., a new genus of the order Actinomycetales. Int. J. Syst. Evol. Microbiol. 50:1163–1171.

Tamura, T., L. Zhiheng, Z. Yamei, and K. Hatano. 2000b. Actinoalloteichus cyanogriseus gen. nov., sp. nov. Int. J. Syst. Evol. Microbiol. 50:1435–1440.

Taptykova, S. D., L. P. Terekhova, and V. M. Adanin. 1990. Menaquinone composition in Nocardiopsis spp. and some Actinomadura spp. [in Russian]. Mikrobiologiya 59:650–655.

Thomson, J. A. 1993. Molecular biology of xylan degradation. FEMS Microbiol. Rev. 104:65–82.

Tomita, K., Y. Hoshino, and T. Miyaki. 1993. Kibdelosporangium albatum sp. nov., producer of the antiviral antibiotics cycloviracins. Int. J. Syst. Bacteriol. 43:297–301.

Trigo, C., and A. S. Ball. 1994. Is the solubilized product from the degradation of lignocellulose by actinomycetes a precursor of humic substances? Microbiology 140:3145–3152.

Tsujibo, H., T. Sato, M. Inui, H. Yamamoto, and Y. Ina-mori. 1988. Intracellular accumulation of phenazine antibiotics produced by an alkaliphilic actinomycete. I. Taxonomy, isolation and identification of the phen-azine antibiotics. Agr. Biol. Chem. 52:301–306.

Tsujibo, H., K. Mijamoto, T. Hasegawa, and Y. Inamori. 1990a. Amino acid composition and partial sequences of two types of alkaline serine proteases from Nocardiopsis dassonvillei subsp. prasina OPC-210. Agr. Biol. Chem. 54:2177–2179.

Tsujibo, H., K. Mijamoto, T. Hasegawa, and Y. Inamori. 1990b. Purification and characterization of two types of alkaline serine proteases produced by an alkaliphilic actinomycete. J. Appl. Bacteriol. 69:520–529.

Tsujibo, H., T. Sakamoto, K. Miyamoto, G. Kusano, M. Ogura, T. Hasegawa, and Y. Inamori. 1990c. Isolation of cytotoxic substance, kalafungin from an alkaliphilic actinomycete, Nocardiopsis dassonvillei subsp. prasina. Chem. Pharm. Bull. (Tokyo) 38:2299–2300.

Tsujibo, H., T. Sakamoto, N. Nishino, T. Hasegawa, and Y. Inamori. 1990d. Purification and properties of three types of xylanases produced by an alkaliphilic actinomycete. J. Appl. Bacteriol. 69:398–405.

Tsujibo, H., T. Sakamoto, K. Miyamoto, T. Hasegawa, M. Fujimoto, and Y. Inamori. 1991. Amino acid compositions and partial sequences of xylanases from a new subspecies, Nocardiopsis dassonvillei subsp. alba OPC-18. Agric Biol Chem. 55:2173–2174.

Tul'skaya, E. M., G. M. Streshinskaya, I. B. Naumova, A. S. Shashkov, and L. P. Terekhova. 1993. A new structural type of teichoic acid and some chemotaxonomic criteria of two species Nocardiopsis dassonvillei and Nocardiopsis antarcticus:1993. Arch. Microbiol. 160:299–305.

Tul'skaya, E. M., A. S. Shashkov, L. I. Evtushenko, and I. B. Naumova. 2000. Cell wall teichoic acids of Nocardiopsis prasina VKM Ac-1880T. Microbiology (Moscow) 69:48–50.

Tuncer, M., A. Ball, A. Rob, and M. Wilson. 1999. Optimization of extracellular lignocellulytic enzyme production by a thermophilic actinomycete Thermomonospora fusca BD25. Enz. Microbiol. Technol. 25:38–47.

Uchida, K., and K. Aida. 1977. Acyl type of bacterial cell wall: Its simple identification by colorimetric method. J. Gen. Appl. Microbiol. 23:249–260.

Vacca, J. G., and W. D. Bellamy. 1976. Classification and morphological properties of a high temperature, cellulolytic actinomycete. In: Abstracts of the Annual meeting of the American Society for Microbiology. 117.

Vandamme, P., B. Pot, M. Gillis, P. de Vos, K. Kersters, and J. Swings. 1996. Polyphasic taxonomy, a consensus approach to bacterial systematics. Microbiol. Rev. 60:407–438.

Van den Borgard, H. G. G. Van den Ende, P. C. Van Loon, and L. J. Van Griensven. 1993. Mushroom worker's lung: Serologic reactions to thermophilic actinomycetes present in the air of compost tunnels. Mycopathologia 122:21–28.

Vasanthi, V., and S. L. Hoti. 1992. Microbial flora in gut of Culex quinquefasciatus breeding in cess pits. SE Asian J. Trop. Med. Pub. Health 23:312–317.

Ventosa, A., J. J. Nieto, and A. Oren. 1998. Biology of moderately halophilic aerobic bacteria. Microbiol. Molec. Biol. Rev. 62:504–544.

Voelskow, H. 1988. Extrazelluläre mikrobielle Polysaccharide: Biopolymere mit interessanten Anwendungsmöglichkeiten. Forum Mikrobiologie 6:273–283.

Waksman, S. A., W. W. Umbert, and T. S. Gordon. 1939. Thermophilic actinomycetes and fungi in soils and composts. Soil. Sci. 47:37–61.

Wang, Y., Z. S. Zhang, J. S. Ruan, Y. M. Wang, and S. M. Ali. 1999. Investigation of actinomycete diversity in the trop-ical rainforests of Singapore. J. Ind. Microbiol. Biotechnol. 23:178–187.

Warwick, S., T. Bowen, H. McVeigh, and T. M. Embley. 1994. A phylogenetic analysis of the family Pseudonocardiaceae and the genera Actinokineospora and Saccharothrix with 16S rRNA sequences and a proposal to combine the genera Amycolata and Pseudonocardia in an emended genus Pseudonocardia. Int. J. Syst. Bacteriol. 44:293–299.

Wellington, E. M. H., and S. T. Williams. 1978. Preserva-tion of actinomycete inoculum in frozen glycerol. Mi-crobios Lett. 6:151–157.

Williams, S. T., G. P. Sharpels, and R. M. Bradshaw. 1974. Spore formation in Actinomadura dassonvillei (Brocq–Rousseau) Lechevalier and Lechevalier. J. Gen. Microbiol. 84:415–419.

Williams, S. T., and E. M. H. Wellington. 1981. The genera Actinomadura, Actinopolyspora, Excellospora, Microbispora, Micropolyspora, Microtetraspora, Nocardiopsis, Saccharopolyspora, and Pseudonocardia. In: A. Balows, H. G. Trüper, M. Dworkin, W. Harder, and K.-H. Schleifer (Eds.) The Prokaryotes. Springer-Verlag. Berlin, Germany. 2103–2117.

Williams, S. T., M. Goodfellow, and G. Alderson. 1989. Genus Streptomyces Waksman and Henrici 1943, 339AL. In: S. T. Williams, M. E. Sharpe, and J. G. Holt (Eds.) Bergey's Manual of Systematic Bacteriology. Williams and Wilkins. Baltimore, MD. 4:2452–2492.

Willscher, S., and K. Bosecker. 2001. Studies on the leaching behaviour of heterotrophic microorganisms isolated from basic slag dump. In: V. S. T. Ciminelli and O. Garcia Jr. (Eds.) Biohydrometallurgy: Fundamentals, Technology and Sustainable Development, Part B. Elsevier. Amsterdam, The Netherlands. 513–521.

Wong, K. K., L. Tan, and J. N. Saddler. 1988. Multiplicity of β-1,4-xylanase in microorganisms. Microbiol. Rev. 52:305–317.

Wood, W. A. 1985. Useful biodegradation of cellulose. Ann. Phythochem. Soc. Eur. 26:295–309.

Xu, L.-H., Y.-Q. Tiang, Y. F. Zhang, L.-X. Zhao, and C.-L. Jiang. 2001. Streptomyces thermogriseus, a new species of the genus Streptomyces from soil, lake and hot spring. Int. J. Syst. Bacteriol. 48:1089–1093.

Yamada, Y., K. Aoki, and Y. Tahara. 1982a. The structure of the hexahydrogenated isoprenoid side-chain menaquinone with nine isoprene units isolated from Actinomadura madurae. J. Gen. Appl. Microbiol. 28:321–329.

Yamada, Y., C. F. Hou, J. Sasaki, Y. Tahara, and H. Yosh-ioka. 1982b. The structure of the octahydrogenated iso-prenoid side-chain menaquinone with nine isoprene units isolated from Streptomyces albus. J. Gen. Appl. Microbiol. 28:519–529.

Yamashita, T., M. Imoto, K. Isshiki, T. Sawa, H. Naganawa, S. Kurasawa, B.-Q. Zhu, and K. Umezawa. 1988. Isolation of a new indole alkaloid, pendolmycin, from Nocardiopsis. J. Nat. Prod. 51:1184–1187.

Yassin, A. F., E. A. Galinski, A. Wohlfahrth, K.-D. Jahnke, K. P. Schaal, and H. G. Trüper. 1993. A new actinomycete species, Nocardiopsis lucentensis sp. nov. Int. J. Syst. Bacteriol. 43:266–271.

Yassin, A. F., F. A. Rainey, J. Burghardt, D. Gierth, J. Ungerechts, I. Lux, P. Seifert, C. Bal., and K. P. Schaal. 1997. Description of Nocardiopsis synnemataformans sp. nov. elevation of Nocardiopsis alba subsp. prasina to Nocardiopsis prasina comb. nov., and designation of Nocardiopsis antarctica and Nocardiopsis alborubida as later subjective synonyms of Nocardiopsis dassonvillei. Int. J. Syst. Bacteriol. 47:983–988.

Yoshida, M., K. Mastubara, T. Kudo, and K. Horikoshi. 1991. Actinopolyspora mortivallis sp. nov., a moderately halophilic actinomycete. Int. J. Syst. Bacteriol. 41:15–20.

Zhang, Z., Y. Wang, and J. Ruan. 1997. A proposal to revive the genus Kitasatospora (Omura, Takahashi, Iwai, and Tanaka 1982). Int. J. Syst. Bacteriol. 47:1048–1054.

Zhang, Z., Y. Wang, and J. Ruan. 1998. Reclassification of Thermomonospora and Microtetraspora. Int. J. Syst. Bacteriol. 48:411–422.

Zhang, S., B. K. Barr, and D. B. Wilson. 2000a. Effects of noncatalytic residue mutations on substrate specific ligand binding of Thermobifida fusca endocellulase Cel6A. Eur. J. Biochem. 267:244–252.

Zhang, S., D. C. Irwin, and D. B. Wilson. 2000b. Site-directed mutation of noncatalytic residues of Thermobifida fusca exocellulase Cel6B. Eur. J. Biochem. 267:3101–3115.

Zhang, Z., T. Kudo, Y. Nakajima, and Y. Wang. 2001. Clarification of the relationship between the members of the family Thermomonosporaceae on the basis of 16S rDNA, 16S-23S rRNA internal transcribed spacer and 23S rDNA sequences and chemotaxonomic analyses. Int. J. Syst. Evol. Microbiol. 51:373–383.

Zippel, M., and M. Neigenfind. 1988. Preservation of streptomycetes. J. Gen. Microbiol. 34:7–14.

Prokaryotes (2006) 3:796–818
DOI: 10.1007/0-387-30743-5_30

CHAPTER 1.1.15

Corynebacterium—Nonmedical

WOLFGANG LIEBL

Introduction

Corynebacterium is a genus within the actino-mycetes subphylum of Gram-positive eubacteria (Stackebrandt et al., 1997). These rod-shaped bacteria have DNA with a high G+C content and irregular ("coryneform") cell morphology. The phylogeny and taxonomy of corynebacteria (see Phylogeny and Taxonomy) have been changed many times, but modern methods of chemotax-onomy and molecular systematics now provide adequate means for the reliable identification and classification of *Corynebacterium* species.

Currently, there are more than 50 validated species of *Corynebacterium*. A large number of these (30) was described in the last ten years. Most *Corynebacterium* species are considered to be of medical significance and are therefore treated in chapter Corynebacterium—Medical in the second edition, whereas the following will be treated:

C. ammoniagenes (other name: *Brevibacterium ammoniagenes* [basonym]; Cooke and Keith, 1927; Collins, 1987b)

C. amycolatum (Collins et al., 1988)

C. callunae (Lee and Good, 1963; Yamada and Komagata, 1972a; Yamada and Komagata, 1972b)

C. casei (Brennan et al., 2001)

C. flavescens (Barksdale et al., 1979)

C. glutamicum (other names: *Brevibacterium divaricatum* [subjective synonym], *Corynebacterium lilium* [subjective synonym]; Kinoshita et al., 1958; Abe et al., 1967; see Liebl et al., 1991)

C. mooreparkense (Brennan et al., 2001)

C. terpenotabidum (formerly *Arthrobacter* spec.; Ikeguchi et al., 1988; Takeuchi et al., 1999)

C. variabile (formerly *C. variabilis*, nom. corrig. Int. J. Syst. Bacteriol. 48:1073; other names: *Arthrobacter variabilis* [basonym], *Caseobacter polymorphus* [subjective synonym]; Müller, 1961; Collins, 1987a; Collins, 1989b)

C. vitaeruminis (formerly *C. vitarumen*, nom. corrig. Int. J. Syst. Bacteriol. 47:908; other name: *Brevibacterium vitarumen* (basonym); Bechdel et al., 1928; Lanéelle et al., 1980)

In addition to the species listed above, certain bacteria frequently encountered in the relevant literature will be discussed, i.e., "*C. lilium*," "*Brevibacterium flavum*," "*Brevibacterium lactofermentum*" and *Brevibacterium divaricatum*. None of these nomenclatural "Brevibacterium" species is a true member of the genus *Brevibacterium*. Numerous data exist (Abe et al., 1967; Minnikin et al., 1978; Suzuki et al., 1981; W. Liebl et al., unpublished data) indicating their close relatedness if not identity with *C. glutamicum*. Indeed, the type strains of *B. divaricatum* and *C. lilium* and some representative strains of "*B. flavum*" and "*B. lactofermentum*" were transferred to *Corynebacterium glutamicum* (Liebl et al., 1991).

These organisms treated in this chapter were originally isolated from very diverse habitats. Some of them are industrially important bacteria used in large-scale biotechnological applications. The characteristic features of the genus *Corynebacterium* as outlined by Collins and Cummins (1986b) are also true for the bacteria of this section: Gram-positive (sometimes unevenly stained), nonsporeforming, nonmotile, not acid-fast, straight or slightly curved rods, ovals or clubs, often with metachromatic granules, often typically V-shaped arrangement of cells (see Fig. 1), facultatively anaerobic to aerobic, catalase positive, chemo-organotrophic, peptidoglycan directly crosslinked of the type A1 γ (Schleifer and Kandler, 1972) with *meso*-diaminopimelic acid (*meso*-A_2pm) as the cross-linking amino acid, predominant cell wall sugars are arabinose and galactose, mycolic acids (corynemycolic acids = short-chain α-substituted-β-hydroxy acids with 22–36 carbon atoms) present except in *C. amycolatum*, straight-chain saturated or monounsaturated fatty acids, 10-methyl-branched chain acids may be present, and eight and/or nine isoprene unit dihydrogenated menaquinones. It is important to note that members of *Corynebacterium* and related bacteria have an unusual cell envelope structure when compared to other Gram-positive bacteria. Including the reclassified and new species of *Corynebacterium*, the genus has a DNA base composition covering a wide range of approxi-

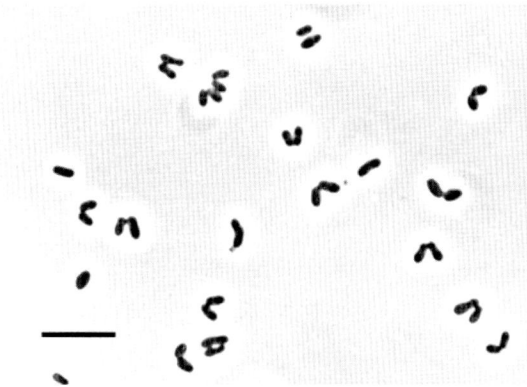

Fig. 1. Phase contrast microphotograph of *Corynebacterium glutamicum* cells grown in BMCG synthetic broth for 16 h. Bar = 10 μm.

mately 46–71 mol% G+C, but most species have between 51 and 68 mol% G+C.

Phylogeny and Taxonomy

Corynebacteria belong to the large group of Gram-positive bacteria with a high G+C content of their DNA (generally above 50 mol%, with a few exceptions) and represent one of the main lines of descent within the domain Bacteria. This group was described on the basis of 16S-rRNA cataloguing as part of the actinomycetes subdivision or subphylum of Gram-positive eubacteria (Stackebrandt and Woese, 1981). Chemotaxonomic studies comparing the cell wall composition (peptidoglycan structure and occurrence of mycolic acids) and lipid profiles suggested that the genera *Mycobacterium*, *Nocardia* and *Rhodococcus* are close relatives of *Corynebacterium* and led to the proposal to combine the four into the so-called "CMN-group" (Barksdale, 1970). Later, it was suggested to place these genera together with *Caseobacter* (later transferred to *Corynebacterium*; Collins et al., 1989b) into a group called "*Mycobacteriaceae*" (Barksdale, 1981; Jones and Collins, 1986). A different classification concept considered the genera *Corynebacterium* and *Mycobacterium* separately and combined the other mycolate-containing, cell wall type IV actinomycetes into the family *Nocardiaceae* (see Goodfellow, 1992). In principle, the grouping of taxa based on the chemotaxonomic markers just mentioned (chemotype IV and mycolic acids) proved to be correct. This classification concept withstood rigorous phylogenetic analysis, i.e., 16S rDNA sequence comparison, and now it is clear that the CMN-group, which from today's standpoint encompasses the genera *Corynebacterium*, *Dietzia*, *Gordona*, *Mycobacterium*, *Nocardia*, *Rhodococcus*, *Tsuka-*

murella and the mycolateless *Turicella*, forms a robust monophyletic taxon (Pascual et al., 1995; Ruimy et al., 1995).

In 1997, a new hierarchal classification system was proposed for the actinomycete subphylum (Stackebrandt et al., 1997). In this comprehensive system, the delineation of taxa is based solely on 16S rRNA/rDNA sequence data. The organisms and the families they belong to are placed into a phylogenetically meaningful framework of higher taxa under the umbrella of the new class *Actinobacteria*. In this classification concept, the genera *Corynebacterium* and *Turicella* form the family *Corynebacteriaceae*, one of six families (*Corynebacteriaceae*, *Dietziaceae*, *Gordoniaceae*, *Mycobacteriaceae*, *Nocardiaceae* and *Tsukamurellaceae*) within the suborder *Corynebacterinae* within the order *Actinomycetales* within the subclass *Actinobacteridae* within the class *Actinobacteria*. A pattern of 16S rDNA signatures characteristic for the family *Corynebacteriaceae* was described by Stackebrandt et al. (1997).

In the course of the decades after creation of the genus *Corynebacterium* by Lehmann and Neumann (1896), many bacterial isolates were assigned to the genus, often for reasons not justifiable from a modern taxonomic point of view. Therefore, the genus *Corynebacterium* traditionally comprised an extremely diverse collection of microorganisms, accommodated in one group mainly on the basis of their common cell morphology, staining properties and respiratory metabolism. Obviously, the sole use of these criteria could not result in a homogenous group of organisms. Chemotaxonomic markers later helped to redraw the borderline of the genus *Corynebacterium*, and subsequently the transfer of various species from *Corynebacterium* to other genera, as well as the inclusion of species previously placed elsewhere, was published (Collins and Bradbury, 1986a; Collins and Cummins, 1986b; Collins, 1987a; Collins, 1987b; Collins et al., 1988; Collins et al., 1989a). More recently, a number of new species of *Corynebacterium* were described, i.e., 30 new species were validly published between 1991 and 2001. Most of them were isolated from clinical and veterinary specimens. This fact indicates that the genus presumably contains many opportunistic pathogens. None of the plant pathogenic coryneform bacteria listed in the eighth edition of *Bergey's Manual of Determinative Bacteriology* (Cummins et al., 1974), which were treated as members of the genus *Corynebacterium* for many years, is a true *Corynebacterium* sensu stricto (see Collins and Bradbury, 1986a). This notion was supported mainly by the accumulation of chemotaxonomic data (i.e., cell wall composition, mycolic acid content and mol% G+C) arguing against the retention of phytopathogenic "corynebacteria" in the genus and finally led to the reclassification of these bacteria in other genera, mainly in *Curtobacterium* and *Clavibacter* (Collins and Jones, 1982a; Collins and Jones, 1983; Collins et al., 1981; Collins et al., 1982c; Goodfellow, 1984b; Goodfellow, 1984a; Davis et al., 1984). These genera are now classified in a different suborder (*Micrococcinae*) of the order *Actinomycetales* than *Corynebacterium* (suborder *Corynebacterinae*; see Stackebrandt et al., 1997).

Comparative 16S rDNA sequence analysis of the *Actinomycetales* taxa with cell wall chemotype IV and mycolic acids revealed that these taxa, including *Corynebacterium*, *Dietzia*, *Gordona*, *Mycobacterium*, *Nocardia*, *Rhodococcus*, and *Tsukamurella*, and the mycolate-less *Turicella* form a natural suprageneric group (Pascual et al., 1995; Ruimy et al., 1995). Within this group, the species of the genus *Corynebacterium* apparently form a monophyletic association, although the genus exhibits considerable phylogenetic depth. Interestingly, the mycolic acid-less species *C. amycolatum* and the single species of *Turicella*, *T. otitidis* (Funke et al., 1994), which also lacks mycolic acids, were found to form deep branches close to the periphery of the genus *Corynebacterium* (Pascual et al., 1995). However, this phylogenetic position of *Turicella* was not confirmed by Ruimy et al. (1995), who reported that the genus appeared as a branch included within the *Corynebacterium* unit and suggested the standing of *Turicella* as a separate genus be re-evaluated.

Despite the improvements in the phylogenetic affiliation and taxonomy of *Corynebacterium* species, it should be kept in mind that the genus is phenotypically still very diverse and contains both aerobes and facultative anaerobes, both mycolate- and non-mycolate-containing species, both tuberculostearic acid- and non-tuberculostearic acid-containing species, and both medically important species and saprophytic, nonpathogenic species. Today, more than 50 validated species of *Corynebacterium* are known, many of which (30) were described in the last ten years. A 16S rDNA sequence-based phylogenetic tree of the species belonging to the genus is shown in Fig. 2.

In the literature, the term "coryneform bacteria" (the origin of this expression and its introduction into the literature was examined by Keddie, 1978) was previously sometimes used as a supergeneric group designation. From today's point of view, this practice is troublesome because the term traditionally was applied to a diverse collection of bacteria of various genera (e.g., *Arthrobacter*, *Brevibacterium*, *Cellulomonas*, *Corynebacterium*, and

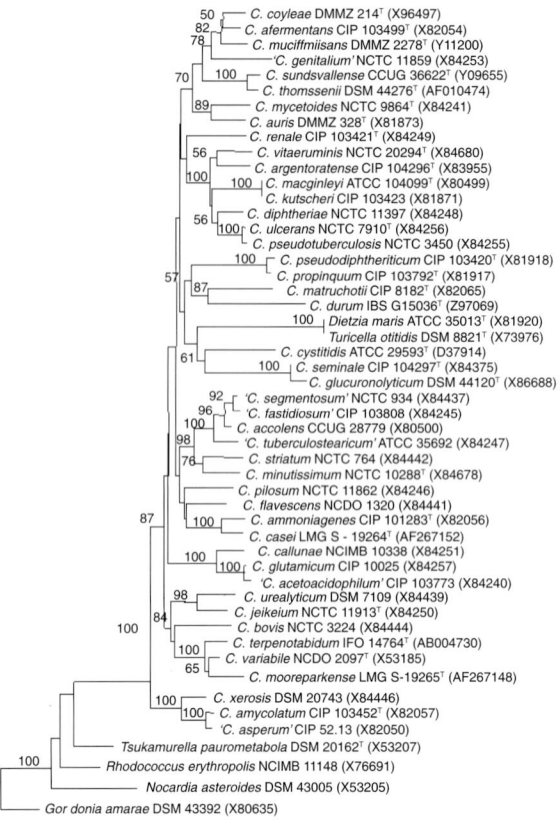

Fig. 2. Unrooted tree of the phylogenetic relationships among the species of the genus *Corynebacterium*. The tree, which is based on 16S rDNA sequence data and was constructed by the neighbour-joining method, was kindly supplied by Dr. T. M. Cogan, Dairy Products Research Centre, Fermoy, County Cork, Ireland. Bootstrap values, expressed as percentages of 1000 replications, are given at the branch points.

Mycobacterium, to name just a few; see Keddie, 1978 and Goodfellow and Minnikin, 1981). The rather confusing situation is best documented by the concept of the chapter entitled "Coryneform Group of Bacteria" in the eighth edition of *Bergey's Manual of Determinative Bacteriology* (Rogosa et al., 1974), where, as Barksdale (1981) points out, *Corynebacterium* is placed "along with a polytypic array of bacteria whose common property would seem to be pleomorphism." The advanced classification methods available now have led to a sophisticated picture of the interrelationships between the genera comprising the coryneform organisms (see Stackebrandt et al., 1997). The term "coryneform" is taxonomically not satisfactory and should be avoided where possible. It may nevertheless be a useful expression if used as a purely descriptive term of bacterial morphology.

Habitat

The bacteria discussed in this chapter were originally isolated from a broad variety of habitats, including soil, feces, dairy products, animal skin, and plant material (Liebl, 1992a; Table 1). *Arthrobacter variabilis* and *Caseobacter polymorphus*, both now accommodated in the new species *Corynebacterium variabile* (Collins, 1987a; Collins et al., 1989b), were originally isolated from animal fodder and cheese, respectively (Müller, 1961; Crombach, 1978). In a numerical taxonomic study of cheese-smear coryneform bacteria isolated from the rind of different cheese varieties, Seiler (1986) found that about 20% of the white and yellow coryneform isolates clustered in a phenon closely resembling the "*Brevibacterium ammoniagenes*" group of Seiler (1983a). Recent work dealing with the microflora of smear-ripened cheeses resulted in the description of two new species of *Corynebacterium*, i.e., *C. mooreparkense* and *C. casei* (Brennan et al., 2001). Other cheese-smear isolates were classified as *C. ammoniagenes* and *C. variabile* (Eliskases-Lechner and Ginzinger, 1995; Valdes-Stauber et al., 1997). Strains classified as "*Brevibacterium ammoniagenes*" (*C. ammoniagenes*) and *C. glutamicum* with numerical taxonomy methods were isolated from piggery wastes by Seiler and Hennlich (1983b). Interestingly, significant numbers of bacteria displaying properties characteristic of the genus *Corynebacterium* have been isolated from marine samples (Bousfield, 1978). However, these isolates were not differentiated to the species level and they were not investigated with molecular systematics methods. Nevertheless, it seems that nonmedical corynebacteria are widely disseminated in nature, although there are no data available concerning the numbers present in different habitats.

Isolation

Nonmedical corynebacteria can be isolated from a variety of different habitats (Liebl, 1992a; see above and Table 1). However, to the authors' knowledge, no selective media or enrichment procedures have been described which are specifically devised for this group of organisms.

Most strains normally grow well if cultivated aerobically at 30°C in standard peptone-yeast extract media like the *Corynebacterium* medium (DSM, Deutsche Sammlung von Mikroorganismen und Zellkulturen GmbH, 1998 catalogue of strains, medium no. 53) shown below, although growth on very rich media such as brain heart infusion (Difco) is generally faster and more abundant.

Corynebacterium Medium

Casein peptone, tryptic digest	10 g
Yeast extract	5 g
Glucose	5 g
NaCl	5 g
(Agar	15 g)
Distilled water	1,000 ml
pH 7.2–7.4	

For *C. ammoniagenes*, a chemically defined medium was described by Nara et al. (1969a). For *C. glutamicum*, synthetic media, e.g., BMCG-broth (Liebl et al., 1989b; see recipe below), have been designed which support abundant growth of this species. The addition of certain substances such as 0.1% citrate or low concentrations (10^{-5} M) of some dihydroxyphenolic compounds (catechol and protocatechuate) has been shown to greatly stimulate growth of *C. glutamicum* in synthetic broth, presumably by assisting in the assimilation of iron by this organism (Von der Osten et al., 1989; Liebl et al., 1989b).

BMCG Synthetic Broth:

$(NH_4)_2SO_4$	7 g
Distilled water	850 ml
10 × M9 solution	100 ml

Autoclave 20 min., 121°C, then add aseptically:

200 × Salt solution	5 ml
Trace element solution	2 ml
1 M $CaCl_2$	0.05 ml
Vitamin stock solution	1 ml
20% Glucose	50 ml
10 mM Catechol	1 ml

(To prepare 10 × M9 stock, add 60 g of Na_2HPO_4, 30 g of KH_2PO_4, 5 g of NaCl, and 10 g of NH_4Cl to 1,000 ml of distilled H_2O, and adjust to pH 7.3; to prepare 200 × salt solution, add 80 g of $MgSO_4 \cdot 7H_2O$, 4 g of $FeSO_4 \cdot 7H_2O$, 0.4 g of $MnSO_4 \cdot H_2O$, and 5 g of NaCl to 1,000 ml of distilled H_2O; to prepare trace element solution, add 88 mg of $Na_2B_4O_7 \cdot 10H_2O$, 40 mg of $(NH_4)_6Mo_7O_{24} \cdot 4H_2O$, 10 mg of $ZnSO_4 \cdot 7H_2O$, 270 mg of $CuSO_4 \cdot 5H_2O$, 7.2 mg of $MnCl_2 \cdot 4H_2O$, and 870 mg of $FeCl_3 \cdot 6H_2O$ to 1,000 ml of distilled H_2O; to prepare vitamin stock solution, add 1 mg of biotin and 10 mg of thiamine HCl per 1 ml of distilled H_2O; the 200 × salt solution, trace element solution and 1 M $CaCl_2$ are autoclaved separately, and the vitamin and glucose solutions are filter sterilized; the catechol stock is adjusted to a neutral pH, sterilized by filtration, stored in aliquots at −20°C and added aseptically to the medium just before inoculation.)

Corynebacteria are generally exacting in their nutritional requirements. It is noteworthy that all strains of glutamic acid-producing corynebacteria (*C. glutamicum* and similar bacteria) are dependent upon the presence of biotin in the growth medium, and some additionally require

Table 1. Differential characteristics of saprophytic *Corynebacterium* species.

	C. ammoniagenes	C. amycolatum	C. callunae	C. casei	C. flavescens	C. glutamicum	C. mooreparkense	C. terpenotabidum	C. variabile	C. vitarumen
Acid from:										
Glucose	ND	+	+	+	+	+	+	-	-	+
Galactose	ND	-	-	-	+	-	-	-	ND	ND
Lactose	ND	-	-	-	-	-	-	ND	ND	-
Maltose	ND	+	+	-	-	+	-	ND	ND	+
Sucrose	ND	ND	+	-	-	+	-	ND	-	+
Trehalose	ND	ND	ND	ND	-	+	-	ND	ND	+
Tyrosine hydrolysis	+	-	-	-	ND	-	+	-	-	ND
Esculin hydrolysis	ND	-	+	-	ND	-	+	-	-	+
Hippurate hydrolysis	+	ND	-	ND	-	+	ND	ND	ND	+
Nitrate → nitrite	+	ND	+	+	-	+	-	-	+	-
Starch hydrolysis	-	+	-	ND	-	-	ND	-	-	+
Urease	+	ND	+	-	-	+	+	+	+	+
Voges Proskauer	ND	+	ND	ND	+	-	ND	+	ND	+
10-Methyl-octadecanoic acid[a] present	+	ND	-	ND	-	-	ND	+	+	+
Mycolic acids present	+	-	+	+	+	+	+	+	+	+
Mol% G+C	53.7–55.8	61.0	51.2	51	58.3	55.5–57.5	60	67.5	65.0	64.8
Habitats	Feces of infants, pig manure, and smear-ripened cheese	Human skin	Heather	Smear-ripened cheese	Dairy products	Soil, animal feces, vegetables, and fruits	Smear-ripened cheese	Soil	Animal fodder, and smear-ripened cheese	Rumen of cow

Symbols: +, most or all strains display the property; –, most or all strains do not display the property; and ND, no data available or conflicting data or variable.

[a]Alternate name: tuberculostearic acid.

Data from Abe et al., 1967; Barksdale et al., 1979; Brennan et al., 2001; Collins, 1987a,b, 1989; Collins and Cummins, 1986; Collins et al., 1988; Lanéelle et al., 1980; Seiler, 1983; Takeuchi et al., 1999; and Yamada and Komagata, 1972.

thiamine or *p*-aminobenzoic acid (Abe et al., 1967).

Identification

MORPHOLOGY For *C. glutamicum* cells grown in various media and sampled after different incubation periods, we have not encountered problems in recognizing the morphological features typical of *Corynebacterium*, i.e., somewhat irregular rod-shaped cells, often arranged in V-formations, and so forth. However, morphological differences occur depending on the media used and the culture age (Fig. 1). Therefore, for the investigation of the species described in this chapter and to have reproducible culture conditions for reasons of comparison, it may be useful to employ EYGA medium when examining morphological features as recommended by Cure and Keddie (1973).

EYGA Medium:

K_2HPO_4	1.1 g
KH_2PO_4	0.86 g
$CaCl_2$	0.025 g
$MgSO_4 \cdot 7H_2O$	0.2 g
NaCl	0.1 g
$(NH_4)_2SO_4$	0.5 g
Yeast extract	1 g
Glucose	1 g
Vitamin B_{12}	0.002 mg
Trace element solution	3 ml
Agar	12 g
Distilled water	1,000 ml
Adjust to pH 6.8	

(To prepare trace element solution, add 5 g of EDTA, 2.2 g of $ZnSO_4 \cdot 7H_2O$, 0.57 g of $MnSO_4 \cdot 4H_2O$, 0.5 g of $FeSO_4 \cdot 7H_2O$, 0.161 g of $CoCl_2 \cdot 6H_2O$, 0.157 g of $CuSO_4 \cdot 5H_2O$, and 0.151 g of $Na_2MoO_4 \cdot 2H_2O$ to 600 ml of distilled water.)

IDENTIFICATION OF *CORYNEBACTERIUM* SENSU STRICTO AND DIFFERENTIATION FROM OTHER GENERA A bacterial isolate which fulfills the following criteria can be considered to be a member of the *Corynebacterium* sensu stricto: facultatively anaerobic or aerobic, coryneform morphology, cell wall containing *meso*-A_2pm, arabinose and galactose and short-chain mycolic acids with 22–36 carbon atoms, and 51–65 mol% G+C. Using these criteria only, certain rhodococci may be confused with *Corynebacterium*, although their mol% G+C (approximate range for *Rhodococcus* is 60–69%) is generally higher than that found for most corynebacteria. Also, rhodococci are considered to be aerobes, whereas most representatives of *Corynebacterium* are facultatively anaerobic.

Mycolic acids (2-alkyl-branched-3-hydroxy-acids) are valuable chemotaxonomic markers for *Corynebacterium* and related taxa (Minnikin et

al., 1978). They are unique lipophilic components of the cell envelopes of the genera *Corynebacterium*, *Gordona*, *Mycobacterium*, *Nocardia*, *Rhodococcus* and *Tsukamurella*. The most convenient method for the determination of mycolic acids (see Keddie and Jones, 1981 for an evaluation of the methods available) is the acid methanolysis of bacterial cells and subsequent thin layer chromatographic analysis of the resulting mycolic acid methyl esters (Minnikin et al., 1975; Minnikin et al., 1978).

The presence of mycolic acids is a criterion that allows the genera mentioned above to be separated from all other bacteria of similar coryneform morphology. Also, mycolic acid-containing genera can be differentiated to a certain degree by examining the size of these compounds (see Collins and Cummins, 1986b). However, part of *Rhodococcus* overlaps with *Corynebacterium* with respect to the mycolate sizes (Collins et al., 1982b), which may pose problems. With the exception of *C. amycolatum*, which does not contain mycolic acids (Collins et al., 1988), all bacterial species discussed in this chapter have mycolic acids with 26–38 carbon atoms (Minnikin et al., 1978; Lanéelle et al., 1980).

DIFFERENTIATION AT THE SPECIES LEVEL Biochemical characteristics that may be useful for differentiating saprophytic *Corynebacterium* species are listed in Table 1.

Preservation

Strains of *Corynebacterium* grown on agar slants or plates can be kept at 4°C for at least 4–6 weeks. Collins and Cummins (1986b) reported that cultures grown in chopped meat medium (without glucose) will remain viable for many months if stored in the dark at room temperature. Long-term storage can be achieved by adding 20% glycerol or 7% dimethylsulfoxide (DMSO) to liquid cultures and freezing at −70°C. Lyophilization is also a reliable means of long-term preservation. For this purpose, we harvest 10 ml of overnight cultures grown in the *Corynebacterium* medium described above, resuspend the cells in 1 ml of 10% skim milk (Difco) and, upon lyophilization of 0.2 ml aliquots, store them over silica gel in sealed glass tubes.

Cell Wall Characteristics

The chemical nature of the major constituents of the cell walls of corynebacteria and related genera has been the subject of many studies (Minnikin et al., 1978; Collins et al., 1982b; Minnikin and O'Donnell, 1984; Daffe and Draper, 1998). The peptidoglycan is directly crosslinked (A1γ-type; Schleifer and Kandler, 1972) with *meso*-diaminopimelic acid (*meso*-A$_2$pm). An arabinogalactan polysaccharide, which is partially esterified by a special class of lipids, i.e., the mycolic acids, is covalently linked to the peptidoglycan. The arabinan and galactan segments of the *Corynebacterium* arabinogalactan also contain significant amounts of mannose and glucose substituents. Additionally, high- and low-molecular-mass glucan, arabinomannan, lipoglycans, and a protein surface layer are present in the cell wall of corynebacteria (Peyret et al., 1993; Sutcliffe, 1995; Soual-Hoebeke et al., 1999; Puech et al., 2001).

Interestingly, molecular biological, biochemical and ultrastructural studies recently showed that the architecture and function of the cell wall of the Gram-positive corynebacteria quite closely resemble those of the Gram-negative cell envelope, although the molecular details are strikingly different. Typical Gram-negative bacteria are characterized by an outer membrane which is a lipid bilayer composed of phospholipids and lipopolysaccharides. In *Corynebacterium* and related Gram-positive bacteria, it appears that the covalently linked mycolic acids, together with free mycolates (trehalose dicorynemycolates and trehalose monocorynemycolates) and phospholipids, form an outer membrane-like diffusion barrier (Fig. 3). Evidence that this barrier could represent a lipid bilayer was obtained with freeze-etch electron microscopy techniques that revealed a main fracture plane within the cell wall of corynemycolate-containing strains of corynebacteria. The mycolate-less cells of *C. amycolatum* strains, on the other hand, exhibited a fracture plane only within the cytoplasmic membrane (Puech et al., 2001). Mutants defective in protein components involved in extracytoplasmic lipid metabolism (mycoloyl transferases) displayed a decreased mycolate

content and an altered cell wall permeability, which points to an important role of the mycolic acids for the outer membrane-like barrier (Jackson et al., 1999; Puech et al., 2000).

The hypothesis of a closed outer membrane bilayer in corynebacterial cells (Puech et al., 2001; Fig. 3) demands a specific mechanism(s) for the transport of hydrophilic solutes across this barrier. Indeed, channel-forming activity in the cell wall of *C. glutamicum* was described by Niederweis et al. (1995), indicating a hydrophilic pathway through the mycolic acid-containing layer—an observation made previously also in mycobacteria. Recently, pore-forming proteins of corynebacteria, nocardiae and mycobacteria have been described. In contrast to the typical trimeric porins of the Gram-negative outer membranes, the porins of *Corynebacterium* strains are low-molecular-mass peptides that can form ion-permeable channels (Mukhopadhyay et al., 1997; Lichtinger et al., 1998; Lichtinger et al., 1999; Kartmann et al., 1999; Rieß et al., 1999). These recent findings about the cell wall structure have important implications for many aspects of corynebacterial biology, including solute uptake and excretion, cell surface properties, susceptibility to antibacterial compounds, etc. They also explain why corynebacteria are relatively poor in the secretion of proteins (Liebl and Sinskey, 1988; Liebl et al., 1989c).

Genetics

The industrial interest in corynebacteria has led to the development of cloning systems for these bacteria, mainly of *C. glutamicum* and closely related organisms. As a prerequisite for the construction of cloning vectors, numerous corynebacterial strains were screened for the presence of plasmids. Plasmids were detected in strains of medically significant and nonmedical corynebac-

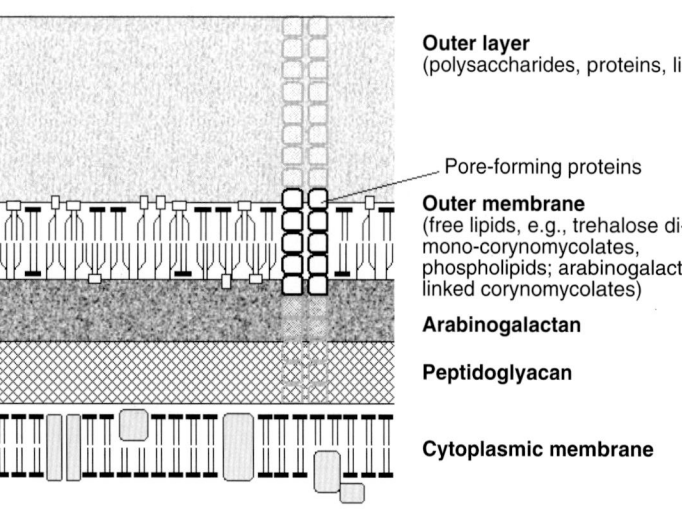

Outer layer
(polysaccharides, proteins, lipids)

Pore-forming proteins

Outer membrane
(free lipids, e.g., trehalose di- and mono-corynomycolates, phospholipids; arabinogalactan-linked corynomycolates)

Arabinogalactan

Peptidoglyacan

Cytoplasmic membrane

Fig. 3. Hypothetical model of the cell envelope of typical mycolic acid-containing corynebacteria. The model as described by Puech et al. (2001) is based on the chemical analysis of cell walls, biochemical and biophysical characterization of cell wall proteins, and ultrastructural studies including transmission electron microscopy and freeze-etch electron microscopy, mainly of corynebacteria and mycobacteria. The bilayer nature of the outer membrane, and the depicted organization of pore-forming proteins spanning all layers of the cell envelope, are speculative (see text).

teria, i.e., *C. diphtheriae, C. glutamicum,* "*B. lactofermentum,*" *C. callunae,* "*C. lilium,*" "*C. melassecola,*" *C. renale* and *C. xerosis*, with sizes ranging between about 1.4 and 95 kilobases (Miwa et al., 1984; Katsumata et al., 1984; Sandoval et al., 1984; Sandoval et al., 1985; Yoshihama et al., 1985; Serwold-Davis et al., 1987; Sonnen et al., 1991; Takeda et al., 1990; Nath and Deb, 1995; Deb and Nath, 1999). Interestingly, certain plasmids of *Brevibacterium linens* function in *Corynebacterium* species (Ankri et al., 1996a). Apart from a few plasmids carrying antibiotic resistance determinants (Katsumata et al., 1984; Tauch et al., 1995), most of the known corynebacterial plasmids are cryptic. The *C. glutamicum* plasmid pCG4 was reported to carry an integron (Nesvera et al., 1998). A multitude of cloning vectors for amino acid-producing corynebacteria have been constructed, most of which are shuttle vectors based on cryptic corynebacterial plasmids (Martin et al., 1987; Archer et al., 1989; Eikmanns et al., 1991; Jakoby et al., 1999).

Several methods for the introduction of recombinant plasmid molecules in corynebacteria are available. Protoplast transformation, a method established for use with various other Gram-positive host organisms, was developed for amino acid-producing corynebacteria simultaneously by different groups (Katsumata et al., 1984; Santamaria et al., 1984; Yoshihama et al., 1985). In these transformation systems, not genuine (spherical) protoplasts are used, but osmotically sensitive cells that retain rod-shaped morphology. Electroporation was shown to be a more convenient and much more efficient method for transformation of *C. glutamicum* (Liebl et al., 1989a; Haynes and Britz, 1989; Dunican and Shivnan, 1989; Bonamy et al., 1990). Finally, the conjugal transfer of mobilizable shuttle plasmids from Gram-negative *Escherichia coli* to various corynebacterial strains was described as a further method for gene transfer into *Corynebacterium* (Schäfer et al., 1990). Besides introducing and establishing autonomously replicating multicopy plasmids in corynebacterial cells, methods are available for the integration, deletion or replacement of DNA sequences in the chromosome of *Corynebacterium*. For this purpose, typically, nonreplicative vectors carrying DNA segments homologous to specific regions of the chromosome are mobilized or electroporated into *Corynebacterium* cells, and chromosomal integration occurs via homologous recombination (Schwarzer and Pühler, 1991; Reyes et al., 1991; Schäfer et al., 1994b). In addition, a method for the chromosomal integration of replicative plasmid DNA was described by Ikeda and Katsumata (1998). Finally, chromosomal integration of vectors via

short homologous sequences of 8–12 bp, presumably by virtue of a RecA-independent illegitimate recombination mechanism, was reported to occur at low frequencies (Mateos et al., 1996).

Numerous observations made during transformation experiments suggest the wide dissemination of restriction/modification systems in corynebacteria (e.g., Follettie and Sinskey, 1986; Katsumata et al., 1984; Bonamy et al., 1990; Vertes et al., 1993c; Vertes et al., 1993b; Ankri et al., 1996b). The problems posed by these systems for cloning experiments can be overcome by the use of restriction-deficient host strains (Liebl et al., 1989a; Bonnassie et al., 1990; Liebl and Schein, 1990; Schäfer et al., 1990; Schäfer et al., 1994a). In some strains, inactivation of the host cell restriction system by treatment of the cells with stress factors such as heat (Schäfer et al., 1994a; van der Rest et al., 1999), or the use of unmethylated DNA for transformation (Ankri et al., 1996b), has been shown to be useful to avoid host cell restriction of incoming DNA.

Polyethylene glycol (PEG)-induced protoplast fusion has been used to obtain hybrid organisms from mutant strains of amino acid-producing corynebacteria (Kaneko and Sakaguchi, 1979; Karasawa et al., 1986) and may have potential in the breeding of improved industrial amino acid producers. However, since it is a rather nonspecific method, its usefulness in basic research is limited.

Phages of several amino acid-producing corynebacteria (*C. glutamicum, C. lilium, B. lactofermentum, B. flavum*) have been described (Oki and Ogata, 1968a; Patek et al., 1985; Trautwetter et al., 1987; Sonnen et al., 1990). Certain phages have been used in transfection experiments with protoplasts of amino acid-producing corynebacteria to optimize protoplast transformation protocols (Katsumata et al., 1984; Sanchez et al., 1986; Yeh et al., 1985). Phage-mediated transduction in *B. flavum* was reported by Momose et al. (1976). A cosmid vector for glutamic acid-producing bacteria has been constructed that was transduced at low frequency through phage infection (Miwa et al., 1985). However, presently too little is known about the biology of the bacteriophages of saprophytic corynebacteria to allow their exploitation as tools in genetic research.

Most genes introduced into amino acid-producing corynebacteria by recombinant DNA methods are homologous genes from these bacteria themselves, encoding enzymes involved in amino acid biosynthesis or deregulated derivatives thereof. Using this strategy, one hopes to obtain improved production strains by raising the activity of rate-limiting enzymes and thus removing bottlenecks of amino acid biosynthesis. On the other hand, heterologous expression signals are readily utilized by corynebacteria, making these organisms versatile hosts for recombinant gene expression. Heterologous antibiotic resistance genes of Gram-negative (*Escherichia coli*) and Gram-positive origin (*Staphylococcus aureus, Enterococcus faecalis, Streptomyces acrimycini* and *Streptomyces hygroscopicus*) were observed to function in *C. glutamicum* (*B. lactofermentum*; Ozaki et al., 1984; Batt et al., 1985; Yoshihama et al., 1985; Santamaria et al., 1984; Santamaria et al., 1987). Additionally, a number of other heterologous genes have been expressed in these bacteria, e.g., lactose utilization genes from *Escherichia coli* (Brabetz et al., 1991; Brabetz et al., 1993), threonine biosynthetic genes from *E. coli* (Patek et al., 1989), ovine γ-interferon (Billman-Jacobe et al., 1994), proteases (Billman-Jacobe et al., 1995), an α-amylase gene from *Bacillus amyloliquefaciens* (Smith et al., 1986), exoglucanase and endoglucanase A genes from *Cellulomonas fimi* (Paradis et al., 1987), and the genes encoding thermonuclease from *Staphylococcus aureus* and lipase from *Staphylococcus hyicus* (Liebl et al., 1989c). The extracellular enzymes encoded by the *Cellulomonas* and *Staphylococcus* genes are secreted to the culture medium by the corynebacterial hosts (Paradis et al., 1987; Liebl et al., 1989c; Liebl et al., 1992a). Furthermore, the *E. coli* lac, lacUV5, tac, and trp promoters are functional in *C. glutamicum* ("*B. lactofermentum*"; Morinaga et al., 1987; Brabetz et al., 1991; Eikmanns et al., 1991; Liebl et al., 1992b). The broad recognition of heterologous expression signals by *C. glutamicum* contrasts with the situation found with other Gram-positive cloning hosts, e.g., *Bacillus subtilis*, where genes of Gram-negative origin are generally not efficiently expressed due to more stringent requirements of the respective Gram-positive transcription/translation apparatus (McLaughlin et al., 1981; Moran et al., 1982; Graves and Rabinowitz, 1986). Patek et al. (1996) undertook a detailed survey of promoter sequences from *C. glutamicum*. This study showed that a corynebacterial promoter consensus can be deduced which is similar to the consensus pattern recognized by the major RNA polymerase σ-factor (σ^{70}) of *E. coli*. The conserved −35 and −10 sequences of *C. glutamicum* promoters were found to be tttGcca.a and ggTA.aaT, respectively, with the core consensus hexamer motifs being ttGcca and TA.aaT, respectively. The −10 region appears to play the predominant role in corynebacterial promoter recognition, whereas the −35 hexamer motif is less conserved than in other bacteria (Patek et al., 1996). This conclusion is in accordance with the observation that a mutation in the −10 region of the *E. coli* lac promoter from TATGTT to TATATT improved

the efficiency of this promoter in *C. glutamicum* (Brabetz et al., 1991).

The biotechnological importance of amino acid-producing corynebacteria (see below) also has led to a pronounced interest in the genetic capacity of *C. glutamicum*. As a consequence, the nucleotide sequence of the genome of *C. glutamicum* has been determined several times by consortia of research groups and different companies in the past few years. Nevertheless, as of 2001, no complete genome sequence is available in the publicly accessible sequence databases. The genome size is about 3.08 Mbp (Bathe et al., 1996). Various transposable elements are known to exist in the genome of *C. glutamicum* strains. Five insertion (IS) elements have been detected in different strains (IS family affiliation and *C. glutamicum* strain(s) in parentheses): IS*1206* (IS*3* family; ATCC21086; Bonamy et al., 1994), IS*31831* (ISL*3* family; ATCC31831; Vertes et al., 1994), IS*13869* (ISL*3* family; ATCC13869; Correia et al., 1996), IS*Cg1* (ISL*3* family; type strain ATCC13032; Jäger et al., 1995), and IS*Cg2* (IS*30* family; type strain ATCC13032, ATCC13869, and ATCC14020; Quast et al., 1999). It is noteworthy in this context that transposons bearing antibiotic resistance determinants from other *Corynebacterium* species are capable of transposition in *C. glutamicum* (Tauch et al., 1995; Tauch et al., 1998).

Physiology

Little effort has been devoted to the metabolism of saprophytic corynebacteria. Only amino acid-producing corynebacteria, particularly *C. glutamicum* (including "*Brevibacterium flavum*"), have been studied in more detail. In "*B. flavum*," the uptake of glucose and fructose seems to be catalyzed by specific phosphoenolpyruvate:sugar phosphotransferase (PTS) systems (Mori and Shiio, 1987a; Mori and Shiio, 1987b), although conflicting data have been presented by Marauska et al. (1981). The main routes for hexose breakdown in amino acid-producing corynebacteria are the Embden-Meyerhof-Parnas and the pentose phosphate pathways followed by further oxidation through the tricarboxylic acid cycle (Fig. 4). Maltose and sucrose are cleaved intracellularly in "*B. flavum*" by maltase and invertase, respectively. The catabolism of mannitol is initiated by mannitol dehydrogenase. Ribose and gluconate are assimilated via the pentose phosphate pathway (Mori and Shiio, 1987b). It has been demonstrated that after internalization into *C. glutamicum* cells, acetate is utilized by acetate kinase (Ack) and phosphotransacetylase (Pta; Reinscheid et al., 1999), the key enzymes of the glyoxylate pathway, which is necessary as an anaplerotic

Fig. 4. Schematic of metabolic pathways leading to glutamate and lysine in *Corynebacterium glutamicum*. Key enzymes discussed in the text are: ASK, aspartate kinase; DDH, diaminopimelate dehydrogenase; DDS, dihydrodipicolinate synthase; HDH, homoserine dehydrogenase; ODH, 2-oxoglutarate dehydrogenase; PPC, phosphoenolpyruvate carboxylase; PYC, pyruvate carboxylase; and SUT, succinyl transferase.

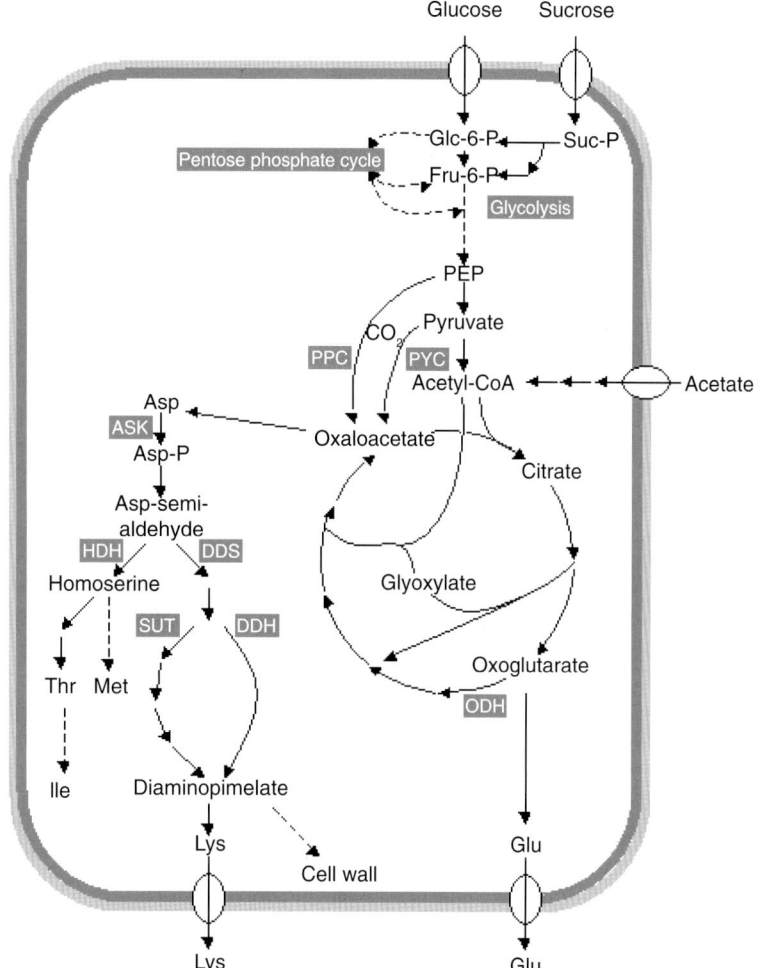

sequence for the assimilation of this C_2 compound (Shiio et al., 1959). The glyoxylate cycle enzymes isocitrate lyase (AceA) and malate synthase (AceB) are allosterically controlled by intermediates of central metabolism, and their synthesis is regulated at the transcriptional level (Reinscheid et al., 1994a; Reinscheid et al., 1994b). Finally, amino acid-producing corynebacteria are able to assimilate various organic acids and ethanol (Oki et al., 1968b; Yamada and Komagata, 1972b).

Corynebacterium glutamicum and similar bacteria are used for the fermentative production of amino acids, with L-glutamic acid and L-lysine being the most important as far as quantity is concerned. Since the biosynthetic pathways in *C. glutamicum* and similar bacteria are basically the same as found in other organisms, the question arises which pecularities make these bacteria superior amino acid excreters. Some features of the metabolism related to the high-level amino acid production by *C. glutamicum* and related organisms are discussed below.

As a soil bacterium, *C. glutamicum* needs to cope with osmotic changes caused by variations in the availability of water. Also in industrial amino acid production processes, this organism is confronted with significant variations in the osmolality of the surrounding media (high sugar concentration at the beginning and high amino acid concentration towards the end of the process). To this end, Krämer and coworkers have characterized four secondary carriers for the uptake of different compatible solutes (i.e., betaine, proline and ectoine), and for the rapid efflux of solutes, they have characterized osmoregulated channels, which (under hypo-osmotic conditions) may be mechanosensitive ion channels in the cytoplasmic membrane (see Peter et al. [1998]; Ruffert et al., 1997; Ruffert et al., 1999).

Applications

AMINO ACID FERMENTATIONS Certain saprophytic corynebacteria have a long tradition as industrial microorganisms in biotechnological production processes. The most prominent example is the production of amino acids by *C. glutamicum* and other similar bacteria. The worldwide production of amino acids amounts to more than 10^9 kg per year. Two of the most important amino acids in terms of quantity, i.e., L-glutamic acid and L-lysine, are produced fermentatively on a large scale by *C. glutamicum*. *Corynebacterium glutamicum* (synonym "*Micrococcus glutamicus*") was first isolated in 1957 by Kinoshita while searching for an efficient glutamic acid-producing microorganism among bacterial isolates (Kinoshita and Nakayama, 1978). Subsequently, a large number of glutamic acid producers were isolated and classified in different genera as various different species, e.g., "*Brevibacterium flavum*," "*B. lactofermentum*," "*B. roseum*," "*B. divaricatum*," "*C. lilium*," "*C. herculis*," "*Microbacterium ammoniaphilum*," etc. However, the separate species status of these organisms is in serious doubt (see introduction to this chapter), and some of these organisms carry invalid names, while others have been reclassified as *C. glutamicum* (Liebl et al., 1991).

Worldwide production of monosodium glutamate (MSG), which has a meaty taste and is used mainly as a flavoring agent, was estimated at approximately 6–8×10^8 kg (Leuchtenberger, 1996; Demain, 2000). Under certain culture conditions, *C. glutamicum* excretes remarkably large amounts of glutamic acid into the growth medium (>100 g per liter). In typical fermentations, the molar yield of glutamate from sugar is 0.50–0.75 (Kinoshita and Nakayama, 1978; Demain, 2000). It has been known for a long time that glutamate excretion is coupled with alterations in the cell membrane (Takinami et al., 1968; Hoischen and Krämer, 1990). However, in contrast to previous long-held beliefs (Shiio et al., 1963; Kimura, 1963; Demain and Birnbaum, 1968), glutamate excretion does not occur as a consequence of mere leakiness of the cells via relaxed membrane permeability (Hoischen and Krämer, 1989; Krämer, 1994a; Krämer, 1994b). Also, glutamate excretion by producing *C. glutamicum* cells is not achieved by the reverse action of the glutamate uptake system present in this organism, as was suggested by Clement et al. (Clement et al., 1984; Clement et al., 1986), since glutamate uptake proceeds via a primary active system that is not reversible (Krämer, 1994a; Kronemeyer et al., 1995). Instead, a specific transport system responsible for glutamate excretion by *C. glutamicum* under biotin limitation was described by Hoischen and Krämer (1989). This efficient efflux system for glutamate, which is energy-dependent and functions against an existing chemical gradient (Gutman et al., 1992), is thought to be a key player in glutamic acid overproduction.

In industrial glutamic acid fermentations, membrane alterations and glutamate excretion are often brought about by biotin limitation in the culture broth. The concentration of biotin for optimal glutamate accumulation is approximately 1 µg/liter (Kinoshita and Tanaka, 1972; Hoischen and Krämer, 1989). *Corynebacterium glutamicum* and similar amino acid-producing bacteria are biotin auxotrophs, and biotin limitation leads to suboptimal growth and gives rise to alterations in the lipid content and composition of the cell membrane by interfering with fatty acid synthesis (Takinami et al., 1968; Hoischen and Krämer, 1990). It is, however, not clear how biotin starvation actually triggers glutamate excretion. With an experimental setup which employs monitoring the lipid composition and glutamate excretion activity, upon adding biotin to biotin-limited (glutamate-producing) cells, it was observed that changes in glutamate excretion can be provoked on a much shorter time scale than significant alterations in the lipid content and composition take place (Hoischen and Krämer, 1990). Therefore, it was concluded that the decreased lipid content of the membrane, although a necessary prerequisite, is not sufficient for high glutamate efflux activity by the carrier (Hoischen and Krämer, 1990). Possible components of biotin-dependent acyl-CoA carboxylase activities have been found (DtsR1, DtsR2 and AccBC) which are necessary for the synthesis of fatty acids and mycolates and which are likely to be involved in the chain of events leading to membrane alterations and glutamate efflux under biotin limitation (Kimura et al., 1996; Kimura et al., 1997; Jäger et al., 1996; Eggeling and Sahm, 1999).

For the use of cheap substrates rich in biotin, e.g., cane molasses, and for glutamic acid fermentation, other processes that function independent of biotin limitation have been developed. The addition of penicillin during the logarithmic growth phase leads to effective glutamate excretion. This treatment does not provoke cell lysis but causes a rapid, 97–99.5% decrease in cell viability (Kinoshita, 1985). Alternative methods include the addition of fatty acid derivatives such as polyoxyethylene sorbitan monooleate (Tween 60) during growth or the use of oleic acid or glycerol-requiring auxotrophs which are kept under oleic acid or glycerol limitation, respectively, during fermentation (Kinoshita and Nakayama, 1978). In small-scale experiments, high glutamate excretion activity was also brought about by the addition of amine surfactants (Dupperay et al., 1992) or by adding local anesthetics like tetracaine at concentrations known to alter the membrane fluidity (Krämer, 1994a).

Another factor long suspected to be involved in the triggering of glutamate production is the activity of oxoglutarate dehydrogenase. However, the traditional idea that oxoglutarate dehydrogenase is absent in *C. glutamicum* is clearly not true (Shiio and Ujigawa-Takeda, 1980). This fact is in accordance with results showing that 1) *C. glutamicum* can grow well on L-glutamate, and 2) when the glyoxylate cycle is not active, cells are still able to grow rapidly, with complete oxidation of glucose to carbon dioxide. The oxoglutarate dehydrogenase complex appears to have an unusual structure in *C. glutamicum* (Usuda et al., 1996). Interestingly, under conditions of glutamate excretion, the activity of this enzyme is reduced by 90%, which favors glutamate production by the enzyme competing for the same substrate as oxoglutarate dehydrogenase and glutamate dehydrogenase. Inactivation of odhA, the gene for one subunit of the oxoglutarate dehydrogenase in the wild type of *C. glutamicum*, leads to high-level glutamate excretion, even without biotin limitation (Kawahara et al., 1997). A comprehensive, recent overview of factors linked with the still puzzling phenomenon of glutamate efflux was given by Eggeling and Sahm (1999).

Another amino acid that is produced fermentatively with *C. glutamicum* and similar bacteria is L-lysine (estimated annual production 2.5 (10^8 kg). In industrial production plants, the concentration of lysine in the fermentation broth can reach up to 120–170 g/liter at the end of the process (Eggeling and Sahm, 1999; Demain, 2000), with molar yields of L-lysine from glucose of 0.25–0.35. L-Lysine is an essential amino acid, which is used mainly as an animal feed supplement, and far smaller amounts are produced for the pharmaceutical and cosmetics industries. The sequence of biosynthetic reactions leading to the production of L-lysine from aspartate is part of a branched pathway that also produces threonine, isoleucine, and methionine, all members of the so-called "aspartate family of amino acids" (Fig. 4). Unusually, the biosynthesis of DL-diaminopimelate, which is needed as the precursor of L-lysine as well as for cell wall peptidoglycan synthesis, occurs via a split pathway in *C. glutamicum*. The common part of these pathway variants consists of the activation and subsequent reduction of aspartate to aspartate semialdehyde by aspartate kinase (LysC) and aspartate semialdehyde dehydrogenase (Asd), followed by conversion of aspartate semialdehyde to piperideine dicarboxylate via dihydrodipicolinate with dihydrodipicolinate synthase (DapA) and dihydrodipicolinate reductase (DapB). Both the succinylase and the dehydrogenase pathway are present for the conversion of piperideine dicarboxylate to DL-diaminopimelate (Schrumpf et al., 1991). Whereas the latter branch is a one-step reaction with D-diaminopimelate dehydrogenase (Ddh), the succinylase branch proceeds via four steps with succinylated intermediates. The dehydrogenase

variant is energetically less expensive than the succinylase variant. However, owing to the low affinity of the dehydrogenase towards ammonium, the dehydrogenase branch is not operational at low ammonium concentrations, and under these conditions the succinylase branch must be used for DL-diaminopimelate synthesis (Sonntag et al., 1993). Although not essential for L-lysine synthesis, inactivation of the dehydrogenase leads to reduced lysine production (Schrumpf et al., 1991). It is clear from metabolic flux analyses that lysine-producing *C. glutamicum* strains can use both pathways simultaneously. However, while strain MH20-22B grown in continuous culture synthesized 26% of lysine via the dehydrogenase reaction, strain ATCC21253 grown in batch culture displayed a higher proportion of lysine synthesis (63%) through this route (Marx et al., 1996; Wittmann and Heinzle, 2001). Although strain differences may play a role, the main reason for this discrepancy probably lies in the different cultivation conditions. It is known that the flux partitioning between the two pathways of DL-diaminopimelate synthesis depends on the supply of ammonium, leading to a relative contribution of the dehydrogenase pathway to lysine synthesis between 0 and 72% (Sonntag et al., 1993; Eggeling and Sahm, 1999).

While glutamic acid production by *C. glutamicum* can be achieved simply by adjusting the critical culture parameters (biotin limitation, addition of penicillin, etc., see above), leading to a "metabolic overflow" situation and active glutamate efflux, the high-yield fermentation of lysine by this organism requires the use of mutated or genetically modified strains. To understand the reasons that may make amino acid-producing corynebacteria so well suited for the overproduction of L-lysine, it is important to realize the relative simplicity of regulation found in this multibranched biosynthetic pathway when compared with the situation found in other bacteria like *E. coli*. There is no regulation of the synthesis of the lysine-specific enzymes within the biosynthetic routes of the aspartate family in *C. glutamicum* (Yeh et al., 1988). Homoserine dehydrogenase, the enzyme located at the branching point leading to threonine, isoleucine and methionine, is repressed by methionine (Nara et al., 1961; Cremer et al., 1988). Inhibition of enzyme activity in the lysine biosynthetic sequence was only found for aspartate kinase (Nakayama et al., 1966; Cremer et al., 1988), the first step of the entire aspartate family branched pathway. In contrast to *E. coli*, which contains three different aspartate kinase isoenzymes (two of which are bifunctional proteins displaying aspartate kinase plus homoserine dehydrogenase activity; Cohen, 1983) that are each regulated by a different end product, *C. glutamicum* has only a single aspartate kinase. Feedback inhibition of this enzyme occurs by the concerted action of lysine plus threonine. This type of regulation represents a means for overall control of the synthesis of the amino acids of both major branches of the pathway. The flow of aspartate semialdehyde to either the diaminopimelate/lysine or the methionine/threonine/isoleucine branch is regulated solely at the level of homoserine dehydrogenase, the initial enzyme of the latter branch, by repression of the enzyme by methionine and sensitive feedback inhibition by threonine. Additionally, homoserine dehydrogenase activity is inhibited 50% by L-isoleucine (Cremer et al., 1988). There is no regulation of dihydrodipicolinate synthase, the other enzyme at the branching point competing for aspartate semialdehyde and the initial enzyme of the diaminopimelate/lysine-specific part of the pathway. Accumulation of threonine efficiently shuts down the methionine/threonine/isoleucine branch by sensitive feedback inhibition of homoserine dehydrogenase, directing the flow of aspartate semialdehyde into the lysine branch. Excess lysine in turn can lead to the decrease of carbon flow into the whole biosynthetic pathway via the multivalent inhibition of aspartate kinase by lysine plus threonine. The rather simple regulation strategies found in amino acid-producing corynebacteria and the lack of additional and differentially regulated isoenzymes at crucial nodes of the biosynthetic pathways (for example, as noted above, *E. coli* has three distinct and separately regulated aspartate kinase isoenzymes) may explain the relative ease of obtaining high-level lysine-producing *C. glutamicum* strains. Because of the lack of regulation of the initial enzymes of the diaminopimelate/lysine branch, a relatively simple way to obtain a lysine overproducer is to genetically knock out homoserine dehydrogenase, which leads to the shut-down of the methionine/threonine/isoleucine branch. In 1958, Kinoshita et al. isolated such homoserine- or both methionine plus threonine-requiring mutants of *C. glutamicum* that produced L-lysine from carbohydrate (Kinoshita et al., 1958). Such threonine- and methionine-requiring strains produce lysine if threonine is supplemented at a limiting concentration, because a low intracellular threonine level bypasses the concerted feedback inhibition of aspartate kinase. A different strategy used to obtain lysine-producing strains employs the selection of regulatory mutants that display an increased resistance to antimetabolites. Isolation of *C. glutamicum* mutants resistant to the lysine analogues S-(2-aminoethyl)-L-cysteine (AEC) or O-(2-aminoethyl)-L-serine ("oxalysine") leads to strains with increased lysine production due to the release of feedback inhibition of the first step of lysine synthesis from aspartate. These mutants contain a deregulated aspartate kinase that is no

longer subject to the concerted feedback inhibition by lysine plus threonine (Shiio and Sano, 1969; Sano and Shiio, 1970; Schrumpf et al., 1992; Jetten et al., 1995; Leuchtenberger, 1996). Industrial strains used for the production of amino acids often have combinations of auxotrophic and regulatory mutations (see Leuchtenberger, 1996).

L-Lysine excretion, like glutamate excretion, is via active transport and involves a specific carrier. However, in this case the mechanism is secondary active transport using a $(2OH^-)$-lysine symporter (Bröer and Krämer, 1991; Krämer, 1994a; Krämer, 1994b). The lysine export carrier LysE is needed during growth on lysine-containing peptides, to avoid the accumulation of lysine in the cell. In the presence of lysyl-alanine, the intracellular lysine concentration in a LysE-deficient mutant can reach more than 1 M, and growth inhibition is observed (Vrljic et al., 1996). Strikingly, the lysine secretion system of certain production strains has been shown to be regulated differently and to display kinetic properties different from the wild type (Bröer et al., 1993), indicating the importance of efficient lysine export for its fermentative overproduction.

Much attention has been paid to the central metabolism of *C. glutamicum*, which provides the precursors of amino acid biosynthesis, particularly in the past few years using modern techniques of metabolic flux analysis. During both L-glutamic acid and L-lysine production from glucose, the major flow of carbon is channeled through the pentose phosphate pathway (Dominguez et al., 1998; Wittmann and Heinzle, 2001). For example, during the lysine-production phase of strain ATCC 21253 grown in batch culture, only 24.4% of the incoming glucose was routed directly into glycolysis while 71% of the incoming carbon entered the central metabolism via the oxidative part of the pentose phosphate pathway, thus supplying ample amounts of NADPH (together with the isocitrate dehydrogenase reaction of the citric acid cycle, even an excess of NADPH) for lysine biosynthesis (Wittmann and Heinzle, 2001).

An important factor for the overproduction of L-glutamate as well as L-lysine is the replenishment of intermediates of the citric acid cycle withdrawn for their synthesis, i.e., 2-oxoglutarate and oxaloacetate, respectively. To this end, *C. glutamicum* possesses two enzymes which ensure the continuous supply of oxaloacetate to the citric acid cycle during growth on glucose: phosphoenolpyruvate (PEP) carboxylase and pyruvate carboxylase (Eikmanns et al., 1989; Jetten et al., 1994; Mori and Shiio, 1985a; Mori and Shiio, 1985b; Ozaki and Shiio, 1969; Peters-Wendisch et al., 1996, 1997, 1998; Fig. 4). Phosphoenolpyruvate carboxykinase, another activity with anaplerotic potential present in *C. glutamicum*, is completely inhibited in the direction of oxaloacetate synthesis by low concentrations of ATP (Jetten and Sinskey, 1993) and therefore is thought to function mainly in gluconeogenesis rather than in anaplerosis. Finally, the glyoxylate cycle, which is used as an anaplerotic sequence by some bacteria during growth on glucose, is needed for anaplerosis during growth of *C. glutamicum* on acetate but is also not required for anaplerosis during glucose consumption since its activity is tightly downregulated in glucose-grown cells (Reinscheid et al., 1994b), and mutants of a lysine-producing strain defective in the glyoxylate cycle enzyme isocitrate dehydrogenase are not affected in their growth on glucose or in their lysine productivity (Peters-Wendisch et al., 1996). Also, PEP carboxylase-negative mutants of *C. glutamicum* grow normally and are able to produce lysine on various media (Gubler et al., 1994; Peters-Wendisch et al., 1993; Peters-Wendisch et al., 1996), which demonstrates that this enzyme is dispensable for *C. glutamicum* and its loss can be compensated by pyruvate carboxylase. Pyruvate carboxylase mutants, on the other hand, although able to grow on glucose and acetate, cannot grow on lactate or pyruvate, indicating that pyruvate carboxylase is the crucial anaplerotic activity during growth on these C_3-compounds (Peters-Wendisch et al., 1997). If both the PEP carboxylase and the pyruvate carboxylase activities are deleted, *C. glutamicum* can no longer grow on glucose (Peters-Wendisch et al., 1998). Interestingly, recent metabolic flux analyses of glutamate and lysine producers revealed that in addition to the anaplerotic carboxylation, also decarboxylation of C_4 units happens to a certain extent in vivo, leading to a continuous depletion of these C_4 intermediates from the citric acid cycle. Thus, there is a significant backward flux from C_4 units of the citric acid cycle to C_3 units of glycolysis, implying a waste of ATP via a "futile cycle" at the anaplerotic node (Marx et al., 1996; Wittmann and Heinzle, 2001). Under conditions of glutamate production, however, the exchange flux between C4 unit generation and decarboxylation is largely reduced (see Eggeling and Sahm, 1999).

Besides glutamate and lysine producers, *C. glutamicum* strains have been constructed which hyperproduce other amino acids, e.g., threonine, isoleucine, tyrosine, phenylalanine, or tryptophan (Kinoshita and Nakayama, 1978; Kinoshita, 1985; Katsumata and Ikeda, 1993; Sahm et al., 1995; Leuchtenberger, 1996). Traditionally, strains suited for amino acid production have been obtained by multiple rounds of classical random mutagenesis and screening procedures. Although this approach has proved to be very efficient and has led to amino acid producers with remarkable productivities, further improvement is presently sought with more rational strategies and by the application of recombinant DNA technology with the aid of the transformation systems now available for corynebacteria. One example in this field is the construction of tryptophan-hyperproducing *C. glutamicum* strains, yielding 58 g/liter tryptophan by 1) deregulating and/or overexpressing the terminal pathways leading to tryptophan and serine (a substrate of the final reaction in the tryptophan pathway) and by 2) low-copy amplification of the transketolase gene to modify the pentose phosphate pathway for an improved supply of the aromatic biosynthesis precursor erythrose 4-phosphate (Ikeda and Katsumata, 1999). In parallel to using genetic engineering for strain improvement of corynebacteria, the focus of research is being extended to include additional aspects besides merely the sequence of biosynthesis reactions leading to a specific amino acid. In this way, one hopes to come to a broad understanding of fluxes and flux control in *Corynebacterium* central metabolism and of the regulation of expression of genes of importance for amino acid hyperproduction. To this end, metabolic flux analysis, using either in-vivo-NMR or MALDI-TOF mass spectrometry as an analytical tool, is being used to quantify intracellular flux distributions, to compare key fluxes in wild-type and mutant strains, and to develop mathematical models for the *C. glutamicum* central metabolism (Vallino and Stephanopoulos, 1993; Marx et al., 1996; Park et al., 1997; Wittmann and Heinzle, 2001). In addition, due to the availability of the *C. glutamicum* genome sequence, methods of postgenome-analysis, i.e., transcript profiling with DNA chip technology (Loos et al., 2001) or proteome analysis, are used to detect possible bottlenecks at the level of the regulation, biosynthesis and stability of enzymes important for amino acid production. Coupling of these modern analytical approaches with the recombinant DNA technology available (extrachromosomal and chromosomal) is expected to lead to the specific design of further improved strains.

NUCLEOTIDE FERMENTATIONS AND OTHER APPLICATIONS Saprophytic corynebacteria are also used in the fermentative production of nucleotides, which are of interest primarily as flavor-enhancing additives in foods (Komata, 1976). Mutant strains of *C. glutamicum* have been selected which excrete the purine ribonucleoside 5′-monophosphates 5′-inosinic acid (IMP), 5′-xanthylic acid (XMP), and 5′-guanylic acid (GMP; Demain, 1978). Similarly, adenine-requiring mutants of *C. ammoniagenes* (previously designated *Brevibacterium ammoniagenes*, later transferred to the genus *Corynebacterium* by Collins, 1987b) were found to accumulate IMP in fermentations (Nara et al., 1967). Remarkably, intact wild-type cells of *C. ammoniagenes* are able to convert hypoxanthine to IMP, guanine to GMP,

GDP and GTP, and adenine to AMP, ADP and ATP (Nara et al., 1967, 1968a; Tanaka et al., 1968) by "salvage synthesis" reactions. In the case of IMP production, particularly the concentration of manganese ions (Mn^{2+}) was found to be of importance, as very low as well as excessive amounts of Mn^{2+} negatively affected IMP accumulation (Furuya et al., 1970; Nara et al., 1968b). In the presence of nonlimiting amounts of manganese, nucleotide production can be achieved by the addition of certain antibiotics or surfactants or the use of mutant strains that accumulate IMP under excess manganese conditions (Nara et al., 1969b; Furuya et al., 1969). Improved *C. ammoniagenes* mutants producing more than 20 g per liter of IMP were selected. These were leaky adenine-auxotrophic, manganese-insensitive, and guanine-auxotrophic, and they displayed increased IMP-excretion characteristics (Teshiba and Furuya, 1982; Teshiba and Furuya, 1984). A different method for IMP production is fermentative production of inosine with *C. ammoniagenes* mutants (which are also leaky adenine-auxotrophic and guanine-auxotrophic) followed by chemical phosphorylation of the nucleoside with phosphoryl chloride ($POCl_3$). Inosine productivity by *C. ammoniagenes* strains can exceed 30 g per liter (Kotani et al., 1978). Mori et al. (1997) recently described a new process for IMP production, in which inosine fermentation by *C. ammoniagenes* is followed by enzymatic phosphorylation via addition of recombinant guanosine/inosine kinase-overexpressing *E. coli* cells. During this coupled process, which is carried out in the presence of detergents for cell permeabilization, the *C. ammoniagenes* cells regenerate the ATP needed for the kinase reaction (Mori et al., 1997). Analogous coupled systems were used for the production of sugar nucleotides like UDP-galactose or UDP-N-acetylglucosamine (Koizumi et al., 1998; Tabata et al., 2000). For production of GMP, xanthosine 5'-monophosphate (XMP) produced fermentatively by an appropriate *C. ammoniagenes* mutant strain can be converted to GMP by addition of overexpressed XMP aminase (Fujio et al., 1997). *Corynebacterium ammoniagenes* also has been used for the production of ATP from adenine (Fujio and Furuya, 1983). In addition to the production of nucleotides from purine bases, *C. ammoniagenes* was found to be suited for the fermentative production of nicotinamide adenine dinucleotide (NAD) when cultured in medium containing adenine and nicotinic acid or nicotinamide. Nicotinic acid mononucleotide was produced if only nicotinic acid or nicotinamide was present (Nakayama et al., 1968). Finally, Koizumi et al. (2000) reported high-level production (more than 15 g per liter) of the vitamin riboflavin (vitamin B_2) by a metabolically engineered recombinant strain of *C. ammoniagenes* with enhanced activities of riboflavin biosynthetic enzymes.

Literature Cited

Abe, S., K. Takayama, and S. Kinoshita. 1967. Taxonomical studies on glutamic acid-producing bacteria. J. Gen. Appl. Microbiol. 13:279–301.

Ankri, S., I. Bouvier, O. Reyes, F. Predali, and G. Leblon. 1996a. A Brevibacterium linens pRBL1 replicon functional in Corynebacterium glutamicum. Plasmid 36:36–41.

Ankri, S., O. Reyes, and G. Leblon. 1996b. Electrotransformation of highly restrictive corynebacteria with synthetic DNA. Plasmid 35:62–66.

Archer, J. A. C., M. T. Follettie, and A. J. Sinskey. 1989. Biology of Corynebacterium glutamicum: a molecular approach. In: C. L. Hershberger, S. W. Queener, and G. Hegeman (Eds.) Genetics and Molecular Biology of Industrial Microorganisms. American Society for Microbiology. Washington DC, 27–33.

Barksdale, L. 1970. Corynebacterium diphtheriae and its relatives. Bacteriol. Rev. 34:378–422.

Barksdale, L., M. A. Lanéelle, M. C. Pollice, J. Asselineau, M. Welby, and M. V. Norgard. 1979. Biological and chemical basis for the reclassification of Microbacterium flavum Orla-Jensen as Corynebacterium flavescens nom. nov. Int. J. Syst. Bacteriol. 29:222–233.

Barksdale, L. 1981. The genus Corynebacterium. In: M. P. Starr, H. Stolp, H. G. Trüper, A. Balows, and H. G. Schlegel (Eds.) The Prokaryotes. Springer-Verlag. New York, NY. II:1827–1837.

Bathe, B., J. Kalinowski, and A. Pühler. 1996. A physical and genetic map of the Corynebacterium glutamicum ATCC 13032 chromosome. Molec. Gen. Genet. 252:255–265.

Batt, C. A., W. S. Shanabruch, and A. J. Sinskey. 1985. Expression of pAM 1 tetracycline resistance gene in Corynebacterium glutamicum: Segregation of antibiotic resistance due to intramolecular recombination. Biotechnol. Lett. 7:717–722.

Bechdel, S. I., H. E. Honeywell, R. A. Dutcher, and M. H. Knutsen. 1928. Synthesis of vitamin B in the rumen of the cow. J. Biol. Chem. 80:231–238.

Billman-Jacobe, H., A. L. M. Hodgson, M. Lightowlers, P. R. Wood, and A. J. Radford. 1994. Expression of ovine gamma interferon in Escherichia coli and Corynebacterium glutamicum. Appl. Environ. Microbiol. 60:1641–1645.

Billman-Jacobe, H., L. Wang, A. Kortt, D. Stewert, and A. J. Radford. 1995. Expression and secretion of heterologous proteases by Corynebacterium glutamicum. Appl. Environ. Microbiol. 61:1610–1613.

Bonamy, C., A. Guyonvarch, O. Reyes, F. David, and G. Leblon. 1990. Interspecies electro-transformation in corynebacteria. FEMS MIcrobiol. Lett. 66:263–270.

Bonamy, C., J. Labarre, O. Reyes, and G. Leblon. 1994. Identification of IS1206, a Corynebacterium glutamicum IS3-related insertion sequence and phylogenetic analysis. Molec. Microbiol. 14:571–581.

Bonnassie, S., J. Oreglia, A. Trautwetter, and A. M. Sicard. 1990. Isolation and characterization of a restriction and modification deficient mutant of Brevibacterium lactofermentum. FEMS Microbiol. Lett. 72:143–146.

Bousfield, I. J. 1978. The taxonomy of coryneform bacteria from the marine environment. *In:* I. J. Bousfield and A. G. Callely (Eds.) Coryneform Bacteria. Academic Press. London, UK. 217–233.

Brabetz, W., W. Liebl, and K. H. Schleifer. 1991. Studies on the utilization of lactose by Corynebacterium glutamicum bearing the lactose operon of Escherichia coli. Arch. Microbiol. 155:607–612.

Brabetz, W., W. Liebl, and K. H. Schleifer. 1993. Lactose permease of Escherichia coli catalyses active β-galactoside transport in a Gram-positive bacterium. J. Bacteriol. 175:7488–7491.

Brennan, N. M., R. Brown, M. Goodfellow, A. C. Ward, T. P. Beresford, P. J. Simpson, P. F. Fox, and T. M. Cogan. 2001. Corynebacterium mooreparkense sp. nov. and Corynebacterium casei sp. nov., isolated from the surface of a smear-ripened cheese. Int. J. Syst. Evol. Microbiol. 51:843–852.

Bröer, S., and R. Krämer. 1991. Lysine secretion by Corynebacterium glutamicum. 2: Energetics and mechanism of the transport system. Eur. J. Biochem. 202:137–143.

Bröer, S., R. Eggeling, and R. Krämer. 1993. Strains of Corynebacterium glutamicum with different lysine productivities may have different lysine excretion systems. Appl. Environ. Microbiol. 59:316–321.

Clement, Y., B. Escoffier, M. C. Trombe, and G. Lanéelle. 1984. Is glutamate excreted by its uptake system in Corynebacterium glutamicum? A working hypothesis. J. Gen Microbiol. 130:2589–2594.

Clement, Y., and G. Lanéelle. 1986. Glutamate excretion mechanism in Corynebacterium glutamicum: Triggering by biotin starvation or by surfactant addition. J. Gen. Microbiol. 132:925–929.

Cohen, G. N. 1983. The common pathway to lysine, methionine, and threonine. *In:* K. M. Herrmann and R. L. Somerville (Eds.) Amino Acids: Biosynthesis and Genetic Regulation. Addison-Wesley Publishing. Reading, MA. 147–171.

Collins, M. D., D. Jones, and R. M. Kroppenstedt. 1981. Reclassification of Corynebacterium ilicis (Mandel, Guba and Litsky) in the genus Arthrobacter, as Arthrobacter ilicis comb. nov. Zentralbl. Bakteriol. Mikrobiol. Hyg. Abt. I., Orig. C 2:318–323.

Collins, M. D., and D. Jones. 1982a. Taxonomic studies on Corynebacterium beticola (Abdou). J. Appl. Bacteriol. 52:229–233.

Collins, M. D., M. Goodfellow, and D. E. Minnikin. 1982b. A survey of the structures of mycolic acids in Corynebacterium and related taxa. J. Gen. Microbiol. 128:129–149.

Collins, M. D., D. Jones, and R. M. Kroppenstedt. 1982c. In validation of the publication of new names and combinations previously effectively published outside the IJSB List No. 9. Int. J. Syst. Bacteriol. 32:384–385.

Collins, M. D., and D. Jones. 1983. Reclassification of Corynebacterium flaccumfaciens, Corynebacterium betae, Corynebacterium oortii and Corynebacterium poinsettiae in the genus Curtobacterium, as Curtobacterium flaccumfaciens comb. nov. J. Gen Microbiol. 129:3545–3548.

Collins, M. D., and J. F. Bradbury. 1986a. Plant pathogenic species of Corynebacterium. *In:* P. H. A. Sneath, N. S. Mair, M. E. Sharpe, and J. G. Holt (Eds.) Bergey's Manual of Systematic Bacteriology. Williams and Wilkins. Baltimore, MD. 2:1276–1283.

Collins, M. D., and C. S. Cummins. 1986b. Genus Corynebacterium Lehmann and Neumann 1896. *In:* P. H. A. Sneath, N. S. Mair, M. E. Sharpe, and J. G. Holt (Eds.) Bergey's Manual of Systematic Bacteriology. Williams and Wilkins. Baltimore, MD. 2:1266–1276.

Collins, M. D. 1987a. Transfer of Arthrobacter variabilis (Müller) to the genus Corynebacterium, as Corynebacterium variabilis comb. nov. Int. J. Syst. Bacteriol. 37:287–288.

Collins, M. D. 1987b. Transfer of Brevibacterium ammoniagenes (Cooke and Keith) to the genus Corynebacterium, as Corynebacterium ammoniagenes comb. nov. Int. J. Syst. Bacteriol. 37:442–443.

Collins, M. D., R. A. Burton, and D. Jones. 1988. Corynebacterium amycolatum, sp. nov., a new mycolic acid-less Corynebacterium species from human skin. FEMS Microbiol. Lett. 49:349–352.

Collins, M. D., J. Smida, M. Dorsch, and E. Stackebrandt. 1989a. Tsukamurella gen. nov. harboring Corynebacterium paurometabolum and Rhodococcus aurantiacus. Int. J. Syst. Bacteriol. 38:385–391.

Collins, M. D., J. Smida, and E. Stackebrandt. 1989b. Phylogenetic evidence for the transfer of Caseobacter polymorphus (Crombach) to the genus Corynebacterium. Int. J. Syst. Bacteriol. 39:7–9.

Cooke, J. V., and H. R. Keith. 1927. A type of urea-splitting bacterium found in the human intestinal tract. J. Bacteriol. 13:315–319.

Correia, A., A. Pisabarro, J. M. Castro, and J. F. Martin. 1996. Cloning and characterization of an IS-like element present in the genome of Brevibacterium lactofermentum ATCC13869. Gene 170:91–94.

Cremer, J., C. Treptow, L. Eggeling, and H. Sahm. 1988. Regulation of enzymes of lysine biosynthesis in Corynebacterium glutamicum. J. Gen. Microbiol. 134:3221–3229.

Crombach, W. H. J. 1978. Caseobacter polymorphus gen. nov., sp. nov., a coryneform bacterium from cheese. Int. J. Syst. Bacteriol. 28:354–366.

Crombach, W. H. J. 1986. Genus Caseobacte 1978. *In:* P. H. A. Sneath, N. S. Mair, M. E. Sharpe, and J. G. Holt (Eds.) Bergey's Manual of Systematic Bacteriology. Williams and Wilkins. Baltimore, MD. 2:1318–1319.

Cummins, C. S., R. A. Lelliott, and M. Rogosa. 1974. Genus I Corynebacterium Lehmann and Neumann 1896, 350. *In:* R. E. Buchanan, and N. E. Gibbons (Eds.) Bergey's Manual of Determinative Bacteriology, 8th ed. Williams and Wilkins. Baltimore, MD. 602–617.

Cure, G. L., and R. M. Keddie. 1973. Methods for the morphological examination of aerobic coryneform bacteria. *In:* R. G. Board, and D. N. Lovelock (Eds.) Sampling: Microbiological Monitoring of Environments. Academic Press. New York, NY. Society for Applied Bacteriology Technical Series 7:123–135.

Daffe, M., and P. Draper. 1998. The envelope layers of mycobacteria with reference to their pathogenicity. Adv. Microb. Physiol. 39:131–203.

Davis, M. J., A. G. Gillespie Jr., A. K. Vidaver, and R. W. Harris. 1984. Clavibacter: a new genus containing some phytopathogenic coryneform bacteria, including Clavibacter xyli subsp. xyli sp. nov. subsp. nov. and Clavibacter xyli subsp. cynodontis subsp. nov., pathogens that cause ratoon stunting disease of sugar cane and Bermudagrass stunting disease. Int. J. Syst. Bacteriol. 34:107–117.

Deb, J. K., and N. Nath. 1999. Plasmids of corynebacteria. FEMS Microbiol. Lett. 175:11–20.

Demain, A. L., and J. Birnbaum. 1968. Alteration of permeability for the release of metabolites from the microbial cell. Curr. Top. Microbiol. Immunol. 46:1–25.

Demain, A. L. 1978. Production of nucleotides by microorganisms. *In:* A. H. Rose (Ed.) Primary Products of Metabolism. Academic Press. London, UK. 187–208.

Demain, A. L. 2000. Microbial biotechnology. Trends Biotechnol. 18:26–31.

Dominguez, H., C. Rollin, A. Guyonvarch, J. L. Guerquin-Kern, M. Cocaign-Bousquet, and A. D. Lindley. 1998. Carbon-flux distribution in the central metabolic pathways of Corynebacterium glutamicum during growth on fructose. Eur. J. Biochem. 254:96–102.

Dunican, L. K., and E. Shivnan. 1989. High frequency transformation of whole cells of amino acid producing coryneform bacteria using high voltage electroporation. Bio/Technology 7:1067–1070.

Dupperay, F., D. Jezequel, A. Ghazi, L. Letellier, and E. Shechter. 1992. Excretion of glutamate from Corynebacterium glutamicum triggered by amine surfactants. Biochim. Biophys. Acta 1103:250–258.

Eggeling, L., and H. Sahm. 1999. L-Glutamate and L-lysine: Traditional products with impetuous developments. Appl. Microbiol. Biotechnol. 52:146–153.

Eikmanns, B. J., M. T. Follettie, M. U. Griot, and A. J. Sinskey. 1989. The phosphoenolpyruvate carboxylase gene of Corynebacterium glutamicum: Molecular cloning, nucleotide sequence, and expression. Molec. Gen. Genet. 218:330–339.

Eikmanns, B. J., E. Kleinertz, W. Liebl, and H. Sahm. 1991. A family of Corynebacterium glutamicum/E. coli vectors for gene cloning, controlled gene expression, and promoter probing. Gene 102:93–98.

Eliskases-Lechner, F., and W. Ginzinger. 1995. The bacterial flora of surface ripened cheese with special regard to coryneforms. Lait 7:571–584.

Follettie, M. T., and A. J. Sinskey. 1986. Recombinant DNA technology for Corynebacterium glutamicum. Food Technol. 40:88–94.

Fujio, T., and A. Furuya. 1983. Production of ATP from adenine by Brevibacterium (Corynebacterium) ammoniagenes. J. Ferment. Technol. 61:261–267.

Fujio, T., T. Nishi, S. Ito, and A. Maruyama. 1997. High level expression of XMP aminase in Escherichia coli and its application for the industrial production of 5′-guanylic acid. Biosci. Biotechnol. Biochem. 61:840–845.

Funke, G., S. Stubbs, M. Altwegg, A. Carlotti, and M. D. Collins. 1994. Turicella otitidis gen. nov., spec. nov., a coryneform bacterium isolated from patients with otitis media. Int. J. Syst. Bacteriol. 44:270–273.

Furuya, A., S. Abe, and S. Kinoshita. 1969. Accumulation of 5′inosinic acid by a manganese-insensitive mutant of Brevibacterium ammoniagenes. Appl. Microbiol. 18:977–984.

Furuya, A., S. Abe, and S. Kinoshita. 1970. Effects of manganese and adenine on 5′ inosinic acid accumulation by a mutant of Brevibacterium ammoniagenes. Agr. Biol. Chem. 34:210–221.

Goodfellow, M., and D. E. Minnikin. 1981. Introduction to the coryneform bacteria. *In:* M. P. Starr, H. Stolp, H. G. Trüper, A. Balows and H. G. Schlegel (Eds.) The Prokaryotes. Springer-Verlag. New York, NY. II:1811–1826.

Goodfellow, M. 1984a. In validation of the publication of new names and combinations previously effectively published outside the IJSB List No. 10. Int. J. Syst. Bacteriol. 34:503–504.

Goodfellow, M. 1984b. Reclassification of Corynebacterium fascians (Tilford) Dowson in the genus Rhodococcus, as Rhodococcus fascians comb. nov. Syst. Appl. Microbiol. 5:225–229.

Goodfellow, M. 1992. The family Nocardiaceaea. *In:* A. Balows, H. G. Trüper, M. Dworkin, W. Harder, and K.-H. Schleifer (Eds.) The Prokaryotes, 2nd ed. Springer-Verlag. New York, NY. II:1188–1213.

Graves, M. C., and J. C. Rabinowitz. 1986. In vivo and in vitro transcription of the Clostridium pasteurianum ferredoxin gene. J. Biol. Chem. 261:11409–11415.

Gubler, M., S. M. Park, M. Jetten, G. Stephanopoulos, and A. J. Sinskey. 1994. Effects of phosphoenolpyruvate carboxylase-deficiency on metabolism and lysine production in Corynebacterium glutamicum. Appl. Microbiol. Biotechnol. 40:857–863.

Gutman, M., C. Hoischen, and R. Krämer. 1992. Carrier-mediated glutamate secretion by Corynebacterium glutamicum under biotin limitation. Biochim. Biophys. Acta 1112:115–123.

Haynes, J. A., and M. L. Britz. 1989. Electrotransformation of Brevibacterium lactofermentum and Corynebacterium glutamicum: growth in tween 80 increases transformation frequencies. FEMS Microbiol. Lett. 61:329–334).

Hoischen, C., and R. Krämer. 1989. Evidence for an efflux carrier system involved in the secretion of glutamate by Corynebacterium glutamicum. Arch. Microbiol. 151:342–347.

Hoischen, C., and R. Krämer. 1990. Membrane alteration is necessary but not sufficient for effective glutamate secretion in Corynebacterium glutamicum. J. Bacteriol. 172:3409–3416.

Ikeda, M., and R. Katsumata. 1998. A novel system with positive selection for the chromosomal integration of replicative plasmid DNA in Corynebacterium glutamicum. Microbiology 144:1863–1868.

Ikeda, M., and R. Katsumata. 1999. Hyperproduction of tryptophan by Corynebacterium glutamicum with the modified pentose phosphate pathway. Appl. Environ. Microbiol. 65:2497–2502.

Ikeguchi, N., T. Nihira, A. Kishimoto, and Y. Yamada. 1988. Oxidative pathway from squalene to geranylacetone in Arthrobacter sp. strain Y-11. Appl. Environ. Microbiol. 54:381–385.

Jackson, M., C. Raynaud, M.-A. Laneele, C. Guilhot, C. Laurent-Winter, D. Ensergueix, B. Gicquel, and M. Daffe. 1999. Inactivation of the antigen 85C gene profoundly affects the mycolate content and alters the permeability of the Mycobacterium tuberculosis cell envelope. Molec. Microbiol. 31:1573–1587.

Jäger, W., A. Schäfer, J. Kalinowski, and A. Pühler. 1995. Isolation of insertion elements from Gram-positive Brevibacterium, Corynebacterium and Rhodococcus strains using the Bacillus subtilis sacB gene as a positive selection marker. FEMS Microbiol. Lett. 126:1–6.

Jäger, W., P. G. Peters-Wendisch, J. Kalinowski, and A. Pühler. 1996. A Corynebacterium glutamicum gene encoding a two-domain protein similar to biotin carboxylases and biotin-carboxyl-carrier proteins. Arch. Microbiol. 166:76–82.

Jakobi, M., C.-E. Ngouoto-Nkili, and A. Burkovski. 1999. Construction and application of new Corynebacterium glutamicum vectors. Biotechnol. Tech. 13:437–441.

Jetten, M. S. M., and A. J. Sinskey. 1993. Characterization of phosphoenolpyruvat carboxykinase from Corynebacterium glutamicum. FEMS Microbiol. Lett. 111:183–188.

Jetten, M. S. M., G. A. Pitoc, M. T. Follettie, and A. J. Sinskey. 1994. Regulation of phospho(enol)pyruvate- and oxaloacetate-converting enzymes in Corynebacterium glutamicum. Appl. Microbiol. Biotechnol. 41:47–52.

Jetten, M. S. M., M. T. Follettie, and A. J. Sinskey. 1995. Effect of different levels of aspartokinase on the lysine production of Corynebacterium lactofermentum. Appl. Microbiol. Biotechnol. 43:76–82.

Jones, D., and M. D. Collins. 1986. Irregular, nonsporing Gram-positive rods. *In:* P. H. A. Sneath, N. S. Mair, M. E. Sharpe, and J. G. Holt (Eds.) Bergey's Manual of Systematic Bacteriology. Williams and Wilkins. Baltimore, MD. 2:1261–1266.

Kaneko, H., and K. Sakaguchi. 1979. Fusion of protoplasts and genetic recombination of Brevibacterium flavum. Agric. Biol. Chem. 43:1007–1013.

Karasawa, M., O. Tosaka, S. Ikeda, and H. Yoshii. 1986. Application of protoplast fusion to the development of L-threonine and L-lysine producers. Agric. Biol. Chem. 50:339–346.

Kartmann, B., Stengler, S., and M. Niederweis. 1999. Porins in the cell wall of Mycobacterium tuberculosis. J. Bacteriol. 181:6543–6546.

Katsumata, R., A. Ozaki, T. Oka, and A. Furuya. 1984. Protoplast transformation of glutamate-producing bacteria with plasmid DNA. J. Bacteriol. 159:306–311.

Katsumata, R., and M. Ikeda. 1993. Hyperproduction of tryptophan in Corynebacterium glutamicum by pathway engineering. Bio/Technology 11:921–925.

Kawahara, Y., K. Takahasi-Fuke, E. Shimizu, T. Nakamatsu, and S. Nakamori. 1997. Relationship between the glutamate production and activity of the 2-oxoglutarate dehydrogenase in Brevibacterium lactofermentum. Biosci. Biotechnol. Biochem. 61:1109–1112.

Keddie, R. M., and G. L. Cure. 1977. The cell wall composition and distribution of free mycolic acids in named strains of coryneform bacteria and in isolates from various natural sources. J. Appl. Bacteriol. 42:229–252.

Keddie, R. M. 1978. What do we mean by coryneform bacteria? *In:* I. J. Bousfield and A. G. Callely (Eds.) Coryneform Bacteria. Academic Press. London, UK. 1–12.

Keddie, R. M., and D. Jones. 1981. Saprophytic, aerobic coryneform bacteria. *In:* M. P. Starr, H. Stolp, H. G. Trüper, A. Balows, and H. G. Schlegel (Eds.) The Prokaryotes. Springer-Verlag. New York, NY. II:1838–1878.

Kimura, M. 1963. The effect of biotin on the amino acid biosynthesis by Micrococcus glutamicus. J. Gen. Appl. Microbiol. 9:205–212.

Kimura, E., C. Abe, Y. Kawahara, and T. Nakamatsu. 1996. Molecular cloning of a novel gene, dtsR, which rescues the detergent sensitivity of a mutant derived from Brevibacterium lactofermentum. Biosci. Biotechnol. Biochem. 60:1565–1570.

Kimura, E., C. Abe, Y. Kawahara, T. Nakamatsu, and H. Tokuda. 1997. A dtsR gene-disrupted mutant of Brevibacterium lactofermentum requires fatty acids for growth and efficiently produces L-glutamate in the presence of an excess of biotin. Biochem. Biophys. Res. Commun. 234:157–161.

Kinoshita, S., K. Nakayama, and S. Kitada. 1958. L-lysine production using microbial auxotroph. J. Gen. Appl. Microbiol. 4:128–129.

Kinoshita, S., and K. Tanaka. 1972. Glutamic acid. *In:* K. Yamada, S. Kinoshita, T. Tsunoda, and K. Aida (Eds.) The Microbial Production of Amino Acids. Kodansha/John Wiley and Sons. Tokyo, Japan. New York, NY. 263–324.

Kinoshita, S., and K. Nakayama. 1978. Amino acids. *In:* A. H. Rose (Ed.) Primary Products of Metabolism. Academic Press. London, UK. 209–261.

Kinoshita, S. 1985. Glutamic acid bacteria. *In:* A. L. Demain, and N. A. Solomon (Eds.) Biology of Industrial Microorganisms. Benjamin/Cummins. London, UK. 115–142.

Koizumi, S., T. Endo, K. Tabata, and A. Ozaki. 1998. Large-scale production of UDP-galactose and globotriose by coupling metabolically engineered bacteria. Nature Biotechnology 16:847–850.

Koizumi, S., Y. Yonetani, A. Maruyama, and S. Teshiba. 2000. Production of riboflavin by metabolically engineered Corynebacterium ammoniagenes. Appl. Microbiol. Biotechnol. 53:674–679.

Komata, Y. 1976. Utilization in foods. *In:* K. Ogata, S. Kinoshita, T. Tsunoda, and K. Aida (Eds.) Microbial Production of Nucleic Acid-related Substances. Kodansha/John Wiley and Sons. Tokyo, Japan. New York, NY. 299–319.

Kotani, Y., K. Yamaguchi, F. Kato, and A. Furuya. 1978. Inosine accumulation by mutants of Brevibacterium ammoniagenes. Strain improvement and culture conditions. Agric. Biol. Chem. 42:399–405.

Krämer, R. 1994a. Secretion of amino acids by bacteria: physiology and mechanism. FEMS Microbiol. Rev. 13:75–94.

Krämer, R. 1994b. Systems and mechanisms of amino acid uptake and excretion in prokaryotes. Arch. Microbiol. 162:1–13.

Kronemeyer, W., N. Peekhaus, R. Krämer, H. Sahm, and L. Eggeling. 1995. Structure of the gluABCD cluster encoding the glutamate uptake system of Corynebacterium glutamicum. J. Bacteriol. 177:1152–1158.

Lanéelle, M. A., J. Asselineau, M. Welby, M. V. Norgard, T. Imaeda, M. C. Pollice, and L. Barksdale. 1980. Biological and chemical basis for the reclassification of Brevibacterium vitarumen (Bechdel et al.) Breed (Approved lists nov. and as Brevibacterium liquefaciens Okabayashi and Musuo (Approved lists nov. and as Corynebacterium vitarumen (Bechdel et al. 1980) as Corynebacterium liquefaciens (Okabayashi and Masuo) comb. J. Syst. Bacteriol. 30:539–546.

Lee, W. H., and R. C. Good. 1963. Amino acid synthesis. US Patent. 3087863.

Lehmann, K. B., and R. Neumann. 1896. Atlas und Grundriss der Bakteriologie und Lehrbuch der speciellen bakteriologischen Diagnostik, 1st ed.. Lehmann. Munich, Germany.

Leuchtenberger, W. 1996. Amino acids—technical production and use. *In:* H.-J. Rehm, G. Reed, A. Pühler, and P. Stadler (Eds.) Biotechnology, 2nd ed.. Wiley/VCH. Weinheim, Germany. 6:465–502.

Lichtinger, T., A. Burkovski, M. Niederweis, R. Krämer, and R. Benz. 1998. Biochemical and biophysical characterization of the cell wall porin of Corynebacterium glutamicum: The channel is formed by a low molecular mass polypeptide. Biochemistry 37:15024–15032.

Lichtinger, T., B. Heym, E. Maier, H. Eichner, S. T. Cole, and R. Benz. 1999. Evidence for a small anion-selective channel in the cell wall of Mycobacterium bovis BCG besides a wide cation-selective pore. FEBS Lett. 454:349–355.

Liebl, W., and A. J. Sinskey. 1988. Molecular cloning and nucleotide sequence of a gene involved in the production of extracellular DNase by Corynebacterium glutamicum. *In:* A. T. Ganesan and J. A. Hoch (Eds.) Genetics and Biotechnology of Bacilli. Academic Press. San Diego, CA. 2:383–388.

Liebl, W., A. Bayerl, B. Schein, U. Stillner, and K.-H. Schleifer. 1989a. High efficiency electroporation of intact Corynebacterium glutamicum cells. FEMS Microbiol. Lett. 65:299–304.

Liebl, W., R. Klamer, and K.-H. Schleifer. 1989b. Requirement of chelating compounds for the growth of Corynebacterium glutamicum in synthetic media. Appl. Microbiol. Biotechnol. 32:205–210.

Liebl, W., K.-H. Schleifer, and A. J. Sinskey. 1989c. Secretion of heterologous proteins by Corynebacterium glutamicum. *In:* L. O. Butler, C. Harwood, and B. E. B. Moseley (Eds.) Genetic Transformation and Expression. Intercept. Andover, UK. 553–559.

Liebl, W., and B. Schein. 1990. Isolation of restriction deficient mutants of Corynebacterium glutamicum. *In:* D. Behrens and P. Krämer (Eds.) Dechema Biotechnology Conferences VCH. Weinheim, Germany. 4:323–327.

Liebl, W., M. Ehrmann, W. Ludwig, and K. H. Schleifer. 1991. Transfer of Brevibacterium divaricatum DSM 20297T, "Brevibacterium flavum" DSM 20411, "Brevibacterium lactofermentum" DSM 20412 and DSM 1412, and Corynebacterium lilium DSM 20137T to Corynebacterium glutamicum and their distinction by rRNA gene restriction patterns. Int. J. Syst. Bacteriol. 41:255–260.

Liebl, W. 1992a. Corynebacterium: Non-medical. *In:* A. Balows, H. G. Trüper, M. Dworkin, W. Harder, and K.-H. Schleifer (Eds.) The Prokaryotes, 2nd ed. Springer-Verlag. New York, NY. II:1157–1171.

Liebl, W., A. J. Sinskey, and K. H. Schleifer. 1992b. Expression, secretion, and processing of staphylococcal nuclease by Corynebacterium glutamicum. J. Bacteriol. 174:1854–1861.

Loos, A., C. Glanemann, L. B. Willis, X. M. O'Brien, P. A. Lessard, R. Gerstmeier, S. Guillouet, and A. J. Sinskey. 2001. Development and validation of Corynebacterium gutamicum DNA microarrays. Appl. Environ. Microbiol. 67:2310–2318.

Marauska, D. F., M. P. Ruklish, and N. I. Galynina. 1981. Energy-dependence of glucose transport in Brevibacterium flavum. Mikrobiologiya 50:763–768.

Martin, J. F., R. Santamaria, H. Sandoval, G. del Real, L. M. Mateos, J. A. Gil, and A. Aguilar. 1987. Cloning systems in amino acid-producing corynebacteria. Bio/Technology 5:137–146.

Martin, J. F. 1989. Molecular genetics of amino acid-producing corynebacteria. *In:* S. Banmberg, I. Hunter, and M. Rhodes (Eds.) Society for General Microbiology Symposium. Cambridge University Press. Cambridge, UK. 44:25–59.

Marx, A., A. A. de Graaf, W. Wiechert, L. Eggeling, and H. Sahm. 1996. Determination of the fluxes in the central metabolism of Corynebacterium glutamicum by nuclear magnetic resonance spectroscopy combined with metabolite balancing. Biotechnol. Bioeng. 49:111–129.

Mateos, L. M., A. Schäfer, J. Kalinowski, J. F. Martin, and A. Pühler. 1996. Integration of narrow-host-range vectors from Escherichia coli into the genomes of amino acid-

producing corynebacteria after intergeneric conjugation. J. Bacteriol. 178:5768–5775.

McLaughlin, J. R., C. L. Murray, and J. C. Rabinowitz. 1981. Unique features in the ribosome binding site sequence of the Gram-positive Staphylococcus aureus α-lactamase gene. J. Biol. Chem. 256:11283–11291.

Minnikin, D. E., L. Alshamaony, and M. Goodfellow. 1975. Differentiation of Mycobacterium, Nocardia and related taxa by thin-layer chromatographic analysis of whole cell methanolysates. J. Gen. Microbiol. 88:200–204.

Minnikin, D. E., M. Goodfellow, and M. D. Collins. 1978. Lipid composition in the classification and identification of coryneform and related taxa. *In:* I. J. Bousfield and A. G. Callely (Eds.) Coryneform Bacteria. Academic Press. London, UK. 85–160.

Minnikin, D. E., and A. G. O'Donnell. 1984. Actinomycete envelope lipid and peptidoglycan composition. *In:* M. Goodfellow, M. Mordarski, and S. T. Williams (Eds.) The Biology of Actinomycetes. Academic Press. London, UK. 337–388.

Miwa, K., H. Matsui, M. Terabe, S. Nakamori, K. Sano, and H. Momose. 1984. Cryptic plasmids in glutamic acid producing bacteria. Agric. Biol. Chem. 48:2901–2903.

Miwa, K., H. Matsui, M. Terabe, K. Ito, K. Ishida, H. Takagi, S. Nakamori, and K. Sano. 1985. Construction of novel shuttle vectors and a cosmid vector for the glutamic acid-producing bacteria Brevibacterium lactofermentum and Corynebacterium glutamicum. Gene 39:281–286.

Momose, H., S. Miyashiro, and M. Oba. 1976. On the transducing phages in glutamic acid producing bacteria. J. Gen. Appl. Microbiol. 22:119–129.

Moran Jr., C. P., N. Lang, S. F. J. LeGrice, G. Lee, M. Stephens, A. L. Sonnenshein, J. Pero, and R. Losick. 1982. Nucleotide sequences that signal the initiation of transcription and translation in Bacillus subtilis. Molec. Gen. Genet. 186:339–346.

Mori, M., and I. Shiio. 1985a. Purification and some properties of phosphoenolpyruvate carboxylase from Brevibacterium flavum and its aspartate-overproducing mutant. J. Biochem. 97:1119–1128.

Mori, M., and I. Shiio. 1985b. Synergistic inhibition of phosphoenolpyruvate carboxylase by aspartate and 2-oxoglutarate in Brevibacterium flavum. J. Biochem. 98:1621–1630.

Mori, M., and I. Shiio. 1987a. Phosphoenolpyruvate sugar phosphotransferase systems and sugar metabolism in Brevibacterium flavum. Agric. Biol. Chem. 51:2671–2678.

Mori, M., and I. Shiio. 1987b. Pyruvate formation and sugar metabolism in an amino acid-producing bacterium, Brevibacterium flavum. Agric. Biol. Chem. 51:129–138.

Mori, H., A. Iida, T. Fujio, and S. Teshiba. 1997. A novel process of inosine 5'-monophosphate production using overexpressed guanosine/inosine kinase. Appl. Microbiol. Biotechnol. 48:693–698.

Morinaga, Y., M. Tsuchiya, K. Miwa, and K. Sano. 1987. Expression of Escherichia coli promoters in Brevibacterium lactofermentum using the shuttle vector pEB003. J. Biotechnol. 5:305–312.

Mukhopadhyay, S., D. Basu, and P. Chakrabarti. 1997. Characterization of a porin from Mycobacterium smegmatis. J. Bacteriol. 179:6205–6207.

Müller, G. 1961. Mikrobiologische Untersuchungen über die "Futterverpilzung durch Selbsterhitzung III". Mitteilung: Ausführliche Beschreibung neuer Bakterienspecies. Zentralbl. Bakteriol. Parasitenkd. Infektionskr. Hyg. Abt. 2, Orig. Reihe A 114:520–537.

Nakayama, K., H. Tanaka, H. Hagino, and S. Kinoshita. 1966. Studies on lysine fermentation. V: Concerted feedback inhibition of aspartokinase and the absence of lysine inhibitionon aspartic semialdehyde-pyruvate condensation in Micrococcus glutamicus. Agric. Biol. Chem. 30:611–616.

Nakayama, K., Z. Sato, H. Tanaka, and S. Kinoshita. 1968. Production of NAD and nicotinic acid mononucleotide with Brevibacterium ammoniagenes. Agric. Biol. Chem. 32:1331–1336.

Nara, T., H. Samejima, C. Fujita, M. Ito, K. Nakayama, and S. Kinoshita. 1961. L-Homoserine fermentation. VI: Effect of threonine and methionine on L-homoserine in dehydrogenase in Micrococcus glutamicus 534-Col47. Agric. Biol. Chem. 25:532–541.

Nara, T., M. Misawa, and S. Kinoshita. 1967. Production of 5'-Inosinic acid by an adenine auxotroph of Brevibacterium ammoniagenes. Agric. Biol. Chem. 31:1351–1356.

Nara, T., M. Misawa, and S. Kinoshita. 1968a. Fermentative production of 5'-purine ribonucleotides by Brevibacterium ammoniagenes. Agric. Biol. Chem. 32:561–567.

Nara, T., M. Misawa, and S. Kinoshita. 1968b. Pantothenate, thiamine and manganese in 5'-purine ribonucleotide production by Brevibacterium ammoniagenes. Agric. Biol. Chem. 32:1153–1161.

Nara, T., T. Komuro, M. Misawa, and S. Kinoshita. 1969a. Growth responses of Brevibacterium ammoniagenes. Agric. Biol. Chem. 33:1030–1036.

Nara, T., M. Misawa, T. Komuro, and S. Kinoshita. 1969b. Effect of antibiotics and surface-active agents on 5'-purinenucleotide production by Brevibacterium ammoniagenes. Agric. Biol. Chem. 33:1198–1204.

Nath, N., and J. K. Deb. 1995. Partial characterization of small plasmids of Corynebacterium renale. Plasmid 34:229–233.

Nesvera, J., J. Hochmannova, and M. Patek. 1998. An integron of class 1 is present on the plasmid pCG4 from Gram-positive bacterium Corynebacterium glutamicum. FEMS Microbiol. Lett. 169:391–395.

Niederweis, M., E. Maier, T. Lichtinger, R. Benz, and R. Krämer. 1995. Identification of channel-forming activity in the cell wall of Corynebacterium glutamicum. J. Bacteriol. 177:5716–5718.

Oki, T., and K. Ogata. 1968a. Size and morphology of Brevibacterium phages. Agric. Biol. Chem. 32:241–248.

Oki, T., Y. Sayama, Y. Nishimura, and A. Ozaki. 1968b. L-Glutamic acid formation by microorganisms from ethanol. Agric. Biol. Chem. 32:119–120.

Ozaki, H., and I. Shiio. 1969. Regulation of the TCA and glyoxylate cycles in Brevibacterium flavum. J. Biochem. 66:297–311.

Ozaki, A., R. Katsumata, T. Oka, and A. Furuya. 1984. Functional expression of the genes of Escherichia coli in gram-positive Corynebacterium glutamicum. Molec. Gen. Genet 196:175–178.

Paradis, F. W., R. A. I. Warren, D. G. Kilburn, and R. C. Miller Jr. 1987. The expression of Cellulomonas fimi cellulase genes in Brevibacterium lactofermentum. Gene 61:199–206.

Park, S. M., C. Shaw-Reid, A. J. Sinskey, and G. Stephanopoulos. 1997. Elucidation of anaplerotic pathways in Corynebacterium glutamicum via 13C-NMR spectros-

copy and GC-MS. Appl. Microbiol. Biotechnol. 47:430–440.

Pascual, C., P. A. Lawson, J. A. E. Farrow, M. N. Gimenez, and M. D. Collins. 1995. Phylogenetic analysis of the genus Corynebacterium based on 16S rRNA gene sequences. Int. J. Syst. Bacteriol. 45:724–728.

Patek, M., J. Ludvik, O. Benada, J. Hochmannova, J. Nesvera, V. Krumphanzl, and M. Bucko. 1985. New bacteriophage-like particles in Corynebacterium glutamicum. Virology 140:360–363.

Patek, M., O. Navratil, J. Hochmannova, J. Nesvera, and J. Hubacek. 1989. Expression of the threonine operon from Escherichia coli in Brevibacterium flavum and Corynebacterium glutamicum. Biotechnol. Lett. 11:231–236.

Patek, M., B. Eikmanns, J. Patek, and H. Sahm. 1996. Promoters from Corynebacterium glutamicum: Cloning, molecular analysis and search for a consensus motif. Microbiology 142:1297–1309.

Peter, H., B. Weil, A. Burkovski, R. Krämer, and S. Morbach. 1998. Corynebacterium glutamicum is equipped with four secondary carriers for compatible solutes: Identification, sequencing, and characterization of the proline/ectoine uptake system, ProP, and the ectoine/proline/glycine betaine carrier, EctP. J. Bacteriol. 180:6005–6012.

Peters-Wendisch, P. G., B. J. Eikmanns, G. Thierbach, B. Bachmann, and H. Sahm. 1993. Phosphoenolpyruvate carboxylase in Corynebacterium glutamicum is dispensable for growth and lysine production. FEMS Microbiol. Lett. 112:269–274.

Peters-Wendisch, P. G., V. F. Wendisch, A. A. de Graaf, B. J. Eikmanns, and H. Sahm. 1996. C_3-Carboxylation as an anaplerotic reaction in phosphoenolpyruvate carboxylase-deficient Corynebacterium glutamicum. Arch. Microbiol. 165:387–396.

Peters-Wendisch, P. G., V. F. Wendisch, S. Paul, B. J. Eikmanns, and H. Sahm. 1997. Pyruvate carboxylase as an anaplerotic enzyme in Corynebacterium glutamicum. Microbiology 143:1095–1103.

Peters-Wendisch, P. G., C. Kreutzer, J. Kalinowski, M. Patek, H. Sahm, and B. J. Eikmanns. 1998. Pyruvate carboxylase from Corynebacterium glutamicum: Characterization, expression and inactivation of the pyc gene. Microbiology 144:915–927.

Peyret, J.-L., N. Bayan, G. Joliff, T. Gulik-Krzywicki, L. Mathieu, E. Schechter, and G. Leblon. 1993. Characterization of the cspB gene encoding PS2, an ordered surface-layer protein in Corynebacterium glutamicum. Molec. Microbiol. 9:97–109.

Puech, V., N. Bayan, K. Salim, G. Leblon, and M. Daffe. 2000. Characterization of the in vivo acceptors of the mycoloyl residues transferred by the corynebacterial PS1 and the related mycobacterial antigens 85. Molec. Microbiol. 35:1026–1041.

Puech, V., M. Chami, A. Lemassu, M.-A. Laneelle, B. Schiffler, P. Gounon, N. Bayan, R. Benz, and M. Daffe. 2001. Structure of the cell envelope of corynebacteria: Importance of the non-covalently bound lipids in the formation of the cell wall permeability barrier and fracture plane. Microbiology 147:1365–1382.

Quast, K., B. Bathe, A. Pühler, and J. Kalinowski. 1999. The Corynebacterium glutamicum insertion sequence ISCg2 prefers conserved target sequences located adjacent to genes involved in aspartate and glutamate metabolism. Molec. Gen. Genet. 262:568–578.

Reinscheid, D. J., B. J. Eikmanns, and H. Sahm. 1994a. Characterization of the isocitrate lyase gene from Corynebacterium glutamicum and biochemical analysis of the enzyme. J. Bacteriol. 176:3474–3483.

Reinscheid, D. J., B. J. Eikmanns, and H. Sahm. 1994b. Malate synthase from Corynebacterium glutamicum: Sequence analysis of the egene and biochemical characterization of the enzyme. Microbiology 140:3099–3108.

Reinscheid, D. J., S. Schnicke, D. Rittmann, U. Zahnow, H. Sahm, and B. J. Eikmanns. 1999. Cloning, sequence analysis, expression and inactivation of the Corynebacterium glutamicum pta-ack operon encoding phosphotransacetylase and acetate kinase. Microbiology 145:503–513.

Reyes, O., A. Guyonvarch, C. Bonamy, V. Salti, F. David, and G. Leblon. 1991. "Integron"-bearing vectors: A method suitable for stable chromosomal integration in highly restrictive corynebacteria. Gene 107:61–68.

Rieß, F. G., T. Lichtinger, R. Cseh, A. F. Yassin, K. P. Schaal, and R. Benz. 1999. The cell wall porin of Nocardia farcinica: Biochemical identification of the channel-forming protein and biophysical characterization of the channel properties. Molec. Microbiol. 29:139–150.

Rogosa, M., C. S. Cummins, R. A. Lelliott, and R. M. Keddie. 1974. Coryneform group of bacteria. In: R. E. Buchanan and N. E. Gibbons (Eds.) Bergey's Manual of Determinative Bacteriology, 8th ed. Williams and Wilkins. Baltimore, MD. 599–632.

Ruffert, S., C. Lambert, H. Peter, V. F. Wendisch, and R. Krämer. 1997. Efflux of compatible solutes in Corynebacterium glutamicum mediated by osmoregulated channel activity. Eur. J. Biochem. 247:572–580.

Ruffert, S., C. Berrier, R. Krämer, and A. Ghazi. 1999. Identification of mechanosensitive ion channels in the cytoplasmic membrane of Corynebacterium glutamicum. J. Bacteriol. 181:1673–1676.

Ruimy, R., P. Riegel, P. Boiron, V. Boivin, H. Monteil, and R. Christen. 1995. A phylogeny of the genus Corynebacterium deduced from analysis of small-subunit ribosomal DNA sequences. Int. J. Syst. Bacteriol. 45:740–746.

Sahm, H., L. Eggeling, B. Eikmanns, and R. Krämer. 1995. Metabolic design in amino acid producing bacterium Corynebacterium glutamicum. FEMS Microbiol. Rev. 16:243–252.

Sanchez, F., M. A. Penalva, C. Patino, and V. Rubio. 1986. An efficient method for the introduction of viral DNA into Brevibacterium lactofermentum protoplasts. J. Gen. Microbiol. 132:1767–1770.

Sandoval, H., A. Aguilar, D. Paniagua, and J. F. Martin. 1984. Isolation and physical characterization of plasmid pCC1 from Corynebacterium callunae and construction of hybrid derivatives. Appl. Microbiol. Biotechnol. 19:409–413.

Sandoval, H., G. Del Real, L. M. Mateos, A. Aguilar, and J. F. Martin. 1985. Screening of plasmids in non-pathogenic corynebacteria. FEMS Microbiol. Lett. 27:93–98.

Sano, K., and I. Shiio. 1970. Microbial production of L-lysine III. Production by mutants resistant to S-(2-aminoethyl)-L-cysteine. J. Gen. Appl. Microbiol. 16:373–391.

Sano, K., K. Ito, K. Miwa, and S. Nakamori. 1987. Amplification of the phosphoenol pyruvate carboxylase gene of Brevibacterium lactofermentum to improve amino acid production. Agric. Biol. Chem. 51:597–599.

Santamaria, R., J. A. Gil, J. M. Mesas, and J. F. Martin. 1984. Characterization of an endogenous plasmid and development of cloning vectors and a transformation system

in Brevibacterium lactofermentum. J. Gen. Microbiol. 130:2237–2246.

Santamaria, R. I., J. F. Martin, and J. A. Gil. 1987. Identification of a promoter sequence in the plasmid pUL340 of Brevibacteriium lactofermentum and construction of new cloning vectors for corynebacteria containing two selectable markers. Gene 56:199–208.

Schäfer, A., J. Kalinowski, R. Simon, A. H. Seep-Feldhaus, and A. Pühler. 1990. High-frequency conjugal transfer from Gram-negative Escherichia coli to various Gram-positive coryneform bacteria. J. Bacteriol. 172:1663–1666.

Schäfer, A., A. Schwarzer, J. Kalinowski, and A. Pühler. 1994a. Cloning and characterization of a DNA region encoding a stress-sensitive restriction system from Corynebacterium glutamicum ATCC 13032 and analysis of its role in intergeneric conjugation with Escherichia coli. J. Bacteriol. 176:7309–7319.

Schäfer, A., A. Tauch, W. Jäger, J. Kalinowski, G. Thierbach, and A. Pühler. 1994b. Small mobilizable multi-purpose cloning vectors derived from the Escherichia coli plasmids pK18 and pK19: Selection of defined deletions in the chromosome of Corynebacterium glutamicum. Gene 145:69–73.

Schleifer, K. H., and O. Kandler. 1972. Peptidoglycan types of bacterial cell walls and their taxonomic implications. Bacteriol. Rev. 35:407–477.

Schrumpf, B., A. Schwarzer, J. Kalinowski, A. Pühler, L. Eggeling, and H. Sahm. 1991. A functionally split pathway for lysine synthesis in Corynebacterium glutamicum. J. Bacteriol. 173:4510–4516.

Schrumpf, B., L. Eggeling, and H. Sahm. 1992. Isolation and prominent characteristics of an L-lysine hyperproducing strain of Corynebacterium glutamicum. Appl. Microbiol. Biotechnol. 37:566–571.

Schwarzer, A., and A. Pühler. 1991. Manipulation of Corynebacterium glutamicum by gene disruption and replacement. Bio/Technology 9:84–87.

Seiler, H. 1983a. Identification key for coryneform bacteria derived by numerical taxonomic studies. J. Gen. Microbiol. 129:1433–1471.

Seiler, H., and W. Hennlich. 1983b. Characterization of coryneform bacteria in piggery wastes. Syst. Appl. Microbiol. 4:132–140.

Seiler, H. 1986. Identification of cheese-smear coryneform bacteria. J. Dairy Res. 53:439–449.

Serwold-Davis, T. M., N. Groman, and M. Rabin. 1987. Transformation of Corynebacterium diphtheriae, Corynebacterium ulcerans, Corynebacterium glutamicum and Escherichia coli with C. diphtheria plasmid pNG2. Proc. Natl. Acad. Sci. USA 84:4964–4968.

Shiio, I., S. I. Otsuka, and T. Tsunoda. 1959. Glutamic acid formation from glucose by bacteria. I: Enzymes of the Embden-Meyerhof-Parnas pathway, the Krebs-cycle, and the glyoxylate bypass in cell extracts of Brevibacterium flavum No. 2247. J. Biochem. 46: 1303–1311.

Shiio, I., S. I. Otsuka, and M. Takahashi. 1961. Significance of a-ketoglutaric dehydrogenase on the glutamic acid formation in Brevibacterium flavum. J. Biochem. 51:164–165.

Shiio, I., S. I. Otsuka, and N. Katsuya. 1963. Cellular permeability and extracellular formation of glutamic acid in Brevibacterium flavum. J. Biochem. 53:333–340.

Shiio, I., and K. Sano. 1969. Microbial production of L-lysine. II: Production by mutants sensitive to threonine or methionine. J. Gen. Appl. Microbiol. 15:267–287.

Shiio, I., and K. Ujigawa-Takeda. 1980. Presence and regulation of a-ketoglutarate dehydrogenase complex in a glutamate-producing bacterium, Brevibacterium flavum. Agric. Biol. Chem. 44:1897–1904.

Skerman, V. B. D., V. McGowan, and P. H. A. Sneath. 1980. Approved lists of bacterial names. Int. J. Syst. Bacteriol. 30:225–420.

Smith, M. D., J. L. Flickinger, D. W. Lineberger, and B. Schmidt. 1986. Protoplast transformation in coryneform bacteria and introduction of an a-amylase gene from Bacillus amyloliquefaciens into Brevibacterium lactofermentum. Appl. Environ. Microbiol. 51:634–639.

Sonnen, H., J. Schneider, and H. J. Kutzner. 1990. Characterization of GA1, an inducible phage particle from Brevibacterium flavum. J. Gen. Microbiol. 136:567–571.

Sonnen, H., G. Thierbach, S. Kautz, J. Kalinowski, J. Schneider, A. Pühler, and H. Kutzner. 1991. Characterization of pGA1, a new plasmid from Corynebacterium glutamicum LP-6. Gene 107:69–74.

Sonntag, K., L. Eggeling, A. A. De Graaf, and H. Sahm. 1993. Flux partitioning in the split pathway of lysine synthesis in Corynebacterium glutamicum: Quantification by 13C- and 1H-NMR spectroscopy. Eur. J. Biochem. 213:1325–1331.

Soual-Hoebeke, E., C. de Sousa-D'Auria, M. Chami, M.-F. Baucher, A. Guyonvarch, N. Bayan, K. Salim, and G. Leblon. 1999. S-layer protein production by Corynebacterium strains is dependent on the carbon source. Microbiology 145:3399–3408.

Stackebrandt, E., and C. R. Woese. 1981. The evolution of prokaryotes. *In:* M. J. Carlile, J. F. Collins, and B. E. B. Moseley (Eds.) Molecular and Cellular Aspects of Microbial Evolution. Cambridge University Press. Cambridge, UK. 1–31.

Stackebrandt, E., F. A. Rainey, and N. L. Ward-Rainey. 1997. Proposal for a new hierarchal classification system, Actinobacteria classis nov. Int. J. Syst. Bacteriol. 47:479–491.

Stanek, J. L., and G. D. Roberts. 1974. Simplified approach to identification of aerobic actinomycetes by thin-layer chromatography. Appl. Microbiol. 28:226–231.

Sutcliffe, I. C. 1995. Identification of a lipoarabinomannan-like lipoglycan in Corynebacterium matruchotii. Arch. Oral Biol. 40:1119–1124.

Suzuki, K., T. Kaneko, and K. Komagata. 1981. Deoxyribonucleic acid homologies among coryneform bacteria. Int. J. Syst. Bacteriol. 31:131–138.

Tabata, K., S. Koizumi, T. Endo, and A. Ozaki. 2000. Production of UDP-N-acetyglucosamine by coupling metabolically engineered bacteria. Biotechnol. Lett. 22:479–483.

Takeda, Y., M. Fuji, I. Nakajyo, T. Nishimura, and S. Issiki. 1990. Isolation of a tetracycline resistance plasmid from a glutamate producing corynebacterium, Corynebacterium melassecola. J. Ferment. Bioeng. 70:177–179.

Takeuchi, M., T. Sakane, T. Nihira, Y. Yamada, and K. Imai. 1999. Corynebacterium terpenotabidum sp. nov., a bacterium capable of degrading squalene. Int. J. Syst. Bacteriol. 49:223–229.

Takinami, K., H. Yoshii, Y. Yamada, H. Okada, and K. Kinoshita. 1968. Control of L-glutamic acid fermentation by biotin and fatty acid. Amino Acid Nucl. Acid 18:120–160.

Tanaka, H., Z. Sato, K. Nakayama, and S. Kinoshita. 1968. Formation of ATP, GTP, and their related substances by Brevibacterium ammoniagenes. Agric. Biol. Chem. 32:721–726.

Tauch, A., F. Kassing, J. Kalinowski, and A. Pühler. 1995. The erythromycin resistance gene of Corynebacterium xerosis R-plasmid p-TP10 also carrying chloramphenicol, kanamycin, and tetracycline resistance is capable of transposition in Corynebacterium glutamicum. Plasmid 33:168–179.

Tauch, A., Z. Zheng, A. Pühler, and J. Kalinowski. 1998. Corynebacterium stratium chloramphenicol resistance transposon Tn5564: Genetic organization and transposition in Corynebacterium glutamicum. Plasmid 40:126–139.

Teshiba, S., and A. Furuya. 1982. Mechanisms of 5'-inosinic acid accumulation by permeability mutants of Brevibacterium ammoniagenes I. Genetical improvement of 5'-IMP productivity of a permeability mutant of B. ammoniagenes. Agric. Biol. Chem. 46:2257–2263.

Teshiba, S., and A. Furuya. 1984. Mechanisms of 5'-inosinic acid accumulation by permeability mutants of Brevibacterium ammoniagenes IV. Excretion mechanisms of 5'-IMP. Agric. Biol. Chem. 48:1311–1317.

Trautwetter, A., C. Blanco, and A. M. Sicard. 1987. Structural characteristics of the Corynebacterium lilium bacteriophage CL$_{31}$. J. Virol. 61:1540–1545.

Usuda, Y., N. Tujimoto, C. Abe, Y. Asakura, E. Kimura, Y. Kawahara, O. Kurahashi, and H. Matsui. 1996. Molecular cloning of the Corynebacterium glutamicum (Brevibacterium lactofermentum AJ12036) odhA gene encoding a novel type of 2-oxoglutarate dehydrogenase. Microbiology 142:3347–3354.

Valdes-Stauber, N., S. Scherer, and H. Seiler. 1997. Identification of yeast and coryneform bacteria from the surface microflora of brick cheese. Int. J. Food. Microbiol. 34:115–129.

Vallino, J. J., and G. Stephanopoulos. 1993. Metabolic flux distributions in Corynebacterium glutamicum during growth and lysine overproduction. Biotechnol. Bioeng. 41:633–646.

van der Rest, M. E., C. Lange, and D. Molenaar. 1999. A heat shock following electroporation induces highly efficient transformation of Corynebacterium glutamicum with xenogeneic plasmid DNA. Appl. Microbiol. Biotechnol. 52:541–545.

Vertes, A. A., Harakeyama, K., Inui, M., Kobayashi, M., Kurusu, Y., and H. Yukawa. 1993a. Presence of mrr- and mcr-like restriction systems in coryneform bacteria. Res. Microbiol. 144:181–185.

Vertes, A. A., K. Harakeyama, M. Inui, M. Kobayashi, Y. Kurusu, and H. Yukawa. 1993b. Replacement recombination in coryneform bacteria: High efficiency integration requirement for non-methylated plasmid DNA. Biosci. Biotech. Biochem. 57:2036–2038.

Vertes, A. A., M. Inui, M. Kobayashi, Y. Kurusu, and H. Yukawa. 1993c. Presence of mrr- and mcr-like restriction systems in coryneform bacteria. Res. Microbiol. 144:181–185.

Vertes, A. A., M. Inui, M. Kobayashi, Y. Kurusu, and H. Yukawa. 1994. Isolation and characterization of IS31831, a transposable element from Corynebacterium glutamicum. Molec. Microbiol. 11:739–746.

Von der Osten, C. H., C. Gioannetti, and A. J. Sinskey. 1989. Design of a defined medium for growth of Corynebacterium glutamicum in which citrate facilitates iron uptake. Biotechnol. Lett. 11:11–16.

Vrljic, M., H. Sahm, and L. Eggeling. 1996. A new type of transporter with a new type of cellular function: L-lysine export in Corynebacterium glutamicum. Molec. Microbiol. 22:815–826.

Wittmann, C., and E. Heinzle. 2001. Application of MALDITOF MS to lysine-producing Corynebacterium glutamicum. A novel approach for metabolic flux analysis. Eur. J. Biochem. 268:2441–2455.

Yamada, K., and K. Komagata. 1972a. Taxonomic studies on coryneform bacteria: Classification of coryneform bacteria. J. Gen. Appl. Microbiol. 18:417–431.

Yamada, K., and K. Komagata. 1972b. Taxonomic studies on coryneform bacteria. IV: Morphological, cultural, biochemical, and physiological characteristics. J. Gen. Appl. Microbiol. 18:399–416.

Yeh, P., J. Oreglia, and A. M. Sicard. 1985. Transfection of Corynebacterium lilium protoplasts. J. Gen. Microbiol. 131:3179–3183.

Yeh, P., A. M. Sicard, and A. J. Sinskey. 1988. General organization of the genes specifically involved in the diaminopimelate-lysine biosynthetic pathway of Corynebacterium glutamicum. Molec. Gen. Genet. 212:105–111.

Yoshihama, M., K. Higashiro, E. A. Rao, M. Akedo, W. G. Shanabruch, M. T. Follettie, G. C. Walker, and A. J. Sinskey. 1985. Cloning vector system for Corynebacterium glutamicum. J. Bacteriol. 162:591–597.

Prokaryotes (2006) 3:819–842
DOI: 10.1007/0-387-30743-5_31

CHAPTER 1.1.16

The Genus *Corynebacterium*—Medical

ALEXANDER VON GRAEVENITZ AND KATHRYN BERNARD

Phylogeny and Taxonomy

The genus *Corynebacterium* is a member of the class Actinobacteria (high G+C Gram-positive bacilli), subclass Actinobacteridae, order Actinomycetales, suborder Corynebacterineae and family Corynebacteriaceae, which also includes the genus *Turicella* (Stackebrandt et al., 1997). This designation is based on extensive 16S rRNA gene sequencing data which demonstrated that the genus *Corynebacterium* is most closely related to non-acid-fast, partially acid-fast or acid-fast genera of the suborder Corynebacterineae, which also includes the families (genus/genera) Dietziaceae (*Dietzia*), Gordoniaceae (*Gordonia* and *Skermania*), Mycobacteriaceae (*Mycobacterium*), Nocardiaceae (*Nocardia* and *Rhodococcus*), Tsukamurellaceae (*Tsukamurella*) and Williamsiaceae (*Williamsia*) (Funke et al., 1994; Pascual et al., 1995; Ruimy et al., 1995; Chun et al., 1997; Stackebrandt et al., 1997; Kämpfer et al., 1999).

Currently, the genus *Corynebacterium* consists of 59 validly described species, of which 35 species as well as two taxon groups are considered to be medically relevant and will be discussed in this chapter. These include *C. accolens, C. afermentans, C. amycolatum, C. appendicis, C. argentoratense, C. aurimucosum, C. auris, C. confusum, C. coyleae, C. diphtheriae, C. durum, C. falsenii, C. freneyi, C. glucuronolyticum, C. imitans, C. jeikeium, C. kroppenstedtii, C. lipophiloflavum, C. macginleyi, C. matruchotii, C. minutissimum, C. mucifaciens, C. mycetoides, C. propinquum, C. pseudodiphtheriticum, C. pseudotuberculosis, C. riegelii, C. simulans, C. singulare, C. striatum, C. sundsvallense, C. thomssenii, C. ulcerans, C. urealyticum* and *C. xerosis,* as well as Coryneform CDC groups G and F-1. A "black-pigmented *Corynebacterium* species" associated with human disease, but not as yet validated, is also described briefly here.

Valid *Corynebacterium* species not known to be isolated from diseased humans and most commonly recovered from soil, animals, the environment or foods are described further in another chapter (Corynebacterium–Nonmedical in this Volume). The following 12 mostly nonhuman taxa have an internet or literature presence, but have not been validated: *C. acetoacidophilum, C. cervicis, C. crenatum, C. fastidiosum, C. genitalium, C. melassecola, C. nephridii, C. nigricans* (the "black-pigmented *Corynebacterium*"), *C. pseudogenitalium, C. segmentosum, C. thermoaminogenes* and *C. tuberculostearicum.*

The core description of the genus *Corynebacterium* has been emended substantively over the years, on the basis of comprehensive 16S rRNA gene sequence analysis for taxa with specific chemotaxonomic, morphological and phenotypic traits. These traits include: 1) the G+C content, which varies from 46 mol% (*C. kutscheri*) to 74 mol% (*C. auris*) (Funke et al., 1995b), but generally ranges from 51 to 65 mol%, indicating wide diversity within this genus (Collins and Cummins, 1986); 2) the cell wall, which contains *meso*-diaminopimelic acid (*m*-DAPA) as the diamino acid; and 3) short-chain mycolic acids called "corynemycolates" with 22–36 carbon atoms (Collins and Cummins, 1986) in all validly described species except *C. amycolatum* (Collins et al., 1988) and *C. kroppenstedtii* (Collins et al., 1998). Degradation products of corynemycolates, if present in abundance, may be observed to co-elute with cellular fatty acids (CFAs) of *Corynebacterium* species when analyzed using the Sherlock system (MIDI, Newark, DE; Bernard et al., 1991; Funke et al., 1995a). The cell wall of *Corynebacterium* species also contains arabinose and galactose (arabinogalactan; Collins and Cummins, 1986); 4) straight-chained saturated and unsaturated CFAs with significant volumes of palmitic (C16:0), oleic (C18:1 ω 9c) and stearic (C18:0) acids as the main CFAs. Tuberculostearic acid (C10 Me18:0) may be found in small volumes in lipophilic species (Bernard et al., 1991; von Graevenitz et al., 1991) but also in some non-lipophilic ones such as *C. minutissimum* and *C. confusum*. A large volume of C16:1 ω 7c is found in *C. diphtheriae, C. ulcerans* and *C. pseudotuberculosis* but not in any other *Corynebacterium* species to date and so may be useful as a rapid identifier (Bernard et al., 1991); 5) dihydrogenated menaquinones with eight and/or nine isoprene units (Collins and Cummins, 1986). Noncovalently bound lipids play a key role in the formation of the cell wall permeability barrier and fracture plane (Puech et al., 2001; see the chapter on Corynebacterium—Nonmedical in this Volume).

Gram staining of corynebacteria shows short or medium length, slightly curved, Gram-positive rods with non-parallel or "palisading" sides and widened ends, giving some organisms a club shape. However, some species differ from that classic morphology in demonstrating longer (*C. durum*) or thinner rods with "bulges" (*C. sundsvallense*) or "whiphandles" (*C. matruchotii*; see the section on Identification in this Chapter).

The genus *Corynebacterium* includes fermentative and oxidative species as well as species that neither ferment nor oxidize. All are catalase-positive, and medically relevant species are nonmotile. Twelve species or biovars to date are lipophilic, i.e., they grow poorly at 35–37°C in 24 h or longer on standard laboratory media but show enhanced growth in 48–72 h on sheep blood or brain heart infusion broth enriched with a lipid such as 0.1–1.0% Tween 80 (Riegel et al., 1994) or serum added to carbohydrate broth (Hollis and Weaver, 1981). Lipophilism has been serendipitously observed in *C. accolens* when enhanced growth was observed in the presence of *Staphylococcus aureus*, which supplied exogenous enrichment (Neubauer et al., 1991).

Habitat

The medically important corynebacteria are commensals and/or pathogens in man and animals. As commensals, they can be found on the skin or on mucous membranes (see the sections on Ecology and Individual Species in this Chapter). Table 1 lists corynebacteria recovered from animals.

Isolation

Sampling for corynebacteria follows the usual procedures for medical bacteria. Special precautions are not necessary. Their maintenance is assured in the usual semi-solid transport media.

Although corynebacteria grow on the usual blood and chocolate agars (but not on enteric media) when incubated aerobically, growth of lipophilic species (Table 2) is strongly enhanced by the addition of 0.1–1% Tween 80 (Riegel

Table 1. *Corynebacteria* occurring in animals.

Species	Occurrence (animal species)	Location or disease	References
C. amycolatum	Cattle	Mastitis	Hommez et al., 1999
C. auriscanis	Dog	Otitis and suppurations	Collins et al., 1999b
C. bovis	Cattle	Mastitis and abscesses	Watts et al., 2000
C. camporealensis[a]	Sheep	Mastitis	Femandez-Garayzabal et al., 1998
C. cystitidis	Cattle	Prepuce pyelonephritis	Yanagawa et al., 1986
C. diphtheriae	Cattle	Dermatitis	Corboz et al., 1996
		Mastitis	Greathead et al., 1963
		Wound infection	Henricson et al., 2000
C. glucuronolyticum	Pig	Genital tract	Devriese et al., 2000
C. kutscheri	Mouse, rat, vole, guinea pig, and hamster(?)	Latent abscesses in immunocompromised hosts	Boot et al., 1994
C. mastitidis	Sheep	Mastitis	Femandez-Garayzabal et al., 1997
C. minutissimum	Cattle	Mastitis	Hommez et al., 1999
C. phocae	Seal	Nasal cavity	Pascual et al., 1998
C. pilosum	Cattle	Urogenital tract	Hiramune et al., 1988
C. pseudotuberculosis	Sheep, goat, and horse	Lymphadenitis, lymphangitis, septicemia, and pneumonia	Brown and Olander, 1987
	Cattle	Mastitis	Watts et al., 2000
	Pig	Vagina	Takahashi et al., 1997
C. ulcerans	Cattle	Mastitis	Hommez et al., 1999
	Horse	Respiratory tract	Lipsky et al., 1982
	Monkey	Respiratory infections and bite wounds	
C. urealyticum	Dog	Encrusted cystitis	Gomez et al., 1995

[a]correct: C. camporealense.

Table 2. The medically relevant *Corynebacterium* spp.

Corynebacterium species	Metabolic process	Lipophilism	Nitrate reduction	Urease	Esculin hydrolysis	PYR/ PAL	Acid production from:					CAMP reaction	Other traits
							Glucose	Maltose	Sucrose	Mannitol	Xylose		
C. accolens	F	+	+	-	-	V/-	+	-	V	V	-	-	Can be mannose +
C. afermentans subsp. *afermentans*	O	-	-	-	-	+/+	-	-	-	-	-	V	None
C. afermentans subsp. *lipophilum*	O	+	-	-	-	+/+	-	-	-	-	-	V	None
C. amycolatum	F	-	V	V	-	+/+	+	V	V	-	-	-	Most O/129 resistant, PA[a] detected, lacks corynemycolates, and grows at 42°C, not 20°C
C. appendicis	F	+	-	+	-	+/+	(+)	(+)	-	-	-	ND	Slowly reactive with sugars, and TBSA detected
C. argentoratense	F	-	-	-	-	+/V	+	-	-	-	-	-	Chymotrypsin may be positive, PA detected, and fructose +
C. aurimucosum	F	-	-	-	-	+/-	+	+	+	-	-	ND	Yellowish, adherent colonies
C. auris	O	-	-	-	-	+/+	-	-	-	-	-	+	Dry, slightly adherent to agar
C. bovis[b]	F	+	-	-	-	-/+	+	-	-	-	-	-	TBSA positive, fructose +, and can be oxidase +
C. confusum	F	-	+	-	-	+/+	(+)	-	-	-	-	-	Tyrosine negative, PA detected, and TBSA +
C. coyleae	F	-	-	-	-	+/+	(+)	-	-	-	-	+	Ribose, fructose, and mannose +
C. diphtheriae biovar *gravis*	F	-	+	-	-	-/-	+	+	-	-	-	-	Glycogen +, and PA detected
C. diphtheriae biovar *intermedius*	F	+	+	-	-	-/-	+	+	-	-	-	-	PA detected and lipophilic
C. diphtheriae biovar *mitis* and *belfanti*	F	-	+/-[c]	-	-	-/-	+	+	-	-	-	-	Glycogen -, and PA detected
C. durum	F	-	+	(V)	(V)	+/-	+	+	+	V	-	-	Sticky colonies, PA detected, long rods with 'bulges" in Gram stained preparations, galactose +, and PYRa -

(*Continued*)

Table 2. Continued

Corynebacterium species	Metabolic process	Lipophilism	Nitrate reduction	Urease	Esculin hydrolysis	PYR/PAL	Glucose	Maltose	Sucrose	Mannitol	Xylose	CAMP reaction	Other traits
							Acid production from:						
C. falsenii	F	–	–	(+)	–	(+)/+	(+)	V	–	–	–	–	Yellowish colonies
C. freneyi	F	–	V	–	–	+/+	+	+	+	–	–	ND	α-glucosidase +, and grows at 20°C and 42°C
C. glucuronolyticum	F	–	V	V	V	+/V	+	V	+	–	V	+	β-glucuronidase and leucine arylamidase +, and PA detected
C. imitans	F	–	–	–	–	(+)/+	+	+	(+)	–	–	+	Tyrosine –, O/129 resistant, and PA not detected
C. jeikeium	O	+	–	–	–	+/+	+	V	–	–	–	–	Fructose –
C. kroppenstedtii	F	(+)	–	–	+	+/–	+	V	+	–	–	–	Lacks mycolic acids, TBSA +, PA detected
C. lipophiloflavum	O	+	–	–	–	+/+	–	–	–	–	–	–	Yellow colonies
C. macginleyi	F	+	+	–	–	–/+	+	–	+	V	–	–	None
C. matruchotii	F	–	+	–	V	+/+	+	+	+	–	–	–	"Whip handles" in Gram stained preparations, PA detected, and PYRa +
C. minutissimum	F	–	–	–	–	+/+	+	+	V	V	–	–	Tyrosine +, and PA not detected
C. mucifaciens	O	–	–	–	–	+/+	+	–	V	–	–	–	Mucoid, yellowish colonies
C. mycetoides^d	O	–	–	–	–	unk/unk	–	–	–	–	–	unk	Described as yellowish colonies
C. propinquum	O	–	+	–	–	V/V	–	–	–	–	–	–	Tyrosine +
C. pseudodiphtheriticum	O	–	+	+	–	+/V	–	–	–	–	–	–	None
C. pseudotuberculosis	F	–	V	+	–	–/V	+	+	V	–	–	REV	PA detected
C. riegelii	F	–	–	+	–	V/V	–	(+)	–	–	–	–	None
C. sanguinis	F	–	–	–	–	+/+	(+)	+	+	–	–	–	Yellowish colonies
C. singulare	F	–	–	+	–	+/+	+	+	+	–	–	–	Tyrosine +
C. simulans^e	F	–	+	–	–	V/+	+	–	+	–	–	–	Reduces nitrite
C. striatum	F	–	+	–	–	+/+	+	–	V	–	–	V	Tyrosine +
C. sundsvallense	F	–	–	+	–	V/V	+	+	+	–	–	–	Sticky colonies, and "bulges" in Gram stained preparations

												Comment
C. thomssenii	F	−	−	+	−	+/+	+	+	−	+	−	N-Acetyl-β-glucosaminidase +, and adherent
C. ulcerans	F	−	−	+	−	−/+	+	+	−	+	REV	Glycogen +, and PA detected
C. urealyticum	O	+	−	+	−	+/V	−	−	−	−	−	None
C. xerosis	F	−	V	−	−	+/+	+	+	+	+	−	O/129 susceptible, and PA not detected
CDC group F-1	F	+	V	+	−	+/−	+	+	−	+	−	None
CDC group G	F	+	V	−	−	+/+	V	V	−	V	−	Fructose +
"black Corynebacterium"[f]	F	−	−	−	−	V/V	+	(+)	−	+	−	Black pigmented colonies, and adherent

Symbols: +, positive; −, negative; V, variable; (), delayed or weak reaction; and ND, no data.

Abbreviations: PYZ, pyrazinamidase; PAL, alkaline phosphatase; CAMP reaction, Christie-Atkins-Munch-Petersen reaction; F, fermentative; O, oxidative; TBSA, tuberculostearic acid; REV, CAMP inhibition reaction; PYRa, pyrrolidonyl arylamidase (using API Coryne, API ZYM [Biomerieux]) or equivalent; and unk, unknown.

[a]PA detected, propionic acid detected as a glucose fermentation product.

[b]*C. bovis* blood culture isolate was also ONPG positive, oxidase positive, weakly maltose positive but negative by API Coryne; propionic acid was not detected; β-galactosidase was not observed using two methods (API Coryne and API Zym); API Coryne code obtained 0101104 (K. A. Bernard et al., 2002).

[c]*C. diphtheriae* biovar *mitis* is nitrate reductase + and *C. diphtheriae* biovar *belfanti* is nitrate reductase −.

[d]*C. mycetoides* originally recovered from human skin ulcer. The reference strain of this species (NCTC 9864 = ATCC 43995) has been the only strain recovered and cited in the literature since its validation in 1982. Data here are from Hollis and Weaver, 1981.

[e]*C. simulans*, a strong nitrite reducer at low and high concentrations, may appear to be nitrate reduction negative unless further tested using zinc dust; one strain was catalase negative (Bernard et al., 2002).

[f]Shukla et al. (2001) found 1 strain to be pyrazinamidase and alkaline phosphatase positive but Bernard et al. (2002) found strains to be negative for these enzymes, using an API Coryne strip, giving rise to the code 0000125.

From Hollis and Weaver (1981); Collins et al. (1982a, b); Collins and Jones (1983); Collins et al. (1982a, b); Collins and Cummins (1986); Jackman et al. (1987); Collins et al. (1988, 1998, 1999); Neubauer et al. (1991); Hollis (1992); Riegel et al. (1993a, b, 1994, 1995a–d); Funke et al. (1995a,b, 1996a,c, 1997a–e, 1998a, 1999a,b, submitted); Zinkemagel et al. (1996); Wauters et al. (1998); Sjöden et al. (1998); Zimmermann et al. (1998); Funke and Bernard (1999d); Wattiau et al. (2000); Reynaud et al. (2001); Rassoulian-Barrett (2001); Shukla et al. (2001); Yassin et al. (2002a, b); and K. A. Bernard et al. (unpublished data).

et al., 1994; see the section on Phylogeny and Taxonomy). Nonfermentative species will show very limited growth under anaerobic conditions. Media (such as colistin-nalidixic acid blood agar) that are mostly inhibitory for Gram-negative bacteria may be used to select corynebacteria. Fosfomycin may also be added to blood agar either as a disk (Wirsing von König et al., 1988) or as fosfomycin-sodium succinate (100–200 mg + 12.5 mg of glucose-6-phosphate per liter; Hiramune et al., 1988; von Graevenitz et al., 1998c).

In addition, various selective media are available for *C. diphtheriae*, such as Tinsdale agar (Tinsdale, 1947), cystine-tellurite blood agar (Frobisher, 1937), Hoyle's lysed blood tellurite agar (Hoyle, 1941), and a liquid enrichment medium (Calalb et al., 1961). They are based on the ability of tellurium to inhibit most other bacteria except Gram-positive cocci, corynebacteria, and yeasts, which however produce smaller colonies on tellurite media than does *C. diphtheriae*. Furthermore, *C. diphtheriae* reduces tellurium salts to tellurium, which accumulates in the colonies, rendering them black. On Tinsdale agar, *C. diphtheriae* and the closely related species, *C. ulcerans* and *C. pseudotuberculosis*, also produce gray haloes due to the action of cystinase on L-cystine which yields hydrogen sulfide. Nonselective media for the isolation of *C. diphtheriae*, such as Loeffler's or Pai's slants (MacFaddin, 1985), favor its growth over other organisms and yield a characteristic microscopic and colonial morphology (see the section on *C. diphtheriae* in this Chapter).

For other corynebacteria (e.g., *C. jeikeium* and *C. urealyticum*), selective media based on "natural" resistance to certain antimicrobials have been devised as well, e.g., blood agar containing gentamicin (Tompkins et al., 1982), ticarcillin + fosfomycin + 5-fluorouracil (Wichmann et al., 1984) with cefotaxime added (De Briel et al., 1991), or polymyxin B + aztreonam + amphotericin B (Zapardiel et al., 1998) with fosfomycin added (Garcia-Bravo et al., 1997).

Detection of corynebacteria by polymerase chain reaction (PCR) methods will be covered in the discussion of individual species (see Individual Species).

Identification

Stains

Gram stains of corynebacteria from routine media reveal a typical coryneform morphology, i.e., of straight to slightly curved, nonsporeforming, non-acid-fast Gram-positive rods, often with tapered ends, sometimes club-shaped or ellipsoidal, arranged in angular or palisade formations

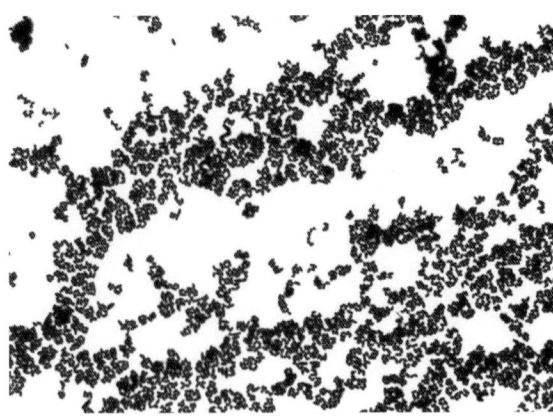

Fig. 1. Gram stain of a 24-h culture of *C. amycolatum* ATCC 49368 on sheep blood agar.

(Fig. 1; see the section on Phylogeny and Taxonomy). Staining may be uneven. These features may, however, be observed in any of the "coryneforms" (e.g., *Actinomyces* and *Arcanobacterium*) and have, therefore, little specificity for true corynebacteria. A few species have characteristic morphologies (e.g., *C. durum*, *C. sundsvallense*, *C. matruchotii*; see the section on Individual Species in this Chapter). When a special stain (e.g., Loeffler's methylene blue; MacFaddin, 1985) on *C. diphtheriae* growing on Loeffler's or Pai's slants is applied, characteristic slender rods in V- or L-shaped arrangements are seen, with metachromatic (polyphosphate) granules at one or both ends or in the center that appear reddish within the blue cells. When Neisser's stain (MacFaddin, 1985) is used, metachromatic granules will stain brownish-black, and the cells stain yellow. On other media and in other corynebacterial species (e.g., *C. pseudodiphtheriticum* and *C. imitans*), metachromatic granules are poorly, if ever, seen.

Culture Media

Colonial morphologies of corynebacteria are not characteristic except: 1) lipophilic species form very small colonies on unsupplemented blood or chocolate agar, and 2) *C. amycolatum*, *C. xerosis* and *C. matruchotii* show mainly rough colony types. Biovars of *C. diphtheriae* may be distinguishable on cystine-tellurite blood agar, with biovar *gravis* (and *C. ulcerans*) showing large gray to black colonies with a granular, radially striated surface and crenated margins and biovar *mitis* showing intensely black colonies with a glossy convex surface and regular margins (see the section on *C. diphtheriae* in this Chapter). This biovar may also cause a slight β-hemolysis. The rare biovar *intermedius* is lipophilic. A presumptive diagnosis of *C. diphtheriae* may also be made from colonial morphology on Tinsdale agar (see the section on Isolation in this Chapter).

Biochemical Reactions

Corynebacteria cannot be speciated in the clinical laboratory without testing for biochemical reactions. While all are catalase-positive, individual species either do or do not ferment carbohydrates (which can be assessed on triple sugar iron agar or Kligler's agar). Further indispensable reactions are those for lipophilism, nitrate reduction, urease production, esculin hydrolysis, acid production from various sugars (at least from glucose, maltose, sucrose, xylose and mannitol), and the CAMP (Christie-Atkins-Munch-Petersen) reaction, which tests synergistic hemolysis with a β-hemolysin-producing strain of *S. aureus* (von Graevenitz and Funke, 1996). One may also include pyrazinamidase and alkaline phosphatase (von Graevenitz et al., 1998a), susceptibility to the vibriostatic agent O/129, growth at 20°C, nitrite reduction, fermentation of glucose at 42°C, as well as other hydrolytic, fermentation, and assimilation reactions (Renaud et al., 1996; Früh et al., 1998; Hommez et al., 1999; Wattiau et al., 2000).

Miniaturized systems for identification are available as well, such as API (RAPID) Coryne (API bioMerieux, La Balme-les Grottes, France), which is based on pre-formed enzyme reactions and acid formation from sugars. Its second version correctly identified, after 24 h, 93.5% of the corynebacterial species included in the database; however, in 64%, additional tests were necessary, mainly to differentiate *C. amycolatum* from *C. striatum* (Funke et al., 1997e). The Biolog system (Biolog, Hazelwood, CA, United States), based on utilization reactions, correctly identified (in 24 h) 38% of corynebacterial strains to species and an additional 29% to genus level only (Lindenmann et al., 1995). Fifty-four percent of *C. bovis* strains were correctly identified with this system (Watts et al., 2000). The RapID CB Plus system (bioMerieux), based on utilization and single substrate enzymatic tests, correctly identified 81.2% of corynebacterial strains to species and an additional 7.2% to genus level in 24 h (Funke et al., 1998d).

Chemotaxonomic Tests

In some instances, chemotaxonomic tests may be necessary to differentiate corynebacteria from other genera. Among them, cellular fatty acid analysis (Bernard et al., 1991; von Graevenitz et al., 1991) and determination of metabolic fatty acids (Reddy and Kao, 1978; Früh et al., 1998; Bernard et al., 2002) are particularly valuable.

Patterns of the former are almost genus-specific (see the section on Phylogeny and Taxonomy in this Chapter). Metabolic acids are not species-specific, but certain acids are only observed in certain species (e.g., propionic acid in *C. diphtheriae*, *C. ulcerans*, *C. amycolatum* and *C. glucuronolyticum* and succinic acid in these species as well as in *C. minutissimum*, *C. pseudotuberculosis*, *C. renale*, *C. striatum* and CDC Group G; Frueh et al., 1998; Table 2). Patterns of corynemycolic acid as obtained by high pressure liquid chromatography have been found specific for five species (De Briel et al., 1992), but routine diagnostic use of corynemycolic acid patterns has not been investigated or recommended. Recently, Fourier-transform infrared spectroscopy has been used for identification, with >87.3% correct results at the species level (Oberreuter et al., 2002).

Molecular Tests and Typing

16S rRNA gene sequencing is now widely used for speciation. In one study, the MicroSeq 500 16S bacterial sequencing kit (Perkin-Elmer Biosystems, Foster City, CA, United States) which sequences the 527-bp 5′ end of the 16S rRNA gene gave results for the genus concordant with phenotypic tests in all of 42 isolates but was correct for species only in 64.3% of strains (Tang et al., 2000).

Amplified rDNA restriction analysis (ARDRA) with the enzymes *Alu* I, *Cfo*I and *Rsa*I was able to separate most corynebacterial species except a few non fermentative ones (Vaneechoutte et al., 1995). Ribotyping using various restriction enzymes has been applied successfully to identification of corynebacterial species (Riegel et al., 1993a; Riegel et al., 1993b; Björkroth et al., 1999), as has analysis of the 16S–23S rRNA intragenic spacer region in single species (Aubel et al., 1997), pulsed field gel electrophoresis (PFGE; Connor et al., 2000), and whole-cell protein profiling (Sjöden et al., 1998), although gel-to-gel variation may be considerable.

For strain typing, amplified fragment length polymorphism (AFLP; De Zoysa and Efstratiou, 2000) has been seen as an alternative to ribotyping (Gruner et al., 1992). Although it could not distinguish three ribotypes of *C. diphtheriae*, it was able to discriminate three other ribotypes further. Other methods used for typing this species are PFGE (De Zoysa et al., 1995) and random amplification of polymorphic DNA (RAPD) (De Zoysa and Efstratiou, 1999), multilocus enzyme electrophoresis (MEE), and single strain conformation polymorphism (SSCP; Popovic et al., 1999). Ribotyping, however, is considered the standard for *C. diphtheriae* by the

European Laboratory Working Group on Diphtheria (Popovic et al., 2000). Molecular typing methods, mostly ribotyping, have also been used to differentiate between isolates of *C. jeikeium* and *C. urealyticum* (Soto et al., 1991; see the section on Epidemiology in this Chapter).

Preservation

Well-grown cultures of corynebacteria may be kept on Loeffler's, Pai's, or blood agar slants at room temperature for months to years (Mitscherlich and Marth, 1984). Alternatively, they may be lyophilized in the usual manner.

Physiology

Toxins

Exotoxins may be formed by strains of *C. diphtheriae*, *C. ulcerans* and *C. pseudotuberculosis* and are recognized virulence factors. A review of the diphtheria toxin (DT) and its formation has appeared recently (Holmes, 2000). Target organs are the heart (myocardium) and nerve tissues.

The DT (a 58-kDa protein with 535 amino acids) consists of three domains of known sequence: 1) the N-terminal C domain, responsible for adenosine diphosphate (ADP) ribosylation of elongation factor (EF)-2; 2) the central T domain, responsible for insertion into membranes at an acidic pH, channel formation, and translocation of the C domain across the endosomal membrane into the cytosol; and 3) the C-terminal R domain, responsible for binding of DT to the receptor, i.e., the heparin-binding epidermal growth factor (HB-EGF) precursor on susceptible cells. Proteolytic "nicking" results in fragments A (corresponding to the C domain) and B (corresponding to the T and R domains), which however remain connected by a single disulfide bond. Binding is followed by receptor-mediated endocytosis of DT. The acidic pH induces a conformational change and allows the central domain to be inserted into membranes and to form a channel through which translocation of the A fragment occurs across the endosomal membrane into the cytosol. The A fragment, having become enzymatically active following reduction of the disulfide bond, catalyzes the transfer of ADP-ribose from nicotinamide adenine dinucleotide (NAD) to EF-2, thus inactivating the latter and thereby inhibiting protein synthesis (chain elongation on the ribosome). The acceptor for the ribosylation is diphthamide, a post-translationally modified histidine.

The DT inhibits protein synthesis in cell-free extracts from all animal species but is toxic to intact cells of certain mammals only, probably because of the presence of high-affinity receptors. It is synthesized by strains that are lysogenic for the bacteriophage β, which carries the structural gene *tox*. Thus, toxigenicity is determined by lysogenic conversion, and the *tox* determinant of corynephage β is the structural gene for DT.

Nontoxigenic strains either do not have detectable *tox*-related sequences or have such sequences that (probably as a result of a point mutation) do not encode functional DT. Also, rare strains with no infective corynephages have been observed (Cianciotto and Groman, 1997). The virulence of such strains to date has not been explained. All *tox*-bearing corynephages (except the δ phage), including the mutant ones, share extensive homology and have similar restriction fragment length polymorphism (RFLP) patterns with the β-converting phage (Buck et al., 1985). Chromosomal integration occurs in analogy to λ lysogenization of *Escherichia coli* (Buck and Groman, 1981). Two sites (*attB* 1 and *attB* 2), located on the bacterium, serve to attach the β-prophage with its attachment site *attP* on the chromosome. The *tox* gene is located adjacent to this site.

Addition of iron to growth media for *C. diphtheriae* inhibits toxin production (Pappenheimer and Johnson, 1936) owing to the diphtheria toxin repressor DtxR, a 226-amino-acid polypeptide encoded by *dtxR* and acting at the level of transcription of the DT gene. DtxR binds DNA by a sequence-specific mechanism and functions as an iron-dependent regulatory protein that represses DT synthesis, corynebacterial siderophore, and other components of iron uptake. A possible utilization of heme or hemoglobin as iron source is regulated by the *hmuO* gene (Schmitt, 1997). Specific recombination between *attP* and the *attB* attachment sites may also occur in *C. ulcerans* and *C. pseudotuberculosis*, which will then produce DT. Other corynebacterial species, even though they may show *attB*-related sequences (Cianciotto et al., 1986), have never been observed to produce DT.

The traditional test for the detection of DT uses two guinea pigs, both of them injected intraperitoneally with a broth culture of the bacterium, but only one of them protected by antitoxin. Toxigenic strains will cause death in the unprotected animal within 1–3 days, with characteristically swollen and hemorrhagic adrenal glands upon autopsy. This test has now been replaced by a variety of in vitro methods, which have recently been reviewed (Efstratiou et al., 2000). The widely used immunoprecipitation (Elek) test (Elek, 1949) has recently been modified in various ways (Engler et al., 1997; Reinhardt et al., 1998) to overcome its dependency on the quality of the reagents and the time needed for completion.

PCR amplification of a 246-bp fragment of the DT gene (Pallen, 1991; Mikhailovich et al., 1995) has also found acceptance but only detects DT gene sequences of both fragments, not active DT itself, and as such would not be a determinative test. New immunological tests such as enzyme immunoassay (EIA; Engler and Efstratiou, 2000), immunoblotting (Hallas et al., 1990), ADP ribosylation of EF-2 (Hauser et al., 1993), and immunochromatography (Engler et al., 2002) only detect fragment A but have at least a shorter time frame than the Elek test. The Vero cell bioassay (Miyamura et al., 1974) is specific for active DT but requires specialized tissue culture facilities and more time than the other tests. Concordance between the in vivo, cytotoxicity and modified Elek tests has been found very high, i.e., over 99% (Efstratiou et al., 1998).

A second toxin found only in *C. ulcerans* and *C. pseudotuberculosis* is phospholipase D (PLD), which hydrolyzes sphingomyelin in mammalian cell membranes, thereby releasing choline (Soucek et al., 1971). The genes and gene products of the two species show 80% DNA sequence homology and 87% amino acid sequence homology, respectively (Cuevas and Songer, 1993), and their PLD has more than 60% DNA and amino acid sequence homology with the PLD of *Arcanobacterium haemolyticum* (McNamara et al., 1995). PLD can be detected by inhibition of the CAMP test.

Other Potential Virulence Factors

Potential adhesins such as hemagglutinins, hydrophobins, sugar residues, and enzymes with *trans*-sialidase activity have been described in *C. diphtheriae* (Mattos-Guaraldi et al., 2000). Extracellular slime production was found in central nervous system shunts colonized by *C. jeikeium*, *C. amycolatum* and nonspeciated corynebacteria (Bayston et al., 1994). In *C. urealyticum*, however, adherence does not seem to be a predictor for virulence (Marty et al., 1991).

Plasmids

Plasmids have thus far been found in *C. renale* (Nath and Deb, 1995), *C. diphtheriae* and *C. jeikeium*. In six epidemic erythromycin- and clindamycin-resistant strains of *C. diphtheriae*, a single plasmid of 9.6 MDa was detected containing the *ermCd* gene, which encoded an inducible rRNA methylase (Roberts et al., 1992). In a study of 39 "JK type" corynebacteria, 23 possessed plasmids, forming six groups on restriction endonuclease analysis. Four groups with a plasmid DNA of over 6.4 kbp were associated with a bacteriocin-like substance but not with antimicrobial resistance (Kerry-Williams and Noble, 1986; Pitcher et al., 1990). Sequencing of the multiresistance plasmid pTP10 from a strain of *C. striatum* has provided genetic information on the mechanisms of resistance to 16 antimicrobials (Tauch et al., 2000).

Ecology

Medically important corynebacteria are commensals and potential pathogens in man and animals. Some data are available on their survival in the inanimate environment, as well as on the species distribution in the normal flora.

Corynebacterium diphtheriae strains from carriers may be toxigenic or nontoxigenic and may be found in the throat, in the nose, or on the skin (see the section on *C. diphtheriae* in this Chapter). While carriage in the upper respiratory tract is now rare among healthy populations in nonepidemic areas, it may become prevalent during epidemics of throat diphtheria (Mikhailovich et al., 1995). Skin carriage, mostly of nontoxigenic strains, is in the United States and Western Europe nowadays associated with homelessness and intravenous drug abuse but has been recently discovered in other populations as well (Funke et al., 1999a; see the section on *C. diphtheriae*). The organism is also able to survive in the inanimate environment of patients with diphtheria, albeit rarely (Larsson et al., 1987), and has survived in milk as well (Kersten, 1909).

On healthy skin, *C. jeikeium* has been detected in intertriginous areas (Larson et al., 1986a; Larson et al., 1986b), and large-colony diphtheroids as well as unspecified lipophilic diphtheroids were found on the same sites, in the anterior nares, and in the axilla (Aly and Maybach, 1977). There was an inverse correlation between the occurrence of these lipophilic strains and of *C. jeikeium* in healthy as well as diseased individuals, but the frequency and density of *C. jeikeium* in diseased individuals were higher (Larson et al., 1986b). This organism is also able to survive in the inanimate environments such as air and surfaces in the rooms of patients (Quinn et al., 1984; Telander et al., 1988). The same is true for *C. urealyticum* (Nieto et al., 1996), which may also be isolated from the intact skin of hospitalized patients (Soriano et al., 1988), and for *C. striatum* (Brandenburg et al., 1996). *Corynebacterium minutissimum* may also be part of the normal skin flora (Pitcher, 1978).

Corynebacteria belonging to the normal flora of mucous membranes have only been speciated recently (Funke et al., 1997f; Gordts et al., 2000). In one study of throat cultures of two different populations, *C. durum* was found as the most frequent corynebacterium, followed by CDC Group G, *C. jeikeium* and *C. pseudodiphtheriti-*

cum (von Graevenitz et al., 1998c). The normal conjunctival flora, on the other hand, contains only lipophilic corynebacteria; in one series, they were CDC Group G, *C. macginleyi*, *C. afermentans* subsp. *lipophilum*, *C. accolens* and *C. jeikeium* (von Graevenitz et al., 2001).

Of the veterinary corynebacteria, *C. renale*, *C. pilosum* and *C. cystitidis* are able to survive in soil (Hayashi et al., 1985).

Epidemiology

Except for *C. diphtheriae*, little is known about the spread of corynebacteria. Diphtheria bacilli are known to spread via droplets from the respiratory tract. In the Western hemisphere, however, spread is often from skin lesions via contact or injection paraphernalia (Gruner et al., 1994; see the section on *C. diphtheriae*). Occurrence of a clone of nontoxigenic isolates, first reported from Switzerland (Gruner et al., 1992), was later reported from Northern Germany as well (Funke et al., 1999a). Skin transmission, in fact, seems to be more efficient than throat transmission (MacGregor, 2000). Epidemic transmission from milk, reported in the early literature (Henry, 1920), seems to have been due to milkers' hand contamination and has not been reported lately. Molecular typing of *C. jeikeium* (Kerry-Williams and Noble, 1986; Khabbaz et al., 1986; Pitcher et al., 1990) and of *C. urealyticum* (Soto et al., 1991) has yielded conflicting evidence as to whether patient-to-patient transmission occurred. Such spread has been documented for *C. striatum*, however (Leonard et al., 1994; Brandenburg et al., 1996).

Individual Species

With some exceptions (such as *C. diphtheriae* and closely related taxa), corynebacteria appear to be infrequent human pathogens. Taxonomy and disease potential of these bacteria have so far only sporadically been investigated and thus are still only partially understood. Some species isolated from humans are represented in the literature by only a single strain or a small number of strains and thus lack sufficient depths to demonstrate a possible heterogeneity of phenotypic reactions, which is a less than ideal situation (Christensen et al., 2001). A comprehensive review of medically relevant coryneforms, including a description of older/provisional taxonomic designations, was published in 1997 (Funke et al., 1997f); this background will not be reiterated here. The species are presented in alphabetical order. Colony description is based on at least 24-h growth at 35–37°C on 5% sheep

blood agar. Biochemical reactions commonly used for identification are listed in Table 2. Additional tests, including chemotaxonomic and molecular genetic studies, may be required in a few instances.

Corynebacterium accolens

Corynebacterium accolens is a lipophilic *Corynebacterium* species recovered from eye specimens, including from the healthy human conjunctiva (von Graevenitz et al., 2001), ears, nose, and the oropharynx (Neubauer et al., 1991). Native aortic and mitral valve endocarditis due to this agent has been described (Claeys et al., 1996). *Corynebacterium accolens* resembles phenotypically CDC group G, except for certain enzyme reactions. Strains to date have been susceptible to a broad spectrum of antibiotics, whereas CDC group G is often multiresistant (Funke et al., 1996b).

Corynebacterium afermentans

Corynebacterium afermentans consists of two subspecies found to be relatively nonreactive in substrates, *C. afermentans* subsp. *afermentans* and *C. afermentans* subsp. *lipophilum* (Table 2), and to have a nonfermentative metabolism (Riegel et al., 1993b). Both subspecies may demonstrate a positive CAMP reaction.

Corynebacterum afermentans subsp. *afermentans* is part of normal human skin flora and has been isolated mainly from blood cultures (Riegel et al., 1993b). Sepsis in a neurological patient has been reported (Kumari et al., 1997). *Corynebacterium afermentans* subsp. *afermentans* is difficult to differentiate from *C. auris* and *Turicella otitidis* solely by phenotypic means and even by a larger set of substrates such as that found in the Biolog (Biolog, Inc., Hayward, CA) or other assimilation systems (Funke et al., 1994; Renaud et al., 1996). These taxa may be differentiated by 16S rRNA gene sequencing and chemotaxonomically, as both *C. afermentans* and *C. auris* contain corynemycolates, but *T. otitidis* does not.

Corynebacterium afermentans subsp. *lipophilum* has been isolated from blood cultures as well as from superficial wounds (Riegel et al., 1993a) but may also be recovered from the healthy conjunctiva (von Graevenitz et al., 2001). It has been reported to be an agent of severe disease, including prosthetic valve endocarditis (Sewell et al., 1995) and disseminated infection associated with abscess formation (Dykhuizen et al., 1995). As the name suggests, this is a lipophilic species with smooth, rounded colonies <0.5 mm in diameter after 24 h of incubation. *Corynebacterium afermentans* subsp. *lipophilum* grows poorly on nonlipid containing media when compared to other phenotypically nonreactive taxa that grow luxuriantly in 24 h (*C. auris, C. afermentans* subsp. *afermentans* and *T. otitidis*), which assists with identification.

Corynebacterum amycolatum

This species was originally recovered as a commensal from normal human skin flora (Collins et al., 1988), but its role in human disease was not further elucidated until the mid-1990's. It is acknowledged to be the most frequently encountered *Corynebacterium* species derived from human clinical material (Funke et al., 1997f). Recent reports have attributed a role in fatal sepsis in a premature infant (Berner et al., 1997), a cardioverter-lead electrode infection (Vaneechoutte et al., 1998), septic arthritis (Clarke et al., 1999), catheter-related infection, surgical wound infection, pilonidal cyst formation (Esteban et al., 1999), sepsis in neutropenic patients (de Miguel et al., 1999), bacteremia (Oteo et al., 2001), and native valve endocarditis (Knox and Holmes, 2002). *Corynebacterium amycolatum* strains are usually multidrug resistant (de Miguel-Martinez et al., 1996; Funke et al., 1996b; Lagrou et al., 1998; Troxler et al., 2001; Knox and Holmes, 2002). This species has not been recovered from throat swabs of healthy individuals (von Graevenitz et al., 1998a). *Corynebacterium amycolatum* is one of two corynebacterial species lacking corynemycolates in its cell wall (Collins et al., 1988), is metabolically fermentative and is very diverse in its reaction to standard phenotypic tests, which has been the basis for misidentification as *C. xerosis, C. striatum* or *C. minutissimum* (Zinkernagel et al., 1996; Wauters et al., 1998). Gram stain of this bacterium is shown in Fig. 1. All *C. amycolatum* strains produce propionic acid as major end product of glucose metabolism. Acyl phosphatidylglycerol is a major phospholipid in *C. amycolatum* in contrast to other *Corynebacterium* spp., in which other phospholipids are predominant (Collins et al., 1988).

Corynebacterium appendicis

This species was recently described on the basis of a single strain recovered from an abdominal swab of a patient with appendicitis and abscess formation. It is a urease-positive, lipophilic *Corynebacterium* species that slowly ferments glucose and maltose (Yassin et al., 2002a; Table 2).

Corynebacterium argentoratense

On the basis of phylogenetic studies, this species is closest to *C. diphtheriae* and the related species

C. ulcerans and *C. pseudotuberculosis*, but lacks the diphtheria *tox* gene (Riegel et al., 1995a). To date, *C. argentoratense* has been isolated from the human throat (Riegel et al., 1995a; von Graevenitz et al., 1998c) as well as from a blood culture (Bernard et al., 2002). The CAMP-negative *C. argentoratense* phenotypically resembles *C. coyleae* strains, which are CAMP positive (Table 2). Like *C. diphtheriae* and closely related species, *C. argentoratense* produces propionic acid as a product of glucose fermentation (Bernard et al., 2002). *Corynebacterium argentoratense* is the only medically relevant *Corynebacterium* species that exhibits α-chymotrypsin in the API Zym (bioMérieux) system; however, a blood culture isolate was not observed to produce that enzyme (Bernard et al., 2002).

Corynebacterium aurimucosum

This species has been recently proposed for two strains, one recovered from a human blood culture (Yassin et al., 2002b). As a unique characteristic on sheep blood agar plates, it has, after a 24-h incubation, colonies that are yellowish, sticky and about 1–2 mm in diameter, but on trypticase soy agar without blood supplement, it lacks pigment and is described as slimy. Phenotypically, this species is most similar to *C. minutissimum* except for a lack of alkaline phosphatase activity.

Corynebacterum auris

This species was originally recovered from the ears of pediatric patients with ear infections (Funke et al., 1995b). On the basis of colonial, chemotaxonomic, and genetic studies, *C. auris* strains were found to be discernible from other minimally reactive taxa (e.g., *C. afermentans* subsp. *afermentans* and *Turicella otitidis*; Funke et al., 1994; Funke et al., 1995b), which they otherwise closely resemble phenotypically (Table 2). *Corynebacterium auris* strains are CAMP test positive. Increased minimum inhibitory concentrations (MICs) for β-lactam antibiotics in *C. auris* strains have been observed, but the mechanism is unknown (Funke et al., 1996b).

Corynebacterium bovis

One case study exists that implicates the lipophilic *Corynebacterium* species *C. bovis* as a rare causative agent of human disease (Vale and Scott, 1977), but otherwise this species is an agent of bovine mastitis, a significant disease in cattle (Watts et al., 2000). Human isolates would now presumably have been assigned to one of a number of other lipophilic *Corynebacterium* species if polyphasic methods and recent identification schemes had been available. In a 1997 review, it was mentioned that in fact this species had not been definitively recovered for many years from human clinical material (Funke et al., 1997f). Recently, however, a human blood culture isolate of *C. bovis* was identified on the basis of a polyphasic approach, including phenotypic, chemotaxonomic and genotypic characteristics (Bernard et al., 2002). Microbiologists should therefore be cognizant of the possibility that this bovine disease agent may on rare occasions be recovered from human specimens.

Corynebacterium confusum

This species has been isolated from patients with foot infections, a clinical blood sample (Funke et al., 1998b), and from a breast abscess (CAMP positive isolate; Bernard et al., 2002). *Corynebacterium confusum* reduces nitrate but characteristically will only slowly and weakly ferment glucose and is poorly reactive in most substrates. A small volume of tuberculostearic acid (TBSA) is observed on CFA analysis. This species may require additional testing such as for tyrosine hydrolysis and CAMP reaction to separate it from *C. argentoratense*, *C. coyleae* (which is CAMP positive) and *C. propinquum* (Table 2).

Corynebacterium coyleae

This species has been primarily isolated from normally sterile body fluids including blood cultures from patients with fever of unknown origin (postsurgically or in HIV infections), as well as from genitourinary specimens (Funke et al., 1997d; Bernard et al., 2002). Characteristically, it is CAMP positive and slowly ferments glucose and ribose, but few other sugars (Table 2).

Corynebacterium diphtheriae

Classical respiratory diphtheria, caused by toxigenic strains of *C. diphtheriae* or, rarely, by the closely related species *C. ulcerans*, was one of the most common causes of death among children in the pre-vaccine era (Golaz et al., 2001). Symptoms include a sore throat, low-grade fever, swelling of the neck, and a membrane on the tonsils or throat, which can be associated with asphyxiation if breathing is obstructed. Use of toxoid vaccine has virtually eliminated diphtheria in developed countries and significantly reduced it in developing countries (Golaz et al., 2001). Cutaneous infections are the second classically recognized DT-induced manifestation. An extensive outbreak of diphtheria in Russia and in newly independent states of the former Soviet Union was directly attributed to 1) declining immunity among adults who had received all or

most childhood immunizations but missed having booster shots, 2) persistence of circulating *C. diphtheriae* strains in the population, and 3) dramatic socioeconomic changes (Efstratiou and George, 1996; Popovic et al., 2000). Only a massive immunization program of 70 million people halted the outbreak (Golaz et al., 2000). Developing countries continue to be at considerable risk (Gilbert, 1997), but waning immunity among populations in developed countries have put those people at risk as well, especially when traveling to countries lacking a universal immunization program (Marston et al., 2001).

A growing body of reports continues to document the emergence of newly recognized manifestations of *C. diphtheriae* disease in countries where it had been thought to be eradicated or substantively diminished. Serious diseases caused by nontoxigenic *C. diphtheriae* and closely related species, including splenic abscesses, bacteremia, septic arthritis, and endocarditis (Tiley et al., 1993; Gruner et al., 1994; Efstratiou and George, 1996; Hogg et al., 1996; Funke et al., 1999a; Belko et al., 2000) have been described (see the section on Epidemiology in this Chapter). Pharyngitis has been reported in homosexual, institutional, and military populations (Wilson, 1995; Gilbert, 1997; Reacher et al., 2000). Increased risk for significant infection exists for homeless (Harnisch et al., 1989), intravenous drug abusers (Gruner et al., 1994), and aboriginal populations (Hogg et al., 1996).

Corynebacterium diphtheriae is divided into four biovars–*gravis*, *mitis*, *belfanti* and *intermedius* (see Identification). Biovar differentiation is recommended (Efstratiou and George, 1999; Efstratiou et al., 2000), although biovars cannot be assigned subspecies status nor is their determination a relevant epidemiologic tracking method (Riegel et al., 1995d). Originally, biovars were defined by differences in colony morphology and biochemical tests. However, only the lipophilic biovar *intermedius* may be identified on the basis of colonial morphology (Coyle et al., 1993b). Other *C. diphtheriae* biotypes produce indistinguishable, larger white or opaque colonies after 24 h on sheep blood agar. Biovar *intermedius* occurs only rarely in clinical infections, and *C. diphtheriae* biovar *belfanti* strains rarely harbor the diphtheria toxin gene.

Biochemically, *C. diphtheriae* as well as closely related species, *C. pseudotuberculosis* and *C. ulcerans*, produces cystinase (detected by using Tinsdale medium or equivalent media) and lacks pyrazinamidase (Efstratiou and George, 1999; Efstratiou et al., 2000) and, by these tests, may be discerned from other species (Table 2). Propionic acid is produced as an end product of glucose metabolism (Estrangin et al., 1987; Früh et al., 1998). *Corynebacterium diphtheriae* strains differ from all other coryneform bacteria (except *C. pseudotuberculosis* and *C. ulcerans*) in their CFA patterns by the presence of a large volume of C16:1 ω 7c (Bernard et al., 1991).

Corynebacterium durum

First recovered from respiratory tract specimens, this species has been further isolated from the gingiva, blood cultures, and abscesses (Riegel et al., 1997a; Rassoulian-Barrett et al., 2001). This is the most frequent *Corynebacterium* species isolated from throat swabs of healthy persons (von Graevenitz et al., 1998c). Gram staining of aerobic cultures shows long and filamentous rods with occasional "bulges" but not *C. matruchotii*-like "whiphandles." Differentiation between these two species may be difficult (Table 2), and in the past, some *C. durum* strains were almost assuredly misidentified as *C. matruchotii* when solely phenotypic methods were used (Hollis and Weaver, 1981). *Corynebacterium matruchotii* produces pyrrolidonyl arylamidase, whereas *C. durum* does not (Rassoulian-Barrett et al., 2001). Most *C. durum* strains hydrolyze esculin and ferment mannitol, which is otherwise unusual among *Corynebacterium* spp. (Rassoulian-Barrett et al., 2001). Both *C. matruchotii* and *C. durum* produce propionic acid as a fermentation product (Bernard et al., 2002).

Corynebacterium falsenii

Isolates have been recovered to date from blood culture and cerebral spinal fluid (Sjöden et al., 1998; Bernard et al., 2002). The original description of this species suggested that colonies after a 24-h incubation are whitish, smooth and glistening with entire edges, but with time (72–120 h), exhibit a yellowish pigment that increases in intensity. This phenomenon has otherwise only been rarely observed among *Corynebacterium* species. This species demonstrates a weak urease reaction, with other distinguishing features described in Table 2.

Corynebacterium freneyi

Isolates of this recently described species were derived from pus from a toe, varicose ulcers, and a subcutaneous abscess fistula (Renaud et al., 2001). Biochemically, these strains were nearly indistinguishable from *C. xerosis* and also similar to *C. amycolatum*. Both *C. xerosis* and *C. freneyi* produce α-glucosidase (Table 2). The two species, although closely related phylogenetically by 16S rRNA gene sequencing, could otherwise be differentiated by less commonly used biochemical tests, by 16S–23S intergenic spacer region analysis, and by DNA-DNA hybridization.

Corynebacterium glucuronolyticum

This species was originally recovered from male urogenital specimens but has also been recovered from blood cultures, peritoneal fluid, and dialysis fluid (Funke et al., 1995a; Bernard et al., 2002). *Corynebacterium seminale* (Riegel et al., 1995b) is a junior synonym of *C. glucuronolyticum* (Devriese et al., 2000). This species is the only medical *Corynebacterium* species that produces β-glucuronidase as well as reacts positively in a wide variety of biochemical tests. Urease, if produced, is abundant and its detection is rapid (<10 min; Table 2). Reactions for strains recovered from animals may differ from those recovered from diseased humans (Devriese et al., 2000). These bacteria may exhibit resistance to tetracycline, macrolides and lincosamides (Funke et al., 1996b). Culture-negative prostatic fluids were found to contain 16S rRNA gene sequences homologous with sequences derived for *C. glucuronolyticum*, suggesting that this agent may play a role in prostatitis (Tanner et al., 1999).

Corynebacterium imitans

Person-to-person transmission of this species was suggested when the isolate originally from the throat of a seriously ill child suspected to be suffering from pharyngeal diphtheria appeared as well in three ill adult contacts (Funke et al., 1997a). These strains, upon initial recovery, were described as being "atypical *C. diphtheriae*" because they expressed a number of phenotypic reactions very similar to those found for that species. Subsequent polyphasic analyses defined the group of isolates as a new species, *C. imitans* (Funke et al., 1997a; Table 2). Strains since recovered from blood cultures consistent genotypically and phenotypically with *C. imitans* did not produce propionic acid (unlike *C. diphtheriae*) and lacked the unique CFA composition associated with *C. diphtheriae* and closely related species (Bernard et al., 1991; Bernard et al., 2002). *Corynebacterium imitans* did not produce DT as assessed by the modified Elek test, by *tox* gene detection with PCR, and by use of an immunochromatographic strip (Funke et al., 1997a; Engler et al., 2002).

Corynebacterium jeikeium

This species, a lipophilic corynebacterium species associated with skin flora (Jackman et al., 1987), is one of the most frequently encountered *Corynebacterium* species in clinical specimens (Funke et al., 1997f; Lagrou et al., 1998). This bacterium can cause serious prosthetic joint infection (von Graevenitz et al., 1998b), endocarditis (Jackman et al., 1987; Ross et al., 2001), bacteremia in bone marrow transplant patients with a Hickman catheter (Wang et al., 2001) and in neutropenic patients (Zinner et al., 1999), pneumonia and bacteremia associated with significant disease (Funke et al., 1997f), as well as otitis media (de Miguel-Martinez, 1999). Nosocomial transmission of this bacterium has been suggested (Pitcher et al., 1990). Most strains of *C. jeikeium* exhibit multiresistance to antibiotics but remain sensitive to vancomycin (Riegel et al., 1996; Lagrou et al., 1998). A novel glycopeptide, BI 397, exhibited significant activity against *C. jeikeium* in vitro (Jones et al., 2001). DNA-DNA hybridization studies have shown that *C. jeikeium* includes at least four genomospecies: two that are multiresistant and two that have low penicillin and gentamicin MICs but could not be differentiated phenotypically (Riegel et al., 1994). *Corynebacterium jeikeium* is phenotypically closest to CDC group G, but may be differentiated from that taxon by inability to grow anaerobically and to oxidize fructose (Riegel et al., 1994).

Corynebacterium kroppenstedtii, *C. lipophiloflavum* and *C. macginleyi*

Corynebacterium kroppenstedtii was originally recovered from a sputum sample of a patient with pulmonary disease (Collins et al., 1998), with additional strains being isolated from a breast abscess, an open lung biopsy, as well as sputum (Bernard et al., 2002). It is a lipophilic corynebacterium, the second species in the genus to lack cell wall mycolates, and one of the few *Corynebacterium* species exhibiting esculin hydrolysis.

The description of *C. lipophiloflavum* is based on characteristics of one strain recovered from vaginal discharge from a patient with bacterial vaginosis (Funke et al., 1997b). That strain was found to exhibit an intensely yellow pigment, urease production, but relative substrate non-reactivity.

Corynebacterium macginleyi, initially described as isolates from ocular specimens (Riegel et al., 1995c), has since been recovered from diseased (Funke et al., 1998c; Joussen et al., 2000) as well as healthy eye specimens (von Graevenitz et al., 2001). These bacteria are lipophilic, with reactions as shown in Table 2.

Corynebacterium matruchotii

The association of this species, formerly *Bacterionema matruchotii* (Collins, 1982b), part of the flora of the oral cavity, with calculus and plaque deposits (Siqueira et al., 2000; Rassoulian-Barrett et al., 2001) has been much studied but otherwise recovery of this species from diseased humans has been rare. Gram-stained *Corynebacterium matruchotii* has a most remarkable appearance, which consists of "whips" of long filamentous bacteria with "whiphandles" formed from a short coccobacillus or bacillus attached to one end (Collins, 1982b). A polyphasic study of reference strains of this species has shown that they include some misidentified strains of *C. durum* (Rassoulian-Barrett et al., 2001). One such *C. matruchotii* reference strain, ATCC 43833, has been shown to represent a novel species distinct from *C. matruchotii* and from *C. durum* (Rassoulian-Barrett et al., 2001).

Corynebacterium minutissimum

This species is part of normal human skin flora and was originally thought to be the cause of erythrasma (Collins and Jones, 1983), but that premise has since been refuted (Coyle and Lipsky, 1990). This species recently has been recovered from patients with endophthalmitis (Arsan et al., 1995–1996) and peritonitis in a CAPD patient (Fernandez Giron et al., 1998). Readers are cautioned that antibiotic multiresistant or propionic acid-producing isolates identified as *C. minutissimum* are most likely *C. amycolatum* strains, if polyphasic identification approaches are used (Zinkernagel et al., 1996). Members of this species are usually DNAse producers, hydrolyze tyrosine, are fermentative and, unlike most *Corynebacterium* species, are able to ferment mannitol (Zinkernagel et al., 1996; Table 2).

Corynebacterium mucifaciens

This species has been recovered from blood cultures and other sterile body fluids as well as from abscesses, soft tissues, and a CAPD dialysate (Funke et al., 1997c; Bernard et al., 2002). Its colonies are unusual and distinct among *Corynebacterium* species. They generally exhibit a slight to intense yellow or yellow-brown pigment and are mucoid, somewhat reminiscent of *Rhodococcus equi* colonies (see the chapter on The Phototrophic Alpha-Proteobacteria in Volume 5). The characteristic mucoid appearance was found by electron microscopy to be associated with discernible thin-to-thick cell-connecting filaments, assumed to be a polysaccharide and possibly related to virulence (Funke et al., 1997c). The mucoid appearance is less obvious after extended incubation for 96 h. One blood culture isolate, found to be closest to *C. mucifaciens* by polyphasic study including 16S rRNA gene sequence analysis, was aberrant in that it was not mucoid, not yellow pigmented and thus difficult to identify (Bernard et al., 2002). *Corynebacte-*

rium mucifaciens has an oxidative metabolism (Table 2) and may be separated from *R. equi* by Gram stain appearance, enzyme profile, biochemical reactions, and CFA composition (the TBSA volumes of *C. mucifaciens* being small, in contrast to the very large volumes found in *R. equi*). β-Lactam antibiotics and aminoglycosides show very good activities against *C. mucifaciens* (Funke et al., 1997c).

Corynebacterium mycetoides

This organism was originally isolated in 1942 from skin ulcers of soldiers in the desert of North Africa. The species was later validated (Collins, 1982a). To the best of our knowledge, clinical cases involving this agent have not to date been reported in the literature. Characteristics are shown in Table 2.

Corynebacterium propinquum

Corynebacterium propinquum is the name assigned to encompass the rare human pathogen CDC group ANF-3 (Hollis and Weaver, 1981; Riegel et al., 1993b). This species was thought to be the cause of a case of native valve endocarditis (Petit et al., 1994). It grows well in 24 h, but in biochemical tests other than nitrate reduction and tyrosine hydrolysis, it is relatively nonreactive (Riegel et al., 1993b).

Corynebacterium pseudodiphtheriticum

This species, part of the normal oropharyngeal flora in humans, in recent years has been documented to cause pneumonia in compromised and immunocompromised populations (Izurieta et al., 1997; Gutierrez-Rodero et al., 1999; Martaresche et al., 1999). One case of exudative pharyngitis with a pseudomembrane, which mimicked diphtherial pharyngitis, provoked a significant public health response (Izurieta et al., 1997). This agent has been recovered from keratitis and conjunctivitis (Li and Lal, 2000) and skin infection (Hemsley et al., 1999). Like *C. propinquum*, it is relatively nonreactive in most biochemical tests but does reduce nitrate and hydrolyze urea. *Corynebacterium pseudodiphtheriticum* strains are susceptible to β-lactam antibiotics, but resistance to macrolides and lincosamides has been observed (Gutierrez-Rodero et al., 1999).

Corynebacterium pseudotuberculosis

This species is primarily noted for causing caseous lymphadenitis in sheep, horses, and feral goats; however, it may on rare occasions cause lymphadenitis in humans who have occupational exposure to sheep (Peel et al., 1997). This species is phylogenetically closely related to *C. diphtheriae* and *C. ulcerans* and, like those species, may harbor the diphtheria toxin gene (Pascual et al., 1995; Ruimy et al., 1995; Peel et al., 1997; see the section on Physiology in this Chapter). Like those species, the cell wall of *C. pseudotuberculosis* contains large amounts of the CFA, C16:1 ω 7c (Bernard et al., 1991). Like *C. ulcerans*, *C. pseudotuberculosis* produces urease and, like *Arcanobacterium haemolyticum*, is reverse CAMP positive (Hollis and Weaver, 1981). Molecular-level strain typing has been done using pulsed field gel electrophoresis (PFGE; Connor et al., 2000).

Corynebacterium riegelii

Strains of this species were originally recovered from urinary tract infections of human females (Funke et al., 1998a) but have also been isolated from blood cultures including cord blood (Bernard et al., 2002). *Corynebacterium riegelii* strains exhibit an unusual phenotypic characteristic of fermenting maltose but not glucose or most other sugar substrates. Urease, when present, is both rapidly and strongly positive (Funke et al., 1998a; Table 2).

Corynebacterium sanguinis, C. simulans, and C. singulare

These species have been recently described in the literature. Phenotypic reactions are represented by only a few strains.

Corynebacterium sanguinis has been recovered from blood culture isolates (G. Funke et al., submitted). Characteristically, this species slowly ferments glucose like *C. coyleae* but is CAMP negative. A small volume of TBSA (2–3%) was detected among CFAs.

Corynebacterium simulans was originally discerned from among some *C. striatum*-like isolates (Wattiau et al., 2000) and included isolates from a foot abscess, lymph node biopsy, and a boil. Two additional strains have been characterized from bile and an unusual, repeatedly catalase-negative isolate has been derived from blood culture (Bernard et al., 2002). *Corynebacterium simulans* is closely related phylogenetically to *C. minutissimum*, *C. singulare* and *C. striatum* and is the only species in the genus described to date that reduces nitrite.

Corynebacterium singulare has been recovered from semen and a blood culture (Riegel et al., 1997b). Its biochemical reactions were originally suggestive of *C. minutissimum*, including hydrolysis of tyrosine, except for urease positivity (Table 2). However, polyphasic identification, including phylogenetic analysis, found that these strains were closest to, but discernible from, *C.*

minutissimum. Corynebacterium singulare also does not produce propionic acid as a fermentation product from glucose (Riegel et al., 1997b).

Corynebacterium striatum

This species has been found to be part of the normal human skin flora but has been occasionally recovered from clinical samples in patients with endocarditis (Tattevin et al., 1996; Juurlink et al., 1996; Keijman et al., 2000), meningitis, septicemia, pneumonia, and other invasive infections (Martinez-Martinez et al., 1995; Martinez-Martinez et al., 1997; Weiss et al., 1996). Vertebral osteomyelitis due to this species has been reported (Fernandez-Ayala et al., 2001). Nosocomial transmission of *C. striatum* has been documented (Leonard et al., 1994; Brandenburg et al., 1996). *Corynebacterium striatum* strains hydrolyze tyrosine and are occasionally CAMP reaction positive. Resistance to macrolides and lincosamides due to the presence of an rRNA methylase has been described. *Corynebacterium striatum* may also be resistant to quinolones and tetracyclines (Martinez-Martinez et al., 1996).

Corynebacterium sundsvallense and *C. thomssenii*

Corynebacterium sundsvallense has been isolated from blood cultures, a vaginal swab, and fluid drained from an infected groin (Collins et al., 1999a). Gram staining shows bulges or knobs at the ends of some rods which may resemble those seen in *C. durum* or *C. matruchotii*, but biochemical methods may differentiate those species (Table 2).

Corynebacterium thomssenii was first repeatedly isolated from a patient with pleural effusion (Zimmermann et al., 1998). A second strain was recovered from an environmental specimen (Bernard et al., 2002). Characteristically, *C. thomssenii* is the only *Corynebacterium* species expressing *N*-acetyl-β-glucosaminidase activity. *Corynebacterium sundsvallense* and *C. thomssenii* are phylogenetically more closely related to each other than to all other *Corynebacterium* species and phenotypically most closely resemble each other (Bernard et al., 2002; Table 2).

Corynebacterium ulcerans

Phylogenetically, *C. ulcerans* is most closely related to *C. pseudotuberculosis* and to *C. diphtheriae* (Pascual et al., 1995; Riegel et al., 1995d; Ruimy et al., 1995). This includes the ability to harbor the diphtheria *tox* gene, to express diphtheria toxin, and to show the typical cell wall with significant amounts of C16:1 ω 7c (Bernard et al., 1991). Disease associated with this bacterium is

rare, but if recovered from pseudomembranous material, the disease must be treated as diphtheria (Funke et al., 1997f; Efstratiou and George, 1999; Efstratiou et al., 2000). Cutaneous diphtheria-like disease has been reported associated with *C. ulcerans* (Wagner et al., 2001). Biochemically, these bacteria are similar to *C. pseudotuberculosis*, including reactivity in the reverse CAMP test (Table 2).

Corynebacterium urealyticum

This species has primarily been recovered from urinary tract infections (UTIs) and is especially associated with UTI patient urine with an alkaline pH, resulting in struvite crystals (Funke et al., 1997f; Hertig et al., 2000). This bacterium has been recovered from patients with other serious illnesses including bacteremias associated with pyelonephritis, nephrolithiasis, cancer, and HIV infection (Fernandez-Natal et al., 2001), pericarditis (Ojeda-Vargas et al., 2000), and soft tissue infection (Saavedra et al., 1996). Most biochemical tests are negative except that for urease, which is strongly positive. A PCR-based assay for detection of *C. urealyticum* has been described (Simoons-Smit et al., 2000). Ribotyping has been used to characterize these strains (Nieto et al., 2000). *Corynebacterium urealyticum* is almost always multidrug resistant, but rare penicillin-susceptible strains have also been described (Funke et al., 1997f).

Corynebacterium xerosis

This species was once thought to be a commonly recovered *Corynebacterium* species from clinical material. However, most isolates described in case reports for *C. xerosis* prior to 1996 apparently were most likely misidentified *C. amycolatum* strains, and *C. xerosis* is actually a very rare human pathogen (Funke et al., 1996a). This difficulty in discerning *C. amycolatum* from *C. xerosis* was also found to extend to strains found in commercial culture collections (Coyle et al., 1993a). The newly described species *C. freneyi* is closely related to *C. xerosis* and must be carefully excluded when differentiating these taxa.

Coryneform CDC Group F-1, CDC Group G, and the "Black Pigmented" *Corynebacterium* species

CDC group F-1 bacteria were first described in the Hollis and Weaver scheme (Hollis and Weaver, 1981) and, although genetically distinctive and consistent with the definition of the genus *Corynebacterium*, have not been validated as single or multiple species (Funke et al., 1997f). Phylogenetically, several CDC group F-1 strains

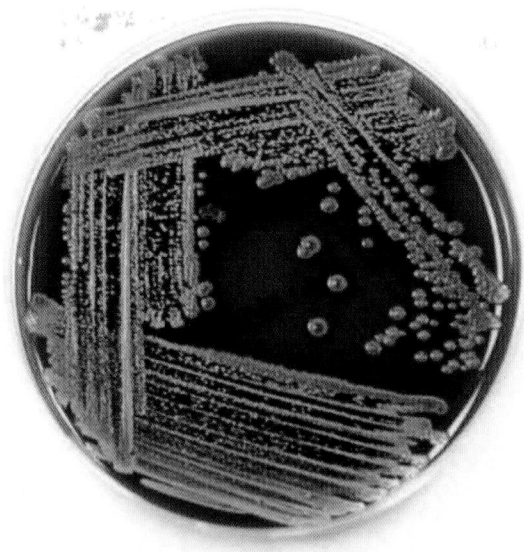

Fig. 2. A 24-h culture of "black pigmented *Corynebacterium* sp." (strain National Microbiology Laboratory [NML] identifier 92-0360) on sheep blood agar.

cluster with "*C. pseudogenitalium*" NCTC 11860 and CDC strain G5911 (Bernard et al., 2000). Isolates are urease producers and lipophilic. CDC group F-1 strains are usually susceptible to penicillin but are often resistant to macrolides.

Coryneform CDC Group G strains also possess features consistent with the genus *Corynebacterium* but could also not be validated as a single or multiple species (Riegel et al., 1995c). Phylogenetically, strains were found to cluster in a clade which includes CDC strain G5840 and "*C. tuberculostearicum*" ATCC 35962 (Bernard et al., 2000). Although phenotypically similar to *C. jeikeium*, these strains can be differentiated by ability to grow anaerobically and to ferment fructose (Riegel et al., 1994; Riegel et al., 1995c). CDC group G isolates are often multiresistant to antibiotics, particularly to macrolides and lincosamides.

Black-pigmented *Corynebacterium* species provisionally called "*C. nigricans*" (Fig. 2) include isolates from a vaginal specimen from a woman with a spontaneous abortion (Shukla et

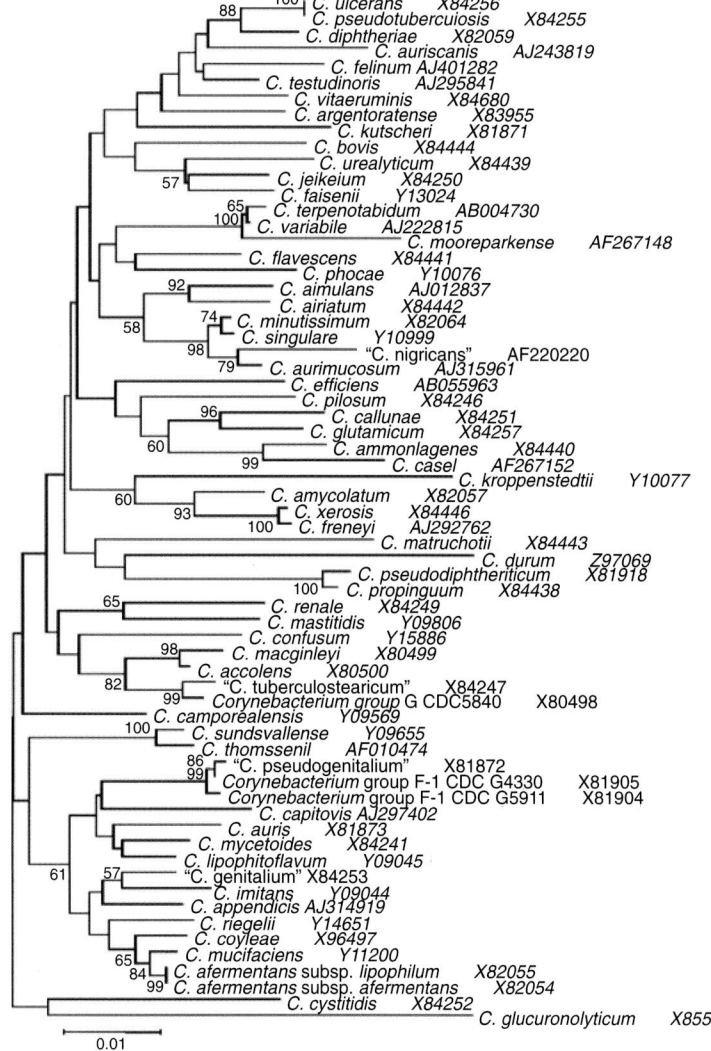

Fig. 3. Unrooted neighbor-joining tree, generated from a distance matrix calculated using the Kimura 2-parameter model of nucleotide substitution. Numbers at nodes represent bootstrap support values (percentage of 1000 resampled datasets that supported the node in the re-calculated tree). Values below 50% have been omitted. Scale bar represents the approximate genetic distance of 0.01 base substitutions per nucleotide pair.

al., 2001) as well as from a vaginal specimen and a vulvar ulcer (Bernard et al., 2002). These strains have genotypic and chemotaxonomic characteristics consistent with the genus *Corynebacterium*. However, phenotypically, they require differentiation from a "black pigmented" *Rothia*-like taxon first referred to as "CDC Group 4" by Hollis (1992), which phylogenetically falls within the *Rothia* genus (K. Bernard, personal observation).

Literature Cited

Aly, R., and H. I. Maibach. 1977. Aerobic microbial flora of intertriginous skin. Appl. Environ. Microbiol. 33:97–100.

Arsan, A. K., S. Sizmaz, S. B. Ozkan, and S. Duman. 1995–1996. Corynebacterium minutissimum endophthalmitis: Management with antibiotic irrigation of the capsular bag. Int. Ophthalmol. 19:313–316.

Aubel, D., F. N. R. Renaud, and J. Freney. 1997. Genomic diversity of several Corynebacterium species identified by amplification of the 16S-23S rRNA gene spacer regions. Int. J. Syst. Bacteriol. 47:767–772.

Bayston, R., C. Compton, and K. Richards. 1994. Production of extracellular slime by coryneforms colonizing hydrocephalus shunts. J. Clin. Microbiol. 32:1705–1709.

Belko, J., D. L. Wessel, and R. Malley. 2000. Endocarditis caused by Corynebcaterium diphtheriae: Case report and review of the literature. Ped. Infect. Dis. J. 19:159–163.

Bernard, K. A., M. Bellefeuille, and E. P. Ewan. 1991. Cellular fatty acid composition as an adjunct in the identification of asporogeneous, aerobic Gram-positive rods. J. Clin. Microbiol. 29:83–89.

Bernard, K., C. Munro, and D. Wiebe. 2000. *In:* Abstracts of the 100th Annual Meeting of the American Society of Microbiology. C106.

Bernard, K. A., C. Munro, D. Wiebe, and E. Ongansoy. 2002. Characteristics of rare or recently described Corynebacterium species recovered from human clinical material in Canada. J. Clin. Microbiol. 40:4375–4381.

Berner, R., K. Pelz, C. Wilhelm, A. Funke, J. U., Leititis, and M. Brandis. 1997. Fatal sepsis caused by Corynebacterium amycolatum in a premature infant. J. Clin. Microbiol. 35:1011–1012.

Björkroth, J., H. Korkeala, and G. Funke. 1999. rRNA gene RFLP as an identifcation tool for Corynebacterium species. Int. J. Syst. Bacteriol. 49:983–989.

Boot, R., H. Thuis, R. Bakker, and J. L. Veenema. 1994. Serological studies of Corynebacterium kutscheri and coryneform bacteria using an enzyme-linked immunosorbent assay (ELISA). Lab. Anim. 29:294–299.

Brandenburg, A., A. van Belkum, C. van Pelt, H. A. Bruining, J. W. Mouton, and H. A. Verbrugh. 1996. Patient-to-patient spread of a single strain of Corynebacterium striatum causing infections in a surgical intensive care unit. J. Clin. Microbiol. 34:2089–2094.

Brown, C. C., and H. J. Olander. 1987. Caseous lymphadenitis of goat and sheep: A review. Vet. Bull. 57:1–12.

Buck, G. A., and N. Groman. 1981. Physical mapping of beta-converting and gamma-nonconverting corynebacteriophage genomes. J. Bacteriol. 148:131–142.

Buck, G. A., R. E. Cross, T. P. Wong, J. Loera, and N. Groman. 1985. DNA relationships among some tox-bearing corynebacteriophages. Infect. Immun. 49:679–684.

Calalb, G., A. Saragea, P. Maximescu, N. Cioroianu, A. Popescu, S. Popa, and A. Mihailescu. 1961. Recherches sur un milieu liquide d'enrichissement pour le diagnostic bacteriologique de la diphtérie. Arch. Roum. Pathol. Exp. 20:95–101.

Christensen, H., M. Bisgaard, W. Frederiksen, R. Muttters, P. Kuhnert, and J. E. Olsen. 2001. Is characterization of a single isolate sufficient for valid publication of a new genus or species? Proposal to modify recommendation 30b of the Bacterial Code (1990 revision). Int. J. Syst. Evol. Microbiol. 51:2221–2225.

Chun, J., L. L. Blackall, S. O. Kang, Y. C. Hah, and M. Goodfellow. 1997. A proposal to reclassify Nocardia pinensis Blackall et al. as Skermania piniformis gen. nov., comb. nov. Int. J. Syst. Bacteriol. 47:127–131.

Cianciotto, N., R. Rappuoli, and N. Groman. 1986. Detection of homology to the beta bacteriophage integration site in a wide variety of Corynebacterium spp. J. Bacteriol. 168:103–106.

Cianciotto, N. P., and N. Groman. 1997. Characterization of bacteriophages from tox-containing, non-toxigenic isolates of Corynebacterium diphtheriae. Microb. Pathog. 22:343–351.

Claeys, G., H. Vanhouteghem, P. Riegel, G. Wauters, R. Hamerlynck, J. Dierick, J. de Witte, and H. Vaneechoutte. 1996. Endocarditis of native aortic and mitral valves due to Corynebacterium accolens: Report of a case and application of phenotypic and genotypic techniques for identification. J. Clin. Microbiol. 34:1290–1292.

Clarke, R., A. Qamruddin, M. Taylor, and H. Panigrahi. 1999. Septic arthritis caused by Corynebacterium amycolatum following vascular graft sepsis. J. Infect. 38:126–127.

Collins, M. D. 1982a. Corynebacterium mycetoides sp. nov., nom. rev. Zbl. Bakteriol. Hyg. 1. Abt. Orig. C3:399–400.

Collins, M. D. 1982b. Reclassification of Bacterionema matruchotii (Mendel) in the genus Corynebacterium, as Corynebacterium matruchotii comb. nov. Zbl. Bakteriol. Hyg. Abt.1 Orig. C3:364–367.

Collins, M. D., and D. Jones. 1983. Corynebacterium minutissimum sp. nov. nom. rev. Int. J. Syst. Bacteriol. 33:870–871.

Collins, M. D., and C. S. Cummins. 1986. Genus Corynebacterium. *In:* P. H. A. Sneath, N. S. Mair, M. E. Sharpe, and J. G. Holt (Eds.)Bergey's Manual of Systematic Bacteriology. Williams and Wilkins. Baltimore, MD. 2:1266–1276.

Collins, M. D., R. A. Burton, and D. Jones. 1988. Corynebacterium amycolatum sp.nov., a new mycolic acid-less Corynebacterium species from human skin. FEMS Microbiol. Lett. 49:349–352.

Collins, M. D., E. Falsen, E. Akervall, B. Sjöden, and A. Alvarez. 1998. Corynebacterium kroppenstedtii sp.nov., a novel corynebacterium that does not contain mycolic acids. Int. J. Syst. Bacteriol. 48:1449–1454.

Collins, M. D., K. A. Bernard, R. A. Hutson, B. Sjöden, A. Nyberg, and E. Falsen. 1999a. Corynebacterium sundsvallense sp.nov., from human clinical specimens. Int. J. Syst. Bacteriol. 49:361–366.

Collins, M. D., L. Hoyles, P. A. Lawson, E. Falsen, R. L. Robson, and G. Foster. 1999b. Phenotypic and phylogenetic characterization of a new Corynebacterium species from dogs: Corynebacterium auriscanis sp. nov. J. Clin. Microbiol. 37:3443–3447.

Connor, K. M., M. M. Quirie, G. Baird, and W. Donachie. 2000. Characterization of United Kingdom isolates of Corynebacterium pseudotuberculosis using pulsed-field gel electrophoresis. J. Clin. Microbiol. 38:2633–2637.

Corboz, L., R. Thoma, U. Braun, and R. Zbinden. 1996. Isolierung von Corynebacterium diphtheriae subsp. belfanti bei einer Kuh mit chronisch-aktiver Dermatitis. Schweiz. Arch. Tierheilk. 138:596–599.

Coyle, M. B., and B. A. Lipsky. 1990. Coryneform bacteria in infectious diseases: Clinical and laboratory aspects. Clin. Microbiol. Rev. 3:227–246.

Coyle, M. B., R. B. Leonard, D. J. Nowowiejski, A. Malekniazi, and D. J. Finn. 1993a. Evidence of multiple taxa within commercially available reference strains of Corynebacterium xerosis. J. Clin. Microbiol. 31:1788–1793.

Coyle, M. B., D. J. Nowowiejski, J. Q. Russell, and N. B. Gorman. 1993b. Laboratory review of reference strains of Corynebacterium diphtheriae indicated mistyped intermedius strains. J. Clin. Microbiol. 31:3060–3062.

Cuevas, W. A., and J. G. Songer. 1993. Arcanobacterium haemolyticum phospholipase D is genetically and functionally similar to Corynebacterium pseudotuberculosis phospholipase D. Infect. Immun. 61:4310–4316.

De Briel, D., J.-C. Langs, G. Rougeron, P. Chabot, and A. LeFaou. 1991. Multiresistant corynebacteria in bacteriuria: A comparative study of Corynebacterium group D2 and Corynebacterium jeikeium. J. Hosp. Infect. 17:35–43.

De Briel, D., F. Couderc, P. Riegel, C. Gallion, J.-C. Langs, and F. Jehl. 1992. Contribution of high-performance liquid chromatography to the identification of some Corynebacterium species by comparison of their corynemycolic acid patterns. Res. Microbiol. 143:191–198.

de Miguel, I., E. Rodriguez, and A. M. Martin. 1999. Corynebacterium amycolatum sepsis in hematologic patients. Enferm. Infecc. Microbiol. Clin. 17:340–341.

de Miguel-Martinez, I., F. Fernandez-Fuertes, A. Ramos-Macias, J. M. Bosch-Benitez, and A. M. Martin-Sanchez. 1996. Sepsis due to multiresistant Corynebacterium amycolatum. Eur. J. Clin. Microbiol. Infect. Dis. 15:617–618.

de Miguel-Martinez, I., A. Ramos-Macias, and A. M. Martin-Sanchez. 1999. Otitis media due to Corynebacterium jeikeium. Eur. J. Clin. Microbiol. Infect. Dis. 18:231–232.

Devriese, L., P. Riegel, J. Hommez, M. Vaneechoutte, T. de Baere, and F. Haesebrouck. 2000. Identification of Corynebacterium glucuronolyticum strains from the urogenital tract of humans and pigs. J. Clin. Microbiol. 38:4657–4659.

De Zoysa, A. S., A. Efstratiou, R. C. George, M. Jahkola, J. Vuopio-Varkila, S. Deshevoi, G. Tseneva, and Y. Rikushin. 1995. Molecular epidemiology of Corynebacterium diphtheriae from Northwestern Russia and surrounding countries studied by using ribotyping and pulsed field gel electrophoresis. J. Clin. Microbiol. 33:1080–1083.

De Zoysa, A. S., and A. Efstratiou. 1999. PCR typing of Corynebacterium diphtheriae by random amplification of polymorphic DNA. J. Med. Microbiol. 48:335–340.

De Zoysa, A. S., and A. Efstratiou. 2000. Use of amplified fragment length polymorphisms for typing Corynebacterium diphtheriae. J. Clin. Microbiol. 38:3843–3845.

Dykhuizen, R. S., G. Douglas, J. Weir, and I. M. Gould. 1995. Corynebacterium afermentans subsp. lipophilum: Multiple abscess formation in brain and liver. Scand. J. Infect. Dis. 27:637–639.

Efstratiou, A., and R. C. George. 1996. Microbiology and epidemiology of diphtheria. Rev. Med. Microbiol. 7:31–42.

Efstratiou, A., K. H. Engler, C. S. Dawes, and D. Sesardic. 1998. Comparison of phenotypic and genotypic methods for detection of diphtheria toxin among isolates of pathogenic corynebacteria. J. Clin. Microbiol. 36:3173–77.

Efstratiou, A., and R. C. George. 1999. Laboratory guidelines for the diagnosis of infections caused by Corynebacterium diphtheriae and C. ulcerans. WHO Comm. Dis. Publ. Hlth. 2:250–257.

Efstratiou, A., K. H. Engler, I. K. Mazurova, T. Glushkevich, J. Vuopio-Varkila, and T. Popovic. 2000. Current approaches to the laboratory diagnosis of diphtheria. J. Infect. Dis. 181, Suppl. 1:138–145.

Elek, S. D. 1949. The plate virulence test for diphtheria. J. Clin. Pathol. 2:250–258.

Engler, K. H., T. Glushkevich, I. Z. Mazurova, R. C. George, and A. Efstratiou. 1997. A modified Elek test for detection of toxigenic corynebacteria in the diagnostic laboratory. J. Clin. Microbiol. 35:495–498.

Engler, K. H., and A. Efstratiou. 2000. Rapid enzyme immunoassay for determination of toxigenicity among clinical isolates of corynebacteria. J. Clin. Microbiol. 38:1385–1389.

Engler, K. H., A. Efstratiou, D. Norn, R. S. Kozlov, I. Selga, T. G. Glushkevich, M. Tam, V. G. Melnikov, I. K. Mazurova, V. E. Kim, G. Y. Tseneva, L. P. Titov, and R. C. George. 2002. Immunochromatographic strip test for rapid detection of diphtheria toxin: Description and multicenter evaluation in areas of low and high prevalance of diphtheria. J. Clin. Microbiol. 40:80–83.

Esteban, J., E. Nieto, R. Calvo, R. R. Fernandez-Robals, P. L. Valero-Guillen, and F. Soriano. 1999. Microbiological characterization and clinical significance of Corynebacterium amycolatum strains. Eur. J. Clin. Microbiol. Infect. Dis. 18:518–521.

Estrangin, E., B. Thiers, and Y. Peloux. 1987. Apport des microméthodes et d'analyse en chromatographie en phase gazeuse des acides carboxyliques issus de la fermentation du glucose dans l'identification des corynébactéries. Ann. Biol. Clin. 45:285–289.

Fernandez-Ayala, M., D. N. Nan, and M. C. Farinas. 2001. Vertebral osteomyelitis due to Corynebacterium striatum. Am. J. Med. 111:167.

Fernandez-Garayzabal, J. F., M. D. Collins, R. A. Hutson, E. Fernandez, R. Monasterio, J. Marco, and L. Dominguez. 1997. Corynebacterium mastitidis sp. nov., isolated from milk of sheep with subclinical mastitis. Int. J. Syst. Bacteriol. 47:1082–1085.

Fernandez-Garayzabal, J. F., M. D. Collins, R. A. Hutson, I. Gonzalez, E. Fernandez, and L. Dominguez. 1998. Corynebacterium camporealensis sp.nov., associated with subclinical mastitis in sheep. Int. J. Syst. Bacteriol. 48:463–468.

Fernandez Giron, F., J. M. Saavedra Martin, M. Benitez Sanchez, F. Fernandez Mora, and E. Rodriguez Gomez. 1998. Corynebacterium minutissimum peritonitis in a CAPD patient. Perit. Dial. Int. 18:345–346.

Fernandez-Natal, I., J. Guerra, M. Alcoba, F. Cachon, and F. Soriano. 2001. Bacteremia caused by multiply resistant

Corynebacterium urealyticum: Six case reports and review. Eur. J. Clin. Microbiol. Infect. Dis. 20:514–517.

Frobisher, M. 1937. Cystine-tellurite agar for C. diphtheriae. J. Infect. Dis. 10:99–105.

Früh, M., A. von Graevenitz, and G. Funke. 1998. Use of second-line biochemical and susceptibility tests for the differential identification of coryneform bacteria. Clin. Microbiol. Infect. 4:332–338.

Funke, G., S. Stubbs, G. E. Pfyffer, M. Marchiani, and M. D. Collins. 1994. Turicella otitidis gen.nov. sp,nov., a coryneform bacterium isolated from patients with otitis media. Int. J. Syst. Bacteriol. 44:270–273.

Funke, G., K. A. Bernard, C. Bucher, G. E. Pfyffer, and M. D. Collins. 1995a. Corynebacterium glucuronolyticum sp.nov., isolated from male patients with genitourinary infections. Med. Microbiol. Lett. 4:204–215.

Funke, G., P. A. Lawson, and M. D. Collins. 1995b. Heterogeneity within Centers for Disease Control and Prevention coryneform group ANF-1 like bacteria and description of Corynebacterium auris sp. nov. Int. J. Syst. Bacteriol. 45:735–739.

Funke, G., P. A. Lawson, K. A. Bernard, and M. D. Collins. 1996a. Most Corynebacterium xerosis strains identified in the routine clinical laboratory correspond to Corynebacterium amycolatum. J. Clin. Microbiol. 34:1124–1128.

Funke, G., V. Pünter, and A. von Graevenitz. 1996b. Antimicrobial susceptibility patterns of some recently established coryneform bacteria. Antimicrob. Agents Chemother. 40:2874–2878.

Funke, G., A. Efstratiou, D. Kuklinska, R. A. Hutson, A. De Zoysa, K. H. Engler, and M. D. Collins. 1997a. Corynebacterium imitans sp. nov., isolated from patients with suspected diphtheria. J. Clin. Microbiol. 35:1978–1983.

Funke, G., R. A. Hutson, M. Hilleringman, W. R. Heizmann, and M. D. Collins. 1997b. Corynebacterium lipophiloflavum sp.nov. isolated from a patient with bacterial vaginosis. FEMS Microbiol. Lett. 150:219–224.

Funke, G., P. A. Lawson, and M. D. Collins. 1997c. Corynebacterium mucifaciens sp. nov., an unusual species from human clinical material. Int. J. Syst. Bacteriol. 47:952–957.

Funke, G., C. Pascual Ramos, and M. D. Collins. 1997d. Corynebacterium coyleae sp.nov. isolated from human clinical specimens. Int. J. Syst. Bacteriol. 47:92–96.

Funke, G., F. N. R. Renaud, J. Freney, and P. Riegel. 1997e. Multicenter evaluation of the updated and extended API (RAPID) Coryne database 2.0. J. Clin. Microbiol. 35:3122–312.

Funke, G., A. von Graevenitz, J. Clarridge, and K. A. Bernard. 1997f. Clinical microbiology of coryneform bacteria. Clin. Microbiol. Rev. 10:3122–3126.

Funke, G., P. A. Lawson, and M. D. Collins. 1998a. Corynebacterium riegelii sp.nov., an unusual species isolated from female patients with urinary tract infections. J. Clin. Microbiol. 36:624–627.

Funke, G., C. R. Osorio, R. Frei, P. Riegel, and M. D. Collins. 1998b. Corynebacterium confusum sp. nov. isolated from human clinical specimens. Int. J. Syst. Bacteriol. 47:92–96.

Funke, G., M. Pagano-Niederer, and W. Bernauer. 1998c. Corynebacterium macginleyi has to date been isolated exclusively from conjunctival swabs. J. Clin. Microbiol. 36:3670–3673.

Funke, G., K. Peters, and M. Aravena-Roman. 1998d. Evaluation of the RapID CB Plus system for identification of coryneform bacteria and Listeria spp. J. Clin. Microbiol. 36:2439–2442.

Funke, G., M. Altwegg, L. Frommelt, and A. von Graevenitz. 1999a. Emergence of related nontoxigenic Corynebacterium diphtheriae biotype mitis strains in Western Europe. Emerg. Infect. Dis. 5:477–480.

Funke, G., and K. A. Bernard. 1999b. Coryneform Grampositive Rods. In: P. R. Murray, E. J. Baron, M. A. Pfaller, F. C. Tenover, and R. H. Yolken (Ed.) Manual of Clinical Microbiology, 7th ed. ASM Press. Washington, DC. 319–345.

Garcia-Bravo, M., J. A. Aguado, J. M. Morales, and A. R. Noriega. 1997. Efficacy of a selective and differential medium for isolating Corynebacterium urealyticum from urine specimens. Clin. Microbiol. Infect. 3:555–558.

Gilbert, L. 1997. Infections with Corynebacterium diphtheriae: Changing epidemiology and clinical manifestations. Comm. Dis. Intell. 21:161–164.

Golaz, A., I. R. Hardy, P. Strebel, K. M. Bisgard, C. Vitek, T. Popovic, and M. Wharton. 2000. Epidemic diphtheria in the newly independent states of the former Soviet Union: Implications for diphtheria control in the United States. J. Infect. Dis. 181, Suppl. 1:S237–S243.

Golaz, A., C. Vitek, T. Popovic, and M. Wharton. 2001. Epidemiology of diphtheria in the 1990s. Clin. Microbiol. Newslett. 23:33–37.

Gomez, A., C. Nombela, J. Zapardiel, and F. Soriano. 1995. An encrusted cystitis caused by Corynebacterium urealyticum in a dog. Austral. Vet. J. 72:72–73.

Gordts, F., S. Halewyck, D. Pierard, L. Kaufman, and P. A. R. Clement. 2000. Micro-biology of the middle meatus: A comparison between normal adults and children. J. Laryngol. Otol. 114:184–188.

Greathead, M. M., and P. J. N. R. Bisschop. 1963. A report on the occurrence of C. diphtheriae in dairy cattle. S. Afr. Med. J. 37:1261–1262.

Gruner, E., P. L. F. Zuber, G. Martinetti Lucchini, A. von Graevenitz, and M. Altwegg. 1992. A cluster of nontoxigenic Corynebacterium diphtheriae infections among Swiss intravenous drug abusers. Med. Microbiol. Lett. 1:160–167.

Gruner, E., M. Opravil, M. Altwegg, and A. von Graevenitz. 1994. Nontoxigenic Corynebacterium diphtheriae isolated from intravenous drug users. Clin. Infect. Dis. 18:94–96.

Gutierrez-Rodero, F., V. Ortiz de la Tabla, C. Martinez, M. M. Masia, A. Mora, C. Escolano, E. Gonzalez, and A. Martin-Hidalgo. 1999. Corynebacterium pseudodiphtheriticum: An easily missed respiratory pathogen in HIV-infected patients. Diagn. Microbiol. Infect. Dis. 33:209–216.

Hallas, G., T. G. Harrison, D. Samuel, and G. Colman. 1990. Detection of diphtheria toxin in culture supernates of Corynebacterium diphtheriae and C. ulcerans by immunoassay with monoclonal antibody. J. Med. Microbiol. 32:247–253.

Harnisch, J. P., E. Tronca, C. M. Nolan, M. Turck, K. K. Holmes. 1989. Diphtheria among alcoholic urban adults. Ann. Intern. Med. 111:71–82.

Hauser, D., M. R. Popoff, M. Kiredjian, P. Bouquet, and F. Bimet. 1993. Polymerase chain reaction assay for diagnosis of potentially toxinogenic Corynebacterium:diphtheriae strains: Correlation with ADP-ribosylation activity assay. J. Clin. Microbiol. 31:2720–2723.

Hayashi, A., R. Yanagawa, and H. Kida. 1985. Survival of Corynebacterium renale, Corynebacterium pilosum, and Corynebacterium cystitidis in soil. Vet. Microbiol. 10:381–386.

Hemsley, C., S. Abraham, and S. Rowland-Jones. 1999. Corynebacterium pseudodiphtheriticum—a skin pathogen. Clin. Infect. Dis. 29(4):938–939.

Henricson, B., M. Segarra, J. Garvin, J. Burns, S. Jenkins, C. Kim, T. Popovic, A. Golaz, and B. Akey. 2000. Toxigenic Corynebacterium diphtheriae associated with an equine wound infection. J. Vet. Diagn. Invest. 12:253–257.

Henry, J. E. 1920. Milk-borne diphtheria: An outbreak traced to infection of a milk handler's finger with B. diphtheriae. J. Am. Med. Assn. 75:1715–1716.

Hertig, A., C. Duvic, Y. Chretien, P. Jungers, J. P. Grunfeld, and P. Rieu. 2000. Encrusted pyelitis of native kidneys. J. Am. Soc. Nephrol. 11:1138–1140.

Hiramune, T., R. Kudoh, N. Kikuchi, and R. Yanagawa. 1988. Selective medium for isolation of urinary corynebacteria and detection of the organisms from bovine vulva and vaginal vestibule using the medium. Jap. J. Vet. Sci. 50:111–114.

Hogg, G. G., J. E. Strachan, L. Huayi, S. A. Beaton, P. M. Robinson, and K. Taylor. 1996. Non-toxigenic Corynebacterium diphtheriae biovar gravis: Evidence for an invasive clone in a south-eastern Australian community. Med. J. Austral. 164:72–75.

Hollis, D. G., and R. E. Weaver. 1981. Gram-positive Organisms: A Guide to Identification. Special Bacteriology Section, Centers for Disease Control. Atlanta, GA.

Hollis, D. G. 1992. Potential New CDC Coryneform Groups. Handout at the 92nd Annual Meeting of the American Society of Microbiology, Washington, DC.

Holmes, R. K. 2000. Biology and molecular epidemiology of diphtheria toxin and the tox gene. J. Infect. Dis. 181, Suppl. 1:156–167.

Hommez, J., L. A. Devriese, M. Vaneechoutte, P. Riegel, P. Butaye, and F. Haese-brouck. 1999. Identification of nonlipophilic corynebacteria isolated from dairy cows with mastitis. J. Clin. Microbiol. 37:954–957.

Hoyle, L. 1941. A tellurite blood agar medium for the rapid diagnosis of diphtheria. Lancet 1:175–176.

Izurieta, H. S., P. M. Strebel, T. Youngblood, D. G. Hollis, and T. Popovic. 1997. Exudative pharyngitis possibly due to Corynebacterium pseudodiphtheriticum, a new challenge in the differential diagnosis of diphtheria. Emerg. Infect. Dis. 3:65–68.

Jackman, P. J. H., D. G. Pitcher, S. Pelczynska, and P. Borman. 1987. Classification of corynebacteria associated with endocarditis (Group JK) as Corynebacterium jeikeium sp. nov. Syst. Appl. Microbiol. 9:83–90.

Jones, R. N., D. J. Biedenbach, D. M. Johnson, M. A. Pfaller. 2001. In vitro evaluation of BI 397, a novel glycopeptide antimicrobial agent. J. Chemother. 13:244–254.

Joussen, A. M., G. Funke, F. Joussen, and G. Herbertz. 2000. Corynebacterium macginleyi: A conjunctiva specific pathogen. Br. J. Ophthalmol. 84:1420–1422.

Juurlink, D. N., A. Borczyk, and A. E. Simor. 1996. Native valve endocarditis due to Corynebacterium striatum. Eur. J. Clin. Microbiol. Infect. Dis. 15:963–965.

Kämpfer, P., M. A. Andersson, F. A. Rainey, R. M. Kroppenstedt, and M. Salkinoja-Salonen. 1999. Williamsia muralis gen. nov., sp. nov., isolated from the indoor environment of a children's day care centre. Int. J. Syst. Bacteriol. 49:681–687.

Keijman, J. M. G., M. R. Luirink, G. Ramsay, and J. A. Jacobs. 2000. Native valve endocarditis due to Corynebacterium striatum. Clin. Microbiol. Newslett. 22:125–127.

Kerry-Williams, S. M., and W. C. Noble. 1986. Plasmids in group JK coryneform bacteria isolated in a single hospital. J. Hosp. Infect. 97:255–263.

Kersten, H. E. 1909. Ueber die Haltbarkeit der Diphtherie- und Paratyphus B-Bazillen in der Milch. Arb. Kaiserl. Gesundheitsamt 30:341–350.

Khabbaz, R. F., J. B. Kaper, M. R. Moody, S. C. Schimpff, and J. H. Tenney. 1986. Molecular epidemiology of group JK Corynebacterium on a cancer ward: Lack of evidence for pateint-to-patient transmission. J. Infect. Dis. 154:95–99.

Knox, K. L., and A. H. Holmes. 2002. Nosocomial endocarditis caused by Corynebacterium amycolatum and other nondiphtheriae corynebacteria. Emerg. Infect. Dis. 8:97–99.

Kumari, P., A. Tyagi, P. Marks, and K. G. Kerr. 1997. Corynebacterium afermentans spp. afermentans in a neurosurgical patient. J. Infect. 35:201–202.

Lagrou, K., J. Verhaegen, M. Janssens, G. Wauters, and L. Verbist. 1998. Prospective study of catalase-positive coryneform organisms in clinical specimens: Identification, clinical relevance, and antibiotic susceptibility. Diagn. Microbiol. Infect. Dis. 30:7–15.

Larson, E. L., K. J. McGinley, A. R. Foglia, G. H. Talbot, and J. J. Leyden. 1986a. Composition and antimicrobic resistance of skin flora in hospitalized and healthy adults. J. Clin. Microbiol. 23:604–608.

Larson, E. L., K. J. McGinley, J. J. Leyden, M. E. Cooley, and G. H. Talbot. 1986b. Skin colonization with antibiotic-resistant (JK group) and antibiotic-sensitive lipophilic diphtheroids in hospitalized and normal adults. J. Infect. Dis. 153:701–706.

Larsson, P., B. Brinkhoff, and L. Larsson. 1987. Corynebacterium diphtheriae: In the environment of carriers and patients. J. Hosp. Infect. 10:282–286.

Leonard, R. B., D. J. Nowowiejski, J. J. Warren, D. J. Finn, and M. B. Coyle. 1994. Molecular evidence of person-to-person transmiussion of a pigmented strain of Corynebacterium striatum in intensive care units. J. Clin. Microbiol. 32:164–169.

Li, A., and S. Lal. 2000. Corynebacterium pseudodiphtheriticum keratitis and conjunctivitis: A case report. Clin. Exp. Ophthalmol. 28:60–61.

Lindenmann, K., A. von Graevenitz, and G. Funke. 1995. Evaluation of the Biolog system for the identification of asporogenous, aerobic Gram-positive rods. Med. Microbiol. Lett. 4:287–296.

Lipsky, B. A., A. C. Goldberger, L. C. Tompkins, and J. J. Plorde. 1982. Infections caused by nondiphtheria corynebacteria. Rev. Infect. Dis. 4:1220–1235.

MacFaddin, J. F. 1985. Media for Isolation-cultivation-identification-maintenance of Medical Bacteria. Williams and Wilkins. Baltimore, MD.

MacGregor, R. R. 2000. Corynebacterium diphtheriae. In: G. L. Mandell, J. E. Bennett, and R. DolinPrinciples and Practice of Infectious Diseases, 5th ed. Churchill Livingstone. Philadelphia, PA. 2190–2197.

Marston, C. K., F. Jamieson, F. Cahoon, G. Lesiak, A. Golaz, M. Reeves, and T. Popovic. 2001. Persistence of a distinct Corynebacterium diphtheriae clonal group within two communities in the United States and Canada where diphtheria is endemic. J. Clin. Microbiol. 39:1586–1590.

Martaresche, C., P. E. Fournier, V. Jacomo, M. Gainnier, A. Boussuge, and M. Drancourt. 1999. A case of Corynebacterium pseudodiphtheriticum nosocomial pneumonia. Emerg. Infect. Dis. 5:722–723.

Martinez-Martinez, L., A. I. Suarez, J. Winstanley, M. C. Ortega, and K. Bernard. 1995. Phenotypic characteristics of 31 strains of Corynebacterium striatum isolated from clinical samples. J. Clin. Microbiol. 33: 2458–2461.

Martinez-Martinez, L., A. Pascual, K. Bernard, and A. I. Suarez. 1996. Antimicrobial susceptibility pattern of Corynebacterium striatum. Antimicrob. Agents Chemother. 40:2671–2672.

Martinez-Martinez, L., A. I. Suarez, J. Rodriguez-Bano, K. Bernard, and M. A. Muniain. 1997. Clinical significance of Corynebacterium striatum isolated from human samples. Clin. Microbiol. Infect. 3:634–639.

Marty, N., L. Agueda, L. Lapchine, D. Clave, S. Henry-Ferry, and G. Chabanon. 1991. Adherence and hemagglutination of Corynebacterium Group D2. Eur. J. Clin. Microbiol. Infect. Dis. 10:20–24.

Mattos-Guaraldi, A. L., L. C. Duarte Formiga, and G. Andrade Pereira. 2000. Cell surface components and adhesion in Corynebacterium diphtheriae. Microb. Infect. 2:1507–1512.

McNamara, P. J., W. A. Cuevas, and J. G. Songer. 1995. Toxic phospholipases D of Corynebacterium pseudotuberculosis, C.ulcerans, and Arcanobacterium haemolyticum: Cloning and sequence homology. Gene 156:113–118.

Mikhailovich, V. M., V. G. Melnikov, I. K. Mazurova, I. K. Wachsmuth, J. D. Wenger, M. Wharton, H. Nakao, and T. Popovic. 1995. Application of PCR for detection of toxigenic Corynebacterium diphtheriae strains isolated during the Russian diphtheria epidemic, 1990 through 1994. J. Clin. Microbiol. 33:3061–3063.

Mitscherlich, E., and E. H. Marth. 1984. Microbial Survival in the Environment. Springer-Verlag. New York, NY. 140–147.

Miyamura, K., S. Nishio, A. Ito, R. Muata, and R. Kono. 1974. Micro cell culture method for determination of diphtheria toxin and antitoxin titre using Vero cells. J. Biol. Stand. 2:189–201.

Nath, N., and J. K. Deb. 1995. Partial characterization of small plasmids of Corynebacterium renale. Plasmid 34:229–233.

Neubauer, M., J. Sourek, M. Ryc, J. Bohacek, M. Mara, and J. Mnukova. 1991. Corynebacterium accolens sp.nov., a Gram-positive rod exhibiting satellitism, from clinical material. Syst. Appl. Microbiol. 14:46–51.

Nieto, E., J. Zapardiel, and F. Soriano. 1996. Environmental contamination by Corynebacterium urealyticum in a teaching hospital. J. Hosp. Infect. 32:78–79.

Nieto, E., A. Vindel, P. L. Valero-Guillen, J. A. Saez-Nieto, and F. Soriano. 2000. Biochemical, antimicrobial susceptibility and genotyping studies on Corynebacterium urealyticum isolates from diverse sources. J. Med. Microbiol. 49:759–763.

Oberreuter, H., H. Seiler, and S. Scherer. 2002. Identification of coryneform bacteria and related texa by Fourier-transform infrared (FT-IR) spectroscopy. Int. J. Syst. Evol. Microbiol. 52:91–100.

Ojeda-Vargas, M., M. A. Gonzalez-Fernandez, D. Romero, A. Cedres, and C. Monzon-Moreno. 2000. Pericarditis caused by Corynebacterium urealyticum. Clin. Microbiol. Infect. 6:560–561.

Oteo, J., B. Aracil, J. Ignacio Alos, and J. Luis Gomez-Garces. 2001. Significant bacteremias by Corynebacterium amycolatum: An emergent pathogen. Enferm. Infecc. Microbiol. Clin. 19:103–106.

Pallen, M. J. 1991. Rapid screening for toxigenic Corynebacterium diphtheriae by the polymerase chain reaction. J. Clin. Pathol. 44:1025–1026.

Pappenheimer, A. M., and S. J. Johnson. 1936. Studies in diphtheria toxin production. I: The effect of iron and copper. Br. J. Exp. Pathol. 17:335–341.

Pascual, C., P. A. Lawson, J. A. E. Farrow, M. Navarro Gimenez, and M. D. Collins. 1995. Phylogenetic analysis of the genus Corynebacterium based on 16S rRNA gene sequences. Int. J. Syst. Bacteriol. 45:724–728.

Pascual, C., G. Foster, N. Alvarez, and M. D. Collins. 1998. Corynebacterium phocae sp.nov., isolated from the common seal (Phoca vitulina). Int. J. Syst. Bacteriol. 48:601–604.

Peel, M. M., G. G. Palmer, A. M. Stacpoole, and T. G. Kerr. 1997. Human lymphadenitis due to Corynebacterium pseudotuberculosis: Report of ten cases from Australia and review. Clin. Infect. Dis. 24:185–191.

Petit, P. L. C., J. W. Bok, J. Thompson, A. G. M. Buiting, and M. B. Coyle. 1994. Native-valve endocarditis due to CDC Coryneform group ANF-3: Report of a case and review of corynebacterial endocarditis. Clin. Infect. Dis. 19:897–901.

Pitcher, D. G. 1978. Aerobic cutaneous coryneforms: Recent taxonomic findings. Br. J. Dermatol. 98:363–370.

Pitcher, D., A. Johnson, F. Allerberger, N. Woodford, and R. George. 1990. An investigation of nosocomial infection with Corynebacterium jeikeium in surgical patients using a ribosomal RNA gene probe. Eur. J. Clin. Microbiol. Infect. Dis. 9:643–648.

Popovic, T., C. Kim, J. Reiss, M. Reeves, H. Nakao, and A. Golaz. 1999. Use of molecular subtyping to document long-term persistence of Corynebacterium diphtheriae in South Dakota. J. Clin. Microbiol. 37: 1092–1099.

Popovic, T., I. K. Mazurova, A. Efstratiou, J. Vuopio-Varkila, M. W. Reeves, A. De Zoysa, T. Glushkevich, and P. Grimont. 2000. Molecular epidemiology of diphtheria. J. Infect. Dis. 181, Suppl. 1:168–177.

Puech, V., M. Chami, A. Lemassu, M. A. Laneelle, B. Schiffler, P. Gounon, B. Bayan, R. Benz, and M. Daffe. 2001. Structure of the cell envelope of corynebacteria: Importance of non-covalently bound lipids in the formation of the cell wall permeability barrier and fracture plane. Microbiology 147:1365–1382.

Quinn, J. P., P. M. Arnow, D. Weil, and J. Rosenbluth. 1984. Outbreak of JK diphtheroid infections associated with environmental contamination. J. Clin. Microbiol. 19:668–671.

Rassoulian-Barrett, S. L., B. T. Cookson, L. C. Carlson, K. A. Bernard, and M. B. Coyle. 2001. Diversity within reference strains of Corynebacterium matruchotii includes Corynebacterium durum and a novel organism. J. Clin. Microbiol. 39:943–948.

Reacher, M., M. Ramsay, J. White, A. De Zoysa, A. Efstratiou, G. Mann, A. Mackay, and R. C. George. 2000. Nontoxigenic Corynebacterium diphtheriae: An emerging pathogen in England and Wales? Emerg. Infect. Dis. 6:640–645.

Reddy, C. A., and M. Kao. 1978. Value of acid metabolic products in identification of certain corynebacteria. J. Clin. Microbiol. 7:428–433.

Reinhardt, D. J., A. Lee, and T. Popovic. 1998. Antitoxin-in-membrane and antitoxin-in-well assays for detection of toxigenic Corynebacterium diphtheriae. J. Clin. Microbiol. 36:207–210.

Renaud, F. N. R., A. Gregory, C. Barreau, D. Aubel, and J. Freney. 1996. Identification of Turicella otitidis isolated from a patient with otorrhea associated with surgery: Differentiation from Corynebacterium afermentans and Corynebacterium auris. J. Clin. Microbiol. 34:2625–2627.

Renaud, F. N. R., D. Aubel, P. Riegel, H. Meugnier, and C. Bollet. 2001. Corynebacterium freneyi sp. nov., a new species of alpha-glucosidase-positive strains related to Corynebacterium xerosis. Int. J. Syst. Evol. Microbiol. 51:1723–1728.

Riegel, P., D. De Briel, G. Prevost, F. Jehl, and H. Monteil. 1993a. Proposal of Corynebacterium propinquum sp.nov. for Corynebacterium group ANF-3 strains. FEMS Microbiol. Lett. 113:229–234.

Riegel, P., D. De Briel, G. Prevost, F. Jehl, and H. Monteil. 1993b. Taxonomic study of Corynebacterium group ANF-1 strains: Proposal of Corynebacterium afermentans sp. nov. containing the subspecies C.afermentans subsp.afermentans subsp. nov. and C. afermentans subsp. lipophilum subsp. nov. Int. J. Syst. Bacteriol. 43:287–92.

Riegel, P., D. De Briel, G. Prevost, F. Jehl, and H. Monteil. 1994. Genomic diversity among Corynebacterium jeikeium strains and comparison with biochemical characteristics. J. Clin. Microbiol. 32:1860–1865.

Riegel, P., R. Ruimy, D. De Briel, G. Prevost, F. Jehl, F. Bimet, R. Christen, and H. Monteil. 1995a. Corynebacterium argentoratense sp. nov., from human throat. Int. J. Syst. Bacteriol. 45:533–537.

Riegel, P., R. Ruimy, D. De Briel, G. Prevost, F. Jehl, F. Bimet, R. Christen, and H. Monteil. 1995b. Corynebacterium seminale sp. nov., a new species associated with genital infections in male patients. J. Clin. Microbiol. 33:2244–2249.

Riegel, P., R. Ruimy, D. De Briel, G. Prevost, F. Jehl, R. Christen, and H. Monteil. 1995c. Genomic diversity and phylogenetic relationships among lipid-requiring diphtheroids from humans and characterization of Corynebacterium macginleyi sp. nov. Int. J. Syst. Bacteriol. 45:128–133.

Riegel, P., R. Ruimy, D. De Briel, G. Prevost, F. Jehl, R. Christen, and H. Monteil. 1995d. Taxonomy of Corynebacterium diphtheriae and related taxa, with recognition of Corynebacterium ulcerans sp. nov., nom. rev. FEMS Microbiol. Lett. 126:271–276.

Riegel, P., R. Ruimy, R. Christen, and H. Monteil. 1996. Species identities and antimicrobial susceptibilities of corynebacteria isolated from various clinical sources. Eur. J. Clin. Microbiol. Infect. Dis. 15:657–662.

Riegel, P., R. Heller, G. Prevost, F. Jehl, and H. Monteil. 1997a. Corynebacterium durum sp. nov., from human clinical specimens. Int. J. Syst. Bacteriol. 47:1107–1111.

Riegel, P., R. Ruimy, F. N. R. Renaud, J. Freney, G. Prevost, F. Jehl, R. Christen, and H. Monteil. 1997b. Corynebacterium singulare sp. nov., a new species for urease-positive strains related to Corynebacterium minutissimum. Int. J. Syst. Bacteriol. 47:1092–1096.

Roberts, M. C., R. B. Leonard, A. Briselden, F. D. Schoenknecht, and M. B. Coyle. 1992. Characterization of antibiotic-resistant Corynebacterium striatum strains. J. Antimicrob. Chemother. 30:463–474.

Ross, M. J., G. Sakoulas, W. J. Manning, W. E. Cohn, and A. Lisbon. 2001. Corynebacterium jeikeium native valve endocarditis following femoral access for coronary angiography. Clin. Infect. Dis. 32:E120–21.

Ruimy, R., P. Riegel, P. Boiron, H. Monteil, and R. Christen. 1995. Phylogeny of the genus Corynebacterium deduced from analyses of small-subunit ribosomal DNA sequences. Int. J. Syst. Bacteriol. 45:740–746.

Saavedra, J. J. N. Rodriguez, A. Fernandez-Jurado, M. D. Vega, L. Pascual, and D. Prados. 1996. A necrotic soft-tissue lesion due to Corynebacterium urealyticum in a neutropenic child. Clin. Infect. Dis. 22:851–852.

Schmitt, M. 1997. Transcription of the Corynebacterium diphtheriae hmuO gene is regulated by iron and heme. Infect. Immun. 65:4634–4641.

Sewell, D. L., M. B. Coyle, and G. Funke. 1995. Prosthetic valve endocarditis caused by Corynebacterium afermentans subsp. lipophilum (CDC coryneform group ANF-1). J. Clin. Microbiol. 33:759–761.

Shukla, S. K., D. N. Vevea, D. N. Frank, N. R. Pace, and K. D. Reed. 2001. Isolation and characterization of a black-pigmented Corynebacterium sp. from a woman with spontaneous abortion. J. Clin. Microbiol. 39:1109–1113.

Simoons-Smit, A. M., P. H. M. Savelkoul, D. W. W. Newling, and C. M. J. Vandenbroucke-Grauls. 2000. Chronic cystitis caused by Corynebacterium urealyticum detected by polymerase chain reaction. Eur. J. Clin. Microbiol. Infect. Dis. 19:949–952.

Siqueira Jr., J. F., I. N. Rocas, R. Souto, M. de Uzeda, and A. P. Colombo. 2000. Checkerboard DNA-DNA hybridization analysis of endodontic infections. Oral Surg. Oral Med. Oral Pathol. Oral Radiol. Endod. 89:744–748.

Sjöden, B., G. Funke, A. Izquierdo, E. Akervall, and M. D. Collins. 1998. Description of some coryneform bacteria from human clinical specimens as Corynebacterium falsenii sp. nov. Int. J. Syst. Bacteriol. 48:69–74.

Soriano, F., and R. Fernandez-Roblas. 1988. Infections caused by antibiotic-resistant Corynebacterium group D2. Eur. J. Clin. Microbiol. Infect. Dis. 7:337–341.

Soto, A., D. G. Pitcher, and F. Soriano. 1991. A numerical analysis of ribosomal RNA gene patterns for typing clinical isolates of Corynebacterium group D2. Epidemiol. Infect. 107:263–272.

Soucek, A., C. Michalic, and A. Souckova. 1971. Identification and characterization of a new enzyme of the group "phospholipase D" isolated from Corynebacterium ovis. Biochem. Biophys. Acta 227:116–128.

Stackebrandt, E., F. A. Rainey, and N. L. Ward-Rainey. 1997. Proposal for a new hierarchic classification system, Actinobacteria classis nov. Int. J. Syst. Bacteriol. 47:479–491.

Takahashi, T., Y. Mori, H. Kobayashi, M. Ochi, N. Kikuchi, and T. Hiramune. 1997. Phylogenetic positions and assignment of swine and ovine corynebacterial isolates based on the 16 rDNA sequence. Microbiol. Immunol. 41:649–655.

Tang, Y., A. von Graevenitz, M. G. Waddington, M. K. Hopkins, D. H. Smith, H. Li, C. P. Kolbert, S. O. Montgomery, and D. H. Persing. 2000. Identification of coryneform bacterial isolates by ribosomal DNA sequence analysis. J. Clin. Microbiol. 38:1676–1678.

Tanner, M. A., D. Shoskes, A. Shahed, and N. R. Pace. 1999. Prevalence of corynebacterial 16S rRNA sequences in patients with bacterial and "nonbacterial" prostatitis. J. Clin. Microbiol. 37:1863–1870.

Tattevin, P., A. Cremieux, C. Muller-Serieys, and C. Carbon. 1996. Native valve endocarditis due to Corynebacterium striatum: First reported case of medical treatment alone. Clin. Infect. Dis. 23:1330–1331.

Tauch, A., S. Krieft, J. Kalinowski, and A. Pühler. 2000. The 51,409-bp R-plasmid pTP10 from the multiresistant clinical isolate Corynebacterium striatum M82B is composed of DNA segments initially identified in soil bacteria and in plant, animal, and human pathogens. Molec. Gen. Genet. 263:1–11.

Telander, B., R. Lerner, J. Palmblad, and O. Ringertz. 1988. Corynebacterium group JK in a hematological ward: Infections, colonization and environmental contamination. Scand. J. Infect. Dis. 20:55–61.

Tiley, S. M., K. R. Kociuba, L. G. Heron, and R. Munro. 1993. Infective endocarditis due to nontoxigenic Corynebacterium diphtheriae: Report of seven cases and review. Clin. Infect. Dis. 16:271–275.

Tinsdale, G. F. W. 1947. A new medium for the isolation and identification of C. diphtheriae based on the production of hydrogen sulfide. J. Pathol. Bacteriol. 59:461–466.

Tompkins, L. S., F. Juffali, and W. E. Stamm. 1982. Use of selective broth enrichment to determine the prevalence of multiply resistant JK corynebacteria on skin. J. Clin. Microbiol. 15:350–351.

Troxler, R., G. Funke, A. von Graevenitz, and I. Stock. 2001. Natural antibiotic susceptibility of recently established coryneform bacteria. Eur. J. Clin. Microbiol. Infect. Dis. 20:315–323.

Vale, J. A., and G. W. Scott. 1977. Corynebacterium bovis as a cause of human disease. Lancet 2:682–684.

Vaneechoutte, M., P. Riegel, D. De Briel, H. Monteil, G. Verschraegen, A. De Rouck, and G. Claeys. 1995. Evaluation of the applicability of amplified rDNA-restriction analysis (ARDRA) to identification of species of the genus Corynebacterium. Res. Microbiol. 146:633–641.

Vaneechoutte, M. D. de Bleser, G. Claeys, G. Verschraegen, T. de Baere, J. Hommez, L. A. Devriese, and P. Riegel. 1998. Cardioverter-lead electrode: Infection due to Corynebacterium amycolatum. Clin. Infect. Dis. 27:1553–1554.

von Graevenitz, A., G. Osterhout, and J. Dick. 1991. Grouping of some clinically relevant Gram-positive rods by automated fatty acid analysis. APMIS 99:147–154.

von Graevenitz, A., and G. Funke. 1996. An identification scheme for rapidly and aerobically growing Gram-positive rods. Zbl. Bakteriol. 284:246–254.

von Graevenitz, A., M. B. Coyle, and G. Funke. 1998a. Corynebacteria and rare coryneforms. In: A. Balows and B. I. Duerden (Eds.) Topley and Wilson's Microbiology and Microbial Infections, 9th ed. Arnold. London, UK. 2:533–548.

von Graevenitz, A., L. Frommelt, V. Pünter-Streit, and G. Funke. 1998b. Diversity of coryneforms found in infections following prosthetic joint insertion and open fractures. Infection 26:36–38.

von Graevenitz, A., V. Pünter-Streit, P. Riegel, and G. Funke. 1998c. Coryneform bacteria in throat cultures of healthy individuals. J. Clin. Microbiol. 36:2087–2088.

von Graevenitz, A., U. Schumacher, and W. Bernauer. 2001. The corynebacterial flora of the normal human conjunctiva is lipophilic. Curr. Microbiol. 42:372–374.

Wagner, J., R. Ignatius, S. Voss, V. Hopfner, S. Ehlers, G. Funke, U. Weber, and H. Hahn. 2001. Infection of the skin caused by Corynebacterium ulcerans and mimicking classical cutaneous diphtheria. Clin. Infect. Dis. 33:1598–1600.

Wang, C. C., D. Mattson, and A. Wald. 2001. Corynebacterium jeikeium bacteremia in bone marrow transplant patients with Hickman catheters. Bone Marrow Transplant. 27:445–449.

Wattiau, P., M. Janssens, and G. Wauters. 2000. Corynebacterium simulans sp. nov., a non-lipophilic, fermentative corynebacterium. Int. J. Syst. Evol. Microbiol. 50:347–353.

Watts, J. L., D. E. Lowery, J. F. Teel, and S. Rossbach. 2000. Identification of Corynebacterium bovis and other coryneforms isolated from bovine mammary glands. J. Dairy Sci. 83:2373–2379.

Wauters, G., B. van Bosterhaut, M. Janssens, and J. Verhaegen. 1998. Identification of Corynebacterium amycolatum and other nonlipophilic fermentative corynebacteria of human origin. J. Clin. Microbiol. 36:1430–1432.

Weiss, K., A. C. Labee, and M. Laverdiere. 1996. Corynebacterium striatum meningitis: Case report and review of an increasingly important Corynebacterium species. Clin. Infect. Dis. 23:1246–1248.

Wichmann, S., C. H. Wirsing von König, E. Becker-Boost, and H. Finger. 1984. Isolation of Corynebacterium group JK from clinical specimens with a semiselective medium. J. Clin. Microbiol. 19:204–206.

Wilson, A. P. R. 1995. The return of Corynebacterium diphtheriae: The rise of non-toxigenic strains. J. Hosp. Infect. 30, Suppl.:S306–S312.

Wirsing von König, C. H., T. Krech, H. Finger, and M. Bergmann. 1988. Use of fosfomycin disks for isolation of diphtheroids. Eur. J. Clin. Microbiol. Infect. Dis. 7:190–193.

Yanagawa, R. 1986. Causative agents of bovine pyelonephritis: Corynebacterium renale, C. pilosum, and C. cystitidis. Progr. Vet. Microbiol. Immunol. 11:158–174.

Yassin, A. F., U. Steiner, and W. Ludwig. 2002a. Corynebacterium appendicis sp. nov. Int. J. Syst. Evol. Microbiol. 52:1165–1169.

Yassin, A. F., U. Steiner, and W. Ludwig. 2002b. Corynebacterium aurimucosum sp. nov. Int. J. Syst. Evol. Microbiol. 52:1001–1005.

Zapardiel, J., E. Nieto, and F. Soriano. 1998. Evaluation of a new selective medium for the isolation of Corynebacterium urealyticum. J. Med. Microbiol. 47:79–83.

Zimmermann, O., C. Spröer, R. M. Kroppenstedt, E. Fuchs, H. G. Köchel, and G. Funke. 1998. Corynebacterium thomssenii sp. nov., a corynebacterium with N-acetyl-beta-glucosaminidase activity from human clinical specimens. Int. J. Syst. Bacteriol. 48:489–494.

Zinkernagel, A. S., A. von Graevenitz, and G. Funke. 1996. Heterogeneity within Corynebacterium minutissimum strains is explained by misidentified Corynebacterium amycolatum strains. Am. J. Clin. Pathol. 106:378–383.

Zinner, S. H. 1999. Changing epidemiology of infections in patients with neutropenia and cancer: Emphasis on Gram-positive and resistant bacteria. Clin. Infect. Dis. 29:490–494.

Prokaryotes (2006) 3:843–888
DOI: 10.1007/0-387-30743-5_32

CHAPTER 1.1.17

The Families Dietziaceae, Gordoniaceae, Nocardiaceae and Tsukamurellaceae

MICHAEL GOODFELLOW AND LUIS ANGEL MALDONADO

Phylogeny

Actinomycetes with *meso*-diaminopimelic acid (*meso*-A$_2$pm), arabinose and galactose in the wall peptidoglycan (wall chemotype IV sensu Lechevalier and Lechevalier, 1970a) belong to two markedly distinct aggregate groups (Goodfellow and Minnikin, 1984b; Goodfellow and Lechevalier, 1989b; Goodfellow et al., 1999). Wall chemotype IV actinomycetes containing mycolic acids (high-molecular-weight 3-hydroxy fatty acids with a long alkyl branch in the two position) are classified in the genera *Corynebacterium*, *Gordonia*, *Mycobacterium*, *Nocardia*, *Rhodococcus*, *Skermania*, *Tsukamurella* and *Williamsia*; their mycolateless counterparts are in the family Pseudonocardiaceae (see The Family Pseudonocardiaceae in the second edition). Mycolic acid-containing strains have many properties in common (Goodfellow and Wayne, 1982a; Goodfellow and Cross, 1984a; Goodfellow and Magee, 1998a; Goodfellow et al., 1998; Goodfellow et al., 1999) and form a monophyletic group in the 16S rRNA gene tree (Mordarski et al., 1980b; Stackebrandt and Woese, 1981; Stackebrandt et al., 1983; Stackebrandt et al., 1997; Goodfellow et al., 1998; Goodfellow et al., 1999). Representatives of the mycolic acid-containing genera form distinct subclades in the 16S rRNA gene tree (Fig. 1).

Mycolic acid-containing actinomycetes, and members of the mycolateless genus *Turicella*, are currently assigned to six families on the basis of 16S rRNA gene similarity values and on the distribution of taxon-specific signature nucleotides (Stackebrandt et al., 1997). All of the families belong to the suborder Corynebacterineae, one of the ten suborders of the order Actinomycetales. The family Corynebacteriaceae contains the genera *Corynebacterium* and *Turicella*; the family Dietziaceae, the genus *Dietzia*; the family Gordoniaceae, the genera *Gordonia*, *Skermania* and *Williamsia*; the family Mycobacteriaceae, the genus *Mycobacterium*; the family Nocardiaceae, the genera *Nocardia* and *Rhodococcus*; and the family Tsukamurellaceae, the genus *Tsukamurella*. The present chapter is restricted to a consideration of the families Dietziaceae, Gordoniaceae, Nocardiaceae and Tsukamurellaceae; the families Corynebacteriaceae (see Corynebacterium—Nonmedical in this Volume and The Genus Corynebacterium—Medical in this Volume) and Mycobacteriaceae (see *Mycobacterium leprae* in this Volume, The Genus Mycobacterium—Medical in this Volume and Mycobacterium leprae in this Volume) are considered elsewhere.

Membership of the families Gordoniaceae, Nocardiaceae and Tsukamurellaceae is currently restricted to actinomycetes with the following characteristics: 1) a peptidoglycan composed of *n*-acetylglucosamine, D-alanine, L-alanine, and D-glutamic acid, with *meso*-A$_2$pm as the diamino acid, and muramic acid in the *n*-glycolated form (Uchida and Aida, 1977; Uchida and Aida, 1979; Uchida and Aida, 1997); 2) a polysaccharide fraction of the wall, rich in arabinose and galactose (whole-organism sugar pattern type A sensu Lechevalier and Lechevalier, 1970b); 3) a phospholipid pattern consisting of diphosphatidylglycerol, phosphatidylethanolamine (a taxonomically significant nitrogenous phospholipid), phosphatidylinositol, and phosphatidylinositol mannosides without phosphatidylcholine or phospholipids containing glucosamine (i.e., phospholipid type 2; Lechevalier et al., 1977; Lechevalier et al., 1981); 4) a fatty acid profile showing major amounts of straight chain, unsaturated fatty acids and tuberculostearic acids (Kroppenstedt, 1985); 5) mycolic acids with 34–78 carbons (Alshamaony et al., 1976a; Alshamaony et al., 1976b; Goodfellow et al., 1978; Goodfellow et al., 1982c); and 6) a DNA G+C content within the range 63–73 mol%. In contrast, strains belonging to the family Dietziaceae have muramic acid in the N-acetylated form and a polar lipid pattern lacking phosphatidylinositol and phosphatidylinositol mannosides (Rainey et al., 1995a).

Some of the major properties of the mycolic acid-containing genera are shown in Fig. 1 and Table 1. Representatives of a number of these taxa have been shown to be closely related using serological techniques (Magnusson, 1976; Pier,

Suprageneric relationships based on 16S rRNA sequence data	Family/genus	Wall chemotype[a]	Whole-organism sugar pattern[b]	Peptidoglycan[c] and muramic acid types[d]		Fatty acid pattern[e]	Major menaquinones[f] (MK)	Phospholipid type[g]	Mol% G+C of DNA
	Dietziaceae								
	Dietzia	IV	A	A1γ	A	1b	−8(H$_2$)	PI	73
	Tsukamurellaceae								
	Tsukamurella	IV	A	A1γ	G	1b	−9	PII	67–78
	Corynebacteriaceae								
	Corynebacterium	IV	A	A1γ	A	1a	−8(H$_2$), −9(H$_2$)	PI	51–63
	Turicella	IV	A	ND	ND	1b	−10, −11	ND	65–72
	Gordoniaceae								
	Gordonia	IV	A	A1γ	G	1b	−9(H$_2$)	PII	60–66
	Williamsia	IV	A	A1γ	G	1b	−9(H$_2$)	PII	64–65
	Skermania	IV	A	A1γ	G	1b	−8(H$_2$, −ωcycl)	PII	67.5
	Nocardiaceae								
	Nocardia	IV	A	A1γ	G	1b	−8(H$_4$, −ωcycl)	PII	64–72
	Rhodococcus	IV	A	A1γ	G	1b	−8(H$_2$)	PII	63–73
	Mycobacteriaceae								
	Mycobacterium	IV	A	A1γ	G	1b	−9(H$_2$)	PII	62–70

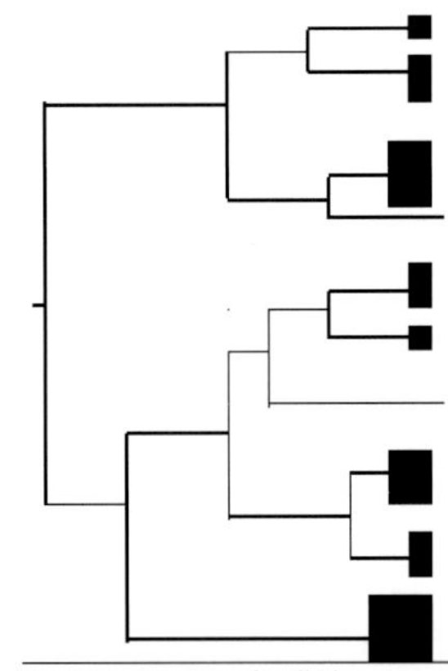

[a] Major constituents of wall chemotypes: I, L-A$_2$pm and glycine; II, meso- A$_2$pm and glycine; III, meso- A$_2$pm; IV, meso-A$_2$pm, arabinose and galactose; V, lysine and ornithine; VI, lysine (with variable presence of aspartic acid and galactose); VIII, diaminobutyric acid and glycine (lysine variable); VIII, ornithine; IX, LL- and mso- A2pm. All wall preparations contain major amounts of alanine, glutamic acid, glucosamine and muramic acid (Lechevalier and Lechevalier, 1970, 1980).

[b] Whole-organism sugar patterns of actinomycetes containing meso- A$_2$pm: A, arabinose and galactose; B, madurose (3-O-methyl-D-galactose); C, no diagnostic sugars; D, arabinose and xylose; ND, not applicable (Lechevalier et al., 1971).

[c] Peptidoglycan classification (Schleifer and Kandler, 1972; Schleifer and Seidl, 1985). Type A, cross-linkage between positions 3 and 4 of two peptide subunits: 1, direct crosslinkage (no interpeptide bridge); 2, crosslinkage by polymerized peptide bridge; 3, crosslinkage by interpeptide bridges consisting of monocarboxylic L-amino acids or glycine, or both; 4, crosslinkage by interpeptide bridges containing a dicarboxylic amino acid; 5, crosslinkage by interpeptide bridges containing a dicarboxylic amino acid and lysine. Peptidoglycan type B, crosslinkage between positions 2 and 4 of two peptide subunits: 1, interpeptide bridge containing a L-diamino acid; 2, interpeptide bridge containing a D-diamino acid. Small Greek letters mak the diversity of amino acids in position 3 of the peptide subunit. Prime (') indicates the replacement of alanine in position 1 in type A peptidoglycan by glycine (Schleifer and Kandler, 1972; Schleifer and Stackebrandt, 1983; Schleifer and Seidl, 1985).

[d] Muramic acid types: A, N-acetylmuramic acid; G, N-glycolymuramic acid (Uchida and Aida, 1977, 1984; Uchida and Seino, 1997).

[e] Fatty acid classification after Kroppenstedt (1985). Numbers refer to the type of fatty acid biosynthetic pathway and letters to types of fatty acids (FA) synthesized. Type 1, pathway generating straight-chain fatty acids, including saturated and unsaturated (FA-type 1a), 10-methyl-branched (FA-type 1b) and cyclopropane fatty acids (FA-type 1c), the latter two being derived from the unsaturated compounds. Type 2, pathway yielding terminally-branched fatty acids (FA-type 2), that is, iso- and anteiso-branched fatty acids. Type 3 encompasses organisms that have complex branched fatty acid types, that is, both 10-methyl-branched (type 1) and iso- and/or anteiso-branched (type 2) fatty acids.

[f] MK-9 (H$_2$), notation for octahydrogenated menaquinone with nine isoprene units.

[g] Characteristic phopholipids: PI, nitrogenous phospholipids absent (with phosphatidylglycerol variable); PII, only phosphatidylethanolamine; PIII, phosphatidylcholine (with phosphatidylethanolamine, phosphatidylmethylethanolamine and phosphatidylglycerol variable; phospholipids containing glucosamine absent); PIV, phospholipids containing glucosamine (with phosphatidylethanolamine variable) and PV, phospholipids containing glucosamine and phosphatidylglycerol (with phosphatidylethanolamine variable); all preparations contain phosphatidylinositol (Lechevaliver et al., 1977, 1981).

Fig. 1. Distribution of chemical markers in genera classified in the suborder Corynebacterineae.

1984). The most comprehensive serological studies have used immunodiffusion techniques, and common precipitinogens have been detected among corynebacteria, gordoniae, mycobacteria, nocardiae and rhodococci (Lind and Ridell, 1976; Lind et al., 1980).

The Family Dietziaceae

DIETZIA Rainey et al. 1995, 35[AL]. This genus was proposed for organisms first known as "*Flavobacterium maris*" (Harrison, 1929) and later as *Rhodococcus maris* (Nesterenko et al., 1982). It currently encompasses three species, *D. maris, D. natronolimnaios* and *D. psychralcaliphilus*, which form a distinct clade in the 16S rRNA Corynebacterineae gene tree (Fig. 2). Dietziae are aerobic, Gram-positive, non-acid-alcohol-fast, non-sporing, catalase-positive actinomycetes, which may form cocci that germinate into short rods or short rod-shaped cells that exhibit snapping division and produce V- forms. Circular, raised or convex, glistening colonies with entire edges are formed on agar media. Pigmentation ranges from orange to coral red. Dietziae are chemoorganotrophics, having an oxidative type of metabolism. In addition to the aforementioned chemical properties, the organism has an A1γ type peptidoglycan (Schleifer and Kandler, 1972) and contains mycolic acids with 34–39 carbons and up to one double bond. Dihydrogenated menaquinone with eight isoprene units is the major isoprenologue. The G+C content of the DNA is 66–73 mol%. Type species: *Dietzia maris* (Nesterenko et al. 1982) Rainey et al. 1995a, 35[AL]. Type strain: ATCC 35013[T] = DSM 43672[T].

The Family Gordoniaceae

GORDONIA (Tsukamura, 1971a) Stackebrandt et al., 1988, 345[AL]. This taxon is the type genus of the family Gordoniaceae. The genus was initially known as *Gordona*; the etymologically correct name, *Gordonia*, was proposed by Stackebrandt et al. (1997). The genus *Gordonia* Tsukamura, 1971a was proposed for some slightly acid-fast actinomycetes isolated from soil and the sputa of patients with pulmonary disease. The three original species, *G. bronchialis, G. rubra* and *G. terrae*, were subsequently reclassified in the revised and re-described genus *Rhodococcus* (Tsukamura, 1974a; Goodfellow and Alderson, 1977).

Table 1. Characteristics of members of the mycolic acid-containing genera.

Characters	Genera								
	1	2	3	4	5	6	7	8	9
Cell morphology	Pleomorphic rods, often club-shaped; commonly in angular and palisade arrangement	Short rods and cocci	Rods and cocci or moderately branching hyphae	Rods, occasionally branched filaments that fragment into rods and coccoid elements	Mycelium which later fragments into rods and cocci	Rods to extensively branched mycelium; the latter fragments into irregular rods and cocci	In early stages of growth (24h), substrate mycelium resembles a pine tree	Rods occurring singly, in pairs or in masses; coccobacillary forms produced	Thin irregular rods or cocci that occur singly or in small clusters
Aerial hyphae	Absent	Absent	Absent	Usually absent[b]	Present	Absent	Present	Absent	Present
Time for visible colonies (days)	1–2	1–3	1–3	2–40	1–5	1–3	9–21	1–3	1–4
Degree of acid-fastness (not necessarily also alcohol fastness)	Sometimes weakly acid-fast	Not acid fast	Often partially acid-fast	Usually strongly acid-fast	Often partially acid-fast	Often partially acid-fast	Not acid-fast	Weak to acid-fast	ND
Strictly aerobic	−	+	+	+	+	+	+	+	+
Mycolic acid type[c]	Single spot	Single spot	Single spot	Multiple spots	Single spot	Single spot	Single spot	Two spots	Single spot
Overall size (number of carbons)	22–38	34–38	46–66	60–90	48–60	30–54	58–64	64–78	50–56
Number of double bonds[d]	0–2	0–1	1–4	1–3	0–3	0–2	2–6	1–6	ND
Fatty acid esters released on pyrolysis (number of carbons)	8–18	ND	16–18	22–26	12–18	12–16	16–20	20–22	ND

Symbols: +, positive; −, negative; and ND, not determined.

[a]Genera: 1, Corynebacterium; 2, Dietzia; 3, Gordonia; 4, Mycobacterium; 5, Nocardia; 6, Rhodococcus; 7, Skermania; 8, Tsukamurella; and 9, Williamsia.

[b]Mycobacterium farcinogenes and Mycobacterium xenopi may occasionally produce aerial hyphae.

[c]Number of mycolic acid spots produced from whole organism methanolysates (Minnikin et al., 1975, 1984a, b; Hamid et al., 1993; Yassin et al., 1997).

[d]In mycobacterial mycolic acids, double bonds may be converted to cyclopropane rings; methyl branches and other oxygen functions may be present (Minnikin et al., 1984a, b; Dobson et al., 1985).

Data taken from Goodfellow et al. (1998a), Kämpfer et al. (1999), and Nishiuchi et al. (2000).

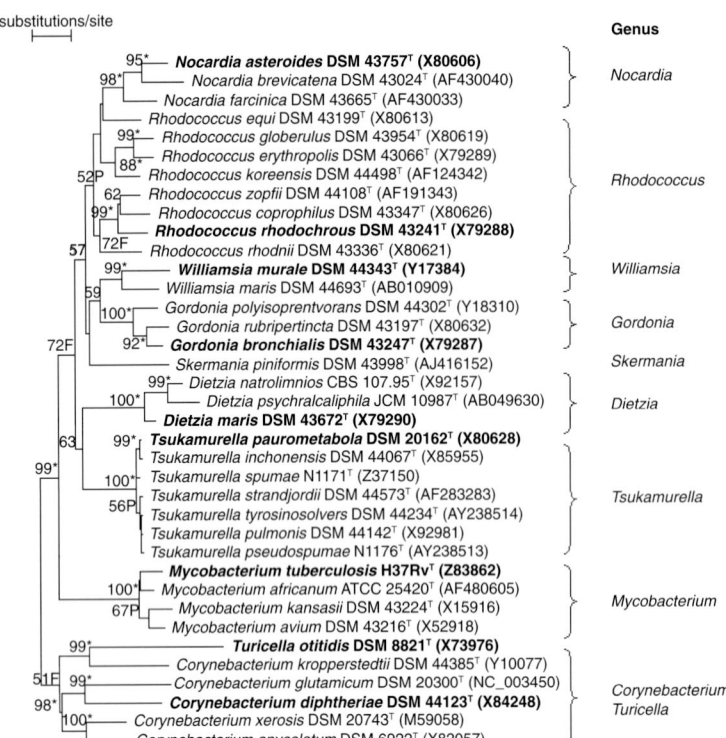

0.1 substitutions/site

Genus

Fig. 2. Neighbor-joining tree based on nearly complete 16S rRNA gene sequences showing the positions of representative strains of genera classified in the suborder Corynebacterineae. Asterisks indicate branches of the tree that were also found using the least-squares, maximum-likelihood, and maximum-parsimony tree-making algorithms. The numbers at the nodes indicate the levels of bootstrap support (%) based on a neighbor-joining analysis of 1000 re-sampled datasets, and only values above 50% are given. The bar indicates 0.1 substitutions per nucleotide position. T, type strain.

Rhodococcal species were eventually assigned to two aggregate groups on the basis primarily of chemical and serological properties (Goodfellow, 1989a). All species originally classified in the genus *Gordonia* contained mycolic acids with between 48 and 66 carbon atoms and major amounts of dihydrogenated menaquinones with nine isoprene units, whereas the remaining strains were characterized by shorter chain mycolic acids (34–52 carbon atoms) and dihydrogenated menaquinones with eight isoprene units as the predominant isoprenologue (Alshamaony et al., 1976b; Collins et al., 1977; Collins et al., 1985b).

The two aggregate groups were also recognized by their antibiotic sensitivity profiles (Goodfellow and Orchard, 1974), delayed skin reaction on sensitized guinea pigs, and polyacrylamide gel electrophoresis of cell extracts (Hyman and Chaparas, 1977). The discovery that the two aggregate groups were phylogenetically distinct led Stackebrandt et al. (1988) to revive the genus *Gordona* Tsukamura 1971a for organisms classified as *R. bronchialis*, *R. rubripertincta*, *R. sputi* and *R. terrae*; DNA homology data indicate that *R. obuensis* Tsukamura 1982 should be considered as a subjective synonym of *R. sputa* (Zakrzewska-Czerwinska et al., 1988).

The genus *Gordonia* currently contains nineteen species which form a distinct clade within the evolutionary radiation encompassed by mycolic acid-containing actinomycetes (Figs. 2

and 3). The taxonomic status of most of these species is underpinned by a wealth of genotypic and phenotypic data (Linos et al., 1999; Kim et al., 2000; Brandão et al., 2001). However, further studies are required to determine the finer relationship between *G. alkalivorans* and *G. nitida*, as the type strains of these species share a 16S rRNA gene similarity of 99.9%, a value which corresponds to two nucleotide differences at 1480 locations (Maldonado et al., 2003).

Gordoniae are aerobic, Gram-positive to Gram-variable, catalase positive, partially acid-alcohol-fast, nonmotile actinomycetes that form short rods and cocci that occur singly, in pairs, in V-shaped arrangements or in short chains. Colonial appearance ranges from convex, shiny and smooth to rough, matt and folded with irregular margins. Pigmentation may be pale yellow, beige or tan through to orange pink or red. The organism has an oxidative type of metabolism, is aryl-sulfatase negative, sensitive to lysozyme, and produces mycobactins. In addition to the chemical properties mentioned earlier, the organism has an A1γ type of peptidoglycan, contains dihydrogenated menaquinones with nine isoprene units as the predominant isoprenologue and has mycolic acids with 46–66 carbons with one to four double bonds. The fatty acid esters released on pyrolysis gas chromatography of mycolic esters contain 16–18 carbon atoms. The G+C content of the DNA is 63–69 mol%. Type species: *Gordonia bronchialis* (Tsukamura, 1971a)

Fig. 3. Neighbor-joining tree based on nearly complete 16S rRNA gene sequences showing relationships between representative strains of the family Gordoniaceae. Asterisks indicate branches of the tree that were also found using the least-squares, maximum-likelihood, and maximum-parsimony tree-making algorithms. The numbers at the nodes indicate the levels of boot-strap support (%) based on a neighbor-joining analysis of 1000 re-sampled datasets, and only values above 50% are shown. F and P indicate branches that were also recovered using the least-squares and maximum-parsimony methods, respectively. The bar indicates 0.1 substitutions per nucleotide position. T, type strain.

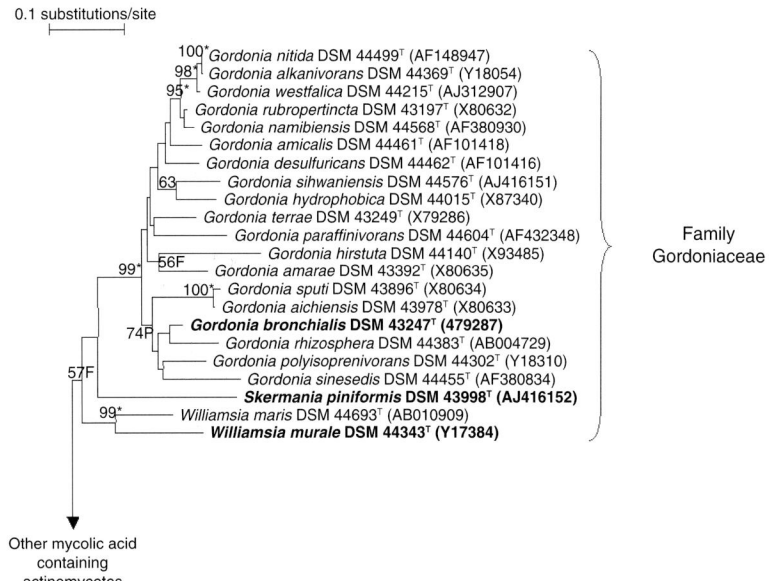

Stackebrandt et al. 1988, 345[AL]. Type strain: ATCC 25592[T] = DSM 43247[T].

SKERMANIA Chun et al., 1997, 129[AL]. This genus contains a single species, *Skermania piniformis*, which falls towards the periphery of the 16S rRNA Gordoniaceae gene tree (Figs. 1 and 3). The taxon was proposed for actinomycetes first known as "Pine Tree-Like Organisms" (PTLO; Blackall et al., 1998) and later as "*Nocardia pinensis*" (Blackall et al., 1989b). The appellation PTLO was given to these organisms as they formed a distinctive tree-like microscopic morphology; branches are acute- rather than right-angled, and those near the apex are shorter than those farther away. The characterization of *S. piniformis* was based on a few strains and a small number of physiological tests, but since then larger numbers of strains with morphology typical of PTLO have been shown to have a unique API ZYM pattern (Soddell and Seviour, 1994). In an extensive numerical taxonomic study, such strains were recovered in a homogeneous cluster which contained the type strain of *S. piniformis* (Soddell and Seviour, 1998).

Skermania are aerobic, Gram-positive, non-acid-fast, nonmotile actinomycetes which form an extensive mycelium that does not fragment in undisturbed culture; secondary branching is rare or absent. Aerial hyphae are not visible to the naked eye, but short branched and unbranched aerial hyphae can be seen microscopically. The microscopic appearance of the organism resembles a pine tree during early stages of growth (24 h). Cells from activated sludge plants or from artificial culture media contain intracellular sudanophilic and polyphosphate inclusions. Colonies (1–2 mm in diameter) form on tryptone-yeast extract-glucose agar within 10–21 days,

although the growth rate is variable. Colonies are orange, opaque, macroscopically dry and friable, and microscopically moist and slimy, have a pasty texture, are difficult to emulsify or sub-culture, and are circular with entire edges. When picked from the agar surface, a whole colony can be removed intact. Strains grow well on media that contain glycerol as a carbon source and asparagine as a nitrogen source. In tryptic yeast extract glucose broth, strains grow as macroscopically visible colonies in a slightly turbid liquid. The maximum and minimum growth temperatures are 31°C and 15°C, respectively. Growth is not enhanced in an atmosphere containing 4% CO_2, nor does growth occur anaerobically, microaerophilically, or in a candle jar. Strains have an oxidative metabolism and are catalase, oxidase, and urease positive.

In addition to the chemical properties mentioned earlier, the organism has an A1γ peptidoglycan and contains mycolic acids with 58–64 carbons with 2–6 double bonds and substantial numbers of unsaturated chains in the 2-position. The fatty acid esters released on pyrolysis gas chromatography of mycolic acids contain 16–20 carbon atoms. Predominant amounts of hexahydrogenated menaquinone with eight isoprene units are produced in which the two end units are cyclized. The G+C content of the DNA is 67.5%, as determined by the thermal denaturation method. Type species: *Skermania piniformis* Chun et al., 1997, 129[AL]. Type strain: DSM 43998[T] = IFO 15059[T].

WILLIAMSIA Kämpfer et al., 1999, 686[AL].

This taxon was proposed to accommodate an aerobic, Gram-positive, non-acid-alcohol-fast, chemoheterotrophic actinomycete that forms short rods and coccoid-like elements. Peritrichous fibrillae distributed over the whole surface of cells are visible by transmission electron microscopy but not in negatively stained preparations. The temperature range for growth is 10–37°C. In addition to the chemical markers mentioned earlier, the organism contains mycolic acids with 50–56 carbon atoms and dihydrogenated menaquinones with nine isoprene units as the predominant isoprenologue. The G+C content of the DNA is 64–65 mol%. Type species: *Williamsia muralis* Kämpfer et al. 1999, 686[AL]. Type strain: DSM 44343[T] = MA140/96[T].

A second species, *W. maris*, has been described for a strain isolated from the Sea of Japan (Stach et al., 2004); this organism was initially considered to be closely related to the genus *Gordonia* (Colquhoun et al., 1998). The type strains of the two *Williamsia* species form a distinct clade towards the margin of the 16S rRNA Gordoniaceae gene tree.

The Family Nocardiaceae

NOCARDIA Trevisan, 1889, 9[AL]. Nocardiae are aerobic, Gram-positive, non-motile catalase-positive actinomycetes that are typically acid-alcohol-fast at some stages of the growth cycle. They form rudimentary to extensively branched substrate hyphae that often fragment in situ or, on mechanical disruption, disrupt into rod-shaped to coccoid, nonmotile elements. Aerial hyphae, at times visible only microscopically, are almost always present. Short-to-long chains of well-to-poorly differentiated conidia may occasionally be found on aerial hyphae and, more rarely, on both aerial and substrate mycelia. Most nocardiae form carotenoid-like pigments that confer various shades of orange, pink, red or yellow to colonies growing on solid culture media. Soluble brown or yellowish diffusible pigments may be produced. Colony morphology is variable; it may be smooth to granular and irregular, wrinkled or heaped. The presence of aerial hyphae often confers a powdery or velvety appearance to the surface of colonies. Nocardiae are chemoorganotrophic and have an oxidative type of metabolism. In addition to the chemical properties already mentioned, the nocardial peptidoglycan is of the A1γ type. The organism contains mycolic acids with 40–64 carbon atoms and up to three double bonds; the fatty acid esters released on pyrolysis gas chromatography of mycolic esters contain 12–18 carbon atoms and may be saturated or unsaturated. The predominant menaquinone corresponds to a hexahydrogenated menaquinone with eight isoprene units in which the end two units are cyclized (i.e., II, III-tetrahydro-ω-[2, 6, 6-trimethylcyclohex-2-en-yimethyl] menaquinone-6). The G+C content of the DNA is 64–72 mol%.

Type species: *Nocardia asteroides* (Eppinger 1891) Blanchard 1896, 856[AL], Opinion 58, Judicial Commission 1985, 538. Type strain: ATCC 19247[T] = DSM 43757[T].

The genus *Nocardia* presently includes 38 species, which form a well-delineated clade in the 16S rRNA Corynebacterineae gene tree (Figs. 1 and 4). Nearly half of these species have been described within the last five years mainly to accommodate organisms isolated from clinical material (Gürtler et al., 2001b; Hamid et al., 2001; Kageyama et al., 2003a; Kageyama et al., 2003b, Yassin et al., 2003) and environmental samples, notably soil (Albuquerque de Barros et al., 2003; Kämpfer et al., 2003; Saintpierre-Bonaccio et al., 2003; Zhang et al., 2004). The taxonomic status of the recently described species is based on a balance set of genotypic and phenotypic criteria. There are good grounds for considering more established species such as *N. brasiliensis*, *N. otitidiscaviarum* and *N. seriola* to be well circumscribed taxa (Gordon and Mihm, 1962; Goodfellow, 1971; Kudo et al., 1988). In contrast, *N. asteroides* has been shown to be heterogeneous on the basis of DNA homology (Mordarski et al., 1977; Mordarski et al., 1978; Bradley et al., 1978b), numerical taxonomic (Tsukamura, 1969; Tsukamura, 1982b; Goodfellow, 1971; Schaal and Reutersberg, 1978b; Orchard and Goodfellow, 1980; Goodfellow et al., 1998), phage sensitivity (Pulverer et al., 1975) and immunological (Pier and Fichtner, 1971; Magnusson, 1976; Kurup and Schribner, 1981; Ridell, 1981) data.

There is evidence that some clinically significant strains identified as *N. asteroides* and *N. farcinica* belong to different species (Wallace et al., 1990; Workman et al., 1998; Roth et al., 2003). One such species, "*N. nova*," is cited as species incertae sedis in the current edition of *Bergey's Manual of Determinative Bacteriology* (Goodfellow and Lechevalier, 1989b). Yano et al. (1990) separated "*N. nova*" from *N. asteroides* and *N. farcinica* strains using DNA-DNA relatedness, mycolic acid, and numerical taxonomic data. *Nocardia* "*nova*" strains also have a distinctive antimicrobial sensitivity pattern (Wallace et al., 1991). Despite the rapid expansion in the number of *Nocardia* species, additional nocardial diversity has still to be described (Wang et al., 1999; Maldonado et al., 2000; Roth et al., 2003).

RHODOCOCCUS (Zopf, 1891) EMEND. Goodfellow et al. 1998, 11[AL]. Rhodococci have a long and checkered taxonomic history, which is described in several review articles (Bousfield and Goodfellow, 1976; Goodfellow and Cross, 1984a; Goodfellow et al., 1998). The genus, with *R. rhodochrous* as the type species, was recognized in the early editions of *Bergey's Manual of Determinative Bacteriology* (Bergey et al., 1923,

Fig. 4. Neighbor-joining tree based on nearly complete 16S rRNA gene sequences showing relationships between representative strains of the family Nocardiaceae. Asterisks indicate branches of the tree that were also found using the least-squares, maximum-likelihood, and maximum-parsimony tree-making algorithms. The numbers at the nodes indicate the levels of bootstrap support (%) based on a neighbor-joining analysis of 1000 re-sampled datasets, and only values above 50% are shown. F and P indicate branches that were also recovered using the least-squares and maximum-parsimony methods, respectively. The bar indicates 0.1 substitutions per nucleotide position. T, type strain.

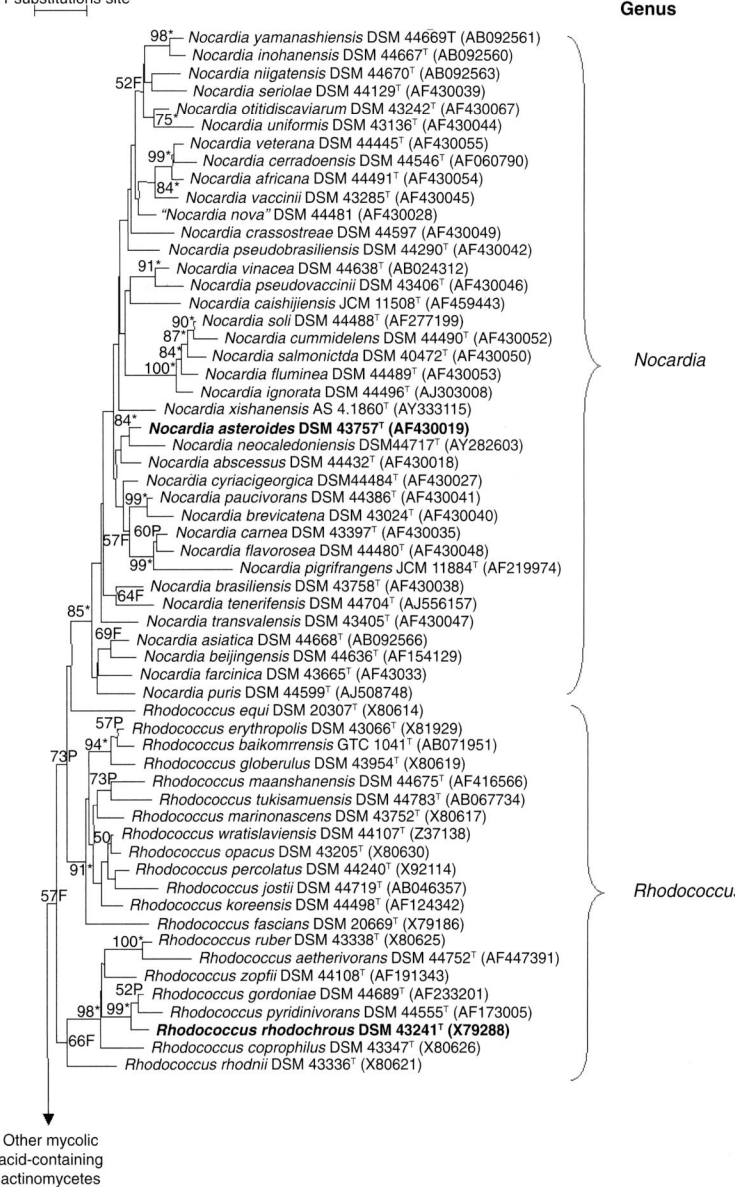

1925, 1930, and 1934). However, rhodococcal species were transferred to the genus *Micrococcus* in the fifth edition of *Bergey's Manual of Determinative Bacteriology* (Bergey et al., 1939) and remained there in the following edition (Breed et al., 1948). Rhodococcal-like strains were also assigned to an assortment of other morphologically defined genera during this time. Indeed, only with the far-reaching changes that occurred in the taxonomy of Gram-positive cocci and mycolic acid-containing strains in light of new developments in bacterial systematics did attention focus on rhodococci.

The epithet *rhodochrous* (Zopf, 1891) was resurrected by Gordon and Mihm (1957) for actinomycetes bearing a multiplicity of generic and specific names but with many properties in common with mycobacteria and nocardiae. The taxon was provisionally assigned to the genus *Mycobacterium* but was later considered to merit generic status. The genus *Rhodococcus* was eventually reintroduced (Tsukamura, 1974a; Goodfellow and Alderson, 1977) and encompasses 16 species in the current edition of *Bergey's Manual of Systematic Bacteriology* (Goodfellow, 1989a). Most of these were circumscribed in numerical taxonomic surveys (Goodfellow and Alderson, 1971; Tsukamura, 1974a; Rowbotham and Cross, 1977b; Goodfellow et al., 1982b; Goodfellow et al., 1982d) and later shown to be homogeneous on chemical and molecular systematic grounds (Minnikin and

Goodfellow, 1980a; Mordarski et al., 1980a; Zakrzewska-Czerwinska et al., 1988).

The reintroduction of the genus *Gordonia* and the recognition of the genus *Dietzia* left the genus *Rhodococcus* as a taxon encompassing aerobic, Gram-positive to Gram-variable, non-motile, catalase-positive actinomycetes that are usually partially acid-alcohol-fast at some stage in the growth cycle and which form rods to extensively branched substrate-mycelia. The growth cycle starts with the coccus or short rod stage, and then different organisms show a more or less complex series of morphological stages: cocci may germinate only into short rods or form filaments with side projections, or show elementary branching, or (in the most differentiated forms) produce branched hyphae. The next generation of cocci or short rods is produced by fragmentation of the rods, filaments and hyphae. Some strains produce sparse, microscopically visible, aerial hyphae that may be branched or form synnemata, which consist of unbranched filaments that coalesce and project upwards.

Rhodococci grow well on standard laboratory media at 30°C although some strains require thiamine. Colonies may be rough, smooth or mucoid, and pigmented buff, cream, yellow, orange or red, though colorless variants occur. Rhodococci are catalase positive, sensitive to lysozyme, arylsulfatase negative, and produce acid from glucose oxidatively. In addition to the chemical properties mentioned earlier, the rhodococcal peptidoglycan belongs to the A1γ type and contains mycolic acids with 30–54 carbons and up to two double bonds. The fatty acid esters released on pyrolysis gas chromatography of mycolic esters have 12–16 carbon atoms. In addition, rhodococci contain dihydrogenated menaquinones with eight isoprene units as the predominant isoprenologue. The G+C content of the DNA is 67–73 mol%. Type species: *Rhodococcus rhodochrous* (Zopf, 1889) Tsukamura 1974, 43[AL]. Type strain: ATCC 13808[T] = DSM 43241[T].

The improved classification of the genus *Rhodococcus* provides a sound basis for the recognition of additional *Rhodococcus* species. The genus currently contains 19 species that form a recognizable group within the evolutionary radiation occupied by mycolic acid containing actinomycetes (Figs. 1 and 4). The taxonomic integrity of rhodococcal species is supported by a combination of genotypic and phenotypic properties (Goodfellow et al., 2004; Jones et al., 2004; Li et al., 2004). Rhodococcal species can be assigned to four 16S rRNA gene subclades (Rainey et al., 1995b; Goodfellow et al., 1998) that have been equated with the rank of species (Goodfellow et al., 1998; McMinn et al., 2000).

The Family Tsukamurellaceae

TSUKAMURELLA Collins et al., 1989, 387[AL]. This taxon was introduced for organisms classified as *Corynebacterium parometabolum* and *Rhodococcus aurantiacus*. *Corynebacterium parometabolum* was proposed by Steinhaus (1941) for bacteria isolated from the mycetome and ovaries of the bedbug (*Cimex lectularis*), but its placement in the genus *Corynebacterium* was questioned (Jones, 1975; Collins and Jones, 1982a). The organism has an A1γ peptidoglycan (Cummins, 1971; Schleifer and Kandler, 1972) but was distinguished from corynebacteria by the presence of long, highly unsaturated, mycolic acids (Collins and Jones, 1982a). A similar series of unsaturated mycolic acids was detected in *Rhodococcus aurantiacus* (Goodfellow et al., 1978; Tomiyasu and Yano, 1984), the generic position of which was also equivocal (Goodfellow, 1989a). This species, first described as *Gordona aurantiaca* (Tsukamura and Mizuno, 1971b), was later reclassified in the genus *Rhodococcus* (Tsukamura, 1974a; Tsukamura and Yano, 1985). In contrast, Goodfellow et al. (1978) considered that *aurantiaca* strains were worthy of generic status as they formed a numerically circumscribed taxon equivalent in rank to the genera *Corynebacterium*, *Mycobacterium*, *Nocardia* and *Rhodococcus* and contained characteristic mycolic acids and unsaturated menaquinones with nine isoprene units. *Corynebacterium parometabolum* was also found to have menaquinones containing unsaturated multiprenyl side chains (Collins and Jones, 1982a).

Corynebacterium parometabolum and *Rhodococcus aurantiacus* were reduced to a single species and reclassified in the genus *Tsukamurella* using 16S rRNA sequence data and the results of the experiments outlined above (Collins et al., 1989). The genus accommodates aerobic, Gram-positive, partially acid-alcohol-fast, nonmotile, nonsporeforming actinomycetes that form straight to slightly curved rods that occur singly, in pairs or in masses; coccobacillary elements also occur. The organism forms white-creamy to orange convex colonies that are dryish but easily emulsified. Tsukamurellae are catalase-positive, are arylsulfatase negative, are resistant to lysozyme, and produce acid from glucose oxidatively. In addition to the chemical properties mentioned earlier, the organism has an A1γ peptidoglycan and highly unsaturated long chain mycolic acids with 64–78 carbon atoms and one to six double bonds. The fatty acids released on pyrolysis mass spectrometry of mycolic esters have 22–26 carbons. Menaquinones are the sole respiratory quinone with unsaturated components with nine isoprene units predominating. The G+C content of the DNA is 67–74 mol%.

Type species: *Tsukamurella paurometabola* (Steinhaus, 1941) Collins et al. 1989[AL]. Type strain: ATCC 8368[T] = DSM 20162[T].

The genus encompasses seven species, which form a distinct clade within the evolutionary radiation occupied by members of the suborder Corynebacterineae. Representatives of these taxa share very high 16S rRNA gene similarity values (>99.5%) but were delineated using DNA-DNA relatedness and phenotypic data (Kattar et al., 2001; Nam et al., 2003). The taxonomic integrity of *T. inchonensis* (Yassin et al., 1995), *T. pseudospumae* (Nam et al., 2004b), *T. pulmonis* (Yassin et al., 1996), *T. spumae* (Nam et al., 2003), *T. strandjordii* (Kattar et al., 2001) and *T. tyrosinosolvens* (Yassin et al., 1998) has been underpinned by the results of an extensive polyphasic taxonomic study (Nam, 2004a). Also, this investigation indicates that *T. paurometabola* is a heterogeneous taxon and that the genus *Tsukamurella* is underspeciated. An additional species, "*T. sunchonensis*" Seong et al. 2003, has yet to be validated.

Habitat

The Family Dietziaceae

Dietzia maris strains have been isolated from soil, the skin and intestinal tract of carp, and deep-sea sediments in the Pacific Ocean (Nesterenko et al., 1982; Rainey et al., 1995a; Colquhoun et al., 1998). In turn, *D. natronolimnaios* was isolated from a moderately saline and alkaline East African soda lake (Duckworth et al., 1998) and *D. psychralcaliphilus* from a drain pool of a fish egg-processing plant (Yumoto et al., 2002). The first documented evidence implicating *D. maris* as an agent of disease involved a catheter-associated bacteremia in an immunocompromised male patient (Bemer-Melchior et al., 1999). The organism has also been isolated from a bone infection (Pidoux et al., 2001).

The Family Gordoniaceae

Gordonia strains are widely distributed in soil and aquatic habitats, including marine sediments (Tsukamura, 1971a; Colquhoun et al., 1998; De los Reyes et al., 1998a; De los Reyes et al., 1998b; Kim et al., 2000). In addition, gordoniae have been isolated from rhizosphere soil (Takeuchi and Hatano, 1998), biofilters (Bendinger et al., 1995), industrial wastewater (Yoon et al., 2000c), wastewater treatment bioreactors (Kim et al. 2003), and activated sludge in aeration tanks of biological sewage-treatment plants (Lemmer and Kroppenstedt, 1984). Large numbers of gordoniae have been isolated from foaming activated sludge plants in various parts of the world (Lechevalier and Lechevalier, 1974; Goodfellow et al., 1996; Goodfellow et al., 1998; Stainsby et al., 2002); *G. amarae* is regarded as a major component of the bacterial community responsible for foaming in such plants (Dhaliwal, 1979; Sezgin et al., 1988; Blackall et al., 1989b; Blackall et al., 1991; De los Reyes et al., 1998a; De los Reyes et al., 1998b). This organism produces a surfactant and cells that are very hydrophobic (Blackall and Marshall, 1989a).

Gordoniae are being increasingly encountered as agents of human infection, but even so their superficial resemblance to more common aerobic actinomycete pathogens, notably mycobacteria and nocardiae, has probably resulted in inadequate diagnosis and hence an underestimation of their clinical significance. *Gordonia bronchialis* and *G. sputi* have been isolated from sputa of patients with pulmonary disease, including bronchiectasis and cavitary pulmonary tuberculosis (Tsukamura, 1971a; Tsukamura, 1978; Tsukamura and Yano, 1985; Guagnani et al., 1998); members of the former taxon caused a nosocomial outbreak of sternal wound infections in patients following coronary artery bypass surgery (Richet et al., 1991) and bacteremia in a patient with a sequestrated lung (Sng et al., 2004). A *Gordonia* strain, related to *G. sputi*, has been implicated in a case of endocarditis in a young woman with a central venous catheter (Lesens et al., 2000). The first recorded case involving *G. rubripertincta* was a lung infection in a young immunocompromised woman where an initial diagnosis was confused with tuberculosis (Hart et al., 1988). Gordoniae, especially *G. terrae*, has been reported to cause central venous catheter related bacteremia (Buchman et al., 1992; Pham et al., 2003), infections of the central nervous system (Drancourt et al., 1994; Drancourt et al., 1997), and cutaneous infections (Martin and Hogan, 1991; Lasker et al., 1992). *Gordonia sputi* has been implicated as a causal agent of mesenteric lymphadenitis of the ileum of a six-year-old pig (Tsukamura et al., 1988).

Skermania piniformis has been found to be abundant in foams on the surface of aeration tanks in activated sludge plants, a feature it has in common with *G. amarae* (Blackall et al., 1989b; Blackall et al., 1989b; Soddell and Seviour, 1994; Soddell and Seviour, 1998). However, unlike *G. amarae* this organism has not been isolated from any other habitat. Little is known about the distribution of *Williamsia*, as only two strains have been isolated to date. *Williamsia muralis* DSM 44343[T] was isolated from indoor building material in the wall of a child day care center in Finland (Kämpfer et al., 1999) and *W. maris* DSM 44693[T] from sediment collected from the Sea of Japan (Stach et al., 2004).

The Family Nocardiaceae

Members of the family Nocardiaceae are common and widely distributed in aquatic and terrestrial habitats, but some are opportunistic pathogens for animals, including humans. Nocardiaceae strains are probably involved in the turnover of organic matter in natural habitats though little is known about their roles therein. Developments in molecular ecology and selective isolation procedures can be expected to yield information on their occurrence, distribution, numbers and activity in natural habitats.

NOCARDIAE AS ANIMAL PATHOGENS. Nocardiae cause several types of suppurative infections in humans (Schaal and Beaman, 1984; Boiron et al., 1993; Boiron et al., 1998; McNeil and Brown, 1994; Goodfellow 1995; Goodfellow 1997; Schaal, 1997; Beaman, 2000a; Beaman, 2002). Infection may occur by inhalation and through contaminated wounds and traumatic implantations. Five forms of disease are recognized, namely, pulmonary nocardiosis; systemic nocardiosis involving two or more body sites; nocardiosis of the central nervous system; extrapulmonary nocardiosis; and cutaneous, subcutaneous and lymphocutaneous nocardiosis (Schaal, 1997). This division is justified from clinical and therapeutic perspectives and is a development of an earlier scheme (Schaal and Beaman, 1984).

Most pulmonary infections in nontropical countries are caused by *N. asteroides*, *N. farcinica* and "*N. nova*"; relatively few are caused by *N. brasiliensis*, *N. otitidiscaviarum*, *N. pseudotuberculosis* and *N. transvalensis*. Hamid et al. (2000) described *N. africana* for strains isolated from sputum samples taken from patients with pulmonary disease at the Chest Unit of Khartoum Teaching Hospital in the Sudan. There have been isolated reports of pulmonary nocardiosis from other tropical countries caused by *N. asteroides*, *N. farcinica*, *N. otitidiscaviarum* and *N. transvalensis* (Osoagbaka et al., 1985; McLeod et al., 1989; Koffi et al., 1998). Cutaneous and subcutaneous infections are caused by *N. asteroides* and *N. brasiliensis* and to a lesser extent by *N. farcinica*, *N. otitidiscaviarum* and *N. transvalensis*. Strains isolated from abscesses have been described as *N. abscessus* (Yassin et al., 2000), *N. puris* (Yassin et al., 2003), and as *N. niigatensis* and *N. yamanashinensis* (Kageyama et al., 2003b). Some nocardiae are agents of actinomycetoma, a localized, chronic and progressive infection of the skin and subcutaneous tissue (Mahgoub and Murray, 1973) which is endemic in many tropical and subtropical regions of the world. The main nocardial agent of this disease is *N. brasiliensis*; *N. asteroides* and *N. otitidiscaviarum* are occasionally involved.

Nocardiae cause infections in a wide range of animals (Beaman and Sugar, 1983; Beaman and Beaman, 1994). Animals susceptible to nocardial infections include birds, cats, cattle, chickens, dogs, ducks, fish, goats, guinea pigs and sheep. Pulmonary and systemic nocardioses, including infections of the brain, are the most frequently recognized conditions, but cutaneous and mycetomatous lesions also occur. In dairy animals, notably cows, nocardial mastitis can be a major problem (Bushnell et al., 1979; Battig et al., 1990; Stark and Anderson, 1992; Manninen et al., 1993; Da Costa et al., 1996). Strains involved in the Canadian epizootic were tentatively identified as *N. farcinica* (Manninen et al., 1993). The most frequently recognized pathogen of animals is *N. asteroides* followed by *N. brasiliensis* and *N. otitidiscaviarum*. *Nocardia crassostreae* has been isolated from infections in oysters (Friedman et al., 1998) and *N. seriolae* from those in fish (Kudo et al., 1988).

NOCARDIAE AS SAPROPHYTES, SYMBIONTS AND PLANT PATHOGENS. Nocardiae are common in soil (Cross et al., 1976); populations up to 7.3×10^4/g dry weight have been found in environmental samples from tropical and temperate regions (Orchard et al., 1977). They also form mutualistic associations with blood-sucking arthropods and occur in aquatic habitats (Cross et al., 1976), where they have been implicated in the biodeterioration of natural rubber joints in water and sewage pipes (Hutchinson et al., 1975; Maldonado et al., 2000). Most isolates from these habitats have been identified as *N. asteroides*, but the application of improved methods for delineating nocardial species has led to novel soil isolates being classified as *N. beijingensis* (Wang et al., 2001), *N. caishijiensis* (Zhang et al., 2003), *N. cerradoensis* (Albuquerque de Barros et al., 2003), *N. cummidelens* (Maldonado et al., 2000), *N. fluminea* (Maldonado et al., 2000), *N. soli* (Maldonado et al., 2000), *N. tenerifensis* (Kämpfer et al., 2004), *N. vinacea* (Kinoshita et al., 2001), and *N. xishanensis* (Zhang et al., 2004). *Nocardia asteroides* and *N. otitidiscaviarum* have been isolated from foaming sewage treatment plants of the activated sludge type (Lechevalier and Lechevalier, 1974; Lechevalier et al., 1977; Blackall et al., 1988; Blackall et al., 1989b). *Nocardia vaccinii*, the only well known plant pathogen, causes galls in blueberry (Demaree and Smith, 1952).

RHODOCOCCI AS SAPROPHYTES, SYMBIONTS AND PATHOGENS. Rhodococci are widely distributed in aquatic and terrestrial habitats. They have been isolated from soil, marine sediments, and from the gut contents of blood-sucking arthropods with which they may form a mutualistic association (Cross et al., 1976; Goodfellow and Aubert, 1980; Colquhoun et al., 1998;

Zhang et al., 2002). *Rhodococcus erythropolis*, *R. rhodochrous* and *R. ruber* have been isolated from soil (Goodfellow and Williams, 1983), *R. marinonascens* from marine sediments (Helmke and Weyland, 1984), and *R. rhodochrous* from foam on the surface of aeration tanks in activated sludge plants (Lemmer and Kroppenstedt, 1984). They have also been recovered from more restricted sources, as exemplified by the isolation of *R. baikonurensis* from air in the Russian space laboratory Mir (Li et al., 2004), *R. gordoniae* from phenol contaminated soil (Jones et al., 2004), *R. jostii* from a medieval grave (Takeuchi et al., 2002), and *R. koreensis* and *R. pyridinivorans* from industrial wastewater (Yoon et al., 2002b; Yoon et al., 2002c).

Rhodococcus coprophilus has been the subject of more extensive ecological studies (Rowbotham and Cross, 1977a). This organism grows on herbivorous dung, but high numbers have also been recorded from grazed pastures and from rivers, streams, and lake muds that receive runoff from land devoted to dairy farming. The coccal survival stage seems to contaminate grass in pastures or hay used during the winter months for fodder and remains viable after ingestion and passage through the rumen. The significant correlation found between the numbers of *R. coprophilus* and fecal streptococci (enterococci) in polluted water led Al-Diwany and Cross (1978) to the view that the organism be considered an indicator of farm animal effluent, a view shared by others (Mara and Oragui, 1981; Oragui and Mara, 1985).

The clinical presentation of rhodococcal infections in humans is influenced by the immune status of the host and the virulence of the infecting organism. The first documented human infection involved a patient suffering from *R. equi* pneumonia who had been given corticosteroid therapy for chronic hepatitis (Golub et al., 1967). Later reports of human infection have underpinned the role of rhodococci as agents of pulmonary disease in severely immunocompromised individuals (McNeil and Brown, 1994). Other rhodococci implicated in pathogenic processes include *R. erythropolis*, *R. globerulus*, *R. luteus*, *R. rhodnii* and *R. rhodochrous* (Alture-Werber et al., 1968; Haburchack et al., 1978; Broughton et al., 1981; Von Below et al., 1991; Cuello et al., 2002).

Rhodococcus equi has long been recognized as an important veterinary pathogen that causes bronchopneumonia, lymphadenitis, and ulcerative enteritis in foals (Barton and Hughes, 1980; Prescott, 1991). Reports of *R. equi* infections in humans, once rare, are increasing in frequency (Scott et al., 1995; Kedlaya et al., 2001; Weinstock and Brown, 2002). Indeed, members of this species have been identified as agents of invasive

pulmonary infections in severely immunosuppressed patients, in particular those with acquired immunodeficiency syndrome (AIDS; McNeil and Brown, 1994; Arlotti et al., 1996). Drancourt et al. (1992) reported 51 infections due to *R. equi*, 20 of which came from human immunodeficiency virus (HIV)-patients and 9 from immunocompetent individuals; they also noted that patients infected with *R. equi* frequently had a history of contact with farm animals.

The Family Tsukamurellaceae

Tsukamurellae have been isolated from activated sludge, arthropods, clinical material and soil (Steinhaus, 1941; Tsukamura and Mizuno, 1971b; Goodfellow et al., 1998; Larkin et al., 1999). They have also been reported to cause catheter-associated sepsis, cutaneous infections, meningitis and pulmonary disease. Respiratory tract colonization occurs in immunocompromised and non-immunocompromised patients, notably those with underlying chronic lung diseases. Tsukamura and Kawakami (1982c) were the first to note the pathogenic role of *Tsukamurella* strains in humans when they repeatedly recovered a pure culture of *T. paurometabola* from the sputa of a patient with a tuberculosis-like disease. This organism has been associated with individual cases of lethal meningitis (Prinz et al., 1985), severe gangrenous tendosynovitis with multiple subcutaneous abscesses (Tsukamura et al., 1988a), and catheter-related bacteremia (Shapiro et al., 1992). Additional cases of catheter-related bacteremia have been attributed to *T. pulmonis* and *T. tyrosinosolvens* (Maertens et al., 1988; Schwartz et al., 2002; Sheridan et al., 2003). Members of these taxa, together with *T. inchonensis*, have been reported to cause lung infections (Yassin et al., 1995; Yassin et al., 1996; Yassin et al., 1997). *Tsukamurella strandjordii* was isolated from blood cultures of a child with acute myelogenous leukemia (Kattar et al., 2001). A pseudoinfection caused by *T. paurometabola* strains was traced back to laboratory contamination (Auerbach et al., 1991).

Isolation and Cultivation

The Family Dietziaceae

Dietzia maris was isolated from soil and the skin and intestinal contents of carp (*Cyprinus carpier*) on mineral salts agar enriched with *n*-alkanes and incubated at 28°C (Nesterenko et al., 1982), and *D. natronolimnaios* was isolated from dilutions of sediment samples made in alkaline broth; the latter was then plated out onto alkaline agar media prior to incubation at 37°C (Duckworth et al., 1998). *Dietzia psychralcaliphila* was isolated from water (6°C, pH 7) obtained from a drain pool of a fish processing plant using a synthetic medium (AT medium) which was supplemented with vaporized *n*-tetradecane as the sole carbon source; inoculated plates were incubated at 4°C for a month (Yumoto et al., 2002).

The Families Gordoniaceae and Tsukamurellaceae

A chemical pretreatment combined with a selective medium has been developed for the isolation of gordoniae and tsukamurellae from sputa and soils (Tsukamura, 1971a; Tsukamura et al., 1988a). Samples and specimens suspended in distilled water (25 ml) are added to an equal volume of NaOH (4%) and liquefied, either by incubation at 37°C for 30 min or by shaking at room temperature for 15–30 min. The digests are inoculated onto Ogawa egg medium (Tsukamura, 1972) and incubated for 4–8 weeks.

For soil samples, soil (5 g) is suspended in distilled water (25 ml) and shaken vigorously in a 300-ml Erlenmeyer flask at room temperature on a reciprocal shaker (stroke: 10 cm, 60 cpm) for 30 min. The suspension is allowed to settle for 10 min, 15 ml of the supernatant is added to an equal volume of NaOH (8%), and the mixture is shaken for 10 min before centrifugation at $500 \times g$ for 15 min. The residue is suspended in 10 ml of a 1% NaH_2PO_4 solution, and 0.002 ml is added to Ogawa egg medium slants, which are incubated for 4–8 weeks.

Colonies growing on the Ogawa egg medium are transferred to fresh slants supplemented with sodium salicylate (0.5 mg/ml), which inhibits *Mycobacterium tuberculosis* (Tsukamura, 1962). Smears, prepared from colonies growing on supplemented slants after 3 weeks of incubation at 37°C, are stained by the Ziehl-Neelsen method and observed by light microscopy. Slightly acidfast, rod-shaped bacteria are typical of gordoniae; those strongly acid-fast are mycobacteria. Gordoniae produce rough, reddish or pinkish colonies on Ogawa egg slants that have been plugged with cotton wool.

Large numbers of gordoniae and tsukamurellae have been isolated on glucose yeast extract agar plates (Gordon and Mihm, 1962) supplemented with cycloheximide and incubated at 30°C for 14 days following inoculation with serially diluted samples collected from foaming activated sludge plants (Goodfellow et al., 1996; Goodfellow et al., 1998). Gordoniae were recognized by their ability to produce rough, grayishpink dry colonies and tsukamurellae by their characteristic deep orange, dry colonies. Prelim-

inary characterization studies showed that the former could be assigned to several novel centers of taxonomic variation within the genus *Gordonia*. Subsequent more exacting studies on representative tsukamurellae led to them being classified into two novel species, namely *T. spumae* and *T. pseudospumae* (Nam et al., 2003; Nam et al., 2004b).

The best media for the isolation of *G. amarae* strains are Czapek's agar supplemented with yeast extract (Higgins and Lechevalier, 1969) and glycerol agar (Gordon and Smith, 1953); these media yield visible colonies in 5–7 days at 38°C (Lechevalier and Lechevalier, 1974; Lechevalier et al., 1976). Similarly, *G. rubripertincta* strains have been isolated on Münz paraffin agar and Winogradsky's nitrate agar supplemented with paraffin (Nesterenko et al., 1978a). Novel gordoniae have been isolated from diverse environmental samples using selective enrichment protocols; organisms isolated in this way have been characterized as *G. alkanivorans* (Kummer et al., 1999), *G. amicalis* (Kim et al., 2000), *G. desulfuricans* (Kim et al., 1999), *G. nitida* (Yoon et al., 2000c), *G. paraffinivorans* (Xue et al., 2003), and *G. sihwensis* (Kim et al., 2003).

The first successful isolation of the relatively slow-growing *S. piniformis* from activated sludge foam and mixed liquor involved the application of a micromanipulatory technique (Blackall et al., 1989b). A Skerman micromanipulator (Skerman, 1968) was used to isolate acute-angled branching filaments from activated sludge samples taken from plants in Queensland, Australia; the filaments were transferred to either yeast glucose agar or tryptone yeast extract agar plates and incubated at a range of temperatures. Visible colonies (1–2 mm in diameter) of *S. piniformis* grew within 10–21 days. The organism has been micromanipulated from activated sludge foam samples taken from various plants in Australia. These findings suggest that the incidence of *S. piniformis* in activated sludge foams has been underreported, as the slow and variable growth of skermaniae means that they are likely to be outcompeted on isolation plates by faster growing mycolic acid-containing actinomycetes, such as *G. amarae* (Soddell and Seviour, 1990; Soddell and Seviour, 1994; Soddell et al., 1992). *Gordonia* and *Tsukamurella* strains, including *G. amarae*, have also been isolated from foaming activated sludge plants by micromanipulation (Tandoi et al., 1992; Seong et al., 1993; Stainsby et al., 2002).

A special semisolid medium, the main nutritive constituents of which were proteose peptone, rabbit serum, gelatin, mineral rabbit kidney, and carbohydrates, were used for the original isolation of *Tsukamurella paurometabola* (Steinhaus, 1941). The organism grew in 24–48 h when subsequently transferred to beef infusion agar. *Tsukamurella strandjordii* and *T. tyrosinosolvens* were obtained directly from blood cultures (Yassin et al., 1997; Kattar et al., 2001), and *T. pulmonis* was obtained from sputum following decontamination with *n*-acetyl-L-cysteine, centrifugation, and cultivation on Lowenstein-Jensen medium (Yassin et al., 1997).

Gordoniae and tsukamurellae grow well on most standard laboratory media, including Bennett (Jones, 1949), brain heart infusion (Difco 0418), modified Sauton supplemented with thiamine (Mordarska et al., 1972), glucose yeast extract (Waksman, 1950) agars, Lowenstein-Jensen medium (BBL 20908), and skermaniae on yeast glucose and tryptone yeast extract agars (Blackall et al., 1989b).

The Family Nocardiaceae

CLASSICAL METHODS OF SELECTIVE ISOLATION FOR NOCARDIAE AND RHODOCOCCI. Classical isolation procedures have depended on the capacity of nocardiae and rhodococci to use hydrocarbons as sole sources of carbon for energy and growth (Cross et al., 1976; Tárnok, 1976). Modifications of Söhngen's paraffin baiting method (Söhngen, 1913) have often been used to isolate *Nocardia* and *Rhodococcus* strains from soil (Portaels, 1976; Schaal and Bickenbach, 1978a) but are of little value in quantitative studies because they merely indicate the presence or absence of the organisms in samples. Many other bacteria and fungi that metabolize paraffin wax frequently outgrow nocardiae and rhodococci on coated glass rods and render isolation impossible. Alternative isolation methods, such as inoculating guinea pig or hamster testicles with soil suspensions supplemented with penicillin and streptomycin (Conti-Diaz et al., 1971) or plating suspensions onto media that contain sodium azide and cholesterol acetate (Farmer, 1962), underestimate populations of nocardiae and rhodococci in natural habitats.

SELECTIVE ISOLATION OF NOCARDIAE. Large numbers of nocardiae have been isolated from soil by plating out soil suspensions onto diagnostic sensitivity test (DST) agar supplemented with antifungal and various combinations of antibacterial antibiotics (Orchard and Goodfellow, 1974; Orchard et al., 1977; Maldonado et al., 2000). Inoculated plates are incubated for up to 21 days at 25°C. Colonies with pink to red stroma, covered to a greater or lesser extent with white aerial hyphae, are characteristic of nocardiae. To date, most isolates have been classified as *N. asteroides*, although laboratory strains of *N. brasiliensis* and *N. otitidiscaviarum* do grow on DST media (Orchard et al., 1977). At least some strains of *N. asteroides* will likely be inhibited by the antibacterial antibiotics in

DST agar (Schaal and Heimerzheim, 1974). A selective medium free of such antibiotics has been used to isolate *N. asteroides* from clinical material (Schaal, 1972).

Sabouraud dextrose agar supplemented with chloramphenicol has been recommended for the recovery of *Nocardia* species from sputum specimens (Ajello and Roberts, 1981), but many nocardiae are inhibited by chloramphenicol (Gutmann et al., 1983). There is encouraging evidence that chemically defined media containing paraffin agar may be useful for the selective isolation of nocardiae from clinical specimens (Shawar et al., 1990). Nonselective media such as brain heart infusion and Sabouraud dextrose agars have been recommended for the isolation of *Nocardia* species from clinical material (Schaal, 1977). Decontamination of respiratory specimens has been shown to be toxic for nocardiae (Murray et al., 1987). However, Hamid et al. (2001) isolated *N. africana* strains on Lowenstein-Jensen slopes that had been inoculated with sputum samples from patients with pulmonary diseases then treated with the digestion-decontamination procedure of Roberts et al. (1991); most of the patients either had not responded to treatment with anti-tubercular drugs or had responded and then relapsed.

Nocardia asteroides and *N. otitidiscaviarum* have been isolated by plating sewage foam onto either Czapek's agar supplemented with yeast extract (0.2%; Higgins and Lechevalier, 1969) or glycerol agar (Gordon and Smith, 1953) and incubating plates at 28°C for 5–7 days. Using nitrite medium (Winogradsky, 1949), Cross et al. (1976) reported high numbers of nocardiae from soils in the Union of Soviet Socialist Republics, but it seems likely that many of their isolates were rhodococci. However, nocardiae are common in soil and have been isolated serendipitously in small numbers on nutrient rich media, as exemplified by the growth of *N. beijingensis* on glucose asparagine agar (Wang et al., 2001), *N. caishijiensis* on Bennett agar (Zhang et al., 2003), *N. cerradoensis* on R5 agar (Albuquerque de Barros et al., 2003), *N. neocaledoniensis* on oatmeal agar (Saintpierre-Bonaccio et al., 2004), *N. tenerifensis* on soil extract agar (Kämpfer et al., 2004), and *N. xishanensis* on modified Sauton agar (Zhang et al., 2004).

Nocardiae grow well on most laboratory media, including brain heart infusion (Difco 0418), nutrient, Sabouraud dextrose, Bennett (Jones, 1949), yeast extract-glucose (Waksman, 1950), modified Sauton (Mordarska et al., 1972), and yeast-malt extract (ISP medium 2; Difco 0770) agars. Media should be incubated aerobically at 25–37°C for up to 3 weeks.

SELECTIVE ISOLATION OF RHODOCOCCI. A heat-pretreatment method combined with a selective medium has been devised to isolate *Rhodococcus coprophilus* from both aquatic and terrestrial habitats (Rowbotham and Cross, 1977a). Thus, environmental samples (2 ml) of cream, milk, or water in 100- × 12-mm glass tubes sealed with silicon rubber bungs are heated in a water bath for 6 min at 55°C prior to further dilution or plating out. Water samples are stored at 4°C before heating. Suspensions of dung, grass, soil (1 : 10), or hay (1 : 50) are homogenized in one-quarter-strength Ringer's solution containing gelatin (0.01%), pH 7.0, before heat pretreatment. Immediately after pretreatment, samples are shaken on a Vortex mixer and portions (0.2 ml) spread onto M3 agar plates (see Media for Isolation and Cultivation in this Chapter), which are incubated at 30°C for 7 days.

Rhodococcus luteus strains were isolated from soil and the skin and intestinal contents of carp on mineral salts agar enriched with *n*-alkanes and incubated at 28°C (Nesterenko et al., 1982); *R. marinonascens* were isolated from marine sediments using a number of rich nutrient media supplemented with seawater and incubated for 8–12 weeks at 18°C (Weyland, 1969), and *R. erythropolis* and *R. rhodochrous* were isolated from mineral salts media supplemented with *m*-cresol or phenol (Gray and Thornton, 1928). Similarly, *R. equi* was isolated from soil, feces, lymph nodes, and the intestinal contents of several animal species using a selective medium (NANAT medium) supplemented with cycloheximide, nalidixic acid, novobiocin, and potassium tellurite (Woolcock et al., 1979; Mutimer and Woolcock, 1980). A selective enrichment broth (TANT broth) containing cycloheximide, nalidixic acid, penicillin, and potassium tellurite incubated at 30°C and used in conjunction with Tinsdale medium (Oxoid) and modified M3 medium has also been used to isolate *R. equi* (Barton and Hughes, 1981).

Rhodococci have been isolated from soil using Czapek's agar (Higgins and Lechevalier, 1969), glycerol agar (Gordon and Smith, 1953), and Winogradsky's nitrate medium (Winogradsky, 1949) and from diseased sweet peas using potato dextrose agar (Tilford, 1936). Novel strains have been isolated from a range of environmental samples using selective enrichment procedures, as illustrated by *R. aetherivorans* (Goodfellow et al., 2004), *R. koreensis* (Yoon et al., 2000a), *R. percolatus* (Briglia et al., 1996), *R. pyridinivorans* (Yoon et al., 2000b) and *R. tukisamuensis* (Matsuyama et al., 2003).

Rhodococci grow well on standard laboratory media such as Bennett (Jones, 1949), brain heart infusion (Difco 0418), glucose yeast extract (Waksman, 1950), modified Sauton plus thiamine (Mordarska et al., 1972), and nutrient

agars and Lowenstein-Jensen medium (BBL 20908).

Media for Isolation and Cultivation

The following are the recipes for those media mentioned in the preceding section that are not commercially available.

Isolation Media

Alkaline Broth (Duckworth et al., 1998)

Glucose	10 g
Peptone (Difco)	0.5 g
Yeast extract (Difco)	0.5 g
K_2HPO_4	0.1 g
$MgSO_4 \cdot 7H_2O$	0.2 g
NaCl	40 g
Na_2CO_3	10 g
Deionized water	1 liter

Autoclave NaCl and Na_2CO_3 separately and add these to the organic components, including 1.5% (w/v) agar, for solid media.

AT Medium (Yumoto et al., 2002)

KNO_3	5.0 g
KH_2PO_4	0.5 g
$MgSO_4 \cdot 7H_2O$	0.5 g
$FeSO_4 \cdot 7H_2O$	0.01 g
$CaCl_2 \cdot 2H_2O$	0.02 g
$MnSO_4 \cdot nH_2O$	0.01 g
$ZnSO_4 \cdot 7H_2O$	0.0005 g
Agar	15 g
100 mM $NaHCO_3/Na_2CO_3$	1 liter

Dissolve the buffer in deionized water and adjust to pH 10. Vaporize *n*-tetradecane, the sole carbon source, onto the surface of the agar plate by inverting the plate over a piece of filter paper that had been soaked in *n*-tetradecane.

Chitin Agar (Weyland, 1969)

Chitin, hydrolyzed and precipitated	10 g
Peptone	0.5 g
Yeast extract	0.1 g
$FePO_4 \cdot H_2O$	0.01 g
Seawater	750 ml
Distilled water	250 ml
Agar	15 g

Adjust to pH 7.5.

Czapek-Dox Agar (Weyland, 1969)

Sucrose	15 g
$NaNO_3$	2 g
$FePO_4 \cdot H_2O$	0.01 g
Magnesium glycerophosphate	0.5 g
Seawater	750 ml
Distilled water	250 ml
Agar	15 g

Adjust to pH 7.6–7.8.

N-GB>DST Agar (Orchard and Goodfellow, 1974)

Supplement diagnostic sensitivity (DST) agar (Oxoid, CM261) with cycloheximide and nystatin (antifungal antibiotics) and various concentrations of chlortetracycline HCl, demethylchlortetracycline HCl, and methacycline HCl (antibacterial antibiotics). Autoclave the DST agar and cycloheximide separately at 15 psi (1.013×10^5 Pa) for 20 min, and filter the remaining antibiotics. Pipet the individual antibiotics separately into Petri dishes and add the basal medium to give the following concentrations (in µg/ml) for four different media:

Medium 1: demethylchlortetracycline HCl (demeclocycline HCl), 5; actidione, 50; and nystatin, 50.

Medium 2: methacycline HCl, 10; actidione, 50; and nystatin, 50.

Medium 3: chlortetracycline HCl (aureomycin), 45; demethylchlortetracycline HCl, 5; actidione, 50; and nystatin, 50.

Medium 4: chlortetracycline HCl, 45; methacycline HCl, 10; actidione, 50; and nystatin, 50.

Media 1 and 2 generally give higher counts of *Nocardia* because of the satisfactory inhibition of other soil bacteria. However, when large mixed populations of bacteria occur, the more selective formulations, media 3 and 4, may be required. The number of unwanted bacteria can also be reduced by heating soil suspensions (2 ml) in a water bath at 55°C for 6 min.

Glucose Extract Agar (Gordon and Mihm, 1962)

Glucose	10 g
Yeast extract	10 g
Agar	15 g
Distilled water	1 liter

Adjust to pH 7.2. Sterilize cycloheximide by membrane filtration and add to the autoclaved and cooled medium to give a final concentration of 50 µl · ml⁻¹.

Glycerol Agar (Gordon and Smith, 1953)

Beef extract	3 g
Glycerol	7%
Peptone	5 g
Agar	15 g
Soil extract	1 liter

Adjust to pH 7.0. Prepare soil extract by autoclaving 1000 g of air-dried soil (that has been sifted through a no. 9 mesh screen) with 2.4 liter of tap water at 121°C for 60 min. Decant the preparation and filter through paper pulp.

M3 Agar (Rowbotham and Cross, 1977a)

KH_2PO_4	0.466 g
Na_2HPO_4	0.732 g
KNO_3	0.01 g
NaCl	0.29 g
$MgSO_4 \cdot 7H_2O$	0.10 g
$CaCO_3$	0.02 g
$FeSO_4 \cdot 7H_2O$	200 µg
$ZnSO_4 \cdot 7H_2O$	180 µg
$MnSO_4 \cdot 7H_2O$	20 µg
Sodium propionate	0.20 g
Agar	18 g
Distilled water	1.0 liter

Adjust to pH 7.0. Filter cycloheximide and thiamine HCl by membrane filtration and add to the autoclaved and cooled medium to give final concentrations of 50 mg/liter and 4.0 mg/liter, respectively.

Mineral Salts Medium (Gray and Thornton, 1928)

K_2HPO_4	1.0 g
$MgSO_4 \cdot 7H_2O$	0.20 g
NaCl	0.10 g
$(NH_4)_2SO_4$ or KNO_3	0.5–1.0 g
$CaCl_2$	0.10 g
$FeCl_3$	0.02 g
Distilled water	1 liter

Inoculate soil (0.5–1 g) into 100 ml of mineral salts solution supplemented (after autoclaving) with m-cresol (0.05%) or phenol (0.05–1.0%). Subculture into fresh flasks containing the same medium, and plate isolates onto appropriate agar media containing m-cresol or phenol.

Mineral Salts Agar (Nesterenko et al., 1982)

KH_2PO_4	0.14 g
KNO_3	1.0 g
$MgSO_4$	0.1 g
NaCl	1.0 g
Na_2HPO_4	0.6 g
Mixture of *n*-alkanes (C-12 to C-22)	20.0 g
Distilled water	500 ml
Tap water	500 ml

Isolate rhodococci on this medium using the method of Yamada et al. (1963).

Mineral Salts Paraffin Agar (Shawar et al., 1990)

KH_2PO_4	3.0 g
K_2HPO_4	1.0 g
NH_4Cl	5.0 g
NH_4NO_3	1.0 g
$FeSO_4$	0.05 g
$MgSO_2 \cdot 7H_2O$	0.05 g
$MnSO_4$	0.05 g
$ZnSO_4$	0.05 g
Bacto agar (Difco)	17 g
Distilled water	1.0 liter

Adjust to pH 7.2. Prepare mineral salts paraffin agar by mixing 9 parts of the carbon-free agar medium with 1 part of paraffin oil (Soybolt viscosity at 37.8°C equals 345–355; J. T. Baker Chemical Co., Phillipsburg, NJ).

Modified Czapek's Agar (Higgins and Lechevalier, 1969)

$NaNO_3$	2.0 g
K_2HPO_4	1.0 g
$MgSO_4 \cdot 7H_2O$	0.5 g
KCl	0.5 g
$FeSO_4$	0.01 g
Sucrose	30.0 g
Yeast extract	2.0 g
Agar	15 g
Distilled water	1.0 liter

Adjust to pH 7.2.

Münz Paraffin Agar (Nesterenko et al., 1978a)

KH_2PO_4	0.14 g
KNO_3	1.0 g
$MgSO_4 \cdot 7H_2O$	0.10 g
NaCl	1.0 g
Na_2HPO_4	0.60 g
Agar	15 g
Distilled water	1.0 liter

Adjust to pH 7.2. Add 10 ml of liquid paraffin to the carbon-free medium after sterilization.

NANAT Medium (Woolcock et al., 1979)

Tryptone soya broth (Oxoid)	30 g
Yeast extract (Oxoid)	1 g
Agar (Difco)	15 g
Distilled water	1 liter

Sterilize at 121°C for 15 min. When cool add cycloheximide, 40 µg/ml; nalidixic acid, 20 µg/ml; and novobiocin, 25 µg/ml. Add potassium tellurite to 0.005%.

Nitrite Medium (Winogradsky, 1949)

NaNO₂	2 g
Na₂CO₄ (anhydrous)	1 g
K₂HPO₄	0.5 g
Agar	15 g
Distilled water	1 liter

Ogawa Egg Medium (Tsukamura, 1972)

Whole chicken eggs	200 ml
Glycerol	6 ml
Malachite green (2%)	6 ml
Sodium glutamate (1%) plus KH₂PO₄ (1%)	100 ml

Adjust to pH 6.8, and sterilize at 90°C for 60 min.

Tellurite-Actidione-Nalidixic Acid-Penicillin (TANT) Selective Enrichment Broth (Barton and Hughes, 1981)

Trypticase soy broth	30 g
Actidione (cycloheximide)	50 µg/ml
Nalidixic acid	20 µg/ml
Potassium tellurite	0.05 g
Penicillin	10 units/ml
Distilled water	1 liter

Inoculate TANT broth (10 ml) with a 1-g environmental sample (e.g., soil or feces) and incubate at 30°C for 6–7 days. Subculture the broth onto Tinsdale (Oxoid) agar supplemented with cycloheximide (50 µg/ml) and onto M3 agar (Rowbotham and Cross, 1977a) modified by the addition of 0.005% potassium tellurite. Colonies of *Rhodococcus equi* appear on these selective media after 4–5 days at 30°C.

Tryptone Yeast Extract Agar (Blackall et al., 1989b)

Glucose	5 g
Tryptone	3 g
Yeast extract	5 g
Distilled water	1 liter

Adjust to pH 7.0 and sterilize at 121°C for 15 min.

Yeast Glucose Agar (Blackall et al., 1989b)

Glucose	10 g
Yeast extract	10 g
Agar	15 g
Distilled water	1 liter

Adjust to pH 7.0 and sterilize at 121°C for 15 min.

Cultivation Media

Bennett Agar (Jones, 1949)

Beef extract	1 g
Glucose	10 g
N-Z amine A (Enzymatic digest of casein)	2 g
Yeast extract	15 g

Adjust to pH 7.3 with NaOH. Alternatively, use modified Bennett agar; in it, glucose is replaced by glycerol (10 g) and N-Z amine A is replaced by Bactocasitone (2 g, Difco).

Glucose-Yeast Extract Agar (Waksman, 1950)

Glucose	10 g
Yeast extract (50% solution)	20 ml

Adjust to pH 6.8.

Modified Sauton Medium (Mordarska et al., 1972)

Asparagine	5.0 g
Casein hydrolysate	2.0 g
Glucose	15.0 g
Sodium citrate	1.5 g
KH₂PO₄	5.0 g
MgSO₄ · 7H₂O	0.5 g
K₂SO₄	0.5 g
Ferric ammonium citrate	Trace
Distilled water	1 liter

Adjust to pH 7.2. Sterilize the glucose separately. For the cultivation of *Rhodococcus* strains, supplement the medium with thiamine (50 mg/1 liter).

Preservation of Cultures

Serial transfer from agar slants of appropriate media (see above) every two months, with storage at 4°C between transfers, is the most convenient method for short-term storage. Lyophilization, storage in liquid nitrogen, or frozen glycerol suspensions can be used for long-term preservation. For lyophilization, cell masses are suspended in a suitable fluid, e.g., 7.5% (w/v) glucose serum or skimmed milk plus 7.5% (w/v) glucose. For storage in liquid nitrogen, the microorganisms are inoculated into small test tubes containing the appropriate medium and incubated until satisfactory growth is visible. The tubes are then closed with cotton wool plugs, dipped in liquid paraffin wax, and placed in a liquid nitrogen container.

Glycerol suspensions are prepared by scraping growth from heavily inoculated agar plates and making heavy suspensions in 3 ml of aqueous glycerol in small vials, which are then stored at –20°C (Wellington and Williams, 1978). The frozen glycerol suspensions serve both as a convenient means of long-term preservation and as a quick source of inoculum. Working inocula are obtained by thawing suspensions at room temperature prior to treating as for broth cultures. After use, glycerol suspensions are promptly frozen and stored again at –20°C.

Identification

Mycolic acid-containing bacteria can be difficult to distinguish from one another and from other actinomycetes using conventional staining and morphological properties. Thus, *Nocardia* strains that lack aerial hyphae cannot always be separated from mycobacteria, rhodococci or "bald" streptomycetes, whereas those producing abundant aerial hyphae can be confused with members of the family Pseudonocardiaceae. Reliable differentiation at the genus level depends to a large extent on the application of chemotaxonomic techniques. Simplified procedures are available for the detection of morphological and staining features (Goodfellow, 1996); cell wall constituents (Hancock, 1994), including the acyl type of muramic acid (Uchida et al., 1999); fatty acids, including mycolic acids (Minnikin et al., 1984b; Embley and Wait, 1994; Goodfellow, 1996); menaquinones (Collins et al., 1985b; Collins, 1994); and polar lipids (Suzuki et al., 1993). Such procedures will be progressively replaced by molecular systematic methods, including the use of nucleic acid probes, oligonucleotide primers and 16S rRNA gene sequencing (Stackebrandt et al., 1997; Laurent et al., 1999; Gürtler and Mayall, 2001a; Monciardini et al., 2002; Stach et al., 2003; Patel et al., 2004).

Identification at the Generic Level

Examination of whole-organism acid or alkaline methanolysates for the presence of mycolic acids is the first stage in the chemical procedure. Mycolic acids vary considerably in structure, ranging from relatively simple mixtures of saturated and unsaturated acids found in corynebacteria to highly complex mixtures characteristic of mycobacteria (Minnikin and Goodfellow, 1980a; Minnikin, 1993; Goodfellow and Magee, 1998a). Qualitative evaluation of mycolic acids can be easily and quickly achieved using a thin-layer chromatographic technique devised by Minnikin et al. (1975). Methanolysates of mycobacteria, other than *M. fallax* and *M. triviate* strains (Dobson et al., 1985), give a multispot pattern of mycolates, those of tsukamurellae give two spots, and those of dietziae, corynebacteria, gordoniae and rhodococci give a single spot, the Rf values of which reflect their chain length and structure (Goodfellow et al., 1976; Minnikin et al., 1980b; Hamid et al., 1993; Minnikin; 1993; Yassin et al., 1995; Yassin et al., 1996; Yassin et al., 1997). Mycolic esters can be positively identified on thin-layer chromatograms by their characteristic immobility when plates are subsequently washed in methanol-water (5:2, v/v; Minnikin et al., 1975). The patterns of mycolic acid types can be used for the identification of mycobacterial species (Goodfellow and Magee, 1998a; Minnikin et al., 1984a; Minnikin et al., 1984c; Minnikin et al., 1985b).

The mycolic acids from mycobacteria are precipitated from ether solution by addition of an equal (Kanetsuna and Bartoli, 1972) or double (Hecht and Causey, 1976) volume of ethanol. Mycolic acids from corynebacteria, gordoniae, nocardiae and rhodococci are not precipitated by

such a procedure, but Hecht and Causey (1976) detected their presence by thin-layer chromatography of the supernatant. A more reliable and sensitive precipitation procedure was introduced by Hamid et al. (1993) to separate mycobacteria from members of other mycolic acid-containing taxa. The basis of the method is the solubility of all mycolic acid methyl esters in acetonitrile/toluene (1:2, v/v) and the insolubility of those from mycobacteria in acetonitrile/toluene (3:2, v/v).

When the presence of mycolic acids has been detected, their esters can be isolated and studied further by pyrolysis gas chromatography, by mass spectrometry of the intact esters, by gas chromatography of trimethylsilyl ether derivatives, and by high-performance liquid chromatography (HPLC) of bromophenacyl derivatives. On pyrolysis, mycolic acid methyl esters yield long-chain methyl esters and aldehydes, which can be analyzed directly by use of gas pyrolysis, whereas the remaining mycolic acid-containing organisms produce esters with fewer carbon atoms (Table 1). Mycolic esters from *G. amarae* release unsaturated C_{16} and C_{18} major components (Lechevalier and Lechevalier, 1974) while production of unsaturated C_{20} and C_{22} esters characterizes the mycolates from *Tsukamurella paurometabola* (Goodfellow et al., 1978). Mycolic acids with unsaturation in the chain in the 2-position have been observed in two nocardial strains, *N. carnea* and *N. vaccinii*, but no details have been reported (Lechevalier and Lechevalier, 1974).

The pyrolysis of mycolates can be observed by mass spectrometry, where the highest peaks in the spectra correspond to anhydromycolates formed by loss of water from the parent molecule (Etémadi, 1967). The overall size of mycolates, their degree of unsaturation, and the nature of both long-alkyl chains may be determined by mass spectrometry. However, the complex mixtures of homologs usually present and the competing fragmentation pathways make interpretation difficult in some cases (Maurice et al., 1971; Alshamaony et al., 1976a; Alshamaony et al., 1976b). The analysis can be taken a stage further by using gas chromatography-mass spectrometry of trimethylsilyl and tertbutyldimethylsilyl derivatives of mycolic acids (Yano et al., 1978; Tomiyasu et al., 1981; Athalye et al., 1984; Pommier and Michel, 1985). This procedure separates mycolic ester derivatives into their homologous components, each of which can be analyzed by mass spectrometry. Capillary gas chromatography-mass spectrometry of trimethylsilyl ester derivatives of mycolic acid methyl esters has been used to distinguish among *Dietzia*, *Gordonia* and *Rhodococcus* strains (Nishiuchi et al., 2000) and to reveal differences in the mycolic acid profiles of *N. asteroides*, *N. farcinica* and *"N. nova"* strains (Butler et al., 1986; Butler et al., 1996; Butler et al., 1999; Baba et al., 1997; Nishiuchi et al., 1999; Butler and Guthertz, 2001).

Different classes of mycobacterial mycolic acids can be separated by HPLC (Qureshi et al., 1978; Steck et al., 1978; Ramos, 1994; Kaneda et al., 1995; Rhodes et al., 2003). This separation is based on the chain lengths, the degree of unsaturation, and other functional groups found in these fatty acids (Svensson et al., 1982). Reverse-phase HPLC of *p*-bromophenacyl esters of mycolic acids may provide a rapid way of distinguishing between mycolic acid-containing genera (Butler et al., 1986). A standardized method for HPLC identification of mycobacteria and an associated database are available (Butler et al., 1996).

Representatives of the genus *Rhodococcus* generally contain longer mycolic acids (34–52 carbon atoms) than those of *Corynebacterium* (22–38 carbon atoms), but an unambiguous distinction between members of the two genera cannot be made by analysis of mycolic acids alone (Collins et al., 1982b). Several rhodococci have mycolic acids that overlap in size with those of corynebacteria (e.g., *R. equi*; Collins et al., 1982b; Barton et al., 1989). Rhodococci are aerobic; although the vast majority of corynebacteria are undoubtedly facultative, obligate aerobic strains do occur which cannot be assigned with confidence to one genus or the other. Criteria that may prove to be of value in the identification of such "difficult strains" include information from DNA base composition and peptidoglycan analyses.

Members of the genus *Rhodococcus* generally have DNA with a higher G+C content (63–73 mol%) than that of *Corynebacterium* (51–63 mol%), and rhodococci contain *n*-glycolyl not acetyl residues in the glycan moiety of their peptidoglycan (Fig. 1). Similarly, the presence of 10-methyloctadecanoic (tuberculostearic) acid and phosphatidylethanolamine can help in the assignment of strains between the two genera. Members of the genus *Rhodococcus* contain tuberculostearic acid whereas corynebacteria, with the exception of *C. bovis*, lack this fatty acid (Lechevalier et al., 1977; Collins et al., 1982c). All other mycolic acid-containing actinomycetes, apart from *M. gordonae*, contain substantial amounts of tuberculostearic acid (Tisdall et al., 1979; Minnikin et al., 1985a). With the exception of corynebacteria and dietziae, all mycolic acid-containing strains contain phosphatidylethanolamine, in addition to diphosphatidylglycerol and the characteristic phosphatidylinositol mannosides (Lechevalier et al., 1977; Lechevalier et al., 1981; Minnikin et al., 1977; Rainey et al., 1995a).

Gordoniae, nocardiae, rhodococci and tsukamurellae can be separated solely on the basis of their predominant menaquinones (Fig. 1). The presence of fully unsaturated menaquinones with nine isoprene units (Goodfellow et al., 1978; Collins and Jones, 1982a) serves to distinguish tsukamurellae from all other mycolic acid-containing actinomycetes. Similarly, nocardiae and skermaniae are characterized by a pseudo-tetrahydrogenated menaquinone with eight isoprene units where the end of the multiprenyl side chain is cyclized (Howarth et al., 1986; Collins et al., 1987; Blackall et al., 1989b; Chun et al., 1997). Gordoniae and rhodococci can be distinguished from one another and from most related strains, as the former contain dihydrogenated menaquinones with nine isoprene units as the predominant isoprenologue and the latter contain the corresponding menaquinone with eight isoprene units (Collins et al., 1977; Collins et al., 1985b). Detailed analytical procedures for determining menaquinone composition have been described (Minnikin et al., 1984d; Kroppenstedt, 1985; Collins, 1985a; Collins et al., 1987). A small-scale integrated chemotaxonomic procedure is available for the detection of wall and lipid markers (O'Donnell et al., 1985).

Susceptibility to bleomycin (2.5 µg/ml), 5-fluorouracil (20 µg/ml), and mitomycin (10 µg/ml) and β-galactosidase activity may prove useful in the differentiation of mycolic acid-containing genera (Tsukamura, 1974b; Tsukamura, 1981a; Tsukamura, 1981b; Tsukamura, 1982a). There is also evidence that rhodococci are more resistant than mycobacteria are to prothionamide (Ridell, 1983) and that they can be distinguished from the latter on the basis of the acid-fast stain, arylsulfatase activity, capacity to use sucrose as a carbon source, and inability to metabolize trimethylamine as a simultaneous carbon and nitrogen source (Tsukamura, 1971a). To separate *Rhodococcus* from *Mycobacterium* and *Nocardia* strains (Ridell and Norlin, 1973) and from *Corynebacterium* (Ridell, 1977) is also possible using diagnostic precipitinogens.

Identification at the Subgeneric Level

Dietzia species can be distinguished from one another using a few phenotypic tests, notably by whether they reduce nitrate, produce H_2S, hydrolyze urea, or utilize citrate, glutamate, mannitol, L-proline or succinate (Yumoto et al., 2002). Similarly, phenotypic tests have been weighted to separate the type strains of the two *Williamsia* species (Stach et al., 2004). Thus, *D. maris* DSM 44693[T], unlike *D. muralis* DSM 44343[T], uses D(+) mannitol, D(+) trehalose and D(+) xylose as sole carbon sources for energy and growth but not adonitol, D(+) mannose or *meso*-inositol. *Gordonia* species (Table 2) and *Tsukamurella* species (Table 3) can be

Table 2. Phenotypic characteristics separating the type strains of Gordonia species.[a]

Characters	Strains																		
	1	2	3	4	5	6	7	8	9	10	11	12	13	14	15	16	17	18	19
Color of colony	Pink/orange	Pink/orange	Tan/white	Red	Brown	Pink	White/light yellow	Tan/white	Orange	Orange	Orange/red	Orange	Pink/orange red	Orange/red	White/tan	Beige	Pink	Pink/orange	Pastel orange
Biochemical tests																			
Esculin hydrolysis	–	–	+	–	–	–	–	+	+	–	+	–	–	–	ND	–	+	+	ND
Allantoin hydrolysis	–	–	+	–	+	–	–	+	+	–	+	–	–	–	ND	–	–	–	ND
Arbutin hydrolysis	+	–	+	–	–	–	–	+	–	–	–	–	–	+	ND	+	+	+	ND
Nitrate reduction	+	+	–	–	+	+	+	+	–	+	+	–	–	–	+	+	+	+	ND
Urea hydrolysis	+	+	+	–	+	–	–	–	+	+	–	+	–	+	ND	+	+	+	ND
Decomposition of (%, w/v)																			
Hypoxanthine (0.4)	–	–	+	–	–	+	–	+	+	–	–	–	–	+	ND	+	+	+	ND
Starch (1)	–	+	+	+	+	+	+	–	–	+	–	+	+	+	ND	+	+	+	ND
Tributyrin (0.1)	–	+	–	–	–	+	–	–	+	+	ND	+	–	+	ND	+	–	–	ND
Tween 80 (1)	+	+	+	–	–	+	+	+	–	+	+	+	–	–	ND	+	–	–	ND
Tyrosine (0.5)	–	–	+	–	–	–	–	+	–	–	–	–	–	–	ND	+	–	+	ND
Uric acid (0.5)	+	–	+	–	+	+	–	+	+	+	–	+	+	+	ND	–	+	+	ND
Xanthine (0.4)	–	–	–	–	+	–	–	–	–	–	–	–	–	–	ND	–	+	+	ND
Growth on sole carbon sources (%, w/v)																			
Arbutin (1)	–	–	+	+	–	+	+	+	–	–	ND	–	+	–	ND	–	–	–	ND
D(+) Cellobiose (1)	–	–	–	–	–	–	+	–	–	–	–	–	+	–	ND	–	–	+	ND
Glycerol (1)	+	–	+	+	+	+	+	+	+	+	+	–	+	+	ND	–	+	+	–
N-acetyl-D-glucosamine (0.1)	–	+	+	–	–	–	+	+	–	–	ND	+	+	+	ND	+	–	–	ND
Adipic acid (0.1)	+	–	+	+	–	+	–	+	+	+	ND	+	+	–	ND	+	+	–	ND
Betaine (0.1)	–	–	–	–	–	–	–	–	–	–	–	–	+	–	ND	–	+	+	ND
Oxalic acid (0.1)	–	–	–	–	–	–	–	–	+	–	–	–	–	–	ND	–	+	+	ND
Propan-1-ol (0.1)	–	–	+	–	–	+	–	+	–	+	ND	–	+	–	ND	+	–	+	ND
Sodium fumarate (1)	–	+	–	–	+	–	–	–	+	+	ND	–	+	+	ND	+	–	+	ND
Growth in the presence of (%, w/v)																			
Oleic acid (0.1)	+	+	+	+	+	+	+	+	+	+	+	+	–	+	ND	+	+	+	ND
Zinc chloride (0.001)	+	+	+	+	+	+	+	+	+	+	+	+	–	+	ND	+	+	+	ND

Symbols and abbreviation: +, positive; –, negative; and ND, not determined.
[a]Strains: 1, G. aichiensis DSM 43978[T]; 2, G. alkanivorans DSM 44369[T]; 3, G. amarae DSM 43392[T]; 4, G. amicalis DSM 44461[T]; 5, G. bronchialis DSM 43247[T]; 6, G. desulfuricans DSM 43247[T]; 7, G. hirsuta DSM 44140[T]; 8, G. hydrophobica DSM 44015[T]; 9, G. namibiensis NCIMB 13780[T]; 10, G. nitida DSM 44499[T]; 11, G. paraffinivorans DSM 44604[T]; 12, G. polyisoprenivorans DSM 44302[T]; 13, G. rhizosphera IFO 16247[T]; 14, G. rubripertincta DSM 43197[T]; 15, G. sihwensis DSM 44576[T]; 16, G. sinesedis DSM 44455[T]; 17, G. sputi DSM 43896[T]; 18, G. terrae DSM 43249[T]; and 19, G. westfalica DSM 44215[T].
Data from Brandão et al. (2000), Linos et al. (2002), Kim et al. (2003), Maldonado et al. (2003), and Xue et al. (2003).

Table 3. Phenotypic properties separating the type strains of *Tsukamurella* species.

Characters	Strains[a]						
	1	2	3	4	5	6	7
Biochemical tests							
Esculin hydrolysis	+	+	+	+	−	+	+
Urea hydrolysis	−	+	−	+	−	+	−
Color of colonies							
Orange/red	+	−	+	−	+	−	−
White/cream	−	+	−	+	−	+	+
Degradation tests							
Hypoxanthine	+	+	+	+	+	−	+
Tyrosine	+	−	+	−	+	−	+
Growth at 10°C	+	−	+	−	−	−	−
Growth on sole carbon sources (1%, w/v)							
D(+) Arabinose	+	+	+	+	−	−	+
L(+) Arabinose	+	−	+	+	+	−	+
D(+) Arabitol	+	−	+	+	+	+	+
D(+) Cellobiose	−	−	−	+	−	−	+
Dulcitol	−	−	−	+	+	−	+
meso-Erythritol	−	−	−	+	+	−	+
D(+) Fructose	+	−	+	+	+	−	+
D(+) Maltose	+	+	+	+	+	−	+
D(+) Mannitol	−	+	−	−	+	+	+
D(+) Melezitose	+	+	+	−	+	−	+
D(+) Melibiose	−	+	−	+	+	+	+
D(+) Ribose	+	+	+	+	+	−	+
D(+) Salicin	−	+	+	+	−	+	+
D(−) Sorbitol	−	+	−	−	+	+	+
D(+) Xylose	−	+	−	+	+	−	+
Resistance to antibiotics (mg · ml^{-1})							
Clindamycin (2)	+	−	+	+	+	+	+
Erythromycin (5)	+	+	−	−	−	+	+
Fusidic acid (10)	+	−	+	+	+	+	+
Tetracycline hydrochloride (10)	+	−	+	+	−	+	+

Symbols: +, present; and −, absent.
[a]Strains: 1, *T. inchonensis* DSM 44067[T]; 2, *T. paurometabola* DSM 21062[T]; 3, *T. pseudospumae* DSM 44118[T]; 4, *T. pulmonis* DSM 44067[T]; 5, *T. spumae* DSM 44113[T]; 6, *T. strandjordii* ATCC BAA176[T]; and 7, *T. tyrosinosolvens* DSM 44234[T].
Data from Nam et al. (2003, 2004).

distinguished using a combination of phenotypic tests.

Nocardia species can be difficult to identify owing to the lack of suitable phenotypic tests. Diagnostic tests recommended for the separation of established nocardial species (Gordon et al., 1978; Goodfellow and Lechevalier, 1989b; Boiron et al., 1993) can be useful but tend to give presumptive identifications. Partial identification of the major causal agents of human infections, notably those causing nocardiosis, can be achieved using a few simple biochemical, degradation and nutritional tests (Table 4) though the time taken from the acquisition of specimens to species identification can take several weeks. The same battery of diagnostic phenotypic tests can be used to distinguish between nocardial species isolated from environmental samples (Table 5). Supplementary tests based upon antibiotic sensitivity and enzyme activity have been recom-

mended for the separation of some *Nocardia* species (Boiron and Provost, 1990a; Boiron and Provost, 1990b; Boiron et al., 1993). An abbreviated battery of tests has been proposed for the identification of medically relevant *Nocardia* species (Kiska et al., 2002), Laboratory specialists need to be aware of the occasional occurrence of atypical strains of *N. asteroides* (Boiron et al., 1990).

Molecular fingerprinting and sequencing procedures provide a rapid, sensitive and effective way of identifying clinically significant nocardiae. The use of the polymerase chain reaction (PCR) coupled with restriction endonuclease analysis of PCR products has provoked some interest for the separation of *Nocardia* species (Lungu et al., 1994; Steingrube et al., 1995; Steingrube et al., 1997; Wilson et al., 1998; Conville et al., 2000), as has ribotyping (Exmelin et al., 1996; Laurent et al., 1996; Isik et al., 2002)

Table 4. Phenotypic properties separating the type strains of *Nocardia* species causing or associated with infections of humans and other animals.

Characters	Strains[a]																				
	1	2	3	4	5	6	7	8	9	10	11	12	13	14	15	16	17	18	19	20	21
Esculin hydrolysis	-	-	ND	+	+	+	+	+	+	ND	ND	+	+	-	+	+	+	+	+	ND	ND
Nitrate reduction	+	+	+	+	+	-	+	+	+	ND	ND	+	+	-	-	ND	+	+	+	-	ND
Urea hydrolysis	+	-	-	+	+	-	+	+	+	+	-	+	+	+	+	ND	+	-	+	+	+
Deposition of (%, w/v)																					
Adenine (0.4)	-	-	-	-	-	-	-	-	-	-	-	-	-	-	+	ND	-	-	-	-	-
Casein (1.0)	-	-	-	-	+	-	-	-	-	-	-	-	-	-	+	-	-	-	+	-	-
Elastin (0.3)	-	-	ND	-	+	-	-	-	-	ND	ND	-	-	-	+	-	-	-	+	-	ND
Hypoxanthine (0.4)	-	-	-	-	+	-	-	-	+	+	+	-	+	-	+	-	-	-	+	-	+
Tyrosine (0.5)	-	-	-	-	+	-	-	-	-	-	-	-	-	+	+	-	+	-	+	-	-
Uric acid (0.5)	-	-	ND	-	-	-	-	ND	-	ND	ND	-	+	-	+	ND	-	-	+	-	ND
Xanthine (0.4)	-	-	-	-	-	-	-	-	-	-	-	-	+	-	-	-	+	-	+	-	-
Growth on sole carbon sources (%, w/v)																					
D(+) Mannitol (1.0)	-	-	ND	+	+	-	-	-	-	ND	ND	+	+	-	+	+	+	-	-	-	ND
α-L-Rhamnose (1.0)	+	-	+	-	-	+	+	-	+	-	-	-	-	-	-	-	-	-	-	+	-
D(+) Sorbitol (1.0)	-	-	-	-	-	-	-	-	-	-	-	-	-	-	+	+	+	ND	-	-	-
D(+) Xylose (1.0)	-	-	ND	-	+	ND	-	-	-	ND	ND	-	-	+	+	-	-	ND	-	ND	ND
Sodium acetate (0.1)	+	-	ND	+	+	+	-	-	+	ND	ND	-	+	+	+	+	+	+	+	-	ND
Sodium citrate (0.1)	+	-	+	+	+	-	-	+	-	-	-	+	+	-	-	+	+	+	-	-	+
Growth at 45°C	-	+	+	+	+		-	+	-	-	-	+	+	-	-	+	+	+	-	-	-

Symbols: +, positive; -, negative; and ND, not determined.
[a]Strain: 1, *N. abscessus* DSM 44432^T; 2, *N. africana* DSM 44491^T; 3, *N. asiatica* JCM 11892^T; 4, *N. asteroides* ATCC 19247^T; 5, *N. brasiliensis* ATCC 19296^T; 6, *N. brevicatena* DSM 43024^T; 7, *N. crasso streae* ATCC 70418^T; 8, *N. cyriacigeorgica* DSM 44484^T; 9, *N. farcinica* ATCC 3318^T; 10, *N. inohanensis* JCM 11891^T; 11, *N. niigatensis* JCM 11894^T; 12, "*N. nova*" JCM 6044; 13, *N. ottidiscaviarum* NCTC 1934^T; 14, *N. paucivorans* DSM 44386^T; 15, *N. pseudobrasiliensis* ATCC 51512^T; 16, *N. puris* DSM 44599^T; 17, *N. salmonicida* JCM 4826^T; 18, *N. seriolae* JCM 3360^T; 19, *N. transvalensis* DSM 43405^T; 20, *N. veterana* DSM 44445^T; and 21, *N. yamanashiensis* JCM 11893^T.
Data from Kageyama et al. (2003a, b) and Saintpierre-Bonaccio et al. (2004).

Table 5. Phenotypic properties separating the type strains of *Nocardia* species causing or associated with infections of humans and other animals.

Characters	1	2	3	4	5	6	7	8	9	10	11	12	13	14	15	16	17
Biochemical tests																	
Esculin hydrolysis	+	–	+	+	+	–	+	+	+	–	+	+	+	+	+	d	+
Nitrate reduction	+	+	+	+	+	–	+	ND	+	+	ND	+	ND	+	+	+	+
Urea hydrolysis	+	+	–	+	+	–	–	ND	+	+	ND	+	ND	+	+	+	+
Decomposition of (% w/v)																	
Adenine (0.4)	–	–	–	–	–	–	–	ND	ND	–	ND	–	ND	–	–	–	–
Casein (1.0)	–	–	–	–	–	–	–	–	–	–	ND	–	ND	–	–	–	–
Elastin (0.3)	–	–	–	–	–	–	–	–	–	–	ND	–	ND	+	–	–	–
Hypoxanthine (0.4)	–	–	–	–	–	–	–	–	–	–	ND	–	ND	+	–	+	–
Tyrosine (0.5)	–	–	–	–	–	–	+	–	+	–	ND	–	ND	+	–	–	–
Uric Acid (0.5)		ND						ND	–	+	ND	–	ND	+	–	+	ND
Xanthine (0.4)	+	–	–	–	–	–	–	–	–	–	ND	–	ND	+	–	–	–
Growth on sole carbon sources (%, w/v)																	
D(+) Mannitol (1.0)	+	–	+	–	–	+	–	+	+	–	ND	–	+	–	+	+	–
α-L-Rhamnose (1.0)	+	+	–	+	–	–	+	–	–	–	–	+	–	–	+	–	–
D(+) Sorbitol (1.0)	+	–	+	+	–	+	–	ND	–	–	ND	–	ND	–	–	+	–
D(+) Xylose (1.0)	+	+	–	–	+	ND	+	–	–	+	ND	+	–	–	+	ND	+
Sodium acetate (0.1)	+	+	+	+	+	+	–	+	–	+	ND	+	+	–	–	ND	–
Sodium citrate (0.1)	+	–	–	–	–	+	+	–	–	–	–	–	+	+	–	d	+
Growth at 45°C	+	–	–	–	–	–	–	+	–	–	ND	–	ND	–	–	–	–

Symbols and abbreviations: +, positive; –, negative; d, doubtful; and ND, not determined.

[a]Strains: 1, *N. beijingensis* IFP 16342[T]; 2, *N. caishijiensis* JCM 11508; 3, *N. carnea* DSM 43397[T]; 4, *N. cerradoensis* DSM 44546[T]; 5, *N. cummidelens* DSM 44490[T]; 6, *N. flavorosea.* JCM 3332[T]; 7, *N. fluminea* DSM 44488[T]; 8, *N. ignorata* DSM 44496[T]; 9, *N. neocaledoniensis* DSM 44717[T]; 10, *N. pigrifrangens* JCM 11884[T]; 11. *N. pseudovaccinii* DSM 43406[T]; 12, *N. soli* DSM 44488T; 13, *N. tenerifensis* DSM 44704[T]; 14, *N. uniformis* JCM 3224[T]; 15, *N. vaccinii* DSM 43285[T]; 16, *N. vinacea* JCM 10988[T]; and 17, *N. xishanensis* JCM 12160[T].

Data from Kämpfer et al. (2004), Saintpierre-Bonaccio et al. (2004), Wang et al. (2004), and Zhang et al. (2004).

and PCR-randomly amplified polymorphic DNA fingerprinting (Isik and Goodfellow, 2002). Isik and his coworkers assigned 47 representatives of 13 validly described species and 4 putatively novel *Nocardia* species to 19 groups; species-specific patterns were recorded for representatives of *N. brasiliensis*, *N. crassostreae*, *N. farcinica*, *N. otitidiscaviarum* and *N. seriola*. Pulsed field gel electrophoresis was used to distinguish between *N. asteroides* strains implicated in a suspected outbreak of nocardiosis (Louie et al., 1997) and *N. farcinica* strains isolated from postoperative infections (Blümel et al., 1998). Partial 16S rRNA gene sequencing has been recommended for the identification of *Nocardia* species (Cloud et al., 2004), as has a digoxigenin-labeled cDNA probe (McNeil et al., 1997). DNA probes have been prepared for the identification of *N. asteroides* using genomic libraries generated for selected strains of *N. asteroides* (Brownell and Belcher, 1990).

Serological tests for the diagnosis of *Nocardia* infections have been developed, but most of them have not reached the desired sensitivity or specificity. Crossreactions of nocardial antigens from culture filtrates or whole-organism extracts from sera from cases of tuberculosis and leprosy have been reported (Shainhaus et al., 1978; Blumer and Kaufman, 1979). Antigens in culture filtrates have been used to define a species-specific antigen for *N. otitidiscaviarium* (Pier and Fichtner, 1971). Tests have also been devised to detect cutaneous hypersensitivity and humoral antibodies for the diagnosis of nocardiosis in humans and animals (Boiron et al., 1993; Bojalil and Zamora, 1993). An enzyme-linked immunosorbent assay (ELISA technique) based on a 55-kDa protein specific to *Nocardia* allowed the detection of antibodies in patients with cutaneous or pulmonary nocardiosis (Angeles and Sugar, 1987; Sugar and Angeles, 1987). A conventional solid-phase ELISA, based on two immunodominant antigens, was used to confirm the diagnosis of *N. brasiliensis* infections in cases of human mycetoma (Salinas-Carmona et al., 1993). The value of the immunoblot technique has been demonstrated in the diagnosis of nocardiosis (Boiron and Provost, 1990c; Boiron and Stynen, 1992a). Monoclonal antibodies developed for use in the localization, purification and characterization of *Nocardia* antigens may serve as diagnostic reagents (Jimenez et al., 1990), as may those raised against a specific 54-kDa protein (Boiron et al., 1992b).

Reliable typing methods are needed to determine the infectious source and mode of transmission of clinically significant nocardiae, particularly since hospital acquired nocardiosis seems to be an emerging problem (Schaal and Lee, 1992; McNeil and Brown, 1994). Antigens in culture filtrates have been used to establish four serotypes within *N. asteroides* and to determine cutaneous hypersensitivity in human patients and cattle infected with *N. asteroides, N. brasiliensis* and *N. otitidiscaviarum* (Pier et al., 1968; Salman et al., 1982). These antigens were used to good effect in epidemiological studies of group infections in humans (Stevens et al., 1981) and cattle (Pier and Fichtner, 1981). Comparisons of plasmid (Jonsson et al., 1986) and restriction fragment length polymorphism (RFLP) profiles (Paterson et al., 1992) have been used to type *N. asteroides* isolates. Strains assigned to the *N. asteroides* complex have been distinguished by ribotyping (Exmelin et al., 1996; Laurent et al., 1996). Pulsed-field gel electrophoresis was used to detect an endemic *N. farcinica* strain responsible for postoperative wound infections, possibly after aerogenic transmission (Blümel et al., 1998).

Members assigned to the four 16S rRNA rhodococcal subclades (Fig. 4) have unique nucleotide signatures (Goodfellow et al., 1998b). Criteria useful for differentiating between members of the *R. erythropolis* and *R. rhodochrous* subclades are shown in Tables 6 and 7, respectively. However, a better panel of phenotypic tests is needed to further improve the identification of rhodococcal species. Rapid enzyme tests show promise in this respect (Mutimer and Woolcock, 1982; Goodfellow et al., 1987; Goodfellow et al., 1990; Boiron et al., 1993; Klatte et al., 1994). PCR primers can be used to distinguish *R. equi* from members of other mycolic acid-containing actinomycetes (Bell et al., 1996). Several PCR-based rapid diagnostic tools are available for the diagnosis and control of pathogenic *R. equi* strains (Takai et al., 1998; Vivrette et al., 2000; Sellon et al., 2001; Arriaga et al., 2002; Ladrón et al., 2003). These methods are based on the IP1/IP2 primer pair designed by Takai et al. (1998) and specific for strains carrying the vap *A+* genotype. A simple modification of this approach allows *R. equi* strains to be classified according to their vap *A*-vap *B* status (Oldfield et al., 2004). The whole-organism protein patterns may also be useful in the identification of *R. equi* (Chirino-Trejo and Prescott, 1987b).

Early detection of *R. equi* disease in foals may be achieved by B-cell epitope mapping of the VapA protein (Vanniasinkan et al., 2001). A comparison has been made of a number of methods used for the diagnosis of *R. equi* pneumonia in foals (Sellon et al., 2001). Ribotyping is a potentially useful epidemiological tool for unravelling relationships between *R. equi* isolates (Lasker et al., 1992); there is evidence that this taxon is heterogeneous (Goodfellow et al., 1982b; Butler et al., 1987; Gotoh et al., 1991; McMinn et al., 2000). PCR amplification of the *Fas-1* gene has been used to detect virulent strains of the plant pathogen *R. fascians* (Stange et al., 1996).

Comprehensive comparative taxonomic studies designed to highlight markers that can be weighted for diagnostic purposes are now possible given the improvements in the classification of *Gordonia, Rhodococcus* and *Tsukamurella* strains. Serological techniques may prove to be of value in distinguishing between species assigned to these genera (Hyman and Chaparas, 1977). Whole cell protein patterns may also be useful in the identification of *R. equi* (Chirino-Trejo and Prescott, 1987).

Pathogenicity

Nocardiae cause a wide range of infections in humans, the most serious of which are actinomycetoma and nocardiosis (Pulverer and Schaal, 1978; Schaal and Beaman, 1984; Schaal, 1997; Pujic and Beaman, 2001; Beaman, 2002). An understanding of the causal agents of these diseases is critical as lesions associated with infections lack distinctive clinical features and may not respond to antimicrobial drugs used to treat common suppurative and septicemic processes. The primary causal agents of actinomycete mycetomas are *Actinomadura* species (see The Family Thermomonosporaceae in this Volume) and *N. asteroides, N. brasiliensis* and *S. somaliensis* (see The Family Streptomycetaceae, Part I: Taxonomy in this Volume and The Family Streptomycetaceae, Part II: Molecular Biology in this Volume). Other nocardial species associated with the disease include *N. farcinica* (Schaal and Lee, 1992), *N. otitidiscaviarum* (Alteras and Feuerman, 1986; Saarinen et al., 2001), *N. pseudobrasiliensis* (Rainey et al., 1996) and *N. transvalensis* (Mirza and Campbell, 1994; Poonwan et al., 1995). Incidence figures for actinomycete mycetomas need to be interpreted with care as the disease is not notifiable and the causal agents can be difficult to identify. Immunologically based diagnostic methods have been used to detect specific antibodies induced by the causal agents, though immunological crossreactions with other pathogenic agents can occur (De Magaldi and Mackenzie, 1990; Salinas-Carmona et al., 1993; Salinas-Carmona, 2001).

Table 6. Phenotypic properties separating the type strains of species classified in the *R. erythropolis* 16S rRNA subclade.

Characters	Strains[a]											
	1	2	3	4	5	6	7	8	9	10	11	12
Morphogenetic sequence	EB-R-C	EB-R-C	H-R-C	EB-R-C	EB-R-C	H-R-C	R-C	H-R-C	H-R-C	H-R-C	EB-R-C	EB-R-C
Biochemical tests												
Esculin hydrolysis	+	−	−	+	−	−	+	+	−	−	−	+
Arbutin hydrolysis	+	−	−	+	−	ND	+	+	+	−	−	+
Urea hydrolysis	+	+	+	+	+	−	+	−	+	+	+	+
Growth on sole carbon sources (1%, w/v)												
L(+) Arabinose	−	−	+	−	w	ND	−	−	−	−	−	+
D(+) Arabitol	−	−	+	+	+	ND	−	+	+	+	+	+
D(+) Cellobiose	−	−	−	−	−	ND	−	+	+	−	−	−
D(+) Galactose	+	ND	+	−	+	−	+	−	+	+	+	+
Glycerol	+	ND	+	+	+	−	+	+	+	+	−	+
meso-Inositol	+	−	+	+	+	+	−	+	+	+	+	+
Inulin	−	−	+	−	+	+	+	−	+	−	−	−
D(+) Lactose	+	−	+	+	+	+	+	−	+	+	+	+
D(+) Maltose	−	−	+	+	+	+	−	−	+	+	−	+
D(+) Mannitol	+	−	+	+	+	+	+	+	+	+	+	+
D(+) Mannose	−	+	+	+	+	ND	+	+	+	+	+	+
D(+) Melibiose	−	−	−	−	+	ND	+	−	+	+	+	+
α-L(−) Rhamnose	−	−	−	−	+	−	w	−	−	−	+	+
D(−) Ribose	+	+	+	+	+	−	+	+	+	+	+	+
D(+) Sorbitol	+	−	+	+	+	−	−	w	+	+	+	+
D(+) Sucrose	+	+	+	+	+	+	+	+	+	+	+	+
D(+) Trehalose	+	+	+	+	+	+	+	−	+	+	+	+
D(+) Turanose	−	−	+	+	+	−	+	−	+	+	+	+
D(+) Xylose	−	+	+	+	+	+	+	−	+	+	−	−

Symbols and abbreviations: +, positive; −, negative; ND, not determined; EB-R-C, elementary branching rod-coccus growth cycle; R-C, rod-coccus growth cycle; and H-R-C, hyphal-rod-coccus growth cycle.

[a]Strains: 1, *R. erythropolis* DSM 43066[T]; 2, *R. baikonurensis* DSM 44587[T]; 3, *R. fascians* DSM 20669[T]; 4, *R. globerulus* DSM 43954[T]; 5, *R. koreensis* DSM 44498[T]; 6, *R. jostii* DSM 44719[T]; 7, *R. maanshanensis* DSM 44675[T]; 8, *R. marinonascens* DSM 43752[T]; 9, *R. opacus* DSM 43205[T]; 10, *R. percolatus* DSM 44240[T]; 11, *R. tukisamuensis* DSM 44783[T]; and 12, *R. wratislaviensis* DSM 44107[T].

Data from Li et al. (2004), Matsuyama et al. (2004), Takeuchi et al. (2002), and Zhang et al. (2002).

Table 7. Phenotypic properties separating the type strains of species classified in the *Rhodococcus rhodochrous* 16S rRNA gene subclade.

Characters	Strains[a]						
	1	2	3	4	5	6	7
Morphogenetic sequences	R-C	H-R-C	EB-R-C	EB-R-C	EB-R-C	H-R-C	H-R-C
Biochemical tests							
Esculin hydrolysis	–	–	–	+	+	–	+
Urea hydrolysis	–	+	–	+	+	–	+
Degradation tests							
DNA	–	–	+	–	+	–	+
Starch	+	+	+	+	–	+	–
Uric acid	–	+	–	–	+	+	+
Growth on sole carbon sources at 1%, w/v							
Arbutin	–	+	+	–	+	–	+
Glycerol	+	–	–	+	+	+	+
meso-Inositol	–	–	–	–	–	–	+
α-Methyl-D-gluconate	–	+	–	–	–	–	+
at 0.1%, w/v							
m-Hydroxybenzoic acid	+	–	+	–	+	–	+
p-Hydroxybenzoic acid	+	–	+	+	+	+	+
Monoethanolamine	+	–	+	–	–	+	+
Sodium benzoate	+	–	+	–	+	+	+
Sodium butyrate	+	–	+	+	–	+	+
Sodium gluconate	+	–	–	–	+	+	+

Symbols and abbreviations: +, positive; –, negative; EB-R-C, elementary branching-rod-coccus cycle; R-C, rod-coccus growth cycle; and H-R-C, hyphal-rod-coccus growth cycle.
[a]Strains: 1, *R. aetherivorans* DSM 44752[T]; 2, *R. coprophilus* DSM 43347[T]; 3, *R. gordoniae* DSM 44689[T]; 4, *R. pyridinovorans* KCTC 0647T; 5, *R. rhodochrous* DSM 43241T; 6, *R. ruber* DSM 43338[T]; and 7, *R. zopfii* DSM 44108[T].
Data from Goodfellow et al. (2004) and Jones et al. (2004).

The clinical picture of actinomycetoma is relatively uniform irrespective of the causal agent. The disease is characterized by abscesses and subcutaneous granulomata and by areas of induration. Sinus tracts, often multiple, may discharge granules that have a characteristic size, shape and color. The granules consist of small colonies of the infective agent surrounded by masses of inflammatory cells. Infections are frequently through the foot, especially in localities where people cannot afford footwear. However, many extrapedal cases are seen involving parts of the body that come into contact with soil. The disease process usually starts at the site of a localized injury such as a puncture wound caused by thorns or splinters. The disease is mainly restricted to tropical and subtropical regions, though mycetomas are occasionally seen in patients from temperate countries. Human mycetoma has been simulated in a mouse model (González-Ochoa, 1973), which has been used to study some aspects of host-parasite relationships (Ortiz-Ortiz et al., 1984).

Nocardiosis usually develops as an opportunistic infection complicating debilitating primary diseases such as diabetes, leukemia, lymphoma and other neoplasms or in patients undergoing immunosuppressive therapeutic procedures (Schaal and Beaman, 1984; McNeil and Brown,

1994; Schaal, 1997; Beaman, 2000a; Beaman, 2002). Accurate diagnosis of the disease depends on the isolation and identification of the causal agent from clinical material, as clinical, radiological and histopathological findings are insufficient for this purpose. Isolation and identification procedures are not straightforward; hence, the true incidence of nocardiosis is masked, a problem which is exacerbated by poor documentation.

Recent increases in the reported frequency of human nocardial infections can be attributed to better clinical microbiological awareness, improved selective isolation and identification procedures, and the extensive use of suppressive drugs; there is also an increasing realization that pulmonary nocardiosis can be misdiagnosed as tuberculosis (Idigbe et al., 1992; Hamid et al., 2001). In France, up to 250 cases of nocardiosis have been estimated annually (Boiron et al., 1990), and corresponding figures from the United States suggest over 1000 cases of the disease per year (Beaman, 1988). Localized cutaneous, subcutaneous and lymphocutaneous nocardioses occur in healthy individuals (Lemar, 1996; Gaude et al., 1999).

Nocardiosis is usually considered to be a late-presenting, community acquired infection though it is becoming increasingly evident that the disease is transmissible. Clusters of patients with *N. asteroides* infections have been reported from heart (Krick et al., 1975; Simpson et al., 1981), liver (Sahathevan et al., 1991) and renal transplant units (Huang et al., 1980; Baddour et al., 1986). Similarly, a cluster of *N. farcinica* postoperative wound infections in patients undergoing cardiac and other vascular surgeries was attributed to environmental factors (Schaal, 1991). Nocardial infections occur in AIDS patients (Kim et al., 1991; Uttamchandani et al., 1994); the suggestion by some clinicians that nocardiosis is a rare complication in AIDS patients has been challenged (Beaman and Beaman, 1994).

The mechanism of pathogenesis and host immunity to nocardial infection has been promoted by Blaine Beaman and his colleagues using murine models (Beaman et al., 1980a; Beaman et al., 1980b; Beaman and Beaman, 1994; Beaman and Beaman, 2000b). Virulent strains of *N. asteroides* are facultative intracellular pathogens that can be grown in a range of cells from humans and experimental animals (Beaman, 1984). The mechanisms of pathogenicity are multiple, complex and not fully understood (Beaman et al., 1992). The virulence of *N. asteroides* strains appears to be influenced by factors that include the age of the culture, its rate of growth, the route of infection, and its capacity to inhibit phagosome-lysozyme fusion, resist oxidative killing mechanisms of phagocytes, and alter lysosomal enzymes within phagocytes. The mechanisms of host resistance to nocardiae are complex and poorly understood (Beaman, 1992). L-Forms have been isolated from humans, but their role in human disease is not known (Beaman, 1982).

In systemic nocardiosis, the primary lesion is usually the lungs, but secondary and often fatal infections may develop in the central nervous system and less frequently in other internal organs. Nocardiae appear to invade the host by direct inhalation of contaminated dust particles or through soil contaminated wounds. Localized cutaneous or subcutaneous infections are usually of a primary nature and are mainly caused by *N. asteroides* or *N. brasiliensis*.

Until recently, dietziae, gordoniae, rhodococci and tsukamurellae were rarely encountered as agents of human infections but are now seen as emerging pathogens. Clinicians need to be aware of the pathogenic potential of these organisms, especially when treating AIDS patients and similarly immunocompromised individuals. *Rhodococcus equi* is known to be an important human pulmonary pathogen of immunocompromised patients, where it causes cavitary pneumonia (Van Etta et al., 1983; Eberzole and Paturzo, 1988; Drancourt et al., 1992; McNeil and Brown, 1994; Scott et al., 1995; Mosser and Hondalus,

1996). A small number of cases in immunocompetent hosts has been reported (Kedlaya et al., 2001).

Rhodococcus equi is pathogenic for foals (Barton and Hughes, 1980; Prescott, 1991). Indeed, the organism was first isolated from the lungs of a foal with chronic suppurating bronchopneumonia (Magnusson, 1923). Pneumonia caused by *R. equi* strains commonly affects foals the ages of 1–6 months with high mortality and has been estimated to account for about 3% of the total number of foal deaths per year (Prescott, 1991). In Ontario it was reported to cause 10% of deaths of foals under 6 months old (Zink et al., 1986). The route of infection is thought to be by inhalation or ingestion through grazing on infected land (Martens et al., 1982; Johnson et al., 1983a; Johnson et al., 1983b).

In the animal host, *R. equi* strains are taken up in macrophages. Equine *R. equi* virulence has a high positive correlation with the expression of the *vapA* gene (Tkachuk-Saad and Prescott, 1991; Haites et al., 1997; Morton et al., 2001), which is located on a circular plasmid, the virulence associated plasmid (VAP; Takai et al., 2000). The *vap* operon of the equine isolate ATCC 33701 is borne within a 27,000-bp "pathogenicity island" and encodes at least four functional genes, *vapA*, *vapC*, *vapD*, *vapE* and *vapF*. Immunoblotting studies (Byrne et al., 2001) show that the gene products, VapA, a cell-surface lipoprotein, and VapC, VapD and VapE, all secreted proteins, are expressed during growth at 37°C but not at 30°C, a result in line with their putative role in virulence. A variant of *vapA*, known as *vapB* (Takai et al., 2000; Byrne et al., 2001), has been identified in some equine isolates from pigs (Takai et al., 1996) and humans (Takai et al., 1994; Takai et al., 1995). A PCR technique allows the selective isolation of *vapA* and *vapB* (Oldfield et al., 2004); these workers showed that the high frequency of *vapA*⁺ amongst equine isolates was significant, thereby confirming the status of *vapA* as a reliable marker of equine virulence.

Oral immunization can protect foals against severe challenge with *R. equi* (Chirino-Trejo et al., 1987a). However, the humoral response to *R. equi* whole-cell antigens does not seem to be important in protection against disease (Chirino-Trejo and Prescott, 1987a). This observation is consistent with the behavior of the organism as a facultative intracellular pathogen (Zink et al., 1985).

Metabolism and Genetics

Early work on the genetics and metabolism of Nocardiaceae strains is difficult to interpret given the uncertain taxonomic status of the tested organisms. Many investigations reported on "nocardiae" can be attributed to strains now classified in the genera *Gordonia* and *Rhodococcus* (Bradley, 1978a; Brownell and Denniston, 1984; Peczynska-Czoch and Mordarski, 1988) or in the family Pseudonocardiaceae (Schupp et al., 1975; Matsushima et al., 1987; Hütter and Eckhardt, 1988). Larkin et al. (1998) have pointed out that many strains assigned to the genus *Rhodococcus* without the aid of 16S rRNA sequence data may belong to other genera, as exemplified by the transfer of *Rhodococcus chlorophenolicus* Apajalahti et al. 1986 to the genus *Mycobacterium* as *Mycobacterium chlorophenolicum* (Apajalahti et al., 1986) Häggblum et al. 1994. Conversely the polychlorinated biphenyl-degrading *Acinetobacter* strain sp. M5 was reclassified as *Rhodococcus globerulus* primarily on the basis of 16S rRNA gene sequence data (Asturias and Timmis, 1993; Asturias et al., 1994).

Clearly, rhodococci and related mycolic acid-containing organisms, notably gordoniae and tsukamurellae, are chemoorganotrophs characteristically exhibiting a remarkable degree of metabolic diversity that allows them to use an immense range of organic compounds as sources of carbon, nitrogen and micronutrients (Tárnok, 1976; Peczynska-Czoch and Mordarski, 1988; Finnerty, 1992; Warhurst and Fewson, 1994; Arenskötter et al., 2004). The list of pathways covered in the review by Warhurst and Fewson (1994), while impressive, probably represents only a fraction of the true extent of the metabolic diversity within this group.

The catabolic potential of gordoniae, nocardiae, rhodococci and tsukamurellae, and probably dietziae and skermaniae, not only includes the capacity to assimilate carbohydrates and proteins but also unusual compounds, such as aliphatic hydrocarbons, aniline, bicyclic and polycyclic hydrocarbons, nitroaromatic compounds, pyridine, and sterols (Tárnok, 1976; Cain, 1981; Janke et al., 1986; Van Ginkel et al., 1987; Peczynska-Czoch and Mordarski, 1988; Williams et al., 1989). Detergents and pesticides, including warfarin, are also modified (Goodfellow and Williams, 1983), and 1,2-epoxyalkanes are produced from 7-alkanes (Fukuhashi et al., 1981). The ability to synthesize complex lipids (Minnikin, 1982; Ioneda, 1988), including mycolic acids, distinguishes dietziae, gordoniae, nocardiae, rhodococci, skermaniae and tsukamurellae from all other bacteria apart from corynebacteria and mycobacteria. Phenol degradation by rhodococci has been examined in batch and continuous culture (Hensel and Straube, 1990; Straube et al., 1990), and oligocarbophily was reported for a rhodococcal strain labeled "*Nocardia corallina*" (Tárnok, 1976).

Rhodococci have frequently been isolated from soils polluted with petroleum (Nesterenko et al., 1978a; Nesterenko et al., 1978b) and have been implicated in the degradation of humic acid (Cross et al., 1976) and lignin-related compounds (Eggeling and Sahm, 1980; Eggeling and Sahm, 1981; Rast et al., 1980) and pesticides (Nagy et al., 1995). Their metabolic versatility includes the ability to break down acrylamide (Arai et al., 1980), aromatic hydrocarbons (Raymond et al., 1971), haloalkanes (Kesseler et al., 1996), and sulfonated azo dyes (Heiss et al., 1992).

Two main types of enzyme are involved in nitrile biotransformations by rhodococci; nitrilases catalyze the direct cleavage of nitriles to yield the corresponding acids and ammonia, whereas nitrile hydratases catalyze the hydration of nitriles to amides (Harper, 1977; Harper, 1985; Nagasawa et al., 1988a; Bunch, 1998). Nagasawa and his colleagues demonstrated the occurrence of a cobalt-induced and cobalt-containing nitrile hydratase in *Rhodococcus rhodochrous* J1; the cells of this organism also produce a nitrilase (Nagasawa et al., 1988b) that has been purified and characterized (Kobayashi et al., 1989a). A novel nitrilase that preferentially catalyzes the hydrolysis of aliphatic nitriles was detected in *R. rhodochrous* K22, a facultative crotononitrile-utilizing actinomycete (Kobayashi et al., 1990); this enzyme is remarkable in its broad substrate specificity for aliphatic nitriles. DNA probes for the α- and β-subunits of the *Rhodococcus* N 774 nitrile hydratase have been used to locate the region on the chromosome that codes for the nitrile hydratase gene (Ikehata et al., 1989). The nucleotide sequence of the nitrile hydratase gene of *R. rhodochrous* J1 has been determined (Kobayashi et al., 1990).

Gordonia strains are a potentially rich source of metabolic diversity. Members of this taxon have the capacity to degrade toxic environment-contaminating compounds, as exemplified by the isolation of strains with hydrocarbon-oxidizing, nitrile-metabolizing, rubber-degrading, and 3-ethylpyridine-degrading pathways (Brandão et al., 2001; Brandão et al., 2003) and novel desulfurizing enzymes (Gilbert et al., 1998; Rhee et al., 1998; Finkel'shtein et al., 1999; Kim et al., 1999).

Early developments in rhodococcal genetics were the subject of several comprehensive reviews (Adams and Brownell, 1976; Brownell and Denniston, 1984; Finnerty, 1992). Recombination was reported in the model organism *R. erythropolis* by Adams and Bradley (1963). Over 60 genetic traits were used in the development of a *R. erythropolis* linkage map, and temperate phages were made available as cloning vectors for establishing a gene cloning system. Genetic recombination was demonstrated between rhodococcal strains labeled "*Nocardia opaca*" and "*Nocardia restricta*." A transferable plasmid described for a strain of the former carries traits allowing for chemolithoautotrophic growth (aut+), including genes for hydrogenase, phosphoribulokinase, and ribulose-bisphosphate-carboxylase production (Reh, 1981a; Reh and Schlegel, 1981b). The "*N. opaca*" strain is able to transfer the aut+ trait to related organisms and to *R. erythropolis*; in the presence of CO_2 and H_2O, it has a generation time of 7 h.

A miniplasmid has been characterized for a rhodococcal strain labeled "*R. corallina*" (Kirby and Usdin, 1985) and *Rhodococcus* genes encoding pigment production were cloned and expressed in *Escherichia coli* (Hill et al., 1989). Plasmid-determined transformation of *cis*-abienol and schareol has been achieved in "*Nocardia restricta*" JTS-162 (Hieda et al., 1982). A generalized transducing bacteriophage is available for *Rhodococcus erythropolis* (Dabbs, 1987).

The need for better genetic tools to characterize and exploit the diverse metabolic activities of members of the families Dietziaceae, Gordoniaceae, Nocardiaceae and Tsukamurellaceae is becoming increasingly evident. Much of the recent work has been focused on rhodococci as illustrated in two excellent reviews (Larkin et al., 1998; Kulakov and Larkin, 2002). These workers note that while considerable progress is being made in rhodococcal genetics, developments have been impeded by difficulties of choosing a model organism, extracting nucleic acids because of the recalcitrance of the rhodococcal cell wall, and the pleomorphic nature of many strains. Apparently, rhodococci exhibit considerable genomic instability.

Rhodococci contain plasmids that range from small cryptic closed circular molecules to large linear plasmids. Linear plasmids have been discovered in many strains and shown to code for a variety of functions such as biphenyl metabolism in *R. erythropolis* strain TA421 (Kosono et al., 1977), hydrogen autotrophy in *R. opacus* strain MR22 (Grzeszik et al., 1997), isopropylbenzene metabolism in *R. erythropolis* strain BD2 (Dabrock et al., 1994), and plant virulence and fasciation genes in *R. fascians* strain D188 (Crespi et al., 1992; Crespi et al., 1994). However, Pisabarro et al. (1998) did not detect any linear replicons in the megabase range in an analysis of several virulent and avirulent strains of *R. fascians*, a difference that may be related to the instability of the rhodococcal genome. Well-characterized circular plasmids are associated with pathogenicity determinants of *R. equi* (Takai et al., 1991; Takai et al., 1993; Takai et al., 1995) and the degradation of organic pollutants (Dabbs, 1998). Circular plasmids have been shown to code for diphenyl metabolism (Masai et al., 1997), chlo-

roalkane degradation (Kulakova et al., 1997), dibenzothiophene desulfurization (Denome et al., 1994; Denise-Larose et al., 1997), and 2-methylaniline metabolism (Schreiner et al., 1991). However, little is known about the replication, stable maintenance, and transfer of rhodococcal plasmids.

Broad-host-range plasmids that can replicate in different actinomycetes, including rhodococci, have yet to be identified (Larkin et al., 1998). This lack of suitable cloning vectors has encouraged several groups to search for DNA fragments of indigenous rhodococcal plasmids, which confer the ability to replicate in particular *Rhodococcus* strains when integrated into *Escherichia coli* replicons. Singer and Finnerty (1988), for instance, constructed the shuttle vector pMVS 301 (10.1 kb) from a 3.8-kb fragment of the cryptic 13.4-kb plasmid pMVS 300 from *Rhodococcus* strain H13-A. Additional cloning vectors have been constructed for different purposes as exemplified by shuttle vectors pK 4 and pRR-6, which have been used to study nitrile conversion by *R. rhodochrous* (Komeda et al., 1996a; Komeda et al., 1996b; Komeda et al., 1997) and to characterize dibenzothiophene desulfurization genes in *R. erythropolis* (Piddington et al., 1995; Li et al., 1996), respectively.

Gene expression and control in *Rhodococcus* are well covered in the review by Larkin et al. (1998). Many rhodococcal genes have been cloned and analyzed with an emphasis directed towards pathways involved in the degradation of aromatic compounds. Sequences are available for operon-like structures that encode the genes for the degradation of biphenyl (Masai et al., 1995), dibenzothiophene desulfurization (Denome et al., 1994), 3-hydroxyphenylpropionic acid (Barnes et al., 1997), isopropylbenzene (Kesseler et al., 1996), and plant virulence loci (Crespi et al., 1994). Enzyme expression in rhodococci is dependent on specific inducers (Komeda et al., 1996a; Komeda et al., 1996b; Barnes et al., 1997; Labbé et al., 1997) or repressors (Li et al., 1996). Transcriptional regulation of the *R. rhodochrous* J1 nitrilase gene (*nit*A) has been analyzed in detail (Komeda et al., 1996b). The transcription of the *nit*A gene was inducible by isovaleronitrile, and a regulatory gene encoding the positive transcriptional regulator *nit*R was essential for the isovaleronitrile dependent induction. In contrast, the dibenzothiophene desulfurization (*dsz*) gene cluster is not inducible but is strongly repressed by sulfate- and sulfur-containing amino acids (Li et al., 1996).

The recent advances in rhodococcal genetics, while welcome, leave plenty of scope for further development. In particular, better cloning methods are needed, as are detailed functional analyses of sequences of regulatory regions, inducers, and associated regulatory processes. Rapid advances can be anticipated on such fronts given the imminent availability of several rhodococcal whole genome sequences.

Genetic recombination and plasmids have been reported in *N. asteroides* (Kasweck et al., 1981; Kasweck and Little, 1982a; Kasweck et al., 1982b) and nocardiophages have been reported for *N. asteroides* (Pulverer et al., 1975; Prauser, 1976; Prauser, 1981b; Andrzejewski et al., 1978), *N. brasiliensis* (Pulverer et al., 1975), *N. carnea* (Williams et al., 1980), *N. otitidiscaviarum*, and *N. vaccinii* (Prauser, 1976). Members of the family Nocardiaceae are generally susceptible to nocardiophages but not strains classified in the families Mycobacteriaceae and Pseudonocardiaceae (Williams et al., 1980; Prauser, 1981a). Conversely, phages that lyse *Amycolatopsis mediterranei* are inactive against nocardiae (Thiemann et al., 1964).

Applications

Nitrile-converting biocatalysts are of industrial interest from the perspectives of treating toxic nitrile- and cyanide-containing wastes (Knowles and Wyatt, 1992) and as agents for the synthesis of chemicals (Liese et al., 2000). Rhodococcal nitrile-converting enzymes are being used commercially as catalysts to convert nitriles to corresponding higher value acids and amides, notably acrylic acid and acrylamide (Beard and Page, 1998; Hughes et al., 1998). These compounds are manufactured by chemical processes in vast amounts and are used to produce polymers for use as dispersants, flocculants and superabsorbants. The chemical production of acrylamide involves the use of a copper catalyst and a process that needs to be carrried out at high temperature under pressure in a nitrogen atmosphere. In contrast, the nitrile hydratase-catalyzed process developed by the Nitto Chemical Industry Company in collaboration with Kyoto University for the production of acrylamide is performed at low temperature under atmospheric pressure using *Rhodococcus* strain N-774 as the biocatalyst. The establishment of an enzymic route to produce acrylamide was the first successful commercial biocatalyst process involving the production of a commodity chemical. The important features of the biocatalysts used by Nitto for acrylamide production have been described by Kobayashi et al. (1992).

Improved biocatalysts for acrylamide production were sought through the Kyoto/Nitto collaboration, a process that led to *Rhodococcus* strain J1 becoming a third-generation biocatalyst for the commercial production of acrylamide from acrylonitrile. The nitrile hydratase of this organism, which is induced when the culture medium is supplemented with cobalt and crotonamide, has been shown to be a versatile and robust enzyme (Nagasawa and Yamada, 1990; Kobayashi et al., 1992; Yamada and Kobayashi, 1996). Given the widespread distribution of rhodococci in the environment, additional representatives of this taxon with similar or better enzymic properties seem likely to be discovered.

Mycolic acid-containing actinomycetes, notably rhodococci, may prove to be a valuable source of novel, commercially significant biosurfactants (Wagner et al., 1976; Gerson and Zajic, 1977; Blackall and Marshall, 1989a; Lang and Philp, 1998; Choi et al., 1999). Rhodococci respond to the presence of alkanes by producing biosurfactants that help them to use hydrophobic compounds as growth substates (McDonald et al., 1981; Wagner et al., 1983; Kurane et al., 1995). Most attention has been focused on trehalose mycolates, notably those from *R. erythropolis* (Kretschmer et al., 1982; Uchida et al., 1989; Lang and Philp, 1998). Surfactants represent very large-scale production chemicals, the demand for which can be expected to increase (Desai and Banat, 1997).

The commercial interest in biosurfactants against chemically synthesized components is due to the increasing environmental concern associated with the continued use of the latter. Rhodococcal surface-active lipids are seen to have some favorable characteristics for certain applications, notably their lipophilicity, nonionic nature, biodegradability, and low toxicity. These compounds can reduce the surface tension of aqueous solutions and the interfacial tension between aqueous and oil phases to levels observed with synthetic surfactants. Their commercial utilization is likely to be where high levels of purity are not required and in the growing environmental remediation industries.

Gordoniae and rhodococci are widespread in the environment through their ability to metabolize a wide range of organic pollutants, many of which are toxic (Dabbs, 1998; Brandão et al., 2001). They are able to survive in polluted habitats under near-starvation conditions (Warhurst and Fewson, 1994; Acharya and Desai, 1997) and have the capacity to degrade diverse hydrocarbons, including halogenated and long chain, as well as numerous substituted aromatic compounds (Tárnok, 1976; Peczynska-Czoch and Mordarski, 1988; Warhurst and Fewson, 1994). The ability of gordoniae and rhodococci to grow in highly polluted environments makes them attractive candidates for the selective removal of unwanted contaminants from commodity products and in bioremediation.

Desulfurization reactions have received a lot of attention because of their potential role in industrial-scale bipurification of coal and crude oil and its distillates by selective removal of contaminating organosulfur. Desulfurization pathways are of prime importance in the development of microbial fuel desulfurization technologies, as a cheap and environmentally friendly alternative to chemical processes. Desulfurizing enzymes remove the sulfur moiety from the organosulfur, leaving the carbon skeleton intact. Two types of desulfurization reaction are recognized. The dibenzothiophene (DBT) specific pathway desulfurizes DBT to inorganic sulfate and 2-hydroxybiphenyl, and the benzothiophene (BTH)-specific pathway desulfurizes BTH to 2- (2′-hydroxyphenyl) ethan-1-ol and probably inorganic sulfate. Considerable interest has been shown in the desulfurizing enzymes produced by *Gordonia* and *Rhodococcus* strains (Oldfield et al., 1998). The DBT-desulfurization pathway was originally identified in *R. erythropolis* strain IGTS8 and the BTH-desulfurization pathway in *Gordonia desulfuricans* strain 213E. Strains of *G. desulfuricans* may find particular application in the desulfurization of diesel fuel, which contains a high proportion of benzothiophene (McFarland et al., 1998).

Most clinically relevant microbial compounds have been natural products or derived from natural products. Among bacteria, actinomycetes have been found to have a unique capacity to produce novel bioactive compounds, notably antibiotics. Early pharmacological screening programs were focused on streptomycetes, but in recent times, the emphasis has switched to other actinomycete genera (Okani and Hotta, 1988; Lazzarini et al., 2000). This change represents a new and important dimension to secondary metabolite discovery, not least because it provides a quantum leap in the number and variety of genetic resources available for exploitation. Mycolic acid-containing actinomycetes have not featured prominently in search and discovery programs, but evidence shows that they may well be worth screening, as exemplified by the ability of the type strain of *Nocardia vinacea* to produce tubelactomicin A, a new 16-membered lactose antibiotic (Kinoshita et al., 2001).

Literature Cited

Acharya, A., and A. J. Desai. 1997. Studies on utilization of acetonitrile by Rhodococcus erythropolis A10. World J. Microbiol. Biotechnol. 13:175–178.

Adams, J. N., and S. G. Bradley. 1963. Recombination events in the bacterial genus Nocardia. Science 140:1392–1394.

Adams, J. M., and G. H. Brownell. 1976. Genetic studies in Nocardia erythropolis. *In:* M. Goodfellow, G. H. Brownell, and J. A. Serrano (Eds.) The Biology of the Nocardiae. Academic Press. London, UK. 225.

Ajello, L., and G. D. Roberts. 1981. Mycetomas. *In:* W. J. Hausler (Ed.) Diagnostic Procedures for Bacterial, Mycotic and Parasitic Infections. American Public Health Association. Washington, DC. 1033.

Albuquerque de Barros, E. V. S., G. P. Manfio, V. Ribeiro Maitan, L. A. Mendes Bataus, S. B. Kim, L. A. Maldonado, and M. Goodfellow. 2003. Nocardia cerradoensis sp. nov., a novel isolate from Cerrado soil in Brazil. Int. J. Syst. Evol. Microbiol. 53:29–33.

Al-Diwany, L. J., and T. Cross. 1978. Ecological studies on nocardioforms and other actinomycetes in aquatic habitats. Zbl. Bakteriol. Suppl. 6:153–160.

Alshamaony, L., M. Goodfellow, and D. E. Minnikin. 1976a. Free mycolic acids as criteria in the classification of Nocardia and the "rhodochrous" complex. J. Gen. Microbiol. 92:188–199.

Alshamaony, L., M. Goodfellow, D. E. Minnikin, and H. Mordarska. 1976b. Free mycolic acids as criteria in the classification of Gordona and the "rhodochrous" complex. J. Gen. Microbiol. 92:183–187.

Alteras, I., and E. J. Feuerman. 1986. The second case of mycetoma due to Nocardia caviae in Israel. Mycopathologia 93:185–187.

Alture-Werber, E., D. O'Hara, and D. B. Louria. 1968. Infections caused by Mycobacterium rhodochrous scotochromogens. Am. Rev. Respir. Dis. 97:694–698.

Andrzejewski, J., G. Müller, E. Röhrscheidt, and D. Pietkiewicz. 1978. Isolation, characterization and classification of Nocardia asteroides bacteriophage. Zbl. Bakteriol. Suppl. 6:319–326.

Angeles, A. M., and A. M. Sugar. 1987. Rapid diagnosis of nocardiosis with enzyme immunoassay. J. Infect. Dis. 155:292–296.

Apajalahti, J. H. A., P. Kärpänoja, and M. S. Salkinoja-Salonen. 1986. Rhodococcus chlorophenolicus sp. nov., a chlorophenol-mineralising actinomycete. Int. J. Syst. Bacteriol. 36:246–251.

Arai, T., S. Kuroda, and I. Watanabe. 1981. Biodegradation of acrylamide monomer by a Rhodococcus strain. Zbl. Bakteriol. 1. Abt., Suppl. 11:297–308.

Arenskütter, M., D. Bröker, and A. Steinbüchel. 2004. Biology of the metabolically diverse genus Gordonia. Appl. Environ. Microbiol. 70:3195–3204.

Arlotti, M., G. Zoboli, G. L. Moseatelli, G. Magnami, R. Maserati, V. Motghi, M. Andreoni, M. Libanore, L. Bonazzi, A. Piscina, and R. Ciammarughi. 1996. Rhodococcus equi infection in HIV-positive subjects: A retrospective analysis of 24 cases. Second. J. Infect. Dis. 28:463–467.

Arnold, M., A. Reittu, A. von Wright, P. J. Martikainen, and M. L. Suihko. 1997. Bacterial degradation of styrene in waste gases using a peat filter. Appl. Microbiol. Biotechnol. 48:738–744.

Arriaga, J. M., N. D. Cohen, J. N. Derr, K. Chaffin, and R. J. Martens. 2002. Detection of Rhodococcus equi by polymerase reaction using species-specific non-proprietary primers. J. Vet. Diagn. Invest. 14:347–353.

Asturias, J. A., and K. N. Timmis. 1993. Three different 2, 3-dihydroxybyphenyl-1, 2 dioxygenase genes in the Gram-positive polychlorobiphenyl-degrading bacterium Rhodococcus globerulus P6. J. Bacteriol. 175:4631–4640.

Athalye, M., W. C. Noble, A. I. Mallet, and D. E. Minnikin. 1984. Gas-chromatography-mass spectrometry of mycolic acids as a tool in the identification of medically important coryneform bacteria. J. Gen. Microbiol. 130:513–519.

Auerbach, S. B., M. M. McNeil, J. M. Brown, B. A. Laskar, and W. R. Jarvis. 1992. Outbreak of pseudoinfection with Tsukamurella paurometabolum traced to laboratory contaminatoin, efficacy of joint epidemiological and laboratory investigation. Clin. Infect. Dis. 14:1015–1022.

Baba, T., Y. Natsuhara, K. Kaneda, and I. Yano. 1997. Granuloma formation activity and mycolic acid composition of mycobacterial cord factor. Cell. Molec. Life Sci. 53:227–232.

Baddour, L. M., V. S. Baselski, M. J. Herr, G. D. Christensen, and A. L. Bisno. 1986. Nocardiosis in recipients of renal transplants: Evidence for nosocomial acquisition. Am. J. Infect. Control. 14:214–219.

Barns, M. R., W. A. Dietz, and P. A. Williams. 1997. A 3-(3-hydroxyphenyl) propionic acid catabolic pathway in Rhodococcus globerulus PWD1: cloning and characterization of the hpp operon. J. Bacteriol. 179:6145–5153.

Barton, M. D., and K. L. Hughes. 1980. Corynebacterium equi: A review. Vet. Bull. 50:65–80.

Barton, M. D., and K. L. Hughes. 1981. Comparison of three techniques for the isolation of Rhodococcus (Corynebacterium) equi from contaminated sources. J. Clin. Microbiol. 13:219–22.

Barton, M. D., M. Goodfellow, and D. E. Minnikin. 1989. Lipid composition in the classification of Rhodococcus equi. Zbl. Bakteriol. 272:154–170.

Battig, U., P. Wegmann, B. Meyer, and J. H. Penseyres. 1990. Nocardia mastitis. I: Clinical signs and diagnoses from 7 individual cases. Schweiz. Arch. Tierheilkde. 132:315–322.

Beaman, B. L., M. E. Gershwin, S. M. Scates, and Y. Ohsugi. 1980a. Immunobiology of germfree mice infected with Nocardia asteroides. Infect. Immunol. 29:733–743.

Beaman, B. L., S. Maslan, S. M. Scates, and J. Rosen. 1980b. Effect of route of inoculation on host resistance to Nocardia. Infect. Immunol. 28:185–189.

Beaman, B. L. 1982. Nocardiosis: Role of the cell wall deficient state of Nocardia. In: G. J. Dominque (Ed.) Cell Wall Defective Bacteria: Basic Principles and Clinical Significance. Addison-Wesley. Reading, MA. 231–255.

Beaman, B. L., and A. M. Sugar. 1983. Interaction of Nocardia in naturally acquired infections in animals. J. Hyg. 91:393–419.

Beaman, B. L. 1984. Actinomycete pathogenesis. In: M. Goodfellow, M. Mordarski, and S. T. Williams (Ed.) The Biology of the Actinomycetes. Academic Press. London, UK. 457–479.

Beaman, B. L. 1988. Nocardia in the etiology of nocardiosis. In: Actinomycetes as Opportunistic Pathogens (Abstracts), 3rd International Symposium of the Research Center for Pathogenic Fungi and Microbial Toxicoses Chiba University Japan. 3–4.

Beaman, B. L. 1992. Nocardia as a pathogen of the brain: Mechanisms of interactions in the murine brain—a review. Gene 115:213–217.

Beaman, B., and L. Beaman. 1994. Nocardia species: host-parasite relationships. Clin. Microbiol. Rev. 7:213–264.

Beaman, B. 2000a. The pathogenesis of Nocardia. In: V. Fischetti, R. Novich, J. Ferreti, D. Portnoy and J. Rood (Eds.) Gram-positive Pathogens. ASM Press. Washington, DC. 594–606.

Beaman, B. L., and L. Beaman. 2000b. Nocardia asteroides as an invasive intracellular pathogen of the brain and lungs. Subcell. Biochem. 33:167–198.

Beaman, B. L. 2002. Nocardiae and agents of actinomycetes. In: Manual of Clinical Laboratory Immunology, 6th ed. ASM Publications. Washington, DC. 511–515.

Beard, T. M., and M. I. Page. 1998. Enantioselective biotransformations using rhodococci. Ant. v. Leeuwenhoek 74:99–106.

Bell, K. S., J. C. Philp, N. Christofi, and D. W. J. Aw. 1996. Identification of Rhodococcus equi using the polymerase chain reaction. Lett. Appl. Microbiol. 23:72–74.

Bemer-Melchior, P., A. Haloun, P. Riegel, and H. Drugeon. 1999. Bacteremia due to Dietzia maris in an immunocompromised patient. Clin. Infect. Dis. 29:1338–1340.

Bendinger, B., F. A. Rainey, R. M. Kroppenstedt, M. Moormann, and S. Klatte. 1995. Gordonia hydrophobica sp. nov., isolated from biofilters for waste gas treatment. Int. J. Syst. Bacteriol. 45:544–548.

Bergey, D. H., F. C. Harrison, R. S. Breed, B. W. Hammer, and F. M. Huntoon (Eds.). 1923. Bergey's Manual of Determinative Bacteriology, 1st ed. Williams and Wilkins. Baltimore, MD.

Bergey, D. H., F. C. Huntoon, R. S. Breed, B. W. Hammer, and F. M. Huntoon (Eds.). 1925. Bergey's Manual of Determinative Bacteriology, 2nd ed. Williams and Wilkins. Baltimore, MD.

Bergey, D. H., F. C. Huntoon, R. S. Breed, B. W. Hammer, and F. M. Huntoon (Eds.). 1930. Bergey's Manual of Determinative Bacteriology, 3rd ed. Williams and Wilkins. Baltimore, MD.

Bergey, D. H., R. S. Breed, B. W. Hammer, F. M. Huntoon, E. G. D. Murray, and F. C. Harrison (Eds.). 1934. Manual of Determinative Bacteriology, 4th ed. Williams and Wilkins. Baltimore, MD.

Bergey, D. H., R. S. Breed, E. G. D. Murray, and A. P. Hitchens. 1939. Bergey's Manual of Determinative Bacteriology, 5th ed. Williams and Wilkins. Baltimore, MD.

Blackall, L. L., A. E. Harbers, P. F. Greenfield, and A. C. Hayward. 1988. Actinomycete scum problems in Australian activated sludge plants. Water Sci. Technol. 20:493–495.

Blackall, L. L., and K. C. Marshall. 1989a. The mechanism of stabilization of actinomycete foams and the prevention of foaming under laboratory conditions. J. Indust. Microbiol. 4:181–188.

Blackall, L. L., J. H. Parlett, A. C. Hayward, D. E. Minnikin, P. F. Greenfield, and A. E. Harbers. 1989b. Nocardia pinensis sp. nov., an actinomycete found in activated sludge foams in Australia. J. Gen. Microbiol. 135:1547–1558.

Blackall, L. L., A. E. Harbers, P. F. Greenfield, and A. C. Hayward. 1991. Foaming in activated sludge plants: A survey in Queensland, Australia and an evaluation of some control strategies. Water Res. 25:313–317.

Blanchard, R. 1896. Parasites végétaux à l'exclusion des bactéries. In: C. Bouchard (Ed.) Traité de pathologie générale. G. Masson. Paris France, 2:811–932.

Blümel, J., E. Blümel, A. F. Yassin, H. Schmidt-Rotte, and K. P. Schaal. 1998. Typing of Nocardia farcinica by pulsed-field gel electrophoresis reveals the endemic strain as source of hospital infections. J. Clin. Microbiol. 36:118–122.

Blumer, S. O., and L. Kaufman. 1979. Microimmunodiffusion test for nocardiosis. J. Clin. Microbiol. 10:308–312.

Boiron, P., and F. Provost. 1990a. Characterisation of Nocardia, Rhodococcus and Gordona species by in vitro susceptibility testing. Zbl. Bakteriol. 274:203–213.

Boiron, P., and F. Provost. 1990b. Enzymatic characterisation of Nocardia species and related bacteria by API ZYM profile. Mycopathologia 110:51–56.

Boiron, P., and F. Provost. 1990c. Use of partially purified 54-kilodalton antigen for diagnosis of nocardiosis by Western blot (immunoblot) assay. J. Clin. Microbiol. 28:328–331.

Boiron, P., C. Lafaurie, A. Rabbache, J. Brown, R. Carteret, and J. Petit. 1990d. Urease-negative Nocardia asteroides causing cutaneous nocardiosis. J. Clin. Microbiol. 28:801–802.

Boiron, P., G. Provost, G. Chevrier, and B. Dupont. 1990e. Review of nocardial infections in France 1987 to 1990. Eur. J. Clin. Microbiol. Infect. Dis. 11:709–714.

Boiron, P., and D. Stynen. 1992a. Immunodiagnosis of nocardiosis. Gene 115:219–222.

Boiron, P., D. Stynen, G. Belkacem, A. Goris, and F. Provost. 1992b. Monoclonal antibodies to a specific 54 kitodalton antigen of Nocardia spp. J. Clin. Microbiol. 30:1033–1035.

Boiron, P., F. Provost, and B. Duport. 1993. Laboratory Methods for the Diagnosis of Nocardiosis. Institut Pasteur. Paris, France.

Boiron, P., R. Locci, M. Goodfellow, S. A. Gumaa, K. Isik, B. Kim, M. M. McNeil, M. C. Salinas-Carmona, and H. Shojaei. 1998. Nocardia, nocardiosis and mycetomas. Med. Mycol. 36 (Suppl.):26–37.

Bojalil, L. F., and A. Zamora. 1963. Precipition and skin tests in the diagnosis of mycetoma due to Nocardia brasiliensis. Proc. Soc. Exp. Biol. Med. 113:40-53.

Bousfield, I. J., and M. Goodfellow. 1976. The "rhodochrous" complex and its relationships with allied taxa. In: M. Goodfellow, G. H. Brownell, and J. A. Serrano (Eds.) The Biology of the Nocardiae. Academic Press. London, UK. 39–65.

Bradley, S. G. 1978a. Physiological aspects of nocardiae. Zbl. Bakteriol. Suppl. 6:287–302.

Bradley, S. G., L. W. Enquist, and H. E. Scribner, 3rd. 1978b. Heterogeneity among deoxyribonucleotide sequences of Actinomycetales. In: E. Freerksen, I. Tárnok, and J. H. Thumin (Eds.) Genetics of the Actinomycetales. Gustav Fischer Verlag. Stuttgart, Germany. 207–224.

Brandão, P. F. B., L. A. Maldonado, A. C. Ward, A. T. Bull, and M. Goodfellow. 2001. Gordonia namibiensis sp. nov., a novel nitrile metabolising actinomycete recovered from an African sand. Syst. Appl. Microbiol. 5:510–515.

Breed, R. S., E. G. D. Murray, and A. P. Hitchens (Eds.). 1948. Bergey's Manual of Determinative Bacteriology, 6th ed. Williams and Wilkins. Baltimore, MD.

Briglia, M., F. A. Rainey, E. Stackebrandt, G. Schraa, and M. S. Salkinoja-Salonen. 1996. Rhodococcus percolatus sp. nov., a bacterium degrading 2, 4, 6-trichlorophenol. Int. J. Syst. Bacteriol. 46:23–30.

Brinkmann, U., and W. Babel. 1996. Simultaneous utilisation of pyridine and fructose by Rhodococcus opacus UFZ B408 without an external nitrogen source. Appl. Environ. Microbiol. 45:217–223.

Broughton, R. A., H. D. Wilson, N. L. Goodman, and J. A. Hedrick. 1981. Septic arthritis and osteomyelitis caused by an organism of the genus Rhodococcus. J. Clin. Microbiol. 13:209–213.

Brownell, G. H., and K. Denniston. 1984. Genetics of nocardioform bacteria. In: M. Goodfellow, M. Mordarski, and S. T. Williams (Eds.) The Biology of the Actinomycetes. Academic Press. London, UK. 201–228.

Brownell, G. H., and K. E. Belcher. 1990. DNA probes for the identification of Nocardia asteroides. J. Clin. Microbiol. 28:2082–2086.

Buchman, A. L., M. M. McNeil, J. M. Brown, B. A. Lasker, and M. E. Ament. 1992. Central venous catheter sepsis caused by unusual Gordona (Rhodococcus species): identification with a digoxigenen-labeled rDNA probe. Clin. Infect. Dis. 15:694–697.

Bunch, A. W. 1998. Biotransformation of nitriles by rhodococci. Ant. v. Leeuwenhoek 74:89–97.

Bushnell, R. B., A. C. Pier, R. E. Fichtner, B. L. Beaman, H. A. Boos, and M. D. Salman. 1979. Clinical and diagnostic aspects of herd problems with nocardial and mycobacterial mastitis. Am. Assc. Vet. Lab. Diag. 22:1–12.

Butler, W. R., D. G. Ahearn, and J. O. Kilburn. 1986. High performance liquid chromatography of mycolic acids as a tool in the identification of Corynebacterium, Nocardia, Rhodococcus, and Mycobacterium species. J. Clin. Microbiol. 23:182–185.

Butler, W. R., O. Kilburn, and G. P. Kubica. 1987. High performance liquid chromatography analysis of mycolic acids as an aid in laboratory identification of Rhodococcus and Nocardia species. J. Clin. Microbiol. 25:2126–2131.

Butler, W. R., M. M. Floyd, V. Silcox, G. Cage, E. Desmond, P. S. Duffey, L. S. Guthertz, W. M. Gross, K. C. Jost Jr., L. S. Ramos, L. Thibert, and N. Warren. 1996. Standardized Method for HPLC Identification of Mycobacteria. Centers for Disease Control and Prevention. Atlanta, GA.

Butler, W. R., and L. S. Guthertz. 2001. Mycolic and analysis by high-performance liquid chromatography for identification of Mycobacterium species. Clin. Microbiol. Rev. 14:704–726.

Byrne, B. A., J. F. Prescott, G. H. Palmer, S. Takai, V. M. Nicholson, D. C. Alperin, and S. A. Hines. 2001. Virulence plasmid of Rhodococcus equi contains inducible gene family encoding several proteins. Infect. Immun. 69:650–656.

Cain, R. B. 1981. Regulation of aromatic and hydroaromatic catabolic pathways in nocardioform actinomycetes. Zbl. Bakteriol. Suppl. 11:335–354.

Chirino-Trejo, J. M., J. F. Prescott. 1987a. Antibody response of horses to Rhodococcus equi antigens. Can. J. Vet. Res 51:301–305.

Chirino-Trejo, J. M., and J. F. Prescott. 1987b. Polyacrylamide gel electrophoresis of whole cell preparations of Rhodococcus equi. Can. J. Vet. Res. 51:297–300.

Chirino-Trejo, J. M., J. F. Prescott, and J. A. Yager. 1987c. Protection of foals against experimental Rhodococcus equi pneumonia by oral immunization. Can. J. Vet Res. 51:444–447.

Choi, K. S., S. H. Kim, and T. H. Lee. 1999. Purification and characterization of biosurfactant from Tsukamurella sp. 26A. J. Microbiol. Biotechnol. 9:32–38.

Chun, J., L. L. Blackall, S.-O. Kang, Y. C. Hah, and M. Goodfellow. 1997. A proposal to reclassify Nocardia pinensis Blackall et al., as Skermania piniformis gen. nov., comb. nov. Int. J. Syst. Evol. Microbiol. 47:127–131.

Cloud, J. L., P. S. Conville, A. Croft, D. Harmsen, F. G. Witebsky, and K. C. Carroll. 2004. Evaluation of partial 16S ribosomal DNA sequencing for identification of Nocardia species by using the microSeq 500 system with an expanded database. J. Clin Microbiol. 42:578–584.

Collins, M. D., T. Pirouz, M. Goodfellow, and D. E. Minnikin. 1977. Distribution of menaquinones in actinomycetes and corynebacteria. J. Gen. Microbiol. 100:221–230.

Collins, M. D., and D. Jones. 1982a. Lipid composition of Corynebacterium paurometabolum (Steinhaus). FEMS Microbiol. Lett. 13:13–16.

Collins, M. D., M. Goodfellow, and D. E. Minnikin. 1982b. A survey of the structure of mycolic acids in Corynebacterium and related taxa. J. Gen. Microbiol. 128:129–149.

Collins, M. D., M. Goodfellow, and D. E. Minnikin. 1982c. Fatty acid composition of some mycolic acid-containing coryneform bacteria. J. Gen. Microbiol. 128:2503–2509.

Collins, M. D. 1985a. Isoprenoid quinone analyses in bacterial classification and identification. In: M. Goodfellow and D. E. Minnikin (Eds.) Chemical Methods in Bacterial Systematics. Academic Press. London, UK. 267–287.

Collins, M. D., M. Goodfellow, D. E. Minnikin, and G. Alderson. 1985b. Menaquinone composition of mycolic acid-containing actinomycetes and some sporoactinomycetes. J. Appl. Bacteriol. 58:77–86.

Collins, M. D., O. W. Howarth, E. Grund, and R. M. Kroppenstedt. 1987. Isolation and structural determination of new members of the vitamin K series in Nocardia brasiliensis. FEMS Microbiol. Lett. 41:35–39.

Collins, M. D., J. Smida, M. Dorsch, and E. Stackebrandt. 1989. Tsukamurella gen. nov. harboring Corynebacterium paurometabolum and Rhodococcus aurantiacus. Int. J. Syst. Bacteriol. 38:385–391.

Collins, M. D. 1994. Isoprenoid quinones. In: M. Goodfellow and A. G. O'Donnell (Eds.) Chemical Methods in Prokaryotic Systematics. John Wiley. Chichester, UK. 265–309.

Colquhoun, J. A., J. Mexson, M. Goodfellow, A. C. Ward, K. Horikoshi, and A. T. Bull. 1998. Novel rhodococci and other mycolate actinomycetes from the deep sea. Ant. v. Leeuwenhoek 74:27–40.

Conti-Diaz, I. A., E. Gezuele, E. Civila, and J. E. Mackinnon. 1971. Termotolerancia y acción patógena de cepas de Nocardia asteroides aislades de fuentas naturales. Revta Urug. Patol. Clin. Microbiol. 9:232–241.

Conville, P. S., S. H. Fischer, C. P. Cartwright, and F. G. Witebsky. 2000. Identification of Nocardia species by restriction endonuclease analysis of an amplified portion of the 16S rRNA gene. J. Clin. Microbiol. 38:158–164.

Crespi, M., E. Messens, A. B. Caplan, M. Vanmontagu, and J. Desomer. 1992. Fasciation induction by the phytopathogen Rhodococcus fascians depends upon a linear plasmid encoding a cytokinin synthase gene. EMBO J. 11:795–804.

Crespi, M., D. Vereecke, W. Temmerman, M. Vanmontagu, and J. Desomer. 1994. The fas operon of Rhodococcus fascians encodes new genes required for the efficient fasciation of host plants. J. Bacteriol. 176:2492–2501.

Cross, T., T. J. Rowbotham, E. N. Mishustin, E. Z. Tepper, F. Antoine-Portaels, K. P. Schaal, and H. Bickenbach. 1976. The ecology of nocardioform actinomycetes. In: M. Goodfellow, G. H. Brownell, and J. A. Serrano (Eds.) The Biology of the Nocardiae. Academic Press. London, UK. 337–371.

Cuello, O. H., M. J. Caorlin, V. E. Reviglio, L. Carvajal, C. P. Juarez, E. Palacio de Guerra, and J. D. Luna. 2002. Rhodococcus globerulus keratitis after laser in situ keratomilensis. J. Cataract. Refract. Surg. 28:2235–2237.

Cummins, C. S. 1971. Cell wall composition in Corynebacterium bovis and some other corynebacteria. J. Bacteriol. 105:1227–1228.

Dabbs, E. R. 1987. A generalized transducing bacteriophage for Rhodococcus erythropolis. Molec. Gen. Genet. 206:116–120.

Dabbs, E. R. 1998. Cloning of genes that have environmental and clinical importance for rhodococci and related bacteria. Ant. v. Leeuwenhoek 74:155–168.

Dabrock, B., M. Kesseler, B. Averhoff, and G. Gottschalk. 1994. Identification and characterization of a transmissible linear plasmid from Rhodococcus erythropolis BD2 that encodes isopropylbenzene and trichloroethane catabolism. Appl. Environ. Microbiol. 60:853–860.

Da Cost, E. O., A. R. Ribeiro, E. T. Watanabe, R. B. Pardo, J. B. Silva, and R. B. Sanches. 1996. An increased incidence of mastitis caused by Prototheca species and Nocardia species on a farm in São Paulo, Brazil. Vet. Res. Comm. 20:237–241.

De los Reyes, M. F., F. L. De los Reyes, 3rd, M. Hernandez, and L. Raskin. 1998a. Identification and quantification of Gordonia amarae strains in activated sludge schemes using comparative rRNA sequence analysis and phylogenetic hybridization probes. Water Sci. Technol. 37:521–5256.

De los Reyes, M. F., F. L. De los Reyes, 3rd, M. Hernandez, and L. Raskin. 1998b. Quantification of Gordonia amarae strains in foaming activated sludge and anaerobic digester systems with oligonucleotide hybridization probes. Appl. Environ. Microbiol. 64:2503–2512.

De Magaldi, S. W., and D. W. Mackenzie. 1990. Comparison of antigens from agents of actinomycetoma by immunodiffusion and electrophoresis procedures. J. Med. Vet. Mycol. 28:363–371.

Demaree, J. B., and N. R. Smith. 1952. Nocardia vaccinii n. sp. causing galls on blueberry plants. Phytopathology 42:249–252.

Denise-Larose, C., D. Labbe, H. Bergeron, A. M. Jones, C. W. Greer, J. Alhawari, M. J. Grossman, B. M. Sankey, and P. C. K. Lau. 1997. Conservation of plasmid-encoded dibenzothiophene desulfurisation genes in several rhodococci. Appl. Environ. Microbiol. 63:2915–2919.

Denome, S. A., E. S. Olsen, and K. D. Young. 1993. Identification and cloning of genes involved in specific desulfurization of dibenzothiophene by Rhodococcus sp. strain IGTS8. Appl. Environ. Microbiol. 59:2837–2843.

Denome, S. A., C. Oldfield, L. J. Nash, and K. D. Young. 1994. Characterization of the desulfurization genes from Rhodococcus sp. strain IGTS8. J. Bacteriol. 176:6707–6716.

Desai, J. D., and I. M. Banat. 1997. Microbial production of surfactants and their commercial potential. Microbiol. Molec. Biol. Rev. 61:47–64.

Dhaliwal, B. S. 1979. Nocardia amarae and activated sludge foaming. J. Water Pollut. Cont. Fed. 51:340–350.

Dobson, G., D. E. Minnikin, S. M. Minnikin, J. H. Parlett, M. Goodfellow, M. Ridell, and M. Magnusson. 1985. Systematic analysis of complex mycobacterial lipids. In: M. Goodfellow and D. E. Minnikin (Eds.) Chemical Methods in Bacterial Systematics. Academic Press. London, UK. 237–265.

Drancourt, M., E. Bonnet, H. Gallais, Y. Peloux, and D. Ravult. 1992. Rhodococcus equi infection in patients with AIDS. J. Infect. 24:123–131.

Drancourt, M., M. M. McNeil, J. M. Brown, B. A. Lasker, M. Maurin, M. Choux, and D. Ravult. 1994. Brain abscess due to Gordonia terrae in an immunocompromised child: Case report and review of infections caused by G. terrae. Clin. Infect. Dis. 19:258–262.

Drancourt, M., J. Pelletier, A. Ali Cherif, and D. Raoult. 1997. Gordonia terrae central nervous system infection in an immunocompetent patient. J. Clin. Microbiol. 35:379–382.

Duckworth, A. W., S. Grant, W. D. Grant, B. E. Jones, and D. Meyer. 1998. Dietzia natronolimnaios sp. nov., a new species of the genus Dietzia isolated from an East African soda lake. Extremophiles 2:359–366.

Eberzole, L. L., and J. L. Paturzo. 1988. Endophthalinitis caused by Rhodococcus equi Prescott serotype 4. J. Clin. Microbiol. 26:1221–1222.

Eggeling, J., and H. Sahm. 1980. Degradation of coniferyl alcohol and other lignin-related aromatic compounds by Nocardia sp. DSM 1069. Arch. Mikrobiol. 126:141–148.

Eggeling, L., and H. Sahm. 1981. Degradation of lignin-related aromatic compounds by Nocardia spec. DSM 1069 and the specificity of demethylation. Zbl. Bakteriol. Suppl. 11:361–366.

Embley, T. M., and R. Wait. 1994. Structural lipids of eubacteria. In: M. Goodfellow and A. G. O'Donnell (Eds.) Chemical Methods in Prokaryotic Systematics. John Wiley. Chichester, UK. 121–161.

Engelhardt, G., P. R. Wallnöfer, and H. G. Rast. 1976. Metabolism of o-phthalic acid by different Gram-negative and Gram-positive soil bacteria. Arch. Microbiol. 109:109–114.

Engelhardt, G., H. G. Rast, and P. R. Wallnöfer. 1979. Degradation of aromatic carboxylic acids by Nocardia spec. DSM 43251. FEMS Microbiol. Lett. 5:245–251.

Eppinger, H. 1891. Über eine nener pathogene Cladothrix und eine durch sie hervorgerufene Pseudotuberculosis (Cladothrichia). Beitr. Pathol. Anat. Allgem. Pathol. 9:287–328.

Etémadi, A. H. 1967. The use of pyrolysis gas chromatography and mass spectrometry in the study of the structure of mycolic acids. J. Gas Chromatogr. 5:447–456.

Exmelin, L., B. Mulbruny, M. Vergnaud, F. Provost, P. Boiron, and C. Morel. 1996. Molecular study of nosocomial nocardiosis outbreaks involving heart transplant recipients. J. Clin. Microbiol. 34:1014–1016.

Farmer, R. 1962. Influence of various chemicals in the isolation of Nocardia from soil. Proc. Acad. Sci. 43:254–256.

Finkel'shtein, Z. I., B. P. Baskunov, E. L. Golovlev, and L. A. Golovieva. 1999. Desulphurization of 4, 6-dimethyldibenzothiophene and dibenzothiophene by Gordonia aichiensis 51 [in Russian]. Mikrobioluja 68:187–190.

Finnerty, W. R. 1992. The biology and the genetics of the genus Rhodococcus. Am. Rev. Microbiol. 46:193–218.

Friedman, C. S., B. L. Beaman, J. Chun, M. Goodfellow, A. Gee, and R. P. Hedrick. 1998. Nocardia crassostreae sp. nov., the causal agent of nocardiosis in Pacific oysters. Int. J. Syst. Bacteriol. 48:237–246.

Fukuhashi, K., A. Taoka, S. Uchida, I. Karube, and S. Suzuki. 1981. Production of 1,2-epoxyalkanes from 1-alkanes by Nocardia corallina B-276. Eur. J. Appl. Microbiol. Biotechnol. 12:39–45.

Gaude, G. S., B. M. Hamashetter, A. S. Bagga, and R. Chatterji. 1999. Clinical profile of pulmonary nocardiosis. Ind. J. Chest Dis. All. Sci. 41:153–157.

Gerson, D. F., and J. E. Zajic. 1977. Bitumen extraction from tar sands with microbial surfactants. The Oil Sands of Canada-Venezuela. CIM Special Publication No. 17:705–710.

Gilbert, S. C., J. Morton, S. Buchanan, C. Oldfield, and A. McRoberts. 1998. Isolation of a unique benzothiophene desulphurizing bacterium, Gordonia sp. strain ZI3ET (NCIMB 40816), and characterization of the desulphurization pathway. Microbiology 144:2545–2553.

Golub, D., G. Falk, and W. W. Spink. 1967. Lung abscesses due to Corynebacterium equi: Report of first human infection. Assoc. Intern. Med. 66:1174–1177.

Gonzáles-Ochoa, A. 1973. Virulence of nocardiae. Can. J. Microbiol. 19:901–904.

Goodfellow, M. 1971. Numerical taxonomy of some nocardioform bacteria. J. Gen. Microbiol. 69:33–80.

Goodfellow, M., and V. A. Orchard. 1974. Antibiotic sensitivity of some nocardioform bacteria and its value as a criterion for taxonomy. J. Gen. Microbiol. 83:375–387.

Goodfellow, M., M. D. Collins, and D. E. Minnikin. 1976. Thin-layer chromatographic analysis of mycolic acid and other long-chain components in whole-organism methanolysates of coryneform and related taxa. J. Gen. Microbiol. 96:351–358.

Goodfellow, M., and G. Alderson. 1977. The actinomycete genus Rhodococcus: a home for the "rhodochrous" complex. J. Gen. Microbiol. 100:99–122.

Goodfellow, M., P. A. B. Orlean, M. D. Collins, L. Alshamaony, and D. E. Minnikin. 1978. Chemical and numerical taxonomy of strains received as Gordona aurantiaca. J. Gen. Microbiol. 109:57–68.

Goodfellow, M., and E. Aubert. 1980. Characterization of rhodococci from the intestinal tract of Rapa Nui cockroaches. In: G. L. Nogrady (Ed.) Microbiology of Easter Island. Sovereign Press. Oakville, Canada. 2:231–240.

Goodfellow, M., and L. G. Wayne. 1982a. Taxonomy and nomenclature. In: C. Ratledge and J. L. Stanford (Eds.) The Biology of the Mycobacteria. Academic Press. London, UK. 1:471–521.

Goodfellow, M., A. R. Beckham, and M. D. Barton. 1982b. Numerical classification of Rhodococcus equi and related actinomycetes. J. Appl. Bacteriol. 53:199–207.

Goodfellow, M., D. E. Minnikin, C. Todd, G. Alderson, S. M. Minnikin, and M. D. Collins. 1982c. Numerical and chemical classification of Nocardia amarae. J. Gen. Microbiol. 128:1283–1297.

Goodfellow, M., C. R. Weaver, and D. E. Minnikin. 1982d. Numerical classification of some rhodococci, corynebacteria and related organisms. J. Gen. Microbiol. 128:731–735.

Goodfellow, M., and S. T. Williams. 1983. Ecology of actinomycetes. Ann. Rev. Microbiol. 37:189–216.

Goodfellow, M., and T. Cross. 1984a. Classification. In: M. Goodfellow, M. Mordarski, and S. T. Williams (Eds.) The Biology of the Actinomycetes. Academic Press. London, UK. 7–164.

Goodfellow, M., and D. E. Minnikin. 1984b. A critical evaluation of Nocardia and related taxa. In: L. Ortiz-Ortiz, L. F. Bojalil, and V. Yakoleff (Eds.) Biological, Biochemical and Biomedical Aspects of Actinomycetes. Academic Press. Orlando, FL. 588–596.

Goodfellow, M., E. G. Thomas, and A. L. James. 1987. Characterisation of rhodococci using peptide hydrolase substrates based on 7-amino-4-methylcoumarin. FEMS Microbiol. Lett. 44:349–355.

Goodfellow, M. 1989a. Genus Rhodococcus Zopf 1891, 28[AL]. In: S. T. Williams, M. E. Sharpe, and J. G. Holt (Eds.) Bergey's Manual of Systematic Bacteriology. Williams and Wilkins. Baltimore MD, 4:2362–2371.

Goodfellow, M., and M. P. Lechevalier. 1989b. Genus Nocardia Trevisan 1889, 9[AL]. *In:* S. T. Williams, M. E. Sharpe, and J. G. Holt (Eds.) Bergey's Manual of Systematic Bacteriology. Williams and Wilkins. Baltimore, MD. 4:2350–2361.

Goodfellow, M., and A. G. O'Donnell. 1989c. Search and discovery of industrially significant actinomycetes. *In:* S. Baumberg, M. Rhodes, and I. Hunter (Eds.) Microbial Products: New Approaches. Cambridge University Press. Cambridge, UK. 343–383.

Goodfellow, M., E. G. Thomas, A. C. Ward, and A. L. James. 1990. Classification and identification of rhodococci. Zbl. Bakteriol. 298:299–315.

Goodfellow, M. 1995. Actinomycetes: Actinomyces, Actinomadura, Nocardia, Streptomyces and related taxa. *In:* J. P. Collee, J. P. Duguid, A. G. Fraser, B. P. Marmion, and A. Simmons (Eds.) Mackie & McCartney Practical Medical Microbiology. Churchill Livingston. Edinburgh, UK.

Goodfellow, M., R. J. Davenport, F. M. Stainsby, and T. P. Curtis. 1996. Actinomycete diversity associated with foaming in activated sludge plants. J. Indust. Microbiol. 17:268–280.

Goodfellow, M. 1997. The actinomycetes: Actinomyces, Nocardia and related genera. *In:* A. Balows and B. I. Duerden (Eds.) Topley and Wilson's Microbiology and Microbial Infections, 9th ed. Edward Arnold. London, UK. 463–489.

Goodfellow, M., and J. G. Magee. 1998a. Taxonomy of mycobacteria. *In:* P. R. J. Gangadharam and P. A. Jenkins (Eds.) Mycobacterial Basic Aspects. Chapman and Hall. New York, MD. 1–71.

Goodfellow, M., G. Alderson, and J. Chun. 1998b. Rhodococcal systematics: problems and developments. Ant. v. Leeuwenhoek 74:3–20.

Goodfellow, M., F. M. Stainsby, R. J. Davenport, J. Chun, and T. Curtis. 1998c. Activated sludge foaming: The true extent of actinomycete diversity. Water Res. Technol. 37:511–519.

Goodfellow, M., K. Isik, and E. Yates. 1999. Actinomycete systematics: an unfinished synthesis. Nova Acta Leopoldina NF80(312):47–82.

Goodfellow, M., A. L. Jones, L. A. Maldonado, and J. Salanitro. 2004. Rhodococcus aetherivorans sp. nov., a new species that contains methyl t-butyl ether-degrading actinomycetes. Syst. Appl. Microbiol. 27:61–65.

Gordon, R. E., and M. M. Smith. 1953. Rapidly growing acid fast bacteria. I: Species description of Mycobacterium phlei Lehmann and Neumann and Mycobacterium smegmatis (Trevisan) Lehmann and Neumann. J. Bacteriol. 66:41–48.

Gordon, R. E., and J. E. Mihm. 1957. A comparative study of some new strains received as nocardiae. J. Bacteriol. 73:15–27.

Gordon, R. E., and J. E. Mihm. 1962. Identification of Nocardia caviae (Erikson) comb. nov. Ann. New York Acad. Sci. 98:628–636.

Gordon, R. E., S. K. Mishra, and D. A. Barnett. 1978. Some bits and pieces of the genus Nocardia, N. carnea, N. vaccinii, N. transvalensis, N. orientalis and N. aerocolonigenes. J. Gen. Microbiol. 108:69–78.

Gotoh, K., M. Mitsuyama, S. Imaizumi, and I. Yano. 1991. Mycolic acid-containing glycolipid as a possible virulence factor of R. equi for mice. Microbiol. Immunol. 35:175–185.

Gray, P. H. H., and H. G. Thornton. 1928. Soil bacteria that decompose certain aromatic compounds. Zbl. Bakteriol. Parasitkde. Abt. II 73:74–96.

Grzeszik, C., M. Lubbers, M. Reh, and H. G. Schlegel. 1997. Genes encoding the NAD-reducing hydrogenase of Rhodococcus opacus MRII. Microbiology 143:1271–1286.

Guagnani, H. C., I. C. Unaogu, F. Provost, and P. Boiron. 1998. Pulmonary infection due to Nocardiopsis dassonvillei, Gordonia sputi, Rhodococcus rhodochrous and Micromonospora sp. in Nigeria and literature review. J. Mycol. Med. 8:21–25.

Gürtler, V., and B. C. Mayall. 2001a. Genomic approaches to typing, taxonomy and evolution of bacterial isolates. Int. J. Syst. Microbiol. 51:3–16.

Gürtler, V., R. Smith, B. C. Mayall, G. Potter-Reinemann, E. Stackebrandt, and R. M. Kroppenstedt. 2001b. Nocardia veterana sp. nov., isolated from human bronchial lavage. Int. J. Syst. Evol. Microbiol. 51:933–936.

Gutmann, L., F. W. Goldstein, M. D. Kitzis, B. Hautefort, C. Darmon, and J. F. Acar. 1983. Susceptibility of Nocardia asteroides to 46 antibiotics, including 22 β-lactams. Antimicrob. Agents Chemother. 23:248–251.

Haburchack, D. R., B. Jeffrey, J. W. Higbee, and E. D. Everett. 1978. Infections caused by rhodochrous. Am. J. Med. 65:298–302.

Häggblom, M., L. J. Nohynek, N. J. Palleroni, K. Kronqvist, E. L. Nurmiaoho-Lassila, M. S. Salkinoja-Salonen, S. Klatte, and R. M. Kroppenstedt. 1994. Transfer of polychlorophenol-degrading Rhodococcus chlorophenolicus (Apajalahti et al., 1996). Syst. Bacteriol. 44:485–493.

Haites, R. E., G. Muscatello, A. P. Begg, and G. F. Browning. 1997. Prevalence of the virulence-associated gene of Rhodococcus equi in isolates from infected foals. J. Clin. Microbiol. 35:1642–1644.

Hamid, M. E., D. E. Minnikin, and M. Goodfellow. 1993. A simple chemical test to distinguish mycobacteria from other mycolic-acid-containing actinomycetes. J. Gen. Microbiol. 139:2203–2213.

Hamid, M. E., L. Maldonado, G. S. Sharaf Eldin, M. A. Mohamed, N. S. Saeed, and M. Goodfellow. 2001. Nocardia africana sp. nov., a new pathogen isolated from patients with pulmonary infections. J. Clin. Pathol. 39:625–630.

Hancock, I. C. 1994. Analysis of cell wall constituents of Gram-positive bacteria. *In:* M. Goodfellow and A. G. O'Donnell (Eds.) Chemical Methods in Prokaryotic Systematics. John Wiley. Chichester, UK. 63–84.

Harper, D. B. 1977. Microbiol metabolism of aromatic nitriles. Biochem. J. 165:309–319.

Harper, D. B. 1985. Characterization of a nitrilase from Nocardia sp. (rhodochrous group) NCIB 11215, using p-hydroxybenzonitrile as sole carbon source. Int. J. Biochem. 17:677–683.

Harrison, F. C. 1929. The discoloration of halibut. Can. J. Res. 1:214–239.

Hart, D. H. L., M. M. Peel, J. H. Andrew, and J. G. W. Burrdon. 1988. Lung infection caused by rhodococcus. Austral. NZ. J. Med. 18:790–791.

Hecht, S. T., and W. A. Causey. 1976. Rapid methods for the detection and identification of mycolic acids in aerobic actinomycetes and related bacteria. J. Clin. Microbiol. 4:284–287.

Heiss, G. S., B. Gowan, and E. R. Dabbs. 1992. Cloning of DNA from a Rhodococcus strain conferring the ability to decolonize sulphonated azo-dyes. FEMS Microbiol. Lett. 99:221–226.

Helmke, E., and H. Weyland. 1984. Rhodococcus marinonascens sp. nov., an actinomycete from the sea. Int. J. Syst. Bacteriol. 34:127–138.

Hensel, J., and G. Straube. 1990. Kinetic studies of phenol degradation by Rhodococcus sp. II: Continuous cultivation. Ant. v. Leeuwenhoek 57:33–36.

Hieda, T., Y. Mikami, Y. Obi, and T. Kisaka. 1982. Plasmid-determined transformation of cis-abienol and sclareol in Nocardia restricta JTS-162. Agric. Biol. Chem. 46:305–306.

Higgins, M. L., and M. P. Lechevalier. 1969. Poorly lytic bacteriophage from Dactylosporangium thailandensis (Actinomycetales). J. Virol. 3:210–216.

Hill, R., S. Hart, N. Illing, R. Kirby, and D. R. Woods. 1989. Cloning and expression of Rhodococcus genes encoding pigment production in Escherichia coli. J. Gen. Microbiol. 135:1507–1513.

Houang, E. T., I. S. Lovett, F. D. Thompson, A. R. Harrison, A. M. Joekes, and M. Goodfellow. 1980. Nocardia asteroides: A transmissible disease. J. Hosp. Infect. 1:31–40.

Howarth, O. W., E. Grund, R. M. Kroppenstedt, and M. D. Collins. 1986. Structural determination of a new naturally occurring cyclic vitamin K. Biochem. Biophys. Res. Commun. 140:916–923.

Hughes, J., Y. C. Armitage, and K. C. Symes. 1998. Application of the whole cell rhodococcal biocatalysts in acrylic polymer manufacture. Ant. v. Leeuwenhoek. 74:107–118.

Hutchinson, M., J. W. Ridgway, and T. Cross. 1975. Biodeterioration of rubber in contact with water, sewage and soil. In: D. W. Lovelock and R. J. Gilbert (Eds.) Microbial Aspects of the Deterioration of Materials. Academic Press. London, UK. 187–202.

Hütter, R. A., and T. Eckhardt. 1988. Genetic manipulation. In: M. Goodfellow, S. T. Williams, and M. Mordarski (Eds.) Actinomycetes in Biotechnology. Academic Press. London, UK. 89–184.

Hyman, I. S., and S. D. Chaparas. 1977. A comparative study of the "rhodochrous" complex and related taxa by delayed type skin reactions on guinea pigs and by polyacrylamide gel electrophoresis. J. Gen. Microbiol. 100:363–371.

Idigbe, E. G., C. Onobogu, and E. K. John. 1992. Human pulmonary nocardiosis. Microbios 69:163–170.

Ikehata, O., M. Nishiyama, S. Horinouchi, and T. Beppu. 1989. Primary structure of nitrile hydratase deduced from the nucleotide sequence of a Rhodococcus species and its expression in Escherichia coli. Eur. J. Biochem. 181:563-570.

Ioneda, T. 1988. Biochemical and physiological aspects of mycolic acids and mycolyl derivatives. In: Y. Okami, T. Beppu, and H. Ogawara (Eds.) Biology of Actinomycetes '88. Japan Scientific Societies Press. Tokyo, Japan. 463–468.

Isik, K., and M. Goodfellow. 2002. Differentiation of Nocardia species by PCR-randomly amplified polymorphic DNA fingerprinting. Syst. Appl. Microbiol. 25:60–67.

Isik, K., N. Sahin, E. Karpitas, and M. Goodfellow. 2002. Typing of some clinically significant Nocardia strains using a digoxigenin-labelled rDNA gene probe. Turkish J. Biol. 26:1–8.

Janke, D., B. Schukat, and H. Prauser. 1986. Screening among nocardioform bacteria strains able to degrade aniline and monochloranilines. J. Basic Microbiol. 6:341–350.

Judicial Commission. 1985. Opinion 58: Confirmation of the types in the Approved Lists as nomenclatural types including recognition of Nocardia asteroides (Eppinger 1891) Blanchard 1896 and Pasteuria multicida (Lehmann and Neumann, 1899) Rosenbusch and Mer-chant 1939 as the respective type species of the genera Nocardia and Pasteurella and the rejection of the species name Pasteurella gallicida (Burrill 1883) Buchanan 1925. Int. J. Syst. Bacteriol. 35:538.

Jiménez, T., A. M. Diaz, and H. Zlotnik. 1990. Monoclonal antibodies to N. asteroides and N. brasiliensis antigens. J. Clin. Microbiol. 28:87–91.

Johnson, J. A., J. F. Prescott, and K. J. Markham. 1983a. The pathology of experimental Corynebacterium equi infection in foals following intrabronchial challenge. Vet. Pathol. 20:440–449.

Johnson, J. A., J. F. Prescott, and K. J. Markham. 1983b. The pathology of experimental Corynebacterium equi infection in foals following intragastric challenge. Vet. Pathol. 20:450–459.

Jones, K. L. 1949. Fresh isolates of actinomycetes in which the presence of sporangeous aerial mycelia is a fluctuating characteristic. J. Bacteriol. 57:141–145.

Jones, D. 1975. A numerical taxonomic study of coryneform and related bacteria. J. Gen. Microbiol. 87:52–96.

Jones, A. L., J. M. Brown, V. Mishra, J. D. Perry, A. G. Steigerwalt, and M. Goodfellow. 2004. Rhodococcus gordoniae sp. nov., an actinomycete isolated from clinical material and phenol-contaminated soil. Int. J. Syst. Evol. Microbiol. 54:407–411.

Jonsson, S., R. J. Wallace Jr., I. S. Hull, and D. M. Musher. 1986. Recurrent Nocardia pneumonia in an adult with chronic granulamatous disease. Am. Rev. Respir. Dis. 123:932–934.

Kageyama, A., N. Poonwan, K. Yazawa, Y. Mikami, and K. Nishimura. 2003a. Nocardia asiatica sp. nov., isolated from patients with nocardiosis in Japan and clinical specimens from Thailand. Int. J. Syst. Evol. Microbiol. 54:125–130.

Kageyama, A., K. Yazawa, K. Nishimura, and Y. Mikami. 2003b. Nocardia inohanensis sp. nov., Nocardia yamanashiensis sp. nov. and Nocardia niigatensis sp.nov., isolated from clinical specimens. Int. J. Syst. Evol. Microbiol. 54:563–569.

Kämpfer, P., M. A. Andersson, F. A. Rainey, R. M. Kroppenstedt, and M. Salkinoja-Salonen. 1999. Williamsia muralis gen. nov., sp. nov., isolated from the indoor environment of a children's day care centre. Int. J. Syst. Bacteriol. 49:681–687.

Kämpfer, P., S. Buczolito, U. Jäckel, I. Grun-Wollny, and H.-J. Busse. 2003. Nocardia tenerifensis sp. nov. Int. J. Syst. Evol. Microbiol. 54:381–383.

Kaneda, K., S. Imaizumi, and I. Yano. 1995. Distribution of C22-, C24- and C26-alpha-unit-containing mycolic acid homologues in mycobacteria. Microbiol. Immun. 39:563–570.

Kanetsuna, F., and A. Bartoli. 1972. A simple chemical method to differentiate Mycobacterium from Nocardia. J. Gen. Microbiol. 70:209–212.

Kasweck, K. L., M. L. Little, and S. G. Bradley. 1981. Characteristics of plasmids in Nocardia asteroides. Actin. Rel. Org. 16:57–63.

Kasweck, K. L., and M. L. Little. 1982a. Genetic recombination in Nocardia asteroides. J. Bacteriol. 149:403–406.

Kasweck, K. L., M. L. Little, and S. G. Bradley. 1982b. Plasmids in mating strains of Nocardia asteroides. Dev. Indust. Microbiol. 23:279–286.

Kattar, M. M., B. T. Cookson, L. C. Carlson, S. K. Stiglich, A. A. Schwartz, T. T. Nguyen, R. Daza, C. K. Wallis, S. L. Yarfitz, and M. B. Coyle. 2001. Tsukamurella strandjordae sp. nov., a proposed new species causing sepsis. J. Clin. Microbiol. 39:1467–1476.

Kayser, K. J., B. A. Bielaga-Jones, K. Jackowski, O. Odusan, and J. J. Kilbane, 2nd. 1993. Utilization of organosulfur compounds by axenic and mixed cultures of Rhodococcus rhodochrous IGTS8. J. Gen. Microbiol. 139:3123–3129.

Kedlaya, I., M. B. Ing, and S. S. Wong. 2001. Rhodococcus equi infections in immunocompetent hosts: A case report and review. Clin. Infect. Dis. 32:e39–e347.

Kesseler, M., E. R. Dabbs, B. Averhoff, and G. Göttschalk. 1996. Studies on the isopropylbenzene 2, 3-dioxygenase and the 3-isopropylcatechol 2, 3-dioxygenase genes encoded by the linear plasmid of Rhodococcus erythropolis BD2. Microbiology 142:3241–3251.

Kim, J., G. Y. Minamoto, and M. H. Grieco. 1991. Nocardial infection as a complication of AIDS: A report of six cases and review. Rev. Infect. Dis. 13:624–629.

Kim, S. B., R. Brown, C. Oldfield, S. C. Gilbert, and M. Goodfellow. 1999. Gordonia desulfuricans sp. nov., a benzothiophene-desulphurizing actinomycete. Int. J. Syst. Bacteriol. 49:1845–1851.

Kim, S. B., R. Brown, C. Oldfield, S. C. Gilbert, S. Ilarionov, and M. Goodfellow. 2000a. Gordonia amicalis sp. nov., a novel dibenzothiophene-desulphurizing actinomycete. J. Clin. Microbiol. 50:2031–2036.

Kim, S. H., E. J. Lim, S. O. Lee, J. D. Lee, and T. H. Lee. 2000b. Purification and characterization of biosurfactants from Nocardia sp. L-417. Biotechnol. Appl. Biochem. 31:249–253.

Kim, K. K., S. L. Lee, R. M. Kroppenstedt, E. Stackebrandt, and S. T. Lee. 2003. Gordonia sihwensis sp. nov., a novel nitrate-reducing bacterium isolated from a wastewater treatment bioreactor. Int. J. Syst. Evol. Microbiol. 53:1427–1433.

Kinoshita, N., Y. Hamma, M. Igarashi, S. Ikeno, M. Hori, and M. Hamada. 2001. Nocardia vinacea sp. nov. Actinomycetology 15:1–5.

Kirby, R., and K. Usdin. 1985. The isolation and restriction mapping of a miniplasmid from the actinomycete Nocardia corallina. FEMS Microbiol. Lett. 27:57–59.

Kiska, D. L., K. Hicks, and D. J. Pettit. 2002. Identification of medically relevant Nocardia species with an abbreviated battery of tests. J. Clin. Microbiol. 40:1346–1351.

Klatte, S., R. M. Kroppenstedt, and F. A. Rainey. 1994. Rhodococcus opacus sp. nov., an unusual nutriionally versatile Rhodococcus species. Syst. Appl. Microbiol. 17:355–360.

Klatte, S., R. M. Kroppenstedt, P. Schumann, K. Altendorf, and F. A. Rainey. 1996. Gordonia hirsuta sp. nov. Int. J. Syst. Bacteriol. 46:876–880.

Knowles, C. J., and J. W. Wyatt. 1992. The degradation of cyanide and nitriles. In: F. C. Fry, G. M. Gadd, R. A. Herbert, C. W. Jones, and J. A. Waton-Craik (Eds.) Microbial Control of Pollution. Cambridge University Press. Cambridge, UK. 113–128.

Kobayashi, M., T. Nagasawa, and H. Yamada. 1989a. Nitrilase of Rhodococcus rhodochrous J1: Purification and characterization. Eur. J. Biochem. 182:349–356.

Kobayashi, M., T. Nagasawa, N. Yanaka, and H. Yamada. 1989b. Nitrilase-catalyzed production of p-aminobenzoic acid from p-aminobenzonitrile with Rhodococcus rhodochrous J1. Biotechnol. Lett. 11:27–30.

Kobayashi, M., N. Yanaka, T. Nagasawa, and H. Yamada. 1990. Purification and characterization of a novel nitrilase of Rhodococcus rhodochrous K22 that acts on aliphatic nitriles. J. Bacteriol. 172(9): 4807–4815.

Kobayashi, M., T. Nagasawa, and H. Yamada. 1992. Enzymatic synthesis of acrylamide: A success story not yet over. TIBTECH 10:402–408.

Koffi, N., E. Aka-Danguy, A. Ngom, B. Kouassi, B.-A. Yaya, and M. Dasso. 1998. Prévalence de la nocardiose pulmonaire en zone d'endémie tuberculeuse. Rev. Mal. Respir. 15:643–647.

Komeda, H., M. Kobayashi, and S. Shimizu. 1996a. A novel gene-cluster including the Rhodococcus rhodochrous J1 nhlBA genes encoding a low-molecular—mass nitrile hydratase (1-nhase) induced by its reaction product. J. Biol. Chem. 271:15796–15802.

Komeda, H., M. Kobayashi, and S. Shimizu. 1996b. Characterization of the gene cluster of high molecular mass nitrile hydratase (L-WHase) induced by its reaction product in Rhodococcus rhodochrous J1. Proc. Natl. Acad. Sci. USA 93:4267–4272.

Komeda, H., M. Kobayashi, and S. Shimizu. 1997. A novel transporter involved in cobalt uptake. Proc. Natl. Acad. Sci. USA 94:36–41.

Kosono, S., M. Maeda, F. Fuji, H. Arai, and T. Kudo. 1997. Three of the seven bphC genes of Rhodococcus erythropolis TA421, isolated from a termite ecosystem, are located on an indigenous plasmid associated with biphenyl degradation. Appl. Environ. Microbiol. 63:3282–3285.

Kretschmer, A., H. Bock, and F. Wagner. 1982. Chemical and physical characterization of interfacial-active lipids from Rhodococcus erythropolis grown on n-alkanes. Appl. Environ. Microbiol. 44:864–870.

Krick, J. A., E. B. Stinson, and J. S. Remington. 1975. Nocardia infection in heart transplant patients. Intern. Med. 82:18–26.

Kroppenstedt, R. M. 1985. Fatty acid and menaquinone analysis of actinomycetes and related organisms. In: M. Goodfellow and D. E. Minnikin (Eds.) Chemical Methods in Bacterial Systematics. Academic Press. London, UK. 173–199.

Kudo, T., K. Hatai, and A. Seino. 1988. Nocardia seriolae sp. nov. causing nocardiosis of cultured fish. Int. J. Syst. Bacteriol. 38:173–178.

Kulakov, L. A., and M. J. Larkin. 2002. Genetic organization of Rhodococcus. In: A. Danchin (Ed.) Genomics of GC-rich Gram-positive Bacteria. Caister Academic Press. Wymondham, UK. 15-64.

Kulakova, A. N., M. J. Larkin, and L. A. Kulakova. 1997. The plasmid-located haloalkane dehydrogenase gene from Rhodococcus rhodochrous NCIMB 13064. Microbiology 143:109–115.

Kummer, C., P. Schumann, and E. Stackebrandt. 1999. Gordonia alkanivorans sp. nov., isolated from tar contaminated soil. Int. J. Syst. Bacteriol. 49:1513–1522.

Kurane, R., T. Suzuki, and Y. Takahara. 1979. Removal of phthalate esters by activated sludge inoculated with a strain of Nocardia erythropolis. Agric. Biol. Chem. 43:421–427.

Kurane, R., T. Suzuki, and Y. Takahara. 1980. Metabolic pathway of phthalate esters by Nocardia erythropolis. Agric. Biol. Chem. 44:523–527.

Kurane, R., T. Suzuki, and S. Fukuoka. 1984. Purification and some properties of a phthalate ester hydrolyzing enzyme from Nocardia erythropolis. Appl. Microbiol. Biotechnol. 20:378–383.

Kurane, R., K. Hatamochi, T. Kakuno, M. Kiyohara, T. Tajima, M. Hirano, and Y. Tamiguchi. 1995. Chemical structure of lipid bioflocculant produced by Rhodococcus erythropolis. Biosci. Biotechnol. Biochem. 59:1652–1656.

Kurup, V. P., and G. H. Schribner. 1981. Antigenic relationship among Nocardia asteroides immunotypes. Microbios 31:25–30.

Labbé, D., J. Garnon, and P. C. K. Lau. 1997. Characterization of the genes encoding a receptor-like histidine kinase and a cognate response regulator from a biphenyl/polychlorobiphenyl-degrading bacterium, Rhdodococcus sp. strain M5. J. Bacteriol. 179:2772–2776.

Lacey, J., and M. Goodfellow. 1975. A novel actinomycete from sugar-cane bagasse, Saccharopolyspora hirsuta, gen. et sp. nov. J. Gen. Microbiol. 88:75–85.

Ladrón, N., M. Fernandez, J. Agüero, B. G. Zorn, J. A. Vazquez-Boland, and J. Navas. 2003. Rapid identification of Rhodococcus equi by a PCR assay targeting the choE gene. J. Clin. Microbiol. 41:3241–3245.

Lang, S., and J. C. Philp. 1998. Surface lipids in rhodococci. Ant. v. Leeuwenhoek 74:59–70.

Larkin, M. J., R. De Mot, L. A. Kulakov, and I. Nagy. 1998. Applied aspects of Rhodococcus genetics. Ant. v. Leeuwenhoek 74:133–153.

Larkin, J. A., L. Lit, J. Sinnott, T. Wills, and A. Szentwanyi. 1999. Infection of a knee prosthesis with Tsukamurella species. Southern Med. J. 92:831–832.

Lasker, B. A., J. M. Brown, and M. M. McNeil. 1992. Identification and epidemiological typing of clinical and environmental isolates of the genus Rhodococcus with use of a digoxigenin-labelled rDNA gene probe. Clin. Infect. Dis. 15:223–253.

Laurent, F., A. Carlotti, P. Boiron, J. Villard, and J. Freney. 1996. Ribotyping: A tool for the taxonomy and identification of the Nocardia asteroides complex species. J. Clin. Microbiol. 34:1079–1082.

Laurent, F. J., F. Provost, and B. Boiron. 1999. Rapid identification of clinically relevant Nocardia species to genus level by 16S rRNA gene PCR. J. Clin. Microbiol. 37:99–102.

Lazzarini, A., L. Cavaletti, G. Toppo, and F. Marinelli. 2000. Rare genera of actinomycetes as potential producers of new antibiotics. Ant. v. Leeuwenhoek 78:399–405.

Lechevalier, H. A., and M. P. Lechevalier. 1970a. Chemical composition as a criterion in the classification of aerobic actinomycetes. Int. J. Syst. Bacteriol. 20:435–443.

Lechevalier, M. P., and H. A. Lechevalier. 1970b. A critical evaluation of the genera of aerobic actinomycetes. In: H. Prauser (Ed.) The Actinomycetales. Gustav Fischer Verlag. Jena, Germany. 393–405.

Lechevalier, M. P., A. C. Horan, and H. A. Lechevalier. 1971. Lipid composition in the classification of nocardiae and mycobacteria. J. Bacteriol. 105:313–318.

Lechevalier, M. P., and H. A. Lechevalier. 1974. Nocardia amarae sp. nov., an actinomycete common in foaming activated sludge. Int. J. Syst. Bacteriol. 24:278–288.

Lechevalier, M. P., C. De Biévre, and H. A. Lechevalier. 1977a. Chemotaxonomy of aerobic actinomycetes: phospholipid composition. Biochem. Syst. Ecol. 5:249–260.

Lechevalier, H. A., M. P. Lechevalier, P. E. Wyszkowski, and F. Mariat. 1977b. Actinomycetes found in sewage-treatment plants of the activated sludge type. In: T. Arai (Ed.) Actinomycetes: The Boundary Micro-organisms. Toppan. Tokyo, Japan. 227–247.

Lechevalier, M. P., and H. A. Lechevalier. 1980. The chemotaxonomy of actinomycetes. In: A. Dietz and D. Thayer (Eds.) Actinomycete Taxonomy. Society for Industrial Microbiology. Arlington VA, Special Publication 6:227–291.

Lechevalier, M. P., A. E. Stern, and H. A. Lechevalier. 1981. Phospholipids in the taxonomy of actinomycetes. Zbl. Bakteriol. Suppl. 11:111–116.

Lemmer, H., and R. M. Kroppenstedt. 1984. Chemotaxonomy and physiology of some actinomycetes isolated from scumming activated sludge. Syst. Appl. Microbiol. 5:124–135.

Lerner, P. I. 1996. Nocardiosis. Clin. Infect. Dis. 22:891–905.

Lesens, O., Y. Hansman, P. Riegel, R. Heller, M. Benaissa-Djellouli, M. Martinot, H. Petit, and D. Christianson. 2000. Bacteremia and endocarditis caused by a Gordonia species in a patient with a central venous catheter. Emerg. Infect. Dis. 6:382–385.

Li, M. Z., C. H. Squires, D. J. Monticello, and J. D. Childs. 1996. Genetic analysis of the dsz promoter and associated regulatory regions of Rhodococcus erythropolis IGTS8. J. Bacteriol. 178:6409–6418.

Li, Y., Y. Kawamura, N. Fujiwara, T. Naka, H. Liu, X. Huang, K. Kubayashi, and T. Ezaki. 2004. Rothia aeria sp. nov., Rhodococcus baikonurensis sp. nov. and Arthrobacter russicus sp. nov., isolated from air in the Russian space laboratory Mir. Int. J. Syst. Evol. Microbiol. 54:827–835.

Liese, A., K. Seelbach, and C. Wandrey. 2000. Industrial Biotransformations. Wiley-VCH. Weinheim, Germany.

Lind, A., and M. Ridell. 1976. Serological relationships between Nocardia, Mycobacterium, Corynebacterium and the "rhodochrous" taxon. In: M. Goodfellow, G. H. Brownell, and J. A. Serrano (Eds.) The Biology of the Nocardiae. Academic Press. London, UK. 220–235.

Lind, A., Ö. Ouchterlony, and M. Ridell. 1980. Mycobacterial antigens. In: G. Meissner and Schimiedel (Eds.) Infektionskrankheiten und ihre Erreger, Band 4: Mycobakterien und mykobakterielle Krankheiten. Fischer Verlag. Jena, Germany. 275–303.

Lind, A., and M. Ridell. 1983. Immunological classification: Immunodiffusion and immunoelectrophoresis. In: G. Kubica and L. G. Wayne (Eds.) The Mycobacteria: A Sourcebook. Marcel Dekker. New York, NY. 67–82.

Linos, A., A. Steinbüchel, C. Spröer, and R. M. Kroppenstedt. 1999. Gordonia polyisoprenivorans sp. nov., a rubber-degrading actinomycete isolated from an automobile tyre. Int. J. Syst Bacteriol. 49:1785–1791.

Linos, A., M. M. Berekaa, R. Reichelt, U. Keller, J. Schmitt, H. C. Fleming, R. M. Kroppenstedt, and A. Steinbüchel. 2000. Biodegradation of cis- 1, 4-polyisoprene rubbers by distinct actinomycetes: Microbial strategies and detailed surface analysis. Appl. Environ. Microbiol. 66:1639–1645.

Linos, A., M. M. Berekaa, A. Steinbüchel, K. K. Kim, C. Sproer, and R. M. Kroppenstedt. 2002. Gordonia westfalica sp. nov., a novel rubber degrading actinomycete. Int. J. Syst. Evol. Microbiol. 52:1133–1139.

Louie, L., M. Louie, and A. E. Simor. 1997. Investigation of a pseudo-outbreak of Nocardia asteroides infection by pulsed-field gel electrophoresis and randomly amplified polymorphic DNA PCR. J. Clin. Microbiol. 35:1582–1584.

Lungu, O., P. D. Latta, I. Weitzman, and S. Silverstein. 1994. Differentiation of Nocardia from rapidly growing Mycobacterium species by PCR-RFLP analysis. Diagn. Microbiol. Infect. Dis. 18:13–18.

Maertins, J., P. Wattiau, J. Verhaegen, M. Boogaerts, L. Verbist, and G. Waut. 1998. Catheter-related bacteremia due to Tsukamurella pulmonis. Clin. Microbiol. Infect. 4:51–53.

Magnusson, H. 1923. Spezifische infektioese pneumonie bien Fohlen: Eis neuer Eitererregen beun Pferd. Arch. Wiss. Prakt. Tierheilkde. 50:22–28.

Magnusson, M. 1976. Sensitin tests in Nocardia taxonomy. In: M. Goodfellow, G. H. Brownell, and J. A. Serrano (Eds.) The Biology of the Nocardiae. Academic Press. London, UK. 236–265.

Mahgoub, E. S., and I. G. Murray. 1973. Mycetoma. Heinemann. London, UK.

Maldonado, L., J. V. Hookey, A. C. Ward, and M. Goodfellow. 2000. The Nocardia salmonicida clade, including descriptions of Nocardia cummideless sp. nov., Nocardia fluminea sp. nov. and Nocardia soli sp. nov. Ant. v. Leeuwenhoek 78:367–377.

Maldonado, L. A., F. M. Stainsby, A. C. Ward, and M. Goodfellow. 2003. Gordonia sinesedis sp. nov., a novel soil isolate. Ant. v. Leeuwenhoek. 83:75–80.

Manninen, K. I., R. A. Smith, and L. O. Kim. 1993. Highly presumptive identification of bacterial isolates associated with the recent Canada-wide mastitis epizootic as Nocardia farcinica. Can. J. Microbiol. 39:635–641.

Mara, D. D., and J. I. Oragui. 1981. Occurrence of Rhodococcus coprophilus and associated actinomycetes in faeces, sewage, and fresh water. Appl. Environ. Microbiol. 42:1037–1042.

Martens, R. J., R. A. Fiske, and H. W. Renshaw. 1982. Experimental subacute foal pneumonia inducible by aerosol administration of Corynebacterium equi. Equine Vet. 14:111–116.

Martin, J., and D. J. Hogan. 1991. Rhodococcus infection of the skin with lymphadenitis in a nonimmunocompromised girl. J. Am. Acad. Dermatol. 24:328–332.

Masai, E., A. Yamada, J. M. Healey, T. Hatta, K. Kimbara, M. Fukuda, and Y. Yano. 1995. Characterisation of biphenyl catabolic genes of Gram-positive polychlorinated biphenyl degrader Rhodococcus strain RHA1. Appl. Environ. Microbiol. 61:2079–2085.

Masai, E., K. Sugiyama, N. Iwashita, S. Shimizu, J. E. Hauschild, T. Hatta, K. Kimbara, K. Yano, and M. Fukuda. 1997. The bph DEF metacleavage pathway genes involved in biphenyl/polychlorinated biphenyl degradation are located on a linear plasmid and separated from the initial bph ABC genes in Rhodococcus sp. strain RHA1. Gene 187:141–149.

Matsushima, P., M. A. McHenney, and R. H. Baltz. 1987. Efficient transformation of Amycolatopsis orientalis (Nocardia orientalis) protoplasts by Streptomyces plasmids. J. Bacteriol. 169:2298–2300.

Matsuyama, H., I. Yumoto, T. Kudo, and O. Shida. 2004. Rhodococcus tukisamuensis sp. nov., isolated from soil. Int. J. Syst. Microbiol. 53(Pt 5):1333–1337.

Maurice, M. T., M. J. Vacheron, and G. Michel. 1971. Isolément d'acides nocardiques de plusieurs espèces de Nocardia. Chem. Phys. Lipids 7:9–18.

McDonald, C. R., D. G. Cooper, and J. Zajic. 1981. Surface active lipids from Nocardia erythropolis grown on hydrocarbons. Appl. Environ. Microbiol. 41:117–123.

McFarland, B. L., D. J. Boron, W. Deever, J. A. Meyer, A. R. Johnson, and R. M. Atlas. 1998. Biocatalytic sulphur-removal from fuels: applicability for producing low-sulphur gaseline. Crit. Rev. Microbiol. 24:99–147.

McLeod, T., P. Neil, and V. J. Robertson. 1989. Pulmonary diseases in Zimbabwe, Central Africa. Trans. R. Soc. Trop. Med. Hyg. 83:694–697.

McMinn, E. J., G. Alderson, H. I. Dodson, M. Goodfellow, and A. C. Ward. 2000. Genomic and phenomic differentiation of Rhodococcus equi and related strains. Ant. v. Leeuwenhoek 78:331–340.

McNeil, M., and J. Brown. 1994. The medically important actinomycetes: Epidemiology and microbiology. Clin. Microbiol. Rev. 7:357–417.

McNeill, M. M., S. Ray, P. E. Kozarsky, and J. M. Brown. 1997. Nocardia farcinica pneumonia in a previously healthy woman: species characterization with use of a digoxigenen-labelled cDNA probe. Infect. Dis. 25:933–934.

Minnikin, D. E., L. Alshamaony, and M. Goodfellow. 1975. Differentiation of Mycobacterium, Nocardia and related taxa by thin-layer chromatographic analysis of whole-organism methanolysates. J. Gen. Microbiol. 88:200–204.

Minnikin, D. E., P. V. Patel, L. Alshamaony, and M. Goodfellow. 1977. Polar lipid composition in the classification of Nocardia and related bacteria. Int. J. Syst. Bacteriol. 27:104–117.

Minnikin, D. E., and M. Goodfellow. 1980a. Lipid composition in the classification and identification of acid-fast bacteria. In: M. Goodfellow and R. G. Board (Eds.) Microbiological Classification and Identification. Academic Press. London, UK. 189–256.

Minnikin, D. E., I. G. Hutchinson, A. B. Caldicott, and M. Goodfellow. 1980b. Thin-layer chromatography of methanolysates of mycolic acid-containing bacteria. J. Chromatogr. 188:221–233.

Minnikin, D. E. 1982. Lipids: Complex lipids, their chemistry, biosynthesis and roles. In: C. Ratledge and J. L. Stanford (Eds.) The Biology of the Mycobacteria. Academic Press. London, UK. 1:95–184.

Minnikin, D. E., S. M. Minnikin, I. G. Hutchinson, M. Goodfellow, and J. M. Grange. 1984a. Mycolic acid patterns of representative strains of Mycobacterium fortuitum, "Mycobacterium peregrinum" and Mycobacterium smegmatis. J. Gen. Microbiol. 130:363–367.

Minnikin, D. E., S. M. Minnikin, A. G. O'Donnell, and M. Goodfellow. 1984b. Extraction of mycobacterial mycolic acids and other long-chain compounds by an alkaline methanolysis procedure. J. Microbiol. Meth. 2:243–249.

Minnikin, D. E., S. M. Minnikin, J. H. Parlett, M. Goodfellow, and M. Magnusson. 1984c. Mycolic acid patterns of some species of Mycobacterium. Arch. Microbiol. 139:225–231.

Minnikin, D. E., A. G. O'Donnell, M. Goodfellow, G. Alderson, M. Athalye, A. Schaal, and J. H. Parlett. 1984d. An integrated procedure for the extraction of isoprenoid quinones and polar lipids. J. Microbiol. Meth. 2:233–241.

Minnikin, D. E., G. Dobson, M. Goodfellow, P. Draper, and M. Magnusson. 1985a. Quantitative comparison of the mycolic and fatty acid composition of Mycobacterium leprae and Mycobacterium gordonae. J. Gen. Microbiol. 131:2013–2021.

Minnikin, D. E., S. M. Minnikin, J. H. Parlett, and M. Goodfellow. 1985b. Mycolic acid patterns of some rapidly-growing species of Mycobacterium. Zbl. Bakteriol. Hyg. A259:446–460.

Minnikin, D. E. 1993. Mycolic acids. In: K. D. Mukherjee and N. Weber (Eds.) CRC Handbook of Chromatography. CRC Press. Boca Raton, FL. 339–347.

Mirza, S. H., and C. Campbell. 1994. Mycetoma caused by Nocardia transvalensis. J. Clin. Pathol. 47:85–86.

Monciardini, P., M. Sasio, L. Cavaletti, C. Chiocchini, and S. Donadio. 2002. New PCR primers for the selective amplification of 16S rDNA from different groups of actinomycetes. FEMS Microbiol. Ecol. 42:419–429.

Mordarska, H., M. Mordarski, and M. Goodfellow. 1972. Chemotaxonomic characters and classification of some nocardioform bacteria. J. Gen. Microbiol. 71: 77–86.

Mordarski, M., K. P. Schaal, K. Szyba, G. Pulverer, and A. Tkacz. 1977. Interrelation of Nocardia asteroides and related taxa as indicated by deoxyribonucleic acid reassociation. Int. J. Syst. Bacteriol. 27:66–70.

Mordarski, M., K. P. Schaal, A. Tkacz, G. Pulverer, K. Szyba, and M. Goodfellow. 1978. Deoxyribonucleic acid base composition and homology studies on Nocardia. Zbl. Bakteriol. Suppl. 6:91–97.

Mordarski, M., M. Goodfellow, K. Szyba, A. Tkacz, G. Pulverer, and K. P. Schaal. 1980a. Deoxyribonucleic acid reassociation in the classification of the genus Rhodococcus. Int. J. Syst. Bacteriol. 30:521–527.

Mordarski, M., M. Goodfellow, A. Tkacz, G. Pulverer, and K. P. Schaal. 1980b. Ribosomal ribonucleic acid similarities in the classification of Rhodococcus and related taxa. J. Gen. Microbiol. 118:313–319.

Morton, A. C., A. P. Begg, G. A. Anderson, S. Takai, C. Lämmler, and G. F. Browning. 2001. Epidemiology of Rhodococcus equi strains on thoroughbred horse farms. Appl. Environ. Microbiol. 67:2167–2175.

Mosser, D. M., and M. K. Hondulas. 1996. Rhodococcus equi: An emerging opportunistic pathogen. Trends Microbiol. 4:29–33.

Murray, P. R., R. L. Heeren, and A. C. Niles. 1987. Effect of decontamination procedures on recovery of Nocardia spp. J. Clin. Microbiol. 25:2010–2011.

Mutimer, M. D., and J. B. Woolcock. 1980. Corynebacterium equi in cattle and pigs. Vet. Q. 2:25–27.

Mutimer, M. D., and J. B. Woolcock. 1982. APIZYM for identification of Corynebacterium equi. Zbl. Bakteriol. Hyg. I. Abt. Orig. C3:410–415.

Nagasawa, T., M. Kobayashi, and H. Yamada. 1988a. Optimum culture conditions for the production of benzonitrilase by Rhodococcus rhodochrous J1. Arch. Microbiol. 150:89–94.

Nagasawa, T., K. Takeuchi, and H. Yamada. 1988b. Occurrence of a cobalt-induced and cobalt-containing nitrile hydratase in Rhodococcus rhodochrous J1. Biochem. Biophys. Res. Commun. 155:1008–1016.

Nagasawa, T., and H. Yamada. 1989. Microbial transformations of nitriles. TIBTECH 7:153–158.

Nagasawa, T., and H. Yamada. 1990. Large-scale bioconversion of nitriles into useful amides and acids. In: D. A. Abramowicz (Ed.) Biocatalysis. Van Nostrand Reinhold. New York, NY. 277–318.

Nagy, I., F. Campernolle, K. Ghys, J. Vanderleiden, and R. de Mot. 1995. A single cytochrome P-450 system is involved in degradation of the herbicide EPTC and altrazine by Rhodococcus sp. N186/21. Appl. Environ. Microbiol. 61:2056–2060.

Nam, S.-W., J. Chun, S. Kim, W. Kim, J. Zakrzewska-Czerwinska, and M. Goodfellow. 2003. Tsukamurella spumae sp. nov., a novel actinomycete associated with foaming in activated sludge plants. Syst. Appl. Microbiol. 26:367–375.

Nam, S.-W. 2004a. Tsukamurella Systematics Revisited [PhD thesis]. University of Newcastle. Newcastle upon Tyne, UK.

Nam, S.-W., W. Kim, J. Chun, and M. Goodfellow. 2004b. Tsukamurella pseudospumae sp. nov., a novel actinomycete from activated sludge foam. Int. J. Syst. Evol. Microbiol. 54:1209–1212.

Nesterenko, O. A., S. A. Kasumova, and S. I. Kvasnikov. 1978a. Microorganisms of the genus Nocardia and the "rhodochrous" group in the soils of the Ukranian S.S.R. USSR Microbiology 47:699–703.

Nesterenko, O. A., S. A. Kasumova, and S. I. Kvasnikov. 1978b. Microorganisms of the Nocardia genus and the "rhodochrous" group in soils of the Ukranian S.S.R. Mikrobiologija 47:866–870.

Nesterenko, O. A., T. M. Nogina, S. A. Kasumova, E. I. Kvasnikov, and S. G. Batrakov. 1982. Rhodococcus luteus nom. nov. and Rhodococcus maris nom. nov. Int. J. Syst. Bacteriol. 32:1–14.

Nishiuchi, Y., T. Baba, H. H. Hotta, and I. Yano. 1999. Mycolic acid analysis in Nocardia species: The mycolic acid compositions of Nocardia asteroides, N. farcinica and N. nova. J. Microbiol. Meth. 37:111–122.

Nishiuchi, Y., T. Baba, and I. Yano. 2000. Mycolic acids from Rhodococcus, Gordonia, and Dietzia. J. Microbiol. Meth. 40:1–9.

O'Donnell, A. G., D. E. Minnikin, and M. Goodfellow. 1985. Integrated lipid and wall analysis of actinomycetes. In: M. Goodfellow and D. E. Minnikin (Eds.) Chemical Methods in Bacterial Systematics. Academic Press. London, UK. 131–143.

Okami, Y., and K. Hotta. 1988. Search and discovery of new antibiotics. In: M. Goodfellow, S. T. Williams, and M. Mordarski (Eds.) Actinomycete Biology. Academic Press. London, UK. 33–67.

Oldfield, C., O. Pogrebinsky, J. Simmonds, E. Olson, and C. F. Kulpa. 1997. Elucidation of the metabolic pathway for dibenzothiophene desulphurisation by Rhodococcus sp. strain IGTS8. Microbiology 163:2961–2973.

Oldfield, C., N. T. Wood, S. C. Gilbert, F. D. Murray, and F. R. Faure. 1998. Desulphurisation of benzothiophene and dibenzothiophene by actinomycete organisms belonging to the genus Rhodococcus and related taxa. Ant. v. Leeuwenhoek 74:119–132.

Oldfield, C., H. Bonella, L. Renwick, H. I. Dodson, G. Alderson, and M. Goodfellow. 2004. Rapid determination of vapA/vapB genotype in Rhodococcus equi using a differential polymerase chain reaction method. Ant. v. Leeuwenhoek 85:317-326.

Oragui, J. L., and D. D. Mara. 1985. Fecal streptococci, Rhodococcus coprophilus and bifidobacteria as specific indicator organisms of fecal pollution. J. Appl. Bacteriol. 59:v–vi.

Orchard, V. A., and M. Goodfellow. 1974. The selective isolation of Nocardia from soil using antibiotics. J. Gen. Microbiol. 85:160–162.

Orchard, V. A., M. Goodfellow, and S. T. Williams. 1977. Selective isolation and occurrence of nocardiae in soil. Soil Biol. Biochem. 9:233–238.

Orchard, V. A., and M. Goodfellow. 1980. Numerical classification of some named strains of Nocardia asteroides and related isolates from soil. J. Gen. Microbiol. 118:295–312.

Ortiz-Ortiz, L., E. I. Melendro, and C. Conde. 1984. Host-parasite relationship in infections due to Nocardia brasiliensis. In: L. Ortiz-Ortiz, L. F. Bojalil, and V. Yakoleff

(Eds.) Biological, Biochemical and Biomedical Aspects of Actinomycetes. Academic Press. Orlando, FL. 119–133.

Osoagbaka, O. U., and A. N. U. Njoku-Obianu. 1985. Nocardiosis in pulmonary diseases in parts of Nigeria. I: Preliminary observations of five cases. J. Trop. Med. Hyg. 88:367–377.

Pagilla, K. R., A. Sood, and H. Kim. 2001. Gordona (Nocardia) amarae foaming due to biosurfactant production. In: V. Tandoi, R. Passino, and C. M. Blundo (Eds.) Proceedings of the IWA International Specialised Conference on Microorganisms in Activated Sludge and Bioprocesses, Rome, Italy. 402–408.

Patel, J. B., R. J. Wallace Jr., B. A. Brown-Elliott, T. Taylor, C. Imperatrice, D. G. B. Leonard, R. W. Wilson, L. Mann, K. C. Just, and I. Nachamkin. 2004. Sequence based identification of aerobic actinomycetes. J. Clin. Microbiol. 42:2530–2540.

Paterson, J. E., K. Chapin-Robertson, S. Waycott, P. Farrel, A. McGeer, M. M. McNeil, and S. C. Edberg. 1992. Pseudoepidemic of Nocardia asteroides associated with a mycobacterial culture system. J. Clin. Microbiol. 30:1357–1360.

Peczynska-Czoch, W., and M. Mordarski. 1988. Actinomycete enzymes. In: M. Goodfellow, S. T. Williams, and M. Mordarski (Eds.) Actinomycetes in Biotechnology. Academic Press. London, UK. 219–283.

Pham, A. S., I. De, K. V. Rolston, J. J. Rarrand, and X. Y. Han. 2003. Catheter-related bacteremia caused by a nocardioform actinomycete Gordonia terrae. Clin. Infect. Dis. 36:524–527.

Piddington, C. S., B. R. Kovacevich, and J. Rambosek. 1995. Sequence and molecular characterization of a DNA region encoding the dibenzothiophene desulpurization operon of Rhodococcus sp. strain IGTS8. Appl. Environ. Microbiol. 61:468–475.

Pidoux, O., J.-N. Argenson, V. Jacomo, and M. Drancourt. 2001. Molecular identification of a Dietzia maris hip prosthesis infection isolate. J. Clin. Microbiol. 39:2634–2636.

Pier, A. C., J. R. Thurston Jr., and A. B. Larson. 1968. A diagnostic antigen for nocardiosis: comparative tests in cattle with nocardiosis and mycobacteriosis. Am. J. Vet. Res. 29:397–403.

Pier, A. C., and R. E. Fichtner. 1971. Serologic typing of Nocardia asteroides by immunodiffusion. Am. Rev. Respir. Dis. 103:698–707.

Pier, A. C., and R. E. Fichtner. 1981. Distribution of serotypes of Nocardia asteroides from animal, human and environmental sources. J. Clin. Microbiol. 13:548–553.

Pier, A. C. 1984. Serological relationships among aerobic actinomycetes in human and animal diseases. In: L. Ortiz-Ortiz, L. F. Bojalil, and V. Yakoloeff (Eds.) Biological, Biochemical and Biomedical Aspects of Actinomycetes. Academic Press. Orlando, FL. 135-142.

Pisabarro, A., A. Correia, and J. F. Martin. 1998. Pulsed-field gel electrophoresis analysis of the genome of Rhodococcus fascians: genome size and linear and circular replicon composition in virulent and avirulent strains. Curr. Microbiol. 36:302–308.

Pommier, M. T., and G. Michel. 1985. Occurrence of corynemycolic acids in strains of Nocardia otitidiscaviarum. J. Gen. Microbiol. 131:2637–2641.

Poonwan, N., M. Kusum, Y. Mikami, K. Yazawa, Y. Tanaka, T. Gonoi, T. Hasegawa, and K. Kanyama. 1995. Pathogenic Nocardia isolated from clinical specimens including those of AIDS patients in Thailand. Eur. J. Epidemiol. 11:507–512.

Portaels, F. 1976. Isolation and distribution of nocardiae in the Bas-Zaire. Ann. Soc. Belge Méd. Trop. 56:73–83.

Prauser, H. 1976. Host-phage relationships in nocardioform organisms. In: M. Goodfellow, G. H. Brownell, and J. A. Serrano (Eds.) The Biology of the Nocardiae. Academic Press. London, UK. 266–284.

Prauser, H. 1981a. Nocardioform organisms: General characterisation and taxonomic relationships. Zbl. Bakteriol. Suppl. 11:17–24.

Prauser, H. 1981b. Taxon specificity of lytic actinophages that do not multiply in the cells affected. Zbl. Bakteriol. Suppl. 11:87–92.

Prescott, J. F. 1991. Rhodococcus equi: an animal and human pathogen. Clin. Microbiol. Rev. 4:20–34.

Prinz, G., E. Bán, S. Fekete, and Z. Szabó. 1985. Meningitis caused by Gordona aurantiaca (Rhodococcus aurantiacus). J. Clin. Microbiol. 22:472–474.

Pujic, P., and B. L. Beaman. 2001. Actinomyces and Nocardia. In: M. Sussman (Ed.) Molecular Medical Microbiology. Academic Press. Washington, DC. 937-960.

Pulverer, G., H. Schütt-Gerowitt, and K. P. Schaal. 1975. Bacteriophages of Nocardia asteroides. Med. Microbiol. Immunol. 161:113–122.

Pulverer, G. H., and K. P. Schaal. 1978. Pathogenicity and medical importance of aerobic and anaerobic actinomycetes. Zbl. Bakteriol. Suppl. 6:417–427.

Qureshi, N., K. Takayama, H. C. Jordi, and H. K. Schnoes. 1978. Characterisation of the purified components of a new homologous series of α-mycolic acids from Mycobacterium tuberculosis H37 Ra. J. Biol. Chem. 253:5411–5417.

Rainey, F. A., J. Burghardt, R. M. Kroppenstedt, S. Klatte, and E. Stackebrandt. 1995a. Phylogenetic analysis of the genera Rhodococcus and Nocardia and evidence for the evolutionary origins of the genus Nocardia from within the radiation of Rhodococcus species. Microbiology 141:523–528.

Rainey, F. A., S. Klatte, R. M. Kroppenstedt, and E. Stackebrandt. 1995b. Dietzia, a new genus including Dietzia maris comb. nov., formerly Rhodococcus maris. Int. J. Syst. Bacteriol. 45:32–36.

Ramos, L. S. 1994. Characterization of mycobacteria species by HPLC and pattern recognition. J. Chromatogr. Sci. 32:219–227.

Rast, H. G., G. Engelhardt, W. Diegler, and P. R. Wallhoffer. 1980. Bacterial degradation of model compounds for lignin and chlorophenol derived lignin bound residues. FEMS Microbiol. Lett. 8:259–263.

Raymond, R. L., V. W. Jamison, and J. O. Hudson. 1971. Hydrocarbon cooxidation in microbial systems. Lipids 6:453–457.

Reh, M. 1981a. Chemolithoautotrophy as an autonomous and transferable property of Nocardia opaca lb. Zbl. Bakteriol. Suppl. 11:577–583.

Reh, M., and H. G. Schlegel. 1981b. Hydrogen autotrophy as a transferable genetic character of Nocardia opaca lb. J. Gen. Microbiol. 126:327–336.

Rhee, S. K., J. H. Chang, Y. K. Chang, and H. O. Chang. 1998. Desulfurization of dibenzothiophene and diesel oils by newly isolated Gordonia strain CYKS1. Appl. Environ. Microbiol. 64:2327–2332.

Rhodes, M. W., H. Kator, S. Kotob, P. van Berkum, I. Kaattari, W. Vogelbein, F. Quinn, M. M. Floyd, W. R. Butler, and A. Ottinger. 2003. Mycobacterium shottsii sp. nov., a slow-growing species isolated from Chesapeake Bay striped bass (Morone saxatilis). Int. J. Syst. Evol. Microbiol. 53:421–424.

Richet, H., P. C. Craven, J. M. Brown, B. A. Lasker, C. D. Cox, M. M. McNeil, A. D. Tice, W. R. Jarvis, and O. C. Tablan. 1991. A cluster of Rhodococcus (Gordona) bronchialis sternal-wound infections after coronary—artery bypass surgery. N. Engl. J. Med. 324:104–109.

Ridell, M., and M. Norlin. 1973. Serological study of Nocardia by using mycobacterial precipitation reference systems. J. Bacteriol. 113:1–7.

Ridell, M. 1977. Studies on corynebacterial precipitinogens common to mycobacteria, nocardiae and rhodococci. Int. Arch. Allergy Appl. Immunol. 55:468–475.

Ridell, M. 1981. Immunodiffusion studies of some Nocardia strains. J. Gen. Microbiol. 123:69–74.

Ridell, M. 1983. Sensitivity to capreomycin and prothionamide in strains of Mycobacterium, Nocardia, Rhodococcus and related taxa for taxonomical purposes. Zbl. Bakteriol. Hyg. A255:309–316.

Ristau, E., and F. Wagner. 1983. Formation of novel anionic trehalose tetraesters from Rhodococcus erythropolis under growth-limiting conditions. Biotechnol. Lett. 5:95–100.

Roberts, G. D., E. W. Koneman, and Y. K. Kim. 1991. Mycobacterium. In: A. Balows, W. J. Hausler Jr., K. L. Herrmann, H. D. Isenberg, and H. J. Shadomy (Eds.) Manual of Clinical Microbiology, 4th ed. American Society for Microbiology. Washington, DC. 304–309.

Roth, A., S. Andrees, R. M. Kroppenstedt, D. Harmsen, and H. Mauch. 2003. Phylogeny of the genus Nocardia based on reassessed 16S rRNA gene sequences reveals underspeciation and division of strains classified as Nocardia asteroides into three established species and two unnamed taxons. J. Clin. Microbiol. 41:851–856.

Rowbotham, T. J., and T. Cross. 1977a. Ecology of Rhodococcus coprophilus and associated actinomycetes in freshwater and agricultural habitats. J. Gen. Microbiol. 100:231–240.

Rowbotham, T. J., and T. Cross. 1977b. Rhodococcus coprophilus sp. nov.: An aerobic nocardioform actinomycete belonging to the "rhodochrous" complex. J. Gen. Microbiol. 100:123–138.

Ruimy, R., P. Riegel, A. Carlotti, P. Boiron, G. Bernardine, H. Monteil, R. J. Wallace Jr., and R. Christen. 1996. Nocardia pseudobrasiliensis sp. nov., a new species of Nocardia which groups bacterial strains previously identified as Nocardia brasiliensis and associated with invasive disease. Int. J. Syst. Bacteriol. 46:259–264.

Saarinen, K. A., G. G. Lestringant, J. Czechowski, and P. M. Frossard. 2001. Cutaneous nocardiosis of the chest wall and pleura—10 year consequences of a hand mycetoma. Dermatology 202:131–133.

Sahathevan, M., F. A. H. Harvey, G. Forbes, J. O'Grady, A. Gimson, S. Bragman, R. Jensen, J. Philpott-Howard, R. Williams, and M. W. Casewell. 1991. Epidemiology, bacteriology and control of an outbreak of Nocardia asteroides infection in a liver unit. J. Hospt. Infect., Suppl. A. 18:472–480.

Saintpierre-Bonaccio, D., L. A. Maldonado, H. Amir, R. Pineau, and M. Goodfellow. 2004. Nocardia neocaledoniensis sp. nov., a novel actinomycete isolated from a New-Caledonian brown hypermagnesian ultramafic soil. Int. J. Syst. Evol. Microbiol. 54:599–603.

Salanitro, J. P., L. A. Diaz, M. P. Williams, and H. L. Wisniewski. 1994. Isolation of a bacterial culture that degrades a methyl-t-butyl ether. Appl. Environ. Microbiol. 60:2593–2596.

Salinas-Carmona, M. C., O. Welsh, and S. M. Casillas. 1993. Enzyme-linked immunosorbent assay for serological diagnosis of Nocardia brasiliensis and clinical correlation with mycetoma infections. J. Clin. Microbiol. 31:2901–2906.

Salinas-Carmona, M. C. 2001. Anti-Nocardia brasiliensis antibodies in patients with actinomycetoma and their clinical usefulness [in Spanish]. Gaceta. Med. Mex. 137:1–8.

Salman, M. D., R. B. Bushnell, and A. C. Pier. 1982. Determination of sensitivity and specificity of the Nocardia asteroides skin test for detection of bovine mammary infections caused by Nocardia asteroides and Nocardia caviae. Am. J. Vet. Res. 43:332–335.

Schaal, K. P. 1972. Zur mikrobiologischen Diagnostik der Nocardiose. Zbl. Bakteriol. Parasitkde. Abt. I, Orig. Reihe A 220:242–246.

Schaal, K. P., and H. Heimerzheim. 1974. Mikrobiologische Diagnose und Therapie der Lungennocardiose. Mykosen 17:313–319.

Schaal, K. P. 1977. Nocardia, Actinomadura and Streptomyces. In: A. von Graevenitz (Ed.) CRC Handbook Series in Chemical Laboratory Sciences, Section E., Volume 1: Chemical Microbiology. CRC Press. Cleveland, OH.

Schaal, K. P., and H. Bickenbach. 1978a. Soil occurrence of pathogenic nocardiae. Zbl. Bakteriol. Suppl. 6:429–434.

Schaal, K. P., and H. Reutersberg. 1978b. Numerical taxonomy of Nocardia asteroides. Zbl. Bakteriol. Suppl. 6:53–62.

Schaal, K. P., and B. L. Beaman. 1984. Chemical significance of actinomycetes. In: M. Goodfellow, M. Mordarski, and S. T. Williams (Eds.) The Biology of Actinomycetes. Academic Press. London, UK. 389–424.

Schaal, K. P., and H. J. Lee. 1992. Actinomycete infections in humans—a review. Gene 115:201–211.

Schaal, K. P. 1997. Actinomycoses, Actinobacillosis and related diseases. In: W. Hausler and M. Lursman (Eds.) Topley and Wilson's Microbiology and Microbial Infections 9th ed. Edward Arnold. London, UK. 3.

Schleifer, K.-H., and O. Kandler. 1972. Peptidoglycan types of bacterial cell walls and their taxonomic implications. Bacteriol. Rev. 36:407–477.

Schleifer, K.-H., and E. Stackebrandt. 1983. Molecular systematics of prokaryotes. Ann. Rev. Microbiol. 37:143–187.

Schleifer, K.-H., and P. H. Seidl. 1985. Chemical composition of structure of murein. In: M. Goodfellow and D. E. Minnikin (Eds.) Chemical Methods in Bacterial Systematics. Academic Press. London, UK. 201–209.

Schreiner, A., K. Fuchs, F. Lottspeich, H. Poth, and F. Lingens. 1991. Degradation of 2-methylaniline in Rhodococcus-rhodochrous-cloning and expression of 2 clustered catechol 2, 3-dioxygenase genes from strain CTM. J. Gen. Microbiol. 137:2041–2048.

Schupp, T., R. Hütter, and D. A. Hopwood. 1975. Genetic recombination in Nocardia mediterranei. J. Bacteriol. 121:128–135.

Schwartz, M. A., S. R. Tabet, A. C. Collier, C. K. Wallis, L. C. Carlson, T. T. Nguyen, M. M. Kattar, and M. B. Coyle. 2002. Central venous catheter-related bacteremia due to Tsukamurella species in immunocompromised host: a case series and review of the literature. Clin. Infect. Dis. 35:72–77.

Scott, M. A., B. S. Graham, R. Verall, R. Dixon, W. Schaffner, and K. T. Thom. 1995. Rhodococcus equi—an increasingly recognised opportunistic pathogen: Report of 12 cases and review of 65 cases in the literature. Am. J. Clin. Pathol. 103:649–655.

Sellon, D. C., T. E. Besser, S. L. Vivrette, and R. S. McConnico. 2001. Comparison of nucleic acid amplification, serology, and microbiologic culture for diagnosis of Rhodococcus equi pneumonia in foals. J. Clin. Microbiol. 39:1289–1293.

Seong, C. N., Y. S. Kim, K. S. Baik, S. D. Lee, Y. C. Hah, S. B. Kim, and M. Goodfellow. 1999. Mycolic acid-containing actinomycetes associated with activated sludge foam. J. Microbiol. 37:66–72.

Seong, C. N., Y. S. Kim, K. S. Baik, S. K. Choi, M. B. Kim, S. B. Kim, and M. Goodfellow. 2003. Tsukamurella sunchonensis sp. nov., a bacterium associated with foam in activated sludge. J. Microbiol. 41:83–88.

Sezgin, M., M. P. Lechevalier, and P. R. Karr. 1988. Isolation and identification of actinomycetes present in activated sludge scum. Water Sci. Technol. 20:257–263.

Shainhaus, J. Z., A. C. Pier, and D. A. Stevens. 1978. Complement fixation antibody test for human nocardiosis. J. Clin. Microbiol. 8:516–519.

Shapiro, C. L., R. H. Haft, N. M. Gantz, G. V. Doern, J. C. Christenson, R. O'Brien, J. C. Overall, B. A. Brown, and R. J. Wallace Jr. 1992. Tsukamurella paurometabolum, a novel pathogen causing catheter-related bacteremia in patients with cancer. Clin. Infect. Dis. 14:200–203.

Shawar, R. M., D. G. Moore, and M. T. Larocco. 1990. Cultivation of Nocardia spp. on chemically defined media for selective recovery of isolates from clinical specimens. J. Clin. Microbiol. 28:508–512.

Sheridan, E. A. S., S. Warwick, A. Chan, M. D. Antonia, M. Koliou, and A. Sefton. 2003. Tsukamurella tyrosinosolvens intravascular catheter infection identified using 16S ribosomal DNA sequencing. Clin. Infect. Dis. 36:69–70.

Simpson, G. L., E. B. Stinson, E. J. Septimus, and E. Rosenblum. 1981. Nocardial infections in the immunocompromised host, a detailed study in a defined population. Rev. Infect. Dis. 3:492–507.

Singer, M. E. V., and W. R. Finnerty. 1988. Construction of an Escherichia coli-Rhodococcus shuttle vector and plasmid transformation in Rhodococcus species. J. Bacteriol. 170:638–645.

Skerman, V. B. D. 1968. A new type of micromanipulator and microforge. J. Gen. Microbiol. 54:287–297.

Sng, L.-H., T. H. Koh, R. S. Toney, M. Floyd, W. R. Butler, and B. H. Tan. 2004. Bacteremia caused by Gordonia bronchialis in a patient with sequestrated lung. J. Clin. Microbiol. 42:2870–2871.

Soddell, J. A., and R. J. Seviour. 1990. Microbiology of foaming in activated sludge plants. J. Appl. Bacteriol. 69:145–176.

Soddell, J. A., G. Knight, W. Strachan, and R. J. Seviour. 1992. Nocardioforms, not Nocardia foams. Water Sci. Technol. 26:455–460.

Soddell, J. A., and R. J. Seviour. 1994. Incidence and morphological variability of Nocardia pinensis in Australian activated sludge plants. Water Res. 28:2343–2351.

Soddell, J. A., and R. J. Seviour. 1998. Numerical taxonomy of Skermania piniformis and related isolates from activated sludge. J. Appl. Microbiol. 84:272–284.

Söhngen, N. L. 1913. Benzin, Petroleum, Paraffinöl und Paraffin als Kohlenstoff und Energiequelle für Mikroben. Zbl. Bakt. Parasitkde., Abt. 2 37:595–609.

Stach, J. E. M., L. A. Maldonado, A. C. Ward, M. Goodfellow, and A. T. Bull. 2003. New primers for the class Actinobacteria: application tomarine and terrestrial environments. Environ. Microbiol. 5:828–841.

Stach, J. E. M., L. A. Maldonado, A. C. Ward, A. T. Bull, and M. Goodfellow. 2004. Williamsia maris sp. nov., a novel actinomycete isolated from the Sea of Japan. Int. J. Syst. Evol. Microbiol. 54:191–194.

Stackebrandt, E., and C. R. Woese. 1981. Towards a phylogeny of the actinomycetes and related organisms. Curr. Microbiol. 5:197–202.

Stackebrandt, E., W. Ludwik, E. Seewaldt, and K.-H. Schleifer. 1983. Phylogeny of sporeforming members of the order Actinomycetales. Int. J. Syst. Bacteriol. 33:173–180.

Stackebrandt, E., J. Smida, and M. D. Collins. 1988. Evidence of phylogenetic heterogeneity within the genus Rhodococcus: Revival of the genus Gordona (Tsukamura). J. Gen. Appl. Microbiol. 34:341–348.

Stackebrandt, E., F. A. Rainey, and N. L. Ward-Rainey. 1997. Proposal for a new hierarchic classification system, Actinobacteria classis nov. Int. J. Syst. Bacteriol. 47:479–491.

Stainsby, F. M., J. Soddell, R. Seviour, J. Upton, and M. Goodfellow. 2002. Dispelling the Nocardia amarae myth: A phylogenetic and phenotypic study of mycolic acid-containing actinomycetes isolated from activated sludge foam. Water Sci. Technol. 46:81–90.

Stange, R. R., D. Jeffares, C. Young, D. B. Scott, J. R. Eason, and P. E. Jameson. 1996. PCR amplification of the Fas-1 gene for the detection of virulent strains of Rhodococcus fascians. Plant Pathol. 45:407–417.

Stark, D. A., and N. G. Anderson. 1990. A case control study of Nocardia mastitis in Ontario dairy herds. Can. Vet. J. 31:197–201.

Steck, P. A., B. A. Schwartz, M. S. Rosendahl, and G. R. Gray. 1978. Mycolic acids—a reinvestigation. J. Biol. Chem. 253:5625–5629.

Steingrube, V. A., J. L. Gilson, B. A. Brown, Y. Zhang, R. W. Wilson, M. Rajagopalan, and R. J. Wallace Jr. 1995. PCR amplification and restriction endonuclease analysis of a 65-kilodallon heat shock protein gene sequence for taxonomic separation of rapidly growing mycobacteria. J. Clin. Microbiol. 35:149–153.

Steingrube, V. A., R. W. Wilson, B. A. Brown, K. C. Jost Jr., Z. Blacklock, J. L. Gilson, and R. C. Wallace Jr. 1997. Rapid identification of clinically significant species and taxa of aerobic actinomycetes, including Actinomadura, Gordonia, Nocardia, Rhodococcus, Streptomyces and Tsukamurella isolates, by DNA amplification and restriction analysis. J. Clin. Microbiol. 35:817–822.

Steinhaus, E. A. 1941. A study of the bacteria associated with thirty species of insects. J. Bacteriol. 42:757–790.

Stevens, D. A., A. C. Pier, B. L. Beaman, P. A. Morozumi, I. S. Lovett, and E. Huang. 1981. Laboratory evaluation of an outbreak of nocardiosis in immunocompromised hosts. Am. J. Med. 71:928–934.

Straube, G., J. Hensel, C. Niedan, and E. Straube. 1990. Kinetic studies of phenol degradation by Rhodococcus sp P1. Ant. v. Leeuwenhoek 57:29–32.

Sugar, A. M., and A. M. Angeles. 1987. Identification of a common immunodominant protein in culture filtrate of three Nocardia species and use in etiologic diagnosis of mycetoma. J. Clin. Microbiol. 25:2278–2280.

Suzuki, K., M. Goodfellow, and A. G. O'Donnell. 1993. Cell envelopes and classification. In: M. Goodfellow and A. G. O'Donnell (Eds.) Handbook of New Bacterial Systematics. Academic Press. London, UK. 195–250.

Svensson, L., L. Sisfontes, G. Nyborg, and R. Blanstrand. 1982. High performance liquid chromatography and glass capillary gas chromatography of geometric and positional isomers of long chain monounsaturated fatty acids. Lipids 17:50–59.

Takai, S., T. Sekizaki, T. Ozawa, T. Sugawara, Y. Watanabe, and S. Tsubaki. 1991. Association between a large plasmid and 15-kitodalton and 17-kilodalton antigens in virulent Rhodococcus equi. Infect. Immun. 59:4056–4060.

Takai, S., Y. Watanabe, T. Ikeda, T. Ozawa, S. Matsukura, Y. Tamada, S. Tsubaki, and T. Sekizaki. 1993. Virulence associated plasmids in Rhodococcus equi. J. Clin. Microbiol. 31:1726–1729.

Takai, S., Y. Sasaki, T. Ikeda, T. Uchida, S. Tsubaki, and T. Sekizaki. 1994. Virulence of Rhodococcus equi from patients with and without AIDS. J. Clin. Microbiol. 32:457–460.

Takai, S., Y. Imai, N. Fukunaga, Y. Uchida, K. Kamisawa, Y. Sasaki, S. Tsubaki, and T. Sekizaki. 1995. Identification of virulence-associated antigen and plasmids in Rhodococcus equi from patients with AIDS. J. Infect. Dis. 172:1306–1311.

Takai, S., N. Fukunaga, S. Ochiai, Y. Imasi, Y. Sasaki, S. Tsubaki, and T. Seizaki. 1996. Identification of intermediately virulent Rhodococcus equi isolates from pigs. J. Clin. Microbiol. 34:1034–1037.

Takai, S., C. Vigo, H. Ikushima, T. Higuchi, S-T. Hagiwara, S. Hashikura, Y. Sasaki, S. Tsulaki, T. Anazi, and M. Kanada. 1998. Detection of virulent Rhodococcus equi in tracheal aspirate samples by polymerase chain reaction for rapid diagnosis of R. equi pneumonia in foals. Vet. Microbiol. 61:51–58.

Takai, S., S. A. Hines, T. Sekizaki, V. Nicholson, D. A. Alperin, M. Osaki, D. Takamatsu, M. Nakamura, K. Suzuki, N. Ogino, T. Kakuda, H. Dan, and J. F. Prescott. 2000. DNA sequence and comparison of virulence plasmids from Rhodococcus equi ATCC 33701 and 103. Infect. Immun. 68:6840–6847.

Takeuchi, M., and K. Hatano. 1998. Gordonia rhizosphera sp. nov., isolated from the mangrove rhizosphere. Int. J. Syst. Bacteriol. 48:907–912.

Takeuchi, M., K. Hatano, I. Sedlácek, and Z. Pácova. 2002. Rhodococcus jostii sp. nov., isolated from a medieval grave. Int. J. Syst. Evol. Microbiol. 52:409–413.

Tandoi, V., L. L. Blackall, and N. Caravaglio. 1992. Isolation by micromanipulation of actinomycetes in scum producing activated sludge. Ann. Microbiol. Enzymol. 42:217–225.

Tárnok, I. 1976. Metabolism in nocardiae and related bacteria. In: M. Goodfellow, G. H. Brownell, and J. A. Serrano (Eds.) The Biology of the Nocardiae. Academic Press. London, UK. 451–500.

Thiemann, J. E., C. Hengeller, A. Virgilio, O. Buelli, and G. Licciardello. 1964. Rifampicin 33: Isolation of actinophages active on Streptomyces mediterranei and characterisation of phage-resistant strains. Appl. Microbiol. 12:261–268.

Tilford, P. E. 1936. Fasciation of sweet peas caused by Phytomonas fascians n. sp. J. Agric. Res. 53:383–394.

Tisdall, P. A., G. D. Roberts, and J. P. Anhalt. 1979. Identification of clinical isolates with gas-liquid chromatography alone. J. Clin. Microbiol. 10:506–514.

Tkachuk-Saad, O., and J. F. Prescott. 1991. Rhodococcus equi plasmids: Isolation and partial characterisation. J. Clin. Microbiol. 29:2696–2700.

Tomioka, N., H. Uchiyama, and O. Yagi. 1994. Cesium accumulation and growth characteristics of Rhodococcus erythropolis CS98 and Rhodococcus sp. strain CS402. Appl. Environ. Microbiol. 60:2227–2231.

Tomiyasu, I., S. Toriyama, I. Yano, and M. Masui. 1981. Changes in molecular species composition of nocardomycolic acids in Nocardia rubra by the growth temperature. Chem. Phys. Lipids 28:41–54.

Tomiyasu, I., and I. Yano. 1984. Separation and analysis of novel polyunsaturated mycolic acids from a psychrophilic acid-fast bacterium, Gordona aurantiaca. Eur. J. Biochem. 139:173–180.

Trevisan, V. 1989. Generiele Specie delle Batteriacee. Zanaboni and Gabuzzi. Milan, Italy.

Tsukamura, M. 1962. Differentiation of Mycobacterium tuberculosis from other mycobacteria by sodium salicylate susceptibility. Am. Rev. Respir. Dis. 86:81–83.

Tsukamura, M. 1969. Numerical taxonomy of the genus Nocardia. J. Gen. Microbiol. 56:265–287.

Tsukamura, M. 1971a. Proposal of a new genus, Gordona, for slightly acid-fast organisms occurring in sputa of patients with pulmonary disease and in soil. J. Gen. Microbiol. 68:15–26.

Tsukamura, M., and S. Mizuno. 1971b. A new species Gordona aurantiaca occurring in sputa of patients with pulmonary disease. (in Japanese). Kekkaku 46:93–98.

Tsukamura, M. 1972. An improved selective medium for atypical mycobacteria. Japanese J. Microbiol. 16:243–246.

Tsukamura, M. 1974a. A further numerical taxonomic study of the rhodochrous group. Japanese Microbiol. 18:37–44.

Tsukamura, M. 1974b. Differentiation of the Mycobacterium rhodochrous group from nocardiae by β-galactosidase activity. J. Gen. Microbiol. 80:553–555.

Tsukamura, M. 1978. Numerical classification of Rhodococcus (formerly Gordona) organisms recently isolated from sputa of patients: description of Rhodococcus sputi Tsukamura sp. nov. Int. J. Syst. Bacteriol. 28:169–181.

Tsukamura, M. 1981a. Differentiation between the genera Mycobacterium, Rhodococcus and Nocardia by susceptibility to 5-fluorouracil. J. Gen. Microbiol. 125:205–208.

Tsukamura, M. 1981b. Tests from susceptibility to mitomycin C as aids in differentiating the genus Rhodococcus from the genus Nocardia and for differentiating Mycobacterium fortuitum and Mycobacterium chelonei from other rapidly growing mycobacteria. Microbiol. Immunol. 25:1197–1199.

Tsukamura, M. 1982a. Differentiation between the genera Rhodococcus and Nocardia and between species of the genus Mycobacterium by susceptibility to bleomycin. J. Gen. Microbiol. 128:2385–2388.

Tsukamura, M. 1982b. Numerical analysis of the taxonomy of nocardiae and rhodococci. Division of Nocardia asteroides sensu stricto into two species and descriptions of Nocardia paratuberculosis sp. nov. Tsukamura (formerly Kyoto-1 group of Tsukamura), Nocardia nova sp. nov. Tsukamura, Rhodococcus chubuensis sp.nov. Tsukamura and Rhodococcus obuensis sp.nov. Tsukamura. Microbiol. Immunol. 26:1101–1119.

Tsukamura, M., and K. Kawakami. 1982c. Lung infection caused by Gordona aurantiaca (Rhodococcus aurantiacus). J. Clin. Microbiol. 16:604–607.

Tsukamura, M., and I. Yano. 1985. Rhodococcus sputi sp. nov., nom. rev., and Rhodococcus aurantiacus sp. nov., nom. rev. Int. J. Syst. Bacteriol. 35:364–368.

Tsukamura, M., K. Hikosaka, K. Nishimura, and S. Hara. 1988a. Severe progressive subcutaneous abscesses and necrotizing tenosynovitis caused by Rhodococcus aurantiacus. J. Clin. Microbiol. 26:201–205.

Tsukamura, M., C. Komatsuzaki, R. Sakai, K. Kaneda, T. Kudo, and A. Seino. 1988b. Mesenteric lymphadenitis of swine caused by Rhodococcus sputi. J. Clin. Microbiol. 26:155–157.

Uchida, K., and K. Aida. 1977. Acyl type of bacterial cell wall: its simple identification by colorimetric method. J. Gen. Appl. Microbiol. 23:249–260.

Uchida, K., and K. Aida. 1979. Taxonomic significance of cell-wall acyl type in Corynebacterium-Mycobacterium-Nocardia group by a glycolate test. J. Gen. Appl. Microbiol. 25:169–183.

Uchida, Y., R. Tsuchiya, M. Chino, J. Hirano, and T. Tabuchi. 1989. Extracellular accumulation of mono- and di-succinoyl trehalose lipids by a strain of Rhodococcus erythropolis grown on n-alkanes. Agric. Biol. Chem. 53:757–763.

Uchida, K., and K. Aida. 1997. Intra- and intergen relationships of various actinomycete strains based on the acyl types of the muramyl residue in cell wall peptidoglycans examined in a glycolate test. Int. J. Syst. Bacteriol. 47:182–190.

Uchida, K., T. Kudo, K. Suzuki, and T. Nakase. 1999. A new rapid method of glycolate test by diethyl ether extraction, which is applicable to a small amount of bacterial cells of less than one milligram. J. Gen. Appl. Microbiol. 45:49–56.

Uttamchandani, R. B., G. L. Daikos, and R. R. Reyes. 1994. Nocardiosis in 30 patients with advanced human immunodeficiency virus infection: clinical features and outcome. Clin. Infect. Dis. 18:348–353.

Van Etta, L. L., G. A. Filice, R. M. Ferguson, and D. N. Gerding. 1983. Corynebacterium equi: A review of 12 cases of human infection. Rev. Infect. Dis. 5:1012–1018.

Van Ginkel, C. G., H. G. J. Welten, S. Hartmans, and J. A. M. de Bont. 1987. Metabolism of trans-2-butene and butane in Nocardia TB1. J. Gen. Microbiol. 133:1713–1720.

Vanniasinkam, T., M. D. Barton, and M. W. Heuzenroeder. 2001. B-Cell epitope mapping of the Vap A protein of Rhodococcus equi: Implications for early detection of R. equi disease in foals. J. Clin. Microbiol. 39:1633–1637.

Vivrette, S. L., D. C. Sellon, and D. S. Gibbons. 2000. Clinical application of a polymerase chain reaction assay in the diagnosis of pneumonia caused by R. equi in a horse. J. Am. Vet. Med. Ass. 217:1348–1350.

Von Below, H., C. M. Wilk, K. P. Schaal, and G. O. H. Naumann. 1991. Rhodococcus luteus and Rhodococcus erythropolis chronic endophthalmitis after lens implantation. Am. J. Opthamol. 112:596–597.

Wagner, F., W. Lindorfer, and W. Schultz. 1976. Verfahren zur Verbesserung des Ausbeute bei Gewinnung von Erdol durch Wasserfluten. German Patent 24. 19–267.

Wagner, F., U. Behrendt, H. Bock, A. Kretschmer, S. Lang, and C. Syldatk. 1983. Production and chemical characterisation of surfactants from Rhodococcus erythropolis and Pseudomonas sp. MUB grown on hydrocarbons. In: J. E. Zajic, D. G. Cooper, T. R. Jack, and N. Kosaric (Eds.) Microbial Enhanced Oil Recovery. Penn Well. Tulsa, OK. 55–60.

Waksman, S. A. 1950. The actinomycetes: Their nature, occurrence, activities and importance. Ann. Crypt. Phytopathol. 9:1–230.

Wallace Jr., R. J., M. Tsukamura, and B. A. Brown. 1990. Cefataxine resistant Nocardia asteroides strains are isolates of the controversial species Nocardia farcinica. J. Clin. Microbiol. 28:2726–2732.

Wang, Y., Z. Zhang, J. Ruan, Y. Wang, and S. Ali. 1999. Investigations of actinomycete diversity in the tropical rainforests of Singapore. J. Clin. Microbiol. 23:178–187.

Wang, L., Y. Zhang, Z. Lu, Y. Shi, Z. Liu, L. Maldonado, and M. Goodfellow. 2001. Nocardia beijingensis sp. nov., a novel isolate from soil. Int. J. Syst. Evol. Microbiol. 51:1783–1788.

Wang, L., Y. Zhang, Y. Huang, L. A. Maldonado, Z. Liu, and M. Goodfellow. 2004. Nocardia pigrifrangens sp. nov., a novel actinomycete isolated from a contaminated plate. Int. J. Syst. Evol. Microbiol. 54(Pt 5):1683–1686.

Warhurst, A. W., and C. A. Fewson. 1994. Biotransformations catalyzed by the genus Rhodococcus. Crit. Rev. Biotechnol. 14:29–73.

Weinstock, D. M., and A. E. Brown. 2002. Rhodococcus equi: an emerging pathogen. Clin. Infect. Dis. 34:1379–1385.

Wellington, E. M. H., and S. T. Williams. 1978. Preservation of actinomycete inoculum in frozen glycerol. Microbios Lett. 6:151–159.

Weyland, H. 1969. Actinomycetes in North Sea and Atlantic Ocean sediments. Nature 233:851.

Williams, S. T., E. M. H. Wellington, and L. S. Tipler. 1980. The taxonomic implications of the reactions of representative Nocardia strains to actinophage. J. Gen. Microbiol. 119:173–178.

Williams, D. R., P. W. Trudgill, and D. G. Taylor. 1989. Metabolism of 1,8-cineole by a Rhodococcus species: Ring cleavage reactions. J. Gen. Microbiol. 135:1957–1967.

Wilson, R. W., V. A. Steingrube, B. A. Brown, and R. J. Wallace Jr. 1998. Clinical application of PCR-restriction enzyme pattern analysis for rapid identification of aerobic actinomycete isolates. J. Clin. Microbiol. 36:148–152.

Winogradsky, S. 1949. Microbiologie du sol. Masson. Paris, France.

Woolcock, J. B., A. M. T. Farmer, and M. D. Mutimer. 1979. Selective medium for Corynebacterium equi isolation. J. Clin. Microbiol. 9:640–642.

Workman, M. R., J. Philpott-Howard, M. Yates, D. Deighton, and M. W. Casewell. 1998. Identification and antibiotic susceptibility of Nocardia farcinica and Nocardia nova in the U.K. J. Med. Microbiol. 47:85–90.

Xue, Y., X. Sun, P. Zhou, R. Liu, F. Liang, and Y. Ma. 2003. Gordonia paraffinivorans sp. nov., a hydrocarbon-degrading actinomycete isolated from an oil-producing well. Int. J. Syst. Evol. Microbiol. 53:1643–1646.

Yamada, K., J. Takahashi, K. Kobayashi, and Y. Imada. 1963. Studies on utilisation of hydrocarbons by microorganisms. Agric. Biol. Chem. 27:390–395.

Yamada, H., and M. Kobayashi. 1996. Nitrile hydratase and its application to industrial production of acrylamide. Biosci. Biotechnol. Biochem. 60:1391–1400.

Yano, I., K. Kageyama, Y. Ohno, M. Masui, E. Kusunose, M. Kusunose, and N. Akinori. 1978. Separation and analysis of molecular species of mycolic acids in Nocardia and related taxa by gas chromatography mass spectrometry. Biomed. Mass Spectr. 5:14–24.

Yano, I., T. Imaeda, and M. Tsukamura. 1990. Characterisation of Nocardia nova. Int. J. Syst. Microbiol. 40:170–174.

Yassin, A. F., F. A. Rainey, H. Brzezinka, J. Burghardt, H. J. Lee, and K. P. Schaal. 1995. Tsukamurella inchonensis sp. nov. Int. J. Syst. Bacteriol. 45:522–527.

Yassin, A. F., F. A. Rainey, H. Brzezinka, J. Burghardt, M. Rafai, P. Seifert, K. Feldmann, and K. P. Schaal. 1996. Tsukamurella pulmonis sp. nov. Int. J. Syst. Bacteriol. 46:429–436.

Yassin, A. F., F. A. Rainey, J. Burghardt, H. Brzezinka, S. Schmitt, P. Seifert, O. Zimmermann, H. Mauch, D. Gierth, I. Lux, and K. P. Schaal. 1997. Tsukamurella tyrosinosolvens sp. nov. Int. J. Syst. Bacteriol. 47:607–614.

Yassin, A. F., F. A. Rainey, U. Mendrock, H. Brzezinka, and K. P. Schaal. 2000. Nocardia abscessus sp. nov. Int. J. Syst. Evol. Microbiol. 50:1487–1493.

Yassin, A. F., B. Sträubler, P. Schumann, and K. P. Schaal. 2003. Nocardia puris sp. nov. Int. J. Syst. Evol. Microbiol. 53:1595–1599.

Yoon, J. H., Y.-G. Cho, S.-S. Kang, S. B. Kim, S. T. Lee, and Y.-H. Park. 2000a. Rhodococcus koreensis sp. nov., a 2, 4-dinitrophenoldegrading bacterium. Int. J. Syst. Evol. Microbiol. 50:1193–1201.

Yoon, J. H., S.-S. Kang, Y. G. Cho, S. T. Lee, Y. H. Kho, C.-J. Kim, and Y.-H. Park. 2000b. Rhodococcus pyridinivorans sp. nov., a pyridine-degrading bacterium. Int. J. Syst. Evol. Microbiol. 50:2173–2180.

Yoon, J.-H., J. J. Lee, S.-S. Kang, M. Takeuchi, Y. K. Shin, S. T. Lee, K. H. Kang, and Y.-H. Park. 2000c. Gordonia nitida sp. nov., a bacterium that degrades 3-ethylpyridine and 3-methylpyridine. Int. J. Syst. Evol. Microbiol. 50:1203–1210.

Yumoto, I., A. Nakamura, H. Iwata, K. Kojima, K. Kusumoto, Y. Nodosaka, and H. Matsuyama. 2002. Dietzia psychralcaliphila sp. nov., a novel facultatively psychrophilic alkaliphile that grows on hydrocarbons. Int. J. Syst. Evol. Microbiol. 52:85–90.

Zakrzewska-Czerwinska, J., M. Mordarski, and M. Goodfellow. 1988. DNA base composition and homology values in the classification of some Rhodococcus species. J. Gen. Microbiol. 134:2807–2813.

Zhang, J., Y. Zhang, C. Xiao, Z. Liu, and M. Goodfellow. 2002. Rhodococcus maanshanensis sp. nov., a novel actinomycete from soil. Int. J. Syst. Microbiol. 52:2121–2126.

Zhang, J., Z. Liu, and M. Goodfellow. 2003. Nocardia caishijiensis sp. nov., a novel soil actinomycete. Int. J. Syst. Evol. Microbiol. 53:999–1004.

Zhang, J., Z. Liu, and M. Goodfellow. 2004. Nocardia xishanensis sp. nov., a novel actinomycete isolated from soil. Int. J. Syst. Evol. Microbiol. 54(Pt 6):2301–2305.

Zink, M. C., J. A. Yager, J. F. Prescott, and B. N. Wilkie. 1985. In vitro phagocytosis and killing of Corynebacterium equi by alveolar macrophages of foals. Am. J. Vet. Res. 46:2171–2174.

Zink, M. C., J. A. Yager, and W. L. Smart. 1986. Corynebacterium equi infection in horses 1958–1984: A review of 131 cases. Can. Vet. J. 27:213–217.

Zopf, W. 1889. Über das Mikrochemische Verhatten von Fettfarbstoff-haltigen Organen. Z. Wissensch. Mikrosk. 6:172–177.

Zopf, W. 1891. Über Ausscheidung von Fellfarbstoffen (Lipochromen) seitens gewisser Spattpilze. Ber. Deutsch. Bot. Ges. 9:22–28.

Prokaryotes (2006) 3:889–918
DOI: 10.1007/0-387-30743-5_33

CHAPTER 1.1.18

The Genus *Mycobacterium*—Nonmedical

SYBE HARTMANS, JAN A. M. DE BONT AND ERKO STACKEBRANDT

Introduction

Because new *Mycobacterium* species are being described in rapid order and because awareness of the importance of these organisms in the clinical and nonclinical environment is increasing, the editors of *The Prokaryotes* think that a thorough coverage of the rapidly growing species should await extensive comparative studies and a deeper analysis of their biotechnological and ecological role. Several genomes of slowly growing mycobacteria have been or are in the process of being sequenced, e.g., several strains of *Mycobacterium tuberculosis*, *Mycobacterium bovis*, *Mycobacterium leprae*, and the two subspecies of *M. avium*. Only one genome of rapidly growing mycobacteria, i.e., *Mycobacterium smegmatis* MC2155 (see The Institute for Genomic research website), has been sequenced; this approach will certainly be applied to more members of these organisms. For the time being, the reader is referred to the recent descriptions of novel species (Table 2) to obtain information on specific properties and to the original chapter by Hartmans and De Bont (1991) in *The Prokaryotes*. The phylogenetic placement of nonmedical strains next to their medically relevant neighbors is shown in Figs. 1–4.

Mycobacteria are aerobic, acid-fast actinomycetes that usually form slightly curved or straight nonmotile rods (0.2–0.6 × 1.0–10 μm). Branching and mycelium-like growth may take place with fragmentation into rods and coccoid elements. Many species form whitish or cream-colored colonies, but especially among the rapid growers, there are also many bright yellow or orange species containing carotenoid pigments (David, 1984). In some cases, the pigments are only formed in response to light (photochromogenic species), but most pigmented species also form these pigments in the dark (scotochromogenic species).

Mycobacterium is the only genus listed in the Family Mycobacteriaceae in *Bergey's Manual of Systematic Bacteriology* (Wayne and Kubica, 1986), but the genus is considered to be closely related to the other mycolic acid-containing gen-era of cell wall chemotype IV: *Caseobacter*, *Corynebacterium*, *Nocardia* and *Rhodococcus* (Goodfellow and Cross, 1984).

Chemical differentiation of mycobacteria from the other mycolic acid-containing genera is possible by analysis of the fatty acid esters formed upon pyrolysis of the mycolic acid esters, in combination with the identification of the major menaquinone present in the plasma membrane (Table 1). The Ziehl-Neelsen stain for acid fastness, however, remains the most obvious method to quickly identify mycobacteria (Barksdale and Kim, 1977).

Mycobacteria are the causal agents of two important diseases, tuberculosis and leprosy, and there has thus been a significant clinical interest in the two responsible species. This started with the work of Koch (1882), who detected the tubercle bacillus in stained infected tissues. The generic name *Mycobacterium* was introduced by Lehmann and Neumann (1896) to include the tubercle and leprosy bacilli. For many years after this work, isolates other than *M. tuberculosis*, but resembling it in staining characteristics, were described as "atypical mycobacteria." After the discovery in the early 1950s that several of these "atypical mycobacteria" could also produce disease in humans (see Kubica, 1978), it was recognized that identification of these strains was required. The classification of Runyon, separating the mycobacteria into four groups (photochromogens, scotochromogens, nonphotochromogens, and rapid growers) was introduced in the late 1950s as a systematic base for the description of the "atypical mycobacteria" (Wayne, 1984). This division, based on pigmentation and rate of growth, is still of use to the clinical mycobacteriologist (see The Genus Mycobacterium—Medical in this Volume).

The separation of the genus into two major groups on the basis of the growth rate of the individual species forms the basis of mycobacterial taxonomy. Although not exactly following this division, many slow growers are either associated with or the causal agents of human or other animal diseases. Most rapid growers are not known to be associated with human diseases

Table 1. Differential characteristics of the mycolic acid-containing genera of wall chemotype IV.

| Genus | Mycolic acids | | Predominant menaquinone | N-glycolyl in glycan moiety of cell wall |
	Overall size (number of carbons)	Ester pyrolysis products (number of carbons in chain)		
Caseobacter	30–36	14–18	$MK-8(H_2), MK-9(H_2)$[a]	−
Corynebacterium	22–36	8–18	$MK-8(H_2), MK-9(H_2)$	−
Mycobacterium	60–90	22–26	$MK-9(H_2)$	+
Nocardia	44–60	12–18	$MK-8(H_4)$	+
Rhodococcus	34–64	12–18	$MK-8(H_2), MK-9(H_2)$	+
Tuskamurella	64–78	20–22	MK-9	+

and consequently are often considered as non-pathogens, although in a strict sense, "nonpathogens" may not really exist (Tsukamura, 1984). Rapidly growing mycobacteria are common saprophytes in natural habitats but have received much less attention than the clinically more relevant slow-growing species.

Phylogeny and Taxonomy

Taxonomic studies traditionally have relied heavily on morphological characteristics, but subsequently numerical taxonomy and chemotaxonomic characteristics have played an important role in determining taxonomic relationships. Cell wall and mycolic acid analyses have proved to be of great value in separating the actinomycete genera, while numerical taxonomy has had the most impact on the subgeneric level (Goodfellow and Wayne, 1982). DNA hybridization of genomic DNA and, more recently, 16S rRNA similarity data are of course important tools in determining phylogenetic relations on both the generic and subgeneric levels.

Phylogeny

The number of *Mycobacterium* species has increased from about 40 in 1980 (Skerman et al., 1980) to about 110 in 2004. The description of novel species is paralleled by the development of molecular methods and by the increased recognition that slow growing mycobacteria are clinically important and fast-growing mycobacteria are ecologically important. By the end of 1983, there were 52 described species. Only 6 new species were added between 1984 and 1991 and about 4 new species per year between 1992 and 2003. In 2004, 12 new species were described or are in the process of being validated (Table 2).

The often very close relationship among mycobacterial strains and the finding that the slow growing strains possess a single copy of the *rrn* operon only, thus lacking sequence microheterogeneity, made the 16S rRNA gene an ideal target to be used in the differentiation of strains (De Smet et al., 1996; Holberg-Petersen et al., 1999). Many mycobacterial species can be differentiated on the basis of the sequence of the variable stretch between positions 175 and 238 (*Escherichia coli* nomenclature; Böddinghaus et al., 1990a; Stahl and Urbance, 1990; Pitulle et al., 1992). More than in other bacterial genera, small differences are used as a basis for describing species (Patel et al., 2000), determining intraspecific subclusters (Böddinghaus et al., 1990b; Frothingham and Wilson, 1994). A specific database of high sequence quality (the Ribosomal Differentiation of Medical Microorganisms database; RIDOM) was specifically established for routine diagnosis of these organisms (Harmsen et al., 2003). For those mycobacterial taxa that cannot be discriminated on the basis of their 16S rRNA gene sequence, other genes, such as *gyrB* (Kasai et al., 2000; Niemann et al., 2000), *hsp65* (Devallois et al., 1997; Ringuet et al., 1999), and intraspecific spacers (De Smet et al., 1995; Roth et al., 2000b), were targeted. In addition, the presence of differences in the sequence of other genes (Aranaz et al., 2003) was used in the differentiation of taxa highly related to *Mycobacterium tuberculosis* and *Mycobacterium bovis*.

In the early 1990s, several studies were published which covered the phylogenetic structure of the fast- and slow-growing mycobacteria (Smida et al., 1988; Rogall et al., 1990a; Stahl and Urbance, 1990; Pitulle et al., 1992). The data by and large supported the clustering of mycobacteria according to their growth behavior and indicated that the mostly clinical, slow growing strains evolved from their fast growing relatives (Pitulle et al., 1992). The genomic separateness of the two groups was supported by a sequence idiosyncrasy: the majority of slow growing strains contained an insert (i.e., a long helix between nucleotide positions 451 and 482 [*E. coli* nomenclature]), while the fast growers contained a short helix only. However, some slow growers, such as *Mycobacterium simiae* and *Mycobacterium triviale*, branching deeply within

Table 2. *Mycobacterium* species, the year of their description, and growth velocity.

Taxon	Growth Velocity	Author and validation date[a]
Mycobacterium abscessus	Rapid	Kusunoki and Ezaki, 1992
Mycobacterium africanum	Slow	Skerman et al., 1980 (AL)
Mycobacterium agri	Rapid	Tsukamura, 1981
Mycobacterium aichiense	Rapid	Tsukamura, 1981
Mycobacterium alvei	Rapid	Ausina et al., 1992
Mycobacterium asiaticum	Slow	Skerman et al., 1980 (AL)
Mycobacterium aurum	Rapid	Skerman et al., 1980 (AL)
Mycobacterium austroafricanum	Rapid	Tsukamura et al., 1983c
Mycobacterium avium subsp. *avium*	Slow	Skerman et al., 1980 (AL) emend. Thorel et al., 1990
Mycobacterium avium subsp. *paratuberculosis*	Slow	Thorel et al., 1990
Mycobacterium avium subsp. *Silvaticum*	Slow	Thorel et al., 1990
Mycobacterium bohemicum	Slow	Reischl et al., 1998
Mycobacterium bonickei	Rapid	Schinsky et al., 2004
Mycobacterium botniense	Slow	Torkko et al., 2000
Mycobacterium bovis subsp. *bovis*	Slow	Skerman et al., 1980 (AL)
Mycobacterium branderi	Slow	Koukila-Kähkölä et al., 1995
Mycobacterium brisbanense	Rapid	Schinsky et al., 2004
Mycobacterium brumae	Rapid	Luquin et al., 1993
Mycobacterium canariasense	Rapid	Jiménez et al., 2004
Mycobacterium caprae	Slow	Aranaz et al., 2003
Mycobacterium celatum	Slow	Butler et al., 1993
Mycobacterium chelonae subsp. *chelonae*	Rapid	Skerman et al., 1980 (AL)
Mycobacterium chimaerae	Slow	Tortoli et al., 2004
Mycobacterium chitae	Rapid	Skerman et al., 1980 (AL)
Mycobacterium chlorophenolicum	Rapid	Häggblom et al., 1994
Mycobacterium chubuense	Rapid	Tsukamura, 1981
Mycobacterium confluentis	Rapid	Kirschner et al., 1992
Mycobacterium conspicuum	Slow	Springer et al., 1995a
Mycobacterium cookii	Slow	Kazda et al., 1990
Mycobacterium cosmeticum	Rapid	R. C. Cooksey et al., unpublished
Mycobacterium diernhoferi	Rapid	Tsukamura et al. 1983c
Mycobacterium doricum	Cluster rapid, designated slow	Tortoli et al., 2001
Mycobacterium duvalii	Rapid	Skerman et al., 1980 (AL)
Mycobacterium elephantis	Rapid	Shojaei et al., 2000
Mycobacterium fallax	Rapid	Levy-Frebault et al., 1983
Mycobacterium farcinogenes	Rapid	Skerman et al., 1980 (AL)
Mycobacterium flavescens	Rapid	Skerman et al., 1980 (AL)
Mycobacterium fortuitum subsp. *acetamidolyticum*	Rapid	Tsukamura et al., 1986
Mycobacterium fortuitum subsp. *fortuitum*	Rapid	Skerman et al., 1980 (AL)
Mycobacterium frederiksbergense	Rapid	Willumsen et al., 2001
Mycobacterium gadium	Rapid	Skerman et al., 1980 (AL)
Mycobacterium gastri	Slow	Wayne, 1966
Mycobacterium genavense	Slow	Böttger et al., 1993
Mycobacterium gilvum	Rapid	Skerman et al., 1980 (AL)
Mycobacterium goodii	Rapid	Brown et al., 1999
Mycobacterium gordonae	Slow	Skerman et al., 1980 (AL)
Mycobacterium haemophilum	Slow	Skerman et al., 1980 (AL)
Mycobacterium hassiacum	Cluster slow, designated rapid	Schröder et al., 1997
Mycobacterium heckeshornense	Slow	Roth et al., 2000a
Mycobacterium heidelbergense	Slow	Haas et al., 1997
Mycobacterium hiberniae	Slow	Kazda et al., 1993
Mycobacterium hodleri	Rapid	Kleespies et al., 1996
Mycobacterium holsaticum	Rapid	Richter et al., 2002
Mycobacterium houstonense	Rapid	Schinsky et al., 2004
Mycobacterium immunogenum	Rapid	Wilson et al., 2001
Mycobacterium interjectum	Slow	Springer et al., 1993
Mycobacterium intermedium	Slow	Meier et al., 1993
Mycobacterium intracellulare	Slow	Skerman et al., 1980 (AL)
Mycobacterium kansasii	Slow	Skerman et al., 1980 (AL)
Mycobacterium komossense	Rapid	Skerman et al., 1980 (AL)
Mycobacterium kubicae	Slow	Floyd et al., 2000
Mycobacterium lacus	Slow	Turenne et al., 2002

(*Continued*)

Table 2. *Continued*

Taxon	Growth Velocity	Author and validation date[a]
Mycobacterium lentiflavum	Slow	Springer et al., 1996
Mycobacterium	Slow	Skerman et al., 1980 (AL)
Mycobacterium lepraemurium	Slow	Skerman et al., 1980 (AL)
Mycobacterium madagascariense	Rapid	Kazda et al., 1992
Mycobacterium mageritense	Rapid	Domenech et al., 1997
Mycobacterium malmoense	Slow	Skerman et al., 1980 (AL)
Mycobacterium marinum	Slow	Skerman et al., 1980 (AL)
Mycobacterium microti	Slow	Skerman et al., 1980 (AL)
Mycobacterium montefiorense	Slow	Levi et al., 2003
Mycobacterium moriokaense	Rapid	Tsukamura et al., 1986
Mycobacterium mucogenicum	Rapid	Springer et al., 1995
Mycobacterium murale	Rapid	Vuorio et al., 1999
Mycobacterium nebraskense	Slow	Mohamed et al., 2004
Mycobacterium neoaurum	Rapid	Skerman et al., 1980 (AL)
Mycobacterium neworleansense	Rapid	Schinsky et al., 2004
Mycobacterium nonchromogenicum	Slow	Skerman et al., 1980 (AL)
Mycobacterium novocastrense	Rapid	Shojaei et al., 1997
Mycobacterium obuense	Rapid	Tsukamura and Mizuno, 1981
Mycobacterium palustre	Slow	Torkko et al., 2002
Mycobacterium parafortuitum	Rapid	Skerman et al., 1980 (AL)
Mycobacterium parascrofulaceum	Slow	Turenne et al., 2004
Mycobacterium parmense	Slow	Fanti et al., 2004
Mycobacterium peregrinum	Rapid	Kusunoki and Ezaki, 1992
Mycobacterium phlei	Rapid	Skerman et al., 1980 (AL)
Mycobacterium pinnipedii	Slow	Cousins et al., 2003
Mycobacterium porcinum	Rapid	Tsukamura et al., 1983b
Mycobacterium poriferae	Rapid	Padgitt and Moshier, 1987
Mycobacterium psychrotolerans	Rapid	Trujillo et al., 2004
Mycobacterium pulveris	Rapid	Tsukamura et al., 1983a
Mycobacterium rhodesiae	Rapid	Tsukamura, 1981
Mycobacterium saskatchewanense	Slow	Turenne et al., 2004b
Mycobacterium scrofulaceum	Slow	Skerman et al., 1980 (AL)
Mycobacterium senegalense	Rapid	Skerman et al., 1980 (AL)
Mycobacterium septicum	Rapid	Schinsky et al., 2000
Mycobacterium shimoidei	Slow	Tsukamura, 1982
Mycobacterium shottsii	Slow	Rhodes et al., 2003
Mycobacterium simiae	Slow	Skerman et al., 1980 (AL)
Mycobacterium smegmatis	Rapid	Skerman et al., 1980 (AL)
Mycobacterium sphagni	Rapid	Kazda, 1980
Mycobacterium szulgai	Slow	Skerman et al., 1980 (AL)
Mycobacterium terrae	Slow	Skerman et al., 1980 (AL)
Mycobacterium thermoresistibile	Rapid	Skerman et al., 1980 (AL)
Mycobacterium tokaiense	Rapid	Tsukamura, 1981
Mycobacterium triplex	Slow	Floyd et al., 1996
Mycobacterium triviale	Slow	Skerman et al., 1980 (AL)
Mycobacterium tuberculosis	Slow	Skerman et al., 1980 (AL)
Mycobacterium tusciae	Cluster rapid, designated slow	Tortoli et al., 1999
Mycobacterium ulcerans	Slow	Skerman et al., 1980 (AL)
Mycobacterium vaccae	Rapid	Skerman et al., 1980 (AL)
Mycobacterium vanbaalenii	Rapid	Khan et al., 2002
Mycobacterium wolinskyi	Rapid	Brown et al., 1999
Mycobacterium xenopi	Slow	Skerman et al., 1980 (AL)

[a]Skerman et al. (1980 [AL]) refers to the Approved Lists in which the respective descriptions were validated.

their cluster, lacked the long helix, thus showing the signature of their fast-growing ancestors (Pitulle et al., 1992). Since 1990, the vast majority of the species descriptions contained an indication (from 16S rRNA gene sequences) of the phylogenetic position of the novel species. However, the topologies of most phylogenetic dendrograms differed in detail from each other because of the influence of the treeing algorithm used, the selection of reference sequences, and the addition of novel sequences into the database.

The phylogenetic dendrograms (Figs. 2–4) are from a comprehensive neighbor-joining tree based on 16S rRNA gene sequences of all valid and yet-to-be validated species (Fig. 1). The dendrogram includes new sequences, for which the public records indicated incomplete datasets only (i.e., sequences for type strains of *Mycobacterium agri*, AJ429045; *M. confluentis*, AJ634379; *M. moriokaense*, AJ429044; *M. pulveris*, AJ429046; *M. rhodesiae*, AJ429047; and *M. margaritense*, AJ699399). As compared to the dendrograms based on a very limited dataset in 1992, the complete dataset of *Mycobacterium* type strains has not changed significantly: the rapidly growing strains are more deeply branched (Fig. 1, cluster B) and are ancestors of the slow growing strains (Fig. 1, cluster A). However, it should

be noted that the statistical significance is low for the majority of branching points, indicating that the order at which the species branch from each other is not settled and may change within the group to which they belong when new sequence data are added to the database.

SLOW GROWING MYCOBACTERIA The vast majority of slow growing mycobacteria form three clusters with high intracluster relationships (Fig. 1, cluster A, groups A1–A3). Group A1 contains the well known pathogens such as *Mycobacterium tuberculosis*, *Mycobacterium ulcerans*, *Mycobacterium intracellulare*, *Mycobacterium leprae* and their relatives (Fig. 2). These organisms are characterized by a long helix between position 451 and 482 of the 16S rRNA sequence.

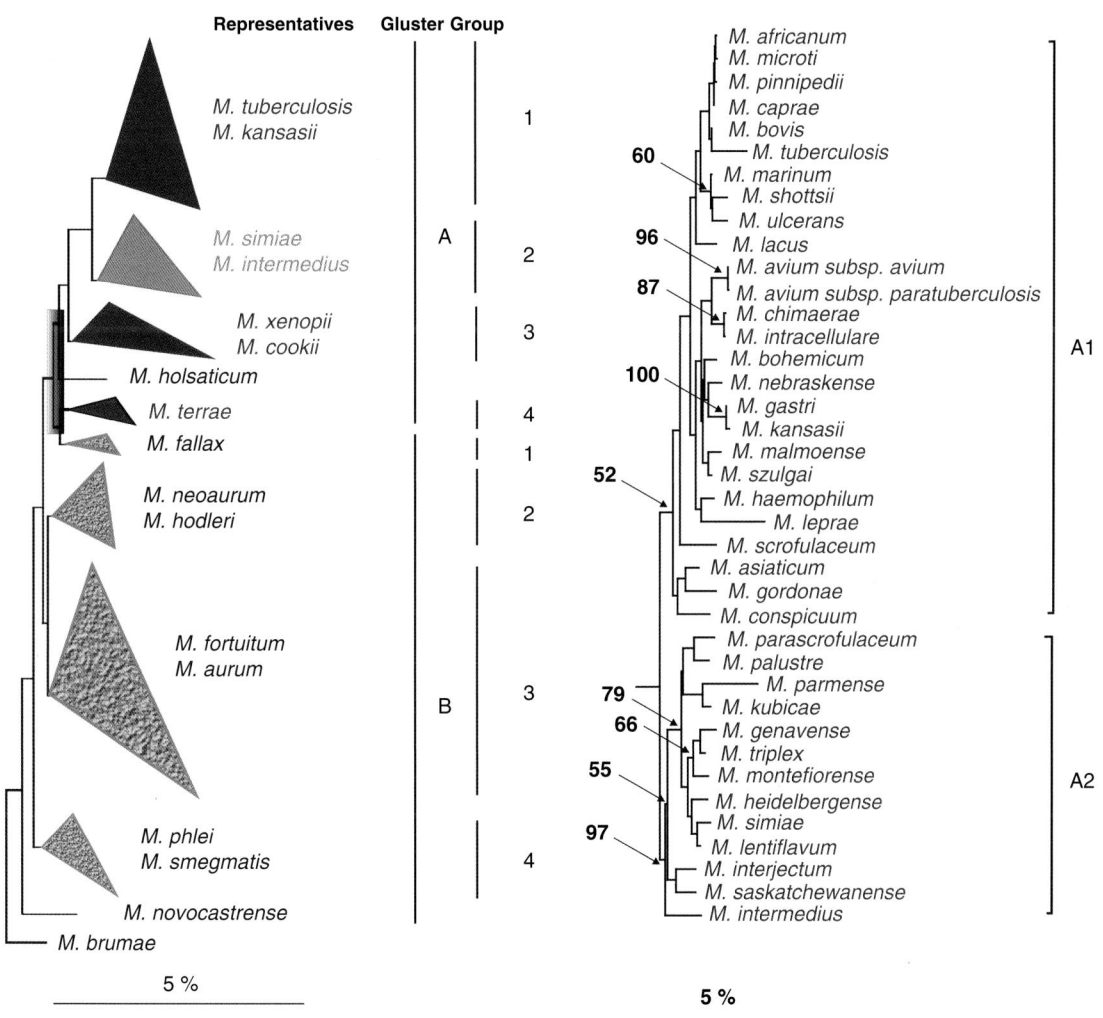

Fig. 1. Schematic graph of a neighbor-joining 16S rRNA gene sequence analysis of type strains of slow growing mycobacterial species. Red triangle: slow growing species containing a long helix between positions 451 and 482; yellow triangle: slow growing species containing a short helix between position 451 and nucleotide position; and gray triangle: fast growing species. Bar: 5% sequence difference.

Fig. 2. Neighbor-joining dendrogram of 16S rRNA gene sequences showing the position of type strains of slow growing mycobacterial species. Numbers at branching points refer to bootstrap values. For references to strain number and 16S rRNA gene sequence accession number, see Table 1. Bar: 5% sequence difference.

Group A2 members, embracing *Mycobacterium simiae*, *Mycobacterium gevanense* and relatives (Fig. 2), however, lack this insert.

Group A3 (Figs. 1 and 3) is interesting from an evolutionary point of view in that this assemblage unites slow growing organisms without the characteristic helix insert (*Mycobacterium shimoidei* and *Mycobacterium triviale*) with other members of this group lacking this insert.

The fourth group of slow growers (Figs. 1 and 3), containing the insert (group A4), occupies a distinctly separate position. The three species *Mycobacterium hiberniae*, *Mycobacterium nonchromogenicum* and *Mycobacterium terrae* cluster next to some fast growing members of *Mycobacterium* (cluster B1). Two species, *Mycobacterium doricum* and *Mycobacterium tusciae*, are defined as slow growing mycobacteria but cluster within the phylogenetic radiation of rapidly growing mycobacteria (Fig. 4).

RAPIDLY GROWING MYCOBACTERIA These organisms form four phylogenetic clusters, three of which (together with two individual lineages defined by *Mycobacterium novocastrense* and *Mycobacterium brumae*) are located at the root of the mycobacterial tree (Figs. 1 and 4). The internal structure allows the definition of four groups (B1–B4), of which only group B4 is well separated from the others. Within group B3, the three species *Mycobacterium abcessus*, *Mycobacterium chelonae* and *Mycobacterium immunogenum* are located at the tip of a longer branch which may indicate that these organisms are subjected to a different evolutionary rate. Interestingly, *Mycobacterium chitae* and *Mycobacterium fallax* (group B1) as well as *Mycobacterium holsaticum* show a closer relatedness to slow growing mycobacteria and thus serve as a bridge between fast and slow growers. Most of the fast growing mycobacterium species have been isolated from environmental samples and some are involved in the degradation of polycyclic aromatic hydrocarbons (e.g., *Mycobacterium vanbaalenii*, *M. hodleri* and *M. frederickbergense*) and pentachlorophenol (*M. chlorophenolicum*). Others were isolated from wound infections (*M. goodie* and *M. wolinskyi*), but their role as pathogens has not been evaluated. The pathogenicity of other fast growing mycobacterial species is well studied (members of cluster B3, e.g., *Mycobacterium fortuitum*, *Mycobacterium septicum* and their relatives). The ecological role of most fast growing mycobacteria has yet to be determined.

Taxonomy

On the basis of chemotaxonomic studies, the genus *Mycobacterium* was placed within the CNM (*Corynebacterium-Nocardia-Mycobacte-*

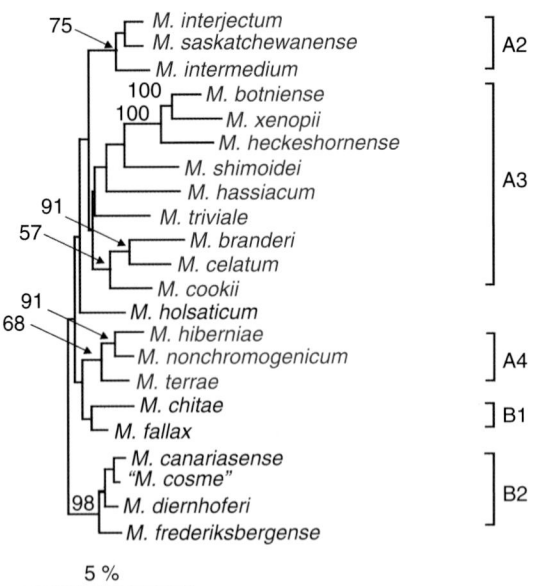

Fig. 3. Neighbor-joining dendrogam of 16S rRNA gene sequences showing the position of type strains of mycobacterial species and the evolution of slow growers (in red) from fast growing ancestors (in blue). Numbers at branching points refer to bootstrap values. For references to strain number and 16S rRNA gene sequence accession number, see Table 1. Bar: 5% sequence difference.

rium) complex. Together with the genera *Caseobacter*, *Rhodococcus* and *Tsukamurella*, these genera all have cell wall chemotype IV (Lechevalier and Lechevalier, 1970) and contain mycolic acids (Collins et al., 1988).

DNA hybridization studies of the genomic DNA have been performed with a number of mycobacteria (Bradley, 1973; Baess, 1982; Lévy-Frébault et al., 1984; Lévy-Frébault et al., 1986b; Garcia and Taberés, 1986; Imaeda et al., 1988; Hurley et al., 1988; Estrada-Garcia et al., 1989) and have generally confirmed the species status already derived from numerical taxonomy studies (Wayne, 1984). More discriminative techniques involving specific DNA probes hybridizing with total DNA (Zainuddin and Dale, 1989; McFadden et al., 1990) or with DNA amplified with the polymerase chain reaction (PCR; Hartskeerl et al., 1989) have been used, but they have focused mainly on the slowly growing species. Sequences of 16S rRNA have also been obtained by direct sequencing of DNA amplified by the PCR method. They were demonstrated to be useful in the differentiation of mycobacteria at the species level (Rogall et al., 1990b). Clearly, the PCR technique will have a significant impact on the identification of mycobacteria in the near future (also see The Genus Mycobacterium—Medical in this Volume). DNA-DNA homology studies can be used to dis-

Fig. 4. Neighbor-joining dendrogram of 16S rRNA gene sequences showing the position of rapidly growing type strains of the genus *Mycobacterium* (in blue). Members in red are slow growing species but they cluster with the fast growers phylogenetically. Numbers at branching points refer to bootstrap values. For references to strain number and 16S rRNA gene sequence accession number, see Table 1. Bar: 5% sequence difference.

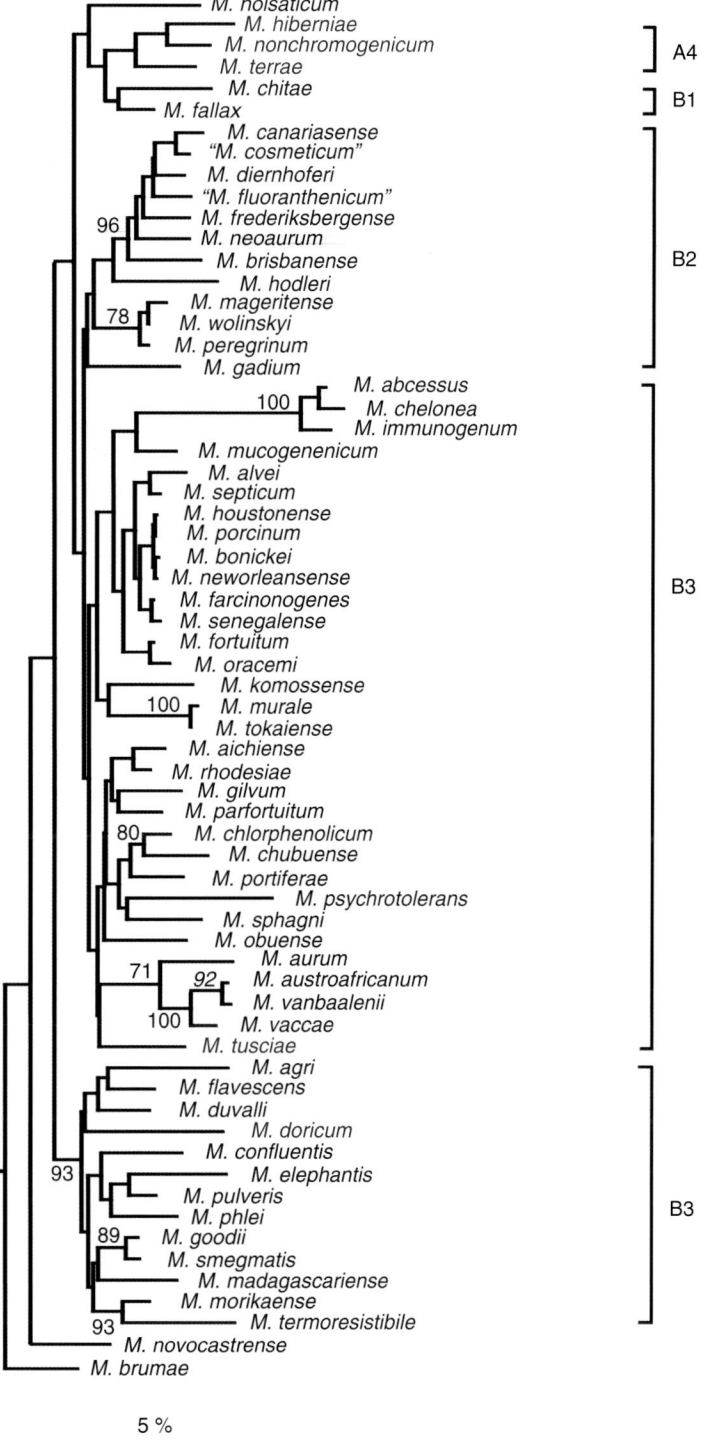

criminate between more closely related taxons, e.g., the different species of a genus. Note however that DNA-DNA hybridization data from different laboratories can only be compared when reference strains are included. DNA homology studies with selected strains of *M. tuberculosis*, *M. bovis*, *M. bovis* BCG (Bacillus Calmette-Guérin), *M. microti* and *M. africanum* revealed that all strains exhibited more than 90% DNA relatedness, whereas DNA relatedness between *M. tuberculosis* and other slowly growing mycobacteria ranged from 9–53% (Imaeda, 1985). Although *M. microti*, *M. tuberculosis*, *M. bovis* and *M. africanum* can be distinguished phenotypically, numerical taxonomy places them all in a cluster distinct from other slowly growing mycobacteria (Wayne and Kubica, 1986). The 16S rRNA sequences of 14 *M. tuberculosis*, two *M. bovis* and two *M. africanum* isolates were also identical (Rogall et al., 1990b). Considering the above, it is anticipated that a proposal will be made to reduce these four species to one species of *M. tuberculosis*, which possibly will be subdivided into subspecies (Wayne and Kubica, 1986).

DNA-DNA hybridization studies have also led to a better understanding of the relationship between *M. paratuberculosis* and the MAIS (*M. avium–intracellulare–scrofulaceum*) complex. The MAIS complex strains have been grouped together on the basis of phenotypic characteristics (Wayne, 1984), but DNA hybridization studies indicate that *M. scrofulaceum* shows little DNA similarity with *M. avium* and *M. intracellulare* (Hurley et al., 1988). The suggestion that *M. avium*, *M. intracellulare* and *M. paratuberculosis* should be grouped together as biovars of a single species is based on these studies (Hurley et al., 1988). The 16S rRNA sequences can however be used to differentiate between representatives of each of these species (Rogall et al., 1990a).

DNA homology studies with rapidly growing mycobacteria have also been performed (Baess, 1982; Lévy-Frébault et al., 1984; Lévy-Frébault et al., 1986b; Garcia and Taberés, 1986). In the study of Baess (1982), *M. chelonae* subsp. *chelonae* and *M. chelonae* subsp. *abscessus* could not be distinguished using the spectrophotometric method, but with the more sensitive S1 nuclease method, the two subspecies were clearly distinct (Lévy-Frébault et al., 1986b). Within the subspecies, homology values were higher than 73% and changes in melting temperature (ΔT_m) less than 2°C, whereas homology values between the *M. chelonae* groups were 26–52% with ΔT_m values of more than 8°C. Both studies indicate that the unrecognized species "*M. perigrinum*," often grouped within the *M. fortuitum* complex, has been revived as an independent species (Table 2).

Habitats

Mycobacteria have been isolated from a very diverse array of biotopes including material of both mammalian and nonmammalian origin, as for instance fresh and salt water, soil, and dust. Some species of the nonpathogenic saprophytes may also occur as opportunistic pathogens. The pathogenic species and their habitats are covered in The Genus Mycobacterium—Medical in this Volume and will only be discussed in this chapter in relation to nonmammalian environments.

Although many *Mycobacterium* species have been isolated from environmental samples, this does not necessarily imply that these strains can all grow under these conditions. Essential in assigning a species to the natural flora of a specific environment is that it should be capable of multiplying actively in these environments (Kazda, 1983). If it lacks this property then it should be regarded as a contaminant. Discriminating between these two possibilities is very difficult because the chance of isolating "contaminating" mycobacteria from environmental samples is quite large as a result of the ability of mycobacteria to survive for very long periods under nongrowth conditions. *Mycobacterium paratuberculosis*, for example, was reported to survive for 252 days in a soil-water slurry (Kazda, 1983).

The most recent reviews on the ecology of mycobacteria are those of Kazda (1983), Collins et al. (1984), and Tsukamura (1984). Kazda (1983) classified the mycobacteria into four groups on the basis of ecologically relevant properties. He distinguished obligate pathogenic, facultative pathogenic, potentially pathogenic, and saprophytic species.

Included in the group of obligate pathogens, which presumably are unable to multiply outside living beings, are the species *M. tuberculosis*, *M. bovis*, *M. africanum*, *M. asiaticum*, *M. malmoense*, *M. microti*, *M. simiae*, *M. szulgai* and *M. haemophilum*. Of these species, *M. tuberculosis* and *M. bovis* have been isolated from wastewater (Kazda, 1983). *Mycobacterum farcinogenes* and *M. shimodei* can also be included in the group of obligate pathogens as they have been isolated from mammalian sources only (Tsukamura, 1982; Wayne and Kubica, 1986).

Kazda assigned *M. leprae*, *M. paratuberculosis* and *M. ulcerans* to the provisional group of the facultative pathogens as their incidence in natural biotopes has not been investigated in detail. Future research may result in the assignment of these species to either the potential or obligate pathogens (Kazda, 1983).

Examples of potential pathogens are *M. avium*, *M. chelonae*, *M. fortuitum*, *M. intracellulare*, *M. kansasii*, *M. marinum*, *M. senegalense*, *M. scrofulaceum* and *M. xenopi* (Kazda, 1983). These species can grow in natural biotopes without losing their pathogenic properties, and they have been isolated from a variety of biotopes (Kazda, 1983; Tsukamura, 1984).

The saprophytic mycobacteria include among others the slowly growing species *M. gordonae*, *M. nonchromogenicum*, *M. triviale*, *M. terrae* and *M. gastri* and most of the rapidly growing species (Table 2), with the exception of *M. chelonae*, *M. fortuitum* and *M. senegalense* and their relatives, which are classified as potential pathogens

(Kazda, 1983). The saprophytic and potentially pathogenic groups (Kazda, 1983) are found in many environments (Tsukamura, 1984) and only occasionally occur as an accompanying flora in pathogenic processes. The ability of organisms from these groups to utilize many different growth substrates and their capacity to survive and multiply under a wide range of environmental conditions apparently enable them to compete successfully with other organisms in many biotopes.

The abundance of the fast-growing species in soils was already demonstrated by Jones and Jenkins (1965). In screening soil samples, 101 acid-fast strains were isolated. Of the 93 isolates studied, five were scotochromogens and all but three of the nonphotochromogens were rapidly growing mycobacteria, forming colonies within 7 days at 25°C.

The most commonly occurring *Mycobacterium* in the environment is probably *M. fortuitum*, as it is often the major *Mycobacterium* species (40–80%) isolated from soil samples (Tsukamura, 1984). Strains of the slowly growing *M. nonchromogenicum* complex (*M. nonchromogenicum*, *M. terrae* and *M. triviale*) and the rapidly growing species *M. aurum*, *M. smegmatis* and *M. agri* have also been isolated frequently from soil samples (Tsukamura, 1984). Isolates from hospital dust were also predominantly *M. fortuitum* (43%), with significant numbers of strains of the *M. nonchromogenicum* complex (25%) and *M. gordonae* (18%), whereas from house dust the majority of the strains isolated belonged to the MAIS (*M. avium–intracellulare–scrofulaceum*) complex (55%), and significant numbers belonged to the *M. nonchromogenicum* complex (23%) and were also *M. gordonae* (11%; Tsukamura, 1984).

The "tap water scotochromogen," *M. gordonae*, is the strain most often isolated from water samples of various origins (Collins et al., 1984). Besides *M. gordonae*, strains of *M. terrae*, *M. phlei* and *M. fortuitum* have also been isolated at a high frequency from various nonmarine aquatic environments (Viallier and Viallier, 1973). Analysis of municipal water supplies and water supplies in hemodialysis centers found strains of *M. fortuitum*, *M. chelonae*, *M. scrofulaceum*, *M. gordonae* and of the *M. avium* and *M. terrae* complexes (Carson et al., 1988). The average number of mycobacteria detected in the municipal water supplies was 74 ± 42 per 100 ml. Marine water samples, usually from coastal waters, predominantly yielded strains of the MAIS complex, although strains of *M. gordonae* and *M. terrae* were also isolated (Collins et al., 1984).

However, when considering studies on the isolation of various *Mycobacterium* species from the environment, note that environmental samples are routinely decontaminated and that the distribution of the species isolated need not necessarily reflect the distribution of these species in the original sample. It may well be that certain species, or cells in a particular physiological state, exhibit a higher- or lower-than-average survival rate during the decontamination procedure. Application of the decontamination procedure therefore may result in an over- or underestimate of the prevalence of a particular species.

Another consequence of routinely applied decontamination procedures is that no reliable data can be obtained on the numbers of viable mycobacteria present in natural environments. The actual numbers present in soil samples may also be considerably higher than anticipated from viable counts of nondecontaminated samples because of the hydrophobic characteristics of the mycobacterial cell wall, which may result in significant bacterial adhesion to surfaces.

Isolation

Mycobacteria are not always readily isolated from natural samples. They grow relatively slowly and are therefore easily overgrown by faster-growing organisms. However, taking advantage of the resistance of mycobacteria to adverse conditions, decontamination procedures and selective media have been developed to increase the efficiency of isolation procedures (Kubica and Good, 1981). Most of the decontamination methods were developed for the isolation of mycobacteria from specimens originating from diseased humans or animals and exploit the resistance of the acid-fast mycobacteria to alkaline and acidic conditions. The most commonly applied decontamination procedures involve treatment of the samples with NaOH (2–4%) for 15–30 min at room temperature or at 37°C (Jenkins et al., 1982). Methods for the selective isolation of mycobacteria from environmental samples have been reviewed by Songer (1981), who, besides reviewing decontamination methods, also discussed numerous selective growth media.

Realize however that no information is available on the number and type(s) of mycobacteria that are lost as a result of the decontamination procedure. Furthermore, the selectivity of the growth conditions (medium composition and incubation temperature) also affects the number and species of mycobacteria that eventually are isolated.

Portaels et al. (1988) have recently compared different decontamination procedures and selective media for the isolation of mycobacteria from soil samples. The best results (low contamination in combination with high positivity rates) were obtained using the following procedure:

Isolation of Mycobacteria from Soil (Portaels et al., 1988)

> Add 0.5 g of soil (wet weight) to 5 ml of sterile trypticase soy broth, shake vigorously, and incubate at 37°C for 5 h. After sedimentation of the soil particles, add 5 ml of malachite green (0.2%), 1 ml of cycloheximide (500 µg per ml), and 5 ml of NaOH (1 M) to the supernatant. After 30 min at room temperature, neutralize the mixture with HCl (1 M) and centrifuge at 2,000 g for 20 min. Inoculate the pellet on Ogawa egg medium containing cycloheximide (500 µg per ml). Incubate at 30°C and inspect every 2 weeks until growth is observed.

Ogawa Egg Medium (Tsukamura et al., 1986b)

> Add to 100 ml of water containing sodium glutamate (1%) and KH_2PO_4 (1%), 200 ml of whole eggs, 6 ml of glycerol, and 6 ml of a malachite green solution (2%). The medium is made up as slopes by heating for 60 min at 90°C.

Using the above method, mycobacteria were isolated from 91% of the soil samples with only 4% of the tubes being contaminated, i.e., exhibiting growth of non-acid-fast microorganisms within 6 months (Portaels et al., 1988). Omission of cycloheximide, malachite green, or both compounds from the egg medium resulted in contamination rates of 14%, 49%, and 63%, respectively.

The decontamination method and incubation conditions may even select for a specific mycobacterial species. *Mycobacterium moriokaense*, for instance, is the predominant species isolated from soil samples using the following procedure:

Isolation of M. moriokaense (Tsukamura et al., 1986b)

> Shake soil sample (20 g) with 100 ml of 0.9% NaCl for 30 min and allow to settle for 15 min. Filter the supernatant and centrifuge the filtrate for 20 min at 700 g. Mix the precipitate in the centrifuge tube with 3 ml of KOH (1%). After 5 min, inoculate 0.02 ml portions of this suspension on slopes of Ogawa egg medium. Incubate the inoculated slopes at 42°C for 7–10 days and subculture the colonies.

The nonpigmented acid-fast colonies isolated using the above procedure were characterized as *M. moriokaense* (80%) and as *M. fortuitum* (20%). The strong selectivity of this procedure is probably due to the combination of the relatively high incubation temperature and the short incubation time. Unfortunately, no mention was made of the numbers of any pigmented acid-fast colonies which might also have grown on the slopes (Tsukamura et al., 1986b).

Another approach for selectively isolating mycobacteria is to use specific carbon sources in simple mineral media. Many years ago, Söhngen (1913) demonstrated that paraffin could be used to enrich mycobacteria from soil and water samples. Likewise, enrichment cultures with ethene as the sole carbon source also exclusively led to the isolation of mycobacteria (Hartmans et al., 1989). Other genera, for instance *Xanthobacter* and *Nocardia* species, may also grow on ethene (Van Ginkel et al., 1987b). However, the very low growth rates of these species with ethene may explain why enrichment cultures with ethene result in the isolation of the faster-growing mycobacteria.

Morpholine (1-oxa-4-azacyclohexane), which for many years was considered nonbiodegradable because of its persistence in waste-water-treatment plants, has recently been demonstrated to be degraded by *Mycobacterium* isolates (Cech et al., 1988; Knapp and Brown, 1988). All strains isolated on morpholine from municipal activated sludge systems and river water were identified as mycobacteria (Knapp and Brown, 1988).

Another approach combines the selectivity of a single carbon source with the selectivity of an antibiotic in enrichment cultures. Gram-positive methanol-utilizing bacteria were isolated from different soil samples with methanol as carbon source in the presence of the antibiotic polymyxin B (Urakami and Yano, 1989). The 14 isolates were all very similar and were identified as mycobacteria closely resembling *M. fortuitum*. The growth rates with methanol, though, varied quite significantly. Enrichment procedures on methanol in the absence of polymyxin B usually result in the isolation of Gram-negative genera.

Although mycobacteria are isolated quite easily from many different environments, they are difficult to quantify, especially in heterogeneous matrices such as soil. As mentioned by Songer (1981), the attachment of mycobacteria to surfaces has not yet been investigated. Considering the hydrophobic character of mycobacteria (Van Loosdrecht et al., 1987), conceivably a large percentage of the mycobacteria present in soil adhere to the surfaces of solids. Biofouling of cellulose diacetate membranes used in reverse-osmosis water purification plants is an example of this characteristic. It has been proposed that the initial step in this process is the attachment of mycobacteria (Ridgway et al., 1984). Therefore, many of the mycobacteria present in environmental samples may be overlooked, especially if interfaces are present.

Cultivation

Mycobacteria can utilize a wide range of carbon compounds. Very often glycerol is designated the preferred carbon since it is utilized as sole source of carbon and energy by all cultivable mycobacteria (Ratledge, 1982b). For clinical isolates, an absolute requirement for carbon dioxide has been reported (Ratledge, 1982b). It has been observed repeatedly that the lag phase of rapidly growing mycobacteria is shortened considerably when liquid cultures are incubated stationary rather than shaken. This effect may be due to a stimulatory effect of carbon dioxide accumulating in the medium of the stationary culture.

Mycobacteria can use a variety of nitrogen sources, including amino acids and ammonium. In many of the frequently used media for the cultivation of mycobacteria, asparagine is the nitrogen source (Jenkins et al., 1982). Many species can also reduce nitrate (Table 3) and use it as a source of nitrogen. Mycobacteria cannot fix nitrogen. The "*M. flavum*" 301 capable of fixing nitrogen (Ratledge, 1982b) is in fact a *Xanthobacter flavus* strain (see The Genus Xanthobacter in Volume 5). Table 3 lists a number of charac-

teristics of rapidly growing species that are useful in identification.

Most mycobacteria do not require any specific growth factors or vitamins in the growth medium. Exceptions are *M. haemophilum*, which requires hemin, and *M. paratuberculosis*, which requires mycobactin, and of course *M. leprae*, which has until now not been cultivated in vitro (Goodfellow and Wayne, 1982).

Problematic in the cultivation of many strains of mycobacteria in liquid culture is the tendency of these organisms to form aggregates and to adhere to the surfaces of growth vessels, probably as a result of the hydrophobic nature of the cell wall. Detergents such as Tween 80 are therefore often added to the growth medium to reduce clumping and to stimulate growth (Ratledge, 1982b). Owing to this tendency of most mycobacteria to form clumps

Table 3. Identifying characteristics of some rapidly growing species of mycobacteria.

Species								Characteristic								
	1	2	3	4	5	6	7	8	9	10	11	12	13	14	15	16
"*M. acetamidolyticum*"	N	+	−	−	+	−	+	+	+	+	−	−	−	−	−	−
M. agri	N	+	+	+	−	+	+	+	v	+	+	−	+	−	−	−
M. aichiense	S		−	+	+	−	v	−	−				−	v	+	−
M. aurum	S	−	−	+	v	−	+	v	v	v	v	−	+	v	+	−
M. austroafricanum	S	−	−	+	+	v	+	+	−	+	−	−	+	−	+	−
M. chelonae																
Subsp. *abscessus*	N	+	−	+	+	v	+	−	−	v	−	−	−	v	−	−
Subsp. *chelonae*	N	−	−	−	+	v	+	−	−	v	−	−	−	v	−	−
M. chitae	N	−	−	+	−	+	+	+	+	+	−	−	−	−	−	−
M. chubuense	S	−	−	+	−	−	+	+	−	+	−	−	+	v	v	v
M. diernhoferi	N	−	−	+	−	−	+	+	+	+	v	−	+	−	+	−
M. duvalii	S		−		−	−	+	+	−	+	−	−	−	−	+	−
M. fallax	N	−	−	−	−	+	+	+	−	−	−	−	−	−	−	−
M. flavescens	S	v	−	−	v	+	+	+	−	v	−	−	−	−	v	v
M. fortuitum	N	v	−	+	+	v	v	+	+	v	+	−	−	v	v	−
M. gadium	S		−	−	−		+			−		−	−			
M. gilvum	S			v	−	v	+	−		+	v	−		−	+	−
M. komossense	S		−	+	−	+	+	−	−	−	−	+	−	+	+	+
M. moriokaense	N	+	−	+	+	+	+	+	−	v	−	−	v	+	+
M. neoaurum	S	v	−	+	+	−	+	v	+	+	+	−	v	+	+	−
M. obuense	S		−	+	+	v	+	−	−	+	+	−	−	v	+	v
M. parafortuitum	P	v	−	+	v	−	v	v	v	+	−	−	v	−	+	−
M. phlei	S	+	+	+	−	+	+	+	v	+	v	−	v	v	+	v
M. porcinum	N	+	−	+	+	v	+	−	+	+	+	+	−	+	+	−
M. poriferae	S	−	−	+	−			−	−	v	−	−	+	+	+	+
M. pulveris	N	+	−	v	−	+	+	+	−	+	−	−	−	−	−	−
M. rhodesiae	S		+	+	+	−	−	−	−	−	−	−	+	−	+	−
M. senegalense	N							+	+	+	+	+			+	
M. smegmatis	N	+	+	+	+	−	−	+	+	+	−	+	+	+	+	+
M. sphagni	S		−	−	−	−	+	−	−	−	−	−	v	+	−
M. thermoresistibile	S	+	+	−	−	+	+	+	−	+	−	−	⫶	−	−	−
M. tokaiense	S		−	+	+	v	+	−	+	+	+	+	+	v	+	+
M. vaccae	P	+	−	+	v	−	v	v	+	+	+	+	v	v	+	−

Symbols: N, nonchromogenic; S, scotochromogenic; P, photochromogenic; +, >85% positive; v, variable; −, <15% positive.
[a]Numbers correspond to the following characteristics: 1, pigmentation; 2, growth at 42°C; 3, growth at 45°C; 4, tolerates 0.2% picric acid; 5, arylsulfatase (3 days); 6, α-esterase; 7, β-esterase; 8, nitrate reduction (24 h); 9, acetamidase; 10, nicotinamidase; 11, allantoinase; 12, succinamidase; 13, xylose utilized; 14, trehalose utilized; 15, mannitol utilized; 16, sorbitol utilized.
Data adapted from Kubica et al. (1972); Saito et al. (1977); Tsukamura (1981); Tsukamura et al. (1981, 1983c, 1986); Tsukamura and Ichiyama (1986); Wayne (1984); Wayne and Kubica (1986); and Padgitt and Moshier (1987).

and to adhere to the surfaces of laboratory fermentors, there have been very few studies of mycobacteria using continuous cultures. One exception was reported by Lowrie et al. (1979). Using a special fermentor in which all stainless steel had been replaced by glass, Teflon, or titanium, *M. bovis* BCG and *M. microti* were grown with 0.08% Tween 80 in the growth medium and by aerating with air containing 5% carbon dioxide. The air supply was introduced in the gas phase of the fermentor, and the stirring rate was kept low to prevent the formation of bubbles.

Some species, however, may be grown in chemostat cultures without having to take any specific precautions. *Mycobacterium phlei* can be grown dispersed in a fermentor (Girbal et al., 1989) without the addition of a detergent, and consequently it has been used quite often for metabolic studies (Ratledge, 1982b). Two other examples of mycobacteria cultivated in continuous culture without detergents are *M. aurum* strain L1 growing on vinyl chloride (Hartmans and De Bont, 1992) and the ethene-utilizing strain E3 growing on ethene (Van Ginkel et al., 1987a).

The mineral salts medium (Wiegant and De Bont, 1980) used in the chemostat studies of *M. aurum* L1 contained 0.2 ml of the Vishniac and Santer (1957) trace element solution per liter. Subsequent chemostat studies using this medium (S. Hartmans and J. A. M. de Bont, unpublished observations) with vinyl chloride as the carbon source revealed that the culture was iron-limited rather than carbon-limited. An improved mineral salts medium with an increased iron content was therefore formulated.

Basal Mineral Salts Medium for the Cultivation of Mycobacteria

K_2HPO_4	1.55 g
$NaH_2PO_4 \cdot 2H_2O$	0.85 g
$(NH_4)_2SO_4$	2.0 g
$MgCl_2 \cdot 6H_2O$	0.1 g
EDTA	10 mg
$ZnSO_4 \cdot 7H_2O$	2 mg
$CaCl_2 \cdot 2H_2O$	1 mg
$FeSO_4 \cdot 7H_2O$	5 mg
$Na_2MoO_4 \cdot 2H_2O$	0.2 mg
$CuSO_4 \cdot 5H_2O$	0.2 mg
$CoCl_2 \cdot 6H_2O$	0.4 mg
$MnCl_2 \cdot 2H_2O$	1 mg
Deionized water	1 liter

Identification

The simplest test for discriminating mycobacteria from other prokaryotes is the Ziehl-Neelsen acid-fast stain. Mycobacteria stained with carbol fuchsin resist decolorization with hydrochloric acid-alcohol. Barksdale and Kim (1977) have termed this property "mycobacterial acid-fastness" or acid-alcohol-fastness, in contrast to the acid-fastness of several other mycolic acid-containing genera, which resist decolorization with dilute mineral acids but which can be decolorized with hydrochloric acid-alcohol. A detailed discussion of the acid-fast stain is presented in the excellent review of Barksdale and Kim (1977).

An alternative method to reliably distinguish mycobacteria from other genera containing mycolic acids (Table 1) is to analyze the mycolic acid methyl esters by thin-layer chromatography (TLC; Minnikin et al., 1980b; Daffé et al., 1983) and to characterize the major menaquinone (Collins et al., 1977). Alternatively, the fatty acid methyl esters formed during pyrolysis gas liquid chromatography of mycolic acid methyl esters can also be analyzed (Kusaka and Mori, 1986).

Identification schemes for mycobacteria are based on a clear distinction between slowly growing and rapidly growing species. Species forming colonies from dilute inocula within 7 days under optimal conditions are classified as rapid growers. Those requiring more than 7 days are designated slow growers. The difference in growth rates is actually quite distinct, with the slow growers usually requiring 10 days or more to form colonies, whereas all rapidly growing species, including the relatively slowly growing *M. flavescens* and *M. thermoresistibile*, form colonies within 4–5 days (Jenkins et al., 1982).

It has been argued that the importance attributed to this characteristic may lead to erroneous results in the identification of mycobacteria, as isolates are usually only compared with known species of the same growth rate category. However, evidence that the majority of the rapidly growing and slowly growing mycobacteria are also clearly separated by 16S RNA studies (Stahl and Urbance, 1990; see Figs 1 and 4) supports the significance usually attributed to this characteristic in identifying mycobacteria.

The identification of slowly growing strains is described in The Genus Mycobacterium—Medical in this Volume as well as in other reviews (Jenkins et al., 1982; Wayne, 1984; Wayne and Kubica, 1986).

Identification of Rapidly Growing Mycobacteria

Identification of rapidly growing mycobacteria can be based on a number of characteristics. Chromogenicity is often used as an initial criterion, primarily of course because of its obvious nature. Numerical studies based on a large number of biochemical and physiological properties have been used to define the presently accepted species, and consequently many of these characteristics can be used in the identification of new isolates. Chemical analyses of cell constituents such as mycolic acids and mycobactins are also very powerful tools in the identification of mycobacteria. DNA analyses including hybridization studies with specific probes and the identification of specific sequences are the most recently developed techniques in mycobacterial identification.

Chromogenicity and Morphology

Chromogenicity is often used as a convenient criterion for identifying or classifying mycobacterial species, but some caution should be taken, as the significance of pigment production in the physiology of mycobacterial species and its taxonomic value are not very well studied (Kaneda et al., 1988). Thus, novel isolates should be compared with the pigmented as well as with the nonpigmented species (e.g., Tsukamura et al., 1986b). When determining chromogenic properties, the effect of light on pigmentation or even on growth itself should also be considered (De Bont et al., 1980).

A description of the morphological characteristics of the different species has been given by Wayne and Kubica (1986), although, according to Tsukamura et al. (1981b), colonial morphology is not very useful in differentiating mycobacterial species.

Biochemical and Physiological Characteristics

A probability matrix for the identification of the slowly growing mycobacteria has been developed on the basis of a large pool of collected data (Wayne et al., 1980). Although many species of rapidly growing mycobacteria also have been subjected to numerical analysis of a large number of biochemical characteristics (Saito et al., 1977; Tsukamura, 1981a; Tsukamura et al., 1981b; Tsukamura et al., 1983c), such a probability matrix has not yet been developed for the rapidly growing species (Wayne, 1984). Identification of the many rapidly growing mycobacterial species that have been proposed during the last 10 years was, however, usually based on studies involving numerical analysis (Kazda, 1980; Tsukamura, 1981a; Tsukamura et al., 1981b; Tsukamura, 1982; Tsukamura et al., 1983a; Tsukamura et al., 1983b; Tsukamura et al., 1983c; Tsukamura et al., 1986b; Lévy-Frébault et al., 1983; Tsukamura and Ichiyama, 1986a; Padgitt and Moshier, 1987). Besides the numerical taxonomic data, additional criteria such as DNA hybridization experiments are generally required to warrant the proposal of a new species (Wayne, 1984).

On the basis of a number of these numerical analysis studies of a large number of strains (Kubica et al., 1972; Saito et al., 1977; Tsukamura, 1981a; Tsukamura et al., 1981b; Tsukamura et al., 1983c; Tsukamura and Ichiyama, 1986a), we have listed several physiological and biochemical characteristics which should be of help in the identification of newly isolated rapidly growing species (Table 3). A detailed description of the methods is given by Vestal (1975). The diagnostic table for rapidly growing species published by Wayne and Kubica (1986) and the references cited therein are also very useful.

Generally, additional tests, such as mycolic acid analysis and DNA-hybridization experiments, are required before an exact identification can be made. Identification of strains of the "*M. parafortuitum* complex," which includes *M. parafortuitum*, *M. diernhoferi*, *M. aurum* and *M. neoaurum*, for example, is sometimes rather difficult. Several strains that clustered with the type strain of *M. neoaurum* in the numerical study of Saito et al. (1977) clustered with the *M. aurum* type strain in a subsequent study by Tsukamura et al. (1983c). In the second study, the above species and the novel species *M. austroafricanum* could be separated satisfactorily. Nevertheless, in view of the similarities observed, the preservation of the *M. parafortuitum* complex containing the five species mentioned above was considered to be justified (Tsukamura et al., 1983c).

The identification of *M. fortuitum* is another example showing that a small number of tests is not sufficient for identifying strains of rapidly growing mycobacteria. This clinically most important rapidly growing species is often identified solely on the basis of its positive 3-day arylsulfatase test. To distinguish *M. fortuitum* from *M. chelonae*, NaCl tolerance, nitrate reduction, and Fe uptake are also often determined (Kubica and Good, 1981). However, as can be seen in Table 3, several other nonphotochromogenic species are also positive in the 3-day arylsulfatase test.

Chemical Analysis of the Mycobacterial Cell

Cell wall and fatty acid analysis has had much impact on the classification of those Gram-positive prokaryotes that have wall chemotype IV (Goodfellow and Cross, 1984), i.e., prokaryotes with a cell wall containing *meso*-diaminopimelic acid (*meso*-A$_2$pm), arabinose, and galactose (Lechevalier and Lechevalier, 1970). In the mycobacterial peptidoglycan, the peptidoglycan residues are *N*-glycolated rather than *N*-acetylated as in most other *meso*-A$_2$pm containing prokaryotes (Brennan, 1988; Table 1). Also very characteristic for mycobacteria is the lipid-rich thick cell envelope containing very long-chain mycolic acids (Draper, 1982; Minnikin and O'Donnell, 1984a; Rastogi et al., 1986; Mamadou et al., 1989; Table 1). The characteristic mycolic acids have received much attention in relation to the identification of mycobacteria, especially at the species level.

Mycolic acids are high-molecular-weight, 3-hydroxy fatty acids substituted with an aliphatic side chain at the C_2 position (Fig. 5). The molecular mass of mycobacterial mycolic acids varies from C_{60} to C_{90} (Table 1). Pyrolysis of the methyl esters of mycobacterial mycolic acids yields long-chain meroaldehydes and fatty acid methyl esters (Fig. 6) with chain lengths of 22–26 carbon atoms (Table 1). This is a very useful characteristic in separating mycobacteria from the other mycolic acid-containing genera, although discrimination between the genera *Tsukamurella* and *Mycobacterium* also requires identification of the major menaquinone (Table 1).

Methanolysates of mycobacterial mycolic acids can be resolved into several classes based on the presence or absence of different functional groups in the longer carbon chain, i.e., R_1 in Fig. 5, of the mycolic acid molecule. Both one-dimensional (Daffé et al., 1983) and two-dimensional (Minnikin et al., 1980b) TLC methods have been used. The least polar compounds are termed "α-mycolic acids" and "α′-mycolic acids." They do not contain any oxygen functions apart from the 3-hydroxy and carboxy functions. In the more polar mycolic acids, the longer chain is substituted with methoxy, keto, epoxy, or carboxy functions (Minnikin, 1982). The various epoxy mycolic acids produce differing characteristic TLC patterns, depending on the method by which the methanolysate (acidic or basic methanolysis) was prepared (Minnikin et al., 1984b).

The predominant α-mycolic acids generally contain 74–82 carbon atoms, whereas the α′-mycolic acids are of a lower molecular weight, with the predominant type containing 60–68 carbon atoms (Kaneda et al., 1988). Five different types of long-chain α-mycolic acids have been separated and identified using argentation TLC, gas chromatography, and mass spectrometry (Minnikin et al., 1984c; Kaneda et al., 1988). These types are I, dicyclopropanoyl; II, monocyclopropanoyl monoenoic; III, methylated monocyclopropanoyl monoenoic; IV, dienoic; and V, methylated dienoic.

The α′-mycolic acids generally contain only one or two double bonds without any cyclopropanoyl functions (Kaneda et al., 1988). Very often (e.g., Collins et al., 1988) the α-mycolic acids of mycobacteria are described as always containing one or two double bonds. This is probably due to the fact that mycolic acids containing a cyclopropanoyl ring and mycolic acids containing a double bond are not separated by TLC. In fact, many species contain only dicyclopropanoyl α-mycolic acids without any double bonds as shown in Table 4, which lists the mycolic acid content by species (Kaneda et al., 1988).

For a number of species, e.g., *M. fortuitum*, *M. smegmatis* (Minnikin et al., 1984b), *M. phlei*, *M. thermoresistibile* (Lévy-Frébault et al., 1986a) and *M. chitae* (Minnikin et al., 1985), the mycolic acids of large numbers of strains have been analyzed, all resulting in a characteristic pattern for each species. Analysis of seven *M. aurum* strains, however, revealed two different mycolic acid patterns (Table 4), with two strains also containing α′-mycolic acids. This discrepancy was also observed by Lanéelle et al. (1988).

The composition of the α- and α′-mycolic acids of a number of strains has been further analyzed by Kaneda et al. (1988) using gas chromatography and mass spectrometry. These authors separated the rapid growers into three groups on the basis of the types of α-mycolic acids present. Group A only contained dienoic α-mycolic acids, group B contained α-mycolic acids containing double bonds and cyclopropanoyl rings, and group C only contained dicyclopropanoyl α-mycolic acids. Group A could be further divided into two groups on the bassis of the number of double bonds in the α′-mycolic acids (Table 4). No α′-mycolic acids were detected in *M. gilvum* by Kaneda et al. (1988), whereas several other authors reported that TLC detected α′-mycolic acids in this species. Unfortunately, *M. fallax* was

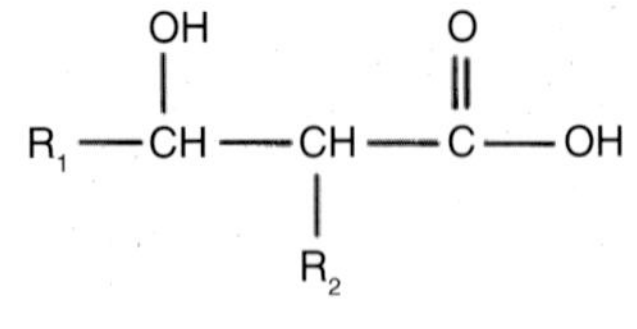

Fig. 5. General formula for mycolic acids. R_1 and R_2 represent variable side chains.

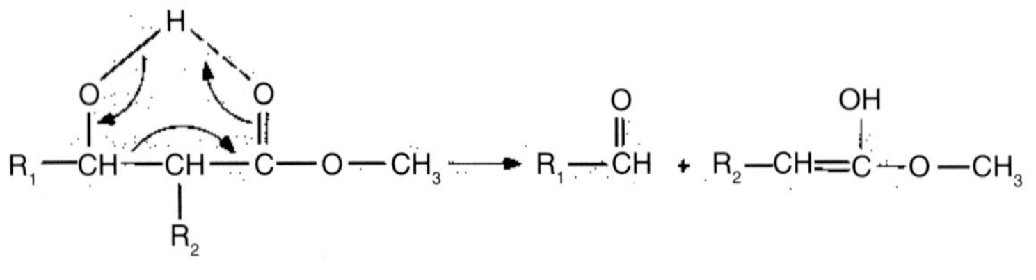

Fig. 6. Rearrangement of the mycolic acid methyl ester during pyrolysis to the meroaldehyde (containing R_1) and the fatty acid methyl ester (containing R_2).

Table 4. Classification of the mycolic acids and pyrolysis products of some of the rapidly growing species of mycobacteria.

Species	Mycolic acids[a]	α-Mycolic acids[b]	α'-Mycolic acids[b]	Ester pyrolysis products[c] 1	2	3
"M. acetamidolyticum"		III, III'	III	24		
M. agri	αα'm	I	IV	24	24	
M. aichiense	αkc				24	
M. aurum	αα'kc/αkc	I/III[a]	IV	22, 24	22/24	22B
M. austroafricanum						
M. chelonae	αα'	III, III'	IV		24	22B
M. chitae	αα'e	III, III'	IV	24		24B
M. chubuense	αα'kc				22, 24	
M. diernhoferi	αkc	I, II, III		22	24	22A
M. duvalii	αα'kc	I	IV	22, 24	22, 24	
M. fallax	α					
M. flavescens	αkc				24	24B
M. fortuitum						
Subsp. *fortuitum*	αα'e	III, III'	III	24	24	24B
Subsp. *peregrinum*	αα'e	III, III'	III	24	24	24B
M. gadium	αkc					
M. gilvum	αα'kc	I			22, 24	22
M. komossense	αkmc/αkc					
M. moriokaense	αkc					
M. neoaurum	αkc				22	22A
M. obuense	αα'kc				22	
M. parafortuitum	αα'kc	I, II, II', III, III'	IV	22	22	22A
M. phlei	αkc	I, II, II'		22, 24	24	24A
M. porcinum	αα'/αe	III, III'	III	24		
M. poriferae	αkc					
M. pulveris		I	IV	22, 24		
M. rhodesiae	αkc	I, II, II', III, III'		22, 24	24	24A
M. senegalense	αα'e			24[c]		
M. smegmatis	αα'e	III, III'	IV	24	24	24B
M. sphagni	αkc				24	
M. thermoresistibile	αα'km	I	IV	24		24B
M. tokaiense	αkc				24	
M. vaccae	αα'kc	I	IV	22, 24	22	24B

[a]Abbreviations: m, methoxy-; k, keto-; c, dicarboxy mycolic acids. Data adapted from Daffé et al. (1983); Minnikin et al. (1984b, 1985); Valero-Guillén and Martin-Luengo (1986); Lévy-Frébault et al. (1986b); Tsukamura et al. (1986); Luquin et al. (1987); Padgitt and Moshier (1987).

[b]Abbreviations: I, dicyclopropanoyl; II, monocyclopropanoyl monoenoic; II', methylated monocyclopropanoyl monoenoic; III, dienoic; III', methylated dienoic; IV, monounsaturated α'-mycolic acids. Data adapted from Kaneda et al. (1988).

[c]Data adapted from: 1, Kaneda et al. (1988); 2, Valero-Guillén and Martin-Luengo (1986); 3, Kusaka and Mori (1986). Chain lengths of the fatty acid methyl esters formed upon pyrolysis: 22A, $C_{22} > 80\%$; 22B, $C_{22} > 50\%$ and $C_{24} > 20\%$; 24A, $C_{24} > 50\%$ and $C_{22} > 20\%$; 24B, $C_{24} > 80\%$.

[d]Type strain data adapted from Lanéelle et al. (1988).

[e]Data adapted from Daffé et al. (1983).

not included in the study of Kaneda et al. (1988), as *M. fallax* was previously reported to contain unique α-mycolic acids containing three double bonds (Lévy-Frébault et al., 1983). Comparison of the data from these authors, however, also reveals a discrepancy in the reported composition of α-mycolic acids of *M. triviale*. Lévy-Frébault et al. (1983) could not detect unsaturated α-mycolic acids in the type strain of *M. triviale*, whereas Kaneda et al. (1988) detected only dienoic α-mycolic acids in the *M. triviale* strain they analyzed.

The chain lengths of the fatty acid esters released upon pyrolysis of mycolic acid methyl esters have also been analyzed for a number of strains (Table 4). Usually only the major chain length is reported. With the exception of *M. aurum* and *M. diernhoferi* (Table 4), no contradictory results were reported. *M. aurum* (ATCC 25793) contained almost equal amounts of C_{22} and C_{24} (Kusaka and Mori, 1986), whereas *M. aurum* (ATCC 25797) mainly contained C_{24} (Valero-Guillén and Martín-Luengo, 1986), while the type strain *M. aurum* (ATCC 23366) only contained C_{22} (Valero-Guillén and Martín-Luengo, 1986; Lanéelle et al., 1988) although it had previously been reported to contain mainly C_{24} (Daffé et al., 1983).

As the analysis of mycobacterial mycolic acid methyl esters by TLC provides a sensitive and relatively easy method to determine mycolic acid patterns, the technique clearly can be of great use in the identification of mycobacteria and in mycobacterial systematics. For a number of species, however, more strains need to be analyzed to confirm the consistency of this characteristic within the species. Especially the situation observed within the *M. aurum* species requires further research, possibly resulting in the reassignment of several strains presently recognized as belonging to this species or to other species or subspecies. These studies should be combined with the data from biochemical numerical studies. The chain length of the fatty acid methyl ester formed during pyrolysis may be of additional use in characterizing new isolates. The detailed characterization of the α-mycolic acids, as was reported by Kaneda et al. (1988), would seem to be less practical in view of the required analytical equipment.

The use of mycobactins (lipid-soluble intracellular siderophores of mycobacteria) as chemotaxonomic markers was proposed in the past. Their use was, however, often hampered by the difficulty of acquiring sufficient amounts of material for analysis. Mycobactin yields of 1–8% (w/w) were reported for *M. smegmatis* using a simple glycerol/asparagine/phosphate medium solidified with agar (Hall and Ratledge, 1982). Subsequently, the same method was used to analyze the mycobactins of a number of different rapidly growing mycobacteria using TLC and high-performance liquid chromatography (HPLC; Hall and Ratledge, 1984). With most strains tested, mycobactin yields of 4–6% (w/w) were obtained. *Mycobacterium aurum*, *M. parafortuitum* and *M. thermoresistibile* did not form detectable amounts of mycobactin under the standard conditions, though more than two-thirds of these strains did yield mycobactins (3–5%) when grown on glucose/yeast extract/agar. The *M. vaccae*, *M. chelonae* subsp. *chelonae,* and *M. komossense* strains tested did not produce detectable amounts of mycobactin on either growth medium. TLC analysis of 32 strains of 15 species of rapidly growing mycobacteria using two different solvent systems revealed that all strains within a single species (except for *M. flavescens*) produced mycobactin with the same R_f value.

TLC analysis of mycobactins is a relatively simple technique, and on the basis of the data presented by Hall and Ratledge (1984), it appears to be a very useful chemotaxonomic character with a high discriminatory power. However, just as in the case of mycolic acid patterns, much larger numbers of strains have to be analyzed to confirm the chemotaxonomic utility of the mycobactin R_f value in identifying mycobacterial species.

Physiology

The physiology of mycobacteria will not be discussed in an exhaustive manner in this chapter as this subject has been dealt with in excellent reviews by Ratledge (Ratledge, 1976; Ratledge, 1982b) and also in the reviews of Ramakrishnan et al. (1972) and Masood et al. (1985). Furthermore, contrary to what was sometimes suspected in the past, mycobacterial physiology differs from that of other aerobic saprophytes only in minor aspects. This section will focus on the areas of mycobacterial physiology that are typical for the genus. The metabolism of unsaturated gaseous hydrocarbons is one area where the role of the mycobacteria is especially important. Another area to be discussed is fatty acid biosynthesis, because mycobacteria clearly differ in this respect from most other prokaryotes. Reserve materials, fatty acid composition, and iron uptake are also discussed.

Catabolic Activities

Mycobacteria are metabolically versatile organisms. They grow not only on common substrates such as sugars, alcohols, and organic acids, but also on a large variety of hydrocarbons including branched-chain, unsaturated, aromatic, and cyclic hydrocarbons (Söhngen, 1913; Lukins and Foster, 1963). Mycobacteria also degrade polycyclic aromatic hydrocarbons, such as pyrene (Heitkamp et al., 1988a; Heitkamp et al., 1988b) and phenanthrene (Guerin and Jones, 1988). Some mycobacteria grow on the simple one-carbon compounds methanol and methylamines (Kato et al., 1988; Urakami and Yano, 1989). In one strain of *M. gastri*, 3-hexulose-6-phosphate synthase was present, indicating formaldehyde incorporation via the ribulose-monophosphate pathway in this strain (Kato et al., 1988). Autotrophic growth on carbon dioxide and hydrogen gas of several strains of *M. smegmatis*, *M. marinum* and *M. fortuitum* was already reported by Lukins and Foster in 1963 (20% of the strains tested). The propene-utilizing strain called "*Mycobacterium* Py1" (De Bont et al., 1980) also grew autotrophically. Under these conditions, ribulose-1,5-bisphosphate carboxylase and a membrane-bound hydrogenase were induced (C. G. van Ginkel and J. A. M. de Bont, unpublished observations). Reports in the older literature on mycobacteria growing on methane should be treated skeptically since such isolates were either lost before a rigorous identification was performed or no longer grew on methane when investigated by others later.

Metabolism of Gaseous Hydrocarbons

In the past, one of the incentives for studying microorganisms degrading gaseous hydrocarbons was the suspected relation between the numbers of these organisms present in soil and the presence of fossil fuel reserves. Such a relationship has, however, not been substantiated (Brisbane and Ladd, 1972).

Growth of mycobacteria with the C_2 hydrocarbons ethane and ethene (ethylene) has been described several times. Using ethane as carbon source, Davis et al. (1956) isolated several types of ethane-utilizing bacteria, with mycobacteria predominating. The isolated mycobacteria could be divided into two groups on the basis of their capacity to grow on complex media. About half of the isolates did not grow on nutrient agar or with glycerol as sole carbon source and were placed in a separate novel species, "*M. paraffinicum.*" The assignment of "*M. paraffinicum*" to the genus *Mycobacterium* is, however, questionable, as it was shown to contain trehalose mycolates of a relatively low molecular weight (Minnikin and Goodfellow, 1980a). Using ethene as carbon source, several mycobacteria were isolated (De Bont, 1976). *Mycobacterium* E20 also grew with ethane. Ethane metabolism was via acetate and ethene metabolism via epoxyethane (De Bont and Harder, 1978). Epoxyethane was further degraded to acetyl-coenzyme (Co)A in a CoA- and nicotinamide adenine dinucleotide (NAD)-dependent reaction (De Bont and Harder, 1978). In contrast to the alkene monooxygenase induced by growth on ethene, the ethane hydroxylase activity of *Mycobacterium* E20 could not be detected in cell-free extracts of ethane-grown cells (De Bont et al., 1979).

Of the short-chain hydrocarbons, the metabolism of propane and its derivatives has received the most attention in the past. The initial step in propane degradation has been the subject of speculation and study for quite some time. Oxidation of both the primary and secondary carbon atoms of propane has been shown to occur (Perry, 1980). *Mycobacterium vaccae* JOB5 has been used in several cases to study three-carbon metabolism (see Perry, 1980). In this strain, acetone and acetol as well as acetate have been detected as intermediates in propane metabolism. These observations contrast with acetone metabolism in several unidentified Gram-positive bacteria that degrade acetone via pyruvate (Taylor et al., 1980).

Mycobacterium Py1, which was isolated with propene as carbon source (De Bont et al., 1980), did not grow on propane or acetone. However, it utilized acetol, which was oxidatively transformed into acetate and formaldehyde by acetol monooxygenase through a Bayer-Villiger type of reaction (Hartmans and de Bont, 1986). Propene metabolism in *Mycobacterium* Py1 proceeds via an initial oxidation to epoxypropane, which is subsequently carboxylated, presumably to acetoacetate.

Fatty Acid Biosynthesis

Mycobacterial lipids have been the subject of numerous studies, often with an emphasis on taxonomic aspects (Ratledge, 1982a; Minnikin, 1982; Brennan, 1988). Besides the characteristic lipid-rich cell wall, mycobacteria also differ from most other prokaryotes in aspects of fatty acid biosynthesis. Fatty acid synthetases are generally divided into two types: the type I "eukaryotic" system and the type II "prokaryotic" system. The type II system, which readily dissociates into separate proteins with discrete catalytic activities, has been detected in all bacteria studied, with the exception of some mycobacteria and corynebacteria. The type I system present in animals and eukaryotic microorganisms has also been found in *M. smegmatis* and *Corynebacterium diphtheriae*.

Fatty acid biosynthesis in *M. smegmatis* has been studied in detail by Bloch and coworkers (Bloch, 1977). They described two fatty acid synthetase (FAS) activities in this species. One is the extensively studied multienzyme complex called "FAS-I," which is probably very much like the FAS of eukaryotes. FAS-I is, however, unique in that it produces both very long (C_{24} and C_{26}) as well as the more common (C_{16} and C_{18}) saturated CoA esters of fatty acids, and its activity is stimulated by certain polysaccharides. The 3-*O*-methylmannose polysaccharide (MMP) and 6-*O*-methylglucose polysaccharide (MGLP) affect the K_m for both acetyl-CoA and malonyl-CoA and can bind palmitoyl-CoA, thus restricting further chain elongation and consequently influencing the bimodal product distribution. The biosynthesis of these polysaccharides has recently been studied by Ballou and coworkers (Weisman and Ballou, 1984a; Weisman and Ballou, 1984b; Kamisango et al., 1987).

The second FAS in *M. smegmatis* studied by Bloch and coworkers is similar to the type II FAS found in other prokaryotes in its requirement for an acyl-carrier protein (ACP). De novo fatty acid synthesis is, however, not observed with the type II FAS of *M. smegmatis*, so that it should actually be regarded as a fatty-acid-elongating system, elongating acyl-CoA esters of C_{16} to C_{28} (Odriozola et al., 1977). The polysaccharides which affect the type I FAS do not affect the type II elongating system of *M. smegmatis*, and FAS II activity is not inhibited by palmitoyl-CoA.

Another fatty-acid-elongation system (FES I), isolated from *M. smegmatis*, requires acetyl-CoA and is apparently ACP-independent (Shimakata et al., 1977). It exhibits optimal activity with the C_8 and C_{10} acetyl-CoA esters. The enzymatic activities of this fatty acid-elongation system, however, under physiological conditions may be involved in the β-oxidation of fatty acids and utilize the stereospecificity of the 3-oxoacyl-CoA reductase, which forms L-hydroxyacyl-CoA esters (Shimakata et al., 1979). L-Hydroxyacyl-CoA esters normally are intermediates in the degradation of fatty acids.

Subsequently, a third elongation system has been described which was apparently also ACP-independent, but which required malonyl-CoA instead of acetyl-CoA. This FES exhibits activity with C_{10} to C_{24} acetyl-CoA esters, with an optimum for stearyl-CoA (Kikuchi and Kusaka, 1982). Also, a very long-chain fatty acid-elongation system was isolated from *M. avium* (Kikuchi et al., 1989), which possibly is involved in the synthesis of mycolic acids. It differs from the ACP-independent, malonyl-CoA-incorporating-elongation system of *M. smegmatis* (Kikuchi and Kusaka, 1982) in its cofactor requirements and its sensitivity towards isoniazid. The authors suggest that isoniazid possibly affects the 3-oxoacyl-CoA and enoyl-CoA reductase activities. This would explain the previously observed effect of isoniazid specifically inhibiting the synthesis of mycolic acids (Winder, 1982).

In contrast to the common bacterial pathway of unsaturated fatty acid formation by elongation of decenoyl-CoA, resulting in the formation of palmitoleic and *cis*-vaccenic ($C_{18:1}\Delta^{11}$) acid, biosynthesis of unsaturated fatty acids in mycobacteria is accomplished by desaturation of stearyl-CoA to oleoyl-CoA, and to a lesser extent palmitoyl-CoA to palmitoleoyl-CoA, by the particulate Δ^9 desaturase (Ratledge, 1982a). A very long-chain, soluble, fatty acid Δ^{15} desaturase was isolated from *M. smegmatis*, which exhibited optimal activity with lignoceroyl-CoA ($C_{24:0}$; Kikuchi and Kusaka, 1986).

The most common mycobacterial branched-chain fatty acid is tuberculostearic acid (D-10-methylstearic acid). It is formed by methylation of oleic acid residues already esterified in phospholipids (Ratledge, 1982a). The methyl group is derived from *S*-adenosylmethionine.

The various different fatty acid synthases and elongating systems that have been found in mycobacteria are of course a reflection of the many different fatty acids present in these organisms. The different constituents will not be discussed in detail here, as excellent reviews are available (Minnikin, 1982; Brennan, 1988).

Reserve Materials

Besides the mycolic acid composition (see Identification), other lipid components of mycobacteria have also been studied. Mycobacteria can contain considerable amounts of triacylglycerols, especially in glycerol-grown cells, resulting in the formation of fat bodies (lipid vacuoles; Brennan, 1988). These lipids may be utilized as a reserve material, though glycogen and trehalose have also been suggested as reserve materials (Ratledge, 1982b). In nitrogen-limited batch cultures of *M. phlei*, both lipid and glycogen accumulation were observed (Antoine and Tepper, 1969). Transfer of these cells to a medium with a high nitrogen content without carbon resulted in restoration of growth and a decrease of the lipid and glycogen content. Similar experiments with *M. smegmatis* (Elbein and Mitchell, 1973) focused on glycogen and trehalose levels. *Mycobacterium smegmatis* grown under nitrogen-limiting conditions had an increased glycogen content, which was rapidly utilized when these cells were transferred to a medium containing sufficient carbon and nitrogen. The trehalose levels were more or less the same under all growth conditions, indicating that under these conditions trehalose is not a reserve material, although the turnover rates were very high (Elbein and Mitchell, 1973).

Starvation experiments performed with *Mycobacterium* sp. strain E3, a "*M. parafortuitum* complex" species, grown under nitrogen or carbon limitation in chemostat cultures have been performed (Habets-Crützen, 1985a). Cells grown under nitrogen limitation contained 10 times more glycogen and only 50% more trehalose and lipid than cells grown under carbon limitation. The reserve materials of these cells were monitored during a 2-day incubation in buffer in the absence of carbon or nitrogen sources. Trehalose levels fell from 2–3% to less than 0.5% within 8 h. The lipid/protein ratio remained constant during the 2-day starvation experiment. Glycogen was only consumed in the cells grown under nitrogen limitation. The 2% glycogen present in cells grown under carbon limitation was not consumed during the 2-day-starvation experiment. Experiments monitoring the NADH-dependent oxidation of propene to 1,2-epoxypropane by starving cells showed that nitrogen-limited-grown cells, containing the higher glycogen levels, produced more of the epoxide (De Haan et al., 1991).

Lipid Composition

The phospholipids of *M. smegmatis* and *M. phlei* have been studied in some detail (Dhariwal et al., 1976). Major phospholipids present in mycobacteria are phosphatidylinositol, phosphatidylethanolamine, diphosphatidylglycerol (cardiolipin), and the phosphatidylinositol mannosides. The phosphatidylinositol mannosides are highly characteristic for actinomycetes and coryneform bacteria (Brennan, 1988).

The phospholipid and fatty acid composition are to a large extent dependent on the culture conditions. The growth temperature and the carbon source as well as the ratio of carbon and nitrogen sources affect the lipid composition (King and Perry, 1975; Dhariwal et al., 1977). Dhariwal et al. (1976) also reported the effect of culture age on the fatty acid composition. The major change they observed was an increase in tuberculostearic acid accompanied by a decrease in oleic acid (18:1) content. This composition is probably a reflection of the biosynthesis of tuberculostearic acid, which is formed by methylation of esterified oleic acid residues. This perhaps also explains why relatively high tuberculostearic acid contents have often been reported, as, very often, cultures were analyzed which were already in the stationary growth phase.

An important aspect which should be emphasized is that these studies were all performed with batch-grown cultures. This implies that the growth conditions and growth rates were rarely constant, thus making it difficult to ascribe the observed changes in fatty acid composition to a specific factor.

More reliable experimental data require the application of chemostat cultures. As discussed in the section on cultivation, the continuous culture of mycobacteria in chemostats has been demonstrated several times. Therefore the regulation of lipid composition, or indeed many other aspects of mycobacterial physiology, should be studied using chemostat cultures.

Iron Uptake

Mycobacterial iron metabolism, especially iron uptake, has been extensively studied by the group of Ratledge (Ratledge, 1982b; Ratledge, 1984). Mycobacteria appear to be unique in producing two different siderophores (exochelins and mycobactins), probably necessitated by the thick lipoidal nature of the cell envelope. Exochelins are extracellular siderophores, which until now have only been poorly characterized. Two types of exochelins, differing in their solubility in organic solvents, have been described (Ratledge, 1982b). The very hydrophobic mycobactins, which are located within the cell envelope, have been studied to a greater extent. The potential for using the mycobactin composition in the identification of rapidly growing mycobacteria (Hall and Ratledge, 1984) is discussed in the section on identification. Recent work has focused on the regulation of the biosynthesis of the different siderophores and other iron-regulated proteins (Sritharan and Ratledge, 1989). For *M.*

neoaurum, these components were coordinately expressed in the presence of low iron concentrations (<0.2 mg/liter). Increasing the iron concentration to 0.5 mg/liter (or more) resulted in repression of the synthesis of all three components (Sritharan and Ratledge, 1989).

Genetics

Our knowledge of the genetics and the molecular biology of mycobacteria lags behind that of thoroughly characterized species such as *Escherichia coli* or *Bacillus subtilis*. This is partly due to the low growth rate of mycobacteria but also to the ineffectiveness of many of the standard molecular biology techniques when applied to the mycobacteria. Recently, however, significant progress has been made, part of which has already been reviewed (Grange, 1982; Hopwood et al., 1988; Konicek et al., 1988).

Organization of Genetic Information

The estimated genome sizes of mycobacteria vary from 3 to 5.5×10^9 daltons (Baess and Mansa, 1978). As a comparison, the genome size of *E. coli* is 2.5×10^9 daltons. Mycobacterial DNA has a high G+C content (66–71 mol% for the strains examined by Baess and Mansa, 1978).

The presence of extrachromosomal DNA (plasmids) has been demonstrated conclusively in several species. Most of the strains studied were, however, slow-growing pathogens (Hopwood et al., 1988). Plasmids have also been described in fast-growing species although no selective markers could be attributed to them (Labidi et al., 1984). A 173-kb plasmid isolated from a *M. scrofulaceum* species (isolated from the environment) encoded mercury and copper resistance (Meissner and Falkinham, 1984; Erardi et al., 1987). This strain contained a total of four plasmids varying in size from 15 to 300 kb. Two smaller mycobacterial plasmids have been studied in more detail: Plasmid pLR7 (15.3 kb) from *M. intracellulare* has been mapped (Crawford and Bates, 1984), and the complete nucleotide sequence of pAL5000, a 4,837-bp plasmid from *M. fortuitum*, has been determined (Rauzier et al., 1988).

The organization and sequences of mycobacterial genes coding for ribosomal RNA (rRNA) have been studied quite intensively. The rRNA can be easily isolated and sequenced, and these sequences have potential as taxonomic markers (Cox and Katoch, 1986) and are used in determining phylogenic relationships (Woese, 1987; also see Identification and Phylogeny and Taxonomy).

Using rRNA probes derived from *E. coli*, hybridization with DNA from *M. phlei* and *M.*

smegmatis revealed that these fast-growing strains contain two rRNA operons, whereas the slow-growing *M. tuberculosis* and *M. intracellulare* appeared to possess only one rRNA operon (Bercovier et al., 1986). The slow-growing *M. bovis* BCG also contains only one set of rRNA genes (Suzuki et al., 1987). It is tempting to speculate that the number of rRNA operons present in the genome forms the genetic basis for the difference in growth rate between the rapidly and slowly growing mycobacteria. *Escherichia coli* grows much faster than the fast-growing mycobacteria and possesses seven rRNA operons (Bercovier et al., 1986).

The rRNA genes of *M. smegmatis* have been studied by restriction analysis (Bercovier et al., 1989), and, like the *M. bovis* BCG rRNA genes (Suzuki et al., 1987), the genes coding for the different rRNAs are organized in operons in the order 16S—23S—5S, as in other eubacteria (Woese, 1987; Clark-Curtiss, 1990).

Genetic Recombination

Papers concerning the transfer of mycobacterial DNA through transduction and conjugation published in the 1970s have been reviewed (Grange, 1982; Konicek et al., 1988) and will not be discussed here. Genetic recombination of mycobacteria by spheroplast fusion to produce genetically modified strains for sterol transformation has also been reported (Jekkel et al., 1989).

However, other methods of genetic recombination are essential for a better genetic characterization of mycobacteria. Methods to efficiently transfer DNA between mycobacteria and *E. coli* and vice versa are consequently an important tool. The approach of Jacobs et al. (1987) to attack this problem was to construct a vector which replicates as a plasmid in *E. coli* and as a phage in mycobacteria. This was achieved by introducing an *E. coli* plasmid replicon into a nonessential region of mycobacteriophage TM4, a temperate phage of *M. avium*. The resulting "phasmid" phAE1 grows as a lytic phage in *M. smegmatis* and replicates as a plasmid in *E. coli*. DNA transfer into *M. smegmatis* was achieved by transfection of protoplasts, and introduction into *E. coli* was done by in vitro packaging with lambda proteins.

Subsequently, efforts were undertaken to use the same approach to construct lysogenic phasmids that would allow the introduction and maintenance of DNA in growing mycobacteria. "Shuttle phasmids" were constructed in a similar manner as described above using the temperate phage L1 which stably lysogenizes *M. smegmatis* by integrating in the chromosome (Snapper et al., 1988). With one of these phasmids (phAE19),

it was possible to lysogenize *M. smegmatis* protoplasts and generate kanamycin- and chloramphenicol-resistant colonies, thus illustrating the possibility of introducing and expressing foreign genes in mycobacteria (Snapper et al., 1988).

A much larger stimulation of research of mycobacterial genetics is expected from the construction of plasmids capable of replicating in both *E. coli* and mycobacteria. Snapper et al. (1988) have constructed such hybrid shuttle plasmids by randomly inserting the *E. coli* plasmid pIJ666, containing an origin of replication and the genes for kanamycin and chloramphenicol resistance in pAL5000 from *M. fortuitum*. Transformation of the pIJ666::pAL5000 library into *M. smegmatis* protoplasts was not successful. Subsequently the high voltage electroporation technique was applied. This method had previously been demonstrated to be useful in the transformation of other Gram-positive bacteria (Chassy and Flickinger, 1987). Electroporation conditions were optimized for the uptake of lytic D29 phage DNA by intact *M. smegmatis* cells, resulting in $>5 \times 10^3$ plaque-forming units per μg DNA. Electroporation of *M. smegmatis* using the optimized procedure with the pIJ666::pAL5000 recombinant library yielded 1–10 kanamycin-resistant transformants per μg DNA. Plasmids isolated from the transformants were used in retransformation experiments yielding kanamycin-resistant *E. coli* and *M. smegmatis* colonies.

Gicquel-Sanzey et al. (1989) used a similar approach to construct the 9.2-kb vector pAL8 by combining pAL5000 with an *E. coli* plasmid and a gene coding for kanamycin resistance. However, transformation of spheroplasts with the pAL8 vector was not successful using conditions under which transformation with the lytic mycobacteriophage D29 resulted in transformation efficiencies of 10^4 to 10^5 per μg DNA. Using the electroporation technique, Gicquel-Sanzey et al. (1989) reported transformation frequencies for *M. smegmatis* of 10 per μg pAL8 DNA. Electroporation of *M. smegmatis* with phage D29 resulted in transformation frequencies of 10^3 per μg DNA, similar to the rate reported by Snapper et al. (1988). Both groups report much higher transformation efficiencies (10^3 per μg DNA) for *M. bovis* BCG. Recently though, high-efficiency-transforming mutants of *M. smegmatis* have been isolated, yielding more than 10^5 transformants per μg DNA (Snapper et al., 1990b). Interestingly, these mutants do not show enhanced transformation frequencies with an integrating vector that recombines into the *M. smegmatis* chromosome. It is suggested that the mutation possibly affects plasmid replication and maintenance in *M. smegmatis*. Using these mutants, an essential replication region of pAL5000 was mapped and the gene coding for the 65-kDa stress-protein antigen of *M. leprae* was expressed (Snapper et al., 1990a).

Deletion experiments with pAL8 have resulted in the construction of a smaller shuttle plasmid of 6.6 kb (pRR3) incorporating only 2.58 kb of pAL5000 (Ranes et al., 1990). The transformation rates of *M. bovis* BCG and a high-transforming mutant of *M. smegmatis* (Snapper et al., 1990b) using the electroporation technique with this plasmid were 10^4 per µg pRR3 DNA.

Transformation by electroporation of *M. aurum* and *M. smegmatis* with the broad-host-range, Gram-negative vector pJRD215 has also been demonstrated (Hermans et al., 1991). In contrast to the constructs described above, this cosmid vector does not contain mycobacterial DNA, i.e., a mycobacterial origin of replication. An advantage of pJRD215 is that it contains the phage lambda *cos* site, allowing the cloning of relatively large DNA fragments in *E. coli* and thereby facilitating the construction of genomic libraries. Expression of pJRD215 in mycobacteria should allow the screening of such libraries by the complementation of mutants. The kanamycin- and streptomycin-resistance genes carried by pJRD215 were both expressed in *M. aurum*. The transformation efficiency, as determined by screening for kanamycin resistance, was rather low at 2×10^2 transformants per µg DNA. Enhancement of the transformation frequency by pretreatment of the cells with isoniazid, as was demonstrated for the transformation of *M. aurum* with pAL8 (Hermans et al., 1990), might also be applied successfully for the transformation of mycobacteria with pJRD215.

Another approach to stably introduce DNA in mycobacteria, using a vector without a mycobacterial origin of replication, requires homologous recombination to take place. This was demonstrated with a shuttle vector that can replicate autonomously in *E. coli* but must integrate into homologous DNA for survival in *M. smegmatis* (Husson et al., 1990). The vector, pY6002, contained an *E. coli* origin of replication and the *pyrF* gene of *M. smegmatis* with the kanamycin-resistance gene *aph* as an insert. The *pyrF*-negative mutants are both uracil auxotrophic and fluorouracil resistant. The transformation frequency of *M. smegmatis* by electroporation with this vector was 10–500 transformants per µg DNA. Integration of pY6002 in the chromosomal DNA of the prototrophic "wild-type" *M. smegmatis* gave two types of kanamycin-resistant recombinants resulting from either a single or double recombination event. Class I recombinants resulting from a single recombination event contained the entire plasmid as well as a functional *pyrF* gene, whereas class II recombinants were uracil auxotrophic and fluorouracil

resistant. In class II recombinants, the plasmid had apparently integrated in the chromosome at the *pyrF* locus and replaced it. Class II transformants could be retransformed with a plasmid containing the intact *pyrF* gene without an *aph* insert (pY6001). Selection for uracil prototrophs gave both classes of transformants. Class I, resulting from a single recombination event, contained both the disrupted and the wild-type *pyrF* gene, while class II kanamycin-sensitive transformants only contained the wild-type *pyrF* gene. Using this technique, the 65-kDa stress-protein antigen of *M. leprae* was expressed in *M. smegmatis* at detectable levels. Two vectors were used with the *M. leprae* gene on a 3.6-kb fragment inserted at different sites of pY6002. The transformation frequencies were comparable to those of pY6002. The two transformants expressing the 65-kDa antigen were kanamycin resistant and fluorouracil sensitive, indicating them to be class I transformants (Husson et al., 1990).

The methods and tools allowing the transformation and molecular genetic manipulation of mycobacteria described above (Jacobs et al., 1987; Snapper et al., 1988; Snapper et al., 1990a; Husson et al., 1990; Hermans et al., 1991) should result in major advances in mycobacterial research in the near future.

Applications

The most obvious applications of mycobacteria are of course associated with the disease tuberculosis. Examples are the BCG (Bacillus Calmette-Guérin) vaccine, which is derived from the *Mycobacterium bovis* BCG strain, and the production of tuberculin, which is extracted from *Mycobacterium* cultures. Tuberculin is used diagnostically in a delayed hypersensitivity type of skin test for the detection of a current or previous infection with *M. tuberculosis*.

In this section we will focus on nonmedical applications. Mycobacteria are used mainly in the area of biocatalysis to perform specific transformation reactions. The justification for using microorganisms or enzymes for a particular transformation is usually the selectivity and specificity (regio- or stereospecificity) exhibited by the biocatalyst (Meijer et al., 1985).

One area in biotechnology that has a large number of commercial applications utilizing mycobacteria is the biotransformation of steroids. These processes involve either the modification of the steroid nucleus of natural or synthetic sterols or the selective degradation of the side chain of naturally occurring sterols such as cholesterol and β-sitosterol (Martin, 1984). The products formed can subsequently be chemically transformed to pharmacologically active sterols (Martin, 1984). Cholesterol degradation by mycobacteria has been known for a long time (Söhngen, 1913) and usually involves simultaneous degradation of both the side chain and the steroid nucleus. Selective modification of only the nucleus or the side chain can be achieved by employing inhibitors in the biotransformation process or by using mutants (Martin, 1984). Wovcha et al. (1978) for example obtained *M. fortuitum* mutants blocked in various steps of sitosterol degradation. With these mutants, intermediates of the steroid-degradation pathway can be produced. Some of these intermediates can be used as substrates for the production of medically useful steroids. Most commercial processes for the selective side-chain cleavage of sterols employ mutant strains of *M. fortuitum* or *M. parafortuitum*, although other *Mycobacterium* species and bacteria of related genera are also used (Martin, 1984).

Many biocatalytic processes comprise only one enzymatic step, using enzyme preparations or whole (permeabilized) cells. An example is the utilization of *M. neoaurum* (ATCC 25795) containing an L-specific aminopeptidase with a very high stereospecificity and broad substrate specificity. Using permeabilized *M. neoaurum* cells, a wide range of L- or D-α-methyl-substituted amino acids can be produced by stereoselective hydrolysis of racemic mixtures of the corresponding amides (Kamphuis et al., 1987).

Another potential application of mycobacteria in the field of biocatalysis is the production of optically active epoxides from alkenes (Habets-Crützen et al., 1985b; Hartmans et al., 1989). Optically pure epoxides form versatile starting materials for the chemical synthesis of optically active pharmaceutical compounds. Screening a number of bacteria from different genera revealed that alkene-grown mycobacteria produce the epoxides examined (1,2-epoxypropane, 1,2-epoxybutane, and 2,3-epoxy-1-chloropropane) in the highest enantiomeric excess (Weijers et al., 1988). However, substantial research efforts will be required to increase the specific activity and the operational stability of the biocatalysts before commercial production of epoxides by the alkene-utilizing mycobacteria can be realized. In this respect, product toxicity and cofactor regeneration have been studied for the production of 1,2-epoxypropane with the ethene-utilizing *Mycobacterium* strain E3 (Habets-Crützen and De Bont, 1985a; Habets-Crützen and De Bont, 1987).

One aspect that has received some attention is the use of organic solvents and immobilization of mycobacteria for application in continuously operated processes (Steinert et al., 1986; Flygare and Larsson, 1987). The advantages of immobilization (i.e., increased reactor productivity or a better operational stability) were however not very evident. One advantage that immobilization of cells can offer is that organic solvents, which can function as a reservoir for substrates and products with a low water solubility, can be used without the problems of cell aggregation and biocatalyst/solvent separation associated with the use of organic solvents with free cells (Linko and Linko, 1985). However, immobilization does not protect the cells against the adverse effects many solvents have on mycobacteria (Brink and Tramper, 1985; Steinert et al., 1986). In the study of Brink and Tramper (1985), complete retention of activity was observed in the presence of dioctyl and didecyl phthalate, but most solvents resulted in decreased biocatalytic activity.

The effect of water-immiscible organic solvents on growth has also been tested with the rapidly growing strain *Mycobacterium* E3 and several other bacteria from different genera (Rezessy-Szabó et al., 1986). From these experiments, *Mycobacterium* E3 was not particularly resistant to organic solvents compared to other bacteria, as for instance *Pseudomonas* species.

Another area of research concerning the potential application of mycobacteria is in environmental biotechnology. Removal of traces of the plant hormone ethene from storage facilities for fruit using immobilized ethene-utilizing mycobacteria is one example (Van Ginkel et al., 1986). Unfortunately, the activity of the ethene-utilizing mycobacteria was very low at the ethene levels (often less than 1 part per million [ppm]) prevailing in the fruit storage facilities. Another example is the application of the vinyl chloride-utilizing *Mycobacterium* strain L1 to remove the carcinogenic vinyl chloride from industrial waste gases (Hartmans and De Bont, 1992). The compound is metabolized by an initial oxidation step to the corresponding epoxide by alkene monooxygenase. Accumulation even of very low concentrations of this reactive intermediate, which can occur because of fluctuations in the vinyl chloride supply, irreversibly inhibited the alkene monooxygenase. As such fluctuations will probably frequently occur, practical application of this strain does not seem realistic.

The use of mycobacteria in the bioremediation of contaminated sediments has also been suggested. Addition of a pyrene-degrading *Mycobacterium* strain to sediments resulted in an enhancement of the mineralization rates of several polycyclic aromatic hydrocarbons. Further investigations were suggested to assess the potential of this strain in the bioremediation of contaminated sediments (Heitkamp and Cerniglia, 1989).

A disadvantage of mycobacteria in general is their relatively low growth rate and hence their low catalytic activities. However, mycobacteria are likely to play an increasingly important role as a source of interesting biocatalytic capacities now that the development of the molecular genetic tools required to increase the expression of desired activities, either in mycobacteria or other hosts, is well under way.

Literature Cited

Antoine, A. D., and B. S. Tepper. 1969. Environmental control of glycogen and lipid content of Mycobacterium phlei. J. Gen. Microbiol. 55:217–226.

Aranaz, A., D. Cousins, A. Mateos, and L. Dominguez. 2003. Elevation of Mycobacterium tuberculosis subsp. caprae Aranaz et al. 1999 to species rank as Mycobacterium caprae comb. nov., sp. nov. Int. J. Syst. Evol. Microbiol. 53:1785–1789.

Ausina, V., M. Luquin, and L. Margarit. 1985. Mycolic acids of Mycobacterium chitae. J. Gen. Microbiol. 131:2237–2239.

Ausina, V., M. Luquin, M. Garcia Barcelo, M. A. Laneelle, V. Levy-Frebault, F. Belda, and G. Prats. 1992. Mycobacterium alvei sp. nov. Int. J. Syst. Bacteriol. 42:529–535.

Baess, I., and B. Mansa. 1978. Determination of genome size and base ratio on deoxyribonucleic acid from mycobacteria. Acta Path. Microbiol. Immunol. Scand. Sect. B 86:309–312.

Baess, I. 1982. Deoxyribonucleic acid relatedness among species of rapidly growing mycobacteria. Acta Path. Microbiol. Immunol. Scand. Sect. B 90:371–175.

Barksdale, L., and K.-S. Kim. 1977. Mycobacterium. Bacteriol. Rev. 41:217–372.

Bercovier, H., O. Kafri, and S. Sela. 1986. Mycobacteria possess a surprisingly small number of ribosomal RNA genes in relation to the size of their genome. Biochem. Biophys. Res. Comm. 136:1136–1141.

Bercovier, H., O. Kafri, D. Kornitzer, and S. Sela. 1989. Cloning and restriction analysis of ribosomal RNA genes from Mycobacterium smegmatis. FEMS Microbiol. Lett. 57:125–128.

Bloch, K. 1977. Control mechanisms for fatty acid synthesis in Mycobacterium smegmatis. Adv. Enzymol. 45:1–84.

Böddinghaus, B., T. Rogall, T. Flohr, H. Blöcker, and E. Böttger. 1990a. Detection and identification of mycobacteria by amplification of rRNA. J. Clin. Microbiol. 28:1751–1759.

Böddinghaus, B., J. Wolters, W. Heikens, and E. C. Böttger. 1990b. Phylogenetic analysis and identification of different serovars of Mycobacterium intracellulare at the molecular level. FEMS Microbiol. Lett. 70:197–203.

Böttger, E. C., B. Hirschel, and M. B. Coyle. 1993. Mycobacterium genavense sp. nov. Int. J. Syst. Bacteriol. 43:841–843.

Bradley, S. G. 1973. Relationships among mycobacteria and nocardiae based upon deoxyribonucleic acid reassociation. J. Bacteriol. 113:645–651.

Brennan, P. J. 1988. Mycobacterium and other actinomycetes. *In:* C. Ratledge and S. G. Wilkinson (Eds.) Microbial Lipids. Academic Press. London, UK. 1:203–298.

Brink, L. E. S., and J. Tramper. 1985. Optimization of organic solvent in multiphase biocatalysis. Biotechnol. Bioengin. 27:1258–1269.

Brisbane, P. G., and J. N. Ladd. 1972. Growth of Mycobacterium paraffinicum on low concentrations of ethane in soils. J. Appl. Bacteriol. 35:659–665.

Brown, B. A., B. Springer, V. A. Steingrube, R. W. Wilson, G. E. Pfyffer, A. M. J. Garcia, M. C. Menendez, B. Rodriguez-Salgado, K. C. Jost Jr., S. H. Chiu, G. O. Onyi, E. C. Böttger, and R. J. Wallace Jr. 1999. Mycobacterium wolinskyi sp. nov. and Mycobacterium goodii sp. nov., two new rapidly growing species related to Mycobacterium smegmatis and associated with human wound infections: a cooperative study from the International Working Group on Mycobacterial Taxonomy. Int. J. Syst. Bacteriol. 49:1493–1511.

Butler, W. R., S. P. O'Connor, M. A. Yakrus, R. W. Smithwick, B. B. Plykaytis, C. W. Moss, M. M. Floyd, C. L. Woodley, J. O. Kilburn, F. S. Vadney, and W. M. Gross. 1993. Mycobacterium celatum sp. nov. Int. J. Syst. Bacteriol. 43:539–548.

Carson, L. A., L. A. Bland, L. B. Cusick, M. S. Favero, G. A. Bolan, A. L. Reingold, and R. B. Good. 1988. Prevalence of nontuberculous mycobacteria in water supplies of hemodialysis centers. Appl. Environ. Microbiol. 54:3122–3125.

Cech, J. S., P. Hartman, M. Slosarek, and J. Chudoba. 1988. Isolation and identification of a morpholine-degrading bacterium. Appl. Environ. Microbiol. 54:619–621.

Chassy, B. M., and J. L. Flickinger. 1987. Transformation of Lactobacillus casei by electroporation. FEMS Microbiol. Lett. 44:173–177.

Clark-Curtiss, J. E. 1990. Genome structure of mycobacteria. *In:* J. J. McFadden (Ed.) Molecular Biology of the Mycobacteria. Academic Press. London, UK. 77–96.

Collins, M. D., T. Pirouz, M. Goodfellow, and D. E. Minnikin. 1977. Distribution of menaquinones in actinomycetes and corynebacteria. J. Gen. Microbiol. 100: 221–230.

Collins, C. H., J. M. Grange, and M. D. Yates. 1984. Mycobacteria in water. J. Appl. Bacteriol. 57:193–211.

Collins, M. D., J. Smida, M. Dorsch, and E. Stackebrandt. 1988. Tsukamurella gen. nov. harboring Corynebacterium paurometabolum and Rhodococcus aurantiacus. Int. J. Syst. Bacteriol. 38:385–391.

Cooksey, R. C., M. A. Yakrus, G. P. Morlock, J. H. DeWaard, and R. W. Butler. Mycobacterium cosmeticum, sp. nov., a novel rapidly growing species. Unpublished A449739 54(Pt 6):2385–2391.

Cousins, D. V., R. Bastida, A. Cataldi, V. Quse, S. Redrobe, S. Dow, P. Duignan, A. Murray, C. Dupont, N. Ahmed, D. M. Collins, W. R. Butler, D. Dawson, D. Rodriguez, J. Loureiro, M. I. Romano, A. Alito, M. Zumarraga, and A. Bernardelli. 2003. Tuberculosis in seals caused by a novel member of the Mycobacterium tuberculosis complex: Mycobacterium pinnipedii sp. nov. Int. J. Syst. Evol. Microbiol. 53:1305–1314.

Cox, R. A., and V. M. Katoch. 1986. Evidence for genetic divergence in ribosomal RNA genes in mycobacteria. FEBS Lett. 195:194–198.

Crawford, J. T., and J. H. Bates. 1984. Restriction endonuclease mapping and cloning of Mycobacterium intracellulare plasmid pLR7. Gene 27:331–333.

Daffé, M., M. A. Lanéelle, C. Asselineau, V. Lévy-Frébault, and H. David. 1983. Intéret taxonomique des acides gras des mycobactéries: proposition d'une méthode d'analyse. Ann. Microbiol. (Paris) 134B:241–256.

David, H. L. 1984. Carotenoid pigments of the mycobacteria. *In:* G. P. Kubica and L. G. Wayne (Eds.) The Mycobacteria: A Source Book, Part A. Marcel Dekker. New York, NY. 537–545.

Davis, J. B., H. H. Chase, and R. L. Raymond. 1956. Mycobacterium paraffinicum n. sp., a bacterium isolated from soil. Appl. Microbiol. 4:310–315.

De Bont, J. A. M. 1976. Oxidation of ethylene by soil bacteria. Ant. v. Leeuwenhoek 42:59–71.

De Bont, J. A. M., and W. Harder. 1978. Metabolism of ethylene by Mycobacterium E20. FEMS Microbiol. Lett. 3:89–93.

De Bont, J. A. M., M. M. Attwood, S. B. Primrose, and W. Harder. 1979. Epoxidation of short-chain alkenes in Mycobacterium E20: The involvement of a specific mono-oxygenase. FEMS Microbiol. Lett. 6:183–188.

De Bont, J. A. M., S. B. Primrose, M. D. Collins, and W. Harder. 1980. Chemical studies on some bacteria which utilize gaseous unsaturated hydrocarbons. J. Gen. Microbiol. 117:97–102.

De Haan, A., M. R. Smith, W. Voorhorst, and J. A. M. de Bont. 1993. The effect of calcium alginate entrapment on the physiology of Mycobacterium sp. strain E3. Applied Microbiology and Biotechnology 38:5:642–648.

De Smet, K. A., I. N. Brown, M. Yates, and J. Ivanyi. 1995. Ribosomal internal transcribed spacer sequences are identical among Mycobacterium avium-intracellulare complex isolates from AIDS patients, but vary among isolates from elderly pulmonary disease patients. Microbiology 141:2739–2747.

De Smet, K. A., T. J. Hellyer, A. W. Khan, I. N. Brown, and J. Ivanyi. 1996. Genetic and serovar typing of clinical isolates of the Mycobacterium avium-intracellulare complex. Tuberc. Lung Dis. 77:71–76.

Devallois, A., K. S. Goh, and N. Rastogi. 1997. Rapid identification of mycobacteria to species level by PCR-restriction fragment length polymorphism analysis of the hsp65 gene and proposition of an algorithm to differentiate 34 mycobacterial species. J. Clin. Microbiol. 35:2969–2973.

Dhariwal, K. R., A. Chander, and T. A. Venkitasubramanian. 1976. Alterations in lipid constituents during growth of Mycobacterium smegmatis CDC 46 and Mycobacterium phlei ATCC 354. Microbios 16:169–182.

Dhariwal, K. R., A. Chander, and T. A. Venkitasubramanian. 1977. Environmental effects on lipids of Mycobacterium phlei ATCC 354. Can. J. Microbiol. 23:7–19.

Domenech, P., M. S. Jiminez, M. C. Menendez, T. J. Bull, S. Samper, A. Manrique, and M. J. Garcia. 1997. Mycobacterium mageritense sp. nov. Int. J. Syst. Bacteriol. 47:535–540.

Draper, P. 1982. The anatomy of mycobacteria. In: C. Ratledge, and J. Stanford (Eds.) The Biology of the Mycobacteria, Volume 1: Physiology, Identification and Classification. Academic Press. London, UK. 9–52.

Elbein, A. D., and M. Mitchell. 1973. Levels of glycogen and trehalose in Mycobacterium smegmatis and the purification and properties of the glycogen synthetase. J. Bacteriol. 113:863–873.

Erardi, F. X., M. L. Failla, and J. O. Falkinham, 3rd. 1987. Plasmid-encoded copper resistance and precipitation by Mycobacterium scrofulaceum. Appl. Environ. Microbiol. 53:1951–1954.

Estrada-Garcia, I. C. E., M. Joseph Colston, and R. A. Cox. 1989. Determination and evolutionary significance of nucleotide sequences near to the 3′-end of 16S ribosomal RNA of mycobacteria. FEMS Microbiol. Lett. 62:285–290.

Fanti, F., E. Tortoli, L. Hall, G. D. Roberts, R. M. Kroppenstedt, I. Dodi, S. Conti, L. Polonelli, and C. Chezzi. 2004. Mycobacterium parmense sp. nov. Int. J. Syst. Evol. Microbiol.

Floyd, M. M., L. S. Guthertz, V. A. Silcox, P. S. Duffey, Y. Jang, E. P. Desmond, J. T. Crawford, and W. R. Butler. 1996. Characterization of an SAV organism and proposal of Mycobacterium triplex sp. nov. J. Clin. Microbiol. 34:2963–2967.

Floyd, M. M., W. M. Gross, D. A. Bonato, V. A. Silcox, R. W. Smithwick, B. Metchock, J. T. Crawford, and W. R. Butler. 2000. Mycobacterium kubicae sp. nov., a slowly growing, scotochromogenic Mycobacterium. Int. J. Syst. Evol. Microbiol. 50:1811–1816.

Flygare, S., and P.-O. Larsson. 1987. Steroid transformation using magnetically immobilized mycobacterium sp. Enz. Microb. Technol. 9:494–499.

Frothingham, R., and K. H. Wilson. 1994. Molecular phylogeny of the Mycobacterium avium complex demonstrates clinically meaningful divisions. J. Infect. Dis. 169:305–312.

Garcia, M. J., and E. Tabarés. 1986. Separation of Mycobacterium gadium from other rapidly growing mycobacteria on the basis of DNA homology and restriction endonuclease analysis. J. Gen. Microbiol. 132:2265–2269.

Gicquel-Sanzey, B., J. Moniz-Pereira, M. Gheorghiu, and J. Rauzier. 1989. Structure of pAL5000, a plasmid from M. fortuitum and its utilization in transformation of mycobacteria. Acta Leprol. 7:208–211.

Girbal, E., R. A. Binot, and P. F. Monsan. 1989. Production, purification, properties and kinetic studies of free and immobilized polyphosphate: glucose-6-phosphotransferase from Mycobacterium phlei. Enz. Microb. Technol. 11:518–527.

Goodfellow, M., and L. G. Wayne. 1982. Taxonomy and nomenclature. In: C. Ratledge and J. Stanford (Eds.) The Biology of the Mycobacteria, Volume 1: Physiology, Identification and Classification. Academic Press. London, UK. 471–521.

Goodfellow, M., and T. Cross. 1984. Classification. In: M. Goodfellow, M. Mordarski, and S. T. Williams (Eds.) The Biology of the Actinomycetes. Academic Press. London, UK. 8–164.

Grange, J. M. 1982. The genetics of mycobacteria and mycobacteriophages. In: C. Ratledge and J. Stanford (Eds.) The Biology of the Mycobacteria, Volume 1: Physiology, Identification and Classification. Academic Press. London, UK. 309–351.

Guerin, W. F., and G. E. Jones. 1988. Mineralization of phenanthrene by a mycobacterium sp. Appl. Environ. Microbiol. 54:937–944.

Haas, W. H., W. R. Butler, P. Kirschner, B. B. Plikaytis, M. B. Coyle, B. Amthor, A. G. Steigerwalt, D. J. Brenner, M. Salfinger, J. T. Crawford, E. C. Böttger, and H. J. Bremer. 1997. A new agent of mycobacterial lymphadenitis in children: Mycobacterium heidelbergense sp. nov. J. Clin. Microbiol. 35:3203–3209.

Habets-Crützen, A. Q. H., and J. A. M. de Bont. 1985a. Inactivation of alkene oxidation by epoxides in alkene- and alkane-grown bacteria. Appl. Microbiol. Biotechnol. 22:428–433.

Habets-Crützen, A. Q. H., S. J. N. Carlier, J. A. M. de Bont, D. Wistuba, V. Schurig, S. Hartmans, and J. Tramper. 1985b. Stereospecific formation of 1,2-epoxypropane, 1,2-epoxybutane and 1-chloro-2,3-epoxypropane by alkene-utilizing bacteria. Enz. Microbiol. Technol. 7:17–21.

Habets-Crützen, A. Q. H., and J. A. M. de Bont. 1987. Effect of co-substrates on 1,2-epoxypropane formation from propene by ethene-utilizing mycobacteria. Appl. Microbiol. Biotechnol. 26:434–438.

Häggblom, M. M., L. J. Nohynek, N. J. Palleroni, K. Kronquist, E. L. Nurmiaho-Lassila, M. S. Salkinoja-Salonen, S. Klatte, and R. M. Kroppenstedt. 1994. Transfer of polychlorophenol-degrading Rhodococcus chlorophenolicus Apajalahti et al. 1986 to the genus Mycobacterium as Mycobacterium chlorophenolicum comb. nov. Int. J. Syst. Bacteriol. 44:485–493.

Hall, R. M., and C. Ratledge. 1982. A simple method for the production of mycobactin, the lipid-soluble siderophore, from mycobacteria. FEMS Microbiol. Lett. 15:133–136.

Hall, R. M., and C. Ratledge. 1984. Mycobactins as chemotaxonomic characters for some rapidly growing mycobacteria. J. Gen. Microbiol. 130:1883–1892.

Harmsen, D., S. Dostal, A. Roth, S. Niemann, J. Rothgänger, M. Sammeth, J. Albert, M. Frosch, and E. L. Richter. 2003. RIDOM: Comprehensive and public sequence database for identification of mycobacterium species. (http://www.biomedcentral.com/1471-2334/3/26). BMC Infect. Dis. 3:26.

Hartmans, S., and J. A. M. de Bont. 1986. Acetol monooxygenase from Mycobacterium Py1 cleaves acetol into acetate and formaldehyde. FEMS Microbiol. Lett. 36:155–158.

Hartmans, S., J. A. M. de Bont, and W. Harder. 1989. Microbial metabolism of short-chain unsaturated hydrocarbons. FEMS Microbiol. Rev. 63:235–264.

Hartmans, S., and J. A. de Bont. 1991. The genus Mycobacterium—nonmedical. In: A. Balows, H. G. Trüper, M. Dworkin., W. Harder, and K.-H. Schleifer (Eds.) The Prokaryotes, 2nd ed. Springer-Verlag. New York, NY pp 1214–1237.

Hartmans, S., and J. A. de Bont. 1992. Aerobic vinyl chloride metabolism in Mycobacterium aurum. Appl. Environ. Microbiol. 58:1220–1226.

Hartskeerl, R. A., M. Y. L. de Wit, and P. R. Klatser. 1989. Polymerase chain reaction for the detection of Mycobacterium leprae. J. Gen. Microbiol. 135:2357–2364.

Heitkamp, M. A., W. Franklin, and C. E. Cerniglia. 1988a. Microbial metabolism of polycyclic aromatic hydrocarbons: Isolation and characterization of a pyrene-degrading bacterium. Appl. Environ. Microbiol. 54:2549–2555.

Heitkamp, M. A., J. P. Freeman, D. W. Miller, and C. E. Cerniglia. 1988b. Pyrene degradation by a mycobacterium sp.: Identification of ring oxidation and ring fission products. Appl. Environ. Microbiol. 54:2556–2565.

Heitkamp, M. A., and C. E. Cerniglia. 1989. Polycyclic aromatic hydrocarbon degradation by a mycobacterium sp. in microcosms containing sediment and water from a pristine ecosystem. Appl. Environ. Microbiol. 55:1968–1973.

Hermans, J., J. G. Boschloo, and J. A. M. de Bont. 1990. Transformation of Mycobacterium aurum by electroporation: The use of glycine, lysozyme and isonicotinic acid hydrazide in enhancing transformation efficiency. FEMS Microbiol. Lett. 72:221–224.

Hermans, J., C. Martin, G. N. M. Huijberts, T. Goosen, and J. A. M. de Bont. 1991. Transformation of Mycobacterium aurum and Mycobacterium smegmatis with the broad-host-range Gram-negative cosmid vector pJRD125. Molec. Microbiol. 5(6):1561–1566.

Holberg-Petersen, M., M. Steinbakk, K. J. Figenschau, E. Jantzen, J. Eng, and K. K. Melby. 1999. Identification of clinical isolates of mycobacterium spp. by sequence analysis of the 16S ribosomal RNA gene: Experience from a clinical laboratory. APMIS 107:231–239.

Hopwood, D. A., T. Kieser, M. J. Colston, and F. I. Lamb. 1988. Molecular biology of mycobacteria. Br. Med. Bull. 44:528–546.

Hurley, S. S., G. A. Splitter, and R. A. Welch. 1988. Deoxyribonucleic acid relatedness of Mycobacterium paratuberculosis to other members of the family Mycobacteriaceae. Int. J. Syst. Bacteriol. 38:143–146.

Husson, R. N., B. E. James, and R. A. Young. 1990. Gene replacement and expression of foreign DNA in mycobacteria. J. Bacteriol. 172:519–524.

Imaeda, T. 1985. Deoxyribonucleic acid relatedness among selected strains of Mycobacterium tuberculosis, Mycobacterium bovis, Mycobacterium bovis BCG, Mycobacterium microti, and Mycobacterium africanum. Int. J. Syst. Bacteriol. 35:147–150.

Imaeda, T., G. Broslawski, and S. Imaeda. 1988. Genomic relatedness among mycobacterial species by nonisotopic blot hybridization. Int. J. Syst. Bacteriol. 38:151–156.

Jacobs, W. R., M. Tuckman, and B. R. Bloom. 1987. Introduction of foreign DNA into mycobacteria using a shuttle phasmid. Nature 327:532–534.

Jekkel, A., E. Csajági, E. Ilköy, and G. Ambrus. 1989. Genetic recombination by spheroplast fusion of sterol-transforming mycobacterium strains. J. Gen. Microbiol. 135:1727–1733.

Jenkins, P. A., S. R. Pattyn, and F. Portaels. 1982. Diagnostic bacteriology. In: C. Ratledge and J. Stanford (Eds.) The Biology of the Mycobacteria, Volume 1: Physiology, Identification and Classification. Academic Press. London, UK. 441–470.

Jiménez, M. S., M. I. Campos-Herrero, D. García, M. Luquín, L. Herrera, and M. J. García. 2004. Mycobacterium canariasense sp. nov. Int. J. Syst. Evol. Microbiol. 54(Pt 5):1729–1734.

Jones, R. J., and D. E. Jenkins. 1965. Mycobacteria isolated from soil. Can. J. Microbiol. 11:127–133.

Kamisango, K.-I., A. Dell, and C. E. Ballou. 1987. Biosynthesis of the mycobacterial O-methylglucose lipopolysaccharide. J. Bacteriol. 262:4580–4586.

Kamphuis, J., C. H. M. Schepers, W. H. J. Boesten, M. J. A. Roberts, P. J. H. Peters, J. A. M. van Balken, E. M. Meijer, and H. E. Schoemaker. 1987. Stereoselective enzymatic hydrolysis of α-disubstituted amino acid amides with an aminopeptidase from Mycobacterium neoaurum. In: O. M. Neijssel, R. R. van der Meer, and K. C. A. M. Luyben (Eds.) Proceedings of the 4th European Congress on Biotechnology. Elsevier Science Publishers. Amsterdam, The Netherlands. 2:164.

Kaneda, K., S. Imaizumi, S. Mizuno, T. Baba, M. Tsukamura, and I. Yano. 1988. Structure and molecular species composition of three homologous series of α-mycolic acids from mycobacterium spp. J. Gen. Microbiol. 134:2213–2229.

Kasai, H., T. Ezaki, and S. Harayama. 2000. Differentiation of phylogenetically related slowly growing mycobacteria by their gyrB sequences. J. Clin. Microbiol. 38:301–308.

Kato, N., N. Miyamoto, M. Shimao, and C. Sakazawa. 1988. 3-Hexulose phosphate synthase from a new facultative methylotroph Mycobacterium gastri MB19. Agric. Biol. Chem. 52:2659–2661.

Kazda, J. F. 1980. Mycobacterium sphagni sp. nov. Int. J. Syst. Bacteriol. 30:77–81.

Kazda, J. F. 1983. The principles of the ecology of mycobacteria. *In:* C. Ratledge and J. Stanford (Eds.) The Biology of the Mycobacteria, Volume 2: Immunological and Environmental Aspects. Academic Press. London UK, 323–415.

Kazda, J., E. Stackebrandt, J. Smida, D. E. Minnikin, M. Daffé, J. H. Parlett, and C. Pitulle. 1990. Mycobacterium cookii sp. nov. Int. J. Syst. Bacteriol. 40:217–223.

Kazda, J., H. J. Müller, E. Stackebrandt, M. Daffe, K. Müller, and C. Pitulle. 1992. Mycobacterium madagascariense sp. nov. Int. J. Syst. Bacteriol. 42:524–528.

Kazda, J., R. Cooney, M. Monaghan, P. J. Quinn, E. Stackebrandt, M. Dorsch, M. Daffé, K. Müller, B. R. Cook, and Z. S. Tarnok. 1993. Mycobacterium hiberniae sp. nov. Int. J. Syst. Bacteriol. 43:352–357.

Khan, A. A., S. J. Kim, D. D. Paine, and C. E. Cerniglia. 2002. Classification of a polycyclic aromatic hydrocarbon-metabolizing bacterium, mycobacterium sp. strain PYR-1, as Mycobacterium vanbaalenii sp. nov. Int. J. Syst. Evol. Microbiol. 52:1997–2002.

Kikuchi, S., and T. Kusaka. 1982. New malonyl-CoA-dependent fatty acid elongation system in Mycobacterium smegmatis. J. Biochem. 92:839–944.

Kikuchi, S., and T. Kusaka. 1986. Isolation and partial characterization of a very long-chain fatty acid desaturation system from the cytosol of Mycobacterium smegmatis. J. Biochem. 99:723–731.

Kikuchi, S., T. Takeuchi, M. Yasui, T. Kusaka, and P. E. Kolattukudy. 1989. A very long-chain fatty acid elongation system in Mycobacterium avium and a possible mode of action of isoniazid on the system. Agric. Biol. Chem. 53:1689–1698.

King, D. H., and J. J. Perry. 1975. The origin of fatty acids in the hydrocarbon-utilizing microorganism Mycobacterium vaccae. Can. J. Microbiol. 21:85–89.

Kirschner, P., A. Teske, K. H. Schröder, R. M. Kroppenstedt, J. Wolters, and E. C. Böttger. 1992. Mycobacterium confluentis sp. nov. Int. J. Syst. Bacteriol. 42:257–262.

Kleespiess, M., R. M. Kroppenstedt, F. A. Rainey, L. E. Webb, and E. Stackebrandt. 1996. Mycobacterium hodleri sp. nov., a new member of the fast-growing mycobacteria capable of degrading polycyclic aromatic hydrocarbons. Int. J. Syst. Bacteriol. 46:683–687.

Knapp, J. S., and V. R. Brown. 1988. Morpholine degradation. Int. Biodet. 24:299–306.

Koch, R. 1882. Die Aetiologie der Tuberkulose. Berl. Klin. Wochenschr. 19:221–230.

Konicek, J., M. Konickova-Radochova, G. Y. Daraselia, and M. Slosarek. 1988. Mycobacteria in the light of modern genetics development. Folia Microbiol. 33:71–19.

Koukila-Kähkölä, P., B. Springer, E. C. Böttger, L. Paulin, E. Jantzen, and M. L. Katila. 1995. Mycobacterium branderi sp. nov., a new potential human pathogen. Int. J. Syst. Bacteriol. 45:549–553.

Kubica, G. P., I. Baess, R. E. Gordon, P. A. Jenkins, J. B. G. Kwapinski, C. McDurmont, S. R. Pattyn, H. Saito, V. Silcox, J. L. Stanford, K. Takeya, and M. Tsukamura. 1972. A co-operative numerical analysis of the rapidly growing mycobacteria. J. Gen. Microbiol. 73:55–70.

Kubica, G. P. 1978. Classification and nomenclature of the mycobacteria. Ann. Microbiol. (Inst. Pasteur) 129A:7–12.

Kubica, G. P., and R. C. Good. 1981. The genus Mycobacterium (except M. leprae). *In:* M. P. Starr, H. Stolp, H. G. Trüper, A. Balows, and H. G. Schlegel (Eds.) The Prokaryotes. Springer-Verlag. Berlin, Germany. 2:1962–1984.

Kusaka, T., and T. Mori. 1986. Pyrolysis gas chromatography-mass spectrometry of mycobacterial mycolic acid methyl esters and its application to the identification of Mycobacterium leprae. J. Gen. Microbiol. 132:3403–3406.

Kusunoki, S., and T. Ezaki. 1992. Proposal of Mycobacterium peregrinum sp. nov., nom. rev., and elevation of Mycobacterium chelonae subsp. abscessus Kubica et al. to species status: Mycobacterium abscessus comb. nov. Int. J. Syst. Bacteriol. 42:240–245.

Labidi, A., C. Dauguet, K. S. Goh, and H. L. David. 1984. Plasmid profiles of Mycobacterium fortuitum complex isolates. Curr. Microbiol. 11:235–340.

Laneelle, M.-A., C. Lacave, M. Daffé, and G. Lanéelle. 1988. Mycolic acids of Mycobacterium aurum. Eur. J. Biochem. 177:631–635.

Lechevalier, M. P., and H. A. Lechevalier. 1970. Chemical composition as a criterion in the classification of aerobic actinomycetes. Int. J. Syst. Bacteriol. 20:435–443.

Lehmann, K. B., and R. Neumann. 1896. Atlas und Grundriss der Bakteriologie und Lehrbuch der speziellen bakteriologischen Diagnostiek. J. F. Lehmann. Munich, Germany.

Levi, M. H., J. Bartell, L. Gandolfo, S. C. Smole, S. F. Costa, L. M. Weiss, L. K. Johnson, G. Osterhout, and L. H. Herbst. 2003. Characterization of Mycobacterium montefiorense sp. nov., a novel pathogenic mycobacterium from moray eels that is related to Mycobacterium triplex. J. Clin. Microbiol. 41:2147–2152.

Lévy-Frébault, V., E. Rafidinarivo, J.-C. Promé, J. Grandry, H. Boisvert, and H. L. David. 1983. Mycobacterium fallax sp. nov. Int. J. Syst. Bacteriol. 33:336–343.

Lévy-Frébault, V., F. Grimont, P. A. D. Grimont, and H. L. David. 1984. Deoxyribonucleic acid relatedness study of Mycobacterium fallax. Int. J. Syst. Bacteriol. 34:423–425.

Lévy-Frébault, V., M. Daffé, E. Restrepo, F. Grimont, P. A. D. Grimont, and H. L. David. 1986a. Differentiation of Mycobacterium thermoresistible from Mycobacterium phlei and other rapidly growing mycobacteria. Ann. Microbiol. (Paris) 137A:143–151.

Lévy-Frébault, V., F. Grimont, P. A. D. Grimont, and H. L. David. 1986b. Deoxyribonucleic acid relatedness study of the Mycobacterium fortuitum-Mycobacterium chelonae complex. Int. J. Syst. Bacteriol. 36:458–460.

Linko, Y.-Y., and P. Linko. 1985. Immobilized biocatalysts in organic synthesis and chemical production. *In:* J. Tramper, H. C. van der Plas, and P. Linko (Eds.) Biocatalysis in Organic Syntheses. Elsevier Science Publishers. Amsterdam, The Netherlands. 159–178.

Lowrie, D. B., V. R. Aber, and P. S. Jackett. 1979. Phagosome-lysosome fusion and cyclic adenosine 3′:5′-monophosphate in macrophages infected with Mycobacterium microti, Mycobacterium bovis BCG or Mycobacterium lepraemurium. J. Gen. Microbiol. 110:431–441.

Lukins, H. B., and J. W. Foster. 1963. Utilization of hydrocarbons and hydrogen by mycobacteria. Z. Allg. Mikrobiol. 3:251–264.

Luquin, M., V. Ausina, V. Vincent-Lévy-Frébault, M. A. Lanéelle, F. Belda, M. Garcia-Barcelo, G. Prats, and M. Daffé. 1993. Mycobacterium brumae sp. nov., a rapidly growing, nonphotochromogenic mycobacterium. Int. J. Syst. Bacteriol. 43:405–413.

Mamadou, D., M.-A. Dupont, and N. Gas. 1989. The cell envelope of Mycobacterium smegmatis: Cytochemistry and architectural implications. FEMS Microbiol. Lett. 61:89–94.

Martin, C. K. A. 1984. Sterols. *In:* K. Kieslich (Ed.) Biotechnology, Volume 6A: Biotransformations. Chemie Verlag. Weinheim, Germany pp. 79–95.

Masood, R., Y. K. Sharma, and T. A. Venkitasubramanian. 1985. Metabolism of mycobacteria. J. Biosci. 7:421–431.

McFadden, J. J., Z. Kunze, and P. Seechurn. 1990. DNA probes for detection and identification. *In:* J. J. McFadden (Ed.) Molecular Biology of the Mycobacteria. Academic Press. London, UK. 139–172.

Meier, A., P. Kirschner, K. H. Schröder, J. Wolters, R. M. Kroppenstedt, and E. C. Böttger. 1993. Mycobacterium intermedium sp. nov. Int. J. Syst. Bacteriol. 43:204–209.

Meijer, E. M., W. H. J. Boesten, H. E. Schoemaker, and J. A. M. van Balken. 1985. Use of biocatalysts in the industrial production of specialty chemicals. *In:* J. Tramper. H. C. van der Plas, and P. Linko (Eds.) Biocatalysis in Organic Syntheses. Elsevier Science Publishers. Amsterdam, The Netherlands. 135–156.

Meissner, P. S., and J. O. Falkinham, 3rd. 1984. Plasmid-encoded mercuric reductase in Mycobacterium scrofulaceum. J. Bacteriol. 157:669–672.

Minnikin, D. E., and M. Goodfellow. 1980a. Lipid composition in the classification and identification of acid-fast bacteria. *In:* M. Goodfellow and R. G. Board (Eds.) Microbiological Classification and Identification. Academic Press. London, UK. 189–256.

Minnikin, D. E., I. G. Hutchinson, A. B. Caldicott, and M. Goodfellow. 1980b. Thin-layer chromatography of methanolysates of mycolic acid-containing bacteria. J. Chromatogr. 188:221–233.

Minnikin, D. E. 1982. Lipids: Complex lipids, their chemistry, biosynthesis, and roles. *In:* C. Ratledge, and J. Stanford (Eds.) The Biology of the Mycobacteria, Volume 1: Physiology, Identification and Classification. Academic Press. London, UK. 95–184.

Minnikin, D. E., and A. G. O'Donnell. 1984a. Actinomycete envelope lipid and peptidoglycan composition. *In:* M. Goodfellow, M. Mordarski, and S. T. Williams (Eds.) The Biology of the Actinomycetes. Academic Press. London, UK. 337–388.

Minnikin, D. E., S. M. Minnikin, I. G. Hutchinson, M. Goodfellow, and J. M. Grange. 1984b. Mycolic acid patterns of representative strains of Mycobacterium fortuitum, "Mycobacterium peregrinum" and Mycobacterium smegmatis. J. Gen. Microbiol. 130:363–367.

Minnikin, D. E., S. M. Minnikin, J. H. Parlett, M. Goodfellow, and M. Magnusson. 1984c. Mycolic acid patterns of some species of mycobacterium. Arch. Microbiol. 139:225–231.

Minnikin, D. E., S. M. Minnikin, J. H. Parlett, and M. Goodfellow. 1985. Mycolic acid patterns of some rapidly-growing species of mycobacterium. Zbl. Bakteriol. Mikrobiol. Hyg., Ser A 259:446–460.

Mohamed, A. M., P. C. Iwen, S. Tarantolo, and S. H. Hinrichs. 2004. Mycobacterium nebraskense sp. nov., a novel slowly growing scotochromogenic species. Unpublished A368456 54(Pt 6):2057–2060.

Niemann, S., D. Harmsen, S. Rüsch-Gerdes, and E. Richter. 2000. Differentiation of clinical Mycobacterium tuberculosis complex isolates by gyrB DNA sequence polymorphism analysis. J. Clin. Microbiol. 38:3231–3234.

Odriozola, J. M., J. A. Ramos, and K. Bloch. 1977. Fatty acid synthetase activity in Mycobacterium smegmatis: Char-

acterization of the acyl carrier protein-dependent elongating system. Biochim. Biophys. Acta 488:207–217.

Padgitt, P. J., and S. E. Moshier. 1987. Mycobacterium poriferae sp. nov., a scotochromogenic, rapidly growing species isolated from a marine sponge. Int. J. Syst. Bacteriol. 37:186–191.

Patel, J. B., D. G. Leonard, X. Pan, J. M. Musser, R. E. Berman, and I. Nachamkin. 2000. Sequence-based identification of mycobacterium species using the MicroSeq 500 16S rDNA bacterial identification system. J. Clin. Microbiol. 38:246–251.

Perry, J. J. 1980. Propane utilization by microorganisms. Adv. Appl. Microbiol. 26:89–115.

Pitulle, C., M. Dorsch, J. Kazda, J. Wolters, and E. Stackebrandt. 1992. Phylogeny of rapidly growing members of the genus Mycobacterium. Int. J. Syst. Bacteriol. 42:337–343.

Portaels, F., A. de Muynck, and M. P. Sylla. 1988. Selective isolation of mycobacteria from soil: A statistical analysis approach. J. Gen. Microbiol. 134:849–855.

Ramakrishnan, T., P. Suryanarayana Murthy, and K. P. Gopinathan. 1972. Intermediary metabolism of mycobacteria. Bacteriol. Rev. 36:65–108.

Ranes, M. G., J. Rauzier, M. Lagranderie, M. Gheorghiu, and B. Gicquel. 1990. Functional analysis of pAL5000, a plasmid from Mycobacterium fortuitum: Construction of a "mini" Mycobacterium-Escherichia coli shuttle vector. J. Bacteriol. 172:2793–2797.

Rastogi, N., C. Frehel, and H. L. David. 1986. Triple-layered structure of mycobacterial cell wall: evidence for the existence of a polysaccharide-rich outer layer in 18 mycobacterial species. Curr. Microbiol. 13:237–242.

Ratledge, C. 1976. The physiology of the mycobacteria. Adv. Microbiol. Physiol. 13:115–244.

Ratledge, C. 1982a. Lipids: Cell composition, fatty acid biosyntheses. *In:* C. Ratledge and J. Stanford (Eds.) The biology of the Mycobacteria, Volume 1: Physiology, Identification and Classification. Academic Press. London, UK. 53–93.

Ratledge, C. 1982b. Nutrition, growth and metabolism. *In:* C. Ratledge and J. Stanford (Eds.) The Biology of the Mycobacteria, Volume 1: Physiology, Identification and Classification. Academic Press. London, UK. 185–271.

Ratledge, C. 1984. Metabolism of iron and other metals by mycobacteria. *In:* G. P. Kubica and L. G. Wayne (Eds.) The Mycobacteria: A Source Book, Part A. Marcel Dekker. New York, NY. 603–627.

Rauzier, J., J. Moniz-Pereira, and J. B. Gicquel-Sanzey. 1988. Complete nucleotide sequence of pAL5000, a plasmid from Mycobacterium fortuitum. Gene 71:315–321.

Reischl, U., S. Emler, Z. Horak, J. Kaustova, R. M. Kroppenstedt, N. Lehn, and L. Naumann. 1998. Mycobacterium bohemicum sp. nov., a new slow-growing scotochromogenic mycobacterium. Int. J. Syst. Bacteriol. 48:1349–1355.

Rezessy-Szabó, J. M., G. N. M. Huijberts, and J. A. M. de Bont. 1986. Potential of organic solvents in cultivating micro-organisms on toxic water-insoluble compounds. *In:* C. Laane, J. Tramper, and M. D. Lilly (Eds.) Biocatalysis in Organic Media. Elsevier Science Publishers. Amsterdam, The Netherlands. 295–302.

Rhodes, M. W., H. Kator, S. Kotob, P. van Berkum, I. Kaattari, W. Vogelbein, F. Quinn, M. M. Floyd, W. R. Butler, and C. A. Ottinger. 2003. Mycobacterium shottsii sp. nov., a slowly growing species isolated from Chesapeake

Bay striped bass Morone saxatilis. Int. J. Syst. Evol. Microbiol. 53:421–424.

Richter, E., S. Niemann, F. O. Gloeckner, G. E. Pfyffer, and S. Rüsch-Gerdes. 2002. Mycobacterium holsaticum sp. nov. Int. J. Syst. Evol. Microbiol. 52:1991–1996.

Ridgway, H. F., M. G. Rigby, and D. G. Argo. 1984. Adhesion of a mycobacterium sp. to cellulose diacetate membranes used in reverse osmosis. Appl. Environ. Microbiol. 47:61–67.

Ringuet, H., C. Akoua-Koffi, S. Honore, A. Varnerot, V. Vincent, P. Berche, J. L. Gaillard, and C. Pierre-Audigier. 1999. hsp65 sequencing for identification of rapidly growing mycobacteria. J. Clin. Microbiol. 37:852–857.

Rogall, T., T. Flohr, and C. Böttger. 1990a. Differentiation of mycobacterium species by direct sequencing of amplified DNA. J. Gen. Microbiol. 136:1915–1920.

Rogall, T., J. Wolters, T. Flohr, and E. C. Böttger. 1990b. Towards a phylogeny and definition of species at the molecular level within the genus Mycobacterium. Int. J. Syst. Bacteriol. 40:323–330.

Roth, A., U. Reischl, N. Schönfeld, L. Naumann, S. Emler, M. Fischer, H. Mauch, R. Loddenkemper, and R. M. Kroppenstedt. 2000a. Mycobacterium heckeshornense sp. nov., a new pathogenic slowly growing mycobacterium sp. causing cavitary lung disease in an immunocompetent patient. J. Clin. Microbiol. 38:4102–4107.

Roth, A., U. Reischl, A. Streubel, L. Naumann, R. M. Kroppenstedt, M. Habicht, M. Fischer, and H. Mauch. 2000b. Novel diagnostic algorithm for identification of mycobacteria using genus-specific amplification of the 16S-23S rRNA gene spacer and restriction endonucleases. J. Clin. Microbiol. 38:1094–1104.

Saito, H., R. E. Gordon, I. Juhlin, W. Käppler, J. B. G. Kwapinski, C. McDurmont, S. R. Pattyn, E. H. Runyon, J. L. Stanford, I. Tarnok, H. Tasaka, M. Tsukamura, and J. Weiszfeiler. 1977. Cooperative numerical analysis of rapidly growing mycobacteria. Int. J. Syst. Bacteriol. 27:75–85.

Schinsky, M. F., M. M. M. cNeil, A. M. Whitney, A. G. Steigerwalt, B. A. Lasker, M. M. Floyd, G. G. Hogg, D. J. Brenner, and J. M. Brown. 2000. Mycobacterium septicum sp. nov., a new rapidly growing species associated with catheter-related bacteraemia. Int. J. Syst. Evol. Microbiol. 50:575–581.

Schinsky, R., E. Morey, A. G. Steigerwalt, M. P. Douglas, R. W. Wilson, M. M. Floyd, W. R. Butler, M. Daneshvar, B. A. Brown-Elliott, R. J. Wallace Jr., M. M. McNeil, D. J. Brenner, and J. M. Brown. 2004. Taxonomic variation in the Mycobacterium fortuitum third biovariant complex: description of Mycobacterium bonickei sp. nov., Mycobacterium houstonense sp. nov., Mycobacterium neworleansense sp. nov., Mycobacterium brisbanense sp. nov. and recognition of Mycobacterium porcinum from human clinical isolates. Int. J. Syst. Evol. Microbiol.

Schröder, K. H., L. Naumann, R. M. Kroppenstedt, and U. Reischl. 1997. Mycobacterium hassiacum sp. nov., a new rapidly growing thermophilic mycobacterium. Int. J. Syst. Bacteriol. 47:86–91.

Shimakata, T., Y. Fujita, and T. Kusaka. 1977. Acetyl-CoA-dependent elongation of fatty acids in Mycobacterium smegmatis. J. Biochem. 82:725–732.

Shimakata, T., Y. Fujita, and T. Kusaka. 1979. Purification and characterization of 3-hydroxyacyl-CoA dehydrogenase from Mycobacterium smegmatis. J. Biochem. 86:1191–1198.

Shojaei, H., M. Goodfellow, J. G. Magee, R. Freemen, F. K. Gould, and C. G. Brignall. 1997. Mycobacterium novocastrense sp. nov., a rapidly growing photochromogenic Mycobacterium. Int. J. Syst. Bacteriol. 47:1205–1207.

Shojaei, H., M. J. G. Agee, R. Freeman, M. Yates, N. U. Horadagoda, and M. Goodfellow. 2000. Mycobacterium elephantis sp. nov., a rapidly growing non-chromogenic Mycobacterium isolated from an elephant. Int. J. Syst. Evol. Microbiol. 50:1817–1819.

Skerman, V. B. D., V. McGowan, and P. H. A. Sneath. 1980. Approved lists of bacterial names. Int. J. Syst. Bacteriol. 30:225–420.

Smida, J., J. Kazda, and E. Stackebrandt. 1988. Molecular evidence for the relationship of Mycobacterium leprae to slow-growing pathogenetic mycobacteria. Int. J. Lepr. 56:449–454.

Snapper, S. B., L. Lugosi, A. Jekkel, R. E. Melton, T. Kieser, B. R. Bloom, and W. R. Jacobs Jr. 1988. Lysogeny and transformation of mycobacteria: Stable expression of foreign genes. Proc. Natl. Acad. Sci. USA 85:6987–6991.

Snapper, S. B., B. R. Bloom, and W. R. Jacobs Jr. 1990a. Molecular genetic approaches to mycobacterial investigation. In: J. J. McFadden (Ed.) Molecular Biology of the Mycobacteria. Academic Press. London, UK. 199–218.

Snapper, S. B., R. E. Melton, S. Mustafa, T. Kieser, and W. R. Jacobs Jr. 1990b. Isolation and characterization of efficient plasmid transformation mutants of Mycobacterium smegmatis. Molec. Microbiol. 4:1911–1919.

Söhngen, N. L. 1913. Benzin, Petroleum, Paraffinöl und Paraffin als Kohlenstoff- und Energiequelle für Mikroben. Zbl. Bakteriol. Parasitenkde. Abt. II 37:595–609.

Songer, J. G. 1981. Methods for selective isolation of mycobacteria from the environment. Can. J. Microbiol. 27:1–7.

Springer, B., P. Kirschner, G. Rost-Meyer, K. H. Schröder, R. M. Kroppenstedt, and E. C. Böttger. 1993. Mycobacterium interjectum, a new species isolated from a patient with chronic lymphadenitis. J. Clin. Microbiol. 31:3083–3089.

Springer, B., E. C. Böttger, P. Kirschner, and R. J. Wallace Jr. 1995a. Phylogeny of the Mycobacterium chelonae-like organism based on partial sequencing of the 16S rRNA gene and proposal of Mycobacterium mucogenicum sp. nov. Int. J. Syst. Bacteriol. 45:262–267.

Springer, B., E. Tortoli, I. Richter, R. Grünewald, S. Rüsch-Gerdes, K. Uschmann., F. Suter, M. D. Collins, R. M. Kroppenstedt, and E. C. Böttger. 1995b. Mycobacterium conspicuum sp. nov., a new species isolated from patients with disseminated infections. J. Clin. Microbiol. 33:2805–2811.

Springer, B., W. K. Wu, T. Bodmer, G. Haase, G. E. Pfyffer, R. M. Kroppenstedt, K. H. Schröder, S. Emler, J. O. Kilburn, P. Kirschner, A. Telenti, M. B. Coyle, and E. C. Böttger. 1996. Isolation and characterization of a unique group of slowly growing mycobacteria: description of Mycobacterium lentiflavum sp. nov. J. Clin. Microbiol. 34:1100–1107.

Sritharan, M., and C. Ratledge. 1989. Co-ordinated expression of the components of iron transport (mycobactin, exochelin and envelope proteins) in Mycobacterium smegmatis. FEMS Microbiol. Lett. 60:183–186.

Stahl, D. A., and J. W. Urbance. 1990. The division between fast- and slow-growing species corresponds to natural relationships among the mycobacteria. J. Bacteriol. 172:116–124.

Steinert, H. J., K. D. Vorlop, and J. Klein. 1986. Steroid side chain cleavage with immobilized living cells in organic solvents. *In:* C. Laane, J. Tramper, and M. D. Lilly (Eds.) Biocatalysis in Organic Media. Elsevier Science Publishers. Amsterdam, The Netherlands. 51–63.

Suzuki, Y., K. Yoshinaga, Y. Ono, A. Nagata, and T. Yamada. 1987. Organization of rRNA genes in Mycobacterium bovis BCG. J. Bacteriol. 169:839–843.

Taylor, D. G., P. W. Trudgill, R. E. Cripps, and P. R. Harris. 1980. The microbial metabolism of acetone. J. Gen. Microbiol. 118:159–170.

Thorel, M. F., M. Krichevsky, and V. V. Lévy-Frébault. 1990. Numerical taxonomy of mycobactin-dependent mycobacteria, emended description of Mycobacterium avium, and description of Mycobacterium avium subsp. avium subsp. nov., Mycobacterium avium subsp. paratuberculosis subsp. nov., and Mycobacterium avium subsp. silvaticum subsp. nov. Int. J. Syst. Bacteriol. 40:254–260.

Torkko, P., S. Suomalainen, E. Iivanainen, M. Suutari, E. Tortoli, L. Paulin, and M. L. Katila. 2000. Mycobacterium xenopi and related organisms isolated from stream waters in Finland and description of Mycobacterium botniense sp. nov. Int. J. Syst. Evol. Microbiol. 50:283–289.

Torkko, P., S. Suomalainen, E. Iivanainen, E. Tortoli, M. Suutari, J. Seppänen, L. Paulin, and M. L. Katila. 2002. Mycobacterium palustre sp. nov., a potentially pathogenic, slowly growing mycobacterium isolated from clinical and veterinary specimens and from Finnish stream waters. Int. J. Syst. Evol. Microbiol. 52:1519–1525.

Tortoli, E., R. M. Kroppenstedt, A. Bartolini, C. Garoli, I. Jan, J. Pawlowski, and S. Emler. 1999. Mycobacterium tusciae sp. nov. Int. J. Syst. Bacteriol. 49:1839–1844.

Tortoli, E., C. Piersimoni, R. M. Kroppenstedt, J. I. Montoya-Burgos, U. Reischl, A. Giacometti, and S. Emler. 2001. Mycobacterium doricum sp. nov. Int. J. Syst. Evol. Microbiol. 51:2007–2012.

Tortoli, E., L. Rindi, M. J. Garcia, P. Chiaradonna, R. Dei, C. Garzelli, R. M. Kroppenstedt, N. Lari, R. Mattei, A. Mariottini, G. Mazzarelli, M. I. Murcia, A. Nanetti, P. Piccoli, and C. Scarparo. 2004. Proposal to elevate the genetic variant MAC-A, included in the Mycobacterium avium complex, to species rank as Mycobacterium chimaerae sp. nov. Int. J. Syst. Evol. Microbiol.

Trujillo, M. E., E. Velázquez, R. M. Kroppenstedt, P. Schumann, R. Rivas, P. F. Mateos, and E. Martínez-Molina. 2004. Mycobacterium psychrotolerans sp. nov., isolated from a pond near an uranium mine. Int. J. Syst. Evol. Microbiol.

Tsukamura, M, Mizuno, S. 1971. Mycobacterium obuense, a rapidly growing scotochromogenic mycobacterium capable of forming a black product from p-aminosalicylate and salicylate. J. Gen. Microbiol. 68:129–134.

Tsukamura, M. 1981a. Numerical analysis of rapidly growing, nonphotochromogenic mycobacteria, including Mycobacterium agri (Tsukamura 1972) Tsukamura sp. nov., nom. rev. Int. J. Syst. Bacteriol. 31:247–258.

Tsukamura, M., S. Mizuno, and S. Tsukamura. 1981b. Numerical analysis of rapidly growing, scotochromogenic mycobacteria, including Mycobacterium obuense sp. nov., nom. rev., Mycobacterium rhodesiae, sp. nov., nom. rev., Mycobacterium aichiense sp. nov., nom. rev., Myco-

bacterium chubuense, sp. nov., nom. rev., and Mycobacterium tokaiense sp. nov., nom. rev. Int. J. Syst. Bacteriol. 31:263–275.

Tsukamura, M. 1982. Mycobacterium shimoidei sp. nov., nom. rev., a lung pathogen. Int. J. Syst. Bacteriol. 32:67–69.

Tsukamura, M., S. Mizuno, and H. Toyama. 1983a. Mycobacterium pulveris sp. nov., a nonphotochromogenic mycobacterium with an intermediate growth rate. Int. J. Syst. Bacteriol. 33:811–815.

Tsukamura, M., H. Nemoto, and H. Yugi. 1983b. Mycobacterium porcinum sp. nov., a porcine pathogen. Int. J. Syst. Bacteriol. 33:162–165.

Tsukamura, M., H. J. van der Meulen, and W. O. K. Grabow. 1983c. Numerical taxonomy of rapidly growing, scotochromogenic mycobacteria of the Mycobacterium parafortuitum complex: Mycobacterium austroafricanum sp. nov. and Mycobacterium diernhoferi sp. nov., nom. rev. Int. J. Syst. Bacteriol. 33:460–469.

Tsukamura, M. 1984. The "non-pathogenic" mycobacteria: Their distribution and ecology in non-living reservoirs. *In:* G. P. Kubica and L. G. Wayne (Eds.) The Mycobacteria: A Source Book, Part B. Marcel Dekker. New York, NY. 1339–1359.

Tsukamura, M., and S. Ichiyama. 1986a. Numerical classification of rapidly growing nonphotochromogenic mycobacteria. Microbiol. Immunol. 30:863–882.

Tsukamura, M., I. Yano, and T. Imaeda. 1986b. Mycobacterium moriokaense sp. nov., a rapidly growing, non Mycobacterium moriokaense sp. photochromogenic mycobacterium. Int. J. Syst. Bacteriol. 36:333–338.

Turenne, C., P. Chedore, J. Wolfe, F. Jamieson, G. Broukhanski, K. May, and A. Kabani. 2002. Mycobacterium lacus sp. nov., a novel slowly growing, non-chromogenic clinical isolate. Int. J. Syst. Evol. Microbiol. 52:2135–2140.

Turenne, C. Y., V. J. Cook, T. V. Burdz, R. J. Pauls, L. Thibert, J. N. Wolfe, and A. Kabani. 2004a. Mycobacterium parascrofulaceum sp. nov., novel slowly growing, scotochromogenic clinical isolates related to Mycobacterium simiae. Int J Syst Evol Microbiol 54(Pt 5):1543–1551.

Turenne, C. Y., L. Thibert, K. Williams, T. V. Burdz, V. J. Cook, J. N. Wolfe, D. W. Cockcroft, and A. Kabani. 2004b. Mycobacterium saskatchewanense sp. nov., a novel slowly growing scotochromogenic species from human clinical isolates related to Mycobacterium interjectum and Accuprobe-positive for Mycobacterium avium complex. Int. J. Syst. Evol. Microbiol.

Urakami, T., and I. Yano. 1989. Methanol-utilizing mycobacterium strains isolated from soil. J. Gen. Appl. Microbiol. 35:125–133.

Valero-Guillén, P. L., and F. Martín-Luengo. 1986. 1-tetradecanol, a new alcohol found in the cell wall of some rapidly growing chromogenic mycobacteria. FEMS Microbiol. Lett. 35:59–63.

Van Ginkel, C. G., H. G. J. Welten, J. A. M. de Bont, and H. A. M. Boerrigter. 1986. Removal of ethene to very low concentrations by immobilized Mycobacterium E3. J. Chem. Technol. Biotechnol. 36:593–598.

Van Ginkel, C. G., A. Q. H. Habets-Crützen, A. R. M. van der Last, and J. A. M. de Bont. 1987a. A description of microbial growth on gaseous alkenes in a chemostat culture. Biotechnol. Bioengin. 30:799–804.

Van Ginkel, C. G., H. G. J. Welten, and J. A. M. de Bont. 1987b. Oxidation of gaseous and volatile hydrocarbons by selected alkene-utilizing bacteria. Appl. Environ. Microbiol. 53:2903–2907.

Van Loosdrecht, M. C. M., J. Lyklema, W. Norde, G. Schraa, and A. J. B. Zehnder. 1987. The role of bacterial cell wall hydrophobicity in adhesion. Appl. Environ. Microbiol. 53:1893–1897.

Vestal, A. L. 1975. Procedures for the Isolation and Identification of Mycobacteria. Center for Disease Control. Atlanta, GA. US Department of Health, Education and Welfare Publication no. (CDC) 76-8230.

Viallier, J., and G. Viallier. 1973. Inventaire des mycobacteries de la nature. Ann. Soc. Belge Med. Trop. 53:361–371.

Vishniac, W., and M. Santer. 1957. The thiobacilli. Bacteriol. Rev. 21:195–213.

Vuorio, R., M. A. Aandersson, F. A. Rainey, R. M. Kroppenstedt, P. Kämpfer, H. J. Busse, M. Viljanen, and M. Salkinoja-Salonen. 1999. A new rapidly growing mycobacterial species, Mycobacterium murale sp. nov., isolated from the indoor walls of a children's day care centre. Int. J. Syst. Bacteriol. 49:25–35.

Wayne, L. G. 1966. Mycobacterial classification. Am. Rev.Respir. Dis. 93:958–959.

Wayne, L. G., E. J. Krichevsky, L. L. Love, R. Johnson, and M. I. Krichevsky. 1980. Taxonomic probability matrix for use with slowly growing mycobacteria. Int. J. Syst. Bacteriol. 30:528–538.

Wayne, L. G. 1984. Mycobacterial speciation. In: G. P. Kubica and L. G. Wayne (Eds.) The Mycobacteria: A Source Book, Part A. Marcel Dekker. New York, NY. 25–65.

Wayne, L. G., and G. P. Kubica. 1986. Genus Mycobacterium Lehmann and Neumann 1896, 363. In: P. H. A. Sneath, N. S. Mair, M. E. Sharpe, and J. G. Holt (Eds.) Bergey's Manual of Systematic Bacteriology. Williams & Wilkins. Baltimore, MD. 2:1436–1457.

Weijers, C. A. G. M., C. G. van Ginkel, and J. A. M. de Bont. 1988. Enantiomeric composition of lower epoxyalkanes produced by methane-, alkane- and alkene-utilizing bacteria. Enz, Microbiol. Technol. 10:214–218.

Weisman, L. S., and C. E. Ballou. 1984a. Biosynthesis of the mycobacterial methylmannose polysaccharide, identification of an α1→4-mannosyltransferase. J. Biol. Chem. 259:3457–3463.

Weisman, L. S., and C. E. Ballou. 1984b. Biosynthesis of the mycobacterial methylmannose polysaccharide, identification of a 3-O-methyltransferase. J. Biol. Chem. 259:3464–3469.

Wiegant, W. M., and J. A. M. de Bont. 1980. A new route for ethylene glycol metabolism in Mycobacterium E44. J. Gen. Microbiol. 120:325–331.

Willumsen, P., U. Karlson, E. Stackebrandt, and R. M. Kroppenstedt. 2001. Mycobacterium frederiksbergense sp. nov., a novel polycyclic aromatic hydrocarbon-degrading mycobacterium species. Int. J. Syst. Evol. Microbiol. 51:1715–1722.

Wilson, R. W., V. A. Steingrube, E. C. Böttger, B. Springer, B. A. Brown-Elliott, V. Vincent, K. C. Jost Jr., Y. Zhang, M. J. Garcia, S. H. Chiu, G. O. Onyi, H. Rossmoor, D. R., Nash, and R. J. Wallace Jr. 2001. Mycobacterium immunogenum sp. nov., a novel species related to Mycobacterium abscessus and associated with clinical disease, pseudo-outbreaks and contaminated metalworking fluids: An international cooperative study on mycobacterial taxonomy. Int. J. Syst. Evol. Microbiol. 51:1751–1764.

Winder, F. G. 1982. Mode of action of the antimycobacterial agents and associated aspects of the molecular biology of the mycobacteria. In: C. Ratledge and J. Stanford (Eds.) The Biology of the Mycobacteria, Volume 1: Physiology, Identification and Classification. Academic Press. London, UK. 353–438.

Woese, C. R. 1987. Bacterial evolution. Microbiol. Rev. 51:221–271.

Wovcha, M. G., F. J. Antosz, J. C. Knight, L. A. Kominek, and T. R. Pyke. 1978. Bioconversion of sitosterol to useful steroidal intermediates by mutants of Mycobacterium fortuitum. Biochim. Biophys. Acta 531:308–321.

Zainuddin, Z. F., and J. W. Dale. 1989. Polymorphic repetitive DNA sequences in Mycobacterium tuberculosis detected with a gene probe from a Mycobacterium fortuitum plasmid. J. Gen. Microbiol. 135:2347–2355.

Prokaryotes (2006) 3:919–933
DOI: 10.1007/0-387-30743-5_34

CHAPTER 1.1.19

The Genus *Mycobacterium*—Medical

BEATRICE SAVIOLA AND WILLIAM BISHAI

Introduction

The genus *Mycobacterium* encompasses a number of medically important species that exact an alarming toll in human morbidity and mortality. Indeed the World Health Organization (WHO) estimates that nearly one third of the world's population (1.8 billion people) are infected with *Mycobacterium tuberculosis*, the cause of tuberculosis (TB) (http://www.who.org). In 1993, the WHO declared a global emergency owing to the fact that TB was epidemic in many areas of the world. Other mycobacterial diseases continue to plague the world's populations as well. *Mycobacterium leprae*, the cause of leprosy, persists in developing countries, and other mycobacteria (ordinarily nonpathogens) have now become threats to individuals infected with the human immunodeficiency virus (HIV).

Mycobacteria have shaped the course of human history. Indeed until 1900, TB was one of the chief causes of death in Europe and the Americas. The nature of the disease, however, remained poorly understood and as a testimonial to this fact its contagiousness was the subject of heated debate well into the 18th and 19th centuries. For many years, psychological and inherited factors were thought to predispose individuals to the disease. However, others thought that a contagious agent was the culprit. This debate ended only after the bacillus *Mycobacterium tuberculosis* was identified as the causative agent of TB. Jean Antoine Villemin in 1868 was the first to transmit TB from man to rabbit (Haas and Haas, 1996). As a military doctor, he had observed that many young healthy military personnel housed in barracks eventually succumbed to TB. He also noted that a disproportionate number of individuals with active TB were prisoners, industrial workers, members of cloistered religious orders as well as military personnel; all of them were housed together with many other people. Villemin speculated that TB was transmissible. He used material (gray and soft tubercles) from humans who had succumbed to TB as well as blood and sputum from tuberculous patients to infect rabbits. The inoculated rabbits indeed

developed pathologic evidence of tuberculosis. Then in 1882, Robert Koch was the first to view *Mycobacterium tuberculosis* through a light microscope. He was then able to grow *M. tuberculosis* in pure culture, infect guinea pigs with the bacilli, and reisolate bacteria from these animals (Adler and Rose, 1996). Thus a new era of mycobacteria research was born.

In Europe and the United States, a sanatorium movement was underway by the 1890s to isolate and cure those patients having TB. Edward Livingston Trudeau started one of the first and most successful of the sanatoria in the United States (Davis, 1996). After providing hospice care for his consumptive brother, Dr. Trudeau was stricken with that same disease. He fled to the Adirondack Mountains believing that fresh air, nutritious food and exercise would restore his health. As hoped, his disease symptoms abated and his health consequently improved. He chose to move permanently to Saranac Lake in the Adirondack Mountains where he bought land, built cottages and started the Adirondack Cottage Sanatorium. Dr. Trudeau immersed himself in research and succeeded in cultivating pure *Mycobacterium tuberculosis*, which he provided to other researchers. Out of the Saranac Lake Laboratory, the Trudeau Institute (http://www.trudeauinstitute.org) for biomedical research was born in 1954. Not until the development of successful antimycobacterial chemotherapy did this and other sanitoria lose their function and close their doors to the public.

In 1944 streptomycin was isolated and identified by Dr. Selman Waksman and his graduate student Albert Schatz (Harris, 1996). It marked the first drug in a line of antibiotics having antimycobacterial properties. In 1952 isonicotinic acid hydrazide (INH) was found to have antimycobacterial properties as well and has since become a mainstay in the antibiotic therapy of TB. With the advent of these new advances in the chemotherapy of TB, victory was declared over the disease; however, this declaration would prove premature. Following an extensive period of decline, the incidence of TB in the United States began to rise again in the late 1980s. The

HIV epidemic combined with an urban drug problem provided fertile ground on which *M. tuberculosis* could flourish anew. Concomitant with this alarming trend was detection of an increasing incidence of drug-resistant *M. tuberculosis*. To combat these problems, the WHO as well as government agencies in countries around the world have increased spending in both the public health forum as well as biomedical research. Today, research on *M. tuberculosis* as well as other members of the genus is providing valuable insight into the basic biology of mycobacterial survival and replication inside the host. Lessons learned through these studies will hopefully provide novel treatments to combat the renewed threat of this age-old enemy.

Taxonomy

The genus *Mycobacterium* comprises a number of Gram-positive aerobic bacteria and is the only member of the family Mycobacteriaceae within the order Actinomycetales. The genus shares an unusually high genomic DNA G+C content (62–70%) and the production of mycolic acids with closely related genera, *Nocardia* and *Corynebacterium*, within Actinomycetales. Phylogenetic trees are available which depict genetic relatedness based on homology of the 16S ribosomal gene sequence. Mycobacteria that have highly homologous rRNA sequences are closely related and are on neighboring branches of the tree (Fig. 1). The mycobacterial phylogenetic tree can be further subdivided into fast and slow growing bacteria. The fast growers form colonies on selective media in less than 7 days and the slow growers, in greater than 7 days. In addition, within the genus *Mycobacterium* a number of species are grouped into complexes (e.g., *M. avium* and *M. tuberculosis* complexes) that include bacterial species that have a high degree of genetic similarity as well as cause similar disease syndromes.

Habitat

Mycobacterium tuberculosis is an obligate pathogen of humans and is rarely identified in other mammals. It is transmitted from person to person, and it has no significant environmental reservoirs. *Mycobacterium bovis*, which causes tuberculosis in humans and in cattle, has a natural reservoir in ruminants. As a consequence, foodstuffs (including cheese and milk originating from these animals) were often contaminated prior to the introduction of current pasteurization and meat inspection procedures. A number of other medically important mycobacteria are found in the environment. Water serves as the habitat for a number of mycobacterial species including *M. marinum*, *M. cheloneae*, *M. fortuitum*, *M. kansasii* and *M. avium*. Although not yet confirmed, stagnant water may be a habitat for *M. ulcerans*. Soil may also harbor mycobacteria including *M. cheloneae*, *M. fortuitum*, and *M. avium*.

Isolation

Medically important mycobacterial species can be isolated and cultured from a number of environmental sources; however, *M. tuberculosis* is rarely isolated from nonhuman sources. As pulmonary tuberculosis is the most frequent form of the disease, sputum is the most common body specimen submitted for culture—although blood, gastric aspirates and biopsy specimens may also be analyzed.

Before sputum samples from patients may be tested for mycobacterial growth, they must be decontaminated to remove rapidly growing upper respiratory flora. Decontamination procedures take advantage of the thick lipid envelope of mycobacteria and this organism's natural resistance to chemical agents. *N*-Acetyl-L-cysteine (0.5–2.0%; NALC) liquefies sputum samples. The compound has minor inhibitory effects on mycobacteria but does allow other antibacterial agents to access and neutralize contaminating organisms. Sodium hydroxide is used to inactivate contaminating bacteria with only modest inhibitory effects against mycobacteria.

Once decontaminated, samples may be used to inoculate either a solid or liquid growth medium. The solid growth media are available as either egg- or agar-based. Both types of solid media contain malachite green dye, which inhibits the growth of many other bacteria. The commonly used Löwenstein-Jensen (LJ) medium is egg-based (Fig. 2), whereas Middlebrook 7H10 and 7H11 are agar-based media. Growth can also be assayed in a liquid medium such as Middlebrook 7H9; however, these media are more problematic because overgrowth by contaminating bacteria is more difficult to detect.

Growth of mycobacteria from clinical specimens is time-consuming because there is often a lag of three to four weeks before sufficient growth is achieved. To shorten growth detection times, the Bactec (Becton-Dickinson) system may be used (Fig. 2). Mycobacteria are grown in 7H9 that contains ^{14}C-labeled palmitic acid as a carbon source. As the bacteria grow they catabolize the radiolabeled carbon source and convert it to gaseous $^{14}CO_2$. A specially designed instrument (Becton Dickinson) samples the amount of ^{14}C above the mycobacterial cultures and converts this information into an index of growth.

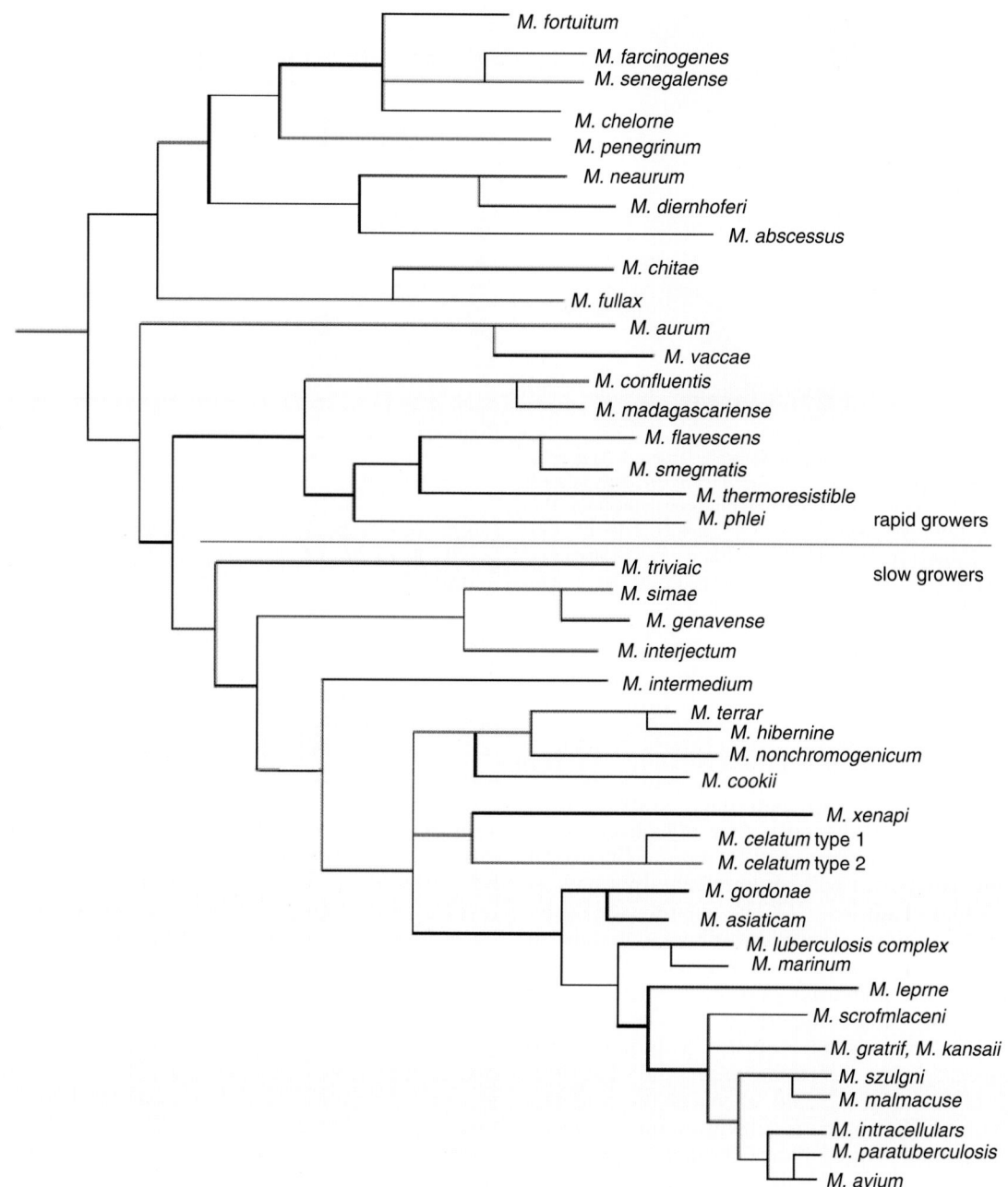

Fig. 1. A phylogenetic tree based on 16S rRNA sequences of the genus *Mycobacterium* depicts closely related species on neighboring branches. Redrawn from Shinnick and Good (1994).

The Bactec method reduces detection time in liquid media to 9–14 days for *M. tuberculosis* (Middleton et al., 1997; Pfyffer et al., 1997). Recently a completely closed system called the mycobacterial growth indicator tube (MGIT) system has been introduced. The tube contains a plastic resin that fluoresces as oxygen levels are depleted by growing bacteria, and an automated fluorescence detector reports growth readings at regular intervals (Saito et al., 1996; Sharp et al., 1996).

Identification

Once mycobacteria have been isolated from either an environmental or a human source, the speciation process can begin. Acid-fast staining

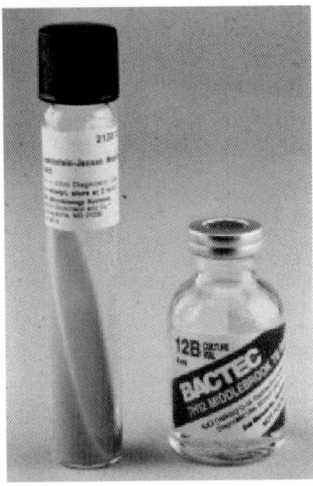

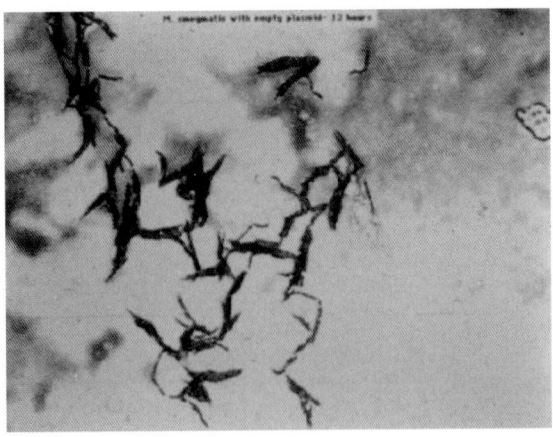

Fig. 3. Kinyoun acid-fast stain of *Mycobacterium smegmatis*. Bacilli appear as small red rods with clumping.

Fig. 2. The growth medium Löwenstein Jensen (left) is used to culture mycobacteria. *Mycobacterium tuberculosis* takes 3 to 4 weeks to produce colonies on this growth medium. The Bactec 12B (right) bottle is used for radiometric detection of mycobacterial growth. Early growth of mycobacteria is detected by the conversion of ^{14}C-labeled palmitate to gaseous ^{14}C-labeled CO_2. An automated sampling needle removes a small amount of gas through a rubber septum each day, and a detector measures the radioactivity obtained. Using the Bactec system, mycobacterial growth may be detected in 7–15 days.

in combination with light microscopy permits an initial rapid diagnosis of a mycobacteriosis. Mycobacteria are acid fast by nature. They form stable complexes with arylmethane dyes such as fuchsin and auramine O. The phenol used in the primary staining procedure allows the stain to penetrate. The mycolic acids in the cell wall act to retain the stain even after the exposure to acid alcohol or strong mineral acids. The resulting acid-fast mycobacteria can be identified microscopically. The Ziehl-Neelsen and Kinyoun methods are acid stains in which the acid-fast bacilli (AFB) appear red against a blue or green background (Fig. 3). Fluorochrome-staining procedures can be used as well. In these procedures, which use auramine O or auramine-rhodamine, the AFB fluoresce yellow to orange. Fluorescence staining is superior as AFB are more easily identified at a lower magnification. Initial microscopic identification of bacteria, while rapid, is not reliable, has a sensitivity that ranges from 22 to 80%, and does not allow species identification of AFB. Therefore, other methods of mycobacterial identification are necessary (Nolte and Metchock, 1995).

A host of biochemical tests can be used to differentiate mycobacterial species. Biochemical tests that distinguish *M. tuberculosis* from other species include the niacin accumulation test, the 68°C catalase test and the nitrate reduction test

(Kent and Kubica, 1985; Nolte and Metchock, 1995). Because *M. tuberculosis* has a blocked pathway for the conversion of free niacin to nicotinic acid mononucleotide, the bacterium accumulates niacin and excretes it into the culture medium. As confirmation, additional biochemical tests (e.g., the nitrate reduction test) must be performed inasmuch as strains of some mycobacterial species other than *M. tuberculosis* also accumulate niacin. *Mycobacterium tuberculosis* is a strongly positive nitrate reducer. Because *M. kansasii, M. szulgai* and *M. fortuitum* also reduce nitrate, an additional test (e.g., heat-stable catalase) should be used. The heat-stable catalase test assays the amount of catalase present in the bacterium after it has been resuspended in a buffer and heated to 68°C. *Mycobacterium tuberculosis* always loses its catalase activity under these conditions.

Once AFB have been observed in culture, most clinical labs use a nucleic acid hybridization test rather than the more laborious biochemical tests to speciate the isolate. The Accuprobe assay is available through Gen-Probe (http://www.gen-probe.com). This assay uses the Hybridization Protection Assay (HPA) where a DNA acridinium-ester-labeled probe hybridizes to a mycobacterial rRNA target sequence. A selection agent cleaves the label away from the unhybridized probe but leaves untouched the probe that is complexed with its target sequence. Light (detected with a luminometer) will be emitted from the DNA probe that is complexed to the rRNA. The advantage of this technique is that it can identify small quantities of mycobacteria due to the many copies of the rRNA target sequence in each mycobacterium. Thus there is an inherent amplification of the signal sequence. Gen-Probe has available probes that are specific for *M. avium, M. intracellulare*, the *M. avium* complex, *M. gordonae*, *M. kansasii* and the *M. tuberculosis* complex.

Nucleic Acid Amplification Tests to Detect and Speciate *M. tuberculosis* Directly from Sputum

Even shorter detection times may be achieved using various commercially available nucleic acid amplification assays (NAA) designed to work using sputum. The United States Food and Drug Administration (USFDA) has approved several commercially available amplification tests for use directly with sputum. These tests include the *Mycobacterium tuberculosis* Direct Test from Gene-Probe (MTD; http://www.gen-probe.com) and the Amplicor (http://www.roche.com/diagnostics/) *Mycobacterium tuberculosis* (MTB) Test. The MTD combines Transcription Mediated Amplification, which can amplify target rRNA sequences, with the HPA described in the previous section. This test is approved by the USFDA for use with sputum samples that are both positive and negative for AFB. The MTB test uses the polymerase chain reaction (PCR) to amplify DNA specific to *M. tuberculosis*. This test is approved by the USFDA for use with sputum samples positive for AFB. The LCx *M. tuberculosis* Test (http://www.abbott.com/research/diagnostics.htm) (Abbott) utilizes a ligase chain reaction where two oligos are hybridized to a target sequence from the mycobacteria and are subsequently ligated. The ligated oligos then serve as templates for additional oligos to hybridize and be ligated, resulting in amplification of the target sequence. At present this test is not approved for use with sputum within the United States. A recent study comparing two NAA methods revealed that MTD and LCx had sensitivities of 98.6 and 100%, respectively, and specificities of 99.4 and 99.3% with smear-positive samples (Wang and Tay, 1999). The Amplicor MTB test was analyzed in clinical trials and had a sensitivity of 95.0% and a specificity of 100% with smear-positive sputum samples (Amplicor, 2000). The NAA tests, while costly, are slightly more sensitive than acid-fast staining procedures. These tests hasten care for infected individuals by quickly identifying mycobacteria within about 25–50% of sputum samples that are negative by acid-fast staining but later culture positive. This aspect of the NAA tests is important especially in view of the recent evidence supporting the transmission of *M. tuberculosis* from patients who did not have acid-fast bacilli in their sputum (Behr et al., 1999). Additionally, in contrast to acid-fast staining, NAA tests provide information about species present within a sputum sample. Thus, *M. tuberculosis*-infected individuals can be isolated and treated more rapidly, preventing the spread of the bacteria. The NAA tests cannot differentiate live from dead bacteria and therefore may not prove helpful for monitoring response to therapy.

Once a bacterial isolate has been identified as *M. tuberculosis*, it can be further subtyped to the strain level. The ability differentiate *M. tuberculosis* on the strain level has enhanced our knowledge of the epidemiology of TB. In fact it has led to the identification of large outbreaks of disease traced back to a single person (Edlin et al., 1992; Valway et al., 1998). Strain typing also has lead to the discovery of unsuspected routes of transmission such as medical equipment (Agerton et al., 1997; Michele et al., 1997), embalming (Sterling et al., 2000), or exposure to childhood TB (Curtis et al., 1999). It has been used, as well, to assess the amount of recent transmission of *M. tuberculosis* in the community, thereby redefining our understanding of the balance between primary active and reactivation TB after a period of latency. Techniques of DNA fingerprinting allow the identification of different *M. tuberculosis* strains. Virtually all *M. tuberculosis* strains have 1–26 copies of an insertion sequence IS*6110*. When a strain's chromosomal DNA is fragmented with restriction enzymes, the number of copies and location of these insertion sequences within the chromosome produce a restriction length polymorphism (RFLP) pattern that characterizes the strain. In addition, the frequency with which IS*6110* moves within the chromosome is low enough to allow identification of specific strains that are the cause of an outbreak. Reasonable identification of a strain can be obtained when the bacterium contains six or more IS*6110* elements. In the event that a strain contains less than six insertion elements, a second probe may be used for increased specificity. This second probe is generally for the polymorphic GC-rich repetitive sequence (PGRS). Distinct strains will have a unique pattern of probe hybridization, thus facilitating identification (Harrington and Bishai, 2000).

Current research in diagnostics for TB is focusing on techniques that are fast, sensitive and specific. A major challenge is to develop inexpensive tests, which may be used in developing countries where TB is abundant. These tests could potentially identify patients with active TB early, thus permitting the isolation of such individuals and preventing the spread of the bacilli. Earlier identification of those individuals that harbor *M. tuberculosis* could possibly lower TB rates in many parts of the developing world.

Physiology

Mycobacteria are straight or slightly curved non-motile rods (approximately 0.2–0.6 μm wide by 1.0–10 μm long; Fig. 4), aerobic, nonsporeformers that grow at times as filaments. Two remarkable bacteriologic features define mycobacteria. First, they have a cell wall that is rich in long-chain fatty acid esters known as "mycolic acids" that are attached to the cell wall through arabinogalactan (Fig. 5). Mycolic acids are chemically related to wax and give the colonies their characteristic waxy appearance and the cells a tendency to clump and resist dispersion. The waxy nature of the coat renders the bacteria not readily stainable (such as by the Gram method) with aniline dyes, though the bacteria are considered to be Gram positive. Mycobacterial mycolic acids do, however, form strong complexes with some dyes that leave them resistant to decolorization with acid alcohol. Therefore the identification of AFB in a sample is suggestive of

mycobacteria (Fig. 3). In addition, this lipid coat renders the mycobacteria impervious to many aseptic solutions and antibiotics. Second, all mycobacteria grow slowly with generation times

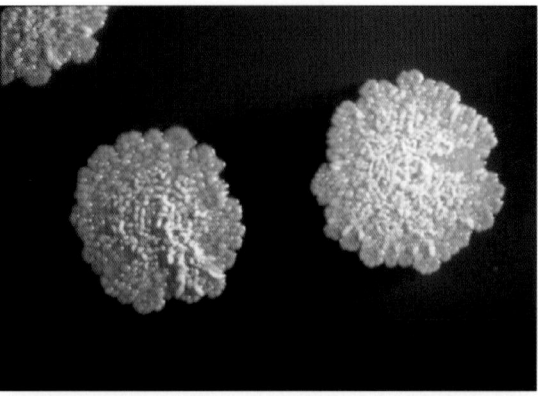

Fig. 4. *Mycobacterium smegmatis* colony morphology after three days of growth from a single bacterium.

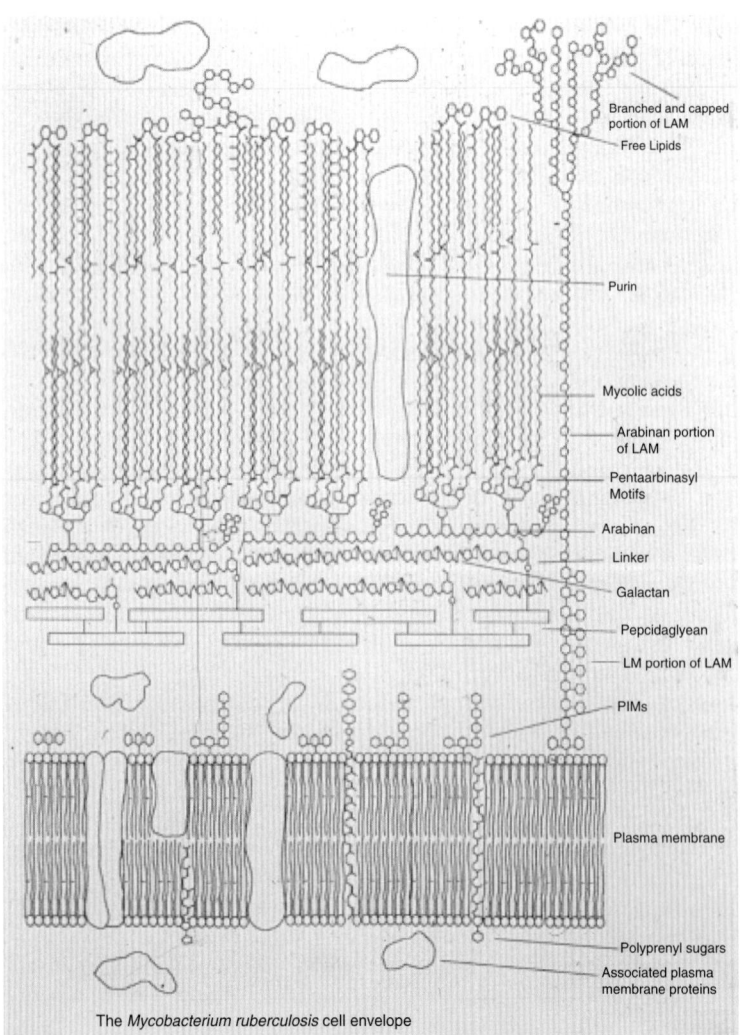

Fig. 5. Schematic representation of the plasma membrane and outer wall of *Mycobacterium tuberculosis* reveals an abundance of mycolic acids. Phosphatidyl-inositol mannosides (PIMs), lipoarabinomannans (LAMs) and lipomannan (LM) portion of LAM are shown as outer wall constituents. From Brennan and Nikaido (1995).

that range from 2 hours for *M. smegmatis* to 12 days for *M. leprae*.

Genetics

The Completed *Mycobacterium tuberculosis* Genome

Genetic research into the virulent nature of *M. tuberculosis* has long been hampered by the organism's slow growth. The generation time of *M. tuberculosis* is 24 hours in contrast to the 20-minute generation time of the common laboratory bacterium *Escherichia coli*. The tendency for mycobacteria to form clumps also leads to significant technical challenges in the research laboratory. As a consequence, traditional genetic screens and selections are difficult to conduct. Therefore, reverse genetic approaches have been emphasized in the study of *M. tuberculosis*. This technique targets specific proteins homologous to proteins of known function and importance in other organisms for study at a molecular level within a host. As a result, the elucidation and analysis of the whole genome sequence of *M. tuberculosis* have become increasingly important.

In 1998, the complete genome sequence of the *M. tuberculosis* strain H37Rv was elucidated (Cole et al., 1998). The completion of the genome sequence was a collaborative effort of the Wellcome Trust and the Institute Pasteur. A web site at the Institute Pasteur shows the completed genome. In addition, The Sanger Center maintains a web site where DNA segments and predicted proteins from the *M. tuberculosis* strain H37Rv may be compared to peptides and DNA from other organisms. The Institute for Genomic Research (TIGR) has completed the sequence for another *M. tuberculosis* strain, CDC1551 (CSU93). At the TIGR web site, one may also compare the chromosomal sequence to other known genes and predicted proteins. In addition, The National Center for Biotechnology Information offers a web site where the *M. tuberculosis* genome may be accessed and the chromosomal sequence may be visualized on a map.

The genome sequence is 4.4 Mbp. Researchers now have at their disposal the information defining every possible drug target, antigen for incorporation into a vaccine, as well as every virulence determinant from *M. tuberculosis*. Annotation of the H37Rv sequence reveals 3,924 open reading frames (Cole et al., 1998). The genome has no easily identifiable pathogenicity islands as other organisms have. However, many repetitive elements were identified, including IS*6110*, which have been used as a tool to identify varying

strains during outbreaks of *M. tuberculosis* (Harrington and Bishai, 2000). The mycobacterium contains within its repertoire genes that could allow the bacterium to adapt to a range of environmental conditions, from growth in a rich broth to survival in host macrophages. These genes include 13 putative σ factors, 140 transcriptional regulators, 32 component regulators that presumably sense environmental signals, and 14 protein kinases or phosphatases. The genome also possesses 250 genes involved in a complex lipid metabolism, which the cell needs to synthesize material for its waxy coat (Cole et al., 1998). Several new classes of putative proteins termed "PE" and "PPE" (each with internal repetitive sequences) were also identified in the genome sequence. Indeed, ten percent of the genome is composed of these new potential coding sequences. The PE or PPE class of proteins has been proposed to be involved in modulation of the host immune response (Cole et al., 1998).

Epidemiology

Tuberculosis causes more deaths than any other infectious organism; 6 million new active cases and 3 million deaths annually are attributed to *M. tuberculosis*. Indeed, rates of TB in the world are predicted to increase 50% each decade. However, the annual incidence of TB in the United States remains low with 8.7 cases/100,000 people. Prior to 1985, TB rates within the United States were steadily declining at approximately 6% per year. In striking contrast to this trend, from 1986 to 1992 the incidence of TB cases rose about 3% per year (Fig. 6). A

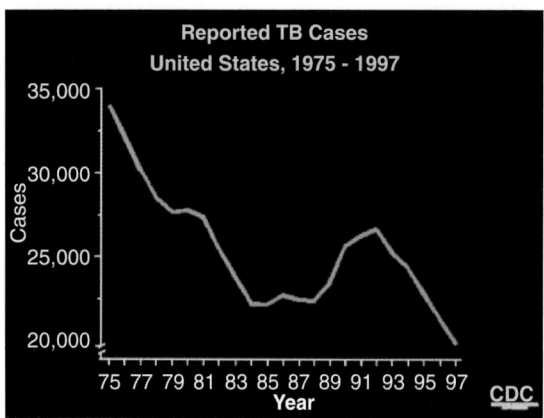

Fig. 6. After many years of decline, reported TB cases in the United States began to rise in the late 1980s. Following public health intervention, TB cases are once again dropping CDC.

maximum of 26,673 cases occurred in 1992. Increased government funding for public health programs and increased use of directly observed therapy (DOT) for tuberculosis have decreased rates since 1992 (Chaisson and Bishai, 1997). As a result, in 1998 the United States Centers for Disease Control and Prevention (CDC) reported 18,361 cases. Again, in 1999 the incidence of new TB cases fell to 17,528 cases.

Despite the fact that the number of cases in the United States is now dropping, there is the new problem of multidrug-resistant TB. The term multidrug-resistant tuberculosis (MDRTB) is used to describe strains that are resistant to two or more of the five first-line anti-TB drugs: isoniazid, rifampin, pyrazinamide, ethambutol and streptomycin. Indeed the incidence of drug resistance has increased in urban settings and among HIV-infected populations. In fact, during 1991, 33% of all TB strains recovered in New York City were MDRTB. The United States rate of MDRTB, however, has now fallen with improved control measures, but the global problem of drug resistance continues. In an international survey between 1994 and 1997, 12% of incident cases were resistant to at least one drug and 7.6% of TB strains were resistant to isoniazid (Pablos-Mendez et al., 1998).

Risk factors for TB include intravenous drug abuse, alcoholism, chronic pulmonary disease, prolonged steroid use, diabetes, renal failure, malnutrition and organ transplant. Infection with HIV is also a risk factor for TB: HIV-infected patients have a 25–50 fold increased risk compared with HIV-negative individuals. A number of social factors also increase a person's risk of TB including institutional living, urban dwellings, poverty and low educational levels (Chaisson and Bishai, 1997).

The incidence of other mycobacterial diseases has increased in recent years. Diseases caused by the *Mycobacterium avium* complex (MAC) were uncommon prior to the AIDS epidemic. Infection with MAC was mostly seen in individuals that were immunocompromised or had underlying lung disease. Indeed, prior to 1981 the incidence of MAC cervical lymphadenitis and MAC pulmonary infection was approximately 300 and 3,000 cases per year, respectively, in the United States. With the onset of the AIDS epidemic, the incidence of disseminated MAC infection jumped in the mid-1990s to approximately 20,000 individuals per year (Chaisson and Bishai, 1997). The MAC rates, however, appear to have peaked in 1997 and have since fallen due to improved antiretroviral therapy for patients infected with HIV (Palella et al., 1998).

Disease

THE *MYCOBACTERIUM TUBERCULOSIS* COMPLEX The *M. tuberculosis* complex is comprised of *M. tuberculosis*, *M. bovis*, *M. microti* and *M. africanum*. Only *M. tuberculosis* and *M. bovis* are a significant source of human disease (http://www.hopkins-tb.org). *Mycobacterium microti* causes disease in voles and was employed as a vaccine in the earlier part of the twentieth century (Wells and Oxon, 1937; Birkhaug, 1946; Wells, 1949; Wells and Wylie, 1954). *Mycobacterium africanum* causes disease in humans but makes up only a minority of cases of pulmonary TB, and its incidence is isolated to Africa.

M. TUBERCULOSIS The main route of infection of *M. tuberculosis* is by person-to-person inhalation of infectious aerosols. Bacilli are transmitted through speaking, coughing, sneezing and singing. Droplets of infectious *M. tuberculosis* are sized in the range of 1–10 μm. After inhalation of the droplets, the bacteria travel to the terminal bronchioles and alveoli where they are phagocytosed by alveolar macrophages. Many of these bacilli are killed in the phagosomes after these fuse with lysosomes and become acidified. *Mycobacterium tuberculosis*, however, is an intracellular pathogen and can efficiently inhibit phagosome-lysosome fusion (Fig. 7). As a result some of the invading *M. tuberculosis* bacilli survive these initial host defenses. Surviving bacilli grow within the macrophages and are released when the macrophages die. Unactivated macrophages arrive from the blood stream and ingest the newly liberated bacilli, which grow symbiotically within the unactivated macrophages for approximately 3 weeks. Eventually the bacilli lyse the macrophages and spill out into the host tissue.

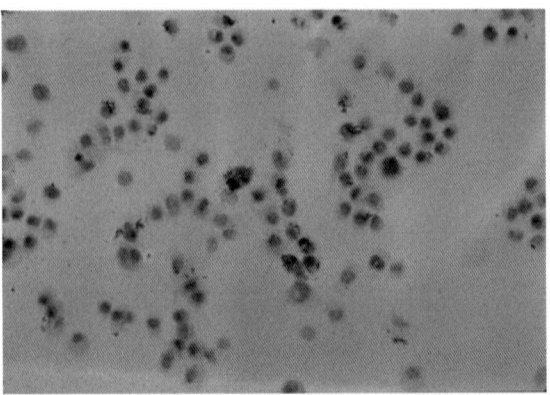

Fig. 7. Macrophages infected with *M. bovis* BCG are visualized by acid-fast staining. Mycobacteria are small red rods within the blue-stained macrophages.

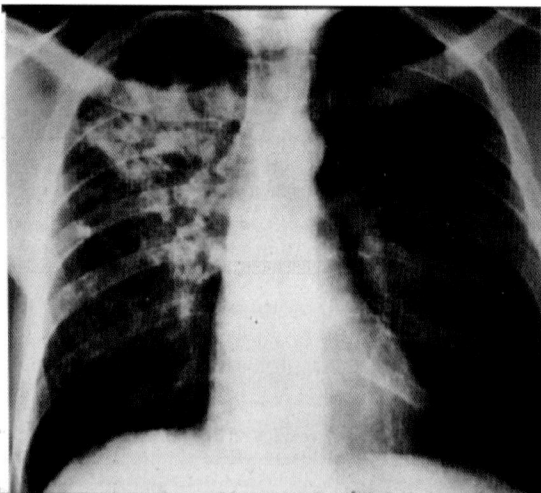

Fig. 8. Cavitary pulmonary tuberculosis in a 36-year old man. Right upper lobe streaky infiltrates and nodules are seen; two cavities are also present near the right apex.

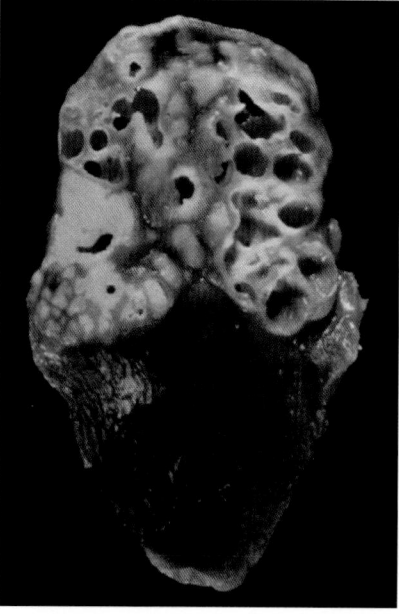

Fig. 9. Lung of a rabbit infected with *M. bovis*. Cavities are abundant within the lung.

In resistant individuals, released bacterial products stimulate strong cell-mediated immunity through Th1 signaling with INF-γ, Il-2, and Il-12. Activated macrophages that can kill *M. tuberculosis*, as well as T cells, are recruited to the periphery of the infectious focus. The bacteria and cellular debris are contained in a tissue structure called "a caseous granuloma." The disease will be halted at the stage where this small granuloma is formed.

In sensitive individuals, cell-mediated immunity is weak, and bacilli continue to multiply. Macrophages arrive to engulf the bacilli while T cells also accumulate. Because the macrophages are inadequately activated, the *M. tuberculosis* bacilli also parasitize the recruited macrophages. Cytotoxic T-cells produce toxic substances, which results in damage of host tissues. This cycle is repeated so that the granuloma enlarges. Eventually, tissue damage may result in cavity formation within the lung as well as in liquefaction of the granuloma (Figs. 8 and 9). *Mycobacterium tuberculosis* bacilli are particularly adept at multiplying in a liquefied cavity from which they are then expelled and spread through coughing and sneezing (Dannenberg, 1993; Dannenberg, 1994).

As they have no environmental reservoirs, *M. tuberculosis* organisms spread from person to person. In addition, fomites are not thought to support the transmission of the disease. Some (25–50%) of the diseased individual's close contacts will become infected with *M. tuberculosis*. However, only a minority (5% of those exposed) is susceptible and will develop primary active disease. In contrast, most humans seem to be naturally resistant to *M. tuberculosis*; these individuals harbor the bacilli and develop only a latent infection. Approximately 95% of infected individuals control the spread of *M. tuberculosis* within the body and disease progression is halted at the stage where a small granuloma is formed. These individuals may have a lifelong latent infection with no symptom other than a positive reaction to the administration of purified protein derivative (PPD) of *M. tuberculosis*, indicating immune memory of *M. tuberculosis* infection. Any process that disrupts the immunocompetence of a latently infected individual, however, may cause active infection to develop. The elderly often have a recrudescence of disease after decades of a latent infection due to their waning immunity. Acquiring a viral infection such as HIV may cause an individual to become immunosuppressed and develop active from latent disease. In fact, HIV-infected individuals have a 4–8% yearly risk for reactivation TB whereas immunocompetent individuals have only a 5–10% lifetime risk (Chaisson and Bishai, 1997).

Disease development due to *M. tuberculosis* infection can be either pulmonary or extrapulmonary. Inasmuch as infection is mainly through the aerosol route, immunocompetent individuals manifest primarily pulmonary disease. Symptoms include cough, chest pain, sputum production, fever, night sweats and hemoptysis in advanced disease. About 15% of TB cases in immunocompetent individuals occur at extrapulmonary sites. Such cases are attributed to reactivation of a latent extrapulmonary focus of infection, because during the initial infection process transient bacteremia may occur. Occasionally hospital and laboratory workers infect soft tissues through accidental injection with *M. tuberculosis*-contaminated syringes. In those cases infection is extrapulmonary but may spread by the blood to the lungs and other organs if the immune system cannot check the growth of the bacilli. Immunocompromised individuals (such as the very young, old and those infected with HIV) are more likely to develop extrapulmonary tuberculosis.

M. Bovis Isolated from cattle, *M. bovis* causes TB in ruminants and occasionally humans. This organism prefers to grow at 37°C and growth of a colony from a single bacillus usually requires 3–4 weeks of culture. Before routine pasteurization was practiced, humans could be infected by drinking *M. bovis*-contaminated milk. In fact milk was an important reservoir of infectious bacilli. However, since pasteurization became commonplace in the United States and in much of the world, the incidence of *M. bovis* infection has decreased considerably. *Mycobacterium bovis* is spread among cattle by an aerosol route and from cattle to humans by either a gastrointestinal or an aerosol route. In humans, *M. bovis* may cause intra-abdominal TB, cervical lymphadenitis (scrofula) or pulmonary TB (Adler and Rose, 1996). *Mycobacterium bovis* was used to derive the well-known attenuated vaccine strain, bacille Calmette Guerin (BCG), which is still used in many parts of the world.

M. Africanum This slow growing mycobacterium, first described in 1969, is infrequently associated with human pulmonary disease. It causes a disease similar to that caused by *M. tuberculosis* and *M. bovis*. Most patients with this disease reside or have resided in Africa. *Mycobacterium africanum* probably spreads by an aerosol route (Adler and Rose, 1996).

The *Mycobacterium Avium* Complex The *M. avium* complex is comprised of three species, *M. avium*, *M. intracellulare* and *M. paratuberculosis*. All three species are common in the environment, as they are isolated from soil, water, food and domestic animals. Categorized as slow growing mycobacteria, these species grow optimally at 37°C and achieve growth in approximately 4–6 weeks (Havlir and Ellner, 2000). Colonies are nonpigmented and are both opaque and domed or flat and transparent (Toosi and Ellner, 1998). *Mycobacterium avium* causes a tuberculosis-like infection in chickens, pigeons and other birds, whereas *M. paratuberculosis* causes Johne's disease in ruminants, which is a contagious enteritis resulting in progressive wasting and eventual death (Fraser et al., 1986). *Mycobacterium*

paratuberculosis has not been shown to cause human disease.

In humans, *M. avium* and *M. intracellulare* can cause pulmonary disease, regional lymphadenitis or disseminated disease. Those individuals who are coinfected with HIV are prone to disseminated disease, whereas those who are not coinfected are likely to have pulmonary disease. Predisposing factors to infection with *M. avium* include underlying lung disease, HIV infection, chronic obstructive pulmonary disease, chronic bronchitis, healed or active TB, pulmonary mycosis and malignancy. Infection occurs by either inhalation or ingestion of the infectious bacilli. Diagnosis in those affected individuals without HIV is made by radiographic evidence of disease, positive sputum cultures and an acid-fast smear. Symptoms are a productive cough, hemoptysis, fever and weight loss (Chaisson and Bishai, 1997).

Disseminated *M. avium* is a late opportunistic infection in those individuals coinfected with HIV. In fact, disseminated MAC generally occurs only in AIDS patients with CD4 cell counts less than $100/mm^3$ (Chaisson and Bishai, 1997). Symptoms of infection are fever, drenching night sweats and weight loss. The hallmark of a disseminated infection is the high circulating levels of MAC bacteremia, which can reach 10^4 bacteria/ml. Organ involvement may be widespread and the tissue burden may reach 10^6 bacteria/g. The gastrointestinal tract is frequently infected, leading to symptoms including nausea, vomiting, watery diarrhea and abdominal pain. Left untreated, the disease will cause progressive clinical deterioration (Havlir and Ellner, 2000).

Mycobacterium Leprae Leprosy is particularly dreaded because it may cause bodily deformities leading to social stigmatization. In centuries past, persons suffering from leprosy were ostracized and forced to live away from the general population, in colonies or leprosaria. Today many patients suffering from leprosy are still concerned about the social stigma associated with leprosy. As of 1999, there were 719,332 leprosy cases registered for treatment. Almost all of these individuals were receiving multidrug therapy. The new case detection rate stayed the same or was increasing with 747,369 new cases detected during 1998 (http://www.who.int/lep). The prevalence of leprosy is not evenly distributed, as much of the disease seen today occurs in the third world. Most cases (72%) are from Asia and Oceania, whereas 18% are from Africa and only 7% are from the Americas (Lockwood and McAdam, 1998).

The genome of the 1–8 μm long and 0.3–0.5 μm wide *M. leprae* bacillus is approximately 3.2 MB (Gelber and Rea, 2000). The sequence has been partially determined and is available for inspection (http://www.sanger.ac.uk/DataSearch). It is speculated that many of the genes essential for the survival of *M. leprae* in the environment have been deleted, and the bacillus has become an obligate parasite of its human host. Indeed to date no one has been able to continuously culture the bacillus in either in vitro conditions or cell culture. It may be true that the mycobacterium has lost genes essential for survival in these ex vivo conditions and thus has an absolute requirement for host factors (Gelber and Rea, 2000). In fact, the only methods to propagate *M. leprae* bacilli are via animal infections with either 9-banded armadillos or severe combined immunodeficiency (SCID) mice. Growth of sufficient bacilli for laboratory manipulation takes approximately one year, inasmuch as the generation time for this organism is approximately 14 days in animal tissues.

Leprosy is a chronic disease with a long incubation period. This fact is affirmed by the observation that children younger than two years do not have leprosy symptoms. In addition, people residing in nonendemic countries who have visited a site with endemic leprosy may develop the disease many years after the initial exposure. Hence the incubation period is estimated to range from 2–12 years. Like those infected with TB, only about 10% of those infected with leprosy go on to develop the disease (Lockwood and McAdam, 1998). Being a human disease, leprosy has no established natural reservoir for infection. *Mycobacterum leprae* is probably transmitted from person to person although the reservoirs and their role in transmission remain controversial issues. In 1898 Schaffer noted that leprosy patients discharged large numbers of acid-fast bacilli (AFB) when coughing, sneezing or speaking normally. In fact, if left untreated a lepromatous leprosy patient may dispel 6.8×10^{10} AFB in a single nose blow. Thus leprosy may to be transmitted via an aerosol route where it goes on to infect the lining of the nose (Lockwood and McAdam, 1998). Though *M. leprae* produces lesions that often involve the extremities and skin, bacilli probably reach these locations by hematogenous spread. In fact people who have leprosy may have granulomata in the lymph nodes, liver, kidney, spleen, bone marrow, adrenals, testes and eyes. Thus the bacilli can spread to remote portions of the body to cause lesions. Although *M. leprae* spreads through the body via a systemic route, it prefers to grow at temperatures below 37°C and therefore has a tropism for the extremities and skin where its growth can flourish.

The outcome of the disease is strongly dependent on the host's immune response. Upon infection, bacilli are engulfed by macrophages. If the macrophages are sufficiently activated to kill the invading bacilli, then the infection will be cleared. Otherwise, the bacilli will go on to replicate within macrophages, lyse the cells, and infect other macrophages (Gelber and Rea, 2000). The host's immune system can take one of two pathways. First, the disease is controlled by strong cell-mediated immunity employing helper T cells of the Th1 lineage. These helper T-cells induce the macrophages to kill the bacilli, and INF-γ and IL-2 are found at the site of infection (Lockwood and McAdam, 1998). Organized granulomata are composed of the macrophages surrounded by the T-cells, resulting in what is known as tuberculoid or paucibacillary leprosy. It is known from the days before antibiotic therapy that tuberculoid leprosy often spontaneously heals (Lockwood and McAdam, 1998). However, tuberculoid leprosy also may result in tissue damage from continuous lymphocyte infiltration into the area of infection. Tuberculoid leprosy has a shorter incubation time, generally 2–5 years, and causes few skin lesions. (Lesions when manifest are hypopigmented and asymmetric.) Nerve damage can occur when granulomata are near small dermal sensory and autonomic nerve fibers (Lockwood and McAdam, 1998). Tuberculoid leprosy is not associated with the presence of stainable *M. leprae* in skin and does not produce upper respiratory signs and symptoms. The reason is probably that strong cell-mediated immunity controls the infection, though the control is at the expense of significant tissue damage. In general, tuberculoid leprosy has a good prognosis with treatment.

The other path of infection occurs in patients who do not have good cell-mediated immunity and good initial killing of the *M. leprae*. In this expression of the disease, named "lepromatous" or "multibacillary" leprosy, bacilli multiply out of control. This form of the disease has a somewhat longer incubation period, 8–12 years. The bacilli may invade Schwann cells, resulting in demyelination of nerves and nerve damage (Lockwood and McAdam, 1998). Granulomata that do form are poorly organized, with high bacterial loads. Bacilli may replicate within dermal cells as evidenced by AFB in the skin. Most (80% of) lepromatous patients have some nasal symptoms due to invasion of the nasal mucosa by *M. leprae*. Occasionally, without treatment, the bacilli may cause so much destruction that the nasal septum collapses. In addition, denervation and loss of pain sensation results in repetitive trauma and injury to limbs. Destruction can occur over vast areas of limbs that in some instances may necessitate amputation.

In between the two extremes or poles, tuberculoid and lepromatous, there is a continuum. These disease states, which are defined pathologically after review of skin biopsy specimens, are referred to as borderline tuberculoid, borderline leprosy and borderline lepromatous. Borderline tuberculoid has some aspects of lepromatous leprosy, whereas borderline lepromatous leprosy has some aspects of tuberculoid leprosy. Borderline leprosy refers to a state in between lepromatous and tuberculoid. Borderline states are extremely unstable and may shift spontaneously between mild and severe disease.

Until recently not much was known about the mechanism by which *M. leprae* interacts with and causes damage to cells of the nervous system. Recent evidence has revealed that *M. leprae* binds to the basal lamina surrounding Schwann cells through an interaction between a 21-kDa surface protein on *M. leprae* and laminin-2 within the basal lamina (Rambukkana et al., 1997; Shimoji et al., 1999). Additionally, *M. leprae* can interact directly with Schwann cells by binding to α-dystroglycan on the Schwann cell surface. This interaction, however, occurs only in the presence of laminin-2 (Rambukkana et al., 1998). These findings have the promise of providing mechanistic information that may aid in the development of improved therapeutics against leprosy.

Mycobacterium Kansasii This slow-growing organism is the second most frequent cause of pulmonary disease by a nontuberculous mycobacteria (Brown and Wallace, 2004). *Mycobacterium kansasii* is found in environmental water, and water is probably the source of human infection (Iseman, 1998). The clinical presentations of pulmonary *M. kansasii* are a chronic cough, low-grade fever, malaise and chest pain. Though clinical manifestations are similar to those caused by *M. tuberculosis*, a chest x-ray usually reveals upper-zone disease with fewer fibronodular regions than found with either *M. tuberculosis* or MAC. Predisposing factors for infection with *M. kansasii* are smoking-induced chronic bronchitis, chronic inorganic dust exposure, and chronic obstructive pulmonary disease (Toosi and Ellner, 1998). Although extrapulmonary infections are rare, *M. kansasii* can cause diffuse cutaneous disease as well as cervical lymphadenitis, especially in children. Patients with AIDS or with impaired cellular immunity may also develop a disseminated infection.

Mycobacterium Marinum *M. marinum* causes a cutaneus disease in humans known as swimming pool granuloma, fish handler's nodule or surfer's nodule. As implied above, water is the major environmental reservoir for this mycobacterial species (Toosi and Ellner, 1998). The bacterium and therefore most cases of infection are mainly found in the coastal areas of North America (Ise-

man, 1998). Humans contract the disease from environmental water, the handling of domestic and wild fish, or the cleaning and maintenance of fish tanks. *Mycobacterium marinum* lesions appear on the extremities of affected individuals, as the bacillus is most adept at growing at a temperature of 32°C (Brown and Wallace, 2004). Interestingly, *M. marinum* is closely related to *M. tuberculosis* on a genetic level. Because *M. marinum* is not transmitted to humans through an aerosol route and is not considered a biosafety level 3 (BSL3) organism, it has been used as a model system for the study of *M. tuberculosis* pathogenesis (Ramakrishnan et al., 2000). In animal tissues, *M. marinum* bacilli infect macrophages and this leads to granuloma formation. As this species of mycobacterium causes disease in both fish and frogs, the nature of its pathogenesis has been investigated in these animal models. Thus, although *M. marinum* is not a major public health problem, research into its basic biology is ongoing due to its genetic similarity with *M. tuberculosis*.

Mycobacterium Scrofulaceum This slowly growing mycobacterium, first isolated in 1956, has an optimal growth temperature of 37°C and, upon initial isolation, can take approximately 4 to 6 weeks to achieve growth. It is a leading cause of scrofula or cervical lymphadenitis in children 1 to 5 years of age. Infrequently it may cause progressive pulmonary disease as well as bone and soft tissue disease and, in some instances, disseminated disease.

Mycobacterium Fortuitum, M. Chelonae and *M. Abscessus* These organisms comprise a group of rapidly growing mycobacteria that are found in soil samples as well as in water; however, their mode of transmission to humans remains unknown (Iseman, 1998). These mycobacteria cause localized infection of the skin, soft tissues and bones. They also may cause pulmonary disease typified by a persistent hacking cough, low-grade fever, chills, malaise and mucopurulent secretions.

Mycobacterium Ulcerans During the first half of the 20th century this mycobacterial disease was recognized in Bairnsdale, Australia. The syndrome produced an ulcer, subsequently named the "Bairnsdale ulcer," or "Buruli ulcer," that was characterized by a chronic progressive ulceration of the skin (Fig. 10). Acid-fast bacilli, later identified as *M. ulcerans*, were eventually cultivated from lesions of infected individuals. Although no bacteria have been cultivated from environmental sources, the use of polymerase chain reaction (PCR) has identified *M. ulcerans* in water sources (Ross et al., 1997). Thus it is

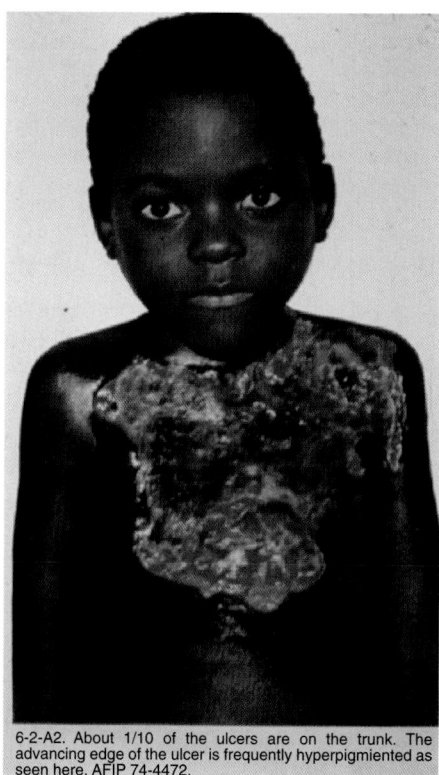

6-2-A2. About 1/10 of the ulcers are on the trunk. The advancing edge of the ulcer is frequently hyperpigmiented as seen here. AFIP 74-4472.

Fig. 10. A young boy infected with *M. ulcerans* has a buruli ulcer covering a portion of his torso. From Connor et al. (1976).

hypothesized that *M. ulcerans* is acquired from the environment via water. Indeed it has been noted that people inhabiting areas close to a body of stagnant water in tropical regions are more at risk for developing a Buruli ulcer (Ross et al., 1997; Johnson et al., 1999). The hallmark of the disease is a skin ulcer that begins as a small nodule but (if left untreated) enlarges to encompass a large surface area. The ulcers are themselves painless, suggesting that there is some nerve involvement. Inflammatory response at the site of colonization by the AFB or the site of ulceration appears to be minimal, even though necrosis of the subcutaneous fat tissue is extensive. In contrast to *M. tuberculosis*, *M. ulcerans* grows as microcolonies extracellularly during human infection. In addition, treatment of the disease is not trivial. Antibiotic therapy has variable efficacy, and therefore the best treatment is surgical excision after the ulcer has been identified at an early nodule stage. Large ulcers require extensive surgery with skin grafting. In extreme cases where muscle tissue is involved, amputation of an affected limb may be required (Iseman, 1998).

It has long been suspected that a toxin mediates the extensive tissue damage associated with the Buruli ulcer (Johnson et al., 1999). The fact that tissue necrosis occurs at sites distal to areas of colonization by *M. ulcerans* seems to imply that a factor that can diffuse to distant tissues is involved. It has recently been discovered that a polyketide toxin named "mycolactone" is responsible for histopathologic changes distal to the site of infection (George et al., 1998; George et al., 1999). This toxin was present in the culture filtrates of growing *M. ulcerans*, indicating that it is secreted into the milieu of the bacterium. When this toxin is injected into guinea pig skin, ulcers similar to those caused by *M. ulcerans* are formed. In addition, exposure of L929 murine fibroblasts to the toxin arrests their cell cycle. Thus, as a toxinogenic microorganism, *M. ulcerans* has a mode of pathogenesis that is very different from that of other members of the genus *Mycobacterium*.

PREVENTION Vaccination with the attenuated *M. bovis* strain bacille Callmette Guerin (BCG) is common in many parts of the world; however, protection rates due to this vaccine vary from 0–80%. Whereas BCG probably provides little protection against adult pulmonary TB, it does appear to protect against childhood disseminated TB (Colditz et al., 1994). An additional problem of vaccination is that (once vaccinated with BCG) individuals will have a positive reaction to the administration of purified protein derivative (PPD) from *M. tuberculosis*, excluding its use as a diagnostic tool. Therefore once a person has been vaccinated, it is impossible to know if they have been infected with *M. tuberculosis* subsequently and therefore if their risk of developing active disease is substantial. Thus in the United States vaccination is not recommended as a measure to control *M. tuberculosis* spread. In the United States and many European nations, it is preferable to periodically administer PPD to persons at high risk of becoming infected with *M. tuberculosis* and to treat prophylactically with antibiotics. Vaccination, however, is useful in reducing miliary and meningeal disease in children in third world countries where tuberculosis rates are high and mass screening is impractical.

Literature Cited

Adler, J., and D. Rose. 1996. Tuberculosis. *In:* W. N. Rom and S. Garay (Eds.) Transmission and Pathogenesis of Tuberculosis. Little, Brown and Company. New York, NY.

Agerton, T., S. Valway, et al. 1997. Transmission of a highly drug-resistant strain (strain W1) of Mycobacterium tuberculosis: Community outbreak and nosocomial transmission via a contaminated bronchoscope. JAMA 278(13):1073–1077.

Amplicor in press. Mycobacterium Tuberculosis Test Insert Information.

Behr, M. A., S. A. Warren, et al. 1999. Transmission of Mycobacterium tuberculosis from patients smear-negative for acid-fast bacilli. Lancet 353(9151):444–449.

Birkhaug, K. 1946. Vaccination with the vole bacillus (Wells). American Review of Tuberculosis 54:41–50.

Brennan, P. J., and H. Nikaido. 1995. The envelope of mycobacteria. Ann. Rev. Biochem. 64:29–63.

Brown, B., and R. Wallace. 2004. Infections due to nontuberculous mycobacteria. In: G. Mandell, J. Bennett, and R. Dolin (Eds.) Principles and Practices of Infectious Diseases. Churchill Livingstone. Philadelphia, PA.

Chaisson, R. E., and W. R. Bishai (Eds.). 1997. Mycobacterium avium and tuberculosis infection: Management in patients with HIV disease. Clinical Care Options for HIV Continuum of Care Series Medical Care Collaborative and Healthcare Communications Group.

Colditz, G. A., T. F. Brewer, et al. 1994. Efficacy of BCG vaccine in the prevention of tuberculosis. JAMA 271(9):698–702.

Cole, S. T., R. Brosch, et al. 1998. Deciphering the biology of Mycobacterium tuberculosis from the complete genome sequence. Nature 393(6685):537–544.

Connor, D., W. Meyers, et al. 1976. Infection by Mycobacterium ulcerans. In: C. Binford and D. Connor (Eds.) Pathology of Tropical and Extraordinary Diseases. Armed Forces Institute of Pathology. Washington DC.

Curtis, A. B., R. Ridzon, et al. 1999. Extensive transmission of Mycobacterium tuberculosis from a child. N. Engl. J. Med. 341(20):1491–1495.

Dannenberg Jr., A. M. 1993. Immunopathogenesis of pulmonary tuberculosis. Hosp. Pract. (Off. Ed.) 28(1):51–58.

Dannenberg Jr., A. M. 1994. Roles of cytotoxic delayed-type hypersensitivity and macrophage-activating cell-mediated immunity in the pathogenesis of tuberculosis. Immunobiology 191(4–5):461–473.

Davis, A. 1996. History of the Sanitorium Movement. In: W. N. Rom and S. Garay (Eds.) Tuberculosis. Little, Brown and Company. New York, NY.

Edlin, B. R., J. I. Tokars, et al. 1992. An outbreak of multi-drug-resistant tuberculosis among hospitalized patients with the acquired immunodeficiency syndrome. N. Engl. J. Med. 326(23):1514–1521.

Fraser, C. M., A. Mays, et al. 1986. The Merk Veterinary Manual. Merk & Co. Rahway, NJ.

Gelber, R., and T. Rea. 2000. Mycobacterium leprae (Leprosy, Hansen's Disease). In: G. Mandell, J. Bennett, and R. Dolin (Eds.) Principles and Practices of Infectious Diseases. Churchill Livingstone. Philadelphia, PA.

George, K. M., L. P. Barker, et al. 1998. Partial purification and characterization of biological effects of a lipid toxin produced by Mycobacterium ulcerans. Infect. Immunol. 66(2):587–593.

George, K. M., D. Chatterjee, et al. 1999. Mycolactone: A polyketide toxin from Mycobacterium ulcerans required for virulence. Science 283(5403):854–857.

Haas, F., and S. Haas. 1996. The origins of Mycobacterium tuberculosis and the notion of its contagiousness. In: W. N. Rom and S. Garay (Eds.) Tuberculosis. Little, Brown and Company. New York, NY.

Harrington, S., and W. Bishai. 2000. Molecular epidemiology and infectious diseases. In: K. Kitzenberg (Ed.) Infectious Disease Epidemiology: Theory and Practice.

Harris, W. 1996. Chemotherapy of tuberculosis: The beginning. In: W. N. Rom and S. Garay (Eds.) Tuberculosis. Little, Brown and Company. New York, NY.

Havlir, D., and J. Ellner. 2000. Mycobacterium avium complex. In: G. Mandell, J. Bennett, and R. Dolin (Eds.) Principles and Practices of Infectious Diseases. Churchill Livingstone. Philadelphia, PA.

Iseman, M. 1998. Nontuberculous mycobacterial infections. S. Gorbach, J. Bartlett, and N. Blacklow. (Eds.) Infectious Diseases. Sauders and Company. Philadelphia, PA.

Johnson, P. D., T. P. Stinear, et al. 1999. Mycobacterium ulcerans—a mini-review. J. Med. Microbiol. 48(6):511–513.

Kent, P. T., and G. P. Kubica. 1985. Public Health Mycobacteriology: A Guide for the Level III Laboratory. Department of Health and Human Services. Atlanta, GA.

Lockwood, D., and K. McAdam. 1998. Leprosy. In: S. Gorbach, J. Bartlett, and N. Blacklow. Infectious Diseases. Saunders and Company. Philadelphia, PA.

Michele, T. M., W. A. Cronin, et al. 1997. Transmission of Mycobacterium tuberculosis by a fiberoptic bronchoscope: Identification by DNA fingerprinting. JAMA 278(13):1093–1095.

Middleton, A. M., M. V. Chadwick, et al. 1997. Evaluation of a commercial radiometric system for primary isolation of mycobacteria over a fifteen-year period. Eur. J. Clin. Microbiol. Infect. Dis. 16(2):166–170.

Nolte, F., and B. Metchock. 1995. Mycobacterium. In: P. Murray, E. Baron, M. Pfaller, F. Tenover, and R. Yolken (Eds.) Manual of Clinical Microbiology. ASM Press. Washington DC.

Pablos-Mendez, A., M. C. Raviglione, et al. 1998. Global surveillance for antituberculosis-drug resistance, 1994–1997: World Health Organization, International Union against Tuberculosis and Lung Disease Working Group on Anti-Tuberculosis Drug Resistance Surveillance. N. Engl. J. Med. 338(23):1641–1649.

Palella Jr., F. J., K. M. Delaney, et al. 1998. Declining morbidity and mortality among patients with advanced human immunodeficiency virus infection: HIV Outpatient Study Investigators. N. Engl. J. Med. 338(13):853–860.

Pfyffer, G. E., C. Cieslak, et al. 1997. Rapid detection of mycobacteria in clinical specimens by using the automated BACTEC 9000 MB system and comparison with radiometric and solid-culture systems. J. Clin. Microbiol. 35(9):2229–2234.

Ramakrishnan, L., N. A. Federspiel, et al. 2000. Granuloma-specific expression of Mycobacterium virulence proteins from the glycine-rich PE-PGRS family. Science 288(5470):1436–1439.

Rambukkana, A., J. L. Salzer, et al. 1997. Neural targeting of Mycobacterium leprae mediated by the G domain of the laminin-alpha2 chain. Cell 88(6):811–821.

Rambukkana, A., H. Yamada, et al. 1998. Role of alpha-dystroglycan as a Schwann cell receptor for Mycobacterium leprae. Science 282(5396):2076–2079.

Ross, B. C., L. Marino, et al. 1997. Development of a PCR assay for rapid diagnosis of Mycobacterium ulcerans infection. J. Clin. Microbiol. 35(7):1696–1700.

Saito, H., Y. Kashiwabara, et al. 1996. Rapid detection of acid-fast bacilli with Mycobacteria Growth Indicator Tube (MGIT). Kekkaku 71(6):399–405.

Sharp, S. E., C. A. Suarez, et al. 1996. Evaluation of the mycobacteria growth indicator tube compared to Septi-

Chek AFB for the detection of mycobacteria. Diagn. Microbiol. Infect. Dis. 25(2):71–75.

Shimoji, Y., V. Ng, et al. 1999. A 21–kDa surface protein of Mycobacterium leprae binds peripheral nerve laminin-2 and mediates Schwann cell invasion. Proc. Natl. Acad. Sci. USA 96(17):9857–9862.

Shinnick, T. M., and R. C. Good. 1994. Mycobacterial taxonomy. Eur. J. Clin. Microbiol. Infect. Dis. 13(11):884–901.

Sterling, T. R., D. S. Pope, et al. 2000. Transmission of Mycobacterium tuberculosis from a cadaver to an embalmer. N. Engl. J. Med. 342(4):246–248.

Toosi, Z., and J. Ellner. 1998. Mycobacterium tuberculosis and other mycobacteria. *In:* S. Gorbach, J. Bartlett, and N. Blacklow (Eds.) Infectious Diseases. Sauders and Company. Philadelphia, PA.

Valway, S. E., M. P. Sanchez, et al. 1998. An outbreak involving extensive transmission of a virulent strain of Mycobacterium tuberculosis. N. Engl. J. Med. 338(10):633–639.

Wang, S. X., and L. Tay. 1999. Evaluation of three nucleic acid amplification methods for direct detection of Mycobacterium tuberculosis complex in respiratory specimens. J. Clin. Microbiol. 37(6):1932–1934.

Wells, A. Q., and D. M. Oxon. 1937. Tuberculosis in wild voles. Lancet 1:1221.

Wells, A. Q. 1949. Vaccination with the murine type of tubercle bacillus (vole bacillus). Lancet 53–55.

Wells, A. Q., and J. A. H. Wylie. 1954. Vaccination against tuberculosis with the vole bacillus. British Medical Bulletin 10:96–100.

Prokaryotes (2006) 3:934–944
DOI: 10.1007/0-387-30743-5_35

CHAPTER 1.1.20

Mycobacterium leprae

THOMAS M. SHINNICK

Although *Mycobacterium leprae* was one of the first bacterial pathogens of humans to be described (Hansen, 1874), progress on understanding the basic biology and pathogenicity of this organism has been greatly hampered by the inability to find a conventional laboratory medium or tissue culture system that can support its growth. Consequently, the only means of propagating this organism at present is by using experimental animals. Furthermore, it has been found that the nine-banded armadillo can be used to produce large numbers of bacilli (Kirchheimer and Storrs, 1971; Storrs, 1971). Relatively little is known, therefore, about the taxonomy, genetics, and biochemistry of this species. A corollary of this is that much of what we do know about *M. leprae* has come from studies of the disease it causes (leprosy or Hansen's disease) and from the animal models. As such, this chapter emphasizes the characteristics and behavior of *M. leprae* in experimental animal model systems and in humans. Several excellent reviews on the clinical aspects, epidemiology, immunology, and pathology of leprosy and on the biochemistry and immunochemistry of *M. leprae* have been published recently, and the reader is referred to these for additional information (Bloom and Godal, 1983; Bloom and Mehra, 1984; Fine, 1982; Gaylord and Brennan, 1987; Hastings, 1986; Hastings and Franzblau, 1988; Jopling and McDougall, 1988; Kaplan and Cohn, 1986; Stewart-Tull, 1982).

Although *M. leprae* can occasionally be found in the body extracellularly, the bacillis appears to be able to replicate only within cells of the host, most commonly in macrophages and Schwann cells (Bloom and Godal, 1983; Kaplan and Cohn, 1986). Hence, *M. leprae* is considered to be an obligate intracellular pathogen. In host cells, the bacilli are found singly or in clumps referred to as *globi* (Cowdry, 1940). The bacilli are straight or slightly curved, Gram-positive, acid-fast, alcohol-fast, nonmotile rods ranging from 1 to 8 µm in length and 0.2 to 0.5 µm in width (Draper, 1983). Acid- and alcohol-fastness refers to the ability of the bacillus to retain the color of certain dyes, usually carbol fuchsin, following treatment with mild acid and alcohol, respectively.

M. leprae has been placed in the genus *Mycobacterium* in the family Actinomycetales based mainly on cell structure, staining properties and chemical composition as well as on the basis of the presence of mycolic acids, antigens characteristic of mycobacteria, and a lipid-rich cell envelope (Draper, 1976; Harboe et al., 1977; Stanford et al., 1975). For example, the *M. leprae* bacillus closely resembles *M. tuberculosis* bacilli in size, morphology, and staining characteristics, although it does stain a little more deeply with carbol fuchsin and the staining is somewhat less acid fast. Recently, analyses of the ribosomal RNA sequences of armadillo-grown *M. leprae* by nucleic acid hybridization techniques (Sela et al., 1989) and by ribosomal RNA sequence comparisons (Smida et al., 1988) revealed that *M. leprae* is closely related to the corynebacteria, nocardia, and mycobacteria, especially to the two slowly growing *Mycobacterium* species *M. avium* and *M. tuberculosis*. Although these observations indicate that *M. leprae* should be classified in the genus *Mycobacterium*, several features distinguish *M. leprae* from other members of the *Mycobacterium* genus. These are: 1) loss of acid-fastness upon extraction with pyridine, although *M. smegmatis*, *M. vaccae*, and *M. phlei* do lose acid fastness after prolonged exposure to pyridine (Fisher and Barksdale, 1971; McCormick and Sanchez, 1979; Skinsnes et al., 1975); 2) ability to oxidize 3,4-dihydroxyphenylalanine (Prabhakaran and Kirchheimer, 1966); 3) replacement of L-alanine with glycine in the linking peptide of peptidoglycan (Draper, 1976); 4) 56% GC content as compared with 65–70% for most *Mycobacterium* species (Clark-Curtiss et al., 1985; Imaeda et al., 1982; Wayne and Gross, 1968); and 5) lack of substantial genomic DNA homology with other *Mycobacterium* species (quite in contrast with the ribosomal rRNA results) (Athwal et al., 1984; Grosskinski et al., 1989). Although these differences are not sufficient to exclude *M. leprae* from the genus *Mycobacterium*, they are sufficient to render the final taxonomic position of *M. leprae* somewhat in doubt.

The Disease

Hansen's disease (leprosy) is a chronic infectious granulomatous disease that primarily affects the peripheral nervous system, skin, and mucous membranes, especially nasal mucosa. In advanced cases, other tissues—including muscle, testes, capillary endothelium, liver, spleen, and bone marrow—can be affected (Desikan and Job, 1968). Although leprosy is not usually fatal in and of itself, 20–30% of patients with untreated or neglected infections develop crippling deformities of hands and feet. A sequela of these deformities is the social stigma that has been historically associated with leprosy. An addition consequence is that suicide is a common cause of death among infected individuals, particularly during episodes of exacerbation of the lesions (erythema nodosum leprosum).

Leprosy has been estimated to afflict 10–12 million individuals world-wide, with most cases being found in tropical and subtropical regions (Sixth Report of the WHO Expert Committee on Leprosy, 1988). Climatic factors do not seem to have a significant impact on the disease, however, since leprosy was widespread in Europe, particularly Norway, in the past centuries (Browne, 1975), and cases have been reported from above the Arctic Circle (Sansarricq, 1981). Within endemic areas, cases of leprosy appear to cluster geographically, with the prevalence of disease exceeding 10 cases per 1000 population in the high prevalence areas (Bloom and Godal, 1983). Although this disease has been cited as the "least infectious" of communicable diseases (McWhirter, 1981), epidemics of leprosy have occurred on some Pacific Islands in which as much as 35% of the population developed the disease over a 20-year period (Wade and Ledowski, 1952).

A feature of the disease that has intrigued clinicians and immunologists is the variety of clinical manifestations that can be found. These range from a single lesion with no detectable bacilli to multiple lesions containing large numbers of bacilli. Bacterial counts up to 5×10^9 organisms per gram of tissue have been found (Collaborative effort, 1975). The variety of disease symptoms is not related to the genetics of the bacterium but rather to the immune responsiveness of the host (Bloom and Mehra, 1984; Kaplan and Cohn, 1986). A clinically and experimentally useful categorization of this disease spectrum is the Ridley-Jopling classification scheme which is based on immunopathologic features of the disease (Ridley and Jopling, 1966). Tuberculoid leprosy (TT) is at one pole of the spectrum and is characterized by one or a few localized skin or nerve lesions, a strong cellular immune response, and a weak humoral immune response. Histologically, one finds well-organized epithelioid cell granulomas with multinucleated giant cells and abundant lymphocytes in TT lesions, but bacilli are usually absent ($<10^5$ bacilli/gram of tissue). At the other extreme is lepromatous leprosy (LL), characterized by numerous small, bilaterally symmetrical skin lesions, a weak or absent cellular immune response, and high antibody titers. Histologically, one observes a foamy macrophage granuloma with few lymphocytes in LL lesions, and the macrophages contain numerous bacilli ($>10^8$ bacilli/gram). Within these extremes are other conditions, including borderline lepromatous (BL; numerous skin lesions, numerous bacilli, granulomas with undifferentiated macrophages and histiocytes), borderline (BB; numerous skin lesions, few acid-fast bacilli, nerve involvement, epithelioid cell granulomas with no giant cells), and borderline tuberculoid (BT; multiple skin lesions, occasional bacilli, nerve involvement, diffuse epithelioid cell granulomas). One additional category, called indeterminant leprosy,

often found when a patient first seeks treatment, usually consists of a single small hypopigmented plaque. Such a lesion may remain indeterminant, regress spontaneously, or progress into a lesion that falls into one of the above categories.

It is generally accepted that *Mycobacterium leprae* is the etiologic agent of leprosy. However, not all of Koch's postulates have been fulfilled for identifying *M. leprae* as the causative agent of human leprosy. That is, a pure culture of *M. leprae* has not been developed from a single bacillus and shown to cause disease. The inability to grow the organism outside of animal hosts is the major stumbling block to completing Koch's postulates. Nonetheless, much evidence has accumulated in support of the role of *M. leprae* as the agent of the disease. The evidence includes the following: 1) *M. leprae* is found in leprosy patients and not in nonleprosy patients; 2) *M. leprae* bacilli are invariably present in lepromatous lesions; 3) leprosy patients display characteristic immune responses to *M. leprae* antigens (Mitsuda and Fernandez reactions; see Jopling and McDougall; 1988); 4) antibodies and T-cells reactive with antigens or epitopes uniquely expressed by *M. leprae* can be isolated from leprosy patients (reviewed in Gaylord and Brennan, 1987); 5) a phenolic glycolipid (PGL-I; Hunter and Brennan, 1981; Hunter et al., 1982) uniquely found in *M. leprae* is also uniquely found in lesions, sera, and urines from leprosy patients (Cho et al., 1983; Cho et al., 1986; Koster et al., 1987; Vemuri et al., 1985; Young et al., 1985); 6) *M. leprae*-specific nucleotide sequences can be found in leprosy lesions (Clark-Curtiss and Docherty, 1989); and 7) drug susceptibility of *M. leprae* in animal models parallels drug efficacy in patients (reviewed in Shepard, 1986).

Several properties of the *M. leprae* bacillus contribute to the features of the disease:

1. *M. leprae* has a generation time of 11–13 days in experimental animals (Levy, 1976; Shepard and McRae, 1971b). Such slow growth might influence the length of time from infection to disease (median interval 2–8 years; Fine, 1982) and the chronic nature of the infection.

2. *M. leprae* grows best at 30°C, which may explain its affinity for cooler parts of the body including skin and nasal mucosa (Shepard, 1965).

3. *M. leprae* infects Schwann cells and is the only bacterial pathogen capable of entering peripheral nerves (Job, 1971; Ridley et al., 1987; Stoner, 1979). These facets play a role in the loss of nerve function and the generation of crippling deformities.

4. *M. leprae* can supress the cellular immune response, perhaps through induction of suppressor T-cells (Bloom and Mehra, 1984;

Mehra et al., 1984) and/or rendering the infected macrophage defective for activation or antigen processing (Desai et al., 1989; Prasad et al., 1987; Sibley and Krahenbuhl, 1988). This may allow the host to tolerate the large bacterial load seen in lepromatous leprosy patients (up to 10^{13} organisms).

5. The unique phenolic glycolipid of *M. leprae*, called PGL-I, is a major component of the cell envelope (2% of total bacterial mass; Hunter and Brennan, 1981), can scavenge hydroxyl radicals and superoxide anions in vitro (Chan et al., 1989) and can inhibit the oxidative response in macrophages that have ingested *M. leprae* (Vachula et al., 1989). Hence, PGL-I may play a role in the ability of the bacteria to survive within macrophages and may also play a role in immunosuppression (Mehra et al., 1984; Prasad et al., 1987).

6. *M. leprae* can survive and multiply within macrophages and Schwann cells. Three possible strategies of intracellular survival have been proposed for *M. leprae*. First, *M. leprae* bacilli might avoid the bactericidal activities of the phagocytic cells by escaping from the phagosome (phagolysosome?) and multiplying in the cytoplasm of the infected cells (Mor, 1983). Second, viable *M. leprae* has been reported to prevent the fusion of phagosomes and lysosomes in murine macrophages (Frehel and Rastogi, 1987; Sibley et al., 1987), but not in Schwann cells (Steinhoff et al., 1989). A similar prevention of phagosome-lysosome fusion is an intracellular survival strategy used by *M. tuberculosis* (d'Arcy Hart, 1982). Third, inside phagosomes *M. leprae* generates around the bacillus a characteristic electron microscope image called an "electron transparent zone (ETZ)" (Draper and Rees, 1970). The ETZ might represent a physical barrier to prevent degradative or bactericidal proteins of the lysosome from reaching the bacterial surface. Regardless of the precise strategy used, one result of evading the bactericidal and degradative activities of the phagocytic cells might be that the *M. leprae* antigens would be less likely to be processed and presented to the immune system. If so, the bacilli might thereby prevent or escape a protective immune response.

Habitat

Leprosy and *M. leprae* have historically been considered to be confined to humans. Indeed, humans are the major host and reservoir of the leprosy bacillus. Humans, however, are not the only possible habitat for *M. leprae*. Naturally acquired, leprosy-like infections have been observed in armadillos (*Dasypus novemcinctus*; Walsh et al., 1975), chimpanzees (*Pan troglodytes*; Donham and Leininger, 1977), and Mangabey monkeys (*Cercocebus torquatas atys*; Meyers et al., 1985). Since the bacilli infecting these animal species are identical to the human pathogen, based on a variety of DNA homology and biochemical studies (Athwal et al., 1984; Clark-Curtiss and Walsh, 1989; Meyers et al., 1985), leprosy should be considered a zoonotic disease (Walsh et al., 1981). Furthermore, the possibility of the transmission of *M. leprae* from animals to humans has been raised by the observation that five armadillo handlers in Texas developed leprosy in the absence of any known contact with a human source of *M. leprae* (Lumpkin et al., 1983).

The isolation of *M. leprae* from soil in Bombay and other environmental sources has also been reported (Kazda, 1981a; Kazda et al., 1986). It is unclear, however, if these reports identify an environmental niche for *M. leprae* or if the isolated bacilli represent "environmental contamination" from *M. leprae*-infected individuals. For example, one possible source of "environmental contamination" might be nasal secretions of lepromatous leprosy patients (see below). Finally, free-living amoebae have been suggested as a potential reservoir for *M. leprae* (Grange and Rowbotham, 1987), although no convincing evidence for the presence or multiplication of *M. leprae* in amoebae has been published.

The mode of transmission of the leprosy bacillus remains unknown. Evidence is accumulating that infection occurs predominantly by way of the respiratory route, although other routes may be responsible for some cases (reviewed in Pallen and McDermott, 1986). Some of the evidence for respiratory transmission includes: 1) the epidemiology of transmission is consistent with spread by a respiratory route (Barton, 1974; Davey and Rees, 1974). 2) The major portal of exit of the leprosy bacillus is the nose. Lepromatous leprosy patients can shed up to 10^8 bacilli per day in nasal discharges, and the bacilli can survive for several days in dried secretions (Davey and Rees, 1974; Shepard, 1962). 3) Bacilli can be aerosolized by coughing and sneezing. 4) The nose has been suggested as a possible site of initial infection in humans (Pallen and McDermott, 1986), and immunodeficient mice can be infected by exposing them to an aerosol containing *M. leprae* (Rees and McDougall, 1976; Chehl et al., 1985). 5) As originally discussed by Koch (1897) and Schaffer (1898), the similarities between tuberculosis and leprosy (e.g., the similarity between the large numbers of *M. leprae* in nasal secretions and the large numbers of *M. tuberculosis* in sputa) are suggestive of an analogous route of transmission.

At present, however, one can not exclude transmission by skin-to-skin contact or by insect vectors. For example, ulcerating skin lesions could be a source of bacilli that might enter a susceptible host through a skin abrasion. Similarly, broken skin might be a portal of entry for bacilli that were aerosolized or otherwise deposited in the environment. With respect to insect vectors, *M. leprae* can be found in flies that have fed on nasal secretions from lepromatous leprosy patients (Greater, 1975; Kirchheimer, 1976), and the transmission of *M. leprae* from humans to mouse footpad by a mosquito vector (*Aedes aegypti*) has been reported (Narayanan et al., 1977). Another possible route is via breast milk since *M. leprae* bacilli can be found in the breast milk of lepromatous patients (Pedley, 1968). The role, if any, of the gastrointestinal tract in the transmission of leprosy is not known. Finally, occasional cases of leprosy may result from accidental inoculation of susceptible individuals by needle prick (Wade, 1948).

Isolation and Propagation

The currently available methods for the isolation and propagation of *M. leprae* are greatly constrained by the lack of an axenic or cell-culture cultivation system. In the laboratory, *M. leprae* has been propagated in both immunocompetent and immunodeficient mice and rats (Colston and Hilson, 1976; Fieldsteel and Levy, 1976; Hilson, 1965; Rees, 1966; Shepard, 1960), hamsters (Binford, 1959), armadillos (Kirchheimer and Storrs, 1971), and in Mangabey, Rhesus, and African green monkeys (Wolf et al., 1985). The features of the growth and pathology of *M. leprae* in each species are somewhat different, and the choice of animal model often depends on exactly what experimental question is being asked (reviewed in Shepard, 1986).

Staining Methods

One consequence of having only animal systems in which to propagate *M. leprae* is that much of the work depends on the enumeration of acid-fast bacilli by direct counting in microscopic fields. The standard staining method is a modified Ziehl-Neelsen technique (e.g., see Jenkins et al., 1982, or Jopling and McDougall, 1988). Briefly, bacilli on a glass slide are covered with a carbol-fuchsin solution and left at room temperature for 20 minutes (cold Ziehl-Neelsen) or heated for 15 minutes over a boiling water bath (hot Ziehl-Neelsen). The slide is washed with water, destained with 1% hydrochloric acid in 70% ethanol, and washed again with water. The sample is then counterstained with 1% methyl-

ene blue, washed in water, and air dried. Several methods for the actual counting of bacilli in the microscope and for using the counts to determine the concentration of bacilli in a sample have been described (Hanks et al., 1964; Shepard and McRae, 1968). The uniformity of staining exhibited by a bacillus is used as an indication of the viability of the bacillus (Hansen and Looft, 1895; Shepard and McRae, 1965a; Waters and Rees, 1962). Rods showing uniform staining are considered to be viable, while bacilli displaying fragmented or granular staining are considered non viable. A "morphologic index" is calculated as the percentage of uniformly staining bacilli and has been used to follow the progress of chemotherapy (Waters and Rees, 1962).

Propagation in Mice

The most frequently used propagation system is growth in the footpad of immunocompetent mice, as was originally described in the landmark paper of Shepard (1960). In this animal model, one typically propagates *M. leprae* by injecting mice subcutaneously with about 5×10^3 bacilli in one of the hind footpads, although as few as 1–10 bacilli are sufficient to establish an infection (Shepard and McRae, 1965a). A typical growth curve displays an initial lag phase followed by logarithmic growth with a doubling time of 11–13 days and a stationary phase with a plateau level of about $1–2 \times 10^6$ bacilli per footpad. The plateau level is thought to be due to development of a cell-mediated immune response to the bacillus. As a consequence, the viability of the *M. leprae* in the footpad decreases dramatically during the plateau phase. Thus, to propagate a strain one should harvest bacilli prior to or early in the plateau phase.

A typical growth curve is accompanied by a typical histopathologic picture. Early in infection one finds single bacilli or small clumps of 2–5 bacilli aligned in a parallel fashion ("cigar packs") in tissue macrophages, muscle, and tendon cells. Later, a cellular infiltrate mainly of macrophages and a few lymphocytes appears, followed by development of a microscopic granuloma with a few large, "foamy" macrophages in the center surrounded by a loose collection of macrophages and lymphocytes. The large macrophages contain numerous bacilli. In late stages, one occasionally observes invasion of the nerves.

In the absence of a T-cell response, multiplication continues to much higher levels. Thus, immunodeficient animals such as thymectomized-irradiated mice (Rees, 1966), neonatally thymectomized Lewis rats (Fieldsteel and Levy; 1976), and nude mice (*nu/nu*) (Colston and Hilson, 1976; Chehl et al., 1983) have been used to propagate *M. leprae*. The minimum infectious dose and generation time during logarithmic growth appear to be the same in the immunodeficient animals as in immunocompetent animals. The bacilli, however, can multiply to up to about 10^9 cells per foot pad in the neonatally thymectomized Lewis rat and about 5×10^{10} cells per footpad in the nude mouse. In the nude mouse one frequently observes spread of the *M. leprae* to other cool sites in the body (e.g., the ears) and invasion into nerves. Overall, the histopathology of the disease in the immunodeficient mice resembles that of human lepromatous leprosy. One drawback to the use of these animals on a routine basis is that they are particularly susceptible to a variety of other infections, and hence, special precautions need to be taken to ensure the survival of the animals for the duration of the experiment (often greater than 1 year).

Propagation in the Armadillo

A second major advance in leprosy research was the development of the nine-banded armadillo as an animal model system for the propagation of *M. leprae* (Kirchheimer and Storrs, 1971; Storrs, 1971). The key here is that in the armadillo the disease resembles human lepromatous leprosy and one can isolate very large numbers of bacilli from each infected armadillo. For example, the bacillary load in lesions can reach 10^{10} organisms per gram of tissue, and the total bacillary load can exceed 10^{12} bacilli per armadillo. Also, the liver and spleen are heavily infected and are easily manipulated sources of bacilli. The availability of the large amounts of *M. leprae* from armadillos opened the way for biochemical and molecular biologic analyses of the bacillus as well as provided a way to produce sufficient *M. leprae* for use in vaccine trials. One technical problem with the use of armadillos is that the *M. leprae* infection is fatal for the armadillo. Hence, to maximize bacillary yield and avoid postmortem contamination of tissues with other bacteria, care must be taken to sacrifice the animal and harvest bacilli just prior to the animal succumbing to the *M. leprae* infection itself. Finally, one must be aware that armadillos may harbor other naturally acquired, fastidious, acid-fast mycobacteria (Kazda, 1981b) and may harbor naturally acquired *M. leprae* (Walsh et al., 1975).

Primate Infection

Naturally occurring *M. leprae* infections have been observed in several primates, and *M. leprae* has been experimentally transmitted to three species of monkeys (Wolf et al., 1985). While the disease in these animals seems to parallel closely disease in humans, their use in the routine isola-

tion and propagation of *M. leprae* is deterred by economic and humane concerns. Nonetheless, the primates may prove to be important model systems for studies on pathology, immunology, and vaccine development.

Maintenance of Viable Cells

Finally, a concern of those working with *M. leprae* is that the bacillus does not store well. Good viability of *M. leprae* in suspensions or in biopsy specimens is maintained for 7–10 days at 0–4°C (Shepard and McRae, 1965b). For long-term storage, one can store the bacilli in liquid nitrogen. Three important points here are: 1) the sample should contain 10% dimethyl sulphoxide or 10% glycerol; 2) slow freezing is required (<1°C/min); and 3) the bacilli rapidly lose viability upon repeated freezing and thawing (Colston and Hilson, 1979; Portaels et al., 1988). For routine laboratory work one usually maintains several stocks of *M. leprae* in continuous serial passage. Of course, one is concerned that serial passaging of *M. leprae*, which typically involves transfer of 10^3 to 10^4 bacilli from mouse to mouse, might lead to changes in the bacillus, e.g., outgrowth of variants. However, some strains have been serially passaged in the Hansen's Disease Laboratory at the Centers for Disease Control (CDC) for over 25 years without any detectable changes in growth pattern (rate and plateau level) or histopathology of infection. Furthermore, by restriction-fragment-length-polymorphism (RFLP) studies, Clark-Curtiss and Walsh (1989) have found that the *M. leprae* bacilli isolated from leprosy patients in India, armadillos in Louisiana, and a naturally infected Mangabey monkey were virtually identical. This exceptional conservation of nucleotide sequence might be related to the long generation time of *M. leprae* or to the presence of a very efficient DNA-damage-repair system. Overall, these observations suggest that *M. leprae* changes very slowly, if at all, during serial passaging.

Isolation from Clinical Specimens

With respect to the isolation of *M. leprae* from a clinical specimen (e.g., biopsy of an LL lesion), the first steps are usually to purify the bacilli and to inject them into the footpad of an immunocompetent mouse such as the BALB/c or CFW strain (reviewed in Shepard, 1986). To purify the bacilli, the usual procedure is to release the bacilli from their intracellular location within a biopsy of infected tissue by mincing the sample, suspending it in Hank's Balanced Salt solution containing 0.1% BSA, and homogenizing it in the presence of 2 to 3-mm-diameter glass beads. To reduce potential contamination with other bacteria, the sample can be treated with 0.5M NaOH for 15 minutes and then neutralized with HCl. Following this treatment, the cells are usually harvested by centrifugation and resuspended in Hank's Balanced Salt solution containing 0.1% BSA. Samples of the bacterial suspension are stained and examined microscopically to determine the number of uniformly staining, acid-fast bacilli. One then injects portions of the samples containing 10^4 or fewer bacilli into the footpads of mice. Tissues from the mice are harvested periodically and examined for the presence and numbers of acid-fast bacilli and the histopathology of the site of infection.

These initial steps allow one to produce sufficient bacilli to begin the process of determining if the isolated organism is *M. leprae* (described below). In addition, the rate of growth in the footpad and the histopathology of the infection in the mouse footpad may be useful in distinguishing *M. leprae* from other mycobacteria (Shepard, 1986). That is, only *M. leprae*, *M. lepraemurium*, *M. marinum*, and *M. ulcerans* of the *Mycobacterium* species grow in the mouse footpad, and each displays a characteristic growth rate and histopathology. For example, *M. marinum* infections appear in about 2 weeks and often cause ulceration at the site of inoculation.

Identification

The lack of an in vitro system for the cultivation of *M. leprae* makes positive identification particularly difficult. Often a presumptive identification of an organism as *M. leprae* is made based simply on two observations—staining characteristics and lack of growth on conventional laboratory media. For example, with clinical specimens, any noncultivable, acid-fast, rod-shaped bacillus isolated from a site displaying a histopathology characteristic of a leprosy lesion is presumed to be *M. leprae*. Although such a definition is adequate for decisions on patient management, it is not particularly useful if one is trying to identify an organism isolated from a source devoid of histopathologic information, such as nasal secretions or the environment. Also, such a definition is clearly insufficient to prove an organism is *M. leprae*. Undoubtedly, the imprecise nature of the definition has contributed to the numerous erroneous claims for the cultivation of *M. leprae* in vitro.

Until recently, only a few rather labor-intensive, time-consuming, and technically difficult tests, such as growth in mouse footpads or presence of *M. leprae*-specific components, were available to confirm the presumptive identification. Fortunately, a variety of relatively simple biochemical, immunological, and molecular-

biological tests is now becoming available that should allow a rapid and definitive identification of *M. leprae*. Some of the potentially useful identification tests are: 1) morphology and staining characteristics; 2) failure to grow on bacteriologic media; 3) growth and histopathology in animal models; 4) drug susceptibility patterns; 5) presence of unique biochemical components or activities; 6) presence of characteristic antigens; and 7) presence of specific nucleotide sequences.

The identification process still starts with an examination of the staining characteristics of the bacilli. *M. leprae* bacilli isolated from lesions are Gram-positive rods. The carbol fuchsin stain is not extractable with acid or alcohol but is extractable with pyridine (Fisher and Barksdale, 1971). Of the other mycobacteria, only *M. smegmatis*, *M. vaccae*, and *M. phlei* display somewhat similar pyridine-extractable acid-fast characteristics (McCormick and Sanchez, 1979; Skinsnes et al., 1975). An important caveat here is that absence of acid-fastness can not be taken as an indication that the organism is not *M. leprae*, since a variety of mycobacteria lose the acid-fast characteristic at various times during their growth.

The second routinely used test is a determination of growth on any of a variety of media used for the propagation of mycobacteria, such as Lowenstein-Jensen, Middlebrook 7H9, or Dubos medium. *M. leprae* is the only *Mycobacterium* species that does *not* grow on at least one type of axenic medium. Of course, one needs to include a control test, such as inoculation into mouse footpads, to ensure that the original preparation did indeed contain viable bacteria.

M. leprae grows in mice, rats, armadillos, and monkeys and produces a characteristic pattern of proliferation and histopathologic changes in each animal (for details, see Shepard, 1986). Hence, growth characteristics can be used for identification purposes. For practical and economic reasons, one usually starts by analyzing growth in mice and then proceeds onto other animal models only as warranted. A key step in this test is careful observation of histopathologic changes. For example, *M. leprae* is the only *Mycobacterium* species to invade peripheral nerves and such invasion is prominent in the later stages of an *M. leprae* infection in the thymectomized-irradiated mouse and in the armadillo. Hence, observation of acid-fast bacilli in nerves is a good indication that the organism is *M. leprae*.

Biochemical Tests

Among the mycobacteria, *M. leprae* has a unique pattern of drug susceptibilities (Shepard, 1971;

Shepard et al., 1983; Hastings and Franzblau, 1988). Thus, one could confirm a presumptive identification by measuring drug susceptibility in: 1) inhibition of in vivo growth in the mouse footpad (Shepard, 1971); 2) inhibition of growth in macrophage culture (Mittal et al., 1983; Ramasesh et al., 1987); or 3) effect of the drug on PGL-I synthesis, palmitate oxidation, or ATP generation in bacilli maintained in a synthetic medium (Franzblau, 1988; Franzblau and Hastings, 1987; Franzblau et al., 1987). Note that the bacilli do not multiply in the systems used by Franzblau and colleagues.

Several identification tests are based on the observation that *M. leprae* contains components, structures, enzymes, and antigens that apparently are not present in any of the other *Mycobacterium* species. For example, enzymatic activities that are apparently unique to *M. leprae* (among the mycobacteria) include the ability to oxidize 3,4-dihydroxyphenylalanine (DOPA) and the synthesis of phenolic glycolipid I (PGL-I). DOPA-oxidase activity can be measured in vitro by following the conversion of DOPA to the pigmented product indole-5,6-quinone (Prabhakaran and Kirchheimer, 1966). This test is not entirely specific for *M. leprae*, since some non-acid-fast bacteria and, more importantly, some mammalian tissues have similar oxidase (phenolase) activities (Prabhakaran, 1967). PGL-I biosynthesis can be followed by measuring the incorporation of ^{14}C-palmitate into PGL-I in an *M. leprae*/macrophage co-culture system (Ramasesh et al., 1987) or in bacilli maintained in a synthetic medium (Franzblau and Hastings, 1987). Since PGL-I is unique to *M. leprae*, this assay is particularly useful. Also, characterization of the susceptibility of the in vitro reaction to various drugs can help confirm the identification.

Characteristic biochemical components of *M. leprae* include: 1) peptidoglycan that contains glycine instead of L-alanine in the linking peptide (Draper, 1976); 2) a GC content of 56 mol% as opposed to GC contents of 65–70 mol% for the other *Mycobacterium* species (Clark-Curtiss et al., 1985; Imaeda et al., 1982; Wayne and Gross, 1968); 3) PGL-I. However, other mycobacteria (e.g., *M. tuberculosis*) have structurally similar phenol-phthiocerol triglycosides (Brennan, 1983; Chatterjee et al., 1989; Daffe et al., 1988; Hunter and Brennan; 1981); and 4) specific mycolic acids (Draper, 1976). Analysis of the mycolic acids may be particularly informative since the high pressure liquid chromatography (HPLC) pattern of mycolic acids produced following alkaline methanolysis seems to be unique for each species of *Mycobacterium*. Indeed, Butler and Kilburn (1988) have devised a relatively simple and rapid scheme based on such HPLC patterns for identifying many of the *Mycobacterium* species.

Immunological Tests

Several antigens (or epitopes?) also appear to be uniquely present in *M. leprae* (reviewed in Gaylord and Brennan, 1987). Particularly useful reagents are monoclonal antibodies directed against the *M. leprae*-specific oligosaccharide of PGL-I (Mehra et al., 1984; Young et al., 1984). Such antibodies have been shown to react only with *M. leprae* and have been used in the serodiagnosis of leprosy (Cho et al., 1986; Koster et al., 1987; Sanchez et al., 1986; Young et al., 1985). Several other monoclonal antibodies are also available that react with other targets that appear to be *M. leprae*-specific (reviewed in Gaylord and Brennan, 1987). One could also assay the reactivity of the unknown with T-cell clones that are directed against *M. leprae*-specific components. Other potential immunological tests include fluorescent-antibody stains (Abe et al., 1980), immunodiffusion assays (Payne et al., 1982), immunoprecipitation assays (Harboe et al., 1978), and the ability to induce a delayed type hypersensitivity reaction in infected individuals similar to that induced by authentic *M. leprae* preparations, e.g. lepromin (Shepard and Guinto, 1963). In any immunological test, one should be aware that *M. leprae* does contain many antigens and epitopes that are cross-reactive with antigens found in one or more of the other *Mycobacterium* species as well as antigens that are cross-reactive with organisms that are evolutionarily very distant, including humans.

Genetic Tests

One can measure the relatedness of two organisms by determining the degree of hybridization between genomic DNAs from the two organisms (Brenner, 1989). Cells from the same species have DNA homologies of greater than 70% and a difference in melting temperature of <5°C. Such genomic DNA:DNA hybridization studies have been reported for *M. leprae* and a variety of other *Mycobacterium* species. *M. leprae* shows little if any homology with the other mycobacteria (Athwal et al., 1984; Grosskinsky et al., 1989). Thus, this sort of test might be useful in identifying *M. leprae*. Unfortunately, such genomic DNA hybridization studies require fairly large amounts of DNA (10–100 μg, 10^9–10^{10} bacilli), not easily obtained for a nonaxenically cultivable bacterium.

One way to circumvent the need for such large amounts of DNA is to use a labelled (e.g., radioactive) nucleic acid probe to assess DNA homologies (reviewed in Tenover, 1989). Using labelled probes, one can reduce the number of organisms needed for analysis to 10^5 to 10^6. Potential hybridization probes (nucleotide sequences that hybridize only with *M. leprae* nucleic acids) for *M. leprae* have been identified and cloned into *E. coli*, which allows easy production of large amounts of the probes. Of particular interest is the work of Clark-Curtiss and colleagues (Clark-Curtiss and Docherty, 1989; Grosskinsky et al., 1989), who have identified a sequence that is present in about 19 copies in the *M. leprae* genome and is not present in other mycobacteria. Using this sequence as a probe and purified DNA, they were able to detect as few as 4×10^3 genome equivalents of *M. leprae* DNA. Furthermore, they were able to detect homologous sequences in skin biopsy samples from lepromatous leprosy patients, which is of potential clinical importance. Incidentally, the *M. leprae* genome appears to be very stable with respect to this sequence in that all *M. leprae* isolates (from patients, armadillos, and monkeys) examined to date have identical hybridization patterns, with the exception of a single difference found in only one of the armadillo isolates (Clark-Curtiss and Walsh, 1989). This observation suggests that the probe should be capable of detecting all *M. leprae* isolates.

Several research groups are currently exploring the potential use of gene amplification techniques such as the polymerase chain reaction (PCR) to detect and identify *M. leprae*. In this technique, one uses oligonucleotide primers to direct the amplification of a particular nucleotide sequence via a bidirectional polymerase cascade reaction (Guatelli et al., 1989; Mullis and Faloona, 1987). The specificity of the reaction is determined by the choice of primers. In some preliminary studies, we have used oligonucleotide primers corresponding to various regions of the gene encoding the 65-kilodalton(kDa) antigen (B. B. Plikaytis and T. M. Shinnick, unpublished observations). Primers representing sequences unique to *M. leprae* direct the amplification of the gene from *M. leprae* but not from the other *Mycobacterium* species tested. The potential power of this technique can be demonstrated by two results. First, using these primers, we can specifically detect as few as 10 copies of the *M. leprae* 65-kDa antigen gene in a sample containing 10^6 copies of the *M. tuberculosis* 65-kDa antigen gene. Second, using crude lysates of *M. leprae*, grown in mouse footpads, we can detect a positive signal in a sample containing as few as 20 bacilli.

These results indicate that the gene amplification technology can be used to amplify specifically sequences from *M. leprae*. Importantly, without optimizing the system, we were able to detect the equivalent of as few as 20 bacilli in a sample—a sensitivity much better than that of the currently available clinical tests. However, much work needs to be done before we can truly assess the potential clinical impact of this technology for leprosy. For example, among other things the system must be optimized with respect to: 1) target sequence and primers; 2) recovery of bacilli and isolation of DNA; 3) reaction conditions; and 4) detection systems. Nonetheless, the degree of sensitivity and specificity displayed by the primers described above is quite encouraging for the potential use of this technology in the rapid detection of small numbers of mycobacteria in clinical specimens and the positive identification of the organism as *M. leprae*.

Summary

Although much has been learned about the structure and composition of the *M. leprae* bacillus, the challenge for the identification and characterization of *M. leprae* still lies with the inability to propagate the organism outside of animal hosts. Until a suitable axenic culture system is developed, one must rely on animal models, hence making identification and characterization quite laborious. Fortunately, there are methods and procedures on the horizon for the analysis of very small numbers of bacilli (PCR) and for the analysis of the metabolic activity of bacilli (Franzblau and Hastings, 1987). These advances will probably facilitate the identification of organisms as *M. leprae* and the characterization of relevant biochemical traits.

Literature Cited

Abe, M., F. Ninagowa, Y. Yoshino, T. Ozawa, K. Saikawa, T. Saito. 1980. Int. J. Lepr. 48:109.

Athwal, R. S., S. S. Deo, T. Imaeda. 1984. Deoxyribonucleic acid relatedness among *Mycobacterium leprae, Mycobacterium tuberculosis*, and selected bacteria by dot blot and spectrophotometric deoxyribonucleic acid hybridization assays. Int. J. Syst. Bacteriol. 34:1136–1141.

Barton, R. F. E. 1974. A clinical study of the nose in leprosy. Lepr. Rev. 45:135–144.

Binford, C. E. 1959. Histiocytic granulomatous mycobacterial lesions produced in the golden hamster (*Cricetus auratus*) inoculated with human leprosy—negative results using other animals. Lab. Invest. 8:901–924.

Bloom, B. R., V. Mehra. 1984. Immunological unresponsiveness in leprosy. Immunol. Rev. 80:5–28.

Bloom, B. R., T. Godal. 1983. Selective primary health care: Strategies for control of disease in the developing world. V. Leprosy. Rev. Infect. Dis. 5:765–780.

Brennan, P. J. 1983. The phthiocerol-containing surface lipids of *Mycobacterium leprae*: a perspective of past and present work. Int. J. Lepr. 51:387–396.

Brenner, D. J. 1989. DNA hybridization for characterization, classification, and identification of bacteria. 75–103. B. Swaminathan and G. Prakash (ed.) Nucleic Acid and Monoclonal Antibody Probes: Applications in Diagnostic Microbiology. Marcel Dekker, Inc. New York.

Browne, S. G. 1975. Some aspects of the history of leprosy: the leprosy of yesteryear. Proc. R. Soc. Med. 68:485–493.

Butler, W. R., J. O. Kilburn. 1988. Identification of major slow-growing pathogenic mycobacteria and *Mycobacterium gordonae* by high-performance liquid chromatography of their mycolic acids. J. Clin. Microbiol. 26:50–53.

Chan, J., T. Fujiwara, P. Brennan, M. McNeil, S. J. Turco, J.-C. Sibille, M. Snapper, P. Aisen, B. R. Bloom. 1989. Microbial glycolipids: Possible virulence factors that scavenge oxygen radicals. Proc. Natl. Acad. Sci. USA. 86:2453–2457.

Chatterjee, D., C. M. Bozic, C. Kinsley, S.-N. P. J. Brennan. 1989. Phenolic glycolipids of *Mycobacterium bovis*: New structures and synthesis of a corresponding seroreactive neoglycoprotein. Infect. Immun. 57:322–330.

Chehl, S., C. K. Job, R. C. Hastings. 1985. The nose: a site for the transmission of leprosy in nude mice. Am. J. Trop. Med. Hyg. 34:1161–1166.

Chehl, S., J. Ruby, C. K. Job, R. C. Hastings. 1983. The growth of *Mycobacterium leprae* in nude mice. Lepr. Rev. 54:283–304.

Cho, S., S. W. Hunter, R. H. Gelber, T. H. Rea, P. J. Brennan. 1986. Quantitation of the phenolic glycolipid of *Mycobacterium leprae* and relevance to glycolipid antigenemia in leprosy. J. Infect. Dis. 153:560–569.

Cho, S., D. L. Yanagihara, S. W. Hunter, R. H. Gelber, P. J. Brennan. 1983. Serological specificity of phenolic glycolipid I from *Mycobacterium leprae* and use in serodiagnosis of leprosy. Infect. Immun. 41:1077–1083.

Clark-Curtis, J. E., M. A. Docherty. 1989. A species-specific repetitive sequence in *Mycobacterium leprae* DNA. J. Infect. Dis. 159:7–15.

Clark-Curtiss, J. E., W. R. Jacobs, M. A. Docherty, L. R. Ritchie, R. Curtiss III. 1985. Molecular analysis of DNA and construction of genomic libraries of *Mycobacterium leprae*. J. Bacteriol. 161:1093–1102.

Clark-Curtiss, J. E., G. P. Walsh. 1989. Conservation of genomic sequences among isolates of *Mycobacterium leprae*. J. Bacteriol. 171:4844–4851.

Collaborative effort of the U.S. Leprosy Panel of the U.S.-Japan Cooperative Medical Sciences Program the Leonard Wood Memorial. 1975. Rifampin therapy of lepromatous leprosy. Am. J. Trop. Med. Hyg. 24:475–484.

Colston, M. J., G. R. F. Hilson. 1976. Growth of *Mycobacterium leprae* and *M. marinum* in congenitally athymic (nude) mice. Nature 262:399–401.

Colston, M. J., G. R. F. Hilson. 1979. The effect of freezing and storage in liquid nitrogen on the viability and growth of *Mycobacterium leprae*. J. Med. Micro. 12:137–142.

Cowdry, E. V. 1940. Cytologic Studies on globi in leprosy. Am. J. Pathol. 16:103–136.

Daffe, M., M. A. Laneelle, C. Lacave, G. Laneelle. 1988. Monoglycosyldiacylphenol-pthiocerol of *Mycobacterium tuberculosis* and *Mycobacterium bovis*. Biochim. Biophys. Acta 958:443–449.

d'Arcy Hart, P. 1982. Lysosome fusion responses of macrophages to infection: behaviour and significance. 437–447. M. L. Karnovsky and L. Bolis (ed.) Phagocytosis: past and future. Academic Press. New York.

Davey, T. F., R. J. W. Rees. 1974. The nasal discharge in leprosy: clinical and bacteriological aspects. Lepr. Rev. 45:121–134.

Desai, S. D., T. J. Birdi, N. H. Antia. 1989. Correlation between macrophage activation and bactericidal function and *Mycobacterium leprae* antigen presentation in macrophages of leprosy patients and normal individuals. Infect. Immun. 57:1311–1317.

Desikan, K. V., C. K. Job. 1968. A review of postmortem findings in 37 cases of leprosy. Int. J. Lepr. 36:31–44.

Donham, K. J., J. R. Leininger. 1977. Spontaneous leprosy-like disease in a chimpanzee. J. Infect. Dis. 136:132–136.

Draper, P. 1976. The cell walls of *Mycobacterium leprae*. Int. J. Lepr. 44:95–98.

Draper, P. 1983. The bacteriology of *Mycobacterium leprae*. Tubercle 64:43–56.

Draper, P., R. J. W. Rees. 1970. Electron-transparent zone of mycobacterium may be a defence mechanism. Nature 228:860–861.

Fieldsteel, A. H., L. Levy. 1976. Neonatally thymectomized Lewis rats infected with *Mycobacterium leprae*: response to primary infection, secondary challenge, and large inocula. Infect. Immun. 14:736–741.

Fine, P. E. M. 1982. Leprosy: The epidemiology of a slow-growing bacterium. Epidemiol. Rev. 4:161–188.

Fisher, C. A., L. Barksdale. 1971. Elimination of the acid-fastness but not the gram positivity of leprosy bacilli after extraction with pyridine. J. Bacteriol. 106:707–708.

Franzblau, S. G. 1988. Oxidation of palmitic acid by *Mycobacterium leprae* in an axenic medium. J. Clin. Microbiol. 26:18–21.

Franzblau, S. G., E. B. Harris, R. C. Hastings. 1987. Axenic incorporation of [U-^{14}C] palmitic acid into phenolic glycolipid I of *Mycobacterium leprae*. FEMS Microbiol. Lett. 48:107–114.

Franzblau, S. G., R. C. Hastings. 1987. Rapid in vitro metabolic screen for antileprosy compounds. Antimicrob. Agents Chemother. 31:780–783.

Frehel, C., N. Rastogi. 1987. *Mycobacterium leprae* surface components intervene in the early phagosome-lysosome fusion inhibition event. Infect. Immun. 55:2916–2921.

Gaylord, H., P. J. Brennan. 1987. Leprosy and the leprosy bacillus: recent developments in characterization of antigens and immunology of the disease. Annu. Rev. Microbiol. 41:645–675.

Grange, J. M., T. J. Rowbotham. 1987. Microbe dependence of *Mycobacterium leprae*: A possible intracellular relationship with protozoa. Int. J. Lepr. 55:565–566.

Greater, J. G. 1975. The fly as potential vector in the transmission of leprosy. Lepr. Rev. 46:279–286.

Grosskinsky, C. M., W. R. Jr. Jacobs, J. E. Clark-Curtiss, B. R. Bloom. 1989. Genetic relationships between *Mycobacterium leprae*, *Mycobacterium tuberculosis*, and candidate leprosy vaccine strains by DNA hybridization: Identification of an *M. leprae*-specific repetitive sequence. Infect. Immun. 57:1535–1541.

Guatelli, J. C., T. R. Gingeras, D. D. Richman. 1989. Nucleic acid amplification in vitro: Detection of sequences with low copy numbers and application to diagnosis of human immunodeficiency virus type 1 infection. Clin. Microbiol. Rev. 2:217–226.

Hanks, J. H., B. R. Chatterjee, M. F. Lechat. 1964. A guide to the counting of mycobacteria in clinical and experimental materials. Int. J. Lepr. 32:156–167.

Hansen, G. A. 1874. Causes of Leprosy. Norsk. Mag. for Laegervidenskaben 4:1–88.

Hansen, G. A., C. Looft. 1895. Leprosy: in its clinical and pathological aspects. Reprinted by John Wright, 1973. Bristol.

Harboe, M., O. Closs, B. Bjorvatn, G. Kronvall, N. H. Axelsen. 1977. Antibody response in rabbits to immunization with *Mycobacterium leprae*. Infect. Immun. 18:792–805.

Hastings, R. C. (ed.). 1986. Leprosy. Churchill Livingstone. Edinburgh, New York.

Hastings, R. C., Franzblau, S. G. 1988. Chemotherapy of leprosy. Annu. Rev. Toxicol. 28:231–45.

Hilson, G. R. F. 1965. Observations on the inoculation of *M. leprae* in the foot pad of the white rat. Int. J. Lepr. 33:662–665.

Hunter, S. W., P. J. Brennan. 1981. A novel phenolic glycolipid from *Mycobacterium leprae* possibly involved in immunogenicity and pathogenicity. J. Bacteriol. 147:728–735.

Hunter, S. W., T. Fujiwara, P. J. Brennan. 1982. Structure and antigenicity of the major specific glycolipid antigen of *Mycobacterium leprae*. J. Biol. Chem. 257:15072–15078.

Imaeda, T., W. F. Kirchheimer, L. Barksdale. 1982. DNA isolated from *Mycobacterium leprae*: genome size, base ratio, and homology with other related bacteria as determined by optical DNA-DNA reassociation. J. Bacteriol. 150:414–417.

Jenkins, P. A., S. R. Pattyn, F. Portaels. 1982. Diagnostic Bacteriology. 441–470. C. Ratledge and J. Stanford (ed.) The Biology of the Mycobacteria, vol. 1. Academic Press. London.

Job, C. K. 1971. Pathology of peripheral nerve lesions in lepromatous leprosy. A light and electron microscopic study. Int. J. Lepr. 39:251–268.

Jopling, W. H., A. C. McDougall. 1988. Handbook of Leprosy. Heinemann Professional Publishing Ltd. Oxford.

Kaplan, G., Z. A. Cohn. 1986. The immunobiology of leprosy. Int. Rev. Exp. Pathol. 28:45–78.

Kazda, J. 1981a. Occurrence of non-cultivatable acid-fast bacilli in the environment and their relationship to *M. leprae*. Lepr. Rev. 52 (Suppl. 1):85–92.

Kazda, J. 1981b. Nine-banded armadillos in captivity: prevention of losses due to parasitic diseases. Some remarks on mycobacteria-free maintenance. Int. J. Lepr. 49:345–346.

Kazda, J., R. Ganapati, C. Revankar, T. M. Buchanan, D. B. Young, L. M. Irgens. 1986. Isolation of environment-derived *Mycobacterium leprae* from the soil in Bombay. Lepr. Rev. 57 (Suppl. 3):201–208.

Kirchheimer, W. F. 1976. The role of arthropods in the transmission of leprosy. Int. J. Lepr. 44:104–107.

Kirchheimer, W. F., E. E. Storrs. 1971. Attempts to establish the armadillo (*Dasypus novemcinctus* Linn.) as a model for the study of leprosy. 1. Report of lepromatoid leprosy in an experimentally infected armadillo. Int. J. Lepr. 39:693–702.

Koch, R. 1897. Die Lepraerkrankungen in Kreise Memel. Abstracted in Baumgartens Jahresbericht 14:428.

Koster, F. T., D. M. Scollard, E. T. Umland, D. B. Fishbein, W. C. Hanley, P. J. Brennan, K. E. Nelson. 1987. Cellular and humoral immune response to a phenolic glycolipid antigen (PGL-I) in patients with leprosy. J. Clin. Microbiol. 25:551–556.

Levy, L. 1976. Studies on the mouse foot pad technique for cultivation of *Mycobacterium leprae*. 3. Doubling time during logarithmic multiplication. Lepr. Rev. 47:103–106.

Lumpkin, L. R., G. F. Fox, J. E. Wolf. 1983. Leprosy in five armadillo handlers. J. Am. Acad. Dermatol. 9:899–903.

McCormick, G. T., R. M. Sanchez. 1979. Pyridine extractability of acid-fastness of *M. leprae*. Int. J. Lepr. 47:495–499.

McWhirter, N. (ed.). 1981. Guiness Book of Records. 28th ed. Trowbridge: Redwood Burn Ltd.

Mehra, V., P. J. Brennan, E. Rada, J. Convit, B. R. Bloom. 1984. Lymphocyte suppression in leprosy induced by unique *M. leprae* glycolipid. Nature 308:194–196.

Meyers, W. M., G. P. Walsh, H. L. Brown, C. H. Binford, G. D. Jr. Imes, T. L. Hadfield, C. J. Schlagel, Y. Fukunishi, P. J. Gerone, R. H. Wolf, B. J. Gormus, L. N. Martin, M. Harboe, T. Imaeda. 1985. Leprosy in a Mangabey monkey—naturally acquired infection. Int. J. Lepr. 53:1–14.

Mittal, A., M. Sathish, P. S. Seshadri, I. Nath. 1983. Rapid radiolabeled-microculture method that uses macrophages for in vitro evaluation of *Mycobacterium leprae* viability and drug susceptibility. J. Clin. Microbiol. 17:704–707.

Mor, N. 1983. Intracellular location of *Mycobacterium leprae* in macrophages of normal and immunodeficient mice and effect of rifampicin. Infect. Immun. 42:802–811.

Mullis, K. B., F. A. Faloona. 1987. Specific synthesis of DNA in vitro via a polymerase-catalyzed chain reaction. Methods in Enzymol. 155:335–350.

Narayanan, E., Sreevetsa, W. F., Kirchheimer, B. M. S., Bedi. 1977. Transfer of leprosy bacilli from patients to mouse foot pads by *Aedes aegypti*. Leprosy in India 49:181–186.

Pallen, M. J., R. D. McDermott. 1986. How might *Mycobacterium leprae* enter the body. Lepr. Rev. 57:289–297.

Payne, S. N., P. Draper, R. J. W. Rees. 1982. Serological activity of purified glycolipid from *Mycobacterium leprae*. Int. J. Lepr. 50:220–221.

Pedley, J. C. 1968. The presence of *M. leprae* in the lumina of the female mammary gland. Lepr. Rev. 39:201.

Portaels, F., K. Fissette, K. De Ridder, P. M. Macedo, A. De Muynck, M. T. Silva. 1988. Effects of freezing and thawing on the viability and the ultrastructure of in vivo grown mycobacteria. Int. J. Lepr. 56:580–587.

Prabhakaran, K. 1967. Oxidation of 3,4-dihydroxyphenylalanine (DOPA) by *Mycobacterium leprae*. Int. J. Lepr. 35:42–51.

Prabhakaran, K., W. F. Kirchheimer. 1966. Use of 3,4-dihydroxyphenylalanine oxidation in the identification of *Mycobacterium leprae*. J. Bacteriol. 92:1267–1268.

Prasad, H. K., R. S. Mishra, I. Nath. 1987. Phenolic glycolipid I of *Mycobacterium leprae* induces general suppression of in vitro concanavalin A responses unrelated to leprosy type. J. Exp. Med. 165:239.

Ramasesh, N., R. C. Hastings, J. L. Krahenbuhl. 1987. Metabolism of *Mycobacterium leprae* in macrophages. Infect. Immun. 55:1203–1206.

Rastogi, N., H. L. David. 1988. Mechanisms of pathogenicity in mycobacteria. Biochemie 70:1101–1120.

Rees, R. J. W. 1966. Enhanced susceptibility of thymectomized and irradiated mice to infection with *Mycobacterium leprae*. Nature 211:657–658.

Rees, R. J. W., A. C. McDougall. 1976. Airborne infection with *Mycobacterium leprae* in mice. Int. J. Lepr. 44:99–103.

Ridley, D. S., W. H. Jopling. 1966. Classification of leprosy according to immunity. A five-group system. Int. J. Lepr. 34:255–273.

Ridley, M. J., M. F. R. Waters, D. S. Ridley. 1987. Events surrounding the recognition of *Mycobacterium leprae* in nerves. Int. J. Lepr. 55:99–108.

Sanchez, G. A., A. Malik, C. Tougne, P. H. Lambert, H. D. Engers. 1986. Simplification and standardization of serodiagnostic tests based on phenolic glycolipid I (PGL-I) antigen. Lepr. Rev. 57 (Suppl. 2):83–93.

Sansarricq, H. 1981. Leprosy in the world today. Lepr. Rev. 51 (Suppl. 1):15–31.

Schaffer, X. 1898. Ueber der Verbreitung der Leprabacillen von den oberen Luftwegen aus. Arch. Dermatol. Syphil. 44:159–174.

Sela, S., J. E. Clark-Curtiss, H. Bercovier. 1989. Characterization and taxonomic implications of the rRNA genes of *Mycobacterium leprae*. J. Bacteriol. 171:70–73.

Shepard, C. C. 1960. The experimental disease that follows the injection of human leprosy bacilli into the footpads of mice. J. Exp. Med. 12:445–454.

Shepard, C. C. 1962. The nasal excretion of *Mycobacterium leprae* in leprosy. Int. J. Lepr. 30:10–18.

Shepard, C. C. 1965. Stability of *Mycobacterium leprae* and temperature optimum for growth. Int. J. Lepr. 33:541–547.

Shepard, C. C. 1971. A survey of drugs with activity against *M. leprae* in mice. Int. J. Lepr. 39:340–348.

Shepard, C. C. 1986. Experimental Leprosy. 269–286. R. C. Hastings (ed.) Leprosy. Churchill Livingstone. Edinburgh.

Shepard, C. C., R. S. Guinto. 1963. Immunological identification of foot-pad isolates as *Mycobacterium leprae* by lepromin reactivity in leprosy patients. J. Exp. Med. 118:195–204.

Shepard, C. C., D. H. McRae. 1965a. *Mycobacterium leprae* in mice: Minimal infectious dose, relationship between staining and infectivity, and effect of cortisone. J. Bacteriol. 89:365–372.

Shepard, C. C., D. H. McRae. 1965b. *Mycobacterium leprae*: Viability at 0°C, 31°C, and during freezing. Int. J. Lepr. 33:316–323.

Shepard, C. C., D. H. McRae. 1968. A method for counting acid-fast bacteria. Int. J. Lepr. 36:78–82.

Shepard, C. C., D. H. McRae. 1971. Hereditary characteristics that varies among isolates of *Mycobacterium leprae*. Infect. Immun. 3:121–126.

Shepard, C. C., R. M. Vanlandingham, L. L. Walker. 1983. Recent studies of antileprosy drugs. Lepr. Rev. 54:23S–30S.

Sibley, L. D., S. G. Franzblau, J. L. Krahenbuhl. 1987. Intracellular fate of *Mycobacterium leprae* in normal and activated macrophages. Infect. Immun. 55:680–685.

Sibley, L. D., J. L. Krahenbuhl. 1988. Defective activation of granuloma macrophages from *Mycobacterium leprae*-infected nude mice. J. Leuk. Biol. 43:60–66.

Sixth Report of the WHO Expert Committee on Leprosy, 1988. World Health Organization Technical Report Series No. 768. Geneva.

Skinsnes, O. K., P. H. C. Chang, E. Matsuo. 1975. Acid-fast properties and pyridine extraction of *M. leprae*. Int. J. Lepr. 43:392–398.

Smida, J., J. Kazda, E. Stackebrandt. 1988. Molecular-genetic evidence for the relationship of *Mycobacterium leprae*

to slow-growing pathogenic mycobacteria. Int. J. Lepr. 56:449–454.

Stanford, J. L., G. A. W. Rook, J. Convit, T. Godal, G. Kronvall, R. J. W. Rees, G. P. Walsh. 1975. Preliminary taxonomic studies of the leprosy bacillus. Brit. J. Exp. Path. 56:579–585.

Steinhoff, U., J. R. Golecki, J. Kazda, S. H. E. Kaufmann. 1989. Evidence for phagosome-lysosome fusion in *Mycobacterium leprae*-infected murine Schwann cells. Infect. Immun. 57:1008–1010.

Stewart-Tull, D. E. S. 1982. *Mycobacterium leprae*—The bacteriologists' enigma. 273–307. C. Ratledge and J. Stanford, (ed.) The biology of mycobacteria, vol. 1. Academic Press. London.

Stoner, G. L. 1979. Importance of the neural predilection of *Mycobacterium leprae* in leprosy. Lancet ii:994–997.

Storrs, E. E. 1971. The nine-banded armadillo: a model for leprosy and other biomedical research. Int. J. Lepr. 39:703–714.

Tenover, F. C. (ed.). 1989. DNA probes for infectious diseases. CRC Press. Boca Raton, FL.

Vachula, M., T. J. Holzer, B. R. Andersen. 1989. Suppression of monocyte oxidative response by phenolic glycolipid I of *Mycobacterium leprae*. J. Immunol. 142:1696–1701.

Vemuri, N., L. Kandke, P. R. Mahadevan, S. W. Hunter, P. J. Brennan. 1985. Isolation of phenolic glycolipid from human lepromatous nodules. Int. J. Lepr. 53:489.

Wade, H. W. 1948. The Michigan inoculation cases. Int. J. Lepr. 16:465–475.

Wade, H. W., V. Ledowski. 1952. The leprosy epidemic at Naura: a review with data on the status since 1937. Int. J. Lepr. 31:34–45.

Walsh, G. P., W. M. Meyers, C. H. Binford, P. J. Gerone, R. H. Wolf, J. R. Leininger. 1981. Leprosy—a zoonosis. Lepr. Rev. 52 (Suppl. 1):77–83.

Walsh, G. P., E. E. Storrs, H. P. Burchfield, W. M. Meyers, C. H. Binford. 1975. Leprosy-like disease occurring naturally in armadillos. J. Reticuloendothel. Soc. 18:347–351.

Waters, M. F. R., R. J. W. Rees. 1962. Changes in the morphology of *Mycobacterium leprae* in patients under treatment. Int. J. Lepr. 30:266–277.

Wayne, L. G., W. M. Gross. 1968. Base composition of deoxyribonucleic acid isolated from mycobacteria. J. Bacteriol. 95:1481–1482.

Wolf, R. H., B. J. Gormus, L. N. Martin, G. B. Baskin, G. P. Walsh, W. M. Meyers, C. H. Binford. 1985. Experimental leprosy in three species of monkeys. Science 227:529–531.

Young, D. B., J. P. Harnisch, J. Knight, T. M. Buchanan. 1985. Detection of phenolic glycolipid I in sera of patients with lepromatous leprosy. J. Infect. Dis. 152:1078.

Young, D. B., S. R. Khanolkar, L. L. Barg, T. M. Buchanan. 1984. Generation and characterization of monoclonal antibodies to the phenolic glycolipid of *Mycobacterium leprae*. Infect. Immun. 43:183–188.

Prokaryotes (2006) 3:945–960
DOI: 10.1007/0-387-30743-5_36

CHAPTER 1.1.21

The Genus *Arthrobacter*

DOROTHY JONES AND RONALD M. KEDDIE

Conn (1928) described a group of bacteria, extremely numerous in certain soils, which were unusual in that they appeared as Gram-negative rods in young cultures and as Gram-positive cocci in older cultures. For these bacteria, Conn (1928) created the species *Bacterium globiforme*, which, as *Arthrobacter globiformis*, was later to become the type species of the genus *Arthrobacter*. The abundance in soil of bacteria similar to Conn's organism, and of other coryneform bacteria, was confirmed later by Jensen (1933, 1934) and Topping (1937, 1938), who, however, referred to them as soil corynebacteria, and by Taylor and Lochhead (1937), who used the name *Bacterium globiforme*. Jensen (1934) considered that these soil bacteria should be classified in the genus *Corynebacterium* because of their morphological resemblance to corynebacteria of animal origin. However, Conn (1947) vigorously opposed this view and created the genus *Arthrobacter* (by reviving an old name), with *A. globiformis* as the type species and with two of Jensen's soil corynebacteria as additional species (Conn and Dimmick, 1947).

In addition to their characteristic morphology and staining reactions, members of the genus *Arthrobacter* were originally described as being highly aerobic, nutritionally nonexacting, and capable of liquefying gelatin slowly (Conn and Dimmick, 1947). These features were chosen mainly to distinguish *Arthrobacter* from *Corynebacterium* as represented by *C. diphtheriae* and similar animal parasitic species. However, because of its poor circumscription (see Gibson, 1953; Jensen, 1952), the genus *Arthrobacter* was not widely accepted until it was included as a member of the family Corynebacteriaceae in the seventh edition of *Bergey's Manual of Determinative Bacteriology* (Breed et al., 1957). But by that time, the genus had been extended to include the two nutritionally exacting species *A. terregens* (Lochhead and Burton, 1953) and *A. citreus* (Sacks, 1954), and shortly afterwards two others were added (Lochhead, 1958a). Indeed, one of Conn's strains of *A. globiformis* was shown subsequently to require biotin for growth (Chan and Stevenson, 1962; Morris, 1960). Thus the concept had developed of *Arthrobacter* as a genus of soil bacteria whose major distinguishing feature was a growth cycle in which the irregular rods in young cultures were replaced by coccoid forms in older cultures; these coccoid forms, when transferred to fresh medium, produced outgrowths ("germinated") to give irregular rods again, and so the cycle was repeated (Fig. 1).

This dependence on morphological features and habitat in the circumscription led to a great deal of confusion in the classification of the genus *Arthrobacter* and thus created considerable problems in the identification of new isolates as arthrobacters. Thus, isolates from soil and, more especially,

This chapter was taken unchanged from the second edition.

those from other habitats have frequently been referred to in the literature as arthrobacters on the basis of morphological features alone, even though they were not necessarily similar to *A. globiformis* in other respects.

It was not surprising that when representatives of the genus were examined by more modern taxonomic methods such as numerical taxonomy (Jones, 1978), various chemotaxonomic techniques (Bowie et al., 1972; Keddie and Cure, 1977, 1978; Minnikin et al., 1978b; Schleifer and Kandler, 1972), and determinations of DNA base ratios (see Skyring and Quadling, 1970; Skyring et al., 1971), it was found to be heterogeneous.

The genus *Arthrobacter* as defined in the eighth edition of *Bergey's Manual* (Keddie, 1974) was heterogeneous, as was noted by Keddie and Jones (1981) in the first edition of *The Prokaryotes*. They referred to *Arthrobacter* in this broad sense as *Arthrobacter* sensu lato (Keddie and Jones, 1981). In *Bergey's Manual of Systematic Bacteriology* (Keddie et al., 1986), the genus was limited to those species which, like the type, *A. globiformis*, contain lysine as the cell wall diamino acid, i.e., *Arthrobacter* sensu stricto (Keddie and Jones, 1981). Thus some species formerly considered to be arthrobacters (Keddie, 1974) have now been removed from the genus.

The two species formerly named *A. terregens* and *A. flavescens*, which contain ornithine as the cell wall diamino acid (Schleifer and Kandler, 1972; Keddie and Cure, 1977), have been transferred to the genus *Aureobacterium* as *Aur. terregens* and *Aur. flavescens* (Collins et al., 1983). The species *Arthrobacter radiotolerans* has now been transferred to the new genus *Rubrobacter* as *R. radiotolerans* (Suzuki et al., 1988).

Although now resolved, the position of the two species *A. simplex* and *A. tumescens* has been more problematical. They were shown by Cummins and Harris (1959) to differ from *A. globiformis* in containing LL-diaminopimelic acid (LL-A$_2$pm) as the cell wall diamino acid, and many other taxonomic differences were detected subsequently (see Keddie et al., 1986, for further details). In the case of *A. simplex*, 16S rRNA cataloging studies showed this species to be only distantly related to *A. globiformis* (Stackebrandt et al., 1980). Conversely, 5S rRNA sequencing indicated that *Pimelobacter simplex* (*A. simplex*—see below) clearly belonged to an "*Arthrobacter*—*Micrococcus*—*Cellulomonas*" subgroup of the coryneform bacteria (Park et al., 1987). While there was general agreement that *A. simplex* and *A. tumescens* should be removed from *Arthrobacter*, there was disagreement about where they should be accommodated. Suzuki and Komagata (1983) created the genus *Pimelobacter* for the LL-A$_2$pm-containing coryneform bacteria and distinguished three species by use of DNA-DNA base-pairing techniques. The first species, *Pimelobacter simplex*, contained most strains of *A. simplex* (and also strains named "*Brevibacterium*

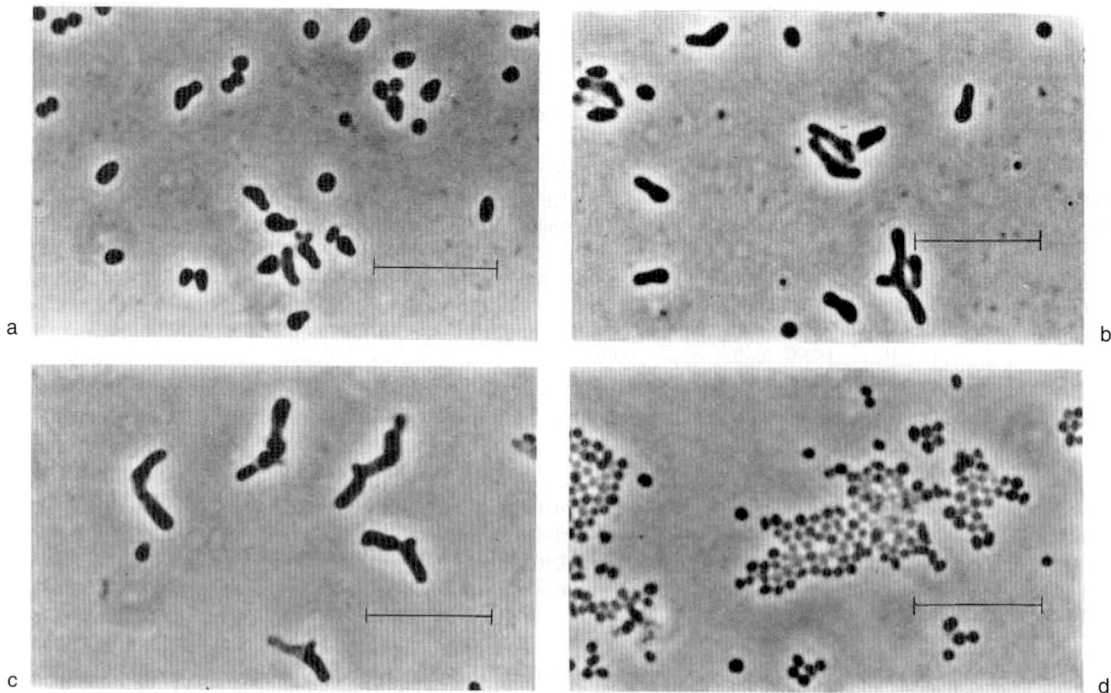

Fig. 1. *Arthrobacter globiformis* (ATCC 8010) grown on medium EYGA at 25°C; the inoculum was of coccoid cells as shown in (d). (a) After 6 h, showing outgrowth of rods from coccoid cells. (b) After 12 h. (c) After 24 h. (d) After 3 days. Bars = 10 μm.

lipolyticum"), the second, *P. tumescens*, was for strains formerly called *A. tumescens*, and a third species, *P. jensenii*, was created for a single strain originally identified as an *A. simplex* strain (Gundersen and Jensen, 1956). The same authors also concluded from their DNA homology studies that *A. simplex* and *A. tumescens* were only distantly related to *Nocardioides albus*, a nocardioform organism. However, O'Donnell et al. (1982) considered that *A. simplex* (but not *A. tumescens*) closely resembled *Nocardioides* species in chemotaxonomic (particularly lipid) characters and proposed that it be transferred to that genus as *Nocardioides simplex*. This view received support from other studies (reviewed by Keddie et al., 1986). In an attempt to resolve the problem, Collins et al. 1989 made a comparative study of the 16S rRNA from the type strains of *Nocardioides albus*, *N. luteus*, *Pimelobacter simplex*, *P. tumescens*, and *P. jensenii* by reverse transcriptase sequencing and compared the results with those from 18 previously studied actinomycetes from 14 different genera. The study confirmed the reclassification of *P. (Arthrobacter) simplex* in the genus *Nocardioides* as *N. simplex* as proposed by O'Donnell et al. (1982) and also showed that *P. jensenii* should be transferred to that genus as *N. jensenii* (Collins et al., 1989). However, *P. (Arthrobacter) tumescens* was so distinct from all the other actinomycete taxa that Collins et al. (1989) proposed its reclassification in a new genus, *Terrabacter*, as *T. tumescens*. The taxonomic status of the single strain of the species *A. duodecadis* remains unresolved. Details of this and some other species named *Arthrobacter*, but now excluded from the genus, are given by Keddie et al. 1986 and for *A. siderocapsulatus* and *A. viscosus* by Collins (1986).

However, Collins (1986) has shown that the lipid composition of "*A. sialophilus*" (Tanenbaum and Flashner, 1977) and the phytopathogen *Agrobacterium pseudotsugae* is con-

sistent with their being members of *Arthrobacter* sensu stricto (Keddie and Jones, 1981), but no formal proposal has been made to include them in the genus.

As presently circumscribed, the genus *Arthrobacter* contains two "groups of species" referred to as the *A. globiformis/A. citreus* group and the *A. nicotianae* group. These groups differ in their peptidoglycan structure, teichoic acid content, and lipid composition (see "Further Identification of Arthrobacters" and Tables 2 and 3 for details). It has been suggested that the genus should be restricted to those bacteria that exhibit the characteristics of the *A. globiformis/A. citreus* group (Minnikin et al., 1978a; Collins and Kroppenstedt, 1983). However, on the basis of DNA-DNA homology studies of representative arthrobacters (Stackebrandt and Fiedler, 1979) and later 16S rRNA cataloging studies, Stackebrandt et al. (1983) concluded that the genus *Arthrobacter* contained two "nuclei," one represented by the *A. globiformis/A. citreus* group of species and the other by the *A. nicotianae* group. This view was adopted by Keddie et al. (1986) and is the one accepted here.

The results of 16S rRNA cataloging studies (Stackebrandt et al., 1980; Stackebrandt and Woese, 1981) indicate that the genus *Arthrobacter* is related to the other coryneform genera, *Aureobacterium, Cellulomonas, Curtobacterium*, and *Microbacterium*, and is more distantly related to *Brevibacterium*. All of these genera are members of the high GC "actinomycete" branch of the Gram-positive eubacteria (Stackebrandt and Woese, 1981). The studies of Stackebrandt et al. (1980) also showed that on a phylogenetic basis the *Arthrobacter* species could not be separated from members of the genus *Micrococcus*. While accepting that the genera *Arthrobacter* and *Micrococcus* are very closely related phylogenetically, we treat them here as distinct taxa for practical purposes.

Table 1. Some characteristics[a] of the *Arthrobacter* species having peptidoglycans of the A3α variation[b] and MK-9 (H₂)[c] as major menaquinones—the *A. globiformis*/*A. citreus* group.[d,e]

Characteristic	A. globiformis[f]	A. crystallopoietes	A. pascens	A. ramosus	A. aurescens	A. histidinolovorans	A. ilicis	A. ureafaciens	A. atrocyaneus	A. oxydans	A. citreus
Peptidoglycan type[b]	Lys-Ala₃	Lys-Ala	Lys-Ala₂	Lys-Ala₄	Lys-Ala-Thr-Ala	Lys-Ala-Thr-Ala	Lys-Ala-Thr-Ala	Lys-Ala-Thr-Ala	Lys-Ser-Ala₂₋₃	Lys-Ser-Thr-Ala	Lys-Thr-Ala₂
Cell wall sugars[g]	Gal, Glu	Gal, Glu	Gal, Glu	Gal, Rha, Man	Gal, (Man)	Gal, Glu	Gal, Rha, Man	Gal, (Man)	Gal, Glu, (Man)	Gal, Glu	Gal
Vitamin requirement; none or biotin only	+	+	+	+	+	+	+[h]	+	+	+	−
Nicotine utilization	−	−	−	−	−	−	−	−	−	+	−
Starch hydrolysis	+	−	+	−	+	−	−	−	+	+	−
Motility	−	−	−	+	−	−	+	−	+	−	+

[a]Most species are represented only by their type strain; therefore, the range of variation within the species is not known. For this reason, data on carbon source utilization tests are not included although studies on a wider range of strains may prove them to be useful. See section on "Soil and Similar Habitats" for a list of near-universal substrates. There are conflicting reports in the literature about the responses of the type strains of some species in some common tests such as nitrate reduction, urea hydrolysis, etc., and therefore these data have been omitted. Other data are from original and/or revised descriptions of species and from Keddie et al. (1966), Yamada and Komagata (1972), and Cure and Keddie, and Robertson and Keddie, unpublished observations.

[b]From Schleifer and Kandler (1972); and Stackebrandt and Fiedler (1979).

[c]From Collins and Jones (1981).

[d]Species are arranged according to peptidoglycan type (Stackebrandt and Fiedler, 1979).

[e]Symbols: +, 90% or more of the strains are positive; −, 90% or more of strains are negative.

[f]Lys, L-lysine; Ala, L-alanine; Thr, L-threonine; Ser, L-serine; Gal, galactose; Glu, glucose; Rha, rhamnose; Man, mannose; ( ), conflicting reports on occurrence.

[g]From Keddie and Cure (1978).

[h]*A. ilicis* grows in a mineral salts-glucose medium only when provided with casamino acids (S. Robertson and R. M. Keddie, unpublished observations).

Table 2. Characters[a] most useful in differentiating the *Arthrobacter* species which have peptidoglycans of the A4α variation[b] and MK-8[c] (or MK-9[d]) as major menaquinones—the *A. nicotianae* group.[e,f]

Characteristics	*A. nicotianae*	*A. protophormiae*	*A. uratoxydans*	*A. sulfureus*
Number of strains studied	6	6	2	3
Peptidoglycan type[b]	Lys-Ala-Glu[g]	Lys-Ala-Glu	Lys-Ala-Glu	Lys-Glu
Major wall sugars[h]	Gal, Glc (one strain)	ND	ND	Gal, Glc (one strain)
Hydrolysis of:				
Starch	+	–	–	–
Casein	+	d	+	–
Utilization of:				
4-Hydroxybenzoate	+	+ (5/6)[i]	–	+
Glyoxylate	d	d	+	–
L-Asparagine	+	+ (5/6)	–	+
L-Arginine	d	+	–	+
L-Histidine	d	+	–	+
D-Xylose	+	– (5/6)	–	–
D-Ribose	+	d	–	d
L-Arabinose	+	+ (5/6)	–	–
D-Galactose	+	+ (5/6)	–	d
L-Rhamnose	–	– (5/6)	+	d
2,3-Butylene glycol	+ (5/6)	–	–	d
Glycerol	+	+	–	–

[a]Adapted from Table 5 in Stackebrandt et al. (1983).
[b]From Schleifer and Kandler (1972); and Stackebrandt et al. (1983).
[c]From Collins and Jones (1981); and Collins and Kroppenstedt (1983).
[d]*A. sulfureus* ("Brevibacterium sulfureum") ATCC 19098 (NCIMB 10355) contains MK-9 as major menaquinone (Collins and Kroppenstedt, 1983); a second strain designated *A. sulfureus* by Stackebrandt et al. (1983), ATCC 15170 (formerly named *A. citreus*), contains similar amounts of MK-9 and MK-10 as major menaquinones (Collins and Kroppenstedt, 1983).
[e]All *A. nicotianae* group strains which have been examined contain teichoic acids (Fiedler and Schäffler, 1987) and contain the polar lipids diphosphatidylglycerol and phosphatidylglycerol, but not phosphatidylinositol (Collins and Kroppenstedt, 1983). According to Stackebrandt et al. (1983) more than 90% of *A. nicotianae* group strains utilize: acetate, propionate, valerate, capronate, heptanoate, caprylate, succinate, DL-malate, citrate, DL-lactate, fumarate, D-gluconate, glycine, L-proline, L-threonine, and L-aspartate; none utilizes adipate, levulinate, and acetamide.
[f]Symbols: see Table 1, footnote [e]. Also: d, 11–89% of strains are positive; ND, no data.
[g]Lys, L-lysine; Ala, L-alanine; Glu, L-glutamic acid; Gal, galactose; Glc, glucose.
[h]From Keddie and Cure (1978).
[i]Fraction of strains giving indicated reaction.

Habitats of *Arthrobacter*

In many ecological studies, isolates have been identified as arthrobacters or described as "arthrobacter-like" simply because they showed the rod-coccus growth cycle and the staining reactions characteristic of the genus. Although the sequence of morphological changes that occurs during the growth cycle is an important distinguishing feature of the genus, it does not occur exclusively in arthrobacters. For example, it is also seen in the genus *Brevibacterium* and in at least some members of the genus *Rhodococcus*. Accordingly, some of the strains described as arthrobacters in the literature cited below may belong to other morphologically similar taxa, especially if they are isolates from habitats other than soil.

Soil and Similar Habitats

Many studies have shown that bacteria of the genus *Arthrobacter* form a numerically important fraction of the indigenous bacterial flora of soils from different parts of the world; they are sometimes the most numerous, single bacterial group recorded in aerobic plate counts (Hagedorn and Holt, 1975b; Holm and Jensen, 1972; Lowe and Gray; 1972; Mulder and Antheunisse; 1963; Skyring and Quadling, 1969; Soumare and Blondeau, 1972). However, both the numbers and the proportions of arthrobacters in "total" counts decrease with increasing soil acidity (Hagedorn and Holt, 1975b; Lowe and Gray, 1972). Among the explanations advanced for their numerical predominance are their extreme resistance to drying (Boylen, 1973; Chen and Alexander, 1973; Labeda et al., 1976; Mulder and Antheunisse, 1963; Robinson et al., 1965) and to starvation (Boylen and Ensign, 1970; Boylen and Mulks, 1978; Zevenhuizen, 1966), factors important in the survival of microorganisms in soil (Gray, 1976). However, the nutritional versatility of the commonly occurring species undoubtedly also plays a part (see below).

Both psychrophilic and psychrotrophic strains of the genus were reported to be the most abun-

Table 3. Characters[a] differentiating *Arthrobacter* from similar genera which either have a rod-coccus growth cycle or have lysine as the cell wall diamino acid.[b]

Genus or species	Mycelium produced	Rod-coccus cycle[c]	Oxygen requirement	Acid from glucose[d]	Cell wall		Major menaquinone[f]
					Diamino acid[e]	Glycine present	
Arthrobacter	–	+	Aerobic	–	Lysine	–	MK-9 (H₂) or MK-8 and/or MK-9
Aureobacterium terregens/Aur. flavescens[g]	–	+	Aerobic	W	Ornithine	+	MK-12 and/or MK-13
Nocardioides simplex[h]	–	+	Aerobic	–	LL-A₂pm	+	MK-8 (H₄)
Terrabacter tumescens[i]	–	+	Aerobic	–	LL-A₂pm	+	MK-8 (H₄)
Microbacterium	–	–	Equivocal	+	Lysine	+	MK-11, MK-12
Renibacterium	–	–	Aerobic	–	Lysine	+	MK-9
Oerskovia	+	–	Facultative	+	Lysine	–	MK-9 (H₄)
Brevibacterium	–	+	Aerobic	–	*meso*-A₂pm	–	MK-8 (H₂)
Rhodococcus	D	D	Aerobic	–	*meso*-A₂pm	–	MK-8 (H₂) or MK-9 (H₂)

[a]Data from Keddie and Cure, 1977, 1978; Keddie and Bousfield, 1980; Sanders and Fryer, 1980; Collins and Jones, 1981; Keddie and Jones, 1981; Lechevalier and Lechevalier, 1981; Collins, 1982; Collins et al., 1989.
[b]Symbols: +, 90% or more of strains are positive; –, 90% or more of strains are negative; W, weak; D, only a proportion of strains studied have the feature cited.
[c]Similar to that in *A. globiformis*.
[d]In peptone-based media.
[e]A₂pm, diaminopimelic acid.
[f]MK-8, MK-9, etc., indicate the number of isoprene units in the menaquinone; (H₂), (H₄), etc., indicate the number of double bonds hydrogenated.
[g]Formerly *Arthrobacter terregens* and *A. flavescens*.
[h]Formerly *A. simplex*.
[i]Formerly *A. tumescens*.

dant and active bacteria in subterranean cave silts (Gounot, 1967), and they also occur in glacier silts (Moiroud and Gounot, 1969). The genus was also represented among isolates from oil brines raised from soil layers some 200–700 m deep (Iizuka and Komagata, 1965). Arthrobacters capable of dissolving aluminium silicates were reported to be common on "karst" rocks, and most were considered to be capable of dinitrogen fixation (Smyk, 1970; Smyk and Ettlinger, 1963). Members of the genus have also been implicated in the growth of manganese nodules in the sea (Ehrlich, 1963, 1968).

There is now much evidence that the predominant soil arthrobacters can use a wide and diverse range of organic substrates as sole or principal sources of carbon and energy (Hagedorn and Holt, 1975a; Keddie, 1974). This nutritional versatility is characteristic of those species which do not require vitamins or other organic growth factors or which require only biotin. Most species now recognized, including *A. globiformis* and new isolates from soil, belong to this nutritional category, with *A. citreus* (type strain) being a notable exception. The detailed carbon nutrition of representatives of a number of species of soil arthrobacters together with 26 unnamed isolates, mainly from soil, was examined by J. D. Owens and R. M. Keddie and G. L. Cure and R. M. Keddie (unpublished observations quoted in Keddie et al., 1986). The following substrates were utilized by some 90% or more of the strains tested: D-xylose, D-glucose, D-mannose, D-galactose, D-fructose, cellobiose, maltose, trehalose, sucrose, raffinose, melezitose, D-gluconolactone, salicin, acetate, propionate, pentanoate, heptanoate, succinate, fumarate, DL-lactate, DL-malate, citrate, pyruvate, oxaloacetate, glycerol, mannitol, *m*-hydroxybenzoate, *p*-hydroxybenzoate, uric acid, glycine, L-α-alanine, D-α-alanine, L-isoleucine, L-threonine, L-lysine, L-arginine, L-aspartate, L-glutamate, L-phenylalanine, L-tyrosine, L-proline, L-histidine, 1,4-butanediamine, agmatine, tyramine, betaine, sarcosine, and creatine. *A. citreus* (type strain) is less nutritionally versatile than the species mentioned above (Keddie, 1974). Other studies have demonstrated the ability of arthrobacters to utilize aromatic compounds (Stevenson, 1967) and nucleic acids and their degradation products (Antheunisse, 1972). The frequent recovery of *Arthrobacter* species from enrichment cultures in which various diverse, organic compounds are supplied as sole carbon sources in simple, mineral media is further evidence of this nutritional versatility. Such compounds include nicotine (Giovanozzi-Sermanni, 1959; Keddie et al., 1966; Sguros, 1955), puromycin amino-nucleoside (Greenberg and Barker, 1962), 2-hydroxypyridine (Ensign and Rittenberg, 1963; Kolenbrander et al., 1976), *n*-alkanes (Klein et al., 1968), lower alcohols (Akiba et al., 1970), choline (Kortstee, 1970), picolinic acid (Tate and Ensign, 1974), and squalene (Yamada et al., 1975).

Arthrobacters have also been shown to degrade herbicides such as disodium endoxo-hexahydrophthalate ("Endothal") (Jensen, 1964) and 2,4-dichlorophenoxyacetate (Cacciari et al., 1971; Loos et al., 1967; Sharpee et al., 1973). Others degrade pesticides, usually by cometabolism, e.g., "diazinon" (Sethunathan and Pathak, 1971) and *m*-chlorobenzoate, the central molecule in many pesticides (Horvath and Alexander, 1970).

Fish and Other Similar Habitats

Coryneform bacteria appear to be common on fish (both marine and freshwater) and on some other seafoods. Many are morphologically similar to arthrobacters (and to *Brevibacterium*) and are frequently referred to by that name. Thus "arthrobacters" have been reported to occur in shark spoilage (referred to as *"Corynebacterium" globiformis* and *C. helvolum*; Wood, 1950), eviscerated freshwater fish (Roth and Wheaton, 1962), fish-pen slime (Chai and Levin, 1975), and Pacific shrimp (Lee and Pfeifer, 1977). Sieburth (1964) reported the isolation of an *Arthrobacter* species from sea water.

However, in the few cases in which coryneform isolates from fish have been examined by suitable modern methods, their relationship to *Arthrobacter* has proved to be much more remote than has been inferred from their morphological similarity, and most, if not all, of them belong to other taxa (see Bousfield, 1978; Crombach, 1974a, 1974b).

Sewage and Similar Habitats

Bacteria identified as *Arthrobacter* species have been isolated from sewage (Nand and Rao, 1972) and from "brewery sewage" (Kaneko et al., 1969). Arthrobacters physiologically similar to those from soil were common in dairy-waste activated sludge (Mulder and Antheunisse, 1963) and were considered to play an important role in the process (Adamse, 1968). A number of the activated-sludge strains were reported by the above authors to decompose phenol. However, subsequent examination of four of these phenol-decomposing strains using chemotaxonomic methods revealed that they were members of the "rhodochrous" taxon *(Rhodococcus)* (Keddie and Cure, 1977). Similarly, although Schefferle (1966) considered that the predominant coryneform bacteria in poultry deep litter could be

placed in the genus *Arthrobacter*, only one of seven of these litter strains subsequently examined by Keddie and Cure (1977) was considered to be a legitimate *Arthrobacter* species. Although many coryneform isolates from aerated, animal-manure slurries are similar in morphology to *Arthrobacter*, only 1 of 16 such isolates examined was identified as *A. globiformis*; most of the remainder were considered to be *"rhodochrous"* strains, and a few were similar to *Brevibacterium linens* (Keddie and Cure, 1977).

Such examples clearly illustrate that new isolates should not be assigned to the genus *Arthrobacter* on the basis of morphology and conventional features alone.

Other Habitats

Coryneform bacteria seem to be relatively common on the aerial surfaces of plants (Austin et al., 1978; Keddie et al., 1966; Mulder et al., 1966) but few have been shown to be legitimate arthrobacters. However, "arthrobacter-like" organisms have been isolated from frozen vegetables (Splittstoesser et al., 1967). Also, two of a number of strains isolated from cauliflower by Lund (1969) were later identified as *A. globiformis* strains by Keddie and Cure (1977), but it is possible that they were contaminants from soil rather than indigenous plant bacteria. However, the organism formerly called *Corynebacterium ilicis*, a pathogen of American holly (Mandel et al., 1961), has been shown to be a legitimate *Arthrobacter* species and has now been transferred to that genus as *A. ilicis* (Collins et al., 1981). The type strain of *A. protophormiae* (formerly *Brevibacterium protophormiae*; Lysenko, 1959) was isolated from an insect, *Protophormia terraenovae*, but other strains were isolated from soil (see Keddie et al., 1986).

Arthrobacters were reported to be numerous in commercial preparations of liquid eggs, probably as a result of contamination from the shells, but they were rarely isolated from turkey giblets (Kraft et al., 1966). Arthrobacters do not appear to have been isolated from clinical sources.

Isolation of Arthrobacters

Arthrobacters have normally been isolated from soil and similar habitats by plating on suitable nonselective ("total count") media, and then picking from a large, random selection of colonies, and identifying as arthrobacters those isolates that show a rod-coccus growth cycle (Holm and Jensen, 1972; Lowe and Gray, 1972; Skyring and Quadling, 1969). This method is only suitable for habitats such as soil in which arthrobacters form an appreciable proportion of the aerobic cultivable population. A further extension of this technique was introduced by Mulder and Antheunisse (1963), who devised what was essentially a method for screening isolates picked from a nonselective medium for those that showed the typical rod-coccus growth cycle of arthrobacters (see also Mulder et al., 1966; Veldkamp, 1965). Hagedorn and Holt (1975b) devised a selective medium said to be suitable for the enumeration of arthrobacters in soil. Thus, when nine different soils were plated on this medium, an average of 74% of the colonies that developed were identified as arthrobacters, and the numbers were similar to those estimated from counts on a nonselective medium (Hagedorn and Holt, 1975b).

However, if identification of the isolates is based only, or largely, on the provision of a rod-coccus growth cycle as was the case in the studies quoted, then bacteria from other genera such as *Brevibacterium* and *Rhodococcus* may be mistaken for arthrobacters. If, however, isolates are screened for the presence of lysine in the cell wall by using a rapid method of cell wall analysis, then this problem is overcome.

As noted above, a number of *Arthrobacter* species have been isolated from enrichment cultures using a variety of organic substrates as sole carbon and energy sources in mineral salts media. However, the primary purpose of such enrichments was not to isolate arthrobacters but to obtain isolates capable of utilizing the particular substrates studied. Accordingly, it is not possible to assess the value of these enrichment methods for the isolation of particular strains of arthrobacters.

Nonselective Media for Isolation of Arthrobacters

The aim of this approach is to use media and conditions of incubation which give the maximum possible counts of soil bacteria capable of growth under aerobic conditions. The media used must therefore contain sufficient amounts of all the organic growth factors and mineral constituents required to provide the diverse nutritional requirements of the indigenous soil bacteria (see Lochhead, 1958b). But, at the same time, media must be sufficiently poor in carbon and energy sources to limit the size of colonies, thereby minimizing antagonistic effects between the components of the population. Various modifications of soil extract agar have been the media most widely used. Lochhead, who pioneered the use of such media, strongly advocated the use of soil extract agar without other additions (Lochhead and Burton, 1956), but for some soils, addition of low concentrations of yeast extract

and glucose can give higher counts (Jensen, 1968). Examples of both kinds of soil extract agars are given below. If necessary, the growth of fungi may be suppressed by incorporating the antibiotics nystatin (50 µg/ml) and cycloheximide (50 µg/ml) in the medium (Williams and Davies, 1965). Other factors important in plating soil samples have been discussed by Jensen (1968). The more important of these are: 1) the soils should be examined within a few hours of sampling; 2) the primary dilution should be dispersed by using a laboratory blender but avoiding heating; 3) a suitable diluent should be used (see below); and 4) plates should be incubated at 25°C for a minimum of 2 weeks.

Jensen (1968) recommends Winogradsky's standard salt solution as a diluent. It has the following composition (Holm and Jensen, 1972):

Winogradsky's Standard Salt Solution

K_2HPO_4	0.25 g
$MgSO_4$	0.125 g
NaCl	0.125 g
$Fe_2(SO_4)_3$	0.0025 g
$MnSO_4$	0.0025 g
Deionized water	1 liter

Adjust pH to 6.5–6.7.

Soil extract and dilute peptone solutions (0.05–0.1%, wt/vol) have also been used successfully but one-fourth-strength Ringer's solution, physiological saline, and tap water are unsuitable (Jensen, 1968). In this context, Owens and Keddie (1969) noted that a chelated mineral salts solution which they devised, but without $(NH_4)_2SO_4$ (mineral base E-N), was a suitable diluent for coryneform bacteria. Mineral base E-N gave slightly better survival of *Arthrobacter globiformis* than a simple salts solution gave and was markedly superior to traditional diluents such as one-fourth-strength Ringer's solution or physiological saline. The preparation of mineral base E-N was described by Cure and Keddie (1973).

Isolation of Arthrobacters Using Soil Extract Agar

Soil Extract (Lochhead and Burton, 1957)

To prepare the soil extract, add 1 kg of soil to 1 liter of tap water and autoclave at 121°C for 20 min; filter and restore volume to 1 liter with tap water. A fertile garden soil usually gives the best results (Jensen, 1968).

Soil Extract Agar for Isolating Arthrobacters (Lochhead and Burton, 1957)

Soil extract (see above), 1 liter; K_2HPO_4, 0.2 g; agar, 15 g; final pH 6.8. Autoclave at 121°C for 20 min. Colonies are picked into tubes of soil extract semisolid medium: soil extract, 1 liter; K_2HPO_4, 0.2 g; yeast extract, 1 g; agar, 3 g; final pH 6.8. Autoclave at 121°C for 20 min.

Soil Extract Agar for Isolating Arthrobacters (Holm and Jensen, 1972)

Soil extract (see above)	400 ml
Tap water	600 ml
Glucose	1 g
Peptone	1 g
Yeast extract	1 g
K_2HPO_4	1 g
Agar	20 g

Adjust pH to 6.5–6.7.

After sterilization and immediately before use, a filter-sterilized solution of cycloheximide is added to the medium to give a final concentration of 40 mg/liter. Before pouring plates, a layer of sterile agar is poured in the bottom of the plates and allowed to solidify. This prevents colonies from spreading between the agar and the bottom of the petri dishes. Colonies are picked onto slants of the same medium.

Isolation and Enumeration of Arthrobacters from Soil (Mulder and Antheunisse, 1963)

Dilutions are prepared and plates poured using the following "poor" medium (g/liter tap water):

$Ca(H_2PO_4)_2$	0.25
K_2HPO_4	1.0
$MgSO_4 \cdot 7H_2O$	0.25
$(NH_4)_2SO_4$	0.25
Casein	1.0
Yeast extract	0.7
Glucose	1.0
Agar	10.0

Adjust pH to 6.9–7.0.

After incubation for 5 days at 25°C, colonies are counted and a large number are transferred to agar slants of the same composition. The slants are incubated for 7 days at 25°C and then examined microscopically. Colonies that consist of coccoid cells are then transferred to a "rich" medium containing (% wt/vol): yeast extract, 0.7; glucose, 1.0; agar, 1.0. The cultures are examined in the exponential phase of growth (usually not more than 24 h), and those showing "germinating" cocci and irregular rods are considered to be "arthrobacters."

Obviously, in the sense used by Mulder and Antheunisse (1963), the term "arthrobacters" refers to all bacteria that show the rod-coccus growth cycle of the genus *Arthrobacter*.

Isolation of Arthrobacters by Selective Media

Selective Medium of Hagedorn and Holt (1975b)

In this method, plate counts are made by spreading 0.1-ml amounts of suitable dilutions over the surface of sterile medium in petri dishes. Peptone solution (0.5%, wt/vol) was used as the diluent by Hagedorn and Holt (1975b). The selective medium has the following composition:

Trypticase soy agar (BBL)	0.4%
Yeast extract (Difco)	0.2%
NaCl	2.0%
Cycloheximide	0.01%
Methyl red (Harleco)	150 µg/ml
Agar	1.5%

The methyl red is filter-sterilized and added aseptically to the autoclaved, cooled medium. The medium is adjusted to the pH of the particular soil being examined.

The selective properties of the medium are said to be unaffected by pH values in the range 5.0–8.5. After incubation for 10 days at 25°C, the plates are counted. Colonies are transferred to slants of trypticase soy agar containing 0.2% yeast extract and examined microscopically for the possession of a morphological growth cycle as described in *Bergey's Manual* (Keddie, 1974). From the results obtained, the authors concluded that 78% of the counts on the selective medium was a suitable approximation of the arthrobacter counts for the soils studied. The authors state that:

The combination of actidione [cycloheximide] at 0.01% and NaCl at 2.0% effectively inhibited all fungi and most streptomycetes, nocardia, and Gram-negative bacteria. The methyl red at 150 µg/ml inhibited other Gram-positive bacteria (bacilli and micrococci) but did not affect the arthrobacters. The pH of the medium, between 5.0 and 8.5, did not affect its selectivity, and the combination of trypticase soy agar at 0.4% and yeast extract at 0.2% gave the highest yield of arthrobacters with the addition of the selective ingredients over the other basal media [tested].

The selective medium gave arthrobacter counts several times higher than those on the nutritionally "poor" medium of Mulder and Antheunisse (1963) for the four soils examined (Hagedorn and Holt, 1975b).

Preservation of Cultures

Stab cultures by loop in TSX semisolid medium (Keddie et al., 1966) or in the soil extract semisolid medium of Lochhead and Burton (1957) (see above) will remain viable for at least three months at room temperature (ca. 20°C) provided they are not allowed to dry out. Cultures may be preserved for longer periods (at least 10 years) by freezing in glass beads at −70°C (Jones et al., 1984). For long-term storage, lyophilization is suitable.

Identification of Arthrobacters

Members of the genus *Arthrobacter* can be recognized by examining the following characters: morphology and staining reactions, oxygen relations, acid production from glucose, and cell wall composition.

The most distinctive feature of arthrobacters is the marked change of form that occurs during the growth cycle on complex media. Stationary phase cultures (usually 2–7 days) are composed entirely or largely of coccoid cells (Fig. 1d) which, on transfer to fresh complex medium, produce one and sometimes two (or occasionally more) outgrowths that give rise to the irregular rods characteristic of exponential phase cultures (Fig. 1a–c). Some of the cells are arranged in V-formations but more complex angular arrangements may also occur. Cells may show primary branching, but true mycelia (showing secondary branching) are not produced. As growth proceeds, the rods become shorter and are eventually replaced by the coccoid forms characteristic of stationary phase cultures (Fig. 1d). For more detailed accounts of *Arthrobacter* morphology and morphogenesis, see Keddie (1974), Luscombe and Gray (1971), Clark (1972), and Duxbury and Gray (1977). Both rod and coccoid forms are Gram positive, but may decolorize readily, and are not acid-fast. The rods are non-motile or motile by one subpolar or a few lateral flagella. They are obligate aerobes. The mode of metabolism is respiratory, never fermentative; little or no acid is formed from sugars in peptone media. The cell wall peptidoglycan contains lysine as diamino acid.

Additional features include the following: they are catalase-positive. The optimum temperature for growth is 25–30°C, and most *Arthrobacter* species grow in the range of about 10°C–35°C; many strains also grow at 5°C, and a few grow at 37°C. Some obligately psychrophilic isolates with a growth range of about –5°C to 20°C that are considered to be a new species of *Arthrobacter* ("*A. glacialis*") were described by Moiroud and Gounot (1969). They do not survive heating at 63°C for 30 min in skim milk. They do not hydrolyze cellulose. DNase is produced and gelatin is usually liquefied (Keddie et al., 1986). The GC content of the DNA of most species is in the range of 59–66 mol% but that of *A. atrocyaneus* is higher, at about 70 mol% (see Keddie et al., 1986, for references).

The extent to which the morphology of arthrobacters changes during the growth cycle is markedly influenced by the nutritional status of the medium (see Clark, 1972; Ensign and Wolfe, 1964; Luscombe and Gray, 1971; Veldkamp et al., 1963); therefore, the medium used for morphological studies must be chosen with care (see Cure and Keddie, 1973). Many different media have been used for this purpose and most are based on soil extract (e.g., see Holm and Jensen, 1972; Lochhead and Burton, 1957). However, such media may give inconsistent results because of the variable nature of soil extract and because some mineral components may be precipitated to different extents during preparation. Medium EYGA (Cure and Keddie, 1973) was devised to overcome such problems and has proved to be satisfactory for examining the morphology of arthrobacters.

Cell wall composition may be determined by one of the rapid methods described by Bousfield et al. (1985) in which cell wall material is prepared by alkali treatment of whole cells.

Further Identification of Arthrobacters

Fifteen species of *Arthrobacter* are recognized in *Bergey's Manual of Systematic Bacteriology* (Keddie et al., 1986). Many of these species were created for single strains that possessed some unusual feature, such as a requirement for a particular growth factor, production of an unusual pigment, the ability to utilize a particular substrate, and so on. Accordingly, such species do not necessarily represent the commonly occurring arthrobacters in the habitats from which they were isolated originally and this may in part explain the common experience that many new soil isolates show little resemblance to the named strains used as reference cultures (see Hagedorn and Holt, 1975a; Keddie et al., 1966; Skyring and Quadling, 1969; Seiler et al., 1980). Also, because most species descriptions are based on single strains, the range of variation within the species is not known. Therefore some of the phenotypic features described for a particular species may be strain rather than species characteristics.

Most *Arthrobacter* species now recognized closely resemble the type species *A. globiformis* in a large number of phenotypic (largely nutritional) characters. For this reason they were referred to as the "globiformis" group of arthrobacters in the first edition of *The Prokaryotes* (Keddie and Jones, 1981) and included the species *A. atrocyaneus*, *A. aurescens*, *A. crystallopoietes*, *A. histidinolovorans*, *A. nicotianae*, *A. oxydans*, *A. pascens*, *A. polychromogenes* (considered a subspecies of *A. oxydans* by Stackebrandt and Fiedler, 1979), *A. ramosus*, and *A. ureafaciens*, together with the species *Brevibacterium sulfureum* (now *A. sulfureum*) and the plant pathogen *Corynebacterium ilicis* (now *A. ilicis*) (see Keddie et al., 1986). A list of substrates utilized by 90% or more of single representatives of these species is given in the section on "Soil and Similar Habitats" above. Detailed nutritional data are not available for the two remaining species now recognized, *A. protophormiae* and *A. uratoxydans* (see Keddie et al., 1986). The remaining species, *A. citreus* (type strain only), is readily distinguished from the other species. It has a complex nutrition and requires biotin, thiamin, nicotinic acid, tyrosine, methionine, cystine, and a siderophore such as ferrichrome or mycobactin for growth (Seidman and Chan, 1969). *A. citreus* also utilizes a much more limited range of compounds as sole or major carbon and energy sources than that used by the other species (Keddie et al., 1986). A number of other strains are listed as *A. citreus* in culture collections but do not have the characteristics of the type strain (ATCC 11624). They were considered to be strains of *A. protophormiae* and *A. uratoxydans* by Stackebrandt et al. (1983) (see Keddie et al., 1986).

However, despite the close phenotypic similarity of most species of *Arthrobacter* (*A. citreus* excepted), two groups within the genus can be distinguished according to their peptidoglycan structure, their lipid composition, and the presence or absence of teichoic acids in the cell walls.

The peptidoglycans of all *Arthrobacter* species contain lysine as diamino acid but more detailed analysis revealed considerable heterogeneity in the cell wall peptidoglycans of the different species. All contain a group A peptidoglycan, i.e., one in which the cross-linkage is between positions 3 and 4 of the peptide subunits. However, the peptide subunits are linked by a number of different interpeptide bridges, depending on the species (Schleifer and Kandler, 1972). Within these numerous different peptidoglycan types, two groups occur, which are referred to as the A3α and A4α variations. In the A3α variation, found in *A. globiformis*, *A. citreus*, and most other species (the "*A. globiformis/A. citreus*" group of Keddie et al., 1986), the interpeptide bridge contains only monocarboxylic acids and/or glycine. In the A4α variation, however, found in *A. nicotianae* and three other species (the "*A. nicotianae*" group of Keddie et al., 1986), the interpeptide bridge always contains a dicarboxylic acid and in most strains also contains alanine (Schleifer and Kandler, 1972). Thus members of the *A. globiformis/A. citreus* group have peptidoglycans with the A3α variation (Schleifer and Kandler, 1972) and contain dihydrogenated menaquinones with nine isoprene units [MK-9(H$_2$)] as their major isoprenoid quinones (Yamada et al., 1976; Collins et al., 1979, 1981; Collins and Jones, 1981). They do not contain teichoic acids in the cell wall (Fiedler and Schäffler, 1987). In contrast, species of the *A. nicotianae* group have peptidoglycans with the A4α variation (Schleifer and Kandler, 1972; Stackebrandt et al., 1983) and, with the exception of *A. sulfureus*, have unsaturated menaquinones with eight isoprene units (MK-8) as major components, although they also contain substantial amounts of unsaturated menaquinones with nine isoprene units (MK-9) as well (Yamada et al., 1976; Collins and Jones, 1981; Collins and Kroppenstedt, 1983). *A. sulfureus* contains either MK-9 as the major menaquinone (Yamada et al., 1976; Collins et al., 1979) or comparable amounts of MK-9 and MK-10 (Collins and Kroppenstedt, 1983). *A. nicotianae* group strains also differ from *A. globiformis/A. citreus* group strains in possessing cell wall teichoic acids

of the poly(glycero- phosphate) type (Fiedler and Schäffler, 1987). Further differences occur between the two groups in the types of polar lipids they contain. All *A. globiformis/A. citreus* group strains contain diphosphatidylglycerol, phosphatidylglycerol, and phosphatidylinositol. With the exception of *A. citreus*, they also contain several glycolipids (Shaw and Stead, 1971; Kostiw et al., 1972; Collins et al., 1981); *A. citreus* contains fewer glycolipids than the other species contain (Collins and Kroppenstedt, 1983).

The major characteristics of the 11 species of the *A. globiformis/A. citreus* group are listed in Table 1. The characteristics of the four species of the *A. nicotianae* group are listed in Table 2. To distinguish the two groups, the peptidoglycan "variation" may be determined by the rapid screening method described by Schleifer and Kandler (1972) (see also Schleifer and Seidl, 1985). Menaquinone composition may be determined by reverse phase partition thin layer chromatography (TLC) or high performance liquid chromatography (HPLC) (Collins et al., 1980, 1983). Polar lipid patterns may be obtained by two-dimensional TLC analysis of free lipid extracts (Collins et al., 1983). Fiedler and Schäffler (1987) state that *A. nicotianae* group strains may be recognized by testing for the presence of lysine, phosphorus, and glycerol in the cell wall. Further differentiation of the species within the *A. globiformis/A. citreus* group and the *A. nicotianae* group depends in large measure on DNA-DNA hybridization techniques (Fiedler and Stackebrandt, 1979; Stackebrandt et al., 1983). *A. citreus* can readily be distinguished but the remaining species share a high degree of phenotypic similarity, and it is not possible to differentiate them by using phenotypic characters. Those characters which may be of value in differentiation are listed in Tables 1 and 2.

REFERENCE STRAINS. *A. globiformis*, ATCC 8010 (NCIMB 8907). *A. nicotianae*, ATCC 15236 (NCIMB 9458). (Note: NCIMB was formerly NCIB.)

Biochemical and Physiological Properties

All species are catalase positive and probably contain cytochromes, though detailed information on the cytochrome content is not available for all species. The cytochrome composition of *A. globiformis* ATCC 4336 (NCIMB 8602) is *bcaa$_3$o* when the strain is in the exponential phase of growth but changes to *bcaa$_3$od* in the stationary phase of growth after the rod-shaped cells become oxygen limited and lose their ability to retain the Gram stain, and coccoid forms appear (Meyer and Jones, 1973; Jones, 1980). A similar observation was made by Faller and Schleifer (1981) when studying the apparent correlation between morphology and cytochrome content in *A. crystallopoietes*. They noted that rod-shaped cells in exponential phase cultures contained large amounts of cytochrome *aa₃* but only traces of cytochrome *d*, whereas coccoid cells in early stationary phase cultures contained relatively larger amounts of cytochrome *d* with lower amounts of cytochrome *aa₃*. However, in keeping with the observations of Meyer and Jones (1973) and Jones (1980), they showed that the increase in cytochrome *d* was not correlated with a change in cell morphology but resulted from the oxygen limitation that occurs in late exponential/early stationary phase cultures.

The fragmentary information available on metabolic pathways in *Arthrobacter* species has been reviewed by Krulwich and Pelliccione (1979). In summary, those few species which have been studied fall into two groups with respect to the pathways of carbohydrate dissimilation utilized. In *A. globiformis*, *A. ureafaciens*, *A. crystallopoietes*, and "*A. pyridinolis*," the primary pathways are the Embden-Meyerhof-Parnas (EMP) pathway and, to a smaller extent, the hexose monophosphate (HMP) pathway. In contrast, *A. pascens* and *A. atrocyaneus* use the Entner-Doudoroff and HMP pathways (Krulwich and Pelliccione, 1979). A list of carbon compounds utilized by a number of species as sole or major carbon sources is given in the section on "Soil and Similar Habitats."

Details of the peptidoglycan structure, lipid composition, menaquinone content, and the presence or absence of teichoic acids of the *Arthrobacter* species now recognized are given in the section "Further Identification of Arthrobacters."

Einck et al. (1973) have summarized the data available on bacteriophages active on *Arthrobacter* species.

Arthrobacters may be distinguished from other coryneforms and morphologically similar bacteria which have lysine in the cell wall or which show a rod-coccus growth cycle by the characters listed in Table 3. The differentiation from *Arthrobacter* of some former species now removed from the genus is also given in Table 3.

Biotechnological Potential

Undoubtedly, soil is the most important habitat of bacteria of the genus *Arthrobacter*. Their abundance in soils of various types and in different geographical locations has been amply dem-

onstrated by many investigators. Their numerical predominance, coupled with the nutritional versatility of the commonly occurring species (Hagedorn and Holt, 1975a; Keddie, 1974), suggests that they may be important agents of mineralization in soil and possibly also in some other habitats. Among the compounds reported to be degraded by arthrobacters are certain herbicides and pesticides, as well as a wide range of naturally occurring and synthetic molecules of various degrees of complexity. Other roles that have been ascribed to at least some *Arthrobacter* species are phytohormone production (Barea et al., 1976; Katznelson and Cole, 1965; Rivière, 1963) and dinitrogen fixation (Cacciari et al., 1971; Smyk, 1970; Smyk and Ettlinger, 1963). Putative *Arthrobacter* species have also been reported to lyse yeast cells (Kitamura et al., 1972) and mycelium of *Fusarium roseum*, a carnation-root pathogen (Morrisey et al., 1976; Szajer and Koths, 1973). In the latter case, the *Arthrobacter* strain investigated produced a chitinase and was considered a possible means of biological control of *Fusarium* diseases.

Products of actual or potential commercial importance obtained from *Arthrobacter* species include glutamic acid (Tanaka and Kimura, 1972; Veldkamp et al., 1963) and α-ketoglutaric acid (Tanaka and Kimura, 1972), although it is likely that in some of the examples mentioned the bacteria concerned were not legitimate *Arthrobacter* species (see Keddie and Cure, 1977, 1978). Also, Veldkamp et al. (1966) described a strain of *A. globiformis* that produced large amounts of riboflavin.

In some species of *Arthrobacter*, the morphological changes that occur during the growth cycle have been shown to be subject to nutritional control, and such species have proved to be useful in the study of bacterial morphogenesis (see review by Clark, 1972).

Literature Cited

Adamse, A. D. 1968. Formation and final composition of the bacterial flora of a dairy waste activated sludge. Water Research 2:665–671.

Akiba, T., Ueyama, H., Seki, M., Fukimbara, T. 1970. Identifications of lower alcohol-utilizing bacteria. Journal of Fermentation Technology 48:323–328.

Antheunisse, J. 1972. Decomposition of nucleic acids and some of their degradation products by microorganisms. Antonie van Leeuwenhoek Journal of Microbiology and Serology 38:311–327.

Austin, B., Goodfellow, M., Dickinson, C. H. 1978. Numerical taxonomy of phylloplane bacteria from *Lolium perenne*. Journal of General Microbiology 104:139–155.

Barea, J. M., Navarro, E., Montoya, E. 1976. Production of plant growth regulators by rhizosphere phosphate-solubilizing bacteria. Journal of Applied Bacteriology 40:129–134.

Bousfield, I. J. 1978. The taxonomy of coryneform bacteria from the marine environment. 217–233. Bousfield, I. J., and Cally, A. G. (ed.) Special publications of the Society for General Microbiology I. Coryneform bacteria. London, Academic Press.

Bousfield, I. J., Keddie, R. M., Dando, T. R., Shaw, S. 1985. Simple rapid methods of cell wall analysis as an aid in the identification of aerobic coryneform bacteria. 221–236. Goodfellow, M., and Minnikin, D. E. (ed.) Chemical methods in bacterial systematics. Society for Applied Bacteriology, Technical Series No. 20. London, Academic Press.

Bowie, I. S., Grigor, M. R., Dunckley, G. G., Loutit, M. W., Loutit, J. S. 1972. The DNA base composition and fatty acid constitution of some Gram-positive pleomorphic soil bacteria. Soil Biology and Biochemistry 4:397–412.

Boylen, C. W. 1973. Survival of *Arthrobacter crystallopoietes* during prolonged periods of extreme desiccation. Journal of Bacteriology 113:33–37.

Boylen, C. W., Ensign, J. C. 1970. Long-term starvation survival of rod and spherical cells of *Arthrobacter crystallopoietes*. Journal of Bacteriology 103:569–577.

Boylen, C. W., Mulks, M. H. 1978. The survival of coryneform bacteria during periods of prolonged nutrient starvation. Journal of General Microbiology 105:323–334.

Breed, R. S., Murray, E. G. D., Smith, N. R. (ed.). 1957. Bergey's manual of determinative bacteriology, 7th ed. Baltimore, MD. Williams & Wilkins.

Cacciari, I., Giovannozzi-Sermanni, G., Grappelli, A., Lippi, D. 1971. Nitrogen fixation by *Arthrobacter* sp. I-Taxonomic study and evidence of nitrogenase activity of two new strains. Annali di Microbiologia ed Enzymologia 21:97–105.

Chai, T. J., Levin, R. E. 1975. Characteristics of heavily mucoid bacterial isolates from fish pen slime. Applied Microbiology 30:450–455.

Chan, E. C. S., Stevenson, I. L. 1962. On the biotin requirement of *Arthrobacter globiformis*. Canadian Journal of Microbiology 8:403–405.

Chen, M., Alexander, M. 1973. Survival of soil bacteria during prolonged desiccation. Soil Biology and Biochemistry 5:213–221.

Clark, J. B. 1972. Morphogenesis in the genus *Arthrobacter*. CRC Critical Reviews in Microbiology 1:521–544.

Collins, M. D. 1982. Lipid composition of *Renibacterium salmoninarum* (Sanders and Fryer). FEMS Microbiology Letters 13:295–297.

Collins, M. D. 1986. Lipid composition of *Arthrobacter siderocapsulatus, A. viscosus,* "*A. oxamicetus,*" "*A. sialophilus,*" "*A. stabilis,*" and "*Agrobacterium pseudotsugene.*" Systematic and Applied Microbiology 8:1–7.

Collins, M. D., Dorsch, M., Stackebrandt, E. 1989. Transfer of *Pimelobacter tumescens* to *Terrabacter* gen. nov. as *Terrabacter tumescens* comb. nov. and of *Pimelobacter jensenii* to *Nocardioides* as *Nocardioides jensenii* comb. nov. International Journal of Systematic Bacteriology 39:1–6.

Collins, M. D., Goodfellow, M., Minnikin, D. E. 1979. Isoprenoid quinones in the classification of coryneform and related bacteria. Journal of General Microbiology 110:127–136.

Collins, M. D., Jones, D. 1981. The distribution of isoprenoid quinone structural types in bacteria and their taxonomic implications. Microbiological Reviews 45:316–354.

Collins, M. D., Jones, D., Keddie, R. M., Kroppenstedt, R. M., Schleifer, K. H. 1983. Classification of some coryneform bacteria in a new genus *Aureobacterium*. Systematic and Applied Microbiology 4:236–252.

Collins, M. D., Jones, D., Kroppenstedt, R. M. 1981. Reclassification of *Corynebacterium ilicis* (Mandel, Guba and Litsky) in the genus *Arthrobacter* as *Arthrobacter ilicis* comb. nov. Zentralblatt für Bakteriologie, Mikrobiologie und Hygiene, Abt. I., Orig. C. 2:318–323.

Collins, M. D., Kroppenstedt, R. M. 1983. Lipid composition as a guide to the classification of some coryneform bacteria containing an A4α type peptidoglycan (Schleifer and Kandler). Systematic and Applied Microbiology 4:95–104.

Collins, M. D., Shah, H. N., Minnikin, D. E. 1980. A note on the separation of natural mixtures of bacterial menaquinones using reverse-phase partition thin-layer chromatography. Journal of Applied Bacteriology 48:277–282.

Conn, H. J. 1928. A type of bacteria abundant in productive soils, but apparently lacking in certain soils of low productivity. New York State Agricultural Experimental Station Technical Bulletin No. 138:3–26.

Conn, H. J. 1947. A protest against the misuse of the generic name *Corynebacterium*. Journal of Bacteriology 54:10.

Conn, H. J., Dimmick, I. 1947. Soil bacteria similar in morphology to *Mycobacterium* and *Corynebacterium*. Journal of Bacteriology 54:291–303.

Crombach, W. H. J. 1974a. Relationships among coryneform bacteria from soil, cheese and sea fish. Antonie van Leeuwenhoek Journal of Microbiology and Serology 40:347–359.

Crombach, W. H. J. 1974b. Morphology and physiology of coryneform bacteria. Antonie van Leeuwenhoek Journal of Microbiology and Serology 40:361–376.

Cummins, C. S., Harris, H. 1959. Taxonomic position of *Arthrobacter*. Nature 184:831–832.

Cure, G. L., Keddie, R. M. 1973. Methods for the morphological examination of aerobic coryneform bacteria. 123–135. Board, R. G., and Lovelock, D. N. (ed.) Sampling-microbiological monitoring of environments. Society for Applied Bacteriology Technical Series 7. New York. Academic Press.

Duxbury, T., Gray, T. R. G. 1977. A microcultural study of the growth of cystites, cocci and rods of *Arthrobacter globiformis*. Journal of General Microbiology 103:101–106.

Ehrlich, H. C. 1963. Bacteriology of manganese nodules. I. Bacterial action on manganese in nodule enrichments. Applied Microbiology 11:15–19.

Ehrlich, H. C. 1968. Bacteriology of manganese nodules. II. Manganese oxidation by cell-free extract from a manganese nodule bacterium. Applied Microbiology 16:197–202.

Einck, K. H., Pattee, P. A., Holt, J. G., Hagedorn, C., Miller, J. A., Berryhill, D. L. 1973. Isolation and characterisation of a bacteriophage of *Arthrobacter globiformis*. Journal of Virology 12:1031–1033.

Ensign, J. C., Rittenberg, S. C. 1963. A crystalline pigment produced from 2-hydroxypyridine by *Arthrobacter crystallopoietes n. sp*. Archiv für Mikrobiologie 47:137–153.

Ensign, J. C., Wolfe, R. S. 1964. Nutritional control of morphogenesis in *Arthrobacter crystallopoietes*. Journal of Bacteriology 87:924–932.

Faller, A. H., Schleifer, K. H. 1981. Effects of growth phase and oxygen supply on the cytochrome composition and morphology of *Arthrobacter crystallopoietes*. Current Microbiology 6:253–258.

Fiedler, F., Schäffler, M. J. 1987. Teichoic acids in cell walls of strains of the *"nicotianae"* group of *Arthrobacter*: a chemotaxonomic marker. Systematic and Applied Microbiology 9:16–21.

Gibson, T. 1953. The taxonomy of the genus *Corynebacterium*. Atti del VI Congresso Internazionale di Microbiologia. Roma. 1:16–20.

Giovanozzi-Sermanni, G. 1959. Una nuova specie di *Arthrobacter* determinante la degradazione della nicotina: *Arthrobacter nicotianae*. II Tabacco. 63:83–86.

Gounot, A. M. 1967. Role biologique des *Arthrobacter* dans les limons souterrains. Annales de I'Institut Pasteur 113:923–945.

Gray, T. R. G. 1976. Survival of vegetative microbes in soil. Symposium of the Society for General Microbiology 26:327–364.

Greenberg, J., Barker, H. A. 1962. A ferrichrome-requiring arthrobacter which decomposes puromycin aminonucleoside. Journal of Bacteriology 83:1163–1164.

Gundersen, K., Jensen, H. L. 1956. A soil bacterium decomposing organic nitro-compounds. Acta Agriculturae Scandinavica 6:100–114.

Hagedorn, C., Holt, J. G. 1975a. A nutritional and taxonomic survey of *Arthrobacter* soil isolates. Canadian Journal of Microbiology 21:353–361.

Hagedorn, C., Holt, J. G. 1975b. Ecology of soil arthrobacters in Clarion-Webster toposequences of Iowa. Applied Microbiology 29:211–218.

Holm, E., Jensen, V. 1972. Aerobic chemoorganotrophic bacteria of a Danish beech forest. Oikos 23:248–260.

Horvath, R. S., Alexander, M. 1970. Cometabolism of *m*-chlorobenzoate by an *Arthrobacter*. Applied Microbiology 20:254–258.

Iizuka, H., Komagata, K. 1965. Microbiological studies on petroleum and natural gas. III. Determination of *Brevibacterium, Arthrobacter, Micrococcus, Sarcina, Alcaligenes,* and *Achromobacter* isolated from oil-brines in Japan. Journal of General and Applied Microbiology 11:1–14.

Jensen, H. L. 1933. Corynebacteria as an important group of soil microorganisms. Proceedings of the Linnean Society of New South Wales 58:181–185.

Jensen, H. L. 1934. Studies on saprophytic mycobacteria and corynebacteria. Proceedings of the Linnean Society of New South Wales 59:19–61.

Jensen, H. L. 1952. The coryneform bacteria. Annual Review of Microbiology 6:77–90.

Jensen, H. L. 1964. Studies on soil bacteria (*Arthrobacter globiformis*) capable of decomposing the herbicide Endothal. Acta Agriculturae Scandinavica 14:193–207.

Jensen, V. 1968. The plate count technique. 158–170. Gray, T. R. G., and Parkinson, D. (ed.) The ecology of soil bacteria. Liverpool, Liverpool University Press.

Jones, C. W. 1980. Cytochrome patterns in classification and identification including their relevance to the oxidase test. 127–138. Goodfellow, M., and Board, R. G. (ed.) Microbial classification and identification. Society for Applied Bacteriology Symposium Series 8. New York Academic Press.

Jones, D. 1978. An evaluation of the contribution of numerical taxonomy to the classification of the coryneform bacteria. 13–46. Bousfield, I. J., and Callely, A. G. (ed.) Coryneform bacteria. Special Publications of the Society for General Microbiology I. London, Academic Press.

Jones, D., Pell, P. A., Sneath, P. H. A. 1984. Maintenance of bacteria on glass beads at –60°C to –76°C. 35–40. Kirsop, B. E., and Snell, J. J. S. (ed.) Maintenance of microorganisms: A manual of laboratory methods. London, Academic Press.

Kaneko, T., Kitamura, K., Yamamoto, Y. 1969. *Arthrobacter luteus* nov. sp. isolated from brewery sewage. Journal of General and Applied Microbiology 15:317–326.

Katznelson, H., Cole, S. E. 1965. Production of gibberellin-like substances by bacteria and actinomycetes. Canadian Journal of Microbiology 11:733–741.

Keddie, R. M. 1974. *Arthrobacter*. 618–625. Buchanan, R. E., and Gibbons, N. E. (ed.) Bergey's manual of determinative bacteriology, 8th ed. Baltimore, Williams & Wilkins.

Keddie, R. M., Bousfield, I. J. 1980. Cell wall composition in the classification and identification of coryneform bacteria. 167–188. Goodfellow, M., and Board, R. G. (ed.) Microbiological classification and identification. Society for Applied Bacteriology Symposium Series No. 8. New York. Academic Press.

Keddie, R. M., Collins, M. D., Jones, D. 1986. Genus *Arthrobacter*. 1288–1301. Sneath, P. H. A., Mair, N. S., Sharpe, M. E., and Holt, J. G. (ed.) Bergey's manual of systematic bacteriology, vol. 2. Baltimore, Williams & Wilkins.

Keddie, R. M., Cure, G. L. 1977. The cell wall composition and distribution of free mycolic acids in named strains of coryneform bacteria and in isolates from various natural sources. Journal of Applied Bacteriology 42:229–252.

Keddie, R. M., Cure, G. L. 1978. Cell wall composition of coryneform bacteria. 47–84. Bousfield, I. J., and Callely, A. G. (ed.) Coryneform bacteria. Special Publications of the Society for General Microbiology I. London, Academic Press.

Keddie, R. M., Jones, D. 1981. Saprophytic, aerobic coryneform bacteria. 1838–1878. Starr, M. P., Stolp, H., Trüper, H. G., Balows, A., and Schlegel, H. G. (ed.) The prokaryotes: A handbook on habitats, isolation and identification of bacteria. Berlin, Springer-Verlag.

Keddie, R. M., Leask, B. G. S., Grainger, J. M. 1966. A comparison of coryneform bacteria from soil and herbage: Cell wall composition and nutrition. Journal of Applied Bacteriology 29:17–43.

Kitamura, K., Kaneko, T., Yamamoto, Y. 1972. Lysis of viable yeast cells by enzymes of *Arthrobacter luteus*. I. Isolation of lytic strain and studies of its lytic activity. Journal of Applied and General Microbiology 18:57–71.

Klein, D. A., Davis, J. A., Casida, L. E. Jr. 1968. Oxidation of *n*-alkanes to ketones by an *Arthrobacter* species. Antonie van Leeuwenhoek Journal of Microbiology and Serology 34:495–503.

Kolenbrander, P. E., Lotong, N., Ensign, J. C. 1976. Growth and pigment production by *Arthrobacter pyridinolis* n. sp. Archives of Microbiology 110:239–245.

Kortstee, G. J. J. 1970. The aerobic decomposition of choline by microorganisms. I. The ability of aerobic organisms, particularly coryneform bacteria, to utilize choline as the sole carbon and nitrogen source. Archiv für Mikrobiologie 71:235–244.

Kostiw, L. L., Boylen, C. W., Tyson, B. J. 1972. Lipid composition of growing and starving cells of *Arthrobacter crystallopoeites*. Journal of Bacteriology 111:103–111.

Kraft, A. A., Ayres, J. C., Torrey, G. S., Salzer, R. H., da Silva, G. A. N. 1966. Coryneform bacteria in poultry, eggs and meat. Journal of Applied Bacteriology 29:161–166.

Krulwich, T. A., Pelliccione, N. J. 1979. Catabolic pathways of coryneforms, nocardias and mycobacteria. Annual Review of Microbiology 33:95–111.

Labeda, D. P., Liu, K. C., Casida, L. E. Jr. 1976. Colonization of soil by Arthrobacter and Pseudomonas under varying conditions of water and nutrient availability as studied by plate counts and transmission electron microscopy. Applied and Environmental Microbiology 31:551–561.

Lechevalier, H. A., Lechevalier, M. P. 1981. Actinomycete genera "in search of a family,". 2118–2123. Starr, M. P., Stolp, H., Trüper, H. G., Balows, A., and Schlegel, H. G. (ed.) The prokaryotes: A handbook on habitats, isolation and identification of bacteria. Berlin, Springer-Verlag.

Lee, J. S., Pfeifer, D. K. 1977. Microbiological characteristics of Pacific shrimp (Pandalus jordani). Applied and Environmental Microbiology 33:853–859.

Lochhead, A. G. 1958a. Two new species of Arthrobacter requiring respectively vitamin B_{12} and the terregens factor. Archiv für Mikrobiologie 31:163–170.

Lochhead, A. G. 1958b. Soil bacteria and growth-promoting substances. Bacteriological Reviews 22:145–153.

Lochhead, A. G., Burton, M. O. 1953. An essential bacterial growth factor produced by microbial synthesis. Canadian Journal of Botany 31:7–22.

Lochhead, A. G., Burton, M. O. 1956. Importance of soil extract for the enumeration and study of soil bacteria. 157–161. Transactions of the 6th International Congress of Soil Science. Paris.

Lochhead, A. G., Burton, M. O. 1957. Qualitative studies of soil micro-organisms. XIV. Specific vitamin requirements of the predominant bacterial flora. Canadian Journal of Microbiology 3:35–42.

Loos, M. A., Roberts, R. N., Alexander, M. 1967. Phenols as intermediates in the decomposition of phenoxyacetates by an Arthrobacter species. Canadian Journal of Microbiology 13:679–690.

Lowe, W. E., Gray, T. R. G. 1972. Ecological studies on coccoid bacteria in a pine forest soil. I. Classification. Soil Biology and Biochemistry 4:459–468.

Lund, B. M. 1969. Properties of some pectolytic, yellow pigmented, Gram-negative bacteria isolated from fresh cauliflowers. Journal of Applied Bacteriology 32:60–67.

Luscombe, B. M., Gray, T. R. G. 1971. Effect of varying growth rate on the morphology of Arthrobacter. Journal of General Microbiology 69:433–434.

Lysenko, O. 1959. The occurrence of species of the genus Brevibacterium in insects. Journal of Insect Pathology 1:34–42.

Mandel, M., Guba, E. F., Litsky, W. 1961. The causal agent of bacterial blight of American holly. Bacteriological Proceedings 61.

Meyer, D. J., Jones, C. W. 1973. Distribution of cytochromes in bacteria: relationship to general physiology. International Journal of Systematic Bacteriology 23:459–467.

Minnikin, D. E., Collins, M. D., Goodfellow, M. 1978a. Menaquinone patterns in the classification of nocardioform and related bacteria. Zentralblatt für Bakteriologie, Parasitenkunde, Infektionskrankheit und Hygiene Abt. 1, Suppl. 6:85–90.

Minnikin, D. E., Goodfellow, M., Collins, M. D. 1978b. Lipid composition in the classification and identification of coryneform and related taxa. 85–160. Bousfield, I. J., and Callely, A. G. (ed.) Special publications of the Society for General Microbiology. I. Coryneform bacteria. London, Academic Press.

Moiroud, A., Gounot, A. M. 1969. Sur une bactérie psychrophile obligatoire isolée de limons glaciaires. Comptes Rendus Hebdomadaires des Seances de l'Academie des Sciences, Serie D 269:2150–2152.

Morris, J. G. 1960. Studies on the metabolism of Arthrobacter globiformis. Journal of General Microbiology 22:564–582.

Morrisey, R. F., Dugan, E. P., Koths, J. S. 1976. Chitinase production by an Arthrobacter sp. lysing cells of Fusarium roseum. Soil Biology and Biochemistry 8:23–28.

Mulder, E. G., Adamse, A. D., Antheunisse, J., Deinema, M. H., Woldendorp, J. W., Zevenhuizen, L. P. T. M. 1966. The relationship between Brevibacterium linens and bacteria of the genus Arthrobacter. Journal of Applied Bacteriology 29:44–71.

Mulder, E. G., Antheunisse, J. 1963. Morphologie, physiologie et écologie des Arthrobacter. Annales de l'Institut Pasteur 105:46–74.

Nand, K., Rao, D. V. 1972. Arthrobacter mysorens—a new species excreting L-glutamic acid. Zentralblatt für Bakteriologie, Parasitenkunde, Infektionskrankheiten und Hygiene, Abt. 2, Orig. 127:324–331.

O'Donnell, A. G., Goodfellow, M., Minnikin, D. E. 1982. Lipids in the classification of Nocardioides: reclassification of Arthrobacter simplex (Jensen) Lochhead in the genus Nocardioides (Prauser) emend. O'Donnell et al. as Nocardioides simplex comb. nov. Archive für Mikrobiologie 133:323–329.

Owens, J. D., Keddie, R. M. 1969. The nitrogen nutrition of soil and herbage coryneform bacteria. Journal of Applied Bacteriology 32:338–347.

Park, Y-H., Hori, H., Suzuki, K.-I., Osawa, S., Komagata, K. 1987. Phylogenetic analysis of the coryneform bacteria by 5S rRNA sequences. Journal of Bacteriology 169:1801–1806.

Rivière, J. 1963. Action des microorganismes de la rhizosphère sur la croissance du blé. II. Isolement et caractérisation des bactéries produisant des phytohormones. Annales de l'Institut Pasteur 105:303–314.

Robinson, J. B., Salonius, P. O., Chase, F. E. 1965. A note on the differential response of Arthrobacter spp. and Pseudomonas spp. to drying in soil. Canadian Journal of Microbiology 11:746–748.

Roth, N. G., Wheaton, R. B. 1962. Continuity of psychrophilic and mesophilic growth characteristics in the genus Arthrobacter. Journal of Bacteriology 83:551–555.

Sacks, L. E. 1954. Observations on the morphogenesis of Arthrobacter citreus, spec. nov. Journal of Bacteriology 67:342–345.

Sanders, J. E., Fryer, J. L. 1980. Renibacterium salmoninarum gen. nov. spec. nov., the causative agent of bacterial kidney disease in salmonid fishes. International Journal of Systematic Bacteriology 30:496–502.

Schefferle, H. E. 1966. Coryneform bacteria in poultry deep litter. Journal of Applied Bacteriology 29:147–160.

Schleifer, K. H., Kandler, O. 1972. Peptidoglycan types of bacterial cell walls and their taxonomic implications. Bacteriological Reviews 36:407–477.

Schleifer, K. H., Seidl, P. H. 1985. Chemical composition and structure of murein. 201–219. Goodfellow, M. and Minnikin, D. E. (ed.) Chemical methods in bacterial systematics. Society for Applied Bacteriology Technical series no. 20. London, Academic Press.

Seidman, P., Chan, E. C. S. 1969. Growth of Arthrobacter citreus in a chemically-defined medium and its requirement for chelating agents with schizokinen activity. Journal of General Microbiology 58:v.

Seiler, H., Braatz, R., Ohmeyer, G. 1980. Numerical cluster analysis of coryneform bacteria from activated sludge. Zentralblatt für Bakteriologie, Parazitenkunde, Infektionskrankheiten und Hygiene Abt 1. Orig C. 1:357–375.

Sethunathan, N., Pathak, M. D. 1971. Development of a diazinon-degrading bacterium in paddy water after repeated application of diazinon. Canadian Journal of Microbiology 17:699–702.

Sguros, P. L. 1955. Microbial transformations of the tobacco alkaloids. I. Cultural and morphological characteristics of a nicotinophile. Journal of Bacteriology 69:28–37.

Sharpee, K. W., Duxbury, J. M., Alexander, M. 1973. 2,4-Dichlorophenoxyacetate metabolism by *Arthrobacter* sp.: Accumulation of a chlorobutenolide. Applied Microbiology 26:445–447.

Shaw, N., Stead, A. 1971. Lipid composition of some species of *Arthrobacter*. Journal of Bacteriology 107:130–133.

Sieburth, J. McN. 1964. Polymorphism of a marine bacterium (*Arthrobacter*) as a function of multiple temperature optima and nutrition. Proceedings of the Symposium on Experimental Marine Ecology. Occasional Publication No. 2:11–16.

Skyring, G. W., Quadling, C. 1969. Soil bacteria: Comparisons of rhizosphere and nonrhizosphere populations. Canadian Journal of Microbiology 15:473–488.

Skyring, G. W., Quadling, C. 1970. Soil bacteria: A principal component analysis and guanine-cytosine contents of some arthrobacter-coryneform soil isolates and of some named cultures. Canadian Journal of Microbiology 16:95–106.

Skyring, G. W., Quadling, C., Rouatt, J. W. 1971. Soil bacteria: Principal component analysis of physiological descriptions of some named cultures of *Agrobacterium*, *Arthrobacter*, and *Rhizobium*. Canadian Journal of Microbiology 17:1299–1311.

Smyk, B. 1970. Fixation of atmospheric nitrogen by the strains of *Arthrobacter*. Zentralblatt für Bakteriologie, Parasitenkunde, Infektionskrankheiten and Hygiene, Abt. 2 Orig. 124:231–237.

Smyk, B. N., Ettlinger, L. 1963. Recherches sur quelque espèces d'arthrobacter fixatrices d'azote isolées des roches karstiques alpines. Annales de L'Institut Pasteur 105:341–348.

Soumare, S., Blondeau, R. 1972. Caractéristiques microbiologiques des sol de la région du nord de la France: Importance des "Arthrobacters." Annales de L'Institut Pasteur 123:239–249.

Splittstoesser, D. F., Wexler, M., White, J., Colwell, R. R. 1967. Numerical taxonomy of Gram-positive and catalase-positive rods isolated from frozen vegetables. Applied Microbiology 15:158–162.

Stackebrandt, E., Fiedler, F. 1979. DNA-DNA homology studies among *Arthrobacter* and *Brevibacterium*. Archives of Microbiology 120:289–295.

Stackebrandt, E., Fowler, V. J., Fiedler, F., Seiler, H. 1983. Taxonomic studies on *Arthrobacter nicotianae* and related taxa. Description of *Arthrobacter uratoxydans* sp. nov. and *Arthrobacter sulfureus* sp. nov. and reclassification of *Brevibacterium protophormiae* as *Arthrobacter protophormiae* comb. nov. Systematic and Applied Microbiology 4:470–486.

Stackebrandt, E., Lewis, B. J., Woese, C. R. 1980. The phylogenetic structure of the coryneform group of bacteria. Zentralblatt für Bakteriologie, Parasitenkunde, Infek-

tionskrankheiten und Hygiene, Abt. 2, Orig. C 1:137–149.

Stackebrandt, E., Woese, C. R. 1981. The evolution of the prokaryotes. 1–3. Carlile, M. J., Collins, J. F., and Moseley, B. E. B. (ed.) Molecular and cellular aspects of microbial evolution. Symposium of the Society for General Microbiology 32. Cambridge, Cambridge University Press.

Stevenson, I. L. 1967. Utilization of aromatic hydrocarbons by *Arthrobacter spp*. Canadian Journal of Microbiology 13:205–211.

Suzuki, K., Collins, M. D., Iijima, E., Komagata, K. 1988. Chemotaxonomic characterization of a radiotolerant bacterium *Arthrobacter radiotolerans*: description of *Rubrobacter radiotolerans* gen. nov., comb. nov. FEMS Microbiology Letters 52:33–40.

Suzuki, K., Komagata, K. 1983. *Pimelobacter* gen. nov.—a new genus of coryneform bacteria with LL-diaminopimelic acid in the cell wall. Journal of General and Applied Microbiology 29:59–71.

Szajer, C., Koths, J. S. 1973. Physiological properties and enzymatic activity of an *Arthrobacter* capable of lysing *Fusarium sp*. Acta Microbiologica Polonica Series B 5:81–86.

Tanaka, K., Kimura, K. 1972. Process for producing L-glutamic acid and alpha-ketoglutaric acid. United States Patent No. 3,642,576.

Tanenbaum, S. W., Flashner, M. 1977. *Arthrobacter sialophilus* sp. nov. a neuramidase-producing coryneform. Canadian Journal of Microbiology 23:1568–1572.

Tate, R. L., Ensign, J. C. 1974. A new species of *Arthrobacter* which degrades picolinic acid. Canadian Journal of Microbiology 20:691–694.

Taylor, C. B., Lochhead, A. G. 1937. A study of *Bacterium globiforme* Conn in soils differing in fertility. Canadian Journal of Research C 15:340–347.

Topping, L. E. 1937. The predominant micro-organisms in soils. I. Description and classification of the organisms. Zentralblatt für Bakteriologie, Parasitenkunde, Infektionskrankheiten und Hygiene, Abt. 2 Orig. 97:289–304.

Topping, L. E. 1938. The predominant micro-organisms in soils. II. The relative abundance of the different types of organisms obtained by plating, and the relation of plate to total counts. Zentralblatt für Bakteriologie, Parasitenkunde, Infektionskrankheiten und Hygiene, Abt. 2 Orig. 98:193–201.

Veldkamp, H. 1965. The isolation of *Arthrobacter*. Zentralblatt für Bakteriologie, Parasitenkunde, Infektionskrankheiten und Hygiene, Abt. 1 Orig. Suppl. 1:265–269.

Veldkamp, H., van den Berg, G., Zevenhuizen, L. P. T. M. 1963. Glutamic acid production by *Arthrobacter globiformis*. Antonie van Leeuwenhoek Journal of Microbiology and Serology 29:35–51.

Veldkamp, H., Venema, P. A. A., Harder, W., Konings, W. N. 1966. Production of riboflavin by *Arthrobacter globiformis*. Journal of Applied Bacteriology 29:107–113.

Williams, S. T., Davies, F. L. 1965. Use of antibiotics for selective isolation and enumeration of actinomycetes in soil. Journal of General Microbiology 38:251–261.

Wood, E. J. F. 1950. The bacteriology of shark spoilage. Australian Journal of Marine and Freshwater Research 1:129–138.

Yamada, K., Komagata, K. 1972. Taxonomic studies on coryneform bacteria. IV. Morphological, cultural, biochemical and physiological characteristics. Journal of General and Applied Microbiology 18:399–416.

Yamada, Y., Motoi, H., Kinoshita, S., Takada, N., Okada, H. 1975. Oxidative degradation of squalene by *Arthrobacter* species. Applied Microbiology 29:400–404.

Yamada, Y., Inouye, G., Tahara, Y., Kondo, K. 1976. The menaquinone system in the classification of coryneform and nocardioform bacteria and related organisms. Journal of General and Applied Microbiology 22:203–214.

Zevenhuizen, L. P. T. M. 1966. Formation and function of the glycogen-like polysaccharide of *Arthrobacter*. Antonie van Leeuwenhoek Journal of Microbiology and Serology 32:356–372.

Prokaryotes (2006) 3:961–971
DOI: 10.1007/0-387-30743-5_37

CHAPTER 1.1.22

The Genus *Micrococcus*

MILOSLAV KOCUR, WESLEY E. KLOOS AND KARL-HEINZ SCHLEIFER

The genus *Micrococcus* consists of Gram-positive spheres occurring in tetrads and in irregular clusters that are usually nonmotile and nonsporeforming. They are catalase positive and usually aerobic with strictly respiratory metabolism. Most species produce carotenoid pigments. The GC content of the DNA ranges from 65 to 75 mol%. There are nine species recognized in the genus (see later, Table 2). The data on GC content of the DNA, chemical cell wall analysis, and a comparative analysis of 16S rRNA sequences indicate that the genus *Micrococcus* is more closely related to the genus *Arthrobacter* than it is to other coccoid genera such as *Staphylococcus* and *Planococcus* (Keddie, 1974; Kloos et al., 1974; Kocur et al., 1971; Stackebrandt and Woese, 1979). For these reasons it cannot be included with the genera *Staphylococcus* and *Planococcus* in the same family Micrococcaceae. Therefore, both the genus *Micrococcus* and the genus *Arthrobacter* should be regarded as closely related, but separate genera.

Habitats

Mammalian skin is now considered as the primary habitat of micrococci. Micrococci are found more consistently and usually in larger populations on mammalian skin than from other sources (Carr and Kloos, 1977; Glass, 1973; Marples, 1965; Noble and Sommerville, 1974; Kloos et al., 1974, 1976).

Human Skin

Human skin is a rich source of micrococci. It was shown that 96% of 115 people living in 18 different states in the USA carried cutaneous populations of micrococci (Kloos et al., 1974). The percentages of individuals carrying various *Micrococcus* species were as follows: *M. luteus*, 90%; *M. varians*, 75%; *M. lylae*, 33%; *M. nishinomiyaensis*, 28%; *M. kristinae*, 25%; *M. roseus*, 15%; *M. sedentarius*, 13%; *M. agilis*, 4%. Populations of *M. luteus* were usually relatively large and this organism was isolated on the average from 51% of the different skin sites sampled of individuals carrying this species. In a temporal study by Kloos and Musselwhite (1975), it was found that micrococci usually constituted from 1 to 20% of the total aerobic bacteria isolated from

the skin of the head, legs, and arms, but less than 1% of those isolated from the high bacterial density areas of the nares and axillae. In the same temporal study, 80% of the individuals carried micrococci on the head, legs, and arms for at least 1 year. Where strain (or clonal) populations could be resolved, it was determined that certain strains of *M. luteus*, *M. varians*, *M. kristinae*, and *M. sedentarius* persisted on specific individuals up to 1 year, which strongly suggested a resident status. Several strains of *M. luteus* that were monitored for longer periods persisted for up to $2^1/_2$ years. Most strains appeared to be more transient and were isolated once or more for periods up to 6 months. In another temporal study of the micrococci of infant skin, Carr and Kloos (1977) reported an increase in the occurrence of micrococci with increasing age up to 10 weeks. Micrococci were rarely isolated from infants less than 1 week of age.

Animal Skin

Micrococci have been isolated from the skin of a variety of mammals, including squirrels of the genus *Sciurus*, rats, raccoons, opossums, horses, swine, cattle, dogs, and various primates (Kloos et al., 1976). The predominant species found on nonhuman mammals studied to date was *M. varians*. *M. luteus* was rarely isolated from nonhuman mammals. Other *Micrococcus* species found on the skin of humans appear to have a narrow host range, as their populations have not yet been discovered on other mammals. Most strains of these specialized species required one or more amino acids and vitamins for growth (Farrior and Kloos, 1975, 1976). The distribution of micrococci on other animals (e.g., birds, reptiles, amphibians, and fish) has not yet been adequately determined.

There are only few data on the distribution of micrococci on fish. Micrococci appear to be numerically diverse on the skin of marine and fresh-water fish, often comprising 1 to 25% of the bacterial population. Similar percentages of micrococci have been found within alimentary tract of different fish (Horsley, 1977). Micrococci predominate in the bacterial population of marine fish captured on the south Australian

This chapter was taken unchanged from the second edition.

coasts, which indicates that this genus is most numerous within these waters (Gillespie and Macrae, 1975). It has been found that micrococci and coryneforms predominate in the bacterial population of sharks (Venkataraman and Sreenivasan, 1953, 1955). The high incidence of micrococci reported on certain fish may be due to their population density in the water or contamination from landing nets or excessive handling by the angler (Horsley, 1977). Micrococci and coryneforms are found among the major groups of bacteria of crustacean shellfish. The percentage composition varies. Williams et al. (1952) found the main groups in the whole Gulf shrimps to be *Acinetobacter (Achromobacter)*, *Micrococcus*, *Pseudomonas*, and *Bacillus*, whereas Sreenivasan (1959) found that prawns in India carried predominately micrococci and coryneforms. The bacterial flora of the Ontario fresh-water fish examined was similar to that reported from marine fish, in which the species *Pseudomonas* and *Micrococcus* were encountered frequently (Evelyn and McDermott, 1961).

Other Sources

With the exception of animal and dairy products, which may be considered as secondary sources of micrococci, most of the sources, such as soil, estuarine mud, marine and fresh water, plants, fomites, dust, and air, contain small isolated populations of micrococci (Aaronson, 1955; Abd-el-Malek and Gibson, 1948; Baird-Parker, 1962; Doeringer and Dugan, 1973; Kitchell, 1962; Pohja, 1960; Streby-Andrews and Kloos, 1971; ZoBell and Upham, 1944).

Soil is not the primary source of micrococci as had been originally assumed. *Micrococcus luteus* cells died relatively quickly when they were added to natural soil. Microscopic observation showed that the cells were being physically destroyed by bacterial predators in the soil. One of the predators was *Streptoverticillium* sp. (Casida, 1980a, 1980b).

Micrococci formed only a limited part of bacterial population of sea water and therefore they have not been studied thoroughly in marine environments (Anderson, 1962; Brisou, 1955; Wood, 1952; ZoBell and Upham, 1944). It has been found that sand taken from public swimming areas at several North Carolina ocean beaches often contained small populations of *M. luteus*, whereas sand taken from remote ocean island beaches, frequented by small numbers of indigenous animals and a few anglers, only occasionally contained populations of *M. varians* and never *M. luteus*, a species common on human skin, but rare on the skin of other animals.

Some micrococci may well survive in beer. Dickscheit (1961) isolated 33 strains of micrococci from samples of beer and also "yellow sarcinas" which were probably strains of *M. kristinae*, *M. varians*, or *M. luteus*. Also Back (1980) isolated several strains from beer which produced a fruity odor. *M. kristinae* formed sediment in beer bottles and changed the flavor of beer.

Isolation

Direct Isolation from Skin Under Nonselective Conditions

Various semiquantitative procedures have been described for isolating aerobic bacteria from human skin (Kloos and Musselwhite, 1975; Pachtman et al., 1954; Smith, 1970; Williamson, 1965; Williamson and Kligman, 1965). The procedure described by Kloos and Musselwhite (1975) has been used for the isolation of micrococci and is suitable for use with human as well as other mammalian skin.

Isolation of Micrococci from Human and Other Mammalian Skin (Kloos and Musselwhite, 1975)

> Sterile cotton swabs were moistened with a detergent containing 0.1% Triton X-100 (Packard) in 0.075M phosphate buffer, pH 7.9 (Williamson, 1965), and rubbed vigorously, with rotation, over approximately 8-cm² sites. Swabbing was performed for 5 s on sites of the forehead, cheek, chin, nares, and axillae that usually contained large populations of bacteria and for 15 s on sites of the arms and legs that usually contained relatively small populations. Swabs taken from the forehead, cheek, chin, external naris, arms, and legs were immediately applied directly on agar media (standard agar plate, 100-mm diameter) by rubbing, with rotation, over the entire surface for two consecutive times. Swabs taken from the anterior nares and axillae were immediately rinsed once in 5 ml of detergent, and the rinse was applied to the surface of agar media. Later, during the course of the study, we observed that adults often contain populations of bacteria on the forehead, cheek, chin, and external nares that were too large to be analyzed by inoculating swabs directly onto media. In these instances, samples taken from a single swab rinse proved to be more satisfactory and produced well isolated colonies.

The Isolating Medium (P agar) of Naylor and Burgi (1956)

Peptone (Difco)	5.0 g
Sodium chloride	5.0 g
Glucose	1.0 g
Agar (Difco)	15.0 g
Distilled water	1 liter

Inoculated agar media are incubated under aerobic conditions at 34°C for 4 days, at which time colonies are counted and recorded according to morphology and pigment. Subcultures (of selected colonies showing *Micrococcus* morphology) are stored at 4°C. For convenience, the original isolation plates could be stored at 4°C for 2 to 3 weeks prior to the isolation of cultures.

This procedure can be modified for use with other mammalian skin as follows.

Isolation of Micrococci from Nonhuman Mammalian Skin (Kloos et al., 1976)

Sites on the body were exposed for swabbing by parting the pelage away from the site area. Swabbing was performed for 5 s on sites in the anterior naris and ventral pouch (opossums only) that usually contained large populations of bacteria and for 15 s on sites of the forehead, forelimbs, hindlimbs, abdomen, and back that usually contained relatively small populations. Swabs taken from the forehead, forelimbs, hindlimbs, abdomen, and back were immediately applied directly on agar media by rubbing, with rotation, over the entire surface for two consecutive times. Swabs taken from the anterior nares and ventral pouch were immediately rinsed once in 5 ml of detergent and then applied to the surface of agar media. (The authors of this chapter have found that rinsing in detergent is also necessary for swabs taken from the face, back, abdomen, and perineum of many Old World monkeys and great apes.)

Direct Isolation from Skin Under Selective Conditions

For shipping agar plates between collecting points or for use with nonhuman mammals, P agar was supplemented with the mold inhibitor cycloheximide (50 µg/ml) (Kloos et al., 1974; Kloos et al., 1976). Many mammals, because of their close contact with soil and foliage, carry large populations of fungi on their skin. If allowed to grow, fungal contaminants may rapidly cover the surface of unsupplemented P agar plates, making the isolation and characterization of bacteria very difficult. If *Bacillus* species are numerous on skin, it may be necessary to supplement P agar with 7% Nace to inhibit their large or spreading colonies. However, such a high amount of sodium chloride will interfere with distinctive *Micrococcus* species colony morphology and *Micrococcus* colonies will not be easily distinguished from *Staphylococcus* colonies.

Selective medium for isolating micrococci (and corynebacteria) in the presence of populations of staphylococci has been described by Curry and Borovian (1976). The nitrofuran-containing medium, known as FTO agar, permits the growth of micrococci and prevents the growth of staphylococci. It is particularly useful for sampling areas of the skin such as the nares, axillae, and perineum where *Staphylococcus* populations are usually very large.

FTO Agar Medium for Direct, Selective Isolation of Micrococci from Skin (Curry and Borovian, 1976)

Tryptic soy agar (Difco) or Trypticase soy agar (BBL)	40.0 g
Yeast extract	1.0 g
Tween 80	5.0 g
Distilled water	1 liter

After autoclaving and cooling to 48°C, 0.1% of a 0.5 acetone stock of the dye Oil Red 0 and 10% of a 0.05% acetone stock of nitrofuran (Furoxone) are added. To prevent precipitation, the latter is added slowly from a 100-ml graduate into swirling agar. Flasks are then left open or loosely covered in the water bath to allow acetone volatilization prior to pouring and hardening plates on a level surface. The plates can be incubated at 34 to 37°C for 3–4 days for adequate development of *Micrococcus* colonies. To reduce competition and crowding by lipophilic corynebacteria, Tween 80 should be omitted from the formula. FTO agar is conveniently prepared in 700-ml amounts in 1-liter Erlenmeyer flasks.

Cultivation

Most of micrococci grow well on nutrient agar or on P agar (Naylor and Burgi, 1956) at 37°C. The only exceptions are *M. agilis*, which is psychrotrophic and grows best at 22 to 25°C, and *M. halobius*, which requires 5% NaCl in cultivating medium such as nutrient agar or P agar.

Preservation

Cultures of micrococci may be stored on nutrient agar in a refrigerator (5°C) for 3 to 5 months if they are in perfectly sealed tubes. They may also be stored on nutrient agar under liquid paraffin in a refrigerator (5°C) for 1 to 2 years. The most reliable method for long-term preservation is lyophilization, using standard methods (Kirsop and Snell, 1984), or in liquid nitrogen.

Identification

Separation of Staphylococci from Micrococci

The characters that separate staphylococci from micrococci are listed in Table 1. Marked differences in the DNA base and cell wall compositions, fatty acid, and menaquinone patterns permit an accurate separation of the members of the two genera. However, these characters cannot be determined in the routine laboratory; therefore, other characters that can be easily analyzed are also listed in Table 1.

A simple test for the separation of staphylococci from micrococci was described by Schleifer and Kloos (1975). It is based on the ability of staphylococci to produce acid aerobically from glycerol in the presence of 0.4 µ/ml of erythromycin and on their sensitivity to lysostaphin. Recent studies on the isolation of micrococci and staphylococci from dry sausage have indicated that the addition of erythromycin to the glycerol medium is not absolutely necessary (Fischer and

Table 1. Separation of staphylococci from micrococci.

Character	*Staphylococcus*	*Micrococcus*	Reference
Anaerobic fermentation of glucose	+ (±, −)	− (±, +)	Evans and Kloos, 1972
FP agar	No growth	Growth	Rheinbaben and Hadlok, 1981
Bacitracin disk diffusion test	Resistant	Susceptible	Falk and Guering, 1983
Acid from glycerol-erythromycin medium	+ (−)	− (+)	Schleifer and Kloos, 1975
Selective medium containing thiocyanate plus azide	Growth	No growth	Schleifer and Krämer, 1980
Susceptibility to the vibriostatic agent 0/129 (0.5 mg/disc)	Growth	No growth	Bouvet et al., 1982
Resistance to lysostaphin	−	+	Klesius and Schuhardt, 1968; Schleifer and Kloos, 1975
Oxidase test	− (+)	+ (−)	Faller and Schleifer, 1981
Modified benzidine test	− (+)	+ (−)	Faller and Schleifer, 1981
Fructose-1,6-biphosphate aldolase (FBP)	Class I[a]	Class II	Götz et al., 1979; Fischer et al., 1982
DNA base composition (mol% GC)	30–38	66–73	Kocur et al., 1971
Major menaquinones	MK-6 to MK-8	Partially hydrogenated MK-7(H_2) to MK-9(H_2)	Collins and Jones, 1981
Long-chain unbranched fatty acids (C-18:0, C-20:0)	+	−	Schleifer and Kroppenstedt, personal communication
Cell wall composition: Peptidoglycan			
Position 1	Ala	Ala	Schleifer, 1986
Position 3	Lys	Lys	
Main interpeptide bridge	Gly$_{5-6}$, Ala-Gly$_4$, or Gly$_{3-5}$, Ser$_{1-2}$	Peptide subunit Ala$_{3-4}$, Asp-Thr-Ala$_3$, or Ser$_2$-D-Glu	
Teichoic acid	+	−	Schleifer, 1973; Schleifer and Kloos, 1975

S. chromogenes, S. hyicus, and *S. intermedius* contain both class I and II FBP-aldolases; *S. caseolyticus* contains only class II. +, positive; ±, weak; −, negative. Symbols in parentheses denote a character frequency of less than 30%.

Schleifer, 1980). Only strains of *Micrococcus kristinae* and a few strains of *M. roseus* produce small amounts of acid aerobically from glycerol, but these organisms can be easily distinguished from staphylococci by their convex colony profile and characteristic colony pigment (Kloos et al., 1974). Furazolidone agar proved to be a suitable medium for the separation of micrococci from staphylococci (Rheinbaben and Hadlok, 1981).

FP Agar for Separating *Micrococcus* from *Staphylococcus* (Rheinbaben and Hadlok, 1981)

Peptone	10.0 g
Yeast extract	5.0 g
NaCl	5.0 g
Glucose	1.0 g
Agar	12.0 g
Distilled water	1 liter

pH 7.0

After autoclaving and cooling to 48°C, 100 ml of a 0.02% acetone solution of furazolidone are mixed under slow stirring with the basal medium. Before pouring the plates, the flasks are left open or loosely covered in a water bath for 3 to min to allow evaporation of acetone.

Micrococci are susceptible to vibriostatic agent 0/129 (0.5 mg/disc) while staphylococci are resistant to it, which enables one to separate both genera. The recommended medium for 0/129 susceptibility testing is Mueller-Hinton agar (Bouvet et al., 1982).

Differentiation of *Micrococcus* species

Micrococci may be differentiated into nine species by means of tests listed in Table 2. Their pigment production and colony morphology may be used as a simple test for their presumptive identification (Kloos et al., 1974). Certain difficulties may occur in the differentiation of *M. luteus* and *M. lylae*, as both species have several features in common. However, *M. lylae* can be distinguished from *M. luteus* by cream-white or unpigmented colonies, lack of growth on inorganic nitrogen agar, lysozyme resistance, and cell wall peptidoglycan. The most common yellow pigmented species, *M. luteus* and *M. varians*, differ in acid production from glucose, nitrate reduction, lysozyme susceptibility, growth on inorganic nitrogen agar and on Simmons' citrate agar, and oxidase reaction. *M. roseus* differs from other species in having pink colonies, nitrate reduction, and an inability to hydrolyze gelatin.

Table 2. Abbreviated scheme for the differentiation of species of the genus *Micrococcus*.[a]

Species	Major pigment[b]	Water-soluble exopigment	Growth on Simmons citrate agar	Growth on inorganic nitrogen agar	Acetoin	Nitrate reduction	Oxidase	Glucose	Glycerol	Lysozyme susceptibility[c]	Arginine dihydrolase	β-Galactosidase	Growth at 37°C	Peptidoglycan type	Amino sugar in cell wall polysaccharide
M. luteus	Y>CW	–	–	+>±,–	–	–>+	+±	–	–	S	–	–	+	L-Lys-peptide subunit	Mannosamineuronic acid
M. lylae	CW, U	–	–	–	–>±	–>+	+,±	–	–	SR	–	–	+	L-Lys-Asp	Galactosamine
M. varians	Y	–	+>±,–	–>±	±,–	+>±	–±	+	–	R	–	–	+	L-Lys-L-Ala₃₋₄	Galactosamine
M. roseus	PR>OR	–	–>±	–	±>–	+>±	–±	+,±	–	SR-R	–	–	+	L-Lys-L-Ala₃₋₄	Galactosamine
M. agilis	R	–	ND	ND	–	–	+	–	–	R	–	+	–	L-Lys-Thr-L-Ala₃	Glucosamine
M. kristinae	PO	–	–	–	+	–>±	+,±	++	++	R	–	–+	+	L-Lys-L-Ala₃	Glucosamine
M. nishino miyaensis	O	–>±	–	±,–	–>±	+,±,–	+,±	–>±	–	SR-R	–	–	+	L-Lys-L-Ser₂-D-Glu	Galactosamine
M. sedentarius	CW>BY	+,±	–	–	–	–	–	–>±	–	S-SR	+	–	+	Uncertain	–
M. halobius	U	–	ND	ND	–	–	+	+	–	R	–	+	+	ND	ND

A single listed symbol denotes a character frequency of about 70–100%; the notation > denotes "a frequency greater than"; a comma between symbols denotes nearly equal frequency.
CW, cream-white; U, unpigmented; PR, pastel red; OR, orange red; R, red; PO, pale orange; O, orange; BY, buttercup yellow; Y, yellow.
S, susceptible (minimal inhibitory concentration, MIC: below 5 µg/ml); SR, slightly resistant (MIC: 5–50 µg/ml); R, resistant (MIC: above 100 µg/ml).
++, strong positive; +, positive; ±, weak; –, negative; ND, not determined.

Micrococcus agilis differs significantly from other micrococci in several features. It possesses flagella, is psychrophilic, and exhibits β-galactosidase activity. *M. kristinae* is a clearly separated species, too. It produces acid from glucose under anaerobic conditions. From glycerol aerobically, it forms acetoin and hydrolyses esculin. *M. kristinae* produces unique wrinkled growth on purple agar (Difco) containing 1% maltose. The orange-pigmented *M. nishinomiyaensis* may be further distinguished from *M. kristinae*, which forms very pale orange colonies, by growth on 7.5% NaCl agar, acetoin production, and esculin hydrolysis. *M. sedentarius* differs from other *Micrococcus* species by being resistant to penicillin and methicillin, producing often water-soluble exopigment, growing very slowly, and reacting positively to the arginine dihydrolase test. *M. halobius* can be easily separated from other species, as it requires at least 5% NaCl for growth.

Physiological and Biochemical Properties

Detailed studies on physiological and biochemical properties of micrococci are rather scarce. Most of these studies have been performed before a clear separation of micrococci from staphylococci was possible. Therefore, only those studies will be taken into consideration where a classification as a *Micrococcus* is guaranteed.

Growth Requirements

Chemically defined media for the growth of *M. luteus* and for the growth and the pigmentation of *M. roseus* have been devised (Wolin and Naylor, 1957; Grula et al., 1961; Cooney and Thierry, 1966). Strains of *M. luteus* may grow in a defined medium containing pyruvate or glutamate as carbon and energy source, biotin, and mineral salts (Perry and Evans, 1966; Salton, 1964). Initiation of growth of *M. luteus* in a defined medium depends upon the presence of an iron-binding compound, such as phenolic compounds (e.g., catechol) or ferrichrome (Salton, 1964; Walsh et al., 1971).

Metabolic Properties

Most micrococci are strictly aerobic. Carbon-containing compounds, oxidized to carbon dioxide and water, include acetate, lactate, pyruvate, succinate, fructose, galactose, glucose, glycerol, maltose, and sucrose. Variable oxidation occurs of mannitol, sorbitol, arabinose, rhamnose, ribose, xylose, and starch. Dulcitol is not oxidized (Saz and Krampitz, 1954; Rosypal and

Kocur, 1963; Perry and Evans, 1960, 1966). Glucose is metabolized by fructose-1, 6-biphosphate and hexose monophosphate pathways and citric acid enzymes (Dawes and Holmes, 1958; Perry and Evans, 1966; Blevins et al., 1969). They possess, like most eubacteria, a class II D-fructose-1,6-biphosphate adolase (Götz et al., 1979). However, strains belonging to *M. varians* and *M. kristinae* can also grow facultatively anaerobically and produce L-lactic acid from glucose. An NAD-dependent L-lactate dehydrogenase could be found in representatives of both species (Hartinger and Schleifer, unpublished observations). *M. luteus* possesses an NAD-dependent and an NAD-independent malic dehydrogenase which can oxidize malic acid to oxalacetic acid (Cohn, 1956). A bacterial tyramine oxidase which catalyzes the oxidation of tyramine and dopamine was for the first time isolated from *M. luteus* (Sarcina lutea) by Yamada et al. (1967). A phosphoglucomutase, which catalyzes the interconversion of glucose-1-phosphate and glucose-6-phosphate, was purified from *M. luteus* (Hanabusa et al., 1966). The metabolism of exogeneous pyrimidine bases and nucleosides was also investigated in *M. luteus* (Auling et al., 1982; Auling and Moss, 1984). The presence of thymidine kinase and phosphorylase could be demonstrated, whereas uridine phosphorylase occurred in a rather weak activity and uridine kinase was not detectable. In contrast to *E. coli*, the pyrimidine nucleotide phosphorylases are not inducible.

Membrane Composition

The cytoplasmic membrane composition, in particular that of *M. luteus*, has been thoroughly studied (Salton, 1987). The major fatty acid component of *Micrococcus* strains is a methyl-branched C_{15}-saturated acid (Girard, 1971; Onishi and Kamekura, 1972; Thirkell and Gray, 1974; Jantzen et al., 1974; Brooks et al., 1980). Micrococci contain relatively high amounts of long-chain aliphatic hydrocarbons in the range of C_{22} to C_{33} (Tornabene et al., 1970; Morrison et al., 1971; Kloos et al., 1974). Cardiolipin and phosphatidylglycerol are the major phospholipids. Unidentified phospholipids a and b were described by Komura et al. 1975. *Micrococcus* species partially contain hydrogenated menaquinones of the type MK-7, MK-8, and MK-9 (Jeffries, 1969; Jeffries et al., 1969; Yamada et al., 1967).

All studied micrococci contain cytochromes of *a*-, *b*-, *c*-, and *d*- types (Faller et al., 1980). Besides cytochromes, menaquinones, and carotenoids, NADH, malate, lactate, and succinate dehydrogenases are also part of the electron-transport system of *M. luteus* (Erickson and

Parker, 1969; Crowe and Owen, 1983a). The succinate dehydrogenase was purified and cross reactions with specific antisera directed to the membrane-bound enzyme of other micrococci could be shown, but no cross reaction occurred with the enzyme from a wide variety of Gram-positive and Gram-negative bacteria (Crowe and Owen, 1983b). The proton-translocating ATP synthase of *M. luteus* has been thoroughly studied (Salton and Schor, 1974; Schmitt et al., 1978).

It consists of the typical F_0F_1 complex. The cytoplasmic membrane is also the site of enzymes involved in phospholipid, peptidoglycan, and teichuronic biosynthesis (De Siervo and Salton, 1971; Park and Matsuhashi, 1984; Traxler et al., 1982). Membranes of *M. luteus* are unusually rich in mannose; much of this is bound in succinylated lipomannan (Owen and Salton, 1975). Part of the mannose is also found in membrane glycoproteins (Doherty et al., 1982). The occurrence of glycoproteins is a rather unusual feature for eubacteria.

Genetics Properties

DNA Relationships

Transformation studies by Kloos (1969a) and DNA-DNA hybridization studies by Ogasawara-Fujita and Sakaguchi (1976) showed that there is no close genetic or DNA relationship among members of the species *M. luteus*, *M. roseus*, and *M. varians*. Schleifer et al. (1979) demonstrated that DNA homology values between the type strains of *M. luteus* and *M. lylae* were in the range of 40 to 50% at optimal reassociation conditions, whereas values of only 10 to 18% were obtained between these species and *M. kristinae* or *M. varians*. Kloos et al. (1974) showed a significant genetic relationship between *M. luteus* and *M. lylae* in transformation studies. Genetics exchange occurred at about 1 to 5% of homologous values using DNA isolated from *M. lylae* with *M. luteus* auxotrophic recipients. An epigenetic relationship between *M. luteus* and *M. lylae* has been demonstrated by comparative immunological studies of catalase using both double-immunodiffusion tests and quantitative micro-complement fixation assays (Rupprecht and Schleifer, 1977). The nucleotide sequence of a 23S rRNA gene of *M. luteus* was determined (Regensburger et al., 1988a). From sequence comparison to 23S rRNA genes from other bacteria, specific probes could be constructed (Regensburger et al., 1988b). One probe (pAR28) reacted only with *M. luteus* and *M. lylae*, while another probe (pAR27) reacted with all micrococci and arthrobacters.

Transformation

Genetic exchange in the genus *Micrococcus* was first demonstrated by transformation of Fleming's *M. luteus* strain (formerly *M. lysodeikticus*) (Kloos, 1968; Kloos and Schultes, 1969; Mahler and Grossmann, 1968; Okubo and Nakayama, 1968). Following the determination of optimal conditions for transformation (Kloos, 1969b), a procedure for transformation was developed for use in genetic mapping (Kloos and Rose, 1970). Transformation has not been reported in other micrococcal species.

Transformation of *Micrococcus luteus* ATCC 27141 Auxotrophs (Kloos and Rose, 1970)

An 18-h P agar slant culture of the recipient strain was suspended in 1 ml saline and diluted 1/100 in saline. Aliquots of 0.1 ml [about 5×10^6 colony-forming units (CFU)] from the diluted suspension were added to duplicate tubes containing 1 ml defined broth supplemented with 20 µg/ml of the required metabolite (e.g., L-tryptophan, L-histidine, adenosine, etc.) (Kloos, 1969b). Cultures were shaken in a 34°C water bath with a Burrell Wrist-Action Shaker at a setting of 4 for 24 h. Following growth, cells (about $1-3 \times 10^8$ CFU/ml) were centrifuged and resuspended in 1 ml transformation buffer: 50 mM Tris (hydroxymethyl) aminomethane + 1-mM $SrCl_2$, pH 7.0. Donor DNA, 0.1 µg in crosses or 10 µg for the construction of double auxotrophic mutants for three-point crosses, in 0.1 ml saline was added to each tube and the mixtures were shaken in a 30°C water bath at a setting of 4 for 30 min. DNA uptake was terminated by the addition of deoxyribonuclease (5µg/ml) and 5 mM $MgSO_4$ with incubation at 37°C for 15 min. After treatment, cells were centrifuged and resuspended in 1 ml saline. Aliquots of 0.1 ml were taken from an appropriate saline dilution of the original suspension and spread on duplicate defined agar plates. Prototrophs were scored after incubation at 34°C for 72 h. Donor-type cotransformants (on appropriate supplemented defined agar plates) were scored after incubation for 96 h. Using the above procedure, transformation frequencies of 0.1–0.7 percent of recipient CFU could be obtained with prototroph donor DNA.

Competence in transformation has been demonstrated in *M. luteus* strains ATCC 27141 and its relatives (ATCC 15801, PU, UM, WRU) and strains ATCC 540 and ATCC 8673.

Chromosomal Gene Arrangements

Tryptophan and histidine biosynthesis genes have been mapped in *M. luteus* by using two-point "best-fit" and donor-types (contransformation) analyses and three-point transformation (Kane-Falce and Kloos, 1975; Kloos and Rose, 1970). These studies have indicated that at least two histidine genes are closely linked, probably contiguous, to the tryptophan gene cluster containing trpE, trpC, trpB, and trpA genes, in that order. Tryptophan genes are located in the main cluster mentioned above and at least in one other unlinked site containing trpD and trpF. Histidine genes are distributed in at least four regions of the chromosome. The largest cluster contains four genes: his (EAH or F), hisB, hisC, and his (EAH or F). Mapping purine biosynthesis genes by the two-point best-fit method has indicated significant separation of the purine genes purE, purJ, and purC and no linkage of these genes to purH (Mohapatra and Kloos, 1974, 1975).

Plasmid Composition

Plasmids have been detected in a small to moderate percentage (7–55%) of strains in most species of micrococci (Mathis and Kloos, 1984). They have not yet been reported in the species *M. lylae* and *M. sedentarius*. Most strains carrying plasmids exhibit only one or two types, ranging in size from 1 to 100SMDa. Plasmid patterns appear to be slightly more complex in *M. nishinomiyaensis*. Thirty-six percent of strains in this species carried two to three different plasmids. To date, most micrococcal plasmids remain cryptic. Several small plasmids identified in *M. nishinomiyaensis*, *M. roseus*, *M. varians*, and *M. luteus* have been studied with restriction endonucleases and may be suitable for use as cloning vectors following some modification.

Pathogenicity for Humans

The data about possible pathogenicity of micrococci for humans are very poor and controversial. Micrococci, however, may be considered as opportunistic pathogens, particularly in view of the increasing number of immunocompromised patients. Micrococci have been reported to be associated with various infections, especially those of the urinary tract (Meers, et al., 1975; Roberts, 1967; Sellin et al., 1975; Telander and Wallmark, 1975). However, these and similar reports cannot be evaluated because the discussed strains were either not sufficiently taxonomically described or incorrectly identified. As a result of the present classification, all the strains from such infections tested to date have been shown to be staphylococci.

Therefore, only the papers of the last 10 to 15 years describing human infections caused by micrococci can be taken into consideration. A case of septic shock caused by *Micrococcus luteus* was described by Albertson et al. (1978). *M. luteus* was reported in a case of cavitating pneumonia of an immunosuppressed patient (Souhami et al., 1979) and in a case of meningitidis (Fosse et al., 1986). *M. kristinae* has been isolated in pure culture from several types of infections in the U.S.A. *M. sedentarius* has been found to be associated with pitted keratolysis

(Nordstrom et al., 1987). Micrococci have been isolated from blood and surgical specimens of patients associated with heart diseases and septic complications following cardiac surgery. Some of the isolates were probably *M. lylae* strains. There is no clear proof that micrococci are pathogenic for nonimmunocompromised humans. Pathogenicity of micrococci for plants has not yet been described.

Application

Micrococci have been reported to be used in the processing of fermented meat products to improve their color, aroma, flavor, and keeping quality (Kitchell, 1962; Niinivaara and Pohja, 1957; Pohja, 1960). However, based on an updated classification of these organisms, most strains used in processing have proved to be staphylococci, in particular, *S. carnosus*, and occasionally *S. xylosus* (W.E. Kloos, unpublished observations). The true micrococci that have been useful in meat processing have proved to be *Micrococcus varians*. Studies by Fischer and Schleifer (1980) showed that besides *M. varians*, *M. kristinae* could also be isolated from fermented sausage. Species such as *Micrococcus luteus* and *M. varians* have been widely used in industries as assay organisms to test various antibiotics in body fluids, feeds, milk, and pharmaceutical preparations (Bowman, 1957; Coates and Argoudelis, 1971; Grove and Randall, 1955; Kirshbaum and Arret, 1959; Simon and Yin, 1970); the vitamin biotin (Aaronson, 1955); and the enzyme lysozyme (Dickman and Proctor, 1952).

Micrococci are among the few bacteria that synthetize long-chain (C_{21}–C_{34}) aliphatic hydrocarbons. These hydrocarbons may have potential economic value if they can be produced and extracted economically and then processed into useful lubricating oils or other petroleum substitutes (Kloos et al., 1974; Morrison et al. 1971; Tornabene et al., 1970).

Literature Cited

Aaronson, S. 1955. Biotin assay with a coccus, *Micrococcus sodonensis*, nov. sp. J. Bacteriol. 69:67–70.

Abd-el-Malek, Y., Gibson, T. 1948. Studies in the bacteriology of milk. II. The staphylococci and micrococci of milk. J. Dairy Res. 15:249–260.

Albertson, D., Natsions, G. A., Gleckman, R. 1978. Septic shock with *Micrococcus luteus*. Arch. Intern. Med. 138:487–488.

Anderson, J. I. W. 1962. Studies on micrococci isolated from the North Sea. J. Appl. Bacteriol. 25:362–368.

Auling, G., Moss, B. 1984. Metabolism of pyrimidine bases and nucleosides in the coryneform bacteria *Brevibacte-*

rium ammoniagenes and *Micrococcus luteus*. J. Bacteriol. 158:733–736.

Auling, G., Prelle, H., Diekmann, H. 1982. Incorporation of deoxyribonucleosides into DNA of coryneform bacteria and the relevance of deoxyribonucleoside kinases. Eur. J. Biochem. 121:365–370.

Back, W. 1980. Taxonomische Untersuchungen an beerschädlichen Bakterien. Habilitationschrift. Technische Universität. Munich.

Baird-Parker, A. C. 1962. The occurrence and enumeration, according to a new classification, of micrococci and staphylococci in bacon and on human and pig skin. J. Appl. Bacteriol. 25:352–361.

Blevins, W. T., Perry, J. J., Evans, J. B. 1969. Growth and macromolecular biosynthesis by *Micrococcus sodonensis* during the utilization of glucose and lactate. Can. J. Microbiol. 15:383–388.

Bouvet, P., Chatelain, R., Riou, J. Y. 1982. Intérêt du composé vibriostatique 0/129 pour différencier les genres *Staphylococcus* et *Micrococcus*. Ann. Inst. Pasteur 113 B:449–453.

Bowman, F. W. 1957. Test organisms for antibiotic microbial assays. Antibiot. Chemother. 7:639–640.

Brisou, J. 1955. La microbiologie de milieu marin. Editions Médicales Flammarion. Paris.

Brooks, B. W., Murray, R. G. E., Johnson, J. L., Stackebrandt, E., Woese, C. R., Fox, G. E. 1980. Red-pigmented micrococci: A basis for taxonomy. Int. J. Syst. Bacteriol. 30:627–646.

Carr, D. L., Kloos, W. E. 1977. Temporal study of the staphylococci and micrococci of normal infant skin. Appl. Environ. Microbiol. 34:673–680.

Casida, L. E. Jr. 1980a. Death of *Micrococcus luteus* in soil. Appl. Environ. Microbiol. 39:1031–1034.

Casida, L. E. Jr. 1980b. Bacterial predators of *Micrococcus luteus* in soil. Appl. Environ. Microbiol. 39:1035–1041.

Coates, J. H., Argoudelis, A. D. 1971. Microbial transformation of antibiotics: Phosphorylation of clindamycin by *Streptomyces coelicolor* Müller. J. Bacteriol. 108:459–464.

Cohn, D. V. 1956. The oxidation of malic acid by *Micrococcus lysodeikticus*. J. Biol. Chem. 221:413–420.

Collins, M. D., Jones, D. 1981. The distribution of isoprenoid quinone structural types in bacteria and their taxonomic implications. Microbiol. Rev. 45:316–354.

Cooney, J. J., Thierry, O. C. 1966. A defined medium for growth and pigment synthesis of *Micrococcus roseus*. Can. J. Microbiol. 12:83–89.

Crowe, B. A., Owen, P. 1983a. Immunochemical analysis of respiratory-chain components of *Micrococcus luteus (lysodeikticus)*. J. Bacteriol. 153:498–505.

Crowe, B. A., Owen, P. 1983b. Molecular properties of succinate dehydrogenase isolated from *Micrococcus luteus (lysodeikticus)*. J. Bacteriol. 153:1493–1501.

Curry, J. C., Borovian, G. E. 1976. Selective medium for distinguishing micrococci from staphylococci in the clinical laboratory. J. Clin. Microbiol. 4:455–457.

Dawes, E. A., Holmes, W. H. 1958. On the quantitative evaluation of routes of glucose metabolism by the use of radioactive glucose. Biophys. Acta 34:551–552.

De Siervo, A. J., Salton, M. R. J. 1971. Biosynthesis of cardiolipin in the membranes of *Micrococcus lysodeikticus*. Biochim. Biophys. Acta 239:280–292.

Dickman, S. R., Proctor, C. M. 1952. Factors affecting the activity of egg white lysozyme. Arch. Biochem. Biophys. 40:364–372.

Dickscheit, R. 1961. Beiträge zur Physiologie und Systematik der Pediokokken des Bieres. Zentralbl. Bakteriol. Parasitenkd. II. Abt 114:270–284.458–471.

Doeringer, R. H., Dugan, P. R. 1973. Growth relationship between the blue-green alga *Anacystis nidulans* and *Sarcina flava* in mixed culture. Abst. Ann. Meet. Am. Soc. Microbiol. 1973:45.

Doherty, H., Condon, C., Owen, P. 1982. Resolution and in vitro glycosylation of membrane glycoproteins in *Micrococcus luteus (lysodeikticus)*. FEMS Microbiol. Lett. 15:331–336.

Erickson, S. K., Parker, G. L. 1969. The electron-transport system of *Micrococcus luteus (Sarcina lutea)*. Biochim. Biophys. Acta 180:56–62.

Evans, J. B., Kloos, W. E. 1972. Use of shake cultures in a semisolid thioglycolate medium for differentiating staphylococci from micrococci. Appl. Microbiol. 23:326–331.

Evelyn, T. P. T., McDermott, L. A. 1961. Bacteriological studies of fresh-water fish. I. Isolation of aerobic bacteria from several species of Ontario fish. Can. J. Microbiol. 7:375–382.

Falk, D., Guering, S. J. 1983. Differentiation of *Staphylococcus* and *Micrococcus* spp. with the taxo A bacitracin disk. J. Clin. Microbiol. 18:719–721.

Faller, A., Götz, F., Schleifer, K. H. 1980. Cytochrome patterns of staphylococci and micrococci and their taxonomic implications. Zbl. Bakteriol. I. Abt. Orig. C1:26–39.

Faller, A., Schleifer, K. H. 1981. Modified oxidase and benzidine tests for separation of staphylococci from micrococci. J. Clin. Microbiol. 13:1031–1035.

Farrior, J. W., Kloos, W. E. 1975. Amino acid and vitamin requirements of *Micrococcus* species isolated from human skin. Int. J. Syst. Bacteriol. 25:80–82.

Farrior, J. W., Kloos, W. E. 1976. Sulfur amino acid auxotrophy in *Micrococcus* species isolated from human skin. Can. J. Microbiol. 22:1680–1690.

Fischer, S., Luczak, H., Schleifer, K. H. 1982. Improved methods for the detection of class I and class II fructose-1, 6-biphosphate aldolases in bacteria. FEMS Microbiol. Lett. 15:103–108.

Fischer, U., Schleifer, K. H. 1980. Zum Verkommen der Gram-positiven, katalase-positiven Kokken in Rohwurst. Fleischwirtschaft. 60:1046–1051.

Fosse, T., Peloux, Y., Granthil, C., Toga, B., Bertrando, J., Sethian, M. 1986. Meningitis due to *Micrococcus luteus*. Eur. J. Clin. Study Treat. Infect. 13:280–281.

Gillespie, N. C., Macrae, I. C. 1975. The bacterial flora of some Queensland fish and its ability to cause spoilage. J. Appl. Bacteriol. 39:91–100.

Girard, A. E. 1971. A comparative study of the fatty acids of some micrococci. Can. J. Microbiol. 17:1503–1508.

Glass, M. 1973. *Sarcina* species on the skin of the human forearm. Trans. St. John's Hosp. Dermatol. Soc. 59:56–60.

Götz, F., Nürnberger, E., Schleifer, K. H. 1979. Distribution of class I and class II D-fructose-1, 6-biphosphate aldolase in various Gram-positive bacteria. FEMS Microbiol. Lett. 5:253–257.

Grove, D. C., Randall, W. A. 1955. Assay Methods of Antibiotics. Medical Encyclopedia. New York. Antibiotics Monographs:2.

Grula, E. A., Luk, S.-K., Chu, Y.-C. 1961. Chemically defined medium for growth of *Micrococcus lysodeikticus*. Can. J. Microbiol. 7:27–32.

Hanabusa, K., Dougherty, H. W., Del Rio, C., Hashimoto, T., Handler, P. 1966. Phosphoglucomutase. II. Preparation and properties of phosphoglucomutases from *Micrococcus lysodeikticus* and *Bacillus cereus*. J. Biol. Chem. 241:3930–3939.

Horsley, R. W. 1977. A review of the bacterial flora of teleosts and elasmobranchs, including methods for its analysis. J. Fish. Biol. 10:529–553.

Jantzen, E., Bergan, T., Bøvre, K. 1974. Gas chromatography of bacterial whole cell methanolysates. VI. Fatty acid composition of strains within *Micrococcaceae*. Acta Pathol. Microbiol. Scand. Sec. B 82:785–798.

Jeffries, L. 1968. Sensitivity to novobiocin and lysozyme in the classification of *Micrococcaceae*. J. Appl. Bacteriol. 31:436–442.

Jeffries, L. 1969. Menaquinone in the classification of *Micrococcaceae* with observations on the application of lysozyme and novobiocin sensitivity tests. Int. J. Syst. Bacteriol. 19:183–187.

Jeffries, L., Cawthorne, M. A., Harris, M., Cook, B., Diplock, A. T. 1969. Menaquinone determination in the taxonomy of *Micrococcaceae*. J. Gen. Microbiol. 54:365–380.

Kane, C. M., Kloos, W. E. 1970. Transformation of *Sarcina flava* and *Micrococcus flavocyaneus*. Genet. Res., Camb. 15:339–343.

Kane-Falce, C. M., Kloos, W. E. 1975. A genetic and biochemical study of histidine biosynthesis in *Micrococcus luteus*. Genetics 79:361–376.

Keddie, R. M. 1974. *Arthrobacter*. 618–625. R. E. Buchanan and N. E. Gibbons (ed.) Bergey's manual of determinative bacteriology, 8th ed. Williams and Wilkins. Baltimore.

Kirshbaum, A., Arret, B. 1959. Outline of details for assaying the commonly used antibiotics. Ant. Chem. 9:613–617.

Kirsop, B. E., Snell, J. J. S. (ed.). 1984. Maintenance of microorganisms. Academic Press. London.

Kitchell, A. G. 1962. Micrococci and coagulase negative staphylococci in cured meats and meat products. J. Appl. Bacteriol. 25:416–431.

Klesius, P. H., Schuhardt, V. T. 1968. Use of lysostaphin in the isolation of highly polymerized deoxyribonucleic acid and in the taxonomy of aerobic *Micrococcaceae*. J. Bacteriol. 95:739–743.

Kloos, W. E. 1968. Evidence of genetic exchange in *Micrococcus lysodeikticus*. Bacteriol. Proc. 1968:55.

Kloos, W. E. 1969a. Transformation of *Micrococcus lysodeikticus* by various members of the family *Micrococcaceae*. J. Gen. Microbiol. 59:247–255.

Kloos, W. E. 1969b. Factors affecting transformation of *Micrococcus lysodeikticus*. J. Bacteriol 98:1397–1399.

Kloos, W. E., Musselwhite, M. S. 1975. Distribution and persistence of *Staphylococcus* and *Micrococcus* species and other aerobic bacteria on human skin. Appl. Microbiol. 30:381–395.

Kloos, W. E., Rose, N. E. 1970. Transformation mapping of tryptophan loci in *Micrococcus luteus*. Genetics 66:595–605.

Kloos, W. E., Schultes, L. M. 1969. Transformation in *Micrococcus lysodeikticus*. J. Gen. Microbiol. 55:307–317.

Kloos, W. E., Tornabene, T. G., Schleifer, K. H. 1974. Isolation and characterization of micrococci from human skin, including two new species: *Micrococcus lylae* and *Micrococcus kristinae*. Int. J. Syst. Bacteriol. 24:79–101.

Kloos, W. E., Zimmerman, R. J., Smith, R. F. 1976. Preliminary studies on the characterization and distribution of *Staphylococcus* and *Micrococcus* species on animal skin. Appl. Environ. Microbiol. 31:53–59.

Kocur, M., Bergan, T., Mortensen, N. 1971. DNA base composition of Gram-positive cocci. J. Gen. Microbiol. 69:167–183.

Komura, I., Yamada, K., Komagata, K. 1975. Taxonomic significance of phospholipid composition in aerobic Gram-positive cocci. J. Gen. Appl. Microbiol. 21:97–107.

Leifson., E. 1963. Determination of carbohydrate metabolism of marine bacteria. J. Bacteriol. 85:1183–1184.

Mahler, I., Grossman, L. 1968. Transformation of radiation sensitive strains of *Micrococcus lysodeikticus*. Biochem. Biophys. Res. Commun. 32:776–781.

Marples, M. J. 1965. The ecology of the human skin. Charles C. Thomas. Springfield, Illinois.

Mathis, J. N., Kloos, W. E. 1984. Isolation and characterization of *Micrococcus* plasmids. Curr. Microbiol. 10:163–171.

Meers, P. D., Whyte, W., Sandys, G. 1975. Coagulase-negative staphylococci and micrococci in urinary tract infections. J. Clin. Pathol. 28:270–273.

Mohapatra, N., Kloos, W. E. 1974. Biochemical and genetic studies of laboratory purine auxotrophic strains *Micrococcus luteus*. Can. J. Microbiol. 20:1751–1754.

Mohapatra, N., Kloos, W. E. 1975. Biochemical characterization and genetic mapping of purine genes in *Micrococcus luteus*. Genet. Res., Camb. 26:163–171.

Morrison, S. J., Tornabene, T. G., Kloos, W. E. 1971. Neutral lipids in the study of the relationships of members of the family *Micrococcaceae*. J. Bacteriol. 108:353–358.

Naylor, H. B., Burgi, E. 1956. Observations on abortive infections of *Micrococcus lysodeikticus* with bacteriophage. Virology 2:577–593.

Niinivaara, F. P., Pohja, M. S. 1957. Erfahrungen über die Herstellung von Rohwurst mittels einer Bakterienreinkultur. Fleischwirtschaft 9:789–790.

Noble, W. C., Somerville, D. A. 1974. Microbiology of human skin. W. B. Saunders. London.

Nordstrom, K. M., McGinley, K. J., Zechman, J. M., Leyden, J. J. 1987. Similarities between *Dermatophilus congolensis* and *Micrococcus sedentarius*: Identity of the etiologic agent of pitted keratolysis. Abstr. Ann. Meet. Amer. Soc. Microbiol. 244.

Ogasawara-Fujita, N., Sakahuchi, K. 1976. Classification of micrococci on the basis of deoxyribonucleic acid homology. J. Gen. Microbiol. 94:97–106.

Okubo, S., Nakayama, H. 1968. Evidence of transformation in *Micrococcus lysodeikticus*. Biochem. Biophys. Res. Commun. 32:825–830.

Onishi, H., Kamekura, H. 1972. *Micrococcus halobius* sp. n. Int. J. Syst. Bacteriol. 22:233–236.

Owen, P., Salton, M. R. J. 1975. A succinylated mannan in the membrane system of *Micrococcus lysodeikticus*. Biochem. Biophys. Res. Comm. 63:875–880.

Pachtman, E. A., Vicher, E. E., Brunner, M. J. 1954. The bacteriologic flora in seborrhoeic dermatitis. J. Invest. Dermatol. 22:389–397.

Park, W., Matsuhashi, M. 1984. *Staphylococcus aureus* and *Micrococcus luteus* peptidoglycan transglycosylases that are not penicillin-binding proteins. J. Bacteriol. 157:538–544.

Perry, J. J., Evans, J. B. 1960. Oxidative metabolism of lactate and acetate by *Micrococcus sodonensis*. J. Bacteriol. 79:113–118.

Perry, J., Evans, J. B. 1966. Oxidation and assimilation of carbohydrates by *Micrococcus sodonensis*. J. Bacteriol. 91:33–38.

Pohja, M. S. 1960. Micrococci in fermented meat products. Classification and description of 171 different strains. Acta Agralia Fennica. 96:1–80.

Regensburger, A., Ludwig, W., Frank, R., Blöcker, H., Schleifer, K.-H. 1988a. Complete nucleotide sequence of a 23S ribosomal RNA gene from *Micrococcus luteus*. Nucl. Acids, Res. 16:2344.

Regensburger, A., Ludwig, W., Schleifer, K.-H. 1988b. DNA probes with different specificities from a cloned 23S rRNA gene of *Micrococcus luteus*. J. Gen. Microbiol. 134:1197–1204.

Rheinbaben, K. E. v., Hadlok, R. M. 1981. Rapid distinction between micrococci and staphylococci with furazolidone agars. Antonie van Leeuwenhoek 47:41–51.

Roberts, A. P. 1967. *Micrococcaceae* from the urinary tract in pregnancy. J. Clin. Pathol. 20:631–632.

Rosypal, S., Kocur, M. 1963. The taxonomic significance of the oxidation of carbon compounds by different strains of *Micrococcus luteus*. Antonie van Leeuwenhoek 29:313–318.

Rupprecht, M., Schleifer, K. H. 1977. Comparative immunological study of catalases in the genus *Micrococcus*. Arch. Microbiol. 114:61–66.

Salton, M. R. J. 1964. Requirements of dehydroxyphenols for the growth of *Micrococcus lysodeikticus* in synthetic media. Biochim. Biophys. Acta 86:421–422.

Salton, M. R. J. 1987. Bacterial membrane proteins. Microbiol. Sciences 4:100–105.

Salton, M. R. J., Schor, M. T. 1974. Release and purification of *Micrococcus lysodeikticus* ATPase from membranes extracted with n-butanol. Biochim. Biophys. Acta 345:74–82.

Saz, H. J., Krampitz, L. O. 1954. The oxidation of acetate by *Micrococcus lysodeikticus*. J. Bacteriol. 67:409–418.

Schleifer, K. H. 1973. Chemical composition of staphylococcal cell walls. 13–23. Jeljaszewicz (ed.) Staphylococci and staphylococcal infections. Recent progress. S. Karger. Basel.

Schleifer, K. H. 1986. Section 12. Gram-positive cocci. 999–1003.P. H. A. Sneath, N. S. Mair, M. E. Sharpe, J. G. Holt (ed.) Bergey's manual of systematic bacteriology, vol. 2. Williams and Wilkins. Baltimore.

Schleifer, K. H., Heise, W., Meyer, S. A. 1979. Deoxyribonucleic acid hybridization studies among soms micrococci. FEMS Lett. 6:33–36.

Schleifer, K. H., Kloos, W. E. 1975. A simple test system for the separation of staphylococci from micrococci. J. Clin. Microbiol. 1:337.

Schleifer, K. H., Kloos, W. E., Kocur, M. 1981. The genus *Micrococcus*. 1539–1547. M. P. Starr, H. Stolp, H. G. Trüper, A. Balows and H. C. Schlegel (ed.) The prokaryotes: a handbook on habitats, isolation and identification of bacteria. Springer-Verlag. Berlin.

Schleifer, K. H., Krämer, E. 1980. Selective medium for isolating staphylococci. Zentralbl. Bakteriol. Mikrobiol. Hyg. Abt. I Orig. C1:270–280.

Schmitt, M., Rittinghaus, K., Scheurich, P., Schwulera, U., Dose, K. 1978. Immunological properties of membrane-bound adenosine triphosphatiase. Biochim. Biophys. Acta 509:410–418.

Sellin, M. A., Cooke, D. I., Gillespie, W. A., Sylvester, D. G. H., Anderson, J. D. 1975. Micrococcal urinary tract infections in young women. Lancet ii:570–572.

Simon, H. J., Yin, E. J. 1970. Microbioassay of antimicrobial agents. Appl. Microbiol. 19:573–579.

Smith, R. F. 1970. Comparative enumeration of lipophilic and nonlipophilic cutaneous diphtheroids and cocci. Appl. Microbiol. 19:254–258.

Souhami, L., Feld, R., Tuffnell, P. G., Feller, T. 1979. *Micrococcus luteus* pneumonia: A case report and review of the literature. Med. Pediatr. Oncol. 7:309–314.

Sreenivasan, A. 1959. A note on the bacteriology of prawns and their preservation by freezing. J. Sci. Ind. Res. 18C:119.

Stackebrandt, E., Woese, C. R. 1979. A phylogenetic dissection of the family *Micrococcaceae*. Curr. Microbiol. 2:317–322.

Streby-Andrews, M. E., Kloos, W. E. 1971. Amino acid auxotrophy in natural strains of *Micrococcus luteus*. Bacteriol. Proc. 1971:27.

Telander, B., Wallmark, G. 1975. *Micrococcus* subgroup 3-a common cause of acute urinary tract infection in women. Lakartidningen 72:1967.

Thirkell, D., Gray, E. M. 1974. Variation in the lipid fatty acid composition in purified membrane fractions from *Sarcina aurantiaca* in relation to growth phase. Antonie van Leeuwenhoek. J. Microbiol. Serol. 40:71–78.

Tornabene, T. G., Morrison, S. J., Kloos, W. E. 1970. Aliphatic hydrocarbon contents of various members of the family *Micrococcaceae*. Lipids 5:929–937.

Traxler, C. I., Goustin, A. S., Anderson, J. S. 1982. Elongation of teichuronic acid chains by a wall-membrane preparation from *Micrococcus luteus*. J. Bacteriol. 150:649–656.

Venkataraman, R., Sreenivasan, A. 1953. The bacteriology of freshwater fish. Ind. J. Med. Res. 41:385–392.

Venkataraman, R., Sreenivasan, A. 1955. Bacterial flora of fresh shark. Curr. Sci. 11:380–381.

Walsh, B. L., O'Dor, J., Warren, R. A. J. 1971. Chelating agents and the growth of *Micrococcus lysodeikticus*. Can. J. Microbiol. 17:593–597.

Williams, O. B., Rees, H. B., Campbell, L. L. 1952. The bacteriology of Gulf Coast shrimp. I. Experimental procedures and quantitative results. Texas J. Sci. 4:49–52.

Williamson, P. 1965. Quantitative estimation of cutaneous bacteria. 3–11.Maibach, H. I., and Hildick-Smith, G. (ed.) Skin bacteria and their role in infection. McGraw-Hill. New York.

Williamson, P., Kligman, A. M. 1965. A new method for the quantitative investigation of cutaneous bacteria. J. Invest. Dermatol. 45:498–503.

Wolin, H. L., Naylor, H. B. 1957. Basic nutritional requirements of *Micrococcus lysodeikticus*. J. Bacteriol. 74:163–167.

Wood, E. J. F. 1952. The micrococci in marine environments. J. Gen. Microbiol. 6:205.

Yamada, H., Uwajima, T., Kumagai, H., Watanabe, M., Ogata, K. 1967. Crystalline tyramine oxidase from *Sarcina lutea*. Biochem. Biophys. Res. Commun. 27:350–355.

ZoBell, C. E., Upham, C. 1944. A list of marine bacteria including descriptions of sixty new species. Bulletin of Scripps Institute of Oceanography, University of California. 5:239–292.

Prokaryotes (2006) 3:972–974
DOI: 10.1007/0-387-30743-5_38

CHAPTER 1.1.23

Renibacterium

HANS-JÜRGEN BUSSE

Renibacterium salmoninarum is the causative agent of Bacterial Kidney Disease (BKD), a chronic and systemic infection of salmonid fish such as Pacific and Atlantic salmon, grayling (*Thymallus thymallus*), lake trout (*Salvelinus namaycush*), and several other species of trout and char. However, *R. salmoninarum* has also been isolated from several non-salmonid fish species that did not show any signs of the disease (Traxler and Bell, 1988). It is one of the most serious diseases in salmonid farming in the Atlantic and Pacific region (Margolis and Evelyn, 1987; Fryer and Lannan, 1993; Olea et al., 1993). The disease occurs mainly in freshwater, but significant mortality also occurs in saltwater (Banner et al., 1983).

The systemic disease, which is a slowly developing chronic infection, derives its name from off-white bacterial lesions in the kidney. BKD is histopathologically characterized as a systemic diffuse granulomatous inflammation. In later stages many organs may become affected and the body cavity may become filled with fluid. Severely infected fish may show no obvious clinical symptoms or pale gills (indicative of anemia), exophthalmia, abdominal distension, skin blisters filled with clear or turbid fluid, shallow ulcers, hemorrhages and, infrequently, intramuscular cavities (Wood and Yasutake, 1956; Fryer and Sanders, 1981; Bruno, 1986). The bacterium is transmitted horizontally, probably by a fecal-oral route (Balfry et al., 1996), and vertically in the egg (Evelyn et al., 1984; Lee and Gordon, 1987). *Renibacterium salmoninarum* occurs extracellularly and intracellularly (Bell, 1961) and survives and multiplies inside fish macrophages (Young and Chapman, 1978).

Therapeutic treatment of diseased fish with antibiotics such as clindamycin, erythromycin, kitasamycin, penicillin G and spiramycin is only of limited use but effective prophylactically (Austin, 1985).

The genus *Renibacterium* with the single species *Renibacterium salmoninarum* was originally described by Sanders and Fryer (1980) and later the description was emended by Goodfellow et al. (1985). Phylogenetically, *Renibacterium salmoninarum* is next but only moderately related to the genera *Arthrobacter* and *Micrococcus* (Stackebrandt et al., 1988; Grayson et al., 2000a).

Cells of this species are fastidious, slowly growing short rods often occurring in pairs that stain strongly Gram positive. For cultivation, KDM-2 medium is recommended (DSMZ, 2001) and the presence of L-cysteine has been demonstrated to support growth (Daly and Stevenson, 1993). Optimal growth occurs between 15°C and 18°C. All strains studied by Goodfellow et al. (1985) are positive for catalase, phosphatase, acid phosphatase, esterase (C8), α-glucosidase, trypsinase, and degradation of casein, tributyrin and Tween 20, and they are sensitive to carbenicillin and cephaloridine treatment. The quinone system consists mainly of menaquinone MK-9 and minor amounts of MK-10. Present in the complex polar lipid profile is diphosphatidyl glycerol, two major and six or seven uncharacterized glycolipids, and two unknown minor phospholipids. In the fatty acid profile, the major components are $C_{15:0}$ *anteiso*, $C_{15:0}$ *iso* and $C_{17:0}$ *anteiso* (Embley et al., 1983). Major cell wall sugar is glucose; arabinose, mannose and rhamnose are also present. The peptidoglycan type is A3α consisting of alanine, glutamic acid, glycine and the diamino acid lysine (Kusser and Fiedler, 1983). Electron microscopic studies have demonstrated the presence of a capsule in the outer layer (Dubreuil et al., 1990). The galactose-rich polysaccharide (GPS) consists of a heptasaccharide repeating unit composed of D-galactose, D-rhamnose, *N*-acetyl-D-glucosamine and *N*-acetyl-L-fucosamine in a molecular ratio of 5.4 : 1.1 : 1.4 : 1.1 (Sørum et al., 1998; Eq. 1). GPS was found to constitute 60–70% of the dry weight of the protein-free cell wall preparations (Fiedler and Draxl, 1986). The G+C content of genomic DNA is 55.5 mol% (Banner et al., 1991). The structure of the heptasaccharide repeating unit (Sørum et al., 1998) is

$$
\begin{array}{c}
\alpha\text{-D-Rha}p\text{-}(1{\rightarrow}3)\text{-}\alpha\text{-L-Fuc}p\text{NAc-}(1{\rightarrow}3) \\
\text{-}\beta\text{-D-Glc}p\text{NAc} \\
1 \\
\downarrow \\
2 \\
{\rightarrow}3)\text{-}\beta\text{-D-Gal}f\text{-}(1{\rightarrow}6)\text{-}\beta\text{-D-Gal}f\text{-}(1{\rightarrow}3) \\
\text{-}\beta\text{-D-Gal}f\text{-}(1{\rightarrow}6)\text{-}\beta\text{-D-Gal}f\text{-}(1{\rightarrow}
\end{array}
\qquad (1)
$$

The cell surface has been shown to be hydrophobic and to have hemagglutinating properties associated with immunomodulation and virulence (Daly and Stevenson, 1987; Bruno, 1988; Wood and Kaattari, 1996). The hemagglutinin was demonstrated to be identical with the soluble heat stable virulence factor p57 (Daly and Stevenson, 1990). The 57-kDa protein, which also is designated "major soluble antigen (MSA), is predominantly localized on the bacterial cell surface, with significant levels released to the environment. Two copies of the MSA encoding gene (*msa*1 and *msa*2) are found in *R. salmoninarum* and both genes are expressed under in vitro conditions (O'Farrell and Strom, 1999; Rhodes et al., 2002). Another antigenic group has been identified which is characterized by a major 30-kDa protein (Bandín et al., 1992).

MSA is the target antigen in several immunological diagnostic assays and PCR-based methods have been developed to detect specific fragments of the MSA encoding gene or the corresponding mRNA (McIntosh et al., 1996; Chase and Pascho, 1998; Cook and Lynch, 1999). Another PCR-approach has been developed to specifically detect *R. salmoninarum* using reverse transcription and nested PCR amplification of hypervariable regions within the 16S rRNA sequence (Magnusson et al., 1994).

The genome of *R. salmoninarum* has been demonstrated to harbor two rRNA operons, which are designated "*rrnA*" and "*rrnB*" (Grayson et al., 2000a). The intergenic spacer between 16S-23S rRNA coding genes (ITS1) was found to consist of 534 bp and to be highly conserved, showing not more than 4 bp difference among different isolates (Grayson et al., 1999; Grayson et al., 2000b). The length of the 23S-5S rRNA intergenic spacer is conserved as well, consisting of 219 bp, and only a small variation exists in isolates derived from areas of the world relatively isolated from the mainstream intensive salmonid cultivation (Grayson et al., 2000a).

KDM-2 Medium (DSMZ-Medium 435; DSMZ, 2001)

Peptone	10.0 g
Yeast extract	0.5 g
L-Cysteine · HCl × H₂O have a look in the Online catalogue of strains from DSMZ!	1.0 g
Calf serum	100.0 ml
Distilled water	900.0 ml
Adjust pH to 6.5.	

Literature Cited

Austin, B. 1985. Evaluation of antimicrobial compounds for the control of bacterial kidney disease in rainbow trout (Salmo gairdneri) Richardson. J. Fish Dis. 8:209–220.

Balfry, S. K., L. J. Albright, and T. P. T. Evelyn. 1996. Horizontal transfer of Renibacterium salmoninarum among farmed salmonids via the fecal-oral route. Dis. Aquat. Org. 25:1–2.

Bandín, I., Y. Santos, B. Margarinos, J. L. Barja, and A. E. Toranzo. 1992. The detection of two antigenic groups among Renibacterium salmoninarum isolates. FEMS Microbiol. Lett. 94:105–110.

Banner, C. R., J. S. Rohovec, and J. L. Fryer. 1983. Renibacterium salmoninarum as a cause of mortality among chinook salmon in salt water. J. World Maricult. Soc. 14:236–239.

Banner, C. R., J. S. Rohovec, and J. L. Fryer. 1991. A new value for mol percent guanine+cytosine of DNA for the salmonid fish pathogen Renibacterium salmoninarum. FEMS Microbiol. Lett. 79:57–59.

Bell, G. R. 1961. Two epidimics of apparent kidney disease in cultured pink salmon (Oncorhynchus gorbuscha). J. Fish. Res. Board Canada 18:562–559.

Bruno, D. W. 1986. Histopathology of bacterial kidney disease in laboratory infected rainbow trout, Salmo gairdneri Richardson, and Atlantic salmon, Salmo salar L., with reference to naturally infected fish. J. Fish Dis. 9:523–537.

Bruno, D. W. 1988. The relationship between autoagglutination, cell surface hydrophobicity and virulence of the fish pathogen Renibacterium salmoninarum. FEMS Microbiol. Lett. 51:135–140.

Chase, D. M., and R. J. Pascho. 1998. Development of a nested polymerase chain reaction for amplification of a sequence of the p57 gene of Renibacterium salmoninarum that provides a highly sensitive method for detection of the bacterium in salmonid kidney. Dis. Aquat. Org. 34:223–229.

Cook, M., and W. H. Lynch. 1999. A sensitive nested reverse transcriptase PCR assay to detect viable cells of the fish pathogen Renibacterium salmoninarum in Atlantic salmon (Salmo salar L.). Appl. Environ. Microbiol. 65:3042–3047.

Daly, J. G., and R. M. Stevenson. 1987. Hydrophobic and haemagglutinating properties of Renibacterium salmoninarum. J. Gen. Microbiol. 133:3575–3580.

Daly, J. G., and R. M. Stevenson. 1990. Characterization of the Renibacterium salmoninarum haemagglutinin. J. Gen. Microbiol. 136:949–953.

Daly, J. G., and R. M. Stevenson. 1993. Nutrient requirements of Renibacterium salmoninarum on agar and in broth media. Appl. Environ. Microbiol. 59:2178–2183.

Dubreuil, D., R. Lallier, and M. Jacques. 1990. Immunoelectron microscopic demonstration that Renibacterium salmoninarum is encapsulated. FEMS Microbiol. Lett. 66:313–316.

Embley, T. M., M. Goodfellow, D. E. Minnikin, and B. Austin. 1983. Fatty acid, isoprenoid quinone and polar lipid composition in the classification of Renibacterium salmoninarum. J. Appl. Bacteriol. 55:31–37.

Evelyn, T. P. T., J. E. Ketcheson, and L. Prosperi-Porta. 1984. Further evidence for the presence of Renibacterium salmoninarum in salmonid eggs and for the failure of povidine-iodine to reduce the intra-ovum infection in water-hardened eggs. J. Fish Dis. 7:173–182.

Fiedler, F., and R. Draxl. 1986. Biochemical and immunochemical properties of the cell surface of Renibacterium salmoninarum. J. Bacteriol. 168:799–804.

Fryer, J. L., and J. E. Sanders. 1981. Bacterial kidney disease of salmonid fish. Ann. Rev. Microbiol. 35:273–298.

Fryer, J. L., and C. N. Lannan. 1993. The history and current status of Renibacterium salmoninarum, the causative agent of bacterial kidney disease in Pacific salmon. Fish. Res. 17:15–33.

Goodfellow, M., T. M. Embley, and B. Austin. 1985. Numerical taxonomy and emended description of Renibacterium salmoninarum. J. Gen. Microbiol. 131:2739–2752.

Grayson, T. H., L. F. Cooper, F. A. Atienzar, M. R. Knowles, and M. L. Gilpin. 1999. Molecular differentiation of Renibacterium salmoninarum isolates from worldwide locations. Appl. Environ. Microbiol. 65:961–968.

Grayson, T. H., S. M. Alexander, L. F. Cooper, and M. L. Gilpin. 2000a. Renibacterium salmoninarum isolates from different sources possess two highly conserved copies of the rRNA operon. Ant. v. Leeuwenhoek 78:51–61.

Grayson, T. H., F. A. Atienzar, S. M. Alexander, L. F. Cooper, and M. L. Gilpin. 2000b. Molecular diversity of Renibacterium salmoninarum isolates determined by randomly amplified polymorphic DNA analysis. Appl. Environ. Microbiol. 66:435–438.

Kusser, W., and F. Fiedler. 1983. Murein type and polysaccharide composition of cell walls from Renibacterium salmoninarum. FEMS Microbiol. Lett. 20:391–394.

Lee, E. G. H., and M. R. Gordon. 1987. Immunofluorescence screening of Renibacterium salmoninarum in the tissues and eggs of farmed chinook salmon spawners. Aquaculture 65:7–14.

Magnusson, H. B., O. H. Fridjonsson, O. S. Andresson, E. Benediktsdottir, S. Gudmundsdottir, and V. Andresdottir. 1994. Renibacterium salmoninarum, the causative agent of bacterial kidney disease in salmonid fish, detected by nested reverse transcription-PCR of 16S rRNA sequences. Appl. Environ. Microbiol. 60:4580–4583.

Margolis, L., and T. P. T. Evelyn. 1987. Aspects of diseases and parasite problems in cultured salmonids in Canada, with emphasis on the Pacific region, and regulatory measures for their control. In: A. Stenmark and G. Malmberg (Eds.) Parasites and Diseases in Natural Waters and Aquaculture in Nordic Countries. Swedish Museum of Natural History. Stockholm, Sweden. 2–19.

McIntosh, D., P. G. Meaden, and B. Austin. 1996. A simplified PCR-based method for the detection of Renibacterium salmoninarum utilizing preparations of rainbow trout (Occorhynchus mykiss, Walbaum) lymphocytes. Appl. Environ. Microbiol. 62:3929–3932.

O'Farrell, C. L., and M. S. Strom. 1999. Differential expression of the virulence-associated protein p57 and characterization its duplicated gene msa in virulent and attenuated strains of Renibacterium salmoninarum. Dis. Aquat. Org. 38:115–123.

Olea, I., D. W. Bruno, and T. S. Hastings. 1993. Detection of Renibacterium salmoninarum in naturally infected Atlantic salmon, Salmo salar L., and rainbow trout, Oncorhynchus mykiss (Walbaum) using an enzyme-linked immunosorbent assay. Aquaculture 116:99–110.

Rhodes, L. D., A. M. Coady, and M. S. Strom. 2002. Expression of duplicate msa genes in the salmonid pathogen Renibacterium salmoninarum. Appl. Environ. Microbiol. 68:5480–5487.

Sanders, J. E., and J. L. Fryer. 1980. Renibacterium salomoninarum gen. nov., sp. nov., the causative agent of bacterial kidney disease in salmonid fishes. Int. J. Syst. Bacteriol. 40:496–502.

Sørum, U., B. Robertsen, and L. Kenne. 1998. Structural studies of the major polysaccharide in the cell wall of Renibacterium salmoninarum. Carbohydr. Res. 306:305–314.

Stackebrandt, E., U. Wehmeyer, H. Nader, and F. Fiedler. 1988. Phylogenetic relationship of the fish pathogenic Renibacterium salmoninarum to Arthrobacter, Micrococcus and related taxa. FEMS Microbiol. 50:117–120.

Traxler, G., and G. Bell. 1988. Pathogens associated with impounded Pacific herring Clupea harengus pallasi, with emphasis on viral erythrocytic necrosis (VEN) and atypical Aeromonas salmonicida. Dis. Aquat. Org. 5:93–100.

Wood, E. M., and W. T. Yasutake. 1956. Histopathology of kidney disease in fish. Am. J. Pathol. 32:845–857.

Wood, P. A., and S. L. Kaattari. 1996. Enhanced immunogenicity of Renibacterium salmoninarum in chinook salmon after removal of the bacterial cell surface-associated 57 kDa protein. Dis. Aquat. Org. 25:71–79.

Young, C. L., and G. B. Chapman. 1978. Ultrastructural aspects of the causative agent and renal histopathology of bacterial kidney disease in brook trout (Salvelinus fontinalis). J. Fish. Res. Board Canada 35:1234–1248.

Prokaryotes (2006) 3:975–982
DOI: 10.1007/0-387-30743-5_39

CHAPTER 1.1.24

The Genus *Stomatococcus*: *Rothia mucilaginosa*, basonym *Stomatococcus mucilaginosus*

ERKO STACKEBRANDT

The species *Stomatococcus mucilaginosus* (Bergan and Kocur, 1982) was recently assigned to the genus *Rothia* as *Rothia mucilaginosa* (Collins et al., 2000). A literature search in Medline indicated that the name change to *Rothia mucilaginosa* has been used in only a single publication (Kazor et al., 2003); as the majority of mostly clinical reports still refer to the basonym *Stomatococcus mucilaginosus*, this name will be used until the revised classification has been fully adopted by medical microbiologists. Electronic publication will facilitate fast updates of taxonomic rearrangements and subsequent changes.

The rationale for the reclassification of *Stomatococcus mucilaginosus* was based mainly on the topology of the 16S rRNA gene tree and some common phenotypic differences between members of *Rothia* and *S. mucilaginosus*. As a consequence of the classification of a mouse isolate as *Rothia nasimurium* type strain CCUG 35957^T (Collins at al., 2000), *Stomatococcus mucilaginosus* branched equidistantly between *R. dentocariosa* and *R. nasimurium* and was therefore reclassified as a member of *Rothia*. Recently, *Rothia amarae* JCM 11375^T was described (Fan et al., 2002), showing a remote relatedness to type strains of the other *Rothia* species. Though formally correct, the authors (Collins et al., 2000; Fan et al., 2002) failed to provide clear-cut evidence that the reclassification of *Stomatococcus* and the affiliation of strains JCM 11375^T and CCUG 35957^T are justified. The past years have shown the high correlation between phylogenetic clustering and distribution of chemotaxonomic properties, and the delineation of groups of actinobacteria at the generic level was often based on their chemotaxonomic coherency. Studies leading to the description and reclassification of the new *Rothia* species have provided information that is either lacking or only fragmentary. If available, these characters are often not as coherent as expected for members of the same genus. The need for a comparative study on chemotaxonomic and phenotypic properties, including all type strains, is therefore obvious.

Taxonomy

The genus *Stomatococcus*, with *S. mucilaginosus* as its only species, was first described after significant phylogenetic and biochemical differences to otherwise morphologically similar organisms were detected (Bergan and Kocur, 1982). *Stomatococcus mucilaginosus* ("*Staphylococcus salivarius*," Andrewes and Gordon, 1907, or "*Micrococcus mucilaginosus*" Bergan et al., 1970) is defined to contain organisms which are Gram-positive, nonmotile, nonsporeforming, encapsulated, spherical cells and which show either a weakly positive or a negative catalase reaction. Cells are arranged mostly in clusters, and occasionally in pairs or tetrads. Originally, following the tradition of defining higher taxa by morphology, *Stomatococcus*—together with the genera *Micrococcus*, *Staphylococcus* and *Planococcus*—was described to constitute the family Micrococcaceae (Bergan and Kocur, 1982; Schleifer, 1986), although a dissection of this family on phylogenetic grounds had already been proposed by Stackebrandt and Woese (1979). Chemical and phylogenetic evidence showed the species *S. mucilaginosus* is constituted by one line of descent within the order Actinomycetales and thus the genus was included in the emended family Micrococcaceae (Stackebrandt et al., 1997), which also included *Micrococcus*, *Arthrobacter*, *Renibacterium*, *Rothia*, *Kocuria* and *Nesterenkonia*. *Rothia dentocariosa* and *Stomatococcus mucilaginosus* were branching adjacent to each other, having a 16S rRNA gene sequence similarity of 97.7%. *Rothia nasimurium* and *R. amarae* are slightly less closely related to these two species and among themselves have a 95.5–96.8% similarity. Members of *Kocuria* are the phylogenetically nearest neighbors (Fig. 1). All members of *Rothia* and *Stomatococcus* possess the family Micrococcaceae-specific 16S

Table 1. Presence of signature nucleotides of the 16S rRNA gene of *Stomatoccus mucilaginosus* [*Rothia mucilaginosa*], *Rothia* and *Kocuria* species.

Position (*E. coli* numbering)	*S. mucilaginosus, R. dentocariosa,* and "*Rothia aerius*"	*Rothia nasimurium*	*Rothia amarae*	*Kocuria* species
80–89	U-A	C-G	G-C	C-G, G-C
81–88	A-U	A-U	U-A	A-U, A-U
134	A	G	G	G
184–193	U-G	U-G	A-U	A-U
199–218	C-G	G-C	G-C	G-C
438–496	G.G	U-A	U-A	U-A
440–494	U-G	C-G	C-G	C-G
437–497	U-A	U-G	U-G	U-G
611	G	C	A	C
1010–1079	U-G	C-G	U-G	C-G
1251	G	A	A	A
1273	G	U	U	G
1285	U	A	A	A

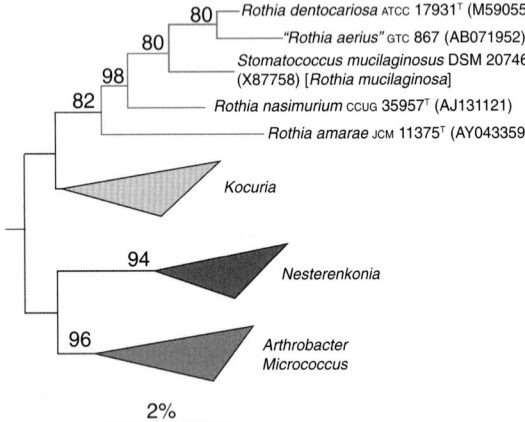

Fig. 1. Phylogenetic tree generated on the basis of almost complete 16S rRNA gene sequences, showing the relationship of *Stomatococcus mucilaginosus* [*Rothia mucilaginosa*] to members of the genus *Rothia* and related genera. The tree was generated by the additive treeing algorithm of De Soete (1983) using the Jukes and Cantor (1969) correction. Bootstrap value was calculated for 500 replicate trees. The bar corresponds to a 2% difference in nucleotide sequences.

rRNA gene signature nucleotides (Stackebrandt et al., 1997). Analysis of the aligned sequences for the presence of species-specific signature nucleotides (Table 1) confirms the phylogenetic branching as shown in Fig. 1, i.e., the isolated position of *S. nasimurium* and *S. amarae* as well as the clustering of *Rothia dentocariosa, S. mucilaginosus,* and the as yet undescribed species "*Rothia aerius*" (Y. Kawamura and T. Ezaki, unpublished observation).

The phenotype of members of *Stomatococcus* and *Rothia* is similar and hardly discriminatory enough to differentiate between species. All members are positive, or at least some strains of a species are positive, for the following characteristics: facultatively anaerobic, nonmotile, acid

production from glucose and trehalose, esculin hydrolysis, and nitrate reduction. All members are negative for the following characteristics: acid production from mannitol and production of sorbitol, urease, and alkaline phosphatase. Other reactions, most of which have not been determined for *R. nasimurium,* are indicated in Table 2. Additional tests for *S. mucilaginosus* and *Rothia* spp. (Gerencser and Bowden, 1986; Collins et al., 2000; Fan et al., 2002) are given in their species description.

Stomatococcus mucilaginosus strains differ from members of *Micrococcus* and catalase-positive staphylococci by failure to grow on media containing 5% NaCl (Bergan and Kocur, 1982; McWhinney et al., 1992; Guermazi et al., 1995). Strains of *Rothia dentocariosa* differ in their ability to grow in 4% NaCl, and none grows in 6% NaCl.

Various commercial systems have been applied to identify and characterize *S. mucilaginosus* strains, including the Biolog identification system (Biolog. Inc., Hayward, CA; Miller et al., 1993), Staph-ident system (bioMerieux Vitek, Hazelwood, MO; Rhoden and Miller, 1995) and API 50 CH system (bioMerieux, Marcy l'Etoile, France; van Tiel, 1995). When compared to conventional identification, the accuracy of the Staph-ident system was 35.3% only, and 41% of strains required an additional test for identification.

Habitats

Stomatococci are normal inhabitants of the mouth and the upper respiratory tract of humans (Bergan et al., 1970), comprising about 3.5% of the predominant cultivable aerobic microflora of the human oral cavity. *Stomatococcus mucilaginosus* is indigenous to the human tongue and pharynx, but it has also been isolated from the nasopharynx, bronchial secretions, and blood (Gordon, 1967; Bergan et al., 1970, Rubin et al., 1978; Pinsky et al., 1989) and from smears of human dental plaque (Bowden, 1969). It may also be among the early colonizers of human teeth (Nyvad and Kilian, 1987). The microbial population has been studied by a culture-independent approach (Kazor et al., 2003) and *S. mucilaginosus* was found among those species most associated with healthy subjects.

Before 1990, it was generally believed that *S. mucilaginosus* was an unlikely pathogen because of its presence as a member of the indigenous microflora of the upper respiratory tract in humans. However, the increasing number of reports on the presence of *S. mucilaginosus* in clinical cases could not be ignored. From 1990 on, the literature treats *S. mucilaginosus* as an opportunistic pathogen (Ascher et al., 1991a; Ascher et al., 1991b). McWhinney et al. (1992) even considered this organism as an important pathogen in severely immunocompromised pediatric patients. Clinicians and microbiologists were asked to be increasingly aware of this pathogen (Kaufhold et al., 1992).

Stomatococcus mucilaginosus has been isolated from patients with severe pericoronitis of mandibular third molars and was found in 71% of patient samples (Peltroche-Llacsahuanga et al., 2000). When caries-reducing sucrose substitutes, e.g., palatinose or leucrose, were metabolized in a batch fermenter, *S. mucilaginosus* fermented these substitutes to produce a water-insoluble polysaccharide. This polysaccharide may be cleaved by the autochthonous flora to produce cariogenic products (glucose and fructose) that another population of the dental flora can use to produce acids (Peltroche-Llacsahuanga et al., 1991).

Stomatococcus mucilaginosus has also been reported in populations of the conjunctival flora of rabbits (8% of specimens; Cooper et al., 2001) and in sickly six-week-old broiler chickens (4% of specimens; Awan and Matsumoto, 1998).

Swab samples from the human tongue or other parts of the oral cavity are inoculated onto blood agar or trypticase soy agar and incubated at 30–37°C. Colonies are usually mucoid, transparent, or whitish and adherent to the agar surface.

Media and Cultivation

Under aerobic conditions, stomatoccoci show good growth within 24–30 h at 30°C on complex media such as brain heart infusion agar (Difco 0418) or casein peptone (tryptic digest)-glucose-yeast extract-NaCl (1:0.5:0.5:0.5%, w/v) broth, pH 7.2–7.4. No growth occurs in media prepared with 5% NaCl or 40% bile (Baird-Parker, 1974).

Preservation

Stock cultures are routinely maintained on agar slants using one of the media indicated above

Table 2. Morphological and phenotypic differences between type strains of *Stomatoccus mucilaginosus* [*Rothia mucilaginosa*] and *Rothia* species.

Properties	*Stomatococcusmuciloaginosus* DSM 20746[T]	*Rothia dentocariosa* ATCC 17931[T]	*Rothia nasimurium* CCUG 35957[T]	*Rothia amarae* JCM 11375[T]
Morphology	Cocci, mostly in clusters, occasionally in pairs and tetrads	Coccoid, diphtheroid, or filamentous	Ovoid cocci	Cocci, single cells, pairs, tetrads, and/or packets
Capsule	+	−	n.d.	n.d.
Catalase	Weak or −	+	+	+
Acid production				
Sucrose	+	+	n.d.	+
Glycerol	d	d[2]	n.d.	+
Maltose	d	+	n.d.	+
Mannose	+	d	n.d.	+
Ribose	n.d.	d	−	+
Salicin	+	+	n.d.	+
Lactose	−	−	+	−
Raffinose	−	−[a]	−	−
Gelatin hydrolysis	+	d	n.d.	+

Symbols: +, positive; and −, negative.
Abbreviations: [T], type strain; CCUG, Culture Collection, University of Göteborg, Dept. of Clinical Bacteriology, Göteborg, Sweden; JCM, Japan Collection of Microorganisms, The Institute of Physical and Chemical Research, Hirosawa, Wako-shi, Japan; n.d., not determined, and d, 11–89% of strains are positive.
[a]See comments of Gerencser and Bowden (1986) in footnote of Table 15.29.

and stored at 4°C. Cultures are transferred to fresh medium every 4 weeks. Liquid cultures can be frozen at –12°C and survive several months. Long-term preservation is done by lyophilization or storage in liquid nitrogen. No information on the viability has been reported.

Identification

The morphological and biochemical tests used for staphylococci and micrococci are recommended (Bergan and Kocur, 1986). Cells are nonmotile and spherical, arranged mostly in clusters, occasionally in pairs or tetrads. A copious capsule is formed (Silva et al., 1977). Oxidase and benzidine tests are negative (Schleifer, 1986). The catalase reaction is negative in about one-half of the strains investigated (Gordon, 1967; Bergan and Kocur, 1986). Cells are resistant to lysostaphin (Schleifer and Kloos, 1975). A variety of sugars is fermented anaerobically. Acid but no gas is produced, although no information on end products is available. *Stomatococcus mucilaginosus* can easily be differentiated from other Gram-positive bacteria when a combination of commercially available identification systems and conventional biochemical tests is combined with adherence to agar surfaces, poor growth on Mueller-Hinton agar, and the presence of a capsule (Kaufhold et al., 1992).

Stomatococcus mucilaginosus is susceptible to micrococcal bacteriophages, most of which are also specifically active on *Micrococcus luteus* (Bauske et al., 1978). *Stomatococcus*-specific phages have as yet not been found. Cytochrome patterns (Stackebrandt et al., 1983) and fatty-acid and polar lipid patterns (Jantzen et al., 1974; Amadi et al., 1988) distinguish stomatococci from micrococci and staphylococci (Lennarz and Talamo, 1966; Faller et al., 1980; Nahaie et al., 1984) but not clearly from other members of *Rothia* (see Table 3). With respect to the presence of unsaturated menaquinones, *S. mucilaginosus* resembles staphylococci and *M. luteus*, but not the other *Micrococcus* species (Amadi et al., 1988); information on the menaqinone composition of *Rothia* species lacks coherency (Table 3). The peptidoglycan type differs among strains of *S. mucilaginosus*, in that the interpeptide bridge consists of L-alanine, D-serine, or glycine (Schleifer and Kandler, 1972; Stackebrandt et al., 1983; Table 3). This type is also found in other members of Micrococcineae, i.e., *Rothia*, *Arthrobacter*, *Kocuria* and *Renibacterium* (see the chapter Introduction to the Taxonomy of Actinobacteria in this Volume). The G+C content of DNA is slightly higher in *S. mucilaginosus* than that of *Rothia* species (Table 3). Despite the variation in peptidoglycan types and the G+C con-

tent of 56–60.5 mol%, all 10 strains of *S. mucilaginosus* included in DNA hybridization studies revealed such a high degree of relatedness (more than 75%) that their allocation to a single species seems justified. The 16S rRNA gene sequence of the type strain contains some *Stomatococcus mucilaginosus*-specific nucleotides (16S rRNA gene positions 139-224: U-A; 453-479: A-U; 467: G; 631: A; 591-648: U-A; 678-713: U-A; 1009-1020: A-U), so that new isolates can be allocated unambiguously to this species. Restriction fragment length polymorphism has been applied to isolates from neutropenic patients with hematological malignancies. These strains were found to be of diverse genetic origin (Granlund et al., 1996).

The type strain of *Stomatococcus mucilaginosus* [*Rothia mucilaginosa*] is CCM 2417^T (=DSM 20746^T = ATCC 25256^T = NCTC 10663^T).

Pathogenicity

Strains of *S. mucilaginosus* have been isolated from the blood cultures of patients with endocarditis, in patients who were intravenous drug abusers (Coudron et al., 1987; Relman et al., 1987), and in patients with pre-existing valvular heart disease (Rubin et al., 1978; Prag et al., 1985), bacteremia (Barlow et al., 1986), and recurrent peritonitis during chronic ambulatory peritoneal dialysis (Ragnaud et al., 1981). Both immunocompetent and immunosuppressed patients are affected. From 1990 on, the number of cases increased in which the organism was isolated from blood, the oral region, and wound sources causing bacteremia and septicemia that, among other factors, developed after endocarditis, surgery, bone marrow transplantation (Cobo et al., 1998), and from infected devices such as indwelling vascular catheters (Mitchell et al., 1990; Ascher et al., 1991a; Kaufhold et al., 1992). More alarming was the finding that *S. mucilaginosus* was also recovered from lower tract infection in a patient with AIDS (Cunniffe et al., 1994), patients with meningitis (Ben Salah et al., 1994; Guermazi et al., 1995; Abraham et al., 1997; Park et al., 1997; Skogen et al., 2001), acute leukemia (Mustafa et al., 1993; Granlund et al., 1996; Trevino et al. 1998; Paci et al., 2000), neutropenia (Oppenheim et al., 1989; Henwick et al., 1993; Gruson et al., 1998), osteomyelitis (Nielsen, 1994), choroid plexus infection (Guermazi et al., 1995), postoperative spondylodiscitis (Bureau-Chalot et al., 2003), and central nervous system infections (Goldman et al., 1998).

All but one isolate from neutropenic patients were sensitive to benzylpenicillin and all were sensitive to vancomycin (McWhinney et al., 1992). In another study, the minimal inhibitory

Table 3. Chemotaxonomic characteristics of *Stomatococcus mucilaginosus* and species of *Rothia.*

Species	*Stomatococcus mucilaginosus* DSM 20746^T	*Rothia dentocariosa* ATCC 17931^T	*Rothia nasimurium* CCUG 35957^T	*Rothia amarae* JCM 11375^T	*Kocuria*	*Nestorenkonia*
Peptidoglycan	L-Lys-L-Ala, or L-Lys-L-Gly(Ser), or L-Lys-Gly[a]	L-Lys-L-Ala$_3$	Lys-Ser-Ala$_2$[b]	Lys-Ala[c]	L-Lys-Ala$_{3-4}$	L-Lys-Gly- L-Glu
Major menaquinone	MK-7	MK-7	MK-8 and MK-7[b]	MK-7 and MK-6(H$_2$)	MK-7(H$_2$) and MK-8(H$_2$)	MK-8 and MK-9
Polar lipids	DPG and PG	DPG and PG	DPG, PG, PI[b]	n.d.	DPG, PG, (PI, PL, and GL)	DPG, PG, PI, PL, and GL
Predominant cellular fatty acid(s)	ai-C$_{15:0}$, i-C$_{16:0}$, ai-C$_{14:0}$, and C$_{16:0}$	ai-C$_{15:0}$, ai-C$_{17:2}$, and C$_{14:0}$	ai-C$_{15:0}$, i-C$_{16:0}$DPG, and PG	ai-C$_{15:0}$ and ai-C$_{17:0}$	ai-C$_{15:0}$ and i-C$_{16:0}$	ai-C$_{15:0}$, ai-C$_{17:0}$, and i-C$_{16:0}$
Mol% G+C	56–60	47–53	56	54.5	66–75	70–72

Abbreviations: MK, menaquinone, MK-6, -7, -8 and -9, menaquinones with 6–9 isoprene side chains, respectively; MK-6(H$_2$), -7(H$_2$), and -8(H$_2$), the same as MK-6, -7 and -8 (but have two hydrogens added to the side chain double bonds); DPG, diphosphatidylglycerol; PG, phosphatidylglycerol; PI, phosphatidylinositol; PL, unidentified phospholipid; GL, unidentified glycolipid, and n.d. not determined.

[a]The peptidoglycan type is indicated as L-Lys-L-Ser-L-Ala by Schleifer and Kandler (1977).
[b]Unpublished data, kindly provided by Peter Schumann.
[c]Fan et al. (2002) report the peptidoglycan type L-Lys-Glu-Ala in the species description, but the presence of lysine and alanine in the text. The code A3*a* refers to the later type.
From Gerencser and Bowden (1986), Collins et al. (2000), Fan et al. (2002) and the chapter Introduction to the Taxonomy of Actinobacteria.

concentration (MIC) at which 55% and 90% of *S. mucilaginosus* isolates from healthy and neutropenic skin and gingiva were respectively inhibited was determined for a broad range of antibiotics (von Eiff et al., 1995). Rifampicin was the most active compound, but ampicillin, imipenem, cefotaxime and clarithromycin were similarly effective. Aminoglycosides were less effective than ß-lactam antibiotics, and only streptomycin could be applied successfully in the study of Chomarat et al. (1991). Cunniffe et al. (1994) reported that an *S. mucilaginosus* isolate was sensitive to rifampicin but resistant to penicillin, co-trimoxazole and ciprofloxaxin. Strains isolated from neutropenic patients who had received ciprofloxacin as part of their prophylactic regime were more resistant to this antibiotic than strains isolated from non-neutropenic patients who had not received the antibiotic (van Tiel et al., 1995). A similar effect was reported by Granlund et al. (1996). Bureau-Chalot et al. (2003) treated *S. mucilaginosus*-caused spondylodiscitis by intravenous treatment with cefotaxime and fosfomycin, followed by oral doses of rifampicin and pristinamycin. Abraham et al. (1997) successfully treated meningitis with a combination of intravenous penicillin G and chloramphenicol and intrathecal vancomycin. Vancomycin was also effective in the study of Granlund et al. (1996). Apparently, different isolates identified as being members of *S. mucilaginosus* react differently to different doses and antibiotics, but *S. mucilaginosus* bacteremia is readily treatable with antibiotics (Ascher et al., 1991a; Ascher et al., 1991b).

Literature Cited

Abraham, J., S. Bilgrami, D. Dorsky, R. L. Edwards, J. Feingold, D. R. Hill, and P. J. Tutschka. 1997. Stomatococcus mucilaginosus meningitis in a patient with multiple myeloma following autologous stem cell transplantation. Bone Marrow Transplant. 19:639–641.

Amadi, E. N., G. Alderson, and D. E. Minnikin. 1988. Lipids in the classification of the genus Stomatococcus. Syst. Appl. Microbiol. 10:111–115.

Andrewes, F. W., and M. H. Gordon. 1907. Report on the biological characters of the staphylococci pathogenic for man. Ann. Rep. Med. Offr. Loc. Govt. 35:543–560.

Ascher, D. P., M. C. Bash, C. Zbick, and C. White. 1991a. Stomatococcus mucilaginosus catheter-related infection in an adolescent with osteosarcoma. South. Med. J. 84:409–410.

Ascher, D. P., C. Zbick, C. White, and G. B. Fischer. 1991b. Infections due to Stomatococcus mucilaginosus: 10 cases and review. Rev. Infect. Dis. 13:1048–1052.

Awan, M. A., and M. Matsumoto. 1998. Heterogeneity of staphylococci and other bacteria isolated from six-week-old boiler chicken. Poult. Sci. 77:944–949.

Baird-Parker, A. C. 1974. Staphylococcus Rosenbach. *In:* R. E. Buchanan and N. E. Gibbons (Eds.) Bergey's Manual of Determinative Bacteriology, 8th ed. Williams and Wilkins. Baltimore, MD. 483–489.

Barlow, J. F., K. A. Vogele, and P. F. Dzintars. 1986. Septicemia with Stomatococcus mucilaginosus. Clin. Microbiol. Newsl. 8:22.

Bauske, R., G. Peters, and G. Pulverer. 1978. Activity spectrum of micrococcal and staphylococcal phages. Zbl. Bakt. Hyg. Abt. Orig. A. 241:24–29.

Ben Salah, H., F. Cockaert, J. Levy, A. Ferster, and C. Devalck. 1994. Stomatococcus mucilaginosus meningitis in immunocompromised child. Arch. Pediatr. 1:813–815.

Bergan, T., K. Bøvre, and B. Hovig. 1970. Priority of Micrococcus mucilaginosus Migula 1900 over Staphylococcus salivarius Andrewes and Gordon 1907 with proposal of a neotype strain. Int. J. Syst. Bacteriol. 20:107–113.

Bergan, T., and M. Kocur. 1982. Stomatococcus mucilaginosus gen. nov. spec. nov., ep. rev., a member of the family Micrococcaceae. Int. J. Syst. Bacteriol. 32:374–377.

Bergan, T., and M. Kocur. 1986. Stomatoccoccus Bergan and Kocur. *In:* P. H. A. Sneath, N. S. Mair, M. E. Sharpe, and J. G. Holt (Eds.) Bergey's Manual of Systematic Bacteriology. Williams and Wilkins. Baltimore, MD. 2:1008–1010.

Bowden, G. H. 1969. The components of cells walls and extracellular slime of four strains of Staphylococcus salivarius isolated from human dental plaque. Arch. Oral. Biol. 14:685–697.

Bureau-Chalot, F., E. Piednoir, A. Bazin, L. Brasme, and O. Bajolet. 2003. Postoperative spondylodiskitis due to Stomatococcus mucilaginosus in an immunocompetent patient. Scand. J. Infect. Dis. 35:146–147.

Chomarat, M., M. G. Vital, and J. P. Flandrois. 1991. Susceptibility to aminoglycosides of 63 strains of Stomatococcus mucilaginosus isolated from sputum. Zentralbl. Bakteriol. 276:63–67.

Cobo, F., J. A. Garcia, M. Jurado, J. C. Alados, and C. Mirando. 1998. Bacteremia caused by Stomatococcus mucilaginosus in a bone marrow transplantation patient. Enferm. Infecc. Microbiol. Clin. 16:150–151.

Collins, M. D., R. A. Hutson, V. Baverud, and E. Falsen. 2000. Characterization of a Rothia-like organism from a mouse: Description of Rothia nasimurium sp. nov. and reclassification of Stomatococcus mucilaginosus as Rothia mucilaginosa comb. nov. Int. J. Syst. Evol. Microbiol. 50:1247–1251.

Cooper, S. C., G. J. McLellan, and A. N. Rycroft. 2001. Conjunctival flora observed in 70 healthy domestic rabbits (Oryctolagus cuniculus). Vet. Rec. 149:232–235.

Coudron, P. E., S. Markowitz, L. B. Mohanty, P. F. Schatzki, and J. M. Payne. 1987. Isolation of Stomatococcus mucilaginosus from drug user with endocarditis. J. Clin. Microbiol. 25:1359–1363.

Cunniffe, J. G., C. Mallia, and P. A. Alcock. 1994. Stomatococcus mucilaginosus lower respiratory tract infection in a patient with AIDS. J. Infect. 29:327–330.

De Soete, G. 1983. A least square algorithm for fitting additive trees to proximity data. Psychometrika 48:621–626.

Faller, A. H., F. Götz;, and K.-H. Schleifer. 1980. Cytochrome patterns of staphylococci and micrococci and their taxonomic implications. Zbl. Bakt. Hyg. I. Abt. Orig. C1:26–39.

Fan, Y., Z. Jin, J. Tong, W. Li, M. Pascial, A. Gamian, Z. Liu, and Y. Huang. 2002. Rothia amarae sp. nov., from sludge of a foul water sewer. Int. J. Syst. Evol. Microbiol. 52:2257–2260.

Gerencser, M. A., and G. H. Bowden. 1986. Rothia. *In:* P. H. A. Sneath, N. S. Mair, M. E. Sharpe, and J. G. Holt (Eds.) Bergey's Manual of Systematic Bacteriology. Williams and Wilkins. Baltimore, MD. 2:1342–1346.

Goldman, M., U. B. Chaudhary, A. Greist, and C. A. Fausel. 1998. Central nervous system infections due to Stomatococcus mucilaginosus in immunocompromised hosts. Clin. Infect. Dis. 27:1241–1246.

Gordon, D. F. 1 967. Reisolation of Staphylococcus salivarius from the human oral cavity. J. Bacteriol. 94:1281–1286.

Granlund, M., M. Linderholm, M. Norgren, C. Olofsson, A. Wahlin, and S. E. Holm. 1996. Stomatococcus mucilaginosus ssepticemia in leucemic patients. Clin. Microbiol. Infect. 2:179–185.

Gruson, D., G. Hilbert, A. Pigneux, F. Vargas, O. Guisset, J. Texier, J. M. Boiron, J. Reiffers, G. Gbikpi-Benissan, and J. P. Cardinaud. 1998. Severe infection caused by Stomatococcus mucilaginosus in a netropenic patient: Case report and review of the literature. Hematol. Cell Ther. 40:167–169.

Guermazi, A., Y. Miaux, and M. Laval-Jeantet. 1995. Imaging of choroid plexus infection by Stomatococcus mucilaginosus in neutropenic patients. Am. J. Neuroradiol. 16:1331–1334.

Henwick, S., M. Koehler, and C. C. Patrick. 1993. Complications of bacteremia due to Stomatococcus mucilaginosus in neutropenic children. Clin. Infect. Dis. 17:667–671.

Jantzen, E., T. Bergan, and K. Bovre. 1974. Gas chromatography of bacterial whole cell methanolysates. VI: Fatty acid composition of strains within Micrococcaceae. Acta Path. Microbiol. Scand Sect. B 82:785–798.

Jukes, T. H., and C. R. Cantor. 1969. Evolution of protein molecules. *In:* H. N. Munro (Ed.) Mammalian Protein Metabolism. Academic Press. New York, NY. 21–132.

Kaufhold, A., R. R. Reinert, and W. Kern. 1992. Bacteremia caused by Stomatococcus mucilaginosus: Report of seven cases and review of the literature. Infection 20:213–220.

Kazor, C. E., P. M. Mitchell, A. M. Lee, L. N. Stokes, W. J. Loesche, F. E. Dewhirst, and B. J. Paster. 2003. Diversity of bacterial populations on the tongue dorsa of patients with halitosis and healthy patients. J. Clin. Microbiol. 41:558–563.

Lennarz, W. J., and B. Talamo. 1966. The chemical characterization and enzymatic synthesis of mannolipids in Micrococcus lysodeikticus. J. Biol. Chem. 241:2702–2719.

Magee, J. T., I. A. Burnett, J. M. Hindmarch, and R. C. Spencer. 1990. Micrococcus and Stomatococcus spp. from human infections. J. Hosp. Infect. 16:67–73.

McWhinney, P. H., C. C. Kibbler, S. H. Gillespie, S. Patel, D. Morrison, A. V. Hoffbrand, and H. G. Prentice. 1992. Stomatococcus mucilaginosus: An emerging pathogen in neutropenic patients. Clin. Infect. Dis. 14:641–646.

Michell, P. S., B. J. Huston, R. N. Jones, L. Holcomb, and F. P. Koontz. 1990. Stomatococcus mucilaginosus bacteremias: Typical case presentations, simplified diagnostic criteria, and a literature review. Diagn. Microbiol. Infect. Dis. 13:521–525.

Miller, J. M., J. W. Biddle, V. K. Quenzer, and J. C. McLaughlin. 1993. Evaluation of Biolog for identificstion of members of the family Micrococcaeae. J. Clin. Microbiol. 31:3170–173.

Mustafa, M. M., L. R. Carlson, and K. Krisher. 1993. Stomatococcus mucilaginosus fatal sepsis in a child with leukaemia. Pediatr. Infect. Dis. J. 12:784–786.

Nahaie, M. R., M. Goodfellow, D. E. Minnikin, and V. Hajek. 1984. Polar lipid and isoprenoid quinone composition in the classification of Staphylococcus. J. Gen. Microbiol. 130:2427–2437.

Nielsen, H. 1994. Vertebral osteomyelitis with Stomatococcus mucilaginosus. Eur. J. Clin. Microbiol. Infect. Dis. 13:775–776.

Nyvad, B., and M. Kilian. 1987. Microbiology of the early colonization of human enamel and root surfaces in vivo. Scand. J. Dent. Res. 95:369–380.

Oppenheim, B. A., N. C. Weightman, and J. Prendeville. 1989. Fatal Stomatococcus mucilaginosus septicaemia in a neutropenic patient. Eur. J. Clin. Microbiol. Infect. Dis. 8:1004–1005.

Paci, C., R. Fanci, C. Casini, P. Pecile, and P. Nicoletti. 2000. Treatment of Stomatococcus mucilaginosus bloodstream infection in two acute leukemia patients, first reported at our cancer center. J. Chemother. 12:536–537.

Park, M. K., J. Khan, F. Stock, and D. R. Lucey. 1997. Successful treatment of Stomatococcus mucilaginosus meningitis with intravenous vancomycin and intravenous ceftriaxone. Clin. Infect. Dis. 24:278.

Peltroche-Llacsahuanga, H., C. J. Hauck, R. Koch, F.Lampert, R. Lutticken, and G. Haagse. 1991. Assessment of acid production by various human oral microorganisms when palatinose or leucrose is utilized. J. Dent. Res. 80:378–384.

Peltroche-Llacsahuanga, H., E. Reichhart, W. Schmitt, R. Lutticken, and G. Haagse. 2000. Investigation of infectious organisms causing pericoronitis of the mandibular third molar. J. Oral Maxillofac. Surg. 58: 611–616.

Pinsky, R. L., V. Piscitelli, and J. E. Petterson. 1989. Endocarditis caused by relatively penicillin-resistant Stomatococcus mucilaginosus. J. Clin. Microbiol. 27: 215–216.

Prag, J., E. Kjoller, and F. Espersen. 1985. Stomatococcus mucilaginosus endocarditis. Eur. J. Clin. Microbiol. 4:422–424.

Ragnaud, J.-M., C. Marceau, M. B. Roche-Bezain, and C. Wone. 1981. Péritonité à rechute à Stomatococcus mucilaginosus chez une malade traitée par dialyse peritonéale continué ambulatoire. Press. Méd. 14:2063.

Relman, D. A., K. Rouff, and M. J. Ferraro. 1987. Stomatococcus mucilaginosus endocarditis in an intravenous drug abuser. J. Infect. Dis. 155:1080–1082.

Rhoden, D. L., and J. M. Miller. 1995. Four-year prospective study of STAPH-IDENT system and conventional method for reference identification of Staphylococcus, Stomatococcus, and Micrococcus spp. J. Clin. Microbiol. 33:96–98.

Rubin, S. J., R. W. Lyons, and A. J. Murcia. 1978. Endocarditis associated with cardial catheterization due to a Gram-positive coccus designated Micrococcus mucilaginosus incertae sedis. J. Clin. Microbiol. 7:546–549.

Schleifer, K.-H., and O. Kandler. 1972. Peptidoglycan types of bacterial cell walls and their taxonomic implications. Bacteriol. Rev. 36:407–477.

Schleifer, K.-H., and W. E. Kloos. 1975. A simple test for the separation of staphylococci from micrococci. J. Clin. Microbiol. 1:337–338.

Schaal, K. P. 1991. The genera Actinomyces, Arcanobacterium and Rothia. The Prokaryotes *In:* E. Balows, H. G. Trüper, M. Dworkin, W. Harder, K.-H. Schleifer, (ed.) Springer-Verlag. New York, NY. USA. 850–905.

Schleifer, K.-H. 1986. Micrococcaceae. *In:* P. H. A. Sneath, N. S. Mair, M. E. Sharpe, and J. G. Holt (Eds.) Bergey's Manual of Systematic Bacteriology. Williams and Wilkins. Baltimore, MD. 2:1003–1035.

Silva, M. T., J. J. Polonia, and M. Kocur. 1977. The fine structure of Micrococcus mucilaginosus. J. Submicrosc. Cytol. 9:53–66.

Skogen, P. G., S. Kolmannskog, and K. Bergh. 2001. Bactericidal activity in cerebrospinal fluid by treating meningitis caused by Stomatococcus mucilaginosus with rifampicin, cefotaxime and vancomycin in a neutropenic child. Clin. Microbiol. Infect. 7:39–42.

Stackebrandt, E., C. R. Woese. 1979. A phylogenetic dissection of the family Micrococcaceae. Curr. Microbiol. 2:317–322.

Stackebrandt, E., C. Scheuerlein, and K.-H. Schleifer. 1983. Phylogenetic and biochemical studies on Stomatococcus mucilaginosus. Syst. Appl. Microbiol. 4:207–217.

Stackebrandt, E., F. A. Rainey, and N. L. Ward-Rainey. 1997. Proposal for a new hierarchic classification system, Actinobacteria classis nov. Int. J. Syst. Bacteriol. 47:479–491.

Trevino, M., A. Garcia-Zabarte, A. Quintas, E. Varala, J. M. Lopez-Paz, A. Jato, C. Garcia-Riestra, and B. J. Regueiro. 1998. Stomatococcus mucilaginosus septicemia in a patient with acute lymphoblastic leukaemia. Eur. J. Clin. Microbiol. Infect. Dis. 17:505–507.

Van Tiel, F. H., B. F. Slangen, H. C. Schouten, and J. A. Jacobs. 1995. Study of Stomatococcus mucilaginosus isolated in a hospital ward using phenotypic characterization. Eur. J. Clin. Microbiol. Infect. Dis. 14:193–198.

Von Eiff, C., M. Hermann, and G. Peters. 1995. Antimicrobial susceptibilities of Stomatococcus mucilaginosus and of Micrococcus spp. Antimicrob. Agents Chemother. 39:268–270.

Prokaryotes (2006) 3:983–1001
DOI: 10.1007/0-387-30743-5_40

CHAPTER 1.1.25

The Family Cellulomonadaceae

ERKO STACKEBRANDT, PETER SCHUMANN AND HELMUT PRAUSER

Introduction

The short history of the family Cellulomonadaceae, suborder Micrococcineae, order Actinomycetales, class Actinobacteria, reflects the changes in the interpretation of sequence based phylogenetic data. When defined originally (Stackebrandt and Prauser, 1991), the family Cellulomonadaceae included the genera *Cellulomonas* (Bergey et al., 1923; emend. Clark, 1952; emend. Stackebrandt et al., 1982), *Oerskovia* (Prauser et al., 1970; emend. Lechevalier, 1972), *Promicromonospora* (Krassilnikov, 1961), and *Jonesia* (Rocourt et al., 1987). The rationale for the establishment of this family was based mainly on comparative analysis of 16S rRNA cataloguing data of the type species of some genera of the order Actinomycetales (Stackebrandt et al., 1980d; Stackebrandt et al., 1983). Phylogenetically, neighboring taxa to Cellulomonadaceae were *Promicromonospora*, *Cellulosimicrobium*, *Rarobacter* (Fig. 1), as well as *Arthrobacter*, *Renibacterium*, *Micrococcus*, *Stomatococcus*, *Dermatophilus* and *Microbacterium* and some other related genera (not shown). Members of the Cellulomonadaceae were phenotypically characterized by a combination of phenotypic characters that distinguish them from the adjacent taxa: 1) from *Microbacterium* and related taxa, by the type A crosslinking of the peptidoglycan moities; 2) from *Dermatophilus* and *Brevibacterium*, by the lack of *meso*-diaminopimelic acid in the peptidoglycan; and 3) from all other members of the order, by the presence of a specific phospholipid type.

The decision to exclude the genus *Jonesia* from the family (Rainey et al., 1995) was due to the change in methodology, i.e., the replacement of 16S rRNA cataloguing by reverse transcriptase sequencing of 16S rRNA and shortly thereafter by analysis of polymerase chain reaction (PCR)-generated 16S rDNA. As a consequence of phylogenetic branching, the order changed and the phylogenetic distinctness of *Jonesia denitrificans* became apparent. *Promicromonospora* was excluded from the family on similar grounds and now constitutes the monogeneric family Promicromonosporaceae (Stackebrandt et al., 1997).

The emended family Cellulomonadaceae, as described by Stackebrandt et al. (1997) on the basis of phylogenetic position and the distribution of signature nucleotides of the 16S rRNA gene of all validly described actinomycete genera, is a member of the suborder Micrococcineae, order Actinomycetales (Stackebrandt et al., 1997). The family then contained the genera *Cellulomonas* and, tentatively, *Rarobacterfaecitabidus*. Shortly afterwards, several new genera were described as new members of the Micrococcineae (Groth et al., 1997a; Groth et al., 1997b; Martin et al., 1997; Groth et al., 1999a, b) and the addition of new 16S rDNA sequences to the database of actinomycete species changed the phylogenetic position of a few taxa and stabilized other taxa. The phylogenetic dendrogram clearly disconnected *Rarobacterfaecitabidus* phylogenetically from the family Cellulomonadaceae. New families were described, i.e., Rarobacteraceae, Bogoriellaceae, Dermacoccaceae, and Sanguibacteraceae (Stackebrandt and Schumann, 2000), leaving Cellulomonadaceae as a family with two genera: *Cellulomonas* and *Oerskovia*.

At present, the family comprises the genera *Cellulomonas* and *Oerskovia* (Stackebrandt et al., 2002), embracing 10 and 4 species, respectively, the type strains of which are listed in Table 1, together with the accession numbers of their 16S rDNAs. Both genera were united under *Cellulomonas* (Stackebrandt et al., 1982) when phylogenetic relatedness based upon 16S rDNA catalogues (Stackebrandt et al., 1980c) was considered more significant than differences in chemotaxonomic properties, such as the amino acid composition of peptidoglycan. In the years following the unification, the high correlation between phylogenetic clustering and chemotaxonomic distinctness became apparent (see chapter "Introduction to the Taxonomy of Actinobacteria") and the reevaluation of phenotypic properties of members of these two genera led to taxonomic consequences: firstly, *Cellulomonascellulans* was excluded from *Cellu-*

lomonas and reclassified as *Cellulosimicrobium-cellulans*, which is a phylogenetic neighbor of *Promicromonospora* (Schumann et al., 2001); secondly, the generic status of *Oerskovia* was reconfirmed and thirdly, *Promicromonosporaenterophila* was reclassified as *Oerskovia enterophila* (Stackebrandt et al., 2002).

The Genus *Cellulomonas*

Introduction

The genus *Cellulomonas* originally described by Bergey et al. (1923) was emended by Clark (1952). The unification of *Cellulomonas* and *Oerskovia* was accompanied by the subsequent

emendation of the genus *Cellulomonas* to accommodate organisms with lysine or ornithine as diagnostic diamino acid of the peptidoglycan (Stackebrandt et al., 1982). As already stated by Stackebrandt and Prauser (1991), "The rational [*sic*] for this union was based on an overestimation of phylogenetic relationships between members of these two genera while, at the same time, neglecting the presence of genus-specific chemotaxonomic properties." The phylogenetic analysis of *Promicromonospora enterophila* (Jáger et al., 1983) and of additional strains of *Cellulomonas* and *Oerskovia* has revealed recently that all these organisms containing lysine in the peptidoglycan clustered together with the type strain of *Oerskoviaturbata* and not with the ornithine containing strains of *Cellulomonas* species. This result supported the view that *Cellulomonas* and *Oerskovia* are separate taxonomic entities that can be differentiated not only by phenotypic characteristics but also phylogenetically after inclusion of additional lysine-containing representatives to 16S rDNA sequence analysis (Stackebrandt et al., 2002).

The type species of the genus *Cellulomonas* is *Cellulomonas flavigena* (Bergey et al., 1923). The species *C. biazotea*, *C. cellasea*, *C. gelida*, *C. fimi* and *C. uda* have been members of the genus since their original description (Bergey et al., 1923). The following species were added later: *C. cellulans* (basonym *Nocardia cellulans*) embracing the misclassified species *C. cartae*, *Brevibacterium lyticum*, *Brevibacterium fermentans* and *Oerskovia xanthineolytica* (Stackebrandt and Kandler, 1980a; Stackebrandt and Keddie, 1986), *C. fermentans* (Bagnara et al., 1985), *C. hominis* (Funke et al., 1995), *C. humilata* (Collins and Pascual, 2000), *C. persica* (Elberson et al., 2000) and *C. iranensis* (Elberson et al., 2000).

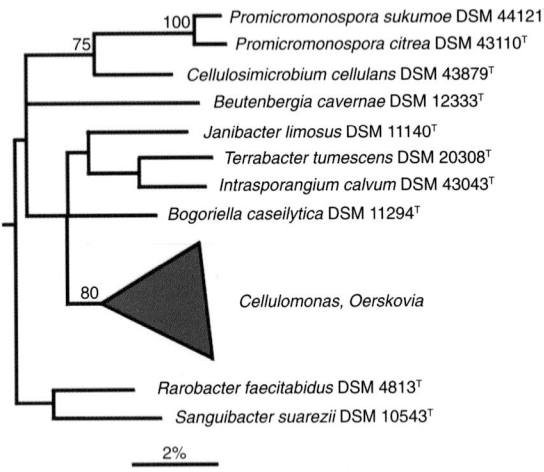

Fig. 1. 16S rDNA relationship of members of *Cellulomonas*, *Oerskovia* and some phylogenetically neighboring taxa (De Soete, 1983). Values refer to bootstrap values (1000 resamplings). Bar = 2% sequence divergence.

Table 1. Type strains of species of *Cellulomonas* and *Oerskovia* and the accession numbers of their 16S rDNA sequences.

Genus	Species	Deposited as	16S rDNA accession numbers
Cellulomonas	*flavigena*	ATCC 482[T], DSM 20109[T]	X83799
	biazotea	ATCC 486[T], DSM 20112[T]	X83802
	uda	ATCC 491[T], DSM 20107[T]	X83801
	fimi	ATCC 484[T], DSM 20113[T]	X79460
	gelida	ATCC 488[T], DSM 20111[T]	X83800
	fermentans	DSM 3133[T]	X83805
	hominis	DSM 9581[T]	X82598
	persica	ATCC 700642[T]	AFO64701
	iranensis	ATCC 700643[T]	AF064702
	humilata	ATCC 25174[T]	X82449
	cellasea	ATCC 487[T], DSM 20118[T]	X83804
Oerskovia	*turbata*	DSM 20577[T]	X83806
	enterophila	DSM 43852[T]	X83807
	jenensis	DSM 46000[T]	AJ314848
	paurometabola	DSM 14281[T]	AJ314851

Abbreviations: DSM, Deutsche Sammlung von Mikroorganismen; and [T], indicates type strain.

As a result of the unification of the genera *Cellulomonas* and *Oerskovia*, *Cellulomonas* (*Oerskovia*) *turbata* was affiliated to the genus *Cellulomonas* (Stackebrandt et al., 1982). The inclusion of representatives of several newly described genera of the suborder Micrococcineae (e.g., Groth et al., 1997a; Groth et al., 1997b; Groth et al., 1999a; Groth et al., 1999b; Martin et al., 1997) and of sequence data of unclassified *Cellulomonas* strains (accession numbers AB023360–AB023367) into phylogenetic analyses based on different treeing algorithms revealed that *C. cellulans* clustered outside the *Cellulomonas* proper. This observation, supported by differences in chemotaxonomic characteristics (e.g., peptidoglycan structure), led to the proposal of the new genus *Cellulosimicrobium* and reclassification of *Cellulomonas cellulans* as *Cellulosimicrobium cellulans*, which is today considered a sister clade of *Promicromonospora* (Schumann et al., 2001).

Phylogeny

Cellulomonas species have been among the first bacteria to be subjected to 16S rDNA cataloguing (Stackebrandt et al., 1980d; Stackebrandt and Woese, 1981), revealing the phylogenetic relatedness to members of *Arthrobacter*, *Micrococcus* and related taxa. This relationship was supported by 5S rDNA analysis performed on *Cellulomonasbiazotea* and a variety of coryneform bacteria (Park et al., 1987).

Analysis of the almost complete 16S rDNA sequence of the type strains reveals a high degree of similarity (above 95% similarity; Table 2, upper right triangle). Only the pair *C. biazotea* and *C. fimi* reveals almost identical sequences (99.7%), while most binary values are below 98%. The phylogenetic position determined by use of different treeing algorithms (distance matrix, De Soete, 1983 and Felsenstein, 1993; and neighbor joining and maximum parsimony, Felsenstein, 1993) under the influence of varying reference organisms reveals that the intrageneric branching order is not stable, which is also reflected by the lack of high bootstrap values. This is especially obvious for *C. hominis*, which changes its position significantly and which can even be found as the most deeply branching member of the family, thus making the genus *Cellulomonas* a non-monophyletic taxon. This phenomenon has already been observed by Funke et al. (1995) in the description of *C. hominis*, which branched adjacent to *Oerskoviaturbata*. Members of *Cellulomonas* and *Oerskovia* can be distinguished by a set of signature nucleotides of their 16S rDNA sequences (Table 3).

Genomic Relatedness

With binary DNA similarity values of less than 60%, the *Cellulomonas* species investigated can be considered genetically well defined. A few species appear closely related as judged from DNA similarity values around 50%, i.e., *C. biazotea* and *C. fimi* and a cluster comprising *C. uda*, *C. flavigena*, *C. persica* and *C. iranensis* (Fig. 2; Table 2, lower left triangle). Similarity values obtained with different hybridization methods as well as reciprocal values may deviate by more than 30%, but this range does not answer the question of the respective strain's affiliation to a different species.

Taxonomy

Cellulomonads are slender irregular rods (ca. 0.4–0.8 μm in diameter), which may vary considerably in length and which may be arranged in V-formation (Fig. 3). On agar media, they may undergo a more or less expressed morphological growth cycle. In this case, colonies resulting appear nocardioform, i.e., they show mycelia-like fringes. The filaments, which may be branching, fragment subsequently into rods (Fig. 4). Nocardioform mycelia-like structures may also appear when filaments penetrate into the agar medium. Week-old cultures are composed mainly of short rods, but a proportion of the cells may be coccoid. Data on the motility of cellulomonads are inconsistent: *C. fermentans*, *C. flavigena*, *C. cellasea*, *C. humilata* and *C. uda* are the only representatives of the genus which have been reported to be nonmotile (Stackebrandt and Kandler, 1979; Schaal, 1986); however, polar multitrichous flagella were detected also for *C. flavigena*, *C. cellasea* and *C. uda* (Thayer, 1984a).

Physiological Properties

Chemotaxis towards cellobiose and hemicellulose hydrolysis products, e.g., cellotriose, D-glucose, xylobiose and D-xylose, as well as other sugars has been observed in *C. gelida* ATCC 486 (Hsing and Canale-Parola, 1992). Two types of separately regulated cellobiose receptors (Cb1 and Cb2) were described, allowing the motile organism to migrate towards plant-derived cellulose and hemicellulose by swimming up concentration gradients of cellobiose and other sugars. While one receptor Cb1 was inducible and bound to cellobiose and xylobiose, receptor Cb2 was synthesized constitutively and bound to cellobiose, cellotriose, xylobiose and D-glucose. In a following paper, the authors (Hsing and Canale-Parola, 1996) described that L-methionine is required for normal cell motility and chemotaxis and that *S*-adenosylmethionine

Table 2. DNA-DNA hybridization similarities (lower left triangle) and 16S rDNA similarities (upper right triangle) among type strains of *Cellulomonas* species.

Species	C. biazotea	C. fimi	C. cellasea	C. gelida	C. iranensis	C. uda	C. flavigena	C. persica	C. humilata	C. hominis	C. fermentans
C. biazotea	100	99.7	97.8	96.0	96.4	96.1	96.6	96.3	96.8	96.0	95.4
C. fimi	52[a], 49[a]	100	97.8	96.1	96.5	96.2	96.6	96.3	96.7	96.0	95.8
C. cellasea	30[a], 35[a]	29[a], 30[a]	100	96.1	96.2	96.0	96.4	96.4	96.2	96.4	95.8
C. gelida	30[a], 31[a]	32[a], 31[a]	26[a], 27[a]	100	98.7	98.4	97.5	98.0	96.1	96.3	94.5
C. iranensis	n.d.	n.d.	n.d.	19[b]	100	98.4	98.1	97.3	96.8	96.5	94.6
C. uda	24[a]	34[a]	27[a], 27[a]	43[a], 45[a], 24[b], 66[b]	39[b]	100	97.0	97.5	96.1	96.1	94.8
C. flavigena	26[a]	27[a]	25[i]	23[a], 17[b], 41[b]	67[b]	24[a], 30[b], 56[b]	100	97.9	96.9	96.6	95.2
C. persica	25[a]	26[a]	n.d.	48[b]	n.d.	54[b]	62[b]	100	95.8	96.9	94.8
C. humilata	n.d.	n.d.	n.d.	n.d.	n.d.	n.d.	n.d.	n.d.	100	96.9	95.4
C. hominis	n.d.	n.d.	n.d.	n.d.	n.d.	n.d.	n.d.	n.d.	n.d.	100	95.7
C. fermentans	n.d.	n.d.	n.d.	n.d.	n.d.	27[c], 40[c]	n.d.	n.d.	n.d.	n.d.	100

Abbreviation: n.d., not determined.
[a]Determined by filter hybridization (Stackebrandt and Kandler, 1979).
[b]Determined by dot blot hybridization (Elberson et al., 2000).
[c]Determined by filter hybridization (Bagnara et al., 1985).

Table 3. Signature nucleotides defining the genera *Cellulomonas* and *Oerskovia*.

16S rDNA position	Members of *Oerskovia* clade	Members of *Cellulomonas* clade
18–917	x-G	x-A
139–224	A-U	U-A
146–176	G-U	G-C
185–192	C-G	R-Y
186–192	U-G	C-G, G-C
601–637	G-C	G-U
602–636	G-U	C-G
612–628	U-A	Y-G
614–626	A-U	G-C, C-G
1120–1153	U-A	C-G
1438–1463	G-U	G-C

Abbreviations: A, adenine; G, guanine; C, cytosine; U, uracil; R, purine (G,A); Y, pyrimidine (C,U); and x, position 18 not determined.
From Stackebrandt et al. (2002).

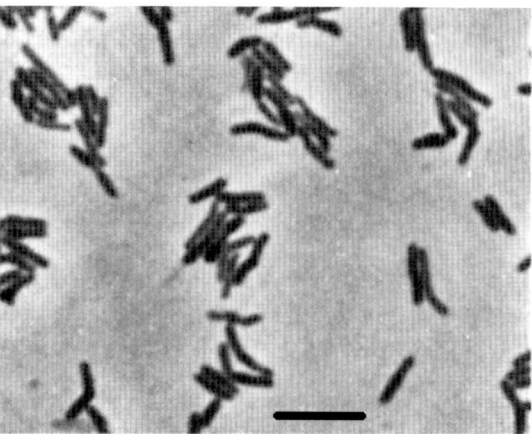

Fig. 3. *Cellulomonas flavigena* CCM 1926 (ATCC482). Coryneform rods from a 28-h slide culture on Bennett's saccharose agar. Bar = 5 µm.

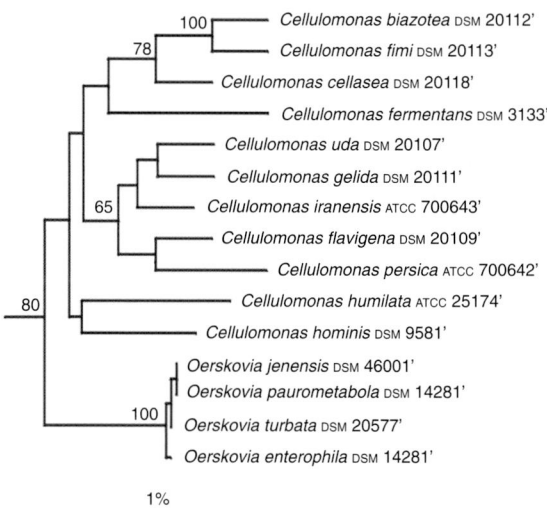

Fig. 2. Intrageneric relationship of *Cellulomonas* and *Oerskovia* species (De Soete, 1983). Values refer to bootstrap values (1000 resamplings). Bar = 1% sequence divergence.

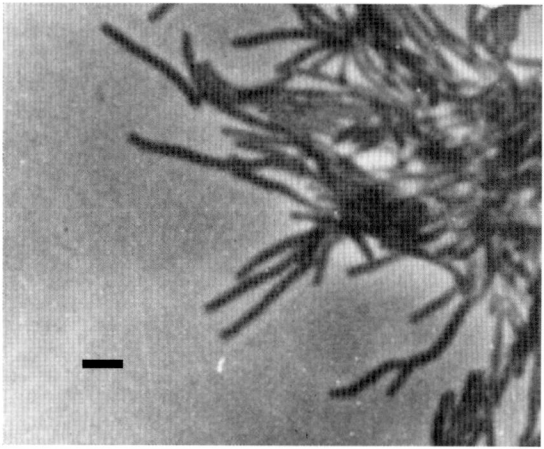

Fig. 4. Nocardioform microcolony of hyphae of *Cellulomonas biazotea* NCIB 8077 on the surface of Bennett's saccharose agar (28-h slide culture). Bar = 10 µm.

is involved in sugar chemotaxis. Methylation occurred posttranslationally, increased upon addition of sugar attractants, and decreased after removal of the stimulating sugars. The cellulolytic enzymes are controlled by catabolic repression, as activity against carboxymethyl cellulose is only low during growth on glucose or cellobiose (Choi et al., 1978; Stoppok et al., 1982).

All cellulomonads are able to grow under aerobic and microaerophilic conditions, and *C. fermentans* (Bagnara et al., 1985) and *C. uda* ATCC 21399 (Dermoun et al., 1988b) grow under strictly anaerobic conditions as well. Glucose uptake of whole cell suspension of a *C. fimi* isolate was two-fold higher under aerobic conditions than under N$_2$ or H$_2$; cellobiose negatively affected glucose uptake (Khanna, 1993).

Glucose, maltose and sucrose are fermented by all species, while a few species fermented one or two of the carbohydrates mannitol, xylose, dextrin, β-methyl-xyloside, rhamnose and gluconate. Anaerobically, resting cells catabolize glucose to various endproducts. Performed on a few species, radiorespirometric studies indicated that glucose is catabolized mainly via the Embden-Meyerhof (EM) pathway and, to some extent, through the hexose monophosphate (HMP) shunt (Stackebrandt and Kandler, 1979). This finding is supported by determination of respective key enzyme activities of these pathways in *C. flavigena* (Kim, 1987) and *C. uda* (Marschoun et al., 1987). As shown with *C. flavigena*, gluconate is catabolized via the Entner-Doudoroff (ED) pathway and HMP shunt (Kim, 1987). Endproducts of aerobic glucose dissimilation are mainly CO$_2$ and either acetate or acetate and L-lactate (Stackebrandt and Kandler, 1980b). Under anaerobic conditions, resting cells produce mainly CO_2, acetate, lactate, and ethanol as well as smaller amounts of succinate and formate. Similar products, although in varying amounts, were reported for *C. fermentans* (Bagnara et al., 1985) and *C. uda* grown under strictly anaerobic conditions (Marschoun et al., 1987).

Resting cultures of all *Cellulomonas* strains investigated show a highly effective symmetric interchange and an asymmetric redistribution of carbon atoms within the hexose molecule (Stackebrandt and Kandler, 1980b). Under energy excess conditions, *C. uda* accumulates glycogen and trehalose, which in turn are degraded at different rates during carbon starvation (Schimz and Overhoff, 1987a; Schimz and Overhoff, 1987b).

Chemotaxonomic Properties

Cellulomonas species have been extensively characterized with respect to chemotaxonomic properties. The diagnostic amino acid in position 3 of the peptide subunit of the peptidoglycan is ornithine with the interpeptide bridge containing either D-aspartic acid (*C. flavigena*, *C. fermentans*, *C. iranensis* and *C. persica*) or D-glutamic acid (all other species; Table 4). Rhamnose is the major diagnostic cell wall sugar in most strains. Menaquinone MK-9(H_4) is the predominant isoprenoid quinone (Collins and Jones, 1981; Collins and Pascual, 2000). The 12-methyltetradecanoic (*ai*-$C_{15:0}$) and hexadecanoic ($C_{16:0}$) acids are the dominant components of fatty acid patterns of cellulomonads, and other branched-chain (e.g., *i*-C15:0, *ai*-C17:0) and straight-chain (e.g., $C_{14:0}$, $C_{15:0}$) fatty acids occur in lower amounts (Funke et al., 1995). No data are available on cellular fatty acids of *C. iranensis*, *C. persica* and *C. humilata*. Phosphatidyglycerol (Lechevalier et al., 1981b) as well as diphosphatidylglycerol and a phosphoglycolipid are the major polar lipids (Minnikin et al., 1979). No data are available on isoprenoid quinones and polar lipids of *C. iranensis*, *C. persica*, *C. humilata* and *C. hominis*. The G+C content of the DNA has been determined in several laboratories using different methods. Values ranging between 71.3 and 76 mol% do not allow differentiation of species (Table 5).

Identification

The individual species can be differentiated by a set of characters that require determination of peptidoglycan, cell wall sugars, end products of glucose degradation, and the utilization of certain sugars and acids (Table 4). Compilations of biochemical reactions are given by Stackebrandt and Kandler (1979), Bagnara et al. (1985), and McHan and Cox (1987). The latter authors propose a simple identification key for most species. Neither genus-specific nor species-specific phages have as yet been detected.

Habitat

The main habitat of cellulomonads appears to be the soil, from which the original cultures were isolated (Kellerman et al., 1913; Kauri and Kushner, 1985; Stackebrandt and Keddie, 1986). This applies also to *Cellulomonas* species described or reclassified recently: *C. persica* and *C. iranensis* were isolated from forest soils (Elberson et al., 2000) and *C. humilata* (basonym *Actinomyces humiferus*) originated from organically rich soils (Collins and Pascual, 2000). Emphasis placed on the cellulolytic activity of these organisms has resulted in the successful isolation of *Cellulomonas* strains from rumen (Lee and Lee, 1986), activated sludge (Ramasamy et al., 1981), and cellulose-enriched environments such as bark and wood (Przybyl, 1979; Deschamps, 1982), coffee cherries (Silva et al., 2000), soils enriched by flax or sisal fibers (Lednicka et al., 2000), and sugar fields (de Leon and Joson, 1980). Cellulolytic culturable bacteria closely related to strains of the genus *Cellulomonas* as revealed by 16S rDNA sequence comparison have been isolated recently from refuse of a landfill (Pourcher et al., 2001) and from an agricultural encatchment (Ulrich and Wirth, 1999). *Cellulomonas hominis* is the first representative of the genus isolated from human clinical samples (cerebrospinal fluid; Funke et al., 1995). Strains of *C. hominis* did not show any cellulolytic activity.

Isolation, Cultivation and Preservation

As described by Stackebrandt and Keddie (1986), a suitable procedure is to enrich cellulomonads in mineral-based medium containing a low concentration (0.05–0.1%) of yeast extract to which filter paper is added as principal carbon source. Isolates are tested on plates containing Avicel, Solka Floc, CF11 cellulose, carboxymethyl cellulose, or phosphoric acid-treated cellulose (Kauri and Kushner, 1985). Cellulose degradation can be visualized owing to the formation of clearing zones. This method is not selective for *Cellulomonas* and isolates must be screened for typical coryneform morphology. Cells are routinely grown in shake flasks at 30°C.

Moderate growth occurs on meat extract agar, peptone agar, or media based on yeast extract or peptone at around neutral pH. Growth-promoting factors in yeast extract are, in part, thiamine and biotin. These factors can be

Table 4. Biochemical reactions of *Cellulomonas* species.

Biochemical reaction	C. biazotea	C. cellasea	C. fermen-tans	C. fimi	C. flavigena	C. gelida	C. hominis	C. humilata	C. iranensis	C. persica	C. uda
Catalase	+	+	−	+	+	+	+	−	n.d.	n.d.	+
Motility	+[b]	−[b]	−	+[b]	+[a,b]	+[b]	+	−	+	+	−[b]
Cellulolytic activity	+	+	+	+	+	+	−	w[c]	+	+	+
Nitrate reduction	+	+	+	+	+	+	+	−	+	+	+
Urease	−	−	−	−	−	−	−	−	+	+	−
Esculin hydrolysis	+	+	+	+	+	+	+	+	n.d.	n.d.	+
Gelatin hydrolysis	+	−	+	+	+	+	+	w	+	w	+
DNase	−	−	+	−	−	+	+	n.d.	+	+	+
Alkaline phosphatase	+	+	−	+	−	−	−	n.d.	n.d.	n.d.	−
Fermentation of											
Glucose	+	+	+	+	+	+	+	+	n.d.	n.d.	+
Maltose	+	+	+	+	+	+	+	+	n.d.	n.d.	+
Sucrose	+	+	+	+	+	+	+	+	n.d.	n.d.	+
Mannitol	−	+	+	−	−	−	+	+	n.d.	n.d.	−
Xylose	+	+	+	+	+	+	−	w	n.d.	n.d.	+
Dextrin	−	−	+	+	+	+	+	+	+	+	+
β-Methyl-xyloside	−	−	+	+	−	−	+	n.d.	n.d.	n.d.	−
Rhamnose	+	−	−	w	−	−	+	+	n.d.	n.d.	−
α-Methyl-mannoside	−	−	−	−	−	−	+	n.d.	n.d.	n.d.	−
Utilization of											
Ribose	−	−	−	−	+	−	−	w[d]	−	−	−
Raffinose	+	−	+	−	−	−	+	w[d]	−	−	−
Acetate	+	+	+	−	+	+	+	n.d.	+	+	+
L-(+)-Lactate	+	+	−	+	−	−	−	n.d.	−	−	−
Gluconate	−	+	−	+	+	−	+	n.d.	−	−	−

Symbols and abbreviations: +, positive; −, negative; w, weakly positive; and n.d., not determined.

[a]Listed negative by Stackebrandt and Kandler (1979).

[b]Possessed polar multitrichous flagella according to Thayer (1984).

[c]Reaction for cellobiose (Schaal, 1986).

[d]Recorded under "acid from" in Table 15.49 by Schaal (1986).

From Stackebrandt and Kandler (1979), Schaal (1986), Funke et al. (1995), and Elberson et al. (2000).

Table 5. Differential chemotaxonomic characteristics of members of the family Cellulomonadaceae and of the type species of the genera *Promicromonospora* and *Cellulosimicrobium*.

Species	Diamino acid of peptidoglycan[a]	Interpeptide bridge of peptidoglycan[a]	Cell wall sugars[b]	G + C content (mol%)[b]
Cellulomonas flavigena	L-Orn	D-Asp	GlcNH$_2$, Rha, Man, and Rib	72.7–74.8
Cellulomonas biazotea	L-Orn	D-Glu	GlcNH$_2$, Rha, Gal, and 6-desoxy-Tal	71.5–75.6
Cellulomonas cellasea	L-Orn	D-Glu	Rha, Man, and 6-desoxy-Tal	75.0
Cellulomonas fermentans	L-Orn	D-Asp	Glc, Rha, and Rib	75.8
Cellulomonas fimi	L-Orn	D-Glu	GlcNH$_2$, Rha, Fuc, and Glc	71.3–72.0
Cellulomonas gelida	L-Orn	D-Glu	GlcNH$_2$ and Glc	72.4–74.4
Cellulomonas hominis	L-Orn	D-Glu	ND	76
Cellulomonas humilata	L-Orn	D-Glu	Rha, Glc, and Fuc	73
Cellulomonas iranensis	L-Orn	D-Asp	GlcNH$_2$, Rha, Glc, and Man	ND
Cellulomonas persica	L-Orn	D-Asp	GlcNH$_2$, Rha, Glc, and Man	ND
Cellulomonas uda	L-Orn	D-Glu	GlcNH$_2$ and Glc	72
Oerskovia turbata	L-Lys	L-Thr←D-Asp	Gal	70.5
Oerskovia enterophila	L-Lys	L-Thr←D-Glu	Rha, Man, Fuc, and GlcNH$_2$	71
Oerskovia jenensis	L-Lys	L-Thr←D-Glu	ND	ND
Oerskovia paurometabola	L-Lys	L-Thr←D-Glu	ND	ND
Promicromonospora citrea	L-Lys	Ala←Glu	Gal	70–75
Cellulosimicrobium cellulans	L-Lys	D-Ser←D-Asp	Rha, Fuc, Gal	74

Abbreviations: Glu, glutamic acid; Ala, alanine; Asp, aspartic acid; Lys, lysine; Orn, ornithine; Ser, serine; Thr, threonine; Rha, rhamnose; Man, mannose; Rib, ribose; Gal, galactose; Tal, talose; Fuc, fucose; Glc, glucose; GlcNH$_2$, glucoseamine; and ND, no data.

[a]Data from Fiedler and Kandler (1973), Stackebrandt et al. (1978), Seidl et al. (1980), Jáger et al. (1983), Bagnara et al. (1985), Evtushenko et al. (1984a), Funke et al. (1995), Collins and Pascual (2000), Elberson et al. (2000), Schumann et al. (2000), Stackebrandt et al. (2002), N. Weiss and P. Schumann (unpublished data).

[b]Data from Stackebrandt and Keddie (1986), Bagnara et al. (1985), Lechevalier and Lechevalier (1986), Kalakoutskii et al. (1986), Funke et al. (1995), Collins and Pascual (2000), Elberson et al. (2000).

supplemented by adding a few drops of a sterile commercially available multivitamin solution (e.g., Multibionta and Merck). *Cellulomonas persica* and *C. iranensis* grow on trypticase soy agar at 30°C (http://www.atcc.org). *Cellulomonas hominis* can be cultured on Columbia agar (Becton Dickinson Microbiology Systems, Cockeysville, MD, United States) with 5% sheep blood in a 5% CO$_2$ atmosphere (Funke et al., 1995) or on trypticase soy yeast extract medium at 37°C (http://www.dsmz.de). *Cellulomonas humilata* can be cultivated on brain heart infusion agar at 30°C (http://www.atcc.org) but does not grow well at 37°C in chemically defined media or in those lacking organic nitrogen and in anaerobic conditions (Collins and Pascual, 2000).

For short-term preservation, stab cultures in semisolid medium should remain viable for several months at room temperature. For long-term maintenance, lyophilization or storage in liquid nitrogen is recommended.

The Glycoside Hydrolases

Of the large variety of hydrolytic starch-, xylan- and cellulose-degrading enzymes detected in cellulomonads, the cellulases are the most salient ones. In a comparative study, *C. biazotea* produced the highest filter-paper cellulase and endo-

glucanase activities, followed by *C. flavigena*, *C. cellasea* and *C. fimi* (Rajoka and Malik, 1997). However, most of the molecular work on cellulase and xylanase genes was determined with *C. fimi* (Table 6). Starch-hydrolyzing activity has been reported in cells of *C. flavigena* (McCarthy and Pembroke, 1988), and the amylase gene product of *Cellulomonas* strain NCIM 2353 has been expressed and characterized (Kumar and Deobakar, 1997).

As summarized by Stackebrandt and Prauser (1991), different kinds of cellulose preparations and derivatives have been tested, including amorphous (Dermoun and Belaich, 1985), swollen, phosphoric-acid-treated (Kauri and Kushner, 1985), microcrystalline (Vladut-Talor et al., 1986; Dermoun and Belaich, 1988a; Poulsen and Petersen, 1988), and carboxymethyl cellulose (CMC). As reported by Kauri and Kushner (1985), degradation of cellulose does not depend on cell-to-fiber contact, but cellulases from three *Cellulomonas* strains were active even when they were physically separated from the fibers. Microcrystalline cellulose (Avicel PH 101, Cellulose MN300, and Whatman grade CC41) appears to be less efficiently attacked than amorphous cellulose (phosphoric-acid-treated Whatman CC41 cellulose; Dermoun and Belaich, 1985). Under anaerobic conditions, celluloses with varying degrees of crystallinity

Table 6. Occurrence of glycoside hydrolases in members of *Cellulomonas*.[a]

Protein	Literature code	Organism	EC	GenBank	SwissProt	Module family
Carbohydrate-binding module						
Cellobiohydrolase A	Cel6B	*Cellulomonas fimi*	3.2.1.91	L25809	P50401	2
Cellobiohydrolase B	Cel48A	*Cellulomonas fimi*	3.2.1.91	L38827	P50899	2
Endo-1,4-glucanase A	Cel6A	*Cellulomonas fimi*	3.2.1.4	M15823	P07984	2
Endo-1,4-glucanase B	Cel9A	*Cellulomonas fimi*	3.2.1.4	M64644	P26225	2
Endo-1,4-glucanase D	Cel5A	*Cellulomonas fimi*	3.2.1.4	L02544	P50400	2
Xylanase (Cex)	Xyn10A	*Cellulomonas fimi*	3.2.1.8	M15824 L11080	P07986 Q59277	2
Xylanase D	Xyn11A	*Cellulomonas fimi*	3.2.1.8 3.1.1.72	X76729	P54865	2
Xylanase (fragment)		*Cellulomonas flavigena* CDBB5321	3.2.1.8	AF338352	Q9AG99	2
Chitinase 63 (Chi63)		*Cellulomonas* sp. GM13	3.2.1.14	AF181718	Q9RB78	2
Endo-1,4-glucanase B	Cel9A	*Cellulomonas fimi*	3.2.1.4	M64644	P26225	3
Endo-1,4-glucanase C	Cel9B	*Cellulomonas fimi*	3.2.1.4	M29707 X57858	P14090	4
Xylanase C	XynC	*Cellulomonas fimi*	3.2.1.8	Z50866	Q59278	9
Xylanase	Xyn10B	"*Cellulomonas pachnodae*"	3.2.1.8	AF120157		22
β-Mannanase	Man26A	*Cellulomonas fimi*	3.2.1.78	AF126471	Q9XCV5	23
Carbohydrate esterase						
Xylanase C	XynC	*Cellulomonas fimi*	3.2.1.8	Z50866	Q59278	4
Xylanase D	Xyn11A	*Cellulomonas fimi*	3.2.1.8 3.1.1.72	X76729	P54865	11
Glycoside hydrolase						
β-Glucosidase		*Cellulomonas fimi*	3.2.1.21	M94865	Q46043	1
β-Mannosidase	Man2A	*Cellulomonas fimi*	3.2.1.25	AF126472		2
β-Glucosidase		*Cellulomonas biazotea*	3.2.1.21	AF005277	O51843	3
Endo-1,4-glucanase D	Cel5A	*Cellulomonas fimi*	3.2.1.4	L02544	P50400	5
Endo-1,4-glucanase B	(cflB)	*Cellulomonas flavigena*	3.2.1.4	AF172345		5
Cellobiohydrolase A	Cel6B	*Cellulomonas fimi*	3.2.1.91	L25809	P50401	5
Endo-1,4-glucanase A	Cel6A	*Cellulomonas fimi*	3.2.1.4	M15823		5
Cellobiohydrolase A	cflA	*Cellulomonas flavigena*	3.2.1.91	AF172344		6
Endo-1,4-glucanase	Cel6A	"*Cellulomonas pachnodae*"	n.d	AF113404		6
Endo-1,4-glucanase		*Cellulomonas uda*	3.2.1.4	M36503	P18336	8
Endo-1,4-glucanase B	Cel9A	*Cellulomonas fimi*	3.2.1.4	M64644	P26225	9
Endo-1,4-glucanase C	Cel9B	*Cellulomonas fimi*	3.2.1.4	M29708 X57858		9
Xylanase (Cex)	Xyn10A	*Cellulomonas fimi*	3.2.1.8	M15824 L11080	P07986 Q59277	10
Xylanase C	XynC	*Cellulomonas fimi*	3.2.1.8	Z50866	Q59278	10
Xylanase D	Xyn11A	*Cellulomonas fimi*	3.2.1.8 3.1.1.72	X76729	P54865	11
Xylanase (fragment)		*Cellulomonas flavigena* CDBB5321	3.2.1.8	AF338352	Q9AG99	11
Xylanase	Xyn11A	"*Cellulomonas pachnodae*"	3.2.1.8	AF120156	Q9RQB8	11
Mannanase	Man26A	*Cellulomonas fimi*	3.2.1.78	AF126471	Q9XCV5	26
Cellobiohydrolase B	Cel48A	*Cellulomonas fimi*	3.2.1.91	L29042 L38827	P50899	48

Abbreviation: EC, enzyme code.

[a]The database contains information on "*Cellulomonas pachnodea*." The taxonomic status of this species has not yet been described.

were metabolized by *C. uda* ATCC 21399 with the same efficiency as by aerobically grown cells, though the growth yield was reduced significantly (Dermoun et al., 1988b). Optimal liquefaction of carboxymethyl cellulose (CMC) gels occurred in a synthetic medium at 40°C at pH of 7.0–7.5 (Thayer et al., 1984b).

As described in the chapter Cellulose-decomposing Bacteria and Their Enzyme Systems in Volume 2 (Bayer et al., 2001), the traditional classification of cellulases and other glycosyl hydrolases, based on substrate type and bonds cleaved by a given enzyme, has recently been changed to take account of the common structural fold and mechanistic themes (Coutinho and Henrissat, 1999; http://www.afmb.cnrs-mrs.fr Carbohydrate-active enzymes server). Different glycosyl hydrolases are grouped into five families:

Glycoside hydrolase, glycosyltransferase, polysaccharide lyase, carbohydrate esterase and carbohydrate-binding module families, three of which are found in cellulomonads. Whereas in the Second edition, the chapter on the family Cellulomonadaceae emphasized the traditional coverage of individual enzymes and their reactions (Stackebrandt and Prauser, 1991), this chapter highlights the type and activity of enzymes. The results published in the past 10 years of cloning, sequencing, mutagenesis and three-dimensional structure studies have been numerous, and even listing the relevant literature would go beyond the limits of this chapter. Rather, the respective proteins occurring in members of *Cellulomonas*, together with the accession numbers of nucleic acid sequences and protein sequences, are listed and will guide the reader to the relevant literature (Table 6). The information has been taken from the website (http://www.afmb.cnrs-mrs.fr/~cazy/CAZY/index.html).

Applications

Cellulomonas strains ATCC 482, ATCC 488, ATCC 491, ATCC 15392 and ATCC 21399 have high protein and essential amino acid values (Dey, 1976), and different strains have been used for single-cell protein production from a variety of waste products, such as sugar cane bagasse (Han et al., 1971; Rodriguez et al., 1993) and rice straw (Han et al., 1971), hemp (Jedar et al., 1987) and ground corn and stalks (Fields et al., 1991). Like other cellulose- and hemicellulose-degrading organisms, *Cellulomonas* strains have been considered potential candidates for waste disposal (Dunlap and Callihan, 1974; Ramasamy et al., 1981) and composting flax and sisal fibers (Lednicka et al., 2000), bagasse (Richard and Peiris, 1981), pith and leaves of sugar cane (Richard and Peiris, 1981; Diaz and Guirola, 1983; Rajoka and Malik, 1986), dried palm oil mill effluent (Agamuthu and Tan, 1985), and shredded newspapers (Rapp et al., 1984) or even as producers of chemicals from low-cost substrates. Mutants of *C. flavigena*, showing elevated xylanase and carboxymethyl cellulase activity, were able to use a larger portion of sugar cane bagasse than their wildtype strains (Ponce-Noyola and de la Torre, 1995; Mayorga-Reyes and Ponce-Noyola, 1998). The isolation of cellulomonads from a landfill of domestic refuse, e.g., *C. fermentans* (Bagnara et al., 1985), *C. hominis*, *C. biazotea*/*C. fimi* and *C. flavigena*/*C. uda* (the latter two pairs of organisms were indistinguishable by numerical analysis) also indicated their potential to degrade solid cellulolytic waste (Pourcher et al., 2001). *Cellulomonas* strain DOT 21, isolated from a domestic refuse (Bichet-Hebe et al.,

1999), has been used to estimate paper degradation by reduction of a whiting fluorescent agent added to white paper prior to the degradation process.

A high number of uncharacterized *Cellulomonas* strains have been isolated from mature coffee cherries of *Coffea arabica* in Brazil (Silva et al., 2000). Under dry conditions, cellulomonads dominated the population of Gram-positive organisms. Their role, however, has not been elucidated.

Cellulomonasflavigena strain HR5, identified by fatty acid analysis, was isolated from agricultural soil in South Korea, contaminated with 4-chlorobenzoate. This plasmid-bearing strain was also able to utilize well 4-bromobenzoic acid, benzoic acid and less well 4-iodobenzoic acid, but not 3-chlorobenzoic acid and 2,4-dichlorophenoxyacetic acid (Yi et al., 2000).

Mixed cultures consisting of *Cellulomonas* sp. ATCC 21399, *Desulfovibrio vulgaris* strain JJ, and *Methanosarcina barkeri* 227 were highly efficient in converting xylan to methane via hydrolysis and acidogenesis (strain ATCC 21399), acetogenesis (strain JJ), and methanogenesis (strain 227; Guyot, 1986). The same *Cellulomonas* strain has also been used in mixed cultures with *Rhodopseudomonas capsulata* to photoevolve molecular hydrogen by the nitrogenase system of the phototrophic strain with cellulose as the sole carbon source (Odom and Wall, 1983).

The ability of cellulomonads to attack cellulose and wheat straw under microaerobic or even anaerobic conditions has been used in mixed cultures to provide nitrogen-fixing strains of *Bacillus macerans* and *Azospirillum brasilense* with energy-yielding products (Halsall and Gibson, 1985; Halsall and Gibson, 1986a; Halsall and Goodchild, 1986b). Good nitrogen-fixing rates have been reported for the pair *Azospirillum brasilense* ATCC 29145 and *Cellulomonas* strain CS117. The latter is a mutant strain selected for its increased production of cellulase and reduced sensitivity to inhibition or repression by accumulated cellobiose and glucose (Haggatt et al., 1978).

Toxicity Evaluation of Cellulomonads

Cellulomonasflavigena ATCC 482, *C. gelida* ATCC 488, *C. uda* ATCC 491, *C. fimi* ATCC 15724 and *Cellulomonas* sp. ATCC 21399 were tested for toxicity by injecting cell extracts as well as viable cells in fertile chicken eggs and some rodents, respectively. Neither were the cell extracts lethal to chick embryos, nor did the viable cells cause generalized or local infections in rats, mice or rabbits. The sexual maturity, fertility and organs of adult rats as well as the vitality of their progeny were not affected by feeding 10% protein from the test strains (Dey and Fields, 1995). The clinical significance of *C. hominis* (Funke et al., 1995) is unknown.

The Genus *Oerskovia*

Introduction

Motile nocardioform bacteria were reported among others by Topping (1937), Ørskov (1938), and Jensen (1953). Erikson (1954) described "Oerskov's motile nocardia, strain 27" as "*Nocardiaturbata*." Prauser (1967) demonstrated phenetic similarities among 12 isolates from different soils and the type strain of "*Nocardiaturbata*." The species, however, differed significantly from authentic *Nocardia* strains: instead of *meso*-diaminopimelic acid (the diagnostic amino acid of the peptidoglycan of nocardiae), lysine was found in strains of "*Nocardiaturbata*," and strains were motile by flagellation of fragmented hyphal parts (Sukapure et al., 1970) and by motile spores (Higgins et al., 1967). These properties formed the basis for the description of "*Nocardiaturbata*" as *Oerskovia turbata* (Prauser et al., 1970), *Oerskovia* being a genus that contained only motile organisms. The genus was emended (Lechevalier, 1972) because the species grew anaerobically on trypticase soy agar as one of 15 media tested, attacked glucose oxidatively and fermentatively, and was catalase-negative when grown under anaerobic conditions.

In the course of the union of the genera *Cellulomonas* and *Oerskovia* the type species of the genus *Oerskovia*, *O. turbata*, as well as a second species, *O. xanthineolytica*, were transferred to the genus *Cellulomonas*. When describing *C. cartae* (ex "*C. cartalyticum*"), Stackebrandt and Kandler (1980a) proposed *O. xanthineolytica* and *Nocardia cellulans* among other taxa as subjective synonyms of *C. cartae*. *Nocardia cellulans* was formally described as *C. cellulans* and consequently the motile strains of *O. xanthineolytica* were combined with nonmotile strains of *C. cellulans* (Metcalf and Brown, 1957), basonym *Nocardia cellulans* (Stackebrandt and Keddie, 1986). Organisms also included in this species were "*Corynebacterium manihot*" (Collard, 1963), *Brevibacterium lyticum* (Takayama et al., 1960), *Brevibacterium fermentans* (Chatelain and Second, 1966), "*Arthrobacter luteus*" (Kaneko et al., 1969), and *C. cartae* (Stackebrandt and Kandler, 1980b). These organisms were united because they shared the peptidoglycan type L-Lys←D-Ser←D-Asp (Seidl et al., 1980; for additional references, see below), were susceptible to *Oerskovia* phages, particularly to those active against *O. xanthineolytica* (Prauser, 1984;

Prauser, 1986), were phenetically similar in numerical taxonomic studies (Bousfield, 1972; Jones, 1975; Seiler et al., 1977; Seiler, 1983), and were genomically closely related as determined by DNA-DNA similarities obtained for *Brevibacterium fermentans*, *C. cellulans*, "*Corynebacterium manihot*," and *O. xanthineolytica* (60–68.1%; Stackebrandt et al., 1980c; Prauser, 1986). Also, comparative 16S rRNA cataloguing revealed a high similarity level for *Nocardia cellulans* and *C. cartae* (Stackebrandt et al., 1980c). *Cellulomonas cellulans* (including the synonym *O. xanthineolytica*) has recently been reclassified as *Cellulosimicrobium cellulans*.

Today, *Oerskovia* and *Cellulomonas* are considered to be well separated, though closely related (Prauser, 1986; Stackebrandt and Prauser, 1991; Funke et al., 1995; Stackebrandt et al., 2002). Supporting arguments are 1) DNA-DNA similarities of only 20 and 26% (Stackebrandt et al., 1980c; Prauser, 1986); 2) 16S rDNA similarities and sets of 16S rDNA signature nucleotides (Stackebrandt et al., 2002) separating members of the two genera; 3) lysine at position 3 of the peptide subunit of peptidoglycan in *Oerskovia* strains, unlike cellulomonads, which have ornithine at this position (Prauser, 1966; Fiedler and Kandler, 1973); 4) insusceptibility of cellulomonads against *O. turbata* phages (Prauser, 1986); 5) differences in the fatty acid patterns (Minnikin et al., 1979), and 6) differences in the formation of mycelia.

Taxonomy

The genus *Oerskovia* has recently been extended by the reclassification of *Promicromonospora enterophila* as *Oerskovia enterophila* and the description of two new species, *O. jenensis* and *O. paurometabola* (Stackebrandt et al., 2002). *Promicromonospora enterophila* has been described for nonmotile strains isolated from feces of a millipede (Jáger et al., 1983). Spores, described to occur in some strains, were killed by heating at 80°C for 5 min (Jáger et al., 1983) and could not be detected in the type strain (Stackebrandt and Prauser, 1991). With respect to the physiological characters (Jáger et al., 1983), *P. enterophila* is closer to *O. turbata* than to the type species *P. citrea*. Furthermore, 16S rDNA analysis places *P. enterophila* next to *O. turbata* (99.6% sequence similarity), and both species share the same peptidoglycan type. *Promicromonospora enterophila* did not group with the type species of *Promicromonospora*, *P. citrea*, and with *P. sukumoe* (Takahashi et al., 1987), which, together with *Cellulosimicrobium cellulans* (Schumann et al., 2001), forms a distinct line of descent within the suborder Micrococcineae. The exclusive susceptibility of *P. enterophila* strains to *Oerskovia* phages showed them to be definitely members of this genus (Prauser, 1986). As a consequence, *P. enterophila* has been classified as *Oerskovia enterophila* (Stackebrandt et al., 2002). The two new species *O. jenensis* and *O. paurometabola* had been listed as strains of *O. turbata*. A polyphasic taxonomic study, including 16S rDNA sequence analysis and DNA-DNA similarity studies, revealed the distinct taxonomic status of these strains within the radiation of the genus *Oerskovia*.

Phylogeny

Analysis of the almost complete 16S rDNA of the type strains and some related strains revealed a high degree of relatedness (above 98% similarity, Table 7). The phylogenetic position of these organisms within the radiation of *Cellulomonas* species and other representatives of the suborder Micrococcineae, determined by different treeing algorithms (distance matrix, De Soete, 1983 and Felsenstein, 1993; neighbor joining and maximum parsimony, Felsenstein, 1993), reveals that the genus *Oerskovia* appears as a sister clade of *Cellulomonas* in most analyses (Fig. 2). As already mentioned in the introduction to this chapter, some phylogenetic trees place *C. hominis* at the root of the *Oerskovia* lineage. The phylogenetic tree (Fig. 1) displays the relative phylogenetic position of these organisms. Members of *Oerskovia* differ from those of *Cellulomonas* in the presence of several signature nucleotides of 16S rDNA (Table 3).

Genomic Relatedness

Discrimination of the type strains of the four *Oerskovia* species by DNA-DNA hybridization reveals similarity values below 70%, indicative of members of the same genomospecies (Wayne et al., 1987). Values above 80% indicate that strain DSM 43878 should be considered a strain of *O. turbata*, and strains DSM 46001 and DSM 46097 can be affiliated to *O. jenensis* (Table 7).

Oerskovia strains are well defined by Ribo-Printer® analysis (Bruce, 1996; Allerberger and Fritschel, 1999) using the restriction enzyme *Pst*I (Stackebrandt et al., 2002).

Taxonomy

Oerskoviae are typical nocardioforms but may show a coryneform appearance depending on the particular strain, the age of the culture, and the external growth conditions. In general, they are characterized by extensively branching vegetative hyphae (ca. 0.5 μm in diameter) that grow on the surface of the agar or penetrate into it. The substrate hyphae fragment into bacillary (Fig. 5) and coccoid or spore-like elements, which can be motile by differently arranged flagella, i.e., subpolar tufts of one to three flagella (Higgins et al., 1967), monotrichous flagella (for short cellular elements), and peritrichous flagella (for longer cells; Sukapure et al., 1970). All elements, motile and nonmotile, resulting from fragmentation may give rise to new mycelia independently of their size. No aerial mycelium is formed. The growth appears bacteroid in smears. Colonies are lemon yellow to whitish. Their consistency is smooth and the surface is glistening with a tendency to dull. Edges show mycelial or at least hyphal character, resembling those of other nocardioforms.

Representatives of the genus *Oerskovia* can be differentiated from members of *Cellulomonas* by the occurrence of L-lysine as diagnostic diamino acid of the peptidoglycan (Table 5). The interpeptide bridge consists of either L-Thr←D-Asp (*O. turbata*) or L-Thr←D-Glu (Table 5).

Table 7. 16S rDNA similarity values (upper right triangle) and DNA-DNA reassociation values (lower left triangle) for type strains of *Oerskovia* spp. and related strains.[a]

Taxon	DSM 20577[T]	DSM 43878	DSM 46000[T]	DSM 46001	DSM 46097	DSM 14281[T]	DSM 43852[T]
O. turbata DSM 20577[T]	100	n.d.	99.9	99.9	99.9	99.9	99.6
DSM 43878	97	100	n.d.	n.d.	n.d.	n.d.	n.d.
O. jenensis DSM 46000[T]	57	n.d	100	100	100	100	99.6
DSM 46001	57	n.d.	94	100	100	100	99.6
DSM 46097	64	67	97	85	100	100	99.6
O. paurometabola DSM 14281[T]	58	65	64	67	75	100	99.6
O. enterophila DSM 43852[T]	55	n.d.	53	49	57	59	100

Abbreviations: See footnote in Table 1; and n.d., not determined.
[a]Average of two measurements.

Table 8. Differentiating phenotypic properties of *Oerskovia* species as determined by Biolog GP.

Characteristics	O. turbata DSM 20577[T], DSM 43878	O. jenensis DSM 46001[T], DSM 46000, DSM 46097	O. enterophila DSM 43852[T]	O. paurometabola DSM 14281[T]
Peptidoglycan type	Lys←Thr ← Asp	Lys←Thr←Glu	Lys←Thr ← Glu	Lys←Thr ← Glu
Motility	+	+	–	–
Utilization of				
Mannan	+	+	+	–
α-Acetyl mannosamine	+	+	+	–
Amygdalin	+	+	+	–
Arbutin	+	+	+	–
Cellobiose	+	+	+	–
D-Fructose	+	+	+	–
DL-Fucose	–	–	+	–
D-Galactose	+	+	+	–
D-Galacturonic acid	–	–	+	–
Gentobiose	+	+	+	–
m-Inositol	–	–	+	–
α-D-Lactose	+	–	+	–
D-Melibiose	+	+	+	–
α-Methyl-D-galactoside	+	+	+	–
Sedoheptulosan	–	–	+	–
Stachyose	–	–	+	–
Acetic acid	+	+	+	–
α-Hydroxybutyric acid	–	–	+	–
γ-Hydroxybutyric acid	–	–	+	–
Lactamide	–	–	+	–
D-Lactic acid methylester	+	+	+	–
L-Lactic acid	+	+	+	–
D-Malic acid	–	–	+	–
L-Asparagine	–	–	+	–
Fructose-6-phosphate	–	–	+	–

Symbols and abbreviations: See footnotes in Tables 1, 4 and 5.

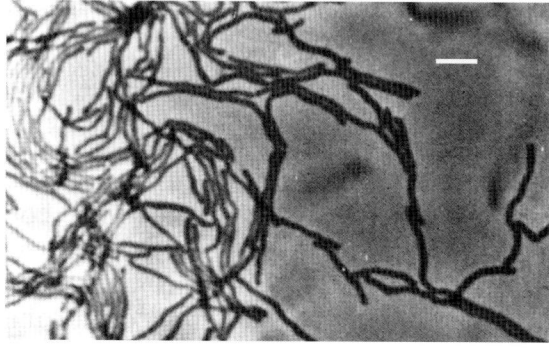

Fig. 5. Nocardioform fragmentation of hyphae of *Oerskovia turbata* NCIB10587 on the agar surface of Bennett's saccharose agar (28-h slide culture). Phase contrast. Bar = 10 μm.

Muramic acid residues of the type species are *N*-acetylated (Uchida and Aida, 1984). The predominating isoprenoid quinone is MK-9(H$_4$). Phosphatidylglycerol, diphosphatidylglycerol and phosphatidylinositol are the major polar lipids (Komura et al., 1975; Lechevalier et al., 1981b). Major fatty acids are *ai*-C$_{15:0}$ and C$_{16:0}$, whereas *i*-C$_{15:0}$, C$_{14:0}$ and *ai*-C$_{17:0}$ occur in smaller amounts (Stackebrandt et al., 2002). Teichoic

acids are lacking (Evtushenko et al., 1984a). The cytochromes belong to the *a, b* and *c* types (Seidl et al., 1980). The G+C content of the type strain of *O. turbata* is 70.5 mol% (Table 5).

The set of phages proposed to be specific of the genus *Oerskovia* (Prauser and Falta, 1968; Prauser, 1984; Prauser, 1986; Stackebrandt and Prauser, 1991) needs to be modified, since the phages O5 (DSM 49112) and O13 (DSM 49139) have host strains recently reclassified as *Cellulosimicrobium cellulans* (DSM 43881 and DSM 46215, respectively). Although no studies on the phage susceptibility of members of the family Cellulomonadaceae have been performed recently, it seems to be reasonable to restrict the set on the *Oerskovia*-specific phages O2 (DSM 49109; host *O. turbata* DSM 43878), O3 (DSM 49138; host *O. jenensis* DSM 46000[T]), and O6 (DSM 49111; host *O. turbata* DSM 20577[T]).

Physiological Properties

Oerskoviae are aerobic to facultatively anaerobic and produce acid from a variety of carbohydrates (Stackebrandt and Kandler, 1980b; Lechevalier and Lechevalier, 1981a). All strains

Table 9. Differentiating phenotypic properties of *Oerskovia* species as determined by API "Coryne."

Organism	*O. turbata*			*O. jenensis*		*O. enterophila*	*O. paurometabola*
	DSM 20577^T	DSM 43878	DSM 46000^T	DSM 46001	DSM 46097	DSM 43852^T	DSM 14281^T
Pyrazinamidase	–	–	–	–	–	+	–
β-Galactosidase	+	+	(+)	(+)	(+)	+	(–)
Urease	–	–	–	–	+	+	+
Gelatin hydrolysis	+	+	–	–	–	–	–
Glycogen fermentation	(+)	–	+	+	+	+	+

Symbols: See footnote in Table 4; (+), weak reaction; and (–), negative reaction.

utilized the following Biolog GP substrates: β-cyclodextrin, dextrin, glycogen, Tween 40, *N*-acetyl glucosamine, D-gluconic acid, α-D-glucose, maltose, maltotriose, mannose, ribose, salicin, sucrose, trehalose, turanose, D-xylose, methyl-pyruvate, glycerol, adenosine, 2-deoxy adenosine, inosine, thymidine, uridine, and adenosine-5′-monophosphate. The following substrates were not utilized by any *Oerskovia* strain: inulin, D-arabitol, melizitose, *p*-hydroxyphenyl acetic acid, α-keto glutaric acid, alanineamide, D-alanine, L-alanine, L-alanyl-glycine, glucose-1-phosphate and glucose-6-phosphate.

API "Coryne" reactions revealed that all strains were positive in the following reactions: oxidase, catalase, nitrate reduction, pyrazimidase, α-glucosidase, *N*-acetyl-β-glucosidase, β-glucosidase, and fermentation of glucose, ribose, xylose, maltose and saccharose. All strains were negative in the following reactions: pyrrolidonyl arylamidase, β-glucuronidase, and fermentation of mannitol and lactose. Distinguishing properties are indicated in Table 9.

Identification

Oerskoviae can readily been identified by morphological, chemotaxonomic and physiological properties indicated in Tables 5, 8 and 9 and Fig. 5.

Habitat

Oerskoviae were randomly and rarely isolated from various types of soils from different geographical regions, composts, decaying plant materials and occasionally from various clinical specimens (Cruickshank et al., 1979; Reller et al., 1975; Sottnek et al., 1977). *Oerskovia enterophila* constitutes the major part of the actinomycete microflora of the intestines and feces of litter-inhabiting millipedes, e.g., *Chromatoiulus projectus* (Dzingov et al., 1982; Jáger et al., 1983; Szabó et al., 1983; Szabó et al., 1986). The species occurs also in the feces of the cave-inhabiting blind isopode *Mesoniscus graniger* (Bodnar et al., 1989). A large homogenous population of facultatively anaerobic *Oerskovia*-type nocardioforms was also found in the gut contents of adult specimens of the common earthworm *Lumbricus polyphemus* (Szabó et al., 1986; Ravasz et al., 1987). None of these oerskoviae could be isolated from the surrounding feeding habitats of the animals. Moreover, the inability of the gut and feces nocardioforms to survive in the natural soil and litter habitats could be demonstrated (Márialigeti et al., 1985). However, among 311 culturable cellulolytic bacterial isolates from soil of an agricultural encatchment, 13 strains were found to show 99.6% 16S rDNA similarity to *Promicromonospora* (*Oerskovia*) *enterophila* (Ulrich and Wirth, 1999).

Isolation, Cultivation and Preservation

Selective media have not been reported for the isolation of strains of the genus. Oerskoviae grow on a wide range of media, as do many other soil bacteria. The main problem in their isolation is the exclusion of organisms that cover large areas of the isolation plates, e.g., swarming bacilli, pseudomonads, and hyphal fungi, as well as the suppression of the numerous streptomycetes, which may be confused at first sight with young stages on the isolation plates and which may possibly antagonize the oerskoviae. Procedures and media used for the isolation of members of the genus *Promicromonospora* (Stackebrandt and Prauser, 1991) can also be recommended for oerskoviae. Tapwater agar (1.5% crude agar in tap water; Lechevalier and Lechevalier, 1989) may also be used to isolate oerskoviae.

Applications

Previous work on *Oerskovia* strains has been performed with *O. xanthineolytica* that has been reclassified recently as *Cellulosimicrobium cellulans* (Schumann et al., 2001). This organism produces inducible extracellular enzymes and rapidly degrades walls of various yeasts (Macmillan et al., 1974). The isolation of α–mannosidase from a strain of *Oerskovia* (Bagiyan et al., 1997) has been reported, but the taxonomic affiliation of this strain remains unclear.

Oerskoviae as Pathogens

The pathogenic potential of oerskoviae was first described by Reller et al. (1975). Strains identified as *O. turbata* were frequently isolated from the blood taken from a patient suffering from endocarditis after homograft replacement of the aortic valve. The source of *O. turbata* remained speculative, although contamination of harvested heart valves with *O. turbata* has been reported (Reller et al., 1975). The occurrence of oerskoviae in various clinical sources was documented (Sottnek et al., 1977) when a large number of motile, Gram-positive, nonsporeforming, yellow-pigmented organisms could be allocated to *O. turbata* and to the former *O. xanthineolytica* (now *Cellulosimicrobium cellulans*). Nine of 31 clinical isolates, collected by the Bacteriology Division, United States Centers for Disease Control (CDC) over a period of 20 years, were identified as *O. turbata*. The source of the *O. turbata* isolates was heart tissues, heart valves, blood and tissues. Since no satisfactory case histories for any of the isolates were available, their clinical significance could not be elucidated.

Vancomycin-sensitive *Oerskovia* strains have been frequently isolated from the commensal bowel flora in humans, where they may cause opportunistic infections. The *vanA* gene, coding for vancomycin and teicoplanin resistance, has been sequenced from the clinical isolate *Oerskovia* strain 892 (Power et al., 1995). The sequence of the plasmid-borne gene was found to be highly similar to that of *Arcanobacterium haemolyticum* and *Enterococcus faecium*. Sialidase (neuraminidase) activity has been detected in culture collection strains of *O. turbata* and a strain isolated from a liver abscess (Müller, 1995). Some of the non-type strains have been reclassified as the new species *O. jenensis* (Stackebrandt et al., 2002). Sialidase activity was not detected in *O. xanthineolytica* (*Cellulosimicrobium cellulans*).

Literature Cited

Agamuthu, P., and E. L. Tan. 1985. Digestion of dried palm oil mill effluent by Cellulomonas species. Microbios Lett. 30:109–113.

Allerberger, F., and S. J. Fritschel. 1999. Use of automated ribotyping of Austrian Listeria monocytogenes isolates to support epidemiological typing. J. Microbiol. Meth. 35:237–244.

Bagiyan, G., E. V. Eneyskaya, A. A. Kulminskaya, A. N. Savel'ev, K. A. Shabalin, and K. N. Neustroev. 1997. The action of αmannosidase from Oerskovia sp. on the mannose-rich O-linked sugar chains of glycoproteins. Eur. J. Biochem. 249:286–292.

Bagnara, C., R. Toci, C. Gaudin, and J. P. Belaich. 1985. Isolation and characterization of a cellulolytic microorganism, Cellulomonas fermentans sp. nov. Int. J. Syst. Bacteriol. 35:502–507.

Bergey, D. H., F. C. Harrison, R. S. Breed, B. W. Hammer, and F. M. Huntoon (Eds.). 1923. Bergey's Manual of Determinative Bacteriology. Williams and Wilkins. Baltimore, MD.

Bichet-Hébé, I., A.-M. Pourcher, L. Sutra, C. Comel, and G. Moguedet. 1999. Detection of a whitening fluorescent agent as an indicator of white paper biodegradation: a new approach to study the kinetics of cellulose hydrolysis by mixed cultures. J. Microbiol. Meth. 37:101–109.

Bodnar, G., I. M. Szabó, and A. Zicsi. 1989. Untersuchungen über die intestinalen Actinomyceten-Gemeinschaften von Mesoniscus graniger Friv./Isopoda. Mémoires de Biospéologie 17:131–136.

Bousfield, I. J. 1972. A taxonomic study of some coryneform bacteria. J. Gen. Microbiol. 71:441–445.

Bruce, J. 1996. Automated system rapidly identifies and characterizes microorganisms in food. Food Technol. 50:77–81.

Chatelain, R., and L. Second. 1966. Taxonomie numerique de quelques Brevibacterium. Ann. Inst. Pasteur 111:630–644.

Choi, W.Y., K. D. Haggett, and N. W. Dunn. 1978. Isolation of a cotton wool degrading strain of Cellulomonas. Mutants with altered ability to degrade cotton wool. Australian J. Biol. Sci. 31:553–564.

Clark, F. E. 1952. The generic classification of the soil corynebacteria. Int. Bull. Bacteriol. Nom. Tax. 2:45–56.

Collard, P. 1963. A species isolated from fermenting cassava roots. J. Appl. Bacteriol. 26:115–116.

Collins, M. D., and D. Jones. 1981. Distribution of isoprenoid quinone structural types in bacteria and their taxonomic implications. Microbiol. Rev. 45:316–354.

Collins, M. D., and C. Pascual. 2000. Reclassification of Actinomyces humiferus (Gledhill and Casida) as Cellulomonas humilata nom. corrig., comb. nov. Int. J. Syst. Evol. Microbiol. 50:661–663.

Coutinho, P. M., and B. Henrissat. 1999. Carbohydrate-active enzymes: an integrated database approach. *In:* H. J. Gilbert, G. J. Davis, B. Henrissat, and B. Svensson (Eds.) Recent Advances in Carbohydrate Bioengineering. Royal Society of Chemistry. Cambridge, UK. 3–12.

Cruickshank, J. G., A. H. Gawler, and C. Shaldon. 1979. Oerskovia species: Rare opportunistic pathogens. J. Med. Microbiol. 12:513–515.

De Leon, C. A., and L. M. Joson. 1980. Conversion of celluloses to protein. Acta Manilana Ser. Natl. Appl. Sci. 19:75–77.

Dermoun, Z., and J. P. Belaich. 1985. Microcalorimetric study of cellulose degradation by Cellulomonas uda ATCC 21399. Biotech. Bioeng. 27:1005–1011.

Dermoun, Z., and J. P. Belaich. 1988a. Crystalline index change in cellulose during aerobic and anaerobic Cellulomonas uda growth. Appl. Microbiol. Biotechnol. 27:399–404.

Dermoun, Z., C. Gaudin, and J. P. Belaich. 1988b. Effects of end-product inhibition of Cellulomonas uda anaerobic growth on cellobiose chemostat culture. J. Bacteriol. 170:2827–2831.

Deschamps, A. M. 1982. Nutritional capacities of bark and wood decaying bacteria with particular emphasis on condensed tannin degrading strains. Eur. J. Path. 12:252–257.

DeSoete, G. 1983. A least squares algorithm for fitting additive trees to proximity data. Psychometrika 48:621–626.

Dey, B. P. 1976. Production, nutritional and toxicological evaluation of Cellulomonas, for protein source (PhD dissertation). University of Missouri-Columbia. Columbia, MO. 1–212.

Dey, B. P., and M. L. Fields. 1995. Toxicity evaluation of strains of Cellulomonas. J. Food Safety 15:265–273.

Diaz, P. L., and H. A. Guirola. 1983. Fermentation study of cellulosic materials of sugarcane by species of the genus Cellulomonas. Rev. Cienc. Biol. 14:283–298.

Dunlap, C E., and C. D. Callihan. 1974. Single cell protein production from cellulosic waste. *In:* H. Yen (Ed.) Recycling and Disposal of Solidwastes: Industrial, Agricultural, Domestic. Ann Harbor Science Publishers. Ann Arbor, MI. 335–347.

Dzingov, A., K. Márialigeti., K. Jáger, E. Contreras, L. Kondics, and I. M. Szabó. 1982. Studies on the microflora of millipedes (Diplopoda). I. A comparison of actinomycetes isolated from surface structures of the exoskeleton and the digestive tract. Pedobiologia 24:1–7.

Elberson, M. A., F. Malekzadeh, M. T. Yazdi, N. Kameranpour, M. R. Noori-Daloii, M. H. Matte, M. Shahamat, R. R. Colwell, and K. R. Sowers. 2000. Cellulomonas persica sp. nov. and Cellulomonas iranensis sp. nov., mesophilic cellulose-degrading bacteria isolated from forest soils. Int. J. Syst. Evol. Microbiol. 50:993–996.

Erikson, D. 1954. Factors promoting cell division in a "soft" mycelial type of Nocardia: Nocardia turbata n. sp. J. Gen. Microbiol. 11:198-208.

Evtushenko, L. I., N. A. Janushkene, G. M. Streshinskaya, I. B. A. Naumova, and N. S. Agre. 1984a. Occurrence of teichoic acids in representatives of the order Actinomycetales. Dokl. Akad. Nauk SSSR 278:237–239.

Evtushenko, L. I., G. F. Levanova, and N. S. Agre. 1984b. Nucleotide composition of DNA and amino acid composition of A4 peptidoglycan in Promicromonospora citrea. Mikrobiologiya 53:519–520.

Felsenstein, J. 1993. PHYLIP (Phylogenetic Inference Package) version 3.51. Department of Genetics, University of Washington. Seattle, WA.

Fiedler, F., and O. Kandler. 1973. Die Mureintypen in der Gattung Cellulomonas Bergey et al. Arch. Microbiol. 89:41–50.

Fields, M. L., S. Tantratian, and R. E. Baldwin. 1991. Production of bacterial and yeast biomass in ground corn cob and ground corn stalk media. J. Food. Prot. 54:117–120.

Funke, G., C. P. Ramos, and M. D. Collins. 1995. Identification of some clinical strains of CDC coryneform group A-3 and A-4 bacteria as Cellulomonas species and proposal of Cellulomonas hominis sp. nov. for some group A-3 strains. J. Clin. Microbiol. 33:2091–2097.

Groth, I., P. Schumann, F. A. Rainey, K. Martin, B. Schuetze, and K. Augsten. 1997a. Bogoriella caseilytica gen. nov., sp. nov., a new alkaliphilic actinomycete from a soda lake in Africa. Int. J. Syst. Bacteriol. 47:788–794.

Groth, I., P. Schumann, F. A. Rainey, K. Martin, B. Schuetze, and K. Augsten. 1997b. Demetria terragena gen. nov., sp. nov., a new genus of actinomycetes isolated from compost soil. Int. J. Syst. Bacteriol. 47:1129–1133.

Groth, I., P. Schumann, K. Martin, B. Schuetze, K. Augsten, I. Kramer, and E. Stackebrandt. 1999a. Ornithinicoccus hortensis gen. nov., sp. nov., a soil actinomycete which contains L-ornithine. Int. J. Syst. Bacteriol. 49:1717–1724.

Groth, I., P. Schumann, B. Schuetze, K. Augsten, I. Kramer, and E. Stackebrandt. 1999b. Beutenbergia cavernae gen. nov., sp. nov., an L-lysine-containing actinomycete isolated from a cave. Int. J. Syst. Bacteriol. 49:1733–1740.

Guyot, J. P. 1986. Role of formate in methanogenesis from xylane by Cellulomonas sp. associated with methanogens and Desulfovibrio vulgaris: inhibition of the aceticlastic reaction. FEMS Microbiol. Lett. 34:149–153.

Haggatt, K. D., W. Y. Choi, and N. W. Dunn. 1978. Mutants of Cellulomonas which produce increased levels of β-glucosidase. Eur. J. Appl. Microbiol. Biotechnol. 6:189–191.

Halsall, D. M., and A. H. Gibson. 1985. Cellulose decomposition and associated nitrogen fixation by mixed cultures of Cellulomonas gelida and Azospirillum species or Bacillus macerans. Appl. Environ. Microbiol. 50:1021–1026.

Halsall, D. M., and A. H. Gibson. 1986a. Comparison of two Cellulomonas strains and their interaction with Azospirillum brasiliense in degradation of wheat straw and associated nitrogen fixation. Appl. Environ. Microbiol. 51:855–861.

Halsall, D. M., and D. J. Goodchild. 1986b. Nitrogen fixation associated with development and localizationof mixed populations of Cellulomonas sp. and Azospirillum brasilense grown on cellulose or wheat straw. Appl. Environ. Microbiol. 51:849–845.

Han, Y. W., C. E. Dunlap, and C. D. Callahan. 1971. Single cell protein from cellulosic waste. Food. Technol. 25:32–34.

Higgins, M. L., M. P. Lechevalier, and H. A. Lechevalier. 1967. Flagellated actinomycetes. J. Bacteriol. 93:1446–1451.

Hsing, W., and E. Canale-Parola. 1992. Cellobiose chemotaxis by the cellulolytic bacterium Cellulomonas gelida. J. Bacteriol. 174:7996–8002.

Hsing, W., and E. Canale-Parola. 1996. A methyl-accepting protein involved in multiple sugar chemotaxis by Cellulomonas gelida. J. Bacteriol. 178:5153–5158.

Jáger, K., K. Márialigeti, M. Hauck, and G. Barabás. 1983. Promicromonospora enterophila sp. nov., a new species of monospore actinomycetes. Int. J. Syst. Bacteriol. 33:525–531.

Jedar, H., A. M. Deschamps, and J. M. Lederbelt. 1987. Production of single cell protein with Cellulomonas sp. on hemstalk wastes. Acta. Biotechnol. 7:103–109.

Jensen, H. L. 1953. Cited from S. A. Waksman (1957), Family II: Actinomycetaceae Buchanan, 1918. *In:* R. S. Breed, E. G. D.Murray, and N. R. Smith (Eds.) Bergey's Manual of Determinative Bacteriology. Williams and Wilkins. Baltimore, MD. 7:713.

Jones, D. 1975. A numerical taxonomic study of coryneform and related bacteria. J. Gen. Microbiol. 87:52–96.

Kalakoutskii, L. V., N. S. Agre, H. Prauser, and L. I. Evtushenko. 1986. Genus Promicromonospora Krasil'nikov, Kalakoutskii & Kirillova 1961, 107[AL]. *In:* P. H. Sneath, N. S. Mair, M. E. Sharpe and J. G. Holt (Eds.) Bergey's Manual of Systematic Bacteriology. William and Wilkins. Baltimore, MD. 1501–1503.

Kaneko, T., K. Kitamura, and Y. Yamamoto. 1969. Arthrobacter luteus nov. sp. isolated from brewery sewage. J. Gen. Appl. Microbiol. 15:317–326.

Kauri, T., and D. J. Kushner. 1985. Role of contact in bacterial degradation of cellulose. FEMS Microbiol. Ecol. 31:301–306.

Kellerman, K. F., F. M. Scales, and N. R. Smith. 1913. Identification and classification of cellulose dissolving bacteria. Zentrabl. Bakteriol. Parasitenk. Infektionskr. Hyg., Abt. II 39:502–522.

Khanna, S. 1993. Glucose uptake by Cellulomonas fimi. World J. Microbiol. Biotechnol. 9:559–561.

Kim, B. H. 1987. Carbohydrate catabolism in cellulolytic strains of Cellulomonas, Pseudomonas, and Nocardia. Kor. J. Microbiol. 25:2833.

Komura, I., K. Yamada, S.-I. Otsuka, and K. Komagata. 1975. Taxonomic significance of phospholipids in coryneform and nocardioform bacteria. J. Gen. Appl. Microbiol. 21:251–261.

Krasilnikov, N. A., L. V. Kalakoutskii, and N. F. Kirillova. 1961. A new genus of ray fungi—Promicromonospora gen. nov. Izv. Akad. Nauk. SSSR Ser. Biol. 1:107–112.

Kumar, N. N., and D. N. Deobagkar. 1997. Expression and characterization of amylase encoded by a gene cloned from Cellulomonas sp. NCIM 2353. World J. Microbiol. Biotechnol. 13:491–496.

Lechevalier, M. P. 1972. Description of a new species, Oerskovia xanthineolytica, and emendation of Oerskovia Prauser et al. Int. J. Syst. Bacteriol. 22:260–264.

Lechevalier, H. A., and M. P. Lechevalier. 1981a. Actinomycete genera "in search of a family." In: M. P. Starr, H. Stolp, H. G. Trüper, A. Balows, and H. G. Schlegel (Eds.) The Prokaryotes. Springer-Verlag. Berlin, Germany. 2118–2123.

Lechevalier, M. P., A. E. Stern, and H. A. Lechevalier. 1981b. Phospholipids in the taxonomy of actinomycetes. In: K. P. Schaal and G. Pulverer (Eds.) Actinomycetes: Proceedings of the Fourth International Symposium on Actinomycete Biology. Gustav Fischer. Stuttgart, Germany. 111–116.

Lechevalier, H. A., and M. P. Lechevalier. 1986. Genus Oerskovia Prauser, Lechevalier & Lechevalier 1970, 534°; emended Lechevalier 1972, 263^AL. In: P. H. A. Sneath, N. S. Mair, M. E. Sharpe and J. G. Holt (Eds.) Bergey's Manual of Systematic Bacteriology. William and Wilkins. Baltimore, MD. 1489–1491.

Lechevalier, H. A., and M. P. Lechevalier. 1989. Genus Oerskovia. In: S. T. Williams, M. E. Sharpe, and J. G. Holt (Eds.) Bergey's Manual of Systematic Bacteriology. Williams and Wilkins. Baltimore, MD. 2379–2382.

Lednicka, D., J. Mergaert, M. C. Cnockaert, and J. Swings. 2000. Isolation and identification of cellulolytic bacteria involved in the degradation of natural cellulosic fibres. Syst. Appl. Microbiol. 23:292–299.

Lee, K. H., and C. Y. Lee. 1986. Characteristics of cellulases from Cellulomonas fimi and conversion of the sawdust into single cell protein. Agricultural Research Soul National University C 10:19–26.

Macmillan, J. D., G. L. Cuffari, T. W. Jeffries, and J. Wilber-Murphy. 1974. Application of yeast cell wall-degrading enzymes for recovery of interacellular products and control of yeast infections. In: Proceedings of the Fourth International Symposium on Yeasts. Vienna, Austria. A13–A14.

Márialigeti, K., E. Contreras, Gy. Barabás, M. Heydrich, and I. M. Szabó. 1985. True intestinal actinomycetes of millipedes (Diplopoda). J. Invertebr. Pathol. 45:120–121.

Marschoun, S., P. Rapp, and F. Wagner. 1987. Metabolism of hexoses and pentoses by Cellulomonas uda under aerobic conditions and during fermentation. Can. J. Microbiol. 33:1024–1031.

Martin, K., P. Schumann, F. A. Rainey, B. Schuetze, and I. Groth. 1997. Janibacter limosus gen. nov., sp. nov., a new actinomycete with meso-diaminopimelic acid in the cell wall. Int. J. Syst. Bacteriol. 47:529–534.

Mayorga-Reyes, L., and T. Ponce-Noyola. 1998. Isolation of a hyperxylanolytic Cellulomonas flavigena mutant growing on continuous culture on sugar cane bagasse. Biotechnol. Lett. 20:443–446.

McCarthy, J. F., and J. T. Pembroke. 1988. The amylase activity associated with Cellulomonas flavigena is cell associated and inducible. Biotechnol. Lett. 10:285–288.

McHan, F., and N. A. Cox. 1987. Differentiation of Cellulomonas species using biochemical tests. Lett. Appl. Microbiol. 4:33–36.

Metcalf, G., and M. Brown. 1957. Nitrogen fixation by new species of Nocardia. J. Gen. Microbiol. 17:567–572.

Minnikin, D. E., M. D. Collins, and M. Goodfellow. 1979. Fatty acid and polar lipid composition in the classification of Cellulomonas, Oerskovia and related taxa. J. Appl. Bacteriol. 47:87–95.

Müller, H. E. 1995. Detection of sialidase activity in Oerskovia (Cellulomonas). Zbl. Bakteriol. 282:13–17.

Odom, J., and J. D. Wall. 1983. Photoproduction of H₂ from cellulose by an anaerobic bacterial coculture. Appl. Environ. Microbiol. 45:1300–1305.

Ørskov, J. 1938. Untersuchungen über Strahlenpilze, reingezüchtet aus dänischen Erdproben. Zbl. Bakt. Hyg., Abt. II 98:344–354.

Park, Y.-H., H. Hori, K.-I. Suzuki, S. Osaa, and K. Komagata. 1987. Phylogenetic analysis of the coryneform bacteria by 5S rRNA sequences. J. Bacteriol. 169:1801–1806.

Ponce-Noyola, T., and M. de la Torre. 1995. Isolation of a high-specific-growth-rate mutant of Cellulomonas flavigena on sugar cane bagasse. Appl. Microbiol. Biotechnol. 42:709–712.

Poulsen, O. M., and L. W. Petersen. 1988. Growth of Cellulomonas sp. ATCC 21399 on different polysaccharides as sole carbon source. Induction of extracellular enzymes. Appl. Microbiol. Biotechnol. 19:480–484.

Pourcher, A.-M., L. Sutra, I. Hébé, G. Moguedet, C. Bollet, P. Simoneau, and L. Gardan. 2001. Enumeration and characterization of cellulolytic bacteria from refuse of a landfill. FEMS Microb. Ecol. 34:229–241.

Power, E. G. M., Y. H. Abdulla, H. G. Talsania, W. Spice, S. Aathithan, and G.-L. French. 1995. VanA genes in vancomycin-resistant clinical isolates of Oerskovia turbata and Arcanobacterium (Corynebacterium) haemolyticum. J. Antimicrob. Chemother. 36:595–606.

Prauser, H. 1967. DAP-freie, gelbe Actinomyceten mit Tendenz zur Beweglichkeit. Zeitschr. Allg. Mikrobiol. 7:81–83.

Prauser, H., and R. Falta. 1968. Phagensensibilität, Zellwand-Zusammensetzung und Taxonomy von Actinomyceten. Zeitschr. Allg. Mikrobiol. 8:39–46.

Prauser, H., M. P. Lechevalier, and H. Lechevalier. 1970. Description of Oerskovia gen. n. to harbor Ørskov's motile nocardia. Appl. Microbiol. 19:534.

Prauser, H. 1984. Phage host ranges in the classification and identification of Gram-positive branched and related bacteria. In: L. Ortiz-Ortiz, L. F. Bojalil, and V. Yakoleff (Eds.) Biological, Biochemical, and Biomedical Aspects of Actinomycetes. Academic Press. Orlando, FL. 617–633.

Prauser, H. 1986. The Cellulomonas, Oersovia, Promicromonospora complex. In: G. Szabó, S. Biro, and M. Goodfellow (Eds.) Biological, Biochemical, and Biomedical Aspects of Actinomycetes, Part B. Akademiai Kiadó. Budapest, Hungary. 527–539.

Przybyl, K. 1979. Bacterial microflora isolated from the bark surface of poplars growing in areas where air pollution is very high. Acta Soc. Bot. Pol. 48:489–496.

Rainey, F. A., N. Weiss, and E. Stackebrandt. 1995. Phylogenetic analysis of the genera Cellulomonas, Promicromonospora, and Jonesia and proposal to exclude the genus Jonesia from the family Cellulomonadaceae. Int. J. Syst. Bacteriol. 45:649–652.

Rajoka, M. I., and K. A. Malik. 1986. Comparison of different strains of Cellulomonas for production of cellulolytic and xylanolytic enzymes from biomass produced on saline lands. Biotechnol. Lett. 8:753–756.

Rajoka, M. I., and K. A. Malik. 1997. Enhanced production of cellulases by Cellulomonas strains grown on different cellulosic residues. Folia Microbiol. 42:59–64.

Ramasamy, K., M. Meyers, J. Bevers, and H. Verachtert. 1981. Isolation and characterization of cellulolytic bacteria from activated sludge. J. Appl. Microbiol. 51: 475–482.

Rapp, P., H. Reng, D. C. Hempel, and F. Wagner. 1984. Cellulose degradation and monitoring of viscosity decrease incultures of Cellulomonas uda grown on printed newspaper. Biotechnol. Bioeng. 26:1167–1175.

Ravasz, K., A. Zicsi, E. Contreras, and I. M. Szabó. 1987. Comparative bacteriological analyses of the fecal matter of different earthworm species. *In:* A. M. B. Pagliai and P. Omodeo (Eds.) On Earthworms: Selected Symposia and Monographs Unione Zoologia Italia. Mucchi. Modena, Italy. 2:389–399.

Reller, L. B., G. L. Maddoux, M. R. Eckman, and G. Pappas. 1975. Bacterial endocarditis caused by Oerskovia turbata. Ann. Int. Med. 83:664–666.

Richard, P. A D., and S. P. Peiris. 1981. The hydrolysis of bagasse hemicellulose by selected strains of Cellulomonas. Biotechnol. Lett. 3:3944.

Rocourt, J., U. Wehmeyer, and E. Stackebrandt. 1987. Transfer of Listeria denitrificans into a new genus Jonesia gen. nov. as Jonesia denitrificans comb. nov. Int. J. Syst. Bacteriol. 37:266–270.

Rodríguez, H., R. Alvarez, and A. Enríques. 1993. Evaluation of different alkali treatments of bagasse pith for cultivation of Cellulomonas sp. World J. Microbiol. Biotechnol. 9:213–215.

Schaal, K. P. 1986. Genus Actinomyces Harz 1877, 133[AL]. *In:* P. H. Sneath, N. S. Mair, M. E. Sharpe, and J. G. Holt (Eds.) Bergey's Manual of Systematic Bacteriology. Williams and Wilkins. Baltimore, MD. 1383–1418.

Schimz, K. L., and B. Overhoff. 1987a. Glycogen, a cytoplasmatic reserve polysaccharide of Cellulomonas sp. (DSM 20108): Its identification, carbon source dependent accumulation, and degradation during starvation. FEMS Microbiol. Lett. 40:325-331.

Schimz, K. L., and B. Overhoff. 1987b. Investigations of the influence of carbon starvation on the carbohydrate storage compounds (trehalose, glycogen), viability, adenylate pool, and adenylate energy charge in Cellulomonas sp. (DSM 20108). FEMS Microbiol. Lett. 40:333–337.

Schumann, P., N. Weiss, and E. Stackebrandt. 2001. Reclassification of Cellulomonas cellulans (Stackebrandt and Keddie, 1986) as Cellulosimicrobium cellulans gen. nov., comb. nov. Int. J. Syst. Evol. Microbiol. 51:1007–1010.

Seidl, P. H., A. H. Faller, R. Loider, and K. H. Schleifer. 1980. Peptidoglycan types and cytochrome patterns of strains of Oerskovia turbata and O. xanthineolytica. Arch. Microbiol. 127:173–178.

Seiler, H., G. Ohmayer, and M. Busse. 1977. Taxonomische Untersuchung an Gram-positiven coryneformen Bakterien unter Anwendung eines EDV-Programms zur Berechnung von Vernetzungsdiagrammen. Zbl. Bakt. Hyg., I. Abt. Orig. A 238:475–488.

Seiler, H. 1983. Identification key for coryneform bacteria derived by numerical taxonomic studies. J. Gen. Microbiol. 129:1433–1471.

Silva, C. F., R. F. Schwan, E. S. Sousa Dias, and A. E. Wheals. 2000. Microbial diversity during maturation and natural processing of coffee cherries of Coffea arabica in Brazil. Int. J. Food Microbiol. 60:251–260.

Sottnek, F. O., J. M. Brown, R. E. Weaver, and G. F. Carroll. 1977. Recognition of Oerskovia species on the clinical laboratory: characterization of 35 isolates. Int. J. Syst. Bacteriol. 27:263–270.

Stackebrandt, E., F. Fiedler, and O. Kandler. 1978. Peptidoglycantyp und Zusammensetzung der Zellwandpolysaccharide von Cellulomonas cartalyticum und einigen coryneformen Organismen. Arch. Microbiol. 117:115–118.

Stackebrandt, E., and O. Kandler. 1979. Taxonomy of the genus Cellulomonas, based on phenotypic characters and deoxyribonucleic acid-deoxyribonucleic acid homology, and proposal of seven neotype strains. Int. J. Syst. Bacteriol. 29:273–282.

Stackebrandt, E., and O. Kandler. 1980a. Cellulomonas cartae sp. nov. Int. J. Syst. Bacteriol. 30:186–188.

Stackebrandt, E., and O. Kandler. 1980b. Fermentation pathway and redistribution of ^{14}C inspecifically labelled glucose in Cellulomonas. Zbl. Bakt. I., Abt. Orig. C1:40–50.

Stackebrandt, E., M. Häringer, and K. H. Schleifer. 1980c. Molecular genetic evidence for the transfer of Oerskovia species into the genus Cellulomonas. Arch. Microbiol. 127:179–185.

Stackebrandt, E., B. J. Lewis, and C. R. Woese. 1980d. The phylogenetic structure of the coryneform group of bacteria. Zbl. Bakt. Hyg. I., Abt. Orig. C1:137–149.

Stackebrandt, E., and C. R. Woese. 1981. Towards a phylogeny of the actinomycetes and related organisms. Curr. Microbiol. 5:197–202.

Stackebrandt, E., H. Seiler, and K. H. Schleifer. 1982. Union of the genera Cellulomonas Bergey et al. and Oerskovia Prauser et al. in a redefined genus Cellulomonas. Zbl. Bakt. Hyg. I., Abt. Orig. C3:401–409.

Stackebrandt, E., W. Ludwig, E. Seewaldt, and K. H. Schleifer. 1983. Phylogeny of sporeforming members of the order Actinomycetales. Int. J. Syst. Bacteriol. 33:173–180.

Stackebrandt, E., and R. M. Keddie. 1986. Genus Cellulomonas Bergey et al. 1923, 154, emend. mut. char. Clark 1952, 50[AL]. *In:* P. H. A. Sneath, N. S. Mair, M. E. Sharpe, and J. G. Holt (Eds.) Bergey's Manual of Systematic Bacteriology. Williams and Wilkins. Baltimore, MD. 1325–1329.

Stackebrandt, E., and H. Prauser. 1991. The family Cellulomonadaceae. *In:* A. Balows, H. G. Trüper, M. Dworkin, W. Harder and K.-H. Schleifer (Eds.) The Prokaryotes. Springer-Verlag. New York, NY. 1323–1345.

Stackebrandt, E., F. A. Rainey, and N. L. Ward-Rainey. 1997. Proposal for a new hierarchic classification system, Actinobacteria classis nov. Int. J. Syst. Bacteriol. 47: 479–491.

Stackebrandt, E., and P. Schumann. 2000. Description of Bogoriellaceae fam. nov., Dermacoccaceae fam. nov., Rarobacteraceae fam. nov. and Sanguibacteraceae fam. nov. and emendation of some families of the suborder Micrococcineae. Int. J. Syst. Evol. Microbiol. 50:1279–1285.

Stackebrandt, E., S. Breymann, U. Steiner, H. Prauser, N. Weiss, and P. Schumann. 2002. Re-evaluation of the status of the genus Oerskovia, reclassification of Promicromonospora enterophila (Jáger et al., 1983) as Oerskovia enterophila comb. nov., and description of Oerskovia jenensis sp. nov. and Oerskovia paurometabola sp. nov. Int. J. Syst. Evol. Microbiol. 52:1105–1111.

Stoppok, W., P. Rapp, and F. Wagner. 1982. Formation, location and regulation of endo-1,4-β-glucanases and β-glucosidases from Cellulomonas uda. Appl. Environ. Microbiol. 44:4–53.

Sukapure, R. S., M. P. Lechevalier, H. Reber, M. L. Higgins, H. A. Lechevalier, and H. Prauser. 1970. Motile nocardoid Actinomycetales. Appl. Microbiol. 19:527–533.

Szabó, I. M., K. Jáger, E. Contreras, K. Márialigeti, A. Dzingov, Gy. Barabás, and M. Pobozsny. 1983. Composition and properties of the external and internal microflora of millipedes (Diplopoda). In: P. Lebrun, H. M. Andre, A. De Medts, C. Gregoire-Wibo, and G. Wauthy (Eds.)Proceedings of the Eighth International Colloquium on Soil Zoology. Dieu-Brichart. Ottignies-Louvain-la-Neuve, France. 197–206.

Szabó, I. M., K. Márialigeti, C. T. Loc, K. Jáger, I. Szabó, E. Contreras, K. Ravasz, M. Heydrich, and E. Palik. 1986. On the ecology of nocardioform intestinal actinomycetes of millipedes (Diplopoda). In: G. Szabó, S. Biró, and M. Goodfellow (Ed.) Biological, Biochemical, and Biomedical Aspects of Actinomycetes, Part B. Akademiai Kiadó. Budapest, Hungary. 701–704.

Takahashi, Y., Y. Tanaka, Y. Iwai, and S. Omura. 1987. Promicromonospora sukumoe sp. nov., a new species of the Actinomycetales. J. Gen. Appl. Microbiol. 33:507–519.

Takayama, K., K. Udagawa, and S. Abe. 1960. Studies on the lytic enzyme produced by Brevibacterium. Part 1. Production of the lytic substance. J. Agric. Chem. Soc. Jap. 34:652–656.

Thayer, D. W. 1984a. Motility and flagellation of cellulomonads. Int. J. Syst. Bacteriol. 34:218–219.

Thayer, D. W., S. V. Lowther, and J. G. Philips. 1984b. Cellulolytic activities of strains of the genus Cellulomonas. Int. J. Syst. Bacteriol. 34:432–438.

Topping, L. E. 1937. The predominant microorganisms in soils. I. Description and classification of the organisms. Zbl. Bakt. II 97:289–304.

Uchida, K., and K. Aida. 1984. An improved method for the glycolate test for simple identification of the acyl type of bacterial cell walls. J. Gen. Appl. Microbiol. 30:131–134.

Ulrich, A., and S. Wirth. 1999. Phylogenetic diversity and population densities of culturable cellulolytic soil bacteria across an agricultural encatchment. Microb. Ecol. 37:238–247.

Vladut-Talor, M., T. Kauri, and D. J. Kushner. 1986. Effects of cellulose on growth, enzyme production, and ultrastructure of a Cellulomonas species. Arch. Microbiol. 144:191–195.

Wayne, L. G., D. J. Brenner, R. R. Colwell, P. A. D. Grimont, O. Kandler, M. I. Krichevsky, L. H. Moore, W. E. C. Moore, R. G. E. Murray, E. Stackebrandt, M. P. Starr, and H. G. Trüper. 1987. Report of the ad hoc committee on reconciliation of approaches to bacterial systematics. Int. J. Syst. Bacteriol. 37:463–464.

Yi, H. R., K.-H. Min, C.-K. Kim, and J.-O. Ka. 2000. Phylogenetic and phenotypic diversity of 4-chlorobenzoate-degrading bacteria isolated from soils. FEMS Microb. Ecol. 31:53–60.

Prokaryotes (2006) 3:1002–1012
DOI: 10.1007/0-387-30743-5_41

CHAPTER 1.1.26

The Family Dermatophilaceae

ERKO STACKEBRANDT

The family Dermatophilaceae embraces two genera, *Dermatophilus* and *Kineosphaera*. *Dermatophilus congolensis* was first described by Van Saceghem (1915) as the causative organism of a disease he named successively as "dermatite granuleuse," "dermatose contagieuse ou impetigo contagieux," and "dermatose dite contagieuse des bovides, impetigo tropical des bovides." This acute, subacute or chronic exudative and proliferative skin disease is prevalent in the humid tropics and subtropics and affects more than thirty animal species and man (Zaria, 1993; Table 1), but animals are also affected in countries of the Northern hemisphere. The disease (also referred to as streptotrichosis, pitted keratolysis, strawberry foot rot, lumpy wool, rain scald or dermatophilosis, depending on the host infected and clinical signs) is economically important, occurring in domestic cattle in tropical regions, sheep in high-rainfall areas, and also wild species (see Zaria [1993] for references). Recently, the genus *Kineosphaera* has been included in the family (Liu et al., 2002). *Kineosphaera limosa*, not reported to be pathogenic, has been isolated from a sequential batch reactor running under alternating anaerobic and aerobic conditions to study biological phosphate removal.

Phylogeny

Austwick (1958) described the family Dermatophilaceae to include the two actinomycete genera *Dermatophilus* and *Geodermatophilus*. The classification of these two genera in a family was based on a single but very unique morphological feature, i.e., division in both transverse and longitudinal planes, which leads to the formation of packets or clusters of cuboid cells or cocci. Considering growth characteristics and developmental stages (Luedemann, 1968), as well as chemical properties (Lechevalier and Lechevalier, 1981; Stackebrandt, 1983), the two previously described species *D. congolensis* and *G. obscurus* differ significantly from each other. It was therefore not surprising that these species

are located at different positions in the dendrogram derived from 16S rRNA cataloguing data. While *G. obscurus*, together with *Frankia*, *Modestobacter* (Mevs et al., 2000) and *Blastococcus* species (Ahrens and Moll, 1970; Urzi et al., 2003), constitutes an individual subline of descent (Stackebrandt et al., 1983; Hahn et al., 1989), *D. congolensis* was found to group by itself within a phylogenetically tight but phenotypically broad cluster, which harbors arthrobacteria, micrococci, cellulomonads and relatives, microbacteria, brevibacteria and others. Consequently, the family Dermatophilaceae was redefined (Stackebrandt and Schumann, 2002) to contain *Dermatophilus* as the only genus, embracing two species, *D. congolens* (Van Saceghem, 1915) and *D. chelonae* (Masters et al., 1995). The type strain of *D. congolensis* has been deposited as ATCC 14637^T = DSM 44180^T = JCM 8106^T = NCTC 13039^T = NRRL B-2350^T. The type strain W16^T of *D. chelonae* is available as ATCC 51576^T = CIP 104541^T = DSM 44178^T = JCM 9706^T.

Analysis of almost complete 16S rRNA gene sequences showed the two type strains of the two described species of *Dermatophilus* to branch within the order Actinomycetales, suborder Micrococcineae. The two type strains share 94.7% sequence similarity, while those with type strains of other genera of the suborder were only 90–92% similar (representatives of *Demetria*, *Dermacoccus*, *Kytococcus*, *Terrabacter* and relatives; see Introduction to the Taxonomy of Actinobacteria in Volume 1 and Liu et al., 2002). Including a varying set of reference organisms, it was noted that the type strains of *D. chelonae* and *D. congolensis* did not consistently group as phylogenetic neighbors. Nevertheless, because of the phylogenetic position, the presence of a defined set of signature nucleotides and common morphological properties the family Dermatophilacea was described (Stackebrandt and Schumann, 2000).

Kineosphaera limosa DSM 14548^T = JCM 11399^T (strain Lpha5^T) was described for a coccoid organism from activated sludge that showed 94% 16S rRNA gene sequence similarity to *Dermatophilus congolensis* DSM44180^T. Unfor-

Table 1. Habitat and geographic region of recent reports on infections by *Dermatophilus* spp.

Habitat	Geographic region	Reference
Camels	Arabia	Gitao et al., 1998a
Camels	Eastern Sudan	Gitao et al., 1998b
Cats	Turkey	Kaya et al., 2000
Dead Beluga whales	St. Lawrence river estuary	Mikaelian et al., 2001
Goats	Tanzania	Msami et al., 2001
Alpine chamois	Italy	De Meneghi et al., 2002
Sheep	Northern Ethiopia	Woldemeskel et al., 2003
Equines		Krüger et al., 1998
Crocodiles	Farms in Australia	Buenviaje et al., 1998
Orangutans	Primate Centre, Germany	Brack et al., 1997
Humans, cattle, buffaloes, goats, horses, and antelopes	India	Pal, 1995
Turtles and tortoises	Zoo, Australia	Masters et al., 1995
Humans	Brazil	Towersey et al., 1993

tunately, the sequence of *D. chelonae* 44178^T was not included in the analysis, even though the sequence of this organism was available at the time of the description of *K. limosa*, giving the impression that the type strain of the latter species branched outside the radiation of the genus *Dermatophilus*. In fact, including sequences of the two type strains of *Dermatophilus*, *K. limosa* DSM 14548^T, and other *Dermatophilus* spp. taken from the EMBL database, it is obvious that *K. limosa* branches within the radiation of the genus *Dermatophilus*. Strain DSM 14548^T clusters with strain Lpha7, another strain from the same activated sludge sample from which strain DSM 14548^T had been recovered (96.9% similarity). Both strains show 95.1–95.9% similarity to *D. chelonae* DSM 44178^T, which indicates a somewhat closer relationship than seen with *D. congolensis* DSM 44180^T (94.0–94.2% similarity). Two other closely related sequences are of the as yet invalid species "*Dermatophilus crocodyli*" (Buenviaje et al., 1998) and a clone obtained from DNA recovered from a microbial community of a laboratory-scale sequence reactor (K. D. Mc Mahon, unpublished observation). The sequence of a strain isolated from a human subgingival plaque defines an individual sublineage (Paster et al., 2001). None of the branches of members of the family Dermatophilaceae is supported by bootstrap values above 70%, and hence the topology of this part of the dendrogram (generated by the additive treeing algorithm of De Soete [1983] on the basis of Jukes and Cantor [1969] correction of binary similarity values) is of little statistical significance.

The presence in strain *Kineosphaera limosa* DSM 14548^T of 16S rRNA gene signature nucleotides defined for members of the genus *Dermatophilus* (Stackebrandt and Schumann, 2000), common chemotaxonomic properties, and the morphology of cells (e.g., forming clusters of cuboid cells or coccoid strains) support the relationship between members of the two genera.

The natural habitat of *D. congolensis* is the diseased tissue of infected animals. It is found in or upon the integument where it causes a chronic or acute exudative dermatitis. This bacterium, which apparently does not grow saprophytically in nature, has been isolated mainly from cattle, sheep, goats and equines and, less commonly, from other sources—including camels, giraffes, gazelles, rodents, monkeys, a polar bear, seals, pigs, humans and even nonmammalian sources, lizards and turtles (Gordon, 1976; Zaria, 1993; Masters et al., 1995; see Table 1 for recent reports). From a medical point of view, *D. congolensis* is unique in that its natural growth cycle is restricted to the living layer of the epidermis. Attempts to find *D. congolensis* in the soil have been unsuccessful (Roberts, 1967b), probably owing to the inability of the organism to survive outside the host environment, especially under dry conditions.

Strains of the second species, *D. chelonae*, have so far only been reported in the study on the original description (Masters et al., 1995). These strains were isolated from an abscess in a tortoise, skin lesions of a turtle, and a nose scab on a snapping turtle, all from the Perth Zoo, Western Australia.

Streptotrichosis in both wild and domestic animals has been reported from virtually every part of the world (Gordon, 1976; Zaria, 1993). As pointed out by Roberts (1981), the disease tends to be more severe in those domestic animals raised under conditions quite different from those of their natural environment, such as European cattle in tropical Africa or Merino sheep in the wetter parts of Australia. Even indigenous African cattle crossbred with exotic breeds are particularly susceptible to streptotrichosis. Cattle imported to Central Africa have only been maintained successfully in an uninfected condition at closely supervised government establishments (Lloyd, 1976). The disease may have an important economic effect in certain countries, and the

1968 Abidjan Conference on Agricultural Research Priorities for Economic Development in Africa has placed streptotrichosis second only to contagious bovine pleuropneumonia on the list of serious bacterial diseases (Doutre and Orue, 1968). Streptotrichosis has been directly responsible for losses in the production of hides, beef and milk, but losses also result from the disease in infected cattle. Lloyd (1976) and Bwangamoi (1976) discuss the effects of streptotrichosis in livestock in Nigeria and East Africa, respectively, estimating that the disease contributes significantly to the reduction of the value of hides (up to 16% for cattle hides), to the decrease in lactation by sucking calves in the field because of lesions on the udder affecting the teats, and to the reduction in meat production caused by either slaughter or death of animals suffering from severe forms of the disease.

A few human isolates are known, most probably as a result of direct contact with infected animals (Kaplan, 1976). One strain, isolated from a wrist infection of an adult male, is on deposit as ATCC 14639.

Kineosphaera limosa Lpha5^T (DSM 14548^T) and its closest relative, strain Lpha7, were isolated only once from an activated sludge reactor, enriched with 2% phosphorous (Liu, 1995; Liu et al., 2000; Liu et al., 2002). Their involvement in the removal of phosphate is unknown. No information is available about pathogenicity of these organisms.

Isolation

The following procedures and descriptions are basically taken from Roberts (1981), supplemented with information from the volume entitled *Dermatophilus Infection in Animals and Man* (Lloyd and Sellers, 1976).

Dermatophilus congolensis is readily isolated from newly formed scabs by chopping or mincing material from the base of the scab in a small volume of sterile nutrient broth or distilled water and then plating on blood agar (5–10% defibrinated sheep blood in a blood agar base). A modified procedure uses Haalstra's method (Haalstra, 1965), growing the isolates on defibrinated 10% human blood agar plates containing 1000 units of polymyxin B sulfate (Pfizer) per ml of medium. To obtain a high yield of good-sized colonies, the plates are incubated at 37°C for about 2 days in an atmosphere of 10% CO_2 or in a candle jar, followed by a day in air. Under a CO_2 atmosphere, the colonies that are formed are grayish, dry and tough and are firmly attached to the agar by invading hyphae and covered by aerial mycelia, as seen under stereoscopic microscope with lateral illumination. On brain-heart-infusion medium, strains have a gray-yellow tint, a wrinkled and shiny surface, and either a rough or smooth consistency, depending on the number of filaments and cocci. During the period in air, the superficial part of the colony is thickened by sporulation events, converting it into a moist, mucoid amber- or cream-colored layer. At this stage, motile zoospores and motile oval or coccoid bodies usually less than a μm in diameter can be observed under the microscope. On agar containing sheep blood, colonies will be surrounded by a zone of complete hemolysis, while no hemolysis occurs with medium containing horse blood.

Isolation of *D. congolensis* from old scabs is difficult because of reduced viability of aged cells or contamination of wetted scabs with other bacteria. Roberts (1981) recommends inoculation of the suspended scab material on the scarified skin of guinea pig or sheep and the use of the resulting scabs for unambiguous identification. Zoospores survive under unfavorable conditions and can resist drying. They can withstand heating at 100°C when dried, and they survive in dry scabs at temperatures between 28 and 31°C for up to 42 months.

Strains of *D. chelonae* were isolated on 9% bovine blood agar (in Oxoid Columbia agar base No. 2; Masters et al., 1995). Growth was observed after 2–3 days of incubation at 37°C in the presence of 10% CO_2 or at 27°C in the ambient atmosphere. Growth also occurred in tryptose-phosphate medium supplemented with 10% ovine serum after 3 days at 37°C when the broth was inoculated with a dense suspension of zoospores.

Kineosphaera limosa was isolated on medium GM1 (Liu et al., 1997) at 30°C at pH 7.0. Medium GM1 contains per litre at pH7.0: glucose 0.5 g, Na-acetate x 3 H_2O 2 g, peptone 0.5 g, yeast extract 0.5 g, KH_2PO_4 0.44 g, $MgSO_4 \times 7H_2O$ 0.5 g, $(NH_4)_2SO_4$ 0.5 g, vitamin solution 10 ml, sludge extract (autoclaved and filtered) 100 ml and agar 16 g.

A lytic phage with species-specific activity was isolated from wool samples infected with *D. congolensis*, collected in Western Australia. This phage reduced the cell numbers of *D. congolensis* on infected wool samples and thus has the potential to be used as a biocontrol agent of dermatophilosis (Patten et al., 1995).

Maintenance Medium

Good growth of *Dermatophilus* strains is achieved in the media recommended by both the American Type Culture Collection (medium 44: brain heart infusion agar, Difco 0418) and the German Culture Collection (medium DSM 65:

brain heart infusion, 18.5 g; glucose, 5.0 g; agar, 12 g; and distilled water, 1 liter; or medium 82: glucose, 4 g; yeast extract, 4 g; malt extract, 10 g; agar, 12 g; and distilled water, 1 liter), 1% tryptone with 0.5% sodium chloride, or on trypticase soy broth agar (medium DSM 535), supplemented with 5% blood.

Kineosphaera limosa is maintained on medium DSM medium 776 (Nakamura et al., 1995), containing in grams per liter: glucose, 0.5; peptone, 0.5; yeast extract, 0.5; Na-glutamate, 0.5; KH_2PO_4, 0.5; $(NH_4)_2SO_4$, 0.1; $MgSO_4 \cdot 7H_2O$, 0.1; and at pH 7.0 and 28°C.

Preservation of Cultures

Well-grown cultures in parafilm-sealed Erlenmeyer flasks may be kept at 4°C for several weeks and frozen at –20°C for several months. For lyophilization, good recovery rates are obtained when fresh materials from colonies on blood agar are suspended in nutrient broth. Vanbreuseghem et al. (1976) recommended the following procedure: the bacterial suspension is made in 2 ml of a mixture of equal volumes of ox serum, a 10% solution of sucrose in distilled water, and a 5% solution of neopeptone (Difco) in distilled water. Viable isolates have been maintained in this state for more than 10 years. Subcultures are made by mixing the lyophilized cells with a solution consisting of 2% glucose and 1% neopeptone in tap water. Freeze dried cultures are stored at 8°C.

Identification

Extensive information is only available for *D. congolensis*. Raised scabs from the skin of an animal should immediately suggest a diagnosis of *Dermatophilus* infection, but it is the morphology, i.e., the stages of the life cycle, that makes identification of this organism relatively easy. Smears of scabs, softened with a few drops of physiological saline, may be stained by the Giemsa, methylene blue, or Grocott silver-methenamine method; according to Gordon (1976), the organisms show up well in Gram-stained smears and in hematoxylin- and eosin-stained sections. The most salient characteristics are the branching mycelium, with formation of transverse rows of cocci.

The morphological features of the life cycle of *D. congolensis* have been extensively described by Van Saceghem (1915), Roberts (1961), Gordon and Edwards (1963), and Gordon (1964). Zoospores germinate by budding to produce a germ tube that extends at its growing tip to elongate into filaments. Transverse septation and branching starts at the other end and consists of the oldest cells of the filament. The structure thickens by division of cells in longitudinal planes. Branches divide at right angles in the same manner as the original filament, giving rise to up to eight parallel rows. In cells enclosed within the epidermis, transverse septation is followed by division exclusively in longitudinal planes, to give each of the original segments of the filament a cross-banded appearance. In liquid growth media, division is not confined so that cuboid packets of cells are formed. When the process of division is complete, the resulting cells separate as fully motile zoospores. These zoospores are enclosed in a thick, irregular mass of capsular matrix that is a degradation product of the cell wall that developed during the divisions of the filament. Zoospores are equipped with from 5 to more than 50 flagella and are isodiametric (Gordon, 1964; Richard et al., 1967; Fig. 2). Certain parts of the life cycle may not occur in artificial culture, depending on the strain investigated and the culture conditions applied (Roberts, 1963). Zoospores of *D. congolensis* are chemotactically attracted to CO_2, which may facilitate their recovery from lesions and epidermis damage (Roberts, 1962). In the presence of 10% CO_2 and at 37°C, good growth is achieved, but aerial hyphae may not be formed and septation and spore formation are delayed. *Dermatophilus congolensis* (Fig. 2D) differs from *D. chelonae* (Fig. 3) in having a thick capsule surrounding mature filaments and fewer flagella per zoospore (Masters et al., 1995).

All members of the family can be identified by their 16S rRNA gene sequence. *Kineosphaera limosa* differs from dermatophili in the lack of formation of a mycelium and of spores in strain 14548[T] (Fig. 4) and in the mol% G+C of DNA (reported to be as high as 71.3 in strain 14548[T], while that of *D. congolensis* DSM 44180[T] is only 57–59; Liu et al., 2002). Other differences occur in the composition of the predominant cellular fatty acids (Table 2). Whether strain DSM 14548[T] shows typical *Dermatophilus*-like morphology when grown under different conditions and whether the discrepancies in the base composition of DNA are confirmed must await further investigations of these properties.

Different methods have been used to characterize strains of *Dermatophilus congolensis* from various geographic regions and hosts. These methods include:

1) Comparison of extracted proteins by sodium dodecyl sulfate polyacrylamide gel electrophoresis (SDS-PAGE) and Western blotting (Makinde and Gyles, 1999). Of different extraction methods used, the separation of lysostaphin-solubilized proteins of whole cells was found suitable for characterization of protein patterns,

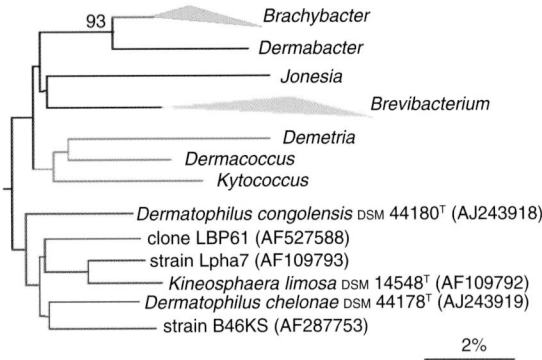

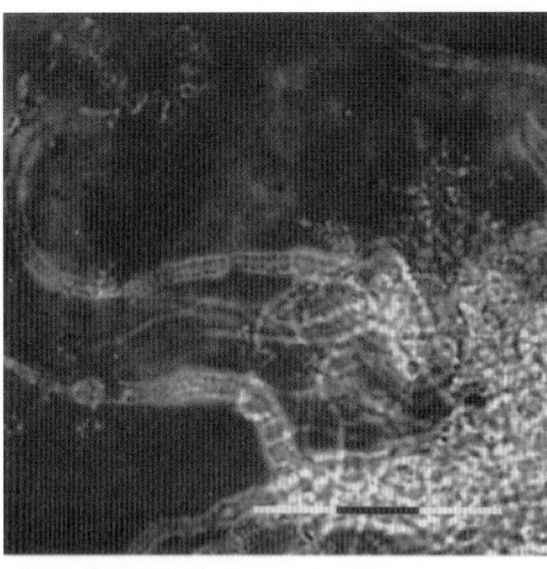

Fig. 1. Phylogenetic tree generated on the basis of 16S rRNA gene sequences showing the nearest neighbors of members of Dermatophilaceae. The tentatively named species "*Dermatophilus crocodyli*" strain TVS 96-490-7b (16S rDNA accession number AF226615) branches with strain LPB61 but was not included in the dendrogram because only 1263 bases are available for comparison. Bootstrap value was calculated for 500 replicate trees. The bar corresponds to a 2% difference in nucleotide sequences.

Fig. 3. Morphology of *Dermatophilus chelonae*. A. Thin capsule (0.5 μm thick) around cells of strain W16[T]. On scale bar, one division = 10 μm. From Masters et al. (1995) with permission.

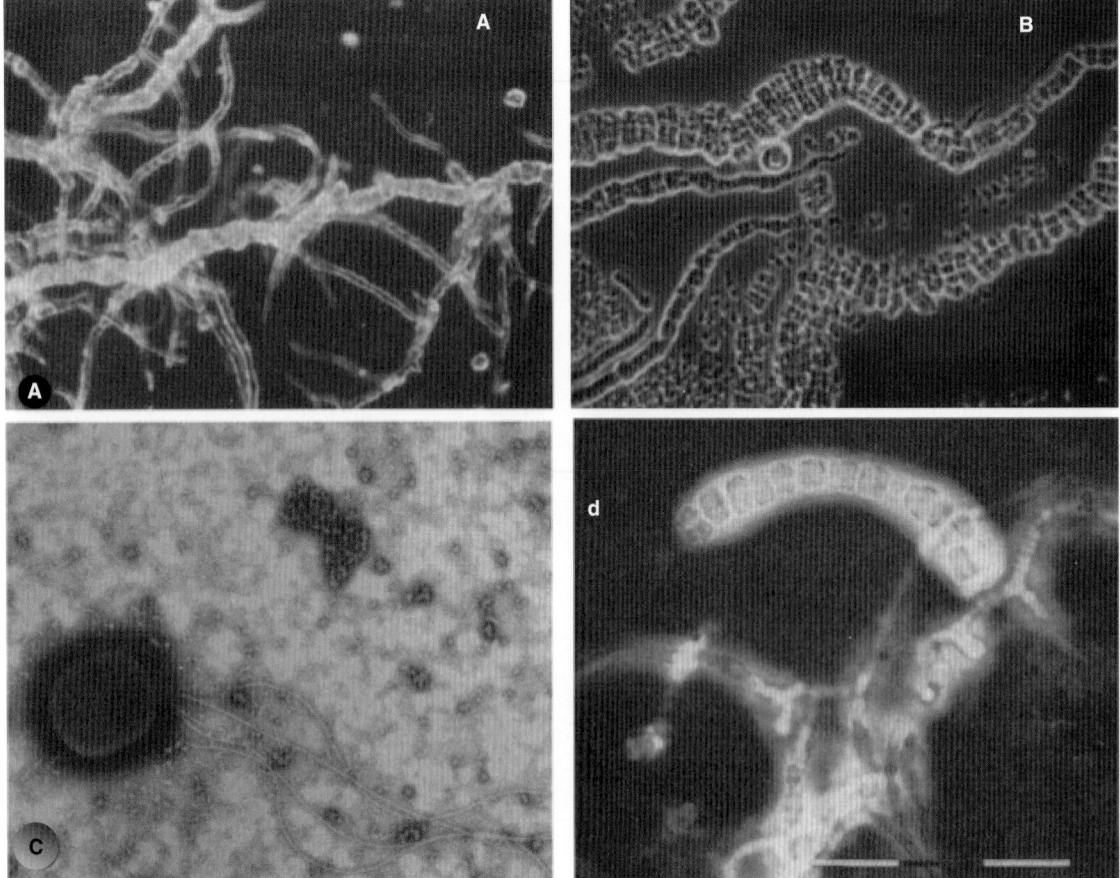

Fig. 2. Morphology of *Dermatophilus congolense*. A) Wet mount of beef infusion-peptone broth culture, showing various stages in development, including fine hyphae and multiseptate filaments that transform into chains of coccal packets. Scale bar = 10 μm. B) Wet mount of broth culture, showing fully segmented hyphae with cubical packets of coccoid spores. Scale bar = 5 μm. C) Negative stain (phosphotungstic acid) of an encapsulated motile spore with tuft of flagella. Scale bar = 10 μm. From Gordon (1964) with permission. D) Thick capsule (2 μm thick) around mature filaments of mammalian strain W15. On scale bar, one division = 10 μm. From Masters et al. (1995) with permission.

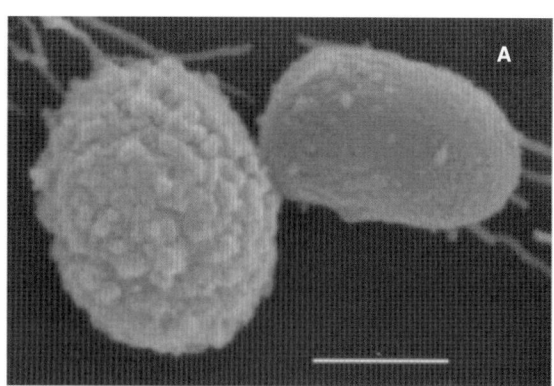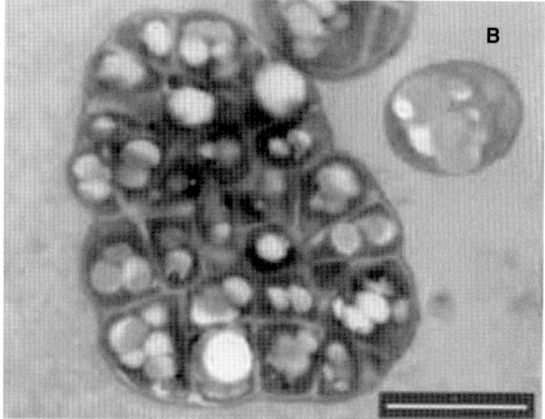

Fig. 4. Morphology of *Kineosphaera limosa* strain Lpha5^T. A) Scanning electron micrograph. B) Transmission electron micrograph, showing intracellular polyhydroxyalkanoate (PHA) granules. Bar = 1.0 μm. From Liu et al. (2002) with permission.

Table 2. Chemotaxonomic characteristics of genera of the family Dermatophilaceae and related genera.

Genus	Diamino acid	Interpeptide bridge	Major menaquinone(s)	Polar lipids	Predominant cellular fatty acids	Mol% G+C
Dermatophilus	*meso*-A$_{2pm}$	None	MK-8(H$_4$)	PG, DPG, PI	C$_{16:0}$, C$_{15:0}$, C$_{14:0}$	57–59
Kineosphaera	*meso*-A$_{2pm}$	None	MK-8(H$_4$)	PG, DPG, PI, PE, PC	C$_{16:0}$, C$_{17:1}$, C$_{18:1}$, C$_{17:0}$	71.3
Dermacoccus	L-Lys	L-Ser$_{1-2}$-D-Glu or L-Ser$_{1-2}$-L-Ala-D-Glu	MK-8(H$_2$)	PI, PG, DPG	*i*-C$_{17:0}$, *ai*-C$_{17:0}$, *i*-C$_{17:1}$	66–71
Kytococcus	L-Lys	D-Glu$_2$	MK-8, MK-9, MK-10	PI, PG, DPG	*ai*-C$_{17:0}$, C$_{17:0}$, *i*-C$_{17:1}$	68–69
Demetria	L-Lys	L-Ser-D-Asp	MK-8(H$_4$)	PI, PE, PG, DPG, PL	C$_{18:1}$, C$_{18:0}$, C$_{17:0}$, *ai*-C$_{17:0}$	66

Abbreviations: A$_{2pm}$, diaminopimelic acid; C$_{14:0}$, tetradecanoic acid; C$_{15:0}$, pentadecanoic acid; C$_{16:0}$, hexadecanoic acid; *i*-C$_{16:0}$, 14-methylpentadecanoic acid; C$_{17:0}$, heptadecanoic acid; *i*-C$_{17:0}$, 15-methylhexadecenoic acid; *ai*-C$_{17:0}$, 14-methylhexadecanoic acid; C$_{17:1}$, heptadecanoic acid; *i*-C$_{17:1}$, 15-methylhexadecenoic acid; C$_{18:0}$, octadecanoic acid; C$_{18:1}$, octadecenoic acid; MK-8(H$_2$) and MK-8(H$_2$) are partially saturated menaquinone with one or two of eight isoprene units hydrogenated, respectively; MK-9 and MK-10 are unsaturated menaquinones with nine or ten isoprene units, respectively; PI, phosphatidylinositol; PG, phosphatidylglycerol; PC, phosphatidylcholine; DPG, diphosphatidylglycerol; PE, phosphatidylethanolamine; and PL, unidentified phospholipid.
Data from Liu et al. (2000) and Stackebrandt and Schumann (2002) and Introduction to the Taxonomy of Actinobacteria.

leading to the recognition of different electropherotypes. Western blotting of different proteins, separated by PAGE with sera from dermatophilosis-infected cattle, demonstrated antigenic relationship between all isolates, leading to the recognition of a prominent common band at 28 kDa.

2) Multilocus enzyme electrophoresis analysis was performed on mainly Australian isolates of *D. congolensis* from warm- and cold-blooded animals (Trott et al., 1995). On the basis of 16 enzyme loci examined, the *Chelonid* isolates grouped differently from the other isolates (these strains were later described as *D. chelonae*). The remaining strains formed two groups that were as distant as those generally found for different species. In each group several subclusters emerged, and in each group α- and β-hemolytic strains were represented. Most of the

non-Australian strains clustered with Australian strains, and most subclusters contain strains from different geographic regions. In conclusion, it was stated that the genetic diversity of *D. congolensis* strains was substantial, confirming results of DNA restriction enzyme analysis.

3) Comparison of riboprints from different *D. congolensis* isolates from one herd in Chad led to the recognition of at least five different restriction fragment length polymorphism patterns, which may partly explain the failure to immunize against dermatophilosis in the field (Faibra, 1993).

4) Analysis of proteases in culture filtrates from *D. congolensis* isolates from different farm animals by substrate (azucasein). SDS-PAGE revealed that bovine isolates contained more proteolytic activity than that of ovine isolates. Serine protease inhibitors had an inhibitory

effect while other classes of proteases had no effect on activity, suggesting that the *Dermatophilus* isolates express a number of alkaline serine proteases (Ambrose et al., 1998).

5) Different DNA isolation protocols were tested for random amplified polymorphic DNA genotyping for isolated of *D. congolensis*. The pattern obtained indicated that genotypic variation between isolates correlated with host species but not with geographic location. Bovine and equine strains appeared more similar to each other than to the ovine strains (Larrasa et al., 2002).

6) Restriction analysis of DNA using the enzymes *Apa*I, *Bam*HI, and *Pvu*II produced profiles that differentiated between *D. congolensis* and *D. chelona*e (Masters et al., 1995).

7) Totally, 120 isolates of *D congolensis* from equine dermatophilosis were subjected to phenotypic characterization by API ZYM™ tests, a conventional physiological test, and analysis of membrane protein profiles (Krüger et al. 1998). These tests revealed high similarity in the data obtained, and minor differences did not correlate with the geographic origin of the isolates. Comparison of the API ZYM™ data with those of Hermoso de Mendoza et al. (1993) revealed similar test profiles. Leucine-arylamidase reacted strongest with the type strain, but esterase (C4), esterase lipase (C8), valine arylamidase, and acid phosphatase also gave distinct reactions with all strains in both studies.

As pointed out in the introduction, dermatophili and *Geodermatophilus obscurus* show striking morphological similarities in certain stages of their life cycles. On the other hand, their differing incompatible ecological niches, together with marked differences in their chemotaxonomic features, colonial pigmentation, location and number of flagella, DNA base composition, mode of branching, and pathogenicity make it easy to clearly differentiate between members of the two genera.

Morphological and chemotaxonomic properties of Dermatophilaceae members are listed in Table 2. Table 3 lists the morphological and physiological properties of members of the family.

Pathogenicity

In principle, sheep and cattle are protected against infection by *Dermatophilus* by the hair coat and dry surface sebum, while the stratum corneum protects uncornified epidermis (Amakari, 1976; Roberts, 1963; Roberts, 1967b). *Dermatophilus congolensis* itself is not highly pathogenic and a combination of factors, such as malnutrition, damage of the skin's defensive barriers, and rainfall, will facilitate infection through

flies attracted by exudates resulting from inflammation caused by arthropods. As summarized by Zaria (1993), growth of *D. congolensis* is inhibited by commensal bacteria, such as *Bacillus* spp. isolated from the skin of healthy sheep, and by coagulase-negative staphylococci isolated from diseased and disease-free pigs; application of antibiotic-producing staphylococci and bacilli to the skin of animals as means of biological control and bacteriotherapy of *D. congolensis* is under discussion (Zaria, 1993).

Zoospores are the principal agents of infection. Spread of infection involving zoospore transfer from wetted scabs to the animal surface is reviewed by Richard and Shotts (1976). It has been observed that introduction of the pathogen into the dermis usually commences during moist atmospheric periods (Roberts, 1967b; see Oppong, 1976) at the sites of trauma, through wounds caused by the activity of biting arthropods, through the contact of plants and animals, and through the sweat ducts and pilosebaceous structures. The main entrance appears to be through hair follicle sheaths, which rupture and allow the organism to spread into the dermis (Amakari, 1976). The odor of cultures of bacteria, including *D. congolensis*, indeed attracts gravid females of the sheep blow fly *Lucilia cuprina* (Morris et al., 1997). *Dermatophilus congolensis* invades only the living layer of the epidermis. From the nature of the skin surface, especially that of cattle, it was deduced that the zoospores require a humid, possibly windswept environment, with a temperature range of 34.5–40°C. In the slightly acidic environment within the epidermis, the organisms proliferate without competing with other microorganisms. Hyphae can be observed in the upper epidermis, especially in the stratum corneum, and within 24 hours of infection the filaments penetrate perpendicularly as far as the epidermal basement membrane (Roberts, 1965; Roberts, 1967b; Zaria, 1993). It is suggested that *D. congolensis* does not penetrate the cellular exudates which separate the infected tissue from the dermal matrix, so that the basement membrane is a natural barrier to penetration. Nevertheless, organisms were seen in the papillary layer of the dermis in those areas where the basement membrane had disintegrated at sites of trauma or when new epidermis became infected by mycelium from hair follicles.

Soon after infection, the invaded cells of the epidermis are covered by one to several layers of laminated scab consisting of cornified epithelial cells and leukocytic cell debris. The characteristic branching mycelium, as well as filaments and zoospores, can be observed in smears of such scabs. Extensive dermal inflammatory reactions are visible on the skin after scabs are removed. New epidermal tissue, formed as a result of cel-

Table 3. Morphological, cultural and physiological characteristics of members of Dermatophilaceae.

Taxon	Morphology	Culture	Physiology
Dermatophilus congolense[a]	Gram-positive branching filaments which divide by transverse and longitudinal septa to form zoospores (5 or more flagella), which form germ tubes, elongate into filaments	On blood agar colonies rough, often becoming viscous, white to gray at first, becoming orange to yellow. Convex colonies with crateriform center may occur.[d] Hemolytic on media containing sheep blood but not horse blood	Acid from glucose and fructose. No acid from lactose, sucrose, xylose, dulcitol, mannitol, sorbitol or salicin. Gelatin and casein hydrolyzed; urease and catalase is produced. Chondoitinase activity against chondroitin 4-sulfate. No hyaluronidase or elastase activity. Nitrate not reduced. Starch hydrolyzed. Susceptible to penicillin, streptomycin, chloramphenicol, erythromycin and tetracyclines. Resistant to antifungal agents
Dermatophilus chelonae[b]	Gram-positive branching filaments which divide by transverse and longitudinal septa to form zoospores (0–6 flagella), which form germ tubes, elongate into filaments. Thin capsule around mature filaments	On bovine blood agar colonies initially dry and adherent, gray white. After transfer to room temperature growth continues, colonies are adherent, white, raised and umbonate or annelliform, becoming sticky after several days. β-Hemolytic on media containing ovine blood; on horse blood nonhemolytic at 37°C and weakly β-hemolytic at 27°C	Acid from glucose, often from fructose and galactose. No acid from lactose, sucrose, xylose, dulcitol, mannitol, sorbitol, trehalose or salicin. Gelatin and casein hydrolyzed; catalase is produced. Urease negative. Chondoitinase activity against chondroitin 4-sulfate. No hyaluronidase or elastase activity. Nitrate weekly reduced to nitrite. Positive lipase activity on tributyrin agar. Susceptible to penicillin, tetracycline, chloramphenicol and sulfafurazol. Resistant to polymyxin, streptomycin and neomycin
Kineosphaera limosa[c]	Gram-positive motile cocci which grow in pairs and packets; nonspore-forming. Polyhydroxybutyrate and polyhydroxyvalerate positive. No accumulation of phosphate	Colonies light-yellow, irregular and umbonate with undulating margins when cells are grown in medium GM1. Hemolysis not reported	Catalase negative, oxidase positive. Obligately aerobic. Nitrate not reduced. Acid not produced. Growth in the presence of up to 3.0% NaCl (w/v)

[a]Data are from Gordon (1989) and Masters etal. (1995).
[b]Data are from Masters et al. (1995).
[c]Data are from Liu et al. (2002).
[d]Zaria (1993) reports on the great variation in the colonial appearance of this species.

lular response, may rapidly be separated from the dermal matrix by invasion of hyphae from infected neighboring follicles (Roberts, 1981). The formation and replacement of keratinic and infected layers of exudated epidermis lead to the formation of massive scabs. The exudates become hard when dry, assuming a paint-brush appearance when it is matted with hair or wool. As discussed by Roberts (Roberts, 1967b; Roberts, 1981) and Oduye (1976), factors limiting the infection are the cellular response and the dermal reactions that are manifested by infiltration of the exudates with lymphocytes, histiocytes and fibroblasts. In animals previously vaccinated, infection ends earlier than in chronic cases, where the cellular response is ineffective even if it is fully developed (Roberts, 1967b).

Besides strains of *D. congolensis* and *D. chelonae*, some *Dermatophilus* strains from lizards have been reported to be pathogenic, one strain for lizard and mice (Anver et al., 1976), another one for sheep (Simmons et al., 1972). Differences in the lesion caused by these organisms have been reported (Montali et al., 1975).

The pathogenesis of dermatophilosis, however, is not well understood (Zaria, 1993), as virulence factors have not yet been defined. Strains of *D. congolensis* and *D. chelonae* are hemolytic (Skalka and Pospisil, 1992) and produce phospholipases (Masters et al., 1995) and proteolytic enzymes (Gordon, 1964; Roberts, 1965; Ambrose et al., 1998). The characteristics and functions of the proteases are not yet known, and neither is the stage in the life cycle known at which production of these factors is most prominent. The immune response to dermatophilosis (including the role of neutrophils in the killing of spores, the characterization of lymphocyte phenotypes, and the phenomenon of susceptibility to dermatophilosis in animals immunocompromised by skin damage and tick [*Amblyomma variegatum*] infestations) has been described in detail by Ambrose et al. (1999).

Successful vaccination against bovine dermatophilosis in southern Chad, using both whole culture and motile zoospores of infective field strains, has been reported by Provost et al. (1976). Systemic antibiotic treatment, including a penicillin and streptomycin combination, has been used in Israel (Nobel, 1976), as well as in Australia where sheep with extensive crusts were made capable of being shorn (Roberts, 1967a). Vaccination is effective against homologous isolates but only partially effective against the challenge of heterologous strains, indicating that immune response may be isolate-specific (How and Lloyd, 1990). Spores and flagella are two of possibly more immunogenic components (Gogolewski et al., 1992). Under laboratory conditions, erythromycin was found to be the most effective antibiotic (Vanbreuseghem et al., 1976). Chemical agents, such as aluminium sulfate, cresol, and copper salts, have been successfully used to treat dermatophilosis in the United Kingdom (Hart et al., 1976). Results of comparative studies—with antibiotics, antiseptics, and miscellaneous compounds applied parenterally and externally—clearly showed the high curative effects of parenteral treatment using antibiotics (Blanchou, 1976).

Literature Cited

Ahrens, R., and G. Moll. 1970. Ein neues knospendes Bakterium aus der Ostsee. Arch. Mikrobiol. 70:242–265.

Amakari, S. F. 1976. Anatomical location of Dermatophilus congolensis in bovine cutaneous streptotrichosis. *In:* D. H. Lloyd and K. C. Sellers (Eds.) Dermatophilus Infection in Animals and Man. Academic Press. London, UK. 163–171.

Ambrose, N. C., M. S. Mijinyawa, and J. Hermoso de Mendoza. 1998. Preliminary characterisation of extracellular serine proteases of Dermatophilus congolensis isolates from cattle, sheep and horses. Vet. Microbiol. 62:321–335.

Ambrose, N., D. Lloyd, and J. C. Maillard. 1999. Immune responses to Dermatophilus congolensis infections. Parasitol. Today 15:295–300.

Anver, M. R., J. S. Park, and H. G. Rush. 1976. Dermatophilosis in the marble lizard (Calotes mystaceus). Lab. Anim. Sci. 26:817–823.

Austwick, P. K. C. 1958. Cutaneous streptotrichosis, mycotic dermatitis and strawberry foot root and the genus Dermatophilus Van Saceghem. Vet. Rev. Annot. 4:33–48.

Blanchou, J. M. 1976. The treatment of infection by Dermatophilus congolensis with particular reference to the disease in cattle. *In:* D. H. Lloyd and K. C. Sellers (Eds.) Dermatophilus Infection in Animals and Man. Academic Press. London, UK. 246–259.

Brack, M., C. Hochleithner, M. Hochleithner, and W. Zenk. 1997. Suspected dermatophilosis in adult orangutan (Pongo pygmaeus). J. Zoo Wildl. Med. 28:336–341.

Buenviaje, G. N., P. W. Ladds, and Y. Martin. 1998. Pathology of skin diseases in crocodiles. Austral. Vet. J. 76:357–363.

Bwangamoi, O. 1976. Economic aspects of streptotrichosis in livestock in East Africa. *In:* D. H. Lloyd and K. C. Sellers (Eds.) Dermatophilus Infection in Animals and Man. Academic Press. London, UK. 292–296.

De Meneghi, D., E. Ferroglio, E. Bollo, L. L. Vizcaino, A. Moresco, and L. Rossi. 2002. Dermatophilosis of alpine Chamois (Rupicapra rupicapra) in Italy. Schweiz. Arch. Tierheilkd. 144:131–136.

De Soete, G. 1983. A least square algorithm for fitting additive trees to proximity data. Psychometrika 48:621–626.

Doutre, M., and J. Orue. 1968. Current problems of bacterial diseases in West Africa. *In:* Agricultural Research Priorities for Economic Development in Africa: The Abijan Conference. 146–173.

Faibra, D. T. 1993. Heterogenicity among Dermatophilus congolensis isolates demonstrated by restriction fragment length polymorphism. Rev. Elev. Méd. Vet. Pays Trop. (France) 46:253–256.

Gitao, C. G., H. Agab, and A. J. Khalifalla. 1998a. An outbreak of a mixed infection of Dermatophilus congolensis and Microsporum gypseum in camels (Camelus dromedarius) in Saudi Arabia. Rev. Sci. Tech. 17:749–755.

Gitao, C. G., H. Agab, and A. J. Khalifalla. 1998b. Outbreaks of Dermatophilus congolensis infection in camels (Camelus dromedarius) from the Butana region in eastern Sudan. Rev. Sci. Technol. 17:743–748.

Gogolewski, R. P., J. A. Mackintosh, S. C. Wilson, and J. C. Chin. 1992. Immunodominant antigens of zoospores from ovine isolates of Dermatophilus congolensis. Vet. Microbiol. 32:305–318.

Gordon, M. A., and M. R. Edwards. 1963. Micromorphology of Dermatophilus congolensis. J. Bacteriol. 86:1101–1115.

Gordon, M. A. 1964. The genus Dermatophilus. J. Bacteriol. 88:509–522.

Gordon, M. A. 1976. Characterization of Dermatophilus congolensis, its affinities with the actinomycetes and differentiation from Geodermatophilus. *In:* D. H. Lloyd and K. C. Sellers (Eds.) Dermatophilus Infection in Animals and Man. Academic Press. London, UK. 187–201.

Gordon, M. A. 1989. Genus Dermatophilus. *In:* S. T. Williams, M. E. Sharpe, and J. G. Holt (Eds.) Bergey's Manual of Systematic Bacteriology. Williams and Wilkins. New York, NY. 2409–2410.

Haalstra, R. T. 1965. Isolation of Dermatophilus congolensis from skin lesions in the diagnosis of streptotrichosis. Vet. Rec. 77:284–291.

Hahn, D., M. P. Lechevalier, A. Fischer, and E. Stackebrandt. 1989. Evidence for a close phylogenetic relationship between members of the genera Frankia, Geodermatophilus and "Blastococcus" and emendation of the family Frankiaceae. Syst. Appl. Microbiol. 11:236–242.

Hart, C. B. 1976. Dermatophilus infection in the United Kingdom. *In:* D. H. Lloyd and K. C. Sellers (Eds.) Dermatophilus Infection in Animals and Man. Academic Press. London, UK. 77–86.

Hermoso de Mendoza, J., A. Arenas, J. M. Alonsa, J. M. Rey, M. C. Gil, J. M. Anton, and M. Hermoso de Mendoza. 1993. Enzymatic activities of Dermatophilus congolensis measured by API ZYM. Vet. Microbiol. 37:175–179.

How, S. J., and D. H. Lloyd. 1990. The effect of recent vaccination on the dose response to experimental Dermatophilus congolensis infection of rabbits. J. Comp. Pathol. 102:157–163.

Jukes, T. H., and C. R. Cantor. 1969. Evolution of protein molecules. *In:* H. N. Munro (Eds.) Mammalian Protein Metabolism. Academic Press. New York, NY. 21–132.

Kaplan, W. 1976. Dermatophilus in primates. *In:* D. H. Lloyd and K. C. Sellers (Eds.) Dermatophilus Infection in Animals and Man. Academic Press. London, UK. 128–138.

Kaya, O., S. Kirkan, and B. Unal. 2000. Isolation of Dermatophilus congolensis from a cat. J. Vet. Med. B. Infect. Dis. Vet. Public Health 47:155–157.

Krüger, B., U. Siesenop, and K. H. Bohm. 1998. Phenotypic characterization of equine Dermatophilus congolensis field isolates. Berl. Münch. Tierärztl. Wochenschr. 111:374–378.

Larrasa, J., A. Garcia, N. C. Ambrose, A. Parra, M. Hermoso de Mendoza, J. Salazar, J. Rey, and J. Hermoso de Mendoza. 2002. A simple random amplified polymorphic DNA genotyping method for field isolates of Dermatophilus congolensis. J. Vet. Med. B. Infect. Dis. Vet. Public Health. 49:135–141.

Lechevalier, H. A., and M. P. Lechevalier. 1981. Introduction to the order Actinomycetales. *In:* M. P. Starr, H. Stolp, H. G. Trüper, A. Balows, and H. G. Schlegel (Eds.) The Prokaryotes. Springer-Verlag. New York, NY. 1915–1922.

Liu, W. T. 1995. Function, Dynamics, and Diversity of Microbial Population in Anaerobic Aerobic Activated Sludge Processes for Biological Phosphate Removal [PhD thesis]. University of Tokyo. Tokyo, Japan.

Liu, W. T., K. Nakamura, T. Matsuo, and T. Mino. 1997. Internal energy-based competition between polyphosphate- and glycogen-accumulating bacteria oin biological phosporus removal reactor-effect of the P/C feeding ratio. Water Res. 31:1430–1438.

Liu, W. T., K. D. Linning, K. Nakamura, T. Mino, T. Matsuo, and L. J. Forney. 2000. Microbial community changes in biological phosphate-removal systems on altering sludge phosphorus content. Microbiology 146:1099–1107.

Liu, W. T., S. Hanada, T. L. Marsh, Y. Kamagata, and K. Nakamura. 2002. Kineosphaera limosa gen. nov., sp. nov., a novel Gram-positive polyhydroxyalkanoate-accumulating coccus isolated from activated sludge. Int. J. Syst. Evol. Microbiol. 52:1845–1849.

Lloyd, D. H. 1976. The economic effects of bovine streptotricosis. *In:* D. H. Lloyd, and K. C. Sellers (Eds.) Dermatophilus Infection in Animals and Man. Academic Press. London, UK. 284–291.

Luedemann, G. M. 1968. Geodermatophilus, a new genus of the Dermatophilaceae (Actinomycetales). J. Bacteriol. 96:1848–1858.

Makinde, A. A., and C. L. Gyles. 1999. A comparison of extracted proteins of isolates of Dermatophilus congolensis by sodium dodecyl sulphate-polyacrylamide gel electrophoresis and Western blotting [Erratum in Vet Microbiol. 72, 341–342]. Vet. Microbiol. 67:251–262.

Masters, A. M., T. M. Ellis, J. M. Carson, S. S. Sutherland, and A. R. Gregory. 1995. Dermatophilus cheloneae sp. nov., isolated from chelonids in Australia. Int. J. Syst. Bacteriol. 45:50–56.

Mevs, U., E. Stackebrandt, P. Schumann, C. A. Gallikowski, and P. Hirsch. 2000. Modestobacter multiseptatus gen. nov., sp. nov., a budding actinomycete from soils of the Asgard Range (Transantarctic Mountains). Int. J. Syst. Evol. Microbiol. 50:337–346.

Mikaelian, I., J. M. Lapointe, P. Labelle, R. Higgins, M. Paradis, and D. Martineau. 2001. Dermatophilus-like infection in beluga whales, Delphinapterus leucas, from the St. Lawrence estuary. Vet. Dermatol. 12:59–62.

Montali, R. J., E. E. Smith, M. Davenport, and M. Bush. 1975. Dermatophilosis in Australian bearded lizards. J. Am. Vet. Med. Assoc. 167:533–555.

Morris, M. C., M. A. Joyce, A. C. Heath, B. Rabel, and G. W. Delisle. 1997. The response of Lucilia cuprina to odours from sheep, offal and bacterial cultures. Med. Vet. Entomol. 11:58–64.

Msami, H. M., D. Khaschabi, K. Schopf, A. M. Kapaga, and T. Shibahara. 2001. Dermatophilus congolensis infection in goats in Tanzania. Trop. Anim. Health Prod. 33:367–377.

Nakamura, K., A. Hiraishi, Y. Yoshimi, M. Kawaharasaki, K. Masuda, and Y. Kamagata. 1995. Microlunatus phosphovorus gen. nov., sp. nov., a new Gram-positive polyphosphate-accumulating bacterium isolated from activated sludge. Int. J. Syst. Bacteriol. 45:17–22.

Nobel, T. A., U. Klopfer, and F. Neumann. 1976. Cutaneous streptotrichosis (dermatophilosis) of cattle in Israel. *In:* D. H. Lloyd and K. C. Sellers (Eds.) Dermatophilus Infection in Animals and Man. Academic Press. London, UK. 70–76.

Oduye, O. O. 1976. The histopathological changes in natural and experimental Dermatophilus congolensis infection of the bovine skin. *In:* D. H. Lloyd and K. C. Sellers (Eds.)Dermatophilus Infection in Animals and Man. Academic Press. London, UK. 172–181.

Oppong, E. N. W. 1976. The epidemiology of Dermatophilus infection of cattle in the Accra plains of Ghana. *In:* D. H. Lloyd and K. C. Sellers (Eds.) Dermatophilus Infection in Animals and Man. Academic Press. London, UK. 17–32.

Pal, M. 1995. Prevalence in India of Dermatophilus congolensis infection in clinical specimens from animals and humans. Rev. Sci. Tech. 14:857–863.

Paster, B. J., S. K. Boches, J. L. Galvin, R. E. Ericson, C. Lau, V. A. Levanos, A. Sahasrabudhe, and F. E. Dewhist. 2001. Bacterial diversity in human subgingival plaque. J. Bacteriol. 183:3770–3783.

Patten, K. M., D. I. Kurtböke, and D. R. Lindsay. 1995. Isolation of Dermatophilus congolensis phage from the "lumpy wool" of sheep in Western Australia. Lett. Appl. Microbiol. 20:199–203.

Provost, A., M. P. Touade, M. Gouillaume, H. Peleton, and F. Damsou. 1976. Vaccination trials against bovine dermatophilosis in southern Chad. *In:* D. Lloyd and K. C. Sellers (Eds.) Dermatophilus Infection in Animals and Man. Academic Press. London, UK. 260–268.

Richard, J. L., A. E. Ritchie, and A. C. Pier. 1967. Electron microscospic anatomy of motile-phase and germinating cells of Dermatophilus congolensis. J. Gen. Microbiol. 49:23–29.

Richard, J. L., and E. B. Shotts. 1976. Wildlife reservoirs of dermatophilosis. *In:* L. A. Page (Ed.) Wildlife Diseases. Plenum Press. New York, NY. 205–214.

Roberts, D. S. 1961. The life cycle of Dermatophilus dermatonomus, the causal agent of ovine mycotic dermatitis. Austral. J. Exper. Biol. Med. Sci. 39:463–476.

Roberts, D. S. 1962. Chemotactic bahaviour of the infective zoospores of Dermatophilus dermatonomus. Austral. J. Agric. Res. 14:400–411.

Roberts, D. S. 1963. The release and survival of Dermatophilus dermatonomus zoospores. Austral. J. Agric. Res. 14:386–399.

Roberts, D. S. 1965. Penetration and irritation of the skin by Dermatophilus congolensis. Brit. J. Exp. Pathol. 46:635–642.

Roberts, D. S. 1967a. Chemotherapy of epidermal infection with Dermatophilus congolensis. J. Comp. Pathol. 77:129–136.

Roberts, D. S. 1967b. Dermatophilus infection. Vet. Bull. 37:513–521.

Roberts, D. S. 1981. The family Dermatophilaceae. *In:* M. P. Starr, H. Stolp, H. G. Trüper, A. Balows, and H. G. Schlegel (Eds.) The Prokaryotes. Springer-Verlag. Berlin, Germany. 2011–2015.

Simmons, G. C., N. D. Sullivan, and P. E. Green. 1972. Dermatophilosis in a lizard (Amphiborulus barbatus). Austral. Vet. J. 48:465–466.

Skalka, B., and L. Pospisil. 1992. Haemolytic interactions of Dermatophilus congolensis. J. Vet. Med. B 39:139–143.

Stackebrandt, E., R. Kroppenstedt, and V. J. Fowler. 1983. A phylogenetic analysis of the family Dermatophilaceae. J. Gen. Microbiol. 129:1831–1838.

Stackebrandt, E., and P. Schumann. 2000. Description of Bogoriellaceae fam. Nov., Dermacoccaceae fam nov., Rarobacteraceae fam. nov and Sanguibacteraceae fam nov. and emendation of some families of the suborder Micrococcineae. Int. J. Syst. Evol. Microbiol. 50:1279–1285.

Towersey, L., E. C. Martins, A. T. Londero, R. J. Hay, P. J. Soares Filho C. M. Takiya, C. C. Martins, and O. F. Gompertz. 1993. Dermatophilus congolensis human infection. J. Am. Acad. Dermatol. 29:351–354.

Trott, D. J., A. M. Masters, J. M. Carson, T. M. Ellis, and D. J. Hampson. 1995. Genetic analysis of Dermatophilus spp. using multilocus enzyme electrophoresis. Zbl. Bakteriol. 282:24–34.

Urzi, C., P. Salamone, P. Schumann, M. Rohde, and E. Stackebrandt. 2003. Blastococcus saxobsidens sp. nov., isolated from marble and calcareous sandstone in quarries and archeological sites in the Mediterranean Basin, and emended description of the genus Blastococcus Ahrens and Moll 1970 and Blastococcus aggregatus Ahrens and Moll 1970. Int. J. Syst. Evol. Microbiol. 53:253–259.

Vanbreuseghem, R., M. Takashio, M. M. El Nageh, D. Presler, M. Selly, and P. Van Wettere. 1976. Some experimental research on Dermatophilus congolensis. *In:* D. H. Lloyd and K. C. Sellers (Eds.) Dermatophilus Infection in Animals and Man. Academic Press. London, UK. 202–212.

Van Saceghem, R. 1915. Dermatose contagieuse (impetigo contagieux). Bull. Soc. Pathol. Exot. 8:354–359.

Woldemeskel, M., and H. Ashenafi. 2003. Study on skin diseases in sheep from northern Ethiopia. Dtsch. Tierärztl. Wochenschr. 110:20–22.

Zaria, L. T. 1993. Dermatophilus congolensis infection (Dermatophilosis) in animals and man: An update. Comp. Immunol. Microbiol. Infect. Dis. 16:179–222.

Prokaryotes (2006) 3:1013–1019
DOI: 10.1007/0-387-30743-5_42

CHAPTER 1.1.27

The Genus *Brevibacterium*

MATTHEW D. COLLINS

Introduction

The genus *Brevibacterium* was established by Breed (1953), with *B. linens* as the type species, for a number of Gram-positive, short, unbranching, rod-shaped bacteria formerly classified in the genus *Bacterium*. Owing to the poor delimitation of the genus, it soon became a repository for a broad range of organisms with very diverse chemical, biochemical and physiological features. Taxonomic studies, in particular those based on cellular chemical analyses, demonstrated the extreme heterogeneity of the genus and resulted in the reclassification of many Brevibacterium species in other coryneform genera (including *Arthrobacter*, *Aureobacterium*, *Corynebacterium*, *Curtobacterium*, *Exiguobacterium*, *Microbacterium*, *Nocardioides*, *Oerskovia* and *Rhodococcus*; see Jones and Keddie, 1985). The description of the genus was subsequently tightened and restricted to the species *Brevibacterium linens* and *B. iodinum* (formerly *Chromobacterium iodinum*; Collins et al., 1980). In addition to *B. linens* and *B. iodinum*, several other species have been assigned to the genus in recent years viz., *B. casei* (Collins et al., 1983), *B. epidermidis* (Collins et al., 1983), *B. mcbrellneri* (McBride et al., 1993), *B. otitidis* (Pascual et al., 1996) and *B. avium* (Pascual and Collins, 1999). The seven currently described species form a phenotypically coherent group of organisms, and 16S rRNA gene sequencing studies have shown the genus to be phylogenetically distinct from other actinomycete taxa (Pascual et al., 1996). The intrageneric phylogenetic relationships of the seven currently described species of *Brevibacterium* inferred from 16S rRNA are shown in Fig. 1.

General Characteristics

Brevibacteria exhibit a distinct rod-coccus cycle during growth on complex media. Cells from older cultures (3–7 days) are composed of coccoid cells (ca. 0.6–1.0 μm diameter), whereas irregular, slender rods are characteristic of exponential phase cultures. Both rod and coccoid forms are Gram-positive, but some strains and older cultures decolorize readily. Members are considered nonmotile although sparse flagellation has been reported for *B. iodinum* (Colwell et al., 1969). Brevibacteria are nonfastidious, chemoorganotrophic, obligately aerobic (oxidative or indifferent toward sugars), and catalase-positive. All brevibacteria are proteolytic, with most strains hydrolyzing casein, gelatin and milk. Brevibacteria grow well the presence of 8% NaCl, and many strains also grow in 15% NaCl.

The cell-wall peptidoglycan is the directly cross-linked type based on *meso*-diaminopimelic acid (A2pm; Schleifer and Kandler, 1972). The walls of brevibacteria contain complex teichoic acids consisting of neutral sugars, amino sugars and sugar alcohols (Anderton and Wilkinson, 1980; Fiedler et al., 1981). The long-chain cellular fatty acids are primarily of the *anteiso*- and *iso*-methyl branched types, with 12-methyltetradecanoic and 14-methylhexadecanoic acids predominating (Collins et al., 1980; Collins et al., 1983; Bousefield et al., 1983). All brevibacteria contain diphosphatidylglycerol, phosphatidylglycerol and dimannosyldiacylglycerol as their major polar lipids; some strains also synthesize phosphatidylinositol (Collins et al., 1980; Collins et al., 1983). Menaquinones are the sole respiratory quinones detected in the cellular membrane of brevibacteria. Dihydrogenated menaquinones with eight isoprene units (MK-8[II-H$_2$]) are generally the major component (Collins et al., 1980; Collins et al., 1983). The G+C content of the DNA of brevibacteria is 60–67 mol%.

Habitat

Brevibacteria exist in a number of different habitats and especially in those having a high salt concentration. *Brevibacterium linens* is a major component of the microflora of surface-ripened cheese such as Limburger, Romadour, Tilsiter and Stilton (Albert et al., 1944; El-Erian, 1969). *Brevibacterium epidermidis* forms part of the res-

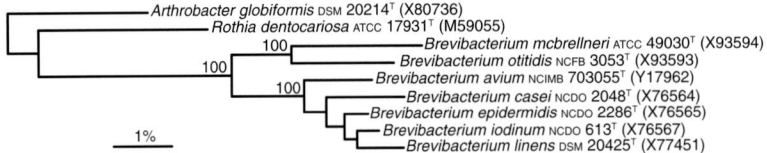

Fig. 1. Dendogram showing intrageneric phylogenetic relationships within the genus *Brevibacterium* based on 16S rRNA. The tree was constructed using the neighbor joining method (Felsenstein, 1989). Bootstrap values (>90%), expressed as a percentage of 500 replications, are given at branching points.

ident flora of the human skin (Pitcher and Noble, 1978), whereas *B. casei* has been isolated from raw milk, cheese curd and cheddar cheese (Sharpe et al., 1976). The natural habitat of this latter species, however, is uncertain. *Brevibacterium casei* has also been isolated from human clinical specimens and it is now thought that most human strains of brevibacteria in fact correspond to this species (Gruner et al., 1994; Funke et al., 1997). Although *B. iodinum* was originally isolated from milk, its natural habitat is unknown (Davis, 1939).

Brevibacteria have been isolated from habitats other than diary products and skin. Orange-pigmented strains considered to be very similar to *B. linens* have been isolated from marine fish (Mulder et al., 1966; Crombach, 1974) and sea-water (Bousefield, 1978). It is pertinent to note that the species *B. linens* is known to be heterogeneous, consisting of at least two DNA-DNA homology groups (Fiedler et al., 1981). Therefore, orange-pigmented brevibacteria may not necessarily correspond to authentic *B. linens*. Strains of brevibacteria also have been isolated from poultry litter (Schefferle, 1966) and pig manure slurry (Keddie and Cure, 1977). *Brevibacterum otitidis*-like organisms have been found in association with sea mammals (G. Foster, C. Pascual, and M. D. Collins, unpublished data). Brevibacteria have also been recovered from bumble-foot lesions of poultry (Mohan, 1981). These latter organisms have recently been classified as a new species, *B. avium* (Pascual and Collins, 1999).

Clinical Significance

Until relatively recently, brevibacteria have not been associated with human infections. There are, however, now several reports of these organisms from human sources, particularly from skin or structures adjacent to skin (Pitcher and Malnick, 1984; Gruner et al., 1994; Funke and Carlotti, 1994). Diseases reported to be due to brevibacteria include osteomyelitis (Neumeister et al., 1993), continuous ambulatory peritoneal dialysis (CAPD) peritonitis (Gruner et al., 1993;

Hummel et al., 1996), and septicemia (Kauko-ranta-Tolvanen et al., 1995; Lina et al., 1994; Reinert et al., 1995). It is now known that organisms formerly referred to as "Centers for Disease Control and Prevention (CDC) groups B-1" and "B-3" from human sources belong to the genus *Brevibacterium* and that most of these correspond to *B. casei* (Gruner et al., 1994). The CDC groups B-1 and B-3 organisms are, however, genetically heterogeneous and there is now evidence of additional unnamed *Brevibacterium* genomospecies having been isolated from human clinical specimens (Gruner et al., 1994). It is pertinent to note that two of the recently described brevibacteria, *B. mcbrellneri* and *B. otitidis*, originated from human sources. *Brevibacterium mcbrellneri* was isolated from infected genital hair of patients with white piedra in association with *Trichosporon beigelii* (McBride et al., 1993), and *B. otitidis* (Pascual et al., 1996) was recovered from human ear infections.

Isolation

Brevibacteria grow well on most peptone-yeast extract based media. Tryptone soya agar supplemented with 4% NaCl (TSAS) is a suitable medium with the following composition (in g/liter): tryptone (Oxoid), 17.0; soya peptone (Oxoid), 3.0; NaCl, 9.0; K_2HPO_4, 2.5; glucose, 2.5; and agar, 12.0. The pH is adjusted to 7.0.

Brevibacteria may be isolated from materials under investigation by streaking or spreading suitable dilutions and incubating at 20–25°C for around 5–7 days. A higher temperature of 30–37°C may be used for skin isolates. For the isolation of *B. linens*, plates should be exposed to light during some period of active growth to enhance pigment formation. A convenient method is to incubate until small colonies are visible and then to remove the plates to a bench exposed to light for the remainder of the period. As the above medium is not selective for brevibacteria, colonies should be examined for a coryneform morphology. Further tests (see below) must be performed to confirm generic and species identification.

Identification

16S rRNA gene sequencing is currently the most reliable means for assigning organisms to the genus and for differentiating between species. There are currently no simple biochemical/physiological criteria for the definitive identification of the genus. Brevibacteria exhibit a marked rod-coccus cycle during growth on complex media. Although morphology is a useful initial test, some members of other Gram-positive genera (e.g., *Arthrobacter*, *Microbacterium*, *Corynebacterium* and *Gordona*) also exhibit this morphology. Chemical tests can, however, be used to distinguish brevibacteria from these and other actinomycete taxa.

The detection of *meso*-A$_2$pm in cell wall or whole-cell hydrolysates serves to distinguish brevibacteria from the majority of other coryneform and actinomycete taxa. Members of the genera Corynebacterium, Gordona and Rhodococcus also contain walls based upon meso-A$_2$pm. Brevibacteria, however, differ from these genera in lacking mycolic acids, possessing predominantly *iso*- and *anteiso*-methyl branched fatty acids, and lacking phosphatidylinositol dimannosides (Collins et al., 1980; Collins et al., 1983). Brevibacteria contain wall teichoic acids (Fiedler et al., 1981; Anderton and Wilkinson, 1980; Anderton and Wilkinson, 1985). These compounds also serve to distinguish brevibacteria from a host of other taxa. Characteristics useful in distinguishing *Brevibacterium* from other actinomycete and coryneform genera are shown in Table 1.

Although 16S rRNA gene sequencing is currently the most reliable means for identifying brevibacteria to the species level, this can also be achieved, albeit to a lesser degree of confidence, using phenotypic criteria. *Brevibacterium linens* is most easily recognized by producing yellow to deep orange smooth colonies (Mulder et al., 1966; Jones et al., 1973), but it is important to bear in mind that the species is genomically heterogeneous. The color reaction of *B. linens* with various chemical reagents has been used as a taxonomic tool for the differentiation of *B. linens* from other coryneform species (Grecz and Dack, 1961), but this is not exclusively selective (Jones et al., 1973). The pigments of *B. linens* have been shown to consist of the aromatic carotenoids 3,3'-dihydroxy-isorenieratene, isorenieratene and 3-hydroxy-isorenieratene (Kohl et al., 1983). In contrast to colonies of *B. linens*, the colonies of *B. iodinum* are characterized by the production of very distinctive purple extracellular crystals of iodinin (Davis, 1939; Clemo and Daglish, 1950; Sneath, 1960). *Brevibacterum mcbrellneri* produces dry friable contoured colonies, whereas the other described species of the genus (viz., *B. avium*, *B. casei*, *B. epidermidis* and *B. otitidis*) generally produce smooth colonies that are whitish-gray in color. To identify brevibacteria to the species level using phenotypic tests, a combination of biochemical and physiological criteria is necessary. Some tests that are useful in differentiating the currently described species of the genus are shown in Table 2.

Production of Bacteriocins and Antimicrobials

Brevibacterium linens has been reported to produce a variety of bacteriocins and antimicrobial substances. Some strains of *B. linens*

Table 1. Differentiation of *Brevibacterium* from other actinomycete and corneform genera.

Genus	Morphology	Wall diamino acid	Arabinogalactan polymer	Mycolic acids	Fatty acid type	Major menaquinone
Brevibacterium	Marked rod-coccus cycle	*meso*-A$_2$pm	–	–	S,A,I	MK-8(H$_2$)
	Marked rod-coccus cycle	Lysine	–	–	S,A,I	MK-9(H$_2$) or MK-8/MK-9
Cellulomonas	Irregular rods, some coccoid forms	L-Omithine	–	–	S,A,I	MK-9(H$_4$)
Clavibacter	Irregular rods	Diaminobutyric	–	–	S,A,I	MK-9,MK-10
Corynebacterium	Irregular rods	*meso*-A$_2$pm	+	+	S,U,(T)	MK-8(H$_2$), MK-9(H$_2$)
Curtobacterium	Irregular rods	D-Omithine	–	–	S,A,I	MK-9
Gordona	Rod-coccus cycle	*meso*-A$_2$pm	+	+	S,U,T	MK-9(H$_2$)
Microbacterium	Irregular rods, some coccoid forms	Lysine	–	–	S,A,I	MK-10, MK-11, MK-12
Rhodococcus	Rods and coccoid forms	*meso*-A$_2$pm	+	+	S,U,T	MK-8(H$_2$)

Symbols: +, present; and –, absent.

Abbreviations: S, straight-chain saturated; A, *anteiso*; I, *iso*; T, tuberculostearic acid; (T), tuberculostearic acid variable; MK-8-12, menaquinones with 8–12 isoprene units; MK-8(H$_2$) and MK-9(H$_2$), dehydrogenated menaquinones with 8 and 9 isoprene units, respectively; and A$_2$pm, diaminopimelic acid.

Table 2. Tests useful for differentiating species of the genus *Brevibacterium*.

Characteristics	B. linens	B. iodinum	B. casei	B. epidermidis	B. mcbrellneri	B. otitidis	B. avium
Colony morphology and consistency	Smooth, creamy	Smooth, creamy	Smooth, creamy	Smooth, creamy	Contoured friable, dry	Smooth, creamy	Smooth, creamy
Pigment	Yellow-orange	Whitish-grayish plus iodinum crystals	Whitish-grayish	Whitish-yellowish	Whitish-beige	Whitish-yellowish	Whitish-grayish
Growth at 37°C	W	+	+	+	+	+	+
Nitrate reduction	+	+	$V^{(-)}$	$V^{(+)}$	−	−	+
α-Glucosidase activity	−	−	V	−	−	−	−
Xanthine hydrolysis	+	+	+	+	V	−	+
Utilization of							
D-Arabinose	−	−	+	−	−	−	−
L-Arabinose	−	−	−	−	−	−	+
D-Arabitol	$V^{(-)}$	−	−	+	−	−	+
Gluconate	−	−	+	+	−	−	−
Mannitol	$V^{(-)}$	−	−	+	−	−	+

Symbols and Abbreviations: +, positive; −, negative; V, variable; $V^{(-)}$, variable with most negative; $V^{(+)}$, variable with most positive; and W, very weak growth.
Data from Jones and Keddie (1986), McBride et al. (1993), Funke and Carlotti (1994), Pascual et al. (1996), Pascual and Collins (1999).

produce a bacteriocin, Linecin A, which is inhibitory to some other *B. linens* strains but not to other species of the genera *Brevibacterium*, *Corynebacterium* and *Micrococcus* (Kato et al., 1991). A bacteriocin designated "Linocin M18," with a broad activity spectrum (active against species of the genera *Arthrobacter*, *Bacillus*, *Corynebacterium*, *Micrococcus* and *Listeria*), has been characterized from a strain of *B. linens* (originating from the surface of red smear cheese; Valdes-Stauber and Scherer, 1994). An antimicrobial substance, designated "Linenscin OC2," which is different from Linecin A and Linocin M18, has been isolated from *B. linens* strain OC2 (Maisnier-Patin and Richard, 1995). Linenscin OC2 is inhibitory towards food-borne pathogens such as *Staphylococcus aureus* and *Listeria monocytogenes*, but not against Gram-negative bacteria. *Brevibacterium linens* strains (isolated from brine used to salt red smear cheese) have also been shown to produce an antimicrobial agent against *Listeria* species (Martin et al., 1995). Similarly, Ryser et al. (1994) isolated from cheese an orange coryneform resembling *B. linens* that produces a bacteriocin-like agent inhibitory for listeriae.

Applications

Brevibacterium linens and other *B. linens*-like organisms are residents of the exterior of surface-ripened soft cheese of the Limburger variety, where they are considered to contribute to the surface color and ripening (Albert et al., 1944; El-Erian, 1969; Eliskases-Lechner and Ginzinger, 1995; Valdès-Stauber et al., 1997). *Brevibacterium linens* and some other brevibacteria produce methanethiol from L-methionine (Sharpe et al., 1976; Sharpe et al., 1977). Methanethiol is an important constituent of the aroma of cheddar cheese, and it is thought that the production of this compound by *B. linens* and related organisms may contribute to the aroma and flavor of surface-ripened cheeses such as Limburger and Romadour. Historically, *B. linens* was considered to be the bacterium solely responsible for ripening of bacterial smear-ripened cheese, but other brevibacteria, micrococci and coryneforms are also of importance.

In addition to the commercial production of *B. linens* for use as a cheese-ripening agent, several patents have been published involving the use of brevibacteria in biotechnological applications, including wastewater treatment, vitamin production, and pharmaceutical processes (Anonymous, 1987, 1988, 1990, 1991; Farrar and Flesher, 1989).

Plasmids and Gene Sequence Information

A multicopy plasmid of size 7.75 kb, designated "pBL100," has been isolated from *B. linens* CECT 75 (Sandoval et al., 1985). A plasmid designated "pBL33" has been isolated from *B. linens* ATCC 9174, which has a size of 7.5 kb and a similar restriction map to that of pBL100 (Kato et al., 1989). Holtz et al. (1992) purified plasmids pBL100 and pBL 33 from *B. linens* ATCC 9174 and DSM 20158. Phenotypic studies indicated resistance to kanamycin is possibly conferred by pBL100 but not by pBL 33. Some strains of *B. linens* do not contain plasmids.

Very little gene sequence information is available on authentic species of the genus *Brevibacterium*. Although many sequence entries are present in the European Molecular Biology Laboratory (EMBL)/GenBank under *Brevibacterium*, the great majority of these derive from species that are not genuine members of the genus. Most available *Brevibacterium* sequence data correspond to that of small subunit rRNA. Extensive complete 16S-23S ribosomal RNA data exist on multiple strains of *B. linens*, although comparative data for other *Brevibacterium* species are not available. There is a paucity of other sequence information, and of that which is available, all pertains to *B. linens* (see Table 3).

Table 3. Gene sequence information on *B. linens*.

Gene/sequence	Accession number(s)
Complete ribosomal RNA operon	U59265-U59268, U62440, U62441, U58866, U62524, U90442, U90443
Insertion sequence (IS)-like element	AFO52055
23S rRNA gene insertion sequence	M85098
Insertion sequence IS5 transposase	AF195203
Tuf gene (elongation factor TU)	X76863
LinM18 gene (linocin M18)	X93588
Plasmid LIM complete sequence	AY004211
Plasmid pBL-A8 gene encoding Rep protein	Y11902
Plasmid pRBL1	U39878
Gene for 1,4-butanediol diacrylate	AB020733
TmpA, a putative efflux protein gene	AF030288

Literature Cited

Albert, J. O., H. F. Long, and B. W. Hammer. 1944. Classification of the organisms important in dairy products IV: Bacterium linens. Agricultural Experimental Station. Iowa Research Bulletin 328:234–259.

Anderton, W. J., and S. G. Wilkinson. 1980. Evidence for the presence of a new class of teichoic acid in the cell wall of bacterium NCTC 9742. J. Gen. Microbiol. 118:343–351.

Anderton, W. J., and S. G. Wilkinson. 1985. Structural studies on a mannitol teichoic acid from the cell wall of bacterium N.C.T.C. 9742. Biochem. J. 226:587–599.

Anonymous inventors. 1987. 3-Amino-1-propionic acid preparation by contacting 3-amino-1-propanol with microbial culture medium, useful for pantothenic acid synthesis. Mitsui from Derwent Info. Ltd. London.

Anonymous inventors. 1988. 1-(4-Methoxy-phenyl)-2-amino-propane, for carbo-styryl derivatives is prepared by culturing microorganism and adding to 1-(4-Methoxy-phenyl)-2-propanone. Tanabe Seiyaku Co. from Derwent Info. Ltd. London.

Anonymous inventors. 1990. Anti-mutagenic agent contains viscous substances produced by Brevibacterium linens as active ingredient. Nippon Oils and Fats Co. Ltd. from Derwent Info. Ltd. London.

Anonymous inventors. 1991. Acetone utilizing microbe belongs to Brevibacterium, used to remove acetone containing gas from organic exhaust gas. Seiko Kakoki Kk, from Derwent Info. Ltd. London.

Bousefield, I. J. 1978. The taxonomy of coryneform bacteria from the marine environment. In: I. J. Bousefield and A. G. G. Callely (Eds.) Coryneform Bacteria. Academic Press. London, 217–233.

Bousefield, I. J., G. L. Smith, T. R. Dando, and G. Hobbs. 1983. Numerical analysis of total fatty acid profiles in the identification of coryneform, nocardioform and some other bacteria. J. Gen. Microbiol. 129:375–394.

Breed, R. S. 1953. The Brevibacteriaceae fam. nov. of order Eubacteriales. In: Riassunti delle Communicazione VI Congresso Internazionale di Microbiologia, Roma. 1:10–15.

Clemo, G. R., and A. F. Daglish. 1950. The phenazine series. Part VIII: The constitution of the pigment of Chromobacterium iodinum. J. Chem. Soc. 1481–1485.

Collins, M. D., D. Jones, R. M. Keddie, and P. H. A. Sneath. 1980. Reclassification of Chromobacterium iodinum (Davis) in a redefined genus Brevibacterium (Breed) as Brevibacterium iodinum nom. rev.; comb. nov. J. Gen. Microbiol. 120:1–10.

Collins, M. D., J. A. E. Farrow, M. Goodfellow, and D. E. Minnikin. 1983. Brevibacterium casei, sp. nov. and Brevibacterium epidermidis sp. nov. Syst. Appl. Microbiol. 4:388–395.

Colwell, R. R., R. V. Citarella, I. Ryman, and G. B. Chapman. 1969. Properties of Pseudomonas iodinum. Can. J. Microbiol. 15:851–857.

Crombach, W. H. J. 1974. Relationships among coryneform bacteria from soil, cheese and sea fish. Ant. v. Leeuwenhoek 40:347–359.

Davis, J. G. 1939. Chromobacterium iodinum (n.sp.). Zentralbl. Bakteriol. Parasitenkd. Infektionskr. Hyg., Abt. II 100:273–276.

El-Erian, A. 1969. Bacteriological Studies on Limburger Cheese (Ph.D. dissertation). Agricultural University, Wageningen. Wageningen, The Netherlands.

Eliskases-Lechner, F., and W. Ginzinger. 1995. The bacterial flora of surface-ripened cheese with special regards to coryneforms. Lait 75:571–584.

Farrar, D., and P. Flesher, inventors. 1989. Dry, low acrylamide monomer containing particulate polyacrylamide production by mixing amidase with coarse aqueous gel polymer particles, absorbing amidase into particles with drying particles. Allied colloids Ltd., Derwent Info. Ltd. London.

Felsenstein, J. 1989. PHYLIP—phylogeny inference package (version 3.2). Cladistics 5:164–166.

Fiedler, F., M. J. Schäffler, and E. Stackebrandt. 1981. Biochemical and nucleic acid hybridization studies on Brevibacterium linens and related strains. Arch. Microbiol. 129:85–93.

Funke, G., and A. Carlotti. 1994. Differentiation of Brevibacterium spp. encountered in clinical specimens. J. Clin. Microbiol. 32:1729–1732.

Funke, G., A. von Graevenitz, J. E. Clarridge, and K. A. Bernard. 1997. Clinical microbiology of coryneform bacteria. Clin. Microbiol. Rev. 10:125–159.

Grecz, N., and G. M. Dack. 1961. Taxonomically significant color reactions of Brevibacterium linens. J. Bacteriol. 82:241–246.

Gruner, E., G. E. Pfyffer, and A. von Graevenitz. 1993. Characterisation of Brevibacterium spp. from clinical specimens. J. Clin. Microbiol. 31:1408–1412.

Gruner, E., A. G. Steigerwalt, D. G. Hollis, R. S. Weyant, R. E. Weaver, C. W. Moss, M. Daneshvar, J. M. Brown, and D. J. Brenner. 1994. Human infection caused by Brevibacterium casei, formerly CDC groups B-1 and B-3. J. Clin. Microbiol. 32:1511–1518.

Holtz, C., N. Domeyer, and B. Kunz. 1992. Occurrence and physical properties plasmids in Brevibacterium linens. Milchwissenschaft 47:705–707.

Hummel, H., W. Bär, K. Dolge-Reetz, H. B. Steinhauer, and G. Funke. 1996. Erfolgreiche Behandlung von durch Brevibacterium casei Verursachten rezidivierenden Peritonitiden bei einem Peritonealdialyse-patienten. Mikrobiologie 6:10–12.

Jones, D., J. Watkins, and S. K. Erickson. 1973. Taxonomically significant colour changes in Brevibacterium linens probably associated with a carotenoid-like pigment. J. Gen. Microbiol. 77:145–150.

Jones, D., and R. M. Keddie. 1985. GENUS Brevibacterium. In: P. H. A. Sneath, N. A. Mair, M. E. Sharpe, and J. G. Holt (Eds.) Bergey's Manual of Systematic Bacteriology. Williams and Wilkins. Baltimore, MD. 2:1301–1313.

Kato, F., N. Hara, K. Matsuyama, K. Hattori, M. Ishii, and A. Murata. 1989. Isolation of plasmids from Brevibacterium. J. Agric. Biol. Chem. 53:879–881.

Kato, F., Y. Eguchi, M. Nakano, T. Oshima, and A. Murata. 1991. Purification and characcretization of Linecin A, a bacteriocin of Brevibacterium linens. J. Agric. Biol. Chem. 55:161–166.

Kaukoranta-Tolvanen, S. S. E., A. Sivonen, A. A. I. Kostiala, P. Hormila, and M. Vaara. 1995. Bacteremia caused by Brevibacterium species in an immunocomprised patient. Eur. J. Clin. Microbiol. Infect. Dis. 14:801–804.

Keddie, R. M., and G. L. Cure. 1977. The cell wall composition and distribution of free mycolic acids in named strains of coryneform bacteria and in isolates from various natural sources. J. Appl. Bacteriol. 42:229–252.

Kohl, W., H. Achenbach, and H. Reichenbach. 1983. The pigments of Brevibacterium linens: Aromatic carotenoids. Phytochem. 22:207–210.

Lina, B., A. Carlotti, V. Lesaint, Y. Devaux, J. Freney, and J. Fleurette. 1994. Persistent bacteremia due to Brevibacterium sp. in an immunocompromised patient. Clin. Infect. Dis. 18:487–488.

Maisnier-Patin, S., and J. Richard. 1995. Activity and purification of Linenscin OC2, an antibacterial substance produced by Brevibacterium linens OC2, an orange cheese coryneform bacterium. Appl. Environ. Microbiol. 61:1847–1852.

Martin, F., K. Friedrich, F. Beyer, and G. Terplan. 1995. Antagonistic effects of strains of Brevibacterium linens against Listeria. Arch. Lebensmittelhygiene 46:7–11.

McBride, M. E., K. M. Ellner, H. S. Black, J. E. Clarridge, and J. E. Wolf. 1993. A new Brevibacterium sp. isolated from infected genital hair of patients with white peidra. J. Med. Microbiol. 39:255–261.

Mohan, K. 1981. Brevibacterium sp. from poultry. Ant. v. Leeuwenhoek 47:449–453.

Mulder, E. G., A. D. Adamse, J. Antheunisse, M. H. Deinema, J. W. Woldendrop, and L. P. T. M. Zevenhuizen. 1966. The relationship between Brevibacterium linens and bacteria of the genus Arthrobacter. J. Appl. Bacteriol. 29:44–71.

Neumeister, B., T. Mandel, E. Gruner, and G. E. Pfyffer. 1993. Brevibacterium species as a cause of osteomyelitis in a neonate. Infection 21:177–178.

Pascual, C., M. D. Collins, G. Funke, and D. G. Pitcher. 1996. Phenotypic and genotypic characterization of two Brevibacterium strains from the human ear: Description of Brevibacterium otitidis sp. nov. Med. Microbiol. Lett. 5:113–123.

Pascual, C., and M. D. Collins. 1999. Description of Brevibacterium avium sp. nov. from poultry. Int. J. Syst. Bacteriol. 49:1527–1530.

Pitcher, D. G., and W. C. Noble. 1978. Aerobic diphtheroids of human skin. In: I. J. Bousefield and A. G. Callely (Eds.) Coryneform Bacteria. Academic Press. London, 265–287.

Pitcher, D. G. and H. Malnick. 1984. Identification of Brevibacterium from clinical sources. J. Clin. Pathol. 37:1395–1398.

Reinert, R. R., N. Schnitzler, G. Haase, R. Lütticken, U. Fabry, K. P. Schaal, and G. Funke. 1995. Recurrent bacteremia due to Brevibacterium casei in an immunocompromised patient. Eur. J. Clin. Microbiol. Infect. Dis. 14:1082–1085.

Ryser, E. T., S. Maisnier-Patin, J. J. Gratadoux, and J. Richard. 1994. Isolation and identification of cheese-smear bacteria inhibitory to Listeria spp. Int. J. Food Microbiol. 21:237–246.

Sandoval, H., G. del Real, L. M. Mateos, A. Aguilar, and J. F. Martin. 1985. Screening of plasmids in non-pathogenic corynebacteria. FEMS Microbiol. Lett. 27:93–98.

Schefferle, H. E. 1966. Coryneform bacteria in poultry deep litter. J. Appl. Bacteriol. 29:147–160.

Schleifer, K. H., and O. Kandler. 1972. Peptidoglycan types of bacterial cell walls and their taxonomic implications. Bacteriol. Rev. 36:407–477.

Sharpe, M. E., B. A. Law, and B. A. Phillips. 1976. Coryneform bacteria producing methanethiol. J. Gen. Microbiol. 94:430–435.

Sharpe, M. E., B. A. Law, B. A. Phillips, and D. G. Pitcher. 1977. Methanethiol production by coryneform bacteria: Strains from dairy and human skin sources and Brevibacterium linens. J. Gen. Microbiol. 101:345–349.

Sneath, P. H. A. 1960. A study of the genus Chromobacterium. Iowa State J. Sci. 34:243–500.

Valdès-Stauber, N., and S. Scherer. 1994. Isolation and characterisation of Linocin M18, a bacteriocin produced by Brevibacterium linens M18. Appl. Environ. Microbiol. 62:1283–1286.

Valdès-Stauber, N., S. Scherer, and H. Seiler. 1997. Identification of yeasts and coryneform bacteria from the surface microflora of brick cheese. Int. J. Food Microbiol. 34:115–129.

Prokaryotes (2006) 3:1020–1098
DOI: 10.1007/0-387-30743-5_43

CHAPTER 1.1.28

The Family Microbacteriaceae

LYUDMILA I. EVTUSHENKO AND MARIKO TAKEUCHI

Introduction

The family Microbacteriaceae (Park et al., 1993; Stackebrandt et al., 1997) embraces a large group of predominantly aerobic Gram-positive bacteria of high G+C content that are distinguished from other actinobacteria by a combination of their unusual B group cell wall peptidoglycan and unsaturated respiratory menaquinones. The bacteria vary in cell morphology, extending from coccoid, small irregular rods to branched fragmenting hyphae. They are spread in various terrestrial and aquatic ecosystems and can be associated with plants, fungi, animals and clinical specimens (Collins and Bradbury, 1992; Funke et al., 1997; Glöckner et al., 2000). Several species and subspecies of the family include either plant pathogens or organisms for which plant pathogenicity has been suggested (Young et al., 1996; Young et al., 2000). The family is a member of the order Actinomycetales, class Actinobacteria (Stackebrandt et al., 1997), and currently harbors 15 genera: *Agreia, Agrococcus, Agromyces, Clavibacter, Cryobacterium, Curtobacterium, Frigoribacterium, Leifsonia, Leucobacter, Microbacterium, Mycetocola, Okibacterium, Plantibacter, Rathayibacter* and *Subtercola*. Furthermore, a number of other microorganisms of the family were reported or can be suggested by 16S rRNA gene sequence analyses to represent separate lines of descent and may be nuclei of new genera. Those include "*Brevibacterium helvolum*" DSM 20419 (Rainey et al., 1994), "*Corynebacterium bovis*" ATCC 13722 (Watts et al., 2001), some copiotrophic bacteria (Iizuka et al., 1998), "*Cryocola antiquus*" (Kochkina et al., 2001; Gavrish et al., 2003), glacial ice microorganisms (Christner, 2002), fresh water ultramicrobacteria (Hahn et al., 2003) and some others. The history, classification, ecology and other information on the organisms of the validly described genera are reviewed in 15 sections, followed by a brief characterization of some bacteria that are waiting for their valid description as new genera.

The current classification system of the family Microbacteriaceae and the genera it comprises is a good example of the use of a classification strategy based on a polyphasic approach to taxonomy (Colwell, 1970). Determination of cell and cell wall components (Schleifer, 1970; Yamada and Komagata, 1970; Yamada and Komagata, 1972a; Yamada and Komagata, 1972b; Schleifer and Kandler, 1972; Yamada et al., 1976; Keddie and Cure, 1977; Keddie and Cure, 1978; Uchida and Aida, 1977; Minnikin et al., 1978; Collins et al., 1979; Collins et al., 1980b; Collins et al., 1982b; Collins et al., 1983c; Collins et al., 1983d; Collins and Jones, 1980a; Collins and Jones, 1981; Collins and Jones, 1983b; Goodfellow and Minnikin, 1985) played the most important role in development of the classification of the bacteria constituting the family Microbacteriaceae in the "nonmolecular" era. Bacteria of the family have group B-type cell wall peptidoglycan (Schleifer and Kandler, 1972), which is characterized by a crosslinkage between the α-carboxyl group of D-glutamic acid in position 2 of the peptide subunit and the C-terminal D-alanine of an adjacent subunit. A diamino acid such as 2,4-diaminobutyric acid (DAB), lysine or ornithine is always present in the interpeptide bridge of this peptidoglycan and can also occupy the third position of the tetrapeptide subunit. In contrast, most other eubacteria have A-type peptidoglycans, in which the crosslinkages bind the ω-amino group of the diamino acid in position 3 of the peptide subunit to the carboxyl group of the C-terminal D-alanine in position 4 of the adjacent peptide subunit (Schleifer and Kandler, 1972). Typical bacteria with the group B-type peptidoglycan, unsaturated respiratory menaquinones, and other associated chemotaxonomic properties clustered together in one or more closely related groups on the basis of early phylogenetic studies, e.g., oligonucleotide cataloging (Stackebrandt et al., 1980), DNA-rRNA cistron similarities (Döpfer et al., 1982), 5S ribosomal RNA sequence similarities (Park et al., 1987), and similarity values of an insertion sequence within the 23S rRNA gene (Roller et al., 1992).

In 1993, Park et al. analyzed the 5S ribosomal RNA sequences for a large set of bacteria with the B-type peptidoglycan along with other actinomycetes and concluded that the sequence similarity and associated chemotaxonomic characteristics provide compelling grounds for classifying all actinomycetes with a group B-type peptidoglycan in a single family. The authors established the family Microbacteriaceae that included the type genus *Microbacterium* and the genera *Agromyces* (Gledhill and Casida, 1969; Zgurskaya et al., 1992), *Curtobacterium* (Yamada and Komagata, 1972b), *Aureobacterium* (Collins et al., 1983c), and *Clavibacter* (Davis et al., 1984b), which had been described at that time. In 1997, Stackebrandt et al. outlined the skeleton of a hierarchic classification of actinobacteria based on analyses of 16S ribosomal RNA and the distribution of signature nucleotides and added the genera *Agrococcus* (Groth et al., 1996) and *Rathayibacter* (Zgurskaya et al., 1993b) to the family Microbacteriaceae. The following pattern of 16S rDNA signature nucleotides was proposed as characteristic of the family: nucleotides at positions 45-396 (U-A), 144-178 (C-G), 258-268 (A-U), 497 (A), 615-625 (A-U), 694 (G), 771-808 (G-C), 839-847 (G-U), 1256 (G), 1310-1327 (A-U), and 1414-1486 (U-A; Stackebrandt et al., 1997). About that time, the genera *Leucobacter* (Takeuchi et al., 1996) and *Cryobacterium* (Suzuki et al., 1997) were added to the family, which has since been enlarged by several new genera: *Frigoribacterium* (Kämpfer et al., 2000), *Leifsonia* (Evtushenko et. al., 2000; Suzuki et al., 1999), *Subtercola* (Männistö et al., 2000), *Mycetocola* (Tsukamoto et al., 2001), *Agreia* (Evtushenko et al., 2001), *Okibacterium* (Evtushenko et al., 2002), and *Plantibacter* (Behrendt et al., 2002). Figure 1 shows the phylogenetic tree based on 16S rRNA gene sequences, indicating the positions of the current 15 genera and species of the family Microbacteriaceae. Typical cultural, morphological, and chemotaxonomic characteristics of the genera are presented in Table 1.

From the Table and the above-mentioned publications dealing with descriptions of the genera and species of the family Microbacteriaceae, it may be generalized that the family embraces mesophilic or psychrophilic bacteria, which are obligately aerobic to facultatively anaerobic, nonsporeforming organisms with varying cell morphology, ranging from coccoid, small irregular rods to branched fragmenting hyphae. All members of the family contain the group B-type peptidoglycan with varying diamino acids (DAB, lysine or ornithine) and varying structures of the polymers. Mycolic acids and arabinogalactan are not present. All strains are characterized by major unsaturated respiratory menaquinones with isoprene units ranging from 9 to 14 and predominant saturated *anteiso-* and *iso*-methyl branched fatty acids. Phosphopids are mainly diphosphatidylglycerol, phosphatidylglycerol, and a variety of glycolipids. The G+C content of the DNA is 65–78 mol%.

The Genus *Agreia*

The genus *Agreia* was proposed for the type species *Agreia bicolorata*, an aerobic coryneform actinobacterium with an unusual peptidoglycan containing both DAB and ornithine as diamino acids (Evtushenko et al., 2001), isolated from plant galls induced by nematodes of the subfamily Anguininae (Evtushenko et al., 1994).

When the description of the genus *Agreia* was in press, the description of a new species similar to *Agreia* in the peptidoglycan and menaquinone compositions was submitted (Behrendt et al., 2002). Unaware of the description of *Agreia*, the authors classified the strains as *Subtercola pratensis* and emended the description of the genus *Subtercola*. Though the 16S rRNA gene similarity between the new and the described *Subtercola* species was high and the branching topology was supported by high bootstrap values, the phylogenetic and chemotaxonomic similarity between *Subtercola pratensis* and *Agreia bicolorata* (Fig. 1; Table 1) became obvious once both species had been published. Recently, *Subtercola pratensis* has been reclassified as *Agreia pratensis* (Schumann et al., 2003).

Phylogeny

Bacteria of the genus *Agreia* form a tight phylogenetic cluster with a 100% bootstrap confidence level. The closest phylogenetic neighbor of the genus *Agreia* is the genus *Subtercola* (Fig. 1). Strains of *Agreia* show high 16S rDNA sequence similarities (up to 97%) with organisms of the genus *Subtercola* which are approximately of the same level as reported for several other closely related genera in the family Microbacteriaceae, e.g., *Clavibacter*, *Rathayibacter* and *Frigoribacterium* (Kämpfer et al., 2000), or in other actinomycete families (Prauser et al., 1997; Labeda et al., 2001; Lee and Hah, 2002).

Taxonomy

Bacteria of the genus *Agreia* are aerobic, catalase-positive, and characterized by round, glistening, opaque, butyrous, sometimes fluid colonies, usually yellow or yellow-orange in young cultures and becoming red-orange or brown-orange in some strains with age. The cells are nonsporeforming, irregular rods; primary branching may be rarely observed. In older cultures, shorter irregular rods usually

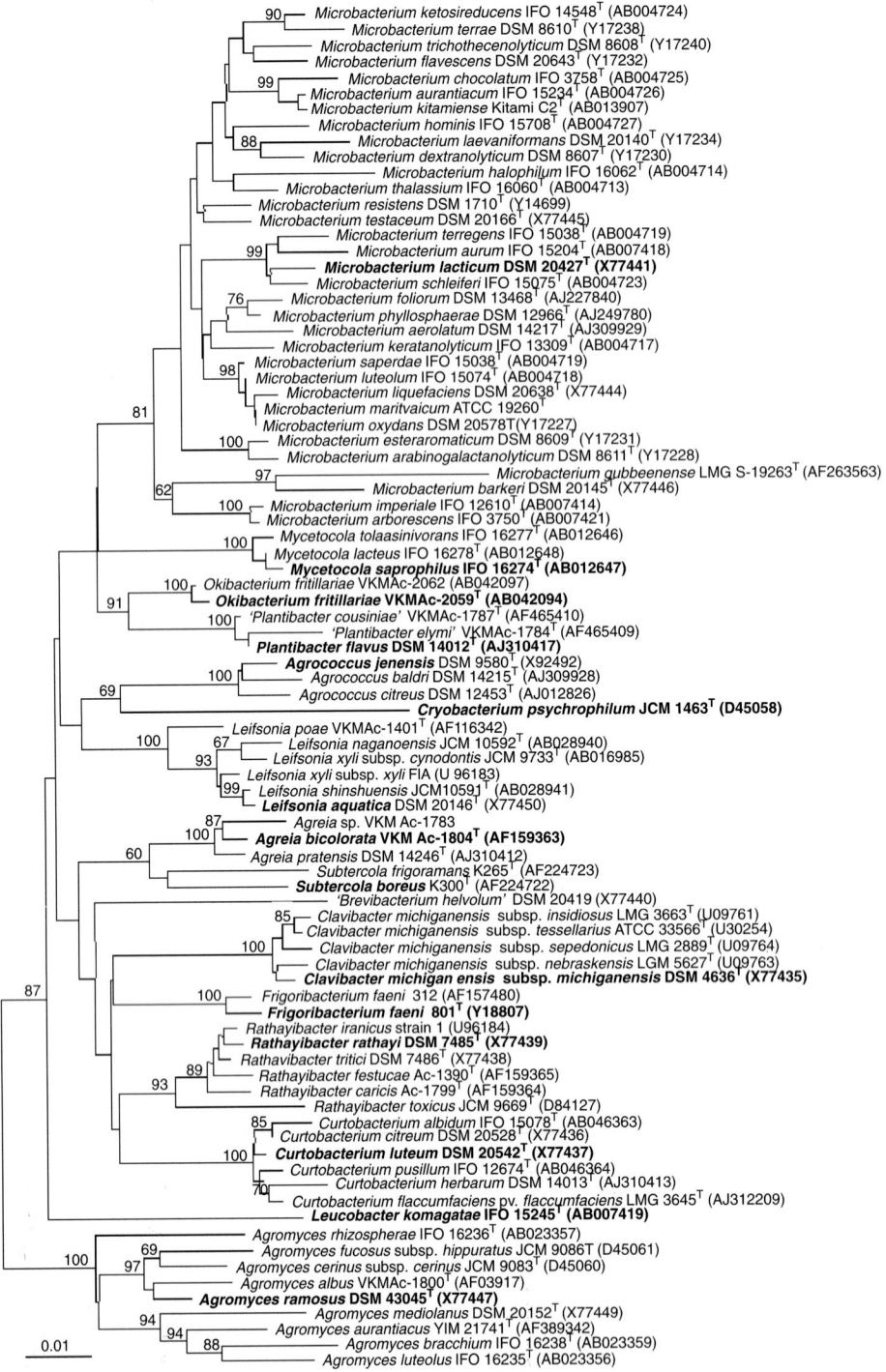

Fig. 1. Phylogenetic tree based on 16S rRNA gene sequence analyses, indicating the positions of 15 genera of the family Microbacteriaceae. The type strains of type species of the genera are given in bold. The tree was generated by the neighbor-joining algorithm (Saitou and Nei, 1987) using Knuc values (Kimura, 1980). The sequence of *Brevibacterium linens* DSM 20425T (X77451) served as an outgroup sequence. The numbers on the tree indicate the percentages of bootstrap samplings (Felsenstein, 1985), derived from 1000 replications. Bar, 1 nucleotide substitution per 100 nucleotides.

Table 1. Differential properties of the genera in the family Microbacteriaceae.

Characteristic	Agreia	Agrococcus	Agromyces	Clavibacter	Cryobacterium	Curto-bacterium	Frigori-bacterium	Leifsonia
Color of colonies	Y/O	W/Y/O	Y/W/O/PG	Y/O/P/WY/BL	P	Y/O/I	PY	Y/WY/W/PB
Morphology	R	C	F/R	R	R[a]	R	R	R
Motility	+/−	−	−	−	−	+/−	+	+
Growth temperature range (°C)	<37	<37	<37	<35	<18	<37	2 ~ 35	<42
Growth optimum (°C)	24 ~ 26	18 ~ 28	26 ~ 30	20 ~ 28	9 ~ 12	24 ~ 26	4 ~ 10 / 24 ~ 26	24 ~ 28
Peptidoglycan amino acids[c]	L-DAB, D-Orn, and Hyg	L-DAB, Asp, and (Thr)[d]	L-DAB	D,L-DAB	L-DAB	D-Orn	L-Lys	D,L-DAB
Peptidoglycan type[e]	B	B2γ	B2γ	B2γ	B2γ	B2β	B2β	B2γ
Acyl type	Acetyl	Acetyl	Acetyl	Acetyl	Acetyl	Acetyl	Acetyl	Acetyl
Major menaquinone, MK-[f]	10	11,12	12	9	10	9	9	11,10/11,12
Predominant fatty acids[h]	a-15:0, i-16:0 and a-17:0	a-15:0, i-16:0, a-17:0	a-15:0, a-17:0, and i-16:0	a-15:0, a-17:0, and i-16:0	a-15:0, a-15:1, i-15:0, and a-17:0	a-15:0, a-17:0, and (Ch-17:0)	a-15:0, i-16:0	a-15:0, a-17:0, and i-16:0
Major 1,1-dimethyl acetals (DMAs)	a-15:0-DMA	−	−	−	−	−	a-15:0-DMA	−
Polar lipids	DPG, PG, and GL	GPG, PG, and GL	DPG, PG, and GL	DPG, PG, and GL	DPG, PG, and GL	DPG, PG, and GL	DPG, PG, and aGL	PG, DPG, and GL
G+C (mol%)	65–67	74–75	69–72	67–78	65	68–71	71	66–71
Habitat/main source of isolation	Plants	Soil and some other sources	Soil	Plants and related sources	Soil	Plants and related sources	Plants, hay dust	Plants, soil, and water

(Continued)

Table 1. *Continued*

Characteristic	Leucobacter	Microbacterium	Mycetocola	Okibacterium	Plantibacter	Rathayibacter	Subtercola
Color of colonies	WB	Y/YW/W/O	Y	Y	Y	Y/RO	PY/BY
Morphology	R	R	R	R	R	R	R[b]
Motility	–	+/–	–	–	–	–	–
Growth temperature range, °C	<37	<42	5 ~ 32	<37	–2 ~ 35	<35	–2 ~ 28
Growth optimum, °C	25	20 ~ 28	25	24 ~ 26	25	24 ~ 28	15 ~ 17
Peptidoglycan amino acids[c]	L-DAB, and GABA	L-Lys / D-Orn, and (Hyg)[d]	Lys	Lys	L-DAB	L-DAB	DAB, and Hyg
Peptidoglycan type[e]	B	B1α, B1β, B2α, B2β	B	B	B2γ	B2γ	B2γ
Acyl type	Acetyl	Glycolyl	Acetyl	Glycolyl	Acetyl	nd	nd
Major menaquinone, MK-[f]	11	11, 12, 13, 14[g]	10	10, 11	10/10.9/10,11	10	9, 10
Predominant fatty acids[h]	a-15:0, a-17:0, i-16:0	a-15:0, a-17:0, i-15:0, and i-16:0	a-15:0, and a-17:0	a-15:0, i-17:0, and i-16:0	a-15:0, i-16:0, and a-17:0	a-15:0, i-16:0, and a-17:0	a-15:0, i-16:0, and a-17:0
Major 1,1-dimethyl acetals, DMA	–	–	–	–	–	–	a-15:0-DMA, and i-16:0-DMA
Polar lipids	DPG, PG, and GL	DPG, PG, GL, and (PGL)	nd	PG, DPG, and GL	PG, DPG, and GL	PG, DPG, (PL), and GL	PG, DPG, PL, and GL
G+C (mol%)	66	65~76	64-65	67	67-70	60-69	64-68
Habitat/main source of isolation	Soil and air	Various	Cultivated mushroom	Plants	Plants and related sources	Plants and related sources	Cold ground water

Symbols and abbreviations: +, positive; –, negative; nd, not determined; BY, bright yellow; I, ivory; O, orange; P, pink; PG, pink gray or pink white; PB, pale brown; PY, pale yellow; RO, rose orange; W, white; WB, whitish brown; WY, whitish yellow; Y, yellow; BL, blue to black somewhat diffusible pigment (indigoidine); C, cocci and coccoid cells; F, fragmenting mycelium; R, irregular rods to short ovoid cells; L-DAB, L-diaminobutyric acid; D-DAB, D-diaminobutyric acid; D-Orn, D-ornithine; L-Lys, L-lysine; Asp, aspartic acid; Thr, threonine; GABA, γ-aminobutyric acid; Hyg, 3-*threo*-hydroxy glutamate; a-15:0-DMA and i-16:0-DMA, 1,1-dimethoxy-*anteiso*-pentadecane and 1,1-dimethoxy-*iso*-hexadecane, respectively; PG, phosphatidylg-lycerol; DPG, diphosphatidylglycerol; PGL, phosphoglycolipid; PL, unknown phospholipid; and GL, glycolipid(s).

[a]Branching occurs in the early growth phase (Suzuki et al., 1997).

[b]Young cells of *S. frigoramans* are frequently swollen at the pole or in the middle (Männistö et al. 2000).

[c]All organisms also contain alanine, glutamic acid and glycine; there are no available data on DAB isomers in *Subtercola* and the configuration of lysine in *Myceticola* and *Okibacterium*.

[d]Parentheses indicate that a compound varies between strains or species.

[e]Peptidoglycan type⁻ is abbreviated as indicated by Schleifer and Kandler (1972).

[f]Menaquinone type is abbreviated as indicated by Collins and Jones (1981).

[g]The distribution of major menaquinone among different species is given in Table 11.

[h]Ch, cyclohexyl fatty acids; other fatty acids are abbreviated as indicated by Collins et al. (1980b).
Data from Yamada and Komagata (1970; 1972a; 1972b), Collins and Jones (1980a), Collins et al. (1980b), Davis et al. (1984b), Zgurskaya et al. (1992; 1993b), Yokota et al. (1993a; 1993b), Groth et al. (1996), Suzuki et al. (1996; 1997; 1999), Takeuchi et al. (1996), Takeuchi and Hatano (1998a; 1998b; 2001), Matsuyama et al. (1999), Schumann et al. (1999), Wieser et al. (1999), Evtushenko et al. (2000; 2001; 2002; submitted), Kämpfer et al. (2000), Männistö et al. (2000), Behrendt et al. (2001; 2002), Tsukamoto et al. (2001), Zlamala et al. (2002a; 2002b), and Li et al. (2003).

predominate; pairs or short chains with diphtheroid arrangements are also observed. A marked rod-coccus growth cycle does not occur. Members of the genus *Agreia* possess an unusual peptidoglycan containing two diamino acids, L-DAB and D-ornithine, and also alanine, glycine, glutamate and hydroxyglutamate. Muramic acid of the peptidoglycan is acetylated. The major menaquinone of respiratory chain is MK-10. The principal phospholipids are phosphatidylglycerol and diphosphatidylglycerol. The major fatty acids are *anteiso*-15:0, *anteiso*-17:0 and *iso*-16:0; 1,1-dimethoxy-*anteiso*-pentadecane occurs in the whole-cell methanolysate. The G+C content of DNA is 65.0–67.0 mol% (Evtushenko et al., 2001; Behrendt et al., 2002; L. I. Evtushenko and M. Takeuchi, unpublished observation).

Currently, the genus contains two validly described species, *Agreia bicolorata* (Evtushenko et al., 2001) and *Agreia pratensis* (Schumann et al., 2003).

Habitat

The data available suggest that the main natural habitat of *Agreia* is plant material. *Agreia bicolorata* was isolated from leaf galls induced by the plant parasitic gall-forming nematode *Heteroanguina graminophila* (the subfamily Anguininae) on its host plant, *Calamagrostis neglecta* (Ehrh.) P. Gaertn. et al. (Evtushenko et al., 1994; Evtushenko et al., 2001). There is some evidence that *Agreia bicolorata* might be a particular nematode associate, like *Rathayibacter* spp. (Riley and Ophel, 1992; Zgurskaya et al., 1993b), carried by nematodes into plants and developing inside plant galls induced by the nematode (Price et al., 1979; Bird, 1981). Strains of *Agreia bicolorata* represent a significant portion of bacterial isolates grown from the leaf gall of narrow reed grass collected in vicinity of Moscow, Russia, and were not observed in the same plants free of nematode infection (Evtushenko et al., 1994; Evtushenko et al., 2001). They also produced abundant polysaccharide slime, which is thought to be one of the characteristic features of nematode-associated bacteria (Bird, 1985). On the other hand, *Agreia bicolorata* may be a secondary settler of the galls (their surface sections), descending from the microbial communities of the phyllosphere or plant tissue. One strain similar to *Agreia bicolorata* was found in stem galls induced by *Anguina agropyri* on *Elymus repens* (L.) Gould collected in the same site (Evtushenko et al., 1994). In addition, yellow-pigmented bacteria of this genus, belonging to other genomic groups, were isolated from the same plants, *Calamagrostis neglecta* and *Elymus repens*, infected by the same nematodes and in other plants without visible nematode infestation

(Dorofeeva et al., 1999; Shramko et al., 1999; L. I. Evtushenko and M. Takeuchi, unpublished observation). The yellow-pigmented strains of *Agreia* were not predominant among other coryneform bacteria revealed by the plating technique. The species *Agreia pratensis* was also isolated from grass (Behrendt et al., 2002; Schumann et al., 2003).

Isolation

Organisms of the genus *Agreia* can be isolated from the grasses and related sources listed in the previous subsection. The approaches described by Dunleavy (1989), Behrendt et al. (1997), and Zinniel et al. (2002) for isolation of bacteria from plants may be applied. The following procedure of preparing a sample for isolation of bacteria from plant galls may be used.

The galls are cut from a plant with a sterile scalpel, treated by a detergent or sterilized or both, washed by sterile water, and pre-incubated in water or in any diluted nutrient media for 1–2 h. Then, the galls are cut into pieces, added to 1–2 ml of 0.85% solution of NaCl (w/v), and ground with a pestle. The obtained suspensions of galls are plated onto an appropriate medium. The strains of *Agreia bicolorata* were first isolated using NWG agar (peptone, 2 g; glucose, 1 g; casein peptone, 1 g; glycerol, 10 ml; wort (a liquid formed by soaking mash in hot water), 100 ml; CaCO$_3$, 5 g; distilled water, 900 ml; agar, 15 g; pH 7.2–7.4) at room temperature (17–24°C; Evtushenko et al., 1994; Evtushenko et al., 2001). Bacteria of this genus can also be isolated and grown on *Corynebacterium* (CB) agar, R2A agar (Difco) and some other media appropriate for isolation and cultivation of plant-associated actinobacteria. The colonies of *Agreia bicolorata* can be recognized owing to their semiliquid consistency and yellow or orange pigment, becoming brown-orange with age on CB agar and some other media. The yellow-pigmented colonies of members of this genus are indistinguishable from colonies of many other coryneform bacteria originating from plant sources.

Corynebacterium (CB) Agar

Glucose	5 g
Yeast extract	5 g
Casamino acids	10 g
NaCl	5 g
Agar	15 g
Water	1 liter
pH 7.2	

R2A Agar (Difco)

Bacto yeast extract	0.5 g
Bacto proteose peptone no. 3	0.5 g
Bacto casamino acids	0.5 g
Bacto dextrose	0.5 g

Soluble starch	0.5 g
Sodium pyruvate	0.3 g
K_2HPO_4	0.3 g
$MgSO_4$	0.05 g
Bacto agar	15 g
Distilled water	1 liter
pH 7.2	

Identification

The unique composition of peptidoglycan with two diamino acids, DAB and ornithine, glutamate plus hydroxyglutamate, alanine and glycine is the most striking chemotaxonomic marker which differentiates *Agreia* from other genera of the family Microbacteriaceae (Table 1). *Agreia bicolorata* is characterized by yellow or orange pigmentation of colonies, becoming brown-orange or brown-red with age, irregular rods (0.4–0.5 × 1.5–2.5 μm), abundant polysaccharide capsule, predominant rhamnose and minor fucose and mannose in the cell wall, and the major menaquinone MK-10, with minor amounts of MK-9. Oxidase reaction with tetramethyl-*p*-phenylenediamine is positive, a wide range of carbohydrate and organic acids is used for the growth, H_2S is produced, and no growth occurs in media with 6% NaCl (Evtushenko et al., 2001). The species *Agreia pratensis* differs from *Agreia bicolorata* by yellow colonies, major menaquinones MK-10 and MK-11, negative oxidase reaction, and some physiological properties (Behrendt et al., 2002).

The Genus *Agrococcus*

The genus *Agrococcus* with the type species *Agrococcus jenensis* was established by Groth et al. (1996). Recently, two other species have been added to the genus (Wieser et al., 1999; Zlamala et al., 2002a).

Phylogeny

Representatives of the genus form a tight phylogenetic cluster with a 100% bootstrap confidence level, linked peripherally to the genus *Cryobacterium* (Suzuki et al., 1997; Fig. 1).

Taxonomy

The genus *Agrococcus* harbors coccoid, non-motile, non-acid-fast, microaerophilic, oxidase-negative bacteria. Their cell wall peptidoglycan contains DAB, alanine, glycine, aspartate, and glutamate; threonine may present in some species (Table 1). No mycolic acids are present. The major menaquinones are MK-11 and MK-12. Diphosphatidylglycerol and phosphatidylg-

lycerol are predominant polar lipids; two unknown glycolipids and some noncharacterized lipids were also detected. Polyamine pattern contains mainly spermine. The main whole-cell sugars glucose and rhamnose and minor amounts of mannose, ribose, galactose and tyvelose are typical for the species *Agrococcus jenensis* and *Agrococcus citreus*. The G+C content of DNA is 74 mol%. The strains grow at 37°C, with no growth at 42°C; they use a wide spectrum of carbon sources for growth and decompose some organic substrates. Usually they tolerate 4% NaCl but not 10% NaCl and are susceptible to many antibiotics but grow in the presence of 300 IU of polymyxin B and 200 μg of sulfonamide (Groth et al., 1996; Altenburger et al., 1996; Wieser et al., 1999; Zlamala et al., 2002a).

The current genus *Agrococcus* includes the following three species: *Agrococcus jenensis* (Groth et al., 1996), *Agrococcus citreus* (Wieser et al., 1999), and *Agrococcus baldri* (Zlamala et al., 2002a). The species differentiation is based on cell wall peptidoglycan composition, polar lipid pattern, fatty acid profile, and a number of physiological properties (Groth et al., 1996; Wieser et al., 1999; Zlamala et al., 2002a) as summarized in the subsection Identification in this Chapter.

Habitat

Little is known about the natural habitat of agrococci, as strains have been obtained from a variety of sources. Strains of *Agrococcus jenensis* were isolated from frozen compost soil in Jena, Germany, and a sandstone surface of the building Alte Pinakothek, Munich, Germany (Groth et al., 1996). Strains of *Agrococcus baldri* were recovered from air of the "Virgilkapelle" in Vienna, Austria (Zlamala et al., 2002a), while *Agrococcus citreus* was isolated from a medieval wall painting in the church of Herberstein castle in Styria, Austria (Wieser et al., 1999). Fritz (2000) found a representative of this genus, strain "*Corynebacterium* cf. *aquaticum*" V4.BO.26 (AJ244681), in the Mediterranean Sea at a depth of 600 m.

Isolation

The dilution plating technique has been applied on nutrient agar containing 2% peptone, PYES agar and some other media to isolate bacteria of the genus *Agrococcus* (Groth et al., 1996; Wieser et al., 1999). Cells of *Agrococcus baldri* were collected from the air using an air sampler and then grown on CasMM-agar (Altenburger et al., 1996) at room temperature (Zlamala et al., 2002a). Organisms of the genus grow well on a number of media, including tryptic soy agar,

PYES agar, CB agar and some others at appropriate temperature conditions, from 18 to 28°C.

Nutrient agar based on 2% peptone (Groth et al., 1996)

Peptone	20 g
Pancreatic digest (meat, fish)	5 g
NaCl	5 g
Agar	12 g
Distilled water	1 liter
pH 7.2–7.4	

PYES agar (Wieser et al., 1999)

Peptone	3 g
Yeast extract	3 g
Na$_2$-succinate	2.3 g
Agar	15 g
Distilled water	1 liter
pH 7.2	

CasMM-agar (Altenburger et al., 1996)

Casein	0.8 g
Yeast extract	0.4 g
K$_2$HPO$_4$	0.6 g
Na$_2$HPO$_4$ · 2H$_2$O	0.5 g
MgSO$_4$ · 7H$_2$O	0.05 g
MgCl$_2$ · 7H$_2$O	0.1 g
KNO$_3$	0.2 g
FeCl$_3$ · 6H$_2$O	0.01 g
Agar	15 g
Distilled water	1 liter
pH 7.0	

Identification

Agrococci can be easily distinguished from other Gram-positive coccoid bacteria by the amino acid composition of peptidoglycan. Spherical shape is the most striking feature that differentiates the *Agrococcus* species from all other genera of the family Microbacteriaceae. Other associated characteristics differentiating the genus from other taxa of the family at the phenotypical level are given in Table 1.

Agrococcus jenensis is differentiated from *Agrococcus citreus* and *Agrococcus baldri* mainly by cell wall peptidoglycan composition and fatty acid profile (Zlamala et al., 2002a). The peptidoglycan of *Agrococcus jenensis* contains alanine, glycine, DAB, aspartate, glutamate and threonine in the molar ratio 3:2:0.7:1:1:1 (data for the type strain; Groth et al., 1996), while no threonine was determined in *Agrococcus citreus*; the molar ratio of alanine, glycine, aspartic and glutamic acids was 3.6:2.1:0.8:0.9:1 (Wieser et al., 1999). Similarly, no threonine was found in *Agrococcus baldri* (Zlamala et al., 2002a). *Agrococcus baldri* differs from the other two species by an unknown glycolipid not detected in *Agrococcus jenensis* or *Agrococcus citreus* (Zlamala et al., 2002a). Some physiological properties useful for discrimination of these three species are listed in Table 2.

The Genus *Agromyces*

Gledhill and Casida (1969) proposed the genus *Agromyces* with the type species *Agromyces ramosus* to accommodate filamentous, branching, fragmenting, nutritionally fastidious, catalase-negative soil isolates. Showing a negative oxidase reaction with benzidine and tetramethyl-*p*-phenylenediamine, the type strain of *Agromyces ramosus* produces cytochromes *b*, *c* and *aa*$_3$,

Table 2. Characteristics that differentiate *Agrococcus* species.

Characteristics	A. jenensis	A. baldri	A. citreus
Threonine in peptidoglycan	+	−	−
Utilization of			
L-Arabinose	−	−	+
D-Cellobiose	−	+	+
Gluconate	−	+	+
D-Mannose	−	+	+
L-Rhamnose	−	+	+
D-Ribose	−	+	+
D-Trehalose	−	+	+
D-Xylose	−	−	+
4-Aminobutyrate	−	+	−
L-Proline	−	+	+
Acid from L-arabinose	+	−	+
Hydrolysis of:			
PNP-β-D-glucopyranoside	+	−	+
Esculin	+	−	+
Tween 20	+	−	+
Growth at 6.5% NaCl	−	d	−

Symbols and abbreviations: +, positive; −, negative; d, variable among strains; and PNP, *p*-nitrophenyl.
Data from Groth et al. (1996) and Zlamala et al. (2002a).

where *aa₃* is the cytochrome oxidase (Jones et al., 1970). The strain was later characterized by a peptidoglycan of the B2γ type (Schleifer and Kandler, 1972; Fiedler and Kandler, 1973a), a major menaquinone MK-12, predominant unsaturated *iso-* and *anteiso*-branched fatty acids, and the polar lipids diphosphatidylglycerol, phosphatidylglycerol, and a glycolipid (Collins, 1982a). In 1992, Zgurskaya et al. described 14 soil isolates that had the morphological and chemotaxonomic characteristics typical of the genus *Agromyces* but grew rapidly on simple media and showed positive catalase and oxidase reactions. The authors classified those into two new species and four subspecies, *Agromyces cerinus* subsp. *cerinus, Agromyces cerinus* subsp. *nitratus, Agromyces fucosus* subsp. *fucosus*, and *Agromyces fucosus* subsp. *hippuratus*, with emendation of the genus description (Zgurskaya et al., 1992). Later, Suzuki et al. (1996) found that the DNA-DNA relatedness between subspecies of *Agromyces fucosus* was 45–47%, lower than the level usually considered to be the border of genomic species (Wayne et al., 1987). Suzuki et al. (1996) described *Agromyces mediolanus*, which included the misclassified bacteria "*Corynebacterium mediolanum*" (Mamoli, 1939), "*Flavobacterium degydrogenans*" (Arnaudi, 1942), and some soil isolates (Aoki et al., 1982), which do not usually produce mycelium. The genus has recently been enlarged by *Agromyces luteolus, Agromyces rhizospherae* and *Agromyces bracchium* isolated from the mangrove rhizosphere (Takeuchi and Hatano, 2001), *Agromyces aurantiacus* found in a Chinese primeval forest soil (Li et al., 2003), and *Agromyces albus* isolated from a plant, *Androsace* sp. (Dorofeeva et al., 2003).

Phylogeny

On the basis of 16S rRNA gene sequencing studies, organisms of *Agromyces* form a phylogenetically coherent group that is clearly distinct from other genera of Microbacteriaceae (Fig. 1).

Taxonomy

The genus *Agromyces* currently harbors nine species and two subspecies:

Agromyces ramosus (Gledhill and Casida, 1969), *Agromyces albus* (Dorofeeva et al., 2003), *Agromyces aurantiacus* (Li et al., 2003), *Agromyces bracchium* (Takeuchi et al., 2001), *Agromyces cerinus* subsp. *cerinus, Agromyces cerinus* subsp. *nitratus, Agromyces fucosus* subsp. *fucosus* and *Agromyces fucosus* subsp. *hippuratus* (Zgurskaya et al., 1992), *Agromyces luteolus, Agromyces rhizospherae* (Takeuchi et al., 2001), and *Agromyces mediolanus* (Suzuki et al., 1996).

Members of the genus are characterized by opaque, paste, entire colonies, usually penetrating into agar. Branching hyphae (width, 0.2–0.8 μm) break up into diphtheroid and rodlike, irregular, nonmotile fragments. Nonmycelial strains occur. No aerial mycelium and spore formation are usually observed, except for *A. aurantiacus* which forms aerial mycelium on glycerol-asparagine agar (Li et al., 2003). Cells are aerobic to microaerophilic. Catalase test and oxidase reaction with tetramethyl-*p*-phenylenediamine are variable for different species. Mesophilic, optimal growth occurs at 24–30°C, and some strains grow at 37°C. Strains utilize a wide range of C and N sources and grow on media containing organic or inorganic nitrogen. The peptidoglycan is of the B2γ type (Schleifer and Kandler, 1972) based upon DAB (mostly L-isomer; Sasaki et al., 1998) and contains amino acids glycine, glutamate, DAB and alanine in a molar ratio close to 1:1:2:1. Galactose and rhamnose occur in the cell wall of most strains; fucose, fructose, glucose, mannose, ribose and tyvelose vary in different species. No mycolic acids are found. The predominant menaquinone is MK-12, and the second most common component is MK-11 or MK-13, depending on the species. Polar lipids are represented by diphosphatidylglycerol, phosphatidylglycerol and glycolipids; *iso-* and *anteiso*-branching fatty acids are predominant and can reach 94% (Zgurskaya et al., 1992; Suzuki et al., 1996; Takeuchi et al., 2001). Cell wall teichoic acids and related anionic polymers were found in some species and subspecies (Naumova et al., 2001). The total polyamine content in some of the *Agromyces* species investigated is lower than that in most other strains of the family Microbacteriaceae and consists of only 0.21–0.28 μmol·g⁻¹, with putrescine as a predominant component (Altenburger et al., 1997).

The species and subspecies differentiation is based on growth, morphological, chemotaxonomic and physiological characteristics as outlined in the subsection Identification in this Chapter.

Habitat

The usual natural habitat of agromycetes is soil of different origin (Gledhill and Casida, 1969; Zgurskaya et al., 1992; Suzuki et al., 1996; Takeuchi and Hatano, 1999; Takeuchi and Hatano, 2001; Li et al., 2003). *Agromyces ramosus* has been reported to be the numerically prevalent organism in soil (Gledhill and Casida, 1969). *Agromyces luteolus, Agromyces rhizospherae* and *Agromyces bracchium* (Takeuchi and Hatano, 2001), along with many kinds of other actinomycetes, were isolated from the rhizosphere of *Bruguiera gymnorhiza* (L.) Savigny and *Sonneratia alba* Sm. in the subtropical and tropical regions of Okinawa, Japan (Hatano, 1997; Takeuchi and Hatano, 1999). Strain *Agromyces mediolanus* ATCC 13990 (the former "*Flavobacterium dehydrogenans*") was recovered from pressed yeast cake (Arnaudi, 1942; Cummins et al., 1974), and *Agromyces albus* was isolated from the above-ground part of a plant *Androsace* sp. (the family Primulaceae), Belgorod region, Russia (Dorofeeva et al., 2003). The microorganisms of the genus were also found in the stalactites, Grotta dei Cervi, Porto Badisco, Italy (Laiz et al., 2000); in more than 500,000-year-old glacial ice, Guliya, China (Christner, 2002); a medieval wall painting (Altenburger et al., 1996); skin and muscle of the Tyrolean iceman (Rollo et al., 2000), and some other sources. Organisms phylogenetically related to *Agromyces* were revealed in casts of the earthworm *Lumbricus rubellus* (Furlong et al., 2002), and a strain tentatively identified as *Agromyces* sp. was isolated from a site highly contaminated with chloroaromatic compounds, although the strain was not capable of trichlorophenol degradation (Maltseva and Oriel, 1996).

Isolation

Many nutrient media, e.g., diluted (1:10) CB agar (Zgurskaya et al., 1992), R agar (Yamada and Komagata, 1972a), AR2 agar (Difco), and some others are suitable for isolation of the fast-growing aerobic species of the genus *Agromyces*. Among other bacteria developing on nutrient agar media, most *Agromyces* strains can preliminarily be recognized owing to their pale-yellow or yellow paste colonies penetrating into agar media. *Agromyces ramosus* is often difficult to isolate because of problems providing the appropriate growth conditions. This species is likely microaerophilic because it lacks catalase and grows best under full oxygen tension, provided that accumulation of H_2O_2 is avoided. Removal of H_2O_2 can be achieved by the addition of catalase, fresh horse blood, or MnO (Jones et al., 1970). *Agromyces ramosus* may be isolated from soil using the dilution frequency procedure described in detail by Casida (1965). The approach is based on the fact that *Agromyces ramosus* occurs in high quantities in many soils. In soils, where *Agromyces ramosus* is not predominant, *Actinomyces humiferus* may be recovered instead of *Agromyces ramosus* (Gledhill and Casida, 1969). *Agromyces ramosus* grows very slowly during the first transfers and might be lost easily. Once adapted, usually growth is without problems on nutritionally rich media. Heart infusion agar at 30°C is a suitable medium for growth (Gledhill and Casida, 1969).

Heart Infusion Agar (Gledhill and Casida, 1969)

Heart infusion broth (Difco)	25 g
Bacto yeast extract (Difco)	5 g
Casitone (Difco)	4 g
Agar	15 g
Distilled water	1 liter
pH 7.0–7.4	

R agar (Yamada and Komagata, 1972a)

Bacto peptone (Difco)	10 g
Bacto yeast extract (Difco)	5 g
Bacto malt extract (Difco)	5 g
Bacto casamino acids (Difco)	5 g
Bacto beef extract (Difco)	2 g
Glycerol	2 g
Tween 80	50 mg
$MgSO_2 \cdot 7H_2O$	1 g
Agar	20 g
Distilled water	1 liter
pH 7.2	

Identification

Novel isolates can be readily identified as members of the genus *Agromyces* by 16S rRNA gene sequence analysis. Most strains can also be referred to the genus by the presence of DAB in the cell wall and a branched fragmenting mycelium (Table 1). Only *Agromyces mediolanus* does not usually produce a mycelium (Suzuki et al., 1996). This species differs from most other DAB-containing genera by the predominant menaquinone MK-12 and nonmotile rod-shaped cells. The major menaquinone MK-12, typical for *Agromyces* species, is also found in members of *Agrococcus* and, together with the second predominant component MK-11, in some species of *Leifsonia* (which are characterized by clearly motile cells; Table 1). In addition, *Agromyces* spp. exclusively possess L-DAB in the peptidoglycan, while *Leifsonia* spp. contain both L- and D- isomers in almost equal proportion (Sasaki et al., 1998). The species and subspecies of the genus *Agromyces* differ from each other in colony color, cell morphology, cell wall sugars, the complex media requirement for growth, the oxidase test with tetramethyl-*p*-phenylenediamine, and some other physiological characteristics (Table 3).

Applications and Properties

Agromyces ramosus ATCC 25173 and several strains assigned to "*Agromyces okinawai*" produced a highly thermally stable immunostimulant with antitumor, antiviral, interferon-inducing and immunity response-potentiating activities (Wakuzaka et al., 1983). Immunoactivating properties were also found for the cell wall polymers of *Agromyces cerinus* (H. I.

Table 3. Characteristics differentiating *Agromyces albus* sp. nov. from other species of the genus *Agromyces*.

Characteristic	A. ramosus	A. albus	A. aurantiacus	A. bracchium	A. cerinus subsp. cerinus	A. cerinus subsp. nitratus	A. fucosus subsp. fucosus	A. fucosus subsp. hippuratus	A. luteolus	A. mediolanus	A. ramosus	A. rhizospherae
Colony color	W	W	O, PG, Y	Y	Y	Y	Y	Y	Y	Y	W	Y
Hyphae	+	+	+	+	+	+	+	+	+	−	+	+
Catalase	−	+	−	+	+	+	+	+	+	+	−	+
Oxidase test	−	+	ND	+	+	+	+	+	+	+	−	d
Microaerophilic	+	−	−	−	−	−	−	−	−	−	+	−
Acid from												
D-Arabinose	+	+	−	−	+	−	+	+	−	−	+	−
D-Galactose	−	+	+	+	+	+	+	+	+	+	−	−
D-Glucose	−	+	−	+	+	+	+	+	+	+	−	+
Inulin	+	+	+	−	−	−	+	−	−	−	+	−
Salicin	−	+	−	+	+	d	+	−	−	+	−	+
D-Sucrose	+	+	+	+	+	+	+	+	d	+	+	−
Hydrolysis												
Casein	ND	+	ND	−	−	−	−	−	−	−	−	−
Starch	−	+	+	+	+	+	+	+	+	−	ND	+
Hypoxanthine	−	+	ND	−	+	+	+	+	−	+	−	−
Xanthine	−	−	ND	−	−	−	−	−	−	+	−	−
Growth, 4% NaCl	d	−	ND	+	−	+	−	−	+	+	d	+
G+C (mol%)	68.9	69.0	72.8	70.0	70.5	70.9	70.6	70.8	71.1	ND	68.9	71.2
Cell wall sugars[a]	Gal, Glu, Man Rha, Xyl	Rha (Gal, Glu, Man)	Rha (Gal, Glu, Man)	Rha (Gal, Glu, and Man)	Gal, Rha, and Tyv (Man)	Gal (Glu, Man, and Rib)	Gal, Rha, and Fuc (Glu, Man)	Gal, and Rha (Man)	Rha, Fruc (Glu, and Man)	ND	Gal, Glu, Man, Rha, and Xyl	Rha (Glu, and Man)
Menaquinones[b]	MK-12 (13)	MK-12 (11)	MK-12	MK-12 (13)	MK-12 (13)	MK-12 (13)	MK-12 (13)	MK-12 (13)	MK-12 (11)	MK-12 (11,10)	MK-12 (13)	MK-12, 11
Source of isolation	Soil	Above-ground part of Androsace sp.	Soil	Rhizosphere of Bruguera gymnorrhiza	Soil	Soil	Soil	Soil	Rhizosphere of Sonneratia alba	Soil	Soil	Rhizosphere of Sonneratia alba

Symbols and abbreviations: +, positive; −, negative; d, weakly positive or variable among tests; ND, not determined; W, white; O, orange; PG, pink gray or pink white; Y, yellow; Fuc, fucose; Gal, Galactose; Glc, glucose; Man, mannose; Rib, ribose; Rha, rhamnose; Tyv, tyvelose; and Xyl, xylose.

[a]Components given in parentheses are present in minor amounts; no available data on sugar amounts in *A. ramosus*.

[b]Menaquinone given in parentheses is present as the second most common component.

Data from Gledhill and Casida (1969), Zgurskaya et al. (1992), Suzuki et al. (1996), Takeuchi and Hatano (2001), Dorofeeva et al. (2003), and Li et al. (2003).

Zgurskaya and L. I. Evtushenko, unpublished observation). Production of 3,6-dideoxy-D-arabinose (tyvelose) was patented for *Agromyces cerinus* subsp. *cerinus* strain (Zgurskaya and Evtushenko, 1996). Production of D-*N*-carbamyl-α-amino acids, e.g., D-*N*-carbamylalanine, by conversion of 5-substituted hydantoin using immobilized *Agromyces* sp. has been reported; the compounds are useful for the production of pharmaceuticals and agrochemicals (Mitsubishi Kaseo Corporation, 1989). An experimental biofilter containing tree bark compost was used to treat raw gas emissions (aldehyde, ketone, furan, thiophene and alkylsulfide compounds, hydrogen sulfide, ammonia and CO_2) from animal rendering. Two strains of *Agromyces*, along with *Corynebacterium* and *Gordonia* strains, were reported among the bacteria present in the biofilter used for waste-gas treatment of an animal rendering plant and pollutant degradation (Bendinger et al., 1990). Some *Agromyces mediolanus* strains assimilate aniline (Aoki et al., 1982; Suzuki et al., 1996). *Agromyces mediolanus* ATCC 14004 and *Agromyces mediolanus* ATCC 13930 are capable of oxidizing steroids (Mamoli, 1939; Arnaudi, 1942), and *Agromyces mediolanus* ATCC 13930 can be used for a microbiological process for the preparation of oxoalkylxanthines (Nesemann et al., 1974). *Agromyces ramosus*, strains ATCC 25173 and PSU 35, attack and destroy cells of *Saccharomyces cerevisiae* and several bacteria, including *Agrobacterium tumefaciens*, *Rhizobium leguminosarum*, *Sinorhizobium meliloti* (formerly, *Rhizobium meliloti*), *Staphylococcus aureus*, *Escherichia coli* and *Pseudomonas fluorescens*. The strains may be manipulated to control levels of *Agrobacterium tumefaciens* and other pathogens without simultaneously depressing the numbers of beneficial organisms in this habitat (Casida, 1983).

Agromyces ramosus was reported to produce the L-form which reverted to a bacterial form when incubated in sterilized soil or in some suitable culture media (Horwitz and Casida, 1978a; Horwitz and Casida, 1978b). Pure cultures of this species can also produce small cells caused by growth on nutritionally limiting agar media (Casida, 1977). *Agromyces ramosus* showed slight activity in oxidizing CO to CO_2; the reaction was proposed to be caused by microbial activity because an appreciable amount of CO was not converted to CO_2 in autoclaved or γ-irradiated soil (Bartholomew and Alexander, 1979). In *Agromyces cerinus* subsp. *cerinus* VKM Ac-1340, a new unusual structure of teichoic acid with arabitol in the main chain was identified (Shashkov et al., 1995). A new polymer of 1,5-poly(ribitol phosphate) with tetrasaccharide substituents was revealed in the cell wall of

Agromyces fucosus subsp. *hippuratus* (Gnilozub et al., 1994).

The Genus *Clavibacter*

The genus *Clavibacter* was established by Davis et al. (1984b) to accommodate several plant pathogenic species and subspecies. These organisms, which were formerly attributed to a separate group of plant pathogens in the genus *Corynebacterium* (Collins and Bradbury, 1986), contain DAB in the cell wall peptidoglycan (type B2γ; Schleifer and Kandler, 1972) and unsaturated menaquinones (Collins and Jones, 1980a). The current genus *Clavibacter* (including one species *Clavibacter michiganensis* with five subspecies) is restricted only by presence of the major menaquinone MK-9, the B2γ type peptidoglycan, and some associated properties (Davis et al., 1984b).

The type strain *Clavibacter michiganensis* subsp. *michiganensis* NCPPB 2581[T] (CIP 104846[T], DSM 46364[T], ICMP 2550[T], JCM 9665[T], LMG 7333), originally described as "*Bacterium michiganense*" (Smith, 1910), was the first documented living bacterium of the genus *Clavibacter sensu stricto*. Later, members of *Clavibacter michiganensis* subsp. *sepedonicus* and *Clavibacter michiganensis* subsp. *insidiosus* originally assigned to the species "*Bacterium sepedonicum*" (Spieckermann and Kotthoff, 1914) and "*Aplanobacter insidiosum*" (McCulloch, 1925) were reported. Following reclassification, the bacteria were allocated to the genus *Corynebacterium* as the separate species *Corynebacterium michiganense*, *Corynebacterium insidiosum* and *Corynebacterium sepedonicum* (Cummins et al., 1974). After half a century, another plant pathogenic species, *Corynebacterium nebraskensis*, was described (Vidaver and Mandel, 1974).

Carlson and Vidaver (1982), who revealed almost identical cellular protein patterns of these organisms and novel strains causing wheat disease, proposed to combine *Corynebacterium michiganense*, *Corynebacterium insidiosum*, *Corynebacterium nebraskense* and *Corynebacterium sepedonicum* into a single species, *Corynebacterium michiganense*. This was consistent with the previous conclusion of Dye and Kemp (1977), based on a high phenotypic similarity of these species. In contrast to Dye and Kemp, who considered the bacteria as pathovars of *Corynebacterium michiganense*, Carlson and Vidaver reclassified them as subspecies, with description of a new subspecies associated with wheat, *Corynebacterium michiganense* subsp. *tessellarium*. The authors noted that the bacterial differences on the basis of colony morphology and pigmentation, overall bacteriocin production (Gross and Vidaver, 1979b), some biochemical reactions, and menaquinone pattern are independent of pathogenic specificity (Carlson and Vidaver, 1982). They also commented on the inappropriateness of the term "pathovar" for avirulent strains and that designation of these organisms as subspecies would allow the rational identification of related saprophytes and avirulent strains. The proposed names were validated as *Corynebacterium michiganense* subsp. *michiganense* (Smith, 1910) Jensen 1934, *Corynebacterium michiganense* subsp. *insidiosum* (McCulloch, 1925) Carlson and Vidaver 1982, *Corynebacterium michiganense* subsp. *nebraskense* (Vidaver and Mandel, 1974) Carlson and Vidaver 1982, *Corynebacterium michiganense* subsp. *sepedonicum* (Spieckermann and Kotthoff, 1914) Carlson and Vidaver 1982, and *Corynebacterium michiganense* subsp. *tessellarium* Carlson and Vidaver 1982.

Unaware of the description of the genus *Clavibacter* and five subspecies of *Clavibacter michiganensis* (Davis et al., 1984b), Collins and Bradbury (1986), summarizing information on DNA-DNA reassociation (Starr et al., 1975; Döpfer et al., 1982), serological specificity (Lazar, 1968), differences in cellular proteins (Carlson and Vidaver, 1982), lipids (Minnikin et al., 1978; Collins and Jones, 1980a), biochemical and physiological features (Jones, 1975; Dye and Kemp, 1977) and some other data, concluded that four subspecies of *Corynebacterium michiganense* (*michiganense, insidiosum, nebraskense* and *sepedonicum*) warrant a separate species status. Although Collins and Bradbury stated that none of these organisms belonged to the genus *Corynebacterium sensu stricto*, they retained the generic name *Corynebacterium* for the purpose of convenience and practicability. With additional data on the differences of the subspecies in allozyme pattern (Riley et al., 1988) and fatty acid composition (Henningson and Gudmestad, 1991) and the results of numerical, chemotaxonomic and DNA-DNA relatedness studies of a large set of plant pathogenic and saprophytic clavibacteria, Zgurskaya (1993a) provided additional evidences that the *Clavibacter michiganensis* subspecies are actually separate species as concluded by Collins and Bradbury (1986). Furthermore, some saprophytic and environmental clavibacteria were found to probably represent novel species of the genus (Zgurskaya, 1993a; Dobrovolskaya et al., 1999; Nazina et al., 2002).

The situation outlined shows that the intrageneric structure of the genus *Clavibacter* needs to be revised in the future, in that the subspecies of *Clavibacter michiganensis* needs to be reclassified as a species.

Phylogeny

Clavibacteria form a tight cluster within the Microbacteriaceae phylogenetic branch and are close to the genus *Frigoribacterium* (Kämpfer et al., 2000), showing up to 97% 16S rDNA sequence similarity to this genus (Fig. 1).

Taxonomy

Currently, the genus *Clavibacter* includes one species, *Clavibacter michiganensis*, with five subspecies (Davis et al., 1984b): *Clavibacter michiganensis* subsp. *michiganensis* (Smith, 1910), *Clavibacter michiganensis* subsp. *insidiosus* (McCulloch, 1925), *Clavibacter michiganensis* subsp. *nebraskensis* (Vidaver and Mandel, 1974), *Clavibacter michiganensis* subsp. *sepedonicus* (Spieckermann and Kotthoff, 1914), and *Clavibacter michiganensis* subsp. *tessellarius* (Carlson and Vidaver, 1982).

Members of the genus *Clavibacter* are characterized by nonsporeforming, nonmotile, irregular rods (ca. $0.5–0.7 \times 0.75–1.5$ μm), which occur singly or in pairs with diphtheroid arrangements. Primary branching is not observed. Cells are aerobic, catalase positive, and usually show a negative or weakly positive oxidase reaction with tetramethyl-*p*-phenylenediamine. Thermal optimum is about 25–28°C. The peptidoglycan type is B2γ [L-DAB] D-glu-D-DAB (Schleifer and Kandler, 1972; Sasaki et al., 1998). Cell wall sugars are represented by rhamnose, galactose and mannose; glucose and fucose also occur in some strains (Davis et al., 1984b; Zgurskaya et al., 1993b). Mycolic acids are not present. The major menaquinone is MK-9; menaquinones MK-7, MK-8 and MK-10 may be detected as minor components. Principal phospholipids are phosphatidylglycerol and diphosphatidylglycerol; three unknown glycolipids are present. Saturated *anteiso*-branched fatty acids *ai*-17:0 and *ai*-15:0 are predominant (Collins and Jones, 1980a; Sasaki et al., 1998). The G+C content of DNA is 69–78 mol% (Yamada and Komagata, 1970; Döpfer et al., 1982). Discrimination of subspecies is based on cultural features, physiological characteristics, and cell wall sugars (see subsection Identification in this Chapter), which are consistent with their grouping on the basis of DNA-DNA relatedness (Döpfer et al., 1982; Zgurskaya, 1993a) and their plant hosts (Vidaver, 1982).

Habitat and Phytopathogenicity

Published data suggest that bacteria of the genus *Clavibacter* are usually associated with different diseased or healthy plants and related sources (Cummins et al., 1974; Vidaver, 1982; Collins and

Bradbury, 1992; Zgurskaya, 1993a; Dobrovolskaya et al., 1999; Sundin and Jacobs, 1999; Behrendt et al., 2002; Zinniel et al., 2002). They can also occur in soil and other environments, including oil fields, glacial and subglacial environments, and a deep subsurface paleosol (Hou et al., 1997; Chandler et al., 1998; Christner, 2002; Nazina et al., 2002).

OCCURRENCE OF PLANT-ASSOCIATED SAPROPHYTIC CLAVIBACTERIA Saprophytic clavibacteria inhabit different plants and related sources worldwide. Along with representatives of *Curtobacterium* and Gram-negative bacteria, they constitute a significant proportion in the microbial communities of grass phyllosphere decaying surface litter after mulching the sward (Behrendt et al., 1997; Behrendt et al., 2002). Numerous clavibacteria, including rose-pigmented strains, were isolated from different plants (*Ammodendron* sp., *Bromus* sp., *Collidonum* sp., *Ephedra* sp., *Gagea* sp., *Salsola* sp., etc.) and related sources collected in desert areas of Middle Asia (Zgurskaya, 1993a; Dobrovolskaya et al., 1999). The pigmented clavibacteria and curtobacteria predominated among highly ultraviolet (UV)-tolerant strains revealed in the phyllosphere of the field-grown peanut, *Arachis hypogeae* L., Texas, United States (Sundin and Jacobs, 1999; Jacobs and Sundin, 2001). Recent data provide evidence that clavibacteria can exist as endophytes (Kado, 1992) living in plant tissues of several agronomic crops without doing substantial harm (Zinniel et al., 2002). However, still little is known about the role of epiphytic and endophytic clavibacteria in the life of the plant. Other endophytic bacteria have been associated with the growth promotion of plants (for the references, see Hallmann et al., 1997). The beneficial effects of bacterial endophytes vary but appear to operate through similar mechanisms (Kloepper et al., 1991). They also may contribute to the control of bacterial and fungal pathogens (Hallmann et al., 1997; Reiter, et al., 2002) and also of plant parasitic nematodes (Hallmann et al., 1997) and insects (Dimock et al., 1988). Several strains of unidentified *Clavibacter* sp. were isolated from seed galls induced by plant parasitic nematode *Anguina agrostis* on *Agrostis* sp. However, clavibacteria were not predominant among other coryneform isolates in bacterium-nematode complexes in the galls and most likely are secondary settlers of the galls, descending from the phyllosphere or healthy plant tissue (L. I. Evtushenko et al., unpublished observation).

PLANT PATHOGENIC *CLAVIBACTER MICHIGANENSIS*

Plant Hosts The main habitats of plant pathogenic *Clavibacter michiganensis* subspecies are considered to be the respective host plants. In nature, each pathogen has been found in one or a few usually related genera of plants (Bradbury, 1986). *Clavibacter michiganensis* subsp. *insidiosus* occurs in lucerne (alfalfa; *Medicago sativa* L.) and a few other pasture legumes (*Trifolium* spp., *Melilotus albus* Medic., *Onobrychis viciifolia* Scop. [syn. *Onobrychis sativa* L.], and *Lotus corniculatus* L.). *Clavibacter michiganensis* subsp. *nebraskensis* is associated with corn (*Zea mays* L.). *Clavibacter michiganensis* subsp. *sepedonicus* inhabits diseased potatoes (*Solanum tuberosum* L.). *Clavibacter michiganensis* subsp. *tessellarius* occurs on wheat (*Triticum aestivum* L.). *Clavibacter michiganensis* subsp. *michiganensis* is limited to solanaceous hosts, principally tomato (*Lycopersicon esculentum* Mill.) and pepper (*Capsicum annuum* L.), but was also revealed in the diseased solanaceous weeds, *Solanum nigrum* L., *Solanum triflorum* Nutt. and *Lycopersicon pimpinellifolium* (L.) Mill. (syn. *Solanum pimpinellifolium* L.) (Vidaver, 1982; Collins and Bradbury, 1986; Vidaver and Davis, 1988). The subspecies *michiganensis*, *sepedonicus* and *insidiosus* were reported to infect some other plants related to their host plants when inoculated (Collins and Bradbury, 1986).

Vascular Diseases Caused by Clavibacteria and Disease Symptoms *Clavibacter michiganensis* with its five subspecies causes a vascular disease with very high titers in a variety of agriculturally important plants and, in fact, may be considered the most important bacterial pathogen (Jahr et al., 1999). The subspecies *sepedonicus* and *michiganensis* are subject to international quarantine regulations (European Union, 1995; Li and De Boer, 1995b). Infection of the host plant usually occurs via wounds, followed by invasion of the xylem vessels, which establishes a systemic vascular disease (Wallis, 1977). After some time, the plants develop wilt and other symptoms and may perish. *Clavibacter michiganensis* subsp. *michiganensis* causes wilt, fruit spot and canker in solanaceous hosts, principally tomato and pepper. The subspecies *insidiosus* causes wilting and stunting in lucerne and a few other pasture legumes. The subspecies *nebraskensis* induces wilt, leaf blight and spot in corn, the subspecies *sepedonicus* is a casuative agent of wilt and tuber rot in potato, while the subspecies *tessellarius* induces a mosaic-like syndrome on wheat (Vidaver, 1982; Vidaver and Davis, 1988). Symptoms of these various hosts, which are often slow to develop, include also decreased vigor, stunting, leaf yellowing and leaf spot, stem or tuber necrosis. Wilt is commonly seen as an early symptom only with *Clavibacter michiganensis* subsp. *michiganensis* infections (Van Alfen,

1989). Environmental factors sometimes crucially affect disease development. Temperature, humidity and pollution can have both synergetic and antagonistic effects on symptom expression (Vidaver, 1982).

An infection at a late stage of plant development resulting in a latent infection is also destructive, as contaminated seeds or, in the case of potato, infected tubers or stems are produced that spread the disease (Nelson, 1982; Jahr et al., 1999).

Pathogenicity Factors Originally, extracellular polysaccharides (EPS) were proposed to be the main pathogenicity factor inducing wilt symptoms by interfering with water transport (for the references, see Vidaver [1982] and Denny [1995]). Strobel and Hess (1968) argued that a mechanism other than plugging exists which causes wilting by the toxic extracellular "glycopeptide" from *Clavibacter michiganensis* subsp. *sepedonicus*. The authors demonstrated that wilt was induced by ultrastructural membrane and cell wall damage rather than interference with water transport. This idea was no longer considered valid, as other workers could not confirm the results (see Vidaver [1982], Van Alfen [1989], and Denny [1995]). However, the toxic "glycopeptides" described by Strobel and Hess (1968) and some others are actually polysaccharide(s) closely associated with polypeptide, and different authors used EPS material of different quality (Denny, 1995). There is no reliable physiological and histological evidence that adequately purified polysaccharides rapidly affect plant membranes during the early stages of pathogenesis (Van Alfen, 1989; Benhamou, 1991; Denny, 1995). Furthermore, Bermpohl et al. (1996), who studied the EPS⁻ mutants of *Clavibacter michiganensis* subsp. *michiganensis* producing only 10% of the amount of EPS of the wildtype, showed that virulence did not change in mutants tested on tomato plants. Paschke and Van Alfen (1993), studying UV-mutants of *Clavibacter michiganensis* subsp. *insidiosus* lacking EPS, also concluded that it was unlikely that EPS of this plant pathogen played an important role in the early stages of disease establishment.

Histological and genetic studies of *Clavibacter michiganensis* suggested that plant cell-wall degrading enzymes contribute to main disease symptoms (Wallis, 1977; Vidaver, 1982; Benhamou, 1991; Meletzus et al., 1993; Jahr et al., 1999). Jahr et al. (2000) presented clear evidence that the 78-kDa CelA protein (746 amino acids) with similarity to endo-β-1,4-glucanases of family A1 cellulases (which is encoded by plasmid pCM1; Meletzus et al., 1993) plays a major role in pathogenicity of

Clavibacter michiganensis subsp. *michiganensis*. Wilt-inducing capability was obtained by endoglucanase expression in plasmid-free, avirulent strains and by complementation of the *celA*-gene-replacement mutant CMM-H4 with the wildtype *celA* gene. Some evidence was obtained that endocellulase is insufficient itself for expression of a virulent phenotype in the tomato test system. It obviously requires an additional gene product, also encoded by plasmid pCM1, which remains to be identified (Jahr et al., 1999). Another pathogenic determinant of *Clavibacter michiganensis* subsp. *michiganensis* is encoded by *pat-1* gene (1500 nucleotides) located on the second plasmid, pCM2 (72 kb). According to sequence homologies, the product of this gene is somewhat homologous to serine proteases from *Lysobacter enzymogenes* and *Streptomyces griseus* (Dreier et al., 1997). In addition, the pathogenic *Clavibacter michiganensis* subsp. *michiganensis* strains were reported to induce a hypersensitive reaction on certain non-host or resistant plants, which is a fast response at the site of pathogen invasion, characterized by a rapid, localized plant cell death (Gitaitis, 1990; Bermpohl et al., 1996; Dangl et al., 1996; Alarcon et al., 1998). The hypersensitivity was also induced when the protein(s) from the culture supernatant of *Clavibacter michiganensis* subsp. *michiganensis* strain 623 is injected into leaves of non-host plants, four-o'clock (*Mirabilis jalapa*) and tobacco (*Nicotiana tabacum*). The properties observed for the necrosis-inducing activity resembled those of harpin and PopA proteins from Gram-negative phytopathogenic bacteria that elicit plant hypersensitive reactions (Alarcon et al., 1998).

Clavibacter michiganensis subsp. *sepedonicus* also produces cellulase (Baer and Gudmestad, 1995), which was shown to play a role in virulence (Metzler et al., 1997; Laine et al., 2000). A cellulase-free mutant as well as a naturally occurring strain that does not contain the pCS1 plasmid was shown to have a markedly reduced virulence on eggplant. Both strains became significantly more virulent after the cellulase gene was introduced into the cells by transformation. The complete nucleotide sequence of the cellulase gene was determined and shown to encode a protein of 727 amino acids with molecular weight of 71.5 kDa. The sequence included a leader sequence for secretion and showed two typical cellulase domains (a catalytic domain and a cellulose binding domain). In addition, the authors identified an unexpected third domain that shows similarity to a plant protein called "expansin," which was believed to interact with cellulose microfibrils during plant cell expansion (Laine et al., 2000). *Clavibacter michiganensis* subsp. *sepedonicus* was also found to produce amylase, which had an effect on virulence, although its effect was less important than that of cellulase. This enzyme probably plays an important role in the rotting of starchy potato tubers (Jahr et al., 1999). The wilting material from this species was also reported to contain 10–30% of nucleic acids, which were inactivated by DNase and RNase in combination (Pearson, 1971; Vidaver, 1982). *Clavibacter michiganensis* subsp. *sepedonicus* elicits a hypersensitive reaction on tobacco, secreting heat stable hypersensitive response-inducing protein(s) (Nissinen et al., 1997).

Southern hybridization experiments revealed that three other subspecies, *Clavibacter michiganensis* subsp. *insidiosus*, *Clavibacter michiganensis* subsp. *tessellarius*, and *Clavibacter michiganensis* subsp. *nebraskensis*, contain sequences hybridizing to the cellulase gene from *Clavibacter michiganensis* subsp. *michiganensis* (Dreier et al., 1995). *Clavibacter michiganensis* subsp. *nebraskensis* was also reported to secrete an anion channel-forming protein (Schurholz et al., 1991). The properties of this voltage-dependent chloride channel protein have only been studied *in vitro*, and the role of this component in disease development remains unclear (Jahr et al., 1999).

The Role of Extracellular Polysaccharides (EPS) Two distinct polymer types represent EPS of *Clavibacter michiganensis* subsp. *michiganensis*, *Clavibacter michiganensis* subsp. *insidiosus* and *Clavibacter michiganensis* subsp. *sepedonicus*. One of them is a low-molecular-weight neutral polymer, and the second is a generally high-molecular-weight acidic polymer that consists of fucose, galactose and glucose, decorated with acetate and pyruvate side groups, or additional mannose (*Clavibacter michiganensis* subsp. *sepedonicus*; for references, see Denny, 1995). It is now clear that EPS of *Clavibacter michiganensis* do not play an important role in the early stages of disease. However, EPS determine multiple and important biological functions and may affect pathogenicity, interfering with water transport. Acid EPS can act as ion exchangers, concentrating minerals and nutrients as well as binding toxic compounds. EPS may also prevent recognition of the pathogen by the plant defense system, block agglutinins or lectins and detoxify phytoalexins or reactive oxygen species (Bradshaw-Rouse et al., 1981; Romeiro et al., 1981; Young and Sequeira, 1986; Kiraly et al., 1997). In addition, EPS mediate adhesion to abiotic and biological surfaces and may also promote infection and colonization of the host plant, a prerequisite for disease development (Melet-

zus et al., 1993; Bermpohl et al., 1996; Saile et al., 1997). Variations in the sugar composition of EPS in *Clavibacter michiganensis* subsp. *michiganensis* were shown to correlate to the inability to colonize tomato plants, indicating that the chemical structure of EPS may play a role in host recognition and the early stages of infection (Bermpohl et al., 1996).

Survival, Sources of Inoculum, and Spreading Plant pathogenic clavibacteria are considered to be poor survivors in soil; survival requires association with diseased plants, residue materials or seeds (Richardson, 1979; Richardson, 1983; Vidaver, 1982; Tsiantos, 1987; Nemeth et al., 1991). However, some bacteria were reported to survive in water for weeks (*Clavibacter michiganensis* subsp. *nebraskensis*) or more than a year (*Clavibacter michiganensis* subsp. *michiganensis*); the survival rate increased with cell concentration and with lowered temperatures (Vidaver, 1977; Vidaver, 1982; Tsiantos, 1986). *Clavibacter michiganensis* subsp. *sepedonicus* can survive for long periods on farm machinery, bins, sacks, kraft paper, and polyethylene plastic (Richardson, 1957; Nelson, 1980) and also in potatos and sugar beets without causing symptoms (Hayward, 1974; Bugbee et al., 1987). Infectious bacteria can be recovered from dried potato stems of infected plants for 26 months and, in case of cv. Russet Burbank, over 5 years (Nelson, 1979; Nelson, 1985). Dried slime from symptomatic tubers especially favors long-term maintenance of the bacterium on surfaces. Freezing temperatures, rather than being deleterious to the bacterium, actually appear to enhance persistence (Slack, 1978; Nelson, 1980). It is suggested that plant pathogenic clavibacteria may spread from plant to plant through all the above sources (Vidaver, 1982). They may also be transmitted by rain, wind, or insects, as occurs for other bacteria inhabiting plants (Lindemann and Upper, 1985; Hirano et al., 1996; Lilley et al., 1997). Hawn (1963) reported that the root-knot nematode, *Ditylenchus dipsaci*, increased the frequency and severity of the bacterial wilt of lucerne caused by *Clavibacter michiganensis* subsp. *insidiosus*.

Isolation

ISOLATION OF PLANT PATHOGENIC CLAVIBACTERIA

Sources and Methods Plant pathogenic subspecies of *Clavibacter michiganensis* can be isolated from both infected and infested plants, particularly if the material is obtained from areas of known infection (Vidaver, 1982). The choice of a suitable piece of tissue is of prime importance to avoid contamination with faster-growing bacteria. Material should be washed with tap water and gently blotted dry with paper toweling. Surface sterilization is usually unnecessary. The tissue should be carefully excised with sterile instruments. Young lesion material, freshly infected material, or the margins of decayed areas should be sought. Direct puncture from fresh tissue by use of a sharp, sterilized dissecting needle onto agar should be successful, as should streaking or plating of comminuted tissues after serial dilution (Vidaver and Davis, 1988). In a wilting plant, the vascular tissue near the base of the wilting sections will increase the recovery rate. The cut tissue is placed into sterile water to release bacteria. Discolorization at the cut vessels is a good indication of releasing. In case of stunting and chlorosis in lucerne or of mosaic in wheat, a vascular infection may occur and recovery of *Clavibacter michiganensis* subsp. *insidiosus* and *Clavibacter michiganensis* subsp. *tessellarius* can be anticipated. In case of cankers and necrotic lesions, the bacteria may be isolated from the moist or dull greasy tissue at the edge of unhealthy tissue. Old necrotic areas, rotten parts, or parts showing advanced stages of disease contain many saprophytes which may overgrow the *Clavibacter* cells, hampering or making impossible direct isolation from such material (Collins and Bradbury, 1992). *Clavibacter michiganensis* subsp. *sepedonicus* is often difficult to isolate from potato tubers because of the presence of many other microorganisms (Vidaver and Davis, 1988).

Isolation Media Primary isolation of plant pathogenic clavibacteria using dilution plating on a suitable selective or nonselective agar media is used to obtain presumptive information on the numbers of different bacteria. Sometimes this method results in the isolation of unexpected or new pathogens (Collins and Bradbury, 1992). Nutrient agar supplemented with 1% glucose or 5% sucrose (Lelliott and Stead, 1987), glucose-yeast-calcium (GYCA) medium (Moffett et al., 1983), NBY medium (Vidaver and Davis, 1988), Doppel's medium (Lelliott and Stead, 1987), CB agar or any other media based on peptone, yeast extract, and glucose may be used as nonselective media for isolation of plant pathogenic clavibacteria.

GYCA Medium (Moffett et al., 1983)

Glucose	5 g
Yeast extract	5 g
Finely ground CaCO₂	41 g
Agar	15 g
Water	1 liter
pH 7.2	

Doppel's Medium (Lelliott and Stead, 1987)

Glucose	10 g
Yeast extract	8 g
Casein hydrolysate (not vitamin free)	8 g
K₂HPO₄	2 g
MgSO₄ · 7H₂O	0.3 g
Agar	12 g
Water	1 liter
pH 7.2	

NBY Medium (Gross and Vidaver, 1979a; Vidaver and Davis, 1988)

Nutrient broth	8 g
Yeast extract	2 g
K₂HPO₄	2 g
KH₂PO₄	0.5 g
Agar	15 g
Water	1 liter
pH 7.2	

Fifty ml of 50% glucose and 1.0 ml of 1M MgSO₄ · 7H₂O are both added after separate filter sterilization.

CNS medium enhances the chance of selectively recovering *Clavibacter michiganensis* subsp. *nebraskensis*, *Clavibacter michiganensis* subsp. *michiganensis* and *Clavibacter michiganensis* subsp. *tessellarius* (Gross and Vidaver, 1979a). The CNS medium is based on NBY medium which is supplemented with 1% lithium chloride and the following antibiotics: nalidixic acid, 25 mg (2.5 ml of freshly made 1% solution in 0.1M NaOH); polymyxin B sulfate, 32 mg (3.2 ml of a 1% aqueous solution of 8,000 USP units/mg, or a total of 256,000 units); cyclohex-imide, 40 mg (4 ml of 1% aqueous solution); and tetrachloro-isophthalonitrile, 0.66 mg (0.625 ml of Bravo 6F diluted 1:50). A slightly different formula of CNS was also published by Vidaver and Davis (1988). Lithium chloride was reported to be transiently toxic to newly isolated *Clavibacter michiganensis* subsp. *nebraskensis*. This can be overcome either by omitting it from the medium or by allowing the plant suspension to stand in a buffer containing sodium for 1–2 h before streaking (Vidaver and Davis, 1988; Zinniel et al., 2002). The medium should also be pretested to determine growth of single colonies in the presence of polymyxin B sulfate preparations, and the concentration of this antibiotic may be reduced (Vidaver and Davis, 1988). SCM agar supports selective isolation of *Clavibacter michiganensis* subsp. *michiganensis* (Fatmi and Schaad, 1988; Vidaver and Davis, 1988). Medium 4m1 (Atlas and Parks, 1993) and semiselective NCP medium (De la Cruz et al., 1992) were proposed for isolation of *Clavibacter michiganensis* subsp. *sepedonicus*. Medium D2 (Kado and Heskett, 1970) was reported to be selective for various coryneforms.

Medium 4m1 (Atlas and Parks, 1993)

Peptone	3 g
Pancreatic digest of casein	3 g
Yeast extract	3 g
Maltose	2 g
Lactose	1 g
Sodium dichromate solution	0.05 g
Agar	15 g
Water	900 ml
pH 7.0–7.2	

Sodium dichromate (0.05 g in 100 ml of distilled water) is sterilized separately and added to sterile medium, cooled to 45–50°C.

ISOLATION OF SAPROPHYTIC CLAVIBACTERIA Saprophytic clavibacteria can be isolated from different plants and some other environmental sources outlined in the subsection Habitat in this Chapter. Early summer, late summer or early autumn (Zinniel et al., 2002; Jacobs and Sundin, 2001) or spring and summer in desert regions (Dobrovolskaya et al., 1999) are suitable sampling periods. Jacobs and Sundin (2001) found that the late summer period was the best sampling time for isolation of UV-tolerant clavibacteria inhabiting field-grown peanut plants. As larger bacterial populations are located on the adaxial surface of leaves (Beattie and Lindow, 1995; Sundin and Jacobs, 1999), this site of leaves is preferable for isolation of epiphytes. To isolate clavibacteria from the phyllosphere of plants, the dilution or leaf impression techniques described by Dickinson et al. (1975), Dunleavy (1989), Behrendt et al. (1997), and Jacobs and Sundin (2001) may be used. Important points are sterile sampling and transport of plants in ice to the laboratory and use of media supplemented with antifungal antibiotics. For isolation of endophytic clavibacteria, the approaches described by Hallmann et al. (1997), Fisher et al. (1992) or Zinniel et al. (2002) may be applied. Surface disinfestation, e.g., with 2% sodium hydrochloride containing 0.1% Tween 20 for 10 s followed by several washes in sterile water or buffer solutions, is an important stage of isolating endophytic bacteria as it significantly reduces (by more than 10,000-fold) external contamination with epiphytic bacteria (Zinniel et al., 2002). Surface detergents, like Tween 20, Tween 80 or Triton X-100, reduce surface tension of the solvent, allowing the disinfectant to reach protected sites and grooves beyond collapsed epidermal cells on the plant surface. Another surface disinfestation technique involves dipping plant tissue into ethanol and flaming of the surface (Dong et al., 1994).

NBY medium, CNS medium (Gross and Vidaver, 1979a), and a modified CNS medium without lithium chloride were successfully used for isolation of the epiphytic and endophytic clavibacteria (Zinniel et al., 2002). They may also be isolated and grown on tryptic soy agar, R2A agar, CB agar, or other media based on peptone, yeast extract, and glucose. *Clavibacter* strains usually grow well at 24–26°C. Incubation at room temperature (18–24°C) in daylight facilitates better formation of pigment that allows one to differentiate some clavibacteria from other bacteria at the earlier stage of investigation.

Identification

GENERIC LEVEL Irregular, nonsporeforming rods isolated from different sources can be easily identified as members of the genus *Clavibacter* by using 16S rDNA sequence analysis and/or some phenotypical properties listed in Table 1. Partial sequence (5'400–500 nucleotides) of 16S rDNA or restriction analysis of amplified 16S rDNA (Lee et al., 1997b; Lee et al., 1997a; Behrendt et al., 2002) is usually discriminatory enough to assign isolates to clavibacteria. Identification based on phenotypic traits involves mainly the detection of DAB in the cell wall and the major menaquinone MK-9. However, as new organisms with the same or similar chemotaxonomic characteristics may occur in different plants or other environments, at least partial 16S rDNA nucleotide sequencing is recommended for determining genus affiliation.

SPECIES LEVEL Cultural, physiological and biochemical features (Table 4) and polyacrylamide gel electrophoresis of soluble proteins (Carlson and Vidaver, 1982) are useful to compare new strains from unusual sources with *Clavibacter michiganensis* subspecies. Fatty acid analysis, rep-polymerase chain reaction (PCR)-mediated genomic fingerprinting (Louws et al., 1998) and Fourier-transform infrared (FT-IR) spectroscopy, which reflects common physical-chemical profiles of bacterial cells (Behrendt et al., 2002; Oberreuter et al., 2002), may also be helpful. Affiliation of saprophytic strains from an unusual host to a valid species requires DNA-DNA relatedness studies or analysis of other genomic characteristics of adequate resolving power (Vandamme et al., 1996; Stackebrandt et al., 2000, 2002; Rossello-Mora and Amann, 2001).

IDENTIFICATION AND DETECTION OF *CLAVIBACTER MICHIGANENSIS* SUBSPECIES *General approaches* The phytopathogenic clavibacteria are relatively easy to identify when isolated from young lesions of the known hosts with characteristic symptoms as a predominant fraction of the microbial community and compared with known pathogens (Vidaver and Davis, 1988). Colony characteristics on NBY agar or other media, growth on

Table 4. Characteristics that differentiate the *Clavibacter michiganensis* subspecies.

Characteristic	*Cl. m.* subsp. *michiganensis*	*Cl. m.* subsp. *insidiosus*	*Cl. m.* subsp. *nebraskensis*	*Cl. m.* subsp. *sepedonicus*	*Cl. m.* subsp. *tesselarius*
Pigment, NBY agar	Y	Y, BL	O (rarely Y)	W	O
Pigment, CB agar	Y	W, YW, BL	O	YW	RO
Cell wall sugars					
Glucose	d	+	+	+	+
Fucose	+	+	+	−	+
Acid from					
Lactose	+	+	+	−	+
Mannose	+	+	+	d	+
L-Ribose	+	+	+	−	+
Inositol	+	−	+[a]	−	+[a]
Sorbitol	−	−	+[b]	−[c]	+
Growth on C-source					
Dulcitol	d	−	+	−	+
D-Rhamnose	+	−	−	−	−
Utilization					
Acetate	+[d]	−	+	+	−
Succinate	+	−	+	+	+
Hydrolysis					
Starch	+	−	+	+	+
Tween 40	+	−	d	+	+
DNA	+	+	+	−	+
H₂S from peptone	+	+[b]	+	−	+
Levan production	−[c]	−[c]	+[e]	−	+[e]
Growth on CNS[f]	+	−	+	−	+
Growth on TTC[f]	+	+	−	−	−
5% NaCl	+	−	+	−	d
0.03% K-Tellurite	d	−	+	−	+
Voges-Proskauer	+[e]	+[e]	−	−	+[e]
G+C (mol%)[f]	73	73	73.5	72	74
Plant host[g]	Tomato and pepper	Lucerne	Corn	Potato	Wheat
Predominant symptoms[g]	Wilt and fruit spot	Wilt and stunting	Wilt and leaf blight	Wilt and tuber rot	Leaf spot

Symbols and abbreviations: +, positive; −, negative; d, different between strains; Y, yellow; O, orange; RO, rose-orange; W, white; YW, yellowish white; and BL, blue to black somewhat diffusible pigment (indigoidine).

[a,b,c,d,e]The opposite results were reported by Zgurskaya (1993a), Dye and Kemp (1977), Davis et al. (1984b), Vidaver and Davis (1988), and Behrendt et al. (2002), respectively, using different methods.

[f]Data from Carlson and Vidaver (1982). Somewhat different values of the G+C content of DNA were reported by other authors.

[g]Data from Vidaver (1982).

Adapted from Dye and Kemp (1977), Carlson and Vidaver (1982), Davis et al. (1984b), Vidaver and Davis (1988), Zgurskaya (1993a), and Behrendt et al. (2002).

CNS agar and TTC agar (glucose, 5 g; peptone, 10 g; casein hydrolysate, 1 g; 2,3,5-triphenyltetrazolium chloride, 0.05 g; agar, 17 g; and water, 1 liter), Gram-positive identification with KOH (Ryu, 1938; Suslow et al., 1982), coryneform morphology (bending type of cell division without a strong tendency for a rod-coccoid cycle), catalase positive and usually oxidase negative reactions, and some other physiological properties (Table 4) are helpful identifying traits. Along with *Clavibacter michiganensis* subspecies, some saprophytic bacteria of other taxa with similar properties may be isolated, especially at the early stages of infection or from old necrotic areas and rotten parts, and an additional confirmation of identification is necessary by using the approaches outlined in the previous subsection. A direct test for virulence on the host or indicator plants or induction of hypersensitive reaction on non-host plants is the appropriate procedure for confirmation or identification of plant pathogenic *Clavibacter michiganensis* (Vidaver and Davis, 1988; Collins and Bradbury, 1992). However, a successful pathogenicity test may prove difficult if the right plants at the right stage of growth are not available (Collins and Bradbury, 1992). In addition, naturally occurring avirulent *Clavibacter michiganensis* subsp. *michiganensis* strains in diseased plants occur (Louws et al., 1998). Identification and confirmation are also possible by different DNA-based techniques or serological tests.

DNA-based methods DNA-based methods offer alternative ways for rapid, highly sensitive, specific detection and identification of plant pathogenic clavibacteria. Originally, DNA hybridization assays to detect and diagnose bacterial ring rot, wilt, and canker in tomato (Johansen et al., 1989; Thompson et al., 1989; Rademaker et al., 1992) were proposed. A PCR technique and specific PCR-primers targeting genomic or plasmid DNAs were developed for detection of *Clavibacter michiganensis* subsp. *sepedonicus* and *Clavibacter michiganensis* subsp. *michiganensis* (Firrao and Locci, 1994; Rademaker and Janse, 1994; Dreier et al., 1995; Hu et al., 1995; Li and De Boer, 1995a; Li and De Boer, 1995b; Slack et al., 1996; Lee et al., 1997a; Mills et al., 1997; Santos et al., 1997; Louws et al., 1998; Jahr et al., 1999; Brown et al., 2002).

A rapid and effective method to identify strains of *Clavibacter michiganensis* subspecies is based on rep-PCR-mediated genomic fingerprinting (Louws et al., 1998). Naturally occurring avirulent *Clavibacter michiganensis* subsp. *michiganensis* strains had rep-PCR fingerprints similar to those of virulent strains of this subspecies (Louws et al., 1998). Nested PCR for ultrasensitive detection of *Clavibacter michiganensis* subsp. *sepedonicus* is based on oligonucleotide primers derived from sequences of the 16S rRNA gene and insertion element IS1121 of this bacterium (Lee et al., 1997a). The method allows detection of very low titers of *Clavibacter michiganensis* subsp. *sepedonicus* in symptom-less potato plants or tubers, which cannot be readily detected by direct PCR (single PCR amplification). Specific detection of *Clavibacter michiganensis* subsp. *sepedonicus* based on amplification of three unique DNA sequences isolated by subtraction hybridization was proposed (Mills et al., 1997). The lowest level of detection of *C. michiganensis* subsp. *sepedonicus* was 100 colony forming units (CFU) per milliliter when cells were added to potato core fluid. Genomic fingerprints of *Clavibacter michiganensis* subsp. *sepedonicus* generated by contour-clamped homogeneous electric field (CHEF) gel electrophoresis of restriction enzyme digested high-molecular weight DNA may be useful to detect genomic variability associated with virulent and avirulent strains of this subspecies (Brown et al., 2002). The sequence information on the endoglucanase (*celA*) gene and the *pat-1* gene encoding serine protease is helpful for detection of *Clavibacter michiganensis* subsp. *michiganensis* (Dreier et al., 1995; Dreier et al., 1997; Jahr et al. 1999; Jahr et al., 2000). A DNA probe derived from the *celA* gene differentiated between *Clavibacter michiganensis* subsp. *michiganensis* and other subspecies. A probe derived from the *pat-*

1 gene is more specific, giving a hybridization signal exclusively with virulent strains of *Clavibacter michiganensis* subsp. *michiganensis*. PCR primers based on the sequence of this gene permit detection of virulent strains of *Clavibacter michiganensis* subsp. *michiganensis* in homogenates prepared from infected plants and naturally contaminated seeds. A positive result was obtained with as few as 2×10^2 bacteria per ml of contaminated plant homogenate (Jahr et al., 1999).

Serological Methods Serological methods are very sensitive and particularly useful for detection of clavibacteria in symptomless infections. Immunofluorescence allows direct observation of cells in the plant host and was reported to be useful for detection of almost all *Clavibacter michiganensis* subspecies (Lelliott and Stead, 1987). The most widely used serological approaches for detection of *Clavibacter michiganensis* subsp. *michiganensis* and *Clavibacter michiganensis* subsp. *sepedonicus* are the enzyme-linked immunosorbent assay (ELISA) and immunofluorescence based on monoclonal antibodies (De Boer and Wieczorek, 1984a; De Boer et al., 1984b; Drennan et al., 1993; Dreier et al., 1995). Occasionally, problems with cross-reacting bacteria may occur (De Boer, 1991; Li and De Boer, 1995b; Slack et al., 1996). Bioassay is sometimes used to complement these tests (Li and De Boer, 1995b). A rhodamine-labelled oligonucleotide probe has been used in conjunction with an indirect immunofluorescence procedure based on a specific monoclonal antibody detected with a fluorescein-labelled conjugate. Simultaneous labelling of bacterial cells with the oligonucleotide and antibody probes was found to allow accurate microscopic identification of single cells of *Clavibacter michiganensis* subsp. *sepedonicus* when isolation and other methods of confirming bacterial identity are hampered or impossible (Li et al., 1997).

Preservation

The principles outlined in the subsection Preservation in this Chapter for *Microbacterium* Species should be followed.

Freeze-drying may sometimes affect virulence of plant pathogens (Servin-Massieu, 1971). Virulence of *Clavibacter michiganensis* subsp. *nebraskensis* and probably other subspecies can be maintained satisfactorily by storage on solid complex media at 6°C for 2 years or by lyophilization. Storage of *Clavibacter michiganensis* subsp. *nebraskensis* in sterile distilled water or phosphate buffer at 6°C or room temperature was unsatisfactory for maintenance of viability and virulence; strains maintained in dried leaves of greenhouse grown plants were viable and virulent up to three months (Vidaver, 1977).

Applications

The physiological and enzymatic activities of wildtypes and mutants of *Clavibacter* spp., including those related to plant pathogenesis, provide evidence that clavibacteria are a potential target for screening metabolites, enzymes and genes to be used in biotechnology, medicine, the food industry, and agriculture. Results obtained during studying the effect of polysaccharides on immunologic factors in animals with *Mycoplasma hominis* infection and histomorphological data showed that polysaccharides of *Clavibacter michiganensis* are immunomodulators (Rudenko et al., 1996). A glycopolymer of *Clavibacter michiganensis* was found to induce γ-interferon (Varbanets et al., 1990a) and acid polysaccharides to induce formation of the tumor necrosis factor and interleukin-1 by peritoneal macrophages of mice (Varbanets et al., 1990b). Extracellular polysaccharides produced by clavibacteria are an easy source of clavan, an L-fucose rich polymer, and some unusual monosaccharides (Vanhooren and Vandamme, 1999; Vanhooren and Vandamme, 2000). L-Fucose can be used in medicine for prevention of tumor cell colonization of the lung, controlling white blood cell formation, rheumatoid arthritis treatment, antigen synthesis for antibody production, and such "cosmeceuticals" as a skin moisturing agent. Clavibacteria produce different biotechnologically applicable enzymes, e.g., oxoglutarate-dehydrogenase used for production of optically active γ-hydroxy-L-glutamic acid (Katsumata and Hashimoto, 1996), specific β-glucosidases (Kiso et al., 2000), and some others. *Clavibacter michiganensis* strain ATZ1 catabolizes triazine herbicides (Seffernick et al., 2000). Cell extracts from this bacterium, which contains a gene homologous to the atrazine chlorohydrolase (AtzA) gene in *Pseudomonas* sp. strain ADP, were able to transform atrazine analogs containing pseudohalide and thiomethyl groups, in addition to the substrates transformed by AtzA from *Pseudomonas* sp. Clavibacteria can also degrade bacterial and synthetic polyesters (e.g., poly-β-hydroxybutyrate, etc.; Mergaert et al., 1995; Mergaert and Swings, 1996). *Clavibacter* strain ALA2 effectively oxidizes linoleic acid from vegetable oils into value-added industrial products. One of the products of the biotransformation, 12,13,17-trihydroxy-9(Z)-octadecenoic acid, was found to inhibit the growth of various plant pathogens and may be used as pesticide (Hou et al., 1997; Hou, 1998; Hou, 1999; Gardner et al., 2000).

Bacteria of the genus *Clavibacter* are also involved in recent studies related to plant disease control. Introduction of disease- and insect resistance into crops and ornamental plants by application of protein derived from a noninfectious strain of *Clavibacter michiganensis* subsp. *sepedonicus* was described (Beer and Butler, 1999). The approach involves applying a hypersensitive response-elicitor protein to plant or plant seed, resulting in a plant with a heritable disease-resistance trait. The treatment induces resistance to different pathogens, including viruses, bacteria, fungi and insects. The possible practical use of clavibacteria is related to their natural endophytic mode of life. Colonization of several crops with endophytic clavibacteria (Zinniel et al., 2002) is a promising tool for delivering enzymes to control plant diseases and other useful products promoting plant growth. This will permit growth of plants with improved capabilities without integration of foreign DNA into the plant genome; this provides an alternative to genetic modification or breeding of plants, which is costly, depends on the plant variety being studied, and may take several years to reach the market (Zupan and Zambryski, 1995; Hallmann et al., 1997). The finding that pathogenic clavibacteria can be converted into endophytes by transformation of plasmid DNA increases the chance of expressing gene products which will prevent infection of host plants by other pathogens (Jahr et al., 1999). A chitinase gene-carrying hybrid plasmid has been constructed that expresses this enzyme when introduced into a plasmid-free strain of *Clavibacter michiganensis* subsp. *michiganensis* (Jahr et al., 1999). Under laboratory conditions, infection of tomato plants with such a strain resulted in a reduced sensitivity to infection of plants by *Fusarium oxysporum* f. sp. *lycopersici*, a fungal pathogen of tomato. Endophytic clavibacteria or plant pathogens converted into endophytes might also be considered as potential vectors of certain foreign genes for construction of transgenic plants.

The Genus *Cryobacterium*

The genus *Cryobacterium*, including the species *Cryobacterium psychrophilum*, was proposed by Suzuki et al. (1997) as a pink, obligately psychrophilic bacterium, which had been previously described by Inoue and Komagata (1976) as "*Curtobacterium psychrophilum*."

The bacterium was isolated from a soil sample from Antarctica by Inoue (1976) and assigned to the genus *Curtobacterium* on the basis of its cell wall amino acid composition, found to be typical of *Curtobacterium* spp., and of its high DNA G+C content (Inoue and Komagata, 1976b). On the basis of *Arthrobacter*-like morphology and low maximum growth temperature, the authors classified it into a new species, "*Curtobacterium psychrophilum*." However, this name was not validly published. In 1997, Suzuki et al. examined the chemotaxonomic features of this strain and found DAB in the cell wall instead of ornithine. Phylogenetic analysis based on 16S ribosomal DNA sequences demonstrated that this bacterium represents a distinct lineage within the radiation of the family Microbacteriaceae, thus representing a new genus (Suzuki et al., 1997).

Phylogeny

The phylogenetic dendrogram (Fig. 1) shows that the genus *Cryobacterium* is distant from all validly described genera of the family Microbacteriaceae, linked peripherally to the genus *Agrococcus* (Groth et al., 1996).

Taxonomy

The only species of the genus, *Cryobacterium psychrophilum*, is characterized by a growth temperature optimum at 9–12°C and no growth at temperatures higher than 18°C. Colony color is pink to reddish-orange. Cells are nonmotile, nonsporeforming, pleomorphic rods. Branching and coccoid forms may occur in the early growth phase and in old culture, respectively. Aerobic, catalase and oxidase are positive. Amino acids of the cell wall peptidoglycan are 2,4-diaminobutyric acid (mainly L-isomer), alanine, glycine, and glutamic acid in the molar ratio consistent with B2γ peptidoglycan (Schleifer and Kandler, 1972). Muramic acid of the peptidoglycan is of acetyl type. Rhamnose and fucose are characteristic cell wall sugars. The main cellular fatty acids are *iso*- and *anteiso*-branched acids. 12-Methyl tetradecenoic acid (*anteiso*-15:1) accounts for 20–26% of the cellular fatty acids in cells grown at 4–10°C. The major menaquinone is MK-10, followed by MK-11 and MK-8. The polar lipids are diphosphatidylglycerol, phosphatidylglycerol, and glycolipid. The G+C content of DNA is approximately 65 mol% (Inoue and Komagata, 1976b; Suzuki et al., 1997; Sasaki et al., 1998).

Habitat

The only strain of *Cryobacterium psychrophilum* was isolated from a soil sample from an area inhabited by plants or animals, collected near Showa Station in Antarctica, the East Ongul Island (the maximum temperature is about +7°C on coastal islands; Inoue, 1976a). The temperature condition of soil from which the strain was isolated, the growth temperature range (<18°C) of the strain, and its fatty acid composition (high

amount of *anteiso*-15:1 used to maintain membrane fluidity at low temperatures; Suzuki et al., 1997), suggest an active state of *Cryobacterium psychrophilum* in soil under permanently low temperatures.

Isolation

Cryobacterium psychrophilum was isolated at 0°C using PYG agar (Inoue and Komagata, 1976b) from a sample stored at 5°C within a few hours after sampling (Inoue, 1976a). Since mesophilic species of the genus may exist in nature, higher temperatures should be tested during the isolation regime. No special media are used for cultivation of heterotrophic microorganisms of the genus *Cryobacterium*. A number of media, including PYG agar (Inoue and Komagata, 1976b) or R agar (Yamada and Komagata, 1972a; Suzuki et al., 1997), are suitable.

PYG agar (Inoue and Komagata, 1976b)

Peptone	1 g
Yeast extract	5 g
Glucose	3 g
Agar	15 g
Water	1 liter
pH 7.2	

Identification

Pink to orange-red colonies, the presence of DAB in cell wall, the major menaquinone MK-10, growth temperature range, and unique cellular fatty acid profile (particularly a significant amount of 12-methyl tetradecenoic acid; Table 1) are good grounds for preliminary identification of *Cryobacterium psychrophilum* at the phenotypic level. Although this fatty acid is also accumulated in cells of some mesophilic strains grown at 10°C, its amount does not exceed 8% (Suzuki et al., 1997). Cell size (0.5–0.7 × 1.0–1.8 μm), cell wall sugar composition, positive DNase test, production of acids from fructose, mannose, glucose, galactose, and sucrose, utilization of lactate, pyruvate, fumarate, and hippurate, low utilization of many other C-sources tested (Inoue and Komagata, 1976b; Suzuki et al., 1997), and probably growth temperature optimum are the basis of its discrimination from other species of the genus, which are yet to be isolated and described.

The Genus *Curtobacterium*

The genus *Curtobacterium* was established by Yamada and Komagata (1972b) for a group of coryneform bacteria that contained ornithine in the cell wall and weakly produced acid from various carbohydrates.

In *Bergey's Manual of Systematic Bacteriology* (Komagata and Suzuki, 1986), five species were included in the genus: *Curtobacterium albidum, Curtobacterium citreum, Curtobacterium luteum, Curtobacterium pusillum* (Yamada and Komagata, 1972b) and *Curtobacterium flaccumfaciens* with pathovars (Collins and Jones, 1983b). Behrendt et al. (2002) have recently described a sixth species, *Curtobacterium herbarum*, which was isolated from the phyllosphere of grasses and the surface litter after mulching the sward. Another plant-associated saprophyte species, *Curtobacterium plantarum*, was proposed by Dunleavy (1989). However, the type strain CL63^T (ATCC 49174^T, DSM 7069^T) of this species has an A-type peptidoglycan, and according to 16S rDNA sequence and fatty acid pattern, should be placed into the genus *Pantoea* (Deutsche Sammlung von Mikroorganismen und Zellkulturen, 1998). As judged from the current knowledge on the occurrence of curtobacteria in different plants (see subsection Habitat in this Chapter), the physiological heterogeneity of strains assigned to *Curtobacterium plantarum*, and the similarity in physiological traits to authentic curtobacteria (Dunleavy, 1989), it might be argued that many strains isolated by Dunleavy were in fact members of the genus *Curtobacterium*.

Since 1922 (Hedges, 1922), organisms comprising the current genus *Curtobacterium* have been assigned to different bacterial genera, e.g., "*Bacterium,*" "*Phytomonas,*" "*Pseudobacterium,*" *Corynebacterium* and *Brevibacterium* (Cummins et al., 1974). By 1972, most of them had been placed within the genera *Brevibacterium* and *Corynebacterium* (Komagata and Iizuka, 1964; Iizuka and Komagata, 1965; Cummins et al., 1974). In an exhaustive taxonomic study of various coryneforms, Yamada and Komagata (1972b) established a new genus *Curtobacterium* for obligately aerobic organisms with ornithine in the cell wall, a high DNA G+C content (66–71 mol%), and weak production of acid from some sugars. The new genus differed from *Cellulomonas*, the other genus known to have ornithine in the cell wall at that time, by the mode of cell division and some physiological properties. The species assigned to the genus *Curtobacterium* included several former *Brevibacterium* spp. (*B. albidum, B. citreum,* "*B. helvolum,*" "*B. insectiphilum*" and *B. luteum* [Komagata and Iizuka, 1964], *B. pusillum* [Iizuka and Komagata, 1965], *B. saperdae* [Lysenko, 1959], *B. testaceum* [Komagata and Iizuka, 1964]), as well as *Corynebacterium flaccumfaciens* (Hedges, 1922; Dowson, 1942) and *Corynebacterium poinsettiae* (Starr and Pirone, 1942), the only plant pathogens with ornithine in their cell walls included in the study of Yamada and Komagata (1972a).

Further investigations (Schleifer and Kandler, 1972; Fiedler et al., 1973b; Yamada et al., 1976; Uchida and Aida, 1977; Collins et al., 1979; Collins et al., 1980b) showed the peptidoglycan structure and menaquinone composition of organisms assigned to this genus were heterogeneous. Consequently, only the non-phytopathogenic species *Curtobacterium albidum, Curtobacterium citreum, Curtobacterium luteum, Curtobacterium pusillum, Curtobacterium saperdae* and *Curtobacterium testaceum* were included in the Approved Lists of Bacterial Names (Skerman et al., 1980). The two latter species were transferred to the genus *Aureobacterium* (Collins et al., 1983c) as they contained *N*-glycolyl residues in the glycan moiety of their walls (Uchida and Aida, 1977), an additional glycine residue in the interpeptide bridge of their peptidoglycan (Schleifer and Kandler, 1972; Collins et al., 1983c), and unusually long menaquinones (MK-11, MK-12; Collins et al., 1980b; Collins et al., 1983c), properties which were in contrast to authentic members of the genus *Curtobacterium* (Table 1). The genera *Aureobacterium* and *Microbacterium* were later united under *Microbacterium* (Takeuchi and Hatano, 1998b).

"*Curtobacterium helvolum*" and "*Cutobacterium insectiphilum,*" proposed by Yamada and Komagata, were not approved (Skerman et al., 1980) and have not been approved since then. Four strains of "*Brevibacterium helvolum*" studied by Yamada and Komagata (1972b) were determined to contain B2γ peptidoglycan based upon DAB (Schleifer and Kandler, 1972). "*Brevibacterium helvolum*" DSMZ 20419 (ATCC 13715), having the same peptidoglycan type (Schleifer and Kandler, 1972; Fiedler et al., 1973b), was shown to represent a separate line of descent within the family Microbacteriaceae and suggested to represent the nucleus of a new genus (Rainey et al., 1994; see the section Other Microorganisms of the Family Microbacteriaceae in this Chapter). Strain "*Brevibacterium helvolum*" ATCC 11822 is a soil isolate originally referred by Jensen (1934) to *Corynebacterium helvolum* but showing morphological and chemotaxonomic characteristics typical of the genus *Arthrobacter* (Yamada and Komagata, 1972b; Fiedler et al., 1973b; Yamada et al., 1976). Other strains labelled "*Brevibacterium helvolum*" are also doubtful members of the genus *Curtobacterium* as the earlier descriptions of this species were contradictory. No phylogenetic or detailed chemotaxonomic data are available with regard to "*Curtobacterium insectiphilum*" AJ 1477 (JCM 1348, IFM AM-23; Yamada and Komagata, 1972b), except for the fatty acid composition which is similar to that of *Curtobacterium* spp. (Suzuki and Komagata, 1983).

"*Brevibacterium insectiphilum*" strain Bhat 146 is phylogenetically closest to *Microbacterium* spp. (Döpfer et al., 1982) and possesses the cell wall peptidoglycan type B2β, [L-Hsr]-D-Glu(Hyg)-Gly-D-Orn (Schleifer and Kandler, 1972), typical of *Microbacterium*.

Besides *Corynebacterium flaccumfaciens* and *Corynebacterium poinsettiae* (Yamada and Komagata, 1972a; Yamada and Komagata, 1972b), two other plant pathogenic species, i.e., *Corynebacterium betae* (Keyworth et al., 1956) and *Corynebacterium oortii* (Saaltink and Maas Geesteranus, 1969), were found to be similar to *Curtobacterium* in chemotaxonomic properties (Schleifer and Kandler, 1972; Yamada et al., 1976; Collins et al., 1979; Collins et al., 1980b) and DNA base ratios (Starr et al., 1975). Also *Corynebacterium betae* was indicated to be similar to *Curtobacterium* spp. on the basis of numerical taxonomic study (Jones, 1975). However, these four plant pathogenic species were listed in the Approved Lists under the generic name *Corynebacterium*. On the basis of nearly identical cellular protein patterns and some other taxonomic characteristics, these four plant pathogens were unified into one species, *Corynebacterium flaccumfaciens* (Carlson and Vidaver, 1982), in accord with the opinion of Dye and Kemp (1977), expressed previously on the basis of high overall phenetic similarity. But in contrast to Dye and Kemp, who treated these four species as pathovars, Carlson and Vidaver (1982) proposed that the traditional names of pathogens in the subspecies be retained and reclassified them into four subspecies. The proposal was substantiated by differences in some biochemical characteristics (Dye and Kemp, 1977), bacteriocin production (Gross and Vidaver, 1979b), preliminary data on low level of DNA-DNA relatedness between two pathovars (Starr et al., 1975), and also host specificity. The subspecies were validated (Moore et al., 1985) as *Corynebacterium flaccumfaciens* subsp. *flaccumfaciens* (Hedges 1922) Dowson 1942; Collins et al. 1979, *Corynebacterium flaccumfaciens* subsp. *betae* (Keyworth et al., 1956; Collins et al., 1979) Carlson and Vidaver, 1982 Collins et al. 1979, *Corynebacterium flaccumfaciens* subsp. *oortii* (Saaltink and Maas Geesteranus, 1969), and *Corynebacterium flaccumfaciens* subsp. *poinsettiae* (Starr and Pirone, 1942; Collins et al., 1979) Carlson and Vidaver 1982.

Later, Collins and Jones (1983b) transferred these plant pathogens to the genus *Curtobacterium* based on the chemotaxonomic data (Schleifer and Kandler, 1972; Yamada et al., 1976; Keddie and Cure, 1978; Collins et al., 1979; Collins et al., 1980b; Collins et al., 1982b). Collins and Jones (1983b) proposed to lower the rank of *Curtobacterium flaccumfaciens* subspecies to the

level of pathovars, taking into account the high degree of DNA-DNA relatedness (Döpfer et al., 1982) and close physiological relationship (Dye and Kemp, 1977). Differences in host specificity and bacterocin production were thus considered insufficient to justify differentiation at the subspecific level (Collins and Jones, 1983b).

Further taxonomic studies are required to elucidate the species structure of the genus *Curtobacterium* and taxonomic ranks of the infraspecies groups comprising *Curtobacterium flaccumfaciens*. Such taxonomic revision is necessary from practical and regulatory points of view. The necessity of such revision is further substantiated by phenotypic differences between the pathovars of *Curtobacterium flaccumfaciens* (Table 6) as well as by information on the isolation of many new *Curtobacterium flaccumfaciens* strains and close relatives from unusual hosts (Raupach and Kloepper, 2000; Araujo et al., 2001) or different non-diseased plants worldwide (Legard et al., 1994; Dobrovolskaya et al., 1999; Elbeltagy et al., 2000; Behrendt et al., 2002; Zinniel et al., 2002; L. I. Evtushenko et al., unpublished observation).

Phylogeny

Members of the genus *Curtobacterium* form a tight (100% bootstrap confidence) phylogenetic cluster within the radiation of bacteria comprising the family Microbacteriaceae (Fig. 1).

Taxonomy

The genus *Curtobacterium* sensu stricto currently harbors the following six species: *Curtobacterium albidum* (Komagata and Iizuka, 1964), *Curtobacterium citreum* (Komagata and Iizuka, 1964), *Curtobacterium luteum* (Komagata and Iizuka, 1964), *Curtobacterium pussilum* (Komagata and Iizuka, 1964) all Yamada and Komagata 1972b, *Curtobacterium flaccumfaciens* (Hedges, 1922) Collins and Jones, 1984, and *Curtobacterium herbarum* (Behrendt et al., 2002).

The characteristics typical of the genus can be summarized as follows. Colonies on nutrient agar are yellow, orange or ivory, shiny, slightly convex and round, with entire margins. Cells are small, irregular, unbranched rods, tending to become shorter in the older culture and multiply by the bending type of cell division. These Gram-positive, nonsporeforming cells usually move by means of lateral flagella and are obligately aerobic, catalase positive, chemoorganotrophic, mesophilic, and capable of growth on a nutrient agar and related media. Growth usually occurs at 5% NaCl but not at 10% NaCl. Acid is produced rather slowly and weakly from carbohy-

drates. The cell wall peptidoglycan based upon
D-ornithine is of the type B2β, [L-homoserine]-
D-Glu-D-Orn (Schleifer and Kandler, 1972). The
glycan moiety of the peptidoglycan contains
acetyl residues (Uchida and Aida, 1977). Mycolic
acids are not present. The predominant fatty
acids are *anteiso*-15:0, *anteiso*-17:0 with substantial amounts of *iso*-16:0, and some others
(Collins et al., 1980b; Stead, 1988). ω-Cyclohexyl
undecanoic acid may occur in some representatives of the genus (Suzuki and Komagata, 1983).
Polar lipids are diphosphatidylglycerol, phosphatidylglycerol, and some glycosyldiacylglycer-
ols; the major respiratory quinone is unsaturated
menaquinone with nine isoprene units (Collins
et al., 1980b). The DNA G+C content is 68.3–75.2
mol% (Yamada and Komagata, 1970; Döpfer et
al., 1982; Behrendt et al., 2002).

The species differentiation is based on growth
characteristics, content of G+C of DNA, fatty
acid composition, and some physiological
properties (see subsection Identification in this
Chapter).

Habitat and Phytopathogenicity

The data available suggest that the main natural
habitat of *Curtobacterium* spp. are different
plants and related sources, although they also
occur in soil and other environments (Iizuka and
Komagata, 1965; Lednicka et al., 2000; Koneva
and Kruglov, 2001). Among validly described
species of the genus, only *Curtobacterium pusillum*, recovered from oil brine in a Japanese oil
field (Iizuka and Komagata, 1965), was found
outside plants. *Curtobacterium albidum*, *Curtobacterium citreum* and *Curtobacterium luteum*
were first isolated from Chinese rice paddies
(Komagata and Iizuka, 1964). *Curtobacterium
herbarum* inhabits the phyllosphere of grasses
and the surface litter after mulching the sward
(Behrendt et al., 2002). *Curtobacterium flaccumfaciens* is the only species of the genus for which
plant pathogenic properties were recorded
(Young et al., 1996).

OCCURRENCE OF PLANT-ASSOCIATED SAPROPHYTIC
CURTOBACTERIA Curtobacteria along with representatives of the genera *Clavibacter*, *Pseudomonas*, *Pantoea* and *Stenotrophomonas* were
reported to be numerically predominant in the
microbial communities of grass phyllosphere and
in decaying surface litter after sward mulching
(Behrendt et al., 1997; Behrendt et al., 2002).
Strains of the genus *Curtobacterium* and *Clavibacter* have been isolated at every sampling and,
in comparison with the Gram-negative bacteria,
often formed a greater fraction. Dunleavy (1989)
found organisms resembling *Curtobacterium*
morphologically and physiologically in leaves of

about 200 species comprising 62 plant families,
including herbs, shrubs and trees from a widespread area in the state of Iowa (United States).
This is consistent with the finding of properly
identified curtobacteria in leaves and seeds of
different plants of the families Asteraceae,
Cypereceae, Poaceae and Lamiaceae and also in
moss in Russia (L. I. Evtushenko et al., unpublished observation). Curtobacteria were also isolated from various plants and related sources
collected in desert areas in Uzbekistan, Turkmenistan and Israel (Dobrovolskaya et al.,
1999). As the isolation methods involved no surface disinfection, it is unclear whether the
saprophytic curtobacteria were epiphytes or
endophytes (Kado, 1992) living in plant tissues
without doing substantive harm or gaining benefit other than residency. *Curtobacterium flaccumfaciens* was found in various plants without
disease symptoms, e.g., in the phyllosphere of
weeds (Vidaver, 1982), grasses (Behrendt et al.,
2002), sugar beet (Thompson et al., 1993), field-
grown peanut (Sundin and Jacobs, 1999), and
spring wheat (Legard et al., 1994).

Zinniel et al. (2002) showed that curtobacteria can exist as endophytes. Endophytic bacteria
identical in 16S rDNA sequences and fatty acid
profiles to *Curtobacterium flaccumfaciens* were
observed within agronomic crops. Endophytes
phylogenetically related to *Curtobacterium citreum* were found in wild and traditionally cultivated rice varieties (Elbeltagy et al., 2000). A
group of endophytic curtobacteria was recovered among other bacteria, which colonized the
foliage, tap roots, and nodules of red clover
plants (*Trifolium pratense* L.; Sturz et al., 1997).
Moreover, in root bacterization experiments,
bacteria identified as *Curtobacterium luteum*
consistently promoted growth of the plant either
individually or in combination with *Rhizobium*
sp. (Sturz et al., 1997). Different strains of the
genus *Curtobacterium* together with *Microbacterium* strains, including *M. testaceum*, were isolated from plant galls induced by plant parasitic
nematodes of the subfamily Anguininae in
seeds, leaves and stems of their host plants
(Evtushenko et al., 1994). In some seed galls of
Agrostis sp., they composed up to 90% bacterial
isolates developed on CB agar (L. I. Evtushenko, unpublished observation). It has not yet
been verified whether these curtobacteria are
carried into the growing points of host plants
like *Rathayibacter* spp. by seed gall nematodes,
Anguina spp. (Price et al., 1979; Riley and
Ophel, 1992), and develop inside plant galls
primordially, or whether they are secondary
settlers of the galls, descending from the phyllosphere or healthy plant tissue. Data regarding
their plant pathogenic properties are also
missing.

PLANT HOSTS AND OCCURRENCE OF PHYTOPATHO-
GENIC *CURTOBACTERIUM FLACCUMFACIENS* The main
habitat of pathovars of *Curtobacterium flaccumfaciens* is considered to be their respective plant
hosts. *Curtobacterium flaccumfaciens* pv. *flaccumfaciens* causes the wilt of beans, i.e., *Phaseo-
lus* species, *Glycine max* (L.) Merr., *Vigna radiata*
(L.) R. Wilczek, *Vigna angularis* (Willd.) Ohwi
and H. Ohashi. *Curtobacterium flaccumfaciens*
pv. *betae* infects red beet (*Beta vulgaris* L.),
inducing wilt and silvering of leaves. *Curtobacterium flaccumfaciens* pv. *oortii* is a causative agent
of wilt and spots of leaves and bulbs of tulips
(*Tulipa* spp.) and *Curtobacterium flaccumfaciens*
pv. *poinsettiae* causes stem canker and leaf spots
of poinsettia (*Euphorbia pulcherrima* Willd. ex
Klotzsch) (Cummins et al., 1974; Hayward, 1974;
Vidaver, 1982; Collins and Bradbury, 1992;
ICMP, 1997). The plant pathogens are considered poor survivors in soil; survival requires
association with diseased plants or residue materials (Vidaver, 1982). Epiphytic survival on, and
natural infection of, weeds has been reported for
Curtobacterium flaccumfaciens pv. *flaccumfaciens* (Schuster and Coyne, 1974; Vidaver, 1982).
Plant pathogenic curtobacteria are usually transmitted by residue materials or seed (Richardson,
1979; Richardson, 1983) or may be transmitted
by rain, wind, or insects, like other bacteria
occurring in plants (Lindemann and Upper,
1985; Hirano et al., 1996; Lilley et al., 1997).

The strains identified as *Curtobacterium flaccumfaciens* have recently been isolated from
macerated leaf tissues of several citrus rootstocks (Araujo et al., 2001) and also shown to be
a control agent for cucumber (Raupach and
Kloepper, 2000). Organisms which resemble
physiologically and biochemically *Curtobacterium flaccumfaciens* pv. *poinsettiae* and *Curtobacterium flaccumfaciens* pv. *flaccumfaciens* have
been found in fasciated mugwort (*Artemisia vulgaris* L.; Battikhi, 2002). However, no genomic
data are available which would support taxonomic affiliation of these organisms.

Pathogenicity Factors Little information is available on the pathogenicity factors of *Curtobacterium flaccumfaciens*. The evidence of enzymatic
activity as a pathogenicity factor of *Curtobacterium flaccumfaciens* is indirect. An unidentified
macerating agent was reported for *Curtobacterium flaccumfaciens* pv. *betae* (Kern and Naef-
Roth, 1971) and will probably also be found in
other subspecies. Earlier histological work on
vascular pathogens, including *Curt. flaccumfaciens* pv. *flaccumfaciens*, is consistent with the
interpretation of enzymatic activity preceding
bacterial appearence (Nelson and Dickey, 1970;
Vidaver, 1982). Electron microscopy of beans
infected with *Curtobacterium flaccumfaciens* pv.

flaccumfaciens showed damage to the xylem and
decomposition of the middle lamella prior to
wilting (Vidaver, 1982). *Curtobacterium flaccumfaciens* possesses β-glucosidase, esterase, peptidases, lipases, etc. and has the high overall
similarity to *Clavibacter michiganense* in many
API ZYM tests and other experimental enzymatic API galleries (the total of 106 tests) (De
Bruyne et al., 1992). In addition, the strain identified as *Curtobacterium flaccumfaciens* on the
basis of the fatty acid and API 20NE profiles was
able to rapidly degrade cellulosic plant fibers *in
vitro* (Lednicka et al., 2000). All these data suggest a possible relatedness between *Clavibacter
michiganensis* and *Curtobacterium flaccumfaciens* in ecological strategies and plant pathogenic properties. Therefore, some aspects of
wilting caused by *Curtobacterium flaccumfaciens*
might be similar to those of *Clavibacter michiganensis* (see the section The Genus *Clavibacter*
in this Chapter).

Isolation

Saprophytic curtobacteria can be isolated
together with clavibacteria from different plants
and some other environmental sources (Dunleavy, 1989; Dobrovolskaya et al., 1999; Jacobs
and Sundin, 2001; Behrendt et al., 2002; Zinniel
et al., 2002). As larger bacterial populations are
located on the adaxial surface of leaves (Beattie
and Lindow, 1995; Sundin and Jacobs, 1999), this
site of leaves is suitable for isolation of epiphytes
by using a leaf impression technique (Dickinson
et al., 1975). The methods described by Dickinson et al. (1975), Dunleavy (1989), Behrendt et
al. (1997), and Jacobs and Sundin (2001) and
discussed in the section The Genus *Clavibacter*
in this Chapter may be used for recovering the
curtobacteria from the plant phyllosphere. For
isolation of endophytic microorganisms, the
approaches outlined by Hallmann et al. (1997)
and Fisher et al. (1992) or described by Zinniel
et al. (2002) should be followed. Surface disinfestation is an important stage of isolating endophytic bacteria as it significantly reduces external
contamination with epiphytic bacteria (Zinniel
et al., 2002). Another surface disinfestation technique involves dipping plant tissue into ethanol
and flamming the surface (Dong et al., 1994).
CNS medium (Vidaver and Starr, 1981; Vidaver
and Davis, 1988) or a modified CNS medium
without lithium chloride may be used for isolation of the epiphytic and endophytic curtobacteria (Zinniel et al., 2002). They may also be
isolated and grown on tryptic soy agar, R2A
agar, NBY agar, PY agar, CB agar or some other
media based on peptone, yeast extract and glucose (Dickinson et al., 1975; Komagata and
Suzuki, 1986; Dunleavy, 1989; Evtushenko et al.,

1994; Behrendt et al., 1997). The strains usually grow well at 26–28°C, but some of them may require lower temperature for growth, 20–24°C. No selective media or special methods for enrichment have been developed for the curtobacteria nonassociated with plants.

To isolate plant pathogens, the choice of suitable diseased material is of prime importance (see the section The Genus *Clavibacter* in this Chapter). Nutrient agar supplemented with 1% glucose was noted to be a satisfactory medium for isolation and growth of plant pathogenic species *Curtobacterium flaccumfaciens* (Lelliott and Stead, 1987). Kado and Heskett (1970) reported that their selective medium D2 gave good growth of one strain each of *Curtobacterium flaccumfaciens* pv. *flaccumfaciens* and *Curtobacterium flaccumfaciens* pv. *poinsettiae*. Growth of the *Curtobacterium flaccumfaciens* pathovars occurs on CNS, but strains of *Curtobacterium flaccumfaciens* pv. *flaccumfaciens* and *Curt. flaccumfaciens* pv. *poinsettiae* may vary in growth on this medium (Vidaver and Starr, 1981).

Identification

GENERIC LEVEL The newly isolated strains can be easily identified as members of the genus *Curtobacterium* by 16S rDNA sequence analysis and/or using some phenotypical properties listed in Table 1. Observation of nonsporeforming, irregular and unbranched rods, along with detection of ornithine in cell wall and the major unsaturated menaquinone MK-9, is sufficient evidence to affiliate isolates to the genus. However, since new undescribed genera with the same or close phenotype may exist in nature, at least partial 16S rDNA nucleotide sequencing or other molecular methods are advisable to support identification. The fatty acid analysis is often used to successfully identify curtobacteria, but such identification should also be confirmed by other methods as bacteria of some genera in the family Microbacteriaceae have similar fatty acid profiles (Collins et al., 1980b; Suzuki and Komagata, 1983; Henningson and Gudmestad, 1991). ω-Cyclohexyl undecanoic acid (ch-C17:0) is a genus-specific property. This fatty acid, up to 51–64% of total cellular fatty acids, was found previously exclusively in *Curtobacterium pusillum* (Suzuki and Komagata, 1983) and has recently also been detected (up to 35%) in some new curtobacteria other than *Curtobacterium pusillum* (L. I. Evtushenko and M. Takeuchi, unpublished observation).

SPECIES LEVEL Species can be discriminated by growth and physiological features. They differ in their ability to hydrolyze starch, gelatin, and casein, to produce acids from some carbohydrates, and in the assimilation of organic acids (Tables 5a and b). As the published data on many physiological properties of the *Curtobacterium* species vary depending on the method applied (Yamada and Komagata, 1972a; Dye and Kemp, 1977; Davis et. al., 1984b; Behrendt et al., 2002), comparative study of new curtobacterial isolates and the type cultures of the known species are prerequisite for identification of the new isolates. Fatty acid analysis, especially detecting ω-cyclohexyl undecanoic acid (Suzuki and Komagata, 1983), and FT-IR spectroscopy (Behrendt et al., 2002; Oberreuter et al., 2002) are helpful approaches. Affiliation of nonpathogenic strains from plants requires DNA-DNA relatedness study or analysis of other adequate genomic characteristics (Vandamme et al., 1996; Stackebrandt et al., 2000, 2002; Rossello-Mora and Amann, 2001).

The phytopathogenic strains of *Curtobacterium flaccumfaciens* and its pathovars are relatively easy to identify when isolated from young lesions of known hosts with characteristic symptoms (Vidaver and Davis, 1988) and compared with known pathogens. Colony morphology, pigmentation, Gram-positive reaction with rapid KOH test (Ryu, 1938; Suslow et al., 1982), palisade cell arrangements in young culture, bending type of cell division without a strong tendency for rod-to-coccoid developmental cycle, some physiological properties, and cell wall sugar composition (Table 6) are helpful traits for identification. Definite identification may need confirmation by pathogenicity and/or serological tests (Vidaver and Davis, 1988), as well as some molecular methods with a taxonomic resolution at the strain-species level. *Curtobacterium flaccumfaciens* pv. *flaccumfaciens* can be identified by using PCR methods (Guimaraes et al., 2001).

Applications

The physiological and enzymatic activities of curtobacteria (Charney, 1966; De Bruyne et al., 1992; Zaczynska et al., 1992; Bark et al., 1993; Lednicka et al., 2000; Froquet et al., 2001) provide evidence that these organisms may be a potential group for screening applicable metabolites and enzymes. *Curtobacterium flaccumfaciens* pv. *flaccumfaciens* can hydrolyze steroids and might be useful in steroid transformation (Charney, 1966). *Curtobacterium flaccumfaciens* pv. *betae* produces glycolipids, which were found to potentiate interferon synthesis by peritoneal cells of BALB/c mice (Zaczynska et al., 1992). *Curtobacterium*-like strains (presumptively identified on the basis of fatty acid methyl esters [FAME] and morphology) isolated from a labo-

Table 5a. Characteristics that differentiate species of the genus *Curtobacterium*.[a]

Characteristics	C. albidum	C. citreum	C. flaccum-faciens[b]	C. herbarum	C. luteum	C. pusillum
Motility	–	+	+	+	+	+
Colony color	I	Y	Y/O	O	Y	LY
DNAse	–	–	w	–	–	–
Hydrolysis of						
Casein	+	–	d	+d[c]	–	–
Esculin	+	+	+	+	–	+
Gelatin	+	–	d	+d[c]	–	+
Starch	–	–	d	–	–	–
Tween 80	+	–	d	+	+	–
Acid from						
Adonitol	–	–	+	–	+	+
Raffinose	+	–	+	+	–	+
L-Rhamnose	+	+	d	–d[c]	+	+
D-Sorbitol	–	–	+	+	–	–
Cyclohexylundecanoic fatty acid[d]	–	–	–	–	–	+
G+C (mol%)[e]	70	70.5	68.3–68.5	71	69.8	69.3

Symbols and abbreviations: +, positive; –, negative; w, weakly positive; d, differs among strains; nd, not determined; colony color on nutrient agar II (SIFIN): I, ivory; Y, yellow; O, orange; and LY, light yellow.
[a]Table is based on the data reported by Behrendt et al. (2002). The type strain of *Curtobacterium plantarum* (Dunleavy, 1989) does not belong to the genus *Curtobacterium* (DSMZ, Catalogue of Strains), and this species is not included in the Table.
[b]Characteristics that differentiate pathovars of the species *C. flaccumfaciens* are given in Table 6.
[c]Reaction of the type strain is positive (+d) or negative (–d).
[d]Data from Suzuki and Komagata (1983).
[e]Data from Yamada and Komagata (1970), exept for the data for *C. herbarum* taken from Behrendt et al. (2002); somewhat higher values of G+C content were reported by Döpfer et al. (1982).

Table 5b. Additional characteristics that differentiate *Curtobacterium albidum*, *C. citreum*, *C. flaccumfaciens*, *C. luteum* and *C. pusillum*.

Characteristics	C. albidum[a]	C. citreum	C. flaccum- faciens[b]	C. luteum[a]	C. pusillum
Colony color	I	DuY	Y/O/P	DaY	PY/GW
Assimilation of					
Lactic acid	+	+	+	–	+
Malic acid	–	+	+	+	–
Fumaric acid	+	+	+	+	d
α-Ketoglutaric acid	–	+	–	–	–
Citric acid	+	d	+	–	–
Glyoxylic acid	+	+	–	–	–
Gluconic acid	+	+	+	+	–

Abbreviations: I, ivory; DuY, dull yellow; DaY, dark yellow; Y, yellow; O, orange; P, pink; PY, pale yellow; GW, grayish white (colony color on nutrient agar); for other definitions see footnote in Table 5a.
[a]Data for type strain only.
[b]Data for *C. flaccumfaciens* pv. *flaccumfaciens* and *C. flaccumfaciens* pv. *poinsettiae* only; characteristics that differentiate all pathovars of the species *C. flaccumfaciens* are given in Table 6.
Data from Yamada and Komagata (1972a; 1972b) and Komagata and Suzuki (1986).

ratory pilot plant were found to accumulate electron-dense granules, which could be identified as polyphosphate granules. The bacteria show adenylate kinase and polyphosphate glucokinase activities (Bark et al., 1993). These features may be applied for a complex process of biological phosphate removal from wastewater. A possible approach for the practical use of curtobacteria is related to their endophytic mode of life. Colonization of several crops with endophytic bacteria (Elbeltagy et al., 2000; Zinniel et al., 2002) sug-

gests that they can be used in future applications, such as delivery of degrading enzymes for controlling of certain plant diseases or other useful products promoting plant growth (Hallmann et al., 1997). This approach would allow growth of plants with improved capabilities without integration of foreign DNA into the plant genome, alternatively to genetic modification or breeding of plants (Zupan and Zambryski, 1995). Endophytic bacteria should be also viewed as potential vectors of foreign genes for construction of

Table 6. Characteristics that differentiate pathovars[a] of the species *Curtobacterium flaccumfaciens*.

Characteristic	*Curt. flac.* pv. *betae*	*Curt. flac.* pv. *laccumfaciens*	*Curt. flac.* pv. *ootrii*	*Curt. Flac.* pv. *poinsettiae*
Colony color, NBY agar	Y	Y/O/P	Y	O
Cell wall sugar				
Galactose	−	+	−	+
Glucose	−	−	+	w
Fucose	+	−	+	+
Hydrolysis of				
Gelatin	−	+[b]	+[b]	−/d[c]
Casein	−	+	+	+/d[c]
DNAse	−	+	−	−
β-Galactosidase (pH 7)	+	d	d	−
Cystin arylamilase (pH 8.5)	d	d	−	−
Acid from L-rhamnose	+	+	+	−[b]
Acid from D-sorbitol	+	−[d]	−[d]	+
Assimilation of				
Acetate	+	+	−	+
Fumarate	−	+	−[e]	+[f]
Propionate	−	−	−	+[e]
KCN-tolerance, 0.0075% (w/v)	+[b]	+	+	−
Production of H₂S	+	+	+	
Max. temperature for growth (°C)	37	37	37	34–35
G+C (mol%)[g]	73.7	72.2	72.2	72.5
Plant host[h]	Beet	Bean	Tulip	Poinsettia
Predominant symptoms[h]	Silvering of leaves; wilt	Wilt	Leaf, bulb spot; wilt	Leaf spot; wilt

Abbreviations: w, weak or questionable; and for other definitions, see footnote in Table 5a.
[a]According to Carlson and Vidaver (1982) and Vidaver and Davis (1988), the pathovars of *Curtobacterium flaccumfaciens* should be classified as separate subspecies (see the text).
[b]The opposite result was reported by Day and Kemp (1977) using different methods.
[c]Different results between strains (Behrendt et al. 2002).
[d,e,f]The opposite results were obtained by Behrendt et al. (2002), Davis et al. (1984b), and Carlson and Vidaver (1982), respectively, using different methods.
[g]Data from Döpfer et al. (1982).
[h]Data from Day and Kemp (1977) and Vidaver (1982).
Data from Day and Kemp (1977), Carlson and Vidaver (1982), Davis et al. (1984), Vidaver and Davis (1988), De Bruyne et al. (1992), and Behrendt et al. (2002).

transgenic plants, although considerably more research is necessary before the practicality of this approach can be assessed.

The Genus *Frigoribacterium*

The genus *Frigoribacterium* with the type species *Frigoribacterium faeni* was proposed by Kämpfer et al. (2000) for psychrophilic strains isolated from airborne hay dust and air inside a museum. Mesophilic organisms of this genus with optimum growth at 26–28°C have recently been isolated from the phyllosphere of fescue (L. I. Evtushenko et al., unpublished observation).

Phylogeny

The strains of the genus *Frigoribacterium* showed highest 16S rRNA gene sequence similarity (up to 97%) with members of the genera *Rathayibacter* and *Clavibacter* (Kämpfer et al., 2000).

Taxonomy

The bacteria form round, convex, pale yellow or yellow colonies on the nutrient agar and can be psychrophilic and mesophilic. Cells are Gram-positive, aerobic, motile, irregular, curved or asymmetric nonsporulating rods. Mycelium is not produced. Catalase test is positive but oxidase reaction varies depending on the strain. The cell wall peptidoglycan contains D-lysine as a diamino acid; the polymer is of type B2β according to Schleifer and Kandler (1972). The glycan moiety of the peptidoglycan contains acetyl residues. The major isoprenoid quinone is menaquinone MK-9. Mycolic acids are not present.

The main fatty acids of *Frigoribacterium faeni* are 12-methyltetradecanoic acid (*anteiso*-15:0), 12-methyl-tetradecenoic acid (*anteiso*-15:1), 14-methyl-pentadecanoic acid (*iso*-16:0) and 14-methyl-hexadecanoic acid (*iso*-17:0), as well as 1,1-dimethoxy-anteisopentadecane (*anteiso*-15:0-DMA). Diphosphatidylglycerol, phosphatidylglycerol and an unknown glycolipid are detected in the polar lipid extracts. The DNA G+C content is approximately 71 mol% (Kämpfer et al., 2000).

Although the DNA-DNA relatedness between strains of the species *Frigoribacterium faeni* was 37–52%, the authors considered them to be different genomovars of a single species as physiological and chemotaxonomic similarity between strains was high. The phenotypic properties related to the species *Frigoribacterium faeni* are outlined in the subsection Identification in this Chapter.

Habitat

Little is known about natural habitats of microorganisms comprising the genus *Frigoribacterium*. The strains were isolated mainly from plants and plant debris. Several psychrophilic strains of *Frigoribacterium faeni* were associated with hay dust in a cattle barn, settled dust in an animal shed, and air inside a museum (Kämpfer et al., 2000). The mesophilic strain of this genus was isolated from the phyllosphere of *Festuca rubra* L. (L. I. Evtushenko et al., unpublished observation).

Isolation and Cultivation

Organisms of the genus *Frigoribacterium* can be isolated from grassess, hay or some related sources using a dilution plating method. Tryptic soy agar, CasMM agar (Altenburger et al., 1996) supplemented with cycloheximide, a tenfold diluted CB agar or R2A agar and incubation at room and lower (4–16°C) temperatures (Kämpfer et al., 2000; L. I. Evtushenko et al., unpublished observation) are suitable isolation conditions. Strains may be recovered from hay dusts as described previously (Andersson et al., 1995; Andersson et al., 1999; Altenburger et al., 1996). Organisms of the genus grow well on a number of media, including tryptic soy agar, PYES agar, CB agar, and some others at appropriate temperature conditions, 4–10°C (*Frigoribacterium faeni*) or 24–26°C (mesophilic strains).

Identification

Isolates affiliated to the family Microbacteriaceae can be readily identified as members of *Frigoribacterium* owing to the presence of lysine in the peptidoglycan and MK-9 as the major menaquinone (Table 1). In the family Microbacteriaceae, only *Okibacterium*, *Mycetocola* and eight species of *Microbacterium* possess cell wall peptidoglycans based upon lysine (Table 7). The major menaquinone MK-9 combined with the acetyl type of muramic acid and/or the G+C content of DNA (71 mol%) clearly discriminate *Frigoribacterium* from *Okibacterium*, which has the same composition and molar ratio of the peptidoglycan amino acids, and also from *Microbacterium* spp. and *Mycetocola* (Table 7). It should be noted that lysine in the cell wall and the major menaquinone MK-9 are also typical of the genus *Jonesia*, family Jonesiaceae (Rainey et al., 1995b; Stackebrandt et al., 1997). However, the latter, besides distant phylogenetic position, clearly differs from *Frigoribacterium* by the presence of serine in the cell wall, G+C content of DNA (56–58 mol%), and some other characteristics (Rainey et al., 1995b).

Helpful identifying traits of the species *Frigoribacterium faeni* are good growth at 0°C and 20°C and optimum temperature of 4–10°C, small cells (0.2–0.3 × 1.0–1.5 μm), negative oxidase reaction, assimilation of a wide variety of sugars, including *N*-acetyl-D-glucosamine, L-arabinose, *p*-arbutin, D-cellobiose, D-fructose, D-galactose, gluconate, D-glucose, D-mannose, D-maltose, L-rhamnose, D-ribose, sucrose, D-trehalose and D-xylose as well as utilization of only a few organic acids as a sole source of carbon (citrate and fumarate). Another diagnostic property of this psychrophilic species is the presence of significant amounts of *anteiso*-15:0-DMA, up to 10–30% of whole-cell fatty substances. The retention time of this compound is almost equal to that of 2-OH-14:0, according to standardized FAME analysis (Kämpfer et al., 2000). This behavior differs from the mode of cold adaptation of many other Gram-positive bacteria, which adapt by changing their ratio of *anteiso*-fatty acids to *iso*-fatty acids (Kaneda, 1991) or accumulate unsaturated fatty acid (*anteiso*-15:1), like *Cryobacterium* and some others (Suzuki et al., 1997). Besides *Frigoribacterium faeni*, DMAs were determined in bacteria of the genera *Subtercola* and *Agreia* (Männistö et al., 2000; Behrendt et al., 2002).

The Genus *Leifsonia*

The genus *Leifsonia* (Evtushenko et al., 2000) was proposed to accommodate bacteria isolated from the root galls on *Poa annua*, the water isolate "*Corynebacterium aquaticum*" (Leifson, 1962) and the plant pathogenic species *Clavibacter xyli* (Davis et al., 1984b) with two subspecies.

The first bacterium of this genus was isolated from distilled water and assigned to a new species "*Corynebacterium aquaticum*" on the basis of morphological and physiological characteristics (Leifson, 1962). Subsequent extensive taxonomic studies revealed DAB in its cell wall peptidoglycan (type B2γ, Schleifer and Kandler,

Table 7. Salient characteristics differentiating the genera *Frigoribacterium*, *Okibacterium*, *Mycetocola* and the *Microbacterium* species containing lysine in the cell wall.

Characteristics	Frigoribacterium	Okibacterium	Mycetocola	M. aurum	M. dextranolyticum M. hominis M. lacticum M. laevaniformans	M. arborescens M. imperiale	M. gubbeenense
Motility	+	−	−	−	−		ND
Murein type[a]	[L-Hsr]-D-Glu-D-Lys	[Hsr]-Glu-Lys	ND	[L-Hsr]-D-Glu (Hyg)-Gly-L-Lys	[L-Lys]-D-Glu (Hyg)-Gly-L-Lys	[L-Hsr]-D-Glu (Hyg)-Gly$_Z$-L-Lys	ND
Amino acids (molar ratio)							
Lysine	1	1	+	1	2	1	+
Glutamic acid	1	1	+	1[b]	1[b]	1[b]	ND
Glycine	1	1	+	2	2	3	ND
Homoserine	+	+	−	+	−	+	ND
Unidentified	−	−	+	−	−		ND
Acyl type	Acetyl	Glycolyl	Acetyl	Glycolyl	Glycolyl	Glycolyl	Glycolyl
Cell wall galactose	ND	−	ND	+	+	+	+
Major MK	9	10,11	10	11,12	11,12	11,12	11,12
G+C (mol%)	71.7	66.6–67.2	63.9–64.7	69.2	68.3–71.2	71.0–71.2	69–72

Symbols and abbreviations: See footnote in Table 1.
[a]Hsr, homoserine; Glu, glutamic acid; Hyg, 3-*threo*-hydroxyglutamic acid; Gly, glycine; Lys, lysine.
Alanine is present in all strains.
[b]Glutamic acid may be substitututed by 3-*threo*-hydroxyglutamic acid.
Data from Collins et al. (1983d), Takeuchi and Hatano (Takeuchi and Hatano, 1998a; Takeuchi and Hatano, 1998b; Kämpfer et al. (2000), Behrendt et al. (2001), Brennan et al. (2001), Tsukamoto et al. (2001), and Evtushenko et al. (2002).

1972), the major menaquinone with 11 and 10 isoprene units (Yamada et al., 1976; Collins et al., 1979; Collins and Jones, 1980a), and some other characteristics (Yamada and Komagata, 1972b; Komura et al., 1975; Collins et al., 1979), which indicated that "*Corynebacterium aquaticum*" was phenotypically close to the group of plant pathogenic bacteria currently belonging to *Clavibacter*, *Rathayibacter* and some other genera of the family Microbacteriaceae. Döpfer et al. (1982), studying DNA-rRNA cistron similarities of coryneform bacteria and related organisms, revealed that "*Corynebacterium aquaticum*" clustered in a homogeneous group together with all other coryneforms containing a peptidoglycan of group B. Subsequently, Park et al. (1987) showed that 5S rRNA sequence of "*Corynebacterium aquaticum*" was the closest to that of *Microbacterium testaceum*, which was the only other member of the current family Microbacteriaceae used in this study.

Phylogenetic analysis based on 16S rDNA analyses (Rainey et al., 1994; Takeuchi and Yokota, 1994) confirmed that "*Corynebacterium aquaticum*" belongs to the phylogenetic cluster composed of actinomycetes of the family Microbacteriaceae, forming a separate subline of descent within the family. The strain was suggested to be the nucleus of a novel genus (Rainey et al., 1994). Plant pathogen *Clavibacter xyli* subsp. *cynodontis* (Davis et al., 1984b) was shown to be a phylogenetic neighbor of "*Corynebacterium aquaticum*" (Takeuchi and Yokota, 1994; Sasaki et al., 1998). On the basis of the phylogenetic data and similarity in salient chemotaxonomic characteristics of "*Corynebacterium aquaticum*" and *Clavibacter xyli* subsp. *cynodontis*, Sasaki et al. (1998) postulated that a new genus should be established to accommodate these organisms. Lee et al. (1997b) showed a high level of phylogenetic similarity between *Clavibacter xyli* subsp. *cynodontis* and *Clavibacter xyli* subsp. *xyli*, which suggested the latter subspecies was another member of this genus.

Later, some coryneform bacteria isolated from the *Poa annua* root gall induced by plant parasitic nematodes of the subfamily Anguininae were shown to be phenotypically and phylogenetically close to "*Corynebacterium aquaticum*" (Dorofeeva et al., 1999). All the above-mentioned bacteria were described as members of the new genus *Leifsonia* (Evtushenko et al., 2000). About that time, Suzuki et al. proposed two other species of this genus, *Leifsonia naganoensis* and *Leifsonia shinshuensis* (Suzuki et al., 1999; Suzuki et al., 2000). The authors also proposed to reclassify *Leifsonia xyli* subsp. *cynodontis* as *Leifsonia cynodontis*, since the type strain of *Leifsonia xyli* subsp. *xyli* is no longer available and this subspecies automatically lost its taxo-

nomic status (Suzuki et al., 1999). However, bacteria of this group exist in nature, they are of importance as a cause of sugarcane stunting disease, and the taxon may be restored in the future, with proposal of the neotype strain. The difference between *Leifsonia xyli* subsp. *cynodontis* 16S rDNA sequence and that of *Leifsonia xyli* subsp. *xyli*, which is about the same level as that found between some other species of the genus (Lee et al., 1997b; Fig. 1), as well as their significant difference in 16S–23S internal transcribed spacer (ITS) sequences (Fegan et al., 1998; Pan et al., 1998), suggests that strains of *Leifsonia xyli* subsp. *xyli* may represent a separate species.

Phylogeny

Members of the genus *Leifsonia* form a tight (100% bootstrap confidence) phylogenetic cluster within the radiation of bacteria comprising the family Microbacteriaceae (Fig. 1).

Taxonomy

The genus *Leifsonia* currently harbors five species and two subspecies:

Leifsonia aquatica (ex Leifson, 1962), *Leifsonia xyli* subsp. *xyli* (Davis et al., 1984b), *Leifsonia xyli* subsp. *cynodontis* (Davis et al., 1984b), *Leifsonia poae* (Evtushenko et al., 2000), *Leifsonia naganoensis* and *Leifsonia shinshuensis* (Suzuki et al., 2000).

Organisms of the genus *Leifsonia* usually form yellow, yellowish-white, white or brownish-white colonies, which are circular, somewhat convex, glistening, opaque, and butyrous. Cells are nonsporeforming, irregular rods, usually fragmenting into shorter rods or coccoid elements. Filaments and primary branching occur in young cultures of some species. Organisms are usually motile, mesophilic, obligately aerobic, and catalase positive. Oxidase test is variable among species. Cell wall peptidoglycan is [L-DAB]D-Glu-D-DAB, B2γ type (Schleifer and Kandler, 1972); both isomers, L-DAB and D-DAB, are usually present in almost equal proportion. The muramic acid is acetylated. The major menaquinones are MK-11 and MK-10 or MK-11 and MK-12. Cell wall sugars consist of a predominant amount of rhamnose and minor amounts of glucose, galactose, and mannose; some species contain fucose. Cell wall teichoic acids and mycolic acids are not present. The principal phospholipids are phosphatidylglycerol and diphosphatidylglycerol. Among the fatty acids, *anteiso*-15:0, *anteiso*-17:0 and *iso*-16:0 predominate. Concentrations of polyamines are low; putrescine is the predominant compound. The DNA G+C content is 66.0–71 mol%. This description is based on the data of Leifson (1962), Fiedler and Kandler (1973a), Yamada et al. (1976), Collins et al. (1979), Collins and Jones (1980a), Döpfer et al. (1982), Davis et al. (1984b), Altenburger et al. (1997), Sasaki et al. (1998), Suzuki et al. (1999), and Evtushenko et al. (2000).

The species differentiation is based on growth, morphological and physiological characteristics, cell wall sugar composition, and menaquinone pattern (see subsection Identification in this Chapter).

Habitat and Phytopathogenicity

Bacteria of the genus *Leifsonia* occur in different environments. The natural habitats of *Leifsonia xyli* and *Leifsonia poae* are plants, where they occur as plant pathogens or saprophytes. *Leifsonia aquatica* has been isolated from distilled water storage tanks and from natural fresh water (Leifson, 1962). Later, representatives of the genus were revealed in bacterioplankton in the Western Mediterranean Sea, at a depth of 80 m (Fritz, 2000). The species *Leifsonia naganoensis*, *Leifsonia shinshuensis* (Suzuki et al., 1999), and some other unidentified organisms of this genus were isolated from soil (Furlong et al., 2002; Janssen et al., 2002; Yanagi et al., 2000).

RATOON STUNTING DISEASE CAUSED BY *LEIFSONIA XYLI* *Leifsonia xyli* subsp. *xyli* invades plant vascular systems, causing ratoon stunting disease in sugarcane (*Saccharum officinarum* L.) in most sugarcane-growing areas of the world (Steindl, 1961; Davis et al., 1984b; Gillaspie and Teackle, 1989). *Leifsonia xyli* subsp. *cynodontis* occurs in Bermuda grass, *Cynodon dactylon* (L.) Pers, and was experimentally shown to cause a disease that does not manifest naturally in this plant (Davis et al., 1980). *Leifsonia xyli* subsp. *xyli* and *Leifsonia xyli* subsp. *cynodontis* are pathogenically distinct, and each induces symptoms in its natural host but not in natural hosts of the other bacterium or in most other plants; each pathogen has been found in nature only in the host in which it was first discovered (Davis et al., 1980; Davis et al., 1984b; Gillaspie and Teackle, 1989). However, both bacteria may inhabit and multiply in the xylem vessels of both hosts and other plants without producing disease symptoms (Steindl, 1961; Davis et al., 1984b; Kostka et al., 1988; Fahey et al., 1991). *Leifsonia xyli* subsp. *xyli* is experimentally transmissible to several gramineous hosts, most of which remain symptomless (Steindl, 1961). The other subspecies, *Leifsonia xyli* subsp. *cynodontis*, or its recombinant with the crystal protein gene of *Bacillus thuringiensis* subsp. *kurstaki*, artificially colonizes more then 80 plant species of 26 families (Kostka et al., 1988; Fahey et al., 1991).

Host plants infected by *Leifsonia xyli* initiate slower growth, look stunted, and may have fewer and thinner stalks, but they show no other external diagnostic symptoms. If plenty of water is available, infected plants may show no stunting and no loss in yield. But infected plants often show internal symptoms. Very young shoots of sugarcane have pinkish discoloration just below the meristematic area of the shoot, above its attachment to the seed piece. In mature canes, symptoms appear as discoloration of individual vascular bundles at the nodes. The pathogen survives in the infected plants and propagative materials such as seeds. The bacteria are also spread by cutting knives and by cultivation and harvesting equipment. They are spread to different countries through infected sugarcane germplasm (Davis et al., 1980).

OCCURRENCE OF OTHER PLANT-ASSOCIATED BACTERIA OF THE GENUS *LEIFSONIA* Strains of *Leifsonia poae* were isolated along with other bacteria from surface-sterilized *Poa annua* root galls induced by the grass root-gall nematode *Subanguina radicicola* (Evtushenko et al., 2000). There is no evidence of whether this species is a plant pathogen or a root endophyte. Also, no direct relationship between *Leifsonia poae* and the nematode *Subanguina radicicola* has been deduced.

Nevertheless, the nematode may play a role in bacterial invasion. Plant parasitic nematodes, penetrating roots in the growing zone, a region also preferably colonized by saprophytic bacteria (Grundler and Wyss, 1995), may create wounds which serve as avenues of entry for the endophytic bacteria. In addition, leakage of plant nutrients supports a higher external population of bacteria, thus facilitating a bigger potential source of endophytic colonizers (Hallmann et al., 1997). *Subanguina radicicola* might also be a nematode vector of *Leifsonia poae* similar to the case described for the *Rathayibacter-Anguina* associations (Price et al., 1979; Riley and Ophel, 1992). Other strains of the genus *Leifsonia* were found in the above-ground parts of grasses without visible nematode infestation (L. I. Evtushenko et al., unpublished observation).

CLINICAL BACTERIA THAT MIGHT BELONG TO THE GENUS *LEIFSONIA* Some strains phenotypically resembling *Leifsonia aquatica* ("*Corynebacterium aquaticum*") were reported to be an agent of meningitis (Beckwith et al., 1986), bacteremia (Kaplan and Israel, 1988; Fischer et al., 1994; Moore and Norton, 1995; Vasseur et al., 1998), urinary tract infections (Tendler and Bottone, 1989), wound infection (Larsson et al., 1996), and continuous ambulatory peritoneal peritonitis (Morris et al., 1986; Casella et al., 1988; Kwon and Lee, 1997). As no data on 16S rDNA sequences or chemotaxonomic characteristics of most such bacteria are available, their taxonomic position remains unclear. Moreover, some clinical bacteria labelled "*Corynebacterium aquaticum*" were shown to be actually *Microbacterium* spp. (Funke et al., 1994; Grove et al., 1999).

Isolation

ISOLATION OF FAST-GROWING SAPROPHYTIC BACTERIA OF THE GENUS *LEIFSONIA* The fast-growing saprophytic organisms of *Leifsonia* can be successfully isolated from plants, soil and other environments on CB agar, R agar (Yamada and Komagata, 1972a), and some related media by the dilution plating technique. For isolation of *Leifsonia aquatica*, the medium originally developed for the culture of *Caulobacter* (peptone, 3 g; yeast extract, 1 g; K₂HPO₄, 1 g; agar, 15 g; water, 1 liter; pH 7.1) may be used (Leifson, 1962). Taking into account that *Leifsonia aquatica* is resistant to many antibiotics (Evtushenko et al., 2000), some of them, e.g., chloramphenicol ($10 \mu g \cdot ml^{-1}$), may be added to the agar medium to reduce development of many other bacteria. *Leifsonia poae* can be isolated from root galls induced by the root-gall nematode *Subanguina radicicola* by the following procedure.

Galls inhabited with bacteria are washed, sterilized by 75% (v/v) ethanol for 1 min, dried and pre-incubated in sterile tap water at 24°C for 2 h. Then, galls are cut into pieces, added to 2 ml of NaCl solution at a concentration of 0.85% (w/v), and ground using a pestle. KOH solution may be added to this ground gall suspension to inhibit most Gram-negative bacteria. One drop of this suspension is plated onto CB agar and incubated at room temperature (18–24°C).

Different media based on glucose, peptone and yeast extract are appropriate for cultivation of the rapidly growing organisms of the genus *Leifsonia* (Suzuki et al., 1999; Evtushenko et al., 2000).

ISOLATION AND CULTIVATION OF *LEIFSONIA XYLI* Certain general rules of isolation of plant pathogens should be followed in the isolation of slowly growing plant pathogens of the genus *Leifsonia* from diseased plants (see the section The Genus *Clavibacter* in this Chapter). The choice of suitable diseased material is of primary importance. The xylem-inhabiting bacteria can be isolated from affected vascular bundles at the nodes, which are usually inhabited by bacteria in the infected plant. Under the microscope, affected bundles appear plugged with bacteria contained in a colored gummy substance. The complex SC

medium (Davis et al., 1980) may be suitable for isolation of both *Leifsonia xyli* subsp. *xyli* and *Leifsonia xyli* subsp. *cynodontis* (Davis et al., 1984b). If this or other nonselective media are used, successful isolation of these two pathogens requires surface sterilization of plant material and other precautions to avoid or reduce the growth of saprophytes (Vidaver and Davis, 1988; Collins and Bradbury, 1992). For isolation of *Leifsonia xyli* subsp. *cynodontis*, a selective complex SCMS medium was developed (Davis and Augustin, 1984a). The SCMS medium is based on SC medium, which is supplemented with the folowing antibiotics: cycloheximide, 50 mg; 3-hydroxy-5-methylisoxazole (hymexazole), 35 mg; colistin methanesulfonate, 10 mg; and polymyxin B sulfate, 50 mg. The antibiotics are dissolved in 10 ml of water, filter sterilized, and added to 1 liter of molten SC agar at 50°C. Visible colonies appear within 7–14 d on this medium, which are round, convex, entire, and nonpigmented or pale yellow. This medium may also be suitable for isolation of its close relative *Leifsonia xyli* subsp. *xyli* (Collins and Bradbury, 1992).

SC medium is used for growing of *Leifsonia xyli* subsp. *cynodontis* and *Leifsonia xyli* subsp. *xyli* (Davis et al., 1984b). Extracts of Bermuda grass and maize, as well as xylem fluid collected from maize plants, enhance the growth of *Leifsonia xyli* subsp. *cynodontis* (Haapalainen et al., 2000). The growth was enhanced even with 1% (v/v) maize xylem sap in a rich culture medium, giving both an increased growth rate and a higher stationary phase cell density (Haapalainen et al., 2000).

SC medium (Davis et al., 1980)

Cornmeal agar	17 g
Papaic digest of soy meal (or phytone peptone or soytone)	8 g
K₂HPO₄	1 g
KH₂PO₄	1 g
MgSO₂ · 7H₂O	0.2 g
Water	1 liter
Bovine hemin chloride	15 mg
Bovine serum albumin (20% solution), fraction V	2 g
Glucose (50% solution)	0.5 g
Cystine (10% solution)	1 g

The 15 mg of bovine hemin chloride is 15 ml of 0.1% bovine hemin chloride in 0.05N NaOH. The bovine serum albumin (20%), glucose (50%) and cystcine (10%) solutions should be filter sterilized and then 10, 1 and 10 ml, respectively, added to the sterilized molten medium at 50°C. The pH is then adjusted to 6.6.

Identification and Detection

GENERIC LEVEL Clearly motile irregular rods, presence of DAB in the cell wall, and major

Table 8. Characterictics that differentiate species of the genus *Leifsonia*.

Characteristics	*L. aquatica*	*L. nagano-ensis*	*L. shinshu-ensis*	*L. poae*	*L. xyli* subsp. *xyli*	*L. xyli* subsp. *cynodontis*
Colony color	Y	W, PB	W, PB	PY, Y	W, PY	PY, Y
Growth on R agar	+	+	+	+	–	–
Visible colonies, day	2	2	2	2	7–10	5–7
Cell width (μm)	0.4–0.7	0.3–0.5	0.3–0.4	0.4–0.8	0.2	0.2–0.3
Cell length (μm)	1.2–2.5	1.8–3.0	2.5–3.0	1.5–8.0	5.0	3.0–6.0
Fucose in the cell wall	+	ND	ND	–	+	+
Utilization of						
Citrate	+	–	–	–	–	+
Gluconate	+	–	–	–	–	–
Propionate	+	–	+	+	–	–
Acid from						
D-Galactose	+	+	–	+	–	–
Salicin	+	+	–	+	–	–
D-Sucrose	w	+	–	+	–	–
Glucose	+	+	–	+	+	+
Mannose	+	+	–	+	w	+
Melibiose	–	+	–	–	–	–
Inulin	–	–	+	–	–	–
Hydrolysis of starch	+	+	+	–	–	+
Growth at NaCl, 5%	+	–	–	–	–	–
Major menaquinone, MK	11, 10	11, 10	11, 12	11, 10	ND	11, 12
G+C (mol%)	70	71	71	69	66	69
Source of isolation	Distilled water	Soil	Soil	Root gall on *Poa annua*	*Saccharum*, interspecific hybrid	*Cynodon dactylon*

Symbols and abbreviations: +, positive; –, negative; w, weakly positive; ND, not determined; W, white; Y, yellow; PB, pale brown; PY, pale yellow; for other definitions see footnote in Table 1.
Data from Davis et al. (1984b), Suzuki et al. (1999), and Evtushenko et al. (2000).

menaquinones MK-11, 10 or MK-11, 12 are the most important phenotypical characteristics allowing the preliminary affiliation of a new mesophilic isolate to the genus *Leifsonia*. However, as new undescribed genera with the same or close characteristics may exist in nature, and weakly motile isolates of *Leifsonia* spp. might be confused with some *Agromyces* species or some others also having DAB in the cell wall (Table 1), at least partial 16S rDNA nucleotide sequencing or restriction analysis of amplified 16S rDNA (Lee et al., 1997b) is necessary to confirm the generic affiliation of the isolate.

SPECIES LEVEL The phenotypic properties given in Table 8 allow affiliation of a fast-growing strain of the genus *Leifsonia* belonging to the species *Leifsonia aquatica*, *Leifsonia naganoensis*, *Leifsonia shinshuensis* and *Leifsonia poae*. Colony color, cell size, menaquinone composition, G+C content of DNA, the presence of fucose in the cell wall, growth in 5% NaCl, and some other physiological features are valuable identifying traits.

The slow-growing plant pathogenic bacteria are relatively easily identified as *Leifsonia xyli* subsp. *xyli* or *Leifsonia xyli* subsp. *cynodontis* at

isolation from sugarcane or Bermuda grass with the typical symptoms of stunting disease. It is usually sufficient to reveal slow-growing, thin, elongated Gram-positive rods showing a tendency to coryneform morphology (Vidaver and Davis, 1988). 16S-23S ITS spacer sequence analysis (Fegan et al., 1998; Pan et al., 1998) is extremely useful for unambiguous identification of *Leifsonia xyli* subsp. *xyli* and *Leifsonia xyli* subsp. *cynodontis*. Some physiological properties (Table 8), cellular protein profiles (Davis et al., 1984b) and 16S rDNA-RFLP patterns (Lee et al., 1997b) are additional helpful identifying traits. Confirmation of identity (and virulence) of pathogens can also be achieved by inoculation of bacteria into a healthy host plant. However, direct testing of virulence is time-consuming and it is often difficult to determine because plant infections are symptomless.

DETECTION OF *LEIFSONIA XYLI* SUBSP. *XYLI* For practical purposes, *Leifsonia xyli* subsp. *xyli* can be detected by using microscopic (Steindl, 1976) and DNA-based approaches (Fegan et al., 1998; Pan et al., 1998) or immunological techniques (Gillaspie and Teackle, 1989; Shine and Comstock, 1993).

DNA-based methods are widely used for diagnosis. This approach is very helpful when symptomless infection or stunting is caused by some other biotic and abiotic factors. The PCR protocols for sensitive and specific detection of *Leifsonia xyli* based on 16S–23S ITS spacer sequences have been reported (Fegan et al., 1998; Pan et al., 1998). The method was successfully applied to the detection of *Leifsonia xyli* subsp. *xyli* in fibrovascular fluid extracted from sugarcane and was sensitive to approximately 22 cells per PCR assay (Fegan et al., 1998). A multiplex PCR test was developed to differentiate this bacterium from *Leifsonia xyli* subsp. *cynodontis* in a single PCR assay (Fegan et al., 1998). *Leifsonia xyli* subsp. *xyli* can also be detected by DNA hybridization using the fragments of genomic DNA as specific hybridization probes, although weak cross-hybridization may also occur with *Leifsonia xyli* subsp. *cynodontis* (Chung et al., 1994).

Immunological techniques, in particular, tissue immunoblotting, are widely used for diagnosis (Gillaspie and Teackle, 1989; Shine and Comstock, 1993). However, they are usually effective in detecting of *Leifsonia xyli* subsp. *xyli* later in the growing season when the bacterium is present at a high titer; these methods can be unreliable in early to mid-season, when a manangement decision must be made about which stalks are to be used as the seed cane for the next crop cycle (Pan et al., 1998). In addition, crossreactivity of polyclonal antisera often reduces the specificity of identification (Shine and Comstock, 1993).

Applications and Properties

Owing to low virulence of *Leifsonia xyli* subsp. *cynodontis* and its ability to be artificially transferred into numerous host plants, use of this genetically engineered bacterium for plant protection from disease and insect damage seems promising (Metzler et al., 1997). The engineered bacteria producing beneficial proteins in plants can be used instead of transforming the plant genome. *Leifsonia xyli* subsp. *cynodontis* was modified to produce the crystal protein of *Bacillus thuringiensis* subsp. *kurstaki* and used to protect maize from the European corn borer (*Ostrinia nubilalis*; Dimock et al., 1988; Lampel et al., 1994). Although no increase in the yield was registered at field testing, the considerable protection of maize was observed (Tomasino et al., 1995). The genetically transformed *Leifsonia xyli* subsp. *cynodontis* was also claimed to synthesize and deliver a bioactive compound (antigen, enzyme, hormone, etc.) to animals. The microorganism was cultured to produce this compound in plants, which when eaten by the

animal could induce a therapeutic, biochemical, or immunological response (Koprowski et al., 1997). Several procedures and components useful for genetic manipulation of *Leifsonia xyli* subsp. *cynodontis* were developed and elaborated (Metzler et al., 1992; Metzler et al., 1997; Taylor et al., 1993; Uratani et al., 1995; Haapalainen et al., 1996; Haapalainen et al., 1998; Brumbley et al., 2002).

A new type of glycoglycerolipids was obtained from strains S361 and S365, identified as *Leifsonia aquatica* ("*Corynebacterium aquaticum*"; Yanagi et al., 2000). A glycolipid from strain S365 (FERM P-17163) has anti-active oxygen disturbance and cell proliferation activities and was claimed to repair cell membranes damaged by active oxygen (Matsufuji et al., 2000). In a cell extract of *Leifsonia aquatica* ("*Corynebacterium aquaticum*") M-13, a new enzyme, levodione reductase (an alcohol dehydrogenase/ reductase), was found (Wada et al., 1999; Yoshisumi et al., 2001). This enzyme catalyzed regio- and stereoselective reduction of levodione to actinol and was highly activated by monovalent cations.

The Genus *Leucobacter*

The genus *Leucobacter* with the type species *Leucobacter komagatae* (Takeuchi et al., 1996) was described as an aerobic, nonsporulating rod-shaped organism with an unusual cell wall peptidoglycan containing γ-aminobutyric acid (GABA). Although this compound has been found in every class of living organism, including bacteria, it has not been reported to constitute a major component of the peptidoglycan in bacteria except for *Leucobacter*. GABA is the major inhibitory neurotransmitter in the central nervous system of invertebrates and vertebrates and causes depressant actions in the brain (Turner and Whittle, 1983; Eldefrawi and Eldefrawi; 1987). The role of GABA in the microorganisms, which lack nervous systems, is unclear. A GABA uptake system with binding characteristics like those of the GABA brain receptor is present in *Pseudomonas fluorescens* (Guthrie et al., 1989) and accumulation of GABA in growing cultures is shown to occur by means of an active transport system in *Saccharomyces cerevisiae* (McKelvey et al., 1990).

Phylogeny

The phylogenetic dendrogram (Fig. 1) shows that *Leucobacter komagatae* IFO 15245[T] represents a distinct lineage within the radiation of Microbacteriaceae.

Taxonomy

Cells of *Leucobacter komagatae* IFO 15245[T] are Gram-positive, nonsporulating, nonmotile, irregular rods that are 0.2–0.3 µm wide and 1.0–1.5 µm long. The optimal temperature for growth is 25°C; no growth occurs at 37 or 4°C. Growth on peptone-yeast agar gives usually low convex, circular, rough, opaque and whitish-brown colonies. Catalase is produced, but oxidase test is negative. The amino acids of the cell wall peptidoglycan are L-alanine, D-glutamic acid, L-DAB, glycine, and an unusual amino acid, GABA; the molar ratio of these compounds was 1.90: 1.00: 0.85: 0.89: 0.72. The acyl type of the cell wall is acetyl. Galactose and glucose are present in the cell wall. The major isoprenoid quinone is menaquinone MK-11. Mycolic acids are not present. The major cellular fatty acids are *anteiso*-15:0, *iso*-16:0 and *anteiso*-17:0. Diphosphatidylglycerol, phosphatidylglycerol, and one unknown glycolipid are present. The G+C content of the DNA is 66 mol%.

The phenotypic properties characteristic of the species are outlined in the subsection Identification in this Chapter.

Habitat

Little is known about the natural habitats of members of the genus *Leucobacter*. The type strain *Leucobacter komagatae* IFO 15245[T] on which the description of the genus was based has been isolated as a contaminant in an L-drying ampoule labelled as "*Pseudomonas riboflavina*" IFO 13584, which was later re-identified as *Devosia riboflavina* (Nakagawa et al., 1996). The ampoule was made in 1973 and maintained in the Institute for Fermentation, Osaka. On the basis of 16S rRNA gene sequence analysis, other isolates have been found to be in close phylogenetic relationship to the genus *Leucobacter*. Strain W5P-3 (AF323267) was isolated from a microbial community oxidizing inorganic sulfide and mercaptans (Duncan et al., 2001), the glacial ice bacterium SB12K-2-1 (AF479358) was recovered from 12,000 year-old glacial ice from Sajama, Bolivia (Christner, 2002), and two bacteria were isolated from a bioreactor designed for treatment of gray water (O. Malseva, personal communication) and from an environmental sample (AB012594; Uemori, 1999).

Identification

The most striking phenotypic characteristic which differentiates *Leucobacter komagatae* from 11 genera of the family Microbacteriaceae that have DAB-based cell wall pepdidoglycans is the presence of GABA in the cell wall (Table 1). Table 9 shows the phenotypic characteristics differentiating the genus *Leucobacter* from the other seven genera, members of which contain L-DAB as major diamino acids in their cell walls like *Leucobacter*. In addition to GABA in the cell wall, the genus *Leucobacter* can be discriminated from these genera by the pigmentation, morphology, optimal growth temperature, menaquinone type, mol% G+C content of DNA, and cell wall sugars. Rhamnose is the major sugar component of the cell wall of all the strains belonging to these genera except for *Agromyces cerinus* subsp. *nitratus*, whereas rhamnose is not found in the cell wall of *Leucobacter komagatae*, its major cell wall sugars being galactose and glucose. The optimal temperature for growth readily separates members of *Leucobacter* from those of *Cryobacterium* and *Subtercola*.

In addition to the characteristics given above and listed in the subsection Taxonomy in this Chapter, the following properties may be useful in assigning a bacterial isolate to *Leucobacter komagatae*: ability to produce urease and hydrogen sulfide; positive methyl red test; negative oxidase, nitrate reductase and Voges-Proskauer reactions; the ability to utilize acetate, D- and L-lactate, malate, succinate, propionate, oxalate, and hippurate; and ability to hydrolyze Tween 20, 40, 60, and 80, and urea. Gelatin, starch, and esculin are not hydrolyzed. Growth occurs in the presence of 5.0% NaCl and at pH 10.0, but not pH 5.0.

The Genus *Microbacterium*

The genus *Microbacterium*, the type genus of the family Microbacteriaceae, was proposed by Orla-Jensen (1919) for Gram-positive, nonspore-forming, rod-shaped bacteria isolated during studies on lactic acid-producing bacteria. Currently, the genus *Microbacterium* harbors 33 species listed in Table 10.

Members of the genus *Microbacterium* were characterized largely by their marked heat resistance, presence in dairies, and production of small amounts of L(+) lactic acid from glucose (Doetsch, 1957; Robinson, 1966a). Four species, *Microbacterium lacticum*, "*Microbacterium flavum*," "*Microbacterium mesentericum*" and "*Microbacterium liquefaciens*" (Orla-Jensen, 1919), were originally included in this genus, while two species, "*Microbacterium thermosphactum*" (McLean and Sulzbacher, 1953) and "*Microbacterium ammoniaphilum*" (Abe et al., 1967), were added later. Subsequent studies showed that the genus *Microbacterium* was heterogeneous. It was revealed that "*M. flavum*," "*M. ammoniaphilum*" and "*M. mesentericum*" are authentic corynebacteria (Keddie et al., 1966; Robinson, 1966b; Rogosa and Keddie, 1974;

Table 9. Diagnostic characteristics of *Leucobacter komagate* and the genera having L-DAB in the peptidoglycan.

Characteristics	Leucobacter	Agreia	Agrococcus	Agromyces	Cryobacterium	Plantibacter	Rathayibacter	Subtercola
Color or colony	WB	Y/O/OB	W/Y/O	Y/W/O/PG	P	Y	Y/O	PY/BY
Cell morphology	Rods	Rods	Cocci	Hyphae, rods	Rods	Rods	Rods	Rods
Optimal growth temp. (°C)	25	24–26	18–28	26–30	9–12	24–26	24–28	15–17
Diagnostic amino acids[a]	L-DAB, and GABA	L-DAB, D-Orn, and Hyg	L-DAB, Asp, and (Thr)	L-DAB	L-DAB	L-DAB	L-DAB	DAB, and Hyg
Major menaquinone, MK-	11	10	11,12	12	10	10/9,10/10,11	10	9,10
DMA	−	+	−	−	−	−	−	+
Cell wall sugars								
Glucose	+	−	+	(+)	−	+	+	(+)
Galactose	+	−	+	+	−	−	(+)	−
Mannose	−	+	(+)	(+)	−	+	+	(+)
Ribose	−	−	(+)	−	−	+	−	(+)
Rhamnose	−	+	+	+[b]	+	+	+	+
Fucose	−	+	−	(+)	+	(+)	−	−
Xylose	−	−	−	(+)	−	−	(+)	+
Tyvelose	−	−	(+)	(+)	−	−	−	−
G+C (mol%)	66	65–67	74–75	69–72	65	67–70	63–72	64–68

Abbreviations: W, white; WB, whitish brown; Y, yellow; O, orange; OB, orange brown; P, pink; PG, pink gray or pink white; PY, pale yellow; L-DAB, L-diaminobutyric acid; GABA, γ-aminobutyric acid; D-Orn, D-ornithine; Hyg, 3-*threo*-hydroxy glutamate; Asp, aspartate; Thr, theonine; parentheses indicate the characteristic varies among different species; and for other definitions, see footnote in Table 1.

[a]Alanine, glutamate and glycine are present in all organisms; there are no available data on the DAB-isomers in *Subtercola*.

[b]*A. cerinus* subsp. *nitratus* does not have rhamnose in the cell wall.

Data from Zgurskaya et al. (1992; 1993b), Evtushenko et al. (Evtushenko et al., 1994; 2002; unpublished), Groth et al. (1996), Suzuki et al. (1996; 1997), Takeuchi et al. (1996), Sasaki et al. (1998), Wieser et al. (1999), Männistö et al. (2000), Takeuchi and Hatano (2001), Behrendt et al. (2002), Zlamala et al. (2002a), and Li et al. (2003).

Table 10. Species in the genus *Microbacterium* and some related information.

	Name	Basonym/former name	Source/habitat	References
1	*M. aerolatum*		Air	Zlamala et al., 2002b
2	*M. arabinogalactanolyticum*	*Aureobacterium arabinogalactanolyticum*	Soil	Kotani et al., 1972
3	*M. arborescens*	"*Bacillus arborescens*"	Water and soil	Frankland and Frankland, 1889
		"*Flavobacterium arborescens*"	River and lake water	Bergey et al., 1923
		[*Microbacterium laevaniformans*]	Sewage	Yokota et al., 1993b
4	*M. aurantiacum*		Corn steep liquor	Yokota et al., 1993b
5	*M. aurum*		Raw domestic sewage/water	Dias et al., 1962b
6	*M. barkeri*	*Aureobacterium barkeri*		Collins et al., 1983c
7	*M. chocolatum*	"*Chromobacterium chocolatum*"	Culture contamination	Clise, 1948
8	*M. dextranolyticum*	*Flavobacterium* sp.	Soil	Kobayashi et al., 1978
9	*M. esteraromaticum*	"*Bacterium esteraromaticum*"	Soil	Omelianski, 1923
		Flavobacterium esteraromaticum	Dairy water	Soppeland, 1924
		"*Flavobacterium suaveolens*"		
		Aureobacterium esteraromaticum		
10	*M. flavescens*	*Arthrobacter flavescens*,	Soil	Lochhead, 1958
		Aureobacterium flavescens		Collins et al., 1983d
11	*M. foliorum*		Phyllosphere of grasses, and the surface litter after mulching the sward	Behrendt et al., 2001
12	*M. gubbeenense*		Surface of Irish farmhouse smear-ripened cheese	Brennan et al., 2001
13	*M. halophilum*		Soil in the mangrove rhizosphere	Takeuchi and Hatano, 1998a
14	*M. hominis*		Lung aspirate	Funke et al., 1995
15	*M. imperiale*	"*Bacterium imperiale*"	Alimentary tract of the imperial moth *Eacles imperialis*	Steinhaus, 1941
		Brevibacterium imperiale		
16	*M. keratanolyticum*	*Aureobacterium keratanolyticum*	Insect	Lysenko, 1959
			Soil	Nakazawa et al., 1975b
				Nakazawa and Suzuki, 1975a
17	*M. ketosireducens*		Soil	Takeuchi and Hatano, 1998a
18	*M. kitamiense*		Wastewater of a sugar-beet factory	Matsuyama et al., 1999

No.	Species (synonyms)	Source	Reference
19	*M. lacticum*	Milk, dairy products and dairy equipment	Orla-Jensen, 1919
20	*M. laevaniformans* "*Corynebacterium laevaniformans*"	Activated sludge, blood	Dias and Bhat, 1962a; Dias, 1963
21	*M. liquefaciens* *Aureobacterium liquefaciens*	Milk, cheese, dairy products, and dairy equipment	Funke et al., 1995; Collins et al., 1983c
22	*M. luteolum* *Aureobacterium luteolum*	Soil	Topping, 1937
23	*M. maritypicum* "*Flavobacterium marinotypicum*"	Sea water and marine mud	ZoBell and Upham, 1944; Weeks, 1974
24	*M. oxydans* *Brevibacterium oxydans*	Contaminated hospital material	Chatelain and Second, 1966
25	*M. phyllosphaerae*	Phyllosphere of grasses, the litter layer after mulching the sward	Behrendt et al., 2001
26	*M. resistens* *Aureobacterium resistens*	Human clinical specimens	Funke et al., 1998
27	*M. saperdae* *Brevibacterium saperdae, Curtobacterium saperdae, Aureobacterium saperdae*	Dead larvae of Elm borer *Saperda caracharias*	Lysenko, 1959; Collins et al., 1983d
28	*M. schleiferi* *Aureobacterium schleiferi*	Soil	Schleifer et al., 1983
29	*M. terrae* *Aureobacaterium terrae*	Soil	Yokota et al., 1993a
30	*M. terregens* *Arthrobacter terregens, Aureobacterium terregens*	Soil	Lochhead and Burton, 1953; Collins et al., 1983d
31	*M. testaceum* *Brevibacterium testaceum, Curtobacterium testaceum, Aureobacterium testaceum*	Rice paddy, soil	Komagata and Iizuka, 1964
32	*M. thalassium*	Soil in the mangrove rhizosphere	Takeuchi and Hatano, 1998a
33	*M.trichothecenolyticum* *A. trichothecenolyticum*	Soil	Nakayama et al., 1980

Barksdale et al., 1979; Collins et al., 1982b), while "*M. thermosphactum*" was reclassified as *Brochothrix thermosphacta* (Sneath and Jones, 1976). "*Microbacterium liquefaciens*," on the other hand, was considered by many investigators to be merely a gelatin-liquefying variety of *Microbacterium lacticum* (Doetsch and Rakosky, 1950; Jayne-Williams and Skerman, 1966). However, *Microbacterium lacticum* contained lysine as cell wall diamino acid (Keddie et al., 1966; Robinson, 1966b), while "*M. liquefaciens*" had ornithine. "*Microbacterium liquefaciens*" was subsequently reclassified in the new genus *Aureobacterium* as *Aureobacterium liquefaciens* (Collins et al., 1983c).

As a result of systematic chemotaxonomic investigations, the genus *Microbacterium* was defined largely in chemical terms and was thus restricted to those organisms which contained a group B type peptidoglycan based upon L-lysine, predominantly methyl-branched fatty acids, long unsaturated menaquinones (MK-10 to MK-12), and a G+C content of 69–75 mol% (Collins et al., 1983d). The genus *Microbacterium* included six species, viz. *Microbacterium lacticum* (the type species; Orla-Jensen, 1919), *Microbacterium arborescens* (Imai et al., 1984), *Microbacterium aurum*, *Microbacterium dextranolyticum* (Yokota et al., 1993b), *Microbacterium imperiale* and *Microbacterium laevaniformans* (Collins et al., 1983d). On the other hand, the genus *Aureobacterium* including six species was proposed by Collins et al. (1983c) to accommodate organisms which contained a group B type peptidoglycan based upon D-ornithine, predominantly methyl-branched fatty acids, long unsaturated menaquinones (MK-10 to MK-12), and a G+C content of 69–75 mol% (Collins et al., 1983c). Later, Yokota et al. (1993a) and Funke et al. (1998) added seven and one species, respectively.

Thus, the genera *Microbacterium* and *Aureobacterium* were separated primarily on the basis of the cell wall peptidoglycan: B1-type peptidoglycan with an L-diamino acid, L-lysine, was typical of the genus *Microbacterium*, and B2-type peptidoglycan with a D-diamino acid, D-ornithine, was characteristic of the *Aureobacterium* species (Schleifer, 1970; Schleifer and Kandler, 1972; Yokota et al., 1993a). However, in other chemotaxonomic features, the organisms of the two genera exhibited very similar profiles, and a comparative 16S rDNA sequence analysis of members of these genera indicated that they were phylogenetically intermixed within a single monophyletic group (Rainey et al., 1994; Takeuchi and Yokota, 1994). The high phylogenetic similarity and the close relationship in physiological and chemotaxonomic features other than the diamino acid in the cell wall provided evidence that the genera *Aureobacterium*

and *Microbacterium* should be unified (Takeuchi and Hatano, 1998b).

The history and origin of current *Microbacterium* species are outlined in the Table 10. As shown, 11 of 33 species, including the type species *Microbacterium lacticum*, have been assigned to the genus *Microbacterium* in their original descriptions, while the other species were initially dispersed among different genera. *Microbacterium esteraromaticum* was originally described as "*Bacterium esteraromaticum*" (Omelianski, 1923) and later re-allocated to the genus *Flavobacterium* as *Flavobacterium esteraromaticum* (Bergey et al., 1930; Weeks, 1974). "*Flavobacterium suaveolens*" (Soppeland, 1924; Bergey and Breed, 1948; Weeks, 1974) resembled *F. esteraromaticum* chemotaxonomically (Takeuchi and Yokota, 1991), and both type strains showed high DNA-DNA hybridization values (Yokota et al., 1993a), leading to the description of *Aureobacterium esteraromaticum*. *Microbacterium saperdae* and *Microbacterium testaceum* were originally assigned to the genus *Brevibacterium* as *Brevibacterium saperdae* (Lysenko, 1959) and *Brevibacterium testaceum* (Komagata and Iizuka, 1964), respectively, and later transferred into the newly established genus *Curtobacterium* by Yamada and Komagata (1972b). Subsequent chemotaxonomic investigations revealed that *B. saperdae* and *B. testaceum* differed from curtobacteria and they were reclassified as members of *Aureobacterium* (Collins et al., 1983c).

Phylogeny

The genus *Microbacterium* accommodates microorganisms with a B group peptidoglycan containing lysine or ornithine, displaying four variations of the peptidoglycan type. In this respect the genus is unique, as most other actinomycete genera are homogeneous with respect to peptidoglycan type. Nevertheless, microbacteria with different peptidoglycans formed a monophyletic cluster on the basis of 16S rRNA gene sequence analysis (Rainey et al., 1994; Takeuchi and Yokota, 1994; Takeuchi and Hatano, 1998b; Brennan et al., 2001), and the phylogenetic relatedness of *Microbacterium* species does not match differences in the configuration of diamino acids in the interpeptide bridge (peptidoglycan type B1 or B2) or in the amino acid at position 3 of the peptide unit in the peptidoglycan (peptidoglycan type B1*a*/B1*b* or B2*a*/B2*b*). The following unique nucleotides of 16S rRNA gene were revealed for all strains of the genus *Microbacterium*: G at position 232 (*Escherichia coli* numbering; Brosius et al., 1978), U at position 279, A at position 780, U-A base pair at position 838–856, and A-U base pair at position 929–1386 (Takeuchi and Hatano, 1998b).

Taxonomy

The current genus *Microbacterium* harbors 33 species (Collins et al., 1983c; Imai et al., 1984; Yokota et al., 1993a; Yokota et al., 1993b; Funke et al., 1998; Takeuchi and Hatano, 1998b; Takeuchi and Hatano, 1998a; Matsuyama et al., 1999; Schumann et al., 1999; Behrendt et al., 2001; Brennan et al., 2001; Zlamala et al. 2002b), listed in Table 10.

Members of the genus have white, whitish yellow, yellow or orange colonies. In young cultures, cells are small irregular short slender rods (ca. 0.4–0.8 μm in diameter and 0.7–4 μm in length) occurring singly or in groups. They become shorter with age (3–7 d), a proportion may be coccoid, but a marked rod-coccus growth cycle does not occur. Some cells are arranged at angles to each other to give V-forms. Primary branching is uncommon. Cells are Gram positive, but older cultures may fail to retain the Gram stain. Cells do not form endospores and may be nonmotile or motile by lateral flagella. Cells are chemoorganotrophic and catalase positive. Metabolism is primarily respiratory but may also be fermentative. Microbacteria may show weak oxidative acid production from some carbohydrates.

The cell wall peptidoglycan contains L-lysine as the diamino acid, type B1α (*Microbacterium dextranolyticum*, *Microbacterium hominis*, *Microbacterium lacticum* and *Microbacterium laevaniformans*; Schleifer and Kandler, 1972; Collins et al., 1983d; Yokota et al., 1993b; Takeuchi and Hatano, 1998a) or type B1b (*Microbacterium arborescens*, *Microbacterium aurum* and *Microbacterium imperiale*; Schleifer and Kandler, 1972; Imai et al., 1984; Yokota et al., 1993b), or has D-ornithine as the diamino acid, type B2a (*Microbacterium keratanolyticum* [Yokota et al., 1993a] and *Microbacterium resistens* [Funke et al., 1998; Behrendt et al., 2001]) or type B2β (all other species; Schleifer and Kandler, 1972; Yokota et al., 1993a; Yokota et al., 1993b; Takeuchi and Hatano, 1998b; Takeuchi and Hatano, 1998a; Behrendt et al., 2001; Table 11). The presence of 3-*threo*-hydroxyglutamic acid in the peptidoglycan, which particularly substitutes glutamic acid at position 2 of the peptide subunit, is typical of microbacteria.

Microbacterium schleiferi contains hydroxyornithine as a diamino acid in its cell wall peptidoglycan (Schleifer et al., 1983), variation B2b of Schleifer and Kandler (Schleifer and Kandler, 1972; Yokota et al., 1993a), and the molar ratio of ornithine plus hydroxyornithine to homoserine to glutamate plus hydroxyglutamate to glycine to alanine to lysine is 1:1:1:2:1:0.4. The

glycan moiety of the cell wall contains both N-glycolyl and N-acetyl residues (Uchida and Aida, 1977). Arabinogalactan is not produced. A variety of sugars (e.g., galactose, glucose, rhamnose, 6-deoxytalose, fucose and xylose) occur in the cell walls; composition varies with strain and species (Keddie and Cure, 1977; Yokota et al., 1993a; Yokota et al., 1993b; Takeuchi and Hatano, 1998b; Takeuchi and Hatano, 1998a). Mycolic acids are not produced. Menaquinones are the sole respiratory quinones. All strains contain very long, unsaturated menaquinones, such as MK-11, MK-12 and MK-13, while MK-10 and MK-14 are minor components (Collins et al., 1980b; Collins et al., 1983c; Collins et al., 1983d; Yokota et al., 1993a; Yokota et al., 1993b). The long-chain cellular fatty acids are mainly of the *anteiso*- and *iso*-methyl branched types, with *anteiso*-15:0, *iso*-16:0 and *anteiso*-17:0 acids predominating. All microbacteria contain diphosphatidylglycerol and phosphatidylglycerol as their major polar lipids, and minor amounts of monomannosyldiacylglycerol, phosphoglycolipid and unidentified glycolipids may also occur (Collins et al., 1983c; Collins et al., 1983d). The G+C content is 66–72 mol%.

The phenotypic delineation of 33 *Microbacterium* species is based on some chemotaxonomic features, such as diamino acid in the cell wall, murein type, major menaquinone and cell wall sugars, combined with growth and physiological features (see subsection Identification in this Chapter and Tables 7, 11 and 12).

Habitat

Members of the genus *Microbacterium* are widely distributed in various environments (Table 10) and can be associated with plants, insects and clinical specimens. The examples presented in the Table 10 and outlined below, while not exhaustive, demonstrate the vast diversity of habitats of microbacteria.

Microbacterium lacticum is usually found in milk, powdered milk, cheese, and on dairy equipment, and it is considered to represent a major part of the thermoduric coryneform bacterial population of such sources (Jayne-Williams and Skerman, 1966; Thomas et al., 1967; Gillies, 1971). *Microbacterium liquefaciens* has also been isolated from milk, cheese, dairy products, and dairy equipment (Collins et al., 1983d). Organisms that phenotypically resemble this species are also found in soil (Topping, 1937) and frozen vegetables (Splittstoesser et al., 1967). *Microbacterium halophilum*, *Microbacterium thalassium* and some other strains considered to be new species of the genus *Microbacterium* were isolated from soils in the mangrove rhizosphere in Okinawa, Japan

Table 11. Chemotaxonomic characteristics of the species in the genus *Microbacterium*.

	Species	G+C content (mol%)	Major menaquinone[a]	Cell wall			Fatty acids[c,d]	Polar lipids[c,d]
				Murein type[b]	Diamino acid	Sugars[c]		
1	*M. aerolatum*	69.3–69.7	MK-12, 13	B2β	Orn	No data	*ai*-C15, *ai*-C17, and *i*-C16	DPG, PG, and GL
2	*M. arabinogalactanolyticum*	69.3	MK-12, 13	B2β	Orn	Gal	*ai*-C15, and *ai*-C17	DPG, PG, and (GL)
3	*M. arborescens*	71	MK-11, 12	B1β	Lys	6dT, Man, and Gal	*ai*-C15, and *ai*-C17	nd
4	*M. aurantiacum*	70.1–70.3	MK-12	B2β	Orn	Rha, Gal, and Fuc	*ai*-C15, *ai*-C17, and *i*-C16	nd
5	*M. aurum*	69.2	MK-11, 12	B1β	Lys	Gal, Glc, and Fuc	*ai*-C15, *ai*-C17, and *i*-C16	DPG, PG, and DMG
6	*M. barkeri*	68.7–74.4	MK-11, 12	B2β	Orn	Rha	*ai*-C15, and *ai*-C17	DPG, PG, and DGG
7	*M. chocolatum*	69.5	MK-12	B2β	Orn	Rha, Gal, (Man, and Xyl)	*ai*-C15, *ai*-C17, and *i*-C16	nd
8	*M. dextranolyticum*	68.3	MK-11, 12	B1α	Lys	Gal, 6dT, Rha, and (Glc)	*ai*-C15, and *ai*-C17	DPG, PG, DMG, and PGL
9	*M. esteraromaticum*	68.8	MK-12, 13	B2β	Orn	Gal, and Glc	*ai*-C15, and *ai*-C17	DPG, PG, and (GL)
10	*M. flavescens*	66.9–70.3	MK-13, 14	B2β	Orn	Gal, Glc, and Rha	*ai*-C15, and *ai*-C17	DPG, PG, and DGG
11	*M. foliorum*	67	MK-12, 11, 10	B1	Lys	Rha, Gal, and Man	*ai*-C15, *ai*-C17, and *i*-C16	DPG, PI, and PGM
12	*M. gubbeenense*	69–75	MK-11, 12	B2β	Orn	Rha, and Gal	*ai*-C15, *ai*-C17, and *i*-C16	nd
13	*M. halophilum*	67.2	MK-11,12,13	B2β	Orn	Man, Gal, and Glc	*ai*-C15, *ai*-C17, *i*-C15, and *i*-C16	nd
14	*M. hominis*	71.2	MK-11, 12	B1α	Lys	Rha, 6dT, Gal, and Man	*ai*-C15, *ai*-C17, and *i*-C16	nd
15	*M. imperiale*	71.0–75.4	MK-11, 12	B1β	Lys	Gal, Man, and Rha	*ai*-C15, and *ai*-C17	DPG, PG, and DMG
16	*M. keratanolyticum*	66.5	MK-12, 13	B2α	Orn	Gal	*ai*-C15, and *ai*-C17	DPG, PG, and (GL)
17	*M. ketosireducens*	69.7–69.8	MK-13	B2β	Orn	No data	*ai*-C15, *ai*-C17, and *i*-C16	nd
18	*M. kitamiense*	69.2	MK-11, 12	B2β	Orn	No data	*ai*-C15, *ai*-C17, and *i*-C16	nd
19	*M. lacticum*	69.0–74.9	MK-11, 12	B1α	Lys	Gal, Rha, and Man	*ai*-C15, *ai*-C17	DPG, PG, and DMG
20	*M. laevaniformans*	70.0–73.7	MK-11, 12	B1α	Lys	Gal, Man, Xyl, and Rha	*ai*-C15, and *ai*-C17	DPG, PG, and DMG
21	*M. liquefaciens*	68.2–72.5	MK-11, 12	B2β	Orn	Rha	*ai*-C15, and *ai*-C17	DPG, PG, and DGG
22	*M. luteolum*	70.4	MK-12	B2β	Orn	Rha, Gal, and (Glc)	*ai*-C15, and *ai*-C17	DPG, PG, and (GL)
23	*M. maritypicum*	71.6	MK-12	B2β	Orn	Gal	*ai*-C15, and *i*-C16	nd
24	*M. oxydans*	70.71	MK-11, 12	B2β	Orn	Rha, Man, Glc, and Gal	*ai*-C15, *ai*-C17, and *i*-C16	DPG, and PG
25	*M. phyllosphaerae*	64	MK-12, 11, 10	B2β	Orn	Gal, and Rha	*ai*-C15, *ai*-C17, *i*-C16, and *i*-C15	nd
26	*M. resistens*	nd	MK-12, 13	B2α	Orn	No data	*ai*-C15, *ai*-C17, and *i*-C16	nd
27	*M. saperdae*	68.5–72.5	MK-11, 12	B2β	Orn	Gal, and Glc	*ai*-C15, and *ai*-C17	DPG, PG, and DGG
28	*M. schleiferi*	66.9	MK-11, 12	B2β	Orn	6dT, Man, and Gal	*ai*-C15, and *ai*-C17	DPG, PG, and (GL)
29	*M. terrae*	70.7	MK-13, 14	B2β	Orn	Rha, Gal, and Glc	*ai*-C15, and *ai*-C17	DPG, PG, and (GL)
30	*M. terregens*	68.2–75.6	MK-12, 13	B2β	Orn	Gal, Rha, and 6dT	*ai*-C15, and *ai*-C17	DPG, PG, and DGG
31	*M. testaceum*	65.4–72.7	MK-11, 10	B2β	Orn	Rha, Man, Gal	*ai*-C15, and *ai*-C17	DPG, PG, and DGG
32	*M. thalassium*	69.1–69.7	MK-11, 12	B2β	Orn	Gal, Glc, (Rha, and Man)	*ai*-C15, *ai*-C17, *i*-C16, and (*i*-C15)	nd
33	*M. trichothecenolyticum*	69.0	MK-12, 13	B2β	Orn	Gal, and Glc	*ai*-C15, and *ai*-C17	DPG, PG and (GL)

Abbreviations: Lys, lysine; Orn, ornithine; Rha, rhamnose; Glc, glucose; Gal, galactose; Man, mannose; 6dT, 6-deoxytalose; Fuc, fucose; Xyl, xylose; *ai*-C15, 12-methyl tetradecanoic acid; *ai*-C17, 14-methyl hexadecanoic acid; *i*-C15, 13-methyl tetradecanoic acid; *i*-C16, 14-methyl pentadecanoic acid; *i*-C17, 14-methyl hexadecanoic acid; PG, phosphatidylglycerol; DPG, diphosphatidylglycerol; PGL, phosphoglycolipid; GL, glycolipid; DMG, dimannosyldiacylglycerol; DGG, diglycosyldiacyl glycerol; PGM, phosphatidylglycerol mannoside; and nd, not determined.

[a]Menaquinone type is abbreviated as indicated by Collins and Jones (1981).

[b]Murein type is abbreviated as indicated by Schleifer and Kandler (1972).

[c]Parentheses indicate that a compound may not be present.

[d]Fatty acid pattern and phospholipid type are abbreviated as in Collins et al. (1980b).

Data from Takeuchi and Yokota (1989; 1993; 1994), Takeuchi et al. (1990), Yokota et al. (1993a; 1993b), Funke et al. (1998), Takeuchi and Hatano (1998a; 1998b), Matsuyama et al. (1999), Schumann et al. (1999), Behrendt et al. (2001), Brennan et al. (2001), and Zlamala et al. (2002b).

Table 12. Growth, morphological and physiological characteristics[a] that differentiate the species in the genus *Microbacterium*.

Species	Color of colony	Motility	Growth at 2% NaCl	Growth at 37°C	GEL (Hydrolysis)	STA	Tween 20	Tween 40	Tween 60	Tween 80	VP	H2S	ADH	ARA	NAG	CIT (Assimilation)	MLT	PAC	FUM	Acid from GLC
1. *M. aerolatum*	Y	−	+	+	−	ND	−	ND	ND	−	−	−	ND	+	+	+	+	+	+	+
2. *M. arabinogalactanolyticum*	YW	−	+	−	+	+	−	−	+	+	−	+	+	+	+	−/+w	+	+	+	−
3. *M. arborescens*	O	+/−	ND	−	+	+	+	−	−	+	−	+	−	+	+	+w	+	−	+	+
4. *M. aurantiacum*	O	−	+	+	−	−	−	+	−	d	−	+	−	d	+w	−	d	−	+	d
5. *M. aurum*	YW	−	+	+	+	+	+	−	+w	−	−	+	−	+	+w	+/−	−	−	−	+
6. *M. barkeri*	Y/W	+	+	+	+	+	+	+	+	+	−	+	+	−	+	+/−	+	−	+	−
7. *M. chocolatum*	O/DO	−	+	+	−	+w	+	+	+	+	−	+	−	+w/−	+w/−	−	−	−	d	−
8. *M. dextranolyticum*	W	−	ND	−	−	−	d	+	+	+	+	+	+	−	−	−	+/−	+	+	+
9. *M. esteraromaticum*	YW/Y	+	−	d	−	+	+	+	+	+	+	−	−	+	−	−	+/−	−	+	+
10. *M. flavescens*	Y	+	+	−	+	+	ND	+	−	−	−	−	−	+	−	−	+/−	−	+w	+
11. *M. foliorum*	Y/LY	+	+	d	+	+w	+	ND	+	+	−	+	−	+	+	d	+	d	ND	+
12. *M. gubbeenense*	Cr	+	+	−	−	−	−	−	−	−	−	+	ND	+	+	+	+	ND	+	+
13. *M. halophilum*	YW	−	+	+	+w	+	+w	−	+	+	−	−	−	−	−	−	−	−	+	+
14. *M. hominis*	YW	−	+	+	−	−	+	−	−	−	+	+	−	+	+	+	−	−	+	+
15. *M. imperiale*	O	+	ND	+	−	+	−	−	−	+	+	+	−	+	+w	+	+	−	+	+
16. *M. keratanolyticum*	Y	+	+	−	+	−	−	−	−	+w	−	+	+w	+w	+	−	−	−	+	−
17. *M. ketosireducens*	Y	−	+	−	+	+	+	+	+	+	−	−	+	+	−	d	−	−	d	+
18. *M. kitamiense*	O	−	−	+	+	+	−	−	−	−	−	−	−	−	−	−	−	−	−	+
19. *M. lacticum*	Y/YW	−	+w	−	−	+	−	−	−	+w	−	−	−	+w	+	−	+	−	+	+
20. *M. laevaniformans*	Y/YW	−	+	+	d	+	−	−	−	−	+	+	−	+	−	d	+/−	−	+	+
21. *M. liquefaciens*	BY/Y	+	+	+	+	−	−	−	−	−	+	+	d	−	+	d	+/−	−	+/−	+
22. *M. luteolum*	YW	−	−	−	+	−	−	−	−	+	−	−	−	−	+	−	+	−	+	+
23. *M. maritypicum*	LY	+	+	+	−	−	+	+	+	−	+	+	−	+	+	+	+	+	−	+
24. *M. oxydans*	Y	+	+	+	+	−	ND	ND	ND	+	+	−	−	+	+w	+	+	−	ND	+
25. *M. phyllosphaerae*	Y	+	+	d	+	+w	ND	ND	+	+	−	+	−	+	+	−	+	d	ND	+
26. *M. resistens*	Y	−	+	+	+	−	ND	ND	ND	−	−	−	−	+	+w	+	−	−	ND	+
27. *M. saperdae*	YW	+	−	−	−	+	+	+	−	+	+	+	−	+	+	−	+	−	−	+
28. *M. schleiferi*	YW	−	+w	+w	−	−	d	+	d	+w	−	−	−	+w	+w	−	+/−	−	+	+
29. *M. terrae*	Y	−	+	−	+	+	d	+	+	+	+	+	−	+	+	−	+/−	−	+	+
30. *M. terregens*	Y	−	+w	−	−	−	+	+	−	+	−	−	−	−	−	−	+	−	+	+
31. *M. testaceum*	O	+	−	+	+	−	−	+	+	+	−	−	−	+	+	+/−	+	+w	+	+
32. *M. thalassium*	Y	−	+	−	+	−	−	+	+	−	−	−	−	−	−	−	−	−	+	−
33. *M. trichotheceno*	Y	−	+w	−	−	+	+	−	−	−	+	+	−	+	+	+/−	−	−	+	+

Symbols and abbreviations: +, positive reaction; +w, weakly positive; −, negative; +/−, different results among strains; ND, not determined; YW, yellow white; Y, yellow; O, orange; DO, dull orange; W, white; LY, light yellow; BY, bright yellow; Cr, cream; GEL, gelatin; STA, starch; H2S, H2S production; VP, Voges-Proskauer test; ADH, arginine dihydrolase; ARA, arabinose; NAG, *N*-acetylglucosamine; CIT, citrate; MLT, malate; PAC, phenyl acetate; FUM, fumarate; and GLC, glucose.

[a]Most tests for characterizing the biochemical profiles of studied strains were performed by the methods described by Cowan (1974), Behrendt et al. (1999) or Takeuchi and Hatano (1998a). API 20NE (bioMérieux) and the API 50CH gallery using the API CHE suspension medium (bioMérieux) were used to determine physiological and biochemical characteristics of some species. Data from references listed in Table 11.

(Takeuchi and Hatano, 1998a; Takeuchi and Hatano, 1999). In the mangrove forests, complex environments have formed under the influence of tide, the influx of fresh water, and high temperature and humidity in the subtropical and tropical climates. The soils in such environments are muddy and are reported to be anoxic, low in nutrients, and to have higher concentrations of heavy metals and higher salinity than terrestrial soils have (Wakushima et al., 1994). Microbacteria were also revealed in the rhizoplane of oilseed rape (Kaiser et al., 2001).

Microbacterium foliorum and *Microbacterium phyllosphaerae* were isolated from grasses and the surface litter after mulching the sward (Behrendt et al., 2001). *Microbacterium testaceum*, first isolated from rice paddy mud (Komagata and Iizuka, 1964), was also found among endophytic bacteria, which reside within plant hosts without causing disease symptoms (Zinniel et al., 2002). Elbeltagy et al. (2000) found endophytic microbacteria and curtobacteria along with some other strains in wild and traditionally cultivated rice varieties. Different strains of the genus *Microbacterium*, including *Microbacterium testaceum*, were isolated from plant galls of grass induced by plant parasitic nematodes of the subfamily Anguininae (Evtushenko et al., 1994; L. I. Evtushenko and M. Takeuchi, unpublished observation). However, microbacteria do not seem to be predominant among other coryneform isolates in bacterium-nematode complexes in the galls. They are probably secondary settlers of the galls, descending from the phylloplane or the healthy plant tissue, and have no relationship with nematodes. In contrast, "*Microbacterium nematophilum*" was reported to be the infectious agent of the nematode *Caenorhabditis elegans*, which fortuitously contaminates the nematode (Hodgkin et al., 2000). The bacteria adhere to the cuticle of susceptible nematodes and induce substantial local swelling of the underlying hypodermal tissue. The induced deformation appears to be part of a survival strategy for the bacteria, as *Caenorhabditis elegans* appears to be their predator (Hodgkin et al., 2000). *Microbacterium imperiale* was isolated from alimentary tract of the imperial moth *Eacles imperialis* (Steinhaus, 1941) and *Microbacterium saperdae* was isolated from dead larvae of Elm borer *Saperda caracharias* (Lysenko, 1959). Recently, some microbacteria were isolated as microbial symbionts of Great Barrier Reef sponges (Burja et al., 1999), and they were also found among cellulolytic bacteria inhabiting the gut of the termite *Zootermopsis angusticollis* (Wenzel et al., 2002).

Free-living, aerobic, copiotrophic ultramicrobacteria (Koch, 1996) that passed through a 0.45-μm membrane filter and had a cell volume of less than 0.3 μm³ were isolated from polluted urban soil (Iizuka et al., 1998). Microbacteria were revealed in the sea mud of the Mariana Trench (Takami et al., 1997) and in the Western Mediterranean Sea (Fritz, 2000). They were found in 12,000 year-old glacial ice from Sajama, Bolivia (Christner, 2002) and in permafrost sediments recovered from a frozen (annual mean temperature –12°C) late Pliocene layer, 1.8–3 million years old, in East Siberia (Karasev et al., 1998; Kochkina et al., 2001). It is interesting to note that many of the permafrost microbacteria were found in samples containing plant residues and were associated with *Rhodococcus* sp. in the two-member cultures isolated on agar media (Kochkina et al., 2001). Phylogenetic and phenotypic data indicate that some of these strains represent new species of the genus *Microbacterium* (L. I. Evtushenko and M. Takeuchi, unpublished observation). Microbacteria were also recovered from condensation water samples of the Russian space station "Mir" (Kawamura et al., 2001) and in a phosphorus-removing bacterial consortium consisting initially of 18 bacterial strains in a sequence batch reactor under alternating aerobic/anaerobic conditions and without additional carbon sources (Hollender et al., 2002).

CLINICAL MICROBACTERIA Several reports have lately indicated the association of *Microbacterium* spp. (*Aureobacterium* spp.) with diseases (Funke et al., 1997). Microbacterial strains resembling organisms of CDC coryneform groups A-4 and A-5 (Hollis and Weaver, 1981) were recovered from different clinical specimens, mostly blood cultures (Funke et al., 1995). Clinically significant strains, presumptively microbacteria, were isolated from patients with endocarditis (Lifshitz et al., 1991), septicemia (Bizette et al., 1995), catheter-related septicemia (Campbell et al., 1994), exogenous endophthalmitis (Barker et al., 1990), acute myelogenous leukemia (Nolte et al., 1996), and some other diseases. *Microbacterium* sp., along with other microorganisms, were detected by molecular methods in advanced noma lesions, although owing to the limited number of available noma subjects it was impossible to associate specific species with the disease (Paster et al., 2002). Hanscom and Maxwell (1979) experimentally demonstrated that a clinical *Microbacterium* strain fulfilled the Koch-Henle postulates for causing endophthalmitis. Most microbacterial infections were reportedly self-limited (Carlotti et al., 1993; Funke et al., 1997). A case of lethal *Microbacterium* bacteremia in an immunocompromised patient was described by Nolte et al. (1996). Saweljew et al. (1996) reported a case of fatal systemic infection of an immunocompetent patient with *Microbacterium* sp. The mode of transmission of most *Microbacterium* strains to humans remains to be elucidated, although the central venous catheter was previously proposed to be a source of bacteremia (Campbell et al., 1994; Grove et al., 1999). The documented case of catheter-related *Microbacterium* bacteremia has recently been reported by Lau et al. (2002).

Some clinical strains were identified or found to be new species. The strain *Microbacterium* sp. CDC group A-4 from lung aspirate was described as the new species named "*Microbacterium hominis*," and *Microbacterium* sp. CDC group A-5 from blood was identified as *Microbacterium laevaniformans* (Takeuchi and Hatano, 1998a). Strains from corneal ulcer and blood culture of a patient with acute myelogenous leukemia were described as the new species *Microbacterium resistens* (Funke et al., 1998; Behrendt et al., 2001). *Microbacterium* sp. isolated repeatedly from blood taken through the lumina of a central venous catheter of a patient with multiple myeloma had a 99.5% identity with *Microbacterium liquefaciens*, but there was a slight phenotypic difference from the description of this species (Grove et al., 1999). A bacterium showing 99.4% 16S rDNA sequence similarity to *Microbacterium oxydans* was revealed in both the blood culture and central catheter tip of a patient with chronic myeloid leukemia (Lau et al., 2002).

Isolation and Cultivation

Since most strains of *Microbacterium lacticum* are thermoduric (survive 63°C for 30 min or 72°C for 15 min in skim milk; Abd-el-Malek and Gibson, 1952; Jayne-Williams and Skerman, 1966), the usual procedure for isolating this organism from dairy products involves plating laboratory-pasteurized (63°C for 30 min) samples onto a suitable, nonselective medium. However, the pasteurization step should be omitted if nonthermoduric strains are also to be isolated. A suitable nonselective medium is yeast extract milk agar (YMA; Harrigan and McCance, 1976). Dias and Bhat (1962a) have described an enrichment procedure for the isolation of *Microbacterium laevaniformans* from raw sewage and activated sludge. Bacteria other than *Microbacterium laevaniformans* are, however, also enriched by this method. A good general-purpose medium for the cultivation of most microbacteria is Corynebacterium (CB) agar containing 0.5% glucose, 1% casein-peptone-tryptic digest, 0.5% yeast extract, 0.5% NaCl and 1.5% agar (pH 7.2). Strains should be incubated aerobically at 28–30°C. *Microbacterium terregens* and *Microbacterium flavescens* reportedly have a requirement for the terregens factor (Lochhead and Burton, 1953; Lochhead, 1958), and these

species can be cultivated on a variety of conventional media supplemented with soil extract. A suitable medium is soil extract nutrient agar. However, most microbacteria (including *Microbacterium terregens* and *Microbacterium flavescens*) grow well on peptone-yeast extract (PY) agar medium containing 1% (w/v) peptone, 0.2% yeast extract, 0.2% NaCl, 0.2% MgSO$_4$ · 7H$_2$O and 1% agar (pH 7.0) or on CB agar.

Soil Extract Nutrient Agar

Lab-Lemco beef extract	1 g
Yeast extract	2 g
Peptone	5 g
NaCl	5 g
Agar	15 g
Soil extract	1 liter
pH 7.0	

Soil extract is prepared by suspending 500–1000 g of soil in 1 liter of tap water and autoclaving at 121°C for 30 min. Clear supernatant is obtained by filtration or centrifugation.

Identification

A classification system based on phylogenetic clustering and the presence of taxon-specific 16S rDNA/RNA signature nucleotides enables new isolate assignment to the genus *Microbacterium* or even to a certain species group. Cell morphology combined with some chemotaxonomic characteristics, mainly the composition of amino acids of peptidoglycan and major menaquinones, is useful for preliminary identification of novel isolates as microbacteria at the phenotypical level (Tables 1 and 7). Besides the genus *Microbacterium*, only three genera in the family, namely, *Frigoribacterium*, *Mycetocola* and *Okibacterium*, have lysine in their peptidoglycan and two genera, *Curtobacterium* and *Agreia*, contain ornithine. Among the genera with lysine-based peptidoglycan, the *Microbacterium* species are distinguished by the presence of 1 or 2 molecules of lysine and 2 or 3 molecules of glycine in the polymer subunit and the major menaquinones MK-11 and MK-12 (Table 7). Microbacteria with ornithine in the cell wall clearly differ from *Curtobacterium* and *Agreia* in the major menaquinones. They can also be separated from *Agreia* by lacking DAB in the cell wall.

Some chemotaxonomic features (Table 11), such as major menaquinone, murein type, diamino acids, and sugar composition of the cell wall, vary among the *Microbacterium* species and may serve to differentiate individual species from others. It is currently not easy to select growth and physiological characteristics that allow unambiguous identification of *Microbacterium* species. However, some of them (Table 12) combined with the chemotaxonomic properties (Tables 7 and 11) permit new isolates of the genus *Microbacterium* to be identified at the species level.

Preservation

Various methods for long-term preservation of microorganisms, including freeze-drying, liquid-drying (L-drying), and freezing at –80°C or at –196°C, are suitable for preservation and maintenance of microbacteria and other organisms of the family. L-drying involves vacuum-drying of samples from the liquid state without freezing, and it is known to be useful for the preservation of microorganisms that are sensitive to freeze-drying (Anner, 1970; Imai and Sakane, 1985; Malik, 1990; Sakane et al., 1992). Iijima and Sakane (1973) improved the L-drying method described by Anner (1970). Potassium phosphate (0.1M) buffer, containing 3% monosodium glutamate, 1.5% ribitol and 0.05% cysteine-HCl (pH 7.0), is a useful protective medium for most bacteria. Two additives, ribitol and cysteine-HCl, prevent mutation induced by desiccation (Sakane et al., 1983; Imai and Sakane, 1985).

Most Gram-positive bacteria including microbacteria and other members of the family Microbacteriaceae dried by the L-drying method usually showed a high survival rate, and these dried cultures appeared to have been preserved for a long time. Sakane and Kuroshima (1997), who examined the viabilities of 59 L-dried bacterial cultures stored 21–24 years at 5°C and compared these viabilities with previous estimates, found that the survival rates of L-dried microorganisms stored 16 years at 5°C were unchanged and that the rate after storage for over 20 years could be estimated by an accelerated storage test requiring 2 weeks at 37°C.

Applications and Properties

Some microbacterial strains show interesting biochemical features, and some organisms play important roles in the chemical industry. *Microbacterium arabinogalactanolyticum* produces *a*-arabinofuranosidase (Kotani et al., 1972). *Microbacterium dextranolyticum* synthesizes dextran-*a*-1,2-debranching enzyme (Kobayashi et al., 1978; Mitsuishi et al., 1979; Mitsuishi et al., 1980). *Microbacterium keratanolyticum* produces keratan sulfate endo-*b*-galactosidase (Nakazawa and Suzuki, 1975a; Nakazawa et al., 1975b). *Microbacterium ketosireducens* reduces 2,5-diketo-D-gluconic acid (Takeuchi and Hatano, 1998a). *Microbacterium trichothecenolyticum* was reported to degrade trichothecene mycotoxins (such as T-2 toxin, diacetoxyscirpenol, neosolaniol, nivalenol, and fusarenon-X) produced by various species of fungi and can occur in agricultural products, including cereals and pulses (Nakayama et al., 1980; Ueno et al., 1983). Consumption of trichothecene-contaminated foods by farm animals and humans leads to mycotoxicosis. Trichothecenes are known to induce hematological disorders such as neutropenia, aplastic anemia, and thrombocytopenia in humans and animals (Froquet et al., 2001). Shafiee et al. (1998) purified, characterized and immobilized an NADPH-dependent enzyme involved in the chiral specific reduction of the keto ester M, an intermediate in the synthesis of an anti-asthma drug, Montelukast, from "*Microbacterium campoquemadoensis.*" Asano and Tanetani (1998) found a thermostable phenylalanine dehydrogenase from a mesophilic *Microbacterium* sp. Desaint et al. (2000) isolated the carbofuran-degrading microbacteria from soil, and Chee-Sanford et al. (2001) reported on the diversity of tetracycline resistance genes in a microbacterium (phylogenetically closest to *Microbacterium oxydans*) found in waste lagoons and in groundwater underlying two swine farms. Unusual microbacteria with arsenic (III)-oxidizing and chromate (VI)-reducing activities were found in sewage and tannery waste (Pattanapipitpaisal et al., 2001; Mokashi and Paknikar, 2002). *Microbacterium esteraromaticum* is an aroma-producing bacterium (Omelianski, 1923; Soppeland, 1924). *Microbacterium barkeri* shows pectinolytic activity (Dias et al., 1962b). *Microbacterium kitamiense* and *Microbacterium phyllosphaerae* produce insoluble and soluble exopolysaccharides (Matsuyama et al., 1999; Behrendt et al., 2001). *Microbacterium laevaniformans* produces levan from sucrose and raffinose (Dias and Bhat, 1962a). *Microbacterium resistens* is resistant to vancomycin (Funke et al., 1998).

In addition to the 33 discussed species (Table 10) another new species, *Microbacterium paraoxydans* with the type strain CF36 (= CCUG 46601 = DSM 15019), has recently been described (Laffineur et al., 2003) and validated. Representatives of the new species were isolated from the blood of a child with acute lymphoblastic leukemia who was perfused with a central venous catheter. The physiological, chemotaxonomic, and genetic characteristics of *Microbacterium paraoxydans* indicated that this species is closely related to *Microbacterium oxydans*. The G+C content of its DNA is 69.9 mol%.

The Genus *Mycetocola*

The genus *Mycetocola*, with the type species *Mycetocola saprophilus*, was established by Tsukamoto et al. (2001) to accommodate tolaasin-detoxifying bacteria, which were isolated from the cultivated mushroom *Pleurotus ostreatus*. Tolaasins, pathogenic toxins (Nutkins et al., 1991; Shirata et al., 1995) produced extracellularly by *Pseudomonas tolaasii* Paine, cause brown blotch disease in economically important cultivated mushrooms, *Pleurotus ostreatus* Kummer and *Agaricus bisporus* Singer (Tolaas, 1915; Suyama and Fujii, 1993).

Phylogeny

Phylogenetically (Fig. 1), the type strains of three species of the genus *Mycetocola* form an independent, well-defined (100% bootstrap confidence) cluster that is usually located close to the root of the genus *Microbacterium* branch.

Taxonomy

Members of the genus *Mycetocola* are Gram-positive, nonmotile, irregular rods, 0.2–0.4 μm wide and 2.0–3.5 μm long. Mycelium and endospores are not produced. Growth on peptone-yeast agar gives usually convex, smooth, yellowish-white colonies. Bacteria are obligately aerobic, and no growth occurs in anaerobic conditions. The optimal growth temperature is 25°C, the maximum temperature for growth is 33°C, and no growth occurs at 4°C. Acid is produced from glucose and various other carbohydrates. Catalase is produced, but oxidase is not. All bacteria of this genus are able to detoxify tolaasins produced by *Pseudomonas tolaasii*. Cell wall peptidoglycan contains lysine as the diamino acid, and the muramic acid of the cell wall is an acetyl type. The major isoprenoid quinone is menaquinone MK-10, and small amounts of MK-9 and MK-11 are present. The major fatty acids are *anteiso*-$C_{15:0}$ and *anteiso*-$C_{17:0}$. The G+C content is 63.9–64.7 mol% (Tsukamoto et al., 2001).

Currently, the genus includes three species, *Mycetocola saprophilus*, *Mycetocola tolaasinivorans* and *Mycetocola lacteus* (Tsukamoto et al., 2001), which show similar physiologial properties. The species differentiation is based on a low DNA-DNA relatedness level, indicating separate genomic species (Wayne et al., 1987), which is consistent with some phenotypical characteristics discussed in the subsection Identification in this Chapter.

Habitat

The data available suggest that the natural habitat of *Mycetocola* is mushrooms. Bacteria of this genus were found to be associated with *Agaricales* mushrooms such as *Pleurotus ostreatus* (Jacq.:Fr) Kummer and *Lepista nuda* (Bull.: Fr.)

Cooke (Tsukamoto et al., 1998; Tsukamoto et al., 2001; Tsukamoto et al., 2002). Some mushrooms undergo blackening by *Pseudomonas tolaasii*, which produces extracellular lipodepsipeptide toxins (i.e., tolaasins; Tolaas, 1915; Suyama and Fujii, 1993) that induce blotch symptoms (Nutkins et al., 1991; Rainey et al., 1993; Shirata et al., 1995; Murata and Magae, 1996; Murata et al., 1998). *Mycetocola* spp. detoxify the tolaasins and suppress the development of the disease (Tsukamoto et al., 1998). *Mycetocola* species are saprophytic but not parasitic or pathogenic to *Pleurotus ostreatus* (Tsukamoto et al., 1998).

Isolation

Bacteria of the genus *Mycetocola* can be found in rotting fruiting bodies of some cultivated or wild Agaricales fungi (e.g., *Pleurotus ostreatus* and *Lepista nuda*). For isolation, the potato-semisynthetic agar (PSA; Wakimoto, 1955) and T-PAF medium (Suyama et al., 1995) can be used.

PSA medium (Wakimoto, 1955)

Sucrose	15 g
Peptone	5 g
Ca(NO$_3$)$_2$ · 4H$_2$O	0.5 g
Na$_2$HPO$_4$ · 12H$_2$O	2 g
Decoction of 300 g of potato tuber slices	1 liter
Agar	15 g
pH 7.0	

T-PAF medium is based on *Pseudomonas* agar F (PAF) supplemented with tolaasin (Suyama et al., 1995). Viable cells of *Mycetocola* form yellow colonies, while *Pseudomonas tolaasii* produces distinctive blue colonies on T-PAF and PSA (Suyama et al., 1995; Wakimoto, 1955). The tolaasin-detoxifying *Mycetocola* spp. can also be isolated by using the potato tuber slice method (Shirata et al., 1995). Microorganisms selected as tolaasin-detoxifying strains are cultured aerobically at 28°C in a peptone-yeast (PY) medium containing 10 g of peptone, 2 g of yeast extract, 2 g of NaCl and 2 g of D-glucose in 1 liter of distilled water (pH 7.2) or other media based on peptone, glucose and yeast extract.

Identification

As shown in Tables 1 and 7, only members of three genera in the family Microbacteriaceae, *Frigoribacterium* (Kämpfer et al., 2000), *Okibacterium* (Evtushenko et al., 2002) and *Microbacterium*, along with the genus *Mycetocola*, have lysine in the cell wall. Among them, only *Mycetocola* and *Frigoribacterium* are characterized by presence of lysine as the diamino acid and acetyl-type muramic acid in the peptidoglycan. Bacteria of these two genera can be distinguished by motility, growth temperature, major menaquinone, and fatty acids (including 1,1-dimethyl acetals), and DNA G+C content. In addition, the ability to detoxify tolaasins has not been reported for *Frigoribacterium* and other Microbacteriaceae.

The three species in the genus *Mycetocola* are very similar phylogenetically and physiologically. To assess whether a new isolate belongs to the known species of *Mycetocola* or to a new species, the following characteristics, along with the source of isolation, may be used: catalase reaction is positive, but oxidase, Voges-Proskauer and methyl red tests are negative; nitrate or nitrate is not reduced; hydrogen sulfide, lecithinase, tyrosinase, and urease are not produced; esculin is hydrolyzed, but arbutin, arginine, and casein are not hydrolyzed; no liquefaction of gelatin and production of indole, 3-ketolactose, 2-ketogluconate and levan from sucrose occur; fumarate is used as a sole carbon source, but asparagine is not used as a sole nitrogen and carbon source; benzoate, butyrate, *m*-hydroxybenzoate, malonate, oxalate, propionate, D- and L-tartarate are not utilized; acid is produced from D-cellobiose, dextrin, D-fructose, D-galactose, D-glucose, glycerol, lactose, maltose, D-mannitol, D-mannose, melibiose, D-ribose, salicin, D-sorbitol, sucrose, trehalose or D-xylose, but not from D- and L-arabinose, D-dulcitol, *m*-inositol, inulin, D-raffinose, L-rhamnose, or starch. Growth occurs in peptone water, but not in Corn's solution, Fermi's solution, and Uschinsky's solution. Maceration of potato tubers and hypersensitive reaction in tobacco leaves are not found (Tsukamoto et al., 2001).

To preliminarily assign a novel isolate to one of the known *Mycetocola* species, the following information is helpful: *Mycetocola lacteus* produces acids from both D-melezitose and erythritol and also utilizes both citrate and Tween 80; *Mycetocola tolaasinivorans* does not produce acid from either D-melezitose or erythritol and does not utilize either citrate or Tween 80. *Mycetocola saprophilus* produces acid from D-melezitose but not from erythritol and utilizes citrate but does not hydrolyze Tween 80. DNA-DNA hybridization tests or analyses of other adequate genomic characteristics (Vandamme et al., 1996; Stackebrandt et al., 2000, 2002; Rossello-Mora and Amann, 2001) may be necessary to support species identification.

Applications and Features

Tolaasin produced by *Pseudomonas tolaasii* disrupts cell membranes by forming a voltage-gated ion channel and acts as a biosurfactant, leading to the cell death of mushroom mycelia (Brodey et al., 1991; Hutchison and Johnstone, 1993). Elimination of contaminants by periodically cleaning facilities with disinfectants such as chlorine and by carefully controlling humidity and temperature during cultivation of mushrooms seems to be effective in controlling the outbreak of disease (Fletcher et al., 1989). Development of efficient biological methods to control disease is important, since, in principle, the use of chemicals or antibiotics should be avoided in mushroom culture. Use of strains of *Mycetocola* could be a unique approach for developing a biocontrol system of cultivated mushrooms, particularly *P. ostreatus*. It is different from the biocontrol system in *Agaricus bisporus* cultivation, where an antagonistic bacterium, i.e., *Pseudomonas fluorescens* (Trevisan) Migula, is allowed to grow on the mushroom casing layer to suppress physically the incoming population of *Pseudomonas tolaasii* (Nair and Fahy, 1972; Nair and Fahy, 1976; Fermor et al., 1991). However, *Pseudomonas fluorescens* is pathogenic to cultivated mushrooms by producing various antifungal agents (Laville et al., 1992; Corbell and Loper, 1995) and is closely related to *Pseudomonas tolaasii* (Thorn and Tsuneda, 1996).

In contrast to *Pseudomonas fluorescens*, strains of *Mycetocola* are saprophytic and markedly reduce the level of tolaasin by live suppressive bacterial cells (Tsukamoto et al., 1998). However, many problems must be solved before the bacterium can be used as a control agent. The major problem is the possibility of side effects of the *Mycetocola* strain on the environment as well as on human health. Another problem is that the effectiveness of *Mycetocola* strains in preventing the development of brown blotch was not as high as the suppression of the tolaasin effect *in vitro*. Reduced disease suppression *in vivo* may be partly due to the decreased population size of strains inoculated on *Pleurotus ostreatus*. This could be attributed to the presence of host components detrimental to the strains or to the limitation of nutrients available to the bacterium on the host mycelia. The suppressive bacterium may be useful in future biocontrol system development and/or the construction of genetically modified edible fungi resistant to the disease caused by *Pseudomonas tolaasii*.

The Genus *Okibacterium*

The genus *Okibacterium* with the single species *Okibacterium fritillariae* was proposed to accommodate aerobic coryneform bacteria isolated from plant seeds (Evtushenko et al., 2002).

phytic bacteria and plant parasite nematodes (Hallmann et al., 1997).

Except for the gumming disease caused by "*Corynebacterium agropyri*" (O'Gara, 1916; Cummins et al., 1974) that might correspond to *Plantibacter* spp., there are no examples of plant pathogenic properties in plantibacteria. Strains of "*Corynebacterium agropyri*" are phenotypically similar to *Rathayibacter* (Cummins et al., 1974; Murray, 1986) and *Plantibacter* strains. "*Corynebacterium agropyri*" CS 35 (T. D. Murray, CA-1), investigated in greater detail, had DAB in the cell wall and clearly differed from *Rathayibacter* in its allozyme pattern (Riley et al., 1988). The strain showed specific adhesion to *Anguina* species, but pathogenicity was not observed (Riley et al., 1988). Like "*Plantibacter elymi*," strains of "*Corynebacterium agropyri*" were associated with wheat grasses and isolated from *Pascopyrum smithii* (Rydb.) A. Love (syn. *Agropyron smithii* Rydb.) or from 30- to 40-year-old herbarium specimens of the same plant and *Elymus trachycaulus* (Link) Gould ex Shinners [syn. *Agropyron trachycaulum* (Link) Malte] and *Elymus lanceolatus* (Scribn. and J. G. Sm.) Gould (syn. *Agropyron riparium* Scribn. and J. G. Sm.) (O'Gara, 1916; Cummins et al., 1974; Murray, 1986).

Isolation

Plantibacter spp. can be isolated from plants and plant-related sources as indicated in the subsection Habitat in this Chapter. The approaches described by Behrendt et al. (1997), Hallmann et al. (1997), and Zinniel et al. (2002) may be applied for isolation (see the section The Genus *Clavibacter* in this Chapter and the subsection Isolation in this Chapter). CNS medium (Vidaver and Davis, 1988) or a modified CNS medium without lithium chloride was successfully used for isolation of endophytic bacteria, including plantibacteria (Zinniel et al., 2002). The procedure of preparing a sample for isolation of bacteria from plant galls is described in the section The Genus *Agreia* in this Chapter. The ground gall suspension is plated onto an appropriate medium, e.g., NWG agar, tenfold diluted CB medium (Evtushenko et al., 1994; Evtushenko et al., 2001), or 2AR agar (Behrendt et al., 1997). Sterilization of gall surface reduces the amount of Gram-negative bacteria, which also develop on isolation media.

Identification

Plantibacteria clearly differ from their phylogenetic neighbor, *Okibacterium*, in DAB-based peptidoglycan and the major menaquinone MK-10, or MK-10 combined with MK-9 (Tables 1 and 9). Phenotypically, plantibacteria resemble members of the genus *Rathayibacter* and might be confused with organisms of this genus, which are also mesophilic, have the same cell size and shape, peptidoglycan of the B2γ type, the major menaquinone MK-10 (Table 1), and association with grasses and nematodes of the subfamily Anguininae (Riley and Ophel, 1992; Zgurskaya et al., 1993b; Dorofeeva et al., 2002). FT-IR spectroscopy, as a physicochemical whole-cell fingerprint technique (Oberreuter et al., 2002), is extremely helpful to cluster members of these genera at the phenotypic level (Behrendt et al., 2002). *Plantibacter* strains differ from *Rathayibacter* spp. in the composition of menaquinones. Plantibacteria produce cytochrome oxidase aa_3, while *Rathayibacter* spp. synthesize mainly cytochrome oxidase bb_3 or both bb_3 and aa_3, depending on growth conditions (Trutko et al., 2003). *Plantibacter* spp. may differ from *Rathayibacter* in the pattern and total amount of polyamines (Altenburger et al., 1997). Besides, *Rathayibacter* and *Plantibacter* species that are both associated with anguinids usually inhabit plant galls induced by different nematodes and, respectively, occur in different plants (Table 13).

The oxidase test, growth at 6% NaCl, composition of menaquinones, the G+C content of DNA, and some other characteristics can be used to identify plantibacteria at the species level. All 12 strains of three species from plant galls are also clearly separated from each other by cell wall sugars, terminal oxidases of respiratory chains, and the FT-IR spectra (L. I. Evtushenko et al., submitted). The species affiliation of strains is consistent with their isolation source, i.e., plant galls on *Agrostis* sp., *Elymus repens* and *Cousinia onopordioides*, induced by highly specialized plant parasitic nematodes, *Anguina agrostis*, *Anguina agropyri* and *Mesoanguina picridis*, respectively (Table 14).

Applications

Colonization of crops with endophytic plantibacteria (Zinniel et al., 2002) suggests that they may be used in future applications, such as for controlling certain plant diseases or promoting plant growth. This approach allows one to obtain plants with improved capabilities without genetic modification of plants (Hallmann et al., 1997). Taking into account the possible role of plantibacters in anguinid-bacterial complexes, it may be supposed that some species possess certain enzymes degrading plant cells and, probably, some antibiotic substances related to their involvement in the nematode-bacterial complexes.

Table 13. Characteristics that differentiate the genera *Plantibacter* and *Rathayibacter*.[a]

Characteristic	Plantibacter	Rathayibacter
Colony color	Yellow	Yellow, pink-orange
Major menaquinone	MK-10 / MK-9, 10 / MK-10,11	MK-10
Cytochrome oxidase	aa_3	bb_3 / bb_3, and aa_3
Quinole oxidase	bo_3 / bb_3 / none	None
Polyamine pattern	Spermine, and 1,3-diaminopropane	Spermine, and spermidine
Polyamine (total amount, $\mu mol \cdot g^{-1}$)	1.3–1.43	4.8–18.0
Plant host/nematode vector[b]	*Agrostis sp.* / *Anguina agrostis*; *Elymus repens* / *Anguina agropyri*; *Cousinia onopordioides* / *Mesoanguina picridis*	*Festuca rubra* / *Anguina graminis*; *Dactylis glomerata* / *Anguina* sp.; *Triticum aestivum* / *Anguina tritici*; *Lolium rigidum* / *Anguina funesta*; *Agrostis avenacea* / *Anguina* sp.; *Polypogon monspeliensis* / *Anguina* sp.; *Carex* sp. / *Heteroanguina caricis* (?)
Other sources	Grass phyllosphere, surface litter after mulching the sward, agronomic crops, and rhizosphere of potato	No data on properly identified rathayibacteria

[a]Members of the genera also differ in FT-IR spectra (Behrendt et al., 2002).
[b]See Tables 14, 15 and the text for details.
Data from Sabet (1954), Gupta and Swarup (1972), Price et al. (1979), Riley (1987), Riley and McKay (1990), Riley and Ophel (1992), Zgurskaya et al. (1993b), Evtushenko et al. (1994; unpublished), Altenburger et al. (1997), McKay et al. (2001), Behrendt et al. (2002), Dorofeeva et al. (2002), Heuer et al. (2002), Zinniel et al. (2002), and Trutko et al. (2003).

Table 14. Characterictics that differentiate *Plantibacter* species.

Characteristic	*P. flavus*	"*P. agrosticola*"	"*P. elymi*"	"*P. cousiniae*"
Fucose in cell wall	ND	–	–	+
Oxidase test	–	+	+	+
Cytochrome oxidase	ND	aa_3	aa_3	aa_3
Quinole oxidase	ND	bb_3	bo_3	–
Heme *O*	ND	–	+	–
Major menaquinone	MK-10,11	MK-9, 10	MK-9, 10	MK-10
Iso-16:0 Me (%)[a]	–	1,53	3,74	2,89
Hydrolysis				
Starch	d	d	+	+
Hypoxanthine	ND	–	d	+
Xanthine	ND	–	d	+
Tween 80	+	+	d	–
Methyl red test	ND	–	d	+
Growth at 6% NaCl	–	d	d	+
G+C (mol%)	68–70	67.2	66.5	67.8
Source	Grass phyllosphere, and surface litter after mulching the sward	*Agrostis* sp.[b]	*Elymus repens*	*Cousinia onopordioides*
Location of gall	ND	Seed	Stem	Leaf
Nematode vector	ND	*Anguina agrostis*	*Anguina agropyri*	*Mesoanguina picridis*

Symbols and abbreviations: +, positive reaction; –, negative; d, different among strains; ND, not determined.
[a]Inferred on the basis of retention time.
[b]Most probably *Agrostis capillaris* L. (S. A. Subbotin, personal communication).
Data from Evtushenko et al. (1994; unpublished), Behrendt et al. (2002), and Trutko et al. (2003).

The Genus *Rathayibacter*

The genus *Rathayibacter* (Zgurskaya et al., 1993b), previously referred to the genus *Clavibacter* (Davis et al., 1984b), was proposed to accommodate some plant pathogenic species containing DAB as diamino acid in the cell wall peptidoglycan and unsaturated menaquinone MK-10.

The first documented species of the genus were *Rathayibacter rathayi* (originally, "*Aplanobacter rathayi*") and *Rathayibacter tritici* (formerly "*Pseudomonas tritici*") described by Smith (1913) and Hutchinson (1917), respectively. The bacteria were reclassified several times and then allocated to the genus *Corynebacterium* as separate species (Cummins et al., 1974). The third species of the genus, "*Corynebacterium iranicum*," was described by Scharif (1961). However, only *Corynebacterium rathayi* was included in the Approved List (Skerman et al., 1980). Subsequently, Carlson and Vidaver (1982) supported the separate species status of *Corynebacterium iranicum* and *Corynebacterium tritici* on the basis of cellular protein analysis, which also corresponded to the phenotypic study results reported by Dye and Kemp (1977), bacteriocin production (Gross and Vidaver, 1979b), and also menaquinone patterns (Collins and Jones, 1980a). Although the organisms differed from *Corynebacterium sensu stricto* in the cell wall peptidoglycan and lipid composition (Schleifer and Kandler, 1972; Fiedler et al., 1973b; Collins and Jones, 1980a), the generic name *Corynebacterium* was retained for the purpose of convenience and practicality and the species were validated under this generic name (Carlson and Vidaver, 1982).

Davis et al. (1984b) established the genus *Clavibacter* for plant pathogenic bacteria with DAB in the cell and, together with *Corynebacterium michiganensis* and the plant pathogens causing ratoon stunt disease (currently *Leifsonia*), reclassified *Corynebacterium rathayi*, *Corynebacterium tritici* and *Corynebacterium iranicum* into *Clavibacter*. Zgurskaya et al. (1993b) proposed that these three species and *Clavibacter* sp., known as the annual raygrass toxicity (ARGT) bacteria (Stynes and Bird, 1980; Stynes and Bird, 1983; Riley, 1987), be reclassified into the newly established genus *Rathayibacter*. The proposal was based on differences between these organisms and *Clavibacter sensu stricto* in predominant menaquinones, morphological and physiological characteristics, DNA-DNA relatedness, and data provided earlier, e.g., on chemotaxonomy (Minnikin et al., 1978; Collins and Jones, 1980a; Collins, 1983a), physiology (Dye and Kemp, 1977), serology (Riley, 1987), and allozyme (Riley et al., 1988) and whole cell protein patterns (Carlson and Vidaver, 1982). The separation was also consistent with the affiliation of individual bacterial groups to host plants (Dye and Kemp, 1977) and associated diseases as well as to nematode vectors (Riley, 1987; Riley and McKay, 1990; Riley and Ophel, 1992). Subsequently, separation of *Rathayibacter* spp. from *Clavibacter* was supported by the analyses of 16S rRNA gene sequences (Rainey et al., 1994; Takeuchi and Yokota, 1994). In 1998, Sasaki et al. reclassified the ARGT bacteria, which had been validly described by Riley and Ophel (1992) as *Clavibacter toxicus*, into *Rathayibacter*. Recently, two new species, *Rathayibacter festucae* and *Rathayibacter caricis*, have been added to the genus (Dorofeeva et al., 2002).

Phylogeny

Members of the genus *Rathayibacter* form a tight cluster within the Microbacteriaceae phylogenetic branch, which is linked to the genus *Curtobacterium* cluster (Fig. 1). The species *Rathayibacter toxicus* branches mostly distant from all other species comprising the genus.

Taxonomy

The genus *Rathayibacter* currently includes six species: *Rathayibacter rathayi* (Smith, 1913), *Rathayibacter tritici* (Carlson and Vidaver, 1982), *Rathayibacter iranicus* (Carlson and Vidaver, 1982) Zgurskaya et al. 1993b, *Rathayibacter toxicus* (Riley and Ophel, 1992) Sasaki et al. 1998, *Rathayibacter caricis*, and *Rathayibacter festucae* (Dorofeeva et al., 2002).

Members of the genus are characterized by nonsporeforming, nonmotile, irregular rods (2.0–2.5 μm × 0.5–0.8 μm), which occur singly or in pairs with diphtheroid arrangements. Primary branching is not observed. In a 5–7 d culture, rods are replaced by cocco-bacillaric forms, single or arranged in pairs, short chains or clumps. Some fresh isolates exhibited a visible rod-coccoid cycle. Strains are aerobic, catalase positive, and usually show a negative or weakly positive oxidase reaction with the tetramethyl-*p*-phenylenediamine test. Thermal optimum is 24–26°C, no growth occurs at 7 or 37°C. Peptidoglycan is of the B2γ type (Schleifer and Kandler, 1972) on the basis of L-2,4-diaminobutyric acid (L-DAB; Sasaki et al., 1998) and contains the amino acids glycine, glutamate, DAB, and alanine in a molar ratio close to 1:1:2:1. Cell wall sugars are glucose, rhamnose, and mannose. Xylose, galactose and fucose may also occur in some species. No mycolic acids are present. The major menaquinone is MK-10, and MK-8, MK-9 and MK-11 may be detected as minor components. Principal phospholipids are phosphatidylglycerol and diphosphatidylglycerol, and traces of another unidentified phospholipid with a low R_f value may be revealed. Saturated *anteiso-* and *iso-*methyl branched fatty acids are predominant (Zgurskaya et al., 1993b; Sasaki et al., 1998; Dorofeeva et al., 2002).

Consistent with their affiliation to host plants and nematode vectors (Table 15), the differentiation between the species is based on cultural, morphological, physiological and chemotaxonomic characteristics.

Table 15. Characteristics differentiating *Rathayibacter* species.

Characteristics	R. caricis	R. festucae	R. rathayi	R. tritici	R. iranicus	R. toxicus
Colony color	**Yellow**	**Rose-orange**	**Yellow**	**Yellow**	**Yellow**	**Yellow**
Visible growth (days)	2	2	2	2	2	4
Cell wall sugars						
Glucose	+	+	+	+	+	+
Galactose	−	−	+	+	−	−
Mannose	+	+	+	+	+	+
Rhamnose	+	+	+	+	+	+
Xylose	+	+	+	+	−	−
Fucose	+	−	−	−	−	−
Methyl red test	+	+	−	−	−	−
Voges-Proskauer	+	+	−	−	−	−
Utilization of						
Dulcitol	+	+	−	−	−	−
meso-Inositol	+	+	−	−	−	−
Inulin	+	+	−	+	+	−
Melibiose	+	+	−	−	−	−
meso-Erythritol	−	+	−	−	−	−
Salicin	+	+	−	+	+	−
Sorbitol	+	+	−	+	−	−
Hydrolysis						
Tween 60	−	+	+	+	+	−
Tween 80	−	+	−	+	+	−
Growth at 5% NaCl	+	−	−	+	−	−
0.03% Tl-Acetate	−	−	+	+	+	−
G+C (mol%)	68	68	67	69	66	60
Source, plant[a]	Carex sp.	Festuca rubra	Dactylis glomerata	Triticum aestivum	Triticum aestivum	Lolium rigidum[b]
Nematode vector[a]	Heteroanguina caricis?[c]	Anguina graminis	Anguina sp.[d]	Anguina tritici	Anguina tritici	Anguina funesta[b]

Symbols: +, positive reaction; and −, negative reaction.
[a]Data from Sabet (1954), Gupta and Swarup (1972), Price et al. (1979), Vidaver (1982), Riley and McKay (1990), and Riley and Ophel (1992).
[b]The *Rathayibacter toxicus*-like strains were isolated from *Agrostis avenacea* and *Polypogon monspeliensis* infested with an *Anguina* sp. which differs from *Anguina funesta* (McKay et al., 2001).
[c]No nematode infestation of *Carex* sp. was observed, but the sedge species are known to be host plants for *Heteroanguina caricis* (Krall, 1991).
[d]*Anguina* sp. parasitizing *Dactilis glomerata* is phylogenetically distant from *Anguina agrostis* sensu stricto and belongs to a new undescribed species (Powers et al., 2001; S. A. Subbotin, personal communication).
Data from Dye and Kemp (1977), Riley and Ophel (1992), Zgurskaya et al. (1992), Sasaki et al. (1998), and Dorofeeva et al. (2002).

Habitat and Phytopathogenicity

The main habitats of *Rathayibacter* spp., most of which are plant pathogens and associated with plant parasitic nematodes of the subfamily Anguininae, are considered to be their host plants and related sources (Table 15). A representative of *Rathayibacter festucae* was found in a leaf gall of fescue (*Festuca rubra* L., the family Poaceae), and *Rathayibacter caricis* was associated with *Carex* sp. belonging to the family Cyperaceae; there are no data on plant pathogenic properties of these two species (Dorofeeva et al., 2002). Several strains identified as *Rathayibacter* sp. were isolated together with *Clavibacter michiganensis* subsp. *michiganensis* (Labadie and Hebraud, 1997). "*Rathayibacter biopuresis*" was obtained from the gut of the marine worm *Notomastus lobatus* (Wong et al., 1996; Wong et al., 1999). There are no phylogenetic data supporting the taxonomic position of these strains.

GUMMING DISEASE CAUSED BY *RATHAYIBACTER* SPECIES The plant pathogens *Rathayibacter tritici*, *Rathayibacter rathayi*, *Rathayibacter iranicus* and *Rathayibacter toxicus* cause a gumming disease, which particularly affects inflorescences and usually is characterized by yellow bacterial slime on developing seedheads, stems and leaves of a plant host (Sabet, 1954; Scharif, 1961; Gupta and Swarup, 1972; Vidaver, 1982; Bradbury, 1986; Riley and Ophel, 1992). *Rathayibacter tritici* and *Rathayibacter iranicus* occur in wheat (*Triticum aestivum* L.), and *Rathayibacter rathayi* is associated with orchard grass (*Dactylis glomerata* L.; Cummins et al., 1974; Carlson and Vidaver, 1982; Collins and Bradbury, 1992). Some weed grasses were reported to be natural hosts of bacteria referred to as "*Rathayibacter tritici*" (Collins and Bradbury, 1986). However, there are no phylogenetic data supporting the taxonomic position of these bacteria. Experimentally, some Poaceae plants have been successfully inoculated by *Rathayibacter tritici* and *Rathayibacter rathayi* using the nematode *Anguina tritici* as vector (Collins and Bradbury, 1986). *Rathayibacter toxicus* inhabits ryegrass (*Lolium rigidum* Gaud.) and probably some related grasses and is responsible for annual rye grass toxicity (Bird and Stynes, 1977; Riley and Ophel, 1992; Johnston et al., 1996). Pathogenicity of cultured *Rathayibacter toxicus* on ryegrass has never been shown, although its association with field-collected bacterial galls has been demonstrated (Price et al., 1979; Riley and Ophel, 1991b).

RATHAYIBACTER-ANGUINA DISEASE COMPLEXES Under natural conditions, the gumming disease is transmitted by vector nematodes of the subfamily Anguininae, which are highly specialized parasites of plants where they induce specific galls in various organs (Gupta and Swarup, 1972; Price, 1979; Bird, 1981; Riley, 1987; Krall, 1991; McKay and Ophel, 1993; Evtushenko et al., 1994). Strains of each *Rathayibacter* species were observed or suggested (Sabet, 1954; Gupta and Swarup, 1972; Bradbury, 1986; Riley, 1987; Riley and McKay, 1990; Dorofeeva et al., 2002) to be associated with their specific nematode vector, namely *Anguina tritici*, *Anguina funesta*, *Anguina graminis* or *Anguina agrostis* (Southey et al., 1990; Krall, 1991; Table 15). It should be noted that *Anguina agrostis* (Steinbuch, 1799), reported to infest several plants, is a heterogeneous species (Southey et al., 1990). "*Anguina agrostis*" parasitizing *Dactylis glomerata* (the host plant of *Rathayibacter rathayi*) belongs to a new undescribed nematode species (Powers et al., 2001; S. A. Subbotin, personal communication). The bacteria transmitted by the nematode *Anguina funesa* (synonymized previously by some authors with *Anguina agrostis*) and cited in some earlier publications as *Corynebacterium rathayi*, *Clavibacter rathayi*, *Corynebacterium* sp., or *Clavibacter* sp. (e.g., Cummins et al., 1974; Bird et al., 1980; Stynes and Bird, 1980; Stynes and Bird, 1983; Bird, 1981; Bird, 1985; Vidaver, 1982; Murray, 1986) are actually strains of *Rathayibacter toxicus* or closely related organisms.

Bacteria assigned to *Rathayibacter toxicus* are not genetically homogeneous (Riley et al., 1988; Riley and Ophel, 1992). Strains from *Agrostis*

avenacea J. F. Gmel. and *Polypogon monspeliensis* (L.) Desf. were indistinguishable from *Rathayibacter toxicus* in colony morphology, serological reactions, and bacteriophage specificity, but were considerably different in allozyme pattern. In addition, they were associated with an *Anguina* sp. which differs from *Anguina funesta*, a vector for *Rathayibacter toxicus* (McKay et al., 1993; Powers et al., 2001).

Rathayibacter rathayi was reported to be transferred experimentally by *Anguina tritici* into *Triticum* sp., a host of *Rathayibacter tritici*, and like *Rathayibacter tritici*, to cause a disease in this plant (Sabet, 1954). *Triticum dicoccum*, *T. durum* and *T. pyramidale* were successfully inoculated by *Rathayibacter tritici* using the nematode *Anguina tritici* (Collins and Bradbury, 1986). *Rathayibater rathayi* and *Rathayibacter tritici* are transmitted experimentally by *Anguina funesta* (the nematode vector for *Rathayibacter toxicus*) and colonize ryegrass without doing harm to the plant (Riley and Yan, 1999).

No visible symptoms of nematode infestation were observed for *Carex* sp. from which *Rathayibacter caricis* has been isolated (Dorofeeva et al., 2002). However, knowledge of habitats and biology of other rathayibacteria and anguinids suggests that *Rathayibacter caricis* is also associated with a nematode of the subfamily Anguininae. The sedge species are known to be host plants for *Heteroanguina caricis*, belonging to the same subfamily Anguininae (Krall, 1991), and this nematode may be the vector for *Rathayibacter caricis*.

Relationship Between Components in the Rathayibacter-Anguina *Disease Complexes* Little is known about the relationship between components in the *Rathayibacter-Anguina* plant pathogenic complexes, including also bacteriophages (Bird et al., 1980; Riley and Gooden, 1991a; Ophel et al., 1993). The available data suggest that this association is of mutual benefit for *Rathayibacter* and nematode and may be regarded as symbiosis. Nematodes carry bacteria on their surface into the growing points of host plants (Price et al., 1979; Bird, 1981; Bird, 1985; Riley and McKay, 1990) and create wounds in plants, providing avenues for the bacteria to enter tissue and ensuring the leakage of plant nutrients, supporting external bacterial colonization (Hallmann et al., 1997). The bacteria inhabit also many galls induced by nematodes in seeds, leaves, stems, or roots (Price, 1979; Bird, 1981; Krall, 1991; Ophel at al., 1993; Evtushenko et al., 1994). The bacteria within desiccated mature galls are resistant to environmental perturbations (isolated from galls after having been dry for several years; Bird, 1981). In turn, nematodes seem to depend in their life cycles on the associated bacteria. The bacteria are supposed to facilitate the development of nematode larvae within the galls and the release of nematodes, destroying gall envelope by fermentative activities at suitable environmental conditions (Bird, 1981; Bird, 1985). As in case of other nematode-associated bacteria (Akhurst, 1982), they may protect nematode larvae from certain bacterial and fungal infections by producing antimicrobial substances, e.g., bacteriocins (Gross and Vidaver, 1979b) or enzymes (De Bruyne et al., 1992). It is unknown whether the toxins produced by *Rathayibacter toxicus* and related bacteria have any effect on the nematode (Agrios, 1997).

Corynetoxin Produced by Rathayibacter Toxicus *and Some Related Bacteria* Corynetoxin is produced by *Rathayibacter toxicus* when the bacterium is infected with a bacteriophage and produces bacteriophage structural proteins (Ophel et al., 1993). The mechanism by which the bacteriophage affects toxin production is unknown. A small bacteriophage genome could not encode both the gene for toxin biosynthesis and the genes for bacteriophage replication and survival. The bacteriophage may encode a regulatory gene that switches on toxin production or a gene encoding a final step in the corynetoxin biosynthetic pathway. Alternatively, it is possible that bacteriophage infection makes cells more permeable to export of toxin (Ophel et al., 1993). Bacteriophages with similar DNA restriction patterns were also isolated from the toxic *Rathayibacter toxicus*-like strains associated with *Anguina* sp., parasitizing *Agrostis avenacea* and *Polypogon monspeliensis* (McKay et al., 1993).

The corynetoxin is chemically and biologically similar to tunicamycin (Vogel et al., 1982; Cheeke, 1995), which inhibits N-linked protein glycosylation (Jago et al., 1983; Crosti et al., 2001). Tunicamycin is active against fungi, yeasts, Gram-positive bacteria, especially *Bacillus* sp., and viruses (Korzybski et al., 1978). Corynetoxin causes defective formation of blood components of the reticulo-endothelial system and is harmful to grazing animals (Finnie and Mukherjee, 1986; Cheeke, 1995; Finnie, 2001). The corynetoxin may play a role in plant pathogenicity. Tunicamycin, together with brefeldin A, has recently been proved able to induce a strong acceleration of the plant cell death, changes in cell and nucleus morphology, and an increase in DNA fragmentation. These compounds, interfering with the activities of endoplasmic reticulum and of Golgi apparatus, strongly induce a form of programmed plant cell death showing apoptotic features (Crosti et al., 2001).

Other Rathayibacter-*Like Bacteria Associated with Anguinid Nematodes* The *Rathayibacter*-like bacteria isolated from plant galls induced by *Mesoanguina pictidis* on *Cousinia onopordialis* Ledeb. (family Asteraceae) and *Anguina agropyri* on *Elymus repens* (L.) Gould (Evtushenko et al., 1994) were later shown to be members of the genus *Plantibacter* (Shramko et al., 1999; Evtushenko et al., submitted). Isolates of "*Corynebacterium agropyri*" (O'Gara, 1916; Cummins et al., 1974; Murray, 1986), isolated from wheat grasses and shown to have specific adhesion to *Anguina* species (Riley and McKay, 1990), are likewise not members of *Rathayibacter* as they differ significantly in the allozyme pattern (Riley et al., 1988), but they might correspond to *Plantibacter* (see the section The Genus *Plantibacter* in this Chapter). The discovery of unidentified bacteria in *Festuca rubra* subsp. *commutata* infested with the nematode *Anguina* sp., showing toxicity for livestock, was reported by Galloway (1961) and Bird (1981). There are no data showing whether the above bacteria belong to *Rathayibacter toxicus*, *Rathayibacter festucae*, or represent new species of *Rathayibacter* or other genera. Similar bacteria may be responsible for the toxicity of *Festuca rubra*, infested with *Anguina* sp. ("*Anguina agrostis*"), for grazing animals, as observed between 1943 and 1949 (Bird, 1981).

Isolation

To isolate *Rathayibacter* spp. from diseased plants, certain common rules on the isolation of many plant pathogenic bacteria should be followed to reduce contamination with other bacteria that also inhabit plants. For successful isolation, bacteria-containing parts of plants (visible bacterial slime) are used. Developing seedheads and leaves are selected in case the plants are determined to be in the earlier phase of bacterial (nematode) infection. The selected piece of tissue should be carefully excised with sterile instruments, placed in sterile water, and left for a period of few minutes to hours to release the bacteria. Plant galls inhabited by bacteria are also a suitable source of isolation. The procedure of preparing a sample for isolation of bacteria from plant galls is described in the section The Genus *Agreia*. The obtained suspensions of bacteria or plant (gall) material are plated onto a suitable medium. *Rathayibacter* spp. can be isolated on the media recommended for isolation of *Clavibacter*. CNS agar medium (see the section The Genus *Clavibacter* in this Chapter) was reported to be helpful for isolating *Rathayibacter tritici* (Gross and Vidaver, 1979a). A number of media based on glucose, peptone, yeast extract, and casamino acids are suitable for growing rathayibacteria. *Rathayibacter toxicus* grows well on 523M agar medium (Riley and Ophel, 1993), CB (Zgurskaya et al., 1993b), and R medium (Yamada and Komagata, 1972a; Sasaki et al., 1998).

Identification

GENERIC LEVEL *Rathayibacter* spp. form a coherent phylogenetic cluster and novel isolates can be easily attributed to the genus on the basis of partial 16S rRNA gene sequences or 16S rDNA restriction analysis (Lee et al., 1997b; Shramko et al., 1999). Rathayibacteria can also be differentiated from most genera of the family Microbacteriaceae at the phenotypic level by the presence of DAB in the cell wall and the major menaquinone MK-10 (Table 1). Phenotypically, the *Rathayibacter* species resemble organisms of the genus *Plantibacter*. They can be readily differentiated from plantibacteria by FT-IR spectra (Behrendt et al., 2002). Menaquinone composition (Behrendt et al., 2002; Evtushenko et al., submitted), amount and pattern of polyamines (Altenburger et al., 1997), and terminal cytochrome oxidases bb_3 or bb_3 combined with aa_3 (Trutko et al., 2003) are additional useful diagnostic traits (Table 13).

SPECIES LEVEL At the species level, *Rathayibacter* strains are relatively easy to identify when isolated from a known plant host with clear disease symptoms and when the nematode vector is known. Species affiliation of strains is achieved by color of colonies, growth rate, cell wall sugar composition, and some physiological properties (Table 15), as well as allozyme profiles (Riley et al., 1988) and 16S rDNA RFLP patterns (Lee et al., 1997b; Shramko et al., 1999).

The immunodiffusion technique is useful to differentiate *Rathyibacter toxicus* from *Rathayibacter tritici*, *Rathayibacter rathayi* and *Rathayibacter iranicus* as well as *Rathayibacter iranicus* from *Rathayibacter tritici* and *Rathayibacter rathayi*, although serum against *Rathayibacter iranicus* gave a reaction with some *Rathayibacter tritici* strains (Riley, 1987). Polyclonal ELISA tests have been applied to detection of *Rathyibacter toxicus* in paddocks and hay without any problems of crossreactivity with other species (Riley, 1987; McKay and Ophel, 1993). However, other toxic strains from *Agrostis avenacea* and *Polypogon monspeliensis*, which might not be members of *Rathayibacter toxicus* (as derived from different allozyme patterns), were reported to be indistinguishable from *Rathyibacter toxicus* in serological reactions (McKay et al., 1993). Plant pathogenicity testing may be difficult, since nematodes are necessary vectors for transmission of infection by *Rathyibacter* pathogens

(Vidaver and Davis et al., 1988), and crossreactions are possible (Sabet, 1954). Cases where rathayibacteria are isolated from an unusual host plant, plant galls induced by anguinids not included in Table 15, or any other unusual source may represent a new species and should be subjected to thorough taxonomic analysis.

Applications

Much of the present annual ryegrass toxicity management has focused on the control of ryegrass (Riley and Yan, 1999). Control by herbicides is often ineffective as it is associated with a rapid increase in herbicide resistance. The recent finding that some strains of *Rathayibater rathayi* and *Rathayibacter tritici* may be transmitted experimentally by *Anguina funesta* (the nematode vector for *Rathayibacter toxicus*) and may colonize ryegrass, displacing *Rathayibacter toxicus*, offers a potential for novel biocontrol (Riley and Yan, 1999). The physiological and enzymatic activities of rathayibacteria (De Bruyne et al., 1992) as well as the ability to produce some antibiotic substances (Gross and Vidaver, 1979b) demonstrate that they may be useful for screening of metabolites and enzymes, which are applicable for biotechnology and pharmaceuticals. A strain identified as *Rathayibacter* sp. produced cephalexin-chloroperoxidase and other enzymes and use of this strain in a new biomethod for producing cefaclor, an antibiotic of the cephalosporin class, was claimed (Chen and Wong, 1997; Wong et al., 1999). Collagenolytic strains identified as *Rathayibacter* sp. produce an enzyme that is able to degrade native collagen, gelatin and probably other proteins from plants, sharing sequence homologies with collagen; it may be of major importance for degradation of the collagen of different plant parasitic nematodes (Labadie and Hebraud, 1997).

The Genus *Subtercola*

The genus *Subtercola*, including *Subtercola boreus* (the type species) and *Subtercola frigoramans*, was established by Männistö et al. (2000) for aerobic psychrophilic coryneform bacteria isolated from cold groundwater. Another species, *Subtercola pratensis* proposed by Behrendt et al. (2002), has recently been reclassified to the genus *Agreia* (Schumann et al., 2003) and is considered in the respective section.

Phylogeny

Organisms of the genus form a distinct phylogenetic branch within the family Microbacteriaceae cluster. The phylogenetic neighbor of the genus

Subtercola as inferred from 16S rRNA gene sequence analysis is the genus *Agreia* (Fig. 1).

Taxonomy

Members of the genus form pale to bright yellow, circular, convex, and smooth colonies. Cells are nonmotile, irregular, thin (0.2–0.3 μm) rods, single or as v-forms. Endospores are not produced. The bacteria grow at temperature 2–28°C, with the optimum at 15–17°C. The characteristic diamino acid of cell wall peptidoglycan is L-DAB and the peptidoglycan type is B2γ. The predominant menaquinones are MK-9 and MK-10. The whole cell sugars are rhamnose, ribose, xylose and mannose or may be glucose, rhamnose and xylose, depending on the strains. The polar lipids are phosphatidylglycerol, diphosphatidylglycerol, one unknown phospholipid, and two glycolipids. The main cellular fatty acids are 12-methyl-tetradecanoic acid (*anteiso*-15:0), 14-methylpentadecanoic acid (*iso*-16:0) and 14-methyl-hexadecanoic acid (*anteiso*-17:0). The whole-cell methanolysates of *Subtercola* sp. contain also 1,1-dimethoxy-*anteiso*-pentadecane (*anteiso*-15:0 DMA) and 1,1-dimethoxy-*iso*-hexadecane (*iso*-16:0 DMA) in amounts comparable to those of fatty acids. The G+C content of DNA is 64–68 mol% (Männistö et al., 2000).

The genus contains two species, *Subtercola boreus* and *Subtercola frigoramans*, which differ in the 16S rRNA gene sequence and in phenotypic characteristics (Männistö et al., 2000; see subsection Identification in this Chapter).

Habitat

Subtercola boreus and *Subtercola frigoramans* were isolated from the groundwater with a low stable temperature of about 7°C, pumped from a shallow aquifer (18 m deep) located under a glacial gravel ridge in Southern Finland. The water was highly humic (with dissolved organic carbon, 4–13 mg/liter) and rich in iron (15 mg/liter; Männistö et al., 2000).

The study of lipid composition in *Subtercola* has revealed an interesting effect of temperature on cellular fatty acid compositions, which may be related to its psychrophilic mode of life. Decreases in temperature influenced both fatty acid and dimethyl acetal (DMA) compositions of the strains. When growth temperature was lowered from 25 to 4°C, the proportion of *anteiso*-15:0 DMA, the main DMA in the whole-cell methanolysates, increased from 10–11 to 14–20%. Accordingly, the *anteiso*-17:0 and *anteiso*-15:0 contents were reduced and the amounts of *anteiso*-17:0 DMA and *iso*-16:0 DMA fell (Männistö et al., 2000). The authors argued that the low growth temperature favored the synthe-

sis of DMAs with shorter carbon chain lengths, a shift from *iso*-branched DMAs to *anteiso*-branched congeners, or both. The biological function of plasmalogens is not known, but Kaufman et al. (1990) showed that membranes of *M. elsdenii* (with a high plasmalogen content) were more ordered than membranes devoid of plasmalogens. This may indicate that changes in the plasmalogen derived 1,1-dimethyl acetal composition may influence the membrane fluidity (Männistö et al., 2000).

Isolation

The strains were isolated on PYGV agar used earlier for isolation of *Ancalomicrobium* and *Prosthecomicrobium* (Staley, 1968) at 7°C. The cultures grow also on trypticase soy (TS) agar or R2A agar (BBL Microbiology Systems, Cockeysville, MD, United States). The optimum growth is at 15–17°C.

Identification

Novel isolates can be preliminarily identified as members of the genus *Subtercola* by their unique combination of chemotaxonomic characteristics (Tables 1 and 9). The most striking properties, which differentiate *Subtercola* from other genera of the family Microbacteriaceae, are the growth temperature optimum, the presence of DAB and hydroxyglutamic acid in the cell wall peptidoglycan, along with fatty acid composition and large amounts of *anteiso*-1,1-dimethoxy-pentadecane (*anteiso*-15:0 DMA) and *iso*-1,1-dimethoxyhexadecane (*iso*-16:0 DMA) in whole-cell methanolysates (Männistö et al., 2000). Bacteria of the genus *Subtercola* differ from their closest phylogenetic neighbor, the genus *Agreia*, by the major menaquinones MK-9 and MK-10, lack of ornithine in the cell wall, and some other properties.

The type strains of the two species of *Subtercola* are significantly different in their 16S rRNA gene sequence (97% similarity), and most nucleotide changes (in about 30 positions) are located at the 5' 400–500 nucleotides (*E. coli* numbering). This difference can be used to preliminarily identify the strains of the species *Subtercola boreus* and *Subtercola frigoramans* at the phylogenetic level, using partial sequences. Strains of the two species can be separated by their cell morphology, cell wall sugar composition, the G+C content of DNA, and physiological features (Table 16). *Subtercola boreus* is characterized by irregular, short, slightly curved rods, 0.2–0.3 × 0.6–1.0 μm, while the rod-shaped cells of *Subtercola frigoramans* are larger and more pleomorphic, 0.3–0.4 × 0.9–1.5 μm. Young cells of *Subtercola frigoramans* can also be swollen at the

Table 16. Characteristics that differentiate two species of the genus *Subtercola*.

Characteristic	S. boreus	S. frigoramans
Assimilation of		
p-Arbutin	+	−
L-Rhamnose	−	+
D-Ribose	−	+
Salicin	+	−
i-Inositol	−	+
Citrate	−	+
3-Hydroxybenzoate	−	+
Acetate	−	+
Propionate	−	+
cis-Aconitate	−	+
Hydrolysis of		
pNP-β-D-glucuronide		−
pNP-β-D-xylopyranoside	+	−
bis-pNP-phosphate	−	+
DNA G+C (mol%)	64	68

Abbreviation: pNP, p-nitrophenyl.

[a]The species *Subtercola pratensis* is considered in the section The Genus *Agreia*. Data from Männistö et al. (2000).

pole or in the middle. The type strain of the species *Subtercola boreus* has rhamnose, ribose and small amounts of xylose and mannose in the cell wall, whereas the type strain of *Subtercola frigoramans* is characterized by glucose, rhamnose and small amounts of xylose. The latter species exceeds *Subtercola boreus* in the spectrum of assimilated carbon sources (Männistö et al., 2000).

Other Microorganisms of the Family Microbacteriaceae

Besides the microorganisms belonging to the 15 validly described genera reviewed in the previous sections, a large diversity of other bacteria of the family Microbacteriaceae with uncertain taxonomic positions has been revealed and described invalidly. In this section, representatives of some bacterial groups of the family are considered, which form separate lines of descent and may be nuclei of new genera. Some of such bacteria have been known for a long time and are well characterized, while others have been discovered only recently and/or only data on their 16S rRNA gene sequences are available from public databases. The list of bacteria with uncertain generic position considered in the section is far from being complete. The data presented aim only to demonstrate additional diversity, features and habitats of microorganisms related to the family Microbacteriaceae which have not been described or were mentioned only in passing in the previous sections.

"Brevibacterium helvolum" Group

"Brevibacterium helvolum" DSM 20419 (X77440) was isolated from butter and has been studied by many authors. The strain has peptidoglycan B2γ, [L-DAB]-D-Glu-D-DAB (Schleifer and Kandler, 1972; Fiedler et al., 1973b), where the proportions of L- and D-DAB are almost equal (Sasaki et al., 1998). The major menaquinone of this strain is MK-9 (Sasaki et al., 1998) and the G+C content is 72.2 mol% (Döpfer et al., 1982). The strain is chemotaxonomically close to the genus *Clavibacter* but distant from it and other genera at the phylogenetic level and supposed to be the nucleus of a new genus (Rainey et al., 1994). Four other strains of "Brevibacterium helvolum," which contain the B2γ peptidoglycan based upon DAB (Schleifer and Kandler, 1972), were isolated from oil-brine in Yabase oil-field, Japan (Komagata and Iizuka, 1964), and assigned to the genus *Curtobacterium* along with some true curtobacteria (Yamada and Komagata, 1972b). These four strains slightly differ from "Brevibacterium helvolum" DSM 20419 in the G+C content (65.6–65.9 mol%; Yamada and Komagata, 1972b) and fatty acid composition, which is rather similar to that of *Curtobacterium* spp. (Suzuki and Komagata, 1983). Although some applicable features for strains labelled "Brevibacterium helvolum" were reported, only "Brevibacterium helvolum" AJ 1458 described by Komagata and Iizuka (1964) seems to be related to the group under discussion. The strain was claimed for the method of microbial production of L-serine (Kubota and Shiro, 1979). Many other publications dealing with the application or possible use of "*Brevibacterium helvolum*" concern the strains which are outside the above group and belong to the genus *Arthrobacter* (e.g., DeLey and Defloor, 1959; Yorifuji et al., 1995; Kim et al., 2000). The bacterium "*Bacterium helvolum*" was reported to be responsible for septicemia, pericarditis and pleurisy (Brisou et al., 1970). However, there are no chemotaxonomic or phylogenetic data supporting its taxonomic position.

"*Corynebacterium bovis*" Group

"*Corynebacterium bovis*" ATCC 13722 (AF300651) from cow manure, "*Brevibacterium equis*" (AJ251780) from horse infections and several organisms from environment samples (e.g., AB012595) also compose a separate phylogenetic cluster. The cluster is linked to the "*Brevibacterium helvolum*" DSM 20419 line of descent with a 99% bootstrap confidence level. Like "*Brevibacterium helvolum*," the strain "*Corynebacterium bovis*" ATCC 13722 possesses peptidoglycan based upon DAB, but the structure of the polymer was determined to be [L-Hsr]-D-Gly-D-DAB (Fiedler and Kandler, 1973a; Schleifer and Seidl, 1985). The G+C content of DNA is 73.7 mol% (Döpfer et al., 1982). Further taxonomic studies are required to elucidate whether "*Brevibacterium helvolum*" and "*Corynebacterium bovis*" with related organisms represent a single genus or two different genera.

Microbacteriaceae Diversity Associated with Plants

Numerous examples demonstrate that plants and related sources are typical natural habitats of many bacteria comprising the family Microbacteriaceae. Those include the organisms of the genera *Agreia*, *Clavibacter*, *Curtobacterium*, *Frigoribacterium*, *Leifsonia*, *Microbacterium*, *Okibacterium*, *Plantibacter* and *Rathayibacter*, which were reviewed in the respective sections. Besides the known and many new species of these genera (as may be suggested by 16S rRNA gene sequences available from public databases), there are organisms that most likely represent new genera, in particular, the strain labelled "*Curtobacterium* sp." VKM Ac-2058 (AB042093) and some related organisms from plants and plant galls induced by plant parasitic nematodes of the subfamily Anguininae. The bacteria are characterized by the presence of DAB in the cell wall and the major menaquinone MK-11 (L. I. Evtushenko and M. Takeuchi, unpublished observation).

Microbacteriaceae from Low Temperature Environments

Recent data demonstrate that many bacteria of the family Microbacteriaceae occur in frozen condition in different environments. Such bacteria were probably active in natural habitats many (500–3000) years ago and survived frozen until isolation. Some of them may be attributed to the current genera on the basis of 16S rDNA sequence analyses, while others represent separate lines of descent and undoubtedly belong to new undescribed genera. The bacteria "*Cryocola antiquus*," preliminarily assigned to the genus *Cryobacterium* sp. (Karasev et al., 1998; Kochkina et al., 2001), were isolated from a sample of permafrost sediments 23 m deep, recovered from the frozen (mean annual temperature –12°C) late Pliocene layer (1.8–3 million years old) in the Kolyma Lowland, East Siberia, Russia. The bacteria form yellow or white, round, glistening, opaque colonies. Optimum growth temperature is 18–20°C, and no growth occurs at 28°C. Cells are aerobic, catalase- and oxidase-positive, nonsporeforming, long irregular rods (0.4 × 3.5 µm), which fragment into shorter motile cells. Cell wall peptydoglycan contains DAB, alanine, glycine, and glutamic acid in the molar ratio consistent with B2γ type peptidoglycan (Schleifer and Kandler, 1972). Cell wall sugars include rhamnose, which is predominant in amount. The major isoprenoid quinones are MK-11 and MK-12, and the major fatty acids are *anteiso*-15:0, *anteiso*-17:0 and *iso*-16:0. The DNA G+C content is approximately 65 mol%. The phylogenetic analyses based on 16S rDNA sequences indicated that the strains are clustered together with the genus *Cryobacterium psychrophilum* with an 85% bootstrap confidence level and exhibit 96.3–96.6% 16S rDNA binary sequence similarity to this species. However, the permafrost bacteria are clearly distinguished phenotypically from *Cryobacterium psychrophilum*, which is pink-colored, nonmotile, does not grow at 18°C, has the major menaquinone MK-10, and produces 12-methyltetradecenoic fatty acid in significant amounts (Suzsuki et al., 1996). Phenotypically, "*Cryocola antiquus*" strains resemble some species of the genus *Leifsonia*; however, the 16S rRNA gene sequence similarities between "*Cryocola antiquus*" and most species of *Leifsonia* are 95.5–96.0%, the exception being *Leifsonia poae* with 97.1% similarity (Gavrish et al., 2003). The phylogenetic and phenotypic data evidently demonstrate that the strains labelled "*Cryocola antiquus*" cannot be assigned unambiguously to either of these genera and should be described as a new genus. Several other bacteria, e.g., strains G50-TB8 (AF479355) and G200-C11

(AF479342), representing a separate line of descent within the family Microbacteriaceae or grouped together with other uncertain bacteria were recovered from 50- to 200-year-old glacial ice from various geographic locations (Christner, 2002).

Recently, Sheridan et al. (2003) has validly described the new genus and species *Rhodoglubus vestalii* to accommodate a psychrophilic, red-pigmented strain LV3^T (=ATCC BAA-534^T =CIP 107482^T; AJ459101) isolated from Antarctic Dry Valley Lake near the McMurdo Ice Shelf. The strain had an optimal growth temperature of 18°C and grew on a variety of media between – 2 and 21°C. Scanning electron micrographs revealed small rods with unusual bulbous protuberances in all phases of growth. The strain reportedly has ornithine in the cell wall and the major menaquinones MK-11 and MK-12. The predominant fatty acids were anteiso-$C_{15:0}$, iso-$C_{16:0}$ and anteiso-$C_{17:0}$; cells grown at -2°C contained significant amounts of anteiso-$C_{15:1}$. Phylogenetic analysis of the 16S rRNA gene sequence indicated that *Rhodoglubus vestalii* was related to, but distinct from, organisms of the genera *Agreia*, *Leifsonia* and *Subtercola*. Alignments of 16S rRNA sequences showed that the sequence of *Rhodoglubus vestalii* LV3^T contained a 13 bp insertion that was found in only a few related sequences. The G+C content of DNA was approximately 62 mol%. (Sheridan et al., 2003).

Two other psychrophilic strains from Antarctica, CMS 76r^T (=DSM 15304^T =CIP 107783^T; AJ438585) and CMS 81y^T (DSM 15303^T =CIP 107785^T; AJ438586), have recently been isolated from a cyanobacterial mat sample from a pond in Wright Valley, McMurdo. The strains were curved, non-motile, and catalase-positive rods. They had DL-2,4-diaminobutyric acid in the cell wall, the major menaquinone MK-11, high amounts of anteiso- and iso-branched fatty acids, and a DNA G+C content of 64–66 mol%. The strains differed from one another in many phenotypic characteristics and exhibited 30% DNA relatedness, thus indicating that they represent two different species (Reddy et al., 2003). Unaware of the description of the genus *Rhodoglubus* (Sheridan et al., 2003), Reddy et al. (2003) have described these strains, which are phylogenetically closely related to *Rhodoglubus*, as novel species of the genus *Leifsonia*, *Leifsonia rubra* and *Leifsonia aurea*.

Copiotrophic Ultramicrobacteria and Related Organisms

Extremely small bacteria, called "ultramicrobacteria," "nanobacteria," and "picobacteria," exist in several taxonomic groups of parasites or pathogens and are known to come from a wide variety of natural environments (Koch, 1996). Iizuka et al. (1998) described the 0.45-μm filtered, free-living, aerobic copiotrophic ultramicrobacteria of the family Microbacteriaceae from urbane soil, some of which are members of the genus *Microbacterium* while others are phylogenetically distant from all known genera of the family. The average cell volume of bacteria, as measured by scanning electron microscope and epifluorescent microscopy with image analysis, was less than 0.3 μm^3 at any growth stage. They have menaquinones MK-11 and MK-12 and varied in colony pigmentation (cream to yellow), cell size, doubling time, motility, catalase reaction, protease activity, nitrate reduction, and acid production from glucose. In addition, cells of strain 12-8 (AB008510) were divided by multiple longitudinal and transverse septa that resulted in formation of cells of different size and shape. Such multiseptation of cells is typical of some actinobacteria, e.g., *Blastococcus* (Ahrens and Moll, 1970), *Geodermatophilus* (Luedemann, 1968; Urzi et al., 2001), *Modestobacter* (Mevs et al., 2000), and some others, but was not reported for organisms of the family Microbacteriaceae previously. No phenotypic data are available for the copiotrophic strain 121 (AB008511), which is grouped together with the recently described psychrophilic bacteria *Rhodoglubus vestalii* (Sheridan et al., 2003), *Leifsonia rubra* and *Leifsonia aurea* (Reddy et al., 2003) into a separate phylogenetic cluster in the family Microbacteriaceae.

Small cell size (ca. 0.2–0.4 μm in diameter and 0.6–2 μm in length), which is consistent with the cell volume <0.3 μm^3, is typical of cells in cultures of a number of species in the family Microbacteriaceae, e.g., in the genera *Agromyces* (Casida, 1977), *Frigoribacterium* (Kämpfer et al., 2000), *Leifsonia* (Davis et al., 1984b; Suzuki et al., 1999), *Microbacterium* (Iizuka et al. 1998), *Mycetocola* (Tsukamoto et al., 2001), and *Subtercola* (Männistö et al., 2000). Such cell size in culture suggests that they may pass through a 0.45-μm membrane filter and be observed among ultramicro-size bacteria *in situ*.

Microbacteriaceae from Freshwater Ecosystems and Marine Bacterioplankton

Members of the class Actinobacteria, including ultramicro-size representatives of Microbacteriaceae, widely occur in marine and freshwater ecosystems (Hiorns et al., 1997; Takami et al., 1997; Fritz et al., 2000; Methe et al., 1998; Pernthaler et al., 1998; Pernthaler et al., 2001; Zwart et al., 1998; Zwart et al., 2002; Crump et al., 1999;

Rappe et al., 1999; Glöckner et al., 2000; Männistö et al., 2000; Hahn et al., 2003). Broad distribution, abundance, and high biomass of actinobacteria in limnic ecosystems were demonstrated using a combination of phylogenetic analyses and the fluorescence in situ hybridization (FISH) technique (Glöckner et al., 2000). Amount of actinobacteria in fresh water habitats can be comparable with that of β-Proteobacteria and even can become a dominant fraction. These two groups were suggested to inhabit separate functional niches: β-Proteobacteria are able to rapidly utilize the main annual input, whereas the actinobacterial group more efficiently consumes the lower levels of organic carbon at low temperature (Glöckner et al., 2000). Many actinobacterial clones obtained from freshwater habitats, e.g., CRE-FL60 (AF141476; Crump et al., 1999), FukuN101 (AJ289982; Glöckner et al., 2000); PRD01a002B (AF289150; Zwart et al., 2002), represent phylogenetically distant lineages that most likely correspond to separate undescribed genera of the family Microbacteriaceae. A number of isolated strains of the freshwater Microbacteriaceae lineages are ultramicro-size bacteria (cell volume, <0.1 μm^3; Hahn et al., 2003). Yellow-pigmented ultramicrobacteria with close 16S rRNA gene sequences (99.4–100%) were obtained from lakes that are ecologically different and geographically separated by great distances. They have clearly different ecophysiological and phenotypic traits and are distinguished by colony morphologies, growth rates, and temperature optima. The strains have a typical Gram-positive cell wall architecture, but the overall shape of cells is unusual for Microbacteriaceae: selenoid, and occasionally cells form rings when no dissociation occurs after the septum is formed. The cell width of strain MWH-Ta1 (AJ507466), a representative of yellow-pigmented isolates, generally ranged from 154.2 to 244.4 nm. The red-pigmented strain MWH-Ta3 (AJ507468) has markedly larger cells and exhibits levels of 16S rRNA similarity of 94.5% to the yellow-pigmented bacteria. Predator-prey experiments showed that at least one of the ultramicro-size isolates is protected against predation by the bacterivorous nanoflagellate *Ochromonas* sp. (Hahn et al., 2003). The finding that members of the class Actinobacteria are abundant in freshwater ecosystems is quite intriguing in the light of the general perception that this group including Microbacteriaceae is typical for terrestrial environments and less common in aquatic ecosystems (Williams et al., 1984; Madigan et al., 1997). Further studies of Microbacteriaceae from freshwater and marine habitats will discover additional diversity of Microbacteriaceae and may provide an opportunity to gain more insight into the evolutionary process of terrestrial and aquatic actinobacteria.

Literature Cited

Abd-el-Malek, Y., and T. Gibson. 1952. Studies in the bacteriology of milk. III: The Corynebacterium of milk. J. Dairy Res. 19:153–159.

Abe, S., K. Takayama, and S. Kinoshita. 1967. Taxonomical studies on glutamic acid-producing bacteria. J. Gen. Appl. Microbiol. 13:279–301.

Agrios, G. 1997. Plant Pathology, 4th ed. Academic Press. Toronto, Canada.

Ahrens, R., and G. Moll. 1970. Ein neues Knospendes Bacterium aus der Ostsee. Arch. Microbiol. 70:243–265.

Akhurst, R. J. 1982. Antibiotic activity of Xenorhabdus spp., bacteria symbiotically associated with insect pathogenic nematodes of the families Heterorhabditidae and Steinernematidae. J. Gen. Microbiol. 128:3061–3065.

Alarcon, C., J. Castro, F. Munoz, P. Arce-Johnson, and J. Delgado. 1998. Protein(s) from the Gram-positive bacterium Clavibacter michiganensis subsp. michiganensis induces a hypersensitive response in plants. Phytopathology 88:306–310.

Altenburger, P., P. Kämpfer, A. Makristathis, W. Lubitz, and H.-J. Busse. 1996. Classification of bacteria isolated from a medieval wall painting. J. Biotechnol. 47:39–52.

Altenburger, P., P. Kämpfer, V. N. Akimov, W. Lubitz, and H.-J. Busse. 1997. Polyamine distribution in actinomycetes with group B peptidoglycan and species of the genera Brevibacterium, Corynebacterium and Tsukamurella. Int. J. Syst. Bacteriol. 47:270–277.

Andersson, M. A., E.-L. Laukkanen, F. Nurmiaho-Lassila, S. Rainey, S. Niemela, and M. Salkinoja-Salonen. 1995. Bacillus thermosphaericus sp. nov., a new thermophilic ureolytic Bacillus isolated from air. Syst. Appl. Microbiol. 18:203–220.

Andersson, M. A., N. Weiss, F. A. Rainey, and M. S. Salkinoja-Salonen. 1999. Dustborne bacteria in animal sheds, schools and children's day care centers. J. Appl. Microbiol. 86:622–634.

Anner, D. I. 1970. A note on the preservation of soil amoebas by drying in vacuo. Cryobiology 7:129–131.

Aoki, K., R. Shinke, and H. Nishira. 1982. Identification of aniline-assimilating bacteria. Agric. Biol. Chem. 46:2563–2570.

Araujo, W. L., W. Maccheroni Jr., C. I. Aguilar-Vildoso, P. A. Barroso, H. O. Saridakis, and J. L. Azevedo. 2001. Variability and interactions between endophytic bacteria and fungi isolated from leaf tissues of citrus rootstocks. Can. J. Microbiol. 47:229–336.

Arnaudi, C. 1942. Flavobacterium degydrogenans (Micrococcus degydrogenans) und seine Fahigkite zur Oxidation von Steroiden sowie Substanzen aus der Sexualhormonreihe. Zentrbl. Bakteriol. Rarasitenkd. Infektionskr. Hyg. Abt. II, Orig. 105:52–366.

Asano, Y., and M. Tanetani. 1998. Thermostable phenylalanine dehydrogenase from a mesophilic Microbacterium sp. strain DM 86-1. Arch. Microbiol. 169:220–224.

Atlas, R. M., and L. C. Parks. 1993. Handbook of microbiological media. CRC Press. Boca Raton, FL.

Baer, D., and N. C. Gudmestad. 1995. In vitro cellulolytic activity of the plant pathogen Clavibacter michiganensis subsp. sepedonicus. Can. J. Microbiol. 41:877–888.

Bark, K., P. Kämpfer, A. Sponner, and W. Dott. 1993. Polyphosphate-dependent enzymes in some coryneform bacteria isolated from sewage sludge. FEMS Microbiol. Lett. 107:133–138.

Barker, C., J. Leitch, N. P. Brenwald, and M. Farrington. 1990. Mixed haematogenous endophthalmitis caused by Candida albicans and CDC fermentative corynebacterium group A-4. British J. Ophthalmol. 74:247–248.

Barksdale, L., M. A. Laneelle, M. C. Pollice, J. Asselineau, M. Welby, and M. V. Norgard. 1979. Biological and chemical basis for the reclassification of Microbacterium flavum Orla-Jensen as Corynebacterium flavescens nom. nov. Int. J. Syst. Bacteriol. 29:222–233.

Bartholomew, G. W., and M. Alexander. 1979. Microbial metabolism of carbon monoxide in culture and in soil. Appl. Environ. Microbiol. 37:932–927.

Battikhi, M. N. 2002. Characterization of a coryneform isolate from fasciated mugwort (Artemisia vulgaris). New Microbiol. 25:187–193.

Beattie, G. A., and S. E. Lindow. 1995. The secret life of foliar bacterial pathogens on leaves. Ann. Rev. Phytopathol. 33:145–172.

Beckwith, D. G., J. A. Jahre, and S. Haggerty. 1986. Isolation of Corynebacterium aquaticum from spinal fluid of an infant with meningitis. J. Clin. Microbiol. 23:375–376.

Beer, S. V., and J. L. Butler. 1999. Introduction of disease-resistance and insect resistance into crops and ornamental plants. Patent WO 9911133.

Behrendt, U., T. Muller, and W. Seyfarth. 1997. The influence of extensification in grassland management on the populations of micro-organisms in the phyllosphere of grasses. Microbiol. Res. 152:75–85.

Behrendt, U., A. Ulrich, and P. Schumann. 2001. Description of Microbacterium foliorum sp. nov. and Microbacterium phyllosphaerae sp. nov., isolated from the phyllospheres of grasses and the surface litter after mulching the sward and reclassification of Aureobacterium resistens (Funke et al. 1998) as Microbacterium resistens comb. nov. Int. J. Syst. Evol. Microbiol. 51:1267–1276.

Behrendt, U., A. Ulrich, P. Schumann, D. Naumann, and K.-I. Suzuki. 2002. Diversity of grass-associated Microbacteriaceae isolated from the phyllosphere and litter layer after mulching the sward; a polyphasic characterization of Subtercola pratensis sp. nov., Curtobacterium herbarum sp. nov. and Plantibacter flavus gen. nov., sp. nov. Int. J. Syst. Evol. Microbiol. 52:1441–1454.

Bendinger, B., R. M. Kroppenstedt, H. Rijnaarts, H. R. Van-Langenhove, R. C. Oberthuer, and K. Altendorf. 1990. Studies on the microbiology and the degradation capacities of a biofilter. Deutsche Gesellschaft fur Chemisches Apparatewesen (DECHEMA). Biotechnol. Conf. 4 Pt. A:529–534.

Benhamou, N. 1991. Cell surface interactions between tomato and Clavibacter michiganensis subsp. michiganensis: localization of some polysaccharides and hydroxyproline-rich glycoproteins in infected host leaf tissue. Physiol. Plant Pathol. 38:15–38.

Bergey, D. H., F. C. Harrison, R. S. Breed, B. W. Hammer, and F. M. Huntoon. 1923. Bergey's Manual of Determinative Bacteriology, 1st ed. Williams and Wilkins. Baltimore, MD.

Bergey, D. H., F. C. Harrison, R. S. Breed, B. W. Hammer, and F. M. Huntoon. 1930. Bergey's Manual of

Determinative Bacteriology, 3rd ed. Williams and Wilkins. Baltimore, MD.

Bergey, D. H., and R. S. Breed. 1948. Genus III. Flavobacterium Bergey et al. In: R. S. Breed, E. G. D. Murray, and A. P. Hitchens (Eds.) Bergey's Manual of Determinative Bacteriology, 6th ed. Williams and Wilkins. Baltimore, MD. 427–442.

Bermpohl, A., J. Dreier, R. Bahro, and R. Eichenlaub. 1996. Exopolysaccharides in the pathogenic interaction of Clavibacter michiganensis subsp. michiganensis with tomato plants. Microbiol Res. 151:391–399.

Bird, A. F., and B. A. Stynes. 1977. The morphology of a Corynebacterium sp. parasitic on annual rye grass. Phytopathology 67:828–830.

Bird, A. F., B. A. Stynes, and W. W. Thompson. 1980. A comparison of nematode and bacteria-colonized galls induced by Anduina agrostis in Lolium rigidum. Phytopathology 70:1104–1109.

Bird, A. F. 1981. The Anguina-coryneform association. In: B. M. Zuckerman and R. A. Rhode (Eds.) Plant Parasitic Nematodes. Academic Press. New York, NY. 3:303–323.

Bird, A. F. 1985. The nature of the adhesion of Corynebacterium rathayi to the cuticle of the infective larva of Anguina agrostis. Int. J. Parasitol. 15:301–308.

Bizette, G. A., S. A. Kemmerly, J. T. Cole, H. B. Bradford, and B. H. Peltier. 1995. Sepsis due to coryneform group A-4 in an immunocompromised host. Clin. Infect. Dis. 21:1334–1336.

Bradbury, J. F. 1986. Guide to Plant Pathogenic Bacteria. Commonwealth Agricultural Bureau (CAB), International Mycological Institute, Kew. London, UK.

Bradshaw-Rouse, J. J., M. H. Whatley, D. L. Coplin, A. Woods, L. Sequeira, and A. Kelman. 1981. Agglutination of Erwinia stuartii strain with a corn agglutinin: correlation with extracellular polysaccharide production and pathogenicity. Appl. Environ. Microbiol. 42:344–350.

Brennan, N. M., R. Brown, M. Goodfellow, A. Ward, T. P. Beresford, N. Vancanneyt, T. M. Cogan, and P. F. Fox. 2001. Microbacterium gubbeenense sp. nov., from the surface of a smear-ripened cheese. Int. J. Syst. Evol. Microbiol. 51:1969–1976.

Brisou, J., F. F. Denis, and A. Nivet. 1970. Septicemia, pericarditis and pleurisy caused by Bacterium helvolum. Bull. Acad. Natl. Med. 154:16–19.

Brodey, C. L., P. B. Rainey, M. Tester, and K. Johnstone. 1991. Bacterial blotch disease of the cultivated mushroom is caused by an ion channel forming lipodepsipeptide toxin. Molec. Plant-Microbe Interact. 4:407–411.

Brosius, J., J. L. Palmer, J. P. Kennedy, and H. F. Noller. 1978. Complete nucleotide sequence of a 16S ribosomal RNA gene from Escherichia coli. Proc. Natl. Acad. Sci., USA 75:4801–4805.

Brown, S. E., A. A. Reilley, D. L. Knudson, and C. A. Ishimaru. 2002. Genomic fingerprinting of virulent and avirulent strains of Clavibacter michiganensis subspecies sepedonicus. Curr. Microbiol. 44:112–119.

Brumbley, S. M., L. A. Petrasovits, R. G. Birch, and P. W. Taylor. 2002. Transformation and transposon mutagenesis of Leifsonia xyli subsp. xyli, causal organism of ratoon stunting disease of sugarcane. Molec. Plant-Microbe Interact. 15:262–268.

Bugbee, W. M., N. C. Ciuclmestad, C. A. Secor, and P. Nolle. 1987. Sugar beet as a symptomless host for Corynebacterium sepedonicum. Phytopathology 77:765–770.

Burja, A. M., N. S. Webster, P. T. Murphy, and R. T. Hill. 1999. Microbial symbionts of Great Barrier Reef Sponges. Mem. Queensland Mus. 44:63–75.

Campbell, P. B., S. Palladino, and J. P. Flexman. 1994. Catheter-related septicemia caused by a vancomycin-resistant coryneform CDC group A-5. Pathology 26:56–58.

Carlotti, A., H. Meugnier, M. T. Pommier, J. Villard, and J. Freney. 1993. Chemotaxonomy and molecular taxonomy of some coryneform clinical isolates. Zentralbl. Bakteriol. 278:23–33.

Carlson, R. R., and A. K. Vidaver. 1982. Taxonomy of Corynebacterium plant pathogens, including a new pathogen of wheat, based on polyacrylamide gel electrophoresis of cellular proteins. Int. J. Syst. Bacteriol. 32:1315–1326.

Casella, P., M. A. Bosoni, and A. Tommasi. 1988. Recurrent Corynebacterium aquaticum peritonitis in a patient undergoing continuous ambulatory peritoneal dialysis. Clin. Microbiol. Newsl. 10:62–63.

Casida, L. E. 1965. Abundant microorganisms in soil. Appl. Microbiol. 13:327–334.

Casida Jr., L. E. 1977. Small cells in pure cultures of Agromyces ramosus and in natural soil. Can. J. Microbiol. 23:214–216.

Casida Jr., L. E. 1983. Integration of Agromyces ramosus with other bacteria in soil. Appl. Environ. Microbiol. 46:881–888.

Chandler, D. P., F. J. Brockman, T. J. Bailey, and J. K. Fredrickson. 1998. Phylogenetic diversity of Archaea and Bacteria in a deep subsurface Paleosol. Microb. Ecol. 36:37–50.

Charney, W. 1966. Transformation of sterols by Corynebacteriaceae. J. Appl. Bacteriol. 29:93–106.

Chatelain, R., and L. Second. 1966. Taxonomie numérique de quelques Brevibacterium. Ann. Inst. Pasteur (Paris) 111:630–644.

Cheeke, P. R. 1995. Endogenous toxins and mycotoxins in forage grasses and their effects on livestock. J Anim. Sci. 73:909–918.

Chee-Sanford, J. C., R. I. Aminov, I. J. Krapac, N. Garrigues-Jeanjean, and R. I. Mackie. 2001. Occurrence and diversity of tetracycline resistance genes in lagoons and groundwater underlying two swine production facilities. Appl. Environ. Microbiol. 67:1494–1502.

Chen, Y., and B. L. Wong. 1997. Producing antibiotic of cephalosporin class. US Patent. 5,695,951.

Christner, B. C. 2002. Recovery of Bacteria from Glacial and Subglacial Environments (Thesis). Ohio State University. Columbus, OH.

Chung, C. H., C. P. Lin, and C. T. Chen. 1994. Development and application of cloned DNA probes for Clavibacter xyli subsp. xyli, the casual agent of sugarcane ratoon stunting. J. Phytopathol. 141:293–301.

Clise, E. H. 1948. Appendix to suborder Eubacteriineae. In: R. S. Breeds, E. G. D. Murray, and A. P. Hitchens (Eds.) Bergey's Manual of Determinative Bacteriology, 6th ed. Williams and Wilkins. Baltimore, MD. 692–693.

Collins, M. D., M. Goodfellow, and D. E. Minnikin. 1979. Isoprenoid quinones in the classification of coryneform bacteria. J. Gen. Microbiol. 110:127–136.

Collins, M. D., and D. Jones. 1980a. Lipids in the classification and identification of coryneform bacteria containing peptidoglycans based on 2,4-diaminobutyric acid. J. Appl. Bacteriol. 48:459–470.

Collins, M. D., M. Goodfellow, and D. E. Minnikin. 1980b. Fatty acid, isoprenoid quinone and polar lipid composition in the classification of Curtobacterium and related taxa. J. Gen. Microbiol. 118:29–37.

Collins, M. D., and D. Jones. 1981. The distribution of isoprenoid quinone structural types in bacteria and their taxonomic implications. Microbiol. Rev. 45:316–354.

Collins, M. D. 1982a. Lipid composition of Agromyces ramosus. FEMS Microbiol. Lett. 14:187–189.

Collins, M. D., M. Goodfellow, and D. E. Minnikin. 1982b. A survey of the structure of mycolic acids in Corynebacterium and related taxa. J. Gen. Microbiol. 128:129–149.

Collins, M. D. 1983a. Cell wall peptidoglycan and lipid composition of the phytopathogen Corynebacterium rathayi (Smith). Syst. Appl. Microbiol. 4:193–198.

Collins, M. D., and D. Jones. 1983b. Reclassification of Corynebacterium flaccumfaciens, Corynebacterium betae, Corynebacterium oortii and Corynebacterium poinsettiae in the genus Curtobacterium, as Curtobacterium flaccumfaciens comb. nov. J. Gen. Microbiol. 129:3545–3548.

Collins, M. D., D. Jones, R. M. Keddie, R. M. Kroppenstedt, and K.-H. Schleifer. 1983c. Classification of some coryneform bacteria in a new genus Aureobacterium. Syst. Appl. Microbiol. 4:236–252.

Collins, M. D., D. Jones, and R. M. Kroppenstedt. 1983d. Reclassification of Brevibacterium imperiale (Steinhaus) and "Corynebacterium laevaniformans" (Dias and Bhat) in a redefined genus Microbacterium (Orla-Jensen), as Microbacterium imperiale comb. nov. and Microbacterium laevaniformans nom. rev., comb. nov. Syst. Appl. Microbiol. 4:65–78.

Collins, M. D., and D. Jones. 1984. Validation of the publication of new names and new combinations previously effectively published outside the IJSB: List No. 14. Int. J. Syst Bacteriol. 34:270–271.

Collins, M. D., and J. F. Bradbury. 1986. Plant pathogenic species of Corynebacterium. In: V. H. A. Sneath, N. A. Mair, M. E. Sharpe, and J. G. Holt (Eds.) Bergey's Manual of Systematic Bacteriology. Williams and Wilkins. Baltimore, MD. 2:1276–1283.

Collins, M. D., and J. F. Bradbury. 1992. The genera Agromyces, Aureobacterium, Clavibacter, Curtobacterium and Microbacterium. In: A. Balow, H. G. Trüper, M. Dworkin, W. Harder and K.-H. Schleifer (Eds.) The Prokaryotes, 2nd ed. Springer-Verlag. Berlin, Germany. 2:1355–1368.

Colwell, R. R. 1970. Polyphasic taxonomy of bacteria. In: H. Iizuka and T. Hasegawa (Ed.) Proceedings of the International Conference on Culture Collections. Tokyo University Press. Tokyo, Japan. 421–436.

Corbell, N., and J. E. Loper. 1995. A global regulation of secondary metabolite production in Pseudomonas fluorescens Pf-5. J. Bacteriol. 177:6230–6236.

Cowan, S. T. 1974. Cowan and Steel's Manual for the Identification of Medical Bacteria. Cambridge University Press. London, UK.

Crosti, P., M. Malerba, and R. Bianchetti. 2001. Tunicamycin and brefeldin A induce in plant cells a programmed cell death showing apoptotic features. Protoplasma 216:31–38.

Crump, B. C., E. V. Armbrust, and J. A. Baross. 1999. Phylogenetic analysis of particle-attached and free-living bacterial communities in the Columbia River, its estuary, and the adjacent coastal ocean. Appl. Env. Microbiol. 65:3192–3204.

Cummins, C. S., R. A. Lelliott, and M. Rogosa. 1974. Genus Corynebacterium. *In:* R. E. Buchanan and N. E. Bibbons (Eds.) Bergey's Manual of Determinative Bacteriology, 8th ed. Williams and Wilkins. Baltimore, MD. 602–617.

Dangl, J. L., R. A. Dietrich, and M. N. Richberg. 1996. Death don't have no mercy: cell death programs in plant–microbe interactions. Plant Cell 8:1793–1807.

Davis, M. J., A. C. Gillaspie Jr., R. W. Harris, and R. H. Lawson. 1980. Ratoon stunting disease of sugarcane: isolation of the causal bacterium. Science 210:1365–1367.

Davis, M. J., and B. J. Augustin. 1984a. Occurrence in Florida of the bacterium that causes Bermuda grass stunting disease. Plant Dis. 68:1095–1097.

Davis, M. J., A. G. Gillaspie, A. K. Vidaver, and R. W. Harris. 1984b. Clavibacter: a new genus containing some phytopathogenic coryneform bacteria, including Clavibacter xyli subsp. xyli sp. nov., subsp. nov. and Clavibacter xyli subsp. cynodontis subsp. nov., pathogens that cause ratoon stunting disease of sugarcane and Bermudagrass stunting disease. Int. J. Syst. Bacteriol. 34:107–117.

De Boer, S. H., and A. Wieczorek. 1984a. Production of monoclonal antibodies to Corynebacterium sepedonicum. Phytopathology 63:1398–1401.

De Boer, S. H., A. Wieczorek, and A. Kummer. 1984b. An ELISA test for bacterial ring rot of potato with a new monoclonal antibody. Plant Dis. 72:874–878.

De Boer, S. H. 1991. Current status and future prospects of bacterial ring rot testing. Am. Potato J. 59:1–8.

De Bruyne, E., J. Swings, and K. Kersters. 1992. Enzymaic relatedness amongst phytopathogenic coryneform bacteria and its potential use for their identification. Syst. Appl. Microbiol. 15:393–401.

De la Cruz, A. R., M. V. Wiese, and N. W. Schaad. 1992. A semiselective agar medim for isolation of Clavibacter michiganensis subsp. sepedonicus from potato tissues. Plant Dis. 76:830–834.

DeLey, J., and J. Defloor. 1959. 2-Ketogluconoreductase in micro-organisms. Biochim. Biophys. Acta 33:47–54.

Denny, T. P. 1995. Involvement of bacterial polysaccharides in plant pathogenesis. Ann. Rev. Phytopathol. 33:173–197.

Desaint, S., A. Hartmann, N. R. Parekh, and J. Fournier. 2000. Genetic diversity of carbofuran-degrading soil bacteria. FEMS Microbiol. Ecol. 34:173–180.

Dias, F. F., and J. V. Bhat. 1962a. A new levan producing bacterium, Corynebacterium laevaniformans. Ant. v. Leeuwenhoek 28:63–72.

Dias, F. F., M. H. Bilimoria, and J. V. Bhat. 1962b. Corynebacterium barkeri nov. spec., a pectinolytic bacterium exhibiting a biotin-folic acid inter-relationship. J. Indian Inst. Sci. 44:59–67.

Dias, F. F. 1963. Studies in the bacteriology of sewage. J. Indian Inst. Sci. 45:36–48.

Dickinson, C. B., B. Austin, and M. Goodfellow. 1975. Quantitative and qualitative studies of phylloplane bacteria from Lolium rigidum. J. Gen. Microbiol. 91:157–166.

Dimock, M. B., R. M. Beach, and P. S. Carlson. 1988. Endophytic bacteria for the delivery of crop protection agents. Biotechnol. Biol. Pestic. 1:88–92.

Dobrovolskaya, T. G., I. Y. Chernov, L. I. Evtushenko, and D. G. Zvyaginzev. 1999. Synecology of saprotrophic bacteria in subtropical deserts [in Russian]. Uspekhi Sovremennoi Biologii 119:151–164.

Doetsch, R. N., and J. Rakosky. 1950. Is there a Microbacterium liquefaciens? Bacteriol. Proceed. G16:38.

Doetsch, R. N. 1957. Genus IV Microbacterium Orla-Jensen. *In:* R. S. Breed, E. G. D. Murray, and N. R. Smith (Eds.) Bergey's Manual of Systematic Bacteriology, 7th ed. Williams and Wilkins. Baltimore, MD. 600–601.

Dong, Z., M. J. Canny, M. E. McCully, M. R. Roboredo, C. F. Cabadilla, E. Ortega, and R. Rodes. 1994. A nitrogen-fixing endophyte of sugarcane stems. Plant Physiol. 105:1139–1147.

Döpfer, H., E. Stackebrandt, and F. Fiedler. 1982. Nucleic acid hybridization studies on Microbacterium, Curtobacterium, Agromyces and related taxa. J. Gen. Microbiol. 128:1697–1708.

Dorofeeva, L. V., P. A. Shramko, S. A. Subbotin, and L. I. Evtushenko. 1999. New genera of bacteria of the family Microbacteriaceae from plant galls induced by nematodes of the Anguininae subfamily: (http://www.zin.ru/conferences/ns/abstracts.htm). Problems of Nematology: Proceedings of the Zoological Institute RAS 280:101–102.

Dorofeeva, L. V., L. I. Evtushenko, V. I. Krausova, A. V. Karpov, S. A. Subbotin, and J. M. Tiedje. 2002. Rathayibacter caricis sp. nov. and Rathayibacter festucae sp. nov. isolated from the phyllosphere of Carex sp. and the leaf gall induced by nematode Anguina graminis on Festuca rubra L., respectively. Int. J. Syst. Microbiol. 52:1917–1923.

Dorofeeva, L. V., V. I. Krausova, L. I. Evtushenko, and J. M. Tiedje. 2003. Agromyces albus sp. nov., isolated from a plant (Androsace sp.). Int. J. Syst. Evol. Microbiol. 53: (http://ijs.sgmjournals.org/misc/pip.shtml).

Dowson, W. J. 1942. On the generic name of the Gram-positive bacterial plant pathogens. Trans. British Mycol. Soc. 25:311–314.

Dreier, J., A. Bermpohl, and R. Eichenlaub. 1995. Southern hybridization and PCR for specific detection of phytopathogenic Clavibacter michiganensis subsp. michiganensis. Phytopathol. 85:462–468.

Dreier, J., D. Meletzus, and R. Eichenlaub. 1997. Characterization of the plasmid encoded virulence region pat-1 of the phytopathogenic Clavibacter michiganensis subsp. michiganensis. Molec. Plant-Microbe Interact. 10:195–206.

Drennan, J. L., A. A. G. Westra, S. A. Slack, L. M. Delserone, A. Collmer, N. C. Gudmestad, and A. E. Oleson. 1993. Comparison of a DNA hybridization probe and ELISA for the detection of Clavibacter michiganensis subsp. sepedonicus in field grown potatoes. Plant Dis. 77:1243–1247.

Deutsche Sammlung von Mikroorganismen und Zellkulturen (DSMZ). 1998. Catalogue of Strains, 6th ed. German Collection of Microorganisms and Cell Cultures. Braunschweig, Germany.

Duncan, K. E., K. L. Sublette, P. A. Rider, A. Stepp, R. R. Beitle, J. A. Conner, and R. Kolhatkar. 2001. Analysis of a microbial community oxidizing inorganic sulfide and mercaptans. Biotechnol. Prog. 17:768–774.

Dunleavy, J. M. 1989. Curtobacterium plantarum sp. nov. is ubiquitous in plant leaves and is seed transmitted in soybean and corn. Int. J. Syst. Bacteriol. 39:240–249.

Dye, D. W., and W. J. Kemp. 1977. A taxonomic study of plant pathogenic Corynebacterium species. NZ. J. Agric. Res. 20:563–582.

Elbeltagy, A., K. Nishioka, H. Suzuko, T. Sato, Y. Sato, H. Morisaki, H. Mistui, and K. Minamisawa. 2000. Isolation and characterization of endophytic bacteria from wild and traditionaly cultivated rice varieties. Soil Sci. Plant Nutr. 46:617–629.

Eldefrawi, A. T., and M. E. Eldefrawi. 1987. Receptors for γ-aminobutyric acid and voltage-dependent chloride channels as targets for drugs and toxicants. FASEB J. 1:262–271.

European Union (EU). 1995. Commission Directive 95/4/EC amendment of 21 Feb. 1995 to the European Community Plant Health Directive (77/93/EEC). Official Journal of the European Communities 44:56–60.

Evtushenko, L. I., L. V. Dorofeeva, T. G. Dobrovolskaya, and S. A. Subbotin. 1994. Coryneform bacteria from plant galls induced by nematodes of the subfamily Anguininae. Russian J. Nematol. 2:99–104.

Evtushenko, L. I., L. V. Dorofeeva, S. A. Subbotin, J. R. Cole, and J. M. Tiedje. 2000. Leifsonia poae gen. nov., sp. nov., isolated from nematode galls on Poa annua and reclassification of "Corynebacterium aquaticum" Leifson 1962 as Leifsonia aquatica (ex Leifson 1962) gen. nov., rev., comb. nov. and Clavibacter xyli Davis et al. 1984 with two subspecies as Leifsonia xyli (Davis et al. 1984) gen. nov., comb. nov. Int. J. Syst. Evol. Microbiol. 50:371–380.

Evtushenko, L. I., L. V. Dorofeeva, T. G. Dobrovolskaya, G. M. Streshinskaya, S. A. Subbotin, and J. M. Tiedje. 2001. Agreia bicolorata en nov., sp. nov., to accommodate actinobacteria isolated from narrow reed grass infected by the nematode Heteroanguina graminosphila. Int. J. Syst. Evol. Microbiol. 51:2073–2079.

Evtushenko, L. I., L. V. Dorofeeva, V. I. Krausova, E. Y. Gavrish, S. G. Yashina, and M. Takeuchi. 2002. Okibacterium fritillariae gen. nov., sp. nov., a novel genus of the family Microbacteriaceae. Int. J. Syst. Evol. Microbial. 52:987–993.

Fahey, J. W., M. B. Dimock, S. F. Tomasino, J. M. Taylor, and P. S. Carlson. 1991. Genetically engineered endophytes as biocontrol agents: a case study from industry. In: J. H. Andrews and S. S. Hirano (Eds) Microbial Ecology of Leaves. Springer-Verlag. London, UK. 401–411.

Fatmi, M., and N. W. Schaad. 1988. Semiselective agar medium for isolation of Clavibacter michiganensis subsp. michiganensis from tomato seed. Phytopathology 18:121–126.

Fegan, M., B. J. Croft, D. S. Teakle, A. C. Hayward, and G. R. Smith. 1998. Sensitive and specific detection of Clavibacter xyli subsp. xyli, causal agent of ratoon stunting disease of sugarcane, with a polymerase chain reaction-based assay. Plant Pathol. 47:495–504.

Felsenstein, J. 1985. Confidence limits on phylogenies: an approach using the bootstrap. Evolution 39:783–791.

Fermor, T. R., M. B. Henry, J. S. Fenlon, M. J. Glenister, S. P. Lincoln, and J. M. Lynch. 1991. Development and application of a biocontrol system for bacterial blotch of the cultivated mushrooms. Crop Prot. 10:271–278.

Fiedler, F., and O. Kandler. 1973a. Die Aminosauresequenz von 2,4-diamino-buttersaure enhaltenden murein bei

verschidenen coryneformen bacterian und Agromyces ramosus. Arch. Microbiol. 89:51–66.

Fiedler, F., K.-H. Schleifer, and O. Kandler. 1973b. Amino acid sequence of the threonine-containing mureins of coryneform bacteria. J. Bacteriol. 113:8–17.

Finnie, J. W., and T. M. Mukherjee. 1986. Ultrastructural changes in the cerebellum of nursling rats given corynetoxin, the aetiological agent of annual ryegrass toxicity. J. Comp. Pathol. 96:205–216.

Finnie, J. W. 2001. Effect of tunicamycin on hepatocytes in vitro. J. Comp. Pathol. 125:318–321.

Firrao, D., and R. Locci. 1994. Identification of Clavibacter michiganensis subsp. sepedonicus using the polymerase chain reaction. Can. J. Microbiol. 40:148–151.

Fischer, R. A., G. Peters, J. Gehrmann, and H. Jurgens. 1994. Corynebacterium aquaticum septicemia with acute lymphoblastic leukemia. Pediatr. Infect. Dis. J. 13:836–837.

Fisher, P. J., O. Petrini, and H. M. L. Scott. 1992. The distibution of some fungal and bacterial endophytes in maize (Sea mais L.). New Phytol. 122:299–411.

Fletcher, J. T., P. F. White, and R. H. Gaze. 1989. Mushrooms: Pest and Disease Control, 2nd ed. Intercept. Andover, UK.

Frankland, G. C., and P. F. Frankland. 1889. Über einige typische Mikroorganismen im Wasser und im Boden. Zentrlbl. Bakt. Parasit. Inf. Med. Mikrobiol. Immunol. 6:373–400.

Fritz, I. 2000. Das Bakterioplankton im Westlichen Mittelmeer: Analyse der taxonomischen Struktur freilebender und partikelgebundener bakterieller Lebensgemeinschaften mit mikrobiologischen und molekularbiologischen Methoden (PhD Thesis). Gemeinsame Naturwissenschaftliche Fakultaet, Technische Universitaet Carolo-Wilhelmina Braunschweig. Braunschweig, Germany.

Froquet, R., Y. Sibiril, and D. Parent-Massin. 2001. Trichothecene toxicity on human megakaryocyte progenitors (CFU-MK). Hum. Exp. Toxicol. 20:84–89.

Funke, G., A. von Graevenitz, and N. Weiss. 1994. Primary identification of Aureobacterium spp. isolated from clinical specimens as "Corynebacterium aquaticum." J. Clin. Microbiol. 32:2686–2691.

Funke, G., E. Falsen, and C. Barreau. 1995. Primary identification of Microbacterium spp. encountered in clinical specimens as CDC coryneform group A-4 and A-5 bacteria. J. Clin. Microbiol. 33:188–192.

Funke, G., A. von Graevenitz, J. E. Clarridge, 3rd, and K. A. Bernard. 1997. Clinical microbiology of coryneform bacteria. Clin. Microbiol. Rev. 10:125–159.

Funke, G., P. A. Lawson, F. S. Nolte, N. Weiss, and M. D. Collins. 1998. Aureobacterium resistens sp. nov., exhibiting vancomycin resistance and teicoplanin susceptibility. FEMS Microbiol. Lett. 158:89–93.

Furlong, M. A., D. R. Singleton, D. C. Coleman, and W. B. Whitman. 2002. Molecular and culture-based analyses of prokaryotic communities from an agricultural soil and the burrows and casts of the earthworm Lumbricus rubellus. Appl. Environ. Microbiol. 68:1265–1279.

Galloway, J. H. 1961. Grass seed nematode poisoning in livestock. J. Am. Vet. Assoc. 139:1212–1214.

Gardner, H. W., C. T. Hou, D. Weisleder, and W. Brown. 2000. Biotransformation of linoleic acid by Clavibacter sp. ALA2: Heterocyclic and heterobicyclic fatty acids. Lipids 35:1055–1060.

Gavrish, E. Yu., S. G. Karasev, N. E. Suzina, D. A. Gilichinsky, and L. I. Evtushenko. 2003. Cryocola antiquus gen. nov., sp. nov., novel actinobacteria isolated from Siberian permafrost. Abstract Book of the 1st FEMS Congress: Ljubljana, Slovenia, 29 June–3 July 2003. Elsevier Delft 172.

Gillaspie Jr., A. G., and D. S. Teackle. 1989. Ratoon stunting disease. In: C. Ricaud, B. T. Egan, A. G. Gillaspie Jr., and C. G. Hudges (Eds.) Diseases of Sugarcane. Elsevier. Amsterdam, The Netherlands. 59–80.

Gillies, A. J. 1971. Significance of thermoduric organisms in Queensland Cheddar cheese. Australian J. Dairy Techonol. 26:145–149.

Gitaitis, R. D. 1990. Induction of a hypersensitive-like reaction in four-o'clock by Clavibacter michiganensis subsp. michiganensis. Plant Dis. 74:58–60.

Gledhill, W. E., and L. E. Casida. 1969. Predominant catalase-negative soil bacteria. III: Agromyces gen. n., microorganisms intermediary to Actinomyces and Nocardia. Appl. Microbiol. 18:340–349.

Glöckner, F. O., E. Zaichikov, N. Belkova, L. Denissova, J. Pernthaler, A. Pernthaler, and R. Amann. 2000. Comparative 16S rRNA analysis of lake bacterioplankton reveals globally distributed phylogenetic clusters including an abundant group of actinobactereria. Appl. Environ. Microbiol. 66:5053–5065.

Gnilozub, V. A., G. M. Streshinskaia, L. I. Evtushenko, I. B. Naumova, and A. S. Shashkov. 1994. 1,5-Poly(ribitol phosphate) with tetrasaccharide substituents in the cell wall of Agromyces fucosus ssp. hippuratus [in Russian]. Biokhimiia 59:1892–1829.

Goodfellow, M., and D. E. Minnikin. 1985. Introduction to chemosystematics. In: M. Goodfellow and D. E. Minnikin (Eds.) Chemical Methods in Bacterial Systematics. Academic Press. London, UK. 1–15.

Gross, D. C., and A. K. Vidaver. 1979a. A selective medium for the isolation of Corynebacterium nebraskense from soil and plant parts. Phytopathology 69:82–87.

Gross, D. C., and A. K. Vidaver. 1979a. Bacteriocins of phytopathogenic Corynebacterium species. Can. J. Microbiol. 25:367–374.

Groth, I., P. Schumann, N. Weiss, K. Martin, and F. A. Rainey. 1996. Agrococcus jenensis gen. nov., sp. nov., a new genus of actinomycetes with diaminobutyric acid in the cell wall. Int. J. Syst. Bacteriol. 46:234–239.

Grove, D. I., V. Der-Haroutian, and R. M. Ratcliff. 1999. Aureobacterium masquerading as "Corynebacterium aquaticum" infection: Case report and review of the literature. J. Med. Microbiol. 48:965–970.

Grundler, F. M. W., and U. Wyss. 1995. Strategies of root parasitism by sedentary plant parasitic nematodes. In: K. Kohmoto, N. S. Singh, and R. P. Singh (Eds.) Pathogenesis and Host Specificity in Plant Diseases, Volume 2: Histopathological, Biochemical, Genetic and Molecular Bases. Elsevier Science. New York, NY. 309–319.

Guimaraes, P. M., S. Palmano, J. J. Smith, M. F. Grossi de Sa, and G. S. Saddler. 2001. Development of a PCR test for the detection of Curtobacterium flaccumfaciens pv. flaccumfaciens. Ant. v. Leeuwenhoek 80:1–10.

Gupta, P., and G. Swarup. 1972. Ear-cockle and yellow ear rot disease of wheat. II: Nematode bacterial association. Nematologia 18:320–324.

Guthrie, G., and C. Nicholson-Guthrie. 1989. acid uptake by a bacterial system with neurotransmitter binding characteristics. Proc. Natl. Acad. Sci. USA 86:7378–7381.

Haapalainen, M., M. Karp, and M. C. Metzler. 1996. Isolation of strong promoters from Clavibacter xyli subsp. cynodontis using a promoter probe plasmid. Biochim. Biophys. Acta 1305:130–134.

Haapalainen, M. L., N. Kobets, E. Piruzian, and M. C. Metzler. 1998. Integrative vector for stable transformation and expression of a beta-1,3-glucanase gene in Clavibacter xyli subsp. cynodontis. FEMS Microbiol. Lett. 162:1–7.

Haapalainen, M., J. Mattinen, and M. C. Metzler. 2000. The growth of a plant-parasitic bacterium, Clavibacter xyli subsp. cynodontis, is enhanced by xylem fluid components. Physiol. Molec. Plant Path. 56:147–155.

Hahn, M. W., H. Lunsdorf, Q. Wu, M. Schauer, M. G. Hofle, J. Boenigk, and P. Stadler. 2003. Isolation of novel ultramicrobacteria classified as Actinobacteria from five freshwater habitats in Europe and Asia. Appl. Environ. Microbiol. 69:1442–1451.

Hallmann, J., A. Quart-Hallmann, W. F. Mahaffee, and J. W. Kloepper. 1997. Bacterial endophytes in agricultural crops. Can. J. Microbiol. 43:895–914.

Hanscom, T., and A. Maxwell. 1979. Corynebacterium endophthalmitis193: Laboratory studies and report of a case treated by vitrectomy. Arch. Ophthalmol. 97:500–502.

Harrigan, W. F., and M. E. McCance. 1976. Laboratory Methods in Food and Dairy Microbiology, revised ed. Academic Press. London, UK.

Hatano, K. 1997. Actinomycete populations in mangrove rhizospheres. IFO Res. Commun. 18:26–31.

Hawn, E. J. 1963. Transmission of bacterial wilt of alfaalfa by Ditylenchus dipsaci (Kuhn). Nematology 9:65–68.

Hayward, A. C. 1974. Latent infections of bacteria. Ann. Rev. Phytopathol. 12:87–97.

Hedges, F. 1922. A bacterial wilt of beans caused by Bacterium flaccumfaciens nov. sp. Science 55:433–434.

Henningson, P. J., and N. C. Gudmestad. 1991. Fatty acid analysis of phytopathogenic coryneform bacteria. J. Gen. Microbiol. 137:427–440.

Heuer, H., R. M. Kroppenstedt, J. Lottmann, G. Berg, and K. Smalla. 2002. Effects of T4 lysozyme release from transgenic potato roots on bacterial rhizosphere communities are negligible relative to natural factors. Appl. Environ. Microbiol. 68:1325–1335.

Hiorns, W. D., B. A. Methe, S. A. Nierzwicki-Bauer, and J. P. Zehr. 1997. Bacterial diversity in Adirondack mountain lakes as revealed by 16S rRNA gene sequences. Appl. Environ. Microbiol. 63:2957–2960.

Hirano, S. S., L. S. Baker, and C. D. Upper. 1996. Raindrop momentum triggers growth of leaf-associated populations of Pseudomonas syringae on field-grown snap bean plants. Appl. Environ. Microbiol. 62:2560–2566.

Hodgkin, J., P. E. Kuwabata, and B. Corneliussen. 2000. A novel bacterial pathogen, Microbacterium nematophilum, induces morphological change in the nematode C. elegans. Curr Biol. 10:1615–1618.

Hollender, J., U. Dreyer, L. Kornberger, P. Kämpfer, and W. Dott. 2002. Selective enrichment and characterization of a phosphorus-removing bacterial consortium from activated sludge. Appl. Microbiol. Biotechnol. 58:106–111.

Hollis, D. G., and R. E. Weaver. 1981. Gram-positive Organisms: A Guide to Identification. Special Bacteriology Section, Centers for Disease Control. Atlanta, GA.

Horwitz, A. H., and L. E. Casida Jr. 1978a. Effects of magnesium, calcium and serum on reversion of stable L-forms. J. Bacteriol. 136:565–599.

Horwitz, A. H., and L. E. Casida Jr. 1978b. Survival and reversion of a stable L form in soil. Agromyces ramosus forms the L-form reverted to a bacterial form when incubated in sterilized soil. Can. J. Microbiol. 24:50–55.

Hou, C. T., W. Brown, D. P. Labeda, T. P. Abbott, and D. Weisleder. 1997. Microbial production of a novel trihydroxy unsaturated fatty acid from linoleic acid. J. Ind. Microbiol. Biotechnol 19:34–38.

Hou, C. T. 1998. New octadec-9-enoyl derivatives and their production. US Patent 5,852,196.

Hou, C. T. 1999. Biotransformation of unsaturated fatty acids to industrial products. In: Abstracts of the 217th ACS National Meeting, Anahaeim, California, 21–25 March 1999. American Chemistry Society. Anaheim, CA. pt. 1:BIOT160.

Hu, X., F. M. Lai, A. S. N. Reddy, and C. A. Ishumaru. 1995. Quantitative detection of Clavibacter michiganensis subsp. sepedonicus by competetive polymarase chain reaction. Phytopathology 85:1468–1473.

Hutchinson, C. M. 1917. A bacterial disease of wheat in the Punjab. Mem. Dep. Agr. India, Bact. Ser. 1:169–175.

Hutchison, M. L., and K. Johnstone. 1993. Evidence for the involvement of the surface active properties of the extracellular toxin tolaasin in the manifestation of brown blotch disease symptoms by Pseudomonas tolaasii on Agaricus bisporus. Physiol. Molec. Plant Pathol. 42:373–384.

International Collection of Micro-organisms from Plants (ICMP). 1997. Catalogue: (http://www.landcareresearch.co.nz/research/biodiversity/fungiprog/icmp97.pdf). ICMP.

Iijima, T., and T. Sakane. 1973. Method for preservation of bacteria and bacteriophages by drying in vacuo. Cryobiology 10:379–385.

Iizuka, H., and K. Komagata. 1965. Microbiological studies on petroleum and natural gas. III: Determination of Brevibacterium, Arthrobacter, Micrococcus, Sarcina, Alcaligenes and Achromobacter isolated from oil-brines in Japan. J. Gen. Appl. Microbiol. 11:1–14.

Iizuka, T., S. Yamanaka, T. Nishiyama, and A. Hiraishi. 1998. Isolation and phylogenetic analysis of aerobic copiotrophic ultramicrobacteria from urban soil. J. Gen. Appl. Microbiol. 44:75–85.

Imai, K., M. Takeuchi, and I. Banno. 1984. Reclassification of "Flavobacterium arborescens" (Frankland and Frankland) Bergey et al. in the genus Microbacterium (Orla-Jensen) Collins et al., as Microbacterium arborescens comb. nov., nom. rev. Curr. Microbiol. 11:281–284.

Imai, K., and T. Sakane. 1985. Preservation of chemolithotrophic bacteria by L-drying. Fundamentals and Applications of Freeze-drying to Biological Materials, Drug and Foods Stuffs. International Institute of Refrigeration, Commission C1. Paris, France. 233–234.

Inoue, K. 1976a. Quantitative ecology of microorganisms of Showa Station in Antarctica and isolation of psychrophiles. J. Gen. Appl. Microbiol. 22:143–150.

Inoue, K., and K. Komagata. 1976b. Taxonomic study on obligately psychrophilic bacteria isolated from Antarctica. J. Gen. Appl. Microbiol. 22:165–176.

Jacobs, J. L., and G. W. Sundin. 2001. Effect of solar UV-B radiation on a phyllosphere bacterial community. Appl. Environ. Microbiol. 67:5488–5496.

Jago, M. V., A. L. Payne, J. E. Peterson, and T. J. Bagust. 1983. Inhibition of glycosylation by corynetoxin, the causative agent of annual ryegrass toxicity: A comparison with tunicamycin. Chem. Biol. Interact. 45:223–234.

Jahr, H., R. Bahro, A. Burger, J. Ahlemeyer, and R. Eichenlaub. 1999. Interactions between Clavibacter michiganensis and its host plants. Environ. Microbiol. 1:113–118.

Jahr, H., J. Dreier, D. Meletzus, R. Bahro, and R. Eichenlaub. 2000. The endo-beta-1,4-glucanase CelA of Clavibacter michiganensis subsp. michiganensis is a pathogenicity determinant required for induction of bacterial wilt of tomato. Molec. Plant-Microbe Interact. 13:703–714.

Janssen, P. H., P. S. Yates, B. E. Grinton, P. M. Taylor, and M. Sait. 2002. Improved culturability of soil bacteria and isolation in pure culture of novel members of the divisions Acidobacteria, Actinobacteria, Proteobacteria and Verrucomicrobia. Appl. Environ. Microbiol. 68:2391–2396.

Jayne-Williams, D. J., and T. M. Skerman. 1966. Comparative studies on coryneform bacteria from milk and dairy sources. J. Appl. Bacteriol. 29:72–92.

Jensen, H. L. 1934. Studies on saprophytic mycobacteria and corynebacteria. Proc. Linn. Soc. New South Wales 59:19–61.

Johansen, I. E., O. F. Rasmussen, and M. Heide. 1989. Specific identification of Clavibacter michiganensis subsp. sepedonicus by DNA-hybridization probes. Phytopathology 79:1019–1023.

Johnston, M. S., S. S. Sutherland, C. C. Constantine, and D. J. Hampson. 1996. Genetic analysis of Clavibacter toxicus, the agent of annual ryegrass toxicity. Epidemiol. Infect. 117:393–400.

Jones, D., J. Watkins, and D. J. Meyer. 1970. Cytochrome composition and effect of catalase or growth of Agromyces ramosus. Nature 226:1249–1250.

Jones, D. 1975. A numerical taxonomic study of coryneform and related bacteria. J. Gen. Microbiol. 87:52–96.

Kado, C., and M. G. Heskett. 1970. Selective media for isolation of Agrobacterium, Corynebacterium, Erwinia, Pseudomonas and Xanthomonas. Phytopathology 60:229–252.

Kado, C. I. 1992. Plant pathogenic bacteria. In: A. Balows, H. G. Trüper, M. Dworkin, W. Harder and K.-H. Schleifer (Eds.) The Prokaryotes, 2nd ed. Springer-Verlag. New York, NY. 1:659–674.

Kaiser, O., A. Puehler, and W. Selbitschka. 2001. Phylogenetic analysis of microbial diversity in the rhizoplane of oilgd rape (Brassica napus cv. Westar) employing cultivation-dependent and cultivation-independent approaches. Microb. Ecol. 42:136–149.

Kämpfer, P., F. A. Rainey, M. A. Andersson, E.-L. Nurmiaho-Lassila, U. Ulrych, H.-J. Busse, N. Weiss, R. Mikkola, and M. Salkinoja-Salonen. 2000. Frigoribacterium faeni gen. nov., sp. nov., a novel psychrophilic genus of the family Microbacteriaceae. Int. J. Syst. Evol. Microbiol. 50:355–363.

Kaneda, T. 1991. Iso- and anteiso-fatty acids in bacteria: biosynthesis, function and taxonomic significance. Microbiol. Rev. 55:288–302.

Kaplan, A., and F. Israel. 1988. Corynebacterium aquaticum infection in a patient with chronic granulomatous disease. Am. J. Med. Sci. 296:57–58.

Karasev, S. G., L. V. Gurina, E. Y. Garvish, V. M. Adanin, D. A. Gilichinsky, and L. I. Evtushenko. 1998. Viable actinobacteria from ancient Siberian permafrost [in Russian]. Earth Cryosphere 11(2):69–75.

Katsumata, R., and S. Hashimoto. 1996. Production of optically active gamma-hydroxy-L-glutamic acid. Patent EP 692539.

Kaufman, A. E., H. Goldfine, O. Narayan, and S. M. Gruner. 1990. Physical studies on the membranes and lipids of plasmalogen-deficient Megasphaera elsdenii. Chem. Phys. Lipids 55:41–48.

Kawamura, Y., Y. Li, H. Liu, X. Huang, Z. Li, and T. Ezaki. 2001. Bacterial population in Russian space station "Mir." Microbiol. Immunol. 45:819–828.

Keddie, R. M., B. G. S. Leask, and J. M. Grainger. 1966. A comparison of coryneform bacteria from soil and herbage: Cell wall composition and nutrition. J. Appl. Bacteriol. 29:17–43.

Keddie, R. M., and G. L. Cure. 1977. The cell wall composition and distribution of free mycolic acids in named strains of coryneform bacteria and in isolates from various natural sources. J. Appl. Bacteriol. 42:229–252.

Keddie, R. M., and G. L. Cure. 1978. Cell wall composition of coryneform bacteria. In: I. J. Bousfield and A. G. Callely (Eds.) Special Publications of the Society for General Microbiology, Volume I: Coryneform Bacteria. Academic Press. London, UK. 47–83.

Kern, H., and S. Naef-Roth. 1971. Phytolysin, ein durch pflanzenpathogene Pilze gebildeter mazierenrender Factor. Phytopathol. Zeitschr. 71:231–246.

Keyworth, W. G., J. Howell, and W. J. Dowson. 1956. Corynebacterium betae (sp. nov.): The casual organisms of silvering disease of red beet. Plant Pathol. 5:88–90.

Kim, Y. H., T. K. Kwon, S. Park, H. S. Seo, J. J. Cheong, C. H. Kim, J. K. Kim, J. S. Lee, and Y. D. Choi. 2000. Trehalose synthesis by sequential reactions of recombinant altooligosyltrehalose synthase and maltooligosyltrehalose trehalohydrolase from Brevibacterium helvolum. Appl. Environ. Microbiol. 66:4620–4624.

Kimura, M. 1980. A simple method for estimating evolutionary rates of base substitutions through comparative studies of nucleotide suquences. J. Molec. Evol. 16:111–120.

Kiraly, Z., H. M. El-Zahaby, and Z. Klement. 1997. Role of extracellular polysaccharides (EPS) slime of plant pathogenic bacteria in protecting cells to reactive oxygen species. J. Phytopath. 145:59–68.

Kiso, T., S. Kitahata, K. Okamoto, S. Miyoshi, and H. Nakano. 2000. Hydrolysis of beta-glucosyl ester linkage of p-hydroxybenzoyl beta-D-glucose, a chemically synthesized glucoside, by beta-glucosidases. J. Ferment. Bioengin. 90:614–618.

Kloepper, J. W., R. M. Zablotowicz, E. M. Tipping, and R. Lifshitz. 1991. Plant growth promotion mediated by bacterial rhizosphere colonozers. In: D. L. Keister and P. B. Cregan (Eds.) The Rhizosphere and Plant Growth. Kluwer Academic Publishers. Dordrecht, The Netherlands. 315–326.

Kobayashi, M., Y. Mitsuishi, and K. Matsuda. 1978. Pronounced hydrolysis of highly branched dextrans with a new type of dextranase. Biochem. Biophys. Res. Commun. 80:306–312.

Koch, A. L. 1996. What size should a bacterium be? A question of scale. Ann. Rev. Micromiol. 50:317–348.

Kochkina, G. A., N. E. Ivanushkina, S. G. Karasev, E. Yu. Gavrish, L. V. Gurina, L. I. Evtushenko, E. V. Spirina, E. A. Vorob'eva, D. A. Gilichinskii, and S. M. Ozerskaya. 2001. Micromycetes and actinobacteria under natural long-term cryoconservation. Microbiologiia 70:412–420.

Komagata, K., and H. Iizuka. 1964. New species of Brevibacterium isolated from rice. J. Agric. Chem. Soc. Japan 38:496–502.

Komagata, K., and K. Suzuki. 1986. Genus Curtobacterium. In: P. H. A. Sneath, N. A. Mair, M. E. Sharpe and J. G. Holt (Eds.) Bergey's Manual of Systematic Bacteriology. Williams and Wilkins. Baltimore, MD. 2:1313–1317.

Komura, I., K. Yamada, S. Otsuka, S., and K. Komagata. 1975. Taxonomic significance of phospholipids in coryneform and nocardioform bacteria. J. Gen. Appl. Microbiol. 21:251–261.

Koneva, N. D., and Yu. V. Kruglov. 2001. Dynamic of the volume and structure of the soil bacterial complex in the presence of azobenzene [in Russian]. Mikrobiologiya 70:552–557.

Koprowski, H., D. C. Hooper, L. J. Conway, F. H. Michaels, A. Modelska, and Z. F. Fu. 1997. Synthesis and delivery of bioactive compound using endophytic microorganism. Patent WO 9714290.

Korzybski, T., Z. Kowszyk-Gindifer, and W. Kurylowicz. 1978. Antibiotics: Origin, Nature, and Properties. American Society of Microbiology. Washington, DC. 1:773–774.

Kostka, S. J., P. W. Reeser, and D. P. Miller. 1988. Experimental host range of Clavibacter xyli subsp. cynodontis (CXC) and a CXC/Bacillus thuringiensis recombinant (CXC/BT). Phytopathology 78:1540–1547.

Kotani, S., T. Kato, T. Matsubara, M. Sakagoshi, and Y. Hirachi. 1972. Inducible enzyme degrading serologically active polysaccharides from mycobacterial and corynebacterial cells. Biken J. 15:1–15.

Krall, E. L. 1991. Wheat and grass nematodes: Anguina, Subanguina, and related genera. In: W. R. Nickle (Ed.) Manual of Agricultural Mematology. Marcel Dekker. New York, NY. 721–760.

Kubota, K., and T. Shiro. 1979. Method of producing L-serine by fermentation. US Patent 3,623,952.

Kwon, Y. J., and D. Y. Lee. 1997. Corynebacterium aquaticum peritonitis in a patient on CAPD. Perit. Dial. Int. 17:98–99.

Labadie, J., and M. Hebraud. 1997. Purification and characterization of a collagenolytic enzyme produced by Rathayibacter sp. strains isolated from cultures of Clavibacter michiganensis subsp. michiganensis. J. Appl. Microbiol. 82:141–148.

Labeda, D. P., K. Hatano, R. M. Kroppenstedt, and T. Tamura. 2001. Revival of the genus Lentzea and proposal for Lechevalieria gen. nov. Int. J. Syst. Evol. Microbiol. 51:1045–1050.

Laffineur, K., Y. Avesani, G. Cornu, J. Charlier, M. Janssens, G. Wauters, and M. Delmee. 2003. Bacteremia due to a novel Microbacterium species in a patient with leukemia and description of Microbacterium paraoxydans sp. nov. J. Clin. Microbiol. 41:2242–2246.

Laine, M. J., M. Haapalainen, T. Wahlroos, K. Kankare, R. Nissinen, S. Kassuwi, and M. C. Metzler. 2000. The cellulase encoded by the native plasmid of Clavibacter michiganensis subsp. sepedonicus plays a role in virulence and contains an expansin-like domain. Physiol. Molec. Plant Path. 57:221–233.

Laiz, L., I. Groth, P. Schumann, F. Zezza, A. Felske, B. Hermosin, and C. Saiz-Jimenez. 2000. Microbiology of the stalactites from Grotta dei Cervi, Porto Badisco, Italy. Int. Microbiol. 3:25–30.

Lampel, J. S., G. L. Canter, M. B. Dimock, J. L. Kelly, J. J. Anderson, and B. B. Uratani. 1994. Integrative cloning, expression and stability of the cryIA(c) gene from Bacillus thuringiensis subsp. kurstaki in a recombinant strain of Clavibacter xyli subsp. cynodontis. Appl. Environ. Microbiol. 60:501–508.

Larsson, P., O. Lundin, and E. Falsen. 1996. "Corynebacterium aquaticum" wound infection after high-pressure water injection into the foot. Scand. J. Infect. Dis. 28:635–636.

Lau, S. K., P. C. Woo, G. K. Woo, and K. Y. Yuen. 2002. Catheter-related Microbacterium bacteremia identified by 16S rRNA gene sequencing. J. Clin. Microbiol. 40:2681–2685.

Laville, J., C. Voisard, C. Keel, M. Maurhofer, G. Defago, and D. Haas. 1992. Global control in Pseudomonas fluorescens mediating antibiotic synthesis and suppression of black root rot of tobacco. Proc. Natl. Acad. Sci. USA 89:1562–1566.

Lazar, I. 1968. Serological relationships of coryneform bacteria. J. Gen. Bacteriol. 52:77–88.

Lednicka, D., J. Mergaert, M. C. Cnockaert, and J. Swings. 2000. Isolation and identification of cellulolytic bacteria involved in the degradation of natural cellulosic fibres. Syst. Appl. Microbiol. 23:292–299.

Lee, I.-M., I. M. Bartoszyk, D. E. Gundersen, B. Mogen, and R. E. Davis. 1997a. Nested PCR for ultrasensitive detection of the potato ring rot bacterium, Clavibacter michiganensis subsp. sepedonicus. Appl. Environ. Microbiol. 63:2625–2630.

Lee, I.-M., I. M. Bartoszyk, D. E. Gundersen-Rindal, and R. E. Davis. 1997b. Phylogeny and classification of bacteria in the genera Clavibacter and Rathayibacter on the basis of 16S rRNA gene sequence analyses. Appl. Environ. Microbiol. 63:2631–2636.

Lee, S. D., and Y. C. Hah. 2002. Proposal to transfer Catellatospora ferruginea and "Catellatospora ishikariense" to Asanoa gen. nov. as Asanoa ferruginea comb. nov. and Asanoa ishikariensis sp. nov., with emended description of the genus Catellatospora. Int. J. Syst. Evol. Microbiol. 52:967–972.

Legard, D. E., M. P. McQuilken, J. M. Whipps, J. S. Fenlon, T. R. Fermor, I. P. Thompson, M. J. Bailey, and J. M. Lynch. 1994. Studies of seasonal changes in the microbial populations on the phyllosphere of spring wheat as a prelude to the release of a genetically modified microorganism. Agricult. Ecosys. Environ. 50:87–101.

Leifson, E. 1962. The bacterial flora of distilled and stored water. III: New species of the genera Corynebacterium, Flavobacterium, Spirillum and Pseudomonas. Int. Bull. Bacteriol. Nomen. Taxon. 12:161–170.

Lelliott, R. A., and D. E. Stead. 1987. Methods for the Diagnosis of Bacterial Diseases of Plants. Blackwell Scientific. Oxford, UK.

Li, X., and S. H. De Boer. 1995a. Comparison of 16S ribosomal RNA genes in Clavibacter michiganensis subspecies with other coryneform bacteria. Can. J. Microbiol. 41:925–929.

Li, X., and S. H. De Boer. 1995b. Selection of polymerase chain reaction primers from an RNA intergenic spacer region for specific detection of Clavibacter michiganensis subsp. sepedonicus. Phytopathology 85:837–842.

Li, X., S. H. De Boer, and L. J. Ward. 1997. Improved microscopic identification of Clavibacter michiganensis subsp. sepedonicus cells by combining in situ hybridization with immunofluorescence. Lett. Appl. Microbiol. 24:431–434.

Li, W.-J., L.-P. Zhang, X.-L. Xu, X.-L. Cui, L.-H. Xu, E. Stackebrandt, and C.-L. Jiang. 2003. Agromyces aurantiacus sp. nov., isolated from a Chinese primeval forest. Int. J. Syst. Evol. Microbiol. 53:303–307.

Lifshitz, A., N. Arber, E. Pras, Z. Samra, J. Pinkhas, and Y. Sidi. 1991. Corynebacterium CDC group A-4 native valve endocarditis. Eur. J. Clin. Microbiol. Infect. Dis. 10:1056–1057.

Lilley, A. K., R. S. Hails, J. S. Cory, and M. J. Bailey. 1997. The dispersal and establishment of pseudomonad populations in the phyllosphere of sugar beet by phytophagous caterpillars. FEMS Microbiol. Ecol. 24:151–157.

Lindemann, J., and C. D. Upper. 1985. Aerial dispersal of epiphytoc bacteria over bean-plants. Appl. Environ. Microbiol. 50:1229–1232.

Lochhead, A. G., and M. O. Burton. 1953. An essential bacterial growth factor produced by microbial synthesis. Can. J. Bot. 31:7–22.

Lochhead, A. G. 1958. Two new species of Arthrobacter requiring respectively vitamin B12 and the terregens factor. Arch. Mikrobiol. 31:163–170.

Louws, F. J., J. Bell, C. M. Medina-Mora, C. D. Smart, D. Opgenorth, C. A. Ishimaru, M. K. Hausbeck, F. J. de Bruijn, and D. W. Fulbright. 1998. Rep-PCR-mediated genomic fingerprinting: a rapid and effective method to identify Clavibacter michiganensis. Phytopathology 88:862–868.

Luedemann, G. M. 1968. Geodermatophilus, a new genus of the Dermatophilaceae (Actinomycetales). J. Bacteriol. 96:1848–1858.

Lysenko, O. 1959. The occurrence of species of the genus Brevibacterium in insects. J. Insect Pathol. 1:34–42.

Madigan, M. T., J. M. Martinko, and J. Parker. 1997. Brock Biology of Microorganisms, 8th ed. Prentice-Hall. Upper Saddle River, NJ.

Malik, K. A. 1990. A simplified liquid-drying method for the preservation of microorganisms sensitive to freezing and freeze-drying. J. Microbiol. Meth. 12:125–132.

Maltseva, O. V., and P. Oriel. 1996. Degradation of 2,4,6-trichlorophenol by an enrichment culture and isolation of a chlorophenol-degrading haloalkaliphilic bacterium. In: Abstracts of the 96th General Meeting of the American Society for Microbiology, New Orleans, LA, 19–23 May. ASM. New Orleans, LA. 417.

Mamoli, L. 1939. Über biochemische Dehydrierungen in der Cortingruppe. Ber. Dtsch. Chem. Ges. 72:1863–1865.

Männistö, M. K., P. Schumann, F. A. Rainey, P. Kämpfer, I. V. Tsitko, M. A. Tiirola, and M. S. Salkinoja-Salonen. 2000. Subtercola boreus gen. nov., sp. nov. and Subtercola frigoramans sp. nov., two new psychrophilic actinobacteria isolated from boreal groundwater. Int. J. Syst. Evol. Microbiol. 50:1731–1739.

Matsufuji, M., K. Nakada, and A. Yoshimito. 2000. Glycoglycerolipid having anti-active oxygen disorder activity and cell proliferation promotion activity. Patent JP 2000 333694.

Matsuyama, H., K. Kawasaki, I. Yumoto, and O. Shida. 1999. Microbacterium kitamiense sp. nov., a new polysaccharide-producing bacterium isolated from the wastewater of a sugar-beet factory. Int. J. Syst. Bacteriol. 49:1353–1357.

McCulloch, L. 1925. Aplanobacter insidiosum n. sp.: The cause of an alfalfa disease. Phytopathology 15:496–497.

McKay, A. C., and K. M. Ophel. 1993. Toxigenic Clavibacter/Anguina associations infecting grass seedheads. Ann. Rev. Phytopathol. 31:151–167.

McKay, A. C., K. M. Ophel, T. B. Reardon, and J. M. Gooden. 1993. Livestock deaths associated with Clavibacter toxicus/Anguina sp. infection in seedheads of Agrostis avenacea and Polypogon monspeliensis. Plant Dis. 77:635–641.

McKelvey, J., R. Rai, and T. G. Cooper. 1990. GABA Transport in Saccharomyces cerevisiae. Yeasts 6:263–270.

McLean, R. A., and W. L. Sulzbacher. 1953. Microbacterium thermosphactum, spec. nov., a nonheat resistant bacterium from fresh pork sausage. J. Bacteriol. 65:428–433.

Meletzus, D., A. Bermpohl, J. Dreier, and R. Eichenlaub. 1993. Evidence for plasmid encoded virulence factors in the phytopathogenic bacterium Clavibacter michiganensis subsp. michiganensis NCPPB 382. J. Bacteriol. 175:2131–2136.

Mergaert, J., A. Wouters, C. Anderson, and J. Swings. 1995. In situ biodegradation of poly(3-hydroxybutyrate) and poly(3-hydroxybutyrate-co-3-hydroxyvalerate) in natural waters. Can. J. Microbiol. 41, Suppl. 1:154–159.

Mergaert, J., and J. Swings. 1996. Biodiversity of microorganisms that degrade bacterial and synthetic polyesters. J. Ind. Microbiol. 17:463–469.

Methe, B. A., W. D. Hiorns, and J. P. Zehr. 1998. Contrasts between marine and freshwater bacterial community composition—analyses of communities in Lake George and six other Adirondack lakes. Limnol. Oceanogr. 43:368–374.

Metzler, M. C., Y. P. Zhang, and T. A. Chen. 1992. Transformation of the Gram-positive bacterium Clavibacter xyli subsp. cynodontis by electroporation with plasmids from the IncP incompatibility group. J. Bacteriol. 174:4500–4503.

Metzler, M. C., M. J. Laine, and S. H. De Boer. 1997. The status of molecular biological research on the plant pathogenic genus Clavibacter. FEMS Microbiol. Lett. 150:1–8.

Mevs, U., E. Stackebrandt, P. Schumann, C. A. Gallikowski, and P. Hirsch. 2000. Modestobacter multiseptatus gen. nov., sp. nov., a budding actinomycete from soils of the Asgard Range (Transantarctic Mountains). Int. J. Syst. Evol. Microbiol. 50:337–346.

Mills, D., B. W. Russell, and J. W. Hanus. 1997. Specific detection of Clavibacter michiganensis subsp. sepedonicus by amplification of three unique DNA sequences isolated by subtraction hybridization. Phytopathology 87:853–861.

Minnikin, D. E., M. Goodfellow, and M. D. Collins. 1978. Lipid composition in the classification and identification of coryneform and related taxa. In: I. J. Bousfield and A. G. Callely (Eds.) Coryneform Bacteria. Academic Press. London, UK. 85–160.

Mitsuishi, Y., M. Kobayashi, and K. Matsuda. 1979. Dextran α-1,2 debranching enzyme from Flavobacterium sp. M-73: Its production and purification. Agric. Biol. Chem. 43:2283–2290.

Mitsuishi, Y., M. Kobayashi, and K. Matsuda. 1980. Dextran α-(1,2)-debranching enzyme from Flavobacterium sp. M-73: Properties and mode of action. Carbohydr. Res. 83:303–313.

Mitsubishi Kasei Corporation (MKC). 1989. Production of D-N-carbamyl-alpha-amino acids. Patent JP 01228489.

Moffett, M. L., P. C. Fahy, and D. Cartwright. 1983. Corynebactcrium. In: P. C. Fahy and G. J. Persley (Eds.) Plant Bacterial Diseases: A Diagnostic Guide. Academic Press. Sydney, Australia. 45–65.

Mokashi, S. A., and K. M. Paknikar. 2002. Microbacterium arsenic (III) oxidizing Microbacterium lacticum and its use in the treatment of arsenic contaminated groundwater. Lett. Appl. Microbiol. 34:258–262.

Moore, W. E. C., E. P. Cato, and L. V. H. Moore. 1985. Index of the bacterial and yeast nomenclatural changes puplished in the International Journal of Systematic Bacteriology since the 1980 Approved Lists of Bacterial names (1 January 1980 to 1 January 1985). Int. J. Syst. Bacteriol. 35:382–407.

Moore, C., and R. Norton. 1995. Corynebacterium aquaticum septicaemia in a neutropenic patient. J. Clin. Pathol. 48:971–972.

Morris, A. J., G. K. Henderson, D. A. Bremner, and J. F. Collins. 1986. Relapsing peritonitis in a patient undergoing continuous ambulatory peritoneal dialysis due to Corynebacterium aquaticum. J. Infect. 13:151–156.

Murata, H., and Y. Magae. 1996. Toxin production in a mushroom pathogenic bacterium, Pseudomonas tolaasii strain PT814 is activated by signals present ina host, Pleurotus ostreatus and those accumulating in the medium in the course of bacterial growth. In: D. J. Royse (Ed.) Mushroom Biology and Mushroom Products. Pennsylvania State University Press. College Park, PA. 483–494.

Murata, H., T. Tsukamoto, and S. Shirata. 1998. rtpA, A gene encoding a bacterial two-component sensor kinase, determines pathogenic traits of Pseudomonas tolaasii, the causal agent of brown blotch disease of a cultivated mushroom, Pleurotus ostreatus. Mycoscience 39:261–271.

Murray, T. D. 1986. Isolation of Corynebacterium agropyri from 30 to 40-year-old herbarium specimens of Agropyron species. Plant Dis. 70:378–380.

Nair, N. G., and P. C. Fahy. 1972. Bacteria antagonistic to Pseudomonas tolaasii and their control of brown blotch of the cultivated mushroom. J. Appl. Bacteriol. 35:439–442.

Nair, N. G., and P. C. Fahy. 1976. Commercial application of biological control of mushroom bacterial blotch. Austral. J. Agric. Res. 27:415–422.

Nakagawa, Y., T. Sakane, and A. Yokota. 1996. Transfer of "Pseudomonas riboflavina" (Foster 1944), a Gram-negative, motile rod with lon-chain 3-hydroxy fatty acids, to Devosia riboflavina gen. nov., sp. nov., nom. rev. Int. J. Syst. Bacteriol. 46:16–22.

Nakayama, K., A. Kato, Y. Ueno, Y. Minoda, and K. Komagata. 1980. Studies on the metabolism of trichothecene mycotoxins. II: Metabolism of T-2 toxin with the soil bacteria. Proc. Jpn. Assoc. Mycotoxicol. 12:30–32.

Nakazawa, K., and S. Suzuki. 1975a. Purification of keratan sulfate-endogalactosidase and its action on keratan sulfates of different origin. J. Biol. Chem. 250:912–917.

Nakazawa, K., N. Suzuki, and S. Suzuki. 1975b. Sequential degradation of keratan sulfate by bacterial enzymes and purification of a sulfatase in the enzymatic system. J. Biol. Chem. 250:905–911.

Naumova, I. B., A. S. Shashkov, E. M. Tul'skaya, G. M. Streshinskaya, Yu. I. Kozlova, N. V. Potekhina, L. I. Evtushenko, and E. Stackebrandt. 2001. Cell wall teichoic acids: structural diversity, species-specificity in the genus Nocardiopsis and chemotaxonomic perspective. FEMS Microbiol. Rev. 25:269–284.

Nazina, T. N., A. A. Grigor'ian, K. F. Sue, D. S. Sokolova, E. V. Novikova, T. P. Turova, A. B. Poltaraus, S. S. Beliaev, and M. V. Ivanov. 2002. Phylogenetic diversity of aerobic saprotrophic bacteria isolated from the Daqing oil field. Mikrobiologiia 71:91–97.

Nelson, P. E., and R. S. Dickey. 1970. Histopathology of plants infected with vascular bacterial pathogens. Ann. Rev. Phytopathol. 8:259–280.

Nelson, G. A. 1979. Persistence of Corynebacterium sepedonicum in soil and buried steams. Am. Potato J. 56:71–77.

Nelson, G. A. 1980. Long-term survival of Corynebacterium sepedonicum on contaminated surfaces and in infected potato stems. Am. Potato J. 57:595–600.

Nelson, G. A. 1982. Corynebacterium sepedonicum in potato: effect of inoculum concentration on ring rot symptoms and latent infection. Can. J. Plant Pathol. 4:129–133.

Nelson, G. A. 1985. Survival of Corynebacterium sepedonicum in potato stems and on surfaces held at freezing and above-freezing temperatures. Am. Potato J. 63:23–28.

Nemeth, J., E. Laszlo, and L. Emody. 1991. Clavibacter michiganensis ssp. insidiosus in lucerne seeds. Bull. OEPP 21:713–718.

Nesemann, G., H. Thurm, and A. Soeder. 1974. Mikrobiologisches Verfahren zur Herstellung von Oxoalkylxanthinen. German Federal Patent DE2302772.

Nissinen, R., F.-M. Lai, M. J. Laine, P. J. Bauer, A. A. Reilley, X. Li, S. H. De Boer, C. A. Ishimaru, and M. C. Metzler. 1997. Clavibacter michiganensis subsp. sepedonicus elicits a hypersensitive response in tobacco and secretes hypersensitive response-inducing protein(s). Phytopathology 87:678–684.

Nolte, F. S., K. E. Arnold, H. Sweat, E. F. Winton, and G. Funke. 1996. Vancomycin-resistant Aureobacterium species cellulitis and bacteremia in a patient with acute myelogenous leukemia. J. Clin. Microbiol. 34:1992–1994.

Nutkins, J. C., R. J. Mortishire-Smith, L. C. Packman, P. B. Brodey, K. Johnstone, and D. H. Williams. 1991. Structure determination of tolaasins, an extracellular lipodepsipeptide produced by the mushroom pathogen Pseudomonas tolaasii Pine. J. Ann. Chem. Soc. 113:2621–2627.

Oberreuter, H., H. Seiler, and S. Scherer. 2002. Identification of coryneform bacteria and related taxa by Fourier-transform infrared (FT-IR) spectroscopy. Int. J. Syst. Evol. Microbiol. 52:91–100.

O'Gara, P. J. 1916. A bacterial disease of western wheat grasses: First acount of the occurrence of a new type of bacterial disease in America. Science (Washington) 42:616–617.

Omelianski, V. L. 1923. Aroma-producing microorganisms. J. Bacteriol. 8:393–419.

Ophel, K. M., A. F. Bird, and A. Kerr. 1993. Assotiation of bacteriophage particles with toxin production by Clavibacter toxicus, the casual agent of annual ryegrass toxicity. Phytopathology 83:676–681.

Orla-Jensen, S. 1919. The Lactic Acid Bacteria. Host & Son. Copenhagen, Denmark.

Pan, Y.-B., M. P. Grisham, D. M. Burner, K. E. Jr. Damann, and Q. Wei. 1998. A polymerase chain reaction protocol for the detection of Clavibacter xyli subsp. xyli, the causal bacterium of sugarcane ratoon stunting disease. Plant Dis. 82:285–290.

Park, Y. H., H. Hori, K.-I. Suzuki, S. Osawa, and K. Komagata. 1987. Phylogenetic analysis of the coryneform bacteria by 5S rRNA sequences. J. Bacteriol. 169:1801–1806.

Park, Y.-H., K.-I. Suzuki, D.-G. Yim, K.-C. Lee, E. Kim, J. Yoon, S. Kim, Y. Kho, M. Goodfellow, and K. Komagata. 1993. Suprageneric classification of peptidoglycan group B actinomycetes by nucleotide sequencing of 5S ribosomal RNA. Ant. v. Leeuwenhoek 64:307–313.

Paschke, M. C., and N. K. Van Alfen. 1993. Extracellular polysaccharide impaired mutants of Clavibacter michiganensis subsp. michiganensis with tomato plants. Microbiol. Res. 151:1–9.

Paster Jr., B. J., W. A. Falkler Jr., C. O. Enwonwu, E. O. Idigbe, K. O., Savage, V. A., Levanos, M. A. Tamer, R. L. Ericson, C. N. Lau, and F. E. Dewhirst. 2002. Prevalent bacterial species and novel phylotypes in advanced noma lesions. J. Clin. Microbiol. 40:2187–2191.

Pattanapipitpaisal, P., N. L. Brown, and L. E. Macaskie. 2001. Chromate reduction and 16S rRNA identification of bacteria isolated from a Cr(VI)-contaminated site. Appl. Microbiol. Biotechnol. 57:257–261.

Pearson, D. 1971. The Biological Activity of Phytotoxic Extracts from Corynebacterium sepedonicum (MS thesis). University of Nebraska. Lincoln, NB.

Pernthaler, J., F. O. Glöckner, A. Unterholzner, A. Alfreider, R. Psenner, and R. Amann. 1998. Seasonal community and population dynamics of pelagic Bacteria and Archaea in a high mountain lake. Appl. Environ. Microbiol. 64:4299–4306.

Pernthaler, J., T. Posch, K. Simek, J. Vrba, A. Pernthaler, F. O. Glöckner, U. Nübel, R. Psenner, and R. Amann. 2001. Predator-specific enrichment of Actinobacteria from a cosmopolitan freshwater clade in mixed continuous culture. Appl. Environ. Microbiol. 67:2145–2155.

Powers, T. O., A. L. Szalanski, P. G. Mullin, T. S. Harris, T. Bertozzi, and J. A. Griesbach. 2001. Identification of seed gall nematodes of agronomic and regulatory concern with PCR-RFLP of ITS1. J. Nematology 33, pp:191–194.

Prauser, H., P. Schumann, F. A. Rainey, R. M. Kroppenstedt, and E. Stackebrandt. 1997. Terracoccus luteus gen. nov., sp. nov., an LL-diaminopimelic acid-containing coccoid actinomycete from soil. Int. J. Syst. Bacteriol. 47:1218–1224.

Price, P. G., J. M. Fisher, and A. Kerr. 1979. On Anguina funesta n. sp. and its association with Corynebacterium sp. in infected Loluim rigidum. Nematologica 25:76–85.

Pukall, R., D. Buntefuss, A. Fruhling, M. Rohde, R. M. Kroppenstedt, J. Burghardt, P. Lebaron, L. Bernard, and E. Stackebrandt. 1999. Sulfitobacter mediterraneus sp. nov., a new sulfite-oxidizing member of the alpha-Proteobacteria. Int. J. Syst. Bacteriol. 49:513–519.

Rademaker, J. I. W., M. Thalen, and J. D. Janse. 1992. Experiences with DNA hybridization using a biotinylated probe against Clavibacter michiganensis subsp. sepedonicus. Meded. Fac. Landbouwwet. Rijksuniv. Gent 57:263–268.

Rademaker, J. I. W., and J. D. Janse. 1994. Detection and identification of Clavibacter michiganensis subsp. sepedonicus and Clavibacter michiganensis subsp. michiganensis by nonradioactive hybridization, polymerase chain reaction and restriction enzyme analysis. Can. J. Microbiol. 40:1007–1018.

Rainey, P. B., C. L. Brodey, and K. Johnstone. 1993. Identification of a gene cluster encoding three high-molecular-weight proteins, which is required for synthesis of tolaasin by the mushroom pathogen Pseudomonas tolaasii. Molec. Microbiol. 8:643–652.

Rainey, F., N. Weiss, H. Prauser, and E. Stackebrandt. 1994. Further evidence for the phylogenetic coherence of actinomycetes with group B-peptidoglycan and evidence for the phylogenetic intermixing of the genera Microbacterium and Aureobacterium as determined by 16S rRNA analysis. FEMS Microbiol. Lett. 118:135–140.

Rainey, F. A., R. U. Ehlers, and E. Stackebrandt. 1995a. Inability of the polyphasic approach to systematics to determine the relatedness of the genera Xenorhabdus and Photorhabdus. Int. J. Syst. Bacteriol. 45:379–381.

Rainey, F. A., N. Weiss, and E. Stackebrandt. 1995b. Phylogenetic analysis of the genera Cellulomonas, Promicromonospora and Jonesia and proposal to exclude the genus Jonesia from the family Cellulomonadaceae. Int. J. Syst. Bacteriol. 45:649–652.

Rappe, M., D. Gordon, K. Vergin, and S. Giovannoni. 1999. Phylogeny of actinobacteria small subunit (SSU) rRNA gene clones recovered from marine bacterioplankton. Syst. Appl. Microbiol. 22:106–112.

Raupach, G. S., and J. W. Kloepper. 2000. Biocontrol of cucumber diseases in the field by plant promoting rhizobacteria with and without methyl bromide fumigation. Plant Dis. 84:1073–1075.

Reddy, G. S. N., J. S. S. Prakash, G. Srinivas, G. I. Matsumoto, and S. Shivaji. 2003. Leifsonia rubra sp. nov. and Leifsonia aurea sp. nov., psychrophiles from a pond in Antarctica. Int. J. Syst. Evol. Microbiol. 53:977–984.

Reiter, B., U. Pfeifer, H. Schwab, and A. Sessitsch. 2002. Response of endophytic bacterial communities in potato plants to infection with Erwinia carotovora subsp. atroseptica. Appl. Environ. Microbiol. 68:2261–2268.

Richardson, L. T. 1957. Quantitative determination of viability of potato ring rot bacteria following storage, heat and gas treatment. Can. J. Bot. 35:647–656.

Richardson, M. J. 1979. An Annotated List of Seed-borne Disease, 3rd ed. Commonwealth Agricultural Bureau (CAB), International Mycological Institute, Kew. London, UK. Phytopathological Paper 23.

Richardson, M. J. 1983. An annotated List of Seed-borne Diseases, 3rd ed., Suppl. 2. International Seed Testing Association. Zurich, Switzerland.

Riley, I. T. 1987. Serological relationships between strains of coryneform bacteria responsible for annual ryegrass toxicity and other plant-pathogenic corynebacteria. Int. J. Syst. Bacteriol. 35:53–159.

Riley, I. T., T. B. Reardon, and A. C. McKay. 1988. Genetic analysis of plant pathogenic bacteria in the genus Clavibacter using allozyme electrophoresis. J. Gen. Microbiol. 134:3025–3030.

Riley, I. T., and A. C. Mc Kay. 1990. Specificity of the adhesion of some plant pathogenic microorganisms to the cuticle of the nematodes in the genus Anguina (Nematoda: Anguinidae). Nematologica 36:90–103.

Riley, I. T., and J. M. Gooden. 1991a. Bacteriophage specific for the Clavibacter spp. associated with annual ryegrass toxicity. Lett. Appl. Microbiol. 12:158–160.

Riley, I. T., and K. M. Ophel. 1991b. Inoculation of Lolium rigidum with Clavibacter sp., the toxigenic bacteria associated with annual ryegrass toxicity. J. Appl. Bacteriol. 71:302–306.

Riley, I. T., and K. M. Ophel. 1992. Clavibacter toxicus sp. nov., the bacterium responsible for annual ryegrass toxicity in Australia. Int. J. Syst. Bacteriol. 42:64–68.

Riley, I., and G. Yan. 1999. Evaluation of non-toxigenic, nematode-vectored Clavibacter spp. for potential biocontrol of Clavibacter toxicus, the bacterium responsible for annual ryegrass toxicity. Australasian Ass. Nematol. 10(2).

Robinson, K. 1966a. Some observations on the taxonomy of the genus Microbacterium. I: Cultural and physiological reactions and heat resistance. J. Appl. Bacteriol. 29:607–615.

Robinson, K. 1966b. Some observations on the taxonomy of the genus Microbacterium. II: Cell wall analysis, gel electrophoresis and serology. J. Appl. Bacteriol. 29:616–624.

Rogosa, M., and R. Keddie. 1974. Genus Microbacterium. In: R. E. Buchanan and N. E. Gibbons (Eds.) Bergey's Manual of Determinative Bacteriology, 8th ed. Williams and Wilkins. Baltimore, MD. 628–629.

Roller, C., W. Ludwig, and K.-H. Schleifer. 1992. Grampositive bacteria with a high G+C content are characterized by a common insertion within their 23S rRNA genes. J. Gen. Microbiol. 138:167–1175.

Rollo, F., S. Luciani, A. Canapa, and I. Marota. 2000. Analysis of bacterial DNA in skin and muscle of the Tyrolean iceman offers new insight into the mummification process. Am. J. Phys. Anthropol. 111:211–219.

Romeiro, R., A. Karr, and R. Goodman. 1981. Isolation of a factor from apple that agglutinates Erwinia amylovora. Plant Physiol. 68:772–777.

Rossello-Mora, R., and R. Amann. 2001. The species concept for prokaryotes. FEMS Microbiol. Rev. 25:39–67.

Rudenko, A. V., and L. D. Varbanets. 1996. The effect of bacterial polysaccharides on the course of a Mycoplasma infection [in Russian]. Mikrobiol. Zh. 58(4):67–73.

Ryu, E. 1938. On the Gram-differentiation of of bacteria by the simplest method. J. Jpn. Soc. Vet. Sci. 17:58–63.

Saaltink, G. J., and P. H. Maas Geesteranus. 1969. A new disease of tulip caused by Corynebacterium oortii nov. sp. Theth. J. Plant Pathol. 75:123–128.

Sabet, A. K. 1954. On the host range and systematic position of the bacteria responsible for the yellow slime diseases of wheat (Ttiticum vulgare Vill.) and cocksfoot grass (Dactilis glomerata L.). Ann. Appl. Biol. 41:606–611.

Saile, E., J. A. McGarvey, M. A. Schell, and T. P. Denny. 1997. Role of extracellular polysaccharide and endoglucanase in root invasion and colonization of tomato plants by Ralstonia solanacearum. Phytopathology 87:1264–1271.

Saitou, N., and M. Nei. 1987. The neighbor-joining method: a new method for reconstructing phylogenetic trees. Molec. Biol. Evol. 4:406–425.

Sakane, T., I. Banno, and T. Iijima. 1983. Compounds protecting L-dried cultures from mutation. IFO Res. Comm. 11:14–24.

Sakane, T., I. Fukuda, T. Itoh, and A. Yokota. 1992. A long-term preservation of halophilic archaebacteria and thermoacidophilic archaebacteria by liquid drying. J. Microbiol. Meth. 16:281–287.

Sakane, T., and K.-I. Kuroshima. 1997. Viabilities of dried cultures of various bacteria after preservation fo over 20 years and their prediction by the accelerated storage test. Microbiol. Cult. Coll. 13:1–7.

Santos, M. S., L. Cruz, P. Norskov, and O. F. Rasmussen. 1997. A rapid and sensitive detection of Clavibacter michiganensis subsp. michiganensis in tomato seeds by polymerase chain reaction. Seed Sci. Technol. 25:581–584.

Sasaki, J., M. Chijimatsu, and K.-I. Suzuki. 1998. Taxonomic significance of 2,4-diaminobutyric acid isomers in the cell wall peptidoglycan of actinomycetes and reclassification of Clavibacter toxicus as Rathayibacter toxicus comb. nov. Int. J. Syst. Bacteriol. 48:403–410.

Saweljew, P., J. Kunkel, A. Feddersen, M. Baumert, J. Baehr, W. Ludwig, S. Bhakdi, and M. Husmann. 1996. Case of fatal systemic infection with an Aureobacterium sp.: Identification of isolate by 16S rRNA gene analysis. J. Clin. Microbiol. 34:1540–1541.

Scharif, G. 1961. Corynebacterium iranicum sp. nov. on wheat (Triticum aestivum) in Iran and a comparative study of it with C. tritici and C. rathayi. Entomol. Phytopathol. Appl. 19:1–4.

Schleifer, K.-H. 1970. Die Mureintypen in der Gattung Microbacterium. Arch. Microbiol. 71:271–282.

Schleifer, K.-H., and O. Kandler. 1972. Peptidoglycan types of bacterial cell walls and their taxonomic implications. Bacteriol. Rev. 36:407–477.

Schleifer, K.-H., I. Hayn, H. P. Seidl, and J. Firl. 1983. Threo-β-hydroxy-ornithine: a natural constituent of the peptidoglycan of Corynebacterium species Co 112. Arch. Microbiol. 134:243–246.

Schleifer, K.-H., and H. P. Seidl. 1985. Chemical composition and structure of murein. In: M. Goodfellow and D. E. Minnikin (Eds.) Chemical Methods in Bacterial Systematics. Academic Press. London, UK. 201–219.

Schumann, P., F. A. Rainey, J. Burghardt, E. Stackebrandt, and N. Weiss. 1999. Reclassification of Brevibacterium oxydans (Chatelain and Second 1966) as Microbacterium oxydans comb. nov. Int. J. Syst. Bacteriol. 49:175–177.

Schumann, P., U. Behrendt, A. Ulrich, and K. Suzuki. 2003. Reclassification of Subtercola pratensis (Behrendt et al. 2002) as Agreia pratensis comb. nov. Int. J. Syst. Evol. Microbiol. 53(Pt 6):2041–2044.

Schurholz, T., M. Wilimzig, E. Katsiou, and R. Eichenlaub. 1991. An anion channel forming activity from the plant pathogenic bacterium Clavibacter michiganensis ssp. nebraskensis. J. Membr. Biol. 123:1–8.

Schuster, M. L., and D. P. Coyne. 1974. Survival mechanisms of phytopathogenic bacteria. Ann. Rev. Phytopathol. 12:199–221.

Seffernick, J. L., G. Johnson, M. J. Sadowsky, and L. P. Wackett. 2000. Substrate specificity of atrazine chlorohydrolase and atrazine-catabolizing bacteria. Appl. Environ. Microbiol. 66:4247–4252.

Servin-Massieu, M. 1971. Effects of freeze-drying and sporulation on microbial variation. Curr. Topics Microbiol. Immunol. 54:119–150.

Shafiee, A., H. Motamedi, and A. King. 1998. Purification, characterization and immobilization of an NADPH-dependent enzyme involved in the chiral specific reduction of the keto ester M., an intermediate in the synthesis of an anti-asthma drug, Montelukast, from Microbacterium campoquemadoensis (MB5614). Appl. Microbiol. Biotechnol. 49:709–717.

Shashkov, A. S., G. M. Streshinskaya, V. A. Gnilozub, L. I. Evtushenko, and I. B. Naumova. 1995. Poly(arabitol phosphate) teichoic acid in the cell wall of Agromyces cerinus subsp. VKM Ac-1340. FEBS Lett. 371:163–166.

Sheridan, P. P., Loveland-Curtze, J., Miteva, V. I., and Brenchley, J. E. 2003. Isolation and characterization of Rhodoglobus vestali gen. nov., sp. nov., a novel psychrophilic organism isolated from an Antarctic Dry Valley Lake. Int. J. Syst. Evol. Microbiol. 53:985–994.

Shine, J. M., and J. C. Comstock. 1993. Digital image analysis system for determining tissue-blot immunoassay results for ratoon stunting disease of sugarcane. Plant. Dis. 77:511–513.

Shirata, A., K. Sugaya, M. Takasugi, and K. Monde. 1995. Isolation and biological activity of toxins produced by a Japanese strain of Pseudomonas tolaasii, the pathogen of bacterial rot of cultivated Oyster mushroom. Ann. Phytopathol Soc. Jpn. 61:493–502.

Shramko, P. A., S. G. Bezborodnikov, L. V. Dorofeeva, and L. I. Evtushenko. 1999. Heterogeneity of restriction profiles of amplified ribosomal DNA of Rathayibacter spp. and related bacteria. Problems of Nematology: Proceedings of the Zoological Institute, RAS 280:103–104.

Skerman, V. B. D., V. McGowan, and P. H. A. Sneath. 1980. Approved lists of bacterial names. Int. J. Syst. Bacteriol. 30:225–420.

Slack, S. A. 1978. Biology and ecology of Corynebacterium sepedonicum. Bacterial Ring Rot Proc. 64:665–670.

Slack, S. A., J. L. Drennan, and A. G. Westra. 1996. Comparison of PCR, ELISA and DNA hybridization for the detection of Clavibacter michiganensis subsp. sepedonicus in field-grown potatoes. Plant Dis. 80:519–524.

Smith, E. F. 1910. A new tomato disease of economic importance. Science (Washington) 31:794–796.

Smith, E. F. 1913. A new type of bacterial disease. Science (Washington) 38:926.

Sneath, P. H., and D. Jones. 1976. Brochothrix, a new genus tentatively placed in the family Lactobacillaceae. Int. J. Syst. Bacteriol. 26:102–104.

Soppeland, L. 1924. Flavobacterium suaveolens, a new species of aromatic bacillus isolated from dairy water. J. Agric. Res. 28:275–276.

Southey, J. F., P. B. Topham, and D. J. F. Brown. 1990. Taxonomy of some species of Anguina Scopoli, 1777 (sensu Brzeski, 1981) forming galls on Gramineae: value of diagnostic characters and present status of nominal species. Rev. Nematol. 13:127–142.

Spieckermann, A., and P. Kotthoff. 1914. Untersuchunger uber die Kartoffelpflanze und ihre Krankheiten. Landhr. Jb. 46:659–732.

Splittstoesser, D. F., M. Wexler, J. White, and R. R. Colweeel. 1967. Numerical taxonomy of Gram-positive and catalase-positive rods isolated from frozen vegetables. Appl. Microbiol. 15:158–162.

Stackebrandt, E., B. J. Lewis, and C. R. Woese. 1980. The phylogenetic structure of the coryneform group of bacteria. Zbl. Bakt., Abt. 1, Orig. C 2:137–149.

Stackebrandt, E., F. Rainey, and N. L. Ward-Rainey. 1997. Proposal for a new hierarchic classification system, Actinobacteria classis nov. Int. J. Syst. Bacteriol. 47:479–491.

Stackebrandt, E. 2000. Defining taxonomic ranks. In: M. Dworkin, S. Falkow, E. Rosenberg, K.-H. Schleifer, and E. Stackebrandt (Eds.) The Prokaryotes, 3rd ed. Springer-Verlag. New York, NY.

Stackebrandt, E., W. Frederiksen, G. M. Garrity, P. A. D. Grimont, P. Kämpfer, M. C. J. Maiden, X. Nesme, R. Rosselló-Mora, J. Swings, H. G. Trüper, L. Vauterin, A. C. Ward, and W. B. Whitman. 2002. Report of the ad hoc committee for the re-evaluation of the species definition in bacteriology. Int. J. Syst. Evol. Microbiol. 52:1043–1047.

Staley, J. T. 1968. Prosthecomicrobium and Ancalomicrobium: new prosthecate freshwater bacteria. J. Bacteriol. 95:1921–1942.

Starr, M. P., and P. P. Pirone. 1942. Phytomonas poinsettiae n. sp., the cause of a bacterial disease of Poinsettia. Phytopathology 32:1076–1081.

Starr, M. P., M. Mandel, and N. Murata. 1975. The phytopathogenic coryneform bacteria in the light of DNA base composition and DNA-DNA segmental homology. J. Gen. Appl. Microbiol. 21:13–26.

Stead, D. E. 1988. A study of the bacteria associated fatty acid profiling. Acta Horticulturae 225:39–46.

Steinbuch, J. G. 1799. Das Grasalchen, Vibrio agrostis, entdeckt unter beschreiben von J. G. Steinbuch. Der Naturforscher 28:233–259.

Steindl, D. R. L. 1961. Ratoon stunting disease. In: J. P. Martin, E. V. Abbot, and C. G. Hughes (Eds.) Sugarcane Diseases of the World. Elsevier. Amsterdam, The Netherlands 1:433–459.

Steindl, D. R. L. 1976. The use of phase contrast microscopy in the identification of ratoon stunting disease. Proc. Qld. Soc. Sugar Cane Technol. 43:71–72.

Steinhaus, E. A. 1941. A study of the bacteria associated with thirty species of insects. J. Bacteriol. 42:757–790.

Strobel, G. A., and W. M. Hess. 1968. Biological activity of a phytotoxic glycopeptide produced by Corynebacterium sepedonicum. Plant Physiol. 43:1673–1688.

Sturz, A. V., B. R. Christie, B. G. Matheson, and J. Nowak. 1997. Biodiversity of endophytic bacteria which colonize red clover nodules, roots, stems and foliage and their influence on host growth. Biol. Fertil. Soils 25:13–19.

Stynes, B. A., and A. F. Bird. 1980. Anguina agrostis, the vector of annual rye grass toxicity in Australia. Nematologica 26:475–490.

Stynes, B. A., and A. F. Bird. 1983. Development of annual ryegrass toxicity. Austral. J. Agric. Res. 34:653–660.

Sundin, G. W., and J. L. Jacobs. 1999. Ultraviolet padiation (UVR) sensitivity analysis and UVR survival strategies of a bacterial community from the phyllosphere of field-grown peanut (Arachis hypogeae L.). Microb. Ecol. 38:27–38.

Suslow, T. V., M. N. Schroth, and M. Isaka. 1982. Application of a rapid method for Gram differentiation of plant pathogenic and saprophytic bacteria without staining. Phytopathology 72:917–918.

Suyama, K., and H. Fujii. 1993. Bacterial disease occurred on cultivated mushroom in Japan. J. Agric. Sci. Tokyo Nogyo Daigaku 38:35–50.

Suyama, K., H. Negishi, and S. Wakimoto. 1995. Selective medium for Pseudomonas tolaasii. Ann. Phytopathol. Soc. Japan 61:255–256.

Suzuki, K., and K. Komagata. 1983. Taxonomic significance of cellular fatty acid composition in some coryneform bacteria. Int. J. Syst. Bacteriol. 33:188–200.

Suzuki, K., J. Sasaki, M. Uramoto, T. Nakase, and K. Komagata. 1996. Agromyces mediolanus sp. nov., nom. rev., comb. nov., a species for "Corynebacterium mediolanum" Mamoli 1939 and for some aniline-assim-

ilating bacteria which contain 2,4-diaminobutyric acid in the cell wall peptidoglycan. Int. J. Syst. Bacteriol. 46:88–93.

Suzuki, K., J. Sasaki, M. Uramoto, T. Nakase, and K. Komagata. 1997. Cryobacterium psychrophilum gen. nov., sp. nov., nom. rev., comb. nov., an obligately psychrophilic actinomycete to accommodate "Curtobacterium psychrophilum" Inoue and Komagata 1976. Int. J. Syst. Bacteriol. 47:474–478.

Suzuki, K., M. Suzuki, J. Sasaki, Y.-H. Park, and K. Komagata. 1999. Leifsonia gen. nov., a genus for 2,4-diaminobutyric acid-containing actinomycetes to accommodate "Corynebacterium aquaticum" Leifson 1962 and Clavibacter xyli subsp. cynodontis Davis et al. 1984. J. Gen. Appl. Microbiol. 45:253–262.

Suzuki, K., M. Suzuki, J. Sasaki, Y.-H. Park, and K. Komagata. 2000. Validation of the publication of new names and new combinations previously effectively published outside the IJSB: List No. 75. Int. J. Syst. Evol. Microbiol. 50:1415–1417.

Szallas, E., C. Koch, A. Fodor, J. Burghardt, O. Buss, A. Szentirmai, K. H. Nealson, and E. Stackebrandt. 1997. Phylogenetic evidence for the taxonomic heterogeneity of Photorhabdus luminescens. Int. J. Syst. Bacteriol. 47:402–407.

Takami, H., A. Inoue, F. Fuji, and K. Horikoshi. 1997. Microbial flora in the deepest sea mud of the Mariana Trench. FEMS Microbiol. Lett. 152:279–285.

Takeuchi, M., and A. Yokota. 1989. Cell-wall polysaccharides in coryneform bacteria. J. Gen. Appl. Microbiol. 35:233–252.

Takeuchi, M., A. Yokota, and A. Misaki. 1990. Comparative structures of the cell-wall polysaccharides of four species of the genus Microbacterium. J. Gen. Appl. Microbiol. 36:255–271.

Takeuchi, M., and A. Yokota. 1991. Reclassification of strains of Flavobacterium-Cytophaga group in IFO culture collection. IFO Res. Commun. 15:83–96.

Takeuchi, M., and A. Yokota. 1993. Evaluation of cell-wall sugar composition as a taxonomic marker of some coryneform bacteria. J. Gen. Appl. Microbiol. 39:505–512.

Takeuchi, M., and A. Yokota. 1994. Phylogenetic analysis of the genus Microbacterium based on 16S rRNA gene sequences. FEMS Microbiol. Lett. 124:11–16.

Takeuchi, M., N. Weiss, P. Schumann, and A. Yokota. 1996. Leucobacter komagatae gen. nov., sp. nov., a new aerobic Gram-positive, nonsporulating rod with 2,4-diaminobutyric acid in the cell wall. Int. J. Syst. Bacteriol. 46:967–971.

Takeuchi, M., and K. Hatano. 1998a. Proposal of six new species in the genus Microbacterium and transfer of Flavobacterium marinotypicum ZoBell and Upham to the genus Microbacterium as Microbacterium maritypicum comb. nov. Int. J. Syst. Bacteriol. 48:973–982.

Takeuchi, M., and K. Hatano. 1998b. Union of the genera Microbacterium Orla-Jensen and Aureobacterium Collins et al. in a redefined genus Microbacterium. Int. J. Syst. Bacteriol. 48:739–747.

Takeuchi, M., and K. Hatano. 1999. Phylogenetic analysis of actinobacteria in the mangrove rhizosphere. IFO Res. Commun. 19:47–62.

Takeuchi, M., and K. Hatano. 2001. Agromyces luteolus sp. nov., Agromyces rhizospherae sp. nov. and Agromyces bracchium sp. nov., from the mangrove rhizosphere. Int. J. Syst. Evol. Microbiol. 51:1529–1537.

Taylor, J., Stearman, R. S., and B. B. Uratani. 1993. Development of a native plasmid as a cloning vector in Clavibacter xyli subsp. cynodontis. Plasmid 29:241–244.

Tendler, C., and E. J. Bottone. 1989. Corynebacterium aquaticum urinary tract infection in a neonate and concepts regarding the role of the organism as a neonatal pathogen. J. Clin. Microbiol. 27:343–345.

Thomas, S. B., R. G. Druce, G. J. Peters, and D. G. Griffiths. 1967. Incidence and significance of thermoduric bacteria in farm milk samples: A reappraisal and review. J. Appl. Bacteriol. 30:265–298.

Thompson, E., J. V. Leary, and W. W. C. Chun. 1989. Specific detection of Clavibacter michiganensis subsp. michiganensis by a homologous DNA probe. Phytopathology 79:311–314.

Thompson, I. P., M. J. Bailey, J. S. Fenlon, T. R. Fermor, A. K. Liley, J. M. Lynch, P. J. McCormack, M. P. McQuilken, K. J. Purdy, P. B. Rainey, and J. M. Whipps. 1993. Quantitative and qualitative seasonal changes in the microbial community from the phyllosphere of sugar beet (Beta vulgaris). Plant and Soil 150:177–191.

Thorn, G., and A. Tsuneda. 1996. Molecular genetic characterization of bacterial isolates causing brown blotch on cultivated mushrooms in Japan. Mycoscience 37:409–416.

Tolaas, A. G. 1915. A bacterial disease of cultivated mushrooms. Phytopathology 5:51–54.

Tomasino, S. F., R. T. Lester, M. R. Dimmock, M. R. Beach, and J. L. Kelly. 1995. Field performnce of Clavibacter xyli subsp. cynodontis expressing the insecticidal protein gene cryIA(c) of Bacillus turingiensis against European corn borer in field corn. Biol. Control 5:442–448.

Topping, L. E. 1937. The predominant microorganisms in soils. I: Description and classification of the organisms. Zbl. Bakteriol. Parasitenkd. Infektionskr. Hyg. Abt. 2, Orig. 97:289–304.

Trutko, S. M., L. I. Evtushenko, L. V. Dorofeeva, M. G. Shlyapnikov, E. Yu. Gavrish, N. E. Suzina, and V. K. Akimenko. 2003. Terminal oxidases in representatives of some different genera of the family Microbacteriaceae [in Russian]. Microbiologiia 53:259–267.

Tsiantos, J. 1986. Survival of phytopathogen Corynebacterium michganense pv. michganense in water. Microbios Lett. 32:69–74.

Tsiantos, J. 1987. Transmission of bacterium Corynebacterium michiganense pv. michiganense by seeds. J. Phytopathol. 119:142–146.

Tsukamoto, T., A. Shirata, and H. Murata. 1998. Isolation of a Gram-positive bacterium effective in suppression of brown blotch disease of cultivated mushrooms, Pleurotus ostreatus and Agaricus bisporus, caused by Pseudomonas tolaasii. Mycoscience 39:273–278.

Tsukamoto, T., M. Takeuchi, O. Shida, H. Murata, and A. Shirata. 2001. Proposal of Mycetocola gen. nov. in the family Microbacteriaceae and three new species, Mycetocola saprophilus sp. nov., Mycetocola tolaasinivorans sp. nov. and Mycetocola lacteus sp. nov., isolated from cultivated ostreatus. Int. J. Syst. Evol. Microbiol. 51:937–944.

Tsukamoto, T., H. Murata, and A. Shirata. 2002. Identification of non-pseudomonad bacteria from fruit bodies of wild agaricales fungi that detoxify tolaasin produced by Pseudomonas tolaasii. Biosci. Biotechnol. Biochem. 66:2201–2208.

Turner, A. J., and S. R. Whittle. 1983. Biochemical dissection of the synapse. Biochem. J. 209:29–41.

Uchida, K., and K. Aida. 1977. Acyltype of bacterial cell wall: its simple identification by colorimetric method. J. Gen. Appl. Microbiol. 23:249–260.

Uemori, K. 1999. Gram positive bacteria containing 2,4-diaminobutyric acid in the cell wall: GenBank (www.ncbi.nlm.nih.gov). NCBI DataBase.

Ueno, Y., K. Nakayama, K. Ishii, F. Tashiro, Y. Minoda, T. Omori, and K. Komagata. 1983. Metabolism of T-2 toxin in Curtobacterium sp. strain 114-2. Appl. Environ. Microbiol. 46:120–127.

Uratani, B. B., S. C. Alcorn, B. H. Tsang, and J. L. Kelly. 1995. Construction of secretion vectors and use of heterologous signal sequences for protein secretion in Clavibacter xyli subsp. cynodontis. Molec. Plant-Microbe Interact. 8:892–828.

Urzi, C., L. Brusetti, P. Salamone, C. Sorlini, E. Stackebrandt, and D. Daffonchio. 2001. Biodiversity of Geodermatophilaceae isolated from altered stones and monuments in the Mediterranean basin. Environ. Microbiol. 7:471–479.

Van Alfen, N. K. 1989. Reassessment of plant wilt toxins. Ann. Rev. Phytopathol. 27:533–530.

Vandamme, P., B. Pot, M. Gills, P. De Vos, K. Kersters, and J. Swings. 1996. Polyphasic taxonomy, a consensus apptoach to bacterial systematics. Microbiol. Rev. 60:407–438.

Vanhooren, P. T., and E. J. Vandamme. 1999. L-fucose: Occurrence, physiological role, chemical, enzymatic and microbial synthesis. J. Chem. Technol. Biotechnol. 74:479–497.

Vanhooren, P. T., and E. J. Vandamme. 2000. Microbial production of clavan, an L-fucose rich exopolysaccharide. Prog. Biotechnol. 17:109–114.

Varbanets, L. D., E. D. Prokop'eva, A. A. Prokop'ev, and S. A. Ketlinskii. 1990a. The induction of tumor necrosis factor and interleukin-1 under exposure to the glycopolymers isolated from Pseudomonas solanacearum and Clavibacter michiganense [in Russian]. Mikrobiol. Zh. 52(5):17–23.

Varbanets, L. D., S. L. Rybalko, S. T. Diadiun, O. S. Brovarskaia, and N. V. Moskalenko. 1990b. Glycopolymers of Clavibacter michiganense and Pseudomonas solanacearum—interferon inducers [in Russian]. Mikrobiol. Zh. 52(4):71–77.

Vasseur, E., V. Broc, and J. J. Cocheton. 1998. Septicemia due to Corynebacterium aquaticum in an HIV-seropositive patient with an implantable chamber catheter [in French]. Presse Med. 27:1476–1477.

Vidaver, A. K., and M. Mandel. 1974. Corynebacterium nebraskense, a new, orange-pigmented phytopathogenic species. Int. J. Syst. Bacteriol. 24:482–485.

Vidaver, A. K. 1977. Maintenance of viability and virulence of Corynebacterium nebraskense. Phytopathology 6:825–827.

Vidaver, A. K., and M. P. Starr. 1981. Phytopathogenic coryneform and related bacteria. In: M. P. Starr, H. Stolp, H. G. Trüper, A. Balows and H. G. Schlegel (Eds.) The Prokaryotes. Springer-Verlag. Berlin, Germany. 2:1879–1887.

Vidaver, A. K. 1982. The plant pathogenic corynebacteria. Ann. Rev. Microbiol. 36:495–517.

Vidaver, A. K., and M. J. Davis. 1988. Coryneform plant pathogens. *In:* N. W. Schaad (Ed.) Laboratory Guide for Identification of Plant Pathogenic Bacteria, 2nd ed. APS Press. St. Paul, MN. 104–113.

Vogel, P., B. A. Stynes, W. Coackley, G. T. Yeoh, and D. S. Petterson. 1982. Glycolipid toxins from parasitised annual ryegrass: a comparison with tunicamycin. Biochem. Biophys. Res. Commun. 105:835–840.

Wada, M., A. Yoshizumi, S. Nakamori, and S. Shimizu. 1999. Purification and characterization of monovalent cation-activated levodione reductase from Corynebacterium aquaticum M-13. Appl. Environ. Microbiol. 65:4399–4403.

Wakimoto, S. 1955. Studies on the multiplication of OP1 phage (Xanthomonas oryzae bacteriophage). 1: One-step growth with experiment under various conditions. Sci. Bull. Fac. Agric. Kyushu. Univ. 15:399–416.

Wakushima, S., S. Kuraishi, and N. Sakurai. 1994. Soil salinity and pH in Japanese mangrove forests and growth of cultivated mangrove plants in different soil conditions. J. Plant Res. 107:39–46.

Wakuzaka, Y., and others. 1983. Immunoactivating agent originated from bacteria belonging to Agromyces genus. Patent JP 58043790.

Wallis, F. M. 1977. Ultrastructural histopathology of tomato plants infected with Corynebacterium michiganense. Physiol. Plant Pathol. 11:333–342.

Watts, J. L., D. E. Lowery, J. F. Teel, C. Ditto, J. S. Horng, and S. Rossbach. 2001. Phylogenetic studies on Corynebacterium bovis isolated from bovine mannary glands. J. Dairy Sci. 84:2419–2423.

Wayne, L. G., D. J. Brenner, R. R. Colwell, P. A. D. Grimont, O. Kandler, M. I. Krichevsky, L. H. Moore, W. E. C. Moore, R. G. E. Murray, E. Stackebrandt, M. P. Starr, and H. G. Trüper. 1987. Report of the ad hoc committee on reconciliation of approaches to bacterial systematics. Int. J. Syst. Bacteriol. 37:463–464.

Weeks, O. B. 1974. Genus Flavobacterium Bergey et al. 1923. *In:* R. E. Buchanan and N. E. Gibbons (Eds.) Bergey's Manual of Determinative Bacteriology, 8th ed. Williams and Wilkins. Baltimore, MD. 357–364.

Wenzel, M., I. Schonig, M. Berchtold, P. Kämpfer, and H. Konig. 2002. Aerobic and facultatively anaerobic cellulolytic bacteria from the gut of the termite Zootermopsis angusticollis. J. Appl. Microbiol. 92:32–40.

Wieser, M., P. Schumann, K. Martin, P. Altenburger, J. Burghardt, W. Lubitz, and H.-J. Busse. 1999. Agrococcus citreus sp. nov., isolated from a medieval wall painting of the chapel of Castle Herberstein (Austria). Int. J. Syst. Bacteriol. 49:1165–1170.

Williams, S. T., S. Lanning, and E. M. H. Wellington. 1984. Ecology of acinomycetes. *In:* M. Goodfellow, M. Mordarski and S. T. Williams (Eds.) The Biology of Actinomycetes. Academic Press. London, UK. 481–528.

Wong, B. L., Y. Q. Shen, and Y. P. Chen. 1996. New isolated Rathayibacter microorganism. Patent WO 9619569.

Wong, B. L., Y. Q. Shen, and Y. P. Chen. 1999. Enzymatic production of cefaclor from cephalexin. US Patent 5,939,299.

Yamada, K., and K. Komagata. 1970. Taxonomic studies on coryneform bacteria. III: DNA base composition of coryneform bacteria. J. Gen. Appl. Microbiol. 16:215–224.

Yamada, K., and K. Komagata. 1972a. Taxonomic studies on coryneform bacteria. IV: Morphological, cultural, bio-chemical and physiological characteristics. J. Gen. Appl. Microbiol. 18:399–416.

Yamada, K., and K. Komagata. 1972b. Taxonomic studies on coryneform bacteria. V: Classification of coryneform bacteria. J. Gen. Appl. Microbiol. 18:417–431.

Yamada, Y., G. Inouye, Y. Tahara, and K. Kondo. 1976. The menaquinone system in the classification of coryneform and nocardioform bacteria and related organisms. J. Gen. Appl. Microbiol. 22:203–214.

Yanagi, H., M. Matsufuji, K. Nakata, Y. Nagamatsu, S. Ohta, and A. Yoshimoto. 2000. A new type of glycoglycerolipids from Corynebacterium aquaticum. Biosci. Biotechnol. Biochem. 64:424–427.

Yokota, A., M. Takeuchi, T. Sakane, and N. Weiss. 1993a. Proposal of six new species in the genus Aureobacterium and transfer of Flavobacterium esteraromaticum Omelianski to the genus Aureobacterium as Aureobacterium esteraromaticum comb. nov. Int. J. Syst. Bacteriol. 43:555–564.

Yokota, A., M. Takeuchi, and N. Weiss. 1993b. Proposal of two new species in the genus Microbacterium: Microbacterium dextranolyticum sp. nov. and Microbacterium aurum sp. nov. Int. J. Syst. Bacteriol. 43:549–554.

Yorifuji, T., M. Kaneoke, T. Okazaki, and E. Shimizu. 1995. Degradation of 2-ketoarginine by guanidinobutyrase in arginine aminotransferase pathway of Brevibacterium helvolum. Biosci. Biotechnol. Biochem. 59:512–513.

Yoshisumi, A., M. Wada, H. Takagi, S. Shimizu, and S. Nakamori. 2001. Cloning, sequence analysis and expression in Escherichia coli of the gene encoding monovalent cation-activated levodione reductase from Corynebacterium aquaticum M-13. Biosci. Biotechnol. Biochem. 65:830–836.

Young, D. H., and L. Sequeira. 1986. Binding of Pseudomonas solanacearum fimbriae to tobacco leaf cell walls and its inhibition by bacterial extracellular polysaccharides. Physiol. Molec. Plant. Pathol. 24:393–402.

Young, J. M., G. S. Saddler, Y. Takikawa, S. H. De Boer, L. Vauterin, L. Gardan, R. I. Gvozdyak, and D. E. Stead. 1996. Names of Plant Pathogenic Bacteria, 1864–1995. International Society for Plant Pathology.

Young, J. M., G. S. Saddler, Y. Takikawa, S. H. De Boer, L. Vauterin, L. Gardan, R. I. Gvozdyak, and D. E. Stead. 2000. New Names, 31 December 2000. International Society for Plant Pathology.

Zaczynska, E., Z. Blach-Olszewska, G. Feldmane, A. Duks, H. Mordarska, I. Bizuniak, M. Cembrzynska-Nowak, and C. Lugowski. 1992. Effect of natural and synthetic immunomodulators on the synthesis of interferon by peritoneal cells of mice. Acta Virol. 36:121–128.

Zgurskaya, H. I., L. I. Evtushenko, V. N. Akimov, H. V. Voyevoda, T. G. Dobrovolskaya, L. V. Lysak, and L. V. Kalakoutskii. 1992. Emended description of the genus Agromyces and description of Agromyces cerinus subsp. cerinus sp. nov., subsp. nov., Agromyces cerinus subsp. nitratus sp. nov., subsp. nov., Agromyces fucosus subsp. fucosus sp. nov., subsp. nov. and Agromyces fucosus subsp. hippuratus sp. nov., subsp. nov. Int. J. Syst. Bacteriol. 42:635–641.

Zgurskaya, H. I. 1993a. Systematics of actinomycetes containing diamonobutyric acid in the cell wall (PhD thesis) [in Russian]. Institute of Biochemistry and Physiology of Microorganims of the RAS. Pushchino, Russia.

Zgurskaya, H. I., L. I. Evtushenko, V. N. Akimov, and L. V. Kalakoutskii. 1993b. Rathayibacter gen. nov., including the species Rathayibacter rathayi comb. nov., Rathayibacter tritici comb. nov., Rathayibacter iranicus comb. nov. and six strains from annual grasses. Int. J. Syst. Bacteriol. 43:143–149.

Zgurskaya, H. I., and L. I. Evtushenko. 1996. Agromyces cerinus subsp. cerinus strain for dideoxyarabinose production. Patent RU 2055887.

Zinniel, D. K., P. Lambrecht, N. B. Harris, Z. Feng, D. Kuczmarski, P. Higley, C. A. Ishimura, A. Arunakumari, R. G. Barletta, and A. K. Vidaver. 2002. Isolation and characterization of endophytic colonizing bacteria from agrononic crops and prairie plants. Appl. Environ. Microbiol. 68:2198–2208.

Zlamala, C., P. Schumann, P. Kämpfer, R. Rossello-More, W. Lubitz, and H.-J. Busse. 2002a. Agrococcus baldri sp. nov., a novel species isolated from the air in the "Virgilkapelle" in Vienna. Int. J. Syst. Evol. Microbiol. 52:1211–1216.

Zlamala, C., P. Schumann, P. Kämpfer, M. Valens, R. Rossello-More, W. Lubitz, and H.-J. Busse. 2002b. Microbacterium aerolatum sp. nov., isolated from the air in the "Virgilkapelle" in Vienna. Int. J. Syst. Evol. Microbiol. 52:1229–1234.

ZoBell, C. E., and H. C. Upman. 1944. A list of marine bacteria including descriptions of sixty new species. Bull. Scripps Inst. Oceanogr. Univ. Calif., Tech. Ser. 5:239–292.

Zupan, J. R., and P. Zambryski. 1995. Transfer of T-Dna from Agrobacterim to the plant cell. Plant Physiol. 107:1041–1104.

Zwart, G., W. D. Hiorns, B. A. Methe, M. P. van Agterveld, R. Huismans, S. C. Nold, J. P. Zehr, and H. J. Laanbroek. 1998. Nearly identical 16S rRNA sequences recovered from lakes in North America and Europe indicate the existence of clades of globally distributed freshwater bacteria. Syst. Appl. Microbiol. 21:546–556.

Zwart, G., W. C. Crump, M. P. Kamst-van Agterveld, F. Hagen, and S. K. Han. 2002. Typical fresh water bacteria: an analysis of available 16S rRNA gene sequences from plankton of lakes and rivers. Aquat. Microb. Ecol. 28:141–155.

Prokaryotes (2006) 3:1099–1113
DOI: 10.1007/0-387-30743-5_44

CHAPTER 1.1.29

The Genus *Nocardioides*

JUNG-HOON YOON AND YONG-HA PARK

The genus *Nocardioides* was proposed by Prauser (1976) for Gram-positive, non-acid-fast, aerobic, and mesophilic nocardioform actinomycetes, developing a mycelium that fragments into irregular rod- to coccus-like elements. Originally, the genus contained two species, i.e., *Nocardioides albus* and *Nocardioides luteus* (Prauser, 1984), both forming well-developed mycelia. The primary mycelium shows abundantly branching hyphae growing on the surface and penetrating into agar media. The hyphae of the aerial mycelium either do not branch or are sparsely and irregularly branched, resembling the structure of the aerial mycelium of nocardiae. Both the primary and the aerial mycelia can break up into short rod-like fragments, which give rise to new mycelia, by extruding one or two hyphae. The inclusion of additional species to the genus *Nocardioides* resulted in the morphologic heterogeneity of the genus. Other species subsequently transferred to *Nocardioides* were *Arthrobacter simplex* (O'Donnell et al., 1982) and *Pimelobacter jensenii* (Suzuki and Komagata, 1983). The genus *Pimelobacter* was originally proposed by Suzuki and Komagata (1983) for *Arthrobacter* species with LL-2,6-diaminopimelic acid in the peptidoglycan. Three *Pimelobacter* species were recognized, namely, *P. jensenii* (previously regarded as an atypical strain of *Arthrobacter simplex*), *P. tumescens* (previously, type strain of *Arthrobacter tumescens*), and *P. simplex* (previously regarded as the type strain of *Arthrobacter simplex* and two strains of "*Brevibacterium lipolyticum*"). As a consequence of a taxonomic reevaluation (Collins et al., 1989b), *Pimelobacter jensenii* was transferred to the genus *Nocardioides* as *N. jensenii*, and *P. tumescens* was reclassified in a new genus, *Terrabacter*, as *T. tumescens* (Collins et al., 1989b). The reclassified *N. simplex* and *N. jensenii* are coryneform actinomycetes that form neither a primary mycelium nor an aerial mycelium.

The number of species belonging to the genus *Nocardioides* has continuously increased. Two more coryneform species, *Nocardioides fastidiosa* (Collins and Stackebrandt, 1989a) and *Nocardioides plantarum* (Collins et al., 1994), were isolated from herbage. *Nocardioides fastidiosa* has MK-9(H₄) as the predominant menaquinone. Consistent with this, the genus *Nocardioides* was known to have MK-8(H₄) or MK-9(H₄) as the predominant menaquinone; however, Tamura and Yokota (1994) subsequently transferred *N. fastidiosa* to the genus *Aeromicrobium* as *A. fastidiosum*. Later, three more *Nocardioides* species, *N. pyridinolyticus*, *N. nitrophenolicus* and *N. aquaticus*, were added to the genus. Strains of these species form neither a primary mycelium nor an aerial mycelium.

Currently, there are eight validly described *Nocardioides* species, namely, *N. albus* (Prauser, 1976), *N. aquaticus* (Lawson et al., 2000), *N. jensenii* (Suzuki and Komagata, 1983; Collins et al., 1989b), *N. luteus* (Prauser, 1984), *N. nitrophenolicus* (Yoon et al., 1999), *N. plantarum* (Collins et al., 1994), *N. pyridinolyticus* (Yoon et al., 1997c), and *N. simplex* (O'Donnell et al., 1982). In addition, there are three invalidly described species, namely, "*N. flavus*" (Ruan and Zhang, 1979), "*N. fulvus*" (Ruan and Zhang, 1979), and "*N. thermolilacinus*" (Lu and Yan, 1983). The type strain of "*N. flavus*" was previously assumed to be a synonym of *N. luteus* (Prauser, 1989). Furthermore, the type strains of "*N. flavus*" and "*N. fulvus*" showed 16S rDNA similarity values of 100% with the type strain of *N. luteus* (Yoon et al., 1998b). Two strains of "*N. thermolilacinus*" were considered to be streptomycetes since they exhibit a *Streptomyces*-like life cycle and are susceptible to *Streptomyces* phages, but not to *Nocardioides* phages (Prauser, 1989). Analysis of 16S rDNA sequence also showed that "*N. thermolilacinus*" is most closely related to the genus *Streptomyces* (Yoon et al., 1998b).

The cell wall peptidoglycan of *Nocardioides* strains is defined by LL-2,6-diaminopimelic acid and glycine (wall chemotype I sensu, Lechevalier and Lechevalier, 1970) and a lack of mycolic acids (Prauser, 1976). The cellular fatty acid profile is complex, containing *iso-* and *anteiso-*branched, straight chain, unsaturated, and 10-methyl fatty acids. The major component is 14-methyl pentadecanoic acid (*iso-*C₁₆:₀)

(O'Donnell et al., 1982; Miller et al., 1991; Park et al., 1999). The predominant isoprenoid quinone is a tetrahydrogenated menaquinone with eight isoprene units (MK-8[H$_4$]; O'Donnell et al., 1982; Tamura and Yokota, 1994; Yoon et al., 1997c; Yoon et al., 1999). The G+C content of the DNA is 67–74 mol%. Phylogenetic analyses based on 16S rDNA sequences show that the genus *Nocardioides* is included as the type genus in the family Nocardioidaceae, which also includes the genera *Actinopolymorpha, Aeromicrobium, Kribbella, Hongia* and *Marmoricola* (Miller et al., 1991; Stackebrandt et al., 1997; Yoon et al., 1998b; Park et al., 1999; Lee et al., 2000; Urzì et al., 2000; Wang et al., 2001). The type species of the genus *Nocardioides* is *N. albus* (Prauser, 1976).

Habitats

Members of the genus *Nocardioides* have been isolated from a variety of habitats, including soils, herbage, an oil shale column, industrial wastewater, and water of hyper-saline lakes. *Nocardioides albus* is widely distributed in soil (see Prauser, 1976 and Prauser, 1989) and has been isolated from a variety of soils from different parts of the world (Prauser, 1976). *Nocardioides luteus* was isolated from soils of subtropical regions (Tille et al., 1978; Prauser, 1984). Strains identified as *N. simplex* have also been isolated from a variety of soils: the type strain of *N. simplex* was isolated from rice soil (Jensen, 1934) and other strains from the soil of the petroleum zone in Japan (Iizuka and Komagata, 1964). More recently, *Nocardioides* strains have been isolated from other sources. *Nocardioides plantarum* was isolated from herbage (Collins et al., 1994). *Nocardioides pyridinolyticus*, which can degrade pyridine, was isolated from the oxic zone of an oil shale column (Yoon et al., 1997c). *Nocardioides nitrophenolicus* was isolated from industrial wastewater in South Korea (Yoon et al., 1999). Recently, *N. aquaticus* was isolated from water of Ekho Lake, a hypersaline, heliothermal and meromictic lake, in the ice-free area of the Vestfold Hills in East Antarctica (Lawson et al., 2000).

Isolation and Cultivation

Nocardioides strains can be isolated by the usual dilution plating technique on solid media suitable for actinomycetes, such as the media by the International Streptomyces Project (Shirling and Gottlieb, 1966), glucose-asparagine agar (Lin-

denbein, 1952), and chitin agar (Prauser, 1976). They generally grow well on other kinds of complex organic media too, such as nutrient agar, trypticase soy agar, and YM agar. No selective media or enrichment procedures have yet been described which are specific for organisms of this genus. *Nocardioides* strains grow on a wide range of media. Most *Nocardioides* strains were isolated between 25 and 30°C, but *N. aquaticus* was isolated at 15°C and its optimal temperature for growth is 16–26°C.

Nocardioides albus strains were isolated by using the agar double-layer method, in which dilutions of soil suspensions are mixed into the melted agar, which is then used for the upper layer (Prauser, 1976). The isolation media were commonly used types, such as inorganic salt-starch agar (Gause et al., 1957), arginine-glycerol-salt agar (El-Nakeeb and Lechevalier, 1963), and dilute oatmeal agar (Prauser, 1976). *Nocardioides simplex* strains have been isolated by using different media. The type strain was isolated on asparagine agar (Jensen, 1934), and two other strains previously classified as members of "*Brevibacterium lipolyticum*" were isolated on kerosene medium and on crude oil medium (Iizuka and Komagata, 1964). *Nocardioides plantarum* was isolated and routinely cultivated on EYPB medium (Cure and Keddie, 1973; Collins et al., 1994). *Nocardioides pyridinolyticus* was isolated on minimal salts medium (Lee et al., 1994) containing 0.1% (v/v) pyridine and 0.4% (w/v) glucose. *Nocardioides nitrophenolicus* was isolated on modified minimal salts medium (Cho et al., 1998b; Yoon et al., 1999) supplemented with *p*-nitrophenol (100 mg/liter) as a source of carbon, nitrogen and energy. The most recently described species, *N. aquaticus*, was isolated on oligotrophic PYGV medium (Staley, 1968) prepared with water taken 1 m below the surface of the Antarctic Lake where it was discovered and was cultivated on PYGV medium prepared with artificial seawater (ASW; Lyman and Fleming, 1940). A variety of procedures, using contact plates, sterile cotton swabs, and porous plastic plugs, was used to isolate heterotrophic bacteria from Spanish caves with Paleolithic rock art, and these showed the existence of a *Nocardioides* isolate (Groth et al., 1999). A new method was elaborated for the selective isolation of actinomycetes from soil, which involves the pre-treatment of soil samples with electric pulses (Bulina et al., 1998). Treatment with 3-ms pulses at a field intensity of 12 kV/cm^2 resulted in a significant increase (threefold on average) in the proportion of rare actinomycetes isolated. Interestingly, treatment at a field intensity of 16 kV/cm^2 resulted in highly selective isolation of representatives of the genus *Nocardioides* (Bulina et al., 1998).

Media for Isolation and Cultivation

The following are the recipes for the media mentioned above that are not commercially available.

Arginine-glycerol-salt Agar (El-Nakeeb and Lechevalier, 1963)

Arginine monohydrochloride	1.0 g
Glycerol	12.5 g
K_2HPO_4 (anhyd.)	1.0 g
NaCl	1.0 g
$MgSO_4 \cdot 7H_2O$	0.5 g
$Fe_2(SO_4)_3 \cdot 6H_2O$	0.01 g
$CuSO_4 \cdot 5H_2O$	0.001 g
$ZnSO_4 \cdot 7H_2O$	0.001 g
$MnSO_4 \cdot H_2O$	0.001 g
Agar	15.0 g
Distilled water	1 liter

Adjust pH to 6.9–7.1

Asparagine Agar (Jensen, 1934)

Dextrose	10.0 g
Asparagine	1.0 g
K_2HPO_4 (anhyd.)	1.0 g
$MgSO_4$ (anhyd.)	0.5 g
NaCl	0.5 g
Agar	20.0 g
Distilled water	1 liter

Adjust pH to 7.0–7.2

Chitin Agar (Prauser, 1976)

KCl	0.5 g
K_2HPO_4 (anhyd.)	1.0 g
$MgSO_4$ (anhyd.)	0.5 g
$FeSO_4$ (anhyd.)	0.01 g
Chitin	1.0 g
Agar	15.0 g
Distilled water	1 liter

Adjust pH to 7.0

Crude Oil Medium (Iizuka and Komagata, 1964)

Crude oil	10.0 g
NH_4NO_3	2.5 g
Na_2HPO_4 (anhyd.)	1.0 g
KH_2PO_4 (anhyd.)	0.5 g
$MgSO_4 \cdot 7H_2O$	0.5 g
$MnCl_2 \cdot 4H_2O$	0.2 g
$CaCO_3$ (anhyd.)	5.0 g
$CaCl_2$ (anhyd.)	trace
$ZnSO_4 \cdot 7H_2O$	trace
$FeSO_4 \cdot 7H_2O$	trace
Agar	15.0 g
Distilled water	1 liter

Adjust pH to 7.2 before sterilization. $CaCO_3$ is added after sterilization.

Dilute Oatmeal Agar (Prauser, 1976)

Oatmeal	3.0 g
KNO_3	0.2 g
K_2HPO_4 (anhyd.)	0.5 g
$MgSO_4$ (anhyd.)	0.2 g
Agar	15.0 g
Distilled water	1 liter

Adjust pH to 7.0

Glucose-asparagine Agar (Lindenbein, 1952)

Glucose	10.0 g
L-Asparagine	0.5 g
K_2HPO_4 (anhyd.)	0.5 g
Agar	20.0 g
Distilled water	1 liter

Adjust pH to 6.8–7.0

Glycerol-asparagine Agar (ISP medium 5; Pridham and Lyons, 1961)

L-Asparagine	1.0 g
Glycerol	10.0 g
K_2HPO_4 (anhyd.)	1.0 g
Trace salts solution (see below)	1.0 ml
Agar	20.0 g
Distilled water	1 liter

Adjust pH to 7.0–7.4

Trace Salts Solution

$FeSO_4 \cdot 7H_2O$	0.1 g
$MnCl_2 \cdot 4H_2O$	0.1 g
$ZnSO_4 \cdot 7H_2O$	0.1 g
Distilled water	100 ml

Inorganic Salts-starch Agar (ISP Medium 4; Gause et al., 1957)

Soluble starch	10.0 g
K_2HPO_4 (anhyd.)	1.0 g
$MgSO_4 \cdot 7H_2O$	1.0 g
NaCl	1.0 g
$(NH_4)_2SO_4$ (anhyd.)	2.0 g
$CaCO_3$ (anhyd.)	2.0 g
Trace salts solution (see below)	1.0 ml
Agar	20.0 g
Distilled water	1 liter

Adjust pH to 7.0–7.4

Trace Salts Solution

$FeSO_4 \cdot 7H_2O$	0.1 g
$MnCl_2 \cdot 4H_2O$	0.1 g
$ZnSO_4 \cdot 7H_2O$	0.1 g
Distilled water	100 ml

Kerosene Medium (Iizuka and Komagata, 1964)

Kerosene	10.0 g
NH_4Cl	2.0 g
Na_2HPO_4 (anhyd.)	1.0 g
KH_2PO_4 (anhyd.)	0.5 g
$MgSO_4 \cdot 7H_2O$	0.5 g
NaCl	2.0 g
Agar	15.0 g
Distilled water	1 liter

Adjust pH to 7.2 before sterilization.

Minimal Salts Medium (Lee et al., 1991)

K_2HPO_4 (anhyd.)	1.0 g
KCl	0.25 g
$MgSO_4 \cdot 7H_2O$	0.25 g
Trace element solution (see below)	1.0 ml
Agar	15 g
Distilled water	1 liter

Adjust pH to 7.2

Trace Element Solution

$FeSO_4 \cdot 7H_2O$	4.0 mg
$MnSO_4 \cdot H_2O$	4.0 mg
Na_2MoO_4 (anhyd.)	0.5 mg
$CaCl_2 \cdot H_2O$	50 mg
NaCl	100 mg
Distilled water	100 ml

Modified Minimal Salts Medium (Cho et al., 1998a; Yoon et al., 1999)

K_2HPO_4 (anhyd.)	1.0 g
$NaH_2PO_4 \cdot 2H_2O$	0.5 g
KCl	0.25 g
$MgSO_4 \cdot 7H_2O$	0.25 g
Trace element solution (see below)	1.0 ml
Agar	15 g
Distilled water	1 liter
Adjust pH to 7.2	

Trace Element Solution

$FeSO_4 \cdot 7H_2O$	4.0 mg
$MnSO_4 \cdot H_2O$	4.0 mg
Na_2MoO_4 (anhyd.)	0.5 mg
$CaCl_2 \cdot H_2O$	50 mg
NaCl	100 mg
Distilled water	100 ml

PYGV Marine Medium (Staley, 1968)

Peptone	0.25 g
Yeast extract	0.25 g
Mineral salt solution (see below)	20.0 ml
Glucose solution (2.5%)	10.0 ml
Vitamin solution (see below)	10.0 ml
Agar	15 g
(Artificial) seawater	960 ml

Mix thoroughly all components, except the glucose and vitamin solutions. After autoclaving and cooling, add aseptically filter-sterilized glucose and vitamin solutions. Adjust pH to 7.5 with sterile KOH, if necessary.

Mineral Salt Solution

$MgSO_4 \cdot 7H_2O$	29.7 g
Nitrilotriacetic acid	10.0 g
$CaCl_2 \cdot 2H_2O$	3.34 g
$FeSO_4 \cdot 7H_2O$	99.0 mg
$Na_2MoO_4 \cdot 2H_2O$	13.0 mg
Metals 44 (see below)	50.0 ml
Distilled water	950 ml

Add nitrilotriacetic acid to 500 ml of distilled water. Dissolve by adjusting pH to 6.5 with KOH. Add remaining components and distilled water. Readjust pH to 7.2.

Vitamin Solution

Biotin	2.0 mg
Folic acid	2.0 mg
Pyridoxine · HCl	10.0 mg
Riboflavin	5.0 mg
Thiamine × HCl	5.0 mg
Nicotinamide	5.0 mg
Calcium pantothenate	5.0 mg
Cyanocobalamin	0.1 mg
p-Aminobenzoic acid	5.0 mg
Distilled water	1 liter

Metals 44

EDTA	250.0 mg
$ZnSO_4 \cdot 7H_2O$	1095.0 mg
$FeSO_4 \cdot 7H_2O$	500.0 mg
$MnSO_4 \cdot H_2O$	154.0 mg
$CuSO_4 \cdot 5H_2O$	39.2 mg
$Co(NO_3)_2 \cdot 6H_2O$	24.8 mg
$Na_2B_4O_7 \cdot 10H_2O$	17.7 mg
Distilled water	100 ml

Yeast Extract-malt Extract Agar (ISP Medium 2)

Yeast extract	4.0 g
Malt extract	10.0 g
Glucose	4.0 g
Agar	20.0 g
Distilled water	1 liter
Adjust pH to 7.3	

Preservation

For short-term preservation, serial transfer from agar slants of appropriate media is recommended. Agar slants can be kept at 4°C for at least 2 months. For long-term preservation, lyophilization and storage in liquid nitrogen or in frozen glycerol suspensions are suitable. For lyophilization, the cell mass is suspended in an appropriate fluid, such as 20% (w/v) skim milk. For storage in liquid nitrogen, the cell mass is inoculated into cryo-tubes containing an appropriate fluid, such as 20% glycerol. Glycerol suspension is prepared by suspending the cell mass in aqueous glycerol in the appropriate vial or tube. The vial or tube is stored at –20 or –70°C.

Identification

Members of the genus *Nocardioides* can be distinguished from all other actinomycetes by using a combination of chemical properties and phylogenetic analysis based on 16S rRNA gene sequences. However, within this genus, morphological features vary greatly and may be insufficient to differentiate it reliably from other genera of the family Nocardioidaceae or from other actinomycetes.

Morphology

Nocardioides albus and *N. luteus* form well-developed mycelia (Prauser, 1976; Prauser, 1984). The hyphae of the primary mycelium, 0.5–0.8 μm in diameter, grow on the surface or penetrate into agar media, branch abundantly, and are irregularly septated. Fragmentation of the hyphae begins in the old parts of the colonies, and as it proceeds, additional cross-walls are formed. The hyphae of the aerial mycelium, 0.6–

1.0 µm in diameter, either do not branch or are sparsely and irregularly branched, thus resembling those of nocardiae (Fig. 1). The aerial hyphae break up into rod- or coccus-like fragments, which are more regular than those of the primary mycelium. The fragments of both the primary and the aerial mycelia give rise to new mycelia by extruding new hyphae. Other *Nocardioides* species, namely, *N. simplex*, *N. jensenii*, *N. plantarum*, *N. pyridinolyticus*, *N. nitrophenolicus* and *N. aquaticus*, form neither primary mycelium nor aerial mycelium. The cells of *Nocardioides* species, which form no mycelia, are rods or cocci. These cells show rod-to-coccus morphogenesis from the early exponential growth phase to the stationary growth phase. Figure 2 shows this rod-to-coccus morphogenesis of *N. pyridinolyticus* KCTC 0074BP^T as the culture ages. Cells of *N. simplex* and *N. jensenii* are irregularly rod-shaped in the early phase of growth and coccoid in old cultures. After 1 day of cultivation on YM agar, the cells of the type strains are 1.0–1.2 × 1.5–6.0 µm and 0.6–0.8 × 3.0–7.0 µm, respectively. Cells of *N. plantarum* are short rods or cocci. Cells of *N. pyridinolyticus* are 0.5–0.6 µm wide and 1.2–1.6 µm long in the expo-

nential phase of growth. Cells of *N. nitrophenolicus* are 0.5–0.8 µm wide and 1.0–3.0 µm long in the exponential phase of growth. Cells of *N. aquaticus* are cocci or short rods, measuring 0.9–1.0 µm in diameter in old cultures. Cells either

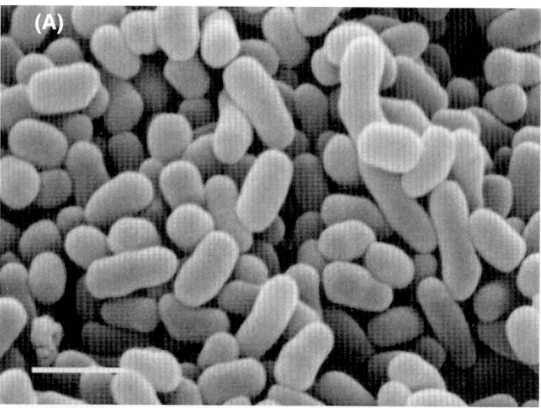

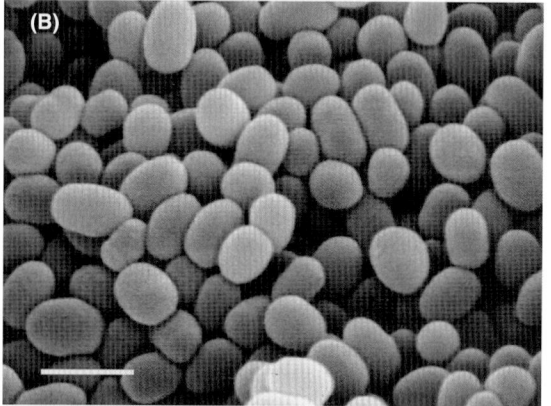

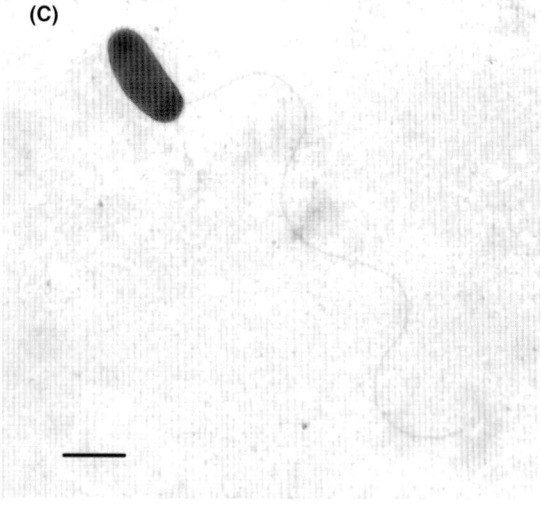

Fig. 2. *Nocardioides pyridinolyticus* (KCTC 0074BPT). Scanning electron micrographs A) and B) and transmission electron micrograph C) of cells grown on nutrient agar. Bar = 1 mm. A) Early growth phase (3-d cultivation on nutrient agar); B) stationary growth phase (7-d cultivation on nutrient agar); and C) early growth phase (3-d cultivation on nutrient agar).

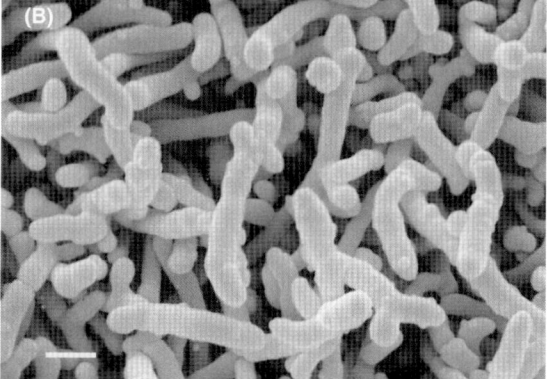

Fig. 1. *Nocardioides albus* (KCTC 9186T). Scanning electron micrographs of aerial mycelia on glucose-asparagine agar. Bar = 1 mm.

are nonmotile or possess a single polar flagellum. The flagellum of *N. pyridinolyticus* KCTC 0074BP^T is shown in Fig. 2.

Colonies of *N. albus* and *N. luteus* appear 1 or 2 days after incubation at 28°C and develop relatively quickly. The primary mycelium of *N. albus* is white, whitish, to faintly cream-colored on oatmeal agar, yeast extract-malt extract agar, glucose-asparagine agar and other agars; in aged cultures, it is whitish to faintly brownish. The aerial mycelium of *N. albus* is thin, dense and chalky. Colonies lacking an aerial mycelium appear pasty, smooth to wrinkled and dull to bright. The aerial mycelium may cover the primary mycelium totally, may be produced only in patches or at the margins of colonies, or may be absent, depending on strain, culture conditions, and the procedures and the duration of maintenance. The formation of aerial mycelium is usually good on yeast extract-malt extract agar, oatmeal agar and glucose-asparagine agar and fair on inorganic salts-starch agar (ISP 4 agar) and glycerol-asparagine agar. On complex organic medium, the aerial mycelium does not form. The colonies do not produce soluble pigments. Faint brownish soluble pigments may occur in aged cultures and reddish brown pigments may occur on media containing tyrosine as sole amino acid. In *N. luteus*, the primary mycelium is yellow on oatmeal agar and other media, varying to orange-yellow in aged cultures. The aerial mycelium on oatmeal agar is thin, dense and cream-colored if well developed, otherwise white. Colonies lacking an aerial mycelium appear pasty, smooth to wrinkled and dull to bright. No soluble pigments are produced, except for a reddish-brown pigment on media containing tyrosine as sole amino acid. Colonies of other *Nocardioides* species are smooth, glossy and raised or convex. Their colors are creamy, yellowish white or dull orange.

The morphological heterogeneity within the genus *Nocardioides* makes it difficult to differentiate the genus from others genera of the family Nocardioidaceae, as well as from other actinomycetes. Of the genera belonging to the same family, the genera *Kribbella* and *Hongia* have morphologies similar to those of *N. albus* and *N. luteus*; the genera *Aeromicrobium* and *Marmoricola* have morphologies similar to those of other *Nocardioides* species. The genus *Marmoricola* shows no rod/coccus life cycle.

Chemotaxonomy

The genus *Nocardioides* is easily differentiated from all other actinomycetes on the basis of the diaminopimelic acid type in the cell wall peptidoglycan, menaquinone profile, and the cellular

fatty acid profile (Park et al., 1999; Wang et al., 2001; Table 1). The genus *Nocardioides* is one of a few genera that have LL-2,6-diaminopimelic acid (LL-DAP) as the diagnostic diamino acid in the cell wall peptidoglycan (Prauser, 1976). This feature, together with glycine in the cell wall peptidoglycan, indicates that this genus is wall chemotype I (Lechevalier and Lechevalier, 1970). This chemical marker distinguishes the genus *Nocardioides* from many other actinomycete genera that possess other types of diamino acids in the cell wall peptidoglycan. Fewer than 20 validly described genera contain LL-DAP in the cell wall peptidoglycan (Maszenan et al., 1999; Lee et al., 2000; Urzì et al., 2000; Yoon and Park, 2000; Wang et al., 2001). The genus *Nocardioides* can be distinguished from most of these on the basis of menaquinone profile (Table 2). The genus *Nocardioides* contains tetrahydrogenated menaquinone with eight isoprenoid units (MK-[8H$_4$]) as the predominant menaquinone, a characteristic shared, along with LL-DAP in the cell wall, only with the genera *Marmoricola* (Urzì et al., 2000), *Terrabacter* (Collins et al., 1989b), and *Terracoccus* (Prauser et al., 1997). The genus *Nocardioides* can be differentiated from these by cellular fatty acid profile.

The cellular fatty acid profiles of *Nocardioides* species are similar, containing *iso-* and *anteiso-* branched, straight chain, unsaturated, and 10-methyl fatty acids (O'Donnell et al., 1982; Miller et al., 1991; Park et al., 1999). The major fatty acid characterizing the genus is *iso*-C$_{16:0}$ (14-methyl pentadecanoic acid). This profile clearly distinguishes it from the genera *Marmoricola*, *Terrabacter* and *Terracoccus*. The *Marmoricola* profile is characterized by the presence of C$_{16:0}$ and C$_{18:1}$ (Urzì et al., 2000). The *Terracoccus* profile has *iso*-C$_{15:0}$ and *anteiso*-C$_{15:0}$ as the major fatty acids (Prauser et al., 1997). The genus *Terrabacter* has shown slightly varying results. O'Donnell et al. (1982) and Park et al. (1999) concluded that major fatty acids are C$_{18:1}$, C$_{16:0}$ and *iso*-C$_{15:0}$; Miller et al. (1991) concluded they are *anteiso*-C$_{15:0}$, C$_{18:1}$ and C$_{17:1}$; Schumann et al. (1997) concluded they are *iso*-C$_{15:0}$ and *iso*-C$_{14:0}$. These differences may be due to differences in cultivation conditions and in preparation methods. The genera *Terrabacter* and *Terracoccus* have been shown to be phylogenetically distantly related to other members of the family Nocardioidaceae (Prauser et al., 1997; Yoon et al., 1998b; Fig. 3).

The phospholipid profiles of *Nocardioides* species are variable. All species so far studied contain phosphatidylglycerol, diphosphatidylglycerol, and other unidentified phospholipids (O'Donnell et al., 1982; Collins et al., 1989b; Lawson et al., 2000). Phosphatidylinositol is

Table 1. Differential characteristics of the genus *Nocardioides* and other genera of the family Nocardioidaceae.

Characteristic	*Nocardioides*	*Aeromicrobium*	*Kribbella*	*Hongia*	*Marmoricola*	*Actinopolymorpha*
Cell morphology	Hyphae, rods, and cocci	Rods	Hyphae	Hyphae	Cocci	Pleomorphism to hyphae
Predominant menaquinone(s)	MK-8(H_4)	MK-9(H_4)	MK-9(H_4)	MK-9(H_4)	MK-8(H_4)	MK-9(H_4), MK-9(H_6), MK-9(H_8), and MK-10(H_4)
Polar lipids	PG, and DPG (PL, PG-OH)	PE, and PG	PC	PC, PG, DPG, and PI	PG, DPG, and PI	PI, PIM, DPG, and PG
Major fatty acid(s)	iso-$C_{16:0}$	10-methyl-$C_{18:0}$, $C_{18:1}ω9c$, and $C_{16:0}$	anteiso-$C_{15:0}$, iso-$C_{16:0}$ or iso-$C_{14:0}$	anteiso-$C_{15:0}$, and iso-$C_{16:0}$	$C_{16:0}$, and $C_{18:1}$	ND
Peptidoglycan interpeptide bridge	Gly	Gly	Gly	Gly	Gly	ND
G+C content (mol%)	67–74	71–73	68–70	71	72	70
Signature nucleotides at						
positions 370:391	G:C	G:C	C:G	C:G	G:C	C:G
positions 602:636	G:T	G:T	G:T	G:T	G:T	A:T

Abbreviations, DPG, diphosphatidylglycerol; PC, phosphatidylcholine; PE, phosphatidylethanolamine; PG, phosphatidylglycerol; PG-OH, phosphatidylglycerol containing 2-hydroxy fatty acids; PI, phosphatidylinositol; PIM, phosphatidylinositol mannosides; PL, unknown phospholipid(s); iso-$C_{16:0}$, 14-methyl pentadecanoic acid; and ND, not determined.
Data taken from Prauser (1976); Prauser (1984); Miller et al. (1991); Suzuki and Komagata (1983); Tamura and Yokota (1994); Park et al. (1999); Lee et al. (2000); Urzì et al. (2000); Wang et al. (2001).

Table 2. Differential characteristics of *Nocardioides* species.

Characteristics	*N. albus*	*N. luteus*	*N. simplex*	*N. jensenii*	*N. plantarum*	*N. pyridinilyticus*	*N. nitrophenolicus*	*N. aquaticus*
Cell morphology	Hyphae	Hyphae	Rods, and cocci	Rods, and cocci	Short rods, and cocci	Rods, and cocci	Rods, and cocci	Rods, and cocci
Mycelium structure	+	+						–
Cell size (μm)	0.5–1.0	0.5–1.0	1.0–1.2 × 1.5–6.0	0.6–0.8 × 3.0–7.0	ND	0.5–0.6 × 1.2–1.6	0.5–0.8 × 1.0–3.0	0.9 × 1.0
Motility	–	ND	+	–	–	–	+	–
Colony color	Whitish to faintly brownish	Yellow to orange-yellow or cream	Yellowish white	Yellowish white	ND	Cream	Yellowish white	Dull orange
Temperature range for growth (°C)	18–37	15–37	10–37	ND	5–30	20–40	15–40	3–33.5
Optimal temperature for growth (°C)	28	28	26–37	ND	25	35	30	16–26
Nitrate reduction	ND	ND	ND	+	–	+	ND	+
Hydrolysis of								
Esculin	+	+	ND	ND	ND	+	+	ND
Casein	+	ND	+	ND	+	ND	+	+
Starch	+	+	ND	ND	–	ND	+	+
Gelatin	ND	ND	+	ND	+	+	ND	+
Tween 80	ND	ND	ND	ND	+	ND	+	+
Urea	ND	ND	–	+	–	ND	+	ND
Utilization of								
L-Arabinose	+	+	ND	ND	–	–	–	ND
D-Cellobiose	ND	ND	ND	ND	+	+	–	ND
Utilization of								
D-Fructose	+	+	ND	ND	+	+	+	ND
D-Galactose	+	ND	ND	ND	ND	+	–	+
D-Glucose	ND	+	+	+	+	+	+	+
Glycerol	–	–	ND	ND	ND	–	–	ND
Inositol	ND	ND	ND	ND	+	+	–	ND
Lactose	ND	ND	ND	ND	ND	+	–	ND
Maltose	+	+	–	ND	+	+	–	ND
D-Mannitol	ND	ND	ND	ND	ND	+	–	ND
D-Mannose	ND	ND	ND	ND	+	+	w	ND
Melezitose	ND	ND	ND	ND	+	ND	–	ND
D-Melibiose	ND	ND	ND	ND	ND	–	–	ND
D-Raffinose	v	–	–	+	ND	–	–	ND
L-Rhamnose	+	+	ND	ND	+	+	+	ND
D-Ribose	ND	ND	ND	+	ND	+	+	ND
Sucrose	v	+	+	ND	+	+	+	ND
D-Trehalose	ND	ND	ND	ND	ND	+	–	ND
D-Xylose	+	+	–	–	+	+	+	ND
G+C content (mol%)	67	68	72–74	69	69	73	71	69
Isolation source	Soil	Soil	Soil	Soil	Herbage	Oil shale column	Industrial wastewater	Saline lake, Antarctica

Symbols and abbreviations: +, positive reaction; –, negative reaction; v, variable; w, weakly positive reaction; and ND, not determined.
Data taken from Collins et al. (1989b); Collins et al. (1994); Lawson et al. (2000); Prauser (1976); Prauser (1984); Prauser (1989); Suzuki and Komagata (1983); Yoon et al. (1997); Yoon et al. (1999).

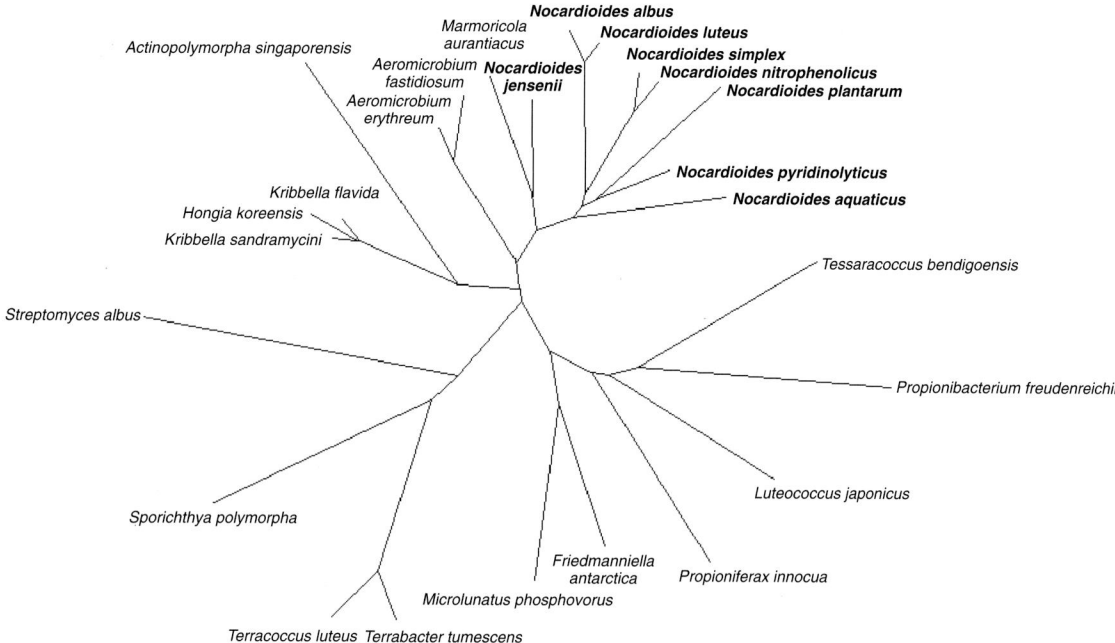

Fig. 3. Phylogenetic tree based on 16S rRNA gene sequences showing the positions of *Nocardioides* species and the representatives of other related taxa. Scale bar represents 0.01 substitutions per nucleotide position.

variably present in *Nocardioides* species. *Nocardioides jensenii* and *N. aquaticus* contain phosphatidylinositol, but *N. albus* and *N. simplex* do not. *Nocardioides albus*, *N. jensenii* and *N. simplex* contain phosphatidylglycerol containing 2-hydroxy fatty acids. Polyamines were analyzed for classification of coryne- and nocardioform actinomycetes with LL-DAP, including five *Nocardioides* species (Busse and Schumann, 1999). Most *Nocardioides* strains have high concentrations of the diamines putrescine and cadaverine (Busse and Schumann, 1999). This similarity is consistent with the phylogenetic homogeneity of *Nocardioides* species (Yoon et al., 1998b). The type strain of *N. plantarum* was found to contain a relatively low concentration of polyamine, and the predominant diamine was cadaverine. When two strains of *N. albus* (IMET 7807[T] and IMET 7815) were grown on a mineral medium (Lindenbein, 1952), cadaverine was again the predominant component. Most strains of the genus *Nocardioides*, when grown in R medium, contain high concentration of putrescine in addition to cadaverine, and this can differentiate them from the genus *Aeromicrobium*.

Molecular Systematics

The DNA G+C content of *Nocardioides* species is 67–74 mol% as determined from the thermal denaturation (T_m) method of Garvie (1978) and Marmur and Doty (1962) or the HPLC method of Tamaoka and Komagata (Tamaoka and Komagata, 1984; Tables 1 and 2). The 16S rRNA gene sequences have been determined for all *Nocardioides* species (Collins et al., 1989b; Collins et al., 1994; Yoon et al., 1998b; Lawson et al., 2000). Levels of nucleotide similarity are 93.5–99.0%. Intraspecific phylogenetic relationships based on 16S rRNA gene sequence similarities have been studied in *N. albus* and *N. simplex* (Yoon et al., 1998b). For 18 strains assigned to *N. albus*, the mean nucleotide similarity is 99.5 ± 0.5% (Yoon et al., 1998b). For *N. simplex*, eight strains have been investigated. Five strains, including the type strain, share an identical 16S rDNA sequence (Yoon et al., 1998b). However, the other three strains (ATCC 13260, 19565, and 19566) exhibited only 90.2–90.7% 16S rDNA similarity to the five closely related *N. simplex* strains and 89.9–92.2% similarity with the type strains of other *Nocardioides* species (Yoon et al., 1998b). These three strains are more related to the genus *Rhodococcus*, especially *R. erythropolis*, than to the genus *Nocardioides* (Yoon et al., 1998b). The results obtained in chemotaxonomic analysis are consistent with this, and the three strains have hence been reclassified as members of the genus *Rhodococcus* (Yoon et al., 1997a).

Studies of 16S rDNA sequence also showed the exact taxonomic positions of some invalid

Nocardioides species (Yoon et al., 1998b). Two strains (71-N54^T = IFO 14396^T and 71-N82 = IFO 14397; Ruan and Zhang, 1979) assigned to "*N. flavus*" and one strain (71-N86^T = JCM 3335^T; Ruan and Zhang, 1979) assigned to "*N. fulvus*" show 100% nucleotide similarity with *N. luteus*. The type strain (71-N54^T) of "*N. flavus*" was assumed to be a synonym of *N. luteus* by Prauser (1989). However, another strain (65-N86 = IFO 14399) assigned to "*N. fulvus*" shows low-level 16S rRNA gene nucleotide similarity (91.4%) with the type strain of "*N. fulvus*." It, together with an unclassified *Nocardioides* sp. ATCC 39419, has been classified as a member of a new genus, *Kribbella* (Park et al., 1999). Two strains (T505^T = IFO 14335^T and T511 = IFO 14336; Lu and Yan, 1983) assigned to "*N. thermolilacinus*" show low levels of 16S rRNA gene similarity (88.9–90.9%) with other members of the genus *Nocardioides*. Phylogenetic analysis based on 16S rRNA gene sequences reveals that they fall within the radiation of the cluster comprising *Streptomyces* species (Yoon et al., 1998b). These two strains also display a streptomyces-like life cycle and are susceptible to *Streptomyces* phages but not to *Nocardioides* phages (Prauser, 1989). Therefore, they are regarded as members of the genus *Streptomyces*.

Phylogenetic analysis based on 16S rRNA gene sequences shows that all *Nocardioides* species, except for *N. jensenii*, form distinct phylogenetic lineages within the radiation of the cluster encompassed by the genus *Nocardioides* (Fig. 3). *Nocardioides jensenii* was found to form a phylogenetic cluster with *Marmoricola aurantiacus* (Urzì et al., 2000; Fig. 3). The genus *Nocardioides* forms a distinct monophyletic group within the evolutionary radiation encompassed by the family Nocardioidaceae, which also includes the genera *Actinopolymorpha* (Wang et al., 2001), *Aeromicrobium* (Miller et al., 1991), *Hongia* (Lee et al., 2000), *Kribbella* (Park et al., 1999), and *Marmoricola* (Urzì et al., 2000). Levels of 16S rRNA gene similarity between *Nocardioides* species and members of other genera belonging to the family Nocardioidaceae are in the range of 89.9–94.3%. Stackebrandt et al. (1997) proposed 16S rRNA gene signature nucleotides characteristic of the family Nocardioidaceae at the following positions: 66-103 (G-C), 328 (C), 370-391 (G-C), 407-435 (A-U), 602-636 (G-U), 658-748 (U-A), 686 (U), 780 (G), 787 (A), 819 (U), 825-875 (G-C), and 1409-1491 (C-G). However, some newly proposed genera within the family have variations at a few of these positions. The genera *Actinopolymorpha*, *Hongia* and *Kribbella* have a C-G pair at positions 370-391 (Wang et al., 2001; Table 1). The genus *Actinopolymorpha* has this plus an A-T pair at positions 602-636 (Wang et al., 2001; Table 1).

Some other genes, such as the 16S-23S rRNA gene's internally transcribed spacer (ITS) region and the ribonuclease P (RNase P) RNA gene, have also been used for elucidating genetic relationships between *Nocardioides* species and between the genus *Nocardioides* and other related taxa (Yoon et al., 1998a; Yoon et al., 2000). In 33 strains assigned to the genus *Nocardioides*, the 16S-23S ITS sequences range from 328 bp to 539 bp. *Nocardioides* strains contain no tRNA sequences in their 16S-23S ITSs, as shown previously for some actinomycete species (Baylis and Bibb, 1988; Suzuki et al., 1988; Pernodet et al., 1989; Normand et al., 1992; Kim et al., 1993; Yoon et al., 1997b). In the type strains of validly described *Nocardioides* species, the 16S-23S ITSs are highly divergent in size and sequence. Their lengths range from 328 bp (*N. nitrophenolicus*) to 514 bp (*N. albus*). Between these type strains, the mean nucleotide similarity is 68.1 ± 16.8%, and this is low enough to distinguish between the species. The most distant relationship was between *N. jensenii* and *N. simplex*, for which the nucleotide similarity was 51.3%. Interestingly, the *N. jensenii* sequence was more similar to those of the genera *Aeromicrobium* and *Kribbella* than to those of other *Nocardioides* species. The 16S-23S ITS of *N. luteus* is only 1 bp different from those of strains 71-N54^T and 71-N82, assigned to "*N. flavus*," and it is identical to that of "*N. fulvus*" strain 71-N86^T, confirming the close relationship deduced from 16S rDNA sequence analysis. Intraspecific relationships were studied in *N. albus* and *N. simplex* (Yoon et al., 1998a). In 18 strains assigned to *N. albus*, the 16S-23S ITS sequences ranged from 468 bp to 533 bp, and the mean intraspecific nucleotide similarity value was 92.1±7.9%. In *N. simplex* strains, the sequences were 386 bp in all but one strain (NCIMB 12919), which has two types of 16S-23S ITS sequences (387 and 388 bp). The levels of nucleotide similarity range from 97.7 to 100%.

RNase P is a ribonuclease that cleaves the 5′ leader sequences of tRNA precursors and is a ribonucleoprotein composed of an RNA molecule of about 400 nucleotides and a protein subunit of about 120 amino acids (Pace and Brown, 1995). RNase P RNA is encoded by a single-copy gene (*rnp B*) in Gram-positive bacteria (Haas et al., 1996) and its size is relatively conserved (350–400 bp in bacteria; Pace and Brown, 1995). Recently, this gene has been evaluated as a new phylogenetic marker (Cho et al., 1998b; Yoon and Park, 2000). RNase P RNA genes have been sequenced to investigate the relationships between the genus *Nocardioides* and other taxa, as well as the relationships between *Nocardioides* species (Yoon and Park, 2000). Among validly described *Nocardioides* species, the sequences are relatively divergent, when compared with those of 16S rDNA. Among the type strains of those species, the mean level of sequence similarity is 85.8 ± 8.2%, and if gaps are included, it is approximately 76.6 ± 12.5%. The topology of the phylogenetic trees of 16S rRNA gene sequences and RNase P RNA gene sequences (Yoon et al., 1998b; Yoon and Park, 2000) was mostly consistent.

In a phylogenetic analysis based on RNase P RNA gene sequences, the type strains of *N. albus* and *N. luteus* are phylogenetic neighbors, with nucleotide similarity of 94%. The type strain (71-N54^T) of "*N. flavus*" has a sequence identical to that of *N. luteus*. The type strain (71-N86^T) of "*N. fulvus*" differs from *N. luteus* by only one base pair. This supports previous data showing that two strains assigned to "*N. flavus*" and one strain assigned to "*N. fulvus*" could be classified as members of *N. luteus*. *Nocardioides simplex* forms the closest phylogenetic cluster with *N. nitrophenolicus*, and the nucleotide similarity between their type strains is 94.7%. Other *Nocardioides* species, *N. jensenii*, *N. plantarum* and *N. pyridinolyticus*, each form distinct phylogenetic lineages. However, in a phylogenetic analysis including other taxa, *Nocardioides* species did not form a monophyletic cluster. In particular, *N. pyridinolyticus* and *N. jensenii* formed lineages that were separate from the cluster that contained the type species of the genus *Nocardioides*, *N. albus*, and most other *Nocardioides* species.

By aligning 16S rDNA sequences of *Nocardioides* species, species-specific sequences were identified (Park et al., 1998) and were used for design of species-specific primers. These primers are: 5′-ACCGGATACGACAACCGATT-3′ (*Escherichia coli*, positions 174-193) for *N. albus* plus *N. luteus*; 5′-TGTAAACCTCTTTCAGC GGG-3′ (*E. coli*, positions 427-446) for *N. jensenii*; 5′-AGGCGGTTCTCTGGCAATGTT-3′ (*E. coli*, positions 729-749) for *N. plantarum*; and 5′-ATAGGGGTCTCTTTGATACT-3′ (*E. coli*, positions 1016-1038) for *N. simplex*. These form the basis of a rapid multiplex polymerase chain reaction (PCR) test to identify *Nocardioides* species. In this test, all the species-specific primers and one universal reverse primer are mixed in a single reaction tube, and the species is identified by the size of PCR product. When representative strains of *Nocardioides* species are studied, the PCR products are clearly distinguished by their sizes in agarose gel electrophoresis (Fig. 4). In the multiplex PCR analysis, products of the type strains of *N. albus* and *N. luteus* appear at the same positions. The type strains of *N. jensenii*, *N. plantarum*, and *N. simplex* give PCR products of 1099, 827 and 529 bp, respectively. Products from additional strains previously identified as *N.*

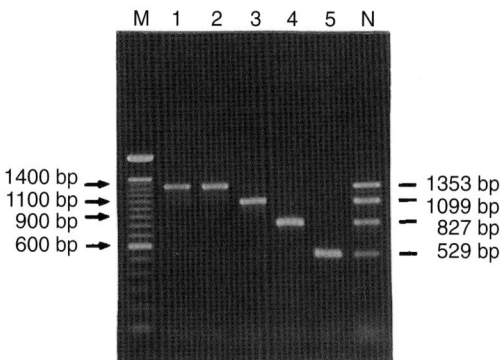

Fig. 4. Agarose gel electrophoresis of amplification products obtained by multiplex PCR from representative strains of the genus *Nocardioides*. Lanes: M, 100 bp DNA ladder (Gibco BRL, Gaithersburg, MD, USA); 1, *N. albus* KCTC 9186^T; 2, *N. luteus* KCTC 9575^T; 3, *N. jensenii* KCTC 9134^T; 4, *N. plantarum* NCIMB 12834^T; 5, *N. simplex* KCTC 9106T; N, species-specific PCR products obtained from four primer sets electrophoresed together.

albus and *N. simplex* appear at the same positions as those of the type strains of corresponding species. Thus the multiplex PCR assay rapidly identifies some *Nocardioides* species and could probably be applied to other recently added *Nocardioides* species, by design of species-specific primers.

Applications

Only a few *Nocardioides* strains produce antibiotics (Ogawara et al., 1985; Dellweg et al., 1988; Sanglier et al., 1993). Strain ATCC 39419, which produces a novel anti-tumor antibiotic, sandramycin, was originally assigned to *Nocardioides* (Matson and Bush, 1989), but has recently been classified as a member of a new genus *Kribbella* (Park et al., 1999). *Nocardioides* strains can perform chemical and enzymatic modifications of complex compounds (Nobile and Belleville, 1958; Behrend and Heesche-Wagner, 1999; Patel et al., 2000) and produce some industrially useful enzymes (Hanson et al., 1994; Jun et al., 1994; Masson et al., 1995; Nishimoto et al., 1996; Siddiqui et al., 2000). Some strains of *N. simplex* oxidize cholesterols (Arima et al., 1969; Nagasawa et al., 1969). Many *Nocardioides* strains play an important role in bioremediation. They degrade a variety of pollutants, including alkanes (Iizuka and Komagata, 1964; Hamamura and Arp, 2000; Hamamura et al., 2001), pyridine (Lee et al., 1994; Yoon et al., 1997c), phenols (Rajan et al., 1996; Cho et al., 1998b; Cho et al., 2000; Ebert et al., 1999; Männistö et al., 1999; Männistö et al., 2001; Yoon et al., 1999), phenanthrene (Iwabuchi et al.,

1998c), and others (Golovleva et al., 1990). The type strain of *N. jensenii* can degrade dinitro-*o*-cresol herbicides (Gundersen and Jensen, 1956), and other *Nocardioides* strains can also degrade herbicides (Travkin et al., 1997; Topp et al., 2000). Some enzymes that degrade pollutants have been characterized (Travkin et al., 1997; Travkin et al., 1999; Iwabuchi et al., 1998c; Saito et al., 1999), and those which degrade phenanthrene have been studied in detail (Iwabuchi and Harayama, 1997; Iwabuchi and Harayama, 1998a; Iwabuchi and Harayama, 1998b; Saito et al., 1999). In the phenanthrene degradation pathway, the structure of the ring cleavage product of an intermediate (1-hydroxy-2-naphthoate) has been determined (Adachi et al., 1999). The gene encoding the enzyme phenanthrene dioxygenase has been cloned, sequenced, and expressed in a functional form in *E. coli* (Saito et al., 2000). Consequently, *Nocardioides* strains may have a role in bioremediation and detoxification of contaminated regions (Männistö et al., 1999; Männistö et al., 2001; Topp et al., 2000).

Acknowledgments. This work was supported by grants HSS0310134 and an NRL research program (grants M10104000294-01J000012800 and NLW0070111) from the Ministry of Science and Technology (MOST) of South Korea. We are grateful to Dr. Hyun Woo Oh for kindly providing electron micrographs.

Literature Cited

Adachi, K., T. Iwabuchi, H. Sano, and S. Harayama. 1999. Structure of the ring cleavage of 1-hydroxy-2-naphthoate, an intermediate of the phenanthrene-degradative pathway of Nocardioides sp. strain KP7. J. Bacteriol. 181:757–763.

Arima, K., M. Nagasawa, M. Bae, and G. Tamura. 1969. Microbial transformation of sterols. Part 1: Decomposition of cholesterol by microorganisms. Agric. Biol. Chem. 33:1636–1643.

Baylis, H. A., and M. J. Bibb. 1988. Transcriptional analysis of the 16S rRNA gene of the rrnD gene set of Streptomyces coelicolor A3(2). Molec. Microbiol. 2:569–579.

Behrend, C., and K. Heesche-Wagner. 1999. Formation of hydride-meisenheimer complexes of picric acid (2,4,6-trinitrophenol) and 2,4-dinitrophenol during mineralization of picric acid by Nocardioides sp. strain CB 22-2. Appl. Environ. Microbiol. 65:1372–1377.

Bulina, T. I., L. P. Terekhova, and M. V. Tiurin. 1998. Use of electric impulses for selective isolation of actinomycetes from soil. Mikrobiologiia 67:556–560.

Busse, H.-J., and P. Schumann. 1999. Polyamine profiles within genera of the class Actinobacteria with LL-diaminopimelic acid in the peptidoglycan. Int. J. Syst. Bacteriol. 49:179–184.

Cho, M., J.-H. Yoon, S.-B. Kim, and Y.-H. Park. 1998a. Application of the ribonuclease P (RNase P) RNA gene sequence for phylogenetic analysis of the genus Saccharomonospora. Int. J. Syst. Bacteriol. 48:1223–1230.

Cho, Y.-G., J.-H. Yoon, Y.-H. Park, and S. T. Lee. 1998b. Simultaneous degradation of p-nitrophenol and phenol by a newly isolated Nocardioides sp. J. Gen. Appl. Microbiol. 44:303–309.

Cho, Y.-G., S. K. Rhee, and S. T. Lee. 2000. Influence of phenol on biodegradation of p-nitrophenol by freely suspended and immobilized Nocardioides sp. NSP41. Biodegradation 11:21–28.

Collins, M. D., and E. Stackebrandt. 1989a. Molecular taxonomic studies on some LL-diaminopimelic acid-containing coryneforms from herbage: description of Nocardioides fastidiosa sp. nov. FEMS Microbiol. Lett. 57:289–294.

Collins, M. D., M. Dorsch, and E. Stackebrandt. 1989b. Transfer of Pimelobacter tumescens to Terrabacter gen. nov. as Terrabacter tumescens comb. nov. and of Pimelobacter jensenii to Nocardioides as Nocardioides jensenii comb. nov. Int. J. Syst. Bacteriol. 39:1–6.

Collins, M. D., S. Cockcroft, and S. Wallbanks. 1994. Phylogenetic analysis of a new LL-diaminopimelic acid-containing coryneform bacterium from herbage, Nocardioides plantarum sp. nov. Int. J. Syst. Bacteriol. 44:523–526.

Cure, G. L., and R. M. Keddie. 1973. Methods for the morphological examination of aerobic coryneform bacteria. Soc. Appl. Bacteriol. Tech. 7:123–135.

Dellweg, H., J. Kurz, W. Pflüger, M. Schedel, G. Vobis, and C. Wünsche. 1988. Rodaplutin, a new peptidylnucleoside from Nocardioides albus. J. Antibiot. 41:1145–1147.

Ebert, S., P.-G. Rieger, and H.-J. Knackmuss. 1999. Function of coenzyme F_{420} in aerobic catabolism of 2,4,6-trinitrophenol and 2,4-dinitrophenol by Nocardioides simplex FJ2-1A. J. Bacteriol. 181:2669–2674.

El-Nakeeb, M. A., and H. A. Lechevalier. 1963. Selective isolation of aerobic actinomycetes. Appl. Microbiol. 11:75–77.

Garvie, E. I. 1978. Streptococcus raffinolactis Orla-Jensen and Hensen. a group N streptococcus found in raw milk. Int. J. Syst. Bacteriol. 28:190–193.

Gause, G. F., T. P. Preobrazhenska, E. S. Kudrina, N. O. Blinov, I. D. Rjabova, and M. A. Sveshnikova. 1957. Problems Pertaining to the Classification of Actinomycetes-antagonists [in Russian]. Medgiz. Moscow, Russia.

Golovleva, L. A., R. N. Pertsova, L. I. Evtushenko, and B. P. Baskunov. 1990. Degradation of 2,4,5-trichlorophenoxy-acetic acid by a Nocardioides simplex culture. Biodegradation 1:263–271.

Groth, I., R. Vettermann, B. Schuetze, P. Schumann, and C. Saiz-Jimenez. 1999. Actinomycetes in Karstic caves of northern Spain (Altamira and Tito Bustillo). J. Microbiol. Meth. 36:115–122.

Gundersen, K., and H. L. Jensen. 1956. A soil bacterium decomposing organic nitro-compounds. Acta Agric. Scand. 6:100–114.

Haas, E. S., A. B. Banta, J. K. Harris, N. R. Pace, and J. W. Brown. 1996. Structure and evolution of ribonuclease P RNA in Gram-positive bacteria. Nucleic. Acids Res. 24:4775–4782.

Hamamura, N., and D. J. Arp. 2000. Isolation and characterization of alkane-utilizing Nocardioides sp. strain CF8. FEMS Microbiol. Lett. 186:21–26.

Hamamura, N., C. M. Yeager, and D. J. Arp. 2001. Two distinct monooxygenases for alkane oxidation in Nocardioides sp. strain CF8. Appl. Environ. Microbiol. 67:4992–4998.

Hanson, R. L., J. M. Wasylyk, V. B. Nanduri, D. L. Cazzulino, R. N. Patel, and L. J. Szarka. 1994. Site-specific enzymatic hydrolysis of taxanes at C-10 and C-13. J. Biol. Chem. 269:22145–22149.

Iizuka, H., and K. Komagata. 1964. Microbiological studies on petroleum and natural gas. I: Determination of hydrocarbon-utilizing bacteria. J. Gen. Appl. Microbiol. 10:207–221.

Iwabuchi, T., and S. Harayama. 1997. Biochemical and genetic characterization of 2-carboxybenzaldehyde dehydrogenase, an enzyme involved in phenanthrene degradation by Nocardioides sp. strain KP7. J. Bacteriol. 179:6488–6494.

Iwabuchi, T., and S. Harayama. 1998a. Biochemical and genetic characterization of trans-2'-carboxybenzalpyruvate hydratase-aldolase from a phenanthrene-degrading Nocardioides strain. J. Bacteriol. 180:945–949.

Iwabuchi, T., and S. Harayama. 1998b. Biochemical and molecular characterization of 1-hydroxy-2-naphthoate dioxygenase from Nocardioides sp. KP7. J. Biol. Chem. 273:8332–8336.

Iwabuchi, T., Y. Inomata-Yamauchi, A. Katsuta, and S. Harayama. 1998c. Isolation and characterization of marine Nocardioides capable of growing and degrading phenanthrene at 42°C. J. Mar. Biotechnol. 6:86–90.

Jensen, H. L. 1934. Studies on saprophytic mycobacteria and corynebacteria. Proc. Linn. Soc. New South Wales 59:19–61.

Jun, H. K., T. S. Kim, and Y. Yeeh. 1994. Purification and characterization of an extracellular adenosine deaminase from Nocardioides sp. J-326TK. Biotechnol. Appl. Biochem. (Korea) 20:265–277.

Kim, E., H. Kim, S.-P. Hong, K. H. Kang, Y. H. Kho, and Y.-H. Park. 1993. Gene organization and primary structure of a ribosomal RNA gene cluster from Streptomyces griseus subsp. griseus. Gene 132:21–31.

Lawson, P. A., M. D. Collins, P. Schumann, B. J. Tindall, P. Hirsch, and M. Labrenz. 2000. New LL-diaminopimelic acid-containing actinomycetes from hypersaline, heliothermal and meromictic Antarctic Ekho Lake: Nocardioides aquaticus sp. nov. and Friedmanniella lacustris sp. nov. Syst. Appl. Microbiol. 23:219–229.

Lechevalier, M. P., and H. A. Lechevalier. 1970. A critical evaluation of the genera of aerobic actinomycetes. *In:* H. Prauser (Ed.) The Actinomycetales. Gustav Fischer-Verlag. Jena, Germany. 393–405.

Lee, S.-T., S.-B. Lee, and Y.-H. Park. 1991. Characterization of a pyridine-degrading branched Gram-positive bacterium isolated from the anoxic zone of an oil shale column. Appl. Microbiol. Biotechnol. 35:824–829.

Lee S.-T., S.-K. Rhee, and G. M. Lee. 1994. Biodegradation of pyridine by freely suspended and immobilized Pimelobacter sp. Appl. Microbiol. Biotechnol. 41:652–657.

Lee, S. D., S.-O. Kang, and Y. C. Hah. 2000. Hongia gen. nov., a new genus of the order Actinomycetales. Int. J. Syst. Evol. Microbiol. 50:191–199.

Lindenbein, W. 1952. Über einige chemisch interessante Aktinomycetenstämme und ihre Klassifizierung. Arch. Mikrobiol. 17:361–383.

Lu, Y., and X. Yan. 1983. Studies of the classification of thermophilic actinomycetes: Determination of thermo-

philic members of Nocardiaceae. Acta Microbiol. Sinica 23:220–228.

Lyman, J., and R. H. Fleming. 1940. Composition of sea water. J. Mar. Res. 3:134–146.

Männistö, M. K., M. A. Tiirola, M. S. Salkinoja-Salonen, M. S. Kulomaa, and J. A. Puhakka. 1999. Diversity of chlorophenol-degrading bacteria isolated from contaminated boreal groundwater. Arch. Microbiol. 171:189–197.

Männistö, M. K., M. S. Salkinoja-Salonen, and J. A. Puhakka. 2001. In situ polychlorophenol bioremediation potential of the indigenous bacterial community of boreal groundwater. Water Res. 35:2496–2504.

Marmur J., and P. Dorty. 1962. Determination of the base composition of deoxyribonucleic acid from its thermal denaturation temperature. J. Molec. Biol. 5:109–118.

Masson, J.-Y., I. Boucher, W. A. Neugebauer, D. Ramotar, and R. Brzezinski.1995. A new chitosanase gene from a Nocardioides sp. is a third member of glycosyl hydrolase family 46. Microbiology 141:2629–2635.

Maszenan, A. M., R. J. Seviour, B. K. C. Patel, P. Schumann, and G. N. Rees. 1999. Tessaracoccus bendigoensis gen. nov., sp. nov., a Gram-positive coccus occurring in regular packages or tetrads, isolated from activated sludge biomass. Int. J. Syst. Bacteriol. 49:459–468.

Matson, J. A., and J. A. Bush. 1989. Sandramycin, a novel antitumor antibiotic produced by a Nocardioides sp.: Production, isolation, characterization and biological properties. J. Antibiot. 42:1763–1767.

Miller, E. S., C. R. Woese, and S. Brenner. 1991. Description of the erythromycin-producing bacterium Arthrobacter sp. strain NRRL B-3381 as Aeromicrobium erythreum gen. nov., sp. nov. Int. J. Syst. Bacteriol. 41:363–368.

Nagasawa, M., M. Bae, G. Tamura, and K. Arima. 1969. Microbial transformation of sterols. Part II: Cleavage of sterol side chains by microorganisms. Agric. Biol. Chem. 33:1644–1650.

Nishimoto, T., M. Nakano, T. Nakada, H. Chaen, S. Fukuda, T. Sugimoto, M. Kurimoto, and Y. Tsujisaka. 1996. Purification and properties of a novel enzyme, trehalose synthase, from Pimelobacter sp. R48. Biosci. Biotech. Biochem. 60:640–644.

Nobile, A., and N. J. Belleville. 1958. Process for production of dienes by corynebacteria. US Patent. 2,837,464.

Normand, P., B. Cournoyer, P. Simonet, and S. Nazaret. 1992. Analysis of a ribosomal RNA operon in the actinomycete Frankia. Gene 111:119–124.

O'Donnell, A. G., M. Goodfellow, and D. E. Minnikin. 1982. Lipids in the classification of Nocardioides: Reclassification of Arthrobacter simplex (Jensen) Lochhead in the genus Nocardioides (Prauser) emend. O'Donnell et al. as Nocardioides simplex comb. nov. Arch Microbiol. 133:323–329.

Ogawara, H., K. Uchino, T. Akiyama, and S. Watanabe. 1985. New 5'-nucleotidase inhibitors, melanocidin A and melanocidin B. I: Taxonomy, fermentation, isolation and biological properties. J. Antibiot. 38:587–591.

Pace, N. R., and J. W. Brown. 1995. Evolutionary perspective on the structure and function of ribonuclease P, a ribozyme. J. Bacteriol. 177:1919–1928.

Park, Y.-H., J.-H. Yoon, and S. T. Lee. 1998. Application of multiplex PCR using species-specific primers within the 16S rRNA gene for rapid identification of Nocardioides strains. Int. J. Syst. Bacteriol. 48:895–900.

Park, Y.-H., J.-H. Yoon, Y. K. Shin, K.-I. Suzuki, T. Kudo, A. Seino, H.-J. Kim, J.-S. Lee, and S. T. Lee. 1999. Classification of "Nocardioides fulvus" IFO 14399 and Nocardioides sp. ATCC 39419 in Kribbella gen. nov., as Kribbella flavida sp. nov. and Kribbella sandramycini sp. nov. Int. J. Syst. Bacteriol. 49:743–752.

Patel, R. N., A. Banerjee, and V. Nanduri. 2000. Enzymatic acetylation of 10-deacetylbaccatin III to baccatin III by C-10 deacetylase from Nocardioides luteus SC 13913. Enz. Microb. Technol. 27:371–375.

Pernodet, J.-L., F. Boccard, M.-T. Alegre, J. Gagnat, and M. Guérineau. 1989. Organization and nucleotide sequence analysis of a ribosomal RNA gene cluster from Streptomyces ambofaciens. Gene 79:33–46.

Prauser, H. 1976. Nocardioides, a new genus of the order Actinomycetales. Int. J. Syst. Bacteriol. 26:58–65.

Prauser, H. 1984. Nocardioides luteus spec. nov. Zeitschr. Allgem. Microbiol. 24:647–648.

Prauser, H. 1989. Genus Nocardioides Prauser 1976. In: S. T. Williams, M. E. Sharpe, and J. G. Holt (Eds.) Bergey's Manual of Systematic Bacteriology. Williams and Wilkins. Baltimore, MD. 4:2371–2375.

Prauser, H., P. Schumann, F. A. Rainey, R. M. Kroppenstedt, and E. Stackebrandt. 1997. Terracoccus luteus gen. nov., sp. nov., an LL-diaminopimelic acid-containing coccoid actinomycete from soil. Int. J. Syst. Bacteriol. 47:1218–1224.

Pridham, T. G., and A. J. Lyons. 1961. Streptomyces albus (Rossi Doria) Waksman et Henrici: Taxonomic study of strains labeled Streptomyces albus. J. Bacteriol. 81:431–441.

Rajan, J., K. Valli, R. E. Perkins, F. S. Sariaslani, S. M. Barns, A.-L. Reysenbach, S. Rehm, M. Ehringer, and N. R. Pace. 1996. Mineralization of 2,4,6-trinitrophenol (picric acid): characterization and phylogenetic identification of microbial strains. J. Indust. Microbiol. 16:319–324.

Ruan, J. S., and Y. M. Zhang. 1979. Two new species of Nocardioides. Acta. Microbiol. Sinica 19:347–352.

Saito, A., T. Iwabuchi, and S. Harayama. 1999. Characterization of genes for enzymes involved in the phenanthrene degradation in Nocardioides sp. KP7. Chemosphere 38:1331–1337.

Saito, A., T. Iwabuchi, and S. Harayama. 2000. A novel phenanthrene dioxygenase from Nocardioides sp. strain KP7: Expression in Escherichia coli. J. Bacteriol. 182:2134–2141.

Sanglier, J. J., E. M. H. Wellington, V. Behal, H. P. Fiedler, R. E. Ghorbel, C. Finance, M. Hacene, A. Kamoun, C. Kelly, D. K. Mercer, S. Prinzis, and C. Trigo. 1993. Novel bioactive compounds from actinomycetes. Res. Microbiol. 144:661–663.

Schumann, P., H. Prauser, F. A. Rainey, E. Stackebrandt, and P. Hirsch. 1997. Friedmanniella antarctica gen. nov., sp. nov., an LL-diaminopimelic acid-containing actinomycete from antarctic sandstone. Int. J. Syst. Bacteriol. 47:278–283.

Shirling, E. B., and D. Gottlieb. 1966. Methods for characterization of Streptomyces species. Int. J. Syst. Bacteriol. 16:313–340.

Siddiqui, J. A., S. M. Shoeb, S. Takayama, E. Shimizu, and T. Yorifuji. 2000. Purification and characterization of histamine dehydrogenase from Nocardioides simplex IFO 12069. FEMS Microbiol Lett. 189:183–187.

Stackebrandt, E., F. A. Rainey, and N. L. Ward-Rainey. 1997. Proposal for a new hierarchic classification system, Actinobacteria classis nov. Int. J. Syst. Bacteriol. 47:479–491.

Staley, J. T. 1968. Prosthecomicrobium and Ancalomicrobium, new prosthecate fresh water bacteria. J. Bacteriol. 95:1921–1944.

Suzuki, K., and K. Komagata. 1983. Pimelobacter gen. nov., a new genus of coryneform bacteria with LL-diaminopimelic acid in the cell wall. J. Gen. Appl. Microbiol. 29:59–71.

Suzuki, Y., Y. Ono, A. Nagata, and T. Yamada. 1988. Molecular cloning and characterization of an rRNA operon in Streptomyces lividans TK21. J. Bacteriol. 170:1631–1636.

Tamaoka, J., and K. Komagata. 1984. Determination of DNA base composition by reverse-phase high-performance liquid chromatography. FEMS Microbiol. Lett. 25:125–128.

Tamura, T., and A. Yokota. 1994. Transfer of Nocardioides fastidiosa Collins and Stackebrandt 1989 to the genus Aeromicrobium as Aeromicrobium fastidiosum comb. nov. Int. J. Syst. Bacteriol. 44:608–611.

Tille, D., H. Prauser, K. Szyba, and M. Mordarski. 1978. On the taxonomic position of Nocardioides albus Prauser by DNA:DNA-hybridization. Zeitschr. Allgem. Mikrobiol. 18:459–462.

Topp, E., W. M. Mulbry, H. Zhu, S. M. Nour, and D. Cuppels. 2000. Characterization of S-triazine herbicide metabolism by a Nocardioides sp. isolated from agricultural soils. Appl. Environ. Microbiol. 66:3134–3141.

Travkin, V. M., A. P. Jadan, F. Briganti, A. Scozzafava, and L. A. Golovleva. 1997. Characterization of an intradiol dioxygenase involved in the biodegradation of the chlorophenoxy herbicides 2,4-D and 2,4,5-T. FEBS Lett. 407:69–72.

Travkin, V. M., E. V. Linko, and L. A. Golovleva. 1999. Purification and characterization of maleylacetate reductase from Nocardioides simplex 3E utilizing phenoxyalcanoic herbicides 2,4-D and 2,4,5-T. Biochemistry (Moscow) 64:625–630.

Urzì, C., P. Salamone, P. Schumann, and E. Stackebrandt. 2000. Marmoricola aurantiacus gen. nov., sp. nov., a coccoid member of the family Nocardioidaceae isolated from a marble statue. Int. J. Syst. Evol. Microbiol. 50:529–536.

Wang, Y. M., Z. S. Zhang, X. L. Xu, J. S. Ruan, and Y. Wang. 2001. Actinopolymorpha singaporensis gen. nov., sp. nov., a novel actinomycetes from the tropical rainforest of Singapore. Int. J. Syst. Evol. Microbiol. 51:467–473.

Yoon, J.-H., J.-S. Lee, Y. K. Shin, Y.-H. Park, and S. T. Lee. 1997a. Reclassification of Nocardioides simplex ATCC 13260, ATCC 19565, and ATCC 19566 as Rhodococcus erythropolis. Int. J. Syst. Bacteriol. 47:904–907.

Yoon, J.-H., S. T. Lee, S.-B. Kim, M. Goodfellow, and Y.-H. Park. 1997b. Inter- and intraspecific genetic analysis of the genus Saccharomonospora with 16S to 23S ribosomal DNA (rDNA) and 23S to 5S rDNA internally transcribed spacer sequences. Int. J. Syst. Bacteriol. 47:661–669.

Yoon, J.-H., S.-K. Rhee, J.-S. Lee, Y.-H. Park, and S. T. Lee. 1997c. Nocardioides pyridinolyticus sp. nov., a pyridine-degrading bacterium isolated from the oxic zone of an oil shale column. Int. J. Syst. Bacteriol. 47:933–938.

Yoon, J.-H., S. T. Lee, and Y.-H. Park. 1998a. Genetic analyses of the genus Nocardioides and related taxa based on 16S-23S rDNA internally transcribed spacer sequences. Int. J. Syst. Bacteriol. 48:641–650.

Yoon, J.-H., S. T. Lee, and Y.-H. Park. 1998b. Inter- and intraspecific phylogenetic analysis of the genus Nocardioides and related taxa based on 16S rDNA sequences. Int. J. Syst. Bacteriol. 48:187–194.

Yoon, J.-H., Y.-G. Cho, S. T. Lee, K.-I. Suzuki, T. Nakase, and Y.-H. Park. 1999. Nocardioides nitrophenolicus sp. nov., a p-nitrophenol-degrading bacterium. Int. J. Syst. Bacteriol. 49:675–680.

Yoon, J.-H., and Y.-H. Park. 2000. Comparative sequence analyses of the ribonuclease P (RNase P) RNA genes from LL-2,6-diaminopimelic acid-containing actinomycetes. Int. J. Syst. Evol. Microbiol. 50:2021–2029.

Index

A

ABC. *See* ATP-binding cassette
Acetate production
  Cellulomonas, 988
  Propionibacterium, 408–9
Acetate utilization
  Corynebacterium, 804–5
  Methanocalculus, 223
  Methanocorpusculum, 220–22
  Methanoculleus, 217–18
  Methanofollis, 219–20
  Methanogenium, 215–16
  Methanoplanus, 218
  Methanosaeta, 253–54
  Methanosarcina, 248
  Methanosarcinales, 244–45
  Micrococcus, 966
  Streptomyces, 612
Acetyl-CoA synthase,
    Thermococcales, 75
Acetyl-CoA synthetase
  Archaeoglobus, 89
  Desulfurococcales, 65
Acidianus, 25, 28–31, 33, 38
Acidianus ambivalens, 24–25, 31,
    33–35, 40
Acidianus ambivalens Lei1^T, 25
Acidianus brierleyi, 24–25, 29–31, 33
Acidianus brierleyi DSM 1651^T, 25
Acidianus infernus, 24–25, 28–29, 31,
    34
Acidianus infernus So-4a^T, 25
Acidilobus, 56–57, 61
Acidilobus aceticus, 52, 56
Acidimicrobium, 305–6
Acidimicrobium ferrooxidans, 305–6
Acidophiles
  Sulfolobales, 26–27, 29–30
  Thermoplasma, 108
Acidothermaceae, 669–78
  phylogenetic tree of, 670–71
  taxonomy of, 672
Acidothermus, 669–70
  isolation of, 674
  physiology of, 674–75
  preservation and cultivation of,
    674
Acidothermus cellulolyticus, 669, 674
Acrocarpospora, 726
  chemotaxonomic and
    morphological features of,
    727

  identification of, 740–46
  isolation and cultivation of, 735
  phylogeny of, 729
  species characteristics of, 747
Acrocarpospora corrugata, 729,
    734–35
Acrocarpospora macrocephala, 729,
    734
Acrocarpospora pleiomorpha, 729,
    734
Actinoalloteichus, 654
  morphology of, 768–69
Actinobacteria
  Actinobacteridae, 797, 819
  Actinomycetales, 430, 983, 1020
  families and genera of, 298
  phylogenetic tree of, 302
  Propionibacterinaea, 383
  Pseudonocardiniae, 654
  species and genera of, 302–4
  taxonomy of, 297–316
Actinobacteridae, Actinomycetales,
    797, 819
Actinobaculum, 430–39, 470–71,
    493–98
  characteristics of, 493–97
  epidemiology of, 497–98
  habitat and ecology of, 497
  taxonomy of, 493–97
Actinobaculum massiliae, 432,
    493–94, 497
Actinobaculum schaalii, 432, 439,
    447, 493–94, 497
Actinobaculum suis, 430, 432, 436,
    439, 447, 493–94, 497
Actinobaculum urinale, 432, 493–94,
    497
Actinocorallia, 682–714
  enzymatic activities of, 704
  habitats of, 688
  identification of, 694–708
  morphological characteristics of,
    701–2, 768–69
  phylogenetic tree of, 689
  species characteristics of, 705
  taxonomic history of, 695–97
Actinocorallia aurantiaca, 705
Actinocorallia glomerata, 705
Actinocorallia herbida, 691, 705
Actinocorallia libanotica, 705
Actinocorallia longicatena, 705
Actinokineospora, 654–61, 666
  characteristics of, 661

  habitat of, 656
  identification of, 659–60
  morphology of, 768–69
  phylogenetic tree of, 663
Actinokineospora diospyrosa, 654
Actinokineospora globicatena, 654
Actinokineospora inagensis, 654
Actinokineospora riparia, 654, 660
Actinokineospora terrae, 654
Actinomadura, 682–714, 725
  habitats of, 688
  identification of, 694–708
  isolation and cultivation of,
    690–91
  morphological characteristics of,
    701–2, 756, 768–69, 778
  phylogenetic tree of, 689
  physiological characteristics of,
    703
  taxonomic history of, 695–97,
    755–57
Actinomadura carminata, 713
Actinomadura citrea, 688, 713
Actinomadura cremea, 688
Actinomadura dassonvillei, 755, 757.
    See also Nocardiopsis
    dassonvillei
Actinomadura echinospora, 688
Actinomadura flava. *See*
    Lechevalieria flava
Actinomadura formosensis, 691, 699
Actinomadura kijaniata, 691, 713
Actinomadura latina, 688, 708–9
Actinomadura macra, 691, 713
Actinomadura madurae, 688, 690,
    698–700, 708–9, 755
Actinomadura oligospora, 713
Actinomadura pelletieri, 688, 690,
    698, 708–9, 755
Actinomadura rubrobrunea, 684,
    688, 691
Actinomadura rugatobispora, 729
Actinomadura spadix, 688, 691
Actinomadura umbrina, 688
Actinomadura verrucosospora, 688
Actinomadura vinacea, 691
Actinomadura viridilutea, 684, 688,
    691, 713
Actinomadura viridis, 729, 731
Actinomadura yumaensis, 713
Actinomyces, 430–93
  chemotaxonomic properties,
    446–48, 468–69, 494

cultivation of, 464
disease from, 482–83
  in animals, 490–93
  in humans, 483–90
  treatment for, 489–90
ecology of, 479–82
habitats of, 455–57
identification of, 464–79
isolation of, 457–63
morphology of, 442–46, 465–67, 469
  cellular, 442–43
  colony, 442–46, 466–67
physiological properties of, 448–55, 469, 474–75
species of, 431–33
taxonomy of, 439–55
Actinomyces bovis, 430–32, 435–36, 439–40, 444–45, 447–48, 455, 457, 467, 491–92
Actinomyces bowdenii, 431–32, 435, 441, 457, 492
Actinomyces canis, 430–32, 438, 457
Actinomyces cardiffensis, 431–32, 435, 441, 445, 456
Actinomyces catuli, 431, 445, 457, 492
Actinomyces coleocanis, 431–32, 435, 438, 441, 457, 492
Actinomyces denticolens, 431–32, 435, 440, 445, 457
Actinomyces europaeus, 431–32, 435, 438, 441, 447, 456
Actinomyces funkei, 431, 456
Actinomyces georgiae, 431–32, 435, 440, 447, 455, 480
Actinomyces gerencseriae, 431–32, 435, 440–41, 443–45, 447, 454, 456, 479–80, 483, 486–87
Actinomyces graevenitzii, 431–32, 435, 441, 445, 455
Actinomyces hongkongensis, 431–32, 435, 438, 449, 456
Actinomyces hordeovulneris, 431–32, 435, 438, 440–41, 445, 449, 457, 491–92
Actinomyces howellii, 431, 457
Actinomyces hyovaginalis, 431–32, 435, 441, 447, 491
Actinomyces israelii, 431–32, 435–36, 439–45, 447–49, 454–56, 459–61, 467, 478–80, 483, 486–87, 489, 491
Actinomyces marimammalium, 431–32, 435, 438, 441, 457, 492
Actinomyces meyeri, 431–32, 435–36, 440, 445, 449, 456, 465, 467, 479, 487
Actinomyces naeslundii, 431–32, 435, 439–40, 442–45, 448–49, 454–57, 476, 479–83, 486–87, 489, 492
Actinomyces nasicola, 431–32, 435, 438, 441, 456
Actinomyces neuii, 431–32, 435, 438, 447, 456, 488
Actinomyces odontolyticus, 431–32, 435, 440, 445, 455–56, 479–80, 486–87, 489

Actinomyces oricola, 431, 441, 445, 456
Actinomyces radicidentis, 431–32, 435, 441, 445, 456
Actinomyces radingae, 431–32, 435, 438, 440, 447, 456, 487
Actinomyces slackii, 431–32, 435, 440, 457
Actinomyces suimastiditis, 431–32, 435, 457
Actinomyces turicensis, 431–32, 435, 440, 447, 456, 487
Actinomyces urogenitalis, 431–32, 435, 441, 445, 456
Actinomyces vaccimaxillae, 432, 441, 457, 492
Actinomyces viscosus, 431–32, 435–36, 439–40, 443, 445, 447–49, 454–57, 476, 480–83, 486–87, 489, 491–92
Actinomycetaceae, 430–516
  Actinobaculum, 430–39, 470–71, 493–98
  Actinomyces, 430–93
  Arcanobacterium, 430–39, 470–71, 498–505
  clusters of, 433–39
  introduction to, 430
  Mobiluncus, 430–39, 506–16
  phylogenetic tree of, 434
  phylogeny of, 430–39
  Varibaculum, 430–39, 470, 505–6
Actinomycetales
  Actinomycineae, 430, 433
  Corynebacterineae, 797, 819
  Microbacteriaceae, 1020
  Micrococcineae, 307, 983, 1002
  Streptomycetaceae, 605
  Streptosporangineae, 682
Actinomycetes. *See* Actinobacteria
Actinomycin production, *Streptomyces*, 566
Actinomycineae, Actinomycetaceae, 430, 433
Actinoplanaceae, 725
Actinoplanes, 623–49, 725
  characteristics of, 639–42
  chemotaxonomic properties of, 636–39
  ecophysiology of, 623–30
  identification of, 635–39
  isolation of, 630–35, 642–45
  morphology of, 636
Actinoplanes azureus, 648
Actinoplanes brasiliensis, 624–25
Actinoplanes caeruleus, 639, 648
Actinoplanes cyaneus, 639
Actinoplanes ferrugineus, 639, 646
Actinoplanes garbadinensis, 639
Actinoplanes ianthinogenes, 639
Actinoplanes italicus, 639
Actinoplanes minutisporangius, 639
Actinoplanes missouriensis, 624
Actinoplanes philipinensis, 607
Actinoplanes rectilineatus, 639
Actinoplanes teichomyceticus, 639, 648
Actinoplanetes, 623–49
  ecophysiology of, 623–30

identification of, 635–39
isolation of, 630–35, 642–45
Actinopolymorpha, characteristics of, 1105, 1108
Actinopolyspora, morphological characteristics of, 756, 768–69, 778
Actinosynnema, 654–62, 666
  characteristics of, 662
  habitat of, 656
  identification of, 660–61
  morphology of, 768–69
  phylogenetic tree of, 663
Actinosynnema mirum, 655, 661
Actinosynnema pretiosum, 655
Actinosynnema pretiosum subsp. *auranticum*, 655, 660
Actinosynnema pretiosum subsp. *pretiosum*, 660
Actinosynnemataceae
  Actinokineospora, 654–61, 666
  Actinosynnema, 654–62, 666
  chemotaxonomic properties of, 658–59
  identification of, 658–65
  isolation and cultivation of, 656–58
  Lechevalieria, 654–59, 665–66
  Lentzea, 654–59, 663–66
  phylogenetic tree of, 655, 663
  Saccharothrix, 654–59, 661–64, 666
Aerobes
  Agreia, 1021
  Arthrobacter, 945
  Brevibacterium, 1013
  Curtobacterium, 1043
  Desulfurococcales, 52
  Friedmanniella, 391
  Glycomyces, 306
  Gordonia, 846
  Halobacteriales, 143
  Herbidospora, 729
  Microbacteriaceae, 1020
  Microbispora, 730
  Microtetraspora, 731
  Mycobacterium, 889
  Nocardia, 848
  Nonomuraea, 731
  Okibacterium, 1068
  Picrophilus, 109
  Planobispora, 732
  Planomonospora, 733
  Planotetraspora, 734
  Skermania, 847
  Sulfolobales, 26–27, 33
  Tsukamurella, 850–51
  Williamsia, 847–48
Aeromicrobium, characteristics of, 1105, 1108
Aeropyrum, 53, 56–58
Aeropyrum pernix, 53, 57–58, 64–65
Agreia, 1020–26
  characteristics of, 1023–24, 1057, 1078
  habitats of, 1025
  phylogenetic tree of, 1022
Agreia bicolorata, 1021, 1025–26
Agreia pratensis, 1021, 1025–26
Agrococcus, 1020, 1026–27

characteristics of, 1023–24, 1027, 1057
habitats of, 1026
phylogenetic tree of, 1022
Agrococcus baldri, 1026–27
Agrococcus citrea, 1026–27
Agrococcus jenensis, 1026–27
Agromyces, 1020, 1027–31, 1081
characteristics of, 1023–24, 1030, 1057
habitats of, 1028–29
phylogenetic tree of, 1022
Agromyces albus, 1028–31
Agromyces aurantiacus, 1028–31
Agromyces bracchium, 1028–31
Agromyces cerinus subsp. *cerinus*, 1028–31
Agromyces cerinus subsp. *nitratus*, 1028–31
Agromyces fucosus subsp. *fucosus*, 1028–31
Agromyces fucosus subsp. *hippuratus*, 1028–31
Agromyces luteolus, 1028–31
Agromyces mediolanus, 1028–31
Agromyces ramosus, 1027–31
Agromyces rhizospherae, 1028–31
Amino acid fermentation, *Corynebacterium*, 805–9
Amino acid utilization, *Mycobacterium*, 899
Aminoglycosides sensitivity, *Stomatococcus*, 980
Ammonia utilization, *Methanococcus*, 182, 264–65
Ammonium utilization, *Mycobacterium*, 899
Amoxicillin sensitivity, *Bifidobacterium*, 349–50
Amphotericin sensitivity, *Gardnerella*, 366–67
Ampicillin resistance/sensitivity
Gardnerella, 366–67, 367
Halobacteriales, 127
Stomatococcus, 980
Thermoplasmatales, 106–7
Amplified ribosomal DNA restriction analysis, 345
Ampullariella, 623–49, 725
characteristics of, 639–42
chemotaxonomic properties of, 636–39
ecophysiology of, 623–30
identification of, 635–39
isolation of, 630–35, 642–45
morphology of, 636
Ampullariella campanulata, 640
Ampullariella digitata, 640
Ampullariella lobata, 640
Ampullariella regularis, 640, 648
Ampullariella violaceochromogenes, 640
Amycolatopsis, morphological characteristics of, 756, 778
Anaerobes
Anaerobiospirillum, 423
Archaeoglobus, 88
Bifidobacteriaceae, 322
Dermabacter, 308

Desulfurococcales, 30, 52, 56, 64
Gardnerella, 365
Methanobacteriaceae, 236
Methanobacteriales, 231
Methanobacterium, 236
Methanococcales, 259–61
methanogens, 165
Methanomicrobiales, 208
Methanosarcinales, 244
Methanothermus, 241–42
Pyrobaculum, 11
Rothia, 315
Stomatococcus, 315
Sulfolobales, 26–27
Thermococcales, 72
Thermofilum, 13
Thermoplasma, 108
Thermoproteales, 10, 18–19
Thermoproteus, 11
Anaerobiospirillum, 419–26
Anaerobiospirillum succiniproducens, 419–26
Anaerobiospirillum thomasii, 419–26
Animal actinomycosis, *Actinomyces*, 490–92
Animal environment
Actinomyces, 456–57
Arcanobacterium, 500–501
Bifidobacterium, 354–55
Corynebacterium, 820
Dermatophilus, 1003
Nocardia, 852
Animal infections, *Arcanobacterium*, 500–505
Anoxic environment
Archaeoglobus, 82
Methanobacteriales, 231
Methanobrevibacter, 238–40
methanogens, 166–67
Methanosarcinales, 244–46
Antibiotic production
Actinoplanetes, 646–49
Actinosynnemataceae, 666
Brevibacterium, 1015–17
Nocardioides, 1109–10
Nocardiopsis, 783
Rathayibacter, 1077
Streptomyces, 562, 566, 589, 613–14
Antibiotic resistance/sensitivity
Bifidobacterium, 349–50
Corynebacterium, 828–36
Gardnerella, 366–67
Halobacteriales, 127–28
Methanococcus, 264
methanogens, 169, 196–98
Propionibacterium, 410, 413
Streptomyces, 567
Thermoplasmatales, 106–7
Applications
Actinomadura, 713–14
Actinoplanetes, 645–49
Actinosynnemataceae, 665–66
Agromyces, 1029, 1031
Arthrobacter, 955
Bifidobacterium, 355–59
Brevibacterium, 1017
Cellulomonas, 992

Clavibacter, 1040
Corynebacterium, 805–10
Curtobacterium, 1046–48
Desulfurococcales, 65–66
Frankineae, 677–78
Gordonia, 871–72
Halobacteriales, 150–51
Leifsonia, 1055
methanogens, 198–99
Microbacterium, 1066
Micrococcus, 968
Mycetocola, 1068
Mycobacterium, 909–10
Nocardioides, 1109–10
Oerskovia, 996
Plantibacter, 1071
Rathayibacter, 1077
Rhodococcus, 871–72
Streptomyces, 613–14
Streptosporangiaceae, 747–48
Succinivibrionaceae, 426
Sulfolobales, 42
Thermomonospora, 713–14
Thermomonosporaceae, 713–14
Thermoproteales, 20
Arcanobacterium, 430–39, 470–71, 498–505
chemotaxonomic properties of, 494, 499–500
ecology and habitat of, 500–501
genetics of, 505
history of, 498–99
identification of, 501–2
isolation and cultivation of, 501
morphological characteristics, 499–500
pathogenicity, 502–5
physiological properties of, 500
taxonomy of, 499–500
Arcanobacterium bernardiae, 432, 438–39, 447, 499–500, 503–4
Arcanobacterium haemolyticum, 430, 432, 436, 438–39, 447, 499–505, 827, 833
Arcanobacterium hippocoleae, 432, 438–39, 499–500, 503
Arcanobacterium phocae, 432, 438–39, 447, 499–500, 503
Arcanobacterium pluranimalium, 432, 438–39, 447, 499–500, 503
Arcanobacterium pyogenes, 430, 432, 435–36, 438–39, 447, 499–505
Archaea
discovery of, 3–8
Eukaryotes v., 6
Archaeoglobaceae, *Archaeoglobus*, 82
Archaeoglobus, 82–97
characteristics of, 83
ecology and habitat of, 82–83, 88–89
enzymes of, 92–93
genetics of, 86–88
identification of, 83–85
isolation of, 85
metabolism of, 89–97
phylogenetic tree of, 84
phylogeny of, 82

Archaeoglobus fulgidus, 82–97
Archaeoglobus fulgidus 7324, 82, 91, 94
Archaeoglobus fulgidus VC-16, 82–83, 87, 89, 94
Archaeoglobus fulgidus Z, 82
Archaeoglobus lithotrophicus, 82
Archaeoglobus profundus, 82, 85
Archaeoglobus veneficus, 82
Arsenate oxidation
    Pyrobaculum, 11
    Thermoproteales, 10, 18
Arthrobacter, 312–15, 945–55, 961
    biochemical and physiological properties of, 954–55
    characteristics of, 949
    globiformis group, 313
    habitats of, 948–50
    history of, 945–47
    identification of, 952–54
    isolation of, 950–52
    morphology of, 313, 952–53
    nicotianae group, 313
    phylogenetic tree of, 314
    species characteristics of, 947–48
Arthrobacter agilis, 314
Arthrobacter atrocyaneus, 314, 953, 955
Arthrobacter aurescens, 314, 953
Arthrobacter citreus, 314, 945, 949, 953–54
Arthrobacter creatinolyticus, 314–15
Arthrobacter crystallopoietes, 314, 953, 955
Arthrobacter cumminsii, 314–15
Arthrobacter duodecadis, 314, 946
Arthrobacter flavescens. See Microbacterium flavescens
Arthrobacter globiformis, 313–15, 945–46, 949–50, 953–55
Arthrobacter histidinolovorans, 314, 953
Arthrobacter ilicis, 314, 950
Arthrobacter mysorens, 314
Arthrobacter nicotianae, 313–14, 946, 953–54
Arthrobacter nicotinovorans, 314
Arthrobacter oxydans, 314, 953
Arthrobacter pascens, 314, 953, 955
Arthrobacter picolinophilus. See Rhodococcus erythropolis
Arthrobacter polychromogenes, 314, 953
Arthrobacter protophormiae, 314, 950, 953–54
Arthrobacter radiotolerans. See Rubrobacter radiotolerans
Arthrobacter ramosus, 314, 953
Arthrobacter siderocapsulatus, 314–15, 946
Arthrobacter simplex, 945–46. *See also Nocardioides simplex*
Arthrobacter sulfureus, 314, 954
Arthrobacter terregens, 945. *See also Microbacterium terregens*
Arthrobacter tumescens, 945–46. *See also Terrabacter tumescens*
Arthrobacter uratoxydans, 314, 953–54

Arthrobacter ureafaciens, 314, 953, 955
Arthrobacter variabilis. See Corynebacterium variabile
Arthrobacter viscosus, 314, 946
Arthrobacter woluwensis, 314–15
Atmospheric methane, methanogens, 169–70
Atopobium, 306
ATPases
    in Desulfurococcales, 60
    in Sulfolobales, 34
ATP-binding cassette (ABC) transporters, Sulfolobales, 35
Aureomycin sensitivity, Halobacteriales, 127
Autotrophic
    Methanothermobacter, 241

B

Bacitracin resistance/sensitivity
    Bifidobacterium, 349
    Halobacteriales, 127
    methanogens, 197
Bacteremia
    Corynebacterium, 832, 834
    Microbacterium, 1064–65
Bacterial vaginosis, 368–70
    clinical diagnosis of, 362–63
    Mobiluncus, 512–16
Bacteriocin production
    Bifidobacterium, 350, 356
    Brevibacterium, 1015–17
Bacteriophages
    Propionibacterium, 409
    Rathayibacter, 1076
Bacteriorhodopsin, Halobacteriales, 113, 137–38, 145
Barophile, Thermococcales, 71
Bifidobacteriaceae, 322–70
    Bifidobacterium, 322–59
    Gardnerella, 322, 359–70
Bifidobacterium, 322–59
    applications of, 355–59
    ecology and habitats of, 331–33, 350–55
    fermentation characterizations of, 336–41
    genetics of, 350
    identification of, 334–46
    isolation of, 333–34
    morphology of, 334, 336
    pathogenicity of, 355
    phylogeny of, 322–23
    physiology of, 346–48
    taxonomy of, 323–31
Bifidobacterium adolescentis, 324, 327–30, 332, 338–40, 343, 346, 348, 352–54, 368
Bifidobacterium angulatum, 324, 327–30, 338–40, 347
Bifidobacterium animalis, 324, 327–30, 338–40, 342, 346, 348–49
Bifidobacterium asteroides, 322, 324, 327–30, 338–40, 347, 350, 354

Bifidobacterium bifidum, 322, 324, 327–32, 336, 338–40, 343, 346, 348–50, 354, 368
Bifidobacterium boum, 324, 327–30, 338–40, 342
Bifidobacterium breve, 324, 327–30, 332, 337–40, 343, 346, 349–50, 352, 354, 368
Bifidobacterium catenulatum, 324, 327–30, 338–40, 347, 350
Bifidobacterium choerinum, 324, 327–30, 338–40, 348
Bifidobacterium coryneforme, 324, 327–30, 338–40
Bifidobacterium cuniculi, 324, 327–30, 338–40, 348, 354
Bifidobacterium denticolens, 324, 327–31, 337–40, 344, 348, 354
Bifidobacterium dentium, 325, 327–31, 337–40, 348, 350, 354
Bifidobacterium gallicum, 325, 327–30, 338–40
Bifidobacterium gallinarum, 325, 327–30, 338–40, 354
Bifidobacterium indicum, 322, 325, 327–30, 338–40, 347, 350, 354
Bifidobacterium infantis, 325, 327–30, 336–40, 343, 346, 348–50, 352, 354
Bifidobacterium inopinatum, 325, 327–30, 337–40, 344, 348, 354
Bifidobacterium lactis, 325, 327–30, 332, 338–40, 346
Bifidobacterium longum, 325, 327–32, 336, 338–40, 343, 346–47, 350, 352–54, 368
Bifidobacterium magnum, 325, 327–30, 338–40, 348, 354
Bifidobacterium merycicum, 325, 327–30, 338–40
Bifidobacterium minimum, 325, 327–30, 332, 338–40
Bifidobacterium pseudocatenulatum, 325, 327–30, 337–40, 350, 352
Bifidobacterium pseudolongum subsp. *globosum*, 326–31, 338–40, 343, 347–48, 350
Bifidobacterium pseudolongum subsp. *pseudolongum*, 326–30, 338–40
Bifidobacterium pullorum, 326–30, 338–40, 354
Bifidobacterium ruminantium, 326–30, 338–40
Bifidobacterium saeculare, 326–30, 338–40
Bifidobacterium subtile, 326–30, 332, 338–40
Bifidobacterium suis, 326–30, 338–40, 348
Bifidobacterium thermoacidophilum, 327–30, 338–40
Bifidobacterium thermophilum, 326–30, 332, 338–40, 347–48
Biochemical properties
    Arthrobacter, 954–55
    Friedmanniella, 387–90
    Luteococcus, 390–91
    Methanococcales, 268–69

methanogens, 188–98
Methanomicrobiales, 223–26
Micrococcus, 966–67
Microlunatus, 391, 393
Micropruina, 391, 394
Propionimicrobium, 396
Streptomycetaceae, 541
Succinivibrionaceae, 424–26
Sulfolobales, 33–38
Tessaracoccus, 391, 396–97
Biocorrosion, methanogens, 199
Biofilm, *Archaeoglobus*, 89
Biopolymer degradation
   Nocardiopsis, 780, 784
   Thermobifida, 780, 784–85
Bioremediation
   Mycobacterium, 910
   Nocardioides, 1109–10
Biotechnology
   Desulfurococcales, 65–66
   Sulfolobales, 42
   Thermoproteales, 20
Biotin requirement,
   Propionibacterium, 409
"Black pigmented"
   Corynebacterium, 834–36
Blastococcus, 669, 671, 1081
Blood environments
   Atopobium, 306
   Corynebacterium, 829–30, 833–34
   Jonesia, 311
   Sanguibacter, 310–11
   Stomatococcus, 315, 977, 978
   Succinivibrio, 419–20, 422
Bogoriella, 307–8
Bogoriella caseilytica, 307
Bogoriellaceae, *Bogoriella*, 307–8
Bovine environment
   Arcanobacterium, 502–3
   Methanobrevibacter, 238–40
   Methanomicrobiales, 211
   Sanguibacter, 310–11
Bovine mastitis, *Corynebacterium*,
   829–30
Brachybacterium, 308
Brachybacterium alimentarium, 308
Brachybacterium conglomeratum,
   308
Brachybacterium faecium, 308
Brachybacterium nesterenkovii, 308
*Brachybacterium
   paraconglomeratum*, 308
Brachybacterium rhamnosum, 308
Brachybacterium tyrofermentans,
   308
Brevibacterium, 951, 1013–17
   characteristics of, 1013, 1015
   genomics of, 1017
   habitats of, 1013–14
   identification of, 1015
   isolation of, 1014
   phylogenetic tree of, 1014
   species characteristics of, 1016
Brevibacterium avium, 1013–15
Brevibacterium casei, 1013–15
Brevibacterium epidermidis, 1013,
   1015
Brevibacterium helvolum group,
   1020, 1079–80

Brevibacterium iodinum, 1013–15
Brevibacterium linens, 1013–17
Brevibacterium mcbrellneri, 1013–15
Brevibacterium otitidis, 1013–15
Buruli ulcer, *Mycobacterium*, 930–
   31

C

Caldivirga, 11–13, 15–16, 82
Caldivirga maquilingensis, 11–13,
   18–19, 82
Calvin cycle
   Halobacteriales, 144
   Pyrodictium, 64
Cancer prevention, *Bifidobacterium*,
   357
Canine environment,
   Anaerobiospirillum, 420
Carbohydrate utilization
   Actinomyces, 448–55
   Archaeoglobus, 91–94
   Gardnerella, 366
   Halobacteriales, 143–44
Carbon dioxide reduction
   Methanococcales, 257
   Methanosarcina, 248–52
   Methanosarcinaceae, 248
   Sulfolobales, 33
Carbon metabolism
   Archaeoglobus, 95
   methanogens, 191–92
Caries, *Actinomyces*, 488–90
β-Carotene, 135
Carotenoid pigments
   Halobacteriales, 135–36
   light damage and, 135
Casein degradation, *Nocardiopsis*,
   784
Caseobacter, 890, 894
Catalase production
   Arcanobacterium, 501
   Arthrobacter, 953–54
   Brevibacterium, 1013
   Corynebacterium, 796
   Gordonia, 846
   Micrococcus, 961
   Oerskovia, 996
Catellatospora, 623–49
   characteristics of, 639–42
   ecophysiology of, 623–30
   identification of, 635–39
   isolation of, 630–35, 642–45
Catellatospora citrea, 640
Catellatospora ferruginea, 640
Catellatospora matsumotoense, 640
Catellatospora tsunoense, 640
Catenuloplanes, morphology of,
   768–69
Cell envelope lipids
   Actinomyces, 446, 468
   Crenarchaeota, 285
   Methanococcus, 262
   Methanomicrobiales, 225–26
   Methanosarcinales, 244
   Micrococcus, 966–67
   Mycetocola, 1067–68
   Mycobacterium, 901–4

Nocardiopsis, 758–60, 767, 771–73,
   776
   Renibacterium, 972
   Streptomyces, 562–65
Cell envelope structure
   Arthrobacter, 953
   Bifidobacterium, 342–43
   Corynebacterium, 796–97, 801–2
   Methanomicrobiales, 224–25
   Mycobacterium, 924
Cell wall inhibitor resistance,
   Thermoplasmatales, 106–7
Cellulomonadaceae, 983–97
   Cellulomonas, 983–84
   Oerskovia, 983–84
Cellulomonas, 983–93
   characteristics of, 987, 1015
   chemotaxonomic properties of,
     988, 990
   habitats of, 988
   identification of, 988
   isolation of, 988
   phylogenetic tree of, 984, 987
   phylogeny of, 985
   physiological properties of,
     985–88
   species of, 984
   taxonomy of, 985
Cellulomonas biazotea, 983–85, 987,
   990, 992
Cellulomonas cellasea, 984–85, 990
Cellulomonas cellulans, 984–85
Cellulomonas fermentans, 984–85,
   987–88, 992
Cellulomonas fimi, 984–85, 987, 990,
   992
Cellulomonas flavigena, 983–85,
   987–88, 990, 992
Cellulomonas gelida, 984
Cellulomonas hominis, 985, 988,
   992
Cellulomonas humilata, 984–85, 988,
   990
Cellulomonas iranensis, 984–85, 988,
   990
Cellulomonas persica, 984–85, 988,
   990
Cellulomonas uda, 984–85, 987–88,
   991–92
Cellulose decomposition
   Cellulomonas, 990–92
   Micromonospora, 627–29
   Streptomyces, 611
   Thermobifida, 763, 784–85
   Thermomonospora, 712
Cellulosimicrobium,
   chemotaxonomic properties
   of, 990
Cellulosimicrobium cellulans, 990,
   993, 995–96
Cenarchaeum, 286–88
Cenarchaeum symbiosum, 286–88
Cephaloridine resistance,
   Streptomyces, 567
Chemolithoautotroph
   Archaeoglobus, 82
   Ferroplasma, 109
Chemolithoheterotroph,
   Archaeoglobus, 82

Chemotaxonomic properties
    Actinobaculum, 493–94
    Actinomyces, 446–48, 494
    Actinoplanes, 636–39
    Actinosynnemataceae, 658–59
    Ampullariella, 636–39
    Arcanobacterium, 494, 499–500
    Cellulomonas, 988
    Dactylosporangium, 636–39
    Demetria, 1007
    Dermacoccus, 1007
    Dermatophilus, 1007
    Friedmanniella, 387–90
    Kineosphaera, 1007
    Kitasatospora, 540
    Kytococcus, 1007
    Luteococcus, 390–91
    methanogens, 198
    Microlunatus, 391, 393
    Micropruina, 391, 394
    Mobiluncus, 507–9, 512
    Nocardioides, 1104–7
    Pilimelia, 636–39
    Propionibacterium, 410
    Propioniferax, 391, 395
    Propionimicrobium, 396
    Rothia, 979
    Stomatococcus, 979
    Streptacidiphilus, 540
    Streptomyces, 540, 562
    Streptomycetaceae, 540
    Tessarococcus, 391, 396–97
    Thermomonosporaceae, 685–86
    Varibaculum, 505
Chicken environment. See Poultry
        environment
Chitin decomposition
    Micromonospora, 627–29
    Streptomyces, 611
Chitinase production, Arthrobacter,
        955
Chloramphenicol resistance/
        sensitivity
    Bifidobacterium, 349
    Halobacteriales, 127
    methanogens, 197
    Propionibacterium, 410
Cholesterol degradation,
        Mycobacterium, 909
Choriobacteraceae, 306
    Atopobium, 306
    Coriobacterium, 306
    Eggerthella, 306
    Slackia, 306
Chromosomes
    Halobacteriales, 146
    Sulfolobales, 35–36
Ciprofloxacin resistance,
        Stomatococcus, 980
Clarithromycin sensitivity,
        Stomatococcus, 980
Clavibacter, 1020, 1031–40, 1045
    characteristics of, 1015, 1023–24,
        1038
    habitats of, 1032–35
    phylogenetic tree of, 1022
Clavibacter michiganensis, 1031–40,
    1045
Clavibacter michiganensis subsp.
    insidiosus, 1031, 1033–40

Clavibacter michiganensis subsp.
    michiganensis, 1031, 1033–40
Clavibacter michiganensis subsp.
    nebraskensis, 1033, 1035–40
Clavibacter michiganensis subsp.
    sepedonicus, 1031, 1033–40
Clavibacter michiganensis subsp.
    tessellarius, 1033, 1035–40
Clindamycin sensitivity
    Bifidobacterium, 349
    Gardnerella, 366–67
CMN-group, 797–98
    Corynebacterium, 797–98
    Dietza, 797–98
    Gordonia, 797–98
    Mycobacterium, 797–98
    Nocardia, 797–98
    Rhodococcus, 797–98
    Tsukamurella, 797–98
    Turicella, 797–98
Coenzyme F_{420}
    Archaeoglobus, 94, 107
    methanogens, 4, 6
    Methanomicrobiales, 224
Coenzyme F_{430}, of methanogenesis,
    189
Coenzyme M
    Methanobrevibacter, 181
    of methanogenesis, 189
    methanogens, 4, 6
    Methanothermobacter, 241
Coenzyme M reductase,
        Methanomicrobiales, 224
Colistin resistance, Gardnerella,
    367
Colony morphology
    Actinobaculum, 494
    Actinomadura, 690
    Actinomyces, 442–46
    Actinoplanes, 639
    Agreia, 1025–26
    Ampullariella, 640
    Brevibacterium, 1014–15
    Catellatospora, 640
    Clavibacter, 1037–38
    Corynebacterium, 824–25
    Dactylosporangium, 640–41
    Glycomyces, 641
    Gordonia, 846
    Micrococcus, 964
    Micromonospora, 641
    Mycobacterium, 889, 924
    Nocardia, 848
    Nocardioides, 1102–4
    Oerskovia, 994
    Pilimelia, 642
    Rathayibacter, 1075
    Rhodococcus, 850
    Skermania, 847
    Streptomyces, 562, 567
    Thermomonospora, 692
    Tsukamurella, 850–51
Compost environments,
        Promicromonospora, 311
Conjunctivitis, Corynebacterium,
    833
Copiotrophic ultramicrobacteria,
    1081
Coriobacterium, 306
Coriobacterium glomerans, 306

Corynebacteriaceae
    Corynebacterium, 797, 819, 843
    Turicella, 797, 819, 843
Corynebacterineae
    Corynebacteriaceae, 797, 819, 843
    Dietziaceae, 797, 819, 843
    Gordoniaceae, 797, 819, 843
    Mycobacteriaceae, 797, 819, 843,
        889
    Nocardiaceae, 797, 819, 843
    Tsukamurellaceae, 797, 819, 843
    Williamsiaceae, 819
Corynebacterium, 920, 1031–32
    "black pigmented" (See "Black
        pigmented"
        Corynebacterium)
    characteristics of, 845, 860, 890,
        1015
    CMN-group, 797–98
    ecology and habitats of, 799, 820,
        827–28
    epidemiology of, 828
    genetics of, 802–4
    identification of, 801, 824–26
    isolation of, 799–801, 820–24
    medical, 819–36
    morphology of, 801, 819–20
    nonmedical, 796–810
    phylogenetic tree of, 798, 835, 846
    phylogeny and taxonomy, 797–98,
        819–20
    physiology of, 804–5, 826–27
    species characteristics of, 800,
        821–23
Corynebacterium accolens, 819, 828
Corynebacterium acetoacidophilum,
    819
Corynebacterium afermentans, 819,
    828–29
Corynebacterium afermentans subsp.
    afermentans, 828–29
Corynebacterium afermentans subsp.
    lipophilum, 828–29
Corynebacterium ammoniagenes,
    796, 799, 809–10
Corynebacterium amycolatum, 796,
    798, 819, 825, 827, 829,
    831–32, 834
Corynebacterium appendicis, 819,
    829
Corynebacterium argentoratense,
    819, 829–30
Corynebacterium aurimucosum, 819,
    829
Corynebacterium auris, 819, 828–29
Corynebacterium bovis, 829–30
Corynebacterium bovis group, 1020,
    1080
Corynebacterium callunae, 796, 803
Corynebacterium casei, 796, 799
Corynebacterium cervicis, 819
Corynebacterium confusum, 819, 830
Corynebacterium coyleae, 819, 830
Corynebacterium crenatum, 819
Corynebacterium cystitidis, 828
Corynebacterium diphtheriae, 607,
    803, 819, 824–28, 830–31,
    833–34
Corynebacterium durum, 819, 824,
    827, 831–32, 834

Corynebacterium falsenii, 819, 831
Corynebacterium fastidiosum, 819
Corynebacterium flavescens, 796
Corynebacterium freneyi, 819, 831
Corynebacterium genitalium, 819
Corynebacterium glucuronolyticum, 819, 825, 831
Corynebacterium glutamicum, 796, 799, 801–9
Corynebacterium imitans, 819, 824, 831
Corynebacterium jeikeium, 819, 824, 827–28, 831–32
Corynebacterium kroppenstedtii, 819, 832
Corynebacterium kutscheri, 819
Corynebacterium lilium, 803
Corynebacterium lipophiloflavum, 819, 832
Corynebacterium macginleyi, 819, 828, 832
Corynebacterium matruchotii, 819, 824, 831–32, 834
Corynebacterium melassecola, 803, 819
Corynebacterium minutissimum, 819, 825, 829, 832–33
Corynebacterium mooreparkense, 796, 799
Corynebacterium mucifaciens, 819, 832–33
Corynebacterium mycetoides, 819, 833
Corynebacterium nephridii, 819
Corynebacterium nigricans, 819
Corynebacterium pilosum, 828
Corynebacterium propinquum, 819, 830, 833
Corynebacterium pseudodiphtheriticum, 819, 824, 827–28, 833
Corynebacterium pseudogenitalium, 819
Corynebacterium pseudotuberculosis, 819, 824–27, 831, 833–34
Corynebacterium renale, 803, 825, 828
Corynebacterium riegelii, 819, 833
Corynebacterium sanguinis, 833–34
Corynebacterium segmentosum, 819
Corynebacterium seminale, 831
Corynebacterium simulans, 819, 833–34
Corynebacterium singulare, 819, 833–34
Corynebacterium striatum, 819, 825, 828–29, 833–34
Corynebacterium sundsvallense, 819, 824, 834
Corynebacterium terpenotabidum, 796
Corynebacterium thermoaminogenes, 819
Corynebacterium thomssenii, 819, 834
Corynebacterium tuberculostearicum, 819
Corynebacterium ulcerans, 819, 824–27, 830–31, 833–34

Corynebacterium urealyticum, 819, 824, 827, 834
Corynebacterium variabile, 315, 796, 799
Corynebacterium vitaeruminis, 796
Corynebacterium xerosis, 803, 819, 829, 831, 834
Coryneform CDC Group F-1, 834–36
Coryneform CDC Group G, 834–36
Corynetoxin, *Rathayibacter*, 1076
Crenarchaeota, 10, 281
    Cenarchaeum, 286–88
    Desulfurococcales of, 10, 23, 52–66
    low-temperature, 283–84
    phylogenetic and ecological perspectives on, 281–88
    phylogenetic tree of, 282
    physiology of, 285–86
    Sulfolobales of, 10, 23–42, 52
    Thermoproteales, 10–20, 23, 52
Crossiella, 654
Cryobacterium, 1020, 1040–41
    characteristics of, 1023–24, 1057
    habitats of, 1041
    phylogenetic tree of, 1022
Cryobacterium psychrophilum, 1040–41, 1080
Cryocola antiquus, 1020
Cultivation
    Acidothermus, 674
    Acrocarpospora, 735
    Actinobaculum, 497
    Actinocorallia, 691
    Actinomadura, 690–91
    Actinomyces, 464
    Actinosynnemataceae, 656–58
    Arcanobacterium, 501
    Archaeoglobus, 85–87
    Desulfurococcales, 61–64
    Dietziaceae, 853–54
    Frankia, 674
    Geodermatophilus, 674
    Gordoniaceae, 854–55
    Herbidospora, 735
    Methanosarcinales, 246–48
    Microbispora, 736–37
    Micrococcus, 963
    Microtetraspora, 736–37
    Mycobacterium, 898–900
    Nanoarchaeum, 278
    Nocardiaceae, 855–56
    Nocardioides, 1100–1102
    Nocardiopsis, 764–65
    Nonomuraea, 736–37
    Planobispora, 737–38
    Planomonospora, 737–38
    Planotetraspora, 735
    Spirillospora, 691
    Sporichthya, 674
    Stomatococcus, 977
    Streptomonospora, 766
    Streptomycetaceae, 590–92
    Streptosporangium, 736–37
    Sulfolobales, 31–32
    Thermobifida, 765–66
    Thermococcales, 70–71
    Thermomonospora, 692–93
    Thermoplasmatales, 107–8

Thermoproteales, 16–18
    Varibaculum, 505–6
Curie-point Pyroloysis Mass Spectrometry (PyMS), *Streptomyces*, 568
Curtobacterium, 1020, 1041–48
    characteristics of, 1015, 1023–24, 1047–48
    habitats of, 1044–45
    phylogenetic tree of, 1022
Curtobacterium albidum, 1042–44, 1047
Curtobacterium citreum, 1042–44, 1047
Curtobacterium flaccumfaciens, 1042–48
Curtobacterium helvolum, 1042
Curtobacterium herbarum, 1042–44, 1047
Curtobacterium insectiphilum, 1042
Curtobacterium luteum, 1042–44, 1047
Curtobacterium plantarum, 1042
Curtobacterium pusillum, 1042–44, 1047
Curtobacterium saperdae, 1042
Curtobacterium testaceum, 1042
Cycloserine resistance, Halobacteriales, 127
L-Cystine oxidation, Thermoproteales, 10
Cytochromes
    Arthrobacter, 955
    Halobacteriales, 143
    Methanocorpusculum, 220
    Micrococcus, 966–67
    in Sulfolobales, 34
Cytostatic production, *Streptomyces*, 613–14

D

Dactylosporangium, 623–49, 725
    characteristics of, 639–42
    chemotaxonomic properties of, 636–39
    ecophysiology of, 623–30
    identification of, 635–39
    isolation of, 630–35, 642–45
    morphology of, 636
Dactylosporangium aurantiacum, 641
Dactylosporangium fulvum, 641
Dactylosporangium matsuzakiense, 641, 648
Dactylosporangium roseum, 641, 648
Dactylosporangium salmoneum, 641, 648
Dactylosporangium thailandense, 641
Dactylosporangium variesprium, 641, 648
Dactylosporangium vinaceum, 641, 648
Dairy environment
    Arcanobacterium, 502–3
    Arthrobacter, 950
    Brevibacterium, 1014

Brevibacterium helvolum group, 1079
  Corynebacterium, 799
  Microbacterium, 1056, 1061
  Propionibacterium, 401–5
Demetria, 309–10, 1002
  chemotaxonomic properties of, 1007
  pathogenicity of, 1008–10
Demetria terragena, 310
Dermabacter, 308
Dermabacter hominis, 308
Dermabacteraceae, 308
  Brachybacterium, 308
  Dermabacter, 308
Dermacoccaceae, 308–10
  Demetria, 309
  Dermacoccus, 309
  Kytococcus, 309
Dermacoccus, 309–10, 1002
  chemotaxonomic properties of, 1007
Dermacoccus nishinomiyaensis, 309–10
Dermatophilaceae, 308–9, 1002–10
  Dermatophilus, 309, 1002–10
  Geodermatophilus, 309
  Kineosphaera, 1002–10
Dermatophilosis, *Demetria*, 1009–10
Dermatophilus, 309, 669, 671, 1002–10
  characteristics of, 1009
  chemotaxonomic properties of, 1007
  identification of, 1005–8
  isolation of, 1004
  morphology of, 1005–6
  phylogenetic tree of, 1006
  phylogeny of, 1002–4
Dermatophilus chelonae, 309, 1002–10
Dermatophilus congolensis, 309, 1002–10
Dermatophilus crocodyli, 1003
Desulfurococcaceae, 52
  Aeropyrum, 53
  Desulfurococcus, 53
  Ignicoccus, 53
  Staphylothermus, 52–53
  Stetteria, 53
  Sulfophobococcus, 53
  Thermodiscus, 52, 56
  Thermosphaera, 56
Desulfurococcales, 10, 23, 52–66
  biotechnology of, 65–66
  characteristics of, 54–55
  Desulfurococcaceae, 52
  ecology and habitats of, 56, 65
  genetics of, 65
  identification of, 57–61
  isolation of, 57
  morphology of, 30, 52, 56–61
  phylogenetic tree of, 53
  phylogeny of, 52–53
  physiology of, 64–65
  Pyrodictiaceae, 52, 56
  taxonomy of, 53–56
Desulfurococcus, 53, 56–57, 64–65

Desulfurococcus amylolyticus, 53, 57, 65
Desulfurococcus mobilis, 53, 57–58
Desulfurococcus mucosus, 53, 57
Desulfurococcus saccharovorans, 53
Diarrhea prevention, *Bifidobacterium*, 356
Dietzia, 819
  characteristics of, 845, 860
  CMN-group, 797–98
  phylogenetic tree of, 846
Dietzia maris, 844, 851, 853
Dietzia natronolimnaios, 844, 851, 853
Dietzia psychralcaliphilus, 844, 851, 854
Dietziaceae, 797, 819, 843–72
  Dietzia, 797, 819, 843
  habitats of, 851
  identification of, 858–65
  isolation and cultivation of, 853–54
  metabolism and genetics of, 869–72
  phylogeny of, 844
Dimethylsulfoxide (DMSO) reduction, Halobacteriales, 145
Diptheria toxin, *Corynebacterium*, 826–27, 830–31
Disease
  Actinomyces
    in animals, 490–93, 502–3
    in humans, 483–90, 502–3
  Bifidobacterium, 355
  Clavibacter, 1032–35
  Dermatophilus, 1003–10
  Gardnerella, 369–70
  Microbacterium, 1064–65
  Mobiluncus, 513–16
  Mycobacterium, 889, 909, 919–31
  Nocardia, 852, 865–68
  Propionibacterium, 412–14
  Tsukamurella, 853
DMSO. *See* Dimethylsulfoxide
DNA-DNA homology
  Arthrobacter, 945–46
  Bifidobacterium, 329, 344
  Brevibacterium, 1014
  Mycobacterium leprae, 940–41
  Nocardiopsis, 780
  Sulfolobales, 30
DNA-DNA hybridization
  Actinosynnemataceae, 658–59
  Arthrobacter, 954
  Bifidobacterium, 334–35
  Cellulomonas, 986
  Corynebacterium, 831
  Halobacteriales, 142
  methanogens, 174
  Micrococcus, 967
  Mycetocola, 1068
  Mycobacterium, 890, 894–96
  Propionibacterium, 402–3, 406
  Sanguibacter, 310–11
  Stomatococcus, 978
  Streptomyces, 569–71

E

Ecology
  Actinobaculum, 497
  Actinomyces, 479–82
  Arcanobacterium, 500–501
  Archaeoglobus, 88–89
  Bifidobacterium, 350–55
  Corynebacterium, 827–28
  Crenarchaeota, 283–85
  Desulfurococcales, 65
  Frankia, 677
  Frankineae, 677
  Gardnerella, 368
  Halobacteriales, 148–50
  Korarchaeota, 281–88
  Mobiluncus, 508–10
  Streptomyces, 614–15
  of Sulfolobales, 42
  Thermococcales, 69–70
Ecophysiology
  Actinoplanetes, 623–30
  Kitasatospora, 575–84
  Streptacidiphilus, 575–84
  Streptomyces, 575–84
Eggerthella, 306
Eggerthella lenta, 306
Elastin degradation, *Nocardiopsis*, 784
Electron transfer, interspecies, 167
Electrophoresis, *Bifidobacterium*, 341–42
Embden-Meyerhof pathway
  Cellulomonas, 987
  Corynebacterium, 804
  Propionibacterium, 408–9
  Streptomyces, 562
  Thermococcales, 75
Endocarditis
  Corynebacterium, 831–34
  Stomatococcus, 978
Entner-Duodoroff pathway
  Solfolobales, 35
  Thermococcales, 75
  Thermoplasma, 109
Environments. *See specific environments*
Enzymology
  Anaerobiospirillum, 423
  Archaeoglobus, 92–93
  Bifidobacterium, 347–48
  Gardnerella, 366
  Streptomyces, 611
  Thermococcales, 77–79
Epidemiology
  Actinobaculum, 497–98
  Actinomyces, 482–83
  Arcanobacterium, 502–3
  Corynebacterium, 828
  Gardnerella, 368–69
  Mobiluncus, 513
  Mycobacterium, 925–31
Epoxide production, *Mycobacterium*, 910
Erythromycin resistance/sensitivity
  Bifidobacterium, 349
  Halobacteriales, 127
  Propionibacterium, 410

Ethanol production, *Cellulomonas*, 988
Eubacterium exiguum. See Slackia
Eubacterium fossor. See Atopobium
Eubacterium lentum. See Eggerthella lenta
Eukaryotes, Archaea v., 6
Euryarchaeota
  Thermococcales, 69
  Thermoplasmatales, 101
Eye infections, *Actinomyces*, 486–87

F

Fatty acid biosynthesis, *Mycobacterium*, 905–6
Fatty acids
  Actinomycetales, 772
  Agreia, 1025
  Agromyces, 1028
  Cellulomonas, 988
  Corynebacterium, 796, 825
  Cryobacterium, 1041
  Curtobacterium, 1044, 1046
  Frigoribacterium, 1048
  Gordonia, 846
  Kineosphaera, 1005
  Leifsonia, 1051
  Microbacteriaceae, 1081
  Microbacterium, 1060–61
  Micrococcus, 966–67
  Mycobacterium, 901–4
  Nocardioides, 1099, 1104
  Plantibacter, 1070
  Rathayibacter, 1073
  Stomatococcus, 978
  Streptomyces, 567–68
  Subtercola, 1078
  Thermobifida, 777
Feces environments
  Actinomyces, 456–57
  Bifidobacterium, 322–23, 326, 329–31, 351–54
  Corynebacterium, 799
  Corynebacterium bovis group, 1080
  Methanobacterium, 181
  Methanobrevibacter, 181, 238–40
  Methanosphaera, 240–41
  Oerskovia, 996
  Propionibacterium, 404–5
  Rhodococcus, 853
  Streptosporangineae, 734
Fermentation characteristics
  Bifidobacterium, 336–41
  Gardnerella, 366
Fermentation test, 337–41
Ferroplasma, 101–10
Ferroplasma acidarmanus, 103, 109
Ferroplasma acidiphilum, 103, 109
Ferroplasmaceae, *Ferroplasma*, 101
Fish environment
  Arthrobacter, 950
  Brevibacterium, 1014
  Micrococcus, 961–62
  Renibacterium, 315–16, 972
Flagella
  Actinoplanes, 636

Ampullariella, 636
Archaeoglobus, 83
Brevibacterium, 1013
Curtobacterium, 1043
Dactylosporangium, 636
Halobacteriales, 138
Methanocalculus, 223
Methanocaldococcus, 262, 266–67
Methanococcales, 257, 268
Methanococcus, 181–82
Methanocorpusculum, 220–22
Methanoculleus, 216–17
Methanofollis, 219–20
Methanogenium, 209–10, 215–16
methanogens, 198
Methanomicrobiales, 226
Methanospirillum, 222
Methanothermococcus, 265–66
Methanothermus, 241–42
Mobiluncus, 507
Oerskovia, 994
Pilimelia, 636
Planobispora, 732
Planotetraspora, 734
Pyrodictium, 60
Succinimonas, 423
Succinivibrio, 422
Thermococcales, 71
Thermoplasmatales, 106
Thermosphaera, 59–60
Fluorescence, Thermoplasmatales, 106
Fluorescent in situ hybridization, 346
Formate oxidation
  Methanobacteriaceae, 236
  Methanobacteriales, 231
  Methanobacterium, 179–81
  Methanobrevibacter, 238–40
  Methanococcus, 181–82
  Methanocorpusculum, 220–22
  Methanoculleus, 183
  Methanofollis, 219–20
  Methanogenium, 183, 216
  Methanospirillum, 184
  Methanothermobacter, 241
  Methanothermococcus, 265–66
Fortner's method, *Actinomyces*, 458–62
2-D Fractionation, 7
Frankia, 669–72
  characteristics of, 673
  ecology of, 677
  genetics of, 676–77
  isolation of, 672–74
  physiology of, 674–75
Frankia alni, 672
Frankiaceae, 669–78
  phylogenetic tree of, 670–71
  taxonomy of, 672
Frankineae
  Acidothermaceae, 669
  characteristics of, 672
  ecology of, 677
  Frankiaceae, 669
  genetics of, 676–77
  Geodermatophilaceae, 669
  Microspaeraceae, 669
  Sporichthyaceae, 669

taxonomy of, 672
Freshwater environments
  Actinokineospora, 655, 656
  Actinoplanetes, 631–32
  Crenarchaeota, 283–85
  Gordonia, 851
  habitats of, 383
  Kocuria, 315
  Luteococcus, 390
  Methanobacteriales, 231
  Methanobacterium, 236
  Methanoculleus, 183
  Methanogenium, 183
  methanogens, 168
  Methanomicrobiales, 208, 211
  Methanosarcina, 185–86
  Methanosarcinales, 244
  Microbacteriaceae, 1081–82
  Micrococcus, 962
  Micromonospora, 627–29
  Mycobacterium, 896
  Nocardia, 852
  Rhodococcus, 852–53
  sporangiate actinoplanetes, 625–26
  Streptomyces, 583–84, 614
  Subtercola, 1078
  Thermococcales, 69
Friedmanniella, 383–90
  characteristics of, 387–91
  isolation of, 385–86
  morphology of, 385–87, 389, 391
Friedmanniella antarctica, 383, 385–91
Friedmanniella capsulata, 385–91
Friedmanniella lacustris, 385–91
Friedmanniella spumicola, 385–91
Frigoribacterium, 1020, 1048–49, 1067, 1069, 1081
  characteristics of, 1023–24, 1050
  habitats of, 1049
  phylogenetic tree of, 1022
Frigoribacterium faeni, 1049
Fructose production, *Microbispora*, 747
Fructose utilization, *Micrococcus*, 966
Fungicide production, *Streptomyces*, 613–14

G

Galactose utilization
  Actinokineospora, 660
  Actinosynnema, 660
  Lechevalieria, 665
  Methanosaeta, 253
  Micrococcus, 966
  Saccharothrix, 662
Gamma proteobacteria. *See also* Proteobacteria
  Ruminobacter, 423–24
  Succinivibrionaceae, 419, 422
Gardnerella, 322, 328, 331, 343–44, 359–70
  characteristics of, 365
  ecology and habitat of, 360, 368

epidemiology of, 368–69
genetics of, 367
identification of, 362–65
isolation of, 360–62
pathogenicity of, 369–70
phylogeny of, 359
physiology of, 365–67
taxonomy of, 359–60
Gardnerella vaginalis, 322, 328, 344, 360–70
Gas vesicles
  Halobacteriales, 138
  Methanosarcina, 185–86
Gastrointestinal environments
  Bifidobacterium, 350–54
  Coriobacterium, 306
  Dietzia, 851
  Methanobacteriales, 231
  methanogens, 166
  Methanosphaera, 240–41
  Oerskovia, 996
G+C content
  Acidimicrobium, 305
  Actinomyces, 442
  Actinoplanetes, 623
  Agrococcus, 1026
  Archaeoglobus, 87
  Atopobium, 306
  Bifidobacterium, 344
  Brevibacterium helvolum group, 1079
  Coriobacterium, 306
  Corynebacterium, 796–97, 801, 819
  Corynebacterium bovis group, 1080
  Cryobacterium, 1041
  Curtobacterium, 1042, 1044
  Desulfurococcales, 57, 59, 60–61
  Dietzia, 844
  Frigoribacterium, 1049
  Glycomyces, 306
  Gordonia, 846
  Halobacteriales, 142
  Herbidospora, 729
  Jonesia, 311
  Kineosphaera, 1005
  Leifsonia, 1051
  Leucobacter, 1056
  Methanocalculus, 223
  Methanocaldococcus, 266–67
  Methanococcales, 257
  Methanocorpusculum, 220–22
  Methanoculleus, 217
  Methanofollis, 219–20
  Methanoplanus, 218
  Methanospirillum, 222
  Microbacteriaceae, 1021, 1080–81
  Microbacterium, 1061, 1066
  Microbispora, 730
  Micrococcus, 313, 961
  Microtetraspora, 731
  Mobiluncus, 507–8
  Mycetocola, 1068
  Mycobacterium, 907–9, 920
  Mycobacterium leprae, 940
  Nanoarchaeum, 277
  Nocardia, 848
  Nocardioides, 1100, 1107
  Oerskovia, 995

Okibacterium, 1069
Promicromonospora, 311
Propionibacterium, 400, 403
Renibacterium, 316, 972
Rothia, 315
Ruminobacter, 423–24
Skermania, 847
Stomatococcus, 978
Streptomyces, 543
Streptomycetaceae, 538
Streptosporangiaceae, 726
Succinivibrionaceae, 419
Sulfolobales, 30–31
Thermococcales, 76
Thermoplasmatales, 103
Tsukamurella, 850–51
Williamsia, 847–48
Gelatin degradation, *Nocardiopsis*, 784
Genetic manipulation
  Halobacteriales, 147–48
  tools for, 41–42
Genetics
  Actinocorallia, 709–13
  Actinomadura, 709–13
  Arcanobacterium, 505
  Archaeoglobus, 87–88
  Bifidobacterium, 350
  Corynebacterium, 802–4
  Desulfurococcales, 65
  Frankia, 676–77
  Frankineae, 676–77
  Gardnerella, 367
  Halobacteriales, 146–48
  Methanococcales, 269
  Micrococcus, 967–68
  Mycobacterium, 907–9, 925
  Nanoarchaeum, 279
  Spirillospora, 709–13
  Streptosporangiaceae, 746–47
  Sulfolobales, 40–42
  Thermomonospora, 709–13
  Thermomonosporaceae, 709–13
  Thermoplasmatales, 109–10
  Thermoproteales, 19
Genital tract
  Actinobaculum, 497
  Actinomyces, 455–57, 479–80
  Gardnerella, 360, 368
  Mobiluncus, 506
  Propionibacterium, 404–5
Genomics
  Brevibacterium, 1017
  Halobacteriales, 146–47
  Methanococcales, 268
  Streptomyces, 607
  of Sulfolobales, 40–41
Gentamycin resistance/sensitivity
  Bifidobacterium, 349
  Gardnerella, 366–67
  methanogens, 197
Geodermatophilaceae, 669–78
  phylogenetic tree of, 670–71
  taxonomy of, 672
Geodermatophilus, 309, 669–72, 1002, 1081
  characteristics of, 673
  isolation of, 673–74
  physiology of, 674–75

preservation and cultivation of, 674
Geodermatophilus obscurus, 672, 1002
Geodermatophilus obscurus subsp. *amargosae*, 672
Geodermatophilus obscurus subsp. *dictyosporus*, 672
Geodermatophilus obscurus subsp. *obscurus*, 672
Geodermatophilus obscurus subsp. *utahensis*, 672
Geosmin, *Streptomyces*, 607
Geothermal environment
  Acidimicrobium, 305
  Methanococcales, 259
  Methanothermus, 181
Gingival environments
  Atopobium, 306
  Rothia, 315
Gingivitis, *Actinomyces*, 488–90
Gluconeogenesis, Halobacteriales, 144
Glucose fermentation/utilization
  Anaerobiospirillum, 423
  Atopobium, 306
  Cellulomonas, 987
  Coriobacterium, 306
  Corynebacterium, 804–5, 833
  Micrococcus, 966
  Streptomyces, 562
  Succinimonas, 423
  Succinivibrio, 422
Glucose isomerase, Actinoplanetes, 646
Glucosidase, *Bifidobacterium*, 347–48
β-Glucuronidase production, *Corynebacterium*, 831
Glutamate dehydrogenases (GDH), *Streptomyces*, 612
Glutamate production
  Arthrobacter, 955
  Corynebacterium, 805–8
Glutamic acid production. *See* Glutamate production
L-Glutamine production. *See* Glutamate production
Glutamine synthetases (GSs), *Frankia*, 675
Glycomyces, 306, 623–49
  characteristics of, 639–42
  ecophysiology of, 623–30
  identification of, 635–39
  isolation of, 630–35, 642–45
  morphology of, 756
Glycomyces harbinensis, 641, 648
Glycomyces rutgersensis, 641
Glycomyces tenuis, 306
Glycomycineae, *Glycomyces*, 306
Glycoprotein cell walls, Halobacteriales, 128–30
Glycosidases, *Bifidobacterium*, 347–48
Glycoside hydrolases, *Cellulomonas*, 990–92
Gordonia, 819, 843
  characteristics of, 845, 860, 1015
  CMN-group, 797–98

phylogenetic tree of, 846
phylogeny of, 844–47
species characteristics of, 861
Gordonia alkalivorans, 846, 854
Gordonia amarae, 854
Gordonia amicalis, 854
Gordonia bronchialis, 844, 846, 851
Gordonia desulfuricans, 854, 872
Gordonia nitida, 846
Gordonia paraffinivorans, 854
Gordonia rubra, 844
Gordonia rubripertincta, 851
Gordonia sihwensis, 854
Gordonia sputi, 851
Gordonia terrae, 844, 851
Gordoniaceae, 797, 819, 843–72
    Gordonia, 819, 843
    habitats of, 851
    identification of, 858–65
    isolation and cultivation of,
        854–55
    metabolism and genetics of,
        869–72
    phylogenetic tree of, 847
    phylogeny of, 844–48
    Skermania, 819, 843
Gram-negative
    Anaerobiospirillum, 423
    Mobiluncus, 506–7
    Ruminobacter, 424
    Succinimonas, 423
    Succinivibrio, 422
    Succinivibrionaceae, 419
Gram-positive
    Acidimicrobium, 305
    Acrocarpospora, 729
    Actinobaculum, 497
    Actinomyces, 440, 442
    Actinomycetaceae, 430
    Atopobium, 306
    Brevibacterium, 1013, 1015
    Coriobacterium, 306
    Corynebacterium, 796, 802, 819–20
    Curtobacterium, 1043
    Dietzia, 844
    Gordonia, 846
    Kitasatospora, 538
    Microbacteriaceae, 1020
    Microbacterium, 1056, 1061, 1066
    Microbispora, 730
    Micrococcus, 313, 961
    Microlunatus, 390
    Micropruina, 393
    Microtetraspora, 731
    Mobiluncus, 506–7
    Nocardia, 848
    Nocardioides, 1099
    Nocardiopsis, 758
    Nonomuraea, 731
    Planobispora, 732
    Planomonospora, 733
    Planotetraspora, 734
    Propionibacterium, 401
    Propioniferax, 394–95
    Renibacterium, 972
    Rhodococcus, 849–50
    Rubrobacter, 305
    Skermania, 847
    Stomatococcus, 975

Streptacidiphilus, 538
Streptomyces, 538, 543, 564
Streptomycetaceae, 538
Streptosporangiaceae, 725
Tsukamurella, 850–51
Williamsia, 847–48
Groundwater environment
    Methanobacteriales, 231
    Methanomicrobiales, 211

H

Habitats. *See specific habitats*
Haloarcula, 114, 120–21, 137, 142,
        144–45
Haloarcula aidinensis, 117
Haloarcula argentinensis, 137
Haloarcula californiae, 117
Haloarcula hispanica, 136, 148
Haloarcula japonica, 115–16,
        128–29, 144
Haloarcula marismortui, 129, 145,
        151
Haloarcula quadrata, 117
Haloarcula vallismortis, 137, 143–45,
        148
Halobacteriales, 113–51
    ecology and habitats of, 139–40,
        148–50
    genetics of, 146–48
    Halobacteriaceae, 114
    Halobacterium, 113
    identification of, 142
    isolation of, 140–42
    morphology of, 115–16
    phylogenetic tree of, 115
    phylogeny of, 114
    physiology of, 143–46
    taxonomy of, 115–39
Halobacterium, 113–14, 118–19, 140
Halobacterium halobium
        IAM13167, 131
Halobacterium salinarum, 113–14,
        116, 127–28, 131, 134–38, 140,
        142–43, 145–50
Halobaculum, 114, 118–19
Halobaculum gomorrense, 148
Halocin secretion, Halobacteriales,
        149
Halococcus, 114, 122–23, 129–30,
        135, 140, 142, 150
Halococcus morrhuae, 129, 136, 143,
        147
Halococcus saccharolyticus, 127
Halococcus salifodinae, 117
Haloferax, 114, 124–25, 127, 130,
        136, 142, 144, 148, 150
Haloferax alicantei, 117
Haloferax denitrificans, 145
Haloferax gibbonsii, 130, 150
Haloferax mediterranei, 117, 127,
        130, 134, 136, 138–39, 144–45,
        147, 151
Haloferax volcanii, 117, 128–29, 135,
        139–40, 142–43, 145, 147–51
Halogeometricum, 114, 122–23
Halogeometricum borinquense, 116,
        138, 145

Halomethanococcus, 175–76, 179,
        186–87
Halomethanococcus doii, 179,
        186–87
Halophile
    Halobacteriales, 113
    Methanosarcinales, 245
    Nesterenkonia, 315
Halorhodopsin, Halobacteriales,
        113, 137–38
Halorubrum, 114, 118–19
Halorubrum distributum, 128,
        139–40, 144
Halorubrum lacusprofundi, 116, 131,
        139
Halorubrum saccharovorum, 117,
        134, 143–44
Halorubrum sodomense, 137, 145,
        148
Halorubrum vacuolatum, 149
Haloterrigena, 114, 126
Haloterrigena turkmenica, 139
Hansen's disease. *See* Leprosy
Heat shock, Sulfolobales, 37–38
Herbicide degradation
    Arthrobacter, 950, 955
    Clavibacter, 1040
Herbidospora, 726
    chemotaxonomic and
        morphological features of,
        727
    identification of, 740–46
    isolation and cultivation, 735
    morphology of, 768–69
    phylogeny of, 729
Herbidospora cretacea, 729, 734–35
Heteropolysaccharide, methanogens,
        193
Heterotrophs
    Halobacteriales, 143
    Picrophilus, 109
Hexose metabolism
    Bifidobacterium, 346–47
    Corynebacterium, 804
    Propionibacterium, 407–9
Hongia
    characteristics of, 1105, 1108
    morphology of, 768–69
Hot spring environments
    Acidimicrobium, 305
    Desulfurococcales, 56
    methanogens, 168
    Methanothermus, 241–42
    Rubrobacter, 305
    Thermoplasmatales, 103–5
    Thermoproteales, 13
Human actinomycoses,
        Actinomyces, 483–86
Human infections
    Arcanobacterium, 500–505
    Mobiluncus, 512–16
Hydrocarbon degradation
    Mycobacterium, 904–5
    Nocardiopsis, 784
Hydrocarbon synthesis,
        Micrococcus, 968
Hydrothermal environment
    Archaeoglobus, 82
    Desulfurococcales, 65

Methanococcales, 259
Methanococcus, 182
Nanoarchaeum, 275–77
Thermococcales, 69–70
Thermoplasmatales, 104
Hypersaline environment
  Halobacteriales, 113, 139, 148–49
  Methanosarcinales, 245–46
  Obligately methylotrophic cocci, 186–87
Hyperthermophile
  Archaeoglobus, 88
  Desulfurococcales, 52, 65
  Methanocaldococcaceae, 257
  Methanocaldococcus, 266–67
  Methanotorris, 267–68
  Sulfolobales, 23
  Thermococcales, 69
  Thermofilum, 13
  Thermoproteales, 10, 20
Hyperthermus, 56–57, 64
Hyperthermus butylicus, 56, 65

I

Identification
  Actinobaculum, 497
  Actinocorallia, 694–708
  Actinokineospora, 659–60
  Actinomadura, 694–708
  Actinomyces, 464–79
  Actinoplanetes, 635–39
  Actinosynnema, 660–61
  Actinosynnemataceae, 658–65
  Agreia, 1026
  Agrococcus, 1027
  Agromyces, 1029
  Arcanobacterium, 501–2
  Archaeoglobus, 83–85
  Arthrobacter, 952–54
  Bifidobacterium, 334–46
  Brevibacterium, 1015
  Cellulomonas, 988
  Clavibacter, 1037–39
  Corynebacterium, 801, 824–26
  Cryobacterium, 1041
  Curtobacterium, 1046
  Dermatophilus, 1005–8
  Desulfurococcales, 57–61
  Dietziaceae, 858–65
  Frigoribacterium, 1049
  Gardnerella, 362–65
  Gordoniaceae, 858–65
  Halobacteriales, 142
  Kineosphaera, 1005–8
  Kitasatospora, 574
  Lechevalieria, 665
  Leifsonia, 1053–55
  Lentzea, 663–65
  Leucobacter, 1056
  Methanobacteriales, 235–36
  Methanococcales, 261–62
  methanogens, 174–76
  Methanomicrobiales, 213–14
  Microbacterium, 1065–66
  Micrococcus, 963–66
  Mobiluncus, 509, 512–13
  Mycetocola, 1067–68

Mycobacterium, 900–904, 921–23
Mycobacterium leprae, 939–41
Nanoarchaeum, 277–78
Nocardiaceae, 858–65
Nocardioides, 1102
Nocardiopsis, 766–77
Oerskovia, 996
Okibacterium, 1069
Plantibacter, 1071
Rathayibacter, 1077
Saccharothrix, 661–63
Spirillospora, 694–708
Stomatococcus, 978
Streptacidiphilus, 574
Streptomonospora, 779–81
Streptomyces, 574–75
Streptomycetaceae, 574–75
Streptosporangiaceae, 740–46
Subtercola, 1078–79
Sulfolobales, 29–32
Thermobifida, 777–79
Thermococcales, 71–72
Thermomonospora, 694–708
Thermomonosporaceae, 694–708
Thermoplasmatales, 106–7
Thermoproteales, 14–16
Tsukamurellaceae, 858–65
Varibaculum, 506
Ignicoccus, 53, 56–58, 64–65
Ignicoccus islandicus, 53, 56, 58, 64
Ignicoccus pacificus, 53, 56
Immune system enhancement, *Bifidobacterium*, 356–57
Internal transcribed spacer sequence analysis, 344–45
Interspecies electron transfer
  methanogens, 167
  Methanomicrobiales, 212
Intrasporangiaceae, 311–12
  Intrasporangium, 311–12
  Janibacter, 311–12
  Terrabacter, 311–12
  Terracoccus, 311–12
Intrasporangium, 311–12
Intrasporangium calvum, 312
Iron oxidation
  Acidimicrobium, 305
  Archaeoglobus, 97
  Thermoproteales, 10, 19–20
Iron uptake, *Mycobacterium*, 907
Isolation
  Acidothermus, 674
  Acrocarpospora, 735
  Actinobaculum, 497
  Actinocorallia, 691
  Actinomadura, 690–91
  Actinomyces, 457–63
  Actinoplanetes, 630–35, 642–45
  Actinosynnemataceae, 656–57
  Agreia, 1025–26
  Agrococcus, 1026
  Agromyces, 1029
  Arcanobacterium, 501
  Archaeoglobus, 85
  Arthrobacter, 950–52
  Bifidobacterium, 333–34
  Brevibacterium, 1014
  Cellulomonas, 988–90

Clavibacter, 1035–37
Corynebacterium, 799–801, 820–24
Cryobacterium, 1041
Curtobacterium, 1045–46
Dermatophilus, 1004
Desulfurococcales, 57
Frankia, 672–74
Frankineae, 672–74
Friedmanniella, 385–86
Frigoribacterium, 1049
Gardnerella, 360–62
Geodermatophilus, 673
Gordoniaceae, 854–55
Halobacteriales, 140–42
Herbidospora, 735
Kineosphaera, 1004
Leifsonia, 1053
Luteococcus, 390
Methanobacteriales, 231–36
Methanococcales, 259–61
methanogens, 169–73
Methanomicrobiales, 212–14
Microbacterium, 1065
Microbispora, 736–37
Micrococcus, 962–63
Microlunatus, 392
Micropruina, 394
Microtetraspora, 736–37
Mobiluncus, 510–12
Mycetocola, 1067
Mycobacterium, 897–98, 920–21
Mycobacterium leprae, 937–39
Nanoarchaeum, 277
Nocardiaceae, 855–56
Nocardioides, 1100–1102
Nocardiopsis, 764–65
Nonomuraea, 736–37
Oerskovia, 996
Okibacterium, 1069
Planobispora, 737–38
Planomonospora, 737–38
Planotetraspora, 735
Plantibacter, 1071
Propionibacterium, 410–12
Propioniferax, 394–95
Propionimicrobium, 395
Rathayibacter, 1076–77
Spirillospora, 691
Sporichthya, 674
Streptomonospora, 766
Streptomyces, 584–90
Streptosporangium, 736–37
Subtercola, 1078
Succinivibrionaceae, 420–21
Sulfolobales, 28–29
Tessarococcus, 396–97
Thermobifida, 765–66
Thermococcales, 70
Thermomonospora, 692–93
Thermoplasmatales, 105–6
Thermoproteales, 13–14
Varibaculum, 505–6

J

Janibacter, 311–12
Janibacter limosus, 311–12

Jonesia, 311
Jonesia denitrificans, 311
Jonesiaceae, *Jonesia*, 311

K

Kanamycin resistance
  Bifidobacterium, 349
  Halobacteriales, 127
  Methanoplanus, 218
Keratinases, *Streptomyces*, 611
Keratitis, *Corynebacterium*, 833
Kibdellosporangium, morphology of, 756
Kineococcaceae, 306–7
  Kineococcus, 306–7
  Kineosporia, 306–7
Kineococcus, 306–7
Kineococcus aurantiacus, 307
Kineosphaera, 1002–10
  characteristics of, 1009
  chemotaxonomic properties of, 1007
  identification of, 1005–8
  isolation of, 1004
  morphology of, 1005, 1007
  pathogenicity of, 1008–10
  phylogenetic tree of, 1006
  phylogeny of, 1002–4
Kineosphaera limosa, 1002–7, 1009
Kineosporia, 306–7
  morphology of, 768–69
Kitasatospora, 538, 542–43
  characteristics of, 580
  chemotaxonomic properties of, 540
  ecophysiology and habitat of, 575–84
  identification of, 574
  morphology of, 540, 756, 768–69
  physiological properties of, 540
Kitasatospora azatica, 543
Kitasatospora cheerisanenesis, 543
Kitasatospora cineracea, 543
Kitasatospora cochleata, 543
Kitasatospora cystarginea, 543
Kitasatospora griseola, 543
Kitasatospora mediocidica, 543
Kitasatospora niigatensis, 543
Kitasatospora paracochleata, 543
Kitasatospora phosalacinea, 543
Kitasatospora setae, 543
Kocuria, 312, 315, 976
Kocuria kristinae, 315
Kocuria palustris, 315
Kocuria rhizophila, 315
Kocuria rosea, 315
Kocuria varians, 315
Korarchaeota
  concluding remarks on, 288
  introduction to, 281
  phylogenetic tree of, 282
Kribella
  characteristics of, 1105, 1108–9
  morphology of, 768–69
Kutzeria, 654
Kytococcus, 309–10, 1002

chemotaxonomic properties of, 1007
Kytococcus sedentarius, 310

L

Lacrimal canaliculitis, *Actinomyces*, 486–87
β-Lactam antibiotic resistance, *Methanobacterium*, 192–93
Lactate production/utilization
  Archaeoglobus, 91
  Cellulomonas, 988
  Micrococcus, 966
Lactic acid production
  Micrococcus, 966
  Propionibacterium, 409
Lactose intolerance, *Bifidobacterium*, 356
Lechevalieria, 654–59, 665–66
  characteristics of, 665
  habitat of, 656
  identification of, 665
  morphology of, 768–69
  phylogenetic tree of, 663
Lechevalieria aerocolonigenes, 655, 665
Lechevalieria flava, 655, 665
Lecithin degradation, *Nocardiopsis*, 784
Leifsonia, 1020, 1049–55, 1081
  characteristics of, 1023–24, 1054
  habitats of, 1052–53
  phylogenetic tree of, 1022
Leifsonia aquatica, 1051–55
Leifsonia aurea, 1081
Leifsonia cyodontis, 1051
Leifsonia naganoensis, 1051–55
Leifsonia poae, 1051–55
Leifsonia rubra, 1081
Leifsonia shinshuensis, 1051–55
Leifsonia xyli subsp. *cyodontis*, 1051–55
Leifsonia xyli subsp. *xyli*, 1051–55
Lentzea, 654–59, 663–66
  characteristics of, 665
  habitat of, 656
  identification of, 663–65
  morphology of, 768–69
  phylogenetic tree of, 663
Lentzea albida, 655
Lentzea albidocapillata. See Saccharothrix albidocapillata
Lentzea califoriensis, 655
Lentzea violacea, 655
Lentzea waywayandensis, 655
Leprosy
  Arcanobacterium, 501
  description of, 934–36
  Mycobacterium, 889, 919, 928–30
  summary of, 941
Leucobacter, 1020, 1055–53
  characteristics of, 1023–24, 1057
  habitats of, 1056
  phylogenetic tree of, 1022
Leucobacter komagatae, 1055–56
Levan production, *Microbacterium*, 1066

Lignin decomposition
  Micromonospora, 627–29
  Streptomyces, 580–81
  Thermobifida, 784–85
  Thermomonospora, 712
Lincomycin resistance/sensitivity
  Bifidobacterium, 349
  Streptomyces, 567
Lipases, *Streptomyces*, 611
Lipid composition
  Mycobacterium, 906–7
  Subtercola, 1078
Listeria denitrificans. See Jonesia denitrificans
Ljungdahl-Wood pathway, methanogens, 191
Luteococcus, 383–85, 388–90
  characteristics of, 390–91
  isolation of, 390
  morphology of, 389–91
Luteococcus japonicus, 383, 390–91
Luteococcus peritonei, 390–91
Lycopene, 135
L-Lysine production, *Corynebacterium*, 805, 807–8

M

Macrolide resistance, methanogens, 197
L-Malate dehydrogenase, *Archaeoglobus*, 90–91
Maltose degradation
  Cellulomonas, 987
  Corynebacterium, 804, 833
  Micrococcus, 966
  Oerskovia, 996
Mannitol fermentation, *Corynebacterium*, 832
Mannose utilization
  Lechevalieria, 665
  Micrococcus, 967
  Oerskovia, 996
Marine bacterioplankton, 1081–82
Marine environments
  Actinoplanetes, 631–32
  Archaeoglobus, 82
  Arthrobacter, 950
  Brevibacterium, 1014
  Crenarchaeota, 283–85
  Desulfurococcales, 56, 64–65
  Gordonia, 851
  Halobacteriales, 139
  Kocuria, 315
  Methanobacteriales, 231
  Methanobrevibacter, 238–40
  Methanococcales, 257, 259
  Methanococcus, 182
  Methanoculleus, 183
  Methanogenium, 183
  methanogens, 168
  Methanomicrobiales, 208, 211
  Methanoplanus, 184
  Methanosarcina, 185–86
  Methanosarcinales, 244, 245–46
  Microbacteriaceae, 1081–82
  Microbacterium, 1064
  Micrococcus, 962

Micromonospora, 627–29
Mycobacterium, 896
Nanoarchaeum, 276–77
Nocardia, 852
Nocardiopsis, 761, 784
Obligately methylotrophic cocci, 186–87
Rhodococcus, 852–53
Streptomyces, 583–84, 614
Sulfolobales, 25, 28
Thermococcales, 69–70, 76
Thermoplasmatales, 104
Thermoproteales, 13
Marmoricola, 1105
Marsh environment
  Methanomicrobiales, 211
  Methanosarcina, 185–86
Menaquinones
  Acrocarpospora, 729
  Actinokineospora, 660
  Actinomadura, 684–85, 699
  Actinomyces, 446–48
  Actinoplanes, 638–39
  Actinosynnema, 660
  Agreia, 1026
  Agrococcus, 1026
  Agromyces, 1028–29
  Ampullariella, 638–39
  Arthrobacter, 312, 947–48, 954
  Bogoriella, 307–8
  Brachybacterium, 308
  Brevibacterium, 1013
  Brevibacterium helvolum group, 1079
  Clavibacter, 1031–32, 1037
  Corynebacterium, 796, 819–20
  Cryobacterium, 1041
  Curtobacterium, 1046
  Dactylosporangium, 638–39
  Demetria, 310
  Dermabacter, 308
  Dietzia, 844
  Frigoribacterium, 1048–49
  Glycomyces, 306
  Gordonia, 846
  Herbidospora, 729
  Intrasporangium, 312
  Jonesia, 311
  Kineococcaceae, 307
  Kineosporia, 307
  Kytococcus, 310
  Leifsonia, 1049, 1051, 1054
  Leucobacter, 1056
  Microbacteriaceae, 1020–21, 1080–81
  Microbacterium, 1060–61, 1065
  Microbispora, 730
  Micrococcaceae, 312
  Micrococcus, 313
  Mycetocola, 1067–68
  Nesterenkonia, 315
  Nocardia, 848
  Nocardioides, 1100, 1104
  Nocardiopsis, 770, 773
  Okibacterium, 1069
  Pilimelia, 638–39
  Planomonospora, 733
  Planotetraspora, 734
  Plantibacter, 1070

Promicromonospora, 311
Rarobacter, 310
Rathayibacter, 1073, 1077
Renibacterium, 316, 972
Rhodococcus, 850
Rothia, 315
Saccharothrix, 662
Sanguibacter, 310–11
Stomatococcus, 315, 978
Streptomonospora, 781
Streptomyces, 562, 565–66
Streptosporangiaceae, 726
Subtercola, 1078
Thermobifida, 760, 777
Thermomonospora, 699
Tsukamurella, 850–51
Williamsia, 847–48
Meningitis
  Stomatococcus, 978
  Tsukamurella, 853
Mesophile
  Curtobacterium, 1043
  Methanocaldococcaceae, 257
  Methanococcaceae, 257
  Methanococcus, 181–82, 262–63
  Methanoculleus, 216–17
  Microbacteriaceae, 1021
  Microbispora, 730
  Microtetraspora, 731
  Nocardioides, 1099
  Streptomyces, 614
Metabolism
  Actinocorallia, 709–13
  Actinomadura, 709–13
  Archaeoglobus, 89–97
  Methanosarcinales, 244–45
  Micrococcus, 966
  Propionibacterium, 408–9
  Spirillospora, 709–13
  Streptosporangiaceae, 746–47
  Thermococcales, 72–79
  Thermomonospora, 709–13
  Thermomonosporaceae, 709–13
Metallosphaera, 25, 28, 30–31, 33
Metallosphaera hakonensis, 25, 28, 31
Metallosphaera hakonensis HO1–1^T, 25
Metallosphaera prunae, 25, 28–29, 31
Metallosphaera prunae Ron12/II^T, 25
Metallosphaera sedula, 25, 28–31, 34
Metallosphaera sedula TH2^T, 25
Methane, atmospheric, methanogens, 168–69
Methane production
  Methanobacteriales, 236–42
  Methanocorpusculaceae, 220–22
  methanogens, 165
  Methanomicrobiales, 214–20
  Methanosarcinales, 244–45
  Methanospirillaceae, 222
Methanobacillus omelianskii, 4
Methanobacteriaceae, 175, 179–80, 232–34, 236–41
  Methanobacterium, 174–75, 179–81, 232–34, 236–37
  Methanobrevibacter, 174–75, 180–81, 232–34, 236, 238–40

  Methanosphaera, 174–75, 181, 232–34, 236, 240–41
  Methanothermobacter, 232–34, 236, 241
Methanobacteriales, 231–42
  characteristics of, 231
  habitats of, 231
  isolation and characterization of, 231–36
  Methanobacteriaceae, 175, 232–34, 236–41
  Methanothermaceae, 175, 232–34, 241–42
Methanobacterium, 5, 174–75, 179–81, 192, 232–34, 236–37
  characteristics of, 176
Methanobacterium alcaliphilum, 176, 179–81, 232–34, 236
Methanobacterium bryantii, 176, 179–81, 232–34, 236–37
Methanobacterium congolense, 232–34, 237
Methanobacterium espanolae, 176, 179–81, 232–34, 237
Methanobacterium formicicum, 168, 176, 179–81, 232–34, 236
Methanobacterium ivanovii, 176, 179–81, 232–34, 237
Methanobacterium oryzae, 232–34, 237
Methanobacterium palustre, 176, 179–81, 232–34, 237
Methanobacterium ruminatanium, 4
Methanobacterium subterraneum, 232–34, 237
Methanobacterium thermoaggregans, 176, 179–81
Methanobacterium thermoalcaliphilum, 176, 179–81
Methanobacterium thermoautotrophicum, 5, 176, 179–81, 192–93, 196–97
Methanobacterium thermoformicicum, 176, 179–81
Methanobacterium uliginosum, 176, 179–81, 232–34, 237
Methanobacterium wolfei, 176, 179–81
Methanobrevibacter, 168, 173, 174–75, 180–81, 211, 232–34, 236, 238–40
  characteristics of, 176
Methanobrevibacter acididurans, 232–34, 238–39
Methanobrevibacter arboriphilus, 176, 180–81, 232–34, 238
Methanobrevibacter curvatus, 232–34, 238
Methanobrevibacter cuticularis, 232–34, 239
Methanobrevibacter filiformis, 232–34, 239
Methanobrevibacter gottschalkii, 232–34, 239
Methanobrevibacter oralis, 232–34, 239

Methanobrevibacter ruminantum,
    168, 176, 180–81, 232–34, 238
Methanobrevibacter smithii, 176,
    181, 232–34, 239
Methanobrevibacter thaueri, 232–34,
    239–40
Methanobrevibacter woesei, 232–34,
    240
Methanobrevibacter wolinii, 232–34,
    240
Methanocalculus halotolerans,
    109–10, 211, 223
Methanocalculus pumilus, 109–10,
    223
Methanocaldococcaceae, 257–58
    Methanocaldococcus, 258–59,
        266–67
    Methanotorris, 258–59, 266–68
Methanocaldococcus, 258–59,
    261–62, 266–67
Methanocaldococcus fervens,
    258–59, 262, 266
Methanocaldococcus infernus,
    258–59, 266–67
Methanocaldococcus jannaschii,
    258–59, 262, 267–69
Methanocaldococcus vulcanius,
    258–59, 267
Methanococcaceae, 257–59
    Methanococcus, 174–75, 181–82,
        258–59, 262–65
    Methanothermococcus, 258–59,
        265–66
Methanococcales, 257–69
    biochemical and physiological
        properties of, 268–69
    habitat of, 259
    identification of, 261–62
    isolation of, 259–61
    Methanocaldococcaceae, 257–58
    Methanococcaceae, 175, 257–58
    phylogenetic tree of, 258
    phylogeny of, 257
    taxonomy of, 258–59
Methanococcoides, 175–76, 179,
    186–87, 249–52
Methanococcoides burtonii, 249–52
Methanococcoides euhalobius, 179,
    186–87
Methanococcoides methylutens, 179,
    186–87, 249–52
Methanococcus, 173, 181–82, 192,
    258–59, 261–65
    characteristics of, 177, 263
Methanococcus aeolicus, 177,
    181–82, 258–59, 263–64
Methanococcus deltae, 258–59, 264
Methanococcus halophilus, 176, 179,
    186–87
Methanococcus jannaschii, 177,
    181–82, 198
Methanococcus maripaludis, 177,
    181–82, 258–59, 262, 264,
    269
Methanococcus vannielii, 177,
    181–82, 196, 258–59, 263–65
Methanococcus voltae, 177, 181–82,
    192, 195, 198, 258–59, 263,
    265, 269

Methanocorpusculaceae
    Methanocorpusculum, 175–76,
        184–85, 209–10, 220–22
Methanocorpusculum, 175–76,
    184–85, 208–10, 220–22
    characteristics of, 177
Methanocorpusculum aggregans,
    178, 184–85, 208–10, 220–21
Methanocorpusculum bavaricum,
    178, 184–85, 208–10, 220–21
Methanocorpusculum labreanum,
    178, 184–85, 209–11, 220–
    21
Methanocorpusculum parvum, 178,
    184–85, 209–10, 220–21
Methanocorpusculum sinense, 178,
    184–85, 208–10, 221–22
Methanoculleus, 175–76, 182–83,
    208–11, 216–18, 261
    characteristics of, 177
Methanoculleus bourgense, 177,
    182–83, 209–10, 217
Methanoculleus marisnigri, 177,
    182–83, 209–11, 217, 224
Methanoculleus oldenburgensis,
    209–11, 217
Methanoculleus olentangyi, 177,
    182–83, 209–11, 217
Methanoculleus palmolei, 208,
    209–10, 217–18
Methanoculleus thermophilicum,
    177, 182–83, 209–11, 218, 224,
    261
Methanofollis, 208–10, 219–20
Methanofollis liminatans, 208–10,
    219, 224
Methanofollis tationis, 209–11,
    219–20
Methanofuran, of methanogenesis,
    188–89
Methanogenesis
    Archaeoglobus, 83
    biochemistry of, 4
    coenzymes of, 188–89
    Methanobacterium, 236–37
    Methanobrevibacter, 238–40
    Methanocaldococcus, 266–67
    Methanococcus, 263–64
    Methanocorpusculum, 220–22
    Methanoculleus, 183, 216–18
    Methanogenium, 183, 215–16
    methanogens, 165
    Methanolacinia, 215
    Methanomicrobium, 214–15
    Methanoplanus, 218–19
    Methanosarcinales, 244–45
    Methanosphaera, 240–41
    Methanospirillum, 184
    Methanothermobacter, 241
    Methanothermococcus, 265–66
    Methanotorris, 267–68
    pathway of, 189–91
    reactions and free energy changes
        of, 166
Methanogenium, 175–76, 182–83,
    198, 209–12, 215–16, 261
    characteristics of, 177
Methanogenium cariaci, 177, 182–83,
    211, 215–16, 224, 226

Methanogenium frigidum, 211,
    215–16
Methanogenium frittonii, 177,
    182–83, 211, 215–16
Methanogenium liminatans, 177,
    182–83, 219. *See also*
        Methanofollis
Methanogenium organophilum, 177,
    182–83, 211, 216, 224
Methanogenium tationis, 177,
    182–83, 219. *See also*
        Methanofollis
Methanogens, 165–99
    biochemical and physiological
        properties, 188–98
    carbon metabolism, 191–92
    cellular morphology of, 174–79
    characteristics of, 175–88
    habitats of, 166–69
    identification of, 173–76
    isolation of, 169–73
    Methanobacteriales, 175, 231–42
    Methanococcales, 175, 257–69
    methanogenesis, 188–91
    Methanomicrobiales, 175
    Methanosarcinales, 244–54
    molecular biology of, 194–96
    nitrogen and sulfur metabolism,
        192
    phylogenetic tree of, 174
Methanohalobium, 175–76, 179,
    186–87, 249–53
Methanohalobium evestigatum, 179,
    186–87, 249–53
Methanohalophilus, 175–76, 179,
    186–87, 249–53
Methanohalophilus halophilus,
    249–53
Methanohalophilus mahii, 179,
    186–87, 249–53
Methanohalophilus oregonense, 179,
    186–87
Methanohalophilus portucalensis,
    249–53
Methanohalophilus zhilinae, 179,
    186–87
Methanolacinia, 175–76, 184, 209–11,
    215
    characteristics of, 177
Methanolacinia paynteri, 178, 184,
    209–11, 215
Methanolobus, 175–76, 179, 186–87,
    249–52
Methanolobus bombayensis, 249–52
Methanolobus oregonensis, 249–52
Methanolobus taylorii, 249–52
Methanolobus tindarus, 179, 186–87,
    249–52
Methanolobus vulcani, 249–52
Methanomicrobiacaea
    Methanoculleus, 175–76, 182–83,
        209–10, 216–18
    Methanofollis, 209–10, 219–20
    Methanogenium, 175–76, 182–83,
        209–10, 215–16
    Methanolacinia, 175–76, 184,
        209–10, 215
    Methanomicrobium, 175–76, 184,
        209–10, 214–15

Methanoplanus, 175–76, 183–84, 209–10, 218–19
Methanospirillum, 175–76, 184
phylogenetic tree of, 183
Methanomicrobiales, 175–76, 208–26
biochemical and physiological properties of, 223–26
characteristics of, 208–14
habitats of, 208–12
isolation of, 212–14
Methanocorpusculaceae, 175–76, 209–10, 220–22
Methanofollis, 208
Methanomicrobiacaea, 175–76, 209–10, 214–20
Methanosarcinaceae, 175–76
Methanospirillaceae, 209–10, 222
Methanomicrobium, 175–76, 184, 209–11, 214–15
characteristics of, 177
Methanomicrobium mobile, 168, 178, 184, 209–10, 214–15, 226
Methanoplanus, 175–76, 183–84, 209–10, 218–19
characteristics of, 177
Methanoplanus endosymbiosus, 168, 178, 183–84, 209–10, 212, 218
Methanoplanus limicola, 178, 183–84, 209–11, 218–19, 224
Methanoplanus petrolearius, 175–76, 183–84, 209–10, 219
Methanosaeta, 175–76, 179, 187, 244–45, 253–54
characteristics of, 177
Methanosaeta concilii, 179, 187, 195, 249–52, 254
Methanosaeta soehngenii, 179, 187
Methanosaeta strain CALS-1, 179, 187
Methanosaeta thermoacetophila, 179, 187
Methanosaeta thermophila, 249–52, 254
Methanosaetaceae, 244–45, 253–54
Methanosaeta, 244–45, 253–54
Methanosalsum, 249–53
Methanosalsum zhilinae, 249–53
Methanosarcina, 173, 175–76, 185–86, 248–52
characteristics of, 177
Methanosarcina acetivorans, 178, 185–86, 248–52
Methanosarcina baltica, 248–52
Methanosarcina barkeri, 178, 185–86, 192–93, 248–52
Methanosarcina frisia, 178, 185–86
Methanosarcina lacustris, 248–52
Methanosarcina mazeii, 178, 185–86, 193, 248–52
Methanosarcina semesiae, 248–52
Methanosarcina siciliae, 248–52
Methanosarcina thermophila, 178, 185–86, 192, 248–52
Methanosarcina vacuolata, 178, 185–86, 248–52
Methanosarcinaceae, 244–45, 248–53
Halomethanococcus, 175–76
Methanococcoides, 175–76, 249–52
Methanohalobium, 175–76, 249–53

Methanohalophilus, 175–76, 249–53
Methanolobus, 175–76, 249–52
Methanosaeta, 175–76, 179, 187
Methanosalsum, 249–53
Methanosarcina, 175–76, 185–86, 248–52
obligately methylotrophic cocci, 179, 186–87
phylogenetic tree of, 183
Methanosarcinales, 211, 244–54
habitats and metabolism of, 244–46
Methanosaetaceae, 244–45, 253–54
Methanosarcinaceae, 244–45, 248–53
Methanosphaera, 168, 174–75, 181, 232–34, 236, 240–41
characteristics of, 177
Methanosphaera cuniculi, 232–34, 240–41
Methanosphaera stadtmaniae, 177, 181, 232–34, 240
Methanospirillaceae, 209–10, 222
Methanospirillum, 209–10, 222
Methanospirillum, 173, 175–76, 184, 192, 209–10, 222
characteristics of, 177
Methanospirillum hungatei, 178, 184, 195, 198, 209–10, 222, 224–26
Methanothermaceae, 175, 232–34, 241–42
Methanothermus, 174–75, 181, 232–34, 241–42
Methanothermobacter, 232–34, 236, 241
Methanothermobacter defluvii, 232–34, 241
Methanothermobacter marburgensis, 232–34, 241
Methanothermobacter thermoautotrophicus, 232–34, 241
Methanothermobacter thermoflexus, 232–34, 241
Methanothermobacter thermophilus, 232–34, 241
Methanothermobacter wolfeii, 232–34, 241
Methanothermococcus, 258–59, 261, 265–66
Methanothermococcus okinawensis, 258–59, 265
Methanothermococcus thermolithotrophicus, 177, 181–82, 258–59, 265–66
Methanothermus, 174–75, 181, 232–34, 241–42
characteristics of, 177
Methanothermus fervidus, 177, 181, 232–34, 241–42
Methanothermus sociabilis, 177, 181, 232–34, 242
Methanothrix. See Methanosaeta
Methanotorris, 258–59, 261–62, 266–68
Methanotorris igneus, 258–59, 266–69

Metronidazole sensitivity
Gardnerella, 367
Microbacteriaceae, 307, 1020–82
Agreia, 1020–26
Agrococcus, 1020, 1026–27
Agromyces, 1020, 1027–31
Brevibacterium helvolum group, 1079–80
Clavibacter, 1020, 1031–40
copiotrophic ultramicrobacteria, 1081
Corynebacterium bovis group, 1020, 1080
Cryobacterium, 1020, 1040–41
Curtobacterium, 1020, 1041–48
from freshwater, 1081–82
Frigoribacterium, 1020, 1048–49
genera properties of, 1023–24
Leifsonia, 1020, 1049–55
Leucobacter, 1020, 1055–53
of low temperature environments, 1080–81
marine bacterioplankton, 1081–82
Microbacterium, 1020, 1056–66
Mycetocola, 1020, 1066–68
Okibacterium, 1020, 1068–69
phylogenetic tree of, 1022
Plantibacter, 1020, 1069–72
Rathayibacter, 1020, 1072–77
Subtercola, 1020, 1077–79
Microbacterium, 1020, 1049, 1056–67, 1081
characteristics of, 1015, 1023–24, 1050
habitats of, 1061–65
phylogenetic tree of, 1022
species of, 1058–59, 1061–63
Microbacterium arabinogalactanolyticum, 1066
Microbacterium arborescens, 1061
Microbacterium aurum, 1061
Microbacterium barkeri, 1066
Microbacterium campoquemadoensis, 1066
Microbacterium dextranolyticum, 1061, 1066
Microbacterium esteraromaticum, 1060, 1066
Microbacterium flavescens, 315, 1065
Microbacterium foliorum, 1064
Microbacterium hominis, 1061, 1065
Microbacterium imperiale, 1061, 1064
Microbacterium keratanolyticum, 1061, 1066
Microbacterium ketosireducens, 1066
Microbacterium kitamiense, 1066
Microbacterium lacticum, 1060–61
Microbacterium laevaniformans, 1061, 1065–66
Microbacterium liquefaciens, 1065
Microbacterium nematophilum, 1064
Microbacterium oxydans, 1065–66
Microbacterium paraoxydans, 1066
Microbacterium phyllosphaerae, 1064, 1066
Microbacterium resistens, 1061, 1065–66

Microbacterium saperdae, 1060
Microbacterium schleiferi, 1061
Microbacterium terregens, 315, 1065
Microbacterium testaceum, 1044, 1060, 1064
Microbacterium trichothecenolyticum, 1066
Microbispora, 725–26
   chemotaxonomic and morphological features of, 727
   identification of, 740–46
   isolation and cultivation, 736–37
   morphological characteristics of, 778
   phylogeny of, 729–30
   species characteristics of, 741
Microbispora aerata, 741
Microbispora amethystogenes, 730, 741
Microbispora bispora. See *Thermobispora bispora*
Microbispora chromogenes, 730
Microbispora corallina, 734
Microbispora diastatica, 730
Microbispora echinospora. See *Actinomadura rugatobispora*
Microbispora indica, 730
Microbispora karnatakensis, 730
Microbispora mesophila, 700, 730
Microbispora parva, 741
Microbispora rosea, 730, 734
Microbispora thermorosea, 730
Microbispora viridis. See *Actinomadura viridis*
Micrococcaceae, 312–16, 975
   Arthrobacter, 312–15
   Kocuria, 312, 315
   Micrococcus, 312–13
   Nesterenkonia, 312, 315
   Renibacterium, 312, 315–16
   Rothia, 312, 315
   Stomatococcus, 312, 315, 975
Micrococcineae, 307
   Cellulomonadaceae, 983
   Dermatophilaceae, 1002
   Microbacteriaceae, 307
   phylogenetic tree of, 304
Micrococcus, 312–13, 961–68
   genetics of, 967–68
   habitats of, 961–62
   identification of, 963–66
   isolation of, 962–63
   pathogenicity of, 968
   physiological and biochemical properties of, 966–67
   species of, 964–66
Micrococcus agilis, 961, 963–66
Micrococcus halobius, 963–66
Micrococcus kristinae, 961–62, 964–68
Micrococcus luteus, 304, 313, 961–62, 964–68
Micrococcus lylae, 304, 313, 961, 964–68
Micrococcus nishinomiyaensis, 961, 964–66, 968
Micrococcus roseus, 961, 964–68

Micrococcus sedentarius, 961, 964–66, 968
Micrococcus varians, 961–62, 964–68
Microlunatus, 383–85, 388–93
   characteristics of, 391, 393
   isolation of, 392
   morphology of, 390–93
Microlunatus phosphovorus, 383, 390–93
Micromonospora, 623–49, 671, 725
   characteristics of, 639–42
   ecophysiology of, 623–30
   identification of, 635–39
   isolation of, 630–35, 642–45
Micromonospora carbonacea, 642
Micromonospora chalcea, 642
Micromonospora coerulea, 642
Micromonospora echinospora, 642, 648
Micromonospora echinospora subsp. *echinospora*, 642
Micromonospora echinospora subsp. *ferruginea*, 642
Micromonospora echinospora subsp. *pallida*, 642
Micromonospora halophytica, 642
Micromonospora inyonensis, 648
Micromonospora olivasterospora, 642, 648
Micromonospora purpurea, 648
Micromonospora purpureochromogenes, 642
Micromonospora rosaria, 642
Micromonosporaceae, 725
Micropruina, 383–85, 388–89, 391, 393–94
   characteristics of, 391, 394
   isolation of, 394
   morphology of, 391, 394
Micropruina glycogenica, 391, 393–94
Microtetraspora, 725–26
   chemotaxonomic and morphological features of, 727
   identification of, 740–46
   isolation and cultivation, 736–37
   morphology of, 756, 768–69
   phylogeny of, 730–31
   species characteristics of, 743
Microtetraspora fusca, 731, 735–37
Microtetraspora glauca, 699–700, 731, 735–37
Microtetraspora malaysiensis, 735–37
Microtetraspora niveoalba, 731, 735–37
Microtetraspora viridis. See *Actinomadura viridis*
Mobiluncus, 430–39, 506–16
   chemotaxonomic properties of, 507–9, 512
   habitat and ecology of, 508–10
   identification of, 509, 512–13
   isolation of, 510–12
   morphology of, 506–9, 512
   pathogenicity of, 512–16
   physiological properties, 508–9, 512
   taxonomy of, 506–8

Mobiluncus curtisii, 430, 506–16
Mobiluncus mulieris, 506–16
Modestobacter, 1081
Modestobacter multiseptatus, 672
Molecular biology
   methanogens, 194–96
   Nocardioides, 1107–9
   Streptomycetaceae, 605–15
   Sulfolobales, 33–38
   Thermococcales, 76–77
Morphology
   Actinobaculum, 494
   Actinomadura, 684
   Actinomyces, 442–46, 465–67
   Actinoplanes, 636
   Actinoplanetes, 635–36
   Ampullariella, 636
   Anaerobiospirillum, 423, 425
   Arcanobacterium, 499–500
   Arthrobacter, 313, 952–53
   Atopobium, 306
   Bifidobacteriaceae, 322
   Bifidobacterium, 334, 336
   Bogoriella, 307–8
   Brachybacterium, 308
   Brevibacterium, 1013, 1015
   Coriobacterium, 306
   Corynebacterium, 796–97, 801, 819–20
   Curtobacterium, 1043
   Dactylosporangium, 636
   Demetria, 310
   Dermatophilus, 1005–6
   Desulfurococcales, 30, 52, 56–61
   Friedmanniella, 385–87, 389, 391
   Frigoribacterium, 1048
   Gardnerella, 360, 363
   Halobacteriales, 115–16
   Janibacter, 312
   Kineococcaceae, 307
   Kineosphaera, 1005, 1007
   Kineosporia, 307
   Kitasatospora, 540
   Korarchaeota, 283
   Leifsonia, 1051
   Lentzea, 663–65
   Luteococcus, 389–91
   Methanobacterium, 179–81, 232–34, 236–37
   Methanobrevibacter, 181, 238–40
   Methanocalculus, 223
   Methanocaldococcus, 266–67
   Methanococcales, 257
   Methanococcoides, 250–52
   Methanococcus, 181–82, 262–65
   Methanocorpusculum, 185, 220–22
   Methanoculleus, 216–18
   Methanofollis, 219
   Methanogenium, 215–16
   methanogens, 174–76
   Methanohalobium, 250–53
   Methanohalophilus, 250–53
   Methanolacinia, 184, 209, 215
   Methanolobus, 250–52
   Methanomicrobiacaea, 182–83
   Methanomicrobium, 184, 209, 214–15
   Methanoplanus, 183–84, 218–19
   Methanosaeta, 187, 254

Methanosalsum, 250–53
Methanosarcina, 185–86, 248–52
Methanosarcinales, 244
Methanosphaera, 181
Methanospirillum, 184, 209–10, 222
Methanothermobacter, 241
Methanothermococcus, 265–66
Methanotorris, 258–59, 266–68
Microbacteriaceae, 1020–21
Microbacterium, 1061
Micrococcus, 313
Microlunatus, 390–93
Micropruina, 391, 394
Mobiluncus, 506–9, 512
Mycetocola, 1067
Mycobacterium, 901, 924
Nanoarchaeum, 277–78
Nocardioides, 1099, 1102–4
Obligately methylotrophic cocci, 186–87
Okibacterium, 1069
Pilimelia, 636
Plantibacter, 1070
Propionibacterium, 401
Propioniferax, 391, 394–95
Propionimicrobium, 389, 396
Rathayibacter, 1073
Renibacterium, 972–73
Rhodococcus, 850
Rothia, 315
Ruminobacter, 423–25
Saccharothrix, 661–62
Stomatococcus, 315, 975
Streptacidiphilus, 540
Streptomyces, 540, 576–78
Streptomycetaceae, 540
Subtercola, 1078–79
Succinimonas, 423, 425
Succinivibrio, 422, 425
Succinivibrionaceae, 419
Sulfolobales, 23, 30
Terrabacter, 312
Tessaracoccus, 389, 391, 396–97
Thermococcales, 69, 71–72
Thermoplasmatales, 106
Thermoproteales, 14–16
Varibaculum, 505
Motility
    Actinokineospora, 659–60
    Methanococcales, 268
    Methanococcus, 181–82
    Methanocorpusculum, 220–22
    Methanofollis, 219–20
    methanogens, 198
    Methanomicrobiales, 226
    Methanospirillum, 222
    Methanothermus, 241–42
    Mobiluncus, 507
    Oerskovia, 994
    Planobispora, 732
    Succinivibrionaceae, 419
Muralytic enzyme sensitivity,
        Propionibacterium, 409
Murein, *Bifidobacterium*, 342–43
Mushrooms, *Mycetocola*, 1067
Mycelium forming
    Acrocarpospora, 729
    Actinokineospora, 655

Actinomadura, 684
Agromyces, 1028–29
Kitasatospora, 542–43
Lechevalieria, 665
Lentzea, 665
Microbispora, 730
Mycobacterium, 889
Nocardioides, 1099, 1102–4
Nocardiopsis, 767, 783
Planotetraspora, 734
Saccharothrix, 662
Skermania, 847
Spirillospora, 708
Streptomonospora, 781
Streptomyces, 539, 562, 576–78
Streptomycetaceae, 538
Streptoverticillium, 579–80
Thermomonospora, 684, 687
Mycetocola, 1020, 1049, 1066–68
    characteristics of, 1023–24, 1050, 1069
    habitats of, 1067
    phylogenetic tree of, 1022
Mycetocola lacteus, 1067–68
Mycetocola saprophilus, 1066–68
Mycetocola tolaasinivorans, 1067–68
Mycobacteriaceae, 797, 819, 889
    Mycobacterium, 819, 843, 889
Mycobacterium, 819, 843
    characteristics of, 845, 860, 890, 901, 903
    clusters of, 893–94
    CMN-group, 797–98
    epidemiology of, 925–31
    genetics of, 907–9, 925
    habitats of, 896–97, 920
    identification of, 900–904, 921–23
    isolation of, 897–98, 920–21
    medical, 919–31
    morphology of, 901, 924
    nonmedical, 889–910
    pathogenicity, 896–97, 919–31, 934–41
    phylogenetic tree of, 846, 893–95, 921
    phylogeny of, 890–94
    physiology of, 904–7, 924–25
    species of, 891–92, 899
    taxonomy of, 894–96, 920
Mycobacterium abcessus, 894, 930
Mycobacterium africanum, 895–96, 926, 928
Mycobacterium agri, 893, 897
Mycobacterium aurum, 897, 900–904, 908–9
Mycobacterium avium, 889, 896, 905, 908, 920, 922, 926, 928
Mycobacterium bovis, 889–90, 895–96, 899, 907–9, 920, 926, 928, 931
Mycobacterium brumae, 894
Mycobacterium chelonae, 894, 896, 901, 904, 920, 930
Mycobacterium chitae, 894, 902
Mycobacterium chlorophenolicum, 894
Mycobacterium confluentis, 893
Mycobacterium diernhoferi, 901, 903
Mycobacterium doricum, 894

Mycobacterium fallax, 894, 902
Mycobacterium farcinogenes, 896
Mycobacterium flavescens, 900
Mycobacterium fortuitum, 894, 896–97, 902, 904, 907–9, 920, 922, 930
Mycobacterium frederickbergense, 894
Mycobacterium gastri, 896, 904
Mycobacterium gilvum, 902
Mycobacterium goodie, 894
Mycobacterium gordonae, 860, 897, 923
Mycobacterium haemophilum, 896
Mycobacterium hiberniae, 894
Mycobacterium holsaticum, 894
Mycobacterium immunogenum, 894
Mycobacterium intracellulare, 893, 896, 907, 922, 928
Mycobacterium kansasii, 896, 920, 922–23, 930
Mycobacterium komossense, 904
Mycobacterium leprae, 889, 893, 896, 909, 919, 925, 928–30, 934–41
    habitat of, 936–37
    identification of, 939–41
    isolation of, 937–39
    properties of, 935–36
Mycobacterium lepraemurium, 939
Mycobacterium margaritense, 893
Mycobacterium marinum, 896, 904, 920, 930, 939
Mycobacterium microti, 895–96, 899, 926
Mycobacterium moidei, 894
Mycobacterium moriokaense, 893
Mycobacterium neoaurum, 901, 907, 909–10
Mycobacterium nonchromogenicum, 894, 897
Mycobacterium novocastrense, 894
Mycobacterium paraffinicum, 904
Mycobacterium parafortuitum, 901, 906, 909
Mycobacterium paratuberculosis, 896, 904, 928
Mycobacterium perigrinum, 896
Mycobacterium phlei, 897, 900, 902, 906–7, 939
Mycobacterium pulveris, 893
Mycobacterium rhodesiae, 893
Mycobacterium scrofulaceum, 896, 907, 930
Mycobacterium senegalense, 896
Mycobacterium septicum, 894
Mycobacterium shimodei, 896
Mycobacterium simiae, 890, 892, 896
Mycobacterium smegmatis, 889, 897, 902, 904–9–922, 925, 939
Mycobacterium szulgai, 896, 922
Mycobacterium terrae, 894, 896–97
Mycobacterium thermoresistibile, 900, 902, 904
Mycobacterium triviale, 890, 892, 894, 896–97, 902–3
Mycobacterium tuberculosis, 607, 854, 889–90, 893, 895–96, 907, 909, 919–28, 931
Mycobacterium tusciae, 894

Mycobacterium ulcerans, 893, 896, 920, 930–31, 939
Mycobacterium vaccae, 904–5, 939
Mycobacterium vanbaalenii, 894
Mycobacterium wolinskyi, 894
Mycobacterium xenopi, 896
Mycolic acids
    Corynebacterium, 801–2, 819
    Dietzia, 844
    Gordonia, 846
    Mycobacterium, 889, 901–4, 920, 924
    Mycobacterium leprae, 940
    Nocardia, 848
    Rhodococcus, 849–50
    Williamsia, 847–48

N

NADH. *See* Nicotinamide adenine dinucleotide
Nalidixic acid resistance
    Bifidobacterium, 349
    Gardnerella, 367
Nanoarchaeota, *Nanoarchaeum,* 274–79
Nanoarchaeum, 274–79
    cultivation of, 278
    genetics of, 279
    habitats of, 275–77
    identification of, 277–78
    introduction to, 274
    isolation of, 277
    phylogenetic tree of, 275–76
    phylogeny of, 274–75
    physiology of, 278–79
    preservation of, 278
    taxonomy of, 274–75
Nanoarchaeum equitans, 274–79
Nanobacteria, 1081
Natrialba, 114, 122–23
Natrialba asiatica, 131, 135
Natrialba magadii, 131, 149
Natrialba vacuolata, 138
Natrinema, 114, 126
Natronobacterium, 114, 122–23
Natronobacterium gregoryi, 116, 131, 137, 149
Natronobacterium innermongoliae, 117
Natronococcus, 114, 124–25
Natronococcus amylolyticus, 149
Natronococcus occultus, 116, 130, 134, 143, 146, 149
Natronomonas, 114, 120–21
Natronomonas pharaonis, 116, 128, 143, 149
Natronorubrum, 114, 126
Neomycin resistance/sensitivity
    Bifidobacterium, 349
    Halobacteriales, 127
    methanogens, 197
Nesterenkonia, 312, 315
Nesterenkonia halobia, 315
Neutral lipids, Halobacteriales, 135
Neutrophiles
    Desulfurococcales, 30
    Thermococcales, 69

Nicotinamide adenine dinucleotide (NADH) dehydrogenase, Sulfolobales, 33–34
Nicotinamide adenine dinucleotide (NADH) oxidase, *Archaeoglobus,* 88–89
Nitrate oxidation, Thermoproteales, 10, 20
Nitrate reduction
    Arcanobacterium, 501
    Bifidobacterium, 349
    Halobacteriales, 145
    Mycobacterium, 899, 922
Nitrite oxidation, Thermoproteales, 10
Nitrogen fixation
    Frankia, 674–75
    Methanococcus, 269
    Streptomyces, 615
Nitrogen metabolism, methanogens, 192
Nocardia, 819, 843, 920
    characteristics of, 845, 860, 862, 864–65, 890
    CMN-group, 797–98
    morphology of, 756
    phylogenetic tree of, 846, 849
    phylogeny of, 848
    species characteristics of, 863–64
Nocardia africana, 852, 855
Nocardia asteroides, 607, 848, 852, 855, 862, 864–65, 868, 871
Nocardia beijingensis, 852
Nocardia brasiliensis, 848, 852, 855, 864–65, 868, 871
Nocardia caishijiensis, 852, 855
Nocardia carnea, 871
Nocardia cerradoensis, 852, 855
Nocardia crassostreae, 864
Nocardia cummidelens, 852
Nocardia farcinica, 848, 852, 864–65
Nocardia fluminea, 852
Nocardia neocaledoniensis, 852
Nocardia nova, 848, 852
Nocardia otitidiscaviarum, 848, 852, 855, 864–65, 871
Nocardia pseudotuberculosis, 852, 865
Nocardia seriola, 848, 864
Nocardia soli, 852
Nocardia tenerifensis, 852, 855
Nocardia transvalensis, 852, 865
Nocardia vaccinii, 852, 871
Nocardia vinacea, 852, 872
Nocardia xishanensis, 855
Nocardiaceae, 797, 819, 843–72, 1099–1110
    habitats of, 852–53, 1100
    identification of, 858–65
    isolation and cultivation of, 855–56
    metabolism and genetics of, 868–72
    Nocardia, 819, 843
    pathogenicity, 865–68
    phylogeny of, 848–50
    Rhodococcus, 819, 843
Nocardioides
    characteristics of, 1105–6

    habitats of, 1100
    phylogenetic tree of, 1107
Nocardioides albus, 1099–1100, 1102, 1107
Nocardioides aquaticus, 1099–1100, 1103, 1107
Nocardioides flavus, 1099, 1108–9
Nocardioides fulvus, 1099, 1108–9
Nocardioides jensenii, 1099, 1103, 1107–10
Nocardioides luteus, 1099–1100, 1102, 1108–9
Nocardioides nitrophenolicus, 1099–1100, 1103, 1109
Nocardioides plantarum, 1099–1100, 1103, 1107, 1109
Nocardioides pyridinolyticus, 1099–1100, 1103, 1109
Nocardioides simplex, 315, 1099–1100, 1103, 1107–9
Nocardioides thermolilacinus, 1099
Nocardiopsaceae, 682, 754–86
    Nocardiopsis, 754
    Streptomonospora, 754
    Thermobifida, 754
Nocardiopsis
    habitat of, 761–63
    identification of, 766–77
    isolation and cultivation of, 764–65
    morphology of, 756, 768–69, 771
    pathogenicity of, 786
    phylogenetic tree of, 780–81
    phylogeny and taxonomy, 754–60
    physiology of, 767, 783–84
    species characteristics of, 774–76
Nocardiopsis alba, 754, 758, 760, 773, 781–82
Nocardiopsis compostus, 754, 761
Nocardiopsis dassonvillei, 757, 761–62, 770–71, 781–82, 784, 786
Nocardiopsis dassonvillei subsp. *albirubida,* 754, 758, 762, 773, 781–82
Nocardiopsis dassonvillei subsp. *dassonvillei,* 754, 758, 762, 773, 781–82
Nocardiopsis dassonvillei subsp. *prasina,* 762, 784
Nocardiopsis exhalans, 754, 761, 781–82
Nocardiopsis halophila, 754, 762, 773, 781–82
Nocardiopsis halotolerans, 754, 762, 781–82
Nocardiopsis kunsanensis, 754, 762, 773, 781–82
Nocardiopsis listeri, 754, 758, 760, 770, 781–82
Nocardiopsis lucentensis, 754, 758, 762, 781–82
Nocardiopsis prasina, 754, 758, 773, 781–82
Nocardiopsis synnemataformans, 754, 773, 781–82
Nocardiopsis trehalosi, 754, 758, 773, 781–83
Nocardiopsis tropica, 754, 781–82

Nocardiopsis umdischolae, 754, 761, 781–82
Nocardiosis, *Nocardia*, 852, 867–68
Nonomuraea, 726
    chemotaxonomic and morphological features of, 727
    identification of, 740–46
    isolation and cultivation, 736–37
    morphological characteristics of, 778
    morphology of, 756, 768–69
    phylogeny of, 731–32
    species characteristics of, 744
Nonomuraea africana, 731
Nonomuraea angiospora, 731
Nonomuraea dietziae, 731
Nonomuraea fastidiosa, 731
Nonomuraea ferruginea, 731, 736
Nonomuraea flexuosa, 731
Nonomuraea helvata, 731, 736
Nonomuraea longicatena, 731
Nonomuraea polychroma, 731
Nonomuraea pusilla, 731, 736, 741
Nonomuraea recticatena, 731
Nonomuraea roseola, 731, 741
Nonomuraea roseoviolacea subsp. carminata, 731
Nonomuraea roseoviolacea subsp. roseoviolacea, 731
Nonomuraea rubra, 731
Nonomuraea salmonea, 731
Nonomuraea spadix, 736
Nonomuraea spiralis, 732, 736, 741
Nonomuraea terrinata, 732
Nonomuraea turkmeniaca, 732
Novobiocin sensitivity
    Halobacteriales, 127
    Propionibacterium, 410
Nucleotide fermentation, *Corynebacterium*, 809–10
Nutrition
    Bifidobacterium, 348–49
    Gardnerella, 366
    Propionibacterium, 408–9

O

Obligate syntrophy, methanogens, 167
Obligately methylotrophic cocci, 179, 186–87
Oerskovia, 983–84, 993–97
    applications of, 996
    characteristics of, 987
    chemotaxonomic properties of, 990
    habitats of, 996
    history of, 993
    identification of, 996
    isolation, cultivation and preservation of, 996
    pathogenicity of, 997
    phenotypic properties of, 995
    phylogenetic tree of, 984, 987
    phylogeny of, 994
    physiological properties of, 995–96

species of, 984
    taxonomy of, 993–95
Oerskovia enterophila, 984, 993–94, 996
Oerskovia jenensis, 984, 993–95, 997
Oerskovia paurometabola, 984, 993–94
Oerskovia turbata, 984–85, 993–95, 997
Oerskovia xanthineolytica, 993, 996–97
Oil environment
    Brevibacterium helvolum group, 1079
    Curtobacterium, 1044
    Methanobacterium, 236–37
    Methanococcales, 259
    Methanomicrobiales, 211
    Nocardioides, 1100
    Nocardiopsis, 784
    Streptomyces, 614
    Thermococcales, 69
Okibacterium, 1020, 1049, 1067–69
    characteristics of, 1023–24, 1050, 1071
    habitats of, 1069
    phylogenetic tree of, 1022
Okibacterium fritillariae, 1068–69
Oligosaccharides, *Bifidobacterium*, 348–49
Oral cavity
    Actinomyces, 455–57, 480–81
    Bifidobacterium, 354
    Corynebacterium, 832
    Rothia, 315
    Stomatococcus, 315, 977
Osmotic adaptation, Halobacteriales, 145–46
Osteomyelitis, *Brevibacterium*, 1014
Oxygen sensitivity
    Bifidobacterium, 346
    methanogens, 169

P

Paleococcus, 69–70, 72, 77
Paleococcus horikoshii, 70
Paleococcus woesei, 70
Pathogenicity
    Actinobaculum, 498
    Actinomadura, 708–9
    Actinomyces, 455–57, 483–93
    Arcanobacterium, 503–5
    Clavibacter, 1031–35, 1039–40
    Curtobacterium, 1044–45
    Demetria, 1008–10
    Kineosphaera, 1008–10
    Leifsonia, 1052–53
    Micrococcus, 968
    Mobiluncus, 514–16
    Mycobacterium, 896–97, 919–31, 934–41
    Nocardia, 852, 865–68
    Nocardiopsis, 786
    Oerskovia, 997
    Propionibacterium, 412–14
    Rathayibacter, 1072, 1074–76
    Rhodococcus, 852–53, 868

Stomatococcus, 977–80
Thermomonosporaceae, 708–9
Varibaculum, 506
Penicillin resistance/sensitivity
    Bifidobacterium, 349
    Gardnerella, 366–67
    Halobacteriales, 127
    methanogens, 197
    Methanoplanus, 218
    Propionibacterium, 410
    Stomatococcus, 980
    Streptomyces, 567
Pentose phosphate pathway, *Corynebacterium*, 804
Pentothenate requirement, *Propionibacterium*, 409
Peptide degradation, Thermococcales, 72–76
Peptidoglycan cell walls
    Actinomyces, 446–48
    Agreia, 1021, 1025
    Agrococcus, 1026
    Agromyces, 1028
    Arthrobacter, 312, 947–48, 954
    Atopobium, 306
    Bifidobacterium, 342–43
    Bogoriella, 307–8
    Brachybacterium, 308
    Brevibacterium, 1013
    Brevibacterium helvolum group, 1079
    Cellulomonadaceae, 983
    Coriobacterium, 306
    Corynebacterium, 796
    Cryobacterium, 1041
    Curtobacterium, 1044
    Demetria, 309–10
    Dermabacter, 308
    Dermacoccus, 309–10
    Dermatophilus, 309
    Dietzia, 844
    Friedmanniella, 385, 388
    Frigoribacterium, 1048–49
    Glycomyces, 306
    Intrasporangiaceae, 311–12
    Kocuria, 315
    Kytococcus, 309–10
    lack of, 5
    Leifsonia, 1049
    Leucobacter, 1055, 1056
    Methanobacteriaceae, 179
    Methanobacterium, 5
    methanogens, 192–93
    Microbacteriaceae, 1020
    Microbacterium, 1060, 1061
    Micrococcaceae, 312
    Micrococcus, 313, 967
    Mycetocola, 1067–68
    Mycobacterium leprae, 940
    Nesterenkonia, 315
    Nocardioides, 1099, 1104
    Nocardiopsis, 758–60
    Oerskovia, 993
    Okibacterium, 1069
    Plantibacter, 1070
    Promicromonospora, 311
    Rarobacter, 310
    Rathayibacter, 1073
    Rhodococcus, 850

Rubrobacter, 305
Sanguibacter, 310–11
Stomatococcus, 315
Streptomonospora, 779, 781
Streptomyces, 562–65
Streptomycetaceae, 539
Streptosporangium, 728
Subtercola, 1078
Tsukamurella, 850–51
Peptostreptoccus heliothrinreducens.
    See Slackia
Pericarditis, *Brevibacterium*
    helvolum group, 1080
Periodontitis, *Actinomyces*, 488–90
Peritonitis, *Brevibacterium*, 1014
Pesticide degradation, *Arthrobacter*,
    950, 955
PFGE. See Pulsed-field gel
    electrophoresis
pH requirement
    Bifidobacterium, 346
    Gardnerella, 365
    Kitasatospora, 538
    Nocardiopsis, 762–63, 783
    Streptacidiphilus, 538
    Streptomyces, 538, 614
    Thermomonospora, 709, 712
Phages
    Corynebacterium, 803
    Oerskovia, 995
    Streptomyces, 569, 610
Pharyngitis, *Corynebacterium*, 833
Phenanthrene degradation,
    Nocardioides, 1110
Phoborhodopsin. *See* Sensory
    rhodopsin II
Phospholipase D
    Arcanobacterium, 827
    Corynebacterium, 827
Phospholipids
    Actinokineospora, 659–60
    Actinoplanes, 638
    Agreia, 1025
    Ampullariella, 638
    Dactylosporangium, 638
    Herbidospora, 729
    Lechevalieria, 665
    Leifsonia, 1051
    Microbacteriaceae, 1021
    Micrococcus, 966–67
    Nocardioides, 1104, 1107
    Nocardiopsis, 767, 771–73
    Pilimelia, 638
    Rathayibacter, 1073
    Renibacterium, 972
    Saccharothrix, 662
    Streptomonospora, 781
Phylogenetic tree
    Acidothermaceae, 670–71
    Actinobacteria, 302
    Actinocorallia, 689
    Actinokineospora, 663
    Actinomadura, 689
    Actinomycetaceae, 434
    Actinosynnema, 663
    Actinosynnemataceae, 655, 663
    Agrococcus, 1022
    Agromyces, 1022
    Archaeoglobus, 84

Arthrobacter, 314
Brevibacterium, 1014
Cellulomonas, 984, 987
Clavibacter, 1022
Corynebacterium, 798, 835, 846
Crenarchaeota, 282
Cryobacterium, 1022
Curtobacterium, 1022
Dermatophilus, 1006
Desulfurococcales, 53
Dietzia, 846
Frankiaceae, 670–71
Frigoribacterium, 1022
Geodermatophilaceae, 670–71
Gordonia, 846
Gordoniaceae, 847
Halobacteriales, 115
Kineosphaera, 1006
Kitasatospora, 538
Korarchaeota, 282
Lechevalieria, 663
Leifsonia, 1022
Lentzea, 663
Leucobacter, 1022
Methanococcales, 258
methanogens, 174
Methanomicrobiacaea, 183
Methanosarcinaceae, 183
Microbacteriaceae, 1022
Microbacterium, 1022
Micrococcineae, 304
Mycetocola, 1022
Mycobacterium, 846, 893–95, 921
Nanoarchaeum, 275–76
Nocardia, 846, 849
Nocardioides, 1107
Nocardiopsis, 780–81
Oerskovia, 984, 987
Okibacterium, 1022
Plantibacter, 1022
Propionibacteriaceae, 384
Propionibacterium, 405
Pseudonocardiaceae, 655, 663
Pseudonocardiniae, 655
Rathayibacter, 1022
Rhodococcus, 846, 849
Rothia, 976
Saccharothrix, 663
Skermania, 846
Spirillospora, 689
Sporichthyaceae, 670–71
Stomatococcus, 976
Streptacidiphilus, 538
Streptomyces, 538
Streptomycetaceae, 538–39
Streptosporangiaceae, 726
Subtercola, 1022
Succinivibrionaceae, 426
Sulfolobales, 24
Thermococcales, 77
Thermomonospora, 689
Thermomonosporaceae, 689
Thermoplasmatales, 102
of Thermoproteales, 11
Tsukamurella, 846
Williamsia, 846
Phylogeny
    Acidothermaceae, 669–72
    Actinomycetaceae, 430–39

Agreia, 1021
Agrococcus, 1026
Agromyces, 1028
Archaeoglobus, 82
Bifidobacterium, 322–23
Cellulomonas, 985
Clavibacter, 1032
Corynebacterium, 797–98, 819–20
Crenarchaeota, 281–82
Cryobacterium, 1041
Curtobacterium, 1043
Dermatophilus, 1002–4
Desulfurococcales, 52–53
Dietzia, 844
Frankiaceae, 669–72
Frigoribacterium, 1048
Gardnerella, 359
Geodermatophilaceae, 669–72
Gordonia, 844–47
Halobacteriales, 114
Kineosphaera, 1002–4
Korarchaeota, 282–83
Leifsonia, 1051
Leucobacter, 1055
Methanococcales, 257
Microbacterium, 1060–61
Mycetocola, 1067
Mycobacterium, 890–94
Nanoarchaeum, 274–75
Nocardia, 848
Nocardiopsis, 754–60
Oerskovia, 994
Okibacterium, 1069
Plantibacter, 1070
Propionibacterium, 405–8
Rathayibacter, 1073
Rhodococcus, 848–50
Skermania, 847
Sporichthyaceae, 669–72
Streptomonospora, 761
Streptosporangiaceae, 725–34
Subtercola, 1077–78
Succinivibrionaceae, 422
Sulfolobales, 23–24
Thermobifida, 760–61
Thermomonosporaceae, 682–88
Thermoplasmatales, 101
of Thermoproteales, 10
Tsukamurella, 850–51
Varibaculum, 505
Williamsia, 847–48
Physiological properties
    Acidothermus, 674–75
    Actinobaculum, 495–97
    Actinomyces, 448–55
    Arcanobacterium, 500
    Arthrobacter, 954–55
    Bifidobacterium, 346–48
    Cellulomonas, 985–88
    Corynebacterium, 804–5, 826–27
    Crenarchaeota, 285–86
    Desulfurococcales, 64–65
    Frankia, 674–75
    Friedmanniella, 387–90
    Gardnerella, 365–67
    Geodermatophilus, 674–75
    Halobacteriales, 143–46
    Kitasatospora, 540
    Luteococcus, 390–91

Methanococcales, 268–69
methanogens, 188–98
Methanomicrobiales, 223–26
Micrococcus, 966–67
Microlunatus, 391, 393
Micropruina, 391, 394
Mobiluncus, 508–9, 512
Mycobacterium, 904–7, 924–25
Nanoarchaeum, 278–79
Nocardiopsis, 783–84
Oerskovia, 995–96
Propioniferax, 391, 395
Propionimicrobium, 396
Sporichthya, 674–75
Streptacidiphilus, 540
Streptomyces, 540
Streptomycetaceae, 540
Succinivibrionaceae, 424–26
Sulfolobales, 33–38
Tessaracoccus, 391, 396–97
Thermobifida, 784–86
Thermococcales, 72–79
Thermoplasmatales, 108–9
Thermoproteales, 18–19
Varibaculum, 505
Picobacteria, 1081
Picrophilaceae, *Picrophilus*, 101
Picrophilus, 101–10
Picrophilus oshimae, 103
Picrophilus torridus, 103
Pig environment, *Arcanobacterium*,
    501
Pigment
    Acrocarpospora, 729
    Actinoplanes, 639
    Actinosynnema, 660
    Agreia, 1021, 1025–26
    Ampullariella, 640
    Brevibacterium, 1014–15
    Catellatospora, 640
    Corynebacterium, 832
    Cryobacterium, 1040–41
    Curtobacterium, 1043
    Dactylosporangium, 641
    Frigoribacterium, 1048
    Glycomyces, 641
    Gordonia, 846
    Herbidospora, 729
    Leifsonia, 1051
    Lentzea, 665
    Microbacteriaceae, 1080–81, 1082
    Microbacterium, 1061
    Microbispora, 730
    Micromonospora, 641–42
    Mycetocola, 1067
    Mycobacterium, 889
    Nocardia, 848
    Nocardioides, 1104
    Oerskovia, 994
    Okibacterium, 1069
    Pilimelia, 642
    Planotetraspora, 734
    Rhodococcus, 850
    Saccharothrix, 661–62
    Spirillospora, 708
    Streptomonospora, 779
    Streptomyces, 575
    Thermomonospora, 684, 687
    Tsukamurella, 850–51

Pilimelia, 623–49
    characteristics of, 639–42
    chemotaxonomic properties of,
        636–39
    ecophysiology of, 623–30
    identification of, 635–39
    isolation of, 630–35, 642–45
    morphology of, 636
Pilimelia anulata, 642
Pilimelia collumellifera, 642
Pilimelia pallida, 642
Pilimelia terevasa, 642
Planobispora, 726
    chemotaxonomic and
        morphological features of,
        727
    identification of, 740–46
    isolation and cultivation, 737–38
    phylogeny of, 732
    species characteristics of, 746
Planobispora longispora, 732, 735,
    743, 746
Planobispora rosea, 732, 735, 743,
    746
Planomonospora, 726
    chemotaxonomic and
        morphological features of,
        727
    identification of, 740–46
    isolation and cultivation, 737–38
    phylogeny of, 732–33
    species characteristics of, 746
Planomonospora alba, 733, 735
Planomonospora parontospora,
    732–33, 735, 746
Planomonospora parontospora
    subsp. *antibiotica*, 733
Planomonospora parontospora
    subsp. *parontospora*, 733
Planomonospora sphaerica, 733,
    735, 746
Planomonospora venezuelensis,
    732–33, 735, 746
Planotetraspora, 726
    chemotaxonomic and
        morphological features of,
        727
    identification of, 740–46
    isolation and cultivation, 735
    phylogeny of, 733–34
Planotetraspora mira, 733–35, 746
Planotetraspora silvatica, 734–35,
    746
Plant environment
    Actinokineospora, 656
    Actinoplanetes, 632
    Actinosynnema, 656
    Agreia, 1025
    Agromyces, 1028–29
    Arthrobacter, 950
    Clavibacter, 1031–35
    Corynebacterium, 799
    Curtobacterium, 1044–45
    Frigoribacterium, 1049
    Herbidospora, 734
    Kineosporia, 307
    Leifsonia, 1052–53, 1055
    Micrococcus, 962
    Nocardia, 852

Nocardioides, 1100
Oerskovia, 996
Plantibacter, 1070–71
Rathayibacter, 1072, 1074–76
Skermania, 851
Streptomyces, 615
Plantibacter, 1020, 1069–72
    characteristics of, 1023–24, 1057,
        1072
    habitats of, 1070–71
    phylogenetic tree of, 1022
Plantibacter agrosticola, 1070
Plantibacter cousiniae, 1070
Plantibacter elymi, 1070
Plantibacter flavus, 1070
Plantoea, 1042
Plasmids
    Brevibacterium, 1017
    Corynebacterium, 802–3, 827
    Frankineae, 676
    Halobacteriales, 146
    Micrococcus, 968
    Streptomyces, 608–10
    Sulfolobales, 40
Pleurisy, *Brevibacterium helvolum*
    group, 1080
Pneumonia
    Corynebacterium, 832–33
    Rhodococcus, 853
Polar lipids
    Actinomyces, 446–48
    Arthrobacter, 954
    Brevibacterium, 1013
    Halobacteriales, 130–35
    Kineococcaceae, 307
    Kineosporia, 307
    Kitasatospora, 538
    Methanolacinia, 215
    Methanomicrobiales, 225–26
    Microbacterium, 1061
    Micrococcaceae, 312
    Microtetraspora, 731
    Nocardiopsis, 770
    Renibacterium, 972
    Stomatococcus, 978
    Streptacidiphilus, 538
    Streptomyces, 538, 562
    Streptomycetaceae, 538
    Streptosporangiaceae, 726
    Subtercola, 1078
    Thermobifida, 760
Polyketide production,
    Streptomyces, 613–14
Polymyxin resistance
    Bifidobacterium, 349
    Halobacteriales, 127
Polysaccharides
    Clavibacter, 1034–35, 1040
    Streptomyces, 610–11
Poultry environment
    Brachybacterium, 308
    Methanomicrobiales, 211
Prebiotics, *Bifidobacterium*, 358–59
Preservation
    Acidothermus, 674
    Actinomyces, 464
    Actinosynnemataceae, 658
    Arthrobacter, 952
    Bifidobacterium, 346

Clavibacter, 1039–40
Corynebacterium, 801, 826
Dermatophilus, 1005
Desulfurococcales, 61–64
Frankia, 674
Gardnerella, 365
Geodermatophilus, 674
Halobacteriales, 142–43
Kineosphaera, 1005
Methanococcales, 262
Microbacterium, 1066
Micrococcus, 963
Nanoarchaeum, 278
Nocardioides, 1102
Nocardiopsis, 782
Sporichthya, 674
Stomatococcus, 977–78
Streptomonospora, 782
Streptomycetaceae, 592
Succinivibrionaceae, 421–22
Sulfolobales, 32
Thermobifida, 782
Thermomonosporaceae, 694
Thermoplasmatales, 108
Thermoproteales, 18
Probiotics
    Bifidobacterium, 355–58
    Propionibacterium, 410
Promicromonospora, 311, 983–84, 993
Promicromonospora citrea, 311, 990, 993
Promicromonospora enterophila, 311, 993–94
Promicromonospora sukumoe, 311, 993
Promicromonosporaceae, *Promicromonospora*, 311
Propionate production, *Propionibacterium*, 408–9
Propionibacteriaceae, 383–84, 400–401
    Friedmanniella, 383–91
    habitats of, 383
    Luteococcus, 383–85, 388–91
    Microlunatus, 383–85, 388–93
    Micropruina, 383–85, 388–89, 391, 393–94
    phylogenetic tree of, 384
    Propionibacterium, 383–85, 400–414
    Propioniferax, 383–85, 388–89, 391, 394–95
    Propionimicrobium, 383–85, 388–89, 391, 395–96
    species of, 384
    Tessarococcus, 383–85, 388–89, 391, 396–97
Propionibacterinaea, Propionibacteriaceae, 383–84, 400–401
Propionibacterium, 383–85, 400–414
    classical, 401–3
    cutaneous, 403–4
        pathogenic, 412–14
    habitats of, 404–5
    isolation of, 410–12
    metabolism and nutritional requirements of, 408–9

morphology of, 401
phylogenetic tree of, 405
phylogeny of, 405–8
Propionibacterium acidipropionici, 401, 405–6
Propionibacterium acnes, 401, 403–7, 409, 412–13
Propionibacterium australiense, 400–401, 405–7, 409
Propionibacterium avidum, 401, 403–7, 409, 412–13
Propionibacterium cyclohexanicum, 400–401, 405–7
Propionibacterium freudenreichii, 401, 404–7, 409
Propionibacterium granulosum, 401–7, 409, 412
Propionibacterium innocuum. See Propioniferax innocua
Propionibacterium jensenii, 401, 405–6, 409–10
Propionibacterium lymphophilum, 383, 400, 405–6. *See also Propionimicrobium*
Propionibacterium microaerophilum, 400–401, 405–7
Propionibacterium propionicum, 383, 401, 405–7, 409, 413–14
Propionibacterium thoenii, 401, 404–6
Propionic acid fermentation, 408
Propionicins, *Propionibacterium*, 409–10
Propioniferax, 383–85, 388–89, 391, 394–95
    characteristics of, 391, 394–95
    isolation of, 394–95
Propioniferax innocua, 383, 391, 394–95
Propionimicrobium, 383–85, 388–89, 391, 395–96
    characteristics of, 396
    isolation of, 395
    morphology of, 389, 396
Propionimicrobium lymphophilum, 395–96
Proteases, *Streptomyces*, 611
Proteobacteria, γ-subclass, 419, 422
Pseudomonas fluorescens, 1068
Pseudomonas putida, 315
Pseudomonas tolaasii, 1067–68
Pseudomurein
    Methanobacteriaceae, 179
    Methanobacteriales, 231
    methanogens, 192
    Methanothermus, 181
Pseudonocardia
    morphological characteristics of, 778
    morphology of, 756, 768–69
Pseudonocardiaceae, 843
    phylogenetic tree of, 655, 663
Pseudonocardiniae
    Actinosynnemataceae, 654
    phylogenetic tree of, 655
    Pseudonocardiaceae, 654
Psycrophile

Arthrobacter, 948–49, 953
Methanogenium, 211, 215–16
Microbacteriaceae, 1021
Nocardiopsis, 762
Pulsed-field gel electrophoresis (PFGE), 345
Pyrobaculum, 11, 15, 18
Pyrobaculum aerophilum, 11, 13, 15, 18–20, 35
Pyrobaculum arsenaticum, 11, 18–20
Pyrobaculum islandicum, 11, 13, 15, 19, 97
Pyrobaculum oguniense, 11, 13, 18–19
Pyrobaculum organotrophum, 11, 15, 18–19
Pyrococcus, 69, 72, 77
Pyrococcus abyssi, 70–72, 76–77
Pyrococcus endeavori, 71
Pyrococcus furiosus, 70–72, 75–77
Pyrococcus glycovorans, 72, 76
Pyrococcus horikoshii, 71–72, 76–77
Pyrococcus woesei, 70–72, 76
Pyrodictiaceae, 52, 56
    Hyperthermus, 56
    Pyrodictium, 52, 56
    Pyrolobus, 56
Pyrodictium, 52, 56–57, 60–61, 64–65
Pyrodictium abyssi, 56, 60–61
Pyrodictium brockii, 56, 60
Pyrodictium occultum, 56, 61
Pyrolobus, 56–57, 61, 64–65
Pyrolobus fumarii, 56, 61, 64

Q

Quinolone resistance, *Gardnerella*, 367
Quinones, in Sulfolobales, 34

R

Randomly amplified polymorphic DNA polymerase chain reaction (RAPD-PCR), *Streptomyces*, 571–72
Rarobacter, 310–11
Rarobacter faecitabidus, 310
Rarobacter incanus, 310
Rarobacteraceae, *Rarobacter*, 310
Rathayibacter, 1020, 1072–77
    characteristics of, 1023–24, 1057, 1069, 1071–72, 1074
    habitats of, 1074–76
    phylogenetic tree of, 1022
Rathayibacter biopuresis, 1074
Rathayibacter caricis, 1072–75
Rathayibacter festucae, 1072–74, 1076
Rathayibacter iranicus, 1073–73, 1077
Rathayibacter rathayi, 1072–75, 1077
Rathayibacter toxicus, 1073–73, 1075–77
Rathayibacter tritici, 1072–75, 1077
5S rDNA, *Cellulomonas*, 985

16S rDNA
 Acidimicrobium, 306
 Actinobacteria, 300–304
 Actinosynnemataceae, 663
 Arthrobacter, 314
 Bifidobacterium, 322–23
 Bogoriella, 307–8
 Cellulomonadaceae, 983
 Cellulomonas, 985–86
 Clavibacter, 1032, 1037
 Corynebacterium, 797–98
 Cryobacterium, 1041
 Curtobacterium, 1044, 1046
 Dermatophilaceae, 309
 Friedmanniella, 385, 390
 Glycomyces, 306
 Jonesia, 311
 Kineococcaceae, 306–7
 Leifsonia, 1051, 1053–54
 Luteococcus, 390
 Microbacteriaceae, 1080
 Microbacterium, 1065
 Micrococcaceae, 312
 Micrococcineae, 303
 Micropruina, 393–94
 Nocardioides, 1100, 1107–9
 Oerskovia, 994, 996
 Okibacterium, 1069
 Plantibacter, 1070
 Promicromonospora, 311
 Propionibacteriaceae, 383
 Propionibacterium, 383, 400–401,
   405–6
 Propionimicrobium, 395
 Rarobacter, 310
 Rubrobacter, 305
 Saccharothrix, 662–63
 Sanguibacter, 310–11
 Streptomyces, 562
23S rDNA, *Bifidobacterium*, 322–23
Renibacterium, 312, 315–16, 972–73
Renibacterium salmoninarum,
   315–16, 972–73
Respiratory chain, Sulfolobales,
   33–34
Respiratory quinones
 Halobacteriales, 136–37
 Rubrobacter, 305
Respiratory tract, *Stomatococcus*,
   315, 976–77
Restriction fragment length
   polymorphism, 345
Retinal pigments, Halobacteriales,
   137–38
Rhizosphere
 Agromyces, 1028–29
 Nocardiopsis, 761
 Plantibacter, 1070–71
 Streptomyces, 615
Rhodococcus, 584, 819, 843, 894,
   951, 1064
 characteristics of, 845, 860, 865,
   890, 1015
 CMN-group, 797–98
 morphology of, 850
 phylogenetic tree of, 846, 849
 phylogeny of, 848–50
 species characteristics of, 866–67
Rhodococcus aetherivorans, 856

Rhodococcus aurantiacus, 850
Rhodococcus baikonurensis, 853
Rhodococcus coprophilus, 853, 856
Rhodococcus equi, 853, 856, 860,
   865, 868
Rhodococcus erythropolis, 315, 853,
   856, 865–66, 869–72
Rhodococcus fascians, 607, 870
Rhodococcus globerulus, 853
Rhodococcus gordoniae, 853
Rhodococcus jostii, 853
Rhodococcus koreensis, 853, 856
Rhodococcus luteus, 853, 856
Rhodococcus marinonascens, 853,
   856
Rhodococcus pyridinivorans, 853,
   856
Rhodococcus rhodnii, 853
Rhodococcus rhodochrous, 848–50,
   853, 856, 865, 867, 869–70
Rhodococcus ruber, 853
Rhodococcus tukisamuensis, 856
Rhodoglobus vestalii, 1081
Rhodopsins, Halobacteriales, 137
Riboflavin production, *Arthrobacter*,
   955
Ribotyping, 345
Rifampicin resistance/sensitivity
 Halobacteriales, 127
 Stomatococcus, 980
Rothia, 312, 315, 975
 chemotaxonomic properties of,
   979
 phylogenetic tree of, 976
 phylogeny of, 975–76
 species characteristics of, 977
Rothia aerius, 976
Rothia amarae, 976
Rothia dentocariosa, 315, 975–76
Rothia mucilaginosus. See
   Stomatococcus mucilaginosus
Rothia nasimurium, 975–76
5S rRNA
 methanogens, 196
 Streptomyces, 571–73
16S rRNA
 Acrocarpospora, 729
 Actinobacteria, 301–2
 Actinomyces, 440–42, 446
 Actinomycineae, 430, 433, 435,
   437–39
 Agreia, 1021
 Agromyces, 1028
 Archaeoglobus, 83
 Arthrobacter, 314
 Bifidobacterium, 322–23, 343–44
 Brevibacterium, 1013, 1015, 1017
 Cellulomonadaceae, 983
 Corynebacterium, 797, 819, 825,
   831
 Dermatophilaceae, 309
 Dermatophilus, 1002–3
 description of, 4, 7–8
 Desulfurococcales, 52–53, 61
 Dietzia, 844
 Frigoribacterium, 1048
 Gardnerella, 359
 Geodermatophilus, 1002
 Halobacteriales, 115–17, 127, 142

Herbidospora, 729
Kineosphaera, 1002–3, 1005
Kitasatospora, 542–43, 574
Leucobacter, 1056
Methanocaldococcus, 262
Methanococcales, 257–59
Methanococcus, 262
methanogens, 5, 174, 196, 244
Methanomicrobiales, 208
Methanosarcinales, 244
Microbacteriaceae, 1020, 1079–80,
   1082
Microbacterium, 1060, 1065
Microbispora, 730
Micrococcaceae, 312
Micrococcus, 961
Microtetraspora, 731
Mobiluncus, 506–7
Mycobacterium, 890, 892–94, 900
Nanoarchaeum, 274–78
Nocardioides, 1102, 1107–8
Nocardiopsis, 757, 766, 780–81
Rathayibacter, 1077
Renibacterium, 315
Rhodococcus, 850, 865
Skermania, 847
Stomatococcus mucilaginosus,
   975–80
Streptacidiphilus, 543, 574
Streptomonospora, 766
Streptomyces, 562, 571–75
Streptomycetaceae, 539–42, 574
Streptosporangiaceae, 726
Streptosporangium, 728
Subtercola, 1078
Succinivibrionaceae, 422
Sulfolobales, 23–24, 30–31
Terrabacter, 312
Thermobifida, 760, 766
Thermococcales, 69, 76
Thermoplasmatales, 101, 103
 of Thermoproteales, 11–13
Varibaculum, 505
Williamsia, 847–48
23S rRNA
 Brevibacterium, 1017
 Corynebacterium, 825, 831
 Desulfurococcales, 53
 Kitasatospora, 542–43
 methanogens, 196
 Micrococcus, 967
 Nanoarchaeum, 274
 Renibacterium, 315
 Streptomyces, 571–73
 Sulfolobales, 24
Rubrobacter, 305
Rubrobacter radiotolerans, 305, 315
Rubrobacter xylanophilus, 305
Rubrobacteraceae, *Rubrobacter*, 305
Rumen environments
 Bifidobacterium, 326
 Methanocorpusculum, 220–22
 methanogens, 168
 Methanomicrobiales, 208, 211–12
 Methanomicrobium, 215
 Ruminobacter, 420
 Succinimonas, 419–20
 Succinivibrio, 419–20
 Succinivibrionaceae, 419–20

Ruminobacter, 419–26
Ruminobacter amylophilus, 419–26

S

Saccharomonospora, morphological
    characteristics of, 778
Saccharopolyspora erythrea, 607
Saccharopolyspora, morphological
    characteristics of, 756,
    768–69, 778
Saccharothrix, 654–59, 661–64, 666
    characteristics of, 664
    habitat of, 656
    history of, 757
    identification of, 661–63
    morphology of, 756, 768–69
    phylogenetic tree of, 663
Saccharothrix aerocolonigenes. *See*
    Lechevalieria aerocolonigenes
Saccharothrix albidocapillata, 655
Saccharothrix australiensis, 655, 662
Saccharothrix coeruleofusca, 656,
    661–62
Saccharothrix coeruleoviolacea, 656
Saccharothrix espanaensis, 655, 662
Saccharothrix flava. *See*
    Lechevalieria flava
Saccharothrix longispora, 656,
    661–62
Saccharothrix mutabilis, 655, 661
Saccharothrix mutabilis subsp.
    capreolus, 655
Saccharothrix mutabilis subsp.
    mutabilis, 655–56, 662
Saccharothrix syringiae, 656
Saccharothrix tangerinus, 662–63
Saccharothrix violacea. *See Lentzea*
    violacea
Saccharothrix waywayandensis. *See*
    Lentzea waywayandensis
Saline environment
    Methanosarcinales, 244–46
    Obligately methylotrophic cocci,
        186–87
Sanguibacter, 310–11
Sanguibacter inulinus, 310–11
Sanguibacter keddieii, 310–11
Sanguibacter suarezii, 310–11
Sanguibacteraceae, *Sanguibacter*,
    310–11
Secondary alcohol utilization
    Methanobacteriaceae, 236
    Methanobacteriales, 231
    Methanobacterium, 179–81
    Methanococcus, 181–82
    Methanocorpusculum, 185
    Methanoculleus, 183
    Methanogenium, 183
Sediment environments
    Actinoplanetes, 631–32
    Catellatospora, 629–30
    Desulfurococcales, 56
    Methanococcales, 259
    methanogens, 166
    Methanomicrobiales, 208, 211
    Methanosarcinales, 244–46
    Microbacterium, 1064

Micrococcus, 962
Nanoarchaeum, 276–77
Nocardiopsis, 761
Obligately methylotrophic cocci,
    186–87
Rhodococcus, 853
sporangiate actinoplanetes,
    625–26
Streptomyces, 583–84
Williamsia, 851
Seed surface, *Okibacterium*, 1069
Selenate oxidation,
    Thermoproteales, 10
Selenite oxidation,
    Thermoproteales, 10
Selenium requirement,
    Methanocaldococcus, 262
Selenium utilization
    Methanocaldococcus, 266–67
    Methanococcus, 264–65
    Methanothermococcus, 265–66
Sensory rhodopsin I,
    Halobacteriales, 138
Sensory rhodopsin II,
    Halobacteriales, 138
Septicemia
    Brevibacterium, 1014
    Brevibacterium helvolum group,
        1080
    Microbacterium, 1064–65
Serology
    Clavibacter, 1039
    Nocardia, 864–65
    Streptomyces, 568–69
Sewage environments
    Acidimicrobium, 306
    Arthrobacter, 950
    Bifidobacterium, 331–33
    Friedmanniella, 385
    habitats of, 383
    Methanobacteriales, 231
    Methanobacterium, 181
    Methanomicrobiales, 208, 211
    Methanosarcinales, 245–46
    Methanothermobacter, 241
    Nocardia, 852
    Nocardioides, 1100
Sewage treatment, methanogens,
    199
Skermania, 819, 843
    characteristics of, 845, 860
    phylogenetic tree of, 846
    phylogeny of, 847
Skermania amarae, 851
Skermania piniformis, 847, 851, 854
Skermania somaliensis, 865
Skin environment
    Brevibacterium, 1014
    Corynebacterium, 799, 827–28,
        831–34
    Dermacoccus, 309–10
    Dermatophilus, 1002–10
    Dietzia, 851
    habitats of, 383
    Kocuria, 315
    Kytococcus, 310
    Micrococcus, 313, 961–62
    Mycobacterium, 889, 919, 928–30,
        934–41

Propionibacterium, 404–5
Slackia, 306
S-layer
    Desulfurococcales, 57–61
    Halobacteriales, 128
    Methanococcus, 181–82, 263
    Methanogenium, 216
    methanogens, 193
    Methanolacinia, 215
    Methanomicrobiacaea, 182
    Methanomicrobiales, 224–25
    Methanoplanus, 184
    Methanothermus, 181
    Nanoarchaeum, 277–78
    Sulfolobales, 30–31
    Thermoplasmatales, 106
Soil environments
    Acidimicrobium, 306
    Actinocorallia, 688, 691
    Actinokineospora, 654–56
    Actinomadura, 688, 690–91
    Actinoplanetes, 631–32
    Actinosynnema, 656
    Agrococcus, 1026
    Agromyces, 1028–29
    Arthrobacter, 945, 948–50
    Bogoriella, 307–8
    Catellatospora, 629–30
    Cellulomonas, 988
    Corynebacterium, 799, 805
    Crenarchaeota, 284–85
    Cryobacterium, 1040–41
    Curtobacterium, 1044
    Demetria, 310
    Desulfurococcales, 56
    Dietzia, 851
    Frankia, 677
    Geodermatophilus, 673
    Glycomyces, 630
    Gordonia, 851
    habitats of, 383
    Halobacteriales, 139
    Herbidospora, 734
    Kineococcus, 307
    Kitasatospora, 575
    Kocuria, 315
    Lechevalieria, 656
    Leifsonia, 1053
    Lentzea, 656
    Luteococcus, 390
    methanogens, 166
    Microbacterium, 1061
    Microbispora, 734
    Micrococcus, 962
    Micromonospora, 627–29
    Microtetraspora, 734–35
    Mycobacterium, 896, 936
    Nocardia, 852
    Nocardioides, 1100
    Nocardiopsis, 761–63, 783
    Oerskovia, 996
    Planobispora, 735
    Planomonospora, 735
    Plantibacter, 1070–71
    Promicromonospora, 311
    Propionibacterium, 404–5
    Rhodococcus, 852–53
    Saccharothrix, 656
    Spirillospora, 691

sporangiate actinoplanetes, 625–26
Streptacidiphilus, 575
Streptomonospora, 763
Streptomyces, 575, 581–85, 614
Streptosporangineae, 734
Terrabacter, 312
Thermoproteales, 19
Tsukamurella, 853
Solfataric environments
Desulfurococcales, 56, 65
methanogens, 168
Methanothermus, 241–42
Nanoarchaeum, 276–77
Sulfolobales, 25, 28
Thermococcales, 69
Thermoplasmatales, 103–5
Thermoproteales, 13, 19
SOR. *See* Sulfur oxygenase reductase
Spermidine
Methanococcus, 262
Methanothermococcus, 262
Spermine, *Methanocaldococcus*, 262
Sphaerobacteraceae, *Acidimicrobium*, 305–6
Spirillospora, 682–714, 725
habitats of, 688, 690
identification of, 694–708
isolation and cultivation of, 691
morphological characteristics of, 701–2
phylogenetic tree of, 689
species characteristics of, 708
taxonomic history of, 695–97
Spirillospora albida, 684, 688, 690–91, 708
Spirillospora rubra, 688, 690–91, 708
Sporangia forming
Actinoplanes, 639
Actinoplanetes, 623–27
Ampullariella, 640
Catellatospora, 640
Dactylosporangium, 641
Glycomyces, 641
Herbidospora, 729
Microbispora, 729
Micromonospora, 641–42
Microtetraspora, 731
Nocardiopsis, 757
Nonomuraea, 731
Pilimelia, 642
Planobispora, 732
Planomonospora, 732–33
Planotetraspora, 734
Streptomonospora, 761
Streptosporangiaceae, 725
Sporichthya
isolation of, 674
physiology of, 674–75
Sporichthya polymorpha, 669, 674
Sporichthyaceae, 669–78
phylogenetic tree of, 670–71
taxonomy of, 672
Sputum, *Mycobacterium*, 921–23
Staphylococcus carnosus, 968
Staphylococcus, separation of, 963–64
Staphylococcus xylosus, 968

Staphylothermus, 52–53, 56–58, 64–65
Staphylothermus hellenicus, 53, 59
Staphylothermus marinus, 53, 59
Starch decomposition
Cellulomonas, 990
Micrococcus, 966
Nocardiopsis, 784
Streptomyces, 581
Steroid production
Curtobacterium, 1046
Mycobacterium, 909–10
Stetteria, 53, 56–57, 59, 64
Stetteria hydrogenophila, 53
Stomach, *Bifidobacterium*, 354
Stomatococcus, 312, 315, 975–80
chemotaxonomic properties of, 979
cultivation of, 977
habitats of, 976–77
identification of, 978
pathogenicity of, 977–80
phylogenetic tree of, 976
preservation of, 977–78
species characteristics of, 977
taxonomy of, 975–76
Stomatococcus mucilaginosus, 315, 975–80
Streptacidiphilus, 538, 543
characteristics of, 581
ecophysiology and habitat of, 575–84
identification of, 574
morphology of, 540
physiological properties of, 540
Streptacidiphilus albus, 543
Streptacidiphilus carbonis, 543
Streptacidiphilus neutrinimicus, 543
Streptoalloteicus, 654
morphology of, 768–69
Streptomonospora, 754
characteristics of, 767, 776
habitats of, 763
identification of, 779–81
isolation and cultivation of, 766
morphology of, 756, 768–69
phylogeny and taxonomy of, 761
preservation of, 782
Streptomonospora salina, 761, 763, 779, 781
Streptomyces, 538, 543–65, 605–15, 725
chemotaxonomic properties of, 540, 562
classification of
genotypic methods, 570–74
numerical, 543
phenotypic methods, 566–70
developmental growth cycle of, 605–7
ecology and habitat of, 575–84, 614–15
general characteristics of, 543, 562
identification of, 574–75
metabolism of, 612–13
morphological characteristics of, 540, 756, 778
pharmacologically active substances from, 613–14

physiological properties of, 540
plasmids of, 608–10
secreted proteins of, 610–11
species and clusters of, 544–61, 566–67
transposons of, 610
Streptomyces abikoensis, 573
Streptomyces acidiscabies, 615
Streptomyces aerocolonigenes. *See* *Lechevalieria aerocolonigenes*
Streptomyces albidoflavus, 567–68
Streptomyces albidoflavus subsp. *albidoflavus*, 567
Streptomyces albidoflavus subsp. *anulatus*, 567
Streptomyces albidoflavus subsp. *halstedii*, 567
Streptomyces albus, 562, 565, 568
Streptomyces aldosporeus supsp. *albosporeus*, 570
Streptomyces ambofaciens, 607
Streptomyces aminophilus, 570
Streptomyces antibioticus, 562, 565, 581, 607, 615
Streptomyces antimycoticus, 578
Streptomyces anultus, 567–69
Streptomyces ardus, 573
Streptomyces aurantiacus, 570
Streptomyces averilitils, 575
Streptomyces avermitilis, 607, 610
Streptomyces blastmyceticus, 573
Streptomyces bluensis, 578
Streptomyces cacaoi supsp. *cacaoi*, 570
Streptomyces caeruleus, 570
Streptomyces chrysomallus, 607
Streptomyces cinnamoneus, 573
Streptomyces clavuligerus, 612
Streptomyces coelicolor, 562, 568, 570, 573, 575, 605–10, 612
Streptomyces colinus, 612
Streptomyces cyaneus, 568
Streptomyces distallicus, 573
Streptomyces erythraeus, 568
Streptomyces eurocidicus, 573
Streptomyces fradiae, 568, 610
Streptomyces glaucescens, 568
Streptomyces griseocarneus, 573
Streptomyces griseofuscus, 562
Streptomyces griseus, 566–68, 581, 606–7
Streptomyces halstedii, 567–68, 611
Streptomyces hiroshimensis, 573
Streptomyces hygroscopicus, 567, 607
Streptomyces ipomoeae, 615
Streptomyces karnatakensis, 578
Streptomyces lavendulae, 568, 572
Streptomyces levoris, 565
Streptomyces lilacinus, 573
Streptomyces lipmanii, 607
Streptomyces lividans, 562, 569–70, 573, 605, 607–8, 610
Streptomyces luteoreticuli, 573
Streptomyces luteosporeus, 573
Streptomyces mashuensis, 573
Streptomyces mobaraensis, 573
Streptomyces moderatus, 607
Streptomyces netropsis, 573

Streptomyces niveus, 570
Streptomyces orinoci, 573
Streptomyces pactum, 611
Streptomyces parvulus, 607
Streptomyces pristinaespiralis, 612
Streptomyces reticuli, 607, 611
Streptomyces rimosus, 565, 568, 611
Streptomyces rochei, 607–9
Streptomyces scabies, 615
Streptomyces spheroides, 570
Streptomyces stramineus, 573
Streptomyces streptomycinii, 565
Streptomyces thermoautotrophicus,
    615
Streptomyces thermocoprophilus,
    583
Streptomyces thermospinipsorus,
    583
Streptomyces thermoviolaceus, 611
Streptomyces thermovulgaris, 565
Streptomyces thioluteus, 573
Streptomyces torulosus, 578
Streptomyces veticillus, 610
Streptomyces violaceoruber, 568
Streptomyces violaceus, 565, 570
Streptomyces violaceusniger, 568
Streptomyces violatus, 570
Streptomyces virginiae, 572
Streptomyces viridiflavus, 573, 606
Streptomyces virniniae, 572
Streptomycetaceae
    chemotaxonomic properties of,
        540
    history of, 538–42
    identification of, 574–75
    isolation of, 584–90
    Kitasatospora, 538, 542–43
    molecular biology of, 605–15
    Streptacidiphilus, 538, 543
    Streptomyces, 538, 543–65, 605–15
    taxonomy of, 538–92
Streptomycin resistance/sensitivity
    Bifidobacterium, 349
    Halobacteriales, 127
    methanogens, 197
    Propionibacterium, 410
    Stomatococcus, 980
    Thermoplasmatales, 106–7
Streptosporangiaceae, 682, 725–48
    Acrocarpospora, 726
    Herbidospora, 726
    identification of, 740–46
    metabolism and genetics of,
        746–47
    Microbispora, 726
    Microtetraspora, 726
    Nonomuraea, 726
    phylogenetic tree of, 726
    phylogeny of, 725–34
    Planobispora, 726
    Planomonospora, 726
    Planotetraspora, 726
    Streptosporangium, 726
Streptosporangineae
    Nocardiopsaceae, 682, 754
    Streptosporangiaceae, 682, 754
    Thermomonosporaceae, 682,
        754
Streptosporangium, 726

    chemotaxonomic and
        morphological features of,
        727
    identification of, 740–46
    isolation and cultivation, 736–37
    phylogeny of, 726–29
    species characteristics of, 728,
        745
Streptosporangium albidum, 728
Streptosporangium amythystogenes,
    743
Streptosporangium fragile, 741
Streptosporangium roseum, 726–29
Streptosporangium viridogriseum,
    728
Streptosporangium viridogriseum
    subsp. *kofuense*, 728
Streptosporangium viridogriseum
    subsp. *viridogriseum*, 728
Streptotrichosis, *Dermatophilus*, 309,
    1003–4
Streptoverticillium, 579–80, 607
Streptoverticillium cinnamomeum,
    579
Streptoverticillium cinnamomeum
    subsp. *azacolutum*, 579
Streptoverticillium mobaraense, 580
Streptoverticillium netropsis, 579
Streptoverticillium reticulum, 579
Streptoverticillium septatum, 579
Stygiolobus, 25, 28–29
Stygiolobus azoricus, 25, 28–31, 33
Stygiolobus azoricus FC6^T, 25
Subtercola, 1020, 1021, 1077–79
    characteristics of, 1023–24, 1057,
        1079
    habitats of, 1078
    phylogenetic tree of, 1022
Subtercola boreus, 1077–79
Subtercola frigoramans, 1077–79
Subtercola pratensis, 1021, 1077
Succinimonas, 419–26
Succinimonas amylolytica, 419–21,
    423, 425–26
Succinivibrio, 419–26
Succinivibrio dextrinosolvens,
    419–22, 424–26
Succinivibrionaceae, 419–26
    Anaerobiospirillum, 419–26
    biochemical and physiological
        properties, 424–26
    habitats of, 419–20
    isolation of, 420–21
    morphology of, 419
    phylogenetic tree of, 426
    phylogeny of, 422
    Ruminobacter, 419–26
    Succinimonas, 419–26
    Succinivibrio, 419–26
    taxonomy of, 422–24
Sucrose degradation
    Cellulomonas, 987
    Corynebacterium, 804, 805
    Micrococcus, 966
    Oerskovia, 996
Sugar degradation
    Sulfolobales, 35
    Thermococcales, 72–76
    Thermoplasma, 109

Sulfadiazine resistance, *Gardnerella*,
    367
Sulfate reduction, *Archaeoglobus*,
    96
Sulfide requirement,
    Methanocaldococcus, 266–67
Sulfide utilization, *Methanococcus*,
    264
Sulfite oxidation, Thermoproteales,
    10
Sulfolobaceae, 23
Sulfolobales, 10, 23–42, 52
    biotechnology of, 42
    characteristics of, 26–27
    ecology and habitats of, 25, 28, 42
    genetics of, 40–42
    identification of, 29–32
    isolation of, 28–29
    morphology, 23, 30
    phylogenetic tree of, 24
    phylogeny of, 23–24
    physiology of, 32–38
    plasmids of, 40
    SOR in, 34–35
    Sulfolobaceae of, 23
    taxonomy of, 24–27
    viruses of, 38–40
Sulfolobicins, Sulfolobales, 35
Sulfolobus, 23–25, 28–30, 33, 36, 38,
    40
Sulfolobus acidocaldarius, 23–25,
    29–30, 33–35, 37, 41
Sulfolobus acidocaldarius 98–3^T, 25
Sulfolobus islandicus, 24, 28, 35, 37,
    40
Sulfolobus metallicus, 24–25, 29,
    33–34
Sulfolobus shibatae, 24–25, 30, 35,
    37, 41
Sulfolobus solfataricus, 23–25, 30,
    33–37, 41
Sulfolobus tengchongensis, 24
Sulfolobus thuringiensis, 24
Sulfolobus tokodaii, 23–25, 31, 33
Sulfolobus yangmingensis, 24–25,
    31
Sulfonamide resistance/sensitivity
    Gardnerella, 367
    Propionibacterium, 410
Sulfophobococcus, 53, 56–57, 59, 64
Sulfophobococcus zilligii, 53, 65
Sulfur metabolism, methanogens,
    192
Sulfur oxidation
    Sulfolobales, 30, 33–35
    Thermofilum, 13
    Thermoproteales, 10, 18, 20
    Thermoproteus, 11
Sulfur oxygenase reductase (SOR),
    in Sulfolobales, 34–35
Sulfur respiration,
    Desulfurococcales, 64
Sulfurisphaera, 25, 28–30, 33
Sulfurisphaera ohwakuensis, 24–25,
    28, 31, 33
Sulfurisphaera ohwakuensis TA-1^T,
    25
Sulfurococcus, 25, 28, 30–31
Sulfurococcus mirabilis, 25

Sulfurococcus mirabilis INMI AT-59[T], 25
Sulfurococcus yellowstonensis, 25
Sulfurococcus yellowstonensis
    Str6kar[T], 25
Superoxide dismutase (Sod)
    Frankia, 675–76
    Frankineae, 675–76
Swamp environment
    Methanococcales, 259
    Methanomicrobiales, 211
    Methanoplanus, 184

T

Taxonomy
    Acidothermaceae, 672
    Actinobacteria, 297–316
    Actinobaculum, 493–97
    Actinomyces, 439–55
    Agreia, 1021, 1025
    Agrococcus, 1026
    Agromyces, 1028
    Arcanobacterium, 499–500
    Bifidobacterium, 323–31
    Cellulomonas, 985
    Corynebacterium, 797–98, 819–20
    Cryobacterium, 1041
    Curtobacterium, 1043–44
    Desulfurococcales, 53–56
    Frankiaceae, 672
    Frankineae, 672
    Frigoribacterium, 1048–49
    Gardnerella, 359–60
    Geodermatophilaceae, 672
    Halobacteriales, 115–39
    Leifsonia, 1051–52
    Leucobacter, 1056
    Methanococcales, 258–59
    Microbacterium, 1061
    Mobiluncus, 506–8
    Mycetocola, 1067
    Mycobacterium, 894–96, 920
    Nanoarchaeum, 274–75
    Nocardiopsis, 754–60
    Oerskovia, 993–95
    Okibacterium, 1069
    Plantibacter, 1070
    Rathayibacter, 1073
    Sporichthyaceae, 672
    Stomatococcus, 975–76
    Streptomonospora, 761
    Streptomycetaceae, 538–92
    Subtercola, 1078
    Succinivibrionaceae, 422–24
    Sulfolobales, 24–27
    Thermobifida, 760–61
    Thermomonosporaceae, 682–88
    Thermoplasmatales, 101–3
    Thermoproteales, 10–13
    Varibaculum, 505
Teichoic acids
    Brevibacterium, 1013
    Nocardiopsis, 758–60, 767, 773
    Thermobifida, 777
Temperature requirement
    Actinomadura, 684
    Bifidobacterium, 346
    Cryobacterium, 1041

Gardnerella, 365
Nocardiopsis, 762
Nonomuraea, 731
Skermania, 847
Subtercola, 1078
Thermobifida, 763
Thermomonospora, 709, 712
Termite gut
    Methanobrevibacter, 238–40
    Methanomicrobiales, 211–12
    Streptomyces, 614
Terrabacter, 311–12, 1002
Terrabacter tumescens, 312, 315
Terracoccus, 311–12
Terracoccus luteus, 312
Tessarococcus, 383–85, 388–89, 391,
    396–97
    characteristics of, 391, 396–97
    isolation of, 396–97
    morphology of, 389, 391, 396–97
Tessarococcus bendigoensis, 383,
    391, 396–97
Tetracycline resistance/sensitivity
    Bifidobacterium, 349
    Gardnerella, 367
    Halobacteriales, 127
    methanogens, 197
    Methanoplanus, 218
    Propionibacterium, 410
Thalassohaline environments,
    Halobacteriales, 139, 148
Thermobifida, 754
    characteristics of, 767
    habitats of, 763
    identification of, 777–79
    isolation and cultivation of,
        765–66
    morphology of, 756, 768–69,
        778–79
    phylogeny and taxonomy of,
        760–61
    physiology of, 784–86
    species identification, 777–79
Thermobifida alba, 683, 700, 760–61,
    763, 777
Thermobifida fusca, 684, 690, 699,
    713–14, 760–61, 763, 777,
    785–86
Thermobifida mesouviformis,
    760–61
Thermobispora, morphological
    characteristics of, 778
Thermobispora bispora, 729–30
Thermocladium, 11–13, 15
Thermocladium modestius, 13, 18
Thermococcales, 69–79
    characteristics of, 73–74
    ecology of, 69–79
    enzymology of, 77–79
    identification and morphology of,
        69, 71–72
    isolation of, 70
    molecular biology of, 76–77
    Paleococcus, 69
    phylogenetic tree of, 77
    physiology and metabolism of,
        72–79
    Pyrococcus, 69
    Thermococcus, 69
Thermococcus, 69, 72, 77

Thermococcus acidoaminovorans,
    69, 71–72
Thermococcus aegaeicus, 72
Thermococcus aggregans, 71–72
Thermococcus alcaliphilus, 69,
    71–72, 76
Thermococcus barophilus, 71, 76
Thermococcus barossii, 76
Thermococcus celer, 69, 75–76
Thermococcus chitinophagus, 71–72,
    76
Thermococcus fumiculans, 72, 76
Thermococcus hydrothermalis,
    71–72
Thermococcus litoralis, 69, 72, 75–76
Thermococcus peptinophilus, 71–72,
    76
Thermococcus profundus, 72, 76
Thermococcus sibiricus, 69, 76
Thermococcus siculi, 71
Thermococcus stetteri, 72, 76
Thermococcus waiotapuoensis, 72
Thermococcus zilligii, 72, 75–76
Thermocrispum, morphological
    characteristics of, 778
Thermodiscus, 52, 56–57, 59, 64
Thermodiscus maritimus, 56, 58
Thermofilaceae, 10, 13
    Thermofilum, 13
Thermofilum, 13, 16
Thermofilum librum, 13
Thermofilum pendens, 13, 19
Thermomonospora, 682–714
    habitats of, 690
    identification of, 694–708
    isolation and cultivation of,
        692–93
    morphological characteristics of,
        701–2, 768–69, 778
    phylogenetic tree of, 689
    species characteristics of, 706–7
    taxonomic history of, 695–97
Thermomonospora chromogena,
    683, 688, 690–92, 698–99, 709,
    712–13
Thermomonospora curvata, 683,
    688, 690–92, 698–700, 709, 712
Thermomonospora mesophilica. See
    Microbispora mesophila
Thermomonosporaceae, 682–714,
    725
    Actinocorallia, 682–714
    Actinomadura, 682–714
    chemotaxonomic markers of,
        685–86
    pathogenicity of, 708–9
    phylogenetic tree of, 689
    phylogeny and taxonomy of,
        682–88
    Spirillospora, 682–714
    Thermomonospora, 682–714
Thermophile
    Crenarchaeota, 282
    Desulfurococcales, 52, 65
    Methanococcaceae, 257
    Methanococcus, 181–82
    Methanoculleus, 216–17
    Methanogenium, 215
    Methanothermobacter, 241
    Methanothermococcus, 265–66

Methanothermus, 181
Microbispora, 730
Pyrodictium, 64
Pyrolobus, 64
Streptomyces, 583, 589–90, 615
Sulfolobales, 23
Thermobifida, 763, 784–85
Thermococcales, 72
Thermoproteales, 10
Thermoplasma, 101–10
Thermoplasma acidophilum, 101–3, 107, 109–10
Thermoplasma volcanium, 101–3, 109–10
Thermoplasmaceae, *Thermoplasma*, 101
Thermoplasmatales, 101–10
    Ferroplasmaceae, 101
    genetics of, 109–10
    habitat of, 103–5
    identification of, 106–7
    introduction to, 101
    isolation of, 105–6
    morphology of, 106
    phylogenetic tree of, 102
    phylogeny of, 101
    physiology of, 108–9
    Picrophilaceae, 101
    taxonomy of, 101–3
    Thermoplasmaceae, 101
Thermoproteaceae, 10
Thermoproteales, 10–20, 23, 52
    biotechnology of, 20
    characteristics of, 12
    cultivation of, 16–18
    genetics of, 19
    habitat of, 13
    identification of, 14–16
    isolation of, 13–14
    morphology of, 14–16
    phylogenetic tree of, 11
    phylogeny of, 10
    physiology of, 18–19
    taxonomy of, 10–13
    Thermofilaceae of, 10, 13
    Thermoproteaceae of, 10–13
Thermoproteus, 11, 18–19
Thermoproteus neutrophilus, 11, 15, 19
Thermoproteus tenax, 11, 13–15, 19
Thermoproteus uzoniensis, 11
Thermosome, *Pyrodictium*, 60
Thermosphaera, 56–57, 59–60, 64
Thermosphaera aggregans, 56, 60, 65
Thermotoga maritima, 97
Thiosulfate oxidation, Thermoproteales, 10
TMAO. *See* Trimethylamine N-oxide
Tolaasin detoxification, *Mycetocola*, 1067–68

Tonsillitis, *Arcanobacterium*, 502
Toxin production
    Corynebacterium, 826–27
    Gardnerella, 366
Transposons, *Streptomyces*, 610
Triacylglycerols (TAG), *Streptomyces*, 607
Tricarboxylic acid cycle
    Corynebacterium, 804
    Halobacteriales, 143
    methanogens, 192
    Solfolobales, 35
Trichothecenes production, *Microbacterium*, 1066
Trimethoprim resistance, *Gardnerella*, 367
Trimethylamine N-oxide (TMAO) reduction, Halobacteriales, 145
Tsukamurella, 819, 843, 894
    characteristics of, 845, 860, 890
    CMN-group, 797–98
    phylogenetic tree of, 846
    phylogeny of, 850–51
    species characteristics of, 862
Tsukamurella inchonensis, 851, 853
Tsukamurella paurometabola, 851, 853–54
Tsukamurella pseudospumae, 851, 854
Tsukamurella pulmonis, 851, 853, 855
Tsukamurella spumae, 854
Tsukamurella strandjordii, 851, 853, 855
Tsukamurella sunchonensis, 851
Tsukamurella tyrosinosolvens, 851, 853, 855
Tsukamurellaceae, 797, 819, 843–72
    habitats of, 853
    identification of, 858–65
    metabolism and genetics of, 869–72
    phylogeny of, 850–51
    Tsukamurella, 819, 843
Tuberculosis, *Mycobacterium*, 889, 909, 919–28, 931
Tungsten utilization, *Methanocaldococcus*, 266–67
Turicella, CMN-group, 797–98
Turicella otitidis, 798, 828–29
Turkey environment. *See* Poultry environment
Tween 80 degradation, *Nocardiopsis*, 784

U

Ultrastructure
    Bifidobacterium, 347–48

Gardnerella, 367
Urease activity, *Bifidobacterium*, 349
Urease production, *Corynebacterium*, 833
Urinary tract, *Actinobaculum*, 497

V

Vaginal environment
    Bifidobacterium, 354
    Corynebacterium, 832
    Gardnerella, 360, 368
    Mobiluncus, 506, 508–9, 512–15
Vancomycin resistance/sensitivity
    Bifidobacterium, 349
    Gardnerella, 366–67
    Methanoplanus, 218
    Oerskovia, 997
    Thermoplasmatales, 106–7
Varibaculum, 430–39, 470, 505–6
    identification of, 506
    isolation and cultivation of, 505–6
    pathogenicity, 506
    phylogeny and taxonomy of, 505
Varibaculum cambriense, 430, 432, 438, 505–6
Viruses, of Sulfolobales, 38–40
Vitamin B_{12}, Methanomicrobiales, 224

W

Walsby's square bacterium, 116
Waste disposal
    Cellulomonas, 992
    methanogens, 198–99
Williamsia, 819, 843
    characteristics of, 845, 860
    phylogenetic tree of, 846
    phylogeny of, 847–48
Williamsia maris, 848, 851
Williamsia muralis, 848, 851
Williamsiaceae, *Williamsia*, 819

X

Xylan decomposition
    Cellulomonas, 992
    Nocardiopsis, 784
    Streptomyces, 611
    Thermomonospora, 712

Z

Zoospores, *Dermatophilus*, 309